Formulas/Equations

Distance Formula — If $P_1 = (x_1, y_1)$ and $P_2 = (x_2, y_2)$

$$d(P_1, P_2) = \sqrt{}$$

Equation of a Circle — The equation of a circle of radiu... ... (h, k) is

$$(x - h)^2 + (y - k)^2 = r^2$$

Slope Formula — The slope m of the line containing the points $P_1 = (x_1, y_1)$ and $P_2 = (x_2, y_2)$ is

$$m = \frac{y_2 - y_1}{x_2 - x_1} \qquad \text{if } x_1 \neq x_2$$

$$m \text{ is undefined} \qquad \text{if } x_1 = x_2$$

Point–Slope Equation of a Line — The equation of a line with slope m containing the point (x_1, y_1) is

$$y - y_1 = m(x - x_1)$$

Slope–Intercept Equation of a Line — The equation of a line with slope m and y-intercept b is

$$y = mx + b$$

Quadratic Formula — The solutions of the equation $ax^2 + bx + c = 0$, $a \neq 0$, are

$$x = \frac{-b \pm \sqrt{b^2 - 4ac}}{2a}$$

If $b^2 - 4ac > 0$, there are two real unequal solutions.
If $b^2 - 4ac = 0$, there is a repeated real solution.
If $b^2 - 4ac < 0$, there are two complex solutions that are not real.

Geometry Formulas

Circle

r = Radius, A = Area, C = Circumference

$A = \pi r^2 \qquad C = 2\pi r$

Triangle

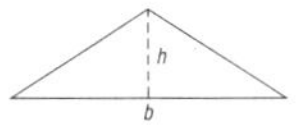

b = Base, h = Altitude (Height), A = area

$A = \frac{1}{2}bh$

Rectangle

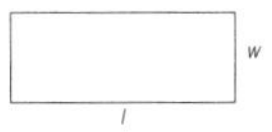

l = Length, w = Width, A = area, P = perimeter

$A = lw \qquad P = 2l + 2w$

Rectangular Box

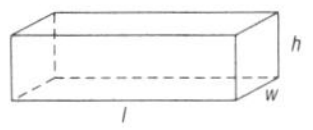

l = Length, w = Width, h = Height, V = Volume

$V = lwh$

Sphere

r = Radius, V = Volume, S = Surface area

$V = \frac{4}{3}\pi r^3 \qquad S = 4\pi r^2$

INSTRUCTOR'S EDITION

SECOND EDITION

College Algebra

Enhanced with Graphing Utilities

Michael Sullivan
Chicago State University

Michael Sullivan, III
Joliet Junior College

Prentice Hall
Upper Saddle River, New Jersey 07458

Executive Acquisitions Editor: Sally Yagan
Editor-in-Chief: Jerome Grant
Marketing Manager: Patrice Lumumba Jones
Supplements Editor/Editorial Assistant: Sara Beth Newell
Editor-in-Chief of Development: Carol Trueheart
Editorial/Production Supervision: Bob Walters
Senior Managing Editor: Linda Mihatov Behrens
Executive Managing Editor: Kathleen Schiaparelli
Assistant Vice President of Production and Manufacturing: David W. Riccardi
Manufacturing Buyer: Alan Fischer
Manufacturing Manager: Trudy Pisciotti
Director of Creative Services: Paul Belfanti
Associate Creative Director: Amy Rosen
Art Director: Maureen Eide
Assistant to Art Director: John Christiana
Art Manager: Gus Vibal
Art Editor: Grace Hazeldine
Interior Designer: Jill Little
Photo Researcher: Kathy Ringrose
Photo Editor: Melinda Reo
Cover Photo: Kids Hiking Together, Red Butte, Aspen, Colorado; Team Russell/Adventure Photo and Film
Art Studio: Academy Artworks

Printed in the United States of America

10 9 8 7 6 5 4 3 2 1

ISBN: 0-13-085303-8

(Student Edition ISBN: 0-13-083335-5)

Prentice-Hall International (UK) Limited, *London*
Prentice-Hall of Australia Pty. Limited, *Sydney*
Prentice-Hall of Canada Inc., *Toronto*
Prentice-Hall Hispanoamericana, S.A., *Mexico*
Prentice-Hall of India Private Limited, *New Delhi*
Prentice-Hall of Japan, Inc., *Tokyo*
Prentice-Hall (Singapore) Pte. Ltd., *Singapore*
Editora Prentice-Hall do Brasil, Ltda., Rio de Janeiro

In Memory of Mary . . .

Wife and Mother

CONTENTS

CHAPTER 7 SEQUENCES; INDUCTION; THE BINOMIAL THEOREM 495

CHAPTER 8 COUNTING AND PROBABILITY 549

CHAPTER 9 CONICS 603

APPENDIX REVIEW 661

PREFACE TO THE INSTRUCTOR

As professors, one at an urban public university and the other at a community college, Michael Sullivan and Michael Sullivan III are aware of the varied needs of College Algebra students, ranging from those having little mathematical background and fear of mathematics courses to those who have had a strong mathematical education and are highly motivated. For some of these students, this will be their last course in mathematics, while others might decide to further their mathematical education. This text is written for both groups. As the author of precalculus, engineering calculus, finite math, and business calculus texts, and as a teacher, Michael understands what students must know if they are to be focused and successful in upper level math courses. However, as a father of four, including the co-author, he also understands the realities of college life. Michael Sullivan III believes passionately in the value of technology as a tool for learning that enhances understanding without sacrificing important skills. Both authors have taken great pains to insure that the text contains solid, student-friendly examples and problems, as well as a clear, seamless writing style.

In the Second Edition

The second edition builds upon a strong foundation by integrating new features and techniques that further enhance student interest and involvement. The elements of the previous edition that have proved successful remain, while many changes, some obvious, others subtle, have been made. The text has been streamlined to increase accessibility. A huge benefit of authoring a successful series is the broad-based feedback upon which improvements and additions are ultimately based. Virtually every change to this edition is the result of thoughtful comments and suggestions made by colleagues and students who have used the previous edition. This feedback has proved invaluable and has been used to make changes that improve the flow and usability of the text. For example, some topics have been moved to better reflect the way teachers approach the course. In other places, problems have been added where more practice was needed. The supplements package has been enhanced through upgrading traditional supplements and adding innovative media components.

Changes to the Second Edition

Specific Organizational Changes

- Chapters 1, 3, and certain Appendix topics have been rewritten and reorganized to now appear in Chapter 1. This format will allow a more efficient use of the text and will improve the flow of the course. Once Chapter 1 is completed, the development of functions proceeds uninterrupted.
- From Chapter 3 to Chapter 1: Solving Equations; Setting up Equations; Applications; Solving Inequalities

- From the Appendix to Chapter 1: Solving Equations; Completing the Square
- From Chapter 1 to Chapter 2: Linear Curve Fitting; Direct Variation (These topics are now treated in the context of a linear function).
- From Chapter 3 to Chapter 4: Complex Numbers; Quadratic Equations with a Negative Discriminate (This section, which may be done after the quadratic equation in Chapter 1, now appears as motivation for the Fundamental Theorem of Algebra). Polynomial and Rational Inequalities (This topic is now handled after the discussion on polynomial and rational functions).
- From Chapter 2 to Chapter 5: One-to-One Functions; Inverse Functions (This topic now appears where it is needed at the start of the Exponential and Logarithmic Function chapter).
- Chapter 2 now appears as Chapters 2 and 3. Included now in Chapter 2 are Quadratic Functions and Models (formerly in Chapter 4).
- The chapter on Systems of Equations and Inequalities appears earlier to accommodate the syllabi of many schools. The section on Nonlinear Equations was moved to the Conics chapter, to make use of the graphs of conics in the solution.
- The chapter on Sequences; Induction; Sets; Probability now consists of two chapters. The last three chapters of the book may be covered in any order now.

Specific Content Changes

- In this edition emphasis is placed on the role of modeling in college algebra. To this end, dedicated sections appear on Linear Functions and Models, Quadratic Functions and Models, Power Functions and Models, Polynomial Functions and Models, and Exponential and Logarithmic Functions and Models. The focus of many of these applications is in the area of business, finance and economics.
- A section on univariate data and obtaining probability from data is new to this edition.
- A more extensive and more useable Appendix is also new. Now reference is made in the text to the Appendix for students who may require more information.
- The use of technology has been updated to include the more powerful features of ZERO(ROOT) VALUE and INTERSECT. The use of TRACE is minimal.

As a result of these changes, this edition will be an improved teaching device for professors and a better learning tool for students.

Acknowledgments

Textbooks are written by authors, but evolve from an idea into final form through the efforts of many people. Special thanks to Don Dellen, who first suggested this book and the other books in this series. Don's extensive contributions to publishing and mathematics are well known; we all miss him dearly.

There are many people we would like to thank for their input, encouragement, patience, and support. They have our deepest thanks and appreciation. We apologize for any omissions . . .

James Africh, *College of DuPage*
Steve Agronsky, *Cal Poly State University*
Dave Anderson, *South Suburban College*
Joby Milo Anthony, *University of Central Florida*
James E. Arnold, *University of Wisconsin-Milwaukee*
Agnes Azzolino, *Middlesex County College*
Wilson P. Banks, *Illinois State University*
Dale R. Bedgood, *East Texas State University*
Beth Beno, *South Suburban College*
Carolyn Bernath, *Tallahassee Community College*
William H. Beyer, *University of Akron*
Richelle Blair, *Lakeland Community College*
Trudy Bratten, *Grossmont College*
William J. Cable, *University of Wisconsin-Stevens Point*
Lois Calamia, *Brookdale Community College*
Roger Carlsen, *Moraine Valley Community College*
John Collado, *South Suburban College*
Denise Corbett, *East Carolina University*
Theodore C. Coskey, *South Seattle Community College*
John Davenport, *East Texas State University*
Duane E. Deal, *Ball State University*
Vivian Dennis, *Eastfield College*
Guesna Dohrman, *Tallahassee Community College*
Karen R. Dougan, *University of Florida*
Louise Dyson, *Clark College*
Paul D. East, *Lexington Community College*
Don Edmondson, *University of Texas-Austin*
Christopher Ennis, *University of Minnesota*
Ralph Esparza, Jr., *Richland College*
Garret J. Etgen, *University of Houston*
W. A. Ferguson, *University of Illinois-Urbana/Champaign*
Iris B. Fetta, *Clemson University*
Mason Flake, *student at Edison Community College*
Merle Friel, *Humboldt State University*
Richard A. Fritz, *Moraine Valley Community College*
Carolyn Funk, *South Suburban College*
Dewey Furness, *Ricke College*
Dawit Getachew, *Chicago State University*
Wayne Gibson, *Rancho Santiago College*
Sudhir Kumar Goel, *Valdosta State University*
Joan Goliday, *Sante Fe Community College*
Frederic Gooding, *Goucher College*
Ken Gurganus, *University of North Carolina*
James E. Hall, *University of Wisconsin-Madison*
Judy Hall, *West Virginia University*
Edward R. Hancock, *DeVry Institute of Technology*
Julia Hassett, *DeVry Institute-Dupage*
Brother Herron, *Brother Rice High School*
Kim Hughes, *California State College-San Bernardino*
Ron Jamison, *Brigham Young University*
Richard A. Jensen, *Manatee Community College*
Sandra G. Johnson, *St. Cloud State University*
Moana H. Karsteter, *Tallahassee Community College*
Arthur Kaufman, *College of Staten Island*
Thomas Kearns, *North Kentucky University*
Keith Kuchar, *Manatee Community College*
Tor Kwembe, *Chicago State University*
Linda J. Kyle, *Tarrant Country Jr. College*
H. E. Lacey, *Texas A & M University*
Christopher Lattin, *Oakton Community College*
Adele LeGere, *Oakton Community College*
Stanley Lukawecki, *Clemson University*
Janice C. Lyon, *Tallahassee Community College*
Virginia McCarthy, *Iowa State University*
Tom McCollow, *DeVry Institute of Technology*
Laurence Maher, *North Texas State University*
Jay A. Malmstrom, *Oklahoma City Community College*
James Maxwell, *Oklahoma State University-Stillwater*

Carolyn Meitler, *Concordia University*
Samia Metwali, *Erie Community College*
Eldon Miller, *University of Mississippi*
James Miller, *West Virginia University*
Michael Miller, *Iowa State University*
Kathleen Miranda, *SUNY at Old Westbury*
A. Muhundan, *Manatee Community College*
Jane Murphy, *Middlesex Community College*
Bill Naegele, *South Suburban College*
James Nymann, *University of Texas-El Paso*
Sharon O'Donnell, *Chicago State University*
Seth F. Oppenheimer, *Mississippi State University*
E. James Peake, *Iowa State University*
Thomas Radin, *San Joaquin Delta College*
Ken A. Rager, *Metropolitan State College*
Elsi Reinhardt, *Truckee Meadows Community College*
Jane Ringwald, *Iowa State University*
Stephen Rodi, *Austin Community College*
Howard L. Rolf, *Baylor University*
Edward Rozema, *University of Tennessee at Chattanooga*
Dennis C. Runde, *Manatee Community College*
John Sanders, *Chicago State University*
Susan Sandmeyer, *Jamestown Community College*
A. K. Shamma, *University of West Florida*
Martin Sherry, *Lower Columbia College*
Anita Sikes, *Delgado Community College*
Timothy Sipka, *Alma College*
Lori Smellegar, *Manatee Community College*
John Spellman, *Southwest Texas State University*
Becky Stamper, *Western Kentucky University*
Judy Staver, *Florida Community College-South*
Neil Stephens, *Hinsdale South High School*
Diane Tesar, *South Suburban College*
Tommy Thompson, *Brookhaven College*
Richard J. Tondra, *Iowa State University*
Marvel Townsend, *University of Florida*
Jim Trudnowski, *Carroll College*
Richard G. Vinson, *University of South Alabama*
Mary Voxman, *University of Idaho*
Darlene Whitkenack, *Northern Illinois University*
Christine Wilson, *West Virginia University*
Carlton Woods, *Auburn University*
George Zazi, *Chicago State University*

Recognition and thanks are due particularly to the following individuals for their valuable assistance in the preparation of this edition: Jerome Grant for his support and commitment; Sally Yagan, for her genuine interest and insightful direction; Bob Walters for his organizational skill as production supervisor; Patrice Lumumba Jones for his innovative marketing efforts; Tony Palermino for his specific editorial comments; the entire Prentice-Hall sales staff for their confidence; and to Katy Murphy and Jon Weerts for checking the answers to all the exercises.

PREFACE TO THE STUDENT

As you begin your study of College Algebra, you may feel overwhelmed by the numbers of theorems, definitions, procedures, and equations that confront you. You may even wonder whether or not you can learn all of this material in the time allotted. These concerns are normal. Do not become frustrated and say "This is impossible." It is possible! Work hard, concentrate, use all the resources available to you, and, above all, don't give up.

For many of you, this may be your last math course; for others, it is just the first in a series of many. Either way, this text was written to help you, the student, master the terminology and basic concepts of College Algebra. These aims have helped to shape every aspect of the book. Many learning aids are built into the format of the text to make your study of the material easier and more rewarding. This book is meant to be a "machine for learning," that can help you focus your efforts and get the most from the time and energy you invest.

About the authors

Michael Sullivan has taught College Algebra courses for over thirty years. He is also the father of four college graduates, including this text's co-author, all of whom called home, from time to time, frustrated and with questions. He knows what you're going through. Michael Sullivan, III, as a fairly recent graduate, experienced, first-hand, learning math using graphing calculators and computer algebra systems. He recognizes that, while this technology has many benefits, it can only be used effectively with an understanding of the underlying mathematics. Having earned advanced degrees in both economics and mathematics, he sees the importance of mathematics in applications.

So we, father and son, have written a text that doesn't overwhelm, or unnecessarily complicate, College Algebra, but at the same time provides you with the skills and practice you need to be successful. Please do not hesitate to contact us through Prentice Hall with any suggestions or comments that would improve this text.

Best Wishes!

Michael Sullivan

Michael Sullivan, III

OVERVIEW

CHAPTER OPENERS AND CHAPTER PROJECTS

Chapter openers use current articles to set up **Chapter Projects.** Many of the concepts that you encounter in this course relate directly to today's headlines and issues. These chapter projects are designed to give you a chance to use math to better understand the world. The list of topics for review will help you in two major ways, ...First, it allows you to review basic concepts immediately before using them in context. Second, it emphasizes the natural building of mathematical concepts throughout the course.

CHAPTER 3

Functions and Their Graphs

Wednesday February 10, 1999
***The Oregonian* "Ship awaits salvage effort"**

COOS BAY—Cleanup crews combed the oil-scarred south coast Tuesday as authorities raced to finish plans to refloat a 639-foot cargo ship mired for six days 150 yards off one of Oregon's most biologically rich beaches.

All day Tuesday, streaks of oil oozed from the cracked hull of the bulk cargo carrier *New Carissa* and spread over six miles of beach. Despite the breached hull, authorities believed they had a better chance of pulling the stricken ship out of beach sands than pumping nearly 400,000 gallons of oil off the ship in the winter surf.

See Chapter Project 1.

Preparing for This Chapter

Before getting started on this chapter, review the following concepts:

- **Slope of a Line (pp. 67–68)**
- **Intervals (pp. 54–55)**
- **Tests for Symmetry of an Equation (p. 23)**
- **Graphs of Certain Equations (Example 2, p. 16; Example 10, p. 23; Example 12, p. 25)**
- **Domain of a Function (pp. 99 and 104)**

Outline

CHAPTER PROJECTS

1. **Oil Spill** An oil tanker strikes a sand bar that rips a hole in the hull of the ship. Oil begins leaking out of the tanker with the spilled oil forming a circle around the tanker. The radius of the circle is increasing at the rate of 2.2 feet per hour.
   (a) Write the area of the circle as a function of the radius, r.
   (b) Write the radius of the circle as a function of time, t.
   (c) What is the radius of the circle after 2 hours? What is the radius of the circle after 2.5 hours?
   (d) Use the result of part (c) to determine the area of the circle after 2 hours and 2.5 hours.
   (e) Determine a function that represents area as a function of time, t.
   (f) Use the result of part (e) to determine the area of the circle after 2 hours and 2.5 hours.
   (g) Compute the average rate of change of the area of the circle from 2 hours to 2.5 hours.
   (h) Compute the average rate of change of the area of the circle from 3 hours to 3.5 hours.
   (i) Based on the results obtained in parts (g) and (h), what is happening to the average rate of change of the area of the circle as time passes?
   (j) If the oil tanker is 150 yards from shore, when will the oil spill first reach the shoreline? (1 yard = 3 feet).
   (k) How long will it be until 6 miles of shoreline is contaminated with oil? (1 mile = 5280 feet)

2. **Economics** For a manufacturer of computer chips, it has been determined that the number Q of chips produced as a function of labor L, in hours, is given by $Q = 8L^{1/2}$. In addition, the cost of production as a function of the number of chips produced per month is $C(Q) = 0.002Q^3 + 0.5Q^2 - Q + 1000$.
   (a) Determine cost as a function of labor.
   (b) If the demand function for the chips is $Q(P) = 5000 - 20P$, where P is the

CLEAR WRITING STYLE

Sullivan's **accessible writing style** is apparent throughout, often utilizing various approaches to the same concept. An author who writes clearly makes potentially difficult concepts intuitive, making class time more productive.

Sometimes it is helpful to think of a function f as a machine that receives as input a number from the domain, manipulates it, and outputs the value. See Figure 3.

Figure 3

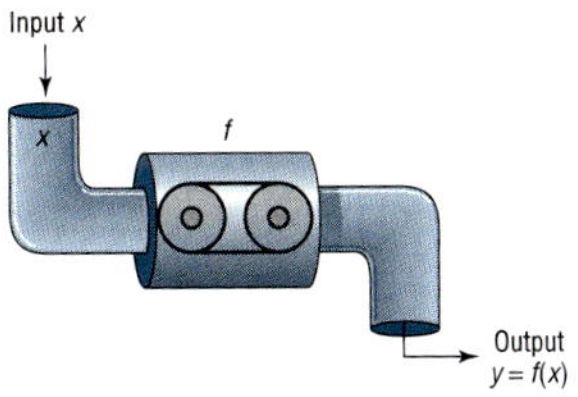

The restrictions on this input/output machine are

1. It only accepts numbers from the domain of the function.
2. For each input, there is exactly one output (which may be repeated for different inputs).

2.1 FUNCTIONS

1. Determine Whether a Relation Represents a Function
2. Find the Value of a Function
3. Find the Domain of a Function
4. Identify the Graph of a Function
5. Obtain Information from or about the Graph of a Function

In Section 1.1, we said that a **relation** is a correspondence between two variables, say x and y. When relations are written as ordered pairs (x, y), we say that x is related to y. Often, we are interested in specifying the type of relation (such as an equation) that might exist between the two variables.

For example, the relation between the revenue R resulting from the sale of x items selling for \$10 each may be expressed by the equation $R = 10x$. If we know how many items have been sold, then we can calculate the revenue by using the equation $R = 10x$. This equation is an example of a *function*.

As another example, suppose that an icicle falls off a building from a height of 64 feet above the ground. According to a law of physics, the distance

LEARNING OBJECTIVES

Begin each section by reading the **Learning Objectives.** These objectives will help you organize your studies and prepare for class. The Learning Objectives also tie the text to MathPro Explorer's Section Objectives.

STEP-BY-STEP EXAMPLES

Step-by-step examples insure that you follow the entire solution process and give you an opportunity to check your understanding of each step.

136 Chapter 2 Linear and Quadratic Functions

SUMMARY **Steps for Graphing a Quadratic Function $f(x) = ax^2 + bx + c, a \neq 0$ by hand**

STEP 1. Determine the vertex, $\left(\frac{-b}{2a}, f\left(\frac{-b}{2a}\right)\right)$.

STEP 2. Determine the axis of symmetry, $x = \frac{-b}{2a}$.

STEP 3. Determine the y-intercept, $f(0)$.

STEP 4. (a) If $b^2 - 4ac > 0$, then the graph of the quadratic function has two x-intercepts, which are found by solving $ax^2 + bx + c = 0$.
(b) If $b^2 - 4ac = 0$, the vertex is the x-intercept.
(c) If $b^2 - 4ac < 0$, there are no x-intercepts.

STEP 5. Determine an additional point if $b^2 - 4ac \leq 0$ by using the y-intercept and the axis of symmetry.

STEP 6. Plot the points and draw the graph.

Applications

Real-world problems may lead to mathematical problems that involve quadratic functions. Two such problems follow.

REAL-WORLD DATA

Real-world data is incorporated into examples and exercise sets to emphasize that mathematics is a tool used to understand the world around us. As you use these problems and examples, you will see the relevance and utility of the skills being covered.

27. Federal Income Tax Two 1998 Tax Rate Schedules are given in the accompanying table. If x equals the amount on Form 1040, line 37, and y equals the tax due, construct a function f for each schedule.

1998 TAX RATE SCHEDULES

SCHEDULE X — IF YOUR FILING STATUS IS SINGLE				SCHEDULE Y-1 — USE IF YOUR FILING STATUS IS MARRIED FILING JOINTLY OR QUALIFYING WIDOW(ER)			
If the amount on Form 1040, line 37, is: Over—	But not over—	Enter on Form 1040, line 38	of the amount over—	If the amount on Form 1040, line 37, is: Over—	But not over—	Enter on Form 1040, line 38	of the amount over—
\$0	\$25,750	_______15%	\$0	\$0	\$43,050	_______15%	\$0
25,750	62,450	\$3,862.50 + 28%	25,750	43,050	104,050	\$6,457.50 + 28%	43,050
62,450	130,250	14,138.50 + 31%	62,450	104,050	158,550	23,537.50 + 31%	104,050
130,250	283,150	35,156.50 + 36%	130,250	158,550	283,150	40,432.50 + 36%	158,550
283,150	_____	90,200.50 + 39.6%	283,150	283,150	_____	85,288.50 + 39.6%	283,150

Solution We begin by choosing the placement of the coordinate axes so that the x-axis coincides with the road surface and the origin coincides with the center of the bridge. As a result, the twin towers will be vertical (height $746 - 220 = 526$ feet above the road) and located 2100 feet from the center. Also, the cable, which has the shape of a parabola, will extend from the towers, open up, and have its vertex at $(0, 0)$. As illustrated in Figure 33, the choice of placement of the axes enables us to identify the equation of the parabola as $y = ax^2$, $a > 0$. We can also see that the points $(-2100, 526)$ and $(2100, 526)$ are on the graph.

Figure 33

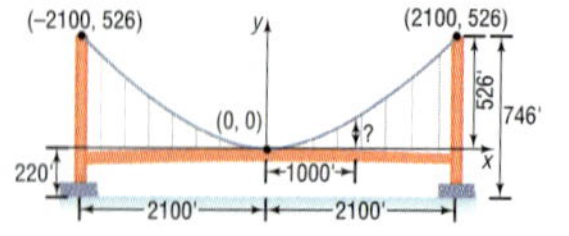

Based on these facts, we can find the value of a in $y = ax^2$.

$$y = ax^2$$

$$526 = a(2100)^2$$

$$a = \frac{526}{(2100)^2}$$

The equation of the parabola is therefore

$$y = \frac{526}{(2100)^2}x^2$$

The height of the cable when $x = 1000$ is

$$y = \frac{526}{(2100)^2}(1000)^2 \approx 119.3 \text{ feet}$$

Thus, the cable is 119.3 feet high at a distance of 1000 feet from the center of the bridge.

NOW WORK PROBLEM **31.**

Fitting a Quadratic Function to Data

3 In Section 2.2, we found the line of best fit for data that appeared to be linearly related. It was noted that data may also follow a nonlinear relation. Figures 34(a) and (b) show scatter diagrams of data that follow a quadratic relation.

Figure 34

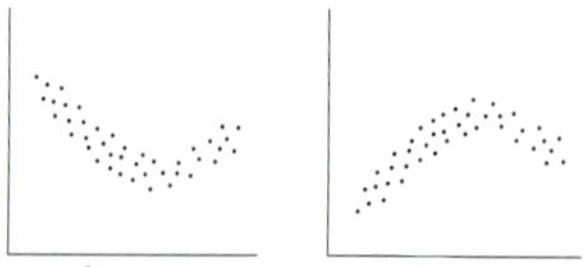

$y = ax^2 + bx + c,\ a > 0$ (a)

$y = ax^2 + bx + c,\ a < 0$ (b)

"NOW WORK" PROBLEMS

Many examples end with **"Now Work Problems."** The problems suggested here are similar to the corresponding examples and provide a great way to check your understanding as you work through the chapter. The solutions to all "Now Work" problems can be found in the back of the text as well as the *Student Solutions Manual.*

31. Suspension Bridge A suspension bridge with weight uniformly distributed along its length has twin towers that extend 75 meters above the road surface and are 400 meters apart. The cables are parabolic in shape and are suspended from the tops of the towers. The cables touch the road surface at the center of the bridge. Find the height of the cables at a point 100 meters from the center. (Assume that the road is level.)

PROCEDURES

Procedures, both algebraic and technical, are clearly expressed throughout the text.

EXAMPLE 3 Solving a Quadratic Equation by Graphing and by Using the Quadratic Formula

Find the real solutions, if any, of the equation $3x^2 - 5x + 1 = 0$.

Solution The equation is in standard form, so we compare it to $ax^2 + bx + c = 0$ to find a, b, and c.

$$3x^2 - 5x + 1 = 0$$
$$ax^2 + bx + c = 0$$

With $a = 3$, $b = -5$, and $c = 1$, we evaluate the discriminant $b^2 - 4ac$.

$$b^2 - 4ac = (-5)^2 - 4(3)(1) = 25 - 12 = 13$$

Since $b^2 - 4ac > 0$, there are two unequal real solutions.

Graphing Solution Figure 18 shows the graph of equation

$$Y_1 = 3x^2 - 5x + 1$$

As expected, we see that there are two x-intercepts: one between 0 and 1, the other between 1 and 2. Using Zero (or Root), twice, the solutions to the equation are 0.23 and 1.43 rounded to two decimal places.

Figure 18

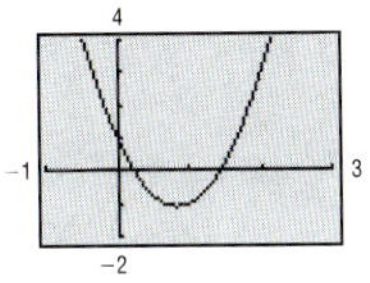

Algebraic Solution We use the quadratic formula with $a = 3$, $b = -5$, $c = 1$, and $b^2 - 4ac = 13$.

$$x = \frac{-b \pm \sqrt{b^2 - 4ac}}{2a} = \frac{5 \pm \sqrt{13}}{6}$$

The solution set is $\left\{\frac{5 - \sqrt{13}}{6}, \frac{5 + \sqrt{13}}{6}\right\}$. These solutions are exact.

NOW WORK PROBLEM 13.

3.1 EXERCISES

In Problems 1–8, match each graph to the function listed whose graph most resembles the one given.

A. Constant function *B. Linear function*
C. Square function *D. Cube function*
E. Square root function *F. Reciprocal function*
G. Absolute value function *H. Greatest-integer function*

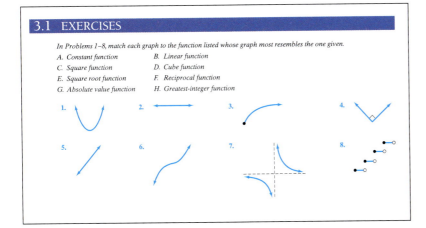

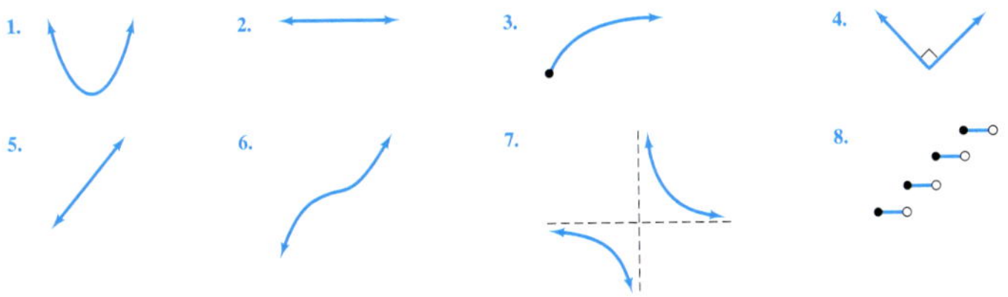

END-OF-SECTION EXERCISES

Sullivan's exercises are unparalleled in terms of thorough coverage and accuracy. Each **end-of-section exercise** set begins with visual and concept based problems, starting you out with the basics of the section. Well-thought-out exercises better prepare you for exams.

28. Employment and the Labor Force The following data represent the civilian labor force (people aged 16 years and older, excluding those serving in the military) and the number of employed people in the United States for the years 1981–1991. Treat the size of the labor force as the independent variable and the number employed as the dependent variable. Both the size of the labor force and the number employed are measured in thousands of people.

Year	Civilian Labor Force	Number Employed
1981	108,670	100,397
1982	110,204	99,526
1983	111,550	100,834
1984	113,544	105,005
1985	115,461	107,150
1986	117,834	109,597
1987	119,865	112,440
1988	121,669	114,968
1989	123,869	117,342
1990	124,787	117,914
1991	125,303	116,877

Source: Business Statistics, 1963–1991, U.S. Department of Commerce, Economics and Statistics Administration, Bureau of Economic Analysis, June 1992.

(a) Use a graphing utility to draw a scatter diagram.
(b) Use a graphing utility to find the line of best fit to the data. Express the solution using function notation.
(c) Interpret the slope.
(d) Predict the number of employed people if the civilian labor force is 122,340 thousand people.

MODELING

Many examples and exercises connect real-world situations to mathematical concepts. Learning to work with **models** is a skill that transfers to many disciplines.

GRAPHING UTILITIES AND TECHNIQUES

Increase your understanding, visualize, discover, explore, and solve problems using a **graphing utility.** Sullivan uses the graphing utility to further your understanding of concepts not to circumvent essential math skills.

Table 8

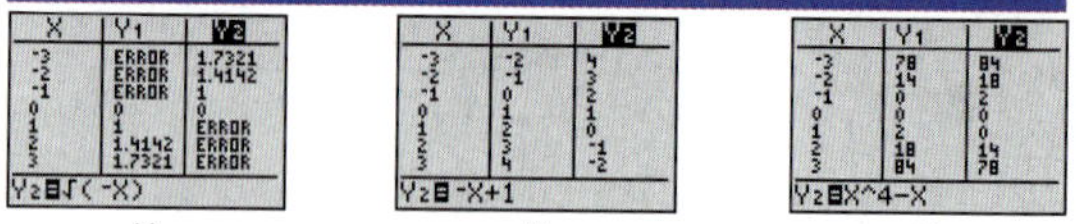

(a) (b) (c)

Figure 34

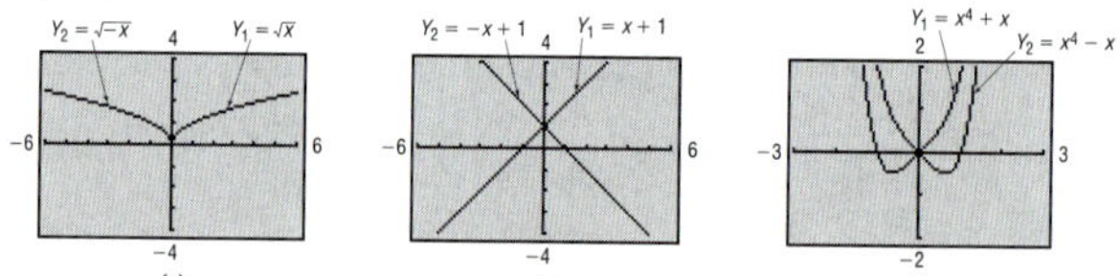

(a) (b) (c)

When the graph of the function $y = f(x)$ is known, the graph of the new function $y = f(-x)$ is the **reflection about the y-axis** of the graph of the function $y = f(x)$.

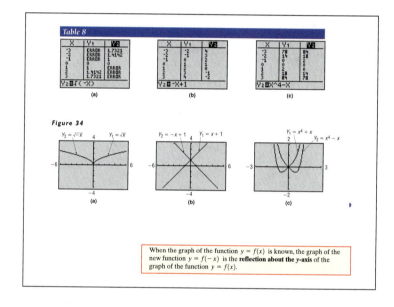

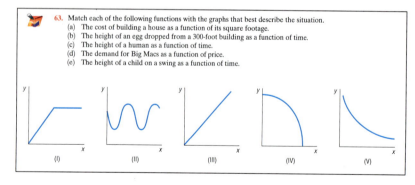

63. Match each of the following functions with the graphs that best describe the situation.
(a) The cost of building a house as a function of its square footage.
(b) The height of an egg dropped from a 300-foot building as a function of time.
(c) The height of a human as a function of time.
(d) The demand for Big Macs as a function of price.
(e) The height of a child on a swing as a function of time.

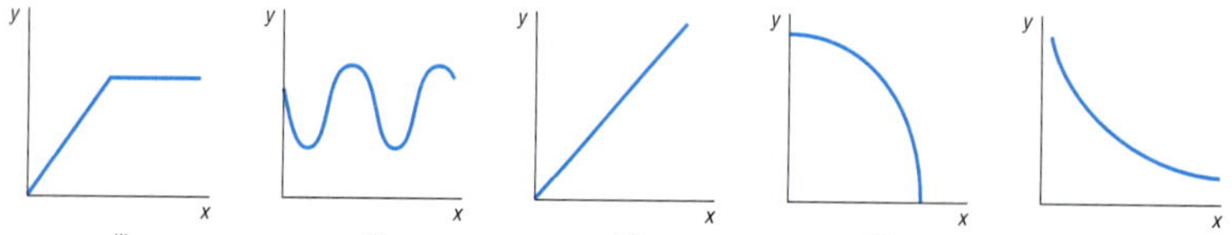

(I) (II) (III) (IV) (V)

DISCUSSION WRITING AND READING PROBLEMS

These **problems** are designed to get you to "think outside the box," therefore fostering an intuitive understanding of key mathematical concepts. In this example, matching the graph to the functions insures that you understand functions at a fundamental level.

CHAPTER REVIEW

LIBRARY OF FUNCTIONS

Linear function
$f(x) = mx + b$ Graph is a line with slope m and y-intercept b.

Constant function
$f(x) = b$ Graph is a horizontal line with y-intercept b (see Figure 14).

Identity function
$f(x) = x$ Graph is a line with slope 1 and y-intercept 0 (see Figure 15).

Square function
$f(x) = x^2$ Graph is a parabola with vortex at (0, 0) (see Figure 16).

Cube function
$f(x) = x^3$ See Figure 17.

Square root function
$f(x) = \sqrt{x}$ See Figure 18.

Reciprocal function
$f(x) = 1/x$ See Figure 19.

Absolute value function
$f(x) = |x|$ See Figure 20.

THINGS TO KNOW

Average rate of change of a function
The average rate of change of f from c to x is

$$\frac{\Delta y}{\Delta x} = \frac{f(x) - f(c)}{x - c}, \quad x \neq c$$

Increasing function
A function f is increasing on an open interval I if, for any choice of x_1 and x_2 in I, with $x_1 < x_2$, we have $f(x_1) < f(x_2)$.

Decreasing function
A function f is decreasing on an open interval I if, for any choice of x_1 and x_2 in I, with $x_1 < x_2$, we have $f(x_1) > f(x_2)$.

Constant function
A function f is constant on an open interval I if, for all choices of x in I, the values of $f(x)$ are equal.

Local maximum
A function f has a local maximum at c if there is an interval I containing c so that, for all $x \neq c$ in I, $f(x) < f(c)$.

Local minimum
A function f has a local minimum at c if there is an interval I containing c so that, for all $x \neq c$ in I, $f(x) > f(c)$.

Even function f
$f(-x) = f(x)$ for every x in the domain ($-x$ must also be in the domain).

Odd function f
$f(-x) = -f(x)$ for every x in the domain ($-x$ must also be in the domain).

HOW TO

Find the average rate of change of a function

Determine algebraically whether a function is even or odd without graphing it

Use a graphing utility to find the local maxima and local minima of a function

Use a graphing utility to determine where a function is increasing or decreasing

Graph certain functions by shifting, compressing, stretching, and/or reflecting (see Table 9)

Find the composite of two functions. Perform operation on functions

Construct functions in applications, including piecewise-defined functions

FILL-IN-THE-BLANK ITEMS

1. The average rate of change of a function equals the _______________ of the secant line containing two points on its graph.
2. A(n) _______________ function f is one for which $f(-x) = f(x)$ for every x in the domain of f; a(n) _______________ function f is one for which $f(-x) = -f(x)$ for every x in the domain of f.
3. Suppose that the graph of a function f is known. Then the graph of $y = f(x - 2)$ may be obtained by a(n) _______________ shift of the graph of f to the _______________ a distance of 2 units.
4. If $f(x) = x + 1$ and $g(x) = x^3$, then _______________ $= (x + 1)^3$.
5. For the two functions f and g, $(g \circ f)(x) =$ _______________.

TRUE/FALSE ITEMS

T F 1. A function f is decreasing on an open interval I if, for any choice of x_1 and x_2 in I, with $x_1 < x_2$, we have $f(x_1) < f(x_2)$.

T F 2. Even functions have graphs that are symmetric with respect to the origin.

T F 3. The graph of $y = f(-x)$ is the reflection about the y-axis of the graph of $y = f(x)$.

T F 4. $f(g(x)) = f(x) \cdot g(x)$.

T F 5. The domain of the composite function $(f \circ g)(x)$ is the same as that of $g(x)$.

REVIEW EXERCISES

Blue problem numbers indicate the authors' suggestions for use in a Practice Test.

In Problems 1 and 2, use the graph of the function f shown on page 210 to find

(a) The domain and range of f

(b) The intervals on which f is increasing, decreasing, or constant

(c) Whether it is even, odd, or neither

(d) The intercepts, if any

CHAPTER REVIEW

The **Chapter Review** helps check your understanding of the chapter materials in several ways. **"Things to Know"** gives a general overview of review topics. The **"How To"** section provides a concept-by-concept listing of the operations you are expected to perform. The **"Review Exercises"** then serve as a chance to practice the concepts presented within the chapter. The review materials are designed to make you, the student, confident in knowing the chapter material.

MULTIMEDIA

Sullivan M@thP@k

An Integrated Learning Environment

Today's textbooks offer a wide variety of ancillary materials to students, from solutions manuals to tutorial software to text-specific Websites. Making the most of all of these resources can be difficult. Sullivan **M@thP@k** helps students get it together. **M@thP@k** seamlessly integrates the following key products into an **integrated learning environment.**

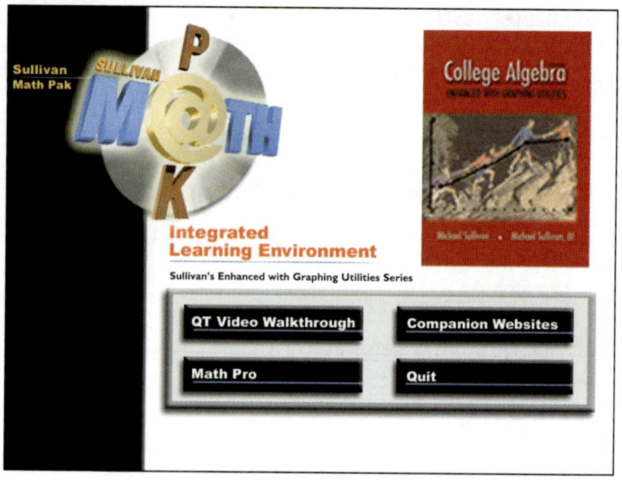

MathPro Explorer 4.0

This hands-on tutorial software reinforces the material learned in class. More than 100 video clips, unlimited practice problems, and interactive step-by-step examples correspond directly to the text's sections and learning objectives. The exploratory component allows students to experiment with mathematical principles on their own in addition to the prescribed exercises.

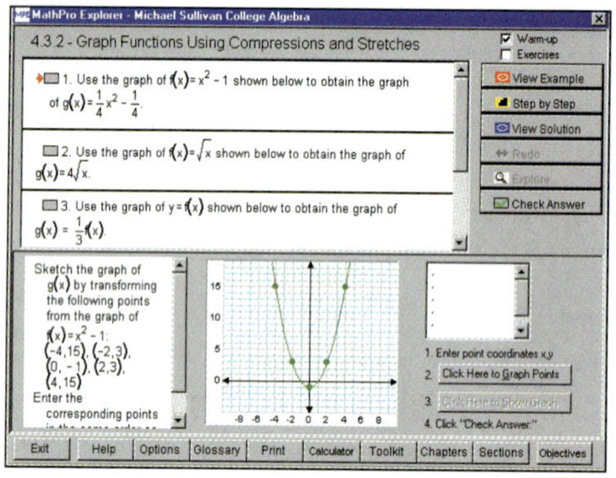

The Sullivan M@thP@k Website

This robust pass-code protected site features quizzes, homework starters, live animated examples, graphing calculator manuals, and much more. It offers the student many, many ways to test and reinforce their understanding of the course material.

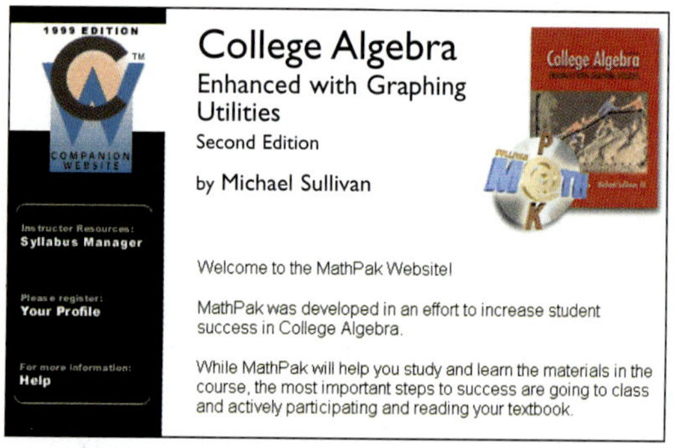

Student Solutions Manual

Written by Michael Sullivan III, co-author of both Sullivan series, and by Katy Murphy, the *Student Solutions Manual* offers thorough solutions that are consistent with the precise mathematics found in the text.

Sullivan M@thP@k.

Helping students *Get it Together.*

ADDITIONAL MEDIA

Sullivan Companion Website

www.prenhall.com/sullivan

This text-specific website beautifully complements the text. Here students can find chapter tests, section-specific links, and PowerPoint downloads in addition to other helpful features.

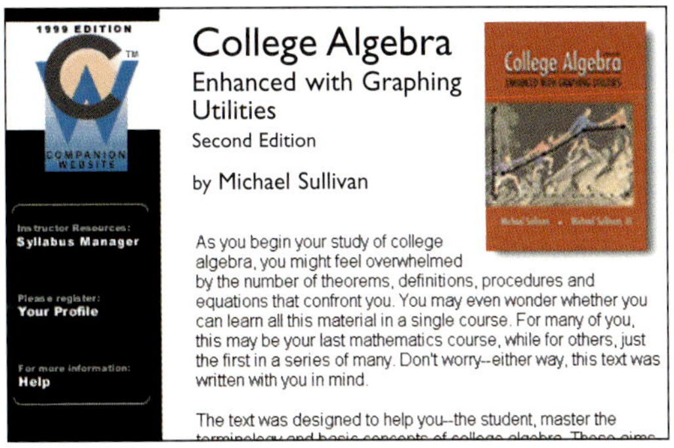

MathPro 4.0 (network version)

This networkable version of MathPro is *free* to adopters.

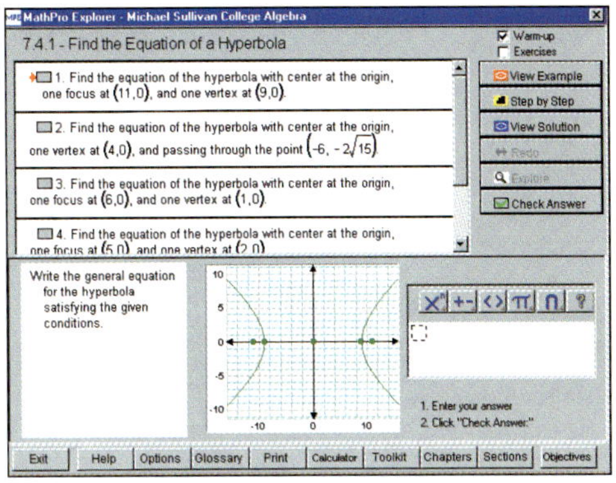

TestPro 4.0

This algorithmically generated testing software features an equation editor and a graphing utility, and offers online testing capabilities. Windows/Macintosh CD-ROM.

SUPPLEMENTS

Student Supplements

Student Solutions Manual

Worked solutions to all odd-numbered exercises from the text and complete solutions for chapter review problems and chapter tests. ISBN: 0-13-085306-2

Lecture Videos

The instructional tapes, in a lecture format, feature worked-out examples and exercises, taken from each section of the text. ISBN: 0-13-040311-3

New York Times Themes of the Times

A *free* newspaper from Prentice Hall and *The New York Times*. Interesting and current articles on mathematics which invite discussion and writing about mathematics.

Mathematics on the Internet

Free guide providing a brief history of the Internet, discussing the use of the World Wide Web, and describing how to find your way within the Internet and how to find others on it.

Instructor Supplements

Instructor's Edition

Includes answers to all exercises at the end of the book. ISBN: 0-13-085303-8

Instructor's Resource Manual

Contains complete step-by-step worked-out solutions to all even-numbered exercises in the textbook. Also included are strategies for using the Collaborative Learning projects found in each chapter. ISBN: 0-13-085307-0

Test Item File

Hard copy of the algorithmic computerized testing materials. ISBN: 0-13-085305-4

LIST OF APPLICATIONS

Economics. *See also* Business; Finance

Education

Electricity

Energy

Engineering. *See also* Architecture

Environment

Finance. *See also* Business; Economics

Geology

Geometry

Marketing. *See also* Business

Medicine

PHOTO CREDITS

CHAPTER 1

Pages 1; 95	Mother and teen daughter check new truck prices	Michelle Bridwell/PhotoEdit

CHAPTER 2

Pages 97; 154	New York Stock Exchange, Blue Monday, 10/27/97	Najlah Feanny/Stock Boston

CHAPTER 3

Pages 155; 212	Grounded freighter *New Carissa* off Coos Bay, Oregon 2/12/99	Reuters/Anthony Bolante/ Archive Photos

CHAPTER 4

Pages 213; 307	School overcrowding in Brooklyn, New York	Ferry/Liaison Agency, Inc.
Page 252	Woodcut of Niccolo Tartaglia	Corbis

CHAPTER 5

Pages 309; 398	Exterior of McDonald's restaurant	Amy C. Etra/PhotoEdit
Page 354	John Napier	The Granger Collection, New York

CHAPTER 6

Pages 401; 493	New York University outdoor graduation	Elena Rooraid/PhotoEdit
Page 461	Arthur Cayley	Corbis

CHAPTER 7

Pages 495; 546	Earth from space, showing North and South America	NASA/GFSC/Tom Stack and Associates
Page 526	Leonardo Fibonacci	The Granger Collection, New York
Page 541	Omar Khayyam, scene from Rubayat	Corbis

CHAPTER 8

Pages 549; 601	Smiling mother hugging her two sons	Myrleen Ferguson/PhotoEdit
Page 576	Blaise Pascal	The Granger Collection, New York

CHAPTER 9

Pages 603; 659	Planet Pluto with its moon Charon	Corbis
Page 652	Illustration of Etienne Bezout	Corbis

APPENDIX

Page 699	Frans Hals "Portrait of Rene Descartes"	Giraudon/Art Resource, N.Y.

CHAPTER

1 Graphs

Big Deals!

Advertisements for cars are bewildering. Is the lowest rate the best rate? Is the longest term the best? What is the 'best' deal?

See Chapter Project 1.

Preparing for This Chapter

Before getting started on this chapter, review the following:

- Real Numbers (Section 1, Appendix)
- Algebra Review (Section 2, Appendix)
- Geometry Review (Section 3, Appendix)

Outline

The idea of using a system of rectangular coordinates dates back to ancient times, when such a system was used for surveying and city planning. Apollonius of Perga, in 200 B.C., used a form of rectangular coordinates in his work on conics, although this use does not stand out as clearly as it does in modern treatments. Sporadic use of rectangular coordinates continued until the 1600s. By that time, algebra had developed sufficiently so that René Descartes (1596–1650) and Pierre de Fermat (1601–1665) could take the crucial step, which was the use of rectangular coordinates to translate geometry problems into algebra problems, and vice versa. This step was supremely important for two reasons. First, it allowed both geometers and algebraists to gain critical new insights into their subjects, which previously had been regarded as separate but now were seen to be connected in many important ways. Second, the insights gained made possible the development of calculus, which greatly enlarged the number of areas in which mathematics could be applied and made possible a much deeper understanding of these areas. With the advent of technology, in particular graphing utilities, we are now able not only to visualize the dual roles of algebra and geometry, but we are also able to solve many problems that required advanced methods before this technology.

1.1 RECTANGULAR COORDINATES; GRAPHING UTILITIES; SCATTER DIAGRAMS

1 Use the Distance Formula

2 Use the Midpoint Formula

3 Draw and Interpret Scatter Diagrams

We locate a point on the real number line* by assigning it a single real number, called the *coordinate of the point.* For work in a two-dimensional plane, we locate points by using two numbers.

We begin with two real number lines located in the same plane: one horizontal and the other vertical. We call the horizontal line the ***x*-axis,** the vertical line the ***y*-axis,** and the point of intersection the **origin *O*.** We assign coordinates to every point on these number lines as shown in Figure 1, using a convenient scale. In mathematics, we usually use the same scale on each axis; in applications, a different scale is often used on each axis.

Figure 1

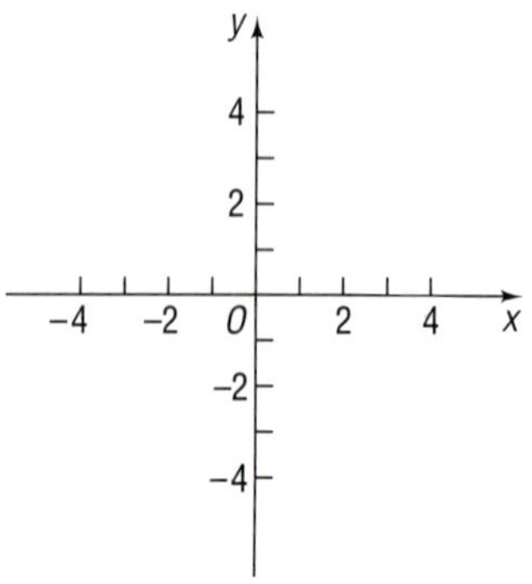

The origin O has a value of 0 on both the x-axis and the y-axis. We follow the usual convention that points on the x-axis to the right of O are associated with positive real numbers, and those to the left of O are associated with negative real numbers. Those on the y-axis above O are associated with positive real numbers, and those below O are associated with negative real numbers. In Figure 1, the x-axis and y-axis are labeled as x and y, respectively, and we have used an arrow at the end of each axis to denote the positive direction.

The coordinate system described here is called a **rectangular** or **Cartesian† coordinate system.** The plane formed by the x-axis and y-axis is some-

*Real numbers are discussed in Section 1 of the Appendix; The real number line is discussed in Section 2.

†Named after René Descartes (1596–1650), a French mathematician, philosopher, and theologian.

Figure 2

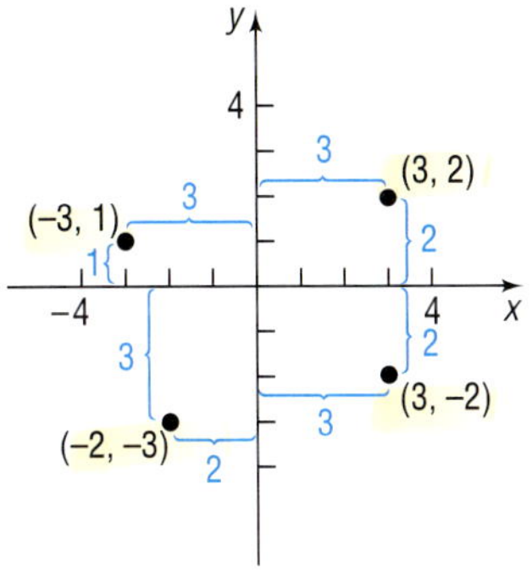

times called the ***xy*-plane,** and the x-axis and y-axis are referred to as the **coordinate axes.**

Any point P in the xy-plane can then be located by using an **ordered pair** (x, y) of real numbers. Let x denote the signed distance of P from the y-axis (*signed* in the sense that, if P is to the right of the y-axis, then $x > 0$, and if P is to the left of the y-axis, then $x < 0$); and let y denote the signed distance of P from the x-axis. The ordered pair (x, y), also called the **coordinates** of P, then gives us enough information to locate the point P in the plane.

For example, to locate the point whose coordinates are $(-3, 1)$, go 3 units along the x-axis to the left of O and then go straight up 1 unit. We **plot** this point by placing a dot at this location. See Figure 2, in which the points with coordinates $(-3, 1)$, $(-2, -3)$, $(3, -2)$, and $(3, 2)$ are plotted.

Figure 3

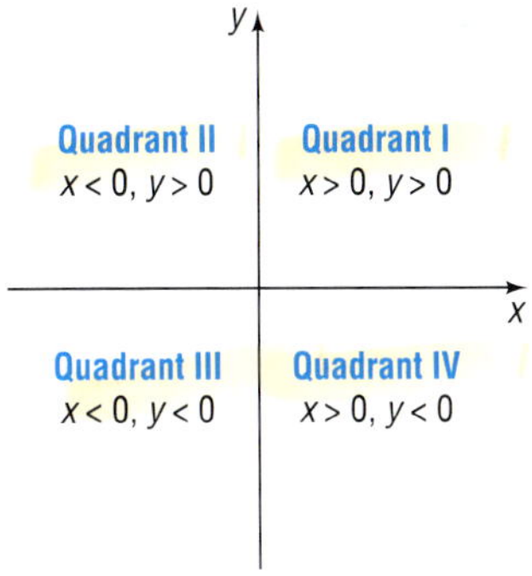

The origin has coordinates $(0, 0)$. Any point on the x-axis has coordinates of the form $(x, 0)$, and any point on the y-axis has coordinates of the form $(0, y)$.

If (x, y) are the coordinates of a point P, then x is called the ***x*-coordinate,** or **abscissa,** of P and y is the ***y*-coordinate,** or **ordinate,** of P. We identify the point P by its coordinates (x, y) by writing $P = (x, y)$. Usually, we will simply say "the point (x, y)" rather than "the point whose coordinates are (x, y)."

The coordinate axes divide the xy-plane into four sections called **quadrants,** as shown in Figure 3. In quadrant I, both the x-coordinate and the y-coordinate of all points are positive; in quadrant II, x is negative and y is positive; in quadrant III, both x and y are negative; and in quadrant IV, x is positive and y is negative. Points on the coordinate axes belong to no quadrant.

NOW WORK PROBLEM 1.

Graphing Utilities

Figure 4

$y = 2x$

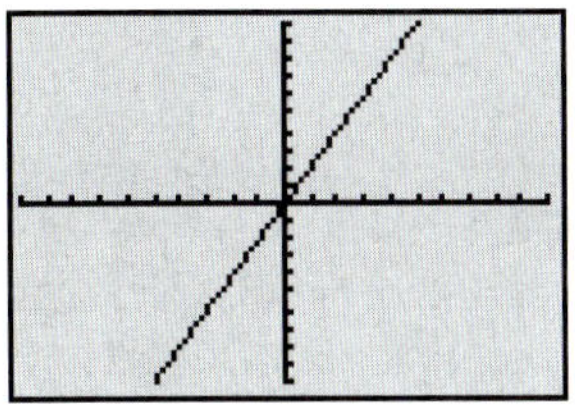

All graphing utilities, that is, all graphing calculators and all computer software graphing packages, graph equations by plotting points on a screen. The screen itself actually consists of small rectangles, called **pixels.** The more pixels the screen has, the better the resolution. Most graphing calculators have 48 pixels per square inch; most computer screens have 32 to 108 pixels per square inch. When a point to be plotted lies inside a pixel, the pixel is turned on (lights up). Thus, the graph of an equation is a collection of pixels. Figure 4 shows how the graph of $y = 2x$ looks on a TI-83 graphing calculator.

The screen of a graphing utility will display the coordinate axes of a rectangular coordinate system. However, you must set the scale on each axis. You must also include the smallest and largest values of x and y that you want included in the graph. This is called **setting the viewing window.**

Figure 5

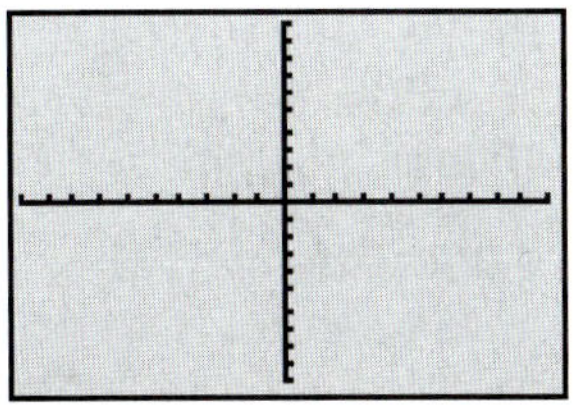

Figure 5 illustrates a typical viewing window.

To select the viewing window, we must give values to the following expressions:

*X*min:	the smallest value of x
*X*max:	the largest value of x
*X*scl:	the number of units per tick mark on the x-axis
*Y*min:	the smallest value of y
*Y*max:	the largest value of y
*Y*scl:	the number of units per tick mark on the y-axis

Figure 6 illustrates these settings and their relation to the Cartesian coordinate system.

Figure 6

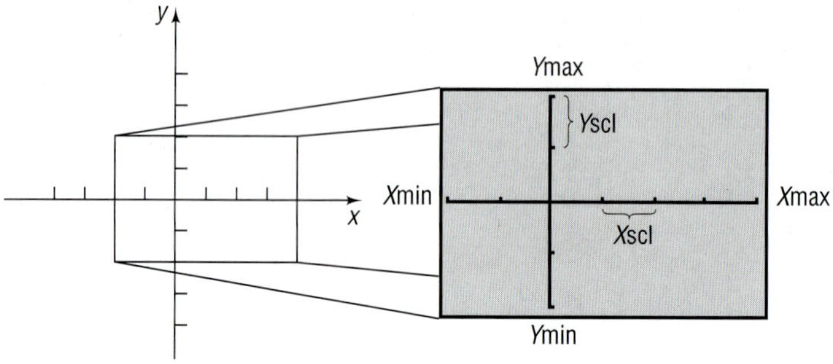

If the scale used on each axis is known, we can determine the minimum and maximum values of x and y shown on the screen by counting the tick marks. Look again at Figure 5. For a scale of 1 on each axis, the minimum and maximum values of x are -10 and 10, respectively; the minimum and maximum values of y are also -10 and 10. If the scale is 2 on each axis, then the minimum and maximum values of x are -20 and 20, respectively; and the minimum and maximum values of y are -20 and 20, respectively.

Conversely, if we know the minimum and maximum values of x and y, we can determine the scales being used by counting the tick marks displayed. We shall follow the practice of showing the minimum and maximum values of x and y in our illustrations so that you will know how the window was set. See Figure 7.

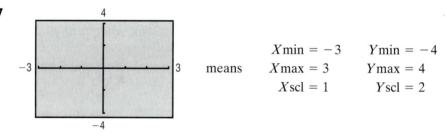

Figure 7

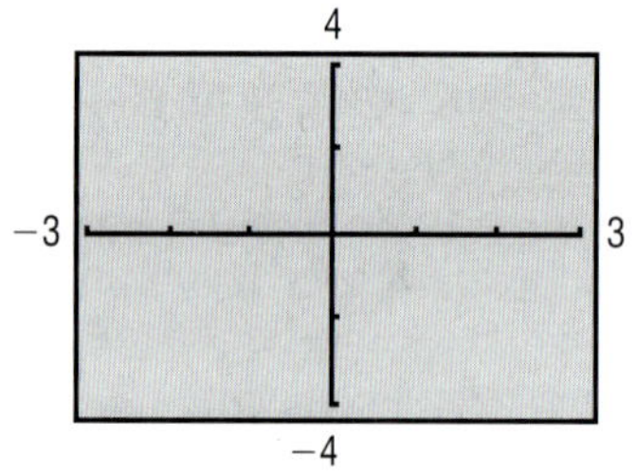

means

$X\text{min} = -3 \qquad Y\text{min} = -4$

$X\text{max} = 3 \qquad Y\text{max} = 4$

$X\text{scl} = 1 \qquad Y\text{scl} = 2$

EXAMPLE 1 Finding the Coordinates of a Point Shown on a Graphing Utility Screen

Find the coordinates of the point shown in Figure 8. Assume the coordinates are integers.

Figure 8

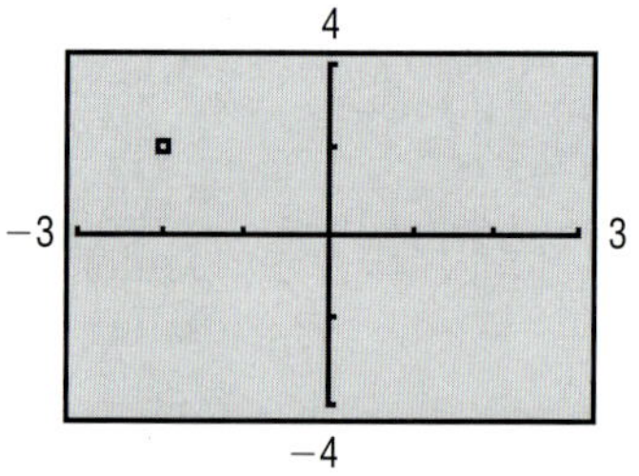

Solution First we note that the viewing window used in Figure 8 is

$$X\text{min} = -3 \qquad Y\text{min} = -4$$
$$X\text{max} = 3 \qquad Y\text{max} = 4$$
$$X\text{scl} = 1 \qquad Y\text{scl} = 2$$

The point shown is 2 tick units to the left on the horizontal axis (scale $= 1$) and 1 tick up on the vertical scale (scale $= 2$). Thus, the coordinates of the point shown are $(-2, 2)$.

NOW WORK PROBLEMS 5 AND 15.

Distance between Points

1 If the same units of measurement, such as inches, centimeters, and so on, are used for both the x-axis and the y-axis, then all distances in the xy-plane can be measured using this unit of measurement.

EXAMPLE 2 Finding the Distance between Two Points

Find the distance d between the points $(1, 3)$ and $(5, 6)$.

Solution First we plot the points $(1, 3)$ and $(5, 6)$ as shown in Figure 9(a). Then we draw a horizontal line from $(1, 3)$ to $(5, 3)$ and a vertical line from $(5, 3)$ to $(5, 6)$, forming a right triangle, as in Figure 9(b). One leg of the triangle is of length 4 and the other is of length 3. By the Pythagorean Theorem,* the square of the distance d that we seek is

$$d^2 = 4^2 + 3^2 = 16 + 9 = 25$$
$$d = 5$$

Figure 9

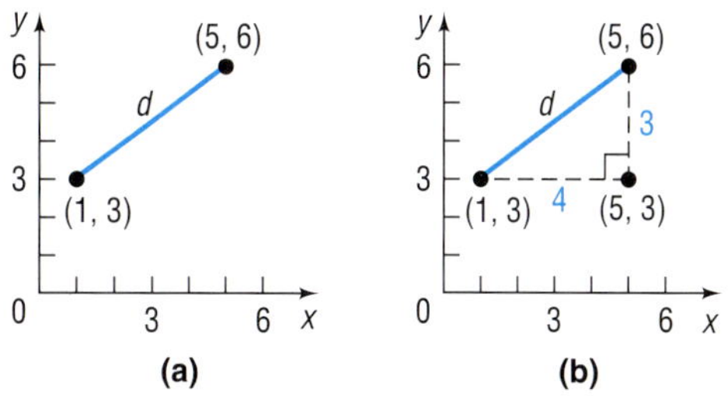

The **distance formula** provides a straightforward method for computing the distance between two points.

THEOREM **Distance Formula**

The distance between two points $P_1 = (x_1, y_1)$ and $P_2 = (x_2, y_2)$, denoted by $d(P_1, P_2)$, is

$$d(P_1, P_2) = \sqrt{(x_2 - x_1)^2 + (y_2 - y_1)^2} \qquad (1)$$

*The Pythagorean Theorem is discussed in Section 3 of the Appendix.

That is, to compute the distance between two points, find the difference of the x-coordinates, square it, and add this to the square of the difference of the y-coordinates. The square root of this sum is the distance. See Figure 10.

Figure 10

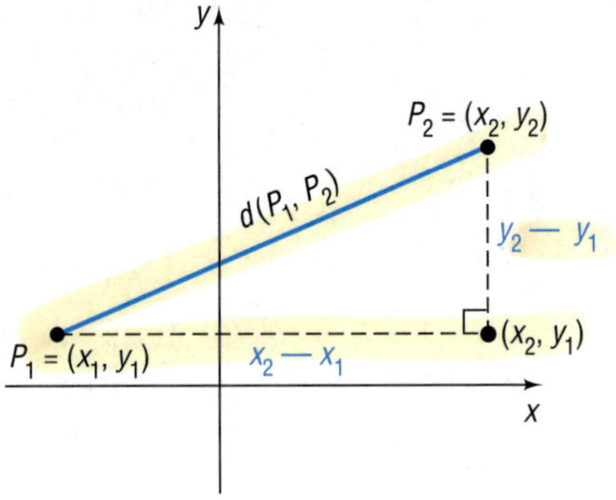

Proof of the Distance Formula Let (x_1, y_1) denote the coordinates of point P_1, and let (x_2, y_2) denote the coordinates of point P_2. Assume that the line joining P_1 and P_2 is neither horizontal nor vertical. Refer to Figure 11(a). The coordinates of P_3 are (x_2, y_1). The horizontal distance from P_1 to P_3 is the absolute value of the difference of the x-coordinates, $|x_2 - x_1|$. The vertical distance from P_3 to P_2 is the absolute value of the difference of the y-coordinates, $|y_2 - y_1|$. See Figure 11(b). The distance $d(P_1, P_2)$ that we seek is the length of the hypotenuse of the right triangle, so, by the Pythagorean Theorem, it follows that

$$\begin{aligned}[d(P_1, P_2)]^2 &= |x_2 - x_1|^2 + |y_2 - y_1|^2 \\ &= (x_2 - x_1)^2 + (y_2 - y_1)^2 \\ d(P_1, P_2) &= \sqrt{(x_2 - x_1)^2 + (y_2 - y_1)^2}\end{aligned}$$

Figure 11

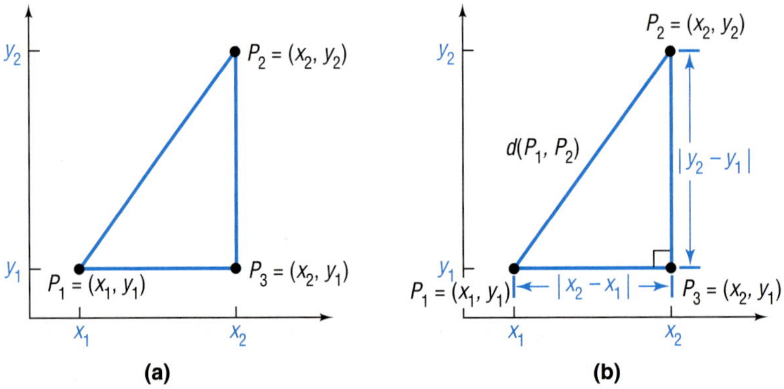

Now, if the line joining P_1 and P_2 is horizontal, then the y-coordinate of P_1 equals the y-coordinate of P_2; that is, $y_1 = y_2$. Refer to Figure 12(a). In this case, the distance formula (1) still works, because, for $y_1 = y_2$, it reduces to

$$d(P_1, P_2) = \sqrt{(x_2 - x_1)^2 + 0^2} = \sqrt{(x_2 - x_1)^2} = |x_2 - x_1|$$

A similar argument holds if the line joining P_1 and P_2 is vertical. See Figure 12(b). Thus, the distance formula is valid in all cases.

Figure 12

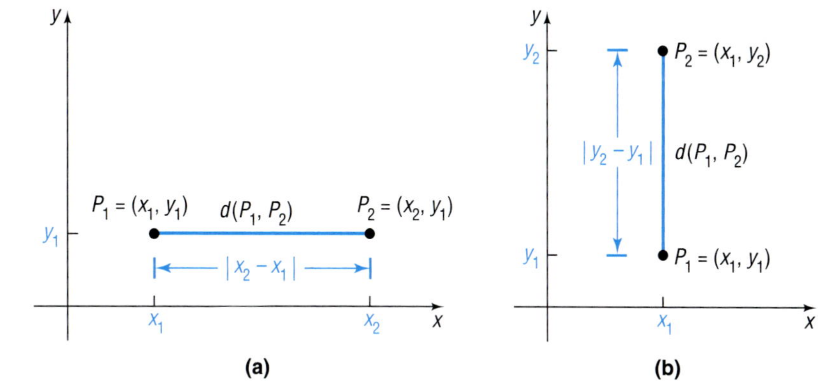

EXAMPLE 3 Finding the Length of a Line Segment

Find the length of the line segment shown in Figure 13.

Figure 13

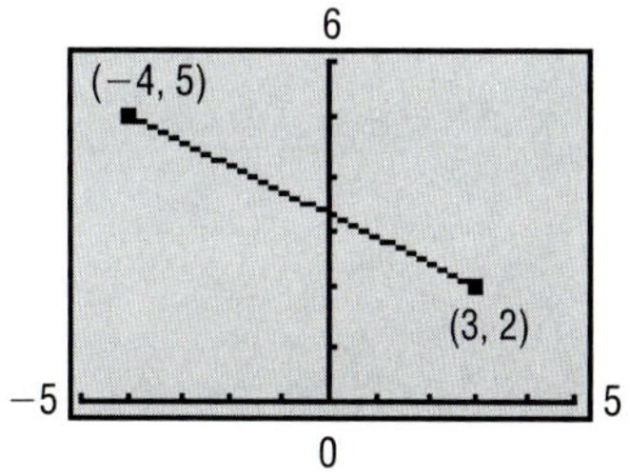

Solution The length of the line segment is the distance between the points $(-4, 5)$ and $(3, 2)$. Using the distance formula (1), the length d is

$$d = \sqrt{[3 - (-4)]^2 + (2 - 5)^2} = \sqrt{7^2 + (-3)^2}$$
$$= \sqrt{49 + 9} = \sqrt{58} \approx 7.62$$

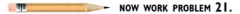

NOW WORK PROBLEM 21.

The distance between two points $P_1 = (x_1, y_1)$ and $P_2 = (x_2, y_2)$ is never a negative number. Furthermore, the distance between two points is 0 only when the points are identical, that is, when $x_1 = x_2$ and $y_1 = y_2$. Also, because $(x_2 - x_1)^2 = (x_1 - x_2)^2$ and $(y_2 - y_1)^2 = (y_1 - y_2)^2$, it makes no difference whether the distance is computed from P_1 to P_2 or from P_2 to P_1; that is, $d(P_1, P_2) = d(P_2, P_1)$.

The introduction to this chapter mentioned that rectangular coordinates enable us to translate geometry problems into algebra problems, and vice versa. The next example shows how algebra (the distance formula) can be used to solve geometry problems.

EXAMPLE 4 Using Algebra to Solve Geometry Problems

Consider the three points $A = (-2, 1)$, $B = (2, 3)$ and $C = (3, 1)$.

(a) Plot each point and form the triangle ABC.
(b) Find the length of each side of the triangle.
(c) Verify that the triangle is a right triangle.
(d) Find the area of the triangle.

Solution

Figure 14

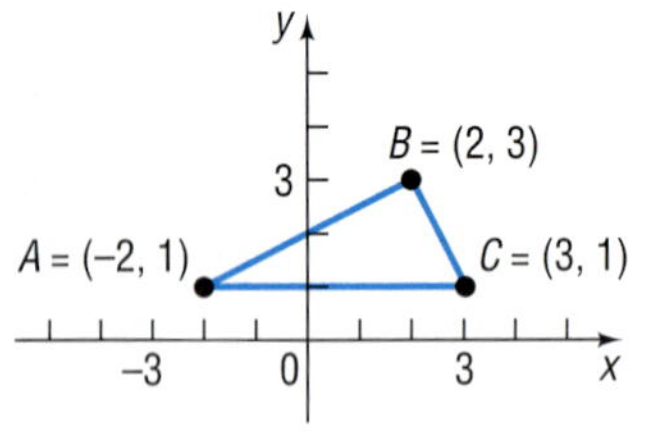

(a) Points A, B, C, and triangle ABC are plotted in Figure 14.

(b) $d(A, B) = \sqrt{[2 - (-2)]^2 + (3 - 1)^2} = \sqrt{16 + 4} = \sqrt{20} = 2\sqrt{5}$
$d(B, C) = \sqrt{(3 - 2)^2 + (1 - 3)^2} = \sqrt{1 + 4} = \sqrt{5}$
$d(A, C) = \sqrt{[3 - (-2)]^2 + (1 - 1)^2} = \sqrt{25 + 0} = 5$

(c) To show that the triangle is a right triangle, we need to show that the sum of the squares of the lengths of two of the sides equals the square of the length of the third side. (Why is this sufficient?) Looking at Figure 14 it seems reasonable to conjecture that the right angle is at vertex B. Thus, we shall check to see whether

$$[d(A, B)]^2 + [d(B, C)]^2 = [d(A, C)]^2$$

We find that

$$\begin{aligned}[d(A, B)]^2 + [d(B, C)]^2 &= (2\sqrt{5})^2 + (\sqrt{5})^2 \\ &= 20 + 5 = 25 = [d(A, C)]^2\end{aligned}$$

so it follows from the converse of the Pythagorean Theorem that triangle ABC is a right triangle.

(d) Because the right angle is at B, the sides AB and BC form the base and altitude of the triangle. Its area is

$$\text{Area} = \frac{1}{2}(\text{Base})(\text{Altitude}) = \frac{1}{2}(2\sqrt{5})(\sqrt{5}) = 5 \text{ square units}$$

NOW WORK PROBLEM **39.**

Midpoint Formula

Figure 15

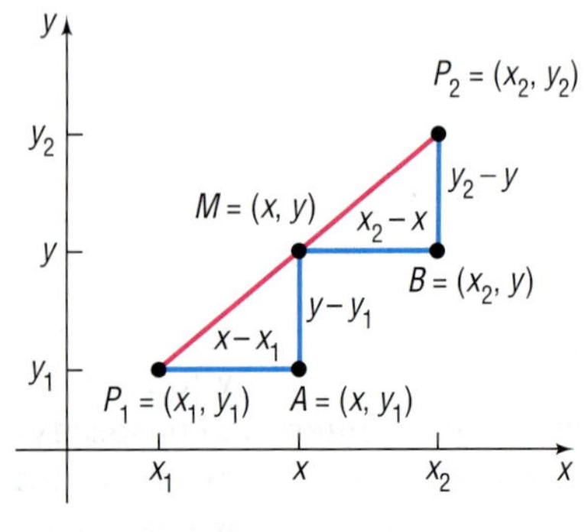

2 We now derive a formula for the coordinates of the **midpoint of a line segment.** Let $P_1 = (x_1, y_1)$ and $P_2 = (x_2, y_2)$ be the end points of a line segment, and let $M = (x, y)$ be the point on the line segment that is the same distance from P_1 as it is from P_2. See Figure 15. The triangles P_1AM and MBP_2 are congruent.* [Do you see why? Angle AP_1M = angle BMP_2,† angle P_1MA = angle MP_2B, and $d(P_1, M) = d(M, P_2)$ is given. So, we have angle–side–angle.] Hence, corresponding sides are equal in length. That is,

*The following statement is a postulate from geometry. Two triangles are congruent if their sides are the same length (SSS), or if two sides and the included angle are the same (SAS), or if two angles and the included sides are the same (ASA).

†Another postulate from geometry states that the transversal $\overline{P_1P_2}$ forms equal corresponding angles with the parallel lines $\overline{P_1A}$ and $\overline{MB}$.

$$x - x_1 = x_2 - x \quad \text{and} \quad y - y_1 = y_2 - y$$
$$2x = x_1 + x_2 \qquad 2y = y_1 + y_2$$
$$x = \frac{x_1 + x_2}{2} \qquad y = \frac{y_1 + y_2}{2}$$

THEOREM

Midpoint Formula

The midpoint (x, y) of the line segment from $P_1 = (x_1, y_1)$ to $P_2 = (x_2, y_2)$ is

$$(x, y) = \left(\frac{x_1 + x_2}{2}, \frac{y_1 + y_2}{2}\right) \tag{2}$$

Thus, to find the midpoint of a line segment, we average the x-coordinates and the y-coordinates of the end points.

EXAMPLE 5 Finding the Midpoint of a Line Segment

Find the midpoint of a line segment from $P_1 = (-5, 3)$ to $P_2 = (3, 1)$. Plot the points P_1 and P_2 and their midpoint. Check your answer.

Solution We apply the midpoint formula (2) using $x_1 = -5$, $x_2 = 3$, $y_1 = 3$, and $y_2 = 1$. Then the coordinates (x, y) of the midpoint M are

$$x = \frac{x_1 + x_2}{2} = \frac{-5 + 3}{2} = -1 \quad \text{and} \quad y = \frac{y_1 + y_2}{2} = \frac{3 + 1}{2} = 2$$

That is, $M = (-1, 2)$. See Figure 16.

Figure 16

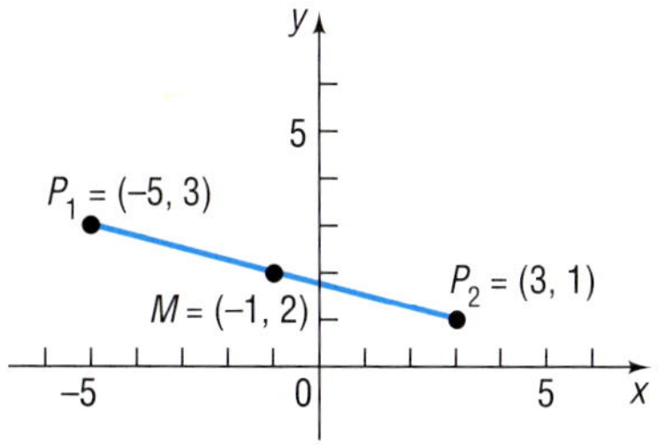

✓CHECK: Because M is the midpoint, we check the answer by verifying that $d(P_1, M) = d(M, P_2)$:

$$d(P_1, M) = \sqrt{[-1 - (-5)]^2 + (2 - 3)^2} = \sqrt{16 + 1} = \sqrt{17}$$
$$d(M, P_2) = \sqrt{[3 - (-1)]^2 + (1 - 2)^2} = \sqrt{16 + 1} = \sqrt{17}$$

NOW WORK PROBLEM **49.**

Scatter Diagrams

3 A **relation** is a correspondence between two variables, say, x and y, and can be written as a set of ordered pairs (x, y). When relations are written as ordered pairs, we say that x is related to y. Often, we are interested in specifying the type of relation (such as an equation) that might exist between two variables. The first step in finding this relation is to plot the ordered pairs using rectangular coordinates. The resulting graph is called a **scatter diagram.**

EXAMPLE 6 Drawing a Scatter Diagram

A farmer collected the following data, which shows crop yields for various amounts of fertilizer used.

Table 1

Plot	1	2	3	4	5	6	7	8	9	10	11	12
Fertilizer, x (Pounds/100 ft^2)	0	0	5	5	10	10	15	15	20	20	25	25
Yield, y (Bushels)	4	6	10	7	12	10	15	17	18	21	23	22
(x, y)	(0, 4)	(0, 6)	(5, 10)	(5, 7)	(10, 12)	(10, 10)	(15, 15)	(15, 17)	(20, 18)	(20, 21)	(25, 23)	(25, 22)

(a) Draw a scatter diagram by hand.

(b) Use a graphing utility to draw a scatter diagram.

(c) Describe what happens to the yield as the number of pounds of fertilizer increases.

Solution (a) To draw a scatter diagram by hand, we plot the ordered pairs listed in Table 1, with the number of pounds of fertilizer as the x-coordinate and the yield (in bushels) as the y-coordinate. See Figure 17(a). Notice that the points in a scatter diagram are not connected.

(b) Figure 17(b) shows a scatter diagram of the data using a graphing utility.

Figure 17

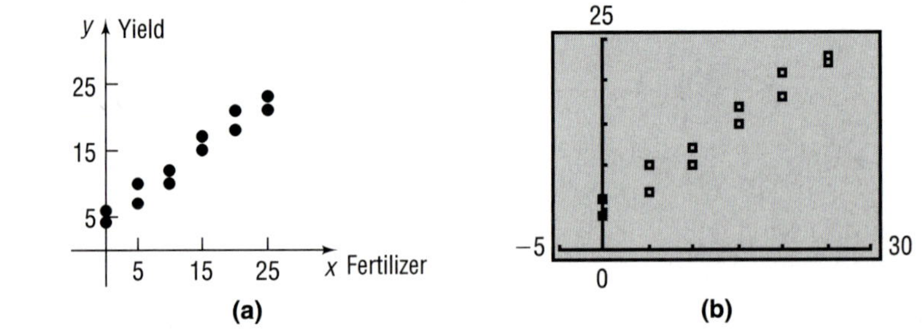

(c) From the scatter diagrams, as the number of pounds of fertilizer increases, we see that the yield also increases.

NOW WORK PROBLEM 65.

1.1 EXERCISES

In Problems 1 and 2, plot each point in the xy-plane. Tell in which quadrant or on what coordinate axis each point lies.

1. (a) $A = (-3, 2)$ (b) $B = (6, 0)$ (c) $C = (-2, -2)$
(d) $D = (6, 5)$ (e) $E = (0, -3)$ (f) $F = (6, -3)$

2. (a) $A = (1, 4)$ (b) $B = (-3, -4)$ (c) $C = (-3, 4)$
(d) $D = (4, 1)$ (e) $E = (0, 1)$ (f) $F = (-3, 0)$

3. Plot the points $(2, 0)$, $(2, -3)$, $(2, 4)$, $(2, 1)$, and $(2, -1)$. Describe the set of all points of the form $(2, y)$, where y is a real number.

4. Plot the points $(0, 3)$, $(1, 3)$, $(-2, 3)$, $(5, 3)$ and $(-4, 3)$. Describe the set of all points of the form $(x, 3)$, where x is a real number.

In Problems 5–8, determine the coordinates of the points shown. Tell in which quadrant each point lies. Assume the coordinates are integers.

5.

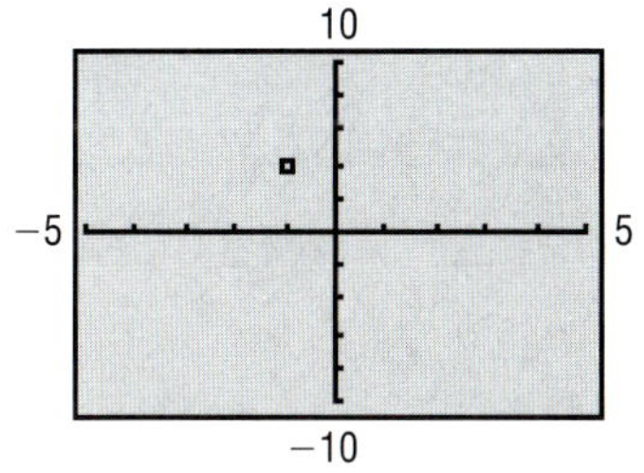

6.

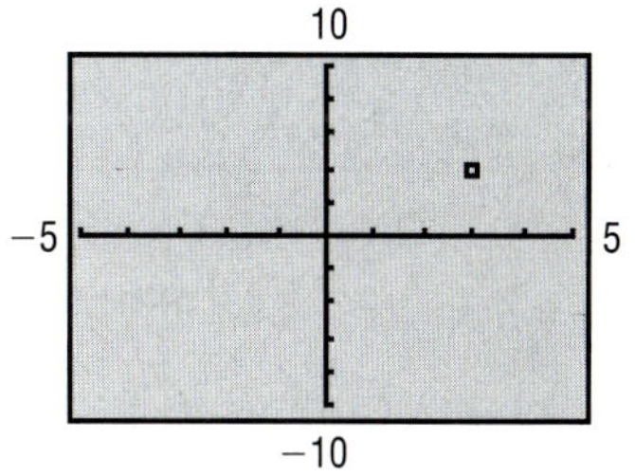

7.

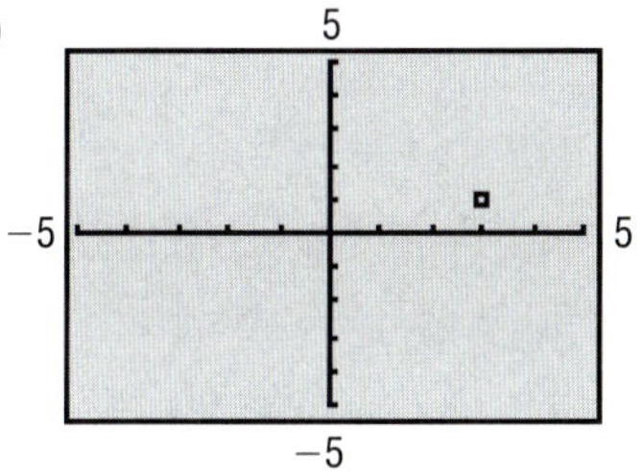

8.

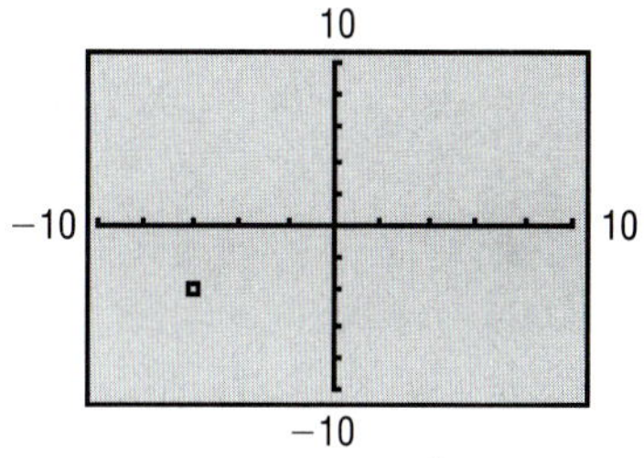

In Problems 9–14, select a setting so that each of the given points will lie within the viewing window.

9. $(-10, 5), (3, -2), (4, -1)$

10. $(5, 0), (6, 8), (-2, -3)$

11. $(40, 20), (-20, -80), (10, 40)$

12. $(-80, 60), (20, -30), (-20, -40)$

13. $(0, 0), (100, 5), (5, 150)$

14. $(0, -1), (100, 50), (-10, 30)$

In Problems 15–20, determine the viewing window used.

15.

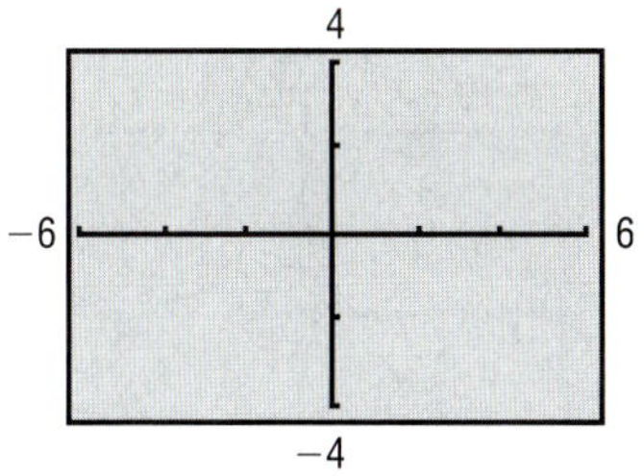

16.

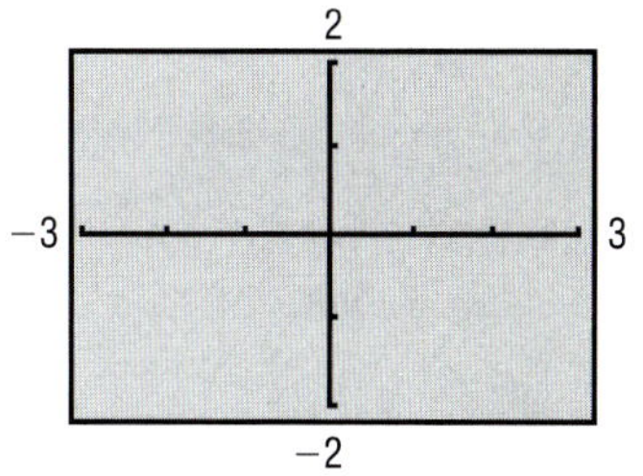

17.

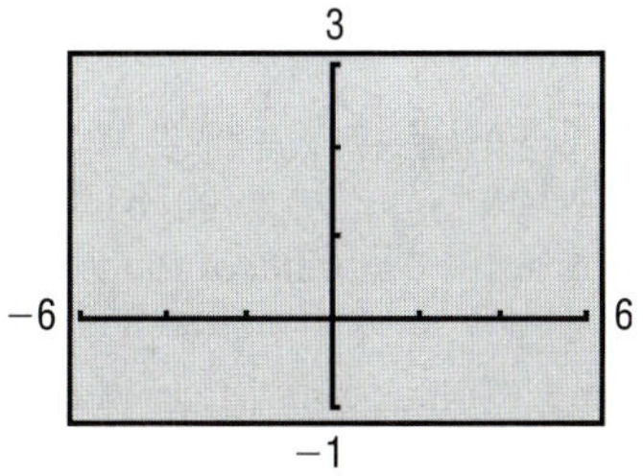

18.

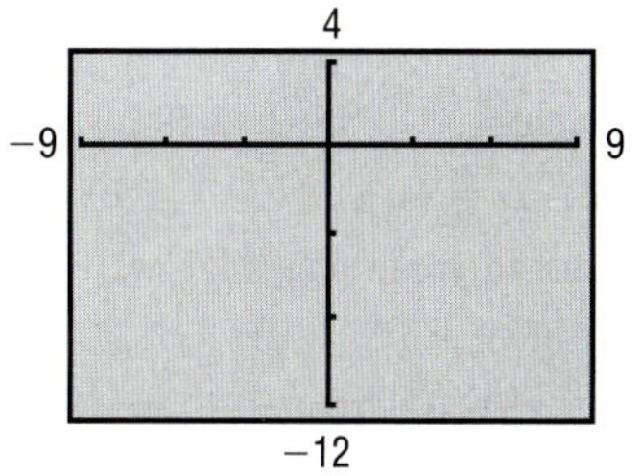

19.

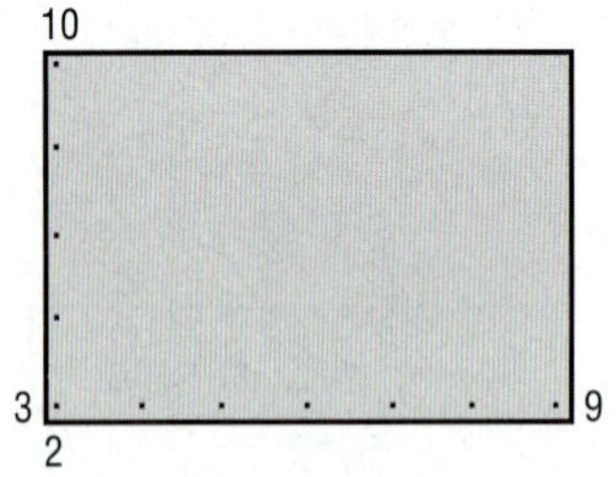

20.

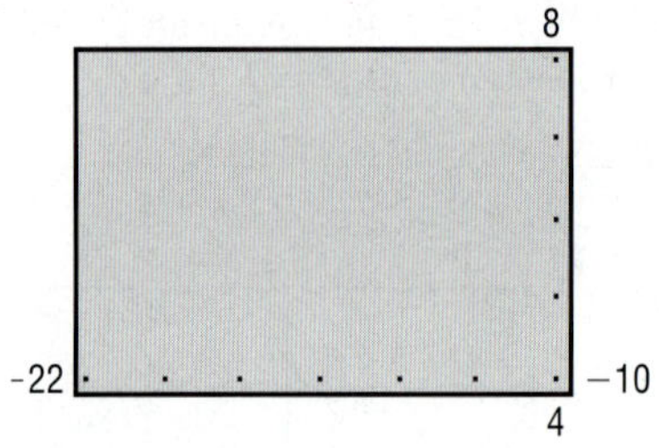

In Problems 21–34, find the distance $d(P_1, P_2)$ between the points P_1 and P_2.

21.

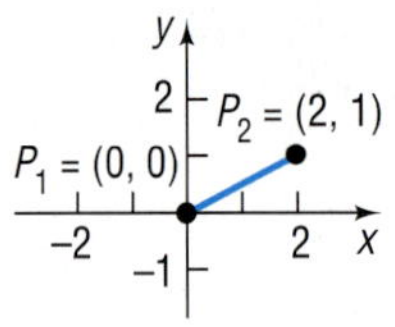

22.

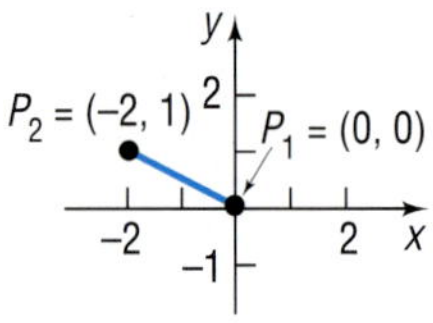

23.

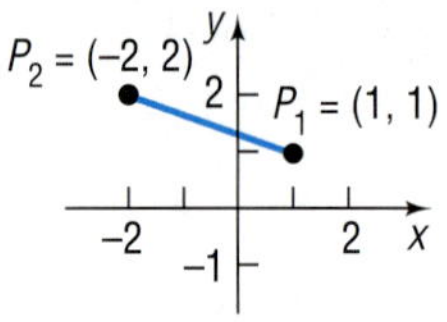

24.

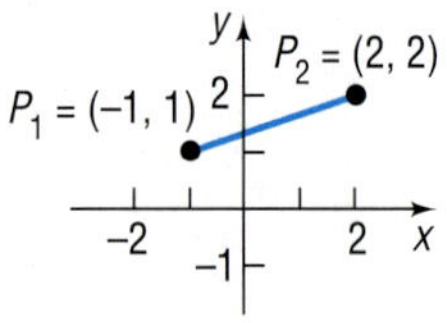

25. $P_1 = (3, -4)$; $P_2 = (5, 4)$

26. $P_1 = (-1, 0)$; $P_2 = (2, 4)$

27. $P_1 = (-3, 2)$; $P_2 = (6, 0)$

28. $P_1 = (2, -3)$; $P_2 = (4, 2)$

29. $P_1 = (4, -3)$; $P_2 = (6, 4)$

30. $P_1 = (-4, -3)$; $P_2 = (6, 2)$

31. $P_1 = (-0.2, 0.3)$; $P_2 = (2.3, 1.1)$

32. $P_1 = (1.2, 2.3)$; $P_2 = (-0.3, 1.1)$

33. $P_1 = (a, b)$; $P_2 = (0, 0)$

34. $P_1 = (a, a)$; $P_2 = (0, 0)$

In Problems 35–38, find the length of the line segment. Assume that the end points of each line segment have integer coordinates.

35.

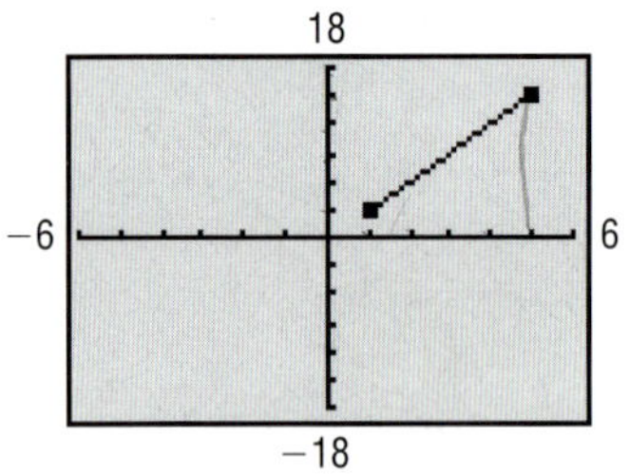

36.

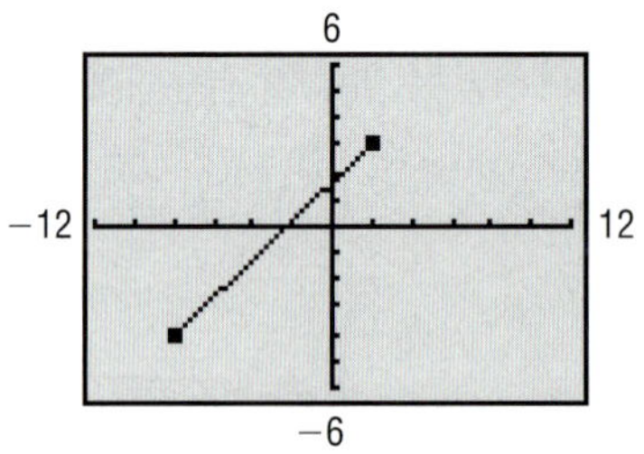

37.

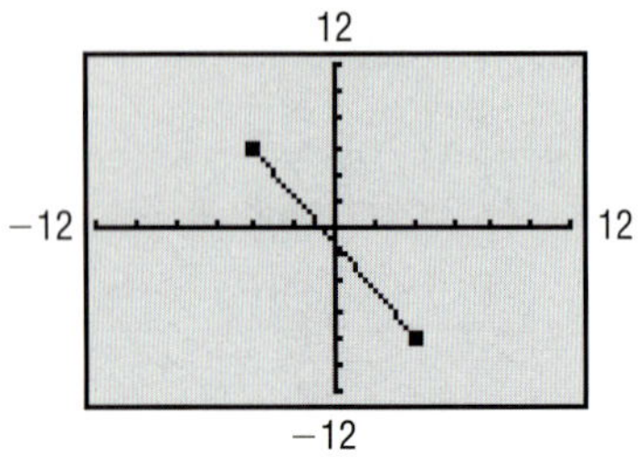

38.

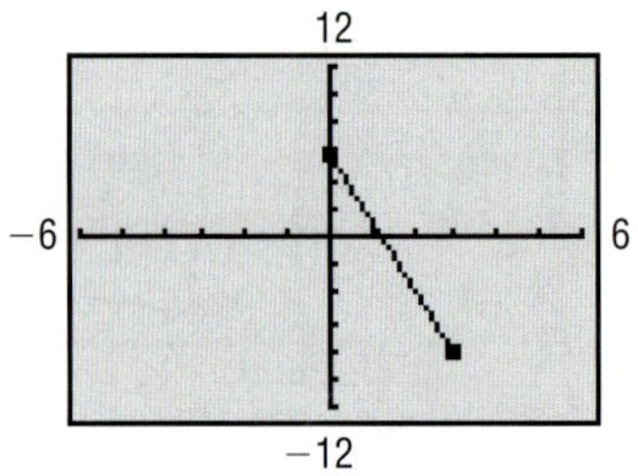

In Problems 39–44, plot each point and form the triangle ABC. Verify that the triangle is a right triangle. Find its area.

39. $A = (-2, 5)$; $B = (1, 3)$; $C = (-1, 0)$

40. $A = (-2, 5)$; $B = (12, 3)$; $C = (10, -11)$

41. $A = (-5, 3)$; $B = (6, 0)$; $C = (5, 5)$

42. $A = (-6, 3)$; $B = (3, -5)$; $C = (-1, 5)$

43. $A = (4, -3)$; $B = (0, -3)$; $C = (4, 2)$

44. $A = (4, -3)$; $B = (4, 1)$; $C = (2, 1)$

45. Find all points having an x-coordinate of 2 whose distance from the point $(-2, -1)$ is 5.

46. Find all points having a y-coordinate of -3 whose distance from the point $(1, 2)$ is 13.

47. Find all points on the x-axis that are 5 units from the point $(4, -3)$.

48. Find all points on the y-axis that are 5 units from the point $(4, 4)$.

In Problems 49–58, find the midpoint of the line segment joining the points P_1 and P_2.

49. $P_1 = (5, -4)$; $P_2 = (3, 2)$

50. $P_1 = (-1, 0)$; $P_2 = (2, 4)$

51. $P_1 = (-3, 2)$; $P_2 = (6, 0)$

52. $P_1 = (2, -3)$; $P_2 = (4, 2)$

53. $P_1 = (4, -3)$; $P_2 = (6, 1)$

54. $P_1 = (-4, -3)$; $P_2 = (2, 2)$

55. $P_1 = (-0.2, 0.3)$; $P_2 = (2.3, 1.1)$

56. $P_1 = (1.2, 2.3)$; $P_2 = (-0.3, 1.1)$

57. $P_1 = (a, b)$; $P_2 = (0, 0)$

58. $P_1 = (a, a)$; $P_2 = (0, 0)$

59. The **medians** of a triangle are the line segments from each vertex to the midpoint of the opposite side (see the figure). Find the lengths of the medians of the triangle with vertices at $A = (0, 0)$, $B = (0, 6)$, and $C = (4, 4)$.

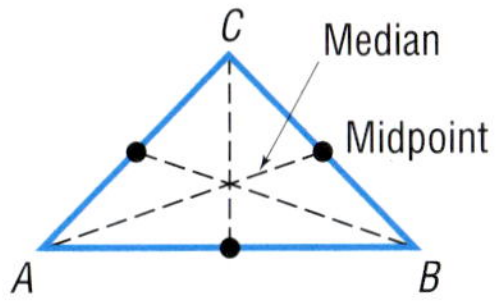

60. An **equilateral triangle** is one in which all three sides are of equal length. If two vertices of an equilateral triangle are (0, 4) and (0, 0), find the third vertex. How many of these triangles are possible?

*In Problems 61–64, find the length of each side of the triangle determined by the three points P_1, P_2, and P_3. State whether the triangle is an isosceles triangle, a right triangle, neither of these, or both. (An **isosceles triangle** is one in which at least two of the sides are of equal length.)*

61. $P_1 = (2, 1)$; $P_2 = (-4, 1)$; $P_3 = (-4, -3)$

62. $P_1 = (-1, 4)$; $P_2 = (6, 2)$; $P_3 = (4, -5)$

63. $P_1 = (-2, -1)$; $P_2 = (0, 7)$; $P_3 = (3, 2)$

64. $P_1 = (7, 2)$; $P_2 = (-4, 0)$; $P_3 = (4, 6)$

65. Prices of Motorola Stock The following data represent the closing price of Motorola, Inc., stock at the end of each month in 1998.

(a) By hand, draw a scatter diagram.

(b) Using a graphing utility, draw a scatter diagram.

(c) How would you describe the trend in these data over time?

Month	Closing Price
January, 1998 ($t = 1$)	$59\frac{9}{16}$
February, 1998 ($t = 2$)	$55\frac{5}{8}$
March, 1998 ($t = 3$)	$60\frac{3}{4}$
April, 1998 ($t = 4$)	$55\frac{3}{4}$
May, 1998 ($t = 5$)	53
June, 1998 ($t = 6$)	$52\frac{9}{16}$
July, 1998 ($t = 7$)	$52\frac{1}{4}$
August, 1998 ($t = 8$)	$42\frac{15}{16}$
September, 1998 ($t = 9$)	$42\frac{7}{8}$
October, 1998 ($t = 10$)	52
November, 1998 ($t = 11$)	$61\frac{7}{8}$
December, 1998 ($t = 12$)	$61\frac{1}{16}$

Courtesy of A. G. Edwards & Sons, Inc.

66. Price of Microsoft Stock The following data represent the closing price of Microsoft Corporation stock at the end of each month in 1998.

(a) By hand, draw a scatter diagram.

(b) Using a graphing utility, draw a scatter diagram.

(c) How would you describe the trend in these data over time?

Month	Closing Price
January, 1998 ($t = 1$)	$74\frac{19}{32}$
February, 1998 ($t = 2$)	$84\frac{3}{4}$
March, 1998 ($t = 3$)	$89\frac{1}{2}$
April, 1998 ($t = 4$)	$90\frac{1}{8}$
May, 1998 ($t = 5$)	$84\frac{13}{16}$
June, 1998 ($t = 6$)	$108\frac{3}{8}$
July, 1998 ($t = 7$)	$109\frac{15}{16}$
August, 1998 ($t = 8$)	$95\frac{15}{16}$
September, 1998 ($t = 9$)	$110\frac{1}{16}$
October, 1998 ($t = 10$)	$105\frac{7}{8}$
November, 1998 ($t = 11$)	122
December, 1998 ($t = 12$)	$138\frac{11}{16}$

Courtesy of A. G. Edwards & Sons, Inc.

67. Government Expenditures The following data represent total federal government expenditures (in billions of dollars) of the United States for the years 1991–1998.

Year	Total Expenditures
1991	1366.1
1992	1381.7
1993	1374.8
1994	1392.8
1995	1407.8
1996	1412.0
1997	1414.0
1998	1441.0

Source: U.S. Office of Management and Budget.

(a) By hand, draw a scatter diagram.
(b) Using a graphing utility, draw a scatter diagram.
(c) How would you describe the trend in these data over time?

68. U.S. Government Debt The following data represent the total debt of the United States federal government (in billions of dollars) for the years 1991–1998.

Year	Total Debt
1991	3598.5
1992	4002.1
1993	4351.4
1994	4643.7
1995	4921.0
1996	5181.9
1997	5369.7
1998	5543.6

Source: U.S. Office of Management and Budget.

(a) By hand, draw a scatter diagram.
(b) Using a graphing utility, draw a scatter diagram.
(c) How would you describe the trend in these data over time?

69. Disposable Personal Income The following data represent per capita disposable personal income x and per capita personal consumption expenditures y for the years 1985–1994 in current dollars (dollars adjusted for inflation).

Per Capita Disposable Income, x	Per Capita Personal Consumption, y
13,258	12,013
13,552	12,336
13,545	12,568
13,890	12,903
14,005	13,025
14,101	13,063
14,003	12,860
14,279	13,110
14,341	13,361
14,696	13,711

Source: U.S. Bureau of Economic Analysis.

(a) By hand, draw a scatter diagram.
(b) Using a graphing utility, draw a scatter diagram.
(c) What happens to per capita personal consumption as per capita disposable income increases?

70. Natural Gas versus Coal The following data represent the amount of energy provided by natural gas and coal measured in trillions of Btu's (British Thermal Units) for the New England states. Let the amount of energy provided by natural gas be x and the amount of energy provided by coal be y.

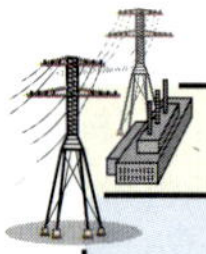

State	Natural Gas, x	Coal, y
Maine	5	22
New Hampshire	17	35
Vermont	8	1
Massachusetts	306	111
Rhode Island	79	0
Connecticut	114	22

Source: U.S. Energy Information Administration.

(a) By hand, draw a scatter diagram.
(b) Using a graphing utility, draw a scatter diagram.
(c) What happens to the amount of energy provided by coal as the amount of energy provided by natural gas increases?

71. **Baseball** A major league baseball "diamond" is actually a square, 90 feet on a side (see the figure). What is the distance directly from home plate to second base (the diagonal of the square)?

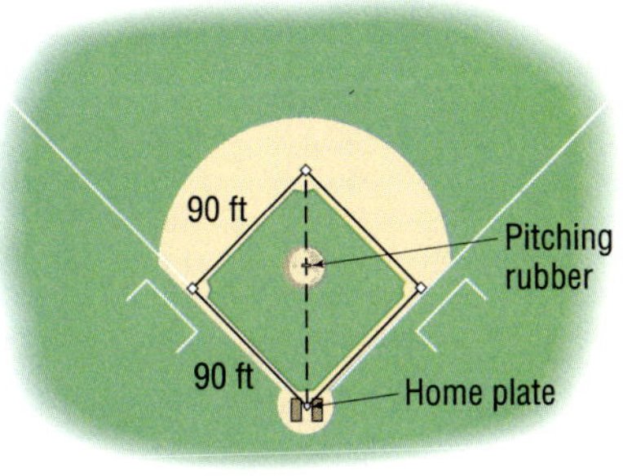

72. **Little League Baseball** The layout of a Little League playing field is a square, 60 feet on a side.* How far is it directly from home plate to second base (the diagonal of the square)?

73. **Baseball** Refer to Problem 71. Overlay a rectangular coordinate system on a major league baseball diamond so that the origin is at home plate, the positive x-axis lies in the direction from home plate to first base, and the positive y-axis lies in the direction from home plate to third base.

(a) What are the coordinates of first base, second base, and third base? Use feet as the unit of measurement.

(b) If the right fielder is located at $(310, 15)$, how far is it from there to second base?

(c) If the center fielder is located at $(300, 300)$, how far is it from there to third base?

*Source: *Little League Baseball, Official Regulations and Playing Rules,* 1998.

74. **Little League Baseball** Refer to Problem 72. Overlay a rectangular coordinate system on a Little League baseball diamond so that the origin is at home plate, the positive x-axis lies in the direction from home plate to first base, and the positive y-axis lies in the direction from home plate to third base.

(a) What are the coordinates of first base, second base, and third base? Use feet as the unit of measurement.

(b) If the right fielder is located at $(180, 20)$, how far is it from there to second base?

(c) If the center fielder is located at $(220, 220)$, how far is it from there to third base?

75. A Dodge Intrepid and a Mack truck leave an intersection at the same time. The Intrepid heads east at an average speed of 30 miles per hour, while the truck heads south at an average speed of 40 miles per hour. Find an expression for their distance apart d (in miles) at the end of t hours.

76. A hot-air balloon, headed due east at an average speed of 15 miles per hour and at a constant altitude of 100 feet, passes over an intersection (see the figure). Find an expression for its distance d (measured in feet) from the intersection t seconds later.

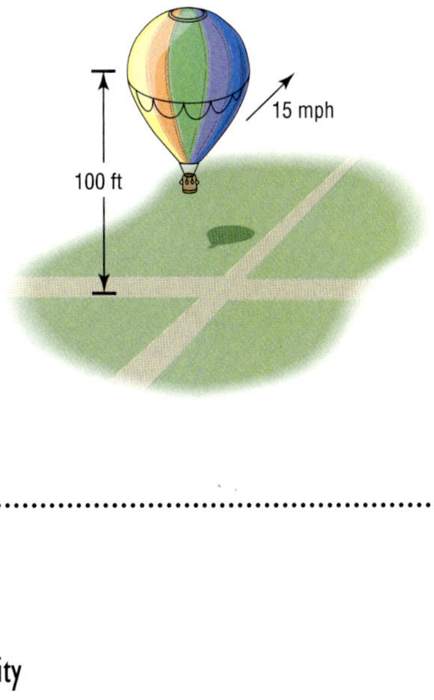

1.2 GRAPHS OF EQUATIONS

1. Graph Equations by Hand
2. Graph Equations Using a Graphing Utility
3. Find Intercepts
4. Test an Equation for Symmetry with Respect to the (a) x-axis, (b) y-axis, and (c) origin

1 Illustrations play an important role in helping us to visualize the relationships that exist between two variable quantities. Figure 18 shows the weekly chart of interest rates for 30-year Treasury bonds from November 29, 1996, through December 31, 1997. For example, the interest rate for January 17, 1997, was about 6.825% (682.5 basis points). Such illustrations are usually

Figure 18
Thirty-year Treasury rates

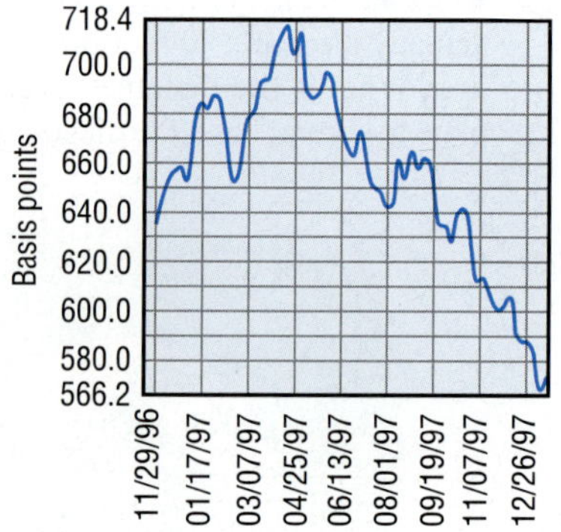

referred to as *graphs.* The **graph of an equation** in two variables x and y consists of the set of points in the xy-plane whose coordinates (x, y) satisfy the equation. Let's look at some examples.

EXAMPLE 1 Graphing an Equation by Hand

Graph the equation: $y = 2x + 5$

Solution We want to find all points (x, y) that satisfy the equation. To locate some of these points (and thus get an idea of the pattern of the graph), we assign some numbers to x and find corresponding values for y:

If	Then	Point on Graph
$x = 0$	$y = 2(0) + 5 = 5$	$(0, 5)$
$x = 1$	$y = 2(1) + 5 = 7$	$(1, 7)$
$x = -5$	$y = 2(-5) + 5 = -5$	$(-5, -5)$
$x = 10$	$y = 2(10) + 5 = 25$	$(10, 25)$

By plotting these points and then connecting them, we obtain the graph of the equation (a *line*), as shown in Figure 19.

Figure 19
$y = 2x + 5$

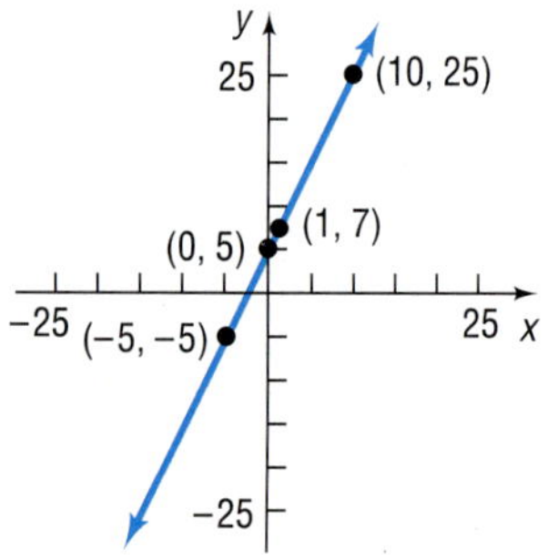

EXAMPLE 2 Graphing an Equation by Hand

Graph the equation: $y = x^2$

Solution Table 2 provides several points on the graph. In Figure 20 we plot these points and connect them with a smooth curve to obtain the graph (a *parabola*).

Table 2

x	$y = x^2$	(x, y)
-4	16	$(-4, 16)$
-3	9	$(-3, 9)$
-2	4	$(-2, 4)$
-1	1	$(-1, 1)$
0	0	$(0, 0)$
1	1	$(1, 1)$
2	4	$(2, 4)$
3	9	$(3, 9)$
4	16	$(4, 16)$

Figure 20

$y = x^2$

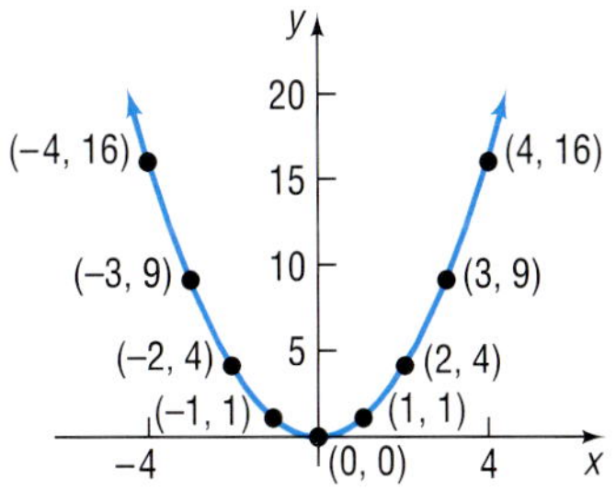

The graphs of the equations shown in Figures 19 and 20 do not show all points. For example, in Figure 19, the point (20, 45) is a part of the graph of $y = 2x + 5$, but it is not shown. Since the graph of $y = 2x + 5$ could be extended out as far as we please, we use arrows to indicate that the pattern shown continues. Thus, it is important when illustrating a graph to present enough of the graph so that any viewer of the illustration will "see" the rest of it as an obvious continuation of what is actually there. This is referred to as a **complete graph.**

So, one way to obtain a complete graph of an equation is to plot a sufficient number of points on the graph until a pattern becomes evident. Then these points are connected with a smooth curve following the suggested pattern. But how many points are sufficient? Sometimes knowledge about the equation tells us. For example, we will learn in Section 1.6 that, if an equation is of the form $y = mx + b$, then its graph is a line. In this case, two points would suffice to obtain the graph.

One of the purposes of this book is to investigate the properties of equations in order to decide whether a graph is complete. Sometimes we shall graph equations by plotting a sufficient number of points on the graph until a pattern becomes evident; then we connect these points with a smooth curve following the suggested pattern. (Shortly, we shall investigate various techniques that will enable us to graph an equation without plotting so many points.) Other times we shall graph equations using a graphing utility.

Using a Graphing Utility to Graph Equations

2 From Examples 1 and 2, we see that a graph can be obtained by plotting points in a rectangular coordinate system and connecting them. Graphing utilities perform these same steps when graphing an equation. For example, the TI-83 determines 95 evenly spaced input values,* uses the equation to determine the output values, plots these points on the screen, and finally (if in the connected mode) draws a line between consecutive points.

*These input values depend on the values of Xmin and Xmax. For example, if $X\text{min} = -10$ and $X\text{max} = 10$, then the first input value will be -10 and the next input value will be $-10 + (10 - (-10))/94 = -9.7872$, and so on.

Most graphing utilities require the following steps in order to obtain the graph of an equation:

Steps for Graphing an Equation Using a Graphing Utility

STEP 1: Solve the equation for y in terms of x.*

STEP 2: Get into the graphing mode of your graphing utility. The screen will usually display $y =$, prompting you to enter the expression involving x that you found in Step 1. (Consult your manual for the correct way to enter the expression; for example, $y = x^2$ might be entered as x^2 or as x*x or as $x\ x^Y\ 2$).

STEP 3: Select the viewing window. Without prior knowledge about the behavior of the graph of the equation, it is common to select the **standard viewing window** initially. The viewing window is then adjusted based on the graph that appears. In this text, the standard viewing window will be

$$X\text{min} = -10 \qquad Y\text{min} = -10$$
$$X\text{max} = 10 \qquad Y\text{max} = 10$$
$$X\text{scl} = 1 \qquad Y\text{scl} = 1$$

STEP 4: Execute.

STEP 5: Adjust the viewing window until a complete graph is obtained.

EXAMPLE 3 Graphing an Equation on a Graphing Utility

Graph the equation $6x^2 + 3y = 36$.

Solution STEP 1: We solve for y in terms of x.

$$6x^2 + 3y = 36$$
$$3y = -6x^2 + 36 \quad \textit{Subtract } 6x^2 \textit{ from both sides of the equation.}$$
$$y = -2x^2 + 12 \quad \textit{Divide both sides of the equation by 3.}$$

STEP 2: From the graphing mode, enter the expression $-2x^2 + 12$ after the prompt $y =$.

STEP 3: Set the viewing window to the standard viewing window.

STEP 4: Execute. The screen should look like Figure 21.

Figure 21

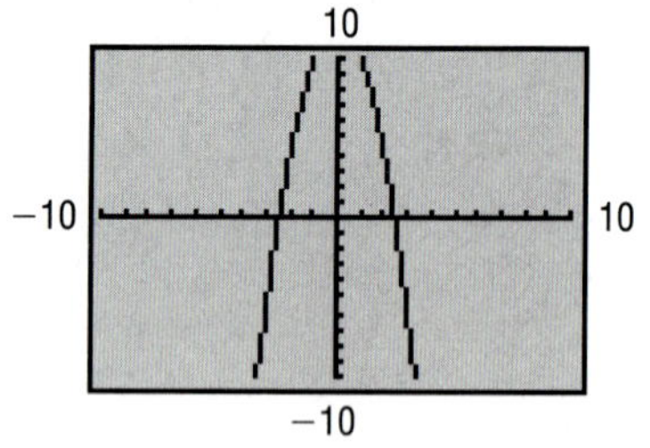

*You may wish to read pp. 681–684 in Section 2 of the Appendix before going on.

STEP 5: The graph of $y = -2x^2 + 12$ is not complete. The value of *Y*max must be increased so that the top portion of the graph is visible. After increasing the value of *Y*max to 12, we obtain the graph in Figure 22.* The graph is now complete.

Figure 22

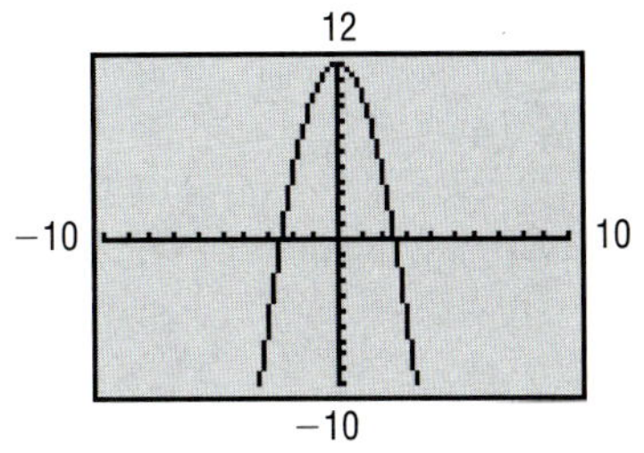

Figure 23

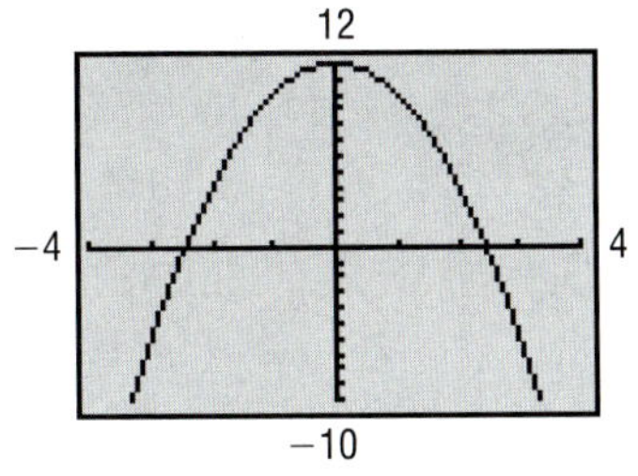

Look again at Figure 22. Although a complete graph is shown, the graph might be improved by adjusting the values of *X*min and *X*max. Figure 23 shows the graph of $y = -2x^2 + 12$ using $X\text{min} = -4$ and $X\text{max} = 4$. Do you think this is a better choice for the viewing window?

NOW WORK PROBLEMS 11(a) AND 11(b).

We said earlier that we would discuss techniques that reduce the number of points required to graph an equation by hand. Two such techniques involve finding *intercepts* and checking for *symmetry*.

Intercepts

3 The points, if any, at which a graph crosses or touches the coordinate axes are called the **intercepts.** See Figure 24. The *x*-coordinate of a point at which the graph crosses or touches the *x*-axis is an ***x*-intercept,** and the *y*-coordinate of a point at which the graph crosses or touches the *y*-axis is a ***y*-intercept.**

Figure 24

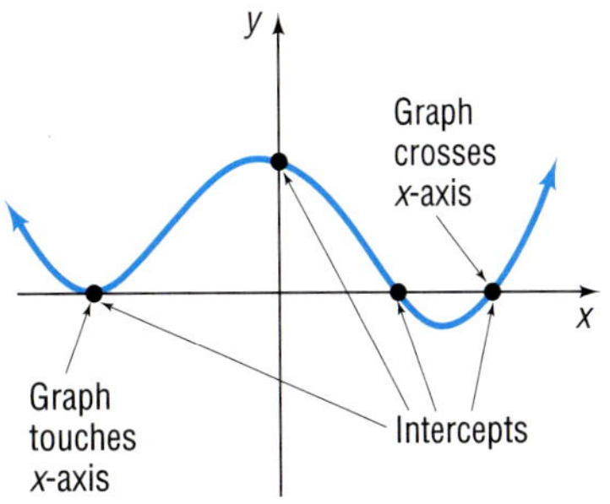

*Some graphing utilities have a ZOOM-FIT feature that determines the appropriate *Y*min and *Y*max for a given *X*min and *X*max. Consult your owner's manual for the appropriate keystrokes.

EXAMPLE 4 Finding Intercepts from a Graph

Find the intercepts of the graph in Figure 25. What are its x-intercepts? What are its y-intercepts?

Figure 25

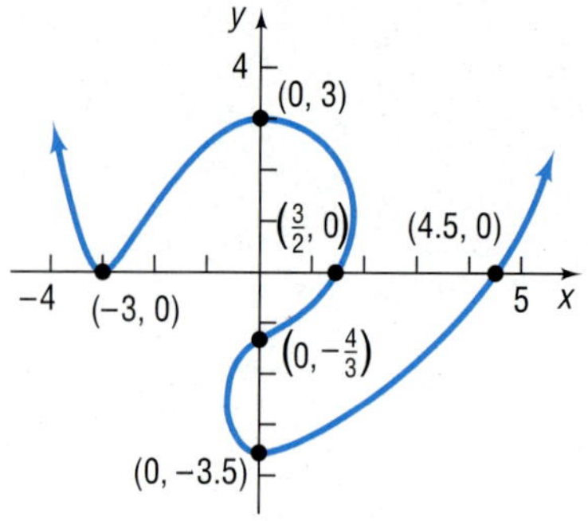

Solution The intercepts of the graph are the points

$$(-3, 0),\quad (0, 3),\quad \left(\frac{3}{2}, 0\right),\quad \left(0, -\frac{4}{3}\right),\quad (0, -3.5),\quad (4.5, 0)$$

The x-intercepts are $-3, \frac{3}{2}, 4.5$; the y-intercepts are $-3.5, -\frac{4}{3}, 3$.

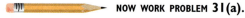

NOW WORK PROBLEM 31(a).

The intercepts of the graph of an equation can be found by using the fact that points on the x-axis have y-coordinates equal to 0, and points on the y-axis have x-coordinates equal to 0.

Procedure for Finding Intercepts

1. To find the x-intercept(s), if any, of the graph of an equation, let $y = 0$ in the equation and solve for x.
2. To find the y-intercept(s), if any, of the graph of an equation, let $x = 0$ in the equation and solve for y.

Because the x-intercepts of the graph of an equation are those x-values for which $y = 0$, they are also called the **zeros** (or **roots**) of the equation.

EXAMPLE 5 Finding Intercepts from an Equation

Find the x-intercept(s) and the y-intercept(s) of the graph of $y = x^2 - 4$.

Solution To find the x-intercept(s), we let $y = 0$ and obtain the equation

$$x^2 - 4 = 0$$
$$(x + 2)(x - 2) = 0 \quad \text{Factor}$$
$$x + 2 = 0 \text{ or } x - 2 = 0 \quad \text{Zero-Product Property}$$
$$x = -2 \text{ or } x = 2$$

The equation has two solutions, -2 or 2. Thus, the x-intercepts (or zeros) are -2 and 2.

To find the y-intercept(s), we let $x = 0$ in the equation:

$$y = x^2 - 4$$
$$y = 0^2 - 4 = -4$$

Thus, the y-intercept is -4.

Table 3

X	Y1
-3	5
-2	0
-1	-3
0	-4
1	-3
2	0
3	5

Y1 ■ X²−4

NOW WORK PROBLEM 69 (LIST THE INTERCEPTS).

Sometimes the TABLE feature of a graphing utility will reveal the intercepts of an equation. (Check your manual to see if your graphing utility has a TABLE feature.) Table 3 shows a table for the equation $y = x^2 - 4$. Can you find the intercepts in the table?

EXAMPLE 6 Using a Graphing Utility to Find Intercepts

Use a graphing utility to find the intercepts of the equation $y = x^3 - 8$.

Solution Figure 26 shows the graph of $y = x^3 - 8$.

Figure 26

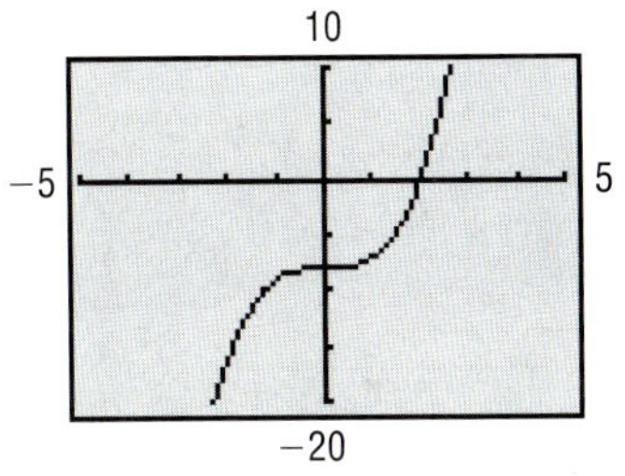

The eVALUEate feature of a TI-83 graphing calculator* accepts as input a value of x and determines the value of y. If we let $x = 0$, we find that the y-intercept is -8. See Figure 27(a).

The ZERO feature of a TI-83 is used to find the x-intercept(s). See Figure 27(b). The x-intercept is 2.

Figure 27

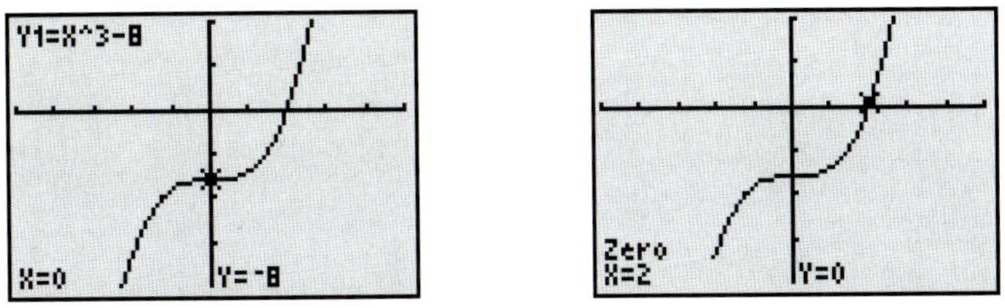

(a) (b)

NOW WORK PROBLEM 69 (LIST THE INTERCEPTS USING A GRAPHING UTILITY).

*Consult your manual to determine the appropriate keystrokes for these features.

Symmetry

4 We have just seen the role that intercepts play in obtaining key points on the graph of an equation. Another helpful tool for graphing equations involves *symmetry,* particularly symmetry with respect to the x-axis, the y-axis, and the origin.

Figure 28

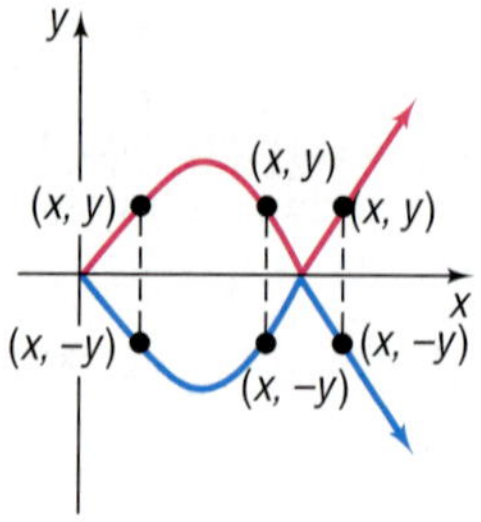

A graph is said to be **symmetric with respect to the x-axis** if, for every point (x, y) on the graph, the point $(x, -y)$ is also on the graph.

Figure 28 illustrates the definition. Notice that when a graph is symmetric with respect to the x-axis the part of the graph above the x-axis is a reflection or mirror image of the part below it, and vice versa.

EXAMPLE 7 Points Symmetric with Respect to the x-Axis

If a graph is symmetric with respect to the x-axis and the point $(3, 2)$ is on the graph, then the point $(3, -2)$ is also on the graph.

Figure 29

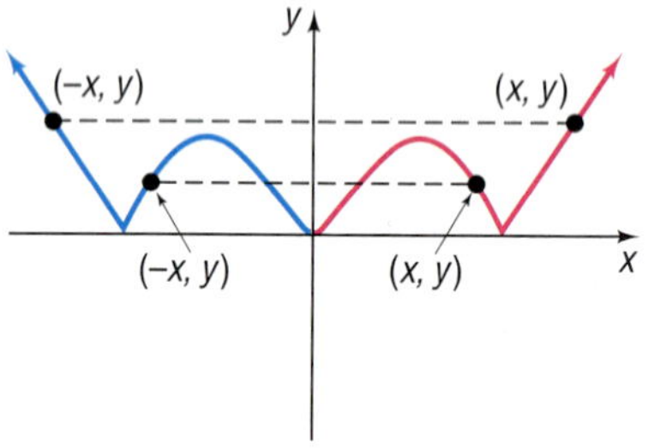

A graph is said to be **symmetric with respect to the y-axis** if, for every point (x, y) on the graph, the point $(-x, y)$ is also on the graph.

Figure 29 illustrates the definition. Notice that when a graph is symmetric with respect to the y-axis the part of the graph to the right of the y-axis is a reflection of the part to the left of it, and vice versa.

EXAMPLE 8 Points Symmetric with Respect to the y-Axis

If a graph is symmetric with respect to the y-axis and the point $(5, 8)$ is on the graph, then the point $(-5, 8)$ is also on the graph.

A graph is said to be **symmetric with respect to the origin** if, for every point (x, y) on the graph, the point $(-x, -y)$ is also on the graph.

Figure 30

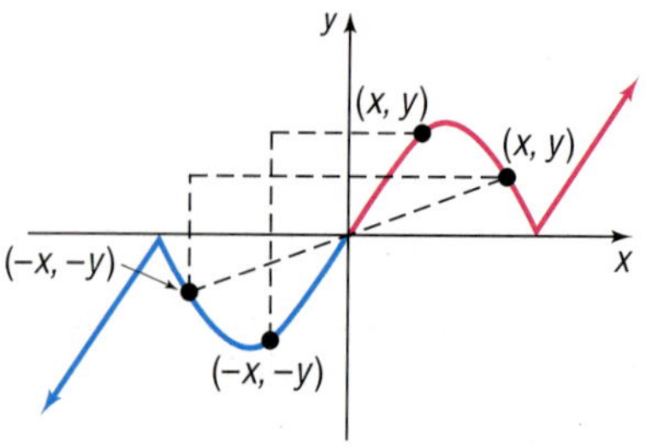

Figure 30 illustrates the definition. Notice that symmetry with respect to the origin may be viewed in two ways:

1. As a reflection about the y-axis, followed by a reflection about the x-axis.
2. As a projection along a line through the origin so that the distances from the origin are equal.

EXAMPLE 9 Points Symmetric with Respect to the Origin

If a graph is symmetric with respect to the origin and the point $(4, 2)$ is on the graph, then the point $(-4, -2)$ is also on the graph.

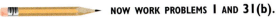

NOW WORK PROBLEMS 1 AND 31(b).

When the graph of an equation is symmetric with respect to a coordinate axis or the origin, the number of points that you need to plot in order to see the pattern is reduced. For example, if the graph of an equation is symmetric with respect to the y-axis, then, once points to the right of the y-axis are plotted, an equal number of points on the graph can be obtained by reflecting them about the y-axis. Thus, before we graph an equation, we first want to determine whether it has any symmetry. The following tests are used for that purpose.

> ***Tests for Symmetry***
>
> To test the graph of an equation for symmetry with respect to the
>
> **x-Axis** Replace y by $-y$ in the equation. If an equivalent equation results, the graph of the equation is symmetric with respect to the x-axis.
>
> **y-Axis** Replace x by $-x$ in the equation. If an equivalent equation results, the graph of the equation is symmetric with respect to the y-axis.
>
> **Origin** Replace x by $-x$ and y by $-y$ in the equation. If an equivalent equation results, the graph of the equation is symmetric with respect to the origin.

EXAMPLE 10 Graphing an Equation by Finding Intercepts and Checking for Symmetry

Graph the equation $y = x^3$ by hand. Find any intercepts and check for symmetry first.

Solution First, we seek the intercepts. When $x = 0$, then $y = 0$; and when $y = 0$, then $x = 0$. Thus, the origin $(0, 0)$ is the only intercept. Now we test for symmetry:

x-Axis: Replace y by $-y$. Since the result, $-y = x^3$, is not equivalent to $y = x^3$, the graph is not symmetric with respect to the x-axis.

y-Axis: Replace x by $-x$. Since the result, $y = (-x)^3 = -x^3$, is not equivalent to $y = x^3$, the graph is not symmetric with respect to the y-axis.

Origin: Replace x by $-x$ and y by $-y$. Since the result, $-y = (-x)^3 = -x^3$, is equivalent to $y = x^3$, the graph is symmetric with respect to the origin.

Table 4

x	$y = x^3$	(x, y)
0	0	(0, 0)
1	1	(1, 1)
2	8	(2, 8)
3	27	(3, 27)

To graph by hand, we use the equation to obtain several points on the graph. Because of the symmetry, we only need to locate points on the graph for which $x \geq 0$. See Table 4. Points on the graph could also be obtained using the TABLE feature on a graphing utility. (Check your manual to see if your graphing utility has this feature.) Using a TI-83, we obtain Table 5. Figure 31 shows the graph.

Table 5

X	Y1
-3	-27
-2	-8
-1	-1
0	0
1	1
2	8
3	27

Y1=X^3

Figure 31
Symmetry with respect to the origin.

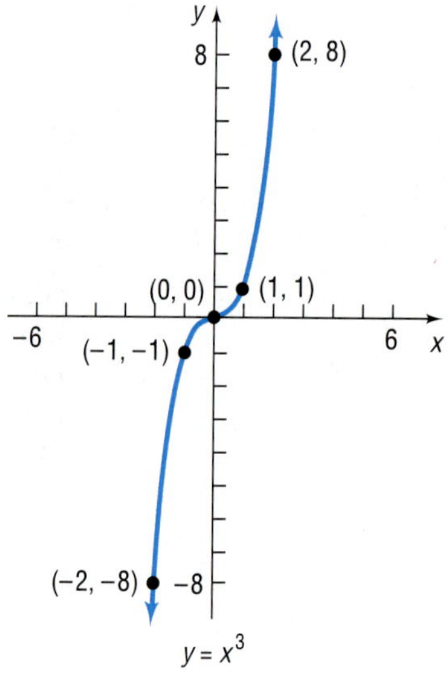

EXAMPLE 11 Graphing an Equation

Graph the equation $x = y^2$. Find any intercepts and check for symmetry first.

Solution

The lone intercept is (0, 0). The graph is symmetric with respect to the x-axis. (Do you see why? Replace y by $-y$). Figure 32 shows the graph drawn by hand.

To graph this equation using a graphing utility, we must write the equation in the form $y = \{\text{expression involving } x\}$. So to graph $x = y^2$ we must solve for y:

$$x = y^2$$
$$y^2 = x$$
$$y = \pm\sqrt{x}$$

Figure 32
Symmetry with respect to the x-axis

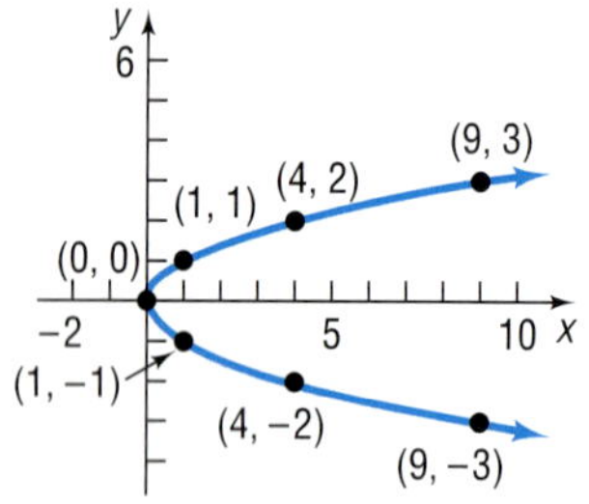

Thus, to graph $x = y^2$, we need to graph both $Y_1 = \sqrt{x}$ and $Y_2 = -\sqrt{x}$ on the same screen. Figure 33 shows the result. Table 6 shows various values of y for a given value of x when $Y_1 = \sqrt{x}$ and $Y_2 = -\sqrt{x}$. Notice that when $x < 0$ we get an error. Can you explain why?

Figure 33

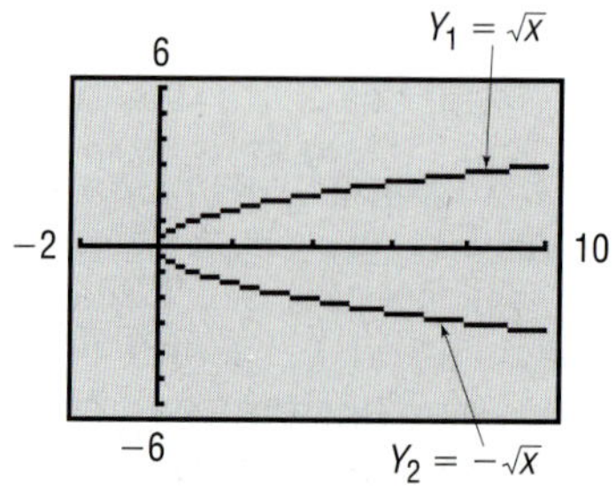

Table 6

X	Y1	Y2
-1	ERROR	ERROR
0	0	0
1	1	-1
2	1.4142	-1.414
3	1.7321	-1.732
4	2	-2
5	2.2361	-2.236

Y1=√(X)

Look again at either Figure 32 or 33. If we restrict y so that $y \geq 0$, the equation $x = y^2$, $y \geq 0$, may be written equivalently as $y = \sqrt{x}$. The portion of the graph of $x = y^2$ in quadrant I is therefore the graph of $y = \sqrt{x}$. See Figure 34.

Figure 34 $y = \sqrt{x}$

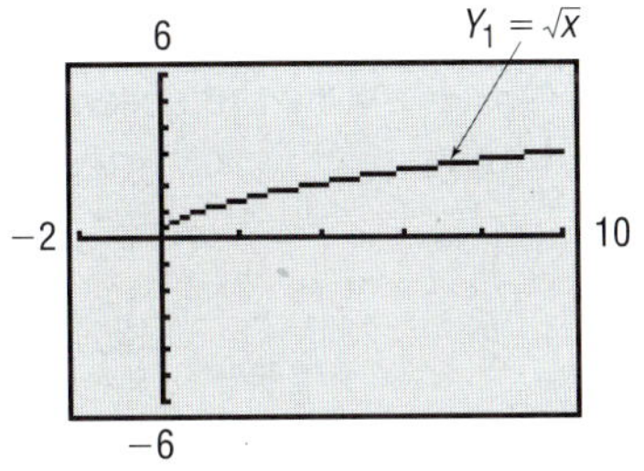

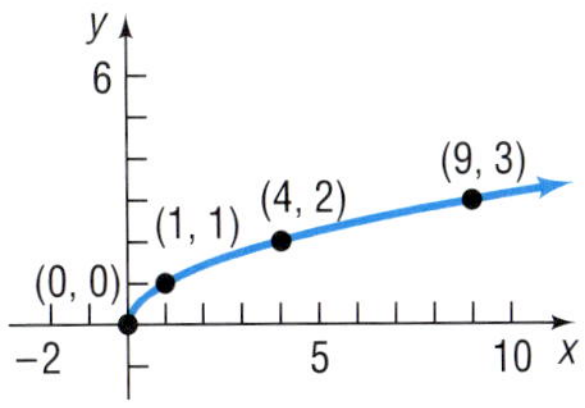

NOW WORK PROBLEM **69.** (TEST FOR SYMMETRY)

EXAMPLE 12 Graphing the Equation $y = 1/x$

Consider the equation: $y = 1/x$

(a) Graph this equation using a graphing utility. Set the viewing window as

$$X\text{min} = -3 \qquad Y\text{min} = -4$$
$$X\text{max} = 3 \qquad Y\text{max} = 4$$
$$X\text{scl} = 1 \qquad Y\text{scl} = 1$$

(b) Use algebra to find any intercepts and test for symmetry.

(c) Draw the graph by hand.

Solution (a) Figure 35 illustrates the graph.

We infer from the graph that there are no intercepts; we may also infer that symmetry with respect to the origin is a possibility. The TRACE feature on a graphing utility can provide further evidence of symmetry with respect to the origin. Using TRACE, we observe that for any ordered pair (x, y) the ordered pair $(-x, -y)$ is also a point on the graph. For example, the points $(0.95744681, 1.0444444)$ and $(-0.95744681, -1.0444444)$ both lie on the graph.

Figure 35

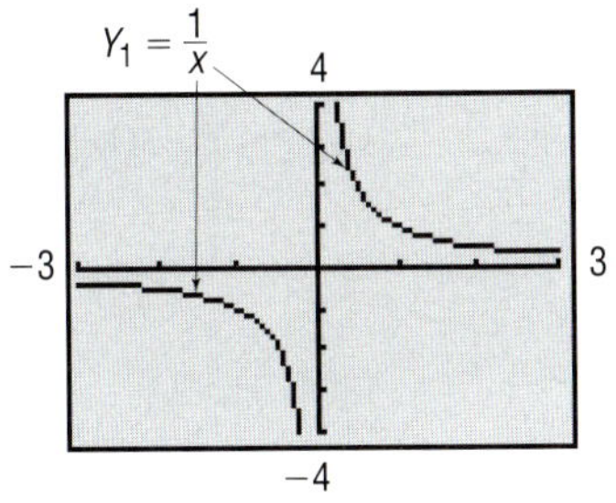

(b) We check for intercepts first. If we let $x = 0$, we obtain a 0 in the denominator, which is undefined. Hence, there is no y-intercept. If we let $y = 0$, we get the equation $1/x = 0$, which has no solution. Hence, there is no x-intercept. Thus, the graph of $y = 1/x$ does not cross the coordinate axes.

Next we check for symmetry:

x-Axis: Replacing y by $-y$ yields $-y = 1/x$, which is not equivalent to $y = 1/x$.

y-Axis: Replacing x by $-x$ yields $y = -1/x$, which is not equivalent to $y = 1/x$.

Origin: Replacing x by $-x$ and y by $-y$ yields $-y = -1/x$, which is equivalent to $y = 1/x$.

The graph is symmetric with respect to the origin. This confirms the inferences drawn in part (a) of the solution.

(c) We can use the equation to obtain some points on the graph. Because of symmetry, we need only find points (x, y) for which x is positive. Also, from the equation $y = 1/x$ we infer that, if x is a large and positive number, then $y = 1/x$ is a positive number close to 0. We also infer that if x is a positive number close to 0 then $y = 1/x$ is a large and positive number. Armed with this information, we can graph the equation. Table 7 and Figure 36 illustrate some of these points and the graph of $y = 1/x$. Observe how the absence of intercepts and the existence of symmetry with respect to the origin were utilized.

Table 7

x	$y = 1/x$	(x, y)
$\frac{1}{10}$	10	$(\frac{1}{10}, 10)$
$\frac{1}{3}$	3	$(\frac{1}{3}, 3)$
$\frac{1}{2}$	2	$(\frac{1}{2}, 2)$
1	1	$(1, 1)$
2	$\frac{1}{2}$	$(2, \frac{1}{2})$
3	$\frac{1}{3}$	$(3, \frac{1}{3})$
10	$\frac{1}{10}$	$(10, \frac{1}{10})$

Figure 36

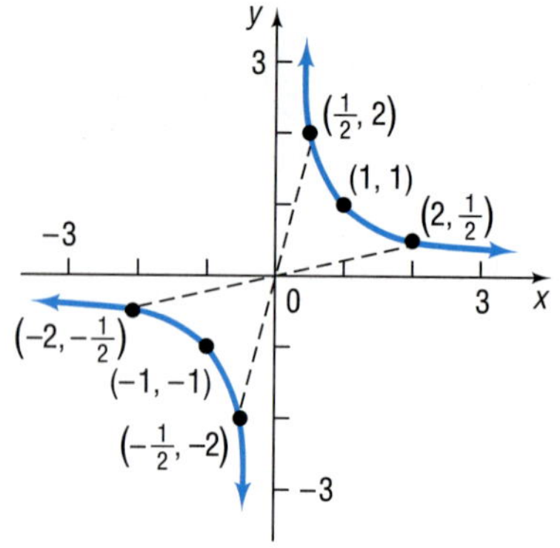

1.2 − 7, 9, 5 19 23 33

1.2 EXERCISES

In Problems 1–10, plot each point. Then plot the point that is symmetric to it with respect to (a) the x-axis; (b) the y-axis; (c) the origin.

1. (3, 4) **2.** (5, 3) **3.** (−2, 1) **4.** (4, −2) **5.** (1, 1)
6. (−1, −1) **7.** (−3, −4) **8.** (4, 0) **9.** (0, −3) **10.** (−3, 0)

In Problems 11–30, graph each equation using the following viewing windows:

(a)	(b)	(c)	(d)
Xmin = −5	Xmin = −10	Xmin = −10	Xmin = −5
Xmax = 5	Xmax = 10	Xmax = 10	Xmax = 5
Xscl = 1	Xscl = 1	Xscl = 2	Xscl = 1
Ymin = −4	Ymin = −8	Ymin = −8	Ymin = −20
Ymax = 4	Ymax = 8	Ymax = 8	Ymax = 20
Yscl = 1	Yscl = 1	Yscl = 2	Yscl = 5

11. $y = x + 2$ **12.** $y = x - 2$ **13.** $y = -x + 2$ **14.** $y = -x - 2$
15. $y = 2x + 2$ **16.** $y = 2x - 2$ **17.** $y = -2x + 2$ **18.** $y = -2x - 2$
19. $y = x^2 + 2$ **20.** $y = x^2 - 2$ **21.** $y = -x^2 + 2$ **22.** $y = -x^2 - 2$
23. $y = 2x^2 + 2$ **24.** $y = 2x^2 - 2$ **25.** $y = -2x^2 + 2$ **26.** $y = -2x^2 - 2$
27. $3x + 2y = 6$ **28.** $3x - 2y = 6$ **29.** $-3x + 2y = 6$ **30.** $-3x - 2y = 6$

In Problems 31–46, the graph of an equation is given.

(a) List the intercepts of the graph.

(b) Based on the graph, tell whether the graph is symmetric with respect to the x-axis, y-axis, and/or origin.

31.

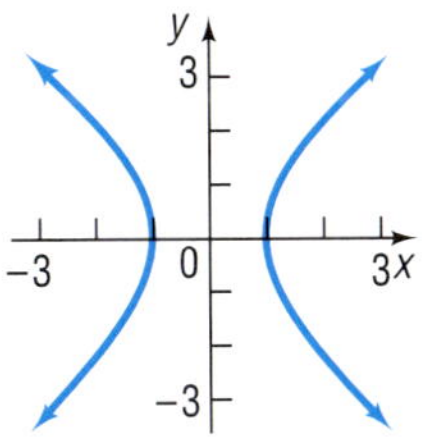

32.

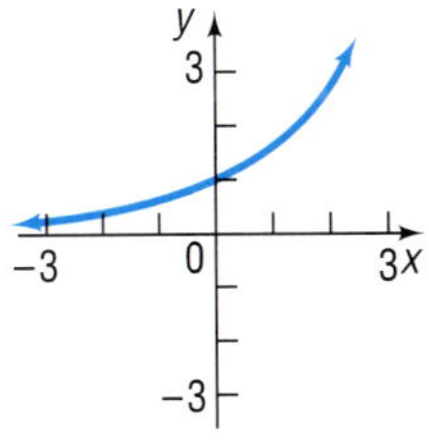

33.

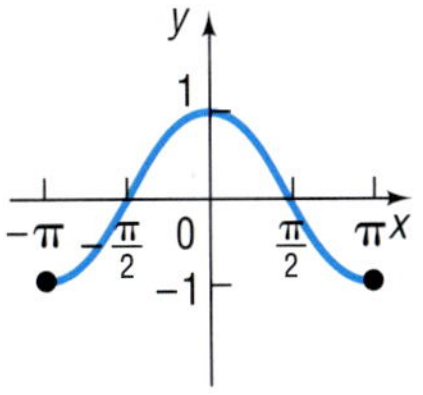

34.

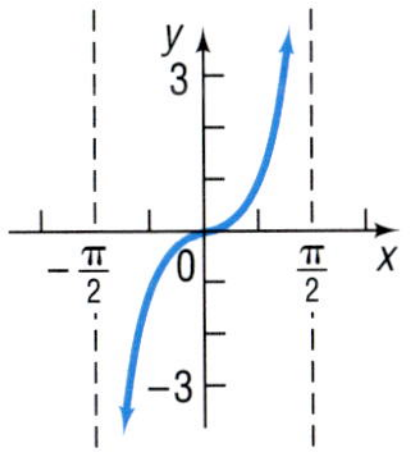

35.

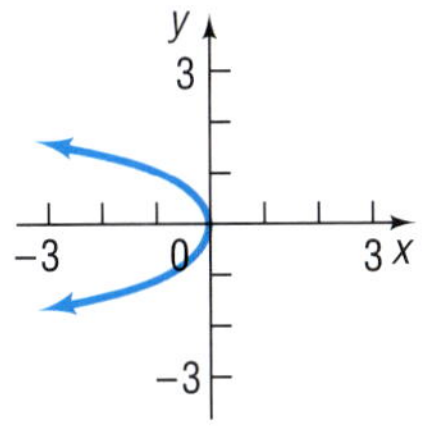

36.

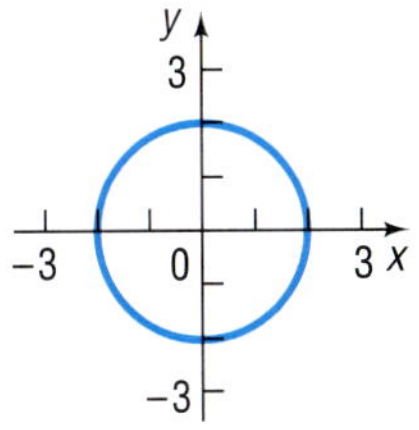

37.

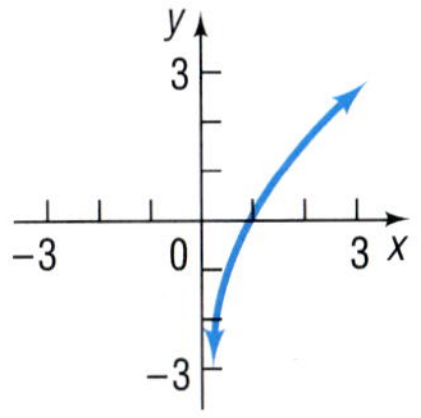

38.

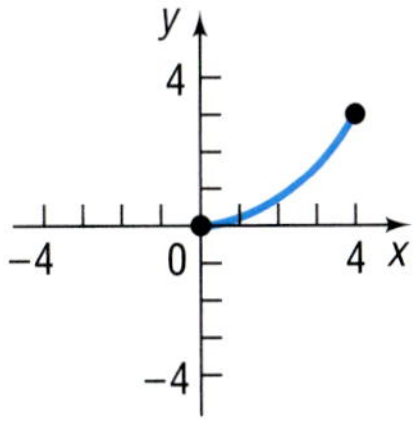

39.

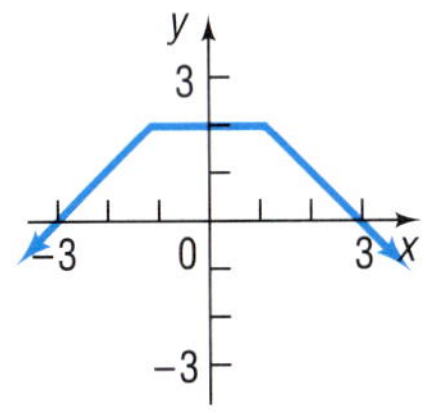

40.

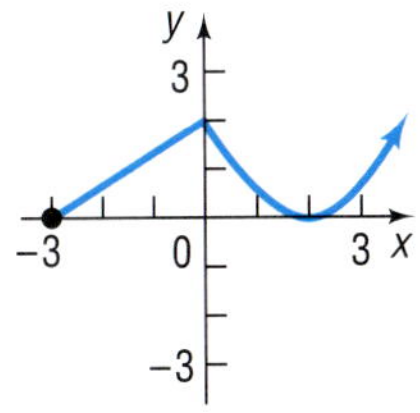

41.

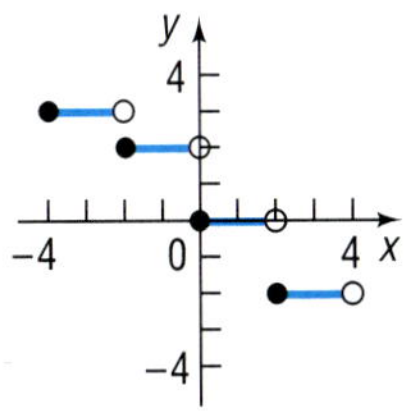

42.

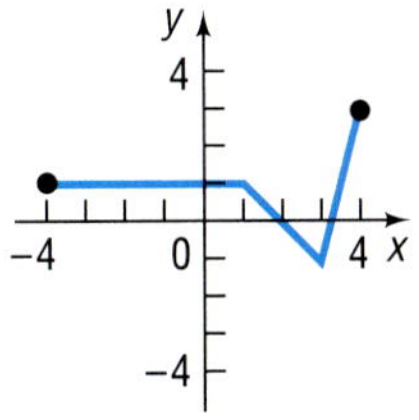

43.

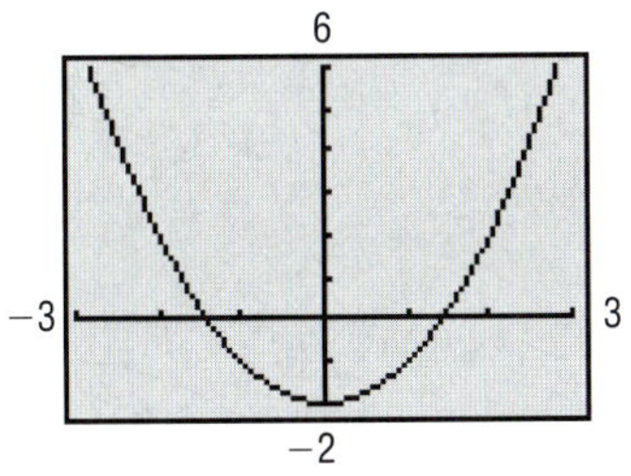

44.

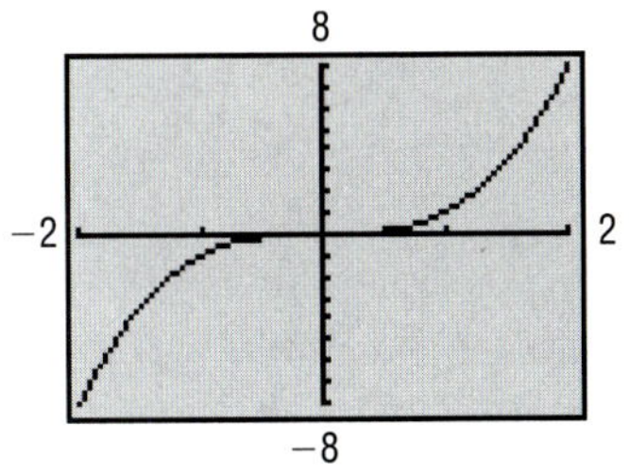

45.

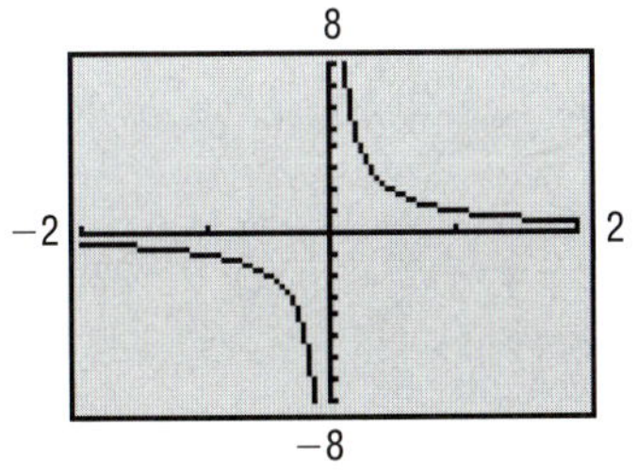

46.

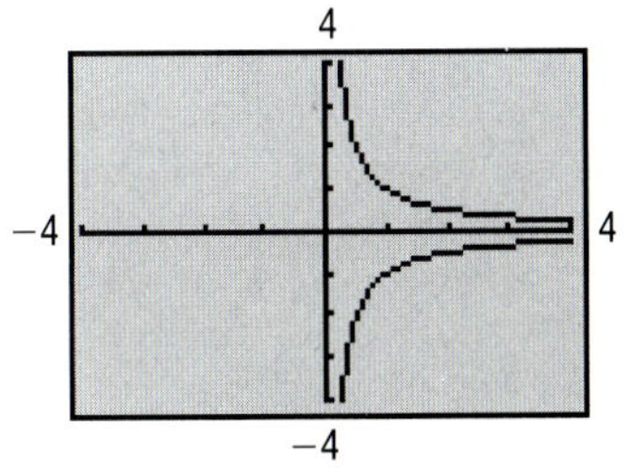

In Problems 47–52, tell whether the given points are on the graph of the equation.

47. Equation: $y = x^4 - \sqrt{x}$
Points: $(0, 0); (1, 1); (-1, 0)$

48. Equation: $y = x^3 - 2\sqrt{x}$
Points: $(0, 0); (1, 1); (1, -1)$

49. Equation: $y^2 = x^2 + 9$
Points: $(0, 3); (3, 0); (-3, 0)$

50. Equation: $y^3 = x + 1$
Points: $(1, 2); (0, 1); (-1, 0)$

51. Equation: $x^2 + y^2 = 4$
Points: $(0, 2); (-2, 2); (\sqrt{2}, \sqrt{2})$

52. Equation: $x^2 + 4y^2 = 4$
Points: $(0, 1); (2, 0); (2, \frac{1}{2})$

53. If $(a, 2)$ is a point on the graph of $y = 5x + 4$, what is a?

54. If $(2, b)$ is a point on the graph of $y = x^2 + 3x$, what is b?

55. If (a, b) is a point on the graph of $2x + 3y = 6$, write an equation that relates a to b.

56. If $(2, 0)$ and $(0, 5)$ are points on the graph of $y = mx + b$, what are m and b?

In Problems 57–60, use the graph on the right.

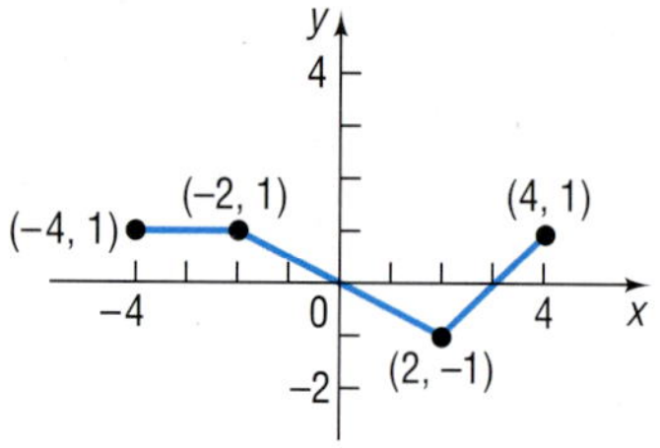

57. Draw the graph to make it symmetric with respect to the x-axis.

58. Draw the graph to make it symmetric with respect to the y-axis.

59. Draw the graph to make it symmetric with respect to the origin.

60. Draw the graph to make it symmetric with respect to the x-axis, y-axis, and origin.

In Problems 61–64, use the graph on the right.

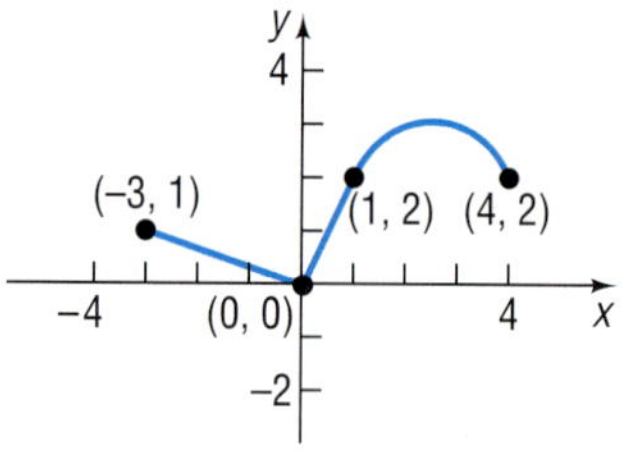

61. Draw the graph to make it symmetric with respect to the x-axis.

62. Draw the graph to make it symmetric with respect to the y-axis.

63. Draw the graph to make it symmetric with respect to the origin.

64. Draw the graph to make it symmetric with respect to the x-axis, y-axis, and origin.

In Problems 65–78, list the intercepts and test for symmetry. Graph each equation using a graphing utility.

65. $x^2 = y$

66. $y^2 = x$

67. $y = 3x$

68. $y = -5x$

69. $x^2 + y - 9 = 0$

70. $y^2 - x - 4 = 0$

71. $9x^2 + 4y^2 = 36$

72. $4x^2 + y^2 = 4$

73. $y = x^3 - 27$

74. $y = x^4 - 1$

75. $y = x^2 - 3x - 4$

76. $y = x^2 + 4$

77. $y = \dfrac{x}{x^2 + 9}$

78. $y = \dfrac{x^2 - 4}{x}$

In Problem 79, you may use a graphing utility, but it is not required.

79. (a) Graph $y = \sqrt{x^2}$, $y = x$, $y = |x|$, and $y = (\sqrt{x})^2$, noting which graphs are the same.
(b) Explain why the graphs of $y = \sqrt{x^2}$ and $y = |x|$ are the same.
(c) Explain why the graphs of $y = x$ and $y = (\sqrt{x})^2$ are not the same.
(d) Explain why the graphs of $y = \sqrt{x^2}$ and $y = x$ are not the same.

80. Make up an equation with the intercepts $(2, 0)$, $(4, 0)$, and $(0, 1)$. Compare your equation with a friend's equation. Comment on any similarities.

81. An equation is being tested for symmetry with respect to the x-axis, the y-axis, and the origin. Explain why, if two of these symmetries are present, then the remaining one must also be present.

82. Draw a graph that contains the points $(-2, -1)$, $(0, 1)$, $(1, 3)$, and $(3, 5)$. Compare your graph with those of other students. Are most of the graphs almost straight lines? How many are "curved"? Discuss the various ways that these points might be connected.

1.3 SOLVING EQUATIONS

1. Solve Equations Using a Graphing Utility
2. Solve Linear Equations
3. Solve Quadratic Equations
4. Solve Equations Involving Absolute Value
5. Solve Radical Equations

Solving Equations Using a Graphing Utility

We shall see as we proceed through this book that some equations can be solved using algebraic techniques that result in exact solutions being obtained. Whenever algebraic techniques exist, we shall discuss them and use them to obtain exact solutions.*

For many equations, though, there are no algebraic techniques that lead to a solution. For such equations, a graphing utility can often be used to investigate possible solutions. When a graphing utility is used to solve an equation, usually *approximate* solutions are obtained. Unless otherwise stated, we shall follow the practice of giving approximate solutions as decimals *rounded to two decimal places.*

1 The ZERO (or ROOT) feature of a graphing utility can be used to find the solutions of an equation when one side of the equation is 0. In using this feature to solve equations, we make use of the fact that the x-intercepts (or zeros) of the graph of an equation are found by letting $y = 0$ and solving the equation for x. Thus, solving an equation for x when one side of the equation is 0 is equivalent to finding where the graph of the corresponding equation crosses or touches the x-axis.

EXAMPLE 1 Using ZERO (or ROOT) to Approximate Solutions of an Equation

Find the solution(s) of the equation $x^2 - 6x + 7 = 0$. Round answers to two decimal places.

Solution The solutions of the equation $x^2 - 6x + 7 = 0$ are the same as the x-intercepts of the graph of $Y_1 = x^2 - 6x + 7$. We begin by graphing the equation.

Figure 37 shows the graph.

Figure 37

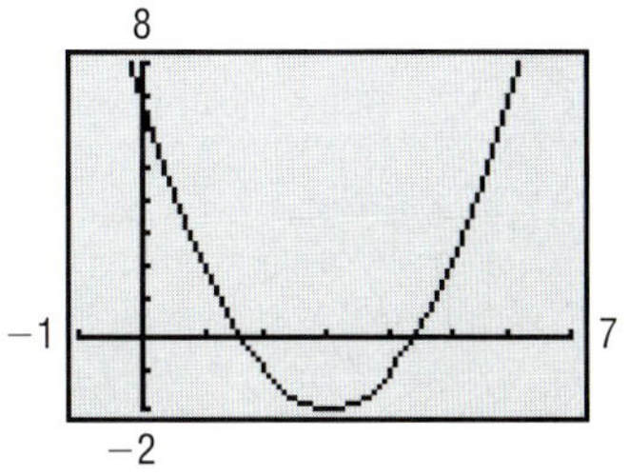

*You may wish to read pp. 681–684, in Section 2 of the Appendix before going on.

From the graph there appear to be two x-intercepts (solutions to the equation): one between 1 and 2; the other between 4 and 5.

Using the ZERO (or ROOT) feature of our graphing utility, we determine that the x-intercepts, and thus the solutions to the equation, are $x = 1.59$ and $x = 4.41$, rounded to two decimal places. See Figure 38(a) and (b).

Figure 38

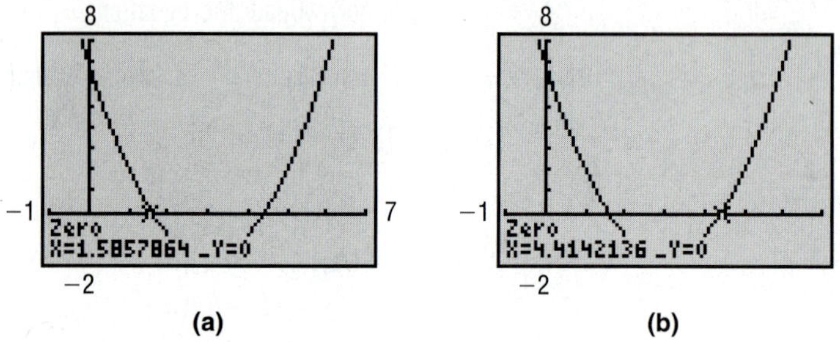

NOW WORK PROBLEM 77.

A second method for solving equations using a graphing utility involves the INTERSECT feature of the graphing utility. This feature is used most effectively when one side of the equation is not 0.

EXAMPLE 2 Using INTERSECT to Approximate Solutions of an Equation

Find the solution(s) to the equation $3(x - 2) = 5(x - 1)$. Round answers to two decimal places.

Solution We begin by graphing each side of the equation as follows: graph $Y_1 = 3(x - 2)$ and $Y_2 = 5(x - 1)$. See Figure 39.

Figure 39

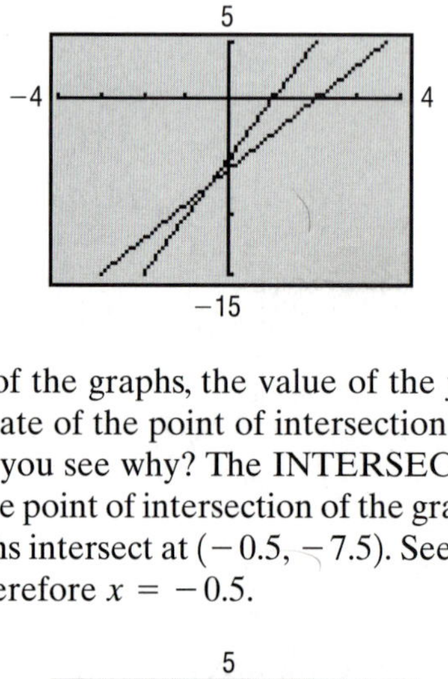

At the point of intersection of the graphs, the value of the y-coordinate is the same. Thus, the x-coordinate of the point of intersection represents the solution to the equation. Do you see why? The INTERSECT feature on a graphing utility determines the point of intersection of the graphs. Using this feature, we find that the graphs intersect at $(-0.5, -7.5)$. See Figure 40. The solution of the equation is therefore $x = -0.5$.

Figure 40

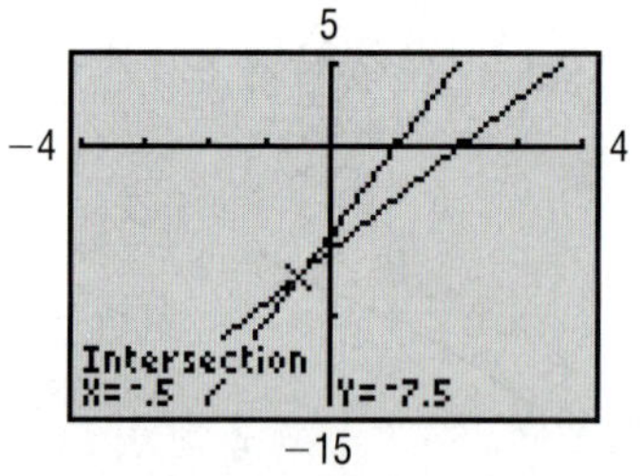

✓CHECK: We can verify our solution by evaluating each side of the equation with $x = -0.5$. See Figure 41. Since the left side of the equation equals the right side of the equation, the solution checks.

Figure 41

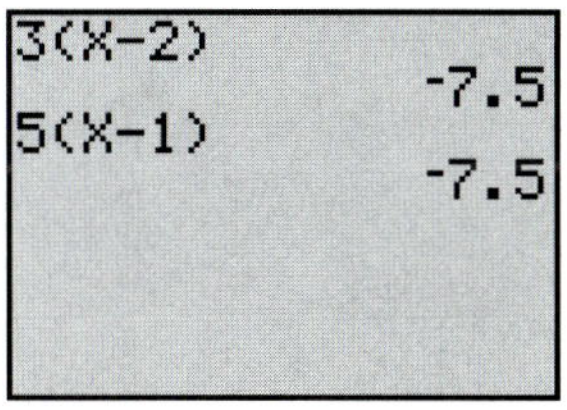

NOW WORK PROBLEM 81.

SUMMARY The steps to follow for approximating solutions of equations are given next.

Steps for Approximating Solutions of Equations Using ZERO (or ROOT)

STEP 1: Write the equation in the form {expression in x} = 0.
STEP 2: Graph Y_1 = {expression in x}.
STEP 3: Use ZERO (or ROOT) to determine each x-intercept of the graph.

Steps for Approximating Solutions of Equations Using INTERSECT

STEP 1: Graph Y_1 = {expression in x on left side of equation}; Y_2 = {expression in x on right side of equation}.
STEP 2: Use INTERSECT to determine each x-coordinate of the points of intersection.

We now discuss equations that can be solved algebraically to obtain exact solutions.

Linear Equations

2 *Linear equations* are equations such as

$$3x + 12 = 0, \qquad -2x + 5 = 0, \qquad 4x - 3 = 0$$

A **linear equation in one variable** is equivalent to an equation of the form

$$ax + b = 0$$

where a and b are real numbers and $a \neq 0$.

Sometimes a linear equation is called a **first-degree equation,** because the left side is a polynomial in x of degree 1.

EXAMPLE 3 Solving a Linear Equation Algebraically

Solve the equation: $3(x-2) = 5(x-1)$

Solution

$$\begin{aligned}
3(x-2) &= 5(x-1) \\
3x - 6 &= 5x - 5 && \textit{Use the Distributive Property.} \\
3x - 6 - 5x &= 5x - 5 - 5x && \textit{Subtract 5x from each side.} \\
-2x - 6 &= -5 && \textit{Simplify.} \\
-2x - 6 + 6 &= -5 + 6 && \textit{Add 6 to each side.} \\
-2x &= 1 && \textit{Simplify.} \\
\frac{-2x}{-2} &= \frac{1}{-2} && \textit{Divide each side by } -2. \\
x &= -\frac{1}{2} && \textit{Simplify.}
\end{aligned}$$

✓CHECK:

$$3(x-2) = 3\left(-\frac{1}{2} - 2\right) = 3\left(-\frac{5}{2}\right) = -\frac{15}{2}$$

$$5(x-1) = 5\left(-\frac{1}{2} - 1\right) = 5\left(-\frac{3}{2}\right) = -\frac{15}{2}$$

Since the two expressions are equal, the solution $x = -\frac{1}{2}$ checks.

Notice that this is the solution obtained in Example 2 using the INTERSECT feature of a graphing utility.

NOW WORK PROBLEM 5.

The next example illustrates the solution of an equation that does not appear to be linear, but leads to a linear equation upon simplification.

EXAMPLE 4 Solving Equations

Solve the equation: $(2x+1)(x-1) = (x+5)(2x-5)$

Graphing Solution Graph $Y_1 = (2x+1)(x-1)$ and $Y_2 = (x+5)(2x-5)$. See Figure 42. Using INTERSECT we find the point of intersection to be (4, 27). The solution of the equation is $x = 4$.

Figure 42

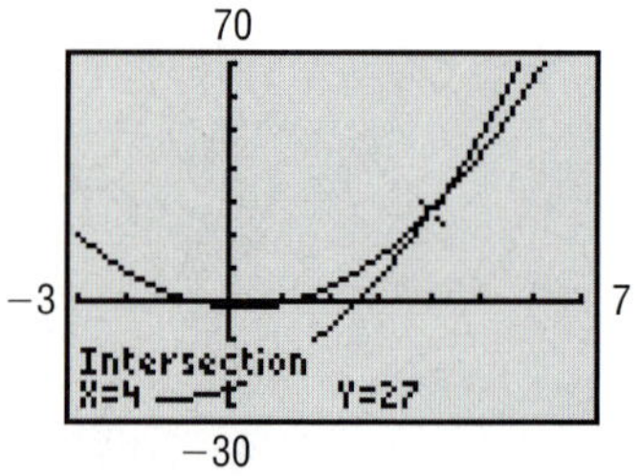

Algebraic Solution

$$\begin{aligned}
(2x + 1)(x - 1) &= (x + 5)(2x - 5) & \\
2x^2 - x - 1 &= 2x^2 + 5x - 25 & &\textit{Multiply and combine like terms.}\\
-x - 1 &= 5x - 25 & &\textit{Subtract } 2x^2 \textit{ from each side.}\\
-x &= 5x - 24 & &\textit{Add 1 to each side.}\\
-6x &= -24 & &\textit{Subtract } 5x \textit{ from each side.}\\
x &= 4 & &\textit{Divide both sides by } -6.
\end{aligned}$$

✓CHECK:
$$(2x + 1)(x - 1) = (8 + 1)(4 - 1) = (9)(3) = 27$$
$$(x + 5)(2x - 5) = (4 + 5)(8 - 5) = (9)(3) = 27$$

Since the two expressions are equal, the solution $x = 4$ checks.

NOW WORK PROBLEM 39.

Quadratic Equations

3

Quadratic equations are equations such as

$$2x^2 + x + 8 = 0, \qquad 3x^2 - 5x + 6 = 0, \qquad x^2 - 9 = 0$$

A general definition is given next.

A **quadratic equation** is an equation equivalent to one of the form

$$ax^2 + bx + c = 0 \tag{1}$$

where a, b, and c are real numbers and $a \neq 0$.

A quadratic equation written in the form $ax^2 + bx + c = 0$ is said to be in **standard form.**

Sometimes, a quadratic equation is called a **second-degree equation,** because the left side is a polynomial of degree 2. We shall discuss two ways of solving quadratic equations: by factoring and by using a graphing utility.*

When a quadratic equation is written in standard form, $ax^2 + bx + c = 0$, it may be possible to factor the expression on the left side as the product of two first-degree polynomials. Then, by setting each factor equal to 0 and solving the resulting linear equations, we obtain the solutions of the quadratic equation.

Let's look at an example.

EXAMPLE 5 Solving a Quadratic Equation by Graphing and by Factoring

Solve the equation: $x^2 = 12 - x$

Graphing Solution

Graph $Y_1 = x^2$ and $Y_2 = 12 - x$. See Figure 43(a). From the graph it appears that there are two points of intersection: one near -4; the other near 3. Using

*A third way, by using the quadratic formula, is discussed in Section 2.3.

INTERSECT, the points of intersection are $(-4, 16)$ and $(3, 9)$, so the solutions of the equation are $x = -4$ and $x = 3$. See Figure 43(b) and (c).

Figure 43

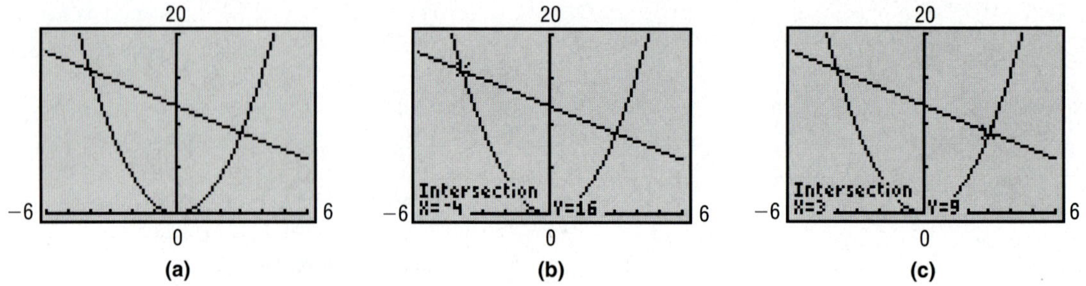

Algebraic Solution We put the equation in standard form by adding $x - 12$ to each side:

$$x^2 = 12 - x$$
$$x^2 + x - 12 = 0$$

The left side of the equation may now be factored as

$$(x + 4)(x - 3) = 0$$

so that

$$x + 4 = 0 \quad \text{or} \quad x - 3 = 0$$
$$x = -4 \qquad\qquad x = 3$$

The solution set is $\{-4, 3\}$.

When the left side factors into two linear equations with the same solution, the quadratic equation is said to have a **repeated solution.** We also call this solution a **root of multiplicity 2,** or a **double root.**

EXAMPLE 6 Solving a Quadratic Equation by Graphing and by Factoring

Solve the equation: $x^2 - 6x + 9 = 0$

Graphing Solution Figure 44 shows the graph of the equation

$$Y_1 = x^2 - 6x + 9$$

From the graph it appears that there is one x-intercept, 3. Using ZERO (or ROOT), we find that the only x-intercept is $x = 3$. Thus, the equation has only the repeated solution 3.

Figure 44

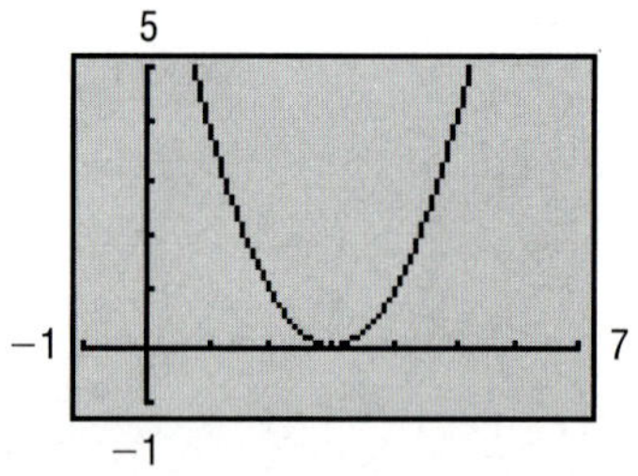

Algebraic Solution This equation is already in standard form, and the left side can be factored:

$$x^2 - 6x + 9 = 0$$
$$(x - 3)(x - 3) = 0$$

so

$$x = 3 \quad \text{or} \quad x = 3$$

This equation has only the repeated solution 3.

NOW WORK PROBLEMS **25** AND **33**.

Equations Involving Absolute Value*

4 Recall that, on the real number line, the absolute value of a equals the distance from the origin to the point whose coordinate is a. For example, there are two points whose distance from the origin is 5 units, -5 and 5. Thus, the equation $|x| = 5$ will have the solution set $\{-5, 5\}$. This leads to the following result:

Equations Involving Absolute Value

If a is a positive real number and if u is any algebraic expression, then

$$|u| = a \quad \text{is equivalent to} \quad u = a \quad \text{or} \quad u = -a \qquad (2)$$

EXAMPLE 7 Solving an Equation Involving Absolute Value

Solve the equation $|x + 4| = 13$.

Graphing Solution For this equation, graph $Y_1 = |x + 4|$ and $Y_2 = 13$ on the same screen and find their point(s) of intersection, if any. See Figure 45. Using the INTERSECT command (twice), we find the points of intersection to be $(-17, 13)$ and $(9, 13)$. The solution set is $\{-17, 9\}$.

Figure 45

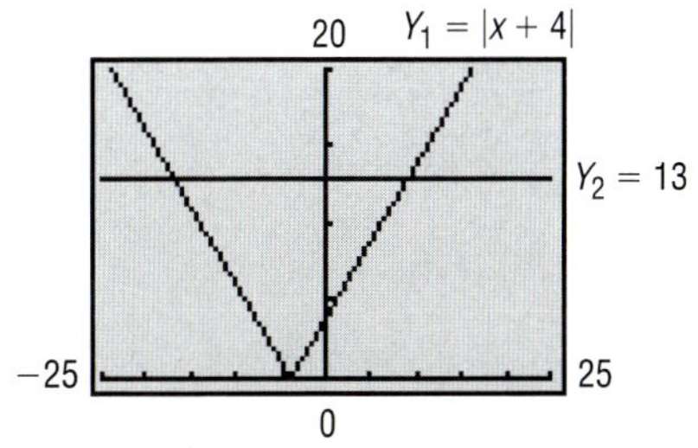

*You may wish to review Absolute Value, pp. 678–679, in Section 2 of the Appendix.

Algebraic Solution This follows the form of equation (2), where $u = x + 4$. Thus, there are two possibilities:

$$x + 4 = 13 \quad \text{or} \quad x + 4 = -13$$
$$x = 9 \qquad\qquad x = -17$$

Thus, the solution set is $\{-17, 9\}$.

NOW WORK PROBLEM 61.

Equations Containing Radicals

5 When the variable in an equation occurs in a square root, cube root, and so on, that is, when it occurs in a radical,* the equation is called a **radical equation.** Sometimes a suitable operation will change a radical equation to one that is linear or quadratic. A commonly used procedure is to isolate the most complicated radical on one side of the equation and then eliminate it by raising each side to a power equal to the index of the radical. Care must be taken, however, because apparent solutions that are not, in fact, solutions of the original equation may result. These are called **extraneous solutions.** Thus, we need to check all answers when working with radical equations.

EXAMPLE 8 Solving a Radical Equation

Find the real solutions of the equation $\sqrt[3]{2x - 4} - 2 = 0$.

Graphing Solution Figure 46 shows the graph of the equation $Y_1 = \sqrt[3]{2x - 4} - 2$. From the graph, we see one x-intercept near 6. Using ZERO (or ROOT), we find that the x-intercept is 6. Thus, the only solution is $x = 6$.

Figure 46

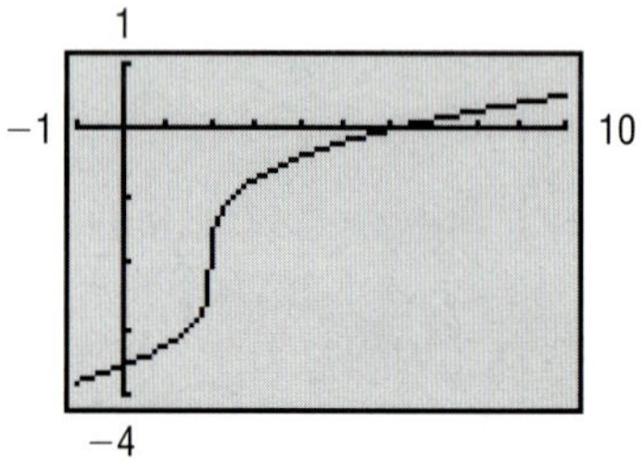

Algebraic Solution The equation contains a radical whose index is 3. We isolate it on the left side:

$$\sqrt[3]{2x - 4} - 2 = 0$$
$$\sqrt[3]{2x - 4} = 2$$

Now raise each side to the third power (the index of the radical is 3) and solve.

$$(\sqrt[3]{2x - 4})^3 = 2^3$$
$$2x - 4 = 8 \quad \text{Simplify}$$
$$2x = 12 \quad \text{Add 4 to each side}$$
$$x = 6 \quad \text{Divide both sides by 2}$$

*A detailed discussion of square roots and radicals may be found in Section 8 of the Appendix.

✓CHECK: $\sqrt[3]{2(6)-4}-2=\sqrt[3]{12-4}-2=\sqrt[3]{8}-2=2-2=0.$

The solution is $x = 6$.

Sometimes, we need to raise each side to a power more than once in order to solve a radical equation algebraically.

EXAMPLE 9 Solving a Radical Equation

Find the real solutions of the equation $\sqrt{2x+3}-\sqrt{x+2}=2$.

Graphing Solution Graph $Y_1=\sqrt{2x+3}-\sqrt{x+2}$ and $Y_2=2$. See Figure 47. From the graph it appears that there is one point of intersection. Using INTERSECT, the point of intersection is (23, 2), so the solution is $x = 23$.

Figure 47

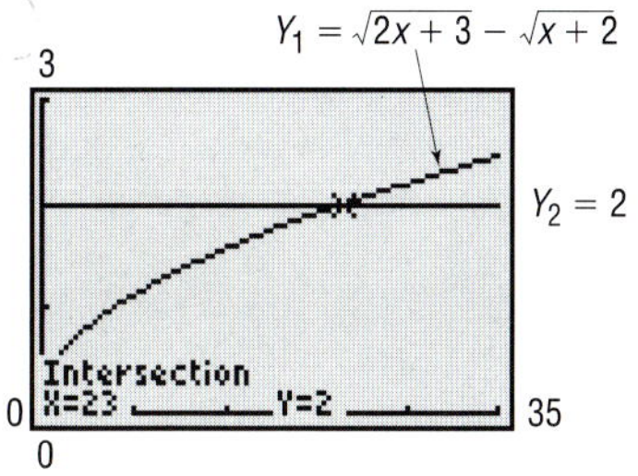

Algebraic Solution First, we choose to isolate the more complicated radical expression (in this case, $\sqrt{2x+3}$) on the left side:

$$\sqrt{2x+3}=\sqrt{x+2}+2$$

Now square both sides (the index of the radical is 2):

$$\begin{aligned}(\sqrt{2x+3})^2 &= (\sqrt{x+2}+2)^2\\ 2x+3 &= (\sqrt{x+2})^2+4\sqrt{x+2}+4\\ 2x+3 &= x+2+4\sqrt{x+2}+4\\ 2x+3 &= x+6+4\sqrt{x+2}\end{aligned}$$

Because the equation still contains a radical, we isolate the remaining radical on the right side and again square both sides.

$$\begin{aligned}x-3 &= 4\sqrt{x+2} &&\\ (x-3)^2 &= 16(x+2) && \textit{Square both sides.}\\ x^2-6x+9 &= 16x+32 && \textit{Remove parentheses.}\\ x^2-22x-23 &= 0 && \textit{Put in standard form.}\\ (x-23)(x+1) &= 0 && \textit{Factor.}\\ x=23 \quad \text{or} \quad x &= -1 &&\end{aligned}$$

The original equation appears to have the solution set $\{-1, 23\}$. However, we have not yet checked.

✓CHECK: $\sqrt{2(23)+3}-\sqrt{23+2}=\sqrt{49}-\sqrt{25}=7-5=2$

$\sqrt{2(-1)+3}-\sqrt{-1+2}=\sqrt{1}-\sqrt{1}=1-1=0$

Thus, the equation has only one solution, 23; the solution -1 is extraneous.

NOW WORK PROBLEMS 49 AND 55.

113, 38, 13, 17, 15, 29, 43

1.3 EXERCISES

In Problems 1–72, solve each equation algebraically. Verify your solution using a graphing utility.

1. $3x + 2 = x + 6$
2. $2x + 7 = 3x + 5$
3. $2t - 6 = 3 - t$
4. $5y + 6 = -18 - y$
5. $6 - x = 2x + 9$
6. $3 - 2x = 2 - x$
7. $3 + 2n = 5n + 7$
8. $3 - 2m = 3m + 1$
9. $2(3 + 2x) = 3(x - 4)$
10. $3(2 - x) = 2x - 1$
11. $8x - (2x + 1) = 3x - 10$
12. $5 - (2x - 1) = 10$
13. $\frac{2}{3}p = \frac{1}{2}p + \frac{1}{3}$
14. $\frac{1}{2} - \frac{1}{3}p = \frac{4}{3}$
15. $0.9t = 0.4 + 0.1t$
16. $0.9t = 1 + t$
17. $\frac{x + 1}{3} + \frac{x + 2}{7} = 5$
18. $\frac{2x + 1}{3} + 16 = 3x$
19. $\frac{2}{y} + \frac{4}{y} = 3$
20. $\frac{4}{y} - 5 = \frac{5}{2y}$
21. $x^2 = 9x$
22. $x^2 = -4x$
23. $x^2 - 25 = 0$
24. $x^2 - 9 = 0$
25. $z^2 + z - 12 = 0$
26. $v^2 + 7v + 12 = 0$
27. $2x^2 - 5x - 3 = 0$
28. $3x^2 + 5x + 2 = 0$
29. $3t^2 - 48 = 0$
30. $2y^2 - 50 = 0$
31. $x(x - 7) + 12 = 0$
32. $x(x + 1) = 12$
33. $4x^2 + 9 = 12x$
34. $25x^2 + 16 = 40x$
35. $6(p^2 - 1) = 5p$
36. $2(2u^2 - 4u) + 3 = 0$
37. $6x - 5 = \frac{6}{x}$
38. $x + \frac{12}{x} = 7$
39. $(x + 7)(x - 1) = (x + 1)^2$
40. $(x + 2)(x - 3) = (x - 3)^2$
41. $x(2x - 3) = (2x + 1)(x - 4)$
42. $x(1 + 2x) = (2x - 1)(x - 2)$
43. $\sqrt{2t - 1} = 1$
44. $\sqrt{3t + 4} = 2$
45. $\sqrt{3t + 1} = -4$
46. $\sqrt{5t + 4} = -3$
47. $\sqrt[3]{1 - 2x} - 3 = 0$
48. $\sqrt[3]{1 - 2x} - 1 = 0$
49. $\sqrt{15 - 2x} = x$
50. $\sqrt{12 - x} = x$
51. $x = 2\sqrt{x - 1}$
52. $x = 2\sqrt{-x - 1}$
53. $3 + \sqrt{3x + 1} = x$
54. $2 + \sqrt{12 - 2x} = x$
55. $\sqrt{2x + 3} - \sqrt{x + 1} = 1$
56. $\sqrt{3x + 7} + \sqrt{x + 2} = 1$
57. $\sqrt{3x + 1} - \sqrt{x - 1} = 2$
58. $\sqrt{3x - 5} - \sqrt{x + 7} = 2$
59. $|2x| = 8$
60. $|3x| = 15$
61. $|2x + 3| = 5$
62. $|3x - 1| = 2$
63. $|1 - 4t| = 5$
64. $|1 - 2z| = 3$
65. $|-2x| = 8$
66. $|-x| = 1$
67. $|-2|x = 4$
68. $|3|x = 9$
69. $\frac{2}{3}|x| = 8$
70. $\frac{3}{4}|x| = 9$
71. $\left|\frac{x}{3} + \frac{2}{5}\right| = 2$
72. $\left|\frac{x}{2} - \frac{1}{3}\right| = 1$

In Problems 73–76, solve each equation. The letters a, b, and c are constants.

73. $ax - b = c, \quad a \neq 0$

74. $1 - ax = b, \quad a \neq 0$

75. $\frac{x}{a} + \frac{x}{b} = c, \quad a \neq 0, \quad b \neq 0, \quad a \neq -b$

76. $\frac{a}{x} + \frac{b}{x} = c, \quad c \neq 0$

In Problems 77–84, use a graphing utility to approximate the real solutions, if any, of each equation rounded to two decimal places.

77. $x^2 - 4x + 2 = 0$

78. $x^2 + 4x + 2 = 0$

79. $x^2 + \sqrt{3}x = 3$

80. $x^2 = 2 - \sqrt{2}x$

81. $\pi x^2 = x + \pi$

82. $\pi x^2 + \pi x = 2$

83. $3x^2 + 8\pi x + \sqrt{29} = 0$

84. $\pi x^2 - 15\sqrt{2}x + 20 = 0$

Problems 85–90 list some formulas that occur in applications. Solve each formula for the indicated variable.

85. **Electricity** $\frac{1}{R} = \frac{1}{R_1} + \frac{1}{R_2}$ for R

86. **Finance** $A = P(1 + rt)$ for r

87. **Mechanics** $F = \frac{mv^2}{R}$ for R

88. **Chemistry** $PV = nRT$ for T

89. **Mathematics** $S = \frac{a}{1 - r}$ for r

90. **Mechanics** $v = -gt + v_0$ for t

91. **Physics: Using Sound to Measure Distance** The depth of a well can sometimes be found by dropping an object into the well and measuring the time elapsed until a sound is heard. If t_1 is the time (measured in seconds) that it takes for the object to strike the bottom of the well, then t_1 will obey the equation $s = 16t_1^2$, where s is the distance (measured in feet). It follows that $t_1 = \sqrt{s}/4$. Suppose that t_2 is the time that it takes for the sound of the impact to reach your ears. Because sound waves are known to travel at a speed of approximately 1100 feet per second, the time t_2 to travel the distance s will be $t_2 = s/1100$. Now $t_1 + t_2$ is the total time that elapses from the moment that the object is dropped to the moment that a sound is heard. Thus, we have the equation

$$\text{Total time elapsed} = \frac{\sqrt{s}}{4} + \frac{s}{1100}$$

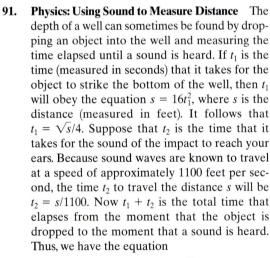

Find the depth of a well if the total time elapsed from dropping a rock to hearing it hit bottom is 4 seconds by graphing

$$Y_1 = \frac{\sqrt{x}}{4} + \frac{x}{1100}, \quad Y_2 = 4$$

for $0 \leq x \leq 300$ and $0 \leq Y_1 \leq 5$ and finding the point of intersection.

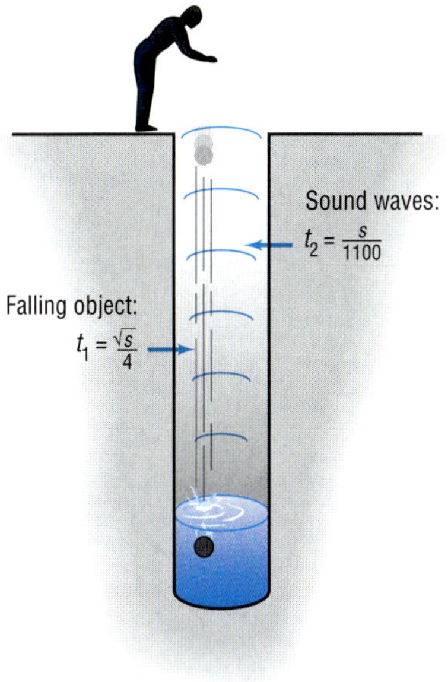

92. Make up a radical equation that has no solution.

93. Make up a radical equation that has an extraneous solution.

94. Discuss the step in the solving process for radical equations that leads to the possibility of extraneous solutions. Why is there no such possibility for linear and quadratic equations?

1.4 SETTING UP EQUATIONS; APPLICATIONS

1 Translate Verbal Descriptions into Mathematical Expressions

2 Set up Applied Problems

3 Solve Interest Problems

4 Solve Mixture Problems

5 Solve Uniform Motion Problems

6 Solve Constant Rate Job Problems

Applied (word) problems do not come in the form "Solve the equation . . . ." Instead, they supply information using words, a verbal description of the real problem. So, to solve applied problems we must be able to translate the verbal description using the language of mathematics. We do this by using variables to represent unknown quantities and then finding relationships (such as equations) that involve these variables. The process of doing all this is called **mathematical modeling.**

Any solution to the mathematical problem must be checked against the mathematical problem, the verbal description, and the real problem. See Figure 48 for an illustration of the modeling process.

Figure 48

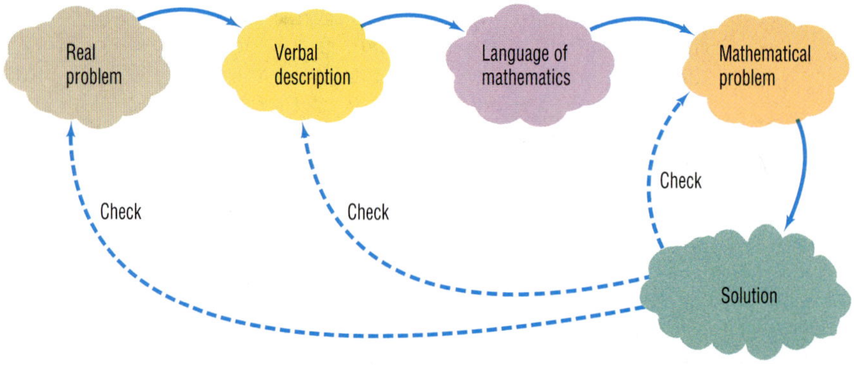

1 Let's look at a few examples that will help you to translate certain words into mathematical symbols.

EXAMPLE 1 Translating Verbal Descriptions into Mathematical Expressions

(a) The area of a rectangle is the product of its length times its width.

Translation: If A is used to represent the area, l the length, and w the width, then $A = lw$.

(b) For uniform motion, the velocity of an object equals the distance traveled divided by the time required.

Translation: If v is the velocity, s the distance, and t the time, then $v = s/t$.

(c) A total of \$5000 is invested, some in stocks and some in bonds. If the amount invested in stocks is x, express the amount invested in bonds in terms of x.

Translation: If x is the amount invested in stocks, then the amount invested in bonds is $5000 - x$, since their sum is $x + (5000 - x) = 5000$.

(d) Let x denote a number.

The number 5 times as large as x is $5x$.

The number 3 less than x is $x - 3$.

The number that exceeds x by 4 is $x + 4$.

The number that, when added to x, gives 5 is $5 - x$.

NOW WORK PROBLEM 1.

2 Always check the units used to measure the variables of an applied problem. In Example 1(a), if l is measured in feet, then w also must be expressed in feet, and A will be expressed in square feet. In Example 1(b), if v is measured in miles per hour, then the distance s must be expressed in miles and the time t must be expressed in hours. It is a good practice to check units to be sure that they are consistent and make sense.

Although each situation has its own unique features, we can provide an outline of the steps to follow in setting up applied problems.

Steps for Setting Up Applied Problems

STEP 1: Read the problem carefully, perhaps two or three times. Pay particular attention to the question being asked in order to identify what you are looking for. If you can, determine realistic possibilities for the answer.

STEP 2: Assign a letter (variable) to represent what you are looking for, and, if necessary, express any remaining unknown quantities in terms of this variable.

STEP 3: Make a list of all the known facts, and translate them into mathematical expressions. These may take the form of an equation (or, later, an inequality) involving the variable. If possible, draw an appropriately labeled diagram to assist you. Sometimes a table or chart helps.

STEP 4: Solve the equation for the variable, and then answer the question.

STEP 5: Check the answer with the facts in the problem. If it agrees, congratulations! If it does not agree, try again.

Let's look at an example.

EXAMPLE 2 Investments

A total of \$18,000 is invested, some in stocks and some in bonds. If the amount invested in bonds is half that invested in stocks, how much is invested in each category?

Solution

STEP 1: We are being asked to find the amount of two investments. These amounts must total \$18,000. (Do you see why?)

STEP 2: If we let x equal the amount invested in stocks, then $18{,}000 - x$ is the amount invested in bonds. [Look back at Example 1(c) to see why.]

STEP 3: We set up a table:

Amount in Stocks	Amount in Bonds	Reason
x	$18{,}000 - x$	Total invested is \$18,000

Since the total amount invested in bonds $(18{,}000 - x)$ is half that in stocks (x), we obtain the equation $18{,}000 - x = \frac{1}{2}x$.

STEP 4:
$$\begin{aligned} 18{,}000 - x &= \tfrac{1}{2}x \\ 18{,}000 &= x + \tfrac{1}{2}x \\ 18{,}000 &= \tfrac{3}{2}x \\ \left(\tfrac{2}{3}\right)18{,}000 &= \left(\tfrac{2}{3}\right)\left(\tfrac{3}{2}x\right) \\ 12{,}000 &= x \end{aligned}$$

Thus, \$12,000 is invested in stocks and \$18,000 − \$12,000 = \$6,000 is invested in bonds.

STEP 5: The total invested is \$12,000 + \$6000 = \$18,000, and the amount in bonds (\$6000) is half that in stocks (\$12,000).

NOW WORK PROBLEM 11.

EXAMPLE 3 Determining an Hourly Wage

Shannon grossed \$435 one week by working 52 hours. Her employer pays time-and-a-half for all hours worked in excess of 40 hours. With this information, can you determine Shannon's regular hourly wage?

Solution STEP 1: We are looking for an hourly wage. Our answer will be in dollars per hour.

STEP 2: Let x represent the regular hourly wage; x is measured in dollars per hour.

STEP 3: We set up a table:

	Hours Worked	Hourly Wage	Salary
Regular	40	x	$40x$
Overtime	12	$1.5x$	$12(1.5x) = 18x$

The sum of regular salary plus overtime salary will equal \$435. From the table, $40x + 18x = 435$.

STEP 4:
$$\begin{aligned} 40x + 18x &= 435 \\ 58x &= 435 \\ x &= 7.50 \end{aligned}$$

Shannon's regular hourly wage is \$7.50 per hour.

STEP 5: Forty hours yields a salary of $40(7.50) = \$300$, and 12 hours of overtime yields a salary of $12(1.5)(7.50) = \$135$, for a total of \$435.

NOW WORK PROBLEM 15.

Interest

3 The next example involves **interest.** Interest is money paid for the use of money. The total amount borrowed (whether by an individual from a bank in the form of a loan or by a bank from an individual in the form of a savings account) is called the **principal.** The **rate of interest,** expressed as a percent, is the amount charged for the use of the principal for a given period of time, usually on a yearly (that is, per annum) basis.

Simple Interest Formula

If a principal of P dollars is borrowed for a period of t years at a per annum interest rate r, expressed as a decimal, the interest I charged is

$$I = Prt \tag{1}$$

Interest charged according to formula (1) is called **simple interest.**

EXAMPLE 4 Finance: Computing Interest on a Loan

Suppose that Juanita borrows \$500 for 6 months at the simple interest rate of 9% per annum. What is the interest that Juanita will be charged on this loan? How much does Juanita owe after 6 months?

Solution The rate of interest is given per annum, so the actual time that the money is borrowed must be expressed in years. The interest charged would be the principal ($P = \$500$) times the rate of interest ($r = 9\% = 0.09$) times the time in years ($t = \frac{6}{12} = \frac{1}{2}$):

$$\text{Interest charged} = I = Prt = (500)(0.09)(\tfrac{1}{2}) = \$22.50$$

Juanita will owe the amount she borrowed, plus interest; that is, she will owe $\$500 + \$22.50 = \$522.50$

EXAMPLE 5 Financial Planning

Candy has \$70,000 to invest and requires an overall rate of return of 9%. She can invest in a safe, government-insured Certificate of Deposit, but it only pays 8%. To obtain 9%, she agrees to invest some of her money in non-insured corporate bonds paying 12%. How much should be placed in each investment to achieve her goals?

Solution STEP 1: The question is asking for two dollar amounts: the principal to invest in the corporate bonds and the principal to invest in the certificate of deposit.

STEP 2: We let x represent the amount (in dollars) to be invested in the bonds. Then $70{,}000 - x$ is the amount that will be invested in the certificate. (Do you see why?)

STEP 3: We set up a table:

	Principal $	Rate	Time yr	Interest $
Bonds	x	$12\% = 0.12$	1	$0.12x$
Certificate	$70{,}000 - x$	$8\% = 0.08$	1	$0.08(70{,}000 - x)$
Total	70,000	$9\% = 0.09$	1	$0.09(70{,}000) = 6300$

Since the total interest from the investments is equal to $0.09(70{,}000) = 6300$, we must have the equation

$$0.12x + 0.08(70{,}000 - x) = 6300$$

(Note that the units are consistent: the unit is dollars on each side.)

STEP 4:
$$\begin{aligned} 0.12x + 5600 - 0.08x &= 6300 \\ 0.04x &= 700 \\ x &= 17{,}500 \end{aligned}$$

Candy should place \$17,500 in the bonds and \$70,000 − \$17,500 = \$52,500 in the certificate.

STEP 5: The interest on the bonds after 1 year is 0.12(\$17,500) = \$2100; the interest on the certificate after 1 year is 0.08(\$52,500) = \$4200. The total annual interest is \$6300, the required amount.

NOW WORK PROBLEM 27.

Mixture Problems

4 Oil refineries sometimes produce gasoline that is a blend of two or more types of fuel; bakeries occasionally blend two or more types of flour for their bread. These problems are referred to as **mixture problems** because they combine two or more quantities to form a mixture.

EXAMPLE 6 Blending Coffees

The manager of a Starbucks store decides to experiment with a new blend of coffee. She will mix some B Grade Columbian coffee that sells for \$5 per pound with some A Grade Arabica coffee that sells for \$10 per pound to get 100 pounds of the new blend. The selling price of the new blend is to be \$7 per pound and there is to be no difference in revenue from selling the new blend versus selling the other types. How many pounds of the B Grade Columbian and A Grade Arabica coffees are required?

Solution Let x represent the number of pounds of the B Grade Columbian coffee. Then $100 - x$ equals the number of pounds of the A Grade Arabica coffee. See Figure 49.

Figure 49

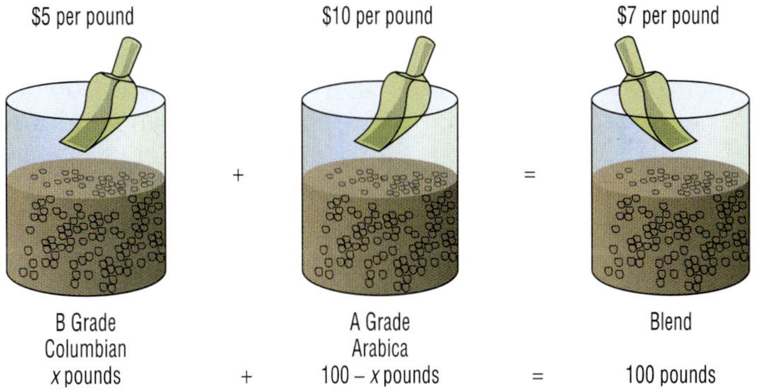

Since there is to be no difference in revenue, we must have

$$\begin{pmatrix}\text{price per}\\ \text{pound}\\ \text{of B Grade}\end{pmatrix}\begin{pmatrix}\text{\# pounds}\\ \text{B Grade}\end{pmatrix} + \begin{pmatrix}\text{price per}\\ \text{pound}\\ \text{of A Grade}\end{pmatrix}\begin{pmatrix}\text{\# pounds}\\ \text{A Grade}\end{pmatrix} = \begin{pmatrix}\text{price per}\\ \text{pound}\\ \text{of blend}\end{pmatrix}\begin{pmatrix}\text{\# pounds}\\ \text{blend}\end{pmatrix}$$

$$5 \cdot x + 10 \cdot (100 - x) = (7)(100)$$

We have the equation

$$\begin{aligned} 5x + 10(100 - x) &= 700 \\ 5x + 1000 - 10x &= 700 \\ -5x &= -300 \\ x &= 60 \end{aligned}$$

The manager should blend 60 pounds of B Grade Columbian coffee with $100 - 60 = 40$ pounds of A Grade Arabica coffee to get the desired blend.

✓CHECK: The 60 pounds of B Grade coffee would sell for $(\$5)(60) = \$300$ and the 40 pounds of A Grade coffee would sell for $(\$10)(40) = \$400$; the total revenue, \$700, equals the revenue obtained from selling the blend, as desired.

NOW WORK PROBLEM **31.**

Uniform Motion

5 Objects that move at a constant velocity are said to be in **uniform motion.** When the average velocity of an object is known, it can be interpreted as its constant velocity. Thus, a bicyclist traveling at an average velocity of 25 miles per hour is in uniform motion.

Uniform Motion Formula

If an object moves at an average velocity v, the distance s covered in time t is given by the formula

$$s = vt \qquad (2)$$

That is, Distance = Velocity · Time.

EXAMPLE 7 Physics: Uniform Motion

Tanya, who is a long-distance runner, runs at an average velocity of 8 miles per hour (mph). Two hours after Tanya leaves your house, you leave in your Honda and follow the same route. If your average velocity is 40 mph, how long will it be before you catch up to Tanya? How far will each of you be from your home?

Solution Refer to Figure 50. We use t to represent the time (in hours) that it takes the Honda to catch up with Tanya. When this occurs, the total time elapsed for Tanya is $t + 2$ hours.

Figure 50

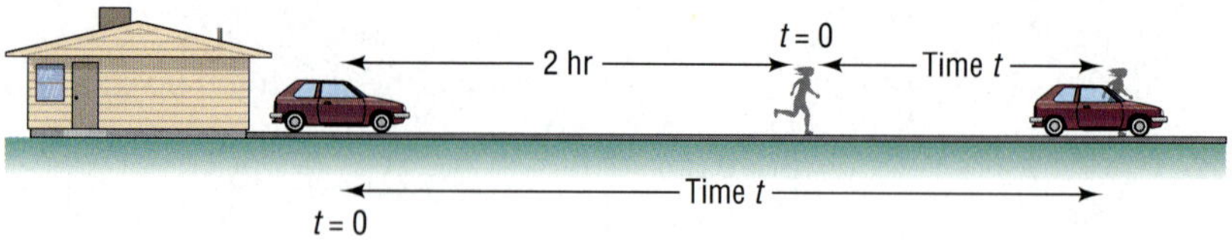

Set up the following table:

	Velocity mph	Time hr	Distance mi
Tanya	8	$t + 2$	$8(t + 2)$
Honda	40	t	$40t$

Since the distance traveled is the same, we are led to the following equation:

$$\begin{aligned} 8(t + 2) &= 40t \\ 8t + 16 &= 40t \\ 32t &= 16 \\ t &= \frac{1}{2} \text{ hour} \end{aligned}$$

It will take the Honda $\frac{1}{2}$ hour to catch up to Tanya. Each of you will have gone 20 miles.

✓CHECK: In 2.5 hours, Tanya travels a distance of $(2.5)(8) = 20$ miles. In $\frac{1}{2}$ hour, the Honda travels a distance of $(\frac{1}{2})(40) = 20$ miles.

EXAMPLE 8 Physics: Uniform Motion

A motorboat heads upstream a distance of 24 miles on the Illinois River, whose current is running at 3 miles per hour. The trip up and back takes 6 hours. Assuming that the motorboat maintained a constant speed relative to the water, what was its speed?

Figure 51

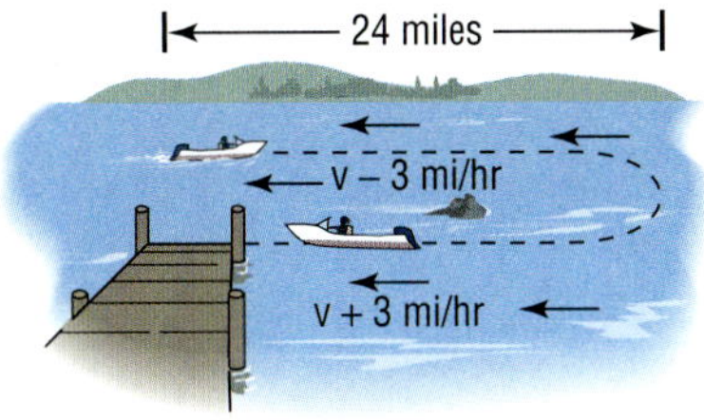

Solution See Figure 51. We use v to represent the constant speed of the motorboat relative to the water. Then the true speed going upstream is $v - 3$ miles per hour, and the true speed going downstream is $v + 3$ miles per hour. Since Distance = Velocity × Time, then Time = Distance/Velocity. We set up a table.

	Velocity mph	Distance mi	Time = Distance/Velocity hr
Upstream	$v - 3$	24	$\dfrac{24}{v-3}$
Downstream	$v + 3$	24	$\dfrac{24}{v+3}$

Since the total time up and back is 6 hours, we have

$$\frac{24}{v-3} + \frac{24}{v+3} = 6$$

$$\frac{24(v+3) + 24(v-3)}{(v-3)(v+3)} = 6 \qquad \textit{Add the rational expressions.*}$$

$$\frac{48v}{v^2 - 9} = 6 \qquad \textit{Simplify.}$$

$$48v = 6(v^2 - 9) \qquad \textit{Clear fractions.}$$

$$8v = v^2 - 9 \qquad \textit{Divide each side by 6.}$$

$$v^2 - 8v - 9 = 0 \qquad \textit{Place in standard form.}$$

$$(v - 9)(v + 1) = 0 \qquad \textit{Factor.}$$

$$v = 9 \quad \text{or} \quad v = -1$$

We discard the solution $v = -1$ mile per hour, so the speed of the motorboat relative to the water is 9 miles per hour.

NOW WORK PROBLEM 37.

Constant Rate Jobs

6 This section involves jobs that are performed at a **constant rate.** Our assumption is that, if a job can be done in t units of time, $1/t$ of the job is done in 1 unit of time. Let's look at an example.

EXAMPLE 9 Working Together to Do a Job

At 10 AM Danny is asked by his father to weed the garden. From past experience, Danny knows that this will take him 4 hours, working alone. His older brother, Mike, when it is his turn to do this job, requires 6 hours. Since Mike wants to go golfing with Danny and has a reservation for 1 PM, he agrees to help Danny. Assuming no gain or loss of efficiency, when will they finish if they work together? Can they make the golf date?

*You may wish to review Rational Expressions in Section 7 of the Appendix.

Solution

	Hours to Do Job	Part of Job Done in 1 Hour
Danny	4	$\frac{1}{4}$
Mike	6	$\frac{1}{6}$
Together	t	$\frac{1}{t}$

We set up the table in the margin. In 1 hour, Danny does $\frac{1}{4}$ of the job, and in 1 hour, Mike does $\frac{1}{6}$ of the job. Let t be the time (in hours) it takes them to do the job together. In 1 hour, then $1/t$ of the job is completed. We reason as follows:

$$\begin{pmatrix}\text{Part done by Danny}\\ \text{in 1 hour}\end{pmatrix} + \begin{pmatrix}\text{Part done by Mike}\\ \text{in 1 hour}\end{pmatrix} = \begin{pmatrix}\text{Part done together}\\ \text{in 1 hour}\end{pmatrix}$$

From the table,

$$\begin{aligned}\frac{1}{4} + \frac{1}{6} &= \frac{1}{t}\\ \frac{3+2}{12} &= \frac{1}{t}\\ \frac{5}{12} &= \frac{1}{t}\\ 5t &= 12\\ t &= \frac{12}{5}\end{aligned}$$

Working together, the job can be done in $\frac{12}{5}$ hours, or 2 hours, 24 minutes. They should make the golf date, since they will finish at 12:24 PM.

NOW WORK PROBLEM 41.

1.4 EXERCISES

In Problems 1–10, translate each sentence into a mathematical equation. Be sure to identify the meaning of all symbols.

1. **Geometry** The area of a circle is the product of the number π times the square of the radius.
2. **Geometry** The circumference of a circle is the product of the number π times twice the radius.
3. **Geometry** The area of a square is the square of the length of a side.
4. **Geometry** The perimeter of a square is four times the length of a side.
5. **Physics** Force equals the product of mass times acceleration.
6. **Physics** Pressure is force per unit area.
7. **Physics** Work equals force times distance.
8. **Physics** Kinetic energy is one-half the product of the mass times the square of the velocity.
9. **Business** The total variable cost of manufacturing x dishwashers is \$150 per dishwasher times the number of dishwashers manufactured.
10. **Business** The total revenue derived from selling x dishwashers is \$250 per dishwasher times the number of dishwashers manufactured.
11. **Finance** A total of \$20,000 is to be invested, some in bonds and some in Certificates of Deposit (CDs). If the amount invested in bonds is to exceed that in CDs by \$2000, how much will be invested in each type of instrument?
12. **Finance** A total of \$10,000 is to be divided up between Yani and Diane, with Diane to receive \$2000 less than Yani. How much will each receive?
13. **Finance** An inheritance of \$900,000 is to be divided among David, Paige, and Dan in the following manner: Paige is to receive $\frac{3}{4}$ of what David gets, while Dan gets $\frac{1}{2}$ of what David gets. How much does each receive?
14. **Sharing the Cost of a Pizza** Carole and Canter agree to share the cost of an \$18 pizza based on

how much each ate. If Carole ate $\frac{2}{3}$ the amount Canter ate, how much should each pay?

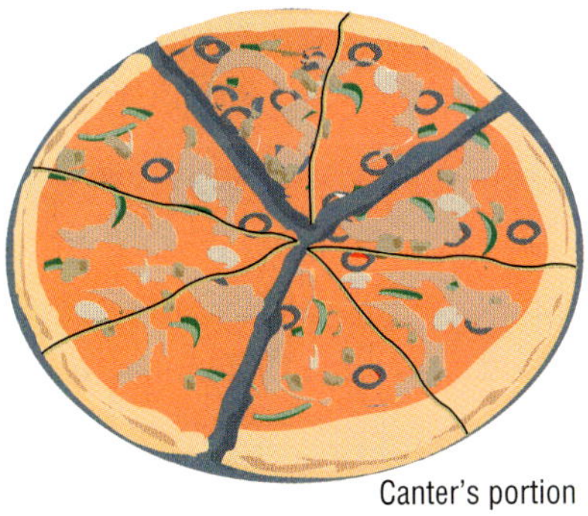

15. **Computing Hourly Wages** Laura, who is paid time-and-a-half for hours worked in excess of 40 hours, had gross weekly wages of $442 for 48 hours worked. What is her regular hourly rate?

16. **Computing Hourly Wages** Amy is paid time-and-a-half for hours worked in excess of 40 hours and double-time for hours worked on Sunday. If Amy had gross weekly wages of $342 for working 50 hours, 4 of which were on Sunday, what is her regular hourly rate?

17. **Football** In an NFL football game, the Bears scored a total of 41 points, including one safety (2 points) and two field goals (3 points each). After scoring a touchdown (6 points), a team is given the chance to score 1 or 2 extra points. The Bears, in trying to score 1 extra point after each touchdown, missed 2 extra points. How many touchdowns did the Bears get?

18. **Basketball** In a basketball game, the Bulls scored a total of 103 points and made three times as many field goals (2 points each) as free throws (1 point each). They also made eleven 3-point baskets. How many field goals did they have?

19. **Geometry** The perimeter of a rectangle is 60 feet. Find its length and width if the length is 8 feet longer than the width.

20. **Geometry** The perimeter of a rectangle is 42 meters. Find its length and width if the length is twice the width.

21. **Computing Grades** Going into the final exam, which will count as two tests, Brooke has test scores of 80, 83, 71, 61, and 95. What score does Brooke need on the final in order to have an average score of 80?

22. **Computing Grades** Going into the final exam, which will count as two-thirds of the final grade, Mike has test scores of 86, 80, 84, and 90. What score does Mike need on the final in order to earn a B, which requires an average score of 80? What does he need to earn an A, which requires an average of 90?

23. **Business: Discount Pricing** A builder of tract homes reduced the price of a model by 15%. If the new price is $125,000, what was its original price? How much can be saved by purchasing the model?

24. **Business: Discount Pricing** A car dealer, at a year-end clearance, reduces the list price of last year's models by 15%. If a certain four-door model has a discounted price of $8000, what was its list price? How much can be saved by purchasing last year's model?

25. **Business: Marking up the Price of Books** A college book store marks up the price that it pays the publisher for a book by 25%. If the selling price of a book is $56.00, how much did the book store pay for this book?

26. **Personal Finance: Cost of a Car** The suggested list price of a new car is $12,000. The dealer's cost is 85% of list. How much will you pay if the dealer is willing to accept $100 over cost for the car?

27. **Financing Planning** Betsy, a recent retiree, requires $6000 per year in extra income. She has $50,000 to invest and can invest in B-rated bonds paying 15% per year or in a Certificate of Deposit (CD) paying 7% per year. How much money should be invested in each to realize exactly $6000 in interest per year?

28. **Financial Planning** After 2 years, Betsy (see Problem 27) finds that she now will require $7000 per year. Assuming that the remaining information is the same, how should the money be reinvested?

29. **Banking** A bank loaned out $12,000, part of it at the rate of 8% per year and the rest at the rate of 18% per year. If the interest received totaled $1000, how much was loaned at 8%?

30. **Banking** Wendy, a loan officer at a bank, has $1,000,000 to lend and is required to obtain an average return of 18% per year. If she can lend at the rate of 19% or at the rate of 16%, how much can she lend at the 16% rate and still meet her requirement?

31. **Blending Teas** The manager of a store that specializes in selling tea decides to experiment with a new blend. She will mix some Earl Gray tea that sells for $5 per pound with some Orange Pekoe tea that sells for $3 per pound to get 100 pounds of the new blend. The selling price of the new blend is to be $4.50 per pound and there is to be no difference in revenue from selling the

new blend versus selling the other types. How many pounds of the Earl Gray tea and Orange Pekoe tea are required?

32. **Business: Blending Coffee** A coffee manufacturer wants to market a new blend of coffee that will cost \$3.90 per pound by mixing two coffees that sell for \$2.75 and \$5 per pound, respectively. What amounts of each coffee should be blended to obtain the desired mixture? ***[Hint: Assume that the total weight of the desired blend is 100 pounds.]***

33. **Business: Mixing Nuts** A nut store normally sells cashews for \$4.00 per pound and peanuts for \$1.50 per pound. But at the end of the month, the peanuts had not sold well, so in order to sell 60 pounds of peanuts, the manager decided to mix the 60 pounds of peanuts with some cashews and sell the mixture for \$2.50 per pound. How many pounds of cashews should be mixed with the peanuts to ensure no change in the profit?

34. **Business: Mixing Candy** A candy store sells boxes of candy containing caramels and cremes. Each box sells for \$12.50 and holds 30 pieces of candy (all pieces are the same size). If the caramels cost \$0.25 to produce and the cremes cost \$0.45 to produce, how many of each should be in a box to make a profit of \$3?

35. **Chemistry: Mixing Acids** How many ounces of pure water should be added to 20 ounces of a 40% solution of muriatic acid to obtain a 30% solution of muriatic acid?

36. **Chemistry: Mixing Acids** How many cubic centimeters of pure hydrochloric acid should be added to 20 cc of a 30% solution of hydrochloric acid to obtain a 50% solution?

37. **Physics: Uniform Motion** A Metra commuter train leaves Union Station in Chicago at 12 noon. Two hours later, an Amtrak train leaves on the same track, traveling at an average speed that is 50 miles per hour faster than the Metra train. At 3 PM, the Amtrak train is 10 miles behind the commuter train. How fast is each going?

38. **Physics: Uniform Motion** Two cars enter the Florida Turnpike at Commercial Boulevard at 8:00 AM, each heading for Wildwood. One car's average speed is 10 miles per hour more than the other's. The faster car arrives at Wildwood at 11:00 AM, $\frac{1}{2}$ hour before the other car. What is the average speed of each car? How far did each travel?

39. **Physics: Uniform Motion** A motorboat can maintain a constant speed of 16 miles per hour relative to the water. The boat makes a trip upstream to a certain point in 20 minutes; the return trip takes 15 minutes. What is the speed of the current? (See the figure.)

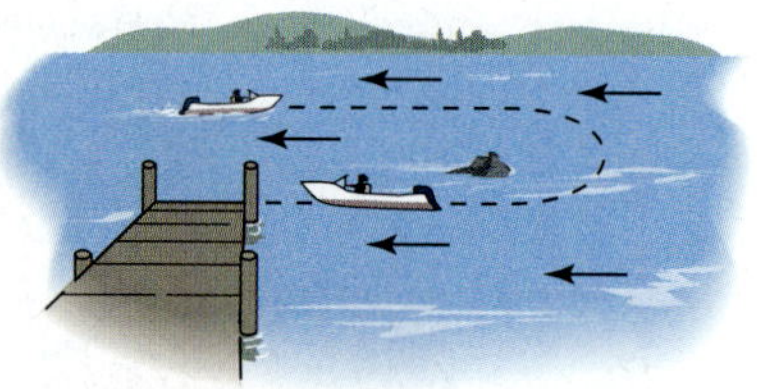

40. **Physics: Uniform Motion** A motorboat heads upstream on a river that has a current of 3 miles per hour. The trip upstream takes 5 hours, while the return trip takes 2.5 hours. What is the speed of the motorboat? (Assume that the motorboat maintains a constant speed relative to the water.)

41. **Working Together on a Job** Trent can deliver his newspapers in 30 minutes. It takes Lois 20 minutes to do the same route. How long would it take them to deliver the newspapers if they work together?

42. **Working Together on a Job** Patrick, by himself, can paint four rooms in 10 hours. If he hires April to help, they can do the same job together in 6 hours. If he lets April work alone, how long will it take her to paint four rooms?

43. **Dimensions of a Window** The area of the opening of a rectangular window is to be 143 square feet. If the length is to be 2 feet more than the width, what are the dimensions?

44. **Dimensions of a Window** The area of a rectangular window is to be 306 square centimeters. If the length exceeds the width by 1 centimeter, what are the dimensions?

45. **Geometry** Find the dimensions of a rectangle whose perimeter is 26 meters and whose area is 40 square meters.

46. **Watering a Field** An adjustable water sprinkler that sprays water in a circular pattern is placed at the center of a square field whose area is 1250 square feet (see the figure). What is the shortest radius setting that can be used if the field is to be completely enclosed within the circle?

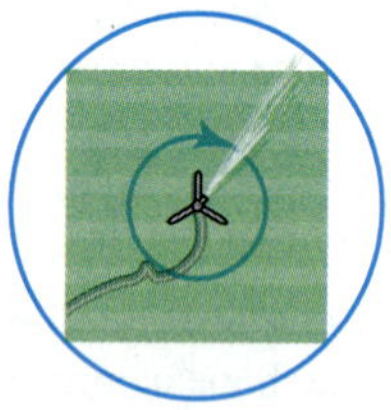

47. **Constructing a Box** An open box is to be constructed from a square piece of sheet metal by removing a square of side 1 foot from each corner and turning up the edges. If the box is to hold 4 cubic feet, what should be the dimensions of the sheet metal?

48. **Constructing a Box** Rework Problem 47 if the piece of sheet metal is a rectangle whose length is twice its width.

49. **Physics: Uniform Motion** A motorboat maintained a constant speed of 15 miles per hour relative to the water in going 10 miles upstream and then returning. The total time for the trip was 1.5 hours. Use this information to find the speed of the current.

50. **Dimensions of a Patio** A contractor orders 8 cubic yards of premixed cement, all of which is to be used to pour a rectangular patio that will be 4 inches thick. If the length of the patio is specified to be twice the width, what will be the patio dimensions? (1 cubic yard = 27 cubic feet)

51. **Enclosing a Garden** A gardener has 46 feet of fencing to be used to enclose a rectangular garden that has a border 2 feet wide surrounding it (see the figure).
    (a) If the length of the garden is to be twice its width, what will be the dimensions of the garden?
    (b) What is the area of the garden?
    (c) If the length and width of the garden were to be the same, what would be the dimensions of the garden?
    (d) What would be the area of the square garden?

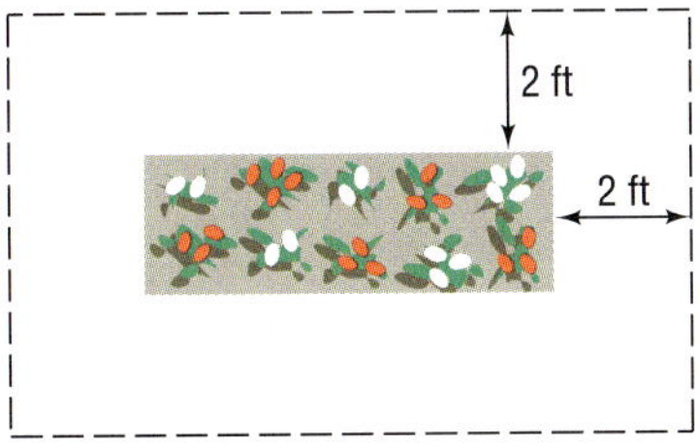

52. **Construction** A pond is enclosed by a wooden deck that is 3 feet wide. The fence surrounding the deck is 100 feet long.
    (a) If the pond is square, what are its dimensions?
    (b) If the pond is rectangular and the length of the pond is three times its width, what are the dimensions of the pond?
    (c) If the pond is circular, what is the diameter of the pond?
    (d) Which pond has the most area?

53. **Constructing a Border around a Pool** A pool in the shape of a circle measures 10 feet across. One cubic yard of concrete is to be used to create a circular border of uniform width around the pool. If the border is to have a depth of 3 inches, how wide will the border be? (1 cubic yard = 27 cubic feet)

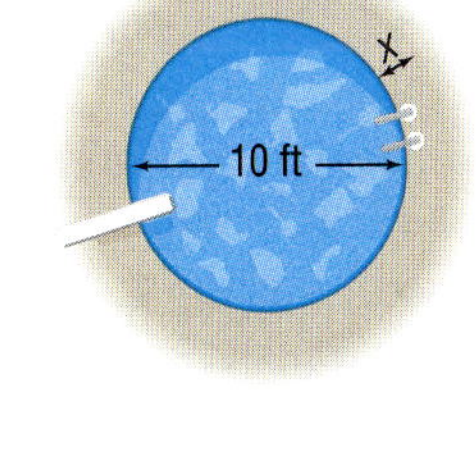

54. **Constructing a Border around a Pool** Rework Problem 53 if the depth of the border is 4 inches.

55. **Constructing a Border around a Garden** A landscaper, who just completed a rectangular flower garden measuring 6 feet by 10 feet, orders 1 cubic yard of premixed cement, all of which is to be used to create a border of uniform width around the garden. If the border is to have a depth of 3 inches, how wide will the border be? (1 cubic yard = 27 cubic feet)

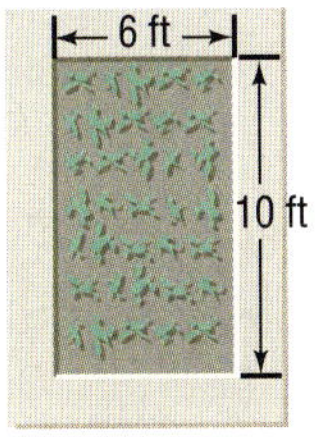

56. **Constructing a Coffee Can** A 39 ounce can of Hills Bros.® coffee requires 188.5 square inches of aluminum. If its height is 7 inches, what is its radius? (The surface area A of a right circular cylinder is $A = 2\pi r^2 + 2\pi rh$, where r is the radius and h is the height.)

57. **Mixing Water and Antifreeze** How much water should be added to 1 gallon of pure antifreeze to obtain a solution that is 60% antifreeze?

58. **Mixing Water and Antifreeze** The cooling system of a certain foreign-made car has a capacity of 15 liters. If the system is filled with a mixture

that is 40% antifreeze, how much of this mixture should be drained and replaced by pure antifreeze so that the system is filled with a solution that is 60% antifreeze?

59. **Cement Mix** A 20 pound bag of Economy brand cement mix contains 25% cement and 75% sand. How much pure cement must be added to produce a cement mix that is 40% cement?

60. **Chemistry: Salt Solutions** How much water must be evaporated from 240 gallons of a 3% salt solution to produce a 5% salt solution?

61. **Reducing the Size of a Candy Bar** A jumbo chocolate bar with a rectangular shape measures 12 centimeters in length, 7 centimeters in width, and 3 centimeters in thickness. Due to escalating costs of cocoa, management decides to reduce the volume of the bar by 10%. To accomplish this reduction, management decides that the new bar should have the same 3 centimeter thickness, but the length and width each should be reduced an equal number of centimeters. What should be the dimensions of the new candy bar?

62. **Reducing the Size of a Candy Bar** Rework Problem 61 if the reduction is to be 20%.

63. **Purity of Gold** The purity of gold is measured in karats, with pure gold being 24 karats. Other purities of gold are expressed as proportional parts of pure gold. Thus, 18 karat gold is $\frac{18}{24}$, or 75% pure gold; 12 karat gold is $\frac{12}{24}$, or 50% pure gold; and so on. How much 12 karat gold should be mixed with pure gold to obtain 60 grams of 16 karat gold?

64. **Chemistry: Sugar Molecules** A sugar molecule has twice as many atoms of hydrogen as it does oxygen and one more atom of carbon than oxygen. If a sugar molecule has a total of 45 atoms, how many are oxygen? How many are hydrogen?

65. **Running a Race** Mike can run the mile in 6 minutes, and Dan can run the mile in 9 minutes. If Mike gives Dan a head start of 1 minute, how far from the start will Mike pass Dan? (See the figure.) How long does it take?

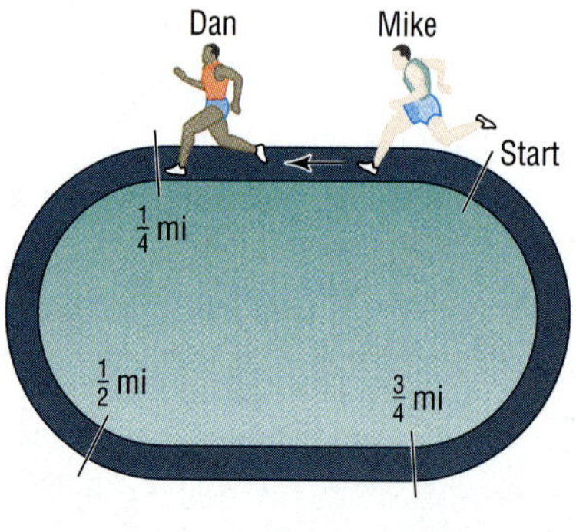

66. **Football** A tight end can run the 100 yard dash in 12 seconds. A defensive back can do it in 10 seconds. The tight end catches a pass at his own 20 yard line with the defensive back at the 15 yard line. (See the figure.) If no other players were nearby, at what yard line will the defensive back catch up to the tight end?
[Hint: At time $t = 0$, the defensive back is 5 yards behind the tight end.]

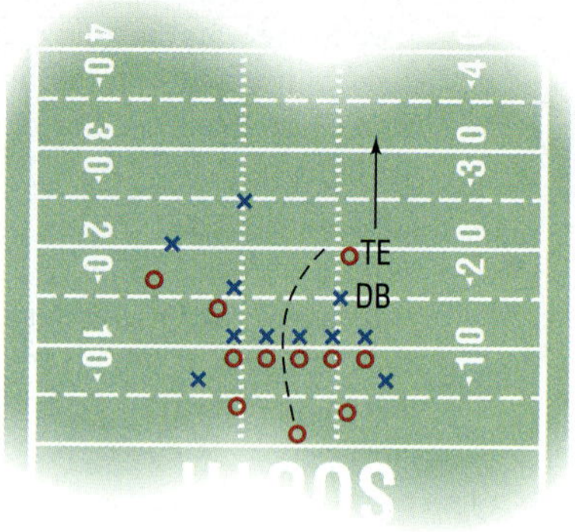

67. **Emptying Oil Tankers** An oil tanker can be emptied by the main pump in 4 hours. An auxiliary pump can empty the tanker in 9 hours. If the main pump is started at 9 AM, when should the auxiliary pump be started so that the tanker is emptied by noon?

68. **Using Two Pumps** A 5 horsepower (hp) pump can empty a pool in 5 hours. A smaller, 2 hp pump empties the same pool in 8 hours. The pumps are used together to begin emptying this pool. After two hours, the 2 hp pump breaks down. How long will it take the larger pump to empty the pool?

69. **Emptying a Tub** A bathroom tub will fill in 15 minutes with both faucets open and the stopper in place. With both faucets closed and the stopper removed, the tub will empty in 20 minutes. How long will it take for the tub to fill if both faucets are open and the stopper is removed?

70. **Range of an Airplane** An air rescue plane averages 300 miles per hour in still air. It carries enough fuel for 5 hours of flying time. If, upon takeoff, it encounters a wind of 30 miles per hour and the direction of the airplane is with the wind in one direction and against it in the other, how far can it fly and return safely? (Assume that the wind remains constant.)

71. **Home Equity Loans** Suppose that you obtain a home equity loan of $100,000 that requires only a monthly interest payment at 10% per annum, with the principal due after 5 years. You

decide to invest part of the loan in a 5 year CD that pays 9% per annum compounded and paid monthly and part in a B+ rated bond due in 5 years that pays 12% per annum compounded and paid monthly. What is the most that you can invest in the CD to ensure that the monthly home equity loan payment is made?

72. **Comparing Olympic Heroes** In the 1984 Olympics, Carl Lewis of the United States won the gold metal in the 100 meter race with a time of 9.99 seconds. In the 1896 Olympics, Thomas Burke, also of the United States, won the gold medal in the 100 meter race in 12.0 seconds. If they ran in the same race repeating their respective times, by how many meters would Lewis beat Burke?

73. **Computing Average Speed** In going from Chicago to Atlanta, a car averages 45 miles per hour, and in going from Atlanta to Miami, it averages 55 miles per hour. If Atlanta is halfway between Chicago and Miami, what is the average speed from Chicago to Miami? Discuss an intuitive solution. Write a paragraph defending your intuitive solution. Then solve the problem algebraically. Is your intuitive solution the same as the algebraic one? If not, find the flaw.

74. **Speed of a Plane** On a recent flight from Phoenix to Kansas City, a distance of 919 nautical miles, the plane arrived 20 minutes early. On leaving the aircraft, I asked the captain, "What was our tail wind?" He replied, "I don't know, but our ground speed was 550 knots." How can you determine if enough information is provided to find the tail wind? If possible, find the tail wind. (1 knot = 1 nautical mile per hour)

75. **Critical Thinking** You are the manager of a clothing store and have just purchased 100 dress shirts for $20.00 each. After 1 month of selling the shirts at the regular price, you plan to have a sale giving 40% off the original selling price. However, you still want to make a profit of $4 on each shirt at the sale price. What should you price the shirts at initially to ensure this? If, instead of 40% off at the sale, you give 50% off, by how much is your profit reduced?

76. **Critical Thinking** Make up a world problem that requires solving a linear equation as part of its solution. Exchange problems with a friend. Write a critique of your friend's problem.

77. **Critical Thinking** Without solving, explain what is wrong with the following mixture problem: How many liters of 25% ethanol should be added to 20 liters of 48% ethanol to obtain a solution of 58% ethanol? Now go through an algebraic solution. What happens?

1.5 SOLVING INEQUALITIES*

1 Use Interval Notation

2 Use Properties of Inequalities

3 Solve Linear Inequalities Algebraically and Graphically

4 Solve Combined Inequalities Algebraically and Graphically

5 Solve Absolute Value Inequalities Algebraically and Graphically

Suppose that a and b are two real numbers and $a < b$. We shall use the notation $a < x < b$ to mean that x is a number *between* a and b. Thus, the expression $a < x < b$ is equivalent to the two inequalities $a < x$ and $x < b$. Similarly, the expression $a \le x \le b$ is equivalent to the two inequalities $a \le x$ and $x \le b$. The remaining two possibilities, $a \le x < b$ and $a < x \le b$, are defined similarly.

Although it is acceptable to write $3 \ge x \ge 2$, it is preferable to reverse the inequality symbols and write instead $2 \le x \le 3$ so that, as you read from left to right, the values go from smaller to larger.

A statement such as $2 \le x \le 1$ is false because there is no number x for which $2 \le x$ and $x \le 1$. Finally, we never mix inequality symbols, as in $2 \le x \ge 3$.

*You may wish to review Inequalities, pp. 676–678, in Section 2 of the Appendix.

Intervals

1 Let a and b represent two real numbers with $a < b$:

A **closed interval,** denoted by **[a, b],** consists of all real numbers x for which $a \leq x \leq b$.

An **open interval,** denoted by **(a, b),** consists of all real numbers x for which $a < x < b$.

The **half-open,** or **half-closed, intervals** are **(a, b],** consisting of all real numbers x for which $a < x \leq b$, and **[a, b),** consisting of all real numbers x for which $a \leq x < b$.

In each of these definitions, a is called the **left end-point** and b the **right end-point** of the interval.

The symbol ∞ (read as "infinity") is not a real number but a notational device used to indicate unboundedness in the positive direction. The symbol $-\infty$ (read as "minus infinity") also is not a real number, but a notational device used to indicate unboundedness in the negative direction. Using the symbols ∞ and $-\infty$, we can define five other kinds of intervals:

$[a, \infty)$	consists of all real numbers x for which $x \geq a$ $(a \leq x < \infty)$
(a, ∞)	consists of all real numbers x for which $x > a$ $(a < x < \infty)$
$(-\infty, a]$	consists of all real numbers x for which $x \leq a$ $(-\infty < x \leq a)$
$(-\infty, a)$	consists of all real numbers x for which $x < a$ $(-\infty < x < a)$
$(-\infty, \infty)$	consists of all real numbers x $(-\infty < x < \infty)$

Note that ∞ (and $-\infty$) is never included as an end-point, since it is not a real number.

Table 8 summarizes interval notation, corresponding inequality notation, and their graphs.

EXAMPLE 1 Writing Inequalities Using Interval Notation

Write each inequality using interval notation.

(a) $1 \leq x \leq 3$ (b) $-4 < x < 0$ (c) $x > 5$ (d) $x \leq 1$

Solution

(a) $1 \leq x \leq 3$ describes all numbers x between 1 and 3, inclusive. In interval notation, we write $[1, 3]$.

(b) In interval notation, $-4 < x < 0$ is written $(-4, 0)$.

(c) $x > 5$ consists of all numbers x greater than 5. In interval notation, we write $(5, \infty)$.

(d) In interval notation, $x \leq 1$ is written $(-\infty, 1]$.

Table 8

Interval	Inequality	Graph
The open interval (a, b)	$a < x < b$	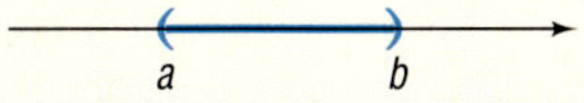
The closed interval $[a, b]$	$a \leq x \leq b$	
The half-open interval $[a, b)$	$a \leq x < b$	
The half-open interval $(a, b]$	$a < x \leq b$	
The interval $[a, \infty)$	$x \geq a$	
The interval (a, ∞)	$x > a$	
The interval $(-\infty, a]$	$x \leq a$	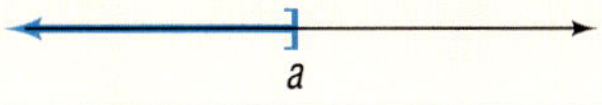
The interval $(-\infty, a)$	$x < a$	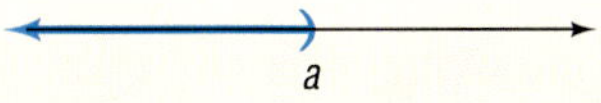
The interval $(-\infty, \infty)$	All real numbers	

EXAMPLE 2 Writing Intervals Using Inequality Notation

Write each interval as an inequality involving x.

(a) $[1, 4)$ (b) $(2, \infty)$ (c) $[2, 3]$ (d) $(-\infty, -3]$

Solution

(a) $[1, 4)$ consists of all numbers x for which $1 \leq x < 4$.
(b) $(2, \infty)$ consists of all numbers x for which $x > 2$ $(2 < x < \infty)$.
(c) $[2, 3]$ consists of all numbers x for which $2 \leq x \leq 3$.
(d) $(-\infty, -3]$ consists of all numbers x for which $x \leq -3$ $(-\infty < x \leq -3)$.

NOW WORK PROBLEMS 1 AND 9.

Properties of Inequalities

2 The product of two positive real numbers is positive, the product of two negative real numbers is positive, and the product of 0 and 0 is 0. Thus, for any real number a, the value of a^2 is 0 or positive; that is, a^2 is nonnegative. This is called the **nonnegative property.**

For any real number a, we have

Nonnegative Property

$$a^2 \geq 0 \tag{1}$$

If we add the same number to both sides of an inequality, we obtain an equivalent inequality. For example, since $3 < 5$, then $3 + 4 < 5 + 4$ or $7 < 9$. This is called the **addition property** of inequalities.

Addition Property of Inequalities

$$\text{If } a < b, \text{ then } a + c < b + c \qquad (2a)$$

$$\text{If } a > b, \text{ then } a + c > b + c \qquad (2b)$$

The addition property states that the sense, or direction, of an inequality remains unchanged if the same number is added to each side. Figure 52 illustrates the addition property (2a). In Figure 52(a), we see that a lies to the left of b. If c is positive, then $a + c$ and $b + c$ each lie c units to the right of a and b, respectively. Consequently, $a + c$ must lie to the left of $b + c$; that is, $a + c < b + c$. Figure 52(b) illustrates the situation if c is negative.

Figure 52

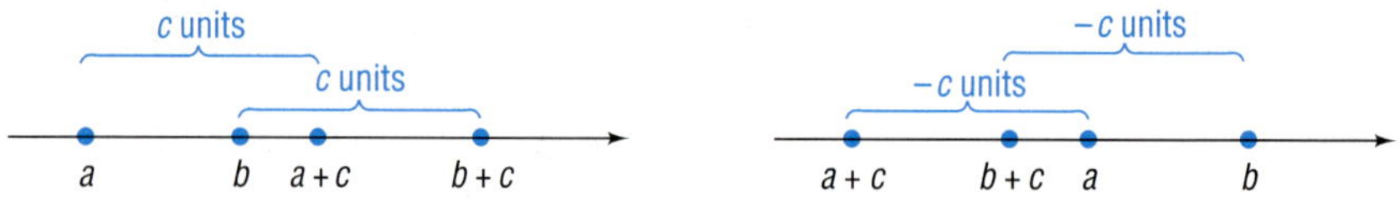

(a) If $a < b$ and $c > 0$, then $a + c < b + c$.

(b) If $a < b$ and $c < 0$, then $a + c < b + c$.

Draw an illustration similar to Figure 52 that illustrates the addition property (2b).

EXAMPLE 3 Addition Property of Inequalities

(a) If $x < -5$, then $x + 5 < -5 + 5$ or $x + 5 < 0$.

(b) If $x > 2$, then $x + (-2) > 2 + (-2)$ or $x - 2 > 0$.

NOW WORK PROBLEM 17.

We will use two examples to arrive at our next property.

EXAMPLE 4 Multiplying an Inequality by a Positive Number

Express as an inequality the result of multiplying each side of the inequality $3 < 7$ by 2.

Solution We begin with

$$3 < 7$$

Multiplying each side by 2 yields the numbers 6 and 14, so we have

$$6 < 14$$

EXAMPLE 5 Multiplying an Inequality by a Negative Number

Express as an inequality the result of multiplying each side of the inequality $9 > 2$ by -4.

Solution We begin with

$$9 > 2$$

Multiplying each side by -4 yields the numbers -36 and -8, so we have

$$-36 < -8$$

Note that the effect of multiplying both sides of $9 > 2$ by the negative number -4 is that the direction of the inequality symbol is reversed.

Examples 4 and 5 illustrate the following general **multiplication properties** for inequalities:

Multiplication Properties for Inequalities

$$\text{If } a < b \text{ and if } c > 0, \text{ then } ac < bc. \qquad \text{If } a < b \text{ and if } c < 0, \text{ then } ac > bc. \tag{3a}$$

$$\text{If } a > b \text{ and if } c > 0, \text{ then } ac > bc. \qquad \text{If } a > b \text{ and if } c < 0, \text{ then } ac < bc. \tag{3b}$$

The multiplication properties state that the sense, or direction, of an inequality *remains the same* if each side is multiplied by a *positive* real number, while the direction is *reversed* if each side is multiplied by a *negative* real number.

EXAMPLE 6 Multiplication Property of Inequalities

(a) If $2x < 6$ then $\frac{1}{2}(2x) < \frac{1}{2}(6)$ or $x < 3$.

(b) If $\dfrac{x}{-3} > 12$, then $-3\left(\dfrac{x}{-3}\right) < -3(12)$ or $x < -36$.

(c) If $-4x > -8$, then $\dfrac{-4x}{-4} < \dfrac{-8}{-4}$ or $x < 2$.

(d) If $-x < 8$, then $(-1)(-x) > (-1)(8)$ or $x > -8$.

NOW WORK PROBLEM 23.

Solving Inequalities

3 An **inequality in one variable** is a statement involving two expressions, at least one containing the variable, separated by one of the inequality symbols, $<$, $\leq$, $>$, or $\geq$. To **solve an inequality** means to find all values of the variable for which the statement is true. These values are called **solutions** of the inequality.

For example, the following are all inequalities involving one variable, x:

$$x + 5 < 8, \qquad 2x - 3 \geq 4, \qquad x^2 - 1 \leq 3, \qquad \frac{x + 1}{x - 2} > 0$$

Two inequalities having exactly the same solution set are called **equivalent inequalities.**

As with equations, one method for solving an inequality is to replace it by a series of equivalent inequalities until an inequality with an obvious solution, such as $x < 3$, is obtained. We obtain equivalent inequalities by applying some of the same operations as those used to find equivalent equations. The addition property and the multiplication properties form the basis for the following procedures.

Procedures That Leave the Inequality Symbol Unchanged

1. Simplify both sides of the inequality by combining like terms and eliminating parentheses:

   Replace $(x + 2) + 6 > 2x + 5(x + 1)$
   by $x + 8 > 7x + 5$

2. Add or subtract the same expression on both sides of the inequality:

   Replace $3x - 5 < 4$
   by $(3x - 5) + 5 < 4 + 5$

3. Multiply or divide both sides of the inequality by the same *positive* expression:

   Replace $4x > 16$ by $\dfrac{4x}{4} > \dfrac{16}{4}$

Procedures That Reverse the Sense or Direction of the Inequality Symbol

1. Interchange the two sides of the inequality:

   Replace $3 < x$ by $x > 3$

2. Multiply or divide both sides of the inequality by the same *negative* expression:

   Replace $-2x > 6$ by $\dfrac{-2x}{-2} < \dfrac{6}{-2}$

To solve an inequality using a graphing utility, we follow these steps:

Steps for Solving Inequalities Graphically

STEP 1: Write the inequality in one of the following forms:

$$Y_1 < Y_2, \quad Y_1 > Y_2, \quad Y_1 \leq Y_2, \quad Y_1 \geq Y_2$$

STEP 2: Graph Y_1 and Y_2 on the same screen.

STEP 3: If the inequality is of the form $Y_1 < Y_2$, determine on what interval Y_1 is below Y_2.

If the inequality is of the form $Y_1 > Y_2$, determine on what interval Y_1 is above Y_2.

If the inequality is not strict ($\leq$ or $\geq$), include the x-coordinates of the points of intersection in the solution.

EXAMPLE 7 Solving an Inequality

Solve the inequality $4x + 7 \geq 2x - 3$, and graph the solution set.

Graphing Solution We graph $Y_1 = 4x + 7$ and $Y_2 = 2x - 3$ on the same screen. See Figure 53. Using the INTERSECT command, we find that Y_1 and Y_2 intersect at $x = -5$. The graph of Y_1 is above that of Y_2, $Y_1 > Y_2$, to the right of the point of intersection. Since the inequality is not strict, the solution set is $\{x|x \geq -5\}$ or, using interval notation, $[-5, \infty)$.

Figure 53

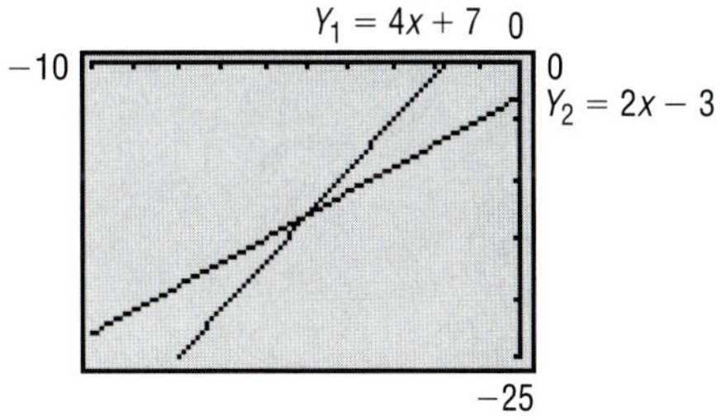

Algebraic Solution

$$4x + 7 \geq 2x - 3$$
$$4x + 7 - 7 \geq 2x - 3 - 7 \quad \textit{Subtract 7 from both sides.}$$
$$4x \geq 2x - 10 \quad \textit{Simplify.}$$
$$4x - 2x \geq 2x - 10 - 2x \quad \textit{Subtract 2x from both sides.}$$
$$2x \geq -10 \quad \textit{Simplify.}$$
$$\frac{2x}{2} \geq \frac{-10}{2} \quad \textit{Divide both sides by 2. (The direction of the inequality symbol is unchanged.)}$$
$$x \geq -5 \quad \textit{Simplify.}$$

Figure 54

$-5 \leq x < \infty$ or $[-5, \infty)$

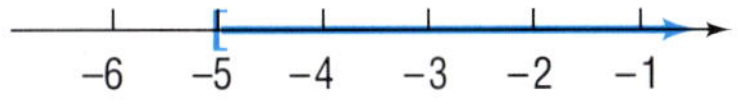

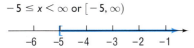

The solution set is $\{x|-5 \leq x < \infty\}$ or, using interval notation, all numbers in the interval $[-5, \infty)$.

See Figure 54 for the graph.

NOW WORK PROBLEM **31.**

EXAMPLE 8 Solving Combined Inequalities

4

Solve the inequality $-5 < 3x - 2 < 1$ and draw a graph to illustrate the solution.

Graphing Solution

To solve a combined inequality, we graph each part: $Y_1 = -5$, $Y_2 = 3x - 2$, and $Y_3 = 1$. We seek the values of x for which the graph of Y_2 is between the graphs of Y_1 and Y_3. See Figure 55. The point of intersection of Y_1 and Y_2 is $(-1, 5)$, and the point of intersection of Y_2 and Y_3 is $(1, 1)$. The inequality is true for all values of x between these two intersection points. Since the inequality is strict, the solution set is $\{x|-1 < x < 1\}$ or, using interval notation, $(-1, 1)$.

Figure 55

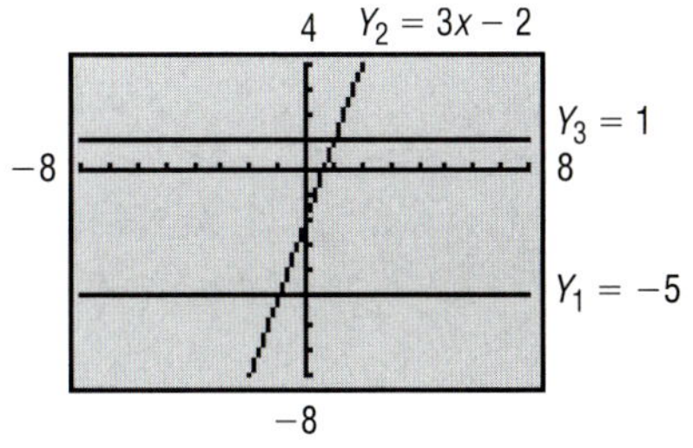

Algebraic Solution Recall that the inequality

$$-5 < 3x - 2 < 1$$

is equivalent to the two inequalities

$$-5 < 3x - 2 \quad \text{and} \quad 3x - 2 < 1$$

We will solve each of these inequalities separately.

$-5 < 3x - 2$		$3x - 2 < 1$
$-5 + 2 < 3x - 2 + 2$	*Add 2 to both sides.*	$3x - 2 + 2 < 1 + 2$
$-3 < 3x$	*Simplify.*	$3x < 3$
$\frac{-3}{3} < \frac{3x}{3}$	*Divide both sides by 3.*	$\frac{3x}{3} < \frac{3}{3}$
$-1 < x$	*Simplify.*	$x < 1$

The solution set of the origin pair of inequalities consists of all x for which

$$-1 < x \quad \text{and} \quad x < 1$$

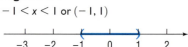

This may be written more compactly as $\{x \mid -1 < x < 1\}$. In interval notation, the solution is $(-1, 1)$. See Figure 56 for the graph.

Figure 56

$-1 < x < 1$ or $(-1, 1)$

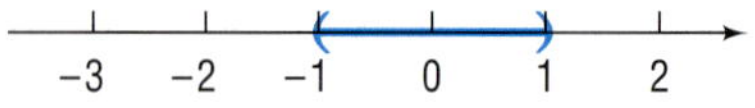

We observe in the preceding process that the two inequalities that we solved required exactly the same steps. A shortcut to solving the original inequality algebraically is to deal with the two inequalities at the same time, as follows:

$$\begin{aligned}
-5 &< 3x - 2 < 1 && \text{Add 2 to each part.}\\
-5 + 2 &< 3x - 2 + 2 < 1 + 2 \\
-3 &< 3x < 3 && \text{Simplify.}\\
\frac{-3}{3} &< \frac{3x}{3} < \frac{3}{3} && \text{Divide each part by 3.}\\
-1 &< x < 1 && \text{Simplify.}
\end{aligned}$$

NOW WORK PROBLEM 51.

5 Let's look at an inequality involving absolute value.

EXAMPLE 9 Solving an Inequality Involving Absolute Value

Solve the inequality: $|x| < 4$

Graphing Solution We graph $Y_1 = |x|$ and $Y_2 = 4$ on the same screen. See Figure 57. Using the INTERSECT command (twice), we find that Y_1 and Y_2 intersect at $x = -4$ and at $x = 4$. The graph of Y_1 is below that of Y_2, $Y_1 < Y_2$, between the points of intersection. Since the inequality is strict, the solution set is $\{x \mid -4 < x < 4\}$ or, using interval notation, $(-4, 4)$.

Figure 57

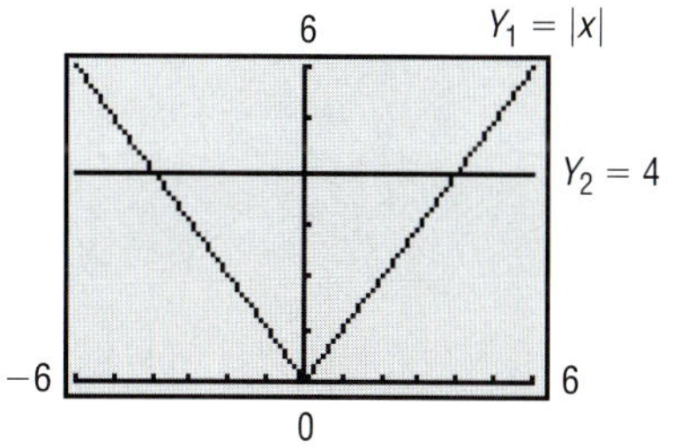

Algebraic Solution

Figure 58

$-4 < x < 4$ or $(-4, 4)$

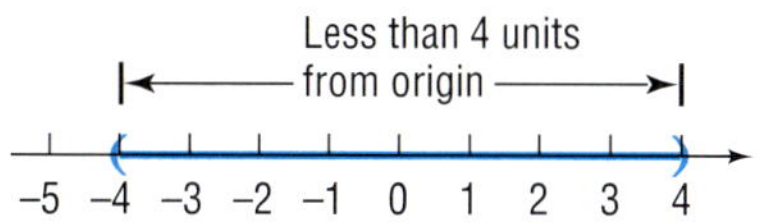

We are looking for all points whose coordinate x is a distance less than 4 units from the origin. See Figure 58 for an illustration. Because any x between -4 and 4 satisfies the condition $|x| < 4$, the solution set consists of all numbers x for which $-4 < x < 4$, that is, all x in $(-4, 4)$.

We are led to the following results:

Inequalities Involving Absolute Value

If a is any positive number and if u is any algebraic expression, then

$\|u\| < a$	is equivalent to	$-a < u < a$	(4)
$\|u\| \le a$	is equivalent to	$-a \le u \le a$	(5)

In other words, $|u| < a$ is equivalent to $-a < u$ and $u < a$.

EXAMPLE 10 Solving an Inequality Involving Absolute Value

Solve the inequality $|2x + 4| \le 3$, and graph the solution set.

Graphing Solution

We graph $Y_1 = |2x + 4|$ and $Y_2 = 3$ on the same screen. See Figure 59. Using the INTERSECT command (twice), we find that Y_1 and Y_2 intersect at $x = -3.5$ and at $x = -0.5$. The graph of Y_1 is below that of Y_2, $Y_1 < Y_2$, between the points of intersection. Since the inequality is not strict, the solution set is $\{x \mid -3.5 \le x \le -0.5\}$ or, using interval notation, $[-3.5, -0.5]$.

Figure 59

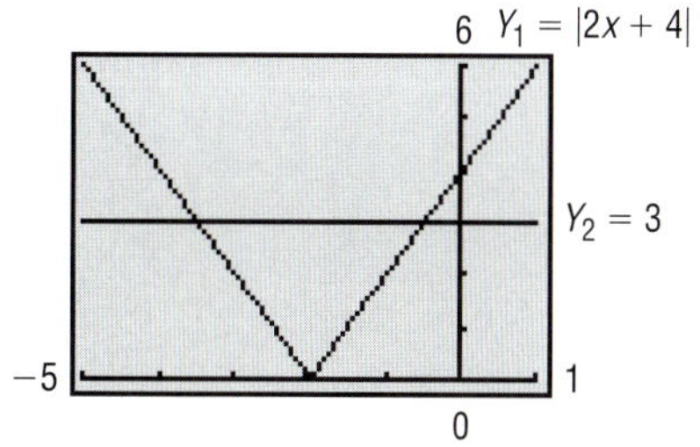

Algebraic Solution

$$\begin{aligned} |2x+4| &\le 3 && \text{This follows the form of statement (5); the expression } u = 2x+4 \text{ is inside the absolute value bars.}\\ -3 \le 2x+4 &\le 3 && \text{Apply statement (5).}\\ -3-4 \le 2x+4-4 &\le 3-4 && \text{Subtract 4 from each part.}\\ -7 \le 2x &\le -1 && \text{Simplify.}\\ \frac{-7}{2} \le \frac{2x}{2} &\le \frac{-1}{2} && \text{Divide each part by 2.}\\ -\frac{7}{2} \le x &\le -\frac{1}{2} && \text{Simplify.} \end{aligned}$$

Figure 60

$-\frac{7}{2} \le x \le -\frac{1}{2}$ or $[-\frac{7}{2}, -\frac{1}{2}]$

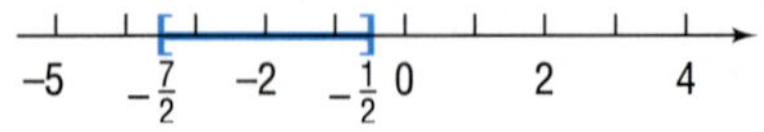

The solution set is $\{x \mid -\frac{7}{2} \le x \le -\frac{1}{2}\}$, that is, all x in $[-\frac{7}{2}, -\frac{1}{2}]$. See Figure 60.

NOW WORK PROBLEM **65.**

EXAMPLE 11 Solving an Inequality Involving Absolute Value

Solve the inequality $|x| > 3$, and graph the solution set.

Graphing Solution

We graph $Y_1 = |x|$ and $Y_2 = 3$ on the same screen. See Figure 61. Using the INTERSECT command (twice), we find that Y_1 and Y_2 intersect at $x = -3$ and at $x = 3$. The graph of Y_1 is above that of Y_2, $Y_1 > Y_2$, to the left of $x = -3$ and to the right of $x = 3$. Since the inequality is strict, the solution set is $\{x \mid -\infty < x < -3 \text{ or } 3 < x < \infty\}$. Using interval notation, the solution is $(-\infty, -3)$ or $(3, \infty)$.

Figure 61

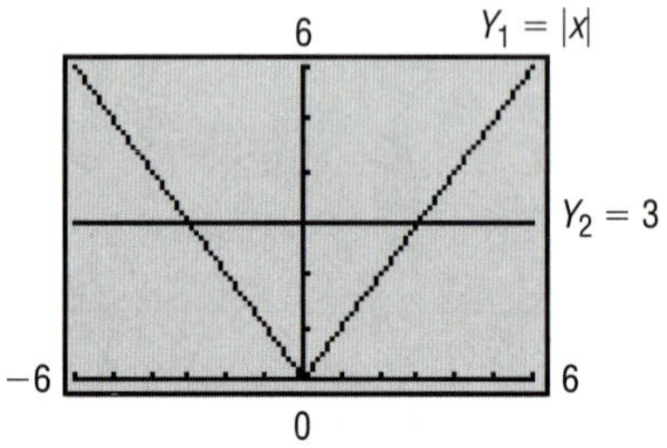

Algebraic Solution

Figure 62

$-\infty < x < -3$ or $3 < x < \infty$; $(-\infty, -3)$ or $(3, \infty)$

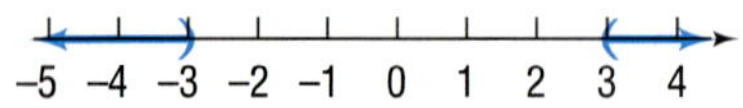

We are looking for all points whose coordinate x is a distance greater than 3 units from the origin. Figure 62 illustrates the situation. We conclude that any x less than -3 or greater than 3 satisfies the condition $|x| > 3$. Consequently, the solution set consists of all numbers x for which $-\infty < x < -3$ or $3 < x < \infty$, that is, all x in $(-\infty, -3)$ or $(3, \infty)$.

Inequalities Involving Absolute Value

If a is any positive number and u is any algebraic expression, then

$\lvert u\rvert > a$	is equivalent to	$u < -a$	or	$u > a$	(6)
$\lvert u\rvert \ge a$	is equivalent to	$u \le -a$	or	$u \ge a$	(7)

EXAMPLE 12 Solving an Inequality Involving Absolute Value

Solve the inequality $|2x - 5| > 3$, and graph the solution set.

Graphing Solution

We graph $Y_1 = |2x - 5|$ and $Y_2 = 3$ on the same screen. See Figure 63. Using the INTERSECT command (twice), we find that Y_1 and Y_2 intersect at $x = 1$ and at $x = 4$. The graph of Y_1 is above that of Y_2, $Y_1 > Y_2$, to the left of $x = 1$ and to the right of $x = 4$. Since the inequality is strict, the solution set is $\{x | -\infty < x < 1 \text{ or } 4 < x < \infty\}$. Using interval notation, the solution is $(-\infty, 1)$ or $(4, \infty)$.

Figure 63

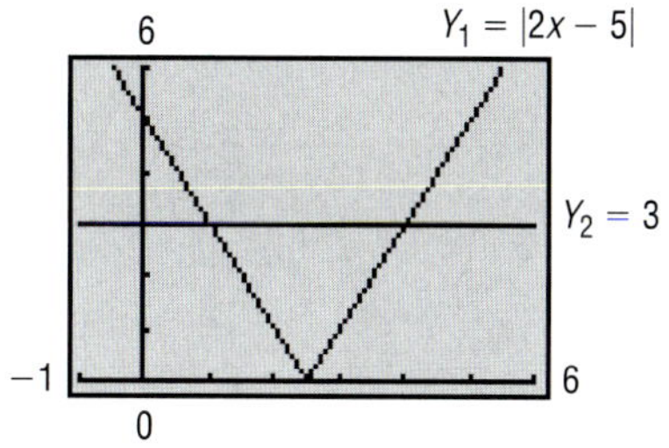

Algebraic Solution

$$|2x - 5| > 3$$ *This follows the form of statement (6); the expression $u = 2x - 5$ is inside the absolute value bars.*

$$2x - 5 < -3 \quad \text{or} \quad 2x - 5 > 3$$ *Apply statement (6).*

$$2x - 5 + 5 < -3 + 5 \quad \text{or} \quad 2x - 5 + 5 > 3 + 5$$ *Add 5 to each part.*

$$2x < 2 \quad \text{or} \quad 2x > 8$$ *Simplify.*

$$\frac{2x}{2} < \frac{2}{2} \quad \text{or} \quad \frac{2x}{2} > \frac{8}{2}$$ *Divide each part by 2.*

$$x < 1 \quad \text{or} \quad x > 4$$ *Simplify.*

The solution set is $\{x | -\infty < x < 1 \text{ or } 4 < x < \infty\}$, that is, all x in $(-\infty, 1)$ or $(4, \infty)$. See Figure 64.

Figure 64
$-\infty < x < 1$ or $4 < x < \infty$;
$(-\infty, 1)$ or $(4, \infty)$

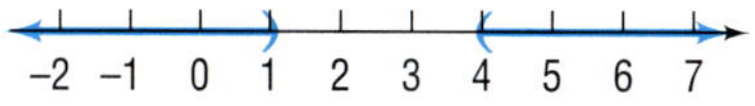

Warning: A common error to be avoided is to attempt to write the solution $x < 1$ or $x > 4$ as $1 > x > 4$, which is incorrect, since there are no numbers x for which $1 > x$ *and* $x > 4$. Another common error is to "mix" the symbols and write $1 < x > 4$, which makes no sense.

NOW WORK PROBLEM 69.

1.5 EXERCISES

In Problems 1–8, write each inequality using interval notation, and illustrate each inequality using the real number line.

1. $0 \le x \le 4$ **2.** $-1 < x < 5$ **3.** $4 \le x < 6$ **4.** $-2 < x < 0$

5. $x \ge 4$ **6.** $x \le 5$ **7.** $x < -4$ **8.** $x > 1$

In Problems 9–16, write each interval as an inequality involving x, and illustrate each inequality using the real number line.

9. $[2, 5]$ **10.** $(1, 2)$ **11.** $(-3, -2)$ **12.** $[0, 1)$

13. $[4, \infty)$ **14.** $(-\infty, 2]$ **15.** $(-\infty, -3)$ **16.** $(-8, \infty)$

In Problems 17–30, fill in the blank with the correct inequality symbol.

17. If $x < 5$, then $x - 5$ ______ 0.

18. If $x < -4$, then $x + 4$ ______ 0.

19. If $x > -4$, then $x + 4$ ______ 0.

20. If $x > 6$, then $x - 6$ ______ 0.

21. If $x \geq -4$, then $3x$ ______ -12.

22. If $x \leq 3$, then $2x$ ______ 6.

23. If $x < 6$, then $-2x$ ______ -12.

24. If $x > -2$, then $-4x$ ______ 8.

25. If $x \geq 5$, then $-4x$ ______ -20.

26. If $x \leq -4$, then $-3x$ ______ 12.

27. If $2x > 6$, then x ______ 3.

28. If $3x \leq 12$, then x ______ 4.

29. If $-\frac{1}{2}x \leq 3$, then x ______ -6.

30. If $-\frac{1}{4}x > 1$, then x ______ -4.

In Problems 31–74, solve each inequality (a) graphically and (b) algebraically. Graph the solution set.

31. $x + 1 < 5$

32. $x - 6 < 1$

33. $1 - 2x \leq 3$

34. $2 - 3x \leq 5$

35. $3x - 7 > 2$

36. $2x + 5 > 1$

37. $3x - 1 \geq 3 + x$

38. $2x - 2 \geq 3 + x$

39. $-2(x + 3) < 8$

40. $-3(1 - x) < 12$

41. $4 - 3(1 - x) \leq 3$

42. $8 - 4(2 - x) \leq -2x$

43. $\frac{1}{2}(x - 4) > x + 8$

44. $3x + 4 > \frac{1}{3}(x - 2)$

45. $\dfrac{x}{2} \geq 1 - \dfrac{x}{4}$

46. $\dfrac{x}{3} \geq 2 + \dfrac{x}{6}$

47. $0 \leq 2x - 6 \leq 4$

48. $4 \leq 2x + 2 \leq 10$

49. $-5 \leq 4 - 3x \leq 2$

50. $-3 \leq 3 - 2x \leq 9$

51. $-3 < \dfrac{2x - 1}{4} < 0$

52. $0 < \dfrac{3x + 2}{2} < 4$

53. $1 < 1 - \frac{1}{2}x < 4$

54. $0 < 1 - \frac{1}{3}x < 1$

55. $(x + 2)(x - 3) > (x - 1)(x + 1)$

56. $(x - 1)(x + 1) > (x - 3)(x + 4)$

57. $x(4x + 3) \leq (2x + 1)^2$

58. $x(9x - 5) \leq (3x - 1)^2$

59. $\dfrac{1}{2} \leq \dfrac{x + 1}{3} < \dfrac{3}{4}$

60. $\dfrac{1}{3} < \dfrac{x + 1}{2} \leq \dfrac{2}{3}$

61. $|2x| < 8$

62. $|3x| < 15$

63. $|3x| > 12$

64. $|2x| > 6$

65. $|x - 2| < 1$

66. $|x + 4| < 2$

67. $|3t - 2| \leq 4$

68. $|2u + 5| \leq 7$

69. $|x - 3| \geq 2$

70. $|x + 4| \geq 2$

71. $|1 - 4x| < 5$

72. $|1 - 2x| < 3$

73. $|1 - 2x| > 3$

74. $|2 - 3x| > 1$

75. Express the fact that x differs from 2 by less than $\frac{1}{2}$ as an inequality involving an absolute value. Solve for x.

76. Express the fact that x differs from -1 by less than 1 as an inequality involving an absolute value. Solve for x.

77. Express the fact that x differs from -3 by more than 2 as an inequality involving an absolute value. Solve for x.

78. Express the fact that x differs from 2 by more than 3 as an inequality involving an absolute value. Solve for x.

79. A young adult may be defined as someone older than 21, but less than 30 years of age. Express this statement using inequalities.

80. Middle-aged may be defined as being 40 or more and less than 60. Express the statement using inequalities.

81. Body Temperature Normal human body temperature is 98.6°F. If a temperature x that differs from normal by at least 1.5° is considered unhealthy, write the condition for an unhealthy temperature x as an inequality involving an absolute value, and solve for x.

82. Household Voltage In the United States, normal household voltage is 115 volts. However, it is not uncommon for actual voltage to differ from normal voltage by at most 5 volts. Express this situation as an inequality involving an absolute value. Use x as the actual voltage and solve for x.

83. Life Expectancy Metropolitan Life Insurance Co. reported that an average 25-year-old male in 1996 could expect to live at least 48.4 more years, and an average 25-year-old female in 1996 could expect to live at least 54.7 more years.

(a) To what age can an average 25-year-old male expect to live? Express your answer as an inequality.
(b) To what age can an average 25-year-old female expect to live? Express your answer as an inequality.
(c) Who can expect to live longer, a male or a female? By how many years?

84. General Chemistry For a certain ideal gas, the volume V (in cubic centimeters) equals 20 times the temperature T (in degrees Celsius). If the temperature varies from 80° to 120°C, inclusive, what is the corresponding range of the volume of the gas?

85. Real Estate A real estate agent agrees to sell a large apartment complex according to the following commission schedule: \$45,000 plus 25% of the selling price in excess of \$900,000. Assuming that the complex will sell at some price between \$900,000 and \$1,100,000, inclusive, over what range does the agent's commission vary? How does the commission vary as a percent of selling price?

86. Sales Commission A used car salesperson is paid a commission of \$25 plus 40% of the selling price in excess of owner's cost. The owner claims that used cars typically sell for at least owner's cost plus \$70 and at most owner's cost plus \$300. For each sale made, over what range can the salesperson expect the commission to vary?

87. Federal Tax Withholding The percentage method of withholding for federal income tax (1998)* states that a single person whose weekly wages, after subtracting withholding allowances, are over \$517, but not over \$1105, shall have \$69.90 plus 28% of the excess over \$517 withheld. Over what range does the amount withheld vary if the weekly wages vary from \$525 to \$600, inclusive?

88. Federal Tax Withholding Rework Problem 87 if the weekly wages vary from \$600 to \$700, inclusive.

89. Electricity Rates Commonwealth Edison Company's summer charge for electricity is 10.494¢ per kilowatt-hour.† In addition, each monthly bill contains a customer charge of \$9.36. If last summer's bills ranged from a low of \$80.24 to a high of \$271.80, over what range did usage vary (in kilowatt-hours)?

90. Water Bills The Village of Oak Lawn charges homeowners \$21.60 per quarter-year plus \$1.70 per 1000 gallons for water usage in excess of 12,000 gallons.‡ In 1995, one homeowner's quarterly bill ranged from a high of \$65.75 to a low of \$28.40. Over what range did water usage vary?

91. Markup of a New Car The markup over dealer's cost of a new car ranges from 12% to 18%. If the sticker price is \$8800, over what range will the dealer's cost vary?

92. IQ Tests A standard intelligence test has an average score of 100. According to statistical theory, of the people who take the test, the 2.5% with the highest scores will have scores of more than 1.96σ above the average, where σ (sigma, a number called the *standard deviation*) depends on the nature of the test. If $\sigma = 12$ for this test and there is (in principle) no upper limit to the score possible on the test, write the interval of possible test scores of the people in the top 2.5%.

93. Computing Grades In your Economics 101 class, you have scores of 68, 82, 87, and 89 on the first four of five tests. To get a grade of B, the average of the first five test scores must be greater than or equal to 80 and less than 90. Solve an inequality to find the range of the score that you need on the last test to get a B.

***Source: Employer's Tax Guide.* Department of the Treasury, Internal Revenue Service, 1998.
†*Source:* Commonwealth Edison Co., Chicago, Illinois, 1998.
‡*Source:* Village of Oak Lawn, Illinois, 1998.

94. **Computing Grades** Repeat Problem 93 if the fifth test counts double.

95. A car that averages 25 miles per gallon has a tank that holds 20 gallons of gasoline. After a trip that covered at least 300 miles, the car ran out of gasoline. What is the range of the amount of gasoline (in gallons) that was in the tank at the start of the trip?

96. Repeat Problem 95 if the same car runs out of gasoline after a trip of no more than 250 miles.

97. **Arithmetic Mean** If $a < b$, show that $a < (a + b)/2 < b$. The number $(a + b)/2$ is called the **arithmetic mean** of a and b.

98. Refer to Problem 97. Show that the arithmetic mean of a and b is equidistant from a and b.

99. **Geometric Mean** If $0 < a < b$, show that $a < \sqrt{ab} < b$. The number $\sqrt{ab}$ is called the **geometric mean** of a and b.

100. Refer to Problems 97 and 99. Show that the geometric mean of a and b is less than the arithmetic mean of a and b.

101. **Harmonic Mean** For $0 < a < b$, let h be defined by

$$\frac{1}{h} = \frac{1}{2}\left(\frac{1}{a} + \frac{1}{b}\right)$$

Show that $a < h < b$. The number h is called the **harmonic mean** of a and b.

102. Refer to Problems 97, 99, and 101. Show that the harmonic mean of a and b equals the geometric mean squared divided by the arithmetic mean.

103. Make up an inequality that has no solution. Make up one that has exactly one solution.

104. The inequality $x^2 + 1 < -5$ has no solution. Explain why.

105. Do you prefer to use inequality notation or interval notation to express the solution to an inequality? Give your reasons. Are there particular circumstances when you prefer one to the other? Cite examples.

106. How would you explain to a fellow student the underlying reason for the multiplication property for inequalities (page 57); that is, the sense or direction of an inequality remains the same if each side is multiplied by a positive real number, while the direction is reversed if each side is multiplied by a negative real number.

1.6 LINES

1. Calculate and Interpret the Slope of a Line
2. Graph Lines
3. Find the Equation of Vertical Lines
4. Use the Point–Slope Form of a Line; Identify Horizontal Lines
5. Find the Equation of a Line from Two Points
6. Write the Equation of a Line in Slope–Intercept Form
7. Identify the Slope and y-intercept of a Line from Its Equation
8. Write the Equation of a Line in General Form
9. Define Parallel and Perpendicular Lines
10. Find Equations of Parallel Lines
11. Find Equations of Perpendicular Lines

In this section we study a certain type of equation that contains two variables, called a *linear equation,* and its graph, a *line.*

Slope of a Line

Figure 65

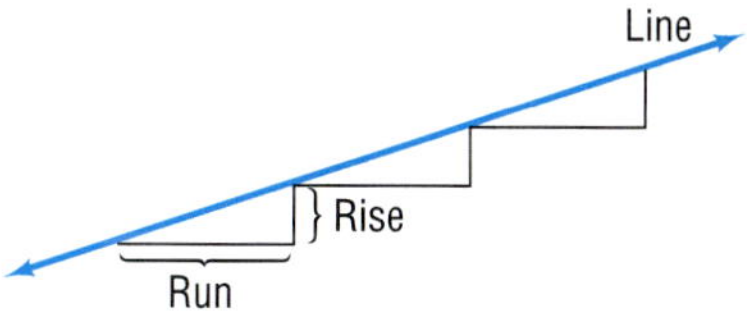

1 Consider the staircase illustrated in Figure 65. Each step contains exactly the same horizontal **run** and the same vertical **rise.** The ratio of the rise to the run, called the *slope,* is a numerical measure of the steepness of the staircase. For example, if the run is increased and the rise remains the same, the staircase becomes less steep. If the run is kept the same, but the rise is increased, the staircase becomes more steep. This important characteristic of a line is best defined using rectangular coordinates.

Let $P = (x_1, y_1)$ and $Q = (x_2, y_2)$ be two distinct points with $x_1 \neq x_2$. The **slope** $\boldsymbol{m}$ of the nonvertical line L containing P and Q is defined by the formula

$$m = \frac{y_2 - y_1}{x_2 - x_1}, \qquad x_1 \neq x_2 \tag{1}$$

If $x_1 = x_2$, L is a **vertical line** and the slope m of L is **undefined** (since this results in division by 0).

Figure 66(a) provides an illustration of the slope of a nonvertical line; Figure 66(b) illustrates a vertical line.

Figure 66

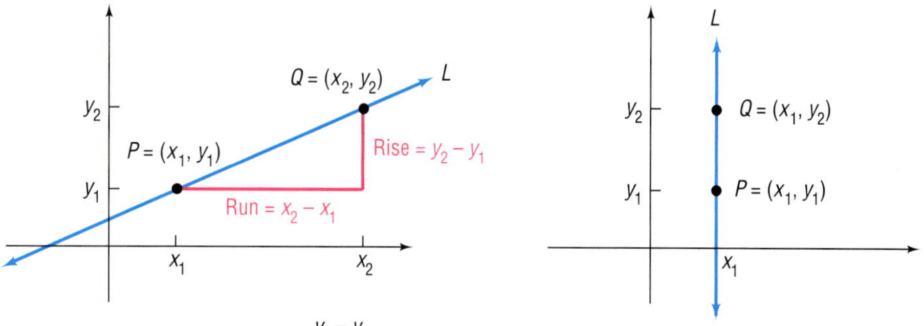

(a) Slope of L is $m = \dfrac{y_2 - y_1}{x_2 - x_1}$ **(b)** Slope is undefined; L is vertical

As Figure 66(a) illustrates, the slope m of a nonvertical line may be viewed as

$$m = \frac{y_2 - y_1}{x_2 - x_1} = \frac{\text{Rise}}{\text{Run}}$$

We can also express the slope m of a nonvertical line as

$$m = \frac{y_2 - y_1}{x_2 - x_1} = \frac{\text{Change in } y}{\text{Change in } x} = \frac{\Delta y}{\Delta x}$$

That is, the slope m of a nonvertical line L measures the amount that y changes as x changes from x_1 to x_2. This is called the **average rate of change** of y with respect to x.

Two comments about computing the slope of a nonvertical line may prove helpful:

1. Any two distinct points on the line can be used to compute the slope of the line. (See Figure 67 for justification.)

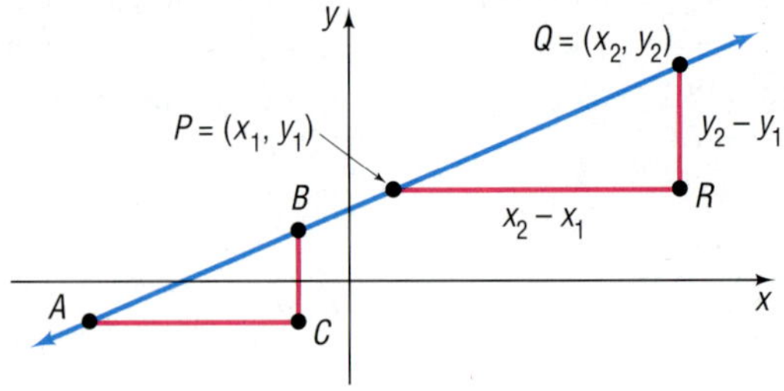

Figure 67

Triangles ABC and PQR are similar (equal angles). Hence, ratios of corresponding sides are proportional. Thus:

$$\text{Slope using } P \text{ and } Q = \frac{y_2 - y_1}{x_2 - x_1} = \text{Slope using } A \text{ and } B = \frac{d(B, C)}{d(A, C)}$$

2. The slope of a line may be computed from $P = (x_1, y_1)$ to $Q = (x_2, y_2)$ or from Q to P because

$$\frac{y_2 - y_1}{x_2 - x_1} = \frac{y_1 - y_2}{x_1 - x_2}$$

EXAMPLE 1 Finding and Interpreting the Slope of a Line Containing Two Points

The slope m of the line containing the points (1, 2) and (5, −3) may be computed as

$$m = \frac{-3 - 2}{5 - 1} = \frac{-5}{4} \quad \text{or as} \quad m = \frac{2 - (-3)}{1 - 5} = \frac{5}{-4} = \frac{-5}{4}$$

For every 4 unit change in x, y will change by −5 units. Thus, if x increases by 4 units, then y will decrease by 5 units.

NOW WORK PROBLEMS 1 AND 7.

Square Screens

To get an undistorted view of slope on a graphing utility, the same scale must be used on each axis. However, most graphing utilities have a rectangular screen. Because of this, using the same interval for both x and y will result in a distorted view. For example, Figure 68 shows the graph of the line $y = x$ connecting the points (−4, −4) and (4, 4).

Figure 68

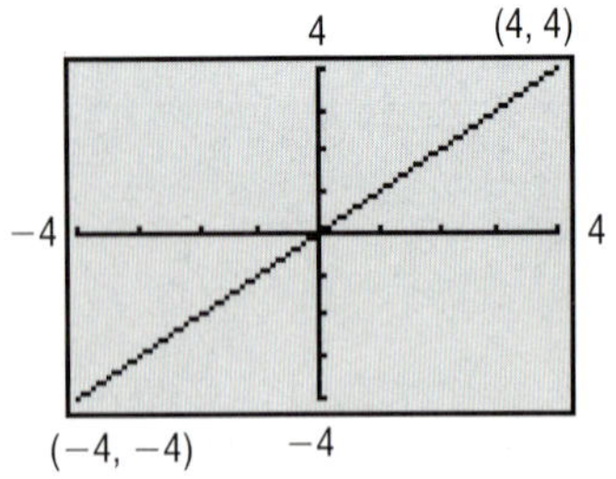

We expect the line to bisect the first and third quadrants, but it doesn't. We need to adjust the selections for Xmin, Xmax, Ymin, and Ymax so that a **square screen** results. On most graphing utilities, this is accomplished by setting the ratio of x to y at 3:2.* In other words,

$$2(X\text{max} - X\text{min}) = 3(Y\text{max} - Y\text{min})$$

Figure 69

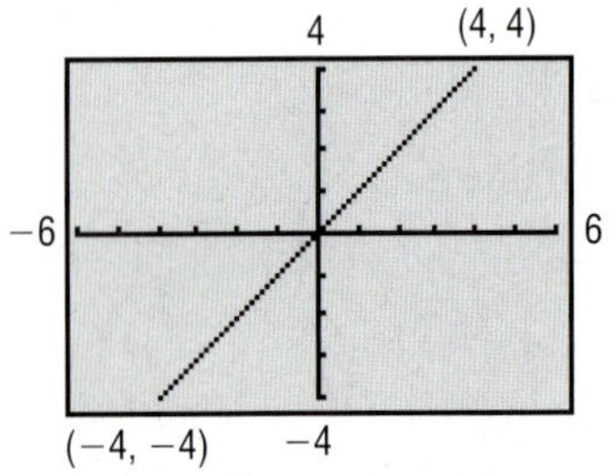

Figure 69 shows the graph of the line $y = x$ on a square screen using a TI-83. Notice that the line now bisects the first and third quadrants. Compare this illustration to Figure 68.

*Most graphing utilities have a feature that automatically squares the viewing window. Consult your owner's manual for the appropriate key strokes.

To get a better idea of the meaning of the slope m of a line, consider the following:

Figure 70

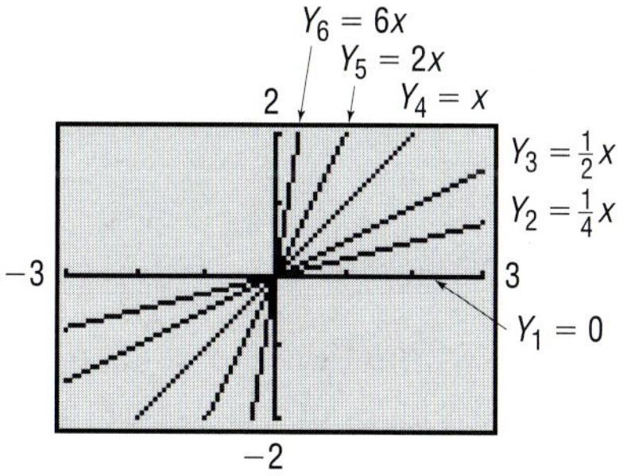

SEEING THE CONCEPT On the same square screen graph the following equations:

$$Y_1 = 0 \quad \textit{Slope of line is 0.}$$
$$Y_2 = \tfrac{1}{4}x \quad \textit{Slope of line is } \tfrac{1}{4}.$$
$$Y_3 = \tfrac{1}{2}x \quad \textit{Slope of line is } \tfrac{1}{2}.$$
$$Y_4 = x \quad \textit{Slope of line is 1.}$$
$$Y_5 = 2x \quad \textit{Slope of line is 2.}$$
$$Y_6 = 6x \quad \textit{Slope of line is 6.}$$

See Figure 70.

Figure 71

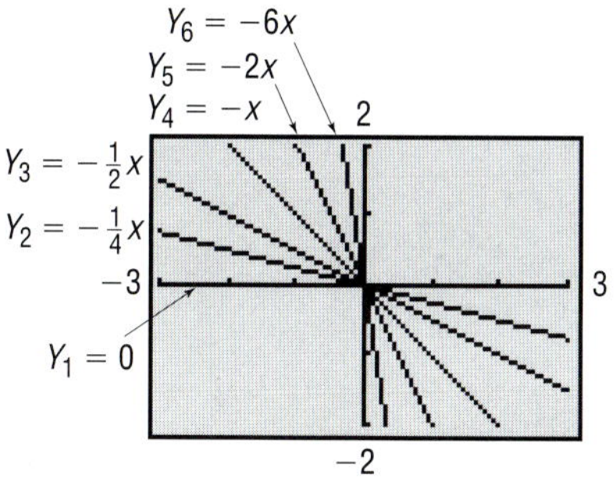

SEEING THE CONCEPT On the same square screen, graph the following equations:

$$Y_1 = 0 \quad \textit{Slope of line is 0.}$$
$$Y_2 = -\tfrac{1}{4}x \quad \textit{Slope of line is } -\tfrac{1}{4}.$$
$$Y_3 = -\tfrac{1}{2}x \quad \textit{Slope of line is } -\tfrac{1}{2}.$$
$$Y_4 = -x \quad \textit{Slope of line is } -1.$$
$$Y_5 = -2x \quad \textit{Slope of line is } -2.$$
$$Y_6 = -6x \quad \textit{Slope of line is } -6.$$

See Figure 71.

Figures 70 and 71 illustrate the following facts:

1. When the slope of a line is positive, the line slants upward from left to right.
2. When the slope of a line is negative, the line slants downward from left to right.
3. When the slope is 0, the line is horizontal.

Figures 70 and 71 also illustrate that the closer the line is to the vertical position, the greater the magnitude of the slope.

2 The next example illustrates how the slope of a line can be used to graph the line.

EXAMPLE 2 Graphing a Line Given a Point and a Slope

Draw a graph of the line that contains the point (3, 2) and has a slope of:

(a) $\frac{3}{4}$ (b) $-\frac{4}{5}$

Figure 72
Slope $= \frac{3}{4}$

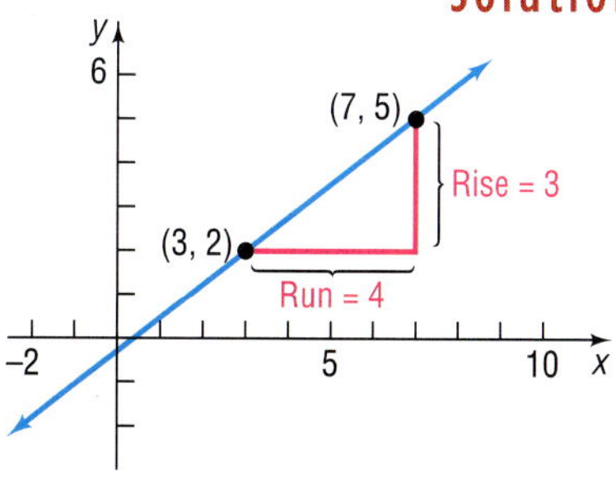

Solution (a) Slope = Rise/Run. The fact that the slope is $\frac{3}{4}$ means that for every horizontal movement (run) of 4 units to the right there will be a vertical movement (rise) of 3 units. If we start at the given point (3, 2) and move 4 units to the right and 3 units up, we reach the point (7, 5). By drawing the line through this point and the point (3, 2), we have the graph. See Figure 72.

(b) The fact that the slope is

$$-\frac{4}{5} = \frac{-4}{5} = \frac{\text{Rise}}{\text{Run}}$$

Figure 73
Slope $= -\frac{4}{5}$

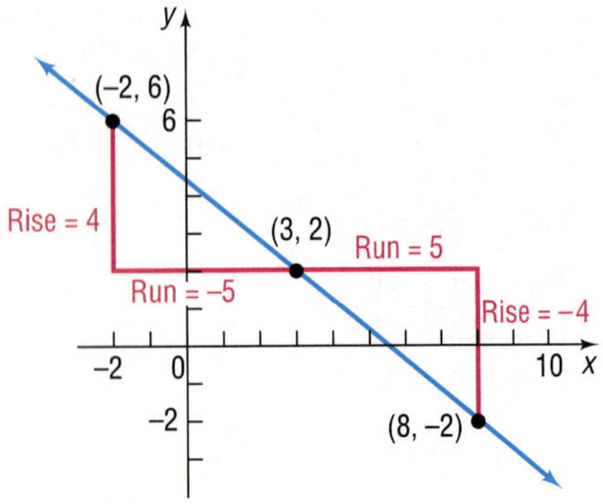

means that for every horizontal movement of 5 units to the right there will be a corresponding vertical movement of -4 units (a downward movement). If we start at the given point $(3, 2)$ and move 5 units to the right and then 4 units down, we arrive at the point $(8, -2)$. By drawing the line through these points, we have the graph. See Figure 73.

Alternatively, we can set

$$-\frac{4}{5} = \frac{4}{-5} = \frac{\text{Rise}}{\text{Run}}$$

so that for every horizontal movement of -5 units (a movement to the left) there will be a corresponding vertical movement of 4 units (upward). This approach brings us to the point $(-2, 6)$, which is also on the graph shown in Figure 73.

NOW WORK PROBLEM 15.

Equations of Lines

3 Now that we have discussed the slope of a line, we are ready to derive equations of lines. As we shall see, there are several forms of the equation of a line. Let's start with an example.

EXAMPLE 3 Graphing a Line

Graph the equation: $x = 3$

Solution Using a graphing utility, we need to express the equation in the form $y = \{\text{expression in } x\}$. But $x = 3$ cannot be put into this form. Consult your manual to determine the methodology required to draw vertical lines. Figure 74(a) shows the graph that you should obtain.

To graph $x = 3$ by hand, we recall that we are looking for all points (x, y) in the plane for which $x = 3$. Thus, no matter what y-coordinate is used, the corresponding x-coordinate always equals 3. Consequently, the graph of the equation $x = 3$ is a vertical line with x-intercept 3 and undefined slope. See Figure 74(b).

Figure 74
$x = 3$

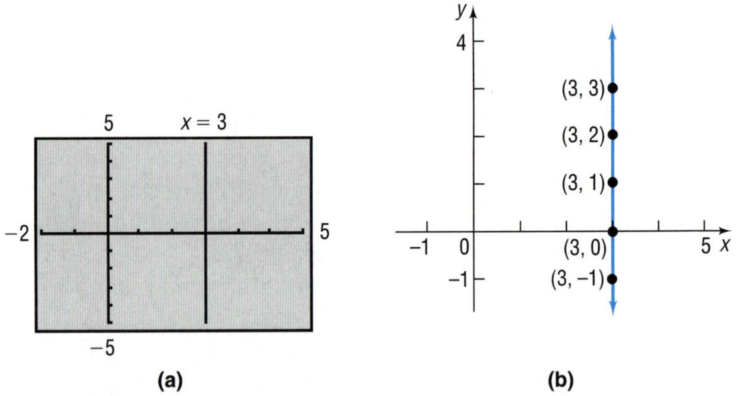

As suggested by Example 3, we have the following result:

THEOREM **Equation of a Vertical Line**

A vertical line is given by an equation of the form

$$x = a$$

where a is the x-intercept.

Figure 75

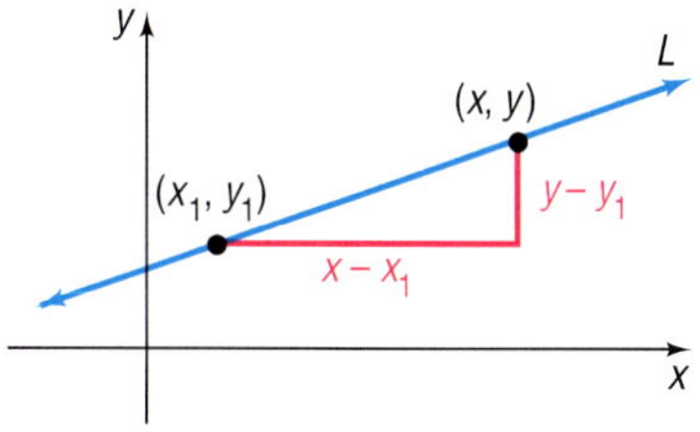

4 Now let L be a nonvertical line with slope m and containing the point (x_1, y_1). See Figure 75. For any other point (x, y) on L, we have

$$m = \frac{y - y_1}{x - x_1} \quad \text{or} \quad y - y_1 = m(x - x_1)$$

THEOREM **Point–Slope Form of an Equation of a Line**

An equation of a nonvertical line of slope m that contains the point (x_1, y_1) is

$$y - y_1 = m(x - x_1) \qquad (2)$$

EXAMPLE 4 Using the Point–Slope Form of a Line

An equation of the line with slope 4 and containing the point (1, 2) can be found by using the point–slope form with $m = 4$, $x_1 = 1$, and $y_1 = 2$.

$$\begin{aligned} y - y_1 &= m(x - x_1) \\ y - 2 &= 4(x - 1) \\ y &= 4x - 2 \end{aligned}$$

See Figure 76.

Figure 76
$y = 4x - 2$

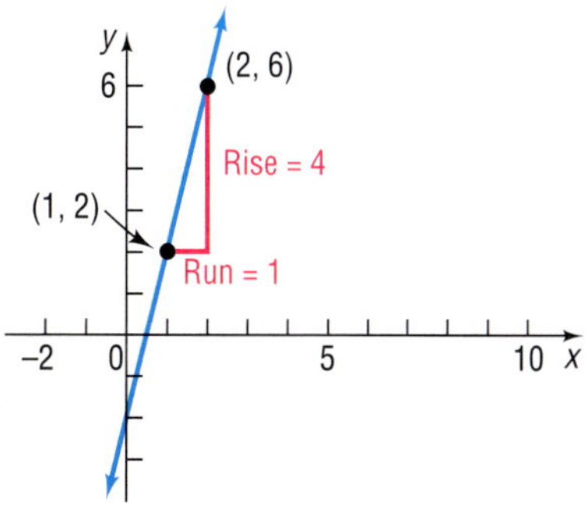

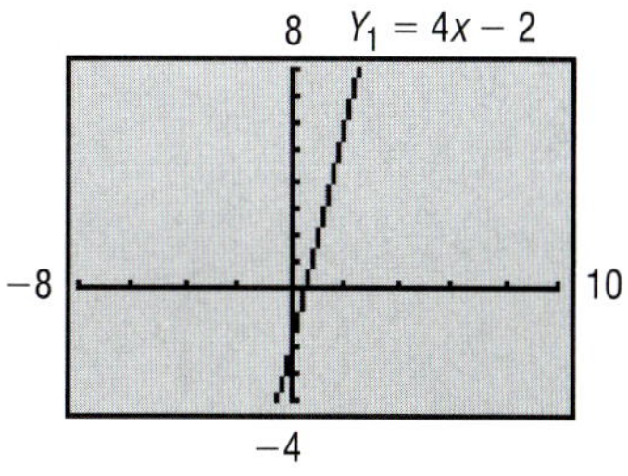

EXAMPLE 5 Finding the Equation of a Horizontal Line

Find an equation of the horizontal line containing the point (3, 2).

Solution The slope of a horizontal line is 0. To get an equation, we use the point–slope form with $m = 0$, $x_1 = 3$, and $y_1 = 2$.

$$\begin{aligned} y - y_1 &= m(x - x_1) \\ y - 2 &= 0 \cdot (x - 3) \\ y - 2 &= 0 \\ y &= 2 \end{aligned}$$

See Figure 77 for the graph.

Figure 77
$y = 2$

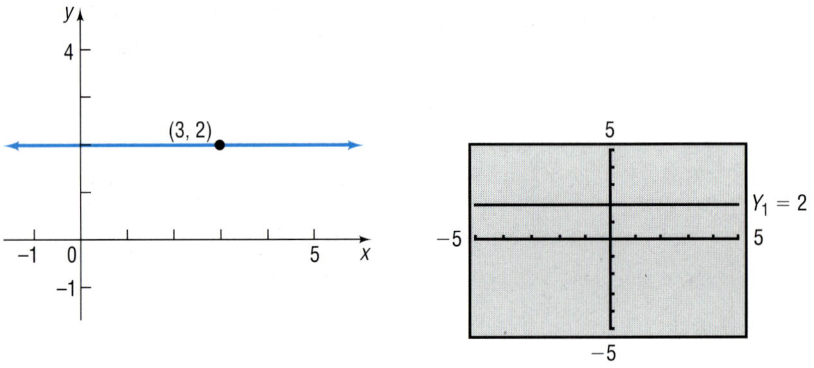

As suggested by Example 5, we have the following result:

THEOREM

Equation of a Horizontal Line

A horizontal line is given by an equation of the form

$$y = b$$

where b is the y-intercept.

EXAMPLE 6 Finding an Equation of a Line Given Two Points

5

Find an equation of the line L containing the points (2, 3) and (−4, 5). Graph the line L.

Solution Since two points are given, we first compute the slope of the line.

$$m = \frac{5 - 3}{-4 - 2} = \frac{2}{-6} = -\frac{1}{3}$$

Figure 78

$y - 3 = -\frac{1}{3}(x - 2)$

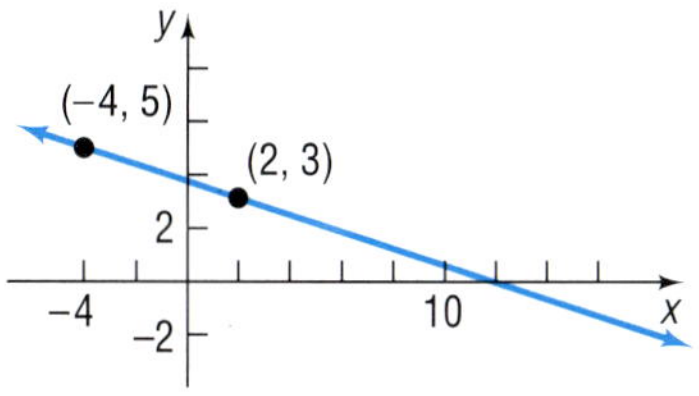

We use the point (2, 3) and the fact that the slope $m = -\frac{1}{3}$ to get the point–slope form of the equation of the line.

$$y - 3 = -\frac{1}{3}(x - 2)$$

See Figure 78 for the graph.

In the solution to Example 6, we could have used the other point, $(-4, 5)$, instead of the point (2, 3). The equation that results, although it looks different, is equivalent to the equation that we obtained in the example. (Try it for yourself.)

6

Another useful equation of a line is obtained when the slope m and y-intercept b are known. In this event, we know both the slope m of the line and a point $(0, b)$ on the line; thus, we may use the point–slope form, equation (2), to obtain the following equation:

$$y - b = m(x - 0) \quad \text{or} \quad y = mx + b$$

THEOREM

Slope–Intercept Form of an Equation of a Line

An equation of a line L with slope m and y-intercept b is

$$y = mx + b \tag{3}$$

Figure 79

$y = mx + 2$

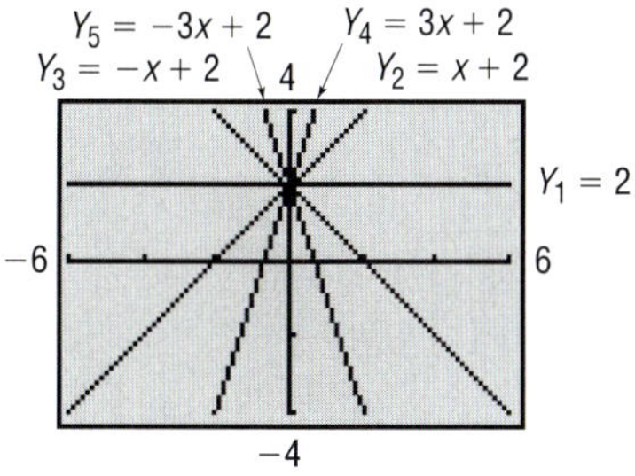

SEEING THE CONCEPT To see the role that the slope m plays, graph the following lines on the same square screen.

$$Y_1 = 2$$
$$Y_2 = x + 2$$
$$Y_3 = -x + 2$$
$$Y_4 = 3x + 2$$
$$Y_5 = -3x + 2$$

See Figure 79. What do you conclude about the lines $y = mx + 2$?

Figure 80

$y = 2x + b$

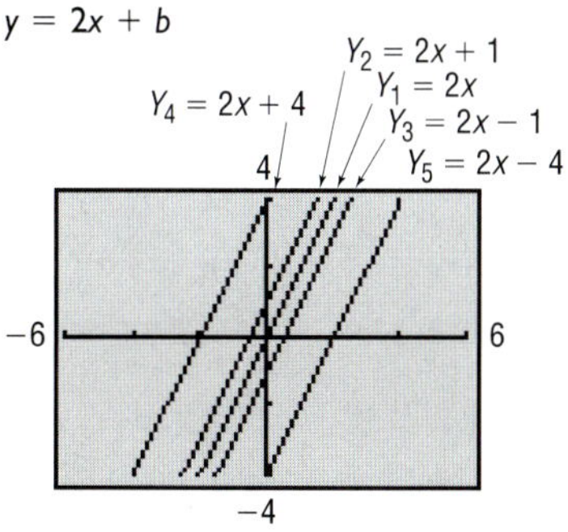

SEEING THE CONCEPT To see the role of the y-intercept b, graph the following lines on the same square screen.

$$Y_1 = 2x$$
$$Y_2 = 2x + 1$$
$$Y_3 = 2x - 1$$
$$Y_4 = 2x + 4$$
$$Y_5 = 2x - 4$$

See Figure 80. What do you conclude about the lines $y = 2x + b$?

7 When the equation of a line is written in slope–intercept form, it is easy to find the slope m and y-intercept b of the line. For example, suppose that the equation of a line is

$$y = -2x + 3$$

Compare it to $y = mx + b$.

$$\underset{y\,=}{y =} \underset{mx}{\underset{\uparrow}{-2x}} + \underset{b}{\underset{\uparrow}{3}}$$

The slope of this line is -2 and its y-intercept is 3.

NOW WORK PROBLEM **55.**

EXAMPLE 7 Finding the Slope and y-Intercept

Find the slope m and y-intercept b of the equation $2x + 4y - 8 = 0$. Graph the equation.

Solution To obtain the slope and y-intercept, we transform the equation into its slope–intercept form. Thus, we need to solve for y.

$$2x + 4y - 8 = 0$$
$$4y = -2x + 8$$
$$y = -\tfrac{1}{2}x + 2 \qquad y = mx + b$$

The coefficient of x, $-\frac{1}{2}$, is the slope, and the y-intercept is 2. We can graph the line in two ways:

1. Use the fact that the y-intercept is 2 and the slope is $-\frac{1}{2}$. Then, starting at the point $(0, 2)$, go to the right 2 units and then down 1 unit to the point $(2, 1)$. See Figure 81.

Figure 81
$2x + 4y - 8 = 0$

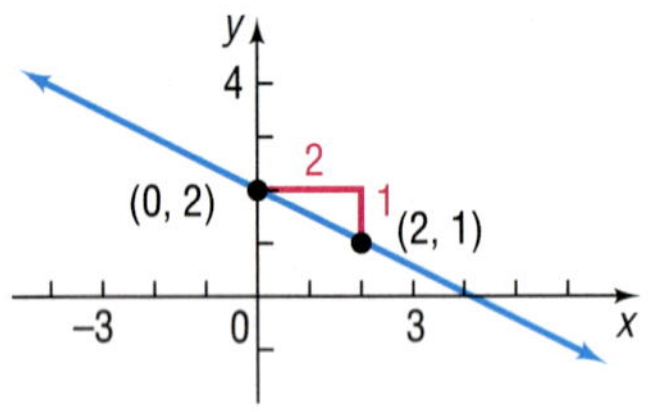

Or:

2. Locate the intercepts. Because the y-intercept is 2, we know that one intercept is $(0, 2)$. To obtain the x-intercept, let $y = 0$ and solve for x. When $y = 0$, we have

$$2x + 4 \cdot 0 - 8 = 0$$
$$2x - 8 = 0$$
$$x = 4$$

Thus, the intercepts are $(4, 0)$ and $(0, 2)$. See Figure 82.

Figure 82
$2x + 4y - 8 = 0$

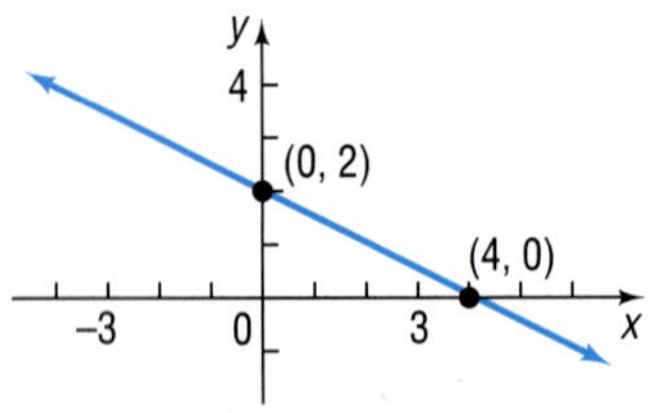

NOW WORK PROBLEM **61.**

8 The form of the equation of the line in Example 7, $2x + 4y - 8 = 0$, is called the *general form.*

The equation of a line L is in **general form** when it is written as

$$Ax + By + C = 0 \qquad (4)$$

where A, B, and C are three real numbers, and A and B are not both 0.

Every line has an equation that is equivalent to an equation written in general form. For example, a vertical line whose equation is

$$x = a$$

can be written in the general form

$$1 \cdot x + 0 \cdot y - a = 0 \quad A = 1, B = 0, C = -a$$

A horizontal line whose equation is

$$y = b$$

can be written in the general form

$$0 \cdot x + 1 \cdot y - b = 0 \quad A = 0, B = 1, C = -b$$

Lines that are neither vertical nor horizontal have general equations of the form

$$Ax + By + C = 0 \quad A \neq 0 \text{ and } B \neq 0$$

Because the equation of every line can be written in general form, any equation equivalent to (4) is called a **linear equation.**

The next example illustrates one way of graphing a linear equation.

EXAMPLE 8 Finding the Intercepts of a Line

Find the intercepts of the line $2x + 3y - 6 = 0$. Graph this line.

Solution To find the point at which the graph crosses the x-axis, that is, to find the x-intercept, we need to find the number x for which $y = 0$. Thus, we let $y = 0$ to get

$$2x + 3(0) - 6 = 0$$
$$2x - 6 = 0$$
$$x = 3$$

The x-intercept is 3. To find the y-intercept, we let $x = 0$ and solve for y.

$$2(0) + 3y - 6 = 0$$
$$3y - 6 = 0$$
$$y = 2$$

The y-intercept is 2.

To graph the line, we use the intercepts. Since the x-intercept is 3 and the y-intercept is 2, we know two points on the line: (3, 0) and (0, 2). Because two points determine a unique line, we do not need further information to graph the line. See Figure 83.

Figure 83

$2x + 3y - 6 = 0$

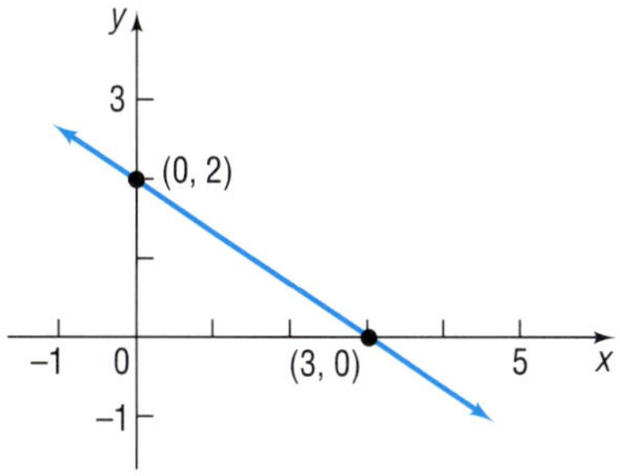

Parallel and Perpendicular Lines

9

When two lines (in the plane) have no points in common, they are said to be **parallel.** Look at Figure 84. There we have drawn two lines and have constructed two right triangles by drawing sides parallel to the coordinate axes. These lines are parallel if and only if the right triangles are similar. (Do you see why? Two angles are equal.) And the triangles are similar if and only if the ratios of corresponding sides are equal.

Figure 84

The lines are parallel if and only if their slopes are equal.

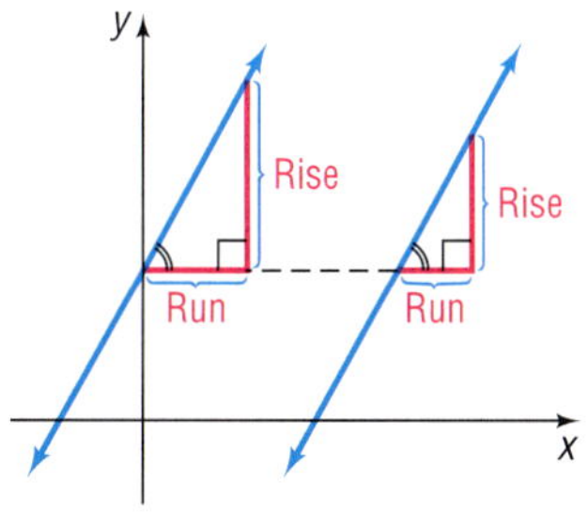

This suggests the following result:

THEOREM Two nonvertical lines are parallel if and only if their slopes are equal and they have different y-intercepts.

The use of the words "if and only if" in the preceding theorem means that actually two statements are being made, one the converse of the other.

If two nonvertical lines are parallel, then their slopes are equal and they have different y-intercepts.

If two nonvertical lines have equal slopes and they have different y-intercepts, then they are parallel.

EXAMPLE 9 Showing That Two Lines Are Parallel

Show that the lines given by the following equations are parallel:

$$L: \quad 2x + 3y - 6 = 0, \qquad M: \quad 4x + 6y = 0$$

Solution To determine whether these lines have equal slopes and different y-intercepts, we write each equation in slope–intercept form:

$$\begin{array}{rlrl} L: & 2x + 3y - 6 = 0 & M: & 4x + 6y = 0 \\ & 3y = -2x + 6 & & 6y = -4x \\ & y = -\tfrac{2}{3}x + 2 & & y = -\tfrac{2}{3}x \\ \multicolumn{2}{r}{\text{Slope} = -\tfrac{2}{3};\ y\text{-intercept} = 2} & \multicolumn{2}{r}{\text{Slope} = -\tfrac{2}{3};\ y\text{-intercept} = 0} \end{array}$$

Because these lines have the same slope, $-\frac{2}{3}$, but different y-intercepts, the lines are parallel. See Figure 85.

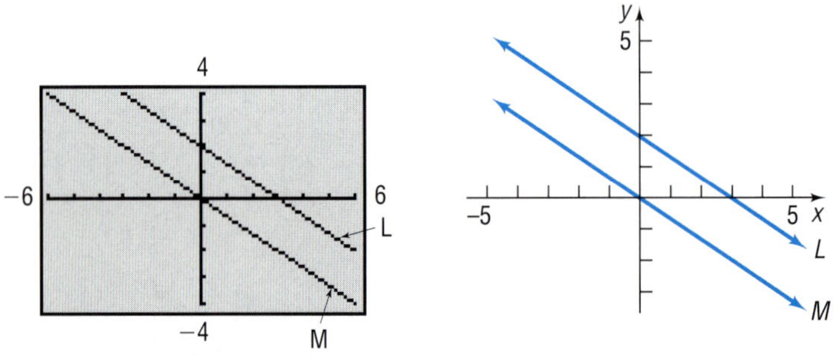

Figure 85
Parallel lines

EXAMPLE 10 Finding a Line That Is Parallel to a Given Line

10

Find an equation for the line that contains the point $(2, -3)$ and is parallel to the line $2x + y - 6 = 0$.

Solution The slope of the line that we seek equals the slope of the line $2x + y - 6 = 0$, since the two lines are to be parallel. Thus, we begin by writing the equation of the line $2x + y - 6 = 0$ in slope–intercept form.

$$\begin{aligned} 2x + y - 6 &= 0 \\ y &= -2x + 6 \end{aligned}$$

The slope is -2. Since the line that we seek contains the point $(2, -3)$, we use the point–slope form to obtain

$$y + 3 = -2(x - 2) \quad \textit{Point–slope form}$$
$$2x + y - 1 = 0 \quad \textit{General form}$$
$$y = -2x + 1 \quad \textit{Slope–intercept form}$$

This line is parallel to the line $2x + y - 6 = 0$ and contains the point $(2, -3)$. See Figure 86

Figure 86

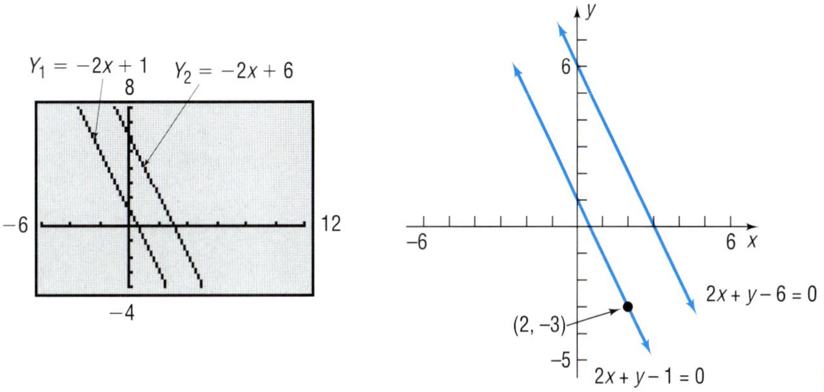

Figure 87
Perpendicular lines

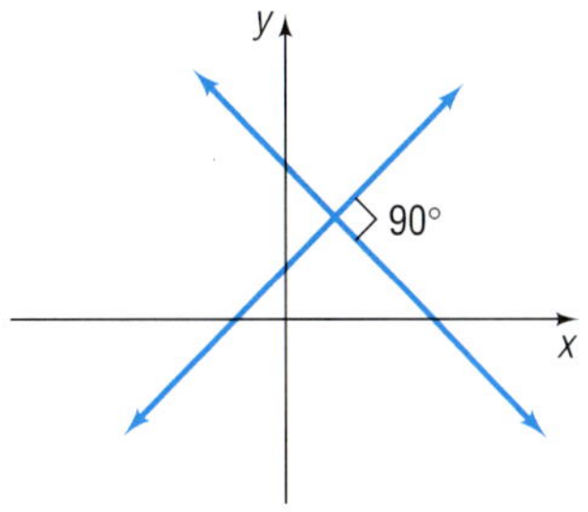

NOW WORK PROBLEM 43.

When two lines intersect at a right angle (90°), they are said to be **perpendicular.** See Figure 87.

The following result gives a condition, in terms of their slopes, for two lines to be perpendicular.

THEOREM Two nonvertical lines are perpendicular if and only if the product of their slopes is -1.

Here we shall prove the "only if" part of the statement:

If two nonvertical lines are perpendicular, then the product of their slopes is -1.

You are asked to prove the "if" part of the theorem; that is:

If two nonvertical lines have slopes whose product is -1, then the lines are perpendicular.

Figure 88

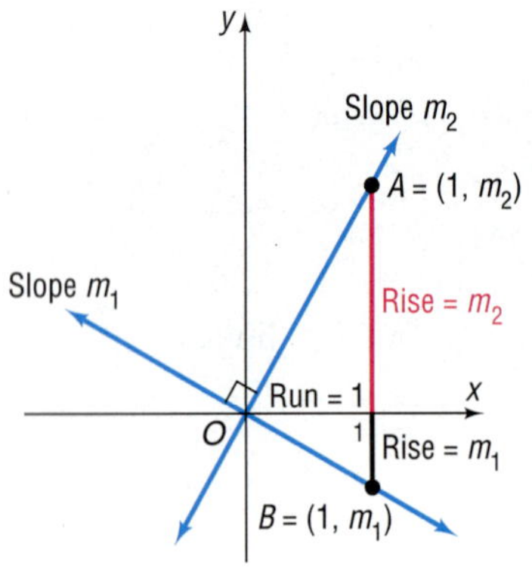

Proof Let m_1 and m_2 denote the slopes of the two lines. There is no loss in generality (that is, neither the angle nor the slopes are affected) if we situate the lines so that they meet at the origin. See Figure 88. The point $A = (1, m_2)$ is on the line having slope m_2, and the point $B = (1, m_1)$ is on the line having slope m_1. (Do you see why this must be true?)

Suppose that the lines are perpendicular. Then triangle OAB is a right triangle. As a result of the Pythagorean Theorem, it follows that

$$[d(O, A)]^2 + [d(O, B)]^2 = [d(A, B)]^2 \tag{5}$$

By the distance formula, we can write each of these distances as

$$\begin{aligned}
[d(O, A)]^2 &= (1 - 0)^2 + (m_2 - 0)^2 = 1 + m_2^2 \\
[d(O, B)]^2 &= (1 - 0)^2 + (m_1 - 0)^2 = 1 + m_1^2 \\
[d(A, B)]^2 &= (1 - 1)^2 + (m_2 - m_1)^2 = m_2^2 - 2m_1m_2 + m_1^2
\end{aligned}$$

Using these facts in equation (5), we get

$$(1 + m_2^2) + (1 + m_1^2) = m_2^2 - 2m_1m_2 + m_1^2$$

which, upon simplification, can be written as

$$m_1m_2 = -1$$

Thus, if the lines are perpendicular, the product of their slopes is -1. ◗

You may find it easier to remember the condition for two nonvertical lines to be perpendicular by observing that the equality $m_1m_2 = -1$ means that m_1 and m_2 are negative reciprocals of each other; that is, either $m_1 = -1/m_2$ or $m_2 = -1/m_1$.

EXAMPLE 11 Finding the Slope of a Line Perpendicular to Another Line

If a line has slope $\frac{3}{2}$, any line having slope $-\frac{2}{3}$ is perpendicular to it. ◗

EXAMPLE 12 Finding the Equation of a Line Perpendicular to a Given Line

11

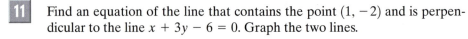

Find an equation of the line that contains the point $(1, -2)$ and is perpendicular to the line $x + 3y - 6 = 0$. Graph the two lines.

Solution We first write the equation of the given line in slope–intercept form to find its slope.

$$\begin{aligned}
x + 3y - 6 &= 0 \\
3y &= -x + 6 \\
y &= -\tfrac{1}{3}x + 2
\end{aligned}$$

The given line has slope $-\frac{1}{3}$. Any line perpendicular to this line will have slope 3. Because we require the point $(1, -2)$ to be on this line with slope 3, we use the point–slope form of the equation of a line.

$$y - (-2) = 3(x - 1) \quad \text{Point–slope form}$$
$$y + 2 = 3(x - 1)$$

This equation is equivalent to the forms

$$3x - y - 5 = 0 \quad \text{General form}$$
$$y = 3x - 5 \quad \text{Slope–intercept form}$$

Figure 89 shows the graphs.

Figure 89

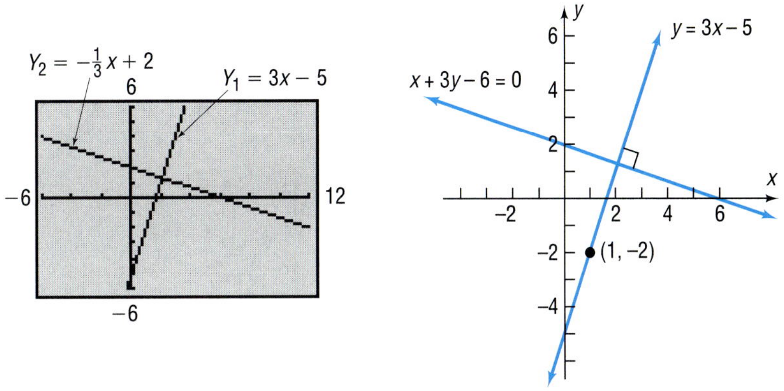

Warning: Be sure to use a square screen when you graph perpendicular lines. Otherwise, the angle between the two lines will appear distorted.

NOW WORK PROBLEM 49.

1.6 EXERCISES

In Problems 1–4, (a) find the slope of the line and (b) interpret the slope.

1. 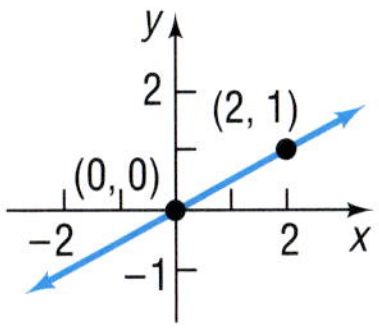

2. 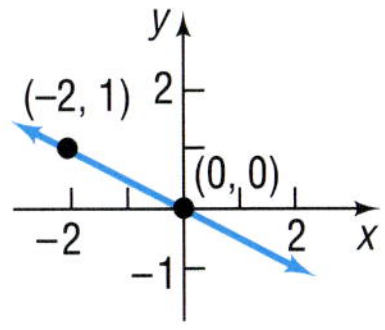

3. 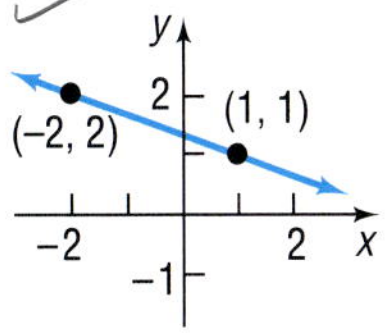

4. 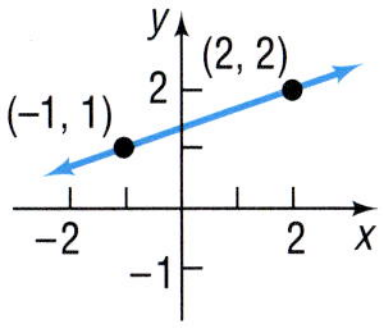

In Problems 5–14, plot each pair of points and determine the slope of the line containing them. Graph the line.

5. $(2, 3); (4, 0)$ **6.** $(4, 2); (3, 4)$ **7.** $(-2, 3); (2, 1)$ **8.** $(-1, 1); (2, 3)$

9. $(-3, -1); (2, -1)$ **10.** $(4, 2); (-5, 2)$ **11.** $(-1, 2); (-1, -2)$ **12.** $(2, 0); (2, 2)$

13. $(\sqrt{2}, 3); (1, \sqrt{3})$ **14.** $(-2\sqrt{2}, 0); (4, \sqrt{5})$

In Problems 15–22, graph, by hand, the line containing the point P and having slope m.

15. $P = (1, 2)$; $m = 3$

16. $P = (2, 1)$; $m = 4$

17. $P = (2, 4)$; $m = \frac{-3}{4}$

18. $P = (1, 3)$; $m = \frac{-2}{5}$

19. $P = (-1, 3)$; $m = 0$

20. $P = (2, -4)$; $m = 0$

21. $P = (0, 3)$; slope undefined

22. $P = (-2, 0)$; slope undefined

In Problems 23–30, find an equation of each line. Express your answer using either the general form or the slope–intercept form of the equation of a line, whichever you prefer.

23.

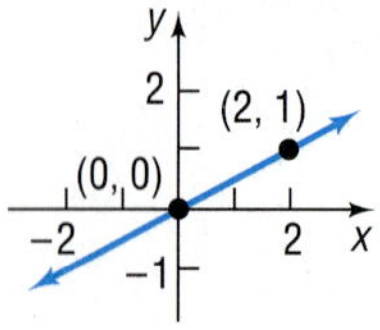

24.

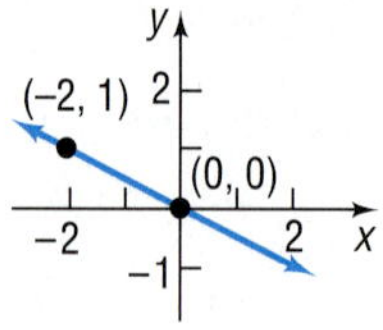

25.

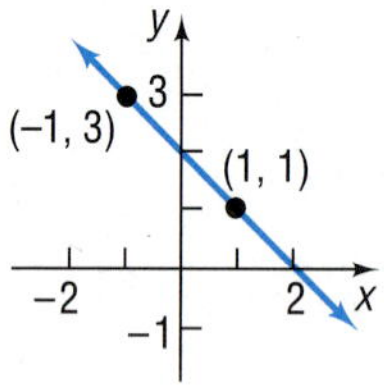

26.

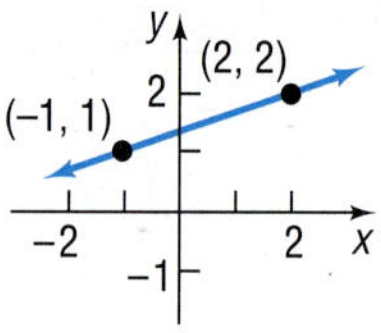

27.

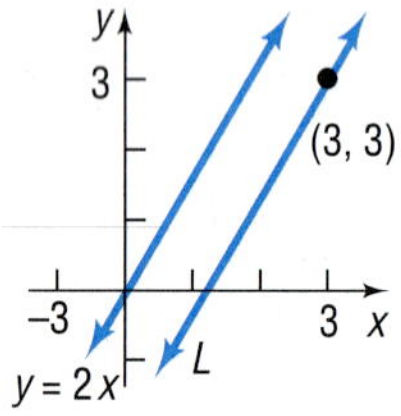

L is parallel to $y = 2x$

28.

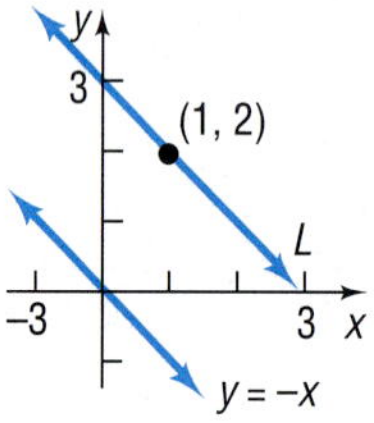

L is parallel to $y = -x$

29.

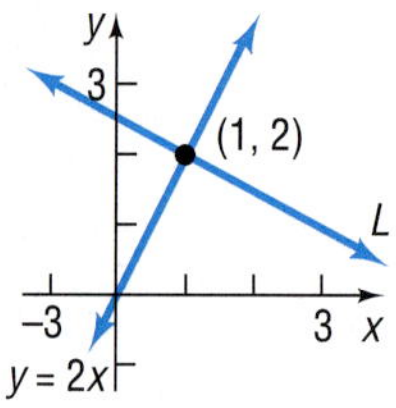

L is perpendicular to $y = 2x$

30.

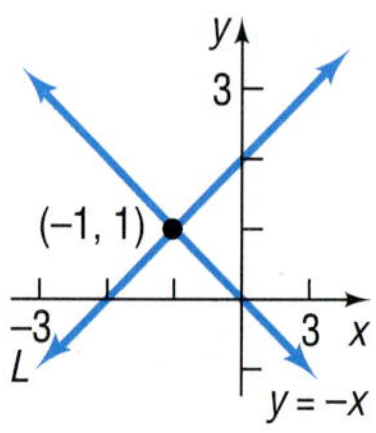

L is perpendicular to $y = -x$

In Problems 31–54, find an equation for the line with the given properties. Express your answer using either the general form or the slope–intercept form of the equation of a line, whichever you prefer.

31. Slope $= 3$; containing the point $(-2, 3)$

32. Slope $= 2$; containing the point $(4, -3)$

33. Slope $= -\frac{2}{3}$; containing the point $(1, -1)$

34. Slope $= \frac{1}{2}$; containing the point $(3, 1)$

35. Containing the points $(1, 3)$ and $(-1, 2)$

36. Containing the points $(-3, 4)$ and $(2, 5)$

37. Slope $= -3$; y-intercept $= 3$

38. Slope $= -2$; y-intercept $= -2$

39. x-intercept $= 2$; y-intercept $= -1$

40. x-intercept $= -4$; y-intercept $= 4$

41. Slope undefined; containing the point $(2, 4)$

42. Slope undefined; containing the point $(3, 8)$

43. Parallel to the line $y = 2x$; containing the point $(-1, 2)$

44. Parallel to the line $y = -3x$; containing the point $(-1, 2)$

45. Parallel to the line $2x - y + 2 = 0$; containing the point $(0, 0)$

46. Parallel to the line $x - 2y + 5 = 0$; containing the point $(0, 0)$

47. Parallel to the line $x = 5$; containing the point $(4, 2)$

48. Parallel to the line $y = 5$; containing the point $(4, 2)$

49. Perpendicular to the line $y = \frac{1}{2}x + 4$; containing the point $(1, -2)$

50. Perpendicular to the line $y = 2x - 3$; containing the point $(1, -2)$

51. Perpendicular to the line $2x + y - 2 = 0$; containing the point $(-3, 0)$

52. Perpendicular to the line $x - 2y + 5 = 0$; containing the point $(0, 4)$

53. Perpendicular to the line $x = 8$; containing the point $(3, 4)$

54. Perpendicular to the line $y = 8$; containing the point $(3, 4)$

In Problems 55–74, find the slope and y-intercept of each line. Graph the line by hand. Check your graph using a graphing utility.

55. $y = 2x + 3$ **56.** $y = -3x + 4$ **57.** $\frac{1}{2}y = x - 1$ **58.** $\frac{1}{3}x + y = 2$ **59.** $y = \frac{1}{2}x + 2$

60. $y = 2x + \frac{1}{2}$ **61.** $x + 2y = 4$ **62.** $-x + 3y = 6$ **63.** $2x - 3y = 6$ **64.** $3x + 2y = 6$

65. $x + y = 1$ **66.** $x - y = 2$ **67.** $x = -4$ **68.** $y = -1$ **69.** $y = 5$

70. $x = 2$ **71.** $y - x = 0$ **72.** $x + y = 0$ **73.** $2y - 3x = 0$ **74.** $3x + 2y = 0$

75. Find an equation of the x-axis. **76.** Find an equation of the y-axis.

In Problems 77–80, match each graph with the correct equation:

(a) $y = x$ *(b)* $y = 2x$ *(c)* $y = x/2$ *(d)* $y = 4x$

77.

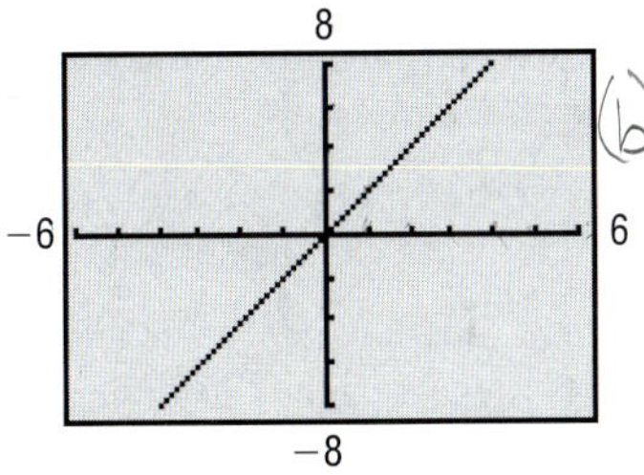

78.

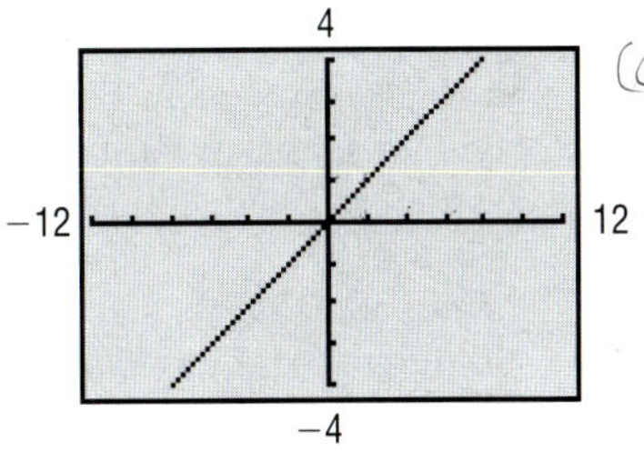

79.

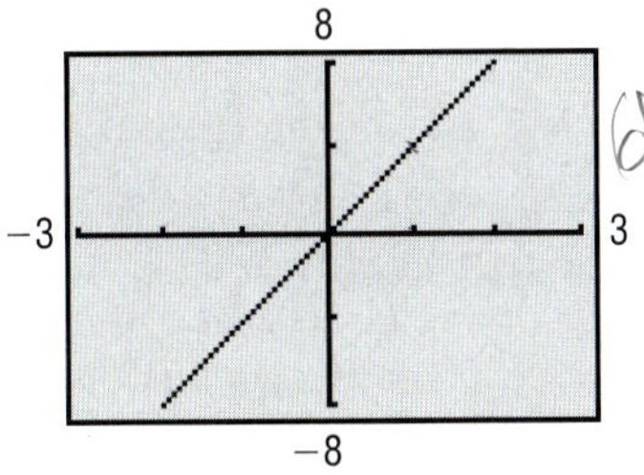

80.

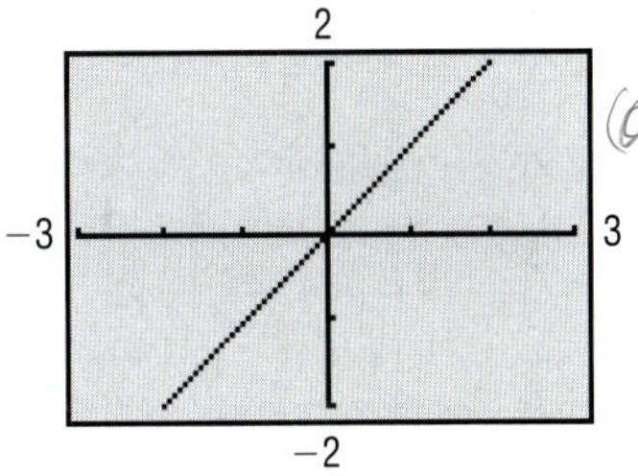

In Problems 81–84, write an equation of each line. Express your answer using either the general form or the slope–intercept form of the equation of a line, whichever you prefer.

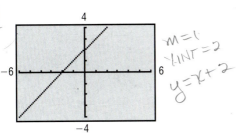

81.

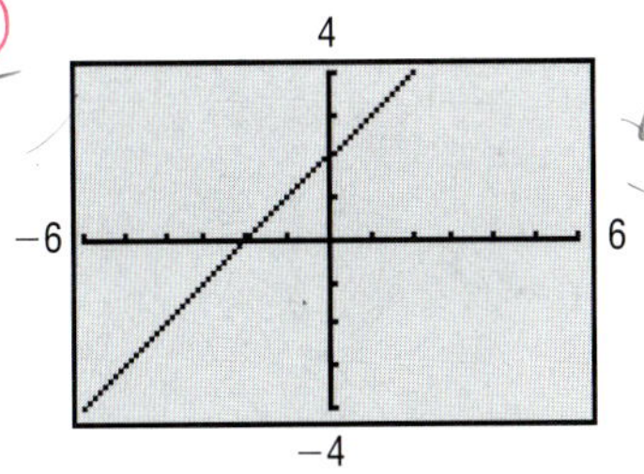

82.

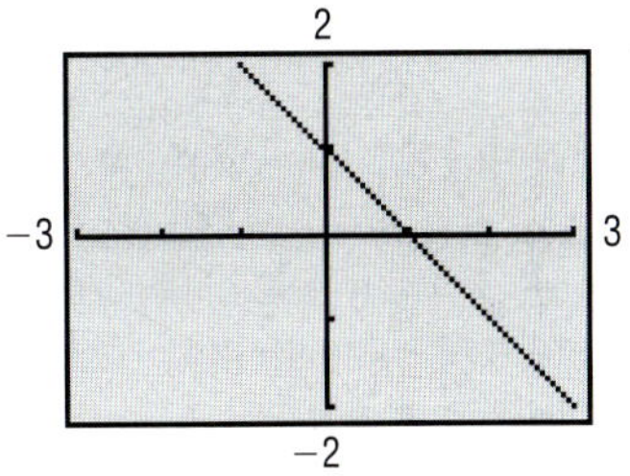

83.

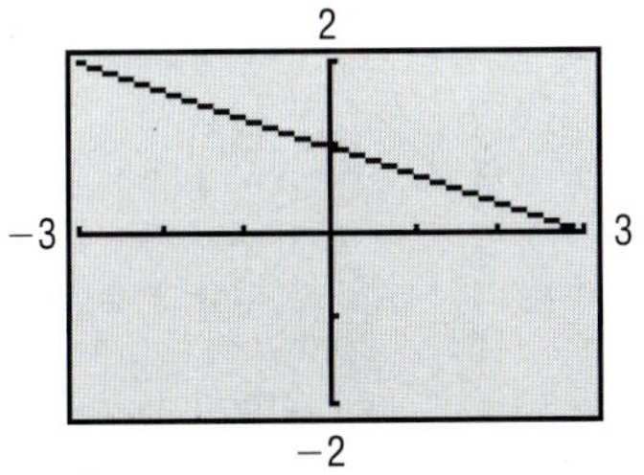

84.

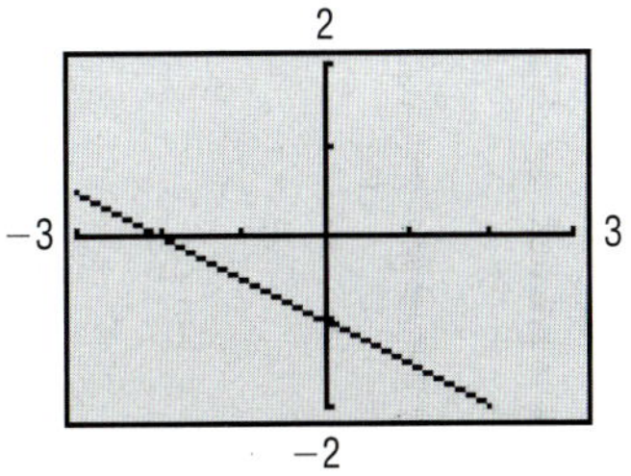

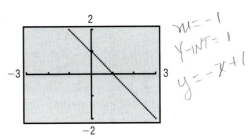

85. Measuring Temperature The relationship between Celsius (°C) and Fahrenheit (°F) degrees of measuring temperature is linear. Find an equation relating °C and °F if 0°C corresponds to 32°F and 100°C corresponds to 212°F. Use the equation to find the Celsius measure of 70°F.

86. Measuring Temperature The Kelvin (K) scale for measuring temperature is obtained by adding 273 to the Celsius temperature.
(a) Write an equation relating K and °C.
(b) Write an equation relating K and °F (see Problem 85).

87. Business: Computing Profit Each Sunday, a newspaper agency sells x copies of certain newspaper for \$1.00 per copy. The cost to the agency of each newspaper is \$0.50. The agency pays a fixed cost for storage, delivery, and so on, of \$100 per Sunday.
(a) Write an equation that relates the profit P, in dollars, to the number x of copies sold. Graph this equation.
(b) What is the profit to the agency if 1000 copies are sold?
(c) What is the profit to the agency if 5000 copies are sold?

88. Business: Computing Profit Repeat Problem 87 if the cost to the agency is \$0.45 per copy and the fixed cost is \$125 per Sunday.

89. Cost of Electricity In 1999, Florida Power and Light Company supplied electricity in the summer months to residential customers for a monthly customer charge of \$5.65 plus 6.543¢ per kilowatt-hour supplied in the month for the first 750 kilowatt-hours used.* Write an equation that relates the monthly charge, C, in dollars, to the number x of kilowatt-hours used in the month. Graph this equation. What is the monthly charge for using 300 kilowatt-hours? For using 750 kilowatt-hours?

C=5.65+.06543(K)
300KWK - \$25.28
750 KWH - \$54.72

90. Show that the line containing the points (a, b) and (b, a) is perpendicular to the line $y = x$. Also show that the midpoint of (a, b) and (b, a) lies on the line $y = x$.

91. The equation $2x - y + C = 0$ defines a **family of lines,** one line for each value of C. On one set of coordinate axes, graph the members of the family when $C = -4$, $C = 0$, and $C = 2$. Can you draw a conclusion from the graph about each member of the family?

92. Rework Problem 91 for the family of lines $Cx + y + 4 = 0$.

*Source: Florida Power and Light Co., Miami, Florida, 1999.

93. Which form of the equation of a line do you prefer to use? Justify your position with an example that shows that your choice is better than another. Give reasons.

94. Can every line be written in slope–intercept form? Explain.

95. Does every line have two distinct intercepts? Explain. Are there lines that have no intercepts? Explain.

96. What can you say about two lines that have equal slopes and equal y-intercepts?

97. What can you say about two lines with the same x-intercept and the same y-intercept? Assume that the x-intercept is not 0.

98. If two lines have the same slope, but different x-intercepts, can they have the same y-intercept?

99. If two lines have the same y-intercept, but different slopes, can they have the same x-intercept? What is the only way that this can happen?

100. The accepted symbol used to denote the slope of a line is the letter m. Investigate the origin of this symbolism. Begin by consulting a French dictionary and looking up the French word *monter.* Write a brief essay on your findings.

101. The term *grade* is used to describe the inclination of a road. How does this term relate to the notion of slope of a line? Is a 4% grade very steep? Investigate the grades of some mountainous roads and determine their slopes. Write a brief essay on your findings.

102. Carpentry Carpenters use the term *pitch* to describe the steepness of staircases and roofs. How does pitch relate to slope? Investigate typical pitches used for stairs and for roofs. Write a brief essay on your findings.

1.7 CIRCLES

1. Graph a Circle by Hand and by Using a Graphing Utility
2. Write the Standard Form of the Equation of a Circle
3. Find the Center and Radius of a Circle from an Equation in General Form

One advantage of a coordinate system is that it enables us to translate a geometric statement into an algebraic statement, and vice versa. Consider, for example, the following geometric statement that defines a circle.

> A **circle** is a set of points in the xy-plane that are a fixed distance r from a fixed point (h, k). The fixed distance r is called the **radius,** and the fixed point (h, k) is called the **center** of the circle.

Figure 90

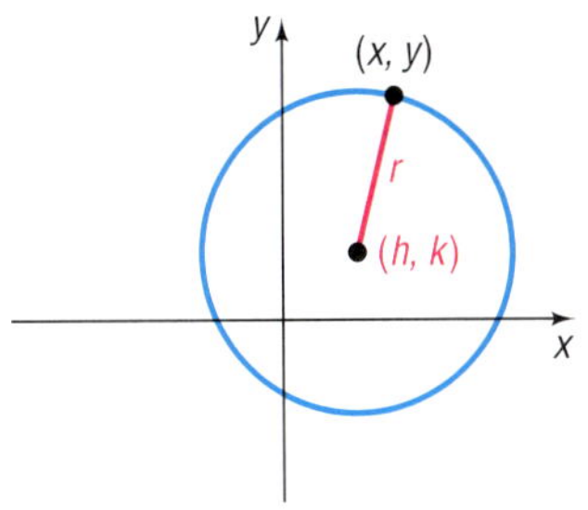

Figure 90 shows the graph of a circle. Is there an equation having this graph? If so, what is the equation? To find the equation, we let (x, y) represent the coordinates of any point on a circle with radius r and center (h, k). Then the distance between the points (x, y) and (h, k) must always equal r. That is, by the distance formula (Section 1.1),

$$\sqrt{(x - h)^2 + (y - k)^2} = r$$

or, equivalently,

$$(x - h)^2 + (y - k)^2 = r^2$$

The **standard form of an equation of a circle** with radius r and center (h, k) is

$$(x - h)^2 + (y - k)^2 = r^2 \qquad (1)$$

1 Conversely, by reversing the steps, we conclude that the graph of any equation of the form of equation (1) is that of a circle with radius r and center (h, k).

EXAMPLE 1 Graphing a Circle

Graph the equation $(x + 3)^2 + (y - 2)^2 = 16$.

Solution The graph of the equation is a circle. To graph a circle on a graphing utility, we must write the equation in the form $y = \{\text{expression involving } x\}$.* Thus, we must solve for y in the equation

$$(x + 3)^2 + (y - 2)^2 = 16$$

$$(y - 2)^2 = 16 - (x + 3)^2 \quad \text{Subtract } (x+3)^2 \text{ from both sides.}$$

$$y - 2 = \pm\sqrt{16 - (x + 3)^2} \quad \text{Take the square root of both sides.}$$

$$y = 2 \pm \sqrt{16 - (x + 3)^2} \quad \text{Add 2 to both sides.}$$

*Some graphing utilities (e.g., TI-82, TI-83, TI-85, and TI-86) have a CIRCLE function that allows the user to enter only the coordinates of the center of the circle and its radius to graph the circle.

Figure 91

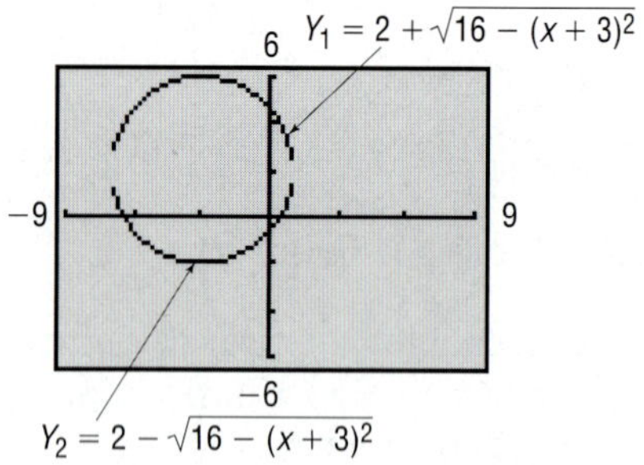

Figure 92

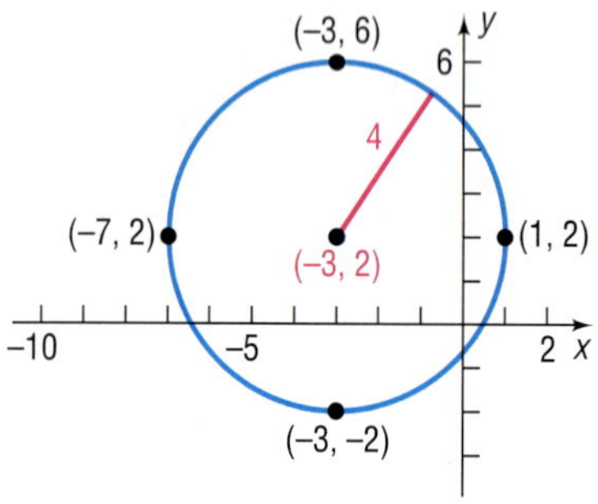

To graph the circle, we graph the top half

$$Y_1 = 2 + \sqrt{16 - (x + 3)^2}$$

and the bottom half

$$Y_2 = 2 - \sqrt{16 - (x + 3)^2}$$

Also, be sure to use a square screen. Otherwise, the circle will appear distorted. Figure 91 shows the graph.

To graph the equation by hand, we first compare the given equation to the standard form of the equation of a circle. The comparison yields information about the circle.

$$\begin{aligned} (x + 3)^2 + (y - 2)^2 &= 16 \\ (x - (-3))^2 + (y - 2)^2 &= 4^2 \\ \uparrow \qquad\quad \uparrow \qquad &\quad \uparrow \\ (x - h)^2 + (y - k)^2 &= r^2 \end{aligned}$$

We see that $h = -3$, $k = 2$, and $r = 4$. Hence, the circle has center $(-3, 2)$ and a radius of 4 units. To graph this circle, we first plot the center $(-3, 2)$. Since the radius is 4, we can locate four points on the circle by going out 4 units to the left, to the right, up, and down from the center. These four points can then be used as guides to obtain the graph. See Figure 92.

NOW WORK PROBLEM 17.

EXAMPLE 2 Writing the Standard Form of the Equation of a Circle

2

Write the standard form of the equation of the circle with radius 3 and center $(1, -2)$.

Solution Using the form of equation (1) and substituting the values $r = 3$, $h = 1$, and $k = -2$, we have

$$(x - h)^2 + (y - k)^2 = r^2$$
$$(x - 1)^2 + (y + 2)^2 = 9$$

NOW WORK PROBLEM 1.

The standard form of an equation of a circle of radius r with center at the origin $(0, 0)$ is

$$x^2 + y^2 = r^2$$

If the radius $r = 1$, the circle whose center is at the origin is called the **unit circle** and has the equation

$$x^2 + y^2 = 1$$

See Figure 93.

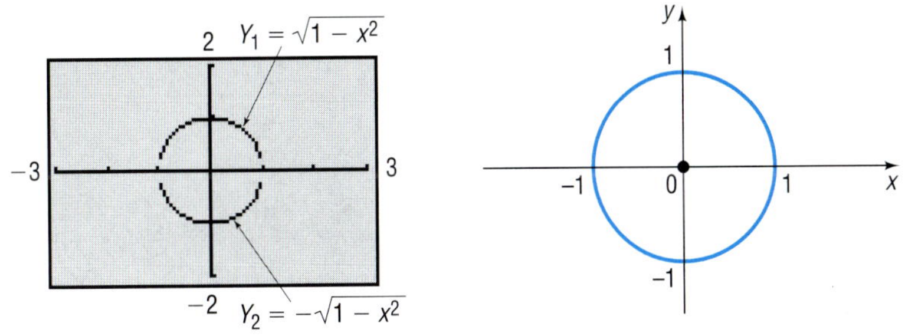

Figure 93
Unit circle $x^2 + y^2 = 1$

If we eliminate the parentheses from the standard form of the equation of the circle obtained in Example 2, we get

$$(x - 1)^2 + (y + 2)^2 = 9$$
$$x^2 - 2x + 1 + y^2 + 4y + 4 = 9$$

which we find, upon simplifying, is equivalent to

$$x^2 + y^2 - 2x + 4y - 4 = 0$$

It can be shown that any equation of the form

$$x^2 + y^2 + ax + by + c = 0$$

has a graph that is a circle, or a point, or has no graph at all. For example, the graph of the equation $x^2 + y^2 = 0$ is the single point (0, 0). The equation $x^2 + y^2 + 5 = 0$, or $x^2 + y^2 = -5$, has no graph, because sums of squares of real numbers are never negative. When its graph is a circle, the equation

$$x^2 + y^2 + ax + by + c = 0$$

is referred to as the **general form of the equation of a circle.**

3 If an equation of a circle is in the general form, we use the method of completing the square to put the equation in standard form so we can identify its center and radius. The idea behind the method of **completing the square** is to "adjust" a second-degree polynomial, $ax^2 + bx + c$, so that it becomes a perfect square, the square of a first-degree polynomial. For example, $x^2 + 6x + 9$ and $x^2 - 4x + 4$ are perfect squares* because

$$x^2 + 6x + 9 = (x + 3)^2 \quad \text{and} \quad x^2 - 4x + 4 = (x - 2)^2$$

How do we adjust the second-degree polynomial? We do it by adding the appropriate number to create a perfect square. For example, to make $x^2 + 6x$ a perfect square, we add 9.

*Perfect squares are discussed in Section 6 of the Appendix.

Let's look at several examples of completing the square when the coefficient of x^2 is 1.

Start	Add	Result
$x^2 + 4x$	4	$x^2 + 4x + 4 = (x + 2)^2$
$x^2 + 12x$	36	$x^2 + 12x + 36 = (x + 6)^2$
$x^2 - 6x$	9	$x^2 - 6x + 9 = (x - 3)^2$
$x^2 + x$	$\frac{1}{4}$	$x^2 + x + \frac{1}{4} = (x + \frac{1}{2})^2$

Do you see the pattern? Provided that the coefficient of x^2 is 1, we complete the square by adding the square of one-half the coefficient of x.

Start	Add	Result
$x^2 + mx$	$\left(\frac{m}{2}\right)^2$	$x^2 + mx + \left(\frac{m}{2}\right)^2 = \left(x + \frac{m}{2}\right)^2$

EXAMPLE 3 Graphing a Circle Whose Equation Is in General Form

3

Graph the equation $x^2 + y^2 + 4x - 6y + 12 = 0$.

Solution We complete the square in both x and y to put the equation in standard form. Group the expression involving x, group the expression involving y, and put the constant on the right side of the equation. The result is

$$(x^2 + 4x) + (y^2 - 6y) = -12$$

Next, complete the square of each expression in parentheses. Remember that any number added on the left side of the equation must be added on the right.

$$(x^2 + 4x + 4) + (y^2 - 6y + 9) = -12 + 4 + 9$$

$$\left(\frac{4}{2}\right)^2 = 4 \qquad \left(\frac{-6}{2}\right)^2 = 9$$

$$(x + 2)^2 + (y - 3)^2 = 1 \quad \textit{Factor.}$$

We recognize this equation as the standard form of the equation of a circle with radius 1 and center $(-2, 3)$.

To graph the equation using a graphing utility, we need to solve for y.

$$(y - 3)^2 = 1 - (x + 2)^2$$
$$y - 3 = \pm\sqrt{1 - (x + 2)^2}$$
$$y = 3 \pm \sqrt{1 - (x + 2)^2}$$

To graph the equation by hand, we use the center $(-2, 3)$ and the radius 1. Figure 94 illustrates the graph.

Figure 94
$x^2 + y^2 + 4x - 6y + 12 = 0$

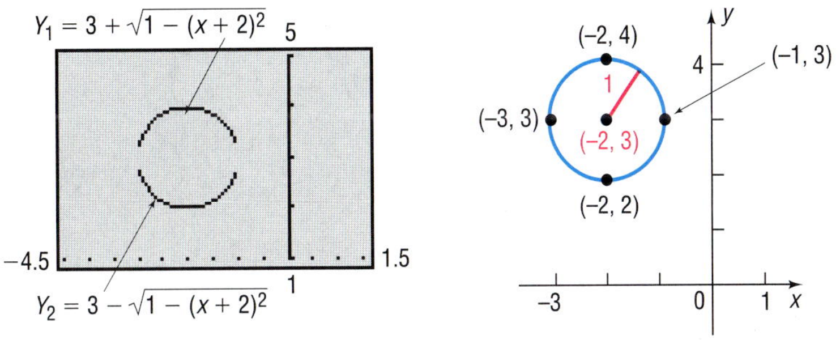

NOW WORK PROBLEM 19.

EXAMPLE 4 Finding the General Equation of a Circle

Find the general equation of the circle whose center is $(1, -2)$ and whose graph contains the point $(4, -2)$.

Solution To find the equation of a circle, we need to know its center and its radius. Here, we know that the center is $(1, -2)$. Since the point $(4, -2)$ is on the graph, the radius r will equal the distance from $(4, -2)$ to the center $(1, -2)$. See Figure 95. Thus,

Figure 95
$x^2 + y^2 - 2x + 4y - 4 = 0$

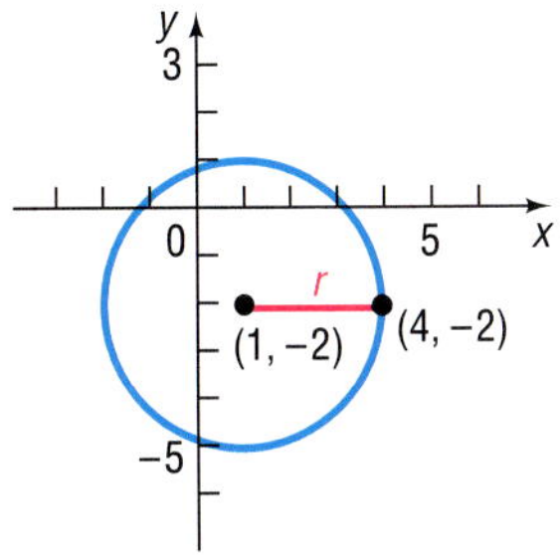

$$r = \sqrt{(4 - 1)^2 + [-2 - (-2)]^2}$$
$$= \sqrt{9} = 3$$

The standard form of the equation of the circle is

$$(x - 1)^2 + (y + 2)^2 = 9$$

Eliminating the parentheses and rearranging terms, we get the general equation

$$x^2 + y^2 - 2x + 4y - 4 = 0$$

Overview

The discussion in Sections 1.6 and 1.7 about lines and circles dealt with two main types of problems that can be generalized as follows:

1. Given an equation, classify it and graph it.
2. Given a graph, or information about a graph, find its equation.

This text deals with both types of problems. We shall study various equations, classify them, and graph them. Although the second type of problem is usually more difficult to solve than the first, in many instances a graphing utility can be used to solve such problems.

1.7 EXERCISES

In Problems 1–4, find the center and radius of each circle. Write the standard form of the equation.

1.
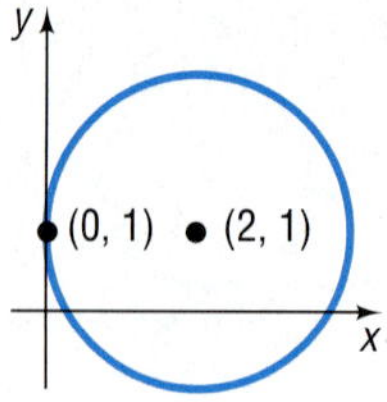

2.
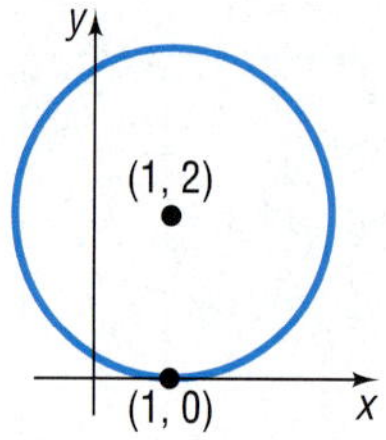

3.
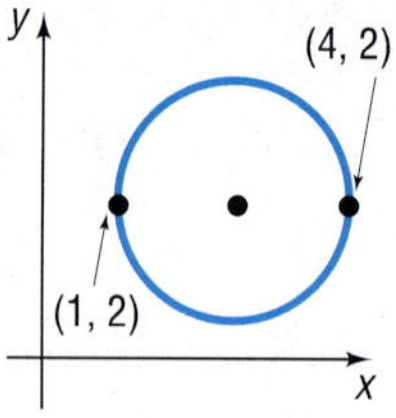

4.
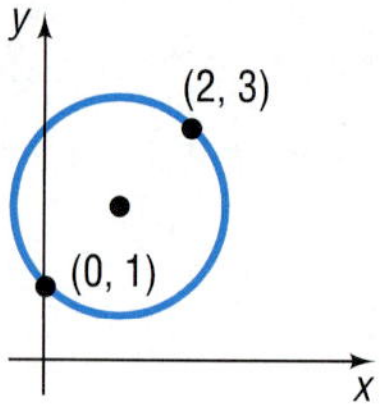

In Problems 5–14, write the standard form of the equation and the general form of the equation of each circle of radius r and center (h, k). By hand, graph each circle.

5. $r = 1;\quad (h, k) = (1, -1)$

6. $r = 2;\quad (h, k) = (-2, 1)$

7. $r = 2;\quad (h, k) = (0, 2)$

8. $r = 3;\quad (h, k) = (1, 0)$

9. $r = 5;\quad (h, k) = (4, -3)$

10. $r = 4;\quad (h, k) = (2, -3)$

11. $r = 2;\quad (h, k) = (0, 0)$

12. $r = 3;\quad (h, k) = (0, 0)$

13. $r = \frac{1}{2};\quad (h, k) = (\frac{1}{2}, 0)$

14. $r = \frac{1}{2};\quad (h, k) = (0, -\frac{1}{2})$

In Problems 15–24, find the center (h, k) and radius r of each circle. By hand, graph each circle.

15. $x^2 + y^2 = 4$

16. $x^2 + (y - 1)^2 = 1$

17. $(x - 3)^2 + y^2 = 4$

18. $(x + 1)^2 + (y - 1)^2 = 2$

19. $x^2 + y^2 + 4x - 4y - 1 = 0$

20. $x^2 + y^2 - 6x + 2y + 9 = 0$

21. $x^2 + y^2 - x + 2y + 1 = 0$

22. $x^2 + y^2 + x + y - \frac{1}{2} = 0$

23. $2x^2 + 2y^2 - 12x + 8y - 24 = 0$

24. $2x^2 + 2y^2 + 8x + 7 = 0$

In Problems 25–30, find the general form of the equation of each circle.

25. Center at the origin and containing the point $(-2, 3)$

26. Center $(1, 0)$ and containing the point $(-3, 2)$

27. Center $(2, 3)$ and tangent to the x-axis.

28. Center $(-3, 1)$ and tangent to the y-axis

29. With end points of a diameter at $(1, 4)$ and $(-3, 2)$

30. With end points of a diameter at $(4, 3)$ and $(0, 1)$

In Problems 31–34, match each graph with the correct equation.

(a) $(x - 3)^2 + (y + 3)^2 = 9$ *(c)* $(x - 1)^2 + (y + 2)^2 = 4$

(b) $(x + 1)^2 + (y - 2)^2 = 4$ *(d)* $(x + 3)^2 + (y - 3)^2 = 9$

31.
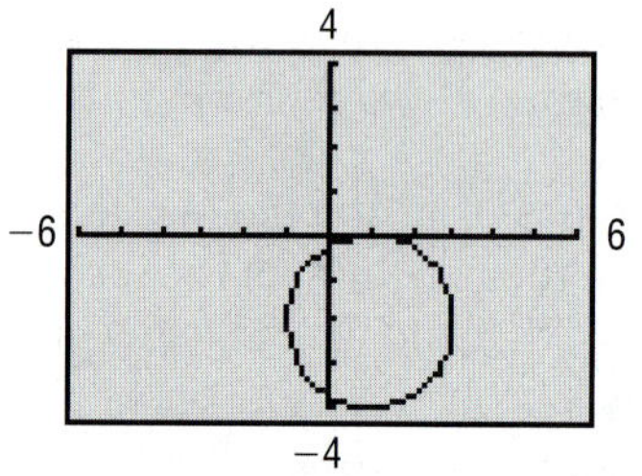

32.
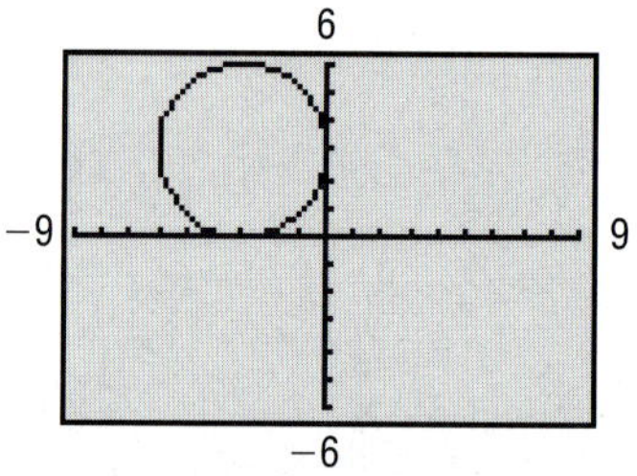

33.

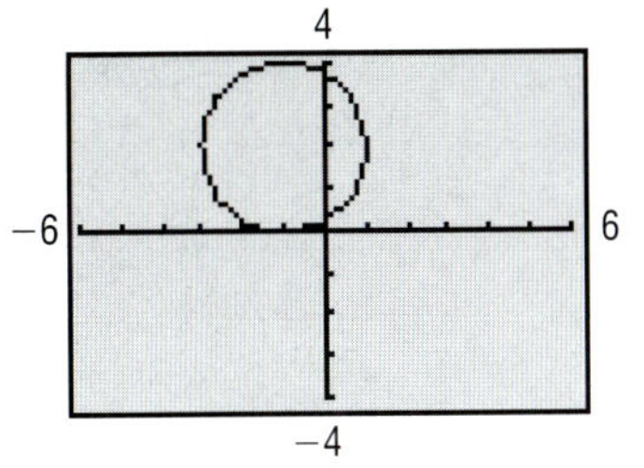

34.

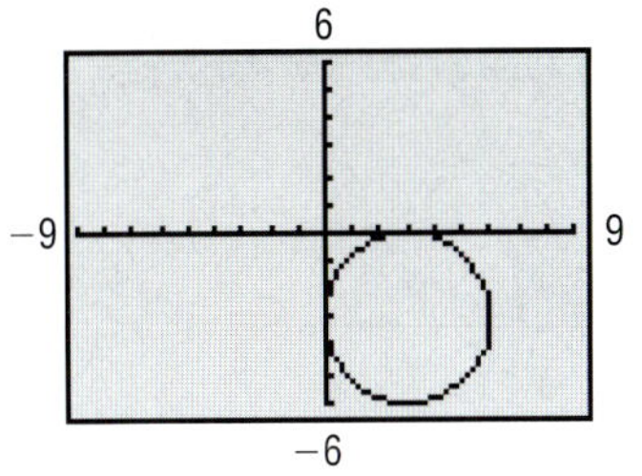

In Problems 35–38, find the standard form of the equation of each circle.

35.

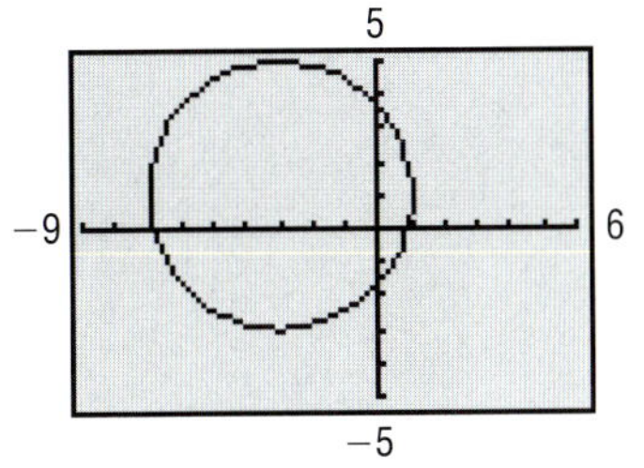

36.

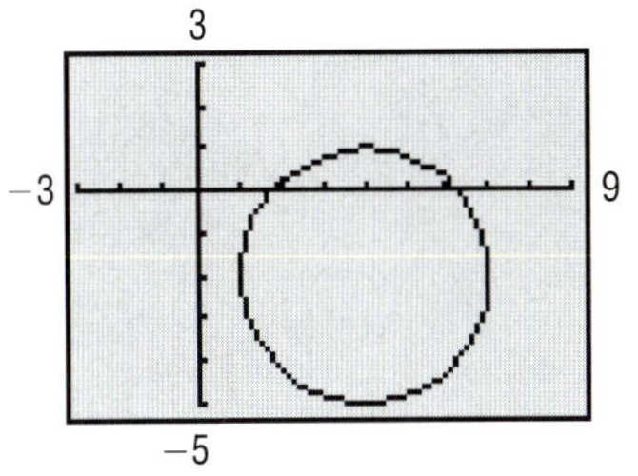

37.

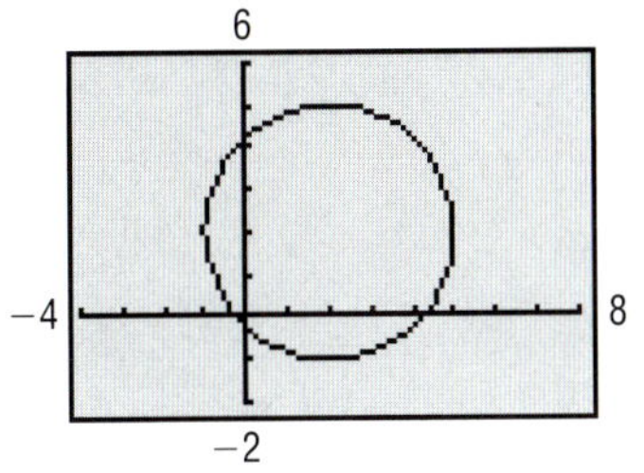

38.

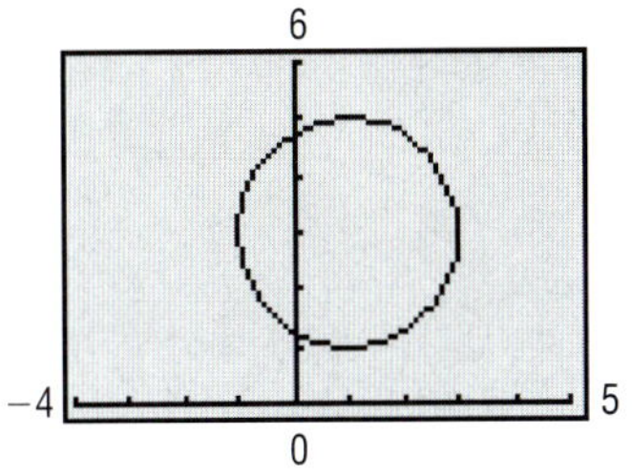

39. The **tangent line** to a circle may be defined as the line that intersects the circle in a single point, called the **point of tangency** (see the figure).

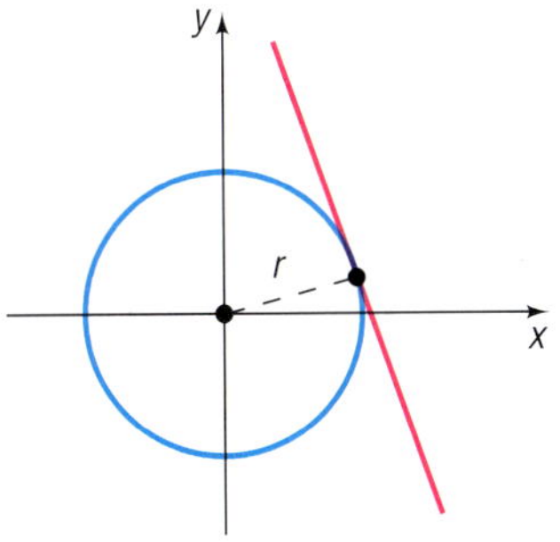

If the equation of the circle is $x^2 + y^2 = r^2$ and the equation of the tangent line is $y = mx + b$, show that:

(a) $r^2(1 + m^2) = b^2$
[Hint: The quadratic equation $x^2 + (mx + b)^2 = r^2$ has exactly one solution.]

(b) The point of tangency is $(-r^2 m/b, r^2/b)$.

(c) The tangent line is perpendicular to the line containing the center of the circle and the point of tangency.

40. The Greek method for finding the equation of the tangent line to a circle used the fact that at any point on a circle the lines containing the center and the tangent line are perpendicular (see Problem 39). Use this method to find an equation of the tangent line to the circle $x^2 + y^2 = 9$ at the point $(1, 2\sqrt{2})$.

41. Use the Greek method described in Problem 40 to find an equation of the tangent line to the circle $x^2 + y^2 - 4x + 6y + 4 = 0$ at the point $(3, 2\sqrt{2} - 3)$.

42. Refer to Problem 39. The line $x - 2y + 4 = 0$ is tangent to a circle at $(0, 2)$. The line $y = 2x - 7$ is tangent to the same circle at $(3, -1)$. Find the center of the circle.

43. Find an equation of the line containing the centers of the two circles

$$x^2 + y^2 - 4x + 6y + 4 = 0 \quad \text{and}$$
$$x^2 + y^2 + 6x + 4y + 9 = 0$$

44. If a circle of radius 2 is made to roll along the x-axis, what is an equation for the path of the center of the circle?

CHAPTER REVIEW

THINGS TO KNOW

Formulas

Distance formula	$d = \sqrt{(x_2 - x_1)^2 + (y_2 - y_1)^2}$
Midpoint formula	$(x, y) = \left(\frac{x_1 + x_2}{2}, \frac{y_1 + y_2}{2}\right)$
Slope	$m = \frac{y_2 - y_1}{x_2 - x_1}$, if $x_1 \neq x_2$; undefined if $x_1 = x_2$
Parallel lines	Equal slopes ($m_1 = m_2$) with different y-intercepts
Perpendicular lines	Product of slopes is -1 ($m_1 \cdot m_2 = -1$)

Equations

Vertical line	$x = a$
Horizontal line	$y = b$
Point–slope form of the equation of a line	$y - y_1 = m(x - x_1)$; m is the slope of the line, (x_1, y_1) is a point on the line
General form of the equation of a line	$Ax + By + C = 0$, A, B, not both 0
Slope–intercept form of the equation of a line	$y = mx + b$; m is the slope of the line, b is the y-intercept
Standard form of the equation of a circle	$(x - h)^2 + (y - k)^2 = r^2$; r is the radius of the circle, (h, k) is the center of the circle
General form of the equation of a circle	$x^2 + y^2 + ax + by + c = 0$
Equation of the unit circle	$x^2 + y^2 = 1$

HOW TO

Use the distance formula
Draw scatter diagrams
Graph equations by plotting points
Graph equations using a graphing utility
Find the intercepts of a graph
Test an equation for symmetry
Solve linear equations in one variable
Solve quadratic equations by factoring
Solve radical equations
Solve equations and inequalities using a graphing utility
Solve equations and inequalities involving absolute value
Solve applied problems
Solve linear inequalities in one variable
Graph lines by hand
Find the slope and intercepts of a line, given the equation
Obtain the equation of a line
Obtain the equation of a circle
Find the center and radius of a circle, given the equation
Graph circles by hand and by using a graphing utility

FILL-IN-THE BLANK ITEMS

1. If (x, y) are the coordinates of a point P in the xy-plane, then x is called the ______________ of P and y is the ______________ of P.
2. If three distinct points P, Q, and R all lie on a line and if $d(P, Q) = d(Q, R)$, then Q is called the ______________ of the line segment from P to R.
3. If for every point (x, y) on a graph the point $(-x, y)$ is also on the graph, then the graph is symmetric with respect to the ______________.
4. To complete the square of the expression $x^2 + 5x$, you would ______________ the number ______________.
5. If $a < 0$, then $|a| =$ ______________.
6. When a quadratic equation has a repeated solution, it is called a(n) ______________ root or a root of ______________ ______________.
7. When an apparent solution does not satisfy the original equation, it is called a(n) ______________ solution.
8. If each side of an inequality is multiplied by a(n) ______________ number, then the direction of the inequality symbol is reversed.
9. The equation $|x^2| = 4$ has two real solutions, ______________ and ______________.
10. The set of points in the xy-plane that are a fixed distance from a fixed point is called a(n) ______________. The fixed distance is called the ______________; the fixed point is called the ______________.
11. The slope of a vertical line is ______________; the slope of a horizontal line is ______________.
12. Two nonvertical lines have slopes m_1 and m_2, respectively. The lines are parallel if ______________; the lines are perpendicular if ______________.

TRUE/FALSE ITEMS

T F 1. The distance between two points is sometimes a negative number.

T F 2. The graph of the equation $y = x^4 + x^2 + 1$ is symmetric with respect to the y-axis.

T F 3. Vertical lines have undefined slope.

T F 4. The slope of the line $2y = 3x + 5$ is 3.

T F 5. Perpendicular lines have slopes that are reciprocals of one another.

T F 6. The radius of the circle $x^2 + y^2 = 9$ is 3.

T F 7. Equations can have no solutions, one solution, or more than one solution.

REVIEW EXERCISES

Blue problem numbers indicate the authors' suggestions for use in a Practice Test.

In Problems 1–28, find all real solutions, if any, of each equation (a) graphically and (b) algebraically.

1. $2 - \frac{x}{3} = 6$
2. $\frac{x}{4} - 2 = 6$
3. $-2(5 - 3x) + 8 = 4 + 5x$
4. $(6 - 3x) - 2(1 + x) = 6x$
5. $\frac{3x}{4} - \frac{x}{3} = \frac{1}{12}$
6. $\frac{4 - 2x}{3} + \frac{1}{6} = 2x$
7. $\frac{x}{x - 1} = \frac{5}{6}, \quad x \neq 1$
8. $\frac{4x - 5}{3 - 7x} = 4, \quad x \neq \frac{3}{7}$
9. $x(1 - x) = -6$
10. $x(1 + x) = 6$
11. $\frac{1}{2}\left(x - \frac{1}{3}\right) = \frac{3}{4} - \frac{x}{6}$
12. $\frac{1 - 3x}{4} = \frac{x + 6}{3} + \frac{1}{2}$
13. $(x - 1)(2x + 3) = 3$
14. $x(2 - x) = 3(x - 4)$
15. $4x + 3 = 4x^2$
16. $5x = 4x^2 + 1$
17. $\sqrt[3]{x - 1} = 2$
18. $\sqrt[4]{1 + x} = 3$

19. $x(x+1) - 2 = 0$
20. $2x^2 - 3x + 1 = 0$
21. $\sqrt{2x-3} + x = 3$
22. $\sqrt{2x-1} = x - 2$
23. $\sqrt{x+1} + \sqrt{x-1} = \sqrt{2x+1}$
24. $\sqrt{2x-1} - \sqrt{x-5} = 3$
25. $|2x+3| = 7$
26. $|3x-1| = 5$
27. $|2-3x| = 7$
28. $|1-2x| = 3$

In Problems 29–38, solve each inequality (a) graphically and (b) algebraically. Graph the solution set.

29. $\dfrac{2x-3}{5} + 2 \le \dfrac{x}{2}$
30. $\dfrac{5-x}{3} \le 6x - 4$
31. $-9 \le \dfrac{2x+3}{-4} \le 7$
32. $-4 < \dfrac{2x-2}{3} < 6$
33. $6 > \dfrac{3-3x}{12} > 2$
34. $6 > \dfrac{5-3x}{2} \ge -3$
35. $|3x+4| < \frac{1}{2}$
36. $|1-2x| < \frac{1}{3}$
37. $|2x-5| \ge 9$
38. $|3x+1| \ge 10$

In Problems 39–48, find an equation of the line having the given characteristics. Express your answer using either the general form or the slope–intercept form of the equation of a line, whichever you prefer.

39. Slope $= -2$; containing the point $(3, -1)$
40. Slope $= 0$; containing the point $(-5, 4)$
41. Slope undefined; containing the point $(-3, 4)$
42. x-intercept $= 2$; containing the point $(4, -5)$
43. y-intercept $= -2$; containing the point $(5, -3)$
44. Containing the points $(3, -4)$ and $(2, 1)$
45. Parallel to the line $2x - 3y + 4 = 0$; containing the point $(-5, 3)$
46. Parallel to the line $x + y - 2 = 0$; containing the point $(1, -3)$
47. Perpendicular to the line $x + y - 2 = 0$; containing the point $(4, -3)$
48. Perpendicular to the line $3x - y + 4 = 0$; containing the point $(-2, 4)$

In Problems 49–54, graph each line using a graphing utility. Find the x-intercept and y-intercept. Also draw each graph by hand, labeling any intercepts.

49. $4x - 5y + 20 = 0$
50. $3x + 4y - 12 = 0$
51. $\dfrac{1}{2}x - \dfrac{1}{3}y + \dfrac{1}{6} = 0$
52. $-\dfrac{3}{4}x + \dfrac{1}{2}y = 0$
53. $\sqrt{2}x + \sqrt{3}y = \sqrt{6}$
54. $\dfrac{x}{3} + \dfrac{y}{4} = 1$

In Problems 55–62, list the x- and y-intercepts and test for symmetry.

55. $2x = 3y^2$
56. $y = 5x$
57. $4x^2 + y^2 = 1$
58. $x^2 - 9y^2 = 9$
59. $y = x^4 + 2x^2 + 1$
60. $y = x^3 - x$
61. $x^2 + x + y^2 + 2y = 0$
62. $x^2 + 4x + y^2 - 2y = 0$

In Problems 63–66, find the center and radius of each circle. Graph each circle using a graphing utility.

63. $x^2 + y^2 - 2x + 4y - 4 = 0$
64. $x^2 + y^2 + 4x - 4y - 1 = 0$
65. $3x^2 + 3y^2 - 6x + 12y = 0$
66. $2x^2 + 2y^2 - 4x = 0$

67. Find the slope of the line containing the points $(7, 4)$ and $(-3, 2)$. What is the distance between these points? What is their midpoint?

68. Find the slope of the line containing the points $(2, 5)$ and $(6, -3)$. What is the distance between these points? What is their midpoint?

69. Show that the points $A = (3, 4)$, $B = (1, 1)$, and $C = (-2, 3)$ are the vertices of an isosceles triangle.

70. Show that the points $A = (-2, 0,)$ $B = (-4, 4)$, and $C = (8, 5)$ are the vertices of a right triangle in two ways:
(a) By using the converse of the Pythagorean Theorem
(b) By using the slopes of the lines joining the vertices

71. Show that the points $A = (2, 5)$, $B = (6, 1)$, and $C = (8, -1)$ lie on a straight line by using slopes.

72. Show that the points $A = (1, 5)$, $B = (2, 4)$, and $C = (-3, 5)$ lie on a circle with center $(-1, 2)$. What is the radius of this circle?

73. The end points of the diameter of a circle are $(-3, 2)$ and $(5, -6)$. Find the center and radius of the circle. Write the general equation of this circle.

74. Find two numbers y such that the distance from $(-3, 2)$ to $(5, y)$ is 10.

75. **Lightning and Thunder** A flash of lightning is seen, and the resulting thunderclap is heard 3 seconds later. If the speed of sound averages 1100 feet per second, how far away is the storm?

76. **Physics: Intensity of Light** The intensity I (in candlepower) of a certain light source obeys the equation $I = 900/x^2$, where x is the distance (in meters) from the light. Over what range of distances can an object be placed from this light source so that the range of intensity of light is from 1600 to 3600 candlepower, inclusive?

77. **Extent of Search and Rescue** A search plane has a cruising speed of 250 miles per hour and carries enough fuel for at most 5 hours of flying. If there is a wind that averages 30 miles per hour and the direction of search is with the wind one way and against it the other, how far can the search plane travel?

78. **Extent of Search and Rescue** If the search plane described in Problem 77 is able to add a supplementary fuel tank that allows for an additional 2 hours of flying, how much farther can the plane extend its search?

79. **Rescue at Sea** A life raft, set adrift from a sinking ship 150 miles offshore, travels directly toward a Coast Guard station at the rate of 5 miles per hour. At the time that the raft is set adrift, a rescue helicopter is dispatched from the Coast Guard station. If the helicopter's average speed is 90 miles per hour, how long will it take the helicopter to reach the life raft?

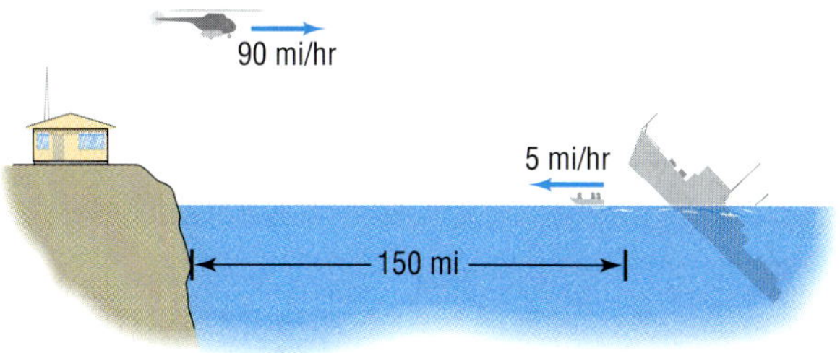

80. **Physics: Uniform Motion** Two bees leave two locations 150 meters apart and fly, without stopping, back and forth between these two locations at average speeds of 3 meters per second and 5 meters per second, respectively. How long is it until the bees meet for the first time? How long is it until they meet for the second time?

81. **Working Together to Get a Job Done** Clarissa and Shawna, working together, can paint the exterior of a house in 6 days. Clarissa by herself can complete this job in 5 days less than Shawna. How long will it take Clarissa to complete the job by herself?

82. **Emptying a Tank** Two pumps of different sizes, working together, can empty a fuel tank in 5 hours. The larger pump can empty this tank in 4 hours less than the smaller one. If the larger one is out of order, how long will it take the smaller one to do the job alone?

83. **Chemistry: Mixing Acids** For a certain experiment, a student requires 100 cubic centimeters of a solution that is 8% HCl. The storeroom has only solutions that are 15% HCl and 5% HCl. How many cubic centimeters of each available solution should be mixed to get 100 cubic centimeters of 8% HCl?

84. **Chemistry: Salt Solutions** How much water must be evaporated from 32 ounces of a 4% salt solution to make a 6% salt solution?

85. **Business: Theater Attendance** The manager of the Coral Theater wants to know whether the majority of its patrons are adults or children. During a week in July, 5200 tickets were sold and the receipts totaled $20,335. The adult admission is $4.75, and the children's admission is $2.50. How many adult patrons were there?

86. **Business: Blending Coffee** A coffee manufacturer wants to market a new blend of coffee that will cost $6 per pound by mixing two coffees that sell for $4.50 and $8 per pound, respectively. What amounts of each coffee should be blended to obtain the desired mixture?
[Hint: Assume that the total weight of the desired blend is 100 pounds.]

87. **Physics: Uniform Motion** A man is walking at an average speed of 4 miles per hour alongside a railroad track. A freight train, going in the same direction at an average speed of 30 miles per hour, requires 5 seconds to pass the man. How long is the freight train? Give your answer in feet.

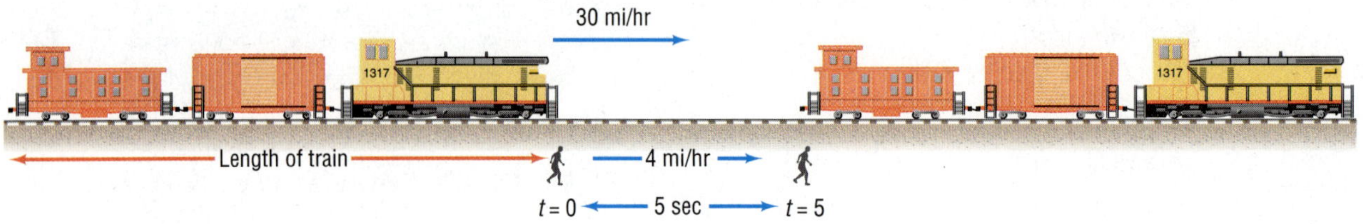

88. One formula stating the relationship between the length l and width w of a rectangle of "pleasing proportion" is $l^2 = w(l + w)$. How should a 4 foot by 8 foot sheet of plasterboard be cut so that the result is a rectangle of "pleasing proportion" with a width of 4 feet?

89. **Business: Determining the Cost of a Charter** A group of 20 senior citizens can charter a bus for a one-day excursion trip for $15 per person. The charter company agrees to reduce the price of each ticket by 10¢ for each additional passenger in excess of 20 who goes on the trip, up to a maximum of 44 passengers (the capacity of the bus). If the final bill from the charter company was $482.40, how many seniors went on the trip, and how much did each pay?

90. A new copying machine can do a certain job in 1 hour less than an older copier. Together they can do this job in 72 minutes. How long would it take the older copier by itself to do the job?

91. **Concentration of Carbon Monoxide in the Air** The following data represent the average concentration of carbon monoxide in parts per million (ppm) in the air for 1987–1993.

Year	Concentration of Carbon Monoxide (ppm)
1987	6.69
1988	6.38
1989	6.34
1990	5.87
1991	5.55
1992	5.18
1993	4.88

Source: U.S. Environmental Protection Agency.

(a) By hand, draw a scatter diagram of the data.
(b) What is the trend in the data? In other words, as time passes, what is happening to the average level of carbon monoxide in the air? Why do you think this is happening?
(c) Treating the year as the x-coordinate and the average level of carbon monoxide as the y-coordinate, use a graphing utility to draw a scatter diagram of the data.
(d) What is the slope of the line joining the points (1987, 6.69) and (1990, 5.87)?
(e) Interpret this slope.
(f) What is the slope of the line joining the points (1990, 5.87) and (1993, 4.88)?
(g) Interpret this slope.
(h) How do you explain the differences among the two slopes obtained?

92. **Value of a Portfolio** The following data represent the value of the Vanguard Index Trust-500 Portfolio for 1993–1997.

Year	Value (Dollars)
1993	40.97
1994	43.83
1995	42.97
1996	57.60
1997	69.17

Source: Vanguard Index Trust, Annual Report 1997.

(a) By hand, draw a scatter diagram of the data.
(b) What is the trend in the data? In other words, as time passes, what is happening to the value of the Vanguard Index Trust-500 Portfolio?
(c) Treating the year as the x-coordinate and the value of the Vanguard Index Trust-500 Portfolio as the y-coordinate, use a graphing utility to draw a scatter diagram of the data.

(d) What is the slope of the line joining the points (1993, 40.97) and (1995, 42.97)?
(e) Interpret this slope.
(f) What is the slope of the line joining the points (1995, 42.97) and (1997, 69.17)?
(g) Interpret this slope.
(h) If you were managing this trust, which of the two slopes would you use to convince someone to invest? Why?

93. Make up four problems that you might be asked to do given the two points $(-3, 4)$ and $(6, 1)$. Each problem should involve a different concept. Be sure that your directions are clearly stated.

94. Describe each of the following graphs. Give justification.
(a) $x = 0$ (b) $y = 0$
(c) $x + y = 0$ (d) $xy = 0$
(e) $x^2 + y^2 = 0$

95. In a 100-meter race, Todd crosses the finish line 5 meters ahead of Scott. To even things up, Todd suggests to Scott that they race again, this time with Todd lining up 5 meters behind the start.
(a) Assuming that Todd and Scott run at the same pace as before, does the second race end in a tie?
(b) If not, who wins?
(c) By how many meters does he win?
(d) How far back should Todd start so that the race ends in a tie?

After running the race a second time, Scott, to even things up, suggests to Todd that he (Scott) line up 5 meters in front of the start.
(e) Assuming again that they run at the same pace as in the first race, does the third race result in a tie?
(f) If not, who wins?
(g) By how many meters?
(h) How far up should Scott start so that the race ends in a tie?

96. Explain the difference between the following three problems. Are there any similarities in their solution?
(a) Write the expression as a single quotient:
$$\frac{x}{x-2} + \frac{x}{x^2-4}$$
(b) Solve: $\frac{x}{x-2} + \frac{x}{x^2-4} = 0$
(c) Solve: $\frac{x}{x-2} + \frac{x}{x^2-4} < 0$

CHAPTER PROJECTS

1. **Financing a Car** Advertisements for cars are bewildering. Suppose you have decided to purchase a car whose list price is \$12,400. You have saved \$2000 to use as a down payment. After shopping around at several dealerships, the offers given in the table are available to you.

	List Price	Dealer Discount	Down Payment	Amount Borrowed	Length of Loan	APR Annual Rate of Interest
Dealer 1	\$12,400	1000	2000	9400	3 years	2.9%
Dealer 1	\$12,400	1000	2000	9400	5 years	5.9%
Dealer 1	\$12,400	1500	2000	8900	4 years	10%
Dealer 2	\$12,400	1500	2000	8900	3 years	9%
Dealer 2	\$12,400	1800	2000	8600	5 years	12%
Dealer 3	\$12,400	400	2000	10,000	2 years	0.5%

CHAPTER PROJECTS *(Continued)*

(a) The first concern that you have is your monthly payment. The formula

$$M = A\left[\frac{1 - (1 + i)^{-n}}{i}\right]^{-1}$$

can be used to compute your monthly payment M. In this formula, A is the amount borrowed, i is the monthly interest rate so $i = \text{APR}/12$, (expressed as a decimal), and n is the length of the loan, in months. Compute the monthly payment for each of the six possibilities.

(b) In each case, how much money do you actually pay for the car?

(c) In each case, how much interest is paid?

(d) Which of the six deals is "best"? Be sure to have reasons!

2. **Economics** An **isocost line** represents the various combinations of two inputs that can be used in a production process while maintaining a constant level of expenditures. The equation for an isocost line is

$$P_L \cdot L + P_K \cdot K = E$$

where

P_L represents the price of labor.
L represents the amount of labor used in the production process.
P_K represents the price of capital (machinery, etc.).
K represents the amount of capital used in the production process.
E represents the total expenditures for the two inputs, labor and capital.

(a) What values of L and K make sense as they are defined above?

(b) Suppose that $P_L = \$400$ per hour and $P_K = \$500$ per hour. If $L = 100$ hours, how many hours of capital, K, may be purchased if expenditures are to be \$100,000?

(c) Suppose that $P_L = \$400$ per hour and $P_K = \$500$ per hour; graph the isocost line if $E = \$100{,}000$ with the number of hours of labor, L, on the x-axis and the number of hours of capital, K, on the y-axis. In which quadrant is your graph located? Why?

(d) Graph the isocost line if E increases to \$120,000. What effect does this have on the graph? What do you think would happen if expenditures were to decrease?

(e) Determine the slope of the isocost line. Interpret its value.

(f) Graph the isocost line with $P_L = \$500$ per hour, $P_K = \$500$ per hour, and $E = \$100{,}000$. Compare this graph with the graph obtained in (c). What effect does the increase in the price of labor have on the graph?

CHAPTER

2 Linear and Quadratic Functions

The Use of Statistics in Making Investment Decisions: Using Beta

One of the most common statistics used by investors is beta, which measures a security or portfolio's movement relative to the S&P 500. When determining beta graphically, returns are plotted on a graph, with the horizontal axis representing the S&P returns and the vertical axis representing the security or portfolio returns; the points that are plotted represent the return of the security versus the return of the S&P 500 at the same time period. A regression line is then drawn, which is a straight line that most closely approximates the points in the graph. Beta represents the slope of that line; a steep slope (45 degrees or greater) indicates that when the stock market moved up or down, the security or portfolio moved up or down on average to the same degree or greater, and a flatter slope indicates that when the stock market moved up or down, the security or portfolio moved to a lesser degree. Since beta measures variability, it is used as a measure of risk; the greater the beta, the greater the risk.

Source: Paul E. Hoffman, ***Journal of the American Association of Individual Investors,*** September, 1991; http://www.aaii.com

See Chapter Project 1.

Preparing for This Chapter

Before getting started on this chapter, review the following concepts:

- **Evaluating Algebraic Expressions (p. 680)**
- **Domain of a Variable (p. 717)**
- **Intercepts (pp. 19–21)**
- **Slope–Intercept Form of a Line (pp. 73–74)**
- **Scatter Diagrams (pp. 9–10)**
- **Solving Quadratic Equations by Factoring (pp. 33–35)**
- **Completing the Square (pp. 85–86)**
- **Solving Equations Using a Graphing Utility (pp. 29–31)**

Outline

Perhaps the most central idea in mathematics is the notion of a *function*. This important chapter deals with what a function is, how to graph linear and quadratic functions, and how these functions are used in applications.

The word *function* apparently was introduced by René Descartes in 1637. For him, a function simply meant any positive integral power of a variable *x*. Gottfried Wilhelm Leibniz (1646–1716), who always emphasized the geometric side of mathematics, used the word *function* to denote any quantity associated with a curve, such as the coordinates of a point on the curve. Leonhard Euler (1707–1783) employed the word to mean any equation or formula involving variables and constants. His idea of a function is similar to the one most often seen in courses that precede calculus. Later, the use of functions in investigating heat flow equations led to a very broad definition, due to Lejeune Dirichlet (1805–1859), which describes a function as a rule or correspondence between two sets. It is his definition that we use here.

2.1 FUNCTIONS

1 Determine Whether a Relation Represents a Function

2 Find the Value of a Function

3 Find the Domain of a Function

4 Identify the Graph of a Function

5 Obtain Information from or about the Graph of a Function

In Section 1.1, we said that a **relation** is a correspondence between two variables, say x and y. When relations are written as ordered pairs (x, y), we say that x is related to y. Often, we are interested in specifying the type of relation (such as an equation) that might exist between the two variables.

For example, the relation between the revenue R resulting from the sale of x items selling for \$10 each may be expressed by the equation $R = 10x$. If we know how many items have been sold, then we can calculate the revenue by using the equation $R = 10x$. This equation is an example of a *function*.

As another example, suppose that an icicle falls off a building from a height of 64 feet above the ground. According to a law of physics, the distance s (in feet) of the icicle from the ground after t seconds is given (approximately) by the formula $s = 64 - 16t^2$. When $t = 0$ seconds, the icicle is $s = 64$ feet above the ground. After 1 second, the icicle is $s = 64 - 16(1)^2 = 48$ feet above the ground. After 2 seconds, the icicle strikes the ground. The formula $s = 64 - 16t^2$ provides a way of finding the distance s when the time t $(0 \leq t \leq 2)$ is prescribed. There is a correspondence between each time t in the interval $0 \leq t \leq 2$ and the distance s. We say that the distance s is a *function* of the time t because

1. There is a correspondence between the set of times and the set of distances.
2. There is exactly one distance s obtained for a prescribed time t in the interval $0 \leq t \leq 2$.

Let's now look at the definition of a function.

Figure 1

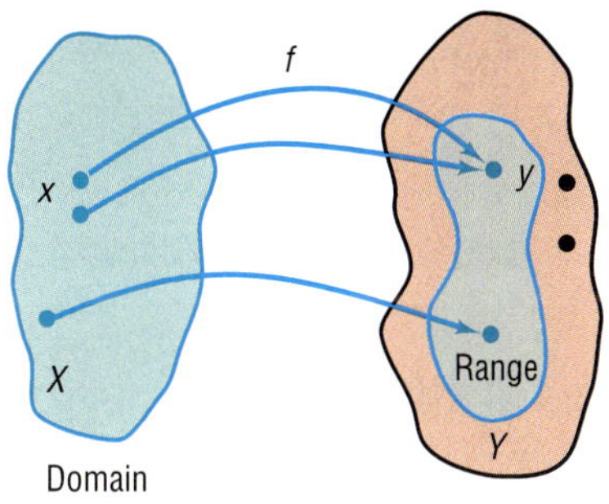

Definition of Function

> Let X and Y be two nonempty sets of real numbers.* A **function** from X into Y is a relation that associates with each element of X a unique element of Y.

The set X is called the **domain** of the function. For each element x in X, the corresponding element y in Y is called the **value** of the function at x, or the **image** of x. The set of all images of the elements of the domain is called the **range** of the function. See Figure 1.

Since there may be some elements in Y that are not the image of some x in X, it follows that the range of a function may be a subset of Y, as shown in Figure 1.

1

Not all relations between two sets are functions.

EXAMPLE 1 Determining Whether a Relation Represents a Function

Determine whether the following relations represent functions.

(a) For this relation, the domain represents the employees of Sara's Pre-Owned Car Mart and the range represents their base salary.

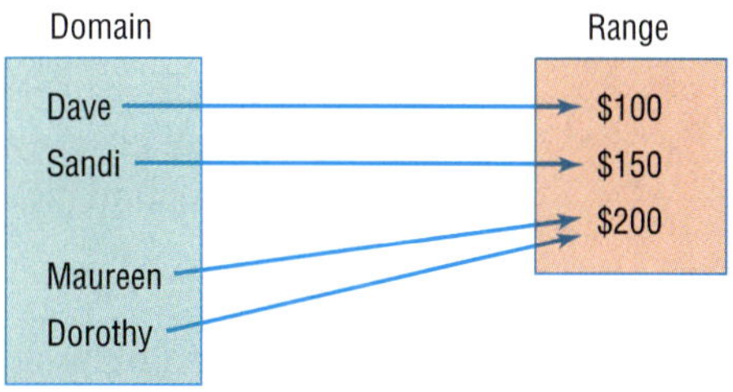

(b) For this relation, the domain represents the employees of Sara's Pre-Owned Car Mart and the range represents their phone number(s).

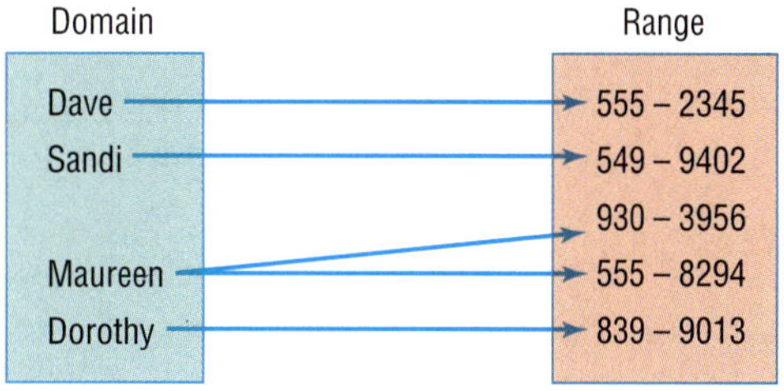

Solution

(a) The relation is a function because each element in the domain corresponds to a unique element in the range. Notice that more than one element in the domain can correspond to the same element in the range.

(b) The relation is not a function because each element in the domain does not correspond to a unique element in the range. Maureen has two telephone numbers; therefore, if Maureen is chosen from the domain, a unique telephone number cannot be assigned to her.

*The two sets X and Y can also be sets of complex numbers (discussed in Section 4.5), and then we have defined a complex function. In the broad definition (due to Lejeune Dirichlet), X and Y can be any two sets.

We may think of a function as a set of ordered pairs (x, y) in which no two distinct pairs have the same first element. The set of all first elements x is the domain of the function, and the set of all second elements y is its range. Thus, there is associated with each element x in the domain a unique element y in the range.

EXAMPLE 2 Determining Whether a Relation Represents a Function

Determine whether each relation represents a function.

(a) $\{(1, 4), (2, 5), (3, 6), (4, 7)\}$
(b) $\{(1, 4), (2, 4), (3, 5), (6, 10)\}$
(c) $\{(-3, 9), (-2, 4), (0, 0), (1, 1), (-3, 8)\}$

Solution (a) This relation is a function because there are no distinct ordered pairs with the same first element.

(b) This relation is a function because there are no distinct ordered pairs with the same first element.

(c) This relation is not a function because there is a first element, -3, that corresponds to two different second elements, 9 and 8.

In Example 2(b), notice that 1 and 2 in the domain each have the same image in the range. This does not violate the definition of a function; two different first elements can have the same second element. A violation of the definition occurs when two ordered pairs have the same first element and different second elements, as in Example 2(c).

NOW WORK PROBLEMS 1 AND 5.

Examples 2(a) and 2(b) demonstrate that a function may be defined by a set of ordered pairs. A function may also be defined by an equation in two variables, usually denoted x and y.

EXAMPLE 3 Example of a Function

Consider the function defined by the equation

$$y = 2x - 5 \qquad 1 \le x \le 6$$

The domain $1 \le x \le 6$ specifies that the number x is restricted to the real numbers from 1 to 6, inclusive. The equation $y = 2x - 5$ specifies that the number x is to be multiplied by 2 and then 5 is to be subtracted from the result to get y. For example, if $x = \frac{3}{2}$, then $y = 2 \cdot \frac{3}{2} - 5 = -2$.

Function Notation

Functions are often denoted by letters such as f, F, g, G, and so on. If f is a function, then for each number x in its domain the corresponding image in the range is designated by the symbol $f(x)$, read as "f of x" or as "f at x." We refer to $f(x)$ as the **value of f at the number x.** Thus, $f(x)$ is the number that results when x is given and the function f is applied; $f(x)$ does *not* mean "f times x." For example, the function given in Example 3 may be written as $y = f(x) = 2x - 5$, $1 \le x \le 6$. Then $f(\frac{3}{2}) = -2$.

Figure 2 illustrates some other functions. Note that in every function illustrated for each x in the domain, there is one value in the range.

Figure 2

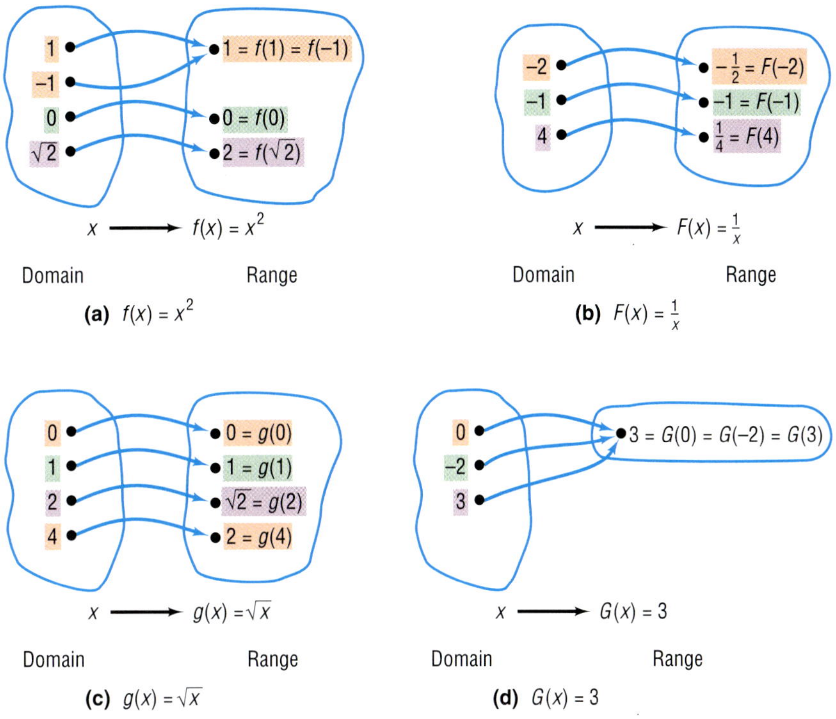

Sometimes it is helpful to think of a function f as a machine that receives as input a number from the domain, manipulates it, and outputs the value. See Figure 3.

Figure 3

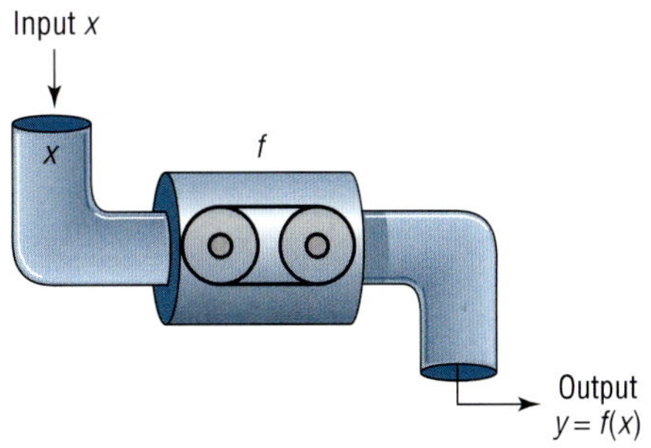

The restrictions on this input/output machine are

1. It only accepts numbers from the domain of the function.
2. For each input, there is exactly one output (which may be repeated for different inputs).

For a function $y = f(x)$, the variable x is called the **independent variable,** because it can be assigned any of the permissible numbers from the domain. The variable y is called the **dependent variable,** because its value depends on x.

Any symbol can be used to represent the independent and dependent variables. For example, if f is the *cube function,* then f can be defined by $f(x) = x^3$ or $f(t) = t^3$ or $f(z) = z^3$. All three functions are the same: Each tells us to cube the independent variable. In practice, the symbols used for the independent and dependent variables are based on common usage, such as using C for cost in business.

2 The variable x is also called the **argument** of the function. Thinking of the independent variable as an argument can sometimes make it easier to find the value of a function. For example, if f is the function defined by $f(x) = x^3$, then f tells us to cube the argument. Thus, $f(2)$ means to cube 2, $f(a)$ means to cube the number a, and $f(x + h)$ means to cube the quantity $x + h$.

EXAMPLE 4 Finding Values of a Function

For the function G defined by $G(x) = 2x^2 - 3x$, evaluate:

(a) $G(3)$ (b) $G(x) + G(3)$ (c) $G(-x)$
(d) $-G(x)$ (e) $G(x + 3)$

Solution (a) We substitute 3 for x in the equation for G to get

$$G(3) = 2(3)^2 - 3(3) = 18 - 9 = 9$$

(b) $G(x) + G(3) = (2x^2 - 3x) + (9) = 2x^2 - 3x + 9$

(c) We substitute $-x$ for x in the equation for G:

$$G(-x) = 2(-x)^2 - 3(-x) = 2x^2 + 3x$$

(d) $-G(x) = -(2x^2 - 3x) = -2x^2 + 3x$

(e) $G(x + 3) = 2(x + 3)^2 - 3(x + 3)$ *Notice the use of parentheses here.*

$$\begin{aligned} G(x+3) &= 2(x + 3)^2 - 3(x + 3) \\ &= 2(x^2 + 6x + 9) - 3x - 9 \\ &= 2x^2 + 12x + 18 - 3x - 9 \\ &= 2x^2 + 9x + 9 \end{aligned}$$

Notice in this example that $G(x + 3) \neq G(x) + G(3)$ and $G(-x) \neq -G(x)$.

NOW WORK PROBLEM 13.

Graphing calculators have special keys that enable you to find the value of certain commonly used functions. For example, you should be able to find the square function $f(x) = x^2$, the square root function $f(x) = \sqrt{x}$, the reciprocal function $f(x) = 1/x = x^{-1}$, and many others that will be discussed later in this book (such as $\ln x$, $\log x$, and so on). Verify the results of Example 5, which follows, on your calculator.

EXAMPLE 5 Finding Values of a Function on a Calculator

(a) $f(x) = x^2$; $f(1.234) = 1.522756$
(b) $F(x) = 1/x$; $F(1.234) = 0.8103727715$
(c) $g(x) = \sqrt{x}$; $g(1.234) = 1.110855526$

Graphing calculators can also be used to evaluate any function that you wish. Figure 4 shows the result obtained in Example 4(a) on a TI-83 graphing calculator with the function to be evaluated, $G(x) = 2x^2 - 3x$, in Y_1.*

Figure 4

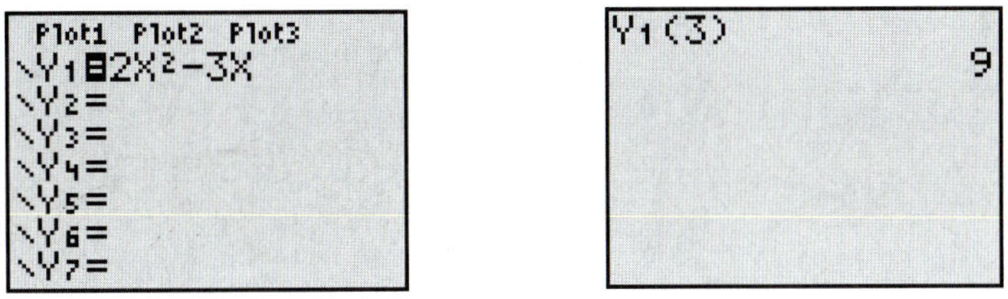

Implicit Form of a Function

In general, when a function f is defined by an equation in x and y, we say that the function f is given **implicitly.** If it is possible to solve the equation for y in terms of x, then we write $y = f(x)$ and say that the function is given **explicitly.** In fact, we usually write "the function $y = f(x)$" when we mean "the function f defined by the equation $y = f(x)$." Although this usage is not entirely correct, it is rather common and should not cause any confusion. For example,

Implicit Form	**Explicit Form**
$3x + y = 5$	$y = f(x) = -3x + 5$
$x^2 - y = 6$	$y = f(x) = x^2 - 6$
$xy = 4$	$y = f(x) = 4/x$

Not all equations in x and y define a function $y = f(x)$. If an equation is solved for y and two or more values of y can be obtained for a given x, then the equation does not define a function. For example, consider the equation $x^2 + y^2 = 1$, which defines the unit circle. If we solve for y, we obtain $y = \pm\sqrt{1 - x^2}$, so two values of y result for numbers x between -1 and 1. Thus, $x^2 + y^2 = 1$ does not define a function.

Comment The explicit form of a function is the form required by a graphing calculator. Now do you see why it is necessary to graph a circle in two "pieces"?

We list next a summary of some important facts to remember about a function f.

SUMMARY OF IMPORTANT FACTS ABOUT FUNCTIONS

1. To each x in the domain of f, there is one and only one image $f(x)$ in the range.
2. f is the symbol that we use to denote the function. It is symbolic of the domain and the equation that we use to get from an x in the domain to $f(x)$ in the range.
3. If $y = f(x)$, then x is called the independent variable or argument of f, and y is called the dependent variable or the value of f at x.

*Consult your owner's manual for the required keystrokes.

Domain of a Function

3 Often, the domain of a function f is not specified; instead, only the equation defining the function is given. In such cases, we agree that the domain of f is the largest set of real numbers for which the value $f(x)$ is a real number. Thus, the domain of f is the same as the domain of the variable x in the expression $f(x)$.

EXAMPLE 6 Finding the Domain of a Function

Find the domain of each of the following functions:

(a) $f(x) = x^2 + 5x$ (b) $g(x) = \dfrac{3x}{x^2 - 4}$ (c) $h(x) = \sqrt{4 - 3x}$

Solution (a) The function tells us to square a number and then add five times the number. Since these operations can be performed on any real number, we conclude that the domain of f is all real numbers.

(b) The function g tells us to divide $3x$ by $x^2 - 4$. Since division by 0 is not defined, the denominator $x^2 - 4$ can never be 0. Thus, x can never equal -2 or 2. The domain of the function g is $\{x | x \neq -2, x \neq 2\}$.

(c) The function h tells us to take the square root of $4 - 3x$. But only non-negative numbers have real square roots. Hence, we require that

$$\begin{aligned} 4 - 3x &\geq 0 \\ -3x &\geq -4 \\ x &\leq \tfrac{4}{3} \end{aligned}$$

The domain of h is $\{x | x \leq \frac{4}{3}\}$ or the interval $(-\infty, \frac{4}{3}]$.

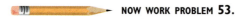

NOW WORK PROBLEM 53.

If x is in the domain of a function f, we shall say that **f is defined at x,** or **$f(x)$ exists.** If x is not in the domain of f, we say that **f is not defined at x,** or **$f(x)$ does not exist.** For example, if $f(x) = x/(x^2 - 1)$, then $f(0)$ exists, but $f(1)$ and $f(-1)$ do not exist. (Do you see why?)

We have not said much about finding the range of a function. The reason is that when a function is defined by an equation it is often difficult to find the range. Therefore, we shall usually be content to find just the domain of a function when only the rule for the function is given. We shall express the domain of a function using inequalities, interval notation, set notation, or words, whichever is most convenient.

The Graph of a Function

In applications, a graph often demonstrates more clearly the relationship between two variables than, say, an equation or table would. For example, Table 1 shows the price per share of Merck stock at the end of each month during 1998.

If we plot the data in Table 1, using the date as the x-coordinate and the price as the y-coordinate, and then connect the points, we obtain Figure 5.

Table 1

Date	Closing Price
1/31/98	115.589
2/28/98	125.621
3/31/98	126.686
4/30/98	119.089
5/31/98	115.63
6/30/98	132.702
7/31/98	122.595
8/31/98	115.029
9/30/98	129.117
10/31/98	134.598
11/30/98	154.591
12/31/98	147.5

Courtesy of A.G. Edwards & Sons, Inc.

Figure 5

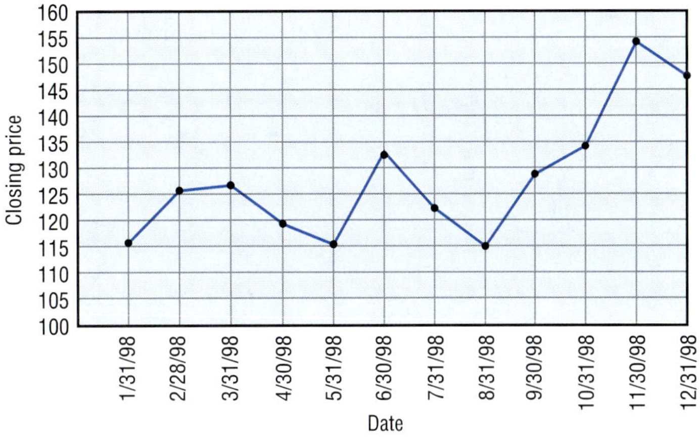

We can see from the graph that the price of the stock was falling during July and August and was rising from September through November. The graph also shows that the lowest price occurred at the end of August, while the highest occurred at the end of November. Equations and tables, on the other hand, usually require some calculations and interpretation before this kind of information can be "seen."

Look again at Figure 5. The graph shows that for each date on the horizontal axis there is only one price on the vertical axis. Thus, the graph represents a function, although the exact rule for getting from date to price is not given.

When a function is defined by an equation in x and y, the **graph** of the function is the graph of the equation, that is, the set of points (x, y) in the xy-plane that satisfies the equation.

4

Not every collection of points in the xy-plane represents the graph of a function. Remember, for a function, each number x in the domain has one and only one image y. Thus, the graph of a function cannot contain two points with the same x-coordinate and different y-coordinates. Therefore, the graph of a function must satisfy the following **vertical-line test.**

THEOREM **Vertical-line Test**

A set of points in the xy-plane is the graph of a function if and only if every vertical line intersects the graph in at most one point.

It follows that, if any vertical line intersects a graph at more than one point, the graph is not the graph of a function.

EXAMPLE 7 Identifying the Graph of a Function

Which of the graphs in Figure 6 are graphs of functions?

Figure 6

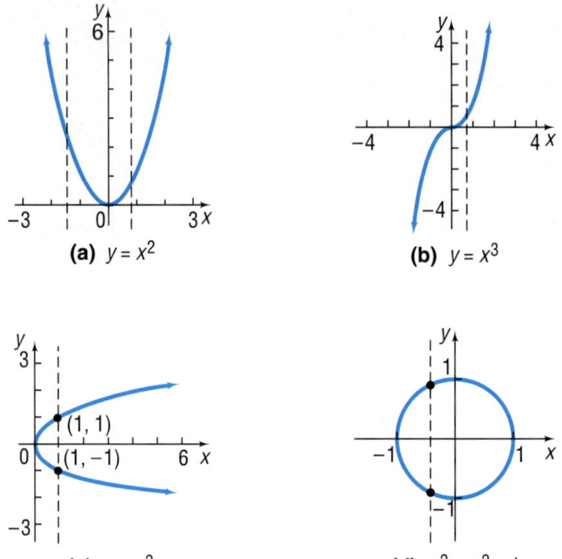

(a) $y = x^2$ **(b)** $y = x^3$ **(c)** $x = y^2$ **(d)** $x^2 + y^2 = 1$

Solution The graphs in Figures 6(a) and 6(b) are graphs of functions, because every vertical line intersects each graph in at most one point. The graphs in Figures 6(c) and 6(d) are not graphs of functions, because a vertical line intersects each graph in more than one point.

NOW WORK PROBLEM 41.

5 If (x, y) is a point on the graph of a function f, then y is the value of f at x, that is, $y = f(x)$. The next example illustrates how to obtain information about a function if its graph is given.

EXAMPLE 8 Obtaining Information from the Graph of a Function

Figure 7

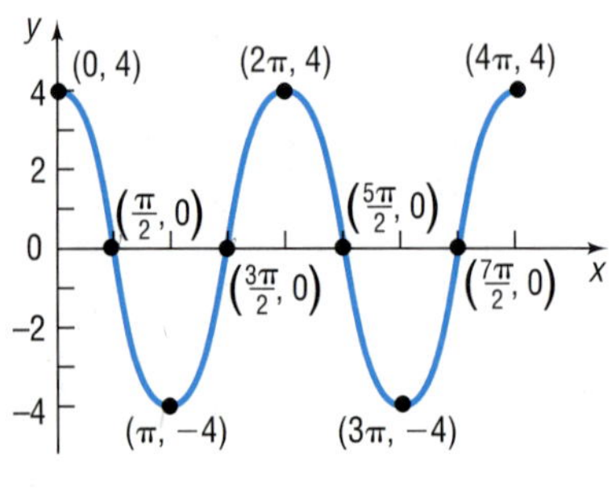

Let f be the function whose graph is given in Figure 7. (The graph of f might represent the distance that the bob of a pendulum is from its "at rest" position. Negative values of y mean that the pendulum is to the left of the "at rest" position, and positive values of y mean that the pendulum is to the right of the "at rest" position.)

(a) What is the value of the function when $x = 0$, $x = 3\pi/2$, and $x = 3\pi$?
(b) What is the domain of f?
(c) What is the range of f?
(d) List the intercepts. (Recall that these are the points, if any, where the graph crosses or touches the coordinate axes.)

Solution (a) Since $(0, 4)$ is on the graph of f, the y-coordinate 4 is the value of f at the x-coordinate 0; that is, $f(0) = 4$. In a similar way, we find that when $x = 3\pi/2$ then $y = 0$; or $f(3\pi/2) = 0$. When $x = 3\pi$, then $y = -4$; or $f(3\pi) = -4$.

(b) To determine the domain of f, we notice that the points on the graph of f will have x-coordinates between 0 and 4π, inclusive; and, for each number x between 0 and 4π, there is a point $(x, f(x))$ on the graph. Thus, the domain of f is $\{x \mid 0 \leq x \leq 4\pi\}$ or the interval $[0, 4\pi]$.

(c) The points on the graph all have y-coordinates between -4 and 4, inclusive; and, for each such number y, there is at least one number x in the domain. Hence, the range of f is $\{y \mid -4 \leq y \leq 4\}$ or the interval $[-4, 4]$.

(d) The intercepts are $(0, 4)$, $(\pi/2, 0)$, $(3\pi/2, 0)$, $(5\pi/2, 0)$, and $(7\pi/2, 0)$.

When the graph of a function is given, its domain may be viewed as the shadow created by the graph on the x-axis by vertical beams of light. Its range can be viewed as the shadow created by the graph on the y-axis by horizontal beams of light. Try this technique with the graph given in Figure 7.

NOW WORK PROBLEMS 37 AND 39.

EXAMPLE 9 Obtaining Information about the Graph of a Function

Consider the function: $f(x) = \dfrac{x}{x + 2}$

(a) Is the point $(1, 1/2)$ on the graph of f?

(b) If $x = -1$, what is $f(x)$? What point is on the graph of f?

(c) If $f(x) = 2$, what is x? What point is on the graph of f?

Solution (a) When $x = 1$, then $f(x) = f(1) = 1/(1 + 2) = 1/3$. The point $(1, 1/2)$ is therefore not on the graph of f. [The point $(1, 1/3)$, however, is].

(b) If $x = -1$, then $f(x) = f(-1) = -1/(-1 + 2) = -1$, so the point $(-1, -1)$ is on the graph of f.

(c) If $f(x) = 2$, then

$$\begin{aligned} \frac{x}{x + 2} &= 2 \\ x &= 2(x + 2) \\ x &= 2x + 4 \\ x &= -4 \end{aligned}$$

The point $(-4, 2)$ is on the graph of f.

NOW WORK PROBLEM 33.

Applications

When we use functions in applications, the domain may be restricted by physical or geometric considerations. For example, the domain of the function f defined by $f(x) = x^2$ is the set of all real numbers. However, if f is used to obtain the area of a square when the length x of a side is known, then we must restrict the domain of f to the positive real numbers, since the length of a side can never be 0 or negative.

EXAMPLE 10 Area of a Circle

Express the area of a circle as a function of its radius.

Solution We know that the formula for the area A of a circle of radius r is $A = \pi r^2$. If we use r to represent the independent variable and A to represent the dependent variable, the function expressing this relationship is

$$A(r) = \pi r^2$$

In this setting, the domain is $\{r|r > 0\}$. (Do you see why?)

NOW WORK PROBLEM 75.

EXAMPLE 11 Getting from an Island to Town

An island is 2 miles from the nearest point P on a straight shoreline. A town is 12 miles down the shore from P.

(a) If a person can row a boat at an average speed of 3 miles per hour and the same person can walk 5 miles per hour, express the time T that it takes to go from the island to town as a function of the distance x from P to where the person lands the boat. See Figure 8.

(b) What is the domain of T?

(c) How long will it take to travel from the island to town if the person lands the boat 4 miles from P?

(d) How long will it take if the person lands the boat 8 miles from P?

Solution (a) Figure 8 illustrates the situation. The distance d_1 from the island to the landing point is the hypotenuse of a right triangle. Using the Pythagorean Theorem, d_1 satisfies the equation

$$d_1^2 = 4 + x^2 \quad \text{or} \quad d_1 = \sqrt{4 + x^2}$$

Figure 8

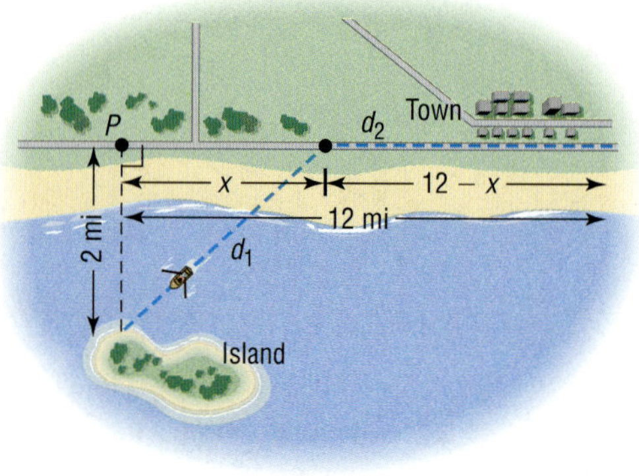

Since the average speed of the boat is 3 miles per hour, the time t_1 that it takes to cover the distance d_1 satisfies

$$d_1 = 3t_1$$

Solving for t_1, we obtain,

$$t_1 = \frac{d_1}{3} = \frac{\sqrt{4 + x^2}}{3}$$

The distance d_2 from the landing point to town is $12 - x$ and the time t_2 that it takes to cover this distance at an average walking speed of 5 miles per hour obey the equation

$$d_2 = 5t_2$$

Solving for t_2, we obtain,

$$t_2 = \frac{d_2}{5} = \frac{12 - x}{5}$$

The total time T of the trip is $t_1 + t_2$. Thus,

$$T(x) = \frac{\sqrt{4 + x^2}}{3} + \frac{12 - x}{5}$$

(b) Since x equals the distance from P to where the boat lands, it follows that the domain of T is $0 \le x \le 12$.

(c) If the boat is landed 4 miles from P, then $x = 4$. The time T that the trip takes is

$$T(4) = \frac{\sqrt{20}}{3} + \frac{8}{5} \approx 3.09 \text{ hours}$$

(d) If the boat is landed 8 miles from P, then $x = 8$. The time T that the trip takes is

$$T(8) = \frac{\sqrt{68}}{3} + \frac{4}{5} \approx 3.55 \text{ hours}$$

NOW WORK PROBLEM 81.

SUMMARY

We list here some of the important vocabulary introduced in this section, with a brief description of each term.

Function	A relation between two sets of real numbers so that each number x in the first set, the domain, has corresponding to it exactly one number y in the second set. A set of ordered pairs (x, y) or $(x, f(x))$ in which no two distinct pairs have the same first element. The range is the set of y values of the function for the x values in the domain. A function f may be defined implicitly by an equation involving x and y or explicitly by writing $y = f(x)$.
Unspecified domain	If a function f is defined by an equation and no domain is specified, then the domain will be taken to be the largest set of real numbers for which the equation defines a real number.

Function notation	$y = f(x)$ f is a symbol for the function. x is the independent variable or argument. y is the dependent variable. $f(x)$ is the value of the function at x, or the image of x.
Graph of a function	The collection of points (x, y) that satisfies the equation $y = f(x)$. A collection of points is the graph of a function provided that every vertical line intersects the graph in at most one point (vertical-line test).

2.1 EXERCISES

In Problems 1–12, determine whether each relation represents a function.

1.

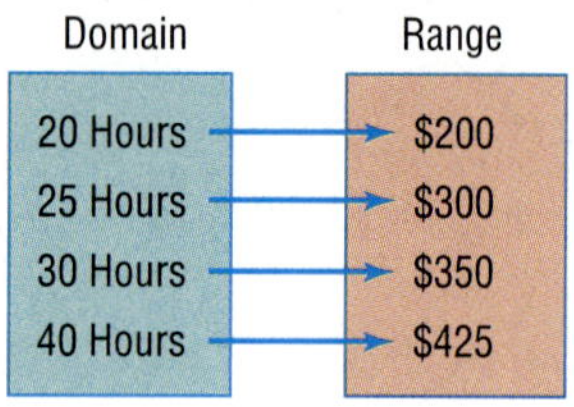

2.

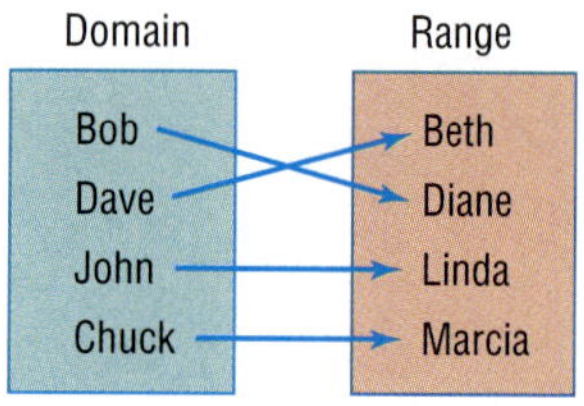

3.

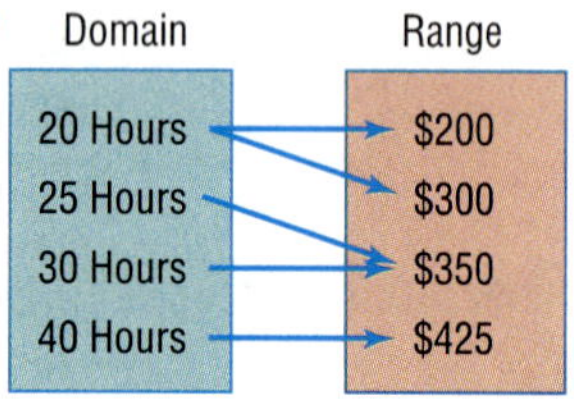

4.

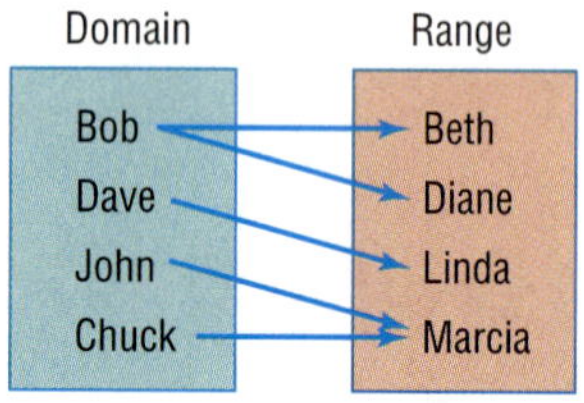

5. $\{(2, 6), (-3, 6), (4, 9), (1, 10)\}$

6. $\{(-2, 5), (-1, 3), (3, 7), (4, 12)\}$

7. $\{(1, 3), (2, 3), (3, 3), (4, 3)\}$

8. $\{(0, -2), (1, 3), (2, 3), (3, 7)\}$

9. $\{(-2, 4), (-2, 6), (0, 3), (3, 7)\}$

10. $\{(-4, 4), (-3, 3), (-2, 2), (-1, 1), (-4, 0)\}$

11. $\{(-2, 4), (-1, 1), (0, 0), (1, 1)\}$

12. $\{(-2, 16), (-1, 4), (0, 3), (1, 4)\}$

In Problems 13–20, find the following values for each function:

(a) $f(0)$; *(b)* $f(1)$; *(c)* $f(-1)$; *(d)* $f(-x)$; *(e)* $-f(x)$; *(f)* $f(x + 1)$

13. $f(x) = -3x^2 + 2x - 4$

14. $f(x) = 2x^2 + x - 1$

15. $f(x) = \dfrac{x}{x^2 + 1}$

16. $f(x) = \dfrac{x^2 - 1}{x + 4}$

17. $f(x) = |x| + 4$

18. $f(x) = \sqrt{x^2 + x}$

19. $f(x) = \dfrac{2x + 1}{3x - 5}$

20. $f(x) = 1 - \dfrac{1}{(x + 2)^2}$

In Problems 21–32, use the graph of the function f given in the figure.

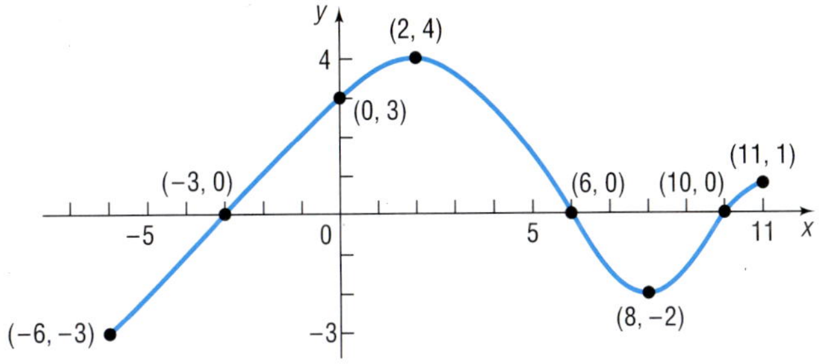

21. Find $f(0)$ and $f(-6)$.

22. Find $f(6)$ and $f(11)$.

23. Is $f(2)$ positive or negative?

24. Is $f(8)$ positive or negative?

25. For what numbers x is $f(x) = 0$?

26. For what numbers x is $f(x) > 0$?

27. What is the domain of f?

28. What is the range of f?

29. What are the x-intercepts?

30. What are the y-intercepts?

31. How often does the line $y = \frac{1}{2}$ intersect the graph?

32. How often does the line $y = 3$ intersect the graph?

In Problems 33–36, answer the questions about the given function.

33. $f(x) = \dfrac{x+2}{x-6}$

(a) Is the point $(3, 14)$ on the graph of f?
(b) If $x = 4$, what is $f(x)$? What point is on the graph of f?
(c) If $f(x) = 2$, what is x? What point is on the graph of f?
(d) What is the domain of f?

34. $f(x) = \dfrac{x^2+2}{x+4}$

(a) Is the point $(1, \frac{3}{5})$ on the graph of f?
(b) If $x = 0$, what is $f(x)$? What point is on the graph of f?
(c) If $f(x) = \frac{1}{2}$, what is x? What points are on the graph of f?
(d) What is the domain of f?

35. $f(x) = \dfrac{2x^2}{x^4+1}$

(a) Is the point $(-1, 1)$ on the graph of f?
(b) If $x = 2$, what is $f(x)$? What point is on the graph of f?
(c) If $f(x) = 1$, what is x? What points are on the graph of f?
(d) What is the domain of f?

36. $f(x) = \dfrac{2x}{x-2}$

(a) Is the point $(\frac{1}{2}, -\frac{2}{3})$ on the graph of f?
(b) If $x = 4$, what is $f(x)$? What point is on the graph of f?
(c) If $f(x) = 1$, what is x? What point is on the graph of f?
(d) What is the domain of f?

In Problems 37–48, determine whether the graph is that of a function by using the vertical-line test. If it is, use the graph to find:

(a) Its domain and range *(b) The intercepts, if any* *(c) Any symmetry with respect to the x-axis, y-axis, or origin*

37.

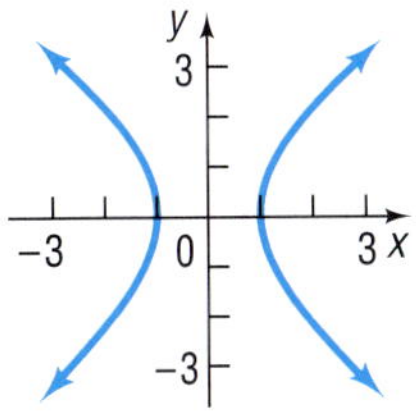

38.

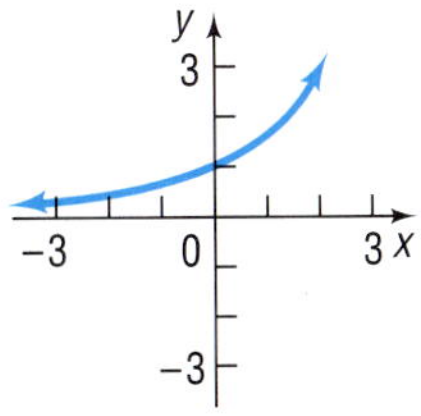

39.

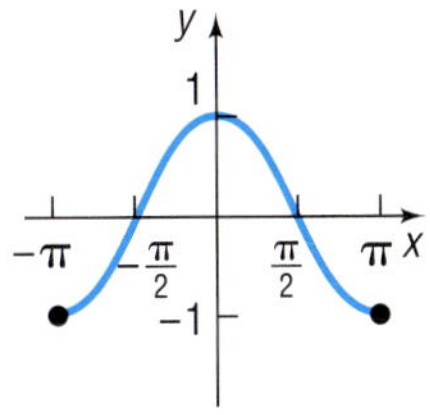

40.

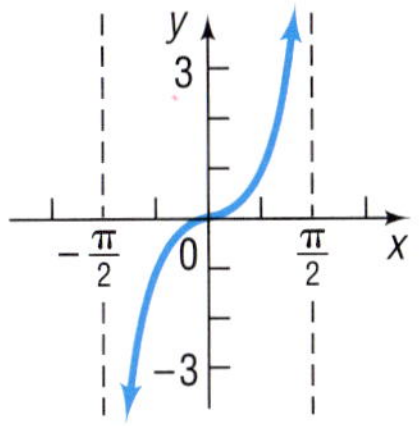

41.

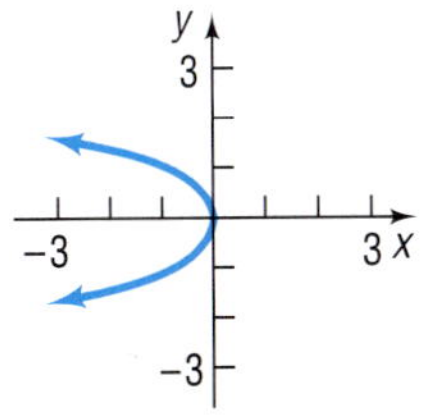

42.

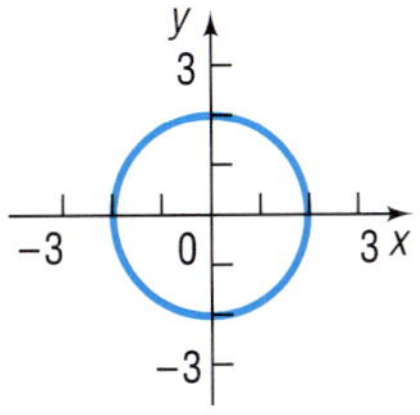

43.

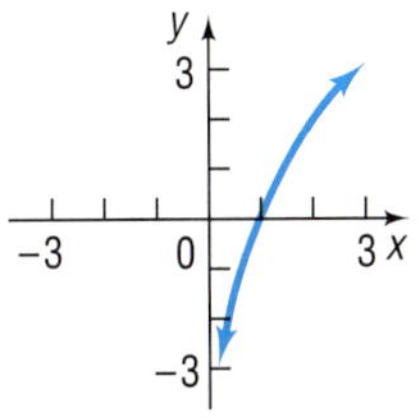

44.

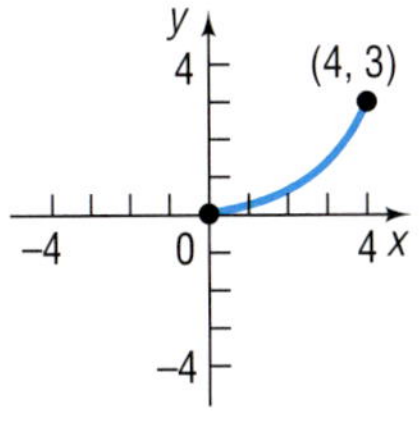

45.

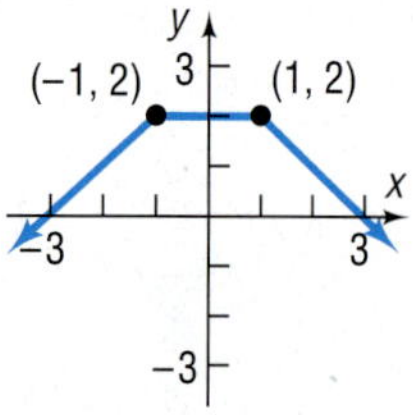

46.

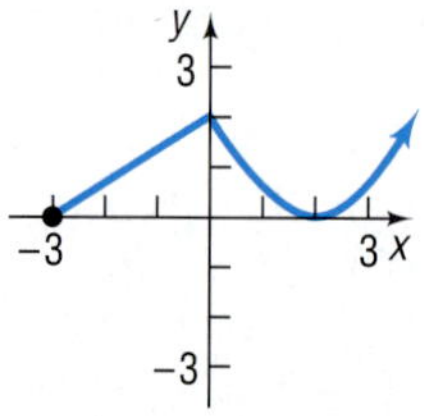

47.

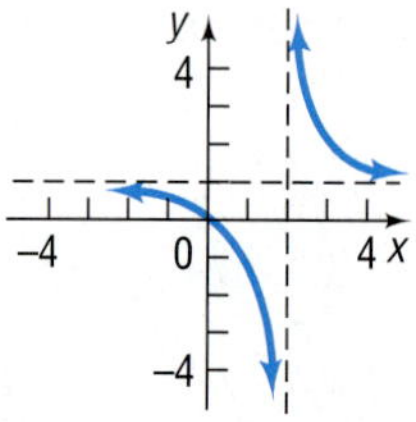

48.

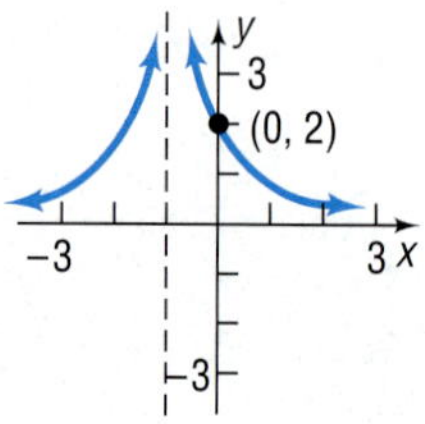

In Problems 49–62, find the domain of each function.

49. $f(x) = 3x + 4$

50. $f(x) = 5x^2 + 2$

51. $f(x) = \dfrac{x}{x^2 + 1}$

52. $f(x) = \dfrac{x^2}{x^2 + 1}$

53. $g(x) = \dfrac{x}{x^2 - 1}$

54. $h(x) = \dfrac{x}{x - 1}$

55. $F(x) = \dfrac{x - 2}{x^3 + x}$

56. $G(x) = \dfrac{x + 4}{x^3 - 4x}$

57. $h(x) = \sqrt{3x - 12}$

58. $G(x) = \sqrt{1 - x}$

59. $f(x) = \dfrac{4}{\sqrt{x - 9}}$

60. $f(x) = \dfrac{x}{\sqrt{x - 4}}$

61. $p(x) = \sqrt{\dfrac{2}{x - 1}}$

62. $q(x) = \sqrt{-x - 2}$

63. Match each of the following functions with the graphs that best describe the situation.
(a) The cost of building a house as a function of its square footage.
(b) The height of an egg dropped from a 300-foot building as a function of time.
(c) The height of a human as a function of time.
(d) The demand for Big Macs as a function of price.
(e) The height of a child on a swing as a function of time.

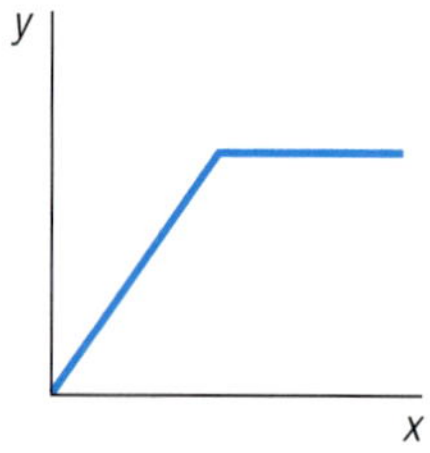

(I)

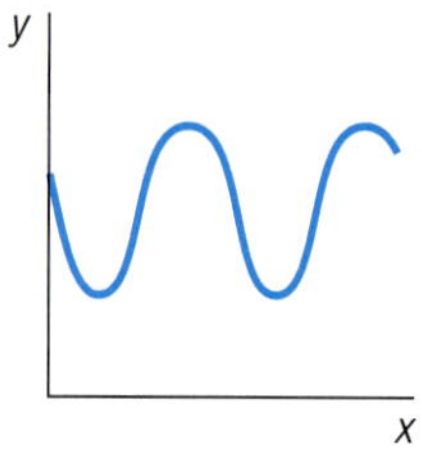

(II)

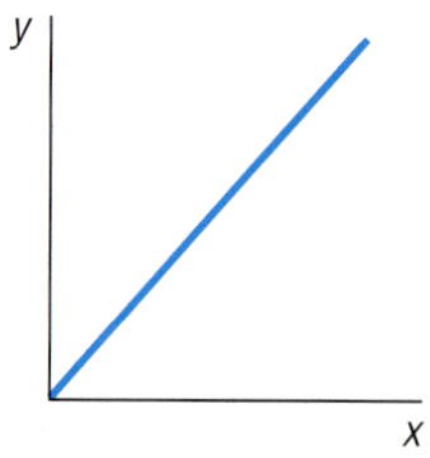

(III)

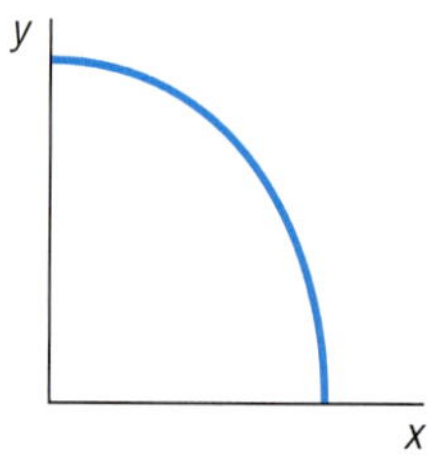

(IV)

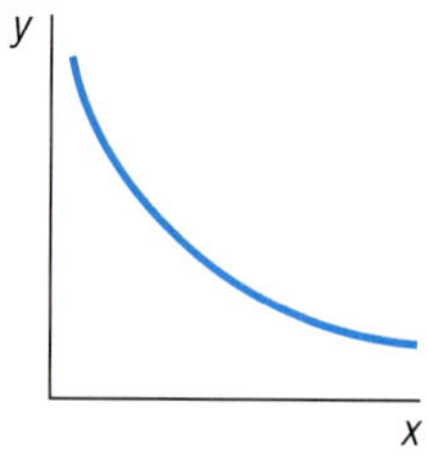

(V)

64. Match each of the following functions with the graph that best describes the situation.
(a) The temperature of a bowl of soup as a function of time.
(b) The number of hours of daylight per day over a two year period.
(c) The population of Florida as a function of time.
(d) The distance of a car traveling at a constant velocity as a function of time.
(e) The height of a golf ball hit with a 7-iron as a function of time.

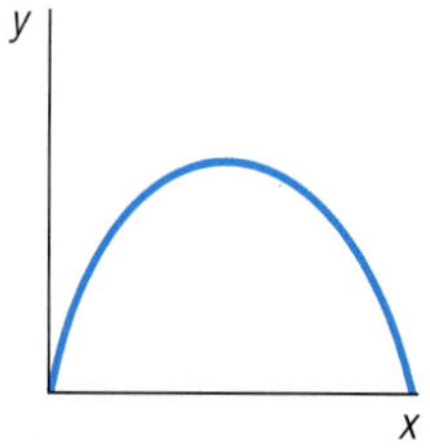

(I)

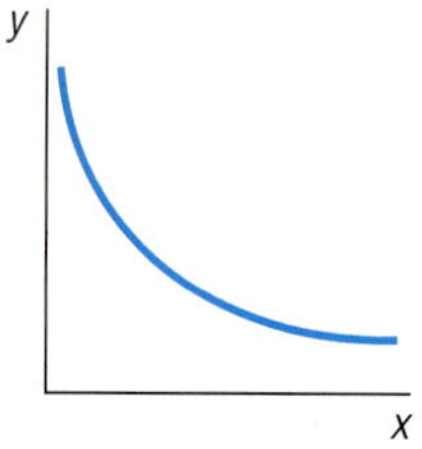

(II)

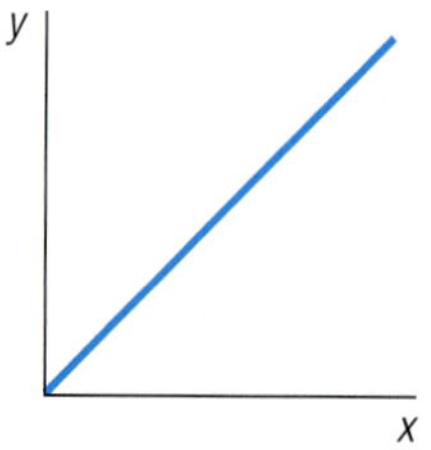

(III)

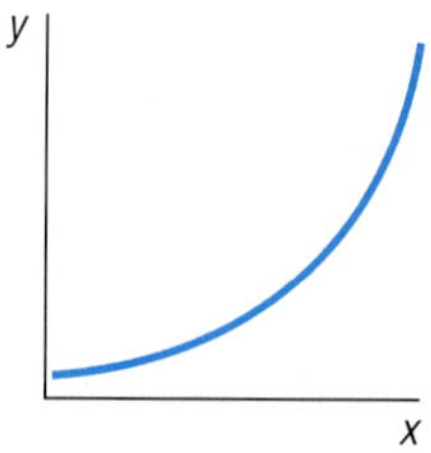

(IV)

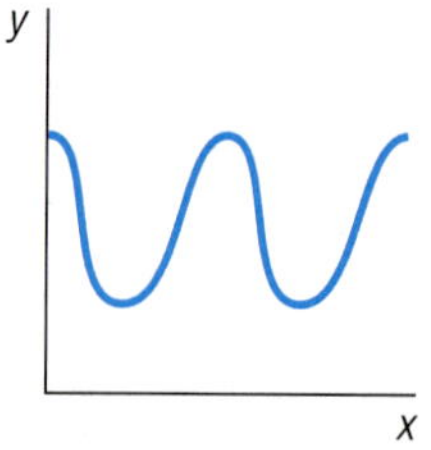

(V)

65. Consider the following scenario: Barbara decides to take a walk. She leaves home, walks 2 blocks in 5 minutes at a constant speed, and realizes that she forgot to lock the door. So Barbara runs home in 1 minute. While at her doorstep, it takes her 1 minute to find her keys and lock the door. Barbara walks 5 blocks in 15 minutes and then decides to jog home. It takes her 7 minutes to get home. Draw a graph of Barbara's distance from home (in blocks) as a function of time.

66. Consider the following scenario: Jayne enjoys riding her bicycle through the woods. At the forest preserve, she gets on her bicycle and rides up a 2,000-foot incline in 10 minutes. She then travels down the incline in 3 minutes. The next 5,000 feet is level terrain and she covers the distance in 20 minutes. She rests for 15 minutes. Jayne then travels 10,000 feet in 30 minutes. Draw a graph of Jayne's distance traveled (in feet) as a function of time.

67. If $f(x) = 2x^3 + Ax^2 + 4x - 5$ and $f(2) = 5$, what is the value of A?

68. If $f(x) = 3x^2 - Bx + 4$ and $f(-1) = 12$, what is the value of B?

69. If $f(x) = (3x + 8)/(2x - A)$ and $f(0) = 2$, what is the value of A?

70. If $f(x) = (2x - B)/(3x + 4)$ and $f(2) = \frac{1}{2}$, what is the value of B?

71. If $f(x) = (2x - A)/(x - 3)$ and $f(4) = 0$, what is the value of A? Where is f not defined?

72. If $f(x) = (x - B)/(x - A)$, $f(2) = 0$, and $f(1)$ is undefined, what are the values of A and B?

73. Effect of Gravity on Earth If a rock falls from a height of 20 meters on Earth, the height H (in meters) after x seconds is approximately

$$H(x) = 20 - 4.9x^2$$

(a) Using a graphing utility, graph $H(x)$.
(b) What is the height of the rock when $x = 1$ second? $x = 1.1$ seconds? $x = 1.2$ seconds? $x = 1.3$ seconds?
(c) When is the height of the rock 15 meters? When is it 10 meters? When is it 5 meters?
(d) When does the rock strike the ground?

74. Effect of Gravity on Jupiter If a rock falls from a height of 20 meters on the planet Jupiter, its height H (in meters) after x seconds is approximately

$$H(x) = 20 - 13x^2$$

(a) Using a graphing utility, graph $H(x)$.
(b) What is the height of the rock when $x = 1$ second? $x = 1.1$ seconds? $x = 1.2$ seconds?
(c) When is the height of the rock 15 meters? When is it 10 meters? When is it 5 meters?
(d) When does the rock strike the ground?

75. Geometry Express the area A of a rectangle as a function of the length x if the length is twice the width of the rectangle.

76. Geometry Express the area A of an isosceles right triangle as a function of the length x of one of the two equal sides.

77. Express the gross salary G of a person who earns \$10 per hour as a function of the number x of hours worked.

78. Tiffany, a commissioned salesperson, earns \$100 base pay plus \$10 per item sold. Express her gross salary G as a function of the number x of items sold.

79. Refer to Example 11. Is there a place to land the boat so that the travel time is least? Do you think this place is closer to town or closer to P? Discuss the possibilities. Give reasons.

80. Refer to Example 11. Graph the function $T = T(x)$. Use TRACE to see how the time T varies as x changes from 0 to 12. What value of x results in the least time?

81. Installing Cable TV MetroMedia Cable is asked to provide service to a customer whose house is located 2 miles from the road along which the cable is buried. The nearest connection box for the cable is located 5 miles down the road (see the figure).

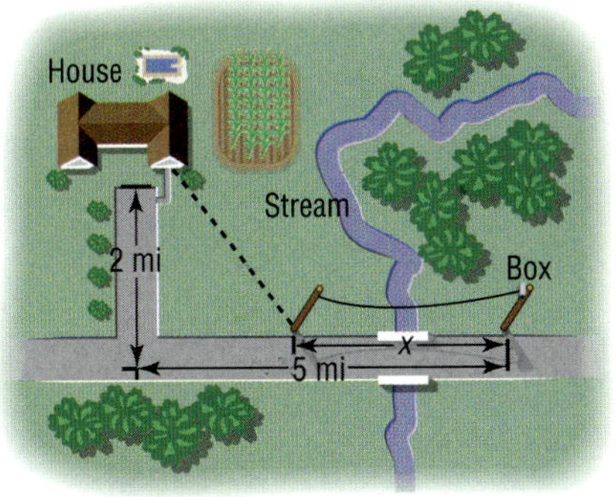

(a) If the installation cost is \$10 per mile along the road and \$14 per mile off the road, express the total cost C of installation as a function of the distance x (in miles) from the connection box to the point where the cable installation turns off the road. Give the domain.

(b) Compute the cost if $x = 1$ mile.
(c) Compute the cost if $x = 3$ miles.
(d) Graph the function $C = C(x)$. Use TRACE to see how the cost C varies as x changes from 0 to 5.
(e) What value of x results in the least cost?

82. Time Required to Go from an Island to a Town An island is 3 miles from the nearest point P on a straight shoreline. A town is located 20 miles down the shore from P. (Refer to Figure 8 for a similar situation.)
(a) If a person has a boat that averages 12 miles per hour and the same person can run 5 miles per hour, express the time T that it takes to go from the island to town as a function of x, where x is the distance from P to where the person lands the boat. Give the domain.
(b) How long will it take to travel from the island to town if you land the boat 8 miles from P?
(c) How long will it take if you land the boat 12 miles from P?
(d) Graph the function $T = T(x)$. Use TRACE to see how the time T varies as x changes from 0 to 20.
(e) What value of x results in the least time?
(f) The least time occurs by heading directly to town from the island. Explain why this solution makes sense.

83. Page Design A page with dimensions of $8\frac{1}{2}$ inches by 11 inches has a border of uniform width x surrounding the printed matter of the page, as shown in the figure.

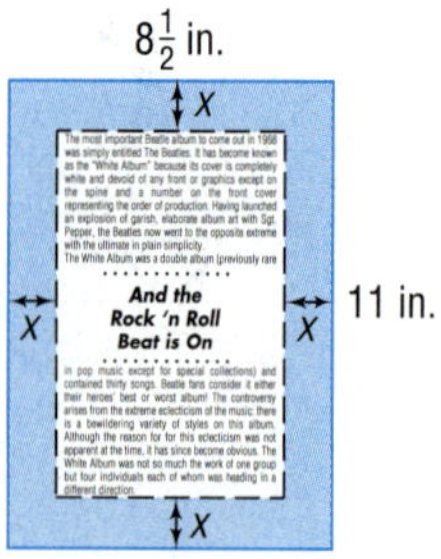

(a) Write a formula for the area A of the printed part of the page as a function of the width x of the border.
(b) Give the domain and range of A.
(c) Find the area of the printed page for borders widths of 1 inch, 1.2 inches, and 1.5 inches.
(d) Graph the function $A = A(x)$.
(e) Use TRACE to determine what margin should be used to obtain an area of 70 square inches and of 50 square inches.

84. Cost of Trans-Atlantic Travel A Boeing 747 crosses the Atlantic Ocean (3000 miles) with an airspeed of 500 miles per hour. The cost C (in dollars) per passenger is given by

$$C(x) = 100 + \frac{x}{10} + \frac{36{,}000}{x}$$

where x is the ground speed (airspeed $\pm$ wind).
(a) What is the cost per passenger for quiescent (no wind) conditions?
(b) What is the cost per passenger with a head wind of 50 miles per hour?
(c) What is the cost per passenger with a tail wind of 100 miles per hour?
(d) What is the cost per passenger with a head wind of 100 miles per hour?
(e) Graph the function $C = C(x)$.
(f) As x varies from 400 to 600 miles per hour, how does the cost vary?

85. Period of a Pendulum The period T (in seconds) of a simple pendulum is a function of its length l (in feet) defined by the equation

$$T(l) = 2\pi\sqrt{\frac{l}{g}}$$

where $g \approx 32.2$ feet per second is the acceleration due to gravity.
(a) Use a graphing utility to graph the function $T = T(l)$.
(b) Use the TRACE function to see how the period T varies as l changes from 1 to 10.
(c) What length should be used if a period of 10 seconds is required?

86. Effect of Elevation on Weight If an object weighs m pounds at sea level, then its weight W (in pounds) at a height of h miles above sea level is given approximately by

$$W(h) = m\left(\frac{4000}{4000 + h}\right)^2$$

(a) If Amy weighs 120 pounds at sea level, how much will she weigh on Pike's Peak, which is 14,110 feet above sea level?
(b) Use a graphing utility to graph the function $W = W(h)$. Use $m = 120$ pounds.
(c) Use the TRACE function to see how weight W varies as h changes from 0 to 5 miles.
(d) At what height will Amy weigh 119.95 pounds?
(e) Does your answer to part (d) seem reasonable?

In Problems 87–94, tell whether the set of ordered pairs (x, y) defined by each equation is a function.

87. $y = x^2 + 2x$

88. $y = x^3 - 3x$

89. $y = \dfrac{2}{x}$

90. $y = \dfrac{3}{x} - 3$

91. $y^2 = 1 - x^2$

92. $y = \pm\sqrt{1 - 2x}$

93. $x^2 + y = 1$

94. $x + 2y^2 = 1$

95. Some functions f have the property that $f(a + b) = f(a) + f(b)$ for all real numbers a and b. Which of the following functions have this property?
(a) $h(x) = 2x$
(b) $g(x) = x^2$
(c) $F(x) = 5x - 2$
(d) $G(x) = 1/x$

96. Draw the graph of a function whose domain is $\{x | -3 \le x \le 8, x \ne 5\}$ and whose range is $\{y | -1 \le y \le 2, y \ne 0\}$. What point(s) in the rectangle $-3 \le x \le 8$, $-1 \le y \le 2$ cannot be on the graph? Compare your graph with those of other students. What differences do you see?

97. Are the functions $f(x) = x - 1$ and $g(x) = (x^2 - 1)/(x + 1)$ the same? Explain.

98. Describe how you would proceed to find the domain and range of a function if you were given its graph. How would your strategy change if, instead, you were given the equation defining the function?

99. How many x-intercepts can the graph of a function have? How many y-intercepts can it have?

100. Is a graph that consists of a single point the graph of a function? Can you write the equation of such a function?

101. Is there a function whose graph is symmetric with respect to the x-axis?

102. Investigate when, historically, the use of function notation $y = f(x)$ first appeared.

2.2 LINEAR FUNCTIONS AND MODELS

1 Graph Linear Functions

2 Distinguish between Linear and Nonlinear Relations

3 Use a Graphing Utility to Find the Line of Best Fit

4 Construct a Model Using Direct Variation

1 In Section 1.6, we introduced lines. In particular, we discussed the slope–intercept form of the equation of a line $y = mx + b$. When we write the slope–intercept form of a line using function notation, we have a *linear function.*

A **linear function** is a function of the form

$$f(x) = mx + b$$

The graph of a linear function is a line with slope m and y-intercept b.

EXAMPLE 1 Graphing a Linear Function

Graph the linear function: $f(x) = -3x + 7$

Solution This is a linear function with slope $m = -3$ and y-intercept 7. To graph this function, we start at $(0, 7)$ and use the slope to find an additional point by moving right 1 unit and down 3 units. See Figure 9.

Figure 9

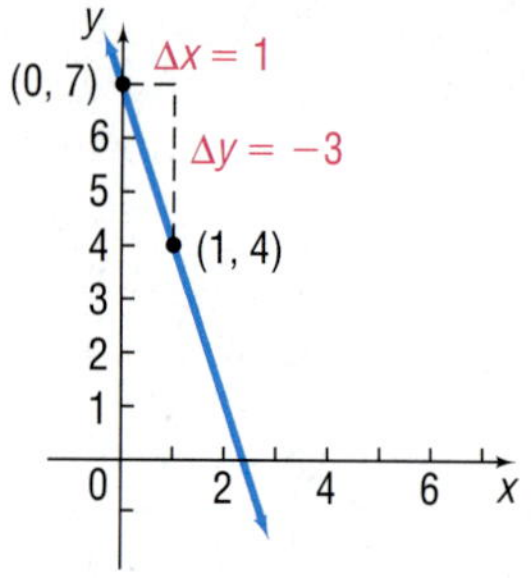

Alternatively, we could have found an additional point by evaluating the function at some $x \neq 0$. For $x = 1$, we find $f(1) = -3(1) + 7 = 4$ and obtain the point $(1, 4)$ on the graph.

NOW WORK PROBLEM 1.

Many real-world situations may be described using linear functions. For example, sometimes we can "fit" a linear function to data.

Fitting a Linear Function to Data

Curve fitting is an area of statistics in which a relation between two or more variables is explained through an equation. For example, the equation $S = \$100{,}000 + 12A$, where S represents sales (in dollars) and A is the advertising expenditure (in dollars), implies that if advertising expenditures were \$0, sales would be $\$100{,}000 + 12(0) = \$100{,}000$; and if advertising expenditures were \$10,000, sales would be $\$100{,}000 + 12(\$10{,}000) = \$220{,}000$. In this model, the variable A is called the predictor (independent) variable, and S is called the response (dependent) variable because, if the level of advertising is known, it can be used to predict sales. Curve fitting is used to find an equation that relates two or more variables, using observed or experimental data.

Steps for Curve Fitting

STEP 1: Obtain data that relate two variables. Then plot ordered pairs of the variables as points to obtain a **scatter diagram.**

STEP 2: Find an equation that describes this relation.

2 Scatter diagrams are used to help us to see the type of relation that exists between two variables. In this text, we will discuss a variety of different relations that may exist between two variables. For now, we concentrate on distinguishing between linear and nonlinear relations. See Figure 10.

Figure 10
Linear Relations

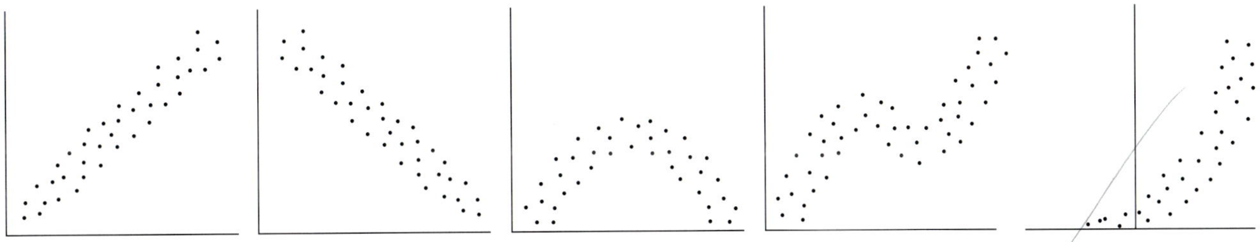

(a) Linear $y = mx + b, m > 0$ **(b)** Linear $y = mx + b, m < 0$ **(c)** Nonlinear **(d)** Nonlinear **(e)** Nonlinear

EXAMPLE 2 Distinguishing between Linear and Nonlinear Relations

Determine whether the relation between the two variables in Figure 11 is linear or nonlinear.

Figure 11

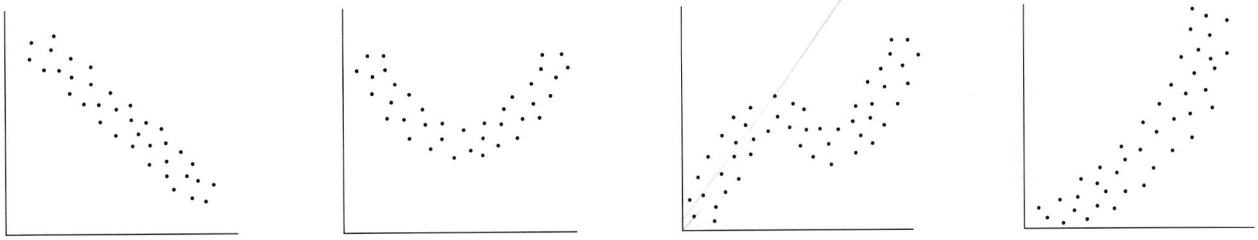

(a) **(b)** **(c)** **(d)**

Solution (a) Linear (b) Nonlinear (c) Nonlinear (d) Nonlinear

We will study linearly related data in this section. Fitting an equation to nonlinear data will be discussed in later chapters.

NOW WORK PROBLEM 9.

Suppose that the scatter diagram of a set of data appears to be linearly related. One way to obtain an equation for such data is to draw a line through two points on the scatter diagram and estimate the equation of the line.

EXAMPLE 3 Finding an Equation for Linearly Related Data

A farmer collected the following data, which show crop yields for various amounts of fertilizer used.

Plot	1	2	3	4	5	6	7	8	9	10	11	12
Fertilizer, X (pounds/100 ft^2)	0	0	5	5	10	10	15	15	20	20	25	25
Yield, Y (bushels)	4	6	10	7	12	10	15	17	18	21	23	22

(a) Draw a scatter diagram of the data.
(b) Select two points from the data and find an equation of the line containing the points.
(c) Graph the line on the scatter diagram.

Solution (a) The data collected indicate that a relation exists between the amount of fertilizer used and crop yield. To draw a scatter diagram, we plot points, using fertilizer as the x-coordinate and yield as the y-coordinate. See Figure 12. From the scatter diagram, it appears that a linear relation with positive slope exists between the amount of fertilizer used and yield.

Figure 12

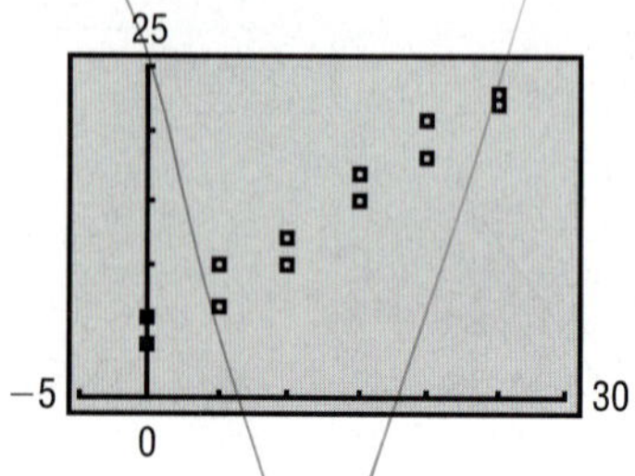

(b) Select two points, say (0, 4) and (25, 23). (You should select your own two points and complete the solution.) The slope of the line joining the points (0, 4) and (25, 23) is

$$m = \frac{23 - 4}{25 - 0} = \frac{19}{25} = 0.76$$

The equation of the line with slope 0.76 and passing through (0, 4) is found using the point–slope form with $m = 0.76$, $x_1 = 0$, and $y_1 = 4$.

$$y - y_1 = m(x - x_1)$$
$$y - 4 = 0.76(x - 0)$$
$$y = 0.76x + 4$$

Figure 13

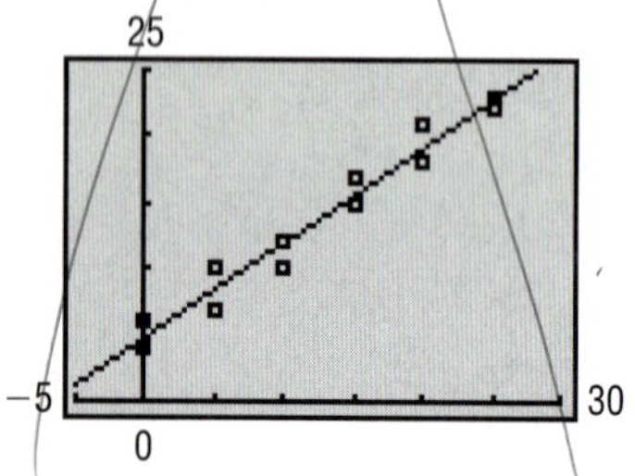

(c) Figure 13 shows the scatter diagram with the graph of the line found in part (b).

NOW WORK PROBLEM 15(a), (b), AND (c).

The line obtained in Example 3 depends on the selection of points, which will vary from person to person. So the line we found might be different from the line you found. Although the line we found in Example 3 appears to "fit" the data well, there may be a line that "fits it better." Do you think your line fits the data better? Is there a line of *best fit*? As it turns out, there is a method for finding the line that best fits linearly related data (called the **line of best fit**).*

Line of Best Fit

EXAMPLE 4 Finding the Line of Best Fit

3

With the data from Example 3, find the line of best fit using a graphing utility.

Solution Graphing utilities contain built-in programs that find the line of best fit for a collection of points in a scatter diagram. (Look in your owner's manual for details on how to execute the program.) Upon executing the LINear RE-

*We shall not discuss in this book the underlying mathematics of lines of best fit. Most books in statistics and many in linear algebra discuss this topic.

Gression program, we obtain the results shown in Figure 14. The output that the utility provides shows us the equation $y = ax + b$, where a is the slope of the line and b is the y-intercept. The line of best fit that relates fertilizer and yield may be expressed as the linear function

$$f(x) = 0.717x + 4.786$$

Figure 15 shows the graph of the line of best fit, along with the scatter diagram.

Figure 14

Figure 15

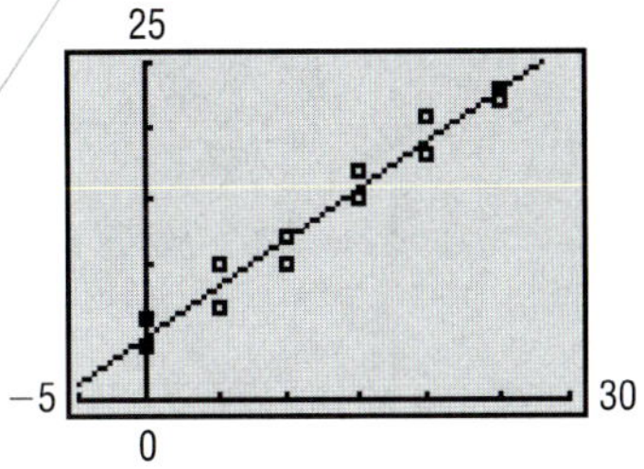

NOW WORK PROBLEM 15(d) AND (e).

Refer back to Example 3. Notice that the data do not represent a function since there are ordered pairs in which the same first element is paired with two different second elements. For example, 0 is matched to both 4 and 6. However, when we find a line that fits these data, we find a function that relates the data. Thus, **curve fitting** is a process whereby we find a functional relationship between two or more variables even though the data, themselves, may not represent a function.

Does the line of best fit appear to be a good fit? In other words, does the line appear to accurately describe the relation between yield and fertilizer?

And just how "good" is this line of best fit? The answers are given by what is called the *correlation coefficient.* Look again at Figure 14. The last line of ourput is $r = 0.98$. This number, called the **correlation coefficient, r,** $0 \leq |r| \leq 1$, is a measure of the strength of the *linear relation* that exists between two variables. The closer that $|r|$ is to 1, the more perfect the linear relationship is. If r is close to 0, there is little or no *linear* relationship between the variables. A negative value of r, $r < 0$, indicates that as x increases y decreases; a positive value of r, $r > 0$, indicates that as x increases y does also. Thus, the data given in Example 3, having a correlation coefficient of 0.98, are strongly indicative of a linear relationship with positive slope.

Direct Variation

When a mathematical model is developed for a real-world problem, it often involves relationships between quantities that are expressed in terms of proportionality.

Force is proportional to acceleration.

For an ideal gas held at a constant temperature, pressure and volume are inversely proportional.

The force of attraction between two heavenly bodies is inversely proportional to the square of the distance between them.

Revenue is directly proportional to sales.

Each of these statements illustrates the idea of **variation,** or how one quantity varies in relation to another quantity. Quantities may vary *directly, inversely,* or *jointly.* We discuss direct variation here.

4

Let x and y denote two quantities. Then y **varies directly** with x, or y is **directly proportional to** x, if there is a nonzero number k such that

$$y = kx$$

Figure 16

$y = kx, k > 0, x \geq 0$

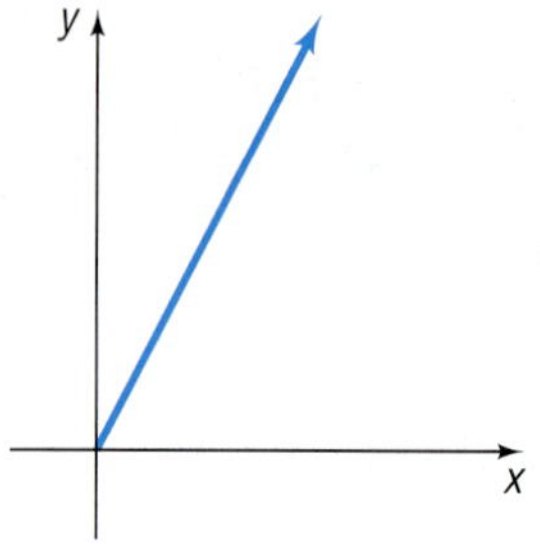

The number k is called the **constant of proportionality.**

Thus, if y varies directly with x, then y is a linear function of x. The graph in Figure 16 illustrates the relationship between y and x if y varies directly with x and $k > 0$, $x \geq 0$. Note that the constant of proportionality is, in fact, the slope of the line.

If we know that two quantities vary directly, then knowing the value of each quantity in one instance enables us to write a formula that is true in all cases.

EXAMPLE 5 Mortgage Payments

The monthly payment p on a mortgage varies directly with the amount borrowed B. If the monthly payment on a 30-year mortgage is \$6.65 for every \$1000 borrowed, find a formula that relates the monthly payment p to the amount borrowed B for a mortgage with the same terms. Then find the monthly payment p when the amount borrowed B is \$120,000.

Solution Because p varies directly with B, we know that

$$p = kB$$

for some constant k. Because $p = 6.65$ when $B = 1000$, it follows that

$$6.65 = k(1000)$$
$$k = 0.00665$$

Figure 17

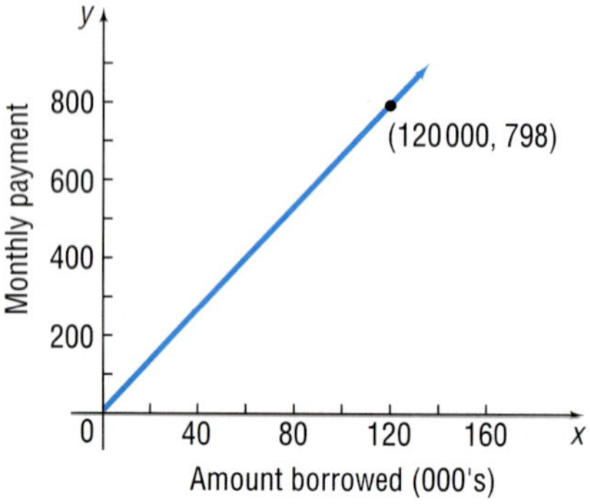

So we have

$$p = 0.00665B$$

We see that p is a linear function of B: $p(B) = 0.00665B$. In particular, when $B = \$120{,}000$, we find that

$$p(120{,}000) = 0.00665(\$120{,}000) = \$798$$

Figure 17 illustrates the relationship between the monthly payment p and the amount borrowed B.

NOW WORK PROBLEM 33.

2.2 EXERCISES

For Problems 1–8, graph each linear function by hand.

1. $f(x) = 2x + 3$
2. $g(x) = 5x - 4$
3. $h(x) = -3x + 4$
4. $p(x) = -x + 6$
5. $f(x) = \frac{1}{4}x - 3$
6. $h(x) = \frac{-2}{3}x + 4$
7. $F(x) = 4$
8. $G(x) = -2$

In Problems 9–14, examine the scatter diagram and determine whether the type of relation, if any, that may exist is linear or nonlinear.

9.

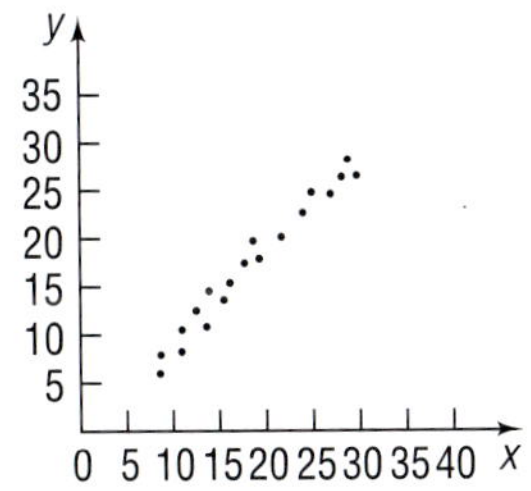

10.

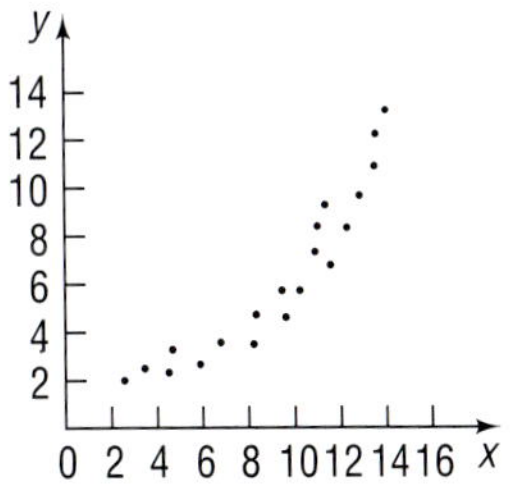

11.

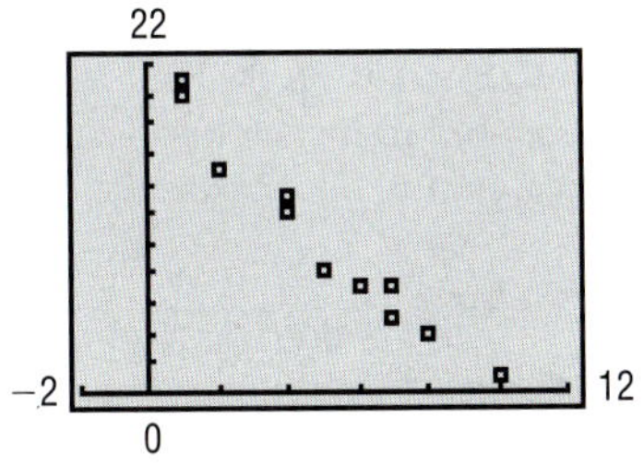

12.

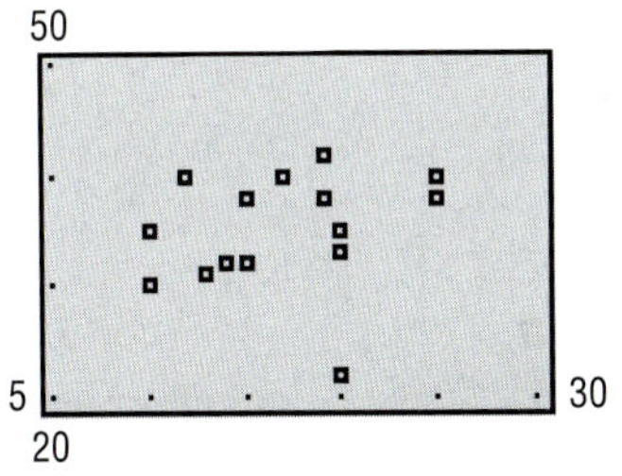

13.

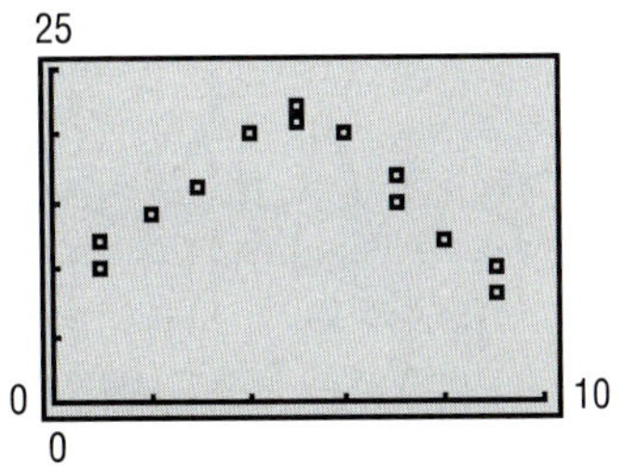

14.

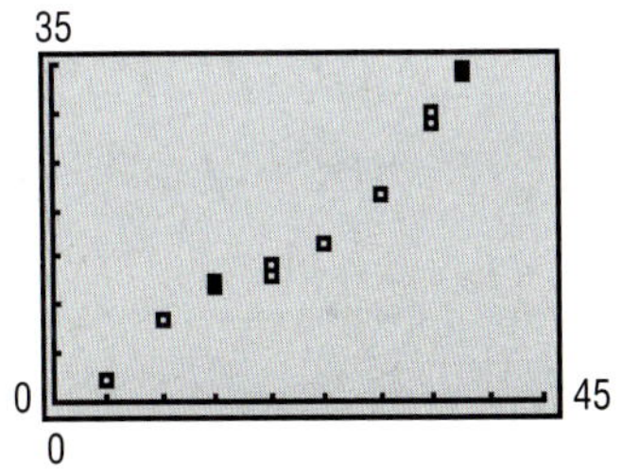

In Problems 15–22:

(a) Draw a scatter diagram.

*(b) Select two points from the scatter diagram and find the equation of the line through the points selected.**

(c) Graph the line found in part (b) on the scatter diagram.

(d) Use a graphing utility to find the line of best fit.

(e) Use a graphing utility to graph the line of best fit on the scatter diagram.

15.

x	3	4	5	6	7	8	9
y	4	6	7	10	12	14	16

16.

x	3	5	7	9	11	13
y	0	2	3	6	9	11

17.

x	−2	−1	0	1	2
y	−4	0	1	4	5

18.

x	−2	−1	0	1	2
y	7	6	3	2	0

19.

x	20	30	40	50	60
y	100	95	91	83	70

20.

x	5	10	15	20	25
y	2	4	7	11	18

21.

x	−20	−17	−15	−14	−10
y	100	120	118	130	140

22.

x	−30	−27	−25	−20	−14
y	10	12	13	13	18

*Answers will vary. We will use the first and last data points in the answer section.

23. Consumption and Disposable Income An economist wishes to estimate a linear function that relates personal consumption expenditures (C) and disposable income (I). Both C and I are in thousands of dollars. She interviews eight heads of households for families of size 4 and obtains the following data:

I (000)	C (000)
20	16
20	18
18	13
27	21
36	27
37	26
45	36
50	39

Let I represent the independent variable and C the dependent variable.

(a) Use a graphing utility to draw a scatter diagram.
(b) Use a graphing utility to find the line of best fit to the data. Express the solution using function notation.
(c) Interpret the slope. The slope of this line is called the **marginal propensity to consume.**
(d) Predict the consumption of a family whose disposable income is $42,000.

24. Marginal Propensity to Save The same economist as in Problem 23 wants to estimate a linear function that relates savings (S) and disposable income (I). Let $S = I - C$ be the dependent variable and I the independent variable. The slope of this line is called the **marginal propensity to save.**

(a) Use a graphing utility to draw a scatter diagram.
(b) Use a graphing utility to find the line of best fit to the data. Express the solution using function notation.
(c) Interpret the slope.
(d) Predict the savings of a family whose income is $42,000.

25. Average Speed of a Car An individual wanted to determine the relation that might exist between speed and miles per gallon of an automobile. Let X be the average speed of a car on the highway measured in miles per hour, and let Y represent the miles per gallon of the automobile. The following data are collected:

X	50	55	55	60	60	62	65	65
Y	28	26	25	22	20	20	17	15

(a) Use a graphing utility to draw a scatter diagram.
(b) Use a graphing utility to find the line of best fit to the data. Express the solution using function notation.
(c) Interpret the slope.
(d) Predict the miles per gallon of a car traveling 61 miles per hour.

26. Height versus Weight A doctor wished to determine whether a relation exists between the height of a female and her weight. She obtained the heights and weights of 10 females aged 18 to 24. Let height be the independent variable X, measured in inches, and weight be the dependent variable Y, measured in pounds.

X	60	61	62	62	64	65	65	67	68	68
Y	105	110	115	120	120	125	130	135	135	145

(a) Use a graphing utility to draw a scatter diagram.
(b) Use a graphing utility to find the line of best fit to the data. Express the solution using function notation.
(c) Interpret the slope.
(d) Predict the weight of a female aged 18 to 24 whose height is 66 inches.

27. Sales Data versus Income The following data represent sales and net income before taxes (both are in billions of dollars) for all manufacturing firms within the United States for 1980–1989. Treat sales as the independent variable and net income before taxes as the dependent variable.

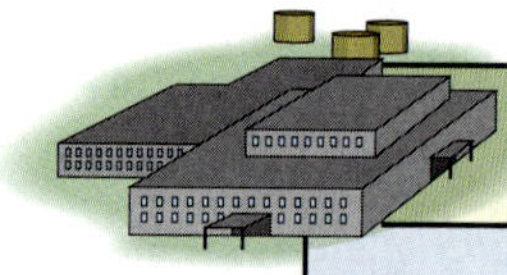

Year	Sales	Net Income Before Taxes
1980	1912.8	92.6
1981	2144.7	101.3
1982	2039.4	70.9
1983	2114.3	85.8
1984	2335.0	107.6
1985	2331.4	87.6
1986	2220.9	83.1
1987	2378.2	115.6
1988	2596.2	154.6
1989	2745.1	136.3

Source: Economic Report of the President, February 1995.

(a) Use a graphing utility to draw a scatter diagram.
(b) Use a graphing utility to find the line of best fit to the data. Express the solution using function notation.
(c) Interpret the slope.
(d) Predict the net income before taxes of manufacturing firms in 1990 if sales are \$2456.4 billion.

28. Employment and the Labor Force The following data represent the civilian labor force (people aged 16 years and older, excluding those serving in the military) and the number of employed people in the United States for the years 1981–1991. Treat the size of the labor force as the independent variable and the number employed as the dependent variable. Both the size of the labor force and the number employed are measured in thousands of people.

Year	Civilian Labor Force	Number Employed
1981	108,670	100,397
1982	110,204	99,526
1983	111,550	100,834
1984	113,544	105,005
1985	115,461	107,150
1986	117,834	109,597
1987	119,865	112,440
1988	121,669	114,968
1989	123,869	117,342
1990	124,787	117,914
1991	125,303	116,877

Source: Business Statistics, 1963–1991, U.S. Department of Commerce, Economics and Statistics Administration, Bureau of Economic Analysis, June 1992.

(a) Use a graphing utility to draw a scatter diagram.
(b) Use a graphing utility to find the line of best fit to the data. Express the solution using function notation.
(c) Interpret the slope.
(d) Predict the number of employed people if the civilian labor force is 122,340 thousand people.

29. Disposable Personal Income Use the data from Problem 69, Section 1.1.
(a) With a graphing utility, find the line of best fit letting per capita disposable income represent the independent variable. Express the solution using function notation.
(b) Interpret the slope.
(c) Predict per capita personal consumption if per capita disposable income is \$14,989.

30. Natural Gas versus Coal Use the data from Problem 70, Section 1.1.
(a) With a graphing utility, find the line of best fit letting the amount of natural gas represent the independent variable. Express the solution using function notation.
(b) Interpret the slope.
(c) Predict the amount of energy provided by coal if the amount of energy provided by natural gas is 67.

31. Demand for Jeans The marketing manager at Levi–Strauss wishes to find a function that relates the demand D of men's jeans and p, the price of the jeans. The following data were obtained based on a price history of the jeans.

Price (\$/Pair), p	Demand (Pairs of Jeans Sold per Day), D
20	60
22	57
23	56
23	53
27	52
29	49
30	44

(a) Does the relation defined by the set of ordered pairs (p, D) represent a function?
(b) Draw a scatter diagram of the data.
(c) Using a graphing utility, find the line of best fit relating price and quantity demanded.
(d) Interpret the slope.
(e) Express the relationship found in part (c) using function notation.
(f) What is the domain of the function?
(g) How many jeans will be demanded if the price is \$28 a pair?

32. Advertising and Sales Revenue A marketing firm wishes to find a function that relates the sales S of a product and A, the amount spent on advertising the product. The data are obtained from past experience. Advertising and sales are measured in thousands of dollars.

Advertising Expenditures, A	Sales, S
20	335
22	339
22.5	338
24	343
24	341
27	350
28.3	351

(a) Does the relation defined by the set of ordered pairs (A, S) represent a function?
(b) Draw a scatter diagram of the data.
(c) Using a graphing utility, find the line of best fit relating advertising expenditures and sales.
(d) Interpret the slope.
(e) Express the relationship found in part (c) using function notation.
(f) What is the domain of the function?
(g) Predict sales if advertising expenditures are \$25,000.

33. **Mortgage Payments** The monthly payment p on a mortgage varies directly with the amount borrowed B. If the monthly payment on a 30-year mortgage is \$6.49 for every \$1000 borrowed, find a linear function that relates the monthly payment p to the amount borrowed B for a mortgage with the same terms. Then find the monthly payment p when the amount borrowed B is \$145,000.

34. **Mortgage Payments** The monthly payment p on a mortgage varies directly with the amount borrowed B. If the monthly payment on a 15-year mortgage is \$8.99 for every \$1000 borrowed, find a linear function that relates the monthly payment p to the amount borrowed B for a mortgage with the same terms. Then find the monthly payment p when the amount borrowed B is \$175,000.

35. **Revenue Function** At the corner Shell station, the revenue R varies directly with the number of gallons of gasoline sold g. If the revenue is \$11.52 when the number of gallons sold is 12, find a linear function that relates revenue R to the number of gallons of gasoline g. Then find the revenue R when the number of gallons of gasoline sold is 10.5.

36. **Cost Function** The cost C of chocolate-covered almonds varies directly with the number of pounds of almonds purchased, A. If the cost is \$23.75 when the number of pounds of chocolate-covered almonds purchased is 5, find a linear function that relates the cost C to the number of pounds of almonds purchased A. Then find the cost C when the number of pounds of almonds purchased is 3.5.

37. **Physics: Falling Objects** The distance s that an object falls is directly proportional to the square of the time t of the fall. If an object falls 16 feet in 1 second, how far will it fall in 3 seconds? How long will it take an object to fall 64 feet?

38. **Physics: Falling Objects** The velocity v of a falling object is directly proportional to the time t of the fall. If, after 2 seconds, the velocity of the object is 64 feet per second, what will its velocity be after 3 seconds?

2.3 QUADRATIC EQUATIONS AND FUNCTIONS

1 Solve Quadratic Equations by Extracting the Root

2 Solve Quadratic Equations by Completing the Square

3 Solve Quadratic Equations Using the Quadratic Formula

4 Determine Whether the Graph of a Quadratic Function Opens Up or Down and Locate Its Vertex

5 Graph Quadratic Functions

Recall that a quadratic equation is an equation equivalent to one of the form

$$ax^2 + bx + c = 0 \qquad (1)$$

where a, b, and c are real numbers and $a \neq 0$. A quadratic equation written in the form of equation (1) is in standard form.

In Section 1.3, we discussed solving quadratic equations by factoring. In this section, we discuss three additional algebraic methods for solving quadratic equations: extracting the root, completing the square, and the quadratic formula. We then use this information to help us graph quadratic functions by hand.

Extracting the Root

1 Suppose that we wish to solve the quadratic equation

$$x^2 = p \tag{2}$$

where $p \geq 0$ is a nonnegative number. We proceed as follows:

$$\begin{aligned} x^2 - p &= 0 && \text{Put in standard form.} \\ (x - \sqrt{p})(x + \sqrt{p}) &= 0 && \text{Factor (over the real numbers).} \\ x = \sqrt{p} \quad \text{or} \quad x &= -\sqrt{p} && \text{Solve.} \end{aligned}$$

We have the following result:

$$\text{If } x^2 = p \text{ and } p \geq 0, \text{ then } x = \sqrt{p} \text{ or } x = -\sqrt{p}. \tag{3}$$

When statement (3) is used, we refer to it as **extracting the root.** Note that if $p > 0$ the equation $x^2 = p$ has two solutions, $x = \sqrt{p}$ and $x = -\sqrt{p}$. We usually abbreviate these solutions as $x = \pm\sqrt{p}$, read as "x equals plus or minus the square root of p." For example, the two solutions of the equation

$$x^2 = 4$$

are

$$x = \pm\sqrt{4}$$

and, since $\sqrt{4} = 2$, we have

$$x = \pm 2$$

The solution set is $\{-2, 2\}$.

EXAMPLE 1 Solving Quadratic Equations by Extracting the Root

Solve each equation.

(a) $x^2 = 5$ (b) $(x - 2)^2 = 16$

Solution (a) We use the result in statement (3) to get

$$\begin{aligned} x^2 &= 5 \\ x &= \pm\sqrt{5} \\ x = \sqrt{5} \quad &\text{or} \quad x = -\sqrt{5} \end{aligned}$$

The solution set is $\{-\sqrt{5}, \sqrt{5}\}$.

(b) We use the result in statement (3) to get

$$(x-2)^2 = 16$$
$$x - 2 = \pm\sqrt{16}$$
$$x - 2 = \sqrt{16} \quad \text{or} \quad x - 2 = -\sqrt{16}$$
$$x - 2 = 4 \qquad\qquad x - 2 = -4$$
$$x = 6 \qquad\qquad x = -2$$

The solution set is $\{-2, 6\}$.

NOW WORK PROBLEM 3.

Completing the Square

2 The next example illustrates how the procedure of completing the square can be used to solve a quadratic equation.

EXAMPLE 2 Solving a Quadratic Equation by Completing the Square

Solve by completing the square: $x^2 + 5x + 4 = 0$

Solution We always begin this procedure by rearranging the equation so that the constant is on the right side.

$$x^2 + 5x + 4 = 0$$
$$x^2 + 5x = -4$$

Since the coefficient of x^2 is 1, we can complete the square on the left side by adding $(\frac{1}{2} \cdot 5)^2 = \frac{25}{4}$. In an equation, whatever we add to the left side must also be added to the right side. Thus, we add $\frac{25}{4}$ to *both* sides.

$$x^2 + 5x + \frac{25}{4} = -4 + \frac{25}{4} \qquad \text{Add } \tfrac{25}{4} \text{ to both sides.}$$
$$\left(x + \frac{5}{2}\right)^2 = \frac{9}{4} \qquad \text{Factor; simplify.}$$
$$x + \frac{5}{2} = \pm\sqrt{\frac{9}{4}} \qquad \text{Apply Statement (3).}$$
$$x + \frac{5}{2} = \pm\frac{3}{2}$$
$$x = -\frac{5}{2} \pm \frac{3}{2}$$
$$x = -\frac{5}{2} + \frac{3}{2} = -1 \quad \text{or} \quad x = -\frac{5}{2} - \frac{3}{2} = -4$$

The solution set is $\{-4, -1\}$.

NOW WORK PROBLEM 7.

The Quadratic Formula

3 We can use the method of completing the square to obtain a general formula for solving the quadratic equation

$$ax^2 + bx + c = 0 \qquad a \neq 0$$

We begin by rearranging the terms as

$$ax^2 + bx = -c$$

Since $a \neq 0$, we can divide both sides by a to get

$$x^2 + \frac{b}{a}x = -\frac{c}{a}$$

Now the coefficient of x^2 is 1. To complete the square on the left side, add the square of $\frac{1}{2}$ the coefficient of x; that is, add

$$\left(\frac{1}{2} \cdot \frac{b}{a}\right)^2 = \frac{b^2}{4a^2}$$

to each side. Then

$$x^2 + \frac{b}{a}x + \frac{b^2}{4a^2} = \frac{b^2}{4a^2} - \frac{c}{a}$$

$$\left(x + \frac{b}{2a}\right)^2 = \frac{b^2 - 4ac}{4a^2} \qquad \frac{b^2}{4a^2} - \frac{c}{a} = \frac{b^2}{4a^2} - \frac{4ac}{4a^2} = \frac{b^2 - 4ac}{4a^2} \tag{4}$$

Provided $b^2 - 4ac \geq 0$, we now can apply the result in statement (3) to get

$$x + \frac{b}{2a} = \pm\sqrt{\frac{b^2 - 4ac}{4a^2}}$$

$$x = -\frac{b}{2a} \pm \frac{\sqrt{b^2 - 4ac}}{2a} = \frac{-b \pm \sqrt{b^2 - 4ac}}{2a}$$

What if $b^2 - 4ac$ is negative? Then equation (4) states that the left expression (a real number squared) equals the right expression (a negative number). Since this occurrence is impossible for real numbers, we conclude that if $b^2 - 4ac < 0$ the quadratic equation has no *real* solution. (We discuss quadratic equations for which the quantity $b^2 - 4ac < 0$ in detail in Section 4.5*.)

We now state the *quadratic formula.*

THEOREM

Quadratic Formula

Consider the quadratic equation

$$ax^2 + bx + c = 0 \qquad a \neq 0$$

If $b^2 - 4ac < 0$, this equation has no real solution.
If $b^2 - 4ac \geq 0$, the real solution(s) of this equation is (are) given by the **quadratic formula:**

$$x = \frac{-b \pm \sqrt{b^2 - 4ac}}{2a} \tag{5}$$

The quantity $\boldsymbol{b^2 - 4ac}$ is called the **discriminant** of the quadratic equation, because its value tells us whether the equation has real solutions. In fact, it also tells us how many solutions to expect.

*Section 4.5 may be covered anytime after completing this section without any loss of continuity.

> ***Discriminant of a Quadratic Equation***
>
> For a quadratic equation $ax^2 + bx + c = 0$:
>
> 1. If $b^2 - 4ac > 0$, there are two unequal real solutions.
> 2. If $b^2 - 4ac = 0$, there is a repeated real solution, a root of multiplicity 2.
> 3. If $b^2 - 4ac < 0$, there is no real solution.

Thus, when asked to find the real solutions, if any, of a quadratic equation, always evaluate the discriminant first to see how many real solutions there are.

EXAMPLE 3 Solving a Quadratic Equation by Graphing and by Using the Quadratic Formula

Find the real solutions, if any, of the equation $3x^2 - 5x + 1 = 0$.

Solution The equation is in standard form, so we compare it to $ax^2 + bx + c = 0$ to find a, b, and c.

$$3x^2 - 5x + 1 = 0$$
$$ax^2 + bx + c = 0$$

With $a = 3$, $b = -5$, and $c = 1$, we evaluate the discriminant $b^2 - 4ac$.

$$b^2 - 4ac = (-5)^2 - 4(3)(1) = 25 - 12 = 13$$

Since $b^2 - 4ac > 0$, there are two unequal real solutions.

Graphing Solution Figure 18 shows the graph of equation

$$Y_1 = 3x^2 - 5x + 1$$

As expected, we see that there are two x-intercepts: one between 0 and 1, the other between 1 and 2. Using ZERO(or ROOT), twice, the solutions to the equation are 0.23 and 1.43 rounded to two decimal places.

Figure 18

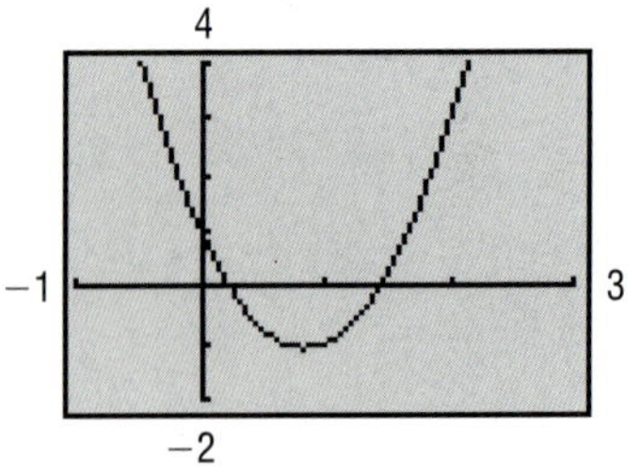

Algebraic Solution We use the quadratic formula with $a = 3$, $b = -5$, $c = 1$, and $b^2 - 4ac = 13$.

$$x = \frac{-b \pm \sqrt{b^2 - 4ac}}{2a} = \frac{5 \pm \sqrt{13}}{6}$$

The solution set is $\left\{\frac{5 - \sqrt{13}}{6}, \frac{5 + \sqrt{13}}{6}\right\}$. These solutions are exact.

NOW WORK PROBLEM 13.

EXAMPLE 4 Solving Quadratic Equations by Graphing and by Using the Quadratic Formula

Find the real solutions, if any, of the equation

$$\tfrac{25}{2}x^2 - 30x + 18 = 0$$

Solution

The equation is given in standard form. However, to simplify the arithmetic, we clear the fractions by multiplying both sides of the equation by 2.

$$\tfrac{25}{2}x^2 - 30x + 18 = 0$$
$$25x^2 - 60x + 36 = 0 \quad \textit{Clear fractions.}$$
$$ax^2 + bx + c = 0 \quad \textit{Compare to standard form.}$$

With $a = 25$, $b = -60$, and $c = 36$, we evaluate the discriminant.

$$b^2 - 4ac = (-60)^2 - 4(25)(36) = 3600 - 3600 = 0$$

The equation has a repeated solution.

Graphing Solution

Figure 19 shows the graph of the equation

$$Y_1 = \frac{25}{2}x^2 - 30x + 18$$

From the graph, we see that the one x-intercept is between 1 and 2. The solution to the equation is 1.2.

Figure 19

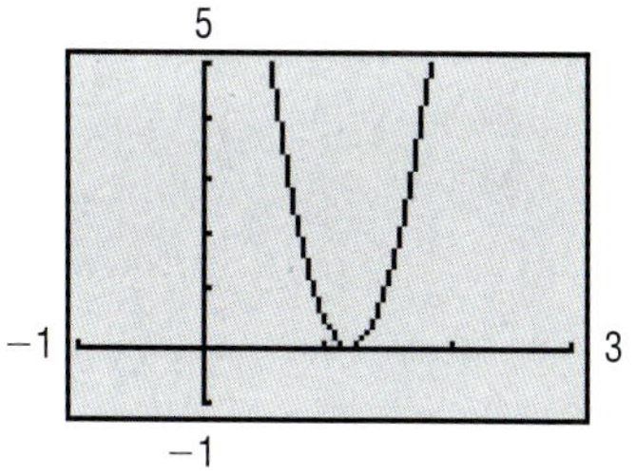

Algebraic Solution

We use $a = 25$, $b = -60$, $c = 36$, and $b^2 - 4ac = 0$ to solve

$$25x^2 - 60x + 36 = 0$$

Using the quadratic formula, we find that

$$x = \frac{-b \pm \sqrt{b^2 - 4ac}}{2a} = \frac{60 \pm \sqrt{0}}{50} = \frac{60}{50} = \frac{6}{5}$$

The repeated solution is $\frac{6}{5}$, which is exact.

EXAMPLE 5 Solving Quadratic Equations by Graphing and by Using the Quadratic Formula

Find the real solutions, if any, of the equation

$$3x^2 + 2 = 4x$$

Solution The equation, as given, is not in standard form.

$$3x^2 + 2 = 4x$$

$$3x^2 - 4x + 2 = 0 \qquad \textit{Put in standard form.}$$

$$ax^2 + bx + c = 0 \qquad \textit{Compare to standard form.}$$

With $a = 3$, $b = -4$, and $c = 2$, we find that

$$b^2 - 4ac = 16 - 24 = -8$$

Since $b^2 - 4ac < 0$, the equation has no real solution.

Graphing Solution We use the standard form of the equation and graph

$$Y_1 = 3x^2 - 4x + 2$$

See Figure 20. We see that there are no x-intercepts, so the equation has no real solution, as expected.

Figure 20

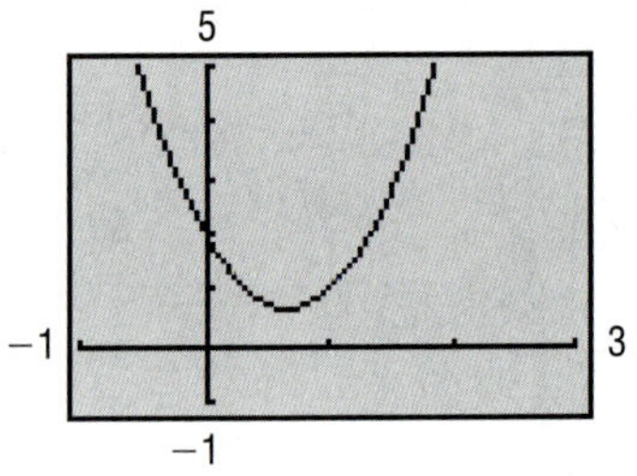

NOW WORK PROBLEM 19.

Quadratic Functions

A **quadratic function** is a function that is defined by a second-degree polynomial in one variable.

A **quadratic function** is a function of the form

$$f(x) = ax^2 + bx + c \tag{6}$$

where a, b, and c are real numbers and $a \neq 0$. The domain of a quadratic function consists of all real numbers.

Many applications require a knowledge of quadratic functions. For example, suppose that Texas Instruments collects the data shown in Table 2 that relate the number of calculators sold at the price p per calculator. Since the price of a product determines the quantity that will be purchased, we treat price as the independent variable.

Table 2

Price per Calculator, p (Dollars)	Number of Calculators, x
60	11,100
65	10,115
70	9,652
75	8,731
80	8,087
85	7,205
90	6,439

A linear relationship between the number of calculators and the price p per calculator may be given by the equation

$$x = 21{,}000 - 150p$$

Then the revenue R derived from selling x calculators at the price p per calculator is

$$\begin{aligned} R &= xp \\ &= (21{,}000 - 150p)p \\ &= -150p^2 + 21{,}000p \end{aligned}$$

Figure 21

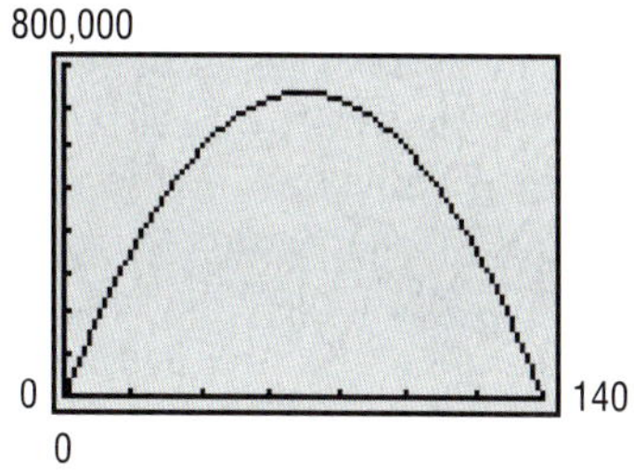

So, the revenue R is a quadratic function of the price p. Figure 21 illustrates the graph of this revenue function, whose domain is $0 \leq p \leq 140$, since both x and p must be nonnegative.

A second situation in which a quadratic function appears involves the motion of a projectile. Based on Newton's second law of motion (force equals mass times acceleration, $F = ma$), it can be shown that, ignoring air resistance, the path of a projectile propelled upward at an inclination to the horizontal is the graph of a quadratic function. See Figure 22 for an illustration.

Figure 22
Path of a cannonball

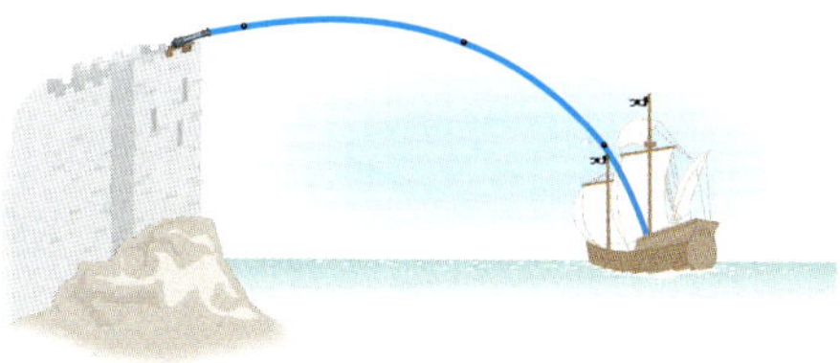

Graphing Quadratic Functions

4 We know how to graph a quadratic function of the form $f(x) = x^2$. Figure 23 shows the graph of three functions of the form $f(x) = ax^2$, $a > 0$, for $a = 1$, $a = \frac{1}{2}$, and $a = 3$. Notice that the larger the value of a, the "narrower" the graph is, and the smaller the value of a, the "wider" the graph is.

Figure 23

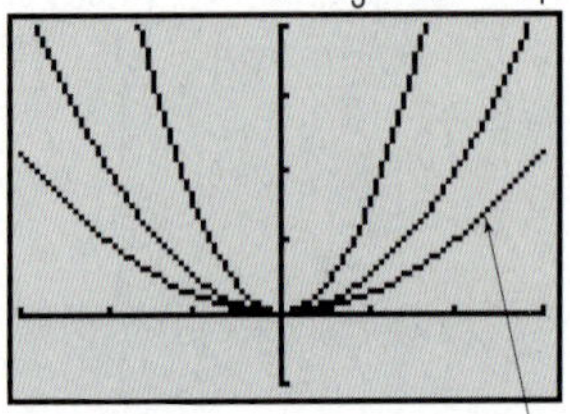

Figure 24

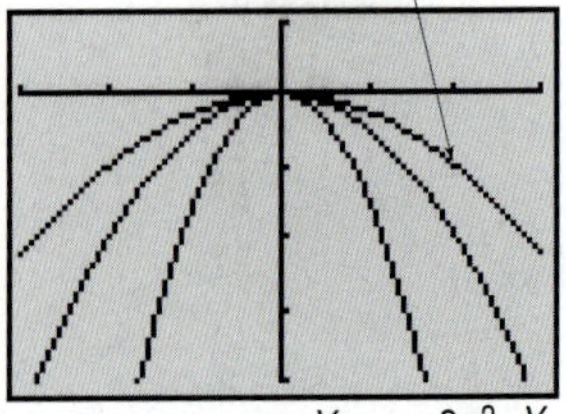

Figure 24 shows the graphs of $f(x) = ax^2$ for $a < 0$. Notice that these graphs are reflections about the x-axis of the graphs in Figure 23. Based on the results of these two figures, we can draw some general conclusions about the graph of $f(x) = ax^2$. First, as $|a|$ increases, the graph becomes "narrower," and as $|a|$ gets closer to zero, the graph gets "wider." Second, if a is positive, then the graph opens "up," and if a is negative, then the graph opens "down."

The graphs in Figures 23 and 24 are typical of the graphs of all quadratic functions, which we call **parabolas.*** Refer to Figure 25, where two parabolas are pictured. The one on the left **opens up** and has a lowest point; the one on the right **opens down** and has a highest point. The lowest or highest point of a parabola is called the **vertex.** The vertical line passing through the vertex in each parabola in Figure 25 is called the **axis of symmetry** (usually abbreviated to **axis**) of the parabola. Because the parabola is symmetric about its axis, the axis of symmetry of a parabola can be used to advantage in graphing the parabola by hand.

Figure 25

Graphs of a quadratic function, $f(x) = ax^2 + bx + c, a \neq 0$

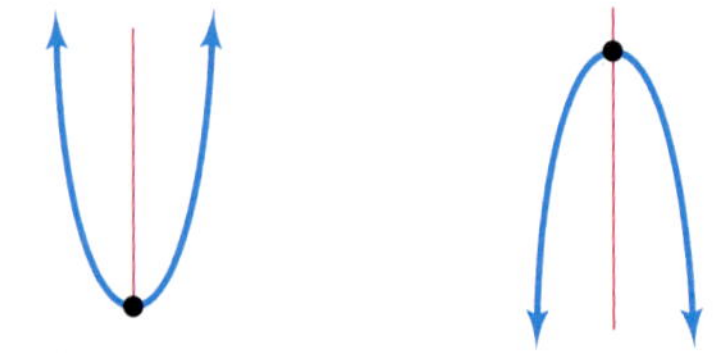

The parabolas shown in Figure 25 are the graphs of a quadratic function $f(x) = ax^2 + bx + c, a \neq 0$. Notice that the coordinate axes are not included in the figure. Depending on the values of a, b, and c, the axes could be placed anywhere. The important fact is that the shape of the graph of a quadratic function will look like one of the parabolas in Figure 25.

A key element in graphing a quadratic function by hand is locating the vertex. To find a formula, we begin with a quadratic function $f(x) = ax^2 + bx + c$, $a \neq 0$, and complete the square in x.

$$f(x) = ax^2 + bx + c \qquad a \neq 0.$$

$$= a\left(x^2 + \frac{b}{a}x\right) + c \qquad \text{Factor out } a \text{ from } ax^2 + bx.$$

$$= a\left(x^2 + \frac{b}{a}x + \frac{b^2}{4a^2}\right) + c - a\left(\frac{b^2}{4a^2}\right) \qquad \text{Complete the square by adding and subtracting } a(b^2/4a^2)\text{. Look closely at this step!}$$

$$= a\left(x + \frac{b}{2a}\right)^2 + c - \frac{b^2}{4a}$$

$$f(x) = a\left(x + \frac{b}{2a}\right)^2 + \frac{4ac - b^2}{4a} \qquad (7)$$

*We shall study parabolas using a geometric definition later in this book.

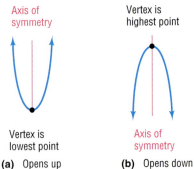

Suppose that $a > 0$. We concentrate on the term $a(x + b/2a)^2$ in (7). If $x = -b/2a$, then this term equals zero. Since $a > 0$, for any other choice of x, this term will be positive. [Do you see why? $a(x + b/2a)^2 > 0$ for all $x \neq -b/2a$.] Thus, in (7), the smallest value of $f(x)$ occurs when $x = -b/2a$. That is, if $a > 0$, the vertex is at $(-b/2a, f(-b/2a))$ and is a minimum point since the parabola opens up.

Similarly, if $a < 0$, the term $a(x + b/2a)^2$ is zero if $x = -b/2a$ and is negative if $x \neq -b/2a$. In this case, the largest value of $f(x)$ occurs when $x = -b/2a$. That is, if $a < 0$, the vertex is at $(-b/2a, f(-b/2a))$ and is a maxium point since the parabola opens down.

We summarize these remarks as follows:

Characteristics of the Graph of a Quadratic Function

$$f(x) = ax^2 + bx + c$$

$$\text{Vertex} = \left(\frac{-b}{2a}, f\left(\frac{-b}{2a}\right)\right) \qquad \text{Axis of symmetry: line } x = \frac{-b}{2a} \qquad (8)$$

Parabola opens up if $a > 0$; the vertex is a minimum point.
Parabola opens down if $a < 0$; the vertex is a maximum point.

EXAMPLE 6 Locating the Vertex without Graphing

Without graphing, locate the vertex and axis of symmetry of the parabola defined by $f(x) = -3x^2 + 6x + 1$. Does it open up or down?

Solution For this quadratic function, $a = -3$, $b = 6$, and $c = 1$. The x-coordinate of the vertex is

$$\frac{-b}{2a} = \frac{-6}{-6} = 1$$

The y-coordinate of the vertex is therefore

$$f\left(\frac{-b}{2a}\right) = f(1) = -3 + 6 + 1 = 4$$

The vertex is located at the point $(1, 4)$. The axis of symmetry is the line $x = 1$. Finally, because $a = -3 < 0$, the parabola opens down.

5 The information we gathered in Example 6, together with the location of the intercepts, usually provides enough information to graph $f(x) = ax^2 + bx + c$, $a \neq 0$, by hand. The y-intercept is the value of f at $x = 0$, that is, $f(0) = c$. The x-intercepts, if there are any, are found by solving the equation

$$f(x) = ax^2 + bx + c = 0$$

This equation has two, one, or no real solutions, depending on whether the discriminant $b^2 - 4ac$ is positive, 0, or negative. Thus, it has corresponding x-intercepts, as follows:

> ***The x-intercepts of a Quadratic Function***
>
> 1. If the discriminant $b^2 - 4ac > 0$, the graph of $f(x) = ax^2 + bx + c$ has two distinct x-intercepts and so will cross the x-axis in two places.
> 2. If the discriminant $b^2 - 4ac = 0$, the graph of $f(x) = ax^2 + bx + c$ has one x-intercept and touches the x-axis at its vertex.
> 3. If the discriminant $b^2 - 4ac < 0$, the graph of $f(x) = ax^2 + bx + c$ has no x-intercept and so will not cross or touch the x-axis.

Figure 26 illustrates these possibilities for parabolas that open up.

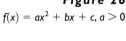

Figure 26
$f(x) = ax^2 + bx + c, a > 0$

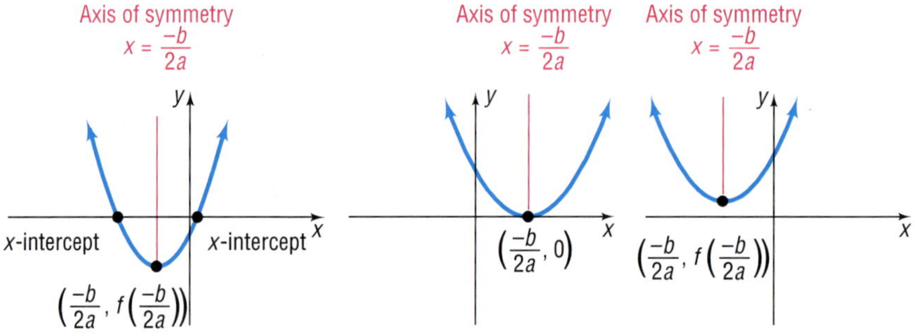

(a) $b^2 - 4ac > 0$ Two x-intercepts

(b) $b^2 - 4ac = 0$ One x-intercept

(c) $b^2 - 4ac < 0$ No x-intercepts

EXAMPLE 7 Graphing a Quadratic Function by Hand Using Its Vertex, Axis, and Intercepts

Use the information from Example 6 and the locations of the intercepts to graph $f(x) = -3x^2 + 6x + 1$.

Solution In Example 6, we found the vertex to be at (1, 4) and the axis of symmetry to be $x = 1$. The y-intercept is found by letting $x = 0$. Thus, the y-intercept is $f(0) = 1$. The x-intercepts are found by solving the equation $f(x) = 0$. This results in the equation

$$-3x^2 + 6x + 1 = 0 \qquad a = -3, b = 6, c = 1$$

The discriminant $b^2 - 4ac = (6)^2 - 4(-3)(1) = 36 + 12 = 48 > 0$, so the equation has two real solutions and the graph has two x-intercepts. Using the quadratic formula, we find that

$$x = \frac{-b + \sqrt{b^2 - 4ac}}{2a} = \frac{-6 + \sqrt{48}}{-6} = \frac{-6 + 4\sqrt{3}}{-6} \approx -0.15$$

and

$$x = \frac{-b - \sqrt{b^2 - 4ac}}{2a} = \frac{-6 - \sqrt{48}}{-6} = \frac{-6 - 4\sqrt{3}}{-6} \approx 2.15$$

The x-intercepts are approximately -0.15 and 2.15.

Figure 27

$f(x) = -3x^2 + 6x + 1$

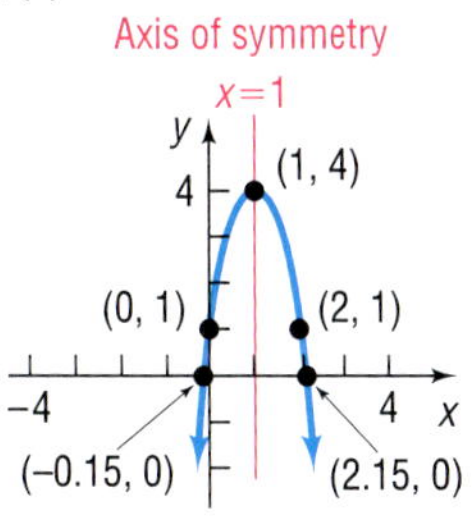

The graph is illustrated in Figure 27. Notice how we used the y-intercept and the axis of symmetry, $x = 1$, to obtain the additional point $(2, 1)$ on the graph. ◗

NOW WORK PROBLEM **37.**

If the graph of a quadratic function has only one x-intercept or none, it is usually necessary to plot an additional point to obtain the graph by hand.

EXAMPLE 8 Graphing a Quadratic Function by Hand Using Its Vertex, Axis, and Intercepts

Graph $f(x) = x^2 - 6x + 9$ by determining whether the graph opens up or down. Find its vertex, axis of symmetry, y-intercept, and x-intercept, if any.

Figure 28

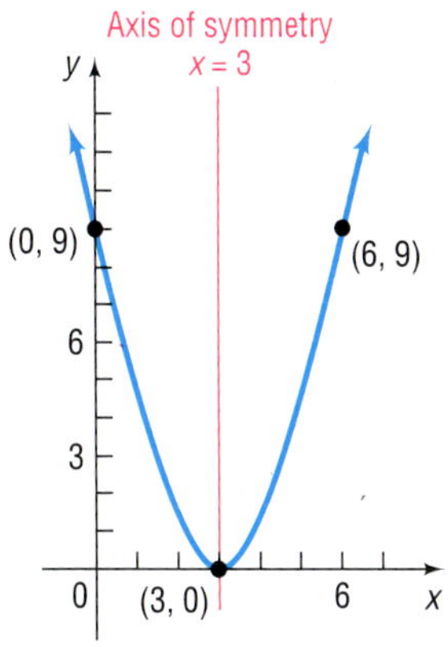

Solution For $f(x) = x^2 - 6x + 9$, we have $a = 1$, $b = -6$, and $c = 9$. Since $a = 1 > 0$, the parabola opens up. The x-coordinate of the vertex is

$$\frac{-b}{2a} = \frac{-(-6)}{2(1)} = 3$$

The y-coordinate of thc vertex is

$$f(3) = (3)^2 - 6(3) + 9 = 0$$

So the vertex is at $(3, 0)$. The axis of symmetry is the line $x = 3$. The y-intercept is $f(0) = 9$. Since the vertex $(3, 0)$ lies on the x-axis, the graph touches the x-axis at the x-intercept. By using the axis of symmetry and the y-intercept at $(0, 9)$, we can locate the additional point $(6, 9)$ on the graph. See Figure 28. ◗

NOW WORK PROBLEM **45.**

EXAMPLE 9 Graphing a Quadratic Function by Hand Using Its Vertex, Axis, and Intercepts

Graph $f(x) = 2x^2 + x + 1$ by determining whether the graph opens up or down. Find its vertex, axis of symmetry, y-intercept, and x-intercepts, if any.

Solution For $f(x) = 2x^2 + x + 1$, we have $a = 2$, $b = 1$, and $c = 1$. Since $a = 2 > 0$, the parabola opens up. The x-coordinate of the vertex is

$$\frac{-b}{2a} = -\frac{1}{4}$$

Figure 29

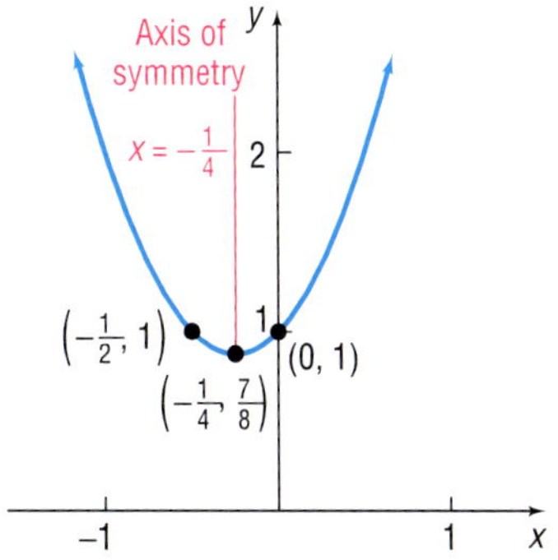

The y-coordinate of the vertex is

$$f\left(-\frac{1}{4}\right) = 2\left(\frac{1}{16}\right) + \left(-\frac{1}{4}\right) + 1 = \frac{7}{8}$$

So the vertex is at $\left(-\frac{1}{4}, \frac{7}{8}\right)$. The axis of symmetry is the line $x = -\frac{1}{4}$. The y-intercept is $f(0) = 1$. The x-intercept(s), if any, obey the equation $2x^2 + x + 1 = 0$. Since the discriminant $b^2 - 4ac = (1)^2 - 4(2)(1) = -7 < 0$, this equation has no real solutions, and therefore the graph has no x-intercepts. We use the point $(0, 1)$ and the axis of symmetry $x = -\frac{1}{4}$ to locate the additional point $\left(-\frac{1}{2}, 1\right)$ on the graph. See Figure 29. ◗

SUMMARY **Steps for Graphing a Quadratic Function $f(x) = ax^2 + bx + c, a \neq 0$ by hand**

STEP 1. Determine the vertex, $\left(\frac{-b}{2a}, f\left(\frac{-b}{2a}\right)\right)$.

STEP 2. Determine the axis of symmetry, $x = \frac{-b}{2a}$.

STEP 3. Determine the y-intercept, $f(0)$.

STEP 4. (a) If $b^2 - 4ac > 0$, then the graph of the quadratic function has two x-intercepts, which are found by solving $ax^2 + bx + c = 0$.

(b) If $b^2 - 4ac = 0$, the vertex is the x-intercept.

(c) If $b^2 - 4ac < 0$, there are no x-intercepts.

STEP 5. Determine an additional point if $b^2 - 4ac \leq 0$ by using the y-intercept and the axis of symmetry.

STEP 6. Plot the points and draw the graph.

Applications

Real-world problems may lead to mathematical problems that involve quadratic functions. Two such problems follow.

EXAMPLE 10 Area of a Rectangle with Fixed Perimeter

A farmer has 50 feet of fence to enclose a rectangular field. Express the area of the field A as a function of the length x of a side.

Solution Consult Figure 30. If the length of the rectangle is x and if w is its width, then the sum of the lengths of the sides is the perimeter, 50.

Figure 30

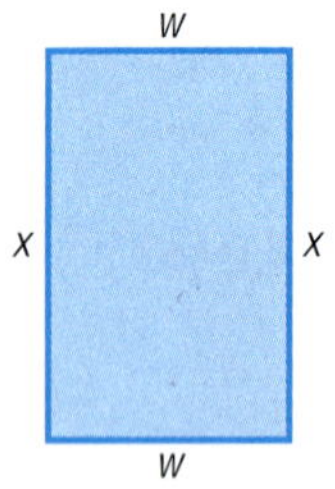

$$x + w + x + w = 50$$
$$2x + 2w = 50$$
$$x + w = 25$$
$$w = 25 - x$$

The area A is length times width, so

$$A = xw = x(25 - x)$$

The area A as a function of x is

$$A(x) = x(25 - x) = -x^2 + 25x$$

Note that we used the symbol A as the dependent variable and also as the name of the function that relates the length x to the area A. As we mentioned earlier, this double usage is common in applications and should cause no difficulties.

EXAMPLE 11 Economics: Demand Equations

In economics, revenue R is defined as the amount of money derived from the sale of a product and is equal to the unit selling price p of the product times the number x of units actually sold. That is,

$$R = xp$$

In economics, the Law of Demand states that p and x are related: As price p increases, the number x of units sold decreases. Suppose that p and x are related by the following

$$p = -\frac{1}{10}x + 20 \qquad 0 \le x \le 200$$

Express the revenue R as a function of the number x of units sold.

Solution Since $R = xp$ and $p = -\frac{1}{10}x + 20$, it follows that

$$R(x) = xp = x\left(-\frac{1}{10}x + 20\right) = -\frac{1}{10}x^2 + 20x$$

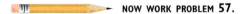

NOW WORK PROBLEM 57.

2.3 EXERCISES

In Problems 1–6, solve each equation by extracting the root. Verify your result using a graphing utility.

1. $x^2 = 16$ **2.** $x^2 = 81$ **3.** $(x + 4)^2 = 9$

4. $(x - 5)^2 = 100$ **5.** $(x - 8)^2 = 17$ **6.** $(x + 1)^2 = 13$

In Problems 7–12, solve each equation by completing the square. Verify your result using a graphing utility.

7. $x^2 + 4x - 21 = 0$ **8.** $x^2 - 6x = 13$ **9.** $x^2 - \frac{1}{2}x = \frac{3}{16}$

10. $x^2 + \frac{2}{3}x = \frac{1}{3}$ **11.** $3x^2 + x - \frac{1}{2} = 0$ **12.** $2x^2 - 3x = 1$

In Problems 13–24, find the real solutions, if any, of each equation. Use the quadratic formula. Verify your results using a graphing utility.

13. $x^2 - 4x + 2 = 0$ **14.** $x^2 + 4x + 2 = 0$ **15.** $x^2 - 4x - 1 = 0$

16. $x^2 + 6x + 1 = 0$ **17.** $2x^2 - 5x + 3 = 0$ **18.** $2x^2 + 5x + 3 = 0$

19. $4y^2 - y + 2 = 0$ **20.** $4t^2 + t + 1 = 0$ **21.** $4x^2 = 1 - 2x$

22. $2x^2 = 1 - 2x$ **23.** $9t^2 - 6t + 1 = 0$ **24.** $4u^2 - 6u + 9 = 0$

In Problems 25–32, match each graph to one of the following functions:

A. $f(x) = x^2 - 1$ *B.* $f(x) = -x^2 - 1$ *C.* $f(x) = x^2 - 2x + 1$ *D.* $f(x) = x^2 + 2x + 1$

E. $f(x) = x^2 - 2x + 2$ *F.* $f(x) = x^2 + 2x$ *G.* $f(x) = x^2 - 2x$ *H.* $f(x) = x^2 + 2x + 2$

25.

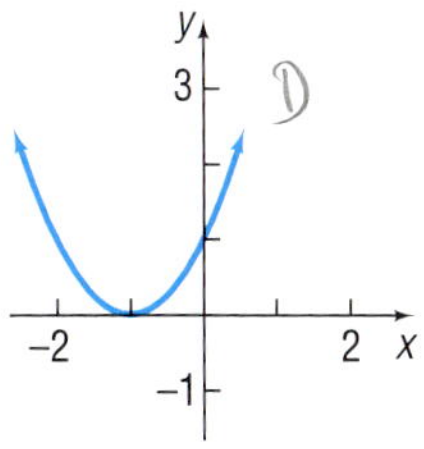

26.

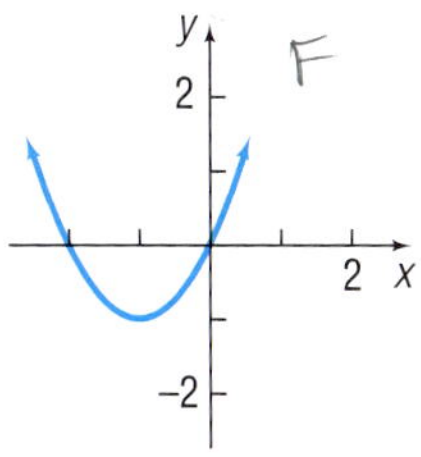

27.

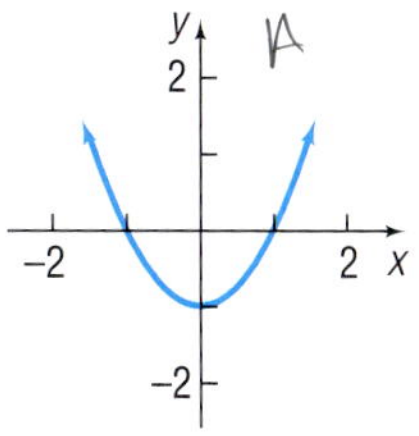

28.

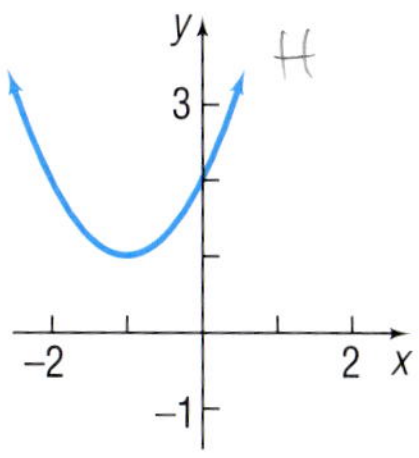

29.

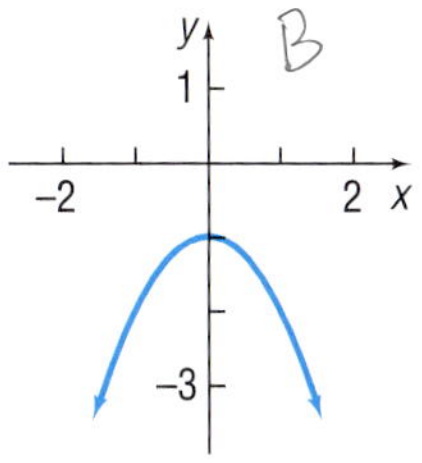

30.

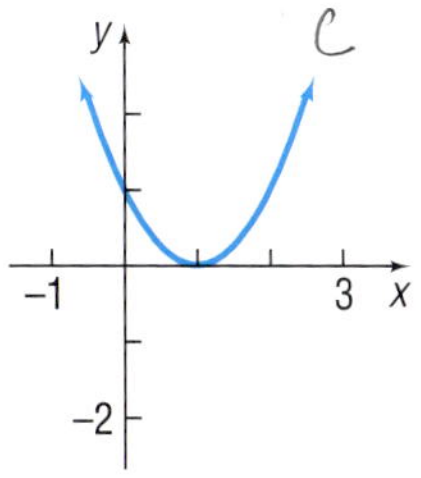

31.

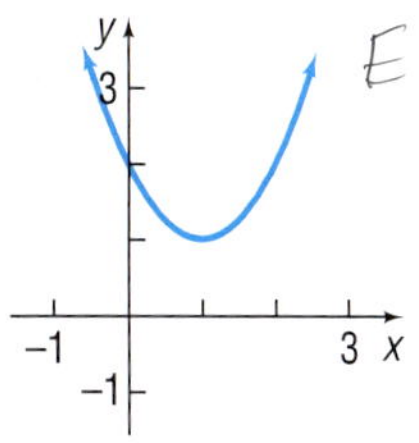

32.

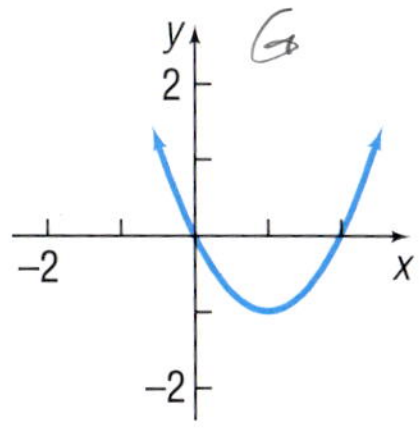

In Problems 33–36, match each graph to one of the following functions.

A. $f(x) = (1/3)x^2 + 2$ B. $f(x) = 3x^2 + 2$

C. $f(x) = x^2 + 5x + 1$ D. $f(x) = -x^2 + 5x + 1$

33.

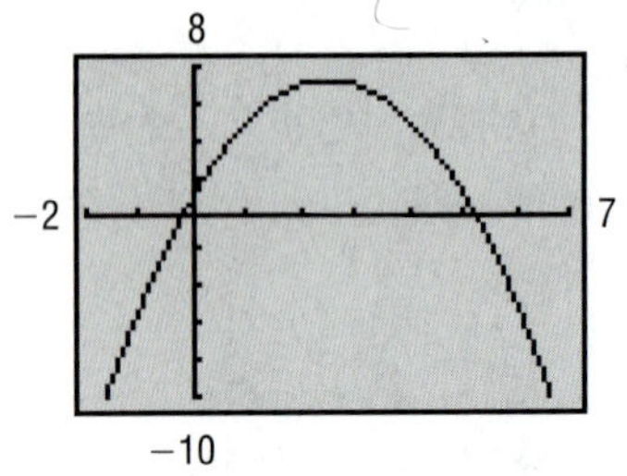

34.

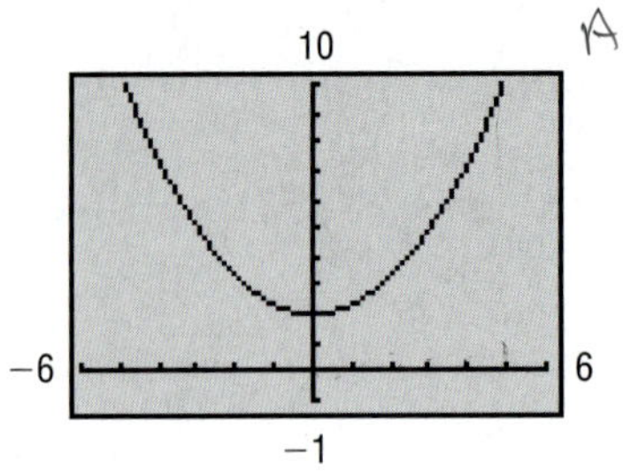

35.

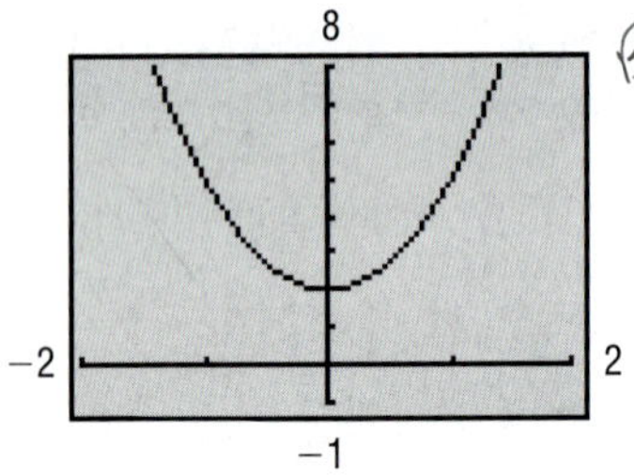

36.

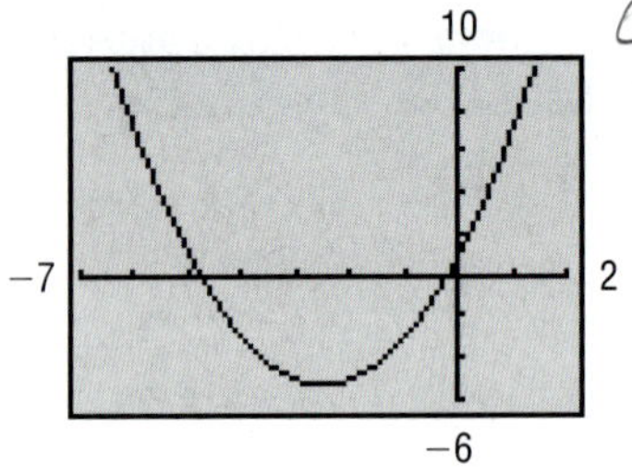

In Problems 37–66, graph each quadratic function by hand by determining whether its graph opens up or down and by finding its vertex, axis of symmetry, y-intercept, and x-intercepts, if any. Verify your results using a graphing utility.

37. $f(x) = x^2 + 2x$ **38.** $f(x) = x^2 - 4x$ **39.** $f(x) = -x^2 - 6x$

40. $f(x) = -x^2 + 4x$ **41.** $f(x) = 2x^2 - 8x$ **42.** $f(x) = 3x^2 + 18x$

43. $f(x) = x^2 + 2x - 8$ **44.** $f(x) = x^2 - 2x - 3$ **45.** $f(x) = x^2 + 2x + 1$

46. $f(x) = x^2 + 6x + 9$ **47.** $f(x) = 2x^2 - x + 2$ **48.** $f(x) = 4x^2 - 2x + 1$

49. $f(x) = -2x^2 + 2x - 3$ **50.** $f(x) = -3x^2 + 3x - 2$ **51.** $f(x) = 3x^2 + 6x + 2$

52. $f(x) = 2x^2 + 5x + 3$ **53.** $f(x) = -4x^2 - 6x + 2$ **54.** $f(x) = 3x^2 - 8x + 2$

55. The graph of the function $f(x) = ax^2 + bx + c$ has vertex at $x = 0$ and passes through the points $(0, 2)$ and $(1, 8)$. Find a, b, and c.

56. The graph of the function $f(x) = ax^2 + bx + c$ has vertex at $x = 1$ and passes through the points $(0, 1)$ and $(-1, -8)$. Find a, b, and c.

57. Demand Equation The price p and the quantity x sold of a certain product obey the demand equation

$$p = -\tfrac{1}{6}x + 100 \qquad 0 \le x \le 600$$

Express the revenue R as a function of x. (Remember, $R = xp$.)

58. Demand Equation The price p and the quantity x sold of a certain product obey the demand equation

$$p = -\tfrac{1}{3}x + 100 \qquad 0 \le x \le 300$$

Express the revenue R as a function of x.

59. Demand Equation The price p and the quantity x sold of a certain product obey the demand equation

$$x = -5p + 100 \qquad 0 \le p \le 20$$

Express the revenue R as a function of x.

60. Demand Equation The price p and the quantity x sold of a certain product obey the demand equation

$$x = -20p + 500 \qquad 0 \le p \le 25$$

Express the revenue R as a function of x.

61. Enclosing a Rectangular Field David has available 400 yards of fencing and wishes to enclose a rectangular area. Express the area A of the rectangle as a function of the width x of the rectangle.

62. Enclosing a Rectangular Field along a River Beth has 3000 feet of fencing available to enclose a rectangular field. One side of the field lies along a river, so only three sides require fencing. Express the area A of the rectangle as a function of x, where x is the length of the side parallel to the river.

63. Make up a quadratic function that opens down and has only one x-intercept. Compare yours with others in the class. What are the similarities? What are the differences?

64. On one set of coordinate axes, graph the family of parabolas $f(x) = x^2 + 2x + c$ for $c = -3$, $c = 0$, and $c = 1$. Describe the characteristics of a member of this family.

65. On one set of coordinate axes, graph the family of parabolas $f(x) = x^2 + bx + 1$ for $b = -4$, $b = 0$, and $b = 4$. Describe the general characteristics of this family.

2.4 QUADRATIC FUNCTIONS AND MODELS

1. Find the Maximum or Minimum Value of a Quadratic Function
2. Use the Maximum or Mininum Value of a Quadratic Function to Solve Applied Problems
3. Use a Graphing Utility to Find the Quadratic Function of Best Fit

Quadratic Models

When a mathematical model leads to a quadratic function, the characteristics of that quadratic function can provide important information about the model. For example, for a quadratic revenue function, we can find the maximum revenue; for a quadratic cost function, we can find the minimum cost.

To see why, recall that the graph of a quadratic function $f(x) = ax^2 + bx + c$ is a parabola with vertex at $(-b/2a, f(-b/2a))$. This vertex is the highest point on the graph if $a < 0$ and the lowest point on the graph if $a > 0$. If the vertex is the highest point ($a < 0$), then $f(-b/2a)$ is the **maximum value** of f. If the vertex is the lowest point ($a > 0$), then $f(-b/2a)$ is the **minimum value** of f.

EXAMPLE 1 Finding the Maximum or Minimum Value of a Quadratic Function

Determine whether the quadratic function

$$f(x) = x^2 - 4x - 5$$

has a maximum or minimum value. Then find the maximum or minimum value.

Solution We compare $f(x) = x^2 - 4x - 5$ to $f(x) = ax^2 + bx + c$. We conclude that $a = 1$, $b = -4$, and $c = -5$. Since $a > 0$, the graph of f opens up, so the vertex is a minimum point. The minimum value occurs at

$$x = \frac{-b}{2a} = \frac{-(-4)}{2(1)} = \frac{4}{2} = 2$$

$\uparrow$ $a = 1, b = -4$

The minimum value is

$$f\left(\frac{-b}{2a}\right) = f(2) = 2^2 - 4(2) - 5 = 4 - 8 - 5 = -9$$

NOW WORK PROBLEM 7.

EXAMPLE 2 Maximizing Revenue

2

The marketing department at Texas Instruments has found that, when certain calculators are sold at a price of p dollars per unit, the revenue R (in dollars) as a function of the price p is

$$R(p) = -150p^2 + 21{,}000p$$

What unit price should be established in order to maximize revenue? If this price is charged, what is the maximum revenue?

Solution The revenue R is

$$R(p) = -150p^2 + 21{,}000p = ap^2 + bp + c$$

The function R is a quadratic function with $a = -150$, $b = 21{,}000$, and $c = 0$. Because $a < 0$, the vertex is the highest point of the parabola. The revenue R is therefore a maximum when the price p is

$$p = \frac{-b}{2a} \underset{\uparrow}{=} \frac{-21{,}000}{2(-150)} = \frac{-21{,}000}{-300} = \$70.00$$

$a = -150,\ b = 21{,}000$

The maximum revenue R is

$$R(70) = -150(70)^2 + 21{,}000(70) = \$735{,}000$$

See Figure 31 for an illustration.

Figure 31
$R(p) = -150p^2 + 21{,}000p$

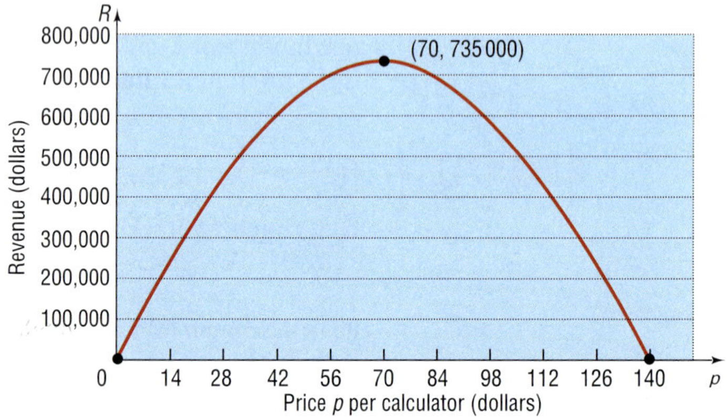

NOW WORK PROBLEM 13.

EXAMPLE 3 Maximizing the Area Enclosed by a Fence

A farmer has 2000 yards of fence to enclose a rectangular field. What are the dimensions of the rectangle that encloses the most area?

Solution The available fence represents the perimeter of the rectangle. If x is the length and w is the width, then

$$2x + 2w = 2000 \qquad (1)$$

The area A of the rectangle is

$$A = xw$$

To express A in terms of a single variable, we solve equation (1) for w and substitute the result in $A = xw$. Then A involves only the variable x. [You could also solve equation (1) for x and express A in terms of w alone. Try it!]

$$2x + 2w = 2000 \qquad \text{Equation (1)}$$
$$2w = 2000 - 2x \qquad \text{Solve for } w.$$
$$w = \frac{2000 - 2x}{2} = 1000 - x$$

Then the area A is

$$A = xw = x(1000 - x) = -x^2 + 1000x$$

Thus, A is a quadratic function of x.

$$A(x) = -x^2 + 1000x \qquad a = -1,\ b = 1000,\ c = 0$$

Since $a < 0$, the vertex is a maximum point on the graph of A. The maximum value occurs at

$$x = \frac{-b}{2a} = -\frac{-1000}{2(-1)} = 500$$

The maximum value of A is

$$A\left(\frac{-b}{2a}\right) = A(500) = -500^2 + 1000(500) = -250{,}000 + 500{,}000 = 250{,}000$$

The most area that can be enclosed by 2000 yards of fence is 250,000 square yards.

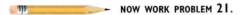

NOW WORK PROBLEM 21.

EXAMPLE 4 Analyzing the Motion of a Projectile

A projectile is fired from a cliff 500 feet above the water at an inclination of 45° to the horizontal, with a muzzle velocity of 400 feet per second. In physics, it is established that the height h of the projectile above the water is given by

$$h(x) = \frac{-32x^2}{(400)^2} + x + 500$$

where x is the horizontal distance of the projectile from the base of the cliff. See Figure 32.

Figure 32

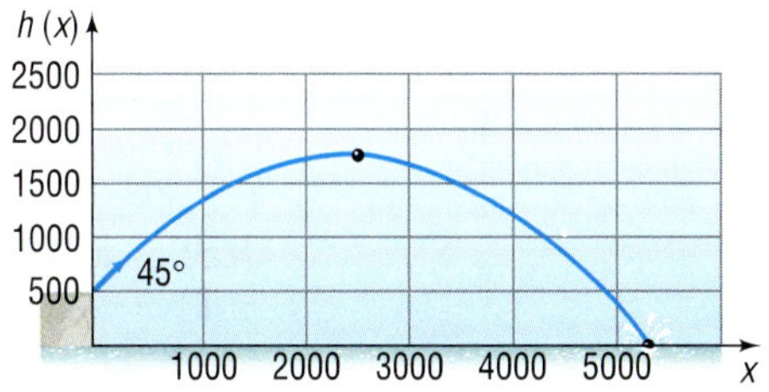

(a) Find the maximum height of the projectile.
(b) How far from the base of the cliff will the projectile strike the water?

Solution (a) The height of the projectile is given by a quadratic function.

$$h(x) = \frac{-32x^2}{(400)^2} + x + 500 = \frac{-1}{5000}x^2 + x + 500$$

We are looking for the maximum value of h. Since the maximum value is obtained at the vertex, we compute

$$x = \frac{-b}{2a} = \frac{-1}{2(-1/5000)} = \frac{5000}{2} = 2500$$

The maximum height of the projectile is

$$h(2500) = \frac{-1}{5000}(2500)^2 + 2500 + 500 = -1250 + 2500 + 500 = 1750 \text{ ft}$$

(b) The projectile will strike the water when the height is zero. To find the distance x traveled, we need to solve the equation

$$h(x) = \frac{-1}{5000}x^2 + x + 500 = 0$$

We use the quadratic formula with

$$b^2 - 4ac = 1 - 4\left(\frac{-1}{5000}\right)(500) = 1.4$$

$$x = \frac{-1 \pm \sqrt{1.4}}{2(-1/5000)} \approx \begin{cases} -458 \\ 5458 \end{cases}$$

We discard the negative solution and find that the projectile will strike the water a distance of about 5458 feet from the base of the cliff.

EXPLORATION Graph

$$h(x) = \frac{-1}{5000}x^2 + x + 500 \qquad 0 \le x \le 5500$$

Use MAXIMUM to find the maximum height of the projectile and use ROOT or ZERO to find the distance from the base of the cliff to where it strikes the water. Compare your results with those obtained in the text. TRACE the path of the projectile. How far from the base of the cliff is the projectile when its height is 1000 ft? 1500 ft?

NOW WORK PROBLEM 29.

EXAMPLE 5 The Golden Gate Bridge

The Golden Gate Bridge, a suspension bridge, spans the entrance to San Francisco Bay. Its 746-foot-tall towers are 4200 feet apart. The bridge is suspended from two huge cables more than 3 feet in diameter; the 90-foot-wide roadway is 220 feet above the water. The cables are parabolic in shape and touch the road surface at the center of the bridge. Find the height of the cable at a distance of 1000 feet from the center.

Solution We begin by choosing the placement of the coordinate axes so that the x-axis coincides with the road surface and the origin coincides with the center of the bridge. As a result, the twin towers will be vertical (height $746 - 220 = 526$ feet above the road) and located 2100 feet from the center. Also, the cable, which has the shape of a parabola, will extend from the towers, open up, and have its vertex at (0, 0). As illustrated in Figure 33, the choice of placement of the axes enables us to identify the equation of the parabola as $y = ax^2$, $a > 0$. We can also see that the points $(-2100, 526)$ and $(2100, 526)$ are on the graph.

Figure 33

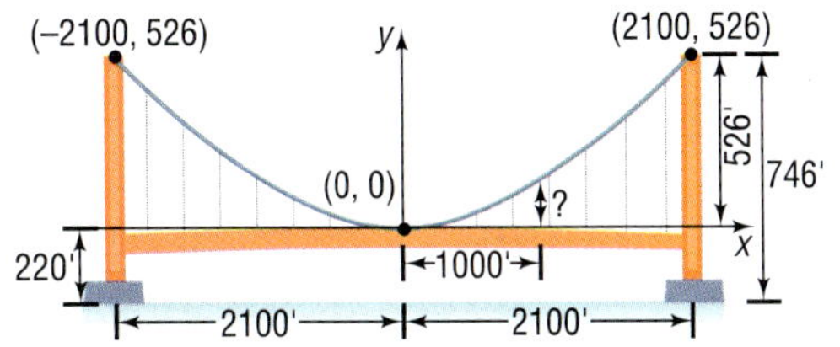

Based on these facts, we can find the value of a in $y = ax^2$.

$$y = ax^2$$
$$526 = a(2100)^2$$
$$a = \frac{526}{(2100)^2}$$

This is because a quadratic of the form y = ax² has vertex at the origin. So, given any point i.e. (2100, 526) we can find a and then sub 1000 for x and get y.

The equation of the parabola is therefore

$$y = \frac{526}{(2100)^2}x^2$$

The height of the cable when $x = 1000$ is

$$y = \frac{526}{(2100)^2}(1000)^2 \approx 119.3 \text{ feet}$$

Thus, the cable is 119.3 feet high at a distance of 1000 feet from the center of the bridge.

NOW WORK PROBLEM **31.**

Fitting a Quadratic Function to Data

3 In Section 2.2, we found the line of best fit for data that appeared to be linearly related. It was noted that data may also follow a nonlinear relation. Figures 34(a) and (b) show scatter diagrams of data that follow a quadratic relation.

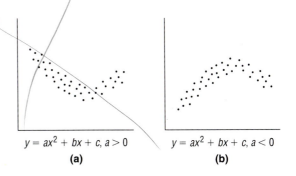

Figure 34

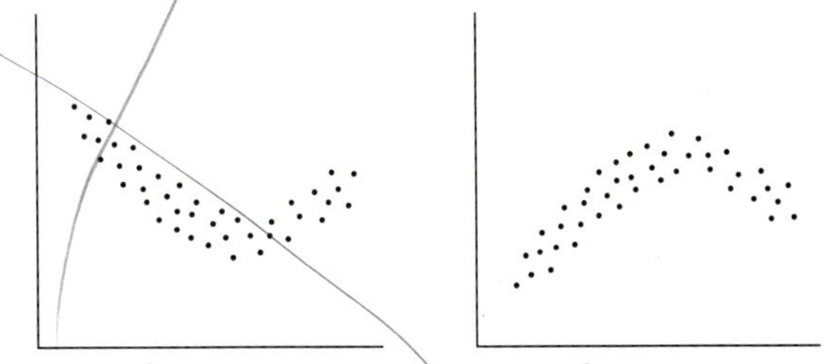

$y = ax^2 + bx + c$, $a > 0$
(a)

$y = ax^2 + bx + c$, $a < 0$
(b)

EXAMPLE 6 Fitting a Quadratic Function to Data

Plot	Fertilizer, x (Pounds/100 ft^2)	Yield, y (Bushels)
1	0	4
2	0	6
3	5	10
4	5	7
5	10	12
6	10	10
7	15	15
8	15	17
9	20	18
10	20	21
11	25	20
12	25	21
13	30	21
14	30	22
15	35	21
16	35	20
17	40	19
18	40	19

Suppose that a farmer collects data, as shown in the table, that show crop yields Y for various amounts of fertilizer used, x.

(a) Draw a scatter diagram of the data. Comment on the type of relation that may exist between the two variables.

(b) The quadratic function of best fit to these data is

$$Y(x) = -0.0171x^2 + 1.0765x + 3.8939$$

Use this function to determine the optimal amount of fertilizer to apply.

(c) Use the function to predict crop yield when the optimal amount of fertilizer is applied.

(d) Use a graphing utility to verify that the function given in part (b) is the quadratic function of best fit.

(e) With a graphing utility, draw a scatter diagram of the data and then graph the quadratic function of best fit on the scatter diagram.

Solution

(a) Figure 35 shows the scatter diagram, from which it appears that the data follow a quadratic relation, with $a < 0$.

Figure 35

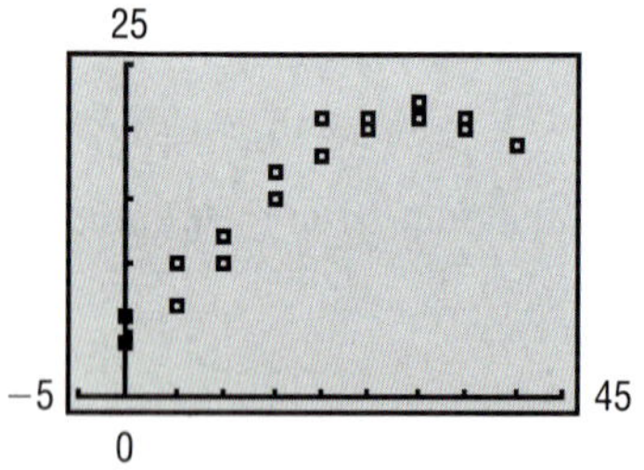

(b) Based on the quadratic function of best fit, the optimal amount of fertilizer to apply is

$$x = \frac{-b}{2a} = \frac{-1.0765}{2(-0.0171)} \approx 31.5 \text{ pounds of fertilizer per 100 square feet}$$

(c) We evaluate the function $Y(x)$ for $x = 31.5$.

$$Y(31.5) = -0.0171(31.5)^2 + 1.0765(31.5) + 3.8939 \approx 20.8 \text{ bushels}$$

If we apply 31.5 pounds of fertilizer per 100 square feet, the crop yield will be 20.8 bushels according to the quadratic function of best fit.

(d) Upon executing the QUADratic REGression program, we obtain the results shown in Figure 36. The output that the utility provides shows us the equation $y = ax^2 + bx + c$. The quadratic function of best fit is $Y(x) = -0.0171x^2 + 1.0765x + 3.8939$, where x represents the amount of fertilizer used and Y represents crop yield.

(e) Figure 37 shows the graph of the quadratic function found in part (d) drawn on the scatter diagram.

Figure 36

Figure 37

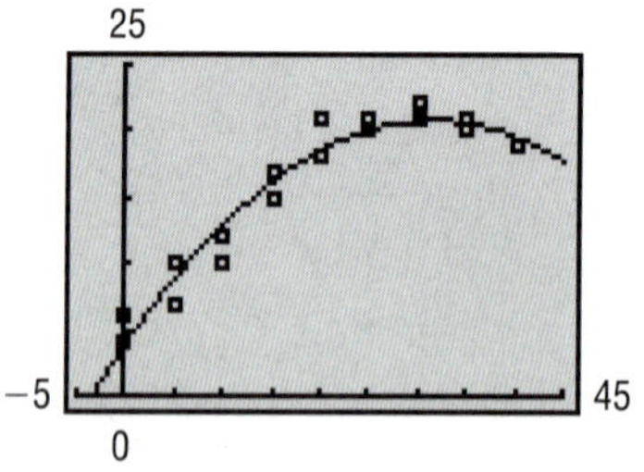

Look again at Figure 36. Notice that the output given by the graphing calculator does not include r, the correlation coefficient. Recall that the correlation coefficient is a measure of the strength of a *linear* relation that exists between two variables. The graphing calculator does not provide an indication of how well the function fits the data in terms of r since the function cannot be expressed as a linear function.

2.4 EXERCISES

In Problems 1–12, determine, without graphing, whether the given quadratic function has a maximum value or a minimum value and then find the value.

1. $f(x) = x^2 + 2$
2. $f(x) = x^2 - 4x$
3. $f(x) = -x^2 - 4x$
4. $f(x) = -x^2 - 6x$
5. $f(x) = 2x^2 + 12x$
6. $f(x) = -2x^2 + 12x$
7. $f(x) = 2x^2 + 12x - 3$
8. $f(x) = 4x^2 - 8x + 3$
9. $f(x) = -x^2 + 10x - 4$
10. $f(x) = -2x^2 + 8x + 3$
11. $f(x) = -3x^2 + 12x + 1$
12. $f(x) = 4x^2 - 4x$

13. Maximizing Revenue Suppose that the manufacturer of a gas clothes dryer has found that, when the unit price is p dollars, the revenue R (in dollars) is

$$R(p) = -4p^2 + 4000p$$

What unit price should be established for the dryer to maximize revenue? What is the maximum revenue?

14. Maximizing Revenue The John Deere company has found that the revenue from sales of heavy-duty tractors is a function of the unit price p that it charges. If the revenue R is

$$R(p) = -\frac{1}{2}p^2 + 1900p$$

what unit price p should be charged to maximize revenue? What is the maximum revenue?

15. Minimizing Cost The daily cost C (in dollars) to produce x portable 12-inch TVs is given by $C(x) = x^2 - 80x + 2000$. How many TVs should be produced to minimize the cost? What is the minimum cost?

16. Minimizing Cost The weekly cost C (in dollars) of manufacturing x cell phones is given by $C(x) = 5x^2 - 200x + 4000$. How many cell phones should be manufactured to minimize the cost? What is the minimum cost?

17. Demand Equation The price p and the quantity x sold of a certain product obey the demand equation

$$p = -\tfrac{1}{6}x + 100 \qquad 0 \le x \le 600$$

(a) Express the revenue R as a function of x. (Remember, $R = xp$.)
(b) What is the revenue if 200 units are sold?
(c) What quantity x maximizes revenue? What is the maximum revenue?
(d) What price should the company charge to maximize revenue?

18. Demand Equation The price p and the quantity x sold of a certain product obey the demand equation

$$p = -\tfrac{1}{3}x + 100 \qquad 0 \le x \le 300$$

(a) Express the revenue R as a function of x.
(b) What is the revenue if 100 units are sold?
(c) What quantity x maximizes revenue? What is the maximum revenue?
(d) What price should the company charge to maximize revenue?

19. Demand Equation The price p and the quantity x sold of a certain product obey the demand equation

$$x = -5p + 100 \qquad 0 \le p \le 20$$

(a) Express the revenue R as a function of x.
(b) What is the revenue if 15 units are sold?
(c) What quantity x maximizes revenue? What is the maximum revenue?
(d) What price should the company charge to maximize revenue?

20. Demand Equation The price p and the quantity x sold of a certain product obey the demand equation

$$x = -20p + 500 \qquad 0 \le p \le 25$$

(a) Express the revenue R as a function of x.
(b) What is the revenue if 20 units are sold?

(c) What quantity x maximizes revenue? What is the maximum revenue?

(d) What price should the company charge to maximize revenue?

21. **Enclosing a Rectangular Field** David has available 400 yards of fencing and wishes to enclose a rectangular area.

(a) Express the area A of the rectangle as a function of the width x of the rectangle.

(b) For what value of x is the area largest?

(c) What is the maximum area?

22. **Enclosing a Rectangular Field** Beth has 3000 feet of fencing available to enclose a rectangular field.

(a) Express the area A of the rectangle as a function of x, where x is the length of the rectangle.

(b) For what value of x is the area largest?

(c) What is the maximum area?

23. **Rectangles with Fixed Perimeter** What is the largest rectangular area that can be enclosed with 400 feet of fencing? What are the dimensions of the rectangle?

24. **Rectangles with Fixed Perimeter** What are the dimensions of a rectangle of a fixed perimeter P that result in the largest area?

25. **Enclosing the Most Area with a Fence** A farmer with 4000 meters of fencing wants to enclose a rectangular plot that borders on a river. If the farmer does not fence the side along the river, what is the largest area that can be enclosed? (See the figure.)

26. **Enclosing the Most Area with a Fence** A farmer with 2000 meters of fencing wants to enclose a rectangular plot that borders on a straight highway. If the farmer does not fence the side along the highway, what is the largest area that can be enclosed?

27. **Enclosing the Most Area with a Fence** A farmer with 10,000 meters of fencing wants to enclose a rectangular field and then divide it into two plots with a fence parallel to one of the sides (see the figure). What is the largest area that can be enclosed?

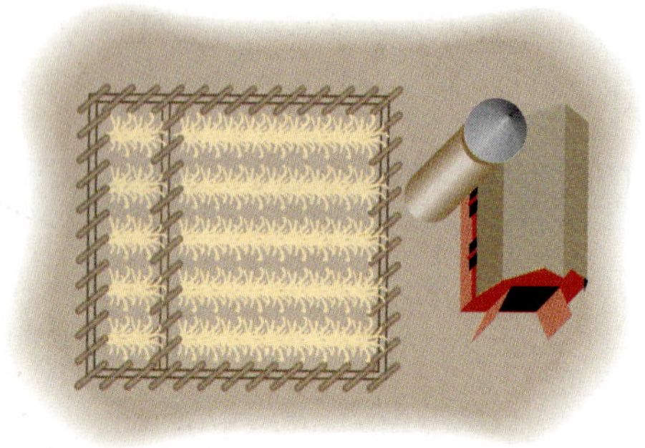

28. **Enclosing the Most Area with a Fence** A farmer with 10,000 meters of fencing wants to enclose a rectangular field and then divide it into three plots with two fences parallel to one of the sides. What is the largest area that can be enclosed?

29. **Analyzing the Motion of a Projectile** A projectile is fired from a cliff 200 feet above the water at an inclination of 45° to the horizontal, with a muzzle velocity of 50 feet per second. The height h of the projectile above the water is given by

$$h(x) = \frac{-32x^2}{(50)^2} + x + 200$$

where x is the horizontal distance of the projectile from the base of the cliff.

(a) How far from the base of the cliff is the height of the projectile a maximum?

(b) Find the maximum height of the projectile.

(c) How far from the base of the cliff will the projectile strike the water?

(d) Using a graphing utility, graph the function h, $0 \le x \le 200$.

(e) Use a graphing utility to verify the solutions found in parts (b) and (c).

(f) When the height of the projectile is 100 feet above the water, how far is it from the cliff?

30. **Analyzing the Motion of a Projectile** A projectile is fired at an inclination of 45° to the horizontal, with a muzzle velocity of 100 feet per second. The height h of the projectile is given by

$$h(x) = \frac{-32x^2}{(100)^2} + x$$

where x is the horizontal distance of the projectile from the firing point.

(a) How far from the firing point is the height of the projectile a maximum?

(b) Find the maximum height of the projectile.
(c) How far from the firing point will the projectile strike the ground?
(d) Using a graphing utility, graph the function h, $0 \leq x \leq 350$.
(e) Use a graphing utility to verify the results obtained in parts (b) and (c).
(f) When the height of the projectile is 50 feet above the ground, how far has it traveled horizontally?

31. **Suspension Bridge** A suspension bridge with weight uniformly distributed along its length has twin towers that extend 75 meters above the road surface and are 400 meters apart. The cables are parabolic in shape and are suspended from the tops of the towers. The cables touch the road surface at the center of the bridge. Find the height of the cables at a point 100 meters from the center. (Assume that the road is level.)

32. **Architecture** A parabolic arch has a span of 120 feet and a maximum height of 25 feet. Choose suitable rectangular coordinate axes and find the equation of the parabola. Then calculate the height of the arch at points 10 feet, 20 feet, and 40 feet from the center.

33. **Constructing Rain Gutters** A rain gutter is to be made of aluminum sheets that are 12 inches wide by turning up the edges 90°. What depth will provide maximum cross-sectional area and hence allow the most water to flow?

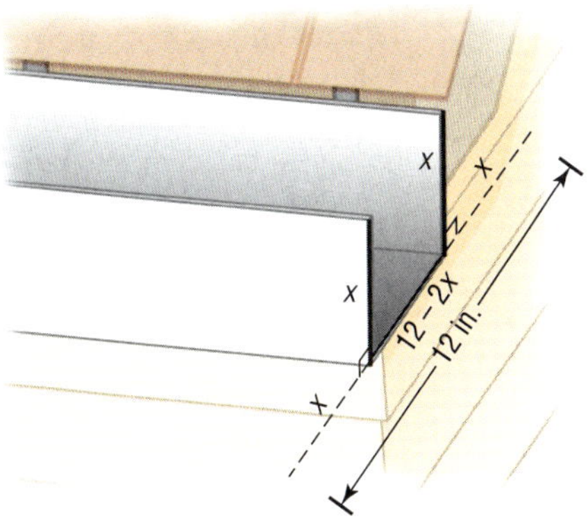

34. **Norman Windows** A Norman window has the shape of a rectangle surmounted by a semicircle of diameter equal to the width of the rectangle (see the figure). If the perimeter of the window is 20 feet, what dimensions will admit the most light (maximize the area)?
[Hint: Circumference of a circle = $2\pi r$; area of a circle = πr^2, where r is the radius of the circle.]

35. **Constructing a Stadium** A track and field playing area is in the shape of a rectangle with semicircles at each end (see the figure). The inside perimeter of the track is to be 1500 meters. What should the dimensions of the rectangle be so that the area of the rectangle is a maximum?

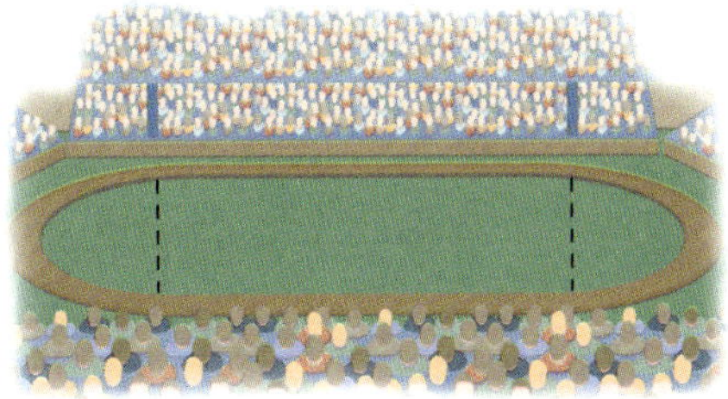

36. **Architecture** A special window has the shape of a rectangle surmounted by an equilateral triangle (see the figure). If the perimeter of the window is 16 feet, what dimensions will admit the most light?
[Hint: Area of an equilateral triangle = $(\sqrt{3}/4)x^2$, where x is the length of a side of the triangle.]

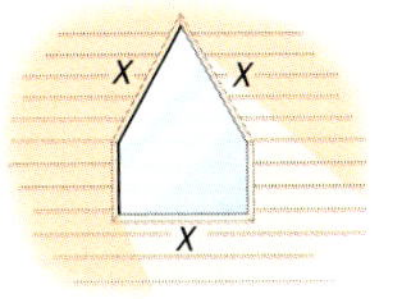

37. **Life Cycle Hypothesis** An individual's income varies with his or her age. The table on p. 148 shows the median income I of individuals of different age groups within the United States for 1995. For each class, let the class midpoint represent the independent variable x. For the class "65 years and older," we will assume that the class midpoint is 69.5.

Age	Class Midpoint, x	Median Income, I
15–24 years	19.5	\$20,979
25–34 years	29.5	\$34,701
35–44 years	39.5	\$43,465
45–54 years	49.5	\$48,058
55–64 years	59.5	\$38,077
65 years and older	69.5	\$19,096

Source: U.S. Census Bureau

(a) Draw a scatter diagram of the data. Comment on the type of relation that may exist between the two variables.

(b) The quadratic function of best fit to these data is

$$I(x) = -42.6x^2 + 3805.5x - 38{,}526$$

Use this function to determine the age at which an individual can expect to earn the most income.

(c) Use the function to predict the peak income earned.

(d) Use a graphing utility to verify that the function given in part (b) is the quadratic function of best fit.

(e) With a graphing utility, graph the quadratic function of best fit on the scatter diagram.

38. Advertising A small manufacturing firm collected the following data on advertising expenditures A (in thousands of dollars) and total revenue R (in thousands of dollars).

Advertising	Total Revenue
20	\$6101
22	\$6222
25	\$6350
25	\$6378
27	\$6453
28	\$6423
29	\$6360
31	\$6231

(a) Draw a scatter diagram of the data. Comment on the type of relation that may exist between the two variables.

(b) The quadratic function of best fit to these data is

$$R(A) = -7.76A^2 + 411.88A + 942.72$$

Use this function to determine the optimal level of advertising for this firm.

(c) Use the function to find the revenue that the firm can expect if it uses the optimal level of advertising.

(d) Use a graphing utility to verify that the function given in part (b) is the quadratic function of best fit.

(e) With a graphing utility, graph the quadratic function of best fit on the scatter diagram.

39. Fuel Consumption The following data represent the average fuel consumption C by cars (in billions of gallons) for the years 1980–1993.

Year, t	Average Fuel Consumption, C
1980	71.9
1981	71.0
1982	70.1
1983	69.9
1984	68.7
1985	69.3
1986	71.4
1987	70.6
1988	71.9
1989	72.7
1990	72.0
1991	70.7
1992	73.9
1993	75.1

Source: U.S. Federal Highway Administration

(a) Draw a scatter diagram of the data. Comment on the type of relation that may exist between the two variables.

(b) The quadratic function of best fit to these data is

$$C(t) = 0.063t^2 - 250t + 248{,}077$$

Use this function to determine the year in which average fuel consumption was lowest.

(c) Use the function to predict the average fuel consumption for 1994.

(d) Use a graphing utility to verify that the function given in part (b) is the quadratic function of best fit.

(e) With a graphing utility, graph the quadratic function of best fit on the scatter diagram.

40. Miles per Gallon An engineer collects data showing the speed s of a Ford Taurus and its average miles per gallon, M. See the table.

Speed, s	Miles per Gallon, M
30	18
35	20
40	23
40	25
45	25
50	28
55	30
60	29
65	26
65	25
70	25

(a) Draw a scatter diagram of the data. Comment on the type of relation that may exist between the two variables.

(b) The quadratic function of best fit to these data is

$$M(s) = -0.018s^2 + 1.93s - 25.34$$

Use this function to determine the speed that maximizes miles per gallon.

(c) Use the function to predict miles per gallon for a speed of 63 miles per hour.

(d) Use a graphing utility to verify that the function given in part (b) is the quadratic function of best fit.

(e) With a graphing utility, graph the quadratic function of best fit on the scatter diagram.

41. Height of a Ball A physicist throws a ball at an inclination of 45° to the horizontal. The following data represent the height of the ball h at the instant it has traveled x feet horizontally.

Distance, x	Height, h
20	25
40	40
60	55
80	65
100	71
120	77
140	77
160	75
180	71
200	64

(a) Draw a scatter diagram of the data. Comment on the type of relation that may exist between the two variables.

(b) The quadratic function of best fit to these data is

$$h(x) = -0.0037x^2 + 1.03x + 5.7$$

Use this function to determine how far the ball will travel before it reaches its maximum height.

(c) Use the function to find the maximum height of the ball.

(d) Use a graphing utility to verify that the function given in part (b) is the quadratic function of best fit.

(e) With a graphing utility, graph the quadratic function of best fit on the scatter diagram.

42. Enrollment in Public Schools The following data represent the enrollment E in all public schools (both elementary and high school) for the academic years 1980–1981 to 1988–1989. Let 1 represent the academic year 1980–1981, 2 the academic year 1981–1982, and so on.

Year, t	Enrollment, E
1	41.5
2	40.8
3	40.1
4	39.6
5	39.1
6	39.1
7	39.6
8	39.8
9	40.1

(a) Draw a scatter diagram of the data. Comment on the type of relation that may exist between the two variables.

(b) The quadratic function of best fit to these data is

$$E(t) = 0.1t^2 - 1.16t + 42.62$$

Use this function to determine when enrollment was lowest.

(c) Use a graphing utility to verify that the function given in part (b) is the quadratic function of best fit.

(d) With a graphing utility, graph the quadratic function of best fit on the scatter diagram.

43. **Chemical Reactions** A self-catalytic chemical reaction results in the formation of a compound that causes the formation ratio to increase. If the reaction rate V is given by

$$V(x) = kx(a - x) \qquad 0 \le x \le a$$

where k is a positive constant, a is the initial amount of the compound, and x is the variable amount of the compound, for what value of x is the reaction rate a maximum?

CHAPTER REVIEW

THINGS TO KNOW

Function

A relation between two sets of real numbers so that each number x in the first set, the domain, has corresponding to it exactly one number y in the second set. The range is the set of y values of the function for the x values in the domain.

x is the independent variable; y is the dependent variable.

A function f may be defined implicitly by an equation involving x and y or explicitly by writing $y = f(x)$.

A function can also be characterized as a set of ordered pairs (x, y) or $(x, f(x))$ in which no two distinct pairs have the same first element.

Function notation

$y = f(x)$

f is a symbol for the function.

x is the argument, or independent variable.

y is the dependent variable.

$f(x)$ is the value of the function at x, or the image of x.

Domain

If unspecified, the domain of a function f is the largest set of real numbers for which $f(x)$ is a real number.

Vertical-line test

A set of points in the plane is the graph of a function if and only if every vertical line intersects the graph in at most one point.

Linear function

$f(x) = mx + b$ Graph is a line with slope m and y-intercept b.

Quadratic equation and quadratic formula

If $ax^2 + bx + c = 0$, $a \neq 0$, and if $b^2 - 4ac \ge 0$, then $x = \dfrac{-b \pm \sqrt{b^2 - 4ac}}{2a}$.

Discriminant

If $b^2 - 4ac > 0$, there are two distinct real solutions.

If $b^2 - 4ac = 0$, there is one repeated real solution.

If $b^2 - 4ac < 0$, there are no real solutions.

Quadratic function

$f(x) = ax^2 + bx + c$ Graph is a parabola that opens up if $a > 0$, and opens down if $a < 0$.

Vertex: $\left(\dfrac{-b}{2a}, f\left(\dfrac{-b}{2a}\right)\right)$

Axis of symmetry: $x = \dfrac{-b}{2a}$

y-intercept: $f(0)$

x-intercept(s): If any, found by solving the equation $ax^2 + bx + c = 0$.

HOW TO

Determine whether a relation represents a function
Find the domain and range of a function from its graph
Find the domain of a function given its equation
Find the equation of the line of best fit
Solve direct variation problems
Solve quadratic equations
Graph quadratic functions
Find the maximum or minimum value of a quadratic function
Solve applied problems

FILL-IN THE BLANK ITEMS

1. If f is a function defined by the equation $y = f(x)$, then x is called the ______________ variable and y is the ______________ variable.
2. A set of points in the xy-plane is the graph of a function if and only if every ______________ line intersects the graph in at most one point of the set.
3. To complete the square of the expression $x^2 + 5x$, you would ______________ the number ______________.
4. The quantity $b^2 - 4ac$ is called the ______________ of a quadratic equation. If it is ______________, the equation has no real solution.
5. For the graph of the linear function $f(x) = mx + b$, m is the ______________ and b is the ______________.
6. The graph of a quadratic function is called a ______________.
7. The function $f(x) = ax^2 + bx + c$, $a \neq 0$, has a minimum value if ______________. The minimum value occurs at $x =$ ______________.

TRUE/FALSE ITEMS

T F 1. Every relation is a function.

T F 2. Vertical lines intersect the graph of a function in no more than one point.

T F 3. The y-intercept of the graph of the function $y = f(x)$ whose domain is all real numbers is $f(0)$.

T F 4. Quadratic equations always have two real solutions.

T F 5. If the discriminant of a quadratic equation is positive, then the equation has two solutions that are negatives of each other.

T F 6. The graph of $f(x) = 2x^2 + 3x - 4$ opens up.

T F 7. The minimum value of $f(x) = -x^2 + 4x + 5$ is $f(2)$.

REVIEW EXERCISES

Blue problem numbers indicate the author's suggestions for use in a Practice Test.

1. Given that f is a linear function, $f(4) = -5$, and $f(0) = 3$, write the equation that defines f.
2. Given that g is a linear function with slope $= -4$ and $g(-2) = 2$, write the equation that defines g.

3. A function f is defined by

$$f(x) = \frac{Ax + 5}{6x - 2}$$

If $f(1) = 4$, find A.

4. A function g is defined by

$$g(x) = \frac{A}{x} + \frac{8}{x^2}$$

If $g(-1) = 0$, find A.

5. Tell which of the following graphs are graphs of functions.

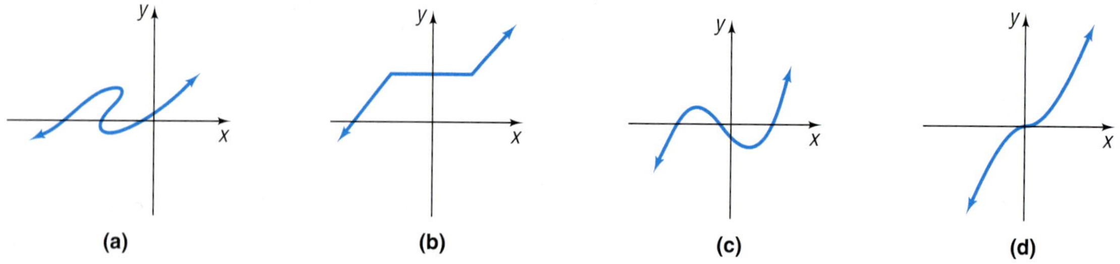

6. Use the graph of the function f shown to find
(a) The domain and range of f
(b) $f(-1)$
(c) The intercepts of f

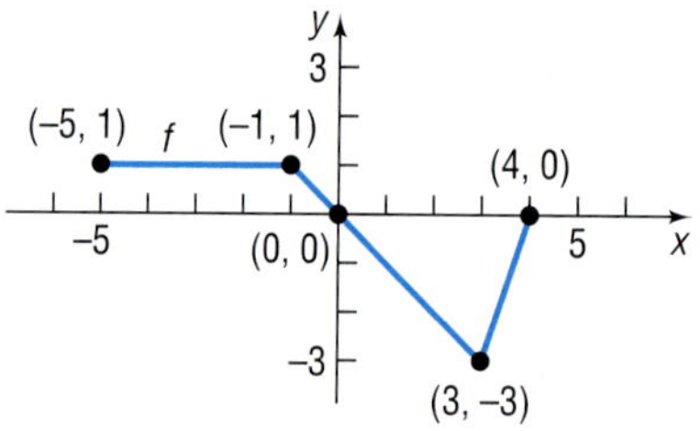

In Problems 7–12, find the following for each function:
(a) $f(-x)$ *(b)* $-f(x)$ *(c)* $f(x+2)$ *(d)* $f(x-2)$

7. $f(x) = \dfrac{3x}{x^2 - 4}$ **8.** $f(x) = \dfrac{x^2}{x + 2}$ **9.** $f(x) = \sqrt{x^2 - 4}$ **10.** $f(x) = |x^2 - 4|$

11. $f(x) = \dfrac{x^2 - 4}{x^2}$ **12.** $f(x) = \dfrac{x^3}{x^2 - 4}$

In Problems 13–20, find the domain of each function.

13. $f(x) = \dfrac{x}{x^2 - 9}$ **14.** $f(x) = \dfrac{3x^2}{x - 2}$ **15.** $f(x) = \sqrt{2 - x}$ **16.** $f(x) = \sqrt{x + 2}$

17. $h(x) = \dfrac{\sqrt{x}}{|x|}$ **18.** $g(x) = \dfrac{|x|}{x}$ **19.** $f(x) = \dfrac{x}{x^2 + 2x - 3}$ **20.** $F(x) = \dfrac{1}{x^2 - 3x - 4}$

In Problems 21–24, graph each linear function.

21. $f(x) = 2x - 5$ **22.** $g(x) = -4x + 7$ **23.** $h(x) = \dfrac{4}{5}x - 6$ **24.** $F(x) = \dfrac{-1}{3}x + 1$

In Problems 25–32, find all real solutions, if any, of each quadratic equation. Verify your results using a graphing utility.

25. $(x + 3)^2 = 100$ **26.** $(x - 5)^2 = 9$ **27.** $4x^2 - 2x - 3 = 0$ **28.** $4x^2 - 6x - 1 = 0$

29. $x^2 + 3x + 1 = 0$ **30.** $x^2 - 4x + 2 = 0$ **31.** $16x^2 + 24x + 9 = 0$ **32.** $-2x^2 + x - 1 = 0$

In Problems 33–42, (a) graph each quadratic function by hand by determining whether its graph opens up or down and by finding its vertex, axis of symmetry, y-intercept, and x-intercepts, if any; (b) verify your results using a graphing utility.

33. $f(x) = \frac{1}{4}x^2 - 16$ **34.** $f(x) = -\frac{1}{2}x^2 + 2$ **35.** $f(x) = -4x^2 + 4x$

36. $f(x) = 9x^2 - 6x + 3$ **37.** $f(x) = \frac{9}{2}x^2 + 3x + 1$ **38.** $f(x) = -x^2 + x + \frac{1}{2}$

39. $f(x) = 3x^2 + 4x - 1$ **40.** $f(x) = -2x^2 - x + 4$ **41.** $f(x) = x^2 - 4x + 6$

42. $f(x) = x^2 + 2x - 3$

In Problems 43–48, determine whether the given quadratic function has a maximum value or a minimum value, and then find the value.

43. $f(x) = 3x^2 - 6x + 4$ **44.** $f(x) = 2x^2 + 8x + 5$ **45.** $f(x) = -x^2 + 8x - 4$

46. $f(x) = -x^2 - 10x - 3$ **47.** $f(x) = -3x^2 + 12x + 4$ **48.** $f(x) = -2x^2 + 4$

49. Distance and Time Using the following data, find a function that relates the distance s, in miles, driven by a Ford Taurus and t, the time the Taurus has been driven.

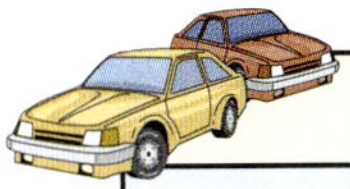

Time (Hours), t	Distance (Miles), s
0	0
1	30
2	55
3	83
4	100
5	150
6	210
7	260
8	300

(a) Does the relation defined by the set of ordered pairs (t, s) represent a function?
(b) Draw a scatter diagram of the data.
(c) Using a graphing utility, find the line of best fit relating time and distance.
(d) Interpret the slope.
(e) Express the relationship found in part (c) using function notation.
(f) What is the domain of the function?
(g) Predict the distance the car is driven after 11 hours.

50. High School versus College GPA An administrator at Southern Illinois University wants to find a function that relates a student's college grade point average G to the high school grade point average x. She randomly selects eight students and obtains the following data:

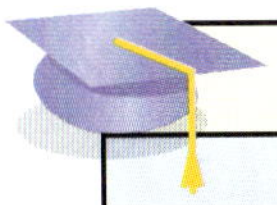

High School GPA, x	College GPA, G
2.73	2.43
2.92	2.97
3.45	3.63
3.78	3.81
2.56	2.83
2.98	2.81
3.67	3.45
3.10	2.93

(a) Does the relation defined by the set of ordered pairs (x, G) represent a function?
(b) Draw a scatter diagram of the data.
(c) Using a graphing utility, find the line of best fit relating high school GPA and college GPA.
(d) Interpret the slope.
(e) Express the relationship found in part (c) using function notation.
(f) What is the domain of the function?
(g) Predict a student's college GPA if her high school GPA is 3.23.

51. Mortgage Payments The monthly payment p on a mortgage varies directly with the amount borrowed B. If the monthly payment on a 30-year mortgage is \$854.00 when \$130,000 is borrowed, find a linear function that relates the monthly payment p to the amount borrowed B for a mortgage with the same terms. Then find the monthly payment p when the amount borrowed B is \$165,000.

52. Revenue Function At the corner Esso station, the revenue R varies directly with the number of gallons of gasoline sold g. If the revenue is \$15.93 when the number of gallons sold is 13.5, find a linear function that relates revenue R to the number of gallons of gasoline g. Then find the revenue R when the number of gallons of gasoline sold is 11.2.

53. Landscaping A landscape engineer has 200 feet of border to enclose a rectangular pond. What dimensions will result in the largest pond?

54. Geometry Find the length and width of a rectangle whose perimeter is 20 feet and whose area is 16 square feet.

55. A rectangle has one vertex on the line $y = 10 - x, x > 0$, another at the origin, one on the positive x-axis, and one on the positive y-axis. Find the largest area A that can be enclosed by the rectangle.

56. Minimizing Cost Callaway Golf Company has determined that the daily cost C of manufacturing x Big Bertha-type gold clubs may be expressed by the quadratic function

$$C(x) = 5x^2 - 620x + 20{,}000$$

(a) How many clubs should be manufactured to minimize the cost?
(b) At this level of production, what is the cost per club?
(c) What are the fixed costs of production (the cost to produce 0 clubs)?

57. **Minimizing Cost** Scott–Jones Publishing Company has found that the cost C for paper, printing, and binding of x textbooks is given by the quadratic function

$$C(x) = 0.003x^2 - 30x + 111{,}800$$

(a) How many books should be manufactured for the cost to be a minimum?
(b) At this level of production, what is the cost per book?
(c) What is the cost to produce one book $(x = 1)$?

58. **Parabolic Arch Bridge** A horizontal bridge is in the shape of a parabolic arch. Given the information shown in the figure, what is the height h of the arch 2 feet from shore?

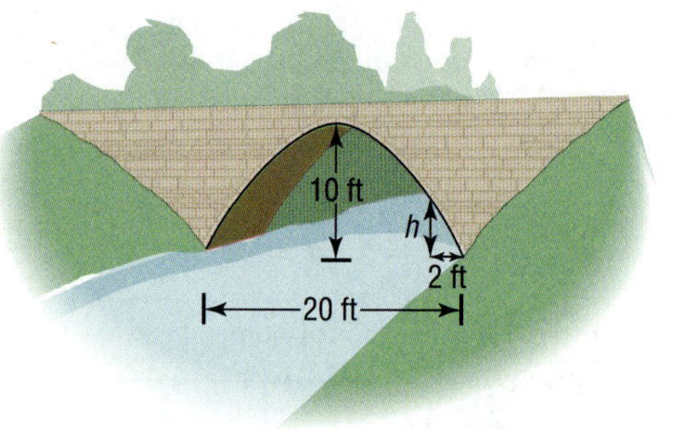

CHAPTER PROJECTS

1. **Analyzing a Stock** The beta, β, of a stock represents the relative risk of a stock compared with a market basket of stocks, such as Standard and Poor's 500 Index of stocks. Beta is computed by finding the slope of the line of best fit between the rate of return of the stock and rate of return of the S&P 500. The rates of return are computed on a weekly basis.

   (a) Find the weekly closing price of your favorite stock and the S&P 500 for 20 weeks. One good source is on the Internet at *http://finance.yahoo.com.*
   (b) Compute the rate of return by computing the weekly percentage change in the closing price of your stock and the weekly percentage change in the S&P 500 using the following formula:

$$\text{Weekly \% change} = \frac{P_2 - P_1}{P_1}$$

   where P_1 is last week's price and P_2 is this week's price.
   (c) Using a graphing utility, find the line of best fit treating the weekly rate of return of the S&P 500 as the independent variable and the weekly rate of return of your stock as dependent variable.
   (d) What is the beta of your stock?
   (e) Compare your result with that of the Value Line Investment Survey found in your library. What might account for any differences?

2. **CBL Experiment** Locate the motion detector on a Calculator Based Laboratory (CBL) or a Calculator Based Ranger (CBR) above a bouncing ball.

   (a) Plot the data collected in a scatter diagram with time as the independent variable.
   (b) Find the quadratic function of best fit for the second bounce.
   (c) Find the quadratic function of best fit for the third bounce.
   (d) Find the quadratic function of best fit for the fourth bounce.
   (e) Compute the maximum height for the second bounce.
   (f) Compute the maximum height for the third bounce.
   (g) Compute the maximum height for the fourth bounce.
   (h) Compute the ratio of the maximum height of the third bounce to the maximum height of the second bounce.
   (i) Compute the ratio of the maximum height of the fourth bounce to the maximum height of the third bounce.
   (j) Compare the results from parts (h) and (i). What do you conclude?

CHAPTER

3 Functions and Their Graphs

Wednesday February 10, 1999
***The Oregonian* "Ship awaits salvage effort"**

COOS BAY—Cleanup crews combed the oil-scarred south coast Tuesday as authorities raced to finish plans to refloat a 639-foot cargo ship mired for six days 150 yards off one of Oregon's most biologically rich beaches.

All day Tuesday, streaks of oil oozed from the cracked hull of the bulk cargo carrier *New Carissa* and spread over six miles of beach. Despite the breached hull, authorities believed they had a better chance of pulling the stricken ship out of beach sands than pumping nearly 400,000 gallons of oil off the ship in the winter surf.

See Chapter Project 1.

Preparing for This Chapter

Before getting started on this chapter, review the following concepts:

- **Slope of a Line (pp. 67–68)**
- **Intervals (pp. 54–55)**
- **Tests for Symmetry of an Equation (p. 23)**
- **Graphs of Certain Equations (Example 2, p. 16; Example 10, p. 23; Example 12, p. 25)**
- **Domain of a Function (pp. 99 and 104)**

Outline

3.1 CHARACTERISTICS OF FUNCTIONS; LIBRARY OF FUNCTIONS

1. Find the Average Rate of Change of a Function
2. Use a Graphing Utility to Locate Local Maxima and Minima
3. Use a Graphing Utility to Determine Where a Function Is Increasing and Decreasing
4. Determine Even or Odd Functions from a Graph
5. Identify Even or Odd Functions from the Equation
6. Graph Certain Important Functions
7. Graph Piecewise-defined Functions

1 Average Rate of Change

In Section 1.6 we said that the slope of a straight line could be interpreted as the average rate of change. Often, we are interested in the rate at which functions change. To find the average rate of change of a function between any two points on its graph, we calculate the slope of the line containing the two points.

EXAMPLE 1 Finding the Average Rate of Change of a Function

The data in Table 1 represent the average cost of tuition and required fees (in dollars) at public four-year colleges.

Table 1

Year	Average Cost of Tuition and Fees ($)
1991	2159
1992	2410
1993	2604
1994	2820
1995	2977
1996	3151
1997	3321

Source: U.S. Center for National Education Statistics

(a) Draw a scatter diagram of the data, treating the year as the independent variable.

(b) Draw a line through the points (1991, 2159) and (1992, 2410) on the scatter diagram found in part (a).

(c) Find the average rate of change of the cost of tuition and required fees from 1991 to 1992.

(d) Draw a line through the points (1995, 2977) and (1996, 3151) on the scatter diagram found in part (a).

(e) Find the average rate of change of the cost of tuition and required fees from 1995 to 1996.

(f) What is happening to the average rate of change as time passes?

Solution

(a) We plot the ordered pairs (1991, 2159), (1992, 2410), and so on, using rectangular coordinates. See Figure 1.

(b) Draw the line through (1991, 2159) and (1992, 2410). See Figure 2.

(c) The average rate of change is found by computing the slope of the line containing the points (1991, 2159) and (1992, 2410).

$$\text{Average rate of change} = \frac{2410 - 2159}{1992 - 1991} = \frac{251}{1} = \$251/\text{year}$$

The average rate of change in the cost of tuition from 1991 to 1992 is $251.

(d) Draw the line through (1995, 2977) and (1996, 3151). See Figure 3.

(e) The average rate of change is found by computing the slope of the line containing the points (1995, 2977) and (1996, 3151).

$$\text{Average rate of change} = \frac{3151 - 2977}{1996 - 1995} = \frac{174}{1} = \$174/\text{year}$$

The average rate of change in the cost of tuition from 1995 to 1996 is $174.

(f) The cost of tuition and fees is increasing over time, but the rate of the increase is declining.

Figure 1

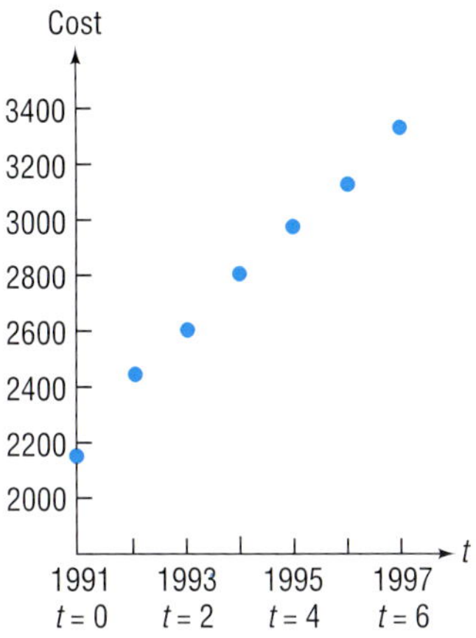

Figure 2

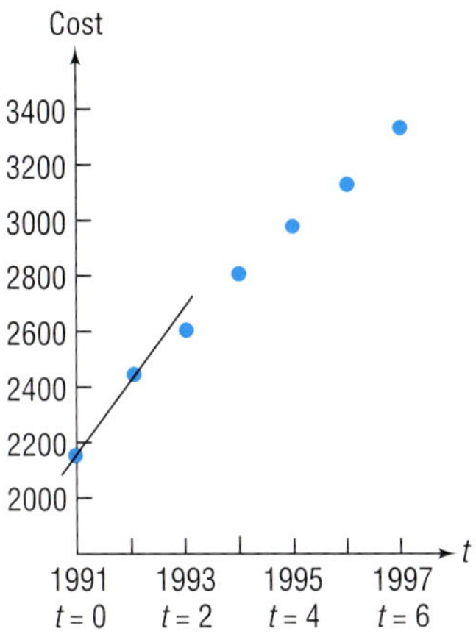

Figure 3

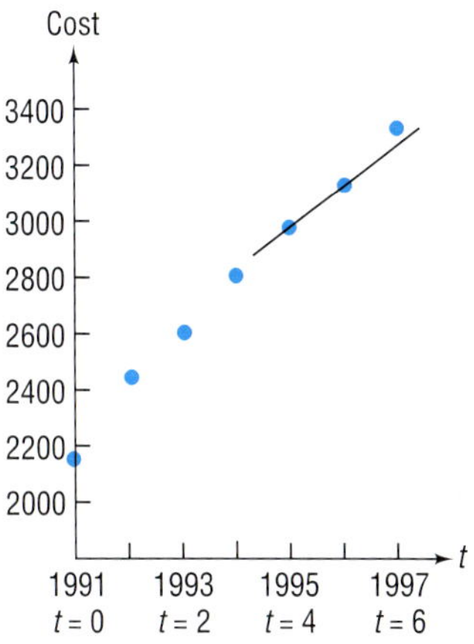

If we know the function C that relates the year to the cost C of tuition and fees, then the average rate of change from 1991 to 1992 may be expressed as

$$\text{Average rate of change} = \frac{C(1992) - C(1991)}{1992 - 1991} = \frac{251}{1} = \$251/\text{year}$$

Expressions like this occur frequently in calculus.

If c is in the domain of a function $y = f(x)$, the **average rate of change of f** from c to x is defined as

$$\text{Average rate of change} = \frac{\Delta y}{\Delta x} = \frac{f(x) - f(c)}{x - c} \qquad x \neq c \qquad (1)$$

This expression is also called the **difference quotient** of f at c.

The average rate of change of a function has an important geometric interpretation. Look at the graph of $y = f(x)$ in Figure 4. We have labeled two points on the graph: $(c, f(c))$ and $(x, f(x))$. The slope of the line containing these two points is

$$\frac{f(x) - f(c)}{x - c}$$

This line is called a **secant line.**

Figure 4

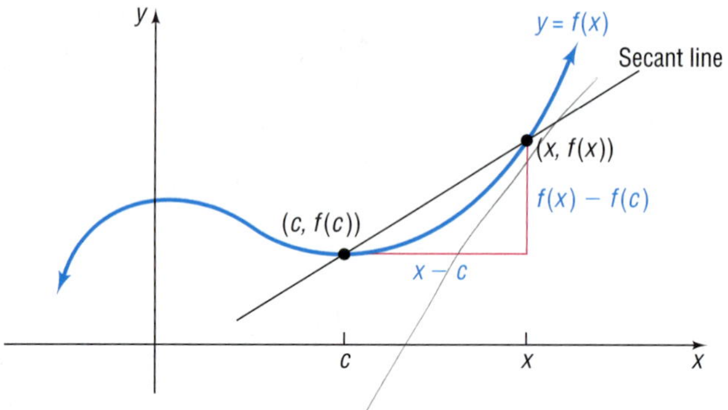

THEOREM **Slope of Secant Line**

The average rate of change of a function equals the slope of the secant line containing two points on its graph.

EXAMPLE 2 Finding the Average Rate of Change of a Function

The quadratic function of best fit for the data in Table 1 where $t = 0$ represents 1991, $t = 1$ represents 1992, and so on, is

$$C(t) = -7.42t^2 + 235.25t + 2168.10$$

(a) Find the average rate of change in tuition and fees from 1991 ($t = 0$) to 1992 ($t = 1$).

(b) Find the average rate of change in tuition and fees from 1991 to t.

Solution (a) For 1991, $t = 0$, and for 1992, $t = 1$. Using expression (1), we find that

$$\text{Average rate of change} = \frac{C(1) - C(0)}{1 - 0}$$

Now

$$C(0) = -7.42(0)^2 + 235.25(0) + 2168.10 = \$2168.10$$

and

$$C(1) = -7.42(1)^2 + 235.25(1) + 2168.10 = -7.42 + 235.25 + 2168.10 = \$2395.93$$

The average rate of change from 1991 to 1992 is

$$\text{Average rate of change} = \frac{C(1) - C(0)}{1 - 0} = \frac{2395.93 - 2168.10}{1} = \$227.83/\text{year}$$

(b) The average rate of change from 1991 $(t = 0)$ to t is

$$\text{Average rate of change} = \frac{C(t) - C(0)}{t - 0} = \frac{(-7.42t^2 + 235.25t + 2168.10) - 2168.10}{t}$$

$$= \frac{-7.42t^2 + 235.25t}{t} = \frac{t(-7.42t + 235.25)}{t} = -7.42t + 235.25 \text{ dollars/year}$$

Using the answer to part (b), we can verify the answer to part (a). Using $t = 1$, we find that

$$\text{Average rate of change} = -7.42(1) + 235.25 = \$227.83/\text{year}$$

NOW WORK PROBLEM **31.**

Increasing and Decreasing Functions

Consider the graph given in Figure 5. If you look from left to right along the graph of the function, you will notice that parts of the graph are rising, parts are falling, and parts are horizontal. In such cases, the function is described as *increasing, decreasing,* and *constant,* respectively.

Figure 5

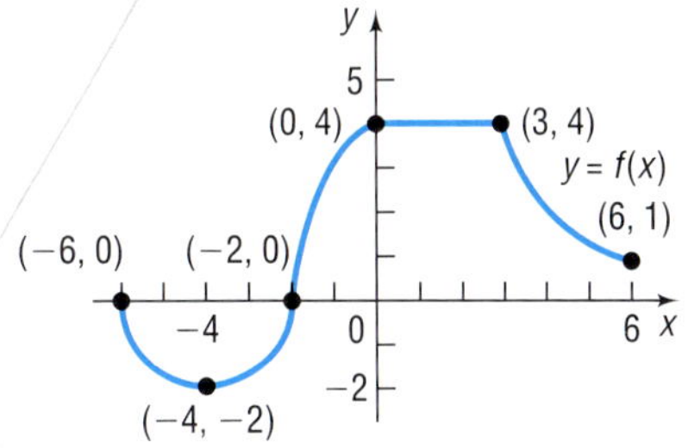

EXAMPLE 3 Determining Where a Function Is Increasing, Decreasing, or Constant

Where is the function in Figure 5 increasing? Where is it decreasing? Where is it constant?

Solution To answer the question of where a function is increasing, where it is decreasing, and where it is constant, we use nonstrict inequalities involving the independent variable x, or we use open intervals* of x-coordinates. The graph in Figure 5 is rising (increasing) from the point $(-4, -2)$ to the point $(0, 4)$, so we conclude that it is increasing on the open interval $(-4, 0)$ (or for $-4 < x < 0$). The graph is falling (decreasing) from the point $(-6, 0)$ to the point $(-4, -2)$ and from the point $(3, 4)$ to the point $(6, 1)$. We conclude that the graph is decreasing on the open intervals $(-6, -4)$ and $(3, 6)$ (or for $-6 < x < -4$ and $3 < x < 6$). The graph is constant on the open interval $(0, 3)$ (or for $0 < x < 3$).

More precise definitions follow:

A function f is **increasing** on an open interval I if, for any choice of x_1 and x_2 in I, with $x_1 < x_2$, we have $f(x_1) < f(x_2)$.

*The open interval (a, b) consists of all real numbers x for which $a < x < b$. Refer to Section 1.5, if necessary.

A function f is **decreasing** on an open interval I if, for any choice of x_1 and x_2 in I, with $x_1 < x_2$, we have $f(x_1) > f(x_2)$.

A function f is **constant** on an open interval I if, for all choices of x in I, the values $f(x)$ are equal.

Thus, the graph of an increasing function goes up from left to right, the graph of a decreasing function goes down from left to right, and the graph of a constant function remains at a fixed height. Figure 6 illustrates the definitions.

Figure 6

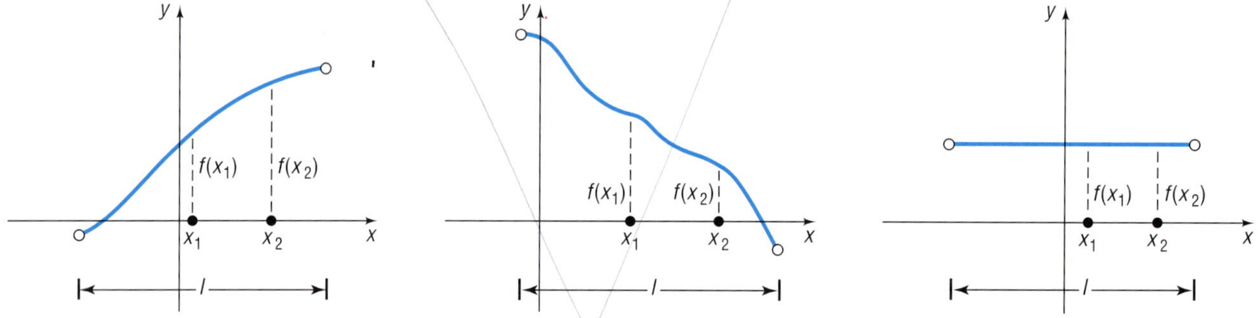

(a) For $x_1 < x_2$ in I, $f(x_1) < f(x_2)$; f is increasing on I

(b) For $x_1 < x_2$ in I, $f(x_1) > f(x_2)$; f is decreasing on I

(c) For all x in I, the values of f are equal; f is constant on I

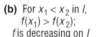

NOW WORK PROBLEM 9(b).

Local Maximum; Local Minimum

2 When the graph of a function is increasing to the left of $x = c$ and decreasing to the right of $x = c$, then at c the value of f is largest. This value is called a *local maximum* of f.

When the graph of a function is decreasing to the left of $x = c$ and is increasing to the right of $x = c$, then at c the value of f is the smallest. This value is called a *local minimum* of f.

Thus, if f has a local maximum at c, then the value of f at c is greater than the values of f near c. If f has a local minimum at c, then the value of f at c is less than the value of f near c. The word *local* is used to suggest that it is only near c that the value $f(c)$ is largest or smallest. See Figure 7

Figure 7

f has a local maximum at x_1 and x_3; f has a local minimum at x_2.

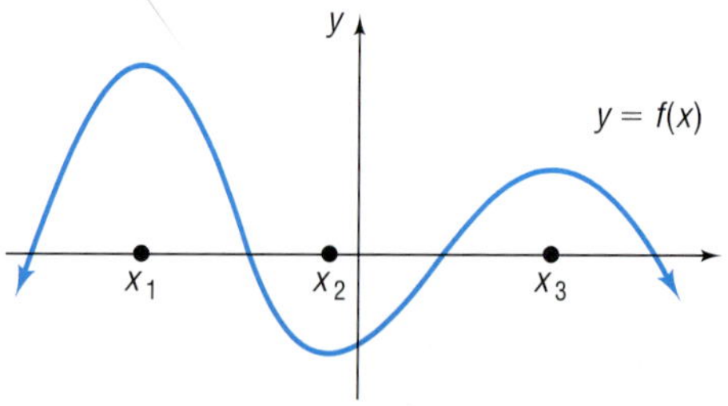

A function f has a **local maximum at c** if there is an interval I containing c so that, for all $x \neq c$ in I, $f(x) < f(c)$. We call $f(c)$ a **local maximum of f.**

A function f has a **local minimum at c** if there is an interval I containing c so that, for all $x \neq c$ in I, $f(x) > f(c)$. We call $f(c)$ a **local minimum of f.**

To locate the exact value at which a function f has a local maximum or a local minimum usually requires calculus. However, a graphing utility may be used to approximate these values by using the MAXIMUM and MINIMUM features.*

EXAMPLE 4 Using a Graphing Utility to Locate Local Maxima and Minima and to Determine Where a Function Is Increasing and Decreasing

(a) Use a graphing utility to graph $f(x) = 6x^3 - 12x + 5$ for $-2 < x < 2$. Determine where f has a local maximum and where f has a local minimum.

(b) Determine where f is increasing and where it is decreasing.

Solution (a) Graphing utilities have a feature that finds the maximum or minimum point of a graph within a given interval. Graph the function f for $-2 < x < 2$. Using MAXIMUM, we find that the local maximum is 11.53 and it occurs at $x = -0.82$, rounded to two decimal places. Using MINIMUM, we find that the local minimum is -1.53 and it occurs at $x = 0.82$, rounded to two decimal places. See Figures 8(a) and (b).

Figure 8

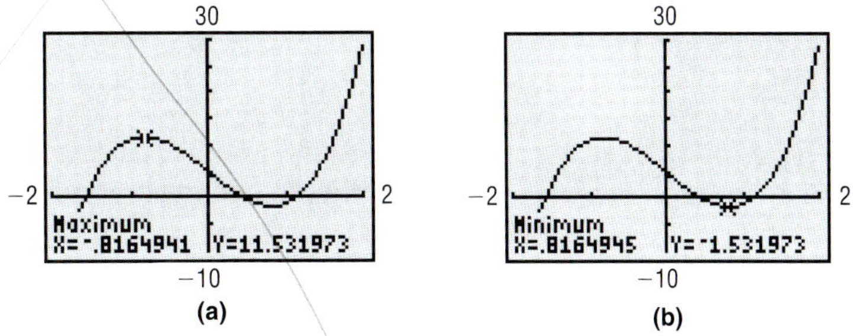

(a) (b)

3 (b) Looking at Figures 8(a) and (b), we see that the graph of f is rising (increasing) from $x = -2$ to $x = -0.82$ and from $x = 0.82$ to $x = 2$, so f is increasing on the open intervals $(-2, -0.82)$ and $(0.82, 2)$ (or for $-2 < x < -0.82$ and $0.82 < x < 2$). The graph is falling (decreasing) from $x = -0.82$ to $x = 0.82$, so f is decreasing on the open interval $(-0.82, 0.82)$ (or for $-0.82 < x < 0.82$).

NOW WORK PROBLEM 67.

*Consult your owner's manual for the appropriate keystrokes.

Even and Odd Functions

4 A function f is even if and only if whenever the point (x, y) is on the graph of f then the point $(-x, y)$ is also on the graph. Algebraically, we define an even function as follows:

A function f is **even** if for every number x in its domain the number $-x$ is also in the domain and

$$f(-x) = f(x)$$

A function f is odd if and only if whenever the point (x, y) is on the graph of f then the point $(-x, -y)$ is also on the graph. Algebraically, we define an odd function as follows:

A function f is **odd** if for every number x in its domain the number $-x$ is also in the domain and

$$f(-x) = -f(x)$$

Refer to Section 1.2, where the tests for symmetry are listed. The following results are then evident.

THEOREM A function is even if and only if its graph is symmetric with respect to the y-axis. A function is odd if and only if its graph is symmetric with respect to the origin.

EXAMPLE 5 Determining Even and Odd Functions from the Graph

Determine whether each graph given in Figure 9 is the graph of an even function, an odd function, or a function that is neither even nor odd.

Figure 9

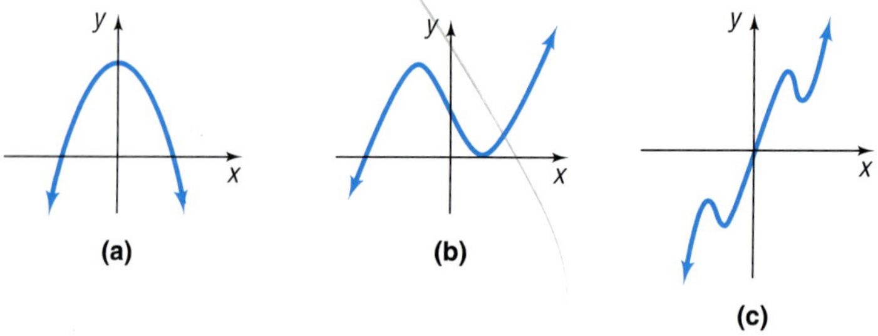

Solution The graph in Figure 9(a) is that of an even function, because the graph is symmetric with respect to the y-axis. The function whose graph is given in Figure 9(b) is neither even nor odd, because the graph is neither symmetric with respect to the y-axis nor symmetric with respect to the origin. The function whose graph is given in Figure 9(c) is odd, because its graph is symmetric with respect to the origin.

NOW WORK PROBLEMS 9(c) AND 13(c).

A graphing utility can be used to conjecture whether a function is even, odd, or neither. As stated, when the graph of an even function contains the point (x, y), it must also contain the point $(-x, y)$. Therefore, if TRACE indicates that both the point (x, y) and the point $(-x, y)$ are on the graph for every x, then we would conjecture that the function is even.*

In addition, the graph of an odd function contains the points $(-x, -y)$ and (x, y). TRACE could be used in the same way to conjecture that the function is odd.*

5 In the next example, we use a graphing utility to conjecture whether a given function is even, odd, or neither. Then we verify our conclusion algebraically.

EXAMPLE 6 Identifying Even and Odd Functions

Determine whether each of the following functions is even, odd, or neither. Then determine whether the graph is symmetric with respect to the y-axis or with respect to the origin.

(a) $f(x) = x^2 - 5$ (b) $g(x) = x^3 - 1$

(c) $h(x) = 5x^3 - x$ (d) $F(x) = |x|$

Solution (a) Graph the function. Use TRACE to determine different pairs of points (x, y) and $(-x, y)$. For example, the point $(1.957447, -1.168402)$ is on the graph. See Figure 10(a). Now move the cursor to -1.957447 and determine the corresponding y-coordinate. See Figure 10(b). Since $(1.957447, -1.168402)$ and $(-1.957447, -1.168402)$ both lie on the graph, we have evidence that the function is even. Repeating this procedure for additional ordered pairs yields similar results. Therefore, we conjecture that the function is even.

Figure 10

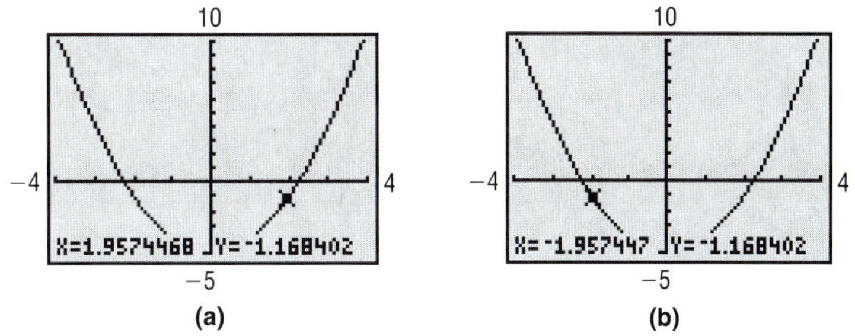

To algebraically verify the conjecture, we replace x by $-x$ in $f(x) = x^2 - 5$. Then

$$f(-x) = (-x)^2 - 5 = x^2 - 5 = f(x)$$

*– Xmin and Xmax must be equal for this to work.

Since $f(-x) = f(x)$, we conclude that f is an even function, and the graph is symmetric with respect to the y-axis.

(b) Graph the function. Using TRACE, the point (1.787234, 4.7087929) is on the graph. See Figure 11(a). Now move the cursor to -1.787234 and determine the corresponding y-coordinate, -6.708793. See Figure 11(b). We conjecture that the function is neither even nor odd since (x, y) does not equal $(-x, y)$ (so it is not even) and (x, y) also does not equal $(-x, -y)$ (so it is not odd).

Figure 11

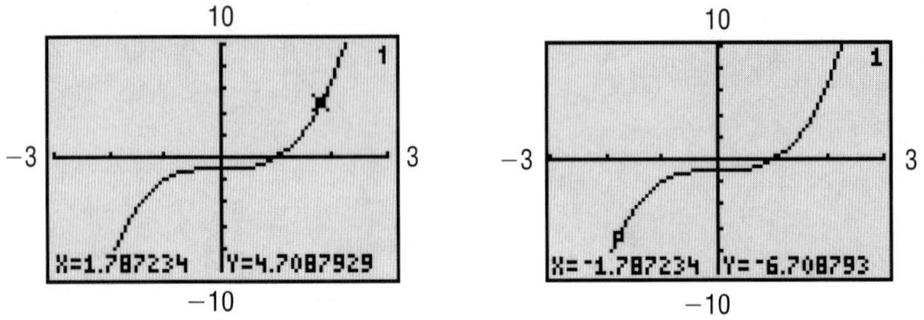

(a) (b)

To algebraically verify the conjecture, we replace x by $-x$. Then

$$g(-x) = (-x)^3 - 1 = -x^3 - 1$$

Since $g(-x) \neq g(x)$ and $g(-x) \neq -g(x) = -(x^3 - 1) = -x^3 + 1$, we conclude that g is neither even nor odd. The graph is not symmetric with respect to the y-axis nor with respect to the origin.

(c) Graph the function. Using TRACE, the point (1.0212766, 4.3047109) is on the graph. See Figure 12(a). We move the cursor to -1.021277 and determine the corresponding y-coordinate, -4.304711. See Figure 12(b). Since (x, y) equals $(-x, -y)$, correct to five decimal places, we have evidence that the function is odd. An examination of other pairs of points leads to the conjecture that the function is odd.

Figure 12

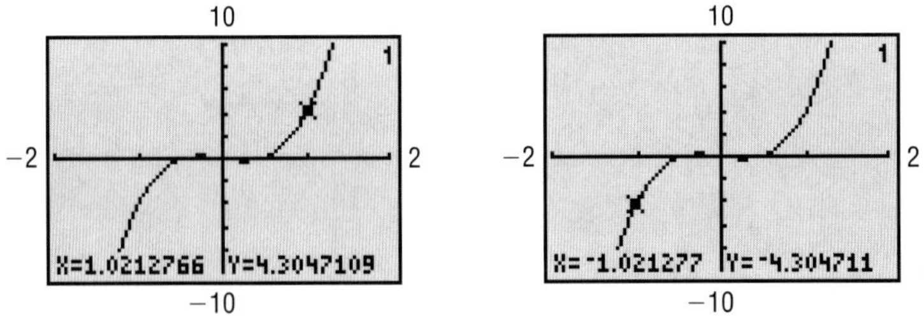

(a) (b)

To algebraically verify the conjecture, we replace x by $-x$ in $h(x) = 5x^3 - x$. Then

$$h(-x) = 5(-x)^3 - (-x) = -5x^3 + x = -(5x^3 - x) = -h(x)$$

Since $h(-x) = -h(x)$, h is an odd function, and the graph of h is symmetric with respect to the origin.

(d) Graph the function. Using TRACE, the point $(-5.744681, 5.7446809)$ is on the graph. See Figure 13(a). We move the cursor to 5.7446809 and determine the corresponding y-coordinate, 5.7446809. See Figure 13(b). Since (x, y) equals $(-x, y)$, we have evidence that the function is even. Repeating this procedure yields similar results. We conjecture that the function is even.

Figure 13

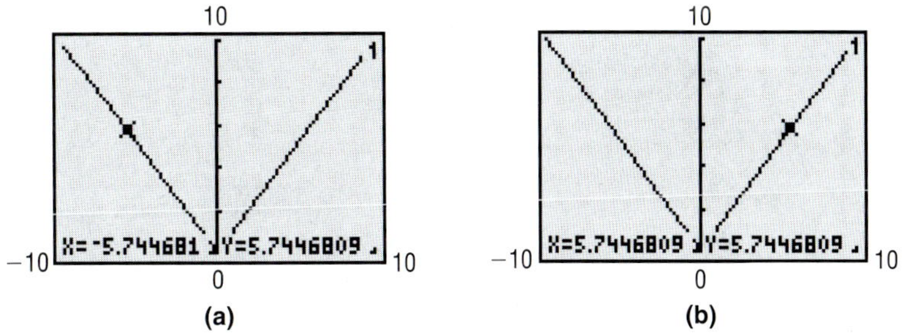

To algebraically verify the conjecture, we replace x by $-x$ in $F(x) = |x|$. Then

$$F(-x) = |-x| = |-1| \cdot |x| = |x| = F(x)$$

Since $F(-x) = F(x)$, F is an even function, and the graph of F is symmetric with respect to the y-axis.

NOW WORK PROBLEM 41.

Library of Functions

6 We now give names to some of the functions that we have encountered. In going through this list, pay special attention to the characteristics of each function, particularly to the shape of each graph. Knowing these graphs will lay the foundation for later graphing techniques.

The first function in this list is one that we discussed in Chapter 2.

Linear Functions

$f(x) = mx + b$	m and b are real numbers

The domain of a **linear function** f consists of all real numbers. The graph of this function is a nonvertical line with slope m and y-intercept b. A linear function is increasing if $m > 0$, decreasing if $m < 0$, and constant if $m = 0$.

Constant Function

$f(x) = b$	b is a real number

Figure 14

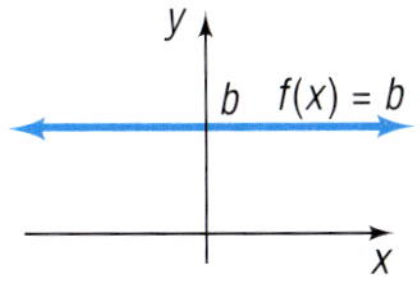

See Figure 14.

A **constant function** is a special linear function ($m = 0$). Its domain is the set of all real numbers; its range is the set consisting of a single number b. Its graph is a horizontal line whose y-intercept is b. The constant function is an even function whose graph is constant over its domain.

Identity Function

$$f(x) = x$$

See Figure 15.

Figure 15

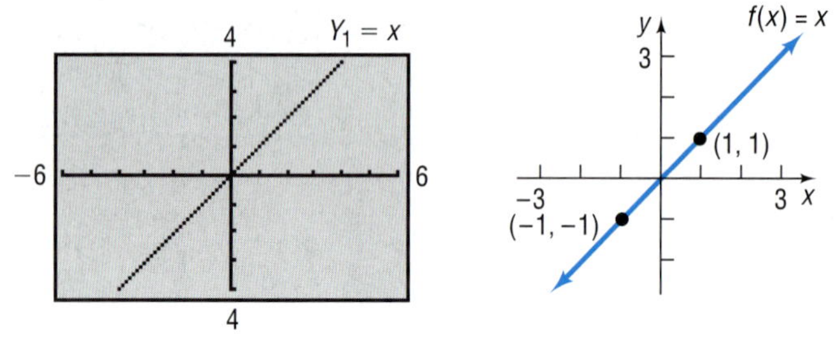

The **identity function** is also a special linear function. Its domain and range are the set of all real numbers. Its graph is a line whose slope is $m = 1$ and whose y-intercept is 0. The line consists of all points for which the x-coordinate equals the y-coordinate. The identity function is an odd function that is increasing over its domain. Note that the graph bisects quadrants I and III.

Square Function

$$f(x) = x^2$$

See Figure 16.

Figure 16

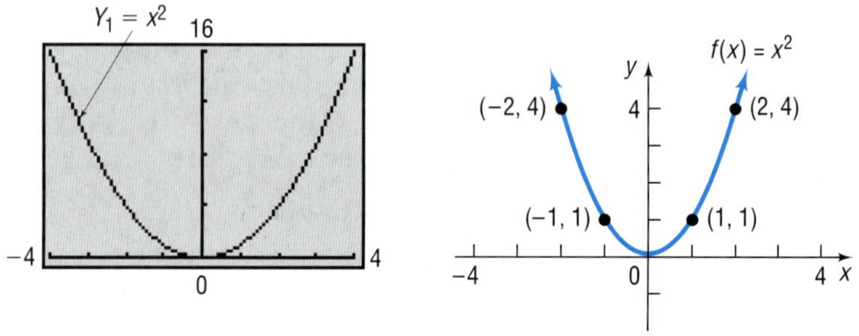

The domain of the **square function** f is the set of all real numbers; its range is the set of nonnegative real numbers. The graph of this function is a parabola, whose vertex is at $(0, 0)$. The square function is an even function that is decreasing on the interval $(-\infty, 0)$ and increasing on the interval $(0, \infty)$.

Cube Function

$$f(x) = x^3$$

See Figure 17.

Figure 17

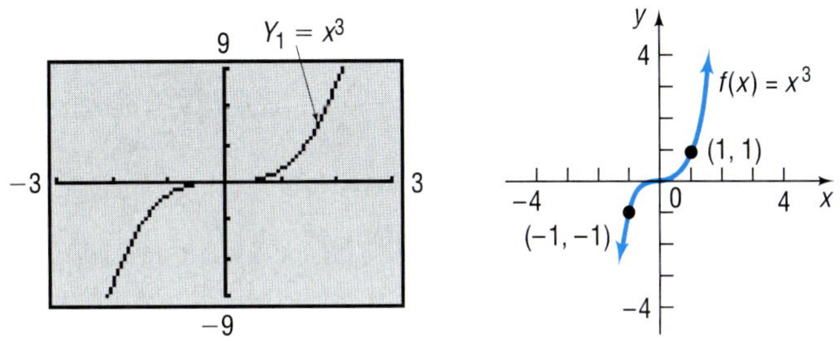

The domain and range of the **cube function** are the set of all real numbers. The intercept of the graph is at $(0, 0)$. The cube function is odd and is increasing on the interval $(-\infty, \infty)$.

Square Root Function

$$f(x) = \sqrt{x}$$

See Figure 18.

Figure 18

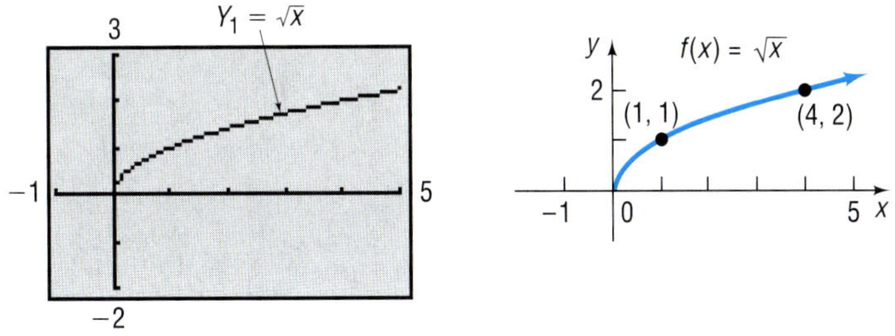

The domain and range of the **square root function** are the set of nonnegative real numbers. The intercept of the graph is at $(0, 0)$. The square root function is neither even nor odd and is increasing on the interval $(0, \infty)$.

Reciprocal Function

$$f(x) = \frac{1}{x}$$

Refer to Example 12, p. 25, for a discussion of the equation $y = 1/x$. See Figure 19.

Figure 19

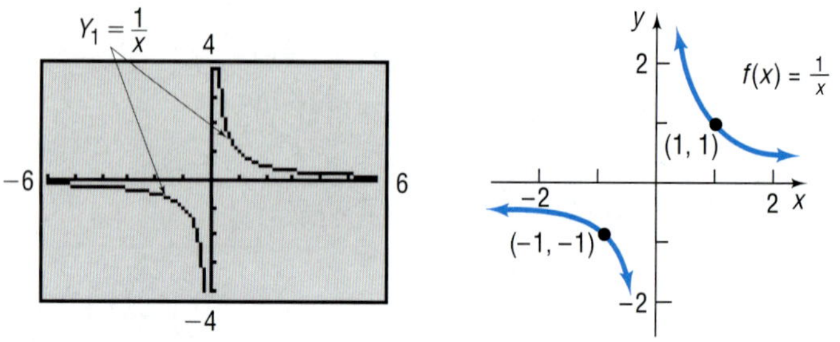

The domain and range of the **reciprocal function** are the set of all nonzero real numbers. The graph has no intercepts. The reciprocal function is decreasing on the intervals $(-\infty, 0)$ and $(0, \infty)$ and is an odd function.

Absolute Value Function

$$f(x) = |x|$$

See Figure 20.

Figure 20

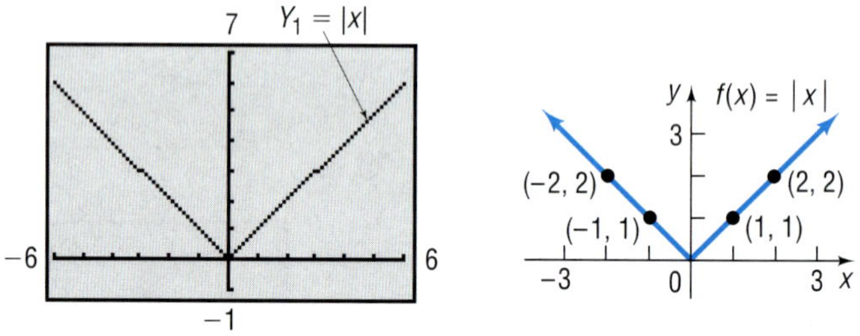

The domain of the **absolute value function** is the set of all real numbers; its range is the set of nonnegative real numbers. The intercept of the graph is at (0, 0). If $x \geq 0$, then $f(x) = x$, and the graph of f is part of the line $y = x$; if $x < 0$, then $f(x) = -x$, and the graph of f is part of the line $y = -x$. The absolute value function is an even function; it is decreasing on the interval $(-\infty, 0)$ and increasing on the interval $(0, \infty)$.

Comment: If your utility has no built-in absolute value function, you can still graph $f(x) = |x|$ by using the fact that $|x| = \sqrt{(x^2)}$.

The notation int(x) stands for the largest integer less than or equal to x. For example,

$$\text{int}(1) = 1 \qquad \text{int}(2.5) = 2 \qquad \text{int}(\tfrac{1}{2}) = 0 \qquad \text{int}(\tfrac{-3}{4}) = -1 \qquad \text{int}(\pi) = 3$$

This type of correspondence occurs frequently enough in mathematics that we give it a name.

Greatest-Integer Function

$$f(x) = \text{int}(x) = \text{greatest integer less than or equal to } x$$

We obtain the graph of $f(x) = \text{int}(x)$ by plotting several points. See Table 2. For values of x, $-1 \leq x < 0$, the value of $f(x) = \text{int}(x)$ is -1; for values of x, $0 \leq x < 1$, the value of f is 0. See Figure 21 for the graph.

Table 2

X	Y1
-1	-1
-.75	-1
-.5	-1
-.25	-1
0	0
.25	0
.5	0

Y1=int(X)

Figure 21

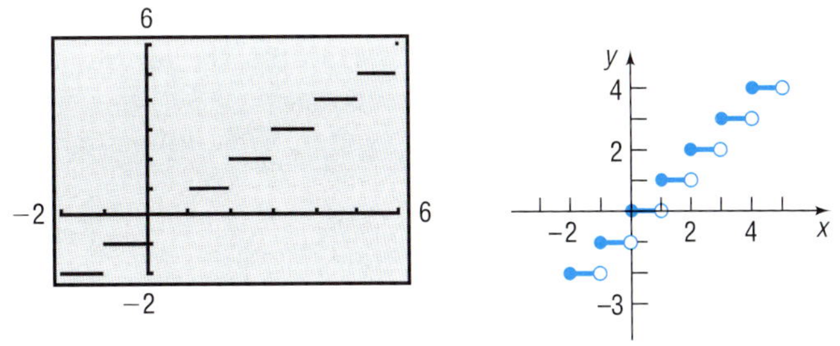

(a) (b)

The domain of the **greatest-integer function** is the set of all real numbers; its range is the set of integers. The y-intercept of the graph is at 0. The x-intercepts lie in the interval $[0, 1)$. The greatest-integer function is neither even nor odd. It is constant on every interval of the form $[k, k + 1)$, for k an integer. In Figure 21(b), we use a solid dot to indicate, for example, that at $x = 1$ the value of f is $f(1) = 1$; we use an open circle to illustrate, for example, that the function does not assume the value of 0 at $x = 1$.

From the graph of the greatest-integer function, we can see why it is also called a **step function.** At $x = 0, x = \pm 1, x = \pm 2$, and so on, this function exhibits what is called a *discontinuity;* that is, at integer values, the graph suddenly "steps" from one value to another without taking on any of the intermediate values. For example, to the immediate left of $x = 3$, the y-coordinates are 2, and to the immediate right of $x = 3$, the y-coordinates are 3.

Comment: When graphing a function, you can choose either the **connected mode,** in which points plotted on the screen are connected, making the graph appear without any breaks, or the **dot mode,** in which only the points plotted appear. When graphing the greatest-integer function with a graphing utility, it is necessary to be in the **dot mode.** This is to prevent the utility from "connecting the dots" when $f(x)$ changes from one integer value to the next. See Figure 22.

Figure 22

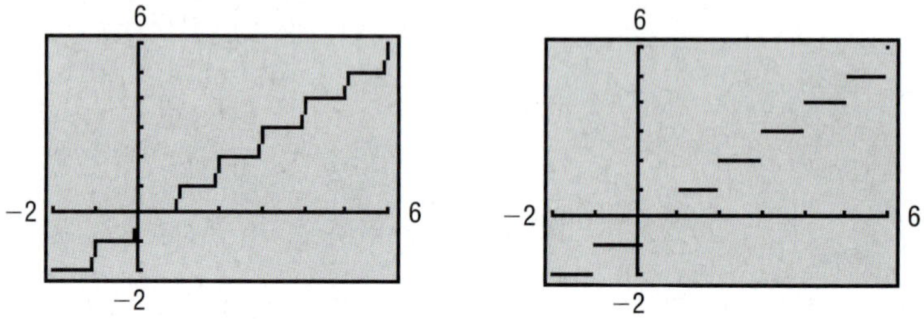

(a) Connected mode

(b) Dot mode

The functions that we have discussed so far are basic. Whenever you encounter one of them, you should see a mental picture of its graph. For example, if you encounter the function $f(x) = x^2$, you should see in your mind's eye a picture like Figure 16.

NOW WORK PROBLEMS 1–8.

Piecewise-defined Functions

7 Sometimes, a function is defined differently on different parts of its domain. For example, the absolute value function $f(x) = |x|$ is actually defined by two equations: $f(x) = x$ if $x \geq 0$ and $f(x) = -x$ if $x < 0$. For convenience, we generally combine these equations into one expression as

$$f(x) = |x| = \begin{cases} x & \text{if } x \geq 0 \\ -x & \text{if } x < 0 \end{cases}$$

When functions are defined by more than one equation, they are called **piecewise-defined** functions.

Let's look at another example of a piecewise-defined function.

EXAMPLE 7 Analyzing a Piecewise-defined Function

For the following function f,

$$f(x) = \begin{cases} -x + 1 & \text{if } -1 \leq x < 1 \\ 2 & \text{if } x = 1 \\ x^2 & \text{if } x > 1 \end{cases}$$

(a) Find $f(0)$, $f(1)$, and $f(2)$.

(b) Determine the domain of f.

(c) Graph f.

(d) Use the graph to find the range of f.

Solution (a) To find $f(0)$, we observe that when $x = 0$ the equation for f is given by $f(x) = -x + 1$. So we have

$$f(0) = -0 + 1 = 1$$

When $x = 1$, the equation for f is $f(x) = 2$. Thus,

$$f(1) = 2$$

When $x = 2$, the equation for f is $f(x) = x^2$. So

$$f(2) = 2^2 = 4$$

(b) To find the domain of f, we look at its definition. We conclude that the domain of f is $\{x | x \geq -1\}$, or $[-1, \infty)$.

(c) On a graphing utility, the procedure for graphing a piecewise-defined function varies depending on the particular utility. In general, you need to enter each piece as a function with a restricted domain. See Figure 23(a).

In graphing piecewise-defined functions, it is usually better to be in the dot mode, since such functions may have breaks (discontinuities).

To graph f on paper, we graph "each piece." Thus, we first graph the line $y = -x + 1$ and keep only the part for which $-1 \leq x < 1$. Then we plot the point $(1, 2)$, because when $x = 1$, $f(x) = 2$. Finally, we graph the parabola $y = x^2$ and keep only the part for which $x > 1$. See Figure 23(b).

Figure 23

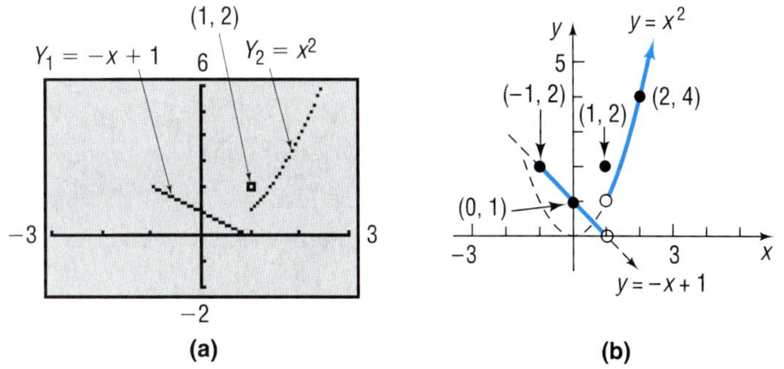

(d) From the graph, we conclude that the range of f is $\{y | y > 0\}$, or $(0, \infty)$.

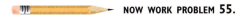

NOW WORK PROBLEM 55.

EXAMPLE 8 Cost of Electricity

In the winter, Commonwealth Edison Company supplies electricity to residences for a monthly customer charge of \$8.91 plus 10.494¢ per kilowatt-hour (kWhr) for the first 400 kWhr supplied in the month and 7.91¢ per kWhr for all usage over 400 kWhr in the month.*

(a) What is the charge for using 300 kWhr in a month?

(b) What is the charge for using 700 kWhr in a month?

(c) If C is the monthly charge for x kWhr, express C as a function of x.

Solution (a) For 300 kWhr, the charge is \$8.91 plus 10.494¢ = \$0.10494 per kWhr. Thus,

$$\text{Charge} = \$8.91 + \$0.10494(300) = \$40.39$$

*Source: Commonwealth Edison Co., Chicago, Illinois, 1997.

(b) For 700 kWhr, the charge is \$8.91 plus 10.494¢ per kWhr for the first 400 kWhr plus 7.91¢ per kWhr for the 300 kWhr in excess of 400. Thus,

$$\text{Charge} = \$8.91 + \$0.10494(400) + \$0.0791(300) = \$74.62$$

(c) If $0 \le x \le 400$, the monthly charge C (in dollars) can be found by multiplying x times \$0.10494 and adding the monthly customer charge of \$8.91. Thus, if $0 \le x \le 400$, then $C(x) = 0.10494x + 8.91$. For $x > 400$, the charge is $0.10494(400) + 8.91 + 0.0791(x - 400)$, since $x - 400$ equals the usage in excess of 400 kWhr, which costs \$0.0791 per kWhr. Thus, if $x > 400$, then

$$\begin{aligned} C(x) &= 0.10494(400) + 8.91 + 0.0791(x - 400) \\ &= 50.89 + 0.0791(x - 400) \\ &= 0.0791x + 19.25 \end{aligned}$$

The rule for computing C follows two equations:

$$C(x) = \begin{cases} 0.10494x + 8.91 & \text{if } 0 \le x \le 400 \\ 0.0791x + 19.25 & \text{if } x > 400 \end{cases}$$

See Figure 24 for the graph.

Figure 24

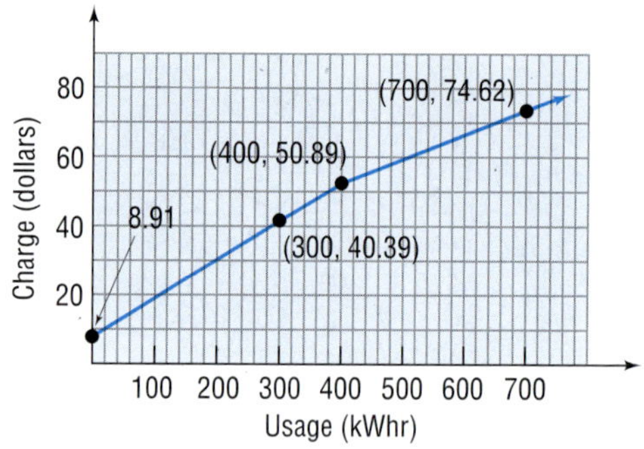

3.1 EXERCISES

In Problems 1–8, match each graph to the function listed whose graph most resembles the one given.

A. Constant function *B. Linear function*
C. Square function *D. Cube function*
E. Square root function *F. Reciprocal function*
G. Absolute value function *H. Greatest-integer function*

1. 

2. 

3. 

4. 

5. 

6. 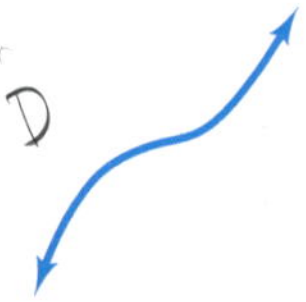

7. 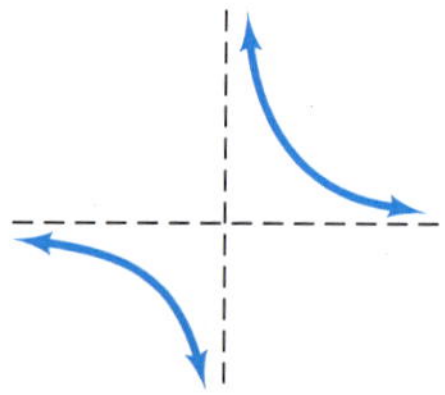

8. 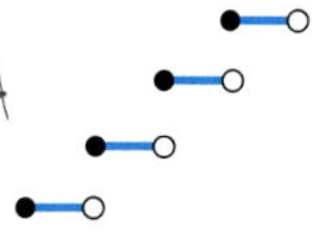

In Problems 9–24, the graph of a function is given. Use the graph to find

(a) Its domain and range

(b) The intervals on which it is increasing, decreasing, or constant

(c) Whether it is even, odd, or neither

(d) The intercepts, if any

9.

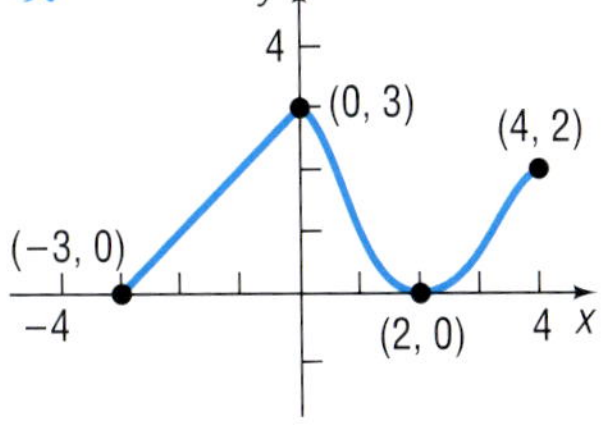

10.

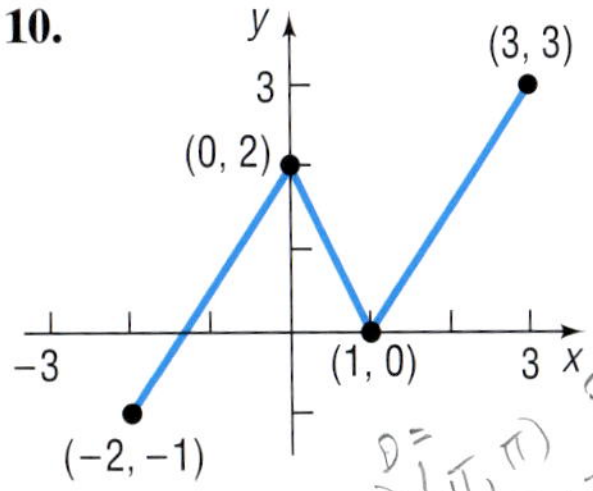

11.

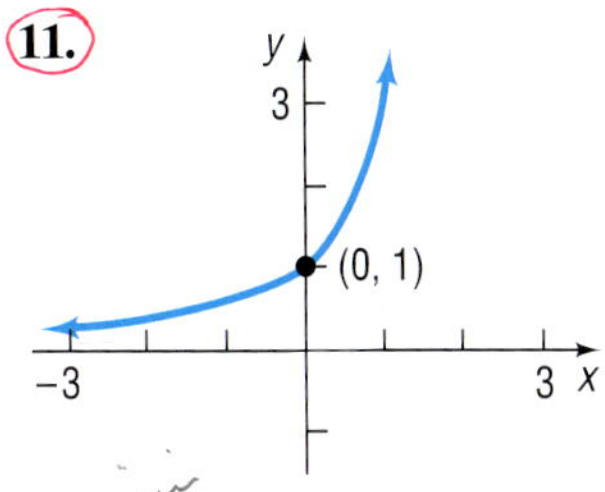

12.

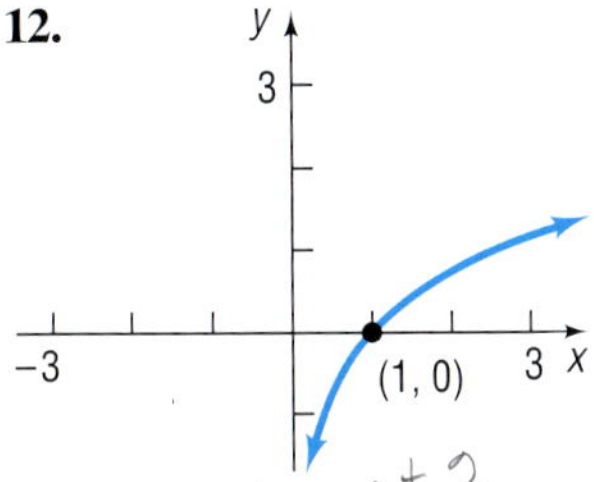

13.

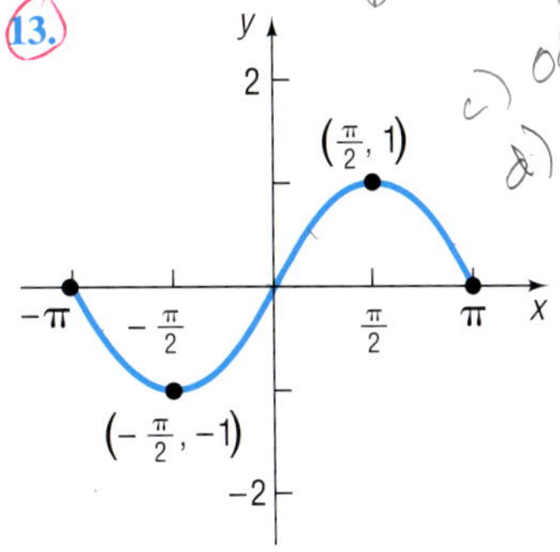

14.

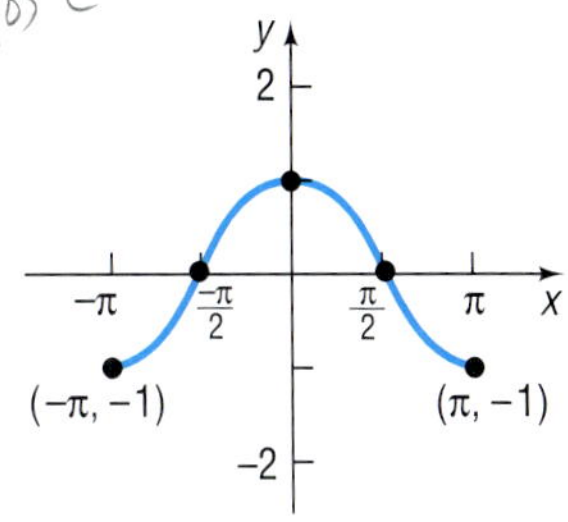

15.

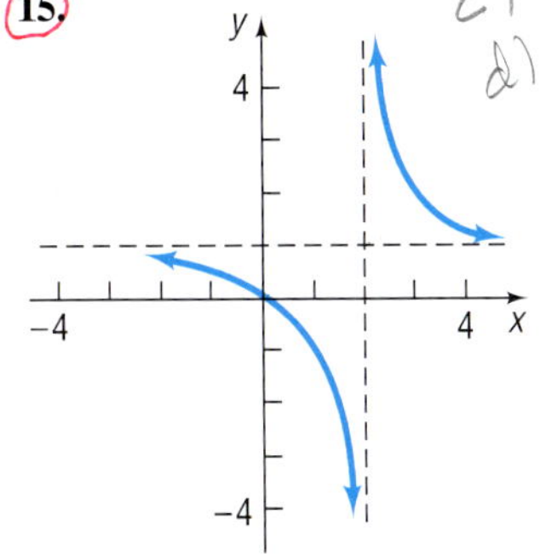

16.

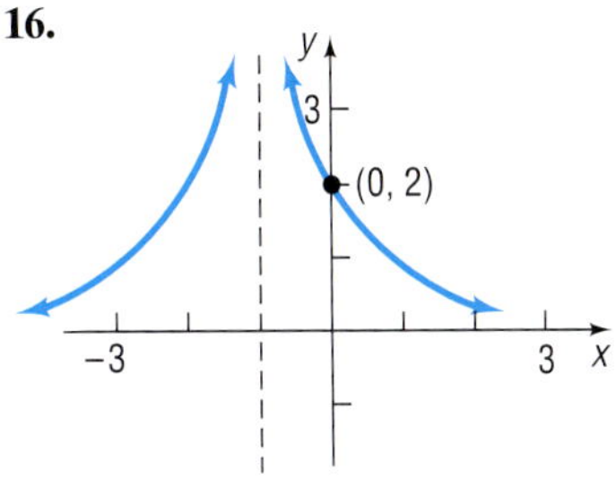

17.

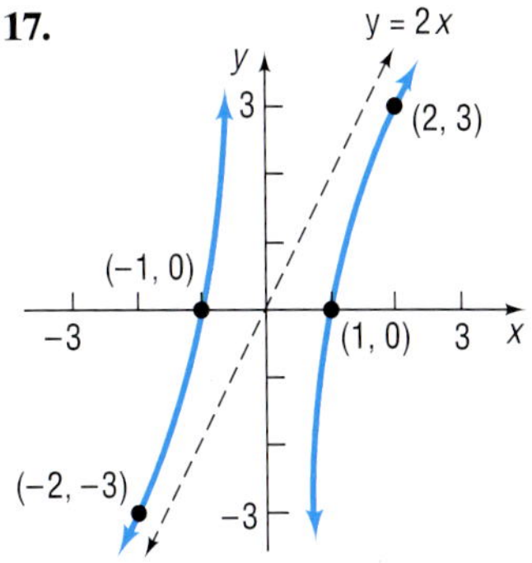

18.

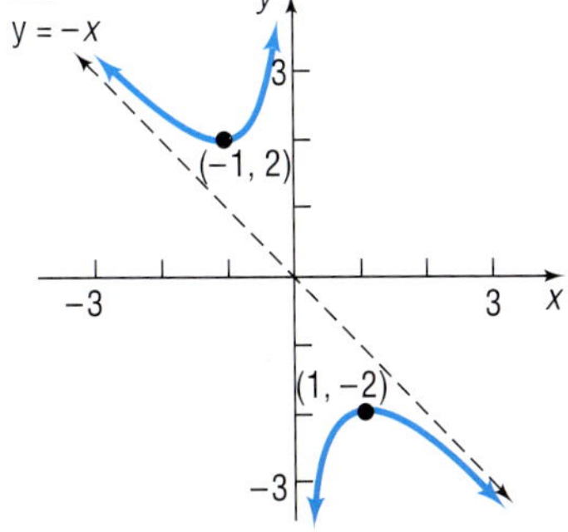

19.

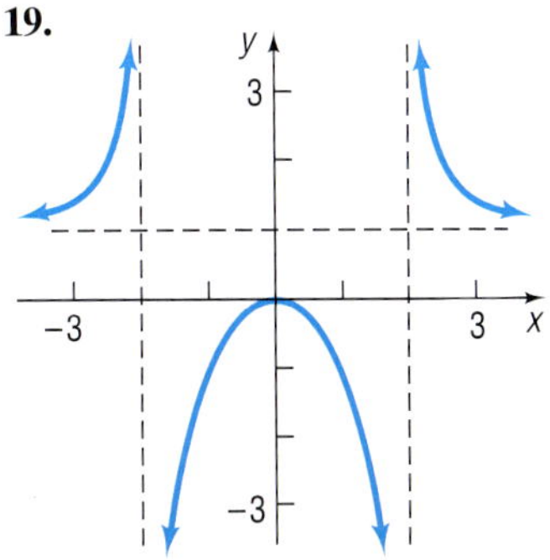

20.

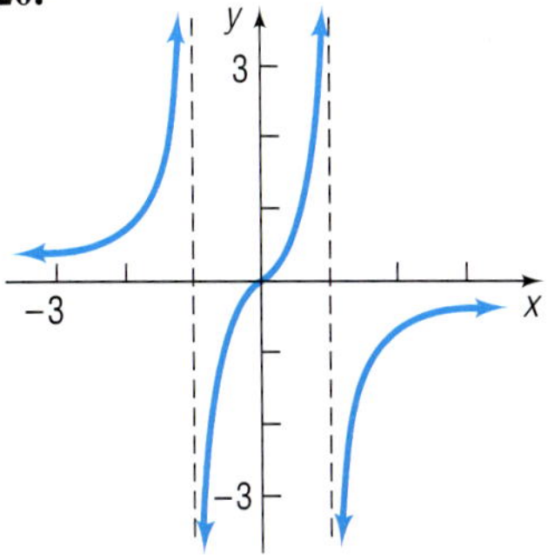

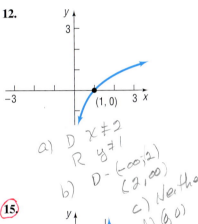

In Problems 21–24, assume that the entire graph is shown.

21.

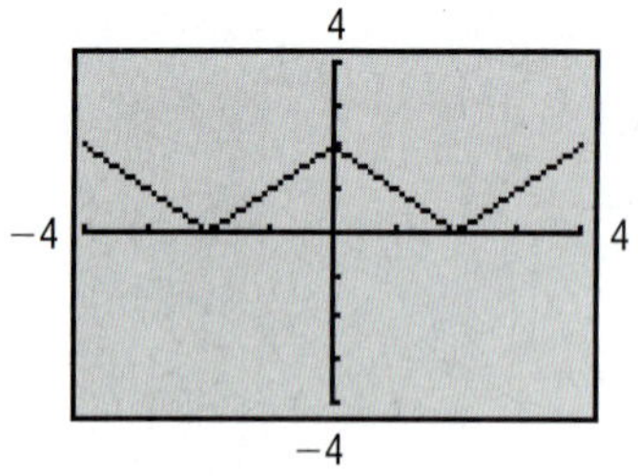

22.

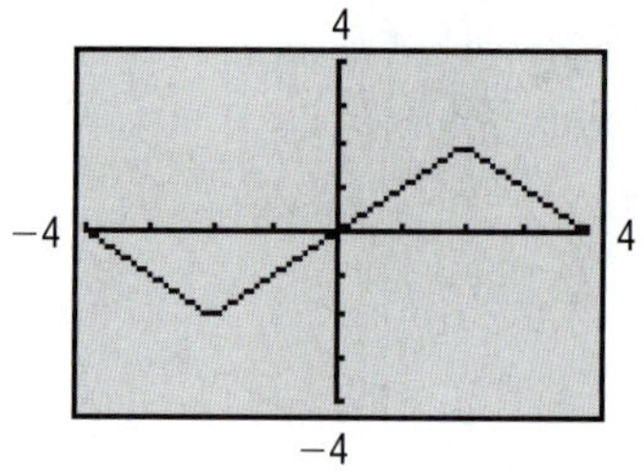

23.

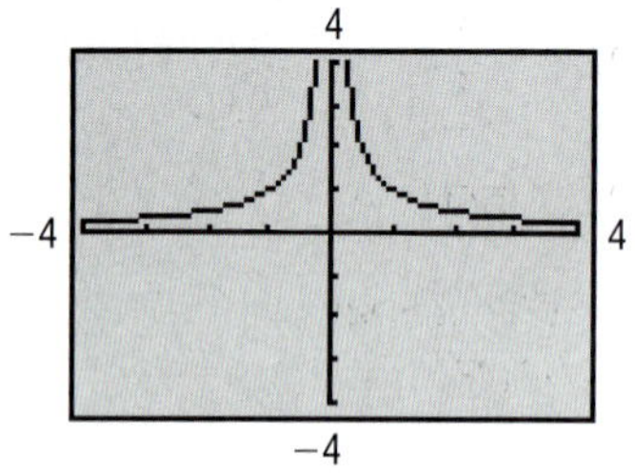

24.

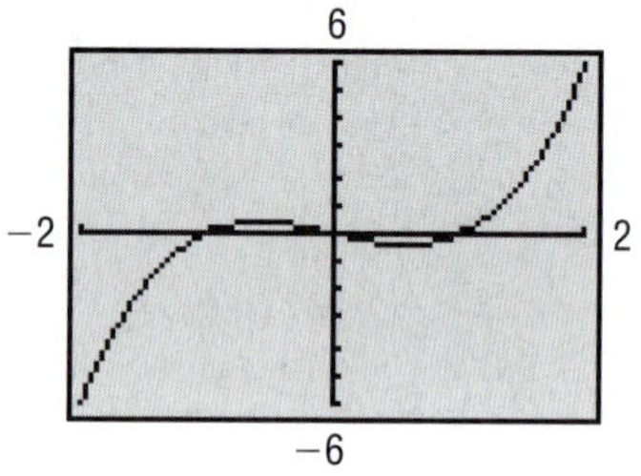

25. If

$$f(x) = \begin{cases} x^2 & \text{if } x < 0 \\ 2 & \text{if } x = 0 \\ 2x + 1 & \text{if } x > 0 \end{cases}$$

find: (a) $f(-2)$ (b) $f(0)$ (c) $f(2)$

26. If

$$f(x) = \begin{cases} x^3 & \text{if } x < 0 \\ 3x + 2 & \text{if } x \ge 0 \end{cases}$$

find: (a) $f(-1)$ (b) $f(0)$ (c) $f(1)$

27. If $f(x) = \text{int}(2x)$, find: (a) $f(1.2)$ (b) $f(1.6)$ (c) $f(-1.8)$

28. If $f(x) = \text{int}(x/2)$, find: (a) $f(1.2)$ (b) $f(1.6)$ (c) $f(-1.8)$

In Problems 29–40, (a) for each function f find the average rate of change of f from 1 to x:

$$\frac{f(x) - f(1)}{x - 1} \qquad x \neq 1$$

(b) Use the result from (a) to compute the average rate of change from $x = 1$ to $x = 2$. Be sure to simplify.

29. $f(x) = 3x$ **30.** $f(x) = -2x$ **31.** $f(x) = 1 - 3x$ **32.** $f(x) = x^2 + 1$

33. $f(x) = 3x^2 - 2x$ **34.** $f(x) = 4x - 2x^2$ **35.** $f(x) = x^3 - x$ **36.** $f(x) = x^3 + x$

37. $f(x) = \dfrac{2}{x + 1}$ **38.** $f(x) = \dfrac{1}{x^2}$ **39.** $f(x) = \sqrt{x}$ **40.** $f(x) = \sqrt{x + 3}$

In Problems 41–52, determine algebraically whether each function is even, odd, or neither. Graph each function and use TRACE to verify your results.

41. $f(x) = 4x^3$ **42.** $f(x) = 2x^4 - x^2$ **43.** $g(x) = 2x^2 - 5$ **44.** $h(x) = 3x^3 + 2$

45. $F(x) = \sqrt[3]{x}$ **46.** $G(x) = \sqrt{x}$ **47.** $f(x) = x + |x|$ **48.** $f(x) = \sqrt[3]{2x^2 + 1}$

49. $g(x) = \dfrac{1}{x^2}$ **50.** $h(x) = \dfrac{x}{x^2 - 1}$ **51.** $h(x) = \dfrac{x^3}{3x^2 - 9}$ **52.** $F(x) = \dfrac{x}{|x|}$

53. How many x-intercepts can a function defined on an interval have if it is increasing on that interval? Explain.

54. How many y-intercepts can a function have? Explain.

In Problems 55–62:

(a) Find the domain of each function.
(b) Locate any intercepts.
(c) Graph each function by hand.
(d) Based on the graph, find the range.
(e) Verify your results using a graphing utility.

55. $f(x) = \begin{cases} 2x & \text{if } x \neq 0 \\ 1 & \text{if } x = 0 \end{cases}$

56. $f(x) = \begin{cases} 3x & \text{if } x \neq 0 \\ 4 & \text{if } x = 0 \end{cases}$

57. $f(x) = \begin{cases} 1 + x & \text{if } x < 0 \\ x^2 & \text{if } x \geq 0 \end{cases}$

58. $f(x) = \begin{cases} 1/x & \text{if } x < 0 \\ \sqrt{x} & \text{if } x \geq 0 \end{cases}$

59. $f(x) = \begin{cases} |x| & \text{if } -2 \leq x < 0 \\ 1 & \text{if } x = 0 \\ x^3 & \text{if } x > 0 \end{cases}$

60. $f(x) = \begin{cases} 3 + x & \text{if } -3 \leq x < 0 \\ 3 & \text{if } x = 0 \\ \sqrt{x} & \text{if } x > 0 \end{cases}$

61. $h(x) = 2\,\text{int}(x)$

62. $f(x) = \text{int}(2x)$

Problems 63–66 require the following definition: ***Secant Line:*** *The slope of the secant line containing the two points $(x, f(x))$ and $(x + h, f(x + h))$ on the graph of a function $y = f(x)$ may be given as*

$$\frac{f(x+h) - f(x)}{(x+h) - x} = \frac{f(x+h) - f(x)}{h}$$

In Problems 63–66, express the slope of the secant line of each function in terms of x and h. Be sure to simplify your answer.

63. $f(x) = 2x + 5$

64. $f(x) = -3x + 2$

65. $f(x) = x^2 + 2x$

66. $f(x) = 1/x$

In Problems 67–74, use a graphing utility to graph each function over the indicated interval and approximate any local maxima and local minima. Determine where the function is increasing and where it is decreasing. Round answers to two decimal places.

67. $f(x) = x^3 - 3x + 2 \quad (-2, 2)$

68. $f(x) = x^3 - 3x^2 + 5 \quad (-1, 3)$

69. $f(x) = x^5 - x^3 \quad (-2, 2)$

70. $f(x) = x^4 - x^2 \quad (-2, 2)$

71. $f(x) = -0.2x^3 - 0.6x^2 + 4x - 6 \quad (-6, 4)$

72. $f(x) = -0.4x^3 + 0.6x^2 + 3x - 2 \quad (-4, 5)$

73. $f(x) = 0.25x^4 + 0.3x^3 - 0.9x^2 + 3 \quad (-3, 2)$

74. $f(x) = -0.4x^4 - 0.5x^3 + 0.8x^2 - 2 \quad (-3, 2)$

75. Revenue from Selling Bikes The following data represent the total revenue that would be received from selling x bicycles at Tunney's Bicycle Shop.

Number of Bicycles, x	Total Revenue, R (Dollars)
0	0
25	28,000
60	45,000
102	53,400
150	59,160
190	62,360
223	64,835
249	66,525

(a) Draw a scatter diagram of the data, treating the number of bicycles produced as the independent variable.
(b) Draw a line through the points (0, 0) and (25, 28,000) on the scatter diagram found in part (a).
(c) Find the average rate of change of revenue from 0 to 25 bicycles.
(d) Interpret the average rate of change found in part (c).
(e) Draw a line through the points (190, 62,360) and (223, 64,835) on the scatter diagram found in part (a).
(f) Find the average rate of change of revenue from 190 to 223 bicycles.
(g) Interpret the average rate of change found in part (f).

76. Cost of Manufacturing Bikes The following data represent the monthly cost of producing bicycles at Tunney's Bicycle Shop.

Number of Bicycles, x	Total Cost of Production, C (Dollars)
0	24,000
25	27,750
60	31,500
102	35,250
150	39,000
190	42,750
223	46,500
249	50,250

(a) Draw a scatter diagram of the data, treating the number of bicycles produced as the independent variable.
(b) Draw a line through the points (0, 24,000) and (25, 27,750) on the scatter diagram found in part (a).
(c) Find the average rate of change of the cost from 0 to 25 bicycles.
(d) Interpret the average rate of change found in part (c).
(e) Draw a line through the points (190, 42,750) and (223, 46,500) on the scatter diagram found in part (a).
(f) Find the average rate of change of the cost from 190 to 223 bicycles.
(g) Interpret the average rate of change found in part (f).

77. Growth of Bacteria The following data represent the population of an unknown bacteria.

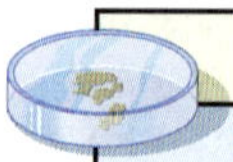

Time (Days)	Population
0	50
1	153
2	234
3	357
4	547
5	839
6	1280

(a) Draw a scatter diagram of the data, treating time as the independent variable.
(b) Draw a line through the points (0, 50) and (1, 153) on the scatter diagram found in part (a).
(c) Find the average rate of change of the population from 0 to 1 days.
(d) Interpret the average rate of change found in part (c).
(e) Draw a line through the points (5, 839) and (6, 1280) on the scatter diagram found in part (a).
(f) Find the average rate of change of the population from 5 to 6 days.
(g) Interpret the average rate of change found in part (f).
(h) What is happening to the average rate of change of the population as time passes?

78. Falling Objects Suppose that you drop a ball from a cliff 1000 feet high. You measure the distance s that the ball has fallen after time t using a motion detector and obtain the following data.

Time, t (Seconds)	Distance, s (Feet)
0	0
1	16
2	64
3	144
4	256
5	400
6	576
7	784

(a) Draw a scatter diagram of the data, treating time as the independent variable.
(b) Draw a line through the points (0, 0) and (2, 64).
(c) Find the average rate of change of the ball from 0 to 2 seconds; that is, find the slope of the line in part (b).
(d) Interpret the average rate of change found in part (c).
(e) Draw a line through the points (5, 400) and (7, 784).
(f) Find the average rate of change of the ball from 5 to 7 seconds, that is, find the slope of the line in part (e).
(g) Interpret the average rate of change found in part (f).
(h) What is happening to the average rate of change of the ball as time passes?

79. **Maximizing the Volume of a Box** An open box with a square base is to be made from a square piece of cardboard 24 inches on a side by cutting out a square from each corner and turning up the sides. (See the illustration.) The volume V of the box as a function of the length x of the side of the square cut from each corner is

$$V(x) = x(24 - 2x)^2$$

Graph V and determine where V is largest.

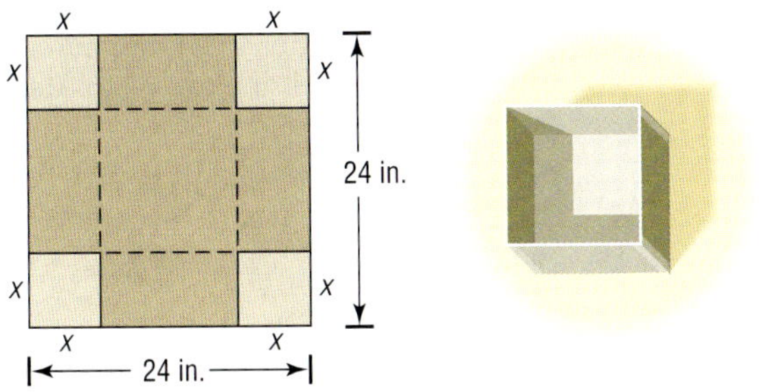

80. **Minimizing the Material Needed to Make a Box** An open box with a square base is required to have a volume of 10 cubic feet. The amount A of material used to make such a box as a function of the length x of a side of the square base is

$$A(x) = x^2 + 40/x$$

Graph A and determine where A is smallest.

81. **Maximum Height of a Ball** The height s of a ball (in feet) thrown with an initial velocity of 80 feet per second from an initial height of 6 feet is given as a function of the time t (in seconds) by

$$s(t) = -16t^2 + 80t + 6$$

(a) Graph s.
(b) Determine the time at which height is maximum.
(c) What is the maximum height?

82. **Minimum Average Cost** The average cost of producing x riding lawn mowers per hour is given by

$$A(x) = 0.3x^2 + 21x - 251 + \frac{2500}{x}$$

(a) Graph A.
(b) Determine the number of riding lawn mowers to produce in order to minimize average cost.
(c) What is the minimum average cost?

83. Graph $y = x^2$. Then on the same screen graph $y = x^2 + 2$, followed by $y = x^2 + 4$, followed by $y = x^2 - 2$. What pattern do you observe? Can you predict the graph of $y = x^2 - 4$? Of $y = x^2 + 5$?

84. Graph $y = x^2$. Then on the same screen graph $y = (x - 2)^2$, followed by $y = (x - 4)^2$, followed by $y = (x + 2)^2$. What pattern do you observe? Can you predict the graph of $y = (x + 4)^2$? Of $y = (x - 5)^2$?

85. Graph $y = |x|$. Then on the same screen graph $y = 2|x|$, followed by $y = 4|x|$, followed by $y = \frac{1}{2}|x|$. What pattern do you observe? Can you predict the graph of $y = \frac{1}{4}|x|$? Of $y = 5|x|$?

86. Graph $y = x^2$. Then on the same screen graph $y = -x^2$. What pattern do you observe? Now try $y = |x|$ and $y = -|x|$. What do you conclude?

87. Graph $y = \sqrt{x}$. Then on the same screen graph $y = \sqrt{-x}$. What pattern do you observe? Now try $y = 2x + 1$ and $y = 2(-x) + 1$. What do you conclude?

88. Graph $y = x^3$. Then on the same screen graph $y = (x - 1)^3 + 2$. Could you have predicted the result?

89. Graph $y = x^2$, $y = x^4$, and $y = x^6$ on the same screen. What do you notice is the same about each graph? What do you notice that is different?

90. Graph $y = x^3$, $y = x^5$, and $y = x^7$ on the same screen. What do you notice is the same about each graph? What do you notice that is different?

91. **Cost of Natural Gas** In January 1997, the Peoples Gas Company had the following rate schedule* for natural gas usage in single-family residences:

Monthly service charge	\$9.00
Per therm service charge	
1st 50 therms	\$0.36375/therm
Over 50 therms	\$0.11445/therm
Gas charge	\$0.3256/therm

(a) What is the charge for using 50 therms in a month?
(b) What is the charge for using 500 therms in a month?
(c) Construct a function that relates the monthly charge C for x therms of gas.
(d) Graph this function.

**Source:* The Peoples Gas Company, Chicago, Illinois.

92. Cost of Natural Gas In January 1997, Northern Illinois Gas Company had the following rate schedule* for natural gas usage in single-family residences:

Monthly customer charge	\$6.00
Distribution charge,	
1st 20 therms	\$0.2012/therm
Next 30 therms	\$0.1117/therm
Over 50 therms	\$0.0374/therm
Gas supply charge	\$0.3113/therm

(a) What is the charge for using 40 therms in a month?

(b) What is the charge for using 202 therms in a month?

(c) Construct a function that gives the monthly charge C for x therms of gas.

(d) Graph this function.

*Source: Northern Illinois Gas Company, Naperville, Illinois.

93. Consider the equation

$$y = \begin{cases} 1 & \text{if } x \text{ is rational} \\ 0 & \text{if } x \text{ is irrational} \end{cases}$$

Is this a function? What is its domain? What is its range? What is its y-intercept, if any? What are its x-intercepts, if any? Is it even, odd, or neither? How would you describe its graph?

94. Define some functions that pass through $(0, 0)$ and $(1, 1)$ are increasing for $x \geq 0$. Begin your list with $y = \sqrt{x}$, $y = x$, and $y = x^2$. Can you propose a general result about such functions?

95. Can you think of a function that is both even and odd?

3.2 GRAPHING TECHNIQUES: TRANSFORMATIONS

1 Graph Functions Using Horizontal and Vertical Shifts

2 Graph Functions Using Compressions and Stretches

3 Graph Functions Using Reflections about the x-Axis or y-Axis

At this stage, if you were asked to graph any of the functions defined by $y = x$, $y = x^2$, $y = x^3$, $y = \sqrt{x}$, $y = |x|$, or $y = 1/x$, your response should be, "Yes, I recognize these functions and know the general shapes of their graphs." (If this is not your answer, review the previous section and Figures 15 through 20.)

Sometimes we are asked to graph a function that is "almost" like one that we already know how to graph. In this section, we look at some of these functions and develop techniques for graphing them. Collectively, these techniques are referred to as **transformations.**

1 Vertical Shifts

EXPLORATION On the same screen, graph each of the following functions:

$$Y_1 = x^2$$
$$Y_2 = x^2 + 1$$
$$Y_3 = x^2 + 2$$
$$Y_4 = x^2 - 1$$
$$Y_5 = x^2 - 2$$

What do you observe?

RESULT Figure 25 illustrates the graphs. You should have observed a general pattern. With $Y_1 = x^2$ on the screen, the graph of $Y_2 = x^2 + 1$ is identical to that of $Y_1 = x^2$, except that it is shifted vertically up 1 unit.

Similarly, $Y_3 = x^2 + 2$ is identical to that of $Y_1 = x^2$, except that it is shifted vertically up 2 units. The graph of $Y_4 = x^2 - 1$ is identical to that of $Y_1 = x^2$, except that it is shifted vertically down 1 unit.

Figure 25

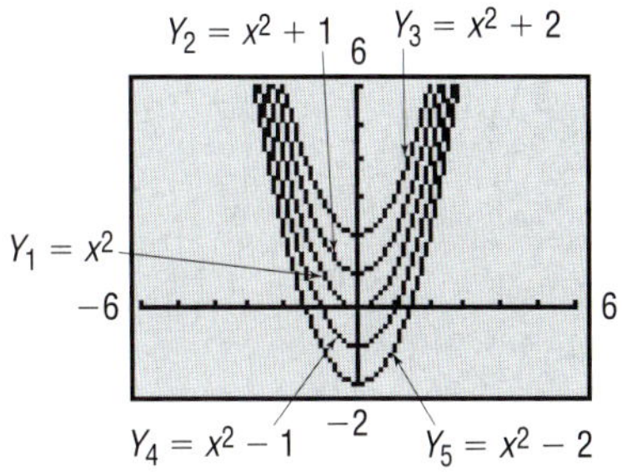

We are led to the following conclusion:

> If a real number c is added to the right side of a function $y = f(x)$, the graph of the new function $y = f(x) + c$ is the graph of f **shifted vertically** up (if $c > 0$) or down (if $c < 0$).

Let's look at an example.

EXAMPLE 1 Vertical Shift Down

Use the graph of $f(x) = x^2$ to obtain the graph of $h(x) = x^2 - 4$.

Solution Table 3 lists some points on the graphs of $f = Y_1$ and $h = Y_2$. Notice each y-coordinate of h is 4 units less than the corresponding y-coordinate of f. The graph of h is identical to that of f, except that it is shifted down 4 units. See Figure 26.

Table 3

X	Y1	Y2
-3	9	5
-2	4	0
-1	1	-3
0	0	-4
1	1	-3
2	4	0
3	9	5

Y2 ■ X²−4

Figure 26

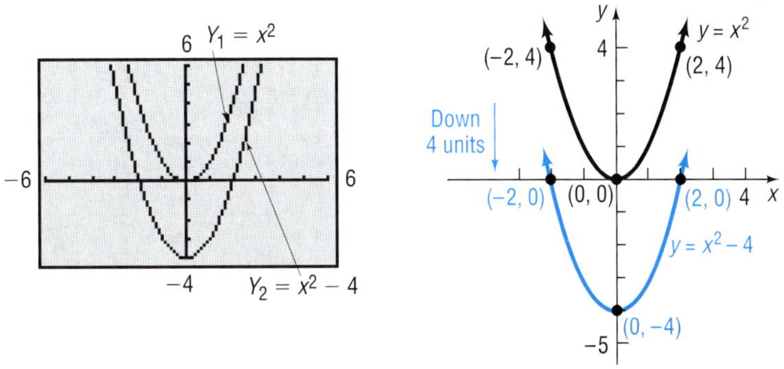

NOW WORK PROBLEM **29.**

Horizontal Shifts

EXPLORATION On the same screen, graph each of the following functions:

$$Y_1 = x^2$$
$$Y_2 = (x - 1)^2$$
$$Y_3 = (x - 3)^2$$
$$Y_4 = (x + 2)^2$$

What do you observe?

Figure 27

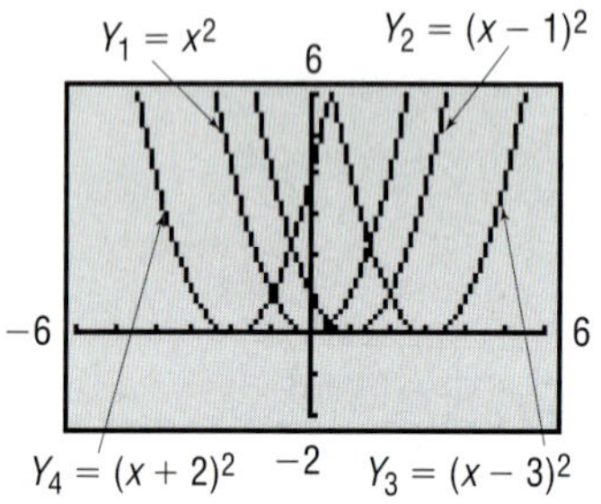

RESULT Figure 27 illustrates the graphs.

You should have observed the following pattern. With the graph of $Y_1 = x^2$ on the screen, the graph of $Y_2 = (x - 1)^2$ is identical to that of $Y = x^2$, except it is shifted horizontally to the right 1 unit. Similarly, the graph of $Y_3 = (x - 3)^2$ is identical to that of $Y_1 = x^2$, except it is shifted horizontally to the right 3 units. Finally, the graph of $Y_4 = (x + 2)^2$ is identical to that of $Y_1 = x^2$, except it is shifted horizontally to the left 2 units.

We are led to the following conclusion.

> If the argument x of a function f is replaced by $x - c$, c a real number, the graph of the new function $g(x) = f(x - c)$ is the graph of f **shifted horizontally** left (if $c < 0$) or right (if $c > 0$).

NOW WORK PROBLEM 35.

In Section 2.3 we graphed quadratic functions by identifying points on the graph of the function, plotting the points, and connecting them with a smooth curve. We can also graph quadratic functions using transformations.

EXAMPLE 2 Graphing a Quadratic Function Using Transformations

Graph the function $f(x) = x^2 + 6x + 4$.

Solution We begin by completing the square on the right side.

$$\begin{aligned} f(x) &= x^2 + 6x + 4 \\ &= (x^2 + 6x) + 4 && \text{Take 1/2 the coefficient of } x \text{ and square the result: } \left(\tfrac{6}{2}\right)^2 = 9. \\ &= (x^2 + 6x + 9) + (4 - 9) && \text{Add and subtract 9.} \\ &= (x + 3)^2 - 5 && \text{Factor.} \end{aligned}$$

We graph f in steps. First, we note that the rule for f is basically a square function, so we begin with the graph of $y = x^2$ as shown in Figure 28(a). Next, to get the graph of $y = (x + 3)^2$, we shift the graph of $y = x^2$ horizontally

3 units to the left. See Figure 28(b). Finally, to get the graph of $y = (x + 3)^2 - 5$, we shift the graph of $y = (x + 3)^2$ vertically down 5 units. See Figure 28(c). Note the points plotted on each graph. Using key points can be helpful in keeping track of the transformation that has taken place.

Figure 28

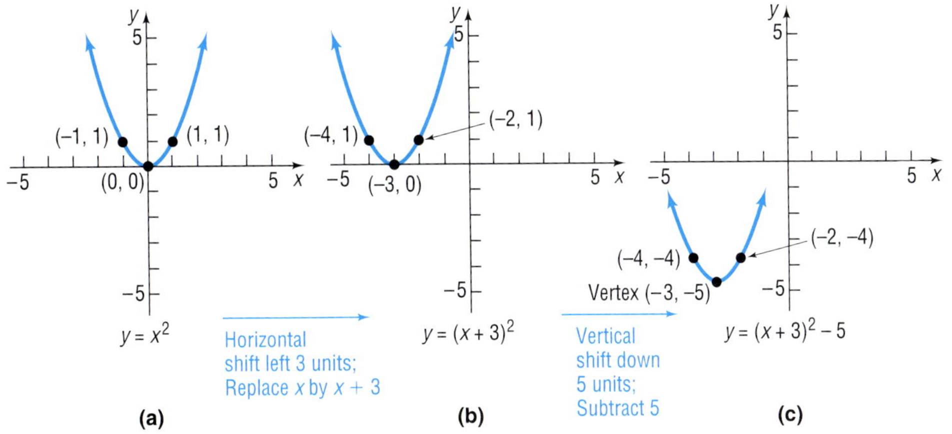

In Example 2, if the vertical shift had been done first, followed by the horizontal shift, the final graph would have been the same. (Try it for yourself.)

We now have two ways to graph a quadratic function:

1. Complete the square and apply shifting techniques (Example 2).
2. Use the results given in display (8) on page 133 in Section 2.3 to find the vertex and axis of symmetry and to determine whether the graph opens up or down. Then locate additional points (such as the y-intercept and x-intercepts, if any) to complete the graph.

NOW WORK PROBLEMS **47** AND **71**.

2 Compressions and Stretches

EXPLORATION On the same screen, graph each of the following functions:

$$Y_1 = |x|$$
$$Y_2 = 2|x|$$
$$Y_3 = 3|x|$$
$$Y_4 = \tfrac{1}{2}|x|$$

Figure 29

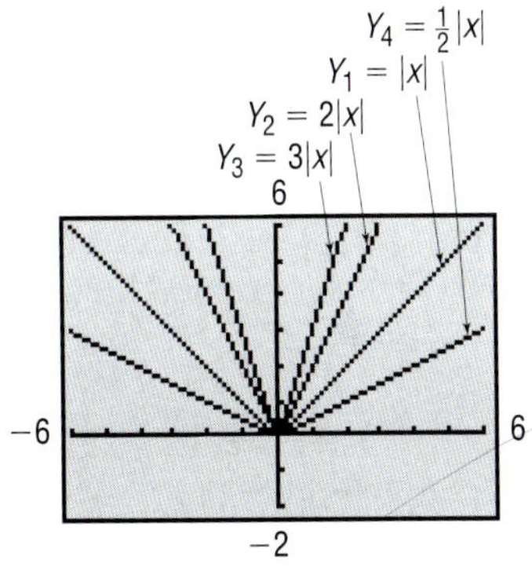

RESULT Figure 29 illustrates the graphs. You should have observed the following pattern. The graphs of $Y_2 = 2|x|$ and $Y_3 = 3|x|$ can be obtained from the graph of $Y_1 = |x|$ by multiplying each y-coordinate of $Y_1 = |x|$ by factors of 2 and 3, respectively. This is sometimes referred to as a vertical *stretch* using factors of 2 and 3.

The graph of $Y_4 = \frac{1}{2}|x|$ can be obtained from the graph of $Y_1 = |x|$ by multiplying each y-coordinate by $\frac{1}{2}$. This is sometimes referred to as a vertical *compression* using a factor of $\frac{1}{2}$. ■

Look at Tables 4 and 5, where $Y_1 = |x|$, $Y_3 = 3|x|$, and $Y_4 = \frac{1}{2}|x|$. Notice that the values for Y_3 in Table 4 are three times the values of Y_1. Therefore, the graph of Y_3 will be vertically *stretched* by a factor of 3. Likewise, the values of Y_4 in Table 5 are half the values of Y_1. Therefore, the graph of Y_4 will be vertically *compressed* by a factor of $\frac{1}{2}$.

Table 4

X	Y1	Y3
-2	2	6
-1	1	3
0	0	0
1	1	3
2	2	6
3	3	9
4	4	12

Y3=3abs(X)

Table 5

X	Y1	Y4
-2	2	1
-1	1	.5
0	0	0
1	1	.5
2	2	1
3	3	1.5
4	4	2

Y4=.5abs(X)

> When the right side of a function $y = f(x)$ is multiplied by a positive number k, the graph of the new function $y = kf(x)$ is a **vertically compressed** (if $0 < k < 1$) or **stretched** (if $k > 1$) version of the graph of $y = f(x)$.

NOW WORK PROBLEM 37.

What happens if the argument x of a function $y = f(x)$ is multiplied by a positive number k creating a new function $y = f(kx)$? To find the answer, we look at the following Exploration.

■ EXPLORATION On the same screen, graph each of the following functions:

$$Y_1 = f(x) = x^2$$
$$Y_2 = f(2x) = (2x)^2$$
$$Y_3 = f\left(\frac{1}{2}x\right) = \left(\frac{1}{2}x\right)^2$$

RESULT You should have obtained the graphs shown in Figure 30. The graph of $Y_2 = (2x)^2$ is the graph of $Y_1 = x^2$ compressed horizontally. Look at Table 6(a). Notice that (1, 1), (2, 4), (4, 16), and (16, 256) are points on the graph of $Y_1 = x^2$. Also, (0.5, 1), (1, 4), (2, 16), and (8, 256) are points on the graph of $Y_2 = (2x)^2$. Thus, the graph of $Y_2 = (2x)^2$ is obtained by multiplying the x-coordinate of each point on the graph of $Y_1 = x^2$ by 1/2. The graph of $Y_3 = \left(\frac{1}{2}x\right)^2$ is the graph of $Y_1 = x^2$ stretched horizontally. Look at Table 6(b). Notice that (0.5, 0.25), (1, 1), (2, 4), and

(4, 16) are points on the graph of $Y_1 = x^2$. Also, (1, 0.25), (2, 1), (4, 4), (8, 16) are points on the graph of $Y_3 = \left(\frac{1}{2}x\right)^2$. Thus, the graph of $Y_3 = \left(\frac{1}{2}x\right)^2$ is obtained by multiplying the x-coordinate of each point on the graph of $Y_1 = x^2$ by a factor of 2.

Figure 30

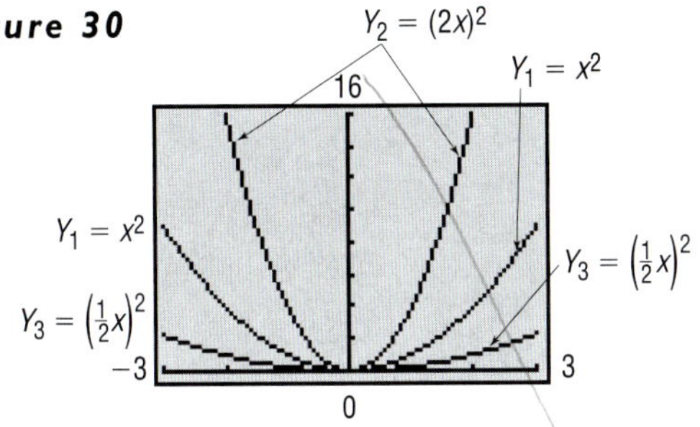

Table 6

(a)

X	Y1	Y2
0	0	0
.5	.25	1
1	1	4
2	4	16
4	16	64
8	64	256
16	256	1024

$Y_2 = (2X)^2$

(b)

X	Y1	Y3
0	0	0
.5	.25	.0625
1	1	.25
2	4	1
4	16	4
8	64	16
16	256	64

$Y_3 = (X/2)^2$

> If the argument of x of a function $y = f(x)$ is multiplied by a positive number k, the graph of the new function $y = f(kx)$ is obtained by multiplying each x-coordinate of $y = f(x)$ by $1/k$. A **horizontal compression** results if $k > 1$ and a **horizontal stretch** occurs if $0 < k < 1$.

Let's look at an example.

EXAMPLE 3 Graphing Using Stretches and Compressions

The graph of $y = f(x)$ is given in Figure 31. Use this graph to find the graphs of

(a) $y = 3f(x)$ (b) $y = f(3x)$

Figure 31

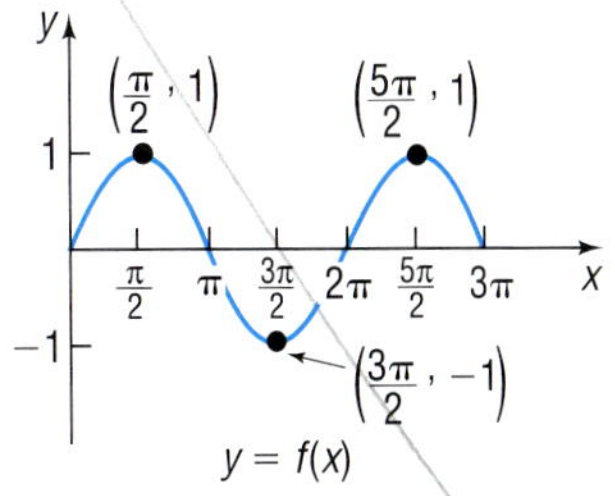

Solution (a) The graph of $y = 3f(x)$ is obtained by multiplying each y-coordinate of $y = f(x)$ by a factor of 3. See Figure 32(a).

(b) The graph of $y = f(3x)$ is obtained from the graph of $y = f(x)$ by multiplying each x-coordinate of $y = f(x)$ by a factor of $\frac{1}{3}$. See Figure 32(b).

Figure 32

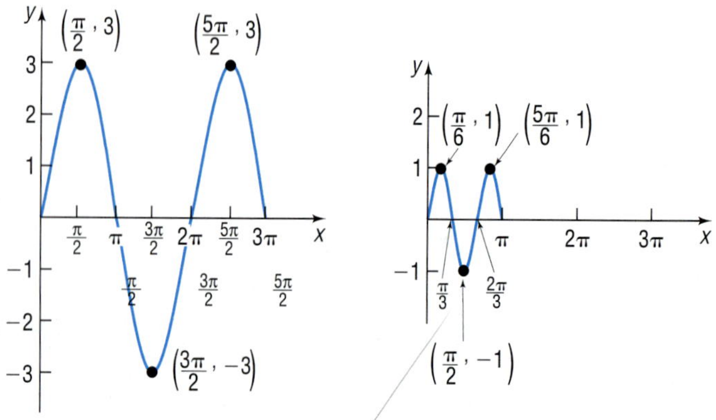

(a) $y = 3f(x)$ **(b)** $y = f(3x)$

NOW WORK PROBLEMS **59(d)** AND **(f)**.

3 Reflections about the x-Axis and the y-Axis

EXPLORATION **Reflection about the x-Axis**

(a) Graph $Y_1 = x^2$, followed by $Y_2 = -x^2$.
(b) Graph $Y_1 = |x|$, followed by $Y_2 = -|x|$.
(c) Graph $Y_1 = x^2 - 4$, followed by $Y_2 = -(x^2 - 4) = -x^2 + 4$.

RESULT See Tables 7(a), (b), and (c) and Figures 33(a), (b), and (c). In each instance, the second graph is the reflection about the x-axis of the first graph.

Table 7

X	Y1	Y2
-3	9	-9
-2	4	-4
-1	1	-1
0	0	0
1	1	-1
2	4	-4
3	9	-9

Y2=-X²

(a)

X	Y1	Y2
-3	3	-3
-2	2	-2
-1	1	-1
0	0	0
1	1	-1
2	2	-2
3	3	-3

Y2=-abs(X)

(b)

X	Y1	Y2
-3	5	-5
-2	0	0
-1	-3	3
0	-4	4
1	-3	3
2	0	0
3	5	-5

Y2=-X²+4

(c)

Figure 33

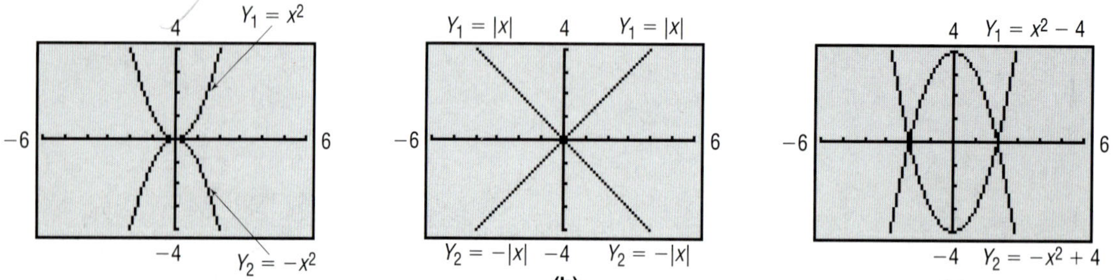

(a) (b) (c)

When the right side of the equation $y = f(x)$ is multiplied by -1, the graph of the new function $y = -f(x)$ is the **reflection about the *x*-axis** of the graph of the function $y = f(x)$.

NOW WORK PROBLEM **43.**

EXPLORATION

(a) Graph $Y_1 = \sqrt{x}$, followed by $Y_2 = \sqrt{-x}$.
(b) Graph $Y_1 = x + 1$, followed by $Y_2 = -x + 1$.
(c) Graph $Y_1 = x^4 + x$, followed by $Y_2 = (-x)^4 + (-x) = x^4 - x$.

RESULT See Tables 8(a), (b), and (c) and Figures 34(a), (b), and (c). In each instance, the second graph is the reflection about the y-axis of the first graph.

Table 8

X	Y1	Y2
-3	ERROR	1.7321
-2	ERROR	1.4142
-1	ERROR	1
0	0	0
1	1	ERROR
2	1.4142	ERROR
3	1.7321	ERROR

Y2=√(-X)

(a)

X	Y1	Y2
-3	-2	4
-2	-1	3
-1	0	2
0	1	1
1	2	0
2	3	-1
3	4	-2

Y2=-X+1

(b)

X	Y1	Y2
-3	78	84
-2	14	18
-1	0	2
0	0	0
1	2	0
2	18	14
3	84	78

Y2=X^4-X

(c)

Figure 34

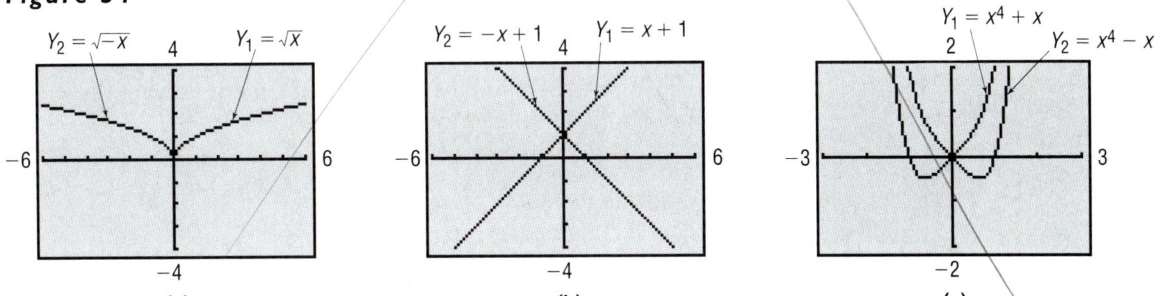

(a) (b) (c)

When the graph of the function $y = f(x)$ is known, the graph of the new function $y = f(-x)$ is the **reflection about the *y*-axis** of the graph of the function $y = f(x)$.

SUMMARY OF GRAPHING TECHNIQUES

Table 9 summarizes the graphing procedures that we have just discussed.

Table 9

To Graph:	Draw the Graph of f and:	Functional Change to $f(x)$
Vertical shifts		
$y = f(x) + c, \quad c > 0$	Raise the graph of f by c units.	Add c to $f(x)$.
$y = f(x) - c, \quad c > 0$	Lower the graph of f by c units.	Subtract c from $f(x)$.
Horizontal shifts		
$y = f(x + c), \quad c > 0$	Shift the graph of f to the left c units.	Replace x by $x + c$.
$y = f(x - c), \quad c > 0$	Shift the graph of f to the right c units.	Replace x by $x - c$.
Compressing or stretching		
$y = kf(x), \quad k > 0$	Multiply each y coordinate of $y = f(x)$ by k.	Multiply $f(x)$ by k.
$y = f(kx), \quad k > 0$	Multiply each x coordinate of $y = f(x)$ by $\frac{1}{k}$.	Replace x by kx.
Reflection about the x-axis		
$y = -f(x)$	Reflect the graph of f about the x-axis.	Multiply $f(x)$ by -1.
Reflection about the y-axis		
$y = f(-x)$	Reflect the graph of f about the y-axis.	Replace x by $-x$.

The examples that follow combine some of the procedures outlined in this section to get the required graph.

EXAMPLE 4 Determining the Function Obtained from a Series of Transformations

Find the function that is finally graphed after the following three transformations are applied to the graph of $y = |x|$.

1. Shift left 2 units.
2. Shift up 3 units.
3. Reflect about the y-axis.

Solution

1. Shift left 2 units: Replace x by $x + 2$ $\quad y = |x + 2|$
2. Shift up 3 units: Add 3 $\quad y = |x + 2| + 3$
3. Reflect about the y-axis: Replace x by $-x$ $\quad y = |-x + 2| + 3$

NOW WORK PROBLEM 25.

EXAMPLE 5 Combining Graphing Procedures

Graph the function $f(x) = \dfrac{3}{x - 2} + 1$.

Solution We use the following steps to obtain the graph of f:

STEP 1: $y = \dfrac{1}{x}$ *Reciprocal function.*

STEP 2: $y = \dfrac{3}{x}$ *Vertical stretch of the graph of* $y = \dfrac{1}{x}$ *by a factor of 3; multiply by 3.*

STEP 3: $y = \dfrac{3}{x-2}$ *Horizontal shift to the right 2 units; replace* x *by* $x - 2$.

STEP 4: $y = \dfrac{3}{x-2} + 1$ *Vertical shift up 1 unit; add 1.*

See Figure 35.

Figure 35

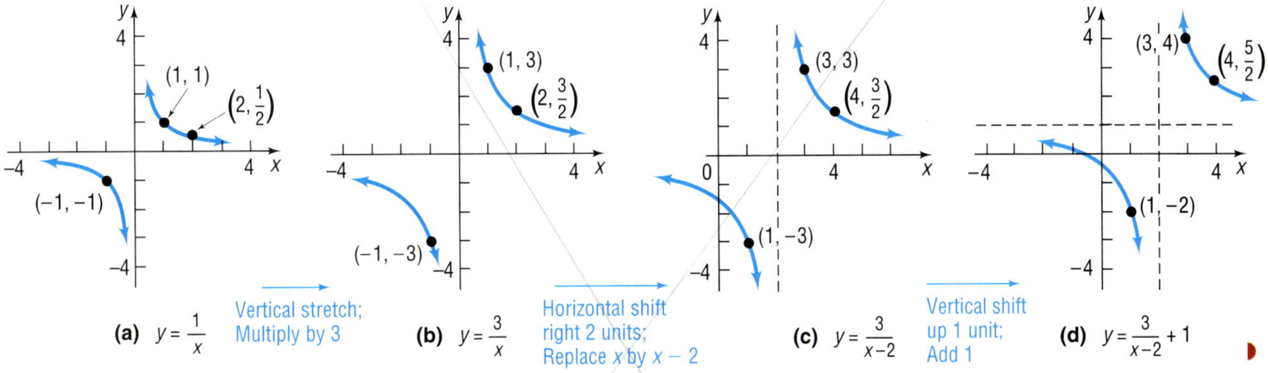

There are other orderings of the steps shown in Example 5 that would also result in the graph of f. For example, try this one:

STEP 1: $y = \dfrac{1}{x}$ *Reciprocal function.*

STEP 2: $y = \dfrac{1}{x-2}$ *Horizontal shift to the right 2 units; replace* x *by* $x - 2$.

STEP 3: $y = \dfrac{3}{x-2}$ *Vertical stretch of the graph of* $y = \dfrac{1}{x-2}$ *by factor of 3; multiply by 3.*

STEP 4: $y = \dfrac{3}{x-2} + 1$ *Vertical shift up 1 unit; add 1.*

NOW WORK PROBLEM 49.

EXAMPLE 6 Combining Graphing Procedures

Graph the function $f(x) = \sqrt{1 - x} + 2$.

Solution We use the following steps to get the graph of $y = \sqrt{1 - x} + 2$.

STEP 1:	$y = \sqrt{x}$	*Square root function.*
STEP 2:	$y = \sqrt{x + 1}$	*Horizontal shift left 1 unit; replace x by $x + 1$.*
STEP 3:	$y = \sqrt{-x + 1} = \sqrt{1 - x}$	*Reflect about y-axis; replace x by $-x$.*
STEP 4:	$y = \sqrt{1 - x} + 2$	*Vertical shift up 2 units; add 2.*

See Figure 36.

Figure 36

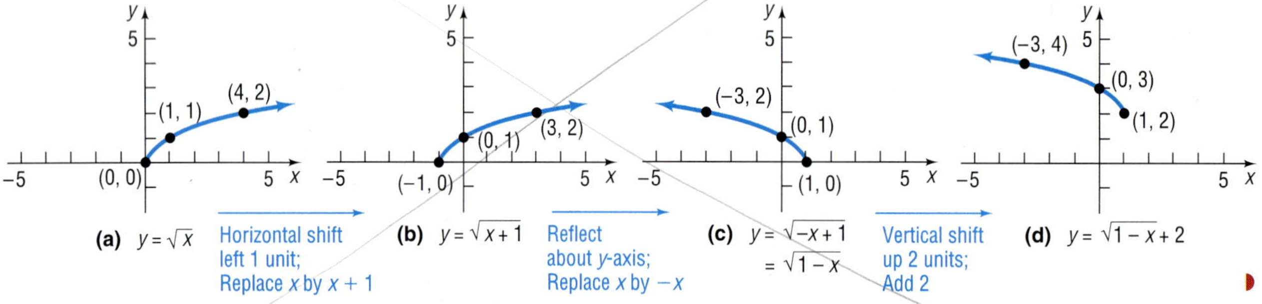

3.2 EXERCISES

In Problems 1–12, match each graph to one of the following functions:

A. $y = x^2 + 2$	B. $y = -x^2 + 2$	C. $y = \lvert x\rvert + 2$	D. $y = -\lvert x\rvert + 2$
E. $y = (x - 2)^2$	F. $y = -(x + 2)^2$	G. $y = \lvert x - 2\rvert$	H. $y = -\lvert x + 2\rvert$
I. $y = 2x^2$	J. $y = -2x^2$	K. $y = 2\lvert x\rvert$	L. $y = -2\lvert x\rvert$

1.
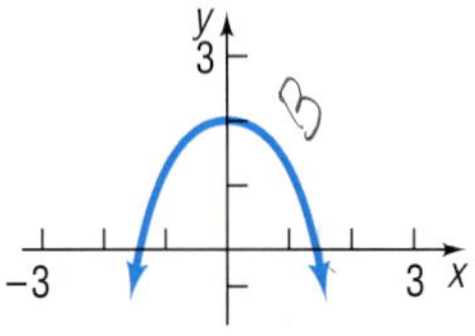

2.
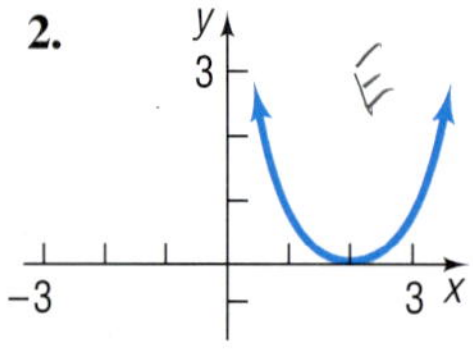

3.
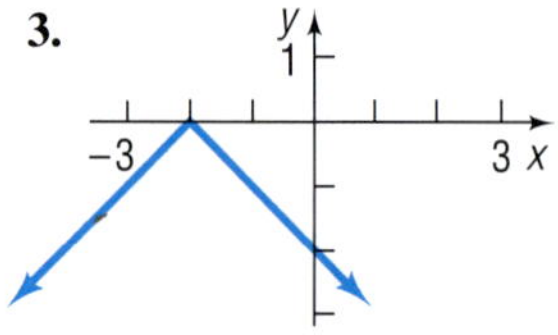

4.
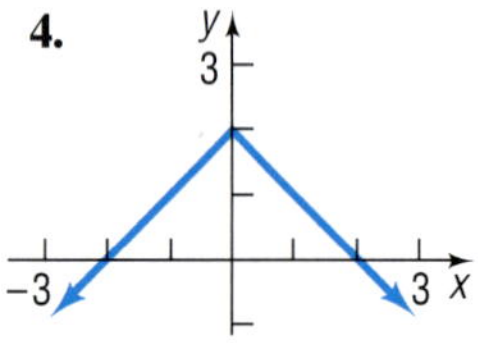

5.
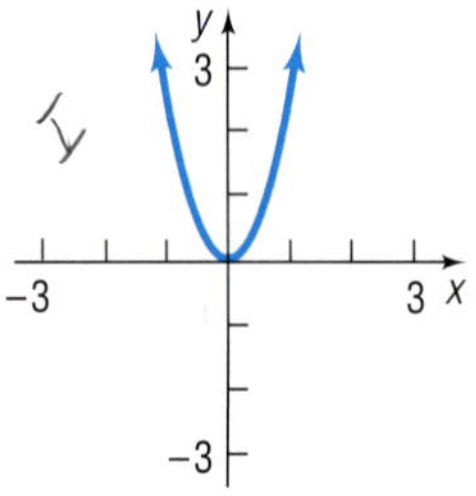

6.
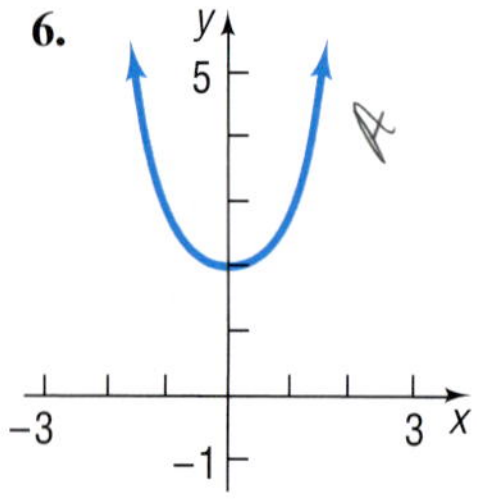

7.
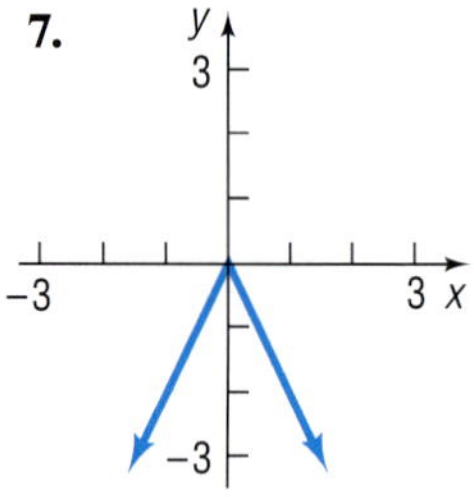

8.
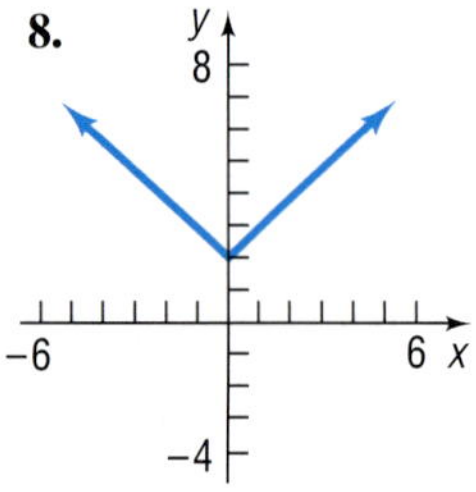

9.
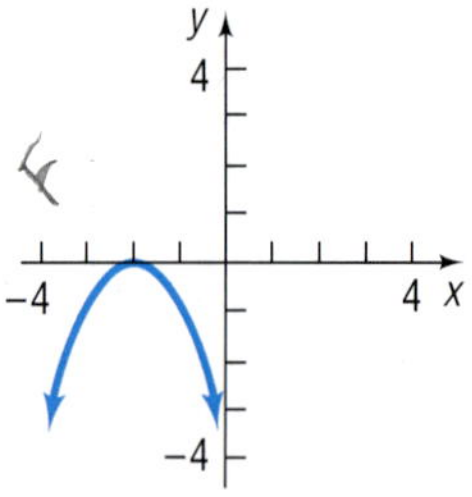

10.
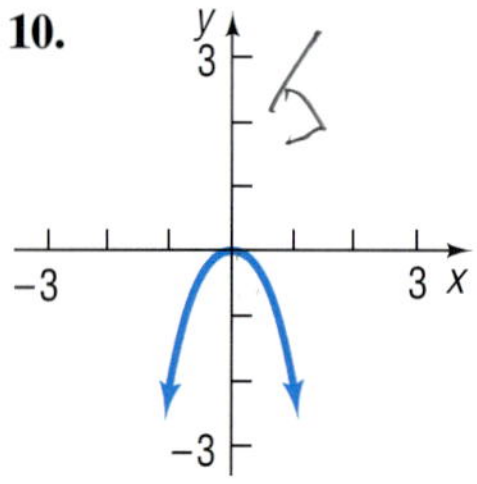

11.
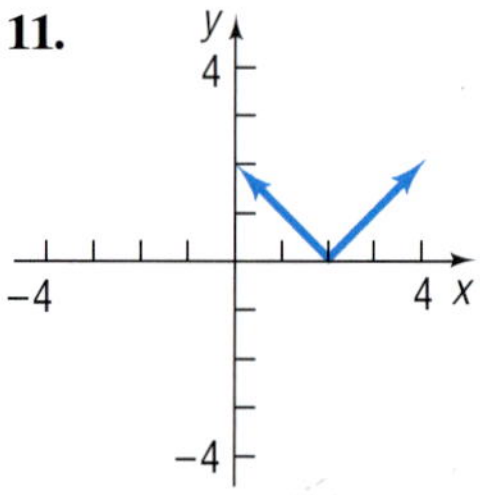

12.
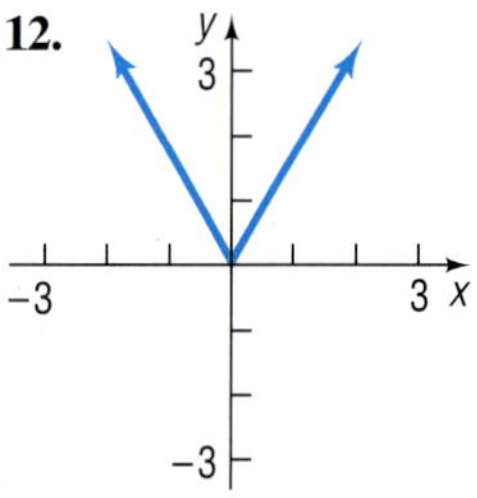

In Problems 13–16, match each graph to one of the following functions:

A. $y = x^3$ *B.* $y = (x + 2)^3$ *C.* $y = -2x^3$ *D.* $y = x^3 + 2$

13.

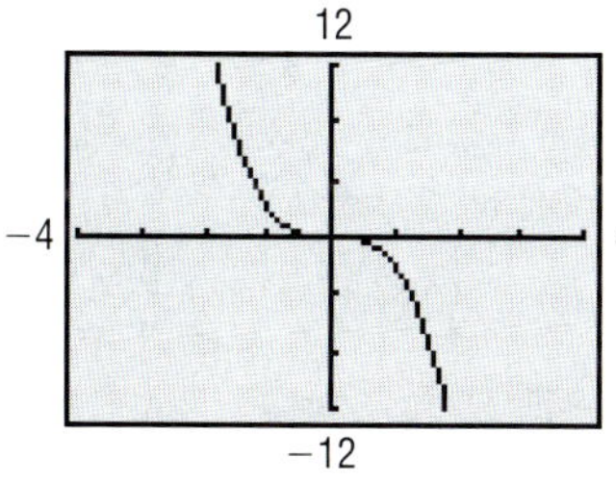

14.

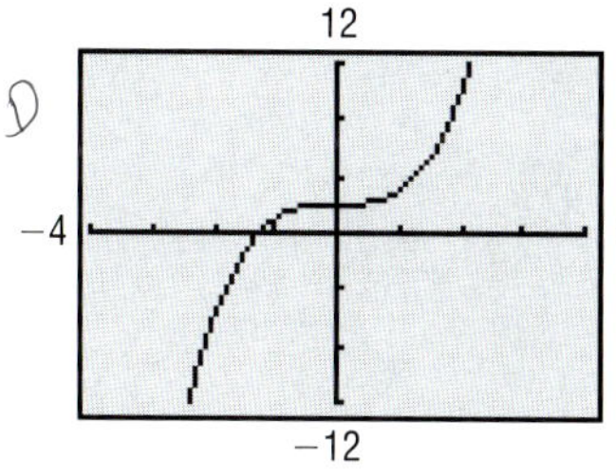

15.

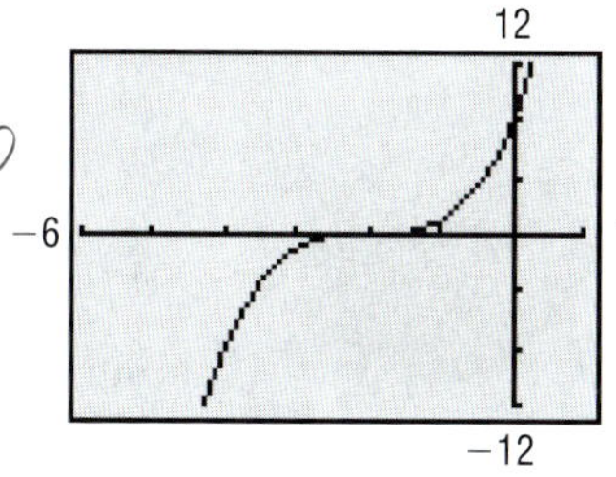

16.

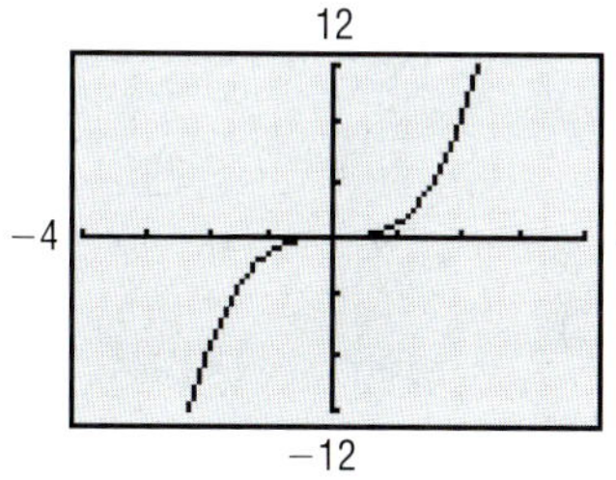

In Problems 17–24, write the function whose graph is the graph of $y = x^3$*, but is*

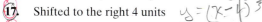

17. Shifted to the right 4 units
18. Shifted to the left 4 units
19. Shifted up 4 units
20. Shifted down 4 units
21. Reflected about the y-axis
22. Reflected about the x-axis
23. Vertically stretched by a factor of 4
24. Horizontally stretched by a factor of 4

In Problems 25–28, find the function that is finally graphed after the following transformations are applied to the graph of $y = \sqrt{x}$.

25. (1) Shift up 2 units
(2) Reflect about the x-axis
(3) Reflect about the y-axis

26. (1) Reflect about the x-axis
(2) Shift right 3 units
(3) Shift down 2 units

27. (1) Reflect about the x-axis
(2) Shift up 2 units
(3) Shift left 3 units

28. (1) Shift up 2 units
(2) Reflect about the y-axis
(3) Shift left 3 units

In Problems 29–58, graph each function by hand using the techniques of shifting, compressing, stretching, and/or reflecting. Start with the graph of the basic function (for example, $y = x^2$*) and show all stages. Verify your answer by using a graphing utility.*

29. $f(x) = x^2 - 1$
30. $f(x) = x^2 + 4$
31. $g(x) = x^3 + 1$
32. $g(x) = x^3 - 1$
33. $h(x) = \sqrt{x - 2}$
34. $h(x) = \sqrt{x + 1}$
35. $f(x) = (x - 1)^3$
36. $f(x) = (x + 2)^3$
37. $g(x) = 4\sqrt{x}$
38. $g(x) = \frac{1}{2}\sqrt{x}$
39. $h(x) = \dfrac{1}{2x}$
40. $h(x) = \dfrac{4}{x}$
41. $f(x) = -|x|$
42. $f(x) = -\sqrt{x}$
43. $g(x) = -\dfrac{1}{x}$
44. $g(x) = -x^3$
45. $h(x) = \text{int}(-x)$
46. $h(x) = \dfrac{1}{-x}$
47. $f(x) = (x + 1)^2 - 3$
48. $f(x) = (x - 2)^2 + 1$
49. $g(x) = \sqrt{x - 2} + 1$
50. $g(x) = |x + 1| - 3$
51. $h(x) = \sqrt{-x} - 2$
52. $h(x) = \dfrac{4}{x} + 2$
53. $f(x) = (x + 1)^3 - 1$
54. $f(x) = 4\sqrt{x - 1}$
55. $g(x) = 2|1 - x|$
56. $g(x) = 4\sqrt{2 - x}$
57. $h(x) = 2\,\text{int}(x - 1)$
58. $h(x) = -x^3 + 2$

In Problems 59–64, the graph of a function f is illustrated. Use the graph of f as the first step toward graphing each of the following functions:

(a) $F(x) = f(x) + 3$ *(b)* $G(x) = f(x + 2)$ *(c)* $P(x) = -f(x)$

(d) $Q(x) = \frac{1}{2}f(x)$ *(e)* $g(x) = f(-x)$ *(f)* $h(x) = f(2x)$

59.

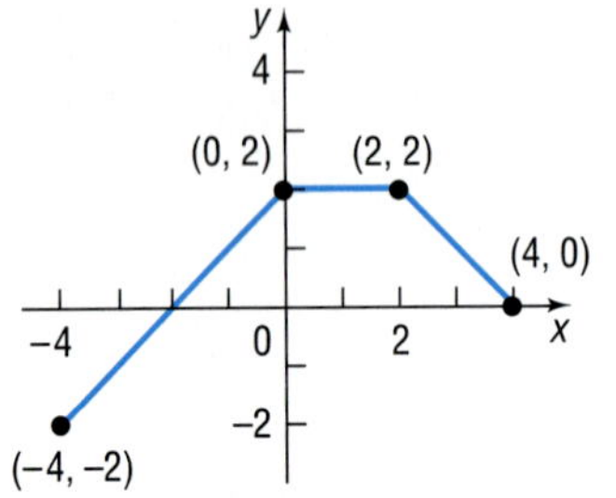

60.

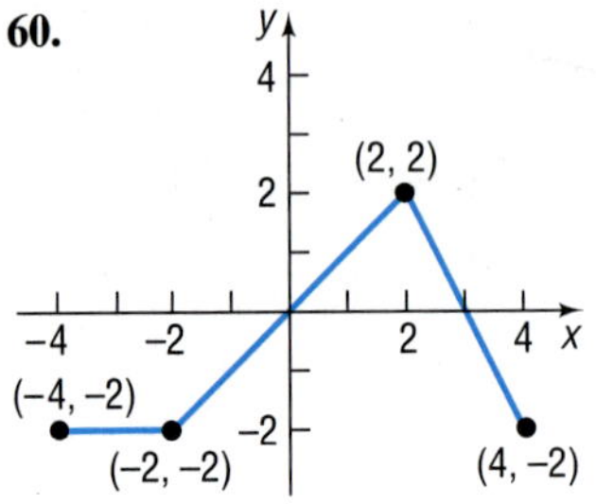

61.

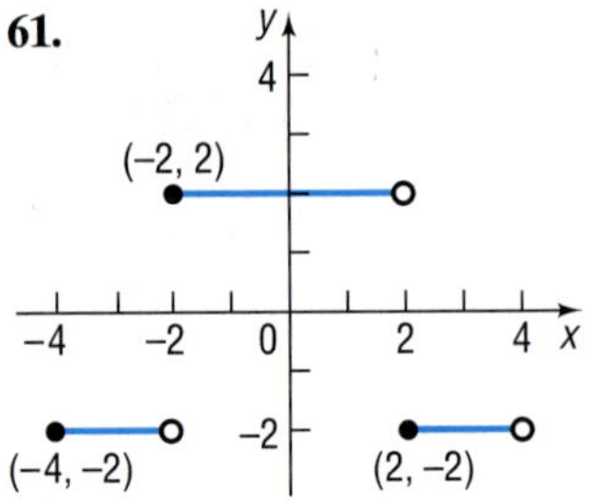

62.

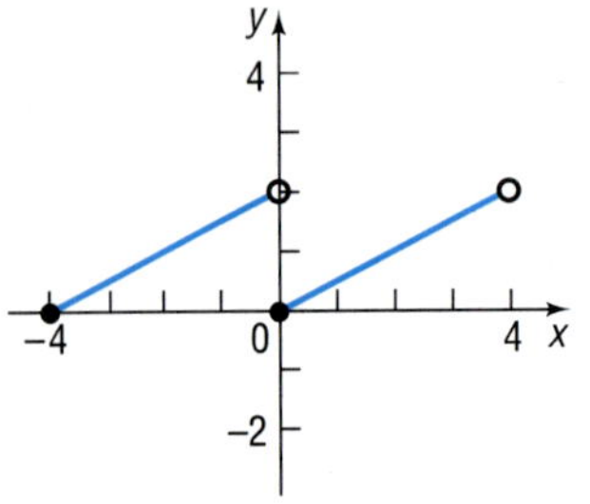

63.

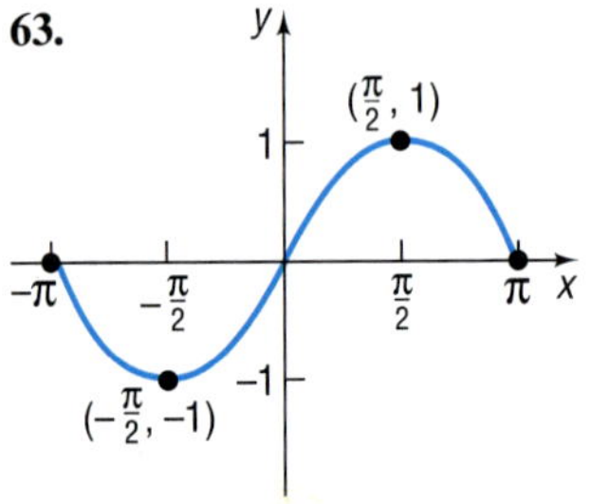

64.

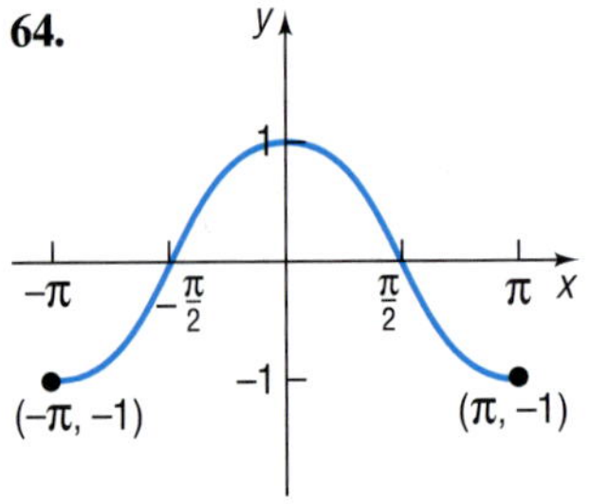

65. Exploration

(a) Use a graphing utility to graph $y = x + 1$ and $y = |x + 1|$.

(b) Graph $y = 4 - x^2$ and $y = |4 - x^2|$.

(c) Graph $y = x^3 + x$ and $y = |x^3 + x|$.

(d) What do you conclude about the relationship between the graphs of $y = f(x)$ and $y = |f(x)|$?

66. Exploration

(a) Use a graphing utility to graph $y = x + 1$ and $y = |x| + 1$.

(b) Graph $y = 4 - x^2$ and $y = 4 - |x|^2$.

(c) Graph $y = x^3 + x$ and $y = |x|^3 + |x|$.

(d) What do you conclude about the relationship between the graphs of $y = f(x)$ and $y = f(|x|)$?

67. The graph of a function f is illustrated in the figure.

(a) Draw the graph of $y = |f(x)|$.

(b) Draw the graph of $y = f(|x|)$.

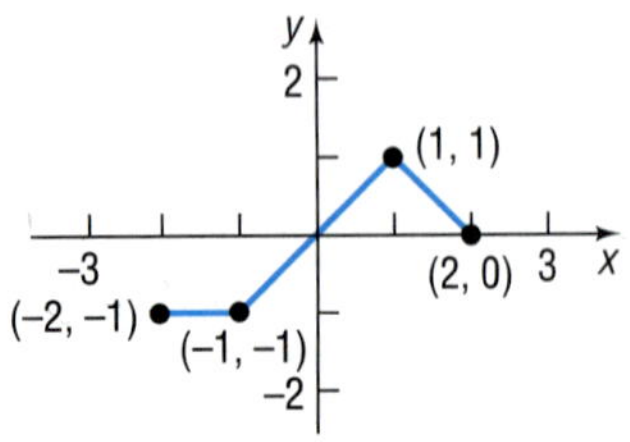

68. The graph of a function f is illustrated in the figure.

(a) Draw the graph of $y = |f(x)|$.

(b) Draw the graph of $y = f(|x|)$.

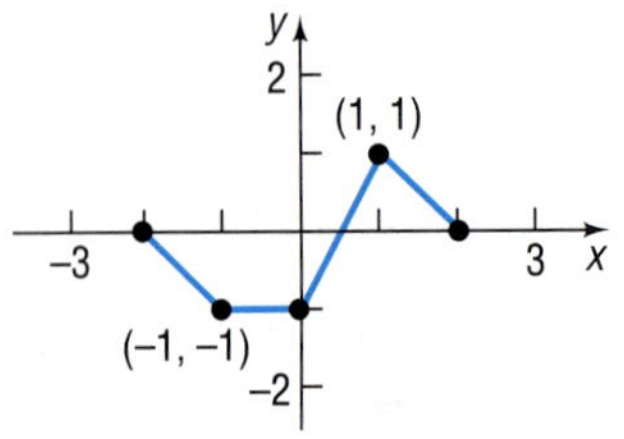

In Problems 69–74, complete the square of each quadratic expression. Then graph each function by hand using the technique of shifting. Verify your results using a graphing utility. (If necessary, refer to Section 1.7 to review Completing the Square.)

69. $f(x) = x^2 + 2x$ **70.** $f(x) = x^2 - 6x$ **71.** $f(x) = x^2 - 8x + 1$

72. $f(x) = x^2 + 4x + 2$ **73.** $f(x) = x^2 + x + 1$ **74.** $f(x) = x^2 - x + 1$

$x^2 - 8x + 16$

75. The equation $y = (x - c)^2$ defines a *family of parabolas*, one parabola for each value of c. On one set of coordinate axes, graph the members of the family for $c = 0$, $c = 3$, and $c = -2$.

76. Repeat Problem 75 for the family of parabolas $y = x^2 + c$.

77. Temperature Measurements The relationship between the Celsius (°C) and Fahrenheit (°F) scales for measuring temperature is given by the equation

$$F = \frac{9}{5}C + 32$$

The relationship between the Celsius (°C) and Kelvin (K) scales is $K = C + 273$. Graph the equation $F = \frac{9}{5}C + 32$ using degrees Fahrenheit on the y-axis and degrees Celsius on the x-axis. Use the techniques introduced in this section to obtain the graph showing the relationship between Kelvin and Fahrenheit temperatures.

78. Period of a Pendulum The period T (in seconds) of a simple pendulum is a function of its length l (in feet) defined by the equation

$$T = 2\pi\sqrt{\frac{l}{g}}$$

where $g \approx 32.2$ feet per second per second is the acceleration of gravity.

(a) Use a graphing utility to graph the function $T = T(l)$.
(b) Now graph the functions $T = T(l + 1)$, $T = T(l + 2)$, and $T = T(l + 3)$.
(c) Discuss how adding to the length l changes the period T.
(d) Now graph the functions $T = T(2l)$, $T = T(3l)$, and $T = T(4l)$.
(e) Discuss how multiplying the length l by factors of 2, 3, and 4 changes the period T.

79. Cigar Company Profits The daily profits of a cigar company from selling x boxes of cigars are given by

$$p(x) = -0.05x^2 + 100x - 2000$$

The government wishes to impose a tax on cigars (sometimes called a *sin tax*) that gives the company the option of either paying a flat tax of \$10,000 per day or a tax of 10% on profits. As Chief Financial Officer (CFO) of the company, you need to decide which tax is the better option for the company.

(a) On the same viewing rectangle, graph $Y_1 = p(x) - 10{,}000$ and $Y_2 = (1 - 0.10)p(x)$.
(b) Based on the graph, which option would you select? Why?
(c) Using the terminology learned in this section, describe each graph in terms of the graph of $p(x)$.
(d) Suppose that the government offered the option of a flat tax of \$4,800 or a tax of 10% on profits. Which option would you select? Why?

3.3 OPERATIONS ON FUNCTIONS; COMPOSITE FUNCTIONS

1 Form the Sum, Difference, Product, and Quotient of Two Functions

2 Form the Composite Function and Find Its Domain

In this section, we introduce some operations on functions. We shall see that functions, like numbers, can be added, subtracted, multiplied, and divided. For example, if $f(x) = x^2 + 9$ and $g(x) = 3x + 5$, then

$$f(x) + g(x) = (x^2 + 9) + (3x + 5) = x^2 + 3x + 14$$

The new function $y = x^2 + 3x + 14$ is called the *sum function* $f + g$. Similarly,

$$f(x) \cdot g(x) = (x^2 + 9)(3x + 5) = 3x^3 + 5x^2 + 27x + 45$$

The new function $y = 3x^3 + 5x^2 + 27x + 45$ is called the *product function* $f \cdot g$.

The general definitions are given next.

If f and g are functions:
Their **sum** $f + g$ is the function defined by

$$(f + g)(x) = f(x) + g(x)$$

The domain of $f + g$ consists of the numbers x that are in the domain of f and in the domain of g.

Their **difference** $f - g$ is the function defined by

$$(f - g)(x) = f(x) - g(x)$$

The domain of $f - g$ consists of the numbers x that are in the domain of f and in the domain of g.

Their **product** $f \cdot g$ is the function defined by

$$(f \cdot g)(x) = f(x) \cdot g(x)$$

The domain of $f \cdot g$ consists of the numbers x that are in the domain of f and in the domain of g.

Their **quotient** f/g is the function defined by

$$\left(\frac{f}{g}\right)(x) = \frac{f(x)}{g(x)}, \qquad g(x) \neq 0$$

The domain of f/g consists of the numbers x for which $g(x) \neq 0$ that are in the domain of f and in the domain of g.

EXAMPLE 1 Operations on Functions

1 Let f and g be two functions defined as

$$f(x) = \sqrt{x + 2} \quad \text{and} \quad g(x) = \sqrt{3 - x}$$

Find the following, and determine the domain in each case.

(a) $(f + g)(x)$ (b) $(f - g)(x)$ (c) $(f \cdot g)(x)$ (d) $(f/g)(x)$

Solution (a) $(f + g)(x) = f(x) + g(x) = \sqrt{x + 2} + \sqrt{3 - x}$

(b) $(f - g)(x) = f(x) - g(x) = \sqrt{x + 2} - \sqrt{3 - x}$

(c) $(f \cdot g)(x) = f(x) \cdot g(x) = (\sqrt{x + 2})(\sqrt{3 - x}) = \sqrt{(x + 2)(3 - x)}$

(d) $\left(\frac{f}{g}\right)(x) = \frac{f(x)}{g(x)} = \frac{\sqrt{x + 2}}{\sqrt{3 - x}} = \frac{\sqrt{(x + 2)(3 - x)}}{3 - x}$

The domain of f consists of all numbers x for which $x \geq -2$; the domain of g consists of all numbers x for which $x \leq 3$. The numbers x common to both these domains are those for which $-2 \leq x \leq 3$. As a result, the numbers x for which $-2 \leq x \leq 3$ comprise the domain of the sum function $f + g$, the difference function $f - g$, and the product function $f \cdot g$. For the quotient function f/g, we must exclude from this set the number 3, because the denominator, g, has the value 0 when $x = 3$. Thus, the domain of f/g consists of all x for which $-2 \leq x < 3$.

NOW WORK PROBLEM 1.

In calculus, it is sometimes helpful to view a complicated function as the sum, difference, product, or quotient of simpler functions. For example,

$F(x) = x^2 + \sqrt{x}$ is the sum of $f(x) = x^2$ and $g(x) = \sqrt{x}$.

$H(x) = (x^2 - 1)/(x^2 + 1)$ is the quotient of $f(x) = x^2 - 1$ and $g(x) = x^2 + 1$.

Composite Functions

2 Consider the function $y = (2x + 3)^2$. If we write $y = f(u) = u^2$ and $u = g(x) = 2x + 3$, then, by a substitution process, we can obtain the original function: $y = f(u) = f(g(x)) = (2x + 3)^2$. This process is called **composition.**

In general, suppose that f and g are two functions, and suppose that x is a number in the domain of g. By evaluating g at x, we get $g(x)$. If $g(x)$ is in the domain of f, then we may evaluate f at $g(x)$ and thereby obtain the expression $f(g(x))$. The correspondence from x to $f(g(x))$ is called a *composite function* $f \circ g$.

Look carefully at Figure 37. Only those x's in the domain of g for which $g(x)$ is in the domain of f can be in the domain of $f \circ g$. The reason is that if $g(x)$ is not in the domain of f then $f(g(x))$ is not defined. Thus, the domain of $f \circ g$ is a subset of the domain of g; the range of $f \circ g$ is a subset of the range of f.

Figure 37

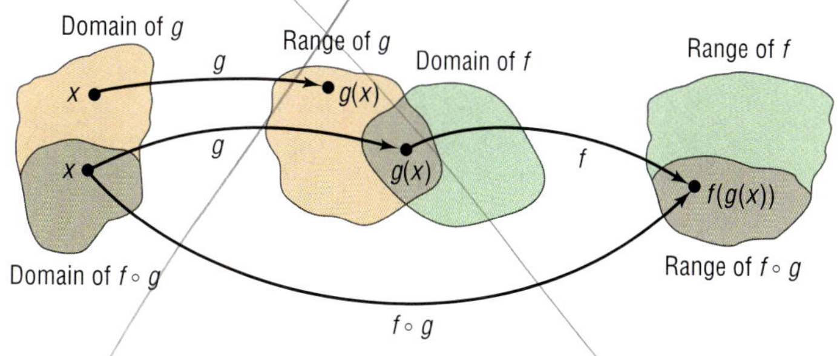

Given two functions f and g, the **composite function,** denoted by $f \circ g$ (read as "f composed with g"), is defined by

$$(f \circ g)(x) = f(g(x))$$

The domain of $f \circ g$ is the set of all numbers x in the domain of g such that $g(x)$ is in the domain of f.

Figure 38 provides a second illustration of the definition. Notice that the "inside" function g in $f(g(x))$ is done first.

Figure 38

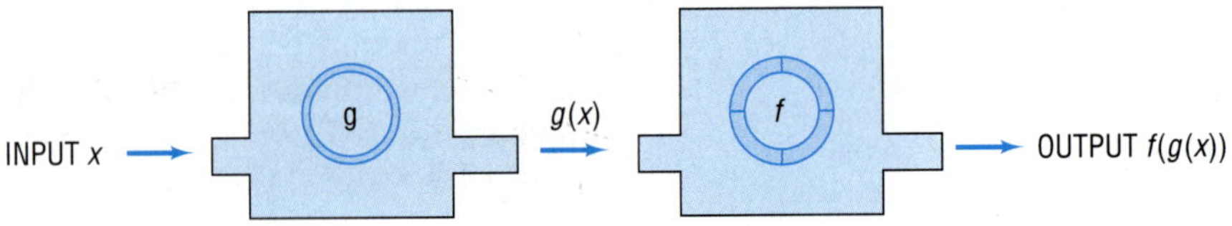

Let's look at some examples.

EXAMPLE 2 Evaluating a Composite Function

Suppose that $f(x) = 2x^2 - 3$ and $g(x) = 4x$. Find

(a) $(f \circ g)(1)$ (b) $(g \circ f)(1)$ (c) $(f \circ f)(-2)$ (d) $(g \circ g)(-1)$

Solution

(a) $(f \circ g)(1) = f(g(1)) = f(4) = 2 \cdot 16 - 3 = 29$

$g(x) = 4x$, $g(1) = 4$; $f(x) = 2x^2 - 3$

(b) $(g \circ f)(1) = g(f(1)) = g(-1) = 4 \cdot (-1) = -4$

$f(x) = 2x^2 - 3$, $f(1) = -1$; $g(x) = 4x$

(c) $(f \circ f)(-2) = f(f(-2)) = f(5) = 2 \cdot 25 - 3 = 47$

$f(-2) = 5$

(d) $(g \circ g)(-1) = g(g(-1)) = g(-4) = 4 \cdot (-4) = -16$

$g(-1) = -4$

Graphing calculators can be used to evaluate composite functions.* Let $Y_1 = f(x) = 2x^2 - 3$ and $Y_2 = g(x) = 4x$. Then, using a TI-83 graphing calculator, $(f \circ g)(1)$ would be found as shown in Figure 39. Notice that this is the result obtained in Example 2(a).

Figure 39

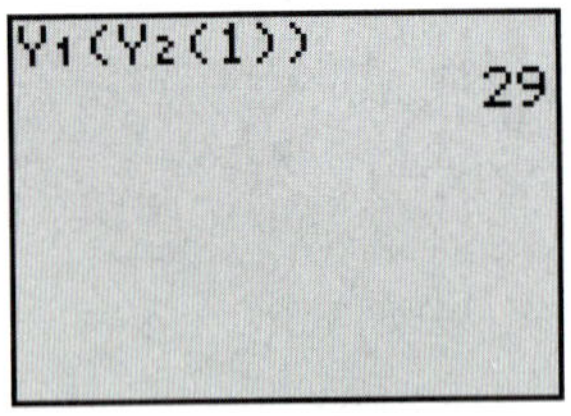

VARS ▸ Select 1 = Function
Select 1 = Y1
Then (Repeat above for Y2
1)) ENTER
29

NOW WORK PROBLEM **13.**

Look back at Figure 37. In determining the domain of the composite function $(f \circ g)(x) = f(g(x))$, keep the following two thoughts in mind about the input x.

1. $g(x)$ must be defined so that any x not in the domain of g will be excluded.
2. $f(g(x))$ must be defined so that any x for which $g(x)$ is not in the domain of f will be excluded.

*Consult your owner's manual for the appropriate keystrokes.

Solution

$$(f \circ g)(x) = f(g(x))$$
$$= f\left(\frac{x+4}{3}\right) \qquad g(x) = \frac{1}{3}(x+4) = \frac{x+4}{3}.$$
$$= 3\left(\frac{x+4}{3}\right) - 4 \qquad \text{Substitute } g(x) \text{ into the rule for } f, f(x) = 3x - 4.$$
$$= x + 4 - 4 = x$$

$$(g \circ f)(x) = g(f(x))$$
$$= g(3x - 4) \qquad f(x) = 3x - 4.$$
$$= \tfrac{1}{3}[(3x - 4) + 4] \qquad \text{Substitute } f(x) \text{ into the rule for } g, g(x) = \tfrac{1}{3}(x+4).$$
$$= \tfrac{1}{3}(3x) = x$$

Thus, $(f \circ g)(x) = (g \circ f)(x) = x$.

In Section 5.1, we shall see that there is an important relationship between functions f and g for which $(f \circ g)(x) = (g \circ f)(x) = x$.

SEEING THE CONCEPT: Using a graphing calculator, let $Y_1 = f(x) = 3x - 4$, $Y_2 = g(x) = \frac{1}{3}(x + 4)$, $Y_3 = f \circ g$, and $Y_4 = g \circ f$. Using the viewing window $-3 \le x \le 3$, $-2 \le y \le 2$, graph only Y_3 and Y_4. What do you see? Use a TABLE or TRACE to verify $Y_3 = Y_4$.

NOW WORK PROBLEM 47.

Calculus Application

Some techniques in calculus require that we be able to determine the components of a composite function. For example, the function $H(x) = \sqrt{x+1}$ is the composition of the functions f and g, where $f(x) = \sqrt{x}$ and $g(x) = x + 1$, because $H(x) = (f \circ g)(x) = f(g(x)) = f(x + 1) = \sqrt{x+1}$.

EXAMPLE 6 Finding the Components of a Composite Function

Find functions f and g such that $f \circ g = H$ if $H(x) = (x^2 + 1)^{50}$.

Solution The function H takes $x^2 + 1$ and raises it to the power 50. A natural way to decompose H is to raise the function $g(x) = x^2 + 1$ to the power 50. Thus, if we let $f(x) = x^{50}$ and $g(x) = x^2 + 1$, then

$$(f \circ g)(x) = f(g(x))$$
$$= f(x^2 + 1)$$
$$= (x^2 + 1)^{50} = H(x)$$

Figure 40

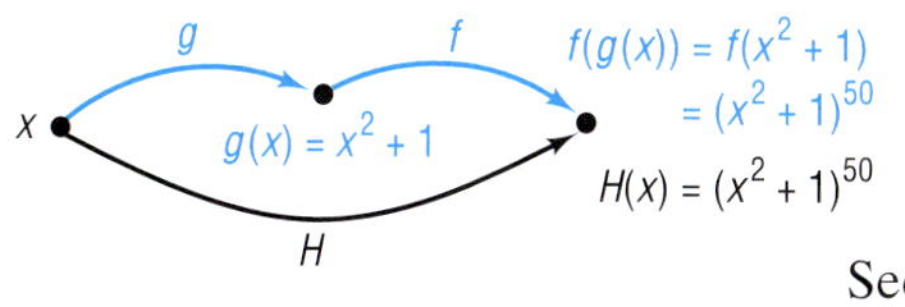

See Figure 40.

Other functions f and g may be found for which $f \circ g = H$ in Example 6. For example, if $f(x) = x^2$ and $g(x) = (x^2 + 1)^{25}$, then

$$(f \circ g)(x) = f(g(x)) = f((x^2 + 1)^{25}) = [(x^2 + 1)^{25}]^2 = (x^2 + 1)^{50}$$

Thus, although the functions f and g found as a solution to Example 6 are not unique, there is usually a "natural" selection for f and g that comes to mind first.

EXAMPLE 7 Finding the Components of a Composite Function

Find functions f and g such that $f \circ g = H$ if $H(x) = 1/(x + 1)$.

Solution Here H is the reciprocal of $g(x) = x + 1$. Thus, if we let $f(x) = 1/x$ and $g(x) = x + 1$, we find that

$$(f \circ g)(x) = f(g(x)) = f(x + 1) = \frac{1}{x+1} = H(x)$$

3.3 EXERCISES

In Problems 1–10, for the given functions f and g, find the following functions and state the domain of each:

(a) $f + g$ (b) $f - g$ (c) $f \cdot g$ (d) f/g

1. $f(x) = 3x + 4;\ g(x) = 2x - 3$
2. $f(x) = 2x + 1;\ g(x) = 3x - 2$
3. $f(x) = x - 1;\ g(x) = 2x^2$
4. $f(x) = 2x^2 + 3;\ g(x) = 4x^3 + 1$
5. $f(x) = \sqrt{x};\ g(x) = 3x - 5$
6. $f(x) = |x|;\ g(x) = x$
7. $f(x) = 1 + \frac{1}{x};\ g(x) = \frac{1}{x}$
8. $f(x) = 2x^2 - x;\ g(x) = 2x^2 + x$
9. $f(x) = \frac{2x+3}{3x-2};\ g(x) = \frac{4x}{3x-2}$
10. $f(x) = \sqrt{x+1};\ g(x) = \frac{2}{x}$
11. Given $f(x) = 3x + 1$ and $(f + g)(x) = 6 - \frac{1}{2}x$, find the function g.
12. Given $f(x) = 1/x$ and $(f/g)(x) = (x + 1)/(x^2 - x)$, find the function g.

In Problems 13–22, for the given functions f and g, find

(a) $(f \circ g)(4)$ (b) $(g \circ f)(2)$ (c) $(f \circ f)(1)$ (d) $(g \circ g)(0)$

Verify your results using a graphing utility.

13. $f(x) = 2x;\ g(x) = 3x^2 + 1$
14. $f(x) = 3x + 2;\ g(x) = 2x^2 - 1$
15. $f(x) = 4x^2 - 3;\ g(x) = 3 - \frac{1}{2}x^2$
16. $f(x) = 2x^2;\ g(x) = 1 - 3x^2$
17. $f(x) = \sqrt{x};\ g(x) = 2x$
18. $f(x) = \sqrt{x+1};\ g(x) = 3x$
19. $f(x) = |x|;\ g(x) = \frac{1}{x^2+1}$
20. $f(x) = |x - 2|;\ g(x) = \frac{3}{x^2+2}$
21. $f(x) = \frac{3}{x^2+1};\ g(x) = \sqrt{x}$
22. $f(x) = x^3;\ g(x) = \frac{2}{x^2+1}$

In Problems 23–30, find the domain of the composite function $f \circ g$.

23. $f(x) = \frac{3}{x-1};\ g(x) = \frac{2}{x}$
24. $f(x) = \frac{1}{x+3};\ g(x) = \frac{-2}{x}$
25. $f(x) = \frac{x}{x-1};\ g(x) = \frac{-4}{x}$
26. $f(x) = \frac{x}{x+3};\ g(x) = \frac{2}{x}$
27. $f(x) = \sqrt{x};\ g(x) = 2x + 3$
28. $f(x) = \sqrt{x-2};\ g(x) = 1 - 2x$
29. $f(x) = \sqrt{x+1};\ g(x) = \frac{2}{x-1}$
30. $f(x) = \sqrt{3-x};\ g(x) = \frac{2}{x-2}$

In Problems 31–46, for the given functions f and g, find

(a) $f \circ g$ (b) $g \circ f$ (c) $f \circ f$ (d) $g \circ g$

State the domain of each composite function.

31. $f(x) = 2x + 3;\ g(x) = 3x$
32. $f(x) = -x;\ g(x) = 2x - 4$
33. $f(x) = 3x + 1;\ g(x) = x^2$
34. $f(x) = x + 1;\ g(x) = x^2 + 4$

35. $f(x) = x^2;\ g(x) = x^2 + 4$

36. $f(x) = x^2 + 1;\ g(x) = 2x^2 + 3$

37. $f(x) = \dfrac{3}{x-1};\ g(x) = \dfrac{2}{x}$

38. $f(x) = \dfrac{1}{x+3};\ g(x) = \dfrac{-2}{x}$

39. $f(x) = \dfrac{x}{x-1};\ g(x) = \dfrac{-4}{x}$

40. $f(x) = \dfrac{x}{x+3};\ g(x) = \dfrac{2}{x}$

41. $f(x) = \sqrt{x};\ g(x) = 2x + 3$

42. $f(x) = \sqrt{x-2};\ g(x) = 1 - 2x$

43. $f(x) = \sqrt{x+1};\ g(x) = \dfrac{2}{x-1}$

44. $f(x) = \sqrt{3-x};\ g(x) = \dfrac{2}{x-2}$

45. $f(x) = ax + b;\ g(x) = cx + d$

46. $f(x) = \dfrac{ax+b}{cx+d};\ g(x) = mx$

In Problems 47–54, show that $(f \circ g)(x) = (g \circ f)(x) = x$.

47. $f(x) = 2x;\ g(x) = \frac{1}{2}x$

48. $f(x) = 4x;\ g(x) = \frac{1}{4}x$

49. $f(x) = x^3;\ g(x) = \sqrt[3]{x}$

50. $f(x) = x + 5;\ g(x) = x - 5$

51. $f(x) = 2x - 6;\ g(x) = \frac{1}{2}(x + 6)$

52. $f(x) = 4 - 3x;\ g(x) = \frac{1}{3}(4 - x)$

53. $f(x) = ax + b;\ g(x) = \dfrac{1}{a}(x - b),\quad a \neq 0$

54. $f(x) = \dfrac{1}{x};\ g(x) = \dfrac{1}{x}$

In Problems 55–60, find functions f and g so that $f \circ g = H$.

55. $H(x) = (2x + 3)^4$

56. $H(x) = (1 + x^2)^3$

57. $H(x) = \sqrt{x^2 + 1}$

58. $H(x) = \sqrt{1 - x^2}$

59. $H(x) = |2x + 1|$

60. $H(x) = |2x^2 + 3|$

61. If $f(x) = 2x^3 - 3x^2 + 4x - 1$ and $g(x) = 2$, find $(f \circ g)(x)$ and $(g \circ f)(x)$.

62. If $f(x) = x/(x - 1)$, find $(f \circ f)(x)$.

63. If $f(x) = 2x^2 + 5$ and $g(x) = 3x + a$, find a so that the graph of $f \circ g$ crosses the y-axis at 23.

64. If $f(x) = 3x^2 - 7$ and $g(x) = 2x + a$, find a so that the graph of $f \circ g$ crosses the y-axis at 68.

65. **Surface Area of a Balloon** The surface area S (in square meters) of a hot-air balloon is given by

$$S(r) = 4\pi r^2$$

where r is the radius of the balloon (in meters). If the radius r is increasing with time t (in seconds) according to the formula $r(t) = \frac{2}{3}t^3$, $t \geq 0$, find the surface area S of the balloon as a function of the time t.

66. **Volume of a Balloon** The volume V (in cubic meters) of the hot-air balloon described in Problem 65 is given by $V(r) = \frac{4}{3}\pi r^3$. If the radius r is the same function of t as in Problem 65, find the volume V as a function of the time t.

67. **Automobile Production** The number N of cars produced at a certain factory in 1 day after t hours of operation is given by $N(t) = 100t - 5t^2$, $0 \leq t \leq 10$. If the cost C (in dollars) of producing N cars is $C(N) = 15{,}000 + 8000N$, find the cost C as a function of the time t of operation of the factory.

68. **Environmental Concerns** The spread of oil leaking from a tanker is in the shape of a circle. If the radius r (in feet) of the spread after t hours is $r(t) = 200\sqrt{t}$, find the area A of the oil slick as a function of the time t.

69. **Production Cost** The price p of a certain product and the quantity x sold obey the demand equation

$$p = -\tfrac{1}{4}x + 100 \qquad 0 \leq x \leq 400$$

Suppose that the cost C of producing x units is

$$C = \frac{\sqrt{x}}{25} + 600$$

Assuming that all items produced are sold, find the cost C as a function of the price p.
[Hint: Solve for x in the demand equation and then form the composite.]

70. **Cost of a Commodity** The price p of a certain commodity and the quantity x sold obey the demand equation

$$p = -\tfrac{1}{5}x + 200 \qquad 0 \leq x \leq 1000$$

Suppose that the cost C of producing x units is

$$C = \frac{\sqrt{x}}{10} + 400$$

Assuming that all items produced are sold, find the cost C as a function of the price p.

71. **Volume of a Cylinder** The volume V of a right circular cylinder of height h and radius r is $V = \pi r^2 h$. If the height is twice the radius, express the volume V as a function of r.

72. **Volume of a Cone** The volume V of a right circular cone is $V = \frac{1}{3}\pi r^2 h$. If the height is twice the radius, express the volume V as a function of r.

3.4 MATHEMATICAL MODELS: CONSTRUCTING FUNCTIONS

1 Construct and Analyze Functions

1 We have already seen examples of real-world problems that lead to linear functions (Section 2.2) and quadratic functions (Section 2.4). Real-world problems may also lead to mathematical models that involve other types of functions.

EXAMPLE 1 Finding the Distance from the Origin to a Point on a Graph

Let $P = (x, y)$ be a point on the graph of $y = x^2 - 1$.

(a) Express the distance d from P to the origin O as a function of x.
(b) What is d if $x = 0$?
(c) What is d if $x = 1$?
(d) What is d if $x = \sqrt{2}/2$?
(e) Use a graphing utility to graph the function $d = d(x)$, $x \geq 0$. Rounded to two decimal places, find the value of x at which d has a local minimum. [This gives the point(s) on the graph of $y = x^2 - 1$ closest to the origin.]

Solution (a) Figure 41 illustrates the graph. The distance d from P to O is

Figure 41

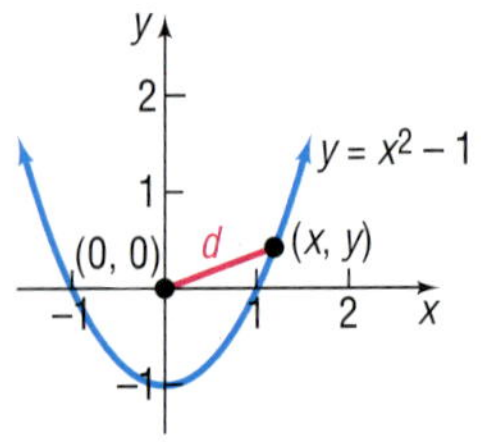

$$d = \sqrt{(x-0)^2 + (y-0)^2} = \sqrt{x^2 + y^2}$$

Since P is a point on the graph of $y = x^2 - 1$, we have

$$d(x) = \sqrt{x^2 + (x^2 - 1)^2} = \sqrt{x^4 - x^2 + 1}$$

Thus, we have expressed the distance d as a function of x.

(b) If $x = 0$, the distance d is

$$d(0) = \sqrt{1} = 1$$

(c) If $x = 1$, the distance d is

$$d(1) = \sqrt{1 - 1 + 1} = 1$$

Figure 42

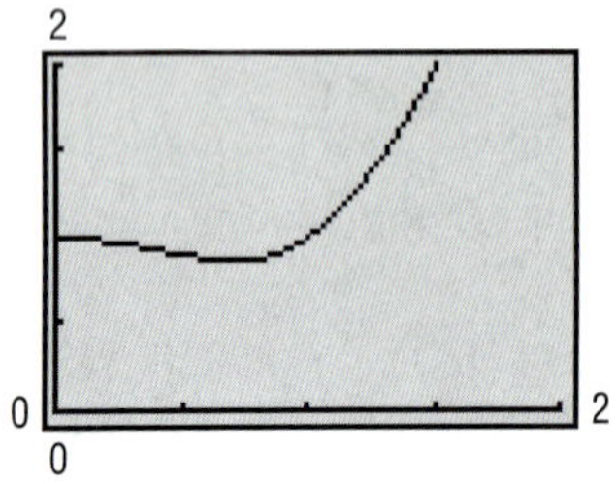

(d) If $x = \sqrt{2}/2$, the distance d is

$$d\left(\frac{\sqrt{2}}{2}\right) = \sqrt{\left(\frac{\sqrt{2}}{2}\right)^4 - \left(\frac{\sqrt{2}}{2}\right)^2 + 1} = \sqrt{\frac{1}{4} - \frac{1}{2} + 1} = \frac{\sqrt{3}}{2}$$

(e) Figure 42 shows the graph of $Y_1 = \sqrt{x^4 - x^2 + 1}$. Using the MINIMUM feature on a graphing utility, we find that when $x \approx 0.71$ the value of d is smallest ($d \approx 0.87$) rounded to two decimal places.

NOW WORK PROBLEM 1.

EXAMPLE 2 Filling a Swimming Pool

A rectangular swimming pool 20 meters long and 10 meters wide is 4 meters deep at one end and 1 meter deep at the other. Figure 43 illustrates a cross-sectional view of the pool. Water is being pumped into the pool at the deep end.

Figure 43

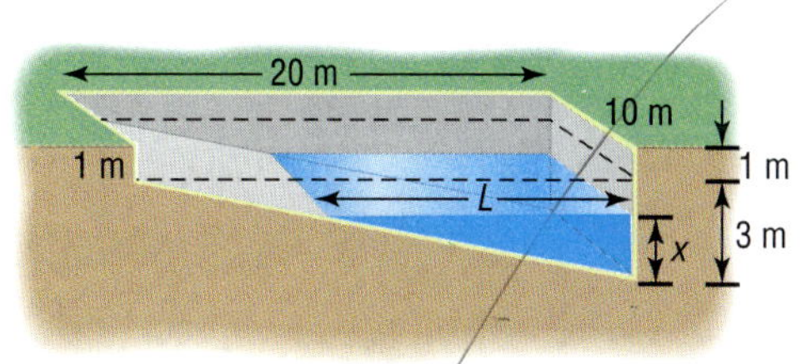

(a) Find a function that expresses the volume V of water in the pool as a function of the height x of the water at the deep end.

(b) Find the volume when the height is 1 meter.

(c) Find the volume when the height is 2 meters.

(d) Use a graphing utility to graph the function $V = V(x)$. At what height is the volume 20 cubic meters? 100 cubic meters?

Solution (a) Let L denote the distance (in meters) measured at water level from the deep end to the short end. Notice that L and x form the sides of a triangle that is similar to the triangle whose sides are 20 meters by 3 meters. Thus, L and x are related by the equation

$$\frac{L}{x} = \frac{20}{3} \quad \text{or} \quad L = \frac{20x}{3} \qquad 0 \le x \le 3$$

The volume V of water in the pool at any time is

$$V = \begin{pmatrix}\text{cross-sectional}\\ \text{triangular area}\end{pmatrix}(\text{width}) = (\tfrac{1}{2}Lx)(10) \quad \text{cubic meters}$$

Since $L = 20x/3$, we have

$$V(x) = \left(\frac{1}{2} \cdot \frac{20x}{3} \cdot x\right)(10) = \frac{100}{3}x^2 \quad \text{cubic meters}$$

(b) When the height x of the water is 1 meter, the volume $V = V(x)$ is

$$V(1) = \frac{100}{3} \cdot 1^2 = 33.3 \text{ cubic meters}$$

(c) When the height x of the water is 2 meters, the volume $V = V(x)$ is

$$V(2) = \frac{100}{3} \cdot 2^2 = \frac{400}{3} = 133.3 \text{ cubic meters}$$

Figure 44

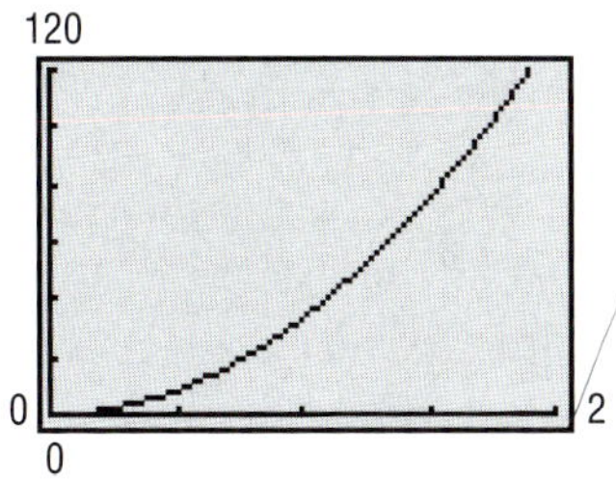

(d) See Figure 44. When $x \approx 0.77$ meter, the volume is 20 cubic meters. When $x \approx 1.73$ meters, the volume is 100 cubic meters.

EXAMPLE 3 Area of an Isosceles Triangle

Consider an isosceles triangle of fixed perimeter p.

(a) If x equals the length of one of the two equal sides, express the area A as a function of x.

(b) What is the domain of A?

(c) Use a graphing utility to graph $A = A(x)$ for $p = 4$.

(d) For what value of x is the area largest?

Solution (a) Look at Figure 45. Since the equal sides are of length x, the third side must be of length $p - 2x$. (Do you see why?) We know that the area A is

$$A = \tfrac{1}{2}(\text{base})(\text{height})$$

To find the height h, we drop the perpendicular to the base of length $p - 2x$ and use the fact that the perpendicular bisects the base. Then, by the Pythagorean Theorem, we have

$$h^2 = x^2 - \left(\frac{p-2x}{2}\right)^2 = x^2 - \frac{1}{4}(p^2 - 4px + 4x^2)$$

$$= px - \frac{1}{4}p^2 = \frac{4px - p^2}{4}$$

$$h = \sqrt{\frac{4px - p^2}{4}} = \frac{\sqrt{p}}{2}\sqrt{4x - p}$$

Figure 45

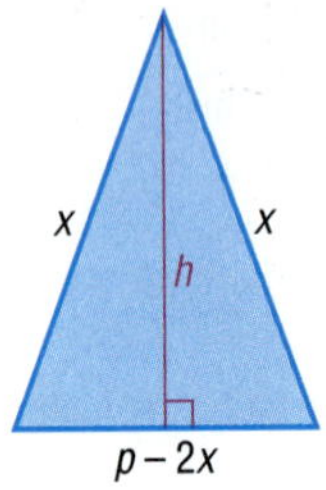

The area A is given by

$$A = \frac{1}{2}\cdot(p - 2x)\frac{\sqrt{p}}{2}\sqrt{4x - p} = \frac{\sqrt{p}}{4}(p - 2x)\sqrt{4x - p}$$

(b) The domain of A is found as follows. Because of the expression $\sqrt{4x - p}$, we require that

$$4x - p > 0$$

$$x > \frac{p}{4}$$

Since $p - 2x$ is a side of the triangle, we also require that

$$p - 2x > 0$$

$$-2x > -p$$

$$x < \frac{p}{2}$$

Thus, the domain of A is $p/4 < x < p/2$, or $(p/4, p/2)$, and we state the function A as

$$A(x) = \frac{\sqrt{p}}{4}(p - 2x)\sqrt{4x - p} \qquad \frac{p}{4} < x < \frac{p}{2}$$

Figure 46

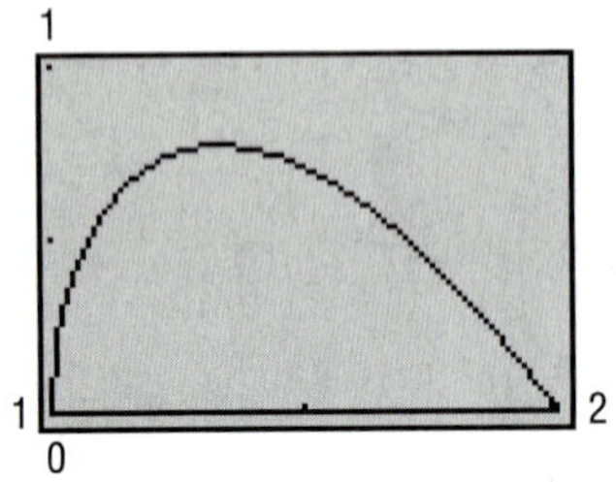

(c) Let $p = 4$. $A(x) = \dfrac{\sqrt{4}}{4}(4 - 2x)\sqrt{4x - 4} = 2(2 - x)\sqrt{x - 1}$, $1 < x < 2$. See Figure 46 for the graph.

(d) The area is largest (approximately 0.77) when $x \approx 1.33$.

NOW WORK PROBLEM 7.

EXAMPLE 4 Minimum Payments and Interest Charged for Credit Cards

(a) Holders of credit cards issued by banks, department stores, oil companies, and so on, receive bills each month that state minimum amounts that must be paid by a certain due date. The minimum due depends on the total amount owed. One such credit card company uses the following rules: For a bill of less than \$10, the entire amount is due. For a bill of at least \$10 but less than \$500, the minimum due is \$10. There is a minimum of \$30 due on a bill of at least \$500 but less than \$1000, a minimum of \$50 due on a bill of at least \$1000 but less than \$1500, and a minimum of \$70 due on bills of \$1500 or more. Find the function f that describes the minimum payment due on a bill of x dollars. Graph f.

(b) The card holder may pay any amount between the minimum due and the total owed. The organization issuing the card charges the card holder interest of 1.5% per month for the first \$1000 owed and 1% per month on any unpaid balance over \$1000. Find the function g that gives the amount of interest charged per month on a balance of x dollars. Graph g.

Solution (a) The function f that describes the minimum payment due on a bill of x dollars is

$$f(x) = \begin{cases} x & \text{if} \quad 0 \le x < 10 \\ 10 & \text{if} \quad 10 \le x < 500 \\ 30 & \text{if} \quad 500 \le x < 1000 \\ 50 & \text{if} \quad 1000 \le x < 1500 \\ 70 & \text{if} \quad 1500 \le x \end{cases}$$

To graph this function f, we proceed as follows. For $0 \le x < 10$, draw the graph of $y = x$; for $10 \le x < 500$, draw the graph of the constant function $y = 10$; for $500 \le x < 1000$, draw the graph of the constant function $y = 30$; and so on. The graph of f is given in Figure 47.

Figure 47

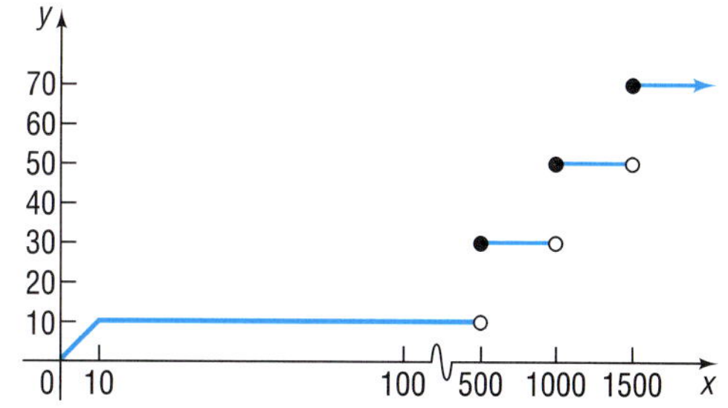

Figure 48

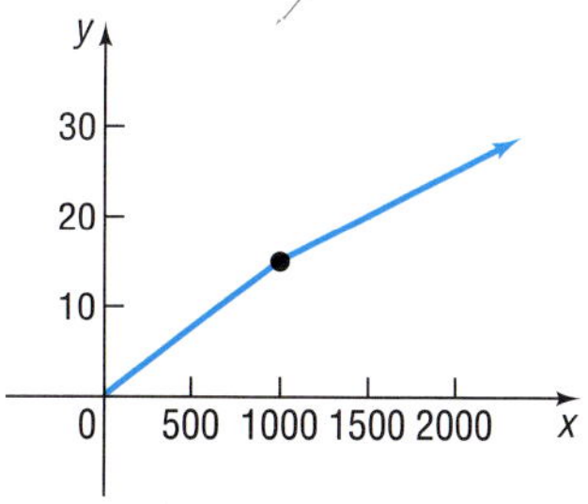

(b) If $g(x)$ is the amount of interest charged per month on a balance of x, then $g(x) = 0.015x$ for $0 \le x \le 1000$. The amount of the unpaid balance above \$1000 is $x - 1000$. If the balance due is $x > 1000$, then the interest is $0.015(1000) + 0.01(x - 1000) = 15 + 0.01x - 10 = 5 + 0.01x$; so

$$g(x) = \begin{cases} 0.015x & \text{if } 0 \le x \le 1000 \\ 5 + 0.01x & \text{if } x > 1000 \end{cases}$$

See Figure 48.

EXAMPLE 5 Making a Playpen*

Figure 49

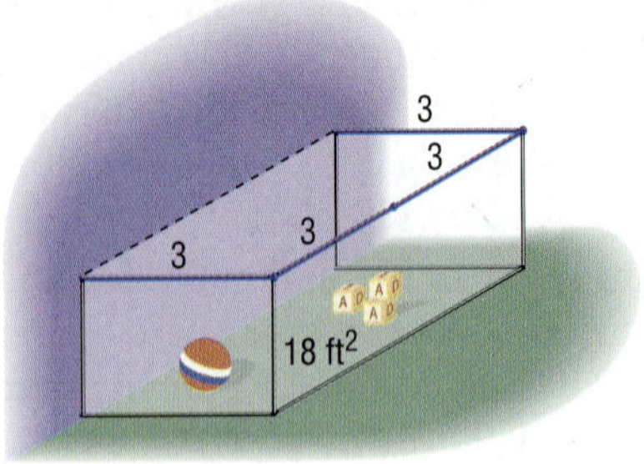

A manufacturer of children's playpens makes a square model that can be opened at one corner and attached at right angles to a wall or, perhaps, the side of a house. If each side is 3 feet in length, the open configuration doubles the available area in which the child can play from 9 square feet to 18 square feet. See Figure 49.

Now suppose that we place hinges at the outer corners to allow for a configuration like the one shown in Figure 50.

(a) Express the area A of this configuration as a function of the distance x between the two parallel sides.
(b) Find the domain of A.
(c) Find A if $x = 5$.
(d) Graph $A = A(x)$. For what value of x is the area largest? What is the maximum area?

Figure 50

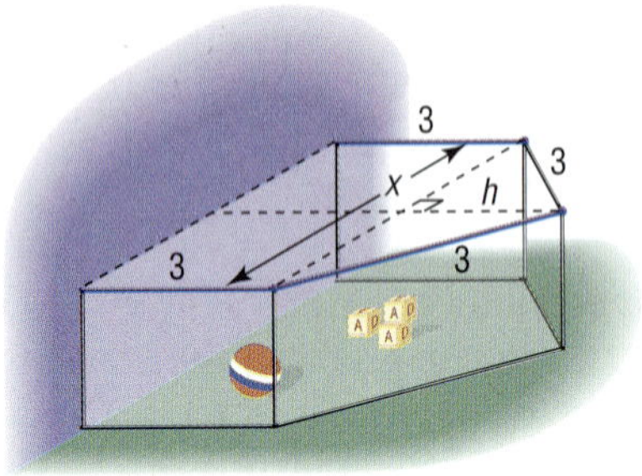

Solution (a) Refer to Figure 50. The area A that we seek consists of the area of a rectangle (with width 3 and length x) and the area of an isosceles triangle (with base x and two equal sides of length 3). The height h of the triangle may be found using the Pythagorean Theorem.

$$h^2 = 3^2 - \left(\frac{x}{2}\right)^2 = 9 - \frac{x^2}{4} = \frac{36 - x^2}{4}$$
$$h = \tfrac{1}{2}\sqrt{36 - x^2}$$

The area A enclosed by the playpen is

$$A = \text{area of rectangle} + \text{area of triangle} = 3x + \tfrac{1}{2}x\left(\tfrac{1}{2}\sqrt{36 - x^2}\right)$$
$$A(x) = 3x + \frac{x\sqrt{36 - x^2}}{4}$$

Thus, we have expressed the area A as a function of x.

(b) To find the domain of A, we note first that $x > 0$, since x is a length. Also, the expression under the radical must be positive, so

$$36 - x^2 > 0$$
$$x^2 < 36$$
$$-6 < x < 6$$

Combining these restrictions, we find that the domain of A is $0 < x < 6$, or $(0, 6)$.

(c) If $x = 5$, the area is

$$A(5) = 3(5) + \frac{5}{4}\sqrt{36 - (5)^2} \approx 19.15 \text{ square feet}$$

Thus, if the width of the playpen is 5 feet, its area is 19.15 square feet.

Figure 51

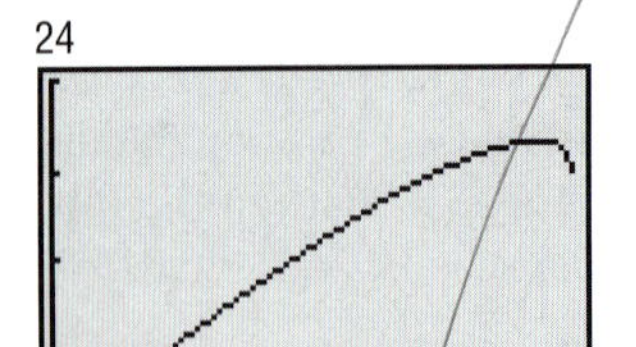

(d) See Figure 51. The maximum area is about 19.82 square feet, obtained when x is about 5.58 feet.

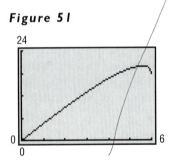

*Adapted from *Proceedings, Summer Conference for College Teachers on Applied Mathematics* (University of Missouri, Rolla), 1971.

3.4 EXERCISES

1. Let $P = (x, y)$ be a point on the graph of $y = x^2 - 8$.
   (a) Express the distance d from P to the origin as a function of x.
   (b) What is d if $x = 0$?
   (c) What is d if $x = 1$?
   (d) Use a graphing utility to graph $d = d(x)$.
   (e) For what values of x is d smallest?

2. Let $P = (x, y)$ be a point on the graph of $y = x^2 - 8$.
   (a) Express the distance d from P to the point $(0, -1)$ as a function of x.
   (b) What is d if $x = 0$?
   (c) What is d if $x = -1$?
   (d) Use a graphing utility to graph $d = d(x)$.
   (e) For what values of x is d smallest?

3. Let $P = (x, y)$ be a point on the graph of $y = \sqrt{x}$.
   (a) Express the distance d from P to the point $(1, 0)$ as a function of x.
   (b) Use a graphing utility to graph $d = d(x)$.
   (c) For what values of x is d smallest?

4. Let $P = (x, y)$ be a point on the graph of $y = 1/x$.
   (a) Express the distance d from P to the origin as a function of x.
   (b) Use a graphing utility to graph $d = d(x)$.
   (c) For what values of x is d smallest?

5. A right triangle has one vertex on the graph of $y = x^3$, $x > 0$, at (x, y), another at the origin, and the third on the positive y-axis at $(0, y)$, as shown in the figure. Express the area A of the triangle as a function of x.

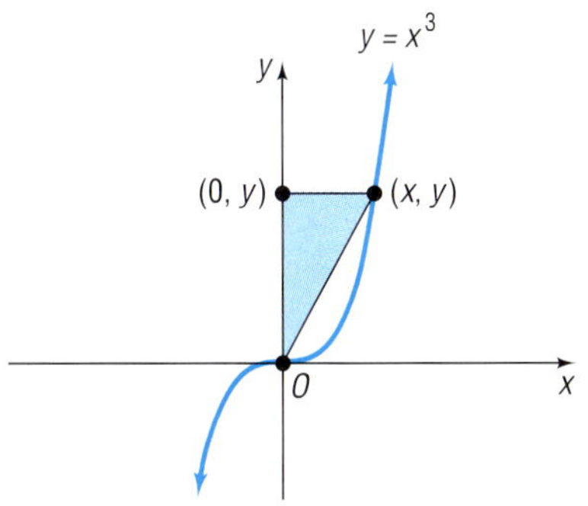

6. A right triangle has one vertex on the graph of $y = 9 - x^2$, $x > 0$, at (x, y), another at the origin, and the third on the positive x-axis at $(x, 0)$. Express the area A of the triangle as a function of x.

7. A rectangle has one corner on the graph of $y = 16 - x^2$, another at the origin, a third on the positive y-axis, and the fourth on the positive x-axis (see the figure).

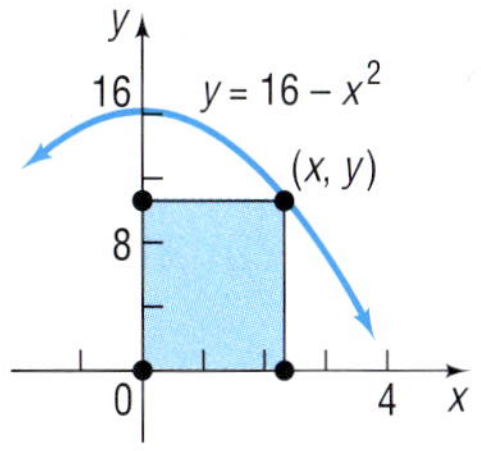

   (a) Express the area A of the rectangle as a function of x.
   (b) What is the domain of A?
   (c) Graph $A = A(x)$. For what value of x is A largest?

8. A rectangle is inscribed in a semicircle of radius 2 (see the figure). Let $P = (x, y)$ be the point in quadrant I that is a vertex of the rectangle and is on the circle.

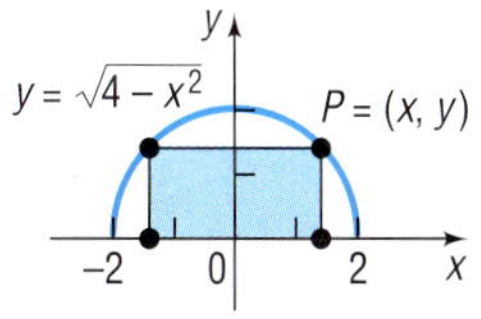

   (a) Express the area A of the rectangle as a function of x.
   (b) Express the perimeter p of the rectangle as a function of x.
   (c) Graph $A = A(x)$. For what value of x is A largest?
   (d) Graph $p = p(x)$. For what value of x is p largest?

9. A rectangle is inscribed in a circle of radius 2 (see the figure). Let $P = (x, y)$ be the point in quadrant I that is a vertex of the rectangle and is on the circle.

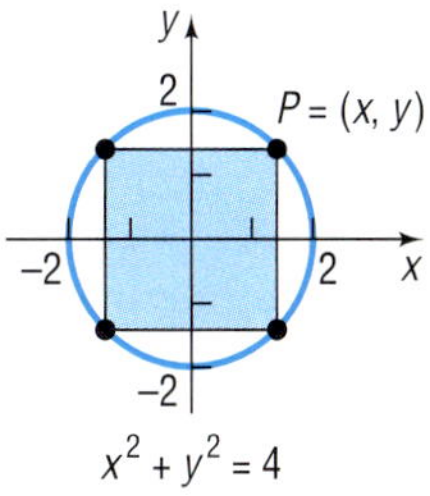

(a) Express the area A of the rectangle as a function of x.
(b) Express the perimeter p of the rectangle as a function of x.
(c) Graph $A = A(x)$. For what value of x is A largest?
(d) Graph $p = p(x)$. For what value of x is p largest?

10. A circle of radius r is inscribed in a square (see the figure).

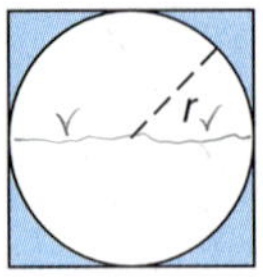

(a) Express the area A of the square as a function of the radius r of the circle.
(b) Express the perimeter p of the square as a function of r.

11. A wire 10 meters long is to be cut into two pieces. One piece will be shaped as a square, and the other piece will be shaped as a circle (see the figure).

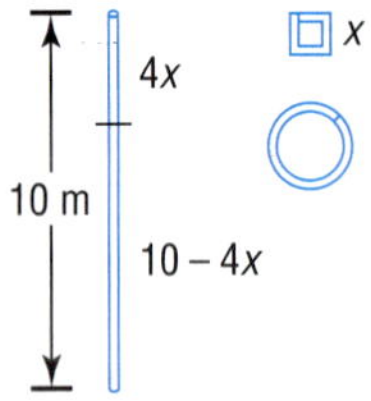

(a) Express the total area A enclosed by the pieces of wire as a function of the length x of a side of the square.
(b) What is the domain of A?
(c) Graph $A = A(x)$. For what value of x is A smallest?

12. A wire 10 meters long is to be cut into two pieces. One piece will be shaped as an equilateral triangle, and the other piece will be shaped as a circle.
(a) Express the total area A enclosed by the pieces of wire as a function of the length x of a side of the equilateral triangle.
(b) What is the domain of A?
(c) Graph $A = A(x)$. For what value of x is A smallest?

13. A wire of length x is bent into the shape of a circle.
(a) Express the circumference of the circle as a function of x.
(b) Express the area of the circle as a function of x.

14. A wire of length x is bent into the shape of a square.
(a) Express the perimeter of the square as a function of x.
(b) Express the area of the square as a function of x.

15. A semicircle of radius r is inscribed in a rectangle so that the diameter of the semicircle is the length of the rectangle (see the figure).

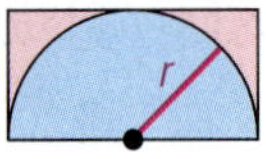

(a) Express the area A of the rectangle as a function of the radius r of the semicircle.
(b) Express the perimeter p of the rectangle as a function of r.

16. An equilateral triangle is inscribed in a circle of radius r. See the figure. Express the circumference C of the circle as a function of the length x of a side of the triangle.
[Hint: First show that $r^2 = x^2/3$.]

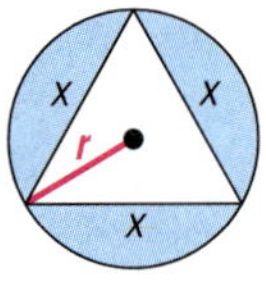

17. An equilateral triangle is inscribed in a circle of radius r. See the figure. Express the area A within the circle, but outside the triangle, as a function of the length x of a side of the triangle.

18. **Cost of Transporting Goods** A trucking company transports goods between Chicago and New York, a distance of 960 miles. The company's policy is to charge, for each pound, \$0.50 per mile for the first 100 miles, \$0.40 per mile for the next 300 miles, \$0.25 per mile for the next 400 miles, and no charge for the remaining 160 miles.
(a) Graph the relationship between the cost of transportation in dollars and mileage over the entire 960-mile route.
(b) Find the cost as a function of mileage for hauls between 100 and 400 miles from Chicago.
(c) Find the cost as a function of mileage for hauls between 400 and 800 miles from Chicago.

19. **Car Rental Costs** An economy car rented in Florida from National Car Rental® on a weekly basis costs \$95 per week. Extra days cost \$24 per day until the day rate exceeds the weekly rate, in which case the weekly rate applies. Find the cost C of renting an economy car as a piecewise-defined function of the number x of days used, where $7 \le x \le 14$. Graph this function.
[Note: Any part of a day counts as a full day.]

20. Rework Problem 19 for a luxury car, which costs \$219 on a weekly basis with extra days at \$45 per day.

21. Two cars leave an intersection at the same time. One is headed south at a constant speed of 30 miles per hour, the other is headed west at a constant speed of 40 miles per hour (see the figure). Express the distance d between the cars as a function of the time t.
[Hint: At $t = 0$, the cars leave the intersection.]

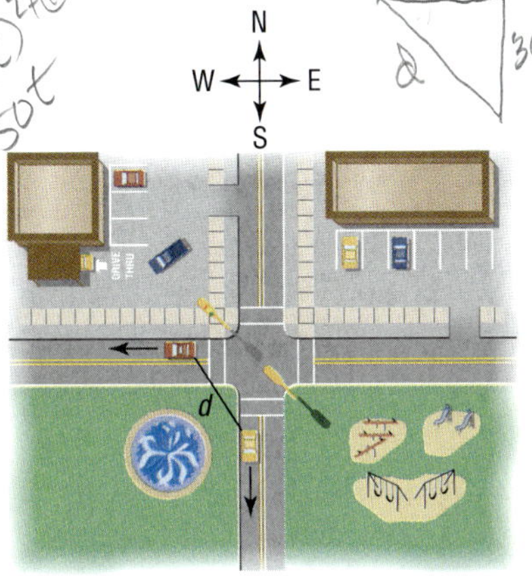

22. Two cars are approaching an intersection. One is 2 miles south of the intersection and is moving at a constant speed of 30 miles per hour. At the same time, the other car is 3 miles east of the intersection and is moving at a constant speed of 40 miles per hour.

(a) Express the distance d between the cars as a function of time t.
[Hint: At $t = 0$, the cars are 2 miles south and 3 miles east of the intersection, respectively.]

(b) Use a graphing utility to graph $d = d(t)$. For what value of t is d smallest?

23. Constructing an Open Box An open box with a square base is to be made from a square piece of cardboard 24 inches on a side by cutting out a square from each corner and turning up the sides (see the figure).

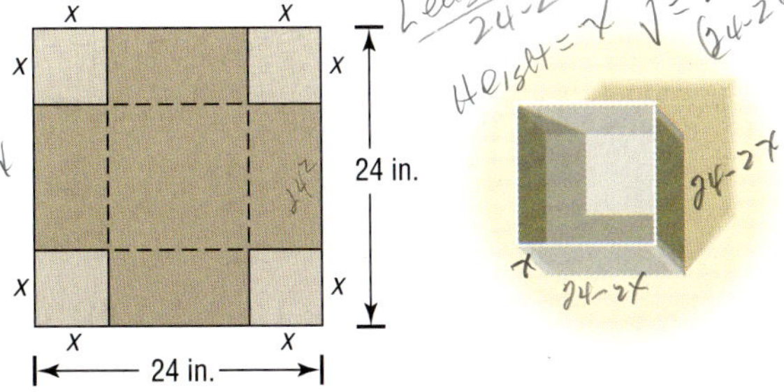

(a) Express the volume V of the box as a function of the length x of the side of the square cut from each corner.

(b) What is the volume if a 3-inch square is cut out?

(c) What is the volume if a 10-inch square is cut out?

(d) Graph $V = V(x)$. For what value of x is V largest?

24. Constructing an Open Box A open box with a square base is required to have a volume of 10 cubic feet.

(a) Express the amount A of material used to make such a box as a function of the length x of a side of the square base.

(b) How much material is required for a base 1 foot by 1 foot?

(c) How much material is required for a base 2 feet by 2 feet?

(d) Graph $A = A(x)$. For what value of x is A smallest?

25. Inscribing a Cylinder in a Sphere Inscribe a right circular cylinder of height h and radius r in a sphere of fixed radius R. See the illustration. Express the volume V of the cylinder as a function of h.
[Hint: $V = \pi r^2 h$. Note also the right triangle.]

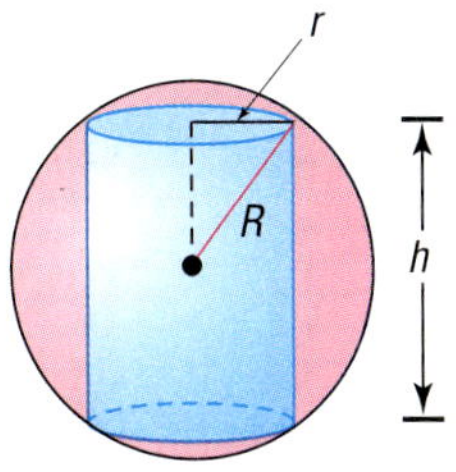

Sphere

26. Inscribing a Cylinder in a Cone Inscribe a right circular cylinder of height h and radius r in a cone of fixed radius R and fixed height H. See the illustration. Express the volume V of the cylinder as a function of r.
[Hint: $V = \pi r^2 h$. Note also the similar triangles.]

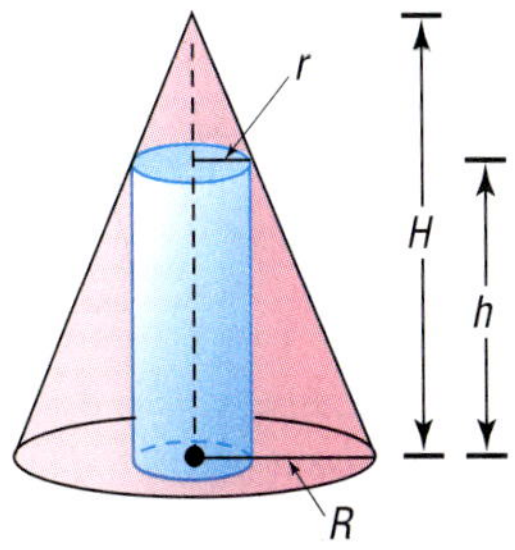

Cone

27. Federal Income Tax Two 1998 Tax Rate Schedules are given in the accompanying table. If x equals the amount on Form 1040, line 37, and y equals the tax due, construct a function f for each schedule.

1998 TAX RATE SCHEDULES

SCHEDULE X — IF YOUR FILING STATUS IS SINGLE				SCHEDULE Y-1 — USE IF YOUR FILING STATUS IS MARRIED FILING JOINTLY OR QUALIFYING WIDOW(ER)			
If the amount on Form 1040, line 37, is: Over—	But not over—	Enter on Form 1040, line 38	of the amount over—	If the amount on Form 1040, line 37, is: Over—	But not over—	Enter on Form 1040, line 38	of the amount over —
\$0	\$25,750	______15%	\$0	\$0	\$43,050	______15%	\$0
25,750	62,450	\$3,862.50 + 28%	25,750	43,050	104,050	\$6,457.50 + 28%	43,050
62,450	130,250	14,138.50 + 31%	62,450	104,050	158,550	23,537.50 + 31%	104,050
130,250	283,150	35,156.50 + 36%	130,250	158,550	283,150	40,432.50 + 36%	158,550
283,150	______	90,200.50 + 39.6%	283,150	283,150	______	85,288.50 + 39.6%	283,150

CHAPTER REVIEW

LIBRARY OF FUNCTIONS

Linear function
$f(x) = mx + b$ Graph is a line with slope m and y-intercept b.

Constant function
$f(x) = b$ Graph is a horizontal line with y-intercept b (see Figure 14).

Identity function
$f(x) = x$ Graph is a line with slope 1 and y-intercept 0 (see Figure 15).

Square function
$f(x) = x^2$ Graph is a parabola with vertex at (0, 0) (see Figure 16).

Cube function
$f(x) = x^3$ See Figure 17.

Square root function
$f(x) = \sqrt{x}$ See Figure 18.

Reciprocal function
$f(x) = 1/x$ See Figure 19.

Absolute value function
$f(x) = |x|$ See Figure 20.

THINGS TO KNOW

Average rate of change of a function
The average rate of change of f from c to x is

$$\frac{\Delta y}{\Delta x} = \frac{f(x) - f(c)}{x - c}, \qquad x \neq c$$

Increasing function
A function f is increasing on an open interval I if, for any choice of x_1 and x_2 in I, with $x_1 < x_2$, we have $f(x_1) < f(x_2)$.

Decreasing function
A function f is decreasing on an open interval I if, for any choice of x_1 and x_2 in I, with $x_1 < x_2$, we have $f(x_1) > f(x_2)$.

Constant function
A function f is constant on an open interval I if, for all choices of x in I, the values of $f(x)$ are equal.

Local maximum
A function f has a local maximum at c if there is an interval I containing c so that, for all $x \neq c$ in I, $f(x) < f(c)$.

Local minimum
A function f has a local minimum at c if there is an interval I containing c so that, for all $x \neq c$ in I, $f(x) > f(c)$.

Even function f
$f(-x) = f(x)$ for every x in the domain ($-x$ must also be in the domain).

Odd function f
$f(-x) = -f(x)$ for every x in the domain ($-x$ must also be in the domain).

HOW TO

Find the average rate of change of a function

Determine algebraically whether a function is even or odd without graphing it

Use a graphing utility to find the local maxima and local minima of a function

Use a graphing utility to determine where a function is increasing or decreasing

Graph certain functions by shifting, compressing, stretching, and/or reflecting (see Table 9)

Perform operation on functions

Find the composite of two functions.

Construct functions in applications, including piecewise-defined functions

FILL-IN-THE-BLANK ITEMS

1. The average rate of change of a function equals the ______________ of the secant line containing two points on its graph.
2. A(n) ______________ function f is one for which $f(-x) = f(x)$ for every x in the domain of f; a(n) ______________ function f is one for which $f(-x) = -f(x)$ for every x in the domain of f.
3. Suppose that the graph of a function f is known. Then the graph of $y = f(x - 2)$ may be obtained by a(n) ______________ shift of the graph of f to the ______________ a distance of 2 units.
4. If $f(x) = x + 1$ and $g(x) = x^3$, then ______________ $= (x + 1)^3$.
5. For the two functions f and g, $(g \circ f)(x) =$ ______________.

TRUE/FALSE ITEMS

T F 1. A function f is decreasing on an open interval I if, for any choice of x_1 and x_2 in I, with $x_1 < x_2$, we have $f(x_1) < f(x_2)$.

T F 2. Even functions have graphs that are symmetric with respect to the origin.

T F 3. The graph of $y = f(-x)$ is the reflection about the y-axis of the graph of $y = f(x)$.

T F 4. $f(g(x)) = f(x) \cdot g(x)$.

T F 5. The domain of the composite function $(f \circ g)(x)$ is the same as that of $g(x)$.

REVIEW EXERCISES

Blue problem numbers indicate the authors' suggestions for use in a Practice Test.

In Problems 1 and 2, use the graph of the function f shown on page 210 to find

(a) The domain and range of f

(b) The intervals on which f is increasing, decreasing, or constant

(c) Whether it is even, odd, or neither

(d) The intercepts, if any

1.

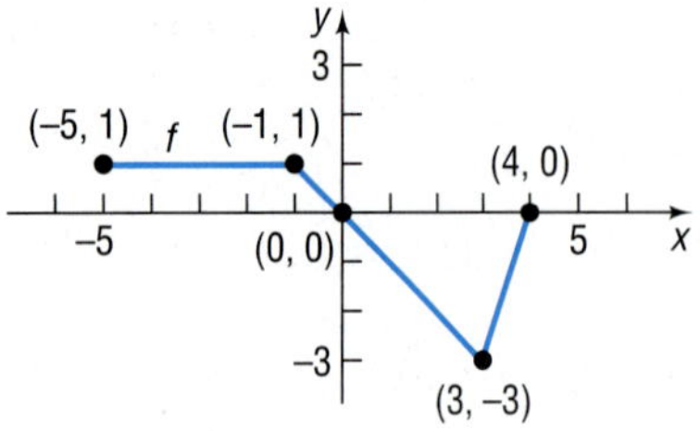

2.

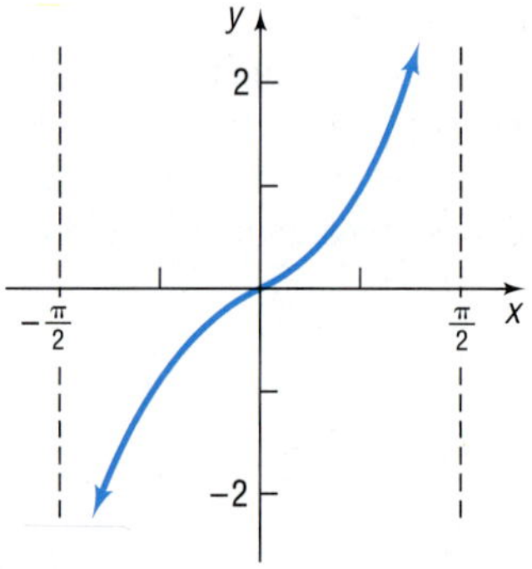

In Problems 3–6, find the average rate of change from 0 to x for each function f. Be sure to simplify.

3. $f(x) = 2 - 5x$ **4.** $f(x) = 2x^2 + 7$ **5.** $f(x) = 3x - 4x^2$ **6.** $f(x) = x^2 - 3x + 2$

In Problems 7–14, determine (algebraically) whether the given function is even, odd, or neither.

7. $f(x) = x^3 - 4x$ **8.** $g(x) = \dfrac{4 + x^2}{1 + x^4}$ **9.** $h(x) = \dfrac{1}{x^4} + \dfrac{1}{x^2} + 1$ **10.** $F(x) = \sqrt{1 - x^3}$

11. $G(x) = 1 - x + x^3$ **12.** $H(x) = 1 + x + x^2$ **13.** $f(x) = \dfrac{x}{1 + x^2}$ **14.** $g(x) = \dfrac{1 + x^2}{x^3}$

In Problems 15–30, graph each function using the techniques of shifting, compressing or stretching, and reflections. Identify any intercepts on the graph. State the domain and, based on the graph, find the range.

15. $F(x) = |x| - 4$ **16.** $f(x) = |x| + 4$ **17.** $g(x) = -|x|$ **18.** $g(x) = \frac{1}{2}|x|$

19. $h(x) = \sqrt{x - 1}$ **20.** $h(x) = \sqrt{x} - 1$ **21.** $f(x) = \sqrt{1 - x}$ **22.** $f(x) = -\sqrt{x}$

23. $F(x) = \begin{cases} x^2 + 4 & \text{if } x < 0 \\ 4 - x^2 & \text{if } x \geq 0 \end{cases}$ **24.** $H(x) = \begin{cases} |1 - x| & \text{if } 0 \leq x \leq 2 \\ |x - 1| & \text{if } x > 2 \end{cases}$ **25.** $h(x) = (x - 1)^2 + 2$

26. $h(x) = (x + 2)^2 - 3$ **27.** $g(x) = (x - 1)^3 + 1$ **28.** $g(x) = (x + 2)^3 - 8$

29. $f(x) = \begin{cases} x & \text{if } 0 < x < 4 \\ 2\sqrt{x} & \text{if } x \geq 4 \end{cases}$ **30.** $f(x) = \begin{cases} 3|x| & \text{if } x < 0 \\ \sqrt{1 - x} & \text{if } 0 \leq x \leq 1 \end{cases}$

In Problems 31–34, use a graphing utility to graph each function over the indicated interval. Approximate any local maxima and local minima. Determine where the function is increasing and where it is decreasing.

31. $f(x) = 2x^3 - 5x + 1 \quad (-3, 3)$ **32.** $f(x) = -x^3 + 3x - 5 \quad (-3, 3)$

33. $f(x) = 2x^4 - 5x^3 + 2x + 1 \quad (-2, 3)$ **34.** $f(x) = -x^4 + 3x^3 - 4x + 3 \quad (-2, 3)$

In Problems 35–40, for the given functions f and g find:

(a) $(f \circ g)(2)$ *(b)* $(g \circ f)(-2)$ *(c)* $(f \circ f)(4)$ *(d)* $(g \circ g)(-1)$

Verify your results using a graphing utility.

35. $f(x) = 3x - 5;\ g(x) = 1 - 2x^2$ **36.** $f(x) = 4 - x;\ g(x) = 1 + x^2$

37. $f(x) = \sqrt{x + 2};\ g(x) = 2x^2 + 1$ **38.** $f(x) = 1 - 3x^2;\ g(x) = \sqrt{4 - x}$

39. $f(x) = \dfrac{1}{x^2 + 4};\ g(x) = 3x - 2$ **40.** $f(x) = \dfrac{2}{1 + 2x^2};\ g(x) = 3x$

In Problems 41–46, find $f \circ g$, $g \circ f$, $f \circ f$, *and* $g \circ g$ *for each pair of functions. State the domain of each.*

41. $f(x) = 2 - x;\ g(x) = 3x + 1$ **42.** $f(x) = 2x - 1;\ g(x) = 2x + 1$

43. $f(x) = 3x^2 + x + 1;\ g(x) = |3x|$ **44.** $f(x) = \sqrt{3x};\ g(x) = 1 + x + x^2$

45. $f(x) = \dfrac{x + 1}{x - 1};\ g(x) = \dfrac{1}{x}$ **46.** $f(x) = \sqrt{x - 3};\ g(x) = \dfrac{3}{x}$

47. For the graph of the function f shown below:

(a) Draw the graph of $y = f(-x)$.
(b) Draw the graph of $y = -f(x)$.
(c) Draw the graph of $y = f(x + 2)$.
(d) Draw the graph of $y = f(x) + 2$.
(e) Draw the graph of $y = 2f(x)$.
(f) Draw the graph of $y = f(3x)$.

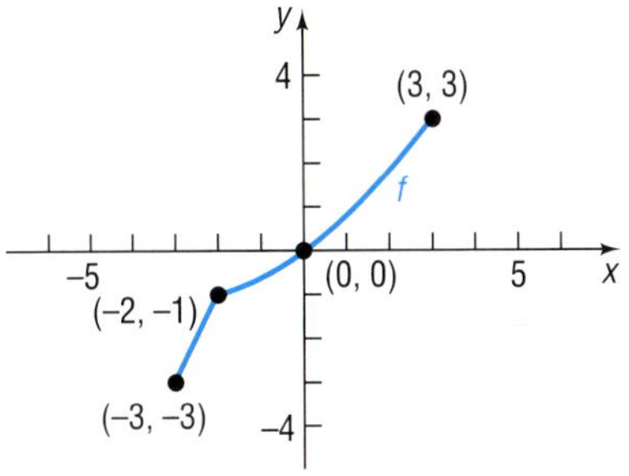

48. Repeat Problem 47 for the graph of the function g shown below.

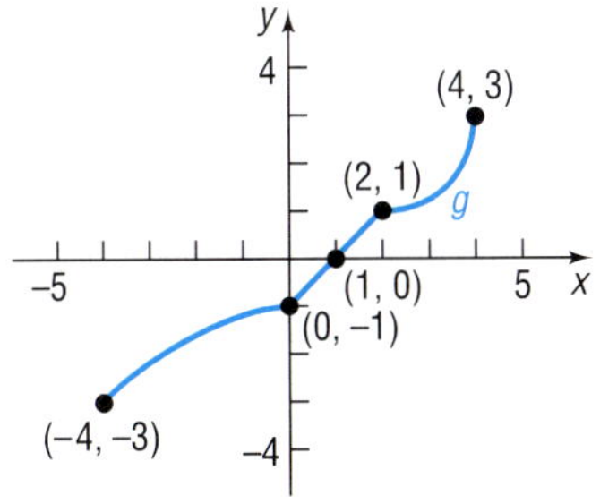

49. Find the point on the line $y = x$ that is closest to the point (3, 1).
[Hint: Find the minimum value of the function $f(x) = d^2$, where d is the distance from (3, 1) to a point on the line.]

50. Find the point on the line $y = x + 1$ that is closest to the point (4, 1).

51. A rectangle has one vertex on the line $y = 10 - x^2$, $x > 0$, another at the origin, one on the positive x-axis, and one on the positive y-axis. Find the largest area A that can be enclosed by the rectangle.

52. Constructing a Closed Box A closed box with a square base is required to have a volume of 10 cubic feet.

(a) Express the amount A of material used to make such a box as a function of the length x of a side of the square base.
(b) How much material is required for a base 1 foot by 1 foot?
(c) How much material is required for a base 2 feet by 2 feet?
(d) Graph $A = A(x)$. For what value of x is A smallest?

53. Spheres The volume V of a sphere of radius r is $V = \frac{4}{3}\pi r^3$; the surface area S of this sphere is $S = 4\pi r^2$. Express the volume V as a function of the surface area S. If the surface area doubles, how does the volume change?

54. Productivity versus Earnings The following data represent the average hourly earnings and productivity (output per hour) of production workers for the years 1986–1995. Let productivity x be the independent variable and average hourly earnings y be the dependent variable.

Productivity	Average Hourly Earnings
94.2	8.76
94.1	8.98
94.6	9.28
95.3	9.66
96.1	10.01
96.7	10.32
100	10.57
100.2	10.83
100.7	11.12
100.8	11.44

Source: Bureau of Labor Statistics.

(a) Draw a scatter diagram of the data.
(b) Draw a line through the points (94.2, 8.76) and (96.7, 10.32) on the scatter diagram found in part (a).
(c) Find the average rate of change of hourly earnings for productivity from 94.2 to 96.7.
(d) Interpret the average rate of change found in part (c).
(e) Draw a line through the points (96.7, 10.32) and (100.8, 11.44) on the scatter diagram found in part (a).
(f) Find the average rate of change of hourly earnings for productivity from 96.7 to 100.8.
(g) Interpret the average rate of change found in part (f).
(h) What is happening to the average rate of change of hourly earnings as productivity increases?

CHAPTER PROJECTS

1. **Oil Spill** An oil tanker strikes a sand bar that rips a hole in the hull of the ship. Oil begins leaking out of the tanker with the spilled oil forming a circle around the tanker. The radius of the circle is increasing at the rate of 2.2 feet per hour.

   (a) Write the area of the circle as a function of the radius, r.

   (b) Write the radius of the circle as a function of time, t.

   (c) What is the radius of the circle after 2 hours? What is the radius of the circle after 2.5 hours?

   (d) Use the result of part (c) to determine the area of the circle after 2 hours and 2.5 hours.

   (e) Determine a function that represents area as a function of time, t.

   (f) Use the result of part (e) to determine the area of the circle after 2 hours and 2.5 hours.

   (g) Compute the average rate of change of the area of the circle from 2 hours to 2.5 hours.

   (h) Compute the average rate of change of the area of the circle from 3 hours to 3.5 hours.

   (i) Based on the results obtained in parts (g) and (h), what is happening to the average rate of change of the area of the circle as time passes?

   (j) If the oil tanker is 150 yards from shore, when will the oil spill first reach the shoreline? (1 yard = 3 feet).

   (k) How long will it be until 6 miles of shoreline is contaminated with oil? (1 mile = 5280 feet)

2. **Economics** For a manufacturer of computer chips, it has been determined that the number Q of chips produced as a function of labor L, in hours, is given by $Q = 8L^{1/2}$. In addition, the cost of production as a function of the number of chips produced per month is $C(Q) = 0.002Q^3 + 0.5Q^2 - Q + 1000$.

   (a) Determine cost as a function of labor.

   (b) If the demand function for the chips is $Q(P) = 5000 - 20P$, where P is the price per chip, find the revenue R as a function of Q. *Hint:* Revenue equals the number of units produced times the price per unit.

   (c) Find revenue as a function of labor.

   (d) Determine profit as a function of labor.

   (e) Using a graphing utility, graph profit as a function of labor. Determine the implied domain of this function.

   (f) Determine the profit-maximizing level of labor.

   (g) Determine the monthly maximum profit of the firm.

   (h) Determine the optimal (profit maximizing) level of output for the firm.

   (i) Determine the optimal price per chip.

CHAPTER

4 Polynomial and Rational Functions

Weber on Vallas Wish List to Ease School Crowding

Vallas said that if a deal can be struck in the coming days, Weber could be reopened as an elementary school in the fall. Schools such as Burbank and other crowded institutions would benefit if such an arrangement were made, he said.

In recent years, Burbank teachers have held classes in the building's storage and loading areas. Tutoring sessions are often held in the foyer of a second-floor lavatory, and as many as four pupils are required to share a locker.

Enrollment is 1,072, nearly 300 more than the school was built to handle. Its heating system has been broken for some time, making it difficult to heat classrooms evenly. (Ray Quintanilla, Chicago Tribune.)

See Chapter Project 1.

Preparing for This Chapter

Before getting started on this chapter, review the following concepts:

- Monomials (Appendix, Section 5, pp. 701–702)
- Polynomials (Appendix, Section 5, pp. 702–704)
- Quadratic Equations (Section 2.3, pp. 124–130)
- Domain of a Function (Section 2.1, p. 104)
- Linear Inequalities (Section 1.5, pp. 57–60)

Outline

4.1 POWER FUNCTIONS AND MODELS

1 Graph Transformations of Power Functions

2 Find the Power Function of Best Fit from Data

Power Functions

A *power function* is a function that is defined by a single monomial.*

A **power function of degree n** is a function of the form

$$f(x) = ax^n \quad (1)$$

where a is a real number, $a \neq 0$, and $n > 0$ is an integer.

Examples of power functions are

$f(x) = 3x$	$f(x) = -5x^2$	$f(x) = 8x^3$	$f(x) = -5x^4$
degree 1	degree 2	degree 3	degree 4

In this section we discuss characteristics of power functions and graph transformations of power functions.

The graph of a power function of degree 1, $f(x) = ax$, is a line with slope a that passes through the origin. The graph of a power function of degree 2, $f(x) = ax^2$, is a parabola, with vertex at the origin, that opens up if $a > 0$ and down if $a < 0$.

If we know how to graph a power function of the form $f(x) = x^n$, then a compression or stretch and, perhaps, a reflection about the x-axis will enable us to obtain the graph of $g(x) = ax^n$, for any real number a. Consequently, we shall concentrate on graphing power functions of the form $f(x) = x^n$.

We begin with power functions of even degree of the form $f(x) = x^n$, $n \geq 2$ and *n even*. The domain of f is the set of all real numbers, and the range is the set of nonnegative real numbers. Such a power function is an even function (do you see why?), and hence its graph is symmetric with respect to the y-axis. Its graph always contains the origin and the points $(-1, 1)$ and $(1, 1)$.

EXPLORATION Using your graphing utility and the viewing window $-2 \leq x \leq 2$, $-4 \leq y \leq 16$, graph the function $Y_1 = f(x) = x^4$. On the same screen, graph $Y_2 = g(x) = x^8$. Now, also on the same screen, graph $Y_3 = h(x) = x^{12}$. What do you notice about the graphs as the magnitude of the exponent increases? Repeat the procedure given above for the viewing window $-1 \leq x \leq 1$, $0 \leq y \leq 1$. What do you notice? See Figures 1(a) and (b).

For large n, it appears that the graph coincides with the x-axis near the origin, but it does not; the graph actually touches the x-axis only at the ori-

*A monomial is the product of a constant times a variable raised to a nonnegative integer power; that is, it is an expression of the form ax^n, $n \geq 0$, an integer. Refer to the Appendix, Section 5, pp. 701–702 for more detail.

Figure 1

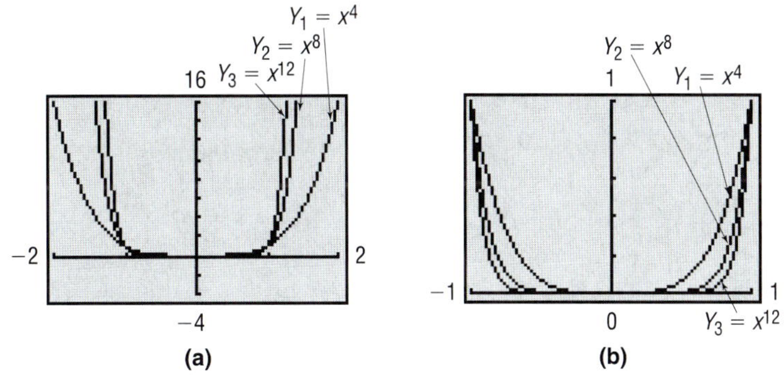

Table 1

X	Y2	Y3
-1	1	1
1	1	1
.5	.00391	2.4E-4
.1	1E-8	1E-12
.01	1E-16	1E-24
.001	1E-24	1E-36
0	0	0

Y2=X^8

gin. See Table 1, where $Y_2 = x^8$ and $Y_3 = x^{12}$; for x close to 0, the values of y are positive and close to 0. Also, for large n, it may appear that for $x < -1$ or for $x > 1$ the graph is vertical, but it is not; it is only increasing very rapidly. If you use TRACE along one of the graphs, these distinctions will be clear.

To summarize:

> ***Characteristics of Power Functions, $y = x^n$, n is an even integer***
>
> 1. The graph is symmetric with respect to the y-axis.
> 2. The domain is the set of all real numbers. The range is the set of nonnegative real numbers.
> 3. The graph always contains the points $(0, 0)$, $(1, 1)$, and $(-1, 1)$.
> 4. As the exponent n increases in magnitude, the graph becomes more vertical when $x < -1$ or $x > 1$, but for x near the origin, the graph tends to flatten out and lie closer to the x-axis.

Now we consider power functions of odd degree of the form $f(x) = x^n$, *n odd*. The domain and range of f are the set of real numbers. Such a power function is an odd function (do you see why?), and hence its graph is symmetric with respect to the origin. Its graph always contains the origin and the points $(-1, -1)$ and $(1, 1)$.

EXPLORATION Using your graphing utility and the viewing window $-2 \le x \le 2$, $-16 \le y \le 16$, graph the function $Y_1 = f(x) = x^3$. On the same screen, graph $Y_2 = g(x) = x^7$ and $Y_3 = h(x) = x^{11}$. What do you notice about the graphs as the magnitude of the exponent increases? Repeat the procedure given above for the viewing window $-1 \le x \le 1$, $-1 \le y \le 1$. What do you notice? The graphs on your screen should look like Figures 2(a) and (b).

Figure 2

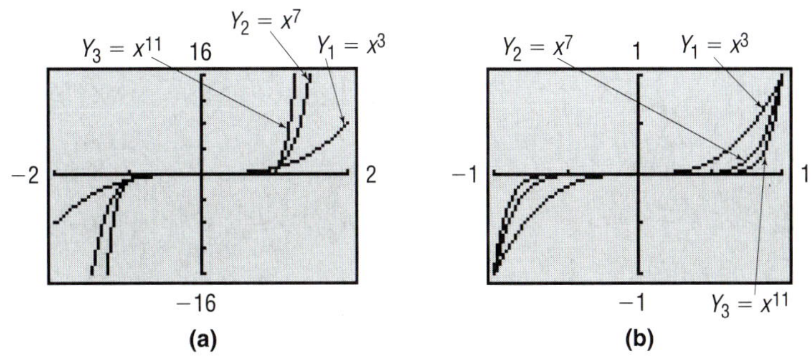

It appears that the graph coincides with the x-axis near the origin, but it does not; the graph actually touches the x-axis only at the origin. Also, it appears that as x increases the graph is vertical, but it is not; it is increasing very rapidly. TRACE along the graphs to verify these distinctions.

To summarize:

> ***Characteristics of Power Functions, $y = x^n$, n is an odd integer***
>
> 1. The graph is symmetric with respect to the origin.
> 2. The domain and range are the set of all real numbers.
> 3. The graph always contains the points $(0, 0)$, $(1, 1)$, and $(-1, -1)$.
> 4. As the exponent n increases in magnitude, the graph becomes more vertical when $x < -1$ or $x > 1$, but for x near the origin the graph tends to flatten out and lie closer to the x-axis.

1

The methods of shifting, compression, stretching, and reflection studied in Section 3.2, when used with the facts just presented, will enable us to graph and analyze a variety of functions that are transformations of power functions.

EXAMPLE 1 Graphing Transformations of Power Functions

Using a graphing utility, show each stage to obtain the graph of $f(x) = 1 - x^5$.

Solution

STAGE 1: Graph $y = x^5$. See Figure 3 and Table 2.

Figure 3

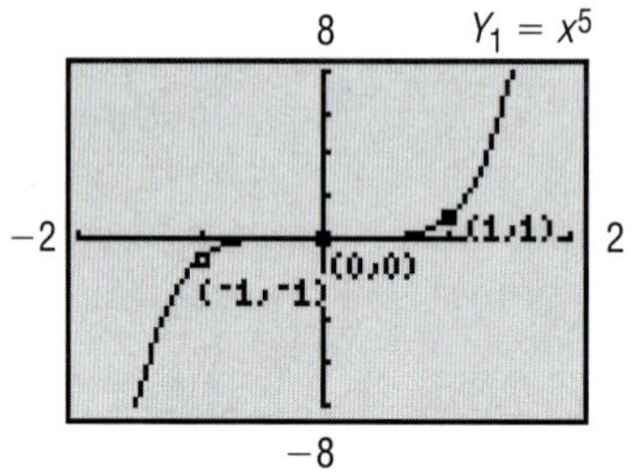

Table 2

X	Y1
-3	-243
-2	-32
-1	-1
0	0
1	1
2	32
3	243

Y1 ■ X^5

STAGE 2: Graph $y = -x^5$ (multiply by -1; reflect about the x-axis). See Figure 4 and Table 3.

Figure 4

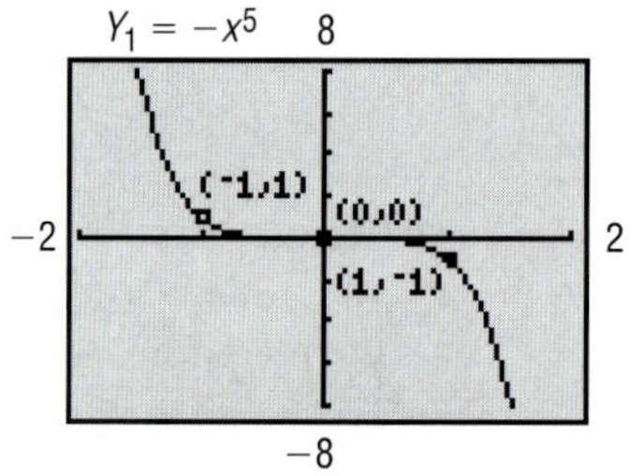

Table 3

X	Y1
-3	243
-2	32
-1	1
0	0
1	-1
2	-32
3	-243

Y1 ■ -X^5

STAGE 3: Graph $y = 1 - x^5$ (add 1; shift up 1 unit). See Figure 5 and Table 4.

Figure 5

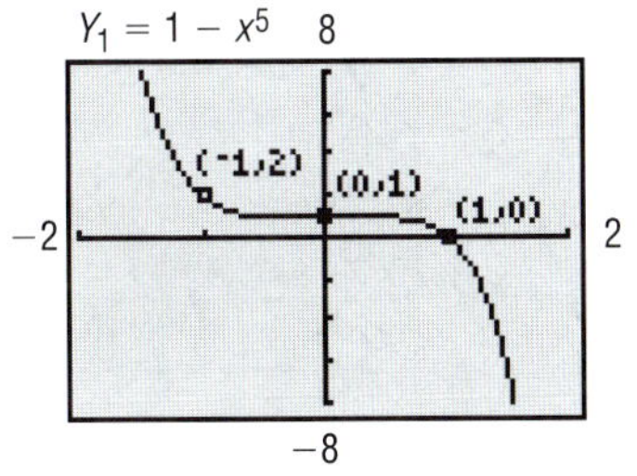

Table 4

X	Y1
-3	244
-2	33
-1	2
0	1
1	0
2	-31
3	-242

Y1=1-X^5

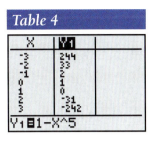

By following key points as we shift, reflect, and so forth, we determine that the intercepts of f are $(0, 1)$ and $(1, 0)$.

EXAMPLE 2 Graphing Transformations of Power Functions

Using a graphing utility, show each stage to obtain the graph of $f(x) = \frac{1}{2}(x - 1)^4$.

Solution STAGE 1: Graph $y = x^4$. See Figure 6 and Table 5.

Figure 6

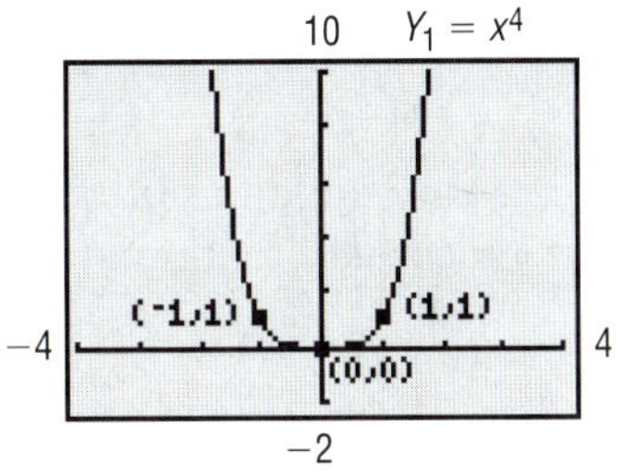

Table 5

X	Y1
-3	81
-2	16
-1	1
0	0
1	1
2	16
3	81

Y1=X^4

STAGE 2: Graph $y = \frac{1}{2}x^4$ (multiply by $\frac{1}{2}$; compress vertically by a factor of $\frac{1}{2}$). See Figure 7 and Table 6.

Figure 7

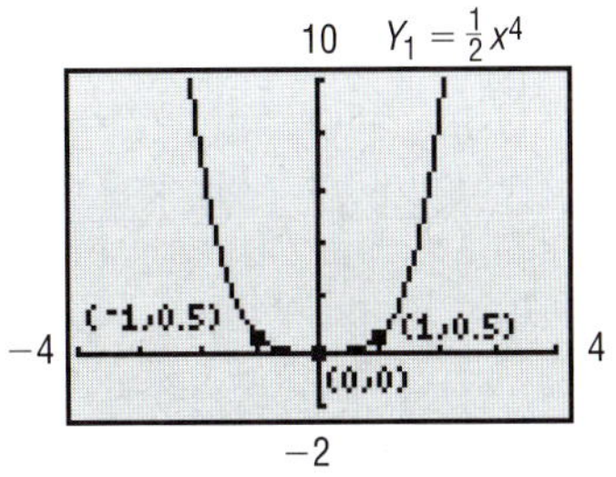

Table 6

X	Y1
-3	40.5
-2	8
-1	.5
0	0
1	.5
2	8
3	40.5

Y1=(1/2)X^4

STAGE 3: Graph $y = \frac{1}{2}(x - 1)^4$ (replace x by $x - 1$; shift right 1 unit). See Figure 8 and Table 7.

Figure 8

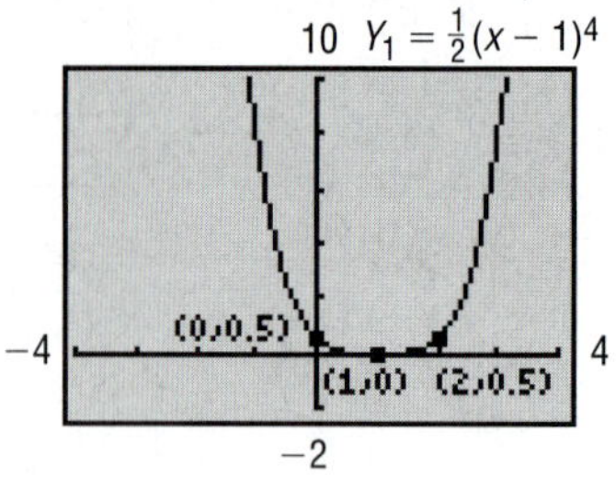

Table 7

The intercepts of f are $(0, \frac{1}{2})$ and $(1, 0)$.

NOW WORK PROBLEM 1.

Modeling Power Functions: Curve Fitting

2 Figure 9 shows a scatter diagram that follows a power function.

Figure 9

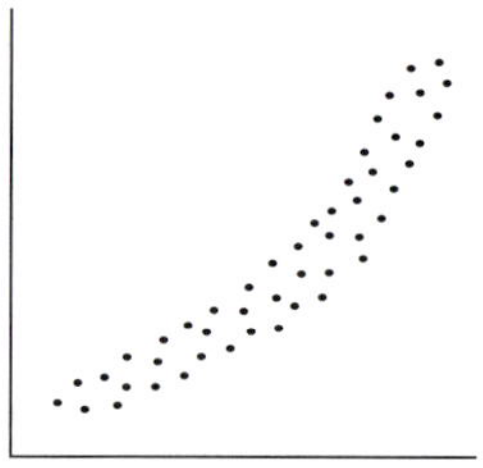

$y = ax^b$, $a > 0$, $b > 1$
$x, y > 0$

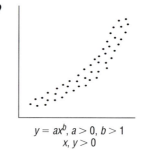

Many situations lead to data that can be modeled using a power function. One such situation involves falling objects.

EXAMPLE 3 Fitting Data to a Power Function

Scott drops a ball from various heights and records the time it takes for the ball to hit the ground. Using a motion detector connected to his graphing calculator, he collects the following data:

Time t (seconds)	1.528	2.015	3.852	4.154	4.625
Distance s (meters)	11.46	19.99	72.41	84.45	104.23

(a) Using a graphing utility, draw a scatter diagram using time t as the independent variable and distance s as the dependent variable.
(b) The power function of best fit to these data is $s(t) = 4.93t^{1.993}$. Predict how long it will take the ball to fall 100 meters.
(c) Graph the power function found in (b) on the scatter diagram.
(d) Physics theory states the distance an object falls is directly proportional to the time squared. Rewrite the model found in (b) so it is of the form

$s = \frac{1}{2}gt^2$, where s is distance, g is acceleration due to gravity, and t is time. What is Scott's estimate of g? (It is known that the acceleration due to gravity is approximately 9.8 meters/sec^2.)

(e) Verify that the power function given in part (b) is the power function of best fit.

Solution (a) After entering the data into the graphing utility, we obtain the scatter diagram shown in Figure 10.

Figure 10

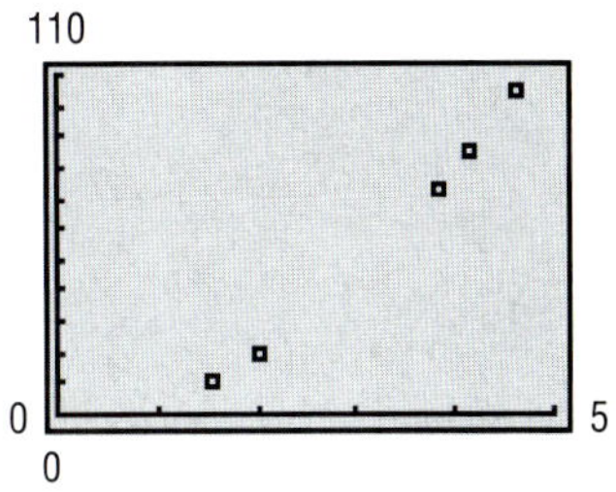

(b) If $s = 100$ meters, then $Y_1 = s(t) = 4.93t^{1.993}$ INTERSECTS $Y_2 = 100$ when $t = 4.53$ seconds, rounded to two decimal places.

(c) Figure 11 shows the graph of the power function on the scatter diagram.

Figure 11

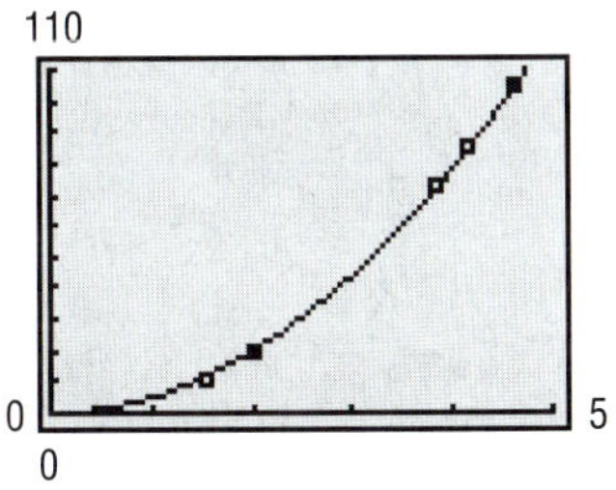

(d) The value of g obeys the equation

$$\frac{1}{2}g = 4.93$$

$$g = 9.86 \text{ meters/sec}^2$$

Scott's estimate for g is 9.86 meters/sec^2.

(e) A graphing utility fits the data in Figure 10 to a power function of the form $y = ax^b$ by using the PoWeR REGression option. See Figure 12.

Figure 12

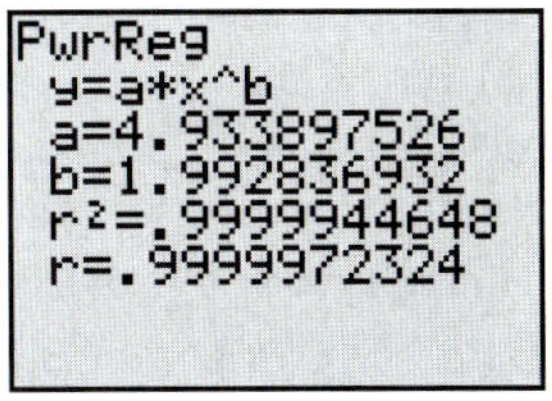

4.1 EXERCISES

In Problems 1–16, using a graphing utility, show the steps required to graph each function.

1. $f(x) = (x + 1)^4$
2. $f(x) = (x - 2)^5$
3. $f(x) = x^5 - 3$
4. $f(x) = x^4 + 2$
5. $f(x) = \frac{1}{2}x^4$
6. $f(x) = 3x^5$
7. $f(x) = -x^5$
8. $f(x) = -x^4$
9. $f(x) = (x - 1)^5 + 2$
10. $f(x) = (x + 2)^4 - 3$
11. $f(x) = 2(x + 1)^4 + 1$
12. $f(x) = \frac{1}{2}(x - 1)^5 - 2$
13. $f(x) = 4 - (x - 2)^5$
14. $f(x) = 3 - (x + 2)^4$
15. $f(x) = -\frac{1}{2}(x - 2)^4 - 1$
16. $f(x) = 1 - 2(x + 1)^5$

17. **CBL Experiment** David is conducting an experiment to estimate the acceleration of an object due to gravity. David takes a ball and drops it from different heights and records the time it takes for the ball to hit the ground. Using an optic laser connected to a stop watch in order to determine the time, he collects the following data:

Time (Seconds)	Distance (Feet)
1.003	16
1.365	30
1.769	50
2.093	70
2.238	80

(a) Draw a scatter diagram using time as the independent variable and distance as the dependent variable.

(b) The power function of best fit to these data is

$$s(t) = 15.9727t^{2.0018}$$

Use this function to predict the time that it will take an object to fall 100 feet.

(c) Graph the function given in part (b) on the scatter diagram.

(d) Verify that the function given in part (b) is the power function of best fit.

(e) Physics theory states the distance an object falls is directly proportional to the time squared. Rewrite the model found in (b) so it is of the form $s = \frac{1}{2}gt^2$, where s is distance, g is the acceleration due to gravity, and t is time. What is David's estimate of g? (It is known that the acceleration due to gravity is approximately 32 feet/sec^2.)

18. **CBL Experiment** Paul, David's friend, doesn't believe David's estimate of acceleration due to gravity is correct. Paul repeats the experiment described in Problem 17 and obtains the following data:

Time (Seconds)	Distance (Feet)
0.7907	10
1.1160	20
1.4760	35
1.6780	45
1.9380	60

(a) Draw a scatter diagram using time as the independent variable and distance as the dependent variable.

(b) The power function of best fit to these data is

$$s(t) = 16.0233t^{1.9982}$$

Use this function to predict the time it will take an object to fall 100 feet.

(c) Graph the function given in part (b) on the scatter digram.

(d) Verify the function given in part (b) is the power function of best fit.

(e) Use the function given in part (b) to estimate g. Why do you think this estimate is different than the one found in part (e) of Problem 17?

19. **Pendulums** The *period* of a pendulum is the time required for one oscillation; the pendulum is usually referred to as *simple* when the angle made to the vertical is less than 5°. An experiment is conducted in which simple pendulums are constructed with different lengths, l, and the corresponding periods T are recorded. The following data are collected:

Length l (in feet)	Period T (in seconds)
1	1.10
2	1.55
3	1.89
4	2.24
5	2.51
6	2.76
7	2.91

(a) Using a graphing utility, draw a scatter diagram of the data with length as the independent variable and period as the dependent variable.

(b) The power function of best fit to these data is

$$T(l) = 1.0948l^{0.5098}$$

Use this function* to predict the period of a simple pendulum whose length is known to be 2.3 feet.

(c) Graph the power function given in part (b) on the scatter diagram.

(d) Verify that the function given in part (b) is the power function of best fit.

*In physics, it is proved that $T = \frac{2\pi}{\sqrt{32}}\sqrt{l}$, provided that friction is ignored. Since $\sqrt{l} = l^{1/2} = l^{0.5}$ and $\frac{2\pi}{\sqrt{32}} \approx 1.1107$, the power function obtained is reasonably close to what is expected.

4.2 POLYNOMIAL FUNCTIONS AND MODELS

1 Identify Polynomials and Their Degree

2 Identify the Zeros of a Polynomial and Their Multiplicity

3 Analyze the Graph of a Polynomial

4 Find the Cubic Function of Best Fit from Data

Polynomial functions are among the simplest expressions in algebra. They are easy to evaluate: only addition and repeated multiplication are required. Because of this, they are often used to approximate other, more complicated functions. In this section, we investigate characteristics of this important class of function.

A **polynomial function** is a function of the form

$$f(x) = a_n x^n + a_{n-1} x^{n-1} + \cdots + a_1 x + a_0 \qquad (1)$$

where $a_n, a_{n-1}, \ldots, a_1, a_0$ are real numbers and n is a nonnegative integer. The domain consists of all real numbers.

1 Thus, a polynomial function is a function whose rule is given by a polynomial in one variable.* The degree of a polynomial function is the degree of the polynomial in one variable.

EXAMPLE 1 Identifying Polynomial Functions

Determine which of the following are polynomial functions. For those that are, state the degree; for those that are not, tell why not.

(a) $f(x) = 2 - 3x^4$ (b) $g(x) = \sqrt{x}$ (c) $h(x) = \dfrac{x^2 - 2}{x^3 - 1}$

(d) $F(x) = 0$ (e) $G(x) = 8$

Solution

(a) f is a polynomial function of degree 4.

(b) g is not a polynomial function. The variable x is raised to the $\frac{1}{2}$ power, which is not a nonnegative integer.

(c) h is not a polynomial function. It is the ratio of two polynomials, and the polynomial in the denominator is of positive degree.

(d) F is the zero polynomial function; it is not assigned a degree.

(e) G is a nonzero constant function, a polynomial function of degree 0.

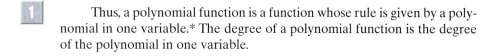

NOW WORK PROBLEMS 1 AND 5.

*A review of polynomials may be found in the Appendix, Section 5, pp. 702–704.

We have already discussed in detail polynomial functions of degrees 0, 1, and 2. See Table 8 for a summary of the characteristics of the graphs of these polynomial functions.

Table 8

Degree	Form	Name	Graph
No degree	$f(x) = 0$	Zero function	The x-axis
0	$f(x) = a_0, \quad a_0 \neq 0$	Constant function	Horizontal line with y-intercept a_0
1	$f(x) = a_1x + a_0, \quad a_1 \neq 0$	Linear function	Nonvertical, nonhorizontal line with slope a_1 and y-intercept a_0
2	$f(x) = a_2x^2 + a_1x + a_0, \quad a_2 \neq 0$	Quadratic function	Parabola: Graph opens up if $a_2 > 0$; graph opens down if $a_2 < 0$

Graphing Polynomials

If you take a course in calculus, you will learn that the graph of every polynomial function is both smooth and continuous. By **smooth,** we mean that the graph contains no sharp corners or cusps; by **continuous,** we mean that the graph has no gaps or holes and can be drawn without lifting pencil from paper. See Figures 13(a) and (b).

Figure 13

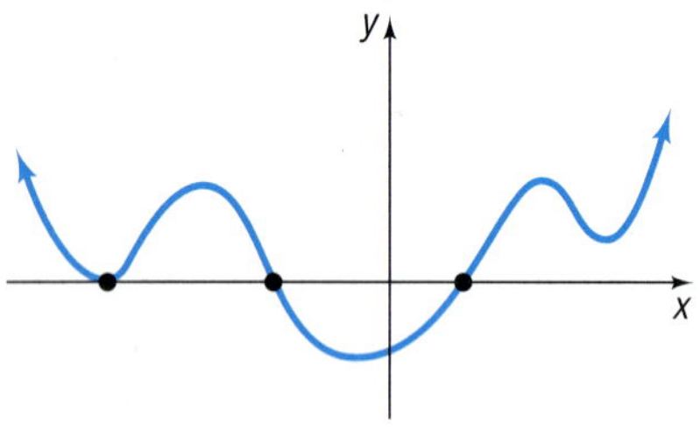

(a) Graph of a polynomial function: smooth, continuous

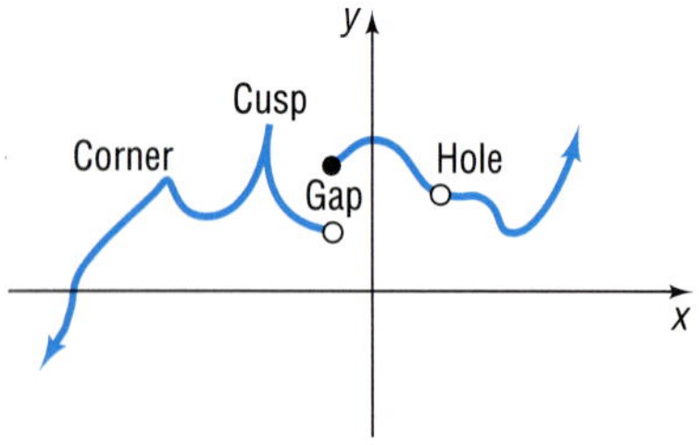

(b) Cannot be the graph of a polynomial function

Figure 14 shows the graph of a polynomial function with four x-intercepts. Notice that at the x-intercepts the graph must either cross the x-axis or touch the x-axis. Consequently, between consecutive x-intercepts the graph is either above the x-axis or below the x-axis. We will make use of this characteristic of the graph of a polynomial shortly.

Figure 14

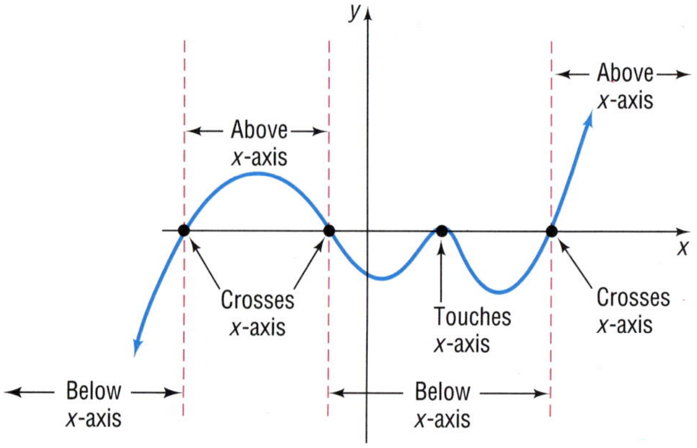

If a polynomial function f is factored completely, it is easy to solve the equation $f(x) = 0$ and locate the x-intercepts of the graph. For example, if $f(x) = (x - 1)^2(x + 3)$, then the solutions of the equation

$$f(x) = (x - 1)^2(x + 3) = 0$$

are easily identified as 1 and -3. Based on this result, we make the following observations:

> If f is a polynomial function and r is a real number for which $f(r) = 0$, then r is called a (real) **zero of *f*,** or **root of *f*.** If r is a (real) zero of f, then
>
> (a) r is an x-intercept of the graph of f.
> (b) $(x - r)$ is a factor of f.

EXAMPLE 2 Finding a Polynomial From Its Zeros

(a) Find a polynomial of degree 3 whose zeros are -3, 2, and 5.
(b) Graph the polynomial found in (a) to verify your result.

Solution (a) If r is a zero of a polynomial f, then $x - r$ is a factor of f. This means that $x - (-3) = x + 3$, $x - 2$, and $x - 5$ are factors of f. As a result, any polynomial of the form

$$f(x) = a(x + 3)(x - 2)(x - 5)$$

where a is any nonzero real number qualifies.

(b) The value of a causes a stretch or compression or reflection, but does not affect the x-intercepts. We choose to graph f with $a = 1$.

$$f(x) = (x + 3)(x - 2)(x - 5) = x^3 - 4x^2 - 11x + 30.$$

Figure 15 shows the graph of f. Notice the x-intercepts are -3, 2, and 5!

Figure 15

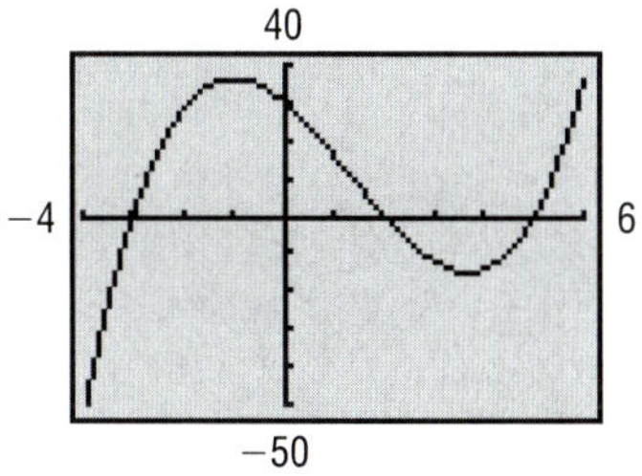

SEEING THE CONCEPT Graph the function found in Example 2 for $a = 2$ and $a = -1$. Does the value of a affect the zeros of f? How does the value of a affect the graph of f?

NOW WORK PROBLEM 11.

2 If the same factor $x - r$ occurs more than once, then r is called a **repeated,** or **multiple, zero of *f*.** More precisely, we have the following definition.

> If $(x - r)^m$ is a factor of a polynomial f and $(x - r)^{m+1}$ is not a factor of f, then r is called a **zero of multiplicity *m* of *f*.**

EXAMPLE 3 Identifying Zeros and Their Multiplicities

For the polynomial

$$f(x) = 5(x - 2)(x + 3)^2\left(x - \frac{1}{2}\right)^4$$

2 is a zero of multiplicity 1
-3 is a zero of multiplicity 2
$\frac{1}{2}$ is a zero of multiplicity 4

In the solution to Example 3 notice that, if you add the multiplicities $(1 + 2 + 4 = 7)$, you obtain the degree of the polynomial.

Suppose that it is possible to factor completely a polynomial function and, as a result, locate all the x-intercepts of its graph (the real zeros of the function). The following example illustrates the role that the multiplicity of the x-intercept plays.

EXAMPLE 4 Investigating the Role of Multiplicity

For the polynomial $f(x) = x^2(x - 2)$:

(a) Find the x- and y-intercepts of the graph.
(b) Using a graphing utility, graph the polynomial.
(c) For each x-intercept, determine whether it is of odd or even multiplicity.

Solution (a) The y-intercept is $f(0) = 0^2(0 - 2) = 0$. The x-intercepts satisfy the equation

$$f(x) = x^2(x - 2) = 0$$

from which we find that

$$x^2 = 0 \quad \text{or} \quad x - 2 = 0$$
$$x = 0 \qquad\qquad x = 2$$

The x-intercepts are 0 and 2.

(b) See Figure 16.

(c) We can see from the factored form of f that 0 is a zero or root of multiplicity 2, and 2 is a zero or root of multiplicity 1; so 0 is of even multiplicity and 2 is of odd multiplicity.

Figure 16

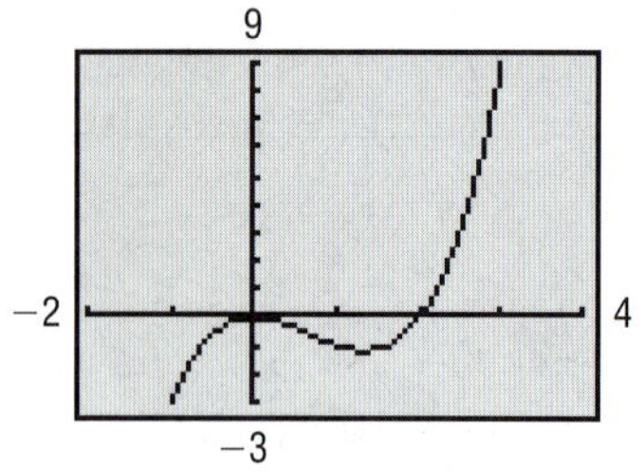

Table 9

X	Y1
-2	-16
-1	-3
0	0
1	-1
2	0
3	9
4	32

Y1=X²(X−2)

We can use TABLE to further analyze the graph. See Table 9. The sign of $f(x)$ is the same on each side of the root 0, so the graph of f just touches the x-axis at 0 (a root of even multiplicity). The sign of $f(x)$ changes from one side of the root 2 to the other, so the graph of f crosses the x-axis at 2 (a root of odd multiplicity). These observations suggest the following result:

If r Is a Zero of Even Multiplicity

Sign of $f(x)$ does not change from one side to the other side of r.	Graph **touches** x-axis at r.

If r Is a Zero of Odd Multiplicity

Sign of $f(x)$ changes from one side to the other side of r.	Graph **crosses** x-axis at r.

NOW WORK PROBLEM 17.

Look again at Figure 16. We can use a graphing utility to determine the graph's local minimum in the interval $0 < x < 2$. After utilizing MINIMUM, we find that the graph's local minimum in the interval $0 < x < 2$ is $(1.33, -1.19)$, rounded to two decimal places. Local minima and local maxima are points where the graph changes direction (i.e., changes from an increasing function to a decreasing function, or vice versa). We call these points **turning points.**

Look again at Figure 16. The graph of $f(x) = x^2(x - 2) = x^3 - 2x^2$, a polynomial of degree 3, has two turning points.

EXPLORATION Graph $Y_1 = x^3$, $Y_2 = x^3 - x$, and $Y_3 = x^3 + 3x^2 + 4$. How many turning points do you see? How does the number of turning points relate to the degree? Graph $Y_1 = x^4$, $Y_2 = x^4 - \frac{4}{3}x^3$, and $Y_3 = x^4 - 2x^2$. How many turning points do you see? How does the number of turning points compare to the degree?

The following theorem from calculus supplies the answer.

THEOREM If f is a polynomial function of degree n, then f has at most $n - 1$ turning points.

One last remark about Figure 16. Notice that the graph of $f(x) = x^2(x - 2)$ looks somewhat like the graph of $y = x^3$. In fact, for very large values of x, either positive or negative, there is little difference.

EXPLORATION Consider the functions Y_1 and Y_2 given below in parts (a), (b), (c). Graph Y_1 and Y_2 on the same viewing window. TRACE for large positive and large negative values of x. What do you notice about the graphs of Y_1 and Y_2 as x becomes very large and positive or very large and negative?

(a) $Y_1 = x^2(x - 2)$; $Y_2 = x^3$
(b) $Y_1 = x^4 - 3x^3 + 7x - 3$; $Y_2 = x^4$
(c) $Y_1 = -2x^3 + 4x^2 - 8x + 10$; $Y_2 = -2x^3$

The behavior of the graph of a function for large values of x, either positive or negative, is referred to as its **end behavior.**

THEOREM **End Behavior**

For large values of x, either positive or negative, the graph of the polynomial

$$f(x) = a_n x^n + a_{n-1}x^{n-1} + \cdots + a_1 x + a_0$$

resembles the graph of the power function

$$y = a_n x^n$$

The following summarizes some features of the graph of a polynomial function.

SUMMARY: GRAPH OF A POLYNOMIAL FUNCTION

$f(x) = a_n x^n + a_{n-1} x^{n-1} + \cdots + a_1 x + a_0, \quad a_n \neq 0$

Degree of the polynomial f: n
Maximum number of turning points: $n - 1$
At zero of even multiplicity: graph of f touches x-axis
At zero of odd multiplicity: graph of f crosses x-axis
Between zeros, graph of f is either above the x-axis or below it.
End Behavior: For large $|x|$, the graph of f behaves like graph of $y = a_n x^n$.

EXAMPLE 5 Analyzing the Graph of a Polynomial Function

3

For the polynomial $f(x) = x^4 - 3x^3 - 4x^2$:

(a) Using a graphing utility, graph f.
(b) Find the x- and y-intercepts.
(c) Determine whether each x-intercept is of odd or even multiplicity.
(d) Find the power function that the graph of f resembles for large values of $|x|$.
(e) Determine the number of turning points on the graph of f.
(f) Determine the local maxima and local minima, if any exist, rounded to two decimal places.

Solution

Figure 17

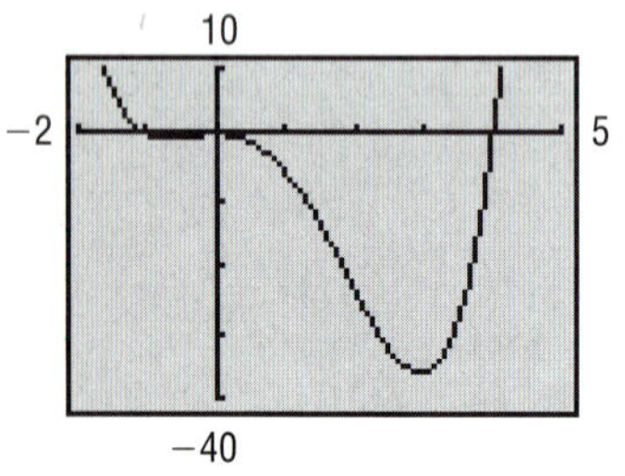

(a) See Figure 17.
(b) The y-intercept is $f(0) = 0$. Factoring, we find that

$$f(x) = x^4 - 3x^3 - 4x^2 = x^2(x^2 - 3x - 4) = x^2(x - 4)(x + 1).$$

We find the x-intercepts by solving the equation

$$f(x) = x^2(x - 4)(x + 1) = 0$$

So

$$x^2 = 0 \quad \text{or} \quad x - 4 = 0 \quad \text{or} \quad x + 1 = 0$$
$$x = 0 \qquad\qquad x = 4 \qquad\qquad x = -1$$

The x-intercepts are -1, 0, and 4.

(c) The intercept 0 is a zero of even multiplicity, 2, so the graph of f will touch the x-axis at 0; 4 and -1 are zeros of odd multiplicity, 1, so the graph of f will cross the x-axis at 4 and -1. Look again at Figure 17. The graph is below the x-axis for $-1 < x < 0$!
(d) The graph of f behaves like $y = x^4$ for large $|x|$.
(e) Since f is of degree 4, the graph can have at most three turning points. From the graph, we see that it has three turning points: one between -1 and 0, one at (0, 0), and one between 2 and 4.
(f) Rounded to two decimal places, the local maximum is (0, 0) and the local minima are $(-0.68, -0.69)$ and $(2.93, -36.10)$.

EXAMPLE 6 Analyzing the Graph of a Polynomial Function

Follow the instructions of Example 5 for the following polynomial:

$$f(x) = x^3 + 2.48x^2 - 4.3155x + 1.484406$$

Solution

(a) See Figure 18.

Figure 18

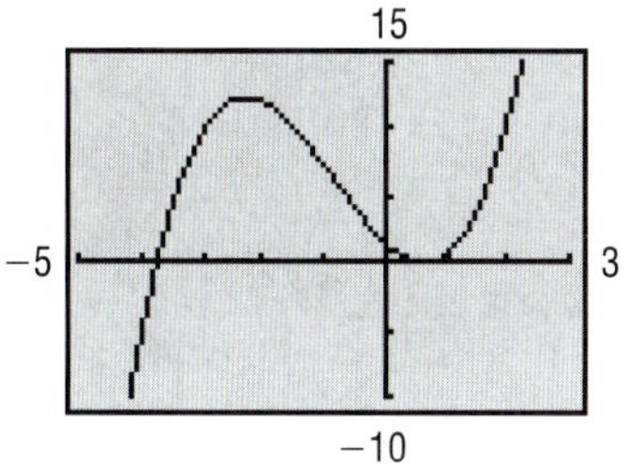

(b) The y-intercept is $f(0) = 1.484406$. In Example 5 we could easily factor $f(x)$ to find the x-intercepts. However, it is not readily apparent how $f(x)$ factors in this example. Therefore, we use a graphing utility and find the x-intercepts to be -3.74 and 0.63.

(c) The x-intercept -3.74 is of odd multiplicity since the graph of f crosses the x-axis at -3.74; the x-intercept 0.63 is of even multiplicity since the graph of f touches the x-axis at 0.63, rounded to two decimal places.

(d) The graph behaves like $y = x^3$ for large $|x|$.

(e) Since f is of degree 3, the graph can have at most two turning points. From the graph we see that it has two turning points: one between -3 and -2, the other at $(0.63, 0)$.

(f) Rounded to two decimal places, the local maximum is $(-2.28, 12.36)$ and the local minimum is $(0.63, 0)$.

NOW WORK PROBLEMS **29** AND **49**.

Modeling Polynomial Functions: Curve Fitting

4 In Section 2.2 we found the line of best fit from data; in Section 2.4, we found the quadratic function of best fit; and in Section 4.1 we found the power function of best fit. It is also possible to find polynomial functions of best fit. However, most statisticians do not recommend finding polynomials of best fit of degree higher than 3.*

Data that follow a cubic relation should look like Figure 19(a) or (b).

Figure 19

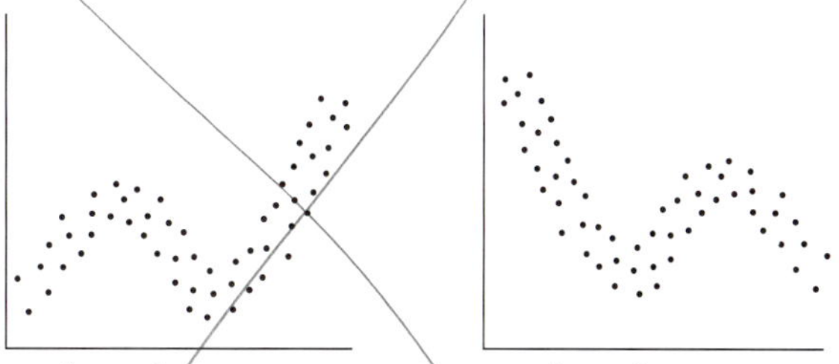

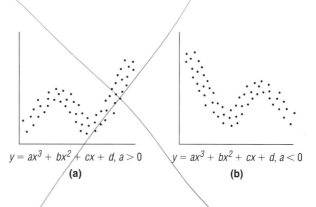

$y = ax^3 + bx^2 + cx + d, a > 0$ (a) $y = ax^3 + bx^2 + cx + d, a < 0$ (b)

*Two points determine a unique linear function. Three noncollinear points determine a unique quadratic function. Four points determine a unique cubic function, and n points will determine a unique polynomial of degree $n - 1$. Therefore, higher-degree polynomials will always "fit" data at least as well as lower-degree polynomials. However, higher-degree polynomials yield highly erratic predictions. Since the ultimate goal of curve fitting is not necessarily to find the model that best fits the data, but instead to explain relationships between two or more variables, polynomial models of degree 3 or less are usually used.

EXAMPLE 7 A Cubic Function of Best Fit

The data in Table 10 represent the number of barrels of refined oil products imported into the United States for 1977–1992, where 1 represents 1977, 2 represents 1978, and so on.

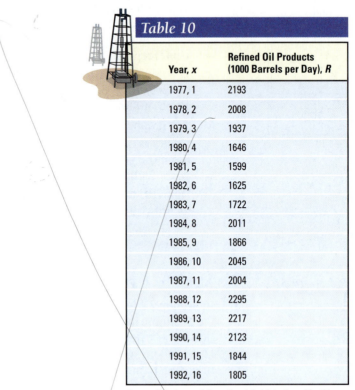

Table 10

Year, x	Refined Oil Products (1000 Barrels per Day), R
1977, 1	2193
1978, 2	2008
1979, 3	1937
1980, 4	1646
1981, 5	1599
1982, 6	1625
1983, 7	1722
1984, 8	2011
1985, 9	1866
1986, 10	2045
1987, 11	2004
1988, 12	2295
1989, 13	2217
1990, 14	2123
1991, 15	1844
1992, 16	1805

Source: U.S. Energy Information Administration, *Monthly Energy Review*, February, 1995.

(a) Draw a scatter diagram of the data. Comment on the type of relation that may exist between the two variables.

(b) The cubic function of best fit is

$$R(x) = -2.4117x^3 + 63.4297x^2 - 453.7614x + 2648.0027,$$

where x represents the year and R represents the number of barrels of refined oil products imported. Use this function to predict the number of barrels of refined oil products imported in 1993 ($x = 17$).

(c) Draw the cubic function of best fit on your scatter diagram.

(d) Verify that the function given in part (b) is the cubic function of best fit.

(e) Do you think the function given in part (b) will be useful in predicting the number of barrels of refined oil products for the year 1999? Why?

Solution (a) Figure 20 shows the scatter diagram.

Figure 20

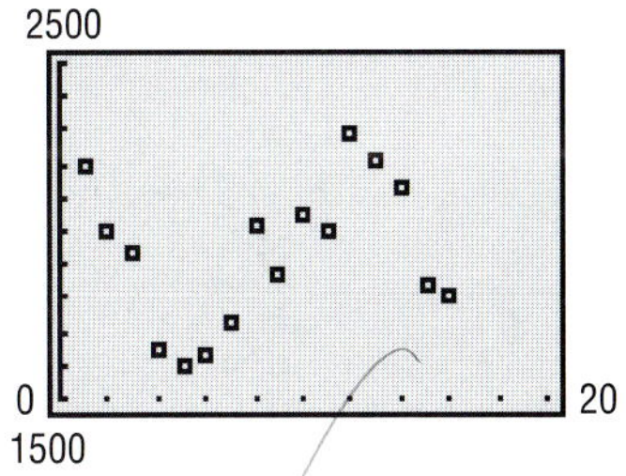

A cubic relation may exist between the two variables.

(b) We evaluate the function $R(x)$ for $x = 17$:

$$R(17) = -2.4117(17)^3 + 63.4297(17)^2 - 453.7614(17) + 2648.0027 \approx 1417$$

So we predict that 1417 barrels of refined oil products will be imported into the United States in 1993.

(c) Figure 21 shows the graph of the cubic function of best fit on the scatter diagram. The function seems to fit the data well.

Figure 21

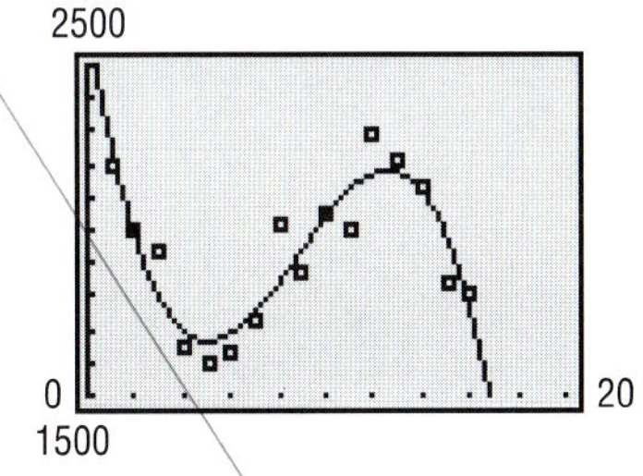

(d) Upon executing the CUBIC REGression program, we obtain the results shown in Figure 22. The output that the utility provides shows us the equation $y = ax^3 + bx^2 + cx + d$.

Figure 22

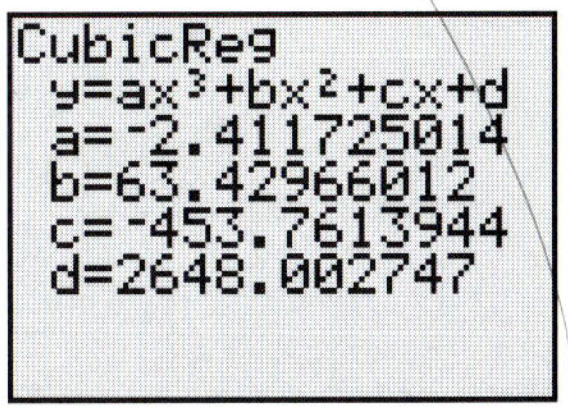
CubicReg
y=ax³+bx²+cx+d
a=-2.411725014
b=63.42966012
c=-453.7613944
d=2648.002747

(e) The graph of the function will continue to decrease for years beyond 1993, indicating that the number of barrels of refined oil imported will continue to decline, eventually reaching 0. This is not a likely scenario. So the function probably will not be useful in predicting the number of barrels of refined oil products imported into the United States in 1999.

4.2 EXERCISES

In Problems 1–10, determine which functions are polynomial functions. For those that are, state the degree. For those that are not, tell why not.

1. $f(x) = 4x + x^3$
2. $f(x) = 5x^2 + 4x^4$
3. $g(x) = \dfrac{1 - x^2}{2}$
4. $h(x) = 3 - \dfrac{1}{2}x$
5. $f(x) = 1 - \dfrac{1}{x}$
6. $f(x) = x(x - 1)$
7. $g(x) = x^{3/2} - x^2 + 2$
8. $h(x) = \sqrt{x}(\sqrt{x} - 1)$
9. $F(x) = 5x^4 - \pi x^3 + \dfrac{1}{2}$
10. $F(x) = \dfrac{x^2 - 5}{x^3}$

In Problems 11–16, form a polynomial whose zeros and degree are given.

11. Zeros: $-1, 1, 3$; degree 3
12. Zeros: $-2, 2, 3$; degree 3
13. Zeros: $-3, 0, 4$; degree 3
14. Zeros: $-4, 0, 2$; degree 3
15. Zeros: $-4, -1, 2, 3$; degree 4
16. Zeros: $-3, -1, 2, 5$; degree 4

In Problems 17–26, for each polynomial function, list each real zero and its multiplicity. Determine whether the graph crosses or touches the x-axis at each x-intercept.

17. $f(x) = 3(x - 7)(x + 3)^2$
18. $f(x) = 4(x + 4)(x + 3)^3$
19. $f(x) = 4(x^2 + 1)(x - 2)^3$
20. $f(x) = 2(x - 3)(x + 4)^3$
21. $f(x) = -2\left(x + \dfrac{1}{2}\right)^2(x^2 + 4)^2$
22. $f(x) = \left(x - \dfrac{1}{3}\right)^2(x - 1)^3$
23. $f(x) = (x - 5)^3(x + 4)^2$
24. $f(x) = (x + \sqrt{3})^2(x - 2)^4$
25. $f(x) = 3(x^2 + 8)(x^2 + 9)^2$
26. $f(x) = -2(x^2 + 3)^3$

In Problems 21–62, for each polynomial function f:

(a) Using a graphing utility, graph f.

(b) Find the x- and y-intercepts.

(c) Determine whether each x-intercept is of odd or even multiplicity.

(d) Find the power function that the graph of f resembles for large values of $|x|$.

(e) Determine the number of turning points on the graph of f.

(f) Determine the local maxima and local minima, if any exist, rounded to two decimal places.

27. $f(x) = (x - 1)^2$
28. $f(x) = (x - 2)^3$
29. $f(x) = x^2(x - 3)$
30. $f(x) = x(x + 2)^2$
31. $f(x) = 6x^3(x + 4)$
32. $f(x) = 5x(x - 1)^3$
33. $f(x) = -4x^2(x + 2)$
34. $f(x) = -\dfrac{1}{2}x^3(x + 4)$
35. $f(x) = x(x - 2)(x + 4)$
36. $f(x) = x(x + 4)(x - 3)$
37. $f(x) = 4x - x^3$
38. $f(x) = x - x^3$
39. $f(x) = x^2(x - 2)(x + 2)$
40. $f(x) = x^2(x - 3)(x + 4)$
41. $f(x) = x^2(x - 2)^2$
42. $f(x) = x^3(x - 3)$
43. $f(x) = x^2(x - 3)(x + 1)$
44. $f(x) = x^2(x - 3)(x - 1)$

45. $f(x) = x(x + 2)(x - 4)(x - 6)$

46. $f(x) = x(x - 2)(x + 2)(x + 4)$

47. $f(x) = x^2(x - 2)(x^2 + 3)$

48. $f(x) = x^2(x^2 + 1)(x + 4)$

49. $f(x) = x^3 + 0.2x^2 - 1.5876x - 0.31752$

50. $f(x) = x^3 - 0.8x^2 - 4.6656x + 3.73248$

51. $f(x) = x^3 + 2.56x^2 - 3.31x + 0.89$

52. $f(x) = x^3 - 2.91x^2 - 7.668x - 3.8151$

53. $f(x) = x^4 - 2.5x^2 + 0.5625$

54. $f(x) = x^4 - 18.5x^2 + 50.2619$

55. $f(x) = x^4 + 0.65x^3 - 16.6319x^2 + 14.209335x - 3.1264785$

56. $f(x) = x^4 + 3.45x^3 - 11.6639x^2 - 5.864241x - 0.69257738$

57. $f(x) = \pi x^3 + \sqrt{2}x^2 - x - 2$

58. $f(x) = -2x^3 + \pi x^2 + \sqrt{3}x + 1$

59. $f(x) = 2x^4 - \pi x^3 + \sqrt{5}x - 4$

60. $f(x) = -1.2x^4 + 0.5x^2 - \sqrt{3}x + 2$

61. $f(x) = -2x^5 - \sqrt{2}x^2 - x - \sqrt{2}$

62. $f(x) = \pi x^5 + \pi x^4 + \sqrt{3}x + 1$

63. Consult the illustration. Which of the following polynomial functions might have this graph? (More than one answer may be possible.)

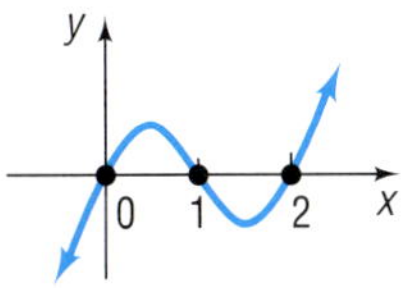

(a) $y = -4x(x - 1)(x - 2)$
(b) $y = x^2(x - 1)^2(x - 2)$
(c) $y = 3x(x - 1)(x - 2)$
(d) $y = x(x - 1)^2(x - 2)^2$
(e) $y = x^3(x - 1)(x - 2)$
(f) $y = -x(1 - x)(x - 2)$

64. Consult the illustration. Which of the following polynomial functions might have this graph? (More than one answer may be possible.)

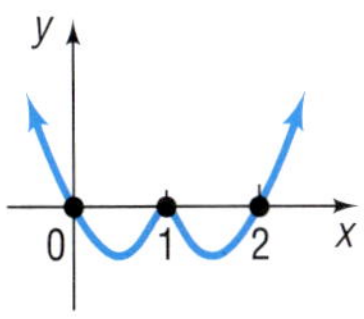

(a) $y = 2x^3(x - 1)(x - 2)^2$
(b) $y = x^2(x - 1)(x - 2)$
(c) $y = x^3(x - 1)^2(x - 2)$
(d) $y = x^2(x - 1)^2(x - 2)^2$
(e) $y = 5x(x - 1)^2(x - 2)$
(f) $y = -2x(x - 1)^2(2 - x)$

65. Consult the illustration. Which of the following polynomial functions might have this graph? (More than one answer may be possible.)

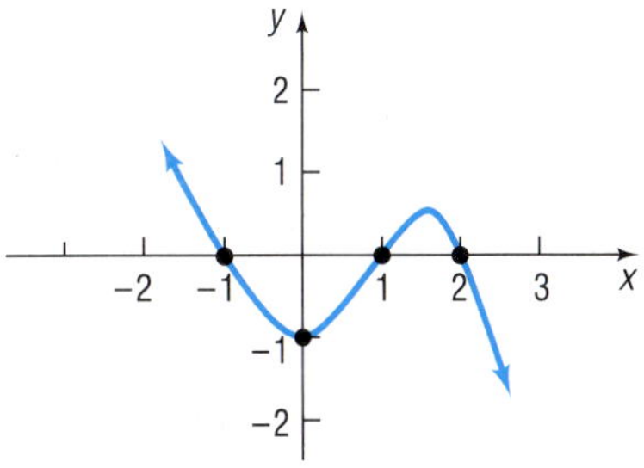

(a) $y = \frac{1}{2}(x^2 - 1)(x - 2)$

(b) $y = -\frac{1}{2}(x^2 - 1)(x - 2)$

(c) $y = (x^2 - 1)\left(1 - \frac{x}{2}\right)$

(d) $y = -\frac{1}{2}(x^2 - 1)^2(x - 2)$

(e) $y = \left(x^2 + \frac{1}{2}\right)(x^2 - 1)(2 - x)$

(f) $y = -(x - 1)(x - 2)(x + 1)$

66. Consult the illustration. Which of the following polynomial functions might have this graph? (More than one answer may be possible.)

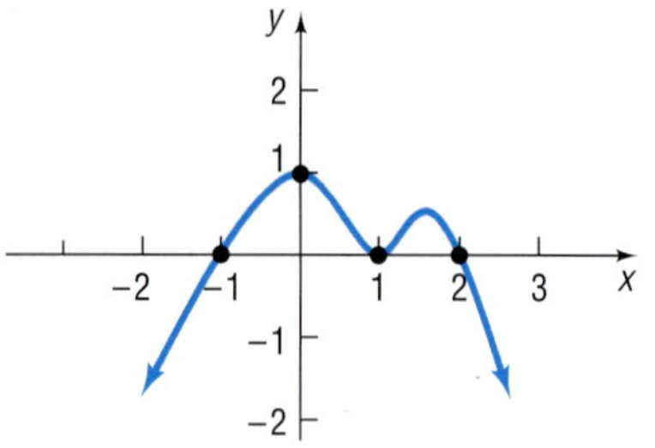

(a) $y = -\frac{1}{2}(x^2 - 1)(x - 2)(x + 1)$

(b) $y = -\frac{1}{2}(x^2 + 1)(x - 2)(x + 1)$

(c) $y = -\frac{1}{2}(x + 1)^2(x - 1)(x - 2)$

(d) $y = (x - 1)^2(x + 1)\left(1 - \frac{x}{2}\right)$

(e) $y = -(x - 1)^2(x - 2)(x + 1)$

(f) $y = -\left(x^2 + \frac{1}{2}\right)(x - 1)^2(x + 1)(x - 2)$

67. Motor Vehicle Thefts The following data represent the number of motor vehicle thefts (in thousands) in the United States for the years 1983–1993, where 1 represents 1983, 2 represents 1984, and so on.

Year, x	Motor Vehicle Thefts, M
1983, 1	1008
1984, 2	1032
1985, 3	1103
1986, 4	1224
1987, 5	1289
1988, 6	1433
1989, 7	1565
1990, 8	1636
1991, 9	1662
1992, 10	1611
1993, 11	1561

Source: U.S. Federal Bureau of Investigation

(a) Draw a scatter diagram of the data. Comment on the type of relation that may exist between the two variables.

(b) The cubic function of best fit to these data is

$$M(x) = -2.4x^3 + 37.3x^2 - 70.4x + 1043.8$$

Use this function to predict the number of motor vehicle thefts in 1994.

(c) Use a graphing utility to verify that the function given in part (b) is the cubic function of best fit.

(d) Graph the cubic function of best fit on the scatter diagram.

(e) Check the prediction of part (b) against actual data. Do you think that the function given in part (b) will be useful in predicting the number of motor vehicle thefts in 1999?

68. Larceny Thefts The following data represent the number of larceny thefts (in thousands) in the United States for the years 1983–1993, where 1 represents 1983, 2 represents 1984, and so on.

Year, x	Larceny Thefts, L
1983, 1	6713
1984, 2	6592
1985, 3	6926
1986, 4	7257
1987, 5	7500
1988, 6	7706
1989, 7	7872
1990, 8	7946
1991, 9	8142
1992, 10	7915
1993, 11	7821

Source: U.S. Federal Bureau of Investigation

(a) Draw a scatter diagram of the data. Comment on the type of relation that may exist between the two variables.

(b) The cubic function of best fit to these data is

$$L(x) = -4.37x^3 + 59.29x^2 - 14.02x + 6578$$

Use this function to predict the number of larceny thefts in 1994.

(c) Use a graphing utility to verify that the function given in part (b) is the cubic function of best fit.

(d) Graph the cubic function of best fit on the scatter diagram.

(e) Check the prediction of part (b) against actual data. Do you think that the function found in part (b) will be useful in predicting the number of larceny thefts in 1999?

69. Cost of Manufacturing The following data represent the cost C of manufacturing Chevy Cavaliers (in thousands of dollars) and the number x of Cavaliers produced.

Number of Cavaliers Produced, x	Cost, C
0	10
1	23
2	31
3	38
4	43
5	50
6	59
7	70
8	85
9	105
10	135

(a) Draw a scatter diagram of the data. Comment on the type of relation that may exist between the two variables.

(b) Find the average rate of change of cost from four to five Cavaliers.

(c) What is the average rate of change of cost from eight to nine Cavaliers?

(d) The cubic function of best fit to these data is

$$C(x) = 0.2x^3 - 2.3x^2 + 14.3x + 10.2$$

Use this function to predict the cost of manufacturing 11 Cavaliers.

(e) Use a graphing utility to verify that the function given in part (d) is the cubic function of best fit.

(f) Graph the cubic function of best fit on the scatter diagram.

(g) Interpret the y-intercept.

70. Cost of Printing The following data represent the weekly cost of printing textbooks C (in thousands) and the number x of texts printed (in thousands).

Number of Text Books, x	Cost, C
0	100
5	128.1
10	144
13	153.5
17	161.2
18	162.6
20	166.3
23	178.9
25	190.2
27	221.8

(a) Draw a scatter diagram of the data. Comment on the type of relation that may exist between the two variables.

(b) Find the average rate of change of cost from 10,000 to 13,000 textbooks.

(c) What is the average rate of change in the cost of producing from 18,000 to 20,000 textbooks?

(d) The cubic function of best fit to these data is

$$C(x) = 0.015x^3 - 0.595x^2 + 9.15x + 98.43$$

Use this function to predict the cost of printing 22,000 texts per week.

(e) Use a graphing utility to verify that the function given in part (d) is the cubic function of best fit.

(f) Graph the cubic function of best fit on the scatter diagram.

(g) Interpret the y-intercept.

71. Can the graph of a polynomial function have no y-intercept? Can it have no x-intercepts? Explain.

72. Write a few paragraphs that provide a general strategy for graphing a polynomial function. Be sure to mention the following: degree, intercepts, and turning points.

73. Make up a polynomial that has the following characteristics: crosses the x-axis at -1 and 4, touches the x-axis at 0 and 2, and is above the x-axis between 0 and 2. Give your polynomial to a fellow classmate and ask for a written critique of your polynomial.

74. Make up two polynomials, not of the same degree, with the following characteristics: crosses the x-axis at -2, touches the x-axis at 1, and is above the x-axis between -2 and 1. Give your polynomials to a fellow classmate and ask for a written critique of your polynomials.

75. The graph of a polynomial function is always smooth and continuous. Name a function studied earlier that is smooth and not continuous. Name one that is continuous, but not smooth.

76. Which of the following statements are true regarding the graph of the cubic polynomial $f(x) = x^3 + bx^2 + cx + d$? (Give reasons for your conclusions.)
(a) It intersects the y-axis in one and only one point.
(b) It intersects the x-axis in at most three points.
(c) It intersects the x-axis at least once.
(d) For $|x|$ very large, it behaves like the graph of $y = x^3$.
(e) It is symmetric with respect to the origin.
(f) It passes through the origin.

4.3 POLYNOMIAL DIVISION

1 Divide Polynomials Using Long Division

2 Divide Polynomials Using Synthetic Division

Long Division

1 The procedure for dividing two polynomials is similar to the procedure for dividing two integers. This process should be familiar to you, but we review it briefly next.

EXAMPLE 1 Dividing Two Integers

Divide 842 by 15.

Solution

$$\begin{array}{rrl} & 56 & \leftarrow \text{Quotient} \\ \text{Divisor} \rightarrow & 15\overline{)842} & \leftarrow \text{Dividend} \\ & \underline{75} & \leftarrow 5 \cdot 15 \text{ (Subtract)} \\ & 92 & \\ & \underline{90} & \leftarrow 6 \cdot 15 \text{ (Subtract)} \\ & 2 & \leftarrow \text{Remainder} \end{array}$$

Thus, $\dfrac{842}{15} = 56 + \dfrac{2}{15}$.

In the long division process detailed in Example 1, the number 15 is called the **divisor,** the number 842 is the **dividend,** the number 56 is the **quotient,** and the number 2 the **remainder.**

To check the answer obtained in a division problem, multiply the quotient by the divisor and add the remainder. The answer should be the dividend.

(Quotient)(Divisor) + Remainder = Dividend

For example, we can check the results obtained in Example 1 as follows:

$$(56)(15) + 2 = 840 + 2 = 842$$

To divide two polynomials, we first must write each polynomial in standard form, that is, in descending powers of the variable. The process then follows a pattern similar to that of Example 1. The next example illustrates the procedure.

EXAMPLE 2 Dividing Two Polynomials

Find the quotient and the remainder when

$$3x^3 + 4x^2 + x + 7 \quad \text{is divided by} \quad x^2 + 1$$

Solution Each polynomial is in standard form. The dividend is $3x^3 + 4x^2 + x + 7$, and the divisor is $x^2 + 1$.

STEP 1: Divide the leading term of the dividend, $3x^3$, by the leading term of the divisor, x^2. Enter the result, $3x$, over the term $3x^3$, as follows:

$$\begin{array}{r}3x \qquad\qquad\qquad\quad \\ x^2+1\overline{)3x^3+4x^2+x+7}\end{array}$$

STEP 2: Multiply $3x$ by $x^2 + 1$ and enter the result below the dividend.

$$\begin{array}{rl}\begin{array}{r}3x \qquad\qquad\qquad\quad \\ x^2+1\overline{)3x^3+4x^2+\ \ x+7} \\ \underline{3x^3 \qquad\quad +3x\ \ \ }\end{array} & \leftarrow 3x\cdot(x^2+1)=3x^3+3x\end{array}$$

↑ *Notice that we align the $3x$ term under the x to make the next step easier.*

STEP 3: Subtract and bring down the remaining terms.

$$\begin{array}{rl}\begin{array}{r}3x \qquad\qquad\qquad\quad \\ x^2+1\overline{)3x^3+4x^2+\ \ x+7} \\ \underline{3x^3 \qquad\quad +3x\ \ \ } \\ 4x^2-2x+7\end{array} & \begin{array}{l}\\ \\ \leftarrow \text{Subtract.} \\ \leftarrow \text{Bring down the } 4x^2 \text{ and the } 7.\end{array}\end{array}$$

STEP 4: Repeat Steps 1–3 using $4x^2 - 2x + 7$ as the dividend.

$$\begin{array}{rl}\begin{array}{r}3x+4 \qquad\qquad\quad \\ x^2+1\overline{)3x^3+4x^2+x+7} \\ \underline{3x^3 \qquad\quad +3x\ \ \ } \\ 4x^2-2x+7 \\ \underline{4x^2 \qquad\ +4} \\ -2x+3\end{array} & \begin{array}{l}\\ \\ \\ \leftarrow \text{Divide } 4x^2 \text{ by } x^2 \text{ to get } 4. \\ \leftarrow \text{Multiply } x^2+1 \text{ by } 4\text{; subtract.} \\ \end{array}\end{array}$$

Since x^2 does not divide $-2x$ evenly (that is, the result is not a monomial), the process ends. The quotient is $3x + 4$, and the remainder is $-2x + 3$.

✓CHECK: (Quotient)(Divisor) + Remainder

$$\begin{aligned} &= (3x + 4)(x^2 + 1) + (-2x + 3) \\ &= 3x^3 + 4x^2 + 3x + 4 + (-2x + 3) \\ &= 3x^3 + 4x^2 + x + 7 = \text{Dividend} \end{aligned}$$

Thus,

$$\frac{3x^3 + 4x^2 + x + 7}{x^2 + 1} = 3x + 4 + \frac{-2x + 3}{x^2 + 1}$$

The next example combines the steps involved in long division.

EXAMPLE 3 Dividing Two Polynomials

Find the quotient and the remainder when

$$x^4 - 3x^3 + 2x - 5 \quad \text{is divided by} \quad x^2 - x + 1$$

Solution In setting up this division problem, it is necessary to leave a space for the missing x^2 term in the dividend.

$$
\begin{array}{rrrrrl}
 & & x^2 - 2x - 3 & & & \leftarrow \text{Quotient} \\
\text{Divisor} \rightarrow & x^2 - x + 1 \,\big) & x^4 - 3x^3 & & + 2x - 5 & \leftarrow \text{Dividend} \\
 & \text{Subtract} \rightarrow & x^4 - x^3 + x^2 & & & \\
 & & -2x^3 - x^2 & & + 2x - 5 & \\
 & \text{Subtract} \rightarrow & -2x^3 + 2x^2 & & - 2x & \\
 & & & -3x^2 & + 4x - 5 & \\
 & \text{Subtract} \rightarrow & & -3x^2 & + 3x - 3 & \\
 & & & & x - 2 & \leftarrow \text{Remainder}
\end{array}
$$

✓CHECK: (Quotient)(Divisor) + Remainder

$$
\begin{aligned}
&= (x^2 - 2x - 3)(x^2 - x + 1) + x - 2 \\
&= x^4 - x^3 + x^2 - 2x^3 + 2x^2 - 2x - 3x^2 + 3x - 3 + x - 2 \\
&= x^4 - 3x^3 + 2x - 5 = \text{Dividend}
\end{aligned}
$$

Thus,

$$\frac{x^4 - 3x^3 + 2x - 5}{x^2 - x + 1} = x^2 - 2x - 3 + \frac{x - 2}{x^2 - x + 1}$$

The process of dividing two polynomials leads to the following result:

THEOREM The remainder after dividing two polynomials is either the zero polynomial or a polynomial of degree less than the degree of the divisor.

NOW WORK PROBLEM 5.

Synthetic Division

2 To find the quotient as well as the remainder when a polynomial function f of degree 1 or higher is divided by $g(x) = x - c$, a shortened version of long division, called **synthetic division,** makes the task simpler.

To see how synthetic division works, we will use long division to divide the polynomial $f(x) = 2x^3 - x^2 + 3$ by $g(x) = x - 3$.

$$
\begin{array}{rrrrl}
 & 2x^2 + 5x + 15 & & & \leftarrow \text{Quotient} \\
x - 3 \,\big) & 2x^3 - x^2 & & + 3 & \\
 & 2x^3 - 6x^2 & & & \\
 & 5x^2 & & & \\
 & 5x^2 & - 15x & & \\
 & & 15x & + 3 & \\
 & & 15x & - 45 & \\
 & & & 48 & \leftarrow \text{Remainder}
\end{array}
$$

✓CHECK: (Divisor)·(Quotient) + Remainder $= (x-3)(2x^2+5x+15)+48$

$$= 2x^3 + 5x^2 + 15x - 6x^2 - 15x - 45 + 48$$
$$= 2x^3 - x^2 + 3$$

The process of synthetic division arises from rewriting the long division in a more compact form, using simpler notation. For example, in the long division above, the terms in color are not really necessary because they are identical to the terms directly above them. With these terms removed, we have

$$
\begin{array}{r|rrrr}
 & 2x^2 & +5x & +15 & \\
\hline
x-3 & 2x^3 & -x^2 & & +3 \\
 & & -6x^2 & & \\
\cline{2-3}
 & & 5x^2 & & \\
 & & & -15x & \\
\cline{3-4}
 & & & 15x & \\
 & & & & -45 \\
\cline{4-5}
 & & & & 48
\end{array}
$$

Most of the x's that appear in this process can also be removed, provided that we are careful about positioning each coefficient. In this regard, we will need to use 0 as the coefficient of x in the dividend, because that power of x is missing. Now we have

$$
\begin{array}{r|rrrr}
 & 2x^2 & +5x & +15 & \\
\hline
x-3 & 2 & -1 & 0 & 3 \\
 & & -6 & & \\
\cline{2-3}
 & & 5 & & \\
 & & & -15 & \\
\cline{3-5}
 & & & 15 & \\
 & & & & -45 \\
\cline{4-5}
 & & & & 48
\end{array}
$$

We can make this display more compact by moving the lines up until the numbers in color align horizontally:

$$
\begin{array}{r|rrrrl}
 & 2x^2 & +5x & +15 & & \text{Row 1} \\
\hline
x-3 & 2 & -1 & 0 & 3 & \text{Row 2} \\
 & & -6 & -15 & -45 & \text{Row 3} \\
\cline{2-5}
 & \bigcirc & 5 & 15 & 48 & \text{Row 4}
\end{array}
$$

Now, if we place the leading coefficient of the quotient (2) in the circled position, the first three numbers in row 4 are precisely the coefficients of the quotient, and the last number in row 4 is the remainder. Thus, row 1 is not really needed, so we can compress the process to three rows, where the bottom row contains the coefficients of both the quotient and the remainder.

$$
\begin{array}{r|rrrrl}
x-3 & 2 & -1 & 0 & 3 & \text{Row 1} \\
 & & -6 & -15 & -45 & \text{Row 2 (subtract)} \\
\cline{2-5}
 & 2 & 5 & 15 & 48 & \text{Row 3}
\end{array}
$$

Recall that the entries in row 3 are obtained by subtracting the entries in row 2 from those in row 1. Rather than subtracting the entries in row 2, we can change the sign of each entry and add. With this modification, our display will look like this:

$$\begin{array}{r|rrrrl} x-3 & 2 & -1 & 0 & 3 & \text{Row 1} \\ & & 6 & 15 & 45 & \text{Row 2 (add)} \\ \hline & 2 & 5 & 15 & 48 & \text{Row 3} \end{array}$$

Notice that the entries in row 2 are three times the prior entries in row 3. Our last modification to the display replaces the $x - 3$ by 3. The entries in row 3 give the quotient and the remainder, as shown next.

$$\begin{array}{r|rrrrl} 3 & 2 & -1 & 0 & 3 & \text{Row 1} \\ & & 6 & 15 & 45 & \text{Row 2 (add)} \\ \hline & 2 & 5 & 15 & 48 & \text{Row 3} \end{array}$$

Quotient: 2 5 15 — Remainder: 48

Let's go through another example step by step.

EXAMPLE 4 Using Synthetic Division to Find the Quotient and Remainder

Use synthetic division to find the quotient and remainder when

$$f(x) = 3x^4 + 8x^2 - 7x + 4 \quad \text{is divided by} \quad g(x) = x - 1$$

Solution

STEP 1: Write the dividend in descending powers of x. Then copy the coefficients, remembering to insert a 0 for any missing powers of x.

$$\begin{array}{rrrrrl} 3 & 0 & 8 & -7 & 4 & \text{Row 1} \end{array}$$

STEP 2: Insert the usual division symbol. Since the divisor is $x - 1$, we insert 1 to the left of the division symbol

$$\begin{array}{r|rrrrrl} 1 & 3 & 0 & 8 & -7 & 4 & \text{Row 1} \end{array}$$

STEP 3: Bring the 3 down two rows, and enter it in row 3:

$$\begin{array}{r|rrrrrl} 1 & 3 & 0 & 8 & -7 & 4 & \text{Row 1} \\ & \downarrow & & & & & \text{Row 2} \\ \hline & 3 & & & & & \text{Row 3} \end{array}$$

STEP 4: Multiply the latest entry in row 3 by 1, and place the result in row 2, but one column over to the right.

$$\begin{array}{r|rrrrrl} 1 & 3 & 0 & 8 & -7 & 4 & \text{Row 1} \\ & & 3 & & & & \text{Row 2} \\ \hline & 3 & & & & & \text{Row 3} \end{array}$$

STEP 5: Add the entry in row 2 to the entry above it in row 1, and enter the sum to row 3.

$$\begin{array}{r|rrrrrl} 1 & 3 & 0 & 8 & -7 & 4 & \text{Row 1} \\ & & 3 & & & & \text{Row 2} \\ \hline & 3 & 3 & & & & \text{Row 3} \end{array}$$

STEP 6: Repeat steps 4 and 5 until no more entries are available in row 1.

$$\begin{array}{r|rrrrrl} 1 & 3 & 0 & 8 & -7 & 4 & \text{Row 1} \\ & & 3 & 3 & 11 & 4 & \text{Row 2 (add)} \\ \hline & 3 & 3 & 11 & 4 & 8 & \text{Row 3} \end{array}$$

STEP 7: The final entry in row 3, an 8, is the remainder; the other entries in row 3 (3, 3, 11, and 4) are the coefficients (in descending order) of a polynomial whose degree is 1 less than that of the dividend; this is the quotient. Thus,

$$\text{Quotient} = 3x^3 + 3x^2 + 11x + 4 \qquad \text{Remainder} = 8$$

✓CHECK: (Divisor)(Quotient) + Remainder

$$= (x - 1)(3x^3 + 3x^2 + 11x + 4) + 8$$
$$= 3x^4 + 3x^3 + 11x^2 + 4x - 3x^3 - 3x^2 - 11x - 4 + 8$$
$$= 3x^4 + 8x^2 - 7x + 4 = \text{dividend}$$

Let's do an example in which all seven steps are combined.

EXAMPLE 5 Using Synthetic Division to Verify a Factor

Use synthetic division to show that $g(x) = x + 3$ is a factor of

$$f(x) = 2x^5 + 5x^4 - 2x^3 + 2x^2 - 2x + 3$$

Solution The divisor is $x + 3 = x - (-3)$, so the row 3 entries will be multiplied by -3, entered in row 2, and added to row 1.

$$\begin{array}{r|rrrrrrl} -3 & 2 & 5 & -2 & 2 & -2 & 3 & \text{Row 1} \\ & & -6 & 3 & -3 & 3 & -3 & \text{Row 2} \\ \hline & 2 & -1 & 1 & -1 & 1 & 0 & \text{Row 3} \end{array}$$

Because the remainder is 0, we have

(Divisor)(Quotient) + Remainder =

$$(x + 3)(2x^4 - x^3 + x^2 - x + 1) = 2x^5 + 5x^4 - 2x^3 + 2x^2 - 2x + 3$$

As we see, $x + 3$ is a factor of $2x^5 + 5x^4 - 2x^3 + 2x^2 - 2x + 3$.

As Example 5 illustrates, the remainder after division gives information about whether the divisor is or is not a factor. We will have more to say about this in the next section.

NOW WORK PROBLEM 39.

4.3 EXERCISES

In Problems 1–26, find the quotient and the remainder. Check your work by verifying that

$$(\textit{Quotient})(\textit{Divisor}) + \textit{Remainder} = \textit{Dividend}$$

1. $4x^3 - 3x^2 + x + 1$ divided by $x + 2$

2. $3x^3 - x^2 + x - 2$ divided by $x + 2$

3. $4x^3 - 3x^2 + x + 1$ divided by $x - 4$

4. $3x^3 - x^2 + x - 2$ divided by $x - 4$

5. $4x^3 - 3x^2 + x + 1$ divided by $x^2 + 2$

6. $3x^3 - x^2 + x - 2$ divided by $x^2 + 2$

7. $4x^3 - 3x^2 + x + 1$ divided by $2x^3 - 1$

8. $3x^3 - x^2 + x - 2$ divided by $3x^3 - 1$

9. $4x^3 - 3x^2 + x + 1$ divided by $2x^2 + x + 1$

10. $3x^3 - x^2 + x - 2$ divided by $3x^2 + x + 1$

11. $4x^3 - 3x^2 + x + 1$ divided by $4x^2 + 1$

12. $3x^3 - x^2 + x - 2$ divided by $3x - 1$

13. $x^4 - 1$ divided by $x - 1$

14. $x^4 - 1$ divided by $x + 1$

15. $x^4 - 1$ divided by $x^2 - 1$

16. $x^4 - 1$ divided by $x^2 + 1$

17. $-4x^3 + x^2 - 4$ divided by $x - 1$

18. $-3x^4 - 2x - 1$ divided by $x - 1$

19. $1 - x^2 + x^4$ divided by $x^2 + x + 1$

20. $1 - x^2 + x^4$ divided by $x^2 - x + 1$

21. $1 - x^2 + x^4$ divided by $1 - x^2$

22. $1 - x^2 + x^4$ divided by $1 + x^2$

23. $x^3 - a^3$ divided by $x - a$

24. $x^3 + a^3$ divided by $x + a$

25. $x^4 - a^4$ divided by $x - a$

26. $x^5 - a^5$ divided by $x - a$

In Problems 27–38, use synthetic division to find the quotient $q(x)$ and remainder R when $f(x)$ is divided by $g(x)$.

27. $f(x) = x^3 - x^2 + 2x + 4;\quad g(x) = x - 2$

28. $f(x) = x^3 + 2x^2 - 3x + 1;\quad g(x) = x + 1$

29. $f(x) = 3x^3 + 2x^2 - x + 3;\quad g(x) = x - 3$

30. $f(x) = -4x^3 + 2x^2 - x + 1;\quad g(x) = x + 2$

31. $f(x) = x^5 - 4x^3 + x;\quad g(x) = x + 3$

32. $f(x) = x^4 + x^2 + 2;\quad g(x) = x - 2$

33. $f(x) = 4x^6 - 3x^4 + x^2 + 5;\quad g(x) = x - 1$

34. $f(x) = x^5 + 5x^3 - 10;\quad g(x) = x + 1$

35. $f(x) = 0.1x^3 + 0.2x;\quad g(x) = x + 1.1$

36. $f(x) = 0.1x^2 - 0.2;\quad g(x) = x + 2.1$

37. $f(x) = x^5 - 1;\quad g(x) = x - 1$

38. $f(x) = x^5 + 1;\quad g(x) = x + 1$

In Problems 39–48, use synthetic division to determine whether $x - c$ is a factor of $f(x)$.

39. $f(x) = 4x^3 - 3x^2 - 8x + 4;\quad x - 2$

40. $f(x) = -4x^3 + 5x^2 + 8;\quad x + 3$

41. $f(x) = 3x^4 - 6x^3 - 5x + 10;\quad x - 2$

42. $f(x) = 4x^4 - 15x^2 - 4;\quad x - 2$

43. $f(x) = 3x^6 + 82x^3 + 27;\quad x + 3$

44. $f(x) = 2x^6 - 18x^4 + x^2 - 9;\quad x + 3$

45. $f(x) = 4x^6 - 64x^4 + x^2 - 15;\quad x + 4$

46. $f(x) = x^6 - 16x^4 + x^2 - 16;\quad x + 4$

47. $f(x) = 2x^4 - x^3 + 2x - 1;\quad x - 1/2$

48. $f(x) = 3x^4 + x^3 - 3x + 1;\quad x + 1/3$

49. When dividing a polynomial by $x - c$, do you prefer to use long division or synthetic division? Does the value of c make a difference to you in choosing? Give reasons.

50. Find the sum of a, b, c, and d if

$$\frac{x^3 - 2x^2 + 3x + 5}{x + 2} = ax^2 + bx + c + \frac{d}{x + 2}$$

4.4 THE REAL ZEROS OF A POLYNOMIAL FUNCTION

1 Utilize the Remainder and Factor Theorems

2 Use the Rational Zeros Theorem to List the Potential Rational Zeros of a Polynomial Function

3 Find the Real Zeros of a Polynomial Function

4 Solve Polynomial Equations

5 Utilize the Theorem: Bounds on Zeros

6 Use the Intermediate Value Theorem

Remainder and Factor Theorems

1 Recall from the previous section that when we divide one polynomial (the dividend) by another (the divisor) we obtain a quotient polynomial and a re-

mainder, the remainder being either the zero polynomial or a polynomial whose degree is less than the degree of the divisor. To check our work, we verify that

$$(\text{Quotient})(\text{Divisor}) + \text{Remainder} = \text{Dividend}$$

This checking routine is the basis for a famous theorem called the **division algorithm* for polynomials,** which we now state without proof.

THEOREM

Division Algorithm for Polynomials

If $f(x)$ and $g(x)$ denote polynomial functions and if $g(x)$ is not the zero polynomial, then there are unique polynomial functions $q(x)$ and $r(x)$ such that

$$\frac{f(x)}{g(x)} = q(x) + \frac{r(x)}{g(x)} \quad \text{or} \quad \underset{\text{dividend}}{f(x)} = \underset{\text{quotient}}{q(x)}\,\underset{\text{divisor}}{g(x)} + \underset{\text{remainder}}{r(x)} \tag{1}$$

where $r(x)$ is either the zero polynomial or a polynomial of degree less than that of $g(x)$.

In equation (1), $f(x)$ is the **dividend,** $g(x)$ is the **divisor,** $q(x)$ is the **quotient,** and $r(x)$ is the **remainder.**

If the divisor $g(x)$ is a first-degree polynomial of the form

$$g(x) = x - c \qquad c \text{ a real number}$$

then the remainder $r(x)$ is either the zero polynomial or a polynomial of degree 0. As a result, for such divisors, the remainder is some number, say R, and we may write

$$f(x) = (x - c)q(x) + R \tag{2}$$

This equation is an identity in x and is true for all real numbers x. In particular, it is true when $x = c$, in which case equation (2) becomes

$$f(c) = (c - c)q(c) + R$$
$$f(c) = R$$

Using this fact in equation (2), we find

$$f(x) = (x - c)q(x) + f(c) \tag{3}$$

*A systematic process in which certain steps are repeated a finite number of times is called an **algorithm.** Thus, long division is an algorithm.

We have now proved the following result, called the **Remainder Theorem.**

REMAINDER THEOREM

Let f be a polynomial function. If $f(x)$ is divided by $x - c$, then the remainder is $f(c)$.

EXAMPLE 1 Using the Remainder Theorem

Find the remainder if $f(x) = x^3 - 4x^2 + 2x - 5$ is divided by

(a) $x - 3$ (b) $x + 2$

Solution (a) We could use long division. However, it is much easier here to use the Remainder Theorem, which says that the remainder is $f(3)$.

$$f(3) = (3)^3 - 4(3)^2 + 2(3) - 5 = 27 - 36 + 6 - 5 = -8$$

The remainder is -8.

(b) To find the remainder when $f(x)$ is divided by $x + 2 = x - (-2)$, we evaluate $f(-2)$.

$$f(-2) = (-2)^3 - 4(-2)^2 + 2(-2) - 5 = -8 - 16 - 4 - 5 = -33$$

The remainder is -33.

EXAMPLE 2 Using Synthetic Division to Find the Value of a Polynomial

Use synthetic division to find the value of $f(x) = -3x^4 + 2x^3 - x + 1$ at $x = -2$; that is, find $f(-2)$.

Solution The Remainder Theorem tells us that the value of a polynomial function at c equals the remainder when the polynomial is divided by $x - c$. This remainder is the final entry of the third row in the process of synthetic division. We want $f(-2)$, so we divide by $x - (-2)$.

$$\begin{array}{r|rrrrr} -2 & -3 & 2 & 0 & -1 & 1 \\ & & 6 & -16 & 32 & -62 \\ \hline & -3 & 8 & -16 & 31 & -61 \end{array}$$

The quotient is $q(x) = -3x^3 + 8x^2 - 16x + 31$; the remainder is $R = -61$. Because the remainder was found to be -61, it follows from the Remainder Theorem that $f(-2) = -61$.

As Example 2 illustrates, we can use the process of synthetic division to find the value of a polynomial function at a number c as an alternative to merely substituting c for x. A graphing utility provides other ways of finding the value of a function, such as using the eVALUEate feature. Consult your manual for details. Then check the result of Example 2.

An important and useful consequence of the Remainder Theorem is the **Factor Theorem.**

FACTOR THEOREM

Let f be a polynomial function. Then $x - c$ is a factor of $f(x)$ if and only if $f(c) = 0$.

The Factor Theorem actually consists of two separate statements:

1. If $f(c) = 0$, then $x - c$ is a factor of $f(x)$.
2. If $x - c$ is a factor of $f(x)$, then $f(c) = 0$.

Thus, the proof requires two parts.

Proof

1. Suppose that $f(c) = 0$. Then, by equation (3), we have
$$f(x) = (x - c)q(x)$$
for some polynomial $q(x)$. That is, $x - c$ is a factor of $f(x)$.
2. Suppose that $x - c$ is a factor of $f(x)$. Then there is a polynomial function q such that
$$f(x) = (x - c)q(x)$$
Replacing x by c, we find that
$$f(c) = (c - c)q(c) = 0 \cdot q(c) = 0$$

This completes the proof.

One use of the Factor Theorem is to determine whether a polynomial has a particular factor.

EXAMPLE 3 Using the Factor Theorem

Use the Factor Theorem to determine whether the function $f(x) = 2x^3 - x^2 + 2x - 3$ has the factor

(a) $x - 1$ (b) $x + 3$

Solution The Factor Theorem states that if $f(c) = 0$ then $x - c$ is a factor.

(a) Because $x - 1$ is of the form $x - c$ with $c = 1$, we find the value of $f(1)$.
$$f(1) = 2(1)^3 - (1)^2 + 2(1) - 3 = 2 - 1 + 2 - 3 = 0$$
By the Factor Theorem, $x - 1$ is a factor of $f(x)$.

(b) To test the factor $x + 3$, we first need to write it in the form $x - c$. Since $x + 3 = x - (-3)$, we find the value of $f(-3)$; we choose to use synthetic division.
$$\begin{array}{r|rrrr} -3 & 2 & -1 & 2 & -3 \\ & & -6 & 21 & -69 \\ \hline & 2 & -7 & 23 & -72 \end{array}$$
Because $f(-3) = -72 \neq 0$, we conclude from the Factor Theorem that $x - (-3) = x + 3$ is not a factor of $f(x)$.

NOW WORK PROBLEM 1.

The next theorem concerns the number of zeros a polynomial function may have. In counting the zeros of a polynomial, we count each zero as many times as its multiplicity.

THEOREM **Number of Zeros**

A polynomial function cannot have more zeros than its degree.

Proof The proof is based on the Factor Theorem. If r is a zero of a polynomial function f, then $f(r) = 0$ and, hence, $x - r$ is a factor of $f(x)$. Thus, each zero corresponds to a factor of degree 1. Because f cannot have more first-degree factors than its degree, the result follows.

Rational Zeros Theorem

2 The next result, called the **Rational Zeros Theorem,** provides information about the rational zeros of a polynomial *with integer coefficients.*

THEOREM **Rational Zeros Theorem**

Let f be a polynomial function of degree 1 or higher of the form

$$f(x) = a_n x^n + a_{n-1}x^{n-1} + \cdots + a_1 x + a_0 \qquad a_n \neq 0, a_0 \neq 0$$

where each coefficient is an integer. If p/q, in lowest terms, is a rational zero of f, then p must be a factor of a_0 and q must be a factor of a_n.

EXAMPLE 4 Listing Potential Rational Zeros

List the potential rational zeros of

$$f(x) = 2x^3 + 11x^2 - 7x - 6$$

Solution Because f has integer coefficients, we may use the Rational Zeros Theorem. First, we list all the integers p that are factors of $a_0 = -6$ and all the integers q that are factors of $a_3 = 2$.

$$p: \quad \pm 1, \pm 2, \pm 3, \pm 6$$
$$q: \quad \pm 1, \pm 2$$

Now we form all possible ratios p/q.

$$\frac{p}{q}: \quad \pm 1, \pm 2, \pm 3, \pm 6, \pm\frac{1}{2}, \pm\frac{3}{2}$$

If f has a rational zero, it will be found in this list, which contains 12 possibilities.

NOW WORK PROBLEM 11.

Be sure that you understand what the Rational Zeros Theorem says: For a polynomial with integer coefficients, *if* there is a rational zero, it is one of those listed. It may be the case that the function does not have any rational zeros.

The Rational Zeros Theorem provides a list of potential rational zeros of a function f. If we graph f, we can get a better sense of the location of the x-intercepts and test to see if they are rational. We can also use the potential rational zeros to select our initial viewing window to graph f and then adjust the window based on the results. The graphs shown throughout the text will be those obtained after setting the final viewing window.

EXAMPLE 5 Finding the Rational Zeros of a Polynomial Function

Continue working with Example 4 to find the rational zeros of

$$f(x) = 2x^3 + 11x^2 - 7x - 6$$

Solution We gather all the information we can about the zeros:

STEP 1: There are at most three zeros.

STEP 2: Now we use the list of potential rational zeros obtained in Example 4: $\pm 1, \pm 2, \pm 3, \pm 6, \pm\frac{1}{2}, \pm\frac{3}{2}$.

We could, of course, test each potential rational zero to see if the value of f there is zero. This is not very efficient. The graph of f will tell us approximately where the real zeros are. So we only need to test those rational zeros that are nearby. Figure 23 shows the graph of f. We see that f has three zeros: one near -6, one between -1 and 0, and one near 1. From our original list of potential rational zeros, we will test:

Figure 23

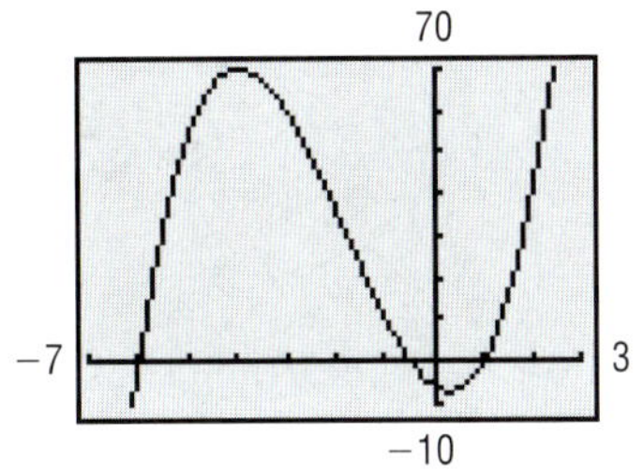

near -6: test -6; between -1 and 0: test $-\frac{1}{2}$; near 1: test 1

$$f(-6) = 0 \qquad f\left(-\frac{1}{2}\right) = 0 \qquad f(1) = 0$$

The three zeros of f are -6, $-\frac{1}{2}$, and 1; each is a rational zero. Since we have found the maximum number of zeros, we are certain that there are no additional zeros off the viewing window.

EXAMPLE 6 Finding the Real Zeros of a Polynomial Function

3

Find the real zeros of $f(x) = 3x^5 - 2x^4 - 15x^3 + 10x^2 + 12x - 8$. Write f in factored form.

Solution We gather all the information we can about the zeros.

STEP 1: There are at most five zeros.

STEP 2: To obtain the list of potential rational zeros, we write the factors p of $a_0 = -8$ and the factors q of $a_5 = 3$.

$$p: \quad \pm 1, \pm 2, \pm 4, \pm 8$$
$$q: \quad \pm 1, \pm 3$$

The potential rational zeros consist of all possible quotients p/q.

$$\frac{p}{q}: \quad \pm 1, \pm 2, \pm 4, \pm 8, \pm\frac{1}{3}, \pm\frac{2}{3}, \pm\frac{4}{3}, \pm\frac{8}{3}$$

Figure 24

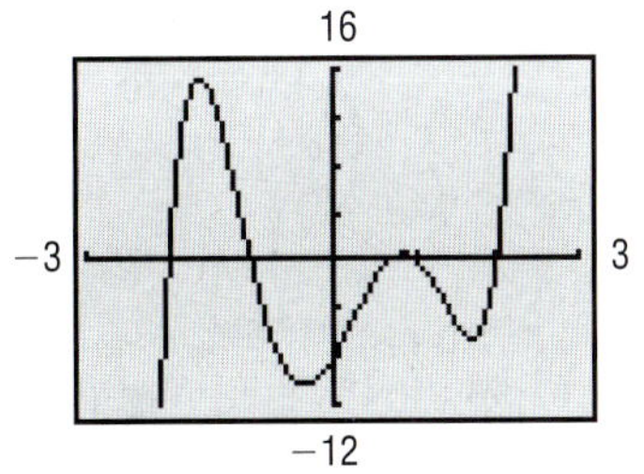

STEP 3: Figure 24 shows the graph of f. The graph has the characteristics that we expect of the given polynomial of degree 5: four turning points, y-intercept -8, and behaves like $y = 3x^5$ for large $|x|$. The graph has five or three x-intercepts: one near -2, one near -1, two (possibly repeated), or none near 1, and one near 2.

Since -2 appears to be a zero and -2 is a potential rational zero, we use synthetic division to determine if $x + 2$ is a factor of f.

$$\begin{array}{r|rrrrrr} -2 & 3 & -2 & -15 & 10 & 12 & -8 \\ & & -6 & 16 & -2 & -16 & 8 \\ \hline & 3 & -8 & 1 & 8 & -4 & 0 \end{array}$$

Since the remainder is 0, we know that -2 is a zero and $x + 2$ is a factor of f. We use the entries in the bottom row of the synthetic division to factor f.

$$\begin{aligned} f(x) &= 3x^5 - 2x^4 - 15x^3 + 10x^2 + 12x - 8 \\ &= (x + 2)(3x^4 - 8x^3 + x^2 + 8x - 4) \end{aligned}$$

Now any solution of the equation $3x^4 - 8x^3 + x^2 + 8x - 4 = 0$ is a zero of f. Because of this we call the equation $3x^4 - 8x^3 + x^2 + 8x - 4 = 0$ a **depressed equation** of f. Since the degree of the depressed equation of f is less than that of the original equation, we work with the depressed equation to find the zeros of f.

We check the potential rational zero -1 next using synthetic division.

$$\begin{array}{r|rrrrr} -1 & 3 & -8 & 1 & 8 & -4 \\ & & -3 & 11 & -12 & 4 \\ \hline & 3 & -11 & 12 & -4 & 0 \end{array}$$

Since the remainder is 0, we know that -1 is a zero and $x + 1$ is a factor of f. Now we can write f as

$$f(x) = (x + 2)(x + 1)(3x^3 - 11x^2 + 12x - 4)$$

We work with the second depressed equation and check the potential rational zero 1 using synthetic division.

$$\begin{array}{r|rrrr} 1 & 3 & -11 & 12 & -4 \\ & & 3 & -8 & 4 \\ \hline & 3 & -8 & 4 & 0 \end{array}$$

Since the remainder is 0, we conclude that 1 is a zero and $x - 1$ is a factor of f. Now we have

$$f(x) = (x + 2)(x + 1)(x - 1)(3x^2 - 8x + 4) \tag{4}$$

The new depressed equation of f, $3x^2 - 8x + 4 = 0$, is a quadratic equation with a discriminant of $b^2 - 4ac = (-8)^2 - 4(3)(4) = 16$. Therefore, this equation has two real solutions, and, in this case, we can find them by factoring.

$$\begin{aligned} 3x^2 - 8x + 4 &= 0 \\ (3x - 2)(x - 2) &= 0 \\ 3x - 2 = 0 \quad \text{or} \quad x - 2 &= 0 \\ x = \frac{2}{3} \qquad\qquad x &= 2 \end{aligned}$$

The zeros of f are $-2, -1, \frac{2}{3}, 1$, and 2.

Based on (4), the factored form of f is

$$\begin{aligned} f(x) &= (x + 2)(x + 1)(x - 1)(3x - 2)(x - 2) \\ &= 3(x + 2)(x + 1)\left(x - \frac{2}{3}\right)(x - 1)(x - 2) \end{aligned}$$

Now we know that the graph of f has two distinct x-intercepts near 1:1 and $\frac{2}{3}$. To see them, we change the viewing rectangle to obtain the graph of f for

Figure 25

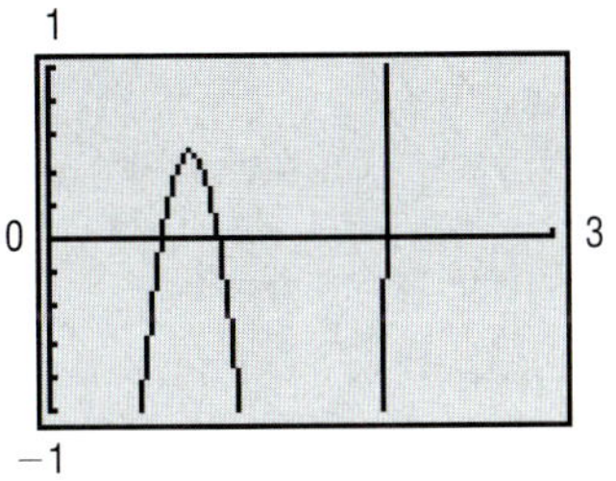

$0 \le x \le 3$. See Figure 25. Now we can see the three distinct positive x-intercepts between 0 and 3.

NOW WORK PROBLEM 29.

4 The procedure outlined in Example 6 for finding the zeros of a polynomial can also be used to solve polynomial equations.

EXAMPLE 7 Solving a Polynomial Equation

Solve the equation $x^5 - 5x^4 + 12x^3 - 24x^2 + 32x - 16 = 0$.

Solution The solutions of this equation are the zeros of the polynomial function

$$g(x) = x^5 - 5x^4 + 12x^3 - 24x^2 + 32x - 16$$

STEP 1: There are at most five real solutions.

STEP 2: Because $a_5 = 1$, the potential rational solutions are the integers $\pm 1, \pm 2, \pm 4, \pm 8$, and ± 16 (the factors of 16).

STEP 3: Figure 26 shows the graph of g. The graph has the characteristics that we expect of this polynomial of degree 5: it behaves like $y = x^5$ for large x, has y-intercept -16, and has two turning points. There is an x-intercept near 1 (where the graph crosses the x-axis) and another near 2 (which may be a root of even multiplicity, since the graph appears to touch the x-axis near 2).

Figure 26

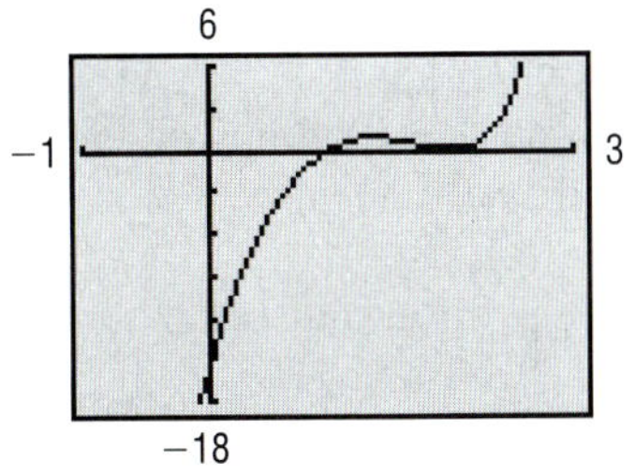

Since 1 appears to be a zero and 1 is a potential rational zero, we use synthetic division to determine if $x - 1$ is a factor of g.

1)	1	−5	12	−24	32	−16
		1	−4	8	−16	16
	1	−4	8	−16	16	0

Since the remainder is 0, we know that 1 is a zero and $x - 1$ is a factor of g. Thus, 1 is a solution to the equation.

We now work with the first depressed equation of g:

$$x^4 - 4x^3 + 8x^2 - 16x + 16 = 0$$

We check the potential rational zero 2 next using synthetic division.

2)	1	−4	8	−16	16
		2	−4	8	−16
	1	−2	4	−8	0

Since the remainder is 0, we conclude that 2 is a zero and $x - 2$ is a factor of g. Thus, 2 is a solution to the equation.

We work with the second depressed equation of g: $x^3 - 2x^2 + 4x - 8 = 0$. Earlier, we conjectured that 2 may be a root of even multiplicity, so we check 2 again.

$$\begin{array}{r|rrrr} 2 & 1 & -2 & 4 & -8 \\ & & 2 & 0 & 8 \\ \hline & 1 & 0 & 4 & 0 \end{array}$$

Since the remainder is 0, 2 is a zero and $x - 2$ is a factor of g. Thus, $x = 2$ is a zero of multiplicity 2.

The third depressed equation is $x^2 + 4 = 0$. This equation has no real solutions.

Thus, the real solutions are 1 and 2 (the latter being a repeated solution).

Based on the solution to Example 7, the polynomial function g can be factored as follows:

$$g(x) = x^5 - 5x^4 + 12x^3 - 24x^2 + 32x - 16 = (x - 1)(x - 2)^2(x^2 + 4)$$

NOW WORK PROBLEM 53.

In Example 7, the quadratic factor $x^2 + 4$ that appears in the factored form of $g(x)$ is called *irreducible*, because the polynomial $x^2 + 4$ cannot be factored over the real numbers. In general, we say that a quadratic factor $ax^2 + bx + c$ is **irreducible** if it cannot be factored over the real numbers, that is, if it is prime over the real numbers.

The polynomial function g of Example 7 has three real zeros and its factored form contains three linear factors and one irreducible quadratic factor. Refer back to the polynomial function f of Example 6. We found that f has five real zeros, so, by the Factor Theorem, its factored form will contain five linear factors.

THEOREM

Every polynomial function (with real coefficients) can be uniquely factored into a product of linear factors and/or irreducible quadratic factors.

We shall prove this result in Section 4.6, and, in fact, we shall draw several additional conclusions about the zeros of a polynomial function. One conclusion is worth noting now. If a polynomial (with real coefficients) is of odd degree, then it must contain at least one linear factor. (Do you see why?) This means it must have at least one real zero.

COROLLARY

A polynomial function (with real coefficients) of odd degree has at least one real zero.

One challenge in using a graphing utility is to set the viewing window so that a complete graph is obtained. The next theorem is a tool that can be used to find bounds on the zeros. This will assure that the function does not have any zeros outside these bounds. Then using these bounds to set *X*min and *X*max assures that all the *x*-intercepts appear in the viewing window.

Bounds on Zeros

5 The search for the real zeros of a polynomial function can be reduced somewhat if *bounds* on the zeros are found. A number M is a **bound** on the zeros of a polynomial if every zero r lies between $-M$ and M, inclusive. That is, M is a bound to the zeros of a polynomial f if

$$-M \leq \text{any zero of } f \leq M$$

THEOREM **Bounds on Zeros**

Let f denote a polynomial function whose leading coefficient is 1:

$$f(x) = x^n + a_{n-1}x^{n-1} + \cdots, + a_1x + a_0$$

A bound M on the zeros of f is the smaller of the two numbers

$$\text{Max}\{1, |a_0| + |a_1| + \cdots + |a_{n-1}|\} \qquad 1 + \text{Max}\{|a_0|, |a_1|, \ldots, |a_{n-1}|\} \quad (5)$$

where Max{ } means "choose the largest entry in { }."

An example will help to make the theorem clear.

EXAMPLE 8 Using the Theorem: Bounds on Zeros

Find a bound to the zeros of each polynomial.

(a) $f(x) = x^5 + 3x^3 - 9x^2 + 5$ (b) $g(x) = 4x^5 - 2x^3 + 2x^2 + 1$

Solution (a) The leading coefficient of f is 1.

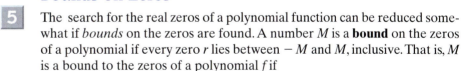

$$f(x) = x^5 + 3x^3 - 9x^2 + 5 \qquad a_4 = 0 \quad a_3 = 3 \quad a_2 = -9 \quad a_1 = 0 \quad a_0 = 5$$

We evaluate the expressions in (5).

$$\text{Max}\{1, |a_0| + |a_1| + \cdots |a_{n-1}|\} = \text{Max}\{1, |5| + |0| + |-9| + |3| + |0|\} = \text{Max}\{1, 17\} = 17$$
$$1 + \text{Max}\{|a_0|, |a_1|, \ldots, |a_{n-1}|\} = 1 + \text{Max}\{|5|, |0|, |-9|, |3|, |0|\} = 1 + 9 = 10$$

The smaller of the two numbers, 10, is the bound. Every zero of f lies between -10 and 10.

(b) First we write g so that its leading coefficient is 1.

$$g(x) = 4x^5 - 2x^3 + 2x^2 + 1 = 4(x^5 - \tfrac{1}{2}x^3 + \tfrac{1}{2}x^2 + \tfrac{1}{4})$$

Next, we evaluate the two expressions in (4) with $a_4 = 0$, $a_3 = -\frac{1}{2}$, $a_2 = \frac{1}{2}$, $a_1 = 0$, $a_0 = \frac{1}{4}$.

$$\text{Max}\{1, |a_0| + |a_1| + \cdots + |a_{n-1}|\} = \text{Max}\{1, |\tfrac{1}{4}| + |0| + |\tfrac{1}{2}| + |-\tfrac{1}{2}| + |0|\} = \text{Max}\{1, \tfrac{5}{4}\} = \tfrac{5}{4}$$

$$1 + \text{Max}\{|a_0|, |a_1|, |a_2|, |a_3|, |a_4|\} = 1 + \text{Max}\{|\tfrac{1}{4}|, |0|, |\tfrac{1}{2}|, |-\tfrac{1}{2}|, |0|\} = 1 + \tfrac{1}{2} = \tfrac{3}{2}$$

The smaller of the two numbers, $\frac{5}{4}$, is the bound. Every zero of g lies between $-\frac{5}{4}$ and $\frac{5}{4}$.

NOW WORK PROBLEM 23.

Steps for Finding the Zeros of a Polynomial

STEP 1: Use the degree of the polynomial to determine the maximum number of zeros.

STEP 2: If the polynomial has integer coefficients, use the Rational Zeros Theorem to identify those rational numbers that are potential zeros.

STEP 3: (a) Using a graphing utility, graph the polynomial.
(b) Conjecture possible zeros based on the graph and the list of potential rational zeros.
(c) Test each potential rational zero.
(d) Find any irrational zeros.

STEP 4: Use the bounds test to be sure that no zeros exist off the viewing window if all the real zeros have not been found.

Let's work one more example that shows these steps.

EXAMPLE 9 Finding the Zeros of a Polynomial

Find all the real zeros of the polynomial function

$$f(x) = x^5 - 1.8x^4 - 17.79x^3 + 31.672x^2 + 37.95x - 8.7121$$

Round answers to two decimal places.

Solution

STEP 1: There are at most five zeros.

STEP 2: Since there are noninteger coefficients, the Rational Zeros Theorem does not apply.

STEP 3: See Figure 27 for the graph of f. We see that f appears to have four x-intercepts: one near -4, one near -1, one between 0 and 1, and one near 3. The x-intercept near 3 might be a zero of even multiplicity since the graph seems to touch the x-axis at the point.

We use the Factor Theorem to determine if -4 and -1 are zeros. Since $f(-4) = f(-1) = 0$, we know that -4 and -1 are zeros. Using ZERO (on ROOT), we find that the remaining zeros are 0.20 and 3.30, rounded to two decimal places.

Figure 27

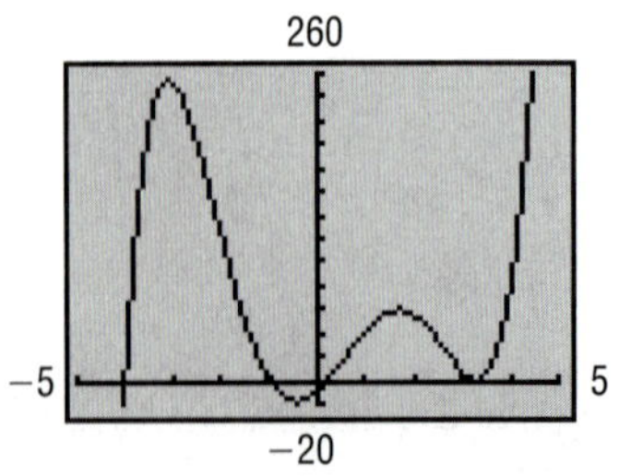

STEP 4: The maximum number of zeros is 5, and we have found 4 zeros, so we will determine the bounds of f to see if there are additional zeros off the viewing window. The leading coefficient of f is 1 with $a_4 = -1.8$, $a_3 = -17.79$, $a_2 = 31.672$, $a_1 = 37.95$, and $a_0 = -8.7121$. We evaluate the expressions in (5).

$$\text{Max}\{1, |-8.7121| + |37.95| + |31.672| + |-17.79| + |-1.8|\} = \text{Max}\{1, 97.9241\} = 97.9241$$
$$1 + \text{Max}\{|-8.7121|, |37.95|, |31.672|, |-17.79|, |-1.8|\} = 1 + 37.95 = 38.95$$

The smaller of the two numbers, 38.95, is the bound. Every zero of f lies between -38.95 and 38.95. Figure 28 shows the graph of f with $X\text{min} = -38.95$ and $X\text{max} = 38.95$. There are no zeros on the graph that have not already been identified. So either 3.30 is a zero of multiplicity 2 or there are two distinct zeros, each of which is 3.30, rounded to two decimal places. [Example 10 provides the answer].

Figure 28

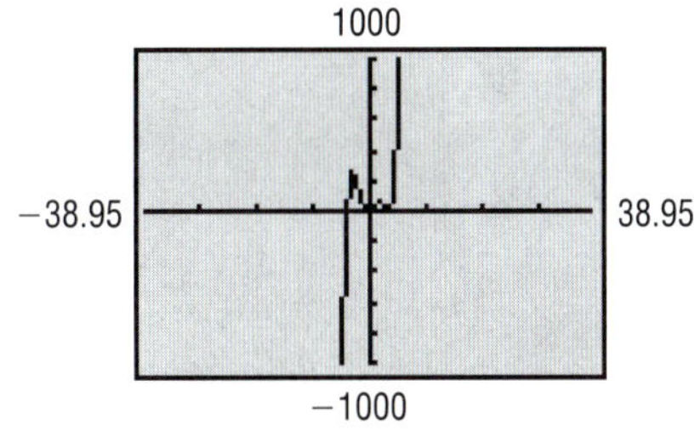

Intermediate Value Theorem

6 The Intermediate Value Theorem requires that the function be *continuous*. Although it requires calculus to explain the meaning precisely, the *idea* of a continuous function is easy to understand. Very basically, a function f is continuous when its graph can be drawn without lifting pencil from paper, that is, when the graph contains no "holes" or "jumps" or "gaps." For example, polynomial functions are continuous.

INTERMEDIATE VALUE THEOREM Let f denote a continuous function. If $a < b$ and if $f(a)$ and $f(b)$ are of opposite sign, then f has at least one zero between a and b.

Although the proof of this result requires advanced methods in calculus, it is easy to "see" why the result is true. Look at Figure 29.

Figure 29
If $f(a) < 0$ and $f(b) > 0$ and if f is continuous, there is at least one x-intercept between a and b.

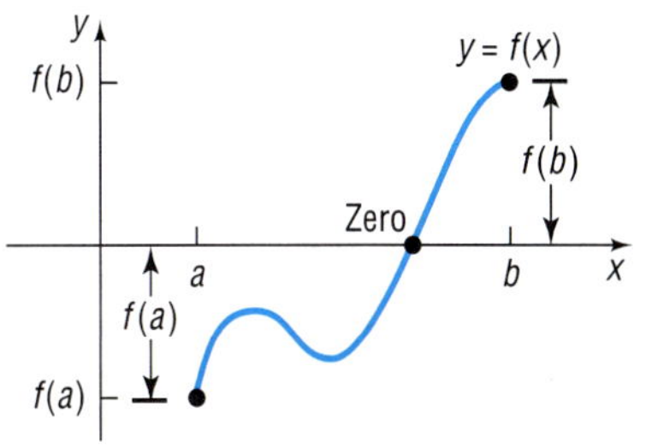

The Intermediate Value Theorem together with the TABLE feature of a graphing utility provide a basis for finding zeros.

EXAMPLE 10 Using the Intermediate Value Theorem and a Graphing Utility to Locate Zeros

Continue working with Example 9 to determine whether there is a repeated zero or two distinct zeros near 3.30.

Solution We use the TABLE feature of a graphing utility. See Table 11. Since $f(3.29) = 0.00956 > 0$ and $f(3.30) = -0.0001 < 0$, by the Intermediate Value Theorem there is a zero between 3.29 and 3.30. Similarly, the table indicates another zero between 3.30 and 3.31. Now we know that the five zeros of f are distinct.

Table 11

X	Y1
3.27	.08567
3.28	.03829
3.29	.00956
3.3	-1E-4
3.31	.0097
3.32	.03936
3.33	.08931

Y1=X^5-1.8X^4-1...

HISTORICAL FEATURE

Formulas for the solution of third- and fourth-degree polynomial equations exist, and, while not very practical, they do have an interesting history.

In the 1500s in Italy, mathematical contests were a popular pastime, and persons possessing methods for solving problems kept them secret. (Solutions that were published were already common knowledge.) Niccolo of Brescia (1500–1557), commonly referred to as Tartaglia ("the stammerer"), had the secret for solving cubic (third-degree) equations, which gave him a decided advantage in the contests. Girolamo Cardano (1501–1576) found out that Tartaglia had the secret and, being interested in cubics, he requested it from Tartaglia. The reluctant Tartaglia hesitated for some time, but finally, swearing Cardano to secrecy with midnight oaths by candlelight, told him the secret. Cardano then published the solution in his book *Ars Magna* (1545), giving Tartaglia the credit but rather compromising the secrecy. Tartaglia exploded into bitter recriminations, and each wrote pamphlets that reflected on the other's mathematics, moral character, and ancestry.

The quartic (fourth-degree) equation was solved by Cardano's student Lodovico Ferrari, and this solution also was included, with credit and this time with permission, in the *Ars Magna*.

Attempts were made to solve the fifth-degree equation in similar ways, all of which failed. In the early 1800s, P. Ruffini, Niels Abel, and Evariste Galois all found ways to show that it is not possible to solve fifth-degree equations by formula, but the proofs required the introduction of new methods. Galois' methods eventually developed into a large part of modern algebra.

4.4 EXERCISES

In Problems 1–10, use the Factor Theorem to determine whether $x - c$ is a factor of $f(x)$.

1. $f(x) = 4x^3 - 3x^2 - 8x + 4; \quad c = 3$

2. $f(x) = -4x^3 + 5x^2 + 8; \quad c = -2$

3. $f(x) = 3x^4 - 6x^3 - 5x + 10; \quad c = 1$

4. $f(x) = 4x^4 - 15x^2 - 4; \quad c = 2$

5. $f(x) = 3x^6 + 2x^3 - 176; \quad c = -2$

6. $f(x) = 2x^6 - 18x^4 + x^2 - 9; \quad c = -3$

7. $f(x) = 4x^6 - 64x^4 + x^2 - 16; \quad c = 4$

8. $f(x) = x^6 - 16x^4 + x^2 - 16; \quad c = -4$

9. $f(x) = 2x^4 - x^3 + 2x - 1; \quad c = \frac{-1}{2}$

10. $f(x) = 3x^4 + x^3 - 3x + 1; \quad c = -\frac{1}{3}$

In Problems 11–22, tell the maximum number of zeros that each polynomial function may have. Then list the potential rational zeros of each polynomial function. Do not attempt to find the zeros.

11. $f(x) = 3x^4 - 3x^3 + x^2 - x + 1$
12. $f(x) = x^5 - x^4 + 2x^2 + 3$
13. $f(x) = x^5 - 6x^2 + 9x - 3$
14. $f(x) = 2x^5 - x^4 - x^2 + 1$
15. $f(x) = -4x^3 - x^2 + x + 2$
16. $f(x) = 6x^4 - x^2 + 2$
17. $f(x) = 3x^4 - x^2 + 2$
18. $f(x) = -4x^3 + x^2 + x + 2$
19. $f(x) = 2x^5 - x^3 + 2x^2 + 4$
20. $f(x) = 3x^5 - x^2 + 2x + 3$
21. $f(x) = 6x^4 + 2x^3 - x^2 + 2$
22. $f(x) = -6x^3 - x^2 + x + 3$

In Problems 23–28, find the bounds to the zeros of each polynomial function.

23. $f(x) = 2x^3 + x^2 - 1$
24. $f(x) = 3x^3 - 2x^2 + x + 4$
25. $f(x) = x^3 - 5x^2 - 11x + 11$
26. $f(x) = 2x^3 - x^2 - 11x - 6$
27. $f(x) = x^4 + 3x^3 - 5x^2 + 9$
28. $f(x) = 4x^4 - 12x^3 + 27x^2 - 54x + 81$

In Problems 29–46, find the real zeros of f. Use the real zeros to factor f.

29. $f(x) = x^3 + 2x^2 - 5x - 6$
30. $f(x) = x^3 + 8x^2 + 11x - 20$
31. $f(x) = 2x^3 - 13x^2 + 24x - 9$
32. $f(x) = 2x^3 - 5x^2 - 4x + 12$
33. $f(x) = 3x^3 + 4x^2 + 4x + 1$
34. $f(x) = 3x^3 - 7x^2 + 12x - 28$
35. $f(x) = x^3 - 8x^2 + 17x - 6$
36. $f(x) = x^3 + 6x^2 + 6x - 4$
37. $f(x) = x^4 + x^3 - 3x^2 - x + 2$
38. $f(x) = x^4 - x^3 - 6x^2 + 4x + 8$
39. $f(x) = 2x^4 + 17x^3 + 35x^2 - 9x - 45$
40. $f(x) = 4x^4 - 15x^3 - 8x^2 + 15x + 4$
41. $f(x) = 2x^4 - 3x^3 - 21x^2 - 2x + 24$
42. $f(x) = 2x^4 + 11x^3 - 5x^2 - 43x + 35$
43. $f(x) = 4x^4 + 7x^2 - 2$
44. $f(x) = 4x^4 + 15x^2 - 4$
45. $f(x) = 4x^5 - 8x^4 - x + 2$
46. $f(x) = 4x^5 + 12x^4 - x - 3$

In Problems 47–52, find the real zeros of f rounded to two decimal places.

47. $f(x) = x^3 + 3.2x^2 - 16.83x - 5.31$
48. $f(x) = x^3 + 3.2x^2 - 7.25x - 6.3$
49. $f(x) = x^4 - 1.4x^3 - 33.71x^2 + 23.94x + 292.41$
50. $f(x) = x^4 + 1.2x^3 - 7.46x^2 - 4.692x + 15.2881$
51. $f(x) = x^3 + 19.5x^2 - 1021x + 1000.5$
52. $f(x) = x^3 + 42.2x^2 - 664.8x + 1490.4$

In Problems 53–62, find the real solutions of each equation.

53. $x^4 - x^3 + 2x^2 - 4x - 8 = 0$
54. $2x^3 + 3x^2 + 2x + 3 = 0$
55. $3x^3 + 4x^2 - 7x + 2 = 0$
56. $2x^3 - 3x^2 - 3x - 5 = 0$
57. $3x^3 - x^2 - 15x + 5 = 0$
58. $2x^3 - 11x^2 + 10x + 8 = 0$
59. $x^4 + 4x^3 + 2x^2 - x + 6 = 0$
60. $x^4 - 2x^3 + 10x^2 - 18x + 9 = 0$
61. $x^3 - \frac{2}{3}x^2 + \frac{8}{3}x + 1 = 0$
62. $x^3 - \frac{2}{3}x^2 + 3x - 2 = 0$

In Problems 63–68, use the Intermediate Value Theorem to show that each function has a zero in the given interval. Approximate the zero rounded to two decimal places.

63. $f(x) = 8x^4 - 2x^2 + 5x - 1;\ [0, 1]$
64. $f(x) = x^4 + 8x^3 - x^2 + 2;\ [-1, 0]$
65. $f(x) = 2x^3 + 6x^2 - 8x + 2;\ [-5, -4]$
66. $f(x) = 3x^3 - 10x + 9;\ [-3, -2]$
67. $f(x) = x^5 - x^4 + 7x^3 - 7x^2 - 18x + 18;\ [1.4, 1.5]$
68. $f(x) = x^5 - 3x^4 - 2x^3 + 6x^2 + x + 2;\ [1.7, 1.8]$

69. **Cost of Manufacturing** In Problem 69 of Section 4.2, you found the cost function for manufacturing Chevy Cavaliers. Use the methods learned in this section to determine how many Cavaliers can be manufactured at a total cost of \$3,000,000. In other words, solve the equation $C(x) = 3000$.

70. **Cost of Printing** In Problem 70 of Section 4.2, you found the cost function for printing textbooks. Use the methods learned in this section to determine how many textbooks can be printed at a total cost of \$200,000. In other words, solve the equation $C(x) = 200$.

71. Find k such that $f(x) = x^3 - kx^2 + kx + 2$ has the factor $x - 2$.

72. Find k such that $f(x) = x^4 - kx^3 + kx^2 + 1$ has the factor $x + 2$.

73. What is the remainder when

$$f(x) = 2x^{20} - 8x^{10} + x - 2$$

is divided by $x - 1$?

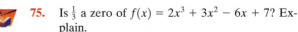

74. What is the remainder when

$$f(x) = -3x^{17} + x^9 - x^5 + 2x$$

is divided by $x + 1$?

75. Is $\frac{1}{3}$ a zero of $f(x) = 2x^3 + 3x^2 - 6x + 7$? Explain.

76. Is $\frac{1}{3}$ a zero of $f(x) = 4x^3 - 5x^2 - 3x + 1$? Explain.

77. Is $\frac{3}{5}$ a zero of $f(x) = 2x^6 - 5x^4 + x^3 - x + 1$? Explain.

78. Is $\frac{2}{3}$ a zero of $f(x) = x^7 + 6x^5 - x^4 + x + 2$? Explain.

79. What is the length of the edge of a cube if, after a slice 1 inch thick is cut from one side, the volume remaining is 294 cubic inches?

80. What is the length of the edge of a cube if its volume could be doubled by an increase of 6 centimeters in one edge, an increase of 12 centimeters in a second edge, and a decrease of 4 centimeters in the third edge?

4.5 COMPLEX NUMBERS; QUADRATIC EQUATIONS WITH A NEGATIVE DISCRIMINANT

1 Add, Subtract, Multiply, and Divide Complex Numbers

2 Solve Quadratic Equations with a Negative Discriminant

One property of a real number is that its square is nonnegative. For example, there is no real number x for which

$$x^2 = -1$$

To remedy this situation, we introduce a number called the **imaginary unit,** which we denote by ***i*** and whose square is -1. Thus,

$$i^2 = -1$$

This should not surprise you. If our universe were to consist only of integers, there would be no number x for which $2x = 1$. This unfortunate circumstance was remedied by introducing numbers such as $\frac{1}{2}$ and $\frac{2}{3}$, the *rational numbers.* If our universe were to consist only of rational numbers, there would be no x whose square equals 2. That is, there would be no number x for which $x^2 = 2$. To remedy this, we introduced numbers such as $\sqrt{2}$ and $\sqrt[3]{5}$, the *irrational numbers.* The *real numbers,* you will recall, consist of the rational numbers and the irrational numbers. Now, if our universe were to consist only of real numbers, then there would be no number x whose square is -1. To remedy this, we introduce a number i, whose square is -1.

In the progression outlined, each time that we encountered a situation that was unsuitable, we introduced a new number system to remedy this situation. And each new number system contained the earlier number system as a subset. The number system that results from introducing the number i is called the **complex number system.**

Complex numbers are numbers of the form $a + bi$, where a and b are real numbers. The real number a is called the **real part** of the number $a + bi$; the real number b is called the **imaginary part** of $a + bi$.

For example, the complex number $-5 + 6i$ has the real part -5 and the imaginary part 6.

When a complex number is written in the form $a + bi$, where a and b are real numbers, we say it is in **standard form.** However, if the imaginary part of a complex number is negative, such as in the complex number $3 + (-2)i$, we agree to write it instead in the form $3 - 2i$.

Also, the complex number $a + 0i$ is usually written merely as a. This serves to remind us that the real numbers are a subset of the complex numbers. The complex number $0 + bi$ is usually written as bi. Sometimes the complex number bi is called a **pure imaginary number.**

1 Equality, addition, subtraction, and multiplication of complex numbers are defined so as to preserve the familiar rules of algebra for real numbers. Thus, two complex numbers are equal if and only if their real parts are equal and their imaginary parts are equal. That is,

Equality of Complex Numbers

$$a + bi = c + di \quad \text{if and only if} \quad a = c \text{ and } b = d \qquad (1)$$

Two complex numbers are added by forming the complex number whose real part is the sum of the real parts and whose imaginary part is the sum of the imaginary parts. That is,

Sum of Complex Numbers

$$(a + bi) + (c + di) = (a + c) + (b + d)i \qquad (2)$$

To subtract two complex numbers, we follow the rule

Difference of Complex Numbers

$$(a + bi) - (c + di) = (a - c) + (b - d)i \qquad (3)$$

EXAMPLE 1 Adding and Subtracting Complex Numbers

(a) $(3 + 5i) + (-2 + 3i) = [3 + (-2)] + (5 + 3)i = 1 + 8i$

(b) $(6 + 4i) - (3 + 6i) = (6 - 3) + (4 - 6)i = 3 + (-2)i = 3 - 2i$

Figure 30

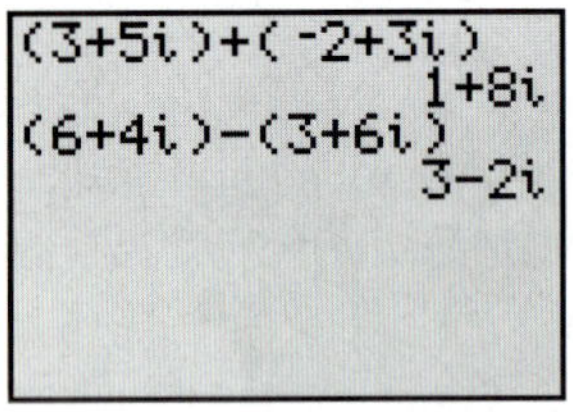

Some graphing calculators have the capability of handling complex numbers.* For example, Figure 30 shows the results of Example 1 using a TI-83 graphing calculator.

NOW WORK PROBLEM 5.

Products of complex numbers are calculated as illustrated in Example 2.

EXAMPLE 2 Multiplying Complex Numbers

$$(5+3i)\cdot(2+7i) = \underset{\substack{\uparrow\\ \text{Distributive property}}}{=} 5\cdot(2+7i) + 3i(2+7i) \underset{\substack{\uparrow\\ \text{Distributive property}}}{=} 10 + 35i + 6i + 21i^2$$

$$\underset{\substack{\uparrow\\ i^2 = -1}}{=} 10 + 41i + 21(-1)$$

$$= -11 + 41i$$

Based on the procedure of Example 2, we define the **product** of two complex numbers by the formula

Product of Complex Numbers

$$(a+bi)\cdot(c+di) = (ac-bd) + (ad+bc)i \qquad (4)$$

Figure 31

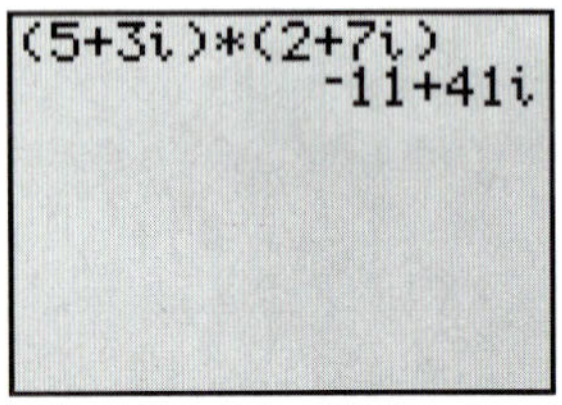

Do not bother to memorize formula (4). Instead, whenever it is necessary to multiply two complex numbers, follow the usual rules for multiplying two binomials, as in Example 2, remembering that $i^2 = -1$. For example,

$$(2i)(2i) = 4i^2 = -4$$
$$(2+i)(1-i) = 2 - 2i + i - i^2 = 3 - i$$

Graphing calculators may also be used to multiply complex numbers. Figure 31 shows the result obtained in Example 2 using a TI-83 graphing calculator.

NOW WORK PROBLEM 11.

Algebraic properties for addition and multiplication, such as the commutative, associative, and distributive properties, hold for complex numbers. Of these, the property that every nonzero complex number has a multiplicative inverse, or reciprocal, requires a closer look.

Conjugates

If $z = a + bi$ is a complex number, then its **conjugate,** denoted by $\bar{z}$, is defined as

$$\bar{z} = \overline{a+bi} = a - bi$$

For example, $\overline{2+3i} = 2 - 3i$ and $\overline{-6-2i} = -6 + 2i$.

*Consult your users manual for the appropriate keystrokes.

EXAMPLE 3 Multiplying a Complex Number by Its Conjugate

Find the product of the complex number $z = 3 + 4i$ and its conjugate $\bar{z}$.

Solution Since $\bar{z} = 3 - 4i$, we have

$$z\bar{z} = (3 + 4i)(3 - 4i) = 9 + 12i - 12i - 16i^2 = 9 + 16 = 25$$

The result obtained in Example 3 has an important generalization.

THEOREM The product of a complex number and its conjugate is a nonnegative real number. That is, if $z = a + bi$, then

$$z\bar{z} = a^2 + b^2 \qquad (5)$$

Proof If $z = a + bi$, then

$$z\bar{z} = (a + bi)(a - bi) = a^2 - (bi)^2 = a^2 - b^2i^2 = a^2 + b^2$$

To express the reciprocal of a nonzero complex number z in standard form, multiply the numerator and denominator by its conjugate $\bar{z}$. That is, if $z = a + bi$ is a nonzero complex number, then

$$\frac{1}{a + bi} = \frac{1}{z} = \frac{1}{z} \cdot \frac{\bar{z}}{\bar{z}} = \frac{\bar{z}}{z\bar{z}} = \frac{a - bi}{(a + bi)(a - bi)}$$

$$= \frac{a - bi}{a^2 + b^2}$$

↑ Use (5).

$$= \frac{a}{a^2 + b^2} - \frac{b}{a^2 + b^2}i$$

EXAMPLE 4 Writing the Reciprocal of a Complex Number in Standard Form

Write $\dfrac{1}{3 + 4i}$ in standard form $a + bi$; that is, find the reciprocal of $3 + 4i$.

Solution The idea is to multiply the numerator and denominator by the conjugate of $3 + 4i$, that is, the complex number $3 - 4i$. The result is

$$\frac{1}{3 + 4i} = \frac{1}{3 + 4i} \cdot \frac{3 - 4i}{3 - 4i} = \frac{3 - 4i}{9 + 16} = \frac{3}{25} - \frac{4}{25}i$$

Figure 32

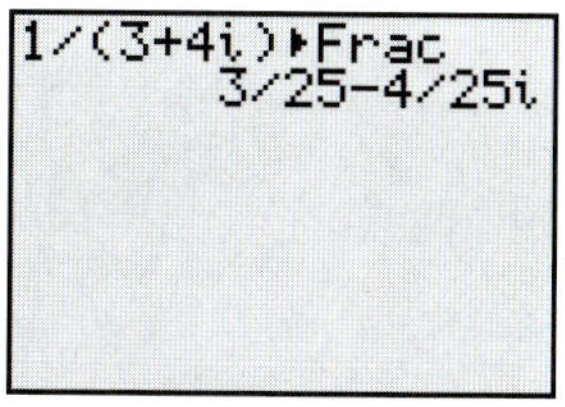

A graphing calculator can be used to verify the result of Example 4. See Figure 32.

To express the quotient of two complex numbers in standard form, we multiply the numerator and denominator of the quotient by the conjugate of the denominator.

EXAMPLE 5 Writing the Quotient of Complex Numbers in Standard Form

Write each of the following in standard form:

(a) $\dfrac{1+4i}{5-12i}$ (b) $\dfrac{2-3i}{4-3i}$

Solution (a) $\dfrac{1+4i}{5-12i} = \dfrac{1+4i}{5-12i} \cdot \dfrac{5+12i}{5+12i} = \dfrac{5+20i+12i+48i^2}{25+144}$

$$= \frac{-43+32i}{169} = \frac{-43}{169} + \frac{32}{169}i$$

(b) $\dfrac{2-3i}{4-3i} = \dfrac{2-3i}{4-3i} \cdot \dfrac{4+3i}{4+3i} = \dfrac{8-12i+6i-9i^2}{16+9} = \dfrac{17-6i}{25} = \dfrac{17}{25} - \dfrac{6}{25}i$

NOW WORK PROBLEM 19.

EXAMPLE 6 Writing Other Expressions in Standard Form

If $z = 2 - 3i$ and $w = 5 + 2i$, write each of the following expressions in standard form:

(a) $\dfrac{z}{w}$ (b) $\overline{z+w}$ (c) $z + \bar{z}$

Solution (a) $\dfrac{z}{w} = \dfrac{z \cdot \overline{w}}{w \cdot \overline{w}} = \dfrac{(2-3i)(5-2i)}{(5+2i)(5-2i)} = \dfrac{10-15i-4i+6i^2}{25+4}$

$$= \frac{4-19i}{29} = \frac{4}{29} - \frac{19}{29}i$$

(b) $\overline{z+w} = \overline{(2-3i)+(5+2i)} = \overline{7-i} = 7 + i$

(c) $z + \bar{z} = (2-3i) + (2+3i) = 4$

The conjugate of a complex number has certain general properties that we shall find useful later.

For a real number $a = a + 0i$, the conjugate is $\bar{a} = \overline{a+0i} = a - 0i = a$. That is,

THEOREM The conjugate of a real number is the real number itself.

Other properties of the conjugate that are direct consequences of the definition are given next. In each statement, z and w represent complex numbers.

THEOREM The conjugate of the conjugate of a complex number is the complex number itself.

$$\overline{(\overline{z})} = z \qquad (6)$$

The conjugate of the sum of two complex numbers equals the sum of their conjugates.

$$\overline{z + w} = \overline{z} + \overline{w} \qquad (7)$$

The conjugate of the product of two complex numbers equals the product of their conjugates.

$$\overline{z \cdot w} = \overline{z} \cdot \overline{w} \qquad (8)$$

We leave the proofs of equations (6), (7), and (8) as exercises.

Powers of i

The **powers of i** follow a pattern that is useful to know.

$$i^1 = i \qquad i^5 = i^4 \cdot i = 1 \cdot i = i$$
$$i^2 = -1 \qquad i^6 = i^4 \cdot i^2 = -1$$
$$i^3 = i^2 \cdot i = -i \qquad i^7 = i^4 \cdot i^3 = -i$$
$$i^4 = i^2 \cdot i^2 = (-1)(-1) = 1 \qquad i^8 = i^4 \cdot i^4 = 1$$

And so on. Thus, the powers of i repeat with every fourth power.

EXAMPLE 7 Evaluating Powers of i

(a) $i^{27} = i^{24} \cdot i^3 = (i^4)^6 \cdot i^3 = 1^6 \cdot i^3 = -i$

(b) $i^{101} = i^{100} \cdot i^1 = (i^4)^{25} \cdot i = 1^{25} \cdot i = i$

EXAMPLE 8 Writing the Power of a Complex Number in Standard Form

Write $(2 + i)^3$ in standard form.

Solution We use the special product formula for $(x + a)^3$.

$$(x + a)^3 = x^3 + 3ax^2 + 3a^2x + a^3$$

Thus,

$$\begin{aligned}(2+i)^3 &= 2^3 + 3\cdot i\cdot 2^2 + 3\cdot i^2\cdot 2 + i^3\\ &= 8 + 12i + 6(-1) + (-i)\\ &= 2 + 11i\end{aligned}$$

NOW WORK PROBLEM 33.

Quadratic Equations with a Negative Discriminant

2 Quadratic equations with a negative discriminant have no real number solution. However, if we extend our number system to allow complex numbers, quadratic equations will always have a solution. Since the solution to a quadratic equation involves the square root of the discriminant, we begin with a discussion of square roots of negative numbers.

If N is a positive real number, we define the **principal square root of $-N$,** denoted by $\sqrt{-N}$, as

$$\sqrt{-N} = \sqrt{N}i$$

where i is the imaginary unit and $i^2 = -1$.

EXAMPLE 9 Evaluating the Square Root of a Negative Number

(a) $\sqrt{-1} = \sqrt{1}i = i$ (b) $\sqrt{-4} = \sqrt{4}i = 2i$

(c) $\sqrt{-8} = \sqrt{8}i = 2\sqrt{2}i$

EXAMPLE 10 Solving Equations

Solve each equation in the complex number system.

(a) $x^2 = 4$ (b) $x^2 = -9$

Solution (a) $x^2 = 4$

$x = \pm\sqrt{4} = \pm 2$

The equation has two solutions, -2 and 2.

(b) $x^2 = -9$

$x = \pm\sqrt{-9} = \pm\sqrt{9}i = \pm 3i$

The equation has two solutions, $-3i$ and $3i$.

NOW WORK PROBLEM 45.

Warning: When working with square roots of negative numbers, do not set the square root of a product equal to the product of the square roots (which can be done with positive numbers). To see why, look at this calculation: We know that $\sqrt{100} = 10$. However, it is also true that $100 = (-25)(-4)$, so

$$10 = \sqrt{100} = \sqrt{(-25)(-4)} = \sqrt{-25}\sqrt{-4} = (\sqrt{25}i)(\sqrt{4}i) = (5i)(2i) = 10i^2 = -10$$

↑ Here is the error.

Because we have defined the square root of a negative number, we now can restate the quadratic formula without restriction.

THEOREM In the complex number system, the solutions of the quadratic equation $ax^2 + bx + c = 0$, where a, b, and c are real numbers and $a \neq 0$, are given by the formula

$$x = \frac{-b \pm \sqrt{b^2 - 4ac}}{2a} \qquad (9)$$

EXAMPLE 11 Solving Quadratic Equations in the Complex Number System

Solve the equation $x^2 - 4x + 8 = 0$ in the complex number system.

Solution Here $a = 1, b = -4, c = 8$, and $b^2 - 4ac = 16 - 4(8) = -16$. Using equation (9), we find that

$$x = \frac{4 \pm \sqrt{-16}}{2} = \frac{4 \pm \sqrt{16}i}{2} = \frac{4 \pm 4i}{2} = 2 \pm 2i$$

The equation has the solution set $\{2 - 2i, 2 + 2i\}$.

✓CHECK:

$$2 + 2i: \quad (2 + 2i)^2 - 4(2 + 2i) + 8 = 4 + 8i + 4i^2 - 8 - 8i + 8$$
$$= 4 - 4 = 0$$
$$2 - 2i: \quad (2 - 2i)^2 - 4(2 - 2i) + 8 = 4 - 8i + 4i^2 - 8 + 8i + 8$$
$$= 4 - 4 = 0$$

Figure 33

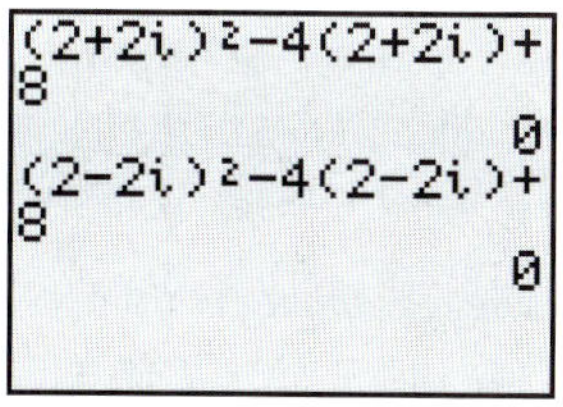

Figure 33 shows the check of the solution using a TI-83 graphing calculator.

NOW WORK PROBLEM **51.**

The discriminant $b^2 - 4ac$ of a quadratic equation still serves as a way to determine the character of the solutions.

Discriminant of a Quadratic Equation

In the complex number system, consider a quadratic equation $ax^2 + bx + c = 0$ with real coefficients.

1. If $b^2 - 4ac > 0$, the equation has two unequal real solutions.
2. If $b^2 - 4ac = 0$, the equation has a repeated real solution, a double root.
3. If $b^2 - 4ac < 0$, the equation has two complex solutions that are not real. The solutions are conjugates of each other.

The third conclusion in the display is a consequence of the fact that if $b^2 - 4ac = -N < 0$ then, by the quadratic formula, the solutions are

$$x = \frac{-b + \sqrt{b^2 - 4ac}}{2a} = \frac{-b + \sqrt{-N}}{2a} = \frac{-b + \sqrt{N}i}{2a} = \frac{-b}{2a} + \frac{\sqrt{N}}{2a}i$$

and

$$x = \frac{-b - \sqrt{b^2 - 4ac}}{2a} = \frac{-b - \sqrt{-N}}{2a} = \frac{-b - \sqrt{N}i}{2a} = \frac{-b}{2a} - \frac{\sqrt{N}}{2a}i$$

which are conjugates of each other.

EXAMPLE 12 Determining the Character of the Solution of a Quadratic Equation

Without solving, determine the character of the solution of each equation.

(a) $3x^2 + 4x + 5 = 0$
(b) $2x^2 + 4x + 1 = 0$
(c) $9x^2 - 6x + 1 = 0$

Solution (a) Here $a = 3$, $b = 4$, and $c = 5$, so $b^2 - 4ac = 16 - 4(3)(5) = -44$. The solutions are complex numbers that are not real and are conjugates of each other.

(b) Here $a = 2$, $b = 4$, and $c = 1$, so $b^2 - 4ac = 16 - 8 = 8$. The solutions are two unequal real numbers.

(c) Here $a = 9$, $b = -6$, and $c = 1$, so $b^2 - 4ac = 36 - 4(9)(1) = 0$. The solution is a repeated real number, that is, a double root.

4.5 EXERCISES

In Problems 1–38, write each expression in the standard form a + bi. Verify your results using a graphing utility.

1. $(2 - 3i) + (6 + 8i)$
2. $(4 + 5i) + (-8 + 2i)$
3. $(-3 + 2i) - (4 - 4i)$
4. $(3 - 4i) - (-3 - 4i)$
5. $(2 - 5i) - (8 + 6i)$
6. $(-8 + 4i) - (2 - 2i)$
7. $3(2 - 6i)$
8. $-4(2 + 8i)$
9. $2i(2 - 3i)$
10. $3i(-3 + 4i)$
11. $(3 - 4i)(2 + i)$
12. $(5 + 3i)(2 - i)$
13. $(-6 + i)(-6 - i)$
14. $(-3 + i)(3 + i)$
15. $\dfrac{10}{3 - 4i}$
16. $\dfrac{13}{5 - 12i}$
17. $\dfrac{2 + i}{i}$
18. $\dfrac{2 - i}{-2i}$
19. $\dfrac{6 - i}{1 + i}$
20. $\dfrac{2 + 3i}{1 - i}$
21. $\left(\dfrac{1}{2} + \dfrac{\sqrt{3}}{2}i\right)^2$
22. $\left(\dfrac{\sqrt{3}}{2} - \dfrac{1}{2}i\right)^2$
23. $(1 + i)^2$
24. $(1 - i)^2$
25. i^{23}
26. i^{14}
27. i^{-15}
28. i^{-23}
29. $i^6 - 5$
30. $4 + i^3$
31. $6i^3 - 4i^5$
32. $4i^3 - 2i^2 + 1$
33. $(1 + i)^3$
34. $(3i)^4 + 1$
35. $i^7(1 + i^2)$
36. $2i^4(1 + i^2)$
37. $i^6 + i^4 + i^2 + 1$
38. $i^7 + i^5 + i^3 + i$

In Problems 39–44, perform the indicated operations and express your answer in the form a + bi.

39. $\sqrt{-4}$
40. $\sqrt{-9}$
41. $\sqrt{-25}$
42. $\sqrt{-64}$
43. $\sqrt{(3 + 4i)(4i - 3)}$
44. $\sqrt{(4 + 3i)(3i - 4)}$

In Problems 45–64, solve each equation in the complex number system. Check your results using a graphing utility.

45. $x^2 + 4 = 0$ **46.** $x^2 - 4 = 0$ **47.** $x^2 - 16 = 0$ **48.** $x^2 + 25 = 0$

49. $x^2 - 6x + 13 = 0$ **50.** $x^2 + 4x + 8 = 0$ **51.** $x^2 - 6x + 10 = 0$ **52.** $x^2 - 2x + 5 = 0$

53. $8x^2 - 4x + 1 = 0$ **54.** $10x^2 + 6x + 1 = 0$ **55.** $5x^2 + 2x + 1 = 0$ **56.** $13x^2 + 6x + 1 = 0$

57. $x^2 + x + 1 = 0$ **58.** $x^2 - x + 1 = 0$ **59.** $x^3 - 8 = 0$ **60.** $x^3 + 27 = 0$

61. $x^4 - 16 = 0$ **62.** $x^4 - 1 = 0$ **63.** $x^4 + 13x^2 + 36 = 0$ **64.** $x^4 + 3x^2 - 4 = 0$

In Problems 65–70, without solving, determine the character of the solutions of each equation. Verify your answer using a graphing utility.

65. $3x^2 - 3x + 4 = 0$ **66.** $2x^2 - 4x + 1 = 0$ **67.** $2x^2 + 3x - 4 = 0$

68. $x^2 + 2x + 6 = 0$ **69.** $9x^2 - 12x + 4 = 0$ **70.** $4x^2 + 12x + 9 = 0$

71. $2 + 3i$ is a solution of a quadratic equation with real coefficients. Find the other solution.

72. $4 - i$ is a solution of a quadratic equation with real coefficients. Find the other solution.

In Problems 73–76, $z = 3 - 4i$ and $w = 8 + 3i$. Write each expression in the standard form $a + bi$.

73. $z + \bar{z}$ **74.** $w - \bar{w}$ **75.** $z\bar{z}$ **76.** $\overline{z - w}$

77. Use $z = a + bi$ to show that $z + \bar{z} = 2a$ and that $z - \bar{z} = 2bi$.

78. Use $z = a + bi$ to show that $(\bar{\bar{z}}) = z$.

79. Use $z = a + bi$ and $w = c + di$ to show that $\overline{z + w} = \bar{z} + \bar{w}$.

80. Use $z = a + bi$ and $w = c + di$ to show that $\overline{z \cdot w} = \bar{z} \cdot \bar{w}$.

81. Explain to a friend how you would add two complex numbers and how you would multiply two complex numbers. Explain any differences in the two explanations.

82. Write a brief paragraph that compares the method used to rationalize the denominator of a rational expression and the method used to write a complex number in standard form.

4.6 COMPLEX ZEROS; FUNDAMENTAL THEOREM OF ALGEBRA

1 Utilize the Conjugate Pairs Theorem to Find the Complex Zeros of a Polynomial

2 Find a Polynomial Function with Specified Zeros

3 Find the Complex Zeros of a Polynomial

In Section 4.4 we found the **real** zeros of a polynomial function. In this section, we will find the **complex** zeros of a polynomial function. Since the set of real numbers is a subset of the set of complex numbers, finding the complex zeros of a function requires finding all zeros of the form $a + bi$. These zeros will be real if $b = 0$.

A variable in the complex number system is referred to as a **complex variable.** A **complex polynomial function** f of degree n is a function of the form

$$f(x) = a_n x^n + a_{n-1} x^{n-1} + \cdots + a_1 x + a_0 \qquad (1)$$

where $a_n, a_{n-1}, \dots, a_1, a_0$ are complex numbers, $a_n \neq 0$, n is a nonnegative integer, and x is a complex variable. Here a_n is called the **leading coefficient** of f. A complex number r is called a (complex) **zero** of f if $f(r) = 0$.

We have learned that some quadratic equations have no real solutions, but that in the complex number system every quadratic equation has a solution, either real or complex. The next result, proved by Karl Friedrich Gauss (1777–1855) when he was 22 years old,* gives an extension to complex polynomials. In fact, this result is so important and useful it has become known as the **Fundamental Theorem of Algebra.**

FUNDAMENTAL THEOREM OF ALGEBRA

Every complex polynomial function $f(x)$ of degree $n \geq 1$ has at least one complex zero.

We shall not prove this result, as the proof is beyond the scope of this book. However, using the Fundamental Theorem of Algebra and the Factor Theorem, we can prove the following result:

THEOREM

Every complex polynomial function $f(x)$ of degree $n \geq 1$ can be factored into n linear factors (not necessarily distinct) of the form

$$f(x) = a_n(x - r_1)(x - r_2) \cdot \ldots \cdot (x - r_n) \tag{2}$$

where $a_n, r_1, r_2, \ldots, r_n$ are complex numbers.

Proof Let

$$f(x) = a_n x^n + a_{n-1}x^{n-1} + \cdots + a_1 x + a_0$$

By the Fundamental Theorem of Algebra, f has at least one zero, say r_1. Then, by the Factor Theorem, $x - r_1$ is a factor, and

$$f(x) = (x - r_1)q_1(x)$$

where $q_1(x)$ is a complex polynomial of degree $n - 1$ whose leading coefficient is a_n. Again, by the Fundamental Theorem of Algebra, the complex polynomial $q_1(x)$ has at least one zero, say r_2. By the Factor Theorem, $q_1(x)$ has the factor $x - r_2$, so

$$q_1(x) = (x - r_2)q_2(x)$$

where $q_2(x)$ is a complex polynomial of degree $n - 2$ whose leading coefficient is a_n. Consequently,

$$f(x) = (x - r_1)(x - r_2)q_2(x)$$

Repeating this argument n times, we finally arrive at

$$f(x) = (x - r_1)(x - r_2) \cdot \ldots \cdot (x - r_n)q_n(x)$$

where $q_n(x)$ is a complex polynomial of degree $n - n = 0$ whose leading coefficient is a_n. Thus, $q_n(x) = a_n x^0 = a_n$, and so

$$f(x) = a_n(x - r_1)(x - r_2) \cdot \ldots \cdot (x - r_n)$$

*In all, Gauss gave four different proofs of this theorem, the first one in 1799 being the subject of his doctoral dissertation.

Complex Polynomials with Real Coefficients

1 We can use the Fundamental Theorem of Algebra to obtain valuable information about the zeros of complex polynomials whose coefficients are real numbers.

CONJUGATE PAIRS THEOREM

Let $f(x)$ be a complex polynomial whose coefficients are real numbers. If $r = a + bi$ is a zero of f, then the complex conjugate $\bar{r} = a - bi$ is also a zero of f.

In other words, for complex polynomials whose coefficients are real numbers, the zeros occur in conjugate pairs.

Proof Let

$$f(x) = a_n x^n + a_{n-1}x^{n-1} + \cdots + a_1 x + a_0$$

where $a_n, a_{n-1}, \ldots, a_1, a_0$ are real numbers and $a_n \neq 0$. If r is a zero of f, then $f(r) = 0$, so

$$a_n r^n + a_{n-1}r^{n-1} + \cdots + a_1 r + a_0 = 0$$

We take the conjugate of both sides to get

$$\overline{a_n r^n + a_{n-1}r^{n-1} + \cdots + a_1 r + a_0} = \bar{0}$$

$$\overline{a_n r^n} + \overline{a_{n-1}r^{n-1}} + \cdots + \overline{a_1 r} + \overline{a_0} = \bar{0}$$ *The conjugate of a sum equals the sum of the conjugates (see Section 4.5).*

$$\overline{a_n}(\bar{r})^n + \overline{a_{n-1}}(\bar{r})^{n-1} + \cdots + \overline{a_1}\,\bar{r} + \overline{a_0} = \bar{0}$$ *The conjugate of a product equals the product of the conjugates.*

$$a_n(\bar{r})^n + a_{n-1}(\bar{r})^{n-1} + \cdots + a_1\bar{r} + a_0 = 0$$ *The conjugate of a real number equals the real number.*

This last equation states that $f(\bar{r}) = 0$; that is, $\bar{r}$ is a zero of f.

The value of this result should be clear. Once we know that, say, $3 + 4i$ is a zero of a polynomial with real coefficients, then we know that $3 - 4i$ is also a zero. This result has an important corollary.

COROLLARY

A complex polynomial f of odd degree with real coefficients has at least one real zero.

Proof Because complex zeros occur as conjugate pairs in a complex polynomial with real coefficients, there will always be an even number of zeros that are not real numbers. Consequently, since f is of odd degree, one of its zeros has to be a real number.

For example, the polynomial $f(x) = x^5 - 3x^4 + 4x^3 - 5$ has at least one zero that is a real number, since f is of degree 5 (odd) and has real coefficients.

EXAMPLE 1 Using the Conjugate Pairs Theorem

A polynomial f of degree 5 whose coefficients are real numbers has the zeros $1, 5i$, and $1 + i$. Find the remaining two zeros.

Solution Since complex zeros appear as conjugate pairs, it follows that $-5i$, the conjugate of $5i$, and $1 - i$, the conjugate of $1 + i$, are the two remaining zeros.

NOW WORK PROBLEM 1.

EXAMPLE 2 Finding a Polynomial Function Whose Zeros Are Given

2

(a) Find a polynomial f of degree 4 whose coefficients are real numbers and that has the zeros $1, 1$, and $-4 + i$.

(b) Graph the polynomial found in part (a) to verify your result.

Solution (a) Since $-4 + i$ is a zero, by the Conjugate Pairs Theorem, $-4 - i$ must also be a zero of f. Because of the Factor Theorem, if $f(c) = 0$, then $x - c$ is a factor of $f(x)$. So we can now write f as

$$f(x) = a(x - 1)(x - 1)[x - (-4 + i)][x - (-4 - i)]$$

where a is any real number. If we let $a = 1$, we obtain

$$\begin{aligned} f(x) &= (x - 1)(x - 1)[x - (-4 + i)][x - (-4 - i)] \\ &= (x^2 - 2x + 1)[x^2 - (-4 + i)x - (-4 - i)x + (-4 + i)(-4 - i)] \\ &= (x^2 - 2x + 1)(x^2 + 4x - ix + 4x + ix + 16 + 4i - 4i - i^2) \\ &= (x^2 - 2x + 1)(x^2 + 8x + 17) \\ &= (x^4 + 8x^3 + 17x^2 - 2x^3 - 16x^2 - 34x + x^2 + 8x + 17) \\ &= x^4 + 6x^3 + 2x^2 - 26x + 17 \end{aligned}$$

Figure 34

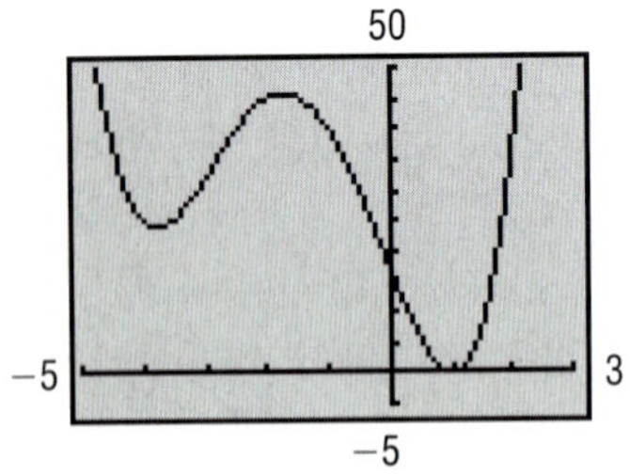

(b) A quick analysis of the polynomial f tells us what to expect:

At most three turning points.
For large $|x|$, the graph will behave like $y = x^4$.
A repeated real zero at 1 so that the graph will touch the x-axis at 1.
The only x-intercept is at 1.

Figure 34 shows the complete graph. (Do you see why? The graph has exactly three turning points.)

EXPLORATION Graph the function found in Example 2 for $a = 2$ and $a = -1$. Does the value of a affect the zeros of f? How does the value of a affect the graph of f?

Now we can prove the theorem we conjectured earlier in Section 4.4

THEOREM Every polynomial function with real coefficients can be uniquely factored over the real numbers into a product of linear factors and/or irreducible quadratic factors.

Proof Every complex polynomial f of degree n has exactly n zeros and can be factored into a product of n linear factors. If its coefficients are real, then those zeros that are complex numbers will always occur as conjugate pairs. As a result, if $r = a + bi$ is a complex zero, then so is $\bar{r} = a - bi$. Consequently, when the linear factors $x - r$ and $x - \bar{r}$ of f are multiplied, we have

$$(x - r)(x - \bar{r}) = x^2 - (r + \bar{r})x + r\bar{r} = x^2 - 2ax + a^2 + b^2$$

This second-degree polynomial has real coefficients and is irreducible (over the real numbers). Thus, the factors of f are either linear or irreducible quadratic factors.

EXAMPLE 3 Finding the Complex Zeros of a Polynomial

3

(a) It is known that $2 + i$ is a zero of $f(x) = x^4 - 8x^3 + 64x - 105$. Find the remaining zeros.

(b) Graph f to verify your results.

Solution (a) Since $2 + i$ is a zero of f and f has real coefficients, then $2 - i$ must also be a zero of f. (Complex zeros appear as conjugate pairs in a polynomial with real coefficients.) So $x - (2 + i)$ and $x - (2 - i)$ are factors of f. Therefore, their product,

$$\begin{aligned}&[x - (2 + i)][x - (2 - i)]\\ &\quad = x^2 - (2 - i)x - x(2 + i) + (2 + i)(2 - i) = x^2 - 4x + 5\end{aligned}$$

is also a factor of f. We use long division to obtain the other factor.

$$\begin{array}{r|l} & x^2 - 4x - 21 \\ \hline x^2 - 4x + 5 & x^4 - 8x^3 \qquad\qquad + 64x - 105 \\ & x^4 - 4x^3 + 5x^2 \\ \hline & -4x^3 - 5x^2 + 64x \\ & -4x^3 + 16x^2 - 20x \\ \hline & -21x^2 + 84x - 105 \\ & -21x^2 + 84x - 105 \\ \hline & 0 \end{array}$$

So $f(x) = (x^2 - 4x + 5)(x^2 - 4x - 21) = (x^2 - 4x + 5)(x - 7)(x + 3)$. By the Factor Theorem, the zeros of f are $-3, 7, 2 + i$, and $2 - i$.

(b) A quick analysis of the polynomial f tells us what to expect.

At most three turning points.
Two x-intercepts: at -3 and 7.
For large $|x|$, the graph will behave like $y = x^4$.

Figure 35

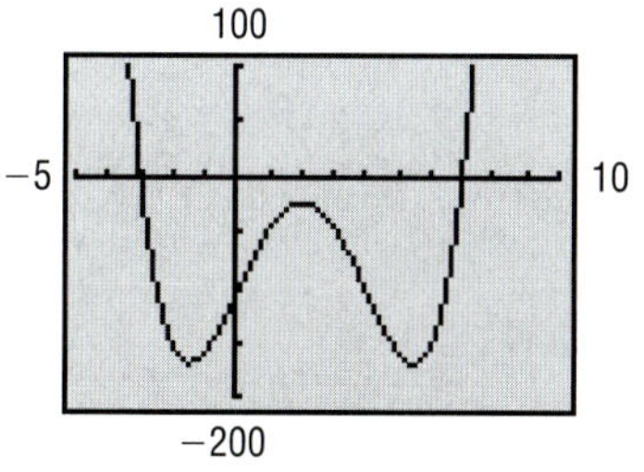

Figure 35 shows the complete graph. (Do you see why? The graph has exactly three turning points.)

NOW WORK PROBLEM 17.

The steps on page 250 in Section 4.4 used to find the real zeros of a polynomial function can also be used to find the complex zeros of a polynomial.

EXAMPLE 4 Finding the Complex Zeros of a Polynomial

Find the complex zeros of the polynomial function

$$f(x) = 3x^4 + 5x^3 + 25x^2 + 45x - 18$$

STEP 1: The degree of f is 4. So f will have four complex zeros.

STEP 2: The Rational Zeros Theorem provides information about the potential rational zeros of polynomials with integer coefficients. For this polynomial (which has integer coefficients), the potential rational zeros are

$$\pm\frac{1}{3},\quad \pm\frac{2}{3},\quad \pm 1,\quad \pm 2,\quad \pm 3,\quad \pm 6,\quad \pm 9,\quad \pm 18$$

Figure 36

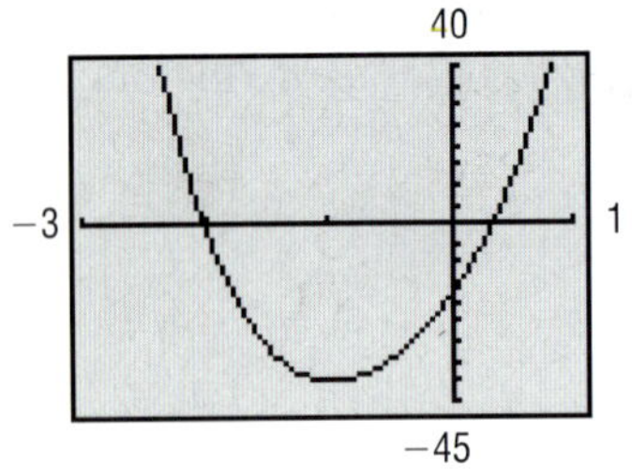

STEP 3: Figure 36 shows the graph of f. The graph has the characteristics that we expect of this polynomial of degree 4: It behaves like $y = 3x^4$ for large $|x|$ and has y-intercept -18. There are x-intercepts near -2 and between 0 and 1. Since $f(-2) = 0$, we know that -2 is a zero of f. We use synthetic division to factor f.

$$\begin{array}{r|rrrrr} -2 & 3 & 5 & 25 & 45 & -18 \\ & & -6 & 2 & -54 & 18 \\ \hline & 3 & -1 & 27 & -9 & 0 \end{array}$$

So $f(x) = (x + 2)(3x^3 - x^2 + 27x - 9)$. From the graph of f and the list of potential rational zeros, it appears that 1/3 may also be a zero of f. We use synthetic division on the depressed equation of f to confirm this.

$$\begin{array}{r|rrrr} 1/3 & 3 & -1 & 27 & -9 \\ & & 1 & 0 & 9 \\ \hline & 3 & 0 & 27 & 0 \end{array}$$

Thus, 1/3 is a zero of f and $x - 1/3$ is therefore a factor. Using the bottom row of the synthetic division, we find that

$$f(x) = (x + 2)\left(x - \frac{1}{3}\right)(3x^2 + 27) = 3(x + 2)\left(x - \frac{1}{3}\right)(x^2 + 9)$$

The factor $x^2 + 9$ does not have any real zeros; its complex zeros are $\pm 3i$. Thus, the complex zeros of $f(x) = 3x^4 + 5x^3 + 25x^2 + 45x - 18$ are -2, 1/3, $3i$, $-3i$.

NOW WORK PROBLEM **27.**

4.6 EXERCISES

In Problems 1–10, information is given about a complex polynomial $f(x)$ whose coefficients are real numbers. Find the remaining zeros of f.

1. Degree 3; zeros: $3, 4 - i$

2. Degree 3; zeros: $4, 3 + i$

3. Degree 4; zeros: $i, 1 + i$

4. Degree 4; zeros: $1, 2, 2 + i$

5. Degree 5; zeros: $1, i, 2i$

6. Degree 5; zeros: $0, 1, 2, i$

7. Degree 4; zeros: $i, 2, -2$

8. Degree 4; zeros: $2 - i, -i$

9. Degree 6; zeros: $2, 2 + i, -3 - i, 0$

10. Degree 6; zeros: $i, 3 - 2i, -2 + i$

In Problems 11–16, form a polynomial $f(x)$ with real coefficients having the given degree and zeros.

11. Degree 4; zeros: $3 + 2i$; 4, multiplicity 2

12. Degree 4; zeros: $i, 1 + 2i$

13. Degree 5; zeros: 2, multiplicity 1; $1 - i$; $1 + i$

14. Degree 6; zeros: $i, 4 - i$; $2 + i$

15. Degree 4; zeros: 3, multiplicity 2; $-i$

16. Degree 5; zeros: 1, multiplicity 3; $1 + i$

In Problems 17–24, use the given zero to find the remaining zeros of each function. Graph the function to verify your results.

17. $f(x) = x^3 - 4x^2 + 4x - 16$; zero: $2i$

18. $g(x) = x^3 + 3x^2 + 25x + 75$; zero: $-5i$

19. $f(x) = 2x^4 + 5x^3 + 5x^2 + 20x - 12$; zero: $-2i$

20. $h(x) = 3x^4 + 5x^3 + 25x^2 + 45x - 18$; zero: $3i$

21. $h(x) = x^4 - 9x^3 + 21x^2 + 21x - 130$; zero: $3 - 2i$

22. $f(x) = x^4 - 7x^3 + 14x^2 - 38x - 60$; zero: $1 + 3i$

23. $h(x) = 3x^5 + 2x^4 + 15x^3 + 10x^2 - 528x - 352$; zero: $-4i$

24. $g(x) = 2x^5 - 3x^4 - 5x^3 - 15x^2 - 207x + 108$; zero: $3i$

In Problems 25–34, find the complex zeros of each polynomial function. Check your results by evaluating the function at each of the zeros.

25. $f(x) = x^3 - 1$

26. $f(x) = x^4 - 1$

27. $f(x) = x^3 - 8x^2 + 25x - 26$

28. $f(x) = x^3 + 13x^2 + 57x + 85$

29. $f(x) = x^4 + 5x^2 + 4$

30. $f(x) = x^4 + 13x^2 + 36$

31. $f(x) = x^4 + 2x^3 + 22x^2 + 50x - 75$

32. $f(x) = x^4 + 3x^3 - 19x^2 + 27x - 252$

33. $f(x) = 3x^4 - x^3 - 9x^2 + 159x - 52$

34. $f(x) = 2x^4 + x^3 - 35x^2 - 113x + 65$

In Problems 35 and 36, tell why the facts given are contradictory.

35. $f(x)$ is a complex polynomial of degree 3 whose coefficients are real numbers; its zeros are $4 + i$, $4 - i$, and $2 + i$.

36. $f(x)$ is a complex polynomial of degree 3 whose coefficients are real numbers; its zeros are $2, i$, and $3 + i$.

37. $f(x)$ is a complex polynomial of degree 4 whose coefficients are real numbers; three of its zeros are $2, 1 + 2i$, and $1 - 2i$. Explain why the remaining zero must be a real number.

38. $f(x)$ is a complex polynomial of degree 4 whose coefficients are real numbers; two of its zeros are -3 and $4 - i$. Explain why one of the remaining zeros must be a real number. Write down one of the missing zeros.

4.7 RATIONAL FUNCTIONS

1 Find the Domain of a Rational Function

2 Determine the Vertical Asymptotes of a Rational Function

3 Determine the Horizontal or Oblique Asymptotes of a Rational Function

4 Analyze the Graph of a Rational Function

5 Solve Applied Problems Involving Rational Functions

Ratios of integers are called *rational numbers.* Similarly, ratios of polynomial functions are called *rational functions.*

A **rational function** is a function of the form

$$R(x) = \frac{p(x)}{q(x)}$$

where p and q are polynomial functions and q is not the zero polynomial. The domain consists of all real numbers except those for which the denominator q is 0.

EXAMPLE 1 Finding the Domain of a Rational Function

1 (a) The domain of $R(x) = \frac{2x^2 - 4}{x + 5}$ consists of all real numbers x except -5.

(b) The domain of $R(x) = \frac{1}{x^2 - 4}$ consists of all real numbers x except -2 and 2.

(c) The domain of $R(x) = \frac{x^3}{x^2 + 1}$ consists of all real numbers.

(d) The domain of $R(x) = \frac{-x^2 + 2}{3}$ consists of all real numbers.

(e) The domain of $R(x) = \frac{x^2 - 1}{x - 1}$ consists of all real numbers x except 1.

It is important to observe that the functions

$$R(x) = \frac{x^2 - 1}{x - 1} \quad \text{and} \quad f(x) = x + 1$$

are not equal, since the domain of R is $\{x \mid x \neq 1\}$ and the domain of f is all real numbers.

NOW WORK PROBLEM 3.

If $R(x) = p(x)/q(x)$ is a rational function and if p and q have no common factors, then the rational function R is said to be in **lowest terms.** For a rational function $R(x) = p(x)/q(x)$ in lowest terms, the zeros, if any, of the numerator are the x-intercepts of the graph of R and so will play a major role in the graph of R. The zeros of the denominator of R [that is, the numbers x, if any, for which $q(x) = 0$], although not in the domain of R, also play a major role in the graph of R. We will discuss this role shortly.

We have already discussed the characteristics of the rational function $f(x) = 1/x$. (Refer to Example 12, page 25). The next rational function we take up is $H(x) = 1/x^2$.

EXAMPLE 2 Graphing $y = \dfrac{1}{x^2}$

Analyze the graph $H(x) = \dfrac{1}{x^2}$.

Solution See Figure 37.

Figure 37

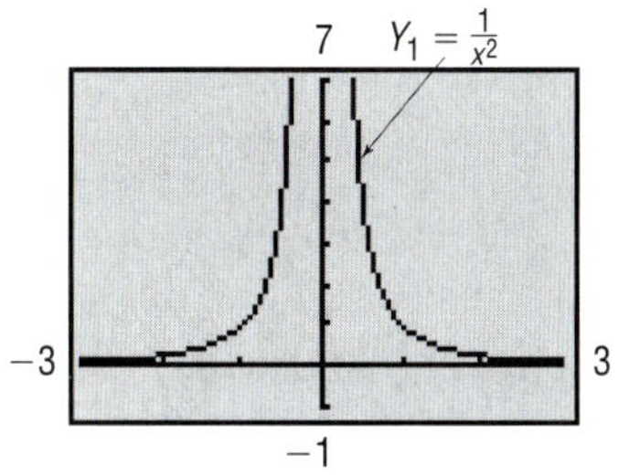

The domain of $H(x) = 1/x^2$ consists of all real numbers x except 0. Thus, the graph has no y-intercept, because x can never equal 0. The graph has no x-intercept because the equation $H(x) = 0$ has no solution. Therefore, the graph of H will not cross either of the coordinate axes.

Because

$$H(-x) = \frac{1}{(-x)^2} = \frac{1}{x^2} = H(x)$$

H is an even function, so its graph is symmetric with respect to the y-axis.

Look again at Figure 37. What happens to the function as the values of x get closer and closer to 0? We use a TABLE to answer the question. See Table 12. There, we see the values of $H(x)$ become larger and larger positive numbers. When this happens, we say that $H(x)$ is **unbounded in the positive direction.** We symbolize this by writing $H(x) \to \infty$ (read as "$H(x)$ **approaches infinity**"). In calculus, **limits** are used to convey these ideas. There we use the symbolism $\lim_{x\to 0} H(x) = \infty$, read as "the limit of $H(x)$ as x approaches 0 is infinity" to mean that $H(x) \to \infty$ as $x \to 0$.

Table 12

X	Y1
.01	10000
.1	100
0	ERROR
1	1
2	.25
10	.01
100	1E-4

Y1 = 1/X²

Look again at Table 12. As $x \to \infty$, the values of $H(x)$ approach 0, the end behavior of the graph, symbolized in calculus by writing $\lim_{x\to\infty} H(x) = 0$.

EXAMPLE 3 Graphing Using Transformations

Using a graphing utility, show each step to graph $R(x) = \dfrac{1}{(x-2)^2} + 1$.

Solution STAGE 1: Graph $y = \dfrac{1}{x^2}$. See Figure 38 and Table 13.

Figure 38

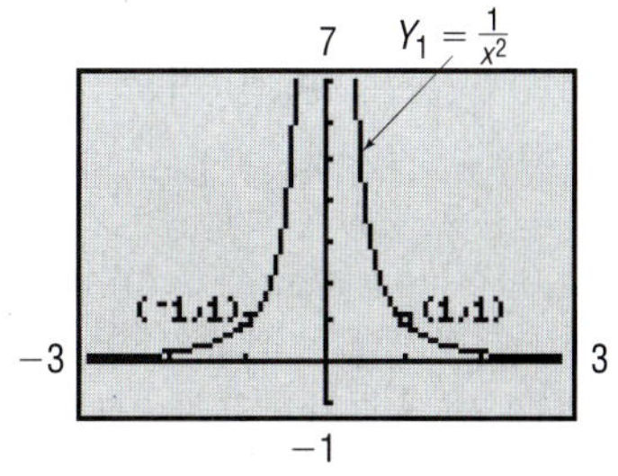

Table 13

X	Y1
-3	.11111
-2	.25
-1	1
0	ERROR
1	1
2	.25
3	.11111

Y1 = 1/X²

STAGE 2: Graph $y = \dfrac{1}{(x-2)^2}$. (Replace x by $x - 2$; shift to the right 2 units.) See Figure 39 and Table 14.

Figure 39

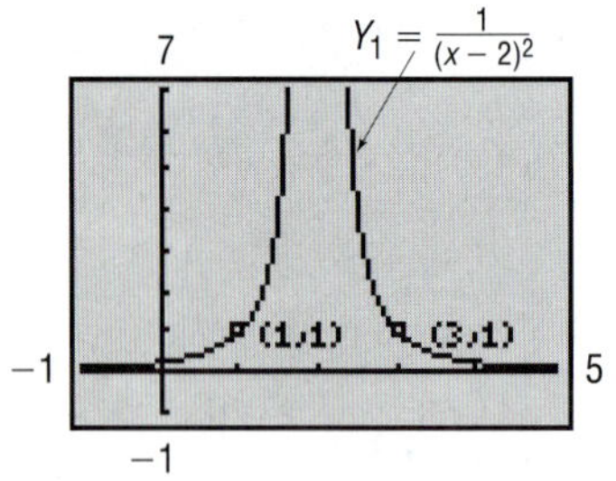

Table 14

X	Y1
-1	.11111
0	.25
1	1
2	ERROR
3	1
4	.25
5	.11111

Y1=1/(X−2)²

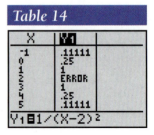

STAGE 3: Graph $y = \dfrac{1}{(x-2)^2} + 1$. (Add 1; shift up 1 unit.) See Figure 40 and Table 15.

Figure 40

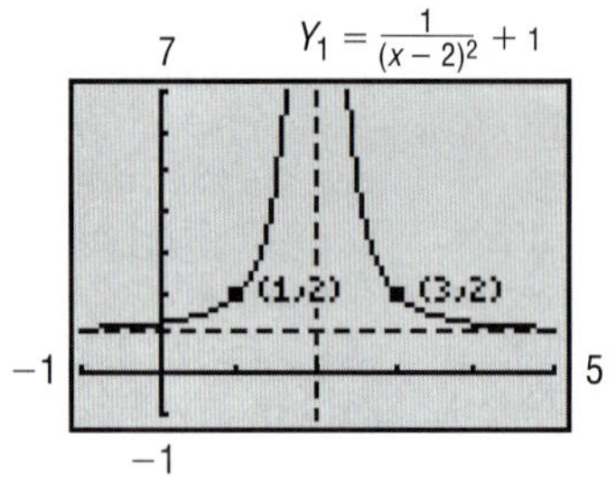

Table 15

X	Y1
-1	1.1111
0	1.25
1	2
2	ERROR
3	2
4	1.25
5	1.1111

Y1=1/(X−2)²+1

NOW WORK PROBLEM 23.

Asymptotes

Notice that the y-axis in Figure 38 is represented by the vertical line $x = 2$ in Figure 40 and the x-axis in Figure 38 is represented by the horizontal line $y = 1$ in Figure 40.

In Figure 40, notice also that as the values of x become more negative, that is, as x becomes **unbounded in the negative direction** ($x \to -\infty$, read as "**x approaches negative infinity**"), the values $R(x)$ approach 1. In fact, we can conclude the following from Figure 40:

1. As $x \to -\infty$, the values $R(x)$ approach 1. $\left[\lim_{x \to -\infty} R(x) = 1\right]$.
2. As x approaches 2, the values $R(x) \to \infty$. $\left[\lim_{x \to 2} R(x) = \infty\right]$.
3. As $x \to \infty$, the values $R(x)$ approach 1. $\left[\lim_{x \to \infty} R(x) = 1\right]$.

Using a graphing utility and the TRACE feature, verify the results discussed above for the graph shown in Figure 40. The vertical line $x = 2$ and the horizontal line $y = 1$ are called *asymptotes* of the graph, which we define as follows:

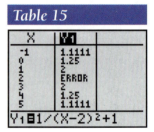

Let R denote a function:

If, as $x \to -\infty$ or as $x \to \infty$, the values of $R(x)$ approach some fixed number L, then the line $y = L$ is a **horizontal asymptote** of the graph of R.

If, as x approaches some number c, the values $|R(x)| \to \infty$, then the line $x = c$ is a **vertical asymptote** of the graph of R.

Even though the asymptotes of a function are not part of the graph of the function, they provide information about how the graph looks. Figure 41 illustrates some of the possibilities.

Figure 41

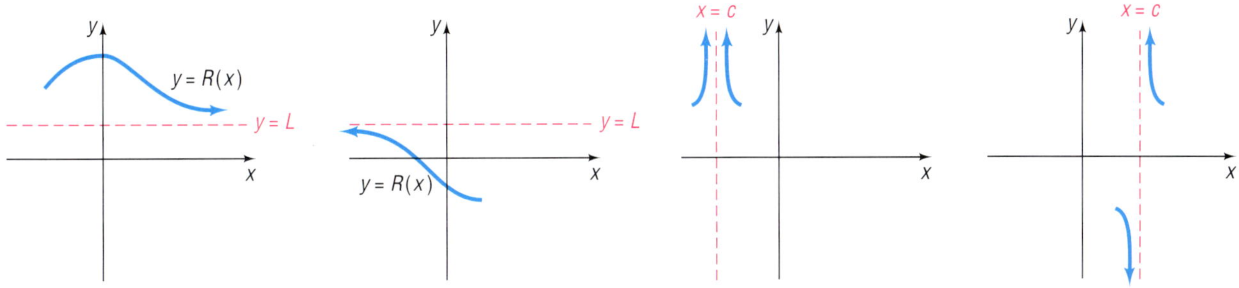

(a) End behavior: As $x \to \infty$, the values of $R(x)$ approach L. That is, the points on the graph of R are getting closer to the line $y = L$; $y = L$ is a horizontal asymptote.

(b) End behavior: As $x \to -\infty$, the values of $R(x)$ approach L. That is, the points on the graph of R are getting closer to the line $y = L$; $y = L$ is a horizontal asymptote.

(c) As x approaches c, the values of $|R(x)| \to \infty$. That is, the points on the graph of R are getting closer to the line $x = c$; $x = c$ is a vertical asymptote.

(d) As x approaches c, the values of $|R(x)| \to \infty$. That is, the points on the graph of R are getting closer to the line $x = c$; $x = c$ is a vertical asymptote.

Figure 42
Oblique asymptote

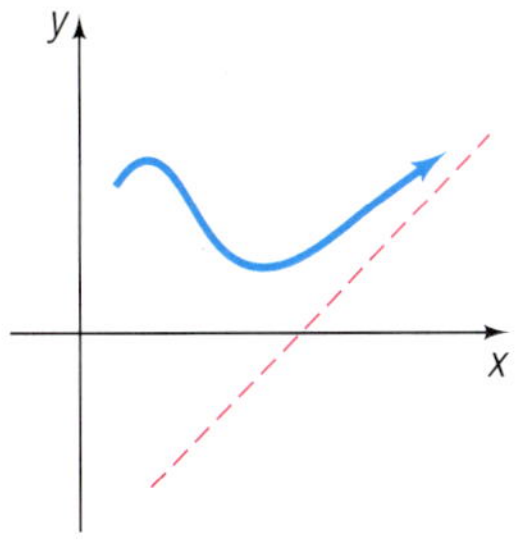

Thus, an asymptote is a line that a certain part of the graph of a function gets closer and closer to but never touches. However, other parts of the graph of the function may intersect a nonvertical asymptote. The graph of the function will never intersect a vertical asymptote. Notice that a horizontal asymptote, when it occurs, describes a certain behavior of the graph as $x \to \infty$ or as $x \to -\infty$, that is, its end behavior. A vertical asymptote, when it occurs, describes a certain behavior of the graph when x is close to some number c.

If an asymptote is neither horizontal nor vertical, it is called **oblique.** Figure 42 shows an oblique asymptote.

Finding Asymptotes

2 The vertical asymptotes, if any, of a rational function $R(x) = p(x)/q(x)$, in lowest terms, are found by factoring the denominator of $q(x)$. Suppose that $x - r$ is a factor of the denominator. Now, as x approaches r, symbolized as

$x \to r$, the values of $x - r$ approach 0, causing the ratio to become unbounded, that is, causing $|R(x)| \to \infty$. Based on the definition, we conclude that the line $x = r$ is a vertical asymptote.

THEOREM **Locating Vertical Asymptotes**

A rational function $R(x) = p(x)/q(x)$, *in lowest terms*, will have a vertical asymptote $x = r$, if $x - r$ is a factor of the denominator q. That is, if r is a real zero of the denominator of a rational function $R(x) = p(x)/q(x)$, in lowest terms, then R will have the vertical asymptote $x = r$.

Warning: If a rational function is not in lowest terms, an application of this theorem may result in an incorrect listing of vertical asymptotes.

EXAMPLE 4 Finding Vertical Asymptotes

Find the vertical asymptotes, if any, of the graph of each rational function.

(a) $R(x) = \dfrac{x}{x^2 - 4}$ (b) $F(x) = \dfrac{x + 3}{x - 1}$ (c) $H(x) = \dfrac{x^2}{x^2 + 1}$

(d) $G(x) = \dfrac{x^2 - 9}{x^2 + 4x - 21}$

Solution

(a) The zeros of the denominator $x^2 - 4$ are -2 and 2. Hence, the lines $x = -2$ and $x = 2$ are the vertical asymptotes of the graph of R.

(b) The only zero of the denominator is 1. Hence, the line $x = 1$ is the only vertical asymptote of the graph of F.

(c) The denominator has no real zeros. Hence, the graph of H has no vertical asymptotes.

(d) Factor $G(x)$ to determine if it is in lowest terms.

$$G(x) = \frac{x^2 - 9}{x^2 + 4x - 21} = \frac{(x + 3)(x - 3)}{(x + 7)(x - 3)} = \frac{x + 3}{x + 7}, \quad x \neq 3$$

The only zero of the denominator of $G(x)$ in lowest terms is -7. Hence, the line $x = -7$ is the only vertical asymptote of the graph of G.

EXPLORATION Graph each of the following rational functions.

$$R(x) = \frac{1}{x - 1} \qquad R(x) = \frac{1}{(x - 1)^2} \qquad R(x) = \frac{1}{(x - 1)^3} \qquad R(x) = \frac{1}{(x - 1)^4}$$

Each has the vertical asymptote $x = 1$. What happens to the value of $R(x)$ as x approaches 1 from the right side of the vertical asymptote? What happens to the value of $R(x)$ as x approaches 1 from the left side of the vertical asymptote? How does the multiplicity of the zero in the denominator affect the graph of R?

3 The procedure for finding horizontal and oblique asymptotes is somewhat more involved. To find such asymptotes, we need to know how the values of a function behave as $x \to -\infty$ or as $x \to \infty$.

If a rational function $R(x)$ is **proper,** that is, if the degree of the numerator is less than the degree of the denominator, then as $x \to -\infty$ or as $x \to \infty$, the values of $R(x)$ approach 0. Consequently, the line $y = 0$ (the x-axis) is a horizontal asymptote of the graph.

THEOREM If a rational function is proper, the line $y = 0$ is a horizontal asymptote of its graph.

EXAMPLE 5 Finding Horizontal Asymptotes

Find the horizontal asymptotes, if any, of the graph of

$$R(x) = \frac{x - 12}{4x^2 + x + 1}$$

Solution The rational function R is proper, since the degree of the numerator, 1, is less than the degree of the denominator, 2. We conclude that the line $y = 0$ is a horizontal asymptote of the graph of R.

To see why $y = 0$ is a horizontal asymptote of the function R in Example 5, we need to investigate the behavior of R for x unbounded. When x is unbounded, the numerator of R, which is $x - 12$, can be approximated by the power function $y = x$, while the denominator of R, which is $4x^2 + x + 1$, can be approximated by the power function $y = 4x^2$.

Thus,

$$R(x) = \frac{x - 12}{4x^2 + x + 1} \approx \frac{x}{4x^2} = \frac{1}{4x} \to 0$$

For x unbounded (↑ at $\approx$); As $x \to -\infty$ or $x \to \infty$ (↑ at $\to 0$)

This shows that the line $y = 0$ is a horizontal asymptote of the graph of R.

If a rational function $R(x)$ is **improper,** that is, if the degree of the numerator is greater than or equal to the degree of the denominator, we must use long division to write the rational function as the sum of a polynomial $f(x)$ plus a proper rational function $r(x)$. That is, we write

$$R(x) = \frac{p(x)}{q(x)} = f(x) + r(x)$$

where $f(x)$ is a polynomial and $r(x)$ is a proper rational function. Since $r(x)$ is proper, then $r(x) \to 0$ as $x \to -\infty$ or as $x \to \infty$. Thus,

$$R(x) = \frac{p(x)}{q(x)} \to f(x) \quad \text{as} \quad x \to -\infty \text{ or as } x \to \infty$$

The possibilities are listed next.

1. If $f(x) = b$, a constant, then the line $y = b$ is a horizontal asymptote of the graph of R.
2. If $f(x) = ax + b$, $a \neq 0$, then the line $y = ax + b$ is an oblique asymptote of the graph of R.
3. In all other cases, the graph of R approaches the graph of f, and there are no horizontal or oblique asymptotes.

The following examples demonstrate these conclusions.

EXAMPLE 6 Finding Horizontal or Oblique Asymptotes

Find the horizontal or oblique asymptotes, if any, of the graph of

$$H(x) = \frac{3x^4 - x^2}{x^3 - x^2 + 1}$$

Solution The rational function H is improper, since the degree of the numerator, 4, is larger than the degree of the denominator, 3. To find any horizontal or oblique asymptotes, we use long division.

$$\begin{array}{r}
3x \phantom{{}-3x^3}+ 3 \phantom{{}+3x+3} \\
x^3 - x^2 + 1 \overline{)\, 3x^4 \phantom{{}-3x^3} - x^2 \phantom{{}+3x}} \\
\underline{3x^4 - 3x^3 \phantom{{}-x^2} + 3x} \\
3x^3 - x^2 - 3x \phantom{{}+3} \\
\underline{3x^3 - 3x^2 \phantom{{}-3x} + 3} \\
2x^2 - 3x - 3
\end{array}$$

Thus,

$$H(x) = \frac{3x^4 - x^2}{x^3 - x^2 + 1} = 3x + 3 + \frac{2x^2 - 3x - 3}{x^3 - x^2 + 1}$$

Then, as $x \rightarrow -\infty$ or as $x \rightarrow \infty$, the remainder will behave as follows:

$$\frac{2x^2 - 3x - 3}{x^3 - x^2 + 1} \approx \frac{2x^2}{x^3} = \frac{2}{x} \rightarrow 0$$

Thus as $x \rightarrow -\infty$ or as $x \rightarrow \infty$, we have $H(x) \rightarrow 3x + 3$. We conclude that the graph of the rational function H has an oblique asymptote $y = 3x + 3$.

EXAMPLE 7 Finding Horizontal or Oblique Asymptotes

Find the horizontal or oblique asymptotes, if any, of the graph of

$$R(x) = \frac{8x^2 - x + 2}{4x^2 - 1}$$

Solution The rational function R is improper, since the degree of the numerator, 2, equals the degree of the denominator, 2. To find any horizontal or oblique asymptotes, we use long division.

$$\begin{array}{r} 2 \\ 4x^2 - 1\overline{)8x^2 - x + 2} \\ \underline{8x^2 \quad\quad - 2} \\ -x + 4 \end{array}$$

Thus,

$$R(x) = \frac{8x^2 - x + 2}{4x^2 - 1} = 2 + \frac{-x + 4}{4x^2 - 1}$$

Then, as $x \to -\infty$ or as $x \to \infty$, the remainder will behave as follows:

$$\frac{-x + 4}{4x^2 - 1} \approx \frac{-x}{4x^2} = \frac{-1}{4x} \to 0$$

Thus, as $x \to -\infty$ or as $x \to \infty$, we have $R(x) \to 2$. We conclude that $y = 2$ is a horizontal asymptote of the graph.

In Example 7, we note that the quotient 2 obtained by long division is the quotient of the leading coefficients of the numerator polynomial and the denominator polynomial $\left(\frac{8}{4}\right)$. This means that we can avoid the long division process for rational functions whose numerator and denominator *are of the same degree* and conclude that the quotient of the leading coefficients will give us the horizontal asymptote.

NOW WORK PROBLEM 39.

EXAMPLE 8 Finding Horizontal or Oblique Asymptotes

Find the horizontal or oblique asymptotes, if any, of the graph of

$$G(x) = \frac{2x^5 - x^3 + 2}{x^3 - 1}$$

Solution The rational function G is improper, since the degree of the numerator, 5, is larger than the degree of the denominator, 3. To find any horizontal or oblique asymptotes, we use long division.

$$\begin{array}{r} 2x^2 - 1 \\ x^3 - 1\overline{)2x^5 - x^3 \quad\quad + 2} \\ \underline{2x^5 \quad\quad - 2x^2 \quad} \\ -x^3 + 2x^2 + 2 \\ \underline{-x^3 \quad\quad\quad + 1} \\ 2x^2 + 1 \end{array}$$

Thus,

$$G(x) = \frac{2x^5 - x^3 + 2}{x^3 - 1} = 2x^2 - 1 + \frac{2x^2 + 1}{x^3 - 1}$$

Then, as $x \to -\infty$ or as $x \to \infty$, the remainder will behave as follows:

$$\frac{2x^2 + 1}{x^3 - 1} \approx \frac{2x^2}{x^3} = \frac{2}{x} \to 0$$

Thus, as $x \to -\infty$ or as $x \to \infty$, we have $G(x) \to 2x^2 - 1$. We conclude that, for large values of x, the graph of G approaches the graph of $y = 2x^2 - 1$. Since $y = 2x^2 - 1$ is not a linear function, G has no horizontal or oblique asymptotes.

We summarize next the procedure for finding horizontal and oblique asymptotes.

Finding Horizontal and Oblique Asymptotes of a Rational Function R

Consider the rational function

$$R(x) = \frac{p(x)}{q(x)} = \frac{a_n x^n + a_{n-1}x^{n-1} + \cdots + a_1 x + a_0}{b_m x^m + b_{m-1}x^{m-1} + \cdots + b_1 x + b_0}$$

in which the degree of the numerator is n and the degree of the denominator is m.

1. If $n < m$, then R is a proper rational function and the graph of R will have the horizontal asymptote $y = 0$ (the x-axis).
2. If $n \geq m$, then R is improper. Here long division is used.
   (a) If $n = m$, the quotient obtained will be a number $L\ (= a_n/b_m)$, and the line $y = L\ (= a_n/b_m)$ is a horizontal asymptote.
   (b) If $n = m + 1$, the quotient obtained is of the form $ax + b$ (a polynomial of degree 1), and the line $y = ax + b$ is an oblique asymptote.
   (c) If $n > m + 1$, the quotient obtained is a polynomial of degree 2 or higher, and R has neither a horizontal nor an oblique asymptote. In this case, for x unbounded, the graph of R will behave like the graph of the quotient.

Note: The graph of a rational function either has one horizontal or one oblique asymptote or else has no horizontal and no oblique asymptote.

Now we are ready to graph rational functions.

Graphing Rational Functions

4 With a graphing utility, the task of graphing a rational function is as simple as entering the function into the utility and pressing graph. However, without appropriate knowledge of the behavior of rational functions, it is dan-

gerous to draw conclusions from the graph. The analysis of a rational function $R(x) = p(x)/q(x)$ requires the following steps:

> ***Analyzing the Graph of a Rational Function***
>
> STEP 1: Find the domain of the rational function.
>
> STEP 2: Locate the intercepts, if any, of the graph. The x-intercepts, if any, of $R(x) = p(x)/q(x)$ satisfy the equation $p(x) = 0$. The y-intercept, if there is one, is $R(0)$.
>
> STEP 3: Test for symmetry. Replace x by $-x$ in $R(x)$. If $R(-x) = R(x)$, there is symmetry with respect to the y-axis; if $R(-x) = -R(x)$, there is symmetry with respect to the origin.
>
> STEP 4: Write R in lowest terms and find the real zeros of the denominator. With R in lowest terms, each zero will give rise to a vertical asymptote.
>
> STEP 5: Locate the horizontal or oblique asymptotes, if any, using the procedure given earlier. Determine points, if any, at which the graph of R intersects these asymptotes.
>
> STEP 6: Graph R.

After graphing, it is important to confirm that the graph is consistent with the conclusions found in Steps 1–5. Let's look at an example.

EXAMPLE 9 Analyzing the Graph of a Rational Function

Analyze the graph of the rational function: $R(x) = \dfrac{x - 1}{x^2 - 4}$.

Solution First, we factor both the numerator and the denominator of R.

$$R(x) = \frac{x - 1}{(x + 2)(x - 2)}$$

R is in lowest terms.

STEP 1: The domain of R is $\{x | x \neq -2, x \neq 2\}$.

STEP 2: We locate the x-intercepts by finding the zeros of the numerator. By inspection, 1 is the only x-intercept. The y-intercept is $R(0) = \frac{1}{4}$.

STEP 3: Because

$$R(-x) = \frac{-x - 1}{x^2 - 4}$$

we conclude that R is neither even nor odd. Thus, there is no symmetry with respect to the y-axis or the origin.

STEP 4: We locate the vertical asymptotes by factoring the denominator: $x^2 - 4 = (x + 2)(x - 2)$. Since R is in lowest terms, the graph of R has two vertical asymptotes: the lines $x = -2$ and $x = 2$.

STEP 5: The degree of the numerator is less than the degree of the denominator, so R is proper and the line $y = 0$ (the x-axis) is a horizontal asymptote of the graph. To determine if the graph of R intersects the horizontal asymptote, we solve the equation $R(x) = 0$. The only solution is $x = 1$, so the graph of R intersects the horizontal asymptote at $(1, 0)$.

STEP 6: The analysis just completed helps us to set the viewing rectangle to obtain a complete graph. Figure 43(a) shows the graph of $R(x) = \dfrac{x - 1}{x^2 - 4}$ in connected mode and Figure 43(b) shows it in dot mode. Notice in Figure 43(a) that the graph has vertical lines at $x = -2$ and $x = 2$. This is due to the fact that, when the graphing utility is in connected mode, it will "connect the dots" between consecutive pixels. We know the graph of R does not cross the lines $x = -2$ and $x = 2$, since R is not defined at $x = -2$ or $x = 2$. Thus, when graphing rational functions, dot mode should be used if extraneous vertical lines appear. You should confirm that all the algebraic conclusions we arrived at in Steps 1–5 are part of the graph. For example, the graph has vertical asymptotes at $x = -2$ and $x = 2$, and the graph has a horizontal asymptote at $y = 0$. The y-intercept is $\frac{1}{4}$ and the x-intercept is 1.

Figure 43

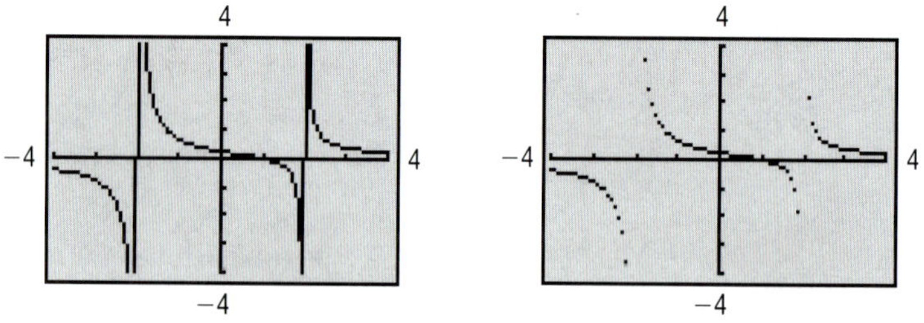

Connected mode
(a)

Dot mode
(b)

NOW WORK PROBLEM 47.

EXAMPLE 10 Analyzing the Graph of a Rational Function

Analyze the graph of the rational function: $R(x) = \dfrac{x^2 - 1}{x}$.

Solution

STEP 1: The domain of R is $\{x | x \neq 0\}$.

STEP 2: The graph has two x-intercepts: -1 and 1. There is no y-intercept, since $x \neq 0$.

STEP 3: Since $R(-x) = -R(x)$, the function is odd and the graph is symmetric with respect to the origin.

STEP 4: Since R is in lowest terms, the graph of $R(x)$ has the line $x = 0$ (the y-axis) as a vertical asymptote.

STEP 5: The rational function R is improper since the degree of the numerator, 2, is larger than the degree of the denominator, 1. To find any horizontal or oblique asymptotes, we use long division.

$$\begin{array}{r} 1x + 0 \\ x\overline{)x^2 - 1} \\ \underline{x^2 } \\ -1 \end{array}$$

The quotient is x, so the line $y = x$ is an oblique asymptote of the graph. To determine whether the graph of R intersects the asymptote $y = x$, we solve the equation $R(x) = x$.

$$R(x) = \frac{x^2 - 1}{x} = x$$
$$x^2 - 1 = x^2$$
$$-1 = 0 \quad \text{Impossible.}$$

Figure 44

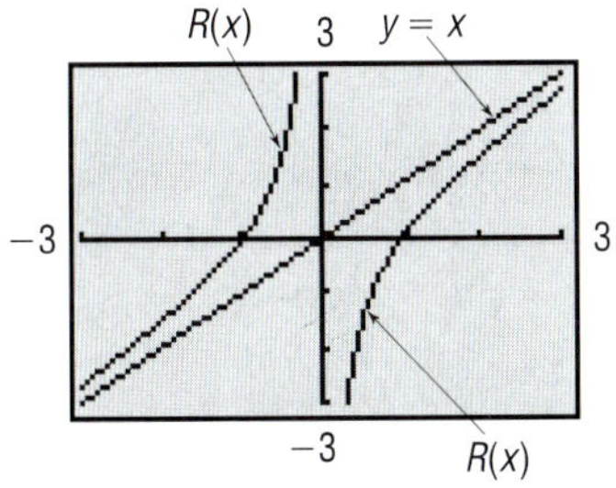

We conclude that the equation $(x^2 - 1)/x = x$ has no solution, so the graph of $R(x)$ does not intersect the line $y = x$.

STEP 6: See Figure 44. We see from the graph that there is no y-intercept and two x-intercepts, -1 and 1. The symmetry with respect to the origin is also evident. We can also see that there is a vertical asymptote at $x = 0$ and an oblique asymptote, $y = x$. Finally, it is not necessary to graph this function in dot mode since no extraneous vertical lines are present.

NOW WORK PROBLEM 55.

EXAMPLE 11 Analyzing the Graph of a Rational Function

Analyze the graph of the rational function $R(x) = \dfrac{3x^2 - 3x}{x^2 + x - 12}$.

Solution We factor R to get

$$R(x) = \frac{3x(x - 1)}{(x + 4)(x - 3)}$$

R is in lowest terms.

STEP 1: The domain of R is $\{x \mid x \neq -4, x \neq 3\}$.

STEP 2: The graph has two x-intercepts: 0 and 1. The y-intercept is $R(0) = 0$.

STEP 3: Because

$$R(-x) = \frac{-3x(-x-1)}{(-x+4)(-x-3)} = \frac{3x(x+1)}{(x-4)(x+3)}$$

we conclude that R is neither even nor odd. Thus, there is no symmetry with respect to the y-axis or the origin.

STEP 4: Since R is in lowest terms, the graph of R has two vertical asymptotes: $x = -4$ and $x = 3$.

STEP 5: Since the degree of the numerator equals the degree of the denominator, the graph has a horizontal asymptote. To find it, we form the quotient of the leading coefficient of the numerator, 3, and the leading coefficient of the denominator, 1. Thus, the graph of R has the horizontal asymptote $y = 3$. To find out whether the graph of R intersects the asymptote, we solve the equation $R(x) = 3$.

$$\begin{aligned} R(x) = \frac{3x^2 - 3x}{x^2 + x - 12} &= 3 \\ 3x^2 - 3x &= 3x^2 + 3x - 36 \\ -6x &= -36 \\ x &= 6 \end{aligned}$$

Thus, the graph intersects the line $y = 3$ only at $x = 6$, and $(6, 3)$ is a point on the graph of R.

STEP 6: Figure 45(a) shows the graph of R in connected mode. Notice the extraneous vertical lines at $x = -4$ and $x = 3$ (the vertical asymptotes). As a result, we shall also graph R in dot mode. See Figure 45(b). We use TRACE to verify that $y = 3$ is a horizontal asymptote by TRACEing for large values of $|x|$ and observing that y gets closer to 3. We can also use TRACE to verify that the graph of R crosses the horizontal asymptote, $y = 3$, at $(6, 3)$.

Figure 45

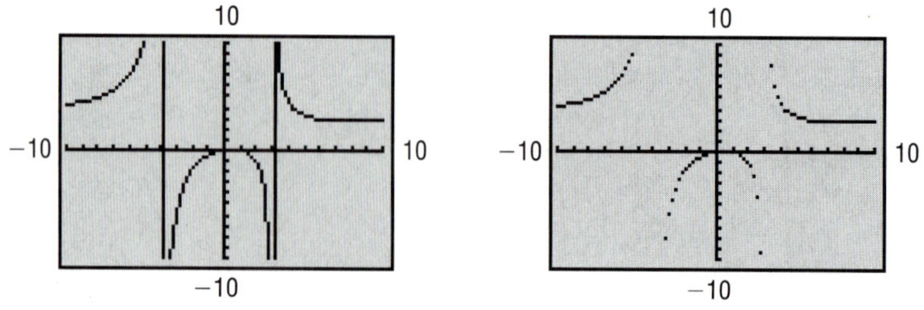

Connected mode
(a)

Dot mode
(b)

Figure 45 does not display the graph between the two x-intercepts, 0 and 1. Nor does it show the graph crossing the horizontal asymptote at $(6, 3)$. To see these parts better, we graph R for $-1 \leq x \leq 2$ [Figure 46(a)] and for $4 \leq x \leq 60$ [Figure 46(b)].

Figure 46

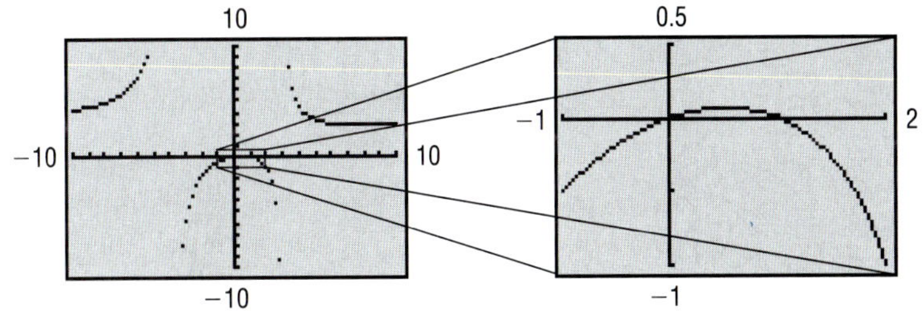

(a)

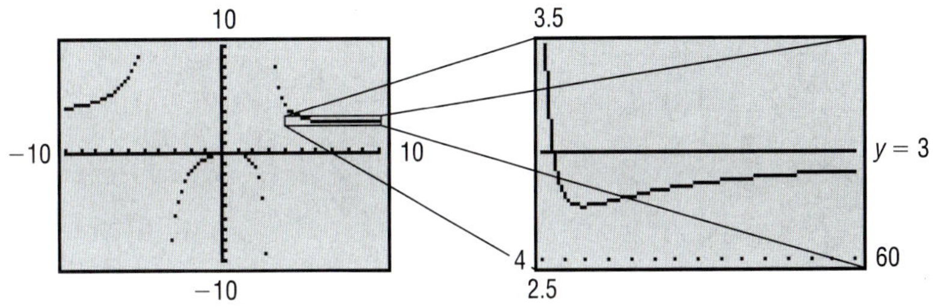

(b)

These graphs now reflect the behavior produced by the analysis. Furthermore, we observe two turning points, one between 0 and 1 and the other to the right of 4. Rounded to two decimal places, these turning points are (0.52, 0.07) and (11.48, 2.75).

EXAMPLE 12 Analyzing the Graph of a Rational Function

Analyze the graph of the rational function

$$R(x) = \frac{0.5x^3 - 2x^2 + 3}{0.1x^3 + x - 3}$$

Solution

STEP 1: The domain of R is $\{x | x \neq 2.09\}$, rounded to two decimal places.

STEP 2: The x-intercepts obey the equation

$$0.5x^3 - 2x^2 + 3 = 0$$

The x-intercepts, rounded to two decimal places, are -1.08, 1.57, and 3.51. The y-intercept is $R(0) = -1$.

STEP 3: Because

$$R(-x) = \frac{-0.5x^3 - 2x^2 + 3}{-0.1x^3 - x - 3}$$

we conclude that R is neither even nor odd. Thus, there is no symmetry with respect to the y-axis or the origin.

STEP 4: We solve the equation

$$0.1x^3 + x - 3 = 0$$

The only solution, rounded to two decimal places, is 2.09. Comparing this result with that of Step 2, we conclude that R is in lowest terms. Thus, the only vertical asymptote is the line $x = 2.09$.

STEP 5: The graph of R has $y = 0.5/0.1 = 5$ as a horizontal asymptote. To see if the graph of R intersects $y = 5$, we solve the equation

$$\frac{0.5x^3 - 2x^2 + 3}{0.1x^3 + x - 3} = 5$$
$$0.5x^3 - 2x^2 + 3 = 0.5x^3 + 5x - 15$$
$$-2x^2 - 5x + 18 = 0$$

The solutions are $x = -4.5$ and $x = 2$ so the points $(-4.5, 5)$ and $(2, 5)$ are on the graph.

STEP 6: The graph of R is shown in Figure 47.

Figure 47

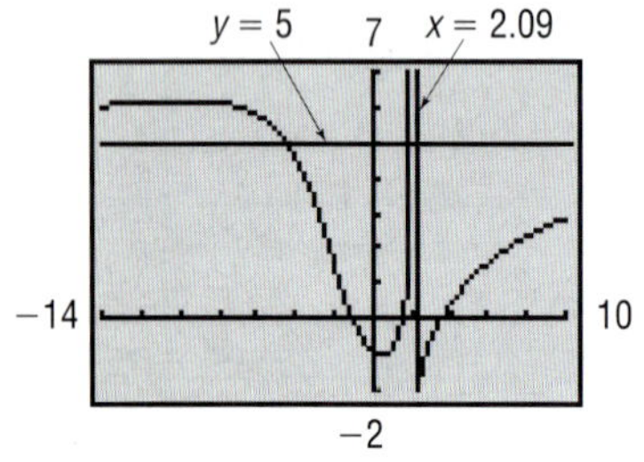

The two turning points, rounded to two decimal places, are at $(-9.55, 6.17)$ and $(0.30, -1.05)$.

EXAMPLE 13 Analyzing the Graph of a Rational Function with a Hole

Analyze the graph of the rational function $R(x) = \dfrac{2x^2 - 5x + 2}{x^2 - 4}$.

Solution We factor R and obtain

$$R(x) = \frac{(2x - 1)(x - 2)}{(x + 2)(x - 2)}$$

In lowest terms,

$$R(x) = \frac{2x - 1}{x + 2}, \; x \neq 2$$

STEP 1: The domain of R is $\{x \mid x \neq -2, x \neq 2\}$.

STEP 2: The graph has one x-intercept: 0.5. The y-intercept is $R(0) = -0.5$.

STEP 3: Because

$$R(-x) = \frac{2x^2 + 5x + 2}{x^2 - 4}$$

we conclude that R is neither even nor odd. Thus, there is no symmetry with respect to the y-axis or the origin.

STEP 4: The graph has one vertical asymptote; $x = -2$, since $x + 2$ is the only factor of the denominator of $R(x)$ *in lowest terms.* However, the rational function is undefined at both $x = 2$ and $x = -2$.

STEP 5: Since the degree of the numerator equals the degree of the denominator, the graph has a horizontal asymptote. To find it, we form the quotient of the leading coefficient of the numerator, 2, and the leading coefficient of the denominator, 1. Thus, the graph of R has the horizontal asymptote $y = 2$. To find whether the graph of R intersects the asymptote, we solve the equation $R(x) = 2$.

$$
\begin{aligned}
R(x) = \frac{2x - 1}{x + 2} &= 2 \\
2x - 1 &= 2(x + 2) \\
2x - 1 &= 2x + 4 \\
-1 &= 4 \qquad \textit{Impossible.}
\end{aligned}
$$

Thus, the graph does not intersect the line $y = 2$.

STEP 6: Figure 48 shows the graph of $R(x)$. Notice that the graph has one vertical asymptote at $x = -2$. Also, the function appears to be continuous at $x = 2$. However, looking at Table 16, we see that R is undefined at $x = 2$. Thus, the graph leads us to the incorrect conclusion that the graph of R is continuous at $x = 2$. In fact, there is a **hole** in the graph where $x = 2$. So, we must be careful when using a graphing utility to analyze rational functions.

Figure 48

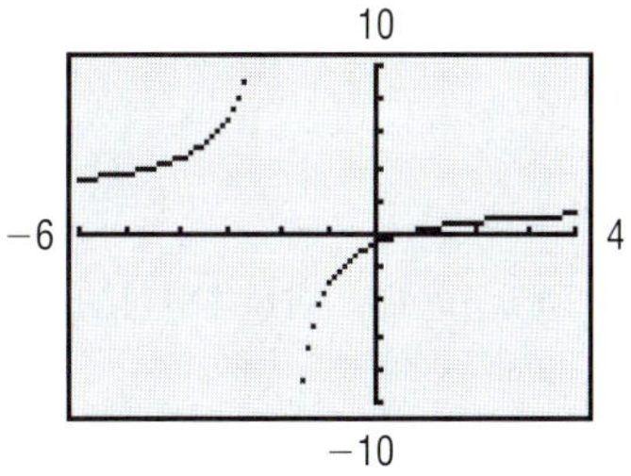

Table 16

X	Y1
-3	7
-2	ERROR
-1	-3
0	-.5
1	.33333
2	ERROR
3	1

Y1=(2X²−5X+2)/(...

As Example 13 shows, the zeros of the denominator of a rational function give rise to either vertical asymptotes or holes on the graph.

Applications Involving Rational Functions

EXAMPLE 14 Finding the Least Cost of a Can

5

Reynolds Metal Company manufactures aluminum cans in the shape of a cylinder with a capacity of 500 cubic centimeters ($\frac{1}{2}$ liter). The top and bottom of the can will be made of a special aluminum alloy that costs 0.05¢ per square centimeter. The sides of the can are to be made of material that costs 0.02¢ per square centimeter.

(a) Express the cost of material for the can as a function of the radius r of the can.
(b) Use a graphing utility to graph the function $C = C(r)$.
(c) What value of r will result in the least cost?
(d) What is this least cost?

Figure 49

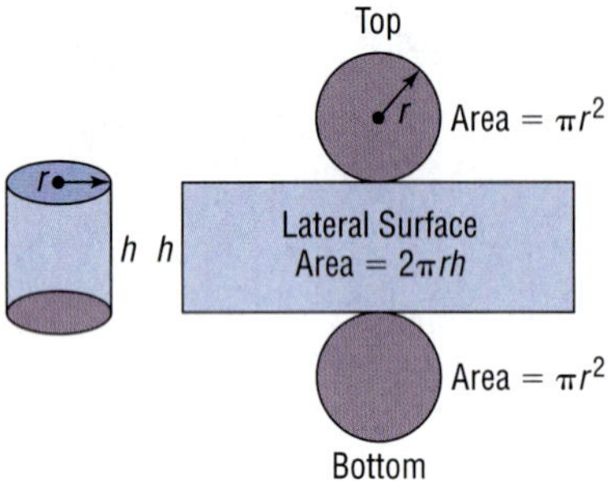

Solution

(a) Figure 49 illustrates the situation. Notice that the material required to produce a cylindrical can of height h and radius r consists of a rectangle of area $2\pi rh$ and two circles, each of area πr^2. The total cost C (in cents) of manufacturing the can is therefore

$$\begin{aligned} C &= \text{Cost of top and bottom} + \text{Cost of side} \\ &= \underbrace{2(\pi r^2)}_{\substack{\text{Total area}\\\text{of top and}\\\text{bottom}}}\ \underbrace{(0.05)}_{\substack{\text{Cost/unit}\\\text{area}}} + \underbrace{(2\pi rh)}_{\substack{\text{Total}\\\text{area of}\\\text{side}}}\ \underbrace{(0.02)}_{\substack{\text{Cost/unit}\\\text{area}}} \\ &= 0.10\pi r^2 + 0.04\pi rh \end{aligned}$$

But we have the additional restriction that the height h and radius r must be chosen so that the volume V of the can is 500 cubic centimeters. Since $V = \pi r^2 h$, we have

Figure 50

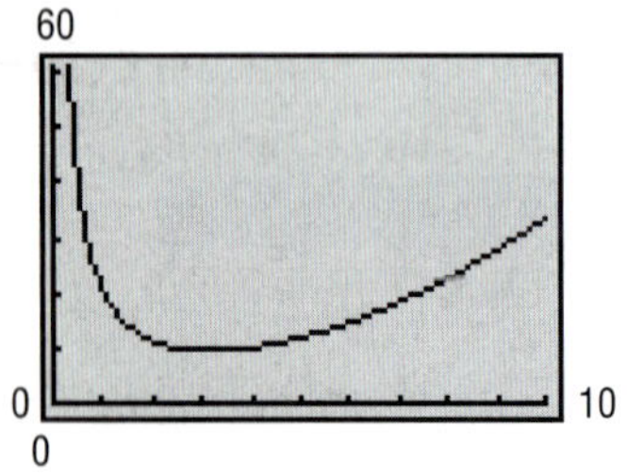

$$500 = \pi r^2 h \quad \text{or} \quad h = \frac{500}{\pi r^2}$$

Thus, the cost C, in cents, as a function of the radius r is

$$C(r) = 0.10\pi r^2 + 0.04\pi r\left(\frac{500}{\pi r^2}\right) = 0.10\pi r^2 + \frac{20}{r} = \frac{0.10\pi r^3 + 20}{r}$$

(b) See Figure 50.
(c) Using the MINIMUM command, the cost is least for a radius of about 3.17 centimeters.
(d) The least cost is 9.47¢.

Figure 51

$y = \frac{k}{x}, k > 0, x > 0$

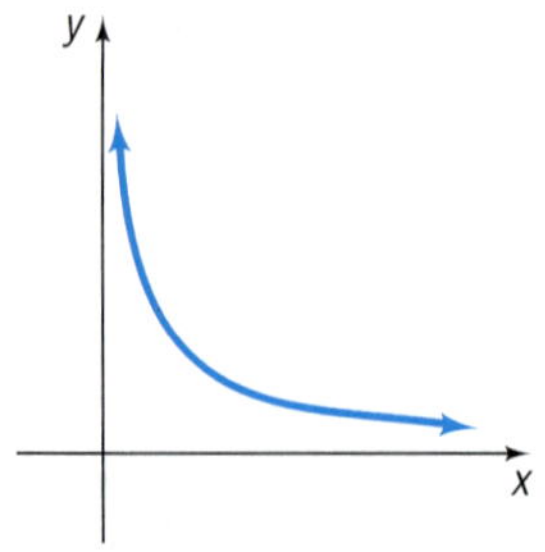

Inverse Variation

Let x and y denote two quantities. Then y **varies inversely** with x, or y is **inversely proportional to** x, if there is a nonzero constant k such that

$$y = \frac{k}{x}$$

The graph in Figure 51 illustrates the relationship between y and x if y varies inversely with x and $k > 0$, $x > 0$.

EXAMPLE 15 Safe Weight That Can Be Supported by a Piece of Pine

The weight W that can be safely supported by a 2-inch by 4-inch piece of lumber varies inversely with its length l. See Figure 52. Experiments indicate that the maximum weight that a 10-foot-long pine 2-by-4 can support is 500 pounds. Write a general formula relating the safe weight W (in pounds) to length l (in feet). Find the maximum weight W that can be safely supported by a length of 25 feet.

Figure 52

Solution Because W varies inversely with l, we know that

$$W = \frac{k}{l}$$

for some constant k. Because $W = 500$ when $l = 10$, we have

$$500 = \frac{k}{10}$$
$$k = 5000$$

Thus, in all cases,

$$W = \frac{5000}{l}$$

In particular, the maximum weight W that can be safely supported by a piece of pine 25 feet in length is

$$W = \frac{5000}{25} = 200 \text{ pounds}$$

Figure 53

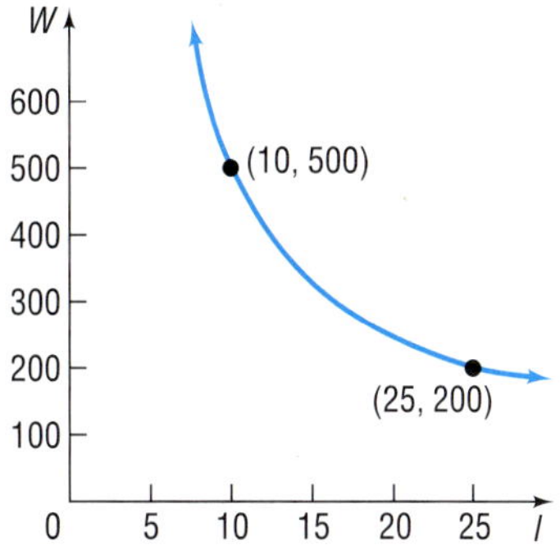

Figure 53 illustrates the relationship between the weight W and the length l.

4.7 EXERCISES

In Problems 1–12, find the domain of each rational function.

1. $R(x) = \dfrac{4x}{x - 3}$ **2.** $R(x) = \dfrac{5x^2}{3 + x}$ **3.** $H(x) = \dfrac{-4x^2}{(x - 2)(x + 4)}$

4. $G(x) = \dfrac{6}{(x + 3)(4 - x)}$ **5.** $F(x) = \dfrac{3x(x - 1)}{2x^2 - 5x - 3}$ **6.** $Q(x) = \dfrac{-x(1 - x)}{3x^2 + 5x - 2}$

7. $R(x) = \dfrac{x}{x^3 - 8}$ **8.** $R(x) = \dfrac{x}{x^4 - 1}$ **9.** $H(x) = \dfrac{3x^2 + x}{x^2 + 4}$

10. $G(x) = \dfrac{x - 3}{x^4 + 1}$ **11.** $R(x) = \dfrac{3(x^2 - x - 6)}{4(x^2 - 9)}$ **12.** $F(x) = \dfrac{-2(x^2 - 4)}{3(x^2 + 4x + 4)}$

In Problems 13–22, use the graph shown to find:

(a) The domain and range of each function

(b) The intercepts, if any

(c) Horizontal asymptotes, if any

(d) Vertical asymptotes, if any

(e) Oblique asymptotes, if any

13.

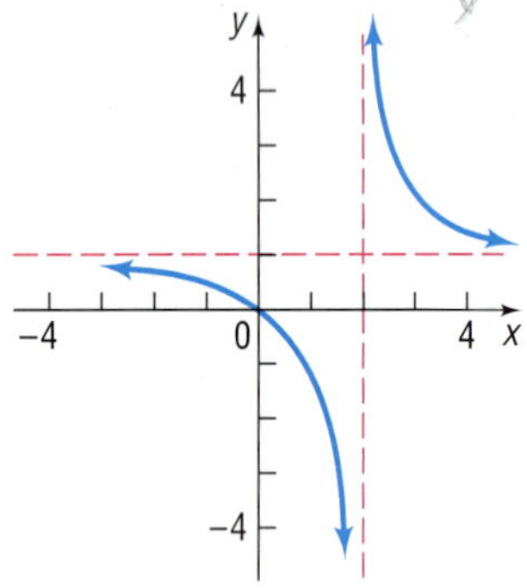

14.

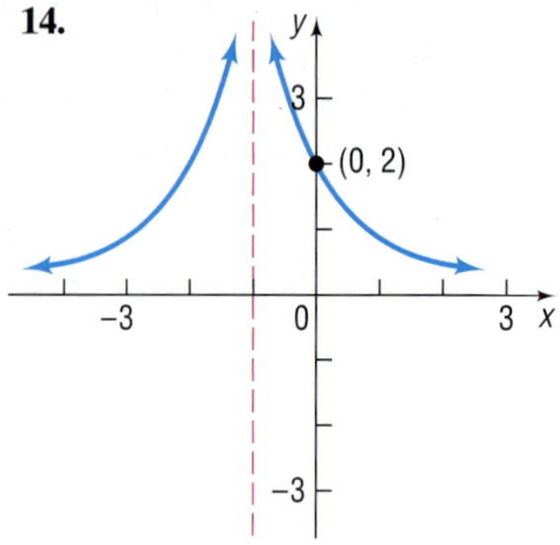

15.

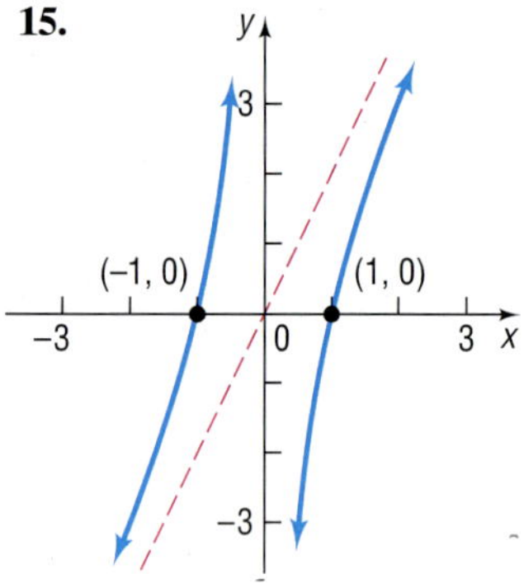

16.

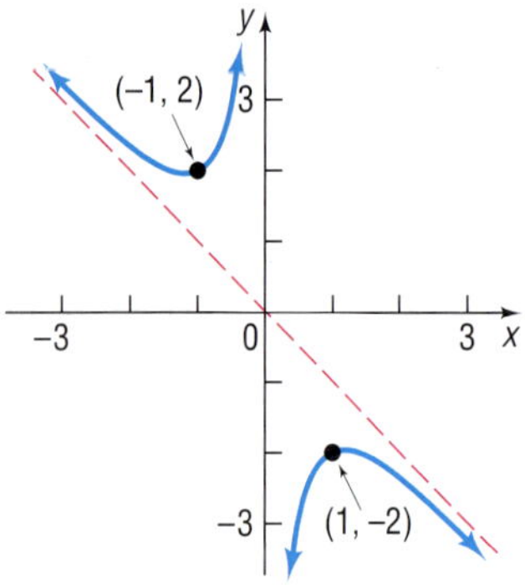

17.

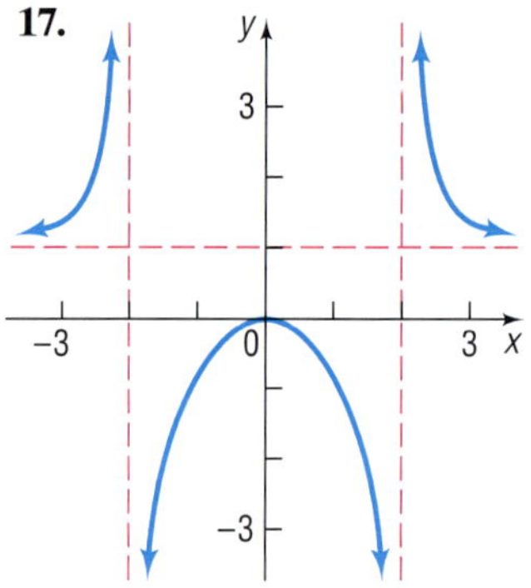

18.

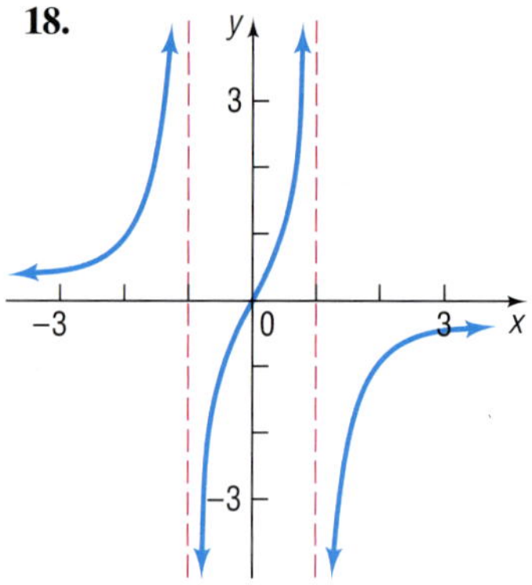

19.

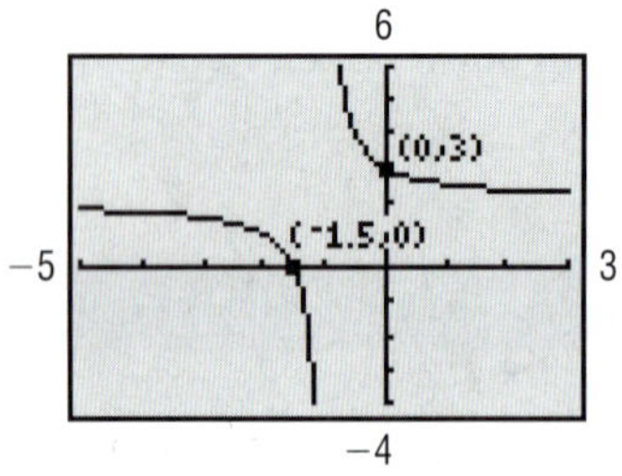

20.

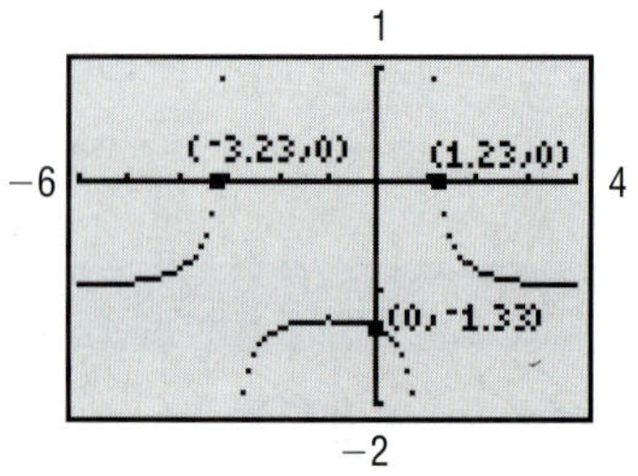

21.

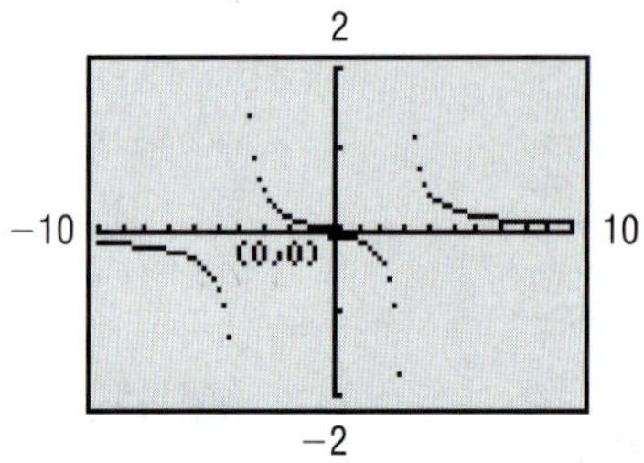

22.

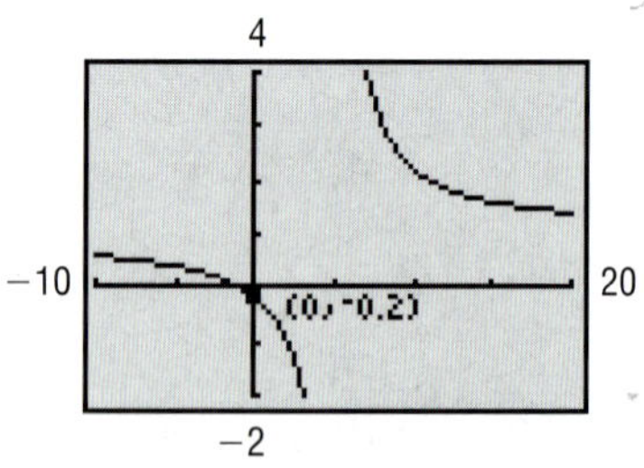

In Problems 23–34, using a graphing utility, show the steps required to graph each rational function using transformations.

23. $R(x) = \dfrac{1}{(x-1)^2}$

24. $R(x) = \dfrac{3}{x}$

25. $H(x) = \dfrac{-2}{x+1}$

26. $G(x) = \dfrac{2}{(x+2)^2}$

27. $R(x) = \dfrac{1}{x^2+4x+4}$

28. $R(x) = \dfrac{1}{x-1} + 1$

29. $F(x) = 1 - \dfrac{1}{x}$

30. $Q(x) = 1 + \dfrac{1}{x}$

31. $R(x) = \dfrac{x^2-4}{x^2}$

32. $R(x) = \dfrac{x-4}{x}$

33. $G(x) = 1 + \dfrac{2}{(x-3)^2}$

34. $F(x) = 2 - \dfrac{1}{x+1}$

In Problems 35–46, find the vertical, horizontal, and oblique asymptotes, if any, of each rational function, without graphing.

35. $R(x) = \dfrac{3x}{x+4}$

36. $R(x) = \dfrac{3x+5}{x-6}$

37. $H(x) = \dfrac{x^4+2x^2+1}{x^2-x+1}$

38. $G(x) = \dfrac{-x^2+1}{x+5}$

39. $T(x) = \dfrac{x^3}{x^4-1}$

40. $P(x) = \dfrac{4x^5}{x^3-1}$

41. $Q(x) = \dfrac{5-x^2}{3x^4}$

42. $F(x) = \dfrac{-2x^2+1}{2x^3+4x^2}$

43. $R(x) = \dfrac{3x^4+4}{x^3+3x}$

44. $R(x) = \dfrac{6x^2+x+12}{3x^2-5x-2}$

45. $G(x) = \dfrac{x^3-1}{x-x^2}$

46. $F(x) = \dfrac{x-1}{x-x^3}$

In Problems 47–90, follow Steps 1 through 6 on page 279 to analyze the graph of each function.

47. $R(x) = \dfrac{x+1}{x(x+4)}$

48. $R(x) = \dfrac{x}{(x-1)(x+2)}$

49. $R(x) = \dfrac{3x+3}{2x+4}$

50. $R(x) = \dfrac{2x+4}{x-1}$

51. $R(x) = \dfrac{3}{x^2-4}$

52. $R(x) = \dfrac{6}{x^2-x-6}$

53. $P(x) = \dfrac{x^4+x^2+1}{x^2-1}$

54. $Q(x) = \dfrac{x^4-1}{x^2-4}$

55. $H(x) = \dfrac{x^3-1}{x^2-9}$

56. $G(x) = \dfrac{x^3+1}{x^2+2x}$

57. $R(x) = \dfrac{x^2}{x^2+x-6}$

58. $R(x) = \dfrac{x^2+x-12}{x^2-4}$

59. $G(x) = \dfrac{x}{x^2-4}$

60. $G(x) = \dfrac{3x}{x^2-1}$

61. $R(x) = \dfrac{3}{(x-1)(x^2-4)}$

62. $R(x) = \dfrac{-4}{(x+1)(x^2-9)}$

63. $H(x) = 4\dfrac{x^2-1}{x^4-16}$

64. $H(x) = \dfrac{x^2+4}{x^4-1}$

65. $F(x) = \dfrac{x^2-3x-4}{x+2}$

66. $F(x) = \dfrac{x^2+3x+2}{x-1}$

67. $R(x) = \dfrac{x^2+x-12}{x-4}$

68. $R(x) = \dfrac{x^2-x-12}{x+5}$

69. $F(x) = \dfrac{x^2+x-12}{x+2}$

70. $G(x) = \dfrac{x^2-x-12}{x+1}$

71. $R(x) = \dfrac{x(x-1)^2}{(x+3)^3}$

72. $R(x) = \dfrac{(x-1)(x+2)(x-3)}{x(x-4)^2}$

73. $R(x) = \dfrac{4x^3-0.5x+2}{2x^3+0.3x^2-1}$

74. $R(x) = \dfrac{3x^4-4x^2+8}{x^4-x^3+2}$

75. $R(x) = \dfrac{3x^3+5x^2-3}{0.1x^4-2x^2+1}$

76. $R(x) = \dfrac{x^2-0.9x+2}{x^3+0.5x^2+1}$

77. $R(x) = \dfrac{0.5x^4-x^3+1}{0.1x^3-\pi x+1}$

78. $R(x) = \dfrac{2x^3+\pi x+3}{-x^2+0.9x+1}$

79. $R(x) = \dfrac{x^2+x-12}{x^2-x-6}$

80. $R(x) = \dfrac{x^2 + 3x - 10}{x^2 + 8x + 15}$

81. $R(x) = \dfrac{6x^2 - 7x - 3}{2x^2 - 7x + 6}$

82. $R(x) = \dfrac{8x^2 + 26x + 15}{2x^2 - x - 15}$

83. $R(x) = \dfrac{x^3 + 2x^2 - 5x - 6}{x^3 + 7x^2 + 7x - 15}$

84. $R(x) = \dfrac{x^3 - 6x^2 - x + 30}{x^3 - 4x^2 + x + 6}$

85. $f(x) = x + \dfrac{1}{x}$

86. $f(x) = 2x + \dfrac{9}{x}$

87. $f(x) = x^2 + \dfrac{1}{x}$

88. $f(x) = 2x^2 + \dfrac{9}{x}$

89. $f(x) = x + \dfrac{1}{x^3}$

90. $f(x) = 2x + \dfrac{9}{x^3}$

91. If the graph of a rational function R has the vertical asymptote $x = 4$, then the factor $x - 4$ must be present in the denominator of R. Explain why.

92. If the graph of a rational function R has the horizontal asymptote $y = 2$, then the degree of the numerator of R equals the degree of the denominator of R. Explain why.

93. Consult the illustration. Which of the following rational functions might have this graph? (More than one answer might be possible.)

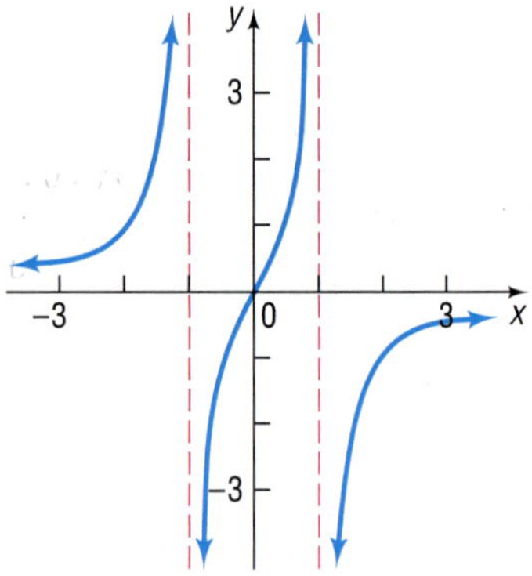

(a) $y = \dfrac{2x}{x^2 - 1}$

(b) $y = \dfrac{-3x}{x^2 - 1}$

(c) $y = \dfrac{x^3}{x^2 - 1}$

(d) $y = \dfrac{x^2 - 1}{-3x}$

(e) $y = \dfrac{-x^3}{(x^2 + 1)(x^2 - 1)}$

(f) $y = \dfrac{-x^2}{(x^2 + 1)(x^2 - 1)}$

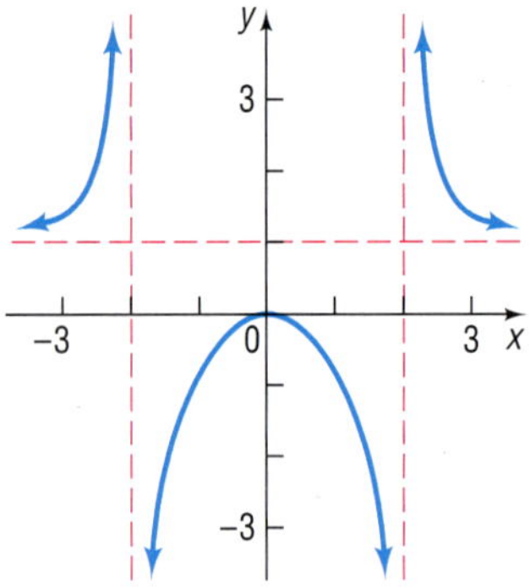

(a) $y = \dfrac{4x^2}{x^2 - 4}$

(b) $y = \dfrac{x}{x^2 - 4}$

(c) $y = \dfrac{x^2}{x^2 - 4}$

(d) $y = \dfrac{x^2(x^2 + 1)}{(x^2 + 4)(x^2 - 4)}$

(e) $y = \dfrac{x^3}{x^2 - 4}$

(f) $y = \dfrac{x^2 - 4}{x^2}$

94. Consult the illustration. Which of the following rational functions might have this graph? (More than one answer may be possible.)

95. Gravity In physics, it is established that the acceleration due to gravity, g, at a height h meters above sea level is given by

$$g(h) = \frac{3.99 \times 10^{14}}{(6.374 \times 10^6 + h)^2}$$

where 6.374×10^6 is the radius of Earth in meters.

(a) What is the acceleration due to gravity at sea level?

(b) The Sears Tower in Chicago, Illinois, is 443 meters tall. What is the acceleration due to gravity at the top of the Sears Tower?

(c) The peak of Mount Everest is 8848 meters above sea level. What is the acceleration due to gravity on the peak of Mount Everest?

(d) Find the horizontal asymptote of $g(h)$.
(e) Using your graphing utility, graph $g(h)$.
(f) Solve $g(h) = 0$. How do you interpret your answer?

96. Population Model A rare species of insect was discovered in the Amazon Rain Forest. In order to protect the species, environmentalists declare the insect endangered and transplant the insects into a protected area. The population of the insect t months after being transplanted is given by P

$$P(t) = \frac{50(1 + 0.5t)}{(2 + 0.01t)}$$

(a) How many insects were discovered? In other words, what was the population when $t = 0$?
(b) What will the population be after 5 years?
(c) Using your graphing utility, graph $P(t)$.
(d) Determine the horizontal asymptote of $P(t)$. What is the largest population that the protected area can sustain?
(e) TRACE $P(t)$ for large values of t to verify your answer to part (d).

97. Drug Concentration The concentration C of a certain drug in a patient's bloodstream t hours after injection is given by

$$C(t) = \frac{t}{2t^2 + 1}$$

(a) Using your graphing utility, graph $C(t)$.
(b) Determine the time at which the concentration is highest.
(c) Find the horizontal asymptote of $C(t)$. What happens to the concentration of the drug as t increases?

98. Drug Concentration The concentration C of a certain drug in a patient's bloodstream t minutes after injection is given by

$$C(t) = \frac{50t}{(t^2 + 25)}$$

(a) Using your graphing utility, graph $C(t)$.
(b) Determine the time at which the concentration is highest.
(c) Find the horizontal asymptote of $C(t)$. What happens to the concentration of the drug as t increases?

99. Average Cost In Problem 69, Exercise 4.2, the cost function for manufacturing Chevy Cavaliers was found to be

$$C(x) = 0.2x^3 - 2.3x^2 + 14.3x + 10.2$$

Economists define the **average cost function** as

$$\overline{C}(x) = \frac{C(x)}{x}$$

(a) Find the average cost function.
(b) What is the average cost of producing six Cavaliers per hour?
(c) What is the average cost of producing nine Cavaliers per hour?
(d) Using your graphing utility, graph the average cost function.
(e) Using your graphing utility, find the number of Cavaliers that should be produced per hour to minimize average cost.
(f) What is the minimum average cost?

100. Average Cost In Problem 70, Exercise 4.2, the cost function for printing textbooks was found to be

$$C(x) = 0.015x^3 - 0.595x^2 + 9.15x + 98.43$$

(a) Find the average cost function (refer to Problem 99).
(b) What is the average cost of printing 13 thousand textbooks per week?
(c) What is the average cost of printing 25 thousand textbooks per week?
(d) Using your graphing utility, graph the average cost function.
(e) Using your graphing utility, find the number of textbooks that should be printed to minimize average cost.
(f) What is the minimum average cost?

101. Minimizing Surface Area United Parcel Service has contracted you to design a closed box with a square base that has a volume of 10,000 cubic inches. See the illustration.

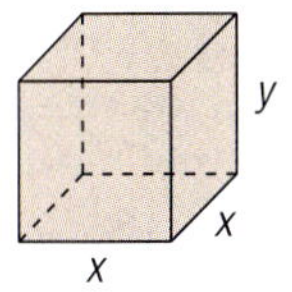

(a) Find a function for the surface area of the box.
(b) Using a graphing utility, graph the function found in part (a).
(c) What is the minimum amount of cardboard that can be used to construct the box?
(d) What are the dimensions of the box that minimize surface area?
(e) Why might UPS be interested in designing a box that minimizes surface area?

102. Minimizing Surface Area United Parcel Service has contracted you to design a closed box with a square base that has a volume of 5000 cubic inches. See the illustration.

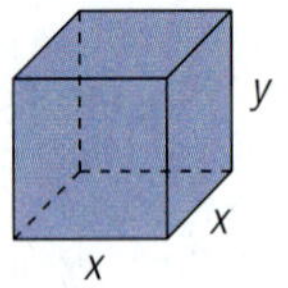

(a) Find a function for the surface area of the box.
(b) Using a graphing utility, graph the function found in part (a).
(c) What is the minimum amount of cardboard that can be used to construct the box?
(d) What are the dimensions of the box that minimize surface area?
(e) Why might UPS be interested in designing a box that minimizes surface area?

103. Cost of a Can A can in the shape of a right circular cylinder is required to have a volume of 500 cubic centimeters. The top and bottom are made of material that costs 6¢ per square centimeter, while the sides are made of material that costs 4¢ per square centimeter.
(a) Express the total cost C of the material as a function of the radius r of the cylinder. (Refer to Figure 49.)
(b) Graph $C = C(r)$. For what value of r is the cost C least?

104. Material Needed to Make a Drum A steel drum in the shape of a right circular cylinder is required to have a volume of 100 cubic feet.
(a) Express the amount A of material required to make the drum as a function of the radius r of the cylinder.
(b) How much material is required if the drum is of radius 3 feet?
(c) Of radius 2 feet?
(d) Of radius 4 feet?
(e) Graph $A = A(r)$. For what value of r is A smallest?

105. Weight of a Body The weight of a body above the surface of Earth varies inversely with the square of the distance from the center of Earth. If a certain body weighs 55 pounds when it is 3960 miles from the center of Earth, how much will it weigh when it is 3965 miles from the center?

106. Weight of a Body The weight of a body varies inversely with the square of its distance from the center of Earth. Assuming that the radius of Earth is 3960 miles, how much would a man weigh at an altitude of 1 mile above Earth's surface if he weighs 200 pounds on Earth's surface?

107. Physics: Vibrating String The rate of vibration of a string under constant tension varies inversely with the length of the string. If a string is 48 inches long and vibrates 256 times per second, what is the length of a string that vibrates 576 times per second?

108. Physics: Vibrating String The rate of vibration of a string under constant tension varies inversely with the length of the string. If a string is 30 centimeters long and vibrates 150 times per second, what is the length of a string that vibrates 220 times per second?

109. Consult the illustration. Make up a rational function that might have this graph. Compare yours with a friend's. What similarities do you see? What differences?

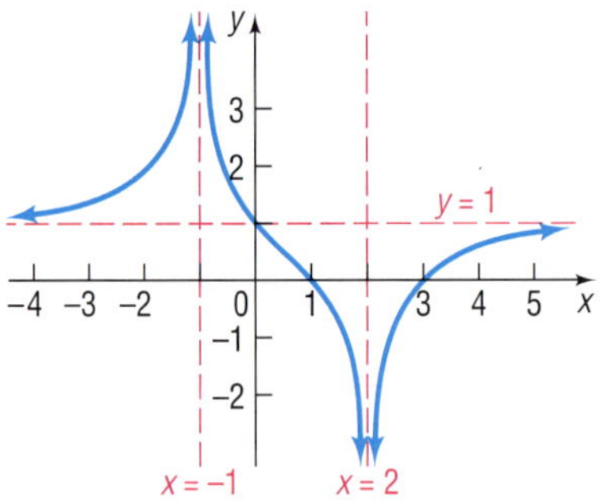

110. Can the graph of a rational function have both a horizontal and an oblique asymptote? Explain.

111. Write a few paragraphs that provide a general strategy for graphing a rational function. Be sure to mention the following: proper, improper, intercepts, and asymptotes.

112. Make up a rational function that has the following characteristics: crosses the x-axis at 3; touches the x-axis at -2; one vertical asymptote, $x = 1$; and one horizontal asymptote, $y = 2$. Give your rational function to a fellow classmate and ask for a written critique of your rational function.

113. Make up a rational function that has $y = 2x + 1$ as an oblique asymptote. Explain the methodology that you used.

4.8 POLYNOMIAL AND RATIONAL INEQUALITIES

1 Solve Polynomial Inequalities Graphically and Algebraically

2 Solve Rational Inequalities Graphically and Algebraically

1 In this section, we consider inequalities that involve polynomials of degree 2 and higher, as well as some that involve rational expressions. We will solve these inequalities using both a graphing approach and an algebraic approach.

To solve polynomial and rational inequalities algebraically, we follow these steps:

Steps for Solving Polynomial and Rational Inequalities Algebraically

STEP 1: Write the inequality so that a polynomial or rational expression f is on the left side and zero is on the right side in one of the following forms:

$$f(x) > 0 \qquad f(x) \geq 0 \qquad f(x) < 0 \qquad f(x) \leq 0$$

For rational expressions, be sure that the left side is written as a single quotient.

STEP 2: Determine the numbers at which the expression f on the left side equals zero, and if the expression is rational, the numbers at which the expression f on the left side is undefined.

STEP 3: Use the numbers found in Step 2 to separate the real number line into intervals.

STEP 4: Select a **test number** in each interval and evaluate f at the test number.

(a) If the value of f is positive, then $f(x) > 0$ for all numbers x in the interval.

(b) If the value of f is negative, then $f(x) < 0$ for all numbers x in the interval.

If the inequality is not strict, include the solutions of $f(x) = 0$ in the solution set.

EXAMPLE 1 Solving Quadratic Inequalities

Solve the inequality $x^2 \leq 4x + 12$, and graph the solution set.

Graphing Solution We graph $Y_1 = x^2$ and $Y_2 = 4x + 12$ on the same screen. See Figure 54 on page 294. Using the INTERSECT command, we find that Y_1 and Y_2 intersect at $x = -2$ and at $x = 6$. The graph of Y_1 is below that of Y_2, $Y_1 < Y_2$, between the points of intersection. Since the inequality is not strict, the solution set is $\{x | -2 \leq x \leq 6\}$ or, using interval notation, $[-2, 6]$.

Figure 54

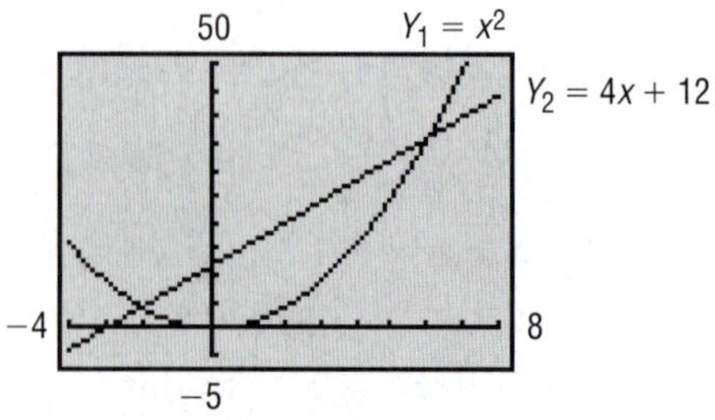

Algebraic Solution

STEP 1: Rearrange the inequality so that 0 is on the right side.

$$x^2 \leq 4x + 12$$
$$x^2 - 4x - 12 \leq 0 \quad \text{Subtract } 4x + 12 \text{ from both sides of the inequality.}$$

STEP 2: Find the zeros of $f(x) = x^2 - 4x - 12$ by solving the equation $x^2 - 4x - 12 = 0$.

$$x^2 - 4x - 12 = 0$$
$$(x + 2)(x - 6) = 0 \quad \text{Factor.}$$
$$x = -2 \quad \text{or} \quad x = 6$$

STEP 3: We use the zeros of f to separate the real number line into three intervals.

$$-\infty < x < -2 \qquad -2 < x < 6 \qquad 6 < x < \infty$$

STEP 4: We select a test number in each interval found in Step 3 and evaluate $f(x) = x^2 - 4x - 12$ at each test number to determine if $f(x)$ is positive or negative. See Table 17.

Table 17

Interval	Test Number	f(x)	Positive/Negative
$-\infty < x < -2$	-3	$f(-3) = (-3)^2 - 4(-3) - 12 = 9$	Positive
$-2 < x < 6$	0	$f(0) = -12$	Negative
$6 < x < \infty$	7	$f(7) = 9$	Positive

Since we want to know where $f(x)$ is negative, we conclude that the solutions are all x such that $-2 < x < 6$. However, because the original inequality is not strict, numbers x that satisfy the equation $x^2 = 4x + 12$ are also solutions of the inequality $x^2 \leq 4x + 12$. Thus, we include -2 and 6. The solution set of the given inequality is $\{x | -2 \leq x \leq 6\}$ or, using interval notation, $[-2, 6]$.

Figure 55
$-2 \leq x \leq 6$

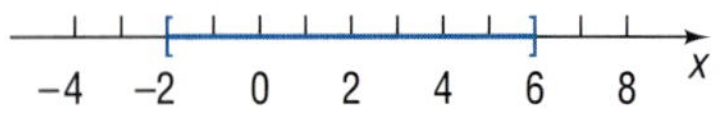

Figure 55 shows the graph of the solution set.

NOW WORK PROBLEMS 3 AND 7.

EXAMPLE 2 Solving a Polynomial Inequality

Solve the inequality $x^4 > x$, and graph the solution set.

Graphing Solution

We graph $Y_1 = x^4$ and $Y_2 = x$ on the same screen. See Figure 56. Using the INTERSECT command, we find that Y_1 and Y_2 intersect at $x = 0$ and at $x = 1$. The graph of Y_1 is above that of Y_2, $Y_1 > Y_2$, to the left of $x = 0$ and to the right of $x = 1$. Since the inequality is strict, the solution set is $\{x | x < 0$ or $x > 1\}$ or, using interval notation, $(-\infty, 0)$ or $(1, \infty)$.

Figure 56

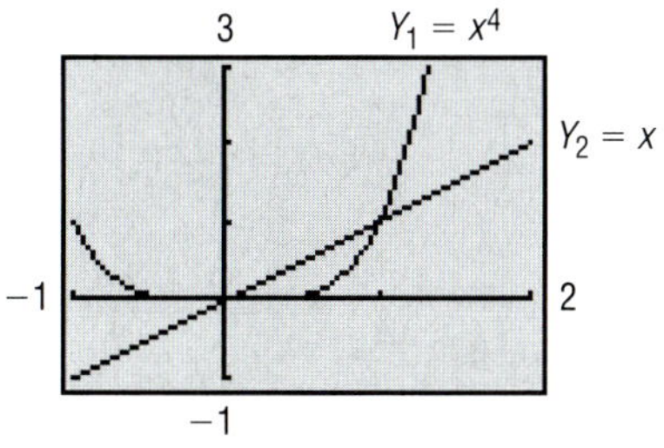

Algebraic Solution

STEP 1: Rearrange the inequality so that 0 is on the right side.

$$x^4 > x$$
$$x^4 - x > 0 \quad \text{Subtract } x \text{ from both sides of the inequality.}$$

STEP 2: Find the zeros of $f(x) = x^4 - x$ by solving $x^4 - x = 0$.

$$x^4 - x = 0$$
$$x(x^3 - 1) = 0 \quad \text{Factor out } x.$$
$$x(x - 1)(x^2 + x + 1) = 0 \quad \text{Factor the difference of two cubes.}$$
$$x = 0 \quad \text{or} \quad x - 1 = 0 \quad \text{or} \quad x^2 + x + 1 = 0 \quad \text{Set each factor equal to zero and solve.}$$
$$x = 0 \quad \text{or} \quad x = 1$$

The equation $x^2 + x + 1 = 0$ has no real solutions. (Do you see why?)

STEP 3: We use the zeros to separate the real number line into three intervals.

$$-\infty < x < 0 \qquad 0 < x < 1 \qquad 1 < x < \infty$$

STEP 4: We select a test number in each interval found in Step 3 and evaluate $f(x) = x^4 - x$ at each test number to determine if $f(x)$ is positive or negative. See Table 18.

Table 18

Interval	Test Number	f(x)	Positive/Negative
$-\infty < x < 0$	-1	$f(-1) = (-1)^4 - (-1) = 2$	Positive
$0 < x < 1$	$1/2$	$f(1/2) = -7/16$	Negative
$1 < x < \infty$	2	$f(2) = 14$	Positive

Since we want to know where $f(x)$ is positive, we conclude that the solutions are numbers x for which $-\infty < x < 0$ or $1 < x < \infty$. Because the original inequality is strict, numbers x that satisfy the equation $x^4 = x$ are not solutions. Thus, the solution set of the given inequality is $\{x | -\infty < x < 0$ or $1 < x < \infty\}$ or, using interval notation, $(-\infty, 0)$ or $(1, \infty)$.

Figure 57 shows the graph of the solution set.

Figure 57

$x < 0$ or $x > 1$

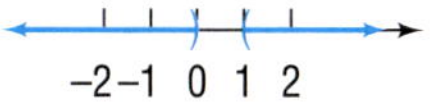

NOW WORK PROBLEM 17.

2 Let's solve a rational inequality.

EXAMPLE 3 Solving a Rational Inequality

Solve the inequality $\frac{4x + 5}{x + 2} \geq 3$, and graph the solution set.

Graphing Solution We first note that the domain of the variable consists of all real numbers except -2. We graph $Y_1 = (4x + 5)/(x + 2)$ and $Y_2 = 3$ on the same screen. See Figure 58. Using the INTERSECT command, we find that Y_1 and Y_2 intersect at $x = 1$. The graph of Y_1 is above that of Y_2, $Y_1 > Y_2$, to the left of $x = -2$ and to the right of $x = 1$. Since the inequality is not strict, the solution set is $\{x | -\infty < x < -2 \text{ or } 1 \leq x < \infty\}$. Using interval notation, the solution set is $(-\infty, -2)$ or $[1, \infty)$.

Algebraic Solution STEP 1: We first note that the domain of the variable consists of all real numbers except -2. We rearrange terms so that 0 is on the right side.

Figure 58

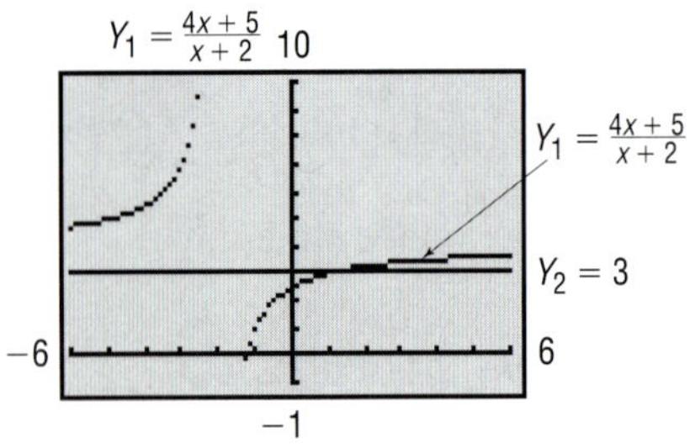

$$\frac{4x + 5}{x + 2} - 3 \geq 0 \quad \textit{Subtract 3 from both sides of the inequality.}$$

STEP 2: Find the zeros and values of x where $f(x) = \frac{4x + 5}{x + 2} - 3$ is undefined. To find these numbers, we must express f as a quotient.

$$f(x) = \frac{4x + 5}{x + 2} - 3 \quad \textit{Least Common Denominator: } x + 2.$$

$$f(x) = \frac{4x + 5}{x + 2} - 3\frac{x + 2}{x + 2} \quad \textit{Multiply } -3 \textit{ by } \frac{x + 2}{x + 2}.$$

$$f(x) = \frac{4x + 5 - 3x - 6}{x + 2} \quad \textit{Distribute } -3.$$

$$f(x) = \frac{x - 1}{x + 2} \quad \textit{Collect like terms.}$$

The zero of f is 1. Also, f is undefined for $x = -2$.

STEP 3: We use the numbers found in Step 2 to separate the real number line into three intervals.

$$-\infty < x < -2 \qquad -2 < x < 1 \qquad 1 < x < \infty$$

STEP 4: We select a test number in each interval found in Step 3 and evaluate $f(x) = \frac{4x + 5}{x + 2} - 3$ at each test number to determine if $f(x)$ is positive or negative. See Table 19.

Table 19

Interval	Test Number	$f(x)$	Positive/Negative
$-\infty < x < -2$	-3	$f(-3) = 4$	Positive
$-2 < x < 1$	0	$f(0) = -1/2$	Negative
$1 < x < \infty$	2	$f(2) = 1/4$	Positive

We want to know where $f(x)$ is positive. We conclude that the solutions are all x such that $-\infty < x < -2$ or $1 < x < \infty$. Because the original inequality is not strict, numbers x that satisfy the equation $\frac{x-1}{x+2} = 0$ are also solutions of the inequality. Since $\frac{x-1}{x+2} = 0$ only if $x = 1$, we conclude that the solution set is $\{x \mid -\infty < x < -2 \text{ or } 1 \le x < \infty\}$ or, using interval notation, $(-\infty, -2)$ or $[1, \infty)$.

Figure 59 shows the graph of the solution set.

Figure 59

$-\infty < x < -2$ or $1 \le x < \infty$

$(-\infty, -2)$ or $[1, \infty)$

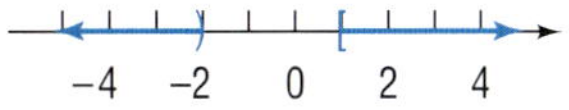

NOW WORK PROBLEM 37.

EXAMPLE 4 Minimum Sales Requirements

Tami is considering leaving her \$30,000 a year job and buying a cookie company. According to the financial records of the firm, the relationship between pounds of cookies sold and profit is exhibited by Table 20.

Table 20

Pounds of Cookies (in Hundreds), x	Profit, P
0	$-20{,}000$
50	-5990
75	412
120	10,932
200	26,583
270	36,948
340	44,381
420	49,638
525	49,225
610	44,381
700	34,220

(a) Draw a scatter diagram of the data in Table 20 with the pounds of cookies sold as the independent variable.

(b) The quadratic function of best fit is $p(x) = -0.31x^2 + 295.86x - 20{,}042.52$. Use this function to determine the number of pounds of cookies that Tami must sell in order for the profits to exceed \$30,000 a year and therefore make it worthwhile for her to quit her job.

(c) Using the function given in part (b), determine the number of pounds of cookies that Tami should sell in order to maximize profits.

(d) Using the function given in part (b), determine the maximum profit that Tami can expect to earn.

(e) Use a graphing utility to verify that the function given in part (b) is the quadratic function of best fit.

Solution (a) Figure 60 shows the scatter diagram.

Figure 60

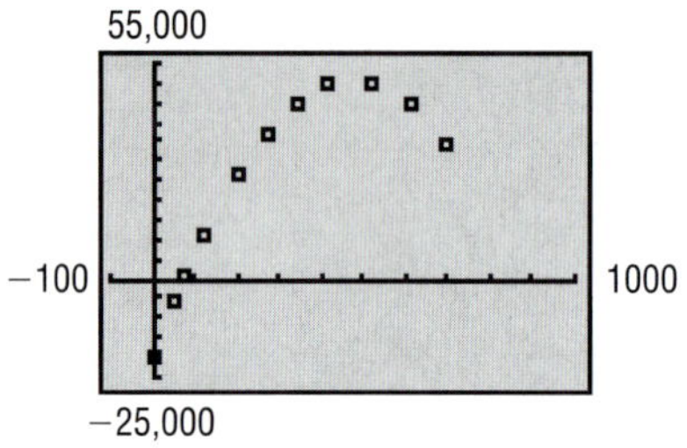

(b) We want profit to exceed \$30,000; therefore, we want to solve the inequality

$$-0.31x^2 + 295.86x - 20{,}042.52 > 30{,}000$$

We graph

$$Y_1 = p(x) = -0.31x^2 + 295.86x - 20{,}042.52 \quad \text{and} \quad Y_2 = 30{,}000$$

on the same screen. See Figure 61.

Figure 61

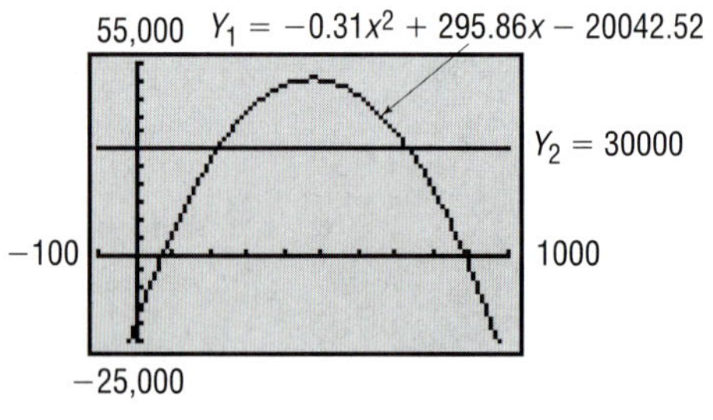

Using the INTERSECT command, we find that Y_1 and Y_2 intersect at $x = 220$ and at $x = 735$. The graph of Y_1 is above that of Y_2, $Y_1 > Y_2$, between the points of intersection. Since the inequality is strict, the solution set is $\{x \mid 220 < x < 735\}$.

(c) The function $p(x)$ given in part (b) is a quadratic function whose graph opens down. The vertex is therefore the highest point. The x-coordinate of the vertex is

$$x = \frac{-b}{2a} = \frac{-295.86}{2(-0.31)} = 477$$

To maximize profit, 477 hundred (47,700) pounds of cookies must be sold.

(d) The maximum profit is

$$p(477) = -0.31(477)^2 + 295.86(477) - 20{,}042.52 = \$50{,}549$$

(e) Using a graphing utility, the quadratic function of best fit is

$$p(x) = -0.31x^2 + 295.86x - 20{,}042.52$$

See Figure 62.

Figure 62

4.8 EXERCISES

In Problems 1–58, solve each inequality (a) graphically and (b) algebraically.

1. $(x-5)(x+2)<0$
2. $(x-5)(x+2)>0$
3. $x^2-4x>0$
4. $x^2+8x>0$
5. $x^2-9<0$
6. $x^2-1<0$
7. $x^2+x>12$
8. $x^2+7x<-12$
9. $2x^2<5x+3$
10. $6x^2<6+5x$
11. $x(x-7)>8$
12. $x(x+1)>20$
13. $4x^2+9<6x$
14. $25x^2+16<40x$
15. $6(x^2-1)>5x$
16. $2(2x^2-3x)>-9$
17. $(x-1)(x^2+x+1)>0$
18. $(x+2)(x^2-x+1)>0$
19. $(x-1)(x-2)(x-3)<0$
20. $(x+1)(x+2)(x+3)<0$
21. $x^3-2x^2-3x>0$
22. $x^3+2x^2-3x>0$
23. $x^4>x^2$
24. $x^4<4x^2$
25. $x^3>x^2$
26. $x^3<3x^2$
27. $x^4>1$
28. $x^3>1$
29. $x^2-7x-8<0$
30. $x^2+12x+32\ge 0$
31. $x^3+x-12\ge 0$
32. $x^3-3x+1\le 0$
33. $x^4-3x^2-4>0$
34. $x^4-5x^2+6<0$
35. $x^3-4\ge 3x^2+5x-3$
36. $x^4-4x\le -x^2+2x+1$
37. $\dfrac{x+1}{x-1}>0$
38. $\dfrac{x-3}{x+1}>0$
39. $\dfrac{(x-1)(x+1)}{x}<0$
40. $\dfrac{(x-3)(x+2)}{x-1}<0$
41. $\dfrac{(x-2)^2}{x^2-1}\ge 0$
42. $\dfrac{(x+5)^2}{x^2-4}\ge 0$
43. $6x-5<\dfrac{6}{x}$
44. $x+\dfrac{12}{x}<7$
45. $\dfrac{x+4}{x-2}\le 1$
46. $\dfrac{x+2}{x-4}\ge 1$
47. $\dfrac{3x-5}{x+2}\le 2$
48. $\dfrac{x-4}{2x+4}\ge 1$
49. $\dfrac{1}{x-2}<\dfrac{2}{3x-9}$
50. $\dfrac{5}{x-3}>\dfrac{3}{x+1}$
51. $\dfrac{2x+5}{x+1}>\dfrac{x+1}{x-1}$

52. $\dfrac{1}{x+2} > \dfrac{3}{x+1}$

53. $\dfrac{x^2(3+x)(x+4)}{(x+5)(x-1)} > 0$

54. $\dfrac{x(x^2+1)(x-2)}{(x-1)(x+1)} > 0$

55. $\dfrac{2x^2-x-1}{x-4} \le 0$

56. $\dfrac{3x^2+2x-1}{x+2} > 0$

57. $\dfrac{x^2+3x-1}{x+3} > 0$

58. $\dfrac{x^2-5x+3}{x-5} < 0$

59. For what positive numbers will the cube of a number exceed 4 times its square?

60. For what positive numbers will the square of a number exceed twice the number?

61. What is the domain of the variable in the expression $\sqrt{x^2-16}$?

62. What is the domain of the variable in the expression $\sqrt{x^3-3x^2}$?

63. What is the domain of the variable in the expression $\sqrt{\dfrac{x-2}{x+4}}$?

64. What is the domain of the variable in the expression $\sqrt{\dfrac{x-1}{x+4}}$?

65. **Physics** A ball is thrown vertically upward with an initial velocity of 80 feet per second. The distance s (in feet) of the ball from the ground after t seconds is $s = 80t - 16t^2$.

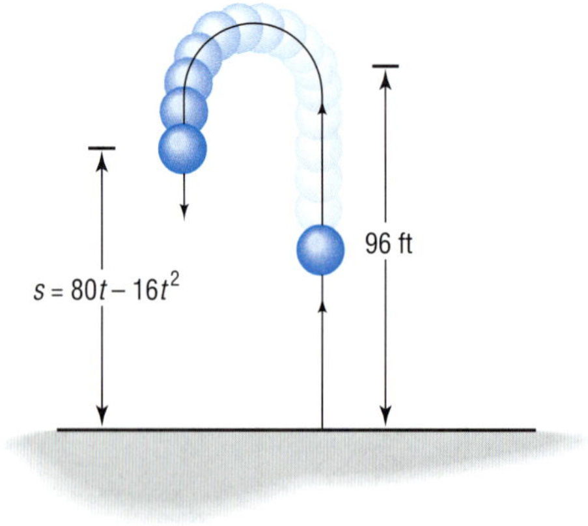

(a) For what time interval is the ball more than 96 feet above the ground? (See the figure.)
(b) Using a graphing utility, graph the relation between s and t.
(c) What is the maximum height of the ball?
(d) After how many seconds does the ball reach the maximum height?

66. **Physics** A ball is thrown vertically upward with an initial velocity of 96 feet per second. The distance s (in feet) of the ball from the ground after t seconds is $s = 96t - 16t^2$.

(a) For what time interval is the ball more than 112 feet above the ground?
(b) Using a graphing utility, graph the relation between s and t.
(c) What is the maximum height of the ball?
(d) After how many seconds does the ball reach the maximum height?

67. **Business** The monthly revenue achieved by selling x wristwatches is figured to be $x(40 - 0.2x)$ dollars. The wholesale cost of each watch is \$32.

(a) How many watches must be sold each month to achieve a profit (revenue − cost) of at least \$50?
(b) Using a graphing utility, graph the revenue function.
(c) What is the maximum revenue that this firm could earn?
(d) How many wristwatches should the firm sell to maximize revenue?
(e) Using a graphing utility, graph the profit function.
(f) What is the maximum profit that this firm can earn?
(g) How many watches should the firm sell to maximize profit?
(h) Provide a reasonable explanation as to why the answers found in parts (d) and (g) differ. Is the shape of the revenue function reasonable in your opinion? Why?

68. **Business** The monthly revenue achieved by selling x boxes of candy is figured to be $x(5 - 0.05x)$ dollars. The wholesale cost of each box of candy is \$1.50.

(a) How many boxes must be sold each month to achieve a profit of at least \$60?
(b) Using a graphing utility, graph the revenue function.
(c) What is the maximum revenue that this firm could earn?
(d) How many boxes of candy should the firm sell to maximize revenue?
(e) Using a graphing utility, graph the profit function.
(f) What is the maximum profit that this firm can earn?

(g) How many boxes of candy should the firm sell to maximize profit?

(h) Provide a reasonable explanation as to why the answers found in parts (d) and (g) differ. Is the shape of the revenue function reasonable in your opinion? Why?

69. Cost of Manufacturing In Problem 69 of Section 4.2, a cubic function of best fit relating the cost C of manufacturing x Chevy Cavaliers in a day was given. Budget constraints will not allow Chevy to spend more than \$97,000 per day. Determine the number of Cavaliers that should be produced in a day.

70. Cost of Printing In Problem 70 of Section 4.2, a cubic function of best fit relating the cost C of printing x textbooks in a week was given. Budget constraints will not allow the printer to spend more than \$170,000 per week. Determine the number of textbooks that should be printed in a week.

71. Mininum Sales Requirements Marissa is thinking of leaving her \$1000 a week job and buying a computer resale shop. According to the financial records of the firm, the profits (in dollars) of the company for different amounts of computers sold and the corresponding profits are as follows:

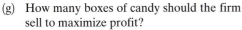

Number of Computers Sold, x	Profit, p
0	−1500
4	−522
7	54
12	775
18	1184
23	1132
29	653

(a) Using a graphing utility, draw a scatter diagram of the data with the number of computers sold as the independent variable.

(b) The quadratic function of best fit is $p(x) = -6.776x^2 + 270.668x - 1500.202$. Use this function to determine the number of computers that Marissa must sell in order for the profits to exceed \$1000 a week and therefore make it worthwhile for her to quit her job.

(c) Using the function given in part (b), determine the number of computers that Marissa should sell in order to maximize profits.

(d) Using the function given in part (b), determine the maximum profit that Marissa can expect to earn.

(e) Use a graphing utility to verify that the function given in part (b) is the quadratic function of best fit.

72. Minimum Sales Requirements Barry is considering the purchase of a gas station. According to the financial records of the gas station, the monthly sales of the gas station (in thousands of gallons of gasoline) and the corresponding profits are as follows:

Gallons of Gasoline (000s), x	Profit, p
50	3947
54	4214
74	4942
92	4838
82	5003
75	4965
100	4521
88	4933
63	4665

(a) Using a graphing utility, draw a scatter diagram of the data with the number of gallons of gasoline sold as the independent variable.

(b) The quadratic function of best fit is $p(x) = -1.17x^2 + 187.23x - 2486.91$. Use this function to determine the number of gallons of gasoline that Barry must sell in order for the profits to exceed \$4000 a month and therefore make it worthwhile for him to quit his job.

(c) Using the function given in part (b), determine the number of gallons of gasoline that Barry should sell in order to maximize profits.

(d) Using the function given in part (b), determine the maximum profit that Barry can expect to earn.

(e) Use a graphing utility to verify that the function given in part (b) is the quadratic function of best fit.

73. Prove that if a, b are real numbers and $a \geq 0$, $b \geq 0$ then

$$a \leq b \quad \text{is equivalent to} \quad \sqrt{a} \leq \sqrt{b}$$

[Hint:** $b - a = (\sqrt{b} - \sqrt{a})(\sqrt{b} + \sqrt{a})$**].

74. Make up an inequality that has no solution. Make up one that has exactly one solution.

75. The inequality $x^2 + 1 < -5$ has no solution. Explain why.

CHAPTER REVIEW

THINGS TO KNOW

Power function	$f(x) = x^n, n \geq 2$ even	Even function: Passes through $(-1, 1), (0, 0), (1, 1)$ Opens up
	$f(x) = x^n, n \geq 3$ odd	Odd function: Passes through $(-1, -1), (0, 0), (1, 1)$ Increasing
Polynomial function	$f(x) = a_n x^n + a_{n-1}x^{n-1} + \cdots + a_1 x + a_0,\ a_n \neq 0$	At most $n - 1$ turning points; End behavior: behaves like $y = a_n x^n$ for large x.
Rational function	$R(x) = \dfrac{p(x)}{q(x)}$	p, q are polynomial functions.

Zeros of a polynomial *f* Numbers for which $f(x) = 0$; these are the x-intercepts of the graph of f.

Remainder Theorem If a polynomial $f(x)$ is divided by $x - c$, then the remainder is $f(c)$.

Factor Theorem $x - c$ is a factor of a polynomial $f(x)$ if and only if $f(c) = 0$.

Rational Zeros Theorem Let f be a polynomial function of degree 1 or higher of the form

$$f(x) = a_n x^n + a_{n-1}x^{n-1} + \cdots + a_1 x + a_0,\ a_n \neq 0,\ a_0 \neq 0$$

where each coefficient is an integer. If p/q, in lowest terms, is a rational zero of f, then p must be a factor of a_0 and q must be a factor of a_n.

Quadratic equation and quadratic formula If $ax^2 + bx + c = 0, a \neq 0$, then $x = \dfrac{-b \pm \sqrt{b^2 - 4ac}}{2a}$.

Discriminant If $b^2 - 4ac > 0$, there are two distinct real solutions.

If $b^2 - 4ac = 0$, there is one repeated real solution.

If $b^2 - 4ac < 0$, there are two distinct complex solutions that are not real; the solutions are conjugates of each other.

Fundamental Theorem of Algebra Every complex polynomial function $f(x)$ of degree $n \geq 1$ has at least one complex zero.

Conjugate Pairs Theorem Let $f(x)$ be a complex polynomial whose coefficients are real numbers. If $r = a + bi$ is a zero of f, then its complex conjugate $\bar{r} = a - bi$ is also a zero of f.

HOW TO

- Graph power functions
- Graph polynomial functions
- Divide polynomials using long division
- Use synthetic division to divide a polynomial by $x - c$
- Find the real zeros of a polynomial by using the Rational Zeros Theorem and depressed equations
- Solve polynomial equations using the Rational Zeros Theorem and depressed equations
- Use the theorem for Bounds on Zeros
- Add, subtract, multiply, and divide complex numbers
- Solve quadratic equations in the complex number system
- Find the complex zeros of a polynomial
- Graph rational functions (see Steps 1 through 6, page 279)
- Solve polynomial and rational inequalities

FILL-IN-THE-BLANK ITEMS

1. In the process of long division,
(Divisor)(Quotient) + ________________ = ________________.
2. When a polynomial function f is divided by $x - c$, the remainder is ________________
3. A polynomial function f has the factor $x - c$ if and only if ________________.
4. A number r for which $f(r) = 0$ is called a(n) ________________ of the function f.
5. The polynomial function $f(x) = x^5 - 2x^3 + x^2 + x - 1$ has at most ________________ real zeros.
6. The possible rational zeros of $f(x) = 2x^5 - x^3 + x^2 - x + 1$ are ________________.
7. The line ________________ is a horizontal asymptote of $R(x) = \dfrac{x^3 - 1}{x^3 + 1}$.
8. The line ________________ is a vertical asymptote of $R(x) = \dfrac{x^3 - 1}{x^3 + 1}$.
9. In the complex number $5 + 2i$, the number 5 is called the ________________ part; the number 2 is called the ________________ part; the number i is called the ________________ ________________.
10. If $3 + 4i$ is a zero of a polynomial of degree 5 with real coefficients, then so is ________________.
11. The equation $|x^2| = 4$ has four solutions: ________________, ________________, ________________, and ________________.

TRUE/FALSE ITEMS

T F 1. Every polynomial of degree 3 with real coefficients has exactly three real zeros.

T F 2. If $2 - 3i$ is a zero of a polynomial with real coefficients, then so is $-2 + 3i$.

T F 3. The graph of $R(x) = \dfrac{x^2}{x - 1}$ has exactly one vertical asymptote.

T F 4. The graph of $f(x) = x^2(x - 3)(x + 4)$ has exactly three x-intercepts.

T F 5. If f is a polynomial function of degree 4 and if $f(2) = 5$, then

$$\frac{f(x)}{x - 2} = p(x) + \frac{5}{x - 2}$$

where $p(x)$ is a polynomial of degree 3.

T F 6. The conjugate of $2 + \sqrt{5}i$ is $-2 - \sqrt{5}i$.

T F 7. A polynomial of degree n with real coefficients has exactly n complex zeros. At most n of them are real numbers.

REVIEW EXERCISES

Blue problem numbers indicate the authors' suggestions for use in a Practice Test.

In Problems 1–6, use a graphing utility to graph each function using transformations (shifting, compressing, stretching, and reflection). Show all the stages.

1. $f(x) = (x + 2)^3$
2. $f(x) = -x^3 + 3$
3. $f(x) = -(x - 1)^4$
4. $f(x) = (x - 1)^4 - 2$
5. $f(x) = (x - 1)^4 + 2$
6. $f(x) = (1 - x)^3$

In Problems 7–14, for each polynomial function f:

(a) Using a graphing utility, graph f.

(b) Find the x- and y-intercepts.

(c) Determine whether each x-intercept is of odd or even multiplicity.

(d) Find the power function that the graph of f resembles for large values of $|x|$.

(e) Determine the number of turning points on the graph of f.

(f) Determine the local maxima and local minima, if any exist, rounded to two decimal places.

7. $f(x) = x(x + 2)(x + 4)$

8. $f(x) = x(x - 2)(x - 4)$

9. $f(x) = (x - 2)^2(x + 4)$

10. $f(x) = (x - 2)(x + 4)^2$

11. $f(x) = x^3 - 4x^2$

12. $f(x) = x^3 + 4x$

13. $f(x) = (x - 1)^2(x + 3)(x + 1)$

14. $f(x) = (x - 4)(x + 2)^2(x - 2)$

In Problems 15–26, discuss each rational function following the six steps on page 279.

15. $R(x) = \dfrac{2x - 6}{x}$

16. $R(x) = \dfrac{4 - x}{x}$

17. $H(x) = \dfrac{x + 2}{x(x - 2)}$

18. $H(x) = \dfrac{x}{x^2 - 1}$

19. $R(x) = \dfrac{x^2 + x - 6}{x^2 - x - 6}$

20. $R(x) = \dfrac{x^2 - 6x + 9}{x^2}$

21. $F(x) = \dfrac{x^3}{x^2 - 4}$

22. $F(x) = \dfrac{3x^3}{(x - 1)^2}$

23. $R(x) = \dfrac{2x^4}{(x - 1)^2}$

24. $R(x) = \dfrac{x^4}{x^2 - 9}$

25. $G(x) = \dfrac{x^2 - 4}{x^2 - x - 2}$

26. $F(x) = \dfrac{(x - 1)^2}{x^2 - 1}$

In Problems 27–30, use synthetic division to find the quotient $q(x)$ and remainder R when $f(x)$ is divided by $g(x)$.

27. $f(x) = 8x^3 - 3x^2 + x + 4;\quad g(x) = x - 1$

28. $f(x) = 2x^3 + 8x^2 - 5x + 5;\quad g(x) = x - 2$

29. $f(x) = x^4 - 2x^3 + x - 1;\quad g(x) = x + 2$

30. $f(x) = x^4 - x^2 + 3x;\quad g(x) = x + 1$

31. List all the potential rational zeros at $f(x) = 12x^8 - x^7 + 6x^4 - x^3 + x - 3$.

32. List all the potential rational zeros of $f(x) = -6x^5 + x^4 + 2x^3 - x + 1$.

In Problems 33–38, follow the steps on page 250 to find all the real zeros of each polynomial function.

33. $f(x) = x^3 - 3x^2 - 6x + 8$

34. $f(x) = x^3 - x^2 - 10x - 8$

35. $f(x) = 4x^3 + 4x^2 - 7x + 2$

36. $f(x) = 4x^3 - 4x^2 - 7x - 2$

37. $f(x) = x^4 - 4x^3 + 9x^2 - 20x + 20$

38. $f(x) = x^4 + 6x^3 + 11x^2 + 12x + 18$

In Problems 39–44, determine the real zeros of the polynomial function. Approximate all irrational zeros rounded to two decimal places.

39. $f(x) = 2x^3 - 11.84x^2 - 9.116x + 82.46$

40. $f(x) = 12x^3 + 39.8x^2 - 4.4x - 3.4$

41. $g(x) = 15x^4 - 21.5x^3 - 1718.3x^2 + 5308x + 3796.8$

42. $g(x) = 3x^4 + 67.93x^3 + 486.265x^2 + 1121.32x + 412.195$

43. $f(x) = 3x^3 + 18.02x^2 + 11.0467x - 53.8756$

44. $f(x) = x^3 - 3.16x^2 - 39.4611x + 151.638$

In Problems 45–48, find the real solutions of each equation.

45. $2x^4 + 2x^3 - 11x^2 + x - 6 = 0$

46. $3x^4 + 3x^3 - 17x^2 + x - 6 = 0$

47. $2x^4 + 7x^3 + x^2 - 7x - 3 = 0$

48. $2x^4 + 7x^3 - 5x^2 - 28x - 12 = 0$

In Problems 49–52, find bounds to the zeros of each polynomial function.

49. $f(x) = x^3 - x^2 - 4x + 2$

50. $f(x) = x^3 + x^2 - 10x - 5$

51. $f(x) = 2x^3 - 7x^2 - 10x + 35$

52. $f(x) = 3x^3 - 7x^2 - 6x + 14$

In Problems 53–56, use the Intermediate Value Theorem to show that each polynomial has a zero in the given interval.

53. $f(x) = 3x^3 - x - 1;\quad [0, 1]$

54. $f(x) = 2x^3 - x^2 - 3;\quad [1, 2]$

55. $f(x) = 8x^4 - 4x^3 - 2x - 1;\quad [0, 1]$

56. $f(x) = 3x^4 + 4x^3 - 8x - 2;\quad [1, 2]$

In Problems 57–66, write each expression in the standard form $a + bi$.

57. $(6 + 3i) - (2 - 4i)$

58. $(8 - 3i) + (-6 + 2i)$

59. $4(3 - i) + 3(-5 + 2i)$

60. $2(1 + i) - 3(2 - 3i)$

61. $\dfrac{3}{3 + i}$

62. $\dfrac{4}{2 - i}$

63. i^{50}

64. i^{29}

65. $(2 + 3i)^3$

66. $(3 - 2i)^3$

In Problems 67–70, information is given abut a complex polynomial $f(x)$ whose coefficients are real numbers. Find the remaining zeros of f.

67. Degree 3; zeros: $4 + i, 6$

68. Degree 3; zeros: $3 + 4i, 5$

69. Degree 4; zeros: $i, 1 + i$

70. Degree 4; zeros: $1, 2, 1 + i$

In Problems 71–84, solve each equation in the complex number system.

71. $x^2 + x + 1 = 0$

72. $x^2 - x + 1 = 0$

73. $2x^2 + x - 2 = 0$

74. $3x^2 - 2x - 1 = 0$

75. $x^2 + 3 = x$

76. $2x^2 + 1 = 2x$

77. $x(1 - x) = 6$

78. $x(1 + x) = 2$

79. $x^4 + 2x^2 - 8 = 0$

80. $x^4 + 8x^2 - 9 = 0$

81. $x^3 - x^2 - 8x + 12 = 0$

82. $x^3 - 3x^2 - 4x + 12 = 0$

83. $3x^4 - 4x^3 + 4x^2 - 4x + 1 = 0$

84. $x^4 + 4x^3 + 2x^2 - 8x - 8 = 0$

In Problems 85–94, solve each inequality (a) graphically and (b) algebraically.

85. $2x^2 + 5x - 12 < 0$

86. $3x^2 - 2x - 1 \geq 0$

87. $\dfrac{6}{x + 3} \geq 1$

88. $\dfrac{-2}{1 - 3x} < 1$

89. $\dfrac{2x - 6}{1 - x} < 2$

90. $\dfrac{3 - 2x}{2x + 5} \geq 2$

91. $\dfrac{(x - 2)(x - 1)}{x - 3} > 0$

92. $\dfrac{x + 1}{x(x - 5)} \leq 0$

93. $\dfrac{x^2 - 8x + 12}{x^2 - 16} > 0$

94. $\dfrac{x(x^2 + x - 2)}{x^2 + 9x + 20} \leq 0$

95. **Relating the Length and Period of a Pendulum** Tom constructs simple pendulums with different lengths, l, and uses a light probe to record the corresponding period, T. The following data are collected:

Length l (in feet)	Period T (in seconds)
0.5	0.79
1.2	1.20
1.8	1.51
2.3	1.65
3.7	2.14
4.9	2.43

(a) Using a graphing utility, draw a scatter diagram of the data with length as the independent variable and period as the dependent variable.

(b) The power function of best fit to these data is

$$T(l) = 1.1094l^{0.4952}$$

Use this function to predict the period of a pendulum whose length is 4 feet.

(c) Graph the power model given in part (b) on the scatter diagram.

(d) Use a graphing utility to verify that the function given in part (b) is the power function of best fit.

96. **AIDS Cases in the United States** The following data represent the cumulative number of reported AIDS cases in the United States from 1990–1997.

Year, t	Number of AIDS Cases, A
1990, 1	193,878
1991, 2	251,638
1992, 3	326,648
1993, 4	399,613
1994, 5	457,280
1995, 6	528,215
1996, 7	594,760
1997, 8	653,253

Source: U.S. Center for Disease Control and Prevention

(a) Using a graphing utility, draw a scatter diagram of the data, treating year as the independent variable.

(b) The cubic function of best fit to these data is

$$A(t) = -212.0t^3 + 2429.1t^2 + 59{,}568.9t + 1{,}300{,}003.1$$

Use this function to predict the cumulative number of AIDS cases reported in the United States in 2000.

(c) Use the function given in part (b) to predict the year in which the cumulative number of AIDS cases reported in the United States reaches 1,000,000.

(c) Draw the cubic function of best fit on the scatter diagram given in part (b).

(d) Use a graphing utility to verify that the function given in part (b) is the cubic function of best fit.

(e) Do you think the function given in part (b) will be useful in predicting the number of AIDS cases in 2010?

97. Design a polynomial function with the following characteristics: degree 6; four real zeros, one of multiplicity 3; y-intercept 3; behaves like $y = -5x^6$ for large values of x. Is this polynomial unique? Compare your polynomial with those of other students. What terms will be the same as everyone else's? Add some more characteristics, such as symmetry or naming the real zeros. How does this modify the polynomial?

98. Design a rational function with the following characteristics: three real zeros, one of multiplicity 2; y-intercept 1; vertical asymptotes $x = -2$ and $x = 3$; oblique asymptote $y = 2x + 1$. Is this rational function unique? Compare yours with those of other students. What will be the same as everyone else's? Add some more characteristics, such as symmetry or naming the real zeros. How does this modify the rational function?

99. The illustration shows the graph of a polynomial function.

(a) Is the degree of the polynomial even or odd?

(b) Is the leading coefficient positive or negative?

(c) Is the function even, odd, or neither?

(d) Why is x^2 necessarily a factor of the polynomial?

(e) What is the minimum degree of the polynomial?

(f) Formulate five different polynomials whose graphs could look like the one shown. Compare yours to those of other students. What similarities do you see? What differences?

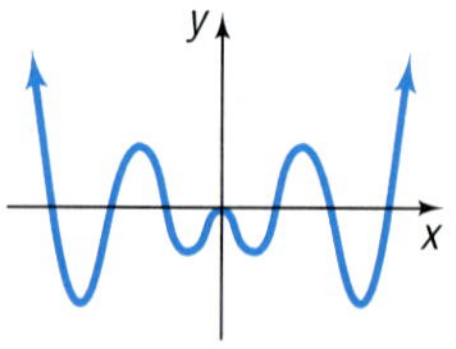

CHAPTER PROJECTS

1. Enrollment in Schools The following data represent the number of students enrolled in grades 9 through 12 in private institutions.

Year	Enrollment (000's)
1975	1300
1980	1339
1981	1400
1982	1400
1983	1400
1984	1400
1985	1362
1986	1336
1987	1247
1988	1206
1989	1193
1990	1137
1991	1125
1992	1163
1993	1191
1994	1236
1995	1260
1996	1297

(a) Using a graphing utility, draw a scatter diagram of the data.

(b) Using a graphing utility, find the cubic function of best fit.

(c) Using a graphing utility, graph the cubic function of best fit on your scatter diagram.

(d) Use the function found in part (b) to predict the number of students enrolled in private institutions for grades 9 through 12 in the year 2000.

(e) Compare the result found in part (d) with that of the U.S. National Center for Education Statistics, whose predictions can be found in the *Statistical Abstract of the United States*. What might account for any differences in your prediction versus that of the U.S. National Center for Education Statistics?

(f) Use the function found in part (b) to predict the year in which enrollment in private institutions for grades 9 through 12 will be 1,400,000. That is, solve $f(x) = 1400$.

CHAPTER PROJECTS *(Continued)*

2. Finding a Function from Its Graph Find a rational function of the form

$$R(x) = \frac{a_n x^n + a_{n-1}x^{n-1} + \cdots + a_0}{b_m x^m + b_{m-1}x^{m-1} + \cdots + b_0}$$

defined by the graph illustrated.

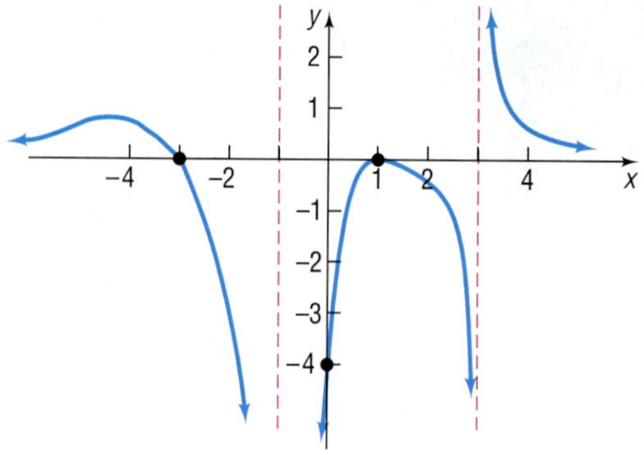

(a) There are two vertical asymptotes. How do they fit?

(b) There are two x-intercepts. Where do they show up in the solution?

(c) Use a graphing utility to check what you have so far. Does it look like the original graph? Notice that 1 and -3 appear to be the only real zeros. What can you tell about the multiplicity of each? Do you need to change the multiplicity of either zero in your solution?

(d) How can you account for the fact that the graph of the function goes to $-\infty$ on both sides of one vertical asymptote and at the other vertical asymptote it goes to $-\infty$ on one side and to ∞ on the other? Can you change the denominator in some way to account for this?

(e) Is the information that you have so far consistent with the horizontal asymptote $y = 0$? If not, how should you adjust your function so that it is consistent?

(f) Is the information that you have so far consistent with the y-intercept? If not, how should you adjust your function so that it is consistent?

(g) Are you finished? How you checked your solution on a graphing utility? Are you satisfied that you have it right?

(h) Are the other functions whose graph might look like the one given? If so, what common characteristics do they have?

CHAPTER

5 Exponential and Logarithmic Functions

The McDonald's Scalding Coffee Case

April 3, 1996

There is a lot of hype about the McDonald's scalding coffee case. No one is in favor of frivolous cases or outlandish results; however, it is important to understand some points that were not reported in most of the stories about the case. McDonald's coffee was not only hot, it was scalding capable of almost instantaneous destruction of skin, flesh and muscle.

Plantiff's expert, a scholar in thermodynamics applied to human skin burns, testified that liquids, at 180 degrees, will cause a full thickness burn to human skin in two to seven seconds. Other testimony showed that as the temperature decreases toward 155 degrees, the extent of the burn relative to that temperature decreases exponentially. Thus, if (the) spill had involved coffee at 155 degrees, the liquid would have cooled and given her time to avoid a serious burn.

Miller, Norman, & Associates, Ltd., Attorneys, Moorhead, MN.

See Chapter Project 1.

Preparing for This Chapter

Before getting started on this chapter, review the following concepts:

- **Exponents (Appendix, Sections 4 and 9)**
- **Functions (Sections 2.1, 3.1, and 3.2)**
- **Simple Interest (p. 43)**
- **Linear Curve Fitting (Section 2.2)**

Outline

5.1 One-to-One Functions; Inverse Functions
5.2 Exponential Functions
5.3 Logarithmic Functions
5.4 Properties of Logarithms
5.5 Logarithmic and Exponential Equations
5.6 Compound Interest
5.7 Growth and Decay
5.8 Exponential, Logarithmic, and Logistic Curve Fitting
Chapter Review
Chapter Projects

Until now, our study of functions has concentrated primarily on polynomial and rational functions. These functions belong to the class of **algebraic functions,** that is, functions that can be expressed in terms of sums, differences, products, quotients, powers, or roots of polynomials. Functions that are not algebraic are termed **transcendental** (they transcend, or go beyond, algebraic functions).

In this chapter, we study two transcendental functions: the *exponential* and *logarithmic functions*. These functions occur frequently in a wide variety of applications, such as biology, chemistry, economics, and psychology.

The chapter begins with a discussion of inverse functions.

5.1 ONE-TO-ONE FUNCTIONS; INVERSE FUNCTIONS

1 Determine Whether a Function Is One-to-One

2 Obtain the Graph of the Inverse Function from the Graph of the Function

3 Find an Inverse Function

In Section 2.1, we said a function f can be thought of as a machine that receives as input a number, say x, from the domain, manipulates it, and outputs the value $f(x)$. The **inverse** of f receives as input a number $f(x)$, manipulates it, and outputs the value x. If the function f is the set of ordered pairs (x, y), then the inverse of f is the set of ordered pairs (y, x).

EXAMPLE 1 Finding the Inverse of a Function

Find the inverse of the following functions.

(a) Let the domain of the function represent the employees of Yolanda's Preowned Car Mart and let the range represent their base salaries.

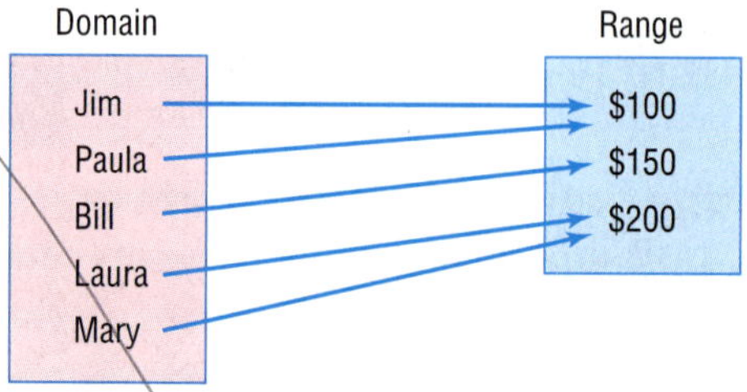

(b) Let the domain of the function represent the employees of Yolanda's Preowned Car Mart and let the range represent their spouse's names.

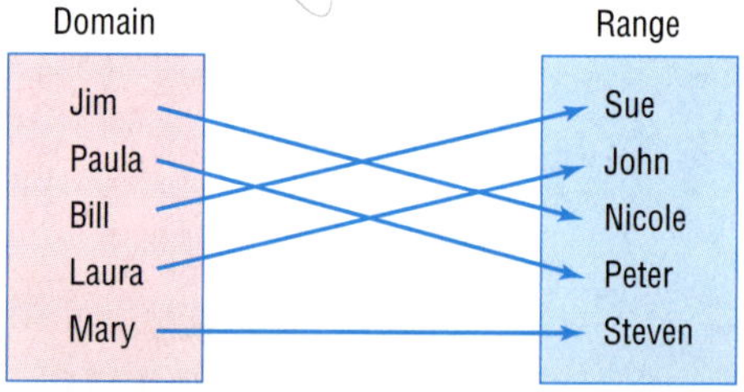

(c) $\{(-3,-27),(-2,-8),(-1,-1),(0,0),(1,1),(2,8),(3,27)\}$
(d) $\{(-3,9),(-2,4),(-1,1),(0,0),(1,1),(2,4),(3,9)\}$

Solution (a) The elements in the domain represent inputs to the function, and the elements in the range represent the outputs. To find the inverse, interchange the elements in the domain with the elements in the range. For example, the function receives as input Bill and outputs \$150. So, the inverse receives an input \$150 and outputs Bill. The inverse of the given function takes the form

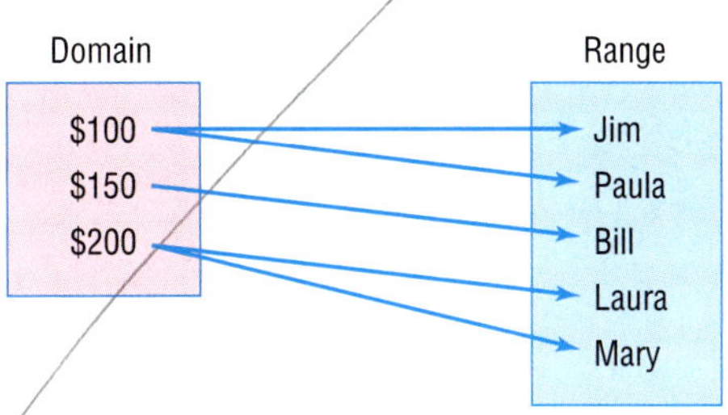

(b) The inverse of the given function is

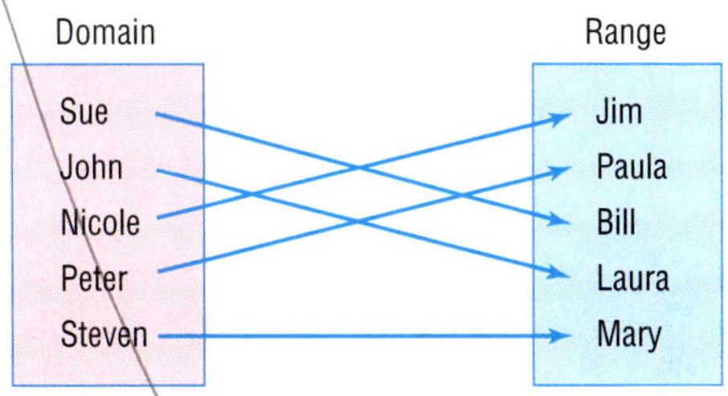

(c) The inverse of the given function is found by interchanging the entries in each ordered pair and so is given by

$$\{(-27,-3),\quad(-8,-2),\quad(-1,-1),\quad(0,0),\quad(1,1),\quad(8,2),\quad(27,3)\}$$

(d) The inverse of the given function is

$$\{(9,-3),\quad(4,-2),\quad(1,-1),\quad(0,0),\quad(1,1),\quad(4,2),\quad(9,3)\}$$

1

We notice that the inverses found in Examples 1(b) and (c) represent functions, since each element in the domain corresponds to a unique element in the range. The inverses found in Examples 1(a) and (d) do not represent functions, since each element in the domain does not correspond to a unique element in the range. Compare the function in Example 1(c) with the function in Example 1(d). For the function in Example 1(c), to every unique x-coordinate there corresponds a unique y-coordinate; for the function in Example 1(d), every unique x-coordinate does not correspond to a unique y-coordinate. Functions where unique x-coordinates correspond to unique y-coordinates are called *one-to-one* functions. So, in order for the inverse of a function f to be a function itself, f must be one-to-one.

A function f is said to be **one-to-one** if, for any choice of numbers x_1 and x_2, $x_1 \neq x_2$, in the domain of f, then $f(x_1) \neq f(x_2)$.

In other words, if f is a one-to-one function, then for each x in the domain of f there is exactly one y in the range, and no y in the range is the image of more than one x in the domain. See Figure 1.

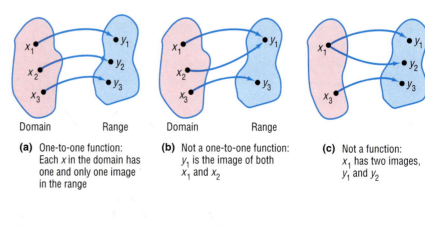

Figure 1

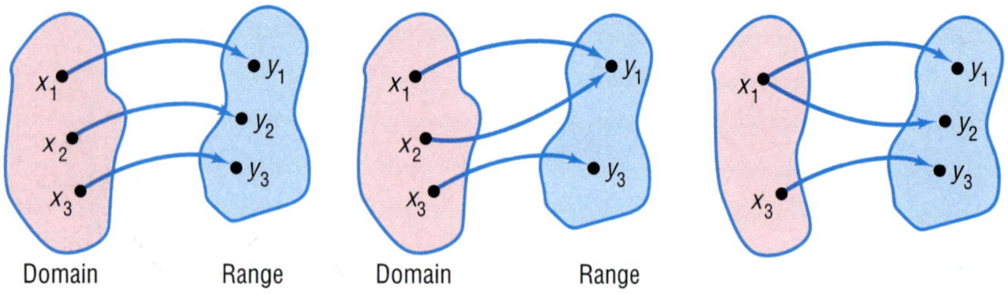

(a) One-to-one function: Each x in the domain has one and only one image in the range

(b) Not a one-to-one function: y_1 is the image of both x_1 and x_2

(c) Not a function: x_1 has two images, y_1 and y_2

NOW WORK PROBLEM 1.

If the graph of a function f is known, there is a simple test, called the **horizontal line test**, to determine whether f is one-to-one.

THEOREM

Horizontal Line Test

If every horizontal line intersects the graph of a function f in at most one point, then f is one-to-one.

Figure 2
$f(x_1) = f(x_2) = h$, but $x_1 \neq x_2$; f is not a one-to-one function.

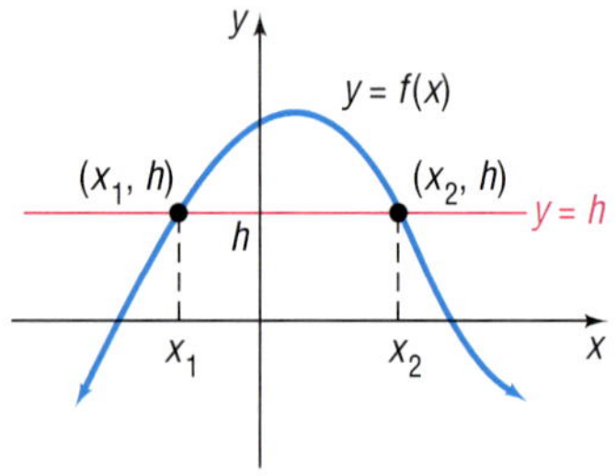

The reason this test works can be seen in Figure 2, where the horizontal line $y = h$ intersects the graph at two distinct points, (x_1, h) and (x_2, h). Since h is the image of both x_1, and x_2, $x_1 \neq x_2$, f is not one-to-one.

EXAMPLE 2 Using the Horizontal Line Test

For each function below, use the graph to determine whether the function is one-to-one.

(a) $f(x) = x^2$ (b) $g(x) = x^3$

Solution (a) Figure 3(a) illustrates the horizontal line test for $f(x) = x^2$. The horizontal line $y = 1$ intersects the graph of f twice, at $(1, 1)$ and at $(-1, 1)$, so f is not one-to-one.

(b) Figure 3(b) illustrates the horizontal line test for $g(x) = x^3$. Because each horizontal line will intersect the graph of g exactly once, it follows that g is one-to-one.

Figure 3

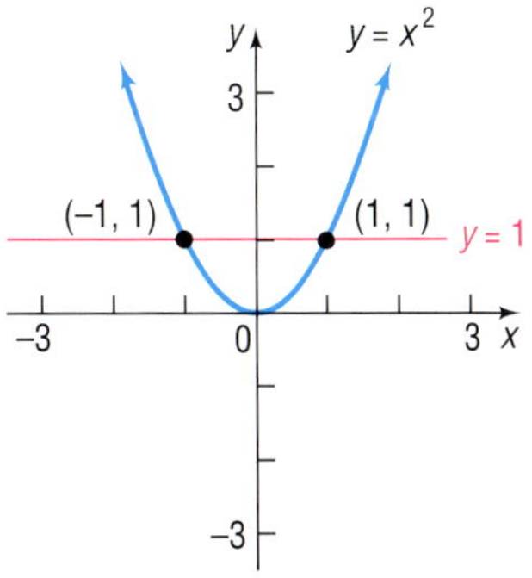

(a) A horizontal line intersects the graph twice; thus, f is not one-to-one

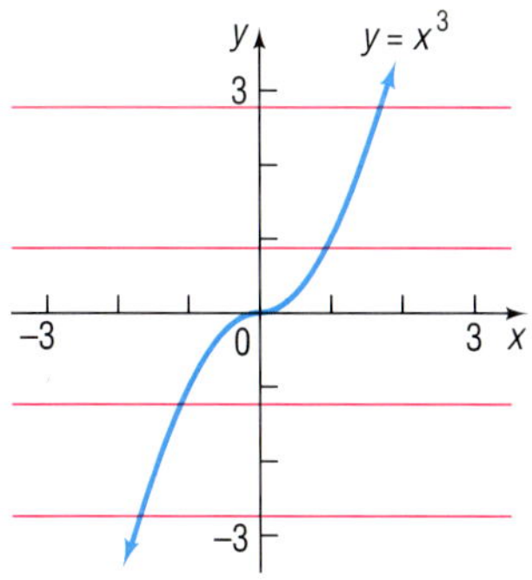

(b) Every horizontal line intersects the graph exactly once; thus, g is one-to-one

NOW WORK PROBLEM 9.

Let's look more closely at the one-to-one function $g(x) = x^3$. This function is an increasing function. Because an increasing (or decreasing) function will always have different y values for unequal x values, it follows that a function that is increasing (decreasing) on its domain is also a one-to-one function.

THEOREM An increasing (decreasing) function is a one-to-one function.

Inverse of a Function $y = f(x)$

For a function $y = f(x)$ to have an inverse *function*, f must be one-to-one. Then for each x in its domain there is exactly one y in its range; furthermore, to each y in the range, there corresponds exactly one x in the domain. The correspondence from the range of f onto the domain of f is, therefore, also a function. It is this function that is the *inverse of f*.

We mentioned in Chapter 2 that a function $y = f(x)$ can be thought of as a rule that tells us to do something to the argument x. For example, the function $f(x) = 2x$ multiplies the argument by 2. An *inverse function* of f undoes whatever f does. For example, the function $g(x) = \frac{1}{2}x$, which divides the argument by 2, is an inverse of $f(x) = 2x$. See Figure 4.

Figure 4

$f(x) = 2x$; $g(x) = \frac{1}{2}x$

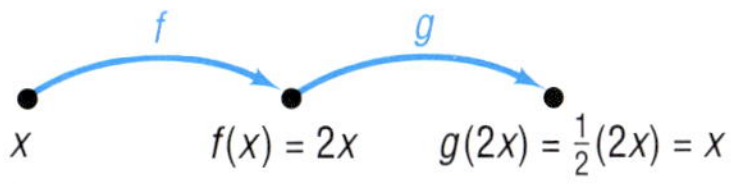

To put it another way, if we think of f as an input/output machine that processes an input x into $f(x)$, then the inverse function reverses this process, taking $f(x)$ back to x. A definition is given next.

Let f denote a one-to-one function $y = f(x)$. The **inverse of f,** denoted by f^{-1}, is a function such that $f^{-1}(f(x)) = x$ for every x in the domain of f and $f(f^{-1}(x)) = x$ for every x in the domain of f^{-1}.

Warning: Be careful! The -1 used in f^{-1} is not an exponent. Thus, f^{-1} does *not* mean the reciprocal of f; it means the inverse of f.

Figure 5

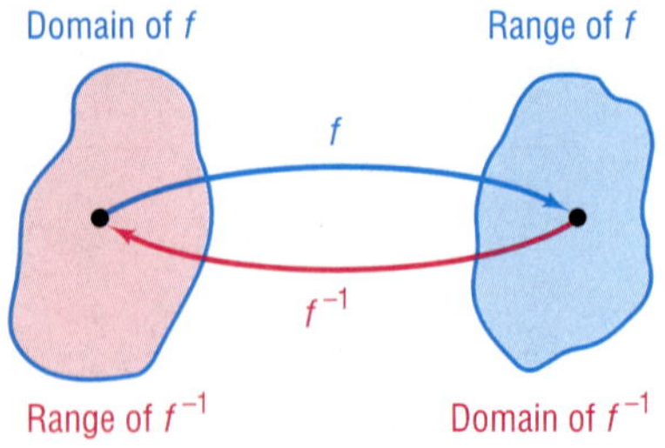

Figure 5 illustrates the definition.

Two facts are now apparent about a function f and its inverse f^{-1}.

Domain of f = Range of f^{-1}	Range of f = Domain of f^{-1}

Look again at Figure 5 to visualize the relationship. If we start with x, apply f, and then apply f^{-1}, we get x back again. If we start with x, apply f^{-1}, and then apply f, we get the number x back again. To put it simply, what f does, f^{-1} undoes, and vice versa:

Input x —Apply f→ $f(x)$ —Apply f^{-1}→ $f^{-1}(f(x)) = x$

Input x —Apply f^{-1}→ $f^{-1}(x)$ —Apply f→ $f(f^{-1}(x)) = x$

In other words,

$$f^{-1}(f(x)) = x \quad \text{and} \quad f(f^{-1}(x)) = x$$

The preceding conditions can be used to verify that a function is, in fact, the inverse of f, as Example 3 demonstrates.

EXAMPLE 3 Verifying Inverse Functions

(a) We verify that the inverse of $g(x) = x^3$ is $g^{-1}(x) = \sqrt[3]{x}$ by showing that

$$g^{-1}(g(x)) = g^{-1}(x^3) = \sqrt[3]{x^3} = x$$

and

$$g(g^{-1}(x)) = g(\sqrt[3]{x}) = (\sqrt[3]{x})^3 = x$$

(b) We verify that the inverse of $h(x) = 3x$ is $h^{-1}(x) = \frac{1}{3}x$ by showing that

$$h^{-1}(h(x)) = h^{-1}(3x) = \tfrac{1}{3}(3x) = x$$

and

$$h(h^{-1}(x)) = h(\tfrac{1}{3}x) = 3(\tfrac{1}{3}x) = x$$

(c) We verify that the inverse of $f(x) = 2x + 3$ is $f^{-1}(x) = \frac{1}{2}(x - 3)$ by showing that

$$f^{-1}(f(x)) = f^{-1}(2x + 3) = \tfrac{1}{2}[(2x + 3) - 3] = \tfrac{1}{2}(2x) = x$$

and

$$f(f^{-1}(x)) = f(\tfrac{1}{2}(x - 3)) = 2[\tfrac{1}{2}(x - 3)] + 3 = (x - 3) + 3 = x$$

EXPLORATION Simultaneously graph $Y_1 = x$, $Y_2 = x^3$, and $Y_3 = \sqrt[3]{x}$ on a square screen, using the viewing rectangle $-3 \le x \le 3$, $-2 \le y \le 2$. What do you observe about the graphs of $Y_2 = x^3$, its inverse $Y_3 = \sqrt[3]{x}$, and the line $Y_1 = x$?

Do you see the symmetry of the graph of Y_2 and its inverse Y_3 with respect to the line $Y_1 = x$?

NOW WORK PROBLEM 21.

2 Geometric Interpretation

For the functions in Example 3(c), we list points on the graph of $f = Y_1$ and on the graph of $f^{-1} = Y_2$ in Table 1.

Table 1

X	Y1
-5	-7
-4	-5
-3	-3
-2	-1
-1	1
0	3
1	5

Y1 = 2X+3

X	Y2
-7	-5
-5	-4
-3	-3
-1	-2
1	-1
3	0
5	1

Y2 = (1/2)(X−3)

Figure 6

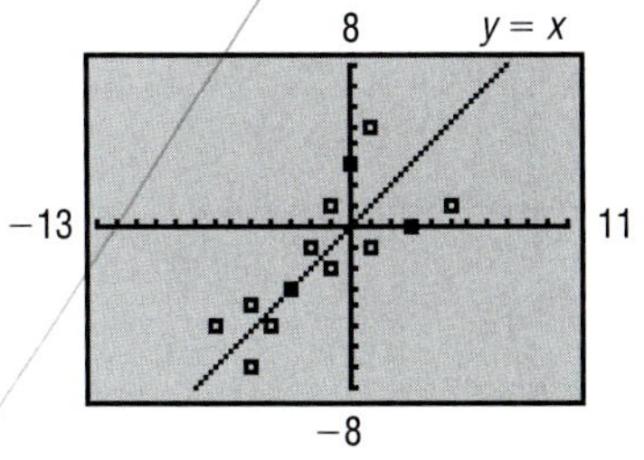

We notice that whenever (a, b) is on the graph of f then (b, a) is on the graph of f^{-1}. Figure 6 shows these points plotted. Also shown there is the graph of $y = x$, which you should observe is a line of symmetry of the points.

Figure 7

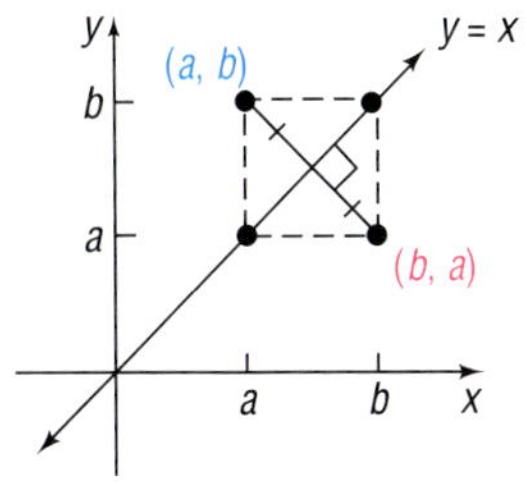

Suppose that (a, b) is a point on the graph of the one-to-one function f defined by $y = f(x)$. Then $b = f(a)$. This means that $a = f^{-1}(b)$, so (b, a) is a point on the graph of the inverse function f^{-1}. The relationship between the point (a, b) on f and the point (b, a) on f^{-1} is shown in Figure 7. The line containing (a, b) and (b, a) is perpendicular to the line $y = x$ and is bisected by the line $y = x$. (Do you see why?) It follows that the point (b, a) on f^{-1} is the reflection about the line $y = x$ of the point (a, b) on f.

THEOREM The graph of a function f and the graph of its inverse f^{-1} are symmetric with respect to the line $y = x$.

Figure 8 illustrates this result. Notice that, once the graph of f is known, the graph of f^{-1} may be obtained by reflecting the graph of f about the line $y = x$.

Figure 8

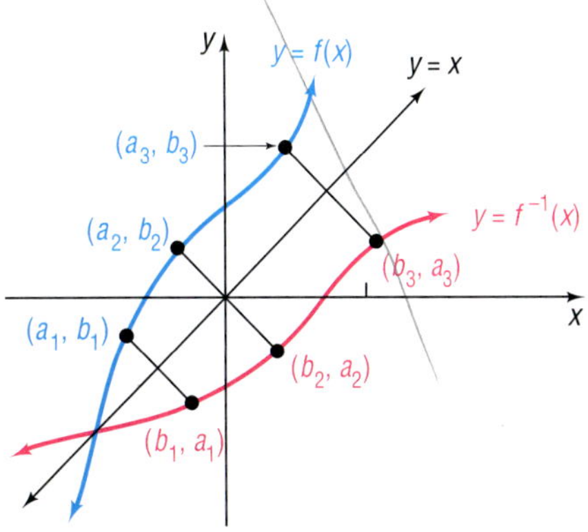

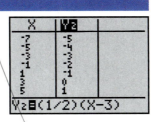

EXAMPLE 4 Graphing the Inverse Function

The graph in Figure 9(a) is that of a one-to-one function $y = f(x)$. Draw the graph of its inverse.

Solution We begin by adding the graph of $y = x$ to Figure 9(a). Since the points $(-2, -1)$, $(-1, 0)$, and $(2, 1)$ are on the graph of f, we know that the points $(-1, -2)$, $(0, -1)$, and $(1, 2)$ must be on the graph of f^{-1}. Keeping in mind that the graph of f^{-1} is the reflection about the line $y = x$ of the graph of f, we can draw f^{-1}. See Figure 9(b).

Figure 9

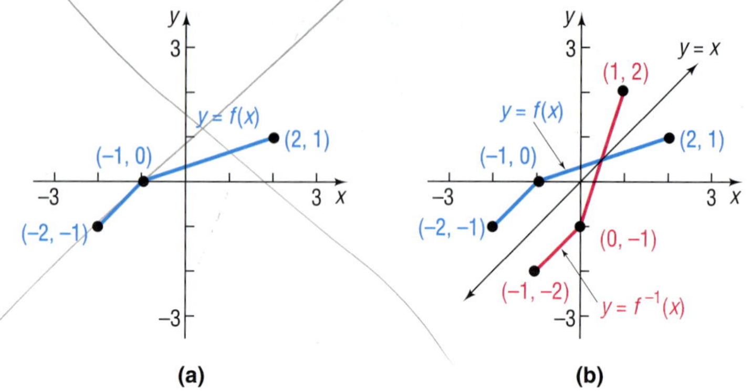

NOW WORK PROBLEM 15.

3 Finding the Inverse Function

The fact that the graph of a one-to-one function f and its inverse are symmetric with respect to the line $y = x$ tells us more. It says that we can obtain f^{-1} by interchanging the roles of x and y in f. Look again at Figure 8. If f is defined by the equation

$$y = f(x)$$

then f^{-1} is defined by the equation

$$x = f(y)$$

The equation $x = f(y)$ defines f^{-1} *implicitly*. If we can solve this equation for y, we will have the *explicit* form of f^{-1}, that is,

$$y = f^{-1}(x)$$

Let's use this procedure to find the inverse of $f(x) = 2x + 3$. (Since f is a linear function and is increasing, we know that f is one-to-one and, so, has an inverse function.)

EXAMPLE 5 Finding the Inverse Function

Find the inverse of $f(x) = 2x + 3$. Also find the domain and range of f and f^{-1}. Graph f and f^{-1} on the same coordinate axes.

Solution In the equation $y = 2x + 3$, interchange the variables x and y. The result,

$$x = 2y + 3$$

is an equation that defines the inverse f^{-1} implicitly. Solving for y, we obtain

$$\begin{aligned} 2y + 3 &= x \\ 2y &= x - 3 \\ y &= \tfrac{1}{2}(x - 3) \end{aligned}$$

The explicit form of the inverse f^{-1} is therefore

$$f^{-1}(x) = \tfrac{1}{2}(x - 3)$$

which we verified in Example 3(c).

Then we find

$$\begin{aligned} \text{Domain } f &= \text{Range } f^{-1} = (-\infty, \infty) \\ \text{Range } f &= \text{Domain } f^{-1} = (-\infty, \infty) \end{aligned}$$

The graphs of $Y_1 = f(x) = 2x + 3$ and its inverse $Y_2 = f^{-1}(x) = \frac{1}{2}(x - 3)$ are shown in Figure 10. Note the symmetry of the graphs with respect to the line $Y_3 = x$.

Figure 10

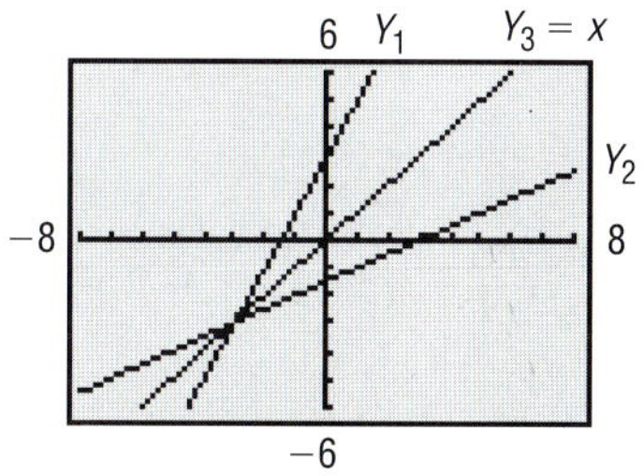

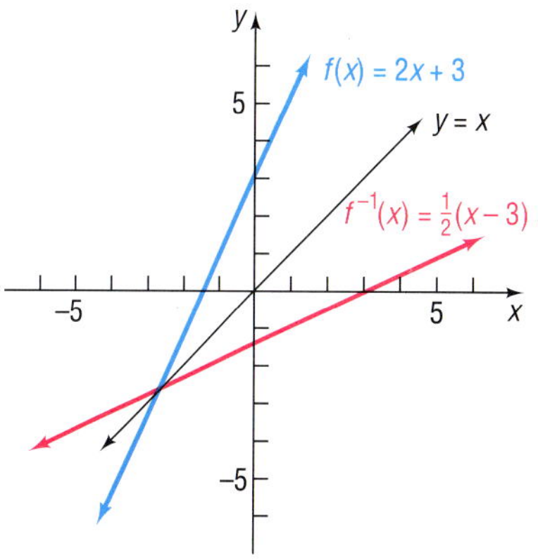

We outline next the steps to follow for finding the inverse of a one-to-one function.

Procedure for Finding the Inverse of a One-to-One Function

STEP 1: In $y = f(x)$, interchange the variables x and y to obtain

$$x = f(y)$$

This equation defines the inverse function f^{-1} implicitly.

STEP 2: If possible, solve the implicit equation for y in terms of x to obtain the explicit form of f^{-1}.

$$y = f^{-1}(x)$$

STEP 3: Check the result by showing that

$$f^{-1}(f(x)) = x \quad \text{and} \quad f(f^{-1}(x)) = x$$

EXAMPLE 6 Finding the Inverse Function

The function

$$f(x) = \frac{2x + 1}{x - 1}, \qquad x \neq 1$$

is one-to-one. Find its inverse and check the result.

Solution

STEP 1: Interchange the variables x and y in

$$y = \frac{2x + 1}{x - 1}$$

to obtain

$$x = \frac{2y + 1}{y - 1}$$

STEP 2: Solve for y.

$$\begin{aligned} x &= \frac{2y + 1}{y - 1} \\ x(y - 1) &= 2y + 1 && \text{Multiply both sides by } y - 1. \\ xy - x &= 2y + 1 && \text{Apply the distributive property.} \\ xy - 2y &= x + 1 && \text{Subtract 2y from both sides; add x to both sides.} \\ (x - 2)y &= x + 1 && \text{Factor.} \\ y &= \frac{x + 1}{x - 2} && \text{Divide by } x - 2. \end{aligned}$$

The inverse is

$$f^{-1}(x) = \frac{x + 1}{x - 2}, \qquad x \neq 2 \qquad \text{Replace } y \text{ by } f^{-1}(x).$$

STEP 3: ✓CHECK:

$$f^{-1}(f(x)) = f^{-1}\left(\frac{2x + 1}{x - 1}\right) = \frac{\frac{2x + 1}{x - 1} + 1}{\frac{2x + 1}{x - 1} - 2} = \frac{2x + 1 + x - 1}{2x + 1 - 2(x - 1)} = \frac{3x}{3} = x$$

$$f(f^{-1}(x)) = f\left(\frac{x + 1}{x - 2}\right) = \frac{2\left(\frac{x + 1}{x - 2}\right) + 1}{\frac{x + 1}{x - 2} - 1} = \frac{2(x + 1) + x - 2}{x + 1 - (x - 2)} = \frac{3x}{3} = x$$

EXPLORATION In Example 6, we found that, if $f(x) = (2x + 1)/(x - 1)$, then $f^{-1}(x) = (x + 1)/(x - 2)$. Compare the vertical and horizontal asymptotes of f and f^{-1}. What did you find? Are you surprised?

NOW WORK PROBLEM 33.

If a function is not one-to-one, then it will have no inverse function. Sometimes, though, an appropriate restriction on the domain of such a function will yield a new function that is one-to-one. Let's look at an example of this common practice.

EXAMPLE 7 Finding the Inverse Function

Find the inverse of $y = f(x) = x^2$ if $x \geq 0$.

Solution The function $f(x) = x^2$ is not one-to-one. [Refer to Example 2(a).] However, if we restrict f to only that part of its domain for which $x \geq 0$, as indicated, we have a new function that is increasing and therefore is one-to-one. As a result, the function defined by $y = f(x) = x^2$, $x \geq 0$, has an inverse function, f^{-1}.

We follow the steps given previously to find f^{-1}.

STEP 1: In the equation $y = x^2$, $x \geq 0$, interchange the variables x and y. The result is

$$x = y^2 \qquad y \geq 0$$

This equation defines (implicitly) the inverse function.

STEP 2: We solve for y to get the explicit form of the inverse. Since $y \geq 0$, only one solution for y is obtained.

$$y = \sqrt{x}$$

So $f^{-1}(x) = \sqrt{x}$.

STEP 3: ✓CHECK: $f^{-1}(f(x)) = f^{-1}(x^2) = \sqrt{x^2} = |x| = x$, since $x \geq 0$

$f(f^{-1}(x)) = f(\sqrt{x}) = (\sqrt{x})^2 = x$

Figure 11 illustrates the graphs of $Y_1 = f(x) = x^2$, $x \geq 0$, and $Y_2 = f^{-1}(x) = \sqrt{x}$.

Figure 11

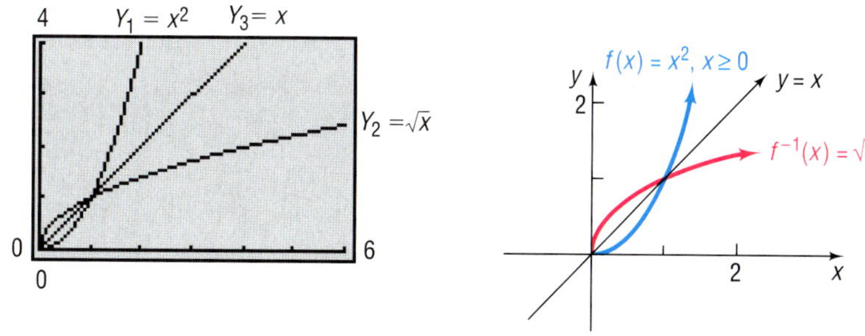

SUMMARY

1. If a function f is one-to-one, then it has an inverse function f^{-1}.
2. Domain f = Range f^{-1}; Range f = Domain f^{-1}.
3. To verify that f^{-1} is the inverse of f, show that $f^{-1}(f(x)) = x$ and $f(f^{-1}(x)) = x$.
4. The graphs of f and f^{-1} are symmetric with respect to the line $y = x$.

5.1 EXERCISES

In Problems 1–8, (a) find the inverse and (b) determine whether the inverse represents a function.

1.

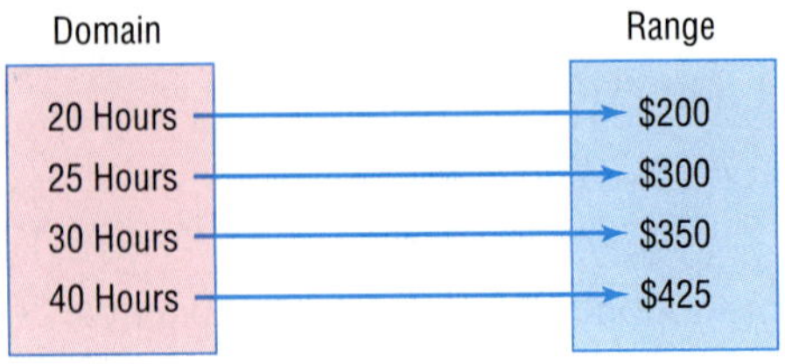

2.

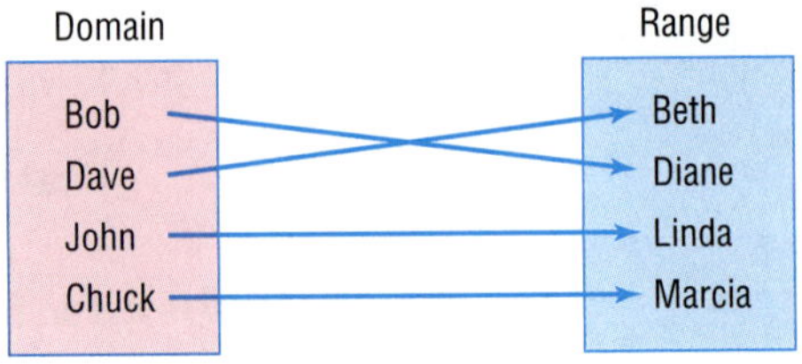

3.

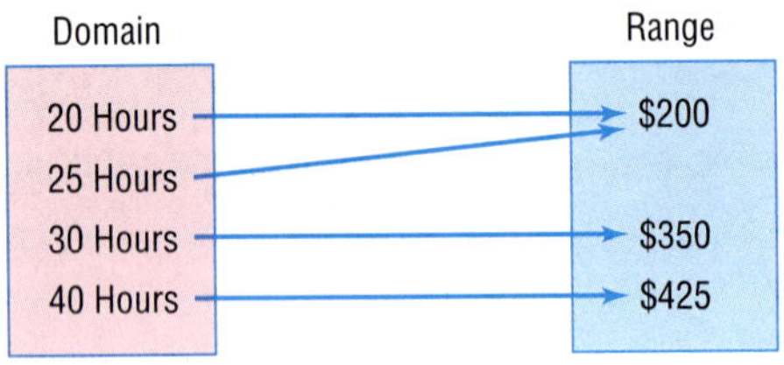

4.

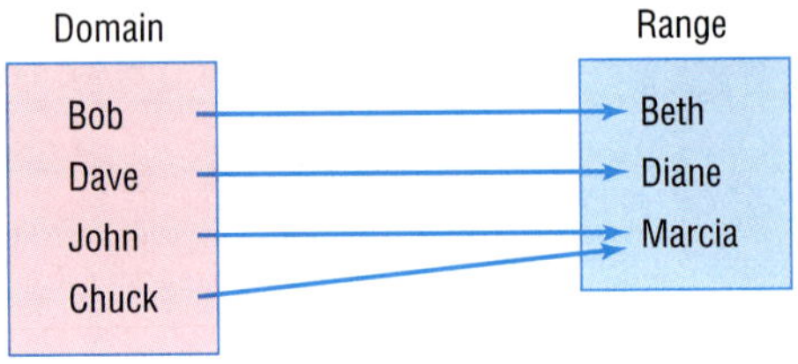

5. $\{(2, 6), (-3, 6), (4, 9), (1, 10)\}$

6. $\{(-2, 5), (-1, 3), (3, 7), (4, 12)\}$

7. $\{(0, 0), (1, 1), (2, 16), (3, 81)\}$

8. $\{(1, 2), (2, 8), (3, 18), (4, 32)\}$

In Problems 9–14, the graph of a function f is given. Use the horizontal line test to determine whether f is one-to-one.

9.

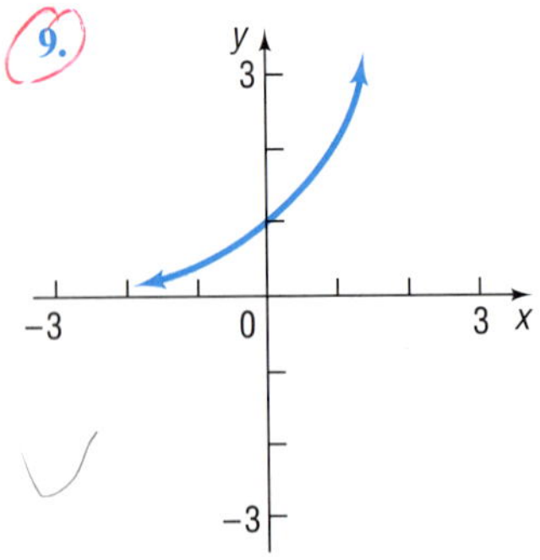

10.

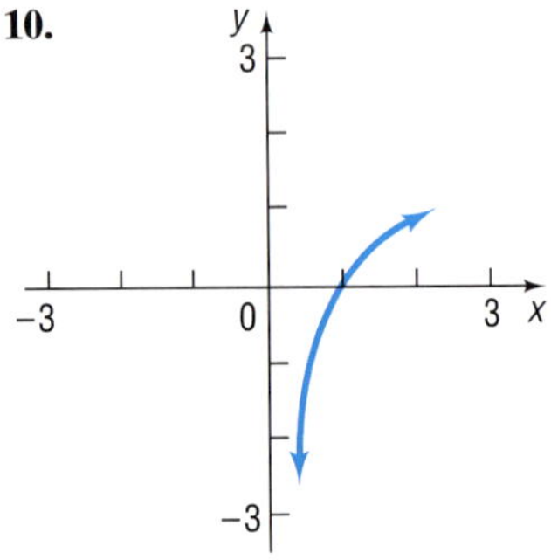

11.

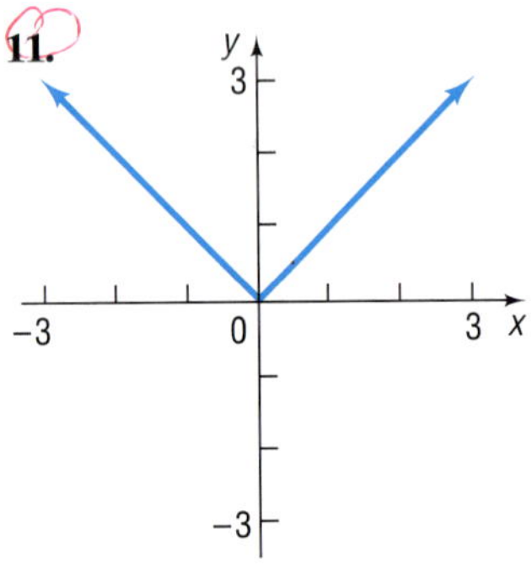

12.

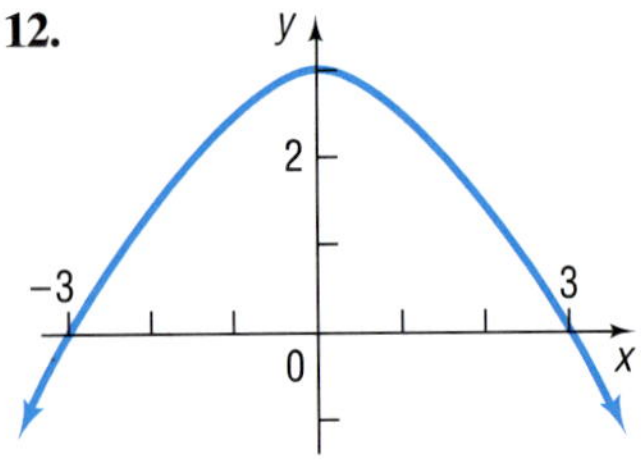

13.

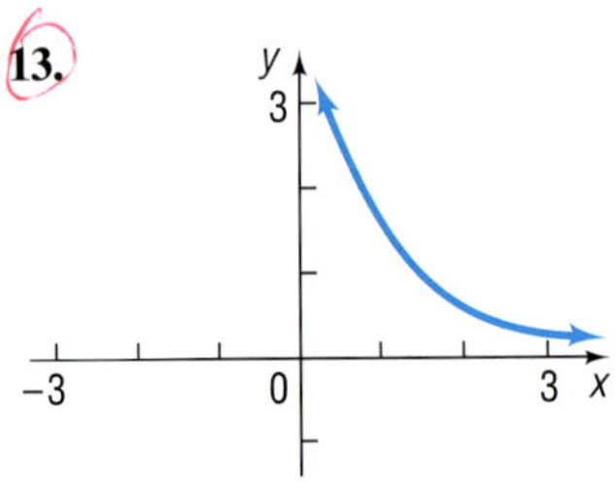

14.

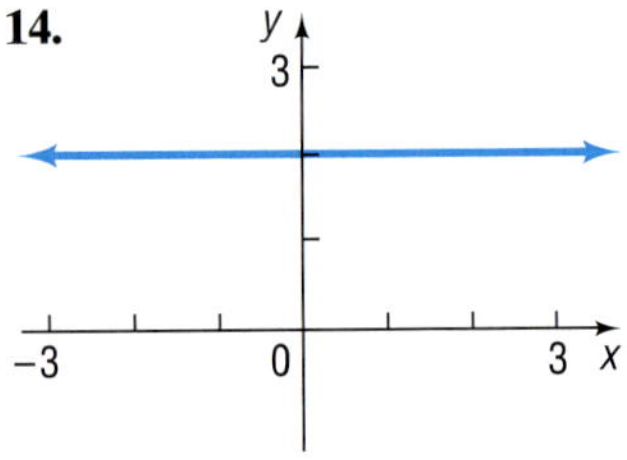

In Problems 15–20, the graph of a one-to-one function f is given. Draw the graph of the inverse function f^{-1}. For convenience (and as a hint), the graph of $y = x$ is also given.

15.

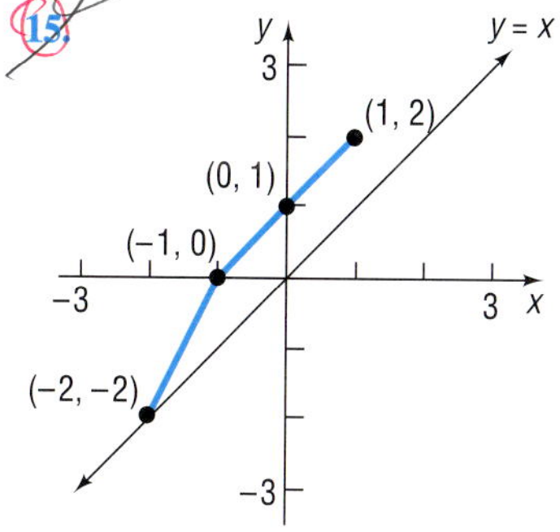

16.

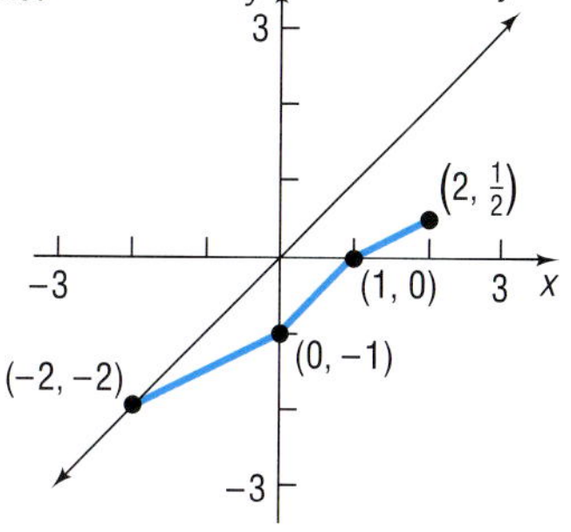

17.

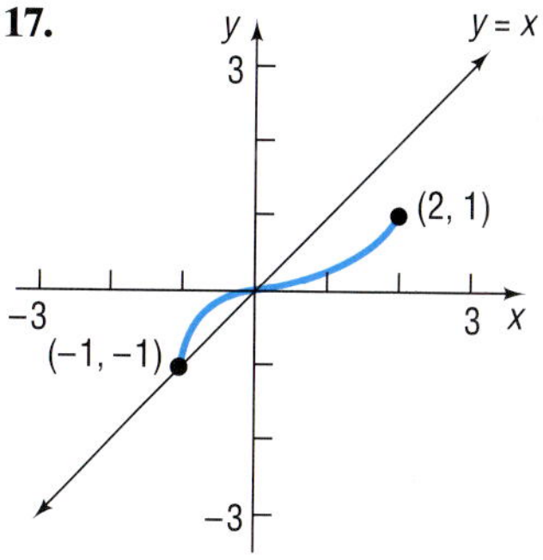

18.

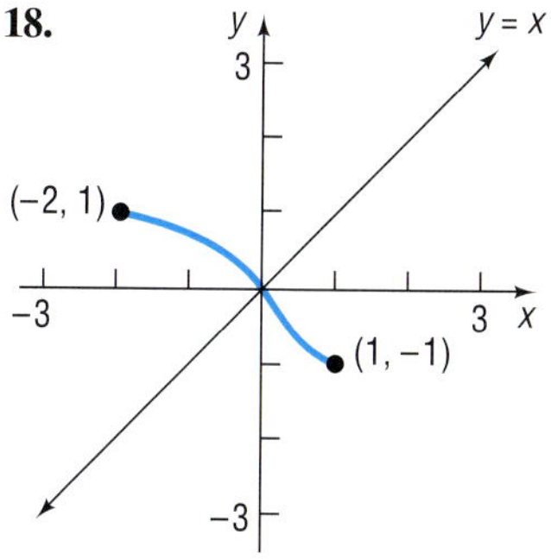

19.

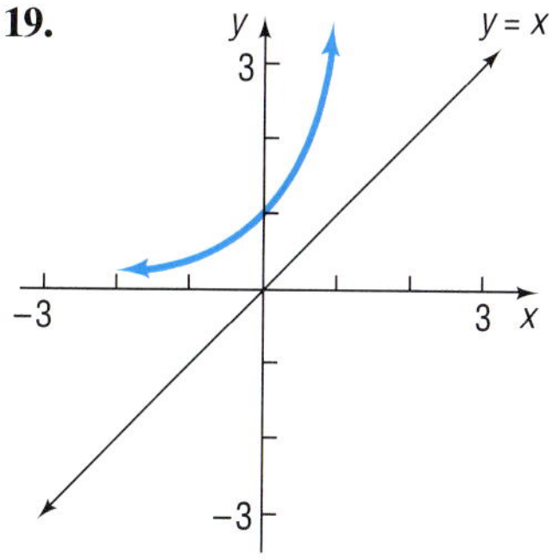

20.

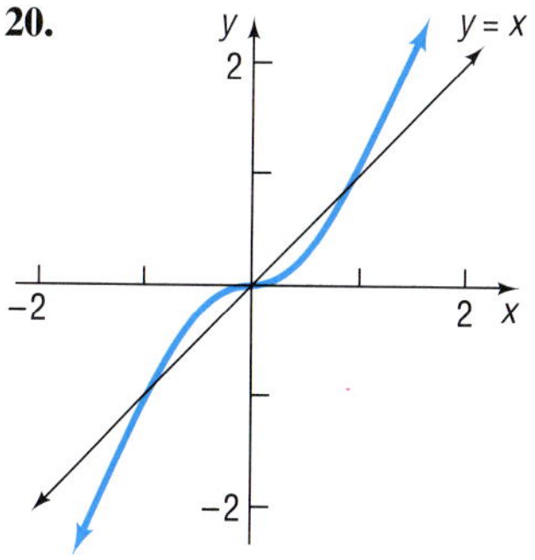

In Problems 21–30, verify that the functions f and g are inverses of each other by showing that $f(g(x)) = x$ and $g(f(x)) = x$. Using a graphing utility, simultaneously graph f, g, and $y = x$ on the same square screen.

21. $f(x) = 3x + 4; \quad g(x) = \frac{1}{3}(x - 4)$

22. $f(x) = 3 - 2x; \quad g(x) = -\frac{1}{2}(x - 3)$

23. $f(x) = 4x - 8; \quad g(x) = \frac{x}{4} + 2$

24. $f(x) = 2x + 6; \quad g(x) = \frac{1}{2}x - 3$

25. $f(x) = x^3 - 8; \quad g(x) = \sqrt[3]{x + 8}$

26. $f(x) = (x - 2)^2; \quad x \geq 2; \quad g(x) = \sqrt{x} + 2, x \geq 0$

27. $f(x) = \frac{1}{x}; \quad g(x) = \frac{1}{x}$

28. $f(x) = x; \quad g(x) = x$

29. $f(x) = \frac{2x + 3}{x + 4}; \quad g(x) = \frac{4x - 3}{2 - x}$

30. $f(x) = \frac{x - 5}{2x + 3}; \quad g(x) = \frac{3x + 5}{1 - 2x}$

In Problems 31–42, the function f is one-to-one. Find its inverse and check your answer. State the domain and range of f and f^{-1}. By hand, graph f, f^{-1}, and $y = x$ on the same coordinate axes. Check your results using a graphing utility.

31. $f(x) = 3x$

32. $f(x) = -4x$

33. $f(x) = 4x + 2$

34. $f(x) = 1 - 3x$

35. $f(x) = x^3 - 1$

36. $f(x) = x^3 + 1$

37. $f(x) = x^2 + 4, \quad x \geq 0$

38. $f(x) = x^2 + 9, \quad x \geq 0$

39. $f(x) = \frac{4}{x}$

40. $f(x) = -\frac{3}{x}$

41. $f(x) = \frac{1}{x - 2}$

42. $f(x) = \frac{4}{x + 2}$

In Problems 43–54, the function f is one-to-one. Find its inverse and check your answer. State the domain and range of f and f^{-1}. Using a graphing utility, simultaneously graph f, f^{-1}, and $y = x$ on the same square screen.

43. $f(x) = \dfrac{2}{3 + x}$

44. $f(x) = \dfrac{4}{2 - x}$

45. $f(x) = (x + 2)^2, \quad x \geq -2$

46. $f(x) = (x - 1)^2, \quad x \geq 1$

47. $f(x) = \dfrac{2x}{x - 1}$

48. $f(x) = \dfrac{3x + 1}{x}$

49. $f(x) = \dfrac{3x + 4}{2x - 3}$

50. $f(x) = \dfrac{2x - 3}{x + 4}$

51. $f(x) = \dfrac{2x + 3}{x + 2}$

52. $f(x) = \dfrac{-3x - 4}{x - 2}$

53. $f(x) = 2\sqrt[3]{x}$

54. $f(x) = \dfrac{4}{\sqrt{x}}$

55. Find the inverse of the linear function $f(x) = mx + b, m \neq 0$.

56. Find the inverse of the function $f(x) = \sqrt{r^2 - x^2}, 0 \leq x \leq r$.

57. Can an even function be one-to-one? Explain.

58. Is every odd function one-to-one? Explain.

59. A function f has an inverse function. If the graph of f lies in quadrant I, in which quadrant does the graph of f^{-1} lie?

60. A function f has an inverse function. If the graph of f lies in quadrant II, in which quadrant does the graph of f^{-1} lie?

61. The function $f(x) = |x|$ is not one-to-one. Find a suitable restriction on the domain of f so that the new function that results is one-to-one. Then find the inverse of f.

62. The function $f(x) = x^4$ is not one-to-one. Find a suitable restriction on the domain of f so that the new function that results is one-to-one. Then find the inverse of f.

63. Temperature Conversion To convert from x degrees Celsius to y degrees Fahrenheit, we use the formula $y = f(x) = \frac{9}{5}x + 32$. To convert from x degrees Fahrenheit to y degrees Celsius, we use the formula $y = g(x) = \frac{5}{9}(x - 32)$. Show that f and g are inverse functions.

64. Demand for Corn The demand for corn obeys the equation $p(x) = 300 - 50x$, where p is the price per bushel (in dollars) and x is the number of bushels produced, in millions. Express the production amount x as a function of the price p.

65. Period of a Pendulum The period T (in seconds) of a simple pendulum is a function of its length l (in feet), given by $T(l) = 2\pi\sqrt{l/g}$, where $g \approx 32.2$ feet per second per second is the acceleration of gravity. Express the length l as a function of the period T.

66. Give an example of a function whose domain is the set of real numbers and that is neither increasing nor decreasing on its domain, but is one-to-one.
[Hint: Use a piecewise-defined function.]

67. Given

$$f(x) = \frac{ax + b}{cx + d}$$

find $f^{-1}(x)$. If $c \neq 0$, under what conditions on a, b, c, and d is $f = f^{-1}$?

68. We said earlier that finding the range of a function f is not easy. However, if f is one-to-one, we can find its range by finding the domain of the inverse function f^{-1}. Use this technique to find the range of each of the following one-to-one functions:

(a) $f(x) = \dfrac{2x + 5}{x - 3}$

(b) $g(x) = 4 - \dfrac{2}{x}$

(c) $F(x) = \dfrac{3}{4 - x}$

69. If the graph of a function and its inverse intersect, where must this necessarily occur? Can they intersect anywhere else? Must they intersect?

70. Can a one-to-one function and its inverse be equal? What must be true about the graph of f for this to happen? Give some examples to support your conclusion.

71. Draw the graph of a one-to-one function that contains the points $(-2, -3)$, $(0, 0)$, and $(1, 5)$. Now draw the graph of its inverse. Compare your graph to those of other students. Discuss any similarities. What differences do you see?

5.2 EXPONENTIAL FUNCTIONS

1. Evaluate Exponential Functions
2. Graph Exponential Functions
3. Define the Number e

1 In the Appendix, Section 9, we give a definition for raising a real number a to a rational power. Based on that discussion, we gave meaning to expressions of the form

$$a^r$$

where the base a is a positive real number and the exponent r is a rational number.

But what is the meaning of a^x, where the base a is a positive real number and the exponent x is an irrational number? Although a rigorous definition requires methods discussed in calculus, the basis for the definition is easy to follow: Select a rational number r that is formed by truncating (removing) all but a finite number of digits from the irrational number x. Then it is reasonable to expect that

$$a^x \approx a^r$$

For example, take the irrational number $\pi = 3.14159\ldots$. Then, an approximation to a^π is

$$a^\pi \approx a^{3.14}$$

where the digits after the hundredths position have been removed from the value for π. A better approximation would be

$$a^\pi \approx a^{3.14159}$$

where the digits after the hundred-thousandths position have been removed. Continuing in this way, we can obtain approximations to a^π to any desired degree of accuracy.

Graphing calculators can easily evaluate expressions of the form a^x as follows. Enter the base a, then press the caret key (^), enter the exponent x, and press enter.

EXAMPLE 1 Using a Calculator to Evaluate Powers of 2

Using a calculator, evaluate:

(a) $2^{1.4}$ (b) $2^{1.41}$ (c) $2^{1.414}$ (d) $2^{1.4142}$ (e) $2^{\sqrt{2}}$

Solution Figure 12 shows the solution to part (a) using a TI-83 graphing calculator.

Figure 12

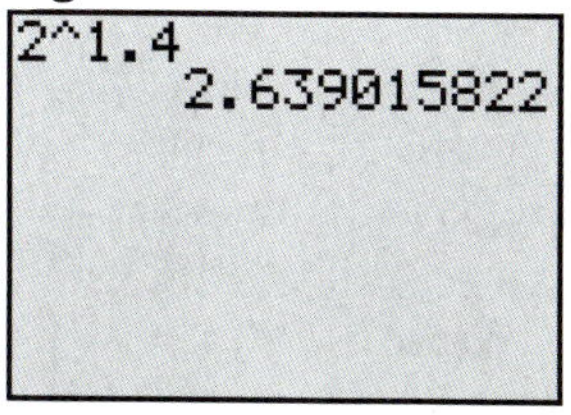

(a) $2^{1.4} \approx 2.639015822$ (b) $2^{1.41} \approx 2.657371628$
(c) $2^{1.414} \approx 2.66474965$ (d) $2^{1.4142} \approx 2.665119089$
(e) $2^{\sqrt{2}} \approx 2.665144143$

NOW WORK PROBLEM 1.

It can be shown that the familiar laws of rational exponents hold for real exponents.

THEOREM

Laws of Exponents

If s, t, a, and b are real numbers with $a > 0$ and $b > 0$, then

$$a^s \cdot a^t = a^{s+t} \qquad (a^s)^t = a^{st} \qquad (ab)^s = a^s \cdot b^s$$
$$1^s = 1 \qquad a^{-s} = \frac{1}{a^s} = \left(\frac{1}{a}\right)^s \qquad a^0 = 1 \tag{1}$$

We are now ready for the following definition:

An **exponential function** is a function of the form

$$f(x) = a^x$$

where a is a positive real number ($a > 0$) and $a \neq 1$. The domain of f is the set of all real numbers.

We exclude the base $a = 1$, because this function is simply the constant function $f(x) = 1^x = 1$. We also need to exclude bases that are negative, because, otherwise, we would have to exclude many values of x from the domain, such as $x = \frac{1}{2}$, $x = \frac{3}{4}$, and so on. [Recall that $(-2)^{1/2} = \sqrt{-2}$, $(-3)^{3/4} = \sqrt[4]{(-3)^3} = \sqrt[4]{-27}$, and so on, are not defined in the system of real numbers.]

Graphs of Exponential Functions

2 First, we graph the exponential function $f(x) = 2^x$.

EXAMPLE 2 Graphing an Exponential Function

Graph the exponential function $f(x) = 2^x$.

Solution The domain of $f(x) = 2^x$ consists of all real numbers. We begin by locating some points on the graph of $f(x) = 2^x$, as listed in Table 2.

Since $2^x > 0$ for all x, the range of f is $(0, \infty)$. From this, we conclude that the graph has no x-intercepts, and, in fact, the graph will lie above the x-axis. As Table 2 indicates, the y-intercept is 1. Table 2 also indicates that as $x \to -\infty$ the value of $f(x) = 2^x$ gets closer and closer to 0. Thus, the x-axis is a horizontal asymptote to the graph as $x \to -\infty$.

To determine the end behavior for x large and positive, look again at Table 2. As $x \to \infty$, $f(x) = 2^x$ grows very quickly, causing the graph of $f(x) = 2^x$ to rise very rapidly. Thus, it is apparent that f is an increasing function and hence is one-to-one.

Figure 13 shows the graph of $f(x) = 2^x$. Notice that all the conclusions given earlier are confirmed by the graph.

Table 2

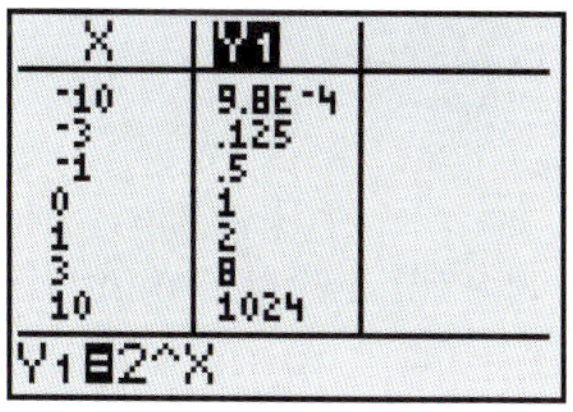

X	Y1
-10	9.8E-4
-3	.125
-1	.5
0	1
1	2
3	8
10	1024

Y1=2^X

Figure 13

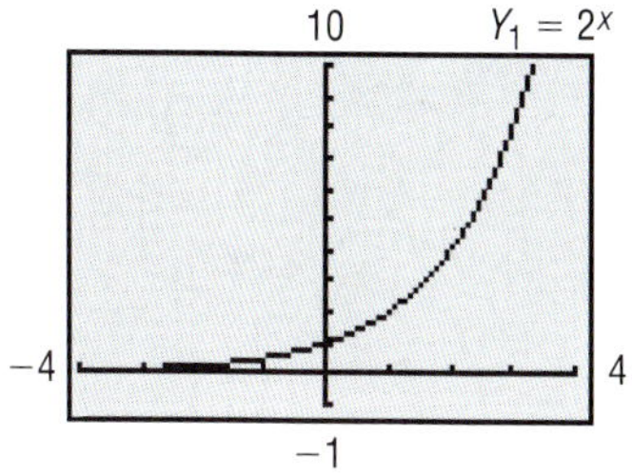

As we shall see, graphs that look like the one in Figure 13 occur very frequently in a variety of situations. For example, look at the graph in Figure 14, which illustrates the closing price of a share of Dell Computer stock. Investors might conclude from this graph that the price of Dell Computer is *behaving exponentially;* that is, the graph exhibits rapid, or exponential, growth. We shall have more to say about situations that lead to exponential growth later in this chapter. For now, we continue to seek properties of the exponential functions.

Figure 14

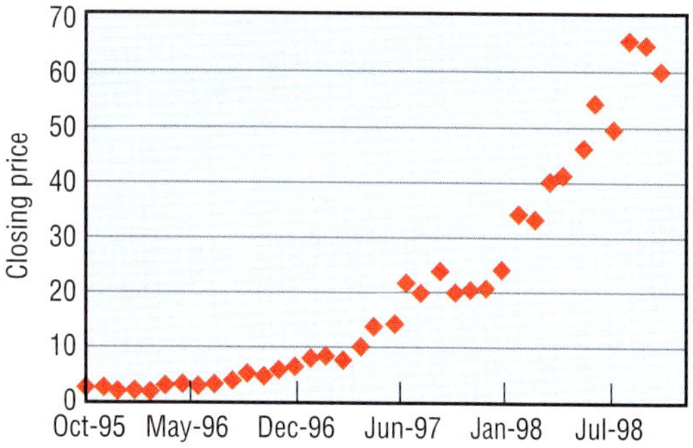

The graph of $f(x) = 2^x$ in Figure 13 is typical of all exponential functions that have a base larger than 1. Such functions are increasing functions and hence are one-to-one. Their graphs lie above the x-axis, pass through the point $(0, 1)$, and thereafter rise rapidly as $x \to \infty$. As $x \to -\infty$, the x-axis $(y = 0)$ is a horizontal asymptote. There are no vertical asymptotes. Finally, the graphs are smooth and continuous, with no corners or gaps.

Figure 15 illustrates the graphs of two more exponential functions whose bases are larger than 1. Notice that for the larger base the graph is steeper when $x > 0$. Figure 16 shows that when $x < 0$ the graph of the equation with the larger base is closer to the x-axis.

Figure 15

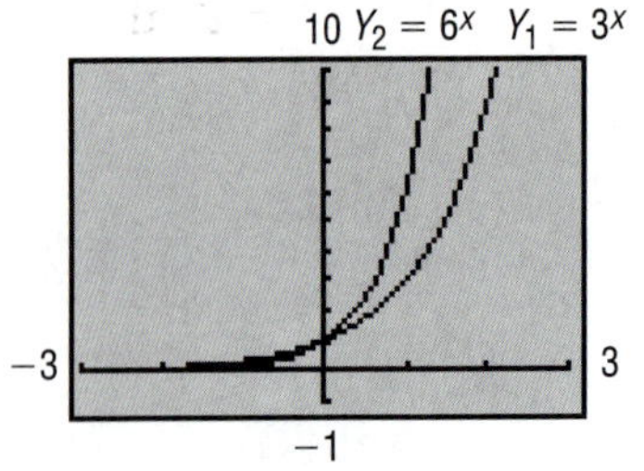

Figure 16

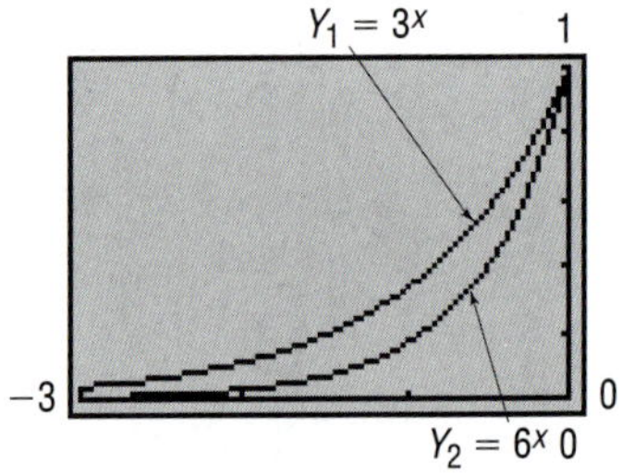

The following display summarizes the information that we have about $f(x) = a^x$, $a > 1$.

Facts About the Graph of an Exponential Function $f(x) = a^x$, $a > 1$

1. The domain is all real numbers; the range is the set of positive real numbers.
2. There are no x-intercepts; the y-intercept is 1.
3. The x-axis ($y = 0$) is a horizontal asymptote as $x \to -\infty$.
4. $f(x) = a^x$, $a > 1$, is an increasing function and is one-to-one.
5. The graph of f contains the points $(0, 1)$ and $(1, a)$.
6. The graph of f is smooth and continuous, with no corners or gaps.

Now we consider $f(x) = a^x$ when $0 < a < 1$.

EXAMPLE 3 Graphing an Exponential Function

Graph the exponential function $f(x) = (\frac{1}{2})^x$.

Solution The domain of $f(x) = (\frac{1}{2})^x$ consists of all real numbers. As before, we locate some points on the graph, as listed in Table 3. Since $(\frac{1}{2})^x > 0$ for all x, the range of f is $(0, \infty)$. Thus, the graph lies above the x-axis and so has no x-intercepts. The y-intercept is 1. As $x \to -\infty$, $f(x) = (\frac{1}{2})^x$ grows very quickly. As $x \to \infty$, the values of $f(x)$ approach 0. Thus, the x-axis ($y = 0$) is a horizontal asymptote as $x \to \infty$. It is apparent that f is a decreasing function and hence is one-to-one. Figure 17 illustrates the graph.

Table 3

X	Y1
-10	1024
-3	8
-1	2
0	1
1	.5
3	.125
10	9.8E-4

Y1=(1/2)^X

Figure 17

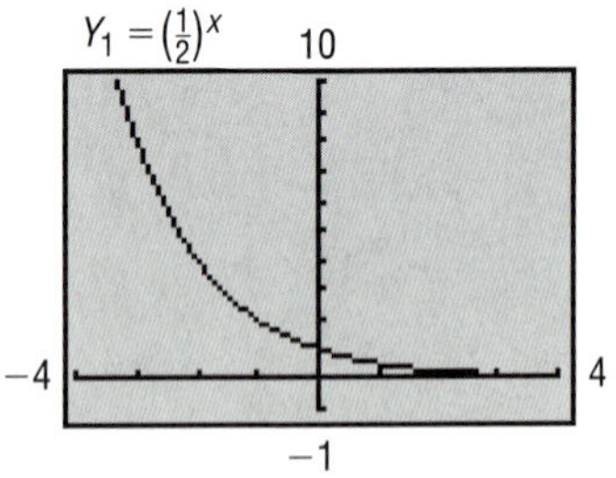

Note that we could have obtained the graph of $y = (\frac{1}{2})^x$ from the graph of $y = 2^x$. If $f(x) = 2^x$, then $f(-x) = 2^{-x} = \frac{1}{2^x} = (\frac{1}{2})^x$. Thus, the graph of $y = (\frac{1}{2})^x = 2^{-x}$ is a reflection about the y-axis of the graph of $y = 2^x$. Compare Figures 13 and 17.

SEEING THE CONCEPT Using a graphing utility, simultaneously graph:

(a) $Y_1 = 3^x, Y_2 = \left(\frac{1}{3}\right)^x$ (b) $Y_1 = 6^x, Y_2 = \left(\frac{1}{6}\right)^x$

Conclude that the graph of $Y_2 = \left(\frac{1}{a}\right)^x$ for $a > 0$ is the reflection about the y-axis of the graph of $Y_1 = a^x$.

The graph of $f(x) = (\frac{1}{2})^x$ in Figure 17 is typical of all exponential functions that have a base between 0 and 1. Such functions are decreasing one-to-one functions. Their graphs lie above the x-axis and pass through the point $(0, 1)$. The graphs rise rapidly as $x \to -\infty$. As $x \to \infty$, the x-axis is a horizontal asymptote. There are no vertical asymptotes. Finally, the graphs are smooth and continuous, with no corners or gaps.

Figure 18 illustrates the graphs of two more exponential functions whose bases are between 0 and 1. Notice that the smaller base results in a graph that is steeper when $x < 0$. Figure 19 shows that when $x > 0$ the graph of the equation with the smaller base is closer to the x-axis.

Figure 18

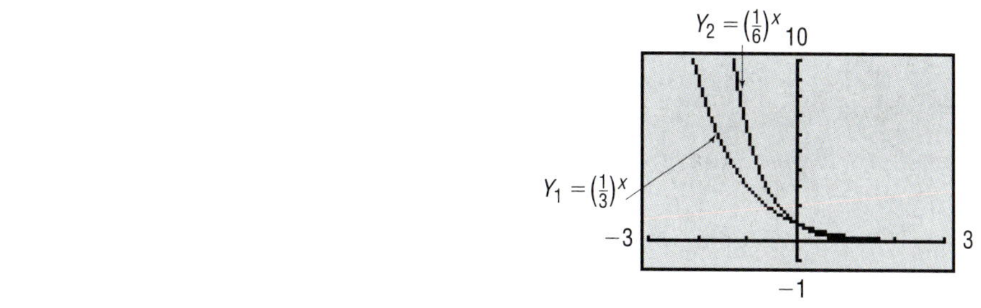

Figure 19

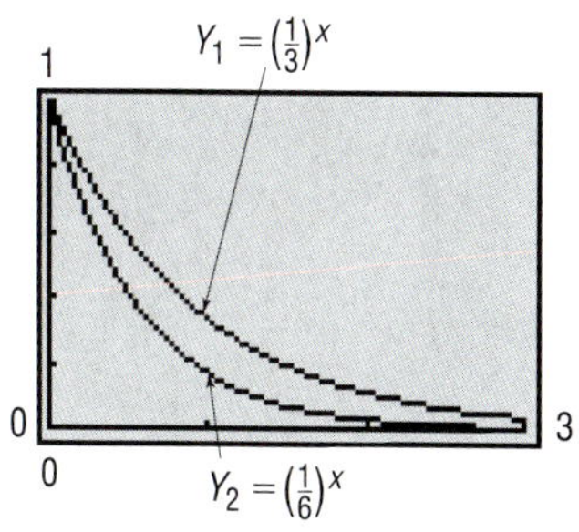

The following display summarizes the information that we have about the function $f(x) = a^x, 0 < a < 1$.

Facts About the Graph of an Exponential Function $f(x) = a^x$, $0 < a < 1$

1. The domain is all real numbers; the range is the set of positive real numbers.
2. There are no x-intercepts; the y-intercept is 1.
3. The x-axis ($y = 0$) is a horizontal asymptote as $x \to \infty$
4. $f(x) = a^x$, $0 < a < 1$, is a decreasing function and is one-to-one.
5. The graph of f contains the points $(0, 1)$ and $(1, a)$.
6. The graph of f is smooth and continuous, with no corners or gaps.

EXAMPLE 4 Graphing Exponential Functions Using Transformations

Graph $f(x) = 2^{-x} - 3$ and determine the domain, range, and horizontal asymptote of f.

Solution Figure 20 shows the various steps.

Figure 20

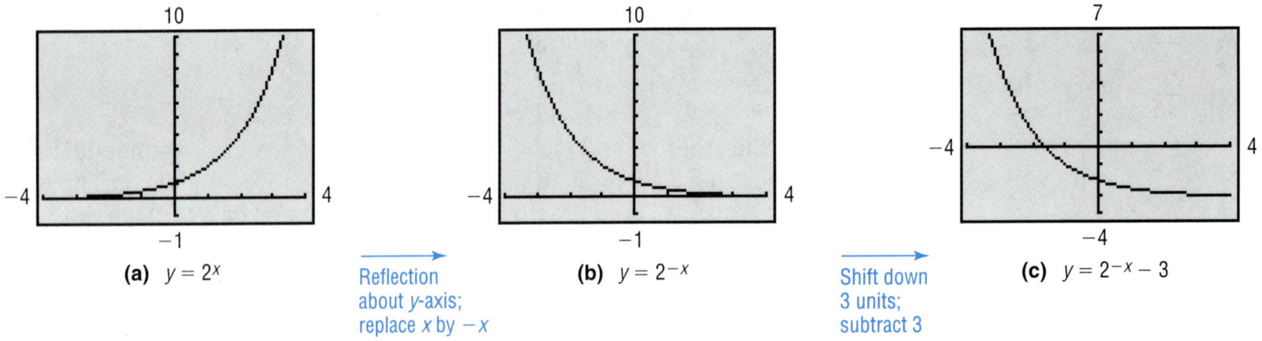

As Figure 20(c) illustrates, the domain of $f(x) = 2^{-x} - 3$ is $(-\infty, \infty)$ and the range is $(-3, \infty)$. The horizontal asymptote of f is the line $y = -3$.

EXAMPLE 5 Graphing Exponential Functions Using Transformations

Graph $f(x) = -(2^{x-3})$ and determine the domain, range, and horizontal asymptote of f.

Solution Figure 21 shows the various steps.

Figure 21

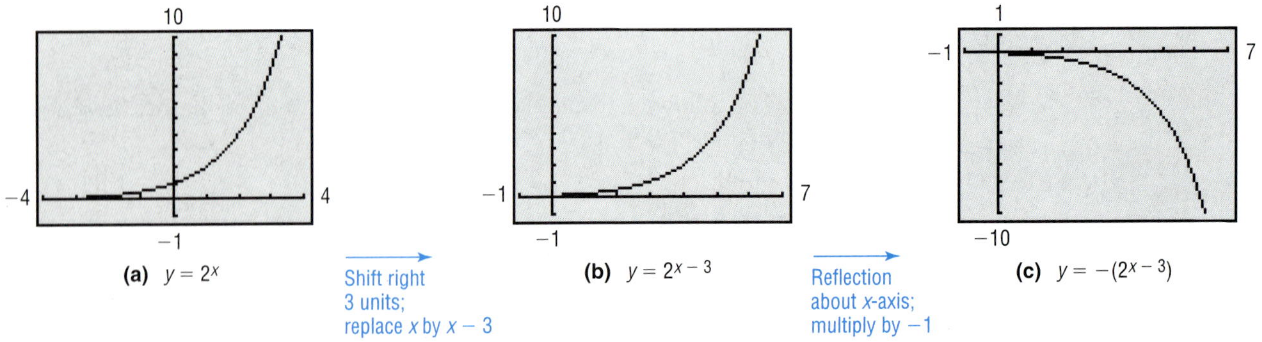

The domain of $f(x) = -(2^{x-3})$ is $(-\infty, \infty)$ and the range is $(-\infty, 0)$. The horizontal asymptote of f is $y = 0$.

NOW WORK PROBLEMS 11 AND 13.

The Base e

3 As we shall see shortly, many problems that occur in nature require the use of an exponential function whose base is a certain irrational number, symbolized by the letter e.

Let's look now at one way of arriving at this important number e.

The **number e** is defined as the number that the expression

$$\left(1 + \frac{1}{n}\right)^n \tag{2}$$

approaches as $n \to \infty$. In calculus, this is expressed using limit notation as

$$e = \lim_{n\to\infty}\left(1 + \frac{1}{n}\right)^n$$

Table 4 illustrates what happens to the defining expression (2) as n takes on increasingly large values. The last number in the last column in the table is correct to nine decimal places and is the same as the entry given for e on your calculator (if expressed correct to nine decimal places).

Table 4

n	$\frac{1}{n}$	$1 + \frac{1}{n}$	$\left(1 + \frac{1}{n}\right)^n$
1	1	2	2
2	0.5	1.5	2.25
5	0.2	1.2	2.48832
10	0.1	1.1	2.59374246
100	0.01	1.01	2.704813829
1,000	0.001	1.001	2.716923932
10,000	0.0001	1.0001	2.718145927
100,000	0.00001	1.00001	2.718268237
1,000,000	0.000001	1.000001	2.718280469
1,000,000,000	10^{-9}	$1 + 10^{-9}$	2.718281827

The exponential function $f(x) = e^x$, whose base is the number e, occurs with such frequency in applications that it is usually referred to as *the* exponential function. Indeed, graphing calculators have the key **e^x** or **exp(x)**, which may be used to evaluate the exponential function for a given value of x. Now use your calculator to find e^x for $x = -2$, $x = -1$, $x = 0$, $x = 1$, and $x = 2$, as we have done to create Table 5. The graph of the exponential function $f(x) = e^x$ is given in Figure 22(a). Since $2 < e < 3$, the graph of $y = e^x$ lies between the graphs of $y = 2^x$ and $y = 3^x$. (See Figure 22(b)).

Table 5

X	Y1
-2	.13534
-1	.36788
0	1
1	2.7183
2	7.3891

Y1=e^(X)

Figure 22

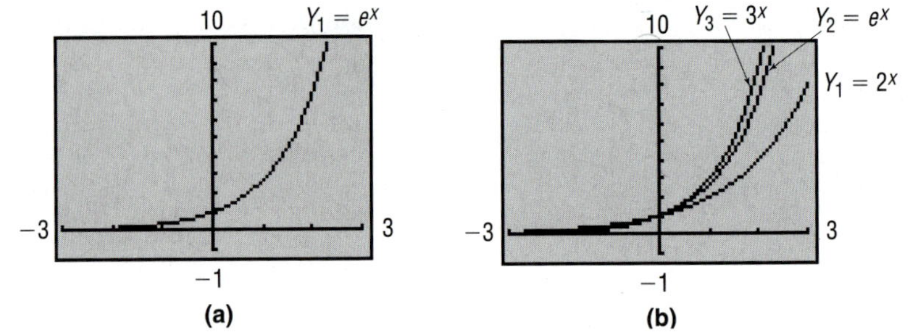

NOW WORK PROBLEM 25.

Many applications involve the exponential functions. Let's look at one.

EXAMPLE 6 Exponential Probability

Between 9:00 PM and 10:00 PM cars arrive at Burger King's drive-thru at the rate of 12 cars per hour (0.2 car per minute). The following formula from statistics can be used to determine the probability that a car will arrive within t minutes of 9:00 PM.

$$F(t) = 1 - e^{-0.2t}$$

(a) Determine the probability that a car will arrive within 5 minutes of 9 PM (that is, before 9:05 PM).

(b) Determine the probability that a car will arrive within 30 minutes of 9 PM (before 9:30 PM).

(c) Graph F using your graphing utility.

(d) Determine how many minutes are needed for the probability to reach 75%.

(e) What value does F approach as t becomes unbounded in the positive direction?

Solution (a) We have $t = 5$. The probability that a car will arrive within 5 minutes is

$$F(5) = 1 - e^{-0.2(5)}$$

We evaluate this expression in Figure 23. We conclude that there is a 63% probability that a car will arrive within 5 minutes.

Figure 23

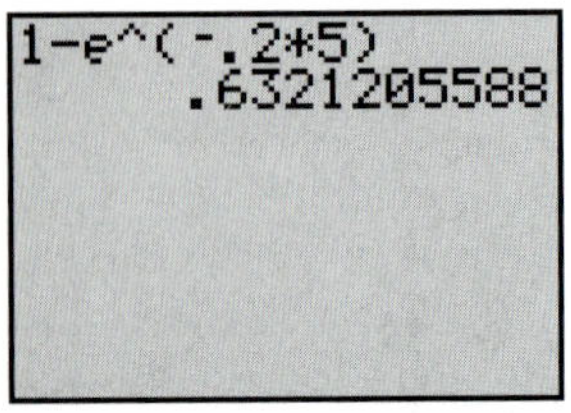

(b) We have $t = 30$. The probability that a car will arrive within 30 minutes is

$$F(30) = 1 - e^{-0.2(30)} \approx 0.9975$$

There is a 99.75% probability that a car will arrive within 30 minutes.

Figure 24

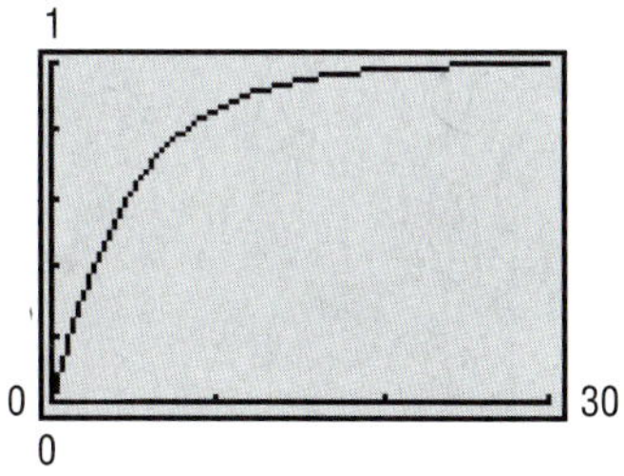

(c) See Figure 24.

(d) Graph $Y_1 = 1 - e^{-0.2x}$, $Y_2 = 0.75$. Using INTERSECT, we find $x = 6.93$. It takes about 7 minutes for the probability to equal 75%. Thus, there is a 75% probability that a car will arrive before 9:07 PM.

(e) As time passes, the probability that a car will arrive increases. The value F that approaches can be found by letting $t \to \infty$. Since $e^{-0.2t} = 1/e^{0.2t}$, it follows that $e^{-0.2t} \to 0$ as $t \to \infty$. Thus, F approaches 1 as t gets large.

NOW WORK PROBLEM **43.**

SUMMARY

Properties of the Exponential Function

$f(x) = a^x, a > 1$	Domain: $(-\infty, \infty)$; Range: $(0, \infty)$; x-intercepts: none; y-intercept: 1; horizontal asymptote: x-axis as $x \to -\infty$; increasing; one-to-one; smooth; continuous. See Figure 13 for a typical graph.
$f(x) = a^x, 0 < a < 1$	Domain: $(-\infty, \infty)$; Range: $(0, \infty)$; x-intercepts: none; y-intercept: 1; horizontal asymptote: x-axis as $x \to \infty$; decreasing; one-to-one; smooth; continuous. See Figure 17 for a typical graph.

5.2 EXERCISES

In Problems 1–10, approximate each number using a calculator. Express your answer rounded to three decimal places.

1. (a) $3^{2.2}$ (b) $3^{2.23}$ (c) $3^{2.236}$ (d) $3^{\sqrt{5}}$

2. (a) $5^{1.7}$ (b) $5^{1.73}$ (c) $5^{1.732}$ (d) $5^{\sqrt{3}}$

3. (a) $2^{3.14}$ (b) $2^{3.141}$ (c) $2^{3.1415}$ (d) 2^{π}

4. (a) $2^{2.7}$ (b) $2^{2.71}$ (c) $2^{2.718}$ (d) 2^{e}

5. (a) $3.1^{2.7}$ (b) $3.14^{2.71}$ (c) $3.141^{2.718}$ (d) π^{e}

6. (a) $2.7^{3.1}$ (b) $2.71^{3.14}$ (c) $2.718^{3.141}$ (d) e^{π}

7. $e^{1.2}$ **8.** $e^{-1.3}$ **9.** $e^{-0.85}$ **10.** $e^{2.1}$

In Problems 11–18, the graph of an exponential function is given. Match each graph to one of the following functions:

A. $y = 3^x$	*B.* $y = 3^{-x}$	*C.* $y = -3^x$	*D.* $y = -3^{-x}$
E. $y = 3^x - 1$	*F.* $y = 3^{x-1}$	*G.* $y = 3^{1-x}$	*H.* $y = 1 - 3^x$

11.

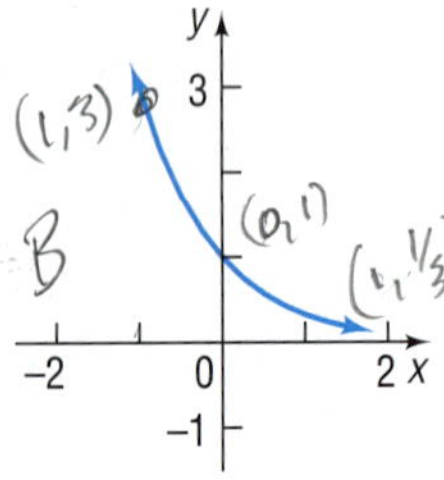

12.

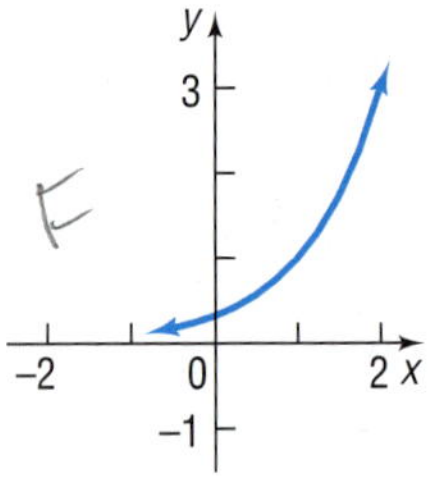

13.

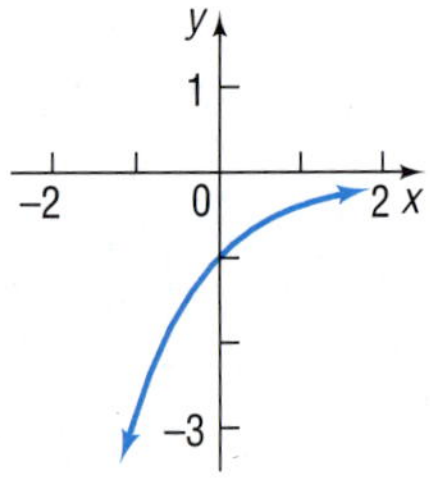

14.

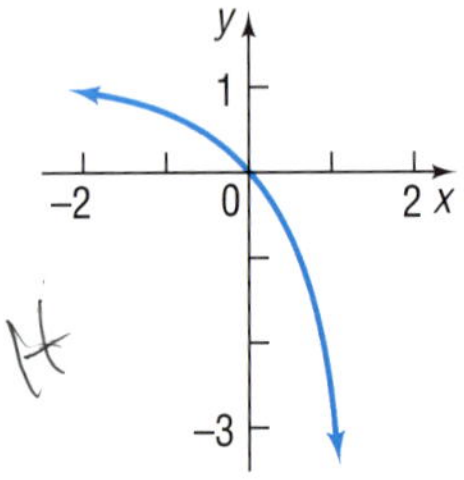

15.

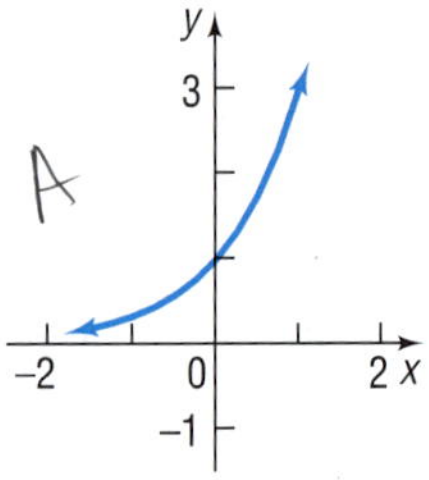

16.

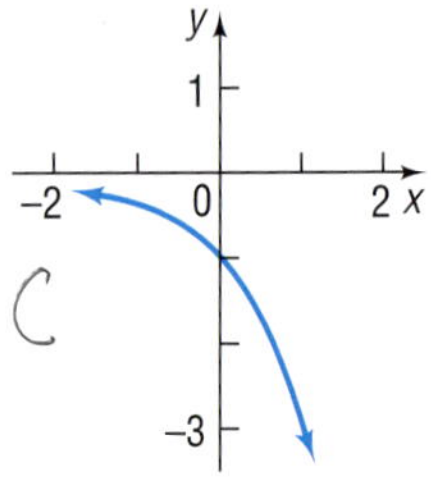

17.

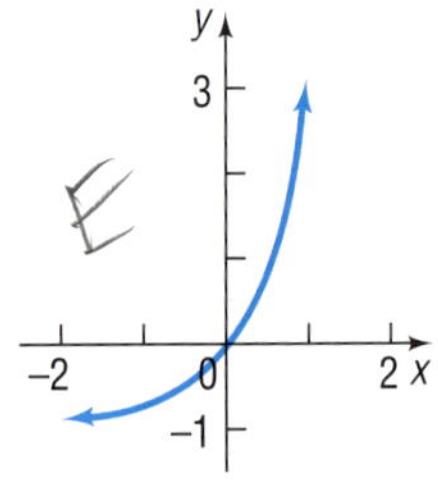

18.

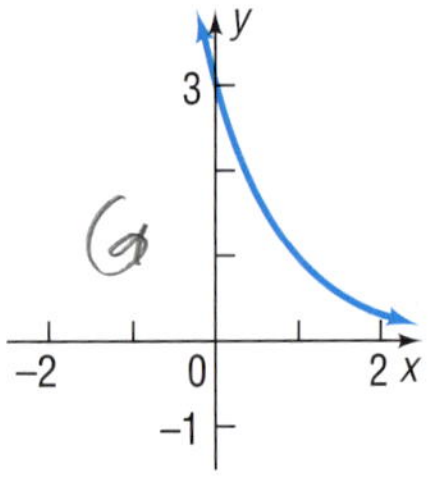

In Problems 19–24, the graph of an exponential function is given. Match each graph to one of the following functions:

A. $y = 4^x$	*B.* $y = 4^{-x}$	*C.* $y = 4^{x-1}$	*D.* $y = 4^x - 1$
E. $y = -4^x$	*F.* $y = 1 - 4^x$		

19.

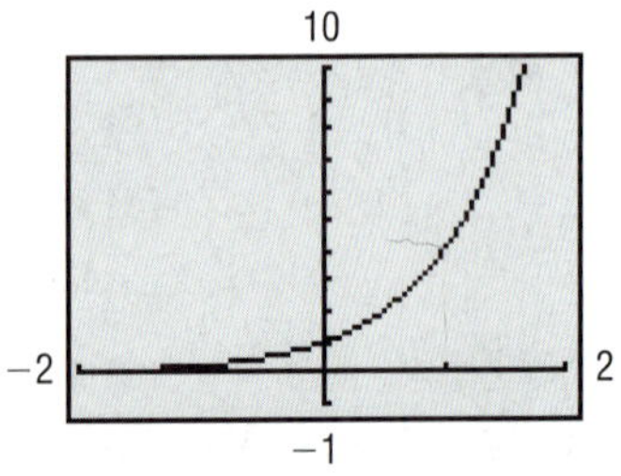

20.

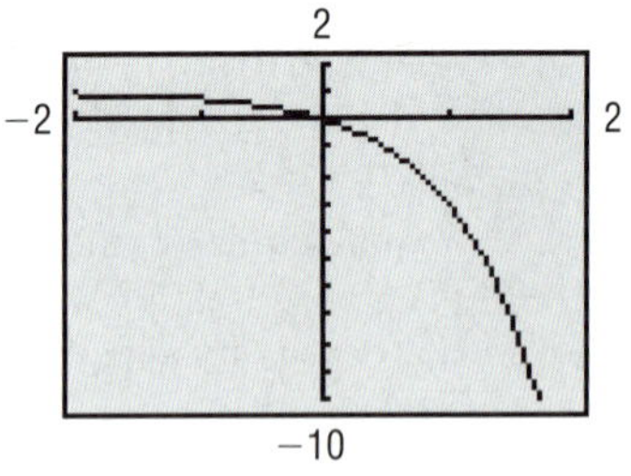

21.

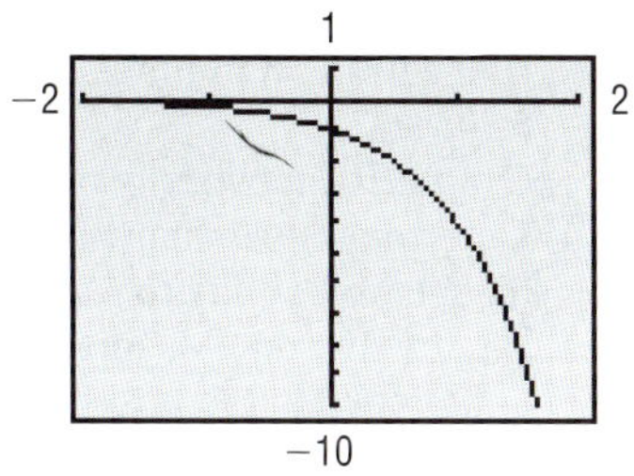

22.

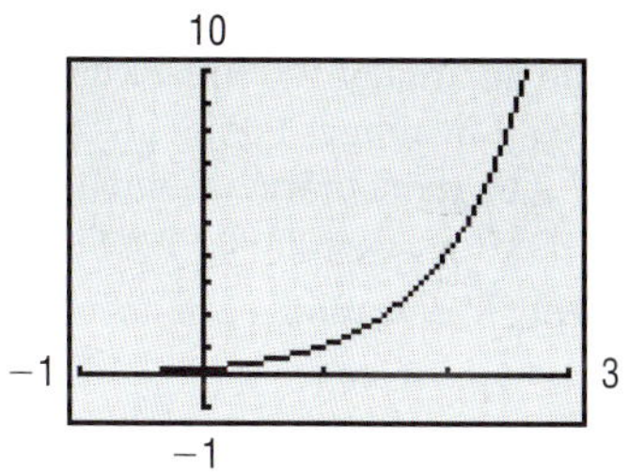

23.

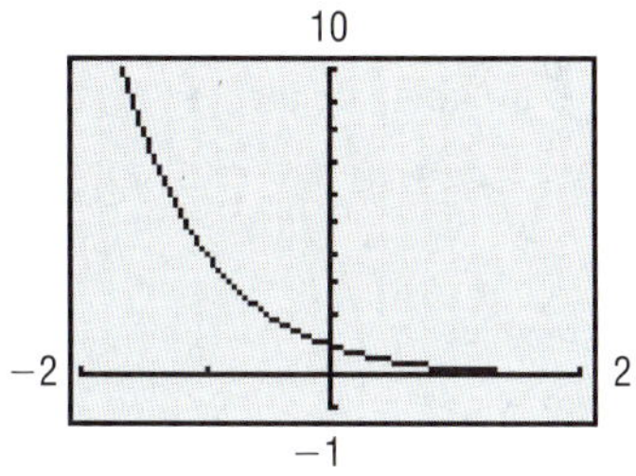

24.

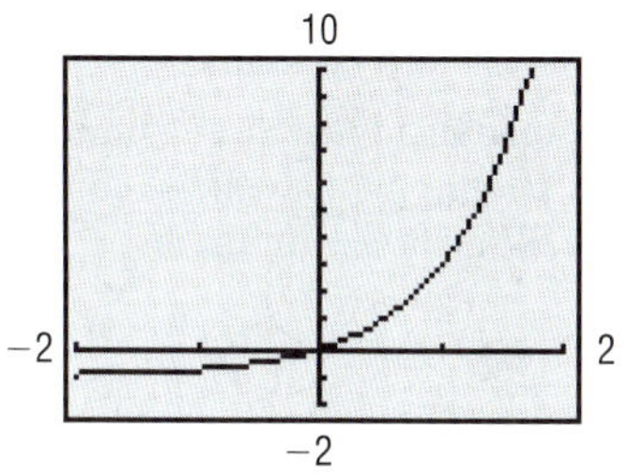

In Problems 25–32, using a graphing utility, show the stages required to graph each function. Determine the domain, range, and horizontal asymptote of each function.

25. $f(x) = e^{-x}$ **26.** $f(x) = -e^{x}$ **27.** $f(x) = e^{x+2}$ **28.** $f(x) = e^{x} - 1$

29. $f(x) = 5 - e^{-x}$ **30.** $f(x) = 9 - 3e^{-x}$ **31.** $f(x) = 2 - e^{-x/2}$ **32.** $f(x) = 7 - 3e^{-2x}$

33. If $4^x = 7$, what does 4^{-2x} equal?

34. If $2^x = 3$, what does 4^{-x} equal?

35. If $3^{-x} = 2$, what does 3^{2x} equal?

36. If $5^{-x} = 3$, what does 5^{3x} equal?

37. Optics If a single pane of glass obliterates 3% of the light passing through it, then the percent p of light that passes through n successive panes is given approximately by the equation

$$p = 100e^{-0.03n}$$

(a) What percent of light will pass through 10 panes?

(b) What percent of light will pass through 25 panes?

38. Atmospheric Pressure The atmospheric pressure p on a balloon or plane decreases with increasing height. This pressure, measured in millimeters of mercury, is related to the number of kilometers h above sea level by the formula

$$p = 760e^{-0.145h}$$

(a) Find the atmospheric pressure at a height of 2 kilometers (over a mile).

(b) What is it at a height of 10 kilometers (over 30,000 feet)?

39. Space Satellites The number of watts w provided by a space satellite's power supply over a period of d days is given by the formula

$$w = 50e^{-0.004d}$$

(a) How much power will be available after 30 days?

(b) How much power will be available after 1 year (365 days)?

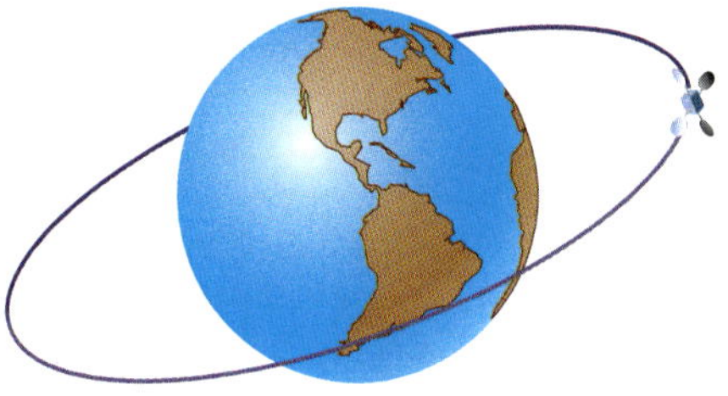

40. Healing of Wounds The normal healing of wounds can be modeled by an exponential function. If A_0 represents the original area of the wound and if A equals the area of the wound after n days, then the formula

$$A = A_0e^{-0.35n}$$

describes the area of a wound on the nth day following an injury when no infection is present to retard the healing. Suppose that a wound initially had an area of 100 square millimeters.

(a) If healing is taking place, how large should the area of the wound be after 3 days?
(b) How large should it be after 10 days?

41. **Drug Medication** The formula

$$D = 5e^{-0.4h}$$

can be used to find the number of milligrams D of a certain drug that is in a patient's bloodstream h hours after the drug has been administered. How many milligrams will be present after 1 hour? After 6 hours?

42. **Spreading of Rumors** A model for the number of people N in a college community who have heard a certain rumor is

$$N = P(1 - e^{-0.15d})$$

where P is the total population of the community and d is the number of days that have elapsed since the rumor began. In a community of 1000 students, how many students will have heard the rumor after 3 days?

43. **Exponential Probability** Between 12:00 PM and 1:00 PM, cars arrive at Citibank's drive-thru at the rate of 6 cars per hour (0.1 car per minute). The following formula from statistics can be used to determine the probability that a car will arrive within t minutes of 12:00 PM.

$$F(t) = 1 - e^{-0.1t}$$

(a) Determine the probability that a car will arrive within 10 minutes of 12:00 PM, (that is, before 12:10 PM).
(b) Determine the probability that a car will arrive within 40 minutes of 12:00 PM (before 12:40 PM).
(c) Graph F using your graphing utility.
(d) Determine how many minutes are needed for the probability to reach 50%?
(e) What value does F approach as t becomes unbounded in the positive direction?

44. **Exponential Probability** Between 5:00 PM and 6:00 PM cars arrive at Jiffy Lube at the rate of 9 cars per hour (0.15 car per minute). The following formula from statistics can be used to determine the probability that a car will arrive within t minutes of 5:00 PM.

$$F(t) = 1 - e^{-0.15t}$$

(a) Determine the probability that a car will arrive within 15 minutes of 5:00 PM (that is, before 5:15 PM).
(b) Determine the probability that a car will arrive within 30 minutes of 5:00 PM (before 5:30 PM).
(c) Graph F using your graphing utility.
(d) Determine how many minutes are needed for the probability to reach 60%.
(e) What value does F approach as t becomes unbounded in the positive direction?

45. **Poisson Probability** Between 5:00 PM and 6:00 PM cars arrive at McDonald's drive-thru at the rate of 20 cars per hour. The following formula from statistics can be used to determine the probability that x cars will arrive between 5:00 PM and 6:00 PM.

$$P(x) = \frac{20^x e^{-20}}{x!}$$

where

$x! = x \cdot (x - 1) \cdot (x - 2) \cdot \cdots \cdot 3 \cdot 2 \cdot 1.$

(a) Determine the probability that $x = 15$ cars will arrive between 5:00 PM and 6:00 PM.
(b) Determine the probability that $x = 20$ cars will arrive between 5:00 PM and 6:00 PM.

46. **Poisson Probability** People enter a line for the *Demon Roller Coaster* at the rate of 4 per minute. The following formula from statistics can be used to determine the probability that x people will arrive within the next minute.

$$P(x) = \frac{4^x e^{-4}}{x!}$$

where

$x! = x \cdot (x - 1) \cdot (x - 2) \cdot \cdots \cdot 3 \cdot 2 \cdot 1.$

(a) Determine the probability that $x = 5$ people will arrive within the next minute.
(b) Determine the probability that $x = 8$ people will arrive within the next minute.

47. **Response to TV Advertising** The percent R of viewers who respond to a television commercial for a new product after t days is found by using the formula

$$R = 70 - 70e^{-0.2t}$$

(a) What percent is expected to respond after 10 days?
(b) What percent has responded after 20 days?
(c) What is the highest percent of people expected to respond?
(d) Graph $R = 70 - 70e^{-0.2t}$, $t > 0$. Use eVALUEate to compare the values of R for $t = 10$ and $t = 20$ to the ones obtained in parts (a) and (b). Determine how many days are required for R to exceed 40%?

48. Profit The annual profit P of a company due to the sales of a particular item after it has been on the market x years is determined to be

$$P = \$100{,}000 - \$60{,}000\left(\tfrac{1}{2}\right)^x$$

(a) What is the annual profit after 5 years?
(b) What is the annual profit after 10 years?
(c) What is the most annual profit that the company can expect from this product?
(d) Graph the profit function. Use eVALUEate to compare the values of P for $x = 5$ and $x = 10$ to the one obtained in parts (a) and (b). Determine how many years it takes before an annual profit of \$65,000 is obtained.

49. Alternating Current in a *RL* Circuit The equation governing the amount of current I (in amperes) after time t (in seconds) in a single RL circuit consisting of a resistance R (in ohms), an inductance L (in henrys), and an electromotive force E (in volts) is

$$I = \frac{E}{R}\left[1 - e^{-(R/L)t}\right]$$

(a) If $E = 120$ volts, $R = 10$ ohms, and $L = 5$ henrys, how much current I_1 is flowing after 0.3 second? After 0.5 second? After 1 second?
(b) What is the maximum current?
(c) Graph this function $I = I_1(t)$, measuring I along the y-axis and t along the x-axis.
(d) If $E = 120$ volts, $R = 5$ ohms, and $L = 10$ henrys, how much current I_2 is flowing after 0.3 second? After 0.5 second? After 1 second?
(e) What is the maximum current?
(f) Graph this function $I = I_2(t)$ on the same viewing window as $I_1(t)$.

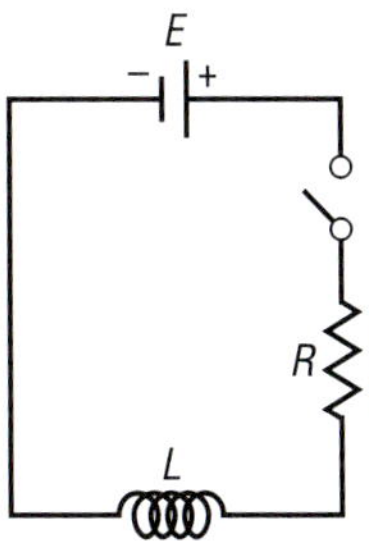

50. Alternating Current in a *RC* Circuit The equation governing the amount of current I (in amperes) after time t (in microseconds) in a single RC circuit consisting of a resistance R (in ohms), a capacitance C (in microfarads), and an electromotive force E (in volts) is

$$I = \frac{E}{R}e^{-t/(RC)}$$

(a) If $E = 120$ volts, $R = 2000$ ohms, and $C = 1.0$ microfarad, how much current I_1 is flowing initially ($t = 0$)? After 1000 microseconds? After 3000 microseconds?
(b) What is the maximum current?
(c) Graph this function $I = I_1(t)$, measuring I along the y-axis and t along the x-axis.
(d) If $E = 120$ volts, $R = 1000$ ohms, and $C = 2.0$ microfarads, how much current I_2 is flowing initially? After 1000 microseconds? After 3000 microseconds?
(e) What is the maximum current?
(f) Graph this function $I = I_2(t)$ on the same viewing window as $I_1(t)$.

51. The *Challenger* Disaster* After the *Challenger*

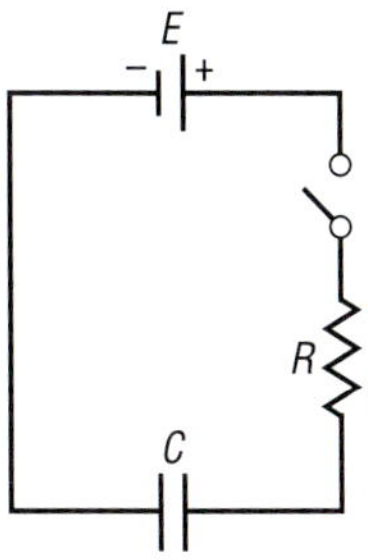

disaster in 1986, a study of the 23 launches that preceded the fatal flight was made. A mathematical model was developed involving the relationship between the Fahrenheit temperature x around the O-rings and the number y of eroded or leaky primary O-rings. The model stated that

$$y = \frac{6}{1 + e^{-(5.085 - 0.1156x)}}$$

where the number 6 indicates the 6 primary O-rings on the spacecraft.

(a) What is the predicted number of eroded or leaky primary O-rings at a temperature of 100°F?

*Linda Tappin, "Analyzing Data Relating to the *Challenger* Disaster," *Mathematics Teacher,* Vol. 87, No. 6, September 1994, pp. 423–426.

(b) What is the predicted number of eroded or leaky primary O-rings at a temperature of 60°F?

(c) What is the predicted number of eroded or leaky primary O-rings at a temperature of 30°F?

(d) Graph the equation. At what temperature is the predicted number of eroded or leaky O-rings 1? 3? 5?

52. Postage Stamps* The cumulative number y of different postage stamps (regular and commemorative only) issued by the U.S. Post Office can be approximated (modeled) by the exponential function

$$y = 78e^{0.025x}$$

where x is the number of years since 1848.

(a) What is the predicted cumulative number of stamps that will have been issued by the year 1995? Check with the Postal Service and comment on the accuracy of using the function.

(b) What is the predicted cumulative number of stamps that will have been issued by the year 1998?

(c) The cumulative number of stamps actually issued by the United States was 2 in 1848, 88 in 1868, 218 in 1888, and 341 in 1908. What conclusion can you draw about using the given function as a model over the first few decades in which stamps were issued?

*David Kullman, "Patterns of Postage-stamp Production," *Mathematics Teacher,* Vol. 85, No. 3, March 1992, pp. 188–189.

53. Another Formula for *e* Use a calculator to compute the value of

$$2 + \frac{1}{2!} + \frac{1}{3!} + \cdots + \frac{1}{n!}$$

for $n = 4, 6, 8$, and 10. Compare each result with e.

[Hint: $1! = 1$, $2! = 2 \cdot 1$, $3! = 3 \cdot 2 \cdot 1$, $n! = n(n-1) \cdot \cdots \cdot (3)(2)(1)$]

54. Another Formula for *e* Use a calculator to compute the various values of the expression Compare the values to e.

$$2 + \cfrac{1}{1 + \cfrac{1}{2 + \cfrac{2}{3 + \cfrac{3}{4 + \cfrac{4}{\text{etc.}}}}}}$$

55. If $f(x) = a^x$, show that

$$\frac{f(x+h) - f(x)}{h} = a^x\left(\frac{a^h - 1}{h}\right)$$

56. If $f(x) = a^x$, show that $f(A + B) = f(A) \cdot f(B)$.

57. If $f(x) = a^x$, show that $f(-x) = \dfrac{1}{f(x)}$.

58. If $f(x) = a^x$, show that $f(\alpha x) = [f(x)]^\alpha$.

59. Historical Problem Pierre de Fermat (1601–1665) conjectured that the function

$$f(x) = 2^{(2^x)} + 1$$

for $x = 1, 2, 3, \ldots$, would always have a value equal to a prime number. But Leonhard Euler (1707–1783) showed that this formula fails for $x = 5$. Use a calculator to determine the prime numbers produced by f for $x = 1, 2, 3, 4$. Then show that $f(5) = 641 \times 6{,}700{,}417$, which is not prime.

60. The bacteria in a 4-liter container double every minute. After 60 minutes the container is full. How long did it take to fill half the container?

61. Explain in your own words what the number e is. Provide at least two applications that require the use of this number.

62. Do you think there is a power function that increases more rapidly than an exponential function whose base is greater than 1? Explain.

5.3 LOGARITHMIC FUNCTIONS

1. Change Exponential Expressions to Logarithmic Expressions
2. Change Logarithmic Expressions to Exponential Expressions
3. Evaluate Logarithmic Functions
4. Determine the Domain of a Logarithmic Function
5. Graph Logarithmic Functions

Recall that a one-to-one function $y = f(x)$ has an inverse function that is defined (implicitly) by the equation $x = f(y)$. In particular, the exponential function $y = f(x) = a^x$, $a > 0$, $a \neq 1$, is one-to-one and hence has an inverse function that is defined implicitly by the equation

$$x = a^y \qquad a > 0, a \neq 1$$

This inverse function is so important that it is given a name, the *logarithmic function.*

The **logarithmic function to the base *a*,** where $a > 0$ and $a \neq 1$, is denoted by $y = \log_a x$ (read as "y is the logarithm to the base a of x") and is defined by

$$y = \log_a x \quad \text{if and only if} \quad x = a^y$$

EXAMPLE 1 Relating Logarithms to Exponents

(a) If $y = \log_3 x$, then $x = 3^y$. Thus, if $x = 9$, then $y = 2$, so $9 = 3^2$ is equivalent to $2 = \log_3 9$.

(b) If $y = \log_5 x$, then $x = 5^y$. Thus, if $x = \frac{1}{5} = 5^{-1}$, then $y = -1$, so $\frac{1}{5} = 5^{-1}$ is equivalent to $-1 = \log_5 (\frac{1}{5})$.

EXAMPLE 2 Changing Exponential Expressions to Logarithmic Expressions

1 Change each exponential expression to an equivalent expression involving a logarithm.

(a) $1.2^3 = m$ (b) $e^b = 9$ (c) $a^4 = 24$

Solution We use the fact that $y = \log_a x$ and $x = a^y$, $a > 0$, $a \neq 1$, are equivalent.

(a) If $1.2^3 = m$, then $3 = \log_{1.2} m$.

(b) If $e^b = 9$, then $b = \log_e 9$.

(c) If $a^4 = 24$, then $4 = \log_a 24$.

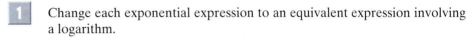

NOW WORK PROBLEM 1.

EXAMPLE 3 Changing Logarithmic Expressions to Exponential Expressions

2 Change each logarithmic expression to an equivalent expression involving an exponent.

(a) $\log_a 4 = 5$ (b) $\log_e b = -3$ (c) $\log_3 5 = c$

Solution (a) If $\log_a 4 = 5$, then $a^5 = 4$.

(b) If $\log_e b = -3$, then $e^{-3} = b$.

(c) If $\log_3 5 = c$, then $3^c = 5$.

NOW WORK PROBLEM 13.

3 To find the exact value of a logarithm, we write the logarithm in exponential notation and use the following fact:

$$\text{If } a^u = a^v, \quad \text{then} \quad u = v. \tag{1}$$

The result (1) is a consequence of the fact that exponential functions are one-to-one.

EXAMPLE 4 Finding the Exact Value of a Logarithmic Function

Find the exact value of:

(a) $\log_2 16$ (b) $\log_3 \frac{1}{3}$ (c) $\log_5 25$

Solution (a) For $y = \log_2 16$, we have the equivalent exponential equation $2^y = 16 = 2^4$, so, by (1), $y = 4$. Thus, $\log_2 16 = 4$.

(b) For $y = \log_3 \frac{1}{3}$, we have $3^y = \frac{1}{3} = 3^{-1}$, so $y = -1$. Thus, $\log_3 \frac{1}{3} = -1$.

(c) For $y = \log_5 25$, we have $5^y = 25 = 5^2$, so $y = 2$. Thus, $\log_5 25 = 2$.

NOW WORK PROBLEM 25.

Domain of a Logarithmic Function

4 The logarithmic function $y = \log_a x$ has been defined as the inverse of the exponential function $y = a^x$. That is, if $f(x) = a^x$, then $f^{-1}(x) = \log_a x$. Based on the discussion given in Section 5.1 on inverse functions, we know that for a function f and its inverse f^{-1}

$$\text{Domain } f^{-1} = \text{Range } f \quad \text{and} \quad \text{Range } f^{-1} = \text{Domain } f$$

Consequently, it follows that

$$\text{Domain of logarithmic function} = \text{Range of exponential function} = (0, \infty)$$
$$\text{Range of logarithmic function} = \text{Domain of exponential function} = (-\infty, \infty)$$

In the next box, we summarize some properties of the logarithmic function:

> $y = \log_a x$ (defining equation: $x = a^y$)
> Domain: $0 < x < \infty$ Range: $-\infty < y < \infty$

Notice that the domain of a logarithmic function consists of the *positive* real numbers. Thus, the argument of a logarithmic function must be greater than zero.

EXAMPLE 5 Finding the Domain of a Logarithmic Function

Find the domain of each logarithmic function.

(a) $F(x) = \log_2(1 - x)$ (b) $g(x) = \log_5\left(\dfrac{1 + x}{1 - x}\right)$ (c) $h(x) = \log_{1/2}|x|$

Solution (a) The domain of F consists of all x for which $(1 - x) > 0$; that is, all $x < 1$, or $(-\infty, 1)$.

(b) The domain of g is restricted to

$$\frac{1 + x}{1 - x} > 0$$

Solve numerator & denominator separately

Solving this inequality, we find that the domain of g consists of all x between -1 and 1, that is, $-1 < x < 1$, or $(-1, 1)$.

(c) Since $|x| > 0$ provided that $x \neq 0$, the domain of h consists of all nonzero real numbers.

NOW WORK PROBLEM **39.**

Graphs of Logarithmic Functions

5 Since exponential functions and logarithmic functions are inverses of each other, the graph of a logarithmic function $y = \log_a x$ is the reflection about the line $y = x$ of the graph of the exponential function $y = a^x$, as shown in Figure 25.

Figure 25

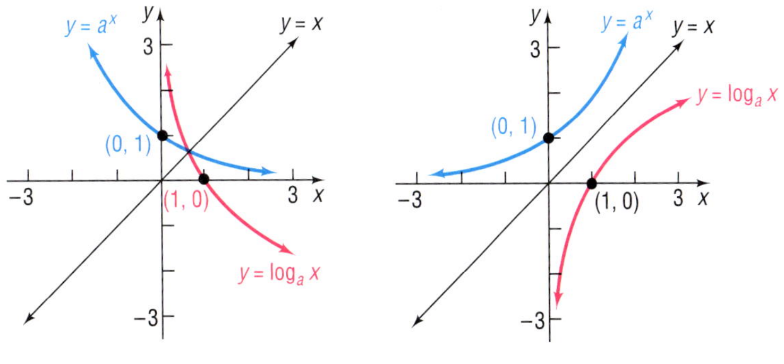

(a) $0 < a < 1$ **(b)** $a > 1$

Facts about the Graph of a Logarithmic Function $f(x) = \log_a x$

1. The x-intercept of the graph is 1. There is no y-intercept.
2. The y-axis ($x = 0$) is a vertical asymptote of the graph.
3. A logarithmic function is decreasing if $0 < a < 1$ and increasing if $a > 1$.
4. The graph is smooth and continuous, with no corners or gaps.

If the base of a logarithmic function is the number e, then we have the **natural logarithm function.** This function occurs so frequently in applications that it is given a special symbol, **ln** (from the Latin, *logarithmus naturalis*). Thus,

$$y = \ln x \quad \text{if and only if} \quad x = e^y$$

Since $y = \ln x$ and the exponential function $y = e^x$ are inverse functions, we can obtain the graph of $y = \ln x$ by reflecting the graph of $y = e^x$ about the line $y = x$. See Figure 26.

Table 6 displays other points on the graph of $f(x) = \ln x$. Notice for $x < 0$ that we obtain an error message. Do you recall why?

Figure 26

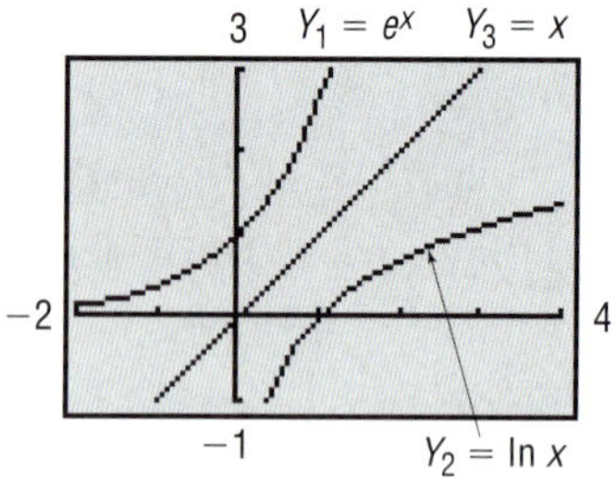

Table 6

X	Y1
-1	ERROR
.5	-.6931
1	0
2	.69315
2.7183	1
3	1.0986

Y1 = ln(X)

EXAMPLE 6 Graphing Logarithmic Functions Using Transformations

Graph $f(x) = -\ln x$ by starting with the graph of $y = \ln x$. Determine the domain, range, and vertical asymptote.

Solution The graph of $f(x) = -\ln x$ is obtained by a reflection about the x-axis of the graph of $f(x) = \ln x$. See Figure 27.

Figure 27

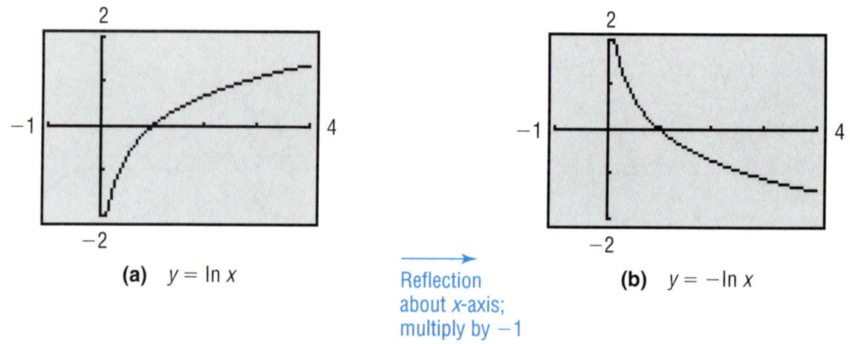

(a) $y = \ln x$ (b) $y = -\ln x$

The domain of $f(x) = -\ln x$ is $(0, \infty)$, the range is $(-\infty, \infty)$, and the vertical asymptote is $x = 0$.

EXAMPLE 7 Graphing Logarithmic Functions Using Transformations

Determine the domain, range, and vertical asymptote of $f(x) = \ln(x + 2)$. Graph $f(x) = \ln(x + 2)$.

Solution The domain consists of all x for which

$$x + 2 > 0 \quad \text{or} \quad x > -2$$

The graph is obtained by applying a horizontal shift to the left 2 units, as shown in Figure 28.

Figure 28

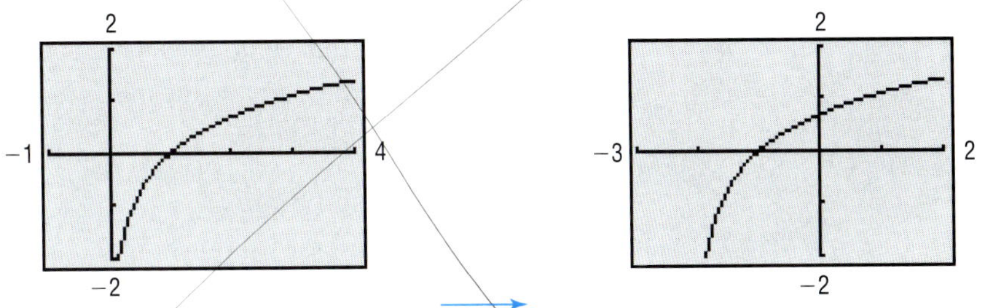

(a) $y = \ln x$ — Shift left 2 units; replace x by $x+2$ — (b) $y = \ln(x + 2)$

The range of $f(x) = \ln(x + 2)$ is $(-\infty, \infty)$, and the vertical asymptote is $x = -2$. [Do you see why? The original vertical asymptote ($x = 0$) is shifted to the left 2 units].

EXAMPLE 8 Graphing Logarithmic Functions Using Transformations

Graph $f(x) = \ln(1 - x)$. Determine the domain, range, and vertical asymptote of f.

Solution The domain consists of all x for which

$$1 - x > 0 \quad \text{or} \quad x < 1$$

To obtain the graph of $y = \ln(1 - x)$, we use the steps illustrated in Figure 29.

Figure 29

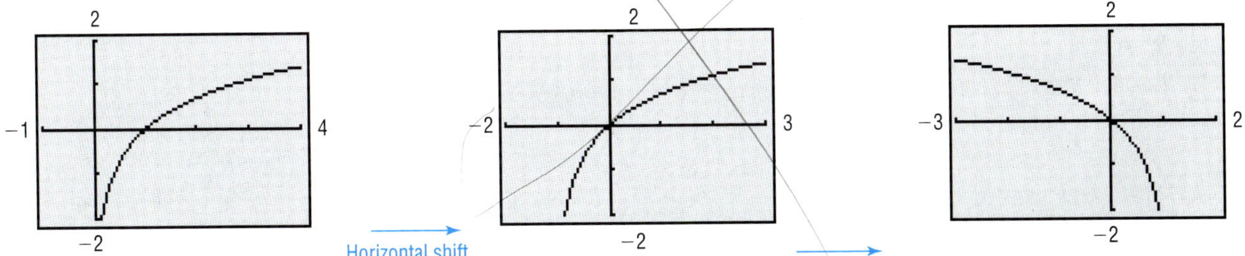

(a) $y = \ln x$ — Horizontal shift left 1 unit; replace x by $x + 1 = 1+x$ — (b) $y = \ln(1 + x)$ — Reflect about y-axis; Replace x by $-x$ — (c) $y = \ln(1 - x)$

The range of $f(x) = \ln(1 - x)$ is $(-\infty, \infty)$, and the vertical asymptote is $x = 1$.

NOW WORK PROBLEM **67.**

EXAMPLE 9 Alcohol and Driving

The concentration of alcohol in a person's blood is measurable. Recent medical research suggests that the risk R (given as a percent) of having an accident while driving a car can be modeled by the equation

$$R = 6e^{kx}$$

where x is the variable concentration of alcohol in the blood and k is a constant.

(a) Suppose that a concentration of alcohol in the blood of 0.04 results in a 10% risk ($R = 10$) of an accident. Find the constant k in the equation. Graph $R = 6e^{kx}$, using this value of k.

(b) Using this value of k, what is the risk if the concentration is 0.17?

(c) Using the same value of k, what concentration of alcohol corresponds to a risk of 100%?

(d) If the law asserts that anyone with a risk of having an accident of 20% or more should not have driving privileges, at what concentration of alcohol in the blood should a driver be arrested and charged with a DUI (Driving Under the Influence)?

Solution (a) For a concentration of alcohol in the blood of 0.04 and a risk of 10%, we let $x = 0.04$ and $R = 10$ in the equation and solve for k.

Figure 30

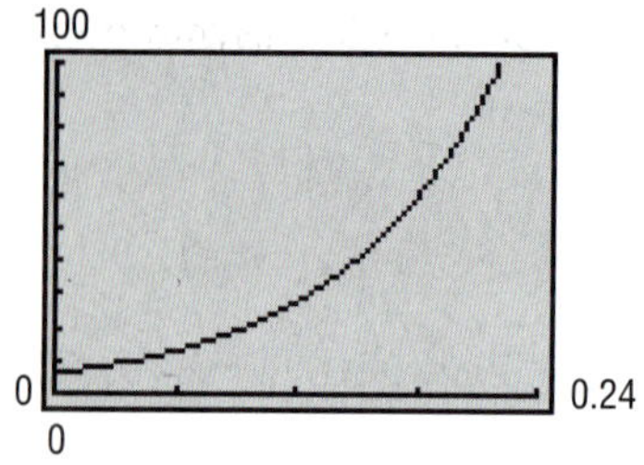

$$R = 6e^{kx}$$

$$10 = 6e^{k(0.04)}$$

$$\frac{10}{6} = e^{0.04k} \qquad \text{Divide both sides by 6.}$$

$$0.04k = \ln\frac{10}{6} = 0.5108256 \qquad \text{Change to a logarithmic expression.}$$

$$k = 12.77 \qquad \text{Solve for } k.$$

See Figure 30 for the graph of $R = 6e^{12.77x}$.

(b) Using $k = 12.77$ and $x = 0.17$ in the equation, we find the risk R to be

$$R = 6e^{kx} = 6e^{(12.77)(0.17)} = 52.6$$

For a concentration of alcohol in the blood of 0.17, the risk of an accident is about 52.6%. Verify this answer using eVALUEate.

(c) Using $k = 12.77$ and $R = 100$ in the equation, we find the concentration x of alcohol in the blood to be

$$R = 6e^{kx}$$

$$100 = 6e^{12.77x}$$

$$\frac{100}{6} = e^{12.77x} \qquad \text{Divide both sides by 6.}$$

$$12.77x = \ln\frac{100}{6} = 2.8134 \qquad \text{Change to a logarithmic expression.}$$

$$x = 0.22 \qquad \text{Solve for } x.$$

For a concentration of alcohol in the blood of 0.22, the risk of an accident is 100%. Verify this answer using a graphing utility.

(d) Using $k = 12.77$ and $R = 20$ in the equation, we find the concentration x of alcohol in the blood to be

$$\begin{aligned} R &= 6e^{kx} \\ 20 &= 6e^{12.77x} \\ \frac{20}{6} &= e^{12.77x} \\ 12.77x &= \ln\frac{20}{6} = 1.204 \\ x &= 0.094 \end{aligned}$$

A driver with a concentration of alcohol in the blood of 0.094 or more should be arrested and charged with DUI. Verify using a graphing utility.

Note: Most states use 0.10 as the blood alcohol content at which a DUI citation is given. A few states use 0.08.

SUMMARY **Properties of the Logarithmic Function**

$f(x) = \log_a x, \quad a > 1$ ($y = \log_a x$ means $x = a^y$)	Domain: $(0, \infty)$; Range: $(-\infty, \infty)$; x-intercept: 1; y-intercept: none; vertical asymptote: $x = 0$ (y-axis); increasing; one-to-one. See Figure 25(b) for a typical graph.
$f(x) = \log_a x, \quad 0 < a < 1$ ($y = \log_a x$ means $x = a^y$)	Domain: $(0, \infty)$; Range: $(-\infty, \infty)$; x-intercept: 1; y-intercept: none; vertical asymptote: $x = 0$ (y-axis); decreasing; one-to-one. See Figure 25(a) for a typical graph.

5.3 EXERCISES

In Problems 1–12, change each exponential expression to an equivalent expression involving a logarithm.

1. $9 = 3^2$ **2.** $16 = 4^2$ **3.** $a^2 = 1.6$ **4.** $a^3 = 2.1$ **5.** $1.1^2 = M$ **6.** $2.2^3 = N$
7. $2^x = 7.2$ **8.** $3^x = 4.6$ **9.** $x^{\sqrt{2}} = \pi$ **10.** $x^\pi = e$ **11.** $e^x = 8$ **12.** $e^{2.2} = M$

In Problems 13–24, change each logarithmic expression to an equivalent expression involving an exponent.

13. $\log_2 8 = 3$ **14.** $\log_3 (\frac{1}{9}) = -2$ **15.** $\log_a 3 = 6$ **16.** $\log_b 4 = 2$
17. $\log_3 2 = x$ **18.** $\log_2 6 = x$ **19.** $\log_2 M = 1.3$ **20.** $\log_3 N = 2.1$
21. $\log_{\sqrt{2}} \pi = x$ **22.** $\log_\pi x = \frac{1}{2}$ **23.** $\ln 4 = x$ **24.** $\ln x = 4$

In Problems 25–36, find the exact value of each logarithm without using a calculator.

25. $\log_2 1$ **26.** $\log_8 8$ **27.** $\log_5 25$ **28.** $\log_3 (\frac{1}{9})$ **29.** $\log_{1/2} 16$ **30.** $\log_{1/3} 9$
31. $\log_{10} \sqrt{10}$ **32.** $\log_5 \sqrt[3]{25}$ **33.** $\log_{\sqrt{2}} 4$ **34.** $\log_{\sqrt{3}} 9$ **35.** $\ln \sqrt{e}$ **36.** $\ln e^3$

$y = \log_{10}\sqrt{10}$ $10^y = 10^{1/2}$

In Problems 37–46, find the domain of each function.

37. $f(x) = \ln(x - 3)$ **38.** $g(x) = \ln(x - 1)$ **39.** $F(x) = \log_2 x^2$ **40.** $H(x) = \log_5 x^3$

41. $h(x) = \log_{1/2}(x^2 - 2x + 1)$ **42.** $G(x) = \log_{1/2}(x^2 - 1)$ **43.** $f(x) = \ln\left(\frac{1}{x + 1}\right)$ **44.** $g(x) = \ln\left(\frac{1}{x - 5}\right)$

45. $g(x) = \log_5\left(\frac{x + 1}{x}\right)$ **46.** $h(x) = \log_3\left(\frac{x}{x - 1}\right)$

$\frac{1}{2}^y = 16$
$2^{-y} = 2^4$
$y = -4$

In Problems 47–50, use a calculator to evaluate each expression. Round your answer to three decimal places.

47. $\ln \dfrac{5}{3}$ **48.** $\dfrac{\ln 5}{3}$ **49.** $\dfrac{\ln (10/3)}{0.04}$ **50.** $\dfrac{\ln (2/3)}{-0.1}$

51. Find a such that the graph of $f(x) = \log_a x$ contains the point $(2, 2)$.

52. Find a such that the graph of $f(x) = \log_a x$ contains the point $(\frac{1}{2}, -4)$.

In Problems 53–60, the graph of a logarithmic function is given. Match each graph to one of the following functions:

A. $y = \log_3 x$ B. $y = \log_3 (-x)$ C. $y = -\log_3 x$ D. $y = -\log_3 (-x)$

E. $y = \log_3 x - 1$ F. $y = \log_3 (x - 1)$ G. $y = \log_3 (1 - x)$ H. $y = 1 - \log_3 x$

53.

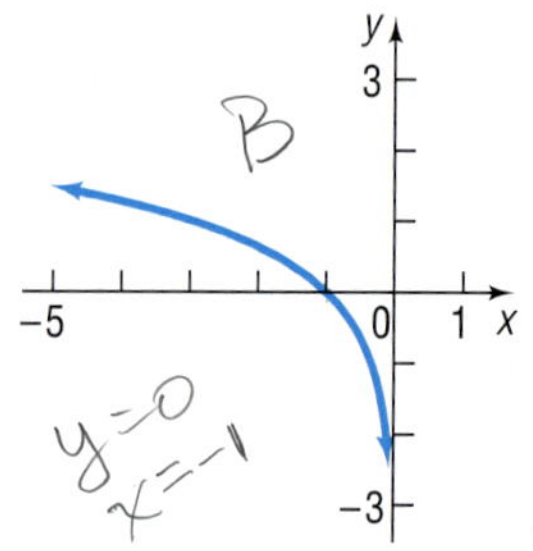

54.

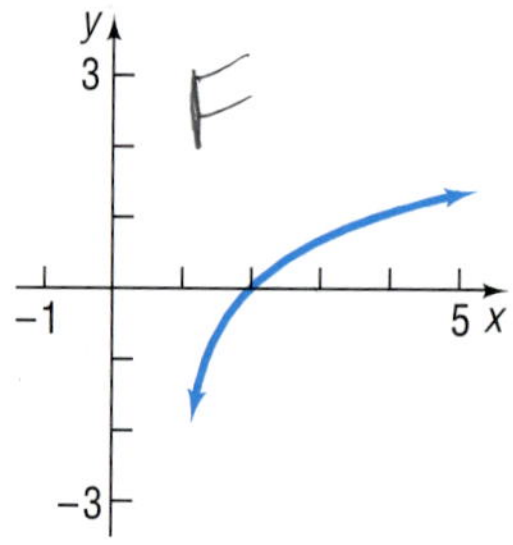

55.

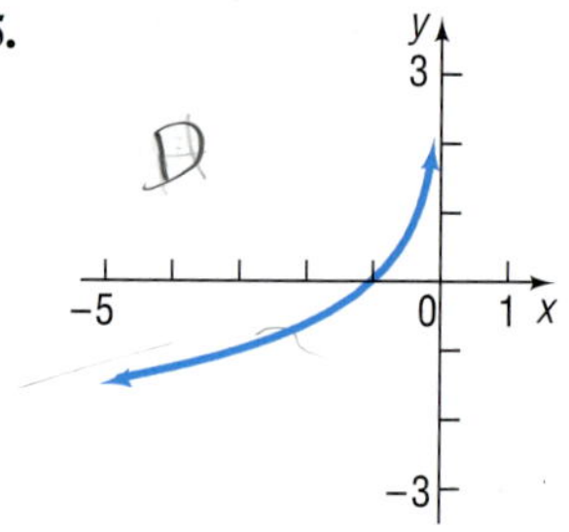

56.

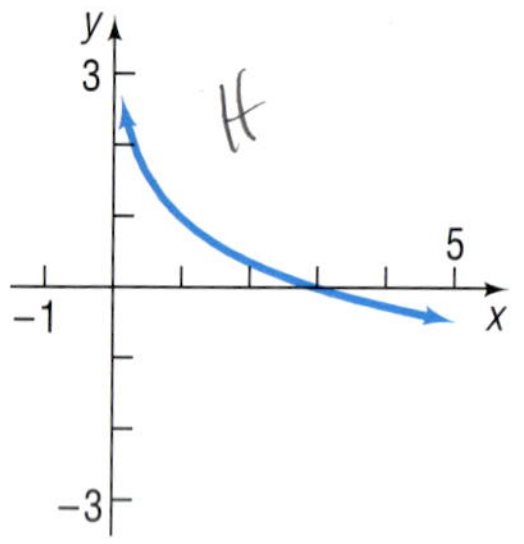

57.

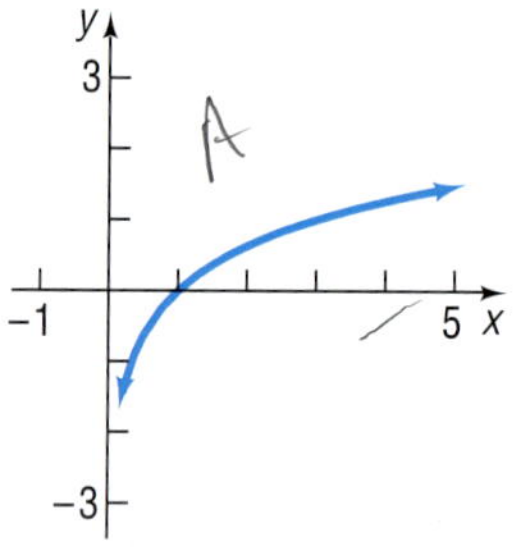

58.

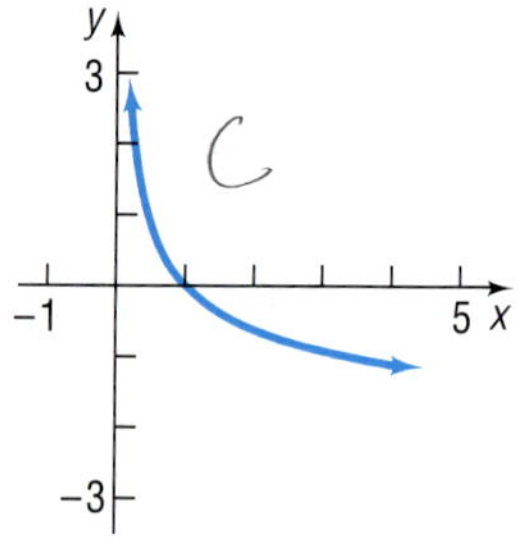

59.

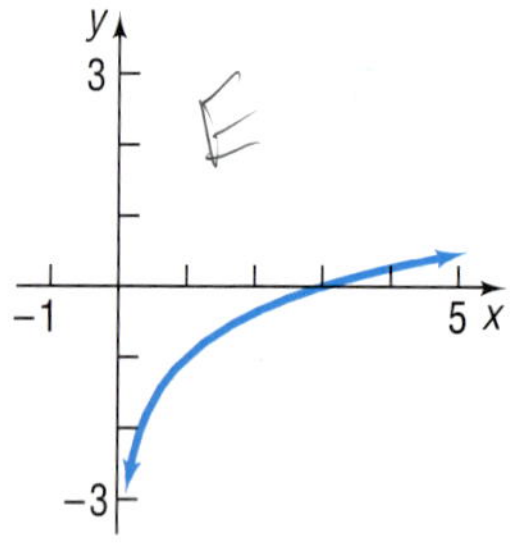

60.

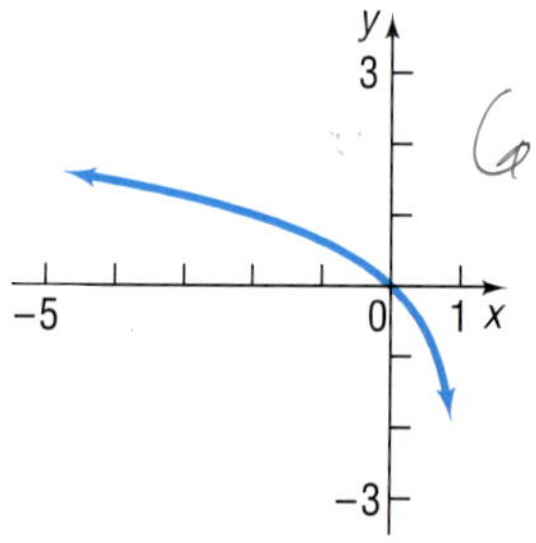

In Problems 61–66, the graph of a logarithmic function is given. Match each graph to one of the following functions:

A. $y = \log_4 x$ B. $y = \log_4 (-x)$ C. $y = \log_4 (x - 1)$

D. $y = -\log_4 x$ E. $y = 1 - \log_4 x$ F. $y = -\log_4 (-x)$

61.

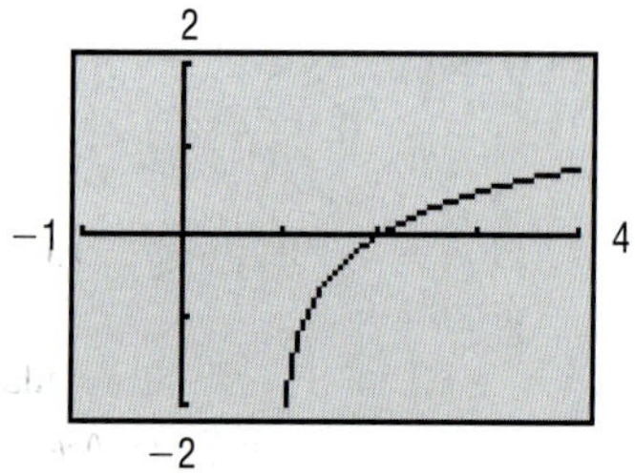

62.

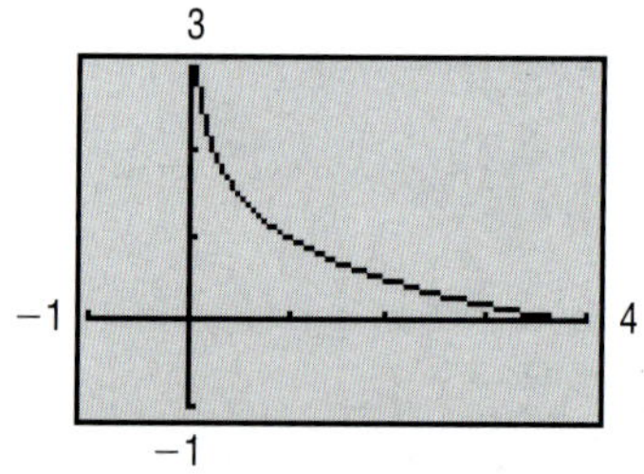

63.

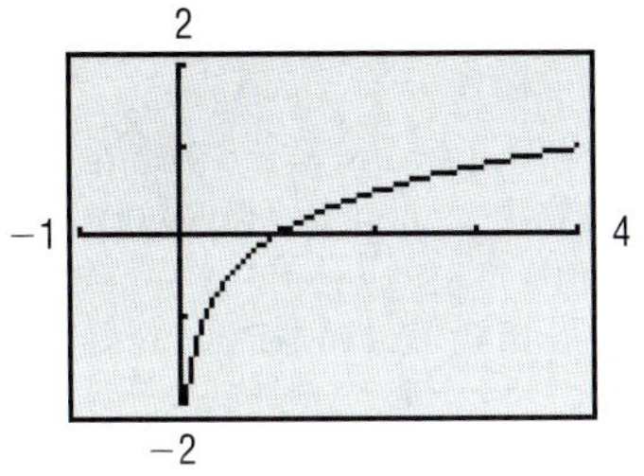

64.

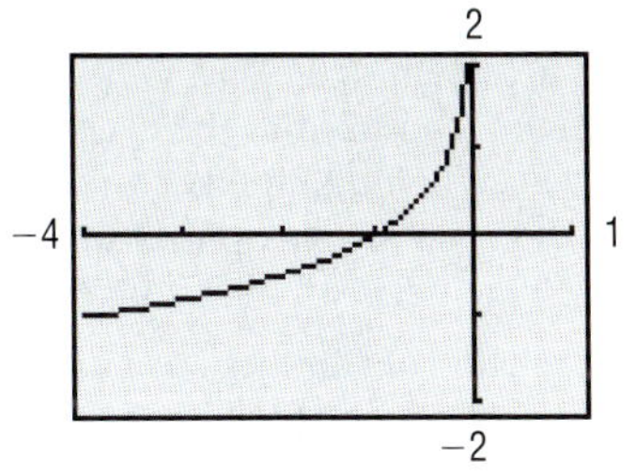

65.

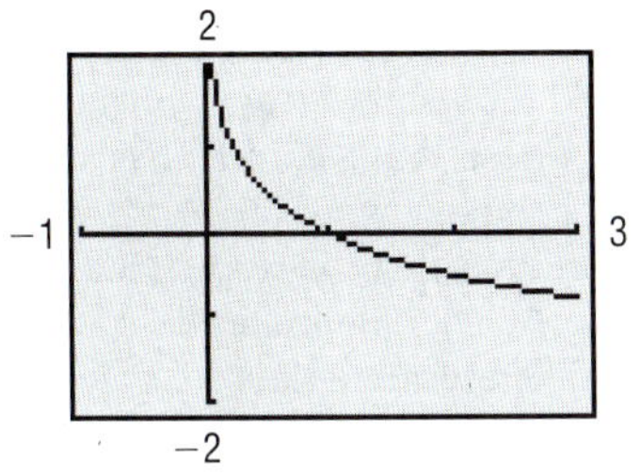

66.

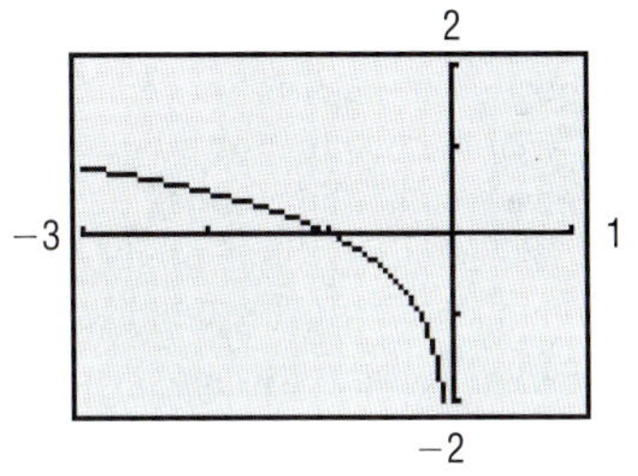

In Problems 67–78, using a graphing utility, show the stages required to graph each function. Determine the domain, range, and vertical asymptote of each function.

67. $f(x) = \ln(x + 4)$ **68.** $f(x) = \ln(x - 3)$ **69.** $f(x) = \ln(-x)$ **70.** $f(x) = -\ln(-x)$

71. $g(x) = \ln 2x$ **72.** $h(x) = \ln \frac{1}{2}x$ **73.** $f(x) = 3 \ln x$ **74.** $f(x) = -2 \ln x$

75. $g(x) = \ln(3 - x)$ **76.** $h(x) = \ln(4 - x)$ **77.** $f(x) = -\ln(x - 1)$ **78.** $f(x) = 2 - \ln x$

79. Optics If a single pane of glass obliterates 10% of the light passing through it, then the percent P of light that passes through n successive panes is given approximately by the equation

$$P = 100e^{-0.1n}$$

(a) How many panes are necessary to block at least 50% of the light?
(b) How many panes are necessary to block at least 75% of the light?

80. Chemistry The pH of a chemical solution is given by the formula

$$\text{pH} = -\log_{10}[H^+]$$

where $[H^+]$ is the concentration of hydrogen ions in moles per liter. Values of pH range from 0 (acidic) to 14 (alkaline).

(a) Find the pH of a 1-liter container of water with 0.0000001 mole of hydrogen ion.
(b) Find the hydrogen ion concentration of a mildly acidic solution with a pH of 4.2.

81. Space Satellites The number of watts w provided by a space satellite's power supply after d days is given by the formula

$$w = 50e^{-0.004d}$$

(a) How long will it take for the available power to drop to 30 watts?
(b) How long will it take for the available power to drop to only 5 watts?

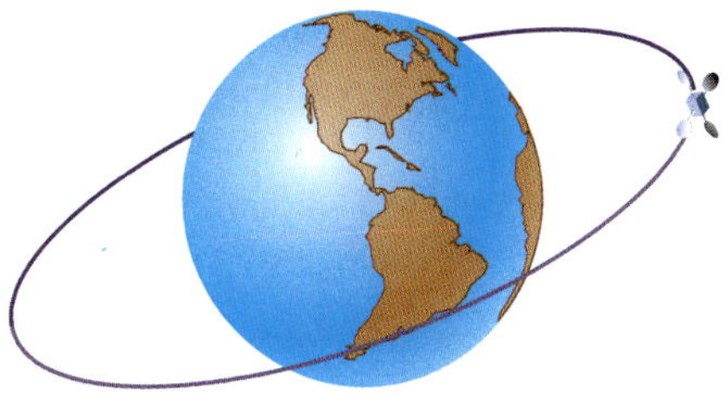

82. Healing of Wounds The normal healing of wounds can be modeled by an exponential function. If A_0 represents the original area of the wound and if A equals the area of the wound after n days, then the formula

$$A = A_0 e^{-0.35n}$$

describes the area of a wound on the nth day following an injury when no infection is present to retard the healing. Suppose that a wound initially had an area of 100 square millimeters.

(a) If healing is taking place, how many days should pass before the wound is one-half its original size?
(b) How long before the wound is 10% of its original size?

83. Drug Medication The formula

$$D = 5e^{-0.4h}$$

can be used to find the number of milligrams D of a certain drug that is in a patient's bloodstream h hours after the drug has been administered. When the number of milligrams reaches 2, the drug is to be administered again. What is the time between injections?

84. **Spreading of Rumors** A model for the number of people N in a college community who have heard a certain rumor is

$$N = P(1 - e^{-0.15d})$$

where P is the total population of the community and d is the number of days that have elapsed since the rumor began. In a community of 1000 students, how many days will elapse before 450 students have heard the rumor?

85. **Current in a *RL* Circuit** The equation governing the amount of current I (in amperes) after time t (in seconds) in a simple RL circuit consisting of a resistance R (in ohms), an inductance L (in henrys), and an electromotive force E (in volts) is

$$I = \frac{E}{R}[1 - e^{-(R/L)t}]$$

If $E = 12$ volts, $R = 10$ ohms, and $L = 5$ henrys, how long does it take to obtain a current of 0.5 ampere? Of 1.0 ampere? Graph the equation.

86. **Learning Curve** Psychologists sometimes use the function

$$L(t) = A(1 - e^{-kt})$$

to measure the amount L learned at time t. The number A represents the amount to be learned, and the number k measures the rate of learning. Suppose that a student has an amount A of 200 vocabulary words to learn. A psychologist determines that the student learned 20 vocabulary words after 5 minutes.
(a) Determine the rate of learning k.
(b) Approximately how many words will the student have learned after 10 minutes?
(c) After 15 minutes?
(d) How long does it take for the student to learn 180 words?

Problems 87–90 use the following discussion: The ***loudness*** *$L(x)$, measured in decibels, of a sound of intensity x, measured in watts per square meter, is defined as $L(x) = 10 \log \frac{x}{I_0}$, where $I_0 = 10^{-12}$ watt per square meter is the least intense sound that a human ear can detect. Determine the loudness, in decibels of each of the following sounds.*

87. **Loudness of Sound** Normal conversation: intensity $x = 10^{-7}$ watt per square meter.

88. **Loudness of Sound** Heavy city traffic: intensity $x = 10^{-3}$ watt per square meter.

89. **Loudness of Sound** Amplified rock music: intensity 10^{-1} watt per square meter.

90. **Loudness of Sound** Diesel truck traveling 40 miles per hour 50 feet away: intensity 10 times that of a passenger car traveling 50 miles per hour 50 feet away whose loudness is 70 decibels.

Problems 91 and 92 use the following discussion: The ***Richter scale*** *is one way of converting seismographic readings into numbers that provide an easy reference for measuring the magnitude M of an earthquake. All earthquakes are compared to a* ***zero-level earthquake*** *whose seismographic reading measures 0.001 millimeter at a distance of 100 kilometers from the epicenter. An earthquake whose seismographic reading measures x millimeters has* ***magnitude*** *$M(x)$ given by*

$$M(x) = \log\left(\frac{x}{x_0}\right)$$

where $x_0 = 10^{-3}$ is the reading of a zero-level earthquake the same distance from its epicenter. Determine the magnitude of the following earthquakes.

91. **Magnitude of an Earthquake** Mexico City in 1985: seismographic reading of 125,892 millimeters 100 kilometers from the center.

92. **Magnitude of an Earthquake** San Francisco in 1906: seismographic reading of 7943 millimeters 100 kilometers from the center.

93. **Alcohol and Driving** The concentration of alcohol in a person's blood is measurable. Suppose that the risk R (given as a percent) of having an accident while driving a car can be modeled by the equation

$$R = 3e^{kx}$$

where x is the variable concentration of alcohol in the blood and k is a constant.

(a) Suppose that a concentration of alcohol in the blood of 0.06 results in a 10% risk ($R = 10$) of an accident. Find the constant k in the equation.

(b) Using this value of k, what is the risk if the concentration is 0.17?

(c) Using the same value of k, what concentration of alcohol corresponds to a risk of 100%?

(d) If the law asserts that anyone with a risk of having an accident of 15% or more should not have driving privileges, at what concentration of alcohol in the blood should a driver be arrested and charged with a DUI?

(e) Compare this situation with that of Example 9. If you were a lawmaker, which situation would you support? Give your reasons.

94. Is there any function of the form $y = x^{\alpha}$, $0 < \alpha < 1$, that increases more slowly than a logarithmic function whose base is greater than 1? Explain.

5.4 PROPERTIES OF LOGARITHMS

1 Write a Logarithmic Expression as a Sum or Difference of Logarithms

2 Write a Logarithmic Expression as a Single Logarithm

3 Evaluate Logarithms Whose Base Is Neither 10 nor e

4 Graph Logarithmic Functions Whose Base Is Neither 10 nor e

Logarithms have some very useful properties that can be derived directly from the definition and the laws of exponents.

EXAMPLE 1 Establishing Properties of Logarithms

(a) Show that $\log_a 1 = 0$. (b) Show that $\log_a a = 1$.

Solution (a) This fact was established when we graphed $y = \log_a x$ (see Figure 25). Algebraically, for $y = \log_a 1$, we have $a^y = 1 = a^0$, so $y = 0$.

(b) For $y = \log_a a$, we have $a^y = a = a^1$, so $y = 1$.

To summarize:

$$\log_a 1 = 0 \qquad \log_a a = 1$$

THEOREM **Properties of Logarithms**

In the properties given next, M and a are positive real numbers, with $a \neq 1$, and r is any real number.

The number $\log_a M$ is the exponent to which a must be raised to obtain M. That is,

$$a^{\log_a M} = M \tag{1}$$

The logarithm to the base a of a raised to a power equals that power. That is,

$$\log_a a^r = r \tag{2}$$

Proof of Property (1) Let $x = \log_a M$. Change this logarithmic expression to the equivalent exponential expression.

$$a^x = M$$

But $x = \log_a M$, so

$$a^{\log_a M} = M$$

Proof of Property (2) Let $x = a^r$. Change this exponential expression to the equivalent logarithmic expression

$$\log_a x = r$$

But $x = a^r$, so

$$\log_a a^r = r$$

EXAMPLE 2 Using Properties (1) and (2)

(a) $2^{\log_2 \pi} = \pi$ (b) $\log_{0.2} 0.2^{-\sqrt{2}} = -\sqrt{2}$ (c) $\ln e^{kt} = kt$

Other useful properties of logarithms are given next.

THEOREM

Properties of Logarithms

In the following properties, M, N, and a are positive real numbers, with $a \neq 1$, and r is any real number.

The Log of a Product Equals the Sum of the Logs

$$\log_a (MN) = \log_a M + \log_a N \quad (3)$$

The Log of a Quotient Equals the Difference of the Logs

$$\log_a \left(\frac{M}{N}\right) = \log_a M - \log_a N \quad (4)$$

$$\log_a \left(\frac{1}{N}\right) = -\log_a N \quad (5)$$

$$\log_a M^r = r \log_a M \quad (6)$$

We shall derive properties (3) and (6) and leave the derivations of properties (4) and (5) as exercises (see Problems 67 and 68).

Proof of Property (3) Let $A = \log_a M$ and let $B = \log_a N$. These expressions are equivalent to the exponential expressions

$$a^A = M \quad \text{and} \quad a^B = N$$

Now

$$\begin{aligned}\log_a(MN) = \log_a(a^A a^B) &= \log_a a^{A+B} && \textit{Law of exponents}\\ &= A + B && \textit{Property (2) of logarithms}\\ &= \log_a M + \log_a N\end{aligned}$$

Proof of Property (6) Let $A = \log_a M$. This expression is equivalent to

$$a^A = M$$

Now

$$\begin{aligned}\log_a M^r = \log_a (a^A)^r &= \log_a a^{rA} && \textit{Law of exponents}\\ &= rA && \textit{Property (2) of logarithms}\\ &= r\log_a M\end{aligned}$$

1 Logarithms can be used to transform products into sums, quotients into differences, and powers into factors. Such transformations prove useful in certain types of calculus problems.

EXAMPLE 3 Writing a Logarithmic Expression as a Sum of Logarithms

Write $\log_a(x\sqrt{x^2+1})$ as a sum of logarithms. Express all powers as factors.

Solution

$$\begin{aligned}\log_a(x\sqrt{x^2+1}) &= \log_a x + \log_a \sqrt{x^2+1} && \textit{Property (3)}\\ &= \log_a x + \log_a (x^2+1)^{1/2}\\ &= \log_a x + \tfrac{1}{2}\log_a(x^2+1) && \textit{Property (6)}\end{aligned}$$

EXAMPLE 4 Writing a Logarithmic Expression as a Difference of Logarithms

Write

$$\log_a \frac{x^2}{(x-1)^3}$$

as a difference of logarithms. Express all powers as factors.

Solution

$$\log_a \frac{x^2}{(x-1)^3} \underset{\uparrow}{=} \log_a x^2 - \log_a(x-1)^3 \underset{\uparrow}{=} 2\log_a x - 3\log_a(x-1)$$

Property (4) *Property (6)*

NOW WORK PROBLEM 13.

EXAMPLE 5 Writing a Logarithmic Expression as a Sum and Difference of Logarithms

Write

$$\log_a \frac{x^3\sqrt{x^2+1}}{(x+1)^4}$$

as a sum and difference of logarithms. Express all powers as factors.

Solution

$$\begin{aligned}
\log_a \frac{x^3\sqrt{x^2+1}}{(x+1)^4} &= \log_a\left(x^3\sqrt{x^2+1}\right) - \log_a (x+1)^4 && \text{Property (4)}\\
&= \log_a x^3 + \log_a \sqrt{x^2+1} - \log_a (x+1)^4 && \text{Property (3)}\\
&= \log_a x^3 + \log_a (x^2+1)^{1/2} - \log_a (x+1)^4 &&\\
&= 3\log_a x + \tfrac{1}{2}\log_a (x^2+1) - 4\log_a (x+1) && \text{Property (6)}
\end{aligned}$$

2 Another use of properties (3) through (6) is to write sums and/or differences of logarithms with the same base as a single logarithm. This skill will be needed to solve certain logarithmic equations discussed in the next section.

EXAMPLE 6 Writing Expressions as a Single Logarithm

Write each of the following as a single logarithm:

(a) $\log_a 7 + 4\log_a 3$

(b) $\frac{2}{3}\log_a 8 - \log_a (3^4 - 8)$

(c) $\log_a x + \log_a 9 + \log_a (x^2+1) - \log_a 5$

Solution (a)

$$\begin{aligned}
\log_a 7 + 4\log_a 3 &= \log_a 7 + \log_a 3^4 && \text{Property (6)}\\
&= \log_a 7 + \log_a 81 &&\\
&= \log_a (7\cdot 81) && \text{Property (3)}\\
&= \log_a 567 &&
\end{aligned}$$

(b)

$$\begin{aligned}
\tfrac{2}{3}\log_a 8 - \log_a (3^4 - 8) &= \log_a 8^{2/3} - \log_a (81 - 8) && \text{Property (6)}\\
&= \log_a 4 - \log_a 73 &&\\
&= \log_a \left(\tfrac{4}{73}\right) && \text{Property (4)}
\end{aligned}$$

(c)

$$\begin{aligned}
\log_a x + \log_a 9 + \log_a (x^2+1) - \log_a 5 &= \log_a 9x + \log_a (x^2+1) - \log_a 5\\
&= \log_a [9x(x^2+1)] - \log_a 5\\
&= \log_a\left[\frac{9x(x^2+1)}{5}\right]
\end{aligned}$$

Warning: A common error made by some students is to express the logarithm of a sum as the sum of logarithms.

$$\log_a (M+N) \quad \text{is not equal to} \quad \log_a M + \log_a N$$

Correct statement $\quad \log_a (MN) = \log_a M + \log_a N \quad$ Property (3)

Another common error is to express the difference of logarithms as the quotient of logarithms.

$$\log_a M - \log_a N \quad \text{is not equal to} \quad \frac{\log_a M}{\log_a N}$$

Correct statement $\quad \log_a M - \log_a N = \log_a\left(\dfrac{M}{N}\right) \quad$ Property (4)

NOW WORK PROBLEM 23.

EXAMPLE 7 Writing a Power Equation as a Linear Equation

Write the equation $y = ax^b$ as a linear equation.

Solution The equation $y = ax^b$ is the model for the power function. Earlier, we obtained a value for the correlation coefficient, r, of this model even though it is not linear. We said this is due to the fact the model can be written in linear form using logarithms. Let's see how.

$$y = ax^b$$
$$\ln y = \ln(ax^b) \quad \text{Take the natural log of both sides.}$$
$$\ln y = \ln a + \ln x^b \quad \text{Property (3)}$$
$$\ln y = \ln a + b \ln x \quad \text{Property (6)}$$
$$Y = \ln a + bX$$

This is an equation in linear form with y-intercept $\ln a$ and slope b.

SEEING THE CONCEPT Redo Example 3 from Section 4.1 by letting the independent variable $x = \ln$ (time) and the dependent variable be $y = \ln$ (distance). Find the line of best fit. What is the slope of your line of best fit? Compare it to the value of b found in Example 3. Why is the value of a found in Example 3 different from the y-intercept of the line of best fit? Compare the values of the correlation coefficient. Are they the same?

There remain two other properties of logarithms that we need to know. They are consequences of the fact that the logarithmic function $y = \log_a x$ is one-to-one.

THEOREM In the following properties, M, N, and a are positive real numbers, with $a \neq 1$:

$$\text{If } M = N, \quad \text{then} \quad \log_a M = \log_a N. \qquad (7)$$
$$\text{If } \log_a M = \log_a N, \quad \text{then} \quad M = N. \qquad (8)$$

Properties (7) and (8) are useful for solving *logarithmic equations,* a topic discussed in the next section.

Using a Calculator to Evaluate and Graph Logarithms with Bases Other Than e or 10

3 Logarithms to the base 10, called **common logarithms,** were used to facilitate arithmetic computations before the widespread use of calculators. (See the Historical Feature at the end of this section.) Natural logarithms, that is, logarithms whose base is the number e, remain very important because they arise frequently in the study of natural phenomena.

Common logarithms are usually abbreviated by writing **log,** with the base understood to be 10, just as natural logarithms are abbreviated by **ln,** with the base understood to be e.

Graphing calculators have both **log** and **ln** keys to calculate the common logarithm and natural logarithm of a number. Let's look at an example to see how to calculate logarithms having a base other than 10 or e.

EXAMPLE 8 Evaluating Logarithms Whose Base Is Neither 10 nor *e*

Evaluate $\log_2 7$.

Solution Let $y = \log_2 7$. Then $2^y = 7$, so

$$
\begin{aligned}
2^y &= 7 \\
\ln 2^y &= \ln 7 && \text{Property (7)} \\
y \ln 2 &= \ln 7 && \text{Property (6)} \\
y &= \frac{\ln 7}{\ln 2} && \text{Solve for } y. \\
&= 2.8074 && \text{Use calculator (} \boxed{\ln} \text{ key).}
\end{aligned}
$$

Example 8 shows how to evaluate a logarithm whose base is 2 by changing to logarithms involving the base. In general, we use the **Change-of-Base Formula.**

THEOREM **Change-of-Base Formula**

If $a \neq 1$, $b \neq 1$, and M are positive real numbers, then

$$\log_a M = \frac{\log_b M}{\log_b a} \tag{9}$$

Proof We derive this formula as follows: Let $y = \log_a M$. Then $a^y = M$, so

$$
\begin{aligned}
\log_b a^y &= \log_b M && \text{Property (7)} \\
y \log_b a &= \log_b M && \text{Property (6)} \\
y &= \frac{\log_b M}{\log_b a} && \text{Solve for } y. \\
\log_a M &= \frac{\log_b M}{\log_b a} && y = \log_a M
\end{aligned}
$$

Since calculators have only keys for $\boxed{\log}$ and $\boxed{\ln}$, in practice, the Change-of-Base Formula uses either $b = 10$ or $b = e$. Thus,

$$\log_a M = \frac{\log M}{\log a} \quad \text{and} \quad \log_a M = \frac{\ln M}{\ln a} \tag{10}$$

EXAMPLE 9 Using the Change-of-Base Formula

Calculate:

(a) $\log_5 89$ (b) $\log_{\sqrt{2}} \sqrt{5}$

Solution (a) $\log_5 89 = \dfrac{\log 89}{\log 5} \approx \dfrac{1.94939}{0.69897} = 2.7889$

or

$$\log_5 89 = \frac{\ln 89}{\ln 5} \approx \frac{4.4886}{1.6094} = 2.7889$$

(b) $\log_{\sqrt{2}} \sqrt{5} = \dfrac{\log \sqrt{5}}{\log \sqrt{2}} = \dfrac{\frac{1}{2}\log 5}{\frac{1}{2}\log 2} \approx 2.3219$

or

$$\log_{\sqrt{2}} \sqrt{5} = \frac{\ln \sqrt{5}}{\ln \sqrt{2}} = \frac{\frac{1}{2}\ln 5}{\frac{1}{2}\ln 2} \approx 2.3219$$

NOW WORK PROBLEM 39.

4 We also use the Change-of-Base Formula to graph logarithmic functions whose base is neither 10 nor *e*.

EXAMPLE 10 Graphing a Logarithmic Function Whose Base Is Neither 10 nor *e*

Use a graphing utility to graph $y = \log_2 x$.

Figure 31

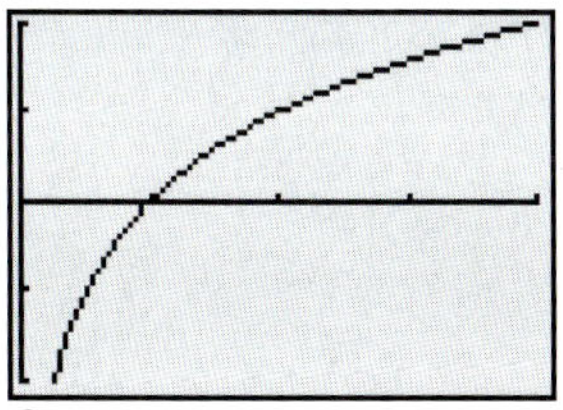

Solution Since graphing utilities only have logarithms with the base 10 or the base *e*, we need to use the Change-of-Base Formula to express $y = \log_2 x$ in terms of logarithms with base 10 or base *e*. We can graph either $y = \ln x/\ln 2$ or $y = \log x/\log 2$ to obtain the graph of $y = \log_2 x$. See Figure 31.

✓CHECK: Verify that $y = \ln x/\ln 2$ and $y = \log x/\log 2$ result in the same graph by graphing each on the same viewing rectangle.

NOW WORK PROBLEM 47.

SUMMARY

Properties of Logarithms

In the list that follows, $a > 0$, $a \neq 1$, and $b > 0$, $b \neq 1$; also, $M > 0$ and $N > 0$.

Definition	$y = \log_a x$ means $x = a^y$
Properties of logarithms	$\log_a 1 = 0$; $\log_a a = 1$
	$a^{\log_a M} = M$; $\log_a a^r = r$
	$\log_a (MN) = \log_a M + \log_a N$
	$\log_a \left(\frac{M}{N}\right) = \log_a M - \log_a N$
	$\log_a \left(\frac{1}{N}\right) = -\log_a N$
	$\log_a M^r = r \log_a M$
Change-of-Base Formula	$\log_a M = \dfrac{\log_b M}{\log_b a}$

HISTORICAL FEATURE

Logarithms were invented about 1590 by John Napier (1550–1617) and Jobst Bürgi (1552–1632), working independently. Napier, whose work had the greater influence, was a Scottish lord, a secretive man whose neighbors were inclined to believe him to be in league with the devil. His approach to logarithms was quite different from ours; it was based on the relationship between arithmetic and geometric sequences (see the chapter on induction and sequences), and not on the inverse function relationship of logarithms to exponential functions (described in Section 5.3). Napier's tables, published in 1614, listed what would now be called *natural logarithms* of sines and were rather difficult to use. A London professor, Henry Briggs, became interested in the tables and visited Napier. In their conversations, they developed the idea of common logarithms, which were published in 1617. Their importance for calculation was immediately recognized, and by 1650 they were being printed as far away as China. They remained an important calculation tool until the advent of the inexpensive handheld calculator about 1972, which has decreased their calculational, but not their theoretical, importance.

A side effect of the invention of logarithms was the popularization of the decimal system of notation for real numbers.

5.4 EXERCISES

In Problems 1–12, suppose that $\ln 2 = a$ *and* $\ln 3 = b$. *Use properties of logarithms to write each logarithm in terms of a and b.*

1. $\ln 6$ **2.** $\ln \frac{2}{3}$ **3.** $\ln 1.5$ **4.** $\ln 0.5$

5. $\ln 2e$ **6.** $\ln\left(\frac{3}{e}\right)$ **7.** $\ln 12$ **8.** $\ln 24$

9. $\ln \sqrt[5]{18}$ **10.** $\ln \sqrt[4]{48}$ **11.** $\log_2 3$ **12.** $\log_3 2$

In Problems 13–22, write each expression as a sum and/or difference of logarithms. Express powers as factors.

13. $\ln(x^2\sqrt{1-x})$ **14.** $\ln(x\sqrt{1+x^2})$ **15.** $\log_2\left(\frac{x^3}{x-3}\right)$ **16.** $\log_5\left(\frac{\sqrt[3]{x^2+1}}{x^2-1}\right)$

17. $\log\left[\frac{x(x+2)}{(x+3)^2}\right]$ **18.** $\log\frac{x^3\sqrt{x+1}}{(x-2)^2}$ **19.** $\ln\left[\frac{x^2-x-2}{(x+4)^2}\right]^{1/3}$ **20.** $\ln\left[\frac{(x-4)^2}{x^2-1}\right]^{2/3}$

21. $\ln\frac{5x\sqrt{1-3x}}{(x-4)^3}$ **22.** $\ln\left[\frac{5x^2\sqrt[3]{1-x}}{4(x+1)^2}\right]$

In Problems 23–32, write each expression as a single logarithm.

23. $3\log_5 u + 4\log_5 v$ **24.** $\log_3 u^2 - \log_3 v$

25. $\log_{1/2}\sqrt{x} - \log_{1/2} x^3$ **26.** $\log_2\left(\frac{1}{x}\right) + \log_2\left(\frac{1}{x^2}\right)$

27. $\ln\left(\frac{x}{x-1}\right) + \ln\left(\frac{x+1}{x}\right) - \ln(x^2-1)$ **28.** $\log\left(\frac{x^2+2x-3}{x^2-4}\right) - \log\left(\frac{x^2+7x+6}{x+2}\right)$

29. $8\log_2\sqrt{3x-2} - \log_2\left(\frac{4}{x}\right) + \log_2 4$ **30.** $21\log_3\sqrt[3]{x} + \log_3 9x^2 - \log_3 25$

31. $2\log_a 5x^3 - \frac{1}{2}\log_a(2x+3)$ **32.** $\frac{1}{3}\log(x^3+1) + \frac{1}{2}\log(x^2+1)$

33. Write the exponential model $y = ab^x$ as a linear model.

34. Is the logarithmic model $y = a + b \ln x$ linear? If so, what is the slope? What is the y-intercept?

In Problems 35–38, find the exact value of each expression.

35. $3^{\log_3 5 - \log_3 4}$ **36.** $5^{\log_5 6 + \log_5 7}$ **37.** $e^{\log_{e^2} 16}$ **38.** $e^{\log_{e^2} 9}$

In Problems 39–46, use the Change-of-Base Formula and a calculator to evaluate each logarithm. Round your answer to three decimal places.

39. $\log_3 21$ **40.** $\log_5 18$ **41.** $\log_{1/3} 71$ **42.** $\log_{1/2} 15$

43. $\log_{\sqrt{2}} 7$ **44.** $\log_{\sqrt{5}} 8$ **45.** $\log_\pi e$ **46.** $\log_\pi \sqrt{2}$

In Problems 47–52, graph each function using a graphing utility and the Change-of-Base Formula.

47. $y = \log_4 x$ **48.** $y = \log_5 x$ **49.** $y = \log_2 (x + 2)$ **50.** $y = \log_4 (x - 3)$

51. $y = \log_{x-1} (x + 1)$ **52.** $y = \log_{x+2} (x - 2)$

In Problems 53–62, express y as a function of x. The constant C is a positive number.

53. $\ln y = \ln x + \ln C$ **54.** $\ln y = \ln(x + C)$ **55.** $\ln y = \ln x + \ln(x + 1) + \ln C$

56. $\ln y = 2 \ln x - \ln(x + 1) + \ln C$ **57.** $\ln y = 3x + \ln C$ **58.** $\ln y = -2x + \ln C$

59. $\ln(y - 3) = -4x + \ln C$ **60.** $\ln(y + 4) = 5x + \ln C$

61. $3 \ln y = \frac{1}{2} \ln(2x + 1) - \frac{1}{3} \ln(x + 4) + \ln C$ **62.** $2 \ln y = -\frac{1}{2} \ln x + \frac{1}{3} \ln(x^2 + 1) + \ln C$

63. Find the value of $\log_2 3 \cdot \log_3 4 \cdot \log_4 5 \cdot \log_5 6 \cdot \log_6 7 \cdot \log_7 8$.

64. Find the value of $\log_2 4 \cdot \log_4 6 \cdot \log_6 8$.

65. Find the value of $\log_2 3 \cdot \log_3 4 \cdot \ldots \cdot \log_n (n + 1) \cdot \log_{n+1} 2$.

66. Find the value of $\log_2 2 \cdot \log_2 4 \cdot \ldots \cdot \log_2 2^n$.

67. Show that $\log_a (M/N) = \log_a M - \log_a N$, where a, M, and N are positive real numbers, with $a \neq 1$.

68. Show that $\log_a (1/N) = -\log_a N$, where a and N are positive real numbers, with $a \neq 1$.

69. Find the domain of $f(x) = \log_a x^2$ and the domain of $g(x) = 2 \log_a x$. Since $\log_a x^2 = 2 \log_a x$, how do you reconcile the fact that the domains are not equal? Write a brief explanation.

5.5 LOGARITHMIC AND EXPONENTIAL EQUATIONS

1 Solve Logarithmic Equations

2 Solve Exponential Equations

3 Solve Logarithmic and Exponential Equations Using a Graphing Utility

Logarithmic Equations

1 Equations that contain terms of the form $\log_a x$, where a is a positive real number, with $a \neq 1$, are often called **logarithmic equations.**

Our practice will be to solve equations, whenever possible, by finding exact solutions using algebraic methods. In such cases, we will also verify the solution obtained by using a graphing utility. When algebraic methods cannot be used, approximate solutions will be obtained using a graphing utility. The reader is encouraged to pay particular attention to the form of equations for which exact solutions are possible.

EXAMPLE 1 Solving a Logarithmic Equation

Solve: $\log_3(4x - 7) = 2$

Solution We can obtain an exact solution by changing the logarithm to exponential form.

$$\begin{aligned} \log_3(4x - 7) &= 2 \\ 4x - 7 &= 3^2 \\ 4x - 7 &= 9 \\ 4x &= 16 \\ x &= 4 \end{aligned}$$

We verify the solution by graphing $Y_1 = \log_3(4x - 7)$ and $Y_2 = 2$ to determine where the graphs intersect. See Figure 32. The graphs intersect at $x = 4$.

Figure 32

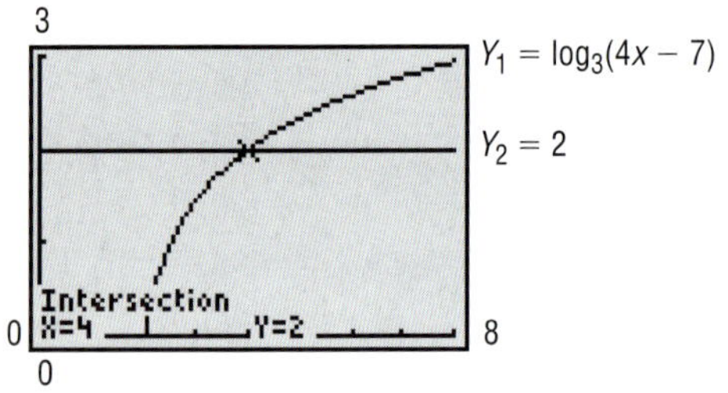

Care must be taken when solving logarithmic equations algebraically. Be sure to check each apparent solution in the original equation and discard any that are extraneous. In the expression $\log_a M$, remember that a and M are positive and $a \neq 1$.

EXAMPLE 2 Solving a Logarithmic Equation

Solve: $2\log_5 x = \log_5 9$

Solution Because each logarithm is to the same base, 5, we can obtain an exact solution as follows:

$$\begin{aligned} 2\log_5 x &= \log_5 9 \\ \log_5 x^2 &= \log_5 9 && \text{Property (6), Section 5.4} \\ x^2 &= 9 && \text{Property (8), Section 5.4} \\ x = 3 \quad &\text{or} \quad x = -3 \end{aligned}$$

Recall that logarithms of negative numbers are not defined, so, in the expression $2\log_5 x$, x must be positive. Therefore, -3 is extraneous and we discard it.

The equation has only one solution, 3.

You should verify for yourself that 3 is the only solution using a graphing utility.

NOW WORK PROBLEM 1.

EXAMPLE 3 Solving a Logarithmic Equation

Solve: $\log_4(x+3) + \log_4(2-x) = 1$

Solution To obtain an exact solution, we need to express the left side as a single logarithm. Then we will change the expression to exponential form.

$$\log_4(x+3) + \log_4(2-x) = 1$$

$$\log_4[(x+3)(2-x)] = 1 \quad \textit{Property (3), Section 5.4}$$

$$(x+3)(2-x) = 4^1 = 4 \quad \textit{Change to an exponential expression.}$$

$$-x^2 - x + 6 = 4 \quad \textit{Simplify.}$$

$$x^2 + x - 2 = 0 \quad \textit{Place the quadratic equation in standard form.}$$

$$(x+2)(x-1) = 0 \quad \textit{Factor.}$$

$$x = -2 \quad \text{or} \quad x = 1 \quad \textit{Zero-Product Property}$$

Since the arguments of each logarithmic expression in the equation are positive for both $x = -2$ and $x = 1$, neither is extraneous. You should verify that both of these are solutions using a graphing utility.

NOW WORK PROBLEM 11.

Exponential Equations

2 Equations that involve terms of the form a^x, $a > 0$, $a \neq 1$, are often referred to as **exponential equations.** Such equations sometimes can be solved by appropriately applying the laws of exponents and equation (1), that is,

$$\text{If } a^u = a^v, \quad \text{then } u = v. \qquad (1)$$

To use equation (1), each side of the equality must be written with the same base.

EXAMPLE 4 Solving an Exponential Equation

Solve: $3^{x+1} = 81$

Solution Since $81 = 3^4$, we can write the equation as

$$3^{x+1} = 81 = 3^4$$

Now we have the same base, 3, on each side, so we can apply (1) to obtain

$$x + 1 = 4$$

$$x = 3$$

We verify the solution by graphing $Y_1 = 3^{x+1}$ and $Y_2 = 81$ to determine where the graphs intersect. See Figure 33. The graphs intersect at $x = 3$.

Figure 33

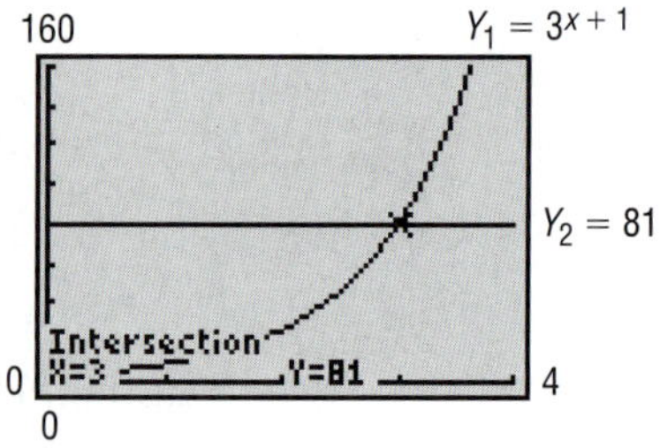

NOW WORK PROBLEM 19.

EXAMPLE 5 Solving an Exponential Equation

Solve: $e^{-x^2} = (e^x)^2 \cdot \dfrac{1}{e^3}$

Solution We use some laws of exponents first to get the same base e on each side.

$$e^{-x^2} = (e^x)^2 \cdot \frac{1}{e^3} = e^{2x} \cdot e^{-3} = e^{2x-3}$$

Now apply (1) to get

$$-x^2 = 2x - 3$$

$$x^2 + 2x - 3 = 0 \quad \textit{Place the quadratic equation in standard form.}$$

$$(x + 3)(x - 1) = 0 \quad \textit{Factor.}$$

$$x = -3 \quad \text{or} \quad x = 1 \quad \textit{Use the Zero-Product Property.}$$

You should verify these solutions using a graphing utility.

EXAMPLE 6 Solving an Exponential Equation

Solve: $4^x - 2^x - 12 = 0$

Solution We note that $4^x = (2^2)^x = 2^{2x} = (2^x)^2$, so the equation is actually quadratic in form, and we can rewrite it as

$$(2^x)^2 - 2^x - 12 = 0 \quad \textit{Let } u = 2^x\textit{; then } u^2 - u - 12 = 0.$$

Now we can factor as usual.

$$(2^x - 4)(2^x + 3) = 0 \qquad (u - 4)(u + 3) = 0$$

$$2^x - 4 = 0 \quad \text{or} \quad 2^x + 3 = 0 \qquad u - 4 = 0 \quad \text{or} \quad u + 3 = 0$$

$$2^x = 4 \qquad 2^x = -3 \qquad u = 2^x = 4 \qquad u = 2^x = -3$$

The equation on the left has the solution $x = 2$, since $2^x = 4 = 2^2$; the equation on the right has no solution, since $2^x > 0$ for all x.

In each of the preceding three examples, we were able to write each exponential expression using the same base, obtaining exact solutions to the equation. When this is not possible, logarithms can sometimes be used to obtain the solution.

EXAMPLE 7 Solving an Exponential Equation

Solve: $2^x = 5$

Solution We write the exponential equation as the equivalent logarithmic equation.

$$2^x = 5$$

$$x = \log_2 5 = \frac{\ln 5}{\ln 2}$$

↑ *Change-of-Base Formula (10), Section 5.4*

Alternatively, we can solve the equation $2^x = 5$ by taking the natural logarithm (or common logarithm) of each side. Taking the natural logarithm,

$$\begin{aligned} 2^x &= 5 \\ \ln 2^x &= \ln 5 \\ x \ln 2 &= \ln 5 && \text{Property (6), Section 5.4} \\ x &= \frac{\ln 5}{\ln 2} \end{aligned}$$

Using a calculator, the solution, rounded to three decimal places, is

$$x = \frac{\ln 5}{\ln 2} \approx 2.322$$

NOW WORK PROBLEM 33.

EXAMPLE 8 Solving an Exponential Equation

Solve: $8 \cdot 3^x = 5$

Solution

$$\begin{aligned} 8 \cdot 3^x &= 5 \\ 3^x &= \tfrac{5}{8} && \text{Solve for } 3^x \\ x &= \log_3\left(\tfrac{5}{8}\right) = \frac{\ln \frac{5}{8}}{\ln 3} && \text{Solve for } x \end{aligned}$$

Using a calculator, the solution, rounded to three decimal places, is

$$x = \frac{\ln\left(\frac{5}{8}\right)}{\ln 3} \approx -0.428$$

EXAMPLE 9 Solving an Exponential Equation

Solve: $5^{x-2} = 3^{3x+2}$

Solution Because the bases are different, we take the natural logarithm of each side and apply appropriate properties of logarithms. The result is an equation in x that we can solve.

$$
\begin{aligned}
5^{x-2} &= 3^{3x+2} \\
\ln 5^{x-2} &= \ln 3^{3x+2} && \textit{Property (7)} \\
(x-2)\ln 5 &= (3x+2)\ln 3 && \textit{Property (6)} \\
(\ln 5)x - 2\ln 5 &= (3\ln 3)x + 2\ln 3 && \textit{Distribute.} \\
(\ln 5)x - (3\ln 3)x &= 2\ln 3 + 2\ln 5 && \textit{Place terms involving x on the left.} \\
(\ln 5 - 3\ln 3)x &= 2\ln 3 + 2\ln 5 && \textit{Factor.} \\
x &= \frac{2(\ln 3 + \ln 5)}{\ln 5 - 3\ln 3} \approx -3.212
\end{aligned}
$$

NOW WORK PROBLEM 41.

Graphing Utility Solutions

3 The techniques introduced in this section only apply to certain types of logarithmic and exponential equations. Solutions for other types are usually studied in calculus, using numerical methods. However, we can use a graphing utility to approximate the solution.

EXAMPLE 10 Solving Equations Using a Graphing Utility

Solve: $\log_3 x + \log_4 x = 4$
Express the solution(s) rounded to two decimal places.

Solution The solution is found by graphing

$$Y_1 = \log_3 x + \log_4 x = \frac{\log x}{\log 3} + \frac{\log x}{\log 4} \quad \text{and} \quad Y_2 = 4$$

(Remember that you must use the Change-of-Base Formula to graph Y_1). Y_1 is an increasing function (do you know why?), and so there is only one point of intersection for Y_1 and Y_2. Figure 34 shows the graphs of Y_1 and Y_2. Using the INTERSECT command, the solution is 11.61, rounded to two decimal places.

Figure 34

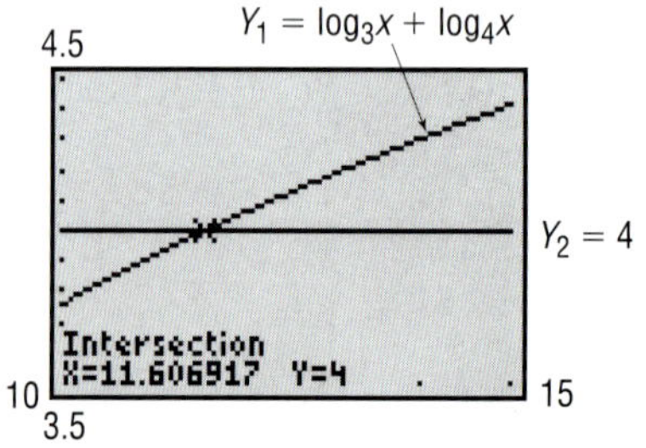

Can you discover an algebraic solution to Example 10?
[Hint: Factor log x from Y_1.]

EXAMPLE 11 Solving Equations Using a Graphing Utility

Solve: $x + e^x = 2$.
Express the solution(s) rounded to two decimal places.

Solution The solution is found by graphing $Y_1 = x + e^x$ and $Y_2 = 2$. Y_1 is an increasing function (do you know why?) and so there is only one point of intersection for Y_1 and Y_2. Figure 35 shows the graphs of Y_1 and Y_2. Using the INTERSECT command, the solution is 0.44 rounded to two decimal places.

Figure 35

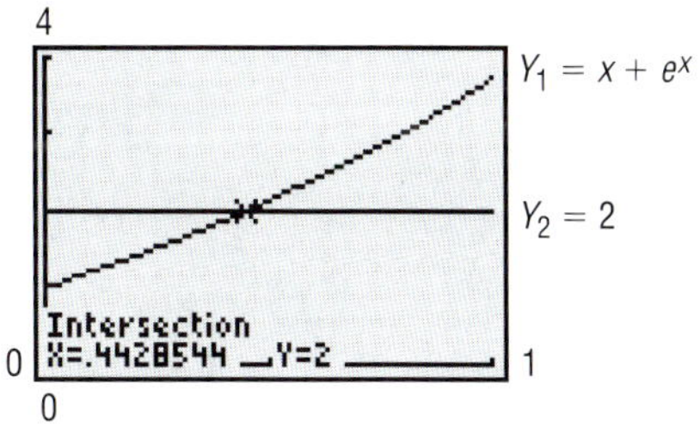

5.5 EXERCISES

In Problems 1–60, solve each equation. Verify your solution using a graphing utility.

1. $\log_2(2x+1) = 3$

2. $\log_3(3x-2) = 2$

3. $\log_3(x^2+1) = 2$

4. $\log_5(x^2+x+4) = 2$

5. $\frac{1}{2}\log_3 x = 2\log_3 2$

6. $-2\log_4 x = \log_4 9$

7. $2\log_5 x = 3\log_5 4$

8. $3\log_2 x = -\log_2 27$

9. $3\log_2(x-1) + \log_2 4 = 5$

10. $2\log_3(x+4) - \log_3 9 = 2$

11. $\log x + \log(x+15) = 2$

12. $\log_4 x + \log_4(x-3) = 1$

13. $\log_x 4 = 2$

14. $\log_x\left(\frac{1}{8}\right) = 3$

15. $\log_3(x-1)^2 = 2$

16. $\log_2(x+4)^2 = 6$

17. $\log_{1/2}(3x+1)^{1/3} = -2$

18. $\log_{1/3}(1-2x)^{1/2} = -1$

19. $2^{2x+1} = 4$

20. $5^{1-2x} = \frac{1}{5}$

21. $3^{x^3} = 9^x$

22. $4^{x^2} = 2^x$

23. $8^{x^2-2x} = \frac{1}{2}$

24. $9^{-x} = \frac{1}{3}$

25. $2^x \cdot 8^{-x} = 4^x$

26. $\left(\frac{1}{2}\right)^{1-x} = 4$

27. $2^{2x} + 2^x - 12 = 0$

28. $3^{2x} + 3^x - 2 = 0$

29. $3^{2x} + 3^{x+1} - 4 = 0$

30. $4^x - 2^x = 0$

31. $4^x = 8$

32. $9^{2x} = 27$

33. $2^x = 10$

34. $3^x = 14$

35. $8^{-x} = 1.2$

36. $2^{-x} = 1.5$

37. $3^{1-2x} = 4^x$

38. $2^{x+1} = 5^{1-2x}$

39. $\left(\frac{3}{5}\right)^x = 7^{1-x}$

40. $\left(\frac{4}{3}\right)^{1-x} = 5^x$

41. $1.2^x = (0.5)^{-x}$

42. $(0.3)^{1+x} = 1.7^{2x-1}$

43. $\pi^{1-x} = e^x$

44. $e^{x+3} = \pi^x$

45. $5(2^{3x}) = 8$

46. $0.3(4^{0.2x}) = 0.2$

47. $400e^{0.2x} = 600$

48. $500e^{0.3x} = 600$

49. $\log_a(x-1) - \log_a(x+6) = \log_a(x-2) - \log_a(x+3)$

50. $\log_a x + \log_a(x-2) = \log_a(x+4)$

51. $\log_{1/3}(x^2+x) - \log_{1/3}(x^2-x) = -1$

52. $\log_4(x^2-9) - \log_4(x+3) = 3$

53. $\log_2 8^x = -3$

54. $\log_3 3^x = -1$

55. $\log_2(x^2+1) - \log_4 x^2 = 1$
[Hint: Change $\log_4 x^2$ to base 2.]

56. $\log_2(3x+2) - \log_4 x = 3$

57. $\log_{16} x + \log_4 x + \log_2 x = 7$

58. $\log_9 x + 3\log_3 x = 14$

59. $(\sqrt[3]{2})^{2-x} = 2^{x^2}$

60. $\log_2 x^{\log_2 x} = 4$

In Problems 61–76, use a graphing utility to solve each equation. Express your answer rounded to two decimal places.

61. $\log_5 x + \log_3 x = 1$

62. $\log_2 x + \log_6 x = 3$

63. $\log_5(x+1) - \log_4(x-2) = 1$

64. $\log_2(x-1) - \log_6(x+2) = 2$

65. $e^x = -x$

66. $e^{2x} = x + 2$

67. $e^x = x^2$

68. $e^x = x^3$

69. $\ln x = -x$

70. $\ln 2x = -x + 2$

71. $\ln x = x^3 - 1$

72. $\ln x = -x^2$

73. $e^x + \ln x = 4$

74. $e^x - \ln x = 4$

75. $e^{-x} = \ln x$

76. $e^{-x} = -\ln x$

5.6 COMPOUND INTEREST

1. Determine the Future Value of a Lump Sum of Money
2. Calculate Effective Rates of Return
3. Determine the Present Value of a Lump Sum of Money
4. Determine the Time Required to Double or Triple a Lump Sum of Money

1 Interest is money paid for the use of money. The total amount borrowed (whether by an individual from a bank in the form of a loan or by a bank from an individual in the form of a savings account) is called the **principal.** The **rate of interest,** expressed as a percent, is the amount charged for the use of the principal for a given period of time, usually on a yearly (that is, per annum) basis.

Simple Interest Formula

If a principal of P dollars is borrowed for a period of t years at a per annum interest rate r, expressed as a decimal, the interest I charged is

$$I = Prt \tag{1}$$

Interest charged according to formula (1) is called **simple interest.**

In working with problems involving interest, we use the term **payment period** as follows:

Annually	Once per year	Monthly	12 times per year
Semiannually	Twice per year	Daily	365 times per year*
Quarterly	Four times per year		

When the interest due at the end of a payment period is added to the principal so that the interest computed at the end of the next payment period is based on this new principal amount (old principal + interest), the interest is said to have been **compounded.** Thus, **compound interest** is interest paid on previously earned interest.

EXAMPLE 1 Computing Compound Interest

A credit union pays interest of 8% per annum compounded quarterly on a certain savings plan. If \$1000 is deposited in such a plan and the interest is left to accumulate, how much is in the account after 1 year?

Solution We use the simple interest formula, $I = Prt$. The principal P is \$1000 and the rate of interest is $8\% = 0.08$. After the first quarter of a year, the time t is $\frac{1}{4}$ year, so the interest earned is

$$I = Prt = (\$1000)(0.08)(\tfrac{1}{4}) = \$20$$

*Most banks use a 360-day "year." Why do you think they do?

The new principal is $P + I = \$1000 + \$20 = \$1020$. At the end of the second quarter, the interest on this principal is

$$I = (\$1020)(0.08)(\tfrac{1}{4}) = \$20.40$$

At the end of the third quarter, the interest on the new principal of $\$1020 + \$20.40 = \$1040.40$ is

$$i = (\$1040.40)(0.08)(\tfrac{1}{4}) = \$20.81$$

Finally, after the fourth quarter, the interest is

$$I = (\$1061.21)(0.08)(\tfrac{1}{4}) = \$21.22$$

Thus, after 1 year the account contains \$1082.43.

The pattern of the calculations performed in Example 1 leads to a general formula for compound interest. To fix our ideas, let P represent the principal to be invested at a per annum interest rate r, which is compounded n times per year. (For computing purposes, r is expressed as a decimal.) The interest earned after each compounding period is the principal times r/n. Thus, the amount A after one compounding period is

$$A = P + P\left(\frac{r}{n}\right) = P\left(1 + \frac{r}{n}\right)$$

After two compounding periods, the amount A, based on the new principal $P(1 + r/n)$, is

$$A = \underbrace{P\left(1 + \tfrac{r}{n}\right)}_{\text{New principal.}} + \underbrace{P\left(1 + \tfrac{r}{n}\right)\left(\tfrac{r}{n}\right)}_{\text{Interest on new principal.}} = P\left(1 + \tfrac{r}{n}\right)\left(1 + \tfrac{r}{n}\right) = P\left(1 + \tfrac{r}{n}\right)^2$$

After three compounding periods,

$$A = P\left(1 + \frac{r}{n}\right)^2 + P\left(1 + \frac{r}{n}\right)^2\left(\frac{r}{n}\right) = P\left(1 + \frac{r}{n}\right)^2\left(1 + \frac{r}{n}\right) = P\left(1 + \frac{r}{n}\right)^3$$

Continuing this way, after n compounding periods (1 year),

$$A = P\left(1 + \frac{r}{n}\right)^n$$

Because t years will contain $n \cdot t$ compounding periods, after t years we have

$$A = P\left(1 + \frac{r}{n}\right)^{nt}$$

THEOREM **Compound Interest Formula**

The amount A after t years due to a principal P invested at an annual interest rate r compounded n times per year is

$$A = P\left(1 + \frac{r}{n}\right)^{nt} \qquad (2)$$

The amount A is typically referred to as the **future value** of the account, while P is called the *present value.*

EXPLORATION To see the effects of compounding interest monthly on an initial deposit of \$1, graph $Y_1 = (1 + \frac{r}{12})^{12x}$ with $r = 0.06$ and $r = 0.12$ for $0 \le x \le 30$. What is the future value of \$1 in 30 years when the interest rate per annum is $r = 0.06$ (6%)? What is the future value of \$1 in 30 years when the interest rate per annum is $r = 0.12$ (12%)? Does doubling the interest rate double the future value?

EXAMPLE 2 Comparing Investments Using Different Compounding Periods

Investing \$1000 at an annual rate of 10% compounded annually, quarterly, monthly, and daily will yield the following amounts after 1 year:

Annual compounding:
$$A = P(1 + r) = (\$1000)(1 + 0.10) = \$1100.00$$

Quarterly compounding:
$$A = P\left(1 + \frac{r}{4}\right)^4 = (\$1000)(1 + 0.025)^4 = \$1103.81$$

Monthly compounding:
$$A = P\left(1 + \frac{r}{12}\right)^{12} = (\$1000)(1 + 0.00833)^{12} = \$1104.71$$

Daily compounding:
$$A = P\left(1 + \frac{r}{365}\right)^{365} = (\$1000)(1 + 0.000274)^{365} = \$1105.16$$

NOW WORK PROBLEM 1.

From Example 2, we can see that the effect of compounding more frequently is that the amount after 1 year is higher: \$1000 compounded 4 times a year at 10% results in \$1103.81; \$1000 compounded 12 times a year at 10% results in \$1104.71; and \$1000 compounded 365 times a year at 10% results in \$1105.16. This leads to the following question: What would happen to the amount after 1 year if the number of times that the interest is compounded were increased without bound?

Let's find the answer. Suppose that P is the principal, r is the per annum interest rate, and n is the number of times that the interest is compounded each year. The amount after 1 year is

$$A = P\left(1 + \frac{r}{n}\right)^n$$

Now suppose that the number n of times that the interest is compounded per year gets larger and larger; that is, suppose that $n \to \infty$. Then

$$A = P\left(1 + \frac{r}{n}\right)^n = P\left[1 + \frac{1}{n/r}\right]^n = P\left(\left[1 + \frac{1}{n/r}\right]^{n/r}\right)^r = P\left[\left(1 + \frac{1}{h}\right)^h\right]^r \quad (3)$$

$$\uparrow \quad h = \frac{n}{r}$$

In (3), as $n \to \infty$, then $h = n/r \to \infty$, and the expression in brackets equals e. [Refer to (2) on p. 329]. Thus, $A \to Pe^r$. Table 7 compares $(1 + r/n)^n$, for large values of n, to e^r for $r = 0.05$, $r = 0.10$, $r = 0.15$, and $r = 1$. The larger n gets the closer $(1 + r/n)^n$ gets to e^r. Thus, no matter how frequent the compounding, the amount after 1 year has the definite ceiling Pe^r.

TABLE 7

	$\left(1+\frac{r}{n}\right)^n$			
	$n = 100$	$n = 1000$	$n = 10{,}000$	e^r
$r = 0.05$	1.0512579	1.05127	1.051271	1.0512711
$r = 0.10$	1.1051157	1.1051654	1.1051703	1.1051709
$r = 0.15$	1.1617037	1.1618212	1.1618329	1.1618342
$r = 1$	2.7048138	2.7169239	2.7181459	2.7182818

When interest is compounded so that the amount after 1 year is Pe^r, we say the interest is **compounded continuously.**

THEOREM **Continuous Compounding**

The amount A after t years due to a principal P invested at an annual interest rate r compounded continuously is

$$A = Pe^{rt} \qquad (4)$$

EXAMPLE 3 Using Continuous Compounding

The amount A that results from investing a principal P of \$1000 at an annual rate r of 10% compounded continuously for a time t of 1 year is

$$A = \$1000e^{0.10} = (\$1000)(1.10517) = \$1105.17$$

NOW WORK PROBLEM 9.

2 The **effective rate of interest** is the equivalent annual simple rate of interest that would yield the same amount as compounding after 1 year. For example, based on Example 3, a principal of \$1000 will result in \$1105.17 at a rate of 10% compounded continuously. To get this same amount using a simple rate of interest would require that interest of \$1105.17 − \$1000.00 = \$105.17 be earned on the principal. Since \$105.17 is 10.517% of \$1000, a simple rate of interest of 10.517% is needed to equal 10% compounded continuously. Thus, the effective rate of interest of 10% compounded continuously is 10.517%.

Based on the results of Examples 2 and 3, we find the following comparisons:

	Annual Rate	Effective Rate
Annual compounding	10%	10%
Quarterly compounding	10%	10.381%
Monthly compounding	10%	10.471%
Daily compounding	10%	10.516%
Continuous compounding	10%	10.517%

NOW WORK PROBLEM 21.

EXAMPLE 4 Computing the Value of an IRA

On January 2, 1996, \$2000 is placed in an Individual Retirement Account (IRA) that will pay interest of 10% per annum compounded continuously. What will the IRA be worth on January 1, 2016?

Solution The amount A after 20 years is

$$A = Pe^{rt} = \$2000e^{(0.10)(20)} = \$14{,}778.11.$$

EXPLORATION How long will it be until $A = \$4000$? \$6000?
[Hint: Graph $Y_1 = 2000e^{0.1x}$ and $Y_2 = 4000$. Use INTERSECT to find x.]

3 When people engaged in finance speak of the "time value of money," they are usually referring to the **present value** of money. The present value of A dollars to be received at a future date is the principal you would need to invest now so that it would grow to A dollars in the specified time period. Thus, the present value of money to be received at a future date is always less than the amount to be received, since the amount to be received will equal the present value (money invested now) *plus* the interest accrued over the time period.

Figure 36

Time is money

We use the compound interest formula (2) to get a formula for present value. If P is the present value of A dollars to be received after t years at a per annum interest rate r compounded n times per year, then, by formula (2),

$$A = P\left(1 + \frac{r}{n}\right)^{nt}$$

To solve for P, we divide both sides by $(1 + r/n)^{nt}$, and the result is

$$\frac{A}{(1 + r/n)^{nt}} = P \quad \text{or} \quad P = A\left(1 + \frac{r}{n}\right)^{-nt}$$

THEOREM

Present Value Formulas

The present value P of A dollars to be received after t years, assuming a per annum interest rate r compounded n times per year, is

$$P = A\left(1 + \frac{r}{n}\right)^{-nt} \tag{5}$$

If the interest is compounded continuously, then

$$P = Ae^{-rt} \tag{6}$$

To prove (6), solve formula (4) for P.

EXAMPLE 5 Computing the Value of a Zero-Coupon Bond

A zero-coupon (noninterest-bearing) bond can be redeemed in 10 years for \$1000. How much should you be willing to pay for it now if you want a return of:

(a) 8% compounded monthly? (b) 7% compounded continuously?

Solution

(a) We are seeking the present value of \$1000. Thus, we use formula (5) with $A = \$1000$, $n = 12$, $r = 0.08$, and $t = 10$.

$$\begin{aligned} P &= A\left(1 + \frac{r}{n}\right)^{-nt} \\ &= \$1000\left(1 + \frac{0.08}{12}\right)^{-12(10)} \\ &= \$450.52 \end{aligned}$$

For a return of 8% compounded monthly, you should pay \$450.52 for the bond.

(b) Here we use formula (6) with $A = \$1000$, $r = 0.07$, and $t = 10$.

$$\begin{aligned} P &= Ae^{-rt} \\ &= \$1000e^{-(0.07)(10)} \\ &= \$496.59 \end{aligned}$$

For a return of 7% compounded continuously, you should pay \$496.59 for the bond.

NOW WORK PROBLEM 11.

EXAMPLE 6 Rate of Interest Required to Double an Investment

4

What annual rate of interest compounded annually should you seek if you want to double your investment in 5 years?

Algebraic Solution

If P is the principal and we want P to double, the amount A will be $2P$. We use the compound interest formula with $n = 1$ and $t = 5$ to find r.

$$\begin{aligned} 2P &= P(1 + r)^5 \\ 2 &= (1 + r)^5 \\ 1 + r &= \sqrt[5]{2} \\ r &= \sqrt[5]{2} - 1 = 1.148698 - 1 = 0.148698 \end{aligned}$$

The annual rate of interest needed to double the principal in 5 years is 14.87%.

Graphing Solution We solve the equation

$$2 = (1 + r)^5$$

for r by graphing the two functions $Y_1 = 2$ and $Y_2 = (1 + x)^5$. The x-coordinate of their point of intersection is the rate r that we seek. See Figure 37. Using the INTERSECT command, we find that the point of intersection of Y_1 and Y_2 is (0.14869835, 2).

Figure 37

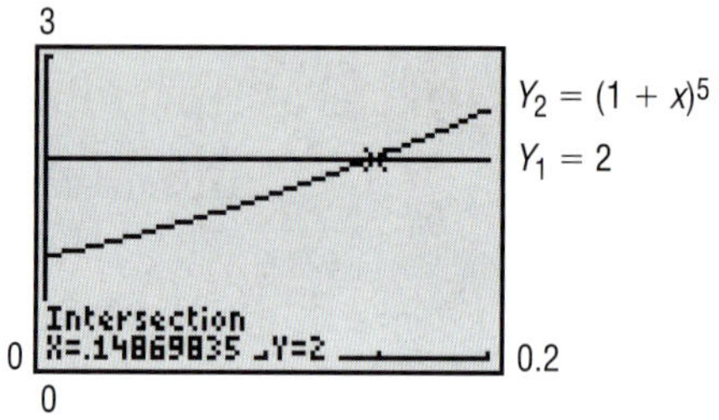

NOW WORK PROBLEM **23.**

EXAMPLE 7 Doubling and Tripling Time for an Investment

(a) How long will it take for an investment to double in value if it earns 5% compounded continuously?

(b) How long will it take to triple at this rate?

Algebraic Solution (a) If P is the initial investment and we want P to double, the amount A will be $2P$. We use formula (4) for continuously compounded interest with $r = 0.05$. Then

$$\begin{aligned} A &= Pe^{rt} \\ 2P &= Pe^{0.05t} \\ 2 &= e^{0.05t} \\ 0.05t &= \ln 2 \\ t &= \frac{\ln 2}{0.05} = 13.86 \end{aligned}$$

It will take about 14 years to double the investment.

Graphing Solution We solve the equation

$$2 = e^{0.05t}$$

for t by graphing the two functions $Y_1 = 2$ and $Y_2 = e^{0.05x}$. Their point of intersection is (13.86, 2). See Figure 38.

Figure 38

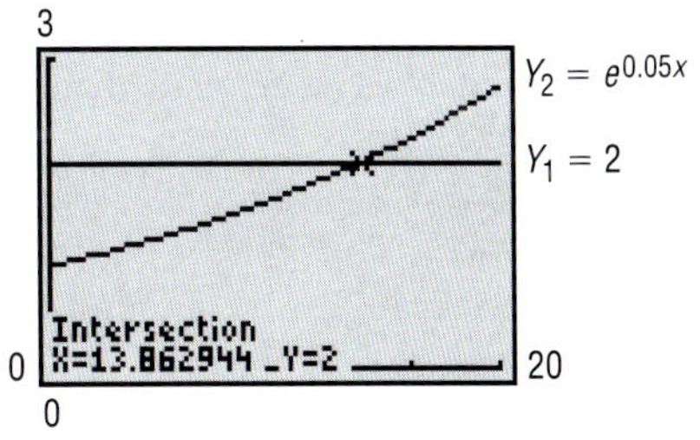

Algebraic Solution

(b) To triple the investment, we set $A = 3P$ in formula (4).

$$\begin{aligned} A &= Pe^{rt} \\ 3P &= Pe^{0.05t} \\ 3 &= e^{0.05t} \\ 0.05t &= \ln 3 \\ t &= \frac{\ln 3}{0.05} = 21.97 \end{aligned}$$

It will take about 22 years to triple the investment.

Graphing Solution

We solve the equation

$$3 = e^{0.05t}$$

for t by graphing the two functions $Y_1 = 3$ and $Y_2 = e^{0.05x}$. Their point of intersection is (21.97, 3). See Figure 39.

Figure 39

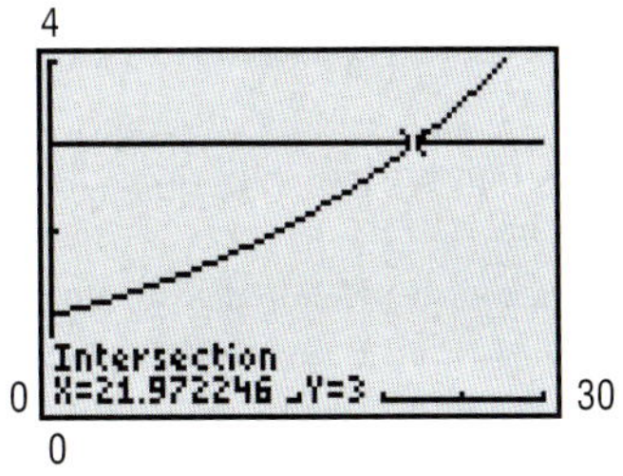

NOW WORK PROBLEM **29.**

5.6 EXERCISES

In Problems 1–10, find the amount that results from each investment.

1. \$100 invested at 4% compounded quarterly after a period of 2 years
2. \$50 invested at 6% compounded monthly after a period of 3 years
3. \$500 invested at 8% compounded quarterly after a period of $2\frac{1}{2}$ years
4. \$300 invested at 12% compounded monthly after a period of $1\frac{1}{2}$ years
5. \$600 invested at 5% compounded daily after a period of 3 years
6. \$700 invested at 6% compounded daily after a period of 2 years
7. \$10 invested at 11% compounded continuously after a period of 2 years
8. \$40 invested at 7% compounded continuously after a period of 3 years
9. \$100 invested at 10% compounded continuously after a period of $2\frac{1}{4}$ years
10. \$100 invested at 12% compounded continuously after a period of $3\frac{3}{4}$ years

In Problems 11–20, find the principal needed now to get each amount; that is, find the present value.

11. To get \$100 after 2 years at 6% compounded monthly
12. To get \$75 after 3 years at 8% compounded quarterly
13. To get \$1000 after $2\frac{1}{2}$ years at 6% compounded daily
14. To get \$800 after $3\frac{1}{2}$ years at 7% compounded monthly
15. To get \$600 after 2 years at 4% compounded quarterly
16. To get \$300 after 4 years at 3% compounded daily
17. To get \$80 after $3\frac{1}{4}$ years at 9% compounded continuously
18. To get \$800 after $2\frac{1}{2}$ years at 8% compounded continuously
19. To get \$400 after 1 year at 10% compounded continuously
20. To get \$1000 after 1 year at 12% compounded continuously
21. Find the effective rate of interest for $5\frac{1}{4}$% compounded quarterly.
22. What interest rate compounded quarterly will give an effective interest rate of 7%?
23. What annual rate of interest is required to double an investment in 3 years? Verify your answer using a graphing utility.
24. What annual rate of interest is required to double an investment in 10 years? Verify your answer using a graphing utility.

In Problems 25–28, which of the two rates would yield the larger amount in 1 year? ***[Hint: Start with a principal of \$10,000 in each instance.]***

25. 6% compounded quarterly or $6\frac{1}{4}$% compounded annually
26. 9% compounded quarterly or $9\frac{1}{4}$% compounded annually
27. 9% compounded monthly or 8.8% compounded daily
28. 8% compounded semiannually or 7.9% compounded daily
29. How long does it take for an investment to double in value if it is invested at 8% per annum compounded monthly? Compounded continuously? Verify your answer using a graphing utility.
30. How long does it take for an investment to double in value if it is invested at 10% per annum compounded monthly? Compounded continuously? Verify your answer using a graphing utility.
31. If Tanisha has \$100 to invest at 8% per annum compounded monthly, how long will it be before she has \$150? If the compounding is continuous, how long will it be?
32. If Angela has \$100 to invest at 10% per annum compounded monthly, how long will it be before she has \$175? If the compounding is continuous, how long will it be?
33. How many years will it take for an initial investment of \$10,000 to grow to \$25,000? Assume a rate of interest of 6% compounded continuously.
34. How many years will it take for an initial investment of \$25,000 to grow to \$80,000? Assume a rate of interest of 7% compounded continuously.
35. What will a \$90,000 house cost 5 years from now if the inflation rate over that period averages 3% compounded annually?
36. Sears charges 1.25% per month on the unpaid balance for customers with charge accounts (interest is compounded monthly). A customer charges \$200 and does not pay her bill for 6 months. What is the bill at that time?

37. Jerome will be buying a used car for $15,000 in 3 years. How much money should he ask his parents for now so that, if he invests it at 5% compounded continuously, he will have enough to buy the car?

38. John will require $3000 in 6 months to pay off a loan that has no prepayment privileges. If he has the $3000 now, how much of it should he save in an account paying 3% compounded monthly so that in 6 months he will have exactly $3000?

39. George is contemplating the purchase of 100 shares of a stock selling for $15 per share. The stock pays no dividends. The history of the stock indicates that is should grow at an annual rate of 15% per year. How much will the 100 shares of stock be worth in 5 years?

40. Tracy is contemplating the purchase of 100 shares of a stock selling for $15 per share. The stock pays no dividends. Her broker says that the stock will be worth $20 per share in 2 years. What is the annual rate of return on this investment?

41. A business purchased for $650,000 in 1994 is sold in 1997 for $850,000. What is the annual rate of return for this investment?

42. Tanya has just inherited a diamond ring appraised at $5000. If diamonds have appreciated in value at an annual rate of 8%, what was the value of the ring 10 years ago when the ring was purchased?

43. Jim places $1000 in a bank account that pays 5.6% compounded continuously. After 1 year, will he have enough money to buy a computer system that costs $1060? If another bank will pay Jim 5.9% compounded monthly, is this a better deal?

44. On January 1, Kim places $1000 in a certificate of deposit that pays 6.8% compounded continuously and matures in 3 months. Then Kim places the $1000 and the interest in a passbook account that pays 5.25% compounded monthly. How much does Kim have in the passbook account on May 1?

45. Will invests $2000 in a bond trust that pays 9% interest compounded semiannually. His friend Henry invests $2000 in a certificate of deposit (CD) that pays $8\frac{1}{2}\%$ compounded continuously. Who has more money after 20 years, Will or Henry?

46. Suppose that April has access to an investment that will pay 10% interest compounded continuously. Which is better: To be given $1000 now so that she can take advantage of this investment opportunity or to be given $1325 after 3 years?

47. Colleen and Bill have just purchased a house for $150,000, with the seller holding a second mortgage of $50,000. They promise to pay the seller $50,000 plus all accrued interest 5 years from now. The seller offers them three interest options on the second mortgage:

(a) Simple interest at 12% per annum
(b) $11\frac{1}{2}\%$ interest compounded monthly
(c) $11\frac{1}{4}\%$ interest compounded continuously

Which option is best; that is, which results in the least interest on the loan?

48. The First National Bank advertises that it pays interest on savings accounts at the rate of 4.25% compounded daily. Find the effective rate if the bank uses (a) 360 days or (b) 365 days in determining the daily rate.

Problems 49–52 involve zero-coupon bonds. A zero-coupon bond is a bond that is sold now at a discount and will pay its face value at some time when it matures; no interest payments are made.

49. A zero-coupon bond can be redeemed in 20 years for $10,000. How much should you be willing to pay for it now if you want a return of:

(a) 10% compounded monthly?
(b) 10% compounded continuously?

50. A child's grandparents are considering buying a $40,000 face value zero-coupon bond at birth so that she will have enough money for her college education 17 years later. If they want a rate of return of 8% compounded annually, what should they pay for the bond?

51. How much should a $10,000 face value zero-coupon bond, maturing in 10 years, be sold for now if its rate of return is to be 8% compounded annually?

52. If Pat pays $12,485.52 for a $25,000 face value zero-coupon bond that matures in 8 years, what is his annual rate of return?

53. Explain in your own words what the term *compound interest* means. What does *continuous compounding* mean?

54. Explain in your own words the meaning of *present value.*

55. Write a program that will calculate the amount after n years if a principal P is invested at $r\%$ per annum compounded quarterly. Use it to verify your answers to Problems 1 and 3.

56. Write a program that will calculate the principal needed now to get the amount A in n years at

$r\%$ per annum compounded daily. Use it to verify your answer to Problem 13.

57. Write a program that will calculate the annual rate of interest required to double an investment in n years. Use it to verify your answers to Problems 23 and 24.

58. Write a program that will calculate the number of years required for an initial investment of x dollars to grow to y dollars at $r\%$ per annum compounded continuously. Use it to verify your answer to Problems 33 and 34.

59. Time to Double or Triple an Investment The formula

$$y = \frac{\ln m}{n \ln\left(1 + \frac{r}{n}\right)}$$

can be used to find the number of years y required to multiply an investment m times when r is the per annum interest rate compounded n times a year.

(a) How many years will it take to double the value of an IRA that compounds annually at the rate of 12%?

(b) How many years will it take to triple the value of a savings account that compounds quarterly at an annual rate of 6%?

(c) Give a derivation of this formula.

60. Time to Reach an Investment Goal The formula

$$y = \frac{\ln A - \ln P}{r}$$

can be used to find the number of years y required for an investment P to grow to a value A when compounded continuously at an annual rate r.

(a) How long will it take to increase an initial investment of \$1000 to \$8000 at an annual rate of 10%?

(b) What annual rate is required to increase the value of a \$2000 IRA to \$30,000 in 35 years?

(c) Give a derivation of this formula.

61. Critical Thinking You have just contracted to buy a house and will seek financing in the amount of \$100,000. You go to several banks. Bank 1 will lend you \$100,000 at the rate of 8.75% amortized over 30 years with a loan origination fee of 1.75%. Bank 2 will lend you \$100,000 at the rate of 8.375% amortized over 15 years with a loan origination fee of 1.5%. Bank 3 will lend you \$100,000 at the rate of 9.125% amortized over 30 years with no loan origination fee. Bank 4 will lend you \$100,000 at the rate of 8.625% amortized over 15 years with no loan origination fee. Which loan would you take? Why? Be sure to have sound reasons for your choice. If the amount of the monthly payment does not matter to you, which loan would you take? Again, have sound reasons for your choice. Use the information in the table to assist you. Compare your final decision with others in the class. Discuss.

	Monthly Payment	Loan Origination Fee
Bank 1	\$786.70	\$1,750.00
Bank 2	\$977.42	\$1,500.00
Bank 3	\$813.63	\$0.00
Bank 4	\$990.68	\$0.00

5.7 GROWTH AND DECAY

1 Find Equations of Populations That Obey the Law of Uninhibited Growth

2 Find Equations of Populations That Obey the Law of Decay

3 Use Newton's Law of Cooling

4 Use Logistic Growth Models

1 Many natural phenomena have been found to follow the law that an amount A varies with time t according to

$$A = A_0 e^{kt} \qquad (1)$$

where A_0 is the original amount ($t = 0$) and $k \neq 0$ is a constant.

If $k > 0$, then equation (1) states that the amount A is increasing over time; if $k < 0$, the amount A is decreasing over time. In either case, when an amount A varies over time according to equation (1), it is said to follow the **exponential law** or the **law of uninhibited growth** $(k > 0)$ **or decay** $(k < 0)$. See Figure 40.

Figure 40

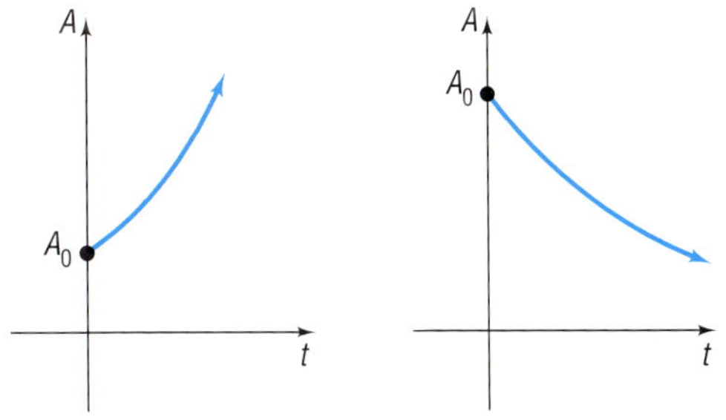

(a) $A(t) = A_0e^{kt}, k > 0$ **(b)** $A(t) = A_0e^{kt}, k < 0$

For example, we saw in Section 5.6 that continuously compounded interest follows the law of uninhibited growth. In this section we shall look at three additional phenomena that follow the exponential law.

Uninhibited Growth

Cell division is a universal process in the growth of many living organisms, such as amoebas, plants, and human skin cells. Based on an ideal situation in which no cells die and no by-products are produced, the number of cells present at a given time follows the law of uninhibited growth. Actually, however, after enough time has passed, growth at an exponential rate will cease due to the influence of factors such as lack of living space and dwindling food supply. The law of uninhibited growth accurately reflects only the early stages of the cell division process.

The cell division process begins with a culture containing N_0 cells. Each cell in the culture grows for a certain period of time and then divides into two identical cells. We assume that the time needed for each cell to divide in two is constant and does not change as the number of cells increases. These new cells then grow, and eventually each divides in two, and so on.

A model that gives the number N of cells in the culture after a time t has passed (in the early stages of growth) is

Uninhibited Growth of Cells

$$N(t) = N_0e^{kt} \qquad k > 0 \tag{2}$$

where N_0 is the initial number of cells and k is a positive constant that represents the growth rate of the cells.

In using formula (2) to model the growth of cells, we are using a function that yields positive real numbers, even though we are counting the number of cells, which must be an integer. This is a common practice in many applications.

EXAMPLE 1 Bacterial Growth

A colony of bacteria increases according to the law of uninhibited growth.

(a) If the number of bacteria doubles in 3 hours, find the function that gives the number of cells in the culture.

(b) How long will it take for the size of the colony to triple?

(c) Using a graphing utility, verify that the population doubles in 3 hours.

(d) Using a graphing utility, approximate the time that it takes for the population to double a second time (that is, increase four times). Does this answer seem reasonable?

Solution (a) Using formula (2), the number N of cells at a time t is

$$N(t) = N_0 e^{kt}$$

where N_0 is the initial number of bacteria present and k is a positive number. We first seek the number k. The number of cells doubles in 3 hours; thus, we have

$$N(3) = 2N_0$$

But $N(3) = N_0 e^{k(3)}$, so

$$N_0 e^{k(3)} = 2N_0$$

$$e^{3k} = 2$$

$$3k = \ln 2 \quad \text{Write the exponential equation as a logarithm.}$$

$$k = \tfrac{1}{3}\ln 2 \approx \tfrac{1}{3}(0.6931) = 0.2310$$

Formula (2) for this growth process is therefore

$$N(t) = N_0 e^{0.2310t}$$

(b) The time t needed for the size of the colony to triple requires that $N = 3N_0$. Thus, we substitute $3N_0$ for N to get

$$3N_0 = N_0 e^{0.2310t}$$

$$3 = e^{0.2310t}$$

$$0.2310t = \ln 3$$

$$t = \frac{1}{0.2310}\ln 3 \approx \frac{1.0986}{0.2310} = 4.756 \text{ hours}$$

It will take about 4.756 hours for the size of the colony to triple.

(c) Figure 41 shows the graph of $Y_1 = 2$ and $Y_2 = e^{0.2310x}$, where x represents the time in hours. Using INTERSECT, it takes 3 hours for the population to double.

Figure 41

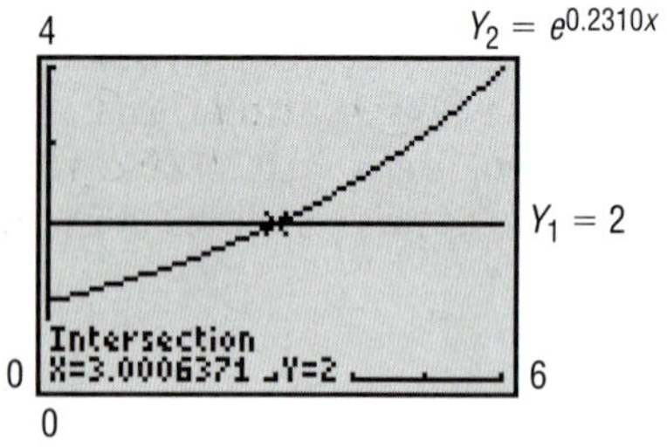

(d) Figure 42 shows the graph of $Y_1 = 4$ and $Y_2 = e^{0.2310x}$, where x represents the time in hours. Using INTERSECT, it takes 6 hours for the population to double a second time.

Figure 42

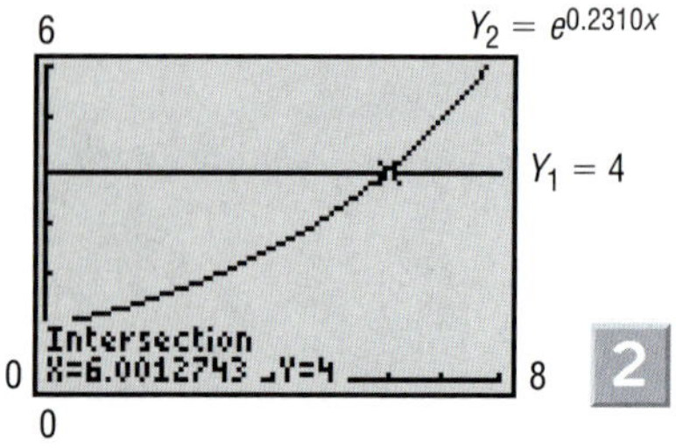

NOW WORK PROBLEM 1.

Radioactive Decay

2

Radioactive materials follow the law of uninhibited decay. Thus, the amount A of a radioactive material present at time t is given by the model

Uninhibited Radioactive Decay

$$A(t) = A_0 e^{kt} \qquad k < 0 \qquad (3)$$

where A_0 is the original amount of radioactive material and k is a negative number that represents the rate of decay.

All radioactive substances have a specific **half-life,** which is the time required for half of the radioactive substance to decay. In **carbon dating,** we use the fact that all living organisms contain two kinds of carbon, carbon 12 (a stable carbon) and carbon 14 (a radioactive carbon, with a half-life of 5600 years). While an organism is living, the ratio of carbon 12 to carbon 14 is constant. But when an organism dies, the original amount of carbon 12 present remains unchanged, whereas the amount of carbon 14 begins to decrease. This change in the amount of carbon 14 present relative to the amount of carbon 12 present makes it possible to calculate when an organism died.

EXAMPLE 2 Estimating the Age of Ancient Tools

Traces of burned wood found along with ancient stone tools in an archaeological dig in Chile were found to contain approximately 1.67% of the original amount of carbon 14.

(a) If the half-life of carbon 14 is 5600 years, approximately when was the tree cut and burned?

(b) Using a graphing utility, graph the relation between the percentage of carbon 14 remaining and time.

(c) Determine the time that elapses until half of the carbon 14 remains. This answer should equal the half-life of carbon 14.

(d) Verify the answer found in part (a).

Solution (a) Using formula (3), the amount A of carbon 14 present at time t is

$$A(t) = A_0 e^{kt}$$

where A_0 is the original amount of carbon 14 present and k is a negative number. We first seek the number k. To find it, we use the fact that after 5600 years half of the original amount of carbon 14 remains, so $A(5600) = \frac{1}{2} A_0$ Thus,

$$\begin{aligned} \frac{1}{2}A_0 &= A_0 e^{k(5600)} \\ \frac{1}{2} &= e^{5600k} \\ 5600k &= \ln\frac{1}{2} \\ k &= \frac{1}{5600}\ln\frac{1}{2} \approx -0.000124 \end{aligned}$$

Formula (3) therefore becomes

$$A(t) = A_0 e^{-0.000124t}$$

If the amount A of carbon-14 now present is 1.67% of the original amount, it follows that

$$\begin{aligned} 0.0167A_0 &= A_0 e^{-0.000124t} \\ 0.0167 &= e^{-0.000124t} \\ -0.000124t &= \ln 0.0167 \\ t &= \frac{1}{-0.000124} \ln 0.0167 \approx 33{,}000 \text{ years} \end{aligned}$$

The tree was cut and burned about 33,000 years ago. Some archaeologists use this conclusion to argue that humans lived in the Americas 33,000 years ago, much earlier than is generally accepted.

(b) Figure 43 shows the graph of $y = e^{-0.000124x}$, where y is the fraction of carbon 14 present and x is the time.

Figure 43

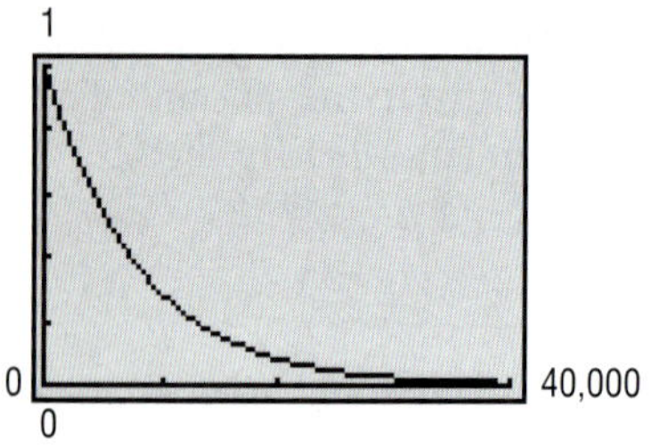

(c) By graphing $Y_1 = 0.5$ and $Y_2 = e^{-0.000124x}$, where x is time, and using INTERSECT, we find that it takes 5590 years until half the carbon 14 remains.

(d) By graphing $Y_1 = 0.0167$ and $Y_2 = e^{-0.000124x}$, where x is time, and using INTERSECT, we find that it takes 33,003 years until 1.67% of the carbon 14 remains.

NOW WORK PROBLEM 3.

Newton's Law of Cooling

3 **Newton's Law of Cooling*** states that the temperature of a heated object decreases exponentially over time toward the temperature of the surrounding medium. That is, the temperature u of a heated object at a given time t can be modeled by the function

Newton's Law of Cooling

$$u(t) = T + (u_0 - T)e^{kt} \qquad k < 0 \tag{4}$$

where T is the constant temperature of the surrounding medium, u_0 is the initial temperature of the heated object, and k is a negative constant.

EXAMPLE 3 Using Newton's Law of Cooling

An object is heated to 100°C (degrees Celsius) and is then allowed to cool in a room whose air temperature is 30°C.

(a) If the temperature of the object is 80°C after 5 minutes, when will its temperature be 50°C?

(b) Using a graphing utility, graph the relation found between the temperature y and time x.

*Named after Sir Isaac Newton (1642–1727), one of the cofounders of calculus.

(c) Using a graphing utility, verify that when $x = 18.6$ then $y = 50$.

(d) Using a graphing utility, determine the elapsed time before the object is 35°C.

(e) What do you notice about y, the temperature, as x, time, increases?

Solution (a) Using formula (4) with $T = 30$ and $u_0 = 100$, the temperature (in degrees Celsius) of the object at time t (in minutes) is

$$u(t) = 30 + (100 - 30)e^{kt} = 30 + 70e^{kt} \tag{5}$$

where k is a negative constant. To find k, we use the fact that $u = 80$ when $t = 5$. Then

$$\begin{aligned} 80 &= 30 + 70e^{k(5)} \\ 50 &= 70e^{5k} \\ e^{5k} &= \frac{50}{70} \\ 5k &= \ln\frac{5}{7} \\ k &= \frac{1}{5}\ln\frac{5}{7} \approx -0.0673 \end{aligned}$$

Formula (5) therefore becomes

$$u(t) = 30 + 70e^{-0.0673t}$$

Now, we want to find t when $u = 50$°C, so

$$\begin{aligned} 50 &= 30 + 70e^{-0.0673t} \\ 20 &= 70e^{-0.0673t} \\ e^{-0.0673t} &= \frac{20}{70} \\ -0.0673t &= \ln\frac{2}{7} \\ t &= \frac{1}{-0.0673}\ln\frac{2}{7} \approx 18.6 \text{ minutes} \end{aligned}$$

Thus, the temperature of the object will be 50°C after about 18.6 minutes.

(b) Figure 44 shows the graph of $y = 30 + 70e^{-0.0673x}$, where y is the temperature and x is the time.

Figure 44

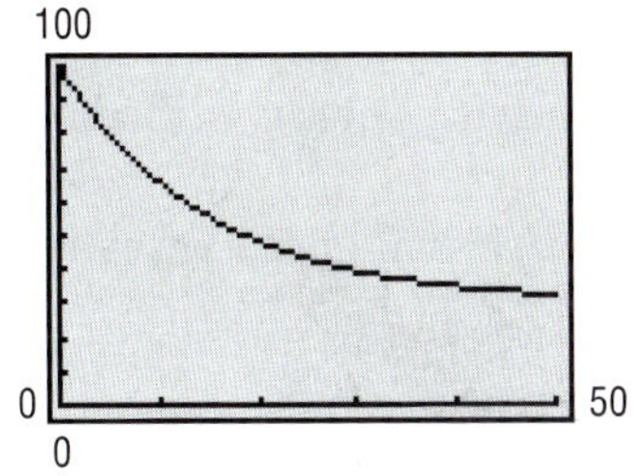

(c) By graphing $Y_1 = 50$ and $Y_2 = 30 + 70e^{-0.0673x}$, where x is time, and using INTERSECT, we find it takes $x = 18.6$ minutes for the temperature to cool to 50°C.

(d) By graphing $Y_1 = 35$ and $Y_2 = 30 + 70e^{-0.0673x}$, where x is time, and using INTERSECT, we find it takes $x = 39.21$ minutes for the temperature to cool to 35°C.

(e) As x increases, the value of $e^{-0.0673x}$ approaches zero, so the value of y approaches 30°C.

Logistic Models

4 The exponential growth model $A(t) = A_0e^{kt}$, $k > 0$, assumes uninhibited growth, meaning that the value of the function grows without limit. Recall we

stated that cell division could be modeled using this function assuming that no cells die and no by-products are produced. However, cell division would eventually be limited by factors such as living space and food supply. The **logistic growth model** is an exponential function that can model situations where the growth of the dependent variable is limited.

Other situations that lead to a logistic growth model include population growth and the sales of a product due to advertising. See Problems 23–26. The logistic growth model is given by the function

Logistic Growth Model

$$P(t) = \frac{c}{1 + ae^{-bt}}$$

where a, b, and c are constants with $c > 0$ and $b > 0$.

The number c is called the **carrying capacity** because the value $P(t)$ approaches c as t approaches infinity; that is $\lim_{t \to \infty} P(t) = c$. Thus, c represents the maximum value that the function can attain.

EXAMPLE 4 Fruit Fly Population

Fruit flies are placed in a half-pint milk bottle with a banana (for food) and yeast plants (for food and to provide a stimulus to lay eggs). Suppose that the fruit fly population after t days is given by

$$P(t) = \frac{230}{1 + 56.5e^{-0.37t}}$$

(a) Using a graphing utility, graph $P(t)$.
(b) What is the carrying capacity of the half-pint bottle? That is, what is $P(t)$ as $t \to \infty$?
(c) How many fruit flies were initially placed in the half-pint bottle?
(d) When will the population of fruit flies be 180?

Figure 45

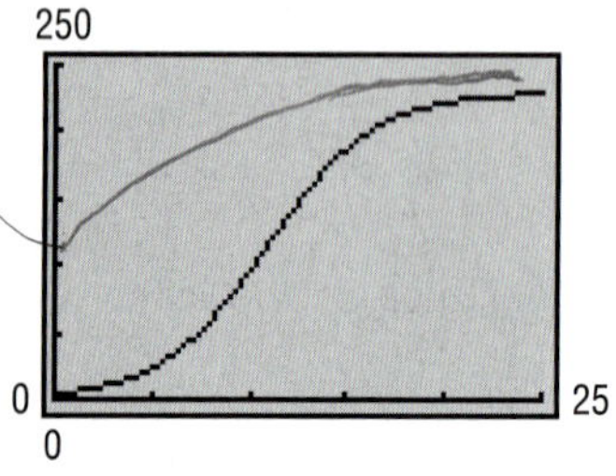

Solution

(a) See Figure 45.
(b) As $t \to \infty$, $e^{-0.37t} \to 0$ and $P(t) \to 230/1$. The carrying capacity of the half-pint bottle is 230 fruit flies.
(c) To find the initial number of fruit flies in the half-pint bottle, we evaluate $P(0)$.

$$P(0) = \frac{230}{1 + 56.5e^{-0.37(0)}}$$
$$= \frac{230}{1 + 56.5}$$
$$= 4$$

So initially there were four fruit flies in the half-pint bottle.

(d) To determine when the population of fruit flies will be 180, we solve the equation

$$
\begin{aligned}
\frac{230}{1 + 56.5e^{-0.37t}} &= 180 \\
230 &= 180(1 + 56.5e^{-0.37t}) \\
1.2778 &= 1 + 56.5e^{-0.37t} && \text{Divide both sides by 180.} \\
0.2778 &= 56.5e^{-0.37t} && \text{Subtract 1 from both sides.} \\
0.0049 &= e^{-0.37t} && \text{Divide both sides by 56.5.} \\
\ln(0.0049) &= -0.37t && \text{Rewrite as a logarithmic expression.} \\
t &\approx 14.4 \text{ days} && \text{Divide both sides by } -0.37.
\end{aligned}
$$

It will take approximately 14.4 days for the population to reach 180 fruit flies.

We could have also solved this problem by graphing $Y_1 = \dfrac{230}{1 + 56.5e^{-0.37t}}$ and $Y_2 = 180$, using INTERSECT to find the solution shown in Figure 46.

Figure 46

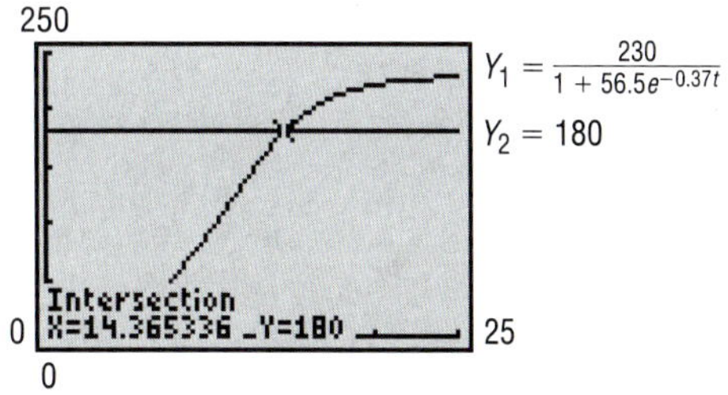

EXPLORATION On the same viewing rectangle, graph $Y_1 = \dfrac{500}{1 + 24e^{-0.03t}}$ and $Y_2 = \dfrac{500}{1 + 24e^{-0.08t}}$. What effect does b have on the logistic growth function?

5.7 EXERCISES

1. **Growth of an Insect Population** The size P of a certain insect population at time t (in days) obeys the function $P(t) = 500e^{0.02t}$.
   (a) After how many days will the population reach 1000? 2000?
   (b) Using a graphing utility, graph the relation between P and t. Verify your answers in part (a) using INTERSECT.

2. **Growth of Bacteria** The number N of bacteria present in a culture at time t (in hours) obeys the function $N(t) = 1000e^{0.01t}$.
   (a) After how many hours will the population equal 1500? 2000?
   (b) Using a graphing utility, graph the relation between N and t. Verify your answers in part (a) using INTERSECT.

3. **Radioactive Decay** Strontium 90 is a radioactive material that decays according to the function $A(t) = A_0e^{-0.0244t}$, where A_0 is the initial amount present and A is the amount present at time t (in years). What is the half-life of strontium 90?

4. **Radioactive Decay** Iodine 131 is a radioactive material that decays according to the function $A(t) = A_0e^{-0.087t}$, where A_0 is the initial amount present and A is the amount present at time t (in days). What is the half-life of iodine 131?

5. (a) Use the information Problem 3 to determine how long it takes for 100 grams of strontium 90 to decay to 10 grams.
   (b) Graph the function $A(t) = 100e^{-0.0244t}$ and verify your answer found in part (a).

6. (a) Use the information in Problem 4 to determine how long it takes for 100 grams of iodine 131 to decay to 10 grams.
   (b) Graph the function $A(t) = 100e^{-0.087t}$ and verify your answer found in part (a).

7. **Growth of a Colony of Mosquitoes** The population of a colony of mosquitoes obeys the law of uninhibited growth. If there are 1000 mosquitoes initially, and there are 1800 after 1 day, what is the size of the colony after 3 days? How long is it until there are 10,000 mosquitoes?

8. **Bacterial Growth** A culture of bacteria obeys the law of uninhibited growth. If 500 bacteria are present initially, and there are 800 after 1 hour, how many will be present in the culture after 5 hours? How long is it until there are 20,000 bacteria?

9. **Population Growth** The population of a southern city follows the exponential law. If the population doubled in size over an 18-month period and the current population is 10,000, what will the population be 2 years from now?

10. **Population Growth** The population of a midwestern city follows the exponential law. If the population decreased from 900,000 to 800,000 from 1993 to 1995, what will the population be in 1997?

11. **Radioactive Decay** The half-life of radium is 1690 years. If 10 grams is present now, how much will be present in 50 years?

12. **Radioactive Decay** The half-life of radioactive potassium is 1.3 billion years. If 10 grams is present now, how much will be present in 100 years? In 1000 years?

13. **Estimating the Age of a Tree** A piece of charcoal is found to contain 30% of the carbon 14 that it originally had.
   (a) When did the tree from which the charcoal came die? Use 5600 years as the half-life of carbon 14.
   (b) Using a graphing utility, graph the relation between the percentage of carbon 14 remaining and time.
   (c) Using INTERSECT, determine the time that elapses until half of the carbon 14 remains.
   (d) Verify the answer found in part (a).

14. **Estimating the Age of a Fossil** A fossilized leaf contains 70% of its normal amount of carbon 14.
   (a) How old is the fossil?
   (b) Using a graphing utility, graph the relation between the percentage of carbon 14 remaining and time.
   (c) Using INTERSECT, determine the time that elapses until half of the carbon 14 remains.
   (d) Verify the answer found in part (a).

15. **Cooling Time of a Pizza** A pizza baked at 450°F is removed from the oven at 5:00 PM into a room that is a constant 70°F. After 5 minutes, the pizza is at 300°F.
   (a) At what time can you begin eating the pizza if you want its temperature to be 135°F?
   (b) Using a graphing utility, graph the relation between temperature and time.
   (c) Using INTERSECT, determine the time that needs to elapse before the pizza is 160°F?
   (d) TRACE the function for large values of time. What do you notice about y, the temperature?

16. **Newton's Law of Cooling** A thermometer reading 72°F is placed in a refrigerator where the temperature is a constant 38°F. If the thermometer reads 60°F after 2 minutes, what will it read after 7 minutes?
   (a) How long will it take before the thermometer reads 39°F?
   (b) Using a graphing utility, graph the relation between temperature and time.

(c) Using INTERSECT, determine the time needed to elapse before the thermometer reads 45°F.
(d) TRACE the function for large values of time. What do you notice about y, the temperature?

17. **Newton's Law of Cooling** A thermometer reading 8°C is brought into a room with a constant temperature of 35°C.
(a) If the thermometer reads 15°C after 3 minutes, what will it read after being in the room for 5 minutes? For 10 minutes?
(b) Graph the relation between temperature and time. TRACE to verify that your answers are correct.
[Hint: You need to construct a formula similar to equation (4).]

18. **Thawing Time of a Steak** A frozen steak has a temperature of 28°F. It is placed in a room with a constant temperature of 70°F. After 10 minutes, the temperature of the steak has risen to 35°F. What will the temperature of the steak be after 30 minutes? How long will it take the steak to thaw to a temperature of 45°F? [See the hint given for Problem 17.] Graph the relation between temperature and time. TRACE to verify that your answer is correct.

19. **Decomposition of Salt in Water** Salt (NaCl) decomposes in water into sodium (NA^+) and chloride (Cl^-) ions according to the law of uninhibited decay. If the initial amount of salt is 25 kilograms and, after 10 hours, 15 kilograms of salt is left, how much salt is left after 1 day? How long does it take until $\frac{1}{2}$ kilogram of salt is left?

20. **Voltage of a Conductor** The voltage of a certain condenser decreases over time according to the law of uninhibited decay. If the initial voltage is 40 volts, and 2 seconds later it is 10 volts, what is the voltage after 5 seconds?

21. **Radioactivity from Chernobyl** After the release of radioactive material into the atmosphere from a nuclear power plant at Chernobyl (Ukraine) in 1986, the hay in Austria was contaminated by iodine 131 (half-life 8 days). If it is all right to feed the hay to cows when 10% of the iodine 131 remains, how long do the farmers need to wait to use this hay?

22. **Pig Roasts** The hotel Bora-Bora is having a pig roast. At noon, the chef put the pig in a large earthen oven. The pig's original temperature was 75°F. At 2:00 PM the chef checked the pig's temperature and was upset because it had reached only 100°F. If the oven's temperature remains a constant 325°F, at what time may the hotel serve its guests, assuming that pork is done when it reaches 175°F?

23. **Proportion of the Population That Owns a VCR** The logistic growth model

$$P(t) = \frac{0.9}{1 + 6e^{-0.32t}}$$

relates the proportion of U.S. households that own a VCR to the year. Let $t = 0$ represent 1984, $t = 1$ represent 1985, and so on.
(a) What proportion of the U.S. households owned a VCR in 1984?
(b) Determine the maximum proportion of households that will own a VCR.
(c) Using a graphing utility, graph $P(t)$.
(d) When will 0.8 (80%) of U.S. households own a VCR?

24. **Market Penetration of Intel's Coprocessor** The logistic growth model

$$P(t) = \frac{0.90}{1 + 3.5e^{-0.339t}}$$

relates the proportion of new personal computers sold at Best Buy that have Intel's latest coprocessor t months after it has been introduced.
(a) What proportion of new personal computers sold at Best Buy will have Intel's latest coprocessor when it's first introduced (that is, at $t = 0$)?
(b) Determine the maximum proportion of new personal computers sold at Best Buy that will have Intel's latest coprocessor.
(c) Using a graphing utility, graph $P(t)$.
(d) When will 0.75 (75%) of new personal computers sold at Best Buy have Intel's latest coprocessor?

25. **Population of a Bacteria Culture** The logistic growth model

$$P(t) = \frac{1000}{1 + 32.33e^{-0.439t}}$$

represents the population of a bacteria after t hours.

(a) Using a graphing utility, graph $P(t)$.
(b) What is the carrying capacity of the environment?
(c) What was the initial amount of bacteria in the population?
(d) When will the amount of bacteria be 800?

26. Population of a Endangered Species Often environmentalists will capture an endangered species and transport the species to a controlled environment where the species can produce offspring and regenerate its population. Suppose that six American Bald Eagles are captured and transported to Montana and set free. Based on experience, the environmentalists expect the population to grow according to the model

$$P(t) = \frac{500}{1 + 83.33e^{-0.162t}}$$

(a) Using a graphing utility, graph $P(t)$.
(b) What is the carrying capacity of the environment?
(c) What is the predicted population of the American Bald Eagle in 20 years?
(d) When will the population be 300?

5.8 EXPONENTIAL, LOGARITHMIC, AND LOGISTIC CURVE FITTING

1 Use a Graphing Utility to Obtain the Exponential Function of Best Fit

2 Use a Graphing Utility to Obtain the Logarithmic Function of Best Fit

3 Use a Graphing Utility to Obtain the Logistic Function of Best Fit

In Section 2.2 we discussed how to find the linear equation of best fit ($y = ax + b$), and in Section 2.4 we discussed how to find the quadratic function of best fit ($y = ax^2 + bx + c$).

In this section we will discuss how to use a graphing utility to find equations of best fit that describe the relation between two variables when the relation is thought to be exponential ($y = ab^x$), logarithmic ($y = a + b \ln x$), or logistic $\left(y = \dfrac{c}{1 + ae^{-bx}}\right)$. As before, we draw a scatter diagram of the data to help to determine the appropriate model to use.

Figure 47 shows scatter diagrams that will typically be observed for the three models. Below each scatter diagram are any restrictions on the values of the parameters.

Figure 47

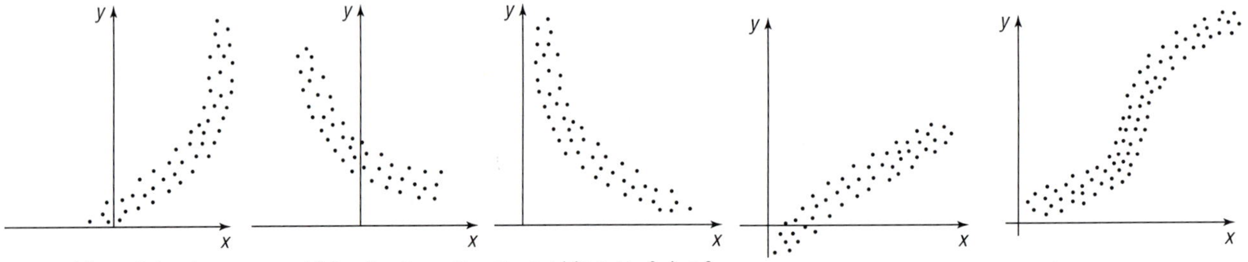

$y = ab^x$, $a > 0$, $b > 1$
Exponential

$y = ab^x$, $0 < b < 1$, $a > 0$
Exponential

$y = a + b \ln x$, $a > 0$, $b < 0$
Logarithmic

$y = a + b \ln x$, $a > 0$, $b > 0$
Logarithmic

$y = \dfrac{c}{1 + ae^{-bx}}$, $a > 0$, $b > 0$, $c > 0$
Logistic

Most graphing utilities have REGression options that fit data to a specific type of curve. Once the data have been entered and a scatter diagram obtained, the type of curve that you want to fit to the data is selected. Then that REGression option is used to obtain the curve of "best fit" of the type selected.

The correlation coefficient r will appear only if the model can be written as a linear expression. Thus, r will appear for the linear, exponential, and logarithmic models, since these models can be written as a linear expression (see Section 5.4). Remember, the closer $|r|$ is to 1, the better the fit.

Let's look at some examples.

Exponential Curve Fitting

1 We saw in Section 5.6 that the future value of money behaves exponentially, and we saw in Section 5.7 that growth and decay models also behave exponentially. The next example shows how data can lead to an exponential model.

EXAMPLE 1 Fitting a Curve to an Exponential Model

Beth is interested in finding a function that explains the closing price of Dell Computer stock at the end of each month. She obtains the following data:

Date	Closing Price	Date	Closing Price
Nov-96	6.3516	Dec-97	21
Dec-96	6.6406	Jan-98	24.8594
Jan-97	8.2656	Feb-98	34.9688
Feb-97	8.8906	Mar-98	33.875
Mar-97	8.4531	Apr-98	40.375
Apr-97	10.4609	May-98	41.2031
May-97	14.0625	Jun-98	46.4062
Jun-97	14.6797	Jul-98	54.2969
Jul-97	21.375	Aug-98	50
Aug-97	20.5156	Sep-98	65.75
Sep-97	24.2188	Oct-98	65.5
Oct-97	20.0312	Nov-98	60.8125
Nov-97	21.0469		

(a) Using a graphing utility, draw a scatter diagram.

(b) Using a graphing utility, fit an exponential model to the data.

(c) Express the function in the form $A = A_0 e^{kt}$.

(d) Graph the exponential function found in part (b) or (c) on the scatter diagram.

(e) Using the solution to part (b) or (c), predict the closing price of Dell Computer stock at the end of December 1998.

(f) Interpret the value of k found in part (c).

Solution (a) Enter the data into the graphing utility, letting 1 represent November, 1996, 2 represent December 1996, and so on. We obtain the scatter diagram shown in Figure 48.

(b) A graphing utility fits the data in Figure 48 to an exponential model of the form $y = ab^x$ by using the EXPonential REGression option. See Figure 49. Thus, $y = ab^x = 6.27284(1.10426)^x$.

Figure 48

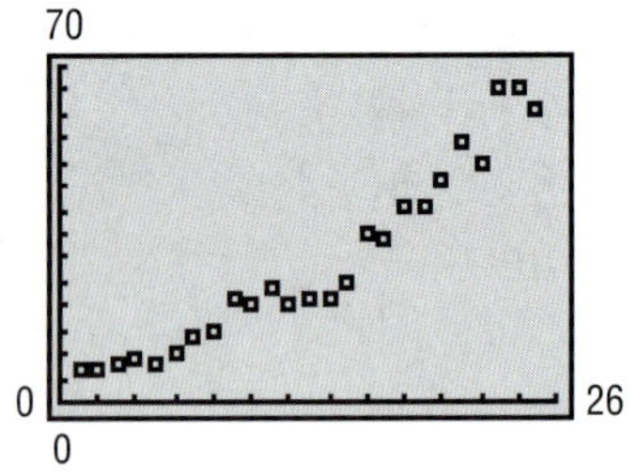

Figure 49

(c) To express $y = ab^x$ in the form $A = A_0e^{kt}$, we proceed as follows:

$$ab^x = A_0e^{kt}$$

We set the coefficients equal to each other and the exponential expressions equal to each other.

$$a = A_0 \qquad b^x = e^{kt} = (e^k)^t$$
$$b = e^k$$

Since $y = ab^x = 6.27284(1.10426)^x$, we find that $a = 6.27284$ and $b = 1.10426$. Thus,

$$a = A_0 = 6.27284 \quad \text{and} \quad b = 1.10426 = e^k$$
$$k = \ln(1.10426) = 0.0992$$

As a result, $A = A_0e^{kt} = 6.27284e^{0.0992t}$.

(d) See Figure 50.

Figure 50

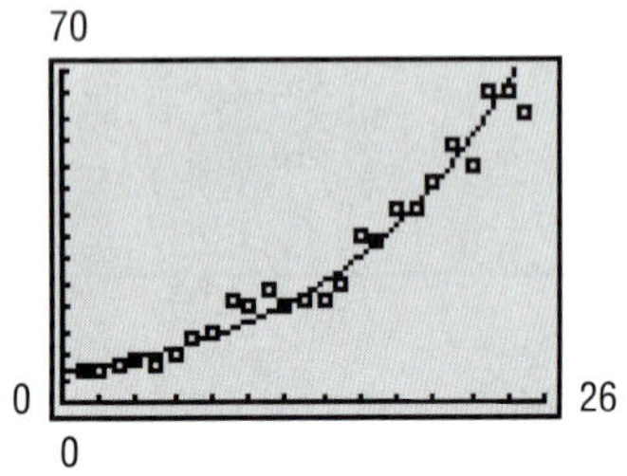

(e) Let $t = 26$ (December 1998) in the function found in part (c). The predicted closing price of Dell Computer stock at the end of December 1998 is

$$A = 6.27284e^{0.0992(26)} = \$82.72$$

(f) The value of k represents the monthly interest rate compounded continuously.

$$A = A_0e^{kt} = 6.27284e^{0.0992t}$$
$$= Pe^{rt} \qquad \text{Equation (4), Section 5.6}$$

Thus, the price of Dell Computer stock has grown at a monthly rate of 9.92% (compounded continuously) between November 1996 and November 1998. The annual rate of continuous growth is $12(9.92\%) = 119.04\%$.

NOW WORK PROBLEM 9.

Logarithmic Curve Fitting

2 Many relations between variables do not follow an exponential model, but, instead, the independent variable is related to the dependent variable using a logarithmic model.

EXAMPLE 2 Fitting a Curve to a Logarithmic Model

Jodi, a meteorologist, is interested in finding a function that explains the relation between the height of a weather balloon (in kilometers) and the atmospheric pressure (measured in millimeters of mercury) on the balloon. She collects the data shown in the table.

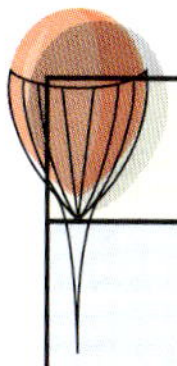

Atmospheric Pressure, p	Height, h
760	0
740	0.184
725	0.328
700	0.565
650	1.079
630	1.291
600	1.634
580	1.862
550	2.235

(a) Using a graphing utility, draw a scatter diagram of the data with atmospheric pressure as the independent variable.

(b) Using a graphing utility, fit a logarithmic model to the data.

(c) Draw the logarithmic function found in part (b) on the scatter diagram.

(d) Use the function found in part (b) to predict the height of the weather balloon if the atmospheric pressure is 560 millimeters of mercury.

Solution (a) After entering the data into the graphing utility, we obtain the scatter diagram shown in Figure 51.

(b) A graphing utility fits the data in Figure 51 to a logarithmic model of the form $y = a + b \ln x$ by using the Logarithm REGression option. See Figure 52. The logarithmic function of best fit to the data is

$$h(p) = 45.7863 - 6.9025 \ln p$$

where h is the height of the weather balloon and p is the atmospheric pressure. Notice that $|r|$ is close to 1, indicating a good fit.

(c) Figure 53 shows the graph of $h(p) = 45.7863 - 6.9025 \ln p$ on the scatter diagram.

(d) Using the function found in part (b), Jodi predicts the height of the weather balloon when the atmospheric pressure is 560 to be

$$h(560) = 45.7863 - 6.9025 \ln 560$$
$$\approx 2.108 \text{ kilometers}$$

Figure 51

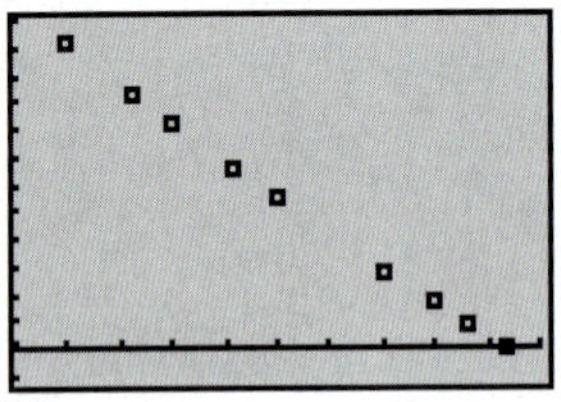

Figure 52

Figure 53

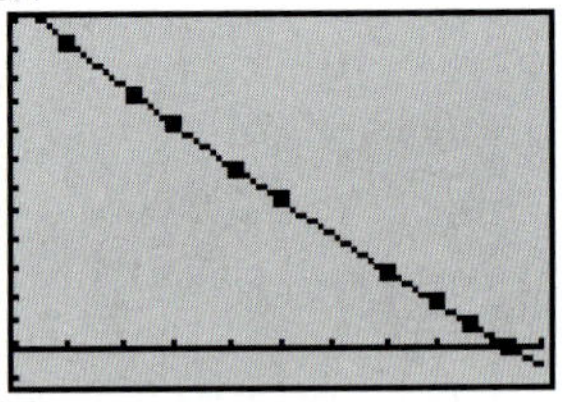

NOW WORK PROBLEM 13.

Logistic Curve Fitting

3 Logistic growth models can be used to model situations where the value of the dependent variable is limited. Many real-world situations conform to this scenario. For example, the population of the human race is limited by the availability of natural resources such as food and shelter. When the value of the dependent variable is limited, a logistic growth model is often appropriate.

EXAMPLE 3 Fitting a Curve to a Logistic Growth Model

The data in the table, obtained from R. Pearl ("The Growth of Population," *Quarterly Review of Biology* 2 (1927): 532–548) represent the amount of yeast biomass after t hours in a culture.

(a) Using a graphing utility, draw a scatter diagram of the data with time as the independent variable.

(b) Using a graphing utility, fit a logistic growth model to the data.

(c) Using a graphing utility, graph the function found in part (b) on the scatter diagram.

Time (in hours)	Yeast Biomass	Time (in hours)	Yeast Biomass
0	9.6	10	513.3
1	18.3	11	559.7
2	29.0	12	594.8
3	47.2	13	629.4
4	71.1	14	640.8
5	119.1	15	651.1
6	174.6	16	655.9
7	257.3	17	659.6
8	350.7	18	661.8
9	441.0		

(d) What is the predicted carrying capacity of the culture?

(e) Use the function found in part (b) to predict the population of the culture at $t = 19$ hours.

Solution (a) See Figure 54.

(b) A graphing utility fits a logistic growth model of the form $y = \dfrac{c}{1 + ae^{-bx}}$ by using the LOGISTIC regression option. See Figure 55. The logistic growth function of best fit to the data is

$$y = \frac{663.0}{1 + 71.6e^{-0.5470x}}$$

where y is the amount of yeast biomass in the culture and x is the time.

(c) See Figure 56.

(d) Based on the logistic growth function found in part (b), the carrying capacity of the culture is 663.

(e) Using the logistic growth function found in part (b), the predicted amount of yeast biomass at $t = 19$ is

$$y = \frac{663.0}{1 + 71.6e^{-0.5470(19)}} = 661.5$$

Figure 54

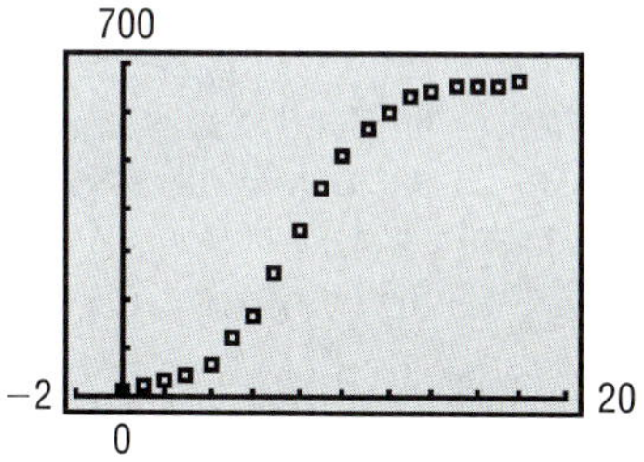

Figure 55

Figure 56

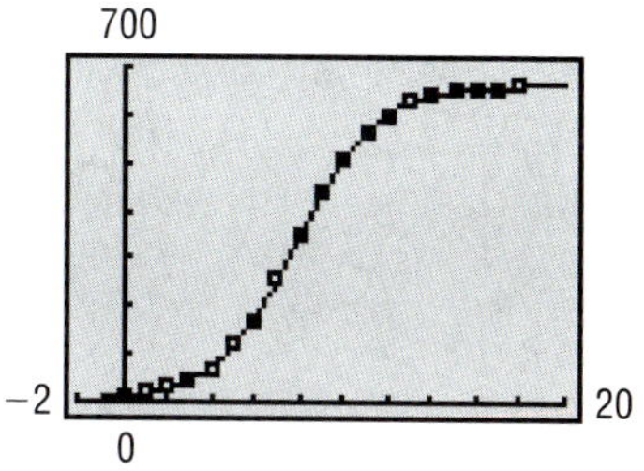

NOW WORK PROBLEM 15.

5.8 EXERCISES

1. **Biology** A certain bacteria initially increases according to the law of uninhibited growth. A biologist collects the following data for this bacteria:
   (a) Using a graphing utility, draw a scatter diagram.
   (b) Using a graphing utility, fit an exponential curve to the data.
   (c) Express the curve in the form $N(t) = N_0 e^{kt}$.
   (d) Graph the exponential function found in part (b) or (c) on the scatter diagram.
   (e) Using the solution to part (b), predict the population at $t = 7$ hours.

Time (Hours)	Population
0	1000
1	1415
2	2000
3	2828
4	4000
5	5656
6	8000

2. **Biology** A colony of bacteria initially increases according to the law of uninhibited growth. A biologist collects the following data for this bacteria:

Time (Days)	Population
0	50
1	153
2	234
3	357
4	547
5	839
6	1280

(a) Using a graphing utility, draw a scatter diagram.
(b) Using a graphing utility, fit an exponential curve to the data.
(c) Express the curve in the form $N(t) = N_0e^{kt}$.
(d) Graph the exponential function found in part (b) or (c) on the scatter diagram.
(e) Using the solution to part (b), predict the population at $t = 7$ days.

3. **Economics and Marketing** A store manager collected the following data regarding price and quantity demanded of shoes:

Price ($/Unit)	Quantity Demanded
79	10
67	20
54	30
46	40
38	50
31	60

(a) Using a graphing utility, draw a scatter diagram with quantity demanded on the x-axis and price on the y-axis.
(b) Using a graphing utility, fit an exponential curve to the data.
(c) Graph the exponential function found in part (b) on the scatter diagram.
(d) Predict the quantity demanded if the price is $60.

4. **Economics and Marketing** A store manager collected the following data regarding price and quantity supplied of dresses:

Price ($/Unit)	Quantity Supplied
25	10
32	20
40	30
46	40
60	50
74	60

(a) Using a graphing utility, draw a scatter diagram with quantity supplied on the x-axis and price on the y-axis.
(b) Using a graphing utility, fit an exponential curve to the data.
(c) Graph the exponential function found in part (b) on the scatter diagram.
(d) Predict the quantity supplied if the price is $45.

5. **Chemistry** A chemist has a 100-gram sample of a radioactive material. He records the amount of radioactive material every week for 6 weeks and obtains the following data:

Week	Weight (in Grams)
0	100.0
1	88.3
2	75.9
3	69.4
4	59.1
5	51.8
6	45.5

(a) Using a graphing utility, draw a scatter diagram with week as the independent variable.
(b) Using a graphing utility, fit an exponential curve to the data.

(c) Express the curve in the form $A(t) = A_0e^{kt}$.
(d) Graph the exponential function found in part (b) or (c) on the scatter diagram.
(e) From the result found in part (b), determine the half-life of the radioactive material.
(f) How much radioactive material will be left after 50 weeks?

6. Chemistry A chemist has a 1000-gram sample of a radioactive material. She records the amount of radioactive material remaining in the sample every day for a week and obtains the following data:

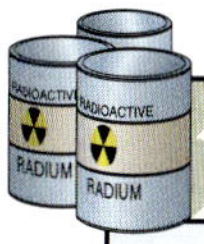

Day	Weight (in Grams)
0	1000.0
1	897.1
2	802.5
3	719.8
4	651.1
5	583.4
6	521.7
7	468.3

(a) Using a graphing utility, draw a scatter diagram with day as the independent variable.
(b) Using a graphing utility, fit an exponential curve to the data.
(c) Express the curve in the form $A(t) = A_0e^{kt}$.
(d) Graph the exponential function found in part (b) or (c) on the scatter diagram.
(e) From the result found in part (b), find the half-life of the radioactive material.
(f) How much radioactive material will be left after 20 days?

7. Finance The following data represent the amount of money an investor has in an investment account each year for 10 years. She wishes to determine the effective rate of return on her investment.

Year	Value of Account
1985	$10,000
1986	$10,573
1987	$11,260
1988	$11,733
1989	$12,424
1990	$13,269
1991	$13,968
1992	$14,823
1993	$15,297
1994	$16,539

(a) Using a graphing utility, draw a scatter diagram with time as the independent variable and the value of the account as the dependent variable.
(b) Using a graphing utility, fit an exponential curve to the data.
(c) Based on the answer in part (b), what was the effective rate of return from this account over the past 10 years?
(d) If the investor plans on retiring in 2020, what will the predicted value of this account be?

8. Finance The following data shows the amount of money an investor has in an investment account each year for 7 years. He wishes to determine the effective rate of return on his investment.

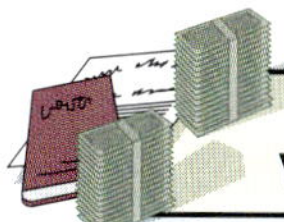

Year	Value of Account
1988	$20,000
1989	$21,516
1990	$23,355
1991	$24,885
1992	$27,434
1993	$30,053
1994	$32,622

(a) Using a graphing utility, draw a scatter diagram with time as the independent variable and the value of the account as the dependent variable.
(b) Using a graphing utility, fit an exponential curve to the data.
(c) Based on the answer to part (b), what was the effective rate of return from this account over the past 7 years?
(d) If the investor plans on retiring in 2020, what will the predicted value of this account be?

9. Microsoft Stock The following data represent the closing price of Microsoft Corporation stock at the end of each month.

Date	Closing Price	Date	Closing Price
Nov-96	19.6094	Dec-97	32.3125
Dec-96	20.6562	Jan-98	37.2969
Jan-97	25.5	Feb-98	42.375
Feb-97	24.375	Mar-98	44.75
Mar-97	22.9219	Apr-98	45.0625
Apr-97	30.375	May-98	42.4062
May-97	31	Jun-98	54.1875
Jun-97	31.5938	Jul-98	54.9688
Jul-97	35.3438	Aug-98	47.9688
Aug-97	33.0469	Sep-98	55.0312
Sep-97	33.0781	Oct-98	52.9375
Oct-97	32.5	Nov-98	61
Nov-97	35.375		

(a) Using a graphing utility, draw a scatter diagram.
(b) Using a graphing utility, fit an exponential model to the data.
(c) Express the function in the form $A(t) = A_0e^{kt}$.
(d) Graph the exponential function found in part (b) or (c) on the scatter diagram.
(e) Using the solution to part (b) or (c), predict the closing price of Microsoft stock at the end of December 1998.

10. Eli Lilly The following data represent the closing price of Eli Lilly stock at the end of each month.

Date	Closing Price	Date	Closing Price
Nov-96	37.2746	Dec-97	68.831
Dec-96	35.5693	Jan-98	66.8538
Jan-97	42.4517	Feb-98	65.2675
Feb-97	42.7457	Mar-98	59.1313
Mar-97	40.2384	Apr-98	68.9865
Apr-97	42.9903	May-98	61.0502
May-97	45.6743	Jun-98	65.8994
Jun-97	53.6858	Jul-98	66.8941
Jul-97	55.4968	Aug-98	65.5973
Aug-97	51.5571	Sep-98	78.1306
Sep-97	59.6264	Oct-98	80.8119
Oct-97	66.0941	Nov-98	89.6875
Nov-97	62.2815		

(a) Using a graphing utility, draw a scatter diagram.
(b) Using a graphing utility, fit an exponential model to the data.
(c) Express the function in the form $A(t) = A_0 e^{kt}$.
(d) Graph the exponential function found in part (b) or (c) on the scatter diagram.
(e) Using the solution to part (b) or (c), predict the closing price of Eli Lilly stock at the end of December 1998.

11. Economics and Marketing The following data represents the price and quantity demanded in 1997 for IBM personal computers at Best Buy.

(a) Using a graphing utility, draw a scatter diagram of the data with price as the dependent variable.
(b) Using a graphing utility, fit a logarithmic model to the data.
(c) Using a graphing utility, draw the logarithmic function found in part (b) on the scatter diagram.
(d) Use the function found in part (b) to predict the number of IBM personal computers that would be demanded if the price were $1650.

Price ($/Computer)	Quantity Demanded
2300	152
2000	159
1700	164
1500	171
1300	176
1200	180
1000	189

12. Economics and Marketing The following data represent the price and quantity supplied in 1997 for IBM personal computers.

Price ($/Computer)	Quantity Supplied
2300	180
2000	173
1700	160
1500	150
1300	137
1200	130
1000	113

(a) Using a graphing utility, draw a scatter diagram of the data with price as the dependent variable.
(b) Using a graphing utility, fit a logarithmic model to the data.
(c) Using a graphing utility, draw the logarithmic function found in part (b) on the scatter diagram.
(d) Use the function found in part (b) to predict the number of IBM personal computers that would be supplied if the price were $1650.

13. Population Model The following data obtained from the U.S. Census Bureau represent the population of the United States. An ecologist is interested in finding a function that describes the population of the United States.

Year	Population
1900	76,212,168
1910	92,228,496
1920	106,021,537
1930	123,202,624
1940	132,164,569
1950	151,325,798
1960	179,323,175
1970	203,302,031
1980	226,542,203
1990	248,709,873

(a) Using a graphing utility, draw a scatter diagram of the data using the year as the independent variable and population as the dependent variable.
(b) Using a graphing utility, fit a logistic model to the data.
(c) Using a graphing utility, draw the function found in part (b) on the scatter diagram.
(d) Based on the function found in part (b), what is the carrying capacity of the United States?
(e) Use the function found in part (b) to predict the population of the United States in 2000.

14. Population Model The following data obtained from the U.S. Census Bureau represent the world population. An ecologist is interested in finding a function that describes the world population.

Year	Population (in Billions)
1981	4.533
1982	4.614
1983	4.695
1984	4.775
1985	4.856
1986	4.941
1987	5.029
1988	5.117
1989	5.205
1990	5.295
1991	5.381
1992	5.469
1993	5.556
1994	5.644
1995	5.732

(a) Using a graphing utility, draw a scatter diagram of the data using year as the independent variable and population as the dependent variable.
(b) Using a graphing utility, fit a logistic model to the data.
(c) Using a graphing utility, draw the function found in part (b) on the scatter diagram.
(d) Based on the function found in part (b), what is the carrying capacity of the world?
(e) Use the function found in part (b) to predict the population of the world in 2000.

15. Population Model The following data obtained from the U.S. Census Bureau represent the population of Illinois. An urban economist is interested in finding a model that describes the population of Illinois.

(a) Using a graphing utility, draw a scatter diagram of the data using year as the independent variable and population as the dependent variable.
(b) Using a graphing utility, fit a logistic model to the data.
(c) Using a graphing utility, draw the function found in part (b) on the scatter diagram.
(d) Based on the function found in part (b), what is the carrying capacity of Illinois?
(e) Use the function found in part (b) to predict the population of Illinois in 2000.

Year	Population
1900	4,821,550
1910	5,638,591
1920	6,485,280
1930	7,630,654
1940	7,897,241
1950	8,712,176
1960	10,081,158
1970	11,110,285
1980	11,427,409
1990	11,430,602

16. Population Model The following data obtained from the U.S. Census Bureau represent the population of Pennsylvania. An urban economist is interested in finding a model that describes the population of Pennsylvania.

Year	Population
1900	6,302,115
1910	7,665,111
1920	8,720,017
1930	9,631,350
1940	9,900,180
1950	10,498,012
1960	11,319,366
1970	11,800,766
1980	11,864,720
1990	11,881,643

(a) Using a graphing utility, draw a scatter diagram of the data using year as the independent variable and population as the dependent variable.
(b) Using a graphing utility, fit a logistic model to the data.
(c) Using a graphing utility, draw the function found in part (b) on the scatter diagram.
(d) Based on the function found in part (b), what is the carrying capacity of Pennsylvania?
(e) Use the function found in part (b) to predict the population of Pennsylvania in 2000.

CHAPTER REVIEW

THINGS TO KNOW

One-to-one function *f*
If $x_1 \neq x_2$, then $f(x_1) \neq f(x_2)$ for any choice of x_1 and x_2 in the domain.

Horizontal line test
If every horizontal line intersects the graph of a function f in at most one point, then f is one-to-one.

Inverse function f^{-1} of f

Domain of f = Range of f^{-1}; Range of f = Domain of f^{-1}

$f^{-1}(f(x)) = x$ and $f(f^{-1}(x)) = x$.

Graphs of f and f^{-1} are symmetric with respect to the line $y = x$.

Properties of the exponential function

$f(x) = a^x, a > 1$ — Domain: $(-\infty, \infty)$; Range: $(0, \infty)$; x-intercepts: none; y-intercept: 1; horizontal asymptote: x-axis as $x \to -\infty$; increasing; one-to-one

See Figure 13 for a typical graph.

$f(x) = a^x, 0 < a < 1$ — Domain: $(-\infty, \infty)$; Range: $(0, \infty)$; x-intercepts: none, y-intercept: 1; horizontal asymptote: x-axis as $x \to \infty$; decreasing; one-to-one

See Figure 17 for a typical graph.

Properties of the logarithmic function

$f(x) = \log_a x, a > 1$
($y = \log_a x$ means $x = a^y$) — Domain: $(0, \infty)$; Range: $(-\infty, \infty)$; x-intercept: 1; y-intercept: none; vertical asymptote: $x = 0$ (y-axis); increasing; one-to-one

See Figure 25(b) for a typical graph.

$f(x) = \log_a x, 0 < a < 1$
($y = \log_a x$ means $x = a^y$) — Domain: $(0, \infty)$; Range: $(-\infty, \infty)$; x-intercept: 1; y-intercept: none; vertical asymptote: $x = 0$ (y-axis); decreasing; one-to-one.

See Figure 25(a) for a typical graph.

Number *e*

Value approached by the expression $\left(1 + \frac{1}{n}\right)^n$ as $n \to \infty$; that is, $\lim_{n\to\infty}\left(1 + \frac{1}{n}\right)^n = e$

Natural logarithm
$y = \ln x$ means $x = e^y$

Properties of logarithms
$\log_a 1 = 0$ $\quad \log_a a = 1$ $\quad a^{\log_a M} = M$ $\quad \log_a a^r = r$

$\log_a (MN) = \log_a M + \log_a N$ $\quad \log_a\left(\frac{M}{N}\right) = \log_a M - \log_a N$ $\quad \log_a\left(\frac{1}{N}\right) = -\log_a N$ $\quad \log_a M^r = r \log_a M$

FORMULAS

Change-of-Base Formula $\quad \log_a M = \dfrac{\log_b M}{\log_b a}$

Compound interest $\quad A = P\left(1 + \dfrac{r}{n}\right)^{nt}$

Continuous compounding	$A = Pe^{rt}$
Present value	$P = A\left(1 + \frac{r}{n}\right)^{-nt}$ or $P = Ae^{-rt}$
Growth and decay	$A(t) = A_0e^{kt}$
Logistic growth	$P(t) = \frac{c}{1 + ae^{-bt}}$

HOW TO

Find the inverse of certain one-to-one functions (see the procedure given on page 317).

Graph f^{-1} given the graph of f.

Graph exponential and logarithmic functions; find their domain, range, and asymptotes.

Solve certain exponential equations.

Solve certain logarithmic equations.

Solve problems involving compound interest.

Solve problems involving growth and decay.

Obtain exponential, logarithmic, and logistic functions of best fit.

FILL-IN-THE-BLANK ITEMS

1. If every horizontal line intersects the graph of a function f at no more than one point, then f is a(n) ______________ function.

2. If f^{-1} denotes the inverse of a function f, then the graphs of f and f^{-1} are symmetric with respect to the line ______________.

3. The graph of every exponential function $f(x) = a^x$, $a > 0$, $a \neq 1$, passes through the two points ______________.

4. If the graph of an exponential function $f(x) = a^x$, $a > 0$, $a \neq 1$, is decreasing, then its base must be less than ______________.

5. If $3^x = 3^4$, then $x =$ ______________.

6. The logarithm of a product equals the ______________ of the logarithms.

7. For every base, the logarithm of ______________ equals 0.

8. If $\log_8 M = \log_5 7/\log_5 8$, then $M =$ ______________.

9. The domain of the logarithmic function $f(x) = \log_a x$ consists of ______________.

10. The graph of every logarithmic function $f(x) = \log_a x$, $a > 0$, $a \neq 1$, passes through the two points ______________.

11. If the graph of a logarithmic function $f(x) = \log_a x$, $a > 0$, $a \neq 1$, is increasing, then its base must be larger than ______________.

12. If $\log_3 x = \log_3 7$, then $x =$ ______________.

TRUE/FALSE ITEMS

T F **1.** If f and g are inverse functions, then the domain of f is the same as the domain of g.

T F **2.** If f and g are inverse functions, then their graphs are symmetric with respect to the line $y = x$.

T F **3.** The graph of every exponential function $f(x) = a^x$, $a > 0$, $a \neq 1$, will contain the points $(0, 1)$ and $(1, a)$.

T F **4.** The graphs of $y = 3^{-x}$ and $y = \left(\frac{1}{3}\right)^x$ are identical.

T F **5.** The present value of \$1000 to be received after 2 years at 10% per annum compounded continuously is approximately \$1205.

T F 6. If $y = \log_a x$, then $y = a^x$.

T F 7. The graph of every logarithmic function $f(x) = \log_a x$, $a > 0$, $a \neq 1$, will contain the points $(1, 0)$ and $(a, 1)$.

T F 8. $a^{\log_M a} = M$, where $a > 0$, $a \neq 1$, $M > 0$.

T F 9. $\log_a (M + N) = \log_a M + \log_a N$, where $a > 0$, $a \neq 1$, $M > 0$, $N > 0$.

T F 10. $\log_a M - \log_a N = \log_a (M/N)$, where $a > 0$, $a \neq 1$, $M > 0$, $N > 0$.

REVIEW EXERCISES

Blue Problem numbers indicate the authors' suggestions for use in a Practice Test.

In Problems 1–6, the function f is one-to-one. Find the inverse of each function and check your answer. Find the domain and range of f and f^{-1}. Use a graphing utility to simultaneously graph f, f^{-1}, and $y = x$ on the same square screen.

1. $f(x) = \dfrac{2x + 3}{5x - 2}$
2. $f(x) = \dfrac{2 - x}{3 + x}$
3. $f(x) = \dfrac{1}{x - 1}$
4. $f(x) = \sqrt{x - 2}$
5. $f(x) = \dfrac{3}{x^{1/3}}$
6. $f(x) = x^{1/3} + 1$

In Problems 7–12, evaluate each expression.

7. $\log_2 \left(\frac{1}{8}\right)$
8. $\log_3 81$
9. $\ln e^{\sqrt{2}}$
10. $e^{\ln 0.1}$
11. $2^{\log_2 0.4}$
12. $\log_2 2^{\sqrt{3}}$

In Problems 13–18, write each expression as a single logarithm.

13. $3 \log_4 x^2 + \dfrac{1}{2} \log_4 \sqrt{x}$
14. $-2 \log_3 \left(\dfrac{1}{x}\right) + \dfrac{1}{3} \log_3 \sqrt{x}$
15. $\ln\left(\dfrac{x - 1}{x}\right) + \ln\left(\dfrac{x}{x + 1}\right) - \ln(x^2 - 1)$
16. $\log (x^2 - 9) - \log (x^2 + 7x + 12)$
17. $2 \log 2 + 3 \log x - \dfrac{1}{2}[\log (x + 3) + \log (x - 2)]$
18. $\dfrac{1}{2} \ln(x^2 + 1) - 4 \ln\dfrac{1}{2} - \dfrac{1}{2}[\ln(x - 4) + \ln x]$

In Problems 19–26, find y as a function of x. The constant C is a positive number.

19. $\ln y = 2x^2 + \ln C$
20. $\ln(y - 3) = \ln 2x^2 + \ln C$
21. $\frac{1}{2} \ln y = 3x^2 + \ln C$
22. $\ln 2y = \ln(x + 1) + \ln(x + 2) + \ln C$
23. $\ln(y - 3) + \ln(y + 3) = x + C$
24. $\ln(y - 1) + \ln(y + 1) = -x + C$
25. $e^{y + C} = x^2 + 4$
26. $e^{3y - C} = (x + 4)^2$

In Problems 27–36, using a graphing utility, show the stages required to graph each function. Determine the domain, range, and any asymptotes.

27. $f(x) = e^{-x}$
28. $f(x) = \ln(-x)$
29. $f(x) = 1 - e^x$
30. $f(x) = 3 + \ln x$
31. $f(x) = 3e^x$
32. $f(x) = \frac{1}{2} \ln x$
33. $f(x) = e^{|x|}$
34. $f(x) = \ln |x|$
35. $f(x) = 3 - e^{-x}$
36. $f(x) = 4 - \ln(-x)$

In Problems 37–56, solve each equation. Verify your result using a graphing utility.

37. $4^{1 - 2x} = 2$
38. $8^{6 + 3x} = 4$
39. $3^{x^2 + x} = \sqrt{3}$
40. $4^{x - x^2} = \frac{1}{2}$
41. $\log_x 64 = -3$
42. $\log_{\sqrt{2}} x = -6$
43. $5^x = 3^{x + 2}$
44. $5^{x + 2} = 7^{x - 2}$
45. $9^{2x} = 27^{3x - 4}$
46. $25^{2x} = 5^{x^2 - 12}$
47. $\log_3 \sqrt{x - 2} = 2$
48. $2^{x + 1} \cdot 8^{-x} = 4$
49. $8 = 4^{x^2} \cdot 2^{5x}$
50. $2^x \cdot 5 = 10^x$
51. $\log_6 (x + 3) + \log_6 (x + 4) = 1$
52. $\log_{10} (7x - 12) = 2 \log_{10} x$
53. $e^{1 - x} = 5$
54. $e^{1 - 2x} = 4$
55. $2^{3x} = 3^{2x + 1}$
56. $2^{x^3} = 3^{x^2}$

In Problems 57–60, use the following result: If x is the atmospheric pressure (measured in millimeters of mercury), then the formula for the altitude $h(x)$ (measured in meters above sea level) is

$$h(x) = (30T + 8000)\log\left(\frac{P_0}{x}\right)$$

where T is the temperature (in degrees Celsius) and P_0 is the atmospheric pressure at sea level, which is approximately 760 millimeters of mercury.

57. **Finding the Altitude of an Airplane** At what height is a Piper Cub whose instruments record an outside temperature of 0°C and a barometric pressure of 300 millimeters of mercury?

58. **Finding the Height of a Mountain** How high is a mountain if instruments placed on its peak record a temperature of 5°C and a barometric pressure of 500 millimeters of mercury?

59. **Atmospheric Pressure Outside an Airplane** What is the atmospheric pressure outside a Boeing 737 flying at an altitude of 10,000 meters if the outside air temperature is −100°C?

60. **Atmospheric Pressure at High Altitudes** What is the atmospheric pressure (in millimeters of mercury) on Mt. Everest, which has an altitude of approximately 8900 meters, if the air temperature is 5°C?

61. **Amplifying Sound** An amplifier's power output P (in watts) is related to its decibel voltage gain d by the formula $P = 25e^{0.1d}$.

(a) Find the power output for a decibel voltage gain of 4 decibels.
(b) For a power output of 50 watts, what is the decibel voltage gain?

62. **Limiting Magnitude of a Telescope** A telescope is limited in its usefulness by the brightness of the star it is aimed at and by the diameter of its lens. One measure of a star's brightness is its *magnitude:* the dimmer the star, the larger its magnitude. A formula for the limiting magnitude L of a telescope, that is, the magnitude of the dimmest star that it can be used to view, is given by

$$L = 9 + 5.1\log d$$

where d is the diameter (in inches) of the lens.
(a) What is the limiting magnitude of a 3.5-inch telescope?
(b) What diameter is required to view a star of magnitude 14?

63. **Product Demand** The demand for a new product increases rapidly at first and then levels off. The percent P of actual purchases of this product after it has been on the market t months is

$$P = 90 - 80\left(\frac{3}{4}\right)^t$$

(a) What is the percent of purchases of the product after 5 months?
(b) What is the percent of purchases of the product after 10 months?
(c) What is the maximum percent of purchases of the product?
(d) How many months does it take before 40% of purchases occur?
(e) How many months before 70% of purchases occur?

64. **Disseminating Information** A survey of a certain community of 10,000 residents shows that the number of residents N who have heard a piece of information after m months is given by the formula

$$m = 55.3 - 6\ln(10{,}000 - N)$$

How many months will it take for half of the citizens to learn about a community program of free blood pressure readings?

65. **Salvage Value** The number of years n for a piece of machinery to depreciate to a known salvage value can be found using the formula

$$n = \frac{\log s - \log i}{\log(1 - d)}$$

where s is the salvage value of the machinery, i is its initial value, and d is the annual rate of depreciation.
(a) How many years will it take for a piece of machinery to decline in value from $90,000 to $10,000 if the annual rate of depreciation is 0.20 (20%)?

(b) How many years will it take for a piece of machinery to lose half of its value if the annual rate of depreciation is 15%?

66. **Funding a College Education** A child's grandparents purchase a $10,000 bond fund that matures in 18 years to be used for her college education. The bond fund pays 4% interest compounded semiannually. How much will the bond fund be worth at maturity?

67. **Funding a College Education** A child's grandparents wish to purchase a bond fund that matures in 18 years to be used for her college education. The bond fund pays 4% interest compounded semiannually. How much should they purchase so that the bond fund will be worth $85,000 at maturity?

68. **Funding an IRA** First Colonial Bankshares Corporation advertised the following IRA investment plans.

Target IRA Plans

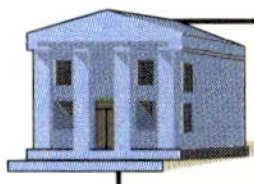

For each $5000 Maturity Value Desired Deposit:	At a Term of:
$620.17	20 Years
$1045.02	15 Years
$1760.92	10 Years
$2967.26	5 Years

(a) Assuming continuous compounding, what was the annual rate of interest that they offered?
(b) First Colonial Bankshares claims that $4000 invested today will have a value of over $32,000 in 20 years. Use the answer found in part (a) to find the actual value of $4000 in 20 years. Assume continuous compounding.

69. **Estimating the Date that a Prehistoric Man Died** The bones of a prehistoric man found in the desert of New Mexico contain approximately 5% of the original amount of carbon 14. If the half-life of carbon 14 is 5600 years, approximately how long ago did the man die?

70. **Temperature of a Skillet** A skillet is removed from an oven whose temperature is 450°F and placed in a room whose temperature is 70°F. After 5 minutes, the temperature of the skillet is 400°F. How long will it be until its temperature is 150°F?

71. **AIDS Infections** The following data represent the cumulative number of HIV/AIDS cases reported worldwide for the years 1980–1995. Let $x = 1$ represent 1980, $x = 2$ represent 1981, and so on.

Year	HIV Infections (Millions)
1980, 1	0.2
1981, 2	0.6
1982, 3	1.1
1983, 4	1.8
1984, 5	2.7
1985, 6	3.9
1986, 7	5.3
1987, 8	6.9
1988, 9	8.7
1989, 10	10.7
1990, 11	13.0
1991, 12	15.5
1992, 13	18.5
1993, 14	21.9
1994, 15	25.9
1995, 16	30.6

Source: Global AIDS Policy Coalition, Harvard School of Public Health, Cambridge, Mass., Private Communication, January, 18, 1996.

(a) Using a graphing utility, draw a scatter diagram of the data using the year as the independent variable and the number of HIV infections as the dependent variable.
(b) Using a graphing utility, determine the logistic function of best fit.
(c) Graph the function found in part (b) on the scatter diagram.
(d) Based on the function found in part (b), what is the maximum number of HIV infections predicted to be?
(e) Using a graphing utility, determine the exponential function of best fit.
(f) Graph the function found in part (e) on the scatter diagram.
(g) Based on the graphs drawn in parts (c) and (f), which model appears to describe the relation between the year and the number of HIV infections reported?

CHAPTER PROJECTS

1. **Hot Coffee** A fast-food restaurant wants a special container to hold coffee. The restaurant wishes the container to quickly cool the coffee from 200°F to 130°F and keep the liquid between 110° and 130°F as long as possible. The restaurant has three containers to select from.

   (1) The CentiKeeper Company has a container that reduces the temperature of a liquid from 200°F to 100°F in 30 minutes by maintaining a constant temperature of 70°F.

   (2) The TempControl Company has a container that reduces the temperature of a liquid from 200°F to 110°F in 25 minutes by maintaining a constant temperature of 60°F.

   (3) The Hot'n'Cold Company has a container that reduces the temperature of a liquid from 200°F to 120°F in 20 minutes by maintaining a constant temperature of 65°F.

You need to recommend which container the restaurant should purchase.

   (a) Use Newton's Law of Cooling to find a function relating the temperature of the liquid over time.

   (b) Graph each function using a graphing utility.

   (c) How long does it take each container to lower the coffee temperature from 200°F to 130°F?

   (d) How long will the coffee temperature remain between 110°F and 130°F? This temperature is considered the optimal drinking temperature.

   (e) Which company would you recommend to the restaurant? Why?

   (f) How might the cost of the container affect your decision?

2. **Depreciation of a New Car** In buying a new car, one consideration might be how well the price of the car holds up over time. Different makes of cars have different depreciation rates. We will discuss one way of computing the depreciation rate for a car. Look up the value of your favorite car for the past 5 years in the NADA ("blue book") available at your local bank, public library, or via the Internet. Let $t = 0$ represent the cost of this car if you purchased it new, $t = 1$ represent the cost of the same car when it is one year old, and so on. Fill in the following table:

Year	Value
$t = 0$	
$t = 1$	
$t = 2$	
$t = 3$	
$t = 4$	
$t = 5$	

   (a) Using a graphing utility, draw a scatter diagram with time as the independent variable and value as the dependent variable.

   (b) Find the exponential function of best fit.

   (c) Express this function in the form $A = A_0e^{rt}$.

   (d) What is the value of A_0? Compare this value to the purchase price of the vehicle.

CHAPTER PROJECTS *(Continued)*

(e) What is the value of r, the depreciation rate?

(f) Compare your value of r to others in your class. Which car has the lowest rate of depreciation? The highest?

(g) How might depreciation factor into your decision regarding a new car purchase?

3. **CBL Experiment** The following data were collected by placing a temperature probe in a cup of hot water, removing the probe, and then recording temperature over time. According to Newton's Law of Cooling, these data should follow an exponential model.

Time	Temperature (F°)
0	175.69
1	173.52
2	171.21
3	169.07
4	166.59
5	164.21
6	161.89
7	159.66
8	157.86
9	155.75
10	153.70
11	151.93
12	150.08
13	148.50
14	146.84
15	145.33
16	143.83
17	142.38
18	141.22
19	140.09
20	138.69
21	137.59
22	136.78
23	135.70
24	134.91
25	133.86

(a) Using a graphing utility, draw a scatter diagram for the data.

(b) Using a graphing utility, fit an exponential model to the data.

(c) Graph the exponential function found in part (b) on the scatter diagram.

(d) Predict how long it will take for the water to reach a temperature of 110°F.

CHAPTER

6 Systems of Equations and Inequalities

Educational attainment

Changes in educational attainment over time indicate changes in the demand for skills and knowledge in the work force, as well as cultural evolution. An increase in the overall level of educational attainment can reflect the increasing emphasis society places on completing high school and college. Completing high school or college is an important educational accomplishment that yields many benefits, such as better job opportunities and higher earnings.

- The educational attainment of 25- to 29-year-olds increased between 1971 and 1997. The percentage of students completing high school rose from 78 to 87 percent; the percentage of high school completers with some college rose from 44 to 65 percent; and the percentage of high school completers with 4 or more years of college rose from 22 to 32 percent.

National Center for Education Statistics, The Condition of Education 1998

See Chapter Project 1.

Preparing for This Chapter

Before getting started on this chapter, review the following concepts:

- **For Sections 6.1, 6.3:** Lines (Section 1.6)
- **For Section 6.6:** Inequalities (Section 1.5)
  Lines (Section 1.6)
- **For Section 6.7:** Identity (p. 681)
  Factoring Polynomials (Appendix, Section 6)
  Proper and Improper Rational Functions (p. 275)

Outline

In this chapter we take up the problem of solving equations and inequalities containing two or more variables. There are various ways to solve such problems:

The *method of substitution* for solving equations in several unknowns goes back to ancient times.

The *method of elimination,* although it had existed for centuries, was put into systematic order by Karl Friedrich Gauss (1777–1855) and by Camille Jordan (1838–1922). This method is now used for solving large systems by computer.

The theory of *matrices* was developed in 1857 by Arthur Cayley (1821–1895), although only later were matrices used as we use them in this chapter. Matrices have become a very flexible instrument, useful in almost all areas of mathematics.

The method of *determinants* was invented by Seki Kōwa) (1642–1708) in 1683 in Japan and by Gottfried Wilhelm von Leibniz (1646–1716) in 1693 in Germany. Both used them only in relation to linear equations. *Cramer's Rule* is named after Gabriel Cramer (1704–1752) of Switzerland, who popularized the use of determinants for solving linear systems.

Section 6.6 introduces *linear programming,* a modern application of linear inequalities to certain types of problems. This topic is particularly useful for students interested in operations research.

Section 6.7, on *partial fraction decomposition,* provides an application of systems of equations. This particular application is one that is used in integral calculus.

6.1 SYSTEMS OF LINEAR EQUATIONS: TWO EQUATIONS CONTAINING TWO VARIABLES

1 Solve Systems of Equations by Substitution

2 Solve Systems of Equations by Elimination

3 Inconsistent Systems and Dependent Equations

We begin with an example.

EXAMPLE 1 Movie Theater Ticket Sales

A movie theater sells tickets for \$8.00 each, with Seniors receiving a discount of \$2.00. One evening the theater took in \$3580 in revenue. If x represents the number of tickets sold at \$8.00 and y the number of tickets sold at the discounted price of \$6.00, write an equation that relates these variables.

Solution Each nondiscounted ticket brings in \$8.00, so x tickets will bring in $8x$ dollars. Similarly, y discounted tickets bring in $6y$ dollars. Since the total brought in is \$3580, we must have

$$8x + 6y = 3580$$

In Example 1, suppose that we also know that 525 tickets were sold that evening. Then we have another equation relating the variables x and y.

$$x + y = 525$$

The two equations

$$\begin{aligned} 8x + 6y &= 3580 \\ x + y &= 525 \end{aligned}$$

form a *system* of equations.

In general, a **system of equations** is a collection of two or more equations, each containing one or more variables. Example 2 gives some samples of systems of equations.

EXAMPLE 2 Examples of Systems of Equations

(a) $\begin{cases} 2x + y = 5 & (1) \\ -4x + 6y = -2 & (2) \end{cases}$ *Two equations containing two variables, x and y.*

(b) $\begin{cases} x + y^2 = 5 & (1) \\ 2x + y = 4 & (2) \end{cases}$ *Two equations containing two variables, x and y.*

(c) $\begin{cases} x + y + z = 6 & (1) \\ 3x - 2y + 4z = 9 & (2) \\ x - y - z = 0 & (3) \end{cases}$ *Three equations containing three variables, x, y, and z.*

(d) $\begin{cases} x + y + z = 5 & (1) \\ x - y = 2 & (2) \end{cases}$ *Two equations containing three variables, x, y, and z.*

(e) $\begin{cases} x + y + z = 6 & (1) \\ 2x + 2z = 4 & (2) \\ y + z = 2 & (3) \\ x = 4 & (4) \end{cases}$ *Four equations containing three variables, x, y, and z.*

We use a brace, as shown, to remind us that we are dealing with a system of equations. We also will find it convenient to number each equation in the system.

A **solution** of a system of equations consists of values for the variables that are solutions of each equation of the system. To **solve** a system of equations means to find all solutions of the system.

For example, $x = 2$, $y = 1$ is a solution of the system in Example 2(a), because

$$2(2) + 1 = 5 \quad \text{and} \quad -4(2) + 6(1) = -2$$

A solution of the system in Example 2(b) is $x = 1$, $y = 2$, because

$$1 + 2^2 = 5 \quad \text{and} \quad 2(1) + 2 = 4$$

Another solution of the system in Example 2(b) is $x = \frac{11}{4}$, $y = -\frac{3}{2}$, which you can check for yourself. A solution of the system in Example 2(c) is $x = 3$, $y = 2$, $z = 1$, because

$$\begin{cases} 3 + 2 + 1 = 6 & (1) \quad x = 3, y = 2, z = 1 \\ 3(3) - 2(2) + 4(1) = 9 & (2) \\ 3 - 2 - 1 = 0 & (3) \end{cases}$$

Note that $x = 3$, $y = 3$, $z = 0$ is not a solution of the system in Example 2(c).

$$\begin{cases} 3 + 3 + 0 = 6 & (1) \quad x = 3, y = 3, z = 0 \\ 3(3) - 2(3) + 4(0) = 3 \neq 9 & (2) \\ 3 - 3 - 0 = 0 & (3) \end{cases}$$

Although these values satisfy equations (1) and (3), they do not satisfy equation (2). Any solution of the system must satisfy *each* equation of the system.

NOW WORK PROBLEM 3.

When a system of equations has at least one solution, it is said to be **consistent;** otherwise, it is called **inconsistent.**

An equation in n variables is said to be **linear** if it is equivalent to an equation of the form

$$a_1x_1 + a_2x_2 + \cdots + a_nx_n = b$$

where $x_1, x_2, \ldots, x_n$ are n distinct variables, $a_1, a_2, \ldots, a_n, b$ are constants, and at least one of the a's is not 0.

Some examples of linear equations are

$$2x + 3y = 2 \qquad 5x - 2y + 3z = 10 \qquad 8x + 8y - 2z + 5w = 0$$

If each equation in a system of equations is linear, then we have a **system of linear equations.** Thus, the systems in Examples 2(a), (c), (d), and (e) are linear, whereas the system in Example 2(b) is nonlinear. In this chapter we shall only solve linear systems (Sections 6.1–6.4). We discuss nonlinear systems in Chapter 9.

Two Linear Equations Containing Two Variables

We can view the problem of solving a system of two linear equations containing two variables as a geometry problem. The graph of each equation in such a system is a straight line. Thus, a system of two equations containing two variables represents a pair of lines. The lines either (1) intersect or (2) are parallel or (3) are **coincident** (that is, identical).

1. If the lines intersect, then the system of equations has one solution, given by the point of intersection. The system is **consistent** and the equations are **independent.**
2. If the lines are parallel, then the system of equations has no solution, because the lines never intersect. The system is **inconsistent.**
3. If the lines are coincident, then the system of equations has infinitely many solutions, represented by the totality of points on the line. The system is **consistent** and the equations are **dependent.**

Figure 1 illustrates these conclusions.

Figure 1

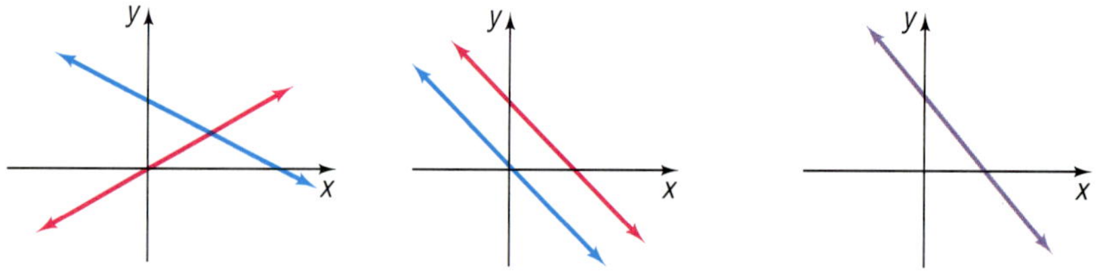

(a) Intersecting lines; system has one solution **(b)** Parallel lines; system has no solution **(c)** Coincident lines; system has infinitely many solutions

EXAMPLE 3 Solving a System of Linear Equations Using a Graphing Utility

Solve: $\begin{cases} 2x + y = 5 & (1) \\ -4x + 6y = 12 & (2) \end{cases}$

Solution First, we solve each equation for y. This is equivalent to writing each equation in slope–intercept form. Equation (1) in slope–intercept form is $Y_1 = -2x + 5$. Equation (2) in slope–intercept form is $Y_2 = \frac{2}{3}x + 2$. Figure 2 shows the graphs using a graphing utility. From the graph in Figure 2, we see that the lines intersect, so the system is consistent and the equations are independent. Using INTERSECT, we obtain the solution (1.125, 2.75).

Figure 2

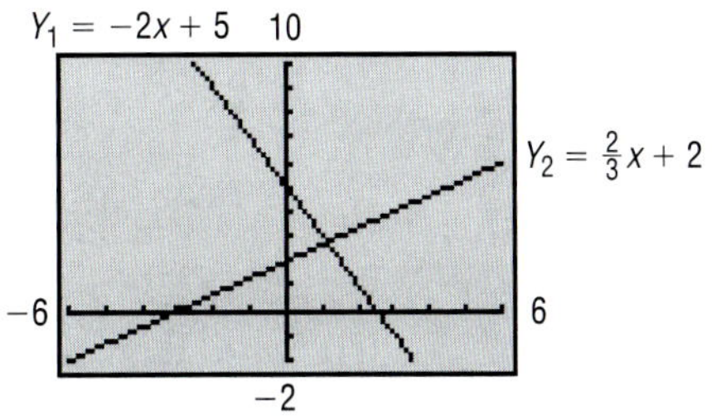

Sometimes, in order to obtain exact solutions, we must use algebraic methods. The first algebraic method that we discuss is the *method of substitution*.

Method of Substitution

1 We illustrate the **method of substitution** by solving the system given in Example 3.

EXAMPLE 4 Solving a System of Linear Equations by Substitution

Solve: $\begin{cases} 2x + y = 5 & (1) \\ -4x + 6y = 12 & (2) \end{cases}$

Solution We solve the first equation for y, obtaining

$$y = 5 - 2x \quad \textit{Subtract 2x from each side of (1)}$$

We substitute this result for y in the second equation. The result is an equation containing just the variable x, which we can then solve for.

$$\begin{aligned} -4x + 6y &= 12 \\ -4x + 6(5 - 2x) &= 12 && y = 5 - 2x. \\ -4x + 30 - 12x &= 12 && \text{Remove parentheses.} \\ -16x &= -18 && \text{Combine like terms and subtract 30 from both sides.} \\ x &= \frac{-18}{-16} = \frac{9}{8} && \text{Divide each side by } -16. \end{aligned}$$

Once we know that $x = \frac{9}{8}$, we can easily find the value of y by **back-substitution,** that is, by substituting $\frac{9}{8}$ for x in one of the original equations. We use the first one.

$$2x + y = 5 \qquad (1)$$
$$2\left(\frac{9}{8}\right) + y = 5 \qquad \text{Let } x = \tfrac{9}{8} \text{ in } (1).$$
$$\frac{9}{4} + y = 5 \qquad \text{Simplify.}$$
$$y = 5 - \frac{9}{4} \qquad \text{Subtract } \tfrac{9}{4} \text{ from both sides.}$$
$$= \frac{20}{4} - \frac{9}{4} = \frac{11}{4}$$

The solution of the system is $x = \frac{9}{8} = 1.125$, $y = \frac{11}{4} = 2.75$.

The method used to solve the system in Example 4 is called **substitution.** The steps to be used are outlined next.

> ***Steps for Solving by Substitution***
>
> STEP 1: Pick one of the equations and solve for one of the variables in terms of the remaining variables.
> STEP 2: Substitute the result in the remaining equations.
> STEP 3: If one equation in one variable results, solve this equation. Otherwise, repeat Step 1 until a single equation with one variable remains.
> STEP 4: Find the values of the remaining variables by back-substitution.
> STEP 5: Check the solution found.

EXAMPLE 5 Solving a System of Linear Equations by Substitution

Solve: $\begin{cases} 3x - 2y = 5 & (1) \\ 5x - y = 6 & (2) \end{cases}$

Solution STEP 1: After looking at the two equations, we conclude that it is easiest to solve for the variable y in equation (2).

$$5x - y = 6$$
$$5x - y - 6 = 0 \qquad \text{Subtract 6 from both sides.}$$
$$5x - 6 = y \qquad \text{Add } y \text{ to both sides.}$$

STEP 2: We substitute this result into equation (1) and simplify.

$$3x - 2y = 5 \qquad (1)$$
$$3x - 2(5x - 6) = 5 \qquad \text{Substitute } y = 5x - 6 \text{ in } (1).$$
$$-7x + 12 = 5 \qquad \text{Remove parentheses and collect like terms.}$$
$$-7x = -7 \qquad \text{Subtract 12 from both sides.}$$
$$x = 1 \qquad \text{Divide both sides by } -7.$$

STEP 3: Because we now have one solution, $x = 1$, we proceed to Step 4.

STEP 4: Knowing $x = 1$, we can find y from the equation

$$y = 5x - 6 = \underset{\substack{\uparrow \\ x = 1}}{5(1)} - 6 = -1$$

STEP 5: ✓CHECK: $\begin{cases} 3(1) - 2(-1) = 3 + 2 = 5 \\ 5(1) - (-1) \;\; = 5 + 1 = 6 \end{cases}$

The solution of the system is $x = 1$, $y = -1$.

NOW USE SUBSTITUTION TO WORK PROBLEM 11.

Method of Elimination

2 A second method for solving a system of linear equations is the *method of elimination.* This method is usually preferred over substitution if substitution leads to fractions or if the system contains more than two variables. Elimination also provides the necessary motivation for solving systems using matrices (the subject of Section 6.3).

The idea behind the method of elimination is to keep replacing the original equations in the system with equivalent equations until a system of equations with an obvious solution is reached. When we proceed in this way, we obtain **equivalent systems of equations.** The rules for obtaining equivalent equations are the same as those studied earlier. However, we may also interchange any two equations of the system and/or replace any equation in the system by the sum (or difference) of that equation and any other equation in the system.

Rules for Obtaining an Equivalent System of Equations

1. Interchange any two equations of the system.
2. Multiply (or divide) each side of an equation by the same nonzero constant.
3. Replace any equation in the system by the sum (or difference) of that equation and a nonzero multiple of any other equation in the system.

An example will give you the idea. As you work through the example, pay particular attention to the pattern being followed.

EXAMPLE 6 Solving a System of Linear Equations by Elimination

Solve: $\begin{cases} 2x + 3y = 1 & (1) \\ -x + y = -3 & (2) \end{cases}$

Solution We multiply each side of equation (2) by 2 so that the coefficients of x in the two equations are negatives of one another. The result is the equivalent system

$$\begin{cases} 2x + 3y = 1 & (1) \\ -2x + 2y = -6 & (2) \end{cases}$$

If we now replace equation (2) of this system by the sum of the two equations, we obtain an equation containing just the variable y, which we can solve for.

$$\begin{cases} 2x + 3y = 1 & (1) \\ -2x + 2y = -6 & (2) \end{cases}$$
$$\begin{aligned} 5y &= -5 && \text{Add (1) and (2)} \\ y &= -1 \end{aligned}$$

We back-substitute by using this value for y in equation (1) and simplify to get

$$\begin{aligned} 2x + 3(-1) &= 1 \\ 2x &= 4 \\ x &= 2 \end{aligned}$$

Thus, the solution of the original system is $x = 2$, $y = -1$. We leave it to you to check the solution.

The procedure used in Example 6 is called the **method of elimination.** Notice the pattern of the solution. First, we eliminated the variable x from the second equation. Then we back-substituted; that is, we substituted the value found for y back into the first equation to find x.

Let's return to the movie theater example (Example 1).

EXAMPLE 7 Movie Theater Ticket Sales

A movie theater sells tickets for \$8.00 each, with Seniors receiving a discount of \$2.00. One evening the theater sold 525 tickets and took in \$3580 in revenue. How many of each type of ticket was sold?

Solution If x represents the number of tickets sold at \$8.00 and y the number of tickets sold at the discounted price of \$6.00, then the given information results in the system of equations

$$\begin{cases} 8x + 6y = 3580 \\ x + y = 525 \end{cases}$$

We use elimination and multiply the second equation by -6 and then add the equations

$$\begin{cases} 8x + 6y = 3580 \\ -6x - 6y = -3150 \end{cases}$$
$$\begin{aligned} 2x &= 430 && \text{Add the equations} \\ x &= 215 \end{aligned}$$

Since $x + y = 525$, then $y = 525 - x = 525 - 215 = 310$. Thus, 215 nondiscounted tickets and 310 Senior discount tickets were sold.

NOW USE ELIMINATION TO WORK PROBLEM 11.

3 The previous examples dealt with consistent systems of equations that had a unique solution. The next two examples deal with two other possibilities that may occur, the first being a system that has no solution.

EXAMPLE 8 An Inconsistent System of Linear Equations

Solve: $\begin{cases} 2x + y = 5 & (1) \\ 4x + 2y = 8 & (2) \end{cases}$

Solution We choose to use the method of substitution and solve equation (1) for y.

$$2x + y = 5$$
$$y = 5 - 2x \quad \text{Subtract } 2x \text{ from each side.}$$

Substituting in equation (2), we get

$$4x + 2y = 8 \quad (2)$$
$$4x + 2(5 - 2x) = 8 \quad y = 5 - 2x$$
$$4x + 10 - 4x = 8 \quad \text{Remove parentheses.}$$
$$0 \cdot x = -2 \quad \text{Combine like terms.}$$

This equation has no solution. Thus, we conclude that the system itself has no solution and is therefore inconsistent.

Figure 3 illustrates the pair of lines whose equations form the system in Example 8. Notice that the graphs of the two equations are lines, each with slope -2; one has a y-intercept of 5, the other a y-intercept of 4. Thus, the lines are parallel and have no point of intersection. This geometric statement is equivalent to the algebraic statement that the system has no solution.

Figure 3

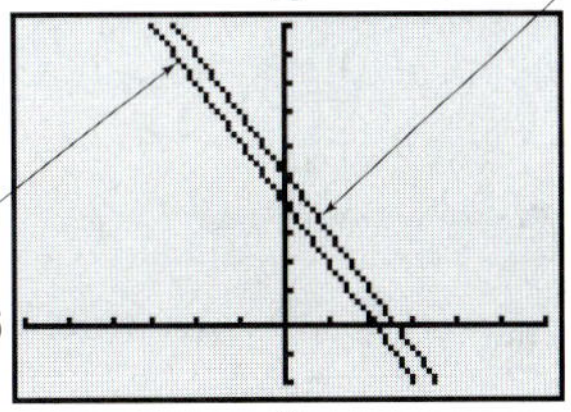

The next example is an illustration of a system with infinitely many solutions.

EXAMPLE 9 Solving a System of Linear Equations with Infinitely Many Solutions

Solve: $\begin{cases} 2x + y = 4 & (1) \\ -6x - 3y = -12 & (2) \end{cases}$

Solution We choose to use the method of elimination:

$$\begin{cases} 2x + y = 4 & (1) \\ -6x - 3y = -12 & (2) \end{cases}$$

$$\begin{cases} 6x + 3y = 12 & (1) \\ -6x - 3y = -12 & (2) \end{cases} \quad \text{Multiply each side of equation (1) by 3.}$$

$$\begin{cases} 6x + 3y = 12 & (1) \\ 0 = 0 & (2) \end{cases} \quad \text{Replace equation (2) by the sum of equations (1) and (2).}$$

The original system is equivalent to a system containing one equation, so the equations are dependent. This means that any values of x and y for which $6x + 3y = 12$ or, equivalently, $2x + y = 4$ are solutions. For example, $x = 2$, $y = 0$; $x = 0$, $y = 4$; $x = -2$, $y = 8$; $x = 4$, $y = -4$; and so on, are solutions. There are, in fact, infinitely many values of x and y for which $2x + y = 4$, so the original system has infinitely many solutions. We will write the solutions of the original systems either as

$$y = 4 - 2x$$

where x can be any real number, or as

$$x = 2 - \frac{1}{2}y$$

where y can be any real number.

Figure 4
$Y_1 = Y_2 = -2x + 4$

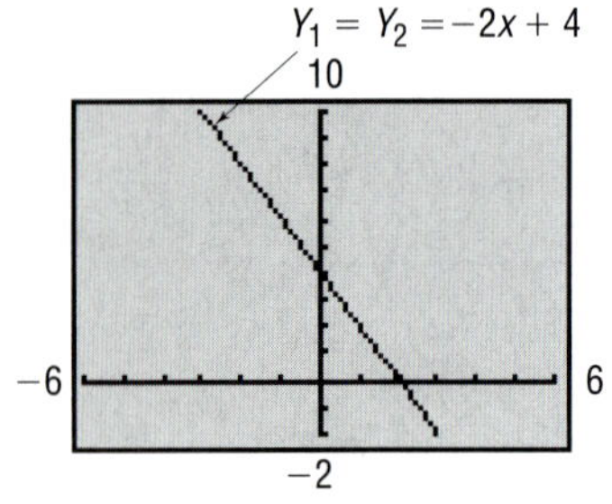

Figure 4 illustrates the situation presented in Example 9. Notice that the graphs of the two equations are lines, each with slope -2 and each with y-intercept 4. Thus, the lines are coincident. Notice also that equation (2) in the original system is just -3 times equation (1), indicating that the two equations are dependent.

For the system in Example 9, we can write down some of the infinite number of solutions by assigning values to x and then finding $y = 4 - 2x$. Thus,

If $x = -2$, then $y = 8$.
If $x = 0$, then $y = 4$.
If $x = 2$, then $y = 0$.

The pairs (x, y) are points on the line in Figure 4.

NOW WORK PROBLEMS 17 AND 21.

6.1 EXERCISES

In Problems 1–8, verify that the values of the variables listed are solutions of the system of equations.

1. $\begin{cases} 2x - y = 5 \\ 5x + 2y = 8 \end{cases}$ $x = 2, y = -1$

2. $\begin{cases} 3x + 2y = 2 \\ x - 7y = -30 \end{cases}$ $x = -2, y = 4$

3. $\begin{cases} 3x - 4y = 4 \\ \frac{1}{2}x - 3y = -\frac{1}{2} \end{cases}$ $x = 2, y = \frac{1}{2}$

4. $\begin{cases} 2x + \frac{1}{2}y = 0 \\ 3x - 4y = -\frac{19}{2} \end{cases}$ $x = -\frac{1}{2}, y = 2$

5. $\begin{cases} x - y = 3 \\ \frac{1}{2}x + y = 3 \end{cases}$ $x = 4, y = 1$

6. $\begin{cases} x - y = 3 \\ -3x + y = 1 \end{cases}$ $x = -2, y = -5$

7. $\begin{cases} 3x + 3y + 2z = 4 \\ x - y - z = 0 \\ 2y - 3z = -8 \end{cases}$ $x = 1, y = -1, z = 2$

8. $\begin{cases} 4x - z = 7 \\ 8x + 5y - z = 0 \\ -x - y + 5z = 6 \end{cases}$ $x = 2, y = -3, z = 1$

In Problems 9–32, solve each system of equations. If the system has no solution, say that it is inconsistent. Use either substitution or elimination. Verify your solution using a graphing utility.

9. $\begin{cases} x + y = 8 \\ x - y = 4 \end{cases}$

10. $\begin{cases} x + 2y = 5 \\ x + y = 3 \end{cases}$

11. $\begin{cases} 5x - y = 13 \\ 2x + 3y = 12 \end{cases}$

12. $\begin{cases} x + 3y = 5 \\ 2x - 3y = -8 \end{cases}$

13. $\begin{cases} 3x = 24 \\ x + 2y = 0 \end{cases}$

14. $\begin{cases} 4x + 5y = -3 \\ -2y = -4 \end{cases}$

15. $\begin{cases} 3x - 6y = 2 \\ 5x + 4y = 1 \end{cases}$

16. $\begin{cases} 2 + 4y = \frac{2}{3} \\ 3x - 5y = -10 \end{cases}$

17. $\begin{cases} 2x + y = 1 \\ 4x + 2y = 3 \end{cases}$

18. $\begin{cases} x - y = 5 \\ -3x + 3y = 2 \end{cases}$

19. $\begin{cases} 2x - y = 0 \\ 3x + 2y = 7 \end{cases}$

20. $\begin{cases} 3x + 3y = -1 \\ 4x + y = \frac{8}{3} \end{cases}$

21. $\begin{cases} x + 2y = 4 \\ 2x + 4y = 8 \end{cases}$

22. $\begin{cases} 3x - y = 7 \\ 9x - 3y = 21 \end{cases}$

23. $\begin{cases} 2x - 3y = -1 \\ 10x + y = 11 \end{cases}$

24. $\begin{cases} 3x - 2y = 0 \\ 5x + 10y = 4 \end{cases}$

25. $\begin{cases} 2x + 3y = 6 \\ x - y = \frac{1}{2} \end{cases}$

26. $\begin{cases} \frac{1}{2}x + y = -2 \\ x - 2y = 8 \end{cases}$

27. $\begin{cases} \frac{1}{2}x + \frac{1}{3}y = 3 \\ \frac{1}{4}x - \frac{2}{3}y = -1 \end{cases}$

28. $\begin{cases} \frac{1}{3}x - \frac{3}{2}y = -5 \\ \frac{3}{4}x + \frac{1}{3}y = 11 \end{cases}$

29. $\begin{cases} 3x - 5y = 3 \\ 15x + 5y = 21 \end{cases}$

30. $\begin{cases} 2x - y = -1 \\ x + \frac{1}{2}y = \frac{3}{2} \end{cases}$

31. $\begin{cases} \dfrac{1}{x} + \dfrac{1}{y} = 8 \\ \dfrac{3}{x} - \dfrac{5}{y} = 0 \end{cases}$

[Hint: Let $u = 1/x$ and $v = 1/y$, and solve for u and v. Then $x = 1/u$ and $y = 1/v$.]

32. $\begin{cases} \dfrac{4}{x} - \dfrac{3}{y} = 0 \\ \dfrac{6}{x} + \dfrac{3}{2y} = 2 \end{cases}$

In Problems 33–38, use a graphing utility to solve each system of equations. Express the solution rounded to two decimal places.

33. $\begin{cases} y = \sqrt{2}x - 20\sqrt{7} \\ y = -0.1x + 20 \end{cases}$

34. $\begin{cases} y = -\sqrt{3}x + 100 \\ y = 0.2x + \sqrt{19} \end{cases}$

35. $\begin{cases} \sqrt{2}x + \sqrt{3}y + \sqrt{6} = 0 \\ \sqrt{3}x - \sqrt{2}y + 60 = 0 \end{cases}$

36. $\begin{cases} \sqrt{5}x - \sqrt{6}y + 60 = 0 \\ 0.2x + 0.3y + \sqrt{5} = 0 \end{cases}$

37. $\begin{cases} \sqrt{3}x + \sqrt{2}y = \sqrt{0.3} \\ 100x - 95y = 20 \end{cases}$

38. $\begin{cases} \sqrt{6}x - \sqrt{5}y + \sqrt{1.1} = 0 \\ y = -0.2x + 0.1 \end{cases}$

39. **Supply and Demand** Suppose that the quantity supplied Q_s and quantity demanded Q_d of T-shirts at a concert is given by the following equations:

$$Q_s = -200 + 50p$$
$$Q_d = 1000 - 25p$$

where p is the price. The equilibrium price of a market is defined as the price at which quantity supplied equals quantity demanded ($Q_s = Q_d$). Find the equilibrium price for T-shirts at this concert. What is the equilibrium quantity?

40. **Supply and Demand** Suppose that the quantity supplied Q_s and quantity demanded Q_d of hot dogs at a baseball game is given by the following equations:

$$Q_s = -2000 + 3000p$$
$$Q_d = 10{,}000 - 1000p$$

where p is the price. The equilibrium price of a market is defined as the price at which quantity supplied equals quantity demanded ($Q_s = Q_d$). Find the equilibrium price for hot dogs at the baseball game. What is the equilibrium quantity?

41. The perimeter of a rectangular floor is 90 feet. Find the dimensions of the floor if the length is twice the width.

42. The length of fence required to enclose a rectangular field is 3000 meters. What are the dimensions of the field if it is known that the difference between its length and width is 50 meters?

43. **Cost of Fast Food** Four large cheeseburgers and two chocolate shakes cost a total of \$7.90. Two shakes cost 15¢ more than one cheeseburger. What is the cost of a cheeseburger? A shake?

44. **Movie Theater Tickets** A movie theater charges \$9.00 for adults and \$7.00 for senior citizens. On a day when 325 people paid an admission, the total receipts were \$2495. How many who paid were adults? How many were seniors?

45. **Mixing Nuts** A store sells cashews for \$5.00 per pound and peanuts for \$1.50 per pound. The manager decides to mix 30 pounds of peanuts with some cashews and sell the mixture for \$3.00 per pound. How many pounds of cashews should be mixed with the peanuts so that the mixture will produce the same revenue as would selling the nuts separately?

46. **Financial Planning** A recently retired couple need \$12,000 per year to supplement their Social Security. They have \$150,000 to invest to obtain this income. They have decided on two investment options: AA bonds yielding 10% per annum and a Bank Certificate yielding 5%.
    (a) How much should be invested in each to realize exactly \$12,000?
    (b) If, after two years, the couple requires \$14,000 per year in income, how should they reallocate their investment to achieve the new amount?

47. **Computing Wind Speed** With a tail wind, a small Piper aircraft can fly 600 miles in 3 hours. Against this same wind, the Piper can fly the same distance in 4 hours. Find the average wind speed and the average airspeed of the Piper.

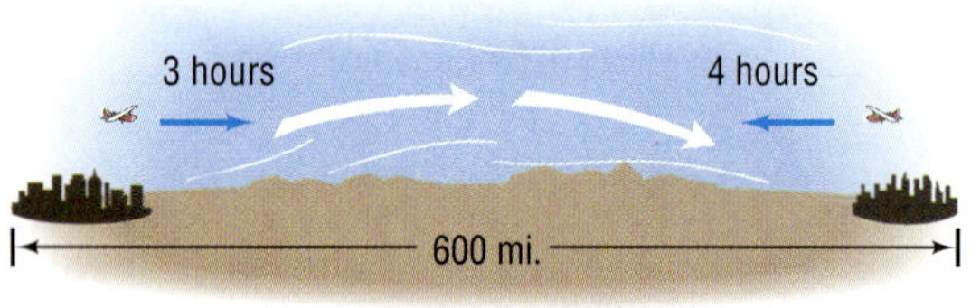
3 hours
4 hours
600 mi.

48. **Computing Wind Speed** The average airspeed of a single-engine aircraft is 150 miles per hour. If the aircraft flew the same distance in 2 hours with the wind as it flew in 3 hours against the wind, what was the wind speed?

49. **Restaurant Management** A restaurant manager wants to purchase 200 sets of dishes. One design costs \$25 per set, while another costs \$45 per set. If she only has \$7400 to spend, how many of each design should be ordered?

50. **Cost of Fast Food** One group of people purchased 10 hot dogs and 5 soft drinks at a cost of \$12.50. A second bought 7 hot dogs and 4 soft drinks at a cost of \$9.00. What is the cost of a single hot dog? A single soft drink?

We paid \$12.50.
How much is one hot dog?
How much is one cola?
HOT DOGS
SODA
We paid \$9.00.
How much is one hot dog?
How much is one cola?
HOT DOGS
SODA

51. **Computing a Refund** The grocery store we use does not mark prices on its goods. My wife went to this store, bought three 1 pound packages of bacon and two cartons of eggs, and paid a total of \$7.45. Not knowing that she went to the store, I also went to the same store, purchased two 1 pound packages of bacon and three cartons of eggs, and paid a total of \$6.45. Now we want to return two 1 pound packages of bacon and two cartons of eggs. How much will be refunded?

52. **Finding the Current of a Stream** Pamela requires 3 hours to swim 15 miles downstream on the Illinois River. The return trip upstream takes 5 hours. Find Pamela's average speed in still water. How fast is the current? (Assume that Pamela's speed is the same in each direction.)

The point at which a company's profits equal zero is called the company's ***break-even point.*** *For Problems 53 and 54, let R represent a company's revenue, let C represent the company's costs, and let x represent the number of units produced and sold each day. Find the firm's break-even point; that is, find x so that $R = C$.*

53. $R = 8x$
    $C = 4.5x + 17{,}500$

54. $R = 12x$
    $C = 10x + 15{,}000$

55. **Curve Fitting** Find real numbers b and c such that the parabola $y = x^2 + bx + c$ passes through the points $(1, 2)$ and $(-1, 3)$.

56. **Curve Fitting** Find real numbers b and c such that the parabola $y = x^2 + bx + c$ passes through the points $(1, 3)$ and $(3, 5)$.

57. Make up a system of two linear equations containing two variables that has
(a) No solution
(b) Exactly one solution
(c) Infinitely many solutions

Give the three systems to a friend to solve and critique.

58. Write a brief paragraph outlining your strategy for solving a system of two linear equations containing two variables.

59. Do you prefer the method of substitution or the method of elimination for solving a system of two linear equations containing two variables? Give reasons.

6.2 SYSTEMS OF LINEAR EQUATIONS: THREE EQUATIONS CONTAINING THREE VARIABLES

1 Solve Systems of Three Equations with Three Variables

2 Inconsistent Systems

3 Dependent Equations

Just as with a system of two linear equations containing two variables, a system of three linear equations containing three variables also has either (1) exactly one solution (a consistent system with independent equations), or (2) no solution (an inconsistent system), or (3) infinitely many solutions (a consistent system with dependent equations).

We can view the problem of solving a system of three linear equations containing three variables as a geometry problem. The graph of each equation in such a system is a plane in space. Thus, a system of three linear equations containing three variables represents three planes in space. Figure 5 illustrates some of the possibilities.

Figure 5

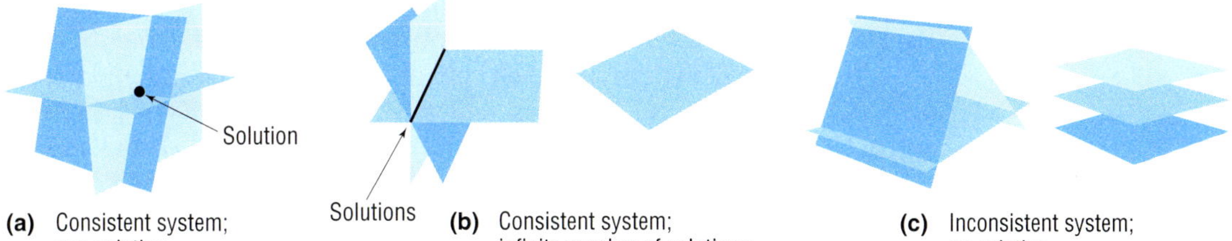

(a) Consistent system; one solution **(b)** Consistent system; infinite number of solutions **(c)** Inconsistent system; no solution

Recall that a **solution** to a system of equations consists of values for the variables that are solutions of each equation of the system. For example, $x = 3$, $y = -1$, $z = -5$ is a solution to the system of equations

$$\begin{cases} x + y + z = -3 & (1) \quad 3 + (-1) + (-5) = -3 \\ 2x - 3y + 6z = -21 & (2) \quad 2(3) - 3(-1) + 6(-5) = 6 + 3 - 30 = -21 \\ -3x + 5y = -14 & (3) \quad -3(3) + 5(-1) = -9 - 5 = -14 \end{cases}$$

because these values of the variables are solutions of each equation.

Typically, when solving a system of three linear equations containing three variables, we use the method of elimination. Recall that the idea behind the method of elimination is to form equivalent equations until a solution is reached. We list the rules for obtaining an equivalent system of equations as a reminder.

> ***Rules for Obtaining an Equivalent System of Equations***
>
> 1. Interchange any two equations of the system.
> 2. Multiply (or divide) each side of an equation by the same nonzero constant.
> 3. Replace any equation in the system by the sum (or difference) of that equation and a nonzero multiple of any other equation in the system.

1 Let's see how elimination works on a system of three equations containing three variables.

EXAMPLE 1 Solving a System of Three Linear Equations with Three Variables

Use the method of elimination to solve the system of equations.

$$\begin{cases} x + y - z = -1 & (1) \\ 4x - 3y + 2z = 16 & (2) \\ 2x - 2y - 3z = 5 & (3) \end{cases}$$

Solution For a system of three equations, we attempt to eliminate one variable at a time, using pairs of equations. Our plan of attack on this system will be to first eliminate the variable x from equations (2) and (3). Next, we will eliminate the variable y from equation (3), leaving only the variable z. Back-substitution can then be used to obtain the values of y and then x.

We begin by multiplying each side of equation (1) by -4 and adding the result to equation (2). (Do you see why? The coefficients on x are now opposites of each other.) We also multiply equation (1) by -2 and add the result to equation (3). Notice that this results in the removal of the x-variable from equations (2) and (3).

$$\begin{array}{rl} -4x - 4y + 4z = 4 & (1) \text{ Multiply by } -4 \\ 4x - 3y + 2z = 16 & (2) \text{ and Add} \\ \hline -7y + 6z = 20 & \end{array}$$

$$\begin{array}{rl} -2x - 2y + 2z = 2 & (1) \text{ Multiply by } -2 \\ 2x - 2y - 3z = 5 & (3) \text{ and Add} \\ \hline -4y - z = 7 & \end{array}$$

$$\begin{cases} x + y - z = -1 & (1) \\ -7y + 6z = 20 & (2) \\ -4y - z = 7 & (3) \end{cases}$$

We now concentrate on equations (2) and (3), treating them as a system of two equations containing two variables. We multiply each side of equation (2) by 4 and each side of equation (3) by -7 and add these equations. The result is the new equation (3).

$$\begin{array}{ll} -7y + 6z = 20 & \textit{Multiply by 4} \\ -4y - z = 7 & \textit{Multiply by } -7 \end{array} \quad \begin{array}{rl} -28y + 24z = & 80 \\ 28y + 7z = & -49 \\ \hline 31z = & 31 \end{array} \textit{ Add} \quad \rightarrow \begin{cases} x + y - z = -1 & (1) \\ -7y + 6z = 20 & (2) \\ 31z = 31 & (3) \end{cases}$$

We now solve equation (3) for z by dividing both sides of the equation by 31.

$$\begin{cases} x + y - z = -1 & (1) \\ -7y + 6z = 20 & (2) \\ z = 1 & (3) \end{cases}$$

Back-substitute $z = 1$ in equation (2) and solve for y.

$$\begin{aligned} -7y + 6z &= 20 && \textit{(2)} \\ -7y + 6(1) &= 20 && z = 1. \\ -7y &= 14 && \textit{Subtract 6 from both sides of the equation.} \\ y &= -2 && \textit{Divide both sides of the equation by } -7. \end{aligned}$$

Finally, we back-substitute $y = -2$ and $z = 1$ in equation (1) and solve for x.

$$\begin{aligned} x + y - z &= -1 && \textit{(1)} \\ x + (-2) - 1 &= -1 && y = -2 \textit{ and } z = 1. \\ x - 3 &= -1 && \textit{Simplify.} \\ x &= 2 && \textit{Add 3 to both sides.} \end{aligned}$$

The solution of the original system is $x = 2$, $y = -2$, $z = 1$. You should verify this solution.

Look back over the solution given in Example 1. Note the pattern of removing one of the variables from two of the equations, followed by solving this system of two equations and two unknowns. Although which variables to remove is your choice, the methodology remains the same for all systems.

NOW WORK PROBLEM 1.

2 The previous example was a consistent system that had a unique solution. The next two examples deal with the two other possibilities that may occur.

EXAMPLE 2 An Inconsistent System of Linear Equations

Solve: $\begin{cases} 2x + y - z = -2 & (1) \\ x + 2y - z = -9 & (2) \\ x - 4y + z = 1 & (3) \end{cases}$

Solution Our plan of attack is the same as in Example 1. We eliminate the variable x from equations (2) and (3); then we eliminate y from equation (3), leaving only the variable z.

We begin by interchanging equations (1) and (2). This is done so that the coefficient of x in (1) is 1. Now it is easier to eliminate the x-variable from equations (2) and (3).

$$\begin{cases} x + 2y - z = -9 & (1) \\ 2x + y - z = -2 & (2) \\ x - 4y + z = 1 & (3) \end{cases}$$

Multiply each side of equation (1) by -2 and add the result to equation (2). Multiply each side of equation (1) by -1 and add the result to equation (3).

$$\begin{array}{rl} -2x - 4y + 2z = 18 & (1) \text{ Multiply by } -2 \\ 2x + y - z = -2 & (2) \text{ and Add} \\ \hline -3y + z = 16 & \end{array} \qquad \begin{array}{rl} -x - 2y + z = 9 & (1) \text{ Multiply by } -1 \\ x - 4y + z = 1 & (3) \text{ and Add} \\ \hline -6y + 2z = 10 & \end{array}$$

$$\begin{cases} x + 2y - z = -9 & (1) \\ -3y + z = 16 & (2) \\ -6y + 2z = 10 & (3) \end{cases}$$

We now concentrate on equations (2) and (3), treating them as system of two equations containing two variables. Multiply each side of equation (2) by -2 and add the result to equation (3).

$$\begin{array}{l} -3y + z = 16 \\ -6y + 2z = 10 \end{array} \quad \text{Multiply by } -2 \quad \begin{array}{r} 6y - 2z = -32 \\ -6y + 2z = 10 \\ \hline 0 = -22 \end{array} \quad \text{Add} \quad \begin{cases} x + 2y - z = -9 & (1) \\ -3y + z = 16 & (2) \\ 0 = -22 & (3) \end{cases}$$

Equation (3) has no solution and thus the system is inconsistent. ◗

3 Now let's look at a system of equations with infinitely many solutions.

EXAMPLE 3 Solving a System of Linear Equations with Infinitely Many Solutions

$$\text{Solve:} \quad \begin{cases} x - 2y - z = 8 & (1) \\ 2x - 3y + z = 23 & (2) \\ 4x - 5y + 5z = 53 & (3) \end{cases}$$

Solution Multiply each side of equation (1) by -2 and add the result to equation (2). Also, multiply each side of equation (1) by -4 and add the result to equation (3).

$$\begin{array}{rl} -2x + 4y + 2z = -16 & (1) \text{ Multiply by } -2 \\ 2x - 3y + z = 23 & (2) \text{ and Add} \\ \hline y + 3z = 7 & \end{array} \qquad \begin{array}{rl} -4x + 8y + 4z = -32 & (1) \text{ Multiply by } -4 \\ 4x - 5y + 5z = 53 & (2) \text{ and Add} \\ \hline 3y + 9z = 21 & \end{array}$$

$$\begin{cases} x - 2y - z = 8 & (1) \\ y + 3z = 7 & (2) \\ 3y + 9z = 21 & (3) \end{cases}$$

Treat equations (2) and (3) as a system of two equations containing two variables, and eliminate the y variable by multiplying both sides of equation (2) by -3 and adding the result to equation (3).

$$\begin{array}{l} y + 3z = 7 \\ 3y + 9z = 21 \end{array} \quad \text{Multiply by } -3 \quad \begin{array}{r} -3y - 9z = -21 \\ 3y + 9z = 21 \\ \hline 0 = 0 \end{array} \quad \text{Add} \quad \begin{cases} x - 2y - z = 8 & (1) \\ y + 3z = 7 & (2) \\ 0 = 0 & (3) \end{cases}$$

The original system is equivalent to a system containing two equations, so the equations are dependent and the system has infinitely many solutions. If we let z represent any real number, then, solving equation (2) for y, we determine that $y = -3z + 7$. Substitute this expression into equation (1) to determine x in terms of z.

$$\begin{aligned} x - 2y - z &= 8 && (1) \\ x - 2(-3z + 7) - z &= 8 && y = -3z + 7 \\ x + 6z - 14 - z &= 8 && \text{Remove parentheses.} \\ x + 5z &= 22 && \text{Combine like terms.} \\ x &= -5z + 22 && \text{Solve for } x. \end{aligned}$$

We will write the solution to the system as

$$\begin{cases} x = -5z + 22 \\ y = -3z + 7 \end{cases}$$

where z can be any real number.

To find specific solutions to the system, choose any value of z and use the equations $x = -5z + 22$ and $y = -3z + 7$ to determine x and y. For example, if $z = 0$, then $x = 22$ and $y = 7$, and if $z = 1$, then $x = 17$ and $y = 4$.

NOW WORK PROBLEM 5

Two points in the Cartesian plane determine a unique line. Given three noncollinear points, we can find the (unique) quadratic function whose graph contains these three points.

EXAMPLE 4 Curve Fitting

Find real numbers a, b, and c so that the graph of the quadratic function $y = ax^2 + bx + c$ contains the points $(-1, -4)$, $(1, 6)$, and $(3, 0)$.

Solution For the point $(-1, -4)$ we have: $-4 = a(-1)^2 + b(-1) + c$ $-4 = a - b + c$

For the point $(1, 6)$ we have: $6 = a(1)^2 + b(1) + c$ $6 = a + b + c$

For the point $(3, 0)$ we have: $0 = a(3)^2 + b(3) + c$ $0 = 9a + 3b + c$

We wish to determine a, b, and c such that each equation is satisfied. That is, we want to solve the following system of three equations containing three variables:

$$\begin{cases} a - b + c = -4 & (1) \\ a + b + c = 6 & (2) \\ 9a + 3b + c = 0 & (3) \end{cases}$$

Figure 6

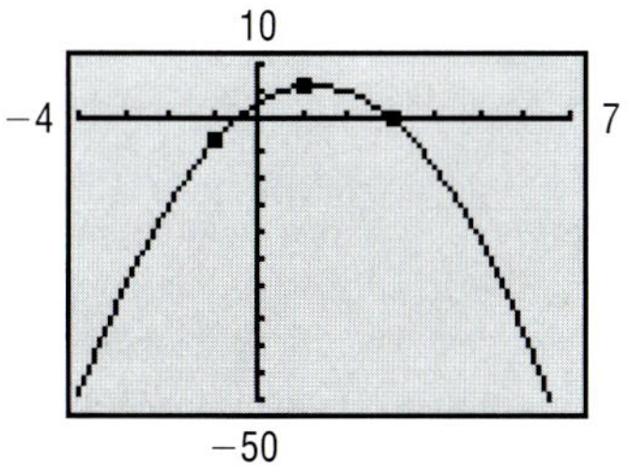

Solving this system of equations, we obtain $a = -2$, $b = 5$, and $c = 3$. So the quadratic function whose graph contains the points $(-1, -4)$, $(1, 6)$, and $(3, 0)$ is

$$y = -2x^2 + 5x + 3 \qquad y = ax^2 + bx + c,\ a = -2, b = 5, c = 3$$

Figure 6 shows the graph of the function along with the three points.

6.2 EXERCISES

In Problems 1–14, solve each system of equations. If the system has no solution, say that it is inconsistent.

1. $\begin{cases} x - y = 6 \\ 2x - 3z = 16 \\ 2y + z = 4 \end{cases}$

2. $\begin{cases} 2x + y = -4 \\ -2y + 4z = 0 \\ 3x - 2z = -11 \end{cases}$

3. $\begin{cases} x - 2y + 3z = 7 \\ 2x + y + z = 4 \\ -3x + 2y - 2z = -10 \end{cases}$

4. $\begin{cases} 2x + y - 3z = 0 \\ -2x + 2y + z = -7 \\ 3x - 4y - 3z = 7 \end{cases}$

5. $\begin{cases} x - y - z = 1 \\ 2x + 3y + z = 2 \\ 3x + 2y = 0 \end{cases}$

6. $\begin{cases} 2x - 3y - z = 0 \\ -x + 2y + z = 5 \\ 3x - 4y - z = 1 \end{cases}$

7. $\begin{cases} x - y - z = 1 \\ -x + 2y - 3z = -4 \\ 3x - 2y - 7z = 0 \end{cases}$

8. $\begin{cases} 2x - 3y - z = 0 \\ 3x + 2y + 2z = 2 \\ x + 5y + 3z = 2 \end{cases}$

9. $\begin{cases} 2x - 2y + 3z = 6 \\ 4x - 3y + 2z = 0 \\ -2x + 3y - 7z = 1 \end{cases}$

10. $\begin{cases} 3x - 2y + 2z = 6 \\ 7x - 3y + 2z = -1 \\ 2x - 3y + 4z = 0 \end{cases}$

11. $\begin{cases} x + y - z = 6 \\ 3x - 2y + z = -5 \\ x + 3y - 2z = 14 \end{cases}$

12. $\begin{cases} x - y + z = -4 \\ 2x - 3y + 4z = -15 \\ 5x + y - 2z = 12 \end{cases}$

13. $\begin{cases} x + 2y - z = -3 \\ 2x - 4y + z = -7 \\ -2x + 2y - 3z = 4 \end{cases}$

14. $\begin{cases} x + 4y - 3z = -8 \\ 3x - y + 3z = 12 \\ x + y + 6z = 1 \end{cases}$

15. **Curve Fitting** Find real numbers a, b, and c such that the graph of the function $y = ax^2 + bx + c$ contains the points $(-1, 4)$, $(2, 3)$, and $(0, 1)$.

16. **Curve Fitting** Find real numbers a, b, and c such that the graph of the function $y = ax^2 + bx + c$ contains the points $(-1, -2)$, $(1, -4)$, and $(2, 4)$.

17. **Electricity: Kirchhoff's Rules** An application of *Kirchhoff's Rules* to the circuit shown results in the following system of equations:

$$\begin{cases} I_2 = I_1 + I_3 \\ 5 - 3I_1 - 5I_2 = 0 \\ 10 - 5I_2 - 7I_3 = 0 \end{cases}$$

Find the currents I_1, I_2, and I_3.*

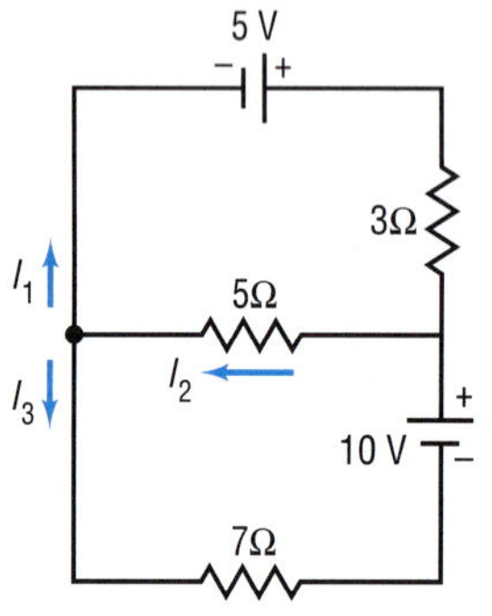

Source: Based on Raymond Serway, *Physics,* 3rd ed. (Philadelphia: Saunders, 1990), Prob. 26, p. 790.

18. **Electricity: Kirchhoff's Rules** An application of Kirchhoff's Rules to the circuit shown results in the following system of equations:

$$\begin{cases} I_3 = I_1 + I_2 \\ 8 = 4I_3 + 6I_2 \\ 8I_1 = 4 + 6I_2 \end{cases}$$

Find the currents I_1, I_2, and I_3.†

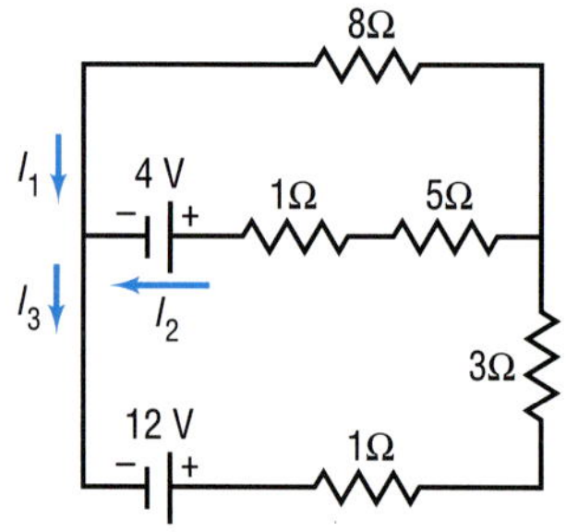

19. **Theater Revenues** A Broadway theater has 500 seats, divided into orchestra, main, and balcony seating. Orchestra seats sell for \$50, main seats for \$35, and balcony seats for \$25. If all the seats are sold, the gross revenue to the theater is

†*Source:* Ibid., Prob. 27, p. 790.

\$17,100. If all the main and balcony seats are sold, but only half the orchestra seats are sold, the gross revenue is \$14,600. How many are there of each kind of seat?

20. **Theater Revenues** A movie theater charges \$8.00 for adults, \$4.50 for children, and \$6.00 for senior citizens. One day the theater sold 405 tickets and collected \$2320 in receipts. There were twice as many children's tickets sold as adult tickets. How many adults, children, and senior citizens went to the theater that day?

21. **Nutrition** A dietician wishes a patient to have a meal that has 66 grams of protein, 94.5 grams of carbohydrates, and 910 milligrams of calcium. The hospital food service tells the dietician that the dinner for today is chicken, corn, and 2% milk. Each serving of chicken has 30 grams of protein, 35 grams of carbohydrates, and 200 milligrams of calcium. Each serving of corn has 3 grams of protein, 16 grams of carbohydrates, and 10 milligrams of calcium. Each glass of milk has 9 grams of protein, 13 grams of carbohydrates, and 300 milligrams of calcium. How many servings of each food should the dietitian provide for the patient?

22. **Investments** Kelly has \$20,000 to invest. As her financial planner, you recommend that she diversify into three investments: Treasury bills that yield 5% simple interest, Treasury bonds that yield 7% simple interest, and corporate bonds that yield 10% simple interest. Kelly wishes to earn \$1390 per year in income. Also, Kelly wants her investment in Treasury bills to be \$3000 more than her investment in corporate bonds. How much money should Kelly place in each investment?

6.3 SYSTEMS OF LINEAR EQUATIONS: MATRICES

1. Write the Augmented Matrix of a System of Linear Equations
2. Write the System from the Augmented Matrix
3. Perform Row Operations on a Matrix
4. Solve Systems of Linear Equations Using Matrices
5. Use a Graphing Utility to Solve a System of Linear Equations

The systematic approach of the method of elimination for solving a system of linear equations provides another method of solution that involves a simplified notation.

Consider the following system of linear equations:

$$\begin{cases} x + 4y = 14 \\ 3x - 2y = 0 \end{cases}$$

If we choose not to write the symbols used for the variables, we can represent this system as

$$\left[\begin{array}{cc|c} 1 & 4 & 14 \\ 3 & -2 & 0 \end{array}\right]$$

where it is understood that the first column represents the coefficients of the variable x, the second column the coefficients of y, and the third column the constants on the right side of the equal signs. The vertical line serves as a reminder of the equal signs. The large square brackets are the traditional symbols used to denote a *matrix* in algebra.

A **matrix** is defined as a rectangular array of numbers,

$$\begin{array}{l} \\ \text{Row } 1 \\ \text{Row } 2 \\ \vdots \\ \text{Row } i \\ \vdots \\ \text{Row } m \end{array} \begin{array}{c} \begin{array}{cccccc} \text{Column } 1 & \text{Column } 2 & & \text{Column } j & & \text{Column } n \end{array} \\ \begin{bmatrix} a_{11} & a_{12} & \cdots & a_{1j} & \cdots & a_{1n} \\ a_{21} & a_{22} & \cdots & a_{2j} & \cdots & a_{2n} \\ \vdots & \vdots & & \vdots & & \vdots \\ a_{i1} & a_{i2} & \cdots & a_{ij} & \cdots & a_{in} \\ \vdots & \vdots & & \vdots & & \vdots \\ a_{m1} & a_{m2} & \cdots & a_{mj} & \cdots & a_{mn} \end{bmatrix} \end{array} \qquad (1)$$

Each number a_{ij} of the matrix has two indexes: the **row index** i and the **column index** j. The matrix shown in display (1) has m rows and n columns. The numbers a_{ij} are usually referred to as the **entries** of the matrix. For example, a_{23} refers to the entry in the second row, third column.

1 Now we will use matrix notation to represent a system of linear equations. The matrices used to represent systems of linear equations are called **augmented matrices.** In writing the augmented matrix of a system, the variables of each equation must be on the left side of the equal sign and the constants on the right side. A variable that does not appear in an equation has a coefficient of 0.

EXAMPLE 1 Writing the Augmented Matrix of a System of Linear Equations

Write the augmented matrix of each system of equations.

(a) $\begin{cases} 3x - 4y = -6 & (1) \\ 2x - 3y = -5 & (2) \end{cases}$ (b) $\begin{cases} 2x - y + z = 0 & (1) \\ x + z - 1 = 0 & (2) \\ x + 2y - 8 = 0 & (3) \end{cases}$

Solution (a) The augmented matrix is

$$\left[\begin{array}{cc|c} 3 & -4 & -6 \\ 2 & -3 & -5 \end{array}\right]$$

(b) Care must be taken that the system be written so that the coefficients of all variables are present (if any variable is missing, its coefficient is 0). Also, all constants must be to the right of the equal sign. Thus, we need to rearrange the given system as follows:

$$\begin{cases} 2x - y + z = 0 & (1) \\ x + z - 1 = 0 & (2) \\ x + 2y - 8 = 0 & (3) \end{cases}$$

$$\begin{cases} 2x - y + z = 0 & (1) \\ x + 0 \cdot y + z = 1 & (2) \\ x + 2y + 0 \cdot z = 8 & (3) \end{cases}$$

The augmented matrix is

$$\left[\begin{array}{ccc|c} 2 & -1 & 1 & 0 \\ 1 & 0 & 1 & 1 \\ 1 & 2 & 0 & 8 \end{array}\right]$$

If we do not include the constants to the right of the equal sign, that is, to the right of the vertical bar in the augmented matrix of a system of equations, the resulting matrix is called the **coefficient matrix** of the system. For the systems discussed in Example 1, the coefficient matrices are

$$\begin{bmatrix} 3 & -4 \\ 2 & -3 \end{bmatrix} \quad \text{and} \quad \begin{bmatrix} 2 & -1 & 1 \\ 1 & 0 & 1 \\ 1 & 2 & 0 \end{bmatrix}$$

NOW WORK PROBLEM 3.

EXAMPLE 2 Writing the System of Linear Equations from the Augmented Matrix

2 Write the system of linear equations corresponding to each augmented matrix.

(a) $\left[\begin{array}{cc|c} 5 & 2 & 13 \\ -3 & 1 & -10 \end{array}\right]$ (b) $\left[\begin{array}{ccc|c} 3 & -1 & -1 & 7 \\ 2 & 0 & 2 & 8 \\ 0 & 1 & 1 & 0 \end{array}\right]$

Solution (a) The matrix has two rows and so represents a system of two equations. The two columns to the left of the vertical bar indicate that the system has two variables. If x and y are used to denote these variables, the system of equations is

$$\begin{cases} 5x + 2y = 13 & (1) \\ -3x + y = -10 & (2) \end{cases}$$

(b) This matrix represents a system of three equations containing three variables. If x, y, and z are the three variables, this system is

$$\begin{cases} 3x - y - z = 7 & (1) \\ 2x \quad + 2z = 8 & (2) \\ \quad y + z = 0 & (3) \end{cases}$$

Row Operations on a Matrix

3 **Row operations** on a matrix are used to solve systems of equations when the system is written as an augmented matrix. There are three basic row operations.

> ***Row Operations***
>
> 1. Interchange any two rows.
> 2. Replace a row by a nonzero multiple of that row.
> 3. Replace a row by the sum of that row and a constant nonzero multiple of some other row.

These three row operations correspond to the three rules given earlier for obtaining an equivalent system of equations. Thus, when a row operation is performed on a matrix, the resulting matrix represents a system of equations equivalent to the system represented by the original matrix.

For example, consider the augmented matrix

$$\left[\begin{array}{cc|c} 1 & 2 & 3 \\ 4 & -1 & 2 \end{array}\right]$$

Suppose that we want to apply a row operation to this matrix that results in a matrix whose entry in row 2, column 1 is a 0. The row operation to use is

Multiply each entry in row 1 by -4 and add the result to the corresponding entries in row 2. (2)

If we use R_2 to represent the new entries in row 2 and we use r_1 and r_2 to represent the original entries in rows 1 and 2, respectively, then we can represent the row operation in statement (2) by

$$R_2 = -4r_1 + r_2$$

Then

$$\left[\begin{array}{cc|c} 1 & 2 & 3 \\ 4 & -1 & 2 \end{array}\right] \underset{\substack{\uparrow \\ R_2 = -4r_1 + r_2}}{\rightarrow} \left[\begin{array}{cc|c} 1 & 2 & 3 \\ -4(1)+4 & -4(2)+(-1) & -4(3)+2 \end{array}\right] = \left[\begin{array}{cc|c} 1 & 2 & 3 \\ 0 & -9 & -10 \end{array}\right]$$

As desired, we now have the entry 0 in row 2, column 1.

EXAMPLE 3 Applying a Row Operation to an Augmented Matrix

Apply the row operation $R_2 = -3r_1 + r_2$ to the augmented matrix

$$\left[\begin{array}{cc|c} 1 & -2 & 2 \\ 3 & -5 & 9 \end{array}\right]$$

Solution The row operation $R_2 = -3r_1 + r_2$ tells us that the entries in row 2 are to be replaced by the entries obtained after multiplying each entry in row 1 by -3 and adding the result to the corresponding entries in row 2. Thus,

$$\left[\begin{array}{cc|c} 1 & -2 & 2 \\ 3 & -5 & 9 \end{array}\right] \underset{\substack{\uparrow \\ R_2 = -3r_1 + r_2}}{\rightarrow} \left[\begin{array}{cc|c} 1 & -2 & 2 \\ -3(1)+3 & (-3)(-2)+(-5) & -3(2)+9 \end{array}\right] = \left[\begin{array}{cc|c} 1 & -2 & 2 \\ 0 & 1 & 3 \end{array}\right]$$

NOW WORK PROBLEM 11.

EXAMPLE 4 Finding a Particular Row Operation

Using the matrix

$$\left[\begin{array}{cc|c} 1 & -2 & 2 \\ 0 & 1 & 3 \end{array}\right]$$

find a row operation that will result in a matrix with a 0 in row 1, column 2.

Solution We want a 0 in row 1, column 2. This result can be accomplished by multiplying row 2 by 2 and adding the result to row 1. That is, we apply the row operation $R_1 = 2r_2 + r_1$.

$$\left[\begin{array}{cc|c} 1 & -2 & 2 \\ 0 & 1 & 3 \end{array}\right] \underset{R_1 = 2r_2 + r_1}{\rightarrow} \left[\begin{array}{cc|c} 2(0) + 1 & 2(1) + (-2) & 2(3) + 2 \\ 0 & 1 & 3 \end{array}\right] = \left[\begin{array}{cc|c} 1 & 0 & 8 \\ 0 & 1 & 3 \end{array}\right]$$

A word about the notation that we have introduced. A row operation such as $R_1 = 2r_2 + r_1$ changes the entries in row 1. Note also that for this type of row operation we change the entries in a given row by multiplying the entries in some other row by an appropriate nonzero number and adding the results to the original entries of the row to be changed.

4 To use row operations to solve a system of linear equations, we transform the augmented matrix of the system and obtain a matrix that is in *echelon form.*

A matrix is in **echelon form** when

1. The entry in row 1, column 1 is a 1, and 0's appear below it.
2. The first nonzero entry in each row after the first row is a 1, 0's appear below it, and it appears to the right of the first nonzero entry in any row above.
3. Any rows that contain all 0's to the left of the vertical bar appear at the bottom.

For example, for a system of three equations containing three variables with a unique solution, the augmented matrix is in echelon form if it is of the form

$$\left[\begin{array}{ccc|c} 1 & a & b & d \\ 0 & 1 & c & e \\ 0 & 0 & 1 & f \end{array}\right]$$

where a, b, c, d, e, and f are real numbers. The last row of the augmented matrix states that $z = f$. We can then determine the value of y using back-substitution with $z = f$ since row 2 represents the equation $y + cz = e$. Finally, x is determined using back-substitution again.

Two advantages of solving a system of equations by writing the augmented matrix in echelon form are the following:

1. The process is algorithmic; that is, it consists of repetitive steps that can be programmed on a computer.
2. The process works on any system of linear equations, no matter how many equations or variables are present.

The next example shows how to write a matrix in echelon form.

EXAMPLE 5 Solving a System of Linear Equations Using Matrices

$$\text{Solve:} \quad \begin{cases} 4x + 3y = 11 & (1) \\ x - 3y = -1 & (2) \end{cases}$$

Solution First, we write the augmented matrix that represents this system.

$$\left[\begin{array}{cc|c} 4 & 3 & 11 \\ 1 & -3 & -1 \end{array}\right]$$

The first step requires getting the entry 1 in row 1, column 1. An interchange of rows 1 and 2 is the easiest way to do this.

$$\left[\begin{array}{cc|c} 1 & -3 & -1 \\ 4 & 3 & 11 \end{array}\right]$$

Next, we want a 0 under the entry 1 in column 1. We use the row operation $R_2 = -4r_1 + r_2$.

$$\left[\begin{array}{cc|c} 1 & -3 & -1 \\ 4 & 3 & 11 \end{array}\right] \underset{R_2 = -4r_1 + r_2}{\rightarrow} \left[\begin{array}{cc|c} 1 & -3 & -1 \\ 0 & 15 & 15 \end{array}\right]$$

Now we want the entry 1 in row 2, column 2. We use $R_2 = \frac{1}{15}r_2$.

$$\left[\begin{array}{cc|c} 1 & -3 & -1 \\ 0 & 15 & 15 \end{array}\right] \underset{R_2 = \frac{1}{15}r_2}{\rightarrow} \left[\begin{array}{cc|c} 1 & -3 & -1 \\ 0 & 1 & 1 \end{array}\right]$$

The second row of the matrix on the right represents the equation $y = 1$. Thus, using $y = 1$, we back-substitute into the equation $x - 3y = -1$ (from the first row) to get

$$\begin{aligned} x - 3(1) &= -1 \quad y = 1 \\ x &= 2 \end{aligned}$$

The solution of the system is $x = 2$, $y = 1$.

NOW WORK PROBLEM 27.

The steps that we used to solve the system of linear equations in Example 5 can be summarized as follows.

Matrix Method for Solving a System of Linear Equations

STEP 1: Write the augmented matrix that represents the system.

STEP 2: Perform row operations that place the entry 1 in row 1, column 1.

STEP 3: Perform row operations that leave the entry 1 in row 1, column 1 unchanged, while causing 0's to appear below it in column 1.

STEP 4: Perform row operations that place the entry 1 in row 2, column 2 and leave the entries in columns to the left unchanged. If it is impossible to place a 1 in row 2, column 2, then proceed to place a 1 in row 2, column 3. Once a 1 is in place, perform row operations to place 0's under it.

STEP 5: Now repeat Step 4, placing a 1 in the next row, but one column to the right. Continue until the bottom row or the vertical bar is reached.

STEP 6: If any rows are obtained that contain only 0's on the left side of the vertical bar, place such rows at the bottom of the matrix.

5 Some graphing utilities have the ability to put an augmented matrix in echelon form. The next example demonstrates this feature using a TI-83 graphing calculator.

EXAMPLE 6 Solving a System of Linear Equations Using Matrices

Solve: $$\begin{cases} x - y + z = 8 & (1) \\ 2x + 3y - z = -2 & (2) \\ 3x - 2y - 9z = 9 & (3) \end{cases}$$

Graphing Solution

The augmented matrix of the system is

$$\left[\begin{array}{rrr|r} 1 & -1 & 1 & 8 \\ 2 & 3 & -1 & -2 \\ 3 & -2 & -9 & 9 \end{array}\right]$$

We enter this matrix into our graphing utility and name it A. See Figure 7(a). Using the REF command on matrix A, we obtain the results shown in Figure 7(b). Since the entire matrix does not fit on the screen, we need to scroll right to see the rest of it. See Figure 7(c).

Figure 7

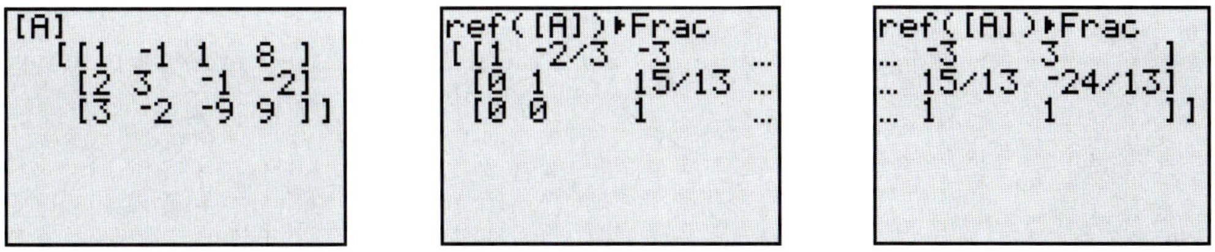

(a) (b) (c)

The system of equations represented by the matrix in echelon form is

$$\begin{cases} x - \frac{2}{3}y - 3z = 3 & (1) \\ y + \frac{15}{13}z = -\frac{24}{13} & (2) \\ z = 1 & (3) \end{cases}$$

Using $z = 1$, we back-substitute to get

$$\begin{cases} x - \frac{2}{3}y - 3(1) = 3 & (1) \\ y + \frac{15}{13}(1) = -\frac{24}{13} & (2) \end{cases} \xrightarrow{\text{Simplify.}} \begin{cases} x - \frac{2}{3}y = 6 & (1) \\ y = \frac{-39}{13} = -3 & (2) \end{cases}$$

Solving the second equation for y, we find that $y = -3$. Back-substituting $y = -3$ into $x - \frac{2}{3}y = 6$, we find that $x = 4$. The solution of the system is $x = 4, y = -3, z = 1$.

Algebraic Solution

STEP 1: The augmented matrix of the system is

$$\left[\begin{array}{ccc|c} 1 & -1 & 1 & 8 \\ 2 & 3 & -1 & -2 \\ 3 & -2 & -9 & 9 \end{array}\right]$$

STEP 2: Because the entry 1 is already present in row 1, column 1, we can go to Step 3.

STEP 3: Perform the row operations $R_2 = -2r_1 + r_2$ and $R_3 = -3r_1 + r_3$. Each of these leaves the entry 1 in row 1, column 1 unchanged, while causing 0's to appear under it.

$$\left[\begin{array}{ccc|c} 1 & -1 & 1 & 8 \\ 2 & 3 & -1 & -2 \\ 3 & -2 & -9 & 9 \end{array}\right] \rightarrow \left[\begin{array}{ccc|c} 1 & -1 & 1 & 8 \\ 0 & 5 & -3 & -18 \\ 0 & 1 & -12 & -15 \end{array}\right]$$

$\uparrow$ $R_2 = -2r_1 + r_2$, $R_3 = -3r_1 + r_3$

STEP 4: The easiest way to obtain the entry 1 in row 2, column 2 without altering column 1 is to interchange rows 2 and 3 (another way would be to multiply row 2 by $\frac{1}{5}$, but this introduces fractions).

$$\left[\begin{array}{ccc|c} 1 & -1 & 1 & 8 \\ 0 & 1 & -12 & -15 \\ 0 & 5 & -3 & -18 \end{array}\right]$$

To get 0's under the 1 in row 2, column 2, perform the row operation $R_3 = -5r_2 + r_3$.

$$\left[\begin{array}{ccc|c} 1 & -1 & 1 & 8 \\ 0 & 1 & -12 & -15 \\ 0 & 5 & -3 & -18 \end{array}\right] \rightarrow \left[\begin{array}{ccc|c} 1 & -1 & 1 & 8 \\ 0 & 1 & -12 & -15 \\ 0 & 0 & 57 & 57 \end{array}\right]$$

$\uparrow$ $R_3 = -5r_2 + r_3$

STEP 5: Continuing, we place a 1 in row 3, column 3 by using $R_3 = \frac{1}{57}r_3$.

$$\left[\begin{array}{ccc|c} 1 & -1 & 1 & 8 \\ 0 & 1 & -12 & -15 \\ 0 & 0 & 57 & 57 \end{array}\right] \rightarrow \left[\begin{array}{ccc|c} 1 & -1 & 1 & 8 \\ 0 & 1 & -12 & -15 \\ 0 & 0 & 1 & 1 \end{array}\right]$$

$\uparrow$ $R_3 = \frac{1}{57}r_3$

Because we have reached the bottom row, the matrix is in echelon form and we can stop.

The system of equations represented by the matrix in echelon form is

$$\begin{cases} x - y + z = 8 & (1) \\ \phantom{x - {}} y - 12z = -15 & (2) \\ \phantom{x - y + 12{}} z = 1 & (3) \end{cases}$$

Using $z = 1$, we back-substitute to get

$$\begin{cases} x - y + 1 = 8 & (1) \\ \phantom{x - {}} y - 12(1) = -15 & (2) \end{cases} \xrightarrow[\text{Simplify.}]{} \begin{cases} x - y = 7 & (1) \\ \phantom{x - {}} y = -3 & (2) \end{cases}$$

Thus, we get $y = -3$, and back-substituting into $x - y = 7$, we find that $x = 4$. The solution of the system is $x = 4$, $y = -3$, $z = 1$.

Notice that the reduced echelon form of the augmented matrix using the graphing utility differs from the reduced echelon form in our algebraic solution, yet both matrices provide the same solution! This is because the two solutions used different row operations to obtain the reduced echelon form. In all likelihood, the two solutions parted ways in Step 4 of the algebraic solution, where we avoided introducing fractions by interchanging rows 2 and 3.

Sometimes, it is advantageous to write a matrix in **reduced row echelon form.** In this form, row operations are used to obtain entries that are 0 above (as well as below) the leading 1 in a row. For example, the echelon form obtained in the algebraic solution to Example 6 is

$$\left[\begin{array}{rrr|r} 1 & -1 & 1 & 8 \\ 0 & 1 & -12 & -15 \\ 0 & 0 & 1 & 1 \end{array}\right]$$

To write this matrix in reduced row echelon form, we proceed as follows:

$$\left[\begin{array}{rrr|r} 1 & -1 & 1 & 8 \\ 0 & 1 & -12 & -15 \\ 0 & 0 & 1 & 1 \end{array}\right] \underset{\substack{\uparrow \\ R_1 = r_2 + r_1}}{\rightarrow} \left[\begin{array}{rrr|r} 1 & 0 & -11 & -7 \\ 0 & 1 & -12 & -15 \\ 0 & 0 & 1 & 1 \end{array}\right] \underset{\substack{\uparrow \\ R_1 = 11r_3 + r_1 \\ R_2 = 12r_3 + r_2}}{\rightarrow} \left[\begin{array}{rrr|r} 1 & 0 & 0 & 4 \\ 0 & 1 & 0 & -3 \\ 0 & 0 & 1 & 1 \end{array}\right]$$

The matrix is now written in reduced row echelon form. The advantage of writing the matrix in this form is that the solution to the system, $x = 4$, $y = -3$, $z = 1$, is readily found, without the need to back-substitute. Another advantage will be seen in Section 6.5, where the inverse of a matrix is discussed.

Figure 8

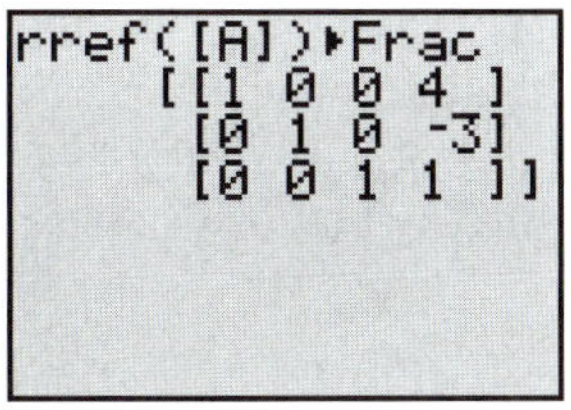

Most graphing utilities also have the ability to put a matrix in reduced row echelon form. Figure 8 shows the reduced row echelon form of the augmented matrix from Example 6 using the RREF command on a TI-83 graphing calculator.

For the remaining examples in this section, we will only provide algebraic solutions to the systems. The reader is encouraged to verify the results using a graphing utility.

NOW WORK PROBLEM 45.

The matrix method for solving a system of linear equations also identifies systems that have infinitely many solutions and systems that are inconsistent. Let's see how.

EXAMPLE 7 Solving a System of Linear Equations Using Matrices

Solve: $\begin{cases} 6x - y - z = 4 & (1) \\ -12x + 2y + 2z = -8 & (2) \\ 5x + y - z = 3 & (3) \end{cases}$

Solution We start with the augmented matrix of the system.

$$\left[\begin{array}{rrr|r} 6 & -1 & -1 & 4 \\ -12 & 2 & 2 & -8 \\ 5 & 1 & -1 & 3 \end{array}\right] \underset{R_1 = -1r_3 + r_1}{\rightarrow} \left[\begin{array}{rrr|r} 1 & -2 & 0 & 1 \\ -12 & 2 & 2 & -8 \\ 5 & 1 & -1 & 3 \end{array}\right] \underset{\substack{R_2 = 12r_1 + r_2 \\ R_3 = -5r_1 + r_3}}{\rightarrow} \left[\begin{array}{rrr|r} 1 & -2 & 0 & 1 \\ 0 & -22 & 2 & 4 \\ 0 & 11 & -1 & -2 \end{array}\right]$$

Obtaining a 1 in row 2, column 2 without altering column 1 can be accomplished only by $R_2 = -\frac{1}{22}r_2$ or by $R_3 = \frac{1}{11}r_3$ and interchanging rows or by $R_2 = \frac{23}{11}r_3 + r_2$. We shall use the first of these.

$$\left[\begin{array}{rrr|r} 1 & -2 & 0 & 1 \\ 0 & -22 & 2 & 4 \\ 0 & 11 & -1 & -2 \end{array}\right] \underset{R_2 = -\frac{1}{22}r_2}{\rightarrow} \left[\begin{array}{rrr|r} 1 & -2 & 0 & 1 \\ 0 & 1 & -\frac{1}{11} & -\frac{2}{11} \\ 0 & 11 & -1 & -2 \end{array}\right] \underset{R_3 = -11r_2 + r_3}{\rightarrow} \left[\begin{array}{rrr|r} 1 & -2 & 0 & 1 \\ 0 & 1 & -\frac{1}{11} & -\frac{2}{11} \\ 0 & 0 & 0 & 0 \end{array}\right]$$

This matrix is in echelon form. Because the bottom row consists entirely of 0's, the system actually consists of only two equations.

$$\begin{cases} x - 2y = 1 & (1) \\ y - \frac{1}{11}z = -\frac{2}{11} & (2) \end{cases}$$

We shall back-substitute the solution for y from the second equation, $y = \frac{1}{11}z - \frac{2}{11}$, into the first equation to get

$$x = 2y + 1 = 2\left(\frac{1}{11}z - \frac{2}{11}\right) + 1 = \frac{2}{11}z + \frac{7}{11}$$

Thus, the original system is equivalent to the system

$$\begin{cases} x = \frac{2}{11}z + \frac{7}{11} & (1) \\ y = \frac{1}{11}z - \frac{2}{11} & (2) \end{cases}$$

where z can be any real number.

Let's look at the situation. The original system of three equations is equivalent to a system containing two equations. This means that any values of x, y, z that satisfy both

$$x = \frac{2}{11}z + \frac{7}{11} \quad \text{and} \quad y = \frac{1}{11}z - \frac{2}{11}$$

will be solutions. For example, $z = 0$, $x = \frac{7}{11}$, $y = -\frac{2}{11}$; $z = 1$, $x = \frac{9}{11}$, $y = -\frac{1}{11}$; and $z = -1$, $x = \frac{5}{11}$, $y = -\frac{3}{11}$ are some of the solutions of the original system. There are, in fact, infinitely many values of x, y, and z for which the two equations are satisfied. That is, the original system has infinitely many solutions. We will write the solution of the original system as

$$\begin{cases} x = \frac{2}{11}z + \frac{7}{11} \\ y = \frac{1}{11}z - \frac{2}{11} \end{cases}$$

where z can be any real number.

We can also find the solution by writing the augmented matrix in reduced row echelon form. Starting with the echelon form, we have

$$\left[\begin{array}{rrr|r} 1 & -2 & 0 & 1 \\ 0 & 1 & -\frac{1}{11} & -\frac{2}{11} \\ 0 & 0 & 0 & 0 \end{array}\right] \underset{R_1 = 2r_2 + r_1}{\rightarrow} \left[\begin{array}{rrr|r} 1 & 0 & -\frac{2}{11} & \frac{7}{11} \\ 0 & 1 & -\frac{1}{11} & -\frac{2}{11} \\ 0 & 0 & 0 & 0 \end{array}\right]$$

The matrix on the right is in reduced row echelon form. The corresponding system of equations is

$$\begin{cases} x - \frac{2}{11}z = \frac{7}{11} & (1) \\ y - \frac{1}{11}z = -\frac{2}{11} & (2) \end{cases}$$

or, equivalently,

$$\begin{cases} x = \frac{2}{11}z + \frac{7}{11} & (1) \\ y = \frac{1}{11}z - \frac{2}{11} & (2) \end{cases}$$

where z can be any real number.

NOW WORK PROBLEM 49.

EXAMPLE 8 Solving a System of Linear Equations Using Matrices

Solve: $\begin{cases} x + y + z = 6 \\ 2x - y - z = 3 \\ x + 2y + 2z = 0 \end{cases}$

Solution We proceed as follows, beginning with the augmented matrix.

$$\left[\begin{array}{rrr|r} 1 & 1 & 1 & 6 \\ 2 & -1 & -1 & 3 \\ 1 & 2 & 2 & 0 \end{array}\right] \underset{\substack{R_2 = -2r_1 + r_2 \\ R_3 = -1r_1 + r_3}}{\rightarrow} \left[\begin{array}{rrr|r} 1 & 1 & 1 & 6 \\ 0 & -3 & -3 & -9 \\ 0 & 1 & 1 & -6 \end{array}\right] \underset{\text{Interchange rows 2 and 3.}}{\rightarrow} \left[\begin{array}{rrr|r} 1 & 1 & 1 & 6 \\ 0 & 1 & 1 & -6 \\ 0 & -3 & -3 & -9 \end{array}\right] \underset{R_3 = 3r_2 + r_3}{\rightarrow} \left[\begin{array}{rrr|r} 1 & 1 & 1 & 6 \\ 0 & 1 & 1 & -6 \\ 0 & 0 & 0 & -27 \end{array}\right]$$

This matrix is in echelon form. The bottom row is equivalent to the equation

$$0x + 0y + 0z = -27$$

which has no solution. Hence, the original system is inconsistent.

NOW WORK PROBLEM 23.

EXAMPLE 9 Nutrition

A dietitian at Cook County Hospital wishes a patient to have a meal that has 65 grams of protein, 95 grams of carbohydrates, and 905 milligrams of calcium. The hospital food service tells the dietitian that the dinner for today is chicken *a la king*, baked potatoes, and 2% milk. Each serving of chicken *a la king* has 30 grams of protein, 35 grams of carbohydrates, and 200 milligrams of calcium. Each serving of baked potatoes contains 4 grams of protein, 33 grams of carbohydrates, and 10 milligrams of calcium. Each glass of milk contains 9 grams of protein, 13 grams of carbohydrates, and 300 milligrams of calcium. How many servings of each food should the dietitian provide for the patient?

Solution Let c, p, and m represent the number of servings of chicken *a la king*, baked potatoes, and milk, respectively. The dietitian wants the patient to have 65 grams of protein. Each serving of chicken *a la king* has 30 grams of protein, so c servings will have $30c$ grams of protein. Each serving of baked potatoes contains 4 grams of protein, so p potatoes will have $4p$ grams of protein. Finally, each glass of milk has 9 grams of protein, so m glasses of milk will have $9m$ grams of protein. The same logic will result in equations for carbohydrates and calcium, and we have the following system of equations:

$$\begin{cases} 30c + 4p + 9m = 65 & \textit{protein equation} \\ 35c + 33p + 13m = 95 & \textit{carbohydrate equation} \\ 200c + 10p + 300m = 905 & \textit{calcium equation} \end{cases}$$

We begin with the augmented matrix and proceed as follows:

$$\left[\begin{array}{ccc|c} 30 & 4 & 9 & 65 \\ 35 & 33 & 13 & 95 \\ 200 & 10 & 300 & 905 \end{array}\right] \underset{R_1 = \left(\frac{1}{30}\right)r_1}{\rightarrow} \left[\begin{array}{ccc|c} 1 & \frac{2}{15} & \frac{3}{10} & \frac{13}{6} \\ 35 & 33 & 13 & 95 \\ 200 & 10 & 300 & 905 \end{array}\right] \underset{\substack{R_2 = -35r_1 + r_2 \\ R_3 = -200r_1 + r_3}}{\rightarrow} \left[\begin{array}{ccc|c} 1 & \frac{2}{15} & \frac{3}{10} & \frac{13}{6} \\ 0 & \frac{85}{3} & \frac{5}{2} & \frac{115}{6} \\ 0 & -\frac{50}{3} & 240 & \frac{1415}{3} \end{array}\right]$$

$$\underset{R_2 = \left(\frac{3}{85}\right)r_2}{\rightarrow} \left[\begin{array}{ccc|c} 1 & \frac{2}{15} & \frac{3}{10} & \frac{13}{6} \\ 0 & 1 & \frac{3}{34} & \frac{23}{34} \\ 0 & -\frac{50}{3} & 240 & \frac{1415}{3} \end{array}\right] \underset{R_3 = \left(\frac{50}{3}\right)r_2 + r_3}{\rightarrow} \left[\begin{array}{ccc|c} 1 & \frac{2}{15} & \frac{3}{10} & \frac{13}{6} \\ 0 & 1 & \frac{3}{34} & \frac{23}{34} \\ 0 & 0 & \frac{4105}{17} & \frac{8210}{17} \end{array}\right] \underset{R_3 = \left(\frac{17}{4105}\right)r_3}{\rightarrow} \left[\begin{array}{ccc|c} 1 & \frac{2}{15} & \frac{3}{10} & \frac{13}{6} \\ 0 & 1 & \frac{3}{34} & \frac{23}{34} \\ 0 & 0 & 1 & 2 \end{array}\right]$$

The matrix is now reduced to echelon form. The final matrix represents the system

$$\begin{cases} c + \frac{2}{15}p + \frac{3}{10}m = \frac{13}{6} & (1) \\ 81p + \frac{3}{34}m = \frac{23}{34} & (2) \\ m = 2 & (3) \end{cases}$$

From (3), we determine that 2 glasses of milk should be served. Back-substitute $m = 2$ into equation (2) to find that $p = \frac{1}{2}$, so $\frac{1}{2}$ of a baked potato should be served. Back-substitute these values into equation (1) and find that $c = 1.5$, so 1.5 servings of chicken *a la king* should be given to the patient to meet the dietary requirements.

6.3 EXERCISES

In Problems 1–10, write the augmented matrix of the given system of equations.

1. $\begin{cases} x - 5y = 5 \\ 4x + 3y = 6 \end{cases}$

2. $\begin{cases} 3x + 4y = 7 \\ 4x - 2y = 5 \end{cases}$

3. $\begin{cases} 2x + 3y - 6 = 0 \\ 4x - 6y + 2 = 0 \end{cases}$

4. $\begin{cases} 9x - y = 0 \\ 3x - y - 4 = 0 \end{cases}$

5. $\begin{cases} 0.01x - 0.03y = 0.06 \\ 0.13x + 0.10y = 0.20 \end{cases}$

6. $\begin{cases} \frac{4}{3}x - \frac{3}{2}y = \frac{3}{4} \\ -\frac{1}{4}x + \frac{1}{3}y = \frac{2}{3} \end{cases}$

7. $\begin{cases} x - y + z = 10 \\ 3x + 2y = 5 \\ x + y + 2z = 2 \end{cases}$

8. $\begin{cases} 5x - y - z = 0 \\ x + y = 5 \\ 2x - 3z = 2 \end{cases}$

9. $\begin{cases} x + y - z = 2 \\ 3x - 2y = 2 \\ 5x + 3y - z = 1 \end{cases}$

10. $\begin{cases} 2x + 3y - 4z = 0 \\ x - 5z + 2 = 0 \\ x + 2y - 3z = -2 \end{cases}$

In Problems 11–20, perform in order (a), followed by (b), followed by (c) on the given augmented matrix.

11. $\left[\begin{array}{rrr|r} 1 & -3 & -5 & -2 \\ 2 & -5 & -4 & 5 \\ -3 & 5 & 4 & 6 \end{array}\right]$ (a) $R_2 = -2r_1 + r_2$ (b) $R_3 = 3r_1 + r_3$ (c) $R_3 = 4r_2 + r_3$

12. $\left[\begin{array}{rrr|r} 1 & -3 & -3 & -3 \\ 2 & -5 & 2 & -4 \\ -3 & 2 & 4 & 6 \end{array}\right]$ (a) $R_2 = -2r_1 + r_2$ (b) $R_3 = 3r_1 + r_3$ (c) $R_3 = 7r_2 + r_3$

13. $\left[\begin{array}{rrr|r} 1 & -3 & 4 & 3 \\ 2 & -5 & 6 & 6 \\ -3 & 3 & 4 & 6 \end{array}\right]$ (a) $R_2 = -2r_1 + r_2$ (b) $R_3 = 3r_1 + r_3$ (c) $R_3 = 6r_2 + r_3$

14. $\left[\begin{array}{rrr|r} 1 & -3 & 3 & -5 \\ 2 & -5 & -3 & -5 \\ -3 & -2 & 4 & 6 \end{array}\right]$ (a) $R_2 = -2r_1 + r_2$ (b) $R_3 = 3r_1 + r_3$ (c) $R_3 = 11r_2 + r_3$

15. $\left[\begin{array}{rrr|r} 1 & -3 & 2 & -6 \\ 2 & -5 & 3 & -4 \\ -3 & -6 & 4 & 6 \end{array}\right]$ (a) $R_2 = -2r_1 + r_2$ (b) $R_3 = 3r_1 + r_3$ (c) $R_3 = 15r_2 + r_3$

16. $\left[\begin{array}{rrr|r} 1 & -3 & -4 & -6 \\ 2 & -5 & 6 & -6 \\ -3 & 1 & 4 & 6 \end{array}\right]$ (a) $R_2 = -2r_1 + r_2$ (b) $R_3 = 3r_1 + r_3$ (c) $R_3 = 8r_2 + r_3$

17. $\left[\begin{array}{rrr|r} 1 & -3 & 1 & -2 \\ 2 & -5 & 6 & -2 \\ -3 & 1 & 4 & 6 \end{array}\right]$ (a) $R_2 = -2r_1 + r_2$ (b) $R_3 = 3r_1 + r_3$ (c) $R_3 = 8r_2 + r_3$

18. $\left[\begin{array}{rrr|r} 1 & -3 & -1 & 2 \\ 2 & -5 & 2 & 6 \\ -3 & -6 & 4 & 6 \end{array}\right]$ (a) $R_2 = -2r_1 + r_2$ (b) $R_3 = 3r_1 + r_3$ (c) $R_3 = 15r_2 + r_3$

19. $\left[\begin{array}{rrr|r} 1 & -3 & -2 & 3 \\ 2 & -5 & 2 & -1 \\ -3 & -2 & 4 & 6 \end{array}\right]$ (a) $R_2 = -2r_1 + r_2$ (b) $R_3 = 3r_1 + r_3$ (c) $R_3 = 11r_2 + r_3$

20. $\left[\begin{array}{rrr|r} 1 & -3 & 5 & -3 \\ 2 & -5 & 1 & -4 \\ -3 & 3 & 4 & 6 \end{array}\right]$ (a) $R_2 = -2r_1 + r_2$ (b) $R_3 = 3r_1 + r_3$ (c) $R_3 = 6r_2 + r_3$

In Problems 21–26, the reduced row echelon form of a system of linear equations is given. Write the system of equations corresponding to the given matrix. Use x, y, or x, y, z as variables. Determine whether the system is consistent or inconsistent. If it is consistent, give the solution.

21. $\left[\begin{array}{rr|r} 1 & 0 & 5 \\ 0 & 1 & -1 \end{array}\right]$

22. $\left[\begin{array}{rr|r} 1 & 0 & -4 \\ 0 & 1 & 0 \end{array}\right]$

23. $\left[\begin{array}{rrr|r} 1 & 0 & 0 & 1 \\ 0 & 1 & 0 & 2 \\ 0 & 0 & 0 & 3 \end{array}\right]$

24. $\left[\begin{array}{rrr|r} 1 & 0 & 0 & 0 \\ 0 & 1 & 0 & 0 \\ 0 & 0 & 0 & 2 \end{array}\right]$

25. $\left[\begin{array}{rrr|r} 1 & 0 & 2 & -1 \\ 0 & 1 & -4 & -2 \\ 0 & 0 & 0 & 0 \end{array}\right]$

26. $\left[\begin{array}{rrr|r} 1 & 0 & 4 & 4 \\ 0 & 1 & 3 & 2 \\ 0 & 0 & 0 & 0 \end{array}\right]$

In Problems 27–62, solve each system of equations using matrices (row operations). If the system has no solution, say that it is inconsistent. Verify your results using a graphing utility.

27. $\begin{cases} x + y = 8 \\ x - y = 4 \end{cases}$

28. $\begin{cases} x + 2y = 5 \\ x + y = 3 \end{cases}$

29. $\begin{cases} x - 5y = -13 \\ 3x + 2y = 12 \end{cases}$

30. $\begin{cases} x + 3y = 5 \\ 2x - 3y = -8 \end{cases}$

31. $\begin{cases} 3x - 6y = 24 \\ 5x + 4y = 12 \end{cases}$

32. $\begin{cases} 2x + 4y = 16 \\ 3x - 5y = -9 \end{cases}$

33. $\begin{cases} 2x + y = 1 \\ 4x + 2y = 6 \end{cases}$

34. $\begin{cases} x - y = 5 \\ -3x + 3y = 2 \end{cases}$

35. $\begin{cases} 2x - 4y = -2 \\ 3x + 2y = 3 \end{cases}$

36. $\begin{cases} 3x + 3y = 3 \\ 4x + 2y = \frac{8}{3} \end{cases}$

37. $\begin{cases} x + 2y = 4 \\ 2x + 4y = 8 \end{cases}$

38. $\begin{cases} 3x - y = 7 \\ 9x - 3y = 21 \end{cases}$

39. $\begin{cases} 2x + 3y = 6 \\ x - y = \frac{1}{2} \end{cases}$

40. $\begin{cases} \frac{1}{2}x + y = -2 \\ x - 2y = 8 \end{cases}$

41. $\begin{cases} 3x - 5y = 3 \\ 15x + 5y = 21 \end{cases}$

42. $\begin{cases} 2x - y = -1 \\ x + \frac{1}{2}y = \frac{3}{2} \end{cases}$

43. $\begin{cases} x - y = 6 \\ 2x - 3z = 16 \\ 2y + z = 4 \end{cases}$

44. $\begin{cases} 2x + y = -4 \\ -2y + 4z = 0 \\ 3x - 2z = -11 \end{cases}$

45. $\begin{cases} x - 2y + 3z = 7 \\ 2x + y + z = 4 \\ -3x + 2y - 2z = -10 \end{cases}$

46. $\begin{cases} 2x + y - 3z = 0 \\ -2x + 2y + z = -7 \\ 3x - 4y - 3z = 7 \end{cases}$

47. $\begin{cases} 2x - 2y - 2z = 2 \\ 2x + 3y + z = 2 \\ 3x + 2y = 0 \end{cases}$

48. $\begin{cases} 2x - 3y - z = 0 \\ -x + 2y + z = 5 \\ 3x - 4y - z = 1 \end{cases}$

49. $\begin{cases} -x + y + z = -1 \\ -x + 2y - 3z = -4 \\ 3x - 2y - 7z = 0 \end{cases}$

50. $\begin{cases} 2x - 3y - z = 0 \\ 3x + 2y + 2z = 2 \\ x + 5y + 3z = 2 \end{cases}$

51. $\begin{cases} 2x - 2y + 3z = 6 \\ 4x - 3y + 2z = 0 \\ -2x + 3y - 7z = 1 \end{cases}$

52. $\begin{cases} 3x - 2y + 2z = 6 \\ 7x - 3y + 2z = -1 \\ 2x - 3y + 4z = 0 \end{cases}$

53. $\begin{cases} x + y - z = 6 \\ 3x - 2y + z = -5 \\ x + 3y - 2z = 14 \end{cases}$

54. $\begin{cases} x - y + z = -4 \\ 2x - 3y + 4z = -15 \\ 5x + y - 2z = 12 \end{cases}$

55. $\begin{cases} x + 2y - z = -3 \\ 2x - 4y + z = -7 \\ -2x + 2y - 3z = 4 \end{cases}$

56. $\begin{cases} x + 4y - 3z = -8 \\ 3x - y + 3z = 12 \\ x + y + 6z = 1 \end{cases}$

57. $\begin{cases} 3x + y - z = \frac{2}{3} \\ 2x - y + z = 1 \\ 4x + 2y = \frac{8}{3} \end{cases}$

58. $\begin{cases} x + y = 1 \\ 2x - y + z = 1 \\ x + 2y + z = \frac{8}{3} \end{cases}$

59. $\begin{cases} x + 2y + z = 1 \\ 2x - y + 2z = 2 \\ 3x + y + 3z = 3 \end{cases}$

60. $\begin{cases} x + 2y - z = 3 \\ 2x - y + 2z = 6 \\ x - 3y + 3z = 4 \end{cases}$

61. $\begin{cases} x + y + z + w = 4 \\ 2x - y + z = 0 \\ 3x + 2y + z - w = 6 \\ x - 2y - 2z + 2w = -1 \end{cases}$

62. $\begin{cases} x + y + z + w = 4 \\ -x + 2y + z = 0 \\ 2x + 3y + z - w = 6 \\ -2x + y - 2z + 2w = -1 \end{cases}$

63. **Curve Fitting** Find the graph of the function $y = ax^2 + bx + c$ that contains the points $(1, 2)$, $(-2, -7)$, and $(2, -3)$.

64. **Curve Fitting** Find the graph of the function $y = ax^2 + bx + c$ that contains the points $(1, -1)$, $(3, -1)$, and $(-2, 14)$.

65. **Curve Fitting** Find the function $f(x) = ax^3 + bx^2 + cx + d$ for which $f(-3) = -112$, $f(-1) = -2$, $f(1) = 4$, and $f(2) = 13$.

66. **Curve Fitting** Find the function $f(x) = ax^3 + bx^2 + cx + d$ for which $f(-2) = -10$, $f(-1) = 3$, $f(1) = 5$, and $f(3) = 15$.

67. **Nutrition** A dietitian at Palos Community Hospital wishes a patient to have a meal that has 78 grams of protein, 59 grams of carbohydrates, and 75 milligrams of vitamin A. The hospital food service tells the dietitian that the dinner for today is salmon steak, baked eggs, and acorn squash. Each serving of salmon steak has 30 grams of protein, 20 grams of carbohydrates, and 2 milligrams of vitamin A. Each serving of baked eggs contains 15 grams of protein, 2 grams of carbohydrates, and 20 milligrams of vitamin A. Each serving of acorn squash contains 3 grams of protein, 25 grams of carbohydrates, and 32 milligrams of vitamin A. How many servings of each food should the dietitian provide for the patient?

68. **Nutrition** A dietitian at General Hospital wishes a patient to have a meal that has 47 grams of protein, 58 grams of carbohydrates, and 630 milligrams of calcium. The hospital food service tells the dietitian that the dinner for today is pork chops, corn on the cob, and 2% milk. Each serving of pork chops has 23 grams of protein, 0 grams of carbohydrates, and 10 milligrams of calcium. Each serving of corn on the cob contains 3 grams of protein, 16 grams of carbohydrates, and 10 milligrams of calcium. Each glass of milk contains 9 grams of protein, 13 grams of carbohydrates, and 300 milligrams of calcium. How many servings of each food should the dietitian provide for the patient?

69. **Financial Planning** Carletta has $10,000 to invest. As her financial consultant, you recommend that she invest in Treasury bills that yield 6%, Treasury bonds that yield 7%, and corporate bonds that yield 8%. Carletta wants to have an annual income of $680, and the amount invested in corporate bonds must be half that invested in Treasury bills. Find the amount in each investment.

70. **Financial Planning** John has $20,000 to invest. As his financial consultant, you recommended that he invest in Treasury bills that yield 5%, Treasury bonds that yield 7%, and corporate bonds that yield 9%. John wants to have an annual income of $1280, and the amount invested in Treasury bills must be two times the amount invested in corporate bonds. Find the amount in each investment.

71. **Production** To manufacture an automobile requires painting, drying, and polishing. Epsilon Motor Company produces three types of cars: the Delta, the Beta, and the Sigma. Each Delta requires 10 hours for painting, 3 hours for drying, and 2 hours for polishing. A Beta requires 16 hours of painting, 5 hours of drying, and 3 hours of polishing, while a Sigma requires 8 hours for painting, 2 hours for drying, and 1 hour for polishing. If the company has 240 hours for painting, 69 hours for drying, and 41 hours for polishing per month, how many of each type of car are produced?

72. **Production** A Florida juice company completes the preparation of its products by sterilizing, filling, and labeling bottles. Each case of orange juice requires 9 minutes for sterilizing, 6 minutes for filling, and 1 minute for labeling Each case of grapefruit juice requires 10 minutes for sterilizing, 4 minutes for filling, and 2 minutes for labeling. Each case of tomato juice requires 12 minutes for sterilizing, 4 minutes for filling, and 1 minute for labeling. If the company runs the cleaning machine for 398 minutes, the filling machine for 164 minutes, and the labeling machine for 58 minutes, how many cases of each type of juice are prepared?

73. **Electricity: Kirchhoff's Rules** An application of Kirchhoff's Rules to the circuit shown results in the following system of equations:

$$\begin{cases} -4 + 8 - 2I_2 = 0 \\ 8 = 5I_4 + I_1 \\ 4 = 3I_3 + I_1 \\ I_3 + I_4 = I_1 \end{cases}$$

Find the currents I_1, I_2, I_3, and I_4.*

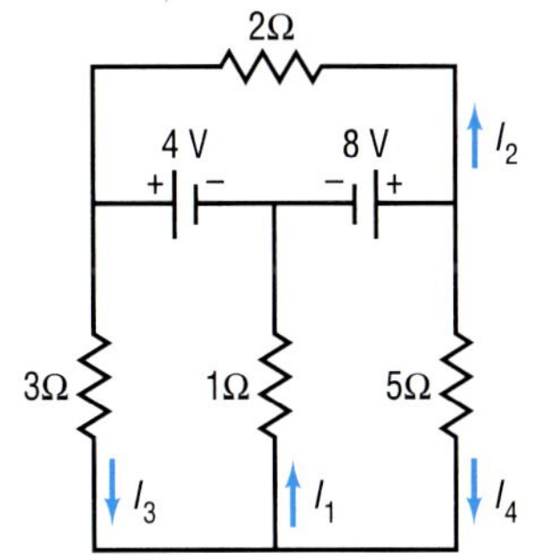

74. **Electricity: Kirchhoff's Rules** An application of Kirchhoff's Rules to the circuit shown results in the following system of equations:

$$\begin{cases} I_1 = I_3 + I_2 \\ 24 - 6I_1 - 3I_3 = 0 \\ 12 + 24 - 6I_1 - 6I_2 = 0 \end{cases}$$

Find the currents I_1, I_2, and I_3.†

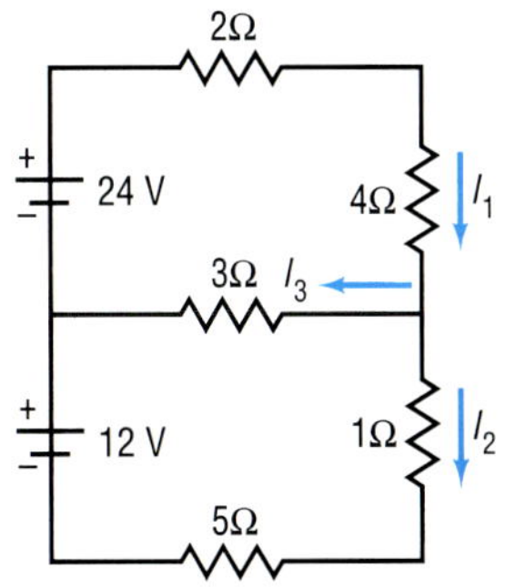

75. Write a brief paragraph or two that outlines your strategy for solving a system of linear equations using matrices.

76. When solving a system of linear equations using matrices, do you prefer to place the augmented matrix in echelon form or in reduced row echelon form? Give reasons for your choice.

77. Make up a system of three linear equations containing three variables that has:
(a) No solution
(b) Exactly one solution
(c) Infinitely many solutions
Give the three systems to a friend to solve and critique.

*Source: Based on Raymond Serway, *Physics,* 3rd ed. (Philadelphia: Saunders, 1990), Prob. 34, p. 79.

†*Source:* Ibid., Prob. 38, p. 791.

6.4 SYSTEMS OF LINEAR EQUATIONS: DETERMINANTS

1 Evaluate 2 by 2 Determinants

2 Use Cramer's Rule to Solve a System of Two Equations, Two Variables

3 Evaluate 3 by 3 Determinants

4 Use Cramer's Rule to Solve a System of Three Equations, Three Variables

5 Know Some Properties of Determinants

1 In the preceding section, we described a method of using matrices to solve a system of linear equations. This section deals with yet another method for solving systems of linear equations; however, it can be used only when the number of equations equals the number of variables. Although the method will work for any system (provided that the number of equations equals the number of variables), it is most often used for systems of two equations containing two variables or three equations containing three variables. This method, called *Cramer's Rule,* is based on the concept of a *determinant.*

2 by 2 Determinants

If a, b, c, and d are four real numbers, the symbol

$$D = \begin{vmatrix} a & b \\ c & d \end{vmatrix}$$

is called a **2 by 2 determinant.** Its value is the number $ad - bc$; that is,

$$D = \begin{vmatrix} a & b \\ c & d \end{vmatrix} = ad - bc \qquad (1)$$

A device that may be helpful for remembering the value of a 2 by 2 determinant is the following:

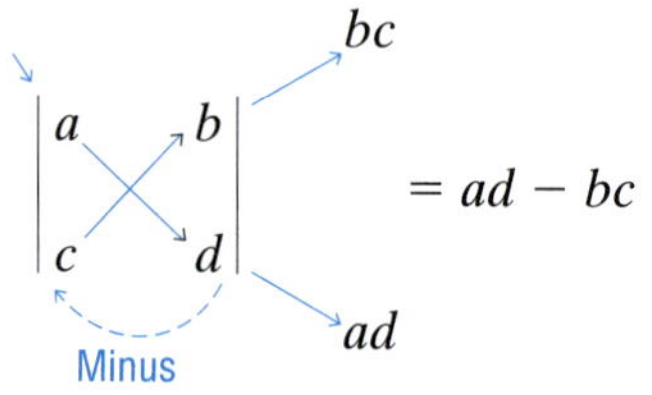

EXAMPLE 1 Evaluating a 2 × 2 Determinant

$$\begin{vmatrix} 3 & -2 \\ 6 & 1 \end{vmatrix} = (3)(1) - (6)(-2) = 3 - (-12) = 15$$

Graphing utilities can also be used to evaluate determinants.

EXAMPLE 2 Using a Graphing Utility to Evaluate a 2 × 2 Determinant

Use a graphing utility to evaluate the determinant from Example 1: $\begin{vmatrix} 3 & -2 \\ 6 & 1 \end{vmatrix}$.

Solution First, we enter the matrix whose entries are those of the determinant into the graphing utility and name it A. Using the determinant command, we obtain the result shown in Figure 9.

Figure 9

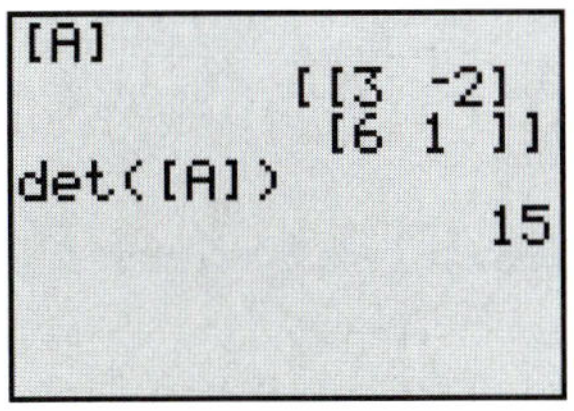

NOW WORK PROBLEM 3.

Let's now see the role that a 2 by 2 determinant plays in the solution of a system of two equations containing two variables. Consider the system

$$\begin{cases} ax + by = s & (1) \\ cx + dy = t & (2) \end{cases} \tag{2}$$

We shall use the method of elimination to solve this system.

Provided $d \neq 0$ and $b \neq 0$, this system is equivalent to the system

$$\begin{cases} adx + bdy = sd & (1) \text{ Multiply by } d. \\ bcx + bdy = tb & (2) \text{ Multiply by } b. \end{cases}$$

On subtracting the second equation from the first equation, we get

$$\begin{cases} (ad - bc)x + 0 \cdot y = sd - tb & (1) \\ \quad bcx \quad + bdy = tb & (2) \end{cases}$$

Now, the first equation can be rewritten using determinant notation.

$$\begin{vmatrix} a & b \\ c & d \end{vmatrix} x = \begin{vmatrix} s & b \\ t & d \end{vmatrix}$$

If $D = \begin{vmatrix} a & b \\ c & d \end{vmatrix} = ad - bc \neq 0$, we can solve for x to get

$$x = \frac{\begin{vmatrix} s & b \\ t & d \end{vmatrix}}{\begin{vmatrix} a & b \\ c & d \end{vmatrix}} = \frac{\begin{vmatrix} s & b \\ t & d \end{vmatrix}}{D} \tag{3}$$

Return now to the original system (2). Provided that $a \neq 0$ and $c \neq 0$, the system is equivalent to

$$\begin{cases} acx + bcy = cs & (1) \text{ Multiply by } c. \\ acx + ady = at & (2) \text{ Multiply by } a. \end{cases}$$

On subtracting the first equation from the second equation, we get

$$\begin{cases} acx + \quad bcy \quad = \quad cs & (1) \\ 0 \cdot x + (ad - bc)y = at - cs & (2) \end{cases}$$

The second equation can now be rewritten using determinant notation.

$$\begin{vmatrix} a & b \\ c & d \end{vmatrix} y = \begin{vmatrix} a & s \\ c & t \end{vmatrix}$$

If $D = \begin{vmatrix} a & b \\ c & d \end{vmatrix} = ad - bc \neq 0$, we can solve for y to get

$$y = \frac{\begin{vmatrix} a & s \\ c & t \end{vmatrix}}{\begin{vmatrix} a & b \\ c & d \end{vmatrix}} = \frac{\begin{vmatrix} a & s \\ c & t \end{vmatrix}}{D} \qquad (4)$$

Equations (3) and (4) lead us to the following result, called **Cramer's Rule.**

THEOREM **Cramer's Rule for Two Equations Containing Two Variables**

The solution to the system of equations

$$\begin{cases} ax + by = s & (1) \\ cx + dy = t & (2) \end{cases} \qquad (5)$$

is given by

$$x = \frac{\begin{vmatrix} s & b \\ t & d \end{vmatrix}}{\begin{vmatrix} a & b \\ c & d \end{vmatrix}}, \qquad y = \frac{\begin{vmatrix} a & s \\ c & t \end{vmatrix}}{\begin{vmatrix} a & b \\ c & d \end{vmatrix}} \qquad (6)$$

provided that

$$D = \begin{vmatrix} a & b \\ c & d \end{vmatrix} = ad - bc \neq 0$$

In the derivation given for Cramer's Rule above, we assumed that none of the numbers a, b, c and d was 0. In Problem 58 you will be asked to complete the proof under the less stringent conditions that $D = ad - bc \neq 0$.

Now look carefully at the pattern in Cramer's Rule. The denominator in the solution (6) is the determinant of the coefficients of the variables.

$$\begin{cases} ax + by = s \\ cx + dy = t \end{cases} \qquad D = \begin{vmatrix} a & b \\ c & d \end{vmatrix}$$

In the solution for x, the numerator is the determinant, denoted by D_x, formed by replacing the entries in the first column (the coefficients of x) in D by the constants on the right side of the equal sign.

$$D_x = \begin{vmatrix} s & b \\ t & d \end{vmatrix}$$

In the solution for y, the numerator is the determinant, denoted by D_y, formed by replacing the entries in the second column (the coefficients of y) in D by the constants on the right side of the equal sign.

$$D_y = \begin{vmatrix} a & s \\ c & t \end{vmatrix}$$

Cramer's Rule then states that, if $D \neq 0$,

$$x = \frac{D_x}{D}, \qquad y = \frac{D_y}{D} \tag{7}$$

EXAMPLE 3 Solving a System of Linear Equations Using Determinants

Use Cramer's Rule, if applicable, to solve the system

$$\begin{cases} 3x - 2y = 4 & (1) \\ 6x + y = 13 & (2) \end{cases}$$

Algebraic Solution

The determinant D of the coefficients of the variables is

$$D = \begin{vmatrix} 3 & -2 \\ 6 & 1 \end{vmatrix} = (3)(1) - (6)(-2) = 15$$

Because $D \neq 0$, Cramer's Rule (7) can be used.

$$x = \frac{D_x}{D} = \frac{\begin{vmatrix} 4 & -2 \\ 13 & 1 \end{vmatrix}}{15} = \frac{30}{15} = 2, \qquad y = \frac{D_y}{D} = \frac{\begin{vmatrix} 3 & 4 \\ 6 & 13 \end{vmatrix}}{15} = \frac{15}{15} = 1$$

The solution is $x = 2$, $y = 1$.

Graphing Solution

Figure 10

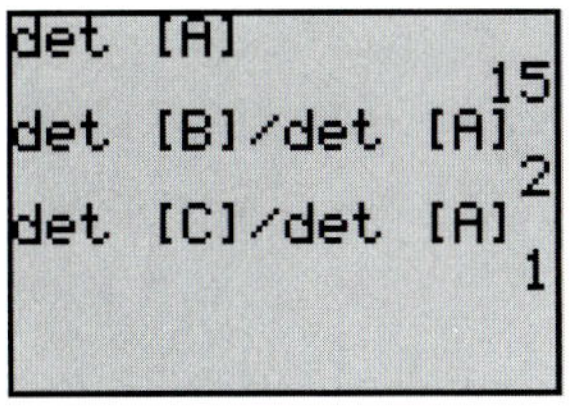

We enter the coefficient matrix into our graphing utility and call it A. Evaluate det$[A]$. Since det$[A] \neq 0$, we can use Cramer's Rule. We enter the matrices D_x and D_y into our graphing utility and call them B and C, respectively. Finally, we find x by calculating det$[B]$/det$[A]$ and y by calculating det$[C]$/det$[A]$. The results are shown in Figure 10.

The solution is $x = 2$, $y = 1$.

In attempting to use Cramer's Rule, if the determinant D of the coefficients of the variables is found to equal 0 (so that Cramer's Rule is not applicable), then the system either is inconsistent or has infinitely many solutions.

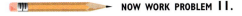

NOW WORK PROBLEM 11.

3 by 3 Determinants

3 In order to use Cramer's Rule to solve a system of three equations containing three variables, we need to define a 3 by 3 determinant.

A **3 by 3 determinant** is symbolized by

$$\begin{vmatrix} a_{11} & a_{12} & a_{13} \\ a_{21} & a_{22} & a_{23} \\ a_{31} & a_{32} & a_{33} \end{vmatrix} \tag{8}$$

in which $a_{11}, a_{12}, \ldots,$ are real numbers.

As with matrices, we use a double subscript to identify an entry by indicating its row and column numbers. For example, the entry a_{23} is in row 2, column 3.

The value of a 3 by 3 determinant may be defined in terms of 2 by 2 determinants by the following formula:

Minus

$$\begin{vmatrix} a_{11} & a_{12} & a_{13} \\ a_{21} & a_{22} & a_{23} \\ a_{31} & a_{32} & a_{33} \end{vmatrix} = a_{11}\begin{vmatrix} a_{22} & a_{23} \\ a_{32} & a_{33} \end{vmatrix} - a_{12}\begin{vmatrix} a_{21} & a_{23} \\ a_{31} & a_{33} \end{vmatrix} + a_{13}\begin{vmatrix} a_{21} & a_{22} \\ a_{31} & a_{32} \end{vmatrix} \tag{9}$$

2 by 2 determinant left after removing row and column containing a_{11}. *2 by 2 determinant left after removing row and column containing a_{12}.* *2 by 2 determinant left after removing row and column containing a_{13}.*

Be sure to take note of the minus sign that appears with the second term—it's easy to forget it! Formula (9) is best remembered by noting that each entry in row 1 is multiplied by the 2 by 2 determinant that remains after the row and column containing the entry have been removed, as follows:

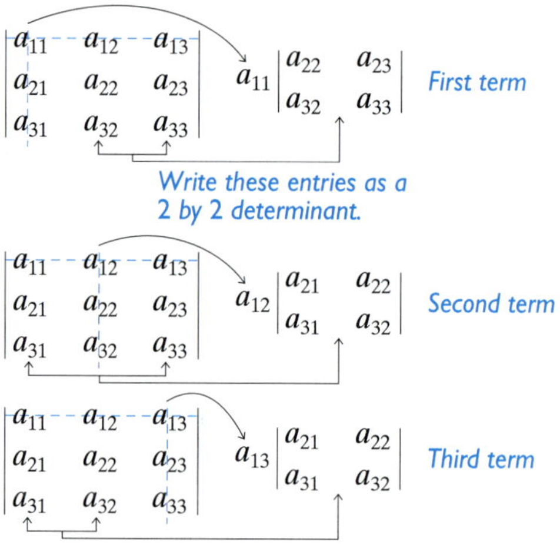

Now insert the minus sign before the middle expression and add.

Minus

$$\begin{vmatrix} a_{11} & a_{12} & a_{13} \\ a_{21} & a_{22} & a_{23} \\ a_{31} & a_{32} & a_{33} \end{vmatrix} = a_{11}\begin{vmatrix} a_{22} & a_{23} \\ a_{32} & a_{33} \end{vmatrix} - a_{12}\begin{vmatrix} a_{21} & a_{23} \\ a_{31} & a_{33} \end{vmatrix} + a_{13}\begin{vmatrix} a_{21} & a_{22} \\ a_{31} & a_{32} \end{vmatrix}$$

Formula (9) exhibits one way to find the value of a 3 by 3 determinant: *by expanding across row 1.* In fact, the expansion can take place across any row or down any column. The terms to be added or subtracted consist of the row (or column) entry times the value of the 2 by 2 determinant that remains after removing the row and column entry. The value of the determinant is found by adding or subtracting the terms according to the following scheme:

$$\begin{bmatrix} (-1)^{1+1} & (-1)^{1+2} & (-1)^{1+3} \\ (-1)^{2+1} & (-1)^{2+2} & (-1)^{2+3} \\ (-1)^{3+1} & (-1)^{3+2} & (-1)^{3+3} \end{bmatrix} = \begin{bmatrix} + & - & + \\ - & + & - \\ + & - & + \end{bmatrix}$$

The exponents are the sum of the row and column of the entry.

For example, if we choose to expand down column 2, we obtain

$$\begin{vmatrix} a_{11} & a_{12} & a_{13} \\ a_{21} & a_{22} & a_{23} \\ a_{31} & a_{32} & a_{33} \end{vmatrix} = -a_{12}\begin{vmatrix} a_{21} & a_{23} \\ a_{31} & a_{33} \end{vmatrix} + a_{22}\begin{vmatrix} a_{11} & a_{13} \\ a_{31} & a_{33} \end{vmatrix} - a_{32}\begin{vmatrix} a_{11} & a_{13} \\ a_{21} & a_{23} \end{vmatrix}$$

↑ *Expand down column 2 $(-, +, -)$.*

If we choose to expand across row 3, we obtain

$$\begin{vmatrix} a_{11} & a_{12} & a_{13} \\ a_{21} & a_{22} & a_{23} \\ a_{31} & a_{32} & a_{33} \end{vmatrix} = a_{31}\begin{vmatrix} a_{12} & a_{13} \\ a_{22} & a_{23} \end{vmatrix} - a_{32}\begin{vmatrix} a_{11} & a_{13} \\ a_{21} & a_{23} \end{vmatrix} + a_{33}\begin{vmatrix} a_{11} & a_{12} \\ a_{21} & a_{22} \end{vmatrix}$$

↑ *Expand across row 3 $(+, -, +)$.*

It can be shown that the value of a determinant does not depend on the choice of the row or column used in the expansion.

EXAMPLE 4 Evaluating a 3×3 Determinant

Find the value of the 3 by 3 determinant: $\begin{vmatrix} 3 & 4 & -1 \\ 4 & 6 & 2 \\ 8 & -2 & 3 \end{vmatrix}$

Solution We choose to expand across row 1.

Remember the minus sign. ↓

$$\begin{vmatrix} 3 & 4 & -1 \\ 4 & 6 & 2 \\ 8 & -2 & 3 \end{vmatrix} = 3\begin{vmatrix} 6 & 2 \\ -2 & 3 \end{vmatrix} - 4\begin{vmatrix} 4 & 2 \\ 8 & 3 \end{vmatrix} + (-1)\begin{vmatrix} 4 & 6 \\ 8 & -2 \end{vmatrix}$$

$$= 3(18 + 4) - 4(12 - 16) + (-1)(-8 - 48)$$

$$= 3(22) - 4(-4) + (-1)(-56)$$

$$= 66 + 16 + 56 = 138$$

We could also find the value of the 3 by 3 determinant in Example 4 by expanding down column 3 (the signs are $+, -, +$):

$$\begin{vmatrix} 3 & 4 & -1 \\ 4 & 6 & 2 \\ 8 & -2 & 3 \end{vmatrix} = (-1)\begin{vmatrix} 4 & 6 \\ 8 & -2 \end{vmatrix} - 2\begin{vmatrix} 3 & 4 \\ 8 & -2 \end{vmatrix} + 3\begin{vmatrix} 3 & 4 \\ 4 & 6 \end{vmatrix}$$

$$= -1(-8 - 48) - 2(-6 - 32) + 3(18 - 16)$$

$$= 56 + 76 + 6 = 138$$

Evaluating 3×3 determinants on a graphing utility follows the same procedure as evaluating 2×2 determinants.

NOW WORK PROBLEM 7.

Systems of Three Equations Containing Three Variables

4 Consider the following system of three equations containing three variables.

$$\begin{cases} a_{11}x + a_{12}y + a_{13}z = c_1 \\ a_{21}x + a_{22}y + a_{23}z = c_2 \\ a_{31}x + a_{32}y + a_{33}z = c_3 \end{cases} \qquad (10)$$

It can be shown that if the determinant D of the coefficients of the variables is not 0, that is, if

$$D = \begin{vmatrix} a_{11} & a_{12} & a_{13} \\ a_{21} & a_{22} & a_{23} \\ a_{31} & a_{32} & a_{33} \end{vmatrix} \neq 0$$

then the unique solution of system (10) is given by

Cramer's Rule for Three Equations Containing Three Variables

$$x = \frac{D_x}{D}, \qquad y = \frac{D_y}{D}, \qquad z = \frac{D_z}{D}$$

where

$$D_x = \begin{vmatrix} c_1 & a_{12} & a_{13} \\ c_2 & a_{22} & a_{23} \\ c_3 & a_{32} & a_{33} \end{vmatrix} \qquad D_y = \begin{vmatrix} a_{11} & c_1 & a_{13} \\ a_{21} & c_2 & a_{23} \\ a_{31} & c_3 & a_{33} \end{vmatrix} \qquad D_z = \begin{vmatrix} a_{11} & a_{12} & c_1 \\ a_{21} & a_{22} & c_2 \\ a_{31} & a_{32} & c_3 \end{vmatrix}$$

The similarity of this pattern and the pattern observed earlier for a system of two equations containing two variables should be apparent.

EXAMPLE 5 Using Cramer's Rule

Use Cramer's Rule, if applicable, to solve the following system:

$$\begin{cases} 2x + y - z = 3 & (1) \\ -x + 2y + 4z = -3 & (2) \\ x - 2y - 3z = 4 & (3) \end{cases}$$

Solution The value of the determinant D of the coefficients of the variables is

$$D = \begin{vmatrix} 2 & 1 & -1 \\ -1 & 2 & 4 \\ 1 & -2 & -3 \end{vmatrix} = 2\begin{vmatrix} 2 & 4 \\ -2 & -3 \end{vmatrix} - 1\begin{vmatrix} -1 & 4 \\ 1 & -3 \end{vmatrix} + (-1)\begin{vmatrix} -1 & 2 \\ 1 & -2 \end{vmatrix}$$

$$= 2(2) - 1(-1) + (-1)(0)$$

$$= 4 + 1 = 5$$

Because $D \neq 0$, we proceed to find the values of D_x, D_y, and D_z.

$$D_x = \begin{vmatrix} 3 & 1 & -1 \\ -3 & 2 & 4 \\ 4 & -2 & -3 \end{vmatrix} = 3\begin{vmatrix} 2 & 4 \\ -2 & -3 \end{vmatrix} - 1\begin{vmatrix} -3 & 4 \\ 4 & -3 \end{vmatrix} + (-1)\begin{vmatrix} -3 & 2 \\ 4 & -2 \end{vmatrix}$$
$$= 3(2) - 1(-7) + (-1)(-2) = 15$$

$$D_y = \begin{vmatrix} 2 & 3 & -1 \\ -1 & -3 & 4 \\ 1 & 4 & -3 \end{vmatrix} = 2\begin{vmatrix} -3 & 4 \\ 4 & -3 \end{vmatrix} - 3\begin{vmatrix} -1 & 4 \\ 1 & -3 \end{vmatrix} + (-1)\begin{vmatrix} -1 & -3 \\ 1 & 4 \end{vmatrix}$$
$$= 2(-7) - 3(-1) + (-1)(-1)$$
$$= -14 + 3 + 1 = -10$$

$$D_z = \begin{vmatrix} 2 & 1 & 3 \\ -1 & 2 & -3 \\ 1 & -2 & 4 \end{vmatrix} = 2\begin{vmatrix} 2 & -3 \\ -2 & 4 \end{vmatrix} - 1\begin{vmatrix} -1 & -3 \\ 1 & 4 \end{vmatrix} + 3\begin{vmatrix} -1 & 2 \\ 1 & -2 \end{vmatrix}$$
$$= 2(2) - 1(-1) + 3(0) = 5$$

As a result,

$$x = \frac{D_x}{D} = \frac{15}{5} = 3 \qquad y = \frac{D_y}{D} = \frac{-10}{5} = -2 \qquad z = \frac{D_z}{D} = \frac{5}{5} = 1$$

The solution is $x = 3$, $y = -2$, $z = 1$.

If the determinant of the coefficients of the variables of a system of three linear equations containing three variables is 0, then Cramer's Rule is not applicable. In such a case, the system either is inconsistent or has infinitely many solutions.

Solving systems of three equations containing three variables using Cramer's Rule on a graphing utility follows the same procedure as that for solving systems of two equations containing two variables.

NOW WORK PROBLEM **29.**

More about Determinants

5 Determinants have several properties that are sometimes helpful for obtaining their value. We list some of them here.

THEOREM The value of a determinant changes sign if any two rows (or any two columns) are interchanged. (11)

Proof for 2 by 2 Determinants

$$\begin{vmatrix} a & b \\ c & d \end{vmatrix} = ad - bc \quad \text{and} \quad \begin{vmatrix} c & d \\ a & b \end{vmatrix} = bc - ad = -(ad - bc)$$

EXAMPLE 6 Demonstrating Theorem (11)

$$\begin{vmatrix} 3 & 4 \\ 1 & 2 \end{vmatrix} = 6 - 4 = 2 \qquad \begin{vmatrix} 1 & 2 \\ 3 & 4 \end{vmatrix} = 4 - 6 = -2$$

THEOREM If all the entries in any row (or any column) equal 0, the value of the determinant is 0. (12)

Proof Merely expand across the row (or down the column) containing the 0's.

THEOREM If any two rows (or any two columns) of a determinant have corresponding entries that are equal, the value of the determinant is 0. (13)

You are asked to prove this result for a 3 by 3 determinant in which the entries in column 1 equal the entries in column 3 in Problem 61.

EXAMPLE 7 Demonstrating Theorem (13)

$$\begin{vmatrix} 1 & 2 & 3 \\ 1 & 2 & 3 \\ 4 & 5 & 6 \end{vmatrix} = 1\begin{vmatrix} 2 & 3 \\ 5 & 6 \end{vmatrix} - 2\begin{vmatrix} 1 & 3 \\ 4 & 6 \end{vmatrix} + 3\begin{vmatrix} 1 & 2 \\ 4 & 5 \end{vmatrix}$$
$$= 1(-3) - 2(-6) + 3(-3)$$
$$= -3 + 12 - 9 = 0$$

THEOREM If any row (or any column) of a determinant is multiplied by a nonzero number k, the value of the determinant is also changed by a factor of k. (14)

You are asked to prove this result for a 3 by 3 determinant using row 2 in Problem 60.

EXAMPLE 8 Demonstrating Theorem (14)

$$\begin{vmatrix} 1 & 2 \\ 4 & 6 \end{vmatrix} = 6 - 8 = -2$$
$$\begin{vmatrix} k & 2k \\ 4 & 6 \end{vmatrix} = 6k - 8k = -2k = k(-2) = k\begin{vmatrix} 1 & 2 \\ 4 & 6 \end{vmatrix}$$

THEOREM

If the entries of any row (or any column) of a determinant are multiplied by a nonzero number k and the result is added to the corresponding entries of another row (or column), the value of the determinant remains unchanged. (15)

In Problem 62, you are asked to prove this result for a 3 by 3 determinant using rows 1 and 2.

EXAMPLE 9 Demonstrating Theorem (15)

$$\begin{vmatrix} 3 & 4 \\ 5 & 2 \end{vmatrix} = \begin{vmatrix} -7 & 0 \\ 5 & 2 \end{vmatrix} = -14$$

Multiply row 2 by −2 and add to row 1.

6.4 EXERCISES

In Problems 1–10, find the value of each determinant (a) by hand and (b) by using a graphing utility.

1. $\begin{vmatrix} 3 & 1 \\ 4 & 2 \end{vmatrix}$ **2.** $\begin{vmatrix} 6 & 1 \\ 5 & 2 \end{vmatrix}$ **3.** $\begin{vmatrix} 6 & 4 \\ -1 & 3 \end{vmatrix}$ **4.** $\begin{vmatrix} 8 & -3 \\ 4 & 2 \end{vmatrix}$ **5.** $\begin{vmatrix} -3 & -1 \\ 4 & 2 \end{vmatrix}$

6. $\begin{vmatrix} -4 & 2 \\ -5 & 3 \end{vmatrix}$ **7.** $\begin{vmatrix} 3 & 4 & 2 \\ 1 & -1 & 5 \\ 1 & 2 & -2 \end{vmatrix}$ **8.** $\begin{vmatrix} 1 & 3 & -2 \\ 6 & 1 & -5 \\ 8 & 2 & 3 \end{vmatrix}$ **9.** $\begin{vmatrix} 4 & -1 & 2 \\ 6 & -1 & 0 \\ 1 & -3 & 4 \end{vmatrix}$ **10.** $\begin{vmatrix} 3 & -9 & 4 \\ 1 & 4 & 0 \\ 8 & -3 & 1 \end{vmatrix}$

In Problems 11–40, solve each system of equations using Cramer's Rule if it is applicable (a) by hand and (b) by using a graphing utility. If Cramer's Rule is not applicable, say so.

11. $\begin{cases} x + y = 8 \\ x - y = 4 \end{cases}$ **12.** $\begin{cases} x + 2y = 5 \\ x - y = 3 \end{cases}$ **13.** $\begin{cases} 5x - y = 13 \\ 2x + 3y = 12 \end{cases}$ **14.** $\begin{cases} x + 3y = 5 \\ 2x - 3y = -8 \end{cases}$

15. $\begin{cases} 3x = 24 \\ x + 2y = 0 \end{cases}$ **16.** $\begin{cases} 4x + 5y = -3 \\ -2y = -4 \end{cases}$ **17.** $\begin{cases} 3x - 6y = 24 \\ 5x + 4y = 12 \end{cases}$ **18.** $\begin{cases} 2x + 4y = 16 \\ 3x - 5y = -9 \end{cases}$

19. $\begin{cases} 3x - 2y = 4 \\ 6x - 4y = 0 \end{cases}$ **20.** $\begin{cases} -x + 2y = 5 \\ 4x - 8y = 6 \end{cases}$ **21.** $\begin{cases} 2x - 4y = -2 \\ 3x + 2y = 3 \end{cases}$ **22.** $\begin{cases} 3x + 3y = 3 \\ 4x + 2y = \frac{8}{3} \end{cases}$

23. $\begin{cases} 2x - 3y = -1 \\ 10x + 10y = 5 \end{cases}$ **24.** $\begin{cases} 3x - 2y = 0 \\ 5x + 10y = 4 \end{cases}$ **25.** $\begin{cases} 2x + 3y = 6 \\ x - y = \frac{1}{2} \end{cases}$ **26.** $\begin{cases} \frac{1}{2}x + y = -2 \\ x - 2y = 8 \end{cases}$

27. $\begin{cases} 3x - 5y = 3 \\ 15x + 5y = 21 \end{cases}$ **28.** $\begin{cases} 2x - y = -1 \\ x + \frac{1}{2}y = \frac{3}{2} \end{cases}$ **29.** $\begin{cases} x + y - z = 6 \\ 3x - 2y + z = -5 \\ x + 3y - 2z = 14 \end{cases}$

30. $\begin{cases} x - y + z = -4 \\ 2x - 3y + 4z = -15 \\ 5x + y - 2z = 12 \end{cases}$ **31.** $\begin{cases} x + 2y - z = -3 \\ 2x - 4y + z = -7 \\ -2x + 2y - 3z = 4 \end{cases}$ **32.** $\begin{cases} x + 4y - 3z = -8 \\ 3x - y + 3z = 12 \\ x + y + 6z = 1 \end{cases}$

33. $\begin{cases} x - 2y + 3z = 1 \\ 3x + y - 2z = 0 \\ 2x - 4y + 6z = 2 \end{cases}$ **34.** $\begin{cases} x - y + 2z = 5 \\ 3x + 2y = 4 \\ -2x + 2y - 4z = -10 \end{cases}$ **35.** $\begin{cases} x + 2y - z = 0 \\ 2x - 4y + z = 0 \\ -2x + 2y - 3z = 0 \end{cases}$

36. $\begin{cases} x + 4y - 3z = 0 \\ 3x - y + 3z = 0 \\ x + y + 6z = 0 \end{cases}$

37. $\begin{cases} x - 2y + 3z = 0 \\ 3x + y - 2z = 0 \\ 2x - 4y + 6z = 0 \end{cases}$

38. $\begin{cases} x - y + 2z = 0 \\ 3x + 2y = 0 \\ -2x + 2y - 4z = 0 \end{cases}$

39. $\begin{cases} \dfrac{1}{x} + \dfrac{1}{y} = 8 \\ \dfrac{3}{x} - \dfrac{5}{y} = 0 \end{cases}$

[Hint: Let $u = 1/x$ and $v = 1/y$ and solve for u and v.]

40. $\begin{cases} \dfrac{4}{x} - \dfrac{3}{y} = 0 \\ \dfrac{6}{x} + \dfrac{3}{2y} = 2 \end{cases}$

In Problems 41–46, solve for x.

41. $\begin{vmatrix} x & x \\ 4 & 3 \end{vmatrix} = 5$

42. $\begin{vmatrix} x & 1 \\ 3 & x \end{vmatrix} = -2$

43. $\begin{vmatrix} x & 1 & 1 \\ 4 & 3 & 2 \\ -1 & 2 & 5 \end{vmatrix} = 2$

44. $\begin{vmatrix} 3 & 2 & 4 \\ 1 & x & 5 \\ 0 & 1 & -2 \end{vmatrix} = 0$

45. $\begin{vmatrix} x & 2 & 3 \\ 1 & x & 0 \\ 6 & 1 & -2 \end{vmatrix} = 7$

46. $\begin{vmatrix} x & 1 & 2 \\ 1 & x & 3 \\ 0 & 1 & 2 \end{vmatrix} = -4x$

In Problems 47–54, use properties of determinants to find the value of each determinant if it is known that

$$\begin{vmatrix} x & y & z \\ u & v & w \\ 1 & 2 & 3 \end{vmatrix} = 4$$

47. $\begin{vmatrix} 1 & 2 & 3 \\ u & v & w \\ x & y & z \end{vmatrix}$

48. $\begin{vmatrix} x & y & z \\ u & v & w \\ 2 & 4 & 6 \end{vmatrix}$

49. $\begin{vmatrix} x & y & z \\ -3 & -6 & -9 \\ u & v & w \end{vmatrix}$

50. $\begin{vmatrix} 1 & 2 & 3 \\ x - u & y - v & z - w \\ u & v & w \end{vmatrix}$

51. $\begin{vmatrix} 1 & 2 & 3 \\ x - 3 & y - 6 & z - 9 \\ 2u & 2v & 2w \end{vmatrix}$

52. $\begin{vmatrix} x & y & z - x \\ u & v & w - u \\ 1 & 2 & 2 \end{vmatrix}$

53. $\begin{vmatrix} 1 & 2 & 3 \\ 2x & 2y & 2z \\ u - 1 & v - 2 & w - 3 \end{vmatrix}$

54. $\begin{vmatrix} x + 3 & y + 6 & z + 9 \\ 3u - 1 & 3v - 2 & 3w - 3 \\ 1 & 2 & 3 \end{vmatrix}$

55. Geometry: Equation of a Line An equation of the line containing the two points (x_1, y_1) and (x_2, y_2) may be expressed as the determinant

$$\begin{vmatrix} x & y & 1 \\ x_1 & y_1 & 1 \\ x_2 & y_2 & 1 \end{vmatrix} = 0$$

Prove this result by expanding the determinant and comparing the result to the two-point form of the equation of a line.

56. Geometry: Collinear Points Using the result obtained in Problem 55, show that three distinct points (x_1, y_1), (x_2, y_2), and (x_3, y_3) are collinear (lie on the same line) if and only if

$$\begin{vmatrix} x_1 & y_1 & 1 \\ x_2 & y_2 & 1 \\ x_3 & y_3 & 1 \end{vmatrix} = 0$$

57. Show that $\begin{vmatrix} x^2 & x & 1 \\ y^2 & y & 1 \\ z^2 & z & 1 \end{vmatrix} = (y - z)(x - y)(x - z)$.

58. Complete the proof of Cramer's Rule for two equations containing two variables.
[Hint: In system (5), page 436, if $a = 0$, then $b \neq 0$ and $c \neq 0$, since $D = -bc \neq 0$. Now show that equation (6) provide a solution of the system when $a = 0$. There are then three remaining cases: $b = 0$, $c = 0$, and $d = 0$.]

59. Interchange columns 1 and 3 of a 3 by 3 determinant. Show that the value of the new determinant is -1 times the value of the original determinant.

60. Multiply each entry in row 2 of a 3 by 3 determinant by the number k, $k \neq 0$. Show that the value of the new determinant is k times the value of the original determinant.

61. Prove that a 3 by 3 determinant in which the entries in column 1 equal those in column 3 has the value 0.

62. Prove that, if row 2 of a 3 by 3 determinant is multiplied by k, $k \neq 0$, and the result is added to the entries in row 1, then there is no change in the value of the determinant.

6.5 MATRIX ALGEBRA

1. Work with Equality and Addition of Matrices
2. Know Properties of Matrices
3. Know How to Multiply Matrices
4. Find the Inverse of a Matrix
5. Solve Systems of Equations Using Inverse Matrices

In Section 6.3, we defined a matrix as an array of real numbers and used an augmented matrix to represent a system of linear equations. There is, however, a branch of mathematics, called **linear algebra,** that deals with matrices in such a way that an algebra of matrices is permitted. In this section, we provide a survey of how this **matrix algebra** is developed.

Before getting started, we restate the definition of a matrix.

A **matrix** is defined as a rectangular array of numbers:

	Column 1	Column 2		Column j		Column n
Row 1	a_{11}	a_{12}	$\cdots$	a_{1j}	$\cdots$	a_{1n}
Row 2	a_{21}	a_{22}	$\cdots$	a_{2j}	$\cdots$	a_{2n}
$\vdots$	$\vdots$	$\vdots$		$\vdots$		$\vdots$
Row i	a_{i1}	a_{i2}	$\cdots$	a_{ij}	$\cdots$	a_{in}
$\vdots$	$\vdots$	$\vdots$		$\vdots$		$\vdots$
Row m	a_{m1}	a_{m2}	$\cdots$	a_{mj}	$\cdots$	a_{mn}

Each number a_{ij} of the matrix has two indexes: the **row index** i and the **column index** j. The matrix shown above has m rows and n columns. The $m \cdot n$ numbers a_{ij} are usually referred to as the **entries** of the matrix. For example, a_{23} refers to the entry in the second row, third column.

Let's begin with an example that illustrates how matrices can be used to conveniently represent an array of information.

EXAMPLE 1 Arranging Data in a Matrix

In a survey of 900 people, the following information was obtained:

200 males	Thought federal defense spending was too high
150 males	Thought federal defense spending was too low
45 males	Had no opinion
315 females	Thought federal defense spending was too high
125 females	Thought federal defense spending was too low
65 females	Had no opinion

We can arrange the above data in a rectangular array as follows:

	Too High	Too Low	No Opinion
Male	200	150	45
Female	315	125	65

or as the matrix

$$\begin{bmatrix} 200 & 150 & 45 \\ 315 & 125 & 65 \end{bmatrix}$$

This matrix has two rows (representing males and females) and three columns (representing "too high," "too low," and "no opinion").

The matrix we developed in Example 1 has 2 rows and 3 columns. In general, a matrix with m rows and n columns is called an $\boldsymbol{m}$ **by** $\boldsymbol{n}$ **matrix.** Thus, the matrix we developed in Example 1 is a 2 by 3 matrix and contains $2 \cdot 3 = 6$ entries. An m by n matrix will contain $m \cdot n$ entries.

If an m by n matrix has the same number of rows as columns, that is, if $m = n$, then the matrix is referred to as a **square matrix.**

EXAMPLE 2 Examples of Matrices

(a) $\begin{bmatrix} 5 & 0 \\ -6 & 1 \end{bmatrix}$ *A 2 by 2 square matrix*

(b) $\begin{bmatrix} 1 & 0 & 3 \end{bmatrix}$ *A 1 by 3 matrix*

(c) $\begin{bmatrix} 6 & -2 & 4 \\ 4 & 3 & 5 \\ 8 & 0 & 1 \end{bmatrix}$ *A 3 by 3 square matrix*

Equality and Addition of Matrices

1 We begin our discussion of matrix algebra by first defining what is meant by two matrices being equal and then defining the operations of addition and subtraction. It is important to note that these definitions require each matrix to have the same number of rows *and* the same number of columns as a prerequisite for equality and for addition and subtraction.

We usually represent matrices by capital letters, such as A, B, C, and so on.

Two m by n matrices A and B are said to be **equal,** written as

$$A = B$$

provided that each entry a_{ij} in A is equal to the corresponding entry b_{ij} in B.

For example,

$$\begin{bmatrix} 2 & 1 \\ 0.5 & -1 \end{bmatrix} = \begin{bmatrix} \sqrt{4} & 1 \\ \frac{1}{2} & -1 \end{bmatrix} \text{ and } \begin{bmatrix} 3 & 2 & 1 \\ 0 & 1 & -2 \end{bmatrix} = \begin{bmatrix} \sqrt{9} & \sqrt{4} & 1 \\ 0 & 1 & \sqrt[3]{-8} \end{bmatrix}$$

$$\begin{bmatrix} 4 & 1 \\ 6 & 1 \end{bmatrix} \neq \begin{bmatrix} 4 & 0 \\ 6 & 1 \end{bmatrix}$$ *Because the entries in row 1, column 2 are not equal*

$$\begin{bmatrix} 4 & 1 & 2 \\ 6 & 1 & 2 \end{bmatrix} \neq \begin{bmatrix} 4 & 1 & 2 & 3 \\ 6 & 1 & 2 & 4 \end{bmatrix}$$ *Because the matrix on the left is 2 by 3 and the matrix on the right is 2 by 4*

Suppose that A and B represent two m by n matrices. We define their **sum** $\boldsymbol{A + B}$ to be the m by n matrix formed by adding the corresponding entries a_{ij} of A and b_{ij} of B. The **difference** $\boldsymbol{A - B}$ is defined as the m by n matrix formed by subtracting the entries b_{ij} in B from the corresponding entries a_{ij} in A. Addition and subtraction of matrices are allowed only for matrices having the same number m of rows and the same number n of columns. Thus, for example, a 2 by 3 matrix and a 2 by 4 matrix cannot be added or subtracted.

Graphing utilities make the sometimes tedious process of matrix algebra easy. Let's compare how a graphing utility adds and subtracts matrices with doing it by hand.

EXAMPLE 3 Adding and Subtracting Matrices

Suppose that

$$A = \begin{bmatrix} 2 & 4 & 8 & -3 \\ 0 & 1 & 2 & 3 \end{bmatrix} \quad \text{and} \quad B = \begin{bmatrix} -3 & 4 & 0 & 1 \\ 6 & 8 & 2 & 0 \end{bmatrix}$$

Find (a) $A + B$ (b) $A - B$

Graphing Solution

Enter the matrices into a graphing utility. Name them $[A]$ and $[B]$. Figure 11 shows the results of adding and subtracting $[A]$ and $[B]$.

Figure 11

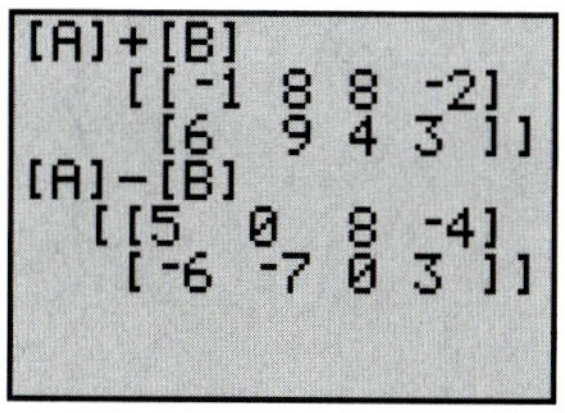

Algebraic Solution (a)

$$A + B = \begin{bmatrix} 2 & 4 & 8 & -3 \\ 0 & 1 & 2 & 3 \end{bmatrix} + \begin{bmatrix} -3 & 4 & 0 & 1 \\ 6 & 8 & 2 & 0 \end{bmatrix}$$

$$= \begin{bmatrix} 2 + (-3) & 4 + 4 & 8 + 0 & -3 + 1 \\ 0 + 6 & 1 + 8 & 2 + 2 & 3 + 0 \end{bmatrix} \quad \text{Add corresponding entries.}$$

$$= \begin{bmatrix} -1 & 8 & 8 & -2 \\ 6 & 9 & 4 & 3 \end{bmatrix}$$

(b)

$$A - B = \begin{bmatrix} 2 & 4 & 8 & -3 \\ 0 & 1 & 2 & 3 \end{bmatrix} - \begin{bmatrix} -3 & 4 & 0 & 1 \\ 6 & 8 & 2 & 0 \end{bmatrix}$$

$$= \begin{bmatrix} 2 - (-3) & 4 - 4 & 8 - 0 & -3 - 1 \\ 0 - 6 & 1 - 8 & 2 - 2 & 3 - 0 \end{bmatrix} \quad \text{Subtract corresponding entries.}$$

$$= \begin{bmatrix} 5 & 0 & 8 & -4 \\ -6 & -7 & 0 & 3 \end{bmatrix}$$

NOW WORK PROBLEM 1.

2 Many of the algebraic properties of sums of real numbers are also true for sums of matrices. Suppose that A, B, and C are m by n matrices. Then matrix addition is **commutative.** That is,

Commutative Property

$$A + B = B + A$$

Matrix addition is also **associative.** That is,

Associative Property

$$(A + B) + C = A + (B + C)$$

Although we shall not prove these results, the proofs, as the following example illustrates, are based on the commutative and associative properties for real numbers.

EXAMPLE 4 Demonstrating the Commutative Property

$$\begin{bmatrix} 2 & 3 & -1 \\ 4 & 0 & 7 \end{bmatrix} + \begin{bmatrix} -1 & 2 & 1 \\ 5 & -3 & 4 \end{bmatrix} = \begin{bmatrix} 2 + (-1) & 3 + 2 & -1 + 1 \\ 4 + 5 & 0 + (-3) & 7 + 4 \end{bmatrix}$$

$$= \begin{bmatrix} -1 + 2 & 2 + 3 & 1 + (-1) \\ 5 + 4 & -3 + 0 & 4 + 7 \end{bmatrix}$$

$$= \begin{bmatrix} -1 & 2 & 1 \\ 5 & -3 & 4 \end{bmatrix} + \begin{bmatrix} 2 & 3 & -1 \\ 4 & 0 & 7 \end{bmatrix}$$

A matrix whose entries are all equal to 0 is called a **zero matrix.** Each of the following matrices is a zero matrix.

$$\begin{bmatrix} 0 & 0 \\ 0 & 0 \end{bmatrix} \quad \textit{2 by 2 square zero matrix} \qquad \begin{bmatrix} 0 & 0 & 0 \\ 0 & 0 & 0 \end{bmatrix} \quad \textit{2 by 3 zero matrix} \qquad \begin{bmatrix} 0 & 0 & 0 \end{bmatrix} \quad \textit{1 by 3 zero matrix}$$

Zero matrices have properties similar to the real number 0. Thus, if A is an m by n matrix and 0 is an m by n zero matrix, then

$$A + 0 = A$$

In other words, the zero matrix is the additive identity in matrix algebra.

We also can multiply a matrix by a real number. If k is a real number and A is an m by n matrix, the matrix kA is the m by n matrix formed by multiplying each entry a_{ij} in A by k. The number k is sometimes referred to as a **scalar,** and the matrix kA is called a **scalar multiple** of A.

EXAMPLE 5 Operations Using Matrices

Suppose that

$$A = \begin{bmatrix} 3 & 1 & 5 \\ -2 & 0 & 6 \end{bmatrix} \qquad B = \begin{bmatrix} 4 & 1 & 0 \\ 8 & 1 & -3 \end{bmatrix} \qquad C = \begin{bmatrix} 9 & 0 \\ -3 & 6 \end{bmatrix}$$

Find (a) $4A$ (b) $\frac{1}{3}C$ (c) $3A - 2B$

Graphing Solution

Enter the matrices $[A]$, $[B]$, and $[C]$ into a graphing utility. Figure 12 shows the required computations.

Figure 12

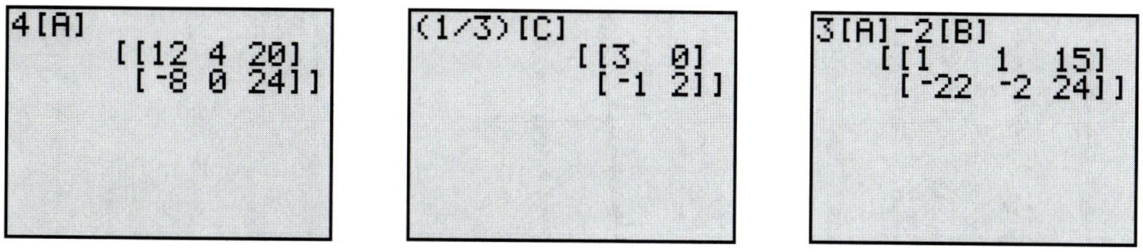

(a) (b) (c)

Algebraic Solution

(a) $$4A = 4\begin{bmatrix} 3 & 1 & 5 \\ -2 & 0 & 6 \end{bmatrix} = \begin{bmatrix} 4 \cdot 3 & 4 \cdot 1 & 4 \cdot 5 \\ 4(-2) & 4 \cdot 0 & 4 \cdot 6 \end{bmatrix} = \begin{bmatrix} 12 & 4 & 20 \\ -8 & 0 & 24 \end{bmatrix}$$

(b) $$\frac{1}{3}C = \frac{1}{3}\begin{bmatrix} 9 & 0 \\ -3 & 6 \end{bmatrix} = \begin{bmatrix} \frac{1}{3} \cdot 9 & \frac{1}{3} \cdot 0 \\ \frac{1}{3}(-3) & \frac{1}{3} \cdot 6 \end{bmatrix} = \begin{bmatrix} 3 & 0 \\ -1 & 2 \end{bmatrix}$$

(c) $3A - 2B = 3\begin{bmatrix} 3 & 1 & 5 \\ -2 & 0 & 6 \end{bmatrix} - 2\begin{bmatrix} 4 & 1 & 0 \\ 8 & 1 & -3 \end{bmatrix}$

$$= \begin{bmatrix} 3 \cdot 3 & 3 \cdot 1 & 3 \cdot 5 \\ 3(-2) & 3 \cdot 0 & 3 \cdot 6 \end{bmatrix} - \begin{bmatrix} 2 \cdot 4 & 2 \cdot 1 & 2 \cdot 0 \\ 2 \cdot 8 & 2 \cdot 1 & 2(-3) \end{bmatrix}$$

$$= \begin{bmatrix} 9 & 3 & 15 \\ -6 & 0 & 18 \end{bmatrix} - \begin{bmatrix} 8 & 2 & 0 \\ 16 & 2 & -6 \end{bmatrix}$$

$$= \begin{bmatrix} 9-8 & 3-2 & 15-0 \\ -6-16 & 0-2 & 18-(-6) \end{bmatrix}$$

$$= \begin{bmatrix} 1 & 1 & 15 \\ -22 & -2 & 24 \end{bmatrix}$$

NOW WORK PROBLEM 5.

We list next some of the algebraic properties of scalar multiplication. Let h and k be real numbers, and let A and B be m by n matrices. Then

Properties of Scalar Multiplication

$$k(hA) = (kh)A$$
$$(k + h)A = kA + hA$$
$$k(A + B) = kA + kB$$

3

Multiplication of Matrices

Unlike the straightforward definition for adding two matrices, the definition for multiplying two matrices is not what we might expect. In preparation for this definition, we need the following definitions:

A **row vector** $\boldsymbol{R}$ is a 1 by n matrix

$$R = [r_1 \quad r_2 \quad \cdots \quad r_n]$$

A **column vector** $\boldsymbol{C}$ is an n by 1 matrix

$$C = \begin{bmatrix} c_1 \\ c_2 \\ \vdots \\ c_n \end{bmatrix}$$

The **product** $\boldsymbol{RC}$ of R times C is defined as the number

$$RC = [r_1 \ r_2 \ \cdots \ r_n]\begin{bmatrix} c_1 \\ c_2 \\ \vdots \\ c_n \end{bmatrix} = r_1c_1 + r_2c_2 + \cdots + r_nc_n$$

Notice that a row vector and a column vector can be multiplied only if they contain the same number of entries.

EXAMPLE 6 The Product of a Row Vector by a Column Vector

If $R = [3 \quad -5 \quad 2]$ and $C = \begin{bmatrix} 3 \\ 4 \\ -5 \end{bmatrix}$, then

$$RC = [3 \quad -5 \quad 2]\begin{bmatrix} 3 \\ 4 \\ -5 \end{bmatrix} = 3 \cdot 3 + (-5)4 + 2(-5)$$
$$= 9 - 20 - 10 = -21$$

Let's look at an application of the product of a row vector by a column vector.

EXAMPLE 7 Using Matrices to Compute Revenue

A clothing store sells men's shirts for \$25, silk ties for \$8, and wool suits for \$300. Last month, the store had sales consisting of 100 shirts, 200 ties, and 50 suits. What was the total revenue due to these sales?

Solution

We set up a row vector R to represent the prices of each item and a column vector C to represent the corresponding number of items sold.

Then

$$\overset{\text{Prices}}{\underset{}{}} R = \begin{matrix} \text{Shirts} & \text{Ties} & \text{Suits} \\ [25 & 8 & 300] \end{matrix} \qquad C = \overset{\text{Number sold}}{\begin{bmatrix} 100 \\ 200 \\ 50 \end{bmatrix}} \begin{matrix} \text{Shirts} \\ \text{Ties} \\ \text{Suits} \end{matrix}$$

The total revenue obtained is the product RC. That is,

$$RC = [25 \ 8 \ 300]\begin{bmatrix} 100 \\ 200 \\ 50 \end{bmatrix}$$
$$= \underbrace{25 \cdot 100}_{\text{Shirt revenue}} + \underbrace{8 \cdot 200}_{\text{Tie revenue}} + \underbrace{300 \cdot 50}_{\text{Suit revenue}} = \underbrace{\$19{,}100}_{\text{Total revenue}}$$

The definition for multiplying two matrices is based on the definition of a row vector times a column vector.

Let A denote an m by r matrix, and let B denote an r by n matrix. The **product** AB is defined as the m by n matrix whose entry in row i, column j is the product of the ith row of A and the jth column of B.

The definition of the product AB of two matrices A and B, in this order, requires that the number of columns of A equal the number of rows of B; otherwise, no product is defined.

$$\underset{m \text{ by } r}{A} \qquad\qquad \underset{r \text{ by } n}{B}$$

Must be same for AB to be defined

AB is m by n.

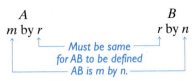

An example will help to clarify the definition.

EXAMPLE 8 Multiplying Two Matrices

Find the product AB if

$$A = \begin{bmatrix} 2 & 4 & -1 \\ 5 & 8 & 0 \end{bmatrix} \quad \text{and} \quad B = \begin{bmatrix} 2 & 5 & 1 & 4 \\ 4 & 8 & 0 & 6 \\ -3 & 1 & -2 & -1 \end{bmatrix}$$

Solution First, we note that A is 2 by 3 and B is 3 by 4, so the product AB is defined and will be a 2 by 4 matrix.

Graphing Solution Enter the matrices A and B into a graphing utility. Figure 13 shows the product AB.

Figure 13

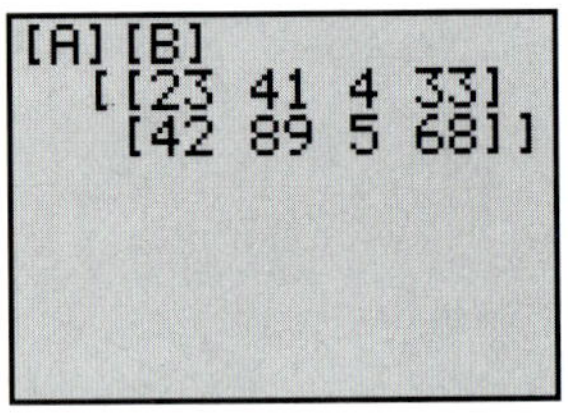

Algebraic Solution Suppose that we want the entry in row 2, column 3 of AB. To find it, we find the product of the row vector from row 2 of A and the column vector from column 3 of B.

Column 3 of B

Row 2 of A

$$[5 \quad 8 \quad 0]\begin{bmatrix} 1 \\ 0 \\ -2 \end{bmatrix} = 5 \cdot 1 + 8 \cdot 0 + 0(-2) = 5$$

So far, we have

Column 3 ↓

$$AB = \begin{bmatrix} \_\_ & \_\_ & \_\_ & \_\_ \\ \_\_ & \_\_ & 5 & \_\_ \end{bmatrix} \leftarrow \text{Row 2}$$

Now, to find the entry in row 1, column 4 of AB, we find the product of row 1 of A and column 4 of B.

Column 4 of B

Row 1 of A

$$[2 \quad 4 \quad -1]\begin{bmatrix} 4 \\ 6 \\ -1 \end{bmatrix} = 2 \cdot 4 + 4 \cdot 6 + (-1)(-1) = 33$$

Continuing in this fashion, we find AB.

$$AB = \begin{bmatrix} 2 & 4 & -1 \\ 5 & 8 & 0 \end{bmatrix} \begin{bmatrix} 2 & 5 & 1 & 4 \\ 4 & 8 & 0 & 6 \\ -3 & 1 & -2 & -1 \end{bmatrix}$$

$$= \begin{bmatrix} \text{Row 1 of } A \text{ times column 1 of } B & \text{Row 1 of } A \text{ times column 2 of } B & \text{Row 1 of } A \text{ times column 3 of } B & \text{Row 1 of } A \text{ times column 4 of } B \\ \text{Row 2 of } A \text{ times column 1 of } B & \text{Row 2 of } A \text{ times column 2 of } B & \text{Row 2 of } A \text{ times column 3 of } B & \text{Row 2 of } A \text{ times column 4 of } B \end{bmatrix}$$

$$= \begin{bmatrix} 2 \cdot 2 + 4 \cdot 4 + (-1)(-3) & 2 \cdot 5 + 4 \cdot 8 + (-1)1 & 2 \cdot 1 + 4 \cdot 0 + (-1)(-2) & 33 \text{ (from earlier)} \\ 5 \cdot 2 + 8 \cdot 4 + 0(-3) & 5 \cdot 5 + 8 \cdot 8 + 0 \cdot 1 & 5 \text{ (from earlier)} & 5 \cdot 4 + 8 \cdot 6 + 0(-1) \end{bmatrix}$$

$$= \begin{bmatrix} 23 & 41 & 4 & 33 \\ 42 & 89 & 5 & 68 \end{bmatrix}$$

NOW WORK PROBLEM 17.

Notice that for the matrices given in Example 8 the product BA is not defined, because B is 3 by 4 and A is 2 by 3. Try calculating BA on a graphing utility. What do you notice?

Another result that can occur when multiplying two matrices is illustrated in the next example.*

EXAMPLE 9 Multiplying Two Matrices

If

$$A = \begin{bmatrix} 2 & 1 & 3 \\ 1 & -1 & 0 \end{bmatrix} \quad \text{and} \quad B = \begin{bmatrix} 1 & 0 \\ 2 & 1 \\ 3 & 2 \end{bmatrix}$$

find (a) AB (b) BA

Solution

(a) $$AB = \underset{2 \text{ by } 3}{\begin{bmatrix} 2 & 1 & 3 \\ 1 & -1 & 0 \end{bmatrix}} \underset{3 \text{ by } 2}{\begin{bmatrix} 1 & 0 \\ 2 & 1 \\ 3 & 2 \end{bmatrix}} = \underset{2 \text{ by } 2}{\begin{bmatrix} 13 & 7 \\ -1 & -1 \end{bmatrix}}$$

(b) $$BA = \underset{3 \text{ by } 2}{\begin{bmatrix} 1 & 0 \\ 2 & 1 \\ 3 & 2 \end{bmatrix}} \underset{2 \text{ by } 3}{\begin{bmatrix} 2 & 1 & 3 \\ 1 & -1 & 0 \end{bmatrix}} = \underset{3 \text{ by } 3}{\begin{bmatrix} 2 & 1 & 3 \\ 5 & 1 & 6 \\ 8 & 1 & 9 \end{bmatrix}}$$

Notice in Example 9 that AB is 2 by 2 and BA is 3 by 3. Thus, it is possible for both AB and BA to be defined, yet be unequal. In fact, even if A and B are both n by n matrices, so that AB and BA are each defined and n by n, AB and BA will usually be unequal.

*For most of the examples that follow, we will multiply matrices by hand. You should verify each result using a graphing utility.

EXAMPLE 10 Multiplying Two Square Matrices

If

$$A = \begin{bmatrix} 2 & 1 \\ 0 & 4 \end{bmatrix} \quad \text{and} \quad B = \begin{bmatrix} -3 & 1 \\ 1 & 2 \end{bmatrix}$$

find (a) AB (b) BA

(a) $$AB = \begin{bmatrix} 2 & 1 \\ 0 & 4 \end{bmatrix}\begin{bmatrix} -3 & 1 \\ 1 & 2 \end{bmatrix}$$
$$= \begin{bmatrix} 2(-3) + 1 \cdot 1 & 2 \cdot 1 + 1 \cdot 2 \\ 0(-3) + 4 \cdot 1 & 0 \cdot 1 + 4 \cdot 2 \end{bmatrix} = \begin{bmatrix} -5 & 4 \\ 4 & 8 \end{bmatrix}$$

(b) $$BA = \begin{bmatrix} -3 & 1 \\ 1 & 2 \end{bmatrix}\begin{bmatrix} 2 & 1 \\ 0 & 4 \end{bmatrix}$$
$$= \begin{bmatrix} (-3)2 + 1 \cdot 0 & (-3)1 + 1 \cdot 4 \\ 1 \cdot 2 + 2 \cdot 0 & 1 \cdot 1 + 2 \cdot 4 \end{bmatrix} = \begin{bmatrix} -6 & 1 \\ 2 & 9 \end{bmatrix}$$

The preceding examples demonstrate that an important property of real numbers, the commutative property of multiplication, is not shared by matrices. Thus in general:

THEOREM Matrix multiplication is not commutative.

NOW WORK PROBLEMS 7 AND 9.

Next we give two of the properties of real numbers that are shared by matrices. Assuming that each product and sum is defined, we have the following:

Associative Property

$$A(BC) = (AB)C$$

Distributive Property

$$A(B + C) = AB + AC$$

The Identity Matrix

For an n by n square matrix, the entries located in row i, column i, $1 \le i \le n$, are called the **diagonal entries.** An n by n square matrix whose diagonal entries are 1's, while all other entries are 0's, is called the **identity matrix I_n.** For example,

$$I_2 = \begin{bmatrix} 1 & 0 \\ 0 & 1 \end{bmatrix} \qquad I_3 = \begin{bmatrix} 1 & 0 & 0 \\ 0 & 1 & 0 \\ 0 & 0 & 1 \end{bmatrix}$$

and so on.

EXAMPLE 11 Multiplication with an Identity Matrix

Let

$$A = \begin{bmatrix} -1 & 2 & 0 \\ 0 & 1 & 3 \end{bmatrix} \quad \text{and} \quad B = \begin{bmatrix} 3 & 2 \\ 4 & 6 \\ 5 & 2 \end{bmatrix}$$

Find (a) AI_3 (b) I_2A (c) BI_2

Solution (a) $AI_3 = \begin{bmatrix} -1 & 2 & 0 \\ 0 & 1 & 3 \end{bmatrix}\begin{bmatrix} 1 & 0 & 0 \\ 0 & 1 & 0 \\ 0 & 0 & 1 \end{bmatrix} = \begin{bmatrix} -1 & 2 & 0 \\ 0 & 1 & 3 \end{bmatrix} = A$

(b) $I_2A = \begin{bmatrix} 1 & 0 \\ 0 & 1 \end{bmatrix}\begin{bmatrix} -1 & 2 & 0 \\ 0 & 1 & 3 \end{bmatrix} = \begin{bmatrix} -1 & 2 & 0 \\ 0 & 1 & 3 \end{bmatrix} = A$

(c) $BI_2 = \begin{bmatrix} 3 & 2 \\ 4 & 6 \\ 5 & 2 \end{bmatrix}\begin{bmatrix} 1 & 0 \\ 0 & 1 \end{bmatrix} = \begin{bmatrix} 3 & 2 \\ 4 & 6 \\ 5 & 2 \end{bmatrix} = B$

Example 11 demonstrates the following property:

Identity Property

If A is an m by n matrix, then

$$I_mA = A \quad \text{and} \quad AI_n = A$$

If A is an n by n square matrix, then $AI_n = I_nA = A$.

Thus, an identity matrix has properties analogous to those of the real number 1. In other words, the identity matrix is a multiplicative identity in matrix algebra.

4 The Inverse of a Matrix

Let A be a square n by n matrix. If there exists an n by n matrix A^{-1}, read "A inverse," for which

$$AA^{-1} = A^{-1}A = I_n$$

then A^{-1} is called the **inverse** of the matrix A.

As we shall soon see, not every square matrix has an inverse. When a matrix A does have an inverse A^{-1}, then A is said to be **nonsingular.** If a matrix A has no inverse, it is called **singular.***

*If the determinant of A is zero, then A is singular. (Refer to Section 6.4.)

EXAMPLE 12 Multiplying a Matrix by Its Inverse

Show that the inverse of

$$A = \begin{bmatrix} 3 & 1 \\ 2 & 1 \end{bmatrix} \quad \text{is} \quad A^{-1} = \begin{bmatrix} 1 & -1 \\ -2 & 3 \end{bmatrix}$$

Solution We need to show that $AA^{-1} = A^{-1}A = I_2$

$$AA^{-1} = \begin{bmatrix} 3 & 1 \\ 2 & 1 \end{bmatrix}\begin{bmatrix} 1 & -1 \\ -2 & 3 \end{bmatrix} = \begin{bmatrix} 3 \cdot 1 + 1(-2) & 3(-1) + 1 \cdot 3 \\ 2 \cdot 1 + 1(-2) & 2(-1) + 1 \cdot 3 \end{bmatrix}$$

$$= \begin{bmatrix} 1 & 0 \\ 0 & 1 \end{bmatrix} = I_2$$

$$A^{-1}A = \begin{bmatrix} 1 & -1 \\ -2 & 3 \end{bmatrix}\begin{bmatrix} 3 & 1 \\ 2 & 1 \end{bmatrix} = \begin{bmatrix} 3 - 2 & 1 - 1 \\ -6 + 6 & -2 + 3 \end{bmatrix} = \begin{bmatrix} 1 & 0 \\ 0 & 1 \end{bmatrix} = I_2$$

We now show one way to find the inverse of

$$A = \begin{bmatrix} 3 & 1 \\ 2 & 1 \end{bmatrix}$$

Suppose that A^{-1} is given by

$$A^{-1} = \begin{bmatrix} x & y \\ z & w \end{bmatrix} \tag{1}$$

where x, y, z, and w are four variables. Based on the definition of an inverse, if, indeed, A has an inverse, we have

$$AA^{-1} = I_2$$

$$\begin{bmatrix} 3 & 1 \\ 2 & 1 \end{bmatrix}\begin{bmatrix} x & y \\ z & w \end{bmatrix} = \begin{bmatrix} 1 & 0 \\ 0 & 1 \end{bmatrix}$$

$$\begin{bmatrix} 3x + z & 3y + w \\ 2x + z & 2y + w \end{bmatrix} = \begin{bmatrix} 1 & 0 \\ 0 & 1 \end{bmatrix}$$

Because corresponding entries must be equal, it follows that this matrix equation is equivalent to four ordinary equations.

$$\begin{cases} 3x + z = 1 \\ 2x + z = 0 \end{cases} \qquad \begin{cases} 3y + w = 0 \\ 2y + w = 1 \end{cases}$$

The augmented matrix of each system is

$$\left[\begin{array}{cc|c} 3 & 1 & 1 \\ 2 & 1 & 0 \end{array}\right] \qquad \left[\begin{array}{cc|c} 3 & 1 & 0 \\ 2 & 1 & 1 \end{array}\right] \tag{2}$$

The usual procedure would be to transform each augmented matrix into reduced echelon form. Notice, though, that the left sides of the augmented matrices are equal, so the same row operations (see Section 6.3) can be used to reduce each one. Thus, we find it more efficient to combine the two augmented matrices (2) into a single matrix, as shown next, and then transform it into reduced row echelon form.

$$\left[\begin{array}{cc|cc} 3 & 1 & 1 & 0 \\ 2 & 1 & 0 & 1 \end{array}\right]$$

Now we attempt to transform the left side into an identity matrix.

$$\left[\begin{array}{cc|cc} 3 & 1 & 1 & 0 \\ 2 & 1 & 0 & 1 \end{array}\right] \underset{\substack{\uparrow \\ R_1 = -1r_2 + r_1}}{\rightarrow} \left[\begin{array}{cc|cc} 1 & 0 & 1 & -1 \\ 2 & 1 & 0 & 1 \end{array}\right]$$

$$\underset{\substack{\uparrow \\ R_2 = -2r_1 + r_2}}{\rightarrow} \left[\begin{array}{cc|cc} 1 & 0 & 1 & -1 \\ 0 & 1 & -2 & 3 \end{array}\right] \tag{3}$$

Matrix (3) is in reduced row echelon form. Now we reverse the earlier step of combining the two augmented matrices in (2) and write the single matrix (3) as two augmented matrices.

$$\left[\begin{array}{cc|c} 1 & 0 & 1 \\ 0 & 1 & -2 \end{array}\right] \text{ and } \left[\begin{array}{cc|c} 1 & 0 & -1 \\ 0 & 1 & 3 \end{array}\right]$$

We conclude from these matrices that $x = 1$, $z = -2$, and $y = -1$, $w = 3$. Substituting these values into matrix (1), we find that

$$A^{-1} = \begin{bmatrix} 1 & -1 \\ -2 & 3 \end{bmatrix}$$

Notice in display (3) that the 2 by 2 matrix to the right of the vertical bar is, in fact, the inverse of A. Also notice that the identity matrix I_2 is the matrix that appears to the left of the vertical bar. These observations and the procedures followed above will work in general.

Procedure for Finding the Inverse of a Nonsingular Matrix

To find the inverse of an n by n nonsingular matrix A, proceed as follows:

STEP 1: Form the matrix $[A|I_n]$.
STEP 2: Transform the matrix $[A|I_n]$ into reduced row echelon form.
STEP 3: The reduced row echelon form of $[A|I_n]$ will contain the identity matrix I_n on the left of the vertical bar; the n by n matrix on the right of the vertical bar is the inverse of A.

In other words, if A is nonsingular, we begin with the matrix $[A|I_n]$ and, after transforming it into reduced row echelon form, we end up with the matrix $[I_n|A^{-1}]$.

Let's look at another example.

EXAMPLE 13 Finding the Inverse of a Matrix

The matrix

$$A = \begin{bmatrix} 1 & 1 & 0 \\ -1 & 3 & 4 \\ 0 & 4 & 3 \end{bmatrix}$$

is nonsingular. Find its inverse.

Graphing Solution Enter the matrix A into a graphing utility. Figure 14 shows A^{-1}.

Figure 14

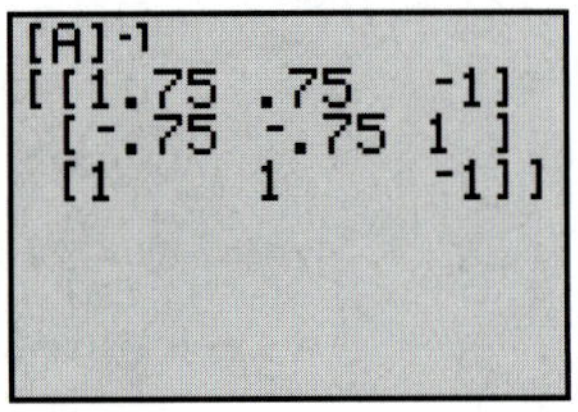

Algebraic Solution First, we form the matrix

$$[A|I_3] = \left[\begin{array}{rrr|rrr} 1 & 1 & 0 & 1 & 0 & 0 \\ -1 & 3 & 4 & 0 & 1 & 0 \\ 0 & 4 & 3 & 0 & 0 & 1 \end{array}\right]$$

Next, we use row operations to transform $[A|I_3]$ into reduced row echelon form.

$$\left[\begin{array}{rrr|rrr} 1 & 1 & 0 & 1 & 0 & 0 \\ -1 & 3 & 4 & 0 & 1 & 0 \\ 0 & 4 & 3 & 0 & 0 & 1 \end{array}\right] \underset{R_2 = r_1 + r_2}{\rightarrow} \left[\begin{array}{rrr|rrr} 1 & 1 & 0 & 1 & 0 & 0 \\ 0 & 4 & 4 & 1 & 1 & 0 \\ 0 & 4 & 3 & 0 & 0 & 1 \end{array}\right] \underset{R_2 = \frac{1}{4}r_2}{\rightarrow} \left[\begin{array}{rrr|rrr} 1 & 1 & 0 & 1 & 0 & 0 \\ 0 & 1 & 1 & \frac{1}{4} & \frac{1}{4} & 0 \\ 0 & 4 & 3 & 0 & 0 & 1 \end{array}\right]$$

$$\underset{\substack{R_1 = -1r_2 + r_1 \\ R_3 = -4r_2 + r_3}}{\rightarrow} \left[\begin{array}{rrr|rrr} 1 & 0 & -1 & \frac{3}{4} & -\frac{1}{4} & 0 \\ 0 & 1 & 1 & \frac{1}{4} & \frac{1}{4} & 0 \\ 0 & 0 & -1 & -1 & -1 & 1 \end{array}\right] \underset{R_3 = -1r_3}{\rightarrow} \left[\begin{array}{rrr|rrr} 1 & 0 & -1 & \frac{3}{4} & -\frac{1}{4} & 0 \\ 0 & 1 & 1 & \frac{1}{4} & \frac{1}{4} & 0 \\ 0 & 0 & 1 & 1 & 1 & -1 \end{array}\right]$$

$$\underset{\substack{R_1 = r_3 + r_1 \\ R_2 = -1r_3 + r_2}}{\rightarrow} \left[\begin{array}{rrr|rrr} 1 & 0 & 0 & \frac{7}{4} & \frac{3}{4} & -1 \\ 0 & 1 & 0 & -\frac{3}{4} & -\frac{3}{4} & 1 \\ 0 & 0 & 1 & 1 & 1 & -1 \end{array}\right]$$

The matrix $[A|I_3]$ is now in reduced row echelon form, and the identity matrix I_3 is on the left of the vertical bar. Hence, the inverse of A is

$$A^{-1} = \begin{bmatrix} \frac{7}{4} & \frac{3}{4} & -1 \\ -\frac{3}{4} & -\frac{3}{4} & 1 \\ 1 & 1 & -1 \end{bmatrix}$$

You can (and should) verify that this is the correct inverse by showing that $AA^{-1} = A^{-1}A = I_3$.

NOW WORK PROBLEM 23.

If transforming the matrix $[A|I_n]$ into reduced row echelon form does not result in the identity matrix I_n to the left of the vertical bar, then A has no inverse. The next example demonstrates such a matrix.

EXAMPLE 14 Showing That a Matrix Has No Inverse

Show that the following matrix has no inverse.

$$A = \begin{bmatrix} 4 & 6 \\ 2 & 3 \end{bmatrix}$$

Graphing Solution Enter the matrix A. Figure 15 shows the result when we try to find its inverse. The ERRor comes about because A is singular.

Figure 15

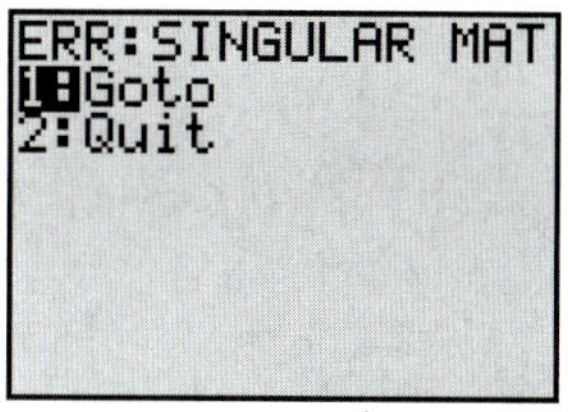

Algebraic Solution Proceeding as in Example 13, we form the matrix

$$[A|I_2] = \left[\begin{array}{cc|cc} 4 & 6 & 1 & 0 \\ 2 & 3 & 0 & 1 \end{array}\right]$$

Then we use row operations to transform $[A|I_2]$ into reduced row echelon form.

$$[A|I_2] = \left[\begin{array}{cc|cc} 4 & 6 & 1 & 0 \\ 2 & 3 & 0 & 1 \end{array}\right] \underset{R_1 = \frac{1}{4}r_1}{\rightarrow} \left[\begin{array}{cc|cc} 1 & \frac{3}{2} & \frac{1}{4} & 0 \\ 2 & 3 & 0 & 1 \end{array}\right] \underset{R_2 = -2r_1 + r_2}{\rightarrow} \left[\begin{array}{cc|cc} 1 & \frac{3}{2} & \frac{1}{4} & 0 \\ 0 & 0 & -\frac{1}{2} & 1 \end{array}\right]$$

The matrix $[A|I_2]$ is sufficiently reduced for us to see that the identity matrix cannot appear to the left of the vertical bar. We conclude that A has no inverse.

SEEING THE CONCEPT Compute the determinant of A in Example 14 using a graphing utility. What is the result? Are you surprised?

NOW WORK PROBLEM 51.

Solving Systems of Linear Equations

5 Inverse matrices can be used to solve systems of equations in which the number of equations is the same as the number of variables.

EXAMPLE 15 Using the Inverse Matrix to Solve a System of Linear Equations

Solve the system of equations: $\begin{cases} x + y = 3 \\ -x + 3y + 4z = -3 \\ 4y + 3z = 2 \end{cases}$

Solution If we let

$$A = \begin{bmatrix} 1 & 1 & 0 \\ -1 & 3 & 4 \\ 0 & 4 & 3 \end{bmatrix} \quad X = \begin{bmatrix} x \\ y \\ z \end{bmatrix} \quad B = \begin{bmatrix} 3 \\ -3 \\ 2 \end{bmatrix}$$

then the original system of equations can be written compactly as the matrix equation

$$AX = B \tag{4}$$

We know from Example 13 that the matrix A has the inverse A^{-1}, so we multiply each side of equation (4) by A^{-1}.

$$\begin{aligned} AX &= B \\ A^{-1}(AX) &= A^{-1}B \\ (A^{-1}A)X &= A^{-1}B && \text{Associative property of multiplication} \\ I_3X &= A^{-1}B && \text{Definition of inverse matrix} \\ X &= A^{-1}B && \text{Property of identity matrix} \end{aligned} \tag{5}$$

Now we use (5) to find $X = \begin{bmatrix} x \\ y \\ z \end{bmatrix}$.

Graphing Solution Enter the matrices A and B into a graphing utility. Figure 16 shows the solution to the system of equations.

Figure 16

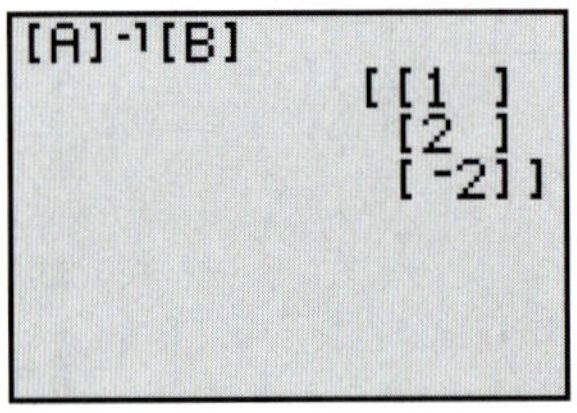

Algebraic Solution

$$X = \begin{bmatrix} x \\ y \\ z \end{bmatrix} = A^{-1}B \underset{\underset{\text{Example 13}}{\uparrow}}{=} \begin{bmatrix} \frac{7}{4} & \frac{3}{4} & -1 \\ -\frac{3}{4} & -\frac{3}{4} & 1 \\ 1 & 1 & -1 \end{bmatrix} \begin{bmatrix} 3 \\ -3 \\ 2 \end{bmatrix} = \begin{bmatrix} 1 \\ 2 \\ -2 \end{bmatrix}$$

Thus, $x = 1$, $y = 2$, $z = -2$.

The method used in Example 15 to solve a system of equations is particularly useful when it is necessary to solve several systems of equations in which the constants appearing to the right of the equal signs change, while the coefficients of the variables on the left side remain the same. See

Problems 31–50 for some illustrations. Be careful; this method can only be used if the inverse exists. If it does not exist, row reduction must be used since the system is either inconsistent or dependent.

HISTORICAL FEATURE

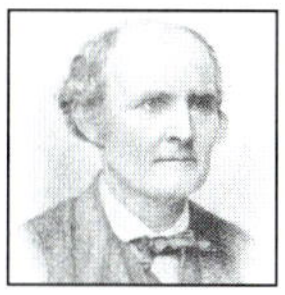

Matrices were invented in 1857 by Arthur Cayley (1821–1895) as a way of efficiently computing the result of substituting one linear system into another (see Historical Problem 2). The resulting system had incredible richness, in the sense that a very wide variety of mathematical systems could be mimicked by the matrices. Cayley and his friend James J. Sylvester (1814–1897) spent much of the rest of their lives elaborating the theory. The torch was then passed to Georg Frobenius (1849–1917), whose deep investigations established a central place for matrices in modern mathematics. In 1924, rather to the surprise of physicists, it was found that matrices (with complex numbers in them) were exactly the right tool for describing the behavior of atomic systems. Today, matrices are used in a wide variety of applications.

HISTORICAL PROBLEMS

1. *Matrices and Complex Numbers* Frobenius emphasized in his research how matrices could be used to mimic other mathematical systems. Here, we mimic the behavior of complex numbers using matrices. Mathematicians call such a relationship an *isomorphism.*

   Complex number $\longleftrightarrow$ Matrix

   $$a + bi \longleftrightarrow \begin{bmatrix} a & b \\ -b & a \end{bmatrix}$$

   Note that the complex number can be read off the top line of the matrix. Thus,

   $$2 + 3i \longleftrightarrow \begin{bmatrix} 2 & 3 \\ -3 & 2 \end{bmatrix} \text{ and } \begin{bmatrix} 4 & -2 \\ 2 & 4 \end{bmatrix} \longleftrightarrow 4 - 2i$$

   (a) Find the matrices corresponding to $2 - 5i$ and $1 + 3i$.
   (b) Multiply the two matrices.
   (c) Find the corresponding complex number for the matrix found in part (b).
   (d) Multiply $2 - 5i$ by $1 + 3i$. The result should be the same as that found in part (c).

   The process also works for addition and subtraction. Try it for yourself.

2. *Cayley's Definition of Matrix Multiplication* Cayley invented matrix multiplication to simplify the following problem:

   $$\begin{cases} u = ar + bs \\ v = cr + ds \end{cases} \qquad \begin{cases} x = ku + lv \\ y = mu + nv \end{cases}$$

(a) Find x and y in terms of r and s by substituting u and v from the first system of equations into the second system of equations.

(b) Use the result of part (a) to find the 2 by 2 matrix A in

$$\begin{bmatrix} x \\ y \end{bmatrix} = A\begin{bmatrix} r \\ s \end{bmatrix}$$

(c) Now look at the following way to do it. Write the equations in matrix form.

$$\begin{bmatrix} u \\ v \end{bmatrix} = \begin{bmatrix} a & b \\ c & d \end{bmatrix}\begin{bmatrix} r \\ s \end{bmatrix} \qquad \begin{bmatrix} x \\ y \end{bmatrix} = \begin{bmatrix} k & l \\ m & n \end{bmatrix}\begin{bmatrix} u \\ v \end{bmatrix}$$

So

$$\begin{bmatrix} x \\ y \end{bmatrix} = \begin{bmatrix} k & l \\ m & n \end{bmatrix}\begin{bmatrix} a & b \\ c & d \end{bmatrix}\begin{bmatrix} r \\ s \end{bmatrix}$$

Do you see how Cayley defined matrix multiplication?

6.5 EXERCISES

In Problems 1–16, use the following matrices to compute the given expression (a) by hand and (b) by using a graphing utility.

$$A = \begin{bmatrix} 0 & 3 & -5 \\ 1 & 2 & 6 \end{bmatrix} \quad B = \begin{bmatrix} 4 & 1 & 0 \\ -2 & 3 & -2 \end{bmatrix} \quad C = \begin{bmatrix} 4 & 1 \\ 6 & 2 \\ -2 & 3 \end{bmatrix}$$

1. $A + B$ **2.** $A - B$ **3.** $4A$ **4.** $-3B$

5. $3A - 2B$ **6.** $2A + 4B$ **7.** AC **8.** BC

9. CA **10.** CB **11.** $C(A + B)$ **12.** $(A + B)C$

13. $AC - 3I_2$ **14.** $CA + 5I_3$ **15.** $CA - CB$ **16.** $AC + BC$

In Problems 17–20, compute each product (a) by hand and (b) by using a graphing utility.

17. $\begin{bmatrix} 2 & -2 \\ 1 & 0 \end{bmatrix}\begin{bmatrix} 2 & 1 & 4 & 6 \\ 3 & -1 & 3 & 2 \end{bmatrix}$

18. $\begin{bmatrix} 4 & 1 \\ 2 & 1 \end{bmatrix}\begin{bmatrix} -6 & 6 & 1 & 0 \\ 2 & 5 & 4 & -1 \end{bmatrix}$

19. $\begin{bmatrix} 1 & 0 & 1 \\ 2 & 4 & 1 \\ 3 & 6 & 1 \end{bmatrix}\begin{bmatrix} 1 & 3 \\ 6 & 2 \\ 8 & -1 \end{bmatrix}$

20. $\begin{bmatrix} 4 & -2 & 3 \\ 0 & 1 & 2 \\ -1 & 0 & 1 \end{bmatrix}\begin{bmatrix} 2 & 6 \\ 1 & -1 \\ 0 & 2 \end{bmatrix}$

In Problems 21–30, each matrix is nonsingular. Find the inverse of each matrix. Be sure to check your answer using a graphing utility (when possible).

21. $\begin{bmatrix} 2 & 1 \\ 1 & 1 \end{bmatrix}$

22. $\begin{bmatrix} 3 & -1 \\ -2 & 1 \end{bmatrix}$

23. $\begin{bmatrix} 6 & 5 \\ 2 & 2 \end{bmatrix}$

24. $\begin{bmatrix} -4 & 1 \\ 6 & -2 \end{bmatrix}$

25. $\begin{bmatrix} 2 & 1 \\ a & a \end{bmatrix}$, $a \neq 0$

26. $\begin{bmatrix} b & 3 \\ b & 2 \end{bmatrix}$, $b \neq 0$

27. $\begin{bmatrix} 1 & -1 & 1 \\ 0 & -2 & 1 \\ -2 & -3 & 0 \end{bmatrix}$

28. $\begin{bmatrix} 1 & 0 & 2 \\ -1 & 2 & 3 \\ 1 & -1 & 0 \end{bmatrix}$

29. $\begin{bmatrix} 1 & 1 & 1 \\ 3 & 2 & -1 \\ 3 & 1 & 2 \end{bmatrix}$

30. $\begin{bmatrix} 3 & 3 & 1 \\ 1 & 2 & 1 \\ 2 & -1 & 1 \end{bmatrix}$

In Problems 31–50, use the inverses found in Problems 21–30 to solve each system of equations by hand.

31. $\begin{cases} 2x + y = 8 \\ x + y = 5 \end{cases}$

32. $\begin{cases} 3x - y = 8 \\ -2x + y = 4 \end{cases}$

33. $\begin{cases} 2x + y = 0 \\ x + y = 5 \end{cases}$

34. $\begin{cases} 3x - y = 4 \\ -2x + y = 5 \end{cases}$

35. $\begin{cases} 6x + 5y = 7 \\ 2x + 2y = 2 \end{cases}$

36. $\begin{cases} -4x + y = 0 \\ 6x - 2y = 14 \end{cases}$

37. $\begin{cases} 6x + 5y = 13 \\ 2x + 2y = 5 \end{cases}$

38. $\begin{cases} -4x + y = 5 \\ 6x - 2y = -9 \end{cases}$

39. $\begin{cases} 2x + y = -3 \\ ax + ay = -a \end{cases}, \quad a \neq 0$

40. $\begin{cases} bx + 3y = 2b + 3 \\ bx + 2y = 2b + 2 \end{cases}, \quad b \neq 0$

41. $\begin{cases} 2x + y = 7/a \\ ax + ay = 5 \end{cases}, \quad a \neq 0$

42. $\begin{cases} bx + 3y = 14 \\ bx + 2y = 10 \end{cases}, \quad b \neq 0$

43. $\begin{cases} x - y + z = 0 \\ -2y + z = -1 \\ -2x - 3y = -5 \end{cases}$

44. $\begin{cases} x + 2z = 6 \\ -x + 2y + 3z = -5 \\ x - y = 6 \end{cases}$

45. $\begin{cases} x - y + z = 2 \\ -2y + z = 2 \\ -2x - 3y = \frac{1}{2} \end{cases}$

46. $\begin{cases} x + 2z = 2 \\ -x + 2y + 3z = -\frac{3}{2} \\ x - y = 2 \end{cases}$

47. $\begin{cases} x + y + z = 9 \\ 3x + 2y - z = 8 \\ 3x + y + 2z = 1 \end{cases}$

48. $\begin{cases} 3x + 3y + z = 8 \\ x + 2y + z = 5 \\ 2x - y + z = 4 \end{cases}$

49. $\begin{cases} x + y + z = 2 \\ 3x + 2y - z = \frac{7}{3} \\ 3x + y + 2z = \frac{10}{3} \end{cases}$

50. $\begin{cases} 3x + 3y + z = 1 \\ x + 2y + z = 0 \\ 2x - y + z = 4 \end{cases}$

In Problems 51–56, by hand, show that each matrix has no inverse.

51. $\begin{bmatrix} 4 & 2 \\ 2 & 1 \end{bmatrix}$

52. $\begin{bmatrix} -3 & \frac{1}{2} \\ 6 & -1 \end{bmatrix}$

53. $\begin{bmatrix} 15 & 3 \\ 10 & 2 \end{bmatrix}$

54. $\begin{bmatrix} -3 & 0 \\ 4 & 0 \end{bmatrix}$

55. $\begin{bmatrix} -3 & 1 & -1 \\ 1 & -4 & -7 \\ 1 & 2 & 5 \end{bmatrix}$

56. $\begin{bmatrix} 1 & 1 & -3 \\ 2 & -4 & 1 \\ -5 & 7 & 1 \end{bmatrix}$

In Problems 57–60, use a graphing utility to find the inverse, if it exists, of each matrix. Round answers to two decimal places.

57. $\begin{bmatrix} 25 & 61 & -12 \\ 18 & -2 & 4 \\ 8 & 35 & 21 \end{bmatrix}$

58. $\begin{bmatrix} 18 & -3 & 4 \\ 6 & -20 & 14 \\ 10 & 25 & -15 \end{bmatrix}$

59. $\begin{bmatrix} 44 & 21 & 18 & 6 \\ -2 & 10 & 15 & 5 \\ 21 & 12 & -12 & 4 \\ -8 & -16 & 4 & 9 \end{bmatrix}$

60. $\begin{bmatrix} 16 & 22 & -3 & 5 \\ 21 & -17 & 4 & 8 \\ 2 & 8 & 27 & 20 \\ 5 & 15 & -3 & -10 \end{bmatrix}$

In Problems 61–64, use the idea behind Example 15 with a graphing utility to solve the following systems of equations. Round answers to two decimal places.

61. $\begin{cases} 25x + 61y - 12z = 10 \\ 18x - 12y + 7z = -9 \\ 3x + 4y - z = 12 \end{cases}$

62. $\begin{cases} 25x + 61y - 12z = 15 \\ 18x - 12y + 7z = -3 \\ 3x + 4y - z = 12 \end{cases}$

63. $\begin{cases} 25x + 61y - 12z = 21 \\ 18x - 12y + 7z = 7 \\ 3x + 4y - z = -2 \end{cases}$

64. $\begin{cases} 25x + 61y - 12z = 25 \\ 18x - 12y + 7z = 10 \\ 3x + 4y - z = -4 \end{cases}$

65. **Computing the Cost of Production** The Acme Steel Company is a producer of stainless steel and aluminum containers. On a certain day, the following stainless steel containers were manufactured: 500 with 10-gallon capacity, 350 with 5-gallon capacity, and 400 with 1-gallon capacity. On the same day, the following aluminum containers were manufactured: 700 with 10-gallon capacity, 500 with 5-gallon capacity, and 850 with 1-gallon capacity.
    (a) Find a 2 by 3 matrix representing the above data. Find a 3 by 2 matrix to represent the same data.
    (b) If the amount of material used in the 10-gallon containers is 15 pounds, the amount used in the 5-gallon containers is 8 pounds, and the amount used in the 1-gallon containers is 3 pounds, find a 3 by 1 matrix representing the amount of material.
    (c) Multiply the 2 by 3 matrix found in part (a) and the 3 by 1 matrix found in part (b) to get a 2 by 1 matrix showing the day's usage of material.
    (d) If stainless steel costs Acme \$0.10 per pound and aluminum costs \$0.05 per pound, find a 1 by 2 matrix representing cost.
    (e) Multiply the matrices found in parts (c) and (d) to determine what the total cost of the day's production was.

66. **Computing Profit** Rizza Ford has two locations, one in the city and the other in the suburbs. In January, the city location sold 400 subcompacts, 250 intermediate-size cars, and 50 station wagons; in February, it sold 350 subcompacts, 100 intermediates, and 30 station wagons. At the suburban location in January, 450 subcompacts, 200 intermediates, and 140 station wagons were sold. In February, the suburban location sold 350 subcompacts, 300 intermediates, and 100 station wagons.
    (a) Find 2 by 3 matrices that summarize the sales data for each location for January and February (one matrix for each month).
    (b) Use matrix addition to obtain total sales for the two-month period.
    (c) The profit on each kind of car is \$100 per subcompact, \$150 per intermediate, and \$200 per station wagon. Find a 3 by 1 matrix representing this profit.
    (d) Multiply the matrices found in parts (b) and (c) to get a 2 by 1 matrix showing the profit at each location.

67. Make up a situation different from any found in the text that can be represented by a matrix.

6.6 SYSTEMS OF LINEAR INEQUALITIES; LINEAR PROGRAMMING

1. Graph a Linear Inequality by Hand
2. Graph a Linear Inequality Using a Graphing Utility
3. Graph a System of Linear Inequalities
4. Set up a Linear Programming Problem
5. Solve a Linear Programming Problem

In Chapter 1, we discussed inequalities in one variable. In this section, we discuss linear inequalities in two variables.

Linear inequalities are inequalities in one of the forms

$$Ax + By < C \qquad Ax + By > C \qquad Ax + By \le C \qquad Ax + By \ge C$$

where A and B are not both zero.

The graph of the corresponding equation of a linear inequality is a line, which separates the xy-plane into two regions, called **half-planes.** See Figure 17.

As shown, $Ax + By = C$ is the equation of the boundary line and it divides the plane into two half-planes, one for which $Ax + By < C$ and the other for which $Ax + By > C$.

Figure 17

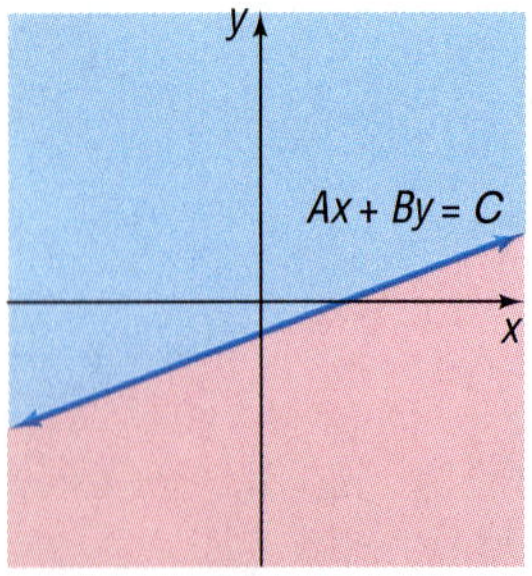

1 A linear inequality in two variables x and y is **satisfied** by an ordered pair (a, b) if, when x is replaced by a and y by b, a true statement results. A **graph of a linear inequality in two variables** x and y consists of all points (x, y) whose coordinates satisfy the inequality.

Let's look at an example.

EXAMPLE 1 Graphing a Linear Inequality

Graph the linear inequality: $3x + y - 6 \le 0$

Solution We begin with the associated problem of the graph of the linear equation

$$3x + y - 6 = 0$$

formed by replacing (for now) the $\le$ symbol with an $=$ sign. The graph of the linear equation is a line. See Figure 18(a). This line is part of the graph of the inequality we seek because the inequality is nonstrict. (Do you see why? We are seeking points for which $3x + y - 6$ is less than *or equal to* 0.)

Now, let's test a few randomly selected points to see whether they belong to the graph of the inequality.

	$3x + y - 6$	Conclusion
$(4, -1)$	$3(4) + (-1) - 6 = 5 > 0$	Does not belong to graph
$(5, 5)$	$3(5) + 5 - 6 = 14 > 0$	Does not belong to graph
$(-1, 2)$	$3(-1) + 2 - 6 = -7 < 0$	Belongs to graph
$(-2, -2)$	$3(-2) + (-2) - 6 = -14 < 0$	Belongs to graph

Look again at Figure 18(a). Notice that the two points that belong to the graph both lie on the same side of the line, and the two points that do not belong to the graph lie on the opposite side. As it turns out, this is always the case. Thus, the graph we seek consists of all points that lie on the same side of the line as do $(-1, 2)$ and $(-2, -2)$, namely, the shaded region in Figure 18(b).

Figure 18

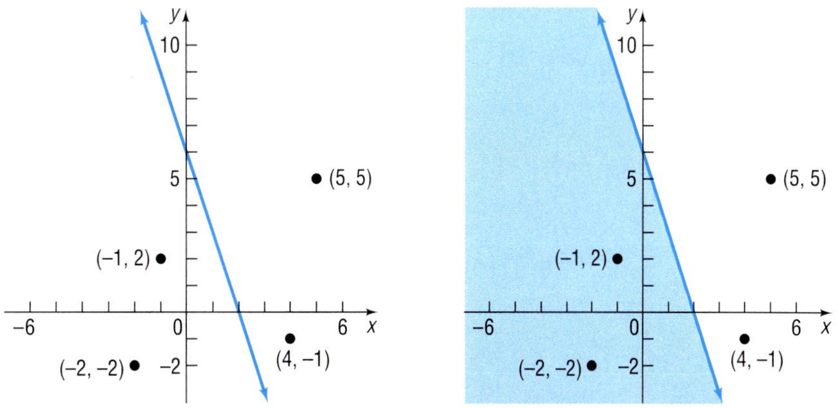

(a) $3x + y - 6 = 0$

(b) Graph of $3x + y - 6 \le 0$

The graph of any linear inequality in two variables may be obtained by graphing the equation corresponding to the inequality, using dashes if the inequality is strict ($<$ or $>$) and solid lines if it is nonstrict ($\le$ or $\ge$). This graph will separate the xy-plane into two half-planes. In each half-plane either all points satisfy the inequality or no points satisfy the inequality. Thus, the use of a single test point is all that is required to determine whether the points of that half-plane are part of the graph or not. The steps to follow are given next.

Steps for Graphing an Inequality By Hand

STEP 1: Replace the inequality symbol by an equal sign and graph the resulting equation. If the inequality is strict, use dashes; if it is nonstrict, use a solid line. This graph separates the xy-plane into two half-planes.

STEP 2: Select a test point P in one of the half-planes.
(a) If the coordinates of P satisfy the inequality, then so do all the points in that half-plane. Indicate this by shading the half-plane.
(b) If the coordinates of P do not satisfy the inequality, then none of the points in that half-plane do, so shade the opposite half-plane.

2 Graphing utilities can also be used to graph linear inequalities. The steps to follow to graph an inequality using a graphing utility are given next.

Steps for Graphing an Inequality Using a Graphing Utility

STEP 1: Replace the inequality symbol by an equal sign and graph the resulting equation.

STEP 2: Select a test point P in one of the half-planes.
(a) Use a graphing utility to determine if the test point P satisfies the inequality. If the test point satisfies the inequality, then so do all the points in this half-plane. Indicate this by using the graphing utility to shade the half-plane.
(b) If the coordinates of P do not satisfy the inequality, then none of the points in that half-plane do, so shade the opposite half-plane.

EXAMPLE 2 Graphing a Linear Inequality Using a Graphing Utility

Use a graphing utility to graph $3x + y - 6 \le 0$.

Solution STEP 1: We begin by graphing the equation $3x + y - 6 = 0$ ($Y_1 = -3x + 6$). See Figure 19.

Figure 19

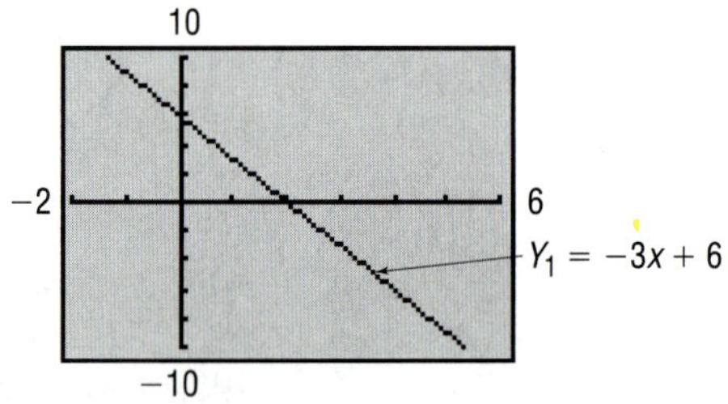

STEP 2: Select a test point in one of the half-planes and determine whether it satisfies the inequality. To test the point $(-1, 2)$, for example, enter $3* - 1 + 2 - 6 \leq 0$. See Figure 20(a). The 1 that appears indicates that the statement entered (the inequality) is true. When the point $(5, 5)$ is tested, a 0 appears, indicating that the statement entered is false. Thus, $(-1, 2)$ is a part of the graph of the inequality and $(5, 5)$ is not, so we shade the half-plane below Y_1. Figure 20(b) shows the graph of the inequality on a TI-83.*

Figure 20

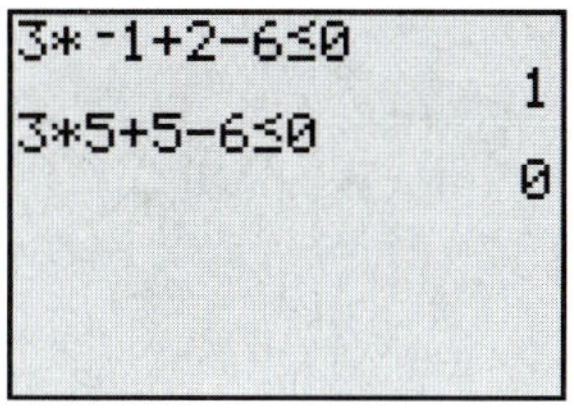

(a)

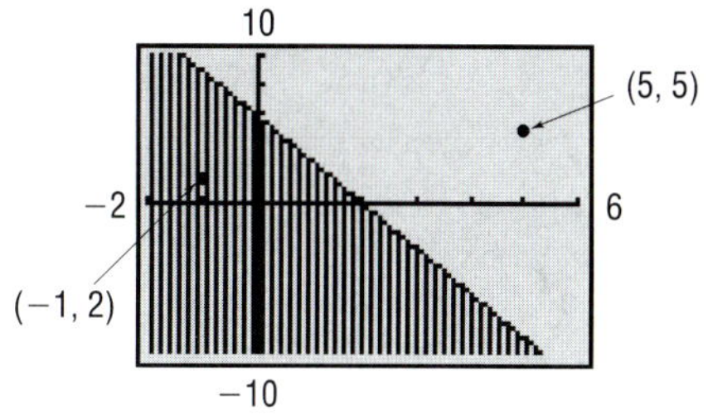

(b)

NOW WORK PROBLEM 5.

EXAMPLE 3 Graphing Linear Inequalities

Graph: (a) $y < 2$ (b) $y \geq 2x$

Solution (a) The graph of the equation $y = 2$ is a horizontal line and is not part of the graph of the inequality, so we use a dashed line. Since $(0, 0)$ satisfies the inequality, the graph consists of the half-plane below the line $y = 2$. See Figure 21.

(b) The graph of the equation $y = 2x$ is a line and is part of the graph of the inequality, so we use a solid line. Using $(3, 0)$ as a test point, we find that it does not satisfy the inequality $[0 < 2 \cdot 3]$. Thus, points in the half-plane on the opposite side of $y = 2x$ satisfy the inequality. See Figure 22.

Figure 21

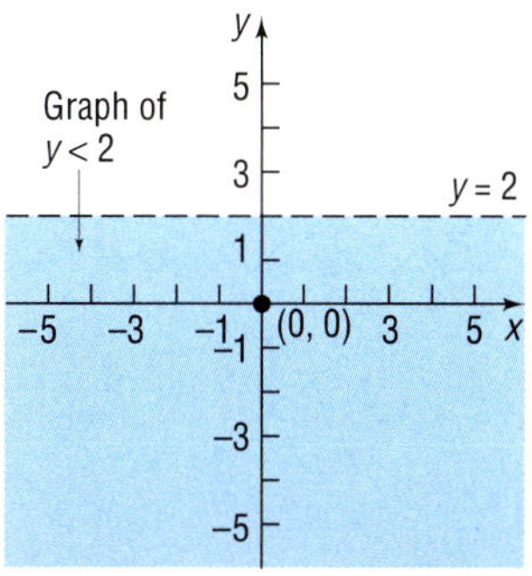

Figure 22

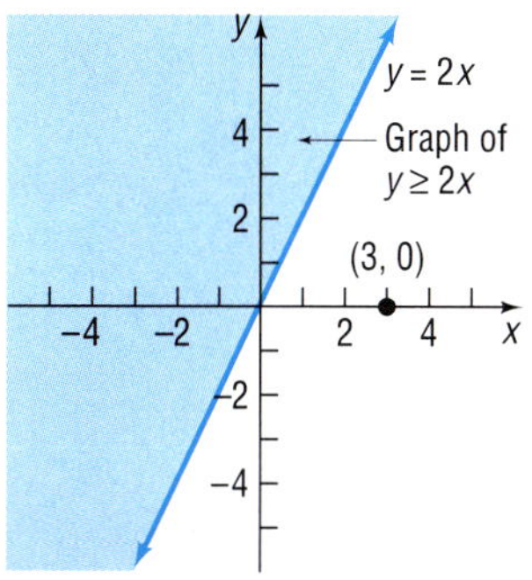

NOW WORK PROBLEM 3.

*Consult your owner's manual for shading techniques.

Systems of Linear Inequalities in Two Variables

3 The **graph of a system of linear inequalities** in two variables x and y is the set of all points (x, y) that simultaneously satisfies *each* of the inequalities in the system. Thus, the graph of a system of linear inequalities can be obtained by graphing each inequality individually and then determining where, if at all, they intersect.

EXAMPLE 4 Graphing a System of Linear Inequalities

Graph the system: $\begin{cases} x + y \geq 2 \\ 2x - y \leq 4 \end{cases}$

Solution First, we graph the inequality $x + y \geq 2$ as the shaded region in Figure 23(a). Next, we graph the inequality $2x - y \leq 4$ as the shaded region in Figure 23(b). Now, superimpose the two graphs, as shown in Figure 23(c). The points that are in both shaded regions [the overlapping, purple region in Figure 23(c)] are the solutions we seek to the system, because they simultaneously satisfy each linear inequality.

Figure 23

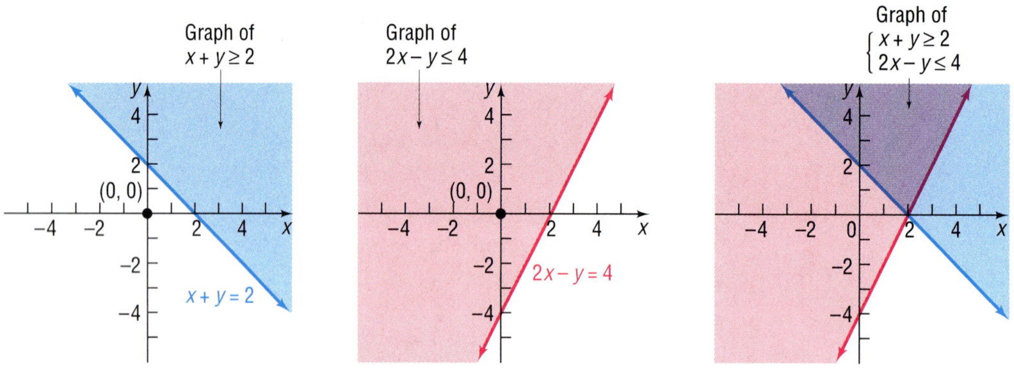

(a) (b) (c)

EXAMPLE 5 Graphing a System of Linear Inequalities Using a Graphing Utility

Graph the system: $\begin{cases} x + y \geq 2 \\ 2x - y \leq 4 \end{cases}$

Solution First, we graph the lines $x + y = 2$ $(Y_1 = -x + 2)$ and $2x - y = 4$ $(Y_2 = 2x - 4)$. See Figure 24.

Figure 24

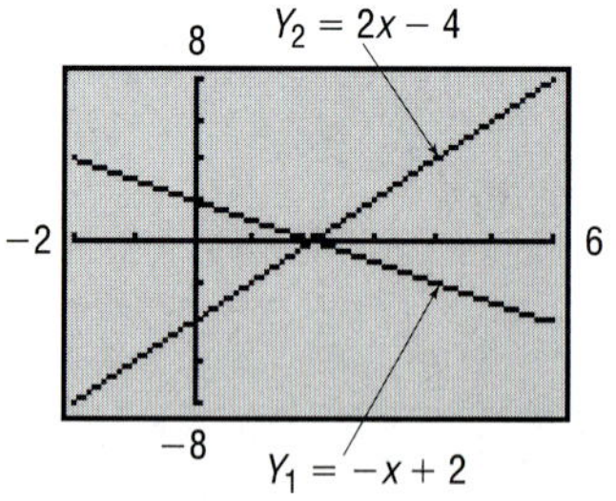

Notice that the graphs divide the viewing window into four regions. We select a test point for each region and determine whether the point makes *both* inequalities true. We choose to test (0, 0), (2, 3), (4, 0), and (2, −2). Figure 25(a) shows that (2, 3) is the only point for which both inequalities are true. Thus, we obtain the graph shown in Figure 25(b).

Figure 25

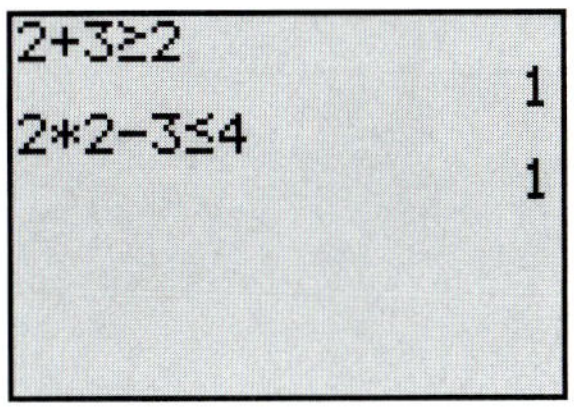

(a)

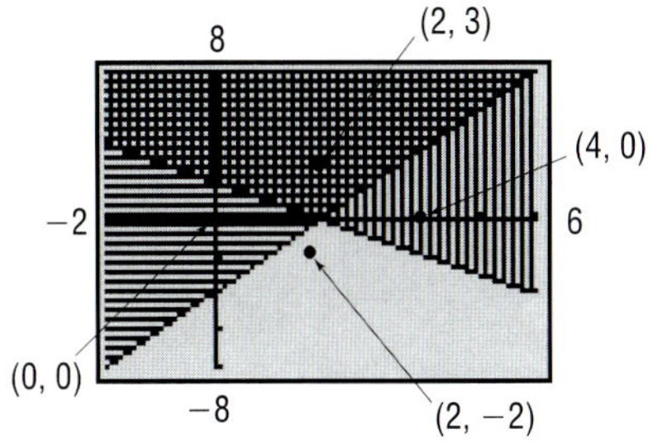

(b)

Notice in Figure 25(b) that the darker region, the graph of the system of inequalities, is the intersection of the graphs of the single inequalities $x + y \geq 2$ and $2x - y \leq 4$. Obtaining Figure 25(b) by this method is sometimes faster than using test points.

NOW WORK PROBLEM 11.

EXAMPLE 6 Graphing a System of Linear Inequalities

Graph the system: $\begin{cases} x + y \leq 2 \\ x + y \geq 0 \end{cases}$

Solution See Figure 26. The overlapping purple shaded region between the two boundary lines is the graph of the system.

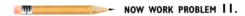

Figure 26

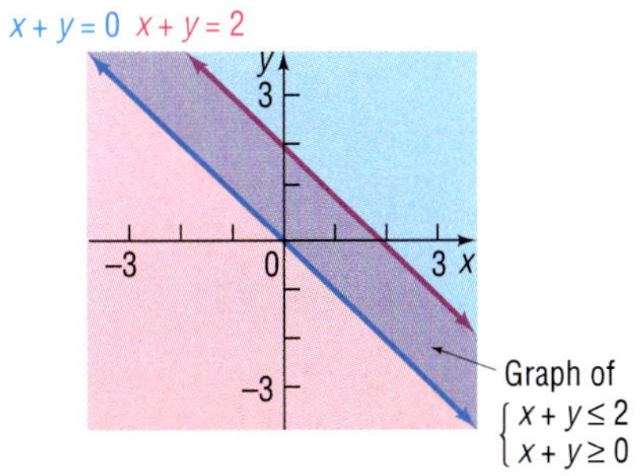

NOW WORK PROBLEM 15.

EXAMPLE 7 Graphing a System of Linear Inequalities

Graph the system: $\begin{cases} 2x - y \geq 2 \\ 2x - y \geq 0 \end{cases}$

Solution See Figure 27. The overlapping purple shaded region is the graph of the system. Note that the graph of the system is identical to the graph of the single inequality $2x - y \geq 2$.

Figure 27

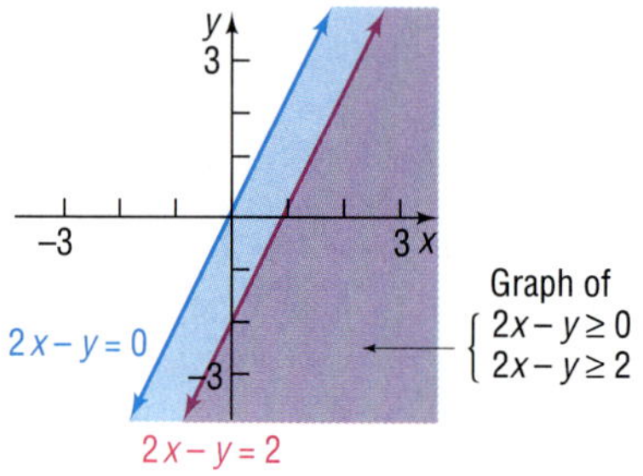

EXAMPLE 8 Graphing a System of Linear Inequalities

Graph the system: $\begin{cases} x + 2y \leq 2 \\ x + 2y \geq 6 \end{cases}$

Solution See Figure 28. Because no overlapping region results, there are no points in the xy-plane that simultaneously satisfy each inequality. Hence, the system has no solution.

Figure 28

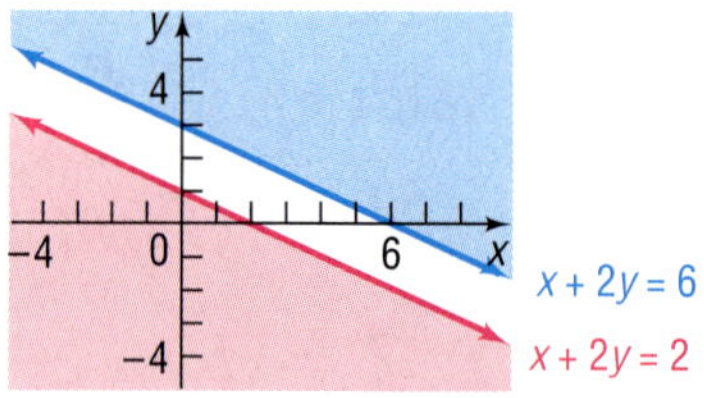

EXAMPLE 9 Graphing a System of Four Linear Inequalities

Graph the system: $\begin{cases} x + y \leq 3 \\ 2x + y \leq 4 \\ x \geq 0 \\ y \geq 0 \end{cases}$

Solution The two inequalities $x \geq 0$ and $y \geq 0$ require that the graph be in quadrant I. Thus, we set our viewing window accordingly. Figure 29 shows the graph of

$x + y = 3$ ($Y_1 = -x + 3$) and $2x + y = 4$ ($Y_2 = -2x + 4$). The graphs divide the viewing window into four regions.

Figure 29

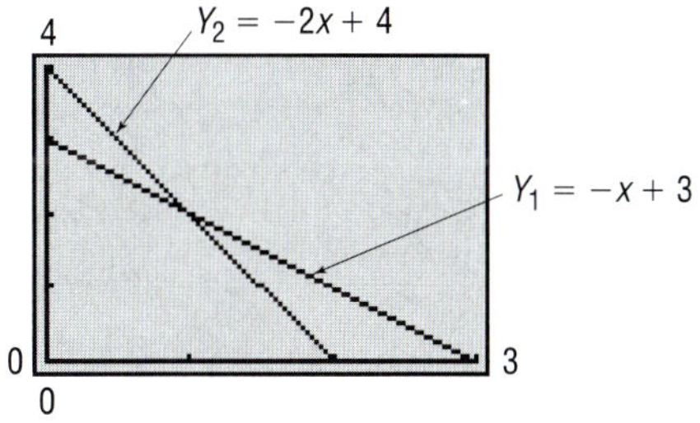

The graph of the inequality $x + y \le 3$ consists of the half-plane below Y_1, and the graph of $2x + y \le 4$ consists of the half-plane below Y_2, so we shade accordingly. See Figures 30(a) and (b).

Figure 30

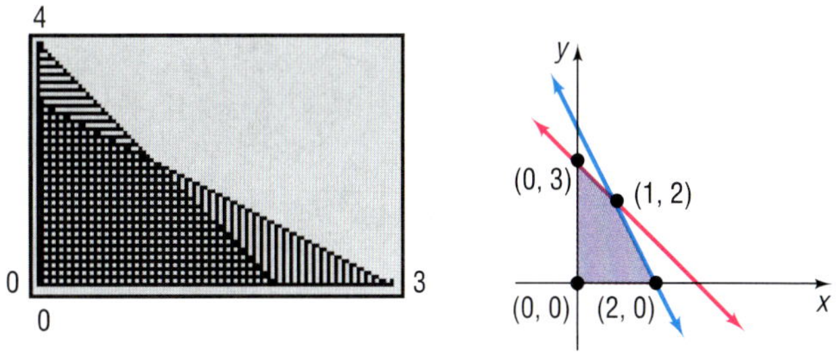

(a) (b)

Notice in Figure 30(b) that those points belonging to the graph that are also points of intersection of boundary lines have been plotted. Such points are referred to as **vertices** or **corner points** of the graph. The corner points (0, 0), (0, 3), (1, 2) and (2, 0) are found by solving the system of equations formed by the two boundary lines. So the corner points are found by solving the following systems:

$$\begin{cases} x = 0 \\ y = 0 \end{cases} \qquad \begin{cases} x + y = 3 \\ x = 0 \end{cases} \qquad \begin{cases} x + y = 3 \\ 2x + y = 4 \end{cases} \qquad \begin{cases} 2x + y = 4 \\ y = 0 \end{cases}$$

Corner point: (0, 0) *Corner point: (0, 3)* *Corner point: (1, 2)* *Corner point: (2, 0)*

The graph of the system of linear inequalities in Figure 30 is said to be **bounded,** because it can be contained within some circle of sufficiently large radius. A graph that cannot be contained in any circle is said to be **unbounded.** For example, the graph of the system of linear inequalities in Figure 27 is unbounded, since it extends indefinitely in a particular direction.

NOW WORK PROBLEM 23.

Linear Programming

4 Historically, linear programming evolved as a technique for solving problems involving resource allocation of goods and materials for the U.S. Air Force during World War II. Today, linear programming techniques are used to solve a wide variety of problems, such as optimizing airline scheduling and

establishing telephone lines. Although most practical linear programming problems involve systems of several hundred linear inequalities containing several hundred variables, we will limit our discussion to problems containing only two variables, because we can solve such problems using graphing techniques.*

EXAMPLE 10 Financial Planning

A retired couple has up to \$25,000 to invest. As their financial adviser, you recommend that they place at least \$15,000 in Treasury bills yielding 6% and at most \$5000 in corporate bonds yielding 9%. How much money should be placed in each investment so that income is maximized?

The problem given here is typical of a *linear programming problem.* The problem requires that a certain linear expression, the income, be maximized. If I represents income, x the amount invested in Treasury bills at 6%, and y the amount invested in corporate bonds at 9%, then

$$I = 0.06x + 0.09y$$

This linear expression is called the **objective function.** Furthermore, the problem requires that the maximum income be achieved under certain conditions or **constraints,** each of which is a linear inequality involving the variables. The linear programming problem given in Example 10 may be restated as

$$\text{Maximize:} \quad I = 0.06x + 0.09y$$

subject to the conditions that

$$\begin{aligned} x \geq 0, \qquad y &\geq 0 \\ x + y &\leq 25{,}000 \\ x &\geq 15{,}000 \\ y &\leq 5000 \end{aligned}$$

In general, every linear programming problem has two components:

1. A linear objective function that is to be maximized or minimized.
2. A collection of linear inequalities that must be satisfied simultaneously.

A **linear programming problem** in two variables x and y consists of maximizing (or minimizing) a linear objective function

$$z = Ax + By$$

where A and B are real numbers, not both 0, subject to certain conditions, or constraints, expressible as linear inequalities in x and y.

*The **simplex method** is a way to solve linear programming problems involving many inequalities and variables. This method was developed by George Dantzig in 1946 and is particularly well suited for computerization. In 1984, Narendra Karmarkar of Bell Laboratories discovered a way of solving large linear programming problems that improves on the simplex method.

To maximize (or minimize) the quantity $z = Ax + By$, we need to identify points (x, y) that make the expression for z the largest (or smallest) possible. But not all points (x, y) are eligible; only those that also satisfy each linear inequality (constraint) can be used. We refer to each point (x, y) that satisfies the system of linear inequalities (the constraints) as a **feasible point.** Thus, in a linear programming problem, we seek the feasible point(s) that maximizes (or minimizes) the objective function.

Let's look again at the linear programming problem in Example 10.

EXAMPLE 11 Analyzing a Linear Programming Problem

Consider the following linear programming problem:

$$\text{Maximize:} \quad I = 0.06x + 0.09y$$

subject to the conditions that

$$\begin{aligned} x \geq 0, \quad & y \geq 0 \\ x + y &\leq 25{,}000 \\ x &\geq 15{,}000 \\ y &\leq 5000 \end{aligned}$$

Graph the constraints. Then graph the objective function for $I = 0, 0.9, 1.35, 1.65$, and 1.8.

Solution Figure 31 shows the graph of the constraints. We superimpose on this graph the graph of the objective function for the given values of I.

For $I = 0$, the objective function is the line $0 = 0.06x + 0.09y$.
For $I = 0.9$, the objective function is the line $0.9 = 0.06x + 0.09y$.
For $I = 1.35$, the objective function is the line $1.35 = 0.06x + 0.09y$.
For $I = 1.65$, the objective function is the line $1.65 = 0.06x + 0.09y$.
For $I = 1.8$, the objective function is the line $1.8 = 0.06x + 0.09y$.

Figure 31

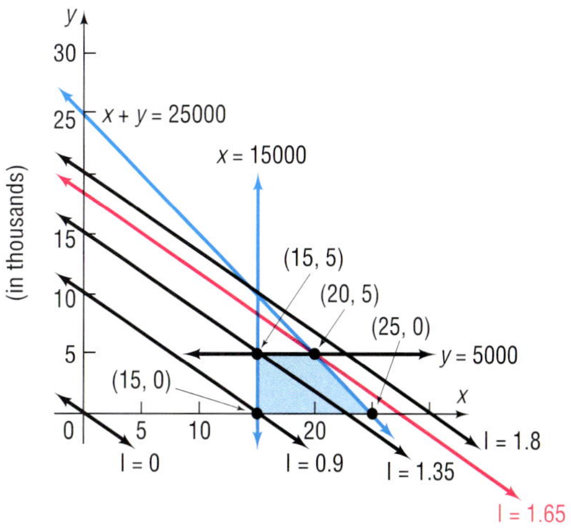

A **solution** to a linear programming problem consists of the feasible point(s) that maximizes (or minimizes) the objective function, together with the corresponding value of the objective function.

5 One condition for a linear programming problem in two variables to have a solution is that the graph of the feasible points be bounded. (Refer to page 471).

If none of the feasible points maximizes (or minimizes) the objective function or if there are no feasible points, then the linear programming problem has no solution.

Consider the linear programming problem stated in Example 11, and look again at Figure 31. The feasible points are the points that lie in the shaded region. For example, (20, 3) is a feasible point, as is (15, 5), (20, 5), (18, 4), and so on. To find the solution of the problem requires that we find a feasible point (x, y) that makes $I = 0.06x + 0.09y$ as large as possible. Notice that, as I increases in value from $I = 0$ to $I = 0.9$ to $I = 1.35$ to $I = 1.65$ to $I = 1.8$, we obtain a collection of parallel lines. Furthermore, notice that the largest value of I that can be obtained while feasible points are present is $I = 1.65$, which corresponds to the line $1.65 = 0.06x + 0.09y$. Any larger value of I results in a line that does not pass through any feasible points. Finally, notice that the feasible point that yields $I = 1.65$ is the point (20, 5), a corner point. These observations form the basis of the following result, which we state without proof.

THEOREM

Location of the Solution of a Linear Programming Problem

If a linear programming problem has a solution, it is located at a corner point of the graph of the feasible points.

If a linear programming problem has multiple solutions, at least one of them is located at a corner point of the graph of the feasible points.

In either case, the corresponding value of the objective function is unique.

We shall not consider here linear programming problems that have no solution. As a result, we can outline the procedure for solving a linear programming problem as follows:

Procedure for Solving a Linear Programming Problem

STEP 1: Write an expression for the quantity to be maximized (or minimized). This expression is the objective function.
STEP 2: Write all the constraints as a system of linear inequalities and graph the system.
STEP 3: List the corner points of the graph of the feasible points.
STEP 4: List the corresponding values of the objective function at each corner point. The largest (or smallest) of these is the solution.

EXAMPLE 12 Solving a Minimum Linear Programming Problem

Minimize the expression

$$z = 2x + 3y$$

subject to the constraints

$$y \le 5 \qquad x \le 6 \qquad x + y \ge 2 \qquad x \ge 0 \qquad y \ge 0$$

Solution The objective function is $z = 2x + 3y$. We seek the smallest value of z that can occur if x and y are solutions of the system of linear inequalities

$$\begin{cases} y \le 5 \\ x \le 6 \\ x + y \ge 2 \\ x \ge 0 \\ y \ge 0 \end{cases}$$

The graph of this system (the feasible points) is shown as the shaded region in Figure 32. We have also plotted the corner points. Table 1 lists the corner points and the corresponding values of the objective function. From the table, we can see that the minimum value of z is 4, and it occurs at the point $(2, 0)$.

Figure 32

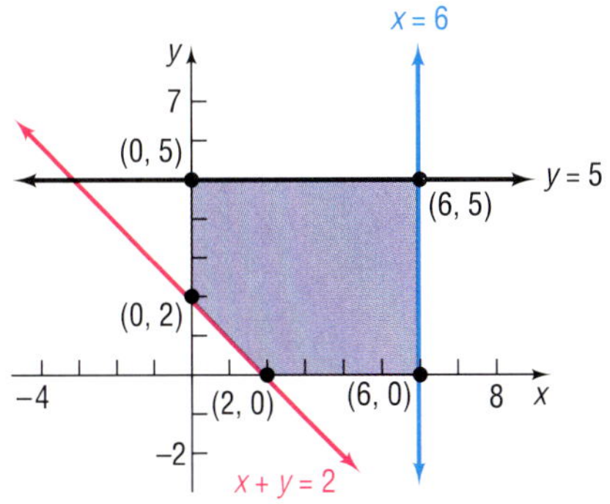

Table 1

Corner Point	Value of the Objective Function
(x, y)	$z = 2x + 3y$
$(0, 2)$	$z = 2(0) + 3(2) = 6$
$(0, 5)$	$z = 2(0) + 3(5) = 15$
$(6, 5)$	$z = 2(6) + 3(5) = 27$
$(6, 0)$	$z = 2(6) + 3(0) = 12$
$(2, 0)$	$z = 2(2) + 3(0) = 4$

NOW WORK PROBLEMS **37** AND **43**.

EXAMPLE 13 Maximizing Profit

At the end of every month, after filling orders for its regular customers, a coffee company has some pure Colombian coffee and some special-blend coffee remaining. The practice of the company has been to package a mixture of the two coffees into 1-pound packages as follows: a low-grade mixture containing 4 ounces of Colombian coffee and 12 ounces of special-blend coffee and a high-grade mixture containing 8 ounces of Colombian and 8 ounces of special-blend coffee. A profit of \$0.30 per package is made on the low-grade mixture, whereas a profit of \$0.40 per package is made on the high-grade mixture. This month, 120 pounds of special-blend coffee and 100 pounds of pure Colombian coffee remain. How many packages of each mixture should be prepared to achieve a maximum profit? Assume that all packages prepared can be sold.

Solution We begin by assigning symbols for the two variables.

$$x = \text{Number of packages of the low-grade mixture}$$
$$y = \text{Number of packages of the high-grade mixture}$$

If P denotes the profit, then

$$P = \$0.30x + \$0.40y$$

This expression is the objective function. We seek to maximize P subject to certain constraints on x and y. Because x and y represent numbers of packages, the only meaningful values for x and y are nonnegative integers. Thus, we have the two constraints

$$x \geq 0 \qquad y \geq 0 \qquad \textit{Nonnegative constraints}$$

We also have only so much of each type of coffee available. For example, the total amount of Colombian coffee used in the two mixtures cannot exceed 100 pounds, or 1600 ounces. Because we use 4 ounces in each low-grade package and 8 ounces in each high-grade package, we are led to the constraint

$$4x + 8y \leq 1600 \qquad \textit{Colombian coffee constraint}$$

Similarly, the supply of 120 pounds, or 1920 ounces, of special-blend coffee leads to the constraint

$$12x + 8y \leq 1920 \qquad \textit{Special-blend coffee constraint}$$

The linear programming problem may be stated as

$$\text{Maximize:} \quad P = 0.3x + 0.4y$$

subject to the constraints

$$x \geq 0 \qquad y \geq 0 \qquad 4x + 8y \leq 1600 \qquad 12x + 8y \leq 1920$$

The graph of the constraints (the feasible points) is illustrated in Figure 33. We list the corner points and evaluate the objective function at each one. In Table 2, we can see that the maximum profit, \$84, is achieved with 40 packages of the low-grade mixture and 180 packages of the high-grade mixture.

Figure 33

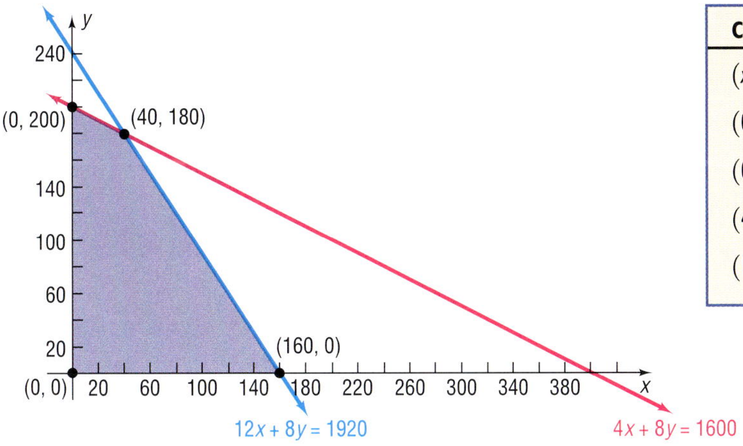

Table 2

Corner Point	Value of Profit
(x, y)	$P = 0.3x + 0.4y$
$(0, 0)$	$P = 0$
$(0, 200)$	$P = 0.3(0) + 0.4(200) = \$80$
$(40, 180)$	$P = 0.3(40) + 0.4(180) = \$84$
$(160, 0)$	$P = 0.3(160) + 0.4(0) = \$48$

NOW WORK PROBLEM 51.

6.6 EXERCISES

In Problems 1–8, graph each inequality by hand.

1. $x \geq 0$ **2.** $y \geq 0$ **3.** $x \geq 4$ **4.** $y \leq 2$

5. $x + y > 1$ **6.** $x + y \leq 9$ **7.** $2x + y \geq 6$ **8.** $3x + 2y \leq 6$

In Problems 9–20, graph each system of inequalities by hand.

9. $\begin{cases} x + y \leq 2 \\ 2x + y \geq 4 \end{cases}$ **10.** $\begin{cases} 3x - y \geq 6 \\ x + 2y \leq 2 \end{cases}$ **11.** $\begin{cases} 2x - y \leq 4 \\ 3x + 2y \geq -6 \end{cases}$ **12.** $\begin{cases} 4x - 5y \leq 0 \\ 2x - y \geq 2 \end{cases}$

13. $\begin{cases} 2x - 3y \leq 0 \\ 3x + 2y \leq 6 \end{cases}$ **14.** $\begin{cases} 4x - y \geq 2 \\ x + 2y \geq 2 \end{cases}$ **15.** $\begin{cases} x - 2y \leq 6 \\ 2x - 4y \geq 0 \end{cases}$ **16.** $\begin{cases} x + 4y \leq 8 \\ x + 4y \geq 4 \end{cases}$

17. $\begin{cases} 2x + y \geq -2 \\ 2x + y \geq 2 \end{cases}$ **18.** $\begin{cases} x - 4y \leq 4 \\ x - 4y \geq 0 \end{cases}$ **19.** $\begin{cases} 2x + 3y \geq 6 \\ 2x + 3y \leq 0 \end{cases}$ **20.** $\begin{cases} 2x + y \geq 0 \\ 2x + y \geq 2 \end{cases}$

In Problems 21–30, graph each system of linear inequalities by hand. Tell whether the graph is bounded or unbounded, and label the corner points.

21. $\begin{cases} x \geq 0 \\ y \geq 0 \\ 2x + y \leq 6 \\ x + 2y \leq 6 \end{cases}$ **22.** $\begin{cases} x \geq 0 \\ y \geq 0 \\ x + y \geq 4 \\ 2x + 3y \geq 6 \end{cases}$ **23.** $\begin{cases} x \geq 0 \\ y \geq 0 \\ x + y \geq 2 \\ 2x + y \geq 4 \end{cases}$ **24.** $\begin{cases} x \geq 0 \\ y \geq 0 \\ 3x + y \leq 6 \\ 2x + y \leq 2 \end{cases}$ **25.** $\begin{cases} x \geq 0 \\ y \geq 0 \\ x + y \geq 2 \\ 2x + 3y \leq 12 \\ 3x + y \leq 12 \end{cases}$

26. $\begin{cases} x \geq 0 \\ y \geq 0 \\ x + y \geq 2 \\ x + y \leq 10 \\ 2x + y \leq 3 \end{cases}$ **27.** $\begin{cases} x \geq 0 \\ y \geq 0 \\ x + y \geq 2 \\ x + y \leq 8 \\ 2x + y \leq 10 \end{cases}$ **28.** $\begin{cases} x \geq 0 \\ y \geq 0 \\ x + y \geq 2 \\ x + y \leq 8 \\ x + 2y \geq 1 \end{cases}$ **29.** $\begin{cases} x \geq 0 \\ y \geq 0 \\ x + 2y \geq 1 \\ x + 2y \leq 10 \end{cases}$ **30.** $\begin{cases} x \geq 0 \\ y \geq 0 \\ x + 2y \geq 1 \\ x + 2y \leq 10 \\ x + y \geq 2 \\ x + y \leq 8 \end{cases}$

In Problems 31–34, write a system of linear inequalities that has the given graph.

31.

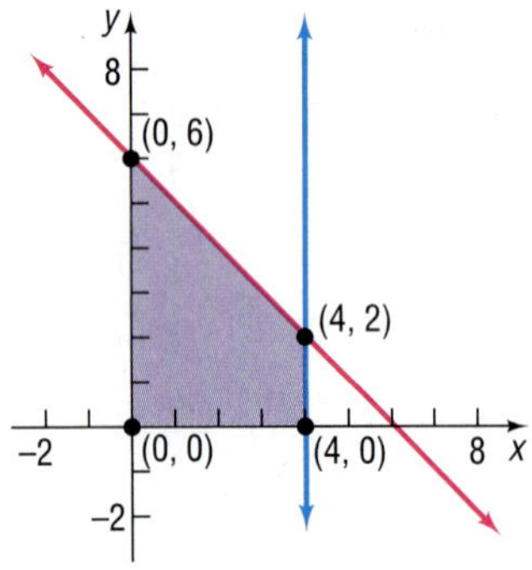

32.

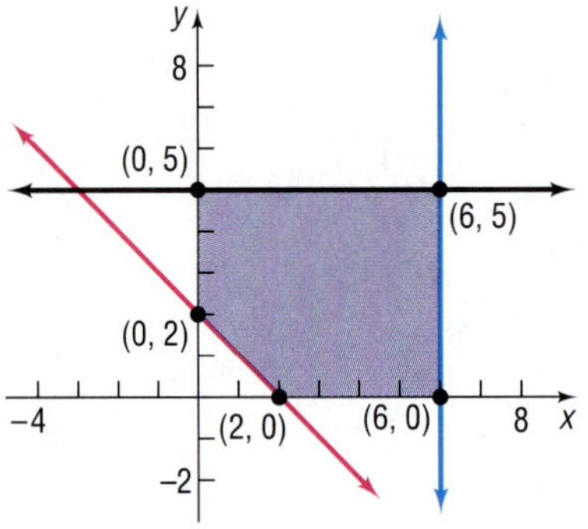

33.

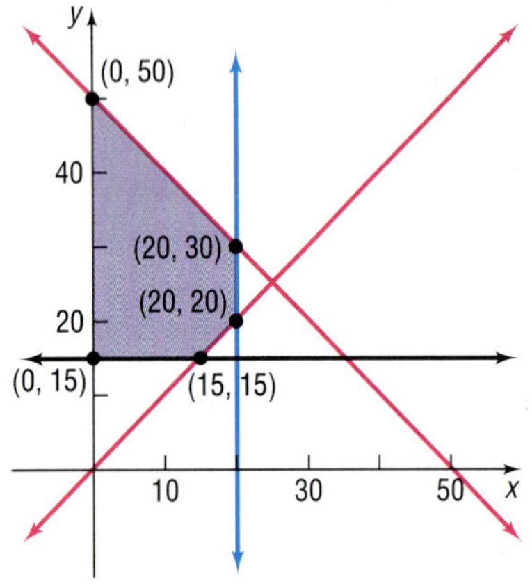

34.

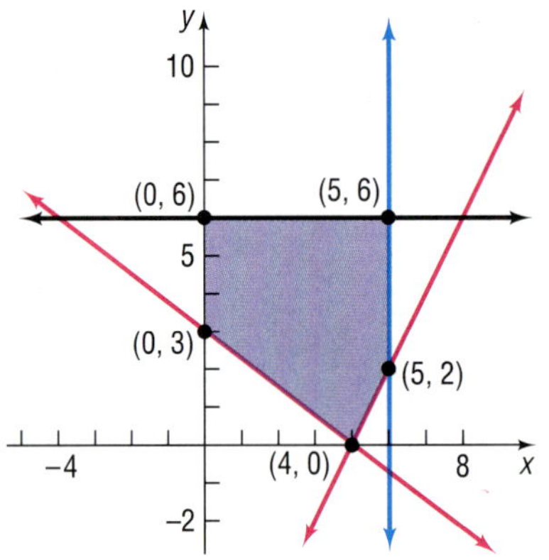

In Problems 35–40, find the maximum and minimum value of the given objective function of a linear programming problem. The figure illustrates the graph of the feasible points.

35. $z = x + y$ **36.** $z = 2x + 3y$ **37.** $z = x + 10y$

38. $z = 10x + y$ **39.** $z = 5x + 7y$ **40.** $z = 7x + 5y$

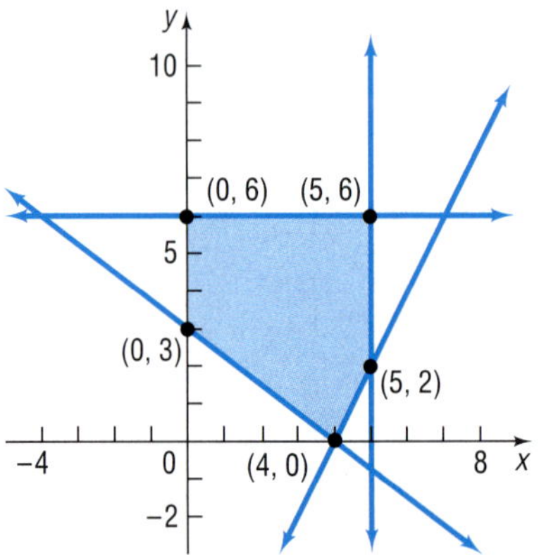

In Problems 41–50, solve each linear programming problem.

41. Maximize $z = 2x + y$ subject to $x \geq 0, \quad y \geq 0, \quad x + y \leq 6, \quad x + y \geq 1$

42. Maximize $z = x + 3y$ subject to $x \geq 0, \quad y \geq 0, \quad x + y \geq 3, \quad x \leq 5, \quad y \leq 7$

43. Minimize $z = 2x + 5y$ subject to $x \geq 0, \quad y \geq 0, \quad x + y \geq 2, \quad x \leq 5, \quad y \leq 3$

44. Minimize $z = 3x + 4y$ subject to $x \geq 0, \quad y \geq 0, \quad 2x + 3y \geq 6, \quad x + y \leq 8$

45. Maximize $z = 3x + 5y$ subject to $x \geq 0, \quad y \geq 0, \quad x + y \geq 2, \quad 2x + 3y \leq 12, \quad 3x + 2y \leq 12$

46. Maximize $z = 5x + 3y$ subject to $x \geq 0, \quad y \geq 0, \quad x + y \geq 2, \quad x + y \leq 8, \quad 2x + y \leq 10$

47. Minimize $z = 5x + 4y$ subject to $x \geq 0, \quad y \geq 0, \quad x + y \geq 2, \quad 2x + 3y \leq 12, \quad 3x + y \leq 12$

48. Minimize $z = 2x + 3y$ subject to $x \geq 0, \quad y \geq 0, \quad x + y \geq 3, \quad x + y \leq 9, \quad x + 3y \geq 6$

49. Maximize $z = 5x + 2y$ subject to $x \geq 0$, $y \geq 0$, $x + y \leq 10$, $2x + y \geq 10$, $x + 2y \geq 10$

50. Maximize $z = 2x + 4y$ subject to $x \geq 0$, $y \geq 0$, $2x + y \geq 4$, $x + y \leq 9$

51. Maximizing Profit A manufacturer of skis produces two types: downhill and cross country. Use the following table to determine how many of each kind of ski should be produced to achieve a maximum profit. What is the maximum profit? What would the maximum profit be if the maximum time available for manufacturing is increased to 48 hours?

	Downhill	Cross Country	Maximum Time Available
Manufacturing time per ski	2 hours	1 hour	40 hours
Finishing time per ski	1 hour	1 hour	32 hours
Profit per ski	\$70	\$50	

52. Farm Management A farmer has 70 acres of land available for planting either soybeans or wheat. The cost of preparing the soil, the workdays required, and the expected profit per acre planted for each type of crop are given in the following table:

	Soybeans	Wheat
Preparation cost per acre	\$60	\$30
Workdays required per acre	3	4
Profit per acre	\$180	\$100

The farmer cannot spend more than \$1800 in preparation costs nor more than a total of 120 workdays. How many acres of each crop should be planted in order to maximize the profit? What is the maximum profit? What is the maximum profit if the farmer is willing to spend no more than \$2400 on preparation?

53. Farm Management A small farm in Illinois has 100 acres of land available on which to grow corn and soybeans. The following table shows the cultivation cost per acre, the labor cost per acre, and the expected profit per acre. The column on the right shows the amount of money available for each of these expenses. Find the number of acres of each crop that should be planted in order to maximize profit.

	Soybeans	Corn	Money Available
Cultivation cost per acre	\$40	\$60	\$1800
Labor cost per acre	\$60	\$60	\$2400
Profit per acre	\$200	\$250	

54. Dietary Requirements A certain diet requires at least 60 units of carbohydrates, 45 units of protein, and 30 units of fat each day. Each ounce of Supplement A provides 5 units of carbohydrates, 3 units of protein, and 4 units of fat. Each ounce of Supplement B provides 2 units of carbohydrates, 2 units of protein, and 1 unit of fat. If Supplement A costs \$1.50 per ounce and Supplement B costs \$1.00 per ounce, how many ounces of each supplement should be taken daily to minimize the cost of the diet?

55. Production Scheduling In a factory, machine 1 produces 8-inch plyers at the rate of 60 units per hour and 6-inch plyers at the rate of 70 units per hour. Machine 2 produces 8-inch plyers at the rate of 40 units per hour and 6-inch plyers at the rate of 20 units per hour. It costs \$50 per hour to operate machine 1, while machine 2 costs \$30 per hour to operate. The production schedule requires that at least 240 units of 8-inch plyers and at least 140 units of 6-inch plyers must be produced during each 10-hour day. Which combination of machines will cost the least money to operate?

56. Farm Management An owner of a fruit orchard hires a crew of workers to prune at least 25 of his 50 fruit trees. Each newer tree requires one hour to prune, while each older tree needs one and a half hours. The crew contracts to work for at least 30 hours and charge \$15 for each newer tree and \$20 for each older tree. To minimize his cost, how many of each kind of tree will the orchard owner have pruned? What will be the cost?

57. Managing a Meat Market A meat market combines ground beef and ground pork in a single package for meat loaf. The ground beef is 75% lean (75% beef, 25% fat) and costs the market \$0.75 per pound. The ground pork is 60% lean and costs the market \$0.45 per pound. The meat loaf must be at least 70% lean. If the market wants to use at least 50 pounds of its available pork, but no more than 200 pounds of its available ground beef, how much ground beef should be mixed with ground pork so that the cost is minimized?

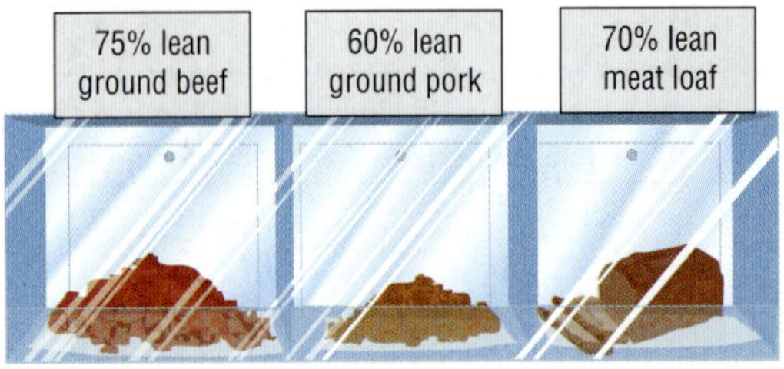

58. **Return on Investment** An investment broker is instructed by her client to invest up to $20,000, some in a junk bond yielding 9% per annum and some in Treasury bills yielding 7% per annum. The client wants to invest at least $8000 in T-bills and no more than $12,000 in the junk bond.
   (a) How much should the broker recommend that the client place in each investment to maximize income if the client insists that the amount invested in T-bills must equal or exceed the amount placed in junk bonds?
   (b) How much should the broker recommend that the client place in each investment to maximize income if the client insists that the amount invested in T-bills must not exceed the amount placed in junk bonds?

59. **Maximizing Profit on Ice Skates** A factory manufactures two kinds of ice skates: racing skates and figure skates. The racing skates require 6 work-hours in the fabrication department, whereas the figure skates require 4 work-hours there. The racing skates require 1 work-hour in the finishing department, whereas the figure skates require 2 work-hours there. The fabricating department has available at most 120 work-hours per day, and the finishing department has no more than 40 work-hours per day available. If the profit on each racing skate is $10 and the profit on each figure skate is $12, how many of each should be manufactured each day to maximize profit? (Assume that all skates made are sold.)

60. **Financial Planning** A retired couple has up to $50,000 to place in fixed-income securities. Their financial advisor suggests two securities to them: one is a AAA bond that yields 8% per annum; the other is a Certificate of Deposit (CD) that yields 4%. After careful consideration of the alternatives, the couple decides to place at most $20,000 in the AAA bond and at least $15,000 in the CD. They also instruct the financial adviser to place at least as much in the CD as in the AAA bond. How should the financial adviser proceed to maximize the return on their investment?

61. **Product Design** An entrepreneur is having a design group produce at least six samples of a new kind of fastener that he wants to market. It costs $9.00 to produce each metal fastener and $4.00 to produce each plastic fastener. He wants to have at least two of each version of the fastener and needs to have all the samples 24 hours from now. It takes 4 hours to produce each metal sample and 2 hours to produce each plastic sample. To minimize the cost of the samples, how many of each kind should the entrepreneur order? What will be the cost of the samples?

62. **Animal Nutrition** Kevin's dog Amadeus likes two kinds of canned dog food. Gourmet Dog costs 40 cents a can and has 20 units of a vitamin complex; the calorie content is 75 calories. Chow Hound costs 32 cents a can and has 35 units of vitamins and 50 calories. Kevin likes Amadeus to have at least 1175 units of vitamins a month and at least 2375 calories during the same time period. Kevin has space to store only 60 cans of dog food at a time. How much of each kind of dog food should Kevin buy each month in order to minimize his cost?

63. **Airline Revenue** An airline has two classes of service: first class and coach. Management's experience has been that each aircraft should have at least 8 but not more than 16 first-class seats and at least 80 but not more than 120 coach seats.
   (a) If management decides that the ratio of first-class to coach seats should never exceed 1:12, with how many of each type seat should an aircraft be configured to maximize revenue?
   (b) If management decides that the ratio of first-class to coach seats should never exceed 1:8, with how many of each type seat should an aircraft be configured to maximize revenue?
   (c) If you were management, what would you do?
   [Hint: Assume that the airline charges $C for a coach seat and $F for a first-class seat; $C > 0$, $F > 0$.

64. **Minimizing Cost** A firm that specializes in raising frying chickens supplements the regular chicken feed with four vitamins. The owner wants the supplemental food to contain at least

50 units of vitamin I, 90 units of vitamin II, 60 units of vitamin III, and 100 units of vitamin IV per 100 ounces of feed. Two supplements are available: supplement A, which contains 5 units of vitamin I, 25 units of vitamin II, 10 units of vitamin III, and 35 units of vitamin IV per ounce; and supplement B, which contains 25 units of vitamin I, 10 units of vitamin II, 10 units of vitamin III, and 20 units of vitamin IV per ounce. If supplement A costs \$0.06 per ounce and supplement B costs \$0.08 per ounce, how much of each supplement should the manager of the farm buy to add to each 100 ounces of feed in order to keep the total cost at a minimum, while still meeting the owner's vitamin specifications?

65. Explain in your own words what a linear programming problem is and how it can be solved.

6.7 PARTIAL FRACTION DECOMPOSITION

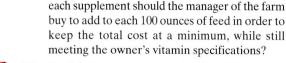

1. Decompose P/Q, where Q has only Nonrepeated Linear Factors
2. Decompose P/Q, where Q has Repeated Linear Factors
3. Decompose P/Q, where Q has only Nonrepeated Irreducible Quadratic Factors
4. Decompose P/Q, where Q has Repeated Irreducible Quadratic Factors

Consider the problem of adding two fractions:

$$\frac{3}{x+4} \quad \text{and} \quad \frac{2}{x-3}$$

The result is

$$\frac{3}{x+4} + \frac{2}{x-3} = \frac{3(x-3) + 2(x+4)}{(x+4)(x-3)} = \frac{5x-1}{x^2+x-12}$$

The reverse procedure, of starting with the rational expression $(5x-1)/(x^2+x-12)$ and writing it as the sum (or difference) of the two simpler fractions $3/(x+4)$ and $2/(x-3)$, is referred to as **partial fraction decomposition,** and the two simpler fractions are called **partial fractions.** Decomposing a rational expression into a sum of partial fractions is important in solving certain types of calculus problems. This section presents a systematic way to decompose rational expressions.

We begin by recalling that a rational expression is the ratio of two polynomials, say, P and $Q \neq 0$. We assume that P and Q have no common factors. Recall also that a rational expression P/Q is called **proper** if the degree of the polynomial in the numerator is less than the degree of the polynomial in the denominator. Otherwise, the rational expression is termed **improper.**

Because any improper rational expression can be reduced by long division to a mixed form consisting of the sum of a polynomial and a proper rational expression, we shall restrict the discussion that follows to proper rational expressions.

The partial fraction decomposition of the rational expression P/Q depends on the factors of the denominator Q. Recall (from Section 4.4) that any polynomial whose coefficients are real numbers can be factored (over the real numbers) into products of linear and/or irreducible quadratic factors. Thus, the denominator Q of the rational expression P/Q will contain only factors of one or both of the following types:

1. *Linear factors* of the form $x - a$, where a is a real number.
2. *Irreducible quadratic factors* of the form $ax^2 + bx + c$, where a, b, and c are real numbers, $a \neq 0$, and $b^2 - 4ac < 0$ (which guarantees that $ax^2 + bx + c$ cannot be written as the product of two linear factors with real coefficients).

1 As it turns out, there are four cases to be examined. We begin with the case for which Q has only nonrepeated linear factors.

CASE 1: Q has only nonrepeated linear factors.

Under the assumption that Q has only nonrepeated linear factors, the polynomial Q has the form

$$Q(x) = (x - a_1)(x - a_2) \cdot \cdots \cdot (x - a_n)$$

where none of the numbers $a_1, a_2, \ldots, a_n$ are equal. In this case, the partial fraction decomposition of P/Q is of the form

$$\frac{P(x)}{Q(x)} = \frac{A_1}{x - a_1} + \frac{A_2}{x - a_2} + \cdots + \frac{A_n}{x - a_n} \qquad (1)$$

where the numbers $A_1, A_2, \ldots, A_n$ are to be determined.

We show how to find these numbers in the example that follows.

EXAMPLE 1 Nonrepeated Linear Factors

Write the partial fraction decomposition of $\dfrac{x}{x^2 - 5x + 6}$

Solution First, we factor the denominator,

$$x^2 - 5x + 6 = (x - 2)(x - 3)$$

and conclude that the denominator contains only nonrepeated linear factors. Then we decompose the rational expression according to equation (1):

$$\frac{x}{x^2 - 5x + 6} = \frac{A}{x - 2} + \frac{B}{x - 3} \qquad (2)$$

where A and B are to be determined. To find A and B, we clear the fractions by multiplying each side by $(x - 2)(x - 3) = x^2 - 5x + 6$. The result is

$$x = A(x - 3) + B(x - 2) \qquad (3)$$

or

$$x = (A + B)x + (-3A - 2B)$$

This equation is an identity in x. Thus, we may equate the coefficients of like powers of x to get

$$\begin{cases} 1 = A + B \\ 0 = -3A - 2B \end{cases}$$

Equate coefficients of x: $1x = (A + B)x$.
Equate coefficients of x^0, the constants: $0x^0 = (-3A - 2B)x^0$.

This system of two equations containing two variables, A and B, can be solved using whatever method you wish. Solving it, we get

$$A = -2 \qquad B = 3$$

Thus, from equation (2), the partial fraction decomposition is

$$\frac{x}{x^2 - 5x + 6} = \frac{-2}{x - 2} + \frac{3}{x - 3}$$

✓CHECK: The decomposition can be checked by adding the fractions.

$$\frac{-2}{x-2} + \frac{3}{x-3} = \frac{-2(x-3) + 3(x-2)}{(x-2)(x-3)} = \frac{x}{(x-2)(x-3)}$$
$$= \frac{x}{x^2 - 5x + 6}$$

NOW WORK PROBLEM 9.

The numbers to be found in the partial fraction decomposition can sometimes be found more readily by using suitable choices for x (which may include complex numbers) in the identity obtained after fractions have been cleared. In Example 1, the identity after clearing fractions, equation (3), is

$$x = A(x - 3) + B(x - 2)$$

If we let $x = 2$ in this expression, the term containing B drops out, leaving $2 = A(-1)$, or $A = -2$. Similarly, if we let $x = 3$, the term containing A drops out, leaving $3 = B$. Thus, as before, $A = -2$ and $B = 3$.

We use this method in the next example.

2

CASE 2: Q has repeated linear factors.

If the polynomial Q has a repeated factor, say $(x - a)^n$, $n \geq 2$ an integer, then, in the partial fraction decomposition of P/Q, we allow for the terms

$$\frac{A_1}{x - a} + \frac{A_2}{(x - a)^2} + \cdots + \frac{A_n}{(x - a)^n}$$

where the numbers $A_1, A_2, \ldots, A_n$ are to be determined.

EXAMPLE 2 Repeated Linear Factors

Write the partial fraction decomposition of $\dfrac{x+2}{x^3-2x^2+x}$.

Solution First, we factor the denominator,

$$x^3 - 2x^2 + x = x(x^2 - 2x + 1) = x(x-1)^2$$

and find that the denominator has the nonrepeated linear factor x and the repeated linear factor $(x-1)^2$. By Case 1, we must allow for the term A/x in the decomposition; and, by Case 2, we must allow for the terms $B/(x-1) + C/(x-1)^2$ in the decomposition.

Thus, we write

$$\frac{x+2}{x^3-2x^2+x} = \frac{A}{x} + \frac{B}{x-1} + \frac{C}{(x-1)^2} \tag{4}$$

Again, we clear fractions by multiplying each side by $x^3 - 2x^2 + x = x(x-1)^2$. The result is the identity

$$x + 2 = A(x-1)^2 + Bx(x-1) + Cx \tag{5}$$

If we let $x = 0$ in this expression, the terms containing B and C drop out, leaving $2 = A(-1)^2$, or $A = 2$. Similarly, if we let $x = 1$, the terms containing A and B drop out, leaving $3 = C$. Thus, equation (5) becomes

$$x + 2 = 2(x-1)^2 + Bx(x-1) + 3x$$

Now, let $x = 2$ (any choice other than 0 or 1 will work as well). The result is

$$\begin{aligned} 4 &= 2(1)^2 + B(2)(1) + 3(2) \\ 2B &= 4 - 2 - 6 = -4 \\ B &= -2 \end{aligned}$$

Thus, we have $A = 2$, $B = -2$, and $C = 3$.

From equation (4), the partial fraction decomposition is

$$\frac{x+2}{x^3-2x^2+x} = \frac{2}{x} + \frac{-2}{x-1} + \frac{3}{(x-1)^2}$$

EXAMPLE 3 Repeated Linear Factors

Write the partial fraction decomposition of $\dfrac{x^3-8}{x^2(x-1)^3}$.

Solution The denominator contains the repeated linear factor x^2 and the repeated linear factor $(x-1)^3$. Thus, the partial fraction decomposition takes the form

$$\frac{x^3-8}{x^2(x-1)^3} = \frac{A}{x} + \frac{B}{x^2} + \frac{C}{x-1} + \frac{D}{(x-1)^2} + \frac{E}{(x-1)^3} \tag{6}$$

As before, we clear fractions and obtain the identity

$$x^3 - 8 = Ax(x-1)^3 + B(x-1)^3 + Cx^2(x-1)^2 + Dx^2(x-1) + Ex^2 \tag{7}$$

Let $x = 0$. (Do you see why this choice was made?) Then

$$\begin{aligned} -8 &= B(-1) \\ B &= 8 \end{aligned}$$

Now let $x = 1$ in equation (7). Then

$$-7 = E$$

Use $B = 8$ and $E = -7$ in equation (7) and collect like terms.

$$\begin{aligned} x^3 - 8 &= Ax(x-1)^3 + 8(x-1)^3 \\ &\quad + Cx^2(x-1)^2 + Dx^2(x-1) - 7x^2 \\ x^3 - 8 - 8(x^3 - 3x^2 + 3x - 1) + 7x^2 &= Ax(x-1)^3 + Cx^2(x-1)^2 + Dx^2(x-1) \\ -7x^3 + 31x^2 - 24x &= x(x-1)[A(x-1)^2 + Cx(x-1) + Dx] \\ x(x-1)(-7x+24) &= x(x-1)[A(x-1)^2 + Cx(x-1) + Dx] \\ -7x + 24 &= A(x-1)^2 + Cx(x-1) + Dx \qquad (8) \end{aligned}$$

We now work with equation (8). Let $x = 0$. Then

$$24 = A$$

Now let $x = 1$ in equation (8). Then

$$17 = D$$

Use $A = 24$ and $D = 17$ in equation (8) and collect like terms.

$$\begin{aligned} -7x + 24 &= 24(x-1)^2 + Cx(x-1) + 17x \\ -24x^2 + 48x - 24 - 17x - 7x + 24 &= Cx(x-1) \\ -24x^2 + 24x &= Cx(x-1) \\ -24x(x-1) &= Cx(x-1) \\ -24 &= C \end{aligned}$$

We now know all the numbers A, B, C, D, and E, so, from equation (6), we have the decomposition

$$\frac{x^3 - 8}{x^2(x-1)^3} = \frac{24}{x} + \frac{8}{x^2} + \frac{-24}{x-1} + \frac{17}{(x-1)^2} + \frac{-7}{(x-1)^3}$$

The method employed in Example 3, although somewhat tedious, is still preferable to solving the system of five equations containing five variables that the expansion of equation (6) leads to.

NOW WORK PROBLEM 15.

3 The final two cases involve irreducible quadratic factors. As mentioned in Section 4.4, a quadratic factor is irreducible if it cannot be factored into linear factors with real coefficients. A quadratic expression $ax^2 + bx + c$ is irreducible whenever $b^2 - 4ac < 0$. For example, $x^2 + x + 1$ and $x^2 + 4$ are irreducible.

CASE 3: Q contains a nonrepeated irreducible quadratic factor.

If Q contains a nonrepeated irreducible quadratic factor of the form $ax^2 + bx + c$, then, in the partial fraction decomposition of P/Q, allow for the term

$$\frac{Ax + B}{ax^2 + bx + c}$$

where the numbers A and B are to be determined.

EXAMPLE 4 Nonrepeated Irreducible Quadratic Factor

Write the partial factor decomposition of $\dfrac{3x-5}{x^3-1}$.

Solution We factor the denominator,

$$x^3 - 1 = (x-1)(x^2+x+1)$$

and find that it has a nonrepeated linear factor $x - 1$ and a nonrepeated irreducible quadratic factor $x^2 + x + 1$. Thus, we allow for the term $A/(x-1)$ by Case 1, and we allow for the term $(Bx + C)/(x^2 + x + 1)$ by Case 3. Hence, we write

$$\frac{3x-5}{x^3-1} = \frac{A}{x-1} + \frac{Bx+C}{x^2+x+1} \tag{9}$$

We clear fractions by multiplying each side of equation (9) by $x^3 - 1 = (x-1)(x^2+x+1)$ to obtain the identity

$$3x - 5 = A(x^2+x+1) + (Bx+C)(x-1) \tag{10}$$

Now let $x = 1$. Then equation (10) gives $-2 = A(3)$, or $A = -\frac{2}{3}$. We use this value of A in equation (10) and simplify.

$$\begin{aligned}
3x - 5 &= -\frac{2}{3}(x^2+x+1) + (Bx+C)(x-1) \\
3(3x-5) &= -2(x^2+x+1) + 3(Bx+C)(x-1) && \text{Multiply each side by 3.} \\
9x - 15 &= -2x^2 - 2x - 2 + 3(Bx+C)(x-1) \\
2x^2 + 11x - 13 &= 3(Bx+C)(x-1) && \text{Collect terms.} \\
(2x+13)(x-1) &= 3(Bx+C)(x-1) && \text{Factor the left side.} \\
2x + 13 &= 3Bx + 3C && \text{Equate coefficients.} \\
2 = 3B &\quad \text{and} \quad 13 = 3C \\
B = \frac{2}{3} &\qquad C = \frac{13}{3}
\end{aligned}$$

Thus, from equation (9), we see that

$$\frac{3x-5}{x^3-1} = \frac{-\frac{2}{3}}{x-1} + \frac{\frac{2}{3}x + \frac{13}{3}}{x^2+x+1}$$

NOW WORK PROBLEM 17.

4

CASE 4: Q contains repeated irreducible quadratic factors.

If the polynomial Q contains a repeated irreducible quadratic factor $(ax^2 + bx + c)^n$, $n \geq 2$, n an integer, then, in the partial fraction decomposition of P/Q, allow for the terms

$$\frac{A_1x + B_1}{ax^2+bx+c} + \frac{A_2x + B_2}{(ax^2+bx+c)^2} + \cdots + \frac{A_nx + B_n}{(ax^2+bx+c)^n}$$

where the numbers $A_1, B_1, A_2, B_2, \ldots, A_n, B_n$ are to be determined.

EXAMPLE 5 Repeated Irreducible Quadratic Factor

Write the partial fraction decomposition of $\dfrac{x^3 + x^2}{(x^2 + 4)^2}$.

Solution The denominator contains the repeated irreducible quadratic factor $(x^2 + 4)^2$, so we write

$$\frac{x^3 + x^2}{(x^2 + 4)^2} = \frac{Ax + B}{x^2 + 4} + \frac{Cx + D}{(x^2 + 4)^2} \qquad (11)$$

We clear fractions to obtain

$$x^3 + x^2 = (Ax + B)(x^2 + 4) + Cx + D$$

Collecting like terms yields

$$x^3 + x^2 = Ax^3 + Bx^2 + (4A + C)x + D + 4B$$

Equating coefficients, we arrive at the system

$$\begin{cases} A = 1 \\ B = 1 \\ 4A + C = 0 \\ D + 4B = 0 \end{cases}$$

The solution is $A = 1, B = 1, C = -4, D = -4$. Hence, from equation (11),

$$\frac{x^3 + x^2}{(x^2 + 4)^2} = \frac{x + 1}{x^2 + 4} + \frac{-4x - 4}{(x^2 + 4)^2}$$

NOW WORK PROBLEM 31.

6.7 EXERCISES

In Problems 1–8, tell whether the given rational expression is proper or improper. If improper, rewrite it as the sum of a polynomial and a proper rational expression.

1. $\dfrac{x}{x^2 - 1}$
2. $\dfrac{5x + 2}{x^3 - 1}$
3. $\dfrac{x^2 + 5}{x^2 - 4}$
4. $\dfrac{3x^2 - 2}{x^2 - 1}$
5. $\dfrac{5x^3 + 2x - 1}{x^2 - 4}$
6. $\dfrac{3x^4 + x^2 - 2}{x^3 + 8}$
7. $\dfrac{x(x - 1)}{(x + 4)(x - 3)}$
8. $\dfrac{2x(x^2 + 4)}{x^2 + 1}$

In Problems 9–42, write the partial fraction decomposition of each rational expression.

9. $\dfrac{4}{x(x - 1)}$
10. $\dfrac{3x}{(x + 2)(x - 1)}$
11. $\dfrac{1}{x(x^2 + 1)}$
12. $\dfrac{1}{(x + 1)(x^2 + 4)}$
13. $\dfrac{x}{(x - 1)(x - 2)}$
14. $\dfrac{3x}{(x + 2)(x - 4)}$
15. $\dfrac{x^2}{(x - 1)^2(x + 1)}$
16. $\dfrac{x + 1}{x^2(x - 2)}$
17. $\dfrac{1}{x^3 - 8}$
18. $\dfrac{2x + 4}{x^3 - 1}$
19. $\dfrac{x^2}{(x - 1)^2(x + 1)^2}$
20. $\dfrac{x + 1}{x^2(x - 2)^2}$
21. $\dfrac{x - 3}{(x + 2)(x + 1)^2}$
22. $\dfrac{x^2 + x}{(x + 2)(x - 1)^2}$
23. $\dfrac{x + 4}{x^2(x^2 + 4)}$
24. $\dfrac{10x^2 + 2x}{(x - 1)^2(x^2 + 2)}$
25. $\dfrac{x^2 + 2x + 3}{(x + 1)(x^2 + 2x + 4)}$
26. $\dfrac{x^2 - 11x - 18}{x(x^2 + 3x + 3)}$

27. $\dfrac{x}{(3x-2)(2x+1)}$ **28.** $\dfrac{1}{(2x+3)(4x-1)}$ **29.** $\dfrac{x}{x^2+2x-3}$

30. $\dfrac{x^2-x-8}{(x+1)(x^2+5x+6)}$ **31.** $\dfrac{x^2+2x+3}{(x^2+4)^2}$ **32.** $\dfrac{x^3+1}{(x^2+16)^2}$

33. $\dfrac{7x+3}{x^3-2x^2-3x}$ **34.** $\dfrac{x^5+1}{x^6-x^4}$ **35.** $\dfrac{x^2}{x^3-4x^2+5x-2}$

36. $\dfrac{x^2+1}{x^3+x^2-5x+3}$ **37.** $\dfrac{x^3}{(x^2+16)^3}$ **38.** $\dfrac{x^2}{(x^2+4)^3}$

39. $\dfrac{4}{2x^2-5x-3}$ **40.** $\dfrac{4x}{2x^2+3x-2}$ **41.** $\dfrac{2x+3}{x^4-9x^2}$

42. $\dfrac{x^2+9}{x^4-2x^2-8}$

CHAPTER REVIEW

THINGS TO KNOW

Systems of equations

Systems with no solutions are inconsistent. Systems with a solution are consistent.

Consistent systems of linear equations have either a unique solution or an infinite number of solutions.

Matrix	Rectangular array of numbers, called entries
m by n matrix	Matrix with m rows and n columns
Identity matrix I	Square matrix whose diagonal entries are 1's, while all other entries are 0's
Inverse of a matrix	A^{-1} is the inverse of A if $AA^{-1} = A^{-1}A = I$
Nonsingular matrix	A matrix that has an inverse

Linear programming problem

Maximize (or minimize) a linear objective function, $z = Ax + By$, subject to certain conditions, or constraints, expressible as linear inequalities in x and y.

Feasible point

A point (x, y) that satisfies the constraints of a linear programming problem

Location of solution

If a linear programming problem has a solution, it is located at a corner point of the graph of the feasible points.

If a linear programming problem has multiple solutions, at least one of them is located at a corner point of the graph of the feasible points.

In either case, the corresponding value of the objective function is unique.

HOW TO

Solve a system of linear equations using the method of substitution

Solve a system of linear equations using the method of elimination

Solve a system of linear equations using matrices

Solve a system of linear equations using determinants

Recognize equal matrices

Add and subtract matrices

Multiply matrices

Find the inverse of a nonsingular matrix

Solve a system of linear equations using the inverse of a matrix

Graph a system of linear inequalities

Find the corner points of the graph of a system of linear inequalities

Solve linear programming problems

Write the partial fraction decomposition of a rational expression

FILL-IN-THE-BLANK ITEMS

1. If a system of equations has no solution, it is said to be _______________.
2. An m by n rectangular array of numbers is called a(n) _______________.
3. Cramer's Rule uses _______________ to solve a system of linear equations.
4. The matrix used to represent a system of linear equations is called a(n) _______________ matrix.
5. A matrix B, for which $AB = I_n$, the identity matrix, is called the _______________ of A.
6. A matrix that has the same number of rows as columns is called a(n) _______________ matrix.
7. In the algebra of matrices, the matrix that has properties similar to the number 1 is called the _______________ matrix.
8. The graph of a linear inequality is called a(n) _______________.
9. A linear programming problem requires that a linear expression, called the _______________ _______________, be maximized or minimized.
10. Each point that satisfies the constraints of a linear programming problem is called a(n) _______________ _______________.
11. A rational function is called _______________ if the degree of its numerator is less than the degree of its denominator.

TRUE/FALSE ITEMS

T F 1. A system of two linear equations containing two unknowns always has at least one solution.

T F 2. The augmented matrix of a system of two equations containing three variables has two rows and four columns.

T F 3. A 3 by 3 determinant can never equal 0.

T F 4. A consistent system of equations will have exactly one solution.

T F 5. Every square matrix has an inverse.

T F 6. Matrix multiplication is commutative.

T F 7. Any pair of matrices can be multiplied.

T F 8. The graph of a linear inequality is a half-plane.

T F 9. The graph of a system of linear inequalities is sometimes unbounded.

T F 10. If a linear programming problem has a solution, it is located at a corner point of the graph of the feasible points.

T F 11. The factors of the denominator of a rational expression are used to arrive at the partial fraction decomposition.

REVIEW EXERCISES

Blue problem numbers indicate the authors' suggestions for use in a Practice Test.

In Problems 1–20, solve each system of equations algebraically using the method of substitution or the method of elimination. If the system has no solution, say that it is inconsistent. Verify your result using a graphing utility.

1. $\begin{cases} 2x - y = 5 \\ 5x + 2y = 8 \end{cases}$

2. $\begin{cases} 2x + 3y = 2 \\ 7x - y = 3 \end{cases}$

3. $\begin{cases} 3x - 4y = 4 \\ x - 3y = \frac{1}{2} \end{cases}$

4. $\begin{cases} 2x + y = 0 \\ 5x - 4y = -\frac{13}{2} \end{cases}$

5. $\begin{cases} x - 2y - 4 = 0 \\ 3x + 2y - 4 = 0 \end{cases}$

6. $\begin{cases} x - 3y + 5 = 0 \\ 2x + 3y - 5 = 0 \end{cases}$

7. $\begin{cases} y = 2x - 5 \\ x = 3y + 4 \end{cases}$

8. $\begin{cases} x = 5y + 2 \\ y = 5x + 2 \end{cases}$

9. $\begin{cases} x - y + 4 = 0 \\ \frac{1}{2}x + \frac{1}{6}y + \frac{2}{5} = 0 \end{cases}$

10. $\begin{cases} x + \frac{1}{4}y = 2 \\ y + 4x + 2 = 0 \end{cases}$

11. $\begin{cases} x - 2y - 8 = 0 \\ 2x + 2y - 10 = 0 \end{cases}$

12. $\begin{cases} x - 3y + \frac{7}{2} = 0 \\ \frac{1}{2}x + 3y - 5 = 0 \end{cases}$

13. $\begin{cases} y - 2x = 11 \\ 2y - 3x = 18 \end{cases}$ **14.** $\begin{cases} 3x - 4y - 12 = 0 \\ 5x + 2y + 6 = 0 \end{cases}$ **15.** $\begin{cases} 2x + 3y - 13 = 0 \\ 3x - 2y = 0 \end{cases}$ **16.** $\begin{cases} 4x + 5y = 21 \\ 5x + 6y = 42 \end{cases}$

17. $\begin{cases} 3x - 2y = 8 \\ x - \frac{2}{3}y = 12 \end{cases}$ **18.** $\begin{cases} 2x + 5y = 10 \\ 4x + 10y = 15 \end{cases}$ **19.** $\begin{cases} x + 2y - z = 6 \\ 2x - y + 3z = -13 \\ 3x - 2y + 3z = -16 \end{cases}$ **20.** $\begin{cases} x + 5y - z = 2 \\ 2x + y + z = 7 \\ x - y + 2z = 11 \end{cases}$

In Problems 21–28, use the following matrices to compute each expression. Verify your result using a graphing utility.

$$A = \begin{bmatrix} 1 & 0 \\ 2 & 4 \\ -1 & 2 \end{bmatrix} \quad B = \begin{bmatrix} 4 & -3 & 0 \\ 1 & 1 & -2 \end{bmatrix} \quad C = \begin{bmatrix} 3 & -4 \\ 1 & 5 \\ 5 & -2 \end{bmatrix}$$

21. $A + C$ **22.** $A - C$ **23.** $6A$ **24.** $-4B$

25. AB **26.** BA **27.** CB **28.** BC

In Problems 29–34, find the inverse of each matrix algebraically, if there is one. If there is not an inverse, say that the matrix is singular. Verify your result using a graphing utility.

29. $\begin{bmatrix} 4 & 6 \\ 1 & 3 \end{bmatrix}$ **30.** $\begin{bmatrix} -3 & 2 \\ 1 & -2 \end{bmatrix}$ **31.** $\begin{bmatrix} 1 & 3 & 3 \\ 1 & 2 & 1 \\ 1 & -1 & 2 \end{bmatrix}$

32. $\begin{bmatrix} 3 & 1 & 2 \\ 3 & 2 & -1 \\ 1 & 1 & 1 \end{bmatrix}$ **33.** $\begin{bmatrix} 4 & -8 \\ -1 & 2 \end{bmatrix}$ **34.** $\begin{bmatrix} -3 & 1 \\ -6 & 2 \end{bmatrix}$

In Problems 35–44, solve each system of equations algebraically using matrices. If the system has no solution, say that it is inconsistent. Verify your result using a graphing utility.

35. $\begin{cases} 3x - 2y = 1 \\ 10x + 10y = 5 \end{cases}$ **36.** $\begin{cases} 3x + 2y = 6 \\ x - y = -\frac{1}{2} \end{cases}$ **37.** $\begin{cases} 5x + 6y - 3z = 6 \\ 4x - 7y - 2z = -3 \\ 3x + y - 7z = 1 \end{cases}$

38. $\begin{cases} 2x + y + z = 5 \\ 4x - y - 3z = 1 \\ 8x + y - z = 5 \end{cases}$ **39.** $\begin{cases} x - 2z = 1 \\ 2x + 3y = -3 \\ 4x - 3y - 4z = 3 \end{cases}$ **40.** $\begin{cases} x + 2y - z = 2 \\ 2x - 2y + z = -1 \\ 6x + 4y + 3z = 5 \end{cases}$

41. $\begin{cases} x - y + z = 0 \\ x - y - 5z - 6 = 0 \\ 2x - 2y + z - 1 = 0 \end{cases}$ **42.** $\begin{cases} 4x - 3y + 5z = 0 \\ 2x + 4y - 3z = 0 \\ 6x + 2y + z = 0 \end{cases}$ **43.** $\begin{cases} x - y - z - t = 1 \\ 2x + y + z + 2t = 3 \\ x - 2y - 2z - 3t = 0 \\ 3x - 4y + z + 5t = -3 \end{cases}$

44. $\begin{cases} x - 3y + 3z - t = 4 \\ x + 2y - z = -3 \\ x + 3z + 2t = 3 \\ x + y + 5z = 6 \end{cases}$

In Problems 45–50, find the value of each determinant algebraically. Verify your result using a graphing utility.

45. $\begin{vmatrix} 3 & 4 \\ 1 & 3 \end{vmatrix}$ **46.** $\begin{vmatrix} -4 & 0 \\ 1 & 3 \end{vmatrix}$ **47.** $\begin{vmatrix} 1 & 4 & 0 \\ -1 & 2 & 6 \\ 4 & 1 & 3 \end{vmatrix}$ **48.** $\begin{vmatrix} 2 & 3 & 10 \\ 0 & 1 & 5 \\ -1 & 2 & 3 \end{vmatrix}$ **49.** $\begin{vmatrix} 2 & 1 & -3 \\ 5 & 0 & 1 \\ 2 & 6 & 0 \end{vmatrix}$ **50.** $\begin{vmatrix} -2 & 1 & 0 \\ 1 & 2 & 3 \\ -1 & 4 & 2 \end{vmatrix}$

In Problems 51–56, use Cramer's Rule, if applicable, to solve each system.

51. $\begin{cases} x - 2y = 4 \\ 3x + 2y = 4 \end{cases}$ **52.** $\begin{cases} x - 3y = -5 \\ 2x + 3y = 5 \end{cases}$ **53.** $\begin{cases} 2x + 3y - 13 = 0 \\ 3x - 2y = 0 \end{cases}$

54. $\begin{cases} 3x - 4y - 12 = 0 \\ 5x + 2y + 6 = 0 \end{cases}$ **55.** $\begin{cases} x + 2y - z = 6 \\ 2x - y + 3z = -13 \\ 3x - 2y + 3z = -16 \end{cases}$ **56.** $\begin{cases} x - y + z = 8 \\ 2x + 3y - z = -2 \\ 3x - y - 9z = 9 \end{cases}$

In Problems 57–62, graph each system of inequalities. Tell whether the graph is bounded or unbounded, and label the corner points.

57. $\begin{cases} -2x + y \le 2 \\ x + y \ge 2 \end{cases}$

58. $\begin{cases} x - 2y \le 6 \\ 2x + y \ge 2 \end{cases}$

59. $\begin{cases} x \ge 0 \\ y \ge 0 \\ x + y \le 4 \\ 2x + 3y \le 6 \end{cases}$

60. $\begin{cases} x \ge 0 \\ y \ge 0 \\ 3x + y \ge 6 \\ 2x + y \ge 2 \end{cases}$

61. $\begin{cases} x \ge 0 \\ y \ge 0 \\ 2x + y \le 8 \\ x + 2y \ge 2 \end{cases}$

62. $\begin{cases} x \ge 0 \\ y \ge 0 \\ 3x + y \le 9 \\ 2x + 3y \ge 6 \end{cases}$

In Problems 63–68, solve each linear programming problem.

63. Maximize $z = 3x + 4y$ subject to $x \ge 0,\ y \ge 0,\ 3x + 2y \ge 6,\ x + y \le 8$

64. Maximize $z = 2x + 4y$ subject to $x \ge 0,\ y \ge 0,\ x + y \le 6,\ x \ge 2$

65. Minimize $z = 3x + 5y$ subject to $x \ge 0,\ y \ge 0,\ x + y \ge 1,\ 3x + 2y \le 12,\ x + 3y \le 12$

66. Minimize $z = 3x + y$ subject to $x \ge 0,\ y \ge 0,\ x \le 8,\ y \le 6,\ 2x + y \ge 4$

67. Maximize $z = 5x + 4y$ subject to $x \ge 0,\ y \ge 0,\ x + 2y \ge 2,\ 3x + 4y \le 12,\ y \ge x$

68. Maximize $z = 4x + 5y$ subject to $x \ge 0,\ y \ge 0,\ 2x + 3y \ge 6,\ x \ge 4,\ 2x + y \le 12$

In Problems 69–78, write the partial decomposition of each rational expression.

69. $\dfrac{6}{x(x-4)}$

70. $\dfrac{x}{(x+2)(x-3)}$

71. $\dfrac{x-4}{x^2(x-1)}$

72. $\dfrac{2x-6}{(x-2)^2(x-1)}$

73. $\dfrac{x}{(x^2+9)(x+1)}$

74. $\dfrac{3x}{(x-2)(x^2+1)}$

75. $\dfrac{x^3}{(x^2+4)^2}$

76. $\dfrac{x^3+1}{(x^2+16)^2}$

77. $\dfrac{x^2}{(x^2+1)(x^2-1)}$

78. $\dfrac{4}{(x^2+4)(x^2-1)}$

79. Find A such that the system of equations has infinitely many solutions.

$$\begin{cases} 2x + 5y = 5 \\ 4x + 10y = A \end{cases}$$

80. Find A such that the system in Problem 79 is inconsistent.

81. Curve Fitting Find the quadratic function $y = ax^2 + bx + c$ that passes through the three points $(0, 1)$, $(1, 0)$, and $(-2, 1)$.

82. Curve Fitting Find the general equation of the circle that passes through the three points $(0, 1)$, $(1, 0)$, and $(-2, 1)$.
[Hint: The general equation of a circle is $x^2 + y^2 + Dx + Ey + F = 0$***.]***

83. Blending Coffee A coffee distributor is blending a new coffee that will cost \$3.90 per pound. It will consist of a blend of \$3.00 per pound coffee and \$6.00 per pound coffee. What amounts of each type of coffee should be mixed to achieve the desired blend?
[Hint: Assume that the weight of the blended coffee is 100 pounds.]

84. Farming A 1000-acre farm in Illinois is used to raise corn and soy beans. The cost per acre for raising corn is \$65 and the cost per acre for soy beans is \$45. If \$54,325 has been budgeted for costs and all the acreage is to be used, how many acres should be allocated for each crop?

85. Cookie Orders A cookie company makes three kinds of cookies, oatmeal raisin, chocolate chip, and shortbread, packaged in small, medium, and large boxes. The small box contains 1 dozen oatmeal raisin and 1 dozen chocolate chip; the medium box has 2 dozen oatmeal raisin, 1 dozen chocolate chip, and 1 dozen shortbread; the large box contains 2 dozen oatmeal raisin, 2 dozen chocolate chip, and 3 dozen shortbread. If you require exactly 15 dozen oatmeal raisin, 10 dozen chocolate chip, and 11 dozen shortbread, how many of each size box should you buy?

86. **Mixed Nuts** A store that specializes in selling nuts has 72 pounds of cashews and 120 pounds of peanuts available. These are to be mixed in 12-ounce packages as follows: a lower-priced package containing 8 ounces of peanuts and 4 ounces of cashews and a quality package containing 6 ounces of peanuts and 6 ounces of cashews.
   (a) Use x to denote the number of lower-priced packages, and use y to denote the number of quality packages. Write a system of linear inequalities that describes the possible number of each kind of package.
   (b) Graph the system and label the corner points.

87. A small rectangular lot has a perimeter of 68 feet. If its diagonal is 26 feet, what are the dimensions of the lot?

88. The area of a rectangular window is 4 square feet. If the diagonal measures $2\sqrt{2}$ feet, what are the dimensions of the window?

89. **Geometry** A certain right triangle has a perimeter of 14 inches. If the hypotenuse is 6 inches long, what are the lengths of the legs?

90. **Geometry** A certain isosceles triangle has a perimeter of 18 inches. If the altitude is 6 inches, what is the length of the base?

91. **Building a Fence** How much fence is required to enclose 5000 square feet by two squares whose sides are in the ratio of 1:2?

92. **Mixing Acids** A chemistry laboratory has three containers of hydrochloric acid, HCl. One container holds a solution with a concentration of 10% HCl, the second holds 25% HCl, and the third holds 40% HCl. How many liters of each should be mixed to obtain 100 liters of a solution with a concentration of 30% HCl? Construct a table showing some of the possible combinations.

93. **Calculating Allowances** Katy, Mike, Danny, and Colleen agreed to do yard work at home for \$45 to be split among them. After they finished, their father determined that Mike deserves twice what Katy gets, Katy and Colleen deserve the same amount, and Danny deserves half of what Katy gets. How much does each receive?

94. **Finding the Speed of the Jet Stream** On a flight between Midway Airport in Chicago and Ft. Lauderdale, Florida, a Boeing 737 jet maintains an airspeed of 475 miles per hour. If the trip from Chicago to Ft. Lauderdale takes 2 hours, 30 minutes and the return flight takes 2 hours, 50 minutes, what is the speed of the jet stream? (Assume that the speed of the jet stream remains constant at the various altitudes of the plane. and that the plane flies with the jetstream one way and against it the other way.)

95. **Constant Rate Jobs** If Bruce and Bryce work together for 1 hour and 20 minutes, they will finish a certain job. If Bryce and Marty work together for 1 hour and 36 minutes, the same job can be finished. If Marty and Bruce work together, they can complete this job in 2 hours and 40 minutes. How long will it take each of them working alone to finish the job?

96. **Maximizing Profit on Figurines** A factory manufactures two kinds of ceramic figurines: a dancing girl and a mermaid, each requiring three processes: molding, painting, and glazing. The daily labor available for molding is no more than 90 work-hours, labor available for painting does not exceed 120 work-hours, and labor available for glazing is no more than 60 work-hours. The dancing girl requires 3 work-hours for molding, 6 work-hours for painting, and 2 work-hours for glazing. The mermaid requires 3 work-hours for molding, 4 work-hours for painting, and 3 work-hours for glazing. If the profit on each figurine is \$25 for dancing girls and \$30 for mermaids, how many of each should be produced each day to maximize profit? If management decides to produce the number of each figurine that maximizes profit, determine which of these processes has excess work-hours assigned to it.

97. **Minimizing Production Cost** A factory produces gasoline engines and diesel engines. Each week the factory is obligated to deliver at least 20 gasoline engines and at least 15 diesel engines. Due to physical limitations, however, the factory cannot make more than 60 gasoline engines nor more than 40 diesel engines. Finally, to prevent layoffs, a total of at least 50 engines must be produced. If gasoline engines cost \$450 each to produce and diesel engines cost \$550 each to produce, how many of each should be produced per week to minimize the cost? What is the excess capacity of the factory; that is, how many of each kind of engine are being produced in excess of the number that the factory is obligated to deliver?

98. Describe four ways of solving a system of three linear equations containing three variables. Which method do you prefer? Why?

CHAPTER PROJECTS

1. **Markov Chains** A **Markov chain** (or process) is one in which future outcomes are determined by a current state. Future outcomes are based on probabilities. The probability of moving to a certain state depends only on the state previously occupied and does not vary with time. An example of a Markov chain would be the maximum education achieved by children based on the highest education attained of their parents, where the states are (1) earned college degree, (2) high-school diploma only, (3) elementary school only. If p_{ij} is the probability of moving from state i to state j, then the **transition matrix** is the $m \times m$ matrix.

$$P = \begin{bmatrix} p_{11} & p_{12} & \cdots & p_{1m} \\ \vdots & \vdots & & \vdots \\ p_{m1} & p_{m2} & \cdots & p_{mm} \end{bmatrix}$$

The following table represents the probabilities of the highest educational level of children based on the highest educational level of their parents.

	Maximum Education that Children Achieve		
Highest Educational Level of Parents	**College**	**High School**	**Elementary**
College	80%	18%	2%
High school	40%	50%	10%
Elementary	20%	60%	20%

For example, the table shows that the probability P_{21} is 40% that parents with a high-school education (row 2) will have children with a college education (column 1).

(a) Convert the percentages to decimals.

(b) What is the transition matrix?

(c) Sum across the rows. What do you notice? Why do you think that you obtained this result?

(d) If P is the transition matrix of a Markov chain, then the (i, j)th entry of P^n (nth power of P) gives the probability of passing from state i to state j in n stages. What is the probability that a grandchild of a college graduate is a college graduate?

(e) What is the probability that the grandchild of a high school graduate finishes college?

(f) The row vector $v^{(0)} = [0.236 \quad 0.581 \quad 0.183]$ represents the proportion of the U.S. population that has college, high school, and elementary school, respectively, as the highest educational level in 1996.* In a Markov chain the probability distribution $v^{(k)}$ after k stages is $v^{(k)} = v^{(0)}P^k$, where P^k is the kth power of the transition matrix. What will be the distribution of highest educational attainment of the grandchildren of the current population?

(g) Calculate $P^3, P^4, P^5, \ldots$. Continue until the matrix does not change. This is called the long-run distribution. What is the long-run distribution of highest educational attainment of the population?

**Source:* U.S. Census Bureau.

2. **Using Matrices to Find the Line of Best Fit** Recall, in Section 2.2, that we found the line of best fit for a set of data that appeared to be linearly related using a graphing utility. The equation was of the form $y = ax + b$, where a was the slope and b was the y-intercept. We now use matrices to find the line of best fit for the following data:

x	3	4	5	6	7	8	9
y	4	6	7	10	12	14	16

We can think of these points as being values of x and y that satisfy the equation

$$b + ax = y$$

where a is the slope and b is the y-intercept. The ordered pairs (x, y) yield a system of equations as follows:

$$\begin{cases} b + 3a = 4 \\ b + 4a = 6 \\ \vdots \\ b + 9a = 16 \end{cases}$$

which can be written in matrix form as follows:

$$\begin{bmatrix} 1 & 3 \\ 1 & 4 \\ 1 & 5 \\ 1 & 6 \\ 1 & 7 \\ 1 & 8 \\ 1 & 9 \end{bmatrix} \begin{bmatrix} b \\ a \end{bmatrix} = \begin{bmatrix} 4 \\ 6 \\ 7 \\ 10 \\ 12 \\ 14 \\ 16 \end{bmatrix}$$

$$AB = Y$$

It can be shown that the solution $\boldsymbol{B} = \begin{bmatrix} b \\ a \end{bmatrix}$ to this system is given by

$$\boldsymbol{B} = (\boldsymbol{A}^T\boldsymbol{A})^{-1}\boldsymbol{A}^T\boldsymbol{Y}$$

where $\boldsymbol{A}^T$ is found by putting the elements in row 1 of $\boldsymbol{A}$ in column 1 of $\boldsymbol{A}^T$, the elements in row 2 of $\boldsymbol{A}$ are inserted in column 2 of $\boldsymbol{A}^T$, and so on. In other words, the rows of $\boldsymbol{A}$ become the columns of $\boldsymbol{A}^T$.

(a) Find $\boldsymbol{A}^T$.

(b) Find $\boldsymbol{B} = \begin{bmatrix} b \\ c \end{bmatrix}$.

(c) Find the line of best fit using the matrix method described above.

(d) Verify your results by using a graphing utility to find the line of best fit.

3. **CBL Experiment** Two students walk toward each other at a *constant* rate. Plotting distance against time on the same viewing window for each student results in a system of linear equations. By finding the intersection point of the two graphs, the time and location where the students pass each other is determined. (Activity 4, Real-World Math with the CBL System, 1994.)

CHAPTER 7

Sequences; Induction; The Binomial Theorem

Will Humans Overwhelm the Earth?

Two hundred years ago the Rev. Thomas Robert Malthus, an English economist and mathematician, anonymously published an essay predicting that the world's burgeoning population would overwhelm the Earth's capacity to sustain it.

Malthus's gloomy forecast, called 'An Essay on the Principle of Population As It Affects the Future Improvement of Society,' was condemned by Karl Marx, Friedrich Engels and many other theorists, and it was still striking sparks last week at a meeting in Philadelphia of the American Anthropological Society. Despite continuing controversy, it was clear that Malthus's conjectures are far from dead.

Among the scores of special conferences organized for the 5,000 participating anthropologists, many touched directly and indirectly on the Malthusian dilemma: Although global food supplies increase arithmetically, the population increases geometrically—a vastly faster rate. The consequence Malthus believed, was that poverty and the misery it imposes will inevitably increase unless the population increase is curbed. ("Will Humans Overwhelm the Earth? The Debate Continues," Malcolm W. Browne, *New York Times*, December 8, 1998.)

See Chapter Project 1.

Preparing for this Chapter

Before getting started on this chapter, review the following concepts:

- **For Section 7.3: Compound Interest** ***(Section 5.6)***

  Rules of Exponents (Appendix, Section 4, p. 693)
- **For Section 7.1: Evaluating Functions (p. 102)**

Outline

This chapter introduces topics that are covered in more detail in courses titled *Discrete Mathematics*. Applications of these topics can be found in the fields of computer science, engineering, business and economics, the social sciences, and the physical and biological sciences.

The chapter may be divided into three independent parts:

Sections 7.1–7.3, Sequences, which are functions whose domain is the set of natural numbers. Sequences form the basis for the *recursively defined functions* and *recursive procedures* used in computer programming.

Section 7.4, Mathematical Induction, a technique for proving theorems involving the natural numbers.

Section 7.5, the Binomial Theorem, a formula for the expansion of $(x + a)^n$, where n is any natural number.

7.1 SEQUENCES

1. Write the First Several Terms of a Sequence
2. Write the Terms of a Sequence Defined by a Recursion Formula
3. Write a Sequence in Summation Notation
4. Find the Sum of a Sequence by Hand and by Using a Graphing Utility

A **sequence** is a function whose domain is the set of positive integers.

Because a sequence is a function, it will have a graph. In Figure 1(a), you will recognize the graph of the function $f(x) = 1/x$, $x > 0$. If all the points on this graph were removed except those whose x-coordinates are positive integers, that is, if all points were removed except $(1, 1), (2, \frac{1}{2}), (3, \frac{1}{3})$, and so on, the remaining points would be the graph of the sequence $f(n) = 1/n$, as shown in Figure 1(b).

Figure 1

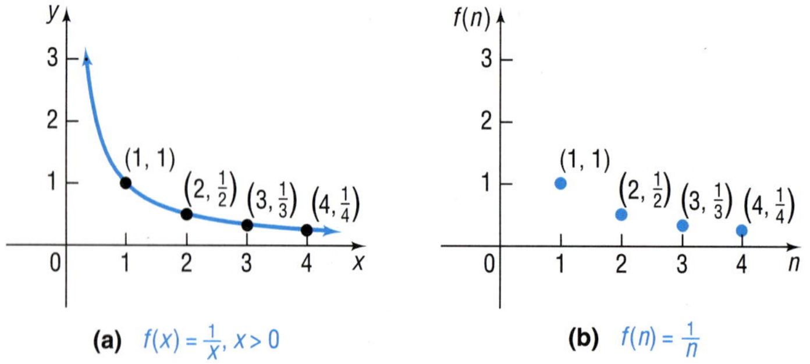

(a) $f(x) = \frac{1}{x}, x > 0$ **(b)** $f(n) = \frac{1}{n}$

A sequence is usually represented by listing its values in order. For example, the sequence whose graph is given in Figure 1(b) might be represented as

$$f(1), f(2), f(3), f(4), \ldots \quad \text{or} \quad 1, \frac{1}{2}, \frac{1}{3}, \frac{1}{4}, \ldots$$

The list never ends, as the ellipsis dots indicate. The numbers in this ordered list are called the **terms** of the sequence.

In dealing with sequences, we usually use subscripted letters, for example, a_1 to represent the first term, a_2 for the second term, a_3 for the third term, and so on. Thus, for the sequence $f(n) = 1/n$, we write

$$a_1 = f(1) = 1, \quad a_2 = f(2) = \frac{1}{2}, \quad a_3 = f(3) = \frac{1}{3}, \quad a_4 = f(4) = \frac{1}{4}, \ldots, \quad a_n = f(n) = \frac{1}{n} \ldots$$

In other words, we usually do not use the traditional function notation $f(n)$ for sequences. For this particular sequence, we have a rule for the nth term, which is $a_n = 1/n$, so it is easy to find any term of the sequence.

When a formula for the nth term of a sequence is known, rather than write out the terms of the sequence, we usually represent the entire sequence by placing braces around the formula for the nth term. For example, the sequence whose nth term is $b_n = (\frac{1}{2})^n$ may be represented as

$$\{b_n\} = \left\{\left(\frac{1}{2}\right)^n\right\}$$

or by

$$b_1 = \frac{1}{2}, \quad b_2 = \frac{1}{4}, \quad b_3 = \frac{1}{8}, \quad \ldots, \quad b_n = \left(\frac{1}{2}\right)^n, \quad \ldots$$

EXAMPLE 1 Writing the First Several Terms of a Sequence

Write down the first six terms of the following sequence and graph it.

$$\{a_n\} = \left\{\frac{n-1}{n}\right\}$$

Figure 2

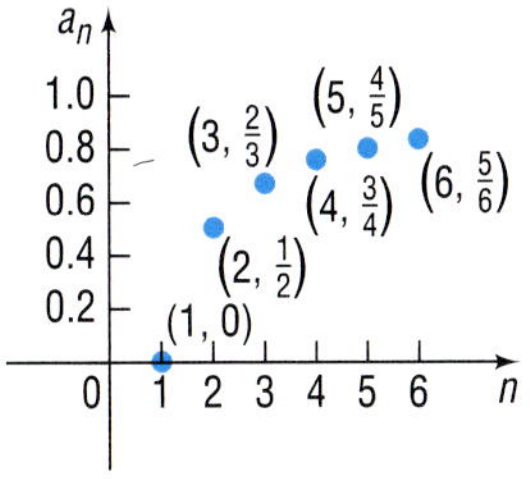

Solution

$$a_1 = 0, \quad a_2 = \frac{1}{2}, \quad a_3 = \frac{2}{3}, \quad a_4 = \frac{3}{4}, \quad a_5 = \frac{4}{5}, \quad a_6 = \frac{5}{6}$$

See Figure 2.

Graphing utilities can be used to write the terms of a sequence and graph them, as the following example illustrates.

EXAMPLE 2 Using a Graphing Utility to Write the First Several Terms of a Sequence

Use a graphing utility to write the first six terms of the following sequence and graph it.

$$\{a_n\} = \left\{\frac{n-1}{n}\right\}$$

Solution Figure 3 shows the sequence generated on a TI-83 graphing calculator. We can see the first few terms of the sequence on the screen. You need to press the right arrow key to scroll right in order to see the remaining terms of the sequence. Figure 4 shows a graph of the sequence. Notice that the first term of the sequence is not visible since it lies on the x-axis. TRACEing the graph will allow you to determine the terms of the sequence.

Figure 3

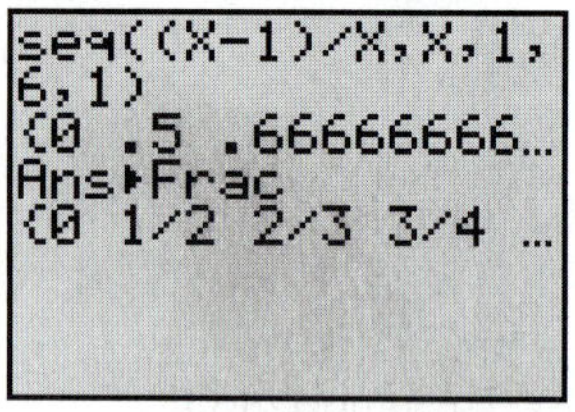

Figure 4

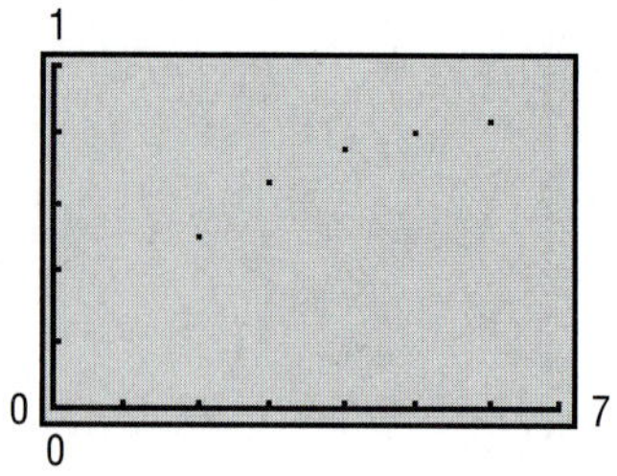

We will usually provide solutions done by hand. The reader is encouraged to verify solutions using a graphing utility.

EXAMPLE 3 Writing the First Several Terms of a Sequence

Write down the first six terms of the following sequence and graph it by hand.

$$\{b_n\} = \left\{(-1)^{n-1}\left(\frac{2}{n}\right)\right\}$$

Solution

$$b_1 = 2, \quad b_2 = -1, \quad b_3 = \frac{2}{3}, \quad b_4 = -\frac{1}{2}, \quad b_5 = \frac{2}{5}, \quad b_6 = -\frac{1}{3}$$

See Figure 5.

Figure 5

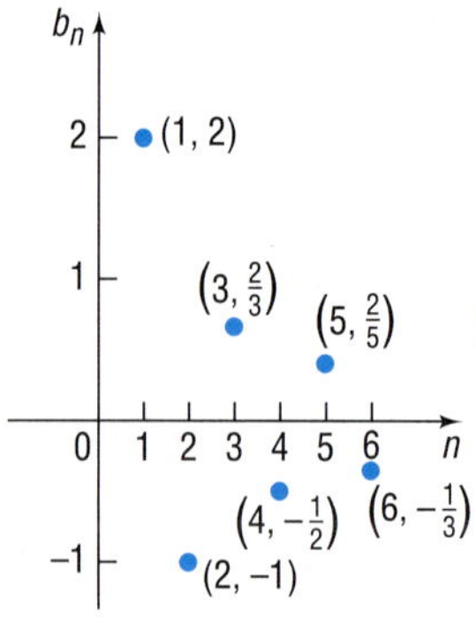

EXAMPLE 4 Writing the First Several Terms of a Sequence

Write down the first six terms of the following sequence and graph it by hand.

$$\{c_n\} = \begin{Bmatrix} n & \text{if } n \text{ is even} \\ 1/n & \text{if } n \text{ is odd} \end{Bmatrix}$$

Solution

$$c_1 = 1, \quad c_2 = 2, \quad c_3 = \frac{1}{3}, \quad c_4 = 4, \quad c_5 = \frac{1}{5}, \quad c_6 = 6$$

See Figure 6.

Figure 6

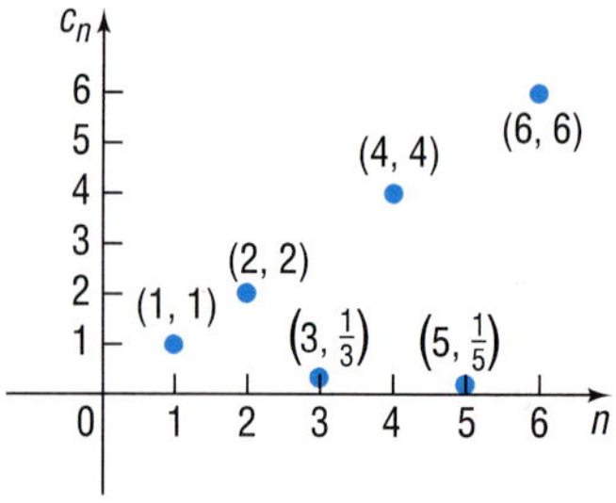

NOW WORK PROBLEMS 3 AND 5.

Sometimes a sequence is indicated by an observed pattern in the first few terms that makes it possible to infer the makeup of the nth term. In the example that follows, a sufficient number of terms of the sequence is given so that a natural choice for the nth term is suggested.

EXAMPLE 5 Determining a Sequence from a Pattern

(a) $e, \frac{e^2}{2}, \frac{e^3}{3}, \frac{e^4}{4}, \ldots$ $\quad a_n = \frac{e^n}{n}$

(b) $1, \frac{1}{3}, \frac{1}{9}, \frac{1}{27}, \ldots$ $\quad b_n = \frac{1}{3^{n-1}}$

(c) $1, 3, 5, 7, \ldots$ $\quad c_n = 2n - 1$

(d) $1, 4, 9, 16, 25, \ldots$ $\quad d_n = n^2$

(e) $1, -\frac{1}{2}, \frac{1}{3}, -\frac{1}{4}, \frac{1}{5}, \ldots$ $\quad e_n = (-1)^{n+1}\left(\frac{1}{n}\right)$

Notice in the sequence $\{e_n\}$ in Example 5(e) that the signs of the terms **alternate.** When this occurs, we use factors such as $(-1)^{n+1}$, which equals 1 if n is odd and -1 if n is even, or $(-1)^n$, which equals -1 if n is odd and 1 if n is even.

NOW WORK PROBLEM 13.

The Factorial Symbol

If $n \geq 0$ is an integer, the **factorial symbol $n!$** is defined as follows:

$$0! = 1 \qquad 1! = 1$$

$$n! = n(n-1) \cdot \ldots \cdot 3 \cdot 2 \cdot 1 \qquad \text{if } n \geq 2$$

For example, $2! = 2 \cdot 1 = 2$, $3! = 3 \cdot 2 \cdot 1 = 6$, $4! = 4 \cdot 3 \cdot 2 \cdot 1 = 24$, and so on. Table 1 lists the values of $n!$ for $0 \leq n \leq 6$.

Table 1

n	0	1	2	3	4	5	6
$n!$	1	1	2	6	24	120	720

Because

$$n! = n\underbrace{(n-1)(n-2) \cdot \ldots \cdot 3 \cdot 2 \cdot 1}_{(n-1)!}$$

we can use the formula

$$n! = n(n-1)!$$

to find successive factorials. For example, because $6! = 720$, we have

$$7! = 7 \cdot 6! = 7(720) = 5040$$

and

$$8! = 8 \cdot 7! = 8(5040) = 40{,}320$$

Comment: Your calculator has a factorial key. Use it to see how fast factorials increase in value. Find the value of 69!. What happens when you try to find 70!? In fact, 70! is larger than 10^{100} (a *googol*), the largest number most calculators can display.

Recursion Formulas

2 A second way of defining a sequence is to assign a value to the first (or the first few) terms and specify the nth term by a formula or equation that involves one or more of the terms preceding it. Sequences defined this way are said to be defined **recursively,** and the rule or formula is called a **recursive formula.**

EXAMPLE 6 Writing the Terms of a Recursively Defined Sequence

Write down the first five terms of the following recursively defined sequence.

$$s_1 = 1, \qquad s_n = 4s_{n-1}$$

Solution The first term is given as $s_1 = 1$. To get the second term, we use $n = 2$ in the formula to get $s_2 = 4s_1 = 4 \cdot 1 = 4$. To get the third term, we use $n = 3$ in the formula to get $s_3 = 4s_2 = 4 \cdot 4 = 16$. To get a new term requires that we know the value of the preceding term. The first five terms are

$$\begin{aligned} s_1 &= 1 \\ s_2 &= 4 \cdot 1 = 4 \\ s_3 &= 4 \cdot 4 = 16 \\ s_4 &= 4 \cdot 16 = 64 \\ s_5 &= 4 \cdot 64 = 256 \end{aligned}$$

Graphing utilities can be used to generate recursively defined sequences.

EXAMPLE 7 Using a Graphing Utility to Write the Terms of a Recursively Defined Sequence

Use a graphing utility to write down the first five terms of the following recursively defined sequence.

$$s_1 = 1, \qquad s_n = 4s_{n-1}$$

Solution First, put the graphing utility into SEQuence mode. Using $Y =$, enter the recursive formula into the graphing utility. See Figure 7(a). Next, set up the viewing window to generate the desired sequence. Finally, graph the recursion relation and use TRACE to determine the terms in the sequence. See Figure 7(b) For example, we see that the fourth term of the sequence is 64.

Note: If your graphing utility is capable of generating tables, the TABLE feature displays the terms in the sequence. See Table 2.

Figure 7

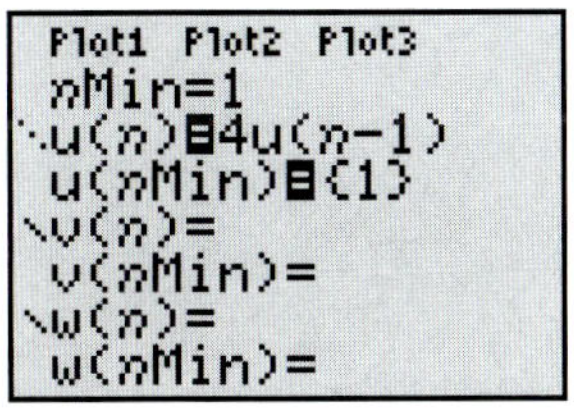

(a)

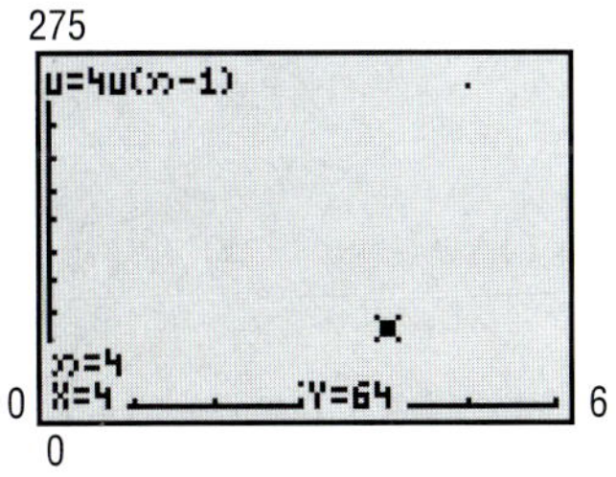

(b)

Table 2

n	$u(n)$
1	1
2	4
3	16
4	64
5	256
6	1024
7	4096

u(n)=4u(n−1)

EXAMPLE 8 Writing the Terms of a Recursively Defined Sequence

Write down the first five terms of the following recursively defined sequence.

$$u_1 = 1, \qquad u_2 = 1, \qquad u_{n+2} = u_n + u_{n+1}$$

Solution We are given the first two terms. To get the third term requires that we know each of the previous two terms. Thus,

$$\begin{aligned} u_1 &= 1 \\ u_2 &= 1 \\ u_3 &= u_1 + u_2 = 1 + 1 = 2 \\ u_4 &= u_2 + u_3 = 1 + 2 = 3 \\ u_5 &= u_3 + u_4 = 2 + 3 = 5 \end{aligned}$$

The sequence defined in Example 8 is called a **Fibonacci sequence,** and the terms of this sequence are called **Fibonacci numbers.** These numbers appear in a wide variety of applications (see Problems 71 and 72).

NOW WORK PROBLEMS 21 AND 29.

EXAMPLE 9 Credit Card Debt

Beth just charged \$5000 on a VISA card that charges 1.5% interest per month on any unpaid balance. She can afford to pay \$100 toward the balance each month. Her balance at the beginning of each month after making a \$100 payment is given by the recursively defined sequence

$$b_1 = \$5000, \qquad b_n = 1.015b_{n-1} - 100$$

(a) Determine Beth's balance after making the first payment; that is, determine b_2, the balance at the beginning of the second month.

(b) Using a graphing utility, graph the recursively defined sequence.

(c) Using a graphing utility, determine when Beth's balance will be below \$4000. How many payments of \$100 have been made?

(d) Using a graphing utility, determine the number of payments it will take until Beth pays off the balance. What is the total of all the payments?

(e) What was Beth's interest expense?

Solution (a) $b_2 = 1.015b_1 - 100 = 1.015(5000) - 100 = \$4975$.

(b) In SEQuence mode, we enter the recursively defined sequence $b_n = 1.015b_{n-1} - 100$ with $b_1 = \$5000$. See Figure 8 for the graph.

Figure 8

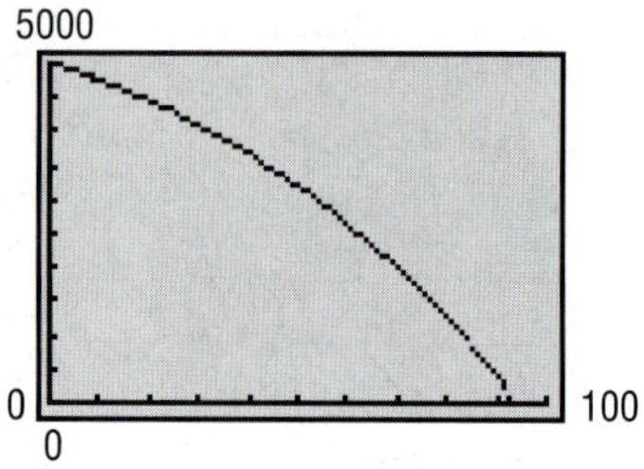

(c) Using the TABLE feature of the TI-83, scroll down until the balance is less than \$4000. See Table 3. (If your graphing utility does not have a TABLE feature, TRACE the graph until the balance is less than \$4000.)

Table 3

n	u(n)
27	4212.2
28	4175.3
29	4138
30	4100
31	4061.5
32	4022.5
33	3982.8

u(n)=1.015u(n−1...

At the beginning of the thirty-third month the balance is less than \$4000. At this time, 32 payments of \$100 each have been made.

(d) Using the TABLE feature, scroll down until the balance is negative. From Table 4, we see that the balance at the beginning of the ninety-fourth month is \$11.01. At this time, 93 payments of \$100 each have been made. The final payment required is \$11.01. Therefore, her total payments were 93(\$100) + \$11.01 = \$9311.01.

Table 4

n	u(n)
89	488.49
90	395.82
91	301.75
92	206.28
93	109.37
94	11.014
95	−88.82

u(n)=1.015u(n−1...

(e) Since Beth's initial balance was \$5000 and her total payments were \$9311.01, the interest expense for this debt was \$9311.01 − \$5000 = \$4311.01

NOW WORK PROBLEM 67.

Summation Notation

3 It is often important to be able to find the sum of the first n terms of a sequence $\{a_n\}$, that is,

$$a_1 + a_2 + a_3 + \cdots + a_n \tag{1}$$

Rather than write down all these terms, we introduce a more concise way to express the sum, called **summation notation.** Using summation notation, we would write the sum (1) as

$$a_1 + a_2 + a_3 + \cdots + a_n = \sum_{k=1}^{n} a_k$$

The symbol Σ (a stylized version of the Greek letter sigma, which is an S in our alphabet) is simply an instruction to sum, or add up, the terms. The integer k is called the **index** of the sum; it tells you where to start the sum and where to end it. Therefore, the expression

$$\sum_{k=1}^{n} a_k \tag{2}$$

is an instruction to add the terms a_k of the sequence $\{a_n\}$ from $k = 1$ through $k = n$. We read expression (2) as "the sum of a_k from $k = 1$ to $k = n$."

EXAMPLE 10 Expanding Summation Notation

Write out each sum.

(a) $\displaystyle\sum_{k=1}^{n} \frac{1}{k}$ (b) $\displaystyle\sum_{k=1}^{n} k!$

Solution (a) $\displaystyle\sum_{k=1}^{n} \frac{1}{k} = 1 + \frac{1}{2} + \frac{1}{3} + \cdots + \frac{1}{n}$ (b) $\displaystyle\sum_{k=1}^{n} k! = 1! + 2! + \cdots + n!$

EXAMPLE 11 Writing a Sum in Summation Notation

Express each sum using summation notation.

(a) $1^2 + 2^2 + 3^2 + \cdots + 9^2$ (b) $1 + \dfrac{1}{2} + \dfrac{1}{4} + \dfrac{1}{8} + \cdots + \dfrac{1}{2^{n-1}}$

Solution (a) The sum $1^2 + 2^2 + 3^2 + \cdots + 9^2$ has 9 terms, each of the form k^2, and starts at $k = 1$ and ends at $k = 9$. Thus,

$$1^2 + 2^2 + 3^2 + \cdots + 9^2 = \sum_{k=1}^{9} k^2$$

(b) The sum

$$1 + \frac{1}{2} + \frac{1}{4} + \frac{1}{8} + \cdots + \frac{1}{2^{n-1}}$$

has n terms, each of the form $1/2^{k-1}$, and starts at $k = 1$ and ends at $k = n$. Thus,

$$1 + \frac{1}{2} + \frac{1}{4} + \frac{1}{8} + \cdots + \frac{1}{2^{n-1}} = \sum_{k=1}^{n} \frac{1}{2^{k-1}}$$

The index of summation need not always begin at 1 or end at n; for example,

$$\sum_{k=0}^{n-1} \frac{1}{2^k} = 1 + \frac{1}{2} + \frac{1}{4} + \cdots + \frac{1}{2^{n-1}}$$

Letters other than k may be used as the index. For example,

$$\sum_{j=1}^{n} j! \quad \text{and} \quad \sum_{i=1}^{n} i!$$

each represent the same sum as the one given in Example 10(b).

NOW WORK PROBLEMS 49 AND 59.

Adding the First n Terms of a Sequence

4 Next, we list some properties of sequences using summation notation. These properties are useful for adding the terms of a sequence.

THEOREM **Properties of Sequences**

If $\{a_n\}$ and $\{b_n\}$ are two sequences and c is a real number, then

1. $\displaystyle\sum_{k=1}^{n} c = c \cdot n$
2. $\displaystyle\sum_{k=1}^{n} ca_k = c\sum_{k=1}^{n} a_k$
3. $\displaystyle\sum_{k=1}^{n} (a_k + b_k) = \sum_{k=1}^{n} a_k + \sum_{k=1}^{n} b_k$
4. $\displaystyle\sum_{k=1}^{n} (a_k - b_k) = \sum_{k=1}^{n} a_k - \sum_{k=1}^{n} b_k$
5. $\displaystyle\sum_{k=1}^{n} a_k = \sum_{k=1}^{j} a_k + \sum_{k=j+1}^{n} a_k$, when $1 < j < n$
6. $\displaystyle\sum_{k=1}^{n} k = \frac{n(n+1)}{2}$
7. $\displaystyle\sum_{k=1}^{n} k^2 = \frac{n(n+1)(2n+1)}{6}$
8. $\displaystyle\sum_{k=1}^{n} k^3 = \left(\frac{n(n+1)}{2}\right)^2$

We shall not prove these properties. The proofs of 1 through 5 are based on properties of real numbers; the proofs of 6 through 8 require mathematical induction, which is discussed in Section 7.4.

EXAMPLE 12 Finding the Sum of a Sequence

Find the sum of each sequence.

(a) $\displaystyle\sum_{k=1}^{5} 3k$ (b) $\displaystyle\sum_{k=1}^{3} (k^3 + 1)$ (c) $\displaystyle\sum_{k=1}^{4} (k^2 - 7k + 2)$

Solution (a) $\sum_{k=1}^{5} 3k = 3\sum_{k=1}^{5} k = 3\left(\frac{5(5+1)}{2}\right) = 3(15) = 45$

Property 2   Property 6

(b) $\sum_{k=1}^{3} (k^3 + 1) = \sum_{k=1}^{3} k^3 + \sum_{k=1}^{3} 1$ *Property 3*

$= \left(\frac{3(3+1)}{2}\right)^2 + 1(3)$ *Properties 1, 8*

$= 36 + 3$

$= 39$

(c) $\sum_{k=1}^{4} (k^2 - 7k + 2) = \sum_{k=1}^{4} k^2 - \sum_{k=1}^{4} 7k + \sum_{k=1}^{4} 2$ *Properties 3, 4*

$= \sum_{k=1}^{4} k^2 - 7\sum_{k=1}^{4} k + \sum_{k=1}^{4} 2$ *Property 2*

$= \frac{4(4+1)(2 \cdot 4 + 1)}{6} - 7\left(\frac{4(4+1)}{2}\right) + 2(4)$

Properties 1, 6, 7

$= 30 - 70 + 8$

$= -32$

EXAMPLE 13 Using a Graphing Utility to Find the Sum of a Sequence

Using a graphing utility, find the sum of each sequence.

(a) $\sum_{k=1}^{5} 3k$ (b) $\sum_{k=1}^{3} (k^3 + 1)$ (c) $\sum_{k=1}^{4} (k^2 - 7k + 2)$

Solution (a) Figure 9 shows the solution using a TI-83 graphing calculator.

Figure 9

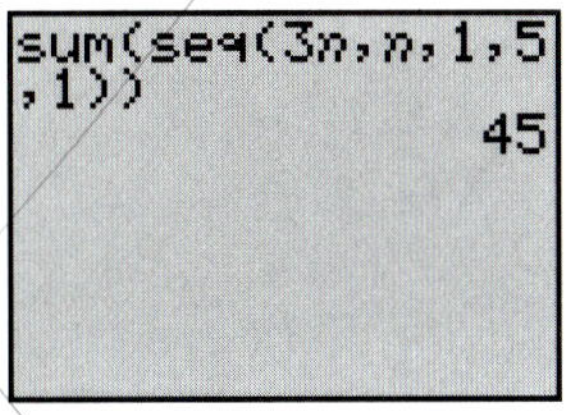

Thus, $\sum_{k=1}^{5} 3k = 45$.

(b) Figure 10 shows the solution using a TI-83 graphing calculator.

Figure 10

Figure 11

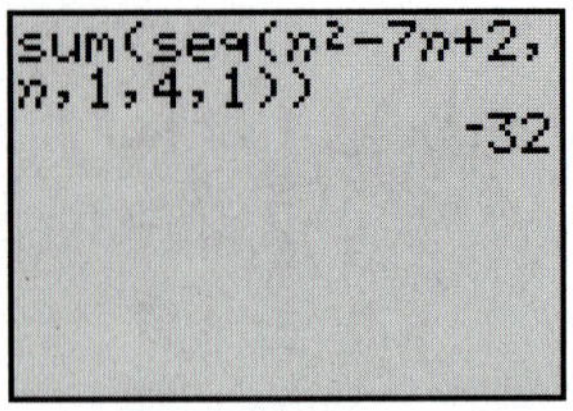

Thus, $\sum_{k=1}^{3}(k^3+1)=39$.

(c) Figure 11 shows the solution using a TI-83 graphing calculator.

Thus, $\sum_{k=1}^{4}(k^2-7k+2)=-32$.

NOW WORK PROBLEM **39.**

7.1 EXERCISES

In Problems 1–12, write down the first five terms of each sequence by hand. Verify your results using a graphing utility.

1. $\{n\}$
2. $\{n^2+1\}$
3. $\left\{\dfrac{n}{n+2}\right\}$
4. $\left\{\dfrac{2n+1}{2n}\right\}$
5. $\{(-1)^{n+1}n^2\}$
6. $\left\{(-1)^{n-1}\left(\dfrac{n}{2n-1}\right)\right\}$
7. $\left\{\dfrac{2^n}{3^n+1}\right\}$
8. $\left\{\left(\dfrac{4}{3}\right)^n\right\}$
9. $\left\{\dfrac{(-1)^n}{(n+1)(n+2)}\right\}$
10. $\left\{\dfrac{3^n}{n}\right\}$
11. $\left\{\dfrac{n}{e^n}\right\}$
12. $\left\{\dfrac{n^2}{2^n}\right\}$

In Problems 13–20, the given pattern continues. Write down the nth term of each sequence suggested by the pattern.

13. $\dfrac{1}{2}, \dfrac{2}{3}, \dfrac{3}{4}, \dfrac{4}{5}, \dots$
14. $\dfrac{1}{1\cdot 2}, \dfrac{1}{2\cdot 3}, \dfrac{1}{3\cdot 4}, \dfrac{1}{4\cdot 5}, \dots$
15. $1, \dfrac{1}{2}, \dfrac{1}{4}, \dfrac{1}{8}, \dots$
16. $\dfrac{2}{3}, \dfrac{4}{9}, \dfrac{8}{27}, \dfrac{16}{81}, \dots$
17. $1, -1, 1, -1, 1, -1, \dots$
18. $1, \dfrac{1}{2}, 3, \dfrac{1}{4}, 5, \dfrac{1}{6}, 7, \dfrac{1}{8}, \dots$
19. $1, -2, 3, -4, 5, -6, \dots$
20. $2, -4, 6, -8, 10, \dots$

In Problems 21–34, a sequence is defined recursively. Write the first five terms by hand. ~~Use a graphing utility to verify your results.~~

21. $a_1=2;\ a_n=3+a_{n-1}$
22. $a_1=3;\ a_n=4-a_{n-1}$
23. $a_1=-2;\ a_n=n+a_{n-1}$
24. $a_1=1;\ a_n=n-a_{n-1}$
25. $a_1=5;\ a_n=2a_{n-1}$
26. $a_1=2;\ a_n=-a_{n-1}$
27. $a_1=3;\ a_n=\dfrac{a_{n-1}}{n}$
28. $a_1=-2;\ a_n=n+3a_{n-1}$
29. $a_1=1;\ a_2=2;\ a_n=a_{n-1}\cdot a_{n-2}$
30. $a_1=-1;\ a_2=1;\ a_n=a_{n-2}+na_{n-1}$
31. $a_1=A;\ a_n=a_{n-1}+d$
32. $a_1=A;\ a_n=ra_{n-1},\ r\neq 0$
33. $a_1=\sqrt{2};\ a_n=\sqrt{2+a_{n-1}}$
34. $a_1=\sqrt{2};\ a_n=\sqrt{a_{n-1}/2}$

In Problems 35–46, find the sum of each sequence. ~~Verify your results using a graphing utility.~~

35. $\sum_{k=1}^{10} 5$
36. $\sum_{k=1}^{20} 8$
37. $\sum_{k=1}^{6} k$
38. $\sum_{k=1}^{4}(-k)$
39. $\sum_{k=1}^{5}(5k+3)$
40. $\sum_{k=1}^{6}(3k-7)$
41. $\sum_{k=1}^{3}(k^2+4)$
42. $\sum_{k=0}^{4}(k^2-4)$
43. $\sum_{k=1}^{6}(-1)^k 2^k$
44. $\sum_{k=1}^{4}(-1)^k 3^k$
45. $\sum_{k=1}^{4}(k^3-1)$
46. $\sum_{k=0}^{3}(k^3+2)$

In Problems 47–56, write out each sum.

47. $\sum_{k=1}^{n} (k+2)$ **48.** $\sum_{k=1}^{n} (2k+1)$ **49.** $\sum_{k=1}^{n} \frac{k^2}{2}$ **50.** $\sum_{k=1}^{n} (k+1)^2$

51. $\sum_{k=0}^{n} \frac{1}{3^k}$ **52.** $\sum_{k=0}^{n} \left(\frac{3}{2}\right)^k$ **53.** $\sum_{k=0}^{n-1} \frac{1}{3^{k+1}}$ **54.** $\sum_{k=0}^{n-1} (2k+1)$

55. $\sum_{k=2}^{n} (-1)^k \ln k$ **56.** $\sum_{k=3}^{n} (-1)^{k+1} 2^k$

In Problems 57–66, express each sum using summation notation.

57. $1 + 2 + 3 + \cdots + 20$

58. $1^3 + 2^3 + 3^3 + \cdots + 8^3$

59. $\frac{1}{2} + \frac{2}{3} + \frac{3}{4} + \cdots + \frac{13}{13+1}$

60. $1 + 3 + 5 + 7 + \cdots + [2(12) - 1]$

61. $1 - \frac{1}{3} + \frac{1}{9} - \frac{1}{27} + \cdots + (-1)^6\left(\frac{1}{3^6}\right)$

62. $\frac{2}{3} - \frac{4}{9} + \frac{8}{27} - \cdots + (-1)^{11+1}\left(\frac{2}{3}\right)^{11}$

63. $3 + \frac{3^2}{2} + \frac{3^3}{3} + \cdots + \frac{3^n}{n}$

64. $\frac{1}{e} + \frac{2}{e^2} + \frac{3}{e^3} + \cdots + \frac{n}{e^n}$

65. $a + (a+d) + (a+2d) + \cdots + (a+nd)$

66. $a + ar + ar^2 + \cdots + ar^{n-1}$

67. Credit Card Debt John has a balance of \$3000 on his Discover card that charges 1% interest per month on any unpaid balance. John can afford to pay \$100 toward the balance each month. His balance at the beginning of each month after making a \$100 payment is given by the recursively defined sequence

$$b_1 = \$3000, \qquad b_n = 1.01b_{n-1} - 100$$

(a) Determine John's balance after making the first payment. That is, determine b_2, the balance at the beginning of the second month.
(b) Using a graphing utility, graph the recursively defined sequence.
(c) Using a graphing utility, determine when John's balance will be below \$2000. How many payments of \$100 have been made?
(d) Using a graphing utility, determine the number of payments that it will take until John pays off the balance. What is the total of all the payments?
(e) What was John's interest expense?

68. Car Loans Phil bought a car by taking out a loan for \$18,500 at 0.5% interest per month. Phil's normal monthly payment is \$434.47 per month, but he decides that he can afford to pay \$100 extra toward the balance each month. His balance each month is given by the recursively defined sequence

$$b_1 = \$18{,}500, \qquad b_n = 1.005b_{n-1} - 534.47$$

(a) Determine Phil's balance after making the first payment. That is, determine b_2, the balance at the beginning of the second month.
(b) Using a graphing utility, graph the recursively defined sequence.
(c) Using a graphing utility, determine when Phil's balance will be below \$10,000. How many payments of \$534.47 have been made?
(d) Using a graphing utility, determine the number of payments it will take until Phil pays off the balance. What is the total of all the payments?
(e) What was Phil's interest expense?

69. Trout Population A pond currently has 2000 trout in it. A fish hatchery decides to add an additional 20 trout each month. In addition, it is known that the trout population is growing 3% per year. The size of the population at the beginning of each month is given by the recursively defined sequence

$$p_1 = 2000, \qquad p_n = 1.03p_{n-1} + 20$$

(a) How many trout are in the pond at the beginning of the second month? That is, what is p_2?
(b) Using a graphing utility, graph the recursively defined sequence.
(c) Using a graphing utility, determine how long will it be before the trout population reaches 5000.

70. Environmental Control The Environmental Protection Agency (EPA) determines that Maple Lake has 250 tons of pollutants as a result of industrial waste and that 10% of the pollutant present is neutralized by solar oxidation every year. The EPA imposes new pollution control laws that result in 15 tons of new pollutant entering the lake each year. The amount of pollutant in the lake each year is given by the recursively defined sequence

$$p_1 = 250, \qquad p_n = 0.9p_{n-1} + 15$$

(a) Determine the amount of pollutant in the lake at the beginning of the second year. That is, determine p_2.
(b) Using a graphing utility, provide a graph of the recursively defined sequence for the next 200 years.
(c) What is the equilibrium level of pollution in Maple Lake? That is, what is $\lim_{n\to\infty} p_n$?

71. Growth of a Rabbit Colony A colony of rabbits begins with one pair of mature rabbits, which will produce a pair of offspring (one male, one female) each month. Assume that all rabbits mature in 1 month and produce a pair of offspring (one male, one female) after 2 months. If no rabbits ever die, how many pairs of mature rabbits are there after 7 months?
[Hint: A Fibonacci sequence models this colony. Do you see why?]

72. Fibonacci Sequence Let

$$u_n = \frac{(1+\sqrt{5})^n - (1-\sqrt{5})^n}{2^n\sqrt{5}}$$

define the nth term of a sequence.
(a) Show that $u_1 = 1$ and $u_2 = 1$.
(b) Show that $u_{n+2} = u_{n+1} + u_n$.
(c) Draw the conclusion that $\{u_n\}$ is a Fibonacci sequence.

73. Pascal's Triangle Divide the triangular array shown (called Pascal's triangle) using diagonal lines as indicated. Find the sum of the numbers in each of these diagonal rows. Do you recognize this sequence?

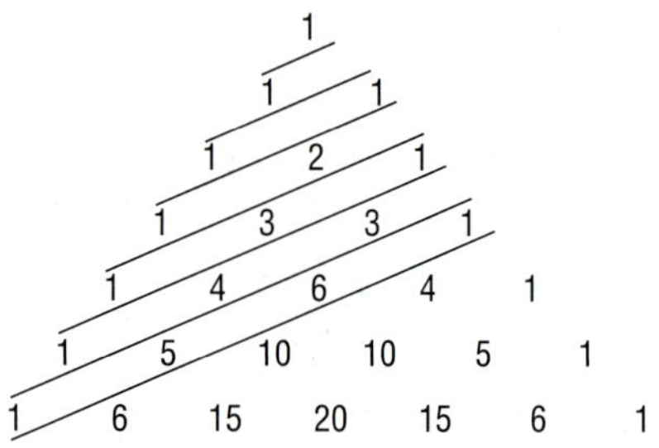

74. Fibonacci Sequence Use the result of Problem 72 to do the following problems:
(a) Write the first 10 terms of the Fibonacci sequence.
(b) Compute the ratio $\dfrac{u_{n+1}}{u_n}$ for the first 10 terms.
(c) As n gets large, what numbers does the ratio approach? This number is referred to as the **golden ratio.** Rectangles whose sides are in this ratio were considered pleasing to the eye by the Greeks. For example, the facade of the Parthenon was constructed using the golden ratio.
(d) Compute the ratio $\dfrac{u_n}{u_{n+1}}$ for the first 10 terms.
(e) As n gets large, what number does the ratio approach? This number is also referred to as the **golden ratio.** This ratio is believed to have been used in the construction of the Great Pyramid in Egypt. The ratio equals the sum of the areas of the four face triangles divided by the total surface area of the Great Pyramid.

75. Investigate various applications that lead to a Fibonacci sequence, such as art, architecture, or financial markets. Write an essay on these applications.

7.2 ARITHMETIC SEQUENCES

1 Determine If a Sequence Is Arithmetic

2 Find a Formula for an Arithmetic Sequence

3 Find the Sum of an Arithmetic Sequence

1 When the difference between successive terms of a sequence is always the same number, the sequence is called **arithmetic.** Thus, an **arithmetic sequence*** may be defined recursively as $a_1 = a$, $a_n - a_{n-1} = d$, or as

*Sometimes called an **arithmetic progression.**

$$a_1 = a, \qquad a_n = a_{n-1} + d \tag{1}$$

where $a = a_1$ and d are real numbers. The number a is the first term, and the number d is called the **common difference.**

Thus, the terms of an arithmetic sequence with first term a and common difference d follow the pattern

$$a, \quad a + d, \quad a + 2d, \quad a + 3d, \ldots$$

EXAMPLE 1 Determining If a Sequence Is Arithmetic

The sequence

$$4, 7, 10, 13, \ldots$$

is arithmetic since the difference of successive terms is 3. The first term is 4, and the common difference is 3.

EXAMPLE 2 Determining If a Sequence Is Arithmetic

Show that the following sequence is arithmetic. Find the first term and the common difference.

$$\{s_n\} = \{3n + 5\}$$

Solution The first term is $s_1 = 3 \cdot 1 + 5 = 8$. The nth and $(n - 1)$st terms of the sequence $\{s_n\}$ are

$$s_n = 3n + 5 \quad \text{and} \quad s_{n-1} = 3(n - 1) + 5 = 3n + 2$$

Their difference is

$$s_n - s_{n-1} = (3n + 5) - (3n + 2) = 5 - 2 = 3$$

Since the difference of two successive terms does not depend on n, the common difference is 3 and the sequence is arithmetic.

EXAMPLE 3 Determining If a Sequence Is Arithmetic

Show that the sequence $\{t_n\} = \{4 - n\}$ is arithmetic. Find the first term and the common difference.

Solution The first term is $t_1 = 4 - 1 = 3$. The nth and $(n - 1)$st terms are

$$t_n = 4 - n \quad \text{and} \quad t_{n-1} = 4 - (n - 1) = 5 - n$$

Their difference is

$$t_n - t_{n-1} = (4 - n) - (5 - n) = 4 - 5 = -1$$

The difference of two successive terms does not depend on n; it always equals the same number, -1. Hence, $\{t_n\}$ is an arithmetic sequence whose common difference is -1.

NOW WORK PROBLEM 3.

2 Suppose that a is the first term of an arithmetic sequence whose common difference is d. We seek a formula for the nth term, a_n. To see the pattern, we write down the first few terms:

$$\begin{aligned} a_1 &= a \\ a_2 &= a + d = a + 1 \cdot d \\ a_3 &= a_2 + d = (a + d) + d = a + 2 \cdot d \\ a_4 &= a_3 + d = (a + 2 \cdot d) + d = a + 3 \cdot d \\ a_5 &= a_4 + d = (a + 3 \cdot d) + d = a + 4 \cdot d \\ &\vdots \\ a_n &= a_{n-1} + d = [a + (n-2)d] + d = a + (n-1)d \end{aligned}$$

We are led to the following result:

THEOREM ***n*th Term of an Arithmetic Sequence**

For an arithmetic sequence $\{a_n\}$ whose first term is a and whose common difference is d, the nth term is determined by the formula

$$a_n = a + (n-1)d \qquad (2)$$

EXAMPLE 4 Finding a Particular Term of an Arithmetic Sequence

Find the thirteenth term of the arithmetic sequence: 2, 6, 10, 14, 18, …

Solution The first term of this arithmetic sequence is $a = 2$, and the common difference is 4. By formula (2), the nth term is

$$a_n = 2 + (n-1)4$$

Hence, the thirteenth term is

$$a_{13} = 2 + 12 \cdot 4 = 50$$

EXPLORATION Use a graphing utility to find the thirteenth term of the sequence given in Example 4. Use it to find the twentieth and fiftieth terms.

EXAMPLE 5 Finding a Recursive Formula for an Arithmetic Sequence

The eighth term of an arithmetic sequence is 75, and the twentieth term is 39. Find the first term and the common difference. Give a recursive formula for the sequence.

Solution By formula (2), we know that $a_n = a + (n-1)d$. As a result,

$$\begin{cases} a_8 = a + 7d = 75 \\ a_{20} = a + 19d = 39 \end{cases}$$

This is a system of two linear equations containing two variables, a and d, which we can solve by elimination. Thus, subtracting the second equation from the first equation, we get

$$-12d = 36$$
$$d = -3$$

With $d = -3$, we find that $a = 75 - 7d = 75 - 7(-3) = 96$. The first term is $a = 96$ and the common difference is $d = -3$. A recursive formula for this sequence is found using (1).

$$a_1 = 96, \qquad a_n = a_{n-1} - 3$$

Based on formula (2), a formula for the nth term of the sequence $\{a_n\}$ in Example 5 is

$$a_n = a + (n-1)d = 96 + (n-1)(-3) = 99 - 3n$$

NOW WORK PROBLEMS 19 AND 25.

EXPLORATION Graph the recursive formula from Example 5, $a_1 = 96$, $a_n = a_{n-1} - 3$, using a graphing utility. Conclude that the graph of the recursive formula behaves like the graph of a linear function. How is d, the common difference, related to m, the slope of a line?

Adding the First n Terms of an Arithmetic Sequence

3 The next result gives a formula for finding the sum of the first n terms of an arithmetic sequence.

THEOREM

Sum of n Terms of an Arithmetic Sequence

Let $\{a_n\}$ be an arithmetic sequence with first term a and common difference d. The sum S_n of the first n terms of $\{a_n\}$ is

$$S_n = \frac{n}{2}[2a + (n-1)d] = \frac{n}{2}(a + a_n) \qquad (3)$$

Proof

$$S_n = a_1 + a_2 + a_3 + \cdots + a_n \qquad \text{Sum of first n terms}$$
$$= a + (a + d) + (a + 2d) + \cdots + [a + (n-1)d] \qquad \text{Formula (2)}$$
$$= \underbrace{(a + a + \cdots + a)}_{n \text{ terms}} + [d + 2d + \cdots + (n-1)d] \qquad \text{Rearrange terms}$$
$$= na + d[1 + 2 + \cdots + (n-1)]$$
$$= na + d\left[\frac{(n-1)n}{2}\right] \qquad \text{Property 6, Section 7.1}$$
$$= na + \frac{n}{2}(n-1)d$$
$$= \frac{n}{2}[2a + (n-1)d] \qquad \text{Factor out } n/2 \qquad (4)$$
$$= \frac{n}{2}[a + a + (n-1)d]$$
$$= \frac{n}{2}(a + a_n) \qquad \text{Formula (2)} \qquad (5)$$

Formula (3) provides two ways to find the sum of the first n terms of an arithmetic sequence. Notice that (4) involves the first term and common difference, while (5) involves the first term and the nth term. Use whichever form is easier.

EXAMPLE 6 Finding the Sum of n Terms of an Arithmetic Sequence

Find the sum S_n of the first n terms of the sequence $\{3n + 5\}$; that is, find

$$8 + 11 + 14 + \cdots + (3n + 5)$$

Solution The sequence $\{3n + 5\}$ is an arithmetic sequence with first term $a = 8$ and the nth term $(3n + 5)$. To find the sum S_n, we use formula (3), as given in (5):

$$S_n = \frac{n}{2}(a + a_n) = \frac{n}{2}[8 + (3n + 5)] = \frac{n}{2}(3n + 13)$$

NOW WORK PROBLEM 33.

EXAMPLE 7 Using a Graphing Utility to Find the Sum of 20 Terms of an Arithmetic Sequence

Use a graphing utility to find the sum S_n of the first 20 terms of the sequence $\{9.5n + 2.6\}$.

Solution Figure 12 shows the results obtained using a TI-83 graphing calculator.

Figure 12

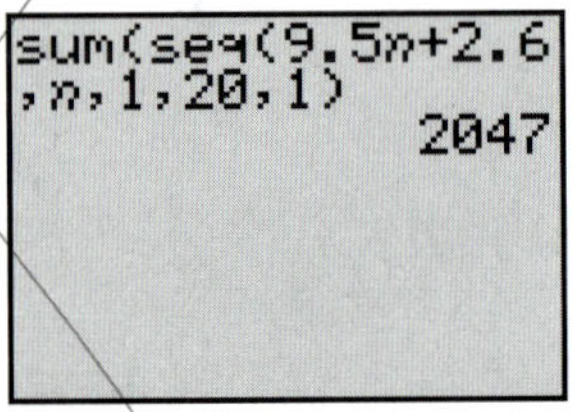

Thus, the sum of the first 20 terms of the sequence $\{9.5n + 2.6\}$ is 2047.

NOW WORK PROBLEM 41.

EXAMPLE 8 Creating a Floor Design

A ceramic tile floor is designed in the shape of a trapezoid 20 feet wide at the base and 10 feet wide at the top. See Figure 13. The tiles, 12 inches by 12 inches, are to be placed so that each successive row contains one less tile than the preceding row. How many tiles will be required?

Figure 13

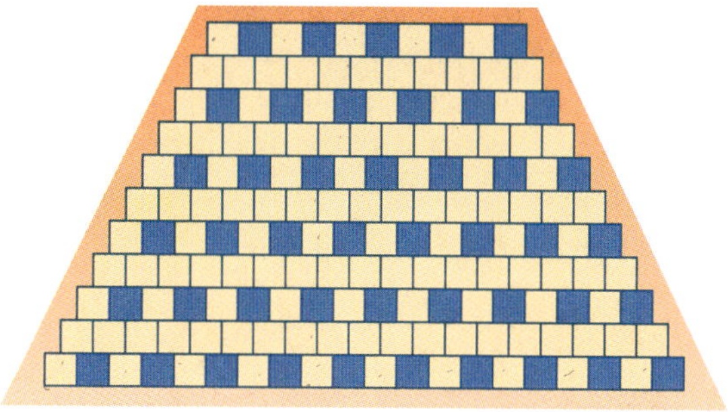

Solution The bottom row requires 20 tiles and the top row, 10 tiles. Since each successive row requires one less tile, the total number of tiles required is

$$S = 20 + 19 + 18 + \cdots + 11 + 10$$

This is the sum of an arithmetic sequence; the common difference is -1. The number of terms to be added is $n = 11$, with the first term $a = 20$ and the last term $a_{11} = 10$. The sum S is

$$S = \frac{n}{2}(a + a_{11}) = \frac{11}{2}(20 + 10) = 165$$

Thus, 165 tiles will be required.

7.2 EXERCISES

In Problems 1–10, an arithmetic sequence is given. Find the common difference and write out the first four terms.

1. $\{n + 4\}$ **2.** $\{n - 5\}$ **3.** $\{2n - 5\}$ **4.** $\{3n + 1\}$ **5.** $\{6 - 2n\}$

6. $\{4 - 2n\}$ **7.** $\left\{\frac{1}{2} - \frac{1}{3}n\right\}$ **8.** $\left\{\frac{2}{3} + \frac{n}{4}\right\}$ **9.** $\{\ln 3^n\}$ **10.** $\{e^{\ln n}\}$

In Problems 11–18, find the nth term of the arithmetic sequence whose initial term a and common difference d are given. What is the fifth term?

11. $a = 2; d = 3$ **12.** $a = -2; d = 4$ **13.** $a = 5; d = -3$

14. $a = 6; d = -2$ **15.** $a = 0; d = \frac{1}{2}$ **16.** $a = 1; d = -\frac{1}{3}$

17. $a = \sqrt{2}; d = \sqrt{2}$ **18.** $a = 0; d = \pi$

In Problems 19–24, find the indicated term in each arithmetic sequence.

19. 12th term of 2, 4, 6, ... **20.** 8th term of $-1, 1, 3, \ldots$

21. 10th term of $1, -2, -5, \ldots$ **22.** 9th term of $5, 0, -5, \ldots$

23. 8th term of $a, a + b, a + 2b, \ldots$ **24.** 7th term of $2\sqrt{5}, 4\sqrt{5}, 6\sqrt{5}, \ldots$

In Problems 25–32, find the first term and the common difference of the arithmetic sequence described. Give a recursive formula for the sequence.

25. 8th term is 8; 20th term is 44

26. 4th term is 3; 20th term is 35

27. 9th term is -5; 15th term is 31

28. 8th term is 4; 18th term is -96

29. 15th term is 0; 40th term is -50

30. 5th term is -2; 13th term is 30

31. 14th term is -1; 18th term is -9

32. 12th term is 4; 18th term is 28

In Problems 33–40, find the sum.

33. $1 + 3 + 5 + \cdots + (2n - 1)$

34. $2 + 4 + 6 + \cdots + 2n$

35. $7 + 12 + 17 + \cdots + (2 + 5n)$

36. $-1 + 3 + 7 + \cdots + (4n - 5)$

37. $2 + 4 + 6 + \cdots + 70$

38. $1 + 3 + 5 + \cdots + 59$

39. $5 + 9 + 13 + \cdots + 49$

40. $2 + 5 + 8 + \cdots + 41$

For Problems 41–46, use a graphing utility to find the sum of each sequence.

41. $\{3.45n + 4.12\}, n = 20$

42. $\{2.67n - 1.23\}, n = 25$

43. $2.8 + 5.2 + 7.6 + \cdots + 36.4$

44. $5.4 + 7.3 + 9.2 + \cdots + 32$

45. $4.9 + 7.48 + 10.06 + \cdots + 66.82$

46. $3.71 + 6.9 + 10.09 + \cdots + 80.27$

47. Find x so that $x + 3$, $2x + 1$, and $5x + 2$ are consecutive terms of an arithmetic sequence.

48. Find x so that $2x$, $3x + 2$, and $5x + 3$ are consecutive terms of an arithmetic sequence.

49. **Drury Lane Theater** The Drury Lane Theater has 25 seats in the first row and 30 rows in all. Each successive row contains one additional seat. How many seats are in the theater?

50. **Football Stadium** The corner section of a football stadium has 15 seats in the first row and 40 rows in all. Each successive row contains two additional seats. How many seats are in this section?

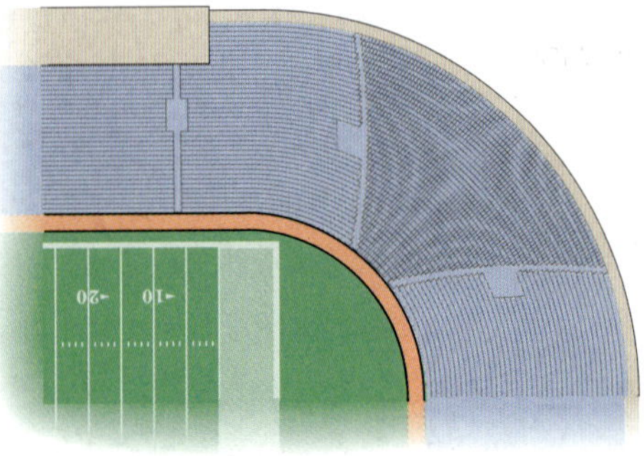

51. **Creating a Mosaic** A mosaic is designed in the shape of an equilateral triangle, 20 feet on each side. Each tile in the mosaic is in the shape of an equilateral triangle, 12 inches to a side. The tiles are to alternate in color as shown in the illustration. How many tiles of each color will be required?

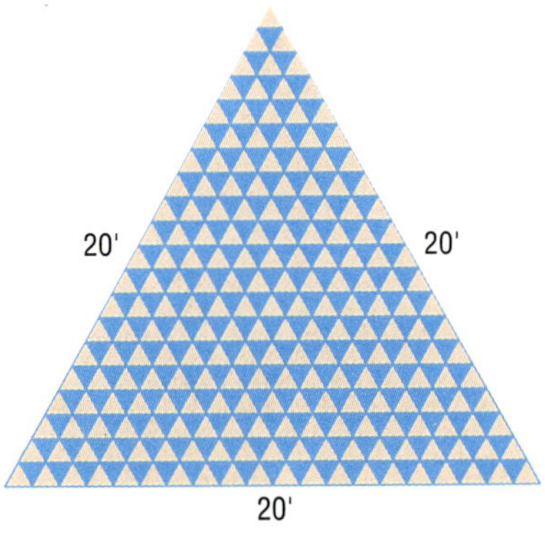

52. **Constructing a Brick Staircase** A brick staircase has a total of 30 steps. The bottom step requires 100 bricks. Each successive step requires two less bricks than the prior step.

(a) How many bricks are required for the top step?

(b) How many bricks are required to build the staircase?

53. **Stadium Construction** How many rows are in the corner section of a stadium containing 2040 seats if the first row has 10 seats and each successive row has 4 additional seats?

54. **Salary** Suppose that you just received a job offer with a starting salary of $35,000 per year and a guaranteed raise of $1400 per year. How many years will it take before your aggregate salary is $617,400?
***Hint:** Your aggregate salary after two years is $35,000 + ($35,000 + $1400).*

55. Make up an arithmetic sequence. Give it to a friend and ask for its twentieth term.

7.3 GEOMETRIC SEQUENCES; GEOMETRIC SERIES

1 Determine If a Sequence Is Geometric

2 Find a Formula for a Geometric Sequence

3 Find the Sum of a Geometric Sequence

4 Solve Annuity Problems

5 Find the Sum of a Geometric Series

1 When the ratio of successive terms of a sequence is always the same nonzero number, the sequence is called **geometric.** Thus, a **geometric sequence*** may be defined recursively as $a_1 = a$, $a_n/a_{n-1} = r$, or as

$$a_1 = a, \qquad a_n = ra_{n-1} \tag{1}$$

where $a_1 = a$ and $r \neq 0$ are real numbers. The number a is the first term, and the nonzero number r is called the **common ratio.**

Thus, the terms of a geometric sequence with first term a and common ratio r follow the pattern

$$a, ar, ar^2, ar^3, \ldots$$

EXAMPLE 1 Determining If a Sequence Is Geometric

The sequence

$$2, 6, 18, 54, 162, \ldots$$

is geometric since the ratio of successive terms is 3 $(\frac{6}{2} = \frac{18}{6} = \cdots = 3)$. The first term is 2, and the common ratio is 3.

EXAMPLE 2 Determining If a Sequence Is Geometric

Show that the following sequence is geometric. Find the first term and the common ratio.

$$\{s_n\} = 2^{-n}$$

Solution The first term is $s_1 = 2^{-1} = \frac{1}{2}$. The nth and $(n - 1)$st terms of the sequence $\{s_n\}$ are

$$s_n = 2^{-n} \quad \text{and} \quad s_{n-1} = 2^{-(n-1)}$$

*Sometimes called a **geometric progression.**

Their ratio is

$$\frac{s_n}{s_{n-1}} = \frac{2^{-n}}{2^{-(n-1)}} = 2^{-n+(n-1)} = 2^{-1} = \frac{1}{2}$$

Because the ratio of successive terms is a nonzero number independent of n, the sequence $\{s_n\}$ is geometric with common ratio $\frac{1}{2}$.

EXAMPLE 3 Determining If a Sequence Is Geometric

Show that the following sequence is geometric. Find the first term and the common ratio.

$$\{t_n\} = \{4^n\}$$

Solution The first term is $t_1 = 4^1 = 4$. The nth and $(n-1)$st terms are

$$t_n = 4^n \quad \text{and} \quad t_{n-1} = 4^{n-1}$$

Their ratio is

$$\frac{t_n}{t_{n-1}} = \frac{4^n}{4^{n-1}} = 4$$

Thus, $\{t_n\}$ is a geometric sequence with common ratio 4.

NOW WORK PROBLEM 3.

2 Suppose that a is the first term of a geometric sequence with common ratio $r \neq 0$. We seek a formula for the nth term a_n. To see the pattern, we write down the first few terms:

$$\begin{aligned}
a_1 &= 1 \cdot a = ar^0 \\
a_2 &= ra = ar^1 \\
a_3 &= ra_2 = r(ar) = ar^2 \\
a_4 &= ra_3 = r(ar^2) = ar^3 \\
a_5 &= ra_4 = r(ar^3) = ar^4 \\
&\vdots \\
a_n &= ra_{n-1} = r(ar^{n-2}) = ar^{n-1}
\end{aligned}$$

We are led to the following result:

THEOREM ***n*th Term of a Geometric Sequence**

For a geometric sequence $\{a_n\}$ whose first term is a and whose common ratio is r, the nth term is determined by the formula

$$a_n = ar^{n-1}, \qquad r \neq 0 \qquad (2)$$

EXAMPLE 4 Finding a Particular Term of a Geometric Sequence

(a) Find the ninth term of the geometric sequence: $10, 9, \frac{81}{10}, \frac{729}{100} \ldots$

(b) Find a recursive formula for this sequence.

Solution (a) The first term of this geometric sequence is $a = 10$ and the common ratio is 9/10. (Use 9/10, or $\frac{81/10}{9} = \frac{9}{10}$, or any two successive terms.) By formula (2), the nth term is

$$a_n = 10\left(\frac{9}{10}\right)^{n-1}$$

Hence, the ninth term is

$$a_9 = 10\left(\frac{9}{10}\right)^{9-1} = 10\left(\frac{9}{10}\right)^8 = 4.3046721$$

(b) The first term in the sequence is 10 and the common ratio is $r = 9/10$. Using formula (1), the recursive formula is $a_1 = 10$, $a_n = \frac{9}{10}a_{n-1}$.

EXPLORATION Use a graphing utility to find the ninth term of the sequence given in Example 4. Use it to find the twentieth and fiftieth terms. Now use a graphing utility to graph the recursive formula found in Example 4, part (b). Conclude that the graph of the recursive formula behaves like the graph of an exponential function. How is r, the common ratio, related to a, the base of the exponential function?

NOW WORK PROBLEMS 25 AND 33.

Adding the First n Terms of a Geometric Sequence

3 The next result gives us a formula for finding the sum of the first n terms of a geometric sequence.

THEOREM **Sum of n Terms of a Geometric Sequence**

Let $\{a_n\}$ be a geometric sequence with first term a and common ratio r. The sum S_n of the first n terms of $\{a_n\}$ is

$$S_n = a\frac{1 - r^n}{1 - r}, \qquad r \neq 0, 1 \qquad (3)$$

Proof

$$S_n = a + ar + \cdots + ar^{n-1} \qquad (4)$$

Multiply each side by r to obtain

$$rS_n = ar + ar^2 + \cdots + ar^n \qquad (5)$$

Now, subtract (5) from (4). The result is

$$S_n - rS_n = a - ar^n$$
$$(1 - r)S_n = a(1 - r^n)$$

Since $r \neq 1$, we can solve for S_n:

$$S_n = a\frac{1 - r^n}{1 - r}$$

EXAMPLE 5 Finding the Sum of n Terms of a Geometric Sequence

Find the sum S_n of the first n terms of the sequence $\left[\left(\frac{1}{2}\right)^n\right]$; that is, find

$$\frac{1}{2} + \frac{1}{4} + \frac{1}{8} + \cdots + \left(\frac{1}{2}\right)^n$$

Solution The sequence $\left[\left(\frac{1}{2}\right)^n\right]$ is a geometric sequence with $a = \frac{1}{2}$ and $r = \frac{1}{2}$. The sum S_n that we seek is the sum of the first n terms of the sequence, so we use formula (3) to get

$$\begin{aligned} S_n = \sum_{k=1}^{n} \left(\frac{1}{2}\right)^k &= \frac{1}{2} + \frac{1}{4} + \frac{1}{8} + \cdots + \left(\frac{1}{2}\right)^n \\ &= \frac{1}{2}\left[\frac{1 - \left(\frac{1}{2}\right)^n}{1 - \frac{1}{2}}\right] \\ &= \frac{1}{2}\left[\frac{1 - \left(\frac{1}{2}\right)^n}{\frac{1}{2}}\right] \\ &= 1 - \left(\frac{1}{2}\right)^n \end{aligned}$$

NOW WORK PROBLEM **39.**

EXAMPLE 6 Using a Graphing Utility to Find the Sum of a Geometric Sequence

Use a graphing utility to find the sum S_n of the first 15 terms of the sequence $\left\{\left(\frac{1}{3}\right)^n\right\}$; that is, find

$$\frac{1}{3} + \frac{1}{9} + \frac{1}{27} + \cdots + \left(\frac{1}{3}\right)^{15}$$

Solution Figure 14 shows the result obtained using a TI-83 graphing calculator.

Figure 14

Thus, the sum of the first 15 terms of the sequence $\left\{\left(\frac{1}{3}\right)^n\right\}$ is 0.4999999652.

NOW WORK PROBLEM **45.**

Annuities

4 In Section 5.6 we developed the compound interest formula, which gives the future value when a fixed amount of money is deposited in an account that pays interest compounded periodically. Often, though, money is invested in small amounts at periodic intervals. An **annuity** is a sequence of equal peri-

odic deposits. The periodic deposits may be made annually, quarterly, monthly, or daily.

When deposits are made at the same time that the interest is credited, the annuity is called **ordinary.** We will only deal with ordinary annuities here. The **amount of an annuity** is the sum of all deposits made plus all interest paid.

Suppose that the interest an account earns is i percent per payment period (expressed as a decimal). For example, if an account pays 12% compounded monthly (12 times a year), then $i = 0.12/12 = 0.01$. If an account pays 8% compounded quarterly (4 times a year), then $i = 0.08/4 = 0.02$. To develop a formula for the amount of an annuity, suppose that \$$P$ is deposited each payment period for n payment periods in an account that earns i percent per payment period. When the last deposit is made at the end of the nth payment period, the first deposit of \$$P$ has earned interest compounded for $n - 1$ payment periods, the second deposit of \$$P$ has earned interest compounded for $n - 2$ payment periods, and so on. Table 5 shows the value of each deposit after n deposits have been made.

Table 5

Deposit	1	2	3	$n-1$	n
Amount	$P(1+i)^{n-1}$	$P(1+i)^{n-2}$	$P(1+i)^{n-3}$	$P(1+i)$	P

The amount A of the annuity is the sum of the amounts shown in Table 5:

$$\begin{aligned} A &= P(1+i)^{n-1} + P(1+i)^{n-2} + \cdots + P(1+i) + P \\ &= P[1 + (1+i) + \cdots + (1+i)^{n-1}] \end{aligned}$$

The expression in brackets is the sum of a geometric sequence with n terms and a common ratio of $(1 + i)$. As a result,

$$\begin{aligned} A &= P[1 + (1+i) + \cdots + (1+i)^{n-2} + (1+i)^{n-1}] \\ &= P\frac{1-(1+i)^n}{1-(1+i)} = P\frac{1-(1+i)^n}{-i} = P\frac{(1+i)^n - 1}{i} \end{aligned}$$

We have established the following result.

THEOREM **Amount of an Annuity**

If P represents the deposit in dollars made at each payment period for an annuity at i percent interest per payment period, the amount A of the annuity after n payment periods is

$$A = P\frac{(1+i)^n - 1}{i} \tag{6}$$

EXAMPLE 7 Determining the Amount of an Annuity

On January 1, 1999, Bill decides to place \$500 into a Roth Individual Retirement Account (IRA) at the end of each quarter for the next 30 years.

What will the value of the IRA be when Bill retires in 30 years if the rate of return of the IRA is assumed to be 10% per annum compounded quarterly?

Solution Let's look at Bill's payment schedule. The account is opened on January 1, 1999, and payments are made at the end of each payment period (quarter), so Bill's first payment of \$500 is made March 31, 1999, his second payment of \$500 is made June 30, 1999, and so on. Since the payments are made at the end of periodic intervals at the same time that interest is credited, this is an ordinary annuity with $n = 4(30) = 120$ quarterly payments of $P = \$500$. The rate of interest per quarter is $i = 0.10/4 = 0.025$. Table 6 shows the payment schedule. The bottom row in Table 6 represents the value of each payment on December 31, 2028 (the date of the last payment). For example, the payment made on March 31, 1999, earns interest for 119 payment periods. The value of this payment on December 31, 2028, is $\$500(1.025)^{119}$.

Table 6

Payment no. (Date)	$n = 1$ (March 31, 1999)	$n = 2$ (June 30, 1999)	$n = 3$ (Sept. 30, 1999)	...	$n = 120$ (Dec. 31, 2028)
Payment	\$500	\$500	\$500	...	\$500
Value of payment on December 31, 2028	$\$500(1.025)^{119}$	$\$500(1.025)^{118}$	$\$500(1.025)^{117}$	...	\$500

So the value of the annuity on December 31, 2028, is the sum of the values of each payment:

$$A = \$500 + \$500(1.025) + \cdots + \$500(1.025)^{118} + \$500(1.025)^{119}$$

Using formula (6) with $P = \$500$, $i = 0.025$, and $n = 120$, we find the value A of the annuity.

$$A = \$500\frac{1.025^{120} - 1}{0.025} = \$367{,}163$$

↑ Formula (6)

We can also represent the amount of an annuity at the time that the nth payment is made using the recursive formula

$$a_1 = P, \qquad a_n = (1 + i)a_{n-1} + P \tag{7}$$

This formula may be explained as follows: the money in the account at the end of the first period, a_1, is $\$P$; the money in the account after $n - 1$ payments (a_{n-1}) earns interest i during the nth period; so when the periodic payment of P dollars is added, the amount after n payments, a_n, is obtained.

Let's use the information provided in Example 7 to get a more concrete understanding of formula (7). Recall that Bill's first payment of \$500 is made on March 31, 1999, his second payment of \$500 is made on June 30,

1999, and so on, until his last payment of \$500 is made on December 31, 2028. Table 7 shows the payment schedule.

Table 7

Payment no. (Date)	$n = 1$ (March 31, 1999)	$n = 2$ (June 30, 1999)	$n = 3$ (Sept. 30, 1999)	...	$n = 120$ (Dec. 31, 2028)
Payment	\$500	\$500	\$500	...	\$500
Value of account on December 31, 2028	$a_1 = \$500$	$a_2 = \$500(1.025) + \$500 = \$1012.50$	$a_3 = \$1012.50(1.025) + \$500 = \$1537.81$	...	$a_{120} = \$357{,}720(1.025) + \$500 = \$367{,}163$

The bottom row of Table 7 represents the value of the account after each payment is made. After the first payment period, the value of the account is \$500. During the second payment period, the \$500 in the account earns $i\%$ interest, so the account value after the second payment period, a_2, is

$$\underbrace{\$500(1.025)}_{a_1\,(1+i)} + \underbrace{\$500}_{P}$$

During the third payment period, the \$1012.50 in the account earns interest, so the account value after the third payment period, a_3, is

$$\underbrace{\$1012.50(1.025)}_{a_2\,(1+i)} + \underbrace{\$500}_{P}$$

This process continues until the 120th payment is made, when the value of the account is \$367,163.

EXAMPLE 8 Determining Information about an Annuity

Use the information given in Example 7 to answer the following questions.

(a) What is the value of the account after the first four payment periods, that is, after the first year?

(b) How long does it take before the value of the account exceeds \$250,000?

(c) What is the value of the account in 30 years?

Solution We use formula (7) with $i = 0.025$, $P = \$500$ and $a_1 = 500$. Then the amount a_n after n payment periods is

$$a_1 = 500, \qquad a_n = (1 + 0.025)a_{n-1} + 500$$

(a) In SEQuence mode on a TI-83, enter the sequence $\{a_n\}$ and create Table 8. After the first payment on March 31, 1999, the value of the account is \$500. After the second payment on June 30, 1999, the value of the account is \$1012.50. After the third payment on September 30, 1999, the value of the account is \$1537.81, and after the fourth payment (on December 31, 1999) the value of the account is \$2076.26.

Table 8

n	u(n)
1	500
2	1012.5
3	1537.8
4	2076.3
5	2628.2
6	3193.9
7	3773.7

u(n)=1.025u(n−1...

(b) Scroll down until the balance is above \$250,000 to obtain Table 9. We conclude that it takes 106 payment periods or 26 years, 6 months for the value of the account to exceed \$250,000.

Table 9

n	u(n)
100	216274
101	222181
102	228236
103	234442
104	240803
105	247323
106	254006

u(n)=1.025u(n−1...

(c) Continue scrolling to obtain Table 10. The amount after 30 years ($n = 4(30) = 120$ payment periods) is \$367,163, the same result obtained in Example 7.

Table 10

n	u(n)
114	313849
115	322196
116	330751
117	339519
118	348507
119	357720
120	367163

u(n)=1.025u(n−1...

A recursive formula can also be used to compute the outstanding balance on a loan, such as a mortgage. The balance on a loan at the time that the nth payment is made is given by the recursive formula

$$a_1 = B, \qquad a_n = (1 + i)a_{n-1} - P \tag{8}$$

where B is the initial balance and P is the periodic payment. Notice that we subtract P dollars in formula (8), reflecting the fact that the payment reduces the balance.

EXAMPLE 9 Mortgage Payments

John and Wanda borrowed \$180,000 at 7% per annum compounded monthly for 30 years to purchase a home. Their monthly payment is determined to be \$1197.54.

(a) Find a recursive formula that represents their balance at the beginning of each month after their payment of \$1197.54 has been made.

(b) Determine their balance after the first payment is made.

(c) When will their balance be below \$170,000?

Solution (a) We use formula (8) with $i = 0.07/12$ and $P = \$1197.54$. Then $a_1 = 180{,}000$, the balance at the beginning of the first month, and

$$a_n = \left(1 + \frac{0.07}{12}\right)a_{n-1} - 1197.54$$

(b) In SEQuence mode on a TI-83, enter the sequence $\{a_n\}$ and create Table 11.

Table 11

n	u(n)
1	180000
2	179852
3	179704
4	179555
5	179405
6	179254
7	179102

u(n)=(1+.07/12)...

After the first payment is made at the beginning of the second month, the balance is $a_2 = \$179{,}852$.

(c) Scroll down until the balance is below \$170,000. See Table 12. After the fifty-eighth payment is made at the beginning of the fifty-ninth month ($n = 59$), the balance is below \$170,000.

Table 12

n	u(n)
53	171067
54	170868
55	170667
56	170465
57	170262
58	170057
59	169852

u(n)=(1+.07/12)...

Geometric Series

An infinite sum of the form

$$a + ar + ar^2 + \cdots + ar^{n-1} + \cdots$$

with first term a and common ratio r, is called an **infinite geometric series** and is denoted by

$$\sum_{k=1}^{\infty} ar^{k-1}$$

5 Based on formula (3), the sum S_n of the first n terms of a geometric series is

$$S_n = a\frac{1 - r^n}{1 - r} = \frac{a}{1 - r} - \frac{ar^n}{1 - r} \qquad (9)$$

If this finite sum S_n approaches a number L as $n \to \infty$, then we call L the **sum of the infinite geometric series,** and we write

$$L = \sum_{k=1}^{\infty} ar^{k-1}$$

THEOREM

Sum of an Infinite Geometric Series

If $|r| < 1$, the sum of the infinite geometric series $\sum_{k=1}^{\infty} ar^{k-1}$ is

$$\sum_{k=1}^{\infty} ar^{k-1} = \frac{a}{1 - r} \qquad (10)$$

Intuitive Proof Since $|r| < 1$, it follows that $|r^n|$ approaches 0 as $n \to \infty$. Then, based on formula (9), the sum S_n approaches $a/(1 - r)$ as $n \to \infty$.

EXAMPLE 10 Finding the Sum of a Geometric Series

Find the sum of the geometric series $2 + \frac{4}{3} + \frac{8}{9} + \cdots$.

Solution The first term is $a = 2$ and the common ratio is

$$r = \frac{\frac{4}{3}}{2} = \frac{4}{6} = \frac{2}{3}$$

Since $|r| < 1$, we use formula (10) to find that

$$2 + \frac{4}{3} + \frac{8}{9} + \cdots = \frac{2}{1 - \frac{2}{3}} = 6$$

NOW WORK PROBLEM 51.

EXPLORATION Use a graphing utility to graph $U_n = 2(\frac{2}{3})^{n-1} + U_{n-1}$ in sequence mode. TRACE the graph for large values of n. What happens to the value of U_n as n increases without bound? What can you conclude about $\sum_{n=1}^{\infty} 2(\frac{2}{3})^{n-1}$?

EXAMPLE 11 Repeating Decimals

Show that the repeating decimal 0.999 ... equals 1.

Solution

$$0.999\ldots = \frac{9}{10} + \frac{9}{100} + \frac{9}{1000} + \cdots$$

Thus, 0.999 ... is a geometric series with first term $\frac{9}{10}$ and common ratio $\frac{1}{10}$. Hence,

$$0.999\ldots = \frac{\frac{9}{10}}{1 - \frac{1}{10}} = \frac{\frac{9}{10}}{\frac{9}{10}} = 1$$

EXAMPLE 12 Pendulum Swings

Figure 15

Initially, a pendulum swings through an arc of 18 inches. See Figure 15. On each successive swing, the length of the arc is 0.98 of the previous length.

(a) What is the length of the arc after 10 swings?
(b) On which swing is the length of the arc first less than 12 inches?
(c) After 15 swings, what total distance will the pendulum have swung?
(d) When it stops, what total distance will the pendulum have swung?

Solution

(a) The length of the first swing is 18 inches. The length of the second swing is 0.98(18) inches; the length of the third swing is $0.98(0.98)(18) = 0.98^2(18)$ inches. The length of arc of the tenth swing is

$$(0.98)^9(18) = 15.007 \text{ inches}$$

(b) The length of arc of the nth swing is $(0.98)^{n-1}(18)$. For this to be exactly 12 inches requires that

$$(0.98)^{n-1}(18) = 12$$
$$(0.98)^{n-1} = \frac{12}{18} = \frac{2}{3}$$
$$n - 1 = \log_{0.98}\left(\frac{2}{3}\right)$$
$$n = 1 + \frac{\ln(\frac{2}{3})}{\ln 0.98} = 1 + 20.07 = 21.07$$

The length of arc of the pendulum exceeds 12 inches on the twenty-first swing and is first less than 12 inches on the twenty-second swing.

(c) After 15 swings, the pendulum will have swung the following total distance L:

$$L = \underset{\text{1st}}{18} + \underset{\text{2nd}}{0.98(18)} + \underset{\text{3rd}}{(0.98)^2(18)} + \underset{\text{4th}}{(0.98)^3(18)} + \cdots + \underset{\text{15th}}{(0.98)^{14}(18)}$$

This is the sum of a geometric sequence. The common ratio is 0.98; the first term is 18. The sum has 15 terms, so

$$L = 18\frac{1 - 0.98^{15}}{1 - 0.98} = 18(13.07) = 235.29 \text{ inches}$$

The pendulum will have swung through 235.29 inches after 15 swings.

(d) When the pendulum stops, it will have swung the following total distance T:

$$T = 18 + 0.98(18) + (0.98)^2(18) + (0.98)^3(18) + \cdots$$

This is the sum of a geometric series. The common ratio is $r = 0.98$; the first term is $a = 18$. The sum is

$$T = \frac{a}{1 - r} = \frac{18}{1 - 0.98} = 900$$

The pendulum will have swung a total of 900 inches when it finally stops.

HISTORICAL FEATURE

Sequences are among the oldest objects of mathematical investigation, having been studied for over 3500 years. After the initial steps, however, little progress was made until about 1600.

Arithmetic and geometric sequences appear in the Rhind papyrus, a mathematical text containing 85 problems copied around 1650 B.C. by the Egyptian scribe Ahmes from an earlier work (see Historical Problem 1). Fibonacci (A.D. 1220) wrote about problems similar to those found in the Rhind papyrus, leading one to suspect that Fibonacci may have had material available that is now lost. This material would have been in the non-Euclidean Greek tradition of Heron (about A.D. 75) and Diophantus (about A.D. 250). One problem, again modified slightly, is still with us in the familiar puzzle rhyme "As I was going to St. Ives . . ." (see Historical Problem 2).

The Rhind papyrus indicates that the Egyptians knew how to add up the terms of an arithmetic or geometric sequence, as did the Babylonians. The rule for summing up a geometric sequence is found in Euclid's *Elements* (book IX, 35, 36), where, like all Euclid's algebra, it is presented in a geometric form.

Investigations of other kinds of sequences began in the 1500s, when algebra became sufficiently developed to handle the more complicated problems. The development of calculus in the 1600s added a powerful new tool, especially for finding the sum of infinite series, and the subject continues to flourish today.

HISTORICAL PROBLEMS

1. *Arithmetic sequence problem from the Rhind payrus (statement modified slightly for clarity)* One hundred loaves of bread are to be divided among five people so that the amounts that they receive form an arithmetic sequence. The first two together receive one-seventh of what the last three receive. How many does each receive? [*Partial answer:* First person receives $1\frac{2}{3}$ loaves.]
2. The following old English children's rhyme resembles one of the Rhind papyrus problems:

   As I was going to St. Ives
   I met a man with seven wives
   Each wife had seven sacks
   Each sack had seven cats
   Each cat had seven kits [kittens]
   Kits, cats, sacks, wives
   How many were going to St. Ives?

(a) Assuming that the speaker and the cat fanciers met by traveling in opposite directions, what is the answer?

(b) How many kittens are being transported?

(c) Kits, cats, sacks, wives; how many?
[Hint: It is easier to include the man, find the sum with the formula, and then subtract 1 for the man.]

7.3 EXERCISES

In Problems 1–10, a geometric sequence is given. Find the common ratio and write out the first four terms.

1. $\{3^n\}$ **2.** $\{(-5)^n\}$ **3.** $\left\{-3\left(\frac{1}{2}\right)^n\right\}$ **4.** $\left\{\left(\frac{5}{2}\right)^n\right\}$ **5.** $\left\{\frac{2^{n-1}}{4}\right\}$

6. $\left\{\frac{3^n}{9}\right\}$ **7.** $\{2^{n/3}\}$ **8.** $\{3^{2n}\}$ **9.** $\left\{\frac{3^{n-1}}{2^n}\right\}$ **10.** $\left\{\frac{2^n}{3^{n-1}}\right\}$

In Problems 11–24, determine whether the given sequence is arithmetic, geometric, or neither. If the sequence is arithmetic, find the common difference; if it is geometric, find the common ratio.

11. $\{n + 2\}$ **12.** $\{2n - 5\}$ **13.** $\{4n^2\}$ **14.** $\{5n^2 + 1\}$ **15.** $\{3 - \frac{2}{3}n\}$

16. $\{8 - \frac{3}{4}n\}$ **17.** $1, 3, 6, 10, \ldots$ **18.** $2, 4, 6, 8, \ldots$ **19.** $\{(\frac{2}{3})^n\}$ **20.** $\{(\frac{5}{4})^n\}$

21. $-1, -2, -4, -8, \ldots$ **22.** $1, 1, 2, 3, 5, 8, \ldots$ **23.** $\{3^{n/2}\}$ **24.** $\{(-1)^n\}$

In Problems 25–32, find the fifth term and the nth term of the geometric sequence whose initial term a and common ratio r are given.

25. $a = 2; r = 3$ **26.** $a = -2; r = 4$ **27.** $a = 5; r = -1$ **28.** $a = 6; r = -2$

29. $a = 0; r = \frac{1}{2}$ **30.** $a = 1; r = -\frac{1}{3}$ **31.** $a = \sqrt{2}; r = \sqrt{2}$ **32.** $a = 0; r = 1/\pi$

In Problems 33–38, find the indicated term of each geometric sequence.

33. 7th term of $1, \frac{1}{2}, \frac{1}{4}, \ldots$ **34.** 8th term of $1, 3, 9, \ldots$

35. 9th term of $1, -1, 1, \ldots$ **36.** 10th term of $-1, 2, -4, \ldots$

37. 8th term of $0.4, 0.04, 0.004, \ldots$ **38.** 7th term of $0.1, 1.0, 10.0, \ldots$

In Problems 39–44, find the sum.

39. $\frac{1}{4} + \frac{2}{4} + \frac{2^2}{4} + \frac{2^3}{4} + \cdots + \frac{2^{n-1}}{4}$ **40.** $\frac{3}{9} + \frac{3^2}{9} + \frac{3^3}{9} + \cdots + \frac{3^n}{9}$

41. $\sum_{k=1}^{n} (\frac{2}{3})^k$ **42.** $\sum_{k=1}^{n} 4 \cdot 3^{k-1}$

43. $-1 - 2 - 4 - 8 - \cdots - (2^{n-1})$ **44.** $2 + \frac{6}{5} + \frac{18}{25} + \cdots + 2\left(\frac{3}{5}\right)^n$

For Problems 45–50, use a graphing utility to find the sum of each geometric sequence.

45. $\frac{1}{4} + \frac{2}{4} + \frac{2^2}{4} + \frac{2^3}{4} + \cdots + \frac{2^{14}}{4}$ **46.** $\frac{3}{9} + \frac{3^2}{9} + \frac{3^3}{9} + \cdots + \frac{3^{15}}{9}$

47. $\sum_{n=1}^{15} (\frac{2}{3})^n$ **48.** $\sum_{n=1}^{15} 4 \cdot 3^{n-1}$

49. $-1 - 2 - 4 - 8 - \cdots - 2^{14}$ **50.** $2 + \frac{6}{5} + \frac{18}{25} + \cdots + 2\left(\frac{3}{5}\right)^{15}$

In Problems 51–60, find the sum of each infinite geometric series.

51. $1 + \frac{1}{3} + \frac{1}{9} + \cdots$

52. $2 + \frac{4}{3} + \frac{8}{9} + \cdots$

53. $8 + 4 + 2 + \cdots$

54. $6 + 2 + \frac{2}{3} + \cdots$

55. $2 - \frac{1}{2} + \frac{1}{8} - \frac{1}{32} + \cdots$

56. $1 - \frac{3}{4} + \frac{9}{16} - \frac{27}{64} + \cdots$

57. $\sum_{k=1}^{\infty} 5\left(\frac{1}{4}\right)^{k-1}$

58. $\sum_{k=1}^{\infty} 8\left(\frac{1}{3}\right)^{k-1}$

59. $\sum_{k=1}^{\infty} 6\left(-\frac{2}{3}\right)^{k-1}$

60. $\sum_{k=1}^{\infty} 4\left(-\frac{1}{2}\right)^{k-1}$

61. Find x so that x, $x + 2$, and $x + 3$ are consecutive terms of a geometric sequence.

62. Find x so that $x - 1$, x, and $x + 2$ are consecutive terms of a geometric sequence.

63. **Retirement** Christine contributes $100 each month to her 401(k). What will be the value of Christine's 401(k) in 30 years if the per annum rate of return is assumed to be 12% compounded monthly?

64. **Saving for a Home** Jolene wants to purchase a new home. Suppose that she invests $400 per month into a mutual fund. If the per annum rate of return of the mutual fund is assumed to be 10% compounded monthly, how much will Jolene have for a down payment after 3 years?

65. **Tax Sheltered Annuity** Don contributes $500 at the end of each quarter to a Tax Sheltered Annuity (TSA). What will the value of the TSA be in 20 years if the per annum rate of return is assumed to be 8% compounded quarterly?

66. **Retirement** Ray, planning on retiring in 15 years, contributes $1000 to an Individual Retirement Account (IRA) semiannually. What will the value of the IRA be when Ray retires if the per annum rate of return is assumed to be 10% compounded semiannually?

67. **Sinking Fund** Scott and Alice want to purchase a vacation home in 10 years and have $50,000 for a down payment. How much should they place in a savings account each month if the per annum rate of return is assumed to be 6% compounded monthly?

68. **Sinking Fund** For a child born in 1996, a 4-year college education at a public university is projected to be $150,000. Assuming an 8% per annum rate of return compounded monthly, how much must be contributed to a college fund every month in order to have $150,000 in 18 years when the child begins college?

69. **Roth IRA** On January 1, 1999, Bob decides to place $500 at the end of each quarter into a Roth Individual Retirement Account.

(a) Find a recursive formula that represents Bob's balance at the end of each quarter if the rate of return is assumed to be 8% per annum compounded quarterly.

(b) How long will it be before the value of the account exceeds $100,000?

(c) What will be the value of the account in 25 years when Bob retires?

70. **Education IRA** On January 1, 1999, John's parents decide to place $45 at the end of each month into an Education IRA.

(a) Find a recursive formula that represents the balance at the end of each month if the rate of return is assumed to be 6% per annum compounded monthly.

(b) How long will it be before the value of the account exceeds $4000?

(c) What will be the value of the account in 16 years when John goes to college?

71. **Home Loan** Bill and Laura borrowed $150,000 at 6% per annum compounded monthly for 30 years to purchase a home. Their monthly payment is determined to be $899.33.

(a) Find a recursive formula for their balance at the beginning of each month after the monthly payment has been made.

(b) Determine Bill and Laura's balance at the beginning of the second month.

(c) Using a graphing utility, create a table showing Bill and Laura's balance at the beginning of each month.

(d) Using a graphing utility, determine the number of months it will take until Bill and Laura's balance is below $140,000.

(e) Using a graphing utility, determine the number of months it will take until Bill and Laura pay off the balance.

(f) Determine Bill and Laura's interest expense when the loan is paid.

(g) Suppose that Bill and Laura decide to pay an additional $100 each month on their loan. Answer parts (a)–(f) under this scenario.

(h) Is it worthwhile for Bill and Laura to pay the additional $100? Explain.

72. **Home Loan** Jodi and Jeff borrowed $120,000 at 6.5% per annum compounded monthly for 30 years to purchase a home. Their monthly payment is determined to be $758.48.

(a) Find a recursive formula for their balance at the beginning of each month after the monthly payment has been made.

(b) Determine Jodi and Jeff's balance at the beginning of the second month.
(c) Using a graphing utility, create a table showing Jodi and Jeff's balance at the beginning of each month.
(d) Using a graphing utility, determine the number of months it will take until Jodi and Jeff's balance is below $100,000.
(e) Using a graphing utility, determine the number of months it will take until Jodi and Jeff pay off the balance.
(f) Determine Jodi and Jeff's interest expense when the loan is paid.
(g) Suppose that Jodi and Jeff decide to pay an additional $100 each month on their loan. Answer parts (a)–(f) under this scenario.
(h) Is it worthwhile for Jodi and Jeff to pay the additional $100? Explain.

73. Multiplier Suppose that, throughout the U.S. economy, individuals spend 90% of every additional dollar that they earn. Economists would say that an individual's **marginal propensity to consume** is 0.90. For example, if Jane earns an additional dollar, she will spend $0.9(1) = \$0.90$ of it. The individual that earns $0.90 (from Jane) will spend 90% of it or $0.81. This process of spending continues and results in an infinite geometric series as follows:

$$1, 0.90, 0.90^2, 0.90^3, 0.90^4, \ldots$$

The sum of this infinite geometric series is called the **multiplier.** What is the multiplier if individuals spend 90% of every additional dollar that they earn?

74. Multiplier Refer to Problem 73. Suppose that the marginal propensity to consume throughout the U.S. economy is 0.95. What is the multiplier for the U.S. economy?

75. Stock Price One method of pricing a stock is to discount the stream of future dividends of the stock. Suppose that a stock pays $\$P$ per year in dividends and, historically, the dividend has been increased $i\%$ per year. If you desire an annual rate of return of $r\%$, this method of pricing a stock states that the price that you should pay is the present value of an infinite stream of payments:

$$\text{Price} = P + P\frac{1+i}{1+r} + P\left(\frac{1+i}{1+r}\right)^2 + P\left(\frac{1+i}{1+r}\right)^3 + \cdots$$

Thus, the price of the stock is the sum of an infinite geometric series. Suppose that a stock pays an annual dividend of $4.00 and, historically, the dividend has been increased 3% per year. You desire an annual rate of return of 9%. What is the most that you should pay for the stock?

76. Stock Price Refer to Problem 75. Suppose that a stock pays an annual dividend of $2.50 and historically, the dividend has increased 4% per year. You desire an annual rate of return of 11%. What is the most that you should pay for the stock?

77. Pendulum Swings Initially, a pendulum swings through an arc of 2 feet. On each successive swing, the length of arc is 0.9 of the previous length.
(a) What is the length of arc after 10 swings?
(b) On which swing is the length of arc first less than 1 foot?
(c) After 15 swings, what total length will the pendulum have swung?
(d) When it stops, what total length will the pendulum have swung?

78. Bouncing Balls A ball is dropped from a height of 30 feet. Each time that it strikes the ground, it bounces up to 0.8 of the previous height.

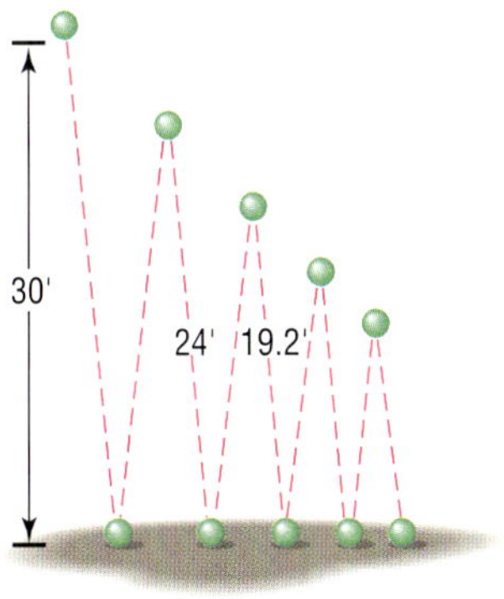

(a) What height will the ball bounce up to after it strikes the ground for the third time?
(b) What is its height after it strikes the ground for the nth time?
(c) How many times does the ball need to strike the ground before its height is less than 6 inches?
(d) What total distance does the ball travel before it stops bouncing?

79. Salary Increases Suppose that you have just been hired at an annual salary of $18,000 and expect to receive annual increases of 5%. What will your salary be when you begin your fifth year?

80. **Equipment Depreciation** A new piece of equipment costs a company \$15,000. Each year, for tax purposes, the company depreciates the value by 15%. What value should the company give the equipment after 5 years?

81. **Critical Thinking** You have just signed a 7-year professional football league contract with a beginning salary of \$2,000,000 per year. Management gives you the following options with regard to your salary over the 7 years.
(1) A bonus of \$100,000 each year
(2) An annual increase of 4.5% per year beginning after 1 year
(3) An annual increase of \$95,000 per year beginning after 1 year
Which option provides the most money over the 7 year period? Which the least? Which would you choose? Why?

82. **A Rich Man's Promise** A rich man promises to give you \$1000 on September 1, 1998. Each day thereafter he will give you $\frac{9}{10}$ of what he gave you the previous day. What is the first date on which the amount that you receive is less than 1¢? How much have you received when this happens?

83. **Grains of Wheat on a Chess Board** In an old fable, a commoner who had just saved the king's life was told he could ask the king for any just reward. Being a shrewd man, the commoner said, "A simple wish, sire. Place one grain of wheat on the first square of a chessboard, two grains on the second square, four grains on the third square, continuing until you have filled the board. This is all I seek." Compute the total number of grains needed to do this to see why the request, seemingly simple, could not be granted. (A chessboard consists of $8 \times 8 = 64$ squares.)

84. Can a sequence be both arithmetic and geometric? Give reasons for your answer.

85. Make up a geometric sequence. Give it to a friend and ask for its twentieth term.

86. Make up two infinite geometric series, one that has a sum and one that does not. Give them to a friend and ask for the sum of each series.

87. If $x < 1$, then $1 + x + x^2 + x^3 + \cdots + x^n + \cdots = 1/(1 - x)$. Make up a table of values using $x = 0.1$, $x = 0.25$, $x = 0.5$, $x = 0.75$, and $x = 0.9$ to compute $1/(1 - x)$. Now determine how many terms are needed in the expansion $1 + x + x^2 + x^3 + \cdots + x^n + \cdots$ before it approximates $1/(1 - x)$ correct to two decimal places. For example, if $x = 0.1$, then $1/(1 - x) = 1/(1 - x) = 10/9 = 1.111\ldots$. The expansion requires three terms.

88. Which of the following choices, *A* or *B*, results in more money?

A: To receive \$1000 on day 1, \$999 on day 2, \$998 on day 3, with the process to end after 1000 days
B: To receive \$1 on day 1, \$2 on day 2, \$4 on day 3, for 19 days

89. You are interviewing for a job and receive two offers:

A: \$20,000 to start with guaranteed annual increases of 6% for the first 5 years
B: \$22,000 to start with guaranteed annual increases of 3% for the first 5 years

Which offer is best if your goal is to be making as much as possible after 5 years? Which is best if your goal is to make as much money as possible over the contract (5 years)?

90. Look at the figure below. What fraction of the square is eventually shaded if the indicated shading process continues indefinitely?

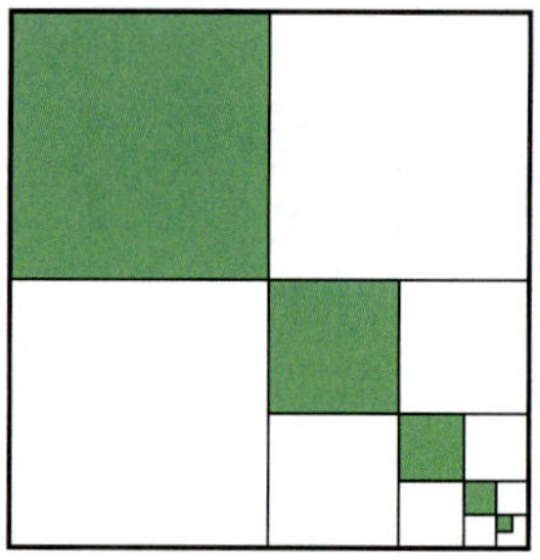

7.4 MATHEMATICAL INDUCTION

1 Prove Statements Using Mathematical Induction

1 *Mathematical induction* is a method for proving that statements involving natural numbers are true for all natural numbers.* For example, the statement "$2n$ is always an even integer" can be proved true for all natural numbers by using mathematical induction. Also, the statement "the sum of the first n positive odd integers equals n^2," that is,

$$1 + 3 + 5 + \cdots + (2n - 1) = n^2 \tag{1}$$

can be proved for all natural numbers n by using mathematical induction.

Before stating the method of mathematical induction, let's try to gain a sense of the power of the method. We shall use the statement in equation (1) for this purpose by restating it for various values of $n = 1, 2, 3, \ldots$:

$n = 1$ The sum of the first positive odd integer is 1^2; $1 = 1^2$.

$n = 2$ The sum of the first 2 positive odd integers is 2^2; $1 + 3 = 4 = 2^2$.

$n = 3$ The sum of the first 3 positive odd integers is 3^2; $1 + 3 + 5 = 9 = 3^2$.

$n = 4$ The sum of the first 4 positive odd integers is 4^2; $1 + 3 + 5 + 7 = 16 = 4^2$.

Although from this pattern we might conjecture that statement (1) is true for any choice of n, can we really be sure that it does not fail for some choice of n? The method of proof by mathematical induction will, in fact, prove that the statement is true for all n.

THEOREM **The Principle of Mathematical Induction**

Suppose that the following two conditions are satisfied with regard to a statement about natural numbers:

CONDITION I: The statement is true for the natural number 1.

CONDITION II: If the statement is true for some natural number k, it is also true for the next natural number $k + 1$.

Then the statement is true for all natural numbers.

We shall not prove this principle. However, we can provide a physical interpretation that will help us to see why the principle works. Think of a collection of natural numbers obeying a statement as a collection of infinitely many dominoes (see Figure 16).

Figure 16

Now, suppose that we are told two facts:

1. The first domino is pushed over.
2. If one of the dominoes falls over, say the kth domino, then so will the next one, the $(k + 1)$st domino.

*Recall from the Appendix, Section 1, that the natural numbers are the numbers 1, 2, 3, 4, ... In other words, the terms *natural numbers* and *positive integers* are synonymous.

Is it safe to conclude that *all* the dominoes fall over? The answer is yes, because if the first one falls (Condition I), then the second one does also (by Condition II); and if the second one falls, then so does the third (by Condition II); and so on.

Now let's prove some statements about natural numbers using mathematical induction.

EXAMPLE 1 Using Mathematical Induction

Show that the following statement is true for all natural numbers n:

$$1 + 3 + 5 + \cdots + (2n - 1) = n^2 \qquad (2)$$

Solution We need to show first that statement (2) holds for $n = 1$. Because $1 = 1^2$, statement (2) is true for $n = 1$. Thus, Condition I holds.

Next, we need to show that Condition II holds. Suppose that we know for some k that

$$1 + 3 + \cdots + (2k - 1) = k^2 \qquad (3)$$

We wish to show that, based on equation (3), statement (2) holds for $k + 1$. Thus, we look at the sum of the first $k + 1$ positive odd integers to determine whether this sum equals $(k + 1)^2$:

$$\begin{aligned}1 + 3 + \cdots + (2k - 1) + (2k + 1) &= \underbrace{[1 + 3 + \cdots + (2k - 1)]}_{= k^2 \text{ by equation (3)}} + (2k + 1)\\ &= k^2 + (2k + 1)\\ &= k^2 + 2k + 1 = (k + 1)^2\end{aligned}$$

Conditions I and II are satisfied; thus, by the Principle of Mathematical Induction, statement (2) is true for all natural numbers.

EXAMPLE 2 Using Mathematical Induction

Show that the following statement is true for all natural numbers n:

$$2^n > n$$

Solution First, we show that the statement $2^n > n$ holds when $n = 1$. Because $2^1 = 2 > 1$, the inequality is true for $n = 1$. Thus, Condition I holds.

Next, we assume, for some natural number k, that $2^k > k$. We wish to show that the formula holds for $k + 1$; that is, we wish to show that $2^{k+1} > k + 1$. Now,

$$2^{k+1} = 2 \cdot 2^k \underset{\substack{\uparrow \\ \text{We know that} \\ 2^k > k.}}{>} 2 \cdot k = k + k \underset{\substack{\uparrow \\ k \geq 1.}}{\geq} k + 1$$

Thus, if $2^k > k$, then $2^{k+1} > k + 1$, so Condition II of the Principle of Mathematical Induction is satisfied. Hence, the statement $2^n > n$ is true for all natural numbers n.

EXAMPLE 3 Using Mathematical Induction

Show that the following formula is true for all natural numbers n:

$$1 + 2 + 3 + \cdots + n = \frac{n(n+1)}{2} \tag{4}$$

Solution First, we show that formula (4) is true when $n = 1$. Because

$$\frac{1(1+1)}{2} = \frac{1(2)}{2} = 1$$

Condition I of the Principle of Mathematical Induction holds.

Next, we assume that formula (4) holds for some k, and we determine whether the formula then holds for $k + 1$. Thus, we assume that

$$1 + 2 + 3 + \cdots + k = \frac{k(k+1)}{2} \quad \text{for some } k \tag{5}$$

Now we need to show that

$$1 + 2 + 3 + \cdots + k + (k+1) = \frac{(k+1)(k+1+1)}{2} = \frac{(k+1)(k+2)}{2}$$

We do this as follows:

$$\begin{aligned}
1 + 2 + 3 + \cdots + k + (k+1) &= \underbrace{[1 + 2 + 3 + \cdots + k]}_{=\frac{k(k+1)}{2} \text{ by equation (5)}} + (k+1) \\
&= \frac{k(k+1)}{2} + (k+1) \\
&= \frac{k^2 + k + 2k + 2}{2} \\
&= \frac{k^2 + 3k + 2}{2} = \frac{(k+1)(k+2)}{2}
\end{aligned}$$

Thus, Condition II also holds. As a result, formula (4) is true for all natural numbers.

NOW WORK PROBLEM 1.

EXAMPLE 4 Using Mathematical Induction

Show that $3^n - 1$ is divisible by 2 for all natural numbers n.

Solution First, we show that the statement is true when $n = 1$. Because $3^1 - 1 = 3 - 1 = 2$ is divisible by 2, the statement is true when $n = 1$. Thus, Condition I is satisfied.

Next, we assume that the statement holds for some k, and we determine whether the statement then holds for $k + 1$. Thus, we assume that $3^k - 1$ is divisible by 2 for some k. We need to show that $3^{k+1} - 1$ is divisible by 2. Now,

$$\begin{aligned}
3^{k+1} - 1 &= 3^{k+1} - 3^k + 3^k - 1 \\
&= 3^k(3 - 1) + (3^k - 1) = 3^k \cdot 2 + (3^k - 1)
\end{aligned}$$

Because $3^k \cdot 2$ is divisible by 2 and $3^k - 1$ is divisible by 2, it follows that $3^k \cdot 2 + (3^k - 1) = 3^{k+1} - 1$ is divisible by 2. Thus, Condition II is also satisfied. As a result, the statement "$3^n - 1$ is divisible by 2" is true for all natural numbers n.

Warning: The conclusion that a statement involving natural numbers is true for all natural numbers is made only after *both* Conditions I and II of the Principle of Mathematical Induction have been satisfied. Problem 27 demonstrates a statement for which only Condition I holds, but the statement is not true for all natural numbers. Problem 28 demonstrates a statement for which only Condition II holds, but the statement is *not* true for any natural number.

7.4 EXERCISES

In Problems 1–26, use the Principle of Mathematical Induction to show that the given statement is true for all natural numbers.

1. $2 + 4 + 6 + \cdots + 2n = n(n + 1)$

2. $1 + 5 + 9 + \cdots + (4n - 3) = n(2n - 1)$

3. $3 + 4 + 5 + \cdots + (n + 2) = \frac{1}{2}n(n + 5)$

4. $3 + 5 + 7 + \cdots + (2n + 1) = n(n + 2)$

5. $2 + 5 + 8 + \cdots + (3n - 1) = \frac{1}{2}n(3n + 1)$

6. $1 + 4 + 7 + \cdots + (3n - 2) = \frac{1}{2}n(3n - 1)$

7. $1 + 2 + 2^2 + \cdots + 2^{n-1} = 2^n - 1$

8. $1 + 3 + 3^2 + \cdots + 3^{n-1} = \frac{1}{2}(3^n - 1)$

9. $1 + 4 + 4^2 + \cdots + 4^{n-1} = \frac{1}{3}(4^n - 1)$

10. $1 + 5 + 5^2 + \cdots + 5^{n-1} = \frac{1}{4}(5^n - 1)$

11. $\dfrac{1}{1 \cdot 2} + \dfrac{1}{2 \cdot 3} + \dfrac{1}{3 \cdot 4} + \cdots + \dfrac{1}{n(n + 1)} = \dfrac{n}{n + 1}$

12. $\dfrac{1}{1 \cdot 3} + \dfrac{1}{3 \cdot 5} + \dfrac{1}{5 \cdot 7} + \cdots + \dfrac{1}{(2n - 1)(2n + 1)} = \dfrac{n}{2n + 1}$

13. $1^2 + 2^2 + 3^2 + \cdots + n^2 = \frac{1}{6}n(n + 1)(2n + 1)$

14. $1^3 + 2^3 + 3^3 + \cdots + n^3 = \frac{1}{4}n^2(n + 1)^2$

15. $4 + 3 + 2 + \cdots + (5 - n) = \frac{1}{2}n(9 - n)$

16. $-2 - 3 - 4 - \cdots - (n + 1) = -\frac{1}{2}n(n + 3)$

17. $1 \cdot 2 + 2 \cdot 3 + 3 \cdot 4 + \cdots + n(n + 1) = \frac{1}{3}n(n + 1)(n + 2)$

18. $1 \cdot 2 + 3 \cdot 4 + 5 \cdot 6 + \cdots + (2n - 1)(2n) = \frac{1}{3}n(n + 1)(4n - 1)$

19. $n^2 + n$ is divisible by 2.

20. $n^3 + 2n$ is divisible by 3.

21. $n^2 - n + 2$ is divisible by 2.

22. $n(n + 1)(n + 2)$ is divisible by 6.

23. If $x > 1$, then $x^n > 1$.

24. If $0 < x < 1$, then $0 < x^n < 1$.

25. $a - b$ is a factor of $a^n - b^n$.
[Hint: $a^{k+1} - b^{k+1} = a(a^k - b^k) + b^k(a - b)$***]***

26. $a + b$ is a factor of $a^{2n+1} + b^{2n+1}$.

27. Show that the statement "$n^2 - n + 41$ is a prime number" is true for $n = 1$, but is not true for $n = 41$.

28. Show that the formula

$$2 + 4 + 6 + \cdots + 2n = n^2 + n + 2$$

obeys Condition II of the Principle of Mathematical Induction. That is, show that if the formula is true for some k it is also true for $k + 1$. Then show that the formula is false for $n = 1$ (or for any other choice of n).

29. Use mathematical induction to prove that if $r \neq 1$ then

$$a + ar + ar^2 + \cdots + ar^{n-1} = a\frac{1 - r^n}{1 - r}$$

30. Use mathematical induction to prove that

$$a + (a + d) + (a + 2d) + \cdots + [a + (n - 1)d] = na + d\frac{n(n - 1)}{2}$$

31. **Geometry** Use mathematical induction to show that the sum of the interior angles of a convex polygon of n sides equals $(n - 2) \cdot 180°$.

32. Extended Principle of Mathematical Induction The Extended Principle of Mathematical Induction states that if Conditions I and II hold, that is,

(I) A statement is true for a natural number j.
(II) If the statement is true for some natural number $k > j$, then it is also true for the next natural number $k + 1$.

then the statement is true for *all* natural numbers $\geq j$.

Use the Extended Principle of Mathematical Induction to show that the number of diagonals in a convex polygon of n sides is $\frac{1}{2}n(n - 3)$. ***[Hint: Begin by showing that the result is true when $n = 4$ (Condition I).]***

33. How would you explain to a friend the Principle of Mathematical Induction?

7.5 THE BINOMIAL THEOREM

1 Evaluate a Binomial Coefficient

2 Expand a Binomial

In the Appendix, Section 5, we listed some special products. Among these were formulas for expanding $(x + a)^n$ for $n = 2$ and $n = 3$. The *Binomial Theorem** is a formula for the expansion of $(x + a)^n$ for n any positive integer. If $n = 1, 2, 3$, and 4, the expansion of $(x + a)^n$ is straightforward:

$$(x + a)^1 = x + a$$ *Two terms, beginning with x^1 and ending with a^1.*

$$(x + a)^2 = x^2 + 2ax + a^2$$ *Three terms, beginning with x^2 and ending with a^2.*

$$(x + a)^3 = x^3 + 3ax^2 + 3a^2x + a^3$$ *Four terms, beginning with x^3 and ending with a^3.*

$$(x + a)^4 = x^4 + 4ax^3 + 6a^2x^2 + 4a^3x + a^4$$ *Five terms, beginning with x^4 and ending with a^4.*

Notice that each expression of $(x + a)^n$ begins with x^n and ends with a^n. As you read from left to right, the powers of x are decreasing, while the powers of a are increasing. Also, the number of terms that appears equals $n + 1$. Notice, too, that the degree of each monomial in the expansion equals n. For example, in the expansion of $(x + a)^3$, each monomial $(x^3, 3ax^2, 3a^2x, a^3)$ is of degree 3. As a result, we might conjecture that the expansion of $(x + a)^n$ would look like this:

$$(x + a)^n = x^n + \_\_ax^{n-1} + \_\_a^2x^{n-2} + \cdots + \_\_a^{n-1}x + a^n$$

where the blanks are numbers to be found. This is, in fact, the case, as we shall see shortly.

First, we need to introduce a symbol.

*The name *binomial* derives from the fact that $x + a$ is a binomial, that is, contains two terms.

The Symbol $\binom{n}{j}$

1 We define the symbol $\binom{n}{j}$, read "n taken j at a time," as follows:

If j and n are integers with $0 \le j \le n$, the **symbol** $\binom{n}{j}$ is defined as

$$\binom{n}{j} = \frac{n!}{j!(n-j)!} \tag{1}$$

Comment: On a graphing calculator, the symbol $\binom{n}{j}$ may be denoted by the key nCr.

EXAMPLE 1 Evaluating $\binom{n}{j}$

Find:

(a) $\binom{3}{1}$ (b) $\binom{4}{2}$ (c) $\binom{8}{7}$ (d) $\binom{65}{15}$

Solution (a) $\binom{3}{1} = \frac{3!}{1!(3-1)!} = \frac{3!}{1!2!} = \frac{3\cdot 2\cdot 1}{1(2\cdot 1)} = \frac{6}{2} = 3$

(b) $\binom{4}{2} = \frac{4!}{2!(4-2)!} = \frac{4!}{2!2!} = \frac{4\cdot 3\cdot 2\cdot 1}{(2\cdot 1)(2\cdot 1)} = \frac{24}{4} = 6$

(c) $\binom{8}{7} = \frac{8!}{7!(8-7)!} = \frac{8!}{7!1!} = \frac{8\cdot \cancel{7!}}{\cancel{7!}\cdot 1!} = \frac{8}{1} = 8$

$8! = 8 \cdot 7!$

Figure 17

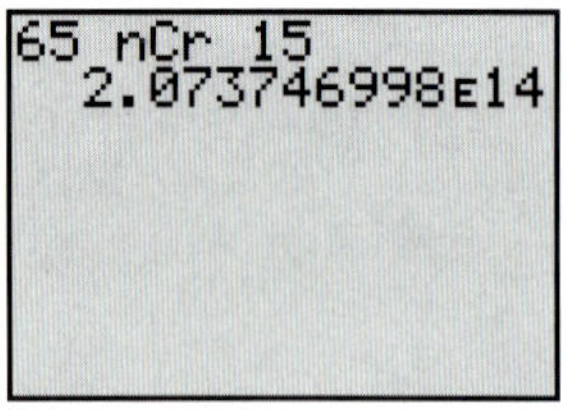

(d) Figure 17 shows the solution using a TI-83 graphing calculator. Thus, $\binom{65}{15} = 2.073746998 \times 10^{14}$.

NOW WORK PROBLEM 1.

Two useful formulas involving the symbol $\binom{n}{j}$ are

$$\binom{n}{0} = 1 \quad \text{and} \quad \binom{n}{n} = 1$$

Proof

$$\binom{n}{0} = \frac{n!}{0!(n-0)!} = \frac{n!}{0!n!} = \frac{1}{1} = 1$$

You are asked to show that $\binom{n}{n} = 1$ in Problem 41 at the end of this section.

Suppose that we arrange the various values of the symbol $\binom{n}{j}$ in a triangular display, as shown next and in Figure 18.

$$\binom{0}{0}$$
$$\binom{1}{0} \quad \binom{1}{1}$$
$$\binom{2}{0} \quad \binom{2}{1} \quad \binom{2}{2}$$
$$\binom{3}{0} \quad \binom{3}{1} \quad \binom{3}{2} \quad \binom{3}{3}$$
$$\binom{4}{0} \quad \binom{4}{1} \quad \binom{4}{2} \quad \binom{4}{3} \quad \binom{4}{4}$$
$$\binom{5}{0} \quad \binom{5}{1} \quad \binom{5}{2} \quad \binom{5}{3} \quad \binom{5}{4} \quad \binom{5}{5}$$

This display is called the **Pascal triangle,** named after Blaise Pascal (1623–1662), a French mathematician.

Figure 18
Pascal triangle

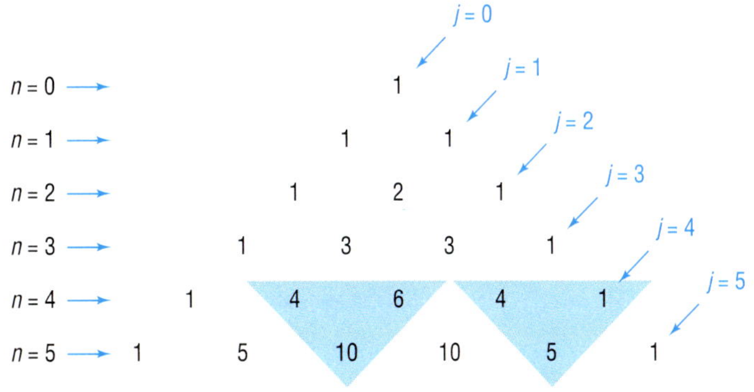

The Pascal triangle has 1's down the sides. To get any other entry, merely add the two nearest entries in the row above it. The shaded triangles in Figure 18 illustrate this feature of the Pascal triangle. Based on this feature, the row corresponding to $n = 6$ is found as follows:

$$\begin{array}{l} n = 5 \rightarrow \quad 1 \quad 5 \quad 10 \quad 10 \quad 5 \quad 1 \\ n = 6 \rightarrow \; 1 \quad 6 \quad 15 \quad 20 \quad 15 \quad 6 \quad 1 \end{array}$$

Later, we shall prove that this addition always works (see the theorem on page 540).

Although the Pascal triangle provides an interesting and organized display of the symbol $\binom{n}{j}$, in practice it is not all that helpful. For example, if you wanted to know the value of $\binom{12}{5}$, you would need to produce 12 rows of the triangle before seeing the answer. It is much faster instead to use the definition (1).

Binomial Theorem

2 Now we are ready to state the **Binomial Theorem.** A proof is given at the end of this section.

THEOREM

Binomial Theorem

Let x and a be real numbers. For any positive integer n, we have

$$(x + a)^n = \binom{n}{0}x^n + \binom{n}{1}ax^{n-1} + \cdots + \binom{n}{j}a^j x^{n-j} + \cdots + \binom{n}{n}a^n$$
$$= \sum_{j=0}^{n}\binom{n}{j}x^{n-j}a^j \qquad (2)$$

Now you know why we needed to introduce the symbol $\binom{n}{j}$; these symbols are the numerical coefficients that appear in the expansion of $(x + a)^n$. Because of this, the symbol $\binom{n}{j}$ is called the **binomial coefficient.**

EXAMPLE 2 Expanding a Binomial

Use the Binomial Theorem to expand $(x + 2)^5$.

Solution In the Binomial Theorem, let $a = 2$ and $n = 5$. Then

$$(x + 2)^5 = \binom{5}{0}x^5 + \binom{5}{1}2x^4 + \binom{5}{2}2^2x^3 + \binom{5}{3}2^3x^2 + \binom{5}{4}2^4x + \binom{5}{5}2^5$$

Use equation (2).

$$= 1 \cdot x^5 + 5 \cdot 2x^4 + 10 \cdot 4x^3 + 10 \cdot 8x^2 + 5 \cdot 16x + 1 \cdot 32$$

Use row $n = 5$ of the Pascal triangle or formula (1) for $\binom{n}{j}$.

$$= x^5 + 10x^4 + 40x^3 + 80x^2 + 80x + 32$$

EXAMPLE 3 Expanding a Binomial

Expand $(2y - 3)^4$ using the Binomial Theorem.

Solution First, we rewrite the expression $(2y - 3)^4$ as $[2y + (-3)]^4$. Now we use the Binomial Theorem with $n = 4$, $x = 2y$, and $a = -3$:

$$[2y + (-3)]^4 = \binom{4}{0}(2y)^4 + \binom{4}{1}(-3)(2y)^3 + \binom{4}{2}(-3)^2(2y)^2 + \binom{4}{3}(-3)^3(2y) + \binom{4}{4}(-3)^4$$

$$= 1 \cdot 16y^4 + 4(-3)8y^3 + 6 \cdot 9 \cdot 4y^2 + 4(-27)2y + 1 \cdot 81$$

Use row $n = 4$ of the Pascal triangle or formula (1) for $\binom{n}{j}$.

$$= 16y^4 - 96y^3 + 216y^2 - 216y + 81$$

In this expansion, note that the signs alternate due to the fact that $a = -3 < 0$.

NOW WORK PROBLEM 17.

EXAMPLE 4 Finding a Particular Coefficient in a Binomial Expansion

Find the coefficient of y^8 in the expansion of $(2y + 3)^{10}$.

Solution We write out the expansion using the Binomial Theorem:

$$(2y+3)^{10} = \binom{10}{0}(2y)^{10} + \binom{10}{1}(2y)^9(3)^1 + \binom{10}{2}(2y)^8(3)^2 + \binom{10}{3}(2y)^7(3)^3 + \binom{10}{4}(2y)^6(3)^4 + \cdots + \binom{10}{9}(2y)(3)^9 + \binom{10}{10}(3)^{10}$$

From the third term in the expansion, the coefficient of y^8 is

$$\binom{10}{2}(2)^8(3)^2 = \frac{10!}{2!8!}\cdot 2^8 \cdot 9 = \frac{10\cdot 9\cdot 8!}{2\cdot 8!}\cdot 2^8\cdot 9 = 103{,}680$$

As this solution demonstrates, we can use the Binomial Theorem to write a particular term in an expansion without writing the entire expansion. Based on the expansion of $(x + a)^n$, the term containing x^j is

$$\binom{n}{n-j}a^{n-j}x^j \qquad (3)$$

For example, we can solve Example 4 by using formula (3) with $n = 10$, $a = 3$, $x = 2y$, and $j = 8$. Then the term containing y^8 is

$$\binom{10}{10-8}3^{10-8}(2y)^8 = \binom{10}{2}\cdot 3^2\cdot 2^8\cdot y^8 = \frac{10!}{2!8!}\cdot 9\cdot 2^8y^8$$
$$= \frac{10\cdot 9\cdot 8!}{2!8!}\cdot 9\cdot 2^8y^8 = 103{,}680y^8$$

EXAMPLE 5 Finding a Particular Term in a Binomial Expansion

Find the sixth term in the expansion of $(x + 2)^9$.

Solution A We expand using the Binomial Theorem until the sixth term is reached:

$$(x+2)^9 = \binom{9}{0}x^9 + \binom{9}{1}x^8\cdot 2 + \binom{9}{2}x^7\cdot 2^2 + \binom{9}{3}x^6\cdot 2^3 + \binom{9}{4}x^5\cdot 2^4 + \binom{9}{5}x^4\cdot 2^5 + \cdots$$

The sixth term is

$$\binom{9}{5}x^4\cdot 2^5 = \frac{9!}{5!4!}\cdot x^4\cdot 32 = 4032x^4$$

Solution B The sixth term in the expansion of $(x + 2)^9$, which has 10 terms total, contains x^4. (Do you see why?) Thus, by formula (3), the sixth term is

$$\binom{9}{9-4}2^{9-4}x^4 = \binom{9}{5}2^5x^4 = \frac{9!}{5!4!}\cdot 32x^4 = 4032x^4$$

NOW WORK PROBLEMS 25 AND 31.

Next we show that the *triangular addition* feature of the Pascal triangle illustrated in Figure 18 always works.

THEOREM If n and j are integers with $1 \le j \le n$, then

$$\binom{n}{j-1} + \binom{n}{j} = \binom{n+1}{j} \tag{4}$$

Proof

$$\begin{aligned}
\binom{n}{j-1} + \binom{n}{j} &= \frac{n!}{(j-1)![n-(j-1)]!} + \frac{n!}{j!(n-j)!} \\
&= \frac{n!}{(j-1)!(n-j+1)!} + \frac{n!}{j!(n-j)!} \\
&= \frac{jn!}{j(j-1)!(n-j+1)!} + \frac{(n-j+1)n!}{j!(n-j+1)(n-j)!} \\
&= \frac{jn!}{j!(n-j+1)!} + \frac{(n-j+1)n!}{j!(n-j+1)!} \\
&= \frac{jn! + (n-j+1)n!}{j!(n-j+1)!} \\
&= \frac{n!(j+n-j+1)}{j!(n-j+1)!} \\
&= \frac{n!(n+1)}{j!(n-j+1)!} = \frac{(n+1)!}{j![(n+1)-j]!} = \binom{n+1}{j}
\end{aligned}$$

Multiply the first term by j/j and the second term by $(n-j+1)/(n-j+1)$.

Now the denominators are equal.

Proof of the Binomial Theorem We use mathematical induction to prove the Binomial Theorem. First, we show that formula (2) is true for $n = 1$:

$$(x+a)^1 = x + a = \binom{1}{0}x^1 + \binom{1}{1}a^1$$

Next we suppose that formula (2) is true for some k. That is, we assume that

$$(x+a)^k = \binom{k}{0}x^k + \binom{k}{1}ax^{k-1} + \cdots + \binom{k}{j-1}a^{j-1}x^{k-j+1} + \binom{k}{j}a^jx^{k-j} + \cdots + \binom{k}{k}a^k \tag{5}$$

Now we calculate $(x+a)^{k+1}$:

$$\begin{aligned}
(x+a)^{k+1} &= (x+a)(x+a)^k = x(x+a)^k + a(x+a)^k \\
&= x\left[\binom{k}{0}x^k + \binom{k}{1}ax^{k-1} + \cdots + \binom{k}{j-1}a^{j-1}x^{k-j+1} + \binom{k}{j}a^jx^{k-j} + \cdots + \binom{k}{k}a^k\right] \\
&\quad + a\left[\binom{k}{0}x^k + \binom{k}{1}ax^{k-1} + \cdots + \binom{k}{j-1}a^{j-1}x^{k-j+1} + \binom{k}{j}a^jx^{k-j} + \cdots + \binom{k}{k-1}a^{k-1}x + \binom{k}{k}a^k\right] \\
&= \binom{k}{0}x^{k+1} + \binom{k}{1}ax^k + \cdots + \binom{k}{j-1}a^{j-1}x^{k-j+2} + \binom{k}{j}a^jx^{k-j+1} + \cdots + \binom{k}{k}a^kx \\
&\quad + \binom{k}{0}ax^k + \binom{k}{1}a^2x^{k-1} + \cdots + \binom{k}{j-1}a^jx^{k-j+1} + \binom{k}{j}a^{j+1}x^{k-j} + \cdots + \binom{k}{k-1}a^kx + \binom{k}{k}a^{k+1} \\
&= \binom{k}{0}x^{k+1} + \left[\binom{k}{1} + \binom{k}{0}\right]ax^k + \cdots + \left[\binom{k}{j} + \binom{k}{j-1}\right]a^jx^{k-j+1} + \cdots + \left[\binom{k}{k} + \binom{k}{k-1}\right]a^kx + \binom{k}{k}a^{k+1}
\end{aligned}$$

Use ↑ equation (5).

Because

$$\binom{k}{0} = 1 = \binom{k+1}{0}, \quad \binom{k}{1} + \binom{k}{0} \underset{\uparrow \atop (4)}{=} \binom{k+1}{1}, \ldots,$$

$$\binom{k}{j} + \binom{k}{j-1} \underset{\uparrow \atop (4)}{=} \binom{k+1}{j}, \ldots, \binom{k}{k} = 1 = \binom{k+1}{k+1}$$

we have

$$(x+a)^{k+1} = \binom{k+1}{0}x^{k+1} + \binom{k+1}{1}ax^k + \cdots + \binom{k+1}{j}a^j x^{k-j+1} + \cdots + \binom{k+1}{k+1}a^{k+1}$$

Thus, Conditions I and II of the Principle of Mathematical Induction are satisfied, and formula (2) is therefore true for all n.

HISTORICAL FEATURE

The case $n = 2$ of the Binomial Theorem, $(a + b)^2$, was known to Euclid in 300 B.C., but the general law seems to have been discovered by the Persian mathematician and astronomer Omar Khayyám (1044?–1123?), who is also well known as the author of the *Rubáiyát,* a collection of four-line poems making observations on the human condition. Omar Khayyám did not state the Binomial Theorem explicitly, but he claimed to have a method for extracting third, fourth, fifth roots, and so on. A little study shows that one must know the Binomial Theorem to create such a method.

The heart of the Binomial Theorem is the formula for the numerical coefficients, and, as we saw, they can be written out in a symmetric triangular form. The Pascal triangle appears first in the books of Yang Hui (about 1270) and Chu Shihchie (1303). Pascal's name is attached to the triangle because of the many applications he made of it, especially to counting and probability. In establishing these results, he was one of the earliest users of mathematical induction.

Many people worked on the proof of the Binomial Theorem, which was finally completed for all n (including complex numbers) by Niels Abel (1802–1829).

7.5 Exercises

In Problems 1–12, evaluate each expression by hand. Use a graphing utility to verify your answer.

1. $\binom{5}{3}$ **2.** $\binom{7}{3}$ **3.** $\binom{7}{5}$ **4.** $\binom{9}{7}$ **5.** $\binom{50}{49}$ **6.** $\binom{100}{98}$

7. $\binom{1000}{1000}$ **8.** $\binom{1000}{0}$ **9.** $\binom{55}{23}$ **10.** $\binom{60}{20}$ **11.** $\binom{47}{25}$ **12.** $\binom{37}{19}$

In Problems 13–24, expand each expression using the Binomial Theorem.

13. $(x + 1)^5$ **14.** $(x - 1)^5$ **15.** $(x - 2)^6$ **16.** $(x + 3)^4$
17. $(3x + 1)^4$ **18.** $(2x + 3)^5$ **19.** $(x^2 + y^2)^5$ **20.** $(x^2 - y^2)^6$
21. $(\sqrt{x} + \sqrt{2})^6$ **22.** $(\sqrt{x} - \sqrt{3})^4$ **23.** $(ax + by)^5$ **24.** $(ax - by)^4$

In Problems 25–38, use the Binomial Theorem to find the indicated coefficient or term.

25. The coefficient of x^6 in the expansion of $(x + 3)^{10}$

26. The coefficient of x^3 in the expansion of $(x - 3)^{10}$

27. The coefficient of x^7 in the expansion of $(2x - 1)^{12}$

28. The coefficient of x^3 in the expansion of $(2x + 1)^{12}$

29. The coefficient of x^7 in the expansion of $(2x + 3)^9$

30. The coefficient of x^2 in the expansion of $(2x - 3)^9$

31. The fifth term in the expansion of $(x + 3)^7$

32. The third term in the expansion of $(x - 3)^7$

33. The third term in the expansion of $(3x - 2)^9$

34. The sixth term in the expansion of $(3x + 2)^8$

35. The coefficient of x^0 in the expansion of $\left(x^2 + \frac{1}{x}\right)^{12}$

36. The coefficient of x^0 in the expansion of $\left(x - \frac{1}{x^2}\right)^9$

37. The coefficient of x^4 in the expansion of $\left(x - \frac{2}{\sqrt{x}}\right)^{10}$

38. The coefficient of x^2 in the expansion of $\left(\sqrt{x} + \frac{3}{\sqrt{x}}\right)^8$

39. Use the Binomial Theorem to find the numerical value of $(1.001)^5$ correct to five decimal places. ***[Hint:** $(1.001)^5 = (1 + 10^{-3})^5$**]***

40. Use the Binomial Theorem to find the numerical value of $(0.998)^6$ correct to five decimal places.

41. Show that $\binom{n}{n} = 1$.

42. Show that, if n and j are integers with $0 \le j \le n$, then

$$\binom{n}{j} = \binom{n}{n - j}$$

Thus, conclude that the Pascal triangle is symmetric with respect to a vertical line drawn from the topmost entry.

43. If n is a positive integer, show that

$$\binom{n}{0} + \binom{n}{1} + \cdots + \binom{n}{n} = 2^n$$

[Hint:** $2^n = (1 + 1)^n$**; now use the Binomial Theorem.]

44. If n is a positive integer, show that

$$\binom{n}{0} - \binom{n}{1} + \binom{n}{2} - \cdots + (-1)^n\binom{n}{n} = 0$$

45. $$\binom{5}{0}\left(\frac{1}{4}\right)^5 + \binom{5}{1}\left(\frac{1}{4}\right)^4\left(\frac{3}{4}\right) + \binom{5}{2}\left(\frac{1}{4}\right)^3\left(\frac{3}{4}\right)^2 + \binom{5}{3}\left(\frac{1}{4}\right)^2\left(\frac{3}{4}\right)^3 + \binom{5}{4}\left(\frac{1}{4}\right)\left(\frac{3}{4}\right)^4 + \binom{5}{5}\left(\frac{3}{4}\right)^5 = ?$$

46. *Stirling's formula* for approximating $n!$ when n is large is given by

$$n! \approx \sqrt{2n\pi}\left(\frac{n}{e}\right)^n\left(1 + \frac{1}{12n - 1}\right)$$

Calculate 12!, 20!, and 25!. Then use Stirling's formula to approximate 12!, 20!, and 25!.

CHAPTER REVIEW

THINGS TO KNOW

Sequence	A function whose domain is the set of positive integers.
Factorials	$0! = 1, 1! = 1, n! = n(n - 1) \cdot \ldots \cdot 3 \cdot 2 \cdot 1$ if $n \ge 2$
Arithmetic sequence	$a_1 = a, a_n = a_{n-1} + d$, where a = first term, d = common difference, $a_n = a + (n - 1)d$
Sum of the first n terms of an arithmetic sequence	$S_n = \frac{n}{2}[2a + (n - 1)d] = \frac{n}{2}(a + a_n)$

Geometric sequence	$a_1 = a, \quad a_n = ra_{n-1}$, where a = first term, r = common ratio, $a_n = ar^{n-1}, \quad r \neq 0$
Sum of the first n terms of a geometric sequence	$S_n = a\dfrac{1 - r^n}{1 - r}, \quad r \neq 0, 1$
Amount of an annuity	$A = P\dfrac{(1 + i)^n - 1}{i}$; $a_1 = P, a_n = (1 + i)a_{n-1} + P$
Infinite geometric series	$a + ar + \cdots + ar^{n-1} + \cdots = \sum_{k=1}^{\infty} ar^{k-1}$
Sum of an infinite geometric series	$\sum_{k=1}^{\infty} ar^{k-1} = \dfrac{a}{1 - r}, \quad \lvert r \rvert < 1$
Principle of Mathematical Induction	Condition I: The statement is true for the natural number 1. Condition II: If the statement is true for some natural number k, it is also true for $k + 1$. Then the statement is true for all natural numbers.
Binomial coefficient	$\dbinom{n}{j} = \dfrac{n!}{j!(n - j)!}$
Pascal triangle	See Figure 18.
Binomial Theorem	$(x + a)^n = \dbinom{n}{0}x^n + \dbinom{n}{1}ax^{n-1} + \cdots + \dbinom{n}{j}a^j x^{n-j} + \cdots + \dbinom{n}{n}a^n$

HOW TO

Write down the terms of a sequence

Use summation notation

Identify an arithmetic sequence

Find the sum of the first n terms of an arithmetic sequence

Identify a geometric sequence

Solve annuity problems

Apply the Binomial Theorem

Find the sum of arithmetic and geometric sequences using a graphing utility

Find the sum of the first n terms of a geometric sequence

Find the sum of an infinite geometric series

Prove statements about natural numbers using mathematical induction

FILL-IN-THE-BLANK ITEMS

1. A(n) ________________ is a function whose domain is the set of positive integers.

2. In a(n) ________________ sequence, the difference between successive terms is always the same number.

3. In a(n) ________________ sequence, the ratio of successive terms is always the same number.

4. The ________________ ________________ is a triangular display of the binomial coefficients.

5. $\dbinom{6}{2} =$ ________________.

TRUE/FALSE ITEMS

T F **1.** A sequence is a function.

T F **2.** For arithmetic sequences, the difference of successive terms is always the same number.

T F **3.** For geometric sequences, the ratio of successive terms is always the same number.

T F **4.** Mathematical induction can sometimes be used to prove theorems that involve natural numbers.

T F **5.** $\binom{n}{j} = \dfrac{j!}{n!(n-j)!}$

T F **6.** The expansion of $(x + a)^n$ contains n terms.

T F **7.** $\sum_{i=1}^{n+1} i = 1 + 2 + 3 + \cdots + n$

REVIEW EXERCISES

Blue problem numbers indicate the authors' suggestions for use in a Practice Test.

In Problems 1–8, write down the first five terms of each sequence.

1. $\left\{(-1)^n\left(\dfrac{n+3}{n+2}\right)\right\}$ **2.** $\{(-1)^{n+1}(2n+3)\}$ **3.** $\left\{\dfrac{2^n}{n^2}\right\}$

4. $\left\{\dfrac{e^n}{n}\right\}$ **5.** $a_1 = 3; \quad a_n = \frac{2}{3}a_{n-1}$ **6.** $a_1 = 4; \quad a_n = -\frac{1}{4}a_{n-1}$

7. $a_1 = 2; \quad a_n = 2 - a_{n-1}$ **8.** $a_1 = -3; \quad a_n = 4 + a_{n-1}$

In Problems 9–20, determine whether the given sequence is arithmetic, geometric, or neither. If the sequence is arithmetic, find the common difference and the sum of the first n terms. If the sequence is geometric, find the common ratio and the sum of the first n terms.

9. $\{n + 5\}$ **10.** $\{4n + 3\}$ **11.** $\{2n^3\}$ **12.** $\{2n^2 - 1\}$

13. $\{2^{3n}\}$ **14.** $\{3^{2n}\}$ **15.** $0, 4, 8, 12, \ldots$ **16.** $1, -3, -7, -11, \ldots$

17. $3, \frac{3}{2}, \frac{3}{4}, \frac{3}{8}, \frac{3}{16}, \ldots$ **18.** $5, -\frac{5}{3}, \frac{5}{9}, -\frac{5}{27}, \frac{5}{81}, \ldots$ **19.** $\frac{2}{3}, \frac{3}{4}, \frac{4}{5}, \frac{5}{6}, \ldots$ **20.** $\frac{3}{2}, \frac{5}{4}, \frac{7}{6}, \frac{9}{8}, \frac{11}{10}, \ldots$

In Problems 21–26, evaluate each sum. Verify your results using a graphing utility.

21. $\sum_{k=1}^{5}(k^2 + 12)$ **22.** $\sum_{k=1}^{3}(k + 2)^2$ **23.** $\sum_{k=1}^{10}(3k - 9)$ **24.** $\sum_{k=1}^{9}(-2k + 8)$

25. $\sum_{k=1}^{7}\left(\dfrac{1}{3}\right)^k$ **26.** $\sum_{k=1}^{10}(-2)^k$

In Problems 27–32, find the indicated term in each sequence (a) by hand and (b) using a graphing utility.

27. 9th term of $3, 7, 11, 15, \ldots$ **28.** 8th term of $1, -1, -3, -5, \ldots$

29. 11th term of $1, \frac{1}{10}, \frac{1}{100}, \ldots$ **30.** 11th term of $1, 2, 4, 8, \ldots$

31. 9th term of $\sqrt{2}, 2\sqrt{2}, 3\sqrt{2}, \ldots$ **32.** 9th term of $\sqrt{2}, 2, 2^{3/2}, \ldots$

In Problems 33–36, find a general formula for each arithmetic sequence.

33. 7th term is 31; 20th term is 96 **34.** 8th term is -20; 17th term is -47

35. 10th term is 0; 18th term is 8 **36.** 12th term is 30; 22nd term is 50

In Problems 37–42, find the sum of each infinite geometric series.

37. $3 + 1 + \frac{1}{3} + \frac{1}{9} + \cdots$ **38.** $2 + 1 + \frac{1}{2} + \frac{1}{4} + \cdots$ **39.** $2 - 1 + \frac{1}{2} - \frac{1}{4} + \cdots$

40. $6 - 4 + \frac{8}{3} - \frac{16}{9} + \cdots$ **41.** $\sum_{k=1}^{\infty} 4\left(\dfrac{1}{2}\right)^{k-1}$ **42.** $\sum_{k=1}^{\infty} 3\left(-\dfrac{3}{4}\right)^{k-1}$

In Problems 43–48, use the Principle of Mathematical Induction to show that the given statement is true for all natural numbers.

43. $3 + 6 + 9 + \cdots + 3n = \frac{3n}{2}(n + 1)$

44. $2 + 6 + 10 + \cdots + (4n - 2) = 2n^2$

45. $2 + 6 + 18 + \cdots + 2 \cdot 3^{n-1} = 3^n - 1$

46. $3 + 6 + 12 + \cdots + 3 \cdot 2^{n-1} = 3(2^n - 1)$

47. $1^2 + 4^2 + 7^2 + \cdots + (3n - 2)^2 = \frac{1}{2}n(6n^2 - 3n - 1)$

48. $1 \cdot 3 + 2 \cdot 4 + 3 \cdot 5 + \cdots + n(n + 2) = \frac{n}{6}(n + 1)(2n + 7)$

In Problems 49–52, expand each expression using the Binomial Theorem.

49. $(x + 2)^5$

50. $(x - 3)^4$

51. $(2x + 3)^5$

52. $(3x - 4)^4$

53. Find the coefficient of x^7 in the expansion of $(x + 2)^9$.

54. Find the coefficient of x^3 in the expansion of $(x - 3)^8$.

55. Find the coefficient of x^2 in the expansion of $(2x + 1)^7$.

56. Find the coefficient of x^6 in the expansion of $(2x + 1)^8$.

57. **Constructing a Brick Staircase** A brick staircase has a total of 25 steps. The bottom step requires 80 bricks. Each successive step requires three less bricks than the prior step.
(a) How many bricks are required for the top step?
(b) How many bricks are required to build the staircase?

58. **Creating a Floor Design** A mosaic tile floor is designed in the shape of a trapezoid 30 feet wide at the base and 15 feet wide at the top. See Figure 13, p. 513. The tiles, 12 inches by 12 inches, are to be placed so that each successive row contains one less tile than the row below. How many tiles will be required?

59. **Retirement Planning** Chris gets paid once a month and contributes \$200 each pay period into his 401(k). If Chris plans on retiring in 20 years, what will the value of his 401(k) be if the per annum rate of return of the 401(k) is 10% compounded monthly?

60. **Retirement Planning** Jacky contributes \$500 every quarter to an IRA. If Jacky plans on retiring in 30 years, what will the value of the IRA be if the per annum rate of return of the IRA is 8% compounded quarterly?

61. **Bouncing Balls** A ball is dropped from a height of 20 feet. Each time it strikes the ground, it bounces up to three-quarters of the previous height.
(a) What height will the ball bounce up to after it strikes the ground for the third time?
(b) What is its height after it strikes the ground for the nth time?
(c) How many times does the ball need to strike the ground before its height is less than 6 inches?
(d) What total distance does the ball travel before it stops bouncing?

62. **Salary Increases** Your friend has just been hired at an annual salary of \$20,000. If she expects to receive annual increases of 4%, what will her salary be as she begins her fifth year?

63. **Home Loan** Mike and Yola borrowed \$190,000 at 6.75% per annum compounded monthly for 30 years to purchase a home. Their monthly payment is determined to be \$1232.34.
(a) Find a recursive formula for their balance at the beginning of each month after the monthly payment has been made.
(b) Determine Mike and Yola's balance at the beginning of the second month.
(c) Using a graphing utility, create a table showing Mike and Yola's balance at the beginning of each month.
(d) Using a graphing utility, determine the number of months it will take until Mike and Yola's balance is below \$100,000.
(e) Using a graphing utility, determine the number of months it will take until Mike and Yola pay off the balance.
(f) Determine Mike and Yola's interest expense when the loan is paid.
(g) Suppose that Mike and Yola decide to pay an additional \$100 each month on their loan. Answer parts (a)–(f) under this scenario.

CHAPTER PROJECTS

1. **Population Growth** The size of the population of the United States essentially depends on its current population, the birth and death rates of the population, and immigration. Suppose that b represents the birth rate of the U.S. population and d represents the death rate of the U.S. population. Then $r = b - d$ represents the growth rate of the population, where r varies from year to year. The U.S. population after n years can be modeled using the recursive function

$$p_n = (1 + r)p_{n-1} + I$$

where I represents net immigration into the United States.

(a) Using data from the National Center for Health Statistics (http://www.fedstats.gov), determine the birth and death rates for all races for the most recent year that data are available. Birth rates are given as the number of live births per 1000 population, while death rates are given as the number of deaths per 100,000 population. Each one must be computed as the number of births (deaths) per individual. For example, in 1990, the birth rate was 16.7 per 1000 and the death rate was 863.8 per 100,000, so $b = 16.7/1000 = 0.0167$, while $d = 863.8/100{,}000 = 0.008638$.

Next, using data from the Immigration and Naturalization Service (http://www.fedstats.gov), determine the immigration to the United States for the same year as the one used to obtain b and d in part (a).

(b) Determine the value of r, the growth rate of the population.

(c) Find a recursive formula for the population of the United States.

(d) Use the recursive formula to predict the population of the United States in the following year. In other words, if data are available up to the year 1994, predict the U.S. population in 1995.

(e) Compare your prediction to actual data.

(f) Do you think the recursive formula found in part (c) will be useful in predicting future populations? Why or why not?

2. **Economics** In many situations, the amount of output a producer manufactures is a function of the price of the product in the prior time period. This is especially true in areas where the producer's output decisions must be made in advance of the date of sale, such as agricultural output, where seeds planted today are sold well into the future and planting decisions must be made based on today's price. We can represent the quantity supplied by the equation $Q_{st} = -a + bP_{t-1}$ (i.e., quantity supplied in time period t depends on the price in period $t - 1$). Since consumers base their decision on today's price, demand is given by the equation $Q_{dt} = c - dP_t$. Equilibrium occurs when $Q_{st} = Q_{dt}$ or $-a + bP_{t-1} = c - dP_t$. Solving for P_t, we find the recursion formula

$$P_t = \frac{a + c - bP_{t-1}}{d} \tag{1}$$

If $b < d$, the system will converge to an equilibrium price, a price where quantity supplied equals quantity demanded and $P_{t-1} = P_t$.

Consider the following model:

$$Q_{st} = -3 + 2P_{t-1} \qquad Q_{dt} = 18 - 3P_t \qquad P_0 = \$2 \text{ (initial price)}$$

The value of b is 2, while the value of d is 3. Therefore, the price will converge to an equilibrium price. To analyze this model, put your graphing utility in WEB format and SEQuence mode.

CHAPTER PROJECTS (*Continued*)

(a) Find a recursive formula for the model using Formula (1).

(b) Graph the recursive formula in WEB format. When graphing in WEB format, the x-axis represents P_{t-1} and the y-axis represents P_t.

(c) TRACE the recursive function. What is the price after one time period ($n = 1$)? Determine Q_{st} and Q_{dt} for this price. They should be equal! This is due to the fact the price will adjust so that $Q_{st} = Q_{dt}$. So, when the price was \$2, quantity supplied was such that the quantity demanded was larger than the amount being supplied. This caused competition among buyers, which caused the price to increase. Now TRACE to the next time period. Explain what happened in the markets from $t = 1$ to $t = 2$.

(d) Find the equilibrium point (if any) for each by TRACEing the function until the price is correct to the nearest penny.

(e) How many time periods does it take until this equilibrium is reached?

(f) Verify that this price is an equilibrium by finding Q_{st} and Q_{dt}. What is the equilibrium quantity supplied and demanded?

CHAPTER

8 Counting and Probability

The Two-Children Problem

PROBLEM: A woman and a man (unrelated) each have two children. At least one of the woman's children is a boy, and the man's older child is a boy. Do the chances that the woman has two boys equal the chances that the man has two boys?

The above problem was posed to Marilyn vos Savant in her column *Ask Marilyn*. Her original answer, based on theoretical probabilities, was that the chances that the woman has two boys are 1 in 3 and the chances that the man has two boys are 1 in 2. This is found by looking at the sample space of two-child families: BB, BG, GB, GG. In the case of the man, we know that his older child is a boy and the sample space reduces to BB and BG. Hence, the probability that he has two boys is 1 out of 2. In the case of the woman, since we only know that she has at least one boy, the sample space reduces to BB, BG, and GB. Thus, her chances of having two boys is 1 out of 3.

The answer about the woman's chances created quite a bit of controversy resulting in many letters which challenged the correctness of her answer (*Parade*, July 27, 1997). Marilyn proposed that readers with exactly two children and at least one boy write in and tell the sex of both their children.

See Chapter Project 1.

Preparing for This Chapter

Outline

This chapter continues with topics that are covered in more detail in courses titled *Finite Mathematics*, *Probability*, or *Discrete Mathematics*. Applications of these topics can be found in the fields of computer science, engineering, business and economics, the social sciences, and the physical and biological sciences.

The first two sections of this chapter deal with techniques and formulas for counting the number of objects in a set, a part of the branch of mathematics called *combinatorics*. These formulas are used in computer science to analyze algorithms and recursive functions and to study stacks and queues. They are also used to determine *probabilities*, the likelihood that a certain outcome of a random experiment will occur, a topic discussed in the last two sections.

8.1 SETS AND COUNTING

1. Find All the Subsets of a Set
2. Find the Intersection and Union of Sets
3. Find the Complement of a Set
4. Count the Number of Elements in a Set

Sets

A **set** is a well-defined collection of distinct objects. The objects of a set are called its **elements.** By **well-defined,** we mean that there is a rule that enables us to determine whether a given object is an element of the set. If a set has no elements, it is called the **empty set,** or **null set,** and is denoted by the symbol $\varnothing$.

Because the elements of a set are distinct, we never repeat elements. Thus, we would never write $\{1, 2, 3, 2\}$; the correct listing is $\{1, 2, 3\}$. Furthermore, because a set is a collection, the order in which the elements are listed is immaterial. Thus, $\{1, 2, 3\}$, $\{1, 3, 2\}$, $\{2, 1, 3\}$, and so on, all represent the same set.

EXAMPLE 1 Writing the Elements of a Set

Write the set consisting of the possible results (outcomes) from tossing a coin twice. Use H for *heads* and T for *tails.*

Solution In tossing a coin twice, we can get heads each time, HH; or heads the first time and tails the second, HT; or tails the first time and heads the second, TH; or tails each time, TT. Because no other possibilities exist, the set of outcomes is

$$\{\text{HH, HT, TH, TT}\}$$

1 If two sets A and B have precisely the same elements, then we say that A and B are **equal** and write $A = B$.

If each element of a set A is also an element of a set B, then we say that A is a **subset** of B and write $A \subseteq B$.

If $A \subseteq B$ and $A \neq B$, then we say that A is a **proper subset** of B and write $A \subset B$.

Thus, if $A \subseteq B$, every element in set A is also in set B, but B may or may not have additional elements. If $A \subset B$, every element in A is also in B, and B has at least one element not found in A.

Finally, we agree that the empty set is a subset of every set; that is,

$$\varnothing \subseteq A \qquad \text{for any set } A$$

EXAMPLE 2 Finding All the Subsets of a Set

Write down all the subsets of the set $\{a, b, c\}$.

Solution To organize our work, we write down all the subsets with no elements, then those with one element, then those with two elements, and finally those with three elements. These will give us all the subsets. Do you see why?

0 Elements	1 Element	2 Elements	3 Elements
$\varnothing$	$\{a\}, \{b\}, \{c\}$	$\{a, b\}, \{b, c\}, \{a, c\}$	$\{a, b, c\}$

NOW WORK PROBLEM 21.

If A and B are sets, the **intersection** of A with B, denoted $A \cap B$, is the set consisting of elements that belong to both A and B. The **union** of A with B, denoted $A \cup B$, is the set consisting of elements that belong to *either* A or B, or both.

EXAMPLE 3 Finding the Intersection and Union of Sets

2 Let $A = \{1, 3, 5, 8\}$, $B = \{3, 5, 7\}$, and $C = \{2, 4, 6, 8\}$. Find:

(a) $A \cap B$ (b) $A \cup B$ (c) $B \cap (A \cup C)$

Solution (a) $A \cap B = \{1, 3, 5, 8\} \cap \{3, 5, 7\} = \{3, 5\}$

(b) $A \cup B = \{1, 3, 5, 8\} \cup \{3, 5, 7\} = \{1, 3, 5, 7, 8\}$

(c) $B \cap (A \cup C) = \{3, 5, 7\} \cap [\{1, 3, 5, 8\} \cup \{2, 4, 6, 8\}]$
$= \{3, 5, 7\} \cap \{1, 2, 3, 4, 5, 6, 8\} = \{3, 5\}$

NOW WORK PROBLEM 5.

3 Usually, in working with sets, we designate a **universal set,** the set consisting of all the elements that we wish to consider. Once a universal set has been designated, we can consider elements of the universal set not found in a given set.

If A is a set, the **complement** of A, denoted $\overline{A}$, is the set consisting of all the elements in the universal set that are not in A.

EXAMPLE 4 Finding the Complement of a Set

If the universal set is $U = \{1, 2, 3, 4, 5, 6, 7, 8, 9\}$, and if $A = \{1, 3, 5, 7, 9\}$, then $\overline{A} = \{2, 4, 6, 8\}$.

Notice that $A \cup \overline{A} = U$ and $A \cap \overline{A} = \varnothing$.

NOW WORK PROBLEM 13.

Figure 1

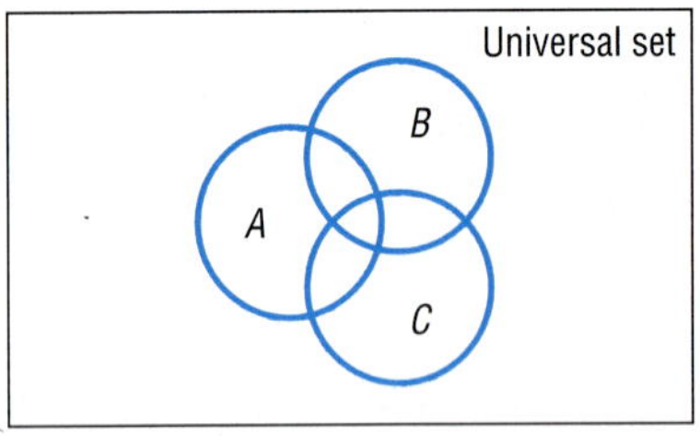

It is often helpful to draw pictures of sets. Such pictures, called **Venn diagrams,** represent sets as circles enclosed in a rectangle, which represents the universal set. Such diagrams often help us to visualize various relationships among sets. See Figure 1.

If we know that $A \subseteq B$, we might use the Venn diagram in Figure 2(a). If we know that A and B have no elements in common, that is, if $A \cap B = \emptyset$, we might use the Venn diagram in Figure 2(b). The sets A and B in Figure 2(b) are said to be **disjoint.**

Figure 2

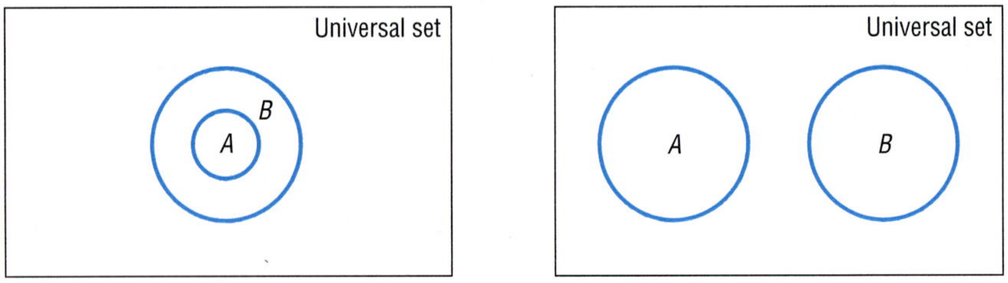

(a) $A \subseteq B$ subset

(b) $A \cap B = \emptyset$ disjoint sets

Figures 3(a), 3(b), and 3(c) use Venn diagrams to illustrate the definitions of intersection, union, and complement, respectively.

Figure 3

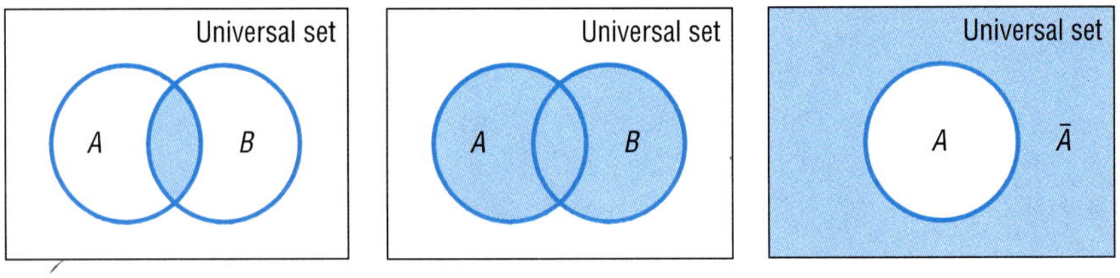

(a) $A \cap B$ intersection

(b) $A \cup B$ union

(c) $\bar{A}$ complement

Counting

4 As you count the number of students in a classroom or the number of pennies in your pocket, what you are really doing is matching, on a one-to-one basis, each object to be counted with the counting numbers $1, 2, 3, \ldots, n$, for some number n. If a set A matched up in this fashion with the set $\{1, 2, \ldots, 25\}$, you would conclude that there are 25 elements in the set A. We use the notation $n(A) = 25$ to indicate that there are 25 elements in the set A.

Because the empty set has no elements, we write

$$n(\emptyset) = 0$$

If the number of elements in a set is a nonnegative integer, we say that the set is **finite.** Otherwise, it is **infinite.** We shall concern ourselves only with finite sets.

From Example 2, we can see that a set with 3 elements has $2^3 = 8$ subsets. This result can be generalized.

> If A is a set with n elements, then A has 2^n subsets.

For example, the set $\{a, b, c, d, e\}$ has $2^5 = 32$ subsets.

EXAMPLE 5 Analyzing Survey Data

In a survey of 100 college students, 35 were registered in College Algebra, 52 were registered in Computer Science I, and 18 were registered in both courses.

(a) How many students were registered in College Algebra or Computer Science I?

(b) How many were registered in neither course?

Solution (a) First, let A = set of students in College Algebra
B = set of students in Computer Science I

Then the given information tells us that

$$n(A) = 35 \qquad n(B) = 52 \qquad n(A \cap B) = 18$$

Figure 4

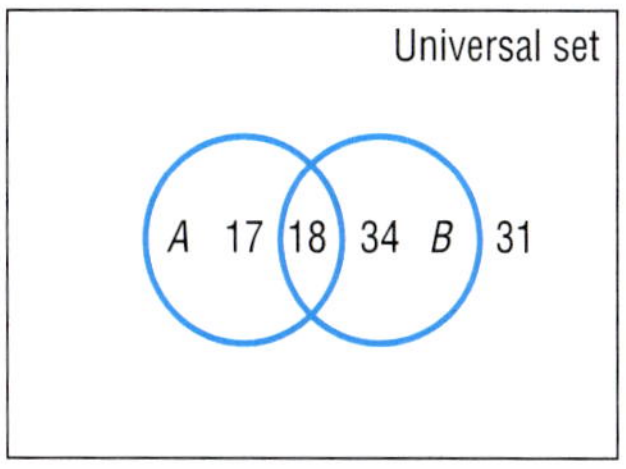

Refer to Figure 4. Since $n(A \cap B) = 18$, we know that the common part of the circles representing set A and set B has 18 elements. In addition, we know that the remaining portion of the circle representing set A will have $35 - 18 = 17$ elements. Similarly, we know that the remaining portion of the circle representing set B has $52 - 18 = 34$ elements. We conclude that $17 + 18 + 34 = 69$ students were registered in College Algebra or Computer Science I.

(b) Since 100 students were surveyed, it follows that $100 - 69 = 31$ were registered in neither course.

NOW WORK PROBLEM 35.

The solution to Example 5 contains the basis for a general counting formula. If we count the elements in each of two sets A and B, we necessarily count twice any elements that are in both A and B, that is, those elements in $A \cap B$. Thus, to count correctly the elements that are in A or B, that is, to find $n(A \cup B)$, we need to subtract those in $A \cap B$ from $n(A) + n(B)$.

THEOREM **Counting Formula**

If A and B are finite sets, then

$$n(A \cup B) = n(A) + n(B) - n(A \cap B) \qquad (1)$$

Refer back to Example 5. Using (1), we have

$$\begin{aligned} n(A \cup B) &= n(A) + n(B) - n(A \cap B) \\ &= 35 + 52 - 18 \\ &= 69 \end{aligned}$$

There are 69 students registered in College Algebra or Computer Science I.

A special case of the counting formula (1) occurs if A and B have no elements in common. In this case, $A \cap B = \varnothing$ so that $n(A \cap B) = 0$.

THEOREM **Addition Principle of Counting**

If two sets A and B have no elements in common, then

$$n(A \cup B) = n(A) + n(B) \qquad (2)$$

We can generalize formula (2).

THEOREM **General Addition Principle of Counting**

If for n sets $A_1, A_2, \ldots, A_n$, no two have elements in common, then

$$n(A_1 \cup A_2 \cup \cdots \cup A_n) = n(A_1) + n(A_2) + \cdots + n(A_n) \qquad (3)$$

EXAMPLE 6 Counting

In 1996 there were 738,028 full-time sworn law-enforcement officers in the United States. Table 1 lists the type of law-enforcement agencies and the corresponding number of full-time sworn officers from each agency.

Table 1

Type of Agency	Number of Full-Time Sworn Officers
Local police	410,956
Sheriff	152,922
State police	54,587
Special police	43,082
Texas constable	1,988
Federal	74,493

Source: Bureau of Justice Statistics.

(a) How many full-time sworn law-enforcement officers in the United States were local police or sheriffs?

(b) How many full-time sworn law-enforcement officers in the United States were local police, sheriffs, or state police?

Solution Let A represent the set of local police, B represent the set of sheriffs, and C represent the set of state police. No two of the sets A, B, and C have elements in common since a single officer cannot be classified in more than one type of agency.

(a) Using formula (2), we have

$$n(A \cup B) = n(A) + n(B) = 410{,}956 + 152{,}922 = 563{,}878$$

There were 563,878 officers that were local police or sheriffs.

(b) Using formula (3), we have

$$n(A \cup B \cup C) = n(A) + n(B) + n(C) = 410{,}956 + 152{,}922 + 54{,}587 = 618{,}465$$

There were 618,465 officers that were local police, sheriffs, or state police.

NOW WORK PROBLEM 39.

8.1 EXERCISES

In Problems 1–10, use $A = \{1, 3, 5, 7, 9\}$, $B = \{1, 5, 6, 7\}$, *and* $C = \{1, 2, 4, 6, 8, 9\}$ *to find each set.*

1. $A \cup B$ **2.** $A \cup C$ **3.** $A \cap B$ **4.** $A \cap C$
5. $(A \cup B) \cap C$ **6.** $(A \cap C) \cup (B \cap C)$ **7.** $(A \cap B) \cup C$ **8.** $(A \cup B) \cup C$
9. $(A \cup C) \cap (B \cup C)$ **10.** $(A \cap B) \cap C$

In Problems 11–20, use U = *universal set* = $\{0, 1, 2, 3, 4, 5, 6, 7, 8, 9\}$, $A = \{1, 3, 4, 5, 9\}$, $B = \{2, 4, 6, 7, 8\}$, *and* $C = \{1, 3, 4, 6\}$ *to find each set.*

11. $\overline{A}$ **12.** $\overline{C}$ **13.** $\overline{A \cap B}$ **14.** $\overline{B \cup C}$ **15.** $\overline{A} \cup \overline{B}$
16. $\overline{B} \cap \overline{C}$ **17.** $\overline{A \cap \overline{C}}$ **18.** $\overline{\overline{B} \cup C}$ **19.** $\overline{A \cup B \cup C}$ **20.** $\overline{A \cap B \cap C}$

21. Write down all the subsets of $\{a, b, c, d\}$.

22. Write down all the subsets of $\{a, b, c, d, e\}$.

23. If $n(A) = 15$, $n(B) = 20$, and $n(A \cap B) = 10$, find $n(A \cup B)$.

24. If $n(A) = 20$, $n(B) = 40$, and $n(A \cup B) = 35$, find $n(A \cap B)$.

25. If $n(A \cup B) = 50$, $n(A \cap B) = 10$, and $n(B) = 20$, find $n(A)$.

26. If $n(A \cup B) = 60$, $n(A \cap B) = 40$, and $n(A) = n(B)$, find $n(A)$.

In Problems 27–34, use the information given in the figure.

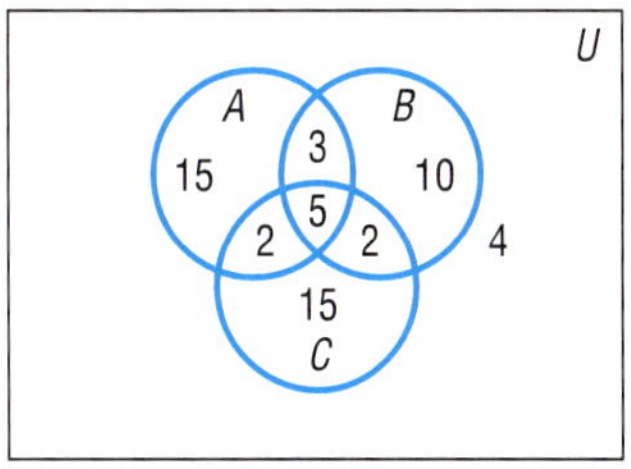

27. How many are in set A?

28. How many are in set B?

29. How many are in A or B?

30. How many are in A and B?

31. How many are in A but not C?

32. How many are not in A?

33. How many are in A and B and C?

34. How many are in A or B or C?

35. Analyzing Survey Data In a consumer survey of 500 people, 200 indicated that they would be buying a major appliance within the next month; 150 indicated that they would buy a car, and 25 said that they would purchase both a major appliance and a car. How many will purchase neither? How many will purchase only a car?

36. Analyzing Survey Data In a student survey, 200 indicated that they would attend Summer Session I and 150 indicated Summer Session II. If 75 students plan to attend both summer sessions and 275 indicated that they would attend neither session, how many students participated in the survey?

37. Analyzing Survey Data In a survey of 100 investors in the stock market,

50 owned shares in IBM
40 owned shares in AT&T
45 owned shares in GE
20 owned shares in both IBM and GE
15 owned shares in both AT&T and GE
20 owned shares in both IBM and AT&T
5 owned shares in all three

(a) How many of the investors surveyed did not have shares in any of the three companies?
(b) How many owned just IBM shares?
(c) How many owned just GE shares?
(d) How many owned neither IBM nor GE?
(e) How many owned either IBM or AT&T but no GE?

38. **Classifying Blood Types** Human blood is classified as either Rh+ or Rh−. Blood is also classified by type: A, if it contains an A antigen; B, if it contains a B antigen; AB, if it contains both A and B antigens; and O, if it contains neither antigen. Draw a Venn diagram illustrating the various blood types. Based on this classification, how many different kinds of blood are there?

39. The following data represent the marital status of males 18 years old and older in March 1997.

Marital Status	Number (in thousands)
Married, spouse present	54,654
Married, spouse absent	3,232
Widowed	2,686
Divorced	8,208
Never married	25,375

Source: Current Population Survey

(a) Determine the number of males 18 years old and older who are married.
(b) Determine the number of males 18 years old and older who are widowed or divorced.
(c) Determine the number of males 18 years old and older who are married, spouse absent, widowed, or divorced.

40. The following data represent the marital status of females 18 years old and older in March 1997.

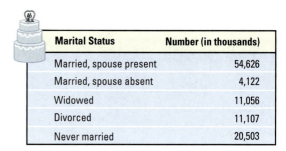

Marital Status	Number (in thousands)
Married, spouse present	54,626
Married, spouse absent	4,122
Widowed	11,056
Divorced	11,107
Never married	20,503

Source: Current Population Survey

(a) Determine the number of females 18 years old and older who are married.
(b) Determine the number of females 18 years old and older who are widowed or divorced.
(c) Determine the number of females 18 years old and older who are married, spouse absent, widowed, or divorced.

41. Make up a problem different from any found in the text that requires the addition principle of counting to solve. Give it to a friend to solve and critique.

42. Investigate the notion of counting as it relates to infinite sets. Write an essay on your findings.

8.2 PERMUTATIONS AND COMBINATIONS

1 Solve Counting Problems Using the Multiplication Principle

2 Solve Counting Problems Using Permutations

3 Solve Counting Problems Using Combinations

1 Counting plays a major role in many diverse areas, such as probability, statistics, and computer science. In this section we shall look at special types of counting problems and develop general formulas for solving them.

We begin with an example that will demonstrate a general counting principle.

EXAMPLE 1 Counting the Number of Possible Meals

The fixed-price dinner at Mabenka Restaurant provides the following choices:

Appetizer:	soup or salad
Entree:	baked chicken, broiled beef patty, baby beef liver, or roast beef au jus
Dessert:	ice cream or cheese cake

How many different meals can be ordered?

Solution Ordering such a meal requires three separate decisions:

Choose an Appetizer	Choose an Entree	Choose a Dessert
2 choices	4 choices	2 choices

Look at the **tree diagram** in Figure 5. We see that, for each choice of appetizer, there are 4 choices of entrees. And for each of these $2 \cdot 4 = 8$ choices, there are 2 choices for dessert. Thus, there are a total of

$$2 \cdot 4 \cdot 2 = 16$$

different meals that can be ordered.

Figure 5

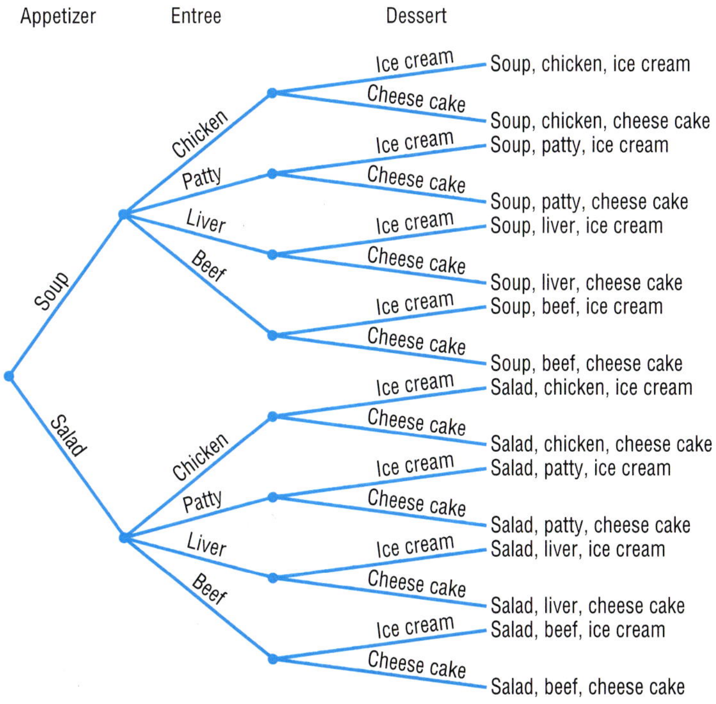

Example 1 illustrates a general counting principle.

THEOREM **Multiplication Principle of Counting**

If a task consists of a sequence of choices in which there are p selections for the first choice, q selections for the second choice, r selections for the third choice, and so on, then the task of making these selections can be done in

$$p \cdot q \cdot r \cdot \ldots$$

different ways.

EXAMPLE 2 Counting Airport Codes

The International Airline Transportation Association (IATA) assigns three-letter codes to represent airport locations. For example, JFK represents Kennedy International in New York. How many different airport codes are possible?

Solution The task consists of making three selections. Each selection requires choosing a letter of the alphabet (26 choices). Thus, by the Multiplication Principle, there are

$$26 \cdot 26 \cdot 26 = 17{,}576$$

different airport codes.

2

In Example 2, we were allowed to repeat a letter. For example, a valid airport code is FLL (Ft. Lauderdale International Airport), in which the letter L appears twice. In the next example, such repetition is not allowed.

EXAMPLE 3 Counting without Repetition

Suppose that we wish to establish a three-letter code using any of the 26 letters of the alphabet, but we require that no letter be used more than once. How many different three-letter codes are there?

Solution The task consists of making three selections. The first selection requires choosing from 26 letters. Because no letter can be used more than once, the second selection requires choosing from 25 letters. The third selection requires choosing from 24 letters. (Do you see why?) By the Multiplication Principle, there are

$$26 \cdot 25 \cdot 24 = 15{,}600$$

different three-letter codes with no letter repeated.

EXAMPLE 4 Birthday Problem

How many ways can 4 people have different birthdays? Assume that there are 365 days in a year.

Solution Once a birthday is selected, that birthday will not be repeated. Using the Multiplication Principle, the first person's birthday can be any one of 365 days, the second person's birthday can be any one of 364 days (we exclude the birthday of the first person), the third person's birthday can be any one of 363 days, and finally the fourth person's birthday can be any one of

362 days. Thus, there are $365 \cdot 364 \cdot 363 \cdot 362 = 17{,}458{,}601{,}160$ ways that four people can have different birthdays.

NOW WORK PROBLEMS 29 AND 47.

Examples 3 and 4 illustrate a type of counting problem referred to as a *permutation*.

A **permutation** is an ordered arrangement of n distinct objects without repetitions. The symbol $P(n, r)$ represents the number of permutations of n distinct objects, taken r at a time, where $r \leq n$.

For example, the question posed in Example 3 asks for the number of ways the 26 letters of the alphabet can be arranged using three nonrepeated letters. The answer is

$$P(26, 3) = 26 \cdot 25 \cdot 24 = 15{,}600$$

To arrive at a formula for $P(n, r)$, we note that the task of obtaining an ordered arrangement of n objects in which only $r \leq n$ of them are used, without repeating any of them, requires making r selections. For the first selection, there are n choices; for the second selection, there are $n - 1$ choices; for the third selection, there are $n - 2$ choices; ...; for the rth selection, there are $n - (r - 1)$ choices. By the Multiplication Principle, we have

$$\begin{aligned} &\quad\ \ \text{1st}\quad\ \text{2nd}\qquad\ \text{3rd}\qquad\qquad\qquad\ r\text{th} \\ P(n, r) &= n \cdot (n - 1) \cdot (n - 2) \cdot \ \dots\ \cdot [n - (r - 1)] \\ &= n \cdot (n - 1) \cdot (n - 2) \cdot \ \dots\ \cdot (n - r + 1) \end{aligned}$$

This formula for $P(n, r)$ can be compactly written using factorial notation.*

$$\begin{aligned} P(n, r) &= n \cdot (n - 1) \cdot (n - 2) \cdot \ \dots\ \cdot (n - r + 1) \\ &= n \cdot (n - 1) \cdot (n - 2) \cdot \ \dots\ \cdot (n - r + 1) \cdot \frac{(n - r) \cdot \ \dots\ \cdot 3 \cdot 2 \cdot 1}{(n - r) \cdot \ \dots\ \cdot 3 \cdot 2 \cdot 1} = \frac{n!}{(n - r)!} \end{aligned}$$

THEOREM **Number of Permutations of *n* Distinct Objects Taken *r* at a Time**

The number of different arrangements of n objects using $r \leq n$ of them, in which

1. the n objects are distinct.
2. once an object is used it cannot be repeated, and
3. order is important.

is given by the formula

$$P(n, r) = \frac{n!}{(n - r)!} \qquad (1)$$

*Recall that $0! = 1, 1! = 1, 2! = 2 \cdot 1, \dots, n! = n(n - 1) \cdot \ \dots\ \cdot 3 \cdot 2 \cdot 1$.

EXAMPLE 5 Computing Permutations

Evaluate: (a) $P(7,3)$ (b) $P(6,1)$ (c) $P(52,5)$

Solution We shall work parts (a) and (b) in two ways.

(a) $P(7,3) = \underbrace{7 \cdot 6 \cdot 5}_{\text{3 factors}} = 210$

or

$$P(7,3) = \frac{7!}{(7-3)!} = \frac{7!}{4!} = \frac{7 \cdot 6 \cdot 5 \cdot \cancel{4!}}{\cancel{4!}} = 210$$

(b) $P(6,1) = \underbrace{6}_{\text{1 factor}} = 6$

or

$$P(6,1) = \frac{6!}{(6-1)!} = \frac{6!}{5!} = \frac{6 \cdot \cancel{5!}}{\cancel{5!}} = 6$$

(c) Figure 6 shows the solution using a TI-83 graphing calculator. Thus, $P(52,5) = 311{,}875{,}200$.

Figure 6

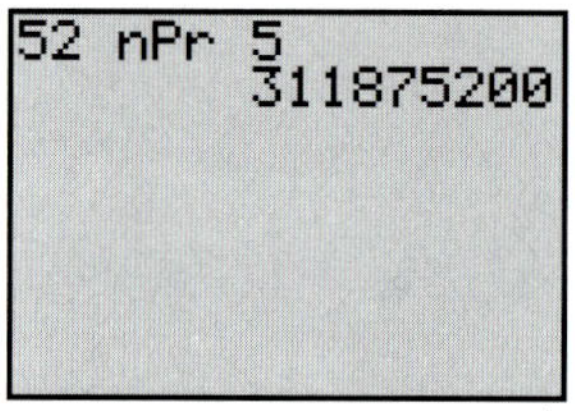

NOW WORK PROBLEM 1.

EXAMPLE 6 Lining Up People

In how many ways can 5 people be lined up?

Solution The 5 people are distinct. Once a person is in line, that person will not be repeated elsewhere in the line; and, in lining up people, order is important. Thus, we have a permutation of 5 objects taken 5 at a time. We can line up 5 people in

$$P(5,5) = \underbrace{5 \cdot 4 \cdot 3 \cdot 2 \cdot 1}_{\text{5 factors}} = 5! = 120 \text{ ways}$$

NOW WORK PROBLEM 31.

Combinations

3 In a permutation, order is important; for example, the arrangements ABC, CAB, BAC, . . . are considered different arrangements of the letters A, B, and C. In many situations, though, order is unimportant. For example, in the card game of poker, the order in which the cards are received does not matter; it is the *combination* of the cards that matters.

A **combination** is an arrangement, without regard to order, of n distinct objects without repetitions. The symbol $C(n, r)$ represents the number of combinations of n distinct objects taken r at a time, where $r \leq n$.

EXAMPLE 7 Listing Combinations

List all the combinations of the 4 objects a, b, c, d taken 2 at a time. What is $C(4, 2)$?

Solution One combination of a, b, c, d taken 2 at a time is

$$ab$$

The object ba is excluded, because order is not important in a combination. The list of all such combinations (convince yourself of this) is

$$ab, \quad ac, \quad ad, \quad bc, \quad bd, \quad cd$$

Thus,

$$C(4, 2) = 6$$

We can find a formula for $C(n, r)$ by noting that the only difference between a permutation and a combination is that we disregard order in combinations. Thus, to determine $C(n, r)$, we need only eliminate from the formula for $P(n, r)$ the number of permutations that were simply rearrangements of a given set of r objects. This can be determined from the formula for $P(n, r)$ by calculating $P(r, r) = r!$. So, if we divide $P(n, r)$ by $r!$, we will have the desired formula for $C(n, r)$:

$$C(n, r) = \frac{P(n, r)}{r!} \underset{\substack{\uparrow \\ \text{Use formula (1).}}}{=} \frac{n!/(n - r)!}{r!} = \frac{n!}{(n - r)!r!}$$

We have proved the following result.

THEOREM **Number of Combinations of n Distinct Objects Taken r at a Time**

The number of different arrangements of n objects using $r \leq n$ of them, in which

1. the n objects are distinct,
2. once an object is used, it cannot be repeated, and
3. order is not important,

is given by the formula

$$C(n, r) = \frac{n!}{(n - r)!r!} \qquad (2)$$

Based on formula (2), we discover that the symbol $C(n, r)$ and the symbol $\binom{n}{r}$ for the binomial coefficients are, in fact, the same. Thus, the Pascal triangle (see Section 7.5) can be used to find the value of $C(n, r)$. However, because it is more practical and convenient, we will use formula (2) instead.

EXAMPLE 8 Using Formula (2)

Use formula (2) to find the value of each expression.

(a) $C(3,1)$ (b) $C(6,3)$ (c) $C(n,n)$ (d) $C(n,0)$ (e) $C(52,5)$

Solution

(a) $C(3, 1) = \dfrac{3!}{(3-1)!1!} = \dfrac{3!}{2!1!} = \dfrac{3 \cdot \cancel{2} \cdot 1}{\cancel{2} \cdot 1 \cdot 1} = 3$

(b) $C(6, 3) = \dfrac{6!}{(6-3)!3!} = \dfrac{6 \cdot 5 \cdot 4 \cdot \cancel{3!}}{3! \cdot \cancel{3!}} = \dfrac{\cancel{6} \cdot 5 \cdot 4}{\cancel{6}} = 20$

(c) $C(n, n) = \dfrac{n!}{(n-n)!n!} = \dfrac{\cancel{n!}}{0!\cancel{n!}} = \dfrac{1}{1} = 1$

(d) $C(n, 0) = \dfrac{n!}{(n-0)!0!} = \dfrac{\cancel{n!}}{\cancel{n!}0!} = \dfrac{1}{1} = 1$

(e) Figure 7 shows the solution using a TI-83 graphing calculator. Thus, $C(52, 5) = 2{,}598{,}960$.

Figure 7

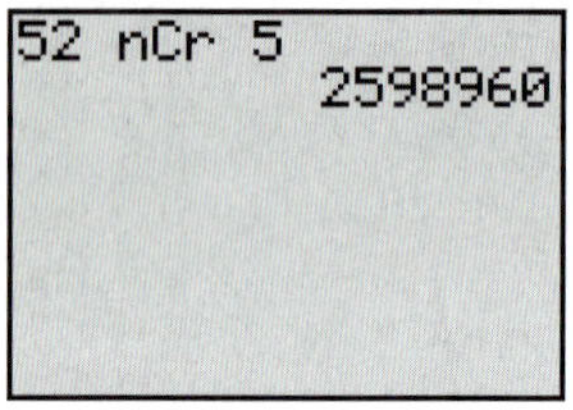

52 MATH ◄ 3 5 ENTER

NOW WORK PROBLEM **9.**

EXAMPLE 9 Forming Committees

How many different committees of 3 people can be formed from a pool of 7 people?

Solution The 7 people are distinct. More important, though, is the observation that the order of being selected for a committee is not significant. Thus, the problem asks for the number of combinations of 7 objects taken 3 at a time.

$$C(7, 3) = \frac{7!}{4!3!} = \frac{7 \cdot 6 \cdot 5 \cdot \cancel{4!}}{\cancel{4!}3!} = \frac{7 \cdot \cancel{6} \cdot 5}{\cancel{6}} = 35$$

EXAMPLE 10 Forming Committees

In how many ways can a committee consisting of 2 faculty members and 3 students be formed if there are 6 faculty members and 10 students eligible to serve on the committee?

Solution The problem can be separated into two parts: the number of ways that the faculty members can be chosen, $C(6, 2)$, and the number of ways that the student members can be chosen, $C(10, 3)$. By the Multiplication Principle, the committee can be formed in

$$C(6, 2) \cdot C(10, 3) = \frac{6!}{4!2!} \cdot \frac{10!}{7!3!} = \frac{6 \cdot 5 \cdot \cancel{4!}}{\cancel{4!}2!} \cdot \frac{10 \cdot 9 \cdot 8 \cdot \cancel{7!}}{\cancel{7!}3!}$$
$$= \frac{30}{2} \cdot \frac{720}{6} = 1800 \text{ ways}$$

NOW WORK PROBLEM **49.**

Permutations with Repetition

Recall that a permutation involves counting *distinct* objects. A permutation in which some of the objects are repeated is called a **permutation with repetition.** Some books refer to this as a **nondistinguishable permutation.**

Let's look at an example.

EXAMPLE 11 Forming Different Words

How many different words can be formed using all the letters in the word REARRANGE?

Solution Each word formed will have 9 letters: 3 R's, 2 A's, 2 E's, 1 N, and 1 G. To construct each word, we need to fill in 9 positions with the 9 letters:

$$\overline{1}\quad \overline{2}\quad \overline{3}\quad \overline{4}\quad \overline{5}\quad \overline{6}\quad \overline{7}\quad \overline{8}\quad \overline{9}$$

The process of forming a word consists of five tasks:

Task 1: Choose the positions for the 3 R's.
Task 2: Choose the positions for the 2 A's.
Task 3: Choose the position for the 2 E's.
Task 4: Choose the position for the 1 N.
Task 5: Choose the position for the 1 G.

Task 1 can be done in $C(9, 3)$ ways. There then remain 6 positions to be filled, so Task 2 can be done in $C(6, 2)$ ways. There remain 4 positions to be filled, so Task 3 can be done in $C(4, 2)$ ways. There remain 2 positions to be filled, so Task 4 can be done in $C(2, 1)$ ways. The last position can be filled in $C(1, 1)$ way. Using the Multiplication Principle, the number of possible words that can be formed is

$$C(9, 3) \cdot C(6, 2) \cdot C(4, 2) \cdot C(2, 1) \cdot C(1, 1) = \frac{9!}{3! \cdot \cancel{6!}} \cdot \frac{\cancel{6!}}{2! \cdot \cancel{4!}} \cdot \frac{\cancel{4!}}{2! \cdot \cancel{2!}} \cdot \frac{\cancel{2!}}{1! \cdot \cancel{1!}} \cdot \frac{\cancel{1!}}{0! \cdot 1!}$$

$$= \frac{9!}{3! \cdot 2! \cdot 2! \cdot 1! \cdot 1!}$$

The form of the answer to Example 11 is suggestive of a general result. Had the letters in REARRANGE each been different, there would have been $P(9, 9) = 9!$ possible words formed. This is the numerator of the answer. The presence of 3 R's, 2 A's, and 2 E's reduces the number of different words, as the entries in the denominator illustrate. We are led to the following result:

THEOREM **Permutations with Repetition**

The number of permutations of n objects of which n_1 are of one kind, n_2 are of a second kind, . . . , and n_k are of a kth kind is given by

$$\frac{n!}{n_1! \cdot n_2! \cdot \ldots \cdot n_k!} \qquad (3)$$

where $n = n_1 + n_2 + \cdots + n_k$.

EXAMPLE 12 Arranging Flags

How many different vertical arrangements are there of 8 flags if 4 are white, 3 are blue, and 1 is red?

Solution We seek the number of permutations of 8 objects, of which 4 are of one kind, 3 of a second kind, and 1 of a third kind. Using formula (3), we find that there are

$$\frac{8!}{4! \cdot 3! \cdot 1!} = \frac{8 \cdot 7 \cdot 6 \cdot 5 \cdot \cancel{4!}}{\cancel{4!} \cdot 3! \cdot 1!} = 280 \text{ different arrangements}$$

NOW WORK PROBLEM 55.

8.2 EXERCISES

In Problems 1–8, find the value of each permutation. Verify your results using a graphing calculator.

1. $P(6, 2)$ **2.** $P(7, 2)$ **3.** $P(5, 5)$ **4.** $P(4, 4)$

5. $P(8, 0)$ **6.** $P(9, 0)$ **7.** $P(8, 3)$ **8.** $P(8, 5)$

In Problems 9–16, use formula (2) to find the value of each combination. Verify your results using a graphing calculator.

9. $C(8, 2)$ **10.** $C(8, 6)$ **11.** $C(6, 4)$ **12.** $C(6, 2)$

13. $C(15, 15)$ **14.** $C(18, 1)$ **15.** $C(26, 13)$ **16.** $C(18, 9)$

17. List all the permutations of 5 objects a, b, c, d, and e taken 3 at a time. What is $P(5, 3)$?

18. List all the permutations of 5 objects a, b, c, d, and e taken 2 at a time. What is $P(5, 2)$?

19. List all the permutations of 4 objects 1, 2, 3, and 4 taken 3 at a time. What is $P(4, 3)$?

20. List all the permutations of 6 objects 1, 2, 3, 4, 5, and 6 taken 3 at a time. What is $P(6, 3)$?

21. List all the combinations of 5 objects a, b, c, d, and e taken 3 at a time. What is $C(5, 3)$?

22. List all the combinations of 5 objects a, b, c, d, and e taken 2 at a time. What is $C(5, 2)$?

23. List all the combinations of 4 objects 1, 2, 3, and 4 taken 3 at a time. What is $C(4, 3)$?

24. List all the combinations of 6 objects 1, 2, 3, 4, 5, and 6 taken 3 at a time. What is $C(6, 3)$?

25. A man has 5 shirts and 3 ties. How many different shirt and tie combinations can he wear?

26. A woman has 3 blouses and 5 skirts. How many different outfits can she wear?

27. Forming Codes How many two-letter codes can be formed using the letters A, B, C, and D? Repeated letters are allowed.

28. Forming Codes How many two-letter codes can be formed using the letters A, B, C, D, and E? Repeated letters are allowed.

29. Forming Numbers How many three-digit numbers can be formed using the digits 0 and 1? Repeated digits are allowed.

30. Forming Numbers How many three-digit numbers can be formed using the digits 0, 1, 2, 3, 4, 5, 6, 7, 8, and 9? Repeated digits are allowed.

31. In how many ways can 4 people be lined up?

32. In how many ways can 5 different boxes be stacked?

33. Forming Codes How many different three-letter codes are there if only the letters A, B, C, D, and E can be used and no letter can be used more than once?

34. Forming Codes How many different four-letter codes are there if only the letters A, B, C, D, E, and F can be used and no letter can be used more than once?

35. Stocks on the NYSE Companies whose stock is listed on the New York Stock Exchange (NYSE) have their company's name represented by either 1, 2, or 3 letters (repetition of letters is allowed). What is the maximum number of companies that can be listed on the New York Stock Exchange?

36. Stocks on the NASDAQ Companies whose stock is listed on the NASDAQ stock exchange have their company's name represented by either 4 or 5 letters (repetition of letters is allowed). What is the maximum number of companies that can be listed on the NASDAQ?

37. Establishing Committees In how many ways can a committee of 4 students be formed from a pool of 7 students?

C(7,4) = 35 ORDER NOT IMP.

38. Establishing Committees In how many ways can a committee of 3 professors be formed from a department having 8 professors?

39. Possible Answers on a True/False Test How many arrangements of answers are possible for a true/false test with 10 questions?

40. Possible Answers on a Multiple-choice Test How many arrangements of answers are possible in a multiple-choice test with 5 questions, each of which has 4 possible answers?

41. How many four-digit numbers can be formed using the digits 0, 1, 2, 3, 4, 5, 6, 7, 8, and 9 if the first digit cannot be 0? Repeated digits are allowed.

42. How many five-digit numbers can be formed using the digits 0, 1, 2, 3, 4, 5, 6, 7, 8, and 9 if the first digit cannot be 0 or 1? Repeated digits are allowed.

43. Arranging Books Five different mathematics books are to be arranged on a student's desk. How many arrangements are possible?

P(5,5) = 120

44. Forming License Plate Numbers How many different license plate numbers can be made using 2 letters followed by 4 digits selected from the digits 0 through 9, if

(a) Letters and digits may be repeated?

(b) Letters may be repeated, but digits may not be repeated?

(c) Neither letters nor digits may be repeated?

45. Stock Portfolios As a financial planner, you are asked to select one stock each from the following groups: 8 DOW stocks, 15 NASDAQ stocks, and 4 global stocks. How many different portfolios are possible?

46. Combination Locks A combination lock has 50 numbers on it. To open it, you turn to a number, then rotate clockwise to a second number, and then counterclockwise to the third number. How many different lock combinations are there?

C(50,3) = 19600

50·49·48 =

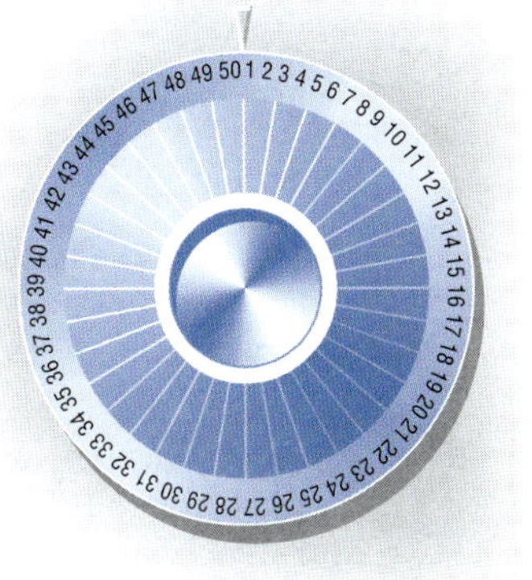

365·364·363 = 48,228,180

47. Birthday Problem How many ways can 3 people have different birthdays? Assume that there are 365 days in a year.

48. Birthday Problem How many ways can 5 people have different birthdays? Assume that there are 365 days in a year.

C(4,2)·C(8,3) = 6·56 = 336

49. A student dance committee is to be formed consisting of 2 boys and 3 girls. If the membership is to be chosen from 4 boys and 8 girls, how many different committees are possible?

50. Baseball Teams A baseball team has 15 members. Four of the players are pitchers, and the remaining 11 members can play any position. How many different teams of 9 players can be formed?

51. The student relations committee of a college consists of 2 administrators, 3 faculty members, and 5 students. There are 4 administrators, 8 faculty members, and 20 students eligible to serve. How many different committees are possible?

52. Football Teams A defensive football squad consists of 25 players. Of these, 10 are linemen, 10 are linebackers, and 5 are safeties. How many different teams of 5 linemen, 3 linebackers, and 3 safeties can be formed?

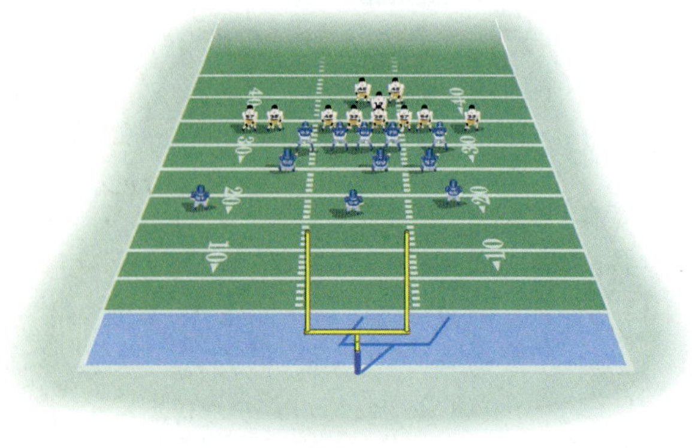

53. **Baseball** In the American Baseball League, a designated hitter may be used. How many batting orders is it possible for a manager to use? (There are 9 regular players on a team.)

54. **Baseball** In the National Baseball League, the pitcher usually bats ninth. If this is the case, how many batting orders are possible for a manager to use?

55. **Forming Words** How many different 9-letter words (real or imaginary) can be formed from the letters in the word ECONOMICS?

56. **Forming Words** How many different 11-letter words (real or imaginary) can be formed from the letters in the word MATHEMATICS?

57. **Senate Committees** The U.S. Senate has 100 members. Suppose that it is desired to place each senator on exactly 1 of 7 possible committees. The first committee has 22 members, the second has 13, the third has 10, the fourth has 5, the fifth has 16, and the sixth and seventh have 17 apiece. In how many ways can these committees be formed?

58. **World Series** In the World Series the American League team (A) and the National League team (N) play until one team wins four games. If the sequence of winners is designed by letters (for example, $NAAAA$ means that the National League team won the first game and the American League won the next four), how many different sequences are possible?

59. **Basketball Teams** A basketball team has 6 players who play guard (2 of 5 starting positions). How many different teams are possible, assuming that the remaining 3 positions are filled and it is not possible to distinguish a left guard from a right guard?

60. **Basketball Teams** On a basketball team of 12 players, 2 only play center, 3 only play guard, and the rest play forward (5 players on a team: 2 forwards, 2 guards, and 1 center). How many different teams are possible, assuming that it is not possible to distinguish left and right guards and left and right forwards?

61. **Selecting Objects** An urn contains 7 white balls and 3 red balls. Three balls are selected. In how many ways can the 3 balls be drawn from the total of 10 balls:
    (a) If 2 balls are white and 1 is red?
    (b) If all 3 balls are white?
    (c) If all 3 balls are red?

62. **Selecting Objects** An urn contains 15 red balls and 10 white balls. Five balls are selected. In how many ways can the 5 balls be drawn from the total of 25 balls:
    (a) If all 5 balls are red?
    (b) If 3 balls are red and 2 are white?
    (c) If at least 4 are red balls?

8.3 PROBABILITY OF EQUALLY LIKELY OUTCOMES

1 Construct Probability Models

2 Compute Probabilities of Equally Likely Outcomes

3 Utilize the Addition Rule to Find Probabilities

4 Utilize the Complement Rule to Find Probabilities

5 Compute Probabilities Using Permutations and Combinations

Probability is an area of mathematics that deals with experiments that yield random results, yet admit a certain regularity. Such experiments do not always produce the same result or outcome, so the result of any one observation is not predictable. However, the results of the experiment over a long period do produce regular patterns that enable us to predict with remarkable accuracy.

EXAMPLE 1 Tossing a Fair Coin

In tossing a fair coin, we know that the outcome is either a head or a tail. On any particular throw, we cannot predict what will happen, but, if we toss the coin many times, we observe that the number of times that a head comes up is approximately equal to the number of times that we get a tail. It seems reasonable, therefore, to assign a probability of $\frac{1}{2}$ that a head comes up and a probability of $\frac{1}{2}$ that a tail comes up.

Probability Models

1 The discussion in Example 1 constitutes the construction of a **probability model** for the experiment of tossing a fair coin once. A probability model has two components: a sample space and an assignment of probabilities. A **sample space** S is a set whose elements represent all the possibilities that can occur as a result of the experiment. Each element of S is called an **outcome.** To each outcome, we assign a number, called the **probability** of that outcome, which has two properties:

1. Each probability is nonnegative.
2. The sum of all the probabilities equals 1.

If a probability model has the sample space

$$S = \{e_1, e_2, \dots, e_n\}$$

where $e_1, e_2, \dots, e_n$ are the possible outcomes, and if $P(e_1), P(e_2), \dots, P(e_n)$ denote the respective probabilities of these outcomes, then

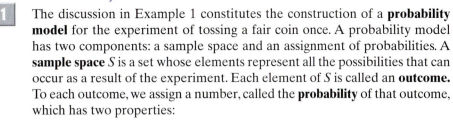

$$P(e_1) \geq 0, P(e_2) \geq 0, \dots, P(e_n) \geq 0 \qquad (1)$$

$$\sum_{i=1}^{n} P(e_i) = P(e_1) + P(e_2) + \cdots + P(e_n) = 1 \qquad (2)$$

EXAMPLE 2 Determining Probability Models

In a bag of M&Ms the candies are colored red, green, blue, brown, yellow, and orange. Suppose that a candy is drawn from the bag and the color is recorded. The sample space of this experiment is {red, green, blue, brown, yellow, orange}. Determine which of the following are probability models.

(a)

Outcome	**Probability**
{red}	0.3
{green}	0.15
{blue}	0
{brown}	0.15
{yellow}	0.2
{orange}	0.2

(b)

Outcome	**Probability**
{red}	0.1
{green}	0.1
{blue}	0.1
{brown}	0.4
{yellow}	0.2
{orange}	0.3

(c)

Outcome	**Probability**
{red}	0.3
{green}	−0.3
{blue}	0.2
{brown}	0.4
{yellow}	0.2
{orange}	0.2

(d)

Outcome	**Probability**
{red}	0
{green}	0
{blue}	0
{brown}	0
{yellow}	1
{orange}	0

Solution

(a) This model is a probability model since all the outcomes have probabilities that are nonnegative and the sum of the probabilities is 1.

(b) This model is not a probability model because the sum of the probabilities is not 1.

(c) This model is not a probability model because $P(\text{green})$ is less than 0. Recall, all probabilities must be nonnegative.

(d) This model is a probability model because all the outcomes have probabilities that are nonnegative, and the sum of the probabilities is 1. Notice that $P(\text{yellow}) = 1$, meaning that this outcome will occur with 100% certainty each time that the experiment is repeated. This means that the entire bag of M&Ms has yellow candies.

NOW WORK PROBLEM 3.

Let's look at an example of constructing a probability model.

EXAMPLE 3 Constructing a Probability Model

An experiment consists of rolling a fair die once.* Construct a probability model for this experiment.

Solution A sample space S consists of all the possibilities that can occur. Because rolling the die will result in one of six faces showing, the sample space S consists of

$$S = \{1, 2, 3, 4, 5, 6\}$$

Figure 8

Because the die is fair, one face is no more likely to occur than another. As a result, our assignment of probabilities is

$$P(1) = \tfrac{1}{6} \qquad P(2) = \tfrac{1}{6}$$
$$P(3) = \tfrac{1}{6} \qquad P(4) = \tfrac{1}{6}$$
$$P(5) = \tfrac{1}{6} \qquad P(6) = \tfrac{1}{6}$$

Now suppose that a die is loaded so that the probability assignments are

$$P(1) = 0, \quad P(2) = 0, \quad P(3) = \frac{1}{3}, \quad P(4) = \frac{2}{3}, \quad P(5) = 0, \quad P(6) = 0$$

This assignment would be made if the die were loaded so that only a 3 or 4 could occur and the 4 is twice as likely as the 3 to occur. This assignment is consistent with the definition since each assignment is nonnegative and the sum of all the probability assignments equals 1.

NOW WORK PROBLEM 19.

*A die is a cube with each face having either 1, 2, 3, 4, 5, or 6 dots on it. See Figure 8.

EXAMPLE 4 Constructing a Probability Model

An experiment consists of tossing a coin. The coin is weighted so that heads (H) is three times as likely to occur as tails (T). Construct a probability model for this experiment.

Solution The sample space S is $S = \{\text{H}, \text{T}\}$. If x denotes the probability that a tail occurs, then

$$P(\text{T}) = x \quad \text{and} \quad P(\text{H}) = 3x$$

Since the sum of the probabilities of the possible outcomes must equal 1, we have

$$\begin{aligned} P(\text{T}) + P(\text{H}) = x + 3x &= 1 \\ 4x &= 1 \\ x &= \frac{1}{4} \end{aligned}$$

Thus, we assign the probabilities

$$P(\text{T}) = \frac{1}{4} \qquad P(\text{H}) = \frac{3}{4}$$

NOW WORK PROBLEM 23.

In working with probability models, the term **event** is used to describe a set of possible outcomes of the experiment. Thus, an event E is some subset of the sample space S. The **probability of an event** E, $E \neq 0$, denoted by $P(E)$, is defined as the sum of the probabilities of the outcomes in E. We can also think of the probability of an event E as the likelihood that the event E occurs. If $E = \varnothing$, then $P(E) = 0$; if $E = S$, then $P(E) = P(S) = 1$.

Equally Likely Outcomes

2 When the same probability is assigned to each outcome of the sample space, the experiment is said to have **equally likely outcomes.***

THEOREM **Probability for Equally Likely Outcomes**

If an experiment has n equally likely outcomes and if the number of ways that an event E can occur is m, then the probability of E is

$$P(E) = \frac{\text{Number of ways that } E \text{ can occur}}{\text{Number of all logical possibilities}} = \frac{m}{n} \tag{3}$$

Thus, if S is the sample space of this experiment, then

$$P(E) = \frac{n(E)}{n(S)} \tag{4}$$

*This is referred to as **classical probability.**

EXAMPLE 5 Calculating Probabilities of Equally Likely Events

Calculate the probability that in a three-child family there are 2 boys and 1 girl. Assume equally likely outcomes.

Solution We begin by constructing a tree diagram to help in listing the possible outcomes of the experiment. See Figure 9, where B stands for boy and G for girl.

Figure 9

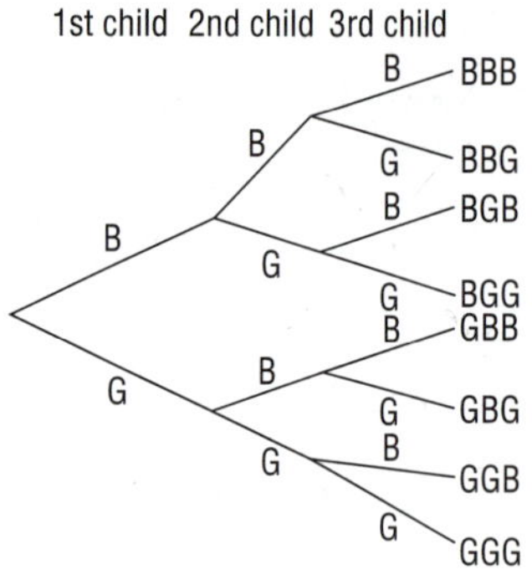

The sample space S of this experiment is

$$S = \{BBB, BBG, BGB, BGG, GBB, GBG, GGB, GGG\}$$

so $n(S) = 8$.

We wish to know the probability of the event E: "having two boys and one girl." From Figure 9, we conclude that $E = \{BBG, BGB, GBB\}$, so $n(E) = 3$. Since the outcomes are equally likely, the probability of E is

$$P(E) = n(E)/n(S) = 3/8.$$

NOW WORK PROBLEM 33.

Compound Probabilities

Thus far, we have calculated probabilities of single events. We will now compute probabilities of multiple events, called **compound probabilities.**

EXAMPLE 6 Computing Compound Probabilities

Consider the experiment of rolling a single fair die. Let E represent the event "roll an odd number" and let F represent the event "roll a 1 or 2."

(a) Write the event E and F.
(b) Write the event E or F.
(c) Compute $P(E)$ and $P(F)$.
(d) Compute $P(E \cap F)$.
(e) Compute $P(E \cup F)$.

Solution The sample space S of the experiment is $\{1, 2, 3, 4, 5, 6\}$, so $n(S) = 6$. Since the die is fair, the outcomes are equally likely. The event E: "roll an odd number," is $\{1, 3, 5\}$, and the event F: "roll a 1 or 2," is $\{1, 2\}$, so $n(E) = 3$ and $n(F) = 2$.

(a) The word "and" in probability means the intersection of two events. The event E and F is

$$E \cap F = \{1, 3, 5\} \cap \{1, 2\} = \{1\}, \qquad n(E \cap F) = 1.$$

(b) The word "or" in probability means the union of the two events. The event E or F is

$$E \cup F = \{1, 3, 5\} \cup \{1, 2\} = \{1, 2, 3, 5\}, \qquad n(E \cup F) = 4.$$

(c) We use formula (4).

$$P(E) = \frac{n(E)}{n(S)} = \frac{3}{6} = \frac{1}{2} \qquad P(F) = \frac{n(F)}{n(S)} = \frac{2}{6} = \frac{1}{3}$$

(d) $P(E \cap F) = \dfrac{n(E \cap F)}{n(S)} = \dfrac{1}{6}$

(e) $P(E \cup F) = \dfrac{n(E \cup F)}{n(S)} = \dfrac{4}{6} = \dfrac{2}{3}$

3 The **Addition Rule** can be used to find the probability of the union of two events.

THEOREM **Addition Rule**

For any two events E and F,

$$P(E \cup F) = P(E) + P(F) - P(E \cap F) \qquad (5)$$

For example, we can use the Addition Rule to find $P(E \cup F)$ in Example 6(e). Then,

$$P(E \cup F) = P(E) + P(F) - P(E \cap F) = \frac{1}{2} + \frac{1}{3} - \frac{1}{6} = \frac{3}{6} + \frac{2}{6} - \frac{1}{6} = \frac{4}{6} = \frac{2}{3}$$

as before.

EXAMPLE 7 Computing Probabilities of Compound Events Using the Addition Rule

If $P(E) = 0.2$, $P(F) = 0.3$, and $P(E \cap F) = 0.1$, find $P(E \cup F)$.

Solution We use the Addition Rule, formula (5).

$$P(E \cup F) = P(E) + P(F) - P(E \cap F) = 0.2 + 0.3 - 0.1 = 0.4$$

A Venn diagram can sometimes be used to obtain probabilities. To construct a Venn diagram representing the information in Example 7, we draw two sets E and F. We begin with the fact that $P(E \cap F) = 0.1$. See Figure 10(a). Then, since $P(E) = 0.2$ and $P(F) = 0.3$, we fill in E with $0.2 - 0.1 = 0.1$ and F with $0.3 - 0.1 = 0.2$. See Figure 10(b). Since $P(S) = 1$, we complete the diagram by inserting $1 - [0.1 + 0.1 + 0.2] = 0.6$. See Figure 10(c). Now it is easy to see, for example, that the probability of F, but not E, is 0.2. Also, the probability of neither E nor F is 0.6.

Figure 10

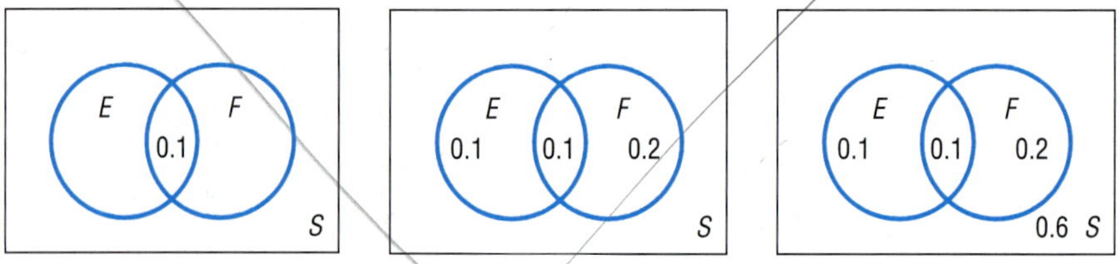

(a) (b) (c)

NOW WORK PROBLEM 41.

If events E and F are disjoint, so that $E \cap F = \varnothing$, we say they are **mutually exclusive.** In this case, $P(E \cap F) = 0$ and the Addition Rule takes the following form:

Mutually Exclusive Events

If E and F are **mutually exclusive events,** then

$$P(E \cup F) = P(E) + P(F) \qquad (6)$$

EXAMPLE 8 Computing Compound Probabilities of Mutually Exclusive Events

If $P(E) = 0.4$ and $P(F) = 0.25$, and E and F are mutually exclusive, find $P(E \cup F)$.

Solution Since E and F are mutually exclusive, we use formula (6).

$$P(E \cup F) = P(E) + P(F) = 0.4 + 0.25 = 0.65$$

NOW WORK PROBLEM 43.

Complements

4 Recall, if A is a set, the complement of A, denoted $\overline{A}$, is the set of all elements in the universal set U not in A. We similarly define the complement of an event.

Complement of an Event

Let S denote the sample space of an experiment and let E denote an event. The **complement of E,** denoted $\overline{E}$, is the set of all outcomes in the sample space S that are not outcomes in the event E.

The complement $\overline{E}$ of an event E in a sample space S has the following two properties:

$$E \cap \overline{E} = \varnothing \qquad E \cup \overline{E} = S$$

Since E and $\overline{E}$ are mutually exclusive, it follows from (6) that

$$P(E \cup \overline{E}) = P(S) = 1 \qquad P(E) + P(\overline{E}) = 1 \qquad P(\overline{E}) = 1 - P(E)$$

Thus, we have the following result.

THEOREM **Computing Probabilities of Complementary Events**

If E represents any event and $\overline{E}$ represents the complement of E, then

$$P(\overline{E}) = 1 - P(E) \tag{7}$$

EXAMPLE 9 Computing Probabilities Using Complements

On the local news the weather reporter stated that the probability of rain is 40%. What is the probability that it will not rain?

Solution The complement of the event "rain" is "no rain." Thus,

$$P(\text{no rain}) = 1 - P(\text{rain}) = 1 - 0.4 = 0.6$$

There is a 60% chance of no rain tomorrow.

NOW WORK PROBLEM 47.

EXAMPLE 10 Birthday Problem

What is the probability that in a group of 10 people at least 2 people have the same birthday? Assume that there are 365 days in a year.

Solution We assume that a person is as likely to be born on one day as another, so we have equally likely outcomes.

We first determine the number of outcomes in the sample space S. There are 365 possibilities for each person's birthday. Since there are 10 people in the group, there are 365^{10} possibilities for the birthdays. [For one person in the group, there are 365 days on which his or her birthday can fall; for two people, there are $(365)(365) = 365^2$ pairs of days; and, in general, using the Multiplication Principle, for n people there are 365^n possibilities.] So,

$$n(S) = 365^{10}$$

We wish to find the probability of the event E: "at least two people have the same birthday." It is difficult to count the elements in this set; it is much easier to count the elements of the complementary event $\overline{E}$: "no two people have the same birthday."

We find $n(\overline{E})$ as follows:

Choose one person at random. There are 365 possibilities for his or her birthday. Choose a second person. There are 364 possibilities for this birthday, if no two people are to have the same birthday. Choose a third person.

There are 363 possibilities left for this birthday. Finally, we arrive at the tenth person. There are 356 possibilities left for this birthday. By the Multiplication Principle, the total number of possibilities is

$$n(\overline{E}) = 365 \cdot 364 \cdot 363 \cdot \ldots \cdot 356.$$

Hence, the probability of event $\overline{E}$ is

$$P(\overline{E}) = \frac{n(\overline{E})}{n(S)} = \frac{365 \cdot 364 \cdot 363 \cdot \ldots \cdot 356}{365^{10}} \approx 0.883$$

The probability of two or more people in a group of 10 people having the same birthday is then

$$P(E) = 1 - P(\overline{E}) = 1 - 0.883 = 0.117$$

The birthday problem can be solved for any group size. The following table gives the probabilities for two or more people having the same birthday for various group sizes. Notice that the probability is greater than $\frac{1}{2}$ for any group of 23 or more people.

	Number of People															
	5	**10**	**15**	**20**	**21**	**22**	**23**	**24**	**25**	**30**	**40**	**50**	**60**	**70**	**80**	**90**
Probability That Two or More Have the Same Birthday	0.027	0.117	0.253	0.411	0.444	0.476	0.507	0.538	0.569	0.706	0.891	0.970	0.994	0.99916	0.99991	0.99999

NOW WORK PROBLEM 65.

5 Probabilities Involving Combinations and Permutations

EXAMPLE 11 Computing Probabilities

Because of a mistake in packaging, 5 defective phones were packaged with 15 good ones. All phones look alike and have equal probability of being chosen. Three phones are selected.

(a) What is the probability that all 3 are defective?
(b) What is the probability that exactly 2 are defective?
(c) What is the probability that at least 2 are defective?

Solution The sample space S consists of the number of ways that 3 objects can be selected from 20 objects, that is, the number of combinations of 20 things taken 3 at a time.

$$n(S) = C(20, 3) = \frac{20!}{17! \cdot 3!} = \frac{20 \cdot 19 \cdot 18}{6} = 1140$$

Each of these outcomes is equally likely to occur.

(a) If E is the event "3 are defective," then the number of elements in E is the number of ways the 3 defective phones can be chosen from the 5 defective phones: $C(5, 3) = 10$. Thus, the probability of E is

$$P(E) = \frac{n(E)}{n(S)} = \frac{10}{1140} \approx 0.0088$$

(b) If F is the event "exactly 2 are defective" and 3 phones are selected, then the number of elements in F is the number of ways to select 2 defective phones from the 5 defective phones and 1 good phone from the 15 good ones. The first of these can be done in $C(5, 2)$ ways and the second in $C(15, 1)$ ways. By the Multiplication Principle, the event F can occur in

$$C(5, 2) \cdot C(15, 1) = \frac{5!}{3! \cdot 2!} \cdot \frac{15!}{14! \cdot 1!} = 10 \cdot 15 = 150 \text{ ways}$$

The probability of F is therefore

$$P(F) = \frac{n(F)}{n(S)} = \frac{150}{1140} \approx 0.1316$$

(c) The event G, "at least two are defective" when 3 are chosen, is equivalent to requiring that either exactly 2 defective are chosen or exactly 3 defective are chosen. That is, $G = E \cup F$. Since E and F are mutually exclusive (it is not possible to select 2 defective phones and, at the same time, select 3 defective phones), we find that

$$P(G) = P(E) + P(F) \approx 0.0088 + 0.1316 = 0.1404$$

NOW WORK PROBLEM 73.

EXAMPLE 12 Tossing a Coin

A fair coin is tossed 6 times.

(a) What is the probability of obtaining exactly 5 heads and one tail?
(b) What is the probability of obtaining between 4 and 6 heads, inclusive?

Solution The number of elements in the sample space S is found using the Multiplication Principle. Each toss results in a head (H) or a tail (T). Since the coin is tossed 6 times, we have

$$n(S) = \underbrace{2 \cdot 2 \cdot \ldots \cdot 2}_{\text{6 tosses}} = 2^6 = 64$$

The outcomes are equally likely since the coin is fair.

(a) Any sequence that contains 5 heads and 1 tail is determined once the position of the 5 heads (or 1 tail) is known. The number of ways that we can position 5 heads in a sequence of 6 slots is $C(6, 5) = 6$. The probability of the event E, exactly 5 heads and one tail, is

$$P(E) = \frac{n(E)}{n(S)} = \frac{C(6,5)}{2^6} = \frac{6}{64} \approx 0.0938$$

(b) Let F be the event: between 4 and 6 heads, inclusive. To obtain between 4 and 6 heads is equivalent to the event: either 4 heads or 5 heads or 6 heads. Since each of these is mutually exclusive (it is impossible to obtain both 4 heads and 5 heads when tossing a coin 6 times), we have

$$\begin{aligned} P(F) &= P(4 \text{ heads or } 5 \text{ heads or } 6 \text{ heads}) \\ &= P(4 \text{ heads}) + P(5 \text{ heads}) + P(6 \text{ heads}) \end{aligned}$$

The probabilities on the right are obtained as in part (a). Thus,

$$P(F) = \frac{C(6,4)}{2^6} + \frac{C(6,5)}{2^6} + \frac{C(6,6)}{2^6} = \frac{15}{64} + \frac{6}{64} + \frac{1}{64} = \frac{22}{64} \approx 0.3438$$

HISTORICAL FEATURE

Set theory, counting, and probability first took form as a systematic theory in the exchange of letters (1654) between Pierre de Fermat (1601–1665) and Blaise Pascal (1623–1662). They discussed the problem of how to divide the stakes in a game that is interrupted before completion, knowing how many points each player needs to win. Fermat solved the problem by listing all possibilities and counting the favorable ones, whereas Pascal made use of the triangle that now bears his name. As mentioned in the text, the entries in Pascal's triangle are equivalent to $C(n, r)$. This recognition of the role of $C(n, r)$ in counting is the foundation of all further developments.

The first book on probability, the work of Christian Huygens (1629–1695), appeared in 1657. In it, the notion of mathematical expectation is explored. This allows the calculation of the profit or loss that a gambler might expect, knowing the probabilities involved in the game (see the Historical Problems that follow).

Although Girolamo Cardano (1501–1576) wrote a treatise on probability, it was not published until 1663 in Cardano's collected works, and this was too late to have any effect on the development of the theory.

In 1713, the posthumously published *Ars Conjectandi* of Jakob Bernoulli (1654–1705) gave the theory the form it would have until 1900. In the current century, both combinatorics (counting) and probability have undergone rapid development due to the use of computers.

A final comment about notation. The notations $C(n, r)$ and $P(n, r)$ are variants of a form of notation developed in England after 1830. The notation $\binom{n}{r}$ for $C(n, r)$ goes back to Leonhard Euler (1707–1783), but is now losing ground because it has no clearly related symbolism of the same type for permutations. The set symbols ∪ and ∩ were introduced by Giuseppe Peano (1858–1932) in 1888 in a slightly different context. The inclusion symbol ⊂ was introduced by E. Schroeder (1841–1902) about 1890. The treatment of set theory in the text is due to George Boole (1815–1864), who wrote $A + B$ for $A \cup B$ and AB for $A \cap B$ (statisticians still use AB for $A \cap B$).

HISTORICAL PROBLEMS

1. *The Problem Discussed by Fermat and Pascal* A game between two equally skilled players, A and B, is interrupted when A needs 2 points to win and B needs 3 points. In what proportion would the stakes be divided?

   [***Note:*** If each play results in 1 point for either player, at most four more plays will decide the game.]

   (a) *Fermat's solution* List all possible outcomes that will end the game to form the sample space (for example, ABA, $ABBB$, etc.). The probabilities for A to win and B to win then determine how the stakes should be divided.

   (b) *Pascal's solution* Use combinations to determine the number of ways that the 2 points needed for A to win could occur in four plays. Then use combinations to determine the number of ways that the 3 points needed for B to win could occur. This is trickier than it looks, since A can win with 2 points in either two plays, three plays, or four plays. Compute the probabilities and compare with the results in part (a).

2. *Huygen's Mathematical Expectation* In a game with n possible outcomes with probabilities $p_1, p_2, \ldots, p_n$, suppose that the *net* winnings are $w_1, w_2, \ldots, w_n$, respectively. Then the mathematical expectation is

$$E = p_1w_1 + p_2w_2 + \cdots + p_nw_n$$

   The number E represents the profit or loss per game in the long run. The following problems are a modification of those of Huygens.

   (a) A fair die is tossed. A gambler wins \$3 if he throws a 6 and \$6 if he throws a 5. What is his expectation?

   [***Note:*** $w_1 = w_2 = w_3 = w_4 = 0$]

   (b) A gambler plays the same game as in part (a), but now the gambler must pay \$1 to play. This means that $w_5 = \$5$, $w_6 = \$2$, and $w_1 = w_2 = w_3 = w_4 = -\$1$. What is the expectation?

8.3 EXERCISES

1. In a probability model, which of the following numbers could be the probability of an outcome: 0, 0.01, 0.35, −0.4, 1, 1.4?

2. In a probability model, which of the following numbers could be the probability of an outcome: 1.5, $\frac{1}{2}$, $\frac{3}{4}$, $\frac{2}{3}$, 0, $-\frac{1}{4}$?

3. Determine whether the following is a probability model.

Outcome	Probability
{1}	0.2
{2}	0.3
{3}	0.1
{4}	0.4

4. Determine whether the following is a probability model.

Outcome	Probability
{Jim}	0.4
{Bob}	0.3
{Faye}	0.1
{Patricia}	0.2

5. Determine whether the following is a probability model.

Outcome	Probability
{Linda}	0.3
{Jean}	0.2
{Grant}	0.1
{Ron}	0.3

6. Determine whether the following is a probability model.

Outcome	Probability
{Lanny}	0.3
{Joanne}	0.2
{Nelson}	0.1
{Rich}	0.5
{Judy}	−0.1

In Problems 7–12, construct a probability model for each experiment.

7. Tossing a fair coin twice

8. Tossing two fair coins once

9. Tossing two fair coins, then a fair die

10. Tossing a fair coin, a fair die, and then a fair coin

11. Tossing three fair coins once

12. Tossing one fair coin three times

In Problems 13–18, use the spinners shown below, and construct a probability model for each experiment.

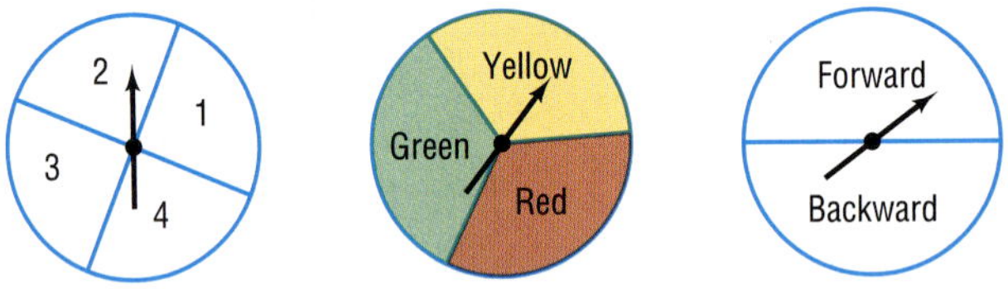

Spinner I Spinner II Spinner III

13. Spin spinner I, then spinner II. What is the probability of getting a 2 or a 4, followed by Red?

14. Spin spinner III, then spinner II. What is the probability of getting Forward, followed by Yellow or Green?

15. Spin spinner I, then II, then III. What is the probability of getting a 1, followed by Red or Green, followed by Backward?

16. Spin spinner II, then I, then III. What is the probability of getting Yellow, followed by a 2 or a 4, followed by Forward?

17. Spin spinner I twice, then spinner II. What is the probability of getting a 2, followed by a 2 or a 4, followed by Red or Green?

18. Spin spinner III, then spinner I twice. What is the probability of getting Forward, followed by a 1 or a 3, followed by a 2 or a 4?

In Problems 19–22, consider the experiment of tossing a coin twice. The table lists six possible assignments of probabilities for this experiment. Using this table, answer the following questions.

19. Which of the assignments of probabilities are consistent with the definition of the probability of an outcome?

20. Which of the assignments of probabilities should be used if the coin is known to be fair?

21. Which of the assignments of probabilities should be used if the coin is known to always come up tails?

22. Which of the assignments of probabilities should be used if tails is twice as likely as heads to occur?

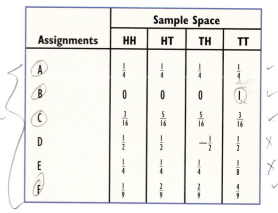

Assignments	Sample Space			
	HH	HT	TH	TT
A	$\frac{1}{4}$	$\frac{1}{4}$	$\frac{1}{4}$	$\frac{1}{4}$
B	0	0	0	1
C	$\frac{3}{16}$	$\frac{5}{16}$	$\frac{5}{16}$	$\frac{3}{16}$
D	$\frac{1}{2}$	$\frac{1}{2}$	$-\frac{1}{2}$	$\frac{1}{2}$
E	$\frac{1}{4}$	$\frac{1}{4}$	$\frac{1}{4}$	$\frac{1}{8}$
F	$\frac{1}{9}$	$\frac{2}{9}$	$\frac{2}{9}$	$\frac{4}{9}$

23. Assigning Probabilities A coin is weighted so that heads is four times as likely as tails to occur. What probability should we assign to heads? to tails?

24. Assigning Probabilities A coin is weighted so that tails is twice as likely as heads to occur. What probability should we assign to heads? to tails?

25. Assigning Probabilities A die is weighted so that an odd-numbered face is twice as likely as an even-numbered face. What probability should we assign to each face?

26. Assigning Probabilities A die is weighted so that a six cannot appear. The other faces occur with the same probability. What probability should we assign to each face?

For Problems 27–30, let the sample space be $S = \{1, 2, 3, 4, 5, 6, 7, 8, 9, 10\}$. *Suppose that the outcomes are equally likely.*

27. Compute the probability of the event $E = \{1, 2, 3\}$.

28. Compute the probability of the event $F = \{3, 5, 9, 10\}$.

29. Compute the probability of the event E: "an even number".

30. Compute the probability of the event F: "an odd number".

For Problems 31 and 32, an urn contains 5 white marbles, 10 green marbles, 8 yellow marbles, and 7 black marbles.

31. If one marble is selected, determine the probability that it is white.

32. If one marble is selected, determine the probability that it is black.

In Problems 33–36, assume equally likely outcomes.

33. Determine the probability of having 3 boys in a three-child family.

34. Determine the probability of having 3 girls in a three-child family.

35. Determine the probability of having 1 girl and 3 boys in a four-child family.

36. Determine the probability of having 2 girls and 2 boys in a four-child family.

For Problems 37–40, two fair die are rolled.

37. Determine the probability that the sum of the two die is 7.

38. Determine the probability that the sum of the two die is 11.

39. Determine the probability that the sum of the two die is 3.

40. Determine the probability that the sum of the two die is 12.

In Problems 41–44, find the probability of the indicated event if $P(A) = 0.25$ and $P(B) = 0.45$.

41. $P(A \cup B)$ if $P(A \cap B) = 0.15$

42. $P(A \cap B)$ if $P(A \cup B) = 0.6$

43. $P(A \cup B)$ if A, B are mutually exclusive

44. $P(A \cap B)$ if A, B are mutually exclusive

45. If $P(A) = 0.60$, $P(A \cup B) = 0.85$ and $P(A \cap B) = 0.05$, find $P(B)$.

46. If $P(B) = 0.30$, $P(A \cup B) = 0.65$ and $P(A \cap B) = 0.15$, find $P(A)$.

47. According to the Federal Bureau of Investigation, in 1997 there was a 25.3% probability of theft involving a motor vehicle. If a victim of theft is randomly selected, what is the probability he or she was not a victim of motor vehicle theft?

48. According to the Federal Bureau of Investigation, in 1997 there was a 5.6% probability of theft involving a bicycle. If a victim of theft is randomly selected, what is the probability he or she was not a victim of bicycle theft?

49. In Chicago, there is a 30% probability that Memorial Day will have a high temperature in the 70s. What is the probability that next Memorial Day will not have a high temperature in the 70s in Chicago?

50. In Chicago, there is a 4% probability that Memorial Day will have a low temperature in the 30s. What is the probability that next Memorial Day will not have a low temperature in the 30s in Chicago?

For Problems 51–54, a golf ball is selected at random from a container. If the container has 9 white balls, 8 green balls, and 3 orange balls, find the probability of each event.

51. The golf ball is white or green.

52. The golf ball is white or orange.

53. The golf ball is not white.

54. The golf ball is not green.

55. On the "Price is Right" there is a game in which a bag is filled with 3 strike chips and 5 numbers. Let's say that the numbers in the bag are 0, 1, 3, 6, and 9. What is the probability of selecting a strike chip or the number 1?

56. Another game on the "Price is Right" requires the contestant to spin a wheel with numbers 5, 10, 15, 20, …, 100. What is the probability that the contestant spins 100 or 30?

Problems 57–60 are based on a consumer survey of annual incomes in 100 households. The following table gives the data.

Income	\$0–9999	\$10,000–19,999	\$20,000–29,999	\$30,000–39,999	\$40,000 or more
Number of households	5	35	30	20	10

57. What is the probability that a household has an annual income of \$30,000 or more?

58. What is the probability that a household has an annual income between \$10,000 and \$29,999, inclusive?

59. What is the probability that a household has an annual income of less than \$20,000?

60. What is the probability that a household has an annual income of \$20,000 or more?

61. Surveys In a survey about the number of TV sets in a house, the following probability table was constructed:

Number of TV sets	0	1	2	3	4 or more
Probability	0.05	0.24	0.33	0.21	0.17

Find the probability of a house having:
(a) 1 or 2 TV sets
(b) 1 or more TV sets
(c) 3 or fewer TV sets
(d) 3 or more TV sets
(e) Less than 2 TV sets
(f) Less than 1 TV set
(g) 1, 2, or 3 TV sets
(h) 2 or more TV sets

62. **Checkout Lines** Through observation it has been determined that the probability for a given number of people waiting in line at the "5 items or less" checkout register of a supermarket is:

Number waiting in line	0	1	2	3	4 or more
Probability	0.10	0.15	0.20	0.24	0.31

Find the probability of:
(a) At most 2 people in line
(b) At least 2 people in line
(c) At least 1 person in line

63. In a certain College Algebra class, there are 18 freshmen and 15 sophomores. Of the 18 freshmen, 10 are male, and of the 15 sophomores, 8 are male. Find the probability that a randomly selected student is:
(a) A freshman or female
(b) A sophomore or male

64. The faculty of the mathematics department at Joliet Junior College is composed of 4 females and 9 males. Of the 4 females, 2 are under the age of 40, and 3 of the males are under age 40. Find the probability that a randomly selected faculty member is:
(a) Female or under age 40
(b) Male or over age 40

65. **Birthday Problem** What is the probability that at least 2 people have the same birthday in a group of 12 people? Assume that there are 365 days in a year.

66. **Birthday Problem** What is the probability that at least 2 people have the same birthday in a group of 35 people? Assume that there are 365 days in a year.

67. **Winning a Lottery** In a certain lottery, there are ten balls, numbered 1, 2, 3, 4, 5, 6, 7, 8, 9, 10. Of these, five are drawn in order. If you pick five numbers that match those drawn in the correct order, you win $1,000,000. What is the probability of winning such a lottery?

$\frac{1}{10}\cdot\frac{1}{9}\cdot\frac{1}{8}\cdot\frac{1}{7}\cdot\frac{1}{6} = 3.3069 \times 10^{-5}$

68. A committee of 6 people is to be chosen at random from a group of 14 people consisting of 2 supervisors, 5 skilled laborers, and 7 unskilled laborers. What is the probability that the committee chosen consists of 2 skilled and 4 unskilled laborers?

69. A fair coin is tossed 5 times.
(a) Find the probability that exactly 3 heads appear.
(b) Find the probability that no heads appear.

70. A fair coin is tossed 4 times.
(a) Find the probability that exactly 1 tail appears.
(b) Find the probability that no more than 1 tail appears.

71. A pair of fair dice is tossed 3 times.
(a) Find the probability that the sum of 7 appears 3 times.
(b) Find the probability that a sum of 7 or 11 appears at least twice.

72. A pair of fair dice is tossed 5 times.
(a) Find the probability that the sum is never 2.
(b) Find the probability that the sum is never 7.

73. Through a mix-up on the production line, 5 defective TVs were shipped out with 25 good ones. If 5 are selected at random, what is the probability that all 5 are defective? What is the probability that at least 2 of them are defective?

74. In a shipment of 50 transformers, 10 are known to be defective. If 30 transformers are picked at random, what is the probability that all 30 are nondefective? Assume that all transformers look alike and have an equal probability of being chosen.

75. In a promotion, 50 silver dollars are placed in a bag, one of which is valued at more than $10,000. The winner of the promotion is given the opportunity to reach into the bag, while blindfolded, and pull out 5 coins. What is the probability that one of the 5 coins is the one valued at more than $10,000?

8.4 ANALYZING UNIVARIATE DATA; OBTAINING PROBABILITIES FROM DATA

1 Classify Data

2 Construct and Interpret Bar Graphs and Pie Charts

3 Construct and Interpret Histograms

4 Compute Probabilities Using Relative Frequencies

5 Simulate Probabilities

1 **Data** are a collection of information about a variable. Data may be a number, a word, or a letter. For example, the miles per gallon obtained by a particular make of car are data that are numeric. The colors of the cars in the South Suburban College student parking lot are data that are words. We usually classify data as either quantitative or qualitative. **Quantitative data** are data that are a numerical measure of some variable. **Qualitative data** are data that describe a nonnumeric characteristic of a variable. So, the miles per gallon for a particular car would be quantitative, while the color of a car is a qualitative variable.

EXAMPLE 1 Classifying Data as Either Quantitative or Qualitative

Determine whether the following sets of data are quantitative or qualitative.

(a) Cost of monthly phone service for households in Jacksonville, Florida

(b) Gender of students enrolled in College Algebra at South Suburban College

(c) Method of payment for tuition bills at Chicago State University for students registering in the current semester

(d) Number of credit card holders who have been late with their payment at least once during the past 12 months

Solution

(a) Quantitative data because the data are numeric

(b) Qualitative data because gender is a nonnumeric characteristic

(c) Qualitative data because the method of payment (credit card, cash, check, etc.) is a nonnumeric characteristic

(d) Quantitative data because the data are numeric

NOW WORK PROBLEM 3.

When data are collected they are generally put into table form and graphs are created from the table. Putting data into the form of a graph allows us to "see" the data. Often a graph will reveal certain characteristics of the data that may not be readily apparent from a table of data. The type of graph drawn will depend on the type of data.

Graphs of Qualitative Data

2 There are two popular methods for graphically displaying a qualitative set of data: bar graphs and pie charts. A bar graph will show the number or the percent of data that are in each category, while a pie chart will show only the percent of data in each category.

EXAMPLE 2 Constructing a Bar Graph

In 1996 there were 738,028 full-time sworn law-enforcement officers in the United States. Table 2 lists the type of law-enforcement agencies and the corresponding number of full-time sworn officers from each agency.

Table 2

Type of Agency	Number of Full-Time Sworn Officers
Local police	410,956
Sheriff	152,922
State police	54,587
Special police	43,082
Texas constable	1,988
Federal	74,493

Source: Bureau of Justice Statistics.

Construct a bar graph for these data.

Solution A horizontal axis is used to indicate the category of agency and a vertical axis is used to represent the number of officers in each category. For each category of agency, we draw rectangles of equal width whose heights represent the number of officers in each category. The rectangles do not touch each other. See Figure 11.

Figure 11

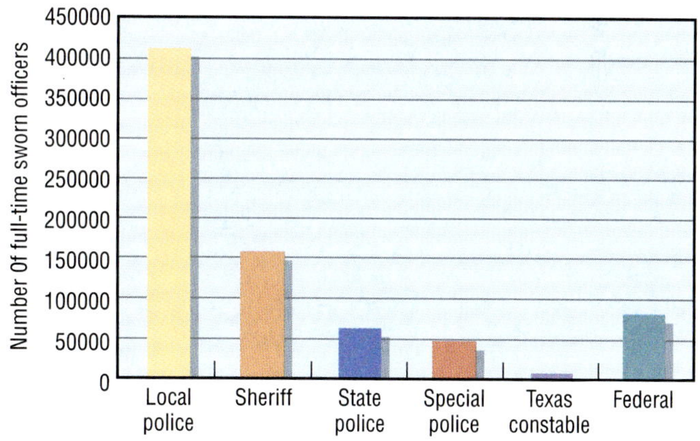

The **relative frequency** of a category is found by dividing the number of items in the category by the total number of items. Relative frequencies can be expressed either as a decimal or as a percent.

EXAMPLE 3 Constructing a Bar Graph of Relative Frequencies

Use the data in Table 2 to construct a bar graph of the relative frequencies.

Solution To construct a bar graph of the relative frequencies, we must first determine the relative frequency of each category. For example, the relative frequency of the category "Local police" is found by dividing the number of local police by the total number of sworn full-time officers. Thus, the relative frequency of this category is $410{,}956/738{,}028 \approx 0.557 = 55.7\%$. Following this procedure for the remaining categories of law enforcement, we obtain Table 3. Notice that the relative frequencies in Table 3 sum to 100%.

Table 3

Category	Number	Relative Frequency
Local police	410,956	0.557 = 55.7%
Sheriff	152,922	0.207 = 20.7%
State police	54,587	0.074 = 7.4%
Special police	43,082	0.058 = 5.8%
Texas constable	1,988	0.003 = 0.3%
Federal	74,493	0.101 = 10.1%
		Sum = 100%

A horizontal axis is used to indicate the category and a vertical axis is used to represent its relative frequency. For each category, we draw rectangles of equal width whose heights represent the relative frequency in each category. The rectangles do not touch each other. See Figure 12.

Figure 12

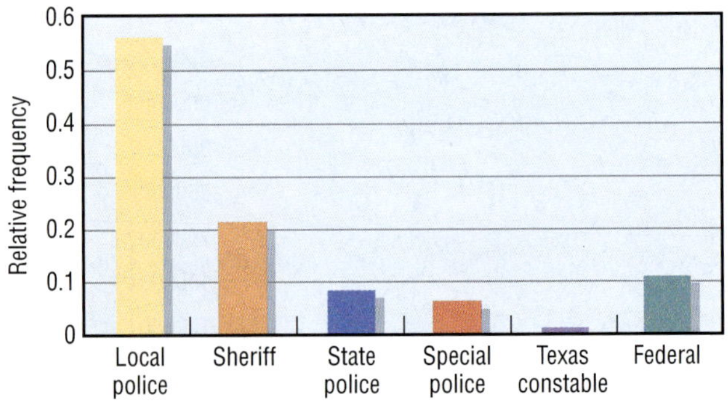

One reason for graphing data is to quickly determine certain information about the data. For example, from Figure 12 we can see that "Local police" has the largest relative frequency of law enforcement.

Qualitative data can also be represented graphically in a pie chart.

EXAMPLE 4 Constructing a Pie Chart

Use the data given in Table 3 to construct a pie chart.

Solution To construct a pie chart, a circle is divided into sectors, one sector for each category of data. The size of each sector is proportional to the total number of sworn officers. Since there are 410,956 local police and 738,028 full-time sworn officers, the relative frequency of this category is $410{,}956/738{,}028 \approx 0.557 = 55.7\%$. Therefore, local police will make up 55.7% of the pie chart. Since a circle has 360°, the degree measure of the sector for this category of officers is $0.557(360°) = 201°$. Following this procedure for the remaining categories of law-enforcement officers, we obtain Table 4.

Table 4

Category	Number of Full-time Sworn Officers	Percent of Total Officers	Degree Measure of Sector*
Local police	410,956	0.557 = 55.7%	201
Sheriff	152,922	0.207 = 20.7%	75
State police	54,587	0.074 = 7.4%	27
Special police	43,082	0.058 = 5.8%	21
Texas constable	1,988	0.003 = 0.3%	1
Federal	74,493	0.101 = 10.1%	36

*The data in column four does not add up to 360 due to rounding.

To construct a pie chart by hand, use a protractor to approximate the angles for each sector. To construct a pie chart with a computer, select a spreadsheet program that has the capability of drawing pie charts. See Figure 13.

Figure 13

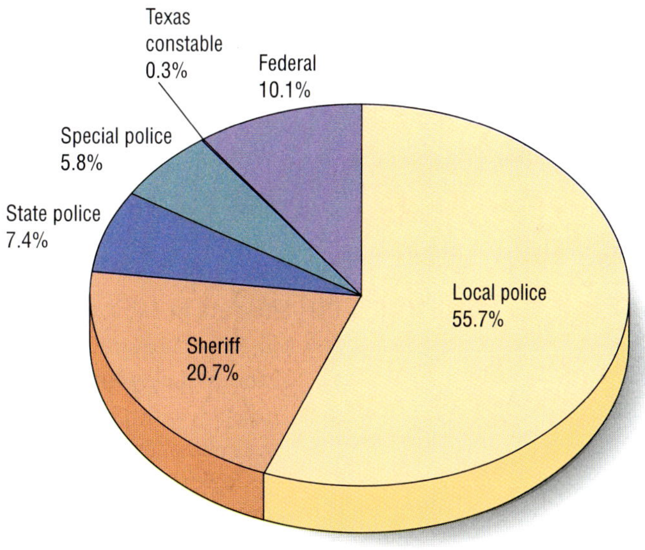

NOW WORK PROBLEM **15(a)–(d).**

3 Graphs of Quantitative Data

Quantitative data can also be organized in frequency tables. The data are divided into **class intervals** of equal width and the number of data that lie in each class interval is recorded. The class intervals do not overlap in order to prevent any data from falling into two different class intervals. The first number in any class interval is called the **lower class limit,** and the last number in any class interval is called the **upper class limit.** The **class width** is the difference between consecutive lower class limits. Data organized in frequency tables allow us to graphically represent quantitative data using *histograms.*

EXAMPLE 5 Drawing a Histogram

Table 5 represents the frequency distribution for the sale price of real estate valued between \$96,000 and \$255,000 in Bridgeview, Illinois, in 1997.

Table 5

Class Interval	Frequency
96–111	9
112–127	13
128–143	10
144–159	9
160–175	4
176–191	2
192–207	1
208–223	1
224–239	2
240–255	5

(a) Determine the number of class intervals.

(b) What is the lower class limit of the first class interval? What is the upper class limit of the first class interval?

(c) Determine the class width.

(d) Draw a histogram of the frequency distribution in Table 5.

Solution (a) The data in Table 5 have 10 class intervals.

(b) The lower class limit of the first class interval is 96. The upper class limit of the first class interval is 111.

(c) The class width is found by finding the difference between consecutive lower class limits. The lower class limit of the second class interval is 112. The lower class limit of the first class interval is 96. The class width is therefore $112 - 96 = 16$.

(d) Histograms are similar to bar charts in that we draw rectangles whose heights equal the frequency of each class interval. The width of each rectangle equals the class width. The bases of the rectangles touch when drawing a histogram. This is because there are no "gaps" in the data. Figure 14 shows a histogram for the data in Table 5.

Figure 14

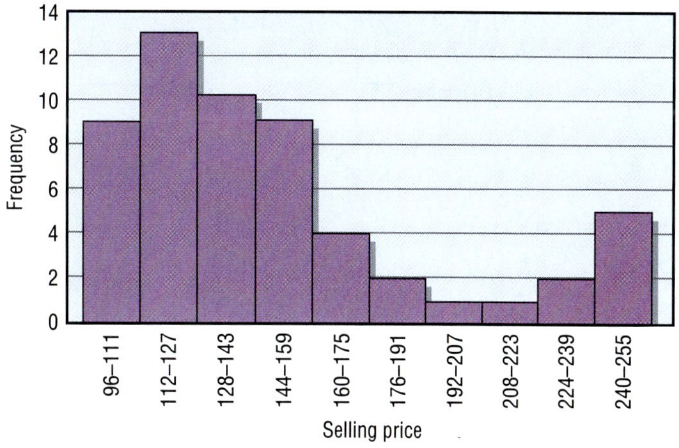

We can also create histograms of the relative frequencies of quantitative data.

EXAMPLE 6 Drawing a Histogram of Relative Frequencies

Use the data in Table 5 to construct a histogram of relative frequencies.

Solution To construct a histogram of relative frequencies, we must first determine the relative frequency of each category. By adding up all the frequencies, we find that the total number of properties sold in Bridgeview, Illinois, in 1997 was 56. The relative frequency is found by dividing the frequency within each class by the total number of properties sold. See Table 6.

Table 6

Class Interval	Frequency	Relative Frequency
96–111	9	$9/56 \approx 0.161$
112–127	13	0.232
128–143	10	0.179
144–159	9	0.161
160–175	4	0.071
176–191	2	0.036
192–207	1	0.018
208–223	1	0.018
224–239	2	0.036
240–255	5	0.089

The histogram of relative frequencies is constructed by drawing rectangles whose heights are the relative frequencies of the class interval. The width of each rectangle equals the class width. Again, the bases of the rectangles touch when drawing a histogram of relative frequencies. See Figure 15.

Figure 15

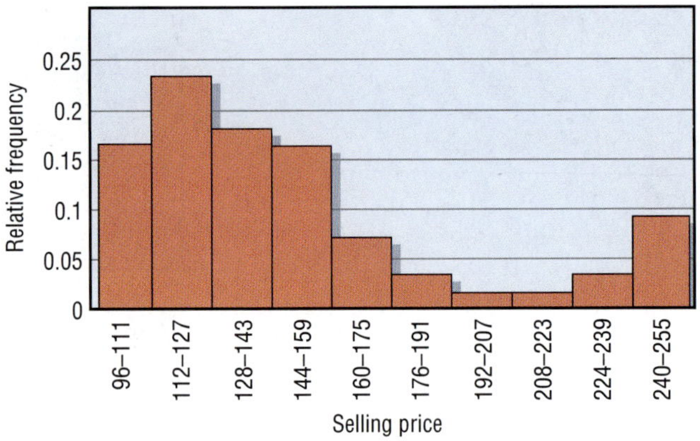

NOW WORK PROBLEM 17(a)–(e).

Probability from Data or Experiments

4 When data are collected and the relative frequency of each category is computed, the relative frequency may be used as the probability. Probabilities assigned in this way are called **empirical probabilities,** because the probability assignment depends on an experiment actually being performed or data actually being collected.

For example, suppose that we revisit the data involving full-time sworn officers in the United States (Table 3, p. 584). Table 7 reproduces the data and shows how probability assignments are made.

Table 7

Category	Relative Frequency	Probability
Local police	0.557 = 55.7%	0.557
Sheriff	0.207 = 20.7%	0.207
State police	0.074 = 7.4%	0.074
Special police	0.058 = 5.8%	0.058
Texas constable	0.003 = 0.3%	0.003
Federal	0.101 = 10.1%	0.101
		Sum = 1.000

Notice in Table 7 that each probability is nonnegative and that the sum of the probabilities is 1, as required.

EXAMPLE 7 Determining the Probability of an Event from an Experiment

A mathematics professor at Joliet Junior College randomly selects 40 currently enrolled students and finds that 25 of them pay their own tuition without assistance. Find the probability that a randomly selected student pays his or her own tuition.

Solution Let E represent the event "pays own tuition." The experiment involves asking students "Do you pay your own tuition?" The experiment is repeated 40 times and the frequency of "yes" responses is 25, so

$$P(E) = \frac{25}{40} = \frac{5}{8} = 0.625$$

There is a 62.5% probability that a randomly selected student pays his or her own tuition.

SEEING THE CONCEPT: Perform the preceding experiment yourself by asking 10 students, "Do you pay your own tuition?" Repeat the experiment three times. Are your probabilities the same? Why might they be different?

Simulation

5 Earlier we discussed the experiment of tossing a coin. We said then that if we assumed that the coin was fair then the outcomes were equally likely. Based on this, we assigned probabilities of 1/2 for heads and 1/2 for tails.

We could also assign probabilities by actually tossing the coin, say 100 times, and recording the frequencies of heads and tails. Perhaps we obtain 47 heads and 53 tails. Then, based on the experiment, we would use empirical probability and assign $P(\text{H}) = 0.47$ and $P(\text{T}) = 0.53$.

Instead of actually physically tossing the coin, we could use simulation; that is, we could use a graphing utility to replicate the experiment of tossing a coin.

EXPLORATION (a) Simulate tossing a coin 100 times. What percent of the time do you obtain heads? (b) Simulate tossing the coin 100 times again. Are the results the same as the first 100 flips? (c) Repeat the simulation by tossing the coin 250 times. Does the percent of heads get closer to $\frac{1}{2}$?

Figure 16

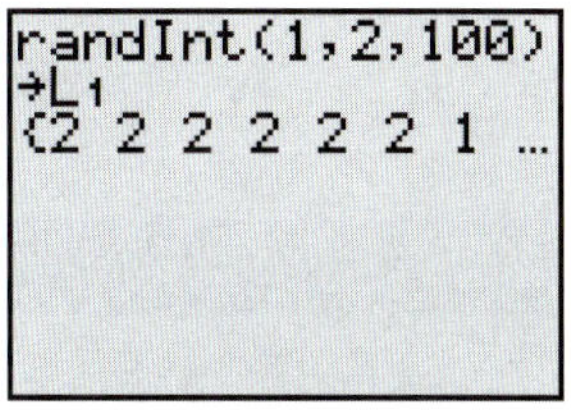

RESULT (a) Using the randInt* feature on a TI-83, we can simulate flipping a coin 100 times by letting 1 represent heads and 2 represent tails, and then storing the random integers in L_1. See Figure 16. Figure 17 shows the number of heads and the number of tails that result from flipping the coin 100 times. Based on this simulation, we would assign probabilities as

$$P(1) = P(\text{heads}) = \frac{48}{100} = 0.48 \quad \text{and} \quad P(2) = P(\text{tails}) = \frac{52}{100} = 0.52$$

*The randInt feature involves a mathematical formula that uses a seed number to generate a sequence of random integers. Consult your owner's manual for setting the seed so that the same random numbers are not generated each time that you repeat this exploration.

(b) Figure 18 shows the number of heads and the number of tails that result from a second simulation of flipping the coin 100 times. Based on this simulation, we would assign probabilities as

$$P(1) = P(\text{heads}) = \frac{47}{100} = 0.47 \quad \text{and} \quad P(2) = P(\text{tails}) = \frac{53}{100} = 0.53$$

Since repetitions of an experiment (simulation) do not necessarily result in the same outcomes, the empirical probabilities from two experiments will generally be different.

(c) Figure 19 shows the number of heads and the number of tails that result from a third simulation of flipping the coin 250 times. Based on this simulation, we would assign probabilities as

$$P(1) = P(\text{heads}) = \frac{124}{250} = 0.496 \quad \text{and} \quad P(2) = P(\text{tails}) = \frac{126}{250} = 0.504$$

Figure 17

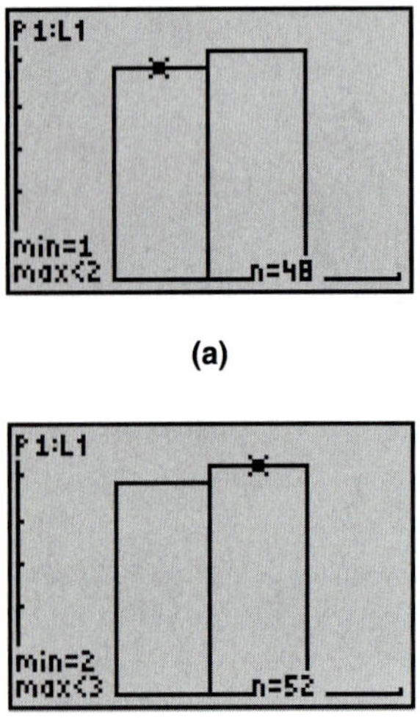

(a) (b)

Figure 18

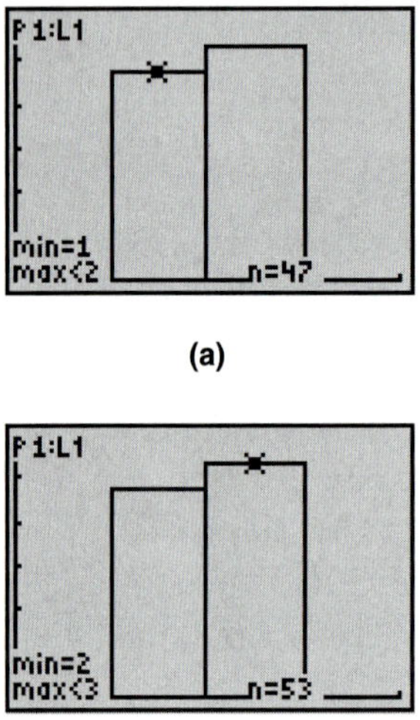

(a) (b)

Figure 19

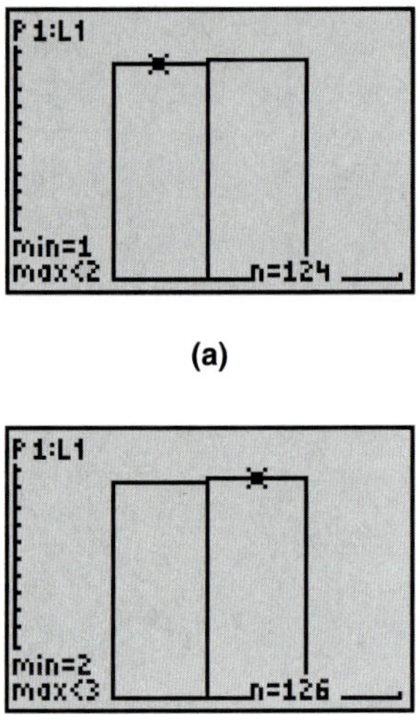

(a) (b)

Notice that the result of flipping the coin more often gives probability assignments that get closer to the ideal of equally likely outcomes. This demonstrates the **Law of Large Numbers,** which states that the more times an experiment involving equally likely outcomes is performed, the closer the empirical probability will come to the probability predicted by equally likely outcomes.

8.4 EXERCISES

For Problems 1–8, classify the data as either qualitative or quantitative.

1. Hair color
2. Number of students enrolled in United States History
3. Gas mileage in the city for a Toyota Camry

4. Brand of stereos a local superstore sells

5. SAT-Verbal scores of the freshman class at the University of Dayton

6. Marital status

7. Bachelor degrees offered by DePaul University

8. Number of hours a bulb burns before burning out

9. **On-Time Performance** The bar chart below represents the overall percentage of reported flight operations arriving on time for 10 different airlines.
(a) Which airline has the highest percentage of on-time flights?
(b) Which airline has the lowest percentage of on-time flights?
(c) What percentage of United Airlines' flights are on time?
(d) What is the probability that a randomly selected Southwest flight is on time?

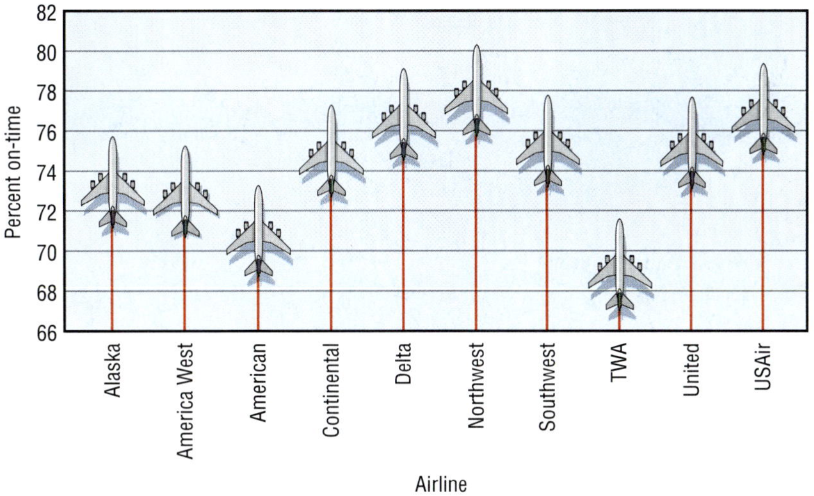

Source: United States Department of Transportation

10. **Income Required for a Loan** The bar chart below shows the minimum annual income required for a $100,000 loan using interest rates available on 12/1/96. Taxes and insurance are assumed to be $230 monthly.
(a) What minimum annual income is needed to qualify for a 5/1 year ARM (Adjustable Rate Mortgage)?
(b) Which loan type requires the most annual income? What minimum annual income is required for this loan type?

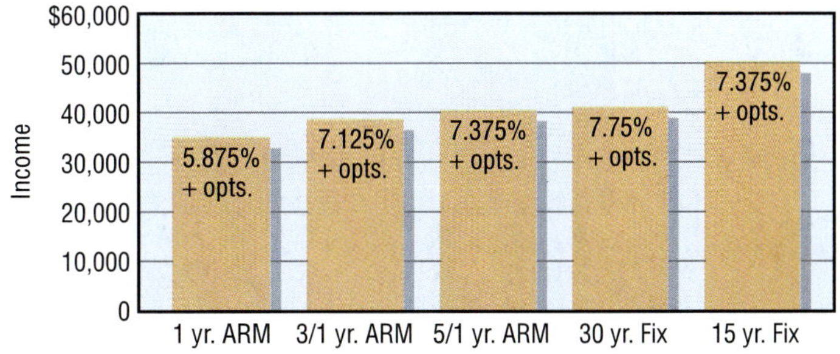

11. Consumer Price Index The Consumer Price Index (CPI) measures inflation. It is calculated by obtaining the prices of a market basket of goods each month. The market basket, along with the percentages of each product, is given in the following pie chart:

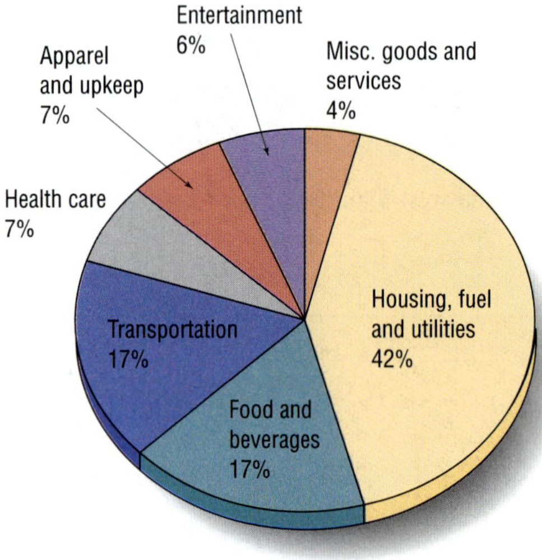

Source: Bureau of Labor Statistics

(a) What is the largest component of the CPI?
(b) What is the smallest component of the CPI?
(c) Senior citizens spend about 14% of their income on health care. Why do you think they feel that the CPI weight for health care is too low?

12. Asset Allocation According to financial planners, an individual's investment mix should change over a person's lifetime. The longer an individual's time horizon, the more the individual should invest in stocks. A financial planner suggested that Jim's retirement portfolio be diversified according to the mix provided in the pie chart below, since Jim has 40 years to retirement.

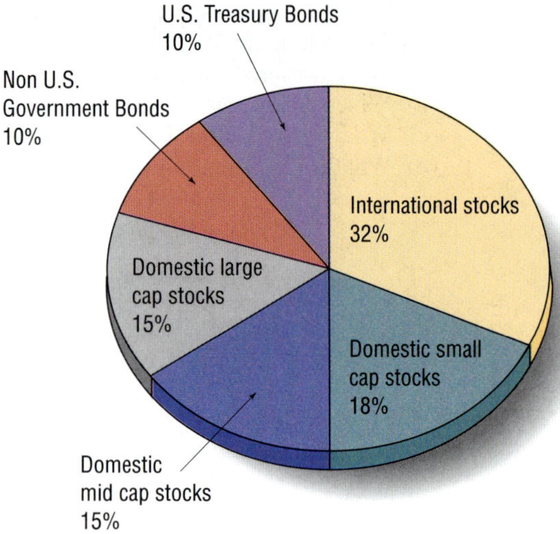

(a) How much should Jim invest in stocks?
(b) How much should Jim invest in bonds?
(c) How much should Jim invest in domestic (U.S.) stocks?
(d) The return on bonds over long periods of time is less than that of stocks. Explain why you think the financial planner recommended what she did for bonds.

13. Licensed Drivers in Florida The histogram below represents the number of licensed drivers between the ages of 20 and 84 in the state of Florida.

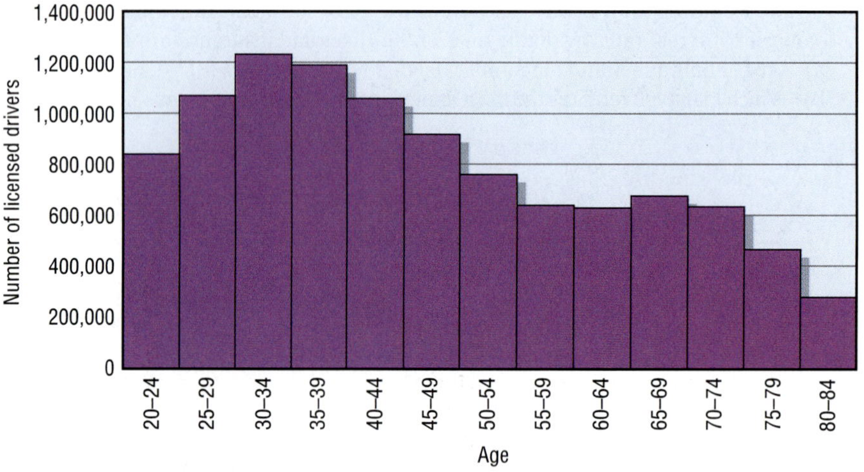

Source: FHWA

(a) Determine the number of class intervals.
(b) What is the lower class limit of the first class interval? What is the upper class limit of the first class interval?
(c) Determine the class width.
(d) How many licensed drivers are 70 to 84 years old?
(e) Which class interval has the most licensed drivers?
(f) Which class interval has the fewest licensed drivers?

14. IQ Scores The histogram below represents the IQ scores of students enrolled in College Algebra at a local university.

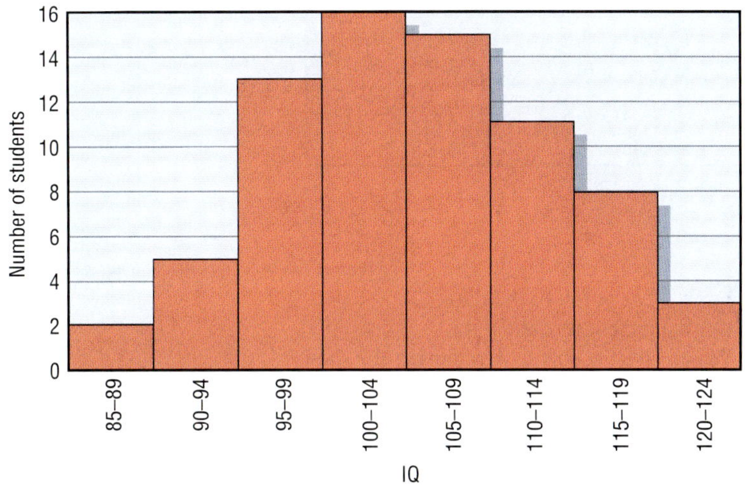

(a) Determine the number of class intervals.
(b) What is the lower class limit of the first class interval? What is the upper limit of the first class interval?
(c) Determine the class width.
(d) How many students have an IQ between 100 and 104?
(e) How many students have an IQ above 110?
(f) How many students are enrolled in College Algebra?

15. Causes of Death The data on the right represent the causes of death for 15–24 year olds in 1995.

(a) Draw a bar graph of the data.
(b) Draw a pie chart of the data.
(c) Which chart seems to summarize the data better?
(d) What was the leading cause of death for 15–24 year olds in 1995?
(e) Draw a bar graph of the relative frequencies.
(f) Construct a probability model for those data.
(g) What is the probability that the cause of death for a randomly selected 15–24 year old was suicide?
(h) What is the probability that the cause of death for a randomly selected 15–24 year old was suicide or malignant neoplasms?
(i) What is the probability that the cause of death for a randomly selected 15–24 year old was not suicide nor malignant neoplasms?

Cause of Death	Number
Accidents and adverse effects	13,532
Homicide and legal intervention	6,827
Suicide	4,789
Malignant neoplasms	1,599
Diseases of heart	964
Human immunodeficiency virus infection	643
Congenital anomalies	425
Chronic obstructive pulmonary diseases	220
Pneumonia and influenza	193
Cerebrovascular diseases	166
All other causes	4,211

Source: National Center for Health Statistics, 1996

16. Licensed Drivers The data below represent the number of licensed drivers in the Great Lake States in 1994.

State	Number of Licensed Drivers
Illinois	7,502,201
Indiana	3,806,329
Michigan	6,601,924
Minnesota	2,705,701
Ohio	7,142,173
Wisconsin	3,554,003

Source: Each state's authorities

(a) Draw a bar graph of the data.
(b) Draw a pie chart of the data.
(c) Which chart seems to summarize the data better?
(d) Which state has the most licensed drivers?
(e) Which state has the fewest licensed drivers?
(f) Draw a bar graph of the relative frequencies.
(g) Construct a probability model for these data.
(h) What is the probability that a randomly selected licensed driver from the Great Lakes States is from Michigan?
(i) What is the probability that a randomly selected licensed driver in the Great Lakes States is from Illinois or Indiana?
(j) What is the probability that a randomly selected licensed driver in the Great Lakes States is not from Illinois or Indiana?

17. Licensed Drivers in Tennessee The following frequency table provides the number of licensed drivers between the ages of 20 and 84 in the state of Tennessee in 1994.

(a) Determine the number of class intervals.
(b) What is the lower class limit of the first class interval? What is the upper class limit of the first class interval?
(c) Determine the class width.
(d) Draw a histogram of the data.
(e) Draw a histogram of the relative frequencies of the data.
(f) Which age group has the most licensed drivers?
(g) Which age group has the fewest licensed drivers?
(h) Construct a probability model for these data.
(i) What is the probability that a randomly selected licensed driver in Tennessee is 45–49?

Age	Number of Licensed Drivers
20–24	345,941
25–29	374,629
30–34	428,748
35–39	439,137
40–44	414,344
45–49	372,814
50–54	292,460
55–59	233,615
60–64	204,235
65–69	181,977
70–74	150,347
75–79	100,068
80–84	50,190

Source: FHWA

18. Licensed Drivers in Hawaii The frequency table below provides the number of licensed drivers between the ages of 20 and 84 in the state of Hawaii in 1994.

Age	Number of Licensed Drivers
20–24	65,951
25–29	78,119
30–34	91,976
35–39	92,557
40–44	87,430
45–49	75,978
50–54	55,199
55–59	39,678
60–64	35,650
65–69	33,885
70–74	26,125
75–79	14,990
80–84	6,952

Source: FHWA

(a) Determine the number of class intervals.
(b) What is the lowest class limit of the first class interval? What is the upper class limit of the first class interval?

(c) Determine the class width.
(d) Draw a histogram of the data.
(e) Draw a histogram of the relative frequencies.
(f) Which age group has the most licensed drivers?
(g) Which age group has the fewest licensed drivers?
(h) Construct a probability model for these data.
(i) What is the probability that a randomly selected licensed driver in Hawaii is 45–49?

19. Undergradute Tuition The data below represent the cost of undergraduate tuition at four-year colleges for 1992–1993 having tuition amounts ranging from $0 through $14,999.

Tuition (Dollars)	Number of 4-Year Colleges
0–999	10
1000–1999	7
2000–2999	45
3000–3999	66
4000–4999	84
5000–5999	84
6000–6999	97
7000–7999	118
8000–8999	138
9000–9999	110
10,000–10,999	104
11,000–11,999	82
12,000–12,999	61
13,000–13,999	34
14,000–14,999	29

Source: The College Board, New York, NY, Annual Survey of Colleges 1992 and 1993.

(a) Determine the number of class intervals.
(b) What is the lower class limit of the first class interval? What is the upper class limit of the first class interval?
(c) Determine the class width.
(d) Draw a histogram of the data.
(e) Draw a histogram of the relative frequencies.
(f) What range of tuition occurs most frequently?
(g) Construct a probability model for these data.
(h) What is the probability that randomly selected four-year college has undergraduate tuition between $11,000 and $11,999?

20. Undergraduate Tuition The following data represent the cost of undergraduate tuition at four-year colleges for 1993–1994 having tuition amounts ranging from $0 through $14,999.

Tuition (Dollars)	Number of 4-Year Colleges
0–999	8
1000–1999	5
2000–2999	28
3000–3999	48
4000–4999	76
5000–5999	65
6000–6999	81
7000–7999	96
8000–8999	112
9000–9999	118
10,000–10,999	106
11,000–11,999	90
12,000–12,999	70
13,000–13,999	59
14,000–14,999	23

Source: The College Board, New York, NY, Annual Survey of Colleges 1992 and 1993.

(a) Determine the number of class intervals.
(b) What is the lower class limit of the first interval? What is the upper limit of the first class interval?
(c) Determine the class width.
(d) Draw a histogram of the data.
(e) Draw a histogram of the relative frequencies.
(f) What range of tuition occurs most frequently?
(g) Construct a probability model for these data.
(h) What is the probability that a randomly selected four-year college has undergraduate tuition between $11,000 and $11,999?

21. The following data represent the marital status of males 18 years old and older in March 1997.

Marital Status	Number (in thousands)
Married, spouse present	54,654
Married, spouse absent	3,232
Widowed	2,686
Divorced	8,208
Never married	25,375

Source: Current Population Survey

(a) Construct a probability model for these data.
(b) Determine the probability that a randomly selected male age 18 years old or older is married, spouse present.
(c) Determine the probability that a randomly selected male age 18 years old or older has never been married.
(d) Determine the probability that a randomly selected male age 18 years old or older is married, spouse present, or married, spouse absent.
(e) What is the probability that the marital status of a randomly selected male 18 years old and older is not "married, spouse present."

22. The following data represent the marital status of females 18 years old and older in March 1997.
(a) Construct a probability model for these data.
(b) Determine the probability that a randomly selected female age 18 years old or older is married, spouse present.
(c) Determine the probability that a randomly selected female age 18 years old or older has never been married.
(d) Determine the probability that a randomly selected female 18 years old or older is married, spouse present, or married, spouse absent.
(e) What is the probability that the marital status of a randomly selected female 18 years old and older is not "married, spouse present."

Marital Status	Number (in thousands)
Married, spouse present	54,626
Married, spouse absent	4,122
Widowed	11,056
Divorced	11,107
Never married	20,503

Source: Current Population Survey

CHAPTER REVIEW

THINGS TO KNOW

Set		Well-defined collection of distinct objects, called elements
Null set	$\varnothing$	Set that has no elements
Equality	$A = B$	A and B have the same elements
Subset	$A \subseteq B$	Each element of A is also an element of B.
Intersection	$A \cap B$	Set consisting of elements that belong to both A and B
Union	$A \cup B$	Set consisting of elements that belong to either A or B, or both
Universal set	U	Set consisting of all the elements that we wish to consider
Complement	$\overline{A}$	Set consisting of elements of the universal set that are not in A
Finite set		The number of elements in the set is a nonnegative integer
Infinite set		A set that is not finite
Counting formula	$n(A \cup B) = n(A) + n(B) - n(A \cap B)$	
Addition Principle		If $A \cap B = \varnothing$, then $n(A \cup B) = n(A) + n(B)$.
Multiplication Principle		If a task consists of a sequence of choices in which there are p selections for the first choice, q selections for the second choice, and so on, then the task of making these selections can be done in $p \cdot q \cdot \ldots$ different ways.
Permutation	$P(n, r) = n(n-1) \cdot \ldots \cdot [n-(r-1)] = \dfrac{n!}{(n-r)!}$	An ordered arrangement of n distinct objects without repetition

Combination	$C(n, r) = \frac{P(n, r)}{r!} = \frac{n!}{(n-r)!r!}$	An arrangement, without regard to order, of n distinct objects without repetition
Permutations with repetition	$\frac{n!}{n_1!n_2!\cdots n_k!}$	The number of permutations of n objects of which n_1 are of one kind, n_2 are of a second kind, ..., and n_k are of a kth kind, where $n = n_1 + n_2 + \cdots + n_k$
Sample space		Set whose elements represent all the logical possibilities that can occur as a result of an experiment
Probability		A number assigned to each outcome of a sample space; the sum of all the probabilities of the outcomes equals 1.
Addition Rule	$P(E \cup F) = P(E) + P(F) - P(E \cap F)$	
Equally likely outcomes	$P(E) = \frac{n(E)}{n(S)}$	The same probability is assigned to each outcome.
Complement of an event	$P(\overline{E}) = 1 - P(E)$	

HOW TO

Find unions, intersections, and complements of sets
Use Venn diagrams to illustrate sets
Recognize a permutation problem
Recognize a combination problem
Solve certain probability problems
Count the elements in a sample space
Draw a tree diagram
Construct probability models
Graph and interpret bar graphs
Graph and interpret pie charts
Draw and interpret histograms

FILL-IN-THE-BLANK ITEMS

1. The ______________ of A with B consists of all elements in either A or B or both; the ______________ of A with B consists of all elements in both A and B.

2. $P(5, 2) =$ ______________; $C(5, 2) =$ ______________.

3. A(n) ______________ is an ordered arrangement of n distinct objects.

4. A(n) ______________ is an arrangement of n distinct objects without regard to order.

5. When the same probability is assigned to each outcome of a sample space, the experiment is said to have ______________ ______________ outcomes.

6. The ______________ of an event E is the set of all outcomes in the sample space S that are not outcomes in the event E.

TRUE/FALSE ITEMS

T F **1.** The intersection of two sets is always a subset of their union.

T F **2.** $P(n, r) = \frac{n!}{r!}$

T F **3.** In a combination problem, order is not important.

T F **4.** In a permutation problem, once an object is used, it cannot be repeated.

T F **5.** The probability of an event can never equal 0.

T F **6.** In a probability model, the sum of all probabilities is 1.

REVIEW EXERCISES

Blue problem numbers indicate the authors' suggestions for use in a Practice Test.

In Problems 1–8, use U = *Universal set* $= \{1, 2, 3, 4, 5, 6, 7, 8, 9\}$, $A = \{1, 3, 5, 7\}$, $B = \{3, 5, 6, 7, 8\}$, *and* $C = \{2, 3, 7, 8, 9\}$ *to find each set.*

1. $A \cup B$ **2.** $B \cup C$ **3.** $A \cap C$ **4.** $A \cap B$

5. $\overline{A} \cup \overline{B}$ **6.** $\overline{B} \cap \overline{C}$ **7.** $\overline{B \cap C}$ **8.** $\overline{A \cup B}$

9. If $n(A) = 8$, $n(B) = 12$, and $n(A \cap B) = 3$, find $n(A \cup B)$.

10. If $n(A) = 12$, $n(A \cup B) = 30$, and $n(A \cap B) = 6$, find $n(B)$.

In Problems 11–16, use the information supplied in the figure:

11. How many are in A?

12. How many are in A or B?

13. How many are in A and C?

14. How many are not in B?

15. How many are in neither A nor C?

16. How many are in B but not in C?

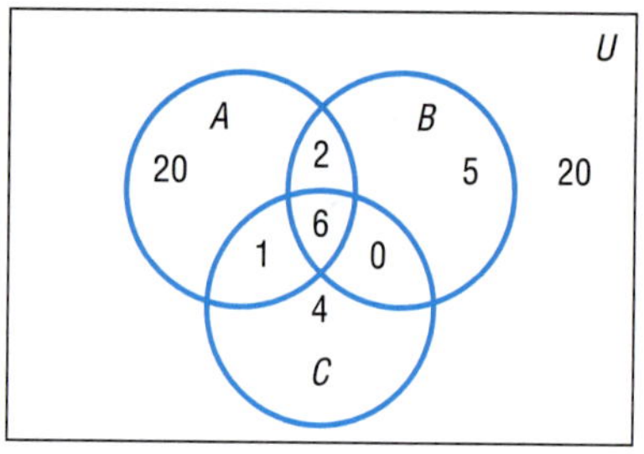

In Problems 17–22, compute the given expression.

17. $5!$ **18.** $6!$ **19.** $P(8, 3)$ **20.** $P(7, 3)$ **21.** $C(8, 3)$ **22.** $C(7, 3)$

23. A clothing store sells pure wool and polyester/wool suites. Each suit comes in 3 colors and 10 sizes. How many suits are required for a complete assortment?

24. In connecting a certain electrical device, 5 wires are to be connected to 5 different terminals. How many different wirings are possible if 1 wire is connected to each terminal?

25. **Baseball** On a given day, the American Baseball League schedules 7 games. How many different outcomes are possible, assuming that each game is played to completion?

26. **Baseball** On a given day, the National Baseball League schedules 6 games. How many different outcomes are possible, assuming that each game is played to completion?

27. If 4 people enter a bus having 9 vacant seats, in how many ways can they be seated?

28. How many different arrangements are there of the letters in the word ROSE?

29. In how many ways can a squad of 4 relay runners be chosen from a track team of 8 runners?

30. A professor has 10 similar problems to put on a test with 3 problems. How many different tests can she design?

31. **Baseball** In how many different ways can the 14 baseball teams in the American League be paired without regard to which team is at home?

32. **Arranging Books on a Shelf** There are 5 different French books and 5 different Spanish books. How many ways are there to arrange them on a shelf if

(a) Books of the same language must be grouped together, French on the left, Spanish on the right?

(b) French and Spanish books must alternate in the grouping, beginning with a French book?

33. **Telephone Numbers** Using the digits 0, 1, 2, . . . , 9, how many 7-digit numbers can be formed if the first digit cannot be 0 or 9 and if the last digit is greater than or equal to 2 and less than or equal to 3? Repeated digits are allowed.

34. Home Choices A contractor constructs homes with 5 different choices of exterior finish, 3 different roof arrangements, and 4 different window designs. How many different types of homes can be built?

35. License Plate Possibilities A license plate consists of 1 letter, excluding O and I, followed by a 4-digit number that cannot have a 0 in the lead position. How many different plates are possible?

36. Using the digits 0 and 1, how many different numbers consisting of 8 digits can be formed?

37. Forming Different Words How many different words can be formed using all the letters in the word MISSING?

38. Arranging Flags How many different vertical arrangements are there of 10 flags, if 4 are white, 3 are blue, 2 are green, and 1 is red?

39. Forming Committees A group of 9 people is going to be formed into committees of 4, 3, and 2 people. How many committees can be formed if

(a) A person can serve on any number of committees?

(b) No person can serve on more than one committee?

40. Forming Committees A group consists of 5 men and 8 women. A committee of 4 is to be formed from this group, and policy dictates that at least 1 woman be on this committee.

(a) How many committees can be formed that contain exactly 1 man?

(b) How many committees can be formed that contain exactly 2 women?

(c) How many committees can be formed that contain at least 1 man?

41. Male Doctorates in Mathematics The data below represent the ethnicity of male doctoral recipients in mathematics in 1995–1996.

Ethnic Group	Number of Doctoral Recipients
Asian, Pacific Islander	318
Black	15
American Indian, Eskimo, Aleut	1
Mexican American, Puerto Rican, or other Hispanic	36
White (non-Hispanic)	527
Unknown	6

Source: 1996 AMS-IMS-MAA Annual Survey

(a) Draw a bar graph of the data.

(b) Draw a bar graph of the relative frequencies.

(c) Draw a pie chart of the data.

(d) Which ethnic group had the most doctoral recipients?

(e) Construct a probability model for the data.

(f) What is the probability that a randomly selected male doctoral recipient is Asian, Pacific Islander?

(g) What is the probability that a randomly selected male doctoral recipient is Asian, Pacific Islander or Mexican American, Puerto Rican, or other Hispanic?

(h) What is the probability that a randomly selected male doctoral recipient is not Asian, Pacific Islander?

42. Female Doctorates in Mathematics The data below represent the ethnicity of female doctoral recipients in mathematics in 1995–1996.

Ethnic Group	Number of Doctoral Recipients
Asian, Pacific Islander	89
Black	4
American Indian, Eskimo, Aleut	0
Mexican American, Puerto Rican, or other Hispanic	7
White (non-Hispanic)	149
Unknown	1

Source: 1996 AMS-IMS-MAA Annual Survey

(a) Draw a bar graph of the data.

(b) Draw a bar graph of the relative frequencies.

(c) Draw a pie chart of the data.

(d) Which ethnic group had the most doctoral recipients?

(e) Construct a probability model for the data.

(f) What is the probability that a randomly selected female doctoral recipient is Asian, Pacific Islander?

(g) What is the probability that a randomly selected female doctoral recipient is Asian, Pacific Islander or Mexican American, Puerto Rican, or other Hispanic?

(h) What is the probability that a randomly selected female doctoral recipient is not Asian, Pacific Islander?

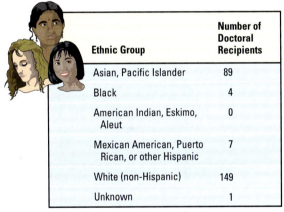

43. Age and DUI in 1980 The data below represent the estimated number of arrests per 100,000 drivers for driving under the influence for individuals 25–64 years of age in 1980.

Age	Estimated Number of Arrests
25–29	1347
30–34	1076
35–39	996
40–44	944
45–49	837
50–54	686
55–59	509
60–64	335

Source: U.S. Bureau of Justice Statistics

(a) What is the class width?
(b) Draw a histogram of the data.
(c) Draw a histogram of the relative frequencies.
(d) Which age group has the highest number of arrests for DUI?
(e) Construct a probability model for the data.
(f) What is the probability that the age of a randomly selected DUI arrest is 35–39?
(g) What is the probability that the age of a randomly selected DUI arrest is 35–39 or 55–59?
(h) What is the probability that the age of a randomly selected DUI arrest is not 35–39?

44. Age and DUI in 1989 The following data represent the estimated number of arrests per 100,000 drivers for driving under the influence for individuals 25–64 years of age in 1989.

Age	Estimated Number of Arrests
25–29	1869
30–34	1486
35–39	1123
40–44	872
45–49	725
50–54	558
55–59	400
60–64	262

Source: U.S. Bureau of Justice Statistics

(a) What is the class width?
(b) Draw a histogram of the data.
(c) Draw a histogram of the relative frequencies.
(d) Which age group has the highest number of arrests for DUI?
(e) Construct a probability model for the data.
(f) What is the probability that the age of a randomly selected DUI arrest is 35–39?
(g) What is the probability that the age of a randomly selected DUI arrest is 35–39 or 55–59?
(h) What is the probability that the age of a randomly selected DUI arrest is not 35–39?

45. Birthday Problem For this problem, assume that the year has 365 days.
(a) How many ways can 18 people have different birthdays?
(b) What is the probability that nobody has the same birthday in a group of 18 people?
(c) What is the probability in a group of 18 people that at least 2 people have the same birthday?

46. Death Rates According to the U.S. National Center for Health Statistics, 32.1% of all deaths in 1994 were due to heart disease.
(a) What is the probability that a randomly selected death in 1994 was due to heart disease?
(b) What is the probability that a randomly selected death in 1994 was not due to heart disease?

47. Unemployment According to the U.S. Bureau of Labor Statistics, 5.4% of the U.S. labor force was unemployed in 1996.
(a) What is the probability that a randomly selected member of the U.S. labor force was unemployed in 1996?
(b) What is the probability that a randomly selected member of the U.S. labor force was not unemployed in 1996?

48. From a box containing three 40-watt bulbs, six 60-watt bulbs, and eleven 75-watt bulbs, a bulb is drawn at random. What is the probability that the bulb is 40 watts? What is the probability that it is not a 75 watt bulb?

49. You have four \$1 bills, three \$5 bills, and two \$10 bills in your wallet. If you pick a bill at random, what is the probability that it will be a \$1 bill?

50. Each of the letters in the word ROSE is written on an index card and the cards are then shuffled. What is the probability that, when the cards are dealt out, they spell the word ROSE?

51. Each of the numbers, $1, 2, \ldots, 100$ is written on an index card and the cards are then shuffled. If a card is selected at random, what is the probability that the number on the card is divisible by 5? What is the probability that the card selected is either a 1 or names a prime number?

52. **Computing Probabilities** Because of a mistake in packaging, a case of 12 bottles of red wine contained 5 Merlot and 7 Cabernet, each without labels. All the bottles look alike and have equal probability of being chosen. Three bottles are selected.

(a) What is the probability that all 3 are Merlot?

(b) What is the probability that exactly 2 are Merlot?

(c) What is the probability that none is a Merlot?

53. **Tossing a Coin** A fair coin is tossed 10 times.

(a) What is the probability of obtaining exactly 5 heads?

(b) What is the probability of obtaining all heads?

(c) Use a graphing utility to simulate this experiment. What is the empirical probability of obtaining exactly 5 heads? All heads?

54. At the Milex tune-up and brake repair shop, the manager has found that a car will require a tune-up with a probability of 0.6, a brake job with a probability of 0.1, and both with a probability of 0.02.

(a) What is the probability that a car requires either a tune-up or a brake job?

(b) What is the probability that a car requires a tune-up but not a brake job?

(c) What is the probability that a car requires neither type of repair?

CHAPTER PROJECTS

1. **Simulation** In the Winter 1998 edition of *Eightysomething!,* Mike Koehler uses simulation to calculate the following probabilities: "A woman and man (unrelated) each have two children. At least one of the woman's children is a boy, and the man's older child is a boy. Do the chances that the woman has two boys equal the chances that the man has two boys?" Go to http://www.ti.com/calc and look up this article. Perform the simulation suggested by Koehler to answer the question.

2. **Surveys** Conduct a survey of 25 students in your school. Ask each student the following questions:

   (a) Do you drive yourself to school?

   (b) Do you pay your own tuition?

   (c) Do you study (i) 0–5 hours per week, (ii) 6–10 hours per week, (iii) 11–15 hours per week, (iv) 16–20 hours per week, (v) more than 20 hours per week?

   (d) Compute the probability that a student (1) drives himself or herself to school, (2) pays his or her own tuition; (3) studies 6–10 hours per week.

   (e) Compare your results with other students in your class. Why might your probabilities differ from the probabilities obtained by other students in the class?

3. **Law of Large Numbers** We can demonstrate the Law of Large Numbers using a graphing calculator such as the TI-83. We will simulate flipping a fair coin 200 times.

   (a) In L_1, store the integers 1–200 using the command $seq(n,n,1,200) \to L_1$.

   (b) In L_2, store the results of flipping a coin 200 times using the command $randInt(0,1,200) \to L_2$. Note that the 0 represents flipping a tail and 1 represents flipping a head.

CHAPTER PROJECTS *(Continued)*

(c) In L_3, store the cumulative number of heads flipped using the command $cumSum(L_2) \rightarrow L_3$.

(d) L_4 will represent the proportion of heads thrown and is found using the command $L_3/L_1 \rightarrow L_4$.

(e) Graph the line $Y_1 = 0.5$ and the data in L_4 versus the data in L_1 using the viewing window:

$$X\text{min} = 0, \quad X\text{max} = 200, \quad Y\text{min} = 0, \quad Y\text{max} = 1$$

(f) What do you observe about the graph of the data as the number of trial flips increases?

CHAPTER

9 Conics

Pluto resumes 'farthest orbit'

WASHINGTON (AP) - Mere days after surviving attacks on its status as a planet, diminutive Pluto is resuming its traditional spot farthest from the sun.

Pluto was on course to cross the orbit of Neptune at 5:08 a.m. Thursday, NASA reported.

Normally the most distant planet from the sun, Pluto has a highly elliptical orbit that occasionally brings it inside the orbit of Neptune.

That last took place on Feb. 7, 1979, and since then Neptune had been the most distant planet.

Now Pluto once again becomes the farthest planet from the sun, where it will remain for 228 years. It takes Pluto 248 years to circle the sun.

(*Source: Daily News Naples,* February 11, 1999.)

See Chapter Project 1.

Preparing for This Chapter

Before getting started on this chapter, review the following concepts:

- **Distance Formula *(p. 5)***
- **Completing the Square *(pp. 85–86)***
- **Intercepts *(pp. 19–21)***
- **Symmetry *(pp. 22–23)***
- **Circles *(Section 1.7)***
- **Systems of Two Linear Equations Containing Two Variables *(Section 6.1)***

Outline

Historically, Apollonius (200 B.C.) was among the first to study *conics* and discover some of their interesting properties. Today, conics are still studied because of their many uses. *Paraboloids of revolution* (parabolas rotated about their axes of symmetry) are used as signal collectors (the satellite dishes used with radar and cable TV, for example), as solar energy collectors, and as reflectors (telescopes, light projection, and so on). The planets circle the Sun in approximately *elliptical* orbits. Elliptical surfaces can be used to reflect signals such as light and sound from one place to another. And *hyperbolas* can be used to determine the positions of ships at sea.

The Greeks used the methods of Euclidean geometry to study conics. We shall use the more powerful methods of analytic geometry, bringing to bear both algebra and geometry, for our study of conics. Thus, we shall give a geometric description of each conic, and then, using rectangular coordinates and the distance formula, we shall find equations that represent conics. We used this same development, you may recall, when we first defined a circle in Section 1.7.

The chapter concludes with a section on systems of nonlinear equations.

9.1 CONICS

1 Know the Names of the Conics

Figure 1

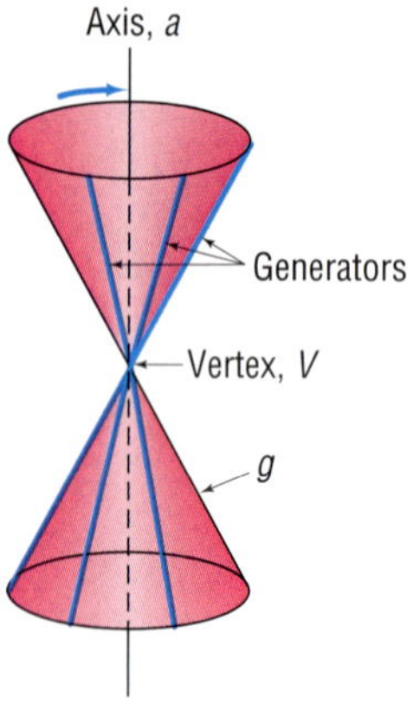

1 The word *conic* derives from the word *cone,* which is a geometric figure that can be constructed in the following way: Let a and g be two distinct lines that intersect at a point V. Keep the line a fixed. Now rotate the line g about a while maintaining the same angle between a and g. The collection of points swept out (generated) by the line g is called a **(right circular) cone.** See Figure 1. The fixed line a is called the **axis** of the cone; the point V is called its **vertex;** the lines that pass through V and make the same angle with a as g are called **generators** of the cone. Thus, each generator is a line that lies entirely on the cone. The cone consists of two parts, called **nappes,** that intersect at the vertex.

Conics, an abbreviation for **conic sections,** are curves that result from the intersection of a (right circular) cone and a plane. The conics we shall study arise when the plane does not contain the vertex, as shown in Figure 2. These conics are **circles** when the plane is perpendicular to the axis of the cone and intersects each generator; **ellipses** when the plane is tilted slightly

Figure 2

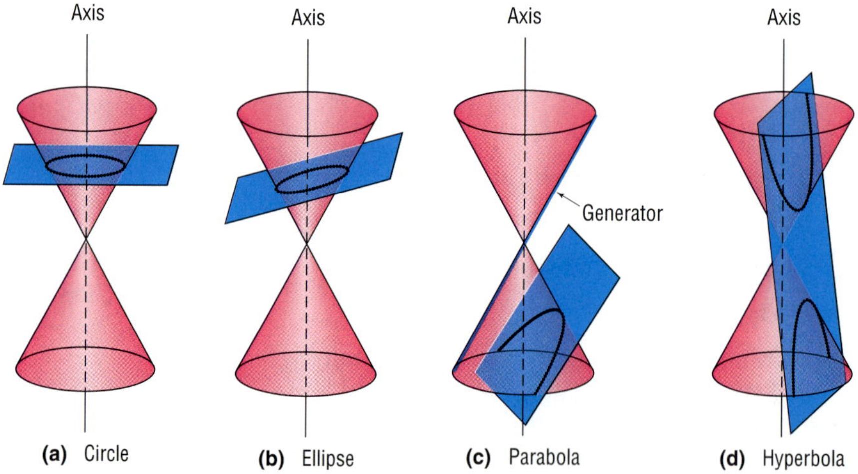

so that it intersects each generator, but intersects only one nappe of the cone; **parabolas** when the plane is tilted further so that it is parallel to one (and only one) generator and intersects only one nappe of the cone; and **hyperbolas** when the plane intersects both nappes.

If the plane does contain the vertex, the intersection of the plane and the cone is a point, a line, or a pair of intersecting lines. These are usually called **degenerate conics.**

9.2 THE PARABOLA

1 Find the Equation of a Parabola

2 Graph Parabolas

3 Discuss the Equation of a Parabola

4 Work with Parabolas with Vertex at (h, k)

5 Solve Applied Problems Involving Properties of Parabolas

We stated earlier (Section 2.3) that the graph of a quadratic function is a parabola. In this section, we begin with a geometric definition of a parabola and use it to obtain an equation.

Figure 3

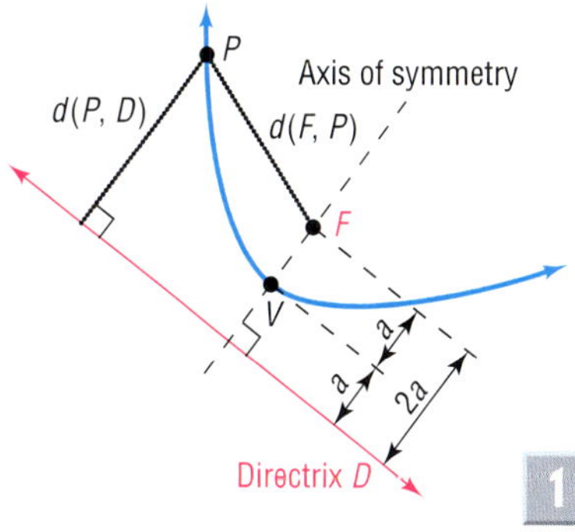

A **parabola** is defined as the collection of all points P in the plane that are the same distance from a fixed point F as they are from a fixed line D. The point F is called the **focus** of the parabola, and the line D is its **directrix.** As a result, a parabola is the set of points P for which

$$d(F, P) = d(P, D) \qquad (1)$$

1

Figure 3 shows a parabola. The line through the focus F and perpendicular to the directrix D is called the **axis of symmetry** of the parabola. The point of intersection of the parabola with its axis of symmetry is called the **vertex** V.

Because the vertex V lies on the parabola, it must satisfy equation (1): $d(F, V) = d(V, D)$. Thus, the vertex is midway between the focus and the directrix. We shall let a equal the distance $d(F, V)$ from F to V. Now we are ready to derive an equation for a parabola. To do this, we use a rectangular system of coordinates positioned so that the vertex V, focus F, and directrix D of the parabola are conveniently located. If we choose to locate the vertex V at the origin $(0, 0)$, then we can conveniently position the focus F on either the x-axis or the y-axis.

Figure 4
$y^2 = 4ax$

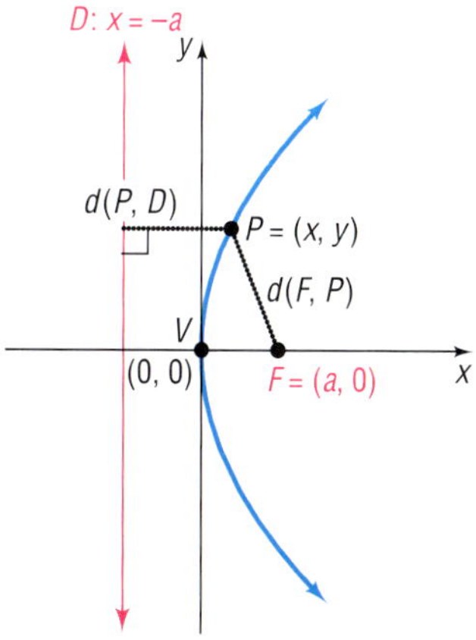

First, we consider the case where the focus F is on the positive x-axis, as shown in Figure 4. Because the distance from F to V is a, the coordinates of F will be $(a, 0)$ with $a > 0$. Similarly, because the distance from V to the directrix D is also a and because D must be perpendicular to the x-axis (since the x-axis is the axis of symmetry), the equation of the directrix D must be $x = -a$. Now, if $P = (x, y)$ is any point on the parabola, then P must obey equation (1):

$$d(F, P) = d(P, D)$$

So we have

$$\begin{aligned} \sqrt{(x-a)^2+y^2} &= |x+a| && \textit{Use the distance formula.}\\ (x-a)^2+y^2 &= (x+a)^2 && \textit{Square both sides.}\\ x^2-2ax+a^2+y^2 &= x^2+2ax+a^2 && \textit{Simplify.}\\ y^2 &= 4ax \end{aligned}$$

THEOREM

Equation of a Parabola; Vertex at (0, 0), Focus at (*a*, 0), *a* > 0

The equation of a parabola with vertex at (0, 0), focus at $(a, 0)$, and directrix $x = -a$, $a > 0$, is

$$y^2 = 4ax \qquad (2)$$

EXAMPLE 1 Finding the Equation of a Parabola

2

Find an equation of the parabola with vertex at (0, 0) and focus at (3, 0). Graph the equation by hand.

Solution The distance from the vertex (0, 0) to the focus (3, 0) is $a = 3$. Based on equation (2), the equation of this parabola is

$$\begin{aligned} y^2 &= 4ax\\ y^2 &= 12x \qquad a = 3. \end{aligned}$$

Figure 5
$y^2 = 12x$

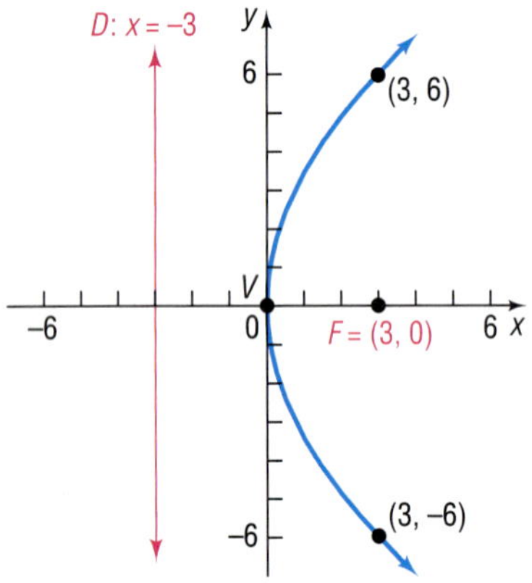

To graph this parabola by hand, it is helpful to plot the two points on the graph above and below the focus. To locate them, we let $x = 3$. Then

$$\begin{aligned} y^2 &= 12x = 12(3) = 36\\ y &= \pm 6 \qquad \textit{Solve for y.} \end{aligned}$$

The points on the parabola above and below the focus are $(3, -6)$ and $(3, 6)$. These points help in graphing the parabola because they determine the "opening." See Figure 5.

In general, the points on a parabola $y^2 = 4ax$ that lie above and below the focus $(a, 0)$ are each at a distance $2a$ from the focus. This follows from the fact that if $x = a$ then $y^2 = 4ax = 4a^2$, or $y = \pm 2a$. The line segment joining these two points is called the **latus rectum;** its length is $4a$.

NOW WORK PROBLEM 17.

EXAMPLE 2 Graphing a Parabola Using a Graphing Utility

Graph the parabola $y^2 = 12x$.

Figure 6
$y^2 = 12x$

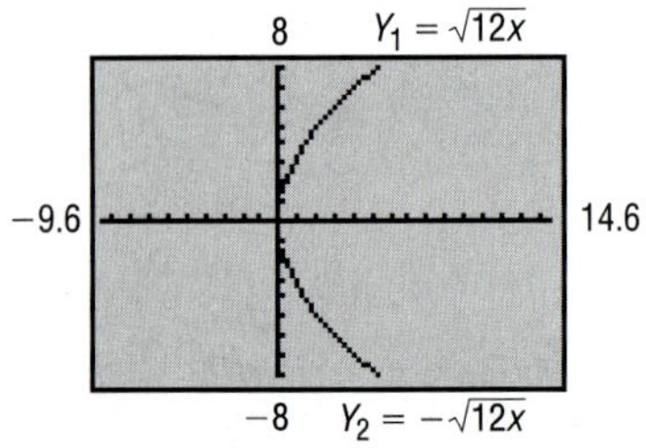

Solution To graph the parabola $y^2 = 12x$, we need to graph the two functions $Y_1 = \sqrt{12x}$ and $Y_2 = -\sqrt{12x}$ on a square screen. Figure 6 shows the graph of $y^2 = 12x$. Notice that the graph fails the vertical line test, so $y^2 = 12x$ is not a function.

By reversing the steps we used to obtain equation (2), it follows that the graph of an equation of the form of equation (2), $y^2 = 4ax$, is a parabola; its vertex is at $(0, 0)$, its focus is at $(a, 0)$, its directrix is the line $x = -a$, and its axis of symmetry is the x-axis.

3 For the remainder of this section, the direction "Discuss the equation" will mean to find the vertex, focus, and directrix of the parabola and graph it.

EXAMPLE 3 Discussing the Equation of a Parabola

Discuss the equation $y^2 = 8x$.

Solution Figure 7(a) shows the graph of $y^2 = 8x$ using a graphing utility. We now proceed to analyze the equation.

The equation $y^2 = 8x$ is of the form $y^2 = 4ax$, where $4a = 8$. Thus, $a = 2$. Consequently, the graph of the equation is a parabola with vertex at $(0, 0)$ and focus on the positive x-axis at $(2, 0)$. The directrix is the vertical line $x = -2$. The two points defining the latus rectum are obtained by letting $x = 2$. Then $y^2 = 16$, or $y = \pm 4$.

Figure 7
$y^2 = 8x$

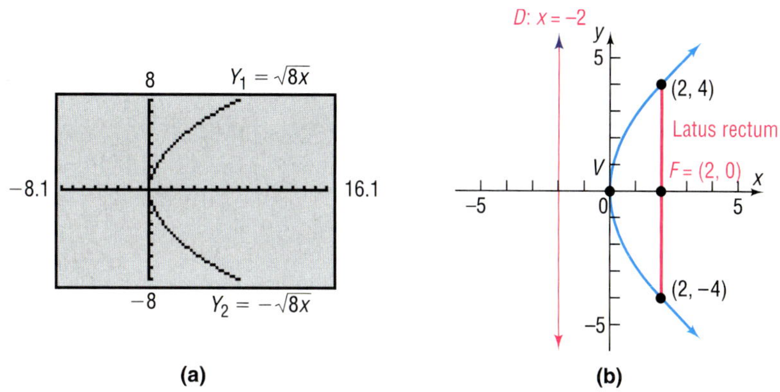

Recall that we arrived at equation (2) after placing the focus on the positive x-axis. If the focus is placed on the negative x-axis, positive y-axis, or negative y-axis, a different form of the equation for the parabola results. The four forms of the equation of a parabola with vertex at $(0, 0)$ and focus on a coordinate axis a distance a from $(0, 0)$ are given in Table 1, and their graphs are given in Figure 8. Notice that each graph is symmetric with respect to its axis of symmetry.

Table 1 Equations of a Parabola: Vertex at (0, 0); Focus on Axis; $a > 0$

Vertex	Focus	Directrix	Equation	Description
$(0, 0)$	$(a, 0)$	$x = -a$	$y^2 = 4ax$	Parabola, axis of symmetry is the x-axis, opens to right
$(0, 0)$	$(-a, 0)$	$x = a$	$y^2 = -4ax$	Parabola, axis of symmetry is the x-axis, opens to left
$(0, 0)$	$(0, a)$	$y = -a$	$x^2 = 4ay$	Parabola, axis of symmetry is the y-axis, opens up
$(0, 0)$	$(0, -a)$	$y = a$	$x^2 = -4ay$	Parabola, axis of symmetry is the y-axis, opens down

Figure 8

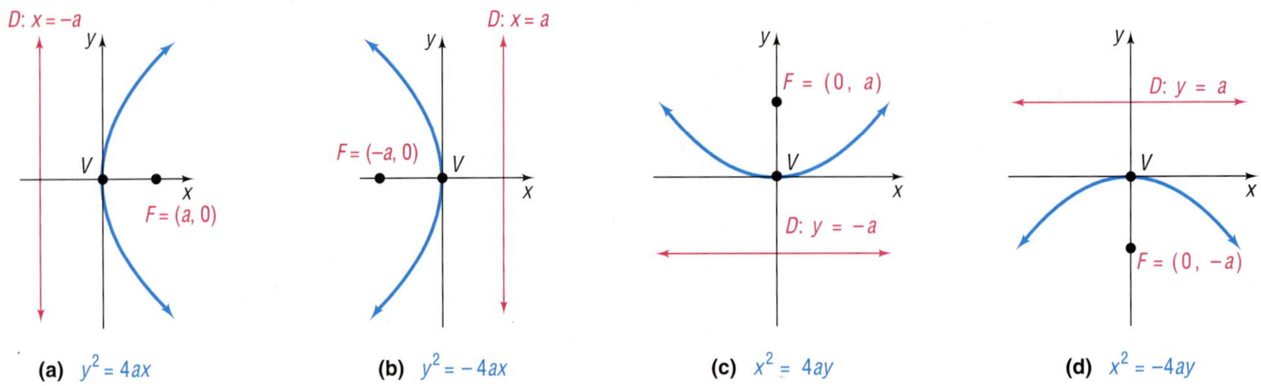

EXAMPLE 4 Discussing the Equation of a Parabola

Discuss the equation $x^2 = -12y$.

Solution Figure 9(a) shows the graph of $x^2 = -12y$ using a graphing utility. We now proceed to analyze the equation.

The equation $x^2 = -12y$ is of the form $x^2 = -4ay$, with $a = 3$. Consequently, the graph of the equation is a parabola with vertex at $(0, 0)$, focus at $(0, -3)$, and directrix the line $y = 3$. The parabola opens down, and its axis of symmetry is the y-axis. To obtain the points defining the latus rectum, let $y = -3$. Then $x^2 = 36$, or $x = \pm 6$. See Figure 9(b).

Figure 9
$x^2 = -12y$

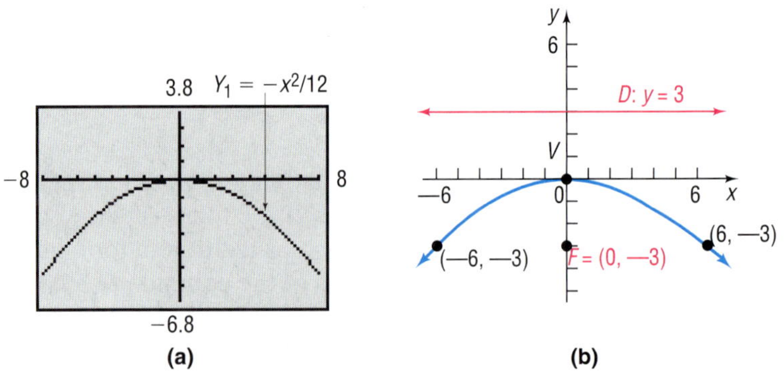

NOW WORK PROBLEM 35.

EXAMPLE 5 Finding the Equation of a Parabola

Find the equation of the parabola with focus at $(0, 4)$ and directrix the line $y = -4$. Graph the equation by hand.

Solution A parabola whose focus is at $(0, 4)$ and whose directrix is the horizontal line $y = -4$ will have its vertex at $(0, 0)$. (Do you see why? The vertex is midway between the focus and the directrix.) Since the vertex is on the positive y-axis, the equation of this parabola is of the form $x^2 = 4ay$, with $a = 4$; that is,

$$x^2 = 4ay = 4(4)y = 16y$$
$$\uparrow$$
$$a = 4$$

Figure 10 shows the graph.

Figure 10
$x^2 = 16y$

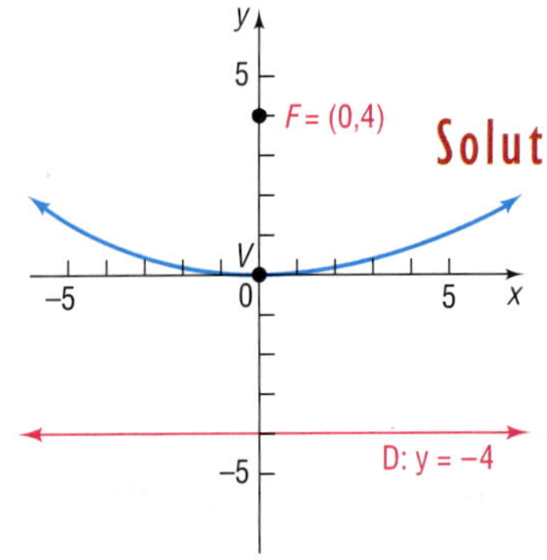

EXAMPLE 6 Finding the Equation of a Parabola

Find the equation of a parabola with vertex at (0, 0) if its axis of symmetry is the x-axis and its graph contains the point $\left(-\frac{1}{2}, 2\right)$. Find its focus and directrix, and graph the equation by hand.

Solution Because the vertex is at the origin, the axis of symmetry is the x-axis, and the graph contains a point in the second quadrant, we see from Table 1 that the form of the equation is

$$y^2 = -4ax$$

Because the point $\left(-\frac{1}{2}, 2\right)$ is on the parabola, the coordinates $x = -\frac{1}{2}$, $y = 2$ must satisfy the equation. Putting $x = -\frac{1}{2}$ and $y = 2$ into the equation, we find that

$$4 = -4a\left(-\tfrac{1}{2}\right)$$
$$a = 2$$

Thus, the equation of the parabola is

$$y^2 = -4(2)x = -8x$$

The focus is at $(-2, 0)$ and the directrix is the line $x = 2$. Letting $x = -2$, we find $y^2 = 16$, or $y = \pm 4$. The points $(-2, 4)$ and $(-2, -4)$ define the latus rectum. See Figure 11.

Figure 11
$y^2 = -8x$

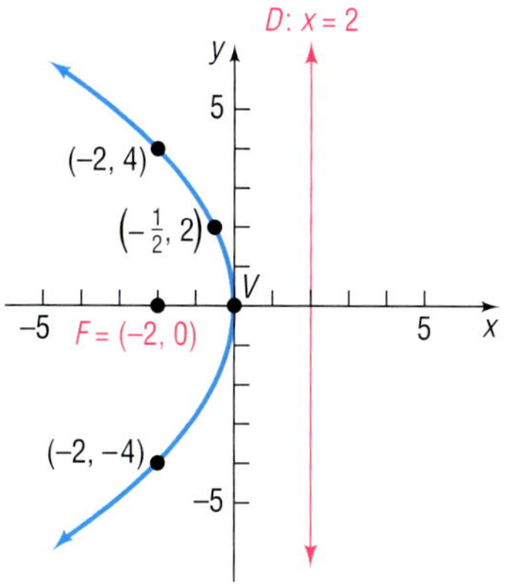

NOW WORK PROBLEM 27.

Vertex at (h, k)

4 If a parabola with vertex at the origin and axis of symmetry along a coordinate axis is shifted horizontally h units and then vertically k units, the result is a parabola with vertex at (h, k) and axis of symmetry parallel to a coordinate axis. The equations of such parabolas have the same forms as those in Table 1, but with x replaced by $x - h$ (the horizontal shift) and y replaced by $y - k$ (the vertical shift). Table 2 gives the forms of the equations of such parabolas. Figure 12(a)–(d) illustrates the graphs for $h > 0$, $k > 0$.

Table 2 *Parabolas with Vertex at (h, k), Axis of Symmetry Parallel to a Coordinate Axis, $a > 0$*

Vertex	Focus	Directrix	Equation	Description
(h, k)	$(h + a, k)$	$x = h - a$	$(y - k)^2 = 4a(x - h)$	Parabola, axis of symmetry parallel to x-axis, opens to right
(h, k)	$(h - a, k)$	$x = h + a$	$(y - k)^2 = -4a(x - h)$	Parabola, axis of symmetry parallel to x-axis, opens to left
(h, k)	$(h, k + a)$	$y = k - a$	$(x - h)^2 = 4a(y - k)$	Parabola, axis of symmetry parallel to y-axis, opens up
(h, k)	$(h, k - a)$	$y = k + a$	$(x - h)^2 = -4a(y - k)$	Parabola, axis of symmetry parallel to y-axis, opens down

Figure 12

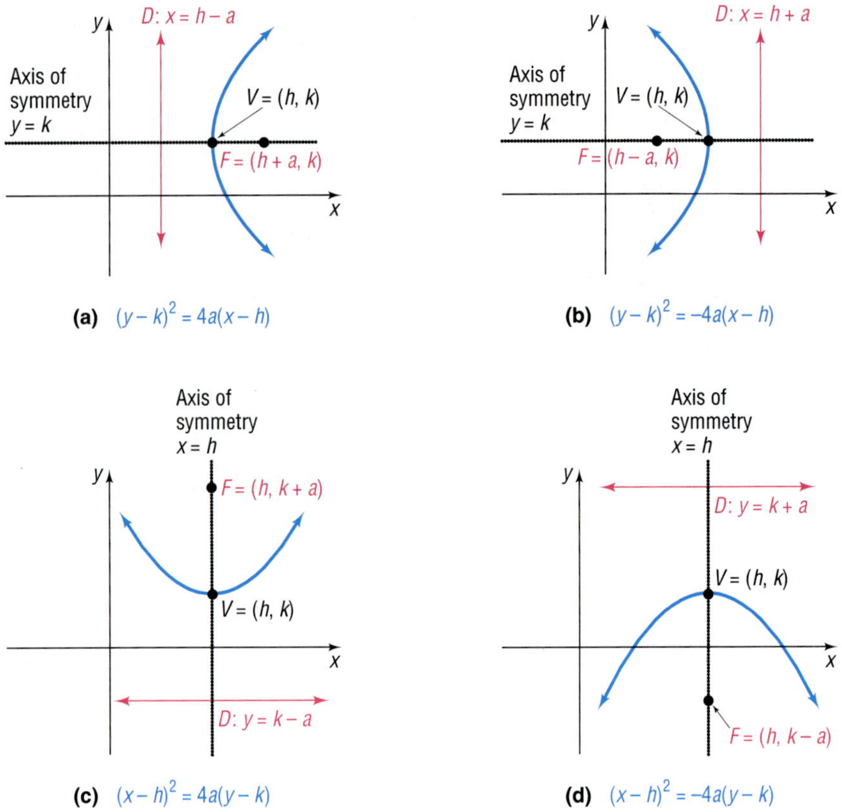

EXAMPLE 7 Finding the Equation of a Parabola, Vertex Not at Origin

Find an equation of the parabola with vertex at $(-2, 3)$ and focus at $(0, 3)$. Graph the equation by hand.

Solution The vertex $(-2, 3)$ and focus $(0, 3)$ both lie on the horizontal line $y = 3$ (the axis of symmetry). The distance a from the vertex $(-2, 3)$ to the focus $(0, 3)$ is $a = 2$. Also, because the focus lies to the right of the vertex, we know that the parabola opens to the right. Consequently, the form of the equation is

$$(y - k)^2 = 4a(x - h)$$

where $(h, k) = (-2, 3)$ and $a = 2$. Therefore, the equation is

$$(y - 3)^2 = 4 \cdot 2[x - (-2)]$$
$$(y - 3)^2 = 8(x + 2)$$

If $x = 0$, then $(y - 3)^2 = 16$. Thus, $y - 3 = \pm 4$, and $y = -1$ or $y = 7$. The points $(0, -1)$ and $(0, 7)$ define the latus rectum; the line $x = -4$ is the directrix. See Figure 13.

Figure 13
$(y - 3)^2 = 8(x + 2)$

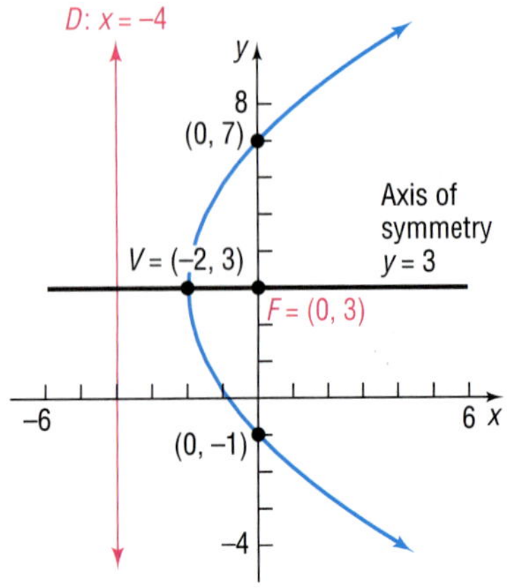

NOW WORK PROBLEM 25.

EXAMPLE 8 Using a Graphing Utility to Graph a Parabola, Vertex Not at Origin

Using a graphing utility, graph the equation $(y - 3)^2 = 8(x + 2)$.

Solution First, we must solve the equation for y.

$$(y-3)^2 = 8(x+2)$$
$$y - 3 = \pm\sqrt{8(x+2)} \quad \textit{Take the square root of each side.}$$
$$y = 3 \pm \sqrt{8(x+2)} \quad \textit{Add 3 to both sides.}$$

Figure 14 shows the graphs of the equations $Y_1 = 3 + \sqrt{8(x+2)}$ and $Y_2 = 3 - \sqrt{8(x+2)}$.

Figure 14

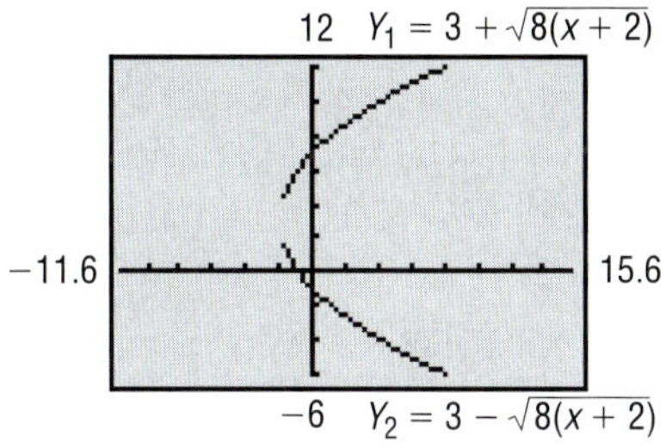

NOW WORK PROBLEM 39.

Polynomial equations define parabolas whenever they involve two variables that are quadratic in one variable and linear in the other. To discuss this type of equation, we first complete the square of the variable that is quadratic.

EXAMPLE 9 Discussing the Equation of a Parabola

Discuss the equation $x^2 + 4x - 4y = 0$.

Solution Figure 15(a) shows the graph of $x^2 + 4x - 4y = 0$ using a graphing utility. We now proceed to analyze the equation.

To discuss the equation $x^2 + 4x - 4y = 0$, we complete the square involving the variable x.

$$x^2 + 4x - 4y = 0$$
$$x^2 + 4x = 4y \quad \textit{Isolate the terms involving x on the left side.}$$
$$x^2 + 4x + 4 = 4y + 4 \quad \textit{Complete the square on the left side.}$$
$$(x + 2)^2 = 4(y + 1) \quad \textit{Factor.}$$

This equation is of the form $(x - h)^2 = 4a(y - k)$, with $h = -2$, $k = -1$, and $a = 1$. The graph is a parabola with vertex at $(h, k) = (-2, -1)$ that opens up. The focus is at $(-2, 0)$, and the directrix is the line $y = -2$. See Figure 15(b).

Figure 15
$x^2 + 4x - 4y = 0$

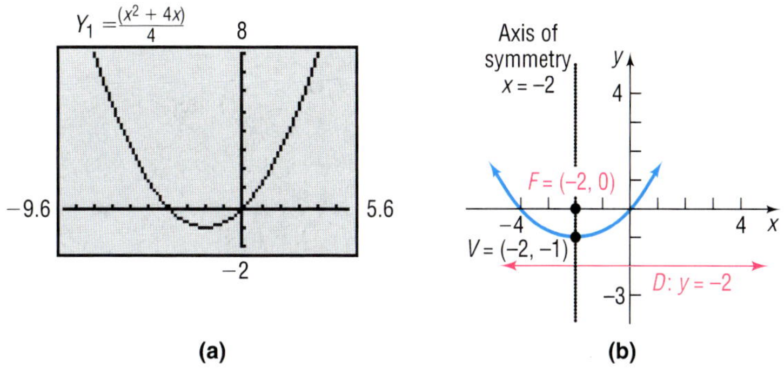

5

Parabolas find their way into many applications. For example, as we discussed in Section 2.4, suspension bridges have cables in the shape of a parabola. Another property of parabolas that is used in applications is their reflecting property.

Figure 16
Searchlight

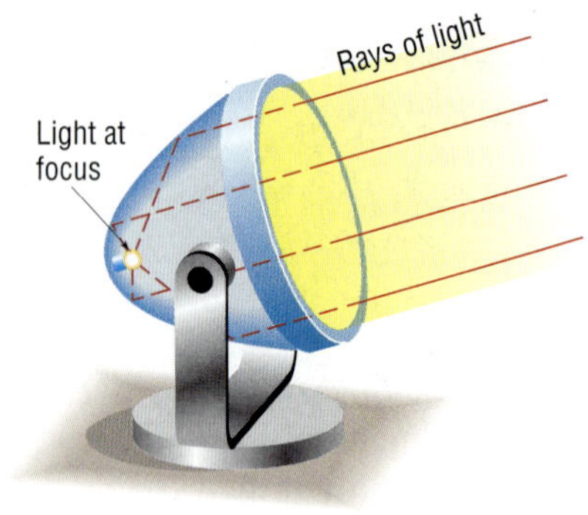

5 Reflecting Property

Suppose that a mirror is shaped like a **paraboloid of revolution,** a surface formed by rotating a parabola about its axis of symmetry. If a light (or any other emitting source) is placed at the focus of the parabola, all the rays emanating from the light will reflect off the mirror in lines parallel to the axis of symmetry. This principle is used in the design of searchlights, flashlights, certain automobile headlights, and other such devices. See Figure 16.

Conversely, suppose that rays of light (or other signals) emanate from a distant source so that they are essentially parallel. When these rays strike the surface of a parabolic mirror whose axis of symmetry is parallel to these rays, they are reflected to a single point at the focus. This principle is used in the design of some solar energy devices, satellite dishes, and the mirrors used in some types of telescopes. See Figure 17.

Figure 17

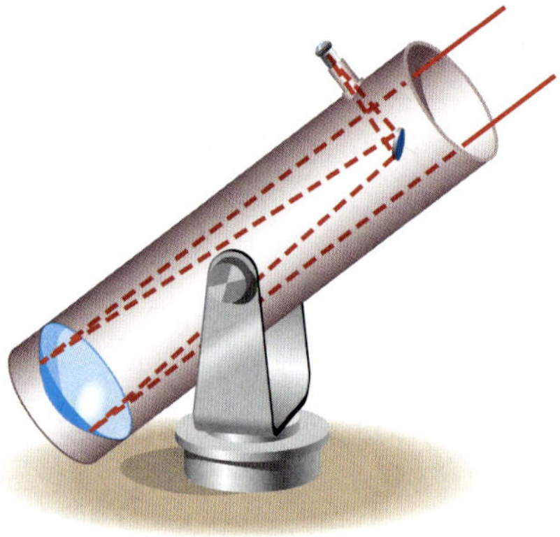

EXAMPLE 10 Satellite Dish

A satellite dish is shaped like a paraboloid of revolution. The signals that emanate from a satellite strike the surface of the dish and are reflected to a single point, where the receiver is located. If the dish is 8 feet across at its opening and is 3 feet deep at its center, at what position should the receiver be placed?

Figure 18

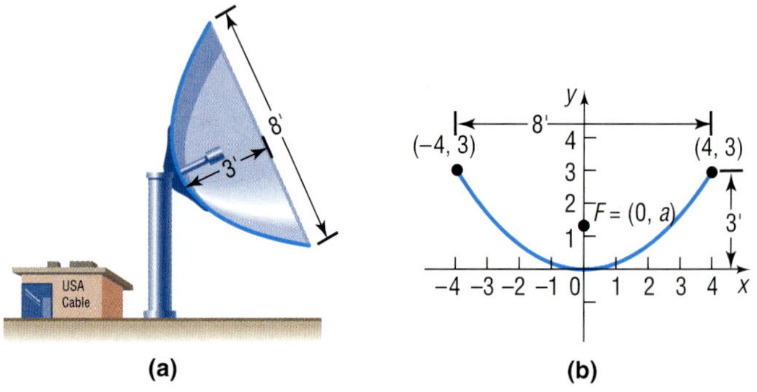

Solution Figure 18(a) shows the satellite dish. We draw the parabola used to form the dish on a rectangular coordinate system so that the vertex of the parabola is at the origin and its focus is on the positive y-axis. See Figure 18(b). The form of the equation of the parabola is

$$x^2 = 4ay$$

and its focus is at $(0, a)$. Since $(4, 3)$ is a point on the graph, we have

$$4^2 = 4a(3)$$
$$a = \frac{4}{3}$$

The receiver should be located $1\frac{1}{3}$ feet from the base of the dish, along its axis of symmetry.

9.2 EXERCISES

In Problems 1–8, the graph of a parabola is given. Match each graph to its equation.

A. $y^2 = 4x$ *B.* $x^2 = 4y$ *C.* $y^2 = -4x$

D. $x^2 = -4y$ *E.* $(y - 1)^2 = 4(x - 1)$ *F.* $(x + 1)^2 = 4(y + 1)$

G. $(y - 1)^2 = -4(x - 1)$ *H.* $(x + 1)^2 = -4(y + 1)$

1.

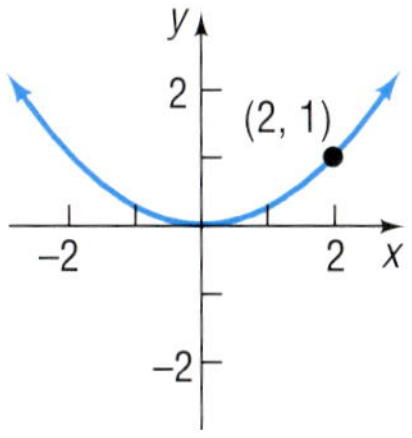

2.

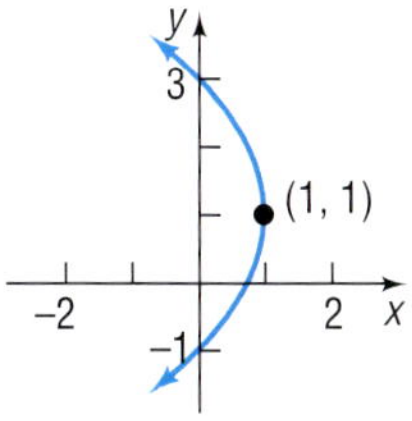

3.

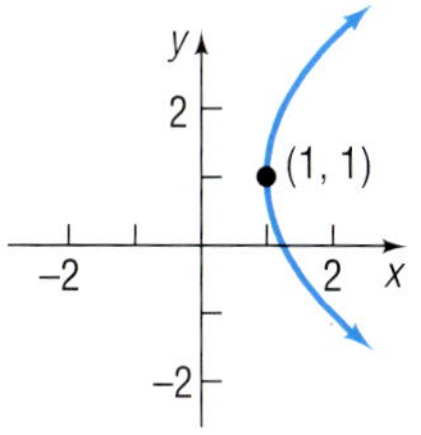

4.

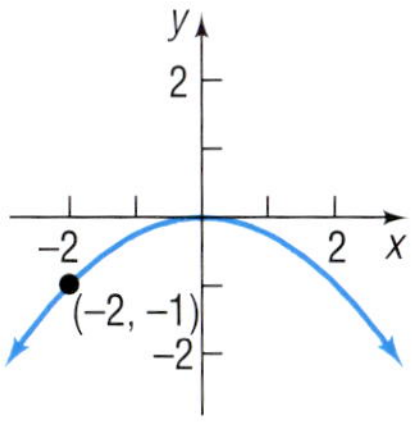

5.

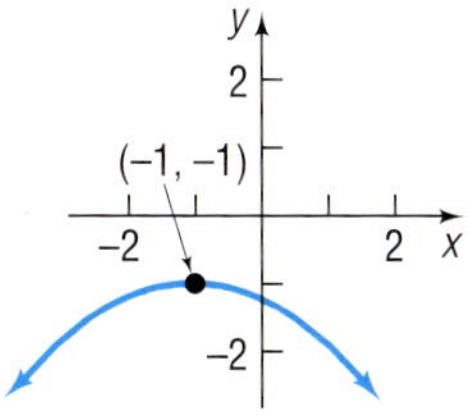

6.

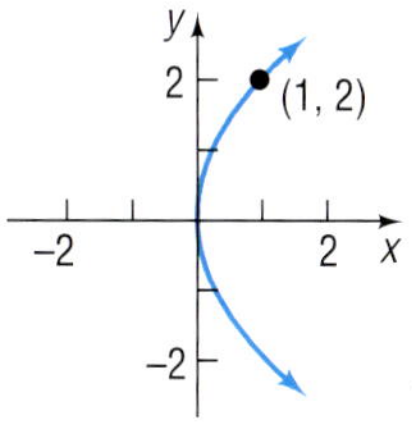

7.

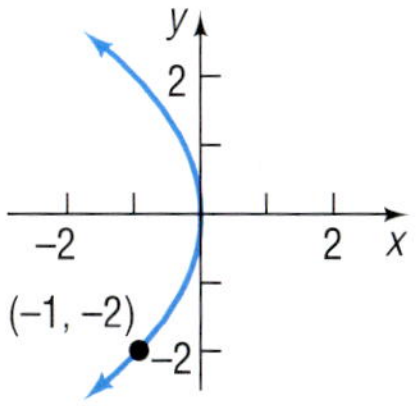

8.

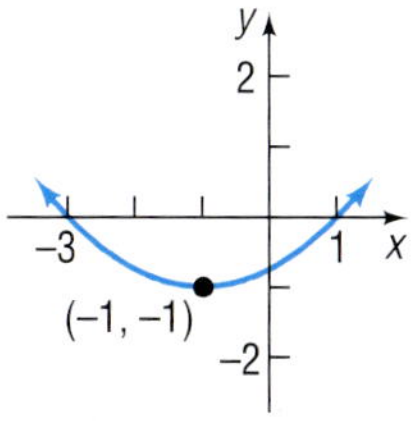

In Problems 9–16, the graph of a parabola is given. Match each graph to its equation.

A. $x^2 = 6y$ *B.* $x^2 = -6y$ *C.* $y^2 = 6x$

D. $y^2 = -6x$ *E.* $(y - 2)^2 = -6(x + 2)$ *F.* $(y - 2)^2 = 6(x + 2)$

G. $(x + 2)^2 = -6(y - 2)$ *H.* $(x + 2)^2 = 6(y - 2)$

9.

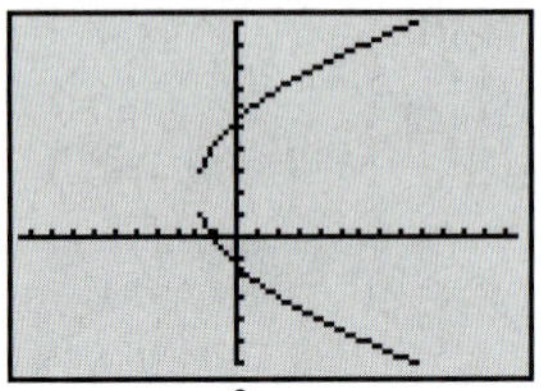

10.

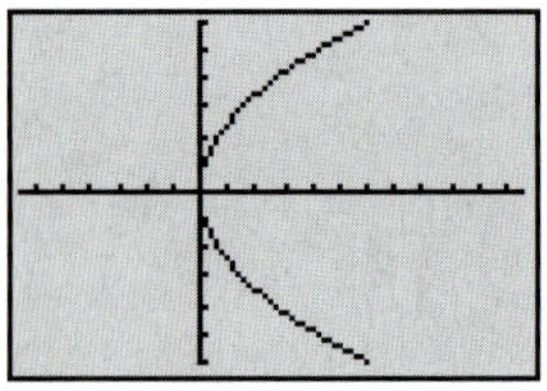

11.

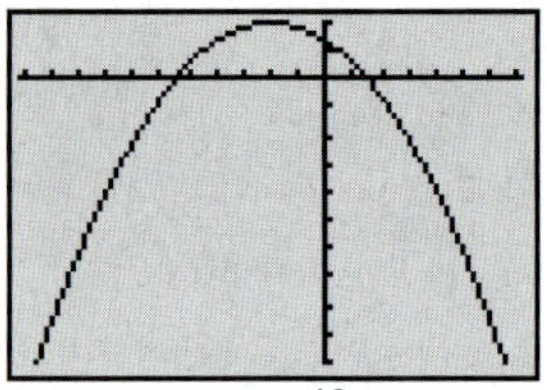

12.

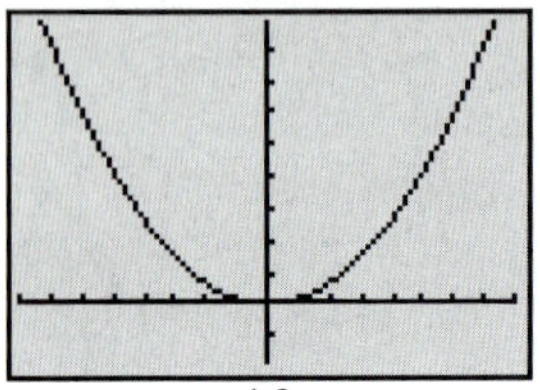

13.

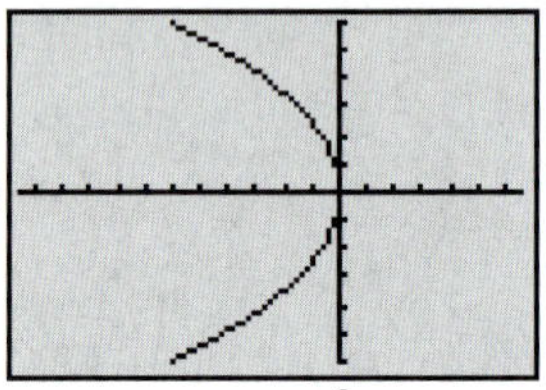

14.

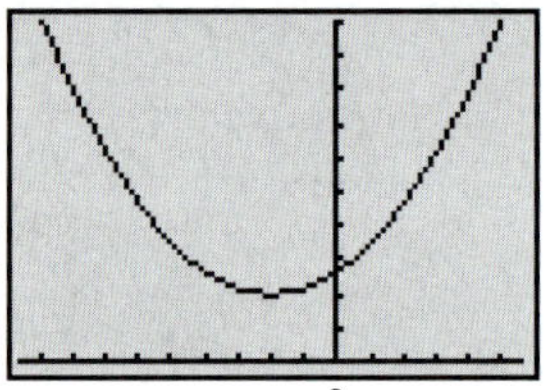

15.

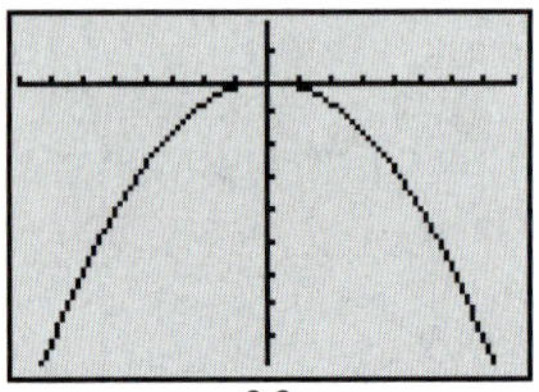

16.

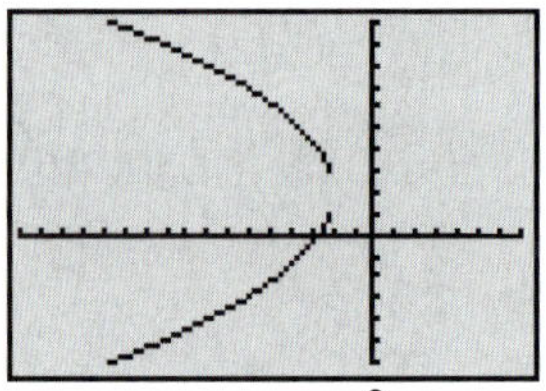

In Problems 17–32, find the equation of the parabola described. Find the two points that define the latus rectum, and graph the equation by hand.

17. Focus at $(4, 0)$; vertex at $(0, 0)$

18. Focus at $(0, 2)$; vertex at $(0, 0)$

19. Focus at $(0, -3)$; vertex at $(0, 0)$

20. Focus at $(-4, 0)$; vertex at $(0, 0)$

21. Focus at $(-2, 0)$; directrix the line $x = 2$

22. Focus at $(0, -1)$; directrix the line $y = 1$

23. Directrix the line $y = -\frac{1}{2}$; vertex at $(0, 0)$

24. Directrix the line $x = -\frac{1}{2}$; vertex at $(0, 0)$

25. Vertex at $(2, -3)$; focus at $(2, -5)$

26. Vertex at $(4, -2)$; focus at $(6, -2)$

27. Vertex at $(0, 0)$; axis of symmetry the y-axis; containing the point $(2, 3)$

28. Vertex at $(0, 0)$; axis of symmetry the x-axis; containing the point $(2, 3)$

29. Focus at $(-3, 4)$; directrix the line $y = 2$

30. Focus at $(2, 4)$; directrix the line $x = -4$

31. Focus at $(-3, -2)$; directrix the line $x = 1$

32. Focus at $(-4, 4)$; directrix the line $y = -2$

In Problems 33–50, find the vertex, focus, and directrix of each parabola. Graph the equation using a graphing utility.

33. $x^2 = 4y$

34. $y^2 = 8x$

35. $y^2 = -16x$

36. $x^2 = -4y$

37. $(y - 2)^2 = 8(x + 1)$

38. $(x + 4)^2 = 16(y + 2)$

39. $(x - 3)^2 = -(y + 1)$

40. $(y + 1)^2 = -4(x - 2)$

41. $(y + 3)^2 = 8(x - 2)$

42. $(x - 2)^2 = 4(y - 3)$

43. $y^2 - 4y + 4x + 4 = 0$

44. $x^2 + 6x - 4y + 1 = 0$

45. $x^2 + 8x = 4y - 8$

46. $y^2 - 2y = 8x - 1$

47. $y^2 + 2y - x = 0$

48. $x^2 - 4x = 2y$

49. $x^2 - 4x = y + 4$

50. $y^2 + 12y = -x + 1$

In Problems 51–58, write an equation for each parabola.

51.

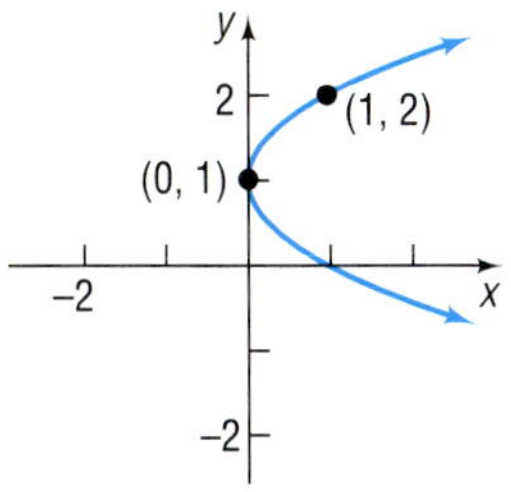

52.

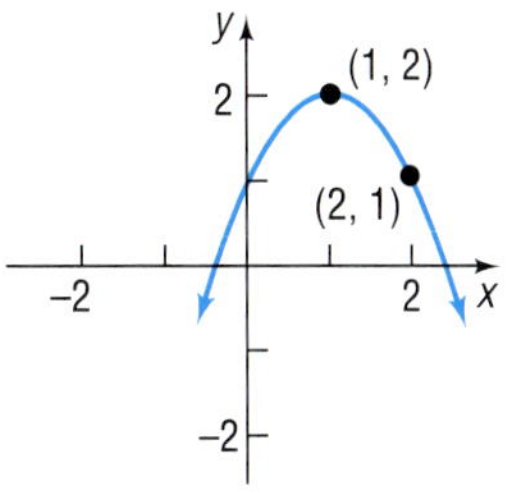

53.

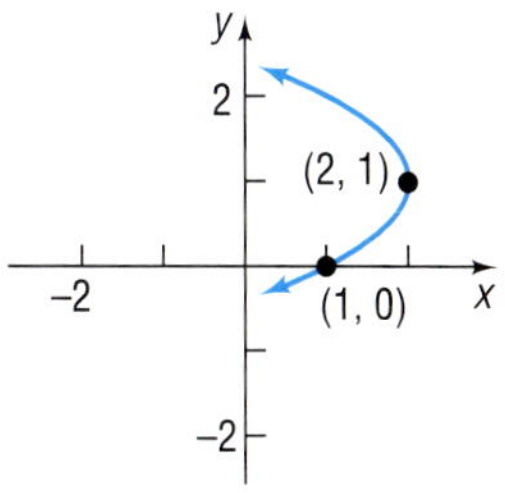

54.

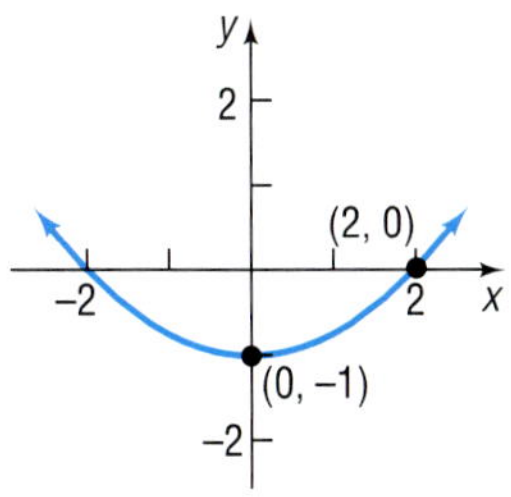

55.

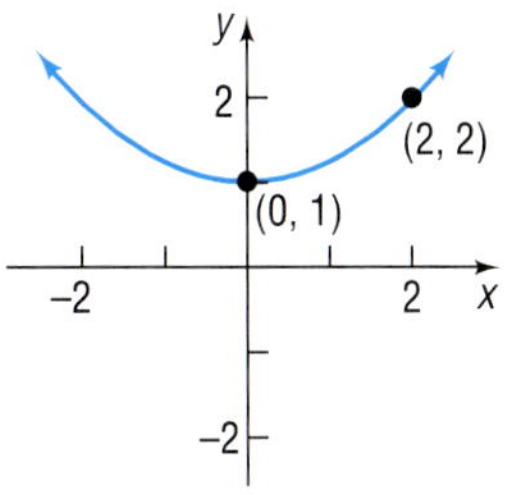

56.

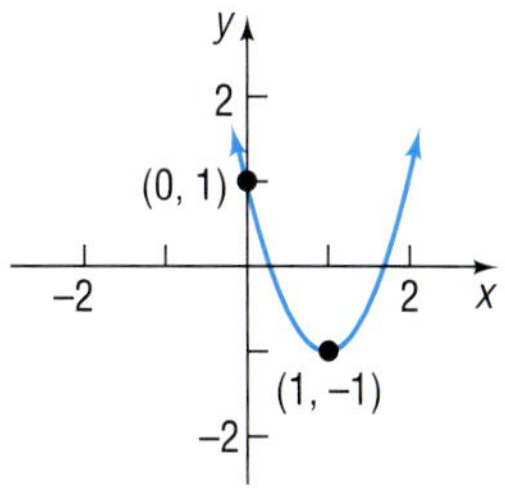

57.

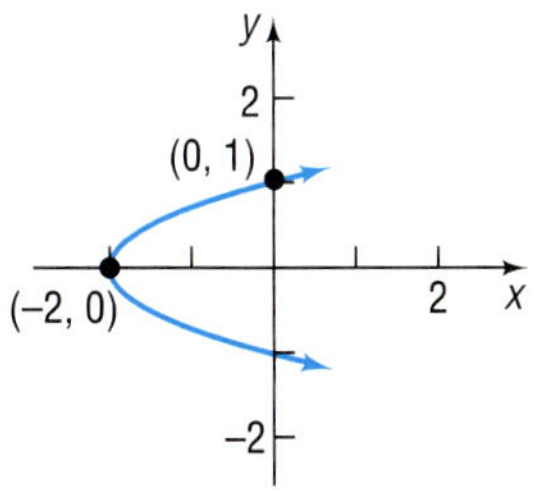

58.

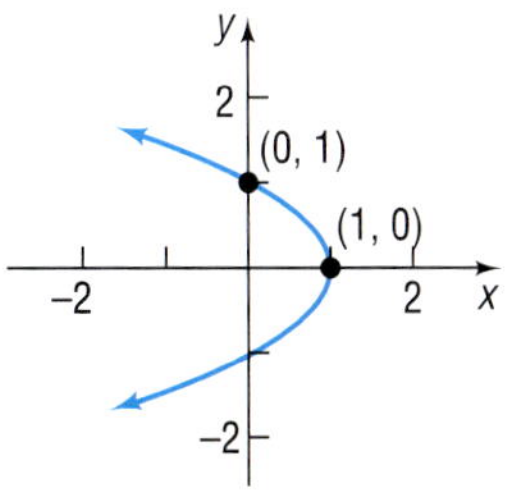

59. **Satellite Dish** A satellite dish is shaped like a paraboloid of revolution. The signals that emanate from a satellite strike the surface of the dish and are reflected to a single point, where the receiver is located. If the dish is 10 feet across at its opening and is 4 feet deep at its center, at what position should the receiver be placed?

60. **Constructing a TV Dish** A cable TV receiving dish is in the shape of a paraboloid of revolution. Find the location of the receiver, which is placed at the focus, if the dish is 6 feet across at its opening and 2 feet deep.

61. **Constructing a Flashlight** The reflector of a flashlight is in the shape of a paraboloid of revolution. Its diameter is 4 inches and its depth is 1 inch. How far from the vertex should the light bulb be placed so that the rays will be reflected parallel to the axis?

62. **Constructing a Headlight** A sealed-beam headlight is in the shape of a paraboloid of revolution. The bulb, which is placed at the focus, is 1 inch from the vertex. If the depth is to be 2 inches, what is the diameter of the headlight at its opening?

63. **Suspension Bridge** The cables of a suspension bridge are in the shape of a parabola, as shown in the figure. The towers supporting the cable are 600 feet apart and 80 feet high. If the cables touch the road surface midway between the towers, what is the height of the cable at a point 150 feet from the center of the bridge?

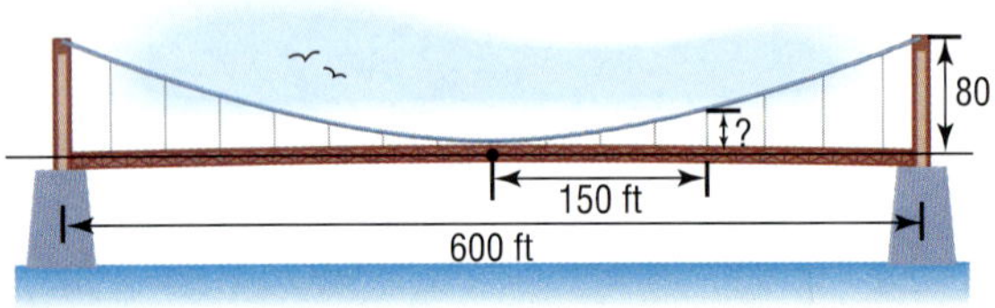

64. **Suspension Bridge** The cables of a suspension bridge are in the shape of a parabola. The towers supporting the cable are 400 feet apart and 100 feet high. If the cables are at a height of 10 feet midway between the towers, what is the height of the cable at a point 50 feet from the center of the bridge?

65. **Searchlight** A searchlight is shaped like a paraboloid of revolution. If the light source is located 2 feet from the base along the axis of symmetry and the opening is 5 feet across, how deep should the searchlight be?

66. **Searchlight** A searchlight is shaped like a paraboloid of revolution. If the light source is located 2 feet from the base along the axis of symmetry and the depth of the searchlight is 4 feet, what should the width of the opening be?

67. **Solar Heat** A mirror is shaped like a paraboloid of revolution and will be used to concentrate the rays of the sun at its focus, creating a heat source. If the mirror is 20 feet across at its opening and is 6 feet deep, where will the heat source be concentrated?

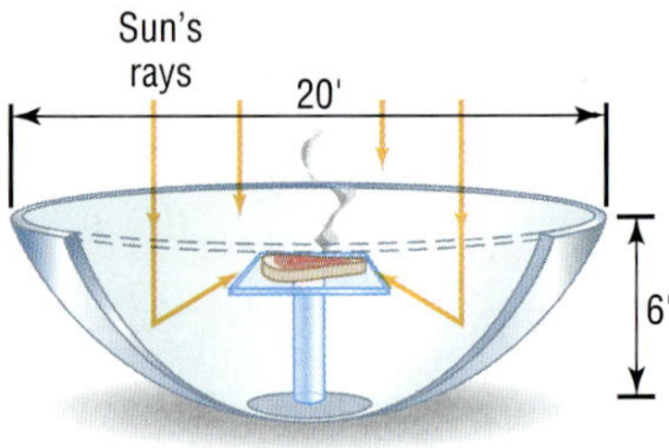

68. **Reflecting Telescope** A reflecting telescope contains a mirror shaped like a paraboloid of revolution. If the mirror is 4 inches across at its opening and is 3 feet deep, where will the collected light be concentrated?

69. **Parabolic Arch Bridge** A bridge is built in the shape of a parabolic arch. The bridge has a span of 120 feet and a maximum height of 25 feet. See the illustration. Choose a suitable rectangular coordinate system and find the height of the arch at distances of 10, 30, and 50 feet from the center.

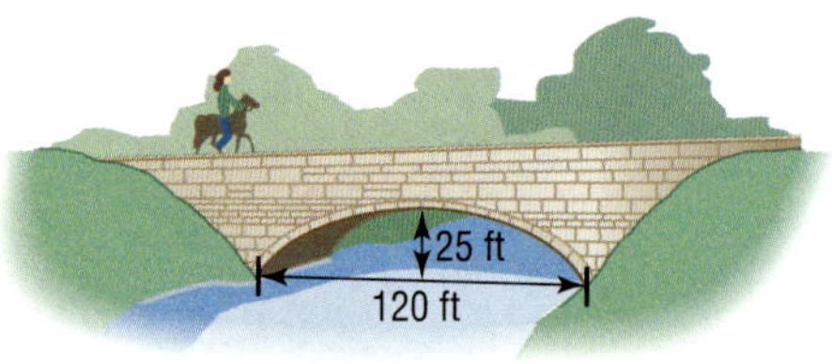

70. **Parabolic Arch Bridge** A bridge is to be built in the shape of a parabolic arch and is to have a span of 100 feet. The height of the arch a distance of 40 feet from the center is to be 10 feet. Find the height of the arch at its center.

71. Show that an equation of the form

$$Ax^2 + Ey = 0 \qquad A \neq 0, E \neq 0$$

is the equation of a parabola with vertex at $(0, 0)$ and axis of symmetry the y-axis. Find its focus and directrix.

72. Show that an equation of the form

$$Cy^2 + Dx = 0 \qquad C \neq 0, D \neq 0$$

is the equation of a parabola with vertex at $(0, 0)$ and axis of symmetry the x-axis. Find its focus and directrix.

73. Show that the graph of an equation of the form

$$Ax^2 + Dx + Ey + F = 0 \qquad A \neq 0$$

(a) Is a parabola if $E \neq 0$.
(b) Is a vertical line if $E = 0$ and $D^2 - 4AF = 0$.
(c) Is two vertical lines if $E = 0$ and $D^2 - 4AF > 0$.
(d) Contains no points if $E = 0$ and $D^2 - 4AF < 0$.

74. Show that the graph of an equation of the form

$$Cy^2 + Dx + Ey + F = 0 \qquad C \neq 0$$

(a) Is a parabola if $D \neq 0$.
(b) Is a horizontal line if $D = 0$ and $E^2 - 4CF = 0$.
(c) Is two horizontal lines if $D = 0$ and $E^2 - 4CF > 0$.
(d) Contains no points if $D = 0$ and $E^2 - 4CF < 0$.

9.3 THE ELLIPSE

1 Find the Equation of an Ellipse

2 Graph Ellipses

3 Discuss the Equation of an Ellipse

4 Work with Ellipses with Center at (h, k)

5 Solve Applied Problems Involving Properties of Ellipses

An **ellipse** is the collection of all points in the plane the sum of whose distances from two fixed points, called the **foci,** is a constant.

Figure 19

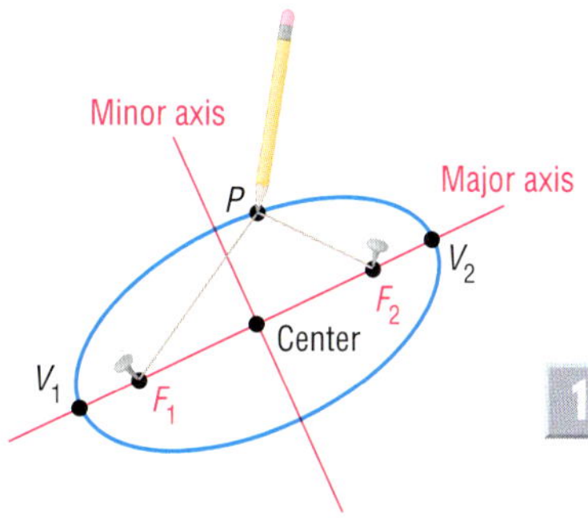

The definition actually contains within it a physical means for drawing an ellipse. Find a piece of string (the length of this string is the constant referred to in the definition). Then take two thumbtacks (the foci) and stick them on a piece of cardboard so that the distance between them is less than the length of the string. Now attach the ends of the string to the thumbtacks and, using the point of a pencil, pull the string taut. Keeping the string taut, rotate the pencil around the two thumbtacks. The pencil traces out an ellipse, as shown in Figure 19.

1

In Figure 19, the foci are labeled F_1 and F_2. The line containing the foci is called the **major axis.** The midpoint of the line segment joining the foci is called the **center** of the ellipse. The line through the center and perpendicular to the major axis is called the **minor axis.**

The two points of intersection of the ellipse and the major axis are the **vertices,** V_1 and V_2, of the ellipse. The distance from one vertex to the other is called the **length of the major axis.** The ellipse is symmetric with respect to its major axis and with respect to its minor axis.

Figure 20
$d(F_1, P) + d(F_2, P) = 2a$

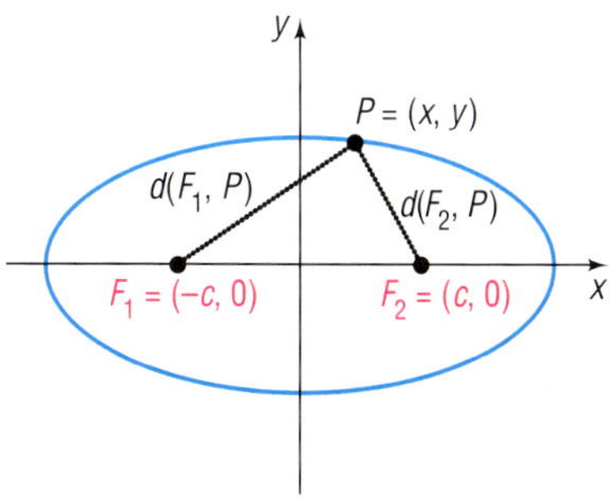

With these ideas in mind, we are now ready to find the equation of an ellipse in a rectangular coordinate system. First, we place the center of the ellipse at the origin. Second, we position the ellipse so that its major axis coincides with a coordinate axis. Suppose that the major axis coincides with the x-axis, as shown in Figure 20. If c is the distance from the center to a focus, then one focus will be at $F_1 = (-c, 0)$ and the other at $F_2 = (c, 0)$. As we shall see, it is convenient to let $2a$ denote the constant distance referred to in the definition. Thus, if $P = (x, y)$ is any point on the ellipse, we have

$$d(F_1, P) + d(F_2, P) = 2a$$ *Sum of the distances from P to the foci equals a constant, 2a.*

$$\sqrt{(x+c)^2 + y^2} + \sqrt{(x-c)^2 + y^2} = 2a$$ *Use the distance formula.*

$$\sqrt{(x+c)^2 + y^2} = 2a - \sqrt{(x-c)^2 + y^2}$$ *Isolate one radical.*

$$(x+c)^2 + y^2 = 4a^2 - 4a\sqrt{(x-c)^2 + y^2} + (x-c)^2 + y^2$$ *Square both sides.*

$$x^2 + 2cx + c^2 + y^2 = 4a^2 - 4a\sqrt{(x-c)^2 + y^2} + x^2 - 2cx + c^2 + y^2$$ *Simplify.*

$$4cx - 4a^2 = -4a\sqrt{(x-c)^2 + y^2}$$ *Isolate the radical.*

$$cx - a^2 = -a\sqrt{(x-c)^2 + y^2}$$ *Divide each side by 4.*

$$c^2x^2 - 2a^2cx + a^4 = a^2[(x-c)^2 + y^2]$$ *Square both sides again.*

$$c^2x^2 - 2a^2cx + a^4 = a^2(x^2 - 2cx + c^2 + y^2)$$ *Simplify.*

$$(c^2 - a^2)x^2 - a^2y^2 = a^2c^2 - a^4$$ *Rearrange the terms.*

$$(a^2 - c^2)x^2 + a^2y^2 = a^2(a^2 - c^2) \qquad (1)$$ *Multiply each side by -1; factor a^2 on the right side.*

To obtain points on the ellipse off the x-axis, it must be that $a > c$. To see why, look again at Figure 20:

$$d(F_1, P) + d(F_2, P) > d(F_1, F_2)$$ *The sum of the lengths of two sides of a triangle is greater than the length of the third side.*

$$2a > 2c$$ *$d(F_1, P) + d(F_2, P) = 2a$; $d(F_1, F_2) = 2c$.*

$$a > c$$

Since $a > c$, we also have $a^2 > c^2$, so $a^2 - c^2 > 0$. Let $b^2 = a^2 - c^2$, $b > 0$. Then $a > b$ and equation (1) can be written as

$$b^2x^2 + a^2y^2 = a^2b^2$$

$$\frac{x^2}{a^2} + \frac{y^2}{b^2} = 1$$ *Divide each side by a^2b^2.*

THEOREM

Equation of an Ellipse; Center at (0, 0); Foci at ($\pm c$, 0); Major Axis along the x-Axis

An equation of the ellipse with center at $(0, 0)$ and foci at $(-c, 0)$ and $(c, 0)$ is

$$\frac{x^2}{a^2} + \frac{y^2}{b^2} = 1 \quad \text{where } a > b > 0 \text{ and } b^2 = a^2 - c^2 \qquad (2)$$

The major axis is the x-axis.

2 As you can verify, the ellipse defined by equation (2) is symmetric with respect to the x-axis, y-axis, and origin.

To find the vertices of the ellipse defined by equation (2), let $y = 0$. The vertices satisfy the equation $x^2/a^2 = 1$, the solutions of which are

Figure 21

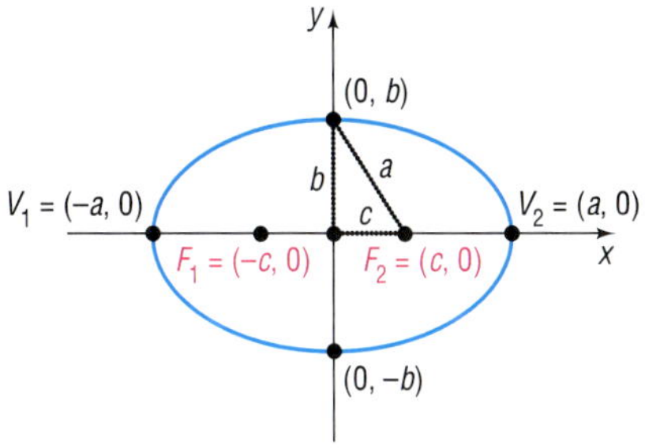

$x = \pm a$. Consequently, the vertices of the ellipse given by equation (2) are $V_1 = (-a, 0)$ and $V_2 = (a, 0)$. The y-intercepts of the ellipse, found by letting $x = 0$, have coordinates $(0, -b)$ and $(0, b)$. These four intercepts, $(a, 0)$, $(-a, 0)$, $(0, b)$, and $(0, -b)$, are used to graph the ellipse by hand. See Figure 21.

Notice in Figure 21 the right triangle formed with the points $(0, 0)$, $(c, 0)$, and $(0, b)$. Because $b^2 = a^2 - c^2$ (or $b^2 + c^2 = a^2$), the distance from the focus at $(c, 0)$ to the point $(0, b)$ is a.

EXAMPLE 1 Finding an Equation of an Ellipse

Find an equation of the ellipse with center at the origin, one focus at $(3, 0)$, and a vertex at $(-4, 0)$. Graph the equation by hand.

Solution The ellipse has its center at the origin and, since the given focus and vertex lie on the x-axis, the major axis is the x-axis. The distance from the center, $(0, 0)$, to one of the foci, $(3, 0)$, is $c = 3$. The distance from the center, $(0, 0)$, to one of the vertices, $(-4, 0)$, is $a = 4$. From equation (2), it follows that

$$b^2 = a^2 - c^2 = 16 - 9 = 7$$

so an equation of the ellipse is

$$\frac{x^2}{16} + \frac{y^2}{7} = 1$$

Figure 22 shows the graph drawn by hand.

Figure 22
$\frac{x^2}{16} + \frac{y^2}{7} = 1$

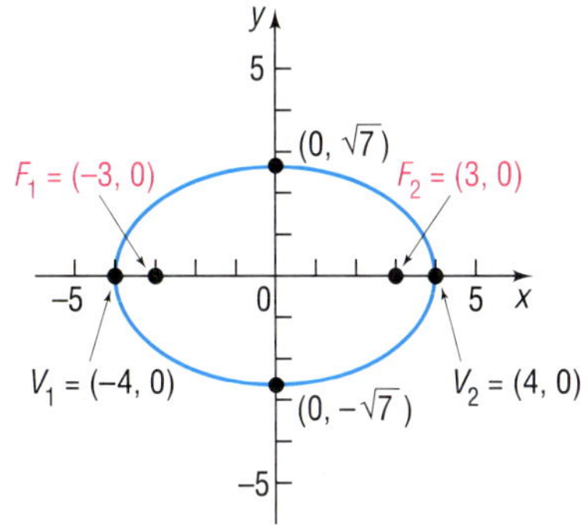

Notice in Figure 22 how we used the intercepts of the equation to graph the ellipse. Following this practice will make it easier for you to obtain an accurate graph of an ellipse when graphing by hand. It also tells you how to set the initial viewing window when using a graphing utility.

EXAMPLE 2 Graphing an Ellipse Using a Graphing Utility

Use a graphing utility to graph the ellipse $\frac{x^2}{16} + \frac{y^2}{7} = 1$.

Solution First, we must solve $\frac{x^2}{16} + \frac{y^2}{7} = 1$ for y.

$$\frac{y^2}{7} = 1 - \frac{x^2}{16} \quad \text{Subtract } \frac{x^2}{16} \text{ from each side.}$$

$$y^2 = 7\left(1 - \frac{x^2}{16}\right) \quad \text{Multiply both sides by 7.}$$

$$y = \pm\sqrt{7\left(1 - \frac{x^2}{16}\right)} \quad \text{Take the square root of each side.}$$

Figure 23

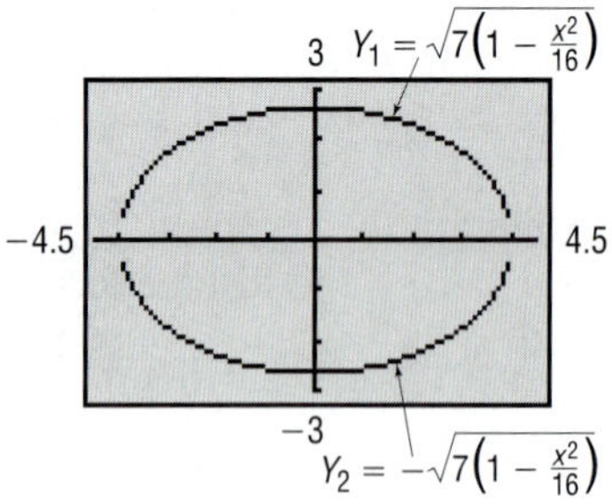

Figure 23* shows the graphs of $Y_1 = \sqrt{7\left(1 - \frac{x^2}{16}\right)}$ and $Y_2 = -\sqrt{7\left(1 - \frac{x^2}{16}\right)}$.

Notice in Figure 23 that we used a square screen. As with circles and parabolas, this is done to avoid a distorted view of the graph.

An equation of the form of equation (2), with $a > b$, is the equation of an ellipse with center at the origin, foci on the x-axis at $(-c, 0)$ and $(c, 0)$, where $c^2 = a^2 - b^2$, and major axis along the x-axis.

3 For the remainder of this section, the direction "Discuss the equation" will mean to find the center, major axis, foci, and vertices of the ellipse and graph it.

EXAMPLE 3 Discussing the Equation of an Ellipse

Discuss the equation $\frac{x^2}{25} + \frac{y^2}{9} = 1$.

Solution Figure 24(a) shows the graph of $\frac{x^2}{25} + \frac{y^2}{9} = 1$ using a graphing utility. We now proceed to analyze the equation. The given equation is of the form of equation (2), with $a^2 = 25$ and $b^2 = 9$. The equation is that of an ellipse with center $(0, 0)$ and major axis along the x-axis. The vertices are at $(\pm a, 0) = (\pm 5, 0)$. Because $b^2 = a^2 - c^2$, we find that

$$c^2 = a^2 - b^2 = 25 - 9 = 16$$

The foci are at $(\pm c, 0) = (\pm 4, 0)$. Figure 24(b) shows the graph drawn by hand.

Figure 24 $\frac{x^2}{25} + \frac{y^2}{9} = 1$

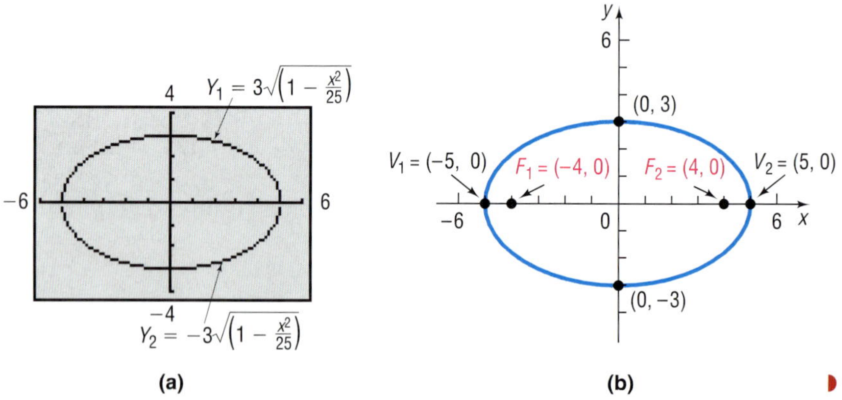

NOW WORK PROBLEMS 9 AND 19.

*The initial viewing window selected was $X\min = -4$, $X\max = 4$, $Y\min = -3$, $Y\max = 3$. Then we used the ZOOM-SQUARE option to obtain the window shown.

If the major axis of an ellipse with center at (0, 0) lies on the y-axis, then the foci are at $(0, -c)$ and $(0, c)$. Using the same steps as before, the definition of an ellipse leads to the following result:

THEOREM **Equation of an Ellipse; Center at (0, 0); Foci at (0, ±c); Major Axis along the y-Axis**

An equation of the ellipse with center at (0, 0) and foci at $(0, -c)$ and $(0, c)$ is

$$\frac{x^2}{b^2} + \frac{y^2}{a^2} = 1 \quad \text{where } a > b > 0 \text{ and } b^2 = a^2 - c^2 \qquad (3)$$

The major axis is the y-axis; the vertices are at $(0, -a)$ and $(0, a)$.

Figure 25 illustrates the graph of such an ellipse. Again, notice the right triangle with the points at (0, 0), $(b, 0)$, and $(0, c)$.

Figure 25

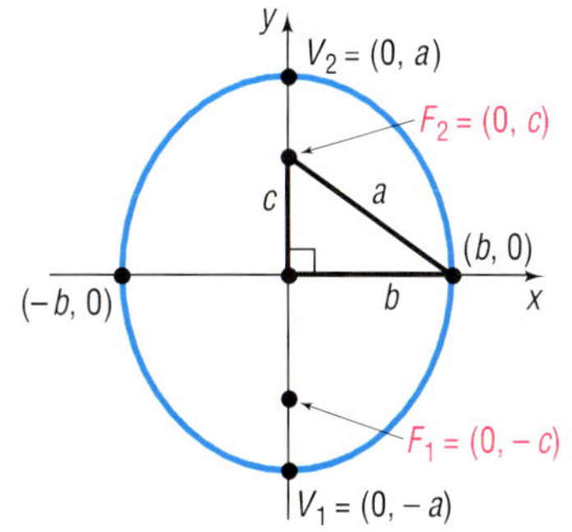

Look closely at equations (2) and (3). Although they may look alike, there is a difference! In equation (2), the larger number, a^2, is in the denominator of the x^2-term, so the major axis of the ellipse is along the x-axis. In equation (3), the larger number, a^2, is in the denominator of the y^2-term, so the major axis is along the y-axis.

EXAMPLE 4 Discussing the Equation of an Ellipse

Discuss the equation $9x^2 + y^2 = 9$.

Solution Figure 26(a) shows the graph of $9x^2 + y^2 = 9$ using a graphing utility. We now proceed to analyze the equation. To put the equation in proper form, we divide each side by 9:

$$x^2 + \frac{y^2}{9} = 1$$

The larger number, 9, is in the denominator of the y^2-term so, based on equation (3), this is the equation of an ellipse with center at the origin and major axis along the y-axis. Also, we conclude that $a^2 = 9$, $b^2 = 1$, and $c^2 = a^2 - b^2 = 9 - 1 = 8$. The vertices are at $(0, \pm a) = (0, \pm 3)$, and the foci are at $(0, \pm c) = (0, \pm 2\sqrt{2})$. The graph, drawn by hand, is given in Figure 26(b).

Figure 26 $x^2 + \frac{y^2}{9} = 1$

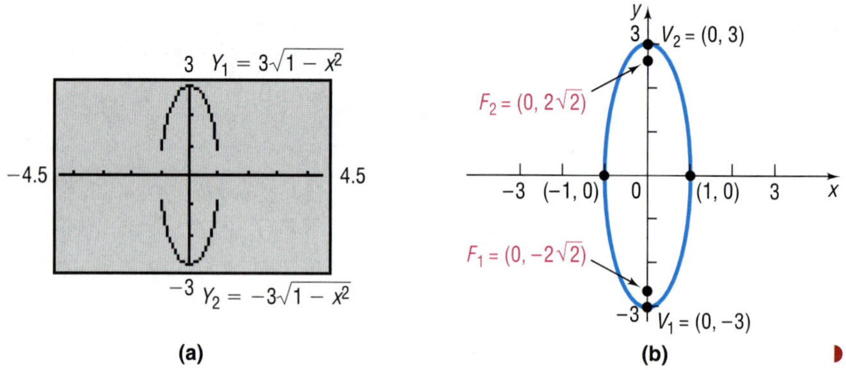

EXAMPLE 5 Finding an Equation of an Ellipse

Find an equation of the ellipse having one focus at $(0, 2)$ and vertices at $(0, -3)$ and $(0, 3)$. Graph the equation by hand.

Solution Because the vertices are at $(0, -3)$ and $(0, 3)$, the center of this ellipse is at their midpoint, the origin. Also, its major axis lies on the y-axis. The distance from the center, $(0, 0)$, to one of the foci, $(0, 2)$, is $c = 2$. The distance from the center, $(0, 0)$, to one of the vertices $(0, 3)$ is $a = 3$. So $b^2 = a^2 - c^2 = 9 - 4 = 5$. The form of the equation of this ellipse is given by equation (3):

$$\frac{x^2}{b^2} + \frac{y^2}{a^2} = 1$$

$$\frac{x^2}{5} + \frac{y^2}{9} = 1$$

Figure 27 $\frac{x^2}{5} + \frac{y^2}{9} = 1$

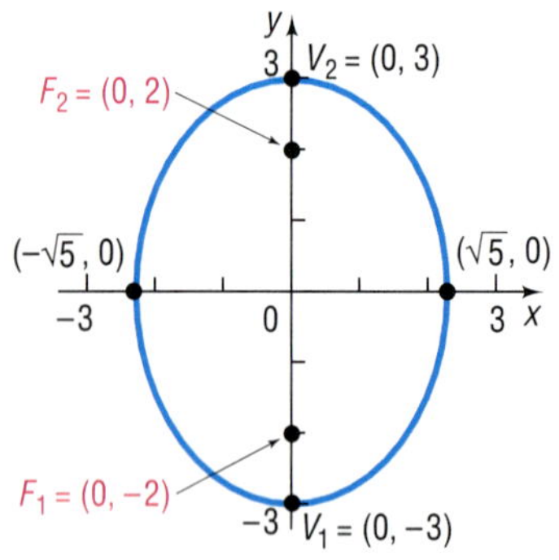

Figure 27 shows the graph.

NOW WORK PROBLEMS 13 AND 21.

The circle may be considered a special kind of ellipse. To see why, let $a = b$ in equation (2) or in equation (3). Then

$$\frac{x^2}{a^2} + \frac{y^2}{a^2} = 1$$

$$x^2 + y^2 = a^2$$

This is the equation of a circle with center at the origin and radius a. The value of c is

$$c^2 = a^2 - b^2 = 0$$

We conclude that the closer the two foci of an ellipse are to the center the more the ellipse will look like a circle.

Center at (h, k)

4 If an ellipse with center at the origin and major axis coinciding with a coordinate axis is shifted horizontally h units and then vertically k units, the result is an ellipse with center at (h, k) and major axis parallel to a coordinate axis. The equations of such ellipses have the same forms as those given in equations (2) and (3), except that x is replaced by $x - h$ (the horizontal shift) and y is replaced by $y - k$ (the vertical shift). Table 3 gives the forms of the equations of such ellipses, and Figure 28 shows their graphs.

Table 3 Ellipses with Center at (h, k) and Major Axis Parallel to a Coordinate Axis

Center	Major Axis	Foci	Vertices	Equation
(h, k)	Parallel to x-axis	$(h + c, k)$	$(h + a, k)$	$\dfrac{(x-h)^2}{a^2} + \dfrac{(y-k)^2}{b^2} = 1,$
		$(h - c, k)$	$(h - a, k)$	$a > b$ and $b^2 = a^2 - c^2$
(h, k)	Parallel to y-axis	$(h, k + c)$	$(h, k + a)$	$\dfrac{(x-h)^2}{b^2} + \dfrac{(y-k)^2}{a^2} = 1,$
		$(h, k - c)$	$(h, k - a)$	$a > b$ and $b^2 = a^2 - c^2$

Figure 28

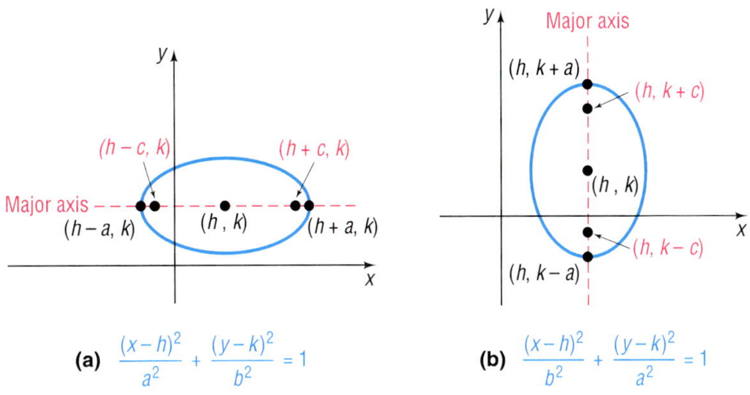

EXAMPLE 6 Finding an Equation of an Ellipse, Center Not at the Origin

Find an equation for the ellipse with center at $(2, -3)$, one focus at $(3, -3)$, and one vertex at $(5, -3)$. Graph the equation by hand.

Solution The center is at $(h, k) = (2, -3)$, so $h = 2$ and $k = -3$. Since the center, focus, and vertex all lie on the line $y = -3$, the major axis is parallel to the x-axis. The distance from the center $(2, -3)$ to a focus $(3, -3)$ is $c = 1$; the distance from the center $(2, -3)$ to a vertex $(5, -3)$ is $a = 3$. Thus, $b^2 = a^2 - c^2 = 9 - 1 = 8$. The form of the equation is

$$\frac{(x-h)^2}{a^2} + \frac{(y-k)^2}{b^2} = 1 \qquad \text{where } h = 2, k = -3, a = 3, b = 2\sqrt{2}$$

$$\frac{(x-2)^2}{9} + \frac{(y+3)^2}{8} = 1$$

Figure 29 $\frac{(x-2)^2}{9} + \frac{(y+3)^2}{8} = 1$

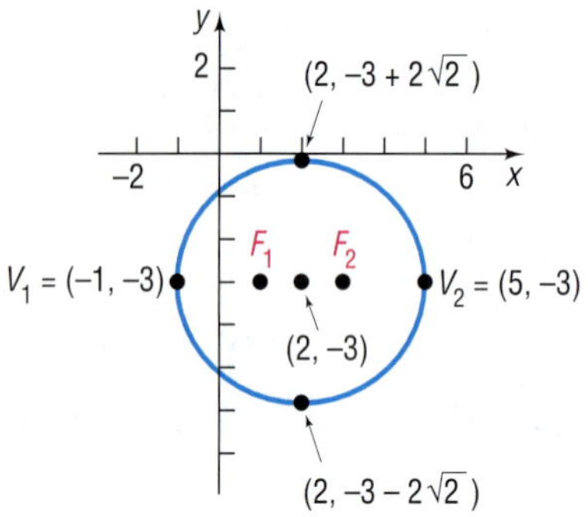

To graph the equation by hand, we first locate the vertices. The major axis is parallel to the x-axis, so the vertices are $a = 3$ units left and right of the center, $(h, k) = (2, -3)$. Therefore, the vertices are

$$V_1 = (2 - 3, -3) = (-1, -3) \quad \text{and} \quad V_2 = (2 + 3, -3) = (5, -3)$$

Since $c = 1$ and the major axis is parallel to the x-axis, the foci are 1 unit left and right of the center. Therefore, the foci are

$$F_1 = (2 - 1, -3) = (1, -3) \quad \text{and} \quad F_2 = (2 + 1, -3) = (3, -3)$$

Finally, we use the value of $b = 2\sqrt{2}$ to find the two points above and below the center:

$$(2, -3 - 2\sqrt{2}) \quad \text{and} \quad (2, -3 + 2\sqrt{2})$$

Figure 29 shows the graph.

EXAMPLE 7 Using a Graphing Utility to Graph an Ellipse, Center Not at the Origin

Using a graphing utility, graph the ellipse: $\frac{(x-2)^2}{9} + \frac{(y+3)^2}{8} = 1$

Solution First, we must solve the equation $\frac{(x-2)^2}{9} + \frac{(y+3)^2}{8} = 1$ for y.

$$\frac{(y+3)^2}{8} = 1 - \frac{(x-2)^2}{9} \qquad \text{Subtract } \tfrac{(x-2)^2}{9} \text{ from each side.}$$

$$(y+3)^2 = 8\left[1 - \frac{(x-2)^2}{9}\right] \qquad \text{Multiply each side by 8.}$$

$$y + 3 = \pm\sqrt{8\left[1 - \frac{(x-2)^2}{9}\right]} \qquad \text{Take the square root of each side.}$$

$$y = -3 \pm \sqrt{8\left[1 - \frac{(x-2)^2}{9}\right]} \qquad \text{Subtract 3 from each side.}$$

Figure 30

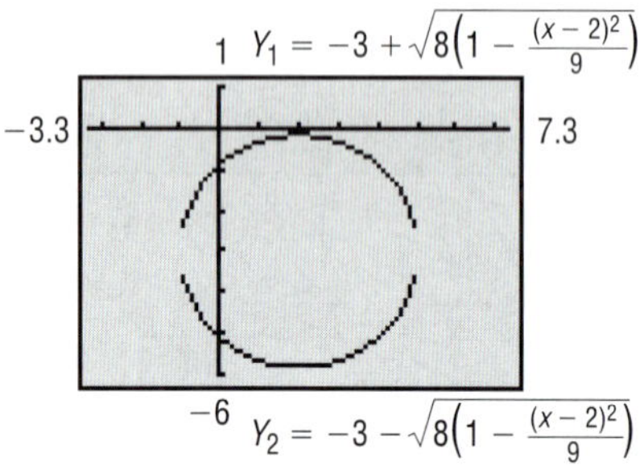

Figure 30 shows the graphs of $Y_1 = -3 + \sqrt{8\left[1 - \frac{(x-2)^2}{9}\right]}$ and $Y_2 = -3 - \sqrt{8\left[1 - \frac{(x-2)^2}{9}\right]}$.

NOW WORK PROBLEM 33.

EXAMPLE 8 Discussing the Equation of an Ellipse

Discuss the equation $4x^2 + y^2 - 8x + 4y + 4 = 0$.

Solution We proceed to complete the squares in x and in y:

$$4x^2 + y^2 - 8x + 4y + 4 = 0$$
$$4x^2 - 8x + y^2 + 4y = -4$$
$$4(x^2 - 2x) + (y^2 + 4y) = -4$$
$$4(x^2 - 2x + 1) + (y^2 + 4y + 4) = -4 + 4 + 4 \qquad \text{Complete each square.}$$
$$4(x - 1)^2 + (y + 2)^2 = 4$$
$$(x - 1)^2 + \frac{(y+2)^2}{4} = 1 \qquad \text{Divide each side by 4.}$$

Figure 31(a) shows the graph of $(x-1)^2 + \frac{(y+2)^2}{4} = 1$ using a graphing utility. We now proceed to analyze the equation.

This is the equation of an ellipse with center at $(1, -2)$ and major axis parallel to the y-axis. Since $a^2 = 4$ and $b^2 = 1$, we have $c^2 = a^2 - b^2 = 4 - 1 = 3$. The vertices are at $(h, k \pm a) = (1, -2 \pm 2)$ or $(1, 0)$ and $(1, -4)$. The foci are at $(h, k \pm c) = (1, -2 \pm \sqrt{3})$ or $(1, -2 - \sqrt{3})$ and $(1, -2 + \sqrt{3})$. Figure 31(b) shows the graph drawn by hand.

Figure 31 $(x-1)^2 + \frac{(y+2)^2}{4} = 1$

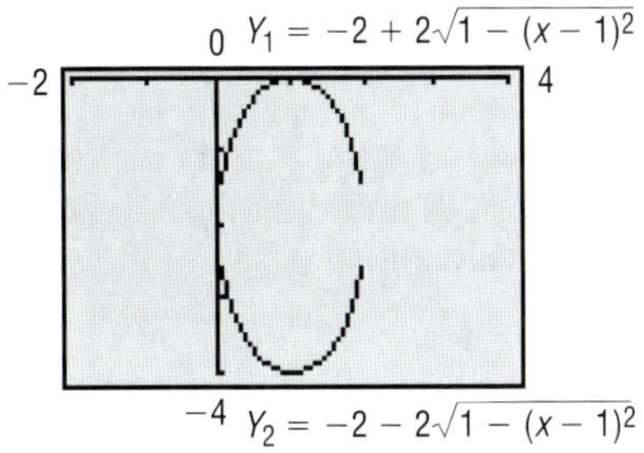

(a)

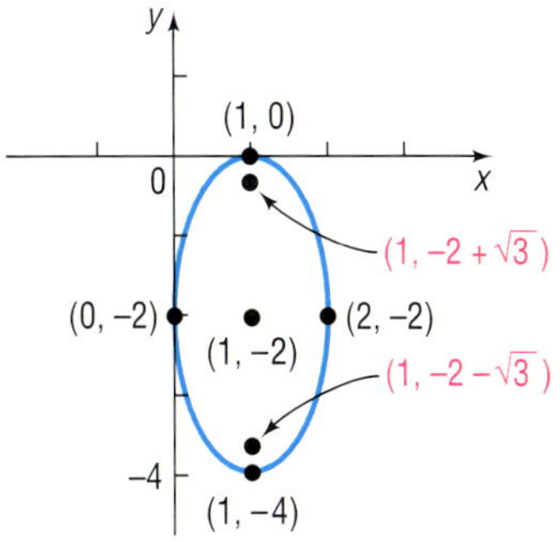

(b)

Applications

5 Ellipses are found in many applications in science and engineering. For example, the orbits of the planets around the Sun are elliptical, with the Sun's position at a focus. See Figure 32.

Figure 32

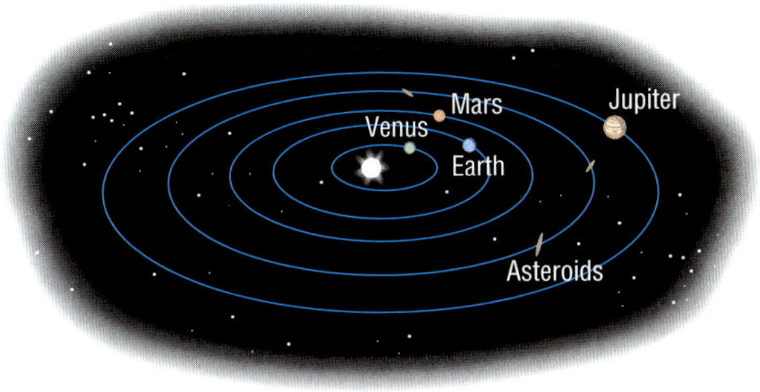

Stone and concrete bridges are often shaped as semielliptical arches. Elliptical gears are used in machinery when a variable rate of motion is required.

Ellipses also have an interesting reflection property. If a source of light (or sound) is placed at one focus, the waves transmitted by the source will reflect off the ellipse and concentrate at the other focus. This is the principle behind "whispering galleries," which are rooms designed with elliptical ceilings. A person standing at one focus of the ellipse can whisper and be heard by a person standing at the other focus, because all the sound waves that reach the ceiling are reflected to the other person.

EXAMPLE 9 Whispering Galleries

Figure 33 shows the specifications for an elliptical ceiling in a hall designed to be a whispering gallery. In a whispering gallery, a person standing at one focus of the ellipse can whisper and be heard by another person standing at the other focus, because all the sound waves that reach the ceiling from one focus are reflected to the other focus. Where in the hall are the foci located?

Figure 33

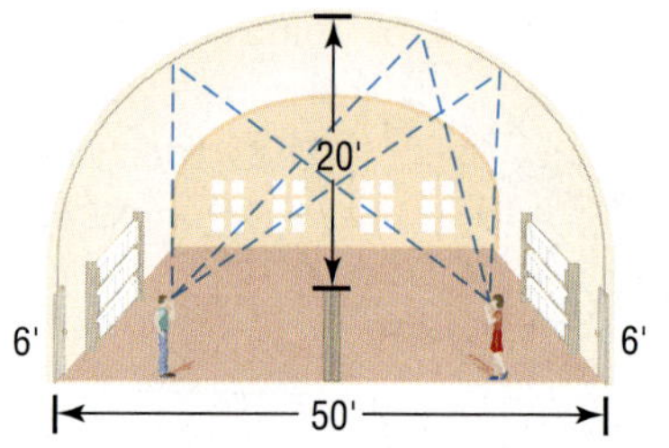

Solution We set up a rectangular coordinate system so that the center of the ellipse is at the origin and the major axis is along the x-axis. See Figure 34. The equation of the ellipse is

$$\frac{x^2}{a^2} + \frac{y^2}{b^2} = 1$$

where $a = 25$ and $b = 20$. Since

$$c^2 = a^2 - b^2 = 25^2 - 20^2 = 625 - 400 = 225$$

we have $c = 15$. Thus, the foci are located 15 feet from the center of the ellipse along the major axis.

Figure 34

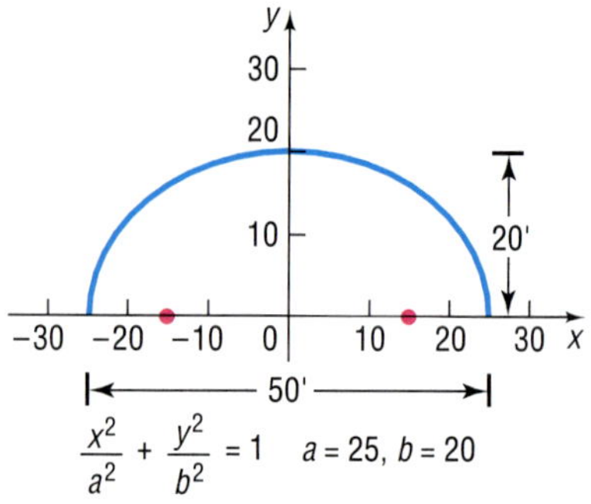

9.3 EXERCISES

In Problems 1–4, the graph of an ellipse is given. Match each graph to its equation.

A. $\frac{x^2}{4} + y^2 = 1$ B. $x^2 + \frac{y^2}{4} = 1$ C. $\frac{x^2}{16} + \frac{y^2}{4} = 1$ D. $\frac{x^2}{4} + \frac{y^2}{16} = 1$

1.

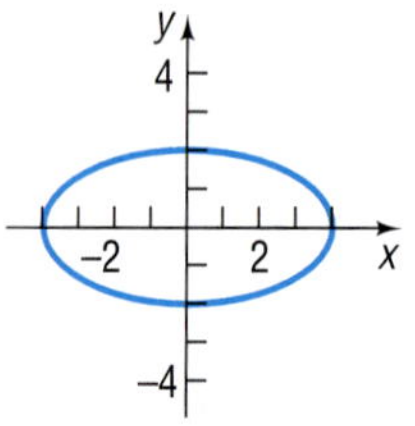

2.

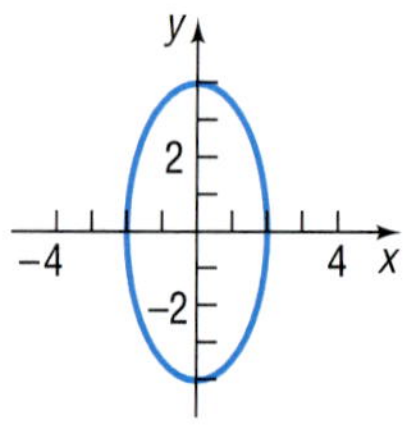

3.

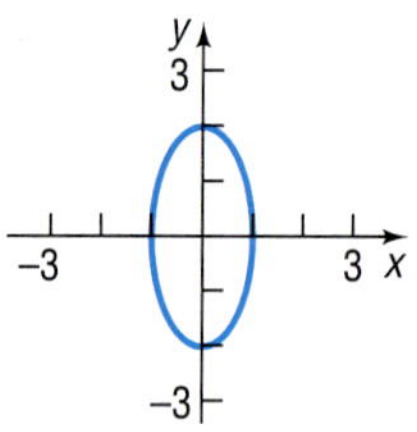

4.

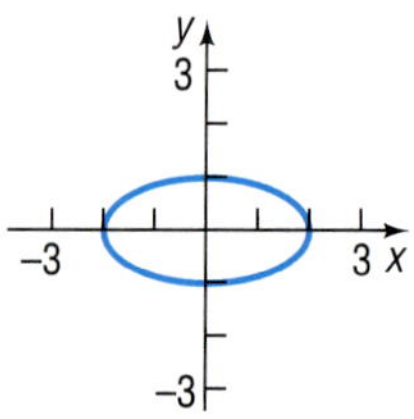

In Problems 5–8, the graph of an ellipse is given. Match each graph to its equation.

A. $\dfrac{(x+1)^2}{4} + \dfrac{(y-1)^2}{9} = 1$ *B.* $\dfrac{(x-1)^2}{4} + \dfrac{(y+1)^2}{9} = 1$

C. $\dfrac{(x-1)^2}{9} + \dfrac{(y+1)^2}{4} = 1$ *D.* $\dfrac{(x+1)^2}{9} + \dfrac{(y-1)^2}{4} = 1$

5.

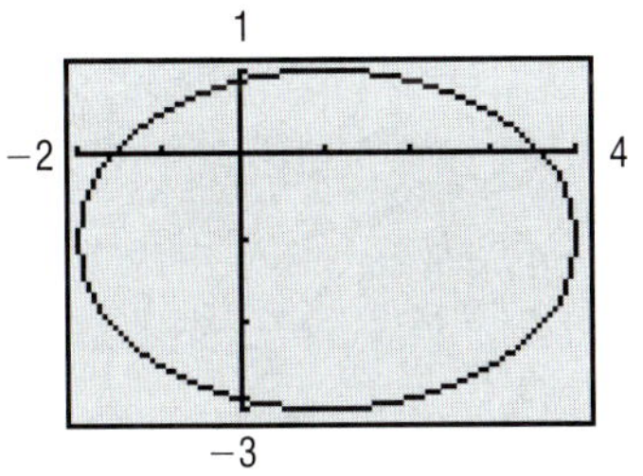

6.

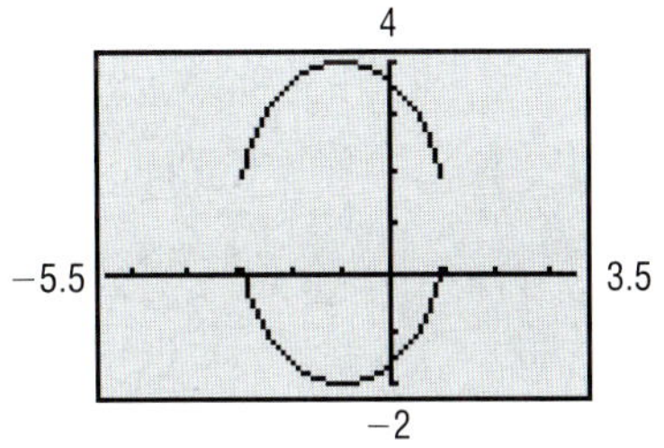

7.

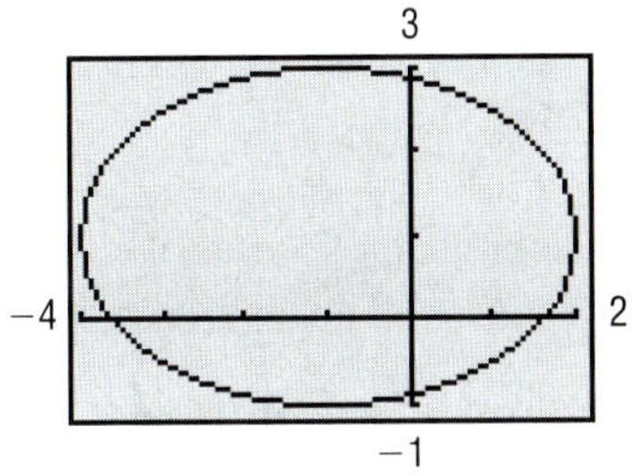

8.

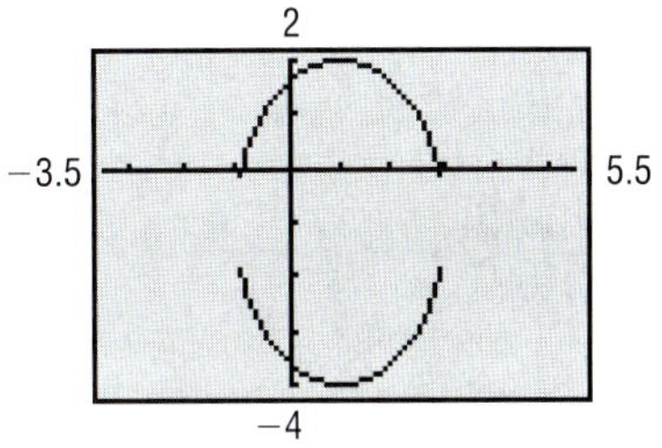

In Problems 9–18, find the vertices and foci of each ellipse. Graph each equation using a graphing utility.

9. $\dfrac{x^2}{25} + \dfrac{y^2}{4} = 1$

10. $\dfrac{x^2}{9} + \dfrac{y^2}{4} = 1$

11. $\dfrac{x^2}{9} + \dfrac{y^2}{25} = 1$

12. $x^2 + \dfrac{y^2}{16} = 1$

13. $4x^2 + y^2 = 16$

14. $x^2 + 9y^2 = 18$

15. $4y^2 + x^2 = 8$

16. $4y^2 + 9x^2 = 36$

17. $x^2 + y^2 = 16$

18. $x^2 + y^2 = 4$

In Problems 19–28, find an equation for each ellipse. Graph the equation by hand.

19. Center at $(0, 0)$; focus at $(3, 0)$; vertex at $(5, 0)$

20. Center at $(0, 0)$; focus at $(-1, 0)$; vertex at $(3, 0)$

21. Center at $(0, 0)$; focus at $(0, -4)$; vertex at $(0, 5)$

22. Center at $(0, 0)$; focus at $(0, 1)$; vertex at $(0, -2)$

23. Foci at $(\pm 2, 0)$; length of the major axis is 6

24. Focus at $(0, -4)$; vertices at $(0, \pm 8)$

25. Foci at $(0, \pm 3)$; x-intercepts are ± 2

26. Foci at $(0, \pm 2)$; length of the major axis is 8

27. Center at $(0, 0)$; vertex at $(0, 4)$; $b = 1$

28. Vertices at $(\pm 5, 0)$; $c = 2$

In Problems 29–32, write an equation for each ellipse.

29.

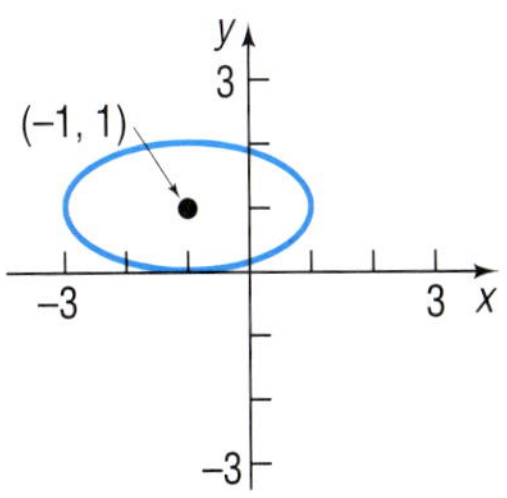

30.

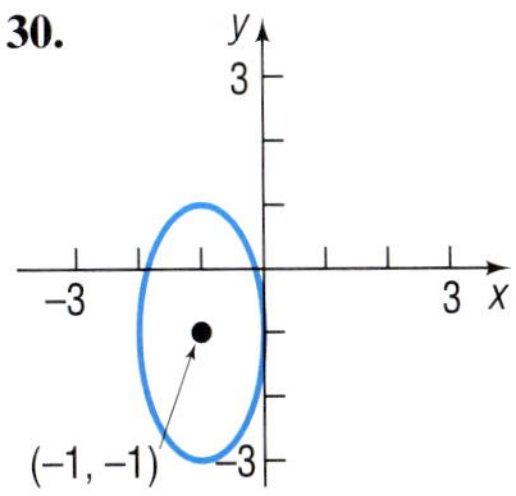

31.

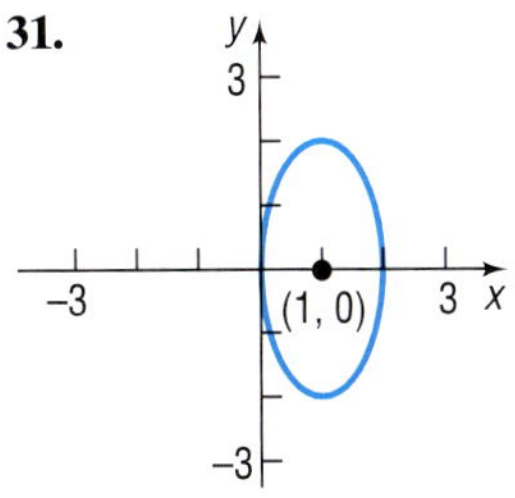

32.

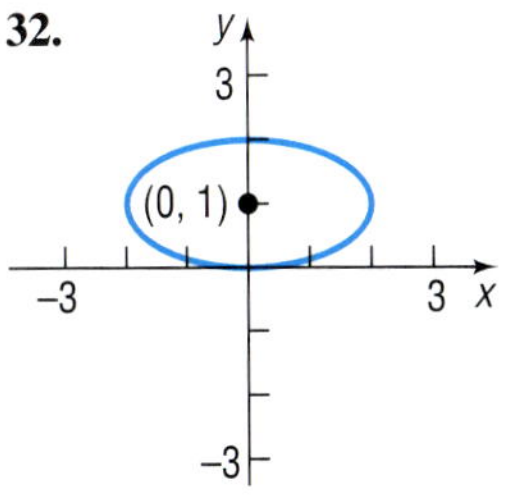

In Problems 33–44, find the center, foci, and vertices of each ellipse. Graph each equation using a graphing utility.

33. $\frac{(x-3)^2}{4} + \frac{(y+1)^2}{9} = 1$

34. $\frac{(x+4)^2}{9} + \frac{(y+2)^2}{4} = 1$

35. $(x+5)^2 + 4(y-4)^2 = 16$

36. $9(x-3)^2 + (y+2)^2 = 18$

37. $x^2 + 4x + 4y^2 - 8y + 4 = 0$

38. $x^2 + 3y^2 - 12y + 9 = 0$

39. $2x^2 + 3y^2 - 8x + 6y + 5 = 0$

40. $4x^2 + 3y^2 + 8x - 6y = 5$

41. $9x^2 + 4y^2 - 18x + 16y - 11 = 0$

42. $x^2 + 9y^2 + 6x - 18y + 9 = 0$

43. $4x^2 + y^2 + 4y = 0$

44. $9x^2 + y^2 - 18x = 0$

In Problems 45–54, find an equation for each ellipse. Graph the equation by hand.

45. Center at $(2, -2)$; vertex at $(7, -2)$; focus at $(4, -2)$

46. Center at $(-3, 1)$; vertex at $(-3, 3)$; focus at $(-3, 0)$

47. Vertices at $(4, 3)$ and $(4, 9)$; focus at $(4, 8)$

48. Foci at $(1, 2)$ and $(-3, 2)$; vertex at $(-4, 2)$

49. Foci at $(5, 1)$ and $(-1, 1)$; length of the major axis is 8

50. Vertices at $(2, 5)$ and $(2, -1)$; $c = 2$

51. Center at $(1, 2)$; focus at $(4, 2)$; contains the point $(1, 3)$

52. Center at $(1, 2)$; focus at $(1, 4)$; contains the point $(2, 2)$

53. Center at $(1, 2)$; vertex at $(4, 2)$; contains the point $(1, 3)$

54. Center at $(1, 2)$; vertex at $(1, 4)$; contains the point $(2, 2)$

In Problems 55–58, graph each function by hand. Use a graphing utility to verify your results.
[Hint: Notice that each function is half an ellipse.]

55. $f(x) = \sqrt{16 - 4x^2}$ **56.** $f(x) = \sqrt{9 - 9x^2}$ **57.** $f(x) = -\sqrt{64 - 16x^2}$ **58.** $f(x) = -\sqrt{4 - 4x^2}$

59. Semielliptical Arch Bridge An arch in the shape of the upper half of an ellipse is used to support a bridge that is to span a river 20 meters wide. The center of the arch is 6 meters above the center of the river (see the figure). Write an equation for the ellipse in which the x-axis coincides with the water level and the y-axis passes through the center of the arch.

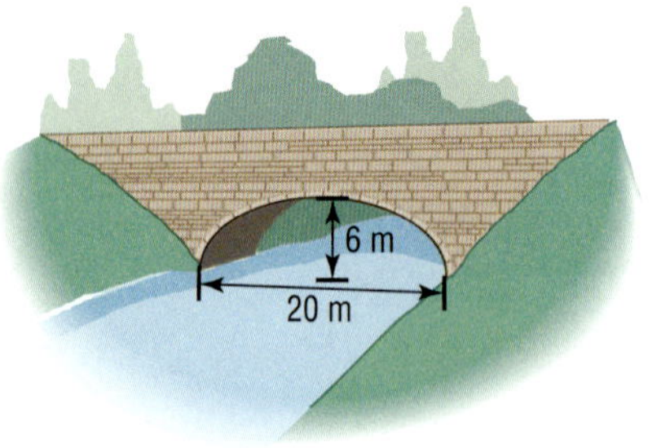

60. Semielliptical Arch Bridge The arch of a bridge is a semiellipse with a horizontal major axis. The span is 30 feet, and the top of the arch is 10 feet above the major axis. The roadway is horizontal and is 2 feet above the top of the arch. Find the vertical distance from the roadway to the arch at 5 foot intervals along the roadway.

61. Whispering Gallery A hall 100 feet in length is to be designed as a whispering gallery. If the foci are located 25 feet from the center, how high will the ceiling be at the center?

62. Whispering Gallery Jim, standing at one focus of a whispering gallery, is 6 feet from the nearest wall. His friend is standing at the other focus, 100 feet away. What is the length of this whispering gallery? How high is its elliptical ceiling at the center?

63. Semielliptical Arch Bridge A bridge is built in the shape of a semielliptical arch. The bridge has a span of 120 feet and a maximum height of 25 feet. Choose a suitable rectangular coordinate system and find the height of the arch at distances of 10, 30, and 50 feet from the center.

64. Semielliptical Arch Bridge A bridge is built in the shape of a semielliptical arch and is to have a span of 100 feet. The height of the arch, at a distance of 40 feet from the center, is to be 10 feet. Find the height of the arch at its center.

65. Semielliptical Arch An arch in the form of half an ellipse is 40 feet wide and 15 feet high at the center. Find the height of the arch at intervals of 10 feet along its width.

66. Semielliptical Arch Bridge An arch for a bridge over a highway is in the form of half an ellipse. The top of the arch is 20 feet above the ground level (the major axis). The highway has four lanes, each 12 feet wide; a center safety strip 8 feet wide; and two side strips, each 4 feet wide. What should the span of the bridge be (the length of its major axis) if the height 28 feet from the center is to be 13 feet?

In Problems 67–70, use the fact that the orbit of a planet about the Sun is an ellipse, with the Sun at one focus. The ***aphelion*** *of a planet is its greatest distance from the Sun and the* ***perihelion*** *is its shortest distance. The* ***mean distance*** *of a planet from the Sun is the length of the semimajor axis of the elliptical orbit. See the illustration.*

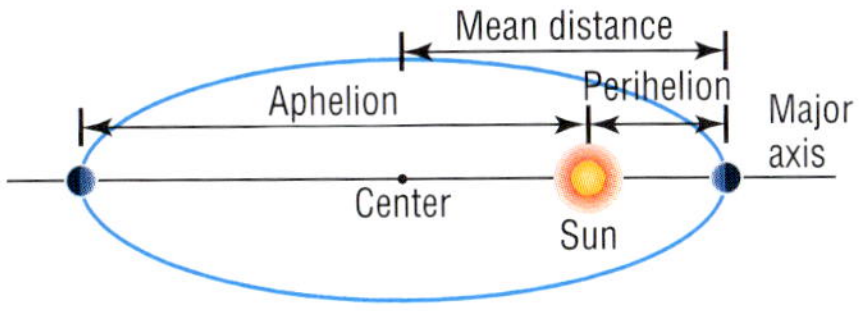

67. Earth The mean distance of Earth from the Sun is 93 million miles. If the aphelion of Earth is 94.5 million miles, what is the perihelion? Write an equation for the orbit of Earth around the Sun.

68. Mars The mean distance of Mars from the Sun is 142 million miles. If the perihelion of Mars is 128.5 million miles, what is the aphelion? Write an equation for the orbit of Mars about the Sun.

69. Jupiter The aphelion of Jupiter is 507 million miles. If the distance from the Sun to the center of its elliptical orbit is 23.2 million miles, what is the perihelion? What is the mean distance? Write an equation for the orbit of Jupiter around the Sun.

70. Pluto The perihelion of Pluto is 4551 million miles and the distance of the Sun from the center of its elliptical orbit is 897.5 million miles. Find the aphelion of Pluto. What is the mean distance of Pluto from the Sun? Write an equation for the orbit of Pluto about the Sun.

71. Consult the figure. A racetrack is in the shape of an ellipse, 100 feet long and 50 feet wide. What is the width 10 feet from the side?

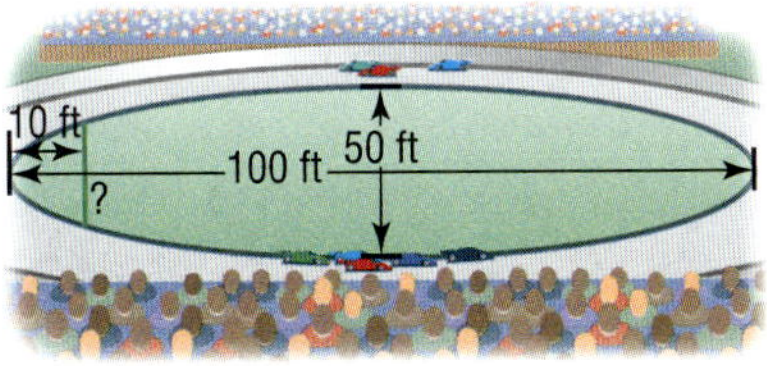

72. A racetrack is in the shape of an ellipse 80 feet long and 40 feet wide. What is the width 10 feet from the side?

73. Show that an equation of the form

$$Ax^2 + Cy^2 + F = 0 \qquad A \neq 0, C \neq 0, F \neq 0$$

where A and C are of the same sign and F is of opposite sign,

(a) Is the equation of an ellipse with center at $(0, 0)$ if $A \neq C$.
(b) Is the equation of a circle with center $(0, 0)$ if $A = C$.

74. Show that the graph of an equation of the form

$$Ax^2 + Cy^2 + Dx + Ey + F = 0 \qquad A \neq 0, C \neq 0$$

where A and C are of the same sign,

(a) Is an ellipse if $(D^2/4A) + (E^2/4C) - F$ is the same sign as A.
(b) Is a point if $(D^2/4A) + (E^2/4C) - F = 0$.
(c) Contains no points if $(D^2/4A) + (E^2/4C) - F$ is of opposite sign to A.

75. The **eccentricity** e of an ellipse is defined as the number c/a, where a and c are the numbers given in equation (2). Because $a > c$, it follows that $e < 1$. Write a brief paragraph about the general shape of each of the following ellipses. Be sure to justify your conclusions.

(a) Eccentricity close to 0
(b) Eccentricity $= 0.5$
(c) Eccentricity close to 1

9.4 THE HYPERBOLA

1. Find the Equation of a Hyperbola with Center at Origin
2. Find the Asymptotes of a Hyperbola
3. Discuss the Equation of a Hyperbola
4. Work with Hyperbolas with Center at (h, k)
5. Solve Applied Problems Involving Properties of Hyperbolas

A **hyperbola** is the collection of all points in the plane the difference of whose distances from two fixed points, called the **foci,** is a constant.

Figure 35

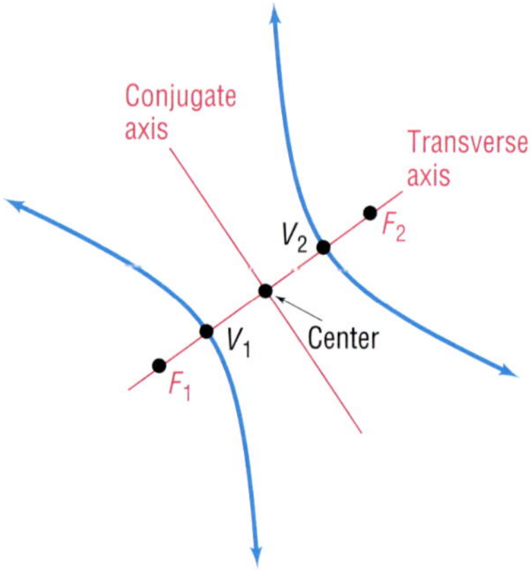

1

Figure 35 illustrates a hyperbola with foci F_1 and F_2. The line containing the foci is called the **transverse axis.** The midpoint of the line segment joining the foci is called the **center** of the hyperbola. The line through the center and perpendicular to the transverse axis is called the **conjugate axis.** The hyperbola consists of two separate curves, called **branches,** that are symmetric with respect to the transverse axis, conjugate axis, and center. The two points of intersection of the hyperbola and the transverse axis are the **vertices,** V_1 and V_2, of the hyperbola.

With these ideas in mind, we are now ready to find the equation of a hyperbola in a rectangular coordinate system. First, we place the center at the origin. Next, we position the hyperbola so that its transverse axis coincides with a coordinate axis. Suppose that the transverse axis coincides with the x-axis, as shown in Figure 36.

If c is the distance from the center to a focus, then one focus will be at $F_1 = (-c, 0)$ and the other at $F_2 = (c, 0)$. Now we let the constant difference of the distances from any point $P = (x, y)$ on the hyperbola to the foci F_1 and F_2 be denoted by $\pm 2a$. (If P is on the right branch, the $+$ sign is used; if P is on the left branch, the $-$ sign is used.) The coordinates of P must satisfy the equation

Figure 36
$d(F_1, P) - d(F_2, P) = \pm 2a$

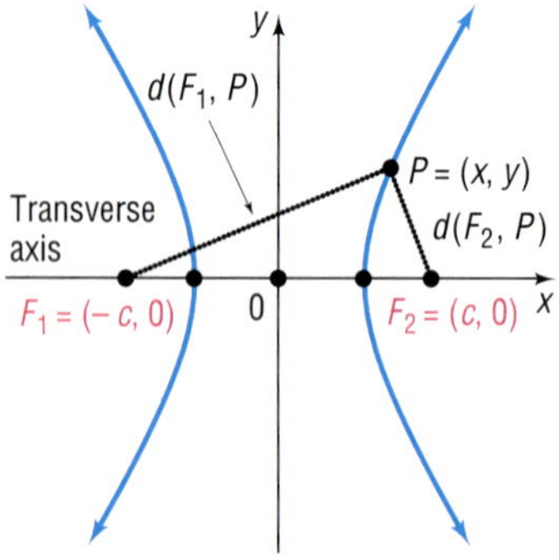

$$d(F_1, P) - d(F_2, P) = \pm 2a$$ *Difference of the distances from P to the foci equals $\pm 2a$.*

$$\sqrt{(x + c)^2 + y^2} - \sqrt{(x - c)^2 + y^2} = \pm 2a$$ *Use the distance formula.*

$$\sqrt{(x + c)^2 + y^2} = \pm 2a + \sqrt{(x - c)^2 + y^2}$$ *Isolate one radical.*

$$(x + c)^2 + y^2 = 4a^2 \pm 4a\sqrt{(x - c)^2 + y^2} + (x - c)^2 + y^2$$ *Square both sides.*

Next, we remove the parentheses:

$$x^2 + 2cx + c^2 + y^2 = 4a^2 \pm 4a\sqrt{(x - c)^2 + y^2} + x^2 - 2cx + c^2 + y^2$$

$$4cx - 4a^2 = \pm 4a\sqrt{(x - c)^2 + y^2}$$ *Isolate the radical.*

$$cx - a^2 = \pm a\sqrt{(x - c)^2 + y^2}$$ *Divide each side by 4.*

$$(cx - a^2)^2 = a^2[(x - c)^2 + y^2]$$ *Square both sides.*

$$c^2x^2 - 2ca^2x + a^4 = a^2(x^2 - 2cx + c^2 + y^2)$$ *Simplify.*

$$c^2x^2 + a^4 = a^2x^2 + a^2c^2 + a^2y^2$$ *Simplify.*

$$(c^2 - a^2)x^2 - a^2y^2 = a^2c^2 - a^4$$ *Rearrange terms.*

$$(c^2 - a^2)x^2 - a^2y^2 = a^2(c^2 - a^2) \qquad (1)$$

To obtain points on the hyperbola off the x-axis, it must be that $a < c$. To see why, look again at Figure 36.

$$
\begin{aligned}
d(F_1, P) &< d(F_2, P) + d(F_1, F_2) && \text{Use triangle } F_1PF_2. \\
d(F_1, P) - d(F_2, P) &< d(F_1, F_2) && P \text{ is on the right branch, so } d(F_1, P) - d(F_2, P) = 2a. \\
2a &< 2c \\
a &< c
\end{aligned}
$$

Since $a < c$, we also have $a^2 < c^2$, so $c^2 - a^2 > 0$. Let $b^2 = c^2 - a^2, b > 0$. Then equation (1) can be written as

$$
\begin{aligned}
b^2x^2 - a^2y^2 &= a^2b^2 \\
\frac{x^2}{a^2} - \frac{y^2}{b^2} &= 1
\end{aligned}
$$

To find the vertices of the hyperbola defined by this equation, let $y = 0$. The vertices satisfy the equation $x^2/a^2 = 1$, the solutions of which are $x = \pm a$. Consequently, the vertices of the hyperbola are $V_1 = (-a, 0)$ and $V_2 = (a, 0)$. Notice that the distance from the center $(0, 0)$ to either vertex is a.

THEOREM

Equation of a Hyperbola; Center at (0, 0); Foci at ($\pm c$, 0); Vertices at ($\pm a$, 0); Transverse Axis along the x-Axis

An equation of the hyperbola with center at $(0, 0)$, foci at $(-c, 0)$ and $(c, 0)$, and vertices at $(-a, 0)$ and $(a, 0)$ is

$$\frac{x^2}{a^2} - \frac{y^2}{b^2} = 1 \quad \text{where } b^2 = c^2 - a^2 \qquad (2)$$

The transverse axis is the x-axis.

As you can verify, the hyperbola defined by equation (2) is symmetric with respect to the x-axis, y-axis, and origin. To find the y-intercepts, if any, let $x = 0$ in equation (2). This results in the equation $y^2/b^2 = -1$, which has no real solution. We conclude that the hyperbola defined by equation (2) has no y-intercepts. In fact, since $x^2/a^2 - 1 = y^2/b^2 \geq 0$, it follows that $x^2/a^2 \geq 1$. Thus, there are no points on the graph for $-a < x < a$. See Figure 37.

Figure 37
$\frac{x^2}{a^2} - \frac{y^2}{b^2} = 1, b^2 = c^2 - a^2$

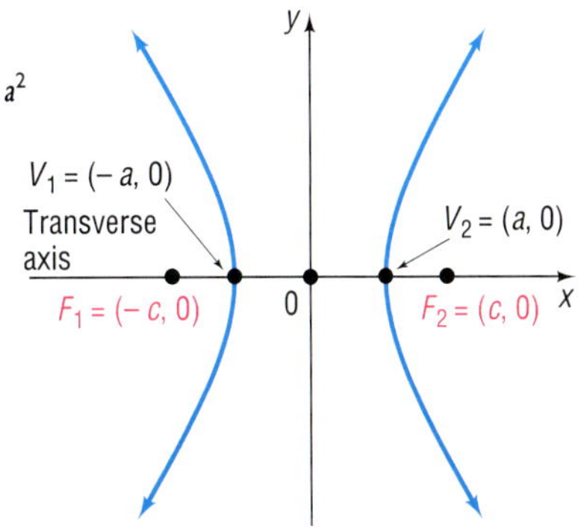

EXAMPLE 1 Finding an Equation of a Hyperbola

Find an equation of the hyperbola with center at the origin, one focus at $(4, 0)$, and one vertex at $(-3, 0)$.

Solution The hyperbola has its center at the origin, and the transverse axis lies on the x-axis. The distance from the center $(0, 0)$ to one focus $(4, 0)$ is $c = 4$. The distance from the center $(0, 0)$ to one vertex $(-3, 0)$ is $a = 3$. From equation (2), it follows that $b^2 = c^2 - a^2 = 16 - 9 = 7$, so an equation of the hyperbola is

$$\frac{x^2}{9} - \frac{y^2}{7} = 1$$

EXAMPLE 2 Using a Graphing Utility to Graph a Hyperbola

Use a graphing utility to graph the hyperbola $\frac{x^2}{9} - \frac{y^2}{7} = 1$.

Solution To graph the hyperbola $\frac{x^2}{9} - \frac{y^2}{7} = 1$, we need to graph the two functions $Y_1 = \sqrt{7}\sqrt{\frac{x^2}{9} - 1}$ and $Y_2 = -\sqrt{7}\sqrt{\frac{x^2}{9} - 1}$. As with graphing circles, parabolas, and ellipses on a graphing utility, we use a square screen setting so that the graph is not distorted. Figure 38 shows the graph of the hyperbola.

Figure 38

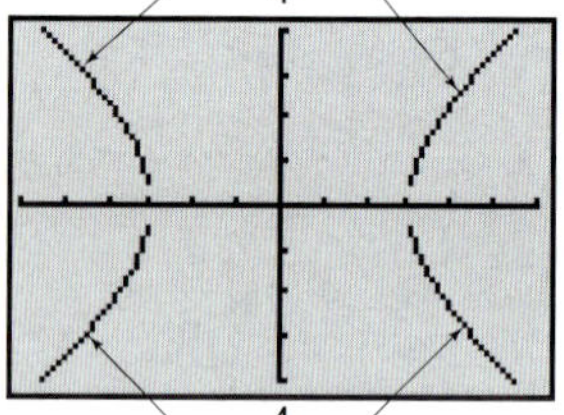

NOW WORK PROBLEM 9.

The next result gives the form of the equation of a hyperbola with center at the origin and transverse axis along the y-axis.

THEOREM

Equation of a Hyperbola; Center at (0, 0); Foci at (0, ±c); Vertices at (0, ±a); Transverse Axis along the y-Axis

An equation of the hyperbola with center at $(0, 0)$, foci at $(0, -c)$ and $(0, c)$, and vertices at $(0, -a)$ and $(0, a)$ is

$$\frac{y^2}{a^2} - \frac{x^2}{b^2} = 1 \quad \text{where } b^2 = c^2 - a^2 \qquad (3)$$

The transverse axis is the y-axis.

Figure 39
$\frac{y^2}{a^2} - \frac{x^2}{b^2} = 1, b^2 = c^2 - a^2$

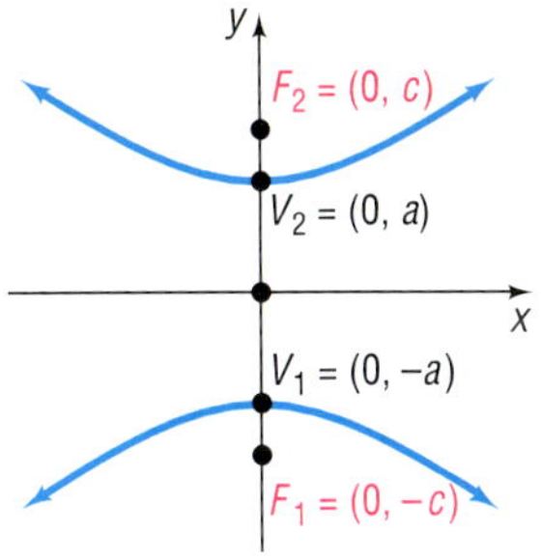

Figure 39 shows the graph of a typical hyperbola defined by equation (3).

An equation of the form of equation (2), $\frac{x^2}{a^2} - \frac{y^2}{b^2} = 1$, is the equation of a hyperbola with center at the origin, foci on the x-axis at $(-c, 0)$ and $(c, 0)$, where $c^2 = a^2 + b^2$, and transverse axis along the x-axis.

An equation of the form of equation (3), $\frac{y^2}{a^2} - \frac{x^2}{b^2} = 1$, is the equation of a hyperbola with center at the origin, foci on the y-axis at $(0, -c)$ and $(0, c)$, where $c^2 = a^2 + b^2$, and transverse axis along the y-axis.

Notice the difference in the forms of equations (2) and (3). When the y^2-term is subtracted from the x^2-term, the transverse axis is the x-axis. When the x^2-term is subtracted from the y^2-term, the transverse axis is the y-axis.

EXAMPLE 3 Finding an Equation of a Hyperbola

Find an equation of the hyperbola having one vertex at $(0, 2)$ and foci at $(0, -3)$ and $(0, 3)$.

Solution Since the foci are at $(0, -3)$ and $(0, 3)$, the center of the hyperbola is at the origin. Also, the transverse axis is along the y-axis. The given information also reveals that $c = 3$, $a = 2$, and $b^2 = c^2 - a^2 = 9 - 4 = 5$. The form of the equation of the hyperbola is given by equation (3):

$$\frac{y^2}{a^2} - \frac{x^2}{b^2} = 1$$

$$\frac{y^2}{4} - \frac{x^2}{5} = 1$$

NOW WORK PROBLEM 13.

Look at the equations of the hyperbolas in Examples 1 and 3. For the hyperbola in Example 1, $a^2 = 9$ and $b^2 = 7$, so $a > b$; for the hyperbola in Example 3, $a^2 = 4$ and $b^2 = 5$, so $a < b$. We conclude that, for hyperbolas, there are no requirements involving the relative sizes of a and b. Contrast this situation to the case of an ellipse, in which the relative sizes of a and b dictate which axis is the major axis. Hyperbolas have another feature to distinguish them from ellipses and parabolas: Hyperbolas have asymptotes.

Asymptotes

2 Recall from Section 4.7 that a horizontal or oblique asymptote of a graph is a line with the property that the distance from the line to points on the graph approaches 0 as $x \to -\infty$ or as $x \to \infty$.

THEOREM **Asymptotes of a Hyperbola**

The hyperbola $\frac{x^2}{a^2} - \frac{y^2}{b^2} = 1$ has the two oblique asymptotes

$$y = \frac{b}{a}x \quad \text{and} \quad y = -\frac{b}{a}x \qquad (4)$$

Proof We begin by solving for y in the equation of the hyperbola:

$$\frac{x^2}{a^2} - \frac{y^2}{b^2} = 1$$
$$\frac{y^2}{b^2} = \frac{x^2}{a^2} - 1$$
$$y^2 = b^2\left(\frac{x^2}{a^2} - 1\right)$$

If $x \neq 0$, we can rearrange the right side in the form

$$y^2 = \frac{b^2x^2}{a^2}\left(1 - \frac{a^2}{x^2}\right)$$
$$y = \pm\frac{bx}{a}\sqrt{1 - \frac{a^2}{x^2}}$$

Now, as $x \to -\infty$ or as $x \to \infty$, the term a^2/x^2 approaches 0, so the expression under the radical approaches 1. Thus, as $x \to -\infty$ or as $x \to \infty$, the value of y approaches $\pm bx/a$; that is, the graph of the hyperbola approaches the lines

$$y = -\frac{b}{a}x \quad \text{and} \quad y = \frac{b}{a}x$$

Thus, these lines are oblique asymptotes of the hyperbola.

Figure 40
$\frac{x^2}{a^2} - \frac{y^2}{b^2} = 1$

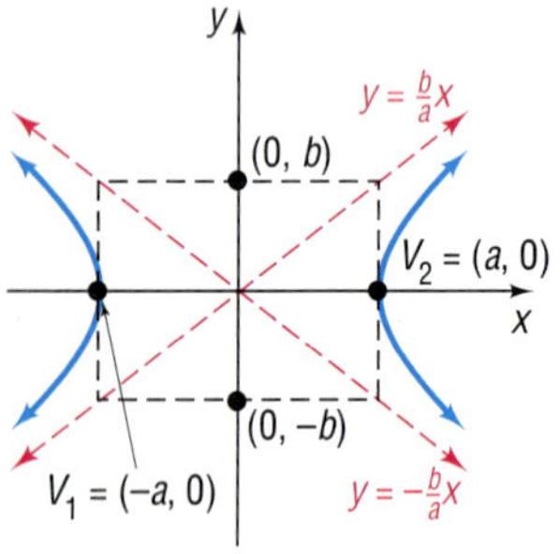

The asymptotes of a hyperbola are not part of the hyperbola, but they do serve as a guide for graphing a hyperbola by hand. For example, suppose that we want to graph the equation

$$\frac{x^2}{a^2} - \frac{y^2}{b^2} = 1$$

We begin by plotting the vertices $(-a, 0)$ and $(a, 0)$. Then we plot the points $(0, -b)$ and $(0, b)$ and use these four points to construct a rectangle, as shown in Figure 40. The diagonals of this rectangle have slopes b/a and $-b/a$, and their extensions are the asymptotes $y = (b/a)x$ and $y = -(b/a)x$ of the hyperbola.

THEOREM

Asymptotes of a Hyperbola

The hyperbola $\frac{y^2}{a^2} - \frac{x^2}{b^2} = 1$ has the two oblique asymptotes

$$y = \frac{a}{b}x \quad \text{and} \quad y = -\frac{a}{b}x \qquad (5)$$

You are asked to prove this result in Problem 64.

3 For the remainder of this section, the direction "Discuss the equation" will mean to find the center, transverse axis, vertices, foci, and asymptotes of the hyperbola and graph it.

EXAMPLE 4 Discussing the Equation of a Hyperbola

Discuss the equation $\frac{y^2}{4} - x^2 = 1$.

Figure 41 $\frac{y^2}{4} - x^2 = 1$

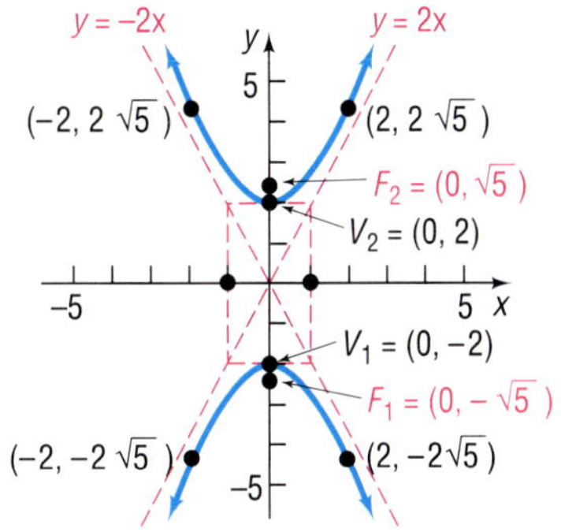

Solution Since the x^2 term is subtracted from the y^2 term, the equation is of the form of equation (3) and is a hyperbola with center at the origin and transverse axis along the y-axis. Also, comparing the above equation to equation (3), we find that $a^2 = 4$, $b^2 = 1$, and $c^2 = a^2 + b^2 = 5$. The vertices are at $(0, \pm a) = (0, \pm 2)$, and the foci are at $(0, \pm c) = (0, \pm\sqrt{5})$. Using (5), the asymptotes are the lines $y = \frac{a}{b}x = 2x$ and $y = -\frac{a}{b}x = -2x$. Form the rectangle containing the points $(0, \pm a) = (0, \pm 2)$ and $(\pm b, 0) = (\pm 1, 0)$. The diagonals of this rectangle are the asymptotes. Now graph the rectangle, the asymptotes, and the hyperbola. See Figure 41.

EXAMPLE 5 Discussing the Equation of a Hyperbola

Discuss the equation $9x^2 - 4y^2 = 36$.

Solution Divide each side of the equation by 36 to put the equation in proper form:

$$\frac{x^2}{4} - \frac{y^2}{9} = 1$$

We now proceed to analyze the equation. The center of the hyperbola is the origin. Since the x^2 term is first in the equation, we know that the transverse axis is along the x-axis and the vertices and foci will lie on the x-axis. Using equation (2), we find $a^2 = 4$, $b^2 = 9$, and $c^2 = a^2 + b^2 = 13$. The vertices are $a = 2$ units left and right of the center at $(\pm a, 0) = (\pm 2, 0)$; the foci are $c = \sqrt{13}$ units left and right of the center at $(\pm c, 0) = (\pm\sqrt{13}, 0)$; and the asymptotes have the equations

$$y = \frac{b}{a}x = \frac{3}{2}x \quad \text{and} \quad y = -\frac{b}{a}x = -\frac{3}{2}x$$

Figure 42(a) shows the graph of $\frac{x^2}{4} - \frac{y^2}{9} = 1$ and its asymptotes, $y = \frac{3}{2}x$ and $y = -\frac{3}{2}x$, using a graphing utility.

To graph the hyperbola by hand, form the rectangle containing the points $(\pm a, 0)$ and $(0, \pm b)$, that is, $(-2, 0)$, $(2, 0)$, $(0, -3)$, and $(0, 3)$. The extensions of the diagonals of this rectangle are the asymptotes. See Figure 42(b) for the graph.

Figure 42 $\frac{x^2}{4} - \frac{y^2}{9} = 1$

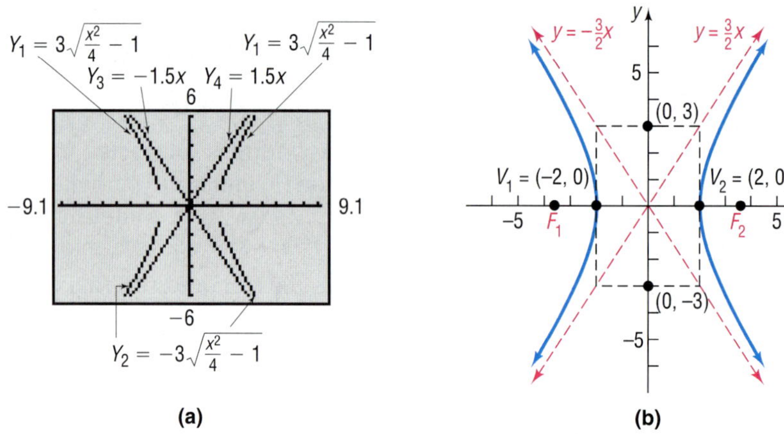

NOW WORK PROBLEM 21.

EXPLORATION Use TRACE to see what happens as x becomes unbounded in the positive direction for both the upper and lower portions of the hyperbola in Example 5.

Center at (h, k)

4

If a hyperbola with center at the origin and transverse axis coinciding with a coordinate axis is shifted horizontally h units and then vertically k units, the result is a hyperbola with center at (h, k) and transverse axis parallel to a coordinate axis. Table 4 gives the forms of the equations of such hyperbolas. See Figure 43 for the graphs.

Table 4 Hyperbolas with Center at (h, k) and Transverse Axis Parallel to a Coordinate Axis

Center	Transverse Axis	Foci	Vertices	Equation	Asymptotes
(h, k)	Parallel to x-axis	$(h \pm c, k)$	$(h \pm a, k)$	$\frac{(x-h)^2}{a^2} - \frac{(y-k)^2}{b^2} = 1, \quad b^2 = c^2 - a^2$	$y - k = \pm\frac{b}{a}(x - h)$
(h, k)	Parallel to y-axis	$(h, k \pm c)$	$(h, k \pm a)$	$\frac{(y-k)^2}{a^2} - \frac{(x-h)^2}{b^2} = 1, \quad b^2 = c^2 - a^2$	$y - k = \pm\frac{a}{b}(x - h)$

Figure 43

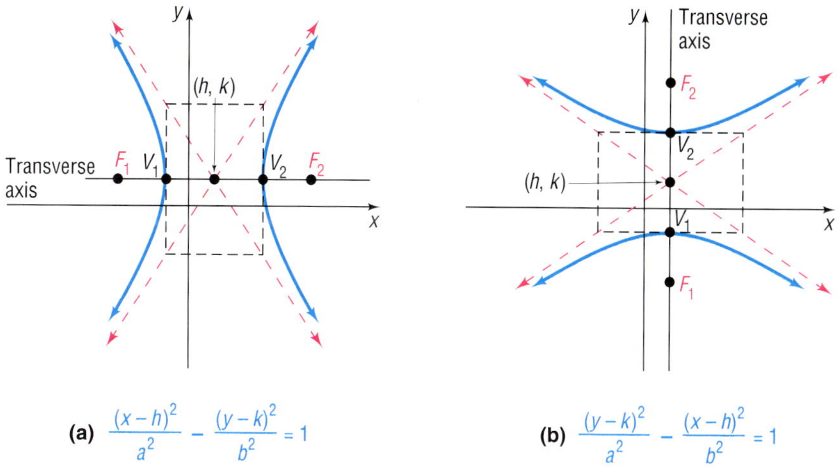

EXAMPLE 6 Finding an Equation of a Hyperbola, Center Not at the Origin

Find an equation for the hyperbola with center at $(1, -2)$, one focus at $(4, -2)$, and one vertex at $(3, -2)$. Graph the equation by hand.

Solution The center is at $(h, k) = (1, -2)$, so $h = 1$ and $k = -2$. Since the center, focus, and vertex all lie on the line $y = -2$, the transverse axis is parallel to the x-axis. The distance from the center $(1, -2)$ to the focus $(4, -2)$ is $c = 3$; the distance from the center $(1, -2)$ to the vertex $(3, -2)$ is $a = 2$. Thus, $b^2 = c^2 - a^2 = 9 - 4 = 5$. The equation is

$$\frac{(x-h)^2}{a^2} - \frac{(y-k)^2}{b^2} = 1$$

$$\frac{(x-1)^2}{4} - \frac{(y+2)^2}{5} = 1$$

See Figure 44.

Figure 44 $\frac{(x-1)^2}{4} - \frac{(y+2)^2}{5} = 1$

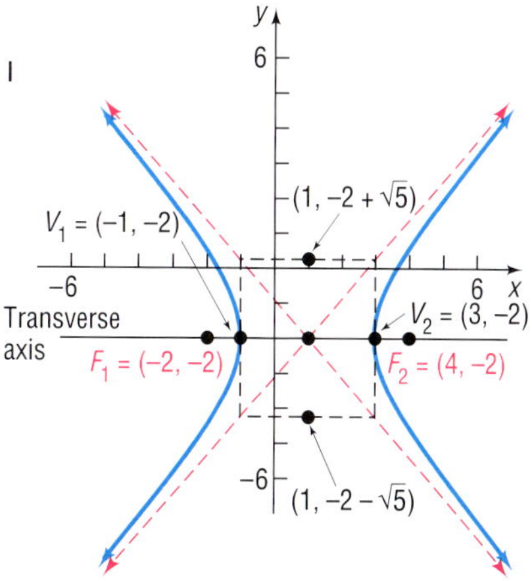

NOW WORK PROBLEM 31.

EXAMPLE 7 Using a Graphing Utility to Graph a Hyperbola, Center Not at the Origin

Use a graphing utility to graph the hyperbola $\dfrac{(x-1)^2}{4} - \dfrac{(y+2)^2}{5} = 1$.

Solution First, we must solve the equation for y:

$$\frac{(x-1)^2}{4} - \frac{(y+2)^2}{5} = 1$$
$$\frac{(y+2)^2}{5} = \frac{(x-1)^2}{4} - 1$$
$$(y+2)^2 = 5\left[\frac{(x-1)^2}{4} - 1\right]$$
$$y + 2 = \pm\sqrt{5\left[\frac{(x-1)^2}{4} - 1\right]}$$
$$y = -2 \pm \sqrt{5\left[\frac{(x-1)^2}{4} - 1\right]}$$

Figure 45

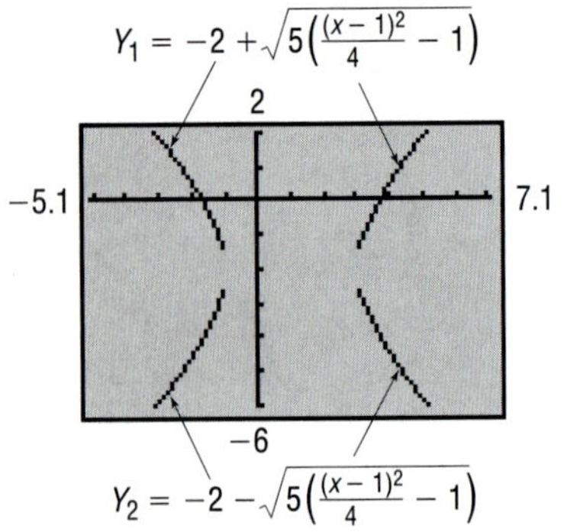

Figure 45 shows the graph of $Y_1 = -2 + \sqrt{5\left[\dfrac{(x-1)^2}{4} - 1\right]}$ and $Y_2 = -2 - \sqrt{5\left[\dfrac{(x-1)^2}{4} - 1\right]}$.

EXAMPLE 8 Discussing the Equation of a Hyperbola

Discuss the equation $-x^2 + 4y^2 - 2x - 16y + 11 = 0$.

Solution We complete the squares in x and in y:

$$-x^2 + 4y^2 - 2x - 16y + 11 = 0$$
$$-(x^2 + 2x) + 4(y^2 - 4y) = -11 \quad \text{Group terms.}$$
$$-(x^2 + 2x + 1) + 4(y^2 - 4y + 4) = -1 + 16 - 11 \quad \text{Complete each square.}$$
$$-(x+1)^2 + 4(y-2)^2 = 4$$
$$(y-2)^2 - \frac{(x+1)^2}{4} = 1 \quad \text{Divide by 4.}$$

Figure 46(a) shows the graph of $(y-2)^2 - \dfrac{(x+1)^2}{4} = 1$ using a graphing utility. We now proceed to analyze the equation.

It is the equation of a hyperbola with center at $(-1, 2)$ and transverse axis parallel to the y-axis. Also, $a^2 = 1$ and $b^2 = 4$, so $c^2 = a^2 + b^2 = 5$. Since the transverse axis is parallel to the y-axis, the vertices and foci are located a and c units above and below the center, respectively. The vertices are at $(h, k \pm a) = (-1, 2 \pm 1)$, or $(-1, 1)$ and $(-1, 3)$. The foci are at

$(h, k \pm c) = (-1, 2 \pm \sqrt{5})$. The asymptotes are $y - 2 = \frac{1}{2}(x + 1)$ and $y - 2 = -\frac{1}{2}(x + 1)$. Figure 46(b) shows the graph drawn by hand.

Figure 46

$$(y-2)^2 - \frac{(x+1)^2}{4} = 1$$

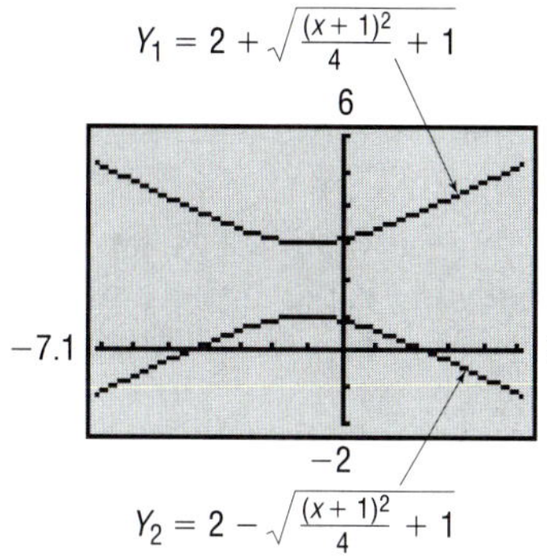

(a)

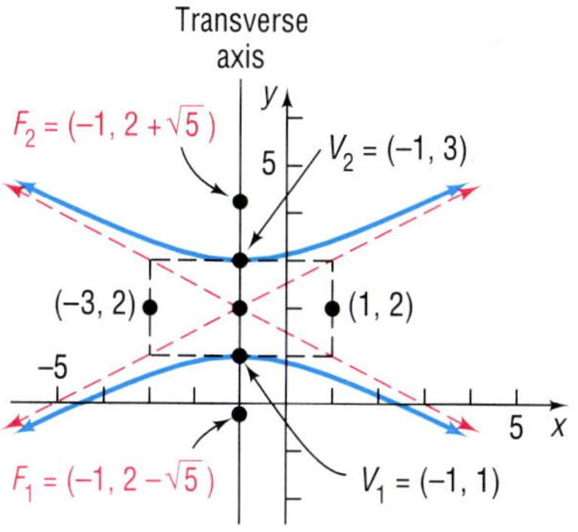

(b)

NOW WORK PROBLEM 45.

Applications

Figure 47

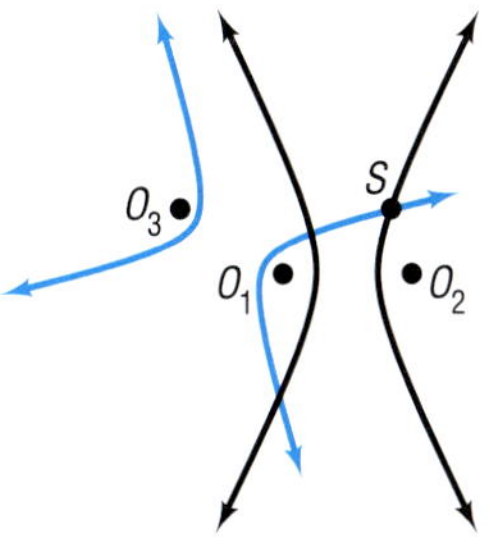

5 Suppose that a gun is fired from an unknown source S. An observer at O_1 hears the report (sound of gun shot) 1 second after another observer at O_2. Because sound travels at about 1100 feet per second, it follows that the point S must be 1100 feet closer to O_2 than to O_1. Thus, S lies on one branch of a hyperbola with foci at O_1 and O_2. (Do you see why? The difference of the distances from S to O_1 and from S to O_2 is the constant 1100.) If a third observer at O_3 hears the same report 2 seconds after O_1 hears it, then S will lie on a branch of a second hyperbola with foci at O_1 and O_3. The intersection of the two hyperbolas will pinpoint the location of S. See Figure 47 for an illustration.

LORAN

Figure 48

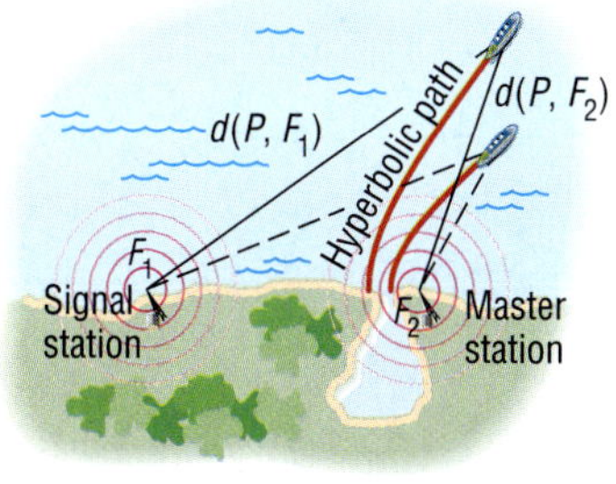

$d(P, F_1) - d(P, F_2)$ = constant

In the LOng RAnge Navigation system (LORAN), a master radio sending station and a secondary sending station emit signals that can be received by a ship at sea. (See Figure 48.) Because a ship monitoring the two signals will usually be nearer to one of the two stations, there will be a difference in the distance that the two signals travel, which will register as a slight time difference between the signals. As long as the time difference remains constant, the difference of the two distances will also be constant. If the ship follows a path corresponding to the fixed time difference, it will follow the path of a hyperbola whose foci are located at the positions of the two sending stations. So for each time difference a different hyperbolic path results, each bringing the ship to a different shore location. Navigation charts show the various hyperbolic paths corresponding to different time differences.

EXAMPLE 9 LORAN

Two LORAN stations are positioned 250 miles apart along a straight shore.

(a) A ship records a time difference of 0.00054 second between the LORAN signals. Set up an appropriate rectangular coordinate system to determine where the ship would reach shore if it were to follow the hyperbola corresponding to this time difference.

(b) If the ship wants to enter a harbor located between the two stations 25 miles from the master station, what time difference should it be looking for?

(c) If the ship is 80 miles offshore when the desired time difference is obtained, what is the approximate location of the ship?
[***Note:*** The speed of each radio signal is 186,000 miles per second.]

Solution

Figure 49

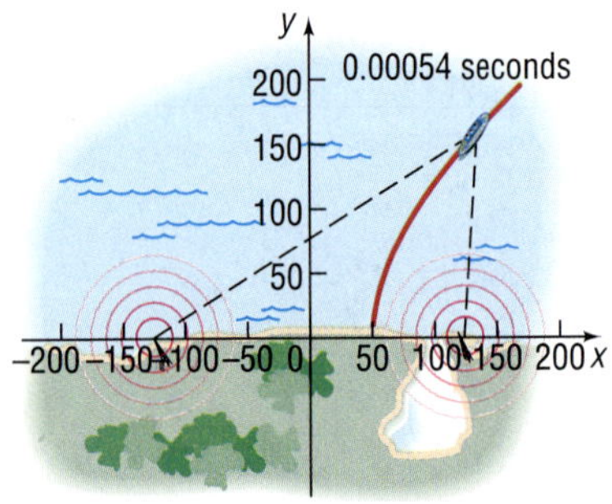

(a) We set up a rectangular coordinate system so that the two stations lie on the x-axis and the origin is midway between them. See Figure 49. The ship lies on a hyperbola whose foci are the locations of the two stations. The reason for this is that the constant time difference of the signals from each station results in a constant difference in the distance of the ship from each station. Since the time difference is 0.00054 second and the speed of the signal is 186,000 miles per second, the difference of the distances from the ship to each station (foci) is

$$\text{Distance} = \text{Speed} \times \text{Time} = 186{,}000 \times 0.00054 \approx 100 \text{ miles}$$

The difference of the distances from the ship to each station, 100, equals $2a$, so $a = 50$ and the vertex of the corresponding hyperbola is at $(50, 0)$. Since the focus is at $(125, 0)$, following this hyperbola the ship would reach shore 75 miles from the master station.

(b) To reach shore 25 miles from the master station, the ship would follow a hyperbola with vertex at $(100, 0)$. For this hyperbola, $a = 100$, so the constant difference of the distances from the ship to each station is 200. The time difference that the ship should look for is

$$\text{Time} = \frac{\text{Distance}}{\text{Speed}} = \frac{200}{186{,}000} = 0.001075 \text{ second}$$

(c) To find the approximate location of the ship, we need to find the equation of the hyperbola with vertex at $(100, 0)$ and a focus at $(125, 0)$. The form of the equation of this hyperbola is

$$\frac{x^2}{a^2} - \frac{y^2}{b^2} = 1$$

where $a = 100$. Since $c = 125$, we have

$$b^2 = c^2 - a^2 = 125^2 - 100^2 = 5625$$

The equation of the hyperbola is

$$\frac{x^2}{100^2} - \frac{y^2}{5625} = 1$$

Figure 50

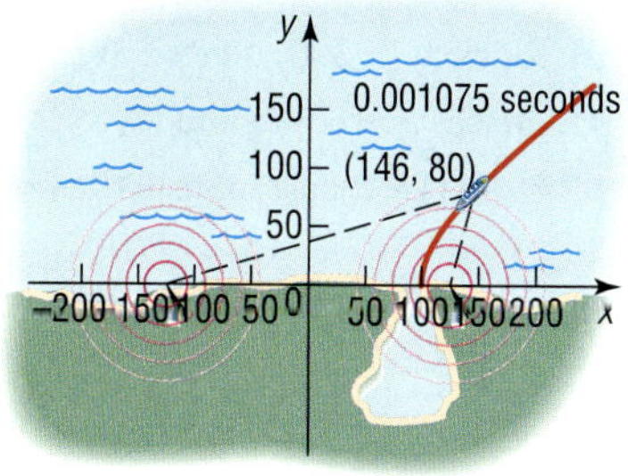

Since the ship is 80 miles from shore, we use $y = 80$ in the equation and solve for x.

$$\frac{x^2}{100^2} - \frac{80^2}{5625} = 1$$

$$\frac{x^2}{100^2} = 1 + \frac{80^2}{5625} = 2.14$$

$$x^2 = 100^2(2.14)$$

$$x = 146$$

The ship is at the position (146, 80). See Figure 50.

9.4 EXERCISES

In Problems 1–4, the graph of a hyperbola is given. Match each graph to its equation.

A. $\frac{x^2}{4} - y^2 = 1$ *B.* $x^2 - \frac{y^2}{4} = 1$ *C.* $\frac{y^2}{4} - x^2 = 1$ *D.* $y^2 - \frac{x^2}{4} = 1$

1.

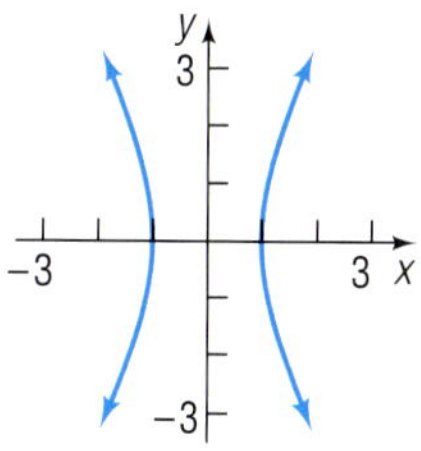

2.

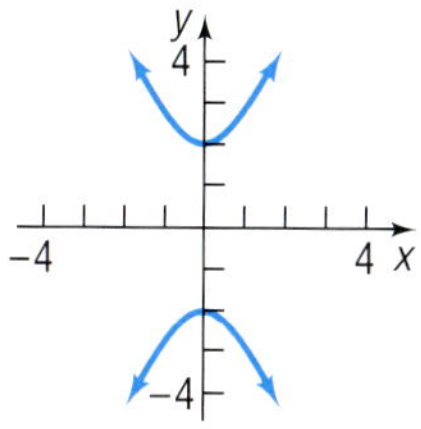

3.

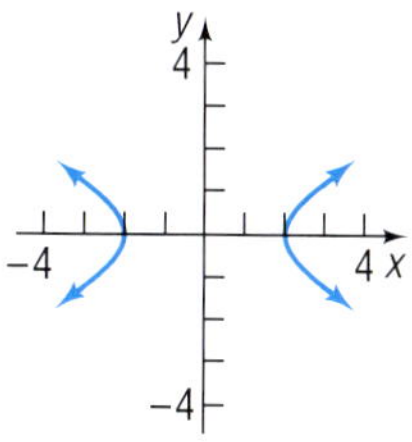

4.

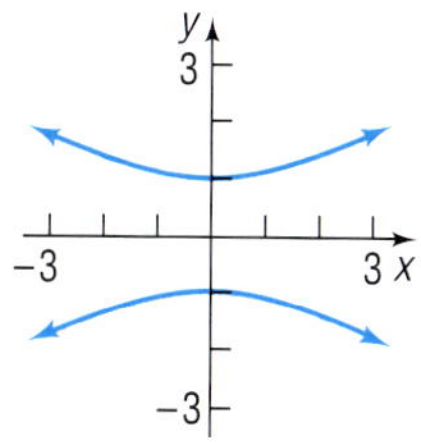

In Problems 5–8, the graph of a hyperbola is given. Match each graph to its equation.

A. $\frac{x^2}{16} - \frac{y^2}{9} = 1$ *B.* $\frac{x^2}{9} - \frac{y^2}{16} = 1$ *C.* $\frac{y^2}{16} - \frac{x^2}{9} = 1$ *D.* $\frac{y^2}{9} - \frac{x^2}{16} = 1$

5.

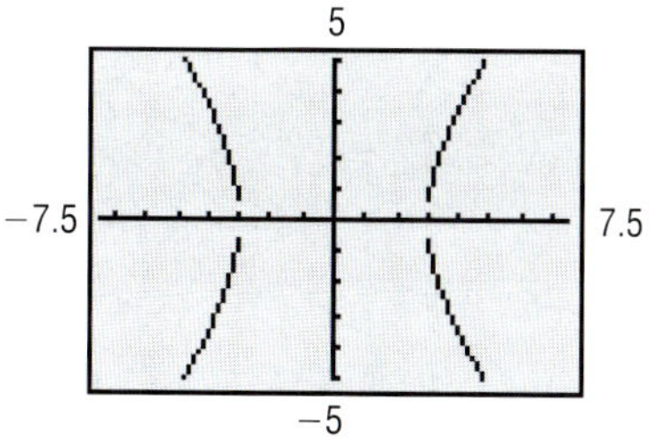

6.

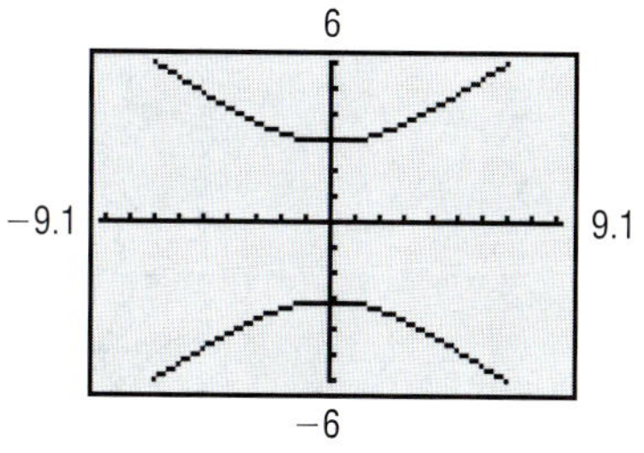

7.

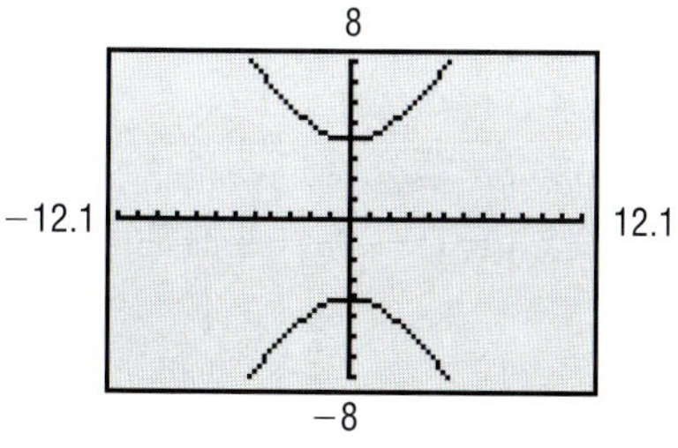

8.

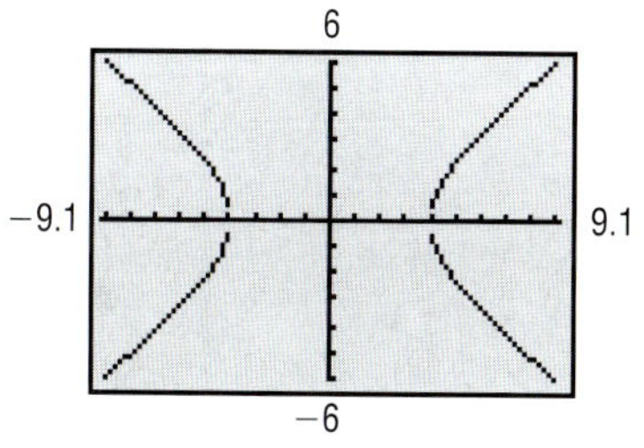

In Problems 9–18, find an equation for the hyperbola described. Graph the equation using a graphing utility.

9. Center at $(0, 0)$; focus at $(3, 0)$; vertex at $(1, 0)$
10. Center at $(0, 0)$; focus at $(0, 5)$; vertex at $(0, 3)$
11. Center at $(0, 0)$; focus at $(0, -6)$; vertex at $(0, 4)$
12. Center at $(0, 0)$; focus at $(-3, 0)$; vertex at $(2, 0)$
13. Foci at $(-5, 0)$ and $(5, 0)$; vertex at $(3, 0)$
14. Focus at $(0, 6)$; vertices at $(0, -2)$ and $(0, 2)$
15. Vertices at $(0, -6)$ and $(0, 6)$; asymptote the line $y = 2x$
16. Vertices at $(-4, 0)$ and $(4, 0)$; asymptote the line $y = 2x$
17. Foci at $(-4, 0)$ and $(4, 0)$; asymptote the line $y = -x$
18. Foci at $(0, -2)$ and $(0, 2)$; asymptote the line $y = -x$

In Problems 19–26, find the center, transverse axis, vertices, foci, and asymptotes. Graph each equation (a) by hand and (b) using a graphing utility.

19. $\dfrac{x^2}{25} - \dfrac{y^2}{9} = 1$
20. $\dfrac{y^2}{16} - \dfrac{x^2}{4} = 1$
21. $4x^2 - y^2 = 16$
22. $y^2 - 4x^2 = 16$
23. $y^2 - 9x^2 = 9$
24. $x^2 - y^2 = 4$
25. $y^2 - x^2 = 25$
26. $2x^2 - y^2 = 4$

In Problems 27–30, write an equation for each hyperbola.

27.
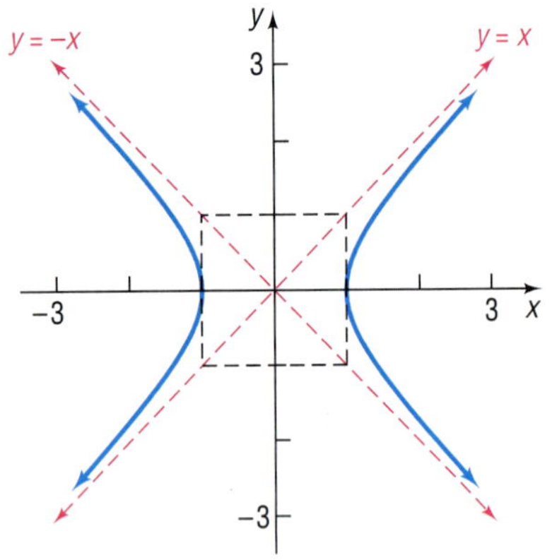

28.
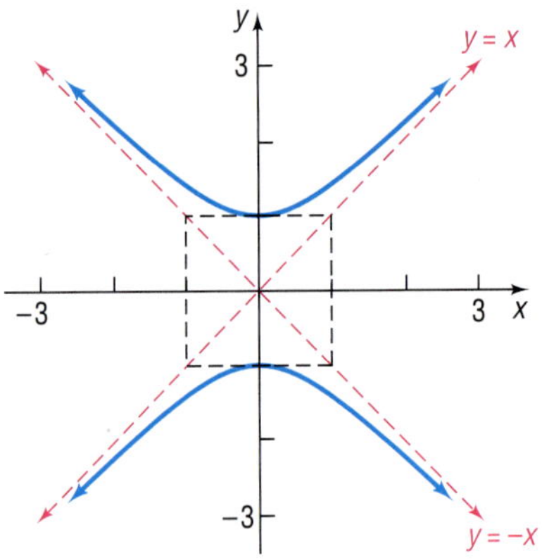

29.
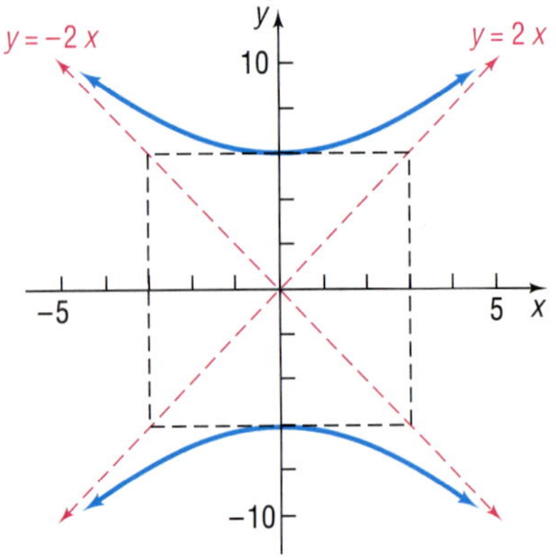

30.
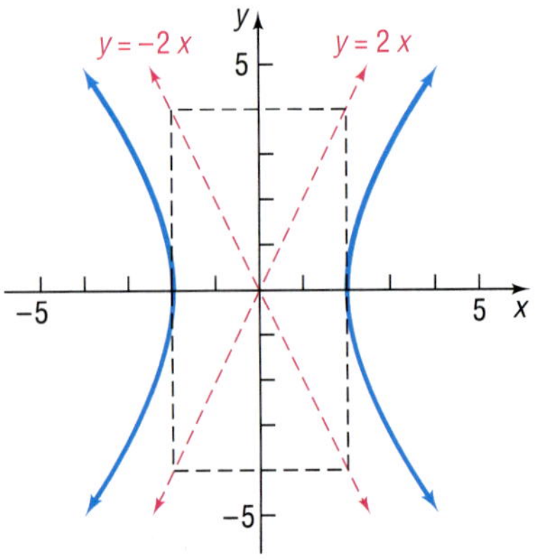

In Problems 31–38, find an equation for the hyperbola described. Graph the equation by hand.

31. Center at $(4, -1)$; focus at $(7, -1)$; vertex at $(6, -1)$
32. Center at $(-3, 1)$; focus at $(-3, 6)$; vertex at $(-3, 4)$
33. Center at $(-3, -4)$; focus at $(-3, -8)$; vertex at $(-3, -2)$
34. Center at $(1, 4)$; focus at $(-2, 4)$; vertex at $(0, 4)$

35. Foci at $(3, 7)$ and $(7, 7)$; vertex at $(6, 7)$

36. Focus at $(-4, 0)$; vertices at $(-4, 4)$ and $(-4, 2)$

37. Vertices at $(-1, -1)$ and $(3, -1)$; asymptote the line $(x - 1)/2 = (y + 1)/3$

38. Vertices at $(1, -3)$ and $(1, 1)$; asymptote the line $(x - 1)/2 = (y + 1)/3$

In Problems 39–52, find the center, transverse axis, vertices, foci, and asymptotes. Graph each equation using a graphing utility.

39. $\dfrac{(x-2)^2}{4} - \dfrac{(y+3)^2}{9} = 1$

40. $\dfrac{(y+3)^2}{4} - \dfrac{(x-2)^2}{9} = 1$

41. $(y - 2)^2 - 4(x + 2)^2 = 4$

42. $(x + 4)^2 - 9(y - 3)^2 = 9$

43. $(x + 1)^2 - (y + 2)^2 = 4$

44. $(y - 3)^2 - (x + 2)^2 = 4$

45. $x^2 - y^2 - 2x - 2y - 1 = 0$

46. $y^2 - x^2 - 4y + 4x - 1 = 0$

47. $y^2 - 4x^2 - 4y - 8x - 4 = 0$

48. $2x^2 - y^2 + 4x + 4y - 4 = 0$

49. $4x^2 - y^2 - 24x - 4y + 16 = 0$

50. $2y^2 - x^2 + 2x + 8y + 3 = 0$

51. $y^2 - 4x^2 - 16x - 2y - 19 = 0$

52. $x^2 - 3y^2 + 8x - 6y + 4 = 0$

In Problems 53–56, graph each function by hand. Verify your answer using a graphing utility.
[Hint: Notice that each function is half a hyperbola.]

53. $f(x) = \sqrt{16 + 4x^2}$

54. $f(x) = -\sqrt{9 + 9x^2}$

55. $f(x) = -\sqrt{-25 + x^2}$

56. $f(x) = \sqrt{-1 + x^2}$

57. LORAN Two LORAN stations are positioned 200 miles apart along a straight shore.
(a) A ship records a time difference of 0.00038 second between the LORAN signals. Set up an appropriate rectangular coordinate system to determine where the ship would reach shore if it were to follow the hyperbola corresponding to this time difference.
(b) If the ship wants to enter a harbor located between the two stations 20 miles from the master station, what time difference should it be looking for?
(c) If the ship is 50 miles offshore when the desired time difference is obtained, what is the approximate location of the ship?
[***Note:*** The speed of each radio signal is 186,000 miles per second.]

58. LORAN Two LORAN stations are positioned 100 miles apart along a straight shore.
(a) A ship records a time difference of 0.00032 second between the LORAN signals. Set up an appropriate rectangular coordinate system to determine where the ship would reach shore if it were to follow the hyperbola corresponding to this time difference.
(b) If the ship wants to enter a harbor located between the two stations 10 miles from the master station, what time difference should it be looking for?
(c) If the ship is 20 miles offshore when the desired time difference is obtained, what is the approximate location of the ship?
[***Note:*** The speed of each radio signal is 186,000 miles per second.]

59. Calibrating Instruments In a test of their recording devices, a team of seismologists positioned two of the devices 2000 feet apart, with the device at point A to the west of the device at point B. At a point between the devices and 200 feet from point B, a small amount of explosive was detonated and a note made of the time at which the sound reached each device. A second explosion is to be carried out at a point directly north of point B.
(a) How far north should the site of the second explosion be chosen so that the measured time difference recorded by the devices for the second detonation is the same as that recorded for the first detonation?
(b) Explain why this experiment can be used to calibrate the instruments.

60. Explain in your own words the LORAN system of navigation.

61. The **eccentricity** e of a hyperbola is defined as the number c/a, where a and c are the numbers given in equation (2). Because $c > a$, it follows that $e > 1$. Describe the general shape of a hyperbola whose eccentricity is close to 1. What is the shape if e is very large?

62. A hyperbola for which $a = b$ is called an **equilateral hyperbola.** Find the eccentricity e of an equilateral hyperbola.
[***Note:*** The eccentricity of a hyperbola is defined in Problem 61.]

63. Two hyperbolas that have the same set of asymptotes are called **conjugate.** Show that the hyperbolas

$$\frac{x^2}{4} - y^2 = 1 \quad \text{and} \quad y^2 - \frac{x^2}{4} = 1$$

are conjugate. Graph each hyperbola on the same set of coordinate axes.

64. Prove that the hyperbola

$$\frac{y^2}{a^2} - \frac{x^2}{b^2} = 1$$

has the two oblique asymptotes

$$y = \frac{a}{b}x \quad \text{and} \quad y = -\frac{a}{b}x$$

65. Show that the graph of an equation of the form

$$Ax^2 + Cy^2 + F = 0 \qquad A \neq 0, C \neq 0, F \neq 0$$

where A and C are of opposite sign, is a hyperbola with center at $(0, 0)$.

66. Show that the graph of an equation of the form

$$Ax^2 + Cy^2 + Dx + Ey + F = 0 \quad A \neq 0, C \neq 0$$

where A and C are of opposite sign,

(a) Is a hyperbola if $(D^2/4A) + (E^2/4C) - F \neq 0$.
(b) Is two intersecting lines if

$$(D^2/4A) + (E^2/4C) - F = 0$$

9.5 SYSTEMS OF NONLINEAR EQUATIONS

1 Solve a System of Nonlinear Equations Using Substitution

2 Solve a System of Nonlinear Equations Using Elimination

1 In Section 6.1 we observed that the solution to a system of linear equations could be found geometrically by determining the point of intersection of the equations in the system. Similarly, when solving systems of nonlinear equations, the solution(s) represent the point(s) of intersection of the equations.

There is no general methodology for solving a system of nonlinear equations by hand. There are times when substitution is best; other times, elimination is best; and there are times when neither of these methods works. Experience and a certain degree of imagination are your allies here.

Before we begin, two comments are in order:

1. If the system contains two variables, then graph them. By graphing each equation in the system, we can get an idea of how many solutions a system has and approximately where they are located.
2. Extraneous solutions can creep in when solving nonlinear systems algebraically, so it is imperative that all apparent solutions be checked.

EXAMPLE 1 Solving a System of Nonlinear Equations

Solve the following system of equations:

$$\begin{cases} 3x - y = -2 & \text{(1) A line.} \\ 2x^2 - y = 0 & \text{(2) A parabola.} \end{cases}$$

Graphing Solution

We use a graphing utility to graph $Y_1 = 3x + 2$ and $Y_2 = 2x^2$. From Figure 51, we see that the system apparently has two solutions. Using INTERSECT, the solutions to the system of equations are $(-0.5, 0.5)$ and $(2, 8)$.

Figure 51

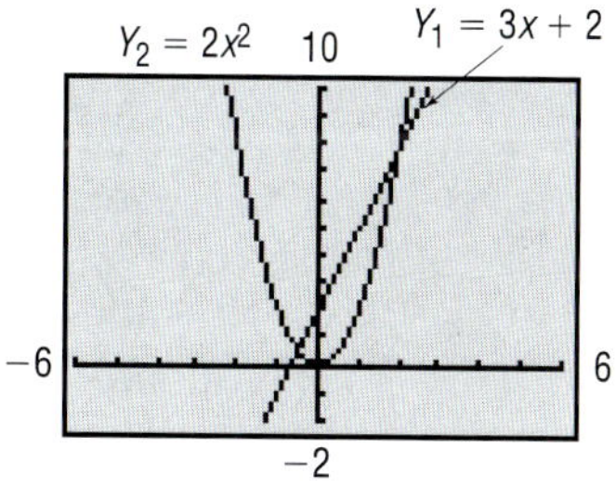

Algebraic Solution Using Substitution

First, we notice that the system contains two variables and that we know how to graph each equation by hand. In Figure 52, we see that the system apparently has two solutions.

Figure 52

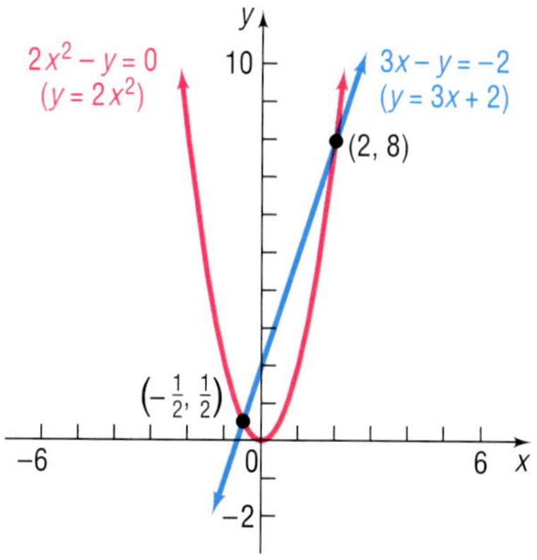

We will use substitution to solve the system. Equation (1) is easily solved for y:

$$\begin{aligned} 3x - y &= -2 \\ y &= 3x + 2 \end{aligned}$$

We substitute this expression for y in equation (2). The result is an equation containing just the variable x, which we can then solve:

$$\begin{aligned} 2x^2 - y &= 0 \\ 2x^2 - (3x + 2) &= 0 && \textit{y = 3x + 2} \\ 2x^2 - 3x - 2 &= 0 && \textit{Simplify.} \\ (2x + 1)(x - 2) &= 0 && \textit{Factor.} \\ 2x + 1 = 0 \quad &\text{or} \quad x - 2 = 0 && \textit{Zero Product Property.} \\ x = -\frac{1}{2} \quad & \quad\quad x = 2 \end{aligned}$$

Using these values for x in $y = 3x + 2$, we find that

$$y = 3\left(-\frac{1}{2}\right) + 2 = \frac{1}{2} \quad \text{or} \quad y = 3(2) + 2 = 8$$

The apparent solutions are $x = -\frac{1}{2}, y = \frac{1}{2}$ and $x = 2, y = 8$.

✓CHECK: For $x = -\frac{1}{2}, y = \frac{1}{2}$:

$$\begin{cases} 3\left(-\frac{1}{2}\right) - \frac{1}{2} = -\frac{3}{2} - \frac{1}{2} = -2 & (1) \\ 2\left(-\frac{1}{2}\right)^2 - \frac{1}{2} = 2\left(\frac{1}{4}\right) - \frac{1}{2} = 0 & (2) \end{cases}$$

For $x = 2, y = 8$:

$$\begin{cases} 3(2) - 8 = 6 - 8 = -2 & (1) \\ 2(2)^2 - 8 = 2(4) - 8 = 0 & (2) \end{cases}$$

Each solution checks. Now we know that the graphs in Figure 52 intersect at $\left(-\frac{1}{2}, \frac{1}{2}\right)$ and at $(2, 8)$.

NOW WORK PROBLEM 11.

2 Our next example illustrates how the method of elimination works for nonlinear systems.

EXAMPLE 2 Solving a System of Nonlinear Equations

Solve: $\begin{cases} x^2 + y^2 = 13 & (1) \text{ A circle.} \\ x^2 - y = 7 & (2) \text{ A parabola.} \end{cases}$

Graphing Solution We use a graphing utility to graph $x^2 + y^2 = 13$ and $x^2 - y = 7$. (Remember that to graph $x^2 + y^2 = 13$ requires two functions, $Y_1 = \sqrt{13 - x^2}$ and $Y_2 = -\sqrt{13 - x^2}$, and a square screen.) From Figure 53 we see that the system apparently has four solutions. Using INTERSECT, the solutions to the system of equations are $(-3, 2)$, $(3, 2)$, $(-2, -3)$, and $(2, -3)$.

Algebraic Solution Using Elimination First, we graph each equation, as shown in Figure 54. Based on the graph, we expect four solutions. By subtracting equation (2) from equation (1), the variable x is eliminated, leaving

$$\begin{cases} x^2 + y^2 = 13 \\ x^2 - y = 7 \end{cases}$$
$$y^2 + y = 6 \quad \text{Subtract.}$$

Figure 53

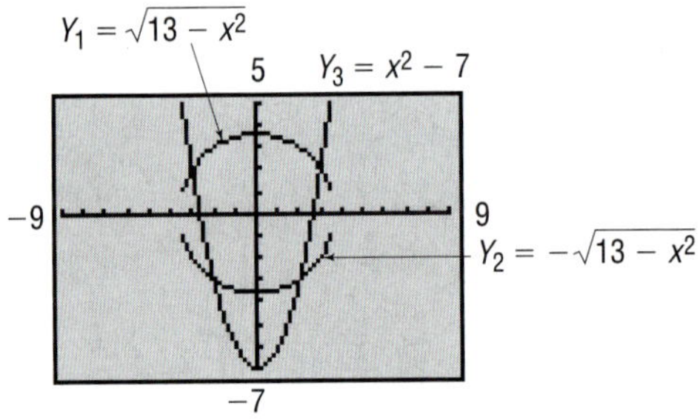

Figure 54

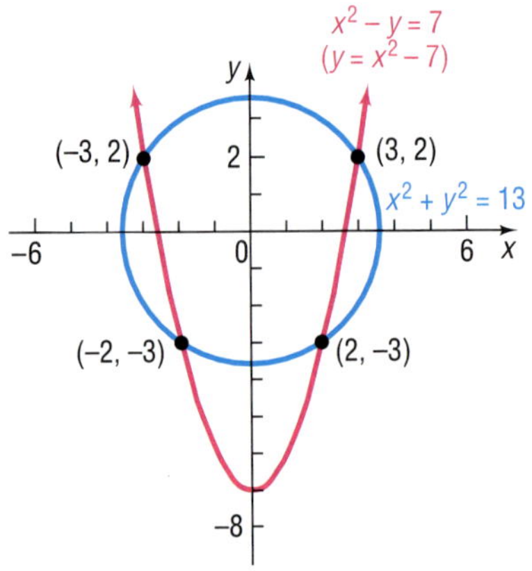

This quadratic equation in y is easily solved by factoring:

$$y^2 + y - 6 = 0$$
$$(y + 3)(y - 2) = 0$$
$$y = -3 \quad \text{or} \quad y = 2$$

We use these values for y in equation (2) to find x. If $y = 2$, then $x^2 = y + 7 = 9$ and $x = 3$ or -3. If $y = -3$, then $x^2 = y + 7 = 4$ and $x = 2$ or -2. Thus, we have four solutions: $x = 3$, $y = 2$; $x = -3$, $y = 2$; $x = 2$, $y = -3$; and $x = -2$, $y = -3$. You should verify that, in fact, these four solutions also satisfy equation (1), so all four are solutions of the system. The four points, $(3, 2)$, $(-3, 2)$, $(2, -3)$, and $(-2, -3)$, are the points of intersection of the graphs. Look again at Figure 54.

NOW WORK PROBLEM 9.

EXAMPLE 3 Solving a System of Nonlinear Equations

Solve: $$\begin{cases} x^2 + x + y^2 - 3y + 2 = 0 & (1) \\ x + 1 + \dfrac{y^2 - y}{x} = 0 & (2) \end{cases}$$

Graphing Solution

First, we multiply equation (2) by x to eliminate the fraction. The result is an equivalent system because x cannot be 0 [look at equation (2) to see why]:

$$\begin{cases} x^2 + x + y^2 - 3y + 2 = 0 & (1) \\ x^2 + x + y^2 - y = 0 & (2) \end{cases}$$

We need to solve each equation for y. First, we solve equation (1) for y:

$$x^2 + x + y^2 - 3y + 2 = 0$$

$$y^2 - 3y = -x^2 - x - 2 \qquad \text{Rearrange so that terms involving } y \text{ are on left side.}$$

$$y^2 - 3y + \frac{9}{4} = -x^2 - x - 2 + \frac{9}{4} \qquad \text{Complete the square involving } y.$$

$$\left(y - \frac{3}{2}\right)^2 = -x^2 - x + \frac{1}{4}$$

$$y - \frac{3}{2} = \pm\sqrt{-x^2 - x + \frac{1}{4}} \qquad \text{Take the square root of both sides.}$$

$$y = \frac{3}{2} \pm \sqrt{-x^2 - x + \frac{1}{4}} \qquad \text{Solve for } y.$$

Now we solve equation (2) for y:

$$x^2 + x + y^2 - y = 0$$

$$y^2 - y = -x^2 - x \qquad \text{Rearrange so that terms involving } y \text{ are on left side.}$$

$$y^2 - y + \frac{1}{4} = -x^2 - x + \frac{1}{4} \qquad \text{Complete the square involving } y.$$

$$\left(y - \frac{1}{2}\right)^2 = -x^2 - x + \frac{1}{4}$$

$$y - \frac{1}{2} = \pm\sqrt{-x^2 - x + \frac{1}{4}} \qquad \text{Take the square root of both sides.}$$

$$y = \frac{1}{2} \pm \sqrt{-x^2 - x + \frac{1}{4}} \qquad \text{Solve for } y.$$

Now graph each equation using a graphing utility. See Figure 55.

Figure 55

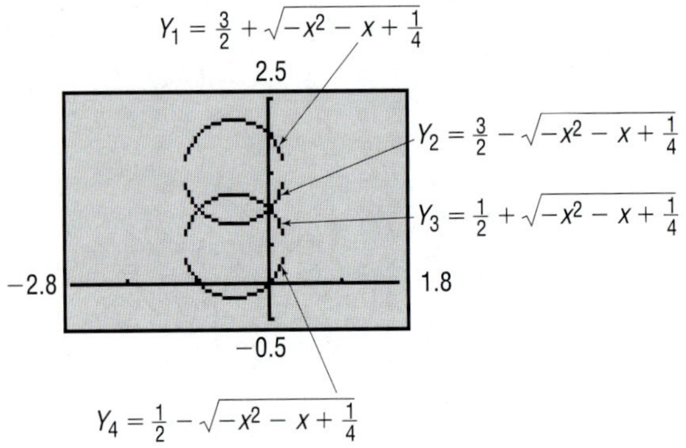

Using INTERSECT, the points of intersection are $(-1, 1)$ and $(0, 1)$. Because x cannot be 0, the value $x = 0$ is extraneous, and we discard it. Thus, the only solution is $x = -1$, $y = 1$.

Algebraic Solution Using Elimination First, we multiply equation (2) by x to eliminate the fraction. The result is an equivalent system because x cannot be 0 [look at equation (2) to see why]:

$$\begin{cases} x^2 + x + y^2 - 3y + 2 = 0 & (1) \\ x^2 + x + y^2 - y = 0 & (2) \end{cases}$$

Now subtract equation (2) from equation (1) to eliminate x. The result is

$$-2y + 2 = 0$$
$$y = 1$$

To find x, we back-substitute $y = 1$ in equation (1):

$$x^2 + x + 1 - 3 + 2 = 0$$
$$x^2 + x = 0$$
$$x(x + 1) = 0$$
$$x = 0 \quad \text{or} \quad x = -1$$

Because x cannot be 0, the value $x = 0$ is extraneous, and we discard it. Thus, the solution is$x = -1$, $y = 1$.

✓CHECK: We now check $x = -1$, $y = 1$:

$$\begin{cases} (-1)^2 + (-1) + 1^2 - 3(1) + 2 = 1 - 1 + 1 - 3 + 2 = 0 & (1) \\ -1 + 1 + \dfrac{1^2 - 1}{-1} = 0 + \dfrac{0}{-1} = 0 & (2) \end{cases}$$

Thus, the only solution to the system is $x = -1$, $y = 1$.

NOW WORK PROBLEMS **25** AND **49**.

EXAMPLE 4 Solving a System of Nonlinear Equations

Solve: $\begin{cases} x^2 - y^2 = 4 & (1)\ \text{A hyperbola.} \\ y = x^2 & (2)\ \text{A parabola.} \end{cases}$

Graphing Solution We graph $Y_1 = x^2$ and $x^2 - y^2 = 4$ in Figure 56. You will need to graph $x^2 - y^2 = 4$ as two functions:

$$Y_2 = \sqrt{x^2 - 4} \quad \text{and} \quad Y_3 = -\sqrt{x^2 - 4}$$

From Figure 56 we see that the graphs of these two equations do not intersect. Thus, the system is inconsistent.

Figure 56

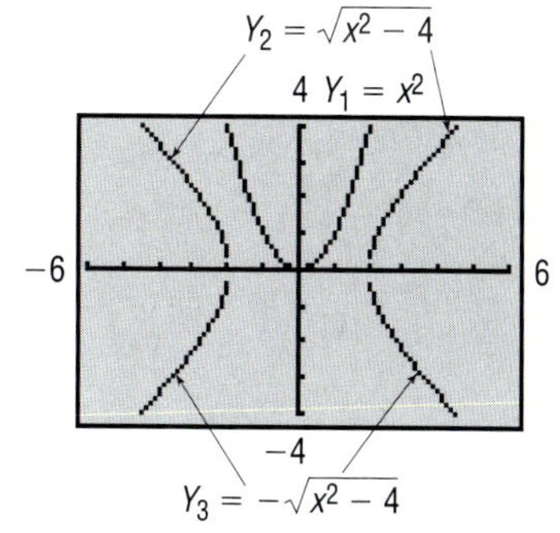

Algebraic Solution

Either substitution or elimination can be used here. We use substitution and replace x^2 by y in equation (1). The result is

$$y - y^2 = 4$$
$$y^2 - y + 4 = 0$$

This is a quadratic equation whose discriminant is $(-1)^2 - 4 \cdot 1 \cdot 4 = -15 < 0$. Thus, the equation has no real solutions, and hence the system is inconsistent.

EXAMPLE 5 Solving a System of Nonlinear Equations

$$\text{Solve:} \quad \begin{cases} 3xy - 2y^2 = -2 & (1) \\ 9x^2 + 4y^2 = 10 & (2) \end{cases}$$

Graphing Solution

To graph $3xy - 2y^2 = -2$, we need to solve for y. In this instance, it is easier to view the equation as a quadratic equation in the variable y.

$$3xy - 2y^2 = -2$$

$$2y^2 - 3xy - 2 = 0 \qquad \textit{Place in standard form.}$$

$$y = \frac{-(-3x) \pm \sqrt{(-3x)^2 - 4(2)(-2)}}{2(2)} \qquad \textit{Use the quadratic formula } a = 2, b = -3x, c = -2.$$

$$y = \frac{3x \pm \sqrt{9x^2 + 16}}{4} \qquad \textit{Simplify.}$$

Figure 57

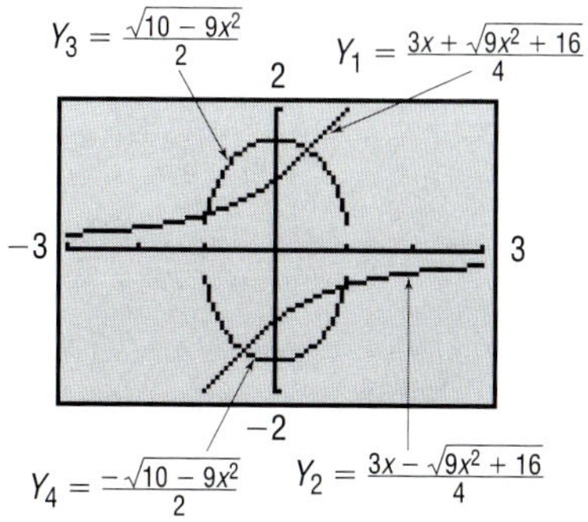

Using a graphing utility, we graph $Y_1 = \dfrac{3x + \sqrt{9x^2 + 16}}{4}$, and $Y_2 = \dfrac{3x - \sqrt{9x^2 + 16}}{4}$. From (2), we graph $Y_3 = \dfrac{\sqrt{10 - 9x^2}}{2}$ and $Y_4 = \dfrac{-\sqrt{10 - 9x^2}}{2}$. See Figure 57.

Using INTERSECT, the solutions to the system of equations are $(-1, 0.5)$, $(0.47, 1.41)$, $(1, -0.5)$, and $(-0.47, -1.41)$, each rounded to two decimal places.

Algebraic Solution We multiply equation (1) by 2 and add the result to equation (2) to eliminate the y^2-terms:

$$\begin{cases} 6xy - 4y^2 = -4 & (1) \\ 9x^2 + 4y^2 = 10 & (2) \end{cases}$$

$$9x^2 + 6xy = 6$$

$$3x^2 + 2xy = 2 \quad \text{Divide each side by 3.}$$

Since $x \neq 0$ (do you see why?), we can solve for y in this equation to get

$$y = \frac{2 - 3x^2}{2x} \qquad x \neq 0$$

Now substitute for y in equation (2) of the system:

$$9x^2 + 4y^2 = 10$$

$$9x^2 + 4\left(\frac{2 - 3x^2}{2x}\right)^2 = 10$$

$$9x^2 + \frac{4 - 12x^2 + 9x^4}{x^2} = 10$$

$$9x^4 + 4 - 12x^2 + 9x^4 = 10x^2$$

$$18x^4 - 22x^2 + 4 = 0$$

$$9x^4 - 11x^2 + 2 = 0 \quad \text{Divide each side by 2.}$$

This quadratic equation (in x^2) can be factored:

$$(9x^2 - 2)(x^2 - 1) = 0$$

$$9x^2 - 2 = 0 \quad \text{or} \quad x^2 - 1 = 0$$

$$x^2 = \tfrac{2}{9} \qquad\qquad x^2 = 1$$

$$x = \pm\frac{\sqrt{2}}{3} \qquad\qquad x = \pm 1$$

To find y, we use the fact that $y = \dfrac{2 - 3x^2}{2x}$ found above:

If $x = \dfrac{\sqrt{2}}{3}$: $\quad y = \dfrac{2 - 3x^2}{2x} = \dfrac{2 - \frac{2}{3}}{2(\sqrt{2}/3)} = \dfrac{4}{2\sqrt{2}} = \sqrt{2}$

If $x = -\dfrac{\sqrt{2}}{3}$: $\quad y = \dfrac{2 - 3x^2}{2x} = \dfrac{2 - \frac{2}{3}}{-2(\sqrt{2}/3)} = \dfrac{4}{-2\sqrt{2}} = -\sqrt{2}$

If $x = 1$: $\quad y = \dfrac{2 - 3x^2}{2x} = \dfrac{2 - 3}{2} = -\dfrac{1}{2}$

If $x = -1$: $\quad y = \dfrac{2 - 3x^2}{2x} = \dfrac{2 - 3}{-2} = \dfrac{1}{2}$

The system has four solutions. Check them for yourself.

NOW WORK PROBLEM **45.**

EXAMPLE 6 Running a Race

In a 50 mile race, the winner crosses the finish line 1 mile ahead of the second place runner and 4 miles ahead of the third place runner. Assuming that each runner maintains a constant speed throughout the race, by how many miles does the second place runner beat the third place runner?

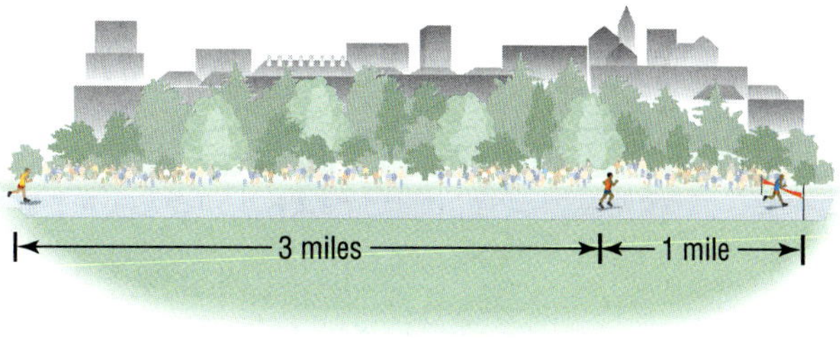

Solution Let v_1, v_2, v_3 denote the speeds of the first, second, and third place runners, respectively. Let t_1 and t_2 denote the times (in hours) required for the first place runner and second place runner to finish the race. Then we have the system of equations

$$\begin{cases} 50 = v_1 t_1 & \text{(1) First place runner goes 50 miles in } t_1 \text{ hours.} \\ 49 = v_2 t_1 & \text{(2) Second place runner goes 49 miles in } t_1 \text{ hours.} \\ 46 = v_3 t_1 & \text{(3) Third place runner goes 46 miles in } t_1 \text{ hours.} \\ 50 = v_2 t_2 & \text{(4) Second place runner goes 50 miles in } t_2 \text{ hours.} \end{cases}$$

We seek the distance of the third place runner from the finish at time t_2. That is, we seek

$$\begin{aligned} 50 - v_3 t_2 &= 50 - v_3\left(t_1 \cdot \frac{t_2}{t_1}\right) \\ &= 50 - (v_3 t_1) \cdot \frac{t_2}{t_1} \\ &= 50 - 46 \cdot \frac{50/v_2}{50/v_1} && \begin{cases} \text{From (3), } v_3 t_1 = 46; \\ \text{from (4), } t_2 = 50/v_2; \\ \text{from (1), } t_1 = 50/v_1. \end{cases} \\ &= 50 - 46 \cdot \frac{v_1}{v_2} \\ &= 50 - 46 \cdot \frac{50}{49} && \text{Form the quotient of (1) and (2).} \\ &\approx 3.06 \text{ miles} \end{aligned}$$

HISTORICAL FEATURE

Recall that, in the beginning of this section, we said imagination and experience are important in solving simultaneous nonlinear equations. Indeed, these kinds of problems lead into some of the deepest and most difficult parts of modern mathematics. Look again at the graphs in Examples 1 and 2 of this section (Figures 51 or 52 and 53 or 54). We see that Example 1 has

two solutions, and Example 2 has four solutions. We might conjecture that the number of solutions is equal to the product of the degrees of the equations involved. This conjecture was indeed made by Étienne Bezout (1730–1783), but working out the details took about 150 years. It turns out that, to arrive at the correct number of intersections, we must count not only the complex intersections, but also the intersections that, in a certain sense, lie at infinity. For example, a parabola and a line lying on the axis of the parabola intersect at the vertex and at infinity. This topic is part of the study of algebraic geometry.

HISTORICAL PROBLEM

1. A papyrus dating back to 1950 B.C. contains the following problem: A given surface area of 100 units of area shall be represented as the sum of two squares whose sides are to each other as $1:\frac{3}{4}$. Solve for the sides by solving the system of equations

$$\begin{cases} x^2 + y^2 = 100 \\ x = \frac{3}{4}y \end{cases}$$

9.5 EXERCISES

In Problems 1–20, use a graphing utility to graph each equation of the system. Then solve the system by finding the intersection points. Express your answer rounded to two decimal places. Also solve each system algebraically.

1. $\begin{cases} y = x^2 + 1 \\ y = x + 1 \end{cases}$

2. $\begin{cases} y = x^2 + 1 \\ y = 4x + 1 \end{cases}$

3. $\begin{cases} y = \sqrt{36 - x^2} \\ y = 8 - x \end{cases}$

4. $\begin{cases} y = \sqrt{4 - x^2} \\ y = 2x + 4 \end{cases}$

5. $\begin{cases} y = \sqrt{x} \\ y = 2 - x \end{cases}$

6. $\begin{cases} y = \sqrt{x} \\ y = 6 - x \end{cases}$

7. $\begin{cases} x = 2y \\ x = y^2 - 2y \end{cases}$

8. $\begin{cases} y = x - 1 \\ y = x^2 - 6x + 9 \end{cases}$

9. $\begin{cases} x^2 + y^2 = 4 \\ x^2 + 2x + y^2 = 0 \end{cases}$

10. $\begin{cases} x^2 + y^2 = 8 \\ x^2 + y^2 + 4y = 0 \end{cases}$

11. $\begin{cases} y = 3x - 5 \\ x^2 + y^2 = 5 \end{cases}$

12. $\begin{cases} x^2 + y^2 = 10 \\ y = x + 2 \end{cases}$

13. $\begin{cases} x^2 + y^2 = 4 \\ y^2 - x = 4 \end{cases}$

14. $\begin{cases} x^2 + y^2 = 16 \\ x^2 - 2y = 8 \end{cases}$

15. $\begin{cases} xy = 4 \\ x^2 + y^2 = 8 \end{cases}$

16. $\begin{cases} x^2 = y \\ xy = 1 \end{cases}$

17. $\begin{cases} x^2 + y^2 = 4 \\ y = x^2 - 9 \end{cases}$

18. $\begin{cases} xy = 1 \\ y = 2x + 1 \end{cases}$

19. $\begin{cases} y = x^2 - 4 \\ y = 6x - 13 \end{cases}$

20. $\begin{cases} x^2 + y^2 = 10 \\ xy = 3 \end{cases}$

In Problems 21–52, solve each system. Use any method you wish.

21. $\begin{cases} 2x^2 + y^2 = 18 \\ xy = 4 \end{cases}$

22. $\begin{cases} x^2 - y^2 = 21 \\ x + y = 7 \end{cases}$

23. $\begin{cases} y = 2x + 1 \\ 2x^2 + y^2 = 1 \end{cases}$

24. $\begin{cases} x^2 - 4y^2 = 16 \\ 2y - x = 2 \end{cases}$

25. $\begin{cases} x + y + 1 = 0 \\ x^2 + y^2 + 6y - x = -5 \end{cases}$

26. $\begin{cases} 2x^2 - xy + y^2 = 8 \\ xy = 4 \end{cases}$

27. $\begin{cases} 4x^2 - 3xy + 9y^2 = 15 \\ 2x + 3y = 5 \end{cases}$

28. $\begin{cases} 2y^2 - 3xy + 6y + 2x + 4 = 0 \\ 2x - 3y + 4 = 0 \end{cases}$

29. $\begin{cases} x^2 - 4y^2 + 7 = 0 \\ 3x^2 + y^2 = 31 \end{cases}$

30. $\begin{cases} 3x^2 - 2y^2 + 5 = 0 \\ 2x^2 - y^2 + 2 = 0 \end{cases}$

31. $\begin{cases} 7x^2 - 3y^2 + 5 = 0 \\ 3x^2 + 5y^2 = 12 \end{cases}$

32. $\begin{cases} x^2 - 3y^2 + 1 = 0 \\ 2x^2 - 7y^2 + 5 = 0 \end{cases}$

33. $\begin{cases} x^2 + 2xy = 10 \\ 3x^2 - xy = 2 \end{cases}$

34. $\begin{cases} 5xy + 13y^2 + 36 = 0 \\ xy + 7y^2 = 6 \end{cases}$

35. $\begin{cases} 2x^2 + y^2 = 2 \\ x^2 - 2y^2 + 8 = 0 \end{cases}$

36. $\begin{cases} y^2 - x^2 + 4 = 0 \\ 2x^2 + 3y^2 = 6 \end{cases}$

37. $\begin{cases} x^2 + 2y^2 = 16 \\ 4x^2 - y^2 = 24 \end{cases}$

38. $\begin{cases} 4x^2 + 3y^2 = 4 \\ 2x^2 - 6y^2 = -3 \end{cases}$

39. $\begin{cases} \dfrac{5}{x^2} - \dfrac{2}{y^2} + 3 = 0 \\ \dfrac{3}{x^2} + \dfrac{1}{y^2} = 7 \end{cases}$

40. $\begin{cases} \dfrac{2}{x^2} - \dfrac{3}{y^2} + 1 = 0 \\ \dfrac{6}{x^2} - \dfrac{7}{y^2} + 2 = 0 \end{cases}$

41. $\begin{cases} \dfrac{1}{x^4} + \dfrac{6}{y^4} = 6 \\ \dfrac{2}{x^4} - \dfrac{2}{y^4} = 19 \end{cases}$

42. $\begin{cases} \dfrac{1}{x^4} - \dfrac{1}{y^4} = 1 \\ \dfrac{1}{x^4} + \dfrac{1}{y^4} = 4 \end{cases}$

43. $\begin{cases} x^2 - 3xy + 2y^2 = 0 \\ x^2 + xy = 6 \end{cases}$

44. $\begin{cases} x^2 - xy - 2y^2 = 0 \\ xy + x + 6 = 0 \end{cases}$

45. $\begin{cases} xy - x^2 + 3 = 0 \\ 3xy - 4y^2 = 2 \end{cases}$

46. $\begin{cases} 5x^2 + 4xy + 3y^2 = 36 \\ x^2 + xy + y^2 = 9 \end{cases}$

47. $\begin{cases} x^3 - y^3 = 26 \\ x - y = 2 \end{cases}$

48. $\begin{cases} x^3 + y^3 = 26 \\ x + y = 2 \end{cases}$

49. $\begin{cases} y^2 + y + x^2 - x - 2 = 0 \\ y + 1 + \dfrac{x - 2}{y} = 0 \end{cases}$

50. $\begin{cases} x^3 - 2x^2 + y^2 + 3y - 4 = 0 \\ x - 2 + \dfrac{y^2 - y}{x^2} = 0 \end{cases}$

51. $\begin{cases} \log_x y = 3 \\ \log_x(4y) = 5 \end{cases}$

52. $\begin{cases} \log_x(2y) = 3 \\ \log_x(4y) = 2 \end{cases}$

In Problems 53–60, use a graphing utility to solve each system of equations. Express the solution(s) rounded to two decimal places.

53. $\begin{cases} y = x^{2/3} \\ y = e^{-x} \end{cases}$

54. $\begin{cases} y = x^{3/2} \\ y = e^{-x} \end{cases}$

55. $\begin{cases} x^2 + y^3 = 2 \\ x^3 y = 4 \end{cases}$

56. $\begin{cases} x^3 + y^2 = 2 \\ x^2 y = 4 \end{cases}$

57. $\begin{cases} x^4 + y^4 = 12 \\ xy^2 = 2 \end{cases}$

58. $\begin{cases} x^4 + y^4 = 6 \\ xy = 1 \end{cases}$

59. $\begin{cases} xy = 2 \\ y = \ln x \end{cases}$

60. $\begin{cases} x^2 + y^2 = 4 \\ y = \ln x \end{cases}$

61. The difference of two numbers is 2 and the sum of their squares is 10. Find the numbers.

62. The sum of two numbers is 7 and the difference of their squares is 21. Find the numbers.

63. The product of two numbers is 4 and the sum of their squares is 8. Find the numbers.

64. The product of two numbers is 10 and the difference of their squares is 21. Find the numbers.

65. The difference of two numbers is the same as their product, and the sum of their reciprocals is 5. Find the numbers.

66. The sum of two numbers is the same as their product, and the difference of their reciprocals is 3. Find the numbers.

67. The ratio of a to b is $\frac{2}{3}$. The sum of a and b is 10. What is the ratio of $a + b$ to $b - a$?

68. The ratio of a to b is $\frac{4}{3}$. The sum of a and b is 14. What is the ratio of $a - b$ to $a + b$?

For the elements given in Problems 69–74, graph each equation by hand and find the point(s) of intersection, if any.

69. Line: $x + 2y = 0$
Circle: $(x - 1)^2 + (y - 1)^2 = 5$

70. Line: $x + 2y + 6 = 0$
Circle: $(x + 1)^2 + (y + 1)^2 = 5$

71. Circle: $(x - 1)^2 + (y + 2)^2 = 4$
Parabola: $y^2 + 4y - x + 1 = 0$

72. Circle: $(x + 2)^2 + (y - 1)^2 = 4$
Parabola: $y^2 - 2y - x - 5 = 0$

73. Equation: $y = \dfrac{4}{x - 3}$
Circle: $x^2 - 6x + y^2 + 1 = 0$

74. Equation: $y = \dfrac{4}{x + 2}$
Circle: $x^2 + 4x + y^2 - 4 = 0$

75. **Geometry** The perimeter of a rectangle is 16 inches and its area is 15 square inches. What are its dimensions?

76. **Geometry** An area of 52 square feet is to be enclosed by two squares whose sides are in the ratio of 2:3. Find the sides of the squares.

77. **Geometry** Two circles have perimeters that add up to 12π centimeters and areas that add up to 20π square centimeters. Find the radius of each circle.

78. **Geometry** The altitude of an isosceles triangle drawn to its base is 3 centimeters, and its perimeter is 18 centimeters. Find the length of its base.

79. **The Tortoise and the Hare** In a 21 meter race between a tortoise and a hare, the tortoise leaves 9 minutes before the hare. The hare, by running at an average speed of 0.5 meter per hour faster than the tortoise, crosses the finish line 3 minutes before the tortoise. What are the average speeds of the tortoise and the hare?

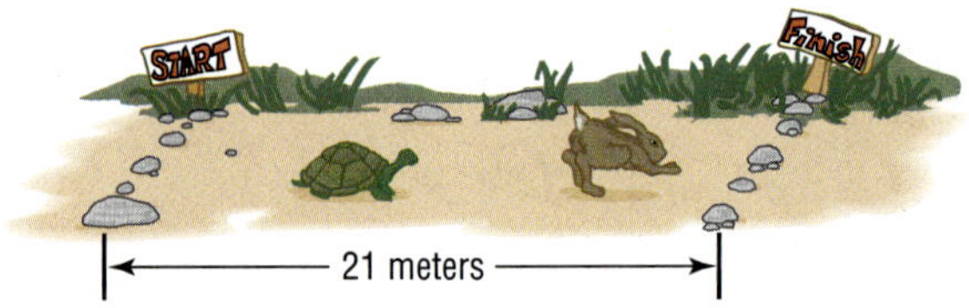

80. **Running a Race** In a 1 mile race, the winner crosses the finish line 10 feet ahead of the second place runner and 20 feet ahead of the third place runner. Assuming that each runner maintains a constant speed throughout the race, by how many feet does the second place runner beat the third place runner?

81. **Constructing a Box** A rectangular piece of cardboard, whose area is 216 square centimeters, is made into an open box by cutting a 2 centimeter square from each corner and turning up the sides. See the figure. If the box is to have a volume of 224 cubic centimeters, what size cardboard should you start with?

82. **Constructing a Cylindrical Tube** A rectangular piece of cardboard, whose area is 216 square centimeters, is made into a cylindrical tube by joining together two sides of the rectangle. (See the figure.) If the tube is to have a volume of 224 cubic centimeters, what size cardboard should you start with?

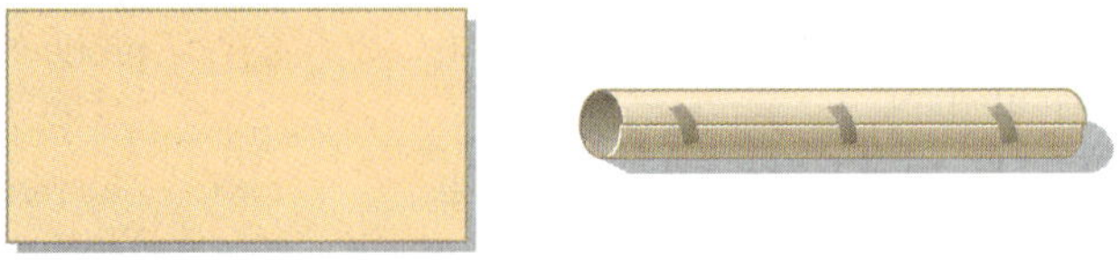

83. Fencing A farmer has 300 feet of fence available to enclose 4500 square feet in the shape of adjoining squares, with sides of length x and y. See the figure. Find x and y.

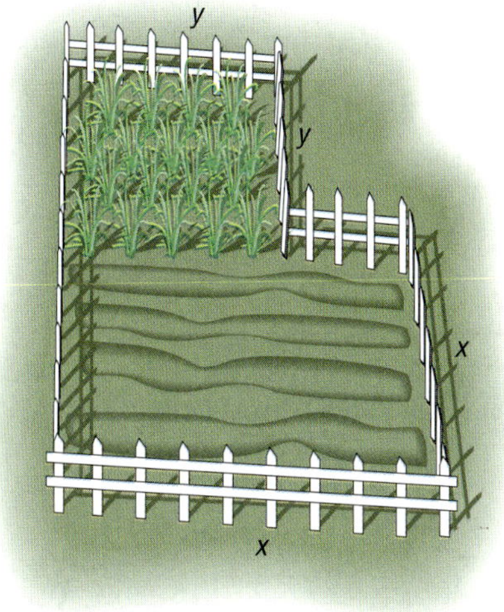

84. Bending Wire A wire 60 feet long is cut into two pieces. Is it possible to bend one piece into the shape of a square and the other into the shape of a circle so that the total area enclosed by the two pieces is 100 square feet? If this is possible, find the length of the side of the square and the radius of the circle.

85. Geometry Find formulas for the length l and width w of a rectangle in terms of its area A and perimeter P.

86. Geometry Find formulas for the base b and one of the equal sides l of an isosceles triangle in terms of its altitude h and perimeter P.

87. Descartes' Method of Equal Roots Descartes' method for finding tangents depends on the idea that, for many graphs, the tangent line at a given point is the *unique* line that intersects the graph at that point only. We will apply his method to find an equation of the tangent line to the parabola $y = x^2$ at the point $(2, 4)$; see the figure. First, we know that the equation of the tangent line must be in the form $y = mx + b$. Using the fact that the point $(2, 4)$ is on the line, we can solve for b in terms of m and get the equation $y = mx + (4 - 2m)$. Now we want $(2, 4)$ to be the *unique* solution to the system

$$\begin{cases} y = x^2 \\ y = mx + 4 - 2m \end{cases}$$

From this system, we get $x^2 = mx + 4 - 2m$ or $x^2 - mx + (2m - 4) = 0$. By using the quadratic formula, we get

$$x = \frac{m \pm \sqrt{m^2 - 4(2m - 4)}}{2}$$

To obtain a unique solution for x, the two roots must be equal; in other words, the discriminant $m^2 - 4(2m - 4)$ must be 0. Complete the work to get m, and write an equation of the tangent line.

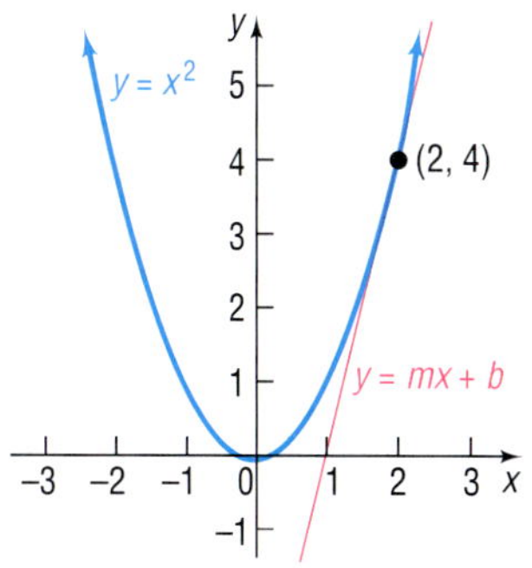

In Problems 88–94, use Descartes' method from Problem 87 to find the equation of the line tangent to each graph at the given point.

88. $x^2 + y^2 = 10$; at $(1, 3)$

89. $y = x^2 + 2$; at $(1, 3)$

90. $x^2 + y = 5$; at $(-2, 1)$

91. $2x^2 + 3y^2 = 14$; at $(1, 2)$

92. $3x^2 + y^2 = 7$; at $(-1, 2)$

93. $x^2 - y^2 = 3$; at $(2, 1)$

94. $2y^2 - x^2 = 14$; at $(2, 3)$

95. If r_1 and r_2 are two solutions of a quadratic equation $ax^2 + bx + c = 0$, then it can be shown that

$$r_1 + r_2 = -\frac{b}{a} \quad \text{and} \quad r_1 r_2 = \frac{c}{a}$$

Solve this system of equations for r_1 and r_2.

96. A circle and a line intersect at most twice. A circle and a parabola intersect at most four times. Deduce that a circle and the graph of a polynomial of degree 3 intersect at most six times. What do you conjecture about a polynomial of degree 4? What about a polynomial of degree n? Can you explain your conclusions using an algebraic argument?

97. Suppose that you are the manager of a sheet metal shop. A customer asks you to manufacture 10,000 boxes, each box being open on top. The boxes are required to have a square base and a 9 cubic foot capacity. You construct the boxes by cutting a square out from each corner of a square piece of sheet metal and folding along the edges.

(a) What are the dimensions of the square to be cut if the area of the square piece of sheet metal is 100 square feet?

(b) Could you make the box using a smaller piece of sheet metal? Make a list of the dimensions of the box for various pieces of sheet metal.

CHAPTER REVIEW

THINGS TO KNOW

Equations of Conics

Parabola	See Tables 1 and 2 (pages 607 and 609).
Ellipse	See Table 3 (page 623).
Hyperbola	See Table 4 (page 636).

HOW TO

Find the vertex, focus, and directrix of a parabola given its equation

Graph a parabola given its equation by hand and by using a graphing utility

Find an equation of a parabola given certain information about the parabola

Find the center, foci, and vertices of an ellipse given its equation

Graph an ellipse given its equation by hand and by using a graphing utility

Find an equation of an ellipse given certain information about the ellipse

Find the center, foci, vertices, and asymptotes of a hyperbola given its equation

Graph a hyperbola given its equation by hand and by using a graphing utility

Find an equation of a hyperbola given certain information about the hyperbola

Solve a system of nonlinear equations

FILL-IN-THE-BLANK ITEMS

1. A(n) ______________ is the collection of all points in the plane such that the distance from each point to a fixed point equals its distance to a fixed line.

2. A(n) ______________ is the collection of all points in the plane the sum of whose distances from two fixed points is a constant.

3. A(n) ______________ is the collection of all points in the plane the difference of whose distances from two fixed points is a constant.

4. For an ellipse, the foci lie on the ______________ axis; for a hyperbola, the foci lie on the ______________ axis.

5. For the ellipse $(x^2/9) + (y^2/16) = 1$, the major axis is along the ______________.

6. The equations of the asymptotes of the hyperbola $(y^2/9) - (x^2/4) = 1$ are ______________ and ______________.

TRUE/FALSE ITEMS

T F **1.** On a parabola, the distance from any point to the focus equals the distance from that point to the directrix.

T F **2.** The foci of an ellipse lie on its minor axis.

T F **3.** The foci of a hyperbola lie on its transverse axis.

T F **4.** Hyperbolas always have asymptotes, and ellipses never have asymptotes.

T F **5.** A hyperbola never intersects its conjugate axis.

T F **6.** A hyperbola always intersects its transverse axis.

REVIEW EXERCISES

Blue problem numbers indicate the authors' suggestions for use in a Practice Test.

In Problems 1–20, identify each equation. If it is a parabola, give its vertex, focus, and directrix; if it is an ellipse, give its center, vertices, and foci; if it is a hyperbola, give its center, vertices, foci, and asymptotes.

1. $y^2 = -16x$

2. $16x^2 = y$

3. $\frac{x^2}{25} - y^2 = 1$

4. $\frac{y^2}{25} - x^2 = 1$

5. $\frac{y^2}{25} + \frac{x^2}{16} = 1$

6. $\frac{x^2}{9} + \frac{y^2}{16} = 1$

7. $x^2 + 4y = 4$

8. $3y^2 - x^2 = 9$

9. $4x^2 - y^2 = 8$

10. $9x^2 + 4y^2 = 36$

11. $x^2 - 4x = 2y$

12. $2y^2 - 4y = x - 2$

13. $y^2 - 4y - 4x^2 + 8x = 4$

14. $4x^2 + y^2 + 8x - 4y + 4 = 0$

15. $4x^2 + 9y^2 - 16x - 18y = 11$

16. $4x^2 + 9y^2 - 16x + 18y = 11$

17. $4x^2 - 16x + 16y + 32 = 0$

18. $4y^2 + 3x - 16y + 19 = 0$

19. $9x^2 + 4y^2 - 18x + 8y = 23$

20. $x^2 - y^2 - 2x - 2y = 1$

In Problems 21–36, obtain an equation of the conic described. Graph the equation by hand.

21. Parabola; focus at $(-2, 0)$; directrix the line $x = 2$

22. Ellipse; center at $(0, 0)$; focus at $(0, 3)$; vertex at $(0, 5)$

23. Hyperbola; center at $(0, 0)$; focus at $(0, 4)$; vertex at $(0, -2)$

24. Parabola; vertex at $(0, 0)$; directrix the line $y = -3$

25. Ellipse; foci at $(-3, 0)$ and $(3, 0)$; vertex at $(4, 0)$

26. Hyperbola; vertices at $(-2, 0)$ and $(2, 0)$; focus at $(4, 0)$

27. Parabola; vertex at $(2, -3)$; focus at $(2, -4)$

28. Ellipse; center at $(-1, 2)$; focus at $(0, 2)$; vertex at $(2, 2)$

29. Hyperbola; center at $(-2, -3)$; focus at $(-4, -3)$; vertex at $(-3, -3)$

30. Parabola; focus at $(3, 6)$; directrix the line $y = 8$

31. Ellipse; foci at $(-4, 2)$ and $(-4, 8)$; vertex at $(-4, 10)$

32. Hyperbola; vertices at $(-3, 3)$ and $(5, 3)$; focus at $(7, 3)$

33. Center at $(-1, 2)$; $a = 3$; $c = 4$; transverse axis parallel to the x-axis

34. Center at $(4, -2)$; $a = 1$; $c = 4$; transverse axis parallel to y-axis

35. Vertices at $(0, 1)$ and $(6, 1)$; asymptote the line $3y + 2x - 9 = 0$

36. Vertices at $(4, 0)$ and $(4, 4)$; asymptote the line $y + 2x - 10 = 0$

37. Find an equation of the hyperbola whose foci are the vertices of the ellipse $4x^2 + 9y^2 = 36$ and whose vertices are the foci of this ellipse.

38. Find an equation of the ellipse whose foci are the vertices of the hyperbola $x^2 - 4y^2 = 16$ and whose vertices are the foci of this hyperbola.

39. Describe the collection of points in a plane so that the distance from each point to the point $(3, 0)$ is three-fourths of its distance from the line $x = \frac{16}{3}$.

40. Describe the collection of points in a plane so that the distance from each point to the point $(5, 0)$ is five-fourths of its distance from the line $x = \frac{16}{5}$.

In Problems 41–50, solve each system of equations algebraically. Verify your result using a graphing utility.

41. $\begin{cases} 2x + y + 3 = 0 \\ x^2 + y^2 = 5 \end{cases}$

42. $\begin{cases} x^2 + y^2 = 16 \\ 2x - y^2 = -8 \end{cases}$

43. $\begin{cases} 2xy + y^2 = 10 \\ 3y^2 - xy = 2 \end{cases}$

44. $\begin{cases} 3x^2 - y^2 = 1 \\ 7x^2 - 2y^2 - 5 = 0 \end{cases}$

45. $\begin{cases} x^2 + y^2 = 6y \\ x^2 = 3y \end{cases}$

46. $\begin{cases} 2x^2 + y^2 = 9 \\ x^2 + y^2 = 9 \end{cases}$

47. $\begin{cases} 3x^2 + 4xy + 5y^2 = 8 \\ x^2 + 3xy + 2y^2 = 0 \end{cases}$

48. $\begin{cases} 3x^2 + 2xy - 2y^2 = 6 \\ xy - 2y^2 + 4 = 0 \end{cases}$

49. $\begin{cases} x^2 - 3x + y^2 + y = -2 \\ \dfrac{x^2 - x}{y} + y + 1 = 0 \end{cases}$

50. $\begin{cases} x^2 + x + y^2 = y + 2 \\ x + 1 = \dfrac{2 - y}{x} \end{cases}$

51. **Mirrors** A mirror is shaped like a paraboloid of revolution. If a light source is located 1 foot from the base along the axis of symmetry and the opening is 2 feet across, how deep should the mirror be?

52. **Parabolic Arch Bridge** A bridge is built in the shape of a parabolic arch. The bridge has a span of 60 feet and a maximum height of 20 feet. Find the height of the arch at distances of 5, 10, and 20 feet from the center.

53. **Semielliptical Arch Bridge** A bridge is built in the shape of a semielliptical arch. The bridge has a span of 60 feet and a maximum height of 20 feet. Find the height of the arch at distances of 5, 10, and 20 feet from the center.

54. **Whispering Gallery** The figure shows the specifications for an elliptical ceiling in a hall designed to be a whispering gallery. Where in the hall are the foci located?

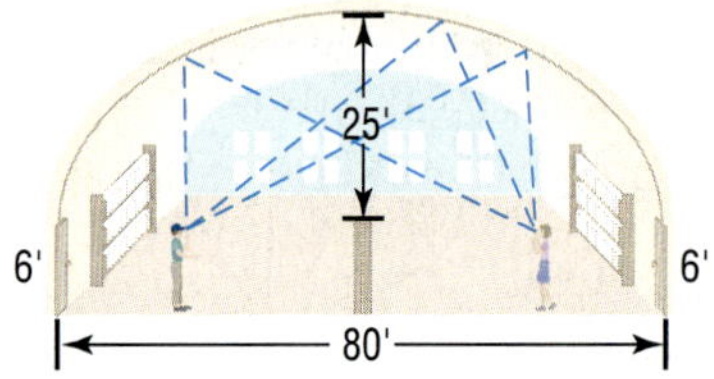

55. **LORAN** Two LORAN stations are positioned 150 miles apart along a straight shore.
   (a) A ship records a time difference of 0.00032 second between the LORAN signals. Set up an appropriate rectangular coordinate system to determine where the ship would reach shore if it were to follow the hyperbola corresponding to this time difference.
   (b) If the ship wants to enter a harbor located between the two stations 15 miles from the master station, what time difference should it be looking for?
   (c) If the ship is 20 miles offshore when the desired time difference is obtained, what is the approximate location of the ship?
   [*Note:* The speed of each radio signal is 186,000 miles per second.]

56. Use the definition of a parabola and the idea of the eccentricity of an ellipse and hyperbola to construct a unifying definition for all three conics. [Refer to Problem 75 in Exercise 9.3 and Problem 61 in Exercise 9.4.]

57. Formulate a strategy for discussing and graphing an equation of the form

$$Ax^2 + Cy^2 + Dx + Ey + F = 0$$

CHAPTER PROJECTS

1. **The Orbits of Neptune and Pluto** The orbit of a planet about the Sun is an ellipse, with the Sun at one focus. The **aphelion** of a planet is its greatest distance from the Sun and the **perihelion** is its shortest distance. The **mean distance** of a planet from the Sun is the length of the semimajor axis of the elliptical orbit. See the illustration.

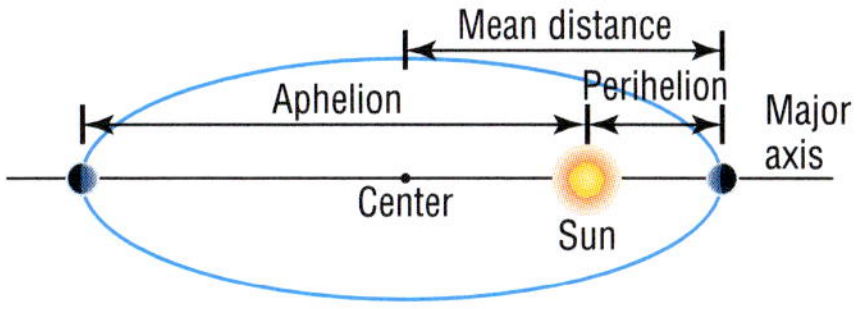

   (a) The aphelion of Neptune is 4532.2×10^6 km and its perihelion is 4458.0×10^6 km. Find the equation for the orbit of Neptune around the Sun.
   (b) The aphelion of Pluto is 7381.2×10^6 km and its perihelion is 4445.8×10^6 km. Find the equation for the orbit of Pluto around the Sun.
   (c) Graph the orbits of Pluto and Neptune on a graphing utility. Do the orbits of the planets intersect?
   (d) The distance from the center (the origin) to the Sun must be the same for both Pluto and Neptune. Therefore, we must shift Pluto's orbit left to get a true representation of the two planet's orbits. The shift amount is equal to Pluto's distance from the center (in the graph in (c)) to the Sun minus Neptune's distance from the center to the Sun. Find the new equation representing the orbit of Pluto.
   (e) Graph the equation for the orbit of Pluto found in (d) along with the equation of the orbit of Neptune.
   (f) Find the point(s) of intersection of the two orbits.
   (g) Do you think two planets ever collide?
   [Hint: Consider space is three dimensional.]

2. **Constructing a Bridge Over the East River** A new bridge is to be constructed over the East River in New York City. The space between the supports needs to be 1050 feet; the height at the center of the arch needs to be 350 feet. Two structural possibilities exist: the support could be in the shape of a parabola or the support could be in the shape of a semiellipse.

   An empty tanker needs a 280 foot clearance to pass beneath the bridge. The width of the channel for each of the two plans must be determined to verify that the tanker can pass through the bridge. (See the illustration on page 660.)

   (a) Determine the equation of a parabola with these characteristics.
   [Hint: Place the vertex of the parabola at the origin to simplify calculations.]
   (b) How wide is the channel that the tanker can pass through?
   (c) Determine the equation of a semiellipse with these characteristics.
   [Hint: Place the center of the semiellipse at the origin to simplify calculations.]
   (d) How wide is the channel that the tanker can pass through?
   (e) If the river were to flood and rise 10 feet, how would the clearances of the two bridges be affected? Does this affect your decision as to which design to choose? Why?

CHAPTER PROJECTS *(Continued)*

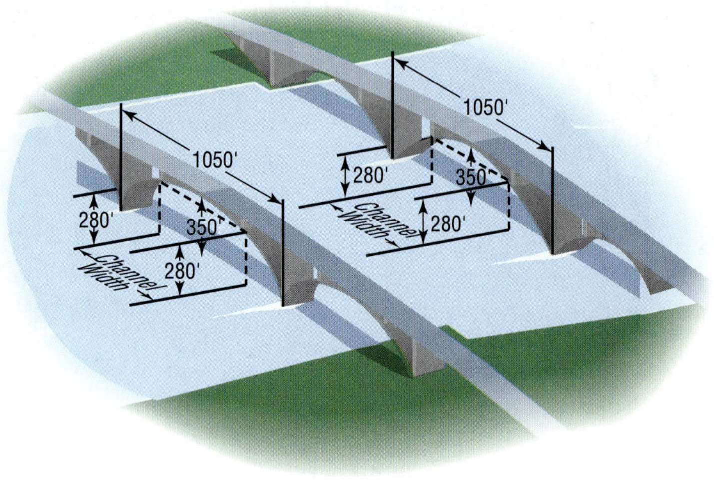

3. **CBL Experiment** The motion of a swinging pendulum is examined by plotting the velocity against the position of the pendulum. The result is the graph of an ellipse. (Activity 17, Real-World Math with the CBL System)

APPENDIX

Review

1 REAL NUMBERS

1 Classify Numbers

2 Evaluate Numerical Expressions

3 Know Properties of Real Numbers

Sets

When we want to treat a collection of similar but distinct objects as a whole, we use the idea of a **set.** For example, the set of *digits* consists of the collection of numbers 0, 1, 2, 3, 4, 5, 6, 7, 8, and 9. If we use the symbol D to denote the set of digits, then we can write

$$D = \{0, 1, 2, 3, 4, 5, 6, 7, 8, 9\}$$

In this notation, the braces { } are used to enclose the objects, or **elements,** in the set. This method of denoting a set is called the **roster method.** A second way to denote a set is to use **set-builder notation,** where the set D of digits is written as

$$D = \{ \quad x \quad | \; x \text{ is a digit}\}$$

Read as "D is the set of all x such that x is a digit."

EXAMPLE 1 Using Set-builder Notation and the Roster Method

(a) $E = \{x | x \text{ is an even digit}\} = \{0, 2, 4, 6, 8\}$

(b) $O = \{x | x \text{ is an odd digit}\} = \{1, 3, 5, 7, 9\}$

In listing the elements of a set, we do not list an element more than once because the elements of a set are distinct. Also, the order in which the elements are listed is not relevant. Thus, for example, $\{2, 3\}$ and $\{3, 2\}$ both represent the same set.

If every element of a set A is also an element of a set B, then we say that A is a **subset** of B. If two sets A and B have the same elements, then we say that A **equals** B. For example, $\{1, 2, 3\}$ is a subset of $\{1, 2, 3, 4, 5\}$, and $\{1, 2, 3\}$ equals $\{2, 3, 1\}$.

Classification of Numbers

1 It is helpful to classify the various kinds of numbers that we deal with as sets. The **counting numbers,** or **natural numbers,** are the set of numbers $\{1, 2, 3, 4, \dots\}$. (The three dots, called an **ellipsis,** indicate that the pattern continues indefinitely.) As their name implies, these numbers are often used to count things. For example, there are 26 letters in our alphabet; there are 100 cents in a dollar. The **whole numbers** are the set of numbers $\{0, 1, 2, 3, \dots\}$, that is, the counting numbers together with 0.

The **integers** are the set of numbers $\{\dots, -3, -2, -1, 0, 1, 2, 3, \dots\}$.

These numbers prove useful in many situations. For example, if your checking account has \$10 in it and you write a check for \$15, you can represent the current balance as $-\$5$.

Notice that the set of counting numbers is a subset of the set of whole numbers. Each time we expand a number system, such as from the whole numbers to the integers, we do so in order to be able to handle new, and usually more complicated, problems. Thus, the integers allow us to solve problems requiring both positive and negative counting numbers, such as profit/loss, height above/below sea level, temperature above/below 0°F, and so on.

But integers alone are not sufficient for *all* problems. For example, they do not answer the question "What part of a dollar is 38 cents?" To answer such a question, we enlarge our number system to include *rational numbers.* For example, $\frac{38}{100}$ answers the question "What part of a dollar is 38 cents?"

A **rational number** is a number that can be expressed as a quotient a/b of two integers. The integer a is called the **numerator,** and the integer b, which cannot be 0, is called the **denominator.** The rational numbers are the set of numbers $\{x \mid x = \frac{a}{b}, \text{ where } a, b \neq 0 \text{ are integers}\}$.

Examples of rational numbers are $\frac{3}{4}, \frac{5}{2}, \frac{0}{4}, -\frac{2}{3}$, and $\frac{100}{3}$. Since $a/1 = a$ for any integer a, it follows that the set of integers is a subset of the set of rational numbers.

Rational numbers may be represented as **decimals.** For example, the rational numbers $\frac{3}{4}$, $\frac{5}{2}$, $-\frac{2}{3}$, and $\frac{7}{66}$ may be represented as decimals by merely carrying out the indicated division:

$$\frac{3}{4} = 0.75 \qquad \frac{5}{2} = 2.5 \qquad -\frac{2}{3} = -0.666\ldots \qquad \frac{7}{66} = 0.1060606\ldots$$

Notice that the decimal representations of $\frac{3}{4}$ and $\frac{5}{2}$ terminate, or end. The decimal representations of $-\frac{2}{3}$ and $\frac{7}{66}$ do not terminate, but they do exhibit a pattern of repetition. For $-\frac{2}{3}$, the 6 repeats indefinitely; for $\frac{7}{66}$, the block 06 repeats indefinitely. It can be shown that every rational number may be represented by a decimal that either terminates or is nonterminating with a repeating block of digits, and vice versa.

On the other hand, there are decimals that do not fit into either of these categories. Such decimals represent **irrational numbers.** Every irrational number may be represented by a decimal that neither repeats nor terminates.

Figure 1

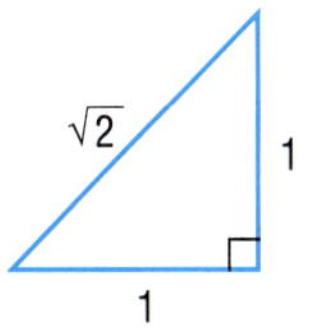

Irrational numbers occur naturally. For example, consider the isosceles right triangle whose legs are each of length 1. See Figure 1. The length of the hypotenuse is $\sqrt{2}$, an irrational number.

Also, the number that equals the ratio of the circumference C to the diameter d of any circle, denoted by the symbol π (the Greek letter pi), is an irrational number. See Figure 2.

Figure 2

$\pi = \frac{C}{d}$

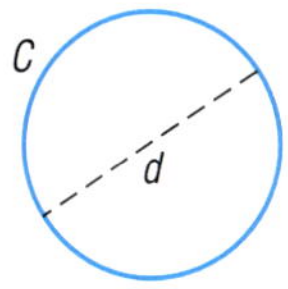

The irrational numbers $\sqrt{2}$ and π have decimal representations that begin as follows:

$$\sqrt{2} = 1.414213\ldots \qquad \pi = 3.14159\ldots$$

In practice, irrational numbers are generally represented by approximations. For example, using the symbol $\approx$ (read as "approximately equal to"), we can write

$$\sqrt{2} \approx 1.4142 \qquad \pi \approx 3.1416$$

Together, the rational numbers and irrational numbers form the set of **real numbers.**

Thus, every decimal may be represented by a real number (either rational or irrational), and every real number may be represented by a decimal. Real numbers in the form of decimals provide a convenient way to measure quantities as they change. Figure 3 shows the relationship of various types of numbers.

Figure 3

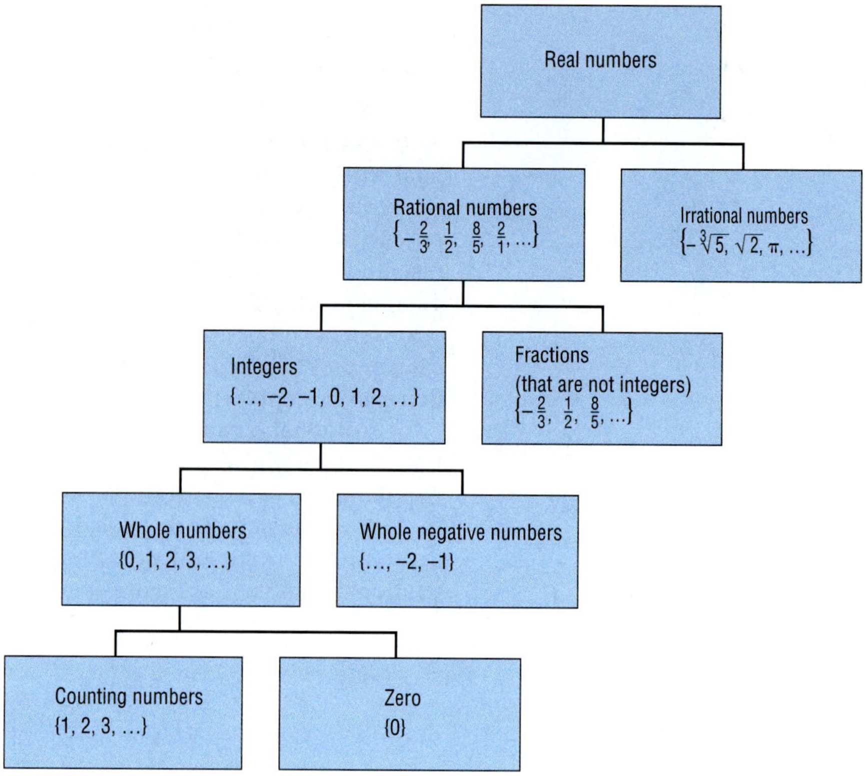

EXAMPLE 2 Classifying the Numbers in a Set

List the numbers in the set

$$\{-3, \tfrac{4}{3}, 0.12, \sqrt{2}, \pi, 2.151515\ldots \text{ (where the block 15 repeats)}, 10\}$$

that are:

(a) Natural numbers (b) Integers (c) Rational numbers
(d) Irrational numbers (e) Real numbers

Solution (a) 10 is the only natural number.

(b) -3 and 10 are integers.

(c) $-3, \frac{4}{3}, 0.12, 2.151515\ldots$, and 10 are rational numbers.

(d) $\sqrt{2}$ and π are irrational numbers.

(e) All the numbers listed are real numbers.

NOW WORK PROBLEM 3.

Operations

In algebra, we use letters such as x, y, a, b, and c to represent numbers. The symbols used in algebra for the operations of addition, subtraction, multiplication, and division are $+$, $-$, $\cdot$, and /. The words used to describe the results of these operations are **sum, difference, product,** and **quotient.** Table 1 summarizes these ideas.

Table 1

Operation	Symbol	Words
Addition	$a + b$	Sum: a plus b
Subtraction	$a - b$	Difference: a less b
Multiplication	$a \cdot b$, $(a) \cdot b$, $a \cdot (b)$, $(a) \cdot (b)$, ab, $(a)b$, $a(b)$, $(a)(b)$	Product: a times b
Division	a/b or $\frac{a}{b}$	Quotient: a divided by b

In algebra, we generally avoid using the multiplication sign $\times$ and the division sign $\div$ so familiar in arithmetic. Notice also that when two expressions are placed next to each other without an operation symbol, as in ab, or in parentheses, as in $(a)(b)$, it is understood that the expressions, called **factors,** are to be multiplied.

The symbol $=$, called an **equal sign** and read as "equals" or "is," is used to express the idea that the number or expression on the left of the equal sign is equivalent to the number or expression on the right.

EXAMPLE 3 Writing Statements Using Symbols

(a) The sum of 2 and 7 equals 9. In symbols, this statement is written as $2 + 7 = 9$.

(b) The product of 3 and 5 is 15. In symbols, this statement is written as $3 \cdot 5 = 15$.

NOW WORK PROBLEM 7.

Order of Operations

2 Consider the expression $2 + 3 \cdot 6$. It is not clear whether we should add 2 and 3 to get 5, and then multiply by 6 to get 30; or first multiply 3 and 6 to get 18, and then add 2 to get 20. To avoid this ambiguity, we have the following agreement.

> We agree that whenever the two operations of addition and multiplication separate three numbers the multiplication operation always will be performed first, followed by the addition operation.

Thus, for $2 + 3 \cdot 6$, we have

$$2 + 3 \cdot 6 = 2 + 18 = 20$$

EXAMPLE 4 Finding the Value of an Expression

Evaluate each expression:

(a) $3 + 4 \cdot 5$ (b) $8 \cdot 2 + 1$ (c) $2 + 2 \cdot 2$

Solution (a) $3 + 4 \cdot 5 = 3 + 20 = 23$ (b) $8 \cdot 2 + 1 = 16 + 1 = 17$

↑ *Multiply first.* ↑ *Multiply first.*

(c) $2 + 2 \cdot 2 = 2 + 4 = 6$

To first add 3 and 4 and then multiply the result by 5, we use parentheses and write $(3 + 4) \cdot 5$. Thus, whenever parentheses appear in an expression, it means "perform the operations within the parentheses first!"

EXAMPLE 5 Finding the Value of an Expression

(a) $(5 + 3) \cdot 4 = 8 \cdot 4 = 32$

(b) $(4 + 5) \cdot (8 - 2) = 9 \cdot 6 = 54$

When we divide two expressions, as in

$$\frac{2 + 3}{4 + 8}$$

it is understood that the division bar acts like parentheses; that is,

$$\frac{2 + 3}{4 + 8} = \frac{(2 + 3)}{(4 + 8)}$$

The following list gives the rules for the order of operations.

> ***Rules for the Order of Operations***
>
> 1. Begin with the innermost parentheses and work outward. Remember that in dividing two expressions the numerator and denominator are treated as if they were enclosed in parentheses.
> 2. Perform multiplications and divisions, working from left to right.
> 3. Perform additions and subtractions, working from left to right.

EXAMPLE 6 Finding the Value of an Expression

Evaluate each expression:

(a) $8 \cdot 2 + 3$ (b) $5 \cdot (3 + 4) + 2$

(c) $\dfrac{2 + 5}{2 + 4 \cdot 7}$ (d) $2 + [4 + 2 \cdot (10 + 6)]$

Solution (a) $8 \cdot 2 + 3 = 16 + 3 = 19$

↑ *Multiply first*

(b) $5 \cdot (3 + 4) + 2 = 5 \cdot 7 + 2 = 35 + 2 = 37$

↑ *Parenthesis first* ↑ *Multiply before adding*

(c) $\dfrac{2+5}{2+4\cdot 7} = \dfrac{2+5}{2+28} = \dfrac{7}{30}$

(d) $2 + [4 + 2\cdot(10+6)] = 2 + [4 + 2\cdot(16)]$
$= 2 + [4 + 32] = 2 + [36] = 38$

NOW WORK PROBLEMS 25 AND 33.

Graphing utilities may be used to find the value of an expression. Figure 4 illustrates the solution to Example 6(a) on a TI-83 graphing calculator. Figure 5 illustrates the solution to Example 6(c). Notice that the answer is expressed as a decimal. The decimal is then converted to a fraction using the ▶FRAC option. Consult your manual to see how your graphing utility does this.

Figure 4

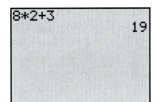

Figure 5

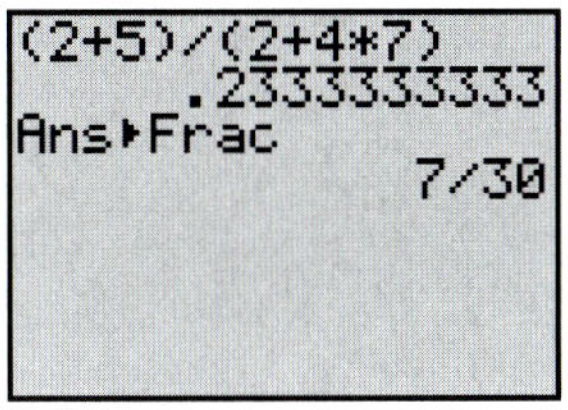

Properties of Real Numbers

3 We have used the equal sign to mean that one expression is equivalent to another. Four important properties of equality are listed next. In this list, a, b, and c represent numbers.

1. The **reflexive property** states that a number always equals itself; that is, $a = a$.
2. The **symmetric property** states that if $a = b$ then $b = a$.
3. The **transitive property** states that if $a = b$ and $b = c$ then $a = c$.
4. The **principle of substitution** states that if $a = b$ then we may substitute b for a in any expression containing a.

Now, let's consider some other properties of real numbers. We begin with an example.

EXAMPLE 7 Commutative Properties

(a) $3 + 5 = 8$
$5 + 3 = 8$
$3 + 5 = 5 + 3$

(b) $2\cdot 3 = 6$
$3\cdot 2 = 6$
$2\cdot 3 = 3\cdot 2$

This example illustrates the **commutative property** of real numbers, which states that the order in which addition or multiplication takes place will not affect the final result.

Commutative Properties

$$a + b = b + a \qquad a \cdot b = b \cdot a \tag{1}$$

Here, and in the properties listed next and on pages 669–672, a, b, and c represent real numbers.

EXAMPLE 8 Associative Properties

(a) $2 + (3 + 4) = 2 + 7 = 9$
$(2 + 3) + 4 = 5 + 4 = 9$
$2 + (3 + 4) = (2 + 3) + 4$

(b) $2 \cdot (3 \cdot 4) = 2 \cdot 12 = 24$
$(2 \cdot 3) \cdot 4 = 6 \cdot 4 = 24$
$2 \cdot (3 \cdot 4) = (2 \cdot 3) \cdot 4$

The way we add or multiply three real numbers will not affect the final result. Thus, expressions such as $2 + 3 + 4$ and $3 \cdot 4 \cdot 5$ present no ambiguity, even though addition and multiplication are performed on one pair of numbers at a time. This property is called the **associative property.**

Associative Property

$$a + (b + c) = (a + b) + c = a + b + c \tag{2a}$$
$$a \cdot (b \cdot c) = (a \cdot b) \cdot c = a \cdot b \cdot c \tag{2b}$$

The next property is perhaps the most important.

Distributive Property

$$a \cdot (b + c) = a \cdot b + a \cdot c \tag{3a}$$
$$(a + b) \cdot c = a \cdot c + b \cdot c \tag{3b}$$

The **distributive property** may be used in two different ways.

EXAMPLE 9 Distributive Property

(a) $2 \cdot (x + 3) = 2 \cdot x + 2 \cdot 3 = 2x + 6$ *Use to remove parentheses.*

(b) $3x + 5x = (3 + 5)x = 8x$ *Use to combine two expressions.*

NOW WORK PROBLEM 53.

The real numbers 0 and 1 have unique properties.

EXAMPLE 10 Identity Properties

(a) $4 + 0 = 0 + 4 = 4$ (b) $3 \cdot 1 = 1 \cdot 3 = 3$

The properties of 0 and 1 illustrated in Example 10 are called the **identity properties.**

Identity Properties

$$0 + a = a + 0 = a \quad a \cdot 1 = 1 \cdot a = a \tag{4}$$

We call 0 the **additive identity** and 1 the **multiplicative identity.**

For each real number a, there is a real number $-a$, called the **additive inverse** of a, having the following property:

Additive Inverse Property

$$a + (-a) = -a + a = 0 \tag{5a}$$

EXAMPLE 11 Finding an Additive Inverse

(a) The additive inverse of 6 is -6, because $6 + (-6) = 0$.
(b) The additive inverse of -8 is $-(-8) = 8$, because $-8 + 8 = 0$.

The additive inverse of a, that is, $-a$, is often called the *negative* of a or the *opposite* of a. The use of such terms can be dangerous because they suggest that the additive inverse is a negative number, which it may not be. For example, the additive inverse of -3, or $-(-3)$, equals 3, a positive number.

For each *nonzero* real number a, there is a real number $1/a$, called the **multiplicative inverse** of a, having the following property:

Multiplicative Inverse Property

$$a \cdot \frac{1}{a} = \frac{1}{a} \cdot a = 1 \qquad \text{if } a \neq 0 \tag{5b}$$

The multiplicative inverse $1/a$ of a nonzero real number a is also referred to as the **reciprocal** of a.

EXAMPLE 12 Finding a Reciprocal

(a) The reciprocal of 6 is $\frac{1}{6}$, because $6 \cdot \frac{1}{6} = 1$.

(b) The reciprocal of -3 is $\frac{1}{-3}$, because $-3 \cdot \frac{1}{-3} = 1$.

(c) The reciprocal of $\frac{2}{3}$ is $\frac{3}{2}$, because $\frac{2}{3} \cdot \frac{3}{2} = 1$.

With these properties for adding and multiplying real numbers, we can now define the operations of subtraction and division as follows:

The **difference** $a - b$, also read "a less b" or "a minus b," is defined as

$$a - b = a + (-b) \tag{6}$$

Thus, to subtract b from a, add the opposite of b to a.

If b is a nonzero real number, the **quotient** a/b, also read as "a divided by b" or "the ratio of a to b," is defined as

$$\frac{a}{b} = a \cdot \frac{1}{b} \quad \text{if } b \neq 0 \tag{7}$$

EXAMPLE 13 Working with Differences and Quotients

(a) $8 - 5 = 8 + (-5) = 3$ (b) $4 - 9 = 4 + (-9) = -5$

(c) $\frac{5}{8} = 5 \cdot \frac{1}{8}$

For any number a, the product of a times 0 is always 0; that is,

Multiplication by Zero

$$a \cdot 0 = 0 \tag{8}$$

For a nonzero number a,

Division Properties

$$\frac{0}{a} = 0 \qquad \frac{a}{a} = 1 \qquad \text{if } a \neq 0 \tag{9}$$

Note: Division by 0 is *not defined.* One reason is to avoid the following difficulty: $\frac{2}{0} = x$ means to find x such that $0 \cdot x = 2$. But $0 \cdot x$ equals 0 for all x, so there is *no* number x such that $\frac{2}{0} = x$.

Rules of Signs

$$a(-b) = -(ab) \qquad (-a)b = -(ab) \qquad (-a)(-b) = ab$$

$$-(-a) = a \qquad \frac{a}{-b} = \frac{-a}{b} = -\frac{a}{b} \qquad \frac{-a}{-b} = \frac{a}{b} \tag{10}$$

EXAMPLE 14 Applying the Rules of Signs

(a) $2(-3) = -(2 \cdot 3) = -6$ (b) $(-3)(-5) = 3 \cdot 5 = 15$

(c) $\dfrac{3}{-2} = \dfrac{-3}{2} = -\dfrac{3}{2}$ (d) $\dfrac{-4}{-9} = \dfrac{4}{9}$ (e) $\dfrac{x}{2} = \dfrac{1}{2} \cdot x$

If c is a nonzero number, then

Cancellation Properties

$$ac = bc \text{ implies } a = b \quad \text{if } c \neq 0$$
$$\frac{ac}{bc} = \frac{a}{b} \quad \text{if } b \neq 0, c \neq 0 \tag{11}$$

EXAMPLE 15 Using the Cancellation Properties

(a) If $2x = 6$, then

$$2x = 6$$
$$2x = 2 \cdot 3 \quad \textit{Factor 6.}$$
$$x = 3 \quad \textit{Cancel the 2's.}$$

(b) $\dfrac{18}{12} = \dfrac{3 \cdot 6}{2 \cdot 6} = \dfrac{3}{2}$

Cancel the 6's.

Zero-Product Property

$$\text{If } ab = 0, \text{ then } a = 0 \text{ or } b = 0, \text{ or both.} \tag{12}$$

EXAMPLE 16 Using the Zero-Product Property

If $2x = 0$, then either $2 = 0$ or $x = 0$. Since $2 \neq 0$, it follows that $x = 0$.

Arithmetic of Quotients

$$\frac{a}{b} + \frac{c}{d} = \frac{ad}{bd} + \frac{bc}{bd} = \frac{ad + bc}{bd} \quad \text{if } b \neq 0, d \neq 0 \qquad (13)$$

$$\frac{a}{b} \cdot \frac{c}{d} = \frac{ac}{bd} \quad \text{if } b \neq 0, d \neq 0 \qquad (14)$$

$$\frac{\frac{a}{b}}{\frac{c}{d}} = \frac{a}{b} \cdot \frac{d}{c} = \frac{ad}{bc} \quad \text{if } b \neq 0, c \neq 0, d \neq 0 \qquad (15)$$

EXAMPLE 17 Adding, Subtracting, Multiplying, and Dividing Quotients

(a) $$\frac{2}{3} + \frac{5}{2} = \frac{2 \cdot 2}{3 \cdot 2} + \frac{3 \cdot 5}{3 \cdot 2} = \frac{2 \cdot 2 + 3 \cdot 5}{3 \cdot 2} = \frac{4 + 15}{6} = \frac{19}{6}$$

By equation (13)

(b) $$\frac{3}{5} - \frac{2}{3} = \frac{3}{5} + \left(-\frac{2}{3}\right) = \frac{3}{5} + \frac{-2}{3}$$

By equation (6) By equation (10)

$$= \frac{3 \cdot 3 + 5 \cdot (-2)}{5 \cdot 3} = \frac{9 + (-10)}{15} = \frac{-1}{15}$$

(c) $$\frac{8}{3} \cdot \frac{15}{4} = \frac{8 \cdot 15}{3 \cdot 4} = \frac{2 \cdot \cancel{4} \cdot \cancel{3} \cdot 5}{\cancel{3} \cdot \cancel{4} \cdot 1} = \frac{2 \cdot 5}{1} = 10$$

By equation (14) By equation (11)

Note: We follow the common practice of using slash marks to indicate cancellations. Slanting the cancellation marks in different directions for different factors, as shown here, is a good practice to follow, since it will help in checking for errors.

(d) $$\frac{\frac{3}{5}}{\frac{7}{9}} = \frac{3}{5} \cdot \frac{9}{7} = \frac{3 \cdot 9}{5 \cdot 7} = \frac{27}{35}$$

By equation (14)

By equation (15)

Note: In writing quotients, we shall follow the usual convention and write the quotient in lowest terms; that is, we write it so that any common factors of the numerator and the denominator have been removed using the cancellation properties, equation (11). Thus,

$$\frac{90}{24} = \frac{15 \cdot \cancel{6}}{4 \cdot \cancel{6}} = \frac{15}{4}$$

$$\frac{24x^2}{18x} = \frac{4 \cdot \cancel{6} \cdot x \cdot \cancel{x}}{3 \cdot \cancel{6} \cdot \cancel{x}} = \frac{4x}{3} \qquad x \neq 0$$

NOW WORK PROBLEMS **35** AND **39**.

Sometimes it is easier to add two fractions using *least common multiples* (LCM). The LCM of two numbers is the smallest number that each has as a common multiple.

EXAMPLE 18 Finding the Least Common Multiple of Two Numbers

Find the least common multiple of 15 and 12.

Solution To find the LCM of 15 and 12, we look at multiples of 15 and 12:

$$15, 30, 45, 60, 75, 90, 105, 120, \ldots$$
$$12, 24, 36, 48, 60, 72, 84, 96, 108, 120, \ldots$$

The *common* multiples are in blue. The *least* common multiple is 60.

EXAMPLE 19 Using the Least Common Multiple to Add Two Fractions

Find: $\frac{8}{15} + \frac{5}{12}$

Solution We use the LCM of the denominators of the fractions and rewrite each fraction using the LCM as a common denominator. The LCM of the denominators (12 and 15) is 60. Rewrite each fraction using 60 as the denominator.

$$\frac{8}{15} + \frac{5}{12} = \frac{8}{15} \cdot \frac{4}{4} + \frac{5}{12} \cdot \frac{5}{5} = \frac{32}{60} + \frac{25}{60} = \frac{32 + 25}{60} = \frac{57}{60}$$

NOW WORK PROBLEM 43.

1 EXERCISES

In Problems 1–6, list the numbers in each set that are (a) Natural numbers, (b) Integers, (c) Rational numbers, (d) Irrational numbers, (e) Real numbers.

1. $A = \{-6, \frac{1}{2}, -1.333\ldots$ (the 3's repeat), $\pi, 2, 5\}$
2. $B = \{-\frac{5}{3}, 2.060606\ldots$ (the block 06 repeats), $1.25, 0, 1, \sqrt{5}\}$
3. $C = \{0, 1, \frac{1}{2}, \frac{1}{3}, \frac{1}{4}\}$
4. $D = \{-1, -1.1, -1.2, -1.3\}$
5. $E = \{\sqrt{2}, \pi, \sqrt{2} + 1, \pi + \frac{1}{2}\}$
6. $F = \{-\sqrt{2}, \pi + \sqrt{2}, \frac{1}{2} + 10.3\}$

In Problems 7–16, write each statement using symbols.

7. The sum of 3 and 2 equals 5.
8. The product of 5 and 2 equals 10.
9. The sum of x and 2 is the product of 3 and 4.
10. The sum of 3 and y is the sum of 2 and 2.
11. The product of 3 and y is the sum of 1 and 2.
12. The product of 2 and x is the product of 4 and 6.
13. The difference x less 2 equals 6.
14. The difference 2 less y equals 6.
15. The quotient x divided by 2 is 6.
16. The quotient 2 divided by x is 6.

In Problems 17–50, evaluate each expression.

17. $9 - 4 + 2$
18. $6 - 4 + 3$
19. $-6 + 4 \cdot 3$
20. $8 - 4 \cdot 2$
21. $4 + 5 - 8$
22. $8 - 3 - 4$
23. $4 + \frac{1}{3}$
24. $2 - \frac{1}{2}$
25. $6 - [3 \cdot 5 + 2 \cdot (3 - 2)]$
26. $2 \cdot [8 - 3(4 + 2)] - 3$
27. $2 \cdot (3 - 5) + 8 \cdot 2 - 1$
28. $1 - (4 \cdot 3 - 2 + 2)$
29. $10 - [6 - 2 \cdot 2 + (8 - 3)] \cdot 2$
30. $2 - 5 \cdot 4 - [6 \cdot (3 - 4)]$
31. $(5 - 3)\frac{1}{2}$
32. $(5 + 4)\frac{1}{3}$
33. $\frac{4 + 8}{5 - 3}$
34. $\frac{2 - 4}{5 - 3}$
35. $\frac{3}{5} \cdot \frac{10}{21}$
36. $\frac{5}{9} \cdot \frac{3}{10}$
37. $\frac{6}{25} \cdot \frac{10}{27}$
38. $\frac{21}{25} \cdot \frac{100}{3}$
39. $\frac{3}{4} + \frac{2}{5}$
40. $\frac{4}{3} + \frac{1}{2}$
41. $\frac{5}{6} + \frac{9}{5}$
42. $\frac{8}{9} + \frac{15}{2}$
43. $\frac{5}{18} + \frac{1}{12}$
44. $\frac{2}{15} + \frac{8}{9}$
45. $\frac{1}{30} - \frac{7}{18}$
46. $\frac{3}{14} - \frac{2}{21}$
47. $\frac{3}{20} - \frac{2}{15}$
48. $\frac{6}{35} - \frac{3}{14}$
49. $\dfrac{\frac{5}{18}}{\frac{11}{27}}$
50. $\dfrac{\frac{5}{21}}{\frac{2}{35}}$

In Problems 51–62, use the distributive property to remove the parentheses.

51. $6(x + 4)$
52. $4(2x - 1)$
53. $x(x - 4)$
54. $4x(x + 3)$
55. $(x + 2)(x + 4)$
56. $(x + 5)(x + 1)$
57. $(x - 2)(x + 1)$
58. $(x - 4)(x + 1)$
59. $(x - 8)(x - 2)$
60. $(x - 4)(x - 2)$
61. $(x + 2)(x - 2)$
62. $(x - 3)(x + 3)$

In Problems 63–72, express each statement as an equation involving the indicated variables and constant.

63. Area of a Rectangle The area A of a rectangle is the product of its length l times its width w.

64. Perimeter of a Rectangle The perimeter P of a rectangle is twice the sum of its length l and its width w.

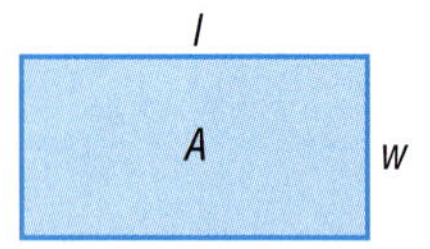

65. Circumference of a Circle The circumference C of a circle is the product of π times its diameter d.

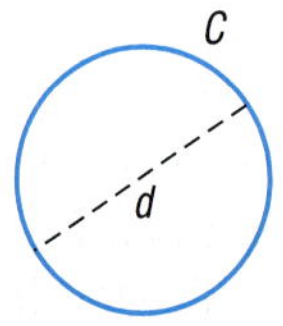

66. Area of a Triangle The area A of a triangle is one-half the product of its base b times its height h.

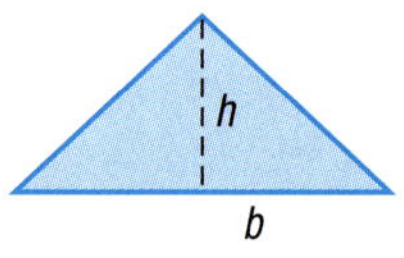

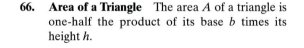

67. Area of an Equilateral Triangle The area A of an equilateral triangle is $\sqrt{3}/4$ times the square of the length x of one side.

68. Perimeter of an Equilateral Triangle The perimeter P of an equilateral triangle is 3 times the length x of one side.

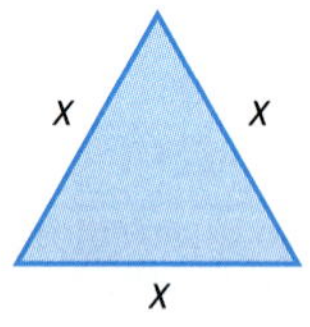

69. **Volume of a Sphere** The volume V of a sphere is the product of $\frac{4}{3}$ times π times the cube of the radius r.

70. **Surface Area of a Sphere** The surface area S of a sphere is the product of 4 times π times the square of the radius r.

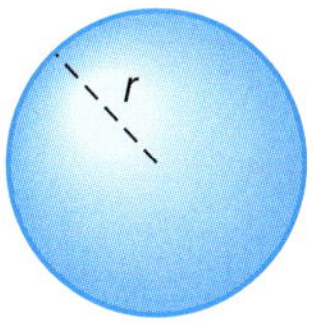

71. **Volume of a Cube** The volume V of a cube is the cube of the length x of a side.

72. **Surface Area of a Cube** The surface area S of a cube is 6 times the square of the length x of a side.

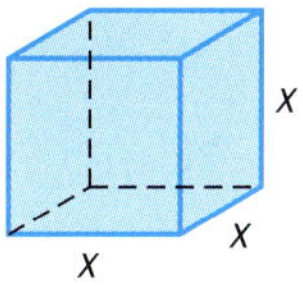

73. Explain to a friend how the distributive property is used to justify the fact that $2x + 3x = 5x$.

74. Explain to a friend why $2 + 3 \cdot 4 = 14$, while $(2 + 3) \cdot 4 = 20$.

75. Explain why $2(3 \cdot 4)$ is not equal to $(2 \cdot 3) \cdot (2 \cdot 4)$.

76. Explain why $\frac{4 + 3}{2 + 5}$ is not equal to $\frac{4}{2} + \frac{3}{5}$.

77. Does $\frac{1}{3}$ equal 0.333? If not, which is larger? By how much?

78. Does $\frac{2}{3}$ equal 0.666? If not, which is larger? By how much?

79. Is subtraction commutative? Support your conclusion with an example.

80. Is subtraction associative? Support your conclusion with an example.

81. Is division commutative? Support your conclusion with an example.

82. Is division associative? Support your conclusion with an example.

83. If $2 = x$, why does $x = 2$?

84. If $x = 5$, why does $x^2 + x = 30$?

85. Are there any real numbers that are both rational and irrational? Are there any real numbers that are neither? Explain your reasoning.

86. Explain why the sum of a rational number and an irrational number must be irrational.

87. What rational number does the repeating decimal $0.9999\ldots$ equal?

2 ALGEBRA REVIEW

1. Graph Inequalities
2. Find Distance on a Real Number Line
3. Evaluate Algebraic Expressions
4. Solve Equations in One Variable

The Real Number Line

The real numbers can be represented by points on a line called the **real number line.** There is a one-to-one correspondence between real numbers and points on a line. That is, every real number corresponds to a point on the line, and each point on the line has a unique real number associated with it.

Pick a point on the line somewhere in the center, and label it O. This point, called the **origin,** corresponds to the real number 0. See Figure 6. The point 1 unit to the right of O corresponds to the number 1. The distance between 0 and 1 determines the **scale** of the number line. For example, the point associated with the number 2 is twice as far from O as 1 is. Notice that an arrowhead on the right end of the line indicates the direction in which the num-

Figure 6
Real number line

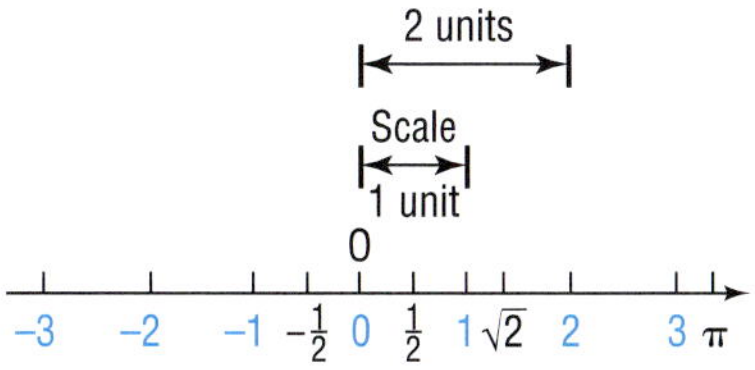

bers increase. Figure 6 also shows the points associated with the irrational numbers $\sqrt{2}$ and π. Points to the left of the origin correspond to the real numbers -1, -2, and so on.

The real number associated with a point P is called the **coordinate** of P, and the line whose points have been assigned coordinates is called the **real number line.**

NOW WORK PROBLEM 1.

The real number line divides the real numbers into three classes, as shown in Figure 7.

Figure 7

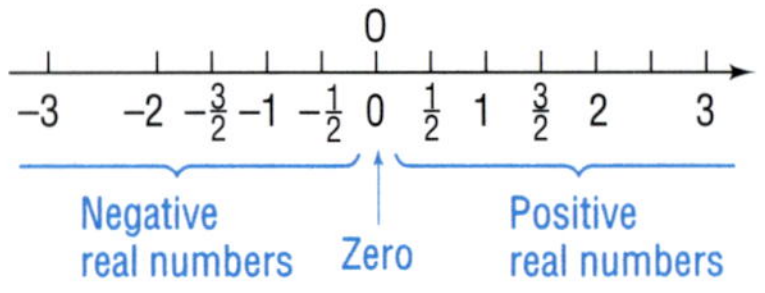

1. The **negative real numbers** are the coordinates of points to the left of the origin O.
2. The real number **zero** is the coordinate of the origin O.
3. The **positive real numbers** are the coordinates of points to the right of the origin O.

Negative and positive numbers have the following multiplication properties:

Multiplication Properties of Positive and Negative Numbers

1. The product of two positive numbers is a positive number.
2. The product of two negative numbers is a positive number.
3. The product of a positive number and a negative number is a negative number.

Inequalities

Figure 8

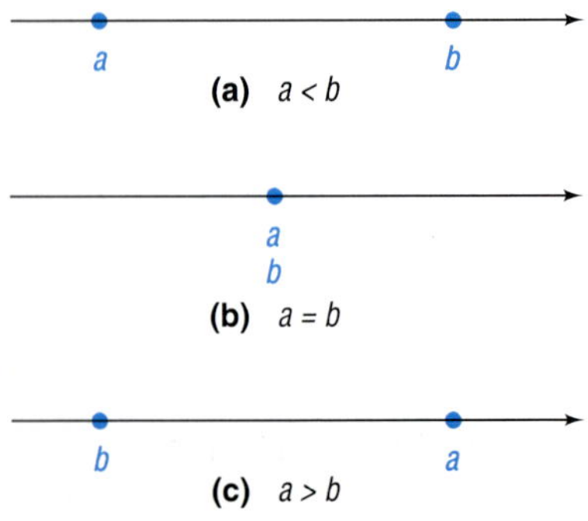

An important property of the real number line follows from the fact that, given two numbers (points) a and b, either a is to the left of b, a equals b, or a is to the right of b. See Figure 8.

If a is to the left of b, we say that "a is less than b" and write $a < b$. If a is to the right of b, we say that "a is greater than b" and write $a > b$. If a equals b, we write $a = b$. If a is either less than or equal to b, we write $a \leq b$. Similarly, $a \geq b$ means that a is either greater than or equal to b. Collectively, the symbols $<$, $>$, $\leq$, and $\geq$ are called **inequality symbols.**

Note that $a < b$ and $b > a$ mean the same thing. Thus, it does not matter whether we write $2 < 3$ or $3 > 2$.

Furthermore, if $a < b$ or if $b > a$, then the difference $b - a$ is positive. Do you see why?

EXAMPLE 1 Using Inequality Symbols

(a) $3 < 7$ (b) $-8 > -16$ (c) $-6 < 0$
(d) $-8 < -4$ (e) $4 > -1$ (f) $8 > 0$

In Example 1(a), we conclude that $3 < 7$ either because 3 is to the left of 7 on the real number line or because the difference $7 - 3 = 4$, a positive real number.

Similarly, we conclude in Example 1(b) that $-8 > -16$ either because -8 lies to the right of -16 on the real number line or because the difference $-8 - (-16) = -8 + 16 = 8$, a positive real number.

Look again at Example 1. Note that the inequality symbol always points in the direction of the smaller number.

Statements of the form $a < b$ or $b > a$ are called **strict inequalities,** while statements of the form $a \leq b$ or $b \geq a$ are called **nonstrict inequalities.** An **inequality** is a statement in which two expressions are related by an inequality symbol. The expressions are referred to as the **sides** of the inequality.

Based on the discussion thus far, we conclude that

$a > 0$	is equivalent to	a is positive
$a < 0$	is equivalent to	a is negative

Thus, we sometimes read $a > 0$ by saying that "a is positive." If $a \geq 0$, then either $a > 0$ or $a = 0$, and we may read this as "a is nonnegative."

NOW WORK PROBLEMS 5 AND 15.

Graphing Inequalities

1 We shall find it useful in later work to graph inequalities on the real number line.

EXAMPLE 2 Graphing Inequalities

(a) On the real number line, graph all numbers x for which $x > 4$.
(b) On the real number line, graph all numbers x for which $x \leq 5$.

Solution (a) See Figure 9. Notice that we use a left parenthesis to indicate that the number 4 is *not* part of the graph.

Figure 9

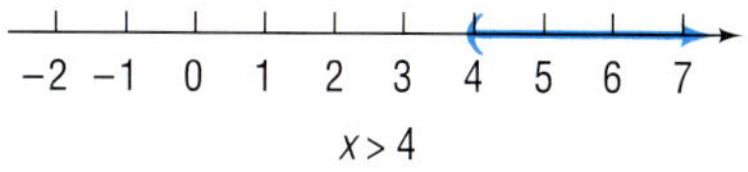

Figure 10

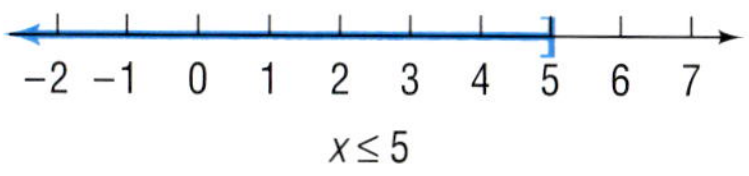

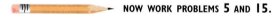

(b) See Figure 10. Notice that we use a right bracket to indicate that the number 5 *is* part of the graph.

Inequalities are often combined.

EXAMPLE 3 Graphing Combined Inequalities

On the real number line, graph all numbers x for which $x > 4$ and $x < 6$.

Solution We first graph each inequality separately, as illustrated in Figure 11(a). Then it is easy to see that the numbers that belong to *both* the graph of $x > 4$ *and* the graph of $x < 6$ are shown in Figure 11(b). For example, 5 is part of the graph because $5 > 4$ and $5 < 6$; 7 is not on the graph because 7 is not less than 6.

Figure 11

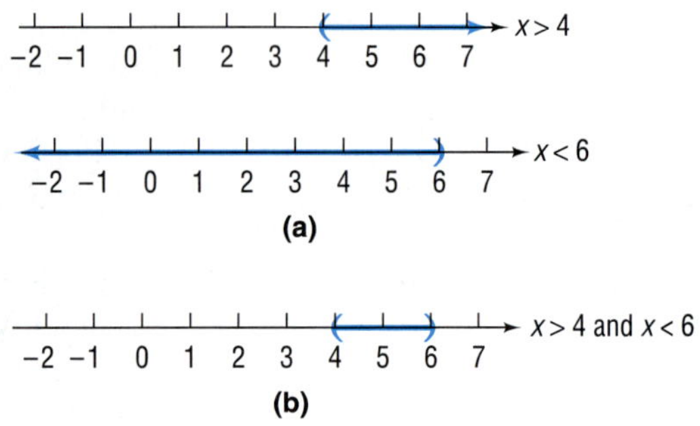

EXAMPLE 4 Graphing Combined Inequalities

On the real number line, graph all numbers x for which $x > 4$ or $x \leq -1$.

Solution See Figure 12(a), where each inequality is graphed separately. The numbers that belong to *either* the graph of $x > 4$ *or* the graph of $x \leq -1$ are graphed in Figure 12(b). For example, 5 is part of the graph because $5 > 4$; 2 is not on the graph because 2 is not greater than 4 nor is 2 less than or equal to -1.

Figure 12

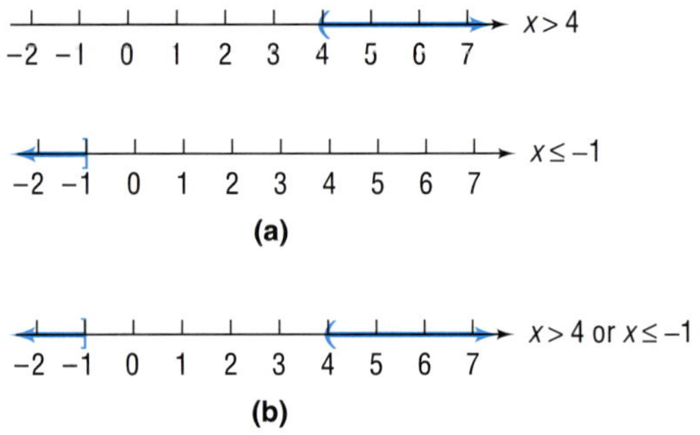

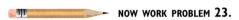

NOW WORK PROBLEM 23.

Absolute Value

Figure 13

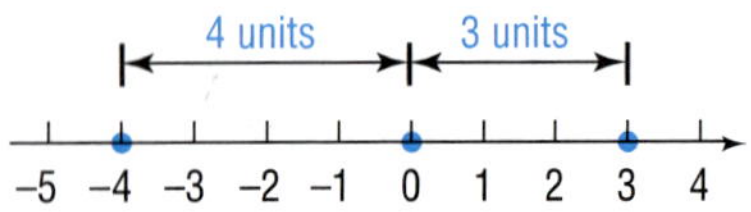

The *absolute value* of a number a is the distance from 0 to a on the number line. For example, -4 is 4 units from 0; and 3 is 3 units from 0. See Figure 13. Thus, the absolute value of -4 is 4, and the absolute value of 3 is 3.

A more formal definition of absolute value is given next.

The **absolute value** of a real number a, denoted by the symbol $|a|$, is defined by the rules

$$|a| = a \quad \text{if } a \geq 0 \qquad \text{and} \qquad |a| = -a \quad \text{if } a < 0$$

For example, since $-4 < 0$, the second rule must be used to get $|-4| = -(-4) = 4$.

EXAMPLE 5 Computing Absolute Value

(a) $|8| = 8$ (b) $|0| = 0$ (c) $|-15| = -(-15) = 15$

2

Look again at Figure 13. The distance from -4 to 3 is 7 units. This distance is the difference $3 - (-4)$, obtained by subtracting the smaller coordinate from the larger. However, since $|3 - (-4)| = |7| = 7$ and $|-4 - 3| = |-7| = 7$, we can use absolute value to calculate the distance between two points without being concerned about which is smaller.

If P and Q are two points on a real number line with coordinates a and b, respectively, the **distance between P and Q,** denoted by $d(P, Q)$, is

$$d(P, Q) = |b - a|$$

Since $|b - a| = |a - b|$, it follows that $d(P, Q) = d(Q, P)$.

EXAMPLE 6 Finding Distance on a Number Line

Let P, Q, and R be points on a real number line with coordinates -5, 7, and -3, respectively. Find the distance

(a) between P and Q (b) between Q and R

Solution (a) $d(P, Q) = |7 - (-5)| = |12| = 12$ (see Figure 14)
(b) $d(Q, R) = |-3 - 7| = |-10| = 10$

Figure 14

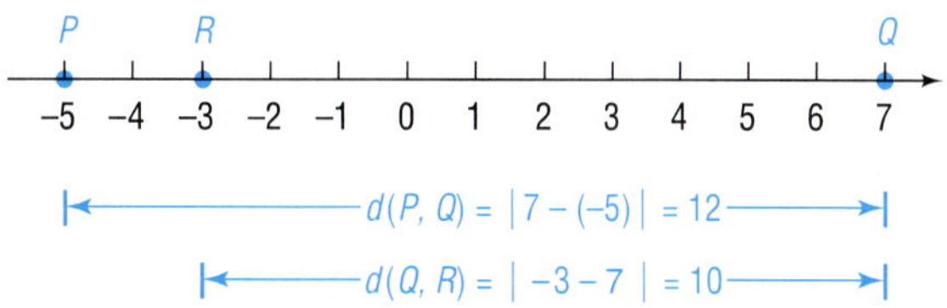

NOW WORK PROBLEM 31.

Constants and Variables

In algebra we use letters such as x, y, a, b, and c to represent numbers. If the letter used is to represent *any* number from a given set of numbers, it is called a **variable.** A **constant** is either a fixed number, such as 5 or $\sqrt{3}$, or a letter that represents a fixed (possibly unspecified) number.

Constants and variables are combined using the operations of addition, subtraction, multiplication, and division to form *algebraic expressions*. Examples of algebraic expressions include

$$x + 3 \qquad \frac{3}{1 - t} \qquad 7x - 2y$$

3 To evaluate an algebraic expression, substitute for each variable its numerical value.

EXAMPLE 7 Evaluating an Algebraic Expression

Evaluate each expression if $x = 3$ and $y = -1$.

(a) $x + 3y$ (b) $5xy$ (c) $\dfrac{3y}{2 - 2x}$ (d) $|-4x + y|$

Solution (a) Substitute 3 for x and -1 for y in the expression $x + 3y$.

$$x + 3y \underset{\substack{\uparrow \\ x = 3,\, y = -1}}{=} 3 + 3(-1) = 3 + (-3) = 0$$

(b) If $x = 3$ and $y = -1$, then

$$5xy = 5(3)(-1) = -15$$

(c) If $x = 3$ and $y = -1$, then

$$\frac{3y}{2 - 2x} = \frac{3(-1)}{2 - 2(3)} = \frac{-3}{2 - 6} = \frac{-3}{-4} = \frac{3}{4}$$

(d) If $x = 3$ and $y = -1$, then

$$|-4x + y| = |-4(3) + (-1)| = |-12 + (-1)| = |-13| = 13$$

Graphing calculators can be used to evaluate algebraic expressions. Figure 15 shows the results of Example 7 using a TI-83.

Figure 15

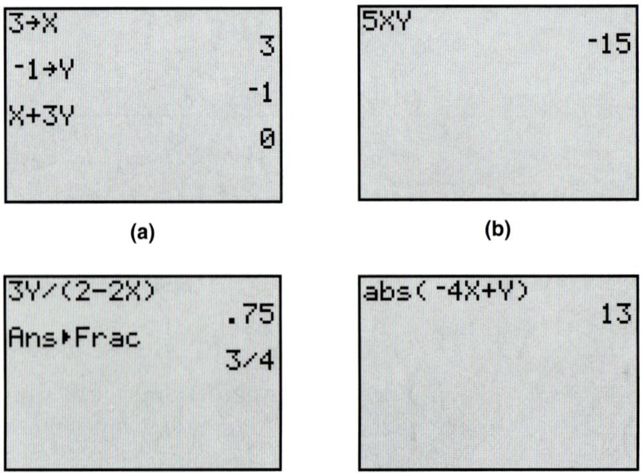

(a) (b) (c) (d)

NOW WORK PROBLEMS 33 AND 41.

Sometimes it is better to simplify an algebraic expression before substituting for the values of the variables.

EXAMPLE 8 Evaluating Algebraic Expressions Involving Exponents*

Evaluate each expression if $m = -2$ and $n = 4$.

(a) $\dfrac{m^2m^{-1}}{n^2}$ (b) $\dfrac{m^5n^3}{m^4}$ (c) $\left(\dfrac{m^{-2}}{n^{-3}}\right)^{-1}$

Solution (a) $\dfrac{m^2m^{-1}}{n^2} = \dfrac{m^{2+(-1)}}{n^2} = \dfrac{m}{n^2} \underset{m=-2,\ n=4}{=} \dfrac{-2}{4^2} = \dfrac{-2}{16} = -\dfrac{1}{8}$

(b) $\dfrac{m^5n^3}{m^4} = m^{5-4}n^3 = mn^3 \underset{m=-2,\ n=4}{=} -2 \cdot (4)^3 = -2 \cdot 64 = -128$

(c) $\left(\dfrac{m^{-2}}{n^{-3}}\right)^{-1} = \dfrac{m^{(-2)(-1)}}{n^{(-3)(-1)}} = \dfrac{m^2}{n^3} \underset{m=-2,\ n=4}{=} \dfrac{(-2)^2}{(4)^3} = \dfrac{4}{64} = \dfrac{1}{16}$

NOW WORK PROBLEM 53.

Equations

4 An **equation in one variable** is a statement in which two expressions, at least one containing the variable, are equal. The expressions are called the **sides** of the equation. Since an equation is a statement, it may be true or false, depending on the value of the variable. The values of the variable, if any, that result in a true statement are called **solutions,** or **roots,** of the equation. To **solve an equation** means to find all the solutions of the equation.

For example, the following are all equations in one variable, x:

$$x + 5 = 9 \qquad x^2 + 5x = 2x - 2 \qquad \frac{x^2 - 4}{x + 1} = 0 \qquad x^2 + 9 = 5$$

The first of these statements, $x + 5 = 9$, is true when $x = 4$ and false for any other choice of x. Thus, 4 is a solution of the equation $x + 5 = 9$. We also say that 4 **satisfies** the equation $x + 5 = 9$.

Sometimes an equation will have more than one solution. For example, the equation

$$x^2 - 4 = 0$$

has either $x = -2$ or $x = 2$ as a solution.

Sometimes we will write the solution of an equation in set notation. This set is called the **solution set** of the equation. For example, the solution set of the equation $x^2 - 9 = 0$ is $\{-3, 3\}$.

Unless indicated otherwise, we will limit ourselves to real solutions. Some equations have no real solution. For example, $x^2 + 9 = 5$ has no real solution, because there is no real number whose square when added to 9 equals 5.

An equation that is satisfied for every choice of the variable for which both sides are defined is called an **identity.** For example, the equation

$$3x + 5 = x + 3 + 2x + 2$$

is an identity, because this statement is true for any real number x.

*A detailed review of integer exponents can be found in Section 4 of this Appendix.

Two or more equations that have precisely the same solutions are called **equivalent equations.** For example, all the following equations are equivalent, because each has only the solution $x = 5$:

$$2x + 3 = 13$$
$$2x = 10$$
$$x = 5$$

These three equations illustrate one method for solving many types of equations: Replace the original equation by an equivalent equation, and continue until an equation with an obvious solution, such as $x = 5$, is reached. The question, though, is "How do I obtain an equivalent equation?" In general, there are five ways to do so.

Procedures That Result in Equivalent Equations

1. Interchange the two sides of the equation:

   Replace $3 = x$ by $x = 3$

2. Simplify the sides of the equation by combining like terms, eliminating parentheses, and so on:

   Replace $(x + 2) + 6 = 2x + 5(x + 1)$
   by $x + 8 = 7x + 5$

3. Add or subtract the same expression on both sides of the equation:

   Replace $3x - 5 = 4$
   by $(3x - 5) + 5 = 4 + 5$

4. Multiply or divide both sides of the equation by the same nonzero expression:

   Replace $3x = 6$
   by $\frac{1}{3} \cdot 3x = \frac{1}{3} \cdot 6$

5. If one side of the equation is 0 and the other side can be factored, then we may use the Zero-Product Property* and set each factor equal to 0:

   Replace $x(x - 3) = 0$
   by $x = 0$ or $x - 3 = 0$

Whenever it is possible to solve an equation in your head, do so. For example:

The solution of $2x = 8$ is $x = 4$.
The solution of $3x - 15 = 0$ is $x = 5$.

Often, though, some rearrangement is necessary.

*The Zero-Product Property says that if $ab = 0$ then $a = 0$ or $b = 0$ or both equal 0.

EXAMPLE 9 Solving an Equation

Solve the equation: $3x - 5 = 4$

Solution We replace the original equation by a succession of equivalent equations.

$$
\begin{aligned}
3x - 5 &= 4 \\
(3x - 5) + 5 &= 4 + 5 && \textit{Add 5 to both sides.} \\
3x &= 9 && \textit{Simplify.} \\
\frac{3x}{3} &= \frac{9}{3} && \textit{Divide both sides by 3.} \\
x &= 3 && \textit{Simplify.}
\end{aligned}
$$

The last equation, $x = 3$, has the single solution 3. All these equations are equivalent, so 3 is the only solution of the original equation $3x - 5 = 4$.

✓CHECK: It is a good practice to check the solution by substituting 3 for x in the original equation.

$$
\begin{aligned}
3x - 5 &= 4 \\
3(3) - 5 &\stackrel{?}{=} 4 \\
9 - 5 &\stackrel{?}{=} 4 \\
4 &= 4
\end{aligned}
$$

The solution checks.

NOW WORK PROBLEM 67.

In the next examples, we use the Zero-Product Property (procedure 5, listed in the box on p. 22).

EXAMPLE 10 Solving Equations by Factoring*

Solve the equations: (a) $x^2 = 4x$ (b) $x^3 - x^2 - 4x + 4 = 0$

Solution (a) We begin by collecting all terms on one side. This results in 0 on one side and an expression to be factored on the other.

$$
\begin{aligned}
x^2 &= 4x \\
x^2 - 4x &= 0 \\
x(x - 4) &= 0 && \textit{Factor.} \\
x = 0 \quad \text{or} \quad x - 4 &= 0 && \textit{Apply the Zero-Product Property.} \\
x = 0 \quad \text{or} \quad x &= 4
\end{aligned}
$$

The solution set is $\{0, 4\}$.

✓CHECK: $x = 0$: $0^2 = 4 \cdot 0$ *So 0 is a solution.*
$x = 4$: $4^2 = 4 \cdot 4$ *So 4 is a solution.*

(b) Do you recall the method of factoring by grouping? We group the terms of $x^3 - x^2 - 4x + 4 = 0$ as follows:

$$(x^3 - x^2) - (4x - 4) = 0$$

*A detailed discussion of factoring may be found in Section 6 of this Appendix.

Factor out x^2 from the first grouping and 4 from the second:

$$x^2(x-1) - 4(x-1) = 0$$

This reveals the common factor $(x - 1)$, so we have

$$(x^2 - 4)(x - 1) = 0$$

$$(x - 2)(x + 2)(x - 1) = 0 \qquad \textit{Factor again.}$$

$$x - 2 = 0 \quad \text{or} \quad x + 2 = 0 \quad \text{or} \quad x - 1 = 0 \qquad \textit{Set each factor equal to 0.}$$

$$x = 2 \qquad x = -2 \qquad x = 1 \qquad \textit{Solve.}$$

The solution set is $\{-2, 1, 2\}$.

✓CHECK: $x = 2$: $2^3 - 2^2 - 4(2) + 4 = 8 - 4 - 8 + 4 = 0$ *2 is a solution.*

$x = -2$: $(-2)^3 - (-2)^2 - 4(-2) + 4 = -8 - (4) + 8 + 4 = 0$ *−2 is a solution.*

$x = 1$: $1^3 - 1^2 - 4(1) + 4 = 1 - 1 - 4 + 4 = 0$ *1 is a solution.*

NOW WORK PROBLEMS 79 AND 87.

2 EXERCISES

1. On the real number line, label the points with coordinates $0, 1, -1, \frac{5}{2}, -2.5, \frac{3}{4}$, and 0.25.
2. Repeat Problem 1 for the coordinates $0, -2, 2, -1.5, \frac{3}{2}, \frac{1}{3}$, and $\frac{2}{3}$.

In Problems 3–12, replace the question mark by $<, >,$ *or* $=$, *whichever is correct.*

3. $\frac{1}{2}$? 0
4. 5 ? 6
5. -1 ? -2
6. -3 ? $-\frac{5}{2}$
7. π ? 3.14
8. $\sqrt{2}$? 1.41
9. $\frac{1}{2}$? 0.5
10. $\frac{1}{3}$? 0.33
11. $\frac{2}{3}$? 0.67
12. $\frac{1}{4}$? 0.25

In Problems 13–20, write each statement as an inequality.

13. x is positive
14. z is negative
15. x is less than 2
16. y is greater than -5
17. x is less than or equal to 1
18. x is greater than or equal to 2
19. x is less than 5 and x is greater than 2
20. y is less than or equal to 2 and y is greater than 0

In Problems 21–28, graph the numbers x, if any, on the real number line.

21. $x \geq -2$
22. $x < 4$
23. $x \geq 4$ and $x < 6$
24. $x > 3$ and $x \leq 7$
25. $x \leq 0$ or $x < 6$
26. $x > 0$ or $x \geq 5$
27. $x \leq -2$ or $x > 1$
28. $x \geq 4$ or $x < -2$

In Problems 29–32, use the real number line below to compute each distance.

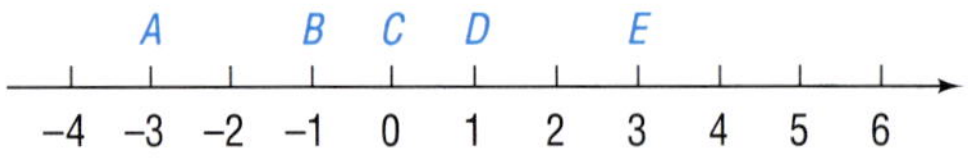

29. $d(D, E)$
30. $d(C, E)$
31. $d(A, E)$
32. $d(D, B)$

In Problems 33–40, evaluate each expression if $x = -2$ *and* $y = 3$. *Check your answer using a graphing utility.*

33. $x + 2y$
34. $3x + y$
35. $5xy + 2$
36. $-2x + xy$
37. $\dfrac{2x}{x - y}$
38. $\dfrac{x + y}{x - y}$
39. $\dfrac{3x + 2y}{2 + y}$
40. $\dfrac{2x - 3}{y}$

In Problems 41–50, find the value of each expression if $x = 3$ *and* $y = -2$. *Check your answers using a graphing utility.*

41. $|x + y|$
42. $|x - y|$
43. $|x| + |y|$
44. $|x| - |y|$
45. $\dfrac{|x|}{x}$
46. $\dfrac{|y|}{y}$
47. $|4x - 5y|$
48. $|3x + 2y|$
49. $||4x| - |5y||$
50. $3|x| + 2|y|$

In Problems 51–60, evaluate each expression if $x = 2$ and $y = -3$. Check your answer using a graphing utility.

51. $3x^0$ **52.** $(3x)^0$ **53.** $\dfrac{x^{-2}y^3}{xy^4}$ **54.** $\dfrac{x^{-2}y}{xy^2}$ **55.** $x^{-1}y^{-1}$

56. $\dfrac{x^{-2}y^{-3}}{x}$ **57.** $\dfrac{x^{-1}}{y^{-1}}$ **58.** $\left(\dfrac{2x}{3}\right)^{-1}$ **59.** $\left(\dfrac{4y}{5x}\right)^{-2}$ **60.** $(x^2y)^{-2}$

In Problems 61–78, solve each equation.

61. $3x + 2 = x$ **62.** $2x + 7 = 3x$ **63.** $2t - 6 = 3 - t$

64. $5y + 6 = -18 - y$ **65.** $6 - x = 2x + 9$ **66.** $3 - 2x = 2 - x$

67. $3 + 2n = 5n + 7$ **68.** $3 - 2m = 3m + 1$ **69.** $2(3 + 2x) = 3(x - 4)$

70. $3(2 - x) = 2x - 1$ **71.** $8x - (3x + 2) = 3x - 10$ **72.** $7 - (2x - 1) = 10$

73. $\frac{3}{2}x + 2 = \frac{1}{2} - \frac{1}{2}x$ **74.** $\frac{1}{3}x = 2 - \frac{2}{3}x$ **75.** $\frac{1}{2}x - 5 = \frac{3}{4}x$

76. $1 - \frac{1}{2}x = 6$ **77.** $\frac{2}{3}p = \frac{1}{2}p + \frac{1}{3}$ **78.** $\frac{1}{2} - \frac{1}{3}p = \frac{4}{3}$

In Problems 79–90, find the real solutions of each equation by factoring.

79. $x^2 - 7x + 12 = 0$ **80.** $x^2 - x - 6 = 0$ **81.** $2x^2 + 5x - 3 = 0$

82. $3x^2 + 5x - 2 = 0$ **83.** $x^3 = 9x$ **84.** $x^4 = x^2$

85. $x^3 + x^2 - 20x = 0$ **86.** $x^3 + 6x^2 - 7x = 0$ **87.** $x^3 + x^2 - x - 1 = 0$

88. $x^3 + 4x^2 - x - 4 = 0$ **89.** $x^3 - 3x^2 - 4x + 12 = 0$ **90.** $x^3 - 3x^2 - x + 3 = 0$

In Problems 91–94, use the formula $C = \frac{5}{9}(F - 32)$ for converting degrees Fahrenheit into degrees Celsius to find the Celsius measure of each Fahrenheit temperature.

91. $F = 32°$ **92.** $F = 212°$ **93.** $F = 77°$ **94.** $F = -4°$

95. Manufacturing Cost The weekly production cost C of manufacturing x watches is given by the formula $C = 4000 + 2x$, where the variable C is in dollars.
(a) What is the cost of producing 1000 watches?
(b) What is the cost of producing 2000 watches?

96. Balancing a Checkbook At the beginning of the month, Mike had a balance of \$210 in his checking account. During the next month, he deposited \$80, wrote a check for \$120, made another deposit of \$25, wrote two checks for \$60 and \$32, and was assessed a monthly service charge of \$5. What was his balance at the end of the month?

97. U.S. Voltage In the United States, normal household voltage is 115 volts. It is acceptable for the actual voltage x to differ from normal by at most 5 volts. A formula that describes this is

$$|x - 115| \le 5$$

(a) Show that a voltage of 113 volts is acceptable.
(b) Show that a voltage of 109 volts is not acceptable.

98. Foreign Voltage In other countries, normal household voltage is 220 volts. It is acceptable for the actual voltage x to differ from normal by at most 8 volts. A formula that describes this is

$$|x - 220| \le 8$$

(a) Show that a voltage of 214 volts is acceptable.
(b) Show that a voltage of 209 volts is not acceptable.

99. Making Precision Ball Bearings The FireBall Company manufactures ball bearings for precision equipment. One of their products is a ball bearing with a stated radius of 3 centimeters (cm). Only ball bearings with a radius within 0.01 cm of this stated radius are acceptable. If x is the radius of a ball bearing, a formula describing this situation is

$$|x - 3| \le 0.01$$

(a) Is a ball bearing of radius $x = 2.999$ acceptable?
(b) Is a ball bearing of radius $x = 2.89$ acceptable?

100. Body Temperature Normal human body temperature is 98.6°F. A temperature x that differs from normal by at least 1.5°F is considered unhealthy. A formula that describes this is

$$|x - 98.6| \ge 1.5$$

(a) Show that a temperature of 97°F is unhealthy.
(b) Show that a temperature of 100°F is not unhealthy.

101. Is there a positive real number "closest" to 0?

102. I'm thinking of a number! It lies between 1 and 10; its square is rational and lies between 1 and 10. The number is larger than π. Correct to two decimal places, name the number. Now think of you own number, describe it, and challenge a fellow student to name it.

103. Write a brief paragraph that illustrates the similarities and differences between "less than" ($<$) and "less than or equal" ($\leq$).

3 GEOMETRY REVIEW

1 Use the Pythagorean Theorem and Its Converse

2 Know Geometry Formulas

In this section we review some topics studied in geometry that we shall need for our study of algebra.

Figure 16

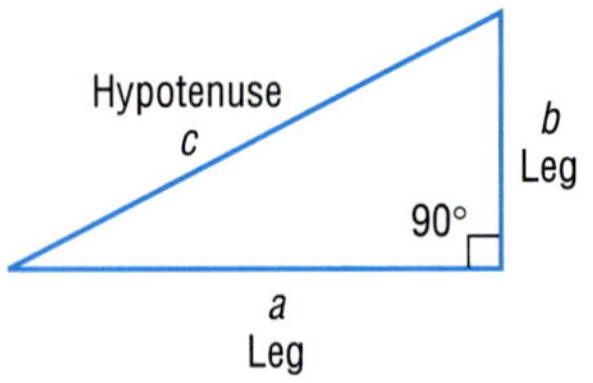

Pythagorean Theorem

1 The *Pythagorean Theorem* is a statement about *right triangles.* A **right triangle** is one that contains a **right angle,** that is, an angle of 90°. The side of the triangle opposite the 90° angle is called the **hypotenuse;** the remaining two sides are called **legs.** In Figure 16 we have used c to represent the length of the hypotenuse and a and b to represent the lengths of the legs. Notice the use of the symbol ⌜ to show the 90° angle. We now state the Pythagorean Theorem.

PYTHAGOREAN THEOREM

In a right triangle, the square of the length of the hypotenuse is equal to the sum of the squares of the lengths of the legs. That is, in the right triangle shown in Figure 16,

$$c^2 = a^2 + b^2 \qquad (1)$$

EXAMPLE 1 Finding the Hypotenuse of a Right Triangle

In a right triangle, one leg is of length 4 and the other is of length 3. What is the length of the hypotenuse?

Solution Since the triangle is a right triangle, we use the Pythagorean Theorem with $a = 4$ and $b = 3$ to find the length c of the hypotenuse. From equation (1), we have

$$\begin{aligned} c^2 &= a^2 + b^2 \\ c^2 &= 4^2 + 3^2 = 16 + 9 = 25 \\ c &= \sqrt{25} = 5 \end{aligned}$$

NOW WORK PROBLEM 3.

The converse of the Pythagorean Theorem is also true.

CONVERSE OF THE PYTHAGOREAN THEOREM

In a triangle, if the square of the length of one side equals the sum of the squares of the lengths of the other two sides, then the triangle is a right triangle. The 90° angle is opposite the longest side.

EXAMPLE 2 Verifying That a Triangle Is a Right Triangle

Show that a triangle whose sides are of lengths 5, 12, and 13 is a right triangle. Identify the hypotenuse.

Solution We square the lengths of the sides.

$$5^2 = 25, \quad 12^2 = 144, \quad 13^2 = 169$$

Notice that the sum of the first two squares (25 and 144) equals the third square (169). Hence, the triangle is a right triangle. The longest side, 13, is the hypotenuse. See Figure 17.

Figure 17

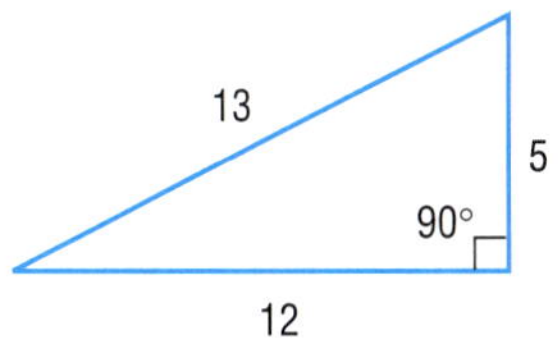

NOW WORK PROBLEM 11.

EXAMPLE 3 Applying the Pythagorean Theorem

The tallest inhabited building in the world is the Sears Tower in Chicago.* If the observation tower is 1450 feet above ground level, how far can a person standing in the observation tower see (with the aid of a telescope)? Use 3960 miles for the radius of Earth. See Figure 18.

[*Note:* 1 mile = 5280 feet]

Figure 18

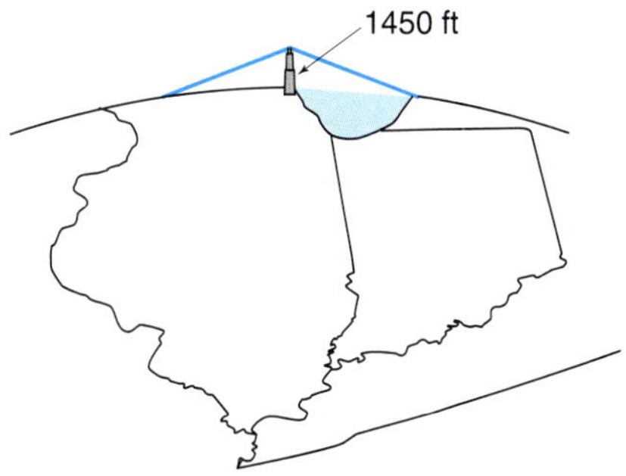

**Source:* Council on Tall Buildings and Urban Habitat (1997): Sears Tower No. 1 for tallest roof (1450 ft) and tallest occupied floor (1431 ft).

Solution From the center of Earth, draw two radii: one through the Sears Tower and the other to the farthest point a person can see from the tower. See Figure 19. Apply the Pythagorean Theorem to the right triangle.

Figure 19

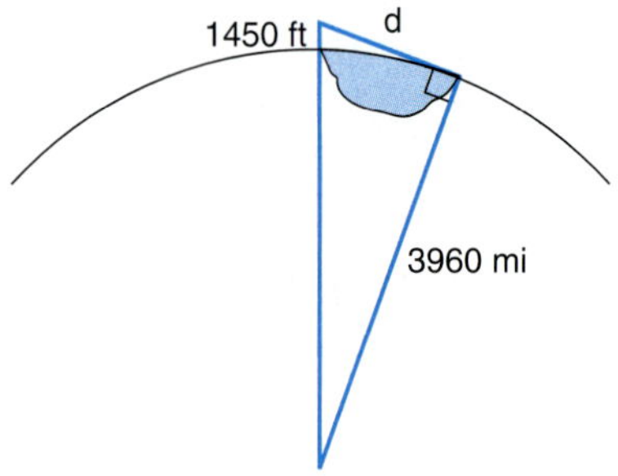

Since 1450 feet = 1450/5280 miles, we have

$$d^2 + (3960)^2 = \left(3960 + \frac{1450}{5280}\right)^2$$

$$d^2 = \left(3960 + \frac{1450}{5280}\right)^2 - (3960)^2 \approx 2175.08$$

$$d \approx 46.64$$

A person can see about 47 miles from the observation tower.

Geometry Formulas

2 Certain formulas from geometry are useful in solving algebra problems. We list some of these formulas next.

For a rectangle of length l and width w,

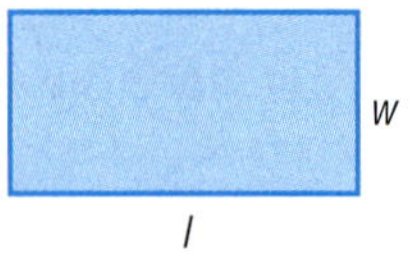

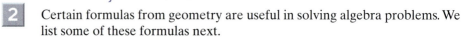

Area = lw Perimeter = $2l + 2w$

For a triangle with base b and altitude h,

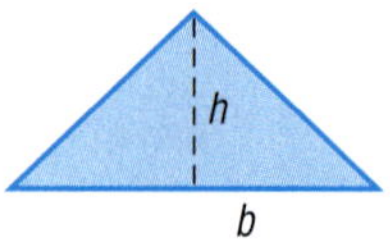

Area = $\frac{1}{2}bh$

For a circle of radius r (diameter $d = 2r$),

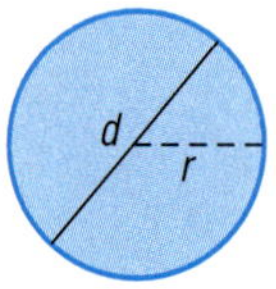

Area = πr^2 Circumference = $2\pi r = \pi d$

For a rectangular box of length l, width w, and height h,

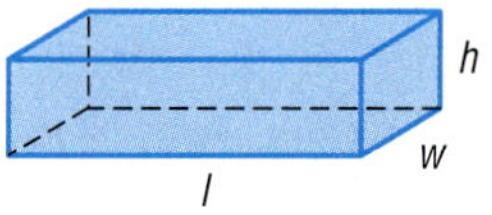

Volume = lwh

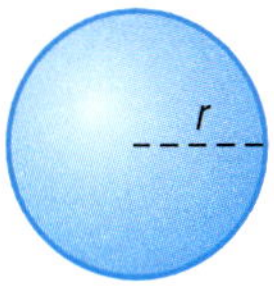

For a sphere of radius r,

$$\text{Volume} = \tfrac{4}{3}\pi r^3 \qquad \text{Surface area} = 4\pi r^2$$

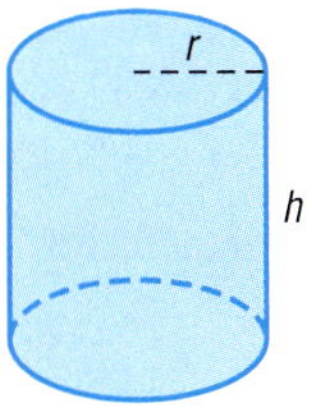

For a right circular cylinder of height h and radius r,

$$\text{Volume} = \pi r^2 h$$

EXAMPLE 4 Using Geometry Formulas

A Christmas tree ornament is in the shape of a semicircle on top of a triangle. How many square centimeters (cm) of copper is required to make the ornament if the height of the triangle is 6 cm and the base is 4 cm?

Figure 20

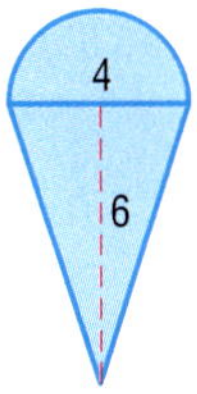

Solution See Figure 20. The amount of copper required equals the shaded area. This area is the sum of the area of the triangle and the semicircle. Thus,

$$\begin{aligned}\text{Area} &= \text{Area of triangle} + \text{Area of semicircle}\\ &= \tfrac{1}{2}bh + \tfrac{1}{2}\pi r^2 = \tfrac{1}{2}(4)(6) + \tfrac{1}{2}\pi 2^2 \quad &&\text{Diameter of semicircle} = 4.\\ &= 12 + 2\pi \approx 18.28 \text{ cm}^2 &&\text{Radius} = 2.\end{aligned}$$

About 18.28 cm^2 of copper is required.

NOW WORK PROBLEM 33.

3 EXERCISES

In Problems 1–6, the lengths of the legs of a right triangle are given. Find the hypotenuse.

1. $a = 5, \quad b = 12$
2. $a = 6, \quad b = 8$
3. $a = 10, \quad b = 24$
4. $a = 4, \quad b = 3$
5. $a = 7, \quad b = 24$
6. $a = 14, \quad b = 48$

In Problems 7–14, the lengths of the sides of a triangle are given. Determine which are right triangles. For those that are, identify the hypotenuse.

7. 3, 4, 5
8. 6, 8, 10
9. 4, 5, 6
10. 2, 2, 3
11. 7, 24, 25
12. 10, 24, 26
13. 6, 4, 3
14. 5, 4, 7
15. Find the area A of a rectangle with length 4 inches and width 2 inches.
16. Find the area A of a rectangle with length 9 centimeters and width 4 centimeters.
17. Find the area A of a triangle with height 4 inches and base 2 inches.
18. Find the area A of a triangle with height 9 centimeters and base 4 centimeters.

19. Find the area A and circumference C of a circle of radius 5 meters.

20. Find the area A and circumference C of a circle of radius 2 feet.

21. Find the volume V of a rectangular box with length 8 feet, width 4 feet, and height 7 feet.

22. Find the volume V of a rectangular box with length 9 inches, width 4 inches, and height 8 inches.

23. Find the volume V and surface area S of a sphere of radius 4 centimeters.

24. Find the volume V and surface area S of a sphere of radius 3 feet.

25. Find the volume V of a right circular cylinder with radius 9 inches and height 8 inches.

26. Find the volume V of a right circular cylinder with radius 8 inches and height 9 inches.

In Problems 27–30, find the area of the shaded region.

27.

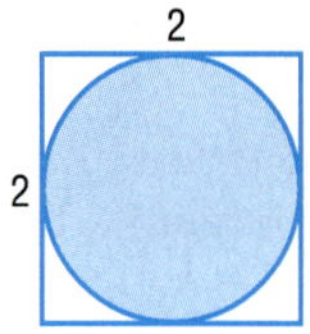

28.

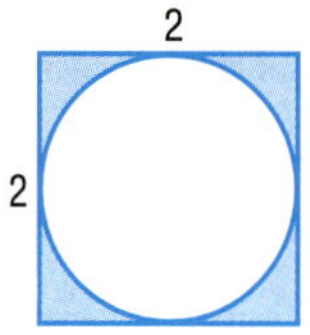

29.

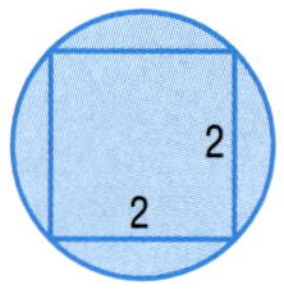

30.

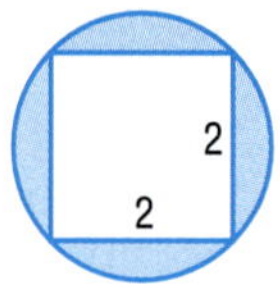

31. How many feet does a wheel with a diameter of 16 inches travel after four revolutions?

32. How many revolutions will a circular disk with a diameter of 4 feet have completed after it has rolled 20 feet?

33. In the figure below, $ABCD$ is a square, with each side of length 6 feet. The width of the border (shaded portion) between the outer square $EFGH$ and $ABCD$ is 2 feet. Find the area of the border.

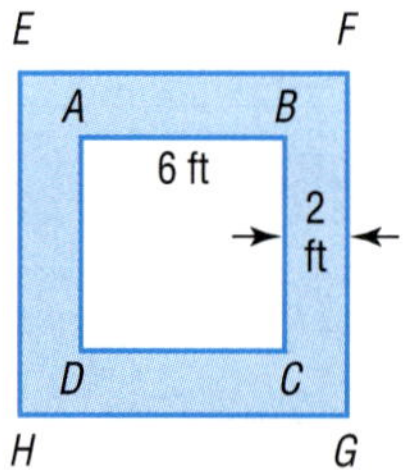

34. Refer to the figure below. Square $ABCD$ has an area of 100 square feet; square $BEFG$ has an area of 16 square feet. What is the area of the triangle CGF?

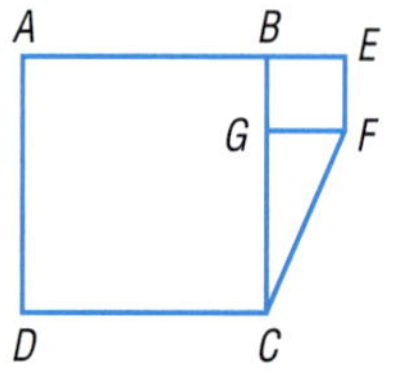

35. Architecture A Norman window consists of a rectangle surmounted by a semicircle. Find the area of the Norman window shown in the illustration. How much wood frame is needed to enclose the window?

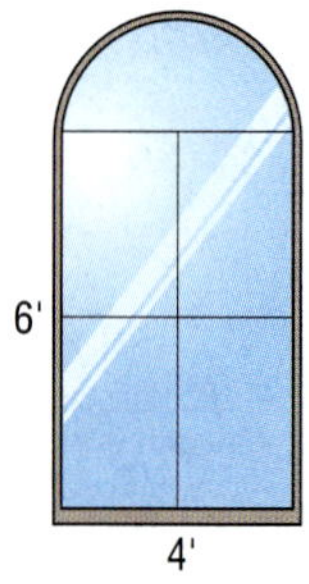

36. Construction A circular swimming pool, 20 feet in diameter, is enclosed by a wooden deck that is 3 feet wide. What is the area of the deck? How much fence is required to enclose the deck?

In Problems 37–39, use the facts that the radius of Earth is 3960 miles and 1 mile = 5280 feet.

37. How Far Can You See? The conning tower of the U.S.S. *Silversides,* a World War II submarine now permanently stationed in Muskegon, Michigan, is approximately 20 feet above sea level. How far can one see from the conning tower?

38. How Far Can You See? A person who is 6 feet tall is standing on the beach in Fort Lauderdale, Florida, and looks out onto the Atlantic Ocean. Suddenly, a ship appears on the horizon. How far is the ship from shore?

39. How Far Can You See? The deck of a destroyer is 100 feet above sea level. How far can a person see from the deck? How far can a person see from the bridge, which is 150 feet above sea level?

40. Suppose that m and n are positive integers with $m > n$. If $a = m^2 - n^2$, $b = 2mn$, and $c = m^2 + n^2$, show that a, b, and c are the lengths of the sides of a right triangle. (This formula can be used to find the sides of a right triangle that are integers, such as 3, 4, 5; 5, 12, 13; and so on. Such triplets of integers are called *Pythagorean triples.*)

41. You have 1000 feet of flexible pool siding and wish to construct a swimming pool. Experiment with rectangular-shaped pools with perimeters of 1000 feet. How do their areas vary? What is the shape of the rectangle with the largest area? Now compute the area enclosed by a circular pool with a perimeter (circumference) of 1000 feet. What would be your choice of shape for the pool? If rectangular, what is your preference for dimensions? Justify your choice. If your only consideration is to have a pool that encloses the most area, what shape should you use?

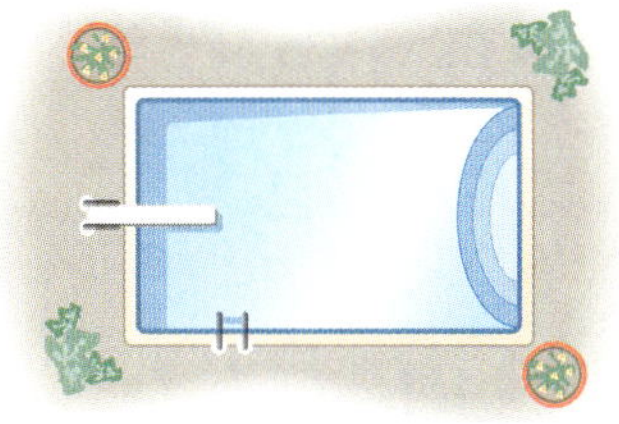

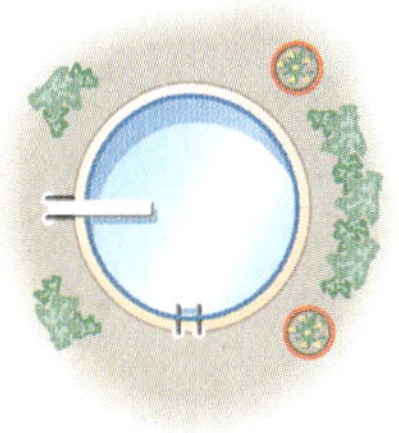

42. The Gibb's Hill Lighthouse, Southampton, Bermuda, in operation since 1846, stands 117 feet high on a hill 245 feet high, so its beam of light is 362 feet above sea level. A brochure states that the light itself can be seen on the horizon about 26 miles distant. Verify the correctness of this information. The brochure further states that ships 40 miles away can see the light and planes flying at 10,000 feet can see it 120 miles away. Verify the accuracy of these statements. What assumption did the brochure make about the height of the ship?

4 INTEGER EXPONENTS

1 Evaluate Expressions Containing Exponents

2 Work with the Laws of Exponents

3 Use a Calculator to Evaluate Exponents

4 Use Scientific Notation

Integer exponents provide a shorthand device for representing repeated multiplications of a real number. For example,

$$3^4 = 3 \cdot 3 \cdot 3 \cdot 3 = 81$$

Additionally, many formulas have exponents. For example,

- The formula for the horsepower rating H of an engine is

$$H = \frac{D^2 N}{2.5}$$

  where D is the diameter of a cylinder and N is the number of cylinders.
- A formula for the resistance R of blood flowing in a blood vessel is

$$R = C\frac{L}{r^4}$$

  where L is the length of the blood vessel, r is the radius, and C is a positive constant.

If a is a real number and n is a positive integer, then the **symbol a^n** represents the product of n factors of a. That is,

$$a^n = \underbrace{a \cdot a \cdot \ldots \cdot a}_{n \text{ factors}} \qquad (1)$$

Here it is understood that $a^1 = a$.

In particular, we have

$$a^1 = a$$
$$a^2 = a \cdot a$$
$$a^3 = a \cdot a \cdot a$$

and so on.

1 In the expression a^n, a is called the **base** and n is called the **exponent,** or **power.** We read a^n as "a raised to the power n" or as "a to the nth power." We usually read a^2 as "a squared" and a^3 as "a cubed."

EXAMPLE 1 Evaluating Expressions Containing Exponents

(a) $2^3 = 2 \cdot 2 \cdot 2 = 8$

(b) $5^2 = 5 \cdot 5 = 25$

(c) $10^1 = 10$

(d) $(-2)^4 = (-2)(-2)(-2)(-2) = 16$

(e) $-2^4 = -(2 \cdot 2 \cdot 2 \cdot 2) = -16$

Notice the difference between Examples 1(d) and 1(e): The exponent applies only to the symbol or parenthetical expression immediately preceding it.

We define a raised to a negative power as follows:

If n is a positive integer and if a is a nonzero real number, then we define

$$a^{-n} = \frac{1}{a^n} \quad \text{if } a \neq 0 \qquad (2)$$

Thus, whenever you encounter a negative exponent, think "reciprocal."

EXAMPLE 2 Evaluating Expressions Containing Negative Exponents

(a) $2^{-3} = \frac{1}{2^3} = \frac{1}{8}$ (b) $x^{-4} = \frac{1}{x^4}$

(c) $\left(\frac{1}{5}\right)^{-2} = \frac{1}{\left(\frac{1}{5}\right)^2} = \frac{1}{\frac{1}{25}} = 25$

NOW WORK PROBLEM 3.

If a is a nonzero number, we define

$$a^0 = 1 \quad \text{if } a \neq 0 \qquad (3)$$

Notice that we do not allow the base a to be 0 in a^{-n} or in a^0.

Laws of Exponents

2 Several general rules can be used when dealing with exponents. The first rule is used when multiplying two expressions that have the same base.

$$\underbrace{3^2 \cdot 3^4}_{\text{Multiplying; Same base}} = \underbrace{(3 \cdot 3)}_{\text{2 factors}}\underbrace{(3 \cdot 3 \cdot 3 \cdot 3)}_{\text{4 factors}} = \underbrace{3 \cdot 3 \cdot 3 \cdot 3 \cdot 3 \cdot 3}_{\text{6 factors}} = \underbrace{3^6}_{\text{Sum of powers 2 and 4; Same base}}$$

Thus, to multiply two expressions having the same base, retain the base and add the exponents.

If a is a real number and m, n are positive integers, then

$$a^m a^n = a^{m+n} \qquad (4)$$

Equation (4) is true whether the integers m and n are positive, negative, or 0. If m, n, or $m + n$ is 0 or negative, then a cannot be 0, as stated earlier.

EXAMPLE 3 Using Equation (4)

(a) $2 \cdot 2^4 = 2^1 \cdot 2^4 = 2^{1+4} = 2^5 = 32$

(b) $(-3)^2(-3)^{-3} = (-3)^{2+(-3)} = (-3)^{-1} = \dfrac{1}{-3}$

(c) $x^{-3} \cdot x^5 = x^{-3+5} = x^2$

(d) $(-y)^{-1}(-y)^{-3} = (-y)^{-1+(-3)} = (-y)^{-4} = \dfrac{1}{(-y)^4} = \dfrac{1}{y^4}$

NOW WORK PROBLEM 11.

Another law of exponents applies when an expression containing a power is itself raised to a power.

$$(2^3)^4 = \underbrace{2^3 \cdot 2^3 \cdot 2^3 \cdot 2^3}_{\text{4 factors}} = \underbrace{\underbrace{(2 \cdot 2 \cdot 2)}_{\text{3 factors}}\underbrace{(2 \cdot 2 \cdot 2)}_{\text{3 factors}}\underbrace{(2 \cdot 2 \cdot 2)}_{\text{3 factors}}\underbrace{(2 \cdot 2 \cdot 2)}_{\text{3 factors}}}_{3 \cdot 4 = 12 \text{ factors}} = 2^{12}$$

Thus, if an expression containing a power is raised to a power, retain the base and multiply the powers.

If a is a real number and m, n are positive integers, then

$$(a^m)^n = a^{mn} \qquad (5)$$

Equation (5) is true whether the integers m and n are positive, negative, or 0. Again, if m or n is 0 or negative, then a must not be 0.

EXAMPLE 4 Using Equation (5)

(a) $[(-2)^3]^2 = (-2)^{3 \cdot 2} = (-2)^6 = 64$

(b) $(3^{-4})^0 = 3^{(-4)(0)} = 3^0 = 1$

(c) $(x^{-3})^2 = x^{-3 \cdot 2} = x^{-6} = \dfrac{1}{x^6}$

(d) $(y^{-1})^{-2} = y^{(-1)(-2)} = y^2$

The next law of exponents involves raising a product to a power.

$$(2 \cdot 5)^3 = (2 \cdot 5)(2 \cdot 5)(2 \cdot 5) = (2 \cdot 2 \cdot 2)(5 \cdot 5 \cdot 5) = 2^3 \cdot 5^3$$

Thus, if a product is raised to a power, the result equals the product of each factor raised to that power.

If a, b are real numbers and n is a positive integer, then

$$(a \cdot b)^n = a^n \cdot b^n \qquad (6)$$

Equation (6) is true whether the integer n is positive, negative, or 0. If n is 0 or negative, neither a nor b can be 0.

EXAMPLE 5 Using Equation (6)

(a) $(2x)^3 = 2^3 \cdot x^3 = 8x^3$

(b) $(-2x)^0 = (-2)^0 \cdot x^0 = 1 \cdot 1 = 1 \qquad x \neq 0$

(c) $(ax)^{-2} = a^{-2}x^{-2} = \frac{1}{a^2} \cdot \frac{1}{x^2} = \frac{1}{a^2x^2} \qquad x \neq 0, a \neq 0$

(d) $(ax)^{-2} = \frac{1}{(ax)^2} = \frac{1}{a^2x^2} \qquad x \neq 0, a \neq 0$

NOW WORK PROBLEM 29.

Equations (4) and (6) both involve products. Two similar laws involve quotients.

If a and b are real numbers and if m and n are integers, then

$$\frac{a^m}{a^n} = a^{m-n} = \frac{1}{a^{n-m}} \qquad \text{if } a \neq 0$$
$$\left(\frac{a}{b}\right)^n = \frac{a^n}{b^n} \qquad \text{if } b \neq 0 \tag{7}$$

EXAMPLE 6 Using Equation (7)

(a) $\frac{2^6}{2^4} = 2^{6-4} = 2^2 = 4$ (b) $\left(\frac{2}{3}\right)^4 = \frac{2^4}{3^4} = \frac{16}{81}$

(c) $\frac{x^{-2}}{x^{-5}} = x^{-2-(-5)} = x^3$ (d) $\left(\frac{5}{2}\right)^{-2} = \frac{1}{\left(\frac{5}{2}\right)^2} = \frac{1}{\frac{5^2}{2^2}} = \frac{1}{\frac{25}{4}} = \frac{4}{25}$

You may have observed the following shortcut for problems like Example 6(d):

$$\left(\frac{a}{b}\right)^{-n} = \left(\frac{b}{a}\right)^n \qquad \text{if } a \neq 0, b \neq 0 \tag{8}$$

EXAMPLE 7 Using Equation (8)

(a) $\left(\frac{2}{3}\right)^{-3} = \left(\frac{3}{2}\right)^3 = \frac{27}{8}$ (b) $\left(\frac{3}{2}\right)^{-3} = \left(\frac{2}{3}\right)^3 = \frac{8}{27}$

NOW WORK PROBLEM 19.

EXAMPLE 8 Using the Laws of Exponents

Write each expression so that all exponents are positive.

(a) $\dfrac{x^5y^{-2}}{(x^3y)^2}$, $x \neq 0, y \neq 0$ (b) $\left(\dfrac{x^{-3}}{3y^{-1}}\right)^{-2}$, $x \neq 0, y \neq 0$

Solution (a) $$\frac{x^5y^{-2}}{(x^3y)^2} = \frac{x^5y^{-2}}{(x^3)^2y^2} = \frac{x^5y^{-2}}{x^6y^2} = \frac{x^5}{x^6}\cdot\frac{y^{-2}}{y^2}$$
$$= x^{5-6}y^{-2-2} = x^{-1}y^{-4} = \frac{1}{x}\cdot\frac{1}{y^4} = \frac{1}{xy^4}$$

(b) $$\left(\frac{x^{-3}}{3y^{-1}}\right)^{-2} = \frac{(x^{-3})^{-2}}{(3y^{-1})^{-2}} = \frac{x^6}{3^{-2}(y^{-1})^{-2}} = \frac{x^6}{\frac{1}{9}y^2} = \frac{9x^6}{y^2}$$

NOW WORK PROBLEM **47.**

Calculator Use

3 Your graphing calculator has the caret [^] key, which is used for computations involving exponents. The next example shows how this key is used.

EXAMPLE 9 Exponents on a Graphing Calculator

Evaluate: $(2.3)^5$

Solution Figure 21 shows the result using a TI-83 graphing calculator.

Figure 21

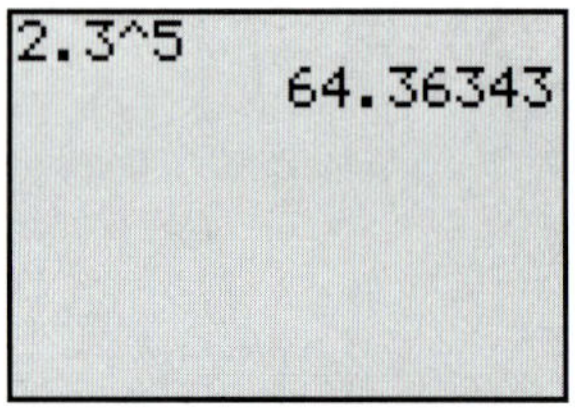

NOW WORK PROBLEM **63.**

Scientific Notation

4 Measurements of physical quantities can range from very small to very large. For example, the mass of a proton is approximately 0.00000000000000000000000000167 kilogram and the mass of Earth is about 5,980,000,000,000,000,000,000,000 kilograms. These numbers obviously are tedious to write down and difficult to read, so we use exponents to rewrite each.

> When a number has been written as the product of a number x, where $1 \leq x < 10$, times a power of 10, it is said to be written in **scientific notation.**

In scientific notation,

$$\text{Mass of a proton} = 1.67 \times 10^{-27} \text{ kilogram}$$
$$\text{Mass of Earth} = 5.98 \times 10^{24} \text{ kilograms}$$

> ***Converting a Decimal to Scientific Notation***
>
> To change a positive number into scientific notation:
>
> 1. Count the number N of places that the decimal point must be moved in order to arrive at a number x, where $1 \leq x < 10$.
> 2. If the original number is greater than or equal to 1, the scientific notation is $x \times 10^N$. If the original number is between 0 and 1, the scientific notation is $x \times 10^{-N}$.

EXAMPLE 10 Using Scientific Notation

Write each number in scientific notation.

(a) 9582 (b) 1.245 (c) 0.285 (d) 0.000561

Solution (a) The decimal point in 9582 follows the 2. Thus, we count

9 5 8 2 .
3 2 1

stopping after three moves because 9.582 is a number between 1 and 10. Since 9582 is greater than 1, we write

$$9582 = 9.582 \times 10^3$$

(b) The decimal point in 1.245 is between the 1 and 2. Since the number is already between 1 and 10, the scientific notation for it is $1.245 \times 10^0 = 1.245$.

(c) The decimal point in 0.285 is between the 0 and the 2. Thus, we count

0 . 2 8 5
1

stopping after one move because 2.85 is a number between 1 and 10. Since 0.285 is between 0 and 1, we write

$$0.285 = 2.85 \times 10^{-1}$$

(d) The decimal point in 0.000561 is moved as follows:

0 . 0 0 0 5 6 1
1 2 3 4

Thus,

$$0.000561 = 5.61 \times 10^{-4}$$

NOW WORK PROBLEM **69.**

EXAMPLE 11 Changing from Scientific Notation to Decimals

Write each number as a decimal.

(a) 2.1×10^4 (b) 3.26×10^{-5} (c) 1×10^{-2}

Solution (a) $2.1 \times 10^4 = 2\ .\ 1\ 0\ 0\ 0 \times 10^4 = 21{,}000$

1 2 3 4

(b) $3.26 \times 10^{-5} = 0\ 0\ 0\ 0\ 0\ 3\ .\ 2\ 6 \times 10^{-5} = 0.0000326$

5 4 3 2 1

(c) $1 \times 10^{-2} = 0\ 0\ 1\ . \times 10^{-2} = 0.01$

2 1

On a graphing calculator, a number such as 3.615×10^{12} is displayed as 3.615E12.

NOW WORK PROBLEM 77.

EXAMPLE 12 Using Scientific Notation

(a) The diameter of the smallest living cell is only about 0.00001 centimeter (cm).* Express this number in scientific notation.

(b) The surface area of Earth is about 1.97×10^8 square miles.† Express the surface area as a whole number.

Solution (a) $0.00001 \text{ cm} = 1 \times 10^{-5} \text{ cm}$ because the decimal point is moved five places and the number is less than 1.

(b) 1.97×10^8 square miles $= 197{,}000{,}000$ square miles.

NOW WORK PROBLEM 85.

SUMMARY

We close this section by summarizing the Laws of Exponents. In the list that follows, a and b are real numbers and m and n are integers. Also, we assume that no denominator is 0 and that all expressions are defined.

Laws of Exponents

$$a^{-n} = \frac{1}{a^n} \qquad a^0 = 1$$

$$a^m a^n = a^{m+n} \qquad (a^m)^n = a^{mn} \qquad (ab)^n = a^n b^n$$

$$\frac{a^m}{a^n} = a^{m-n} = \frac{1}{a^{n-m}} \qquad \left(\frac{a}{b}\right)^n = \frac{a^n}{b^n}$$

*_Powers of Ten,_ Philip and Phylis Morrison.

†_1998 Information Please Almanac._

HISTORICAL FEATURE

Our method of writing exponents originated with René Descartes (1596–1650), although the concept goes back in various forms to the ancient Babylonians. Even after its introduction by Descartes in 1637, the method took a remarkable amount of time to become completely standardized, and expressions like $aaaaa + 3aaaa + 2aaa - 4aa + 2a + 1$ remained common until 1750. The concept of a rational exponent (see Section 9 of this Appendix) was known by 1400, although inconvenient notation prevented the development of any extensive theory. John Wallis (1616–1703), in 1655, was the first to give a fairly complete explanation of negative and rational exponents, and Sir Isaac Newton's (1642–1727) use of them made exponents standard in their current form.

4 EXERCISES

In Problems 1–24, simplify each expression.

1. 4^2 **2.** -4^2 **3.** 4^{-2} **4.** $(-4)^2$ **5.** -4^{-2}

6. $(-4)^{-2}$ **7.** $4^0 \cdot 2^{-3}$ **8.** $(-2)^{-3} \cdot 3^0$ **9.** $2^{-3} + \left(\frac{1}{2}\right)^3$ **10.** $3^{-2} + \left(\frac{1}{3}\right)^2$

11. $3^{-6} \cdot 3^4$ **12.** $4^{-2} \cdot 4^3$ **13.** $\frac{(3^2)^2}{(2^3)^2}$ **14.** $\frac{(2^3)^3}{(2^2)^3}$ **15.** $\left(\frac{2}{3}\right)^{-3}$

16. $\left(\frac{3}{2}\right)^{-2}$ **17.** $\frac{2^3 \cdot 3^2}{2^4 \cdot 3^{-2}}$ **18.** $\frac{3^{-2} \cdot 5^3}{3^2 \cdot 5}$ **19.** $\left(\frac{9}{2}\right)^{-2}$ **20.** $\left(\frac{6}{5}\right)^{-3}$

21. $\frac{2^{-2}}{3}$ **22.** $\frac{3^{-2}}{2}$ **23.** $\frac{-3^{-1}}{2^{-1}}$ **24.** $\frac{-2^{-3}}{-1}$

In Problems 25–56, simplify each expression so that all exponents are positive. Whenever an exponent is negative or 0, we assume that the base does not equal 0.

25. x^0y^2 **26.** $x^{-1}y$ **27.** xy^{-2} **28.** x^0y^4 **29.** $(8x^3)^{-2}$

30. $(-8x^3)^{-2}$ **31.** $-4x^{-1}$ **32.** $(-4x)^{-1}$ **33.** $3x^0$ **34.** $(3x)^0$

35. $\frac{x^{-2}y^3}{xy^4}$ **36.** $\frac{x^{-2}y}{xy^2}$ **37.** $x^{-1}y^{-1}$ **38.** $\frac{x^{-2}y^{-3}}{x}$ **39.** $\frac{x^{-1}}{y^{-1}}$

40. $\left(\frac{2x}{3}\right)^{-1}$ **41.** $\left(\frac{4y}{5x}\right)^{-2}$ **42.** $(x^2y)^{-2}$ **43.** $x^{-2}y^{-2}$ **44.** $x^{-1}y^{-1}$

45. $\frac{x^{-1}y^{-2}z^3}{x^2yz^3}$ **46.** $\frac{3x^{-2}yz^2}{x^4y^{-3}z^2}$ **47.** $\frac{(-2)^3x^4(yz)^2}{3^2xy^3z^4}$ **48.** $\frac{4x^{-2}(yz)^{-1}}{(-5)^2x^4y^2z^{-2}}$ **49.** $\frac{\left(\frac{x}{y}\right)^{-2} \cdot \left(\frac{y}{x}\right)^4}{x^2y^3}$

50. $\frac{\left(\frac{y}{x}\right)^2}{x^{-2}y}$ **51.** $\left(\frac{3x^{-1}}{4y^{-1}}\right)^{-2}$ **52.** $\left(\frac{5x^{-2}}{6y^{-2}}\right)^{-3}$ **53.** $\frac{(xy^{-1})^{-2}}{xy^3}$ **54.** $\frac{(3xy^{-1})^2}{(2x^{-1}y)^3}$

55. $\left(\frac{x}{y^2}\right)^{-2} \cdot (y^2)^{-1}$ **56.** $\frac{(x^2)^{-3}y^3}{(x^3y)^{-2}}$

57. Find the value of the expression $2x^3 - 3x^2 + 5x - 4$ if $x = 2$. If $x = 1$.

58. Find the value of the expression $4x^3 + 3x^2 - x + 2$ if $x = 1$. If $x = 2$.

59. What is the value of $\frac{(666)^4}{(222)^4}$? **60.** What is the value of $(0.1)^3(20)^3$?

In Problems 61–68, use a graphing utility to evaluate each expression. Round your answers to three decimal places.

61. $(8.2)^6$ **62.** $(3.7)^5$ **63.** $(6.1)^{-3}$ **64.** $(2.2)^{-5}$

65. $(-2.8)^6$ **66.** $-(2.8)^6$ **67.** $(-8.11)^{-4}$ **68.** $-(8.11)^{-4}$

In Problems 69–76, write each number in scientific notation.

69. 454.2 **70.** 32.14 **71.** 0.013 **72.** 0.00421

73. 32,155 **74.** 21,210 **75.** 0.000423 **76.** 0.0514

In Problems 77–84, write each number as a decimal.

77. 6.15×10^4 **78.** 9.7×10^3 **79.** 1.214×10^{-3} **80.** 9.88×10^{-4}

81. 1.1×10^8 **82.** 4.112×10^2 **83.** 8.1×10^{-2} **84.** 6.453×10^{-1}

85. Distance from Earth to Its Moon The distance from Earth to the Moon is about 4×10^8 meters.* Express this distance as a whole number.

86. Height of Mt. Everest The height of Mt. Everest is 8872 meters.* Express this height in scientific notation.

87. Wavelength of Visible Light The wavelength of visible light is about 5×10^{-7} meter.* Express this wavelength as a whole number.

88. Diameter of an Atom The diameter of an atom is about 1×10^{-10} meters.* Express this diameter as a whole number.

89. Diameter of Copper Wire The smallest commercial copper wire is about 0.0005 inch in diameter.[+] Express this diameter using scientific notation.

90. Smallest Motor The smallest motor ever made is less than 0.05 centimeter wide.[+] Express this width using scientific notation.

91. World Oil Production In 1996, world oil production averaged 64,000,000 barrels per day.[+] How many barrels on average were produced in the month of April 1996? Express your answer in scientific notation.

92. U.S. Consumption Oil In 1995, the United States consumed 17,640,000 barrels of petroleum products per day.[+] If there are 42 gallons in one barrel, how many gallons did the United States consume per day? Express your answer in scientific notation.

93. Astronomy One light-year is defined by astronomers to be the distance a beam of light will travel in 1 year (365 days). If the speed of light is 186,000 miles per second, how many miles are in a light-year? Express your answer in scientific notation.

94. Astronomy How long does it take a beam of light to reach Earth from the Sun, when the Sun is 93,000,000 miles from Earth? Express your answer in seconds, using scientific notation.

95. Look at the summary box where the Laws of Exponents are given. List them in the order of most importance to you. Write a brief position paper defending your ordering.

96. Write a paragraph to justify the definition given in the text that $a^0 = 1, a \neq 0$.

**Powers of Ten*, Philip and Phylis Morrison.

[+]*1998 Information Please Almanac*

5 POLYNOMIALS

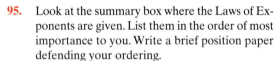

1 Recognize Monomials

2 Recognize Polynomials

3 Add and Subtract Polynomials

4 Multiply Polynomials

5 Know Formulas for Special Products

We have described algebra as a generalization of arithmetic in which letters are used to represent real numbers. From now on, we shall use the letters at the end of the alphabet, such as x, y, and z, to represent variables and the letters at the beginning of the alphabet, such as a, b, and c, to represent constants. Thus, in the expressions $3x + 5$ and $ax + b$, it is understood that x is a variable and that a and b are constants, even though the constants a and b are unspecified. As you will find out, the context usually makes the intended meaning clear.

Now we introduce some basic vocabulary.

A **monomial** in one variable is the product of a constant times a variable raised to a nonnegative integer power. Thus, a monomial is of the form

$$ax^k$$

where a is a constant, x is a variable, and $k \geq 0$ is an integer. The constant a is called the **coefficient** of the monomial. If $a \neq 0$, then k is called the **degree** of the monomial.

EXAMPLE 1 Examples of Monomials

1

Monomial	Coefficient	Degree	
(a) $6x^2$	6	2	
(b) $-\sqrt{2}x^3$	$-\sqrt{2}$	3	
(c) 3	3	0	*Since* $3 = 3 \cdot 1 = 3x^0, x \neq 0$
(d) $-5x$	-5	1	*Since* $-5x = -5x^1$
(e) x^4	1	4	*Since* $x^4 = 1 \cdot x^4$

Now let's look at some expressions that are not monomials.

EXAMPLE 2 Examples of Nonmonomial Expressions

(a) $3x^{1/2}$ is not a monomial, since the exponent of the variable x is $\frac{1}{2}$ and $\frac{1}{2}$ is not a nonnegative integer.

(b) $4x^{-3}$ is not a monomial, since the exponent of the variable x is -3 and -3 is not a nonnegative integer.

NOW WORK PROBLEM 1.

Two monomials with the same variable raised to the same power are called **like terms.** For example, $2x^4$ and $-5x^4$ are like terms. In contrast, the monomials $2x^3$ and $2x^5$ are not like terms.

We can add or subtract like terms using the distributive property. For example,

$$2x^2 + 5x^2 = (2 + 5)x^2 = 7x^2 \quad \text{and} \quad 8x^3 - 5x^3 = (8 - 5)x^3 = 3x^3$$

The sum or difference of two monomials having different degrees is called a **binomial.** The sum or difference of three monomials with three different degrees is called a **trinomial.** For example,

$x^2 - 2$ is a binomial.

$x^3 - 3x + 5$ is a trinomial.

$2x^2 + 5x^2 + 2 = 7x^2 + 2$ is a binomial.

A **polynomial** in one variable is an algebraic expression of the form

$$a_nx^n + a_{n-1}x^{n-1} + \cdots + a_1x + a_0 \qquad (1)$$

where $a_n, a_{n-1}, \ldots, a_1, a_0$ are constants* called the **coefficients** of the polynomial, $n \geq 0$ is an integer, and x is a variable. If $a_n \neq 0$, it is called the **leading coefficient,** and n is called the **degree** of the polynomial.

The monomials that make up a polynomial are called its **terms.** If all the coefficients are 0, the polynomial is called the **zero polynomial,** which has no degree.

Polynomials are usually written in **standard form,** beginning with the nonzero term of highest degree and continuing with terms in descending order according to degree.

EXAMPLE 3 Examples of Polynomials

2

Polynomial	Coefficients	Degree
$3x^2 - 5 = 3x^2 + 0 \cdot x + (-5)$	$3, 0, -5$	2
$8 - 2x + x^2 = 1 \cdot x^2 - 2x + 8$	$1, -2, 8$	2
$5x + \sqrt{2} = 5x^1 + \sqrt{2}$	$5, \sqrt{2}$	1
$3 = 3 \cdot 1 = 3 \cdot x^0$	3	0
0	0	No degree

Although we have been using x to represent the variable, letters such as y or z are also commonly used. Thus,

$3x^4 - x^2 + 2$ is a polynomial (in x) of degree 4.
$9y^3 - 2y^2 + y - 3$ is a polynomial (in y) of degree 3.
$z^5 + \pi$ is a polynomial (in z) of degree 5.

Algebraic expressions such as

$$\frac{1}{x} \quad \text{and} \quad \frac{x^2 + 1}{x + 5}$$

are not polynomials. The first is not a polynomial because $1/x = x^{-1}$ has an exponent that is not a nonnegative integer. Although the second expression is the quotient of two polynomials, the polynomial in the denominator has degree greater than 0, so the expression cannot be a polynomial.

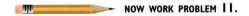

NOW WORK PROBLEM 11.

Adding and Subtracting Polynomials

3 Polynomials are added and subtracted by combining like terms.

*The notation a_n is read as "a sub n." The number n is called a **subscript** and should not be confused with an exponent. We use subscripts in order to distinguish one constant from another when a large or undetermined number of constants is required.

EXAMPLE 4 Adding Polynomials

Find the sum of the polynomials

$$8x^3 - 2x^2 + 6x - 2 \quad \text{and} \quad 3x^4 - 2x^3 + x^2 + x$$

Solution We shall find the sum in two ways.

Horizontal Addition: The idea here is to group the like terms and then combine them.

$$\begin{aligned}(8x^3 - 2x^2 + 6x - 2) &+ (3x^4 - 2x^3 + x^2 + x)\\ &= 3x^4 + (8x^3 - 2x^3) + (-2x^2 + x^2) + (6x + x) - 2\\ &= 3x^4 + 6x^3 - x^2 + 7x - 2\end{aligned}$$

Vertical Addition: The idea here is to vertically line up the like terms in each polynomial and then add the coefficients.

$$\begin{array}{rrrrrr} & x^4 & x^3 & x^2 & x^1 & x^0\\ & & 8x^3 & -\,2x^2 & +\,6x & -\,2\\ (+) & 3x^4 & -\,2x^3 & +\,x^2 & +\,x & \\ \hline & 3x^4 & +\,6x^3 & -\,x^2 & +\,7x & -\,2\end{array}$$

We can subtract two polynomials in either of the previous ways.

EXAMPLE 5 Subtracting Polynomials

Find the difference: $(3x^4 - 4x^3 + 6x^2 - 1) - (2x^4 - 8x^2 - 6x + 5)$

Solution *Horizontal Subtraction*

$$\begin{aligned}(3x^4 - 4x^3 + 6x^2 - 1) &- (2x^4 - 8x^2 - 6x + 5)\\ &= 3x^4 - 4x^3 + 6x^2 - 1 + \underbrace{(-2x^4 + 8x^2 + 6x - 5)}_{\text{Be sure to change the sign of each term in the second polynomial.}}\\ &= (3x^4 - 2x^4) + (-4x^3) + (6x^2 + 8x^2) + 6x + (-1 - 5)\\ &\quad \uparrow \text{ Group like terms.}\\ &= x^4 - 4x^3 + 14x^2 + 6x - 6\end{aligned}$$

Vertical Subtraction: We line up like terms, change the sign of each coefficient of the second polynomial, and add.

$$\begin{array}{rrrrrr} & x^4 & x^3 & x^2 & x^1 & x^0\\ & 3x^4 & -\,4x^3 & +\,6x^2 & & -\,1\\ (-)[& 2x^4 & & -\,8x^2 & -\,6x & +\,5]\\ \hline\end{array} = \begin{array}{rrrrrr} & x^4 & x^3 & x^2 & x^1 & x^0\\ & 3x^4 & -\,4x^3 & +\,6x^2 & & -\,1\\ (+) & -2x^4 & & +\,8x^2 & +\,6x & -\,5\\ \hline & x^4 & -\,4x^3 & +\,14x^2 & +\,6x & -\,6\end{array}$$

The choice of which of these methods to use for adding and subtracting polynomials is left to you. To save space, we shall most often use the horizontal format.

NOW WORK PROBLEM 23.

Multiplying Polynomials

4 Two monomials may be multiplied using the laws of exponents and commutative and associative properties. For example,

$$(2x^3)\cdot(5x^4) = (2\cdot 5)\cdot(x^3\cdot x^4) = 10x^{3+4} = 10x^7$$

Products of polynomials are found by repeated use of the distributive property and the laws of exponents. Again, you have a choice of horizontal or vertical format.

EXAMPLE 6 Multiplying Polynomials

Find the product: $(2x+5)(x^2-x+2)$

Solution *Horizontal Multiplication*

$$(2x+5)(x^2-x+2) = 2x(x^2-x+2) + 5(x^2-x+2)$$

↑ *Distributive property*

$$= (2x\cdot x^2 - 2x\cdot x + 2x\cdot 2) + (5\cdot x^2 - 5\cdot x + 5\cdot 2)$$

↑ *Distributive property*

$$= (2x^3 - 2x^2 + 4x) + (5x^2 - 5x + 10)$$

↑ *Law of exponents*

$$= 2x^3 + 3x^2 - x + 10$$

↑ *Combine like terms*

Vertical Multiplication: The idea here is very much like multiplying a two-digit number by a three-digit number.

$$\begin{array}{rrrrl}
 & & x^2 - & x + 2 & \\
 & & & 2x + 5 & \\ \hline
 & 2x^3 - 2x^2 & + 4x & & \text{This line is } 2x(x^2-x+2). \\
(+) & 5x^2 & - 5x & + 10 & \text{This line is } 5(x^2-x+2). \\ \hline
 & 2x^3 + 3x^2 & - x & + 10 & \text{Sum of the above two lines.}
\end{array}$$

NOW WORK PROBLEM **35.**

Special Products

5 Certain products, which we call **special products,** occur frequently in algebra. We can calculate them easily using the **FOIL** (**F**irst, **O**uter, **I**nner, **L**ast) method of multiplying two binomials:

Outer, First, Inner, Last

$$(ax+b)(cx+d) = ax(cx+d) + b(cx+d)$$

$$= \underbrace{ax\cdot cx}_{\text{First}} + \underbrace{ax\cdot d}_{\text{Outer}} + \underbrace{b\cdot cx}_{\text{Inner}} + \underbrace{b\cdot d}_{\text{Last}}$$

$$= acx^2 + adx + bcx + bd$$

$$= acx^2 + (ad+bc)x + bd$$

EXAMPLE 7 Using FOIL to Find Products of the Form $(x - a)(x + a)$

$$(x - 3)(x + 3) = \underset{F}{x^2} + \underset{O}{3x} - \underset{I}{3x} - \underset{L}{9} = x^2 - 9$$

EXAMPLE 8 Using FOIL to Find the Square of a Binomial

(a) $(x + 2)^2 = (x + 2)(x + 2) = x^2 + 2x + 2x + 4 = x^2 + 4x + 4$
(b) $(x - 3)^2 = (x - 3)(x - 3) = x^2 - 3x - 3x + 9 = x^2 - 6x + 9$

EXAMPLE 9 Using FOIL to Find the Product of Two Binomials

(a) $(x + 3)(x + 1) = x^2 + x + 3x + 3 = x^2 + 4x + 3$
(b) $(2x + 1)(3x + 4) = 6x^2 + 8x + 3x + 4 = 6x^2 + 11x + 4$

NOW WORK PROBLEMS 41 AND 49.

Some special products occur so frequently that they have been given names. The Special Products below are based on Example 7 and 8.

Difference of Two Squares

$$(x - a)(x + a) = x^2 - a^2 \tag{2}$$

Squares of Binomials, or Perfect Squares

$$(x + a)^2 = x^2 + 2ax + a^2 \tag{3a}$$
$$(x - a)^2 = x^2 - 2ax + a^2 \tag{3b}$$

EXAMPLE 10 Using Special Product Formulas

(a) $(x - 5)(x + 5) = x^2 - 5^2 = x^2 - 25$ *Difference of two squares*
(b) $(x + 7)^2 = x^2 + 2 \cdot 7 \cdot x + 7^2 = x^2 + 14x + 49$ *Square of a binomial*

EXAMPLE 11 Using Special Product Formulas

(a) $(2x + 1)^2 = (2x)^2 + 2 \cdot 1 \cdot 2x + 1^2 = 4x^2 + 4x + 1$ *Notice that we used 2x in place of x in formula (3a).*
(b) $(3x - 4)^2 = (3x)^2 - 2 \cdot 3x \cdot 4 + 4^2 = 9x^2 - 24x + 16$ *Replace x by 3x in formula (3b).*

NOW WORK PROBLEM 61.

Let's look at some more examples that lead to general formulas.

EXAMPLE 12 Cubing a Binomial

(a) $(x + 2)^3 = (x + 2)(x + 2)^2 = (x + 2)(x^2 + 4x + 4)$ *Formula (3a)*
$= (x^3 + 4x^2 + 4x) + (2x^2 + 8x + 8)$
$= x^3 + 6x^2 + 12x + 8$

(b) $(x-1)^3 = (x-1)(x-1)^2 = (x-1)(x^2-2x+1)$ *Formula (3b)*

$$= (x^3 - 2x^2 + x) - (x^2 - 2x + 1)$$

$$= x^3 - 3x^2 + 3x - 1$$

Cubes of Binomials, or Perfect Cubes

$$(x+a)^3 = x^3 + 3ax^2 + 3a^2x + a^3 \quad (4a)$$

$$(x-a)^3 = x^3 - 3ax^2 + 3a^2x - a^3 \quad (4b)$$

NOW WORK PROBLEM **79.**

EXAMPLE 13 Forming the Difference of Two Cubes

$$(x-1)(x^2+x+1) = x(x^2+x+1) - 1(x^2+x+1)$$

$$= x^3 + x^2 + x - x^2 - x - 1$$

$$= x^3 - 1$$

EXAMPLE 14 Forming the Sum of Two Cubes

$$(x+2)(x^2-2x+4) = x(x^2-2x+4) + 2(x^2-2x+4)$$

$$= (x^3 - 2x^2 + 4x) + (2x^2 - 4x + 8)$$

$$= x^3 + 8$$

Examples 13 and 14 lead to two more Special Products.

Difference of Two Cubes

$$(x-a)(x^2+ax+a^2) = x^3 - a^3 \quad (5)$$

Sum of Two Cubes

$$(x+a)(x^2-ax+a^2) = x^3 + a^3 \quad (6)$$

Polynomials in Two Variables

A **monomial in two variables** x and y has the form ax^ny^m, where a is a constant, x and y are variables, and n and m are nonnegative integers. The **degree** of a monomial is the sum of the powers of the variables.

For example,

$$2xy^3,\ a^2b^2, \text{ and } x^3y$$

are monomials, each of which has degree 4.

A **polynomial in two variables** x and y is the sum of one or more monomials in two variables. The **degree of a polynomial** in two variables is the highest degree of all the monomials with nonzero coefficients.

EXAMPLE 15 Examples of Polynomials in Two Variables

$3x^2 + 2x^3y + 5$	$\pi x^3 - y^2$	$x^4 + 4x^3y - xy^3 + y^4$
Two variables, degree is 4.	*Two variables, degree is 3.*	*Two variables, degree is 4.*

Multiplying polynomials in two variables is handled in the same way as polynomials in one variable.

EXAMPLE 16 Using a Special Product Formula

To multiply $(2x - y)^2$, use the Squares of Binomials (36) formula with $2x$ instead of x and y instead of a.

$$(2x - y)^2 = (2x)^2 - 2 \cdot 2x \cdot y + y^2 = 4x^2 - 4xy + y^2$$

NOW WORK PROBLEM 73.

5 EXERCISES

In Problems 1–10, tell whether the expression is a monomial. If it is, name the variable(s), the coefficient, and give the degree of the monomial.

1. $2x^3$
2. $-4x^2$
3. $8/x$
4. $-2x^{-3}$
5. $-2xy^2$
6. $5x^2y^3$
7. $8x/y$
8. $-2x^2/y^3$
9. $x^2 + y^2$
10. $3x^2 + 4$

In Problems 11–20, tell whether the expression is a polynomial. If it is, give its degree.

11. $3x^2 - 5$
12. $1 - 4x$
13. 5
14. $-\pi$
15. $3x^2 - \dfrac{5}{x}$
16. $\dfrac{3}{x} + 2$
17. $2y^3 - \sqrt{2}$
18. $10z^2 + z$
19. $\dfrac{x^2 + 5}{x^3 - 1}$
20. $\dfrac{3x^3 + 2x - 1}{x^2 + x + 1}$

In Problems 21–40, add, subtract, or multiply, as indicated. Express your answer as a single polynomial.

21. $(x^2 + 4x + 5) + (3x - 3)$
22. $(x^3 + 3x^2 + 2) + (x^2 - 4x + 4)$
23. $(x^3 - 2x^2 + 5x + 10) - (2x^2 - 4x + 3)$
24. $(x^2 - 3x - 4) - (x^3 - 3x^2 + x + 5)$
25. $(6x^5 + x^3 + x) + (5x^4 - x^3 + 3x^2)$
26. $(10x^5 - 8x^2) + (3x^3 - 2x^2 + 6)$
27. $(x^2 - 3x + 1) + 2(3x^2 + x - 4)$
28. $-2(x^2 + x + 1) + (-5x^2 - x + 2)$
29. $6(x^3 + x^2 - 3) - 4(2x^3 - 3x^2)$
30. $8(4x^3 - 3x^2 - 1) - 6(4x^3 + 8x - 2)$
31. $(x^2 - x + 2) + (2x^2 - 3x + 5) - (x^2 + 1)$
32. $(x^2 + 1) - (4x^2 + 5) + (x^2 + x - 2)$
33. $9(y^2 - 3y + 4) - 6(1 - y^2)$
34. $8(1 - y^3) + 4(1 + y + y^2 + y^3)$
35. $x(x^2 + x - 4)$
36. $4x^2(x^3 - x + 2)$
37. $-2x^2(4x^3 + 5)$
38. $5x^3(3x - 4)$
39. $(x + 1)(x^2 + 2x - 4)$
40. $(2x - 3)(x^2 + x + 1)$

In Problems 41–58, multiply the polynomials using the FOIL method. Express your answer as a single polynomial.

41. $(x + 2)(x + 4)$
42. $(x + 3)(x + 5)$
43. $(2x + 5)(x + 2)$
44. $(3x + 1)(2x + 1)$
45. $(x - 4)(x + 2)$
46. $(x + 4)(x - 2)$
47. $(x - 3)(x - 2)$
48. $(x - 5)(x - 1)$
49. $(2x + 3)(x - 2)$
50. $(2x - 4)(3x + 1)$
51. $(-2x + 3)(x - 4)$
52. $(-3x - 1)(x + 1)$
53. $(-x - 2)(-2x - 4)$
54. $(-2x - 3)(3 - x)$
55. $(x - 2y)(x + y)$
56. $(2x + 3y)(x - y)$
57. $(-2x - 3y)(3x + 2y)$
58. $(x - 3y)(-2x + y)$

In Problems 59–84, multiply the polynomials using the Special Product Formulas.

59. $(x - 7)(x + 7)$
60. $(x - 1)(x + 1)$
61. $(2x + 3)(2x - 3)$
62. $(3x + 2)(3x - 2)$
63. $(x + 4)^2$
64. $(x + 5)^2$
65. $(x - 4)^2$
66. $(x - 5)^2$

67. $(3x + 4)(3x - 4)$
68. $(5x - 3)(5x + 3)$
69. $(2x - 3)^2$
70. $(3x - 4)^2$
71. $(x + y)(x - y)$
72. $(x + 3y)(x - 3y)$
73. $(3x + y)(3x - y)$
74. $(3x + 4y)(3x - 4y)$
75. $(x + y)^2$
76. $(x - y)^2$
77. $(x - 2y)^2$
78. $(2x + 3y)^2$
79. $(x - 2)^3$
80. $(x + 1)^3$
81. $(2x + 1)^3$
82. $(3x - 2)^3$
83. $(x + 2y)^3$
84. $(x - 2y)^3$

85. Explain why the degree of the product of two polynomials equals the sum of their degrees.

86. Explain why the degree of the sum of two polynomials equals the larger of their degrees.

87. Do you prefer adding two polynomials using the horizontal method or the vertical method? Write a brief position paper defending your choice.

88. Do you prefer to memorize the rule for the square of a binomial $(x + a)^2$ or to use FOIL to obtain the product? Write a brief position paper defending your choice.

6 FACTORING POLYNOMIALS

1. Factor the Difference of Two Squares and the Sum and the Difference of Two Cubes
2. Factor Perfect Squares
3. Factor a Second-degree Polynomial: $x^2 + Bx + C$
4. Factor by Grouping
5. Factor a Second-degree Polynomial: $Ax^2 + Bx + C$

Consider the following product:

$$(2x + 3)(x - 4) = 2x^2 - 5x - 12$$

The two polynomials on the left side are called **factors** of the polynomial on the right side. Expressing a given polynomial as a product of other polynomials, that is, finding the factors of a polynomial, is called **factoring.**

We shall restrict our discussion here to factoring polynomials in one variable into products of polynomials in one variable, where all coefficients are integers. We call this **factoring over the integers.**

Any polynomial can be written as the product of 1 times itself or as -1 times its additive inverse. If a polynomial cannot be written as the product of two other polynomials (excluding 1 and -1), then the polynomial is said to be **prime.** When a polynomial has been written as a product consisting only of prime factors, it is said to be **factored completely.** Examples of prime polynomials are

$$2, \quad 3, \quad 5, \quad x, \quad x + 1, \quad x - 1, \quad 3x + 4$$

The first factor to look for in a factoring problem is a common monomial factor present in each term of the polynomial. If one is present, use the distributive property to factor it out.

EXAMPLE 1 Identifying Common Monomial Factors

Polynomial	Common Monomial Factor	Remaining Factor	Factored Form
$2x + 4$	2	$x + 2$	$2x + 4 = 2(x + 2)$
$3x - 6$	3	$x - 2$	$3x - 6 = 3(x - 2)$
$2x^2 - 4x + 8$	2	$x^2 - 2x + 4$	$2x^2 - 4x + 8 = 2(x^2 - 2x + 4)$
$8x - 12$	4	$2x - 3$	$8x - 12 = 4(2x - 3)$
$x^2 + x$	x	$x + 1$	$x^2 + x = x(x + 1)$
$x^3 - 3x^2$	x^2	$x - 3$	$x^3 - 3x^2 = x^2(x - 3)$
$6x^2 + 9x$	$3x$	$2x + 3$	$6x^2 + 9x = 3x(2x + 3)$

Notice that, once all common monomial factors have been removed from a polynomial, the remaining factor is either a prime polynomial of degree 1 or a polynomial of degree 2 or higher. (Do you see why?)

NOW WORK PROBLEM 1.

Special Formulas

1 When you factor a polynomial, first check whether you can use one of the special formulas discussed in the previous section.

Difference of Two Squares	$x^2 - a^2 = (x - a)(x + a)$
Perfect Squares	$x^2 + 2ax + a^2 = (x + a)^2$
	$x^2 - 2ax + a^2 = (x - a)^2$
Sum of Two Cubes	$x^3 + a^3 = (x + a)(x^2 - ax + a^2)$
Difference of Two Cubes	$x^3 - a^3 = (x - a)(x^2 + ax + a^2)$

EXAMPLE 2 Factoring the Difference of Two Squares

Factor completely: $x^2 - 4$

Solution We notice that $x^2 - 4$ is the difference of two squares, x^2 and 2^2. Thus,

$$x^2 - 4 = (x - 2)(x + 2)$$

EXAMPLE 3 Factoring the Difference of Two Cubes

Factor completely: $x^3 - 1$

Solution Because $x^3 - 1$ is the difference of two cubes, x^3 and 1^3, we find that

$$x^3 - 1 = (x - 1)(x^2 + x + 1)$$

EXAMPLE 4 Factoring the Sum of Two Cubes

Factor completely: $x^3 + 8$

Solution Because $x^3 + 8$ is the sum of two cubes, x^3 and 2^3, we have

$$x^3 + 8 = (x + 2)(x^2 - 2x + 4)$$

EXAMPLE 5 Factoring the Difference of Two Squares

Factor completely: $x^4 - 16$

Solution Because $x^4 - 16$ is the difference of two squares, $x^4 = (x^2)^2$ and $16 = 4^2$, we have

$$x^4 - 16 = (x^2 - 4)(x^2 + 4)$$

But $x^2 - 4$ is also the difference of two squares. Thus,

$$x^4 - 16 = (x^2 - 4)(x^2 + 4) = (x - 2)(x + 2)(x^2 + 4)$$

NOW WORK PROBLEMS 11 AND 29.

2 When the first term and third term of a trinomial are both positive and are perfect squares, such as x^2, $9x^2$, 1, and 4, check to see whether the trinomial is a perfect square.

EXAMPLE 6 Factoring Perfect Squares

Factor completely: $x^2 + 6x + 9$

Solution The first term, x^2, and the third term, $9 = 3^2$, are perfect squares. Because the middle term $6x$ is twice the product of x and 3, we have a perfect square.

$$x^2 + 6x + 9 = (x + 3)^2$$

EXAMPLE 7 Factoring Perfect Squares

Factor completely: $9x^2 - 6x + 1$

Solution The first term, $9x^2 = (3x)^2$, and the third term, $1 = 1^2$, are perfect squares. Because the middle term $-6x$ is -2 times the product of $3x$ and 1, we have a perfect square.

$$9x^2 - 6x + 1 = (3x - 1)^2$$

EXAMPLE 8 Factoring Perfect Squares

Factor completely: $25x^2 + 30x + 9$

Solution The first term, $25x^2 = (5x)^2$, and the third term, $9 = 3^2$, are perfect squares. Because the middle term $30x$ is twice the product of $5x$ and 3, we have a perfect square.

$$25x^2 + 30x + 9 = (5x + 3)^2$$

NOW WORK PROBLEMS 21 AND 89.

If a trinomial is not a perfect square, it may be possible to factor it using the technique discussed next.

Factoring a Second-degree Polynomial: $x^2 + Bx + C$

3 The idea behind factoring a second-degree polynomial like $x^2 + Bx + C$ is to see whether it can be made equal to the product of two, possibly equal, first-degree polynomials.

For example, we know that

$$(x + 3)(x + 4) = x^2 + 7x + 12$$

Thus, the factors of $x^2 + 7x + 12$ are $x + 3$ and $x + 4$. Notice the following:

$$x^2 + 7x + 12 = (x + 3)(x + 4)$$

3 and 4 are factors of 12

7 is the sum of 3 and 4

In general, if $x^2 + Bx + C = (x + a)(x + b)$, then $ab = C$ and $a + b = B$.

> To factor a second-degree polynomial $x^2 + Bx + C$, find factors of C whose sum is B. That is, if there are numbers a, b, where $a + b = C$ and $a + b = B$, then
>
> $$x^2 + Bx + C = (x + a)(x + b)$$

EXAMPLE 9 Factoring Trinomials

Factor completely: $x^2 + 7x + 10$

Solution First, determine all possible factors of the constant term 10 and then compute their sums:

Factors of 10	1, 10	−1, −10	2, 5	−2, −5
Sum	11	−11	7	−7

The factors of 10 that add up to 7, the coefficient of the middle term, are 2 and 5. Thus,

$$x^2 + 7x + 10 = (x + 2)(x + 5)$$

EXAMPLE 10 Factoring Trinomials

Factor completely: $x^2 - 6x + 8$

Solution First, determine all possible factors of the constant term 8 and then compute each sum:

Factors of 8	1, 8	−1, −8,	2, 4	−2, −4
Sum	9	−9	6	−6

Since −6 is the coefficient of the middle term,

$$x^2 - 6x + 8 = (x - 2)(x - 4)$$

EXAMPLE 11 Factoring Trinomials

Factor completely: $x^2 - x - 12$

Solution First, determine all possible factors of -12 and then compute each sum:

Factors of -12	$1, -12$	$-1, 12$	$2, -6$	$-2, 6$	$3, -4$	$-3, 4$
Sum	-11	11	-4	4	-1	1

Since -1 is the coefficient of the middle term,

$$x^2 - x - 12 = (x + 3)(x - 4)$$

EXAMPLE 12 Factoring Trinomials

Factor completely: $x^2 + 4x - 12$

Solution The factors -2 and 6 of -12 have the sum 4. Thus,

$$x^2 + 4x - 12 = (x - 2)(x + 6)$$

To avoid errors in factoring, always check your answer by multiplying it out to see if the result equals the original expression.

When none of the possibilities works, the polynomial is prime.

EXAMPLE 13 Identifying Prime Polynomials

Show that $x^2 + 9$ is prime.

Solution First, list the factors of 9 and then compute their sums:

Factors of 9	$1, 9$	$-1, -9$	$3, 3$	$-3, -3$
Sum	10	-10	6	-6

Since the coefficient of the middle term in $x^2 + 9 = x^2 + 0x + 9$ is 0 and none of the sums equals 0, we conclude that $x^2 + 9$ is prime.

Example 13 demonstrates a more general result:

THEOREM Any polynomial of the form $x^2 + a^2$, a real, is prime.

NOW WORK PROBLEMS 35 AND 73.

Factoring by Grouping

4 Sometimes a common factor does not occur in every term of the polynomial, but in each of several groups of terms that together make up the polyno-

mial. When this happens, the common factor can be factored out of each group by means of the distributive property. This technique is called **factoring by grouping.**

EXAMPLE 14 Factoring by Grouping

Factor completely by grouping: $(x^2 + 2)x + (x^2 + 2) \cdot 3$

Solution Notice the common factor $x^2 + 2$. By applying the distributive property, we have

$$(x^2 + 2)x + (x^2 + 2) \cdot 3 = (x^2 + 2)(x + 3)$$

Since $x^2 + 2$ and $x + 3$ are prime, the factorization is complete.

EXAMPLE 15 Factoring by Grouping

Factor completely by grouping: $x^3 - 4x^2 + 2x - 8$

Solution To see if factoring by grouping will work, group the first two terms and the last two terms. Then look for a common factor in each group. In this example, we can factor x^2 from $x^3 - 4x^2$ and 2 from $2x - 8$. The remaining factor in each case is the same, $x - 4$. This means that factoring by grouping will work, as follows:

$$\begin{aligned} x^3 - 4x^2 + 2x - 8 &= (x^3 - 4x^2) + (2x - 8) \\ &= x^2(x - 4) + 2(x - 4) \\ &= (x^2 + 2)(x - 4) \end{aligned}$$

Since $x^2 + 2$ and $x - 4$ are prime, the factorization is complete.

EXAMPLE 16 Factoring by Grouping

Factor completely by grouping: $3x^3 + 4x^2 - 6x - 8$

Solution Here, $3x + 4$ is a common factor of $3x^3 + 4x^2$ and of $-6x - 8$. Hence,

$$\begin{aligned} 3x^3 + 4x^2 - 6x - 8 &= (3x^3 + 4x^2) - (6x + 8) \\ &= x^2(3x + 4) - 2(3x + 4) \\ &= (x^2 - 2)(3x + 4) \end{aligned}$$

Since $x^2 - 2$ and $3x + 4$ are prime (over the integers), the factorization is complete.

NOW WORK PROBLEM 47.

5 Factoring a Second-Degree Polynomial: $Ax^2 + Bx + C$

To factor a second-degree polynomial $Ax^2 + Bx + C$, when $A \neq 1$, follow these steps:

Steps for Factoring $Ax^2 + Bx + C, A \neq 1$

STEP 1: Find the value of AC.
STEP 2: Find the factors of AC that add up to B. That is, find a and b so that $ab = AC$ and $a + b = B$.
STEP 3: Write $Ax^2 + Bx + C = Ax^2 + ax + bx + C$.
STEP 4: Factor this last expression by grouping.

EXAMPLE 17 Factoring Trinomials

Factor completely: $2x^2 + 5x + 3$

Solution Comparing $2x^2 + 5x + 3$ to $Ax^2 + Bx + C$, we find that $A = 2$, $B = 5$, and $C = 3$.

STEP 1: The value of AC is $2 \cdot 3 = 6$.
STEP 2: Determine the factors of $AC = 6$ and compute their sums.

Factors of 6	1, 6	−1, −6	2, 3	−2, −3
Sum	7	−7	5	−5

The factors of 6 that add up to $B = 5$ are 2 and 3.

STEP 3: $2x^2 + 5x + 3 = 2x^2 + 2x + 3x + 3$

STEP 4: Factor by grouping.

$$\begin{aligned} 2x^2 + 2x + 3x + 3 &= (2x^2 + 2x) + (3x + 3) \\ &= 2x(x + 1) + 3(x + 1) \\ &= (2x + 3)(x + 1) \end{aligned}$$

Thus,

$$2x^2 + 5x + 3 = (2x + 3)(x + 1)$$

EXAMPLE 18 Factoring Trinomials

Factor completely: $2x^2 - x - 6$

Solution Comparing $2x^2 - x - 6$ to $Ax^2 + Bx + C$, we find that $A = 2$, $B = -1$, and $C = -6$.

STEP 1: The value of AC is $2 \cdot (-6) = -12$.
STEP 2: Determine the factors of $AC = -12$ and compute their sums.

Factors of −12	1, −12	−1, 12	2, −6	−2, 6	3, −4	−3, 4
Sum	−11	11	−4	4	−1	1

The factors of -12 that add up to $B = -1$ are -4 and 3.

STEP 3: $2x^2 - x - 6 = 2x^2 - 4x + 3x - 6$

STEP 4: Factor by grouping.

$$\begin{aligned} 2x^2 - 4x + 3x - 6 &= (2x^2 - 4x) + (3x - 6) \\ &= 2x(x - 2) + 3(x - 2) \\ &= (2x + 3)(x - 2) \end{aligned}$$

Thus,

$$2x^2 - x - 6 = (2x + 3)(x - 2)$$

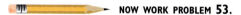

NOW WORK PROBLEM 53.

SUMMARY We close this section with a capsule summary.

Type of Polynomial	Method	Example
Any polynomial	Look for common monomial factors. (Always do this first!)	$6x^2 + 9x = 3x(2x + 3)$
Binomials of degree 2 or higher	Check for a special product:	
	Difference of two squares, $x^2 - a^2$	$x^2 - 16 = (x - 4)(x + 4)$
	Difference of two cubes, $x^3 - a^3$	$x^3 - 64 = (x - 4)(x^2 + 4x + 16)$
	Sum of two cubes, $x^3 + a^3$	$x^3 + 27 = (x + 3)(x^2 - 3x + 9)$
Trinomials of degree 2	Check for a perfect square, $(x \pm a)^2$.	$x^2 + 8x + 16 = (x + 4)^2$
	Follow the STEPS on page 711 or 714.	$6x^2 + x - 1 = (2x + 1)(3x - 1)$
Three or more terms	Grouping	$2x^3 - 3x^2 + 4x - 6 = (2x - 3)(x^2 + 2)$

6 EXERCISES

In Problems 1–10, factor each polynomial by removing the common monomial factor.

1. $3x + 6$ **2.** $7x - 14$ **3.** $ax^2 + a$ **4.** $ax - a$

5. $x^3 + x^2 + x$ **6.** $x^3 - x^2 + x$ **7.** $2x^2 - 2x$ **8.** $3x^2 - 3x$

9. $3x^2y - 6xy^2 + 12xy$ **10.** $60x^2y - 48xy^2 + 72x^3y$

In Problems 11–18, factor the difference of two squares.

11. $x^2 - 1$ **12.** $x^2 - 4$ **13.** $4x^2 - 1$ **14.** $9x^2 - 1$

15. $x^2 - 16$ **16.** $x^2 - 25$ **17.** $25x^2 - 4$ **18.** $36x^2 - 9$

In Problems 19–28, factor the perfect squares.

19. $x^2 + 2x + 1$ **20.** $x^2 - 4x + 4$ **21.** $x^2 + 4x + 4$ **22.** $x^2 - 2x + 1$

23. $x^2 - 10x + 25$ **24.** $x^2 + 10x + 25$ **25.** $4x^2 + 4x + 1$ **26.** $9x^2 + 6x + 1$

27. $16x^2 + 8x + 1$ **28.** $25x^2 + 10x + 1$

In Problems 29–34, factor the sum or difference of two cubes.

29. $x^3 - 27$ **30.** $x^3 + 125$ **31.** $x^3 + 27$

32. $27 - 8x^3$ **33.** $8x^3 + 27$ **34.** $64 - 27x^3$

In Problems 35–46, factor each polynomial.

35. $x^2 + 5x + 6$ **36.** $x^2 + 6x + 8$ **37.** $x^2 + 7x + 6$

38. $x^2 + 9x + 8$ **39.** $x^2 + 7x + 10$ **40.** $x^2 + 11x + 10$

41. $x^2 - 10x + 16$ **42.** $x^2 - 17x + 16$ **43.** $x^2 - 7x - 8$

44. $x^2 - 2x - 8$ **45.** $x^2 + 7x - 8$ **46.** $x^2 + 2x - 8$

In Problems 47–52, factor by grouping.

47. $2x^2 + 4x + 3x + 6$ **48.** $3x^2 - 3x + 2x - 2$ **49.** $2x^2 - 4x + x - 2$

50. $3x^2 + 6x - x - 2$ **51.** $6x^2 + 9x + 4x + 6$ **52.** $9x^2 - 6x + 2x - 3$

In Problems 53–64, factor each polynomial.

53. $3x^2 + 4x + 1$ **54.** $2x^2 + 3x + 1$ **55.** $2z^2 + 5z + 3$

56. $6z^2 + 5z + 1$ **57.** $3x^2 + 2x - 8$ **58.** $3x^2 + 10x + 8$

59. $3x^2 - 2x - 8$ **60.** $3x^2 - 10x + 8$ **61.** $3x^2 + 14x + 8$

62. $3x^2 - 14x + 8$ **63.** $3x^2 + 10x - 8$ **64.** $3x^2 - 10x - 8$

In Problems 65–112, factor completely each polynomial. If the polynomial cannot be factored, say it is prime.

65. $x^2 - 36$ **66.** $x^2 - 9$ **67.** $1 - 4x^2$ **68.** $1 - 9x^2$

69. $x^2 + 7x + 10$ **70.** $x^2 + 5x + 4$ **71.** $x^2 - 10x + 21$ **72.** $x^2 - 6x + 8$

73. $x^2 - 2x + 8$ **74.** $x^2 - 4x + 5$ **75.** $x^2 + 4x + 16$ **76.** $x^2 + 12x + 36$

77. $15 + 2x - x^2$ **78.** $14 + 6x - x^2$ **79.** $3x^2 - 12x - 36$ **80.** $x^3 + 8x^2 - 20x$

81. $y^4 + 11y^3 + 30y^2$ **82.** $3y^3 - 18y^2 - 48y$ **83.** $4x^2 + 12x + 9$ **84.** $9x^2 - 12x + 4$

85. $3x^2 + 4x + 1$ **86.** $4x^2 + 3x - 1$ **87.** $x^4 - 81$

88. $x^4 - 1$ **89.** $x^6 - 2x^3 + 1$ **90.** $x^6 + 2x^3 + 1$

91. $x^7 - x^5$ **92.** $x^8 - x^5$ **93.** $16x^2 + 24x + 9$

94. $9x^2 - 24x + 16$ **95.** $5 + 16x - 16x^2$ **96.** $5 + 11x - 16x^2$

97. $4y^2 - 16y + 15$ **98.** $9y^2 + 9y - 4$ **99.** $1 - 8x^2 - 9x^4$

100. $4 - 14x^2 - 8x^4$ **101.** $x(x + 3) - 6(x + 3)$ **102.** $5(3x - 7) + x(3x - 7)$

103. $(x + 2)^2 - 5(x + 2)$ **104.** $(x - 1)^2 - 2(x - 1)$

105. $(3x - 2)^3 - 27$ **106.** $(5x + 1)^3 - 1$

107. $3(x^2 + 10x + 25) - 4(x + 5)$ **108.** $7(x^2 - 6x + 9) + 5(x - 3)$

109. $x^3 + 2x^2 - x - 2$ **110.** $x^3 - 3x^2 - x + 3$

111. $x^4 - x^3 + x - 1$ **112.** $x^4 + x^3 + x + 1$

113. Show that $x^2 + 4$ is prime. **114.** Show that $x^2 + x + 1$ is prime.

115. Make up a polynomial that factors into a perfect square.

116. Explain to a fellow student what you look for first when presented with a factoring problem. What do you do next?

7 RATIONAL EXPRESSIONS

1 Find the Domain of a Variable

2 Reduce a Rational Expression to Lowest Terms

3 Multiply and Divide Rational Expressions

4 Add and Subtract Rational Expressions

5 Use the Least Common Multiple Method

6 Simplify Mixed Quotients

If we form the quotient of two polynomials, the result is called a **rational expression.** Some examples of rational expressions are

(a) $\frac{x^3 + 1}{x}$ (b) $\frac{3x^2 + x - 2}{x^2 + 5}$ (c) $\frac{x}{x^2 - 1}$ (d) $\frac{xy^2}{(x - y)^2}$

Expressions (a), (b), and (c) are rational expressions in one variable, x, whereas (d) is a rational expression in two variables, x and y.

Rational expressions are described in the same manner as rational numbers. Thus, in expression (a), the polynomial $x^3 + 1$ is called the **numerator,** and x is called the **denominator.** When the numerator and denominator of a rational expression contain no common factors (except 1 and -1), we say that the rational expression is **reduced to lowest terms,** or **simplified.**

1 The polynomial in the denominator of a rational expression cannot be equal to 0 because division by 0 is not defined. Thus, for example, for the expression $\frac{x^3 + 1}{x}$, x cannot take on the value 0.

> The set of values that a variable in an expression may assume is called the **domain** of the variable.

EXAMPLE 1 Finding the Domain of a Variable

The domain of the variable x in the rational expression

$$\frac{5}{x - 2}$$

is $\{x | x \neq 2\}$, since, if $x = 2$, the denominator becomes 0, which is not allowed.

In describing the domain of a variable, we may use either set notation or words, whichever is more convenient.

NOW WORK PROBLEM 1.

2 A rational expression is reduced to lowest terms by factoring completely the numerator and the denominator and canceling any common factors by using the cancellation property.

$$\frac{a\cancel{c}}{b\cancel{c}} = \frac{a}{b} \quad \text{if } b \neq 0, c \neq 0 \qquad (1)$$

EXAMPLE 2 Reducing a Rational Expression to Lowest Terms

Reduce to lowest terms: $\dfrac{x^2 + 4x + 4}{x^2 + 3x + 2}$

Solution We begin by factoring the numerator and the denominator.

$$x^2 + 4x + 4 = (x + 2)(x + 2)$$
$$x^2 + 3x + 2 = (x + 2)(x + 1)$$

Since a common factor, $x + 2$, appears, the original expression is not in lowest terms. To reduce it to lowest terms, we use the cancellation property.

$$\frac{x^2 + 4x + 4}{x^2 + 3x + 2} = \frac{\cancel{(x + 2)}(x + 2)}{\cancel{(x + 2)}(x + 1)} = \frac{x + 2}{x + 1}, \qquad x \neq -2, -1$$

Warning: Apply the cancellation property only to rational expressions written in factored form. Be sure to cancel only common factors!

EXAMPLE 3 Reducing Rational Expressions to Lowest Terms

Reduce each rational expression to lowest terms.

(a) $\dfrac{x^3 - 8}{x^3 - 2x^2}$ (b) $\dfrac{8 - 2x}{x^2 - x - 12}$

Solution (a) $\dfrac{x^3 - 8}{x^3 - 2x^2} = \dfrac{\cancel{(x - 2)}(x^2 + 2x + 4)}{x^2\cancel{(x - 2)}} = \dfrac{x^2 + 2x + 4}{x^2}, \qquad x \neq 0, 2$

(b) $\dfrac{8 - 2x}{x^2 - x - 12} = \dfrac{2(4 - x)}{(x - 4)(x + 3)} = \dfrac{2(-1)\cancel{(x - 4)}}{\cancel{(x - 4)}(x + 3)} = \dfrac{-2}{x + 3}, \quad x \neq -3, 4$

NOW WORK PROBLEM 9.

Multiplying and Dividing Rational Expressions

3 The rules for multiplying and dividing rational expressions are the same as the rules for multiplying and dividing rational numbers. If $\dfrac{a}{b}$ and $\dfrac{c}{d}$, $b \neq 0$, $d \neq 0$, are two rational expressions, then

$$\frac{a}{b} \cdot \frac{c}{d} = \frac{ac}{bd} \qquad \text{if } b \neq 0, d \neq 0 \tag{2}$$

$$\frac{\dfrac{a}{b}}{\dfrac{c}{d}} = \frac{a}{b} \cdot \frac{d}{c} = \frac{ad}{bc} \qquad \text{if } b \neq 0, c \neq 0, d \neq 0 \tag{3}$$

In using equations (2) and (3) with rational expressions, be sure first to factor each polynomial completely so that common factors can be canceled. Leave your answer in factored form.

EXAMPLE 4 Multiplying and Dividing Rational Expressions

Perform the indicated operation and simplify the result. Leave your answer in factored form.

(a) $\dfrac{x^2 - 2x + 1}{x^3 + x} \cdot \dfrac{4x^2 + 4}{x^2 + x - 2}$ (b) $\dfrac{\dfrac{x + 3}{x^2 - 4}}{\dfrac{x^2 - x - 12}{x^3 - 8}}$

Solution (a)

$$\frac{x^2 - 2x + 1}{x^3 + x} \cdot \frac{4x^2 + 4}{x^2 + x - 2} = \frac{(x-1)^2}{x(x^2+1)} \cdot \frac{4(x^2+1)}{(x+2)(x-1)}$$
$$= \frac{(x-1)^2(4)(x^2+1)}{x(x^2+1)(x+2)(x-1)}$$
$$= \frac{4(x-1)}{x(x+2)}, \quad x \neq -2, 0, 1$$

(b)

$$\frac{\dfrac{x + 3}{x^2 - 4}}{\dfrac{x^2 - x - 12}{x^3 - 8}} = \frac{x+3}{x^2-4} \cdot \frac{x^3 - 8}{x^2 - x - 12}$$
$$= \frac{x+3}{(x-2)(x+2)} \cdot \frac{(x-2)(x^2+2x+4)}{(x-4)(x+3)}$$
$$= \frac{(x+3)(x-2)(x^2+2x+4)}{(x-2)(x+2)(x-4)(x+3)}$$
$$= \frac{x^2+2x+4}{(x+2)(x-4)}, \quad x \neq -3, -2, 2, 4$$

NOW WORK PROBLEMS 23 AND 31.

4 Adding and Subtracting Rational Expressions

The rules for adding and subtracting rational expressions are the same as the rules for adding and subtracting rational numbers. Thus, if the denominators of two rational expressions to be added (or subtracted) are equal, we add (or subtract) the numerators and keep the common denominator.

If $\dfrac{a}{b}$ and $\dfrac{c}{b}$ are two rational expressions, then

$$\frac{a}{b} + \frac{c}{b} = \frac{a+c}{b} \qquad \frac{a}{b} - \frac{c}{b} = \frac{a-c}{b} \qquad \text{if } b \neq 0 \qquad (4)$$

EXAMPLE 5 Adding and Subtracting Rational Expressions with Equal Denominators

Perform the indicated operation and simplify the result. Leave your answer in factored form.

(a) $\dfrac{2x^2 - 4}{2x + 5} + \dfrac{x + 3}{2x + 5}, \quad x \neq -\dfrac{5}{2}$ (b) $\dfrac{x}{x - 3} - \dfrac{3x + 2}{x - 3}, \quad x \neq 3$

Solution (a)

$$\frac{2x^2-4}{2x+5}+\frac{x+3}{2x+5}=\frac{(2x^2-4)+(x+3)}{2x+5}$$
$$=\frac{2x^2+x-1}{2x+5}=\frac{(2x-1)(x+1)}{2x+5}$$

(b)

$$\frac{x}{x-3}-\frac{3x+2}{x-3}=\frac{x-(3x+2)}{x-3}=\frac{x-3x-2}{x-3}$$
$$=\frac{-2x-2}{x-3}=\frac{-2(x+1)}{x-3}$$

EXAMPLE 6 Adding Rational Expressions Whose Denominators Are Additive Inverses of Each Other

Perform the indicated operation and simplify the result. Leave your answer in factored form.

$$\frac{2x}{x-3}+\frac{5}{3-x}, \qquad x \neq 3$$

Solution Notice that the denominators of the two rational expressions are different. However, the denominator of the second expression is just the additive inverse of the denominator of the first. That is,

$$3-x=-x+3=-1\cdot(x-3)=-(x-3)$$

Thus,

$$\frac{2x}{x-3}+\frac{5}{3-x} \underset{\uparrow}{=} \frac{2x}{x-3}+\frac{5}{-(x-3)} \underset{\uparrow}{=} \frac{2x}{x-3}+\frac{-5}{x-3}$$

$3-x=-(x-3)$ $\qquad$ $\dfrac{a}{-b}=\dfrac{-a}{b}$

$$=\frac{2x+(-5)}{x-3}=\frac{2x-5}{x-3}$$

NOW WORK PROBLEM **43.**

If the denominators of two rational expressions to be added or subtracted are not equal, we can use the general formulas for adding and subtracting quotients.

$$\frac{a}{b}+\frac{c}{d}=\frac{a\cdot d}{b\cdot d}+\frac{b\cdot c}{b\cdot d}=\frac{ad+bc}{bd} \quad \text{if } b\neq 0, d\neq 0$$
$$\frac{a}{b}-\frac{c}{d}=\frac{a\cdot d}{b\cdot d}-\frac{b\cdot c}{b\cdot d}=\frac{ad-bc}{bd} \quad \text{if } b\neq 0, d\neq 0 \qquad (5)$$

EXAMPLE 7 Adding and Subtracting Rational Expressions with Unequal Denominators

Perform the indicated operation and simplify the result. Leave your answer in factored form.

(a) $\frac{x-3}{x+4} + \frac{x}{x-2}, \quad x \neq -4, 2$ (b) $\frac{x^2}{x^2-4} - \frac{1}{x}, \quad x \neq -2, 0, 2$

Solution (a)

$$\frac{x-3}{x+4} + \frac{x}{x-2} \underset{(5)}{=} \frac{x-3}{x+4}\cdot\frac{x-2}{x-2} + \frac{x+4}{x+4}\cdot\frac{x}{x-2}$$

$$= \frac{(x-3)(x-2) + (x+4)(x)}{(x+4)(x-2)}$$

$$= \frac{x^2-5x+6+x^2+4x}{(x+4)(x-2)} = \frac{2x^2-x+6}{(x+4)(x-2)}$$

(b)

$$\frac{x^2}{x^2-4} - \frac{1}{x} = \frac{x^2}{x^2-4}\cdot\frac{x}{x} - \frac{x^2-4}{x^2-4}\cdot\frac{1}{x} = \frac{x^2(x)-(x^2-4)(1)}{(x^2-4)(x)}$$

$$= \frac{x^3-x^2+4}{(x-2)(x+2)(x)}$$

NOW WORK PROBLEM 53.

Least Common Multiple (LCM)

5 If the denominators of two rational expressions to be added (or subtracted) have common factors, we usually do not use the general rules given by equation (5). Just as with fractions, we apply the **least common multiple (LCM) method.** The LCM method uses the polynomial of least degree that contains each denominator polynomial as a factor.

> ***The LCM Method for Adding or Subtracting Rational Expressions***
>
> The Least Common Multiple Method (LCM) requires four steps:
>
> STEP 1: Factor completely the polynomial in the denominator of each rational expression.
>
> STEP 2: The LCM of the denominator is the product of each of these factors raised to a power equal to the greatest number of times that the factor occurs in the polynomials.
>
> STEP 3: Write each rational expression using the LCM as the common denominator.
>
> STEP 4: Add or subtract the rational expressions using equation (4).

Let's work an example that only requires Steps 1 and 2.

EXAMPLE 8 Finding the Least Common Multiple

Find the least common multiple of the following pair of polynomials:

$$x(x-1)^2(x+1) \quad \text{and} \quad 4(x-1)(x+1)^3$$

Solution STEP 1: The polynomials are already factored completely as

$$x(x-1)^2(x+1) \quad \text{and} \quad 4(x-1)(x+1)^3$$

STEP 2: Start by writing the factors of the left-hand polynomial. (Or you could start with the one on the right.)

$$x(x-1)^2(x+1)$$

Now look at the right-hand polynomial. Its first factor, 4, does not appear in our list, so we insert it.

$$4x(x-1)^2(x+1)$$

The next factor, $x-1$, is already in our list, so no change is necessary. The final factor is $(x+1)^3$. Since our list has $x+1$ to the first power only, we replace $x+1$ in the list by $(x+1)^3$. The LCM is

$$4x(x-1)^2(x+1)^3$$

Notice that the LCM is, in fact, the polynomial of least degree that contains $x(x-1)^2(x+1)$ and $4(x-1)(x+1)^3$ as factors.

NOW WORK PROBLEM 63.

EXAMPLE 9 Using the Least Common Multiple to Add Rational Expressions

Perform the indicated operation and simplify the result. Leave your answer in factored form.

$$\frac{x}{x^2+3x+2} + \frac{2x-3}{x^2-1}, \qquad x \neq -2, -1, 1$$

Solution STEP 1: Factor completely the polynomials in the denominators.

$$x^2+3x+2 = (x+2)(x+1)$$
$$x^2-1 = (x-1)(x+1)$$

STEP 2: The LCM is $(x+2)(x+1)(x-1)$. Do you see why?

STEP 3: Write each rational expression using the LCM as the denominator.

$$\frac{x}{x^2+3x+2} = \frac{x}{(x+2)(x+1)} = \frac{x}{(x+2)(x+1)} \cdot \frac{x-1}{x-1} = \frac{x(x-1)}{(x+2)(x+1)(x-1)}$$

Multiply numerator and denominator by $x-1$ to get the LCM in the denominator.

$$\frac{2x-3}{x^2-1} = \frac{2x-3}{(x-1)(x+1)} = \frac{2x-3}{(x-1)(x+1)} \cdot \frac{x+2}{x+2} = \frac{(2x-3)(x+2)}{(x-1)(x+1)(x+2)}$$

Multiply numerator and denominator by $x+2$ to get the LCM in the denominator.

STEP 4: Now we can add by using equation (4).

$$\frac{x}{x^2+3x+2}+\frac{2x-3}{x^2-1}=\frac{x(x-1)}{(x+2)(x+1)(x-1)}+\frac{(2x-3)(x+2)}{(x+2)(x+1)(x-1)}$$
$$=\frac{(x^2-x)+(2x^2+x-6)}{(x+2)(x+1)(x-1)}$$
$$=\frac{3x^2-6}{(x+2)(x+1)(x-1)}=\frac{3(x^2-2)}{(x+2)(x+1)(x-1)}$$

EXAMPLE 10 Using the Least Common Multiple to Subtract Rational Expressions

Perform the indicated operations and simplify the result. Leave your answer in factored form.

$$\frac{3}{x^2+x}-\frac{x+4}{x^2+2x+1}, \qquad x \neq -1, 0$$

Solution STEP 1: Factor completely the polynomials in the denominators.

$$x^2+x=x(x+1)$$
$$x^2+2x+1=(x+1)^2$$

STEP 2: The LCM is $x(x+1)^2$.

STEP 3: Write each rational expression using the LCM as the denominator.

$$\frac{3}{x^2+x}=\frac{3}{x(x+1)}=\frac{3}{x(x+1)}\cdot\frac{x+1}{x+1}=\frac{3(x+1)}{x(x+1)^2}$$
$$\frac{x+4}{x^2+2x+1}=\frac{x+4}{(x+1)^2}=\frac{x+4}{(x+1)^2}\cdot\frac{x}{x}=\frac{x(x+4)}{x(x+1)^2}$$

STEP 4: Subtract, using equation (4).

$$\frac{3}{x^2+x}-\frac{x+4}{x^2+2x+1}=\frac{3(x+1)}{x(x+1)^2}-\frac{x(x+4)}{x(x+1)^2}$$
$$=\frac{3(x+1)-x(x+4)}{x(x+1)^2}$$
$$=\frac{3x+3-x^2-4x}{x(x+1)^2}$$
$$=\frac{-x^2-x+3}{x(x+1)^2}$$

NOW WORK PROBLEM **73.**

Mixed Quotients

6 When sums and/or differences of rational expressions appear as the numerator and/or denominator of a quotient, the quotient is called a **mixed quotient.*** For example,

*Some texts use the term **complex fraction.**

$$\frac{1+\frac{1}{x}}{1-\frac{1}{x}} \quad \text{and} \quad \frac{\frac{x^2}{x^2-4}-3}{\frac{x-3}{x+2}-1}$$

are mixed quotients. To **simplify** a mixed quotient means to write it as a rational expression reduced to lowest terms. This can be accomplished in either of two ways.

> ***Simplifying a Mixed Quotient***
>
> METHOD 1: Treat the numerator and denominator of the mixed quotient separately, performing whatever operations are indicated and simplifying the results. Follow this by simplifying the resulting rational expression.
>
> METHOD 2: Find the LCM of the denominators of all rational expressions that appear in the mixed quotient. Multiply the numerator and denominator of the mixed quotient by the LCM and simplify the result.

We will use both methods in the next example. By carefully studying each method, you can discover situations in which one method may be easier to use than the other.

EXAMPLE 11 Simplifying a Mixed Quotient

Simplify: $\dfrac{\frac{1}{2}+\frac{3}{x}}{\frac{x+3}{4}}, \quad x \neq -3, 0$

Solution

METHOD 1: First, we perform the indicated operation in the numerator, and then we divide:

$$\frac{\frac{1}{2}+\frac{3}{x}}{\frac{x+3}{4}} \underset{\uparrow}{=} \frac{\frac{1\cdot x+2\cdot 3}{2\cdot x}}{\frac{x+3}{4}} = \frac{\frac{x+6}{2x}}{\frac{x+3}{4}} \underset{\uparrow}{=} \frac{x+6}{2x}\cdot\frac{4}{x+3}$$

Rule for adding quotients — Rule for dividing quotients

$$\underset{\uparrow}{=} \frac{(x+6)\cdot 4}{2\cdot x\cdot(x+3)} = \frac{2\cdot 2\cdot(x+6)}{2\cdot x\cdot(x+3)} = \frac{2(x+6)}{x(x+3)}$$

Rule for multiplying quotients

METHOD 2: The rational expressions that appear in the mixed quotient are

$$\frac{1}{2}, \quad \frac{3}{x}, \quad \frac{x+3}{4}$$

The LCM of their denominators is $4x$. Thus, we multiply the numerator and denominator of the mixed quotient by $4x$ and then simplify.

$$\frac{\frac{1}{2}+\frac{3}{x}}{\frac{x+3}{4}} = \frac{4x\cdot\left(\frac{1}{2}+\frac{3}{x}\right)}{4x\cdot\left(\frac{x+3}{4}\right)} = \frac{4x\cdot\frac{1}{2}+4x\cdot\frac{3}{x}}{\frac{4x\cdot(x+3)}{4}}$$

Multiply by $4x$ — Distributive property in numerator

$$= \frac{2\cdot 2x\cdot\frac{1}{2}+4x\cdot\frac{3}{x}}{\frac{4x\cdot(x+3)}{4}} = \frac{2x+12}{x(x+3)} = \frac{2(x+6)}{x(x+3)}$$

Simplify — Factor

EXAMPLE 12 Simplifying a Mixed Quotient

Simplify: $\dfrac{\frac{x^2}{x-4}+2}{\frac{2x-2}{x}-1}$

Solution We will use Method 1.

$$\frac{\frac{x^2}{x-4}+2}{\frac{2x-2}{x}-1} = \frac{\frac{x^2}{x-4}+\frac{2(x-4)}{x-4}}{\frac{2x-2}{x}-\frac{x}{x}} = \frac{\frac{x^2+2x-8}{x-4}}{\frac{2x-2-x}{x}}$$

$$= \frac{\frac{(x+4)(x-2)}{x-4}}{\frac{x-2}{x}} = \frac{(x+4)(x-2)}{x-4}\cdot\frac{x}{x-2}$$

$$= \frac{(x+4)\cdot x}{x-4}$$

NOW WORK PROBLEM 83.

EXAMPLE 13 Solving an Application in Electricity

Figure 22

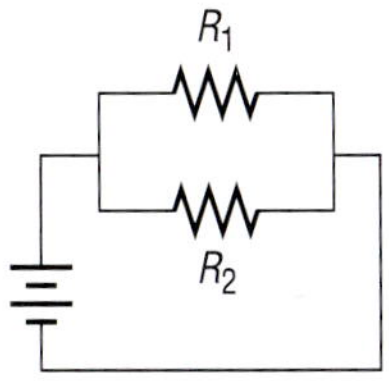

An electrical circuit contains two resistors connected in parallel, as shown in Figure 22. If the resistance of each is R_1 and R_2 ohms respectively, their combined resistance R is given by the formula

$$R = \frac{1}{\frac{1}{R_1}+\frac{1}{R_2}}$$

Express R as a rational expression; that is, simplify the right-hand side of this formula. Evaluate the rational expression if $R_1 = 6$ ohms and $R_2 = 10$ ohms.

Solution We will use Method 2. If we consider 1 as the fraction $\frac{1}{1}$, then the rational expressions in the mixed quotient are

$$\frac{1}{1}, \frac{1}{R_1}, \frac{1}{R_2}$$

The LCM of the denominators is R_1R_2. We multiply the numerator and denominator of the mixed quotient by R_1R_2 and simplify.

$$\frac{1}{\frac{1}{R_1}+\frac{1}{R_2}} = \frac{1 \cdot R_1R_2}{\left(\frac{1}{R_1}+\frac{1}{R_2}\right) \cdot R_1R_2} = \frac{R_1R_2}{\frac{1}{R_1} \cdot R_1R_2 + \frac{1}{R_2} \cdot R_1R_2} = \frac{R_1R_2}{R_2 + R_1}$$

Thus,

$$R = \frac{R_1R_2}{R_2 + R_1}$$

If $R_1 = 6$ and $R_2 = 10$, then

$$R = \frac{6 \cdot 10}{10 + 6} = \frac{60}{16} = \frac{15}{4} \text{ ohms}$$

7 EXERCISES

In Problems 1–8, determine which of the value(s) given below, if any, must be excluded from the domain of the variable in each rational expression:

(a) $x = 3$ *(b)* $x = 1$ *(c)* $x = 0$ *(d)* $x = -1$

1. $\frac{x^2 - 1}{x}$ **2.** $\frac{x^2 + 1}{x}$ **3.** $\frac{x}{x^2 - 9}$ **4.** $\frac{x}{x^2 + 9}$

5. $\frac{x^2}{x^2 + 1}$ **6.** $\frac{x^3}{x^2 - 1}$ **7.** $\frac{x^2 + 5x - 10}{x^3 - x}$ **8.** $\frac{-9x^2 - x + 1}{x^3 + x}$

In Problems 9–22, reduce each rational expression to lowest terms.

9. $\frac{3x + 9}{x^2 - 9}$ **10.** $\frac{4x^2 + 8x}{12x + 24}$ **11.** $\frac{x^2 - 2x}{3x - 6}$ **12.** $\frac{15x^2 + 24x}{3x^2}$

13. $\frac{24x^2}{12x^2 - 6x}$ **14.** $\frac{x^2 + 4x + 4}{x^2 - 16}$ **15.** $\frac{y^2 - 25}{2y^2 - 8y - 10}$ **16.** $\frac{3y^2 - y - 2}{3y^2 + 5y + 2}$

17. $\frac{x^2 + 4x - 5}{x^2 - 2x + 1}$ **18.** $\frac{x - x^2}{x^2 + x - 2}$ **19.** $\frac{x^2 - 4}{x^2 + 5x + 6}$ **20.** $\frac{x^2 + x - 6}{9 - x^2}$

21. $\frac{x^2 + 5x - 14}{2 - x}$ **22.** $\frac{2x^2 + 5x - 3}{1 - 2x}$

In Problems 23–40, perform the indicated operation and simplify the result. Leave your answer in factored form.

23. $\frac{3x + 6}{5x^2} \cdot \frac{x}{x^2 - 4}$ **24.** $\frac{3}{2x} \cdot \frac{x^2}{6x + 10}$ **25.** $\frac{4x^2}{x^2 - 16} \cdot \frac{x - 4}{2x}$

26. $\frac{12}{x^2 - x} \cdot \frac{x^2 - 1}{4x - 2}$ **27.** $\frac{4x - 8}{-3x} \cdot \frac{12}{12 - 6x}$ **28.** $\frac{6x - 27}{5x} \cdot \frac{2}{4x - 18}$

29. $\dfrac{x^2 - 3x - 10}{x^2 + 2x - 35} \cdot \dfrac{x^2 + 4x - 21}{x^2 + 9x + 14}$

30. $\dfrac{x^2 + x - 6}{x^2 + 4x - 5} \cdot \dfrac{x^2 - 25}{x^2 + 2x - 15}$

31. $\dfrac{\dfrac{6x}{x^2 - 4}}{\dfrac{3x - 9}{2x + 4}}$

32. $\dfrac{\dfrac{12x}{5x + 20}}{\dfrac{4x^2}{x^2 - 16}}$

33. $\dfrac{\dfrac{8x}{x^2 - 1}}{\dfrac{10x}{x + 1}}$

34. $\dfrac{\dfrac{x - 2}{4x}}{\dfrac{x^2 - 4x + 4}{12x}}$

35. $\dfrac{\dfrac{4 - x}{4 + x}}{\dfrac{4x}{x^2 - 16}}$

36. $\dfrac{\dfrac{3 + x}{3 - x}}{\dfrac{x^2 - 9}{9x^3}}$

37. $\dfrac{\dfrac{x^2 + 7x + 12}{x^2 - 7x + 12}}{\dfrac{x^2 + x - 12}{x^2 - x - 12}}$

38. $\dfrac{\dfrac{x^2 + 7x + 6}{x^2 + x - 6}}{\dfrac{x^2 + 5x - 6}{x^2 + 5x + 6}}$

39. $\dfrac{\dfrac{2x^2 - x - 28}{3x^2 - x - 2}}{\dfrac{4x^2 + 16x + 7}{3x^2 + 11x + 6}}$

40. $\dfrac{\dfrac{9x^2 + 3x - 2}{12x^2 + 5x - 2}}{\dfrac{9x^2 - 6x + 1}{8x^2 - 10x - 3}}$

In Problems 41–62, perform the indicated operations and simplify the result. Leave your answer in factored form.

41. $\dfrac{x}{2} + \dfrac{5}{2}$

42. $\dfrac{3}{x} - \dfrac{6}{x}$

43. $\dfrac{x^2}{2x - 3} - \dfrac{4}{2x - 3}$

44. $\dfrac{3x^2}{2x - 1} - \dfrac{9}{2x - 1}$

45. $\dfrac{x + 1}{x - 3} + \dfrac{2x - 3}{x - 3}$

46. $\dfrac{2x - 5}{3x + 2} + \dfrac{x + 4}{3x + 2}$

47. $\dfrac{3x + 5}{2x - 1} - \dfrac{2x - 4}{2x - 1}$

48. $\dfrac{5x - 4}{3x + 4} - \dfrac{x + 1}{3x + 4}$

49. $\dfrac{4}{x - 2} + \dfrac{x}{2 - x}$

50. $\dfrac{6}{x - 1} - \dfrac{x}{1 - x}$

51. $\dfrac{4}{x - 1} - \dfrac{2}{x + 2}$

52. $\dfrac{2}{x + 5} - \dfrac{5}{x - 5}$

53. $\dfrac{x}{x + 1} + \dfrac{2x - 3}{x - 1}$

54. $\dfrac{3x}{x - 4} + \dfrac{2x}{x + 3}$

55. $\dfrac{x - 3}{x + 2} - \dfrac{x + 4}{x - 2}$

56. $\dfrac{2x - 3}{x - 1} - \dfrac{2x + 1}{x + 1}$

57. $\dfrac{x}{x^2 - 4} + \dfrac{1}{x}$

58. $\dfrac{x - 1}{x^3} + \dfrac{x}{x^2 + 1}$

59. $\dfrac{x^3}{(x - 1)^2} - \dfrac{x^2 + 1}{x}$

60. $\dfrac{3x^2}{4} - \dfrac{x^3}{x^2 - 1}$

61. $\dfrac{x}{x + 1} + \dfrac{x - 2}{x - 1} - \dfrac{x + 1}{x - 2}$

62. $\dfrac{3x + 1}{x} + \dfrac{x}{x - 1} - \dfrac{2x}{x + 1}$

In Problems 63–70, find the LCM of the given polynomials.

63. $x^2 - 4,\ \ x^2 - x - 2$

64. $x^2 - x - 12,\ \ x^2 - 8x + 16$

65. $x^3 - x,\ \ x^2 - x$

66. $3x^2 - 27,\ \ 2x^2 - x - 15$

67. $4x^3 - 4x^2 + x,\ \ 2x^3 - x^2,\ \ x^3$

68. $x - 3,\ \ x^2 + 3x,\ \ x^3 - 9x$

69. $x^3 - x,\ \ x^3 - 2x^2 + x,\ \ x^3 - 1$

70. $x^2 + 4x + 4,\ \ x^3 + 2x^2,\ \ (x + 2)^3$

In Problems 71–82, perform the indicated operations and simplify the result. Leave your answer in factored form.

71. $\dfrac{x}{x^2 - 7x + 6} - \dfrac{x}{x^2 - 2x - 24}$

72. $\dfrac{x}{x - 3} - \dfrac{x + 1}{x^2 + 5x - 24}$

73. $\dfrac{4x}{x^2 - 4} - \dfrac{2}{x^2 + x - 6}$

74. $\dfrac{3x}{x - 1} - \dfrac{x - 4}{x^2 - 2x + 1}$

75. $\dfrac{3}{(x - 1)^2(x + 1)} + \dfrac{2}{(x - 1)(x + 1)^2}$

76. $\dfrac{2}{(x + 2)^2(x - 1)} - \dfrac{6}{(x + 2)(x - 1)^2}$

77. $\dfrac{x + 4}{x^2 - x - 2} - \dfrac{2x + 3}{x^2 + 2x - 8}$

78. $\dfrac{2x - 3}{x^2 + 8x + 7} - \dfrac{x - 2}{(x + 1)^2}$

79. $\dfrac{1}{x} - \dfrac{2}{x^2 + x} + \dfrac{3}{x^3 - x^2}$

80. $\dfrac{x}{(x - 1)^2} + \dfrac{2}{x} - \dfrac{x + 1}{x^3 - x^2}$

81. $\dfrac{1}{h}\left(\dfrac{1}{x + h} - \dfrac{1}{x}\right)$

82. $\dfrac{1}{h}\left[\dfrac{1}{(x + h)^2} - \dfrac{1}{x^2}\right]$

In Problems 83–92, perform the indicated operations and simplify the result. Leave your answer in factored form.

83. $\dfrac{1 + \dfrac{1}{x}}{1 - \dfrac{1}{x}}$

84. $\dfrac{4 + \dfrac{1}{x^2}}{3 - \dfrac{1}{x^2}}$

85. $\dfrac{x - \dfrac{1}{x}}{x + \dfrac{1}{x}}$

86. $\dfrac{1 - \dfrac{x}{x + 1}}{2 - \dfrac{x - 1}{x}}$

87. $\dfrac{\dfrac{x + 4}{x - 2} - \dfrac{x - 3}{x + 1}}{x + 1}$

88. $\dfrac{\dfrac{x - 2}{x + 1} - \dfrac{x}{x - 2}}{x + 3}$

89. $\dfrac{\dfrac{x - 2}{x + 2} + \dfrac{x - 1}{x + 1}}{\dfrac{x}{x + 1} - \dfrac{2x - 3}{x}}$

90. $\dfrac{\dfrac{2x + 5}{x} - \dfrac{x}{x - 3}}{\dfrac{x^2}{x - 3} - \dfrac{(x + 1)^2}{x + 3}}$

91. $1 - \dfrac{1}{1 - \dfrac{1}{x}}$

92. $1 - \dfrac{1}{1 - \dfrac{1}{1 - x}}$

93. The Lensmaker's Equation The focal length f of a lens with index of refraction n is

$$\frac{1}{f} = (n - 1)\left[\frac{1}{R_1} + \frac{1}{R_2}\right]$$

where R_1 and R_2 are the radii of curvature of the front and back surfaces of the lens. Express f as a rational expression. Evaluate the rational expression for $n = 1.5$, $R_1 = 0.1$ meter, and $R_2 = 0.2$ meter.

94. Electrical Circuits An electrical circuit contains three resistors connected in parallel. If the resistance of each is R_1, R_2, and R_3 ohms, respectively, their combined resistance R is given by the formula

$$\frac{1}{R} = \frac{1}{R_1} + \frac{1}{R_2} + \frac{1}{R_3}$$

Express R as a rational expression. Evaluate R for $R_1 = 5$ ohms, $R_2 = 4$ ohms, and $R_3 = 10$ ohms.

95. The following expressions are called **continued fractions:**

$$1 + \frac{1}{x},\quad 1 + \frac{1}{1 + \dfrac{1}{x}},\quad 1 + \frac{1}{1 + \dfrac{1}{1 + \dfrac{1}{x}}},\quad 1 + \frac{1}{1 + \dfrac{1}{1 + \dfrac{1}{1 + \dfrac{1}{x}}}}, \ldots$$

Each simplifies to an expression of the form

$$\frac{ax + b}{bx + c}$$

Trace the successive values of a, b, and c as you "continue" the fraction. Can you discover the patterns that these values follow? Go to the library and research Fibonacci numbers. Write a report on your findings.

96. Explain to a fellow student when you would use the LCM method to add two rational expressions. Give two examples of adding two rational expressions, one in which you use the LCM and the other in which you do not.

97. Which of the two methods given in the text for simplifying mixed quotients do you prefer? Write a brief paragraph stating the reasons for your choice.

8 SQUARE ROOTS; RADICALS

1 Work with Properties of Square Roots

2 Rationalize the Denominator

3 Simplify *n*th Roots

4 Simplify Radicals

Square Roots

A real number is squared when it is raised to the power 2. The inverse of squaring is finding a **square root.** For example, since $6^2 = 36$ and $(-6)^2 = 36$, the numbers 6 and -6 are square roots of 36.

The symbol $\sqrt{}$, called a **radical sign,** is used to denote the **principal,** or nonnegative, square root. Thus, $\sqrt{36} = 6$.

In general, if a is a nonnegative real number, the nonnegative number b such that $b^2 = a$ is the **principal square root** of a is denoted by $b = \sqrt{a}$.

The following comments are noteworthy:

1. Negative numbers do not have square roots (in the real number system), because the square of any real number is *nonnegative.* For example, $\sqrt{-4}$ is not a real number, because there is no real number whose square is -4.
2. The principal square root of 0 is 0, since $0^2 = 0$. That is, $\sqrt{0} = 0$.
3. The principal square root of a positive number is positive.
4. If $c \geq 0$, then $(\sqrt{c})^2 = c$. For example, $(\sqrt{2})^2 = 2$ and $(\sqrt{3})^2 = 3$.

EXAMPLE 1 Evaluating Square Roots

(a) $\sqrt{64} = 8$ (b) $\sqrt{\dfrac{1}{16}} = \dfrac{1}{4}$ (c) $(\sqrt{1.4})^2 = 1.4$

Examples 1(a) and (b) are examples of **perfect square roots.** Thus, 64 is a **perfect square,** since $64 = 8^2$; and $\frac{1}{16}$ is a perfect square, since $\frac{1}{16} = (\frac{1}{4})^2$.

In general, we have the rule

$$\sqrt{a^2} = |a| \tag{1}$$

Notice the absolute value in equation (1). We need it since the principal square root is nonnegative.

NOW WORK PROBLEM 1.

EXAMPLE 1 Square Roots of Perfect Squares

(a) $\sqrt{(2.3)^2} = |2.3| = 2.3$ (b) $\sqrt{(-2.3)^2} = |-2.3| = 2.3$

(c) $\sqrt{x^2} = |x|$

Properties of Square Roots

1 We begin with the following observation:

$$\sqrt{4 \cdot 25} = \sqrt{100} = 10 \quad \text{and} \quad \sqrt{4}\sqrt{25} = 2 \cdot 5 = 10$$

This suggests the following property of square roots: If a and b are each nonnegative real numbers, then

Product Property of Square Roots

$$\sqrt{ab} = \sqrt{a}\sqrt{b} \qquad (2)$$

When used in connection with square roots, the direction "simplify" means to remove from the square root any perfect squares that occur as factors. We can use equation (2) to simplify a square root that contains a perfect square as a factor, as illustrated by the following examples.

EXAMPLE 3 Simplifying Square Roots

(a) $\sqrt{32} = \sqrt{16 \cdot 2} = \sqrt{16}\sqrt{2} = 4\sqrt{2}$

↑ 16 is a perfect square. ↑ (2)

(b) $\sqrt{5}\sqrt{10} = \sqrt{5 \cdot 10} = \sqrt{50} = \sqrt{25 \cdot 2} = \sqrt{25}\sqrt{2} = 5\sqrt{2}$

↑ (2)

(c) $-3\sqrt{72} = -3\sqrt{36 \cdot 2} = -3\sqrt{36}\sqrt{2} = -3 \cdot 6\sqrt{2} = -18\sqrt{2}$

(d) $\sqrt{75x^2} = \sqrt{(25x^2)(3)} = \sqrt{25x^2}\sqrt{3} = \sqrt{25}\sqrt{x^2}\sqrt{3} = 5|x|\sqrt{3}$

↑ (1)

NOW WORK PROBLEMS 13 AND 27.

Sometimes to multiply two expressions with square roots, we use the distributive property and the product property of square roots. For example,

$$\begin{aligned}\sqrt{3}(4\sqrt{2} - \sqrt{12}) &= \sqrt{3} \cdot 4\sqrt{2} - \sqrt{3} \cdot \sqrt{12} && \text{Distributive Property}\\ &= 4\sqrt{3 \cdot 2} - \sqrt{3 \cdot 12} && \text{Product Property of Square Roots}\\ &= 4\sqrt{6} - \sqrt{36} && \text{Simplify.}\\ &= 4\sqrt{6} - 6\end{aligned}$$

EXAMPLE 4 Multiplying Square Roots

$$\begin{aligned}(\sqrt{5} - 5)(\sqrt{5} + 2) &= \sqrt{5}(\sqrt{5} + 2) - 5(\sqrt{5} + 2)\\ &= \sqrt{5} \cdot \sqrt{5} + \sqrt{5} \cdot 2 - 5 \cdot \sqrt{5} - 5 \cdot 2\\ &= 5 + 2\sqrt{5} - 5\sqrt{5} - 10\\ &= -5 - 3\sqrt{5}\end{aligned}$$

NOW WORK PROBLEM 51.

Another property of square roots is suggested by the following examples:

$$\sqrt{\frac{36}{9}} = \sqrt{4} = 2 \quad \text{and} \quad \frac{\sqrt{36}}{\sqrt{9}} = \frac{6}{3} = 2$$

If a is a nonnegative real number and b is a positive real number, then

Quotient Property of Square Roots

$$\sqrt{\frac{a}{b}} = \frac{\sqrt{a}}{\sqrt{b}} \qquad (3)$$

EXAMPLE 5 Simplifying Square Roots

(a) $\sqrt{\frac{81}{25}} = \frac{\sqrt{81}}{\sqrt{25}} = \frac{9}{5}$

(b) $\frac{\sqrt{24}}{\sqrt{3}} = \sqrt{\frac{24}{3}} = \sqrt{8} = \sqrt{4 \cdot 2} = \sqrt{4}\sqrt{2} = 2\sqrt{2}$

NOW WORK PROBLEM 31.

Rationalizing

2 When square roots occur in quotients, we rewrite the quotient so that the denominator contains no square roots. This process is referred to as **rationalizing the denominator.**

The idea is to multiply by an appropriate expression so that the new denominator contains no square roots. For example:

If Denominator Contains the Factor	Multiply By	To Obtain Denominator Free of Radicals
$\sqrt{3}$	$\sqrt{3}$	$(\sqrt{3})^2 = 3$
$\sqrt{3} + 1$	$\sqrt{3} - 1$	$(\sqrt{3})^2 - 1^2 = 3 - 1 = 2$
$\sqrt{2} - 3$	$\sqrt{2} + 3$	$(\sqrt{2})^2 - 3^2 = 2 - 9 = -7$
$\sqrt{5} - \sqrt{3}$	$\sqrt{5} + \sqrt{3}$	$(\sqrt{5})^2 - (\sqrt{3})^2 = 5 - 3 = 2$

In rationalizing the denominator of a quotient, be sure to multiply both the numerator and the denominator by the expression.

EXAMPLE 6 Rationalizing Denominators

Rationalize the denominator: $\frac{1}{\sqrt{3}}$

Solution The denominator contains the factor $\sqrt{3}$, so we multiply the numerator and denominator by $\sqrt{3}$ to obtain

$$\frac{1}{\sqrt{3}} = \frac{1}{\sqrt{3}} \cdot \frac{\sqrt{3}}{\sqrt{3}} = \frac{\sqrt{3}}{(\sqrt{3})^2} = \frac{\sqrt{3}}{3}$$

EXAMPLE 7 Rationalizing Denominators

Rationalize the denominator: $\dfrac{5}{4\sqrt{2}}$

Solution The denominator contains the factor $\sqrt{2}$, so we multiply the numerator and denominator by $\sqrt{2}$ to obtain

$$\frac{5}{4\sqrt{2}} = \frac{5}{4\sqrt{2}} \cdot \frac{\sqrt{2}}{\sqrt{2}} = \frac{5\sqrt{2}}{4(\sqrt{2})^2} = \frac{5\sqrt{2}}{4 \cdot 2} = \frac{5\sqrt{2}}{8}$$

EXAMPLE 8 Rationalizing Denominators

Rationalize the denominator: $\dfrac{\sqrt{2}}{\sqrt{3}-\sqrt{2}}$

Solution The denominator contains the factor $\sqrt{3}-\sqrt{2}$, so we multiply the numerator and denominator by $\sqrt{3}+\sqrt{2}$ to obtain

$$\frac{\sqrt{2}}{\sqrt{3}-\sqrt{2}} = \frac{\sqrt{2}}{\sqrt{3}-\sqrt{2}} \cdot \frac{\sqrt{3}+\sqrt{2}}{\sqrt{3}+\sqrt{2}} = \frac{\sqrt{2}(\sqrt{3}+\sqrt{2})}{(\sqrt{3})^2-(\sqrt{2})^2}$$

$$= \frac{\sqrt{2}\sqrt{3}+(\sqrt{2})^2}{3-2} = \sqrt{6}+2$$

NOW WORK PROBLEM 69.

*n*th Roots

The **principal *n*th root of a real number *a***, $n \geq 2$ an integer, symbolized by $\sqrt[n]{a}$ is defined as follows:

$$\sqrt[n]{a} = b \quad \text{means} \quad a = b^n$$

where $a \geq 0$ and $b \geq 0$ if $n \geq 2$ is even and a, b are any real numbers if $n \geq 3$ is odd

Notice that if a is negative and n is even then $\sqrt[n]{a}$ is not defined. When it is defined, the principal nth root of a number is unique.

3 The symbol $\sqrt[n]{a}$ for the principal nth root of a is sometimes called a **radical;** the integer n is called the **index,** and a is called the **radicand.** If the index of a radical is 2, we call $\sqrt[2]{a}$ the **square root** of a and omit the index 2 by simply writing $\sqrt{a}$. If the index is 3, we call $\sqrt[3]{a}$ the **cube root** of a.

EXAMPLE 9 Simplifying Principal *n*th Roots

(a) $\sqrt[3]{8} = \sqrt[3]{2^3} = 2$ (b) $\sqrt[3]{-64} = \sqrt[3]{(-4)^3} = -4$

(c) $\sqrt[4]{1/16} = \sqrt[4]{(1/2)^4} = 1/2$ (d) $\sqrt[6]{(-2)^6} = |-2| = 2$

(a), (b), and (c) are examples of **perfect roots.** Thus, 8 and -64 are perfect cube roots, since $8 = 2^3$ and $-64 = (-4)^3$; $\frac{1}{2}$ is a perfect fourth root of $\frac{1}{16}$, since $\frac{1}{16} = (\frac{1}{2})^4$.

Notice the absolute value in Example 9(d). If n is even, the principal nth root must be nonnegative.

In general, if $n \geq 2$ is a positive integer and a is a real number, we have

$$\sqrt[n]{a^n} = a \quad \text{if } n \text{ is odd} \tag{4a}$$
$$\sqrt[n]{a^n} = |a| \quad \text{if } n \text{ is even} \tag{4b}$$

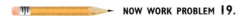

NOW WORK PROBLEM 19.

Radicals provide a way of representing many irrational real numbers. For example, there is no rational number whose square is 2. Using radicals, we can say that $\sqrt{2}$ *is* the positive number whose square is 2.

EXAMPLE 10 Using a Calculator to Approximate Roots

Use a calculator to approximate $\sqrt[5]{16}$.

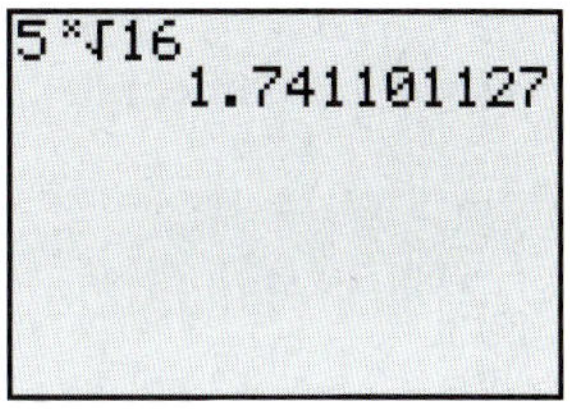

NOW WORK PROBLEM 79.

Properties of Radicals

Let $n \geq 2$ and $m \geq 2$ denote positive integers, and let a and b represent real numbers. Assuming that all radicals are defined, we have the following properties:

Properties of Radicals

$$\sqrt[n]{ab} = \sqrt[n]{a}\sqrt[n]{b} \tag{5a}$$
$$\sqrt[n]{\frac{a}{b}} = \frac{\sqrt[n]{a}}{\sqrt[n]{b}} \tag{5b}$$
$$\sqrt[n]{a^m} = (\sqrt[n]{a})^m \tag{5c}$$
$$\sqrt[m]{\sqrt[n]{a}} = \sqrt[mn]{a} \tag{5d}$$

4 When used in reference to radicals, the direction to "simplify" will mean to remove from the radicals any perfect roots that occur as factors. Let's look at some examples of how the preceding rules are applied to simplify radicals.

EXAMPLE 11 Simplifying Radicals

(a) $\sqrt[3]{16} = \sqrt[3]{8 \cdot 2} = \sqrt[3]{8} \cdot \sqrt[3]{2} = \sqrt[3]{2^3} \cdot \sqrt[3]{2} = 2\sqrt[3]{2}$

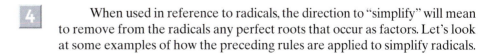

↑ Factor out perfect cube. ↑ (5a)

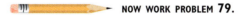

(b) $\sqrt[3]{-16x^4} = \sqrt[3]{-8 \cdot 2 \cdot x^3 \cdot x} = \sqrt[3]{(-8x^3)(2x)}$

↑ *Factor perfect cubes inside radical.* ↑ *Combine perfect cubes.*

$$= \sqrt[3]{(-2x)^3 \cdot 2x} = \sqrt[3]{(-2x)^3} \cdot \sqrt[3]{2x}$$
$$= -2x\sqrt[3]{2x}$$

↑ *(5a)*

EXAMPLE 12 Simplifying Radicals

$$\sqrt[3]{\frac{8x^5}{27}} = \sqrt[3]{\frac{2^3x^3x^2}{3^3}} = \sqrt[3]{\left(\frac{2x}{3}\right)^3 \cdot x^2} = \sqrt[3]{\left(\frac{2x}{3}\right)^3} \cdot \sqrt[3]{x^2} = \frac{2x}{3}\sqrt[3]{x^2}$$

NOW WORK PROBLEM 23.

Two or more radicals can be combined, provided that they have the same index and the same radicand. Such radicals are called **like radicals.**

EXAMPLE 13 Combining Like Radicals

$$-8\sqrt{12} + \sqrt{3} = -8\sqrt{4 \cdot 3} + \sqrt{3}$$
$$= -8 \cdot \sqrt{4}\sqrt{3} + \sqrt{3}$$
$$= -16\sqrt{3} + \sqrt{3} = -15\sqrt{3}$$

EXAMPLE 14 Combining Like Radicals

$$\sqrt[3]{8x^4} + \sqrt[3]{-x} + 4\sqrt[3]{27x} = \sqrt[3]{2^3x^3x} + \sqrt[3]{-1 \cdot x} + 4\sqrt[3]{3^3x}$$
$$= \sqrt[3]{(2x)^3} \cdot \sqrt[3]{x} + \sqrt[3]{-1} \cdot \sqrt[3]{x} + 4\sqrt[3]{3^3} \cdot \sqrt[3]{x}$$
$$= 2x\sqrt[3]{x} - 1 \cdot \sqrt[3]{x} + 12\sqrt[3]{x}$$
$$= (2x + 11)\sqrt[3]{x}$$

NOW WORK PROBLEM 43.

HISTORICAL FEATURE

The radical sign, $\sqrt{\ }$, was first used in print by Coss in 1525. It is thought to be the manuscript form of the letter *r* (for the Latin word *radix* = *root*), although this is not quite conclusively proved. It took a long time for $\sqrt{\ }$ to become the standard symbol for a square root and much longer to standardize $\sqrt[3]{\ }$, $\sqrt[4]{\ }$, $\sqrt[5]{\ }$, and so on. The indexes of the root were placed in every conceivable position, with

$$\sqrt{8}^{\,3}, \quad \sqrt{③8}, \quad \text{and} \quad \sqrt{8}_{3}$$

all being variants for $\sqrt[3]{8}$. The notation $\sqrt{\sqrt{16}}$ was popular for $\sqrt[4]{16}$. By the 1700s, the index had settled where we now put it.

The bar on top of the present radical symbol, as follows,

$$\sqrt{a^2 + 2ab + b^2}$$

is the last survivor of the **vinculum,** a bar placed atop an expression to indicate what we would now indicate with parentheses. For example,

$$a\overline{b + c} = a(b + c)$$

8 EXERCISES

In Problems 1–12, evaluate each perfect root.

1. $\sqrt{25}$
2. $\sqrt{81}$
3. $\sqrt[3]{27}$
4. $\sqrt[3]{125}$
5. $\sqrt[3]{-64}$
6. $\sqrt[3]{-8}$
7. $\sqrt{\dfrac{1}{9}}$
8. $\sqrt[3]{\dfrac{27}{8}}$
9. $\sqrt{25x^4}$
10. $\sqrt[3]{64x^6}$
11. $\sqrt[3]{8(1 + x)^3}$
12. $\sqrt{4(x + 4)^2}$

In Problems 13–38, simplify each expression. Assume that all radicals containing variables are defined.

13. $\sqrt{8}$
14. $\sqrt{27}$
15. $\sqrt{50}$
16. $\sqrt{72}$
17. $\sqrt[3]{16}$
18. $\sqrt[3]{24}$
19. $\sqrt[3]{-16}$
20. $-\sqrt[3]{16}$
21. $\sqrt{\dfrac{25x^3}{9x}}$
22. $\sqrt[3]{\dfrac{x}{8x^4}}$
23. $\sqrt[4]{x^{12}y^8}$
24. $\sqrt[5]{x^{10}y^5}$
25. $\sqrt{36x}$
26. $\sqrt{9x^5}$
27. $\sqrt{3x^2}\sqrt{12x}$
28. $\sqrt{5x}\sqrt{20x^3}$
29. $\dfrac{\sqrt{3xy^3}\sqrt{2x^2y}}{\sqrt{6x^3y^4}}$
30. $\dfrac{\sqrt[3]{x^2y}\sqrt[3]{125x^3}}{\sqrt[3]{8x^3y^4}}$
31. $\sqrt{\dfrac{16y^4}{9x^2}}$
32. $\sqrt{\dfrac{9x^4}{16y^6}}$
33. $(\sqrt{5}\sqrt[3]{9})^2$
34. $(\sqrt[3]{3}\sqrt{10})^4$
35. $\sqrt{\dfrac{2x - 3}{2x^4 + 3x^3}}\sqrt{\dfrac{x}{4x^2 - 9}}$
36. $\sqrt[3]{\dfrac{x - 1}{x^2 + 2x + 1}}\sqrt[3]{\dfrac{(x - 1)^2}{x + 1}}$
37. $\sqrt{\dfrac{x - 1}{x + 1}}\sqrt{\dfrac{x^2 + 2x + 1}{x^2 - 1}}$
38. $\sqrt{\dfrac{x^2 + 4}{x(x^2 - 4)}}\sqrt{\dfrac{4x^2}{x^4 - 16}}$

In Problems 39–48, simplify each expression.

39. $3\sqrt{2} + 4\sqrt{2}$
40. $6\sqrt{5} - 4\sqrt{5}$
41. $-\sqrt{18} + 2\sqrt{8}$
42. $2\sqrt{12} - 3\sqrt{27}$
43. $5\sqrt[3]{2} - 2\sqrt[3]{54}$
44. $9\sqrt[3]{24} - \sqrt[3]{81}$
45. $\sqrt{8x^3} - 3\sqrt{50x}, \quad x \geq 0$
46. $3x\sqrt{9y} + 4\sqrt{25y}, \quad x \geq 0, y \geq 0$
47. $\sqrt[3]{16x^4y} - 3x\sqrt[3]{2xy} + 5\sqrt[3]{-2xy^4}$
48. $8xy - \sqrt{25x^2y^2} + \sqrt[3]{8x^3y^3}, \quad x \geq 0, y \geq 0$

In Problems 49–62, perform the indicated operation and simplify the results.

49. $(3\sqrt{6})(4\sqrt{3})$
50. $(5\sqrt{8})(-3\sqrt{6})$
51. $\sqrt{3}(\sqrt{3} - 4)$
52. $\sqrt{5}(\sqrt{5} + 6)$
53. $3\sqrt{7}(2\sqrt{7} + 3)$
54. $(2\sqrt{6} + 3)(3\sqrt{6})$
55. $(\sqrt{2} - 1)^2, \quad x \geq 0$
56. $(\sqrt{3} + \sqrt{5})^2$
57. $(\sqrt[3]{2} - 1)^3$
58. $(\sqrt[3]{4} + 2)^3$
59. $(2\sqrt{x} - 3)(2\sqrt{x} + 5)$
60. $(4\sqrt{x} - 3)(\sqrt{x} + 3)$
61. $\sqrt{1 - x^2} - \dfrac{1}{\sqrt{1 - x^2}}, \quad -1 < x < 1$
62. $\sqrt{1 - x^2} + \dfrac{x^2}{\sqrt{1 - x^2}}, \quad -1 < x < 1$

In Problems 63–78, rationalize the denominator of each expression.

63. $\dfrac{2}{\sqrt{5}}$ **64.** $\dfrac{\sqrt{3}}{\sqrt{5}}$ **65.** $\dfrac{8}{\sqrt{6}}$ **66.** $\dfrac{5}{\sqrt{10}}$

67. $\dfrac{1}{\sqrt{x}}, \quad x > 0$ **68.** $\dfrac{x}{\sqrt{x^2+4}}$ **69.** $\dfrac{3}{5+\sqrt{2}}$ **70.** $\dfrac{2}{\sqrt{7}-2}$

71. $\dfrac{3}{4+\sqrt{7}}$ **72.** $\dfrac{10}{4-\sqrt{2}}$ **73.** $\dfrac{\sqrt{5}}{2+3\sqrt{5}}$ **74.** $\dfrac{\sqrt{3}}{2\sqrt{3}+3}$

75. $\dfrac{\sqrt{3}-\sqrt{2}}{\sqrt{3}+\sqrt{2}}$ **76.** $\dfrac{\sqrt{5}+\sqrt{3}}{\sqrt{5}-\sqrt{3}}$ **77.** $\dfrac{1}{\sqrt{x}+2}, \quad x \geq 0$ **78.** $\dfrac{1}{\sqrt{x}-3}, \quad x \geq 0, x \neq 9$

In Problems 79–86, use a calculator to approximate each radical. Round your answer to two decimal places.

79. $\sqrt{2}$ **80.** $\sqrt{7}$ **81.** $\sqrt[3]{4}$ **82.** $\sqrt[3]{-5}$

83. $\dfrac{2+\sqrt{3}}{3-\sqrt{5}}$ **84.** $\dfrac{\sqrt{5}-2}{\sqrt{2}+4}$ **85.** $\dfrac{3\sqrt[3]{5}-\sqrt{2}}{\sqrt{3}}$ **86.** $\dfrac{2\sqrt{3}-\sqrt[3]{4}}{\sqrt{2}}$

87. Calculating the Amount of Gasoline in a Tank An Exxon station stores its gasoline in underground tanks that are right circular cylinders lying on their sides. See the illustration. The volume V of gasoline in the tank (in gallons) is given by the formula

$$V = 40h^2\sqrt{\frac{96}{h} - 0.608}$$

where h is the height of the gasoline (in inches) as measured on a depth stick.

(a) If $h = 12$ inches, how many gallons of gasoline are in the tank?

(b) If $h = 1$ inch, how many gallons of gasoline are in the tank?

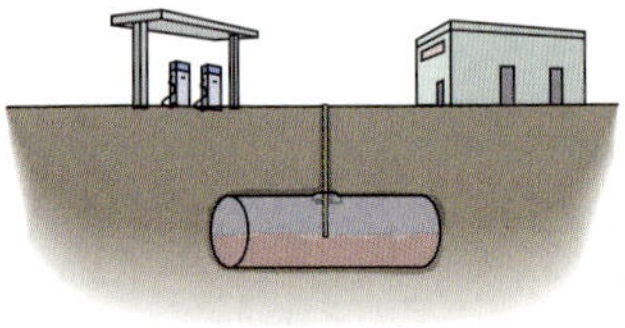

88. Inclined Planes The final velocity v of an object in feet per second (ft/sec) after it slides down a frictionless inclined plane of height h feet is

$$v = \sqrt{64h + v_0^2}$$

where v_0 is the initial velocity (in ft/sec) of the object.

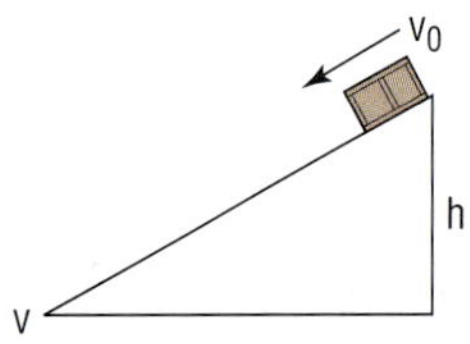

(a) What is the final velocity v of an object that slides down a frictionless inclined plane of height 4 feet? Assume that the initial velocity is 0.

(b) What is the final velocity v of an object that slides down a frictionless inclined plane of height 16 feet? Assume that the initial velocity is 0.

(c) What is the final velocity v of an object that slides down a frictionless inclined plane of height 2 feet with an initial velocity of 4 ft/sec?

In Problems 89–92, use the following information.

Period of a Pendulum The period T, in seconds, of a pendulum of length l, in feet, may be approximated using the formula

$$T = 2\pi\sqrt{l/32}$$

In the following problems, express your answer both as a square root and as a decimal.

89. Find the period T of a pendulum whose length is 64 feet.

90. Find the period T of a pendulum whose length is 16 feet.

91. Find the period T of a pendulum whose length is 8 inches.

92. Find the period T of a pendulum whose length is 4 inches.

93. Give an example to show that $\sqrt{a^2}$ is not equal to a. Use it to explain why $\sqrt{a^2} = |a|$.

9 RATIONAL EXPONENTS

1 Evaluate Expressions with Fractional Exponents

2 Simplify Radicals Using Rational Exponents

Our purpose in this section is to give a definition for "a raised to the power m/n," where a is a real number and m/n is a rational number. However, we want the definition to obey the laws of exponents stated for integer exponents in Appendix Section 4. For example, if the law of exponents $(a^r)^s = a^{rs}$ is to hold, then it must be true that

$$(3^{1/2})^2 = 3^{\frac{1}{2}\cdot 2} = 3^1 = 3$$

That is, $a^{1/n}$ is a number that, when raised to the power n, is a. But this was the definition we gave of the principal nth root of a in Appendix Section 8. Thus,

$$a^{1/2} = \sqrt{a} \qquad a^{1/3} = \sqrt[3]{a} \qquad a^{1/4} = \sqrt[4]{a}$$

and we can state the following definition:

If a is a real number and $n \geq 2$ is an integer, then

$$a^{1/n} = \sqrt[n]{a} \qquad (1)$$

provided that $\sqrt[n]{a}$ exists.

1 **EXAMPLE 1 Writing Expressions Containing Fractional Exponents as Radicals**

(a) $4^{1/2} = \sqrt{4} = 2$ (b) $8^{1/2} = \sqrt{8} = 2\sqrt{2}$

(c) $(-27)^{1/3} = \sqrt[3]{-27} = -3$ (d) $16^{1/3} = \sqrt[3]{16} = 2\sqrt[3]{2}$

Note that if n is even and $a < 0$ then $\sqrt[n]{a}$ and $a^{1/n}$ do not exist.

We now seek a definition for $a^{m/n}$, where m and n are integers containing no common factors (except 1 and -1) and $n \geq 2$. Again, we want the definition to obey the laws of exponents stated earlier. For example,

$$a^{m/n} = a^{m(1/n)} = (a^m)^{1/n} \quad \text{and} \quad a^{m/n} = a^{(1/n)m} = (a^{1/n})^m$$

If a is a real number and m and n are integers containing no common factors, with $n \geq 2$, then

$$a^{m/n} = \sqrt[n]{a^m} = (\sqrt[n]{a})^m \qquad (2)$$

provided that $\sqrt[n]{a}$ exists.

We have two comments about equation (2):

1. The exponent m/n must be in lowest terms and n must be positive.
2. In simplifying the rational expression $a^{m/n}$, either $\sqrt[n]{a^m}$ or $(\sqrt[n]{a})^m$ may be used, the choice depending on which is easier to simplify. Generally, taking the root first, as in $(\sqrt[n]{a})^m$, is easier.

EXAMPLE 2 Using Equation (2)

(a) $4^{3/2} = (\sqrt{4})^3 = 2^3 = 8$
(b) $(-8)^{4/3} = (\sqrt[3]{-8})^4 = (-2)^4 = 16$
(c) $(32)^{-2/5} = (\sqrt[5]{32})^{-2} = 2^{-2} = \frac{1}{4}$
(d) $4^{6/4} = 4^{3/2} = (\sqrt{4})^3 = 2^3 = 8$

NOW WORK PROBLEM 7.

Based on the definition of $a^{m/n}$, no meaning is given to $a^{m/n}$ if a is a negative real number and n is an even integer.

The definitions in equations (1) and (2) were stated so that the Laws of Exponents would remain true for rational exponents. For convenience, we list again the Law of Exponents.

If a and b are real numbers and r and s are rational numbers, then

Laws of Exponents

$$a^r a^s = a^{r+s} \qquad (a^r)^s = a^{rs} \qquad (ab)^r = a^r \cdot b^r$$
$$a^{-r} = \frac{1}{a^r} \qquad \left(\frac{a}{b}\right)^r = \frac{a^r}{b^r} \qquad \frac{a^r}{a^s} = a^{r-s} = \frac{1}{a^{s-r}}$$

where it is assumed that all expressions used are defined.

2 Rational exponents can sometimes be used to simplify radicals.

EXAMPLE 3 Simplifying Radicals Using Rational Exponents

Simplify each expression.

(a) $(\sqrt[4]{7})^2$ (b) $\sqrt[9]{x^3}$ (c) $\sqrt[3]{4}\sqrt{2}$

Solution (a) $(\sqrt[4]{7})^2 = (7^{1/4})^2 = 7^{\frac{1}{4}\cdot 2} = 7^{\frac{1}{2}} = \sqrt{7}$
(b) $\sqrt[9]{x^3} = x^{3/9} = x^{1/3} = \sqrt[3]{x}$
(c) $\sqrt[3]{4}\sqrt{2} = 4^{1/3} \cdot 2^{1/2} = (2^2)^{1/3} \cdot 2^{1/2} = 2^{2/3} \cdot 2^{1/2} = 2^{7/6} = 2 \cdot 2^{1/6} = 2\sqrt[6]{2}$

NOW WORK PROBLEM 21.

The next example illustrates the use of the Laws of Exponents to simplify.

EXAMPLE 4 Simplifying Expressions Containing Rational Exponents

Simplify each expression. Express your answer so that only positive exponents occur. Assume that the variables are positive.

(a) $(x^{2/3}y)(x^{-2}y)^{1/2}$ (b) $\left(\frac{2x^{1/3}}{y^{2/3}}\right)^{-3}$ (c) $\left(\frac{9x^2y^{1/3}}{x^{1/3}y}\right)^{1/2}$

Solution (a)
$$\begin{aligned}(x^{2/3}y)(x^{-2}y)^{1/2} &= (x^{2/3}y)[(x^{-2})^{1/2}y^{1/2}]\\ &= x^{2/3}yx^{-1}y^{1/2}\\ &= (x^{2/3}\cdot x^{-1})(y\cdot y^{1/2})\\ &= x^{-1/3}y^{3/2}\\ &= \frac{y^{3/2}}{x^{1/3}}\end{aligned}$$

(b)
$$\left(\frac{2x^{1/3}}{y^{2/3}}\right)^{-3} = \left(\frac{y^{2/3}}{2x^{1/3}}\right)^3 = \frac{(y^{2/3})^3}{(2x^{1/3})^3} = \frac{y^2}{2^3(x^{1/3})^3} = \frac{y^2}{8x}$$
↑ Equation (8), page 695

(c)
$$\left(\frac{9x^2y^{1/3}}{x^{1/3}y}\right)^{1/2} = \left(\frac{9x^{2-(1/3)}}{y^{1-(1/3)}}\right)^{1/2} = \left(\frac{9x^{5/3}}{y^{2/3}}\right)^{1/2} = \frac{9^{1/2}(x^{5/3})^{1/2}}{(y^{2/3})^{1/2}} = \frac{3x^{5/6}}{y^{1/3}}$$

NOW WORK PROBLEM 43.

9 EXERCISES

In Problems 1–30, simplify each expression.

1. $8^{2/3}$
2. $4^{3/2}$
3. $(-27)^{2/3}$
4. $(-64)^{2/3}$
5. $4^{-3/2}$
6. $(-8)^{-5/3}$
7. $9^{-3/2}$
8. $25^{-5/2}$
9. $\left(\frac{9}{4}\right)^{3/2}$
10. $\left(\frac{27}{8}\right)^{2/3}$
11. $\left(\frac{4}{9}\right)^{-3/2}$
12. $\left(\frac{8}{27}\right)^{-2/3}$
13. $4^{1.5}$
14. $16^{-1.5}$
15. $\left(\frac{1}{4}\right)^{-1.5}$
16. $\left(\frac{1}{9}\right)^{1.5}$
17. $(\sqrt{3})^6$
18. $(\sqrt[3]{4})^6$
19. $(\sqrt{5})^{-2}$
20. $(\sqrt[4]{3})^{-8}$
21. $3^{1/2}\cdot 3^{3/2}$
22. $5^{1/3}\cdot 5^{4/3}$
23. $\frac{7^{1/3}}{7^{4/3}}$
24. $\frac{6^{5/4}}{6^{1/4}}$
25. $2^{1/3}\cdot 4^{1/3}$
26. $9^{1/3}\cdot 3^{1/3}$
27. $\sqrt[4]{3}\cdot\sqrt[4]{27}$
28. $\sqrt[3]{2}\cdot\sqrt[3]{4}$
29. $(\sqrt[4]{2})^{-4}$
30. $(\sqrt[5]{3})^{-5}$

In Problems 31–48, simplify each expression. Express your answer so that only positive exponents occur. Assume that the variables are positive.

31. $(\sqrt[3]{6})^2$
32. $(\sqrt[4]{5})^3$
33. $\sqrt{2}\sqrt[3]{2}$
34. $\sqrt{5}\sqrt[3]{5}$
35. $\sqrt[8]{x^4}$
36. $\sqrt[6]{x^3}$
37. $\sqrt{x^3}\sqrt[4]{x}$
38. $\sqrt[3]{x^2}\sqrt{x}$
39. $x^{3/2}x^{-1/2}$
40. $x^{5/4}x^{-1/4}$
41. $(x^3y^6)^{2/3}$
42. $(x^4y^8)^{5/4}$
43. $(x^2y)^{1/3}(xy^2)^{2/3}$
44. $(xy)^{1/4}(x^2y^2)^{1/2}$
45. $(16x^2y^{-1/3})^{3/4}$
46. $(4x^{-1}y^{1/3})^{3/2}$
47. $\left(\frac{x^{2/5}y^{-1/5}}{x^{-1/3}}\right)^{15}$
48. $\left(\frac{x^{1/2}}{y^2}\right)^4\left(\frac{y^{1/3}}{x^{-2/3}}\right)^3$

ANSWERS

CHAPTER 1 Graphs

1.1 Exercises

1. **(a)** Quadrant II **(b)** Positive x-axis **(c)** Quadrant III **(d)** Quadrant I **(e)** Negative y-axis **(f)** Quadrant IV

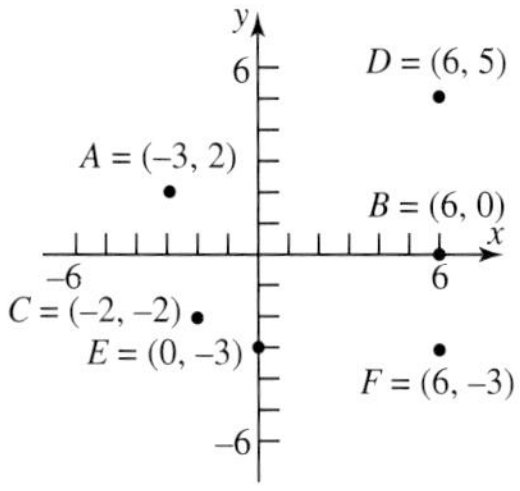

2. **(a)** Quadrant I **(b)** Quadrant III **(c)** Quadrant II **(d)** Quadrant I **(e)** positive y-axis **(f)** negative x-axis

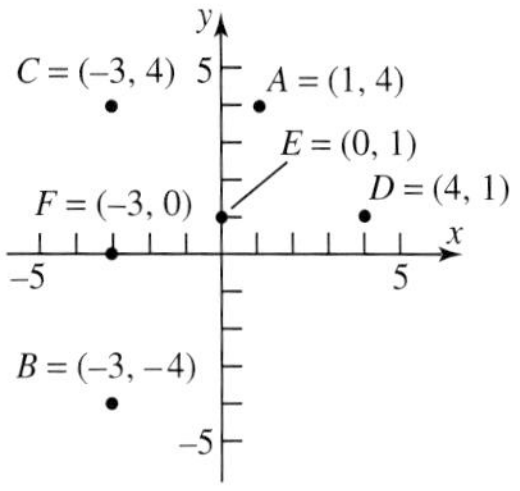

3. The points will be on a vertical line that is 2 units to the right of the y-axis

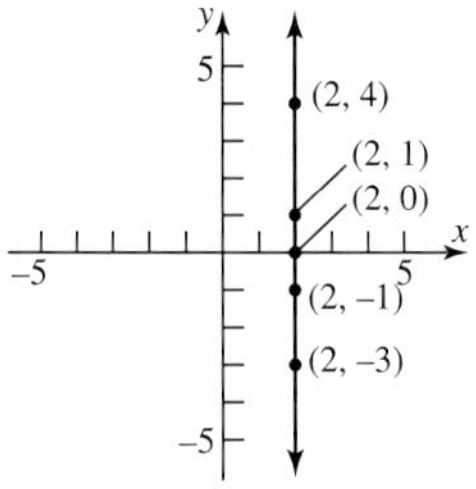

4. The points will be on a horizontal line that is 3 units above the x-axis.

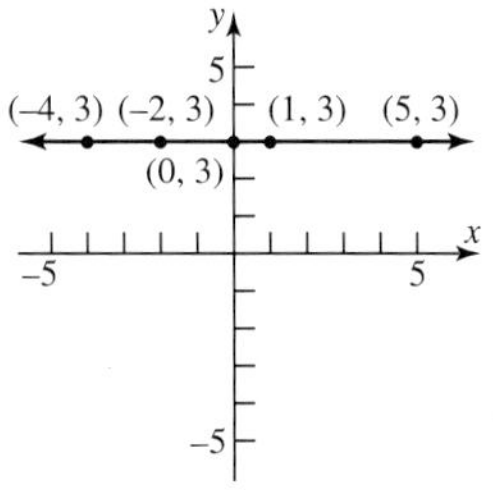

5. $(-1, 4)$ **6.** $(3, 4)$ **7.** $(3, 1)$ **8.** $(-6, -4)$
9. $X\text{min} = -11, X\text{max} = 5, X\text{scl} = 1, Y\text{min} = -3, Y\text{max} = 6, Y\text{scl} = 1$
10. $X\text{min} = -3, X\text{max} = 7, X\text{scl} = 1, Y\text{min} = -4, Y\text{max} = 9, Y\text{scl} = 1$
11. $X\text{min} = -30, X\text{max} = 50, X\text{scl} = 10, Y\text{min} = -90, Y\text{max} = 50, Y\text{scl} = 10$
12. $X\text{min} = -90, X\text{max} = 30, X\text{scl} = 10, Y\text{min} = -50, Y\text{max} = 70, Y\text{scl} = 10$
13. $X\text{min} = -10, X\text{max} = 110, X\text{scl} = 10, Y\text{min} = -10, Y\text{max} = 160, Y\text{scl} = 10$

14. $X\text{min} = -20, X\text{max} = 110, X\text{scl} = 10, Y\text{min} = -10, Y\text{max} = 60, Y\text{scl} = 10$ **15.** $X\text{min} = -6, X\text{max} = 6, X\text{scl} = 2, Y\text{min} = -4, Y\text{max} = 4, Y\text{scl} = 2$ **16.** $X\text{min} = -3, X\text{max} = 3, X\text{scl} = 1, Y\text{min} = -2, Y\text{max} = 2, Y\text{scl} = 1$ **17.** $X\text{min} = -6, X\text{max} = 6, X\text{scl} = 2, Y\text{min} = -1, Y\text{max} = 3, Y\text{scl} = 1$ **18.** $X\text{min} = -9, X\text{max} = 9, X\text{scl} = 3, Y\text{min} = -12, Y\text{max} = 4, Y\text{scl} = 4$ **19.** $X\text{min} = 3, X\text{max} = 9, X\text{scl} = 1, Y\text{min} = 2, Y\text{max} = 10, Y\text{scl} = 2$ **20.** $X\text{min} = -22, X\text{max} = -10, X\text{scl} = 2, Y\text{min} = 4, Y\text{max} = 8, Y\text{scl} = 1$ **21.** $\sqrt{5}$ **22.** $\sqrt{5}$ **23.** $\sqrt{10}$ **24.** $\sqrt{10}$ **25.** $2\sqrt{17}$ **26.** 5 **27.** $\sqrt{85}$ **28.** $\sqrt{29}$ **29.** $\sqrt{53}$ **30.** $5\sqrt{5}$ **31.** 2.625 **32.** 1.92 **33.** $\sqrt{a^2 + b^2}$ **34.** $\sqrt{2}|a|$ **35.** $4\sqrt{10}$ **36.** $\sqrt{149}$ **37.** $2\sqrt{65}$ **38.** $\sqrt{205}$

39. $d(A, B) = \sqrt{13}$
$d(B, C) = \sqrt{13}$
$d(A, C) = \sqrt{26}$
$(\sqrt{13})^2 + (\sqrt{13})^2 = (\sqrt{26})^2$
Area $= \dfrac{13}{2}$ square units

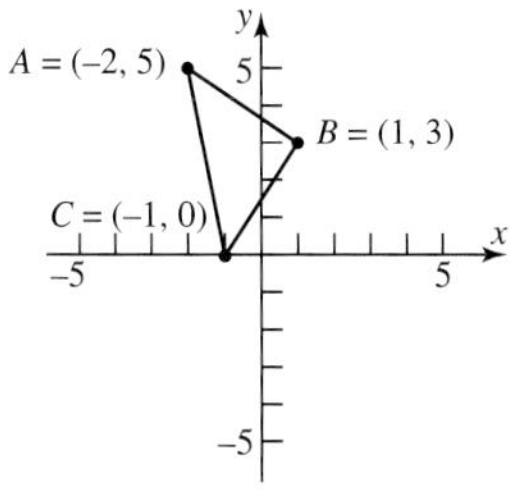

40. $d(A, C) = 20$
$d(A, B) = 10\sqrt{2}$
$d(B, C) = 10\sqrt{2}$
$20^2 = (10\sqrt{2})^2 + (10\sqrt{2})^2$
Area $=$ 100 square units

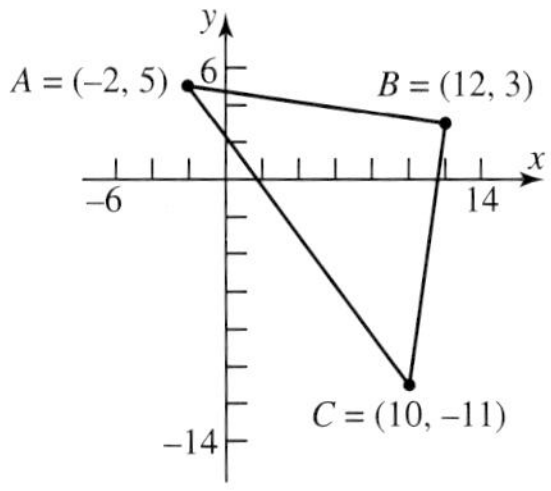

41. $d(A, B) = \sqrt{130}$
$d(B, C) = \sqrt{26}$
$d(A, C) = \sqrt{104}$
$(\sqrt{26})^2 + (\sqrt{104})^2 = (\sqrt{130})^2$
Area $=$ 26 square units

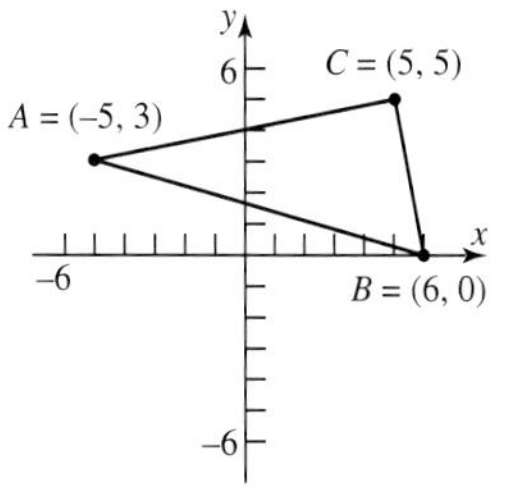

42. $d(A, B) = \sqrt{145}$
$d(A, C) = \sqrt{29}$
$d(B, C) = \sqrt{116}$
$(\sqrt{145})^2 = (\sqrt{29})^2 + (\sqrt{116})^2$
Area $= 29$ square units

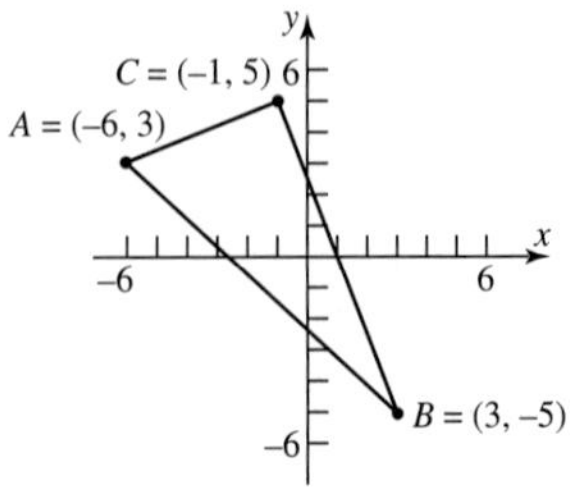

43. $d(A, B) = 4$
$d(A, C) = 5$
$d(B, C) = \sqrt{41}$
$4^2 + 5^2 = 16 + 25 = (\sqrt{41})^2$
Area $= 10$ square units

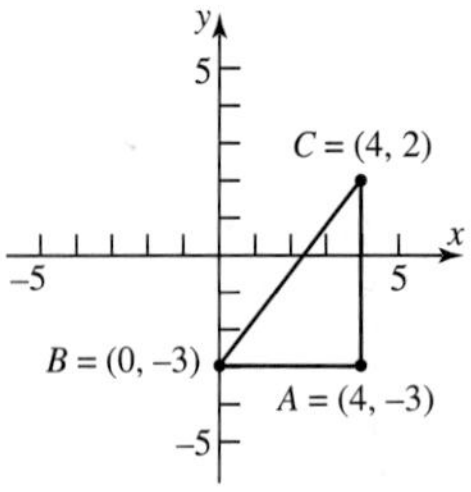

44. $d(A, B) = 4$
$d(A, C) = 2\sqrt{5}$
$d(B, C) = 2$
$(2\sqrt{5})^2 = 4^2 + 2^2$
Area $= 4$ square units

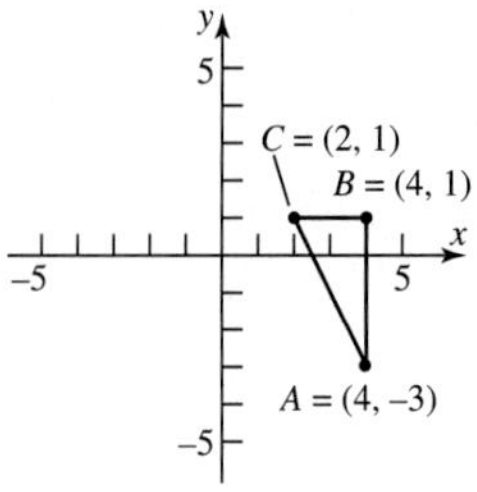

45. $(2, 2); (2, -4)$ **46.** $(13, -3); (-11, -3)$ **47.** $(0, 0); (8, 0)$ **48.** $(0, 1); (0, 7)$ **49.** $(4, -1)$ **50.** $\left(\frac{1}{2}, 2\right)$ **51.** $\left(\frac{3}{2}, 1\right)$ **52.** $\left(3, -\frac{1}{2}\right)$ **53.** $(5, -1)$ **54.** $\left(-1, -\frac{1}{2}\right)$ **55.** $(1.05, 0.7)$ **56.** $(0.45, 1.7)$ **57.** $\left(\frac{a}{2}, \frac{b}{2}\right)$ **58.** $\left(\frac{a}{2}, \frac{a}{2}\right)$ **59.** $\sqrt{17}; 2\sqrt{5}; \sqrt{29}$

60. Two triangles are possible. The third vertex is $(2\sqrt{3}, 2)$ or $(-2\sqrt{3}, 2)$

61. $d(P_1, P_2) = 6; d(P_2, P_3) = 4; d(P_1, P_3) = 2\sqrt{13}$; right triangle

62. $d(P_1, P_2) = \sqrt{53}; d(P_2, P_3) = \sqrt{53}; d(P_1, P_3) = \sqrt{106}$; isosceles right triangle

63. $d(P_1, P_2) = \sqrt{68}; d(P_2, P_3) = \sqrt{34}; d(P_1, P_3) = \sqrt{34}$; isosceles right triangle

64. $d(P_1, P_3) = 5\sqrt{5}; d(P_2, P_3) = 10; d(P_1, P_3) = 5$; right triangle

65. (a)

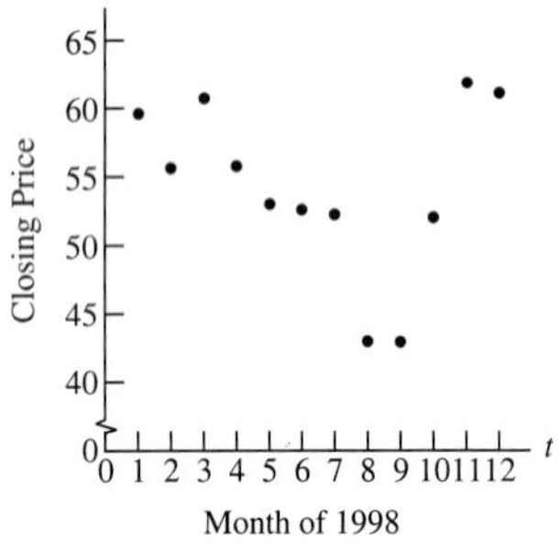

(b)

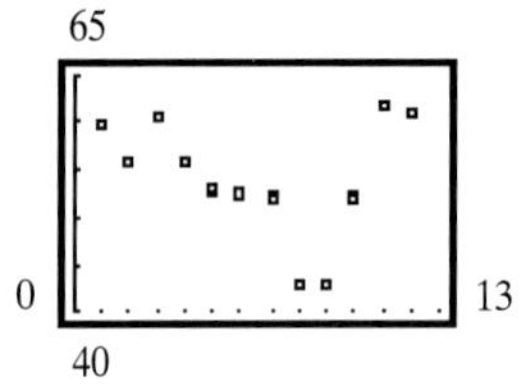

(c) There is no trend apparent. We could say it is U-shaped.

66. (a)

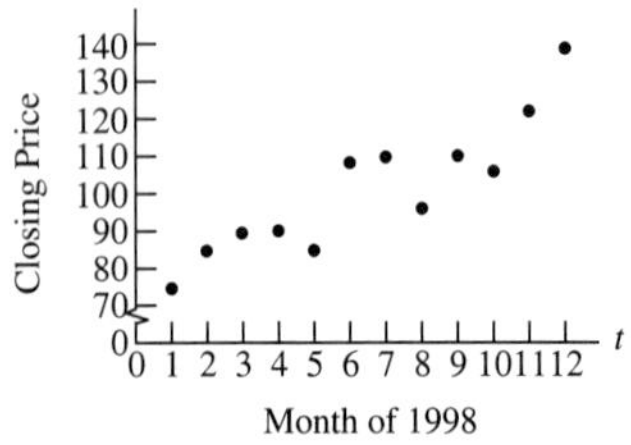

(b)

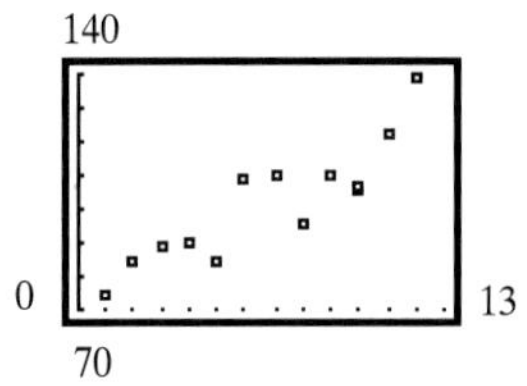

(c) The price of the stock increases with time.

67. (a)

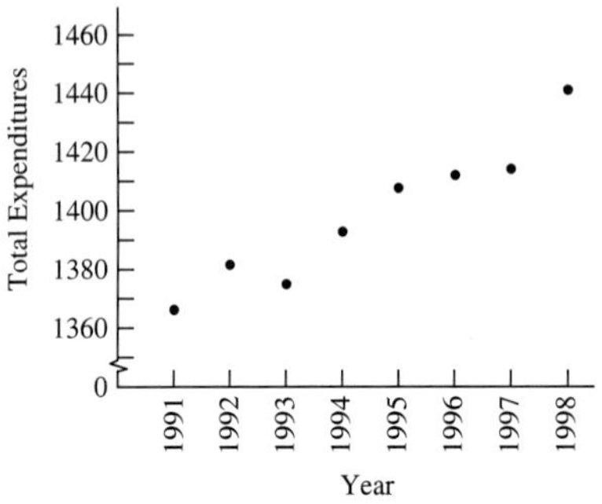

(b)

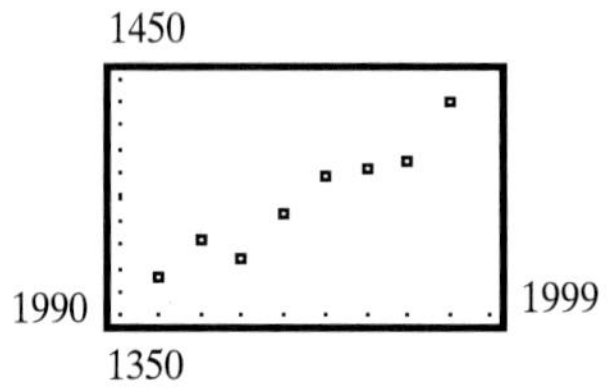

(c) As time passes, total expenditures increase.

68. (a)

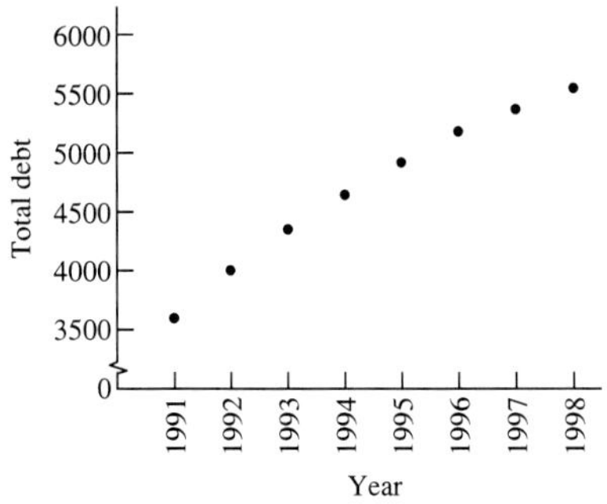

(b)

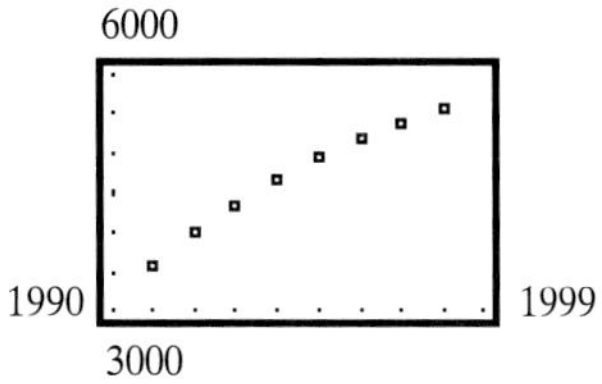

(c) As time passes, total debt increases.

69. (a)

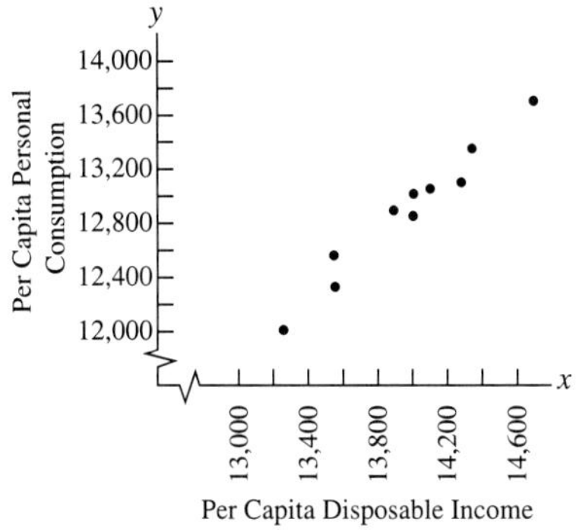

(b)

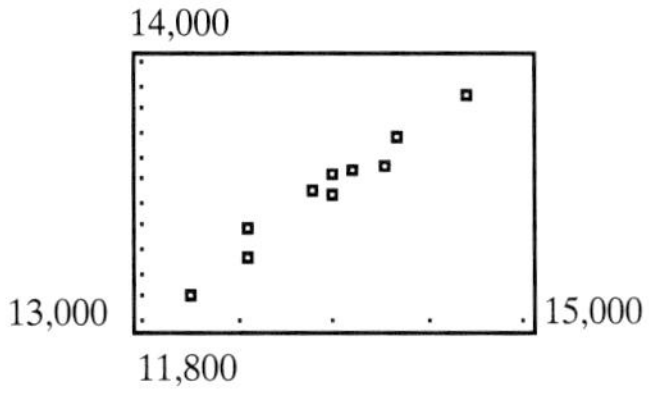

(c) Per capita personal income increases as per capita disposable income increases.

70. (a)

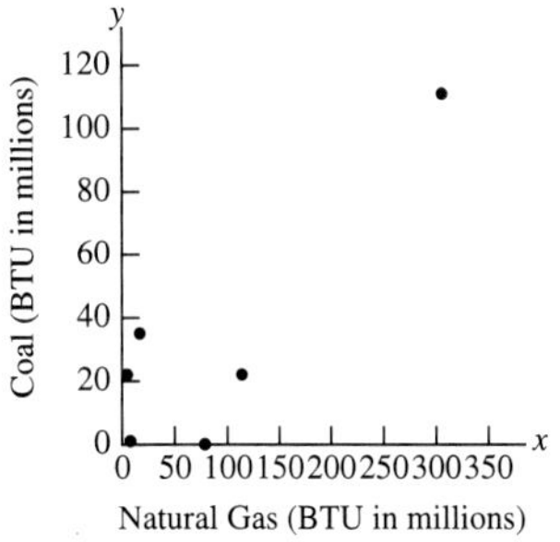

(b)

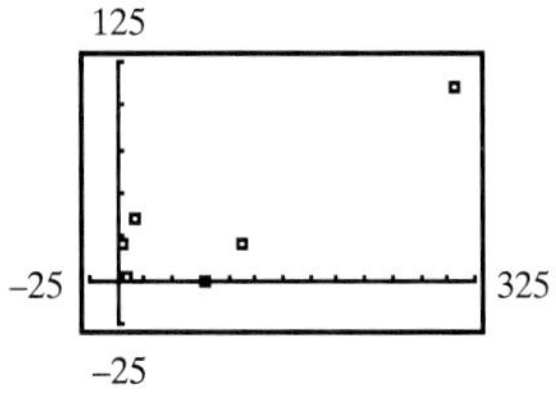

(c) It increases.

71. $90\sqrt{2} \approx 127.28$ ft

72. $60\sqrt{2} \approx 84.9$ ft

73. **(a)** $(90, 0), (90, 90), (0, 90)$ **(b)** 232.4 ft **(c)** 366.2 ft

74. **(a)** $(60, 0), (60, 60), (0, 60)$ **(b)** 126.5 ft **(c)** 272 ft

75. $d = 50t$

76. $\sqrt{10000 + 225t^2}$

1.2 Exercises

1.

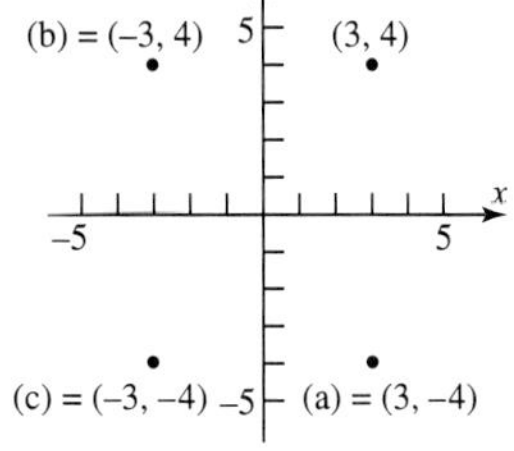

2.

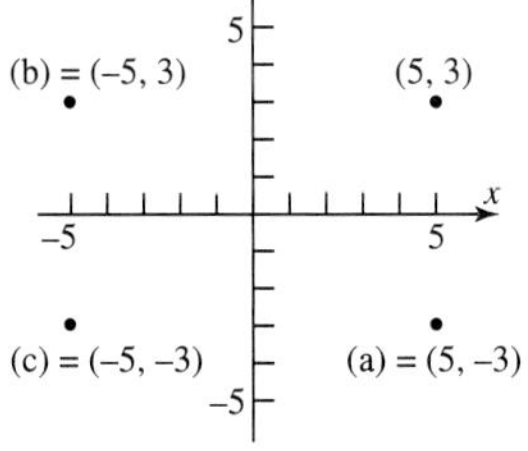

3.

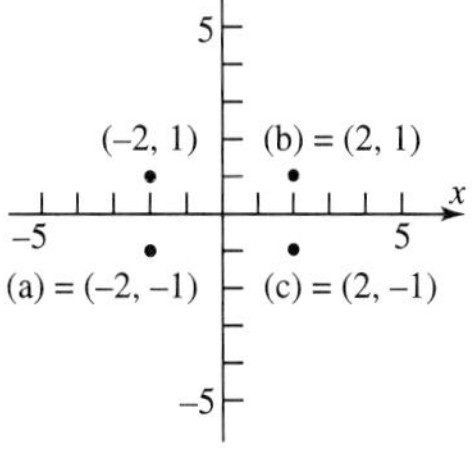

4.

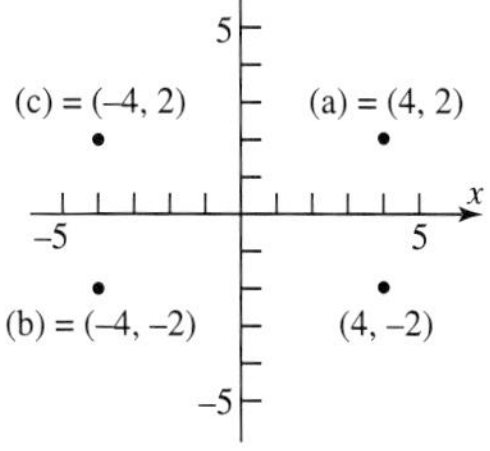

5.

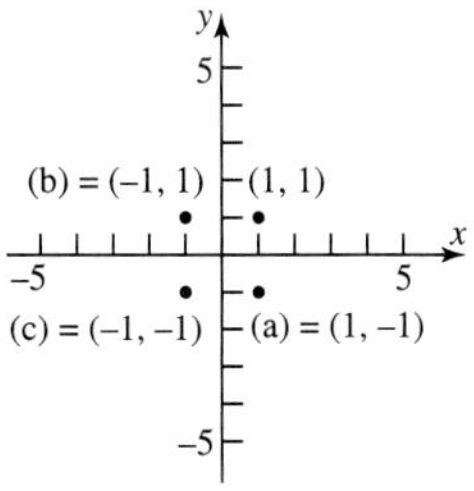

6.

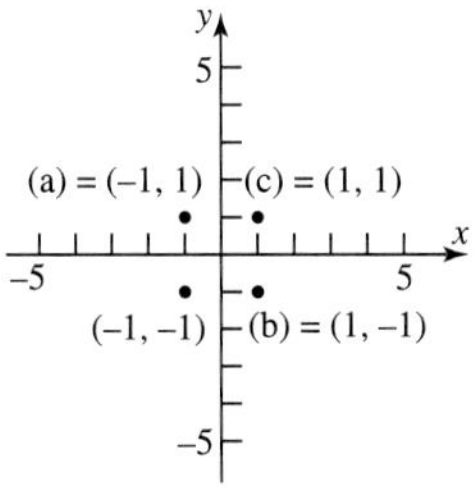

7.

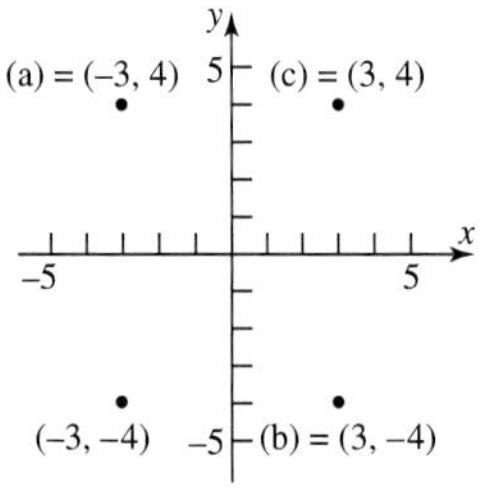

8.

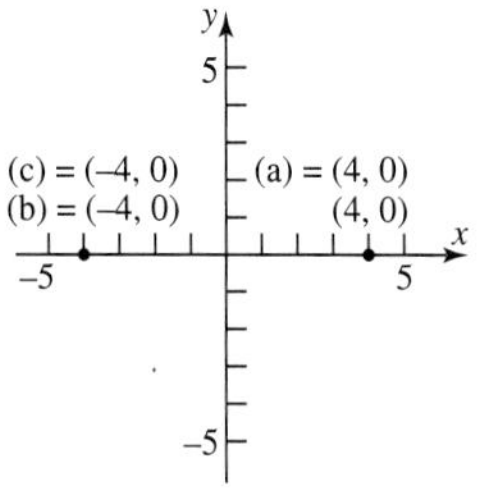

9.

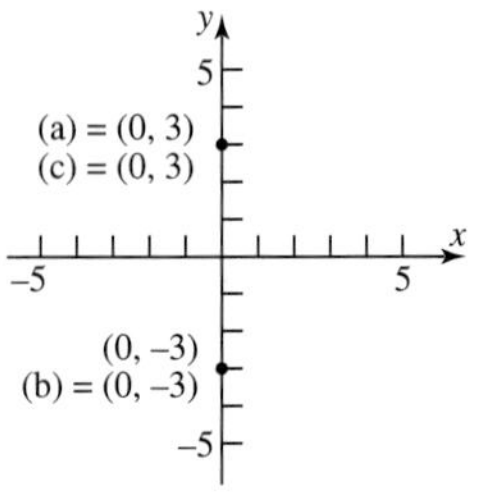

10.

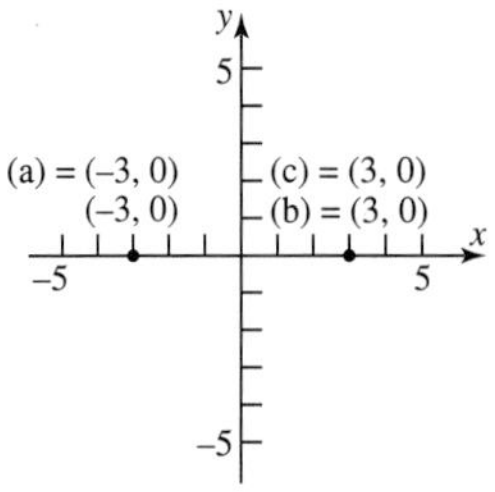

11. (a)

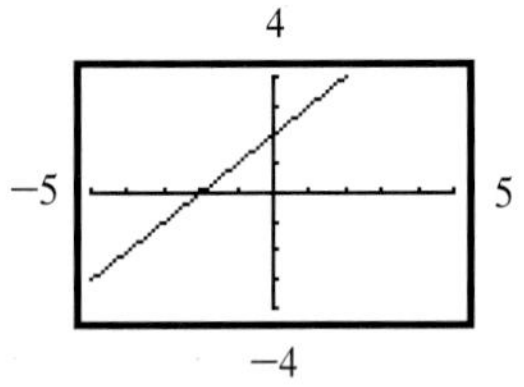

(b)

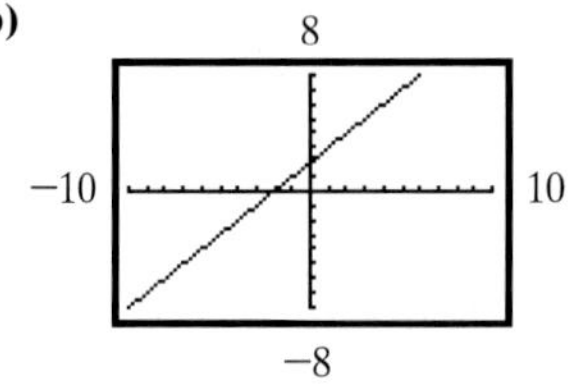

(c)

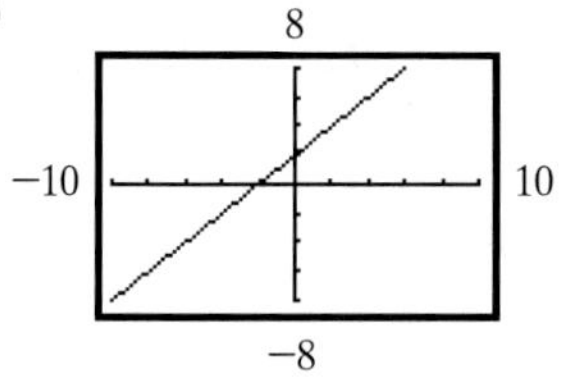

(d)

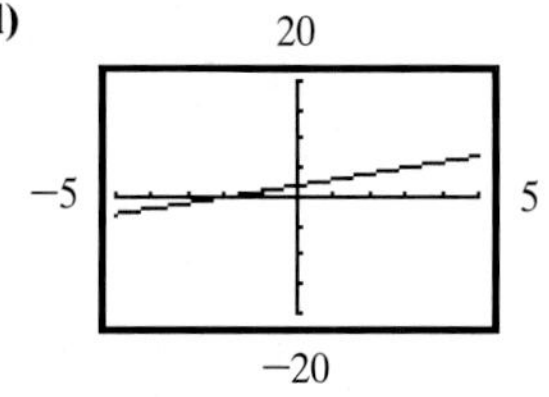

12. (a)

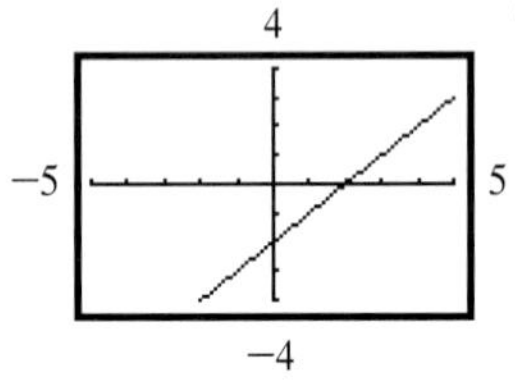

(b)

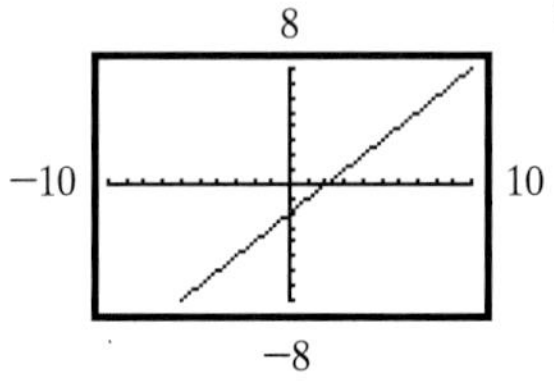

(c)

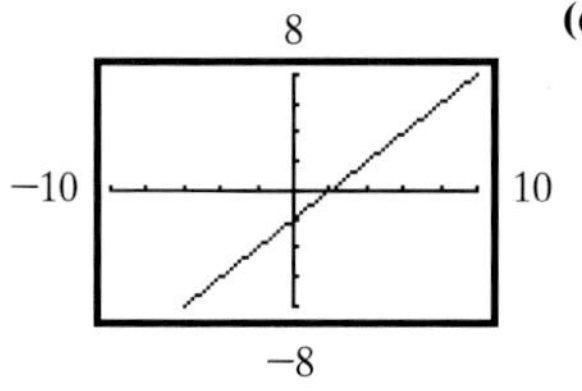

(d)

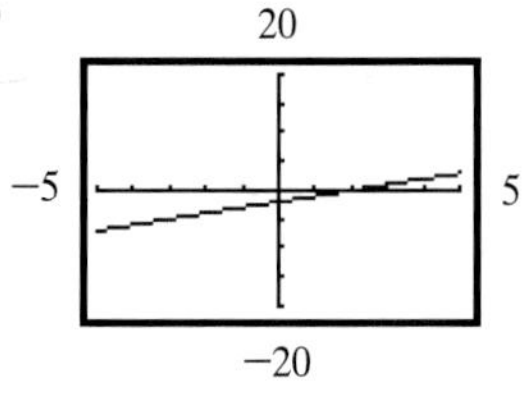

13. (a)

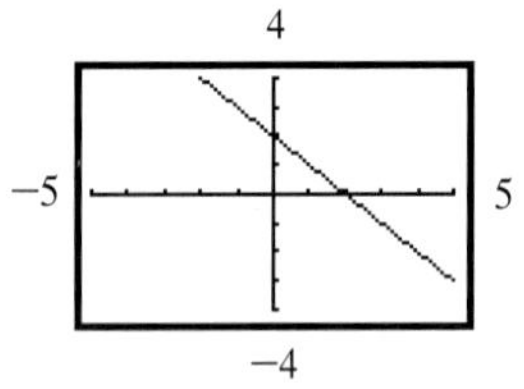

(b)

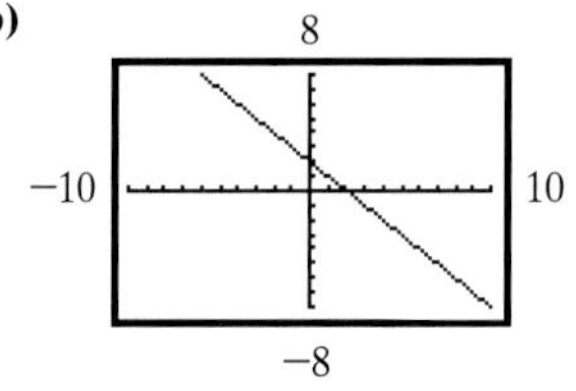

(c)

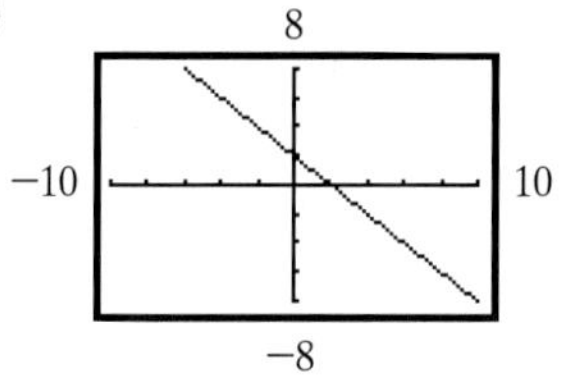

(d)

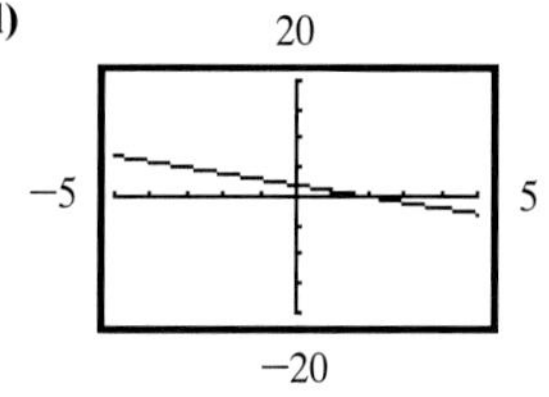

14. (a)

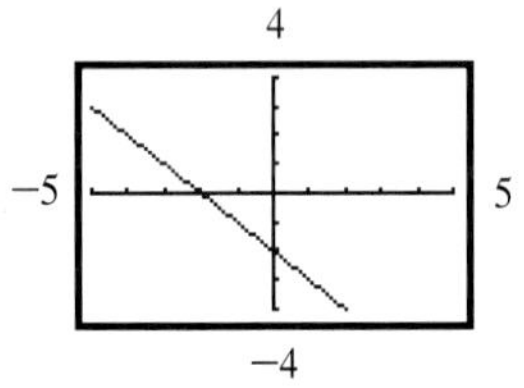

(b)

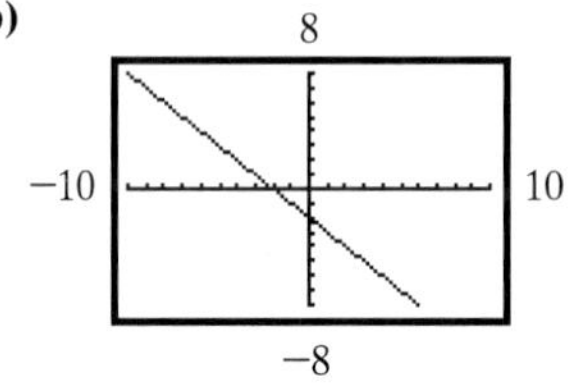

(c)

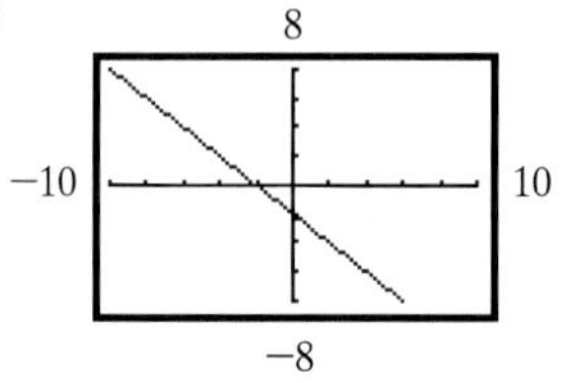

(d)

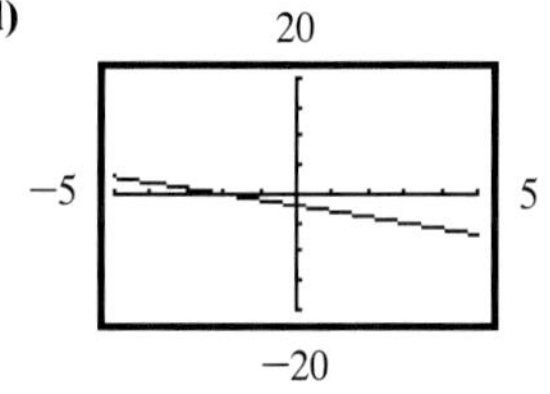

15. (a)

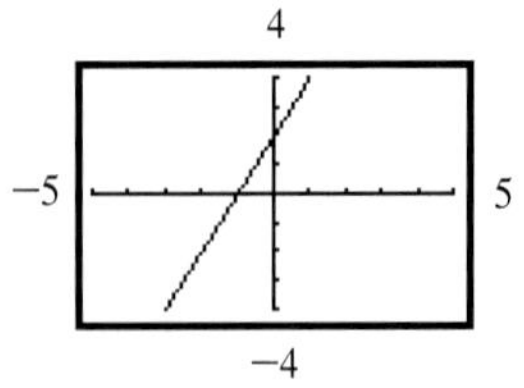

(b)

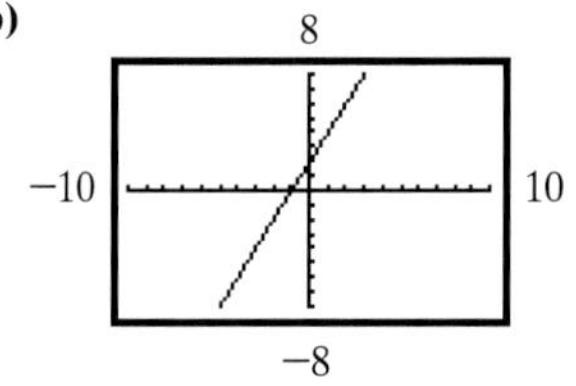

(c)

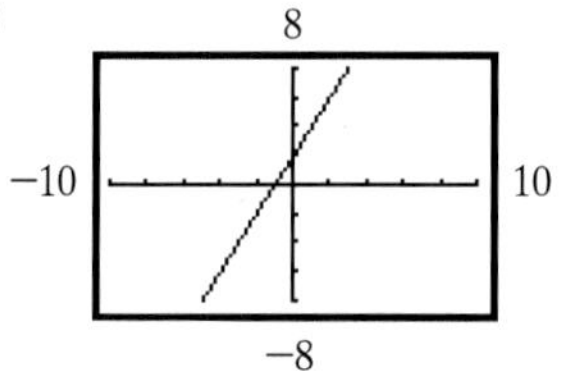

(d)

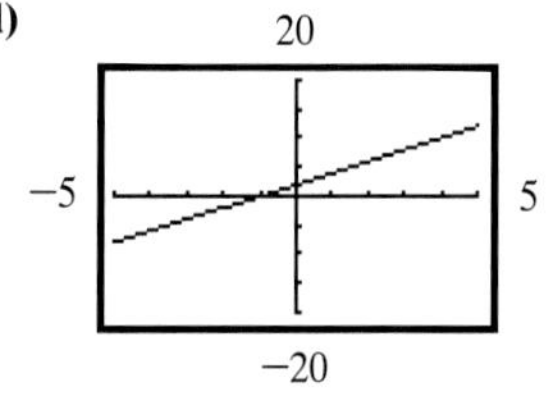

16. (a)

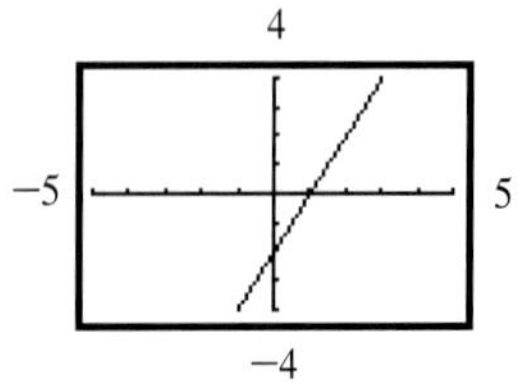

(b)

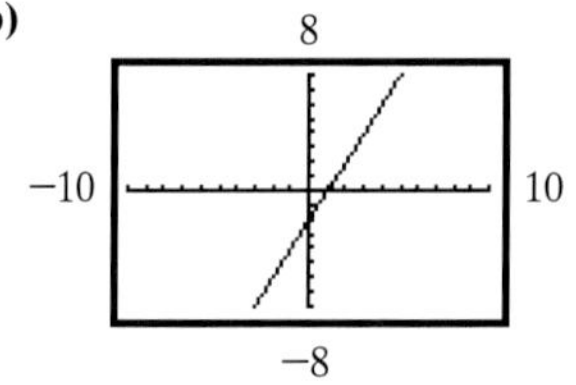

(c)

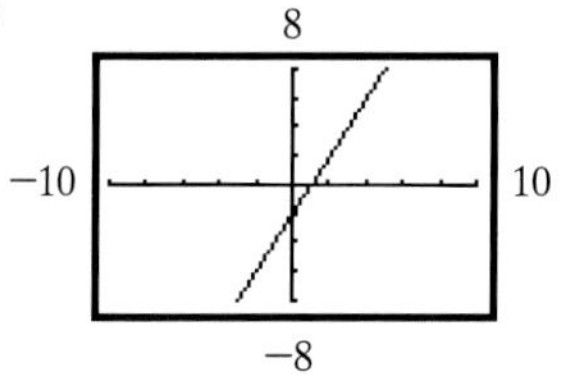

(d)

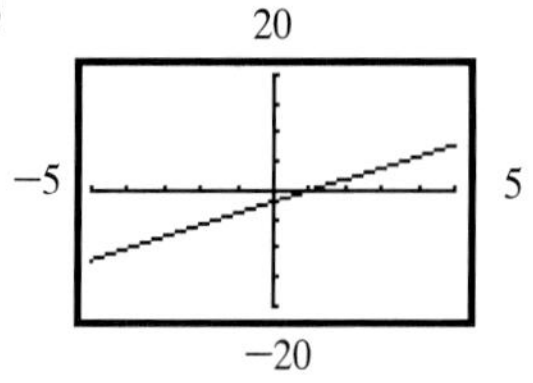

17. **(a)**
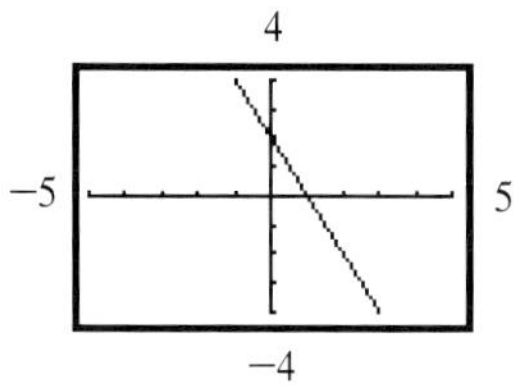

(b)
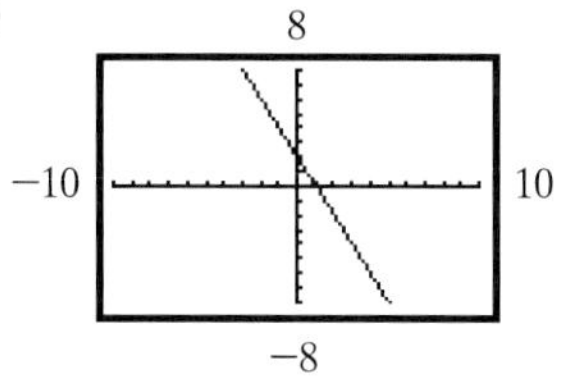

(c)
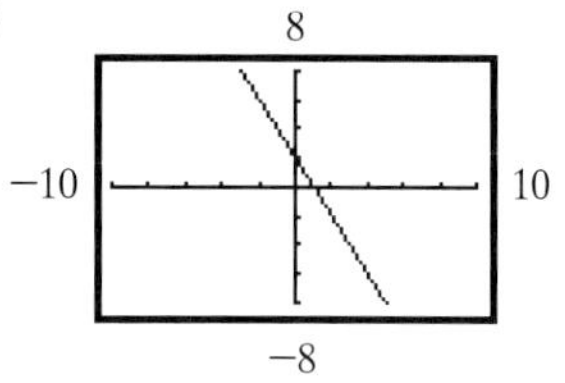

(d)
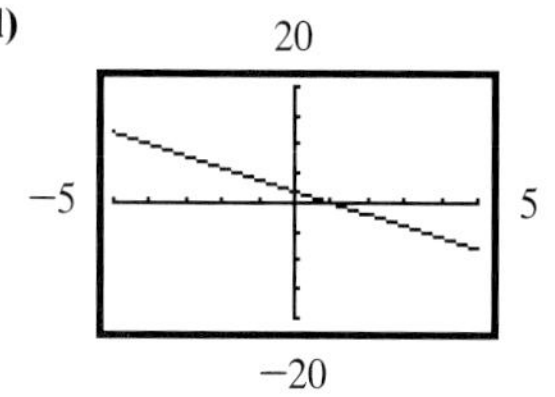

18. **(a)**
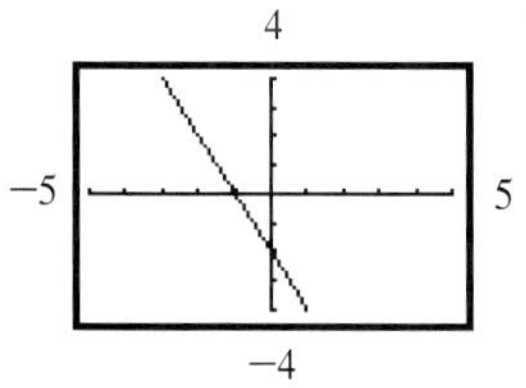

(b)
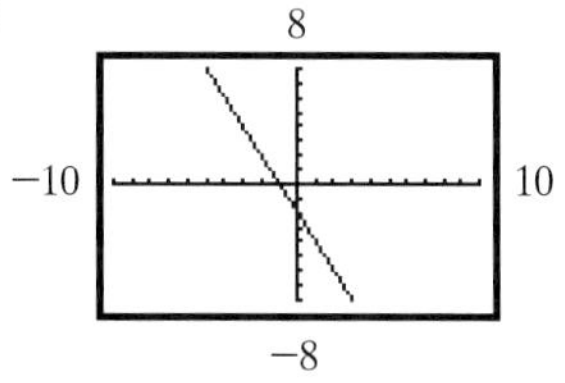

c)
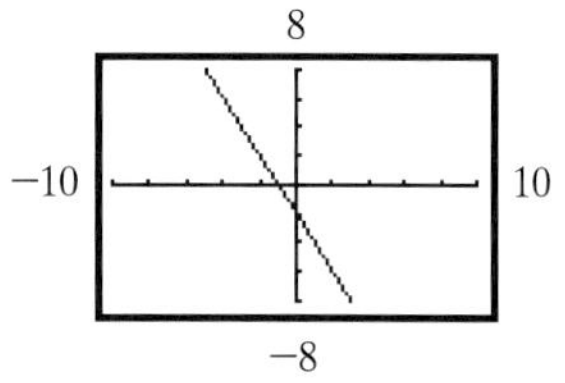

(d)
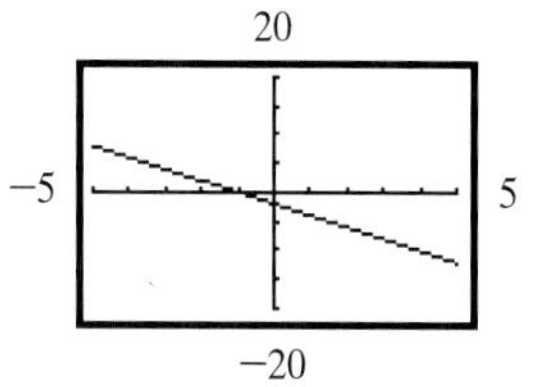

19. **(a)**
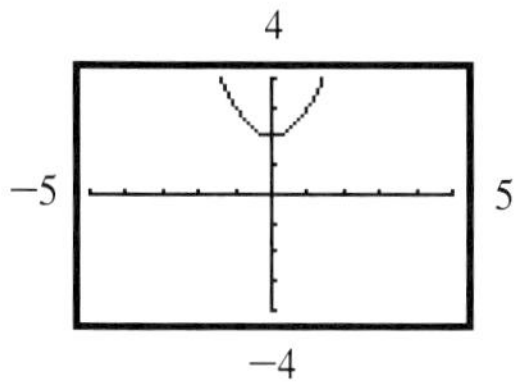

(b)
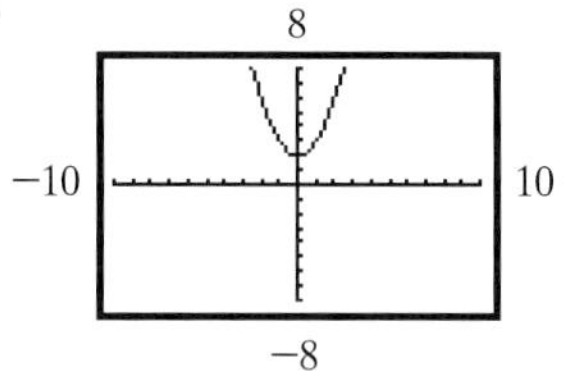

(c)
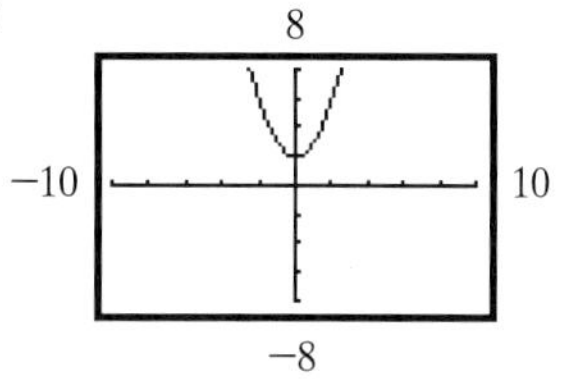

(d)
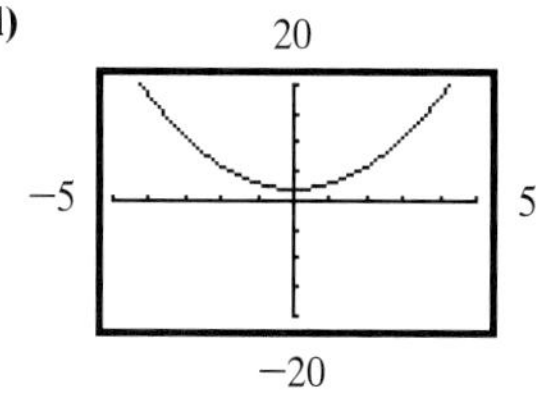

20. **(a)**
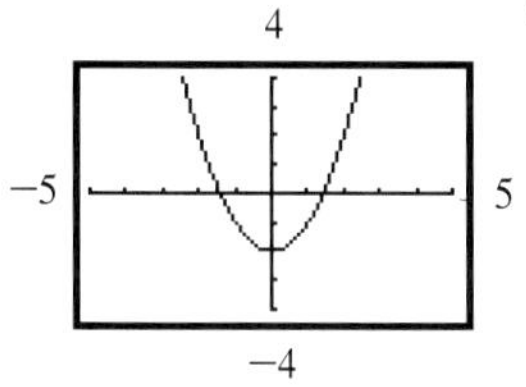

(b)
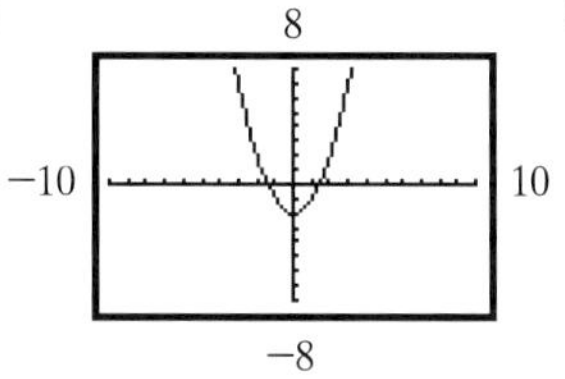

(c)
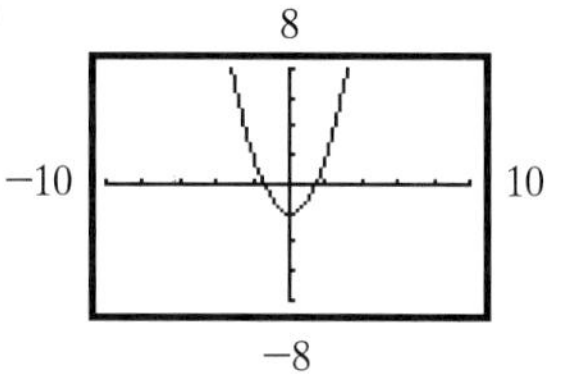

(d)
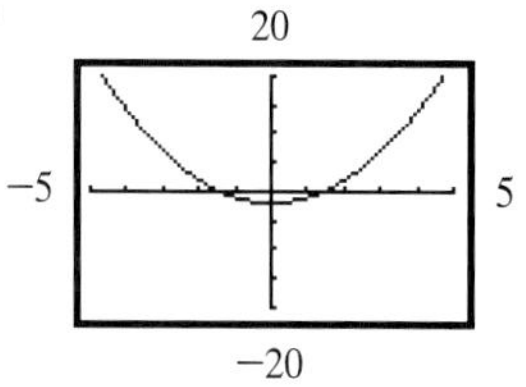

21. **(a)**
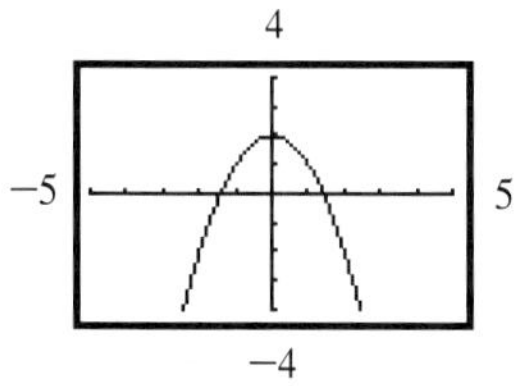

(b)
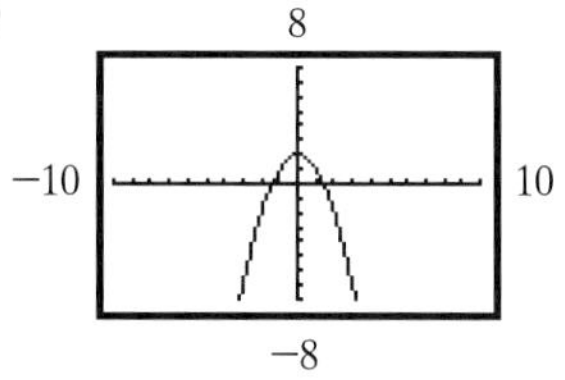

(c)
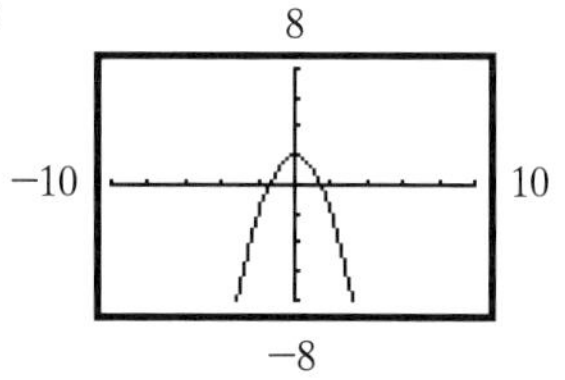

(d)
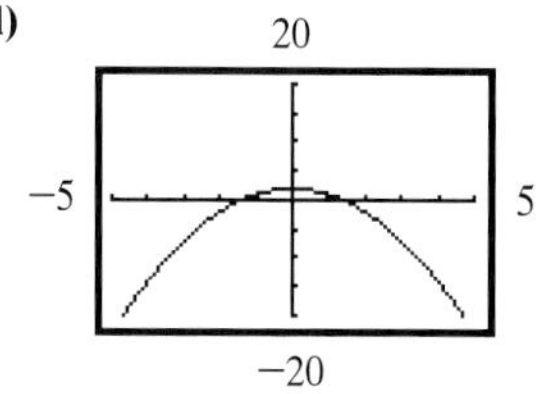

22. **(a)**
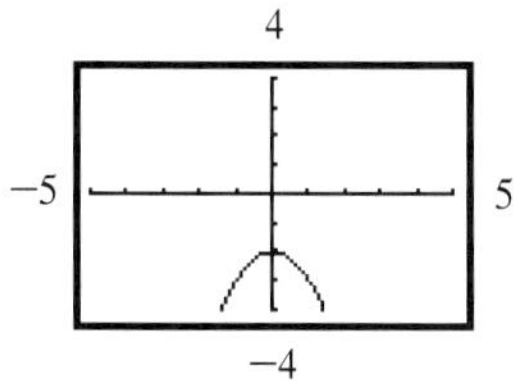

(b)
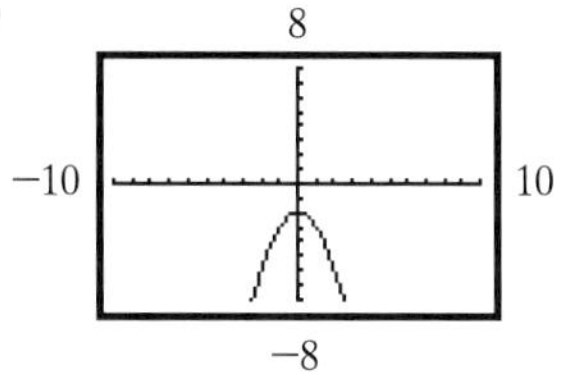

(c)
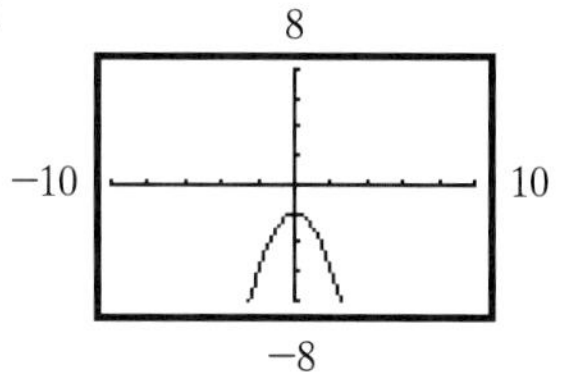

(d)
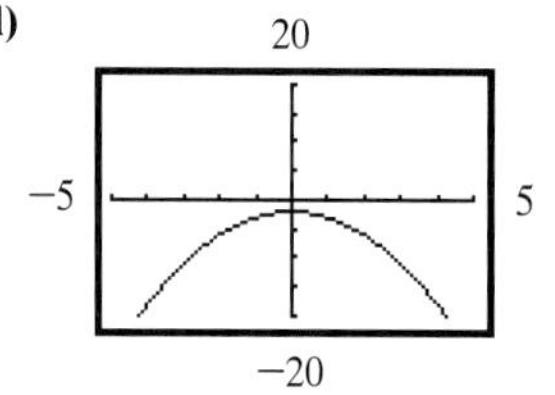

23. **(a)**
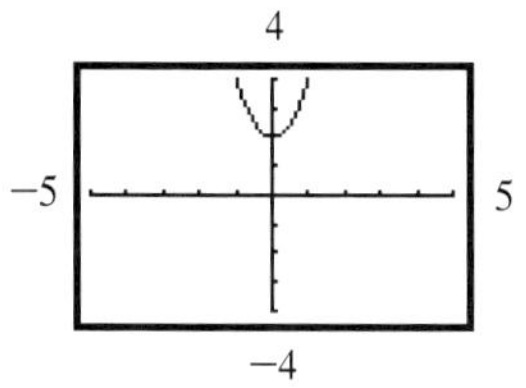

(b)
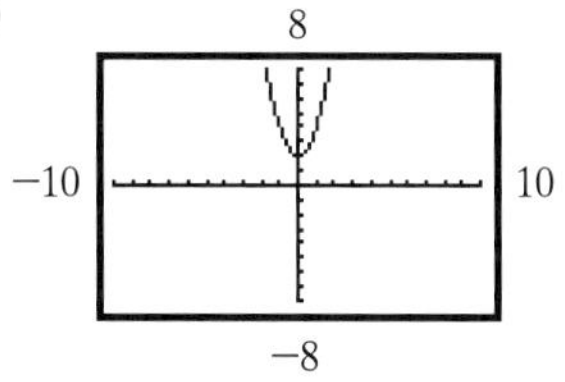

(c)
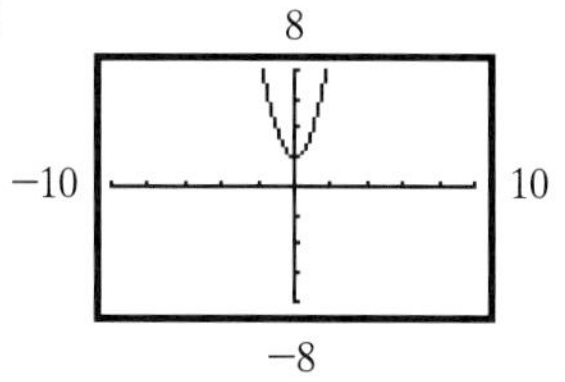

(d)
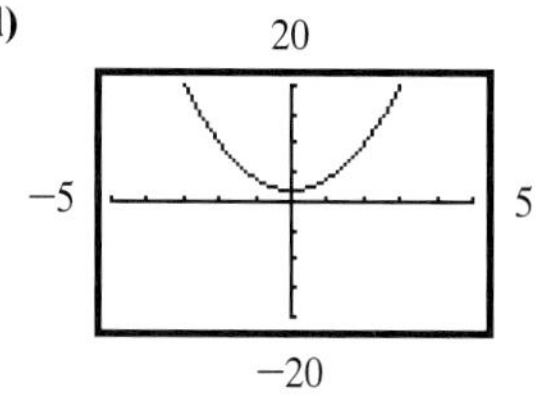

24. **(a)** 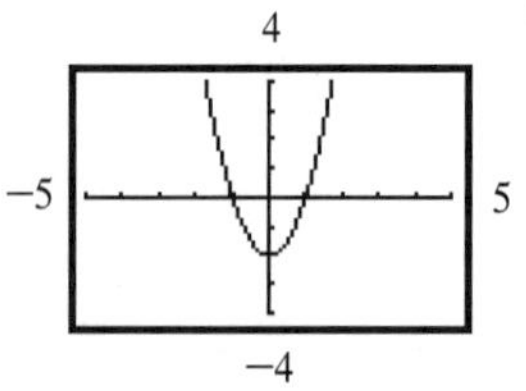

(b) 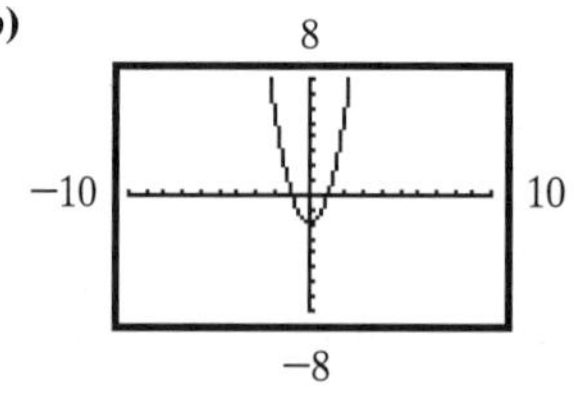

(c) 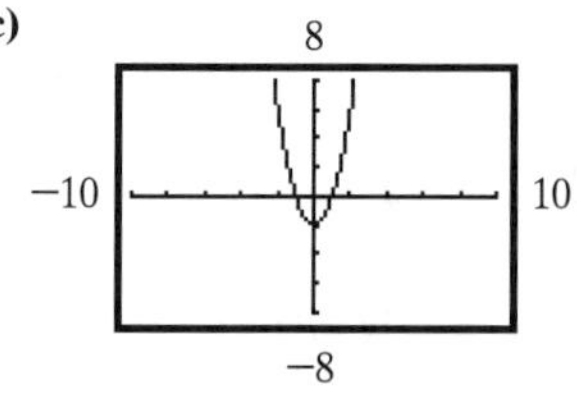

(d) 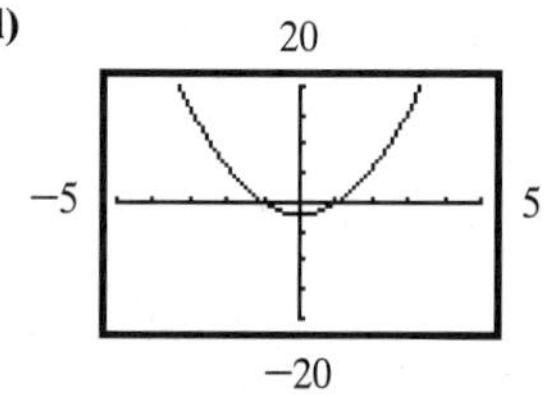

25. **(a)** 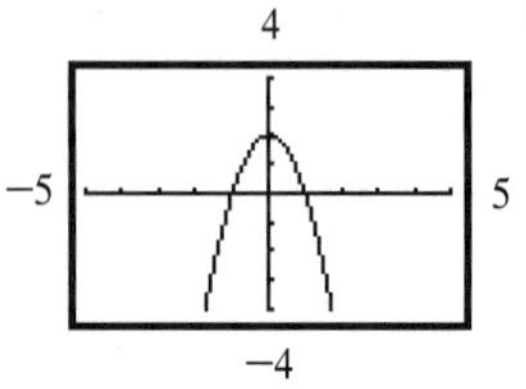

(b) 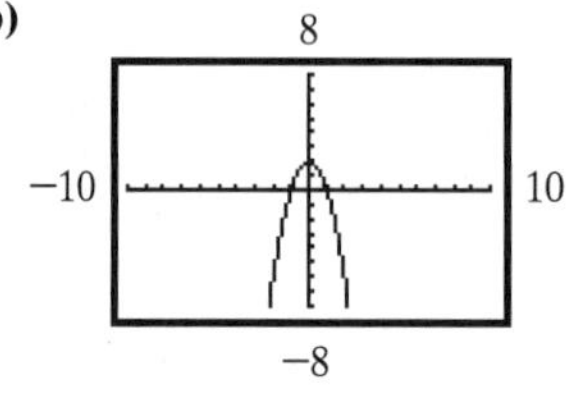

(c) 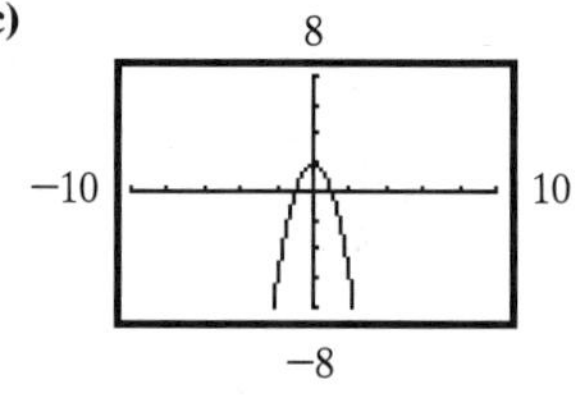

(d) 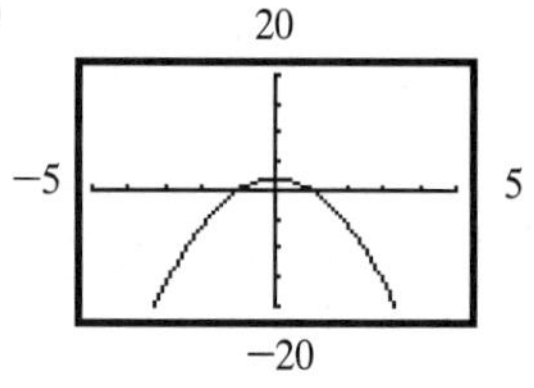

26. **(a)** 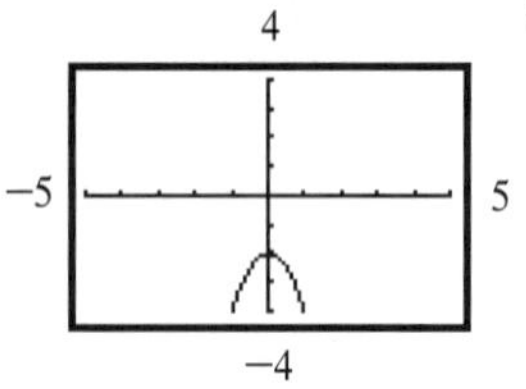

(b) 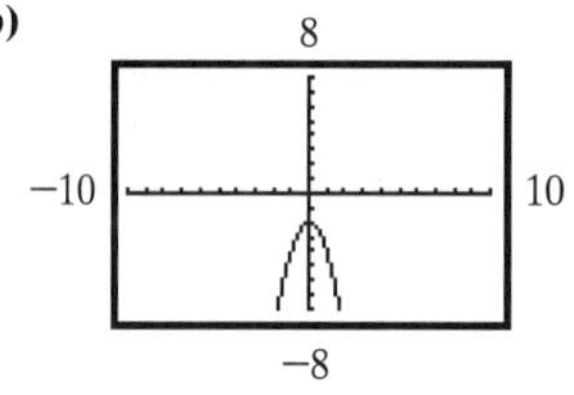

(c) 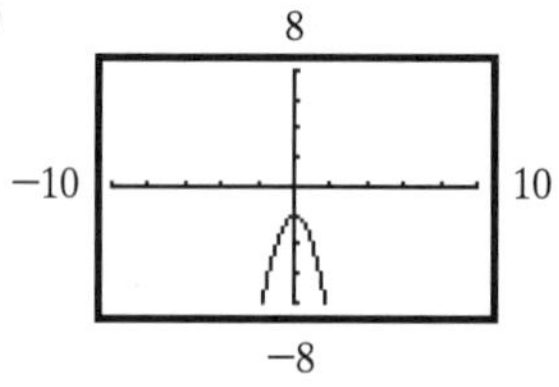

(d) 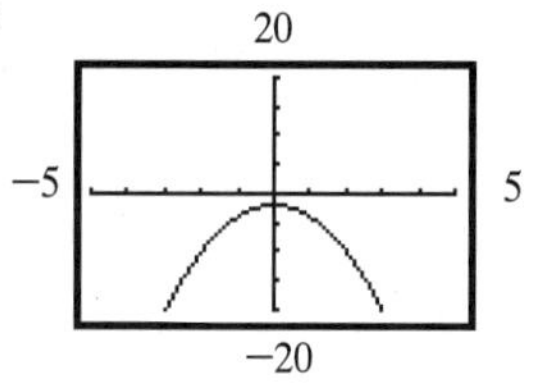

27. **(a)** 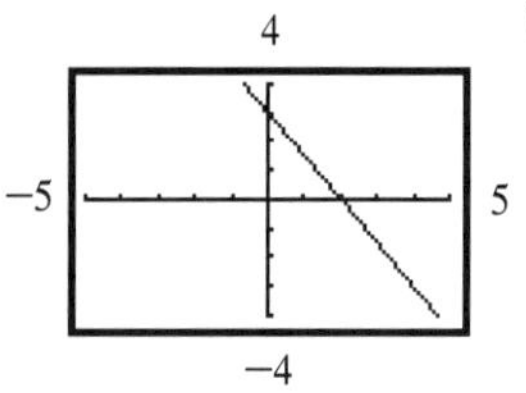

(b) 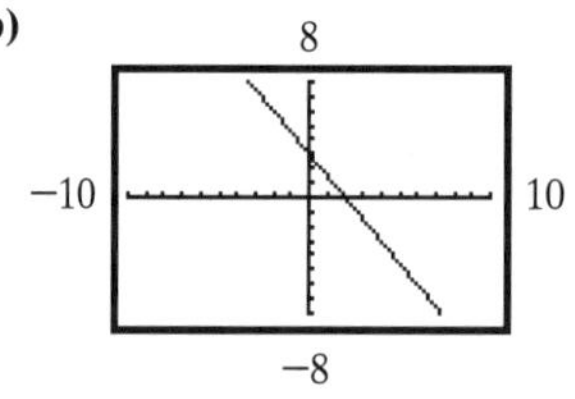

(c) 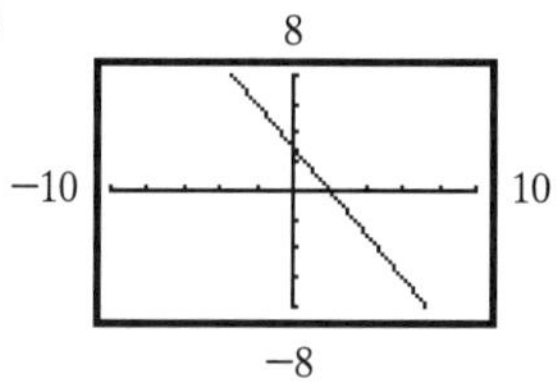

(d) 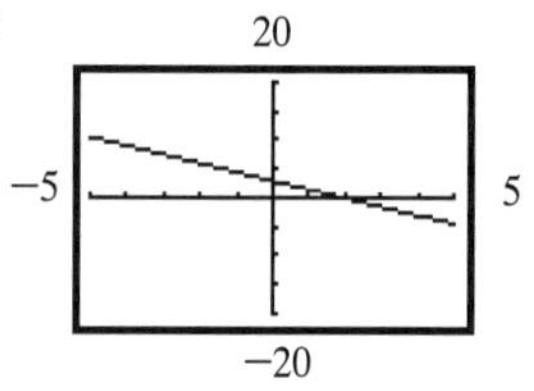

28. **(a)** 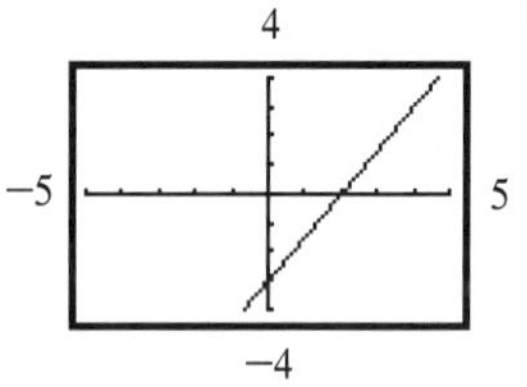

(b) 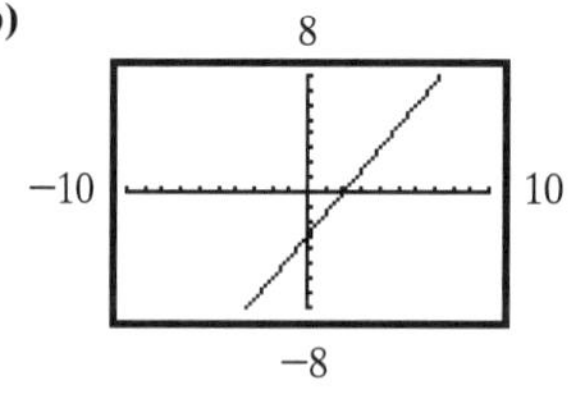

(c) 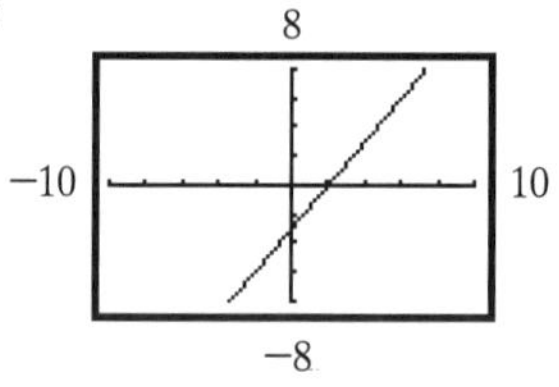

(d) 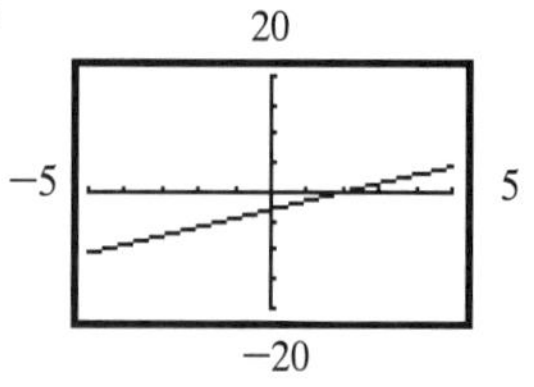

29. **(a)** 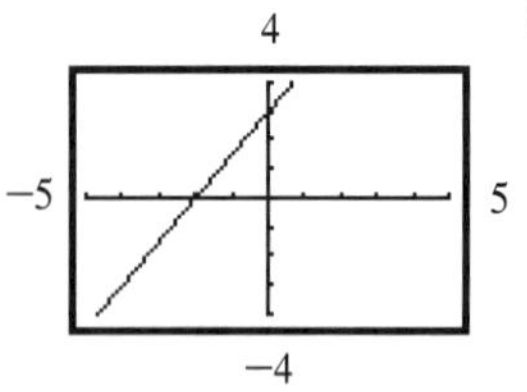

(b) 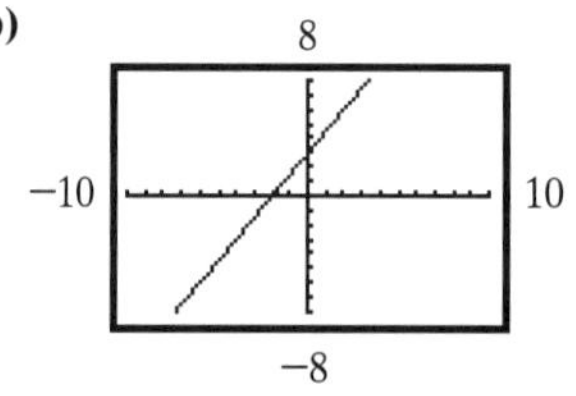

(c) 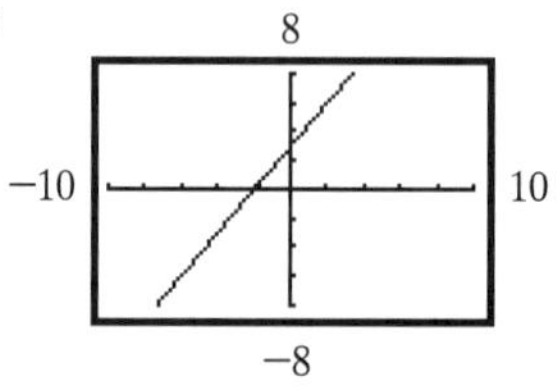

(d) 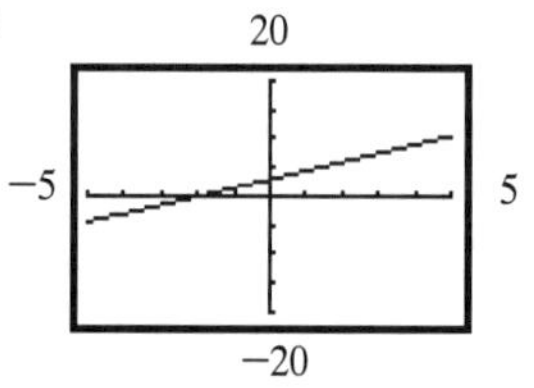

30. **(a)** 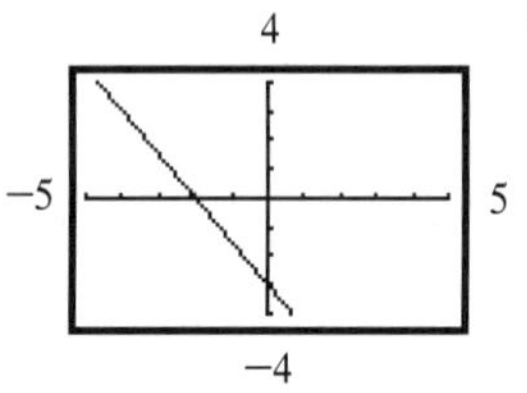

(b) 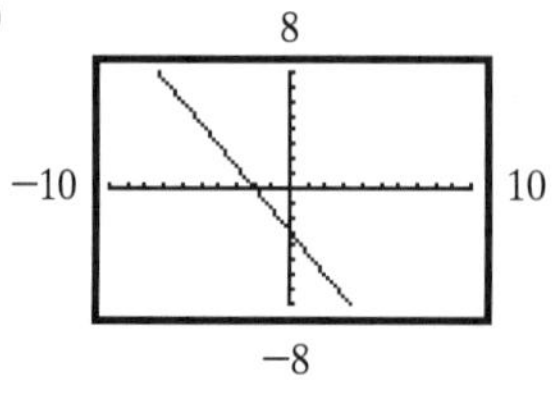

(c) 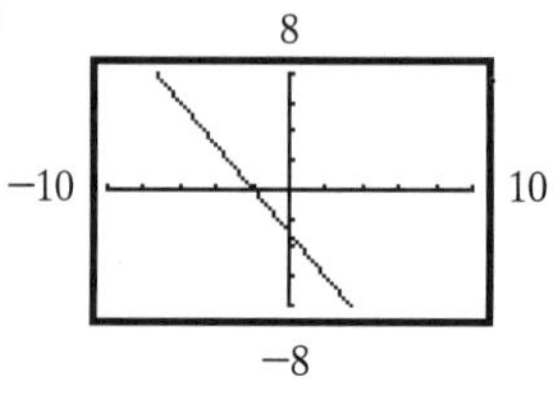

(d) 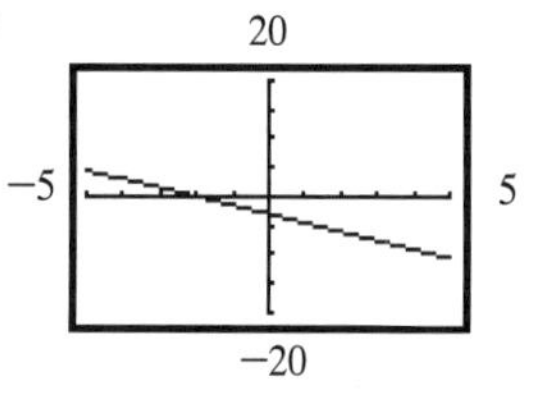

31. **(a)** $(-1, 0), (1, 0)$ **(b)** x-axis, y-axis, origin **32.** **(a)** $(0, 1)$ **(b)** none **33.** **(a)** $\left(-\frac{\pi}{2}, 0\right), (0, 1), \left(\frac{\pi}{2}, 0\right)$ **(b)** y-axis
34. **(a)** $(0, 0)$ **(b)** origin **35.** **(a)** $(0, 0)$ **(b)** x-axis **36.** **(a)** $(-2, 0), (2, 0), (0, -2), (0, 2)$ **(b)** x-axis, y-axis, origin
37. **(a)** $(1, 0)$ **(b)** none **38.** **(a)** $(0, 0)$ **(b)** none **39.** **(a)** $(-3, 0), (0, 2), (3, 0)$ **(b)** y-axis **40.** **(a)** $(-3, 0), (0, 2), (2, 0)$ **(b)** none
41. **(a)** $(x, 0), 0 \leq x < 2$ **(b)** none **42.** **(a)** $(0, 1), (2, 0), (3.25, 0)$ **(b)** none **43.** **(a)** $(-1.5, 0), (0, -2), (1.5, 0)$ **(b)** y-axis
44. **(a)** $(0, 0)$ **(b)** origin **45.** **(a)** none **(b)** origin **46.** **(a)** none **(b)** x-axis **47.** $(0, 0)$ is on the graph.
48. $(0, 0)$ and $(1, -1)$ are on the graph. **49.** $(0, 3)$ is on the graph. **50.** $(0, 1)$ and $(-1, 0)$ are on the graph.

51. $(0, 2)$ and $(\sqrt{2}, \sqrt{2})$ are on the graph. **52.** $(0, 1)$ and $(2, 0)$ are on the graph. **53.** $-\frac{2}{5}$ **54.** 10 **55.** $2a + 3b = 6$ **56.** $b = 5; m = -\frac{5}{2}$

57.

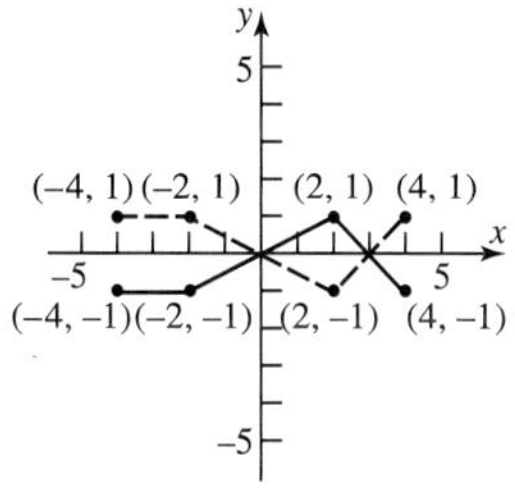

58.

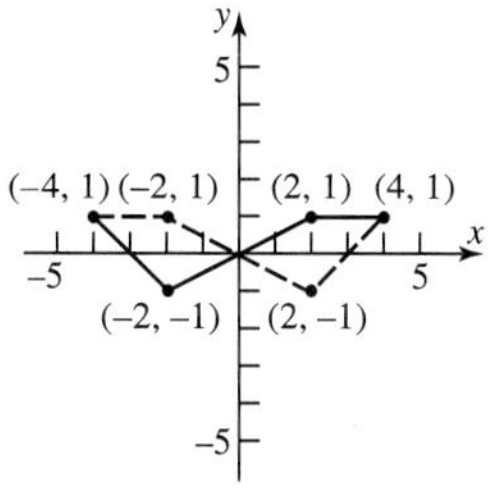

59.

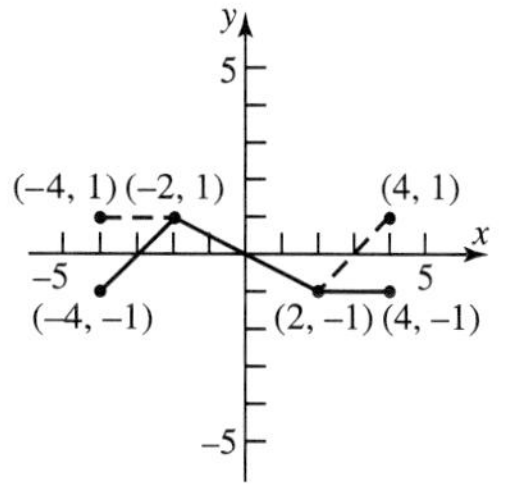

60.

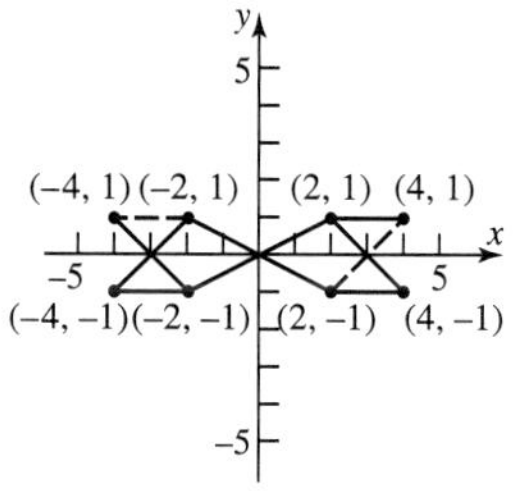

61.

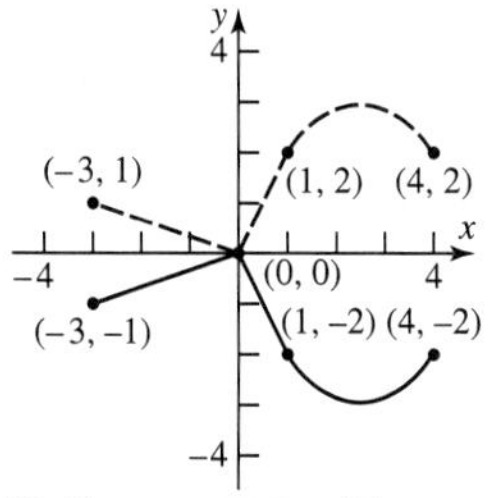

62.

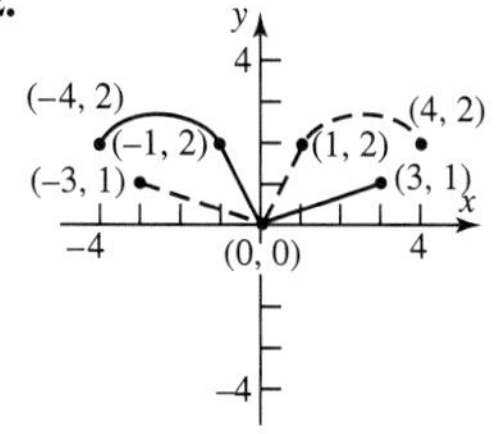

63.

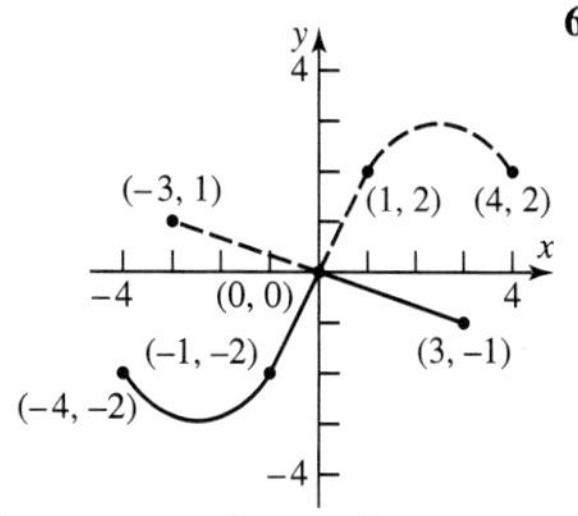

64.

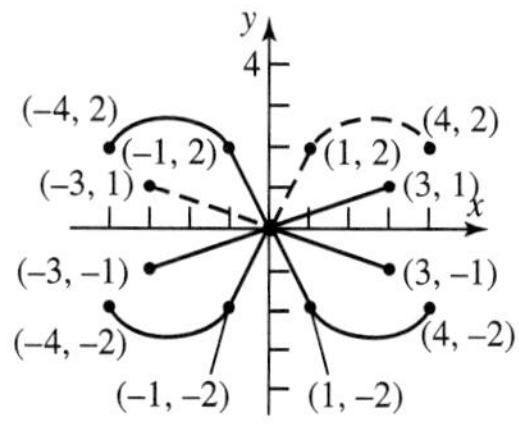

65. $(0, 0)$; symmetric with respect to the y-axis **66.** $(0, 0)$; symmetric with respect to the x-axis
67. $(0, 0)$; symmetric with respect to the origin **68.** $(0, 0)$; symmetric with respect to the origin
69. $(0, 9)$, $(3, 0)$, $(-3, 0)$; symmetric with respect to the y-axis **70.** $(-4, 0)$, $(0, -2)$, $(0, 2)$; symmetric with respect to the x-axis
71. $(-2, 0)$, $(2, 0)$, $(0, -3)$, $(0, 3)$; symmetric with respect to the x-axis, y-axis, and origin
72. $(-1, 0)$, $(1, 0)$, $(0, -2)$, $(0, 2)$; symmetric with respect to the x-axis, y-axis, and origin **73.** $(0, -27)$, $(3, 0)$; no symmetry
74. $(-1, 0)$, $(1, 0)$, $(0, -1)$; symmetric with respect to the y-axis **75.** $(0, -4)$, $(4, 0)$, $(-1, 0)$; no symmetry
76. $(0, 4)$; symmetric with respect to the y-axis **77.** $(0, 0)$; symmetric with respect to the origin
78. $(2, 0)$, $(-2, 0)$; symmetric with respect to the origin

79. **(a)**

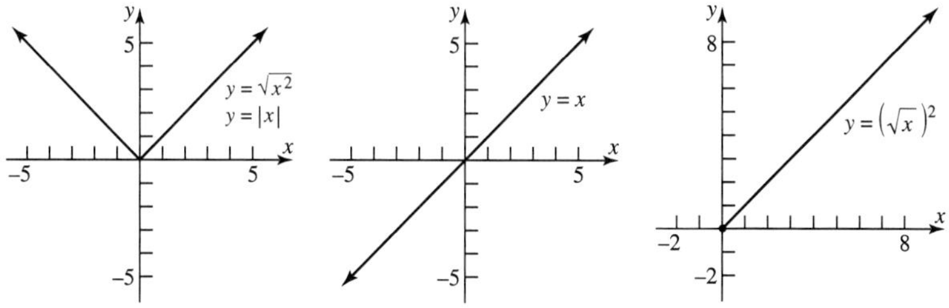

(b) Since $\sqrt{x^2} = |x|$, for all x, the graphs of $y = \sqrt{x^2}$ and $y = |x|$ are the same.

(c) For $y = (\sqrt{x})^2$, the domain of the variable x is $x \geq 0$; for $y = x$, the domain of the variable x is all real numbers. Thus, $(\sqrt{x})^2 = x$ only for $x \geq 0$.

(d) For $y = \sqrt{x^2}$, the range of the variable y is $y \geq 0$; for $y = x$, the range of the variable y is all real numbers. Also, $\sqrt{x^2} = |x|$, which equals x only if $x \geq 0$.

1.3 Exercises

1. 2 **2.** 2 **3.** 3 **4.** −4 **5.** −1 **6.** 1 **7.** $-\frac{4}{3}$ **8.** $\frac{2}{5}$ **9.** −18 **10.** $\frac{7}{5}$ **11.** −3 **12.** −2 **13.** 2 **14.** $\frac{-5}{2}$ **15.** 0.5 **16.** −10 **17.** $\frac{46}{5}$

18. 7 **19.** 2 **20.** $\frac{3}{10}$ **21.** $\{0, 9\}$ **22.** $\{-4, 0\}$ **23.** $\{-5, 5\}$ **24.** $\{-3, 3\}$ **25.** $\{-4, 3\}$ **26.** $\{-4, -3\}$ **27.** $\left\{-\frac{1}{2}, 3\right\}$ **28.** $\left\{-1, -\frac{2}{3}\right\}$

29. $\{-4, 4\}$ **30.** $\{-5, 5\}$ **31.** $\{3, 4\}$ **32.** $\{-4, 3\}$ **33.** $\frac{3}{2}$ **34.** $\frac{4}{5}$ **35.** $\left\{-\frac{2}{3}, \frac{3}{2}\right\}$ **36.** $\left\{\frac{1}{2}, \frac{3}{2}\right\}$ **37.** $\left\{-\frac{2}{3}, \frac{3}{2}\right\}$ **38.** $\{3, 4\}$ **39.** 2 **40.** 3

41. −1 **42.** $\frac{1}{3}$ **43.** 1 **44.** 0 **45.** No real solution **46.** No real solution **47.** −13 **48.** 0 **49.** 3 **50.** 3 **51.** 2 **52.** No real solution

53. 8 **54.** 4 **55.** $\{-1, 3\}$ **56.** −2 **57.** $\{1, 5\}$ **58.** 18 **59.** $\{-4, 4\}$ **60.** $\{-5, 5\}$ **61.** $\{-4, 1\}$ **62.** $\left\{-\frac{1}{3}, 1\right\}$ **63.** $\left\{-1, \frac{3}{2}\right\}$

64. $\{-1, 2\}$ **65.** $\{-4, 4\}$ **66.** $\{-1, 1\}$ **67.** 2 **68.** 3 **69.** $\{-12, 12\}$ **70.** $\{-12, 12\}$ **71.** $\left\{-\frac{36}{5}, \frac{24}{5}\right\}$ **72.** $\left\{-\frac{4}{3}, \frac{8}{3}\right\}$ **73.** $\frac{b + c}{a}$

74. $\frac{1 - b}{a}$ **75.** $\frac{abc}{a + b}$ **76.** $\frac{a + b}{c}$ **77.** $\{0.59, 3.41\}$ **78.** $\{-3.41, -0.59\}$ **79.** $\{-2.80, 1.07\}$ **80.** $\{-2.29, 0.87\}$ **81.** $\{-0.85, 1.17\}$

82. $\{-1.44, 0.44\}$ **83.** $\{-8.16, -0.22\}$ **84.** $\{1.13, 5.62\}$ **85.** $R = \dfrac{R_1R_2}{R_1 + R_2}$ **86.** $r = \dfrac{A - P}{Pt}$ **87.** $R = \dfrac{mv^2}{F}$ **88.** $T = \dfrac{PV}{nR}$
89. $r = \dfrac{S - a}{S}$ **90.** $t = \dfrac{v_0 - v}{g}$ **91.** 229.94 ft

1.4 Exercises

1. $A = \pi r^2$; r = Radius, A = Area **2.** $C = 2\pi r$; r = Radius, C = Circumference **3.** $A = s^2$; A = Area, s = Length of a side
4. $P = 4s$; s = length of a side, P = Perimeter **5.** $F = ma$; F = Force, m = Mass, a = Acceleration
6. $P = \dfrac{F}{A}$; P = Pressure, F = Force, A = Area **7.** $W = Fd$; W = Work, F = Force, d = Distance
8. $K = \dfrac{1}{2}mv^2$; K = Kinetic Energy, m = Mass, v = Velocity **9.** $C = 150x$; C = Total cost, x = number of dishwashers
10. $R = 250x$; R = Total Revenue, x = number of dishwashers **11.** $11,000 will be invested in bonds and $9000 in CDs.
12. Yani will receive $6000 and Diane $4000. **13.** David will receive $400,000, Paige $300,000, and Dan $200,000.
14. Canter pays $10.80 and Carole $7.20. **15.** The regular hourly rate is $8.50 **16.** Leigh's hourly wage is $6.00 per hr.
17. The Bears got 5 touchdowns. **18.** The Bulls had 30 field goals. **19.** The length is 19 ft; the width is 11 ft
20. The length is 14 m; the width is 7 m. **21.** Brooke needs a score of 85. **22.** For a B, Mike needs 78; for an A, he needs 93.
23. The original price was $147,058.82; purchasing the model saves $22,058.82.
24. Its list price was $9411.76; the amount saved was $1411.76. **25.** The bookstore paid $44.80.
26. At $100 over cost, the payment is $10,300. **27.** Invest $31,250 in bonds and $18,750 in CDs.
28. Invest $43,750 in bonds, $6250 in CDs. **29.** $11,600 was loaned out at 8%. **30.** She can lend $333,333.33.
31. Mix 75 lb of Earl Gray tea with 25 lb of Orange Pekoe tea. **32.** 49 lb of coffee I should be mixed with 51 lb of Coffee II.
33. Mix 40 lb of cashews with the peanuts. **34.** Each box should contain 20 carmels and 10 creams. **35.** Add $\dfrac{20}{3}$ oz of pure water.
36. Add 8 cc of pure hydrochloric acid. **37.** The Metra commuter averages 30 mph; the Amtrak averages 80 mph.
38. The average speed of the slower car is 60 mph; the average speed of the faster car is 70 mph. Each traveled 210 mi.
39. The speed of the current is 2.286 mph. **40.** The speed is 9 mph. **41.** Working together, it takes 12 min. **42.** April would take 15 hr.
43. The dimensions are 11 ft by 13 ft **44.** The dimensions are 17 cm by 18 cm. **45.** The dimensions are 5 m by 8 m
46. The shortest radius setting is 25 ft. **47.** The dimensions should be 4 ft by 4 ft. **48.** The dimensions should be 6 ft by 3 ft.
49. The speed of the current is 5 mph. **50.** The patio dimensions will be 36 ft by 18 ft. **51. (a)** The dimensions are 10 ft by 5 ft.
(b) The area is 50 sq ft. **(c)** The dimensions would be 7.5 ft by 7.5 ft. **(d)** The area would be 56.25 sq ft.
52. (a) Its dimensions are 19 ft by 19 ft. **(b)** Its dimensions are 28.5 ft by 9.5 ft. **(c)** The diameter is approximately 25.83 ft.
(d) The circular one has the most. **53.** The border will be 2.71 ft wide. **54.** The border will be 2.13 ft wide.
55. The border will be 2.56 ft wide. **56.** The radius is 3 in. **57.** Add $\dfrac{2}{3}$ gal of water. **58.** 5 liters should be drained.
59. 5 lb must be added. **60.** Evaporate 96 gal of water. **61.** The dimensions should be 11.55 cm by 6.55 cm by 3 cm.
62. The dimensions should be 11.07 cm by 6.07 cm by 3 cm. **63.** 40 g of 12 karat gold should be mixed with 20 g of pure gold.
64. There are 11 atoms of oxygen and 22 atoms of hydrogen.
65. Mike passes Dan $\dfrac{1}{3}$ mi from the start, 2 min from the time Mike started to race.
66. The defensive back catches up to the tight end at the tight end's 45 yard line. **67.** Start the auxiliary pump at 9:45 AM.
68. It will take an hour and 45 min more for the 5 hp pump to empty the pool. **69.** 60 minutes **70.** It can fly 742.5 mi.
71. The most you can invest in the CD is $66,667. **72.** Lewis would beat Burke by 16.75 m. **73.** The average speed is 49.5 mph.
74. The tail wind was 137 knots. **75.** Set the original price at $40. At 50% off, there will be no profit at all.

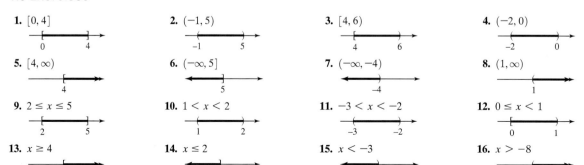

1.5 Exercises

1. $[0, 4]$
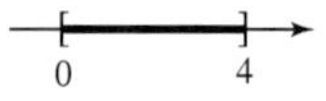

2. $(-1, 5)$
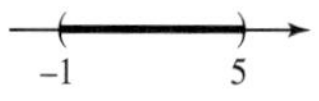

3. $[4, 6)$

4. $(-2, 0)$
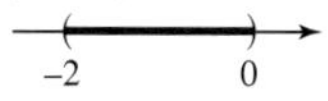

5. $[4, \infty)$
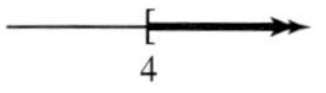

6. $(-\infty, 5]$
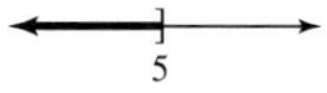

7. $(-\infty, -4)$

8. $(1, \infty)$
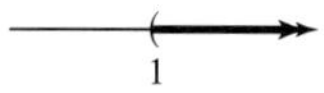

9. $2 \le x \le 5$

10. $1 < x < 2$

11. $-3 < x < -2$

12. $0 \le x < 1$

13. $x \ge 4$
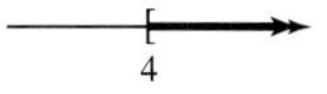

14. $x \le 2$
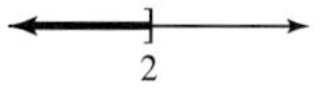

15. $x < -3$
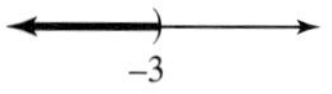

16. $x > -8$
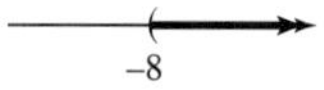

17. $<$ **18.** $<$ **19.** $>$ **20.** $>$ **21.** $\geq$ **22.** $\leq$ **23.** $>$ **24.** $<$ **25.** $<$ **26.** $\geq$ **27.** $>$ **28.** $\leq$ **29.** $\geq$ **30.** $<$

31. $\{x|x < 4\}$ or $(-\infty, 4)$ **32.** $\{x|x < 7\}$ or $(-\infty, 7)$ **33.** $\{x|x \geq -1\}$ or $[-1, \infty)$ **34.** $\{x|x \geq -1\}$ or $[-1, \infty)$

35. $\{x|x > 3\}$ or $(3, \infty)$ **36.** $\{x|x > -2\}$ or $(-2, \infty)$ **37.** $\{x|x \geq 2\}$ or $[2, \infty)$ **38.** $\{x|x \geq 5\}$ or $[5, \infty)$

39. $\{x|x > -7\}$ or $(-7, \infty)$ **40.** $\{x|x < 5\}$ or $(-\infty, 5)$ **41.** $\left\{x \middle| x \leq \frac{2}{3}\right\}$ or $\left(-\infty, \frac{2}{3}\right]$ **42.** $\{x|x \leq 0\}$ or $(-\infty, 0]$

43. $\{x|x < -20\}$ or $(-\infty, -20)$ **44.** $\left\{x \middle| x > -\frac{7}{4}\right\}$ or $\left(-\frac{7}{4}, \infty\right)$ **45.** $\left\{x \middle| x \geq \frac{4}{3}\right\}$ or $\left[\frac{4}{3}, \infty\right)$ **46.** $\{x|x \geq 12\}$ or $[12, \infty)$

47. $\{x|3 \leq x \leq 5\}$ or $[3, 5]$ **48.** $\{x|1 \leq x \leq 4\}$ or $[1, 4]$ **49.** $\left\{x \middle| \frac{2}{3} \leq x \leq 3\right\}$ or $\left[\frac{2}{3}, 3\right]$ **50.** $\{x|-3 \leq x \leq 3\}$ or $[-3, 3]$

51. $\left\{x \middle| -\frac{11}{2} < x < \frac{1}{2}\right\}$ or $\left(-\frac{11}{2}, \frac{1}{2}\right)$ **52.** $\left\{x \middle| -\frac{2}{3} < x < 2\right\}$ or $\left(-\frac{2}{3}, 2\right)$ **53.** $\{x|-6 < x < 0\}$ or $(-6, 0)$

54. $\{x|0 < x < 3\}$ or $(0, 3)$ **55.** $\{x|x < -5\}$ or $(-\infty, -5)$ **56.** $\{x|x < 11\}$ or $(-\infty, 11)$ **57.** $\{x|x \geq -1\}$ or $[-1, \infty)$

58. $\{x|x \leq 1\}$ or $(-\infty, 1]$ **59.** $\left\{x \middle| \frac{1}{2} \leq x < \frac{5}{4}\right\}$ or $\left[\frac{1}{2}, \frac{5}{4}\right)$ **60.** $\left\{x \middle| -\frac{1}{3} < x \leq \frac{1}{3}\right\}$ or $\left(-\frac{1}{3}, \frac{1}{3}\right]$

61. $(-4, 4)$ **62.** $(-5, 5)$ **63.** $(-\infty, -4)$ or $(4, \infty)$ **64.** $(-\infty, -3)$ or $(3, \infty)$

65. $(1, 3)$ **66.** $(-6, -2)$ **67.** $\left[-\frac{2}{3}, 2\right]$ **68.** $[-6, 1]$

69. $(-\infty, 1]$ or $[5, \infty)$ **70.** $(-\infty, -6]$ or $[-2, \infty)$ **71.** $\left(-1, \frac{3}{2}\right)$ **72.** $(-1, 2)$

73. $(-\infty, -1)$ or $(2, \infty)$ **74.** $\left(-\infty, \frac{1}{3}\right)$ or $(1, \infty)$ **75.** $\left|x - 2\right| < \frac{1}{2}$; $\left\{x \middle| \frac{3}{2} < x < \frac{5}{2}\right\}$ **76.** $|x + 1| < 1$; $\{x|-2 < x < 0\}$

77. $|x + 3| > 2$; $\{x|x < -5 \text{ or } x > -1\}$

78. $|x - 2| > 3$; $x < -1$ or $x > 5$ **79.** $21 <$ age < 30

80. $40 \leq$ age < 60 **81.** $|x - 98.6| \geq 1.5$; $x \leq 97.1$°F or $x \geq 100.1$°F **82.** $|x - 115| \leq 5$; $110 \leq x \leq 120$

83. **(a)** Male ≥ 73.4 **(b)** Female ≥ 79.7 **(c)** A female can expect to live at least 6.3 years longer.

84. The volume of the gas ranges from 1600 to 2400 cc, inclusive. **85.** The agent's commission ranges from \$45,000 to \$95,000, inclusive. As a percent of selling price, the commission ranges from 5% to 8.6%, inclusive. **86.** The commission will vary from \$53 to \$145, inclusive.

87. The amount withheld varies from \$72.14 to \$93.14, inclusive. **88.** The amount withheld varies from \$93.14 to \$121.14, inclusive.

89. The usage varies from 675.43 kWhr to 2500.86 kWhr, inclusive. **90.** The water usage varied from 16,000 gal to 37,971 gal.

91. The dealer's cost varies from \$7457.63 to \$7857.14, inclusive. **92.** The people in the top 2.5% have test scores greater than 123.5.

93. You need at least a 74 on the last test. **94.** You need at least a 77 on the last test.

95. The amount of gasoline ranged from 12 to 20 gal, inclusive. **96.** There were 10 gal or less of gasoline at the start of the trip.

97. $\frac{a+b}{2} - a = \frac{a+b-2a}{2} = \frac{b-a}{2} > 0$; therefore, $a < \frac{a+b}{2}$

$b - \frac{a+b}{2} = \frac{2b-a-b}{2} = \frac{b-a}{2} > 0$; therefore, $b > \frac{a+b}{2}$

98. $\frac{a+b}{2} - a = \frac{a+b-2a}{2} = \frac{b-a}{2}$

$b - \frac{a+b}{2} = \frac{2b-a-b}{2} = \frac{b-a}{2}$; thus, $\frac{a+b}{2}$ is equidistant from a and b

99. $(\sqrt{ab})^2 - a^2 = ab - a^2 = a(b-a) > 0$; thus, $(\sqrt{ab})^2 > a^2$ and $\sqrt{ab} > a$
$b^2 - (\sqrt{ab})^2 = b^2 - ab = b(b-a) > 0$; thus $b^2 > (\sqrt{ab})^2$ and $b > \sqrt{ab}$

100. $\frac{a+b}{2} - \sqrt{ab} = \frac{1}{2}[a - 2\sqrt{ab} + b]$

$= \frac{1}{2}(\sqrt{a} - \sqrt{b})^2 > 0$ since $a > b$; thus, $\sqrt{ab} < \frac{a+b}{2}$

101. $h - a = \frac{2ab}{a+b} - a = \frac{ab - a^2}{a+b} = \frac{a(b-a)}{a+b} > 0$; thus, $h > a$

$b - h = b - \frac{2ab}{a+b} = \frac{b^2 - ab}{a+b} = \frac{b(b-a)}{a+b} > 0$; thus $h < b$

102. $\frac{1}{h} = \frac{1}{2}\left(\frac{1}{a} + \frac{1}{b}\right)$

$\frac{2}{h} = \frac{1}{a} + \frac{1}{b} = \frac{a+b}{ab}$

$\frac{h}{2} = \frac{ab}{a+b}$

$h = \frac{2ab}{a+b}$

$h = \frac{(\sqrt{ab})^2}{\frac{1}{2}(a+b)}$

1.6 Exercises

1. (a) $\frac{1}{2}$ **(b)** If x increases by 2 units, y will increase by 1 unit.

2. (a) $-\frac{1}{2}$ **(b)** If x increases by 2 units, y will decrease by 1 unit.

3. (a) $-\frac{1}{3}$ **(b)** If x increases by 3 units, y will decrease by 1 unit.

4. (a) $\frac{1}{3}$ **(b)** If x increases by 3 units, y will increase by 1 unit.

5. Slope $= -\frac{3}{2}$

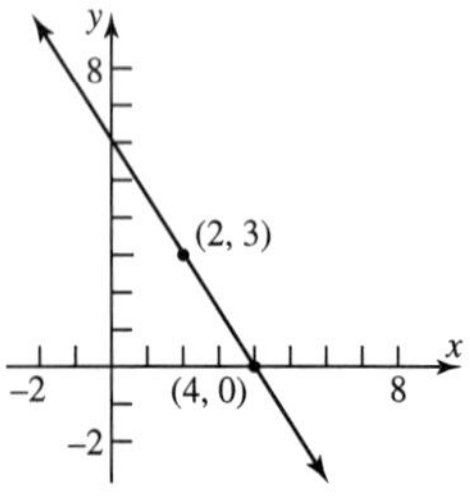

6. Slope $= -2$

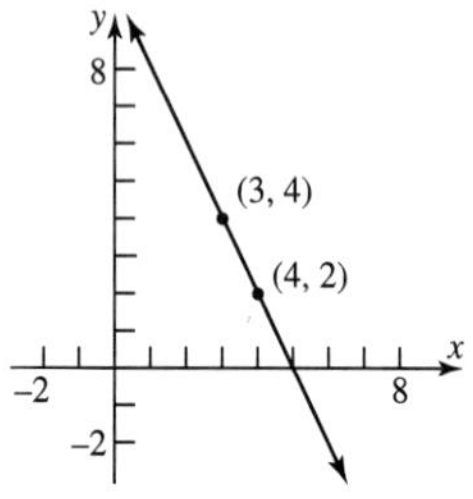

7. Slope $= -\frac{1}{2}$

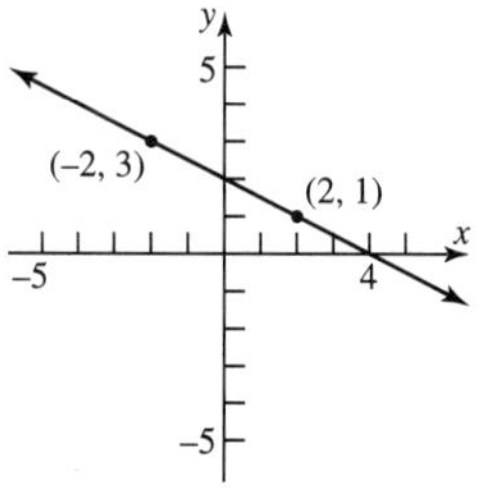

8. Slope $= \frac{2}{3}$

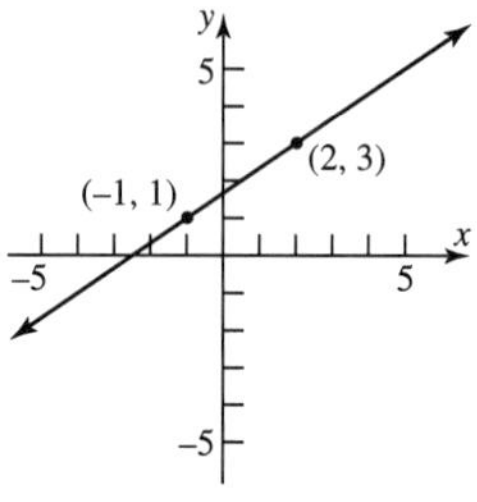

9. Slope $= 0$

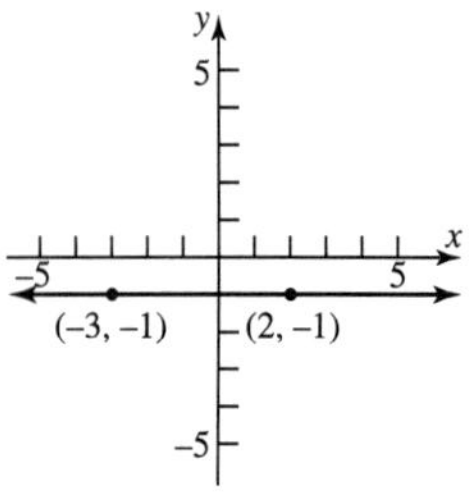

10. Slope $= 0$

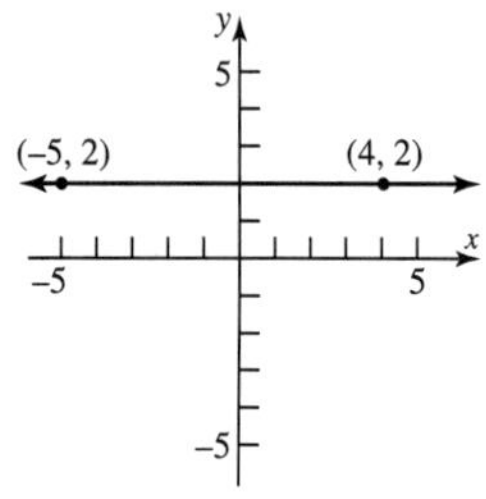

11. Slope undefined

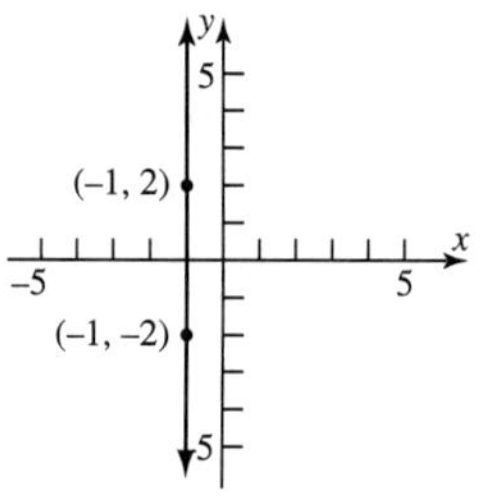

12. Slope undefined

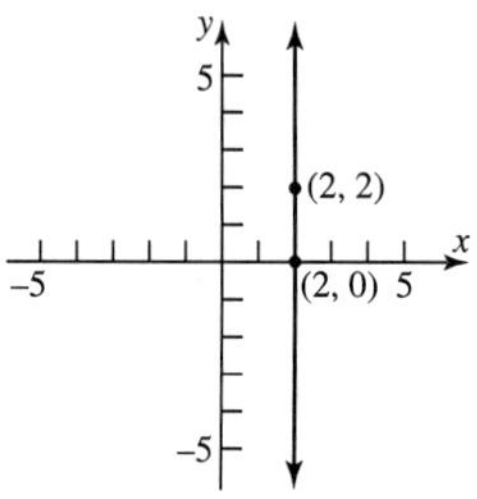

13. Slope $= \dfrac{\sqrt{3}-3}{1-\sqrt{2}} \approx 3.06$

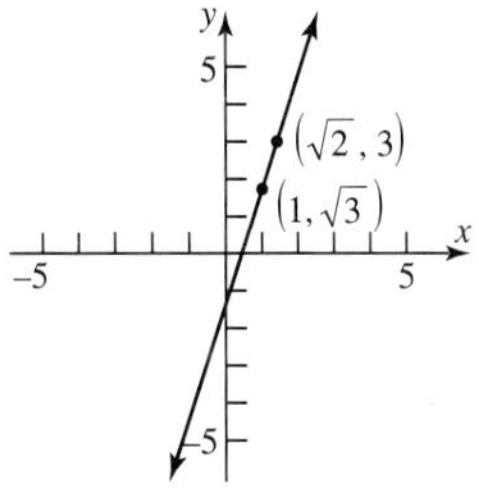

14. Slope $= \dfrac{\sqrt{5}}{4+2\sqrt{2}} \approx 0.33$

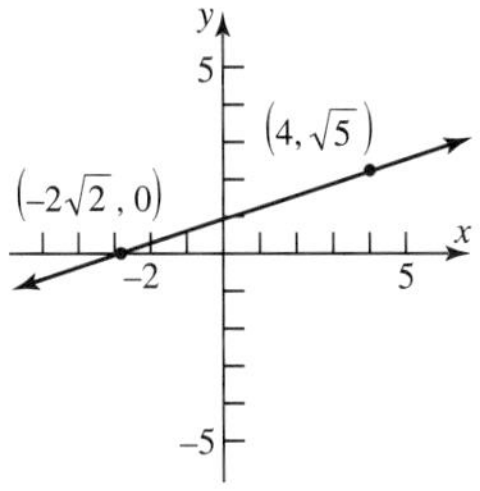

15.

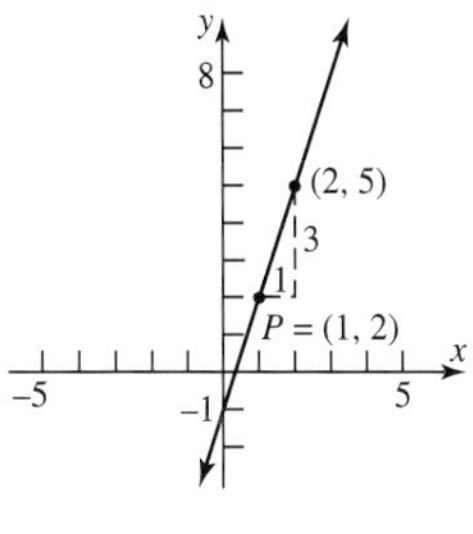

16.

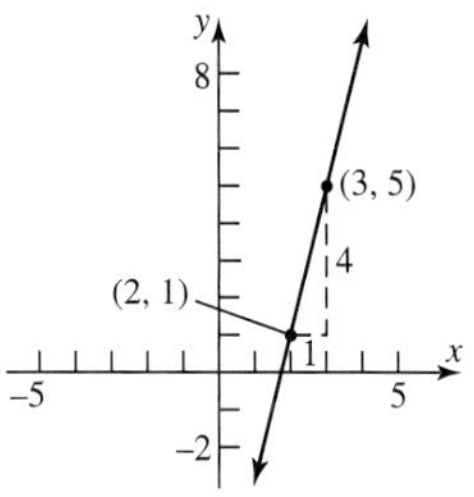

17.

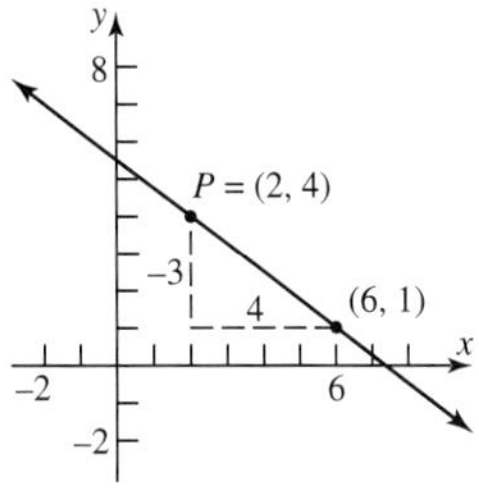

18.

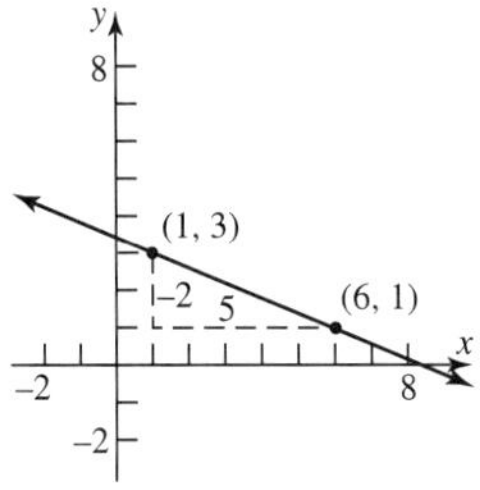

19.

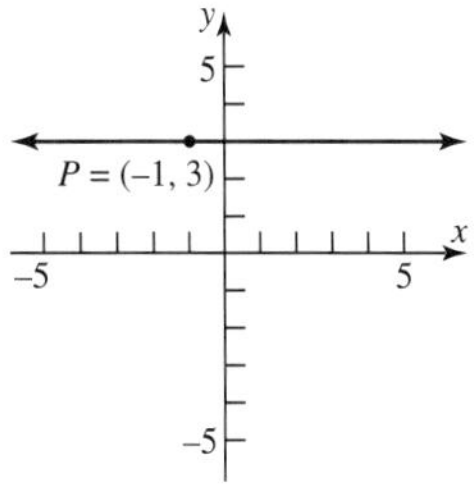

20.

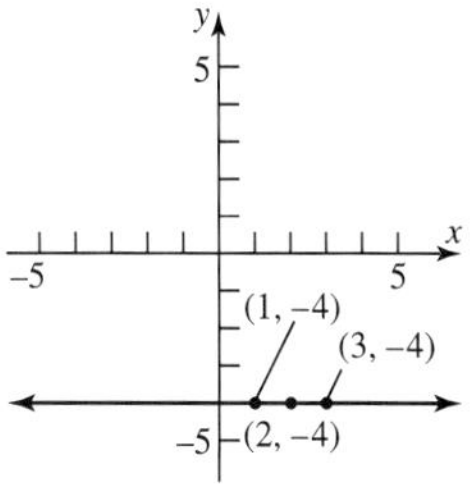

21.

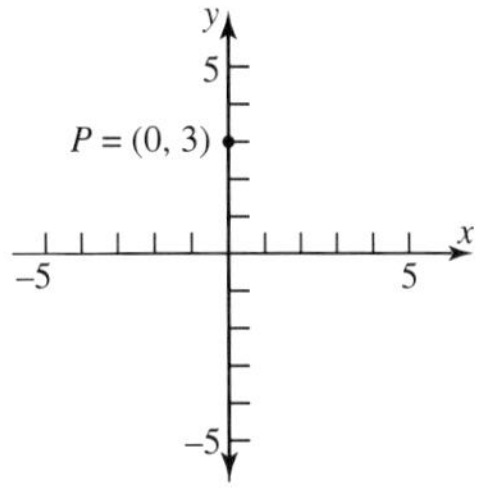

22.

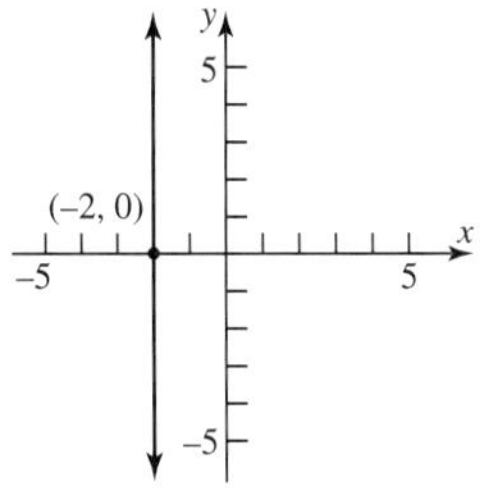

23. $x - 2y = 0$ or $y = \frac{1}{2}x$ **24.** $\frac{1}{2}x + y = 0$ or $y = -\frac{1}{2}x$

25. $x + y = 2$ or $y = -x + 2$

26. $x - 3y + 4 = 0$ or $y = \frac{1}{3}x + \frac{4}{3}$

27. $2x - y - 3 = 0$ or $y = 2x - 3$

28. $x + y - 3 = 0$ or $y = -x + 3$

29. $x + 2y - 5 = 0$ or $y = -\frac{1}{2}x + \frac{5}{2}$

30. $x - y + 2 = 0$ or $y = x + 2$

31. $3x - y + 9 = 0$ or $y = 3x + 9$ **32.** $2x - y - 11 = 0$ or $y = 2x - 11$ **33.** $2x + 3y + 1 = 0$ or $y = -\frac{2}{3}x - \frac{1}{3}$

34. $x - 2y - 1 = 0$ or $y = \frac{1}{2}x - \frac{1}{2}$ **35.** $x - 2y + 5 = 0$ or $y = \frac{1}{2}x + \frac{5}{2}$ **36.** $x - 5y + 23 = 0$ or $y = \frac{1}{5}x + \frac{23}{5}$

37. $3x + y - 3 = 0$ or $y = -3x + 3$ **38.** $2x + y + 2 = 0$ or $y = -2x - 2$ **39.** $x - 2y - 2 = 0$ or $y = \frac{1}{2}x - 1$

40. $x - y + 4 = 0$ or $y = x + 4$ **41.** $x - 2 = 0$; no slope-intercept form **42.** $x - 3 = 0$; no slope-intercept form

43. $2x - y + 4 = 0$ or $y = 2x + 4$ **44.** $3x + y + 1 = 0$ or $y = -3x - 1$ **45.** $2x - y = 0$ or $y = 2x$ **46.** $x - 2y = 0$ or $y = \frac{1}{2}x$

47. $x - 4 = 0$; no slope–intercept form **48.** $y - 2 = 0$ or $y = 2$ **49.** $2x + y = 0$ or $y = -2x$ **50.** $x + 2y + 3 = 0$ or $y = -\frac{1}{2}x - \frac{3}{2}$

51. $x - 2y + 3 = 0$ or $y = \frac{1}{2}x + \frac{3}{2}$ **52.** $2x + y - 4 = 0$ or $y = -2x + 4$ **53.** $y - 4 = 0$; no slope-intercept form

54. $x - 3 = 0$; no slope–intercept form

55. Slope $= 2$; y-intercept $= 3$

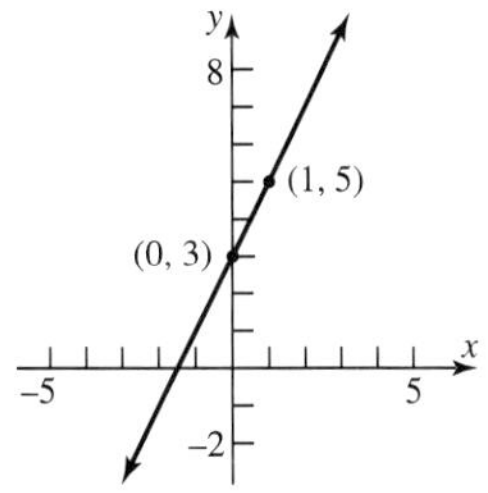

56. Slope $= -3$, y-intercept $= 4$

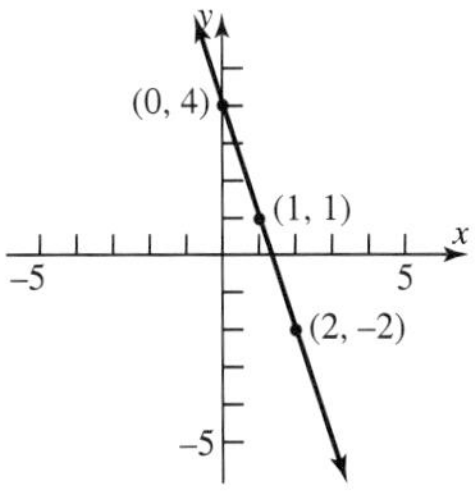

57. $y = 2x - 2$; Slope $= 2$; y-intercept $= -2$

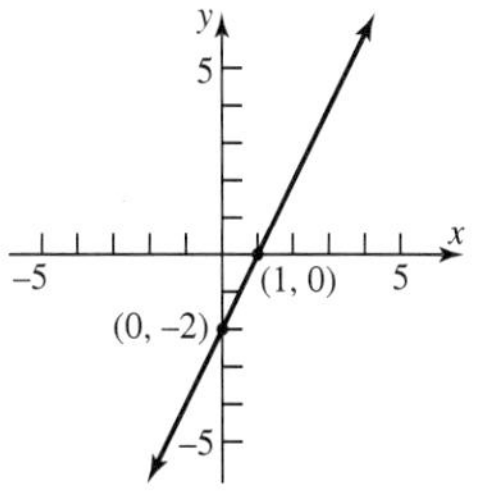

58. $y = -\frac{1}{3}x + 2$; Slope $= -\frac{1}{3}$, y-intercept $= 2$

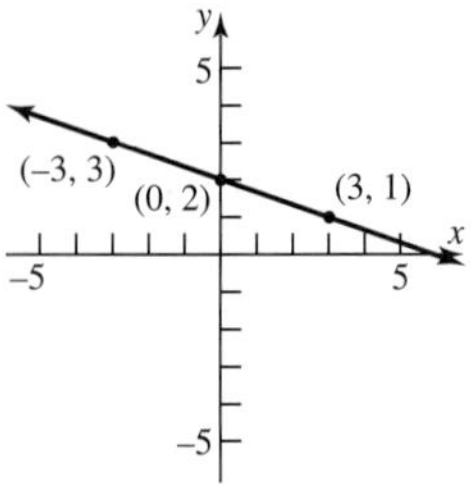

59. Slope $= \frac{1}{2}$; y-intercept $= 2$

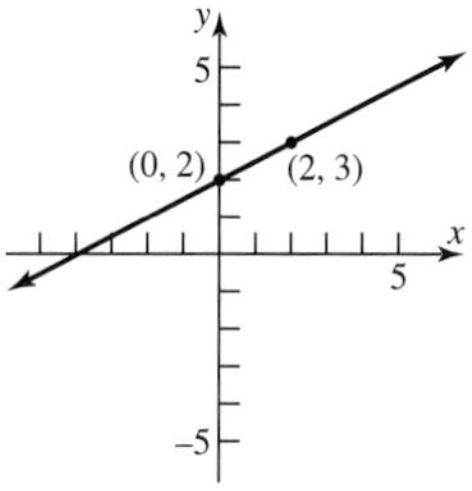

60. Slope $= 2$, y-intercept $= \frac{1}{2}$

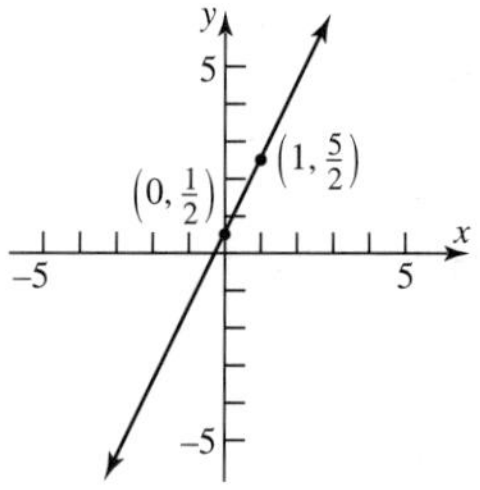

61. $y = -\frac{1}{2}x + 2$; Slope $= -\frac{1}{2}$; y-intercept $= 2$

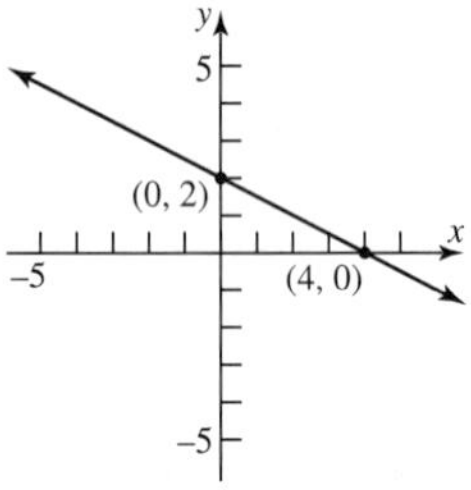

62. $y = \frac{1}{3}x + 2$; Slope $= \frac{1}{3}$, y-intercept $= 2$

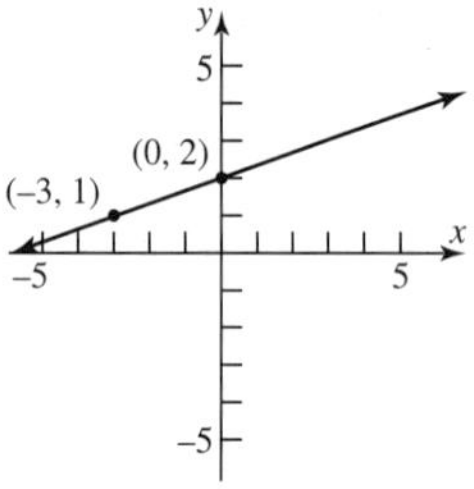

63. $y = \frac{2}{3}x - 2$; Slope $= \frac{2}{3}$; y-intercept $= -2$

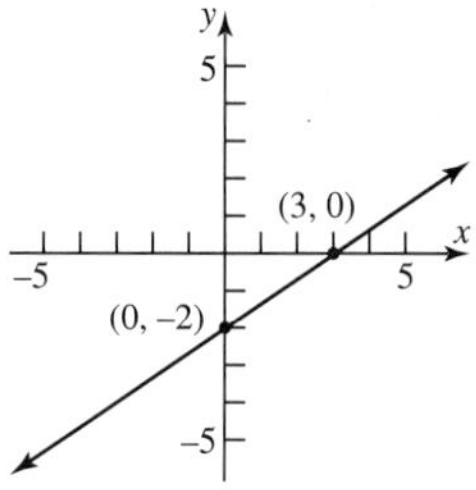

64. $y = -\frac{3}{2}x + 3$; Slope $= -\frac{3}{2}$, y-intercept $= 3$

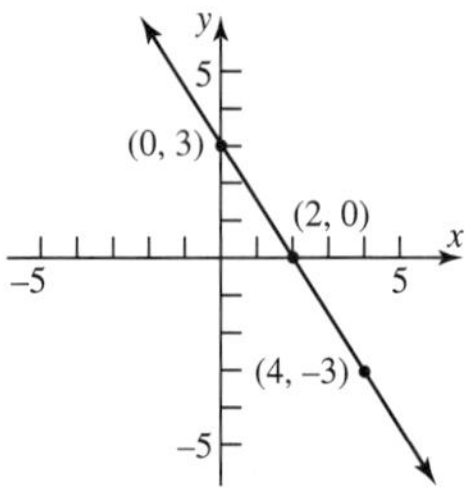

65. $y = -x + 1$; Slope $= -1$; y-intercept $= 1$

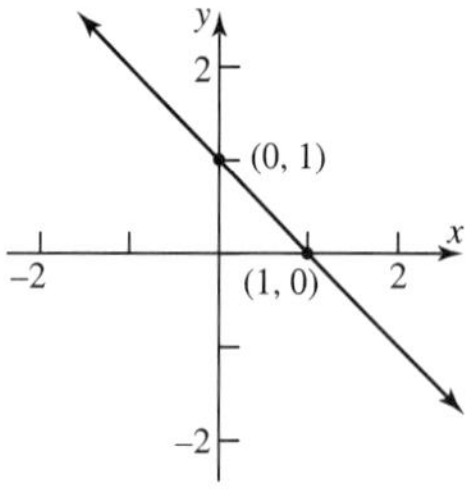

66. $y = x - 2$; Slope $= 1$, y-intercept $= -2$

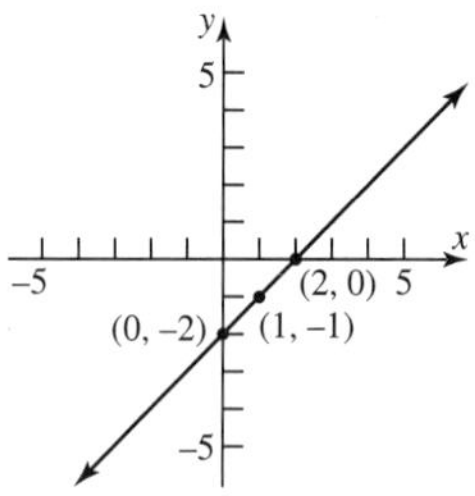

67. Slope undefined; no y-intercept

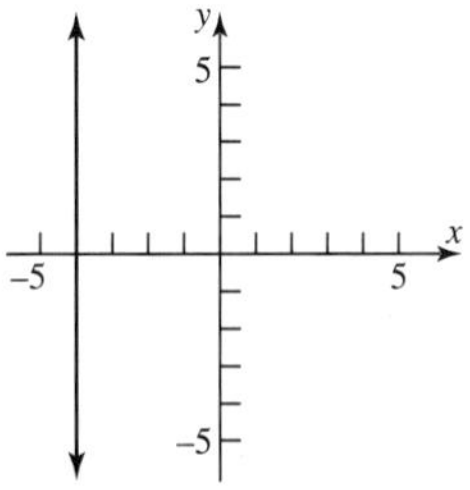

68. Slope $= 0$, y-intercept $= -1$

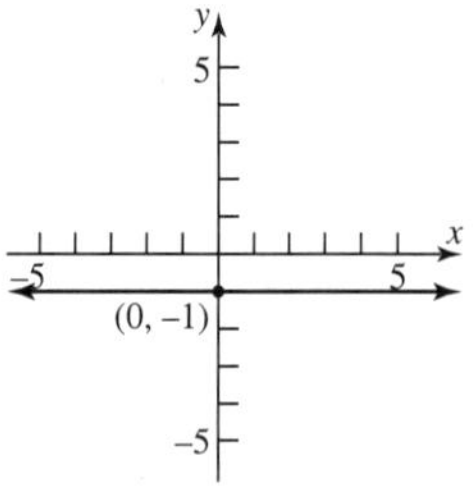

69. Slope $= 0$; y-intercept $= 5$

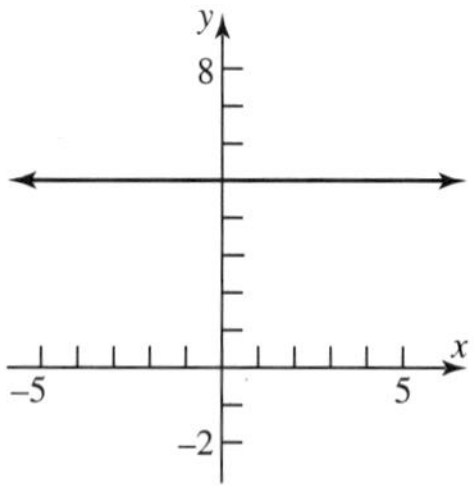

70. Slope undefined; no y-intercept

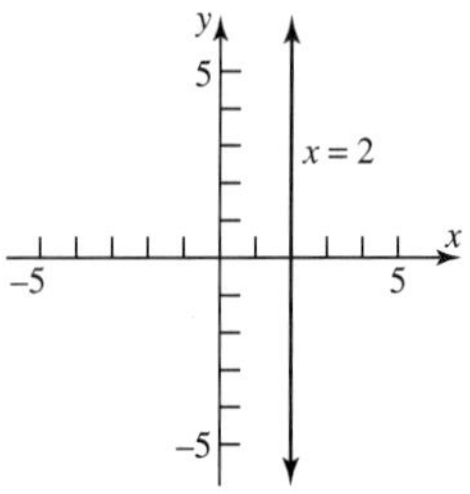

71. $y = x$; Slope $= 1$; y-intercept $= 0$

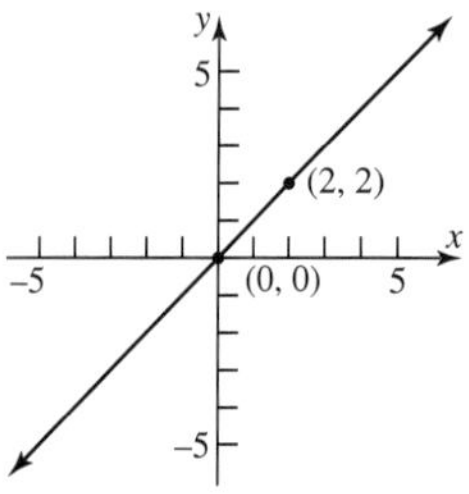

72. $y = -x$, Slope $= -1$, y-intercept $= 0$

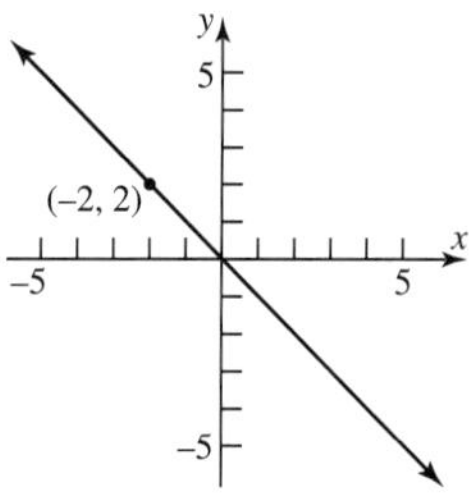

73. $y = \frac{3}{2}x$; Slope $= \frac{3}{2}$; y-intercept $= 0$

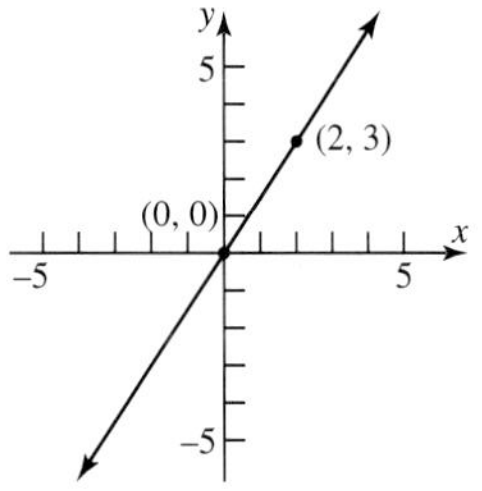

74. $y = -\frac{3}{2}x$, Slope $= -\frac{3}{2}$, y-intercept $= 0$

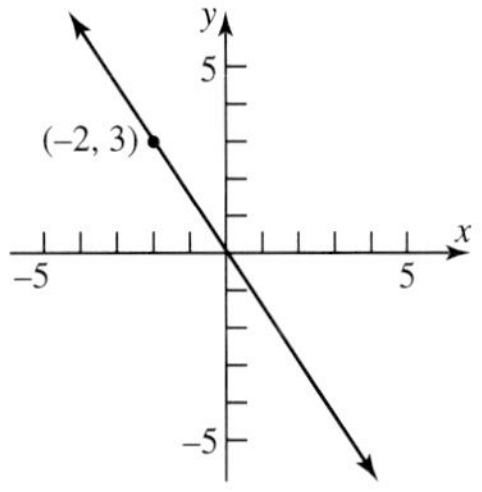

75. $y = 0$ **76.** $x = 0$

77. (b) **78.** (c) **79.** (d) **80.** (a)

81. $x - y + 2 = 0$ or $y = x + 2$

82. $x + y - 1 = 0$ or $y = -x + 1$

83. $x + 3y - 3 = 0$ or $y = -\frac{1}{3}x + 1$

84. $x + 2y + 2 = 0$ or $y = -\frac{1}{2}x - 1$

85. $°C = \frac{5}{9}(°F - 32)$; approximately 21°C **86.** **(a)** $K = °C + 273$ **(b)** $K = \frac{5}{9}°F + \frac{2297}{9}$

87. **(a)** $P = 0.5x - 100$ **(b)** \$400 **(c)** \$2400

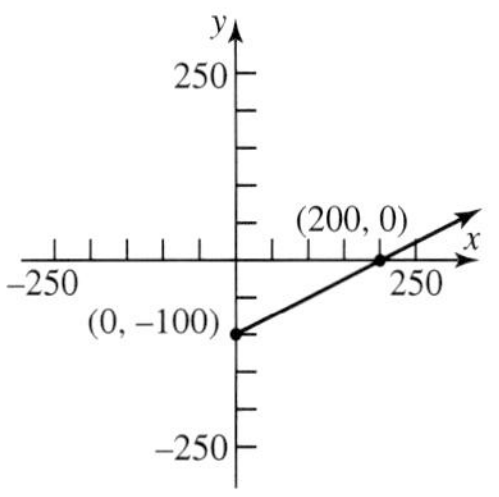

88. **(a)** $P = 0.55x - 125$ **(b)** \$425 **(c)** \$2625

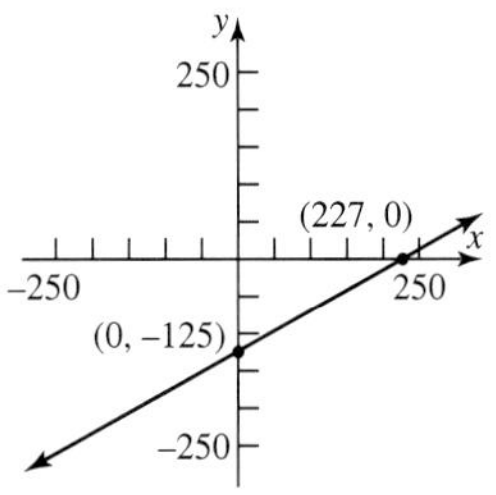

89. $C = 0.06543x + 5.65$; $C = \$25.28$; $C = \$54.72$

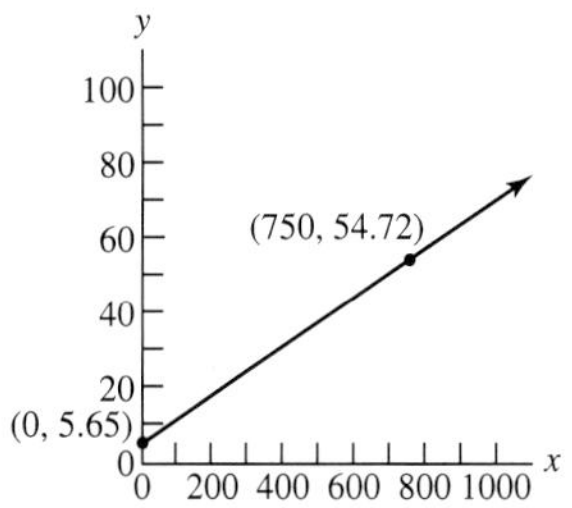

90. Slope from (a, b) to (b, a) is $\frac{a - b}{b - a} = -1$.

Slope of the line $y = x$ is 1.

Since $-1 \cdot 1 = -1$, the line containing the points (a, b) and (b, a) is perpendicular to the line $y = x$.

The midpoint of (a, b) and $(b, a) = \left(\frac{a + b}{2}, \frac{b + a}{2}\right)$.

Since $\frac{b + a}{2} = \frac{a + b}{2}$, the midpoint lies on the line $y = x$.

91. All have the same slope, 2; the lines are parallel.

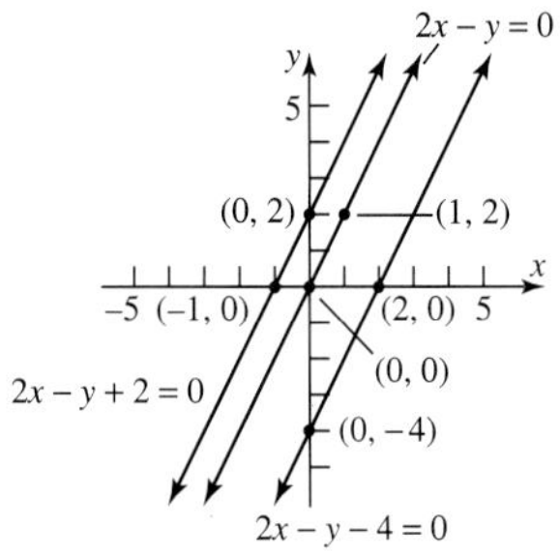

92. The family of lines $Cx + y + 4 = 0$ intersect at the point $(0, -4)$

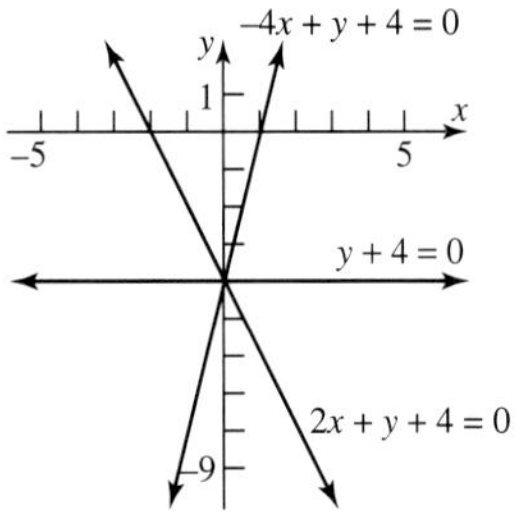

1.7 Exercises

1. Center (2, 1); Radius 2; $(x - 2)^2 + (x - 1)^2 = 4$ **2.** Center (1, 2); Radius $= 2$; $(x - 1)^2 + (y - 2)^2 = 4$

3. Center $\left(\frac{5}{2}, 2\right)$; Radius $\frac{3}{2}$; $\left(x - \frac{5}{2}\right)^2 + (y - 2)^2 = \frac{9}{4}$ **4.** Center (1, 2); Radius $= \sqrt{2}$; $(x - 1)^2 + (y - 2)^2 = 2$

5. $(x-1)^2+(y+1)^2=1$; $x^2+y^2-2x+2y+1=0$

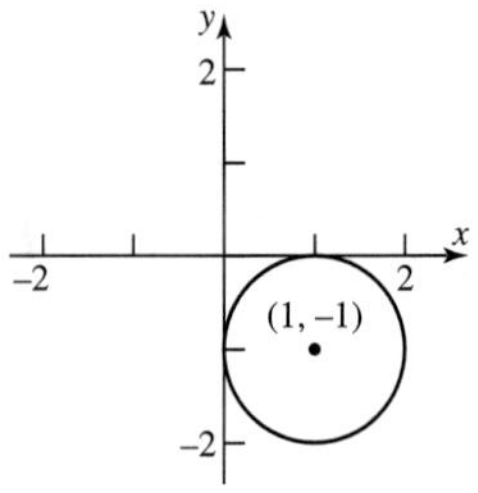

6. $(x+2)^2+(y-1)^2=4$; $x^2+y^2+4x-2y+1=0$

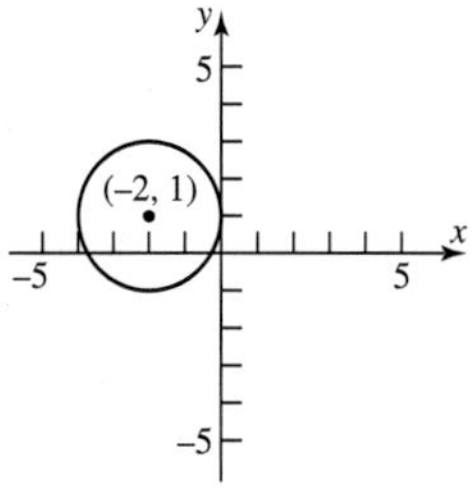

7. $x^2+(y-2)^2=4$; $x^2+y^2-4y=0$

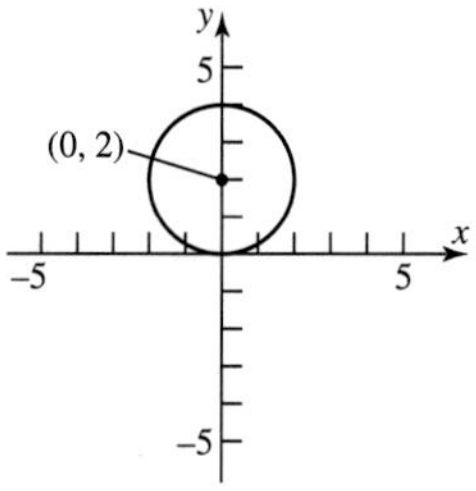

8. $(x-1)^2+y^2=9$; $x^2+y^2-2x-8=0$

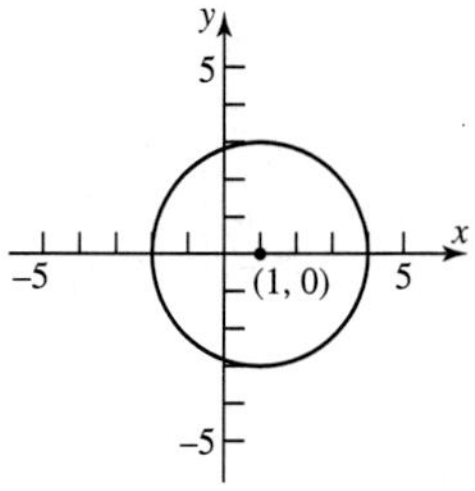

9. $(x-4)^2+(y+3)^2=25$; $x^2+y^2-8x+6y=0$

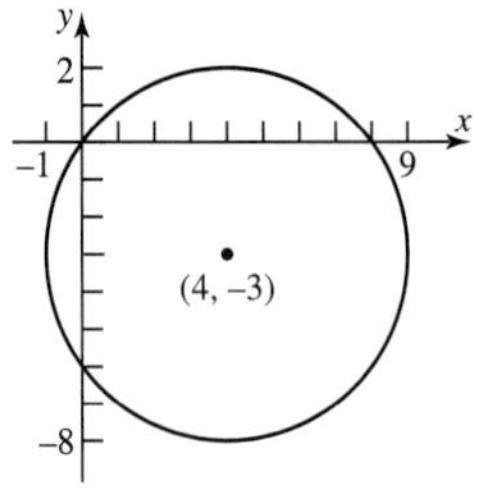

10. $(x-2)^2+(y+3)^2=16$; $x^2+y^2-4x+6y-3=0$

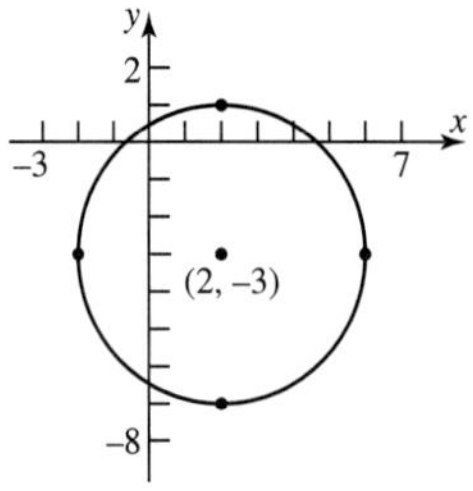

11. $x^2+y^2=4$; $x^2+y^2-4=0$

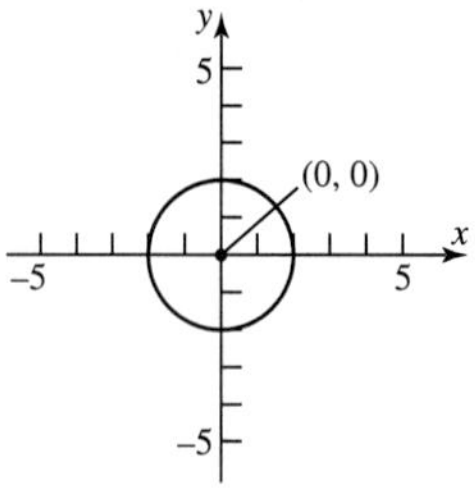

12. $x^2+y^2=9$; $x^2+y^2-9=0$

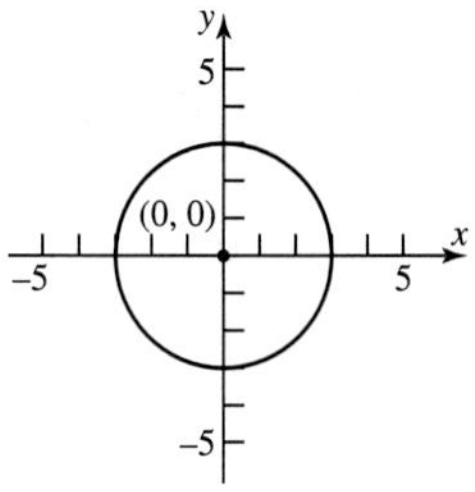

13. $\left(x-\frac{1}{2}\right)^2+y^2=\frac{1}{4}$; $x^2+y^2-x=0$

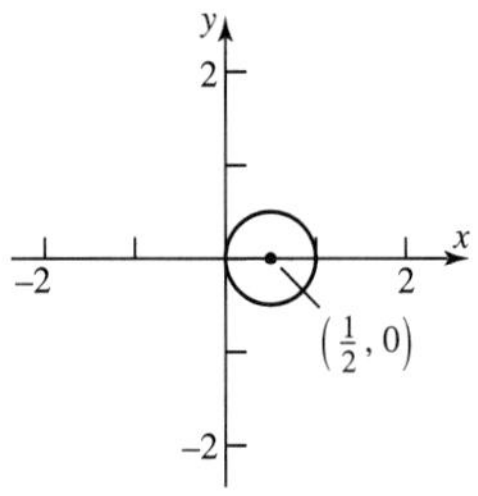

14. $x^2+\left(y+\frac{1}{2}\right)^2=\frac{1}{4}$; $x^2+y^2+y=0$

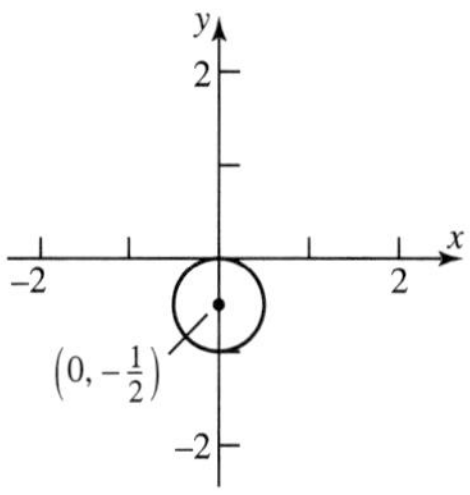

15. $r=2$; $(h,k)=(0,0)$

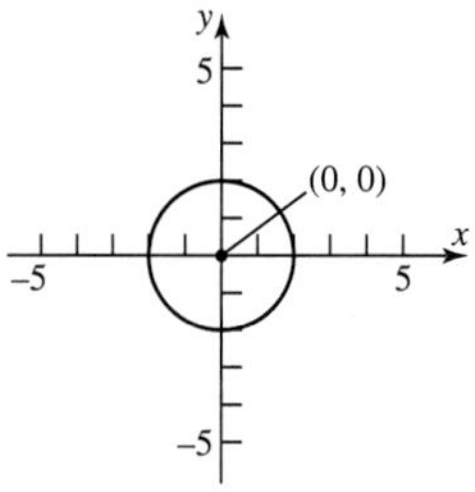

16. $r=1$; $(h,k)=(0,1)$

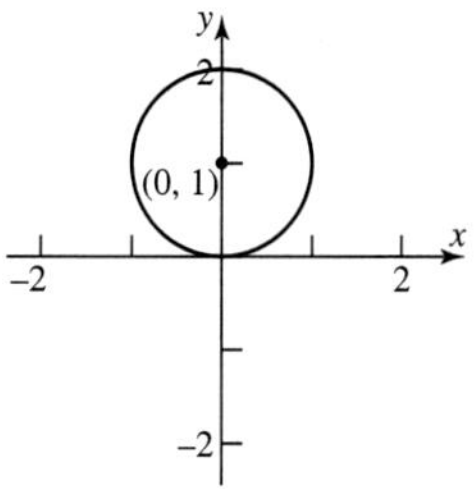

17. $r=2$; $(h,k)=(3,0)$

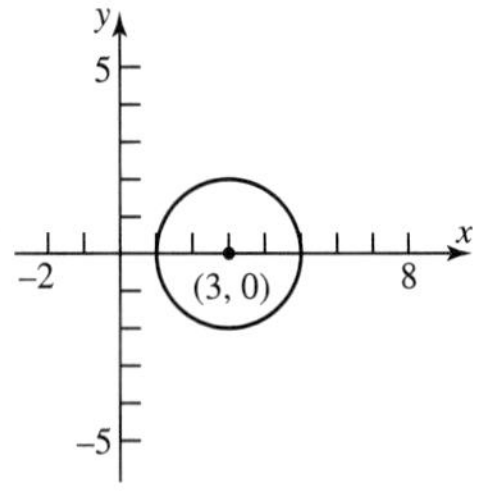

18. $r=\sqrt{2}$; $(h,k)=(-1,1)$

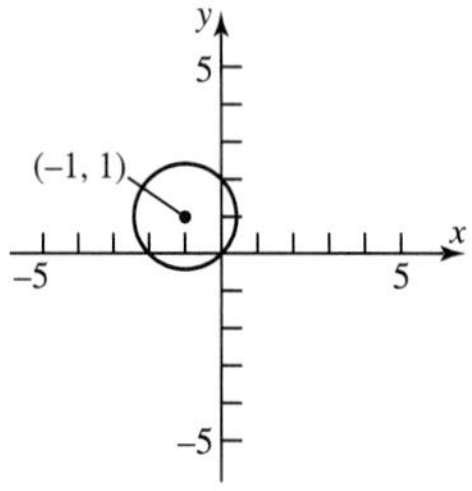

19. $r=3$; $(h,k)=(-2,2)$

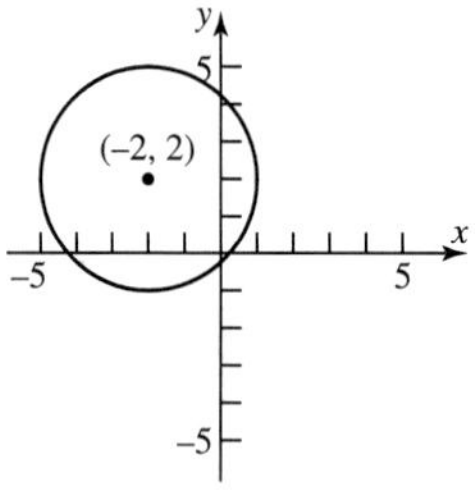

20. $r=1$; $(h,k)=(3,-1)$

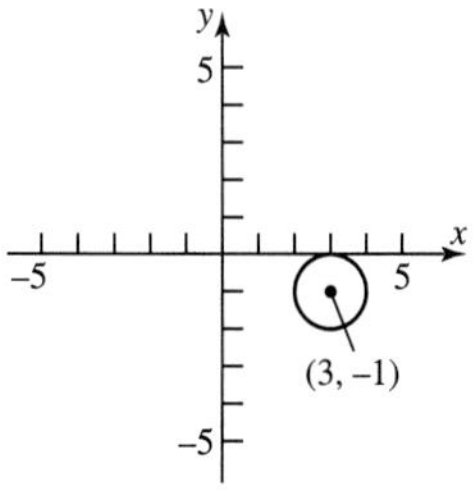

21. $r=\frac{1}{2}$; $(h,k)=\left(\frac{1}{2},-1\right)$

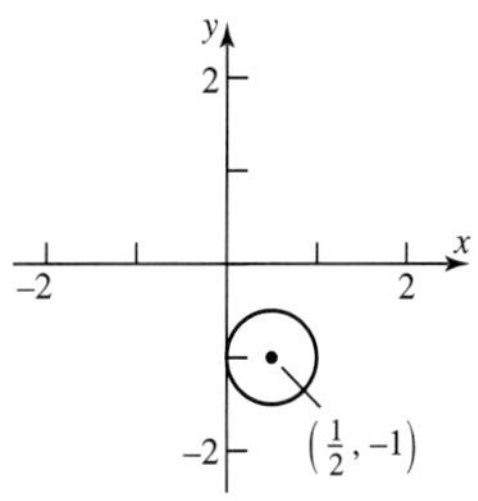

22. $r=1$; $(h,k)=\left(-\frac{1}{2},-\frac{1}{2}\right)$

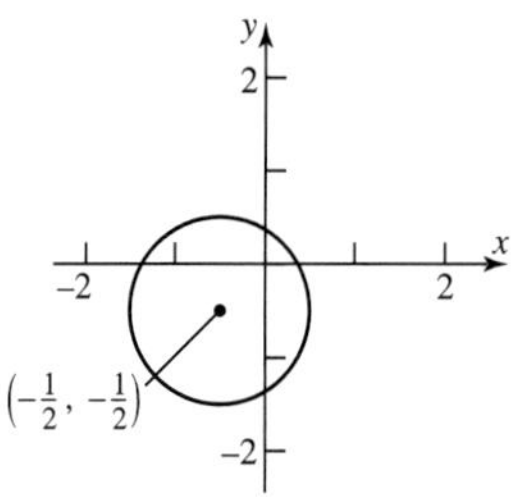

23. $r=5$; $(h,k)=(3,-2)$

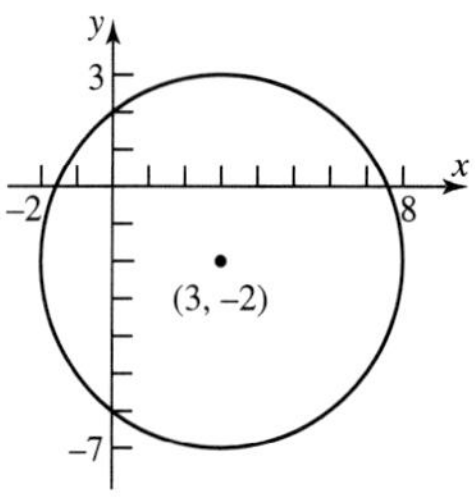

24. $r=\frac{\sqrt{2}}{2}$; $(h,k)=(-2,0)$

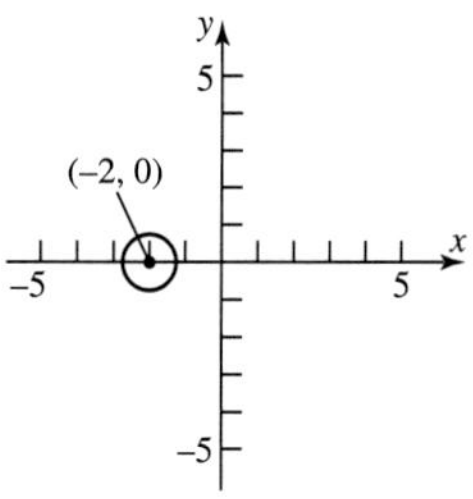

25. $x^2+y^2-13=0$ **26.** $x^2+y^2-2x-19=0$ **27.** $x^2+y^2-4x-6y+4=0$ **28.** $x^2+y^2+6x-2y+1=0$ **29.** $x^2+y^2+2x-6y+5=0$ **30.** $x^2+y^2-4x-4y+3=0$ **31.** (c) **32.** (d) **33.** (b) **34.** (a)

35. $(x + 3)^2 + (y - 1)^2 = 16$ **36.** $(x - 4)^2 + (y + 2)^2 = 9$ **37.** $(x - 2)^2 + (y - 2)^2 = 9$ **38.** $(x - 1)^2 + (y - 3)^2 = 4$

39. (a) $x^2 + (mx + b)^2 = r^2$

$(1 + m^2)x^2 + 2mbx + b^2 - r^2 = 0$

One solution if and only if discriminate $= 0$

$(2mb)^2 - 4(1 + m^2)(b^2 - r^2) = 0$

$-4b^2 + 4r^2 + 4m^2r^2 = 0$

$r^2(1 + m^2) = b^2$

(b) $x = \dfrac{-2mb}{2(1 + m^2)} = \dfrac{-2mb}{2b^2/r^2} = -\dfrac{r^2m}{b}$

$y = m\left(-\dfrac{r^2m}{b}\right) + b = -\dfrac{r^2m^2}{b} + b = \dfrac{-r^2m^2 + b^2}{b} = \dfrac{r^2}{b}$

(c) Slope of tangent line $= m$

Slope of line joining center to point of tangency $= \dfrac{r^2/b}{-r^2m/b} = -\dfrac{1}{m}$

40. $\sqrt{2}x + 4y - 9\sqrt{2} = 0$
41. $\sqrt{2}x + 4y - 11\sqrt{2} + 12 = 0$
42. $(1, 0)$
43. $x + 5y + 13 = 0$
44. $y = 2$

Fill-in-the-Blank Items

1. Abscissa; ordinate **2.** Midpoint **3.** y-axis **4.** Square; $\frac{5}{2}$ **5.** $-a$ **6.** Double; multiplicity two **7.** Extraneous **8.** Negative **9.** $-2; 2$ **10.** Circle; radius; center **11.** Undefined; 0 **12.** $m_1 = m_2$; $m_1 = -\frac{1}{m_2}$, $m_2 \neq 0$

True/False Items

1. F **2.** T **3.** T **4.** F **5.** F **6.** T **7.** T

Review Exercises

1. -12 **2.** 32 **3.** 6 **4.** $\frac{4}{11}$ **5.** $\frac{1}{5}$ **6.** $\frac{9}{16}$ **7.** -5 **8.** $\frac{17}{32}$ **9.** $\{-2, 3\}$ **10.** $\{-3, 2\}$ **11.** $\frac{11}{8}$ **12.** $-\frac{27}{13}$ **13.** $\left\{-2, \frac{3}{2}\right\}$ **14.** $\{-4, 3\}$ **15.** $\left\{-\frac{1}{2}, \frac{3}{2}\right\}$ **16.** $\left\{\frac{1}{4}, 1\right\}$ **17.** 9 **18.** 80 **19.** $\{-2, 1\}$ **20.** $\left\{\frac{1}{2}, 1\right\}$ **21.** 2 **22.** 5 **23.** $\frac{\sqrt{5}}{2}$ **24.** $\{5, 41\}$ **25.** $\{-5, 2\}$ **26.** $\left\{-\frac{4}{3}, 2\right\}$ **27.** $\left\{-\frac{5}{3}, 3\right\}$ **28.** $\{-1, 2\}$

29. $\{x|14 \le x < \infty\}$ or $[14, \infty)$

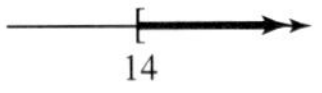

30. $\left\{x \middle| x \ge \frac{17}{19}\right\}$ or $\left[\frac{17}{19}, \infty\right)$

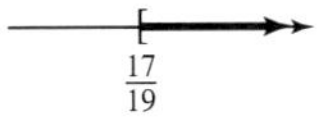

31. $\left\{x \middle| -\frac{31}{2} \le x \le \frac{33}{2}\right\}$ or $\left[-\frac{31}{2}, \frac{33}{2}\right]$

32. $\{x|-5 < x < 10\}$ or $(-5, 10)$

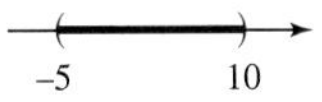

33. $\{x|-23 < x < -7\}$ or $(-23, -7)$

34. $\left\{x \middle| -\frac{7}{3} < x \le \frac{11}{3}\right\}$ or $\left(-\frac{7}{3}, \frac{11}{3}\right]$

35. $\left\{x \middle| -\frac{3}{2} < x < -\frac{7}{6}\right\}$ or $\left(-\frac{3}{2}, -\frac{7}{6}\right)$

36. $\left\{x \middle| \frac{1}{3} < x < \frac{2}{3}\right\}$ or $\left(\frac{1}{3}, \frac{2}{3}\right)$

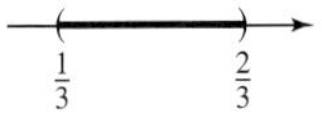

37. $\{x|-\infty < x \le -2$ or $7 \le x < \infty\}$ or $(-\infty, -2]$ or $[7, \infty)$

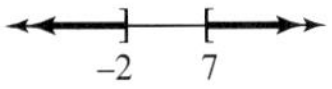

38. $\left\{x \middle| x \le -\frac{11}{3} \text{ or } x \ge 3\right\}$ or $\left(-\infty, -\frac{11}{3}\right]$ or $[3, \infty)$

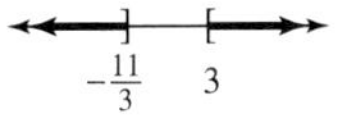

39. $2x + y - 5 = 0$ or $y = -2x + 5$
40. $y - 4 = 0$ or $y = 4$
41. $x + 3 = 0$; no slope-intercept form

42. $\frac{5}{2}x + y - 5 = 0$ or $y = -\frac{5}{2}x + 5$ **43.** $x + 5y + 10 = 0$ or $y = -\frac{1}{5}x - 2$ **44.** $5x + y - 11 = 0$ or $y = -5x + 11$ **45.** $2x - 3y + 19 = 0$ or $y = \frac{2}{3}x + \frac{19}{3}$ **46.** $x + y + 2 = 0$ or $y = -x - 2$ **47.** $-x + y + 7 = 0$ or $y = x - 7$ **48.** $x + 3y - 10 = 0$ or $y = -\frac{1}{3}x + \frac{10}{3}$

49. $4x - 5y + 20 = 0$

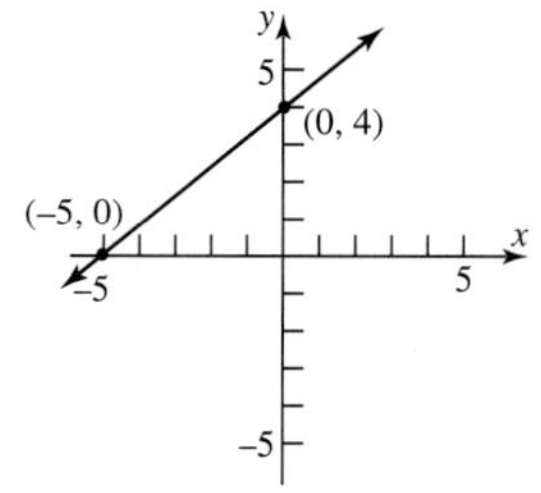

50. $3x + 4y - 12 = 0$

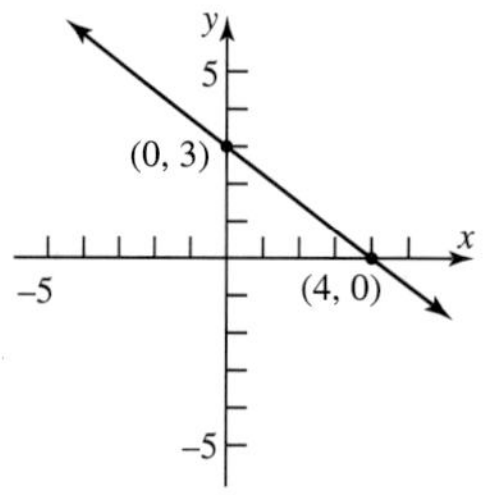

51. $\frac{1}{2}x - \frac{1}{3}y + \frac{1}{6} = 0$

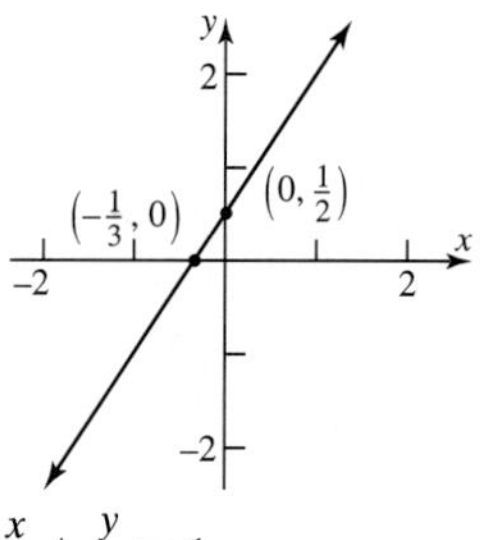

52. $-\frac{3}{4}x + \frac{1}{2}y = 0$

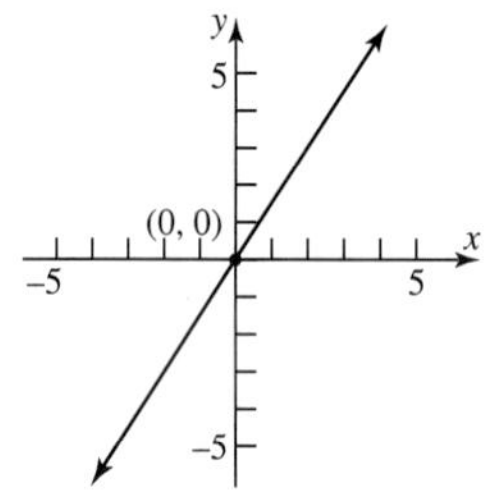

53. $\sqrt{2}x + \sqrt{3}y = \sqrt{6}$

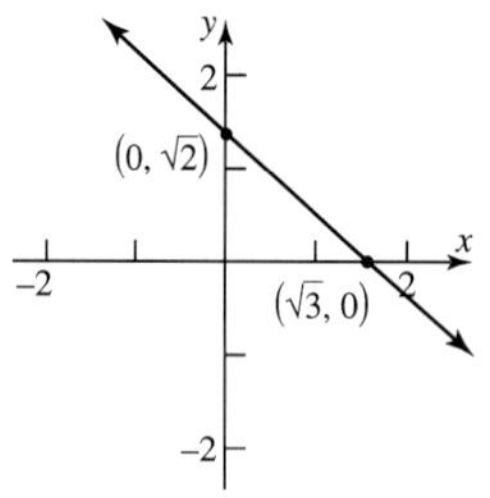

54. $\frac{x}{3} + \frac{y}{4} = 1$

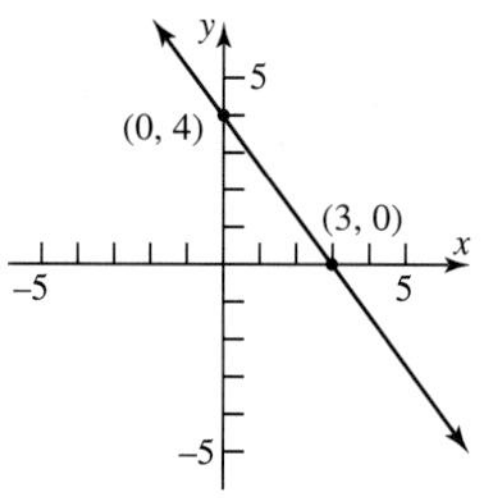

55. Intercept: $(0, 0)$; symmetric with respect to the x-axis **56.** Intercept: $(0, 0)$; symmetric with respect to the origin
57. Intercepts: $\left(\frac{1}{2}, 0\right)$, $\left(-\frac{1}{2}, 0\right)$, $(0, 1)$, $(0, -1)$; symmetric with respect to the x-axis, y-axis, and origin
58. Intercepts: $(-3, 0)$, $(3, 0)$; symmetric with respect to the x-axis, y-axis, and origin.
59. Intercept: $(0, 1)$; symmetric with respect to the y-axis **60.** Intercepts: $(-1, 0)$, $(0, 0)$, $(1, 0)$; symmetric with respect to the origin
61. Intercepts: $(0, 0)$, $(0, -2)$, $(-1, 0)$; no symmetry
62. Intercepts: $(-4, 0)$, $(0, 0)$, $(0, 2)$; no symmetry
63. Center $(1, -2)$; Radius $= 3$ **64.** Center: $(-2, 2)$; Radius $= 3$ **65.** Center $(1, -2)$; Radius $= \sqrt{5}$ **66.** Center: $(1, 0)$; Radius $= 1$

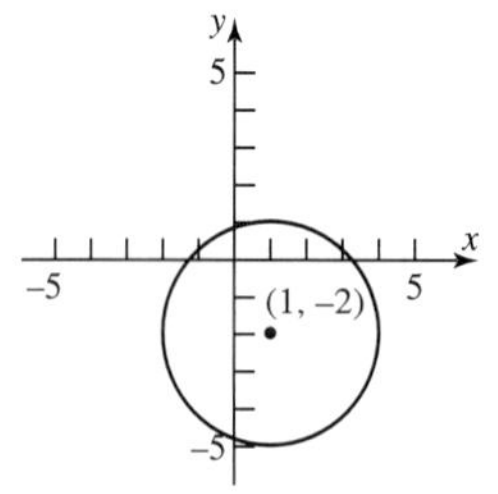

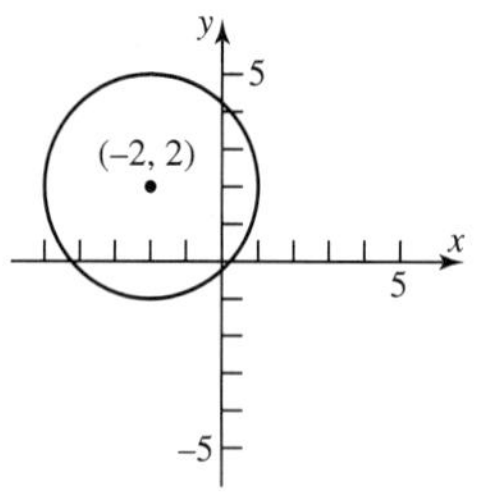

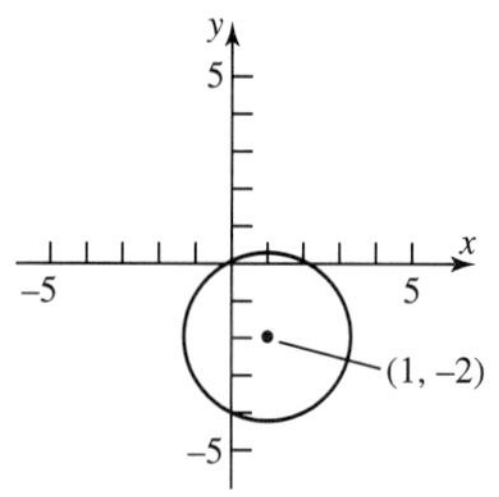

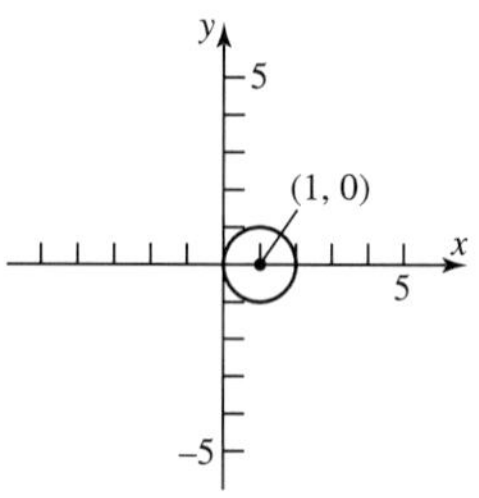

67. Slope $= \frac{1}{5}$; distance $= 2\sqrt{26}$; midpoint $= (2, 3)$ **68.** Slope $= -2$; distance $= 4\sqrt{5}$; midpoint $= (4, 1)$
69. $d(A, B) = \sqrt{13}$; $d(B, C) = \sqrt{13}$
70. (a) $d(A, B) = \sqrt{[-4 - (-2)]^2 + (4 - 0)^2} = \sqrt{4 + 16} = \sqrt{20} = 2\sqrt{5}$
$d(B, C) = \sqrt{[8 - (-4)]^2 + (5 - 4)^2} = \sqrt{144 + 1} = \sqrt{145}$
$d(A, C) = \sqrt{[8 - (-2)]^2 + (5 - 0)^2} = \sqrt{100 + 25} = \sqrt{125} = 5\sqrt{5}$
Since $d[A, B]^2 + d[A, C]^2 = d[B, C]^2$, by the converse of the Pythagorean Theorem, the points A, B, and C are vertices of a right triangle.
(b) Slope of $AB = \frac{4}{-2} = -2$; Slope of $BC = \frac{1}{12}$; Slope of $AC = \frac{5}{10} = \frac{1}{2}$; Since $(-2)\left(\frac{1}{2}\right) = -1$, the lines AB and AC are perpendicular and hence form a right angle.
71. $M_{AB} = -1$; $M_{BC} = -1$
72. $(x + 1)^2 + (y - 2)^2 = r^2$
If $(1, 5)$ lies on the circle, then $r^2 = 13$. Now show that $(2, 4)$ and $(-3, 5)$ lie on the circle $(x + 1)^2 + (y - 2)^2 = 13$
73. Center $(1, -2)$; radius $= 4\sqrt{2}$; $x^2 + y^2 - 2x + 4y - 27 = 0$ **74.** -4 and 8 **75.** The storm is 3300 ft away.
76. The range of distances is from 0.5 m to 0.75 m, $0.5 \leq x \leq 0.75$. **77.** The search plane can go as far as 616 mi.
78. The plane can extend its search 246.4 mi. **79.** The helicopter will reach the life raft in a little less than 1 hr, 35 min.
80. The bees meet for the first time in 18.75 sec. The bees meet for the second time in 37.5 sec.
81. It takes Clarissa 10 days by herself. **82.** It will take about 12 hr, 23 min. **83.** Mix 30cc of 15% HCI with 70cc of 5% HCI.
84. $10\frac{2}{3}$ oz must be evaporated. **85.** There were 3260 adults.

86. The manufacturer should mix 3 parts of \$4.50 coffee and 4 parts \$8.00 coffee. **87.** The freight train is 190.67 ft long.
88. The length should be approximately 6.47 ft. **89.** 36 seniors went on the trip; each one paid \$13.40.
90. It would take the older copier 180 min or 3 hr.

91. (a)

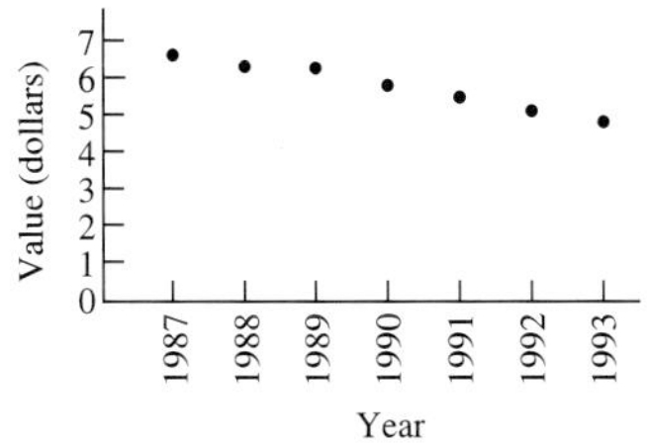

(b) The average level of carbon monoxide in the air is decreasing.

(c)

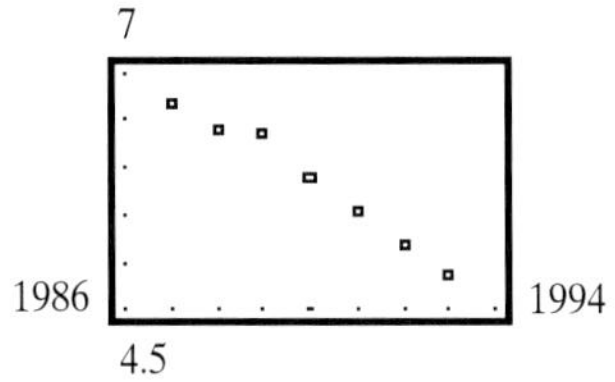

(d) $\dfrac{5.87 - 6.69}{1990 - 1987} = -\dfrac{0.82}{3} \approx -0.27$

(e) Between the years 1987 and 1990, as x increases by 1 year, the level of carbon monoxide decreases by about 0.27 ppm

(f) $-\dfrac{0.99}{3} = -0.33$

(g) As x increases by 1 year, the level of carbon monoxide decreases by 0.33 ppm

92. (a)

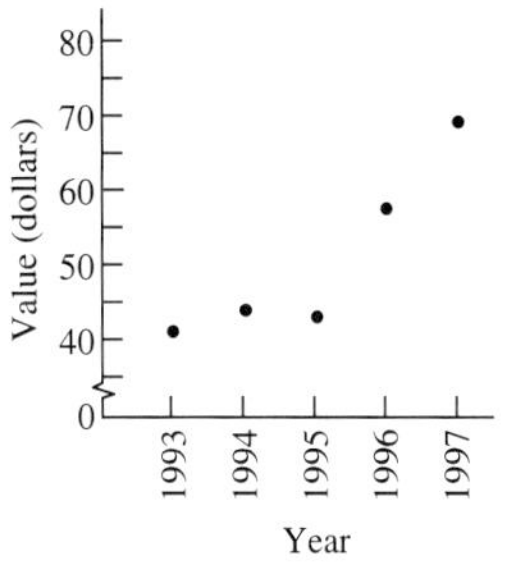

(b) The value is increasing over time.

(c)

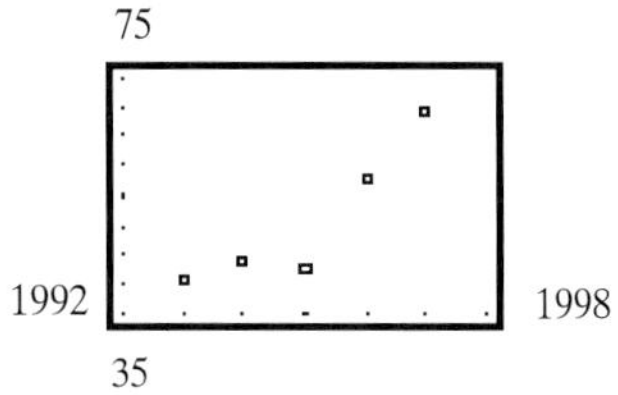

(d) 1

(e) For 1993–1995, the value increased at an average rate of \$1 per year.

(f) 13.1

(g) For 1995–1997, the value increased at an average rate of \$13.10 per year.

94. (a) The y-axis; the value of any x is 0. **(b)** The x-axis; the value of any y is 0. **(c)** The line that passes through the origin with slope -1. **(d)** Either $x = 0$ or $y = 0$; the coordinate axes. **(e)** The unit circle or the circle centered at the origin with radius 1.

95. (a) No **(b)** Todd wins again. **(c)** Todd wins by $\frac{1}{4}$ m. **(d)** Todd should line up 5.26316 m behind the start line. **(e)** Yes

96. (a) is an expression, (b) is an equation, and (c) is an inequality. Simplifying the sum is the first step in solving each of the problems. Solving the equality is one step toward solving the inequality.

CHAPTER 2 Linear and Quadratic Functions

2.1 Exercises

1. Function **2.** Function **3.** Not a function **4.** Not a function **5.** Function **6.** Function **7.** Function **8.** Function **9.** Not a function **10.** Not a function **11.** Function **12.** Function **13. (a)** -4 **(b)** -5 **(c)** -9 **(d)** $-3x^2 - 2x - 4$ **(e)** $3x^2 - 2x + 4$ **(f)** $-3x^2 - 4x - 5$ **14. (a)** -1 **(b)** 2 **(c)** 0 **(d)** $2x^2 - x - 1$ **(e)** $-2x^2 - x + 1$ **(f)** $2x^2 + 5x + 2$ **15. (a)** 0 **(b)** $\frac{1}{2}$ **(c)** $-\frac{1}{2}$ **(d)** $\frac{-x}{x^2+1}$ **(e)** $\frac{-x}{x^2+1}$ **(f)** $\frac{x+1}{x^2+2x+2}$ **16. (a)** $-\frac{1}{4}$ **(b)** 0 **(c)** 0 **(d)** $-\frac{x^2-1}{x-4}$ **(e)** $-\frac{x^2-1}{x+4}$ **(f)** $\frac{x^2+2x}{x+5}$ **17. (a)** 4 **(b)** 5 **(c)** 5 **(d)** $|x| + 4$ **(e)** $-|x| - 4$ **(f)** $|x+1| + 4$ **18. (a)** 0 **(b)** $\sqrt{2}$ **(c)** 0 **(d)** $\sqrt{x^2 - x}$ **(e)** $-\sqrt{x^2 + x}$ **(f)** $\sqrt{x^2 + 3x + 2}$ **19. (a)** $-\frac{1}{5}$ **(b)** $-\frac{3}{2}$ **(c)** $\frac{1}{8}$ **(d)** $\frac{-2x+1}{-3x-5}$ **(e)** $\frac{-2x-1}{3x-5}$ **(f)** $\frac{2x+3}{3x-2}$ **20. (a)** $\frac{3}{4}$ **(b)** $\frac{8}{9}$ **(c)** 0 **(d)** $1 - \frac{1}{x^2 - 4x + 4}$ **(e)** $-1 + \frac{1}{(x+2)^2}$ **(f)** $1 - \frac{1}{(x+3)^2}$ **21.** $f(0) = 3; f(-6) = -3$ **22.** $f(6) = 0; f(11) = 1$ **23.** Positive **24.** Negative **25.** -3, 6 and 10 **26.** $-3 < x < 6$ and $10 < x \le 11$ **27.** $\{x \mid -6 \le x \le 11\}$ **28.** $\{y \mid -3 \le y \le 4\}$ **29.** $(-3, 0), (6, 0), (10, 0)$ **30.** $(0, 3)$ **31.** 3 times **32.** 2 times **33. (a)** No **(b)** $-3; (4, -3)$ **(c)** 14; $(14, 2)$ **(d)** $\{x \mid x \neq 6\}$ **34. (a)** Yes **(b)** $\frac{1}{2}; \left(0, \frac{1}{2}\right)$ **(c)** 0 or $\frac{1}{2}; \left(0, \frac{1}{2}\right)$ and $\left(\frac{1}{2}, \frac{1}{2}\right)$ **(d)** $\{x \mid x \neq -4\}$ **35. (a)** Yes **(b)** $\frac{8}{17}; \left(2, \frac{8}{17}\right)$ **(c)** $-1, 1; (-1, 1), (1, 1)$ **(d)** All real numbers **36. (a)** Yes **(b)** 4; $(4, 4)$ **(c)** $-2; (-2, 1)$ **(d)** $\{x \mid x \neq 2\}$ **37.** Not a function **38.** Function **(a)** Domain: All real numbers; Range: $\{y \mid 0 < y < \infty\}$ **(b)** Intercept: $(0, 1)$ **(c)** None **39.** Function **(a)** Domain: $\{x \mid -\pi \le x \le \pi\}$; Range: $\{y \mid -1 \le y \le 1\}$ **(b)** Intercepts: $\left(-\frac{\pi}{2}, 0\right), \left(\frac{\pi}{2}, 0\right), (0, 1)$ **(c)** y-axis

40. Function **(a)** Domain: $\left\{x \middle| -\frac{\pi}{2} < x < \frac{\pi}{2}\right\}$; Range: All real numbers **(b)** Intercepts: $(0, 0)$ **(c)** Origin
41. Not a function **42.** Not a function
43. Function **(a)** Domain: $\{x|x > 0\}$; Range: All real numbers **(b)** Intercept: $(1, 0)$ **(c)** None
44. Function **(a)** Domain: $\{x|0 \le x \le 4\}$; Range: $\{y|0 \le y \le 3\}$ **(b)** Intercept: $(0, 0)$ **(c)** None
45. Function **(a)** Domain: all real numbers; Range: $\{y|y \le 2\}$ **(b)** Intercept: $(-3, 0), (3, 0), (0, 2)$ **(c)** y-axis
46. Function **(a)** Domain: $\{x|x \ge -3\}$; Range: $\{y|y \ge 0\}$ **(b)** Intercepts: $(-3, 0), (0, 2), (2, 0)$ **(c)** None
47. Function **(a)** Domain: $\{x|x \ne 2\}$; Range: $\{y|y \ne 1\}$ **(b)** Intercept: $(0, 0)$ **(c)** None
48. Function **(a)** Domain: $\{x|x \ne -1\}$; Range: $\{y|y > 0\}$ **(b)** Intercept: $(0, 2)$ **(c)** None
49. All real numbers **50.** All real numbers **51.** All real numbers **52.** All real numbers **53.** $\{x|x \ne -1, x \ne 1\}$ **54.** $\{x|x \ne 1\}$
55. $\{x|x \ne 0\}$ **56.** $\{x|x \ne 0, x \ne -2, x \ne 2\}$ **57.** $\{x|x \ge 4\}$ **58.** $\{x|x \le 1\}$ **59.** $\{x|x > 9\}$ **60.** $\{x|x > 4\}$ **61.** $\{x|x < 1 \text{ or } x \ge 2\}$
62. $\{x|x \le -1 \text{ or } x \ge 2\}$ **63. (a)** III **(b)** IV **(c)** I **(d)** V **(e)** II **64. (a)** II **(b)** V **(c)** IV **(d)** III **(e)** I

65.

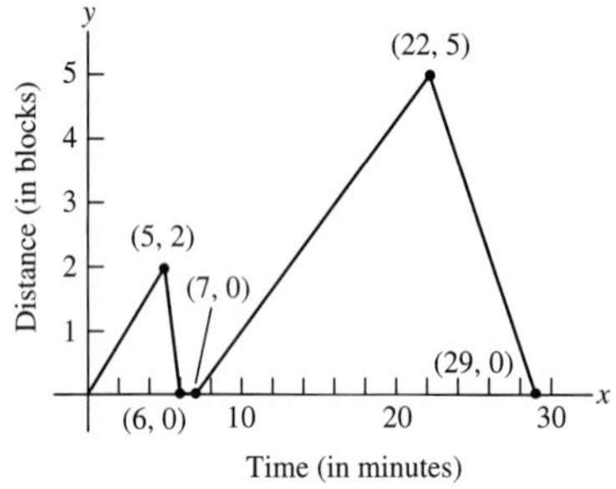

66.

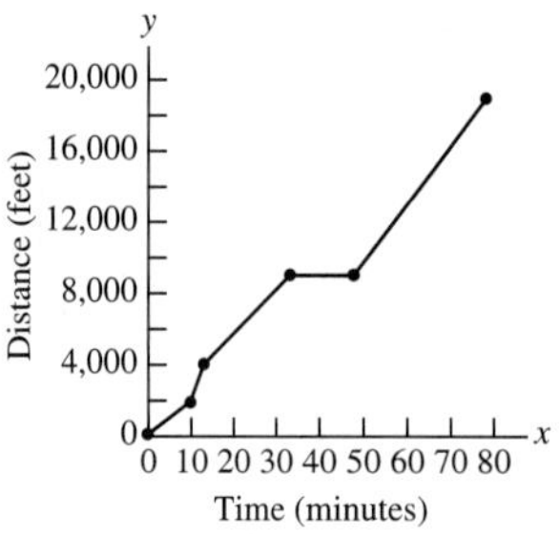

67. $A = -\frac{7}{2}$ **68.** $B = 5$ **69.** $A = -4$
70. $B = -1$ **71.** $A = 8$; undefined at $x = 3$
72. $A = 1, B = 2$
73. (a) 15.1 m, 14.07 m, 12.94 m, 11.72 m
(b) 1.01 sec, 1.42 sec, 1.74 sec **(c)** 2.02 sec

74. (a)

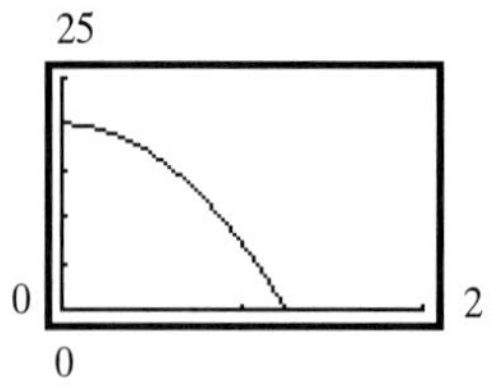

(b) 7m; 4.27m; 1.28 m
(c) 0.62 seconds, 0.88 seconds
(d) 1.07 seconds
(e) 1.24 seconds

75. $A(x) = \frac{1}{2}x^2$
76. $A(x) = \frac{1}{2}x^2$
77. $G(x) = 10x$
78. $G(x) = 100 + 10x$

80. The time is least for $x = 1.50$ mi.

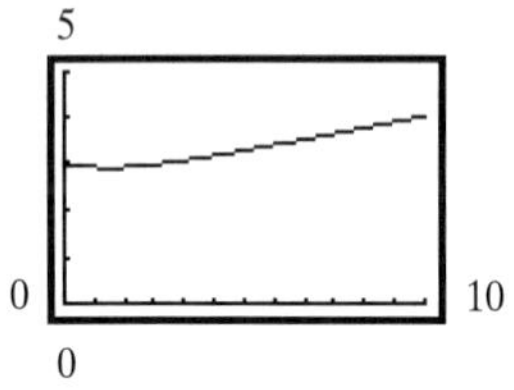

81. (a) $C(x) = 10x + 14\sqrt{x^2 - 10x + 29}, 0 \le x \le 5$
(b) $C(1) = \$72.61$ **(c)** $C(3) = \$69.60$
(d)

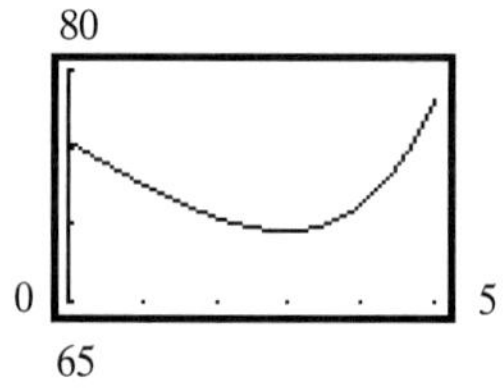

(e) Least cost: $x \approx 2.96$ miles

82. (a) $T(x) = \frac{\sqrt{9 + x^2}}{12} + \frac{20 - x}{5}, 0 \le x \le 20$
(b) 3.11 hrs **(c)** 2.63 hrs
(d)

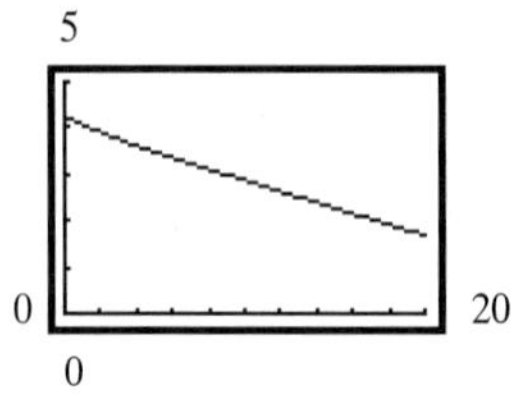

(e) T is least when $x = 20$.

83. (a) $A(x) = (8.5 - 2x)(11 - 2x)$
(b) $0 \le x \le 4.25, 0 \le A \le 93.5$
(c) $A(1) = 58.5 \text{ in}^2, A(1.2) = 52.46 \text{ in}^2, A(1.5) = 44 \text{ in}^2$
(d)

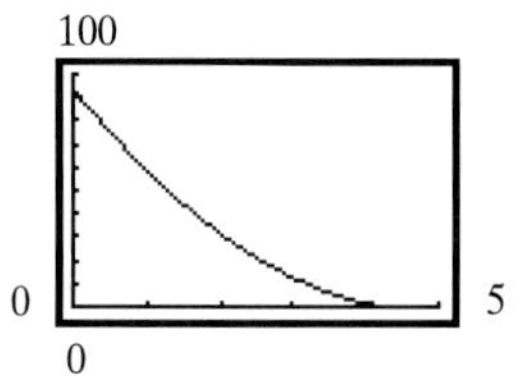

(e) $A(x) = 70$ when $x = 0.64$ in.; $A(x) = 50$ when $x = 1.28$ in.

84. **(a)** \$222 **(b)** \$225 **(c)** \$220 **(d)** \$230
(e)

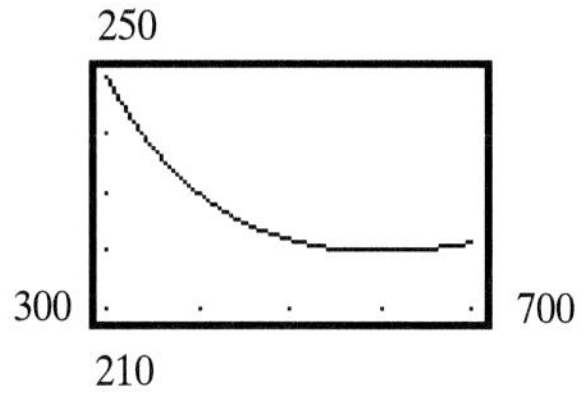

(f) $C(x)$ varies from \$230 to \$220.

86. **(a)** 119.8 lb **(b)**

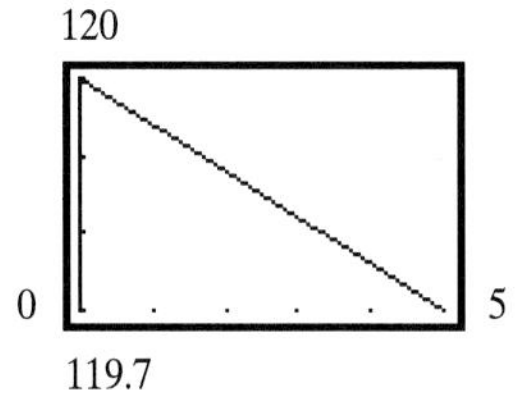

(c) Weight decreases as h approaches 5.
(d) 16.56 mi below sea level.

85. **(a)**

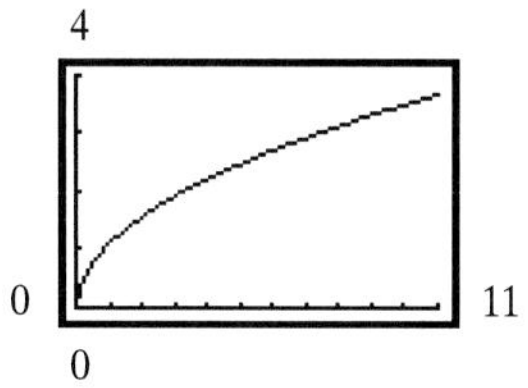

(b) The period of T varies from 1.107 seconds to 3.501 seconds.
(c) 81.56 ft

87. Function **88.** Function **89.** Function **90.** Function
91. Not a function **92.** Not a function **93.** Function
94. Not a function **95.** Only $h(x) = 2x$

2.2 Exercises

1.

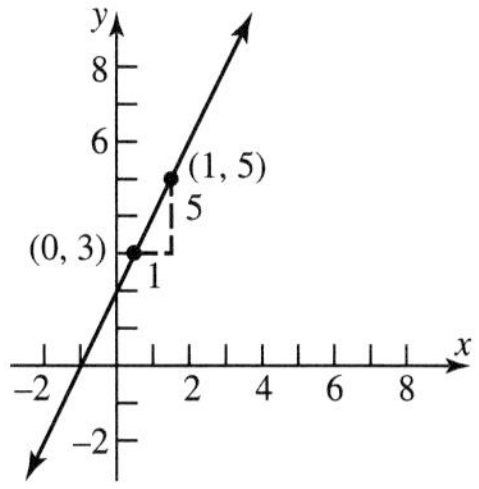

2.

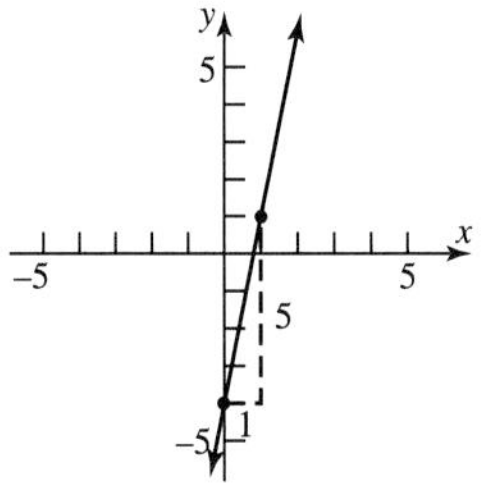

3.

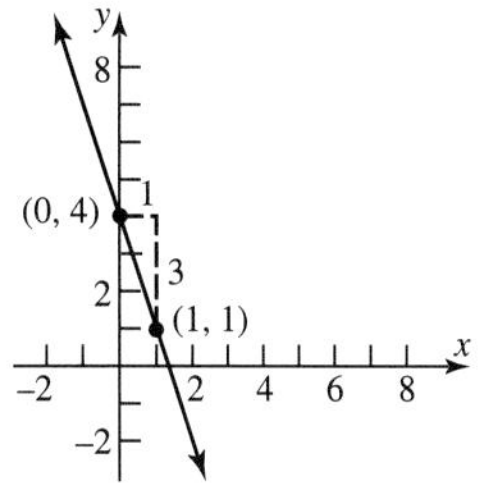

4.

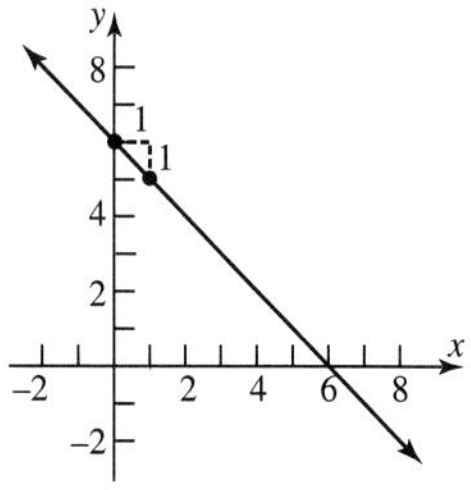

5.

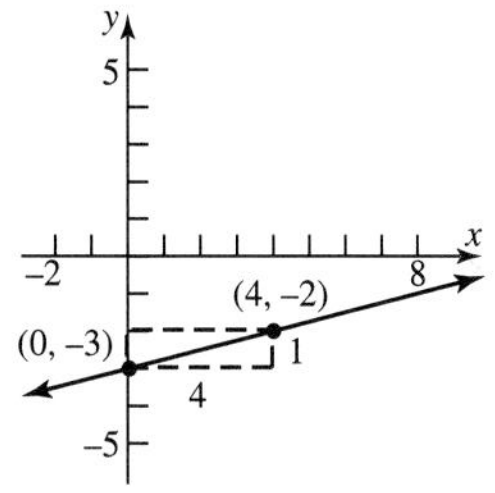

6.

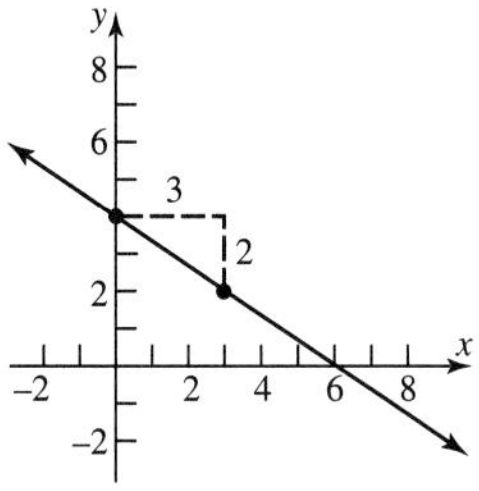

7.

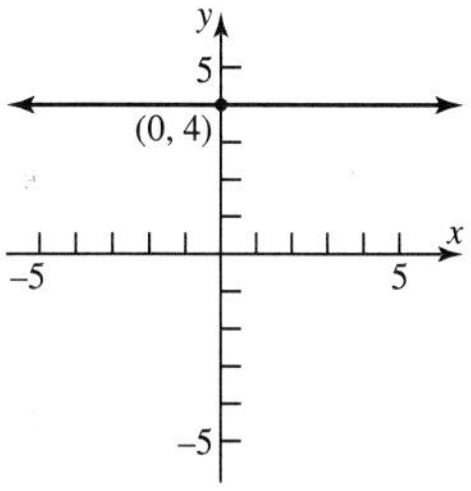

8.

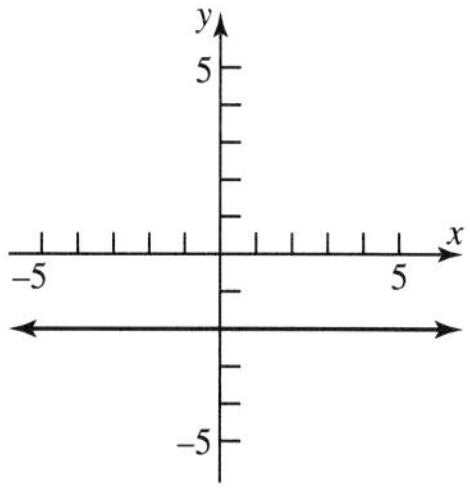

9. Linear relation **10.** Nonlinear relation **11.** Linear relation **12.** Nonlinear relation **13.** Nonlinear relation **14.** Linear relation

15. **(a)**

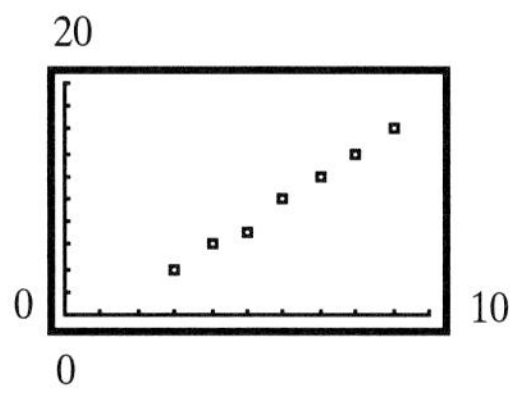

(b) Answers will vary.
Using (4, 6) and (8, 14):
$y = 2x - 2$

(c)

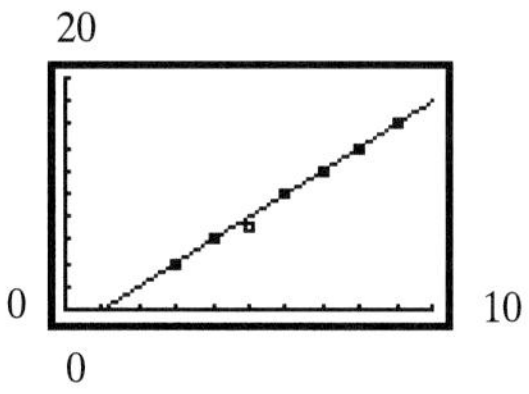

(d) $y = 2.0357x - 2.3571$

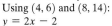

(e)

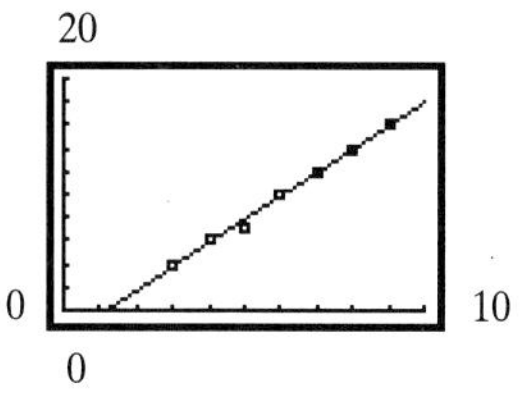

16. (a)

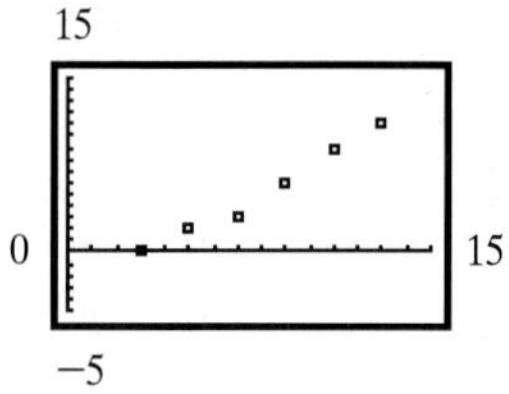

(b) Answers will vary.
Using (5, 2) and (11, 9):
$y = \frac{7}{6}x - \frac{23}{6}$

(c)

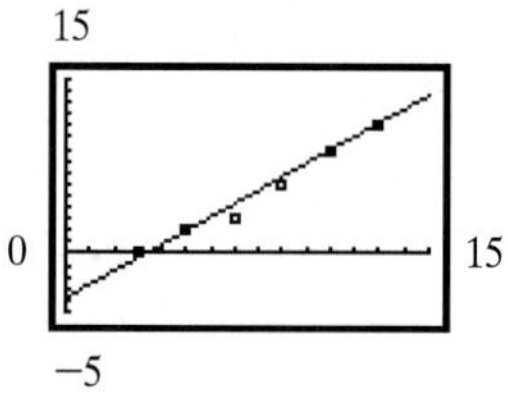

(d) $y = 1.129x - 3.862$

(e)

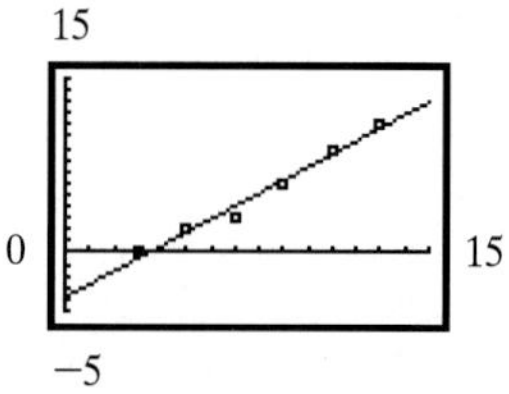

17. (a)

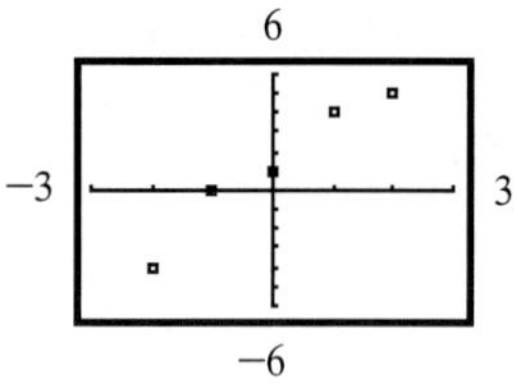

(b) Answers will vary.
Using (−2, −4) and (1, 4):
$y = \frac{8}{3}x + \frac{4}{3}$

(c)

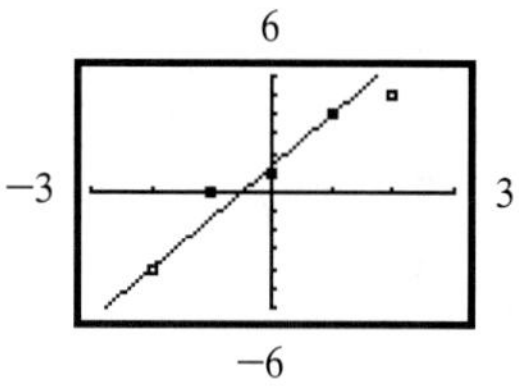

(d) $y = 2.2x + 1.2$

(e)

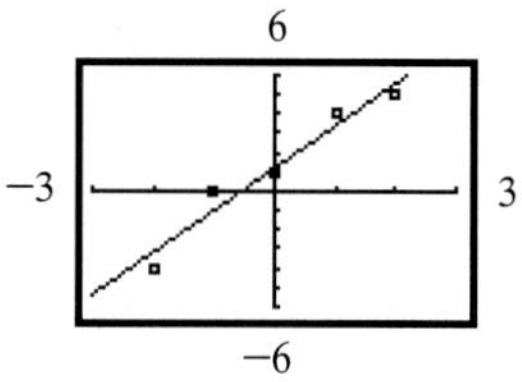

18. (a)

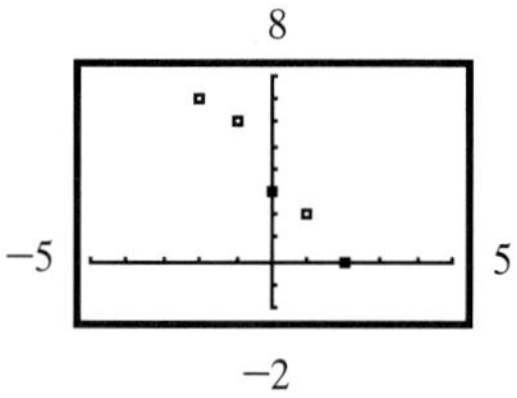

(b) Answers will vary.
Using (−2, 7) and (1, 2):
$y = -\frac{5}{3}x + \frac{11}{3}$

(c)

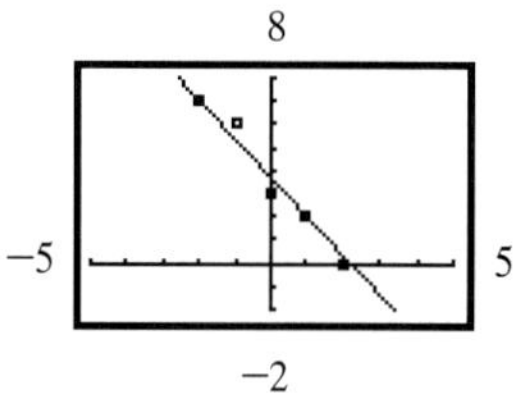

(d) $y = -1.8x + 3.6$

(e)

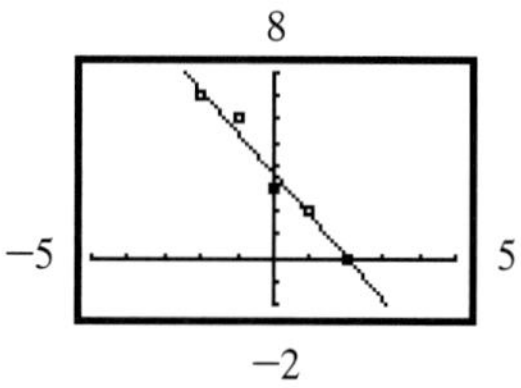

19. (a)

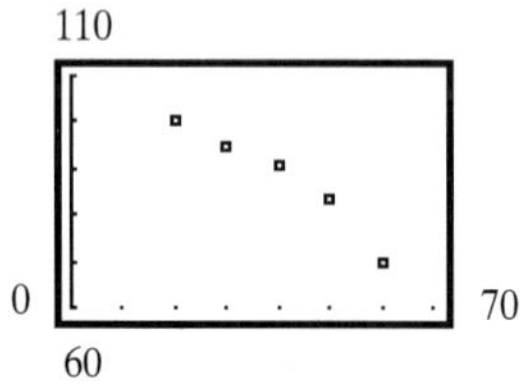

(b) Answers will vary.
Using (30, 95) and (60, 70):
$y = -\frac{5}{6}x + 120$

(c)

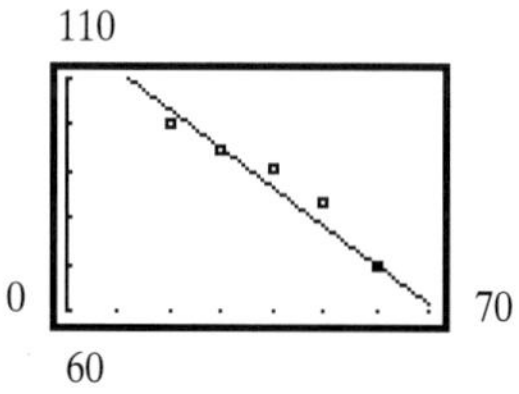

(d) $y = -0.72x + 116.6$

(e)

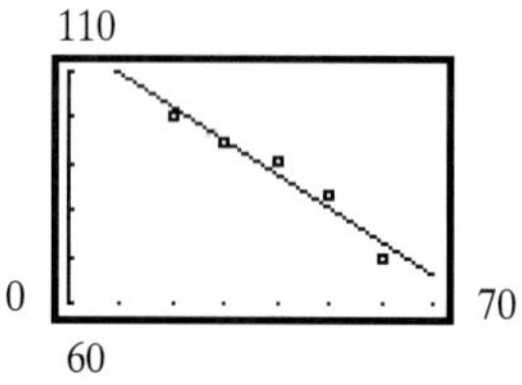

20. (a)

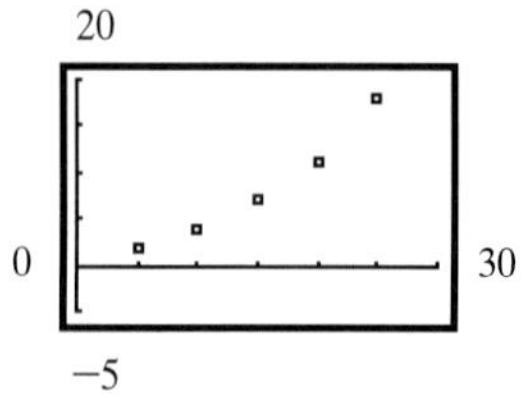

(b) Answers will vary.
Using (10, 4) and (20, 11):
$y = \frac{7}{10}x - 3$

(c)

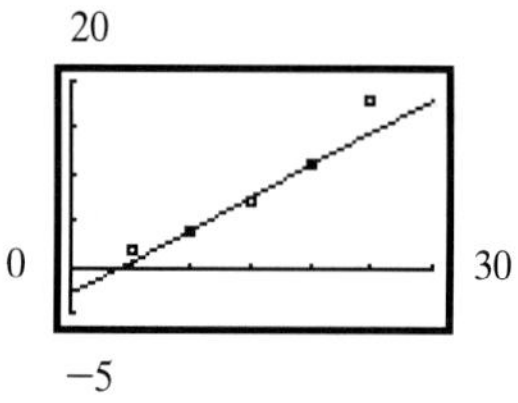

(d) $y = 0.78x - 3.3$

(e)

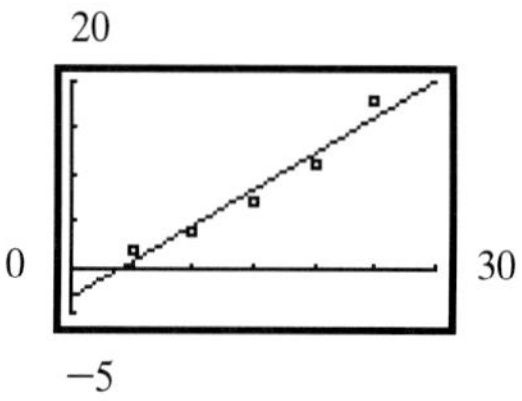

21. (a)

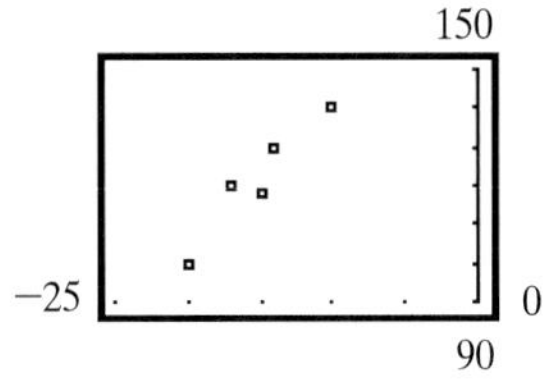

(b) Answers will vary.
Using (−20, 100) and (−15, 118):
$y = \frac{18}{5}x + 172$

(c)

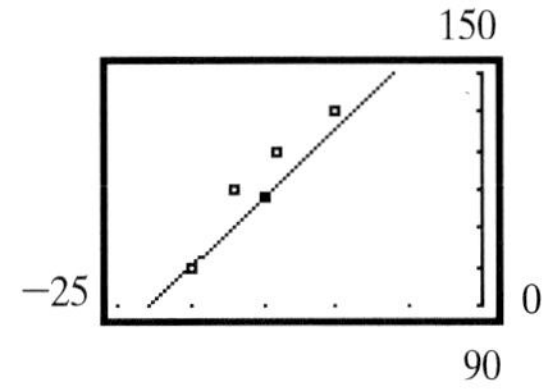

(d) $y = 3.8613x + 180.292$

(e)

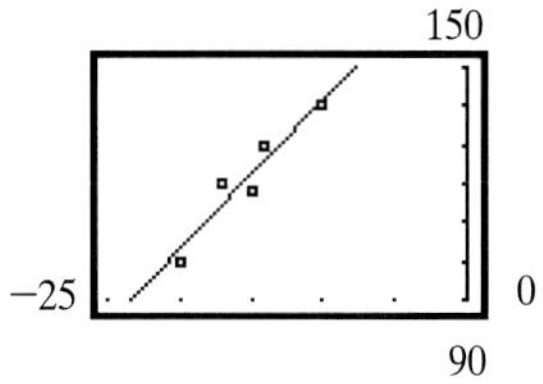

22. (a)

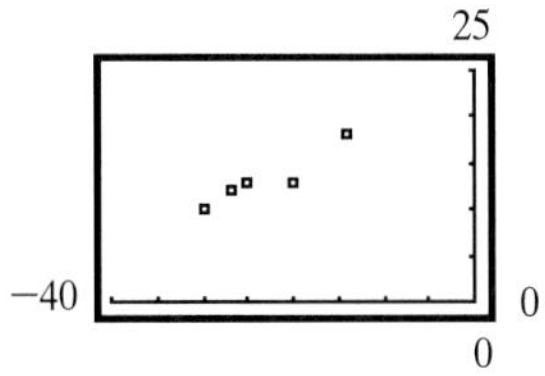

(b) Answers will vary.
Using (−30, 10) and (−14, 18):
$y = \frac{1}{2}x + 25$

(c)

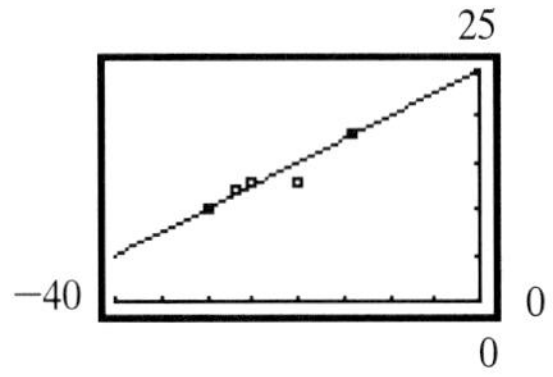

(d) $y = 0.442x + 23.456$

(e)

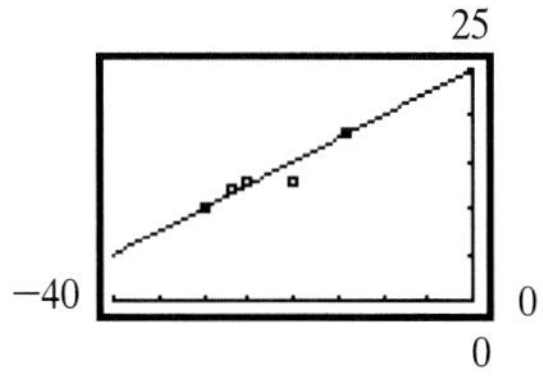

23. (a)

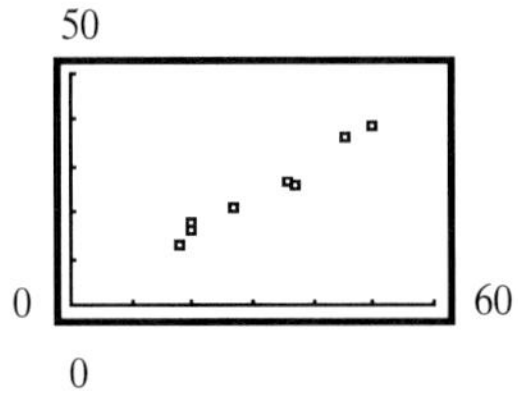

(b) $C(I) = 0.755I + 0.6266$
(c) As disposable income increases by $1000, consumption increases by about $755
(d) $32,337

24. (a)

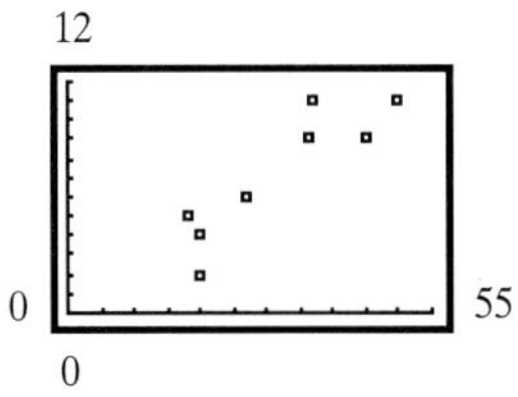

(b) $S(I) = 0.245I - 0.627$
(c) As disposable income increases by $1000, savings increase by $245.
(d) $9663

25. (a)

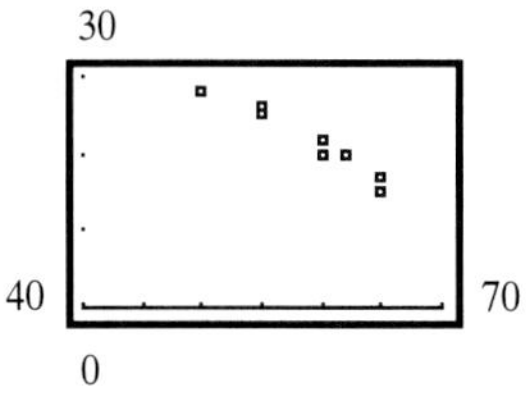

(b) $Y(X) = -0.8265X + 70.3903$
(c) As speed increases by 1 mph, mpg decreases by 0.8265.
(d) 20 mpg

26. (a)

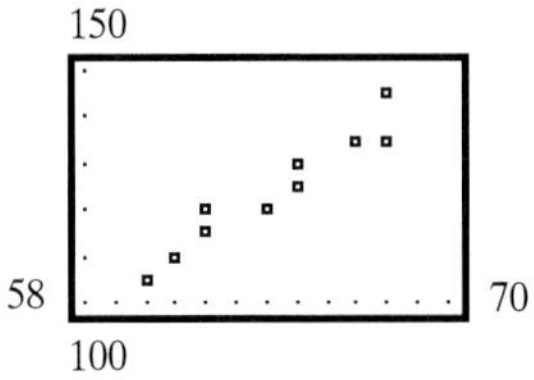

(b) $Y(X) = 4.127X - 140.952$
(c) As height increases by 1 in., weight increases by 4.127 lb.
(d) 131.43 lb

27. (a)

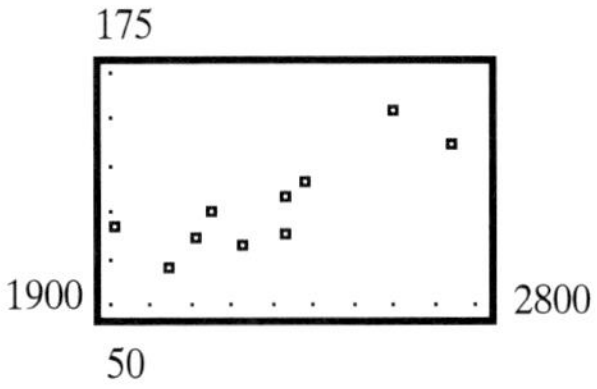

(b) $N(S) = 0.0842S - 88.5776$
(c) As sales increase by $1, NIBT increases by $0.0842
(d) $118.3 billion

28. (a)

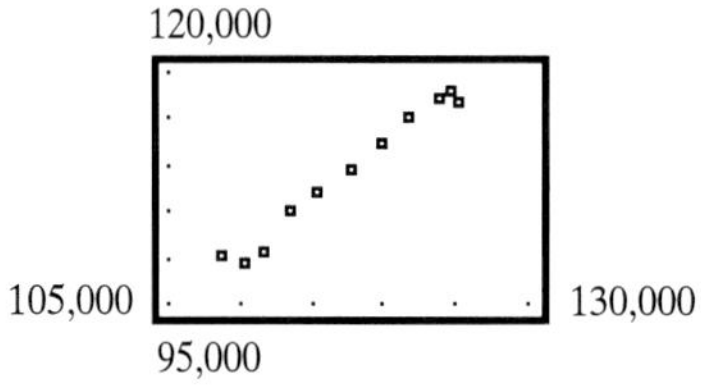

(b) $f(x) = 1.170x - 28{,}230.774$
(c) As the civilian labor force increases by 1000 people, the number of employed people increases by 1170.
(d) 114,907,026

29. (a) $y(x) = 1.13x - 2841.69$
(b) If per capita disposable increases $1, per capita personal consumption increases $1.13.
(c) $14,095.88

30. (a) $f(x) = 0.305x + 4.902$
(b) The amount of energy provided by coal increases by 0.305 trillion BTU for every 1 trillion BTU of energy provided by natural gas.
(c) 25.337 trillion BTUs

31. **(a)** No **(b)**

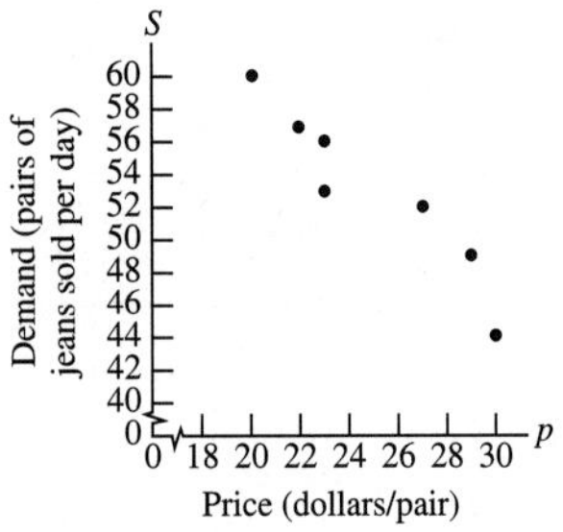

(c) $D = -1.34p + 86.20$
(d) If the price increases \$1, the quantity demanded decreases by 1.6
(e) $D(p) = -1.6p + 92$
(f) $\{p|p > 0\}$
(g) $D(28) = 47.2$; about 47 pairs

32. **(a)** No **(b)**

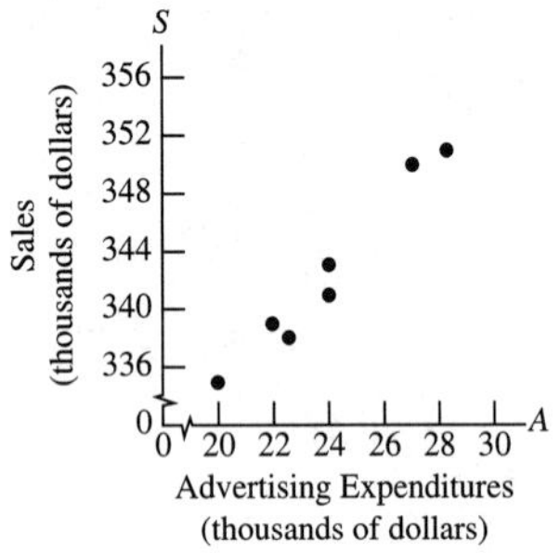

(c) $S = 2.067A + 292.887$
(d) As advertising increases by \$1000, sales increase by \$1928.
(e) $S(A) = 1.928A + 296.446$
(f) $\{A|A \geq 0\}$
(g) \$344,646

33. $p = 0.00649B$; \$941.05 **34.** $p = 0.00899B$; \$1573.25 **35.** $R = 0.96g$; \$10.08 **36.** $C(A) = 4.75A$; \$16.63 **37.** 144 ft; 2 s **38.** 96 ft/s

2.3 Exercises

1. $\{-4, 4\}$ **2.** $\{-9, 9\}$ **3.** $\{-7, -1\}$ **4.** $\{-5, 15\}$ **5.** $\{8 - \sqrt{17}, 8 + \sqrt{17}\}$ **6.** $\{-1 - \sqrt{13}, -1 + \sqrt{13}\}$ **7.** $\{-7, 3\}$ **8.** $\{3 - \sqrt{22}, 3 + \sqrt{22}\}$ **9.** $\left\{-\frac{1}{4}, \frac{3}{4}\right\}$ **10.** $\left\{-1, \frac{1}{3}\right\}$ **11.** $\left\{\frac{-1 - \sqrt{7}}{6}, \frac{-1 + \sqrt{7}}{6}\right\}$ **12.** $\left\{\frac{3 - \sqrt{17}}{4}, \frac{3 + \sqrt{17}}{4}\right\}$ **13.** $\{2 - \sqrt{2}, 2 + \sqrt{2}\}$ **14.** $\{-2 - \sqrt{2}, -2 + \sqrt{2}\}$ **15.** $\{2 - \sqrt{5}, 2 + \sqrt{5}\}$ **16.** $\{-3 - 2\sqrt{2}, -3 + 2\sqrt{2}\}$ **17.** $\left\{1, \frac{3}{2}\right\}$ **18.** $\left\{-\frac{3}{2}, -1\right\}$ **19.** No real solution **20.** No real solution **21.** $\left\{\frac{-1 - \sqrt{5}}{4}, \frac{-1 + \sqrt{5}}{4}\right\}$ **22.** $\left\{\frac{-1 - \sqrt{3}}{2}, \frac{-1 + \sqrt{3}}{2}\right\}$ **23.** $\left\{\frac{1}{3}\right\}$ **24.** No real solution **25.** D **26.** F **27.** A **28.** H **29.** B **30.** C **31.** E **32.** G **33.** D **34.** A **35.** B **36.** C

37.

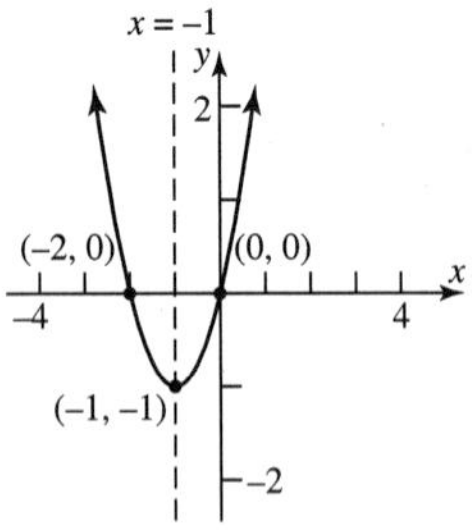

38.

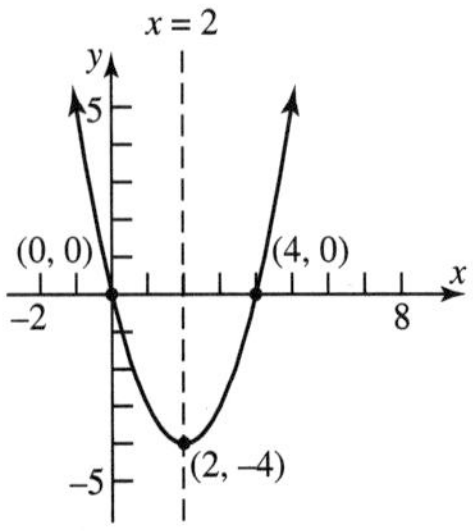

39.

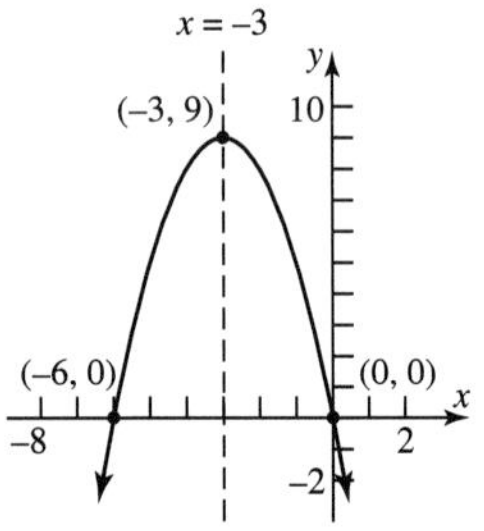

40.

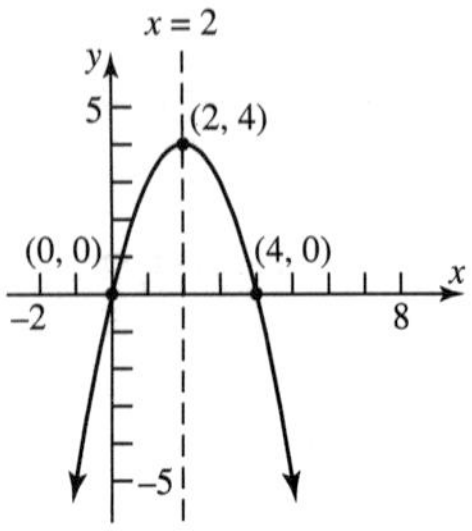

41.

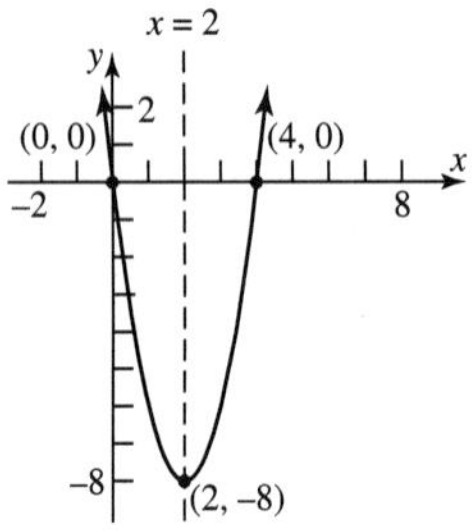

42.

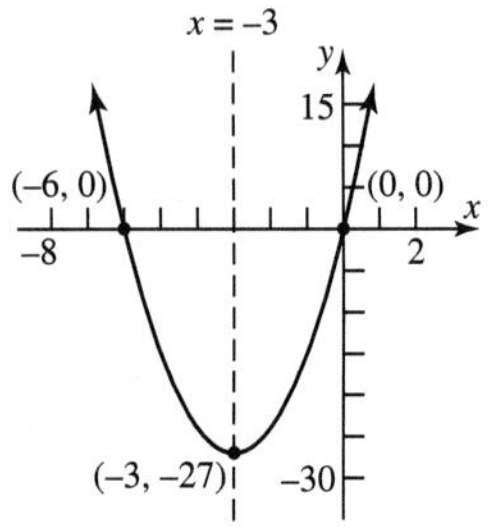

43.

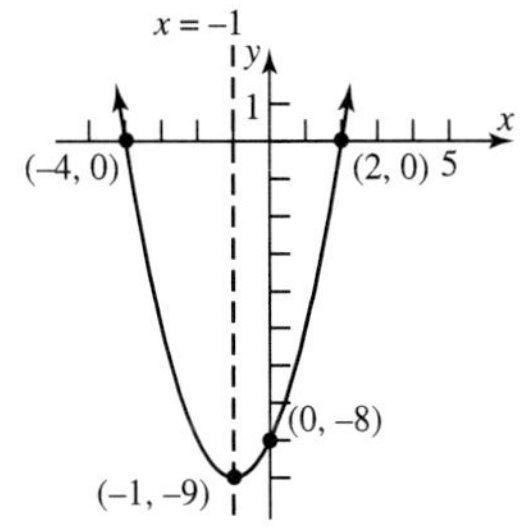

44.

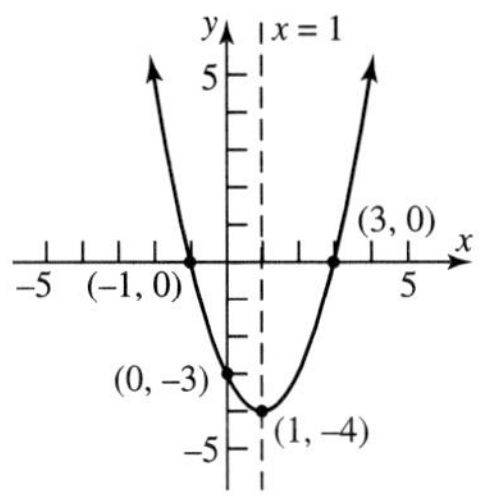

45.

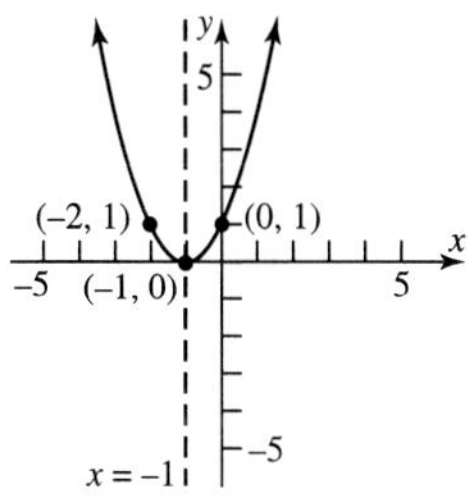

46.

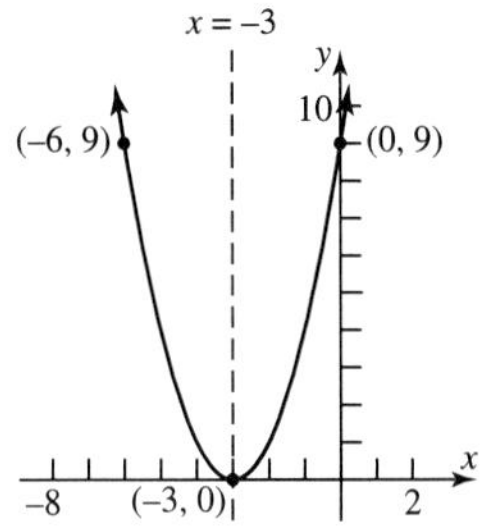

47.

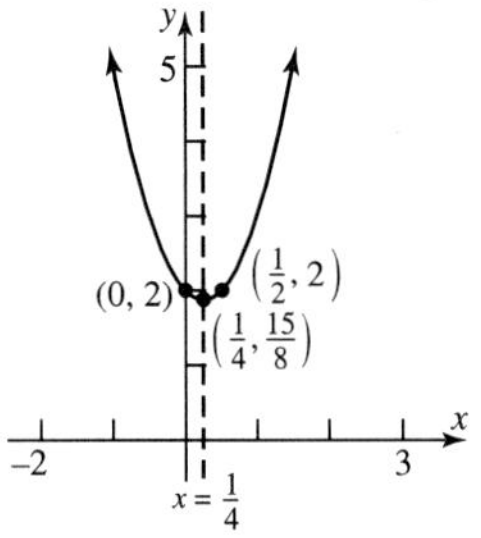

48.

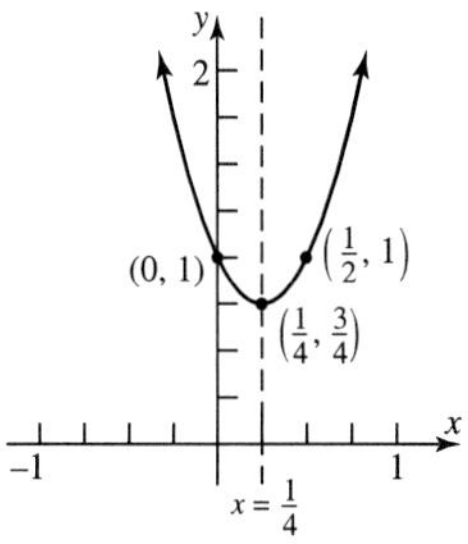

49.

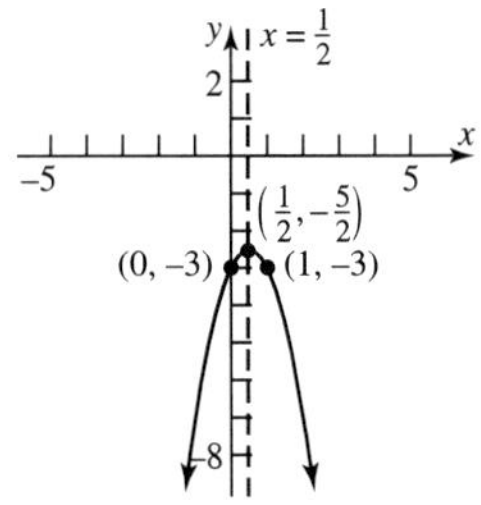

50.

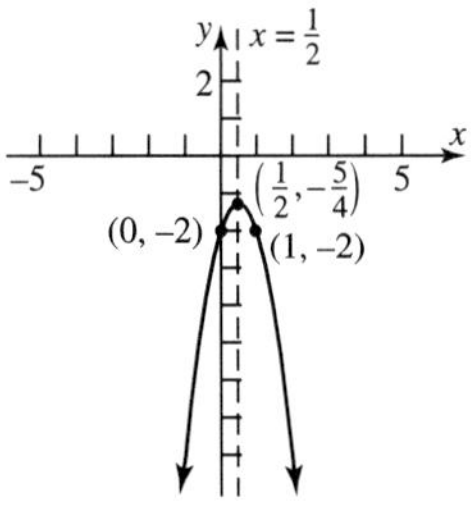

51.

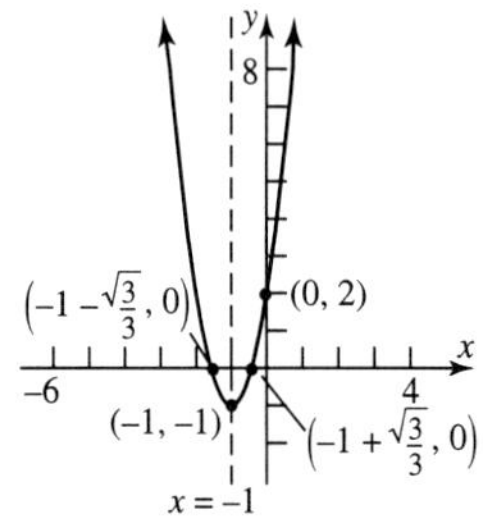

52.

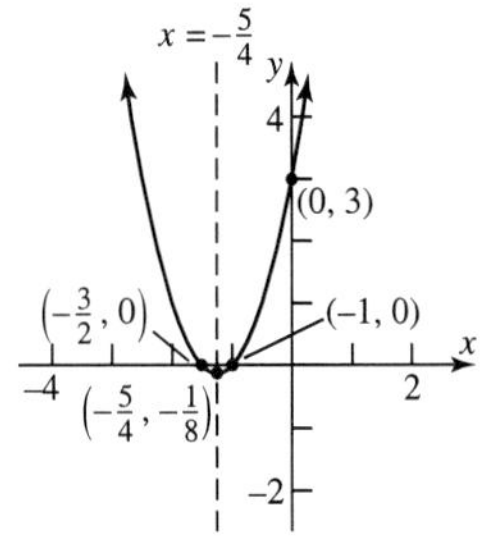

53.

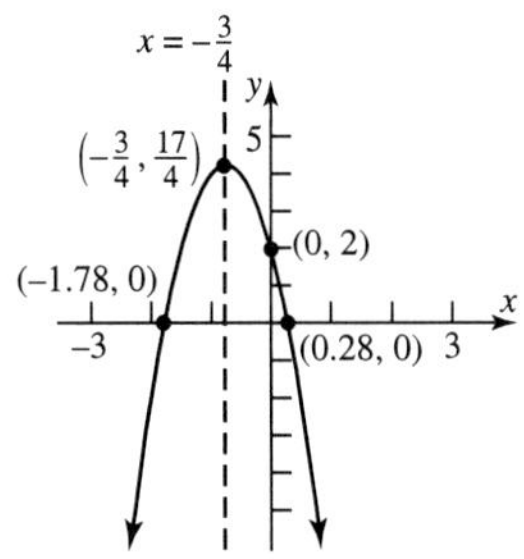

54.

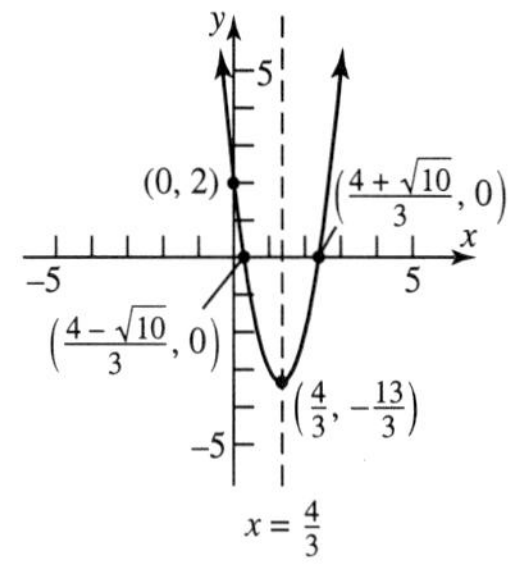

55. $f(x) = 6x^2 + 2$ **56.** $f(x) = -3x^2 + 6x + 1$ **57.** $R(x) = -\frac{1}{6}x^2 + 100x$ **58.** $R(x) = -\frac{1}{3}x^2 + 100x$
59. $R(x) = -\frac{1}{5}x^2 + 20x$ **60.** $R(x) = -\frac{1}{20}x^2 + 25x$ **61.** $A(x) = -x^2 + 200x$ **62.** $A(x) = -\frac{1}{2}x^2 + 1500x$

2.4 Exercises

1. Minimum value; 2 **2.** Minimum value; −4 **3.** Maximum value; 4 **4.** Maximum value; 9 **5.** Minimum value; −18 **6.** Maximum value; 18 **7.** Minimum value; −21 **8.** Minimum value; −1 **9.** Maximum value; 21 **10.** Maximum value; 11 **11.** Maximum value; 13 **12.** Minimum value; −1 **13.** \$500; \$1,000,000 **14.** \$1900; \$1,805,000 **15.** 40; \$400 **16.** 20; \$2000

17. **(a)** $R(x) = -\frac{1}{6}x^2 + 100x$
(b) \$13,333
(c) 300; \$15,000
(d) \$50

18. **(a)** $R(x) = -\frac{1}{3}x^2 + 100x$
(b) \$6667
(c) 150; \$7500
(d) \$50

19. **(a)** $R(x) = -\frac{1}{5}x^2 + 20x$
(b) \$255
(c) 50; \$500
(d) \$10

20. **(a)** $R(x) = -\frac{1}{20}x^2 + 25x$
(b) \$480
(c) 250; \$3125
(d) \$12.50

21. **(a)** $A(x) = -x^2 + 200x$

(b) A is largest when $x = 100$ yd.

(c) 10,000 sq yd

22. **(a)** $A(x) = -x^2 + 1500x$

(b) A is largest when $x = 750$ ft.

(c) 562,500 ft^2

23. 10,000 ft^2; 100 ft by 100 ft

24. $\dfrac{P}{4}$ by $\dfrac{P}{4}$

25. 2,000,000 m^2

26. 500,000 m^2

27. 4,166,666.7 m^2

28. 3,125,000 m^2

29. **(a)** $\dfrac{625}{16} \approx 39$ ft **(b)** 219.5 ft **(c)** 170 ft

(d)

(f) When the height is 100 ft, the projectile is 135.7 ft from the cliff.

30. **(a)** 156.25 ft **(b)** 78.125 ft **(c)** 312.5 ft

(d)

(f) When the height is 50 ft, the projectile has traveled horizontally 62.5 and 250 ft.

31. 18.75 m **32.** $y = -\dfrac{1}{144}x^2 + 25$; 24.3 ft; 22.2 ft; 13.9 ft **33.** 3 in. **34.** Width $= \dfrac{40}{\pi + 4} \approx 5.6$ ft; length ≈ 2.8 ft **35.** $\dfrac{750}{\pi}$ by 375 m

36. $x = \dfrac{16}{6 - \sqrt{3}} \approx 3.75$ ft; other side ≈ 2.38 ft

37. **(a)** Quadratic, $a < 0$.

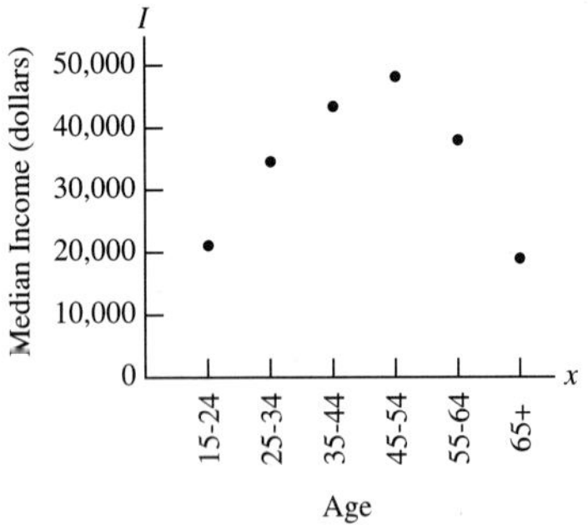

(b) 44.7 years old

(c) \$46,461

(e)

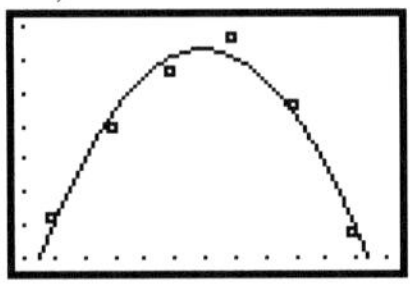

38. **(a)** Quadratic, $a < 0$.

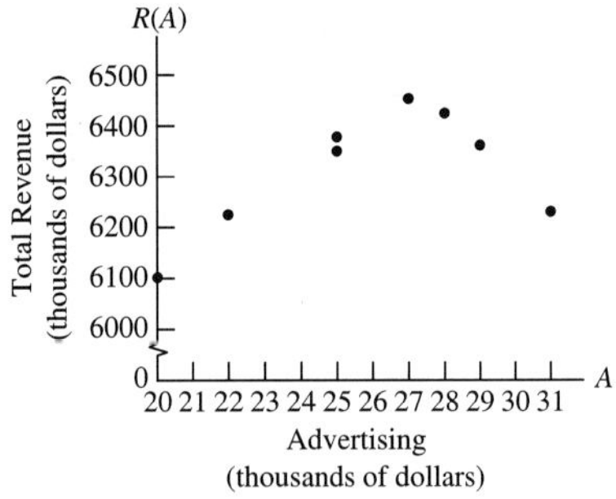

(b) Approximately \$26,500

(c) Approximately \$6,408,000

(e)

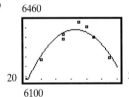

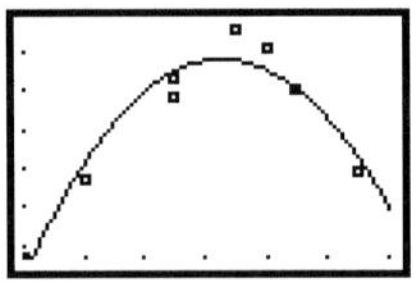

39. **(a)** Quadratic, $a > 0$.

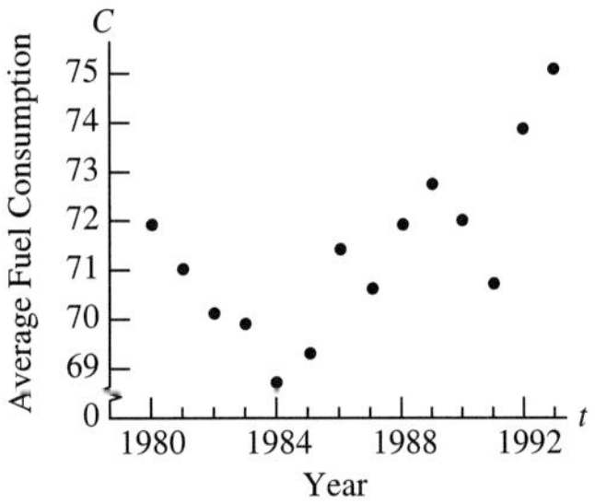

(b) 1984

(c) 67.3 billion gallons

(e)

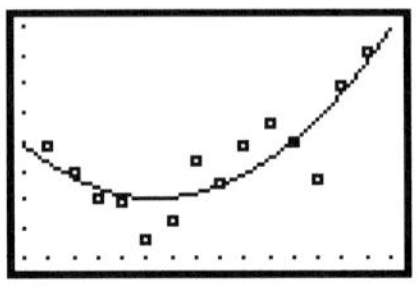

40. **(a)** Quadratic, $a < 0$.

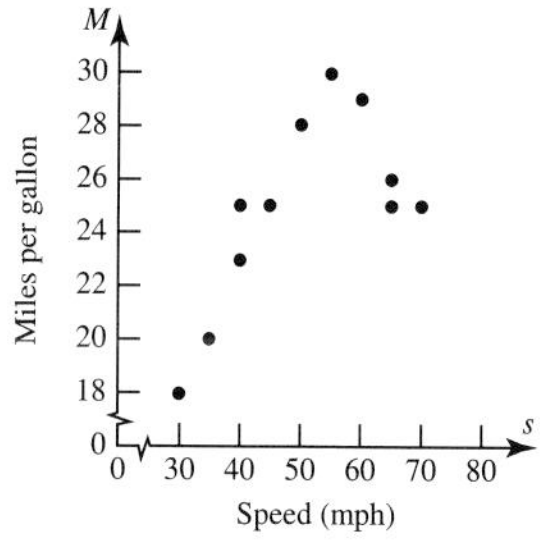

(b) Approximately 53.61 mph
(c) Approximately 24.81 mpg
(e)

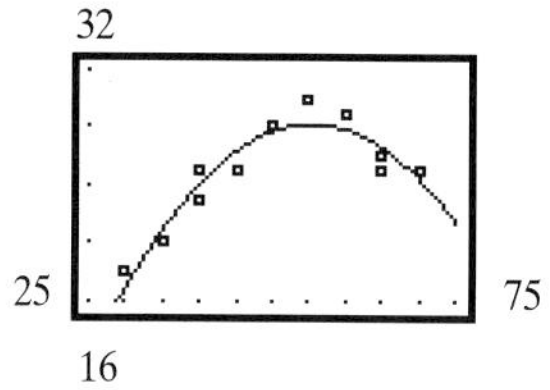

41. **(a)** Quadratic, $a < 0$.

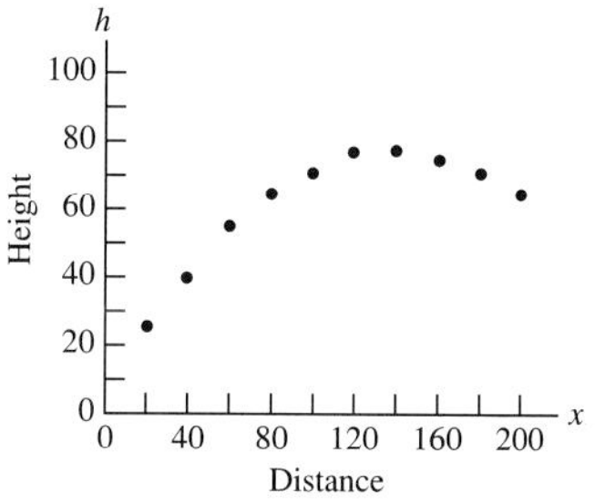

(b) 139.2 ft
(c) 77.4 ft
(e)

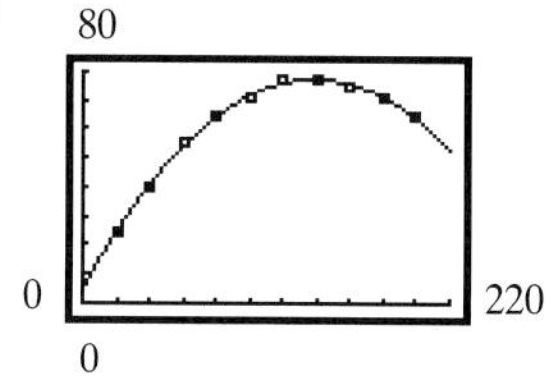

42. **(a)** Quadratic, $a > 0$.

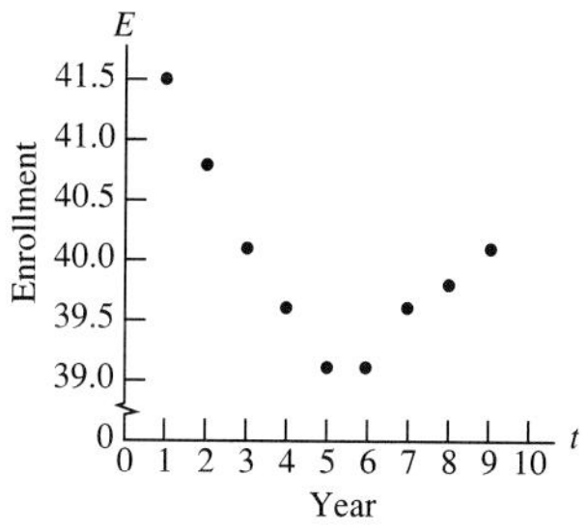

(b) During the academic year 1985–1986.
(d)

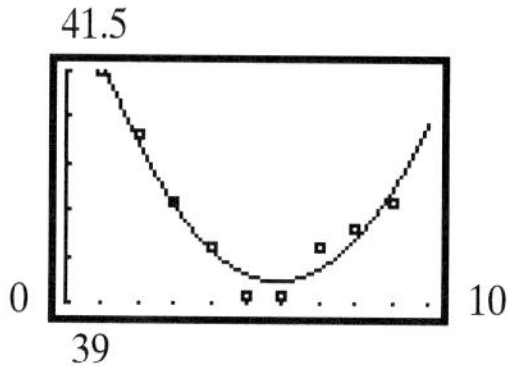

43. $x = \frac{a}{2}$

Fill-in-the-Blank Items

1. independent; dependent **2.** vertical **3.** add; $\frac{25}{4}$ **4.** discriminant; negative **5.** slope; y-intercept **6.** parabola **7.** $a < 0; -\frac{b}{2a}$

True/False Items

1. F **2.** T **3.** T **4.** F **5.** F **6.** T **7.** T

Review Exercises

1. $f(x) = -2x + 3$ **2.** $g(x) = -4x - 6$ **3.** $A = 11$ **4.** $A = 8$ **5.** b, c, d
6. **(a)** Domain: $\{x|-5 \le x \le 4\}$; Range: $\{y|-3 \le y \le 1\}$ **(b)** 1 **(c)** $(0, 0), (4, 0)$
7. **(a)** $f(-x) = \frac{-3x}{x^2 - 4}$ **(b)** $-f(x) = \frac{-3x}{x^2 - 4}$ **(c)** $f(x + 2) = \frac{3x + 6}{x^2 + 4x}$ **(d)** $f(x - 2) = \frac{3x - 6}{x^2 - 4x}$
8. **(a)** $f(-x) = \frac{x^2}{-x + 2}$ **(b)** $-f(x) = \frac{-x^2}{x + 2}$ **(c)** $f(x + 2) = \frac{x^2 + 4x + 4}{x + 4}$ **(d)** $f(x - 2) = \frac{x^2 - 4x + 4}{x}$
9. **(a)** $f(-x) = \sqrt{x^2 - 4}$ **(b)** $-f(x) = -\sqrt{x^2 - 4}$ **(c)** $f(x + 2) = \sqrt{x^2 + 4}$ **(d)** $f(x - 2) = \sqrt{x^2 - 4}$
10. **(a)** $f(-x) = |x^2 - 4|$ **(b)** $-f(x) = -|x^2 - 4|$ **(c)** $f(x + 2) = |x^2 + 4x|$ **(d)** $f(x - 2) = |x^2 - 4x|$
11. **(a)** $f(-x) = \frac{x^2 - 4}{x^2}$ **(b)** $-f(x) = -\frac{x^2 - 4}{x^2}$ **(c)** $f(x + 2) = \frac{x^2 + 4x}{x^2 + 4x + 4}$ **(d)** $f(x - 2) = \frac{x^2 - 4x}{x^2 - 4x + 4}$
12. **(a)** $f(-x) = \frac{-x^3}{x^2 - 4}$ **(b)** $-f(x) = \frac{-x^3}{x^2 - 4}$ **(c)** $f(x + 2) = \frac{x^3 + 6x^2 + 12x + 8}{x^2 + 4x}$ **(d)** $f(x - 2) = \frac{x^3 - 6x^2 + 12x - 8}{x^2 - 4x}$
13. $\{x|x \ne -3, x \ne 3\}$ **14.** $\{x|x \ne 2\}$ **15.** $\{x|x \le 2\}$ **16.** $\{x|x \ge -2\}$ **17.** $\{x|x > 0\}$ **18.** $\{x|x \ne 0\}$ **19.** $\{x|x \ne -3, x \ne 1\}$
20. $\{x|x \ne -1, x \ne 4\}$

21.

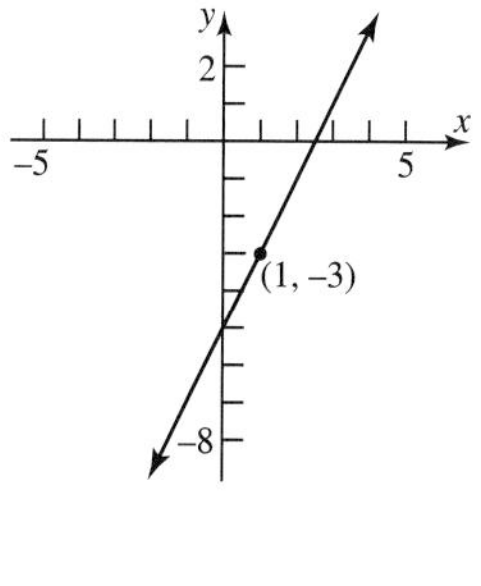

22.

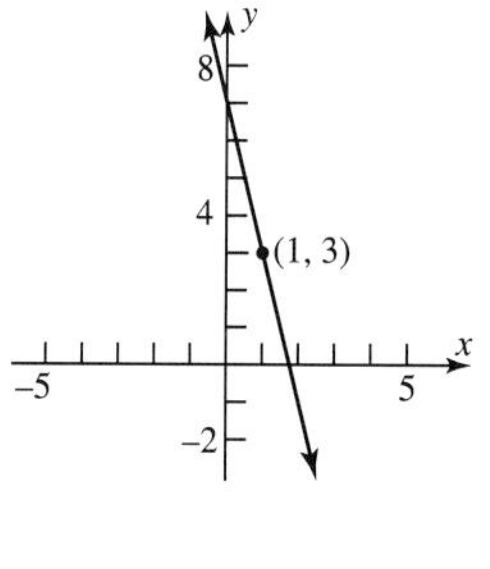

23.

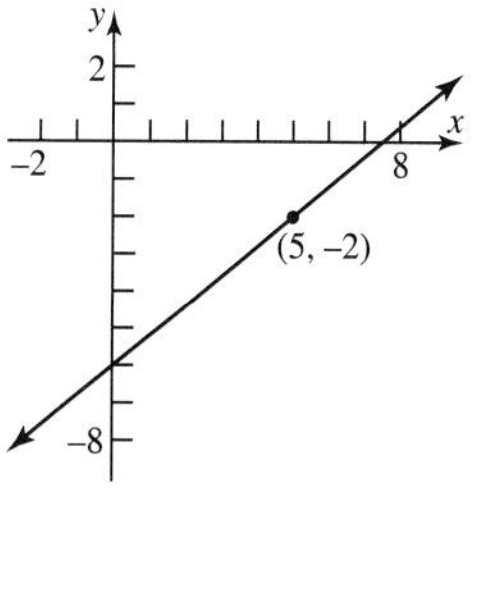

24.

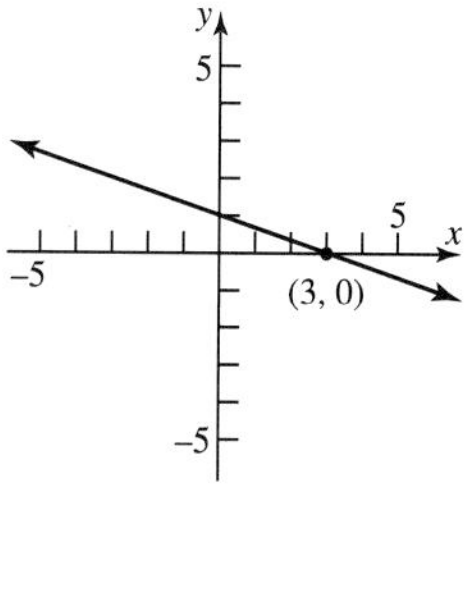

25. $\{-13, 7\}$ **26.** $\{2, 8\}$ **27.** $\left\{\frac{1-\sqrt{13}}{4}, \frac{1+\sqrt{13}}{4}\right\}$ **28.** $\left\{\frac{3-\sqrt{13}}{4}, \frac{3+\sqrt{13}}{4}\right\}$ **29.** $\left\{\frac{-3-\sqrt{5}}{2}, \frac{-3+\sqrt{5}}{2}\right\}$

30. $\{2-\sqrt{2}, 2+\sqrt{2}\}$ **31.** $\left\{-\frac{3}{4}\right\}$ **32.** No real solution

33.

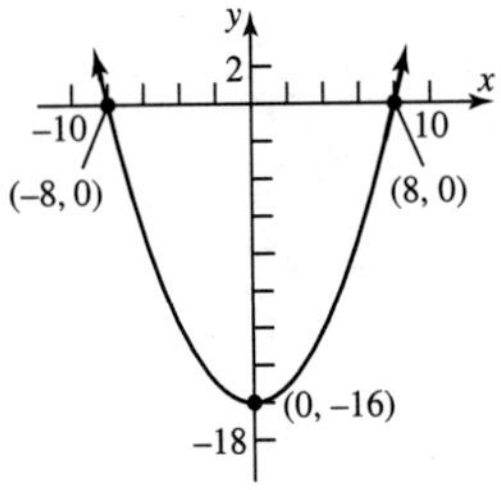

34.

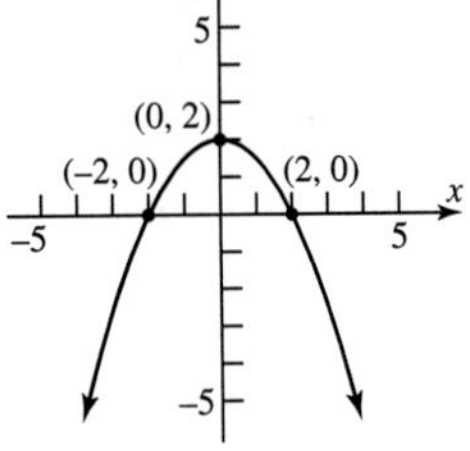

35.

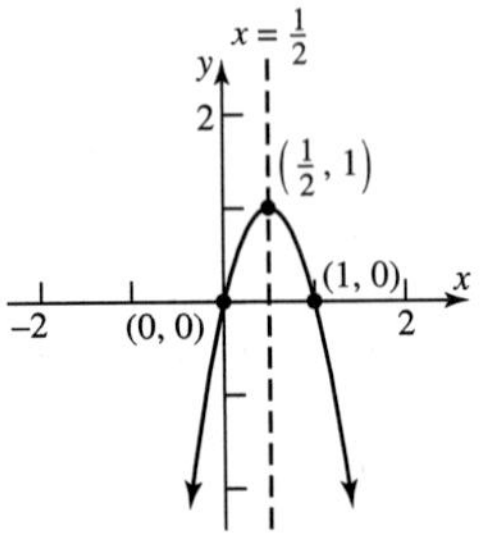

36.

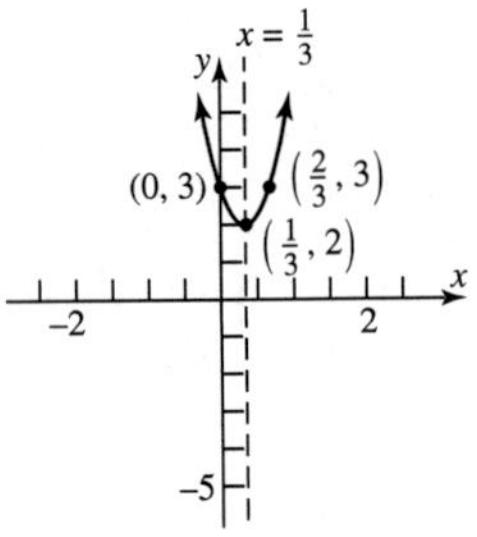

37.

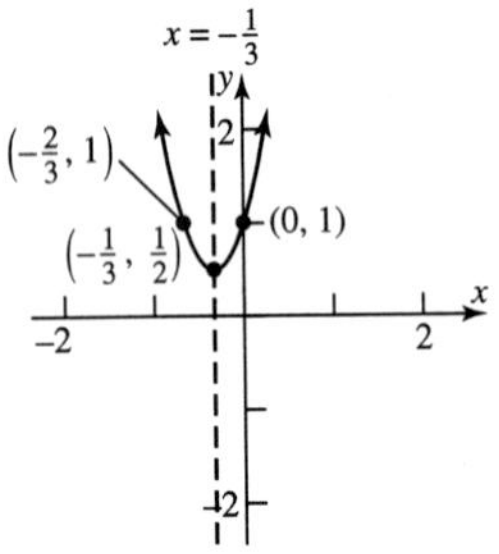

38.

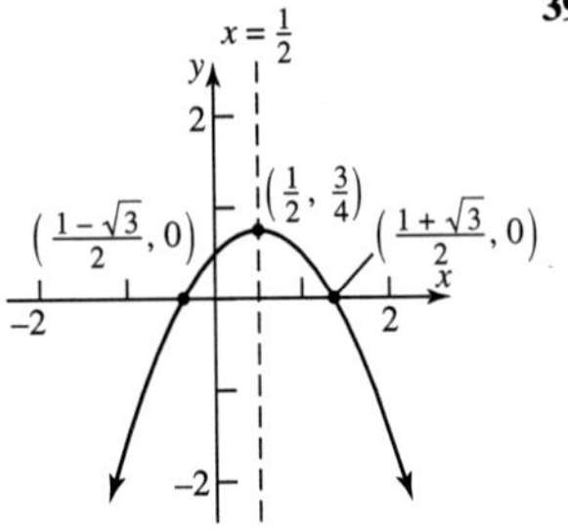

39.

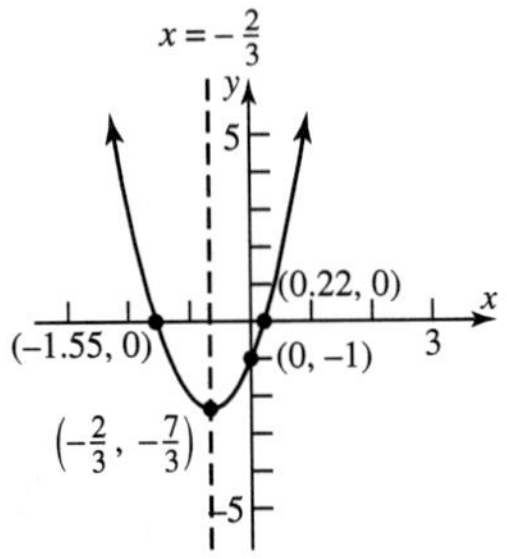

40.

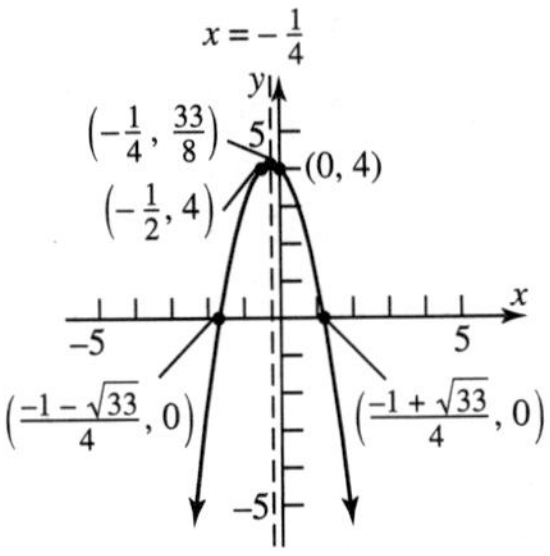

41.

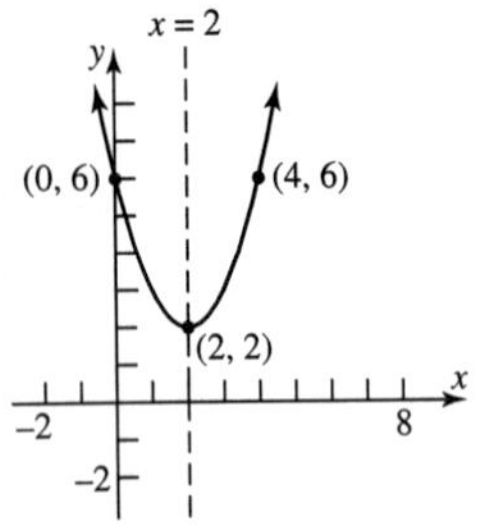

42.

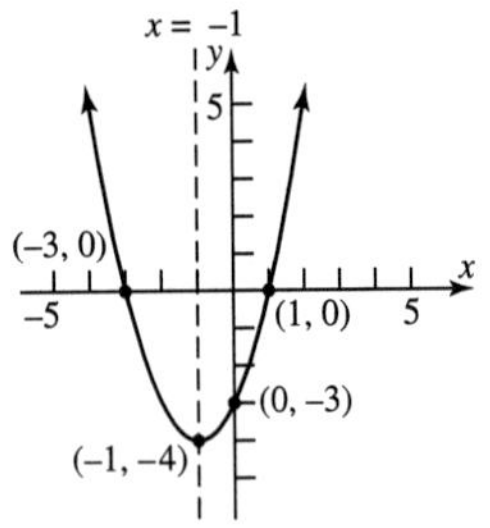

43. Minimum value; 1 **44.** Minimum value; −3
45. Maximum value; 12 **46.** Maximum value; 22
47. Maximum value; 16 **48.** Maximum value; 4

49. **(a)** Yes
(b)

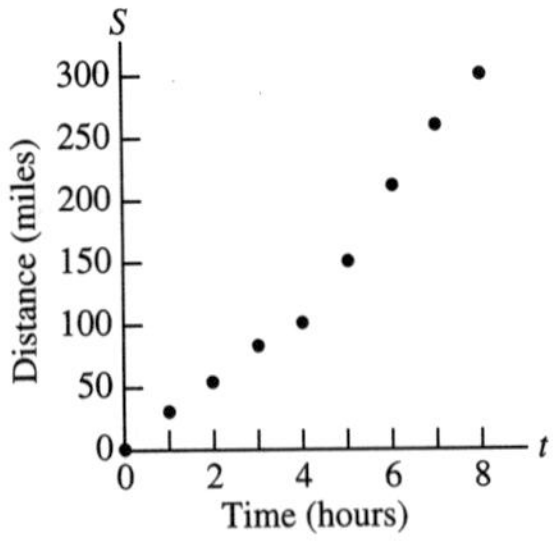

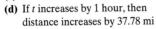

(c) $s = 37.78t - 19.13$
(d) If t increases by 1 hour, then distance increases by 37.78 mi
(e) $s(t) = 37.78t - 19.13$
(f) $\{t | t \geq 0\}$
(g) 396.45

50. **(a)** Yes
(b)

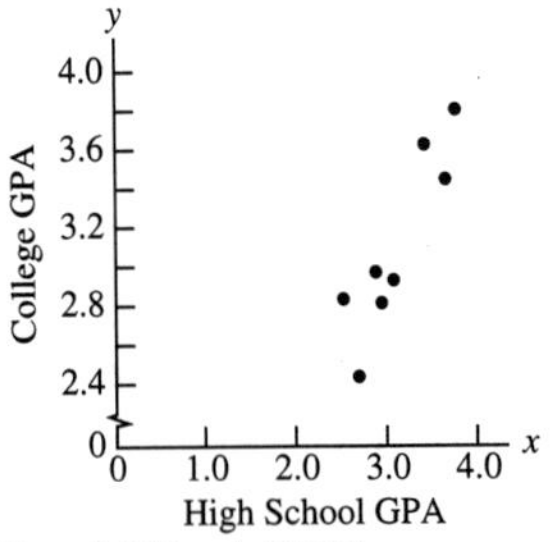

(c) $G = 0.964x + 0.072$
(d) As high school GPA increases 1 point, college GPA increases by 0.964 point.
(e) $G(x) = 0.964x + 0.072$
(f) $\{x | 0 \leq x \leq 4\}$
(g) 3.19

51. $p = 0.00657B$; \$1083.92
52. $R(g) = 1.18g$; \$13.22 **53.** 50 by 50 ft
54. 2 by 8 ft
55. 25 square units
56. **(a)** 62 **(b)** \$12.58 **(c)** \$20,000
57. **(a)** 5000 books **(b)** \$7.36 **(c)** \$111,770
58. 3.6 ft

CHAPTER 3 Functions and Their Graphs

3.1 Exercises

1. C **2.** A **3.** E **4.** G **5.** B **6.** D **7.** F **8.** H **9. (a)** Domain: $\{x \mid -3 \le x \le 4\}$; Range: $\{y \mid 0 \le y \le 3\}$ **(b)** Increasing on $(-3, 0)$ and on $(2, 4)$; Decreasing on $(0, 2)$ **(c)** Neither **(d)** $(-3, 0), (0, 3), (2, 0)$ **10. (a)** Domain: $\{x \mid -2 \le x \le 3\}$; Range: $\{y \mid -1 \le y \le 3\}$ **(b)** Increasing on $(-2, 0)$ and on $(1, 3)$; Decreasing on $(0, 1)$ **(c)** Neither **(d)** $\left(-\frac{4}{3}, 0\right), (0, 2), (1, 0)$ **11. (a)** Domain: all real numbers; Range: $\{y \mid y > 0\}$ **(b)** Increasing on $(-\infty, \infty)$ **(c)** Neither **(d)** $(0, 1)$ **12. (a)** Domain: $\{x \mid x > 0\}$; Range: all real numbers **(b)** Increasing on $(0, \infty)$ **(c)** Neither **(d)** $(1, 0)$ **13. (a)** Domain: $\{x \mid -\pi \le x \le \pi\}$; Range: $\{y \mid -1 \le y \le 1\}$ **(b)** Increasing on $\left(-\frac{\pi}{2}, \frac{\pi}{2}\right)$; Decreasing on $\left(-\pi, -\frac{\pi}{2}\right)$ and on $\left(\frac{\pi}{2}, \pi\right)$ **(c)** Odd **(d)** $(-\pi, 0), (0, 0), (\pi, 0)$ **14. (a)** Domain: $\{x \mid -\pi \le x \le \pi\}$; Range: $\{y \mid -1 \le y \le 1\}$ **(b)** Increasing on $(-\pi, 0)$; Decreasing on $(0, \pi)$ **(c)** Even **(d)** $\left(-\frac{\pi}{2}, 0\right), (0, 1), \left(\frac{\pi}{2}, 0\right)$ **15. (a)** Domain: $\{x \mid x \ne 2\}$; Range: $\{y \mid y \ne 1\}$ **(b)** Decreasing on $(-\infty, 2)$ and on $(2, \infty)$ **(c)** Neither **(d)** $(0, 0)$ **16. (a)** Domain: $\{x \mid x \ne -1\}$; Range: $\{y \mid y > 0\}$ **(b)** Increasing on $(-\infty, -1)$; Decreasing on $(-1, \infty)$ **(c)** Neither **(d)** $(0, 2)$ **17. (a)** Domain: $\{x \mid x \ne 0\}$; Range: all real numbers **(b)** Increasing on $(-\infty, 0)$ and on $(0, \infty)$ **(c)** Odd **(d)** $(-1, 0), (1, 0)$ **18. (a)** Domain: $\{x \mid x \ne 0\}$; Range: $\{y \mid y \le -2 \text{ or } y \ge 2\}$ **(b)** Increasing on $(-1, 0)$ and on $(0, 1)$; Decreasing on $(-\infty, -1)$ and on $(1, \infty)$ **(c)** Odd **(d)** None **19. (a)** Domain: $\{x \mid x \ne -2, x \ne 2\}$; Range: $\{y \mid y \le 0 \text{ or } y > 1\}$ **(b)** Increasing on $(-\infty, -2)$ and on $(-2, 0)$; Decreasing on $(0, 2)$ and on $(2, \infty)$ **(c)** Even **(d)** $(0, 0)$ **20. (a)** Domain: $\{x \mid x \ne -1, x \ne 1\}$; Range: all real numbers **(b)** Increasing on $(-\infty, -1)$ and on $(-1, 1)$ and on $(1, \infty)$ **(c)** Odd **(d)** $(0, 0)$ **21. (a)** Domain: $\{x \mid -4 \le x \le 4\}$; Range: $\{y \mid 0 \le y \le 2\}$ **(b)** Increasing on $(-2, 0)$ and on $(2, 4)$; Decreasing on $(-4, -2)$ and on $(0, 2)$ **(c)** Even **(d)** $(-2, 0), (0, 2), (2, 0)$ **22. (a)** Domain: $\{x \mid -4 \le x \le 4\}$; Range: $\{y \mid -2 \le y \le 2\}$ **(b)** Increasing on $(-2, 2)$; Decreasing on $(-4, -2)$ and on $(2, 4)$ **(c)** Odd **(d)** $(0, 0), (-4, 0), (4, 0)$ **23.** Domain: $\{x \mid -4 \le x < 0 \text{ or } 0 < x \le 4\}$; Range: $\{y \mid 0 < y \le 4\}$ **(b)** Increasing on $(-4, 0)$; Decreasing on $(0, 4)$ **(c)** Even **(d)** None **24. (a)** Domain: $\{x \mid -2 \le x \le 2\}$; Range: $\{y \mid -6 \le y \le 6\}$ **(b)** Increasing on $\left(-2, -\frac{1}{2}\right)$ and $\left(\frac{1}{2}, 2\right)$; decreasing on $\left(-\frac{1}{2}, \frac{1}{2}\right)$ **(c)** Odd **(d)** $(-1, 0), (0, 0), (1, 0)$ **25. (a)** 4 **(b)** 2 **(c)** 5 **26. (a)** -1 **(b)** 2 **(c)** 5 **27. (a)** 2 **(b)** 3 **(c)** -4 **28. (a)** 0 **(b)** 0 **(c)** -1 **29. (a)** 3 **(b)** 3 **30. (a)** -2 **(b)** -2 **31. (a)** -3 **(b)** -3 **32. (a)** $x + 1$ **(b)** 3 **33. (a)** $3x + 1$ **(b)** 7 **34. (a)** $-2x + 2$ **(b)** -2 **35. (a)** $x(x + 1)$ **(b)** 6 **36. (a)** $x^2 + x + 2$ **(b)** 8 **37. (a)** $\frac{-1}{x + 1}$ **(b)** $-\frac{1}{3}$ **38. (a)** $-\frac{x + 1}{x^2}$ **(b)** $-\frac{3}{4}$ **39. (a)** $\frac{1}{\sqrt{x} + 1}$ **(b)** $\frac{1}{\sqrt{2} + 1}$ **40. (a)** $\frac{1}{\sqrt{x + 3} + 2}$ **(b)** $\frac{1}{\sqrt{5} + 2}$ **41.** Odd **42.** Even **43.** Even **44.** Neither **45.** Odd **46.** Neither **47.** Neither **48.** Even **49.** Even **50.** Odd **51.** Odd **52.** Odd **53.** At most one **54.** At most one

55. (a) All real numbers
(b) $(0, 1)$
(c)

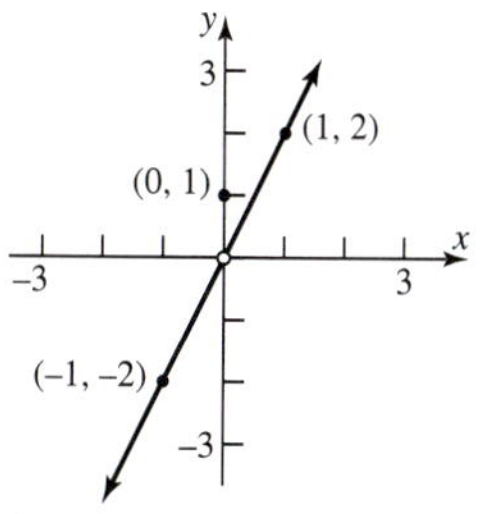

(d) $\{y \mid y \ne 0\}$

56. (a) All real numbers
(b) $(0, 4)$
(c)

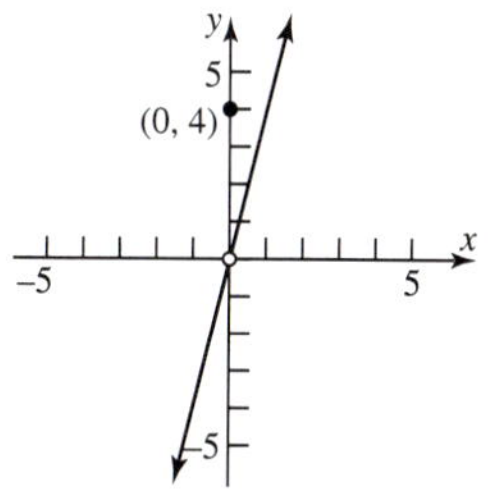

(d) $\{y \mid y \ne 0\}$

57. (a) All real numbers
(b) $(-1, 0), (0, 0)$
(c)

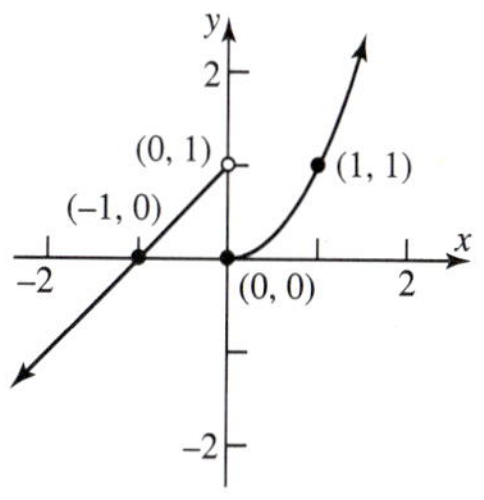

(d) All real numbers

58. (a) All real numbers
(b) $(0, 0)$
(c)

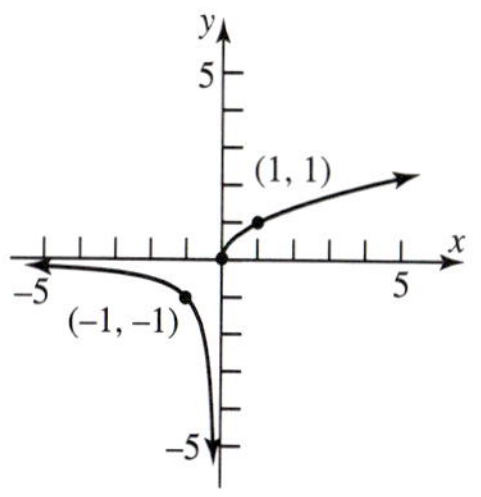

(d) All real numbers

59. **(a)** $\{x|x \geq -2\}$ **(b)** $(0, 1)$ **(c)**

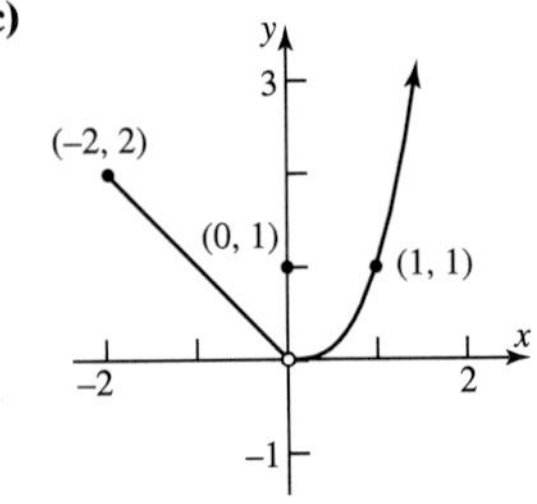

(d) $\{y|y > 0\}$

60. **(a)** $\{x|x \geq -3\}$ **(b)** $(-3, 0), (0, 3)$ **(c)**

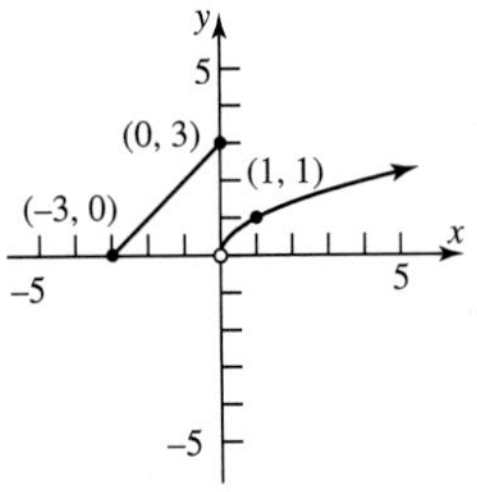

(d) $\{y|y \geq 0\}$

61. **(a)** All real numbers **(b)** $(x, 0)$ for $0 \leq x < 1$ **(c)**

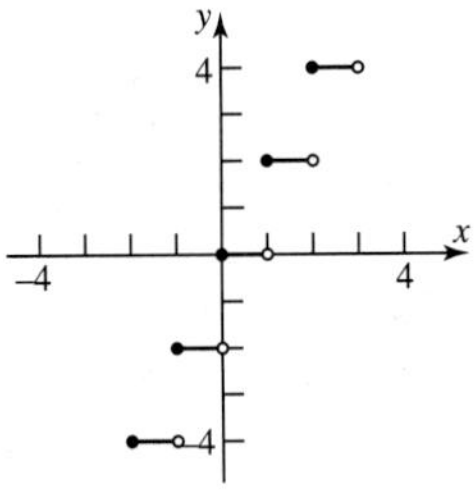

(d) Set of even integers

62. **(a)** All real numbers **(b)** $(x, 0)$ for $0 \leq x < \frac{1}{2}$ **(c)**

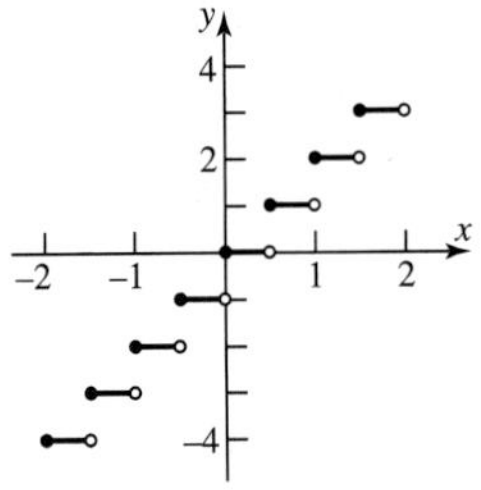

(d) Set of integers

63. 2 **64.** −3 **65.** $2x + h + 2$ **66.** $\dfrac{-1}{(x + h)x}$

67.

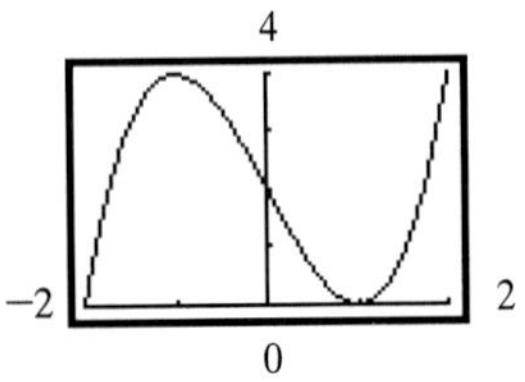

Increasing: $(-2, -1), (1, 2)$
Decreasing: $(-1, 1)$
Local maximum: $(-1, 4)$
Local minimum: $(1, 0)$

68.

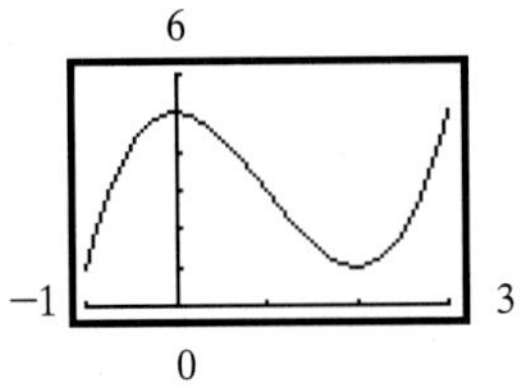

Increasing: $(-1, 0), (2, 3)$
Decreasing: $(0, 2)$
Local maximum: $(0, 5)$
Local minimum: $(2, 1)$

69.

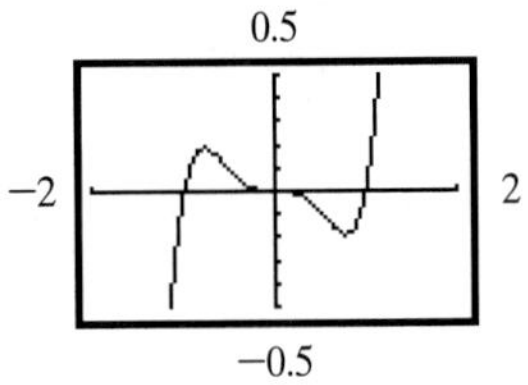

Increasing: $(-2, -0.77), (0.77, 2)$
Decreasing: $(-0.77, 0.77)$
Local maximum: $(-0.77, 0.19)$
Local minimum: $(0.77, -0.19)$

70.

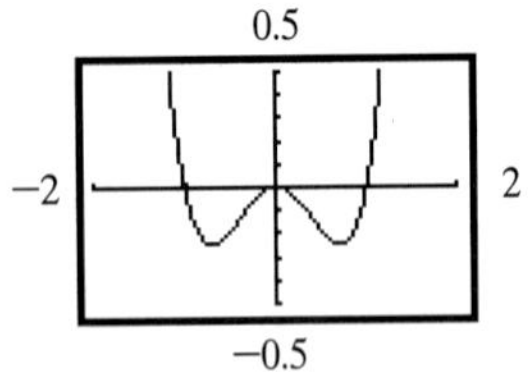

Increasing: $(-0.71, 0), (0.71, 2)$
Decreasing: $(-2, -0.71), (0, 0.71)$
Local maximum: $(0, 0)$
Local minima: $(-0.71, -0.25), (0.71, -0.25)$

71.

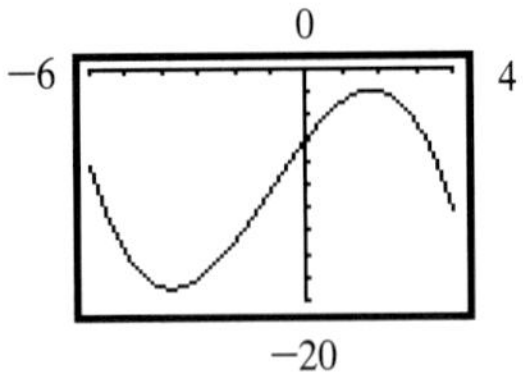

Increasing: $(-3.77, 1.77)$
Decreasing: $(-6, -3.77), (1.77, 4)$
Local maximum: $(1.77, -1.91)$
Local minimum: $(-3.77, -18.89)$

72.

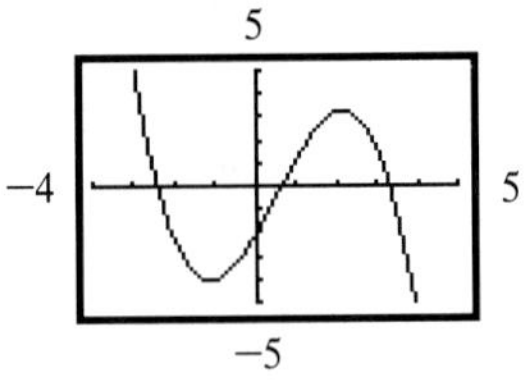

Increasing: $(-1.16, 2.16)$
Decreasing: $(-4, -1.16), (2.16, 5)$
Local maximum: $(2.16, 3.25)$
Local minimum: $(-1.16, -4.05)$

73.

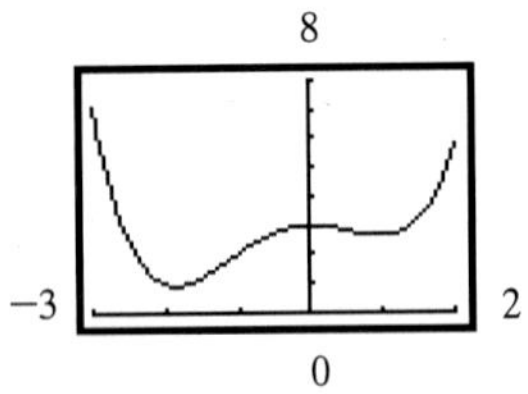

Increasing: $(-1.87, 0), (0.97, 2)$
Decreasing: $(-3, -1.87), (0, 0.97)$
Local maximum: $(0, 3)$
Local minima: $(-1.87, 0.95), (0.97, 2.65)$

74.

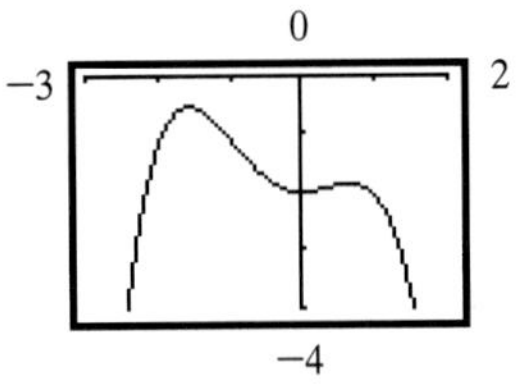

Increasing: $(-3, -1.57), (0, 0.64)$
Decreasing: $(-1.57, 0), (0.64, 2)$
Local maxima: $(-1.57, -0.52), (0.64, -1.87)$
Local minimum: $(0, -2)$

75. (a), (b), (e)

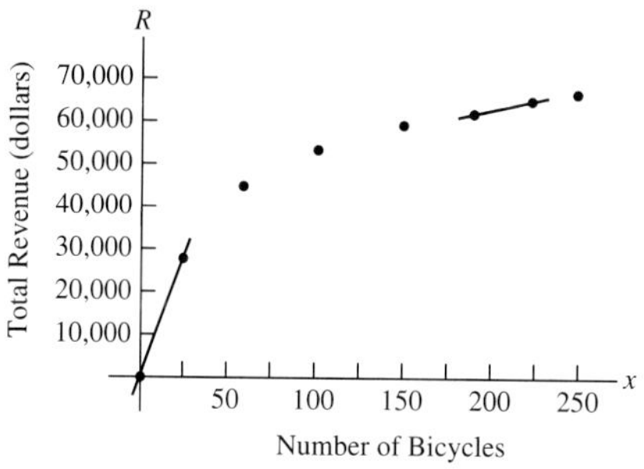

(c) 1120 dollars/bicycle
(d) For each additional bicycle sold between 0 and 25 bicycles, total revenue increases, on average, by \$1120.
(f) 75 dollars/bicycle
(g) For each additional bicycle sold between 190 and 223 bicycles, total revenue increases, on average, by \$75.

76. (a), (b), (e)

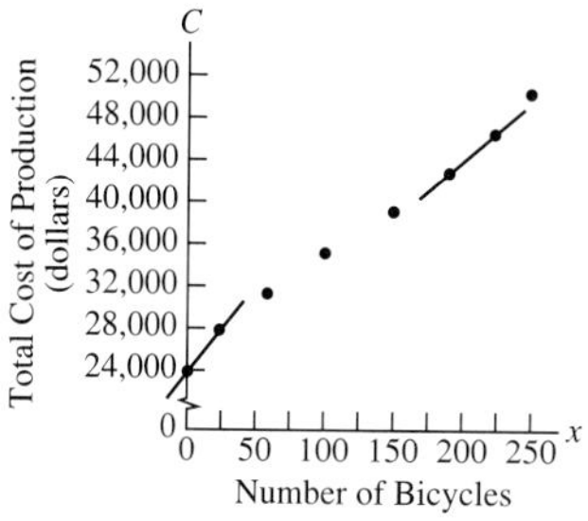

(c) 150 dollars/bicycle
(d) The total cost of producing between 0 and 25 bicycles increases, on average, by \$150 for each additional bicycle manufactured.
(f) 113.64 dollars/bicycle
(g) The total cost of producing between 190 and 223 bicycles increases, on average, by \$113.64 for each additional bicycle manufactured.

77. (a), (b), (e)

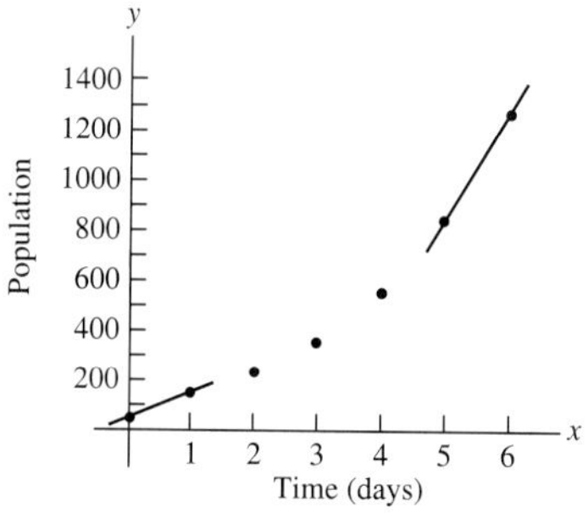

(c) 103 bacteria/day
(d) The population is increasing at an average rate of 103 bacteria per day between day 0 and day 1.
(f) 441 bacteria/day
(g) The population is increasing at an average rate of 441 bacteria per day between day 5 and day 6.
(h) The average rate of change is increasing.

78. (a), (b), (e)

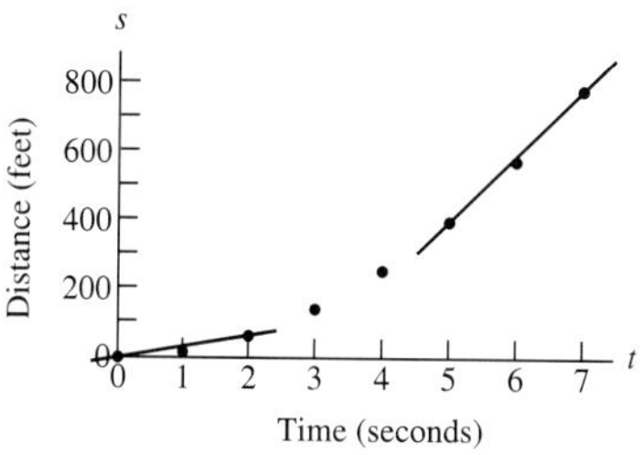

(c) 32 feet/second
(d) The ball is falling at an average speed of 32 feet per second during the first 2 seconds.
(f) 192
(g) Between $t = 5$ seconds and $t = 7$ seconds, the ball is falling at an average speed of 192 feet per second.
(h) The average rate of change is increasing.

79.

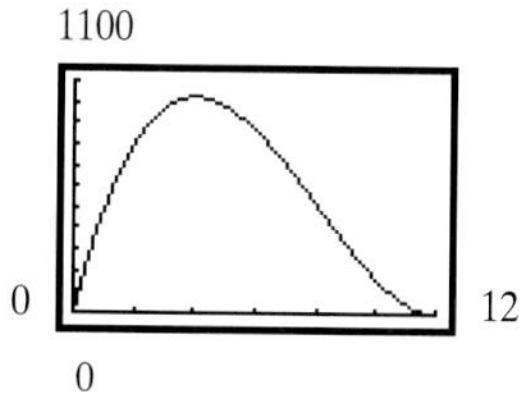

The volume is largest at $x = 4$ inches.

80.

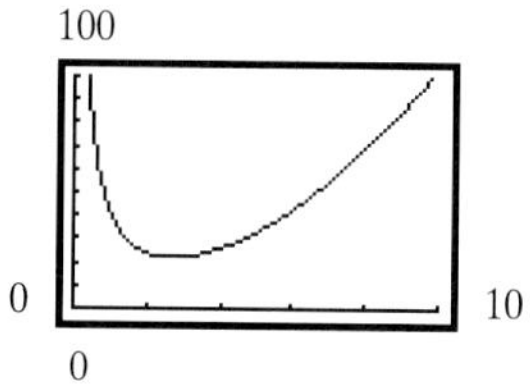

The amount is smallest at $x \approx 2.71$ inches.

81. (a)

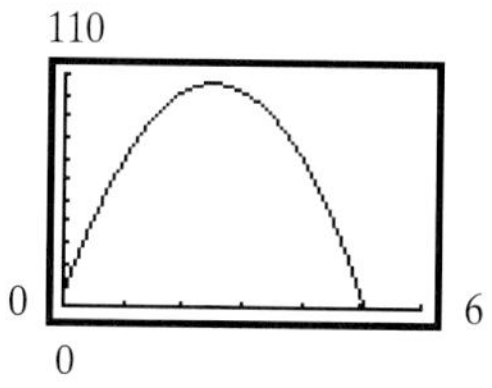

(b) 2.5 seconds
(c) 106 feet

82. (a)

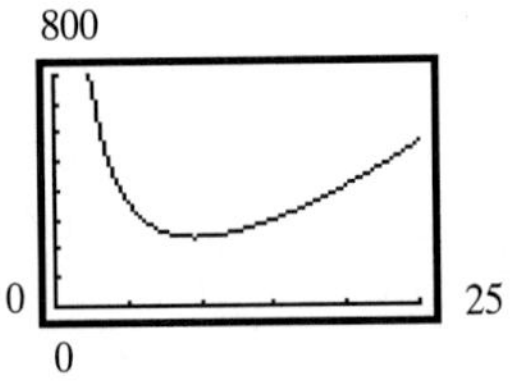

(b) 9.66 lawn mowers per hour
(c) \$238.65

83. Each graph is that of $y = x^2$, but shifted vertically. If $y = x^2 + k, k > 0$, the shift is up k units; if $y = x^2 + k, k < 0$, the shift is down $|k|$ units.
84. Each graph is that of $y = x^2$, but shifted horizontally. If $y = (x - k)^2$ and $k > 0$, the shift is k units to the right. If $y = (x + k)^2$ and $k > 0$, the shift is k units to the left.
85. Each graph is that of $y = |x|$, but either compressed or stretched. If $y = k|x|$ and $k > 1$, the graph is stretched vertically; if $y = k|x|, 0 < k < 1$, the graph is compressed vertically.

86. The graph of $y = -f(x)$ is the reflection about the x-axis of the graph of $y = f(x)$.
87. The graph of $y = f(-x)$ is the reflection about the y-axis of the graph of $y = f(x)$.
88. Shift $y = x^3$ to the right 1 unit and then upward 2 units. **89.** They are all U-shaped and open upward. All three go through the points $(-1, 1)$, $(0, 0)$ and $(1, 1)$. As the exponent increases, the steepness of the curve increases (except near $x = 0$).
90. They are all increasing functions which pass through the points $(-1, -1)$, $(0, 0)$ and $(1, 1)$. As the exponent increases, the steepness of the curve increases (except near $x = 0$).

91. (a) \$43.47 **(b)** \$241.49

(c) $C = \begin{cases} 9 + 0.68935x & \text{if } 0 \le x \le 50 \\ 21.465 + 0.44005x & \text{if } x > 50 \end{cases}$

(d)

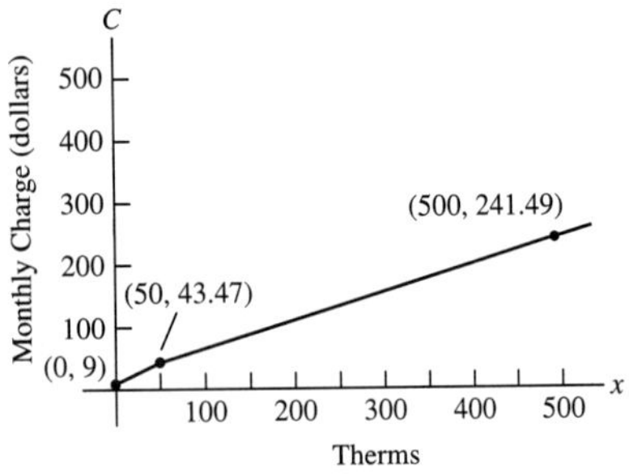

92. (a) \$24.71 **(b)** \$81.94

(c) $C = \begin{cases} 6 + 0.5125x & \text{if } 0 \le x \le 20 \\ 7.79 + 0.423x & \text{if } 20 < x \le 50 \\ 11.505 + 0.3487x & \text{if } x > 50 \end{cases}$

(d)

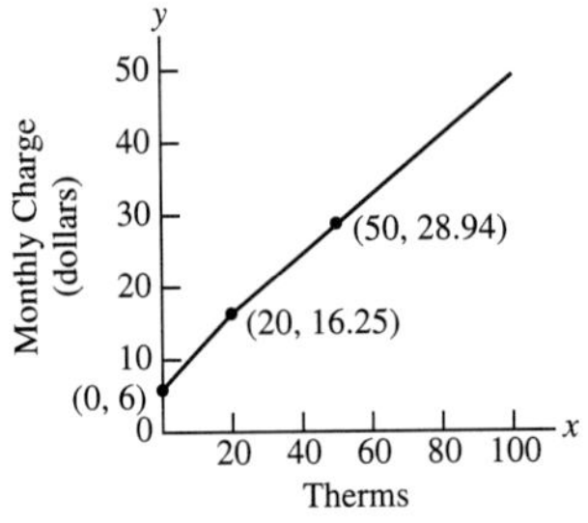

3.2 Exercises

1. B **2.** E **3.** H **4.** D **5.** I **6.** A **7.** L **8.** C **9.** F **10.** J **11.** G **12.** K **13.** C **14.** D **15.** B **16.** A **17.** $y = (x - 4)^3$
18. $y = (x + 4)^3$ **19.** $y = x^3 + 4$ **20.** $y = x^3 - 4$ **21.** $y = -x^3$ **22.** $y = -x^3$ **23.** $y = 4x^3$ **24.** $y = \left(\frac{1}{4}x\right)^3 = \frac{1}{64}x^3$
25. (1) $y = \sqrt{x} + 2$; (2) $y = -(\sqrt{x} + 2)$; (3) $y = -(\sqrt{-x} + 2)$ **26.** (1) $y = -\sqrt{x}$; (2) $y = -\sqrt{x - 3}$; (3) $y = -\sqrt{x - 3} - 2$
27. (1) $y = -\sqrt{x}$; (2) $y = -\sqrt{x} + 2$; (3) $y = -\sqrt{x + 3} + 2$ **28.** (1) $y = \sqrt{x} + 2$; (2) $y = \sqrt{-x} + 2$; (3) $y = \sqrt{-(x + 3)} + 2$

29.

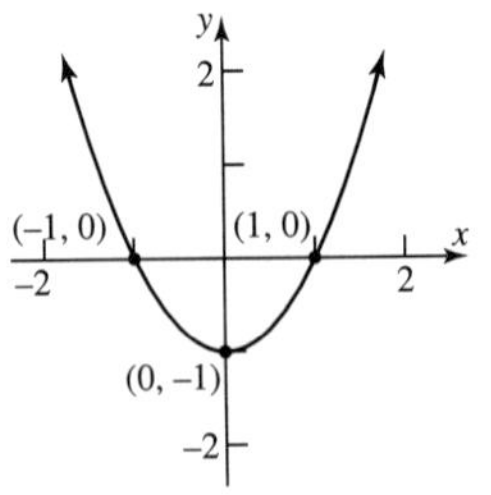

30.

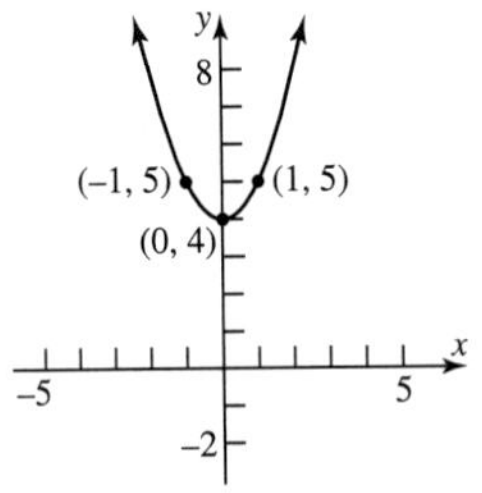

31.

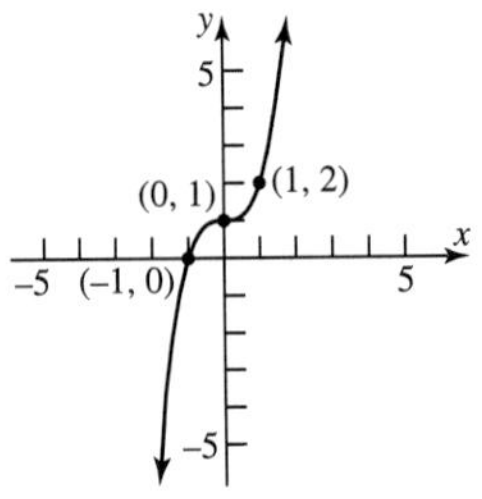

32.

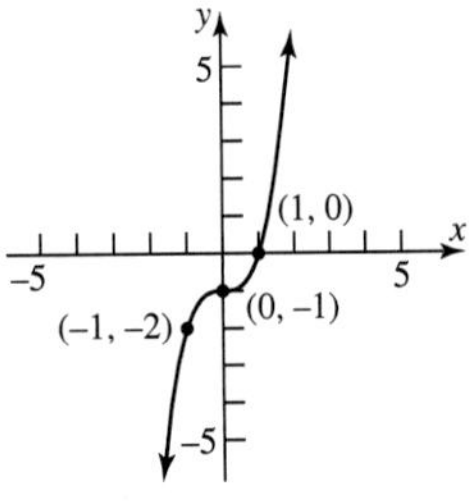

33.

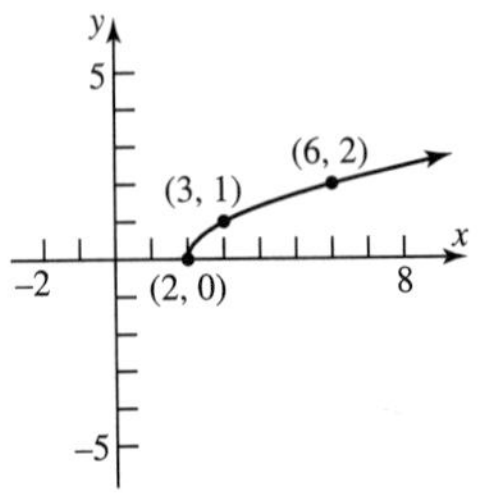

34.

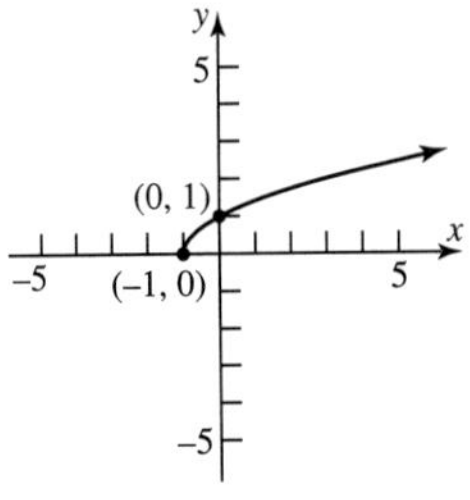

35.

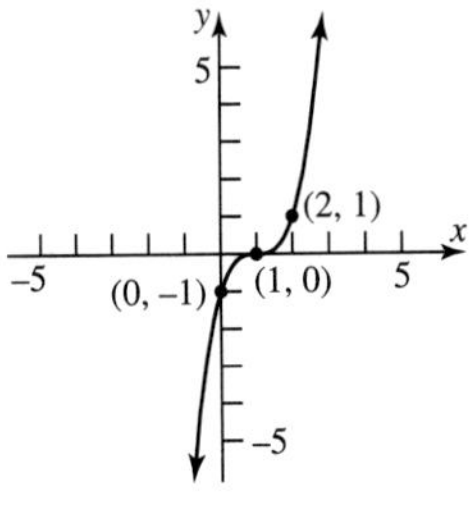

36.

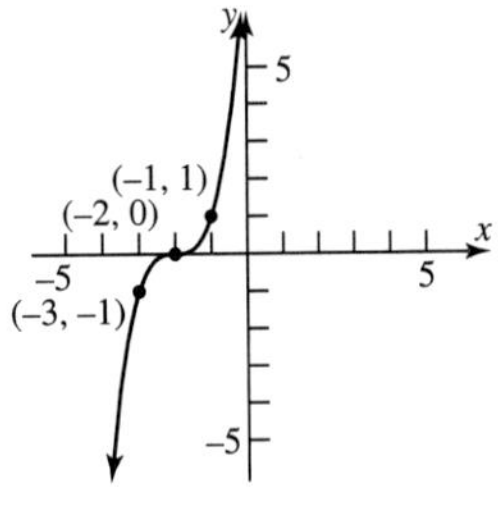

37.

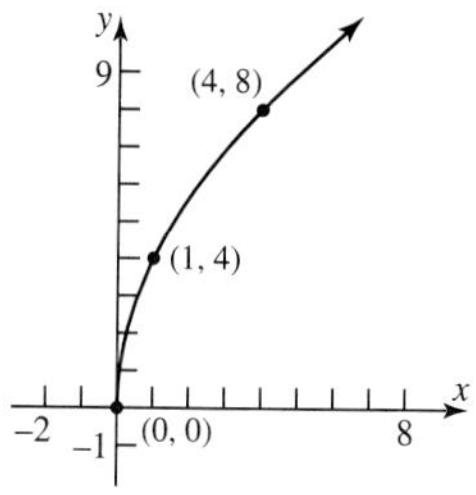

38.

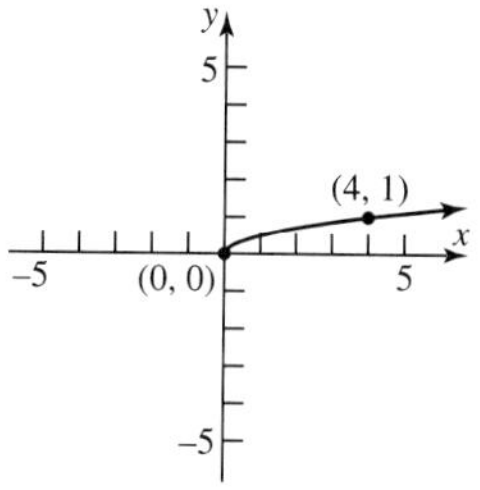

39.

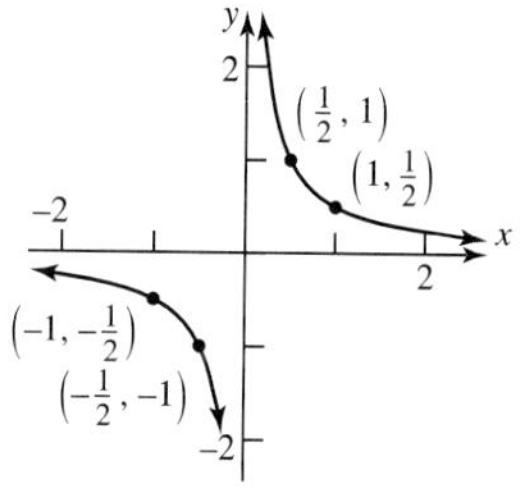

40.

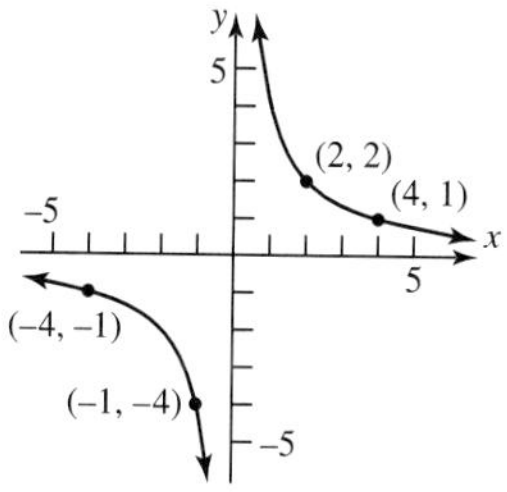

41.

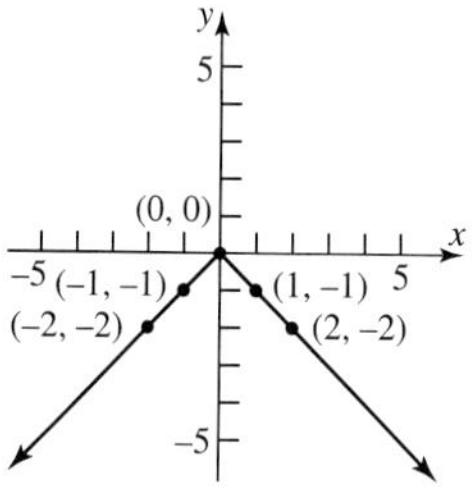

42.

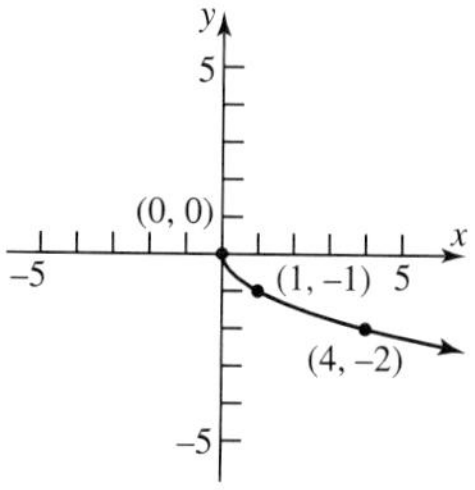

43.

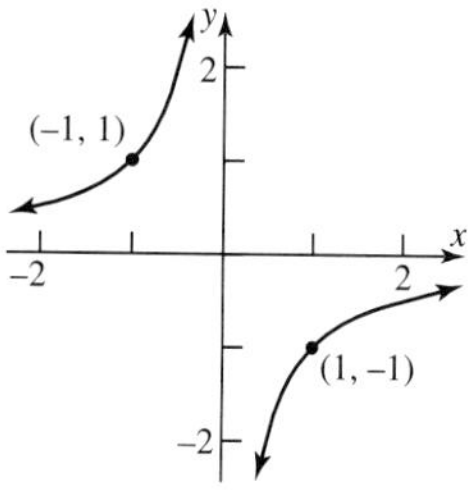

44.

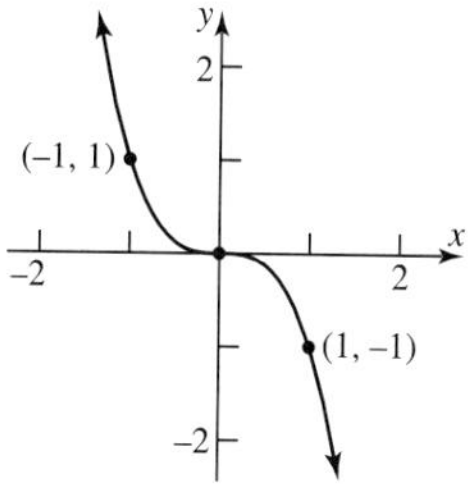

45.

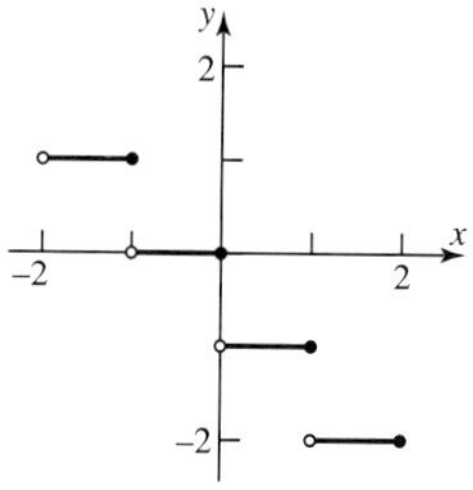

46.

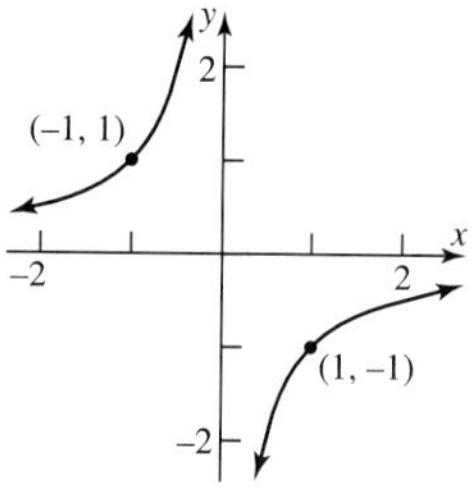

47.

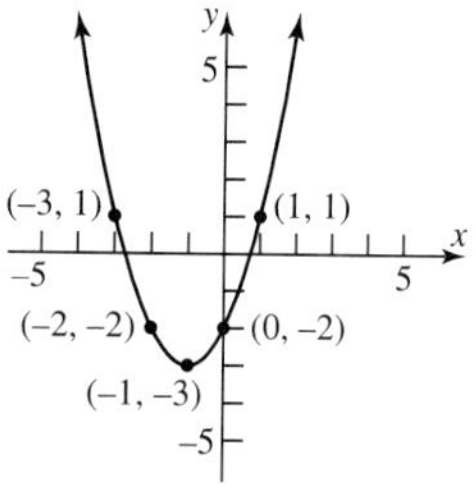

48.

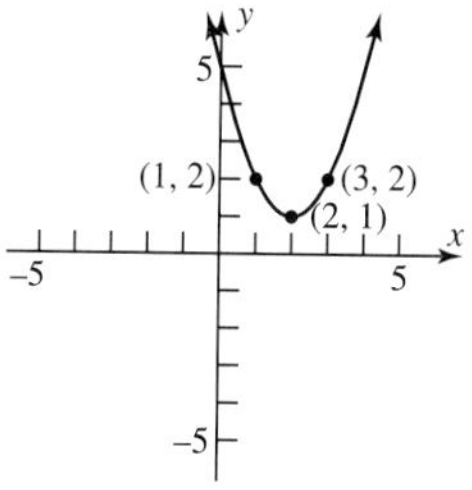

49.

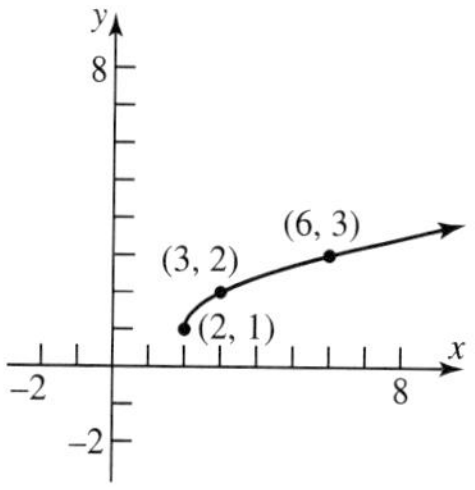

50.

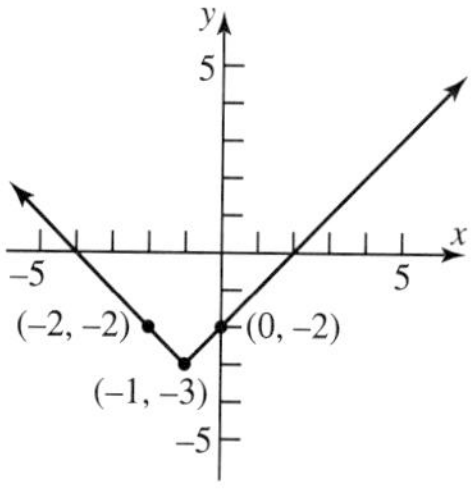

51.

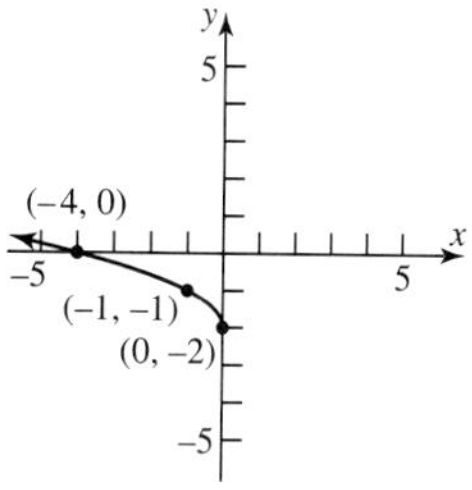

52.

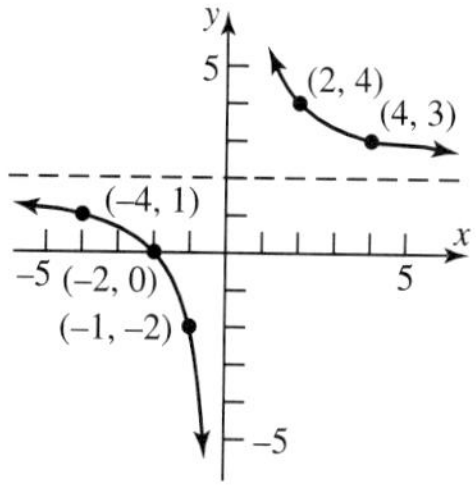

53.

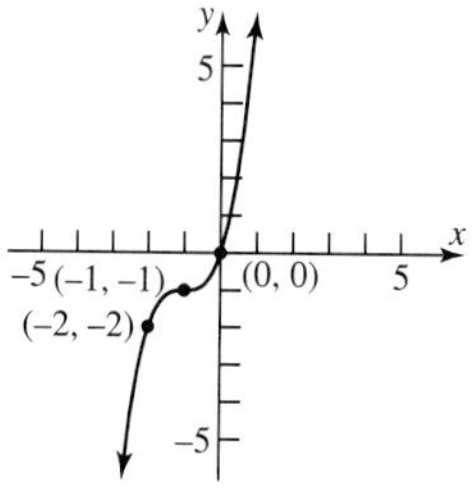

54.

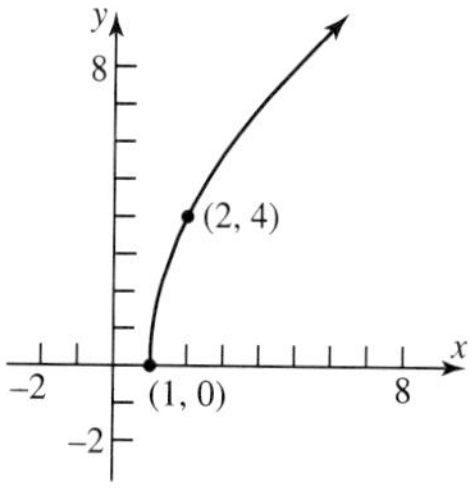

55.

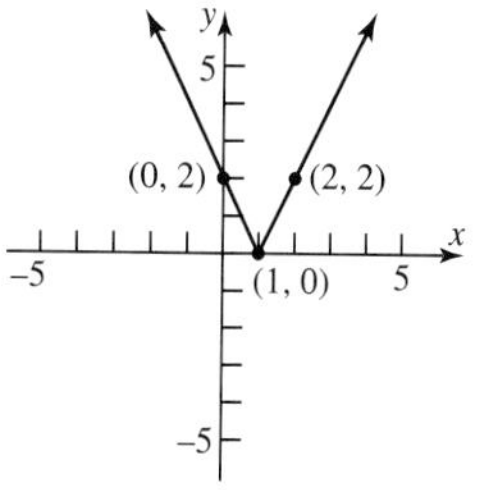

56.

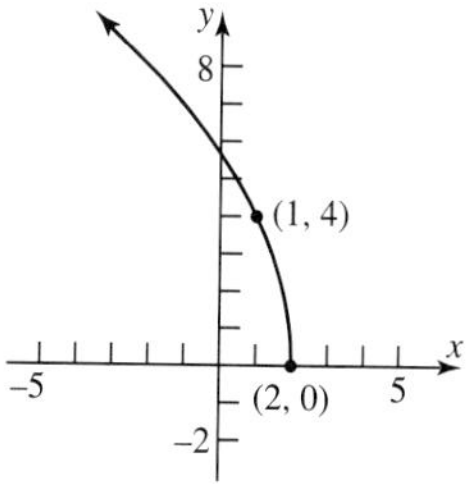

57.

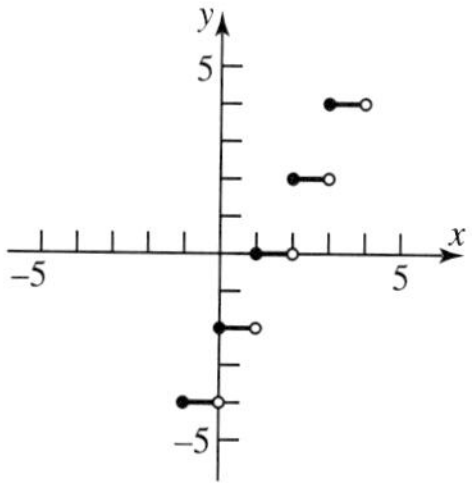

58.

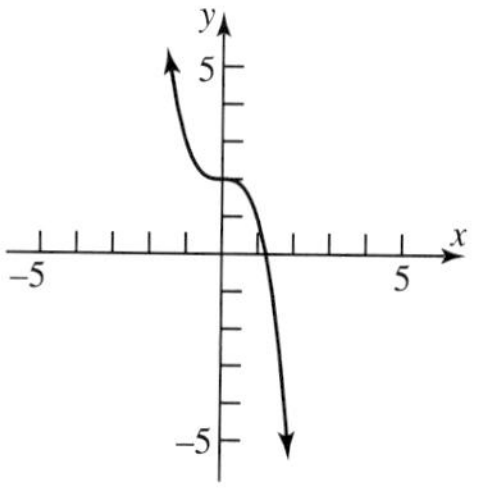

59. (a) $F(x) = f(x) + 3$

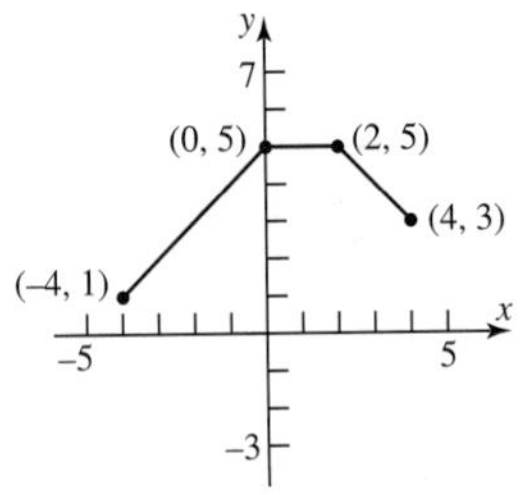

(b) $G(x) = f(x + 2)$

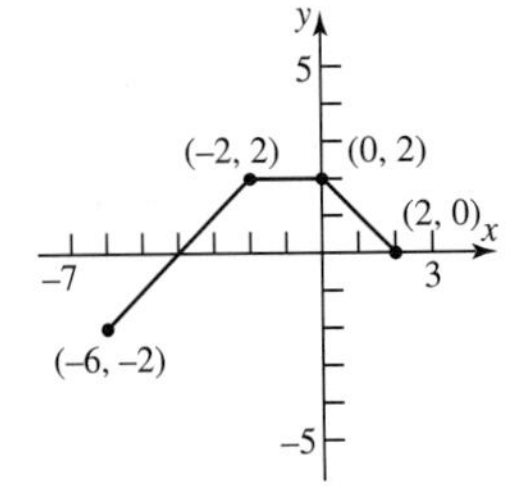

(c) $P(x) = -f(x)$

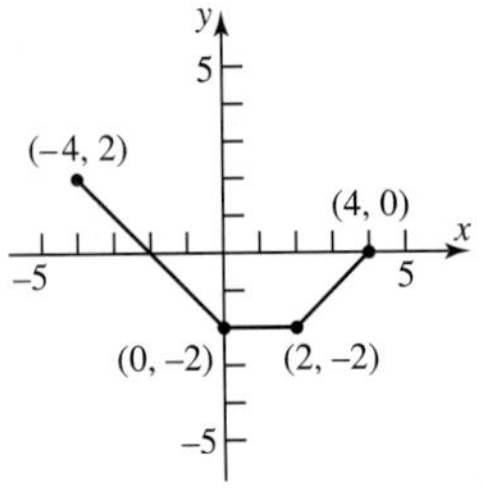

(d) $Q(x) = \frac{1}{2}f(x)$

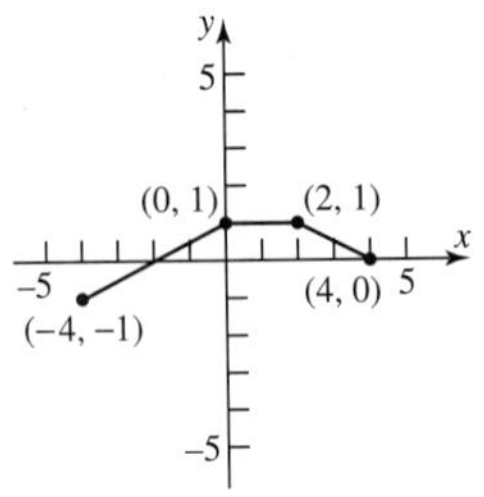

(e) $g(x) = f(-x)$

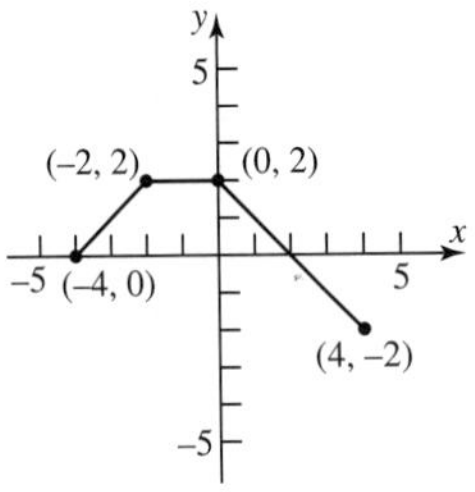

(f) $h(x) = f(2x)$

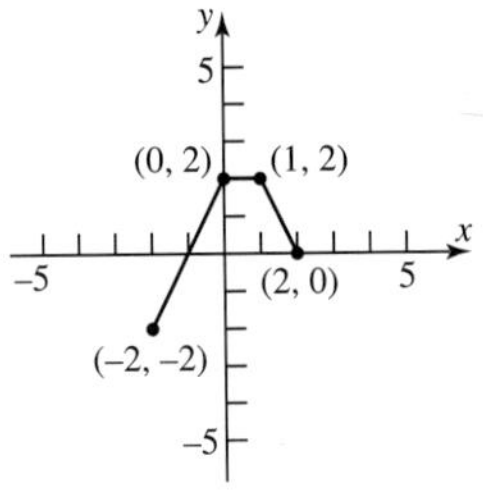

60. (a) $F(x) = f(x) + 3$

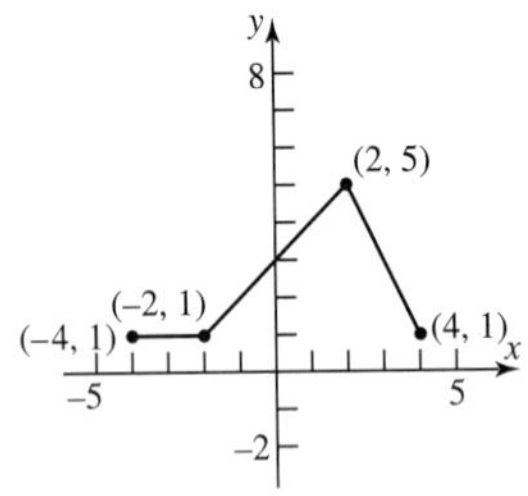

(b) $G(x) = f(x + 2)$

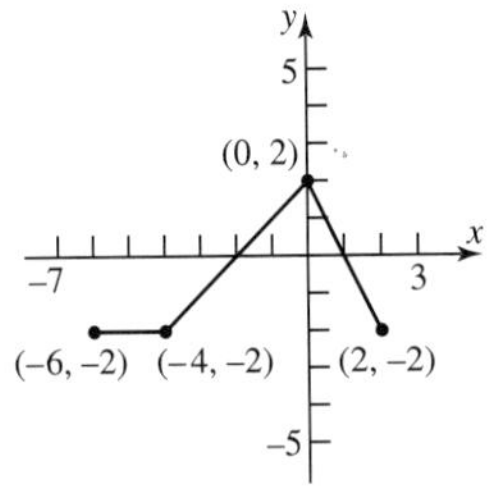

(c) $P(x) = -f(x)$

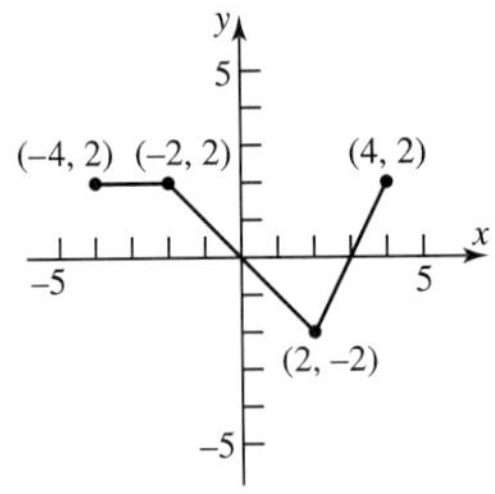

(d) $Q(x) = \frac{1}{2}f(x)$

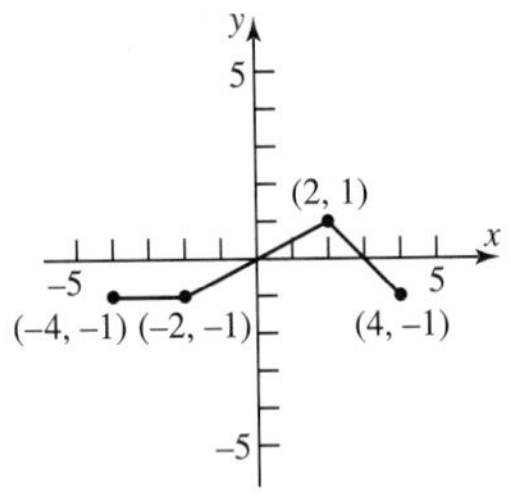

(e) $g(x) = f(-x)$

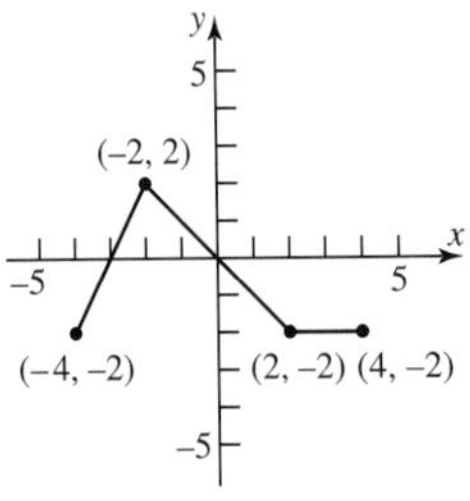

(f) $h(x) = f(2x)$

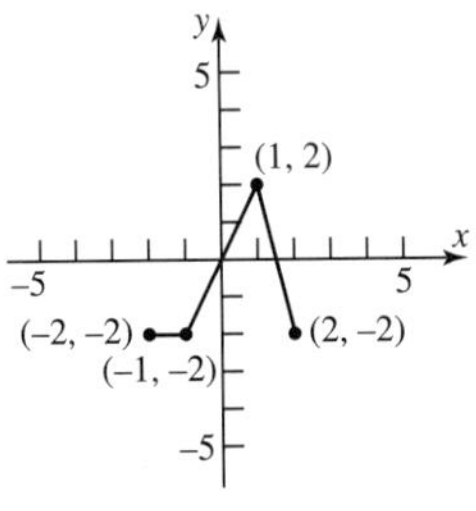

61. (a) $F(x) = f(x) + 3$

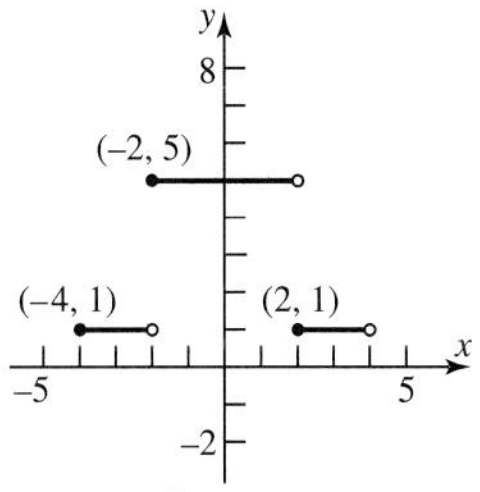

(b) $G(x) = f(x + 2)$

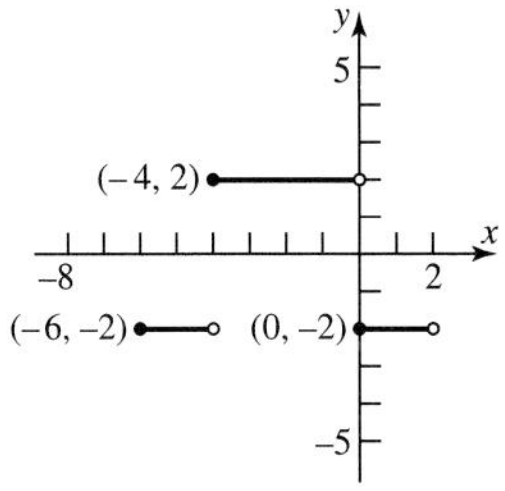

(c) $P(x) = -f(x)$

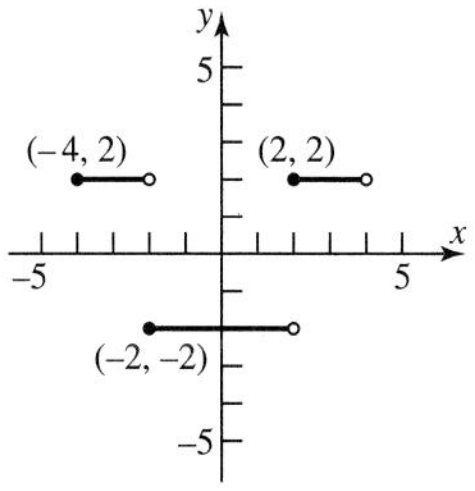

(d) $Q(x) = \frac{1}{2}f(x)$

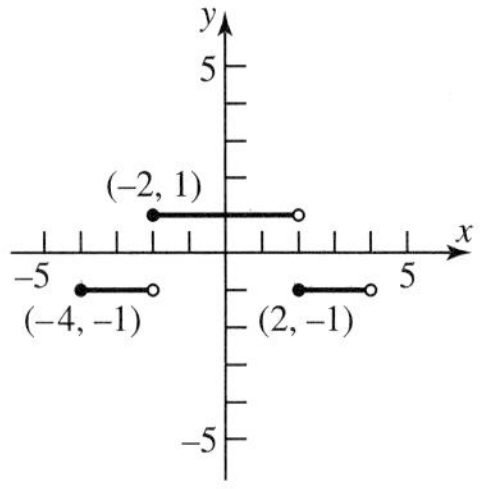

(e) $g(x) = f(-x)$

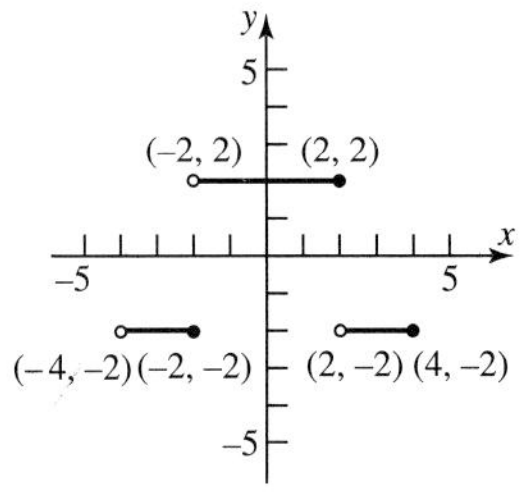

(f) $h(x) = f(2x)$

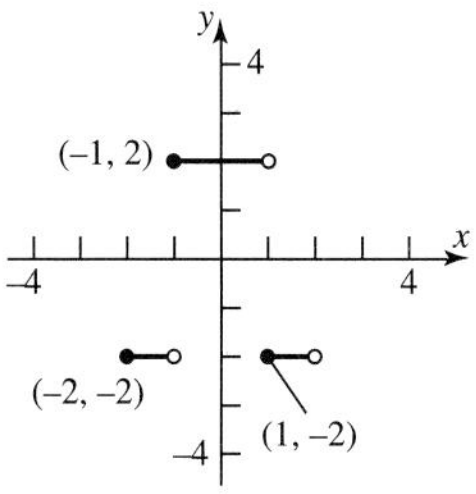

62. (a) $F(x) = f(x) + 3$

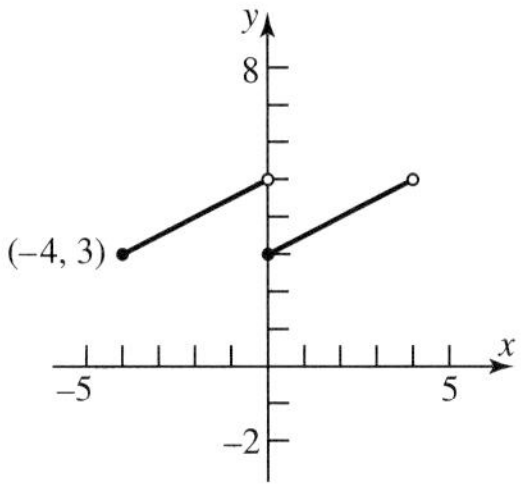

(b) $G(x) = f(x + 2)$

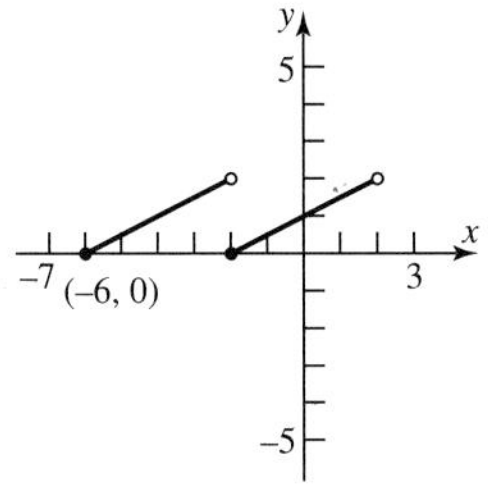

(c) $P(x) = -f(x)$

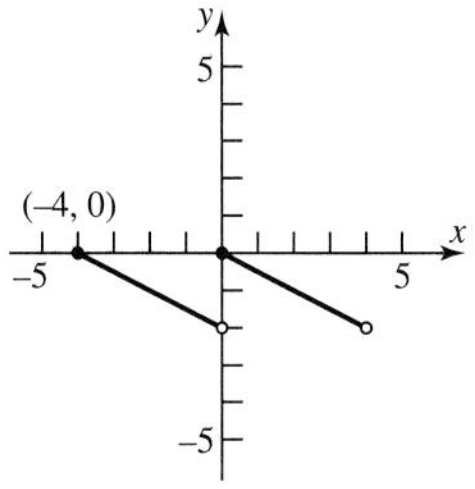

(d) $Q(x) = \frac{1}{2}f(x)$

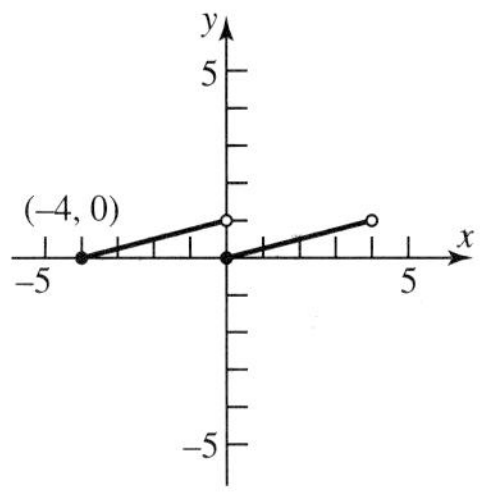

(e) $g(x) = f(-x)$

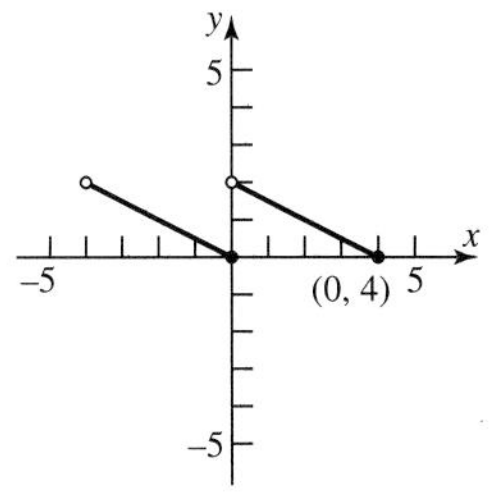

(f) $h(x) = f(2x)$

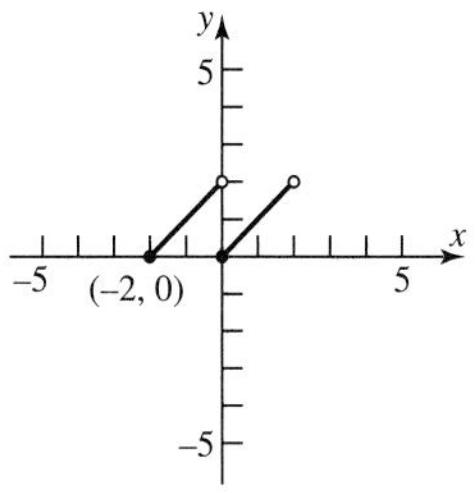

63. (a) $F(x) = f(x) + 3$

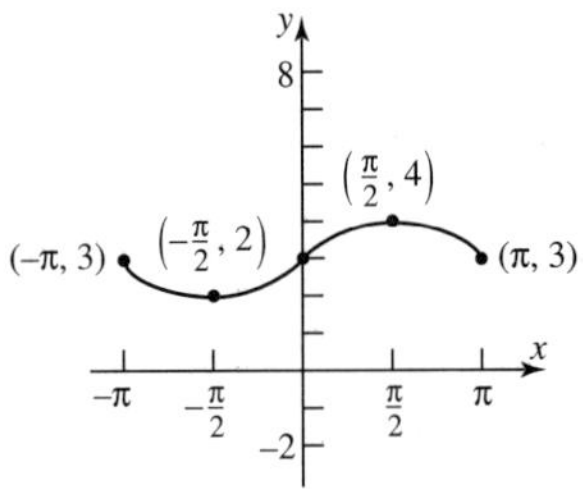

(b) $G(x) = f(x + 2)$

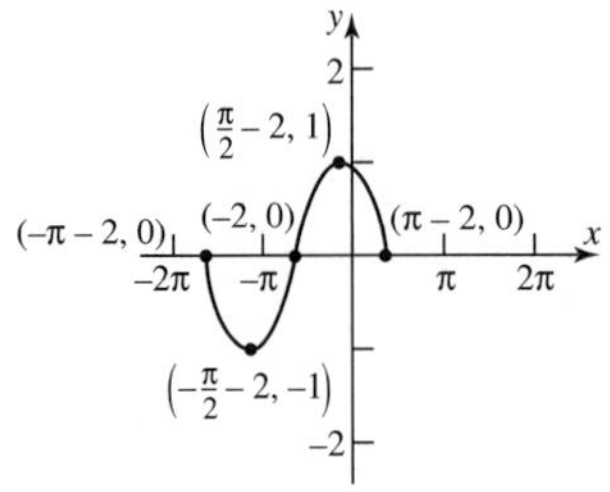

(c) $P(x) = -f(x)$

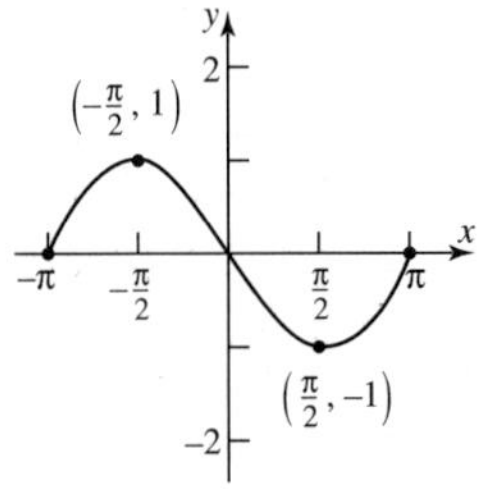

(d) $Q(x) = \frac{1}{2}f(x)$

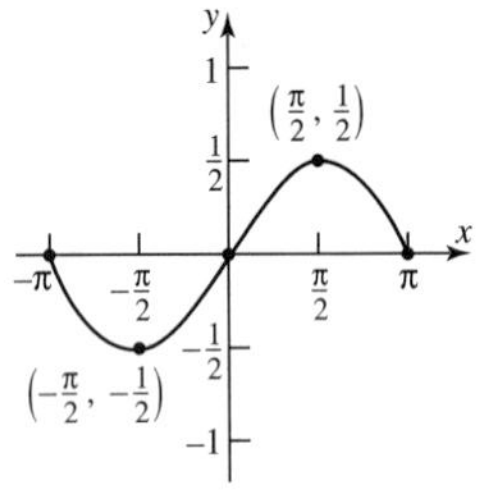

(e) $g(x) = f(-x)$

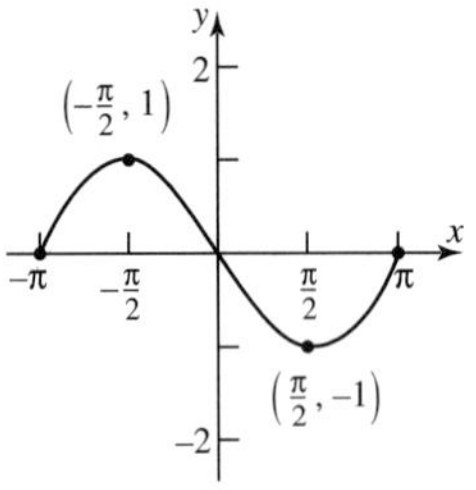

(f) $h(x) = f(2x)$

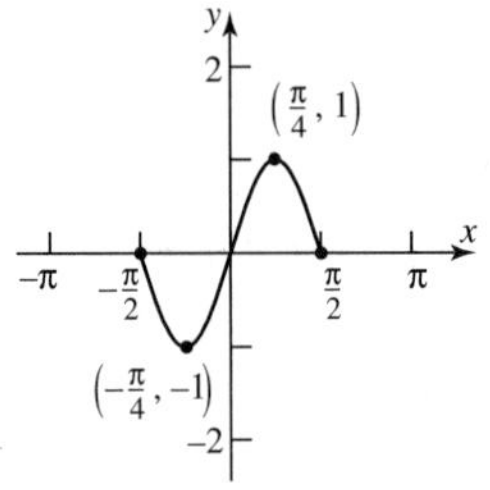

64. (a) $F(x) = f(x) + 3$

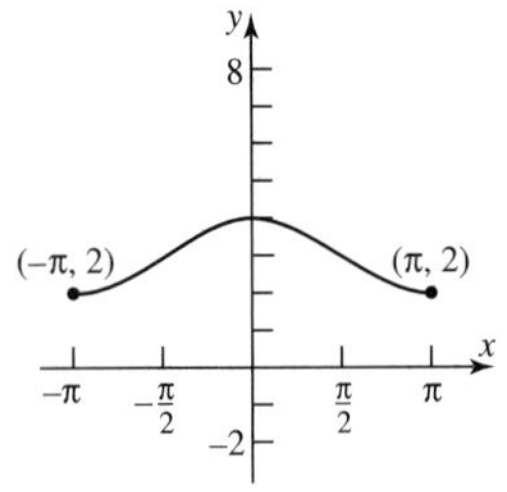

(b) $G(x) = f(x + 2)$

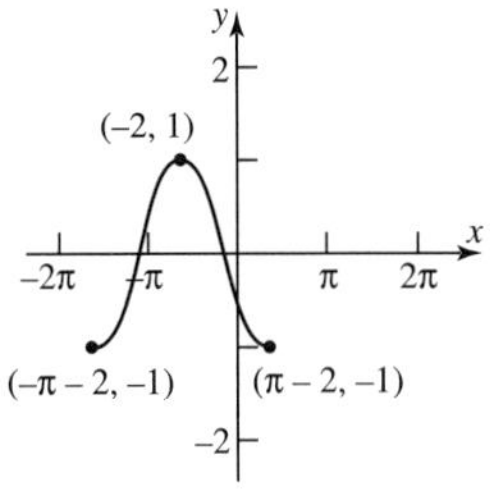

(c) $P(x) = -f(x)$

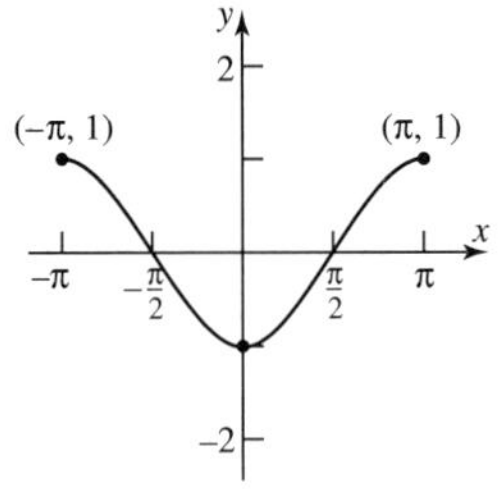

(d) $Q(x) = \frac{1}{2}f(x)$

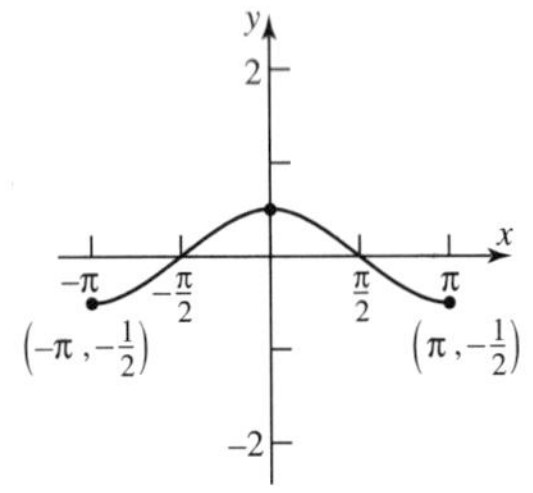

(e) $g(x) = f(-x)$

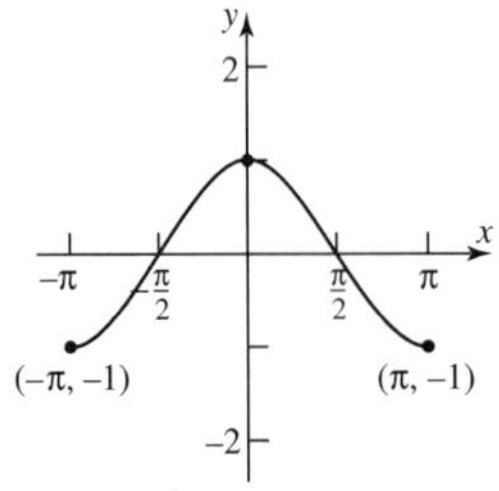

(f) $h(x) = f(2x)$

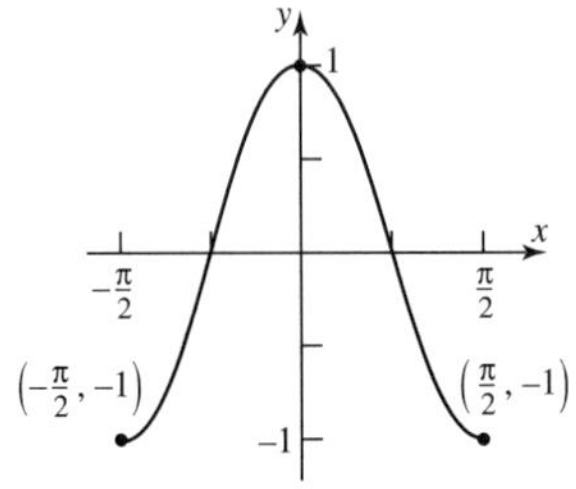

65. (a)

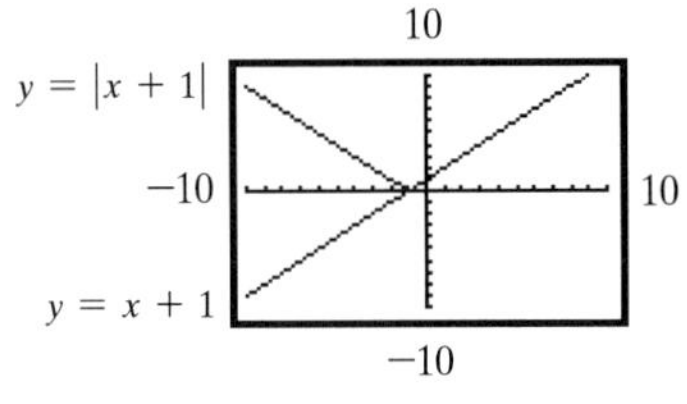

(b)

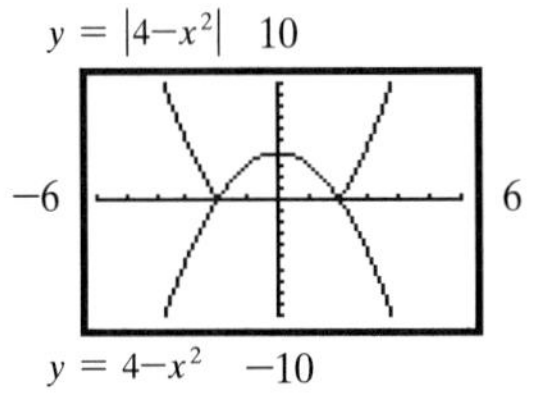

(c)

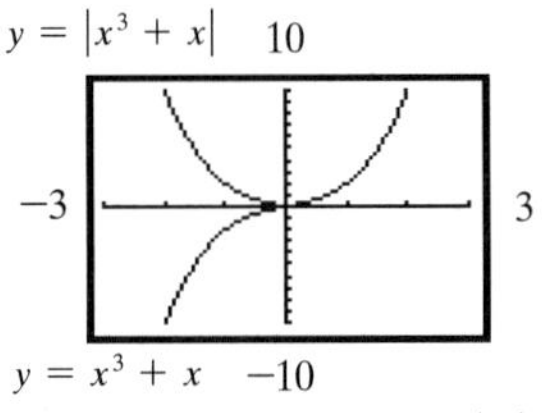

(d) Any part of the graph of $y = f(x)$ that lies below the x-axis is reflected about the x-axis to obtain the graph of $y = |f(x)|$.

66. **(a)**

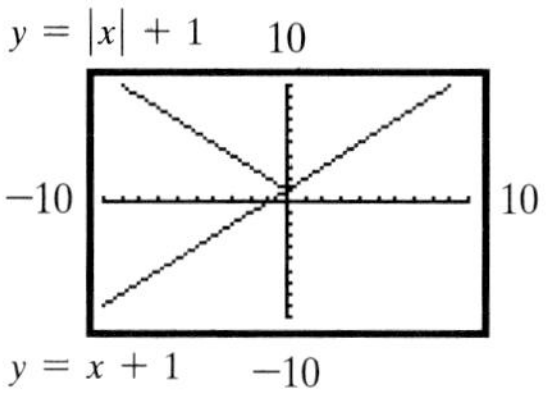

(b)

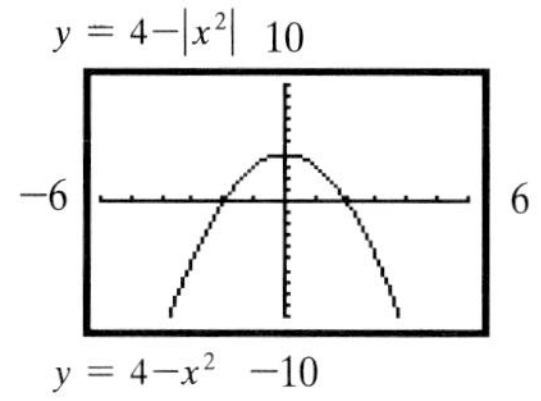

(c)

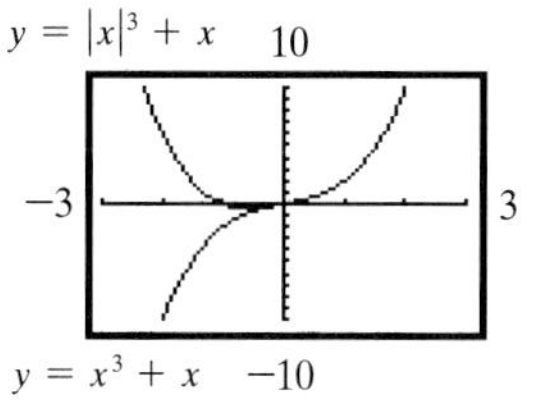

(d) For the graph of $y = f(|x|)$, the graph of $y = f(x)$ to the left of the y-axis is replaced by the reflection about the y-axis of the graph of $y = f(x)$ to the right of the y-axis. To the right of the y-axis, both graphs are the same.

67. **(a)**

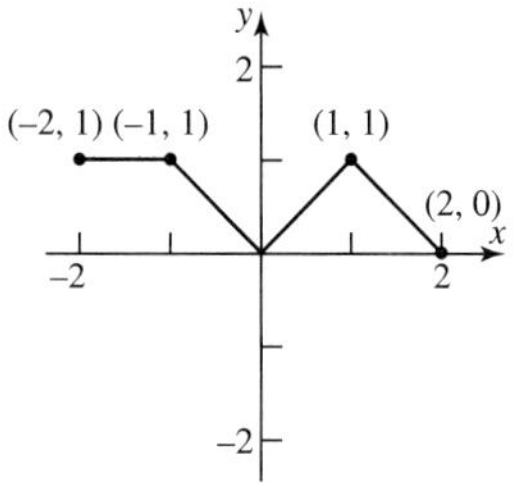

(b)

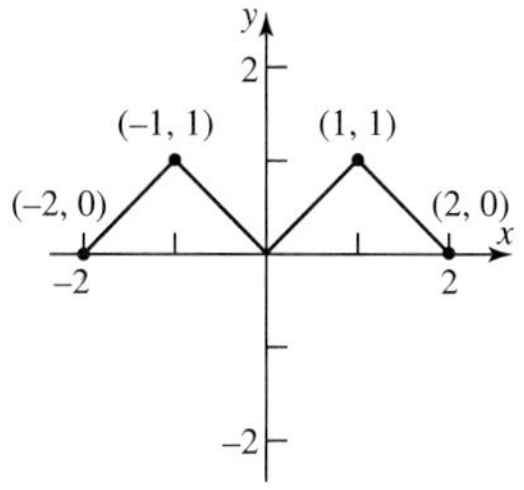

68. **(a)**

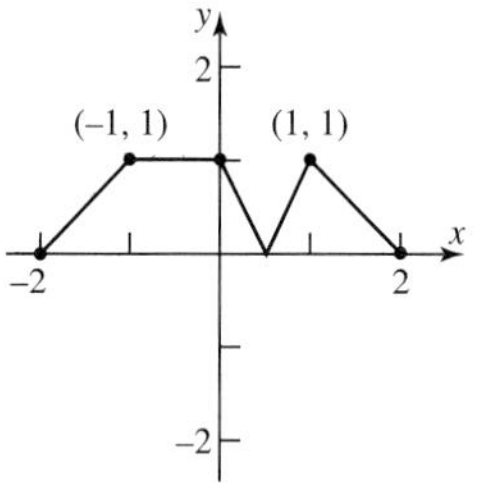

(b)

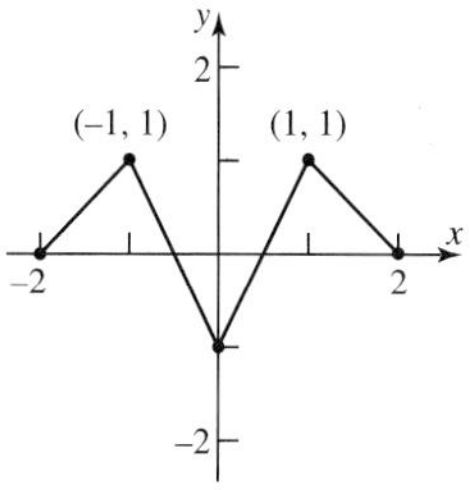

69. $f(x) = (x + 1)^2 - 1$

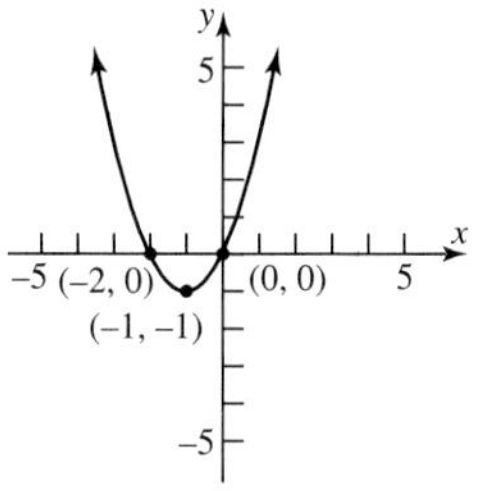

70. $f(x) = (x - 3)^2 - 9$

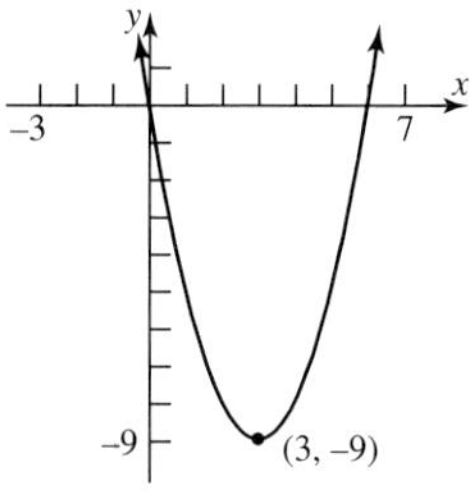

71. $f(x) = (x - 4)^2 - 15$

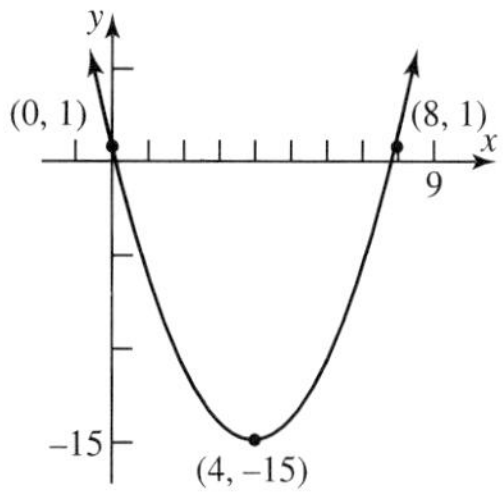

72. $f(x) = (x + 2)^2 - 2$

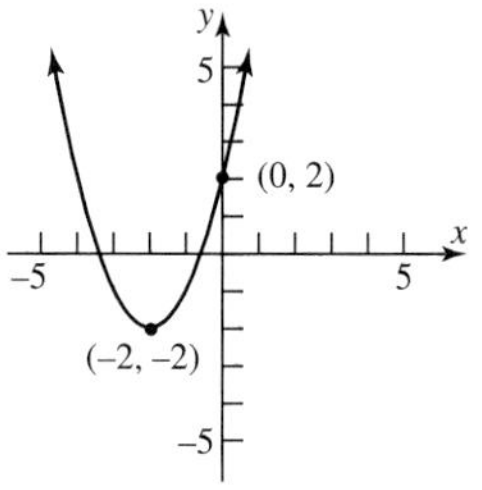

73. $f(x) = \left(x + \frac{1}{2}\right)^2 + \frac{3}{4}$

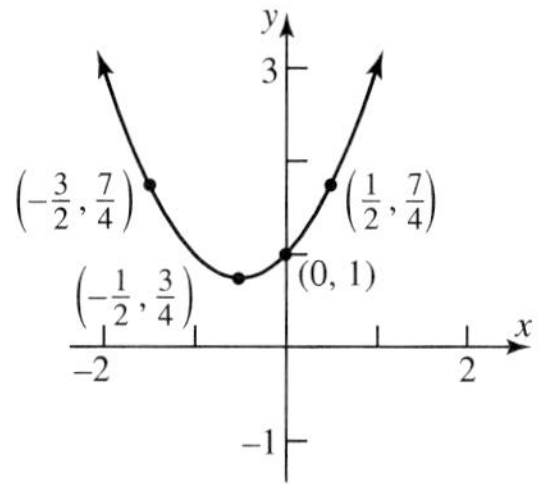

74. $f(x) = \left(x - \frac{1}{2}\right)^2 + \frac{3}{4}$

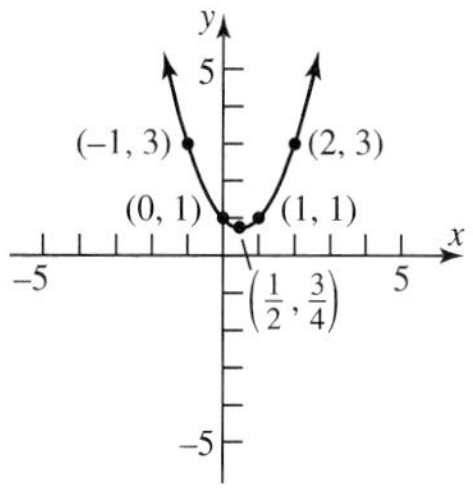

75.

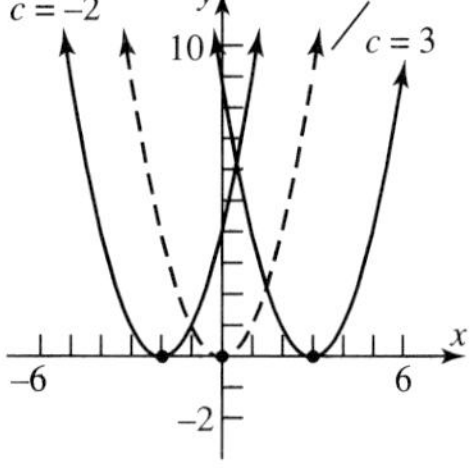

76.

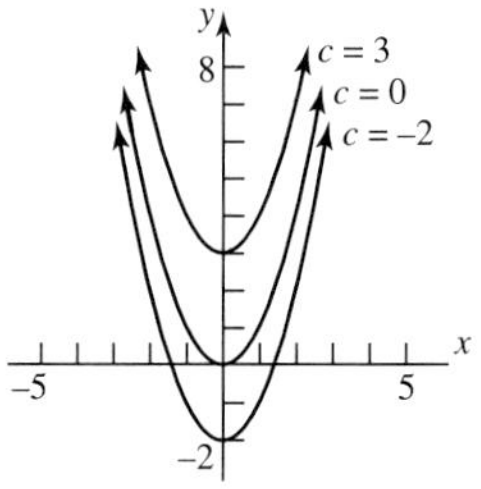

77.

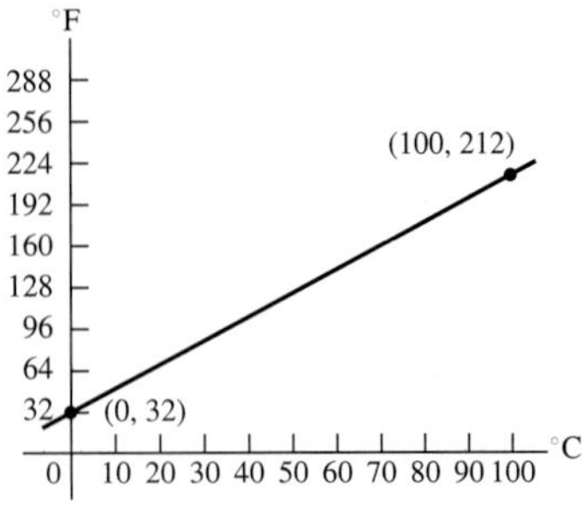

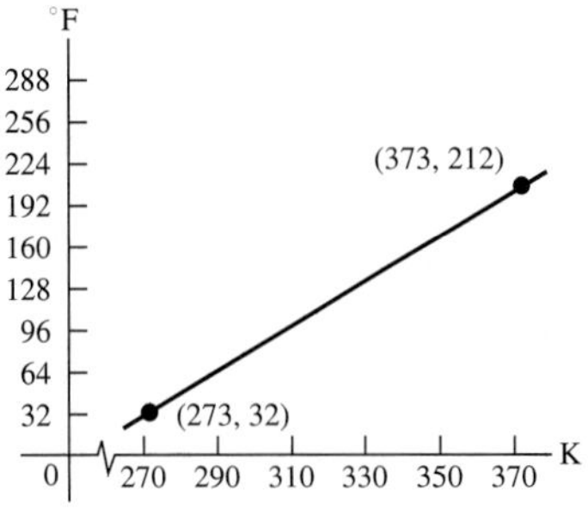

78. (a)

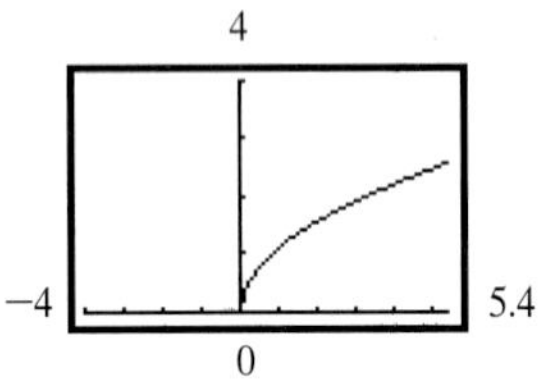

(b)

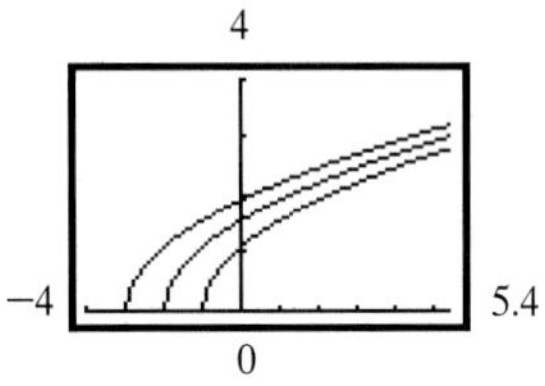

(c) An increase in l increases the period T.

(d)

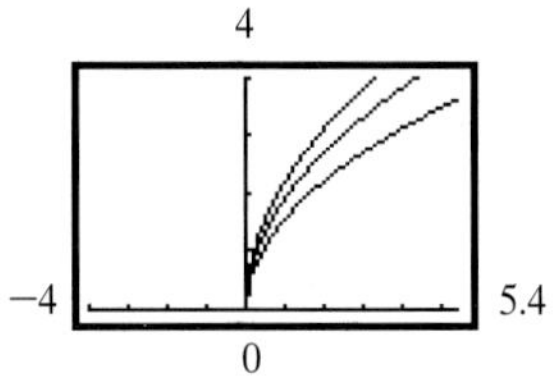

(e) Multiplying the length l by a factor k, increases the period T by a factor of $\sqrt{k}$.

79. (a)

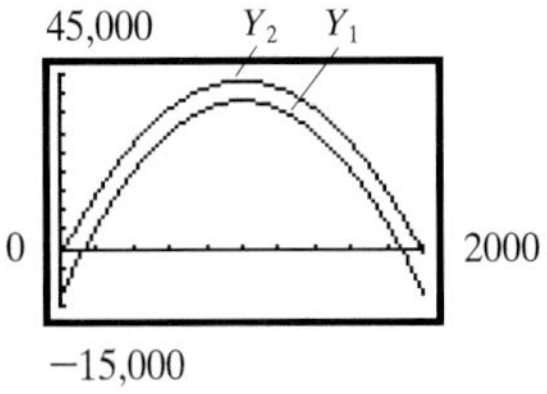

(b) 10% tax
(c) Y_1 is the graph of $p(x)$ shifted down vertically 10,000 units. Y_2 is the graph of $p(x)$ vertically compressed by a factor of 0.9.
(d) 10% tax

3.3 Exercises

1. (a) $(f+g)(x) = 5x + 1$; All real numbers **(b)** $(f-g)(x) = x + 7$; All real numbers **(c)** $(f \cdot g)(x) = 6x^2 - x - 12$; All real numbers **(d)** $\left(\frac{f}{g}\right)(x) = \frac{3x+4}{2x-3}$; $\left\{x \middle| x \neq \frac{3}{2}\right\}$ **2. (a)** $(f+g)(x) = 5x - 1$; All real numbers **(b)** $(f-g)(x) = -x + 3$; All real numbers **(c)** $(f \cdot g)(x) = 6x^2 - x - 2$; All real numbers **(d)** $\left(\frac{f}{g}\right)(x) = \frac{2x+1}{3x-2}$; $\left\{x \middle| x \neq \frac{2}{3}\right\}$ **3. (a)** $(f+g)(x) = 2x^2 + x - 1$; All real numbers **(b)** $(f-g)(x) = -2x^2 + x - 1$; All real numbers **(c)** $(f \cdot g)(x) = 2x^3 - 2x^2$; All real numbers **(d)** $\left(\frac{f}{g}\right)(x) = \frac{x-1}{2x^2}$; $\{x|x \neq 0\}$ **4. (a)** $(f+g)(x) = 4x^3 + 2x^2 + 4$; All real numbers **(b)** $(f-g)(x) = -4x^3 + 2x^2 + 2$; All real numbers **(c)** $(f \cdot g)(x) = 8x^5 + 12x^3 + 2x^2 + 3$; All real numbers **(d)** $\left(\frac{f}{g}\right)(x) = \frac{2x^2+3}{4x^3+1}$; $\left\{x \middle| x \neq -\frac{1}{\sqrt[3]{4}}\right\}$ **5. (a)** $(f+g)(x) = \sqrt{x} + 3x - 5$; $\{x|x \geq 0\}$ **(b)** $(f-g)(x) = \sqrt{x} - 3x + 5$; $\{x|x \geq 0\}$ **(c)** $(f \cdot g)(x) = 3x\sqrt{x} - 5\sqrt{x}$; $\{x|x \geq 0\}$ **(d)** $\left(\frac{f}{g}\right)(x) = \frac{\sqrt{x}}{3x-5}$; $\left\{x \middle| x \geq 0, x \neq \frac{5}{3}\right\}$ **6. (a)** $(f+g)(x) = |x| + x$; All real numbers **(b)** $(f-g)(x) = |x| - x$; All real numbers **(c)** $(f \cdot g)(x) = |x|x$; All real numbers **(d)** $\left(\frac{f}{g}\right)(x) = \frac{|x|}{x}$; $\{x|x \neq 0\}$ **7. (a)** $(f+g)(x) = 1 + \frac{2}{x}$; $\{x|x \neq 0\}$ **(b)** $(f-g)(x) = 1$; $\{x|x \neq 0\}$ **(c)** $(f \cdot g)(x) = \frac{1}{x} + \frac{1}{x^2}$; $\{x|x \neq 0\}$ **(d)** $\left(\frac{f}{g}\right)(x) = x + 1$; $\{x|x \neq 0\}$ **8. (a)** $(f+g)(x) = 4x^2$; All real numbers **(b)** $(f-g)(x) = -2x$; All real numbers **(c)** $(f \cdot g)(x) = 4x^4 - x^2$; All real numbers **(d)** $\left(\frac{f}{g}\right)(x) = \frac{2x^2-x}{2x^2+x}$; $\left\{x \middle| x \neq 0, x \neq -\frac{1}{2}\right\}$ **9. (a)** $(f+g)(x) = \frac{6x+3}{3x-2}$; $\left\{x \middle| x \neq \frac{2}{3}\right\}$ **(b)** $(f-g)(x) = \frac{-2x+3}{3x-2}$; $\left\{x \middle| x \neq \frac{2}{3}\right\}$ **(c)** $(f \cdot g)(x) = \frac{8x^2+12x}{(3x-2)^2}$; $\left\{x \middle| x \neq \frac{2}{3}\right\}$ **(d)** $\left(\frac{f}{g}\right)(x) = \frac{2x+3}{4x}$; $\left\{x \middle| x \neq 0, x \neq \frac{2}{3}\right\}$ **10. (a)** $(f+g)(x) = \sqrt{x+1} + \frac{2}{x}$; $\{x|x \geq -1, x \neq 0\}$ **(b)** $(f-g)(x) = \sqrt{x+1} - \frac{2}{x}$; $\{x|x \geq -1, x \neq 0\}$

(c) $(f \cdot g)(x) = \dfrac{2\sqrt{x+1}}{x}$; $\{x \mid x \geq -1, x \neq 0\}$ **(d)** $\left(\dfrac{f}{g}\right)(x) = \dfrac{x\sqrt{x+1}}{2}$; $\{x \mid x \geq -1, x \neq 0\}$ **11.** $g(x) = 5 - \dfrac{7}{2}x$ **12.** $g(x) = \dfrac{x-1}{x+1}$

13. (a) 98 **(b)** 49 **(c)** 4 **(d)** 4 **14. (a)** 95 **(b)** 127 **(c)** 17 **(d)** 1 **15. (a)** 97 **(b)** $-\dfrac{163}{2}$ **(c)** 1 **(d)** $-\dfrac{3}{2}$ **16. (a)** 4418 **(b)** −191 **(c)** 8 **(d)** −2 **17. (a)** $2\sqrt{2}$ **(b)** $2\sqrt{2}$ **(c)** 1 **(d)** 0 **18. (a)** $\sqrt{13}$ **(b)** $3\sqrt{3}$ **(c)** $\sqrt{\sqrt{2}+1}$ **(d)** 0

19. (a) $\dfrac{1}{17}$ **(b)** $\dfrac{1}{5}$ **(c)** 1 **(d)** $\dfrac{1}{2}$ **20. (a)** $\dfrac{11}{6}$ **(b)** $\dfrac{3}{2}$ **(c)** 1 **(d)** $\dfrac{12}{17}$ **21. (a)** $\dfrac{3}{5}$ **(b)** $\dfrac{\sqrt{15}}{5}$ **(c)** $\dfrac{12}{13}$ **(d)** 0 **22. (a)** $\dfrac{8}{4913}$ **(b)** $\dfrac{2}{65}$

(c) 1 **(d)** $\dfrac{2}{5}$ **23.** $\{x \mid x \neq 0, x \neq 2\}$ **24.** $\left\{x \,\middle|\, x \neq 0, x \neq \dfrac{2}{3}\right\}$ **25.** $\{x \mid x \neq -4, x \neq 0\}$ **26.** $\left\{x \,\middle|\, x \neq 0, x \neq -\dfrac{2}{3}\right\}$ **27.** $\left\{x \,\middle|\, x \geq -\dfrac{3}{2}\right\}$

28. $\left\{x \,\middle|\, x \leq -\dfrac{1}{2}\right\}$ **29.** $\{x \mid x \leq -1 \text{ or } x > 1\}$ **30.** $\left\{x \,\middle|\, x < 2 \text{ or } x \geq \dfrac{8}{3}\right\}$ **31. (a)** $(f \circ g)(x) = 6x + 3$; All real numbers
(b) $(g \circ f)(x) = 6x + 9$; All real numbers **(c)** $(f \circ f)(x) = 4x + 9$; All real numbers **(d)** $(g \circ g)(x) = 9x$; All real numbers
32. (a) $(f \circ g)(x) = 4 - 2x$; All real numbers **(b)** $(g \circ f)(x) = -2x - 4$; All real numbers **(c)** $(f \circ f)(x) = x$; All real numbers
(d) $(g \circ g)(x) = 4x - 12$; All real numbers **33. (a)** $(f \circ g)(x) = 3x^2 + 1$; All real numbers
(b) $(g \circ f)(x) = 9x^2 + 6x + 1$; All real numbers **(c)** $(f \circ f)(x) = 9x + 4$; All real numbers **(d)** $(g \circ g)(x) = x^4$; All real numbers
34. (a) $(f \circ g)(x) = x^2 + 5$; All real numbers **(b)** $(g \circ f)(x) = x^2 + 2x + 5$; All real numbers
(c) $(f \circ f)(x) = x + 2$; All real numbers **(d)** $(g \circ g)(x) = x^4 + 8x^2 + 20$; All real numbers
35. (a) $(f \circ g)(x) = x^4 + 8x^2 + 16$; All real numbers **(b)** $(g \circ f)(x) = x^4 + 4$; All real numbers
(c) $(f \circ f)(x) = x^4$; All real numbers **(d)** $(g \circ g)(x) = x^4 + 8x^2 + 20$; All real numbers
36. (a) $(f \circ g)(x) = 4x^4 + 12x^2 + 10$; All real numbers **(b)** $(g \circ f)(x) = 2x^4 + 4x^2 + 5$; All real numbers
(c) $(f \circ f)(x) = x^4 + 2x^2 + 2$; All real numbers **(d)** $(g \circ g)(x) = 8x^4 + 24x^2 + 21$; All real numbers

37. (a) $(f \circ g)(x) = \dfrac{3x}{2-x}$; $\{x \mid x \neq 0, x \neq 2\}$ **(b)** $(g \circ f)(x) = \dfrac{2(x-1)}{3}$; $\{x \mid x \neq 1\}$ **(c)** $(f \circ f)(x) = \dfrac{3(x-1)}{4-x}$; $\{x \mid x \neq 1, x \neq 4\}$

(d) $(g \circ g)(x) = x$; $\{x \mid x \neq 0\}$ **38. (a)** $(f \circ g)(x) = \dfrac{x}{-2+3x}$; $\left\{x \,\middle|\, x \neq 0, x \neq \dfrac{2}{3}\right\}$ **(b)** $(g \circ f)(x) = -2x - 6$; $\{x \mid x \neq -3\}$

(c) $(f \circ f)(x) = \dfrac{x+3}{3x+10}$; $\left\{x \,\middle|\, x \neq -3, x \neq -\dfrac{10}{3}\right\}$ **(d)** $(g \circ g)(x) = x$; $\{x \mid x \neq 0\}$ **39. (a)** $(f \circ g)(x) = \dfrac{4}{4+x}$; $\{x \mid x \neq -4, x \neq 0\}$

(b) $(g \circ f)(x) = \dfrac{-4(x-1)}{x}$; $\{x \mid x \neq 0, x \neq 1\}$ **(c)** $(f \circ f)(x) = x$; $\{x \mid x \neq 1\}$ **(d)** $(g \circ g)(x) = x$; $\{x \mid x \neq 0\}$

40. (a) $(f \circ g)(x) = \dfrac{2}{2+3x}$; $\left\{x \,\middle|\, x \neq 0, x \neq -\dfrac{2}{3}\right\}$ **(b)** $(g \circ f)(x) = \dfrac{2x+6}{x}$; $\{x \mid x \neq -3, x \neq 0\}$

(c) $(f \circ f)(x) = \dfrac{x}{4x+9}$; $\left\{x \,\middle|\, x \neq -3, x \neq -\dfrac{9}{4}\right\}$ **(d)** $(g \circ g)(x) = x$; $\{x \mid x \neq 0\}$ **41. (a)** $(f \circ g)(x) = \sqrt{2x+3}$; $\left\{x \,\middle|\, x \geq -\dfrac{3}{2}\right\}$
(b) $(g \circ f)(x) = 2\sqrt{x} + 3$; $\{x \mid x \geq 0\}$ **(c)** $(f \circ f)(x) = \sqrt[4]{x}$; $\{x \mid x \geq 0\}$ **(d)** $(g \circ g)(x) = 4x + 9$; All real numbers

42. (a) $(f \circ g)(x) = \sqrt{-1-2x}$; $\left\{x \,\middle|\, x \leq -\dfrac{1}{2}\right\}$ **(b)** $(g \circ f)(x) = 1 - 2\sqrt{x-2}$; $\{x \mid x \geq 2\}$ **(c)** $(f \circ f)(x) = \sqrt{\sqrt{x-2} - 2}$; $\{x \mid x \geq 6\}$

(d) $(g \circ g)(x) = -1 + 4x$; All real numbers **43. (a)** $(f \circ g)(x) = \sqrt{\dfrac{1+x}{x-1}}$; $\{x \mid x \leq -1 \text{ or } x > 1\}$

(b) $(g \circ f)(x) = \dfrac{2}{\sqrt{x+1} - 1}$; $\{x \mid x \geq -1, x \neq 0\}$ **(c)** $(f \circ f)(x) = \sqrt{\sqrt{x+1} + 1}$; $\{x \mid x \geq -1\}$

(d) $(g \circ g)(x) = \dfrac{2(x-1)}{3-x}$; $\{x \mid x \neq 1, x \neq 3\}$ **44. (a)** $(f \circ g)(x) = \sqrt{\dfrac{3x-8}{x-2}}$; $\left\{x \,\middle|\, x < 2 \text{ or } x \geq \dfrac{8}{3}\right\}$

(b) $(g \circ f)(x) = \dfrac{2}{\sqrt{3-x} - 2}$; $\{x \mid x \leq 3, x \neq -1\}$ **(c)** $(f \circ f)(x) = \sqrt{3 - \sqrt{3-x}}$; $\{x \mid -6 \leq x \leq 3\}$

(d) $(g \circ g)(x) = -\dfrac{x-2}{x-3}$; $\{x \mid x \neq 2, x \neq 3\}$ **45. (a)** $(f \circ g)(x) = acx + ad + b$; All real numbers
(b) $(g \circ f)(x) = acx + bc + d$; All real numbers **(c)** $(f \circ f)(x) = a^2x + ab + b$; All real numbers

(d) $(g \circ g)(x) = c^2x + cd + d$; All real numbers **46. (a)** $(f \circ g)(x) = \dfrac{amx + b}{cmx + d}$; $\left\{x \,\middle|\, x \neq -\dfrac{d}{cm}\right\}$

(b) $(g \circ f)(x) = \dfrac{max + mb}{cx + d}$; $\left\{x \,\middle|\, x \neq -\dfrac{d}{c}\right\}$ **(c)** $(f \circ f)(x) = \dfrac{a^2x + ab + bcx + bd}{cax + cb + dcx + d^2}$; $\left\{x \,\middle|\, x \neq -\dfrac{d}{c}, x \neq -\dfrac{cb + d^2}{ca + dc}\right\}$
(d) $(g \circ g)(x) = m^2x$; All real numbers

47. $(f \circ g)(x) = f(g(x)) = f\left(\dfrac{1}{2}x\right) = 2\left(\dfrac{1}{2}x\right) = x$; $(g \circ f)(x) = g(f(x)) = g(2x) = \dfrac{1}{2}(2x) = x$

48. $(f \circ g)(x) = f(g(x)) = f\left(\dfrac{1}{4}x\right) = 4\left(\dfrac{1}{4}x\right) = x$; $(g \circ f)(x) = g(f(x)) = g(4x) = \dfrac{1}{4}(4x) = x$

49. $(f \circ g)(x) = f(g(x)) = f(\sqrt[3]{x}) = (\sqrt[3]{x})^3 = x$; $(g \circ f)(x) = g(f(x)) = g(x^3) = \sqrt[3]{x^3} = x$
50. $(f \circ g)(x) = f(g(x)) = f(x - 5) = (x - 5) + 5 = x$; $(g \circ f)(x) = g(f(x)) = g(x + 5) = (x + 5) - 5 = x$

51. $(f \circ g)(x) = f(g(x)) = f\left(\frac{1}{2}(x+6)\right) = 2\left[\frac{1}{2}(x+6)\right] - 6 = x + 6 - 6 = x;$

$(g \circ f)(x) = g(f(x)) = g(2x - 6) = \frac{1}{2}(2x - 6 + 6) = x$

52. $(f \circ g)(x) = f(g(x)) = f\left(\frac{1}{3}(4 - x)\right) = 4 - 3\left(\frac{1}{3}(4 - x)\right) = x; (g \circ f)(x) = g(f(x)) = g(4 - 3x) = \frac{1}{3}(4 - (4 - 3x)) = x$

53. $(f \circ g)(x) = f(g(x)) = f\left(\frac{1}{a}(x - b)\right) = a\left[\frac{1}{a}(x - b)\right] + b = x; (g \circ f)(x) = g(f(x)) = g(ax + b) = \frac{1}{a}(ax + b - b) = x$

54. $(f \circ g)(x) = f(g(x)) = f\left(\frac{1}{x}\right) = \frac{1}{\frac{1}{x}} = x; (g \circ f)(x) = g(f(x)) = g\left(\frac{1}{x}\right) = \frac{1}{\frac{1}{x}} = x$

55. $f(x) = x^4; g(x) = 2x + 3$ (Other answers are possible.) **56.** $f(x) = x^3; g(x) = 1 + x^2$ (Other answers are possible.)
57. $f(x) = \sqrt{x}; g(x) = x^2 + 1$ (Other answers are possible.) **58.** $f(x) = \sqrt{x}; g(x) = 1 - x^2$ (Other answers are possible.)
59. $f(x) = |x|; g(x) = 2x + 1$ (Other answers are possible.) **60.** $f(x) = |x|; g(x) = 2x^2 + 3$ (Other answers are possible.)

61. $(f \circ g)(x) = 11; (g \circ f)(x) = 2$ **62.** $(f \circ f)(x) = x$ **63.** $-3, 3$ **64.** $-5, 5$ **65.** $S(r(t)) = \frac{16}{9}\pi t^6$ **66.** $V(r(t)) = \frac{32}{81}\pi t^9$

67. $C(N(t)) = 15{,}000 + 800{,}000t - 40{,}000t^2$ **68.** $A(r(t)) = 40{,}000\pi t$ **69.** $C = \frac{2\sqrt{100 - p}}{25} + 600$ **70.** $C = \frac{\sqrt{5(200 - p)}}{10} + 400$

71. $V(r) = 2\pi r^3$ **72.** $V(r) = \frac{2}{3}\pi r^3$

3.4 Exercises

1. (a) $d(x) = \sqrt{x^4 - 15x^2 + 64}$ **(b)** $d(0) = 8$
(c) $d(1) = \sqrt{50} \approx 7.07$
(d)

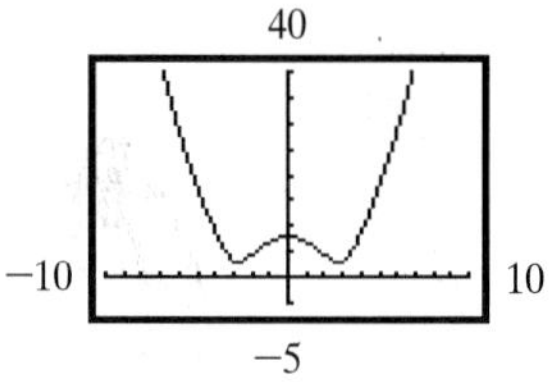

(e) d is smallest when $x \approx -2.74$ or $x \approx 2.74$.

2. (a) $d(x) = \sqrt{x^4 - 13x^2 + 49}$ **(b)** $d(0) = 7$
(c) $d(-1) = \sqrt{37} \approx 6.08$
(d)

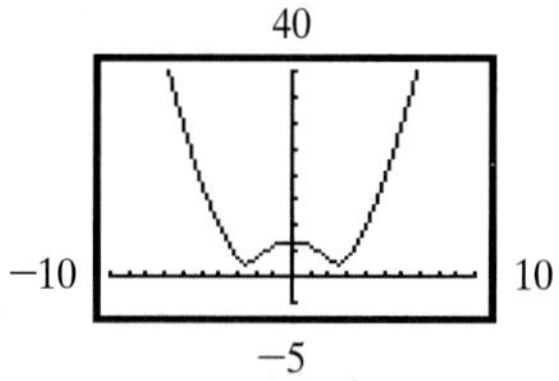

(e) d is smallest when $x \approx -2.55$ or $x \approx 2.55$.

3. (a) $d(x) = \sqrt{x^2 - x + 1}$

(b)

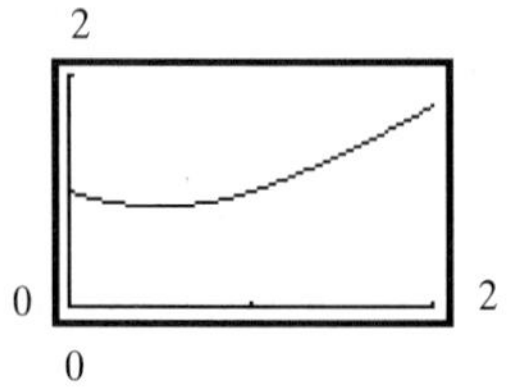

(c) d is smallest when $x = 0.50$.

4. (a) $d(x) = \sqrt{\frac{x^4 + 1}{x^2}}$

(b)

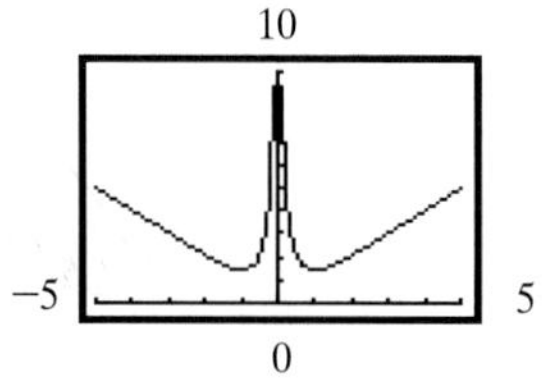

(c) d is smallest when $x = -1$ or $x = 1$.

5. $A(x) = \frac{1}{2}x^4$ **6.** $A(x) = -\frac{1}{2}x^3 + \frac{9}{2}x$

7. (a) $A(x) = x(16 - x^2)$
(b) Domain: $\{x | 0 < x < 4\}$
(c) The area is largest when $x \approx 2.31$.

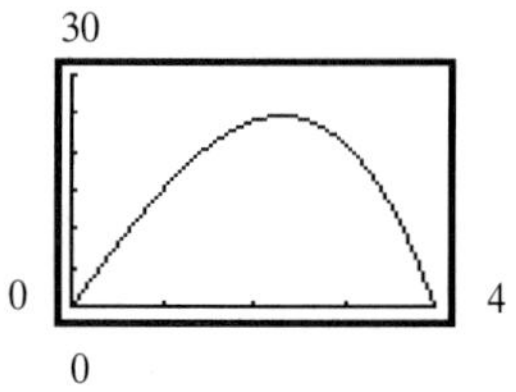

8. (a) $A(x) = 2x\sqrt{4 - x^2}$
(b) $p(x) = 4x + 2\sqrt{4 - x^2}$
(c) A is largest when $x \approx 1.41$. **(d)** p is largest when $x \approx 1.79$.

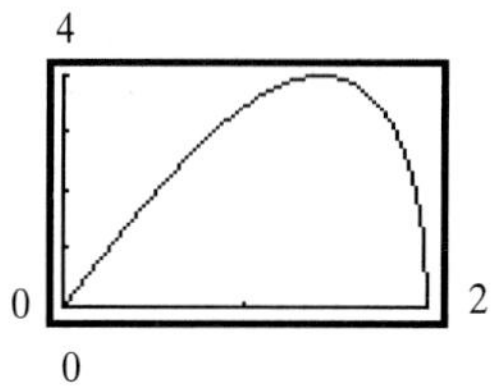

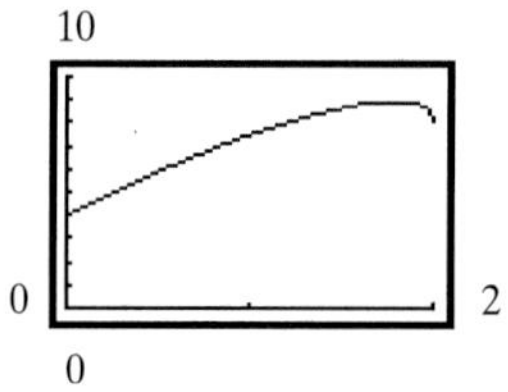

9. **(a)** $A(x) = 4x\sqrt{4 - x^2}$ **(b)** $p(x) = 4x + 4\sqrt{4 - x^2}$
(c) The area is largest when $x \approx 1.41$. **(d)** The perimeter is largest when $x \approx 1.41$.

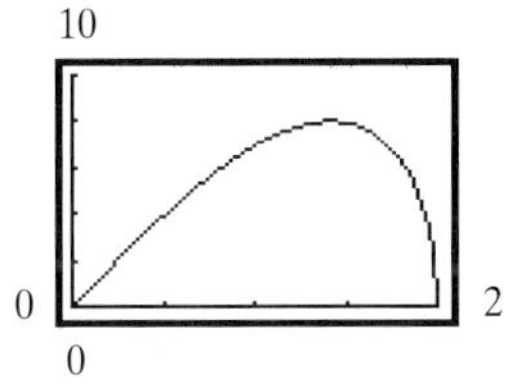

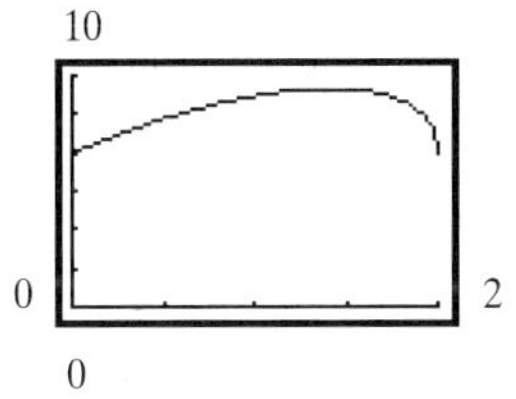

10. **(a)** $A(r) = 4r^2$ **(b)** $p(r) = 8r$

11. **(a)** $A(x) = x^2 + \dfrac{25 - 20x + 4x^2}{\pi}$

(b) Domain: $\{x \mid 0 < x < 2.5\}$

(c) The area is smallest when $x \approx 1.40$ meters.

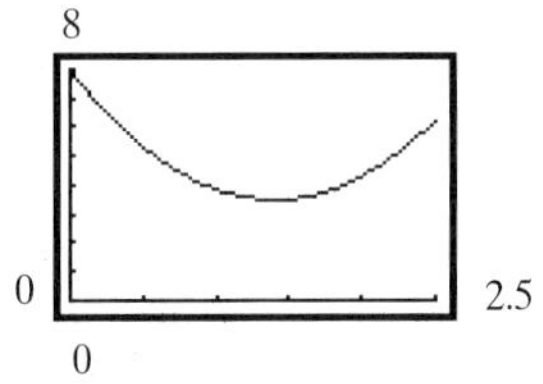

12. **(a)** $A(x) = \dfrac{(10 - 3x)^2}{4\pi} + \dfrac{\sqrt{3}}{4}x^2$

(b) $\left\{x \,\middle|\, 0 < x < \dfrac{10}{3}\right\}$

(c) A is smallest when $x \approx 2.08$.

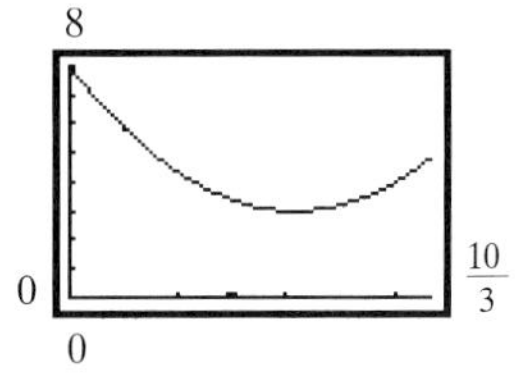

13. **(a)** $C(x) = x$ **(b)** $A(x) = \dfrac{x^2}{4\pi}$ **14.** **(a)** $p(x) = x$ **(b)** $A(x) = \dfrac{x^2}{16}$ **15.** **(a)** $A(r) = 2r^2$ **(b)** $p(r) = 6r$ **16.** $C(x) = \dfrac{2\pi x}{\sqrt{3}}$

17. $A(x) = \left(\dfrac{\pi}{3} - \dfrac{\sqrt{3}}{4}\right)x^2$ **18.** **(a)**

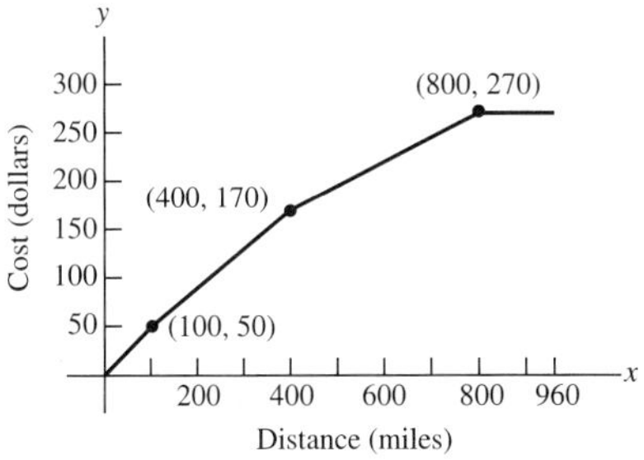

(b) For hauls between 100 and 400 mi, the cost as a function of mileage is $C(x) = 10 + 0.40x$.

(c) For hauls between 400 and 800 mi, the cost as a function of mileage is $C(x) = 70 + 0.25x$.

19. $C = \begin{cases} 95 & \text{if } x = 7 \\ 119 & \text{if } 7 < x \le 8 \\ 143 & \text{if } 8 < x \le 9 \\ 167 & \text{if } 9 < x \le 10 \\ 190 & \text{if } 10 < x \le 14 \end{cases}$

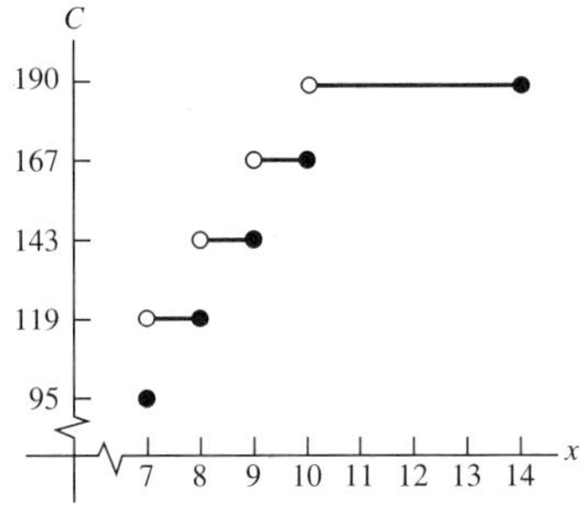

20. $C = \begin{cases} 219 & \text{if } x = 7 \\ 264 & \text{if } 7 < x \le 8 \\ 309 & \text{if } 8 < x \le 9 \\ 354 & \text{if } 9 < x \le 10 \\ 399 & \text{if } 10 < x \le 11 \\ 438 & \text{if } 11 < x \le 14 \end{cases}$

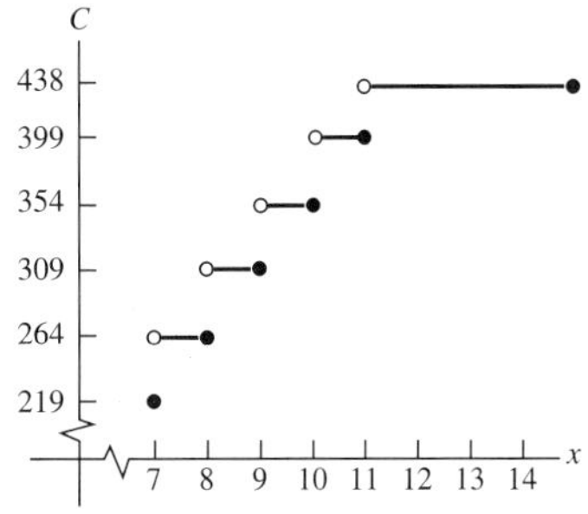

21. $d(t) = 50t$

22. **(a)** $d(t) = \sqrt{2500t^2 - 360t + 13}$
(b) d is smallest when $t = 0.072$ hr.

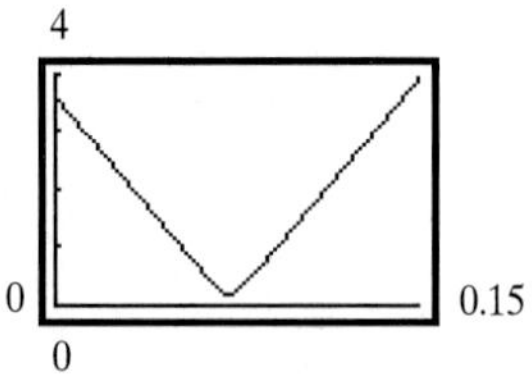

23. **(a)** $V(x) = x(24 - 2x)^2$
(b) 972 in^3 **(c)** 160 in^3
(d) The volume is largest when $x = 4$.

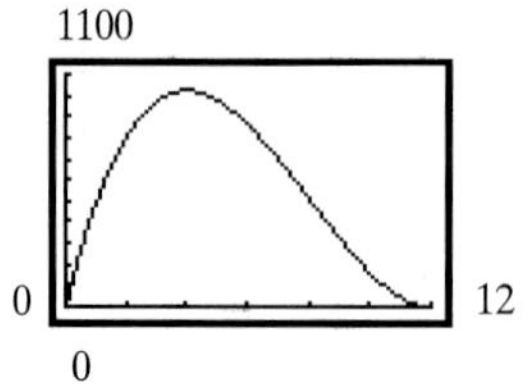

24. **(a)** $A(x) = x^2 + \dfrac{40}{x}$ **(b)** 41 ft^2
(c) 24 ft^2
(d) A is smallest if $x \approx 2.71$ ft

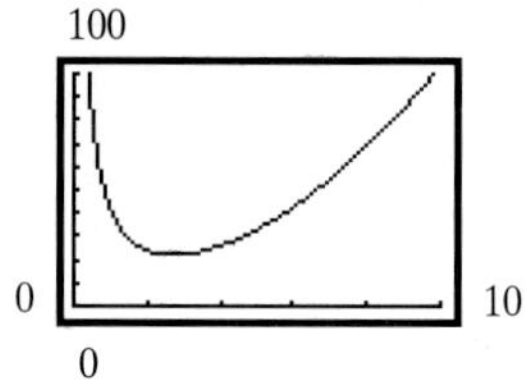

25. $V(h) = \pi h\left(R^2 - \dfrac{h^2}{4}\right)$ **26.** $V(r) = \pi H r^2\left(1 - \dfrac{r}{R}\right)$

27. For schedule X: $f(x) = \begin{cases} 0.15x & \text{if } 0 \le x \le 25{,}750 \\ 368.50 + 0.28(x - 25{,}750) & \text{if } 25{,}750 < x \le 64{,}450 \\ 14{,}138.50 + 0.31(x - 62{,}450) & \text{if } 64{,}450 < x \le 130{,}250 \\ 35{,}156.50 + 0.36(x - 130{,}250) & \text{if } 130{,}250 < x \le 283{,}150 \\ 90{,}200.50 + 0.396(x - 283{,}150) & \text{if } x > 283{,}150 \end{cases}$

For schedule Y-1: $f(x) = \begin{cases} 0.15x & \text{if } 0 \le x \le 43{,}050 \\ 6457.50 + 0.28(x - 43{,}050) & \text{if } 43{,}050 < x \le 104{,}050 \\ 23{,}537.50 + 0.31(x - 104{,}050) & \text{if } 104{,}050 < x \le 158{,}550 \\ 40{,}432.50 + 0.36(x - 158{,}550) & \text{if } 158{,}550 < x \le 283{,}150 \\ 85{,}288.50 + 0.396(x - 283{,}150) & \text{if } x > 283{,}150 \end{cases}$

Fill-in-the-Blank Items

1. Slope **2.** Even;odd **3.** Horizontal; right **4.** $g \circ f$

True/False Items

1. False **2.** False **3.** True **4.** False **5.** False

Review Exercises

1. **(a)** Domain: $[-5, 4]$ or $\{x \mid -5 \le x \le 4\}$
Range: $[-3, 1]$ or $\{y \mid -3 \le y \le 1\}$
(b) Increasing: $(3, 4)$ or $3 < x < 4$
Decreasing: $(-1, 3)$ or $-1 < x < 3$
Constant: $(-5, -1)$ or $-5 < x < -1$
(c) Neither
(d) Intercepts: $(0, 0)$; $(4, 0)$

2. **(a)** Domain: $\left(-\dfrac{\pi}{2}, \dfrac{\pi}{2}\right)$ or $\left\{x \,\middle|\, -\dfrac{\pi}{2} < x < \dfrac{\pi}{2}\right\}$
Range$(-\infty, \infty)$ or all real numbers
(b) Increasing: $\left(-\dfrac{\pi}{2}, \dfrac{\pi}{2}\right)$ or $-\dfrac{\pi}{2} < x < \dfrac{\pi}{2}$
Decreasing: Nowhere
Constant: Nowhere
(c) Odd
(d) Intercept: $(0, 0)$

3. -5 **4.** $2x$ **5.** $3 - 4x$ **6.** $x - 3$ **7.** Odd **8.** Even **9.** Even **10.** Neither **11.** Neither **12.** Neither **13.** Odd **14.** Odd

15.

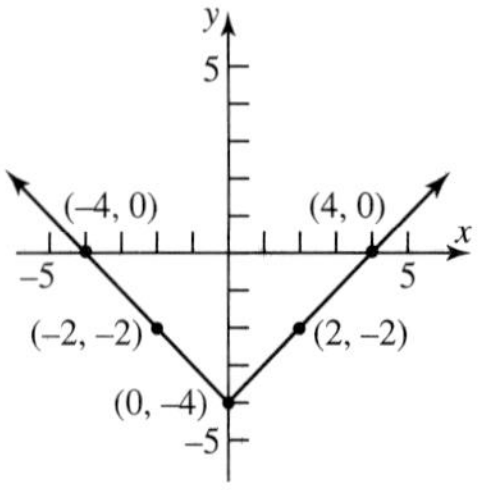

Intercepts: $(-4, 0)$, $(4, 0)$, $(0, -4)$
Domain: all real numbers
Range: $\{y \mid y \ge -4\}$

16.

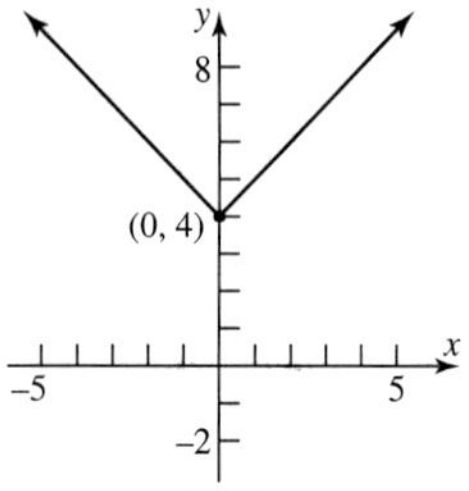

Intercept: $(0, 4)$
Domain: all real numbers
Range: $\{y \mid y \ge 4\}$.

17.

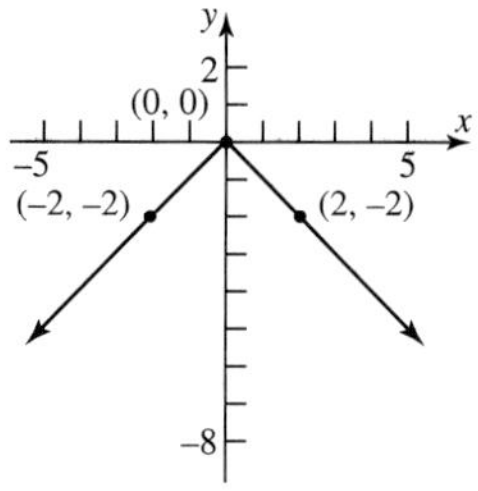

Intercept: $(0, 0)$
Domain: all real numbers
Range: $\{y \mid y \le 0\}$

18.

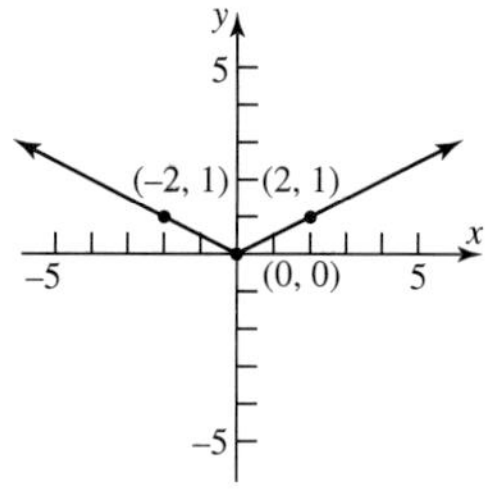

Intercept: $(0, 0)$
Domain: all real numbers
Range: $\{y \mid y \ge 0\}$

19.

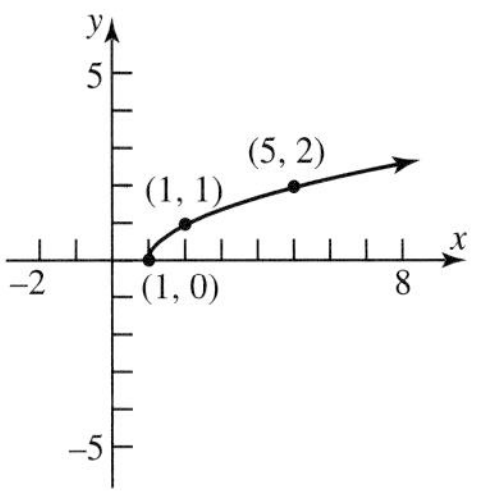

Intercept: (1, 0)
Domain: $\{x|x \geq 1\}$
Range: $\{y|y \geq 0\}$

20.

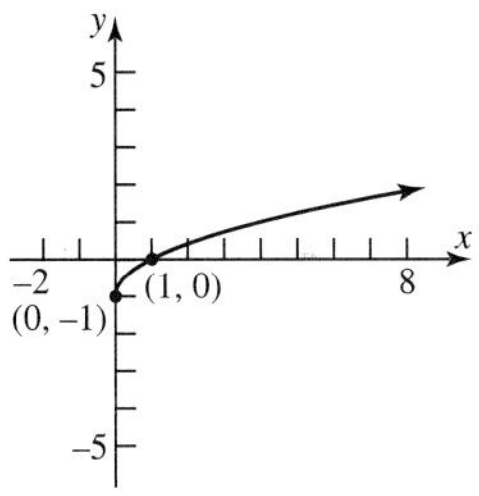

Intercepts: (0, −1), (1, 0)
Domain: $\{x|x \geq 0\}$
Range: $\{y|y \geq -1\}$

21.

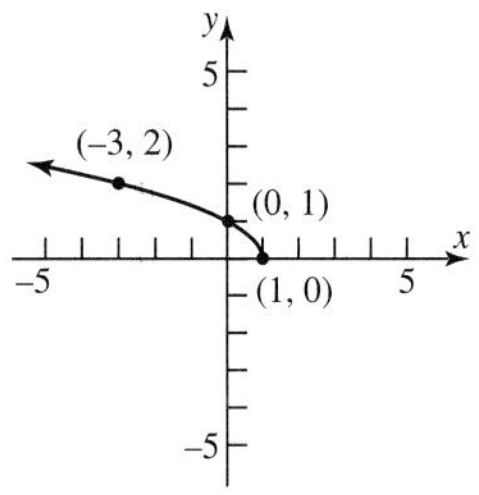

Intercepts: (0, 1), (1, 0)
Domain: $\{x|\ x \leq 1\}$
Range: $\{y|y \geq 0\}$

22.

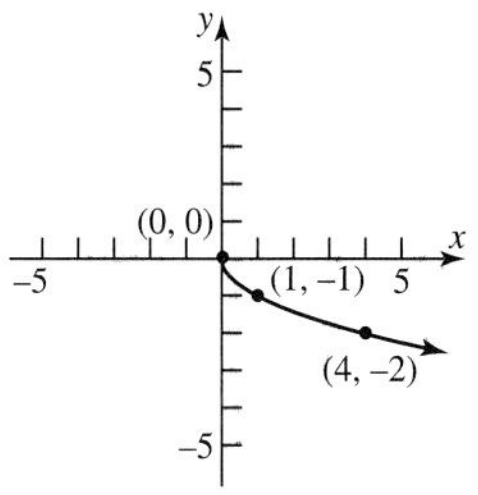

Intercept: (0, 0)
Domain: $\{x|x \geq 0\}$
Range: $\{y|y \leq 0\}$

23.

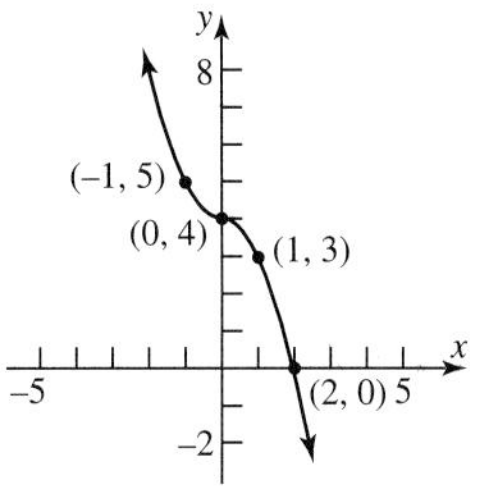

Intercepts: (0, 4), (2, 0)
Domain: all real numbers
Range: all real numbers

24.

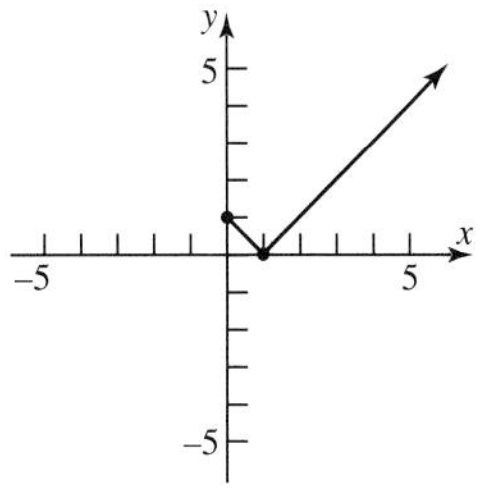

Intercepts: (0, 1), (1, 0)
Domain: $\{x|x \geq 0\}$
Range: $\{y|y \geq 0\}$

25.

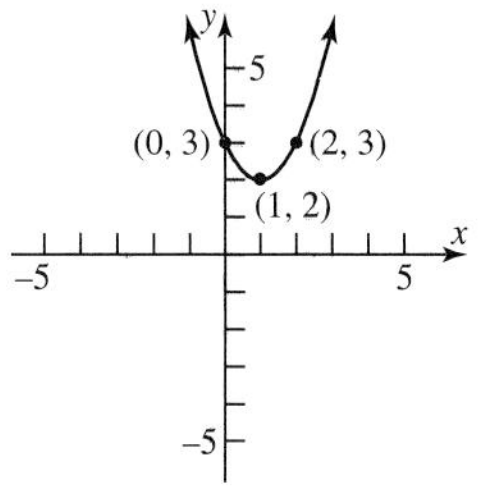

Intercept: (0, 3)
Domain: all real numbers
Range: $\{y|y \geq 2\}$

26.

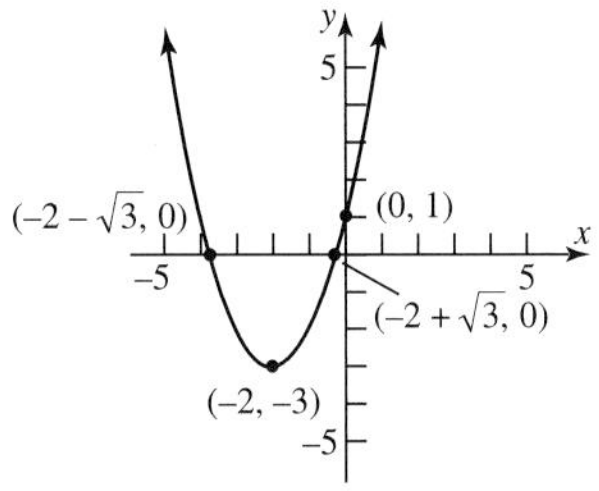

Intercepts: $(-2 - \sqrt{3}, 0)$, $(-2 + \sqrt{3}, 0)$, (0, 1)
Domain: all real numbers
Range: $\{y|y \geq -3\}$

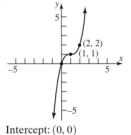

27.

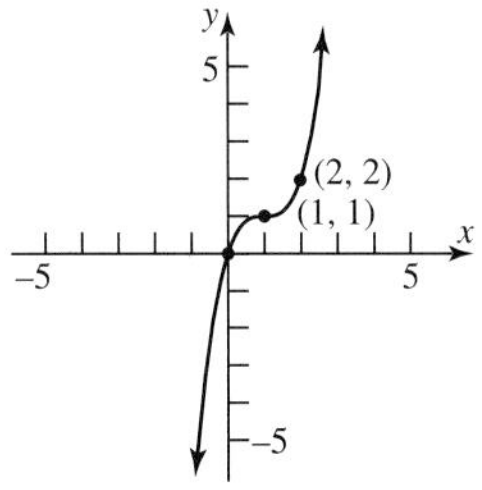

Intercept: (0, 0)
Domain: all real numbers
Range: all real numbers

28.

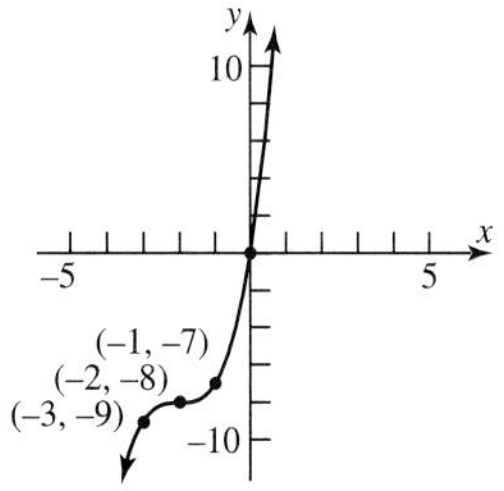

Intercept: (0, 0)
Domain: all real numbers
Range: all real numbers

29.

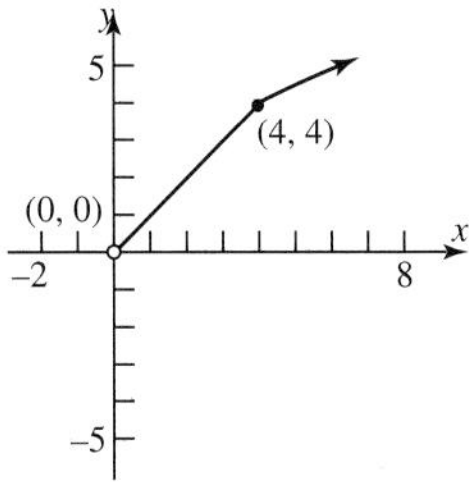

Intercept: none
Domain: $\{x|x > 0\}$
Range: $\{y|y > 0\}$

30.

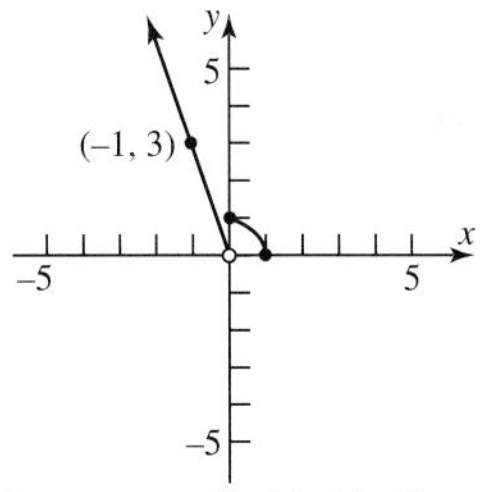

Intercepts: (0, 1), (1, 0)
Domain: $\{x|x \leq 1\}$
Range: $\{y|y \geq 0\}$

31.

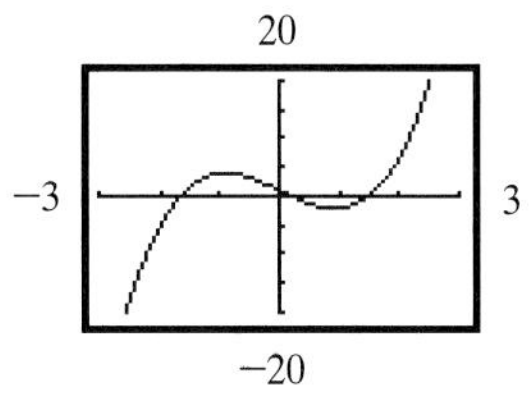

Local maximum: (−0.91, 4.04)
Local minimum: (0.91, −2.04)
Increasing: (−3, −0.91); (0.91, 3)
Decreasing: (−0.91, 0.91)

32.

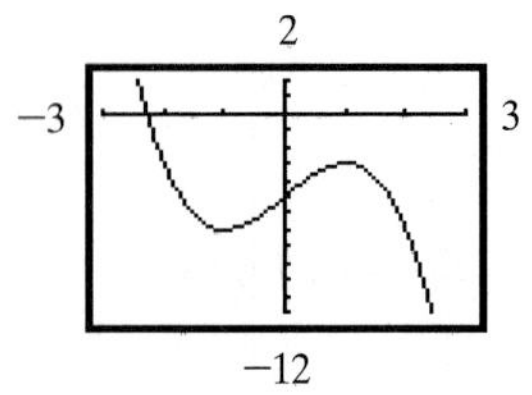

Local maximum: $(1, -3)$
Local minimum: $(-1, -7)$
Increasing: $(-1, 1)$
Decreasing: $(-3, -1), (1, 3)$

33.

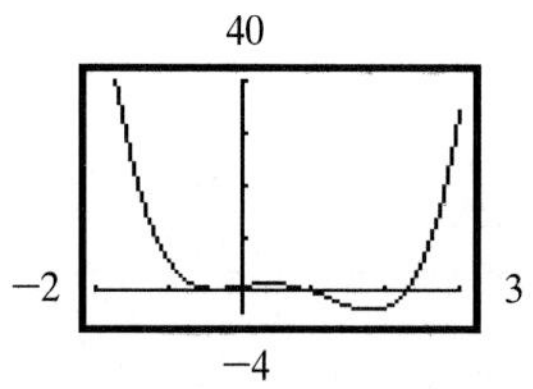

Local maximum: $(0.41, 1.53)$
Local minimum: $(-0.34, 0.54)$; $(1.80, -3.56)$
Increasing: $(-0.34, 0.41)$; $(1.80, 3)$
Decreasing: $(-2, -0.34)$; $(0.41, 1.80)$

34.

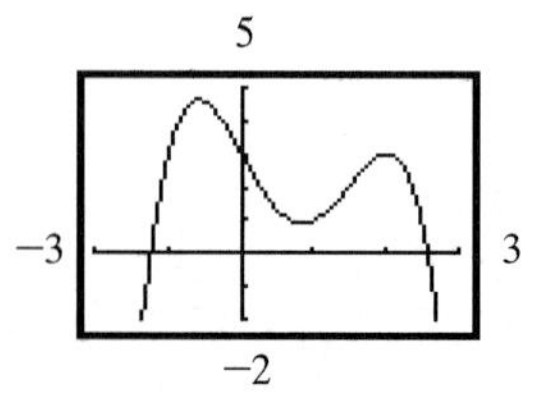

Local maximum: $(-0.59, 4.62), (2, 3)$
Local minimum: $(0.84, 0.92)$
Increasing: $(-2, -0.59), (0.84, 2)$
Decreasing: $(-0.59, 0.84), (2, 3)$

35. (a) -26 **(b)** -241 **(c)** 16 **(d)** -1 **36. (a)** -1 **(b)** 37 **(c)** 4 **(d)** 5 **37. (a)** $\sqrt{11}$ **(b)** 1 **(c)** $\sqrt{\sqrt{6}+2}$ **(d)** 19

38. (a) -5 **(b)** $\sqrt{15}$ **(c)** -6626 **(d)** $\sqrt{4-\sqrt{5}}$ **39. (a)** $\frac{1}{20}$ **(b)** $-\frac{13}{8}$ **(c)** $\frac{400}{1601}$ **(d)** -17 **40. (a)** $\frac{2}{73}$ **(b)** $\frac{2}{3}$ **(c)** $\frac{2178}{1097}$ **(d)** -9

41. $(f \circ g)(x) = 1 - 3x$; All real numbers; $(g \circ f)(x) = 7 - 3x$; All real numbers; $(f \circ f)(x) = x$; All real numbers; $(g \circ g)(x) = 9x + 4$; All real numbers

42. $(f \circ g)(x) = 4x + 1$; All real numbers; $(g \circ f)(x) = 4x - 1$; All real numbers; $(f \circ f)(x) = 4x - 3$; All real numbers; $(g \circ g)(x) = 4x + 3$; All real numbers

43. $(f \circ g)(x) = 27x^2 + 3|x| + 1$; All real numbers; $(g \circ f)(x) = 3|3x^2 + x + 1|$; All real numbers; $(f \circ f)(x) = 3(3x^2 + x + 1)^2 + 3x^2 + x + 2$; All real numbers; $(g \circ g)(x) = 9|x|$; All real numbers

44. $(f \circ g)(x) = \sqrt{3 + 3x + 3x^2}$; All real numbers; $(g \circ f)(x) = 1 + \sqrt{3x} + 3x$; $\{x|x \geq 0\}$; $(f \circ f)(x) = \sqrt{3\sqrt{3x}}$; $\{x|x \geq 0\}$; $(g \circ g)(x) = 3 + 3x + 4x^2 + 2x^3 + x^4$; All real numbers

45. $(f \circ g)(x) = \frac{1+x}{1-x}$; $\{x|\, x \neq 0, x \neq 1\}$; $(g \circ f)(x) = \frac{x-1}{x+1}$; $\{x|\, x \neq -1, x \neq 1\}$; $(f \circ f)(x) = x$; $\{x|x \neq 1\}$; $(g \circ g)(x) = x$; $\{x|x \neq 0\}$

46. $(f \circ g)(x) = \sqrt{\frac{3}{x} - 3}$; $\{x|0 < x \leq 1\}$; $(g \circ f)(x) = \frac{3}{\sqrt{x-3}}$; $\{x|x > 3\}$; $(f \circ f)(x) = \sqrt{\sqrt{x-3} - 3}$; $\{x|x \geq 12\}$; $(g \circ g)(x) = x$; $\{x|x > 0\}$

47. (a)

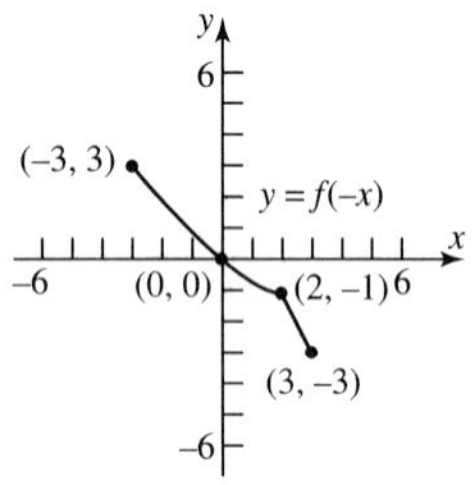

(b)

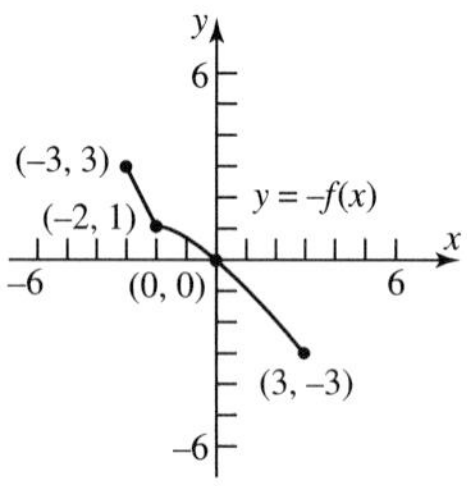

(c)

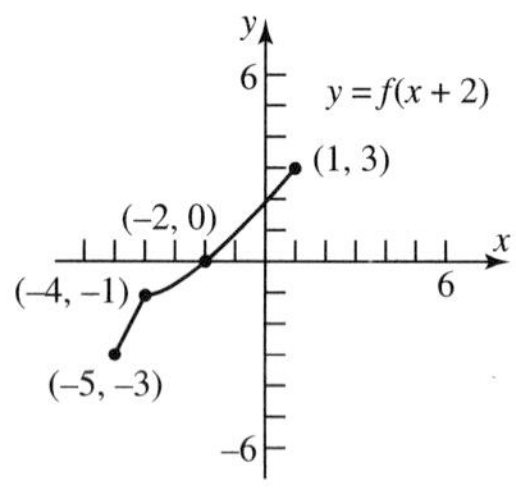

(d)

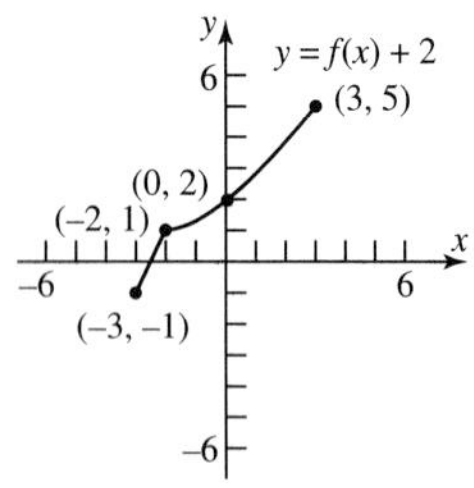

(e)

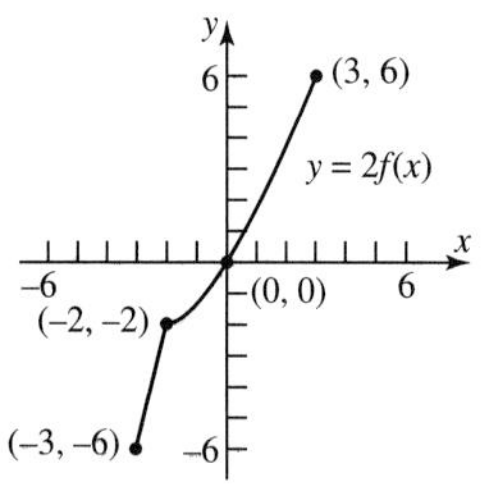

(f)

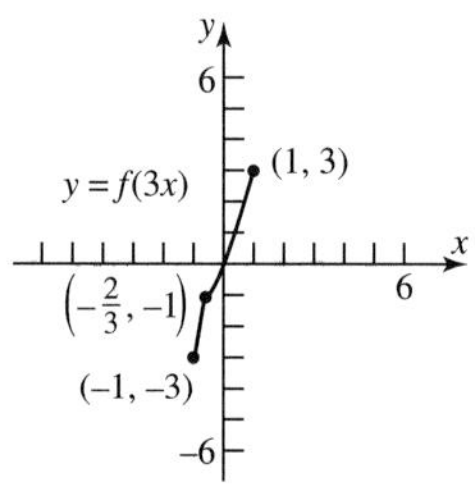

48. (a)

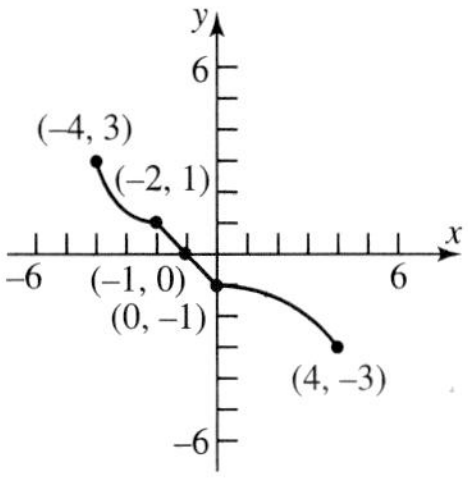

(b)

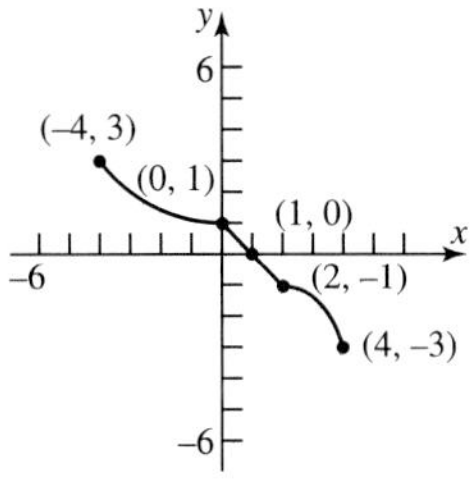

(c)

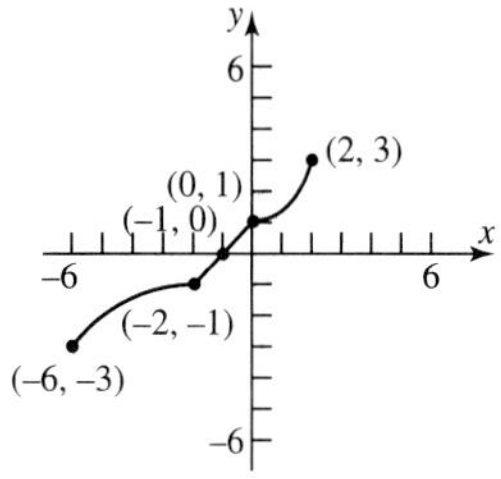

(d)

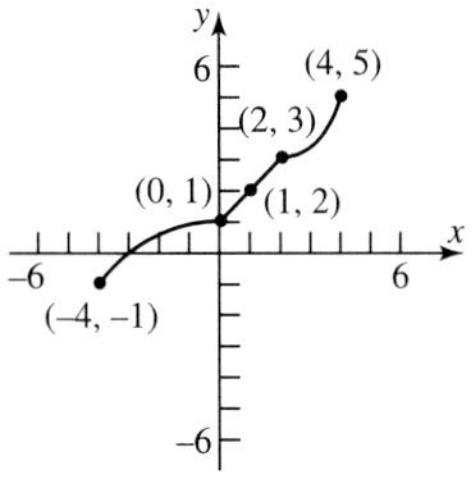

(e)

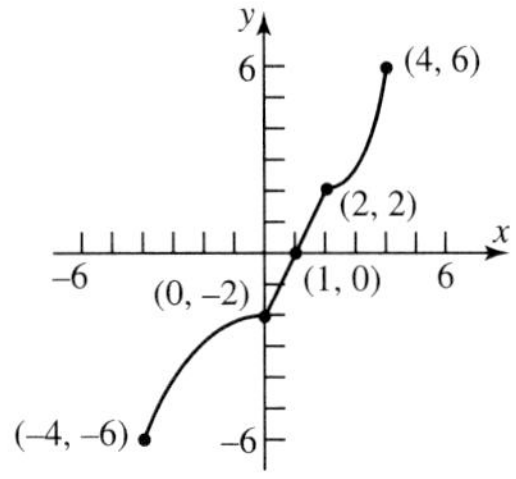

(f)

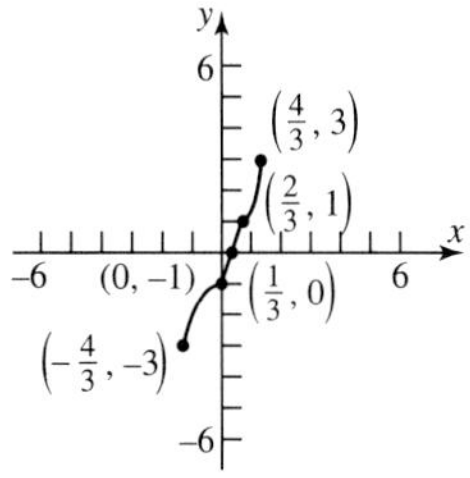

49. $(2, 2)$ **50.** $(2, 3)$ **51.** 25 square units

52. (a) $A(x) = 2x^2 + \dfrac{40}{x}$ **(b)** 42 ft^2 **(c)** 28 ft^2

(d) The area is smallest when $x \approx 2.15$.

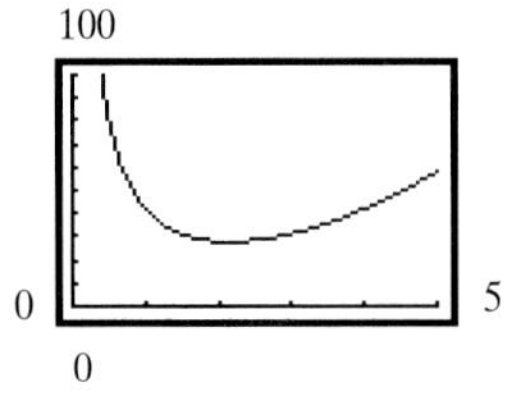

53. (a) $V(S) = \dfrac{S}{6}\sqrt{\dfrac{S}{\pi}}$

(b) $V = \dfrac{1}{3}rS$; if the surface area doubles, the volume doubles.

54. (a), (b), (e)

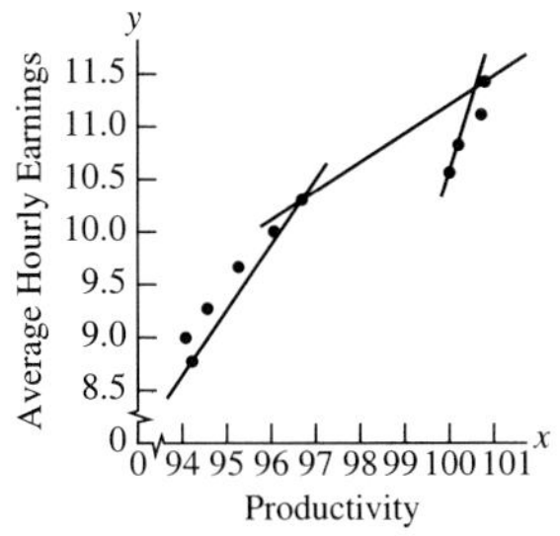

(c) 0.624 dollars/output
(d) For each 1-unit increase in output, earnings increase by an average of \$0.62.
(f) 0.273 dollars/output unit
(g) For each 1-unit increase in output, earnings increased by an average of \$0.27.
(h) It is decreasing.

CHAPTER 4 Polynomial and Rational Functions

4.1 Exercises

1.

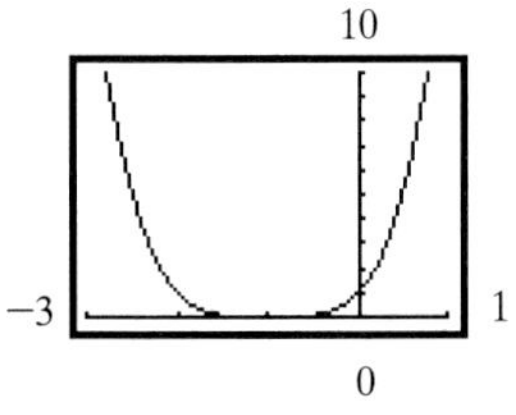

2.

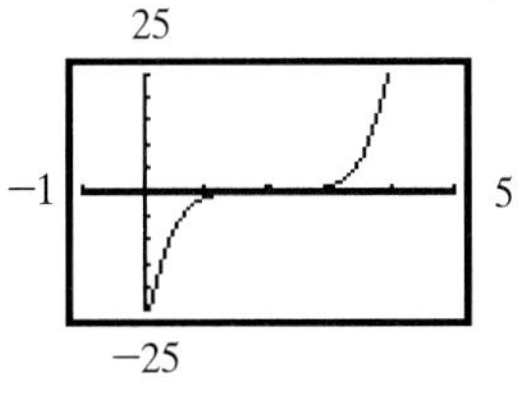

3.

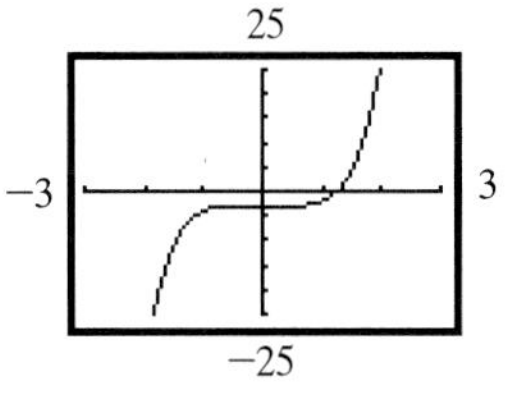

4.

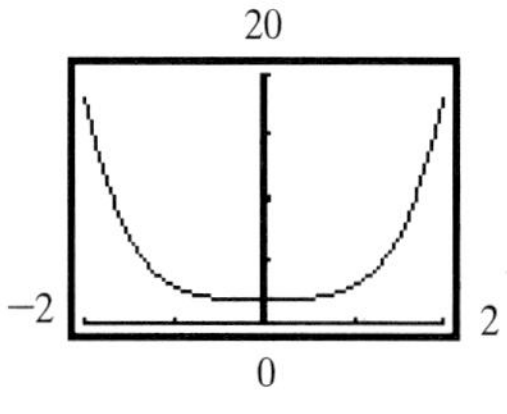

5. 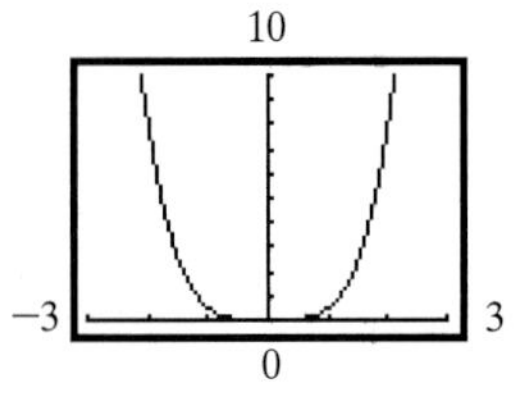

6. 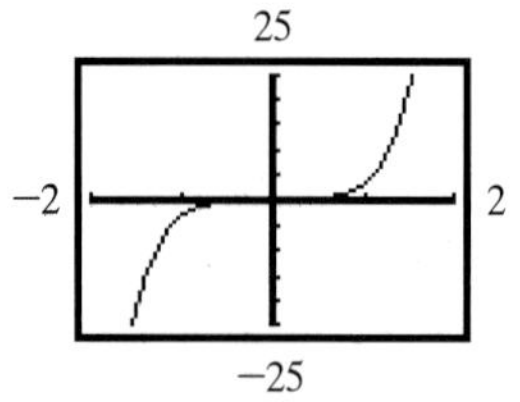

7. 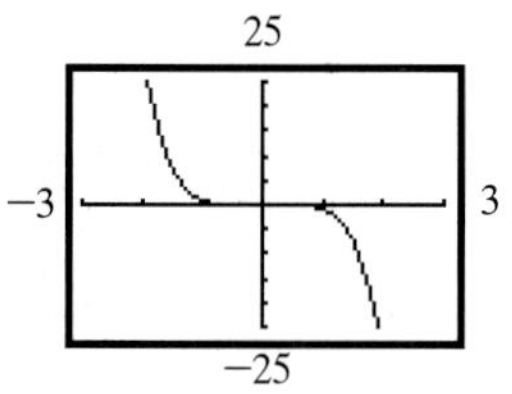

8. 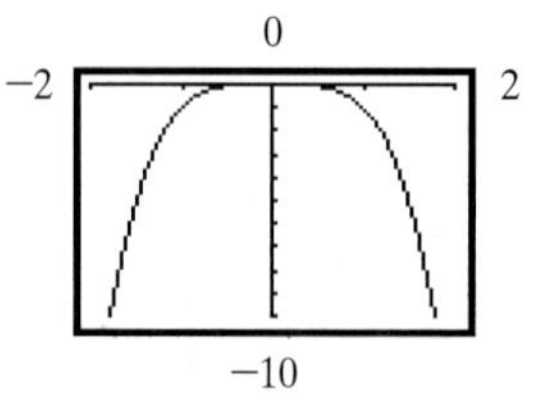

9. 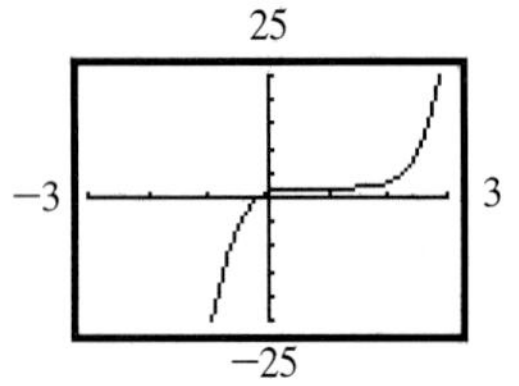

10. 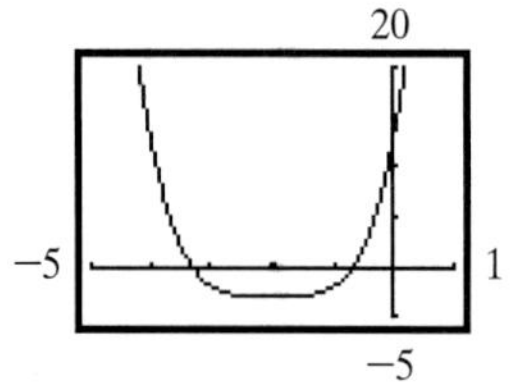

11. 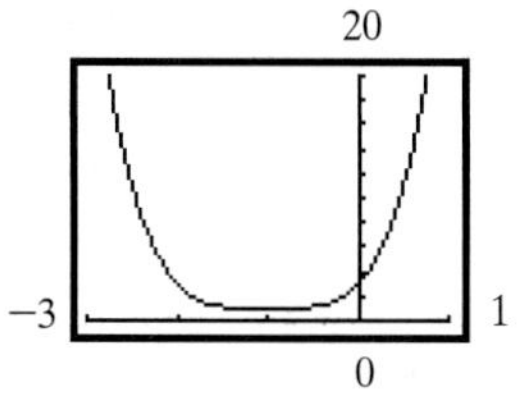

12. 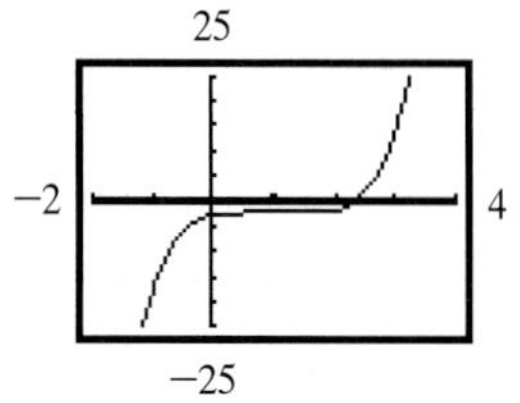

13. 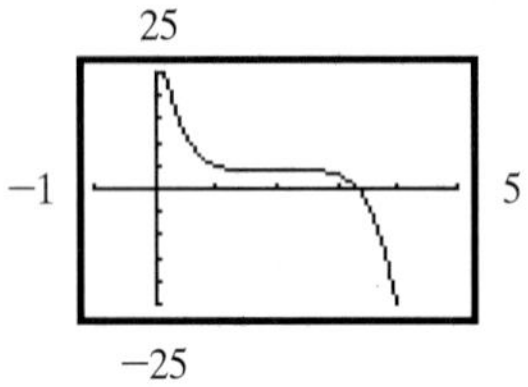

14. 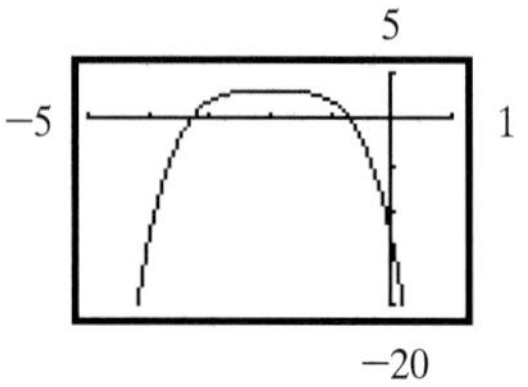

15. 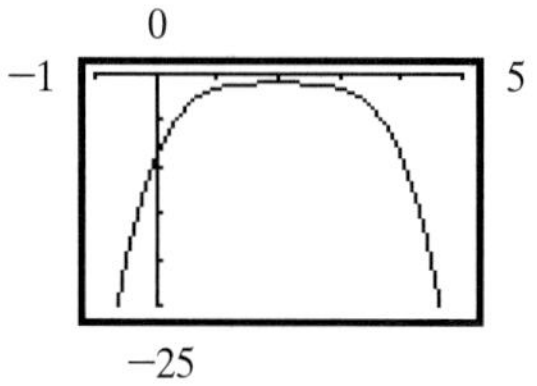

16. 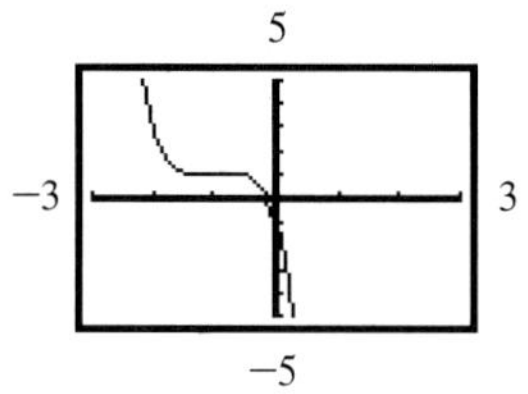

17. **(a)** 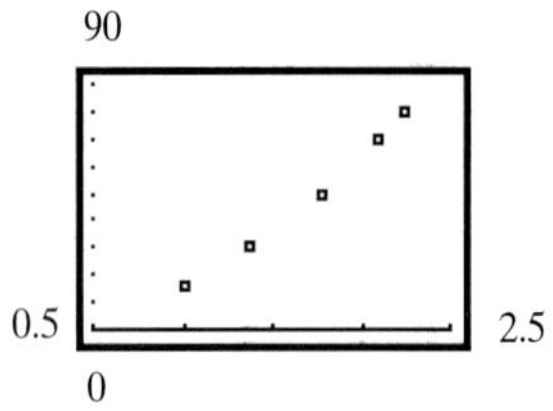

(c) 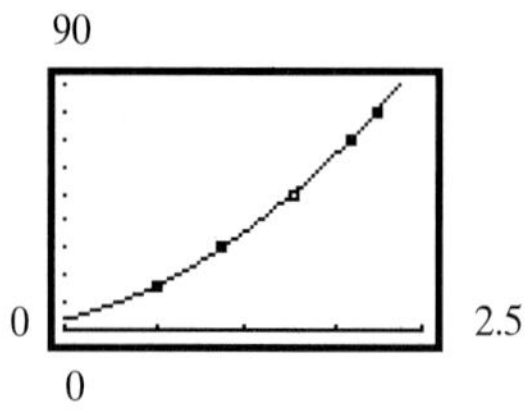

(d) 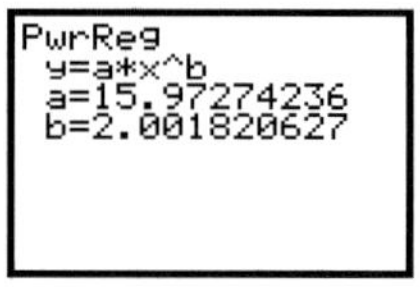

(b) $t \approx 2.5$ sec

(e) $s = \frac{1}{2} \cdot 31.9454t^{2.0018}$, $g \approx 31.9454$ feet/sec^2

18. **(a)** 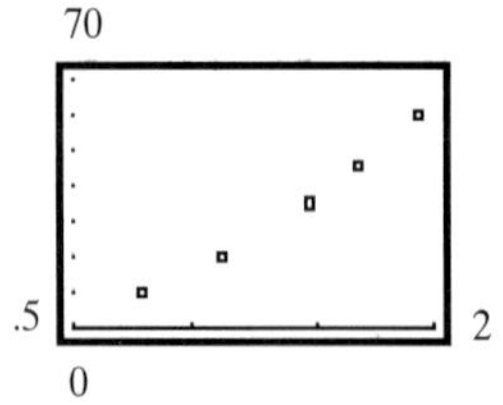

(c) 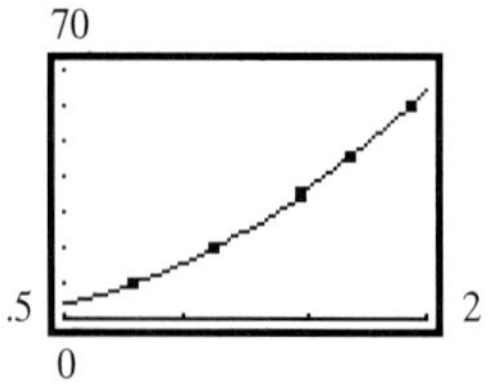

(b) $s \approx 16.0233t^{1.9982}$ **(d)** $t \approx 2.5$ sec

19. **(a)** See (c) **(b)** $T(2.3) = 1.674$ sec **(c)** 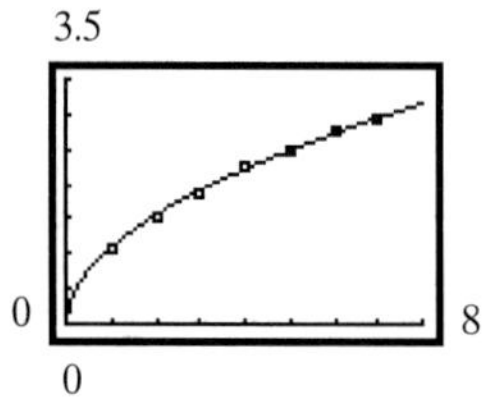

(d) 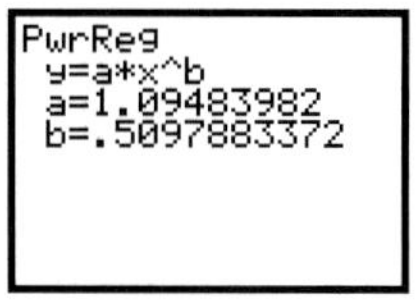

4.2 Exercises

1. Yes; degree 3 **2.** Yes; degree 4 **3.** Yes; degree 2 **4.** Yes; degree 1 **5.** No; x is raised to the -1 power. **6.** Yes; degree 2 **7.** No; x is raised to the $\frac{3}{2}$ power. **8.** No; $h(x) = x - \sqrt{x}$, so x is raised to the $\frac{1}{2}$ power. **9.** Yes; degree 4 **10.** No; $f(x) = x^{-1} - 5x^{-3}$, so x is raised to the -1 power and to the -3 power. **11.** $f(x) = x^3 - 3x^2 - x + 3$ for $a = 1$ **12.** $f(x) = x^3 - 3x^2 - 4x + 12$ for $a = 1$ **13.** $f(x) = x^3 - x^2 - 12x$ for $a = 1$ **14.** $f(x) = x^3 + 2x^2 - 8x$ for $a = 1$ **15.** $f(x) = x^4 - 15x^2 + 10x + 24$ for $a = 1$ **16.** $f(x) = x^4 - 3x^3 - 15x^2 + 19x + 30$ for $a = 1$ **17.** 7, multiplicity 1; -3, multiplicity 2; graph touches the x-axis at -3 and crosses it at 7

18. -4, multiplicity 1; -3, multiplicity 3; graph crosses the x-axis at -4 and -3

19. 2, multiplicity 3; graph crosses the x-axis at 2 **20.** 3, multiplicity 1; -4, multiplicity 3; graph crosses the x-axis at -4 and 3

21. $-\frac{1}{2}$, multiplicity 2; graph touches the x-axis at $-\frac{1}{2}$ **22.** $\frac{1}{3}$, multiplicity 2; 1, multiplicity 3; graph touches the x-axis at $\frac{1}{3}$ and crosses it at 1

23. 5, multiplicity 3; -4, multiplicity 2; graph touches the x-axis at -4 and crosses it at 5

24. $-\sqrt{3}$, multiplicity 2; 2, multiplicity 4; graph touches the x-axis at $-\sqrt{3}$ and at 2

25. No real zeros; graph neither crosses nor touches the x-axis **26.** No real zeros; graph neither crosses nor touches the x-axis

27. (a)

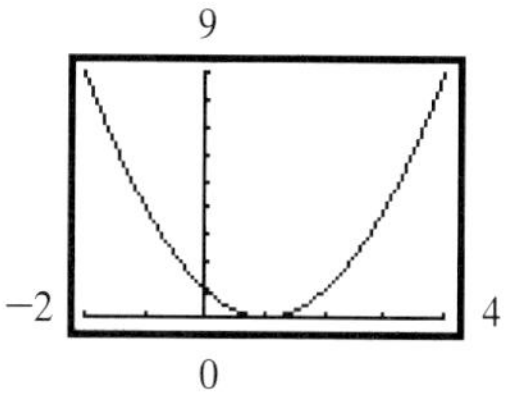

(b) x-intercept: 1; y-intercept: 1
(c) 1: Even
(d) $y = x^2$
(e) 1
(f) Local minimum: (1, 0);
Local maxima: None

28. (a)

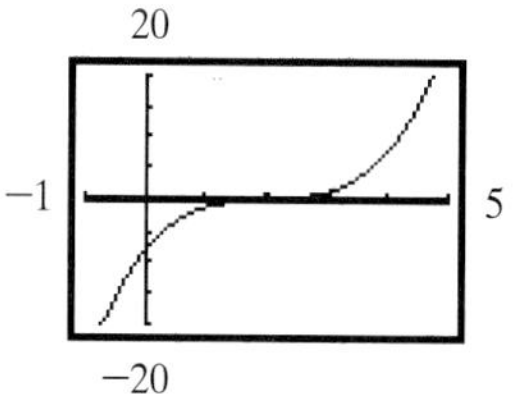

(b) x-intercept: 2; y-intercept: -8
(c) 2: Odd
(d) $y = x^3$
(e) 0
(f) None

29. (a)

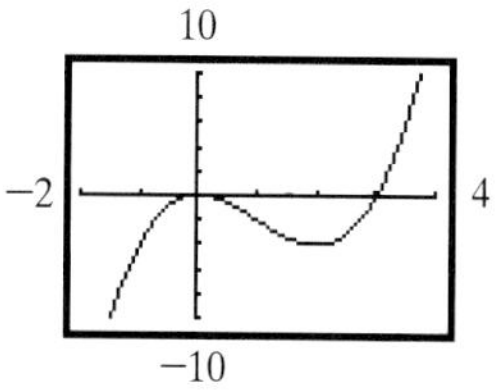

(b) x-intercepts: 0, 3; y-intercept: 0
(c) 0: Even; 3: Odd
(d) $y = x^3$
(e) 2
(f) Local minimum: $(2, -4)$;
Local maximum: (0, 0)

30. (a)

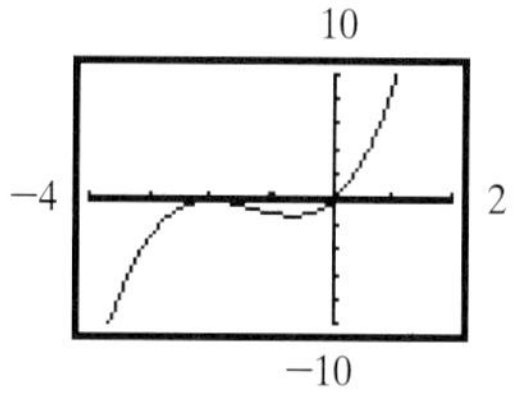

(b) x-intercepts: -2, 0; y-intercept: 0
(c) -2: Even; 0: Odd
(d) $y = x^3$
(e) 2
(f) Local minimum: $(-0.67, -1.19)$;
Local maximum: $(-2, 0)$

31. (a)

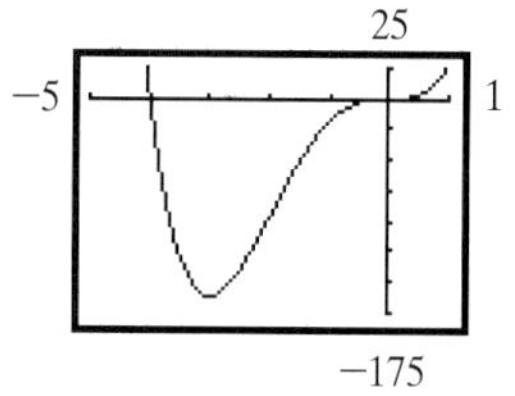

(b) x-intercepts: -4, 0; y-intercept: 0
(c) -4, 0: Odd
(d) $y = 6x^4$
(e) 1
(f) Local minimum: $(-3, -162)$;
Local maxima: None

32. (a)

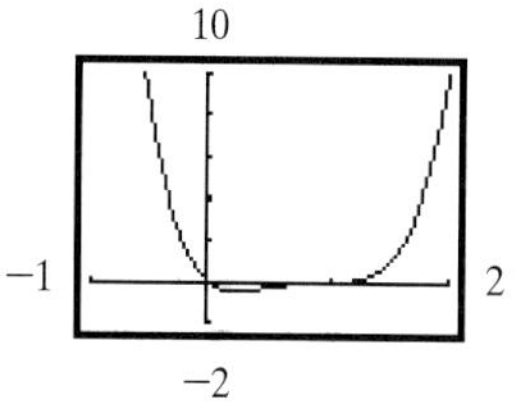

(b) x-intercepts: 0, 1; y-intercept: 0
(c) 0, 1: Odd
(d) $y = 5x^4$
(e) 1
(f) Local minimum: $(0.25, -0.53)$;
Local maxima: None

33. (a)

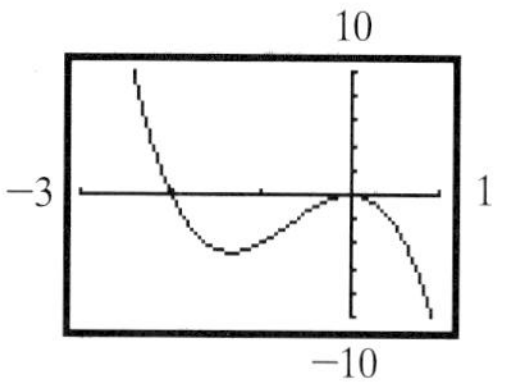

(b) x-intercepts: -2, 0; y-intercept: 0
(c) -2: Odd; 0: Even
(d) $y = -4x^3$
(e) 2
(f) Local minimum: $(-1.33, -4.74)$;
Local maximum: (0, 0)

34. (a)

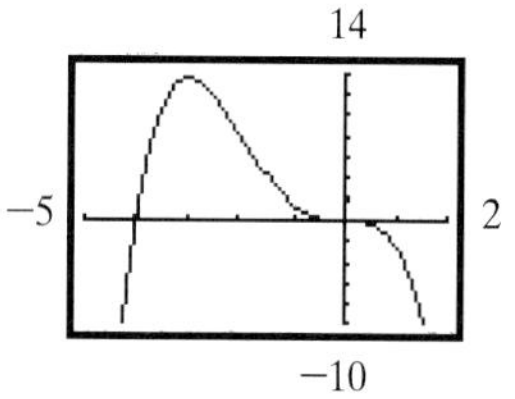

(b) x-intercepts: -4, 0; y-intercept: 0
(c) -4, 0: Odd
(d) $y = -\frac{1}{2}x^4$
(e) 1
(f) Local minima: None;
Local maximum: $(-3, 13.5)$

35. (a)

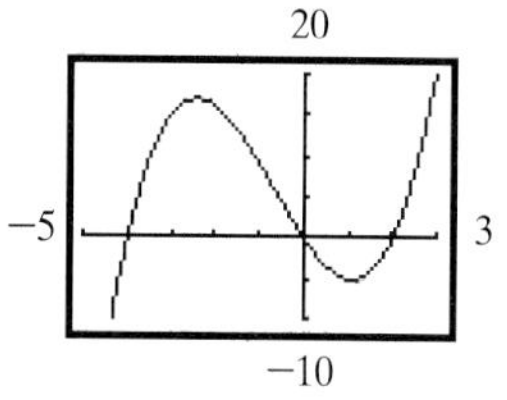

(b) x-intercepts: -4, 0, 2; y-intercept: 0
(c) -4, 0, 2: Odd
(d) $y = x^3$
(e) 2
(f) Local minimum: $(1.10, -5.05)$;
Local maximum: $(-2.43, 16.90)$

36. (a)

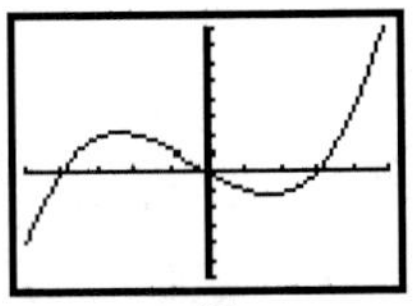

(b) x-intercepts: $-4, 0, 3$; y-intercept: 0
(c) $-4, 0, 3$: Odd
(d) $y = x^3$
(e) 2
(f) Local minimum: $(1.69, -12.60)$; Local maximum: $(-2.36, 20.75)$

37. (a)

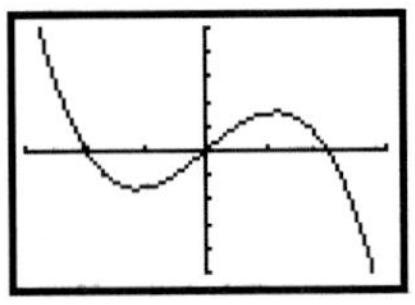

(b) x-intercepts: $-2, 0, 2$; y-intercept: 0
(c) $-2, 0, 2$: Odd
(d) $y = -x^3$
(e) 2
(f) Local minimum: $(-1.15, -3.08)$; Local maximum: $(1.15, 3.08)$

38. (a)

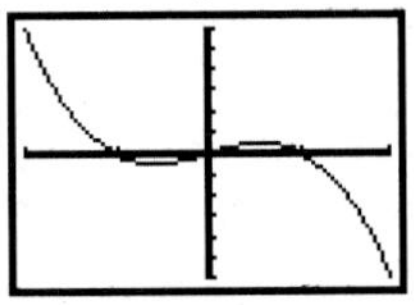

(b) x-intercepts: $-1, 0, 1$; y-intercept: 0
(c) $-1, 0, 1$: Odd
(d) $y = -x^3$
(e) 2
(f) Local minimum: $(-0.58, -0.38)$; Local maximum: $(0.58, 0.38)$

39. (a)

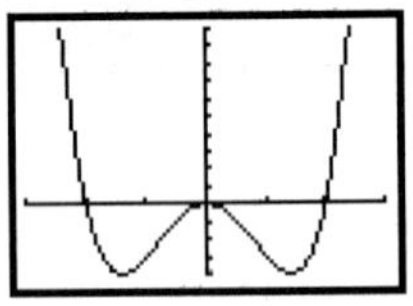

(b) x-intercepts: $-2, 0, 2$; y-intercept: 0
(c) $-2, 2$: Odd; 0: Even
(d) $y = x^4$
(e) 3
(f) Local minima: $(-1.41, -4)$, $(1.41, -4)$; Local maximum: $(0, 0)$

40. (a)

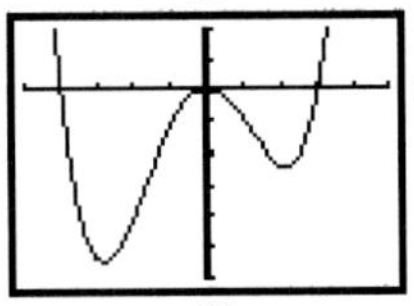

(b) x-intercepts: $-4, 0, 3$; y-intercept: 0
(c) $-4, 3$: Odd; 0: Even
(d) $y = x^4$
(e) 3
(f) Local minima: $(-2.85, -54.64)$, $(2.10, -24.21)$; Local maximum: $(0, 0)$

41. (a)

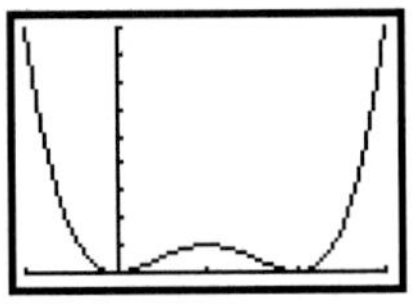

(b) x-intercepts: $0, 2$; y-intercept: 0
(c) $0, 2$: Even
(d) $y = x^4$
(e) 3
(f) Local minima: $(0, 0)$, $(2, 0)$; Local maximum: $(1, 1)$

42. (a)

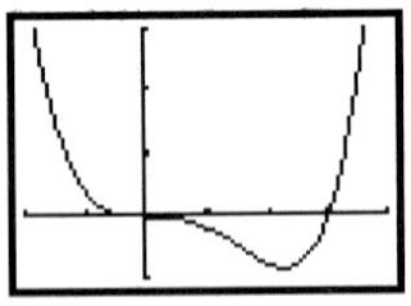

(b) x-intercepts: $0, 3$; y-intercept: 0
(c) $0, 3$: Odd
(d) $y = x^4$
(e) 1
(f) Local minimum: $(2.25, -8.54)$; Local maxima: None

43. (a)

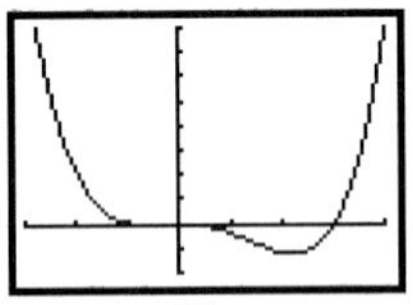

(b) x-intercepts: $-1, 0, 3$; y-intercept: 0
(c) $-1, 3$: Odd; 0: Even
(d) $y = x^4$
(e) 3
(f) Local minima: $(2.19, -12.39)$, $(-0.69, -0.54)$; Local maximum: $(0, 0)$

44. (a)

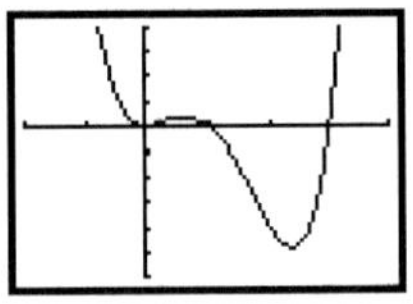

(b) x-intercepts: $0, 1, 3$; y-intercept: 0
(c) 0: Even; $1, 3$: Odd
(d) $y = x^4$
(e) 3
(f) Local minima: $(0, 0)$, $(2.37, -4.85)$; Local maximum: $(0.63, 0.35)$

45. (a)

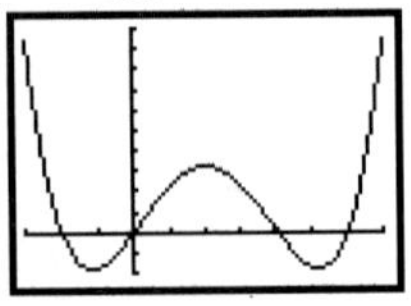

(b) x-intercepts: $-2, 0, 4, 6$; y-intercept: 0
(c) $-2, 0, 4, 6$: Odd
(d) $y = x^4$
(e) 3
(f) Local minima: $(-1.16, -36)$, $(5.16, -36)$; Local maximum: $(2, 64)$

46. (a)

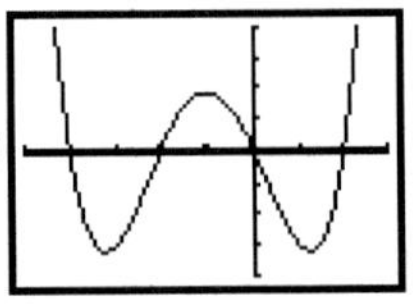

(b) x-intercepts: $-4, -2, 0, 2$; y-intercept: 0
(c) $-4, -2, 0, 2$: Odd
(d) $y = x^4$
(e) 3
(f) Local minima: $(-3.24, -16)$, $(1.27, -16)$; Local maximum: $(-1, 9)$

47. (a)

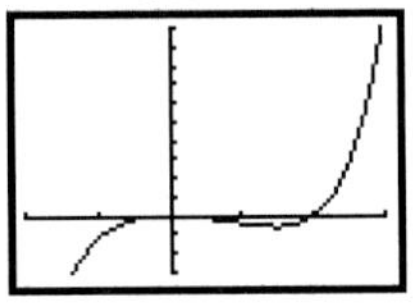

(b) x-intercepts: $0, 2$; y-intercept: 0
(c) 0: Even; 2: Odd
(d) $y = x^5$
(e) 2
(f) Local minimum: $(1.48, -5.91)$; Local maximum: $(0, 0)$

48. (a)

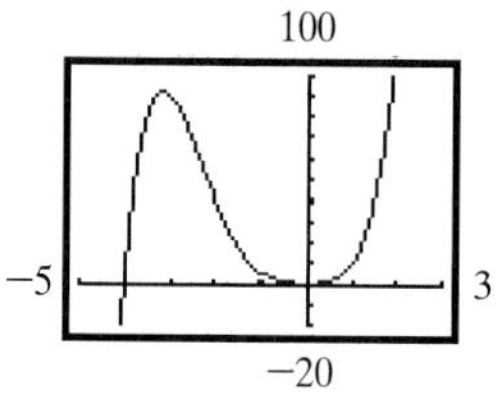

(b) x-intercepts: −4, 0; y-intercept: 0
(c) −4: Odd; 0: Even
(d) $y = x^5$
(e) 2
(f) Local minimum: (0, 0); Local maximum: (−3.17, 92.15)

49. (a)

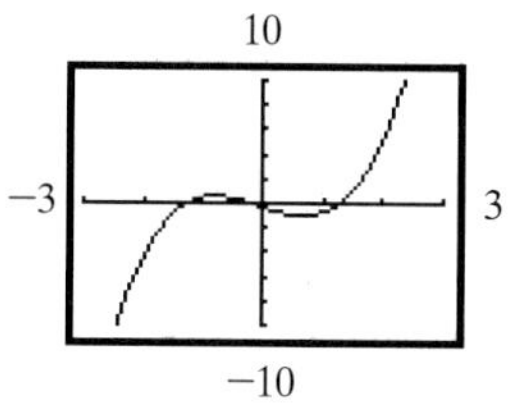

(b) x-intercepts: −1.26, −0.20, 1.26; y-intercept: −0.31752
(c) −1.26, −0.20, 1.26: Odd
(d) $y = x^3$
(e) 2
(f) Local minimum: (0.66, −0.99); Local maximum: (−0.80, 0.57)

50. (a)

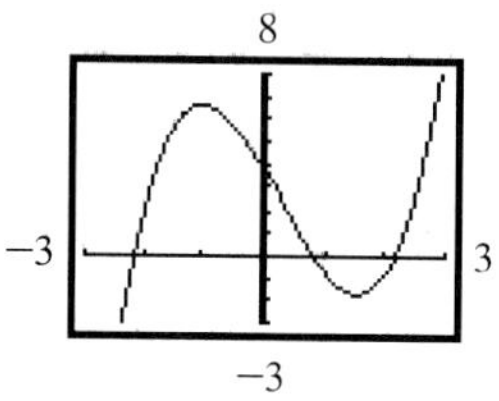

(b) x-intercepts: (−2.16, 0.80, 2.16); y-intercept: 3.73248
(c) −2.16, 0.80, 2.16: Odd
(d) $y = x^3$
(e) 2
(f) Local minimum: (1.54, −1.70); Local maximum: (−1.01, 6.60)

51. (a)

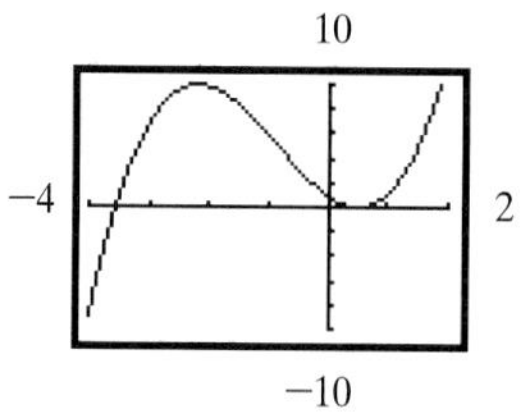

(b) x-intercepts: −3.56, 0.50; y-intercept: 0.89
(c) −3.56: Odd; 0.50: Even
(d) $y = x^3$
(e) 2
(f) Local minimum: (0.50, 0); Local maximum: (−2.21, 9.91)

52. (a)

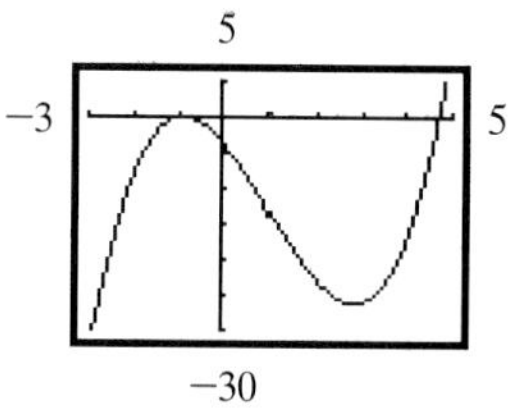

(b) x-intercepts: (−0.9, 4.71); y-intercept: −3.8151
(c) −0.9: Even; 4.71: Odd
(d) $y = x^3$
(e) 2
(f) Local minimum: (2.84, −26.16); Local maximum: (−0.9, 0)

53. (a)

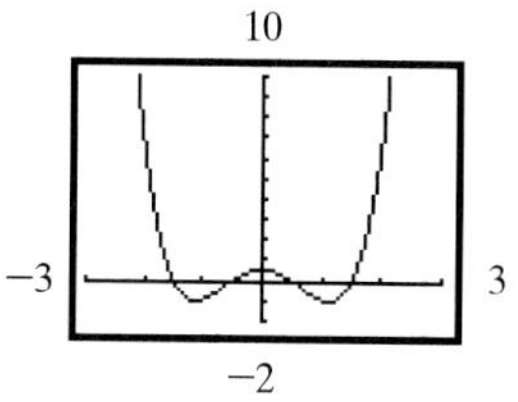

(b) x-intercepts: −1.50, −0.50, 0.50, 1.50; y-intercept: 0.5625
(c) −1.50, −0.50, 0.50, 1.50: Odd
(d) $y = x^4$
(e) 3
(f) Local minima: (−1.12, −1), (1.12, −1); Local maximum: (0, 0.5625)

54. (a)

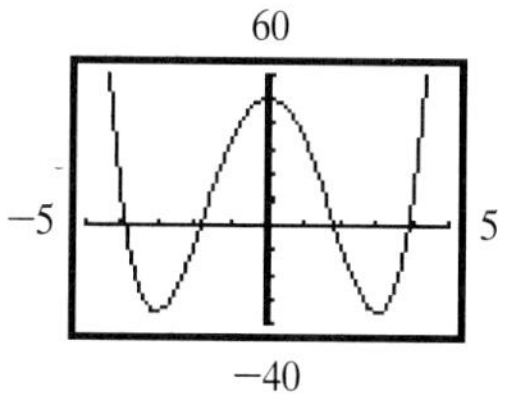

(b) x-intercepts: −3.90, −1.82, 1.82, 3.90; y-intercept: 50.2619
(c) −3.90, −1.82, 1.82, 3.90: Odd
(d) $y = x^4$
(e) 3
(f) Local minima: (−3.04, −35.30), (3.04, −35.30); Local maximum: (0, 50.2619)

55. (a)

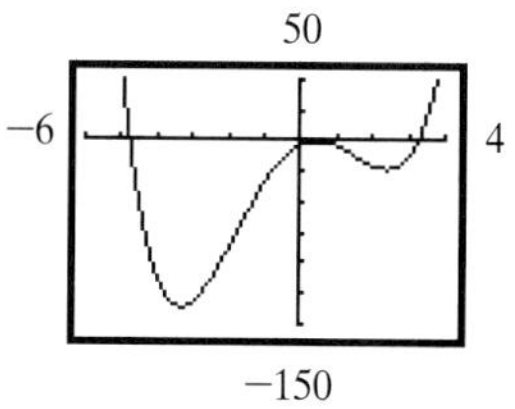

(b) x-intercepts: −4.78, 0.45, 3.23; y-intercept: −3.1264785
(c) −4.78, 3.23: Odd; 0.45: Even
(d) $y = x^4$
(e) 3
(f) Local minima: (−3.32, −135.92), (2.38, −22.67); Local maximum: (0.45, 0)

56. (a)

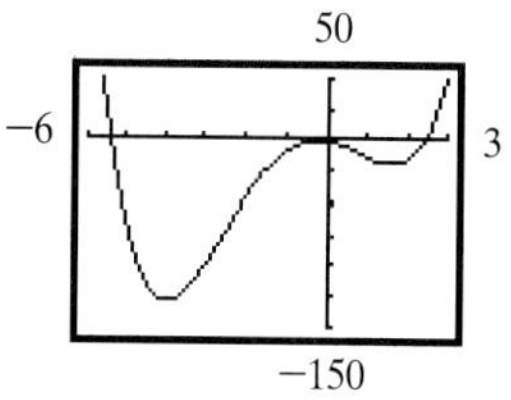

(b) x-intercepts: −5.41, −0.23, 2.42; y-intercept: −0.69257738
(c) −5.41, 2.42: Odd; −0.23: Even
(d) $y = x^4$
(e) 3
(f) Local minima: (−3.97, −128.71), (1.61, −19.25); Local maximum: (−0.23, 0)

57. (a)

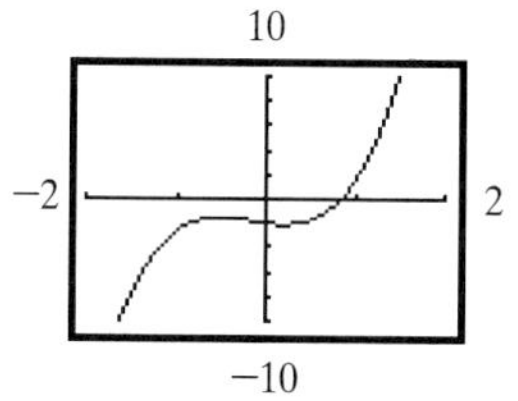

(b) x-intercept: 0.84; y-intercept: −2
(c) 0.84: Odd
(d) $y = \pi x^3$
(e) 2
(f) Local minimum: (0.21, −2.12); Local maximum: (−0.51, −1.54)

58. (a)

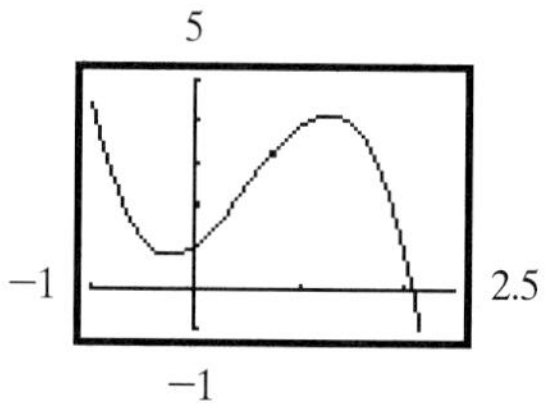

(b) x-intercept: 2.10; y-intercept: 1
(c) 2.10: Odd
(d) $y = -2x^3$
(e) 2
(f) Local minimum: (−0.23, 0.79); Local maximum: (1.27, 4.17)

59. (a)

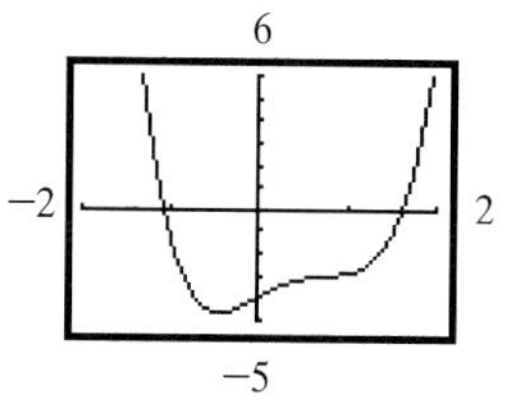

(b) x-intercepts: −1.07, 1.62; y-intercept: −4
(c) −1.07, 1.62: Odd
(d) $y = 2x^4$
(e) 1
(f) Local minimum: (−0.42, −4.64); Local maxima: None

60. (a)

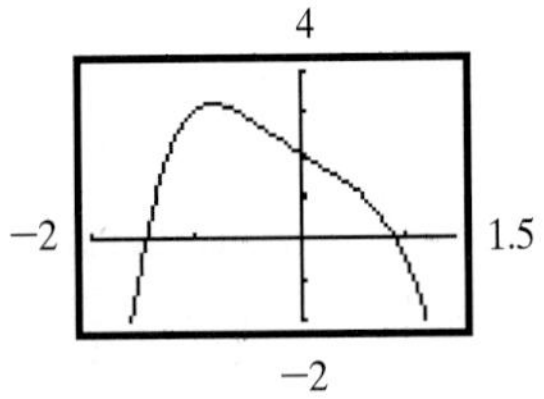

(b) x-intercept: −1.47, 0.91; y-intercept: 2
(c) −1.47, 0.91: Odd
(d) $y = -1.2x^4$
(e) 1
(f) Local minima: None;
Local maximum: (−0.81, 3.21)

61. (a)

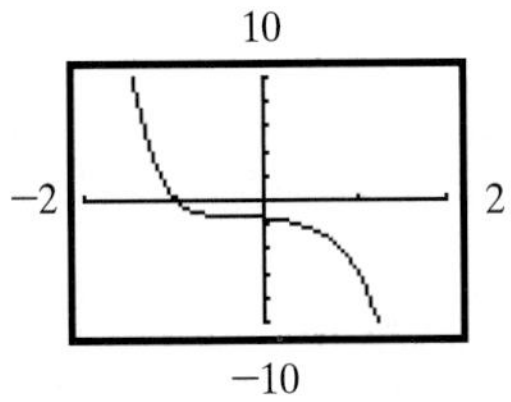

(b) x-intercept: −0.98; y-intercept: $-\sqrt{2}$
(c) −0.98: Odd
(d) $y = -2x^5$
(e) 0
(f) None

62. (a)

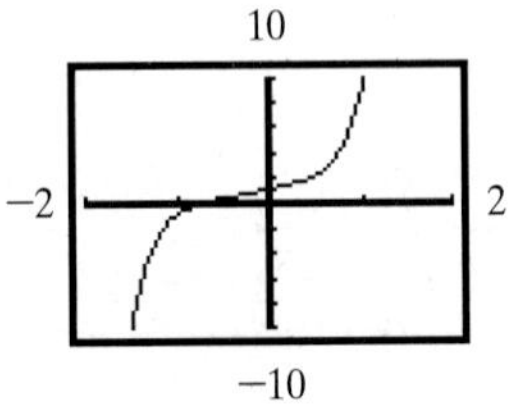

(b) x-intercept: −0.71; y-intercept: 1
(c) −0.71: Odd
(d) $y = \pi x^5$
(e) 0
(f) None

63. c, e, f **64.** c, e, f **65.** b, c, e **66.** d, f

67. (a) Cubic, $a < 0$

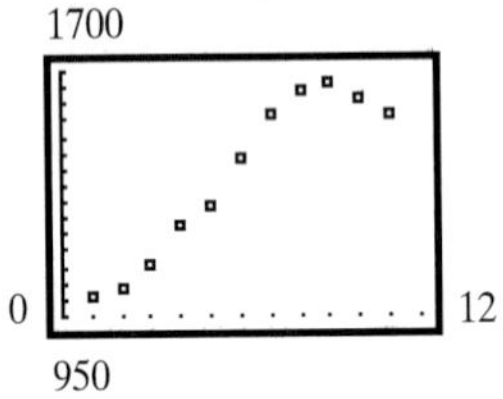

(b) ≈ 1423 thousand motor vehicle thefts

(c)

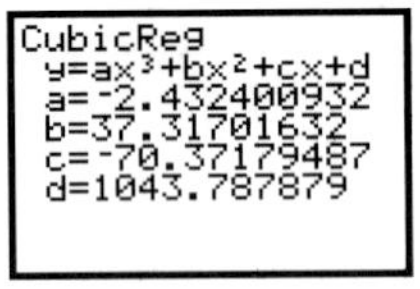

(d)

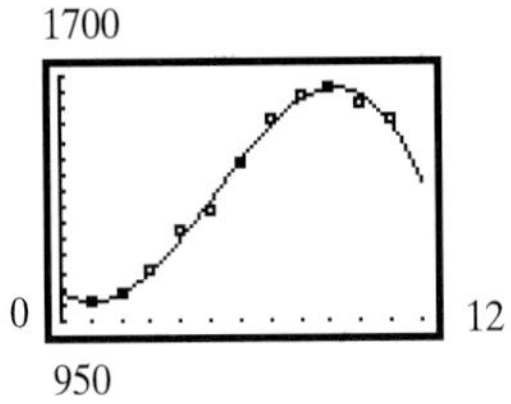

(e) No. It is not likely that motor vehicle thefts will keep decreasing after 1993, which that suggests.

68. (a) Cubic, $a < 0$

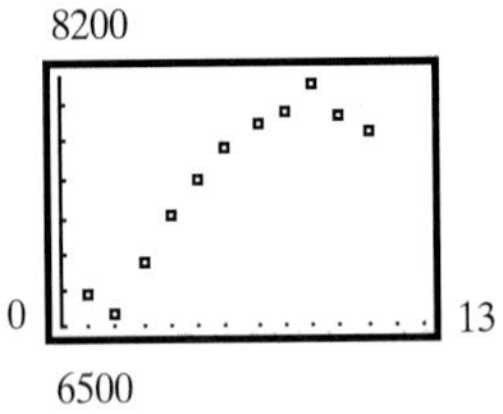

(b) ≈ 7396 thousand larceny thefts (or ≈ 7394 thousand larceny thefts, using unrounded regression function)

(c)

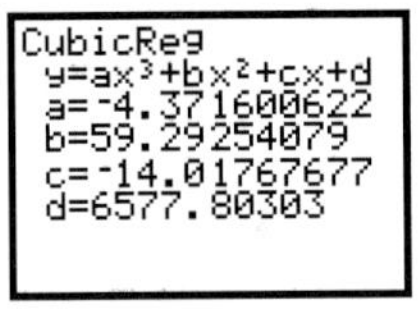

(d)

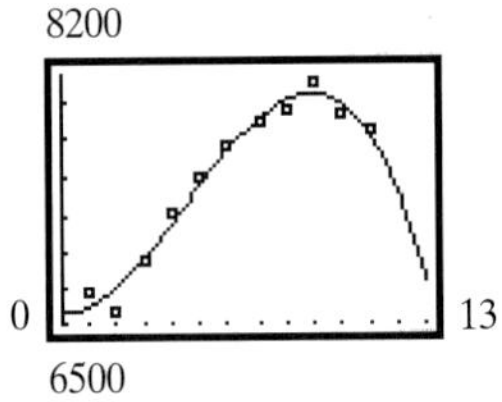

(e) No. It is not likely that larceny thefts will keep decreasing after 1993, which that function suggests.

69. (a) Cubic, $a > 0$

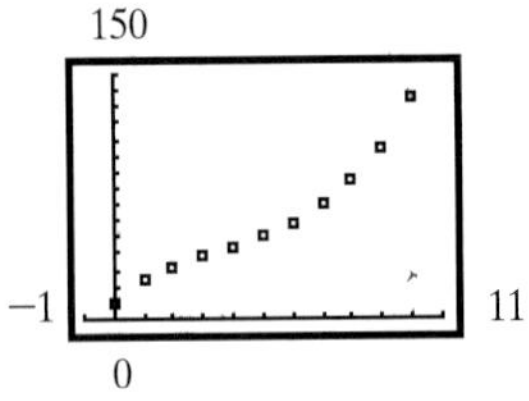

(b) \$7000 per car **(c)** \$20,000 per car
(d) ≈ \$155,400

(e)

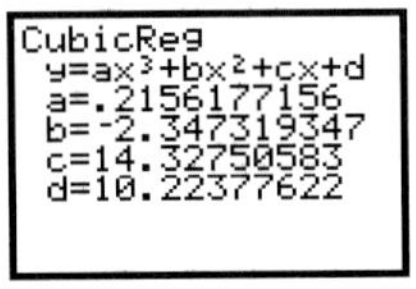

(f)

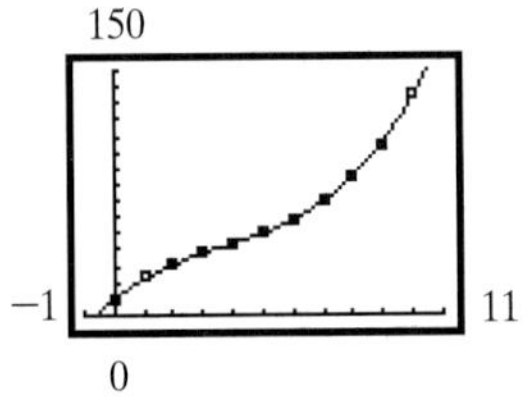

(g) Fixed costs of ≈ \$10,200

70. (a) Cubic, $a > 0$

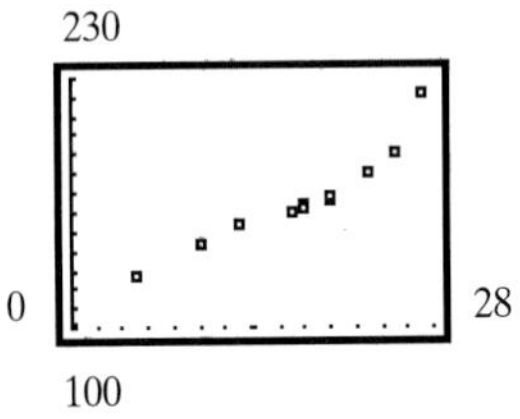

(b) ≈ \$3.17 per textbook
(c) ≈ \$1.85 per textbook
(d) ≈ \$171,470 (or ≈ \$176,295, using unrounded regression function)

(e)

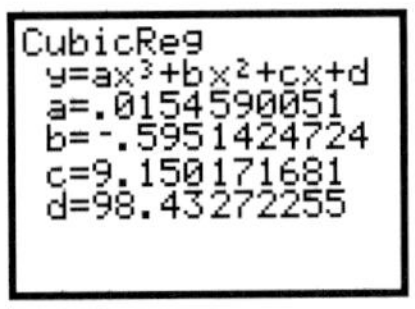

(f)

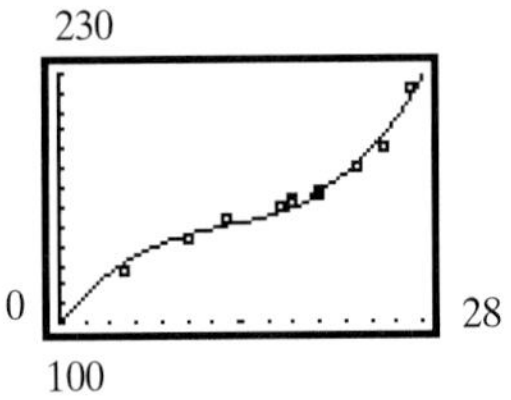

(g) Fixed costs of ≈ \$98,430

71. No, because the domain is $\mathbb{R}$. Yes; for example, $x^2 + 1$ has no x-intercepts. **72.** Degree $n \Rightarrow \le n$ roots and $\le n - 1$ turning points. Roots and multiplicities $\Rightarrow$ crossing/touching x-axis behavior at roots. $a_n x^n \Rightarrow$ end behavior. End behavior and crossing/touching x-axis behavior $\Rightarrow$ where $f > 0$ and $f < 0$. Between any two roots, there is at least one turning point; even roots are turning points; if there are $< n$ roots, there may be extra turning points between roots or outside the outermost roots.
73. $-x^2(x-2)^2(x+1)(x-4) = -x^6 + 7x^5 - 12x^4 - 4x^3 + 16x^2$
74. $(x+2)(x-1)^2 = x^3 - 3x + 2$; $(x+2)(x-1)^2(x^2+1) = x^5 - 2x^3 + 2x^2 - 3x + 2$ **75.** $\text{int}(x)$; $|x|$
76. **(a)** True **(b)** True **(c)** True **(d)** True **(e)** False (unless $b = 0$ and $d = 0$) **(f)** False (unless $d = 0$)

4.3 Exercises

1. Quotient: $4x^2 - 11x + 23$; Remainder: -45 **2.** Quotient: $3x^2 - 7x + 15$; Remainder: -32
3. Quotient: $4x^2 + 13x + 53$; Remainder: 213 **4.** Quotient: $3x^2 + 11x + 45$; Remainder: 178
5. Quotient: $4x - 3$; Remainder: $-7x + 7$ **6.** Quotient: $3x - 1$; Remainder: $-5x$ **7.** Quotient: 2; Remainder: $-3x^2 + x + 3$
8. Quotient: 1; Remainder: $-x^2 + x - 1$ **9.** Quotient: $2x - \frac{5}{2}$; Remainder: $\frac{3}{2}x + \frac{7}{2}$ **10.** Quotient: $x - \frac{2}{3}$; Remainder: $\frac{2}{3}x - \frac{4}{3}$
11. Quotient: $x - \frac{3}{4}$; Remainder: $\frac{7}{4}$ **12.** Quotient: $x^2 + \frac{1}{3}$; Remainder: $-\frac{5}{3}$ **13.** Quotient: $x^3 + x^2 + x + 1$; Remainder: 0
14. Quotient: $x^3 - x^2 + x - 1$; Remainder: 0 **15.** Quotient: $x^2 + 1$; Remainder: 0 **16.** Quotient: $x^2 - 1$; Remainder: 0
17. Quotient: $-4x^2 - 3x - 3$; Remainder: -7 **18.** Quotient: $-3x^3 - 3x^2 - 3x - 5$; Remainder: -6
19. Quotient: $x^2 - x - 1$; Remainder: $2x + 2$ **20.** Quotient: $x^2 + x - 1$; Remainder: $2 - 2x$ **21.** Quotient: $-x^2$; Remainder: 1
22. Quotient: $x^2 - 2$; Remainder: 3 **23.** Quotient: $x^2 + ax + a^2$; Remainder: 0 **24.** Quotient: $x^2 - ax + a^2$; Remainder: 0
25. Quotient: $x^3 + ax^2 + a^2x + a^3$; Remainder: 0 **26.** Quotient: $x^4 + ax^3 + a^2x^2 + a^3x + a^4$; Remainder: 0
27. $q(x) = x^2 + x + 4$; $R = 12$ **28.** $q(x) = x^2 + x - 4$; $R = 5$ **29.** $q(x) = 3x^2 + 11x + 32$; $R = 99$
30. $q(x) = -4x^2 + 10x - 21$; $R = 43$ **31.** $q(x) = x^4 - 3x^3 + 5x^2 - 15x + 46$; $R = -138$ **32.** $q(x) = x^3 + 2x^2 + 5x + 10$; $R = 22$
33. $q(x) = 4x^5 + 4x^4 + x^3 + x^2 + 2x + 2$; $R = 7$ **34.** $q(x) = x^4 - x^3 + 6x^2 - 6x + 6$; $R = -16$
35. $q(x) = 0.1x^2 - 0.11x + 0.321$; $R = -0.3531$ **36.** $q(x) = 0.1x - 0.21$; $R = 0.241$ **37.** $q(x) = x^4 + x^3 + x^2 + x + 1$; $R = 0$
38. $q(x) = x^4 - x^3 + x^2 - x + 1$; $R = 0$ **39.** No; $f(2) = 8$ **40.** No; $f(-3) = 161$ **41.** Yes; $f(2) = 0$ **42.** Yes; $f(2) = 0$
43. Yes; $f(-3) = 0$ **44.** Yes; $f(-3) = 0$ **45.** No; $f(-4) = 1$ **46.** Yes; $f(-4) = 0$ **47.** Yes; $f\left(\frac{1}{2}\right) = 0$ **48.** No; $f\left(-\frac{1}{3}\right) = 2$ **50.** -9

4.4 Exercises

1. No; $f(3) = 61$ **2.** No; $f(-2) = 60$ **3.** No; $f(1) = 2$ **4.** Yes; $f(2) = 0$ **5.** Yes; $f(-2) = 0$ **6.** Yes; $f(-3) = 0$ **7.** Yes; $f(4) = 0$
8. Yes; $f(-4) = 0$ **9.** No; $f\left(-\frac{1}{2}\right) = -\frac{7}{4}$ **10.** No; $f\left(-\frac{1}{3}\right) = 2$ **11.** 4; $\pm 1, \pm\frac{1}{3}$ **12.** 5; $\pm 1, \pm 3$ **13.** 5; $\pm 1, \pm 3$ **14.** 5; $\pm 1, \pm\frac{1}{2}$
15. 3; $\pm 1, \pm 2, \pm\frac{1}{4}, \pm\frac{1}{2}$ **16.** 4; $\pm 1, \pm 2, \pm\frac{1}{2}, \pm\frac{1}{3}, \pm\frac{1}{6}, \pm\frac{2}{3}$ **17.** 4; $\pm 1, \pm 2, \pm\frac{1}{3}, \pm\frac{2}{3}$ **18.** 3; $\pm 1, \pm 2, \pm\frac{1}{2}, \pm\frac{1}{4}$ **19.** 5; $\pm 1, \pm 2, \pm 4, \pm\frac{1}{2}$
20. 5; $\pm 1, \pm 3, \pm\frac{1}{3}$ **21.** 4; , $\pm 1, \pm 2, \pm\frac{1}{6}, \pm\frac{1}{3}, \pm\frac{1}{2}, \pm\frac{2}{3}$ **22.** 3; $\pm 1, \pm 3, \pm\frac{1}{2}, \pm\frac{3}{2}, \pm\frac{1}{3}, \pm\frac{1}{6}$ **23.** -1 and 1 **24.** $-\frac{7}{3}$ and $\frac{7}{3}$ **25.** -12 and 12
26. $-\frac{13}{2}$ and $\frac{13}{2}$ **27.** -10 and 10 **28.** $-\frac{85}{4}$ and $\frac{85}{4}$ **29.** $-3, -1, 2$; $f(x) = (x+3)(x+1)(x-2)$
30. $-5, -4, 1$; $f(x) = (x+5)(x+4)(x-1)$ **31.** $\frac{1}{2}, 3, 3$; $f(x) = 2\left(x - \frac{1}{2}\right)(x-3)^2$ **32.** $-\frac{3}{2}, 2, 2$; $f(x) = (2x+3)(x-2)^2$
33. $-\frac{1}{3}$; $f(x) = 3\left(x + \frac{1}{3}\right)(x^2 + x + 1)$ **34.** $\frac{7}{3}$; $f(x) = (3x-7)(x^2+4)$
35. $3, \frac{5+\sqrt{17}}{2}, \frac{5-\sqrt{17}}{2}$; $f(x) = (x-3)\left(x - \left(\frac{5+\sqrt{17}}{2}\right)\right)\left(x - \left(\frac{5-\sqrt{17}}{2}\right)\right)$
36. $-2, -2+\sqrt{6}, -2-\sqrt{6}$; $f(x) = (x+2)(x+2-\sqrt{6})(x+2+\sqrt{6})$ **37.** $-2, -1, 1, 1$; $f(x) = (x+2)(x+1)(x-1)^2$
38. $-2, -1, 2, 2$; $f(x) = (x+2)(x+1)(x-2)^2$ **39.** $-5, -3, -\frac{3}{2}, 1$; $f(x) = (x+5)(x+3)(2x+3)(x-1)$
40. $-1, -\frac{1}{4}, 1, 4$; $f(x) = (x+1)(4x+1)(x-1)(x-4)$ **41.** $-2, -\frac{3}{2}, 1, 4$; $f(x) = (x+2)(2x+3)(x-1)(x-4)$
42. $-5, 1, \frac{-3+\sqrt{65}}{4}, \frac{-3-\sqrt{65}}{4}$; $f(x) = 2(x+5)(x-1)\left(x + \frac{3-\sqrt{65}}{4}\right)\left(x + \frac{3+\sqrt{65}}{4}\right)$
43. $-\frac{1}{2}, \frac{1}{2}$; $f(x) = (2x+1)(2x-1)(x^2+2)$ **44.** $-\frac{1}{2}, \frac{1}{2}$; $f(x) = (2x+1)(2x-1)(x^2+4)$
45. $2, \frac{\sqrt{2}}{2}, -\frac{\sqrt{2}}{2}$; $f(x) = 4(x-2)\left(x - \frac{\sqrt{2}}{2}\right)\left(x + \frac{\sqrt{2}}{2}\right)\left(x^2 + \frac{1}{2}\right)$

46. $-3, \frac{\sqrt{2}}{2}, -\frac{\sqrt{2}}{2}; f(x) = 4(x+3)\left(x - \frac{\sqrt{2}}{2}\right)\left(x + \frac{\sqrt{2}}{2}\right)\left(x^2 + \frac{1}{2}\right)$ **47.** $-5.9, -0.3, 3$ **48.** $-4.5, -0.7, 2$ **49.** $-3.8, 4.5$ **50.** $-2.3, 1.7$
51. $-43.5, 1, 23$ **52.** $-54.82, 2.76, 9.87$ **53.** $\{-1, 2\}$ **54.** $\left\{-\frac{3}{2}\right\}$ **55.** $\left\{\frac{2}{3}, -1+\sqrt{2}, -1-\sqrt{2}\right\}$ **56.** $\left\{\frac{5}{2}\right\}$ **57.** $\left\{\frac{1}{3}, \sqrt{5}, -\sqrt{5}\right\}$
58. $\left\{-\frac{1}{2}, 2, 4\right\}$ **59.** $\{-3, -2\}$ **60.** $\{1\}$ **61.** $-\frac{1}{3}$ **62.** $\left\{\frac{2}{3}\right\}$ **63.** $f(0) = -1; f(1) = 10$; Zero: 0.22 **64.** $f(-1) = -6; f(0) = 2$; Zero: -0.60
65. $f(-5) = -58; f(-4) = 2$; Zero: -4.05 **66.** $f(-3) = -42; f(-2) = 5$; Zero: -2.17 **67.** $f(1.4) = -0.17536; f(1.5) = 1.40625$; Zero: 1.41 **68.** $f(1.7) = 0.35627; f(1.8) = -1.02112$; Zero: 1.73 **69.** $\approx$ 28 Cavaliers **70.** $\approx$ 26,256 books **71.** $k = 5$ **72.** $k = -\frac{17}{12}$
73. -7 **74.** 1 **75.** No (use the Rational Zeros Theorem) **76.** No (use the Rational Zeros Theorem)
77. No (use the Rational Zeros Theorem) **78.** No (use the Rational Zeros Theorem) **79.** 7 in. **80.** 6 cm or 12 cm

4.5 Exercises

1. $8 + 5i$ **2.** $-4 + 7i$ **3.** $-7 + 6i$ **4.** 6 **5.** $-6 - 11i$ **6.** $-10 + 6i$ **7.** $6 - 18i$ **8.** $-8 - 32i$ **9.** $6 + 4i$ **10.** $-12 - 9i$
11. $10 - 5i$ **12.** $13 + i$ **13.** 37 **14.** -10 **15.** $\frac{6}{5} + \frac{8}{5}i$ **16.** $\frac{5}{13} + \frac{12}{13}i$ **17.** $1 - 2i$ **18.** $\frac{1}{2} + i$ **19.** $\frac{5}{2} - \frac{7}{2}i$ **20.** $-\frac{1}{2} + \frac{5}{2}i$
21. $-\frac{1}{2} + \frac{\sqrt{3}}{2}i$ **22.** $\frac{1}{2} - \frac{\sqrt{3}}{2}i$ **23.** $2i$ **24.** $-2i$ **25.** $-i$ **26.** -1 **27.** i **28.** i **29.** -6 **30.** $4 - i$ **31.** $-10i$ **32.** $3 - 4i$
33. $-2 + 2i$ **34.** 82 **35.** 0 **36.** 0 **37.** 0 **38.** 0 **39.** $2i$ **40.** $3i$ **41.** $5i$ **42.** $8i$ **43.** $5i$ **44.** $5i$ **45.** $\{-2i, 2i\}$ **46.** $\{-2, 2\}$
47. $\{-4, 4\}$ **48.** $\{-5i, 5i\}$ **49.** $\{3 - 2i, 3 + 2i\}$ **50.** $\{-2 + 2i, -2 - 2i\}$ **51.** $\{3 - i, 3 + i\}$ **52.** $\{1 - 2i, 1 + 2i\}$
53. $\left\{\frac{1}{4} - \frac{1}{4}i, \frac{1}{4} + \frac{1}{4}i\right\}$ **54.** $\left\{-\frac{3}{10} - \frac{1}{10}i, -\frac{3}{10} + \frac{1}{10}i\right\}$ **55.** $\left\{-\frac{1}{5} - \frac{2}{5}i, -\frac{1}{5} + \frac{2}{5}i\right\}$ **56.** $\left\{-\frac{3}{13} - \frac{2}{13}i, -\frac{3}{13} + \frac{2}{13}i\right\}$
57. $\left\{-\frac{1}{2} - \frac{\sqrt{3}}{2}i, -\frac{1}{2} + \frac{\sqrt{3}}{2}i\right\}$ **58.** $\left\{\frac{1}{2} - \frac{\sqrt{3}}{2}i, \frac{1}{2} + \frac{\sqrt{3}}{2}i\right\}$ **59.** $\{2, -1 - \sqrt{3}i, -1 + \sqrt{3}i\}$ **60.** $\left\{-3, -\frac{3}{2} - \frac{3\sqrt{3}}{2}i, -\frac{3}{2} + \frac{3\sqrt{3}}{2}i\right\}$
61. $\{-2, 2, -2i, 2i\}$ **62.** $\{-1, 1 -i, i\}$ **63.** $\{-3i, -2i, 2i, 3i\}$ **64.** $\{-1, 1, -2i, 2i\}$ **65.** Two complex solutions.
66. Two unequal real solutions. **67.** Two unequal real solutions. **68.** Two complex solutions. **69.** A repeated real solution.
70. A repeated real solution. **71.** $2 - 3i$ **72.** $4 + i$ **73.** 6 **74.** $6i$ **75.** 25 **76.** $-5 + 7i$
77. $z + \bar{z} = (a + bi) + (a - bi) = 2a; z - \bar{z} = (a + bi) - (a - bi) = 2bi$ **78.** $\bar{\bar{z}} = \overline{\overline{a + bi}} = \overline{a - bi} = a + bi = z$
79. $\overline{z + w} = \overline{(a + bi) + (c + di)} = \overline{(a + c) + (b + d)i} = (a + c) - (b + d)i = (a - bi) + (c - di) = \bar{z} + \bar{w}$
80.
$$\begin{aligned}\overline{z \cdot w} &= \overline{(a + bi) \cdot (c + di)}\\ &= \overline{(ac - bd) + (ad + bc)i}\\ &= (ac - bd) - (ad + bc)i\\ &= (ac - bd) + (-ad - bc)i\\ &= (a - bi) \cdot (c - di)\\ &= \bar{z} \cdot \bar{w}\end{aligned}$$

4.6 Exercises

1. $4 + i$ **2.** $3 - i$ **3.** $-i, 1 - i$ **4.** $2 - i$ **5.** $-i, -2i$ **6.** $-i$ **7.** $-i$ **8.** $2 + i, i$ **9.** $2 - i, -3 + i$ **10.** $-i, 3 + 2i, -2 - i$
11. $f(x) = x^4 - 14x^3 + 77x^2 - 200x + 208; a = 1$ **12.** $f(x) = x^4 - 2x^3 + 6x^2 - 2x + 5; a = 1$
13. $f(x) = x^5 - 4x^4 + 7x^3 - 8x^2 + 6x - 4; a = 1$ **14.** $f(x) = x^6 - 12x^5 + 55x^4 - 120x^3 + 139x^2 - 108x + 85; a = 1$
15. $f(x) = x^4 - 6x^3 + 10x^2 - 6x + 9; a = 1$ **16.** $f(x) = x^5 - 5x^4 + 11x^3 - 13x^2 + 8x - 2; a = 1$ **17.** $-2i, 4$ **18.** $5i, -3$
19. $2i, -3, \frac{1}{2}$ **20.** $-3i, -2, \frac{1}{3}$ **21.** $3 + 2i, -2, 5$ **22.** $1 - 3i, -1, 6$ **23.** $4i, -\sqrt{11}, \sqrt{11}, -\frac{2}{3}$ **24.** $-3i, -3, 4, \frac{1}{2}$
25. $1, -\frac{1}{2} - \frac{\sqrt{3}}{2}i, -\frac{1}{2} + \frac{\sqrt{3}}{2}i$ **26.** $-i, i, -1, 1$ **27.** $2, 3 - 2i, 3 + 2i$ **28.** $-5, -4 + i, -4 - i$ **29.** $-i, i, -2i, 2i$ **30.** $-2i, 2i, -3i, 3i$
31. $-5i, 5i, -3, 1$ **32.** $-3i, 3i, -7, 4$ **33.** $-4, \frac{1}{3}, 2 - 3i, 2 + 3i$ **34.** $-3 - 2i, -3 + 2i, \frac{1}{2}, 5$
35. Zeros that are complex numbers must occur in conjugate pairs; or a polynomial with real coefficients of odd degree must have at least one real zero. **36.** Zeros that are complex numbers must occur in conjugate pairs.
37. If the remaining zero were a complex number, then its conjugate would also be a zero.
38. A missing zero is $4 + i$. If the remaining zero were a complex number, then its conjugate would also be a zero.

4.7 Exercises

1. All real numbers except 3. **2.** All real numbers except -3. **3.** All real numbers except 2 and -4.
4. All real numbers except -3 and 4. **5.** All real numbers except $-\frac{1}{2}$ and 3. **6.** All real numbers except $\frac{1}{3}$ and -2.

7. All real numbers except 2. **8.** All real numbers except −1 and 1. **9.** All real numbers **10.** All real numbers
11. All real numbers except −3 and 3. **12.** All real numbers except −2.
13. **(a)** Domain: $\{x \mid x \neq 2\}$; Range: $\{y \mid y \neq 1\}$ **(b)** $(0, 0)$ **(c)** $y = 1$ **(d)** $x = 2$ **(e)** None
14. **(a)** Domain: $\{x \mid x \neq -1\}$; Range: $\{y \mid y > 0\}$ **(b)** $(0, 2)$ **(c)** $y = 0$ **(d)** $x = -1$ **(e)** None
15. **(a)** Domain: $\{x \mid x \neq 0\}$; Range: all real numbers **(b)** $(-1, 0), (1, 0)$ **(c)** None **(d)** $x = 0$ **(e)** $y = 2x$
16. **(a)** Domain: $\{x \mid x \neq 0\}$; Range: $\{y \mid -\infty < y \leq -2 \text{ or } 2 \leq y < \infty\}$ **(b)** None **(c)** None **(d)** $x = 0$ **(e)** $y = -x$
17. **(a)** Domain: $\{x \mid x \neq -2, x \neq 2\}$; Range: $\{y \mid -\infty < y \leq 0 \text{ or } 1 < y < \infty\}$ **(b)** $(0, 0)$ **(c)** $y = 1$ **(d)** $x = -2, x = 2$ **(e)** None
18. **(a)** Domain: $\{x \mid x \neq -1, x \neq 1\}$; Range: All real numbers **(b)** $(0, 0)$ **(c)** $y = 0$ **(d)** $x = -1, x = 1$ **(e)** None
19. **(a)** Domain: $\{x \mid x \neq -1\}$; Range: $\{y \mid y \neq 2\}$ **(b)** $(-1.5, 0), (0, 3)$ **(c)** $y = 2$ **(d)** $x = -1$ **(e)** None
20. **(a)** Domain: $\{x \mid x \neq -3, x \neq 1\}$; Range: $\left\{y \,\middle|\, -\infty < y \leq -\frac{5}{4} \text{ or } -1 < y < \infty\right\}$ **(b)** $(-3.23, 0), (1.23, 0), (0, -1.33)$ **(c)** $y = -1$ **(d)** $x = -3, x = 1$ **(e)** None
21. **(a)** Domain: $\{x \mid x \neq -4, x \neq 3\}$; Range: All real numbers **(b)** $(0, 0)$ **(c)** $y = 0$ **(d)** $x = -4; x = 3$ **(e)** None
22. **(a)** Domain: $\{x \mid x \neq 5\}$; Range: $\{y \mid y \neq 1\}$ **(b)** $(0, -0.20)$ **(c)** $y = 1$ **(d)** $x = 5$ **(e)** None

23.
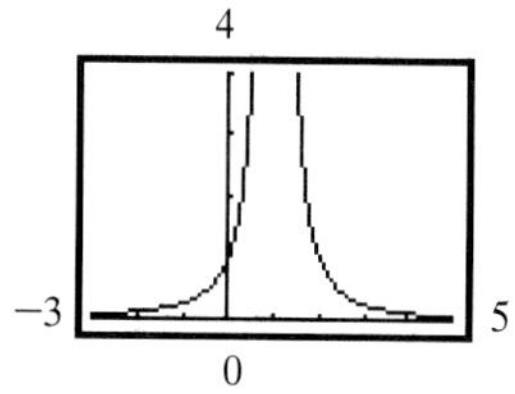

24.
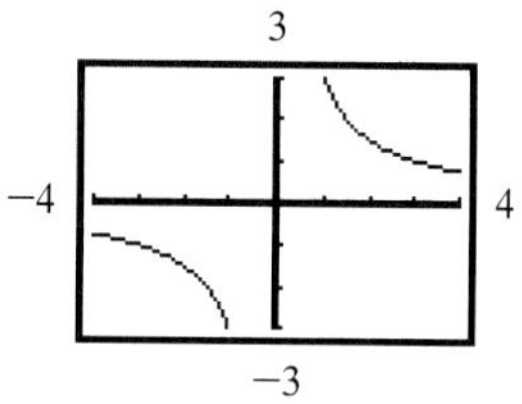

25.
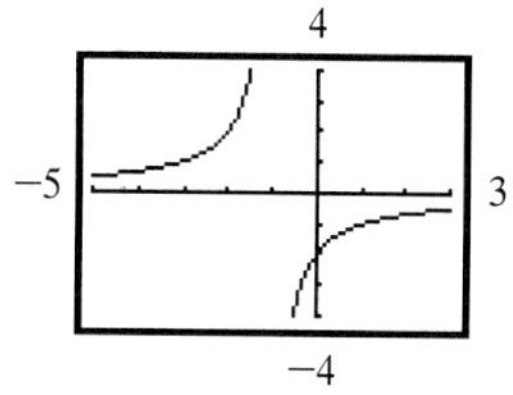

26.
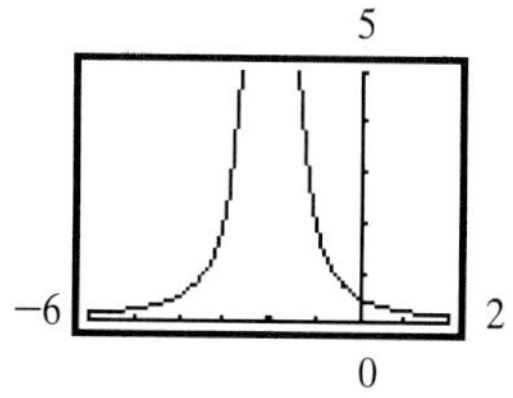

27.
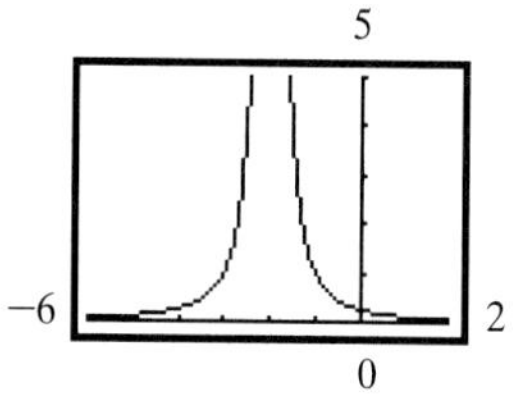

28.
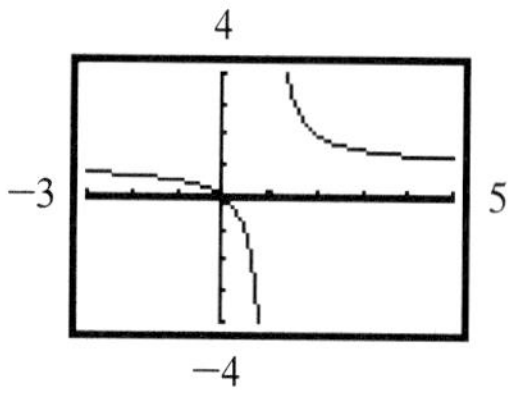

29.
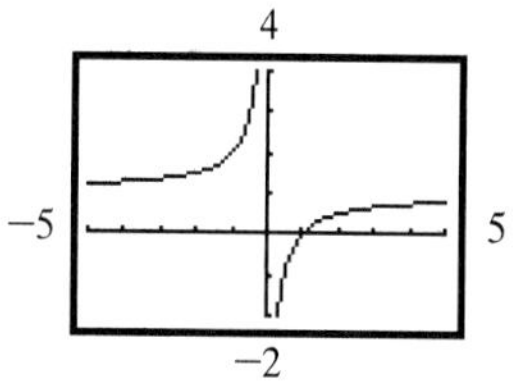

30.
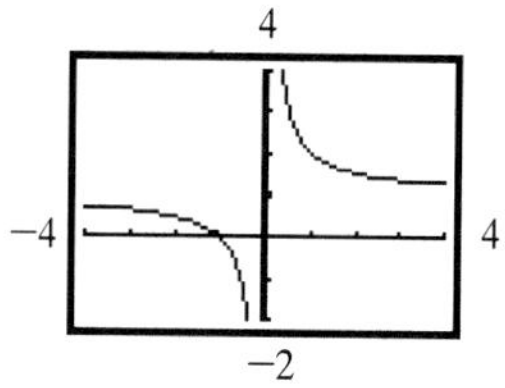

31.
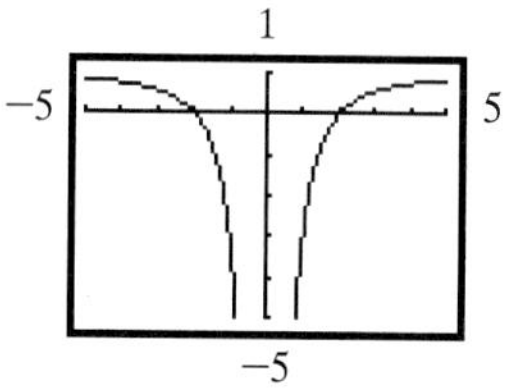

32.
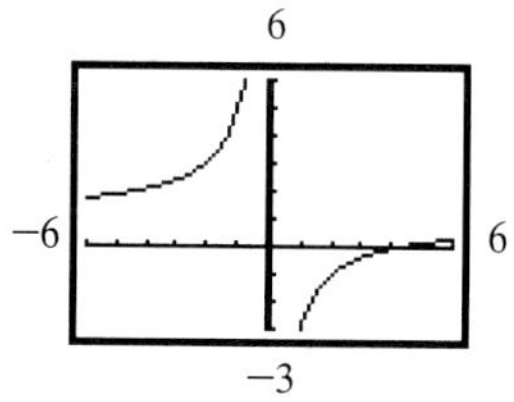

33.
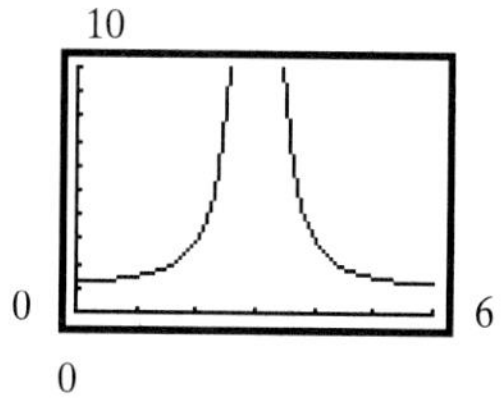

34.
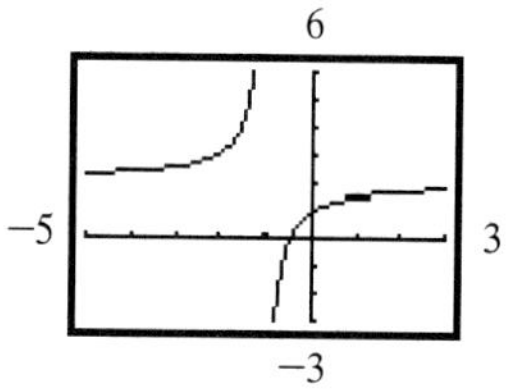

35. Horizontal asymptote: $y = 3$; vertical asymptote: $x = -4$ **36.** Horizontal asymptote: $y = 3$; vertical asymptote: $x = 6$
37. No asymptotes **38.** Oblique asymptote: $y = -x - 5$; vertical asymptote: $x = -5$
39. Horizontal asymptote: $y = 0$; vertical asymptotes: $x = 1, x = -1$ **40.** Vertical asymptote: $x = 1$
41. Horizontal asymptote: $y = 0$; vertical asymptote: $x = 0$ **42.** Horizontal asymptote: $y = 0$; vertical asymptote: $x = 0, x = -2$
43. Oblique asymptote: $y = 3x$; vertical asymptote: $x = 0$ **44.** Vertical asymptotes: $x = -\frac{1}{3}, x = 2$; horizontal asymptote: $y = 2$
45. Oblique asymptote: $y = -(x - 1)$; vertical asymptote: $x = 0$ **46.** Vertical asymptotes: $x = -1, x = 0$; horizontal asymptote: $y = 0$
47. 1. Domain: $\{x \mid x \neq 0, x \neq -4\}$
2. x-intercept: −1; no y-intercept
3. No symmetry
4. Vertical asymptotes: $x = 0, x = -4$
5. Horizontal asymptote: $y = 0$, intersected at $(-1, 0)$
6.
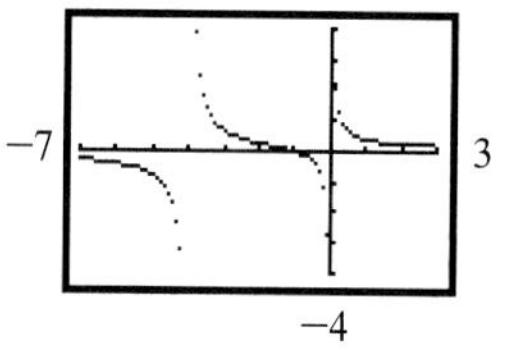

48. 1. Domain: $\{x|x \neq 1, x \neq -2\}$
2. x-intercept: 0; y-intercept: 0
3. No symmetry
4. Vertical asymptotes: $x = 1, x = -2$
5. Horizontal asymptote: $y = 0$, intersected at $(0, 0)$
6.

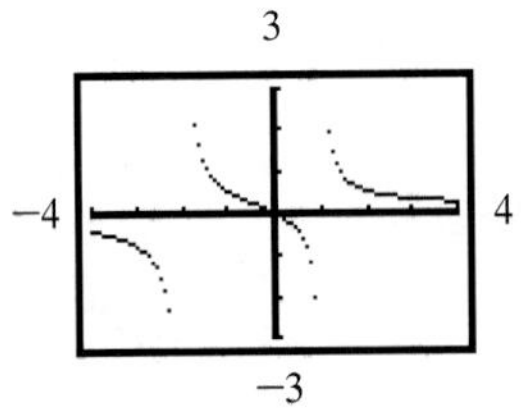

49. 1. Domain: $\{x|x \neq -2\}$
2. x-intercept: -1; y-intercept: $\frac{3}{4}$
3. No symmetry
4. Vertical asymptote: $x = -2$
5. Horizontal asymptote: $y = \frac{3}{2}$, not intersected
6.

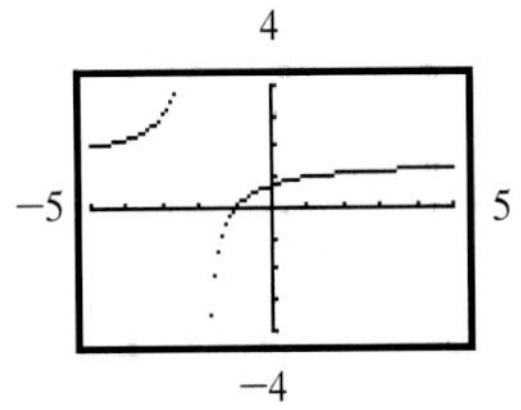

50. 1. Domain: $\{x|x \neq 1\}$
2. x-intercept: -2; y-intercept: -4
3. No symmetry
4. Vertical asymptotes: $x = 1$
5. Horizontal asymptote: $y = 2$, not intersected
6.

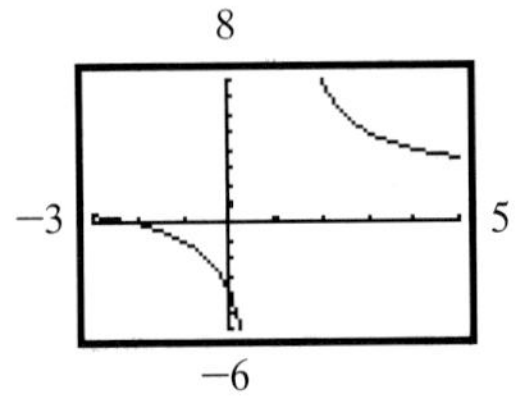

51. 1. Domain: $\{x|x \neq -2, x \neq 2\}$
2. No x-intercept; y-intercept: $-\frac{3}{4}$
3. Symmetric with respect to y-axis
4. Vertical asymptotes: $x = 2, x = -2$
5. Horizontal asymptote: $y = 0$, not intersected
6.

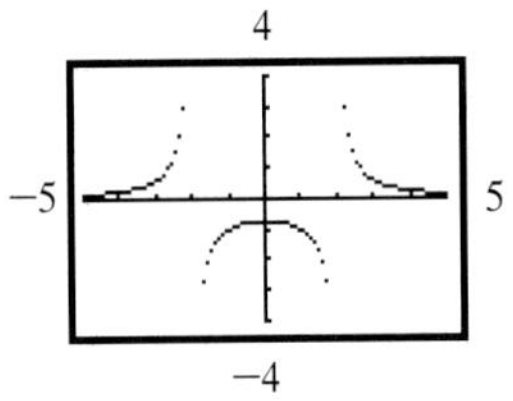

52. 1. Domain: $\{x|x \neq -2, x \neq 3\}$
2. No x-intercept; y-intercept: -1
3. No symmetry
4. Vertical asymptotes: $x = -2, x = 3$
5. Horizontal asymptotes: $y = 0$, not intersected
6.

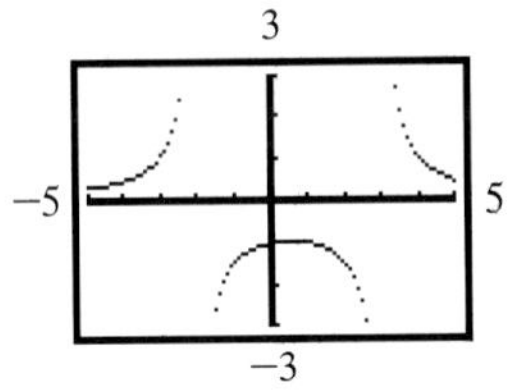

53. 1. Domain: $\{x|x \neq -1, x \neq 1\}$
2. No x-intercept; y-intercept: -1
3. Symmetric with respect to y-axis
4. Vertical asymptotes: $x = -1, x = 1$
5. No horizontal or oblique asymptotes
6.

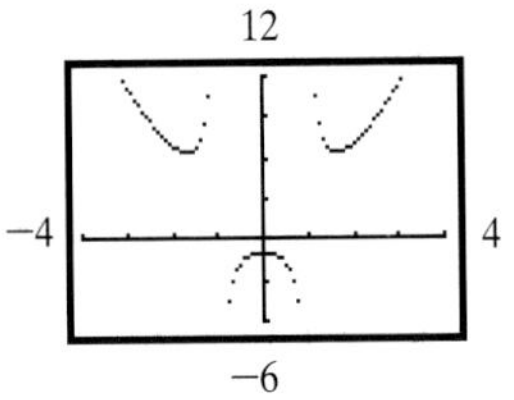

54. 1. Domain: $\{x|x \neq -2, x \neq 2\}$
2. x-intercepts: $1, -1$; y-intercept: $\frac{1}{4}$
3. Symmetric with respect to y-axis
4. Vertical asymptotes: $x = -2, x = 2$
5. No horizontal or oblique asymptotes
6.

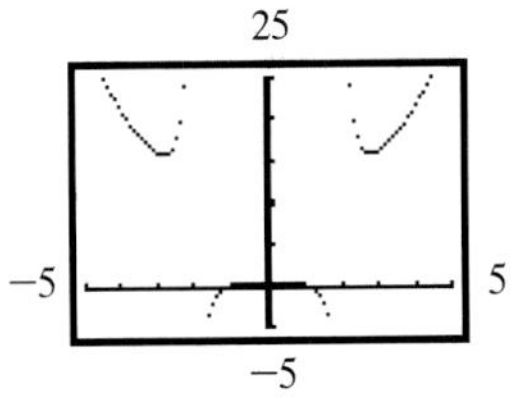

55. 1. Domain: $\{x \mid x \neq -3, x \neq 3\}$
2. x-intercept: 1; y-intercept: $\frac{1}{9}$
3. No symmetry
4. Vertical asymptotes: $x = 3, x = -3$
5. Oblique asymptote: $y = x$, intersected at $\left(\frac{1}{9}, \frac{1}{9}\right)$
6.

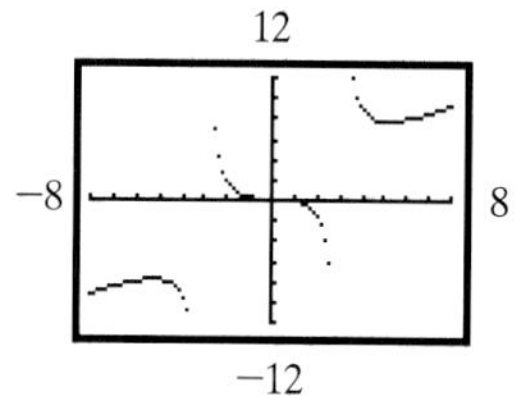

56. 1. Domain: $\{x \mid x \neq 0, x \neq -2\}$
2. x-intercept: −1; no y-intercept
3. No symmetry
4. Vertical asymptotes: $x = 0, x = -2$
5. Oblique asymptote: $y = x - 2$, intersected at $\left(-\frac{1}{4}, -\frac{9}{4}\right)$
6.

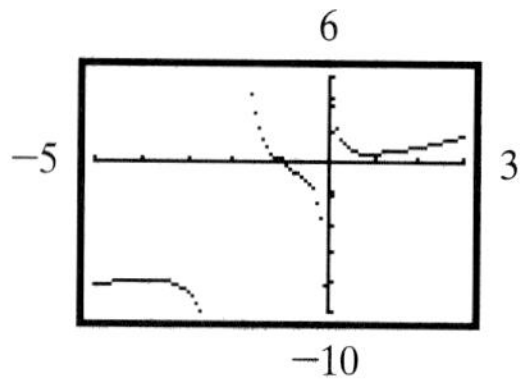

57. 1. Domain: $\{x \neq -3, x \neq 2\}$
2. Intercept: $(0, 0)$
3. No symmetry
4. Vertical asymptotes: $x = 2, x = -3$
5. Horizontal asymptote: $y = 1$, intersected at $(6, 1)$
6.

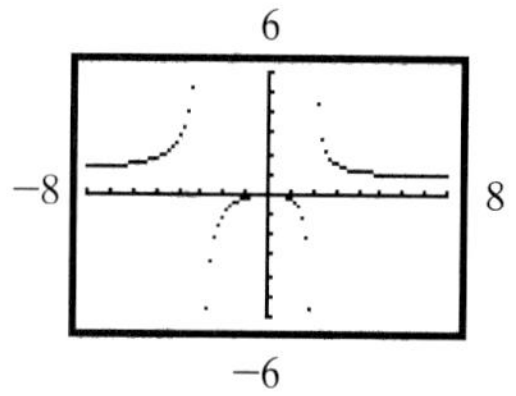

58. 1. Domain: $\{x \mid x \neq -2, x \neq 2\}$
2. x-intercepts: −4, 3; y-intercepts: 3
3. No symmetry
4. Vertical asymptotes: $x = -2, x = 2$
5. Horizontal asymptote: $y = 1$, intersected at $(8, 1)$
6.

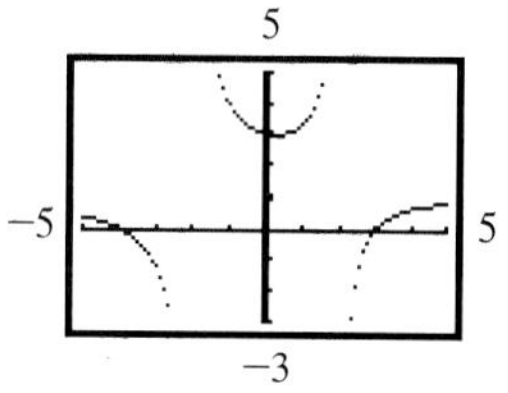

59. 1. Domain: $\{x \mid x \neq -2, x \neq 2\}$
2. Intercept: $(0, 0)$
3. Symmetry with respect to origin
4. Vertical asymptotes: $x = -2, x = 2$
5. Horizontal asymptote: $y = 0$, intersected at $(0, 0)$
6.

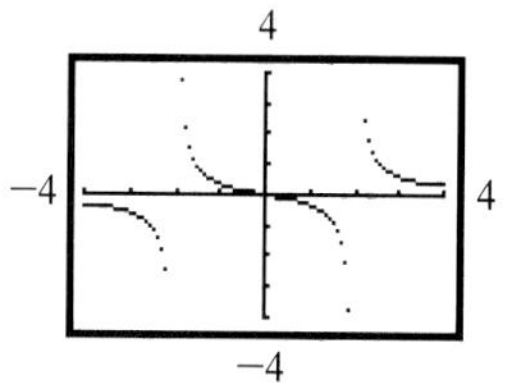

60. 1. Domain: $\{x \mid x \neq -1, x \neq 1\}$
2. Intercept: $(0, 0)$
3. Symmetry with respect to the origin
4. Vertical asymptotes: $x = 1, x = -1$
5. Horizontal asymptote: $y = 0$, intersected at $(0, 0)$
6.

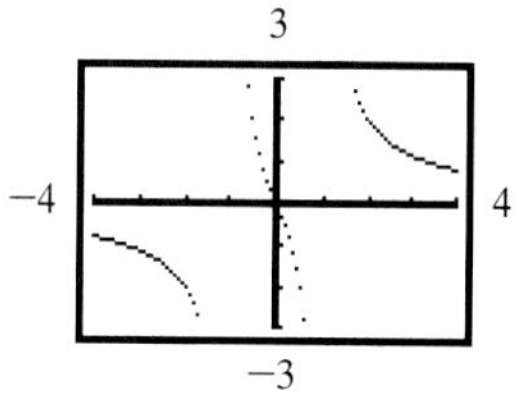

61. 1. Domain: $\{x \mid x \neq 1, x \neq -2, x \neq 2\}$
2. No x-intercept; y-intercept: $\frac{3}{4}$
3. No symmetry
4. Vertical asymptotes: $x = -2, x = 1, x = 2$
5. Horizontal asymptote: $y = 0$, not intersected
6.

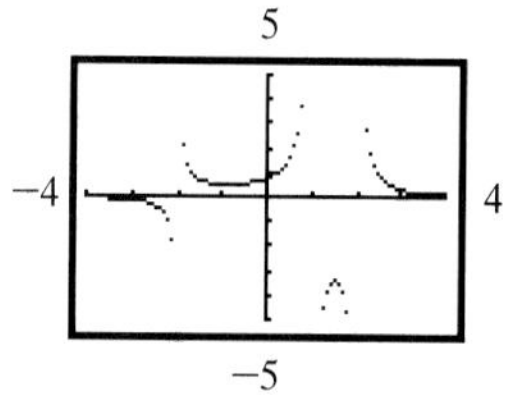

62. 1. Domain: $\{x|x \neq -3, x \neq -1, x \neq 3\}$
2. No x-intercept; y-intercept: $\frac{4}{9}$
3. No symmetry
4. Vertical asymptotes: $x = -1, x = 3, x = -3$
5. Horizontal asymptote: $y = 0$, not intersected
6.

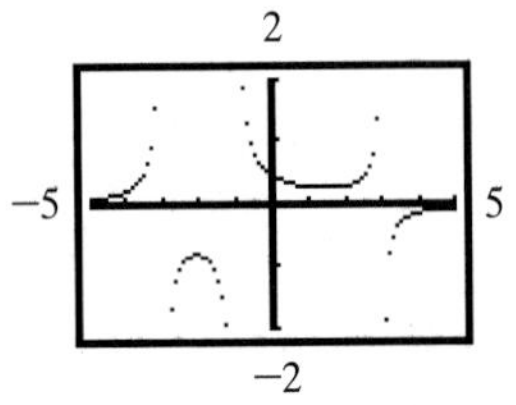

63. 1. Domain: $\{x|x \neq -2, x \neq 2\}$
2. x-intercepts: $-1, 1$; y-intercept: $\frac{1}{4}$
3. Symmetry with respect to y-axis
4. Vertical asymptotes: $x = -2, x = 2$
5. Horizontal asymptote: $y = 0$, intersected at $(-1, 0)$ and $(1, 0)$
6.

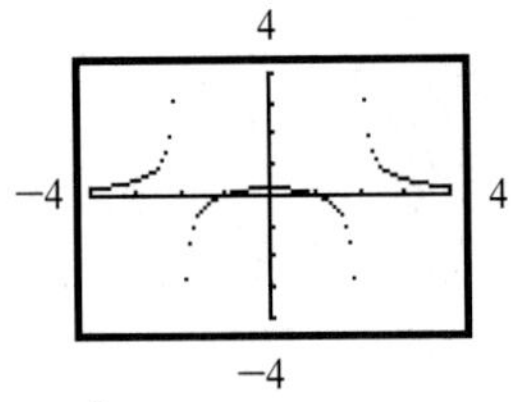

64. 1. Domain: $\{x|x \neq -1, x \neq 1\}$
2. No x-intercept; y-intercept: -4
3. Symmetry about the y-axis
4. Vertical asymptotes: $x = 1, x = -1$
5. Horizontal asymptote: $y = 0$, not intersected
6.

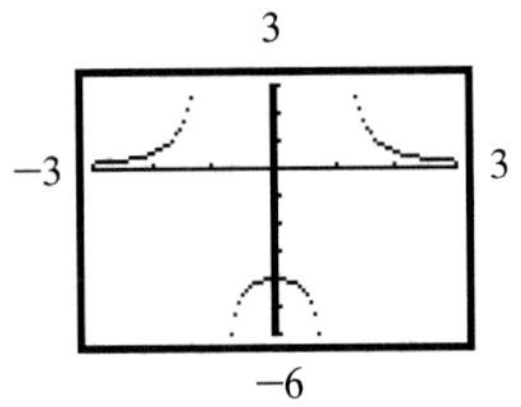

65. 1. Domain: $\{x|x \neq -2\}$
2. x-intercepts: $-1, 4$; y-intercept: -2
3. No symmetry
4. Vertical asymptote: $x = -2$
5. Oblique asymptote: $y = x - 5$, not intersected
6.

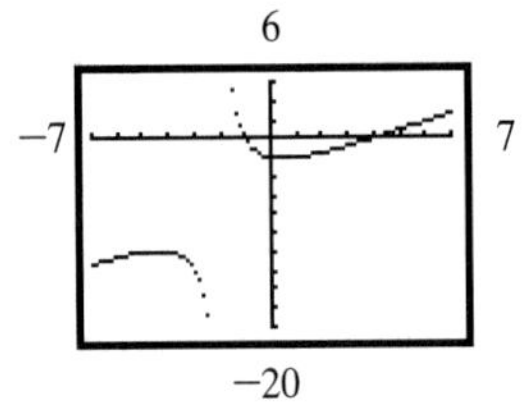

66. 1. Domain: $\{x|x \neq 1\}$
2. x-intercepts: $-2, -1$; y-intercepts: -2
3. No symmetry
4. Vertical asymptote: $x = 1$
5. Oblique asymptote: $y = x + 4$, not intersected
6.

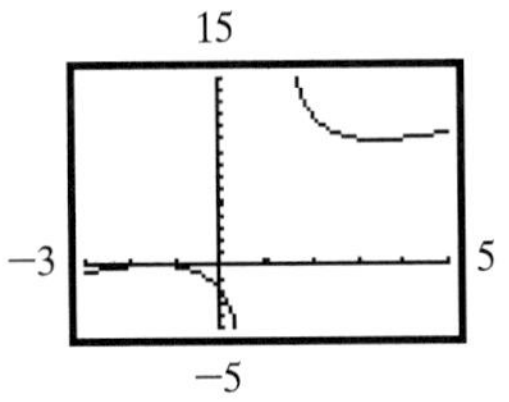

67. 1. Domain: $\{x|x \neq 4\}$
2. x-intercepts: $-4, 3$; y-intercept: 3
3. No symmetry
4. Vertical asymptote: $x = 4$
5. Oblique asymptote: $y = x + 5$, not intersected
6.

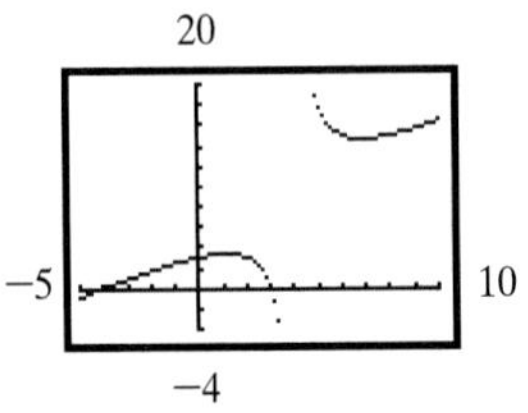

68. 1. Domain: $\{x|x \neq -5\}$
2. x-intercepts: $4, -3$; y-intercept: $-\frac{12}{5}$
3. No symmetry
4. Vertical asymptote: $x = -5$
5. Oblique asymptote: $y = x - 6$, not intersected
6.

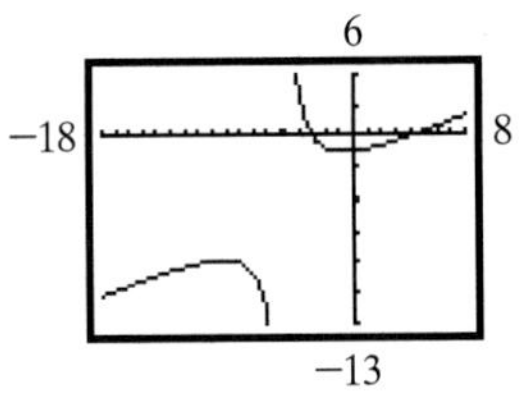

69. 1. Domain: $\{x \mid x \neq -2\}$
2. x-intercepts: $-4, 3$; y-intercept: -6
3. No symmetry
4. Vertical asymptote: $x = -2$
5. Oblique asymptote: $y = x - 1$, not intersected

6.
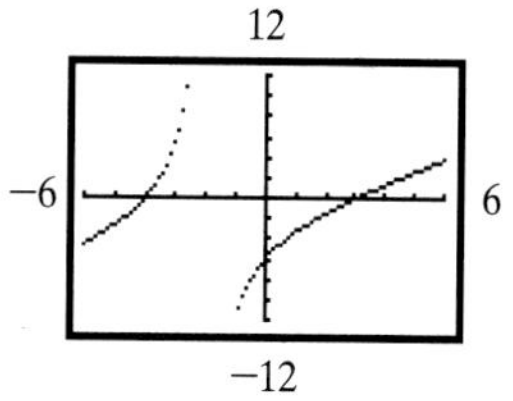

70. 1. Domain: $\{x \mid x \neq -1\}$
2. x-intercepts: $4, -3$; y-intercept: -12
3. No symmetry
4. Vertical asymptote: $x = -1$
5. Oblique asymptote: $y = x - 2$, not intersected

6.
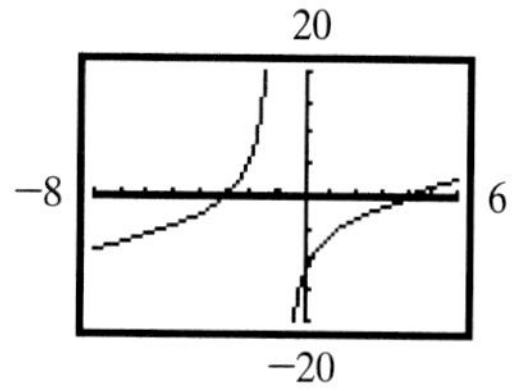

71. 1. Domain: $\{x \mid x \neq -3\}$
2. x-intercepts: $0, 1$; y-intercept: 0
3. No symmetry
4. Vertical asymptote: $x = -3$
5. Horizontal asymptote: $y = 1$, not intersected

6.
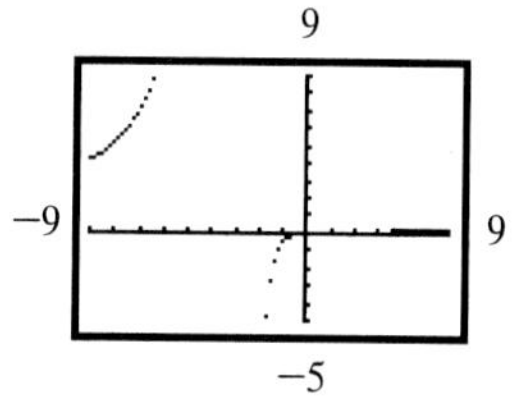

72. 1. Domain: $\{x \mid x \neq 0, x \neq 4\}$
2. x-intercepts: $1, -2, 3$; no y-intercept
3. No symmetry
4. Vertical asymptotes: $x = 0$; $x = 4$
5. Horizontal asymptote: $y = 1$,
intersected at $\left(\frac{7 + \sqrt{33}}{4}, 1\right)$ and $\left(\frac{7 - \sqrt{33}}{4}, 1\right)$

6.
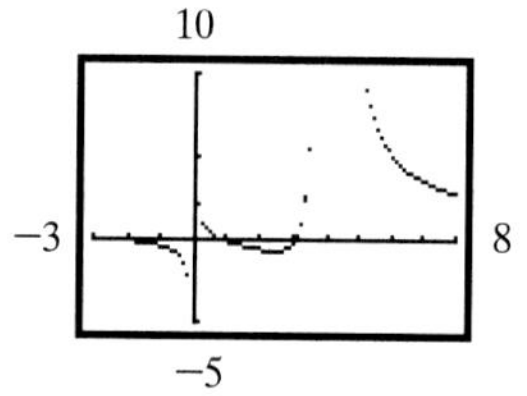

73. 1. Domain: $\{x \mid x \neq 0.73\}$
2. x-intercept: -0.85; y-intercept: -2
3. No symmetry
4. Vertical asymptote: $y = 0.75$
5. Horizontal asymptote: $y = 2$
intersected at $(2.20, 2)$ and $(-3.03, 2)$

6.
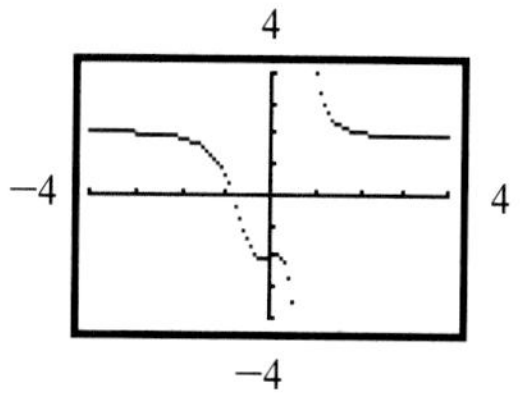

74. 1. Domain: All real numbers
2. No x-intercepts; y-intercept: 4
3. No symmetry
4. No vertical asymptotes
5. Horizontal asymptote: $y = 3$, intersected at $(-0.59, 3)$

6.
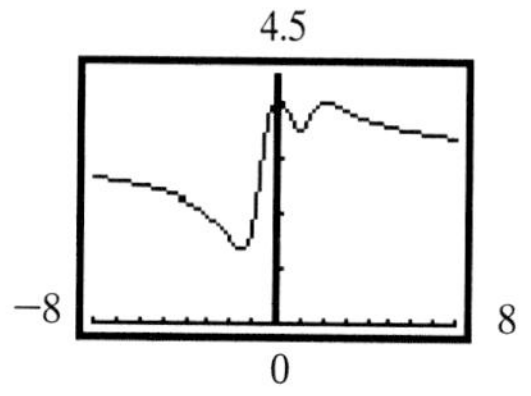

75. 1. Domain: $\{x \mid x \neq -4.41, x \neq -0.72, x \neq 0.72, x \neq 4.41\}$
2. x-intercept: 0.66; y-intercept: -3
3. No symmetry
4. Vertical asymptotes: $x = -4.41$; $x = -0.72$; $x = 0.72$; $x = 4.41$
5. Horizontal asymptote: $y = 0$, intersected at $(0.66, 0)$

6.
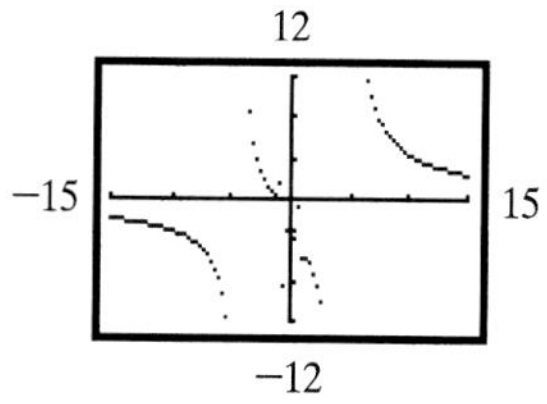

76. 1. Domain: $\{x|x \neq -1.20\}$
2. No x-intercepts; y-intercept: 2
3. No symmetry
4. Vertical asymptotes: $x = -1.20$
5. Horizontal asymptote: $y = 0$, not intersected
6.

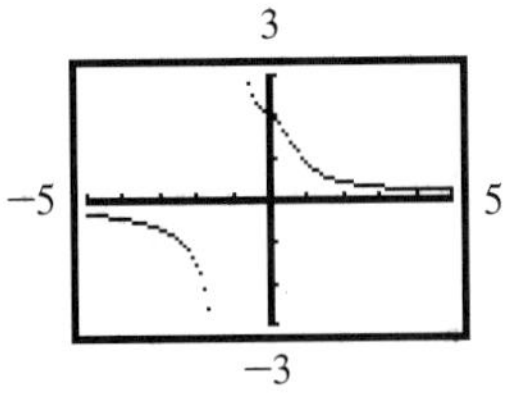

77. 1. Domain: $\{x|x \neq -5.76, x \neq 0.32, x \neq 5.44\}$
2. No x-intercepts; y-intercept: 1
3. No symmetry
4. Vertical asymptotes: $x = -5.76$; $x = 0.32$; $x = 5.44$
5. Oblique asymptotes: $y = 5x - 10$, intersected at $(0.36, -8.21)$ and $(1.96, -0.19)$
6.

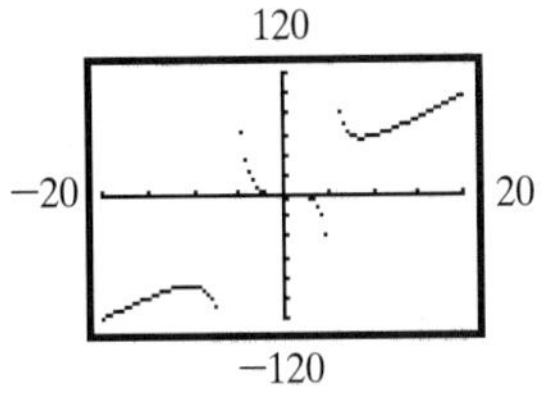

78. 1. Domain: $\{x|x \neq -0.65, x \neq 1.55\}$
2. x-intercept: -0.72; y-intercept: 3
3. No symmetry
4. Vertical asymptotes: $x = -0.65$; $x = 1.55$
5. Oblique asymptotes: $y = -2x - 1.8$, intersected at $(-0.71, -0.38)$
6.

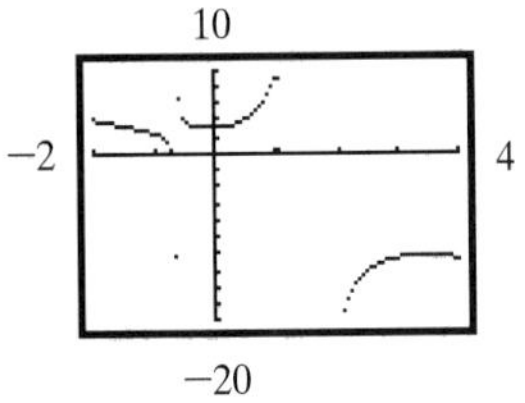

79. 1. Domain: $\{x|x \neq -2, x \neq 3\}$
2. x-intercept: -4; y-intercept: 2
3. No symmetry
4. Vertical asymptote: $x = -2$
5. Horizontal asymptote: $y = 1$, not intersected
6.

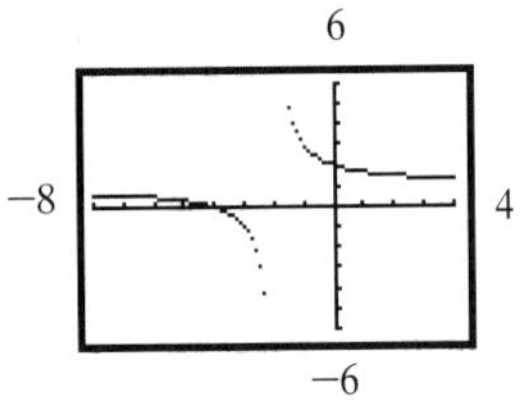

80. 1. Domain: $\{x|x \neq -3, x \neq -5\}$
2. x-intercept: 2; y-intercept: $-\frac{2}{3}$
3. No symmetry
4. Vertical asymptote: $x = -3$
5. Horizontal asymptote: $y = 1$, not intersected
6.

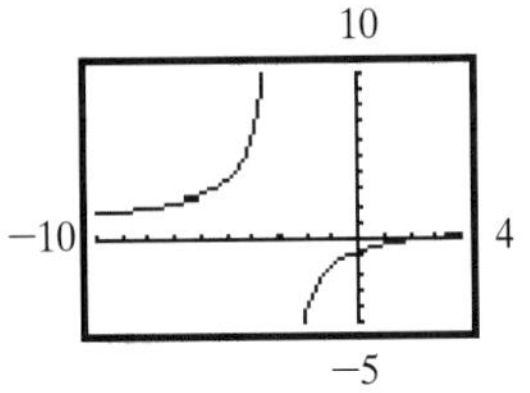

81. 1. Domain: $\left\{x \middle| x \neq \frac{3}{2}, x \neq 2\right\}$
2. x-intercept: $-\frac{1}{3}$; y-intercept: $-\frac{1}{2}$
3. No symmetry
4. Vertical asymptote: $x = 2$
5. Horizontal asymptote: $y = 3$, not intersected
6.

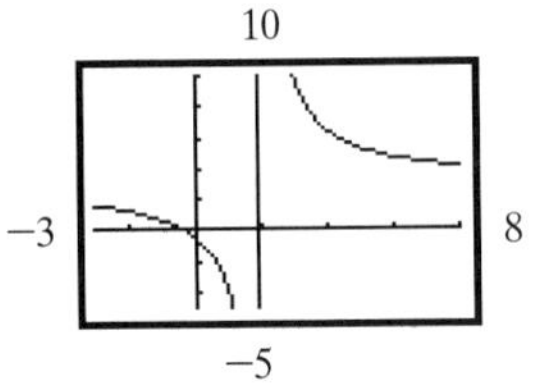

82. 1. Domain: $\left\{x \middle| x \neq -\frac{5}{2}, x \neq 3\right\}$
2. x-intercept: $-\frac{3}{4}$; y-intercept: -1
3. No symmetry
4. Vertical asymptote: $x = 3$
5. Horizontal asymptote: $y = 4$, not intersected
6.

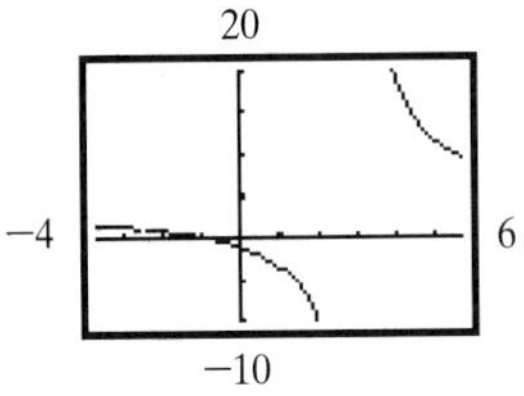

83. 1. Domain: $\{x|x \neq -5, x \neq 1, x \neq -3\}$
2. x-intercept: $-1, 2$; y-intercept: $\frac{2}{5}$
3. No symmetry
4. Vertical asymptotes: $x = -5$; $x = 1$
5. Horizontal asymptote: $y = 1$, intersected at $\left(\frac{3}{5}, 1\right)$
6.

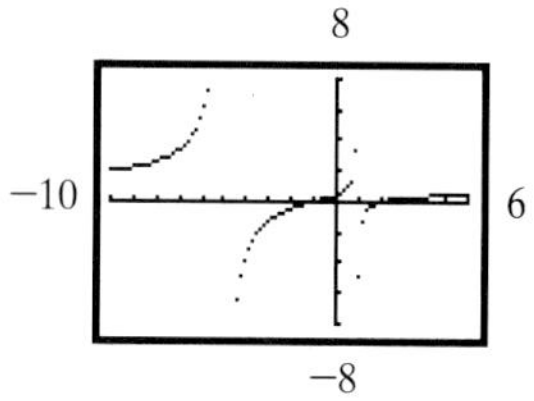

84. 1. Domain: $\{x|x \neq -1, x \neq 2, x \neq 3\}$
2. x-intercepts: $-2, 5$; y-intercept: 5
3. No symmetry
4. Vertical asymptotes: $x = -1$; $x = 2$
5. Horizontal asymptote: $y = 1$, intersected at $(-4, 1)$
6.

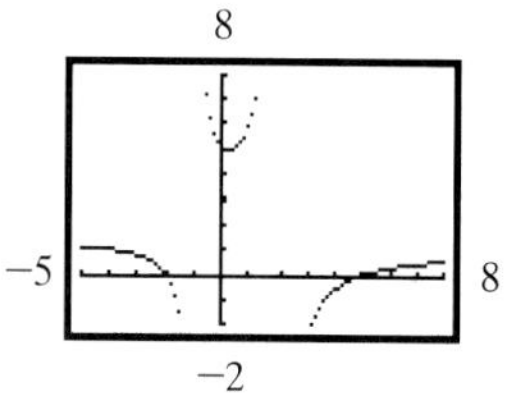

85. 1. Domain: $\{x|x \neq 0\}$
2. No x-intercepts; no y-intercepts
3. Symmetric about the origin
4. Vertical asymptote: $x = 0$
5. Oblique asymptotes: $y = x$, not intersected
6.

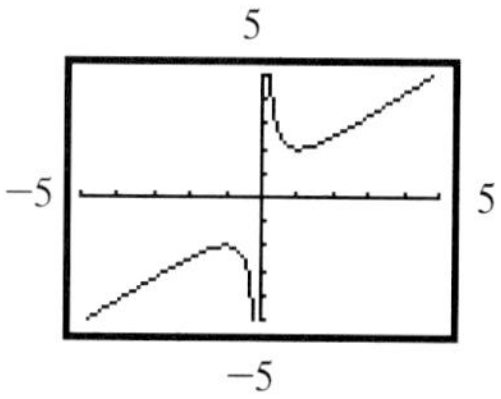

86. 1. Domain: $\{x|x \neq 0\}$
2. No x-intercepts; no y-intercepts
3. Symmetry with respect to the origin
4. Vertical asymptote: $x = 0$
5. Oblique asymptote: $y = 2x$, not intersected
6.

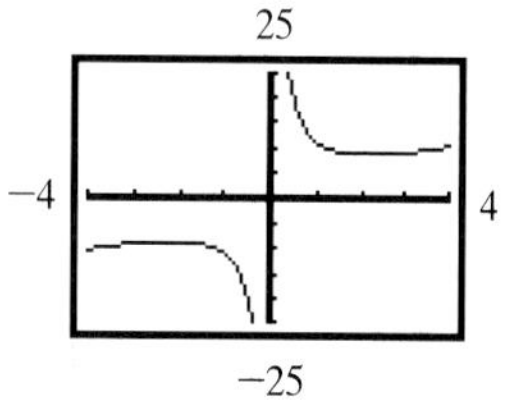

87. 1. Domain: $\{x|x \neq 0\}$
2. x-intercept: -1; no y-intercepts
3. No symmetry
4. Vertical asymptote: $x = 0$
5. No horizontal or oblique asymptotes
6.

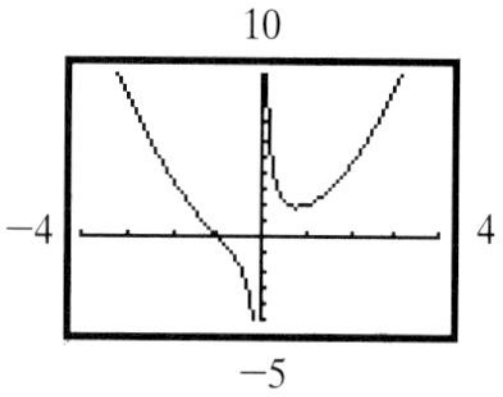

88. 1. Domain: $\{x|x \neq 0\}$
2. x-intercept: -1.65; no y-intercepts
3. No symmetry
4. Vertical asymptote: $x = 0$
5. No horizontal or oblique asymptotes
6.

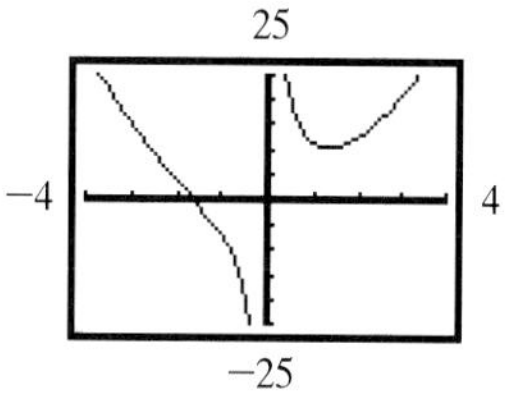

89. 1. Domain: $\{x|x \neq 0\}$
2. No x-intercepts; no y-intercepts
3. Symmetric about the origin
4. Vertical asymptote: $x = 0$
5. Oblique asymptote: $y = x$, not intersected
6.

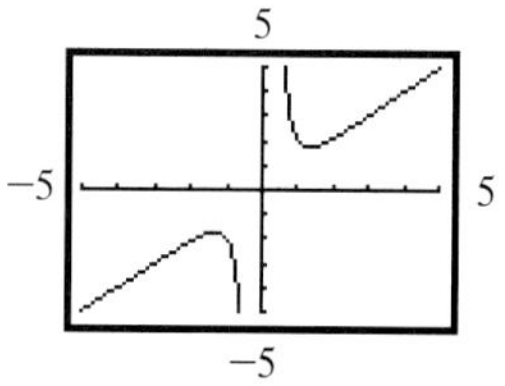

90. 1. Domain: $\{x | x \neq 0\}$
2. No x-intercepts; no y-intercepts
3. Symmetry with respect to the origin
4. Vertical asymptote: $x = 0$
5. Oblique asymptote: $y = 2x$, not intersected
6.

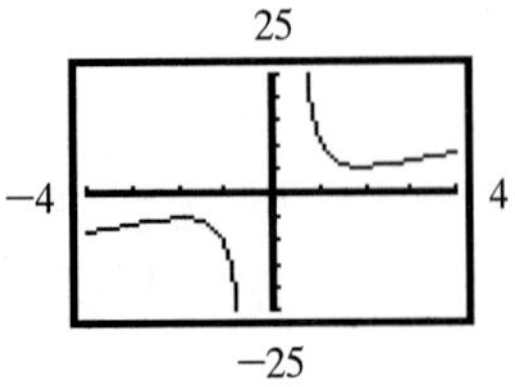

91. 4 must be a zero of the denominator, hence, $x - 4$ must be a factor.
92. If the degree of the numerator were less than the degree of the denominator, the asymptote would be $y = 0$; if it was greater, there would be an oblique asymptote or no asymptote. So, the degrees must be equal. **93.** c, d **94.** b, e

95. (a) 9.82 m/sec^3 **(b)** 9.8195 m/sec^2
(c) 9.7936 m/sec^2 **(d)** h-axis
(e)

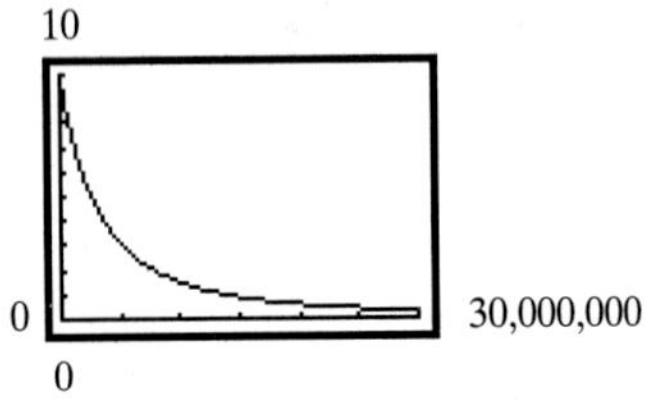

(f) $g(h)$ is never equal to 0, but $g(h) \to 0$ as $h \to \infty$. The further away from sea level you get, the lower the acceleration due to gravity.

96. (a) 25 **(b)** Approximately 596
(c)

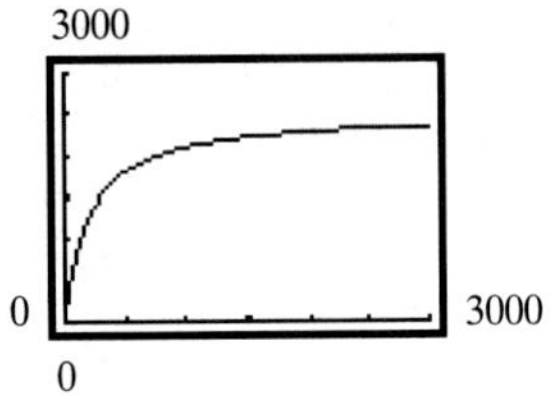

(d) $P = 2500$; 2500

97. (a)

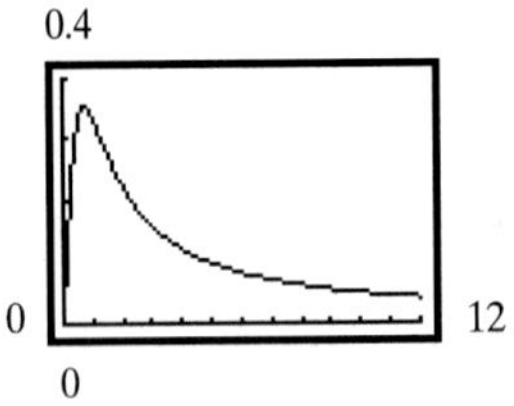

(b) 0.71 hr **(c)** t-axis; $C(t) \to 0$

98. (a)

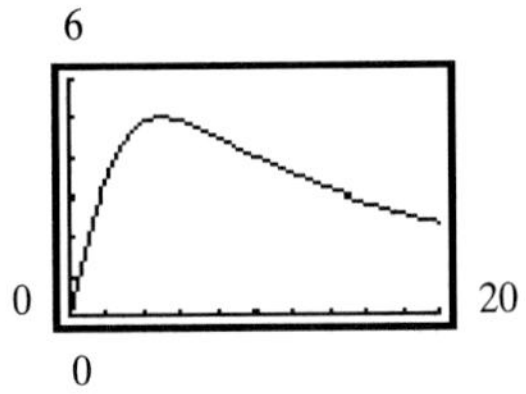

(b) 5 min **(c)** t-axis; $c(t) \to 0$

99. (a) $\overline{C}(x) = \dfrac{0.2x^3 - 2.3x^2 + 14.3x + 10.2}{x}$
(b) $\overline{C}(6) = \$9400$ per car **(c)** $\overline{C}(9) = \$10{,}933$ per car
(d)

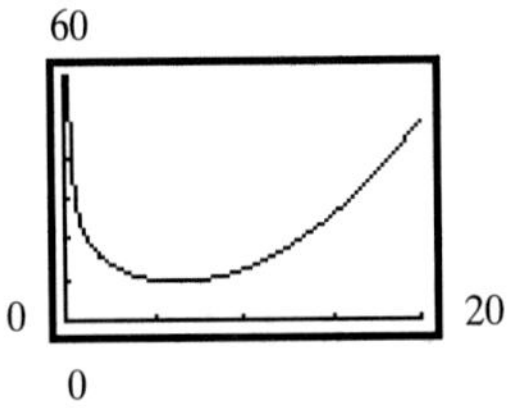

(e) 6 cars **(f)** \$9400 per car

100. (a) $\overline{C}(x) = \dfrac{0.015x^3 - 0.595x^2 + 9.15x + 98.43}{x}$
(b) $\overline{C}(13) = \$11.52$ per book **(c)** $\overline{C}(25) = \$7.59$ per book
(d)

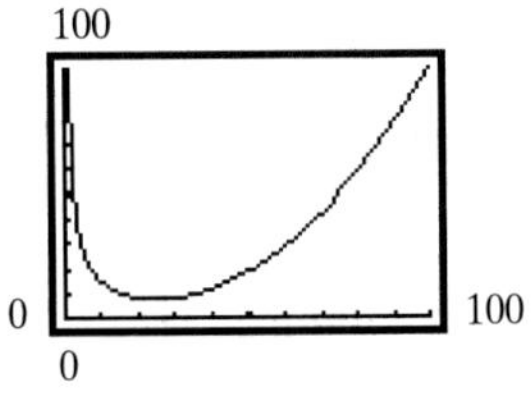

(e) Approximately 25,058 books **(f)** \$7.59 per book

101. (a) $S(x) = 2x^2 + \dfrac{40{,}000}{x}$
(b)

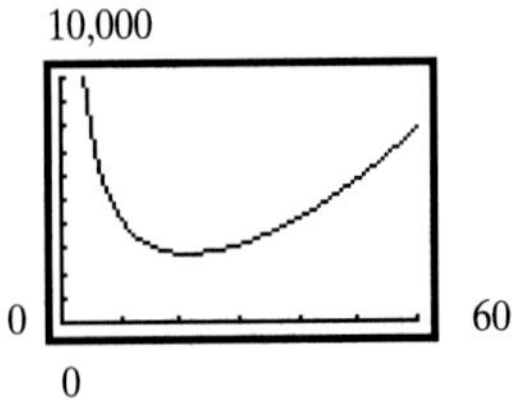

(c) 2784.95 sq in.
(d) 21.54 in. × 21.54 in. × 21.54 in.
(e) Minimizing area of cardboard will minimize their cost.

102. (a) $A(x) = 2x^2 + \dfrac{20{,}000}{x}$
(b)

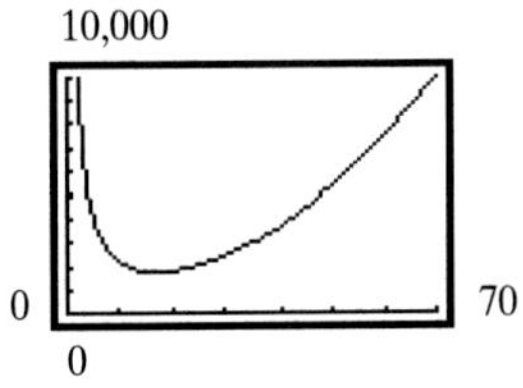

(c) 1754.41 sq in.
(d) 17.10 in. × 17.10 in. × 17.10 in.
(e) Minimizing area of cardboard will minimize their cost.

103. **(a)** $C(r) = 12\pi r^2 + \dfrac{4000}{r}$

(b) $r \approx 3.76$ cm

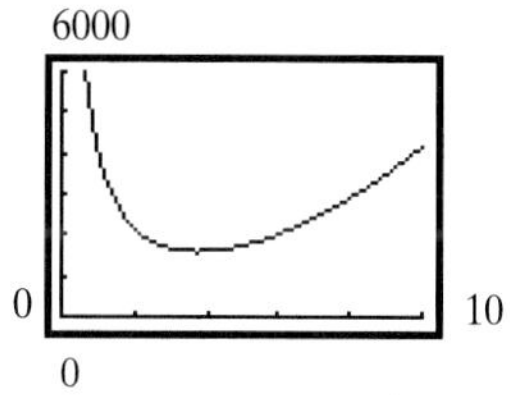

104. **(a)** $A(r) = 2\pi r^2 + \dfrac{200}{r}$ **(b)** Approximately 123.22 sq ft

(c) Approximately 125.13 sq ft **(d)** Approximately 150.53 sq ft

(e)

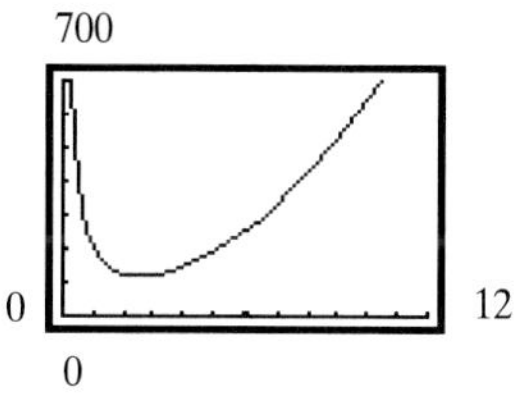

; $r \approx 2.52$ ft

105. 54.86 lb **106.** 199.90 lb **107.** 21.33 in. **108.** 20.45 cm **109.** $\dfrac{(x-1)(x-3)(3x^2+4)}{3(x+1)^2(x-2)^2}$

110. No, because horizontal asymptotes occur when the numerator and denominator have the same degree, whereas oblique asymptotes occur when the degree of the numerator is 1 more than the degree of the denominator.

112. $\dfrac{2(x-3)(x+2)^2}{(x-1)(x^2+1)}$ **113.** $(2x+1) + \dfrac{1}{x} = \dfrac{2x^2+x+1}{x}$

Exercises 4.8

1. $\{x|-2 < x < 5\}$ **2.** $\{x|-\infty < x < -2$ or $5 < x < \infty\}$ **3.** $\{x|-\infty < x < 0$ or $4 < x < \infty\}$ **4.** $\{x|-\infty < x < -8$ or $0 < x < \infty\}$
5. $\{x|-3 < x < 3\}$ **6.** $\{x|-1 < x < 1\}$ **7.** $\{x|-\infty < x < -4$ or $3 < x < \infty\}$ **8.** $\{x|-4 < x < -3\}$ **9.** $\left\{x\middle|-\frac{1}{2} < x < 3\right\}$
10. $\left\{x\middle|-\frac{2}{3} < x < \frac{3}{2}\right\}$ **11.** $\{x|-\infty < x < -1$ or $8 < x < \infty\}$ **12.** $\{x|-\infty < x < -5$ or $4 < x < \infty\}$ **13.** No real solution
14. No real solution **15.** $\left\{x\middle|-\infty < x < -\frac{2}{3} \text{ or } \frac{3}{2} < x < \infty\right\}$ **16.** All real numbers **17.** $\{x|1 < x < \infty\}$ **18.** $\{x|-2 < x < \infty\}$
19. $\{x|-\infty < x < 1$ or $2 < x < 3\}$ **20.** $\{x|-\infty < x < -3$ or $-2 < x < -1\}$ **21.** $\{x|-1 < x < 0$ or $3 < x < \infty\}$
22. $\{x|-3 < x < 0$ or $1 < x < \infty\}$ **23.** $\{x|-\infty < x < -1$ or $1 < x < \infty\}$ **24.** $\{x|-2 < x < 0$ or $0 < x < 2\}$ **25.** $\{x|1 < x < \infty\}$
26. $\{x|-\infty < x < 0$ or $0 < x < 3\}$ **27.** $\{x|-\infty < x < -1$ or $1 < x < \infty\}$ **28.** $\{x|1 < x < \infty\}$ **29.** $\{x|-1 < x < 8\}$
30. $\{x|-\infty < x \le -8$ or $-4 \le x < \infty\}$ **31.** $\{x|2.14 \le x < \infty\}$ **32.** $\{x|-\infty < x \le -1.88$ or $0.35 \le x \le 1.53\}$
33. $\{x|-\infty < x < -2$ or $2 < x < \infty\}$ **34.** $\{x|-\sqrt{3} < x < -\sqrt{2}$ or $\sqrt{2} < x < \sqrt{3}\}$ **35.** $\{x|-1 \le x \le 2-\sqrt{5}$ or $2+\sqrt{5} \le x < \infty\}$
36. $\{x|-0.16 < x < 1.70\}$ **37.** $\{x|-\infty < x < -1$ or $1 < x < \infty\}$ **38.** $\{x|-\infty < x < -1$ or $3 < x < \infty\}$
39. $\{x|-\infty < x < -1$ or $0 < x < 1\}$ **40.** $\{x|-\infty < x < -2$ or $1 < x < 3\}$ **41.** $\{x|-\infty < x < -1$ or $1 < x < \infty\}$
42. $\{x|-\infty < x < -2$ or $2 < x < \infty\}$ **43.** $\left\{x\middle|-\infty < x < -\frac{2}{3} \text{ or } 0 < x < \frac{3}{2}\right\}$ **44.** $\{x|-\infty < x < 0$ or $3 < x < 4\}$
45. $\{x|-\infty < x < 2\}$ **46.** $\{x|4 < x < \infty\}$ **47.** $\{x|-2 < x \le 9\}$ **48.** $\{x|-8 \le x < -2\}$ **49.** $\{x|-\infty < x < 2$ or $3 < x < 5\}$
50. $\{x|-7 < x < -1$ or $3 < x < \infty\}$ **51.** $\{x|-\infty < x < -3$ or $-1 < x < 1$ or $2 < x < \infty\}$ **52.** $\left\{x\middle|-\infty < x < -\frac{5}{2} \text{ or } -2 < x < -1\right\}$
53. $\{x|-\infty < x < -5$ or $-4 < x < -3$ or $1 < x < \infty\}$ **54.** $\{x|-\infty < x < -1$ or $0 < x < 1$ or $2 \le x < \infty\}$
55. $\{x|-\infty < x \le -0.5$ or $1 \le x < 4\}$ **56.** $\left\{x\middle|-2 < x < -1 \text{ or } \frac{1}{3} < x < \infty\right\}$
57. $\left\{x\middle|\frac{(-3-\sqrt{13})}{2} < x < -3 \text{ or } \frac{(-3+\sqrt{13})}{2} < x < \infty\right\}$ **58.** $\left\{x\middle|-\infty < x < \frac{5-\sqrt{13}}{2} \text{ or } \frac{5+\sqrt{13}}{2} < x < 5\right\}$
59. $\{x|4 < x < \infty\}$ **60.** $\{x|2 < x < \infty\}$ **61.** $\{x|-\infty < x \le -4$ or $4 \le x < \infty\}$ **62.** $\{x|x = 0$ or $3 \le x < \infty\}$
63. $\{x|-\infty < x < -4$ or $2 \le x < \infty\}$ **64.** $\{x|-\infty < x < -4$ or $1 \le x < \infty\}$

65. **(a)** The ball is more than 96 feet above the ground from time t between 2 and 3 seconds, $2 < t < 3$.

(b)

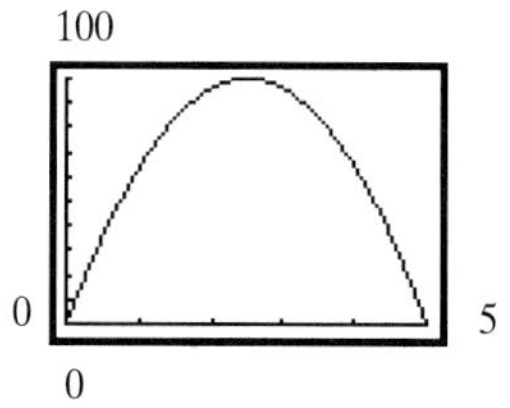

(c) 100 ft **(d)** 2.5 sec

66. **(a)** The ball is more than 112 feet above the ground for time t between 1.59 and 4.41 seconds, $3 - \sqrt{2} < t < 3 + \sqrt{2}$.

(b)

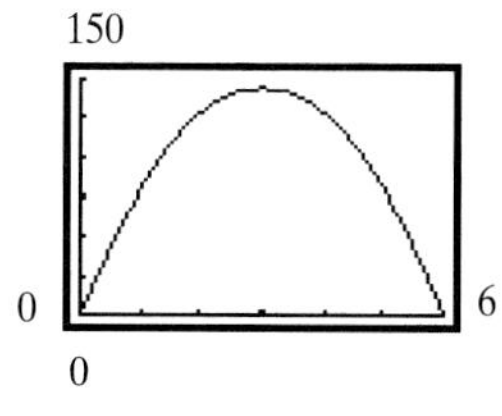

(c) 144 ft **(d)** 3 sec

67. **(a)** For a profit of at least \$50, between 8 and 32 watches must be sold, $8 \le x \le 32$.

(b)

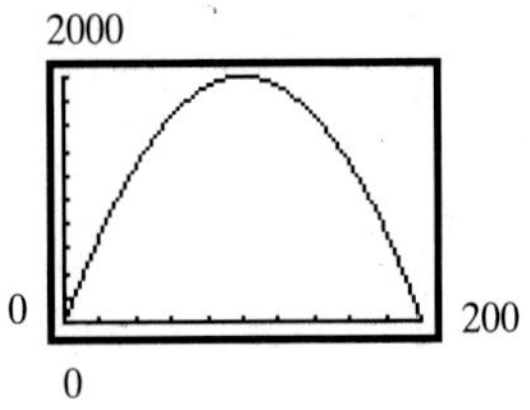

(c) \$2000 **(d)** 100

(e)

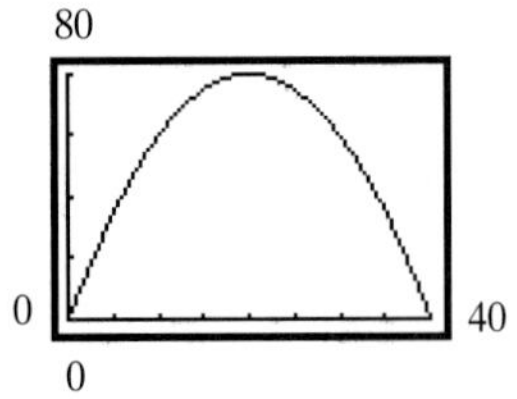

(f) \$80 **(g)** 20

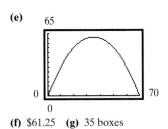

68. **(a)** For a profit of at least \$60, between 30 and 40 boxes must be sold, $30 \le x \le 40$.

(b)

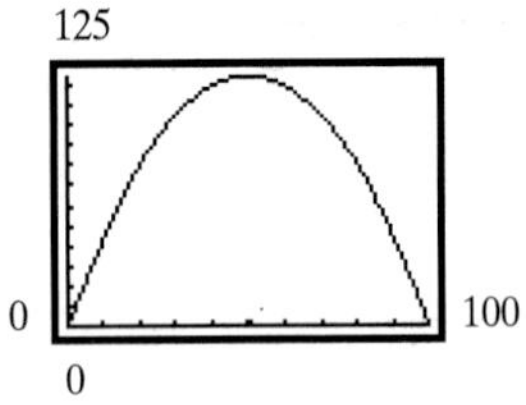

(c) \$125 **(d)** 50 boxes

(e)

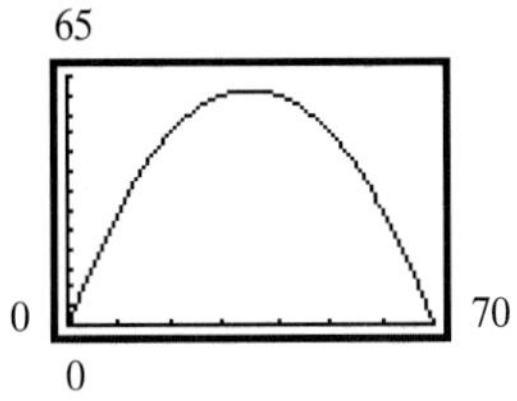

(f) \$61.25 **(g)** 35 boxes

69. Chevy can produce at most 8 Cavaliers in a day, assuming that cars cannot be partially completed in a day.
70. The number of textbooks printed in a week should be at most 21,682 textbooks.

71. **(a)**

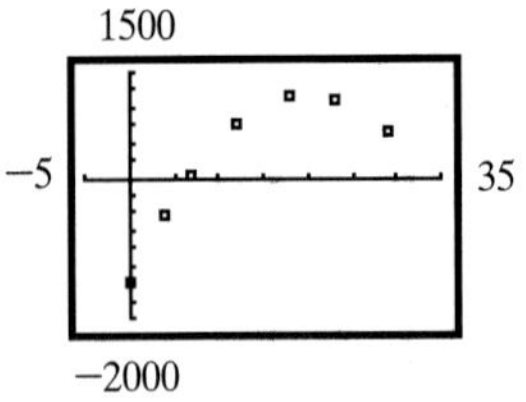

(b) Between 15 and 25 computers.
(c) 20 computers **(d)** \$1202.76
(e)

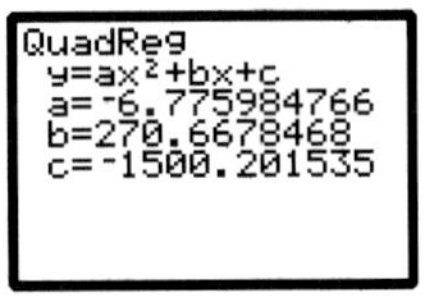

72. **(a)**

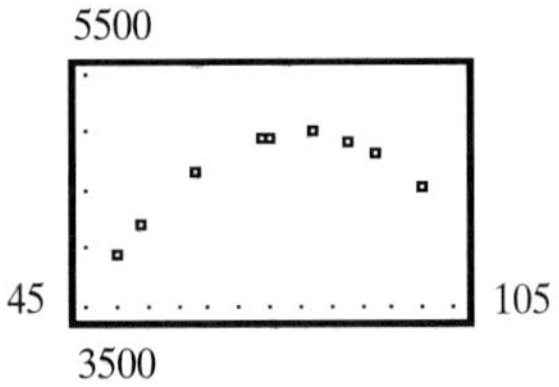

(b) Between 50,727 gallons and 109,299 gallons
(c) 80,013 gal **(d)** \$5003.50
(e)

73. $b - a = (\sqrt{b} - \sqrt{a})(\sqrt{b} + \sqrt{a})$; since $a \ge 0$ and $b \ge 0$, then $\sqrt{a} \ge 0$ and $\sqrt{b} \ge 0$ so that $\sqrt{b} + \sqrt{a} \ge 0$; thus, $b - a \ge 0$ is equivalent to $\sqrt{b} - \sqrt{a} \ge 0$ and $a \le b$ is equivalent to $\sqrt{a} \le \sqrt{b}$
74. $x^2 + 1 < 0; x^2 \le 0$ **75.** $x^2 + 1$ is always positive, hence can never be less than -5.

Fill-in-the-Blank Items

1. Remainder; dividend **2.** $f(c)$ **3.** $f(c) = 0$ **4.** Zero **5.** Five **6.** $\pm 1, \pm\frac{1}{2}$ **7.** $y = 1$ **8.** $x = -1$
9. Real; imaginary; imaginary number **10.** $3 - 4i$ **11.** $2, -2, 2i, -2i$

True/False Items

1. F **2.** F **3.** T **4.** T **5.** T **6.** F **7.** T

Review Exercises

1.

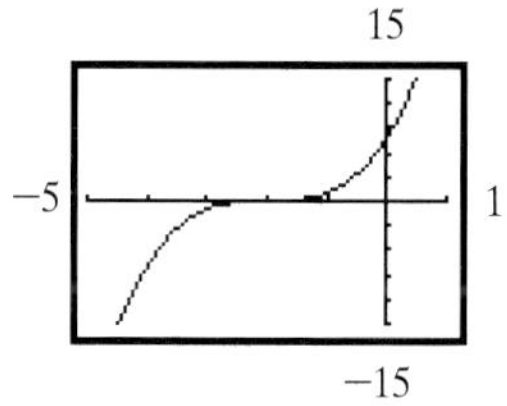

2.

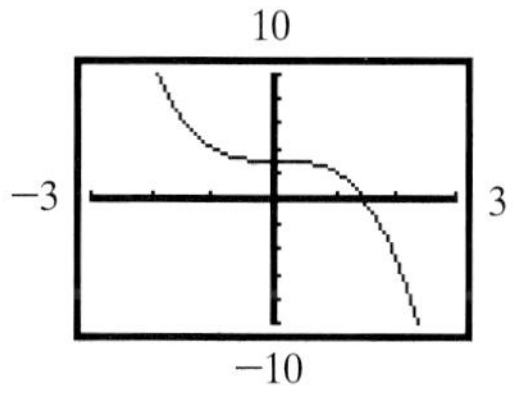

3.

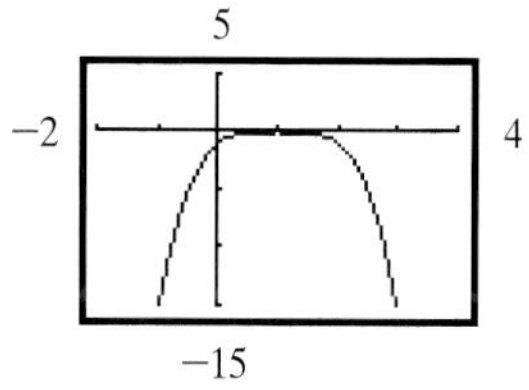

4.

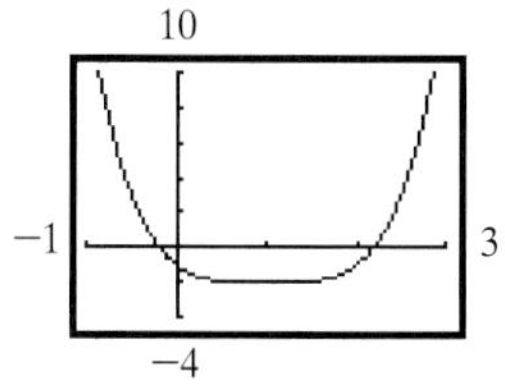

5.

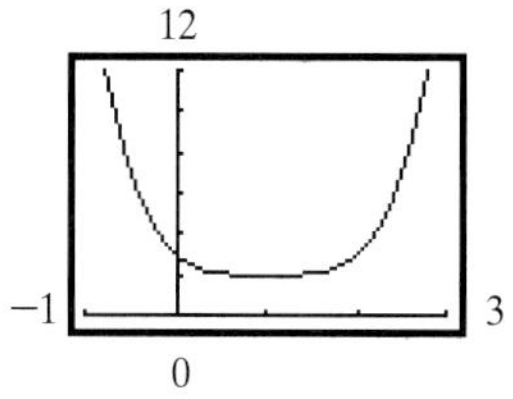

6.

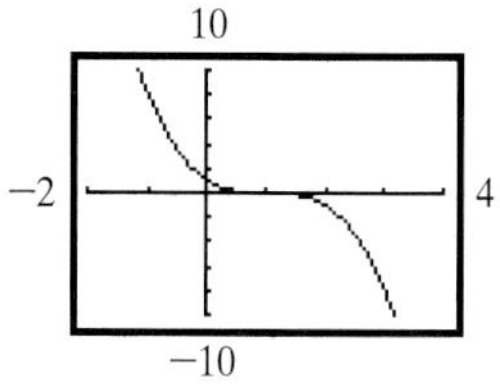

7. (a)

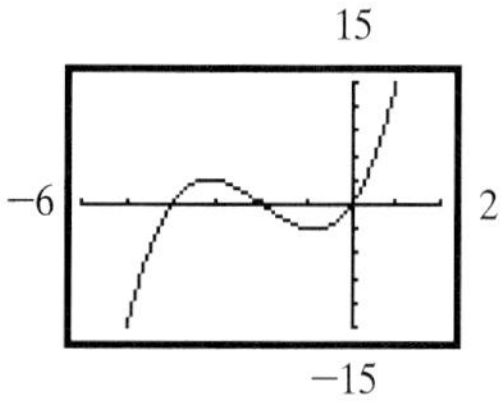

(b) x-intercepts: $-4, -2, 0$; y-intercept: 0
(c) $-4, -2, 0$: Odd
(d) $y = x^3$
(e) 2
(f) Local minimum: $(-0.85, -3.08)$
Local maximum: $(-3.15, 3.08)$

8. (a)

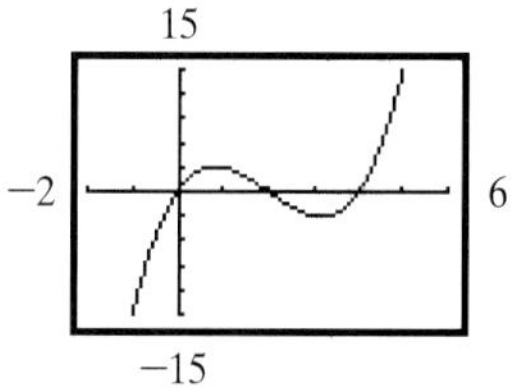

(b) x-intercepts: 0, 2, 4; y-intercept: 0
(c) 0, 2, 4: Odd
(d) $y = x^3$
(e) 2
(f) Local minimum: $(3.15, -3.08)$
Local maximum: $(0.85, 3.08)$

9. (a)

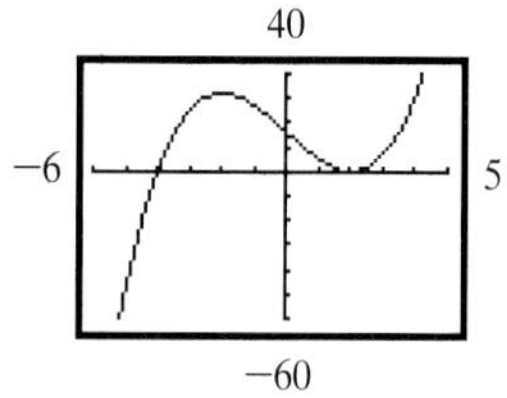

(b) x-intercepts: $-4, 2$; y-intercept: 16
(c) -4: Odd; 2: Even
(d) $y = x^3$
(e) 2
(f) Local minimum: $(2, 0)$
Local maximum: $(-2, 32)$

10. (a)

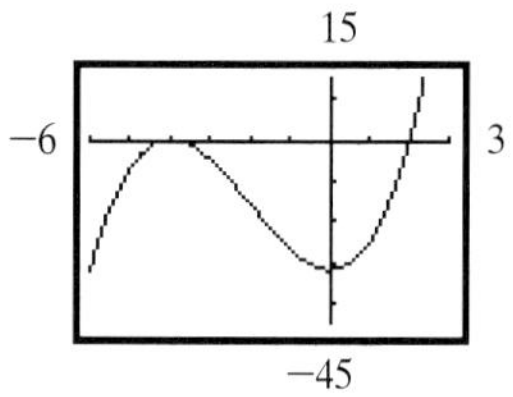

(b) x-intercepts: $-4, 2$; y-intercept: -32
(c) -4: Even; 2: Odd
(d) $y = x^3$
(e) 2
(f) Local minimum: $(0, -32)$
Local maximum: $(-4, 0)$

11. $f(x) = x^3 - 4x^2 = x^2(x - 4)$

(a)

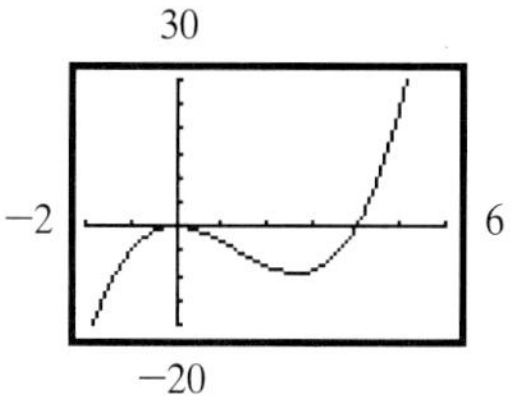

(b) x-intercepts: 0, 4; y-intercept: 0
(c) 0: Even; 4: Odd
(d) $y = x^3$
(e) 2
(f) Local minimum: $(2.67, -9.48)$
Local maximum: $(0.00, 0.00)$

12. (a)

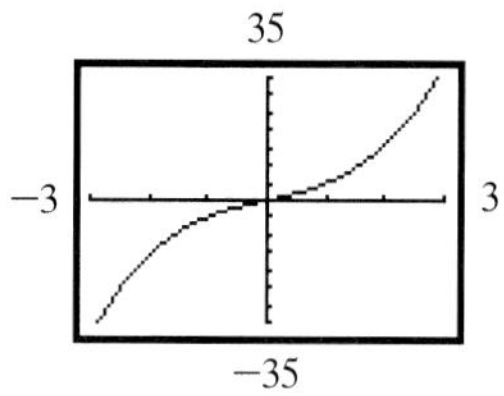

(b) x-intercept: 0; y-intercept: 0
(c) 0: Odd
(d) $y = x^3$
(e) 0
(f) None

13. (a)

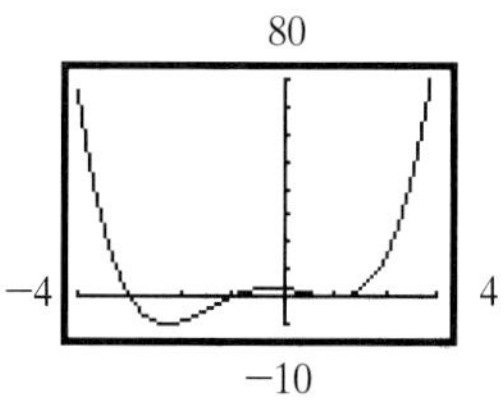

(b) x-intercepts: $-3, -1, 1$; y-intercept: 3
(c) $-3, -1$: Odd; 1: Even
(d) $y = x^4$
(e) 3
(f) Local minima: $(-2.28, -9.91)$, $(1.00, 0.00)$
Local maximum: $(-0.22, 3.23)$

14. (a)

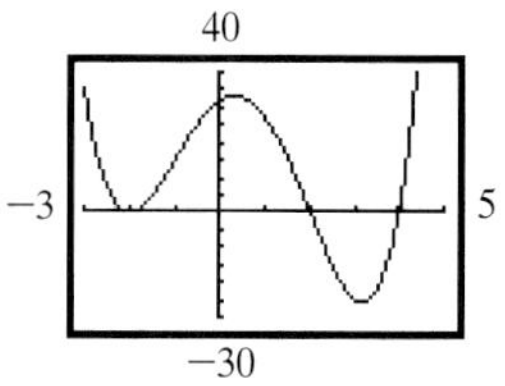

(b) x-intercepts: $-2, 2, 4$; y-intercept: 32
(c) 2, 4: Odd; -2: Even
(d) $y = x^4$
(e) 3
(f) Local minimum: $(-2, 0)$, $(3.19, -25.96)$
Local maximum: $(0.31, 33.28)$

15. 1. Domain: $\{x|x \neq 0\}$
2. x-intercept: 3; no y-intercepts
3. No symmetry
4. Vertical asymptote: $x = 0$
5. Horizontal asymptote: $y = 2$, not intersected
6.

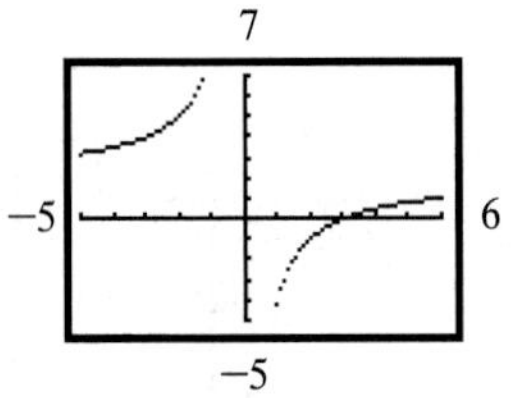

16. 1. Domain: $\{x|x \neq 0\}$
2. x-intercept: 4; no y-intercept
3. No symmetry
4. Vertical asymptote: $x = 0$
5. Horizontal asymptote: $y = -1$, not intersected
6.

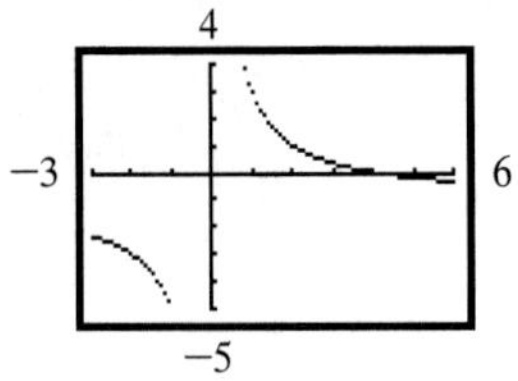

17. 1. Domain: $\{x|x \neq 0, x \neq 2\}$
2. x-intercept: -2; no y-intercept
3. No symmetry
4. Vertical asymptotes: $x = 0$, $x = 2$
5. Horizontal asymptote: $y = 0$, intersected at $(-2, 0)$
6.

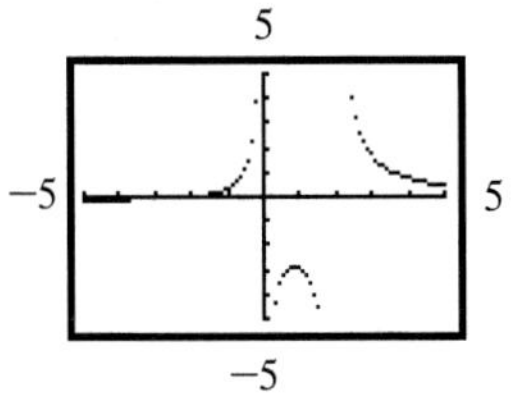

18. 1. Domain: $\{x|x \neq -1, x \neq 1\}$
2. Intercept: $(0, 0)$
3. Symmetric with respect to the origin
4. Vertical asymptotes: $x = -1$ and $x = 1$
5. Horizontal asymptote: $y = 0$, intersected at $(0, 0)$
6.

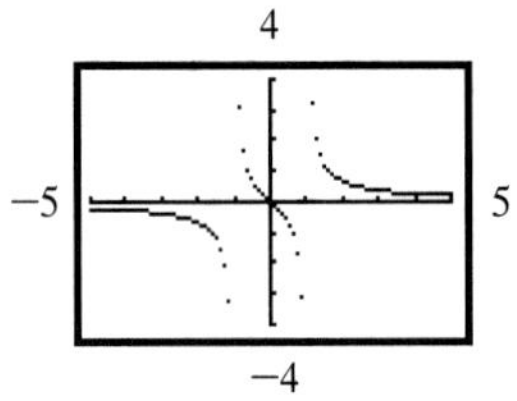

19. 1. Domain: $\{x|x \neq -2, x \neq 3\}$
2. x-intercepts: -3, 2; y-intercept: 1
3. No symmetry
4. Vertical asymptote: $x = -2$, $x = 3$
5. Horizontal asymptote: $y = 1$, intersected at $(0, 1)$
6.

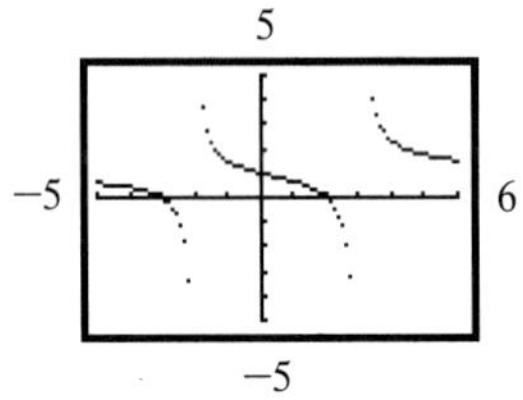

20. 1. Domain: $\{x|x \neq 0\}$
2. x-intercept: 3; no y-intercept
3. No symmetry
4. Vertical asymptote: $x = 0$
5. Horizontal asymptote: $y = 1$, intersected at $\left(\frac{3}{2}, 1\right)$
6.

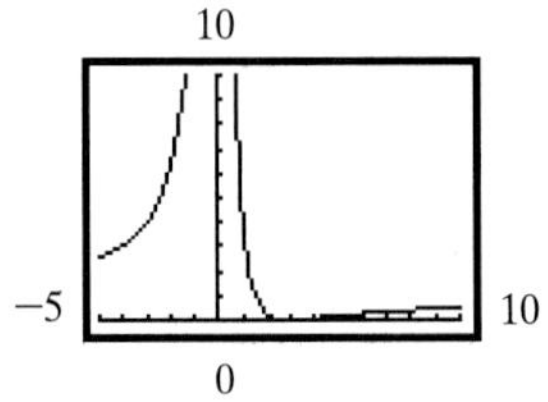

21. 1. Domain: $\{x|x \neq -2, x \neq 2\}$
2. Intercept: $(0, 0)$
3. Symmetric with respect to the origin
4. Vertical asymptotes: $x = -2$, $x = 2$
5. Oblique asymptote: $y = x$, intersected at $(0, 0)$
6.

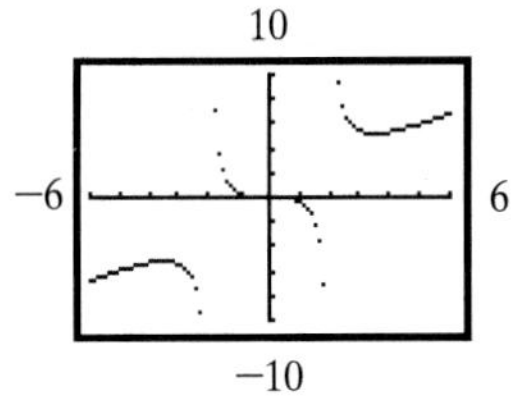

22. 1. Domain: $\{x | x \neq 1\}$
2. Intercept: $(0, 0)$
3. No symmetry
4. Vertical asymptote: $x = 1$
5. Oblique asymptote: $y = 3x + 6$, intersected at $\left(\frac{2}{3}, 8\right)$
6.

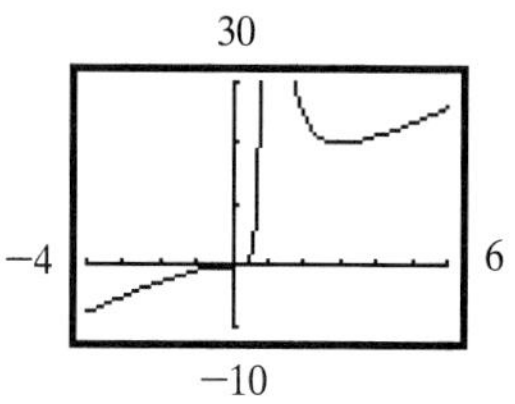

23. 1. Domain: $\{x | x \neq 1\}$
2. Intercept: $(0, 0)$
3. No symmetry
4. Vertical asymptote: $x = 1$
5. No oblique or horizontal asymptote
6.

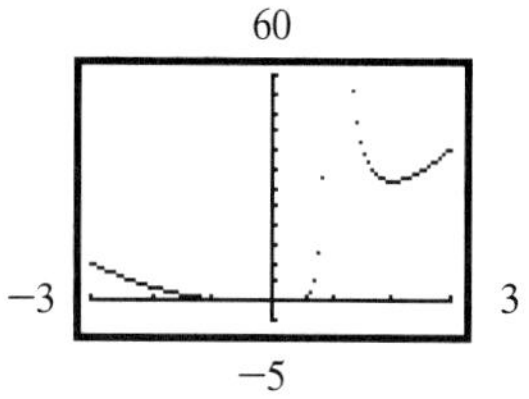

24. 1. Domain: $\{x | x \neq -3, x \neq 3\}$
2. Intercept: $(0, 0)$
3. Symmetric with respect to the y-axis
4. Vertical asymptotes: $x = 3$ and $x = -3$
5. No horizontal asymptotes and no oblique asymptotes
6.

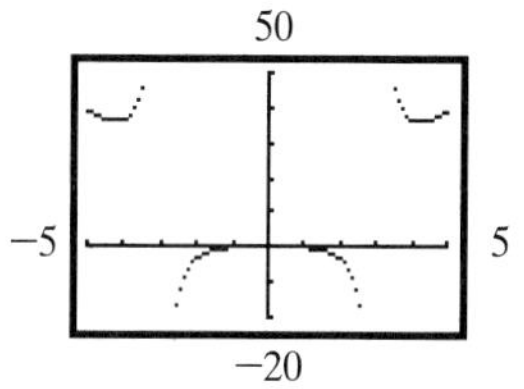

25. 1. Domain: $\{x | x \neq -1, x \neq 2\}$
2. x-intercept: -2; y-intercept: 2
3. No symmetry
4. Vertical asymptote: $x = -1$
5. Horizontal asymptote: $y = 1$, not intersected
6.

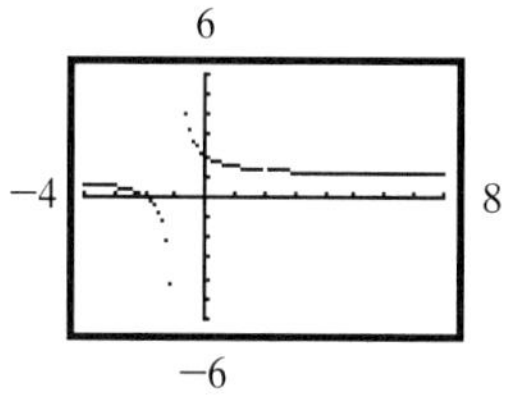

26. 1. Domain: $\{x | x \neq -1, x \neq 1\}$
2. No x-intercepts; y-intercept: -1
3. No symmetry
4. Vertical asymptote: $x = -1$
5. Horizontal asymptote: $y = 1$, not intersected
6.

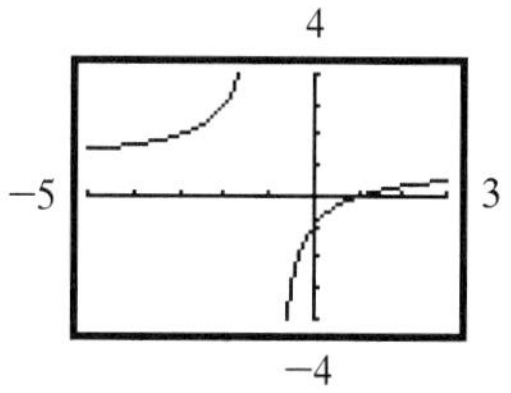

27. $q(x) = 8x^2 + 5x + 6; R = 10$ **28.** $q(x) = 2x^2 + 12x + 19; R = 43$ **29.** $q(x) = x^3 - 4x^2 + 8x - 15; R = 29$
30. $q(x) = x^3 - x^2 + 3; R = -3$ **31.** $\pm\frac{1}{12}, \pm\frac{1}{6}, \pm\frac{1}{4}, \pm\frac{1}{3}, \pm\frac{1}{2}, \pm\frac{3}{4}, \pm 1, \pm\frac{3}{2}; \pm 3$ **32.** $\pm 1, \pm\frac{1}{2}, \pm\frac{1}{3}, \pm\frac{1}{6}$
33. $-2, 1, 4; f(x) = (x + 2)(x - 1)(x - 4)$ **34.** $-1, 4, -2; f(x) = (x + 1)(x - 4)(x + 2)$
35. $\frac{1}{2}$, multiplicity 2; -2; $f(x) = 4\left(x - \frac{1}{2}\right)^2(x + 2)$ **36.** $2, -\frac{1}{2}$, multiplicity 2; $f(x) = 4(x - 2)\left(x + \frac{1}{2}\right)^2$
37. 2, multiplicity 2; $f(x) = (x - 2)^2(x^2 + 5)$ **38.** -3, multiplicity 2; $f(x) = (x + 3)^2(x^2 + 2)$ **39.** $-2.5, 3.1, 5.32$
40. $-3.4, -0.25, 0.33$ **41.** $-11.3, -0.6, 4, 9.33$ **42.** $-10.33, -8.30, -3.56, -0.45$ **43.** $-3.67, 1.33$ **44.** -6.5 **45.** $\{-3, 2\}$ **46.** $\{-3, 2\}$
47. $\left\{-3, -1, -\frac{1}{2}, 1\right\}$ **48.** $\left\{2, -2, -\frac{1}{2}, -3\right\}$ **49.** -5 and 5 **50.** -11 and 11 **51.** $-\frac{37}{2}$ and $\frac{37}{2}$ **52.** $-\frac{17}{3}$ and $\frac{17}{3}$
53. $f(0) = -1; f(1) = 1$ **54.** $f(1) = -2, f(2) = 9$ **55.** $f(0) = -1; f(1) = 1$ **56.** $f(1) = -3, f(2) = 62$ **57.** $4 + 7i$ **58.** $2 - i$
59. $-3 + 2i$ **60.** $-4 + 11i$ **61.** $\frac{9}{10} - \frac{3}{10}i$ **62.** $\frac{8}{5} + \frac{4}{5}i$ **63.** -1 **64.** i **65.** $-46 + 9i$ **66.** $-9 - 46i$ **67.** $4 - i$ **68.** $3 - 4i$
69. $-i, 1 - i$ **70.** $1 - i$ **71.** $\left\{-\frac{1}{2} - \frac{\sqrt{3}}{2}i, -\frac{1}{2} + \frac{\sqrt{3}}{2}i\right\}$ **72.** $\left\{\frac{1 - \sqrt{3}i}{2}, \frac{1 + \sqrt{3}i}{2}\right\}$ **73.** $\left\{\frac{-1 - \sqrt{17}}{4}, \frac{-1 + \sqrt{17}}{4}\right\}$ **74.** $\left\{-\frac{1}{3}, 1\right\}$
75. $\left\{\frac{1}{2} - \frac{\sqrt{11}i}{2}, \frac{1}{2} + \frac{\sqrt{11}i}{2}\right\}$ **76.** $\left\{\frac{1}{2} - \frac{1}{2}i, \frac{1}{2} + \frac{1}{2}i\right\}$ **77.** $\left\{\frac{1}{2} - \frac{\sqrt{23}}{2}i, \frac{1}{2} + \frac{\sqrt{23}}{2}i\right\}$ **78.** $\{-2, 1\}$ **79.** $\{-\sqrt{2}, \sqrt{2}, -2i, 2i\}$

80. $\{-3i, 3i, -1, 1\}$ **81.** $\{-3, 2\}$ **82.** $\{-2, 2, 3\}$ **83.** $\left\{\frac{1}{3}, 1, -i, i\right\}$ **84.** $\{-2, \sqrt{2}, -\sqrt{2}\}$ **85.** $\left\{x \middle| -4 < x < \frac{3}{2}\right\}$

86. $\left\{x \middle| -\infty < x \le -\frac{1}{3} \text{ or } 1 \le x < \infty\right\}$ **87.** $\{x \mid -3 < x \le 3\}$ **88.** $\left\{x \middle| -\infty < x < \frac{1}{3} \text{ or } 1 < x < \infty\right\}$

89. $\{x \mid -\infty < x < 1 \text{ or } 2 < x < \infty\}$ **90.** $\left\{x \middle| -\frac{5}{2} < x \le -\frac{7}{6}\right\}$ **91.** $\{x \mid 1 < x < 2 \text{ or } 3 < x < \infty\}$ **92.** $\{x \mid -\infty < x \le -1 \text{ or } 0 < x < 5\}$

93. $\{x \mid -\infty < x < -4 \text{ or } 2 < x < 4 \text{ or } 6 < x < \infty\}$ **94.** $\{x \mid -\infty < x < -5 \text{ or } -4 < x \le -2 \text{ or } 0 \le x \le 1\}$

95. (a)

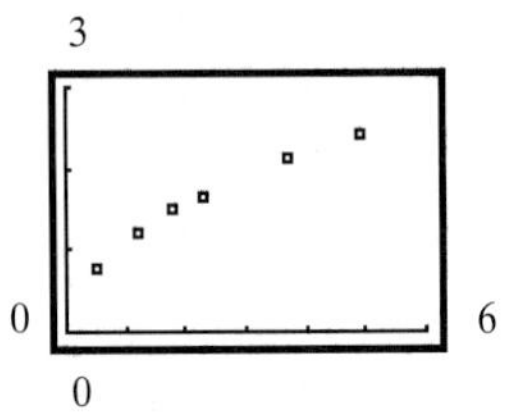

(b) 2.2 sec **(c)**

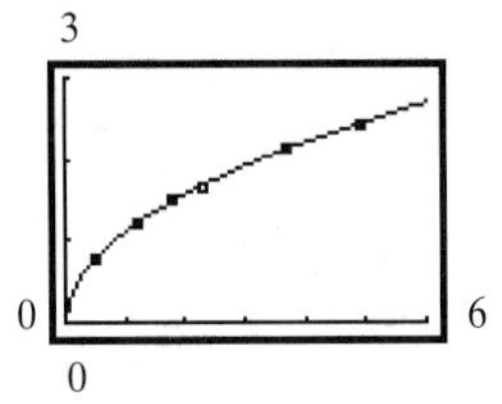

(d)

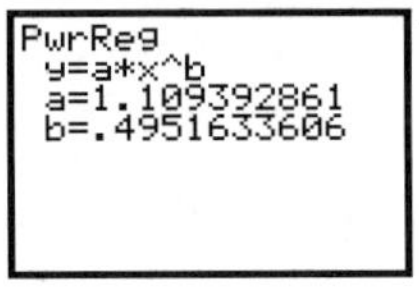

96. (a)

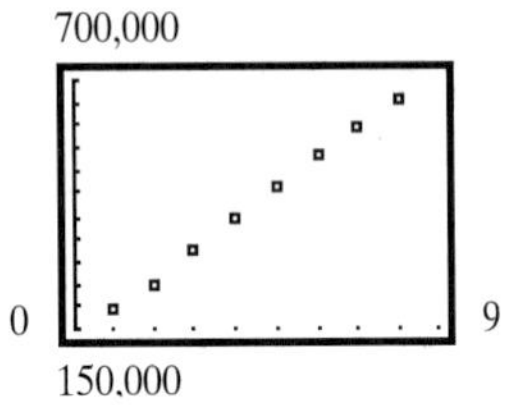

(b) 797,010 (or 797,003, using unrounded regression function)
(c) According to the function, the number of AIDS cases will never reach 1,000,000.

(d)

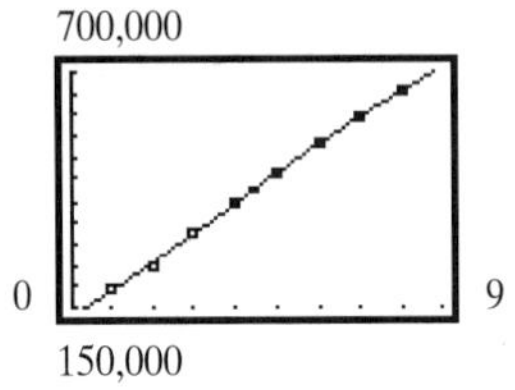

(e)

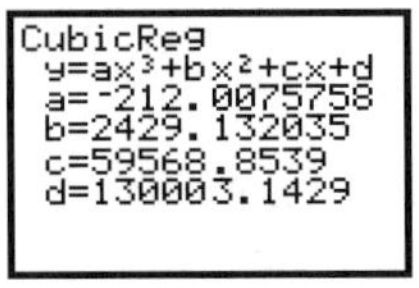

(f) No. This function will start decreasing when $t \approx 14$ and it is not likely that the number of AIDS cases will decrease.

97. $-5(x-1)^3(x-2)(x+1)\left(x-\frac{3}{10}\right)$; This is not unique. **98.** $\dfrac{2\left(x-\frac{3}{2}\right)\left(x+\frac{1}{2}\right)^2(x^2+8)}{(x+2)(x-3)(x^2+1)}$; This is not unique.

99. (a) Even **(b)** Positive **(c)** Even **(d)** $x = 0$ is a root of even multiplicity. **(e)** 8 **(f)** One possibility: $x^2(x^2-1)(x^2-4)(x^2-9)$

CHAPTER 5 Exponential and Logarithmic Functions

5.1 Exercises

1. (a)

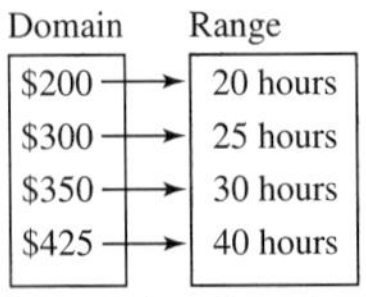

(b) Inverse is a function

2. (a)

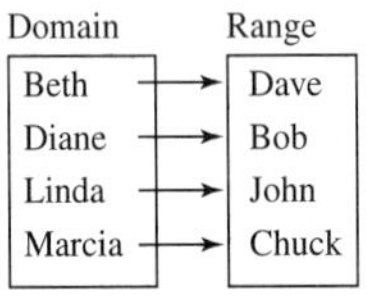

(b) Inverse is a function

3. (a)

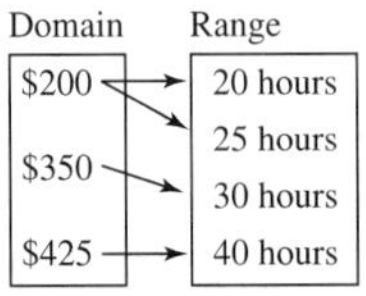

(b) Inverse is not a function

4. (a)

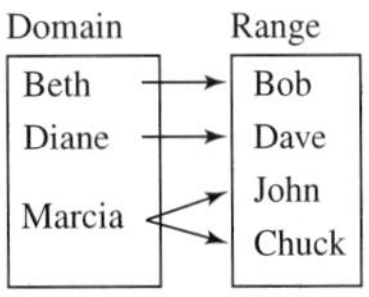

(b) Inverse is not a function

5. (a) $\{(6, 2), (6, -3), (9, 4), (10, 1)\}$ **(b)** Inverse is not a function **6. (a)** $\{(5, -2), (3, -1), (7, 3), (12, 4)\}$ **(b)** Inverse is a function
7. (a) $\{(0, 0), (1, 1), (16, 2), (81, 3)\}$ **(c)** Inverse is a function **8. (a)** $\{(2, 1), (8, 2), (18, 3), (32, 4)\}$ **(b)** Inverse is a function
9. One-to-one **10.** One-to-one **11.** Not one-to-one **12.** Not one-to-one **13.** One-to-one **14.** Not one-to-one

15.

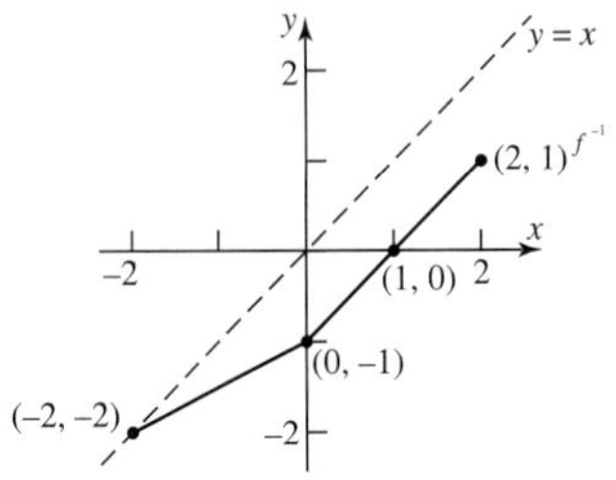

16.

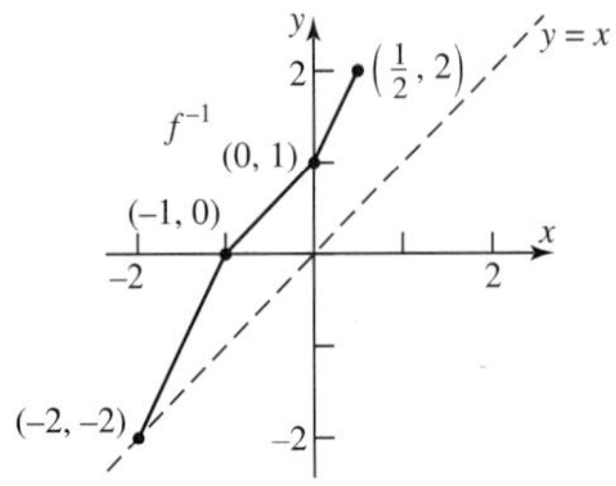

17.

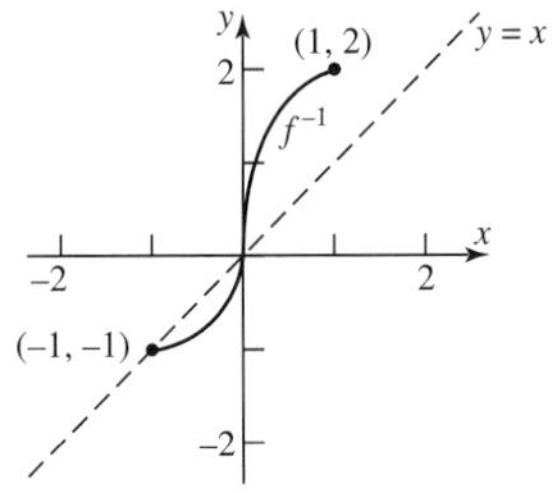

18.

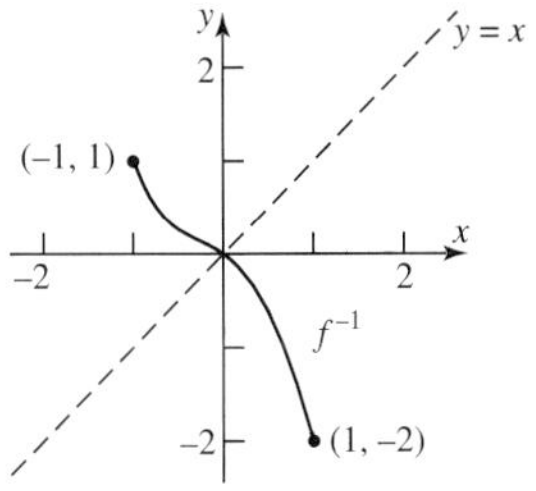

19.

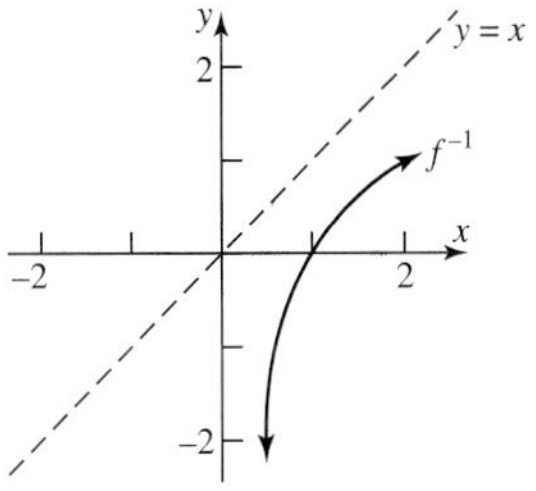

20.

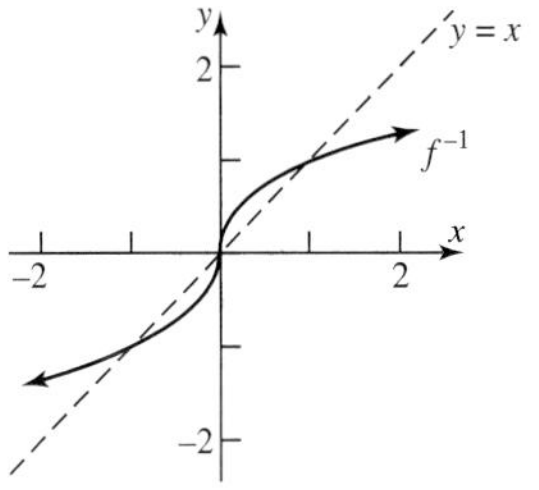

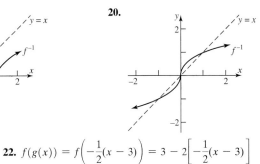

21. $f(g(x)) = f\left(\frac{1}{3}(x-4)\right) = 3\left[\frac{1}{3}(x-4)\right] + 4 = x;$

$g(f(x)) = g(3x+4) = \frac{1}{3}[(3x+4)-4] = x$

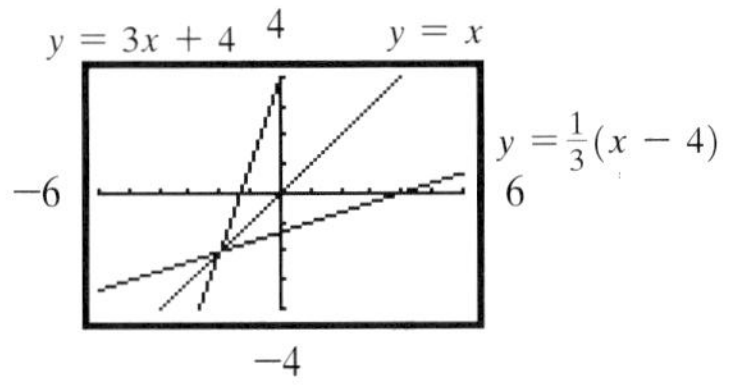

22. $f(g(x)) = f\left(-\frac{1}{2}(x-3)\right) = 3 - 2\left[-\frac{1}{2}(x-3)\right]$

$= 3 + x - 3 = x;$

$g(f(x)) = g(3-2x) = -\frac{1}{2}[(3-2x)-3] = x$

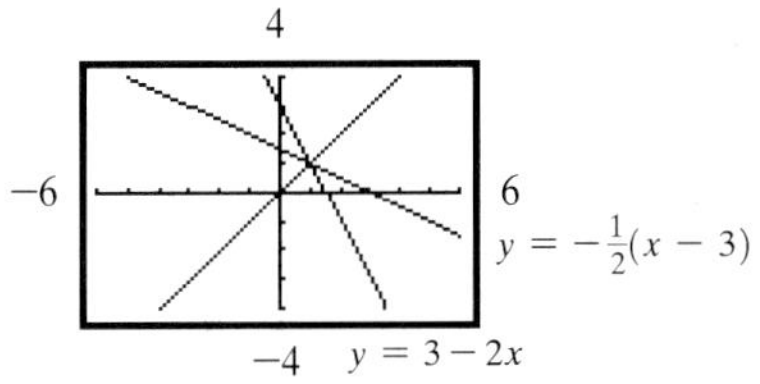

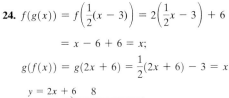

23. $f(g(x)) = 4\left[\frac{x}{4} + 2\right] - 8 = x;$

$g(f(x)) = \frac{4x-8}{4} + 2 = x$

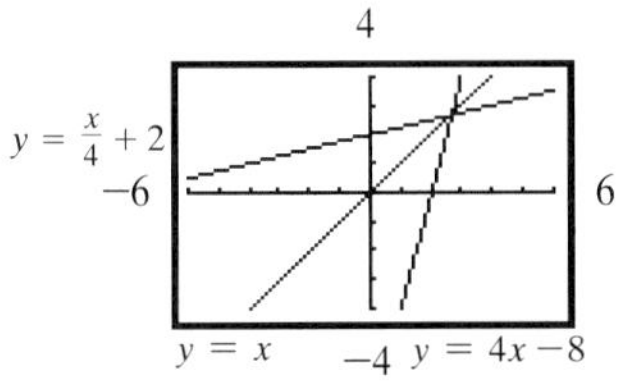

24. $f(g(x)) = f\left(\frac{1}{2}(x-3)\right) = 2\left(\frac{1}{2}x - 3\right) + 6$

$= x - 6 + 6 = x;$

$g(f(x)) = g(2x+6) = \frac{1}{2}(2x+6) - 3 = x$

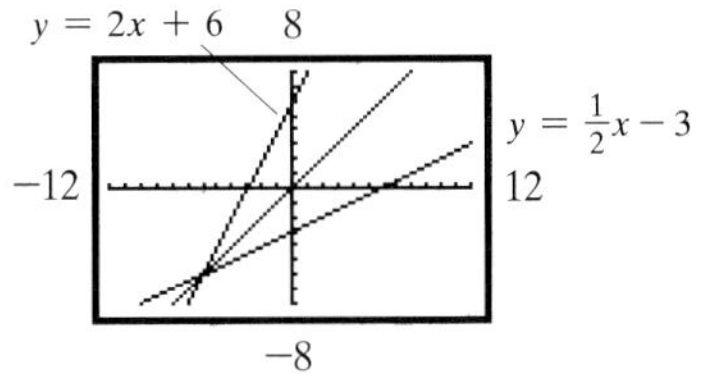

25. $f(g(x)) = (\sqrt[3]{x+8})^3 - 8 = x;$
$g(f(x)) = \sqrt[3]{(x^3-8)+8} = x$

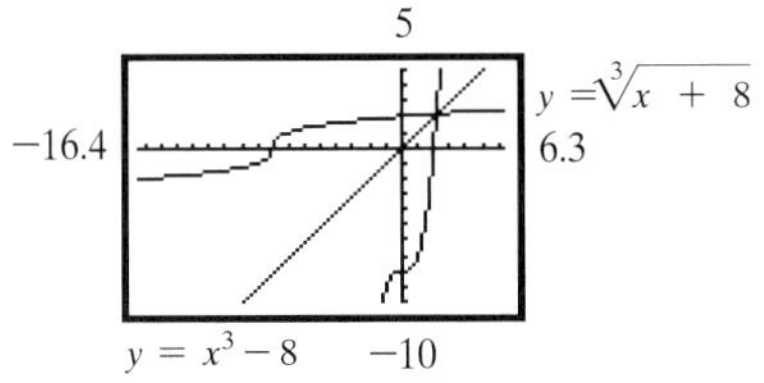

26. $f(g(x)) = f(\sqrt{x}+2) = [(\sqrt{x}+2)-2]^2 = (\sqrt{x})^2 = x$
$g(f(x)) = g((x-2)^2) = \sqrt{(x-2)^2} + 2$
$= |x-2| + 2 = x, x \geq 2$

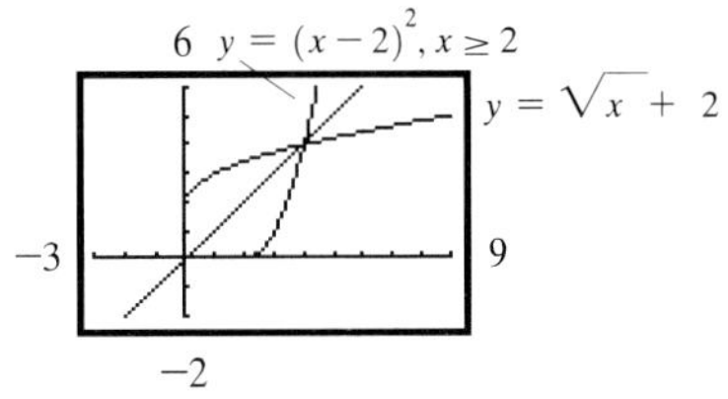

27. $f(g(x)) = \frac{1}{\left(\frac{1}{x}\right)} = x; g(f(x)) = \frac{1}{\left(\frac{1}{x}\right)} = x$

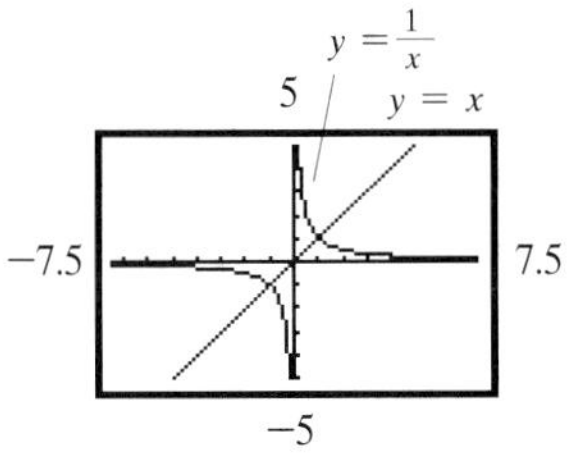

28. $f(g(x)) = f(x) = x; g(f(x)) = g(x) = x$

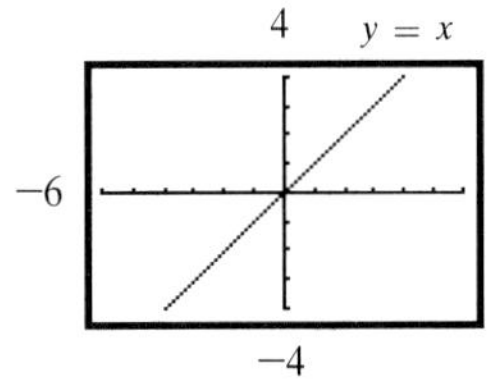

29. $f(g(x)) = \dfrac{2\left(\dfrac{4x-3}{2-x}\right)+3}{\dfrac{4x-3}{2-x}+4} = x;$

$g(f(x)) = \dfrac{4\left(\dfrac{2x+3}{x+4}\right)-3}{2-\dfrac{2x+3}{x+4}} = x$

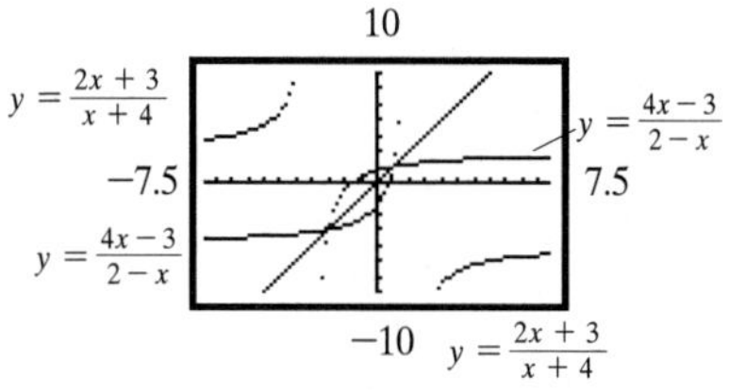

30. $f(g(x)) = f\left(\dfrac{3x+5}{1-2x}\right) = \dfrac{\dfrac{3x+5}{1-2x}-5}{2\left(\dfrac{3x+5}{1-2x}\right)+3} = x;$

$g(f(x)) = g\left(\dfrac{x-5}{2x+3}\right) = \dfrac{3\left(\dfrac{x-5}{2x-3}\right)+5}{1-2\left(\dfrac{x-5}{2x+3}\right)} = x$

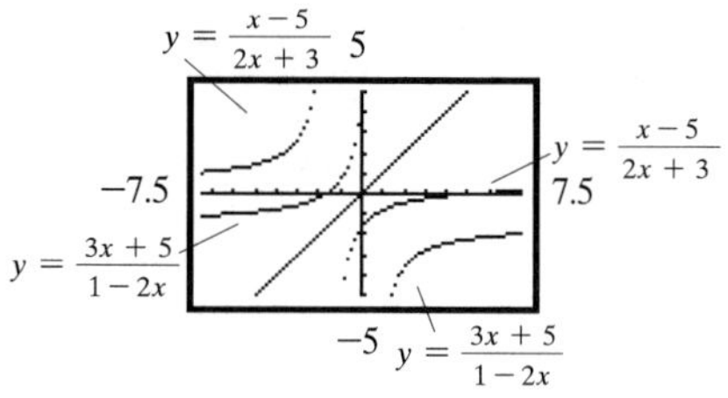

31. $f^{-1}(x) = \dfrac{1}{3}x$

$f(f^{-1}(x)) = 3\left(\dfrac{1}{3}x\right) = x$

$f^{-1}(f(x)) = \dfrac{1}{3}(3x) = x$

Domain f = Range $f^{-1} = (-\infty, \infty)$
Range f = Domain $f^{-1} = (-\infty, \infty)$

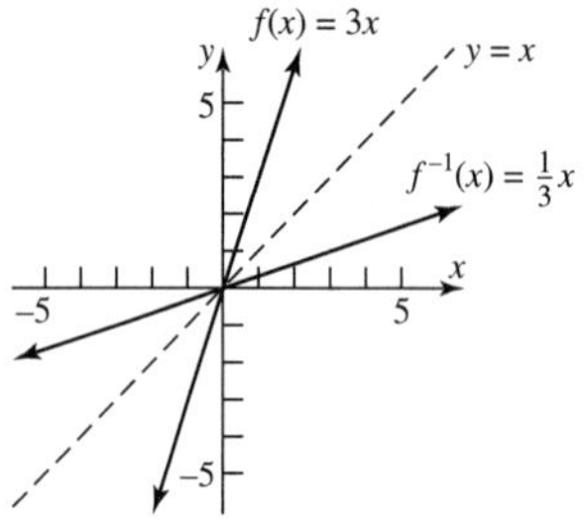

32. $f^{-1}(x) = \dfrac{-x}{4}$

$f(f^{-1}(x)) = -4\left(\dfrac{-x}{4}\right) = x$

$f^{-1}(f(x)) = \dfrac{-(-4x)}{4} = x$

Domain f = Range $f^{-1} = (-\infty, \infty)$
Range f = Domain $f^{-1} = (-\infty, \infty)$

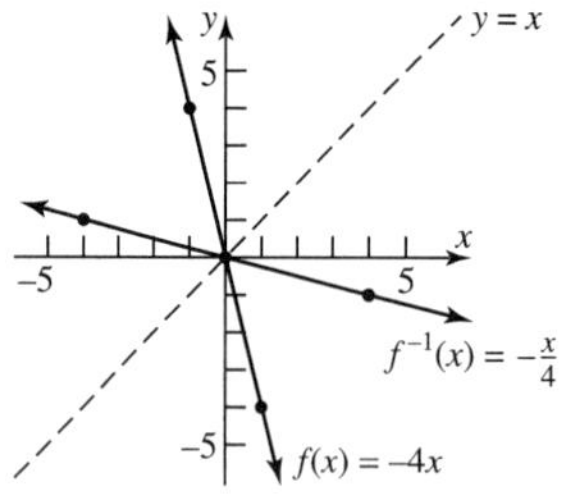

33. $f^{-1}(x) = \dfrac{x}{4} - \dfrac{1}{2}$

$f(f^{-1}(x)) = 4\left(\dfrac{x}{4} - \dfrac{1}{2}\right) + 2 = x$

$f^{-1}(f(x)) = \dfrac{4x+2}{4} - \dfrac{1}{2} = x$

Domain f = Range $f^{-1} = (-\infty, \infty)$
Range f = Domain $f^{-1} = (-\infty, \infty)$

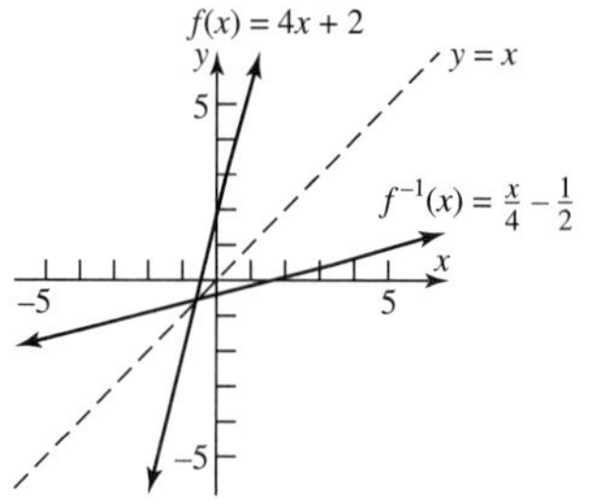

34. $f^{-1}(x) = \dfrac{-x+1}{3}$

$f(f^{-1}(x)) = 1 - 3\left(\dfrac{-x+1}{3}\right) = x$

$f^{-1}(f(x)) = \dfrac{-(1-3x)+1}{3} = x$

Domain f = Range $f^{-1} = (-\infty, \infty)$
Range f = Domain $f^{-1} = (-\infty, \infty)$

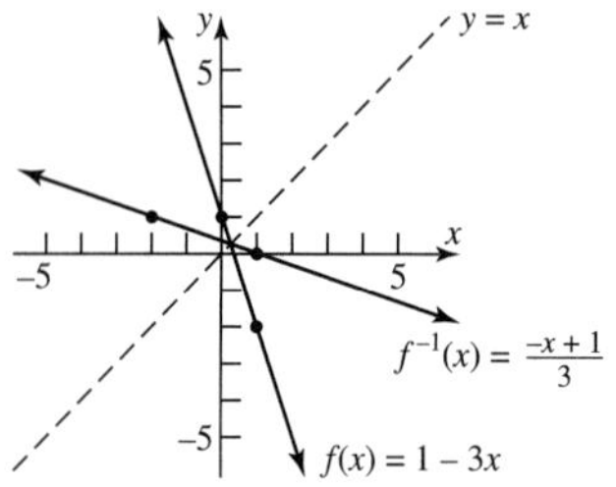

35. $f^{-1}(x) = \sqrt[3]{x+1}$

$f(f^{-1}(x)) = (\sqrt[3]{x+1})^3 - 1 = x$

$f^{-1}(f(x)) = \sqrt[3]{(x^3-1)+1} = x$

Domain f = Range $f^{-1} = (-\infty, \infty)$
Range f = Domain $f^{-1} = (-\infty, \infty)$

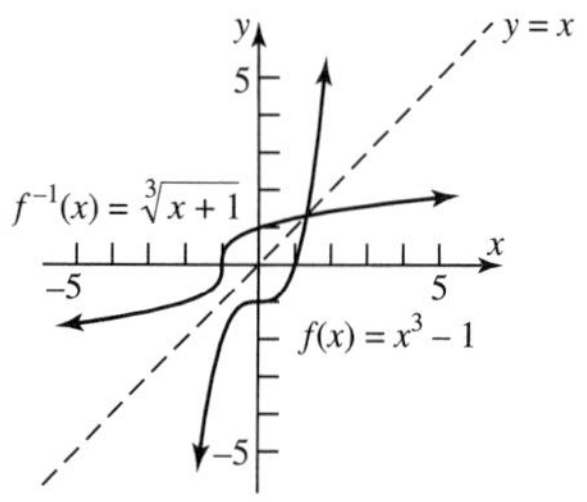

36. $f^{-1}(x) = \sqrt[3]{x-1}$
$f(f^{-1}(x)) = (\sqrt[3]{x-1})^3 + 1 = x$

$f^{-1}(f(x)) = \sqrt[3]{(x^3+1)-1} = x$
Domain f = Range $f^{-1} = (-\infty, \infty)$
Range f = Domain $f^{-1} = (-\infty, \infty)$

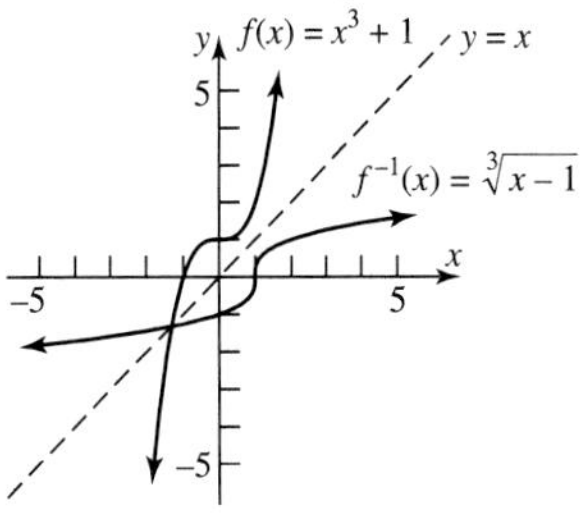

37. $f^{-1}(x) = \sqrt{x-4}$
$f(f^{-1}(x)) = (\sqrt{x-4})^2 + 4 = x$

$f^{-1}(f(x)) = \sqrt{(x^2+4)-4} = \sqrt{x^2} = |x| = x$
Domain f = Range $f^{-1} = [0, \infty)$
Range f = Domain $f^{-1} = [4, \infty)$

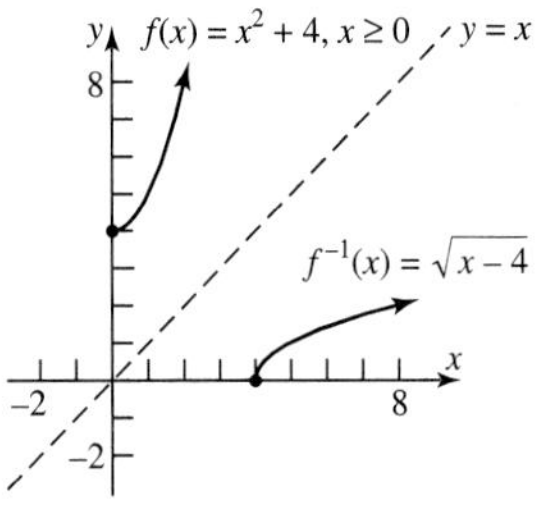

38. $f^{-1}(x) = \sqrt{x-9}$

$f(f^{-1}(x)) = (\sqrt{x-9})^2 + 9 = x$

$f^{-1}(f(x)) = \sqrt{(x^2+9)-9} = \sqrt{x^2} = |x| = x, x \geq 0$

Domain f = Range $f^{-1} = [0, \infty)$
Range f = Domain $f^{-1} = [9, \infty)$

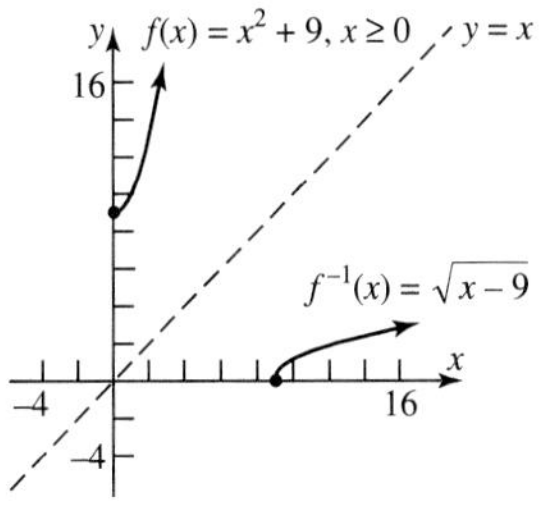

39. $f^{-1}(x) = \dfrac{4}{x}$

$f(f^{-1}(x)) = \dfrac{4}{\frac{4}{x}} = x$

$f^{-1}(f(x)) = \dfrac{4}{\frac{4}{x}} = x$

Domain f = Range f^{-1} = All real numbers except 0
Range f = Domain f^{-1} = All real numbers except 0

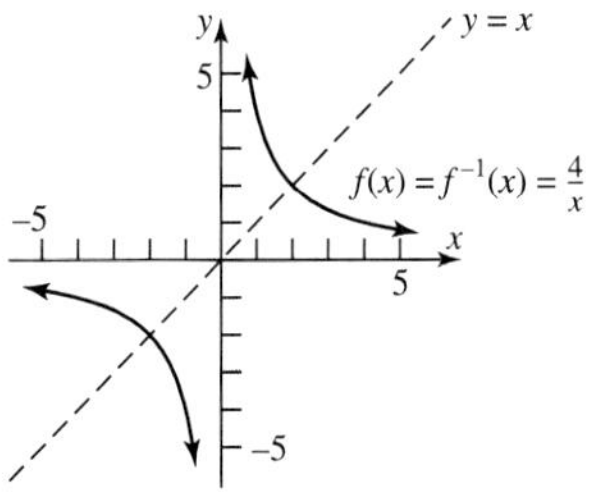

40. $f^{-1}(x) = -\dfrac{3}{x}$

$f(f^{-1}(x)) = -\dfrac{3}{-\frac{3}{x}} = x$

$f^{-1}(f(x)) = -\dfrac{3}{-\frac{3}{x}} = x$

Domain f = Range f^{-1} = All real numbers except 0
Range f = Domain f^{-1} = All real numbers except 0

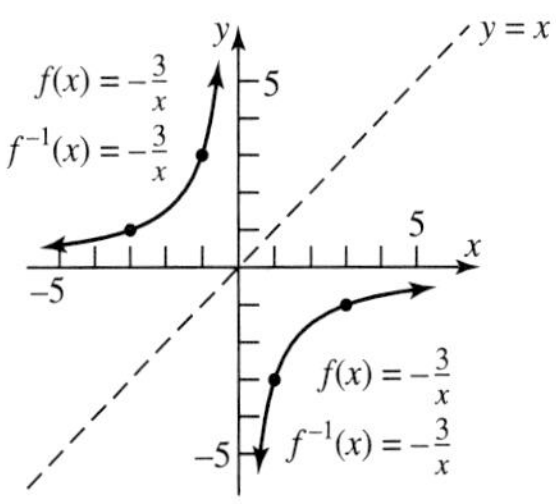

41. $f^{-1}(x) = \dfrac{2x+1}{x}$

$f(f^{-1}(x)) = \dfrac{1}{\frac{2x+1}{x} - 2} = x$

$f^{-1}(f(x)) = \dfrac{2\left(\frac{1}{x-2}\right) + 1}{\frac{1}{x-2}} = x$

Domain f = Range f^{-1} = All real numbers except 2
Range f = Domain f^{-1} = All real numbers except 0

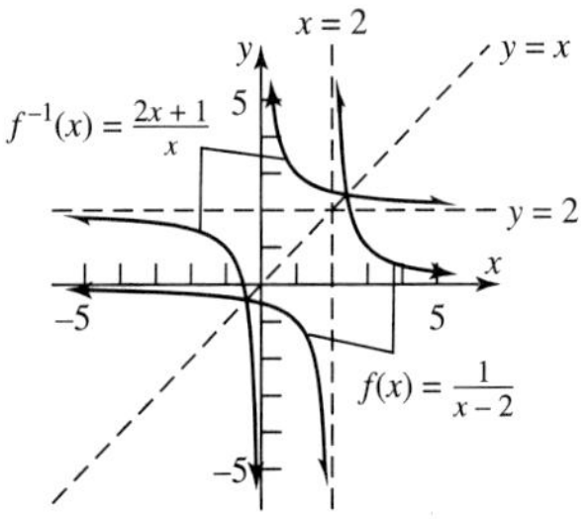

42. $f^{-1}(x) = \dfrac{4 - 2x}{x}$

$$f(f^{-1}(x)) = \frac{4}{\dfrac{4 - 2x}{x} + 2} = x$$

$$f^{-1}(f(x)) = \frac{4 - 2\left(\dfrac{4}{x + 2}\right)}{\dfrac{4}{x + 2}} = x$$

Domain f = Range f^{-1} = All real numbers except -2
Range f = Domain f^{-1} = All real numbers except 0

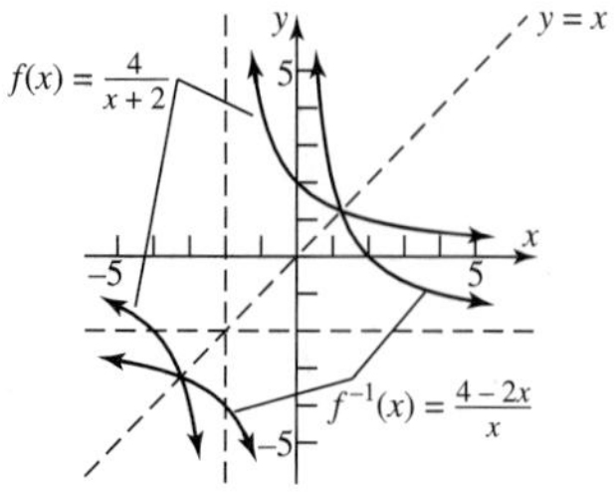

43. $f^{-1}(x) = \dfrac{2 - 3x}{x}$

$$f(f^{-1}(x)) = \frac{2}{3 + \dfrac{2 - 3x}{x}} = x$$

$$f^{-1}(f(x)) = \frac{2 - 3\left(\dfrac{2}{3 + x}\right)}{\dfrac{2}{3 + x}} = x$$

Domain f = Range f^{-1} = All real numbers except -3
Range f = Domain f^{-1} = All real numbers except 0

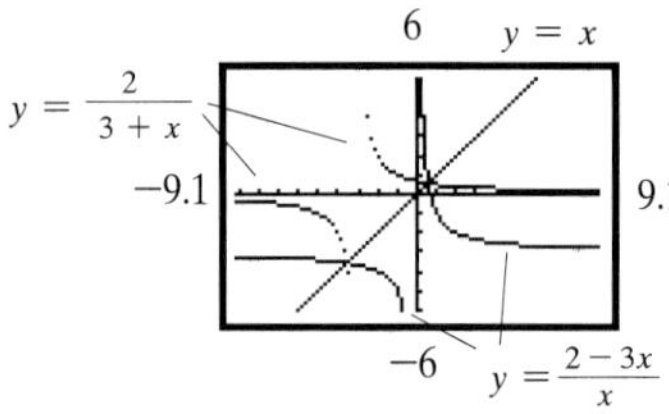

44. $f^{-1}(x) = \dfrac{2x - 4}{x}$

$$f(f^{-1}(x)) = \frac{4}{2 - \left(\dfrac{2x - 4}{x}\right)} = x$$

$$f^{-1}(f(x)) = \frac{2\left(\dfrac{4}{2 - x}\right) - 4}{\dfrac{4}{2 - x}} = x$$

Domain f = Range f^{-1} = All real numbers except 2
Range f = Domain f^{-1} = All real numbers except 0

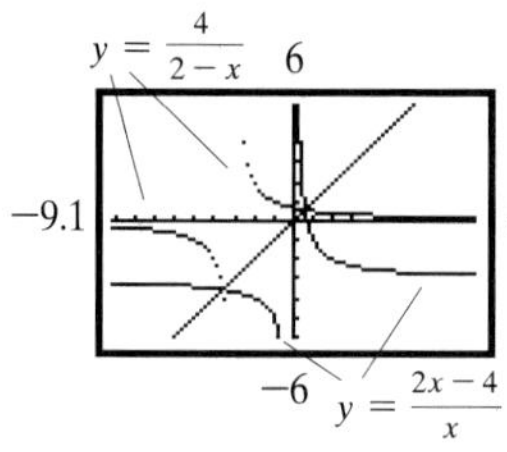

45. $f^{-1}(x) = \sqrt{x} - 2$
$f(f^{-1}(x)) = (\sqrt{x} - 2 + 2)^2 = x$
$f^{-1}(f(x)) = \sqrt{(x + 2)^2} - 2 = |x + 2| - 2 = x, x \geq -2$
Domain f = Range f^{-1} = $[-2, \infty)$
Range f = Domain f^{-1} = $[0, \infty)$

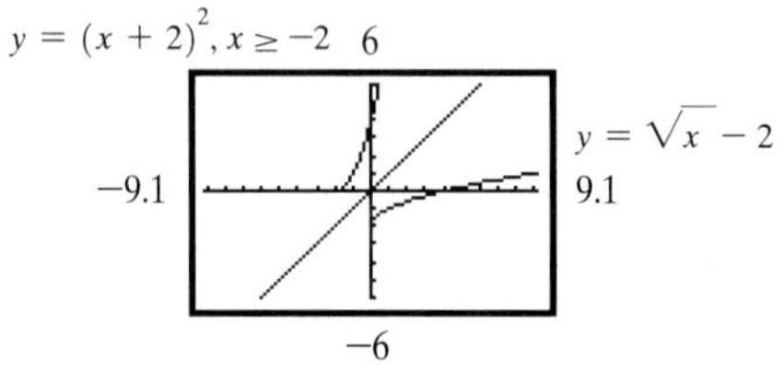

46. $f^{-1}(x) = \sqrt{x} + 1$
$f(f^{-1}(x)) = (\sqrt{x} + 1 - 1)^2 = x$
$f^{-1}(f(x)) = \sqrt{(x - 1)^2} + 1 = |x - 1| + 1 = x, x \geq 1$
Domain f = Range f^{-1} = $[1, \infty)$
Range f = Domain f^{-1} = $[0, \infty)$

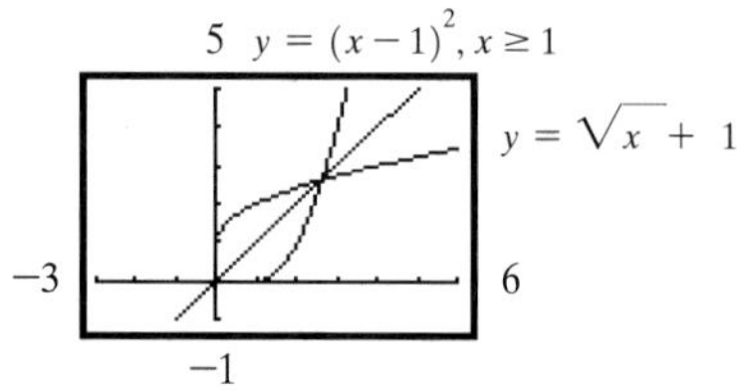

47. $f^{-1}(x) = \dfrac{x}{x-2}$

$$f(f^{-1}(x)) = \frac{2\left(\dfrac{x}{x-2}\right)}{\dfrac{x}{x-2} - 1} = x$$

$$f^{-1}(f(x)) = \frac{\dfrac{2x}{x-1}}{\dfrac{2x}{x-1} - 2} = x$$

Domain f = Range f^{-1} = All real numbers except 1
Range f = Domain f^{-1} = All real numbers except 2

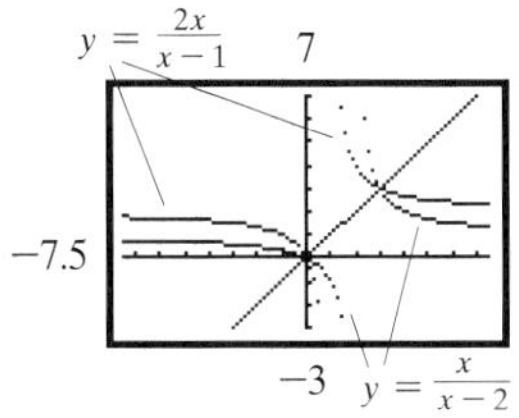

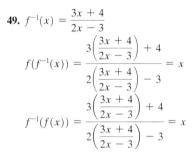

48. $f^{-1}(x) = \dfrac{1}{x-3}$

$$f(f^{-1}(x)) = \frac{3\left(\dfrac{1}{x-3}\right)}{\dfrac{1}{x-3}} = x$$

$$f^{-1}(f(x)) = \frac{1}{\dfrac{3x+1}{x} - 3} = x$$

Domain f = Range f^{-1} = All real numbers except 0
Domain f^{-1} = Range f = All real numbers except 3

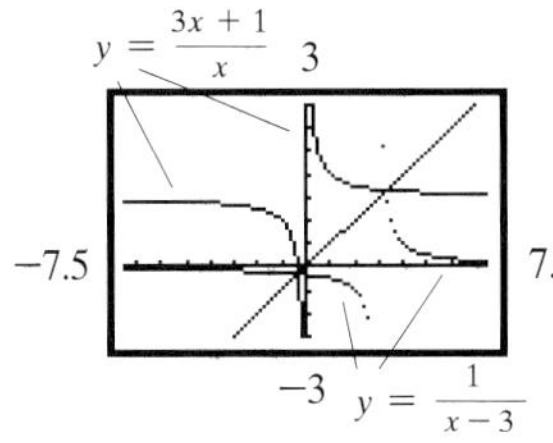

49. $f^{-1}(x) = \dfrac{3x+4}{2x-3}$

$$f(f^{-1}(x)) = \frac{3\left(\dfrac{3x+4}{2x-3}\right) + 4}{2\left(\dfrac{3x+4}{2x-3}\right) - 3} = x$$

$$f^{-1}(f(x)) = \frac{3\left(\dfrac{3x+4}{2x-3}\right) + 4}{2\left(\dfrac{3x+4}{2x-3}\right) - 3} = x$$

Domain f = Range f^{-1} = All real numbers except $\dfrac{3}{2}$

Range f = Domain f^{-1} = All real numbers except $\dfrac{3}{2}$

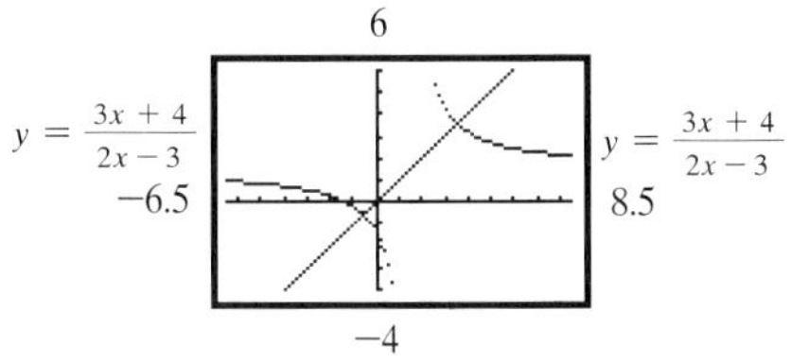

50. $f^{-1}(x) = \dfrac{-4x-3}{x-2}$

$$f(f^{-1}(x)) = \frac{2\left(\dfrac{-4x-3}{x-2}\right) - 3}{\left(\dfrac{-4x-3}{x-2}\right) + 4} = x$$

$$f^{-1}(f(x)) = \frac{-4\left(\dfrac{2x-3}{x+4}\right) - 3}{\left(\dfrac{2x-3}{x+4}\right) - 2} = x$$

Domain f = Range f^{-1} = All real numbers except -4

Range f = Domain f^{-1} = All real numbers except 2

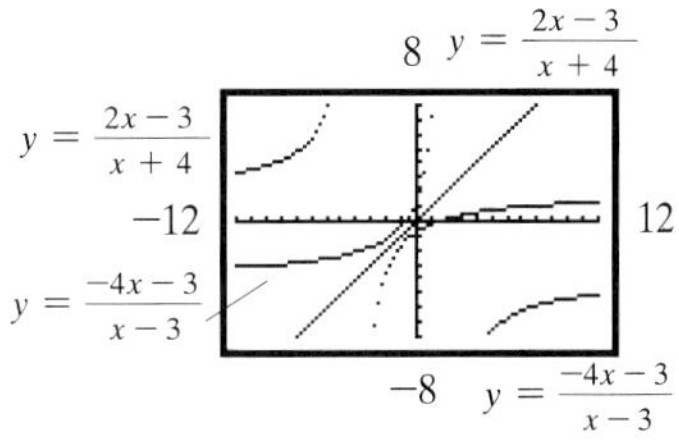

51. $f^{-1}(x) = \dfrac{-2x+3}{x-2}$

$$f(f^{-1}(x)) = \frac{2\left(\dfrac{-2x+3}{x-2}\right)+3}{\dfrac{-2x+3}{x-2}+2} = x$$

$$f^{-1}(f(x)) = \frac{-2\left(\dfrac{2x+3}{x+2}\right)+3}{\dfrac{2x+3}{x+2}-2} = x$$

Domain f = Range f^{-1} = All real numbers except -2
Range f = Domain f^{-1} = All real numbers except 2

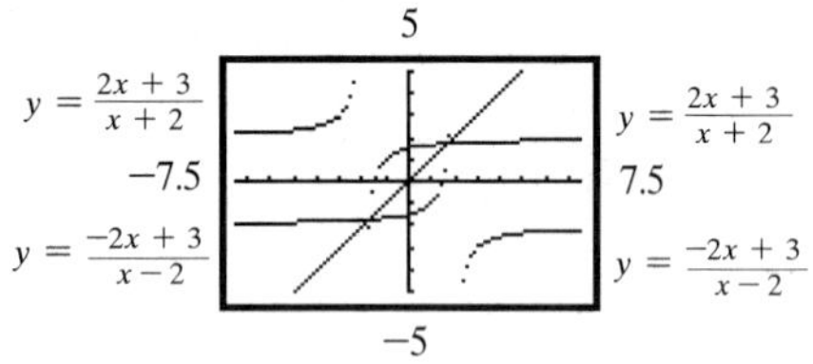

52. $f^{-1}(x) = \dfrac{2x-4}{x+3}$

$$f(f^{-1}(x)) = \frac{-3\left(\dfrac{2x-4}{x+3}\right)-4}{\left(\dfrac{2x-4}{x+3}\right)-2} = x$$

$$f^{-1}(f(x)) = \frac{2\left(\dfrac{-3x-4}{x-2}\right)-4}{\left(\dfrac{-3x-4}{x-2}\right)+3} = x$$

Domain f = Range f^{-1} = All real numbers except 2
Range f = Domain f^{-1} = All real numbers except -3

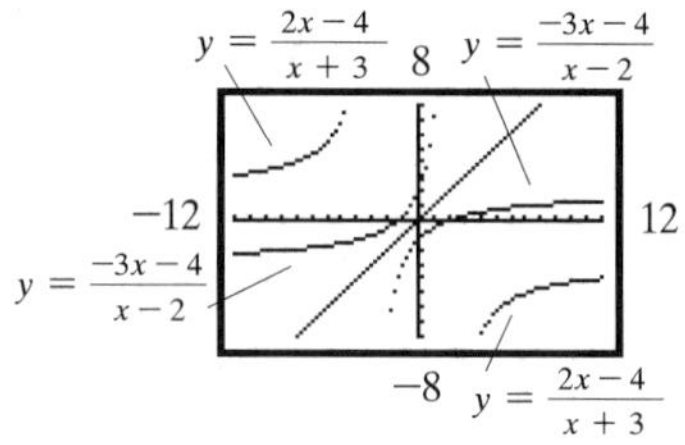

53. $f^{-1}(x) = \dfrac{x^3}{8}$

$$f(f^{-1}(x)) = 2\sqrt[3]{\frac{x^3}{8}} = x$$

$$f^{-1}(f(x)) = \frac{(2\sqrt[3]{x})^3}{8} = x$$

Domain f = Range f^{-1} = $(-\infty, \infty)$
Range f = Domain f^{-1} = $(-\infty, \infty)$

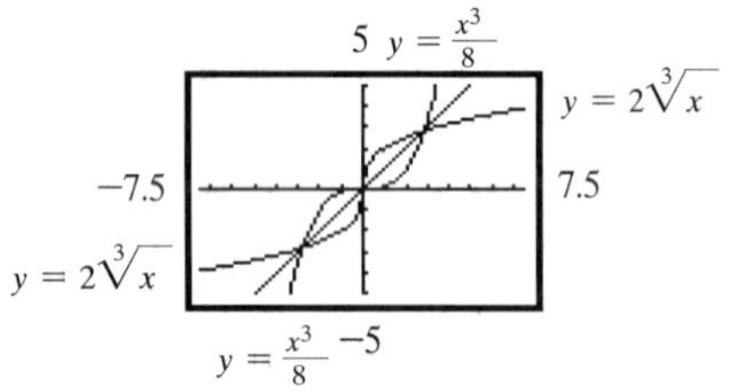

54. $f^{-1}(x) = \dfrac{16}{x^2}, x > 0$

$$f(f^{-1}(x)) = \frac{4}{\sqrt{\dfrac{16}{x^2}}} = \frac{4}{\left|\dfrac{4}{x}\right|} = x, x > 0$$

$$f^{-1}(f(x)) = \frac{16}{\left(\dfrac{4}{\sqrt{x}}\right)^2} = \frac{16}{\dfrac{16}{x}} = x$$

Domain f = Range f^{-1} = $(0, \infty)$
Range f = Domain f^{-1} = $(0, \infty)$

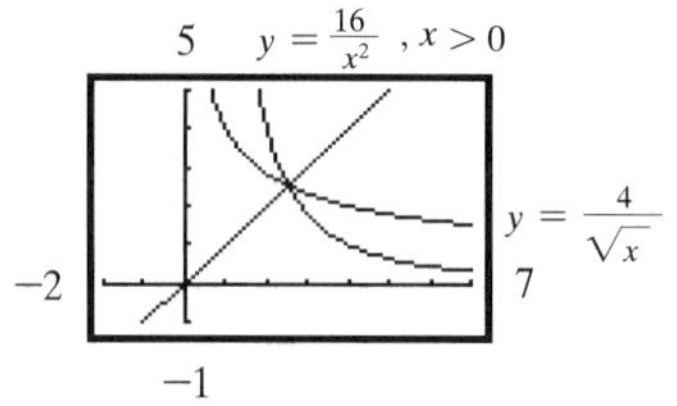

55. $f^{-1}(x) = \dfrac{1}{m}(x - b), m \neq 0$ **56.** $f^{-1}(x) = \sqrt{r^2 - x^2}, 0 \leq x \leq r$ **59.** Quadrant I **60.** Quadrant IV
61. $f(x) = |x|, x \geq 0$, is one-to-one; $f^{-1}(x) = x, x \geq 0$ **62.** $f(x) = x^4, x \geq 0$, is one-to-one; $f^{-1}(x) = \sqrt[4]{x}$
63. $f(g(x)) = \dfrac{9}{5}\left[\dfrac{5}{9}(x - 32)\right] + 32 = x$; $g(f(x)) = \dfrac{5}{9}\left[\left(\dfrac{9}{5}x + 32\right) - 32\right] = x$ **64.** $x(p) = \dfrac{1}{50}(300 - p)$ **65.** $l(T) = \dfrac{gT^2}{4\pi^2}, T > 0$
66. $f(x) = \begin{cases} x \text{ if } x \text{ is rational} \\ -x \text{ if } x \text{ is irrational} \end{cases}$ (Other answers are possible.) **67.** $f^{-1}(x) = \dfrac{-dx + b}{cx - a}$; $f = f^{-1}$ if $a = -d$
68. **(a)** $f^{-1}(x) = \dfrac{3x + 5}{x - 2}$; The range of f is $\{y \mid y \neq 2\}$ **(b)** $g^{-1}(x) = \dfrac{-2}{x - 4}$; The range of g is $\{y \mid y \neq 4\}$
(c) $F^{-1}(x) = \dfrac{4x - 3}{x}$; The range of F is $\{y \mid y \neq 0\}$

5.2 Exercises

1. **(a)** 11.212 **(b)** 11.587 **(c)** 11.664 **(d)** 11.665 **2.** **(a)** 15.426 **(b)** 16.189 **(c)** 16.241 **(d)** 16.242
3. **(a)** 8.815 **(b)** 8.821 **(c)** 8.824 **(d)** 8.825 **4.** **(a)** 6.498 **(b)** 6.543 **(c)** 6.580 **(d)** 6.581
5. **(a)** 21.217 **(b)** 22.217 **(c)** 22.440 **(d)** 22.459 **6.** **(a)** 21.738 **(b)** 22.884 **(c)** 23.119 **(d)** 23.141
7. 3.320 **8.** 0.273 **9.** 0.427 **10.** 8.166 **11.** B **12.** F **13.** D **14.** H **15.** A **16.** C **17.** E **18.** G
19. A **20.** F **21.** E **22.** C **23.** B **24.** D

25.

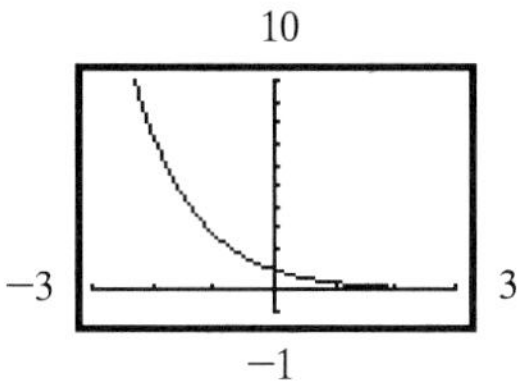

Domain: $(-\infty, \infty)$
Range: $(0, \infty)$
Horizontal asymptote: $y = 0$

26.

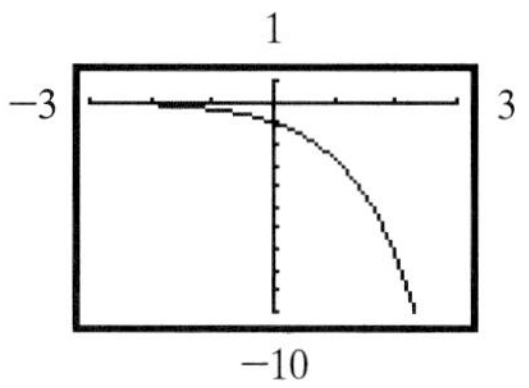

Domain: $(-\infty, \infty)$
Range: $(-\infty, 0)$
Horizontal asymptote: $y = 0$

27.

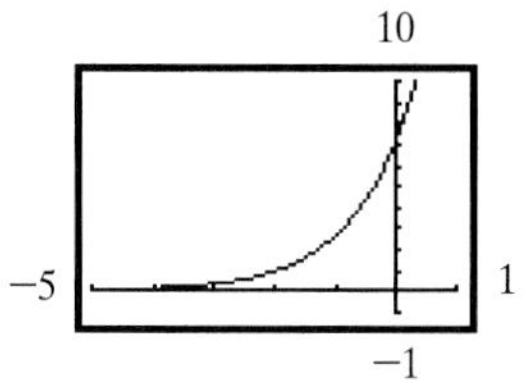

Domain: $(-\infty, \infty)$
Range: $(0, \infty)$
Horizontal asymptote: $y = 0$

28.

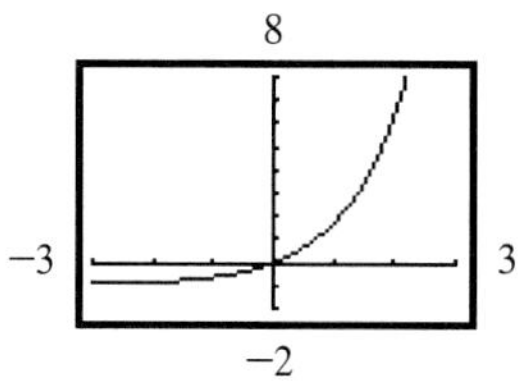

Domain: $(-\infty, \infty)$
Range: $(-1, \infty)$
Horizontal asymptote: $y = -1$

29.

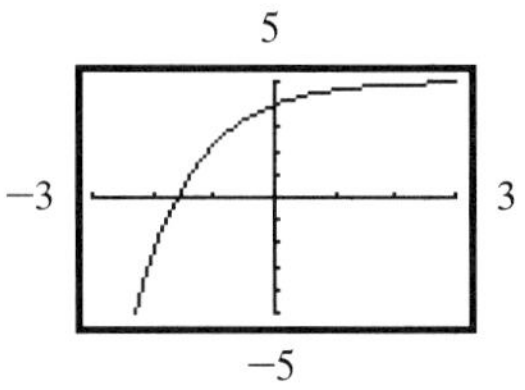

Domain: $(-\infty, \infty)$
Range: $(-\infty, 5)$
Horizontal asymptote: $y = 5$

30.

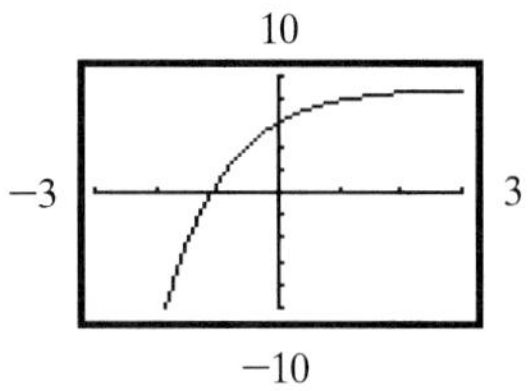

Domain: $(-\infty, \infty)$
Range: $(-\infty, 9)$
Horizontal asymptote: $y = 9$

31.

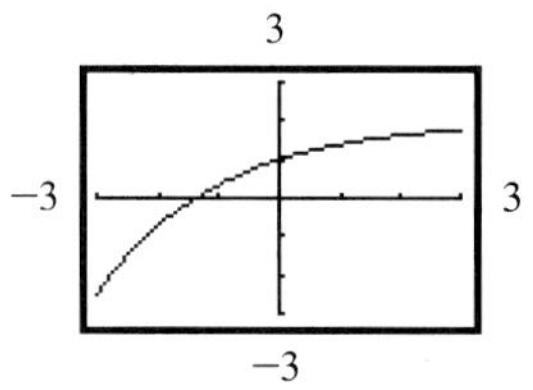

Domain: $(-\infty, \infty)$
Range: $(-\infty, 2)$
Horizontal asymptote: $y = 2$

32.

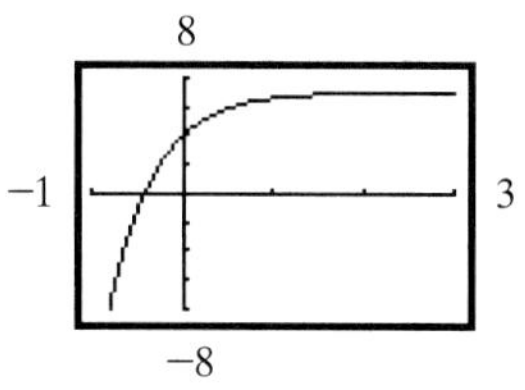

Domain: $(-\infty, \infty)$
Range: $(-\infty, 7)$
Horizontal asymptote: $y = 7$

33. $\frac{1}{49}$ **34.** $\frac{1}{9}$ **35.** $\frac{1}{4}$ **36.** $\frac{1}{27}$ **37.** **(a)** 74% **(b)** 47% **38.** **(a)** 568.68 mm Hg **(b)** 178.27 mm Hg **39.** **(a)** 44 watts **(b)** 11.6 watts
40. **(a)** 34.99 cm^2 **(b)** 3.02 cm^2 **41.** 3.35 milligrams; 0.45 milligrams **42.** About 362 students

43. **(a)** 0.63 **(b)** 0.98
(c)

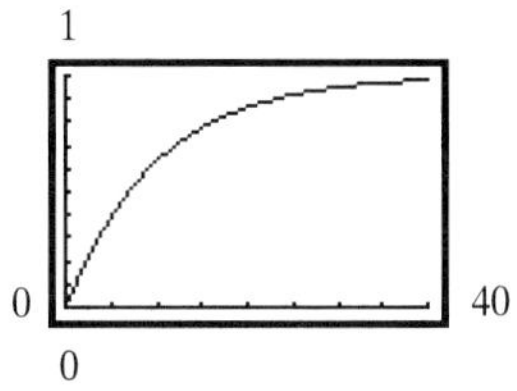

(d) About 7 minutes **(e)** 1

44. **(a)** 89.5% **(b)** 98.9%
(c)

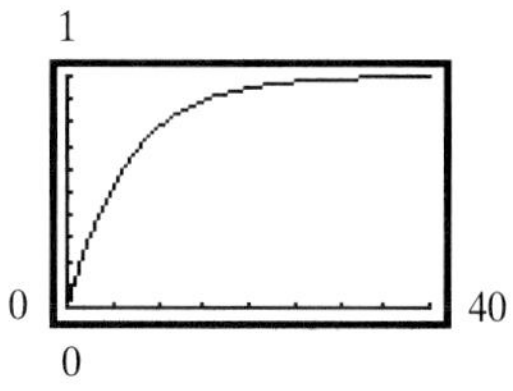

(d) About 6 minutes **(e)** 1

45. **(a)** 5.16% **(b)** 8.88%

46. **(a)** 15.6% **(b)** 3.0%

47. **(a)** 60.5% **(b)** 68.7% **(c)** 70%
(d)

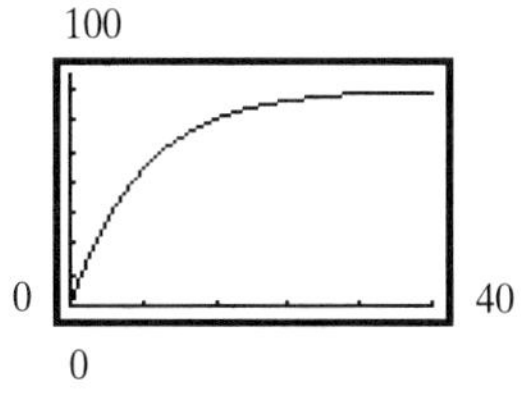

About $4\frac{1}{4}$ days

48. **(a)** \$98,125 **(b)** \$99,941.41 **(c)** \$100,000
(d)

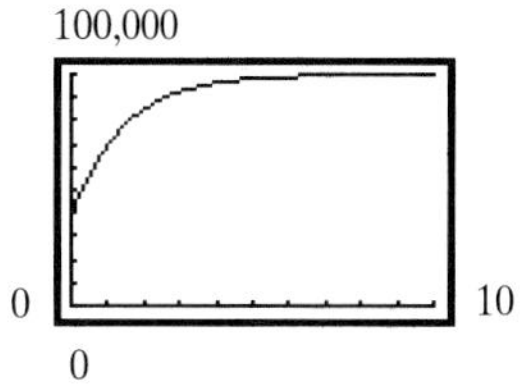

About $\frac{3}{4}$ year

49. **(a)** 5.414 amperes, 7.585 amperes, 10.376 amperes **(b)** 12 amperes
(d) 3.343 amperes, 5.309 amperes, 9.443 amperes **(e)** 24 amperes
(c), **(f)**

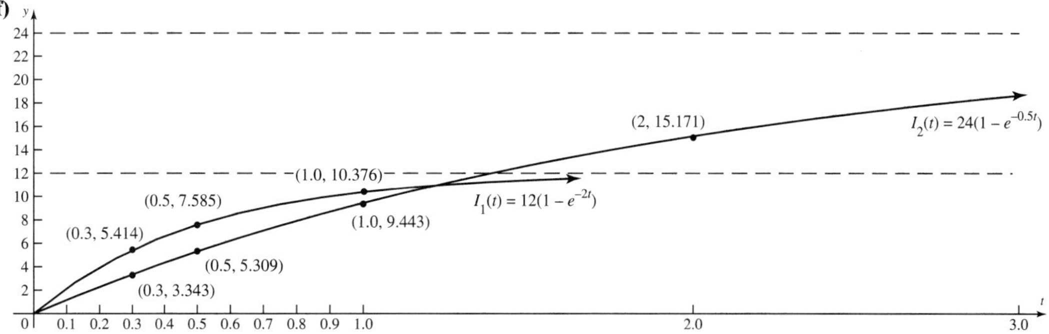

50. **(a)** 0.06 milliamperes, 0.0364 milliamperes, 0.0134 milliamperes **(b)** 0.06 milliamperes
(d) 0.12 milliamperes, 0.0728 milliamperes, 0.0268 milliamperes **(e)** 0.12 milliamperes
(c), (f)

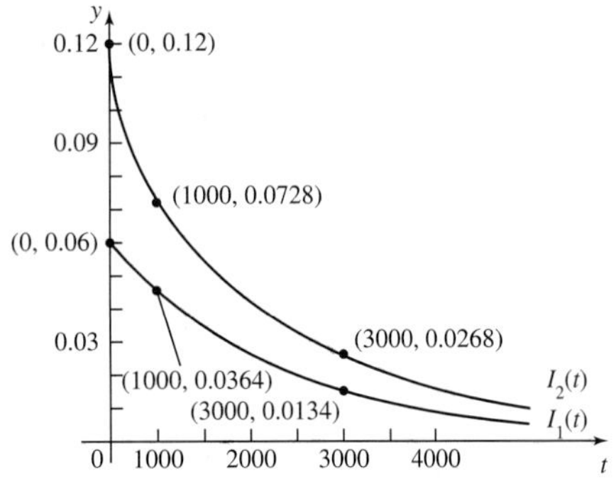

51. **(a)** 9.23×10^{-3}, or about 0
(b) 0.81, or about 1 **(c)** 5.01, or about 5
(d) 57.91°, 43.99°, 30.07°

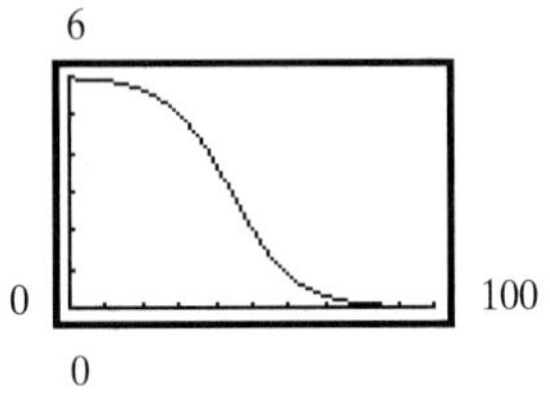

52. **(a)** About 3077 stamps
(b) About 3317 stamps
(c)

Year	Actual	Predicted
1848	2	78
1868	88	129
1888	218	212
1908	341	350

The predictions appear to get better with time.

53. $n = 4: 2.7083; n = 6: 2.7181; n = 8: 2.7182788; n = 10: 2.7182818$
54. 3.0, 2.666666, 2.727272, 2.71698, 2.71845, 2.71826; The expression gets closer to e, 2.71828182846.
55. $\dfrac{f(x+h) = f(x)}{h} = \dfrac{a^{x+h} - a^x}{h} = \dfrac{a^x a^h - a^x}{h} = \dfrac{a^x(a^h - 1)}{h}$ **56.** $f(A+B) = a^{A+B} = a^A \cdot a^B = f(A) \cdot f(B)$
57. $f(-x) = a^{-x} = \dfrac{1}{a^x} = \dfrac{1}{f(x)}$ **58.** $f(\alpha x) = a^{\alpha x} = (a^x)^\alpha = [f(x)]^\alpha$
59. $f(1) = 5, f(2) = 17, f(3) = 257, f(4) = 65{,}537, f(5) = 4{,}294{,}967{,}297 = 641 \times 6{,}700{,}417$ **60.** 59 minutes **62.** No

5.3 Exercises

1. $2 = \log_3 9$ **2.** $2 = \log_4 16$ **3.** $2 = \log_a 1.6$ **4.** $3 = \log_a 2.1$ **5.** $2 = \log_{1.1} M$ **6.** $3 = \log_{2.2} N$ **7.** $x = \log_2 7.2$ **8.** $x = \log_3 4.6$
9. $\sqrt{2} = \log_x \pi$ **10.** $\pi = \log_x e$ **11.** $x = \ln 8$ **12.** $2.2 = \ln M$ **13.** $2^3 = 8$ **14.** $3^{-2} = \dfrac{1}{9}$ **15.** $a^6 = 3$ **16.** $b^2 = 4$ **17.** $3^x = 2$
18. $2^x = 6$ **19.** $2^{1.3} = M$ **20.** $3^{2.1} = N$ **21.** $(\sqrt{2})^x = \pi$ **22.** $\pi^{1/2} = x$ **23.** $e^x = 4$ **24.** $e^4 = x$ **25.** 0 **26.** 1 **27.** 2 **28.** −2
29. −4 **30.** −2 **31.** $\dfrac{1}{2}$ **32.** $\dfrac{2}{3}$ **33.** 4 **34.** 4 **35.** $\dfrac{1}{2}$ **36.** 3 **37.** $\{x \mid x > 3\}$ **38.** $\{x \mid x > 1\}$ **39.** All real numbers except 0
40. $\{x \mid x > 0\}$ **41.** $\{x \mid x \neq 1\}$ **42.** $\{x \mid x < -1 \text{ or } x > 1\}$ **43.** $\{x \mid x > -1\}$ **44.** $\{x \mid x > 5\}$ **45.** $\{x \mid x < -1 \text{ or } x > 0\}$
46. $\{x \mid x < 0 \text{ or } x > 1\}$ **47.** 0.511 **48.** 0.536 **49.** 30.099 **50.** 4.055 **51.** $\sqrt{2}$ **52.** $\sqrt[4]{2}$ **53.** B **54.** F **55.** D **56.** H **57.** A **58.** C
59. E **60.** G **61.** C **62.** E **63.** A **64.** F **65.** D **66.** B

67.

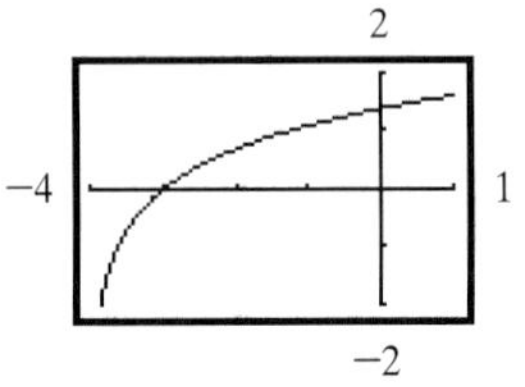

Domain: $(-4, \infty)$
Range: $(-\infty, \infty)$
Vertical asymptote: $x = -4$

68.

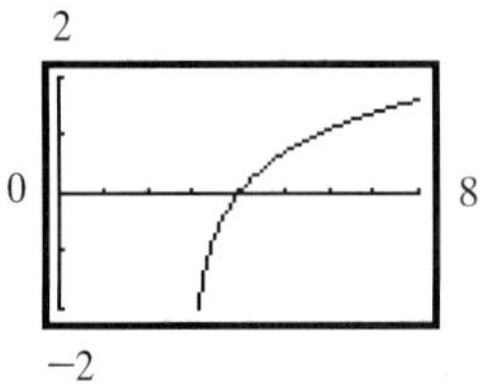

Domain: $(3, \infty)$
Range: $(-\infty, \infty)$
Vertical asymptote: $x = 3$

69.

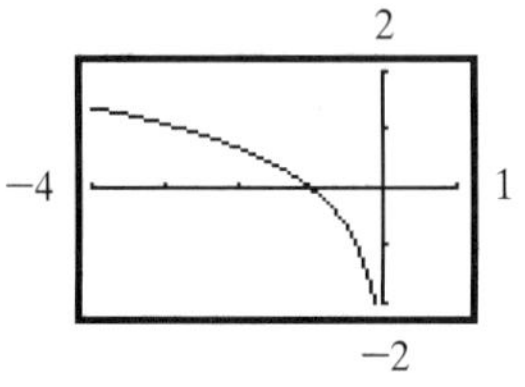

Domain: $(-\infty, 0)$
Range: $(-\infty, \infty)$
Vertical asymptote: $x = 0$

70.

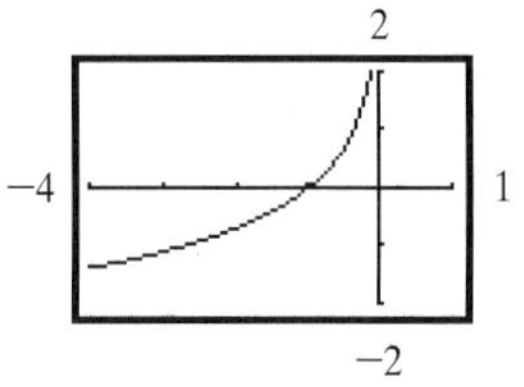

Domain: $(-\infty, 0)$
Range: $(-\infty, \infty)$
Vertical asymptote: $x = 0$

71.

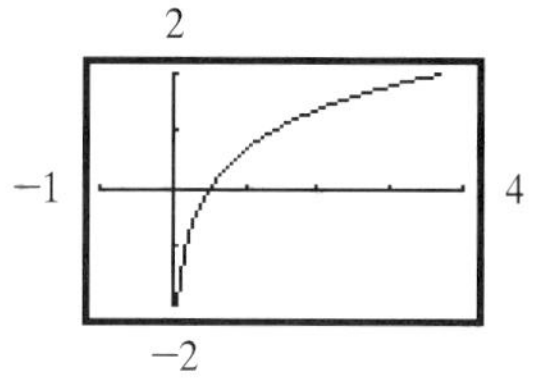

Domain: $(0, \infty)$
Range: $(-\infty, \infty)$
Vertical asymptote: $x = 0$

72.

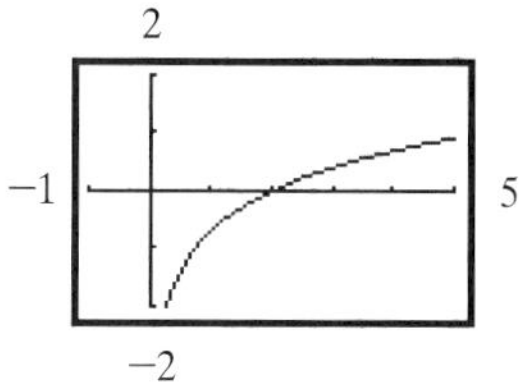

Domain: $(0, \infty)$
Range: $(-\infty, \infty)$
Vertical asymptote: $x = 0$

73.

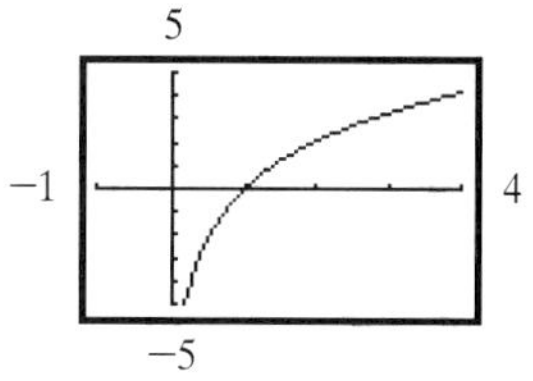

Domain: $(0, \infty)$
Range: $(-\infty, \infty)$
Vertical asymptote: $x = 0$

74.

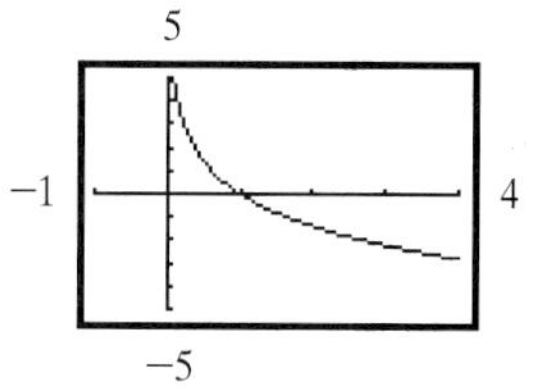

Domain: $(0, \infty)$
Range: $(-\infty, \infty)$
Vertical asymptote: $x = 0$

75.

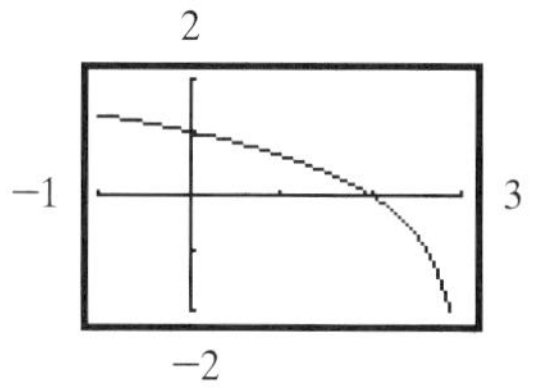

Domain: $(-\infty, 3)$
Range: $(-\infty, \infty)$
Vertical asymptote: $x = 3$

76.

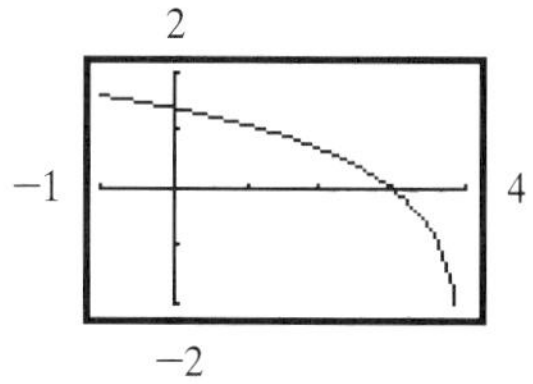

Domain: $(-\infty, 4)$
Range: $(-\infty, \infty)$
Vertical asymptote: $x = 4$

77.

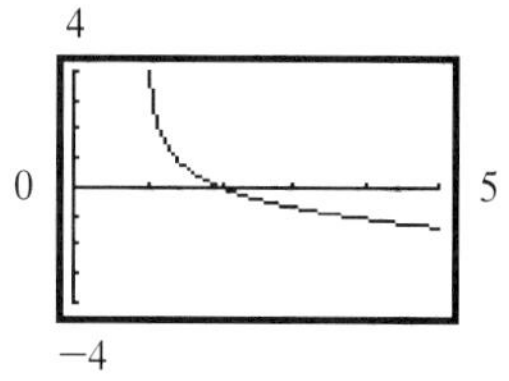

Domain: $(1, \infty)$
Range: $(-\infty, \infty)$
Vertical asymptote: $x = 1$

78.

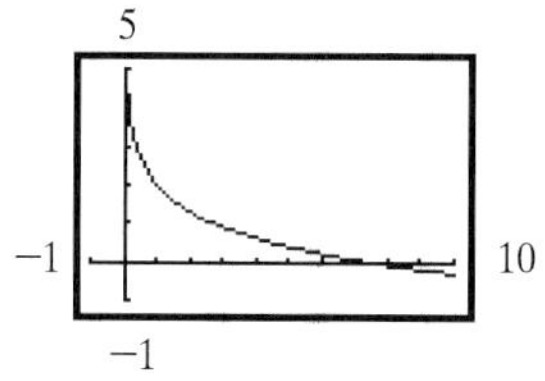

Domain: $(0, \infty)$
Range: $(-\infty, \infty)$
Vertical asymptote: $x = 0$

79. (a) $n \approx 6.93$ so 7 panes are necessary **(b)** $n \approx 13.86$ so 14 panes are necessary **80. (a)** 7 **(b)** 0.0000631 moles per liter
81. (a) $d \approx 127.7$ so it takes about 128 days **(b)** $d \approx 575.6$ so it takes about 576 days
82. (a) $n \approx 1.98$ so it takes about 2 days **(b)** $n \approx 6.58$ so it takes about $6\frac{1}{2}$ days
83. $h \approx 2.29$ so the time between injections is about 2 hours, 17 minutes **84.** $d \approx 3.99$ so it takes about 4 days

85. 0.2695 seconds
0.8959 seconds

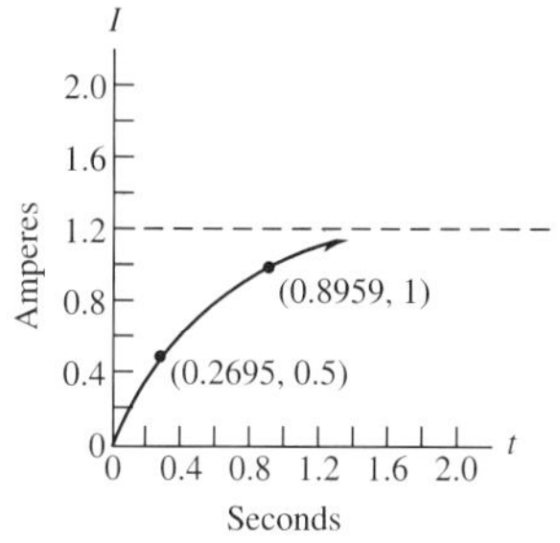

86. (a) 0.0211
(b) 38 words
(c) 54 words
(d) About 1 hour 49 minutes

87. 50 decibels
88. 90 decibels
89. 110 decibels
90. 80 decibels
91. 8.1
92. 6.9

93. (a) $k = 20.07$
(b) 91%
(c) 0.175
(d) 0.08
94. No

5.4 Exercises

1. $a + b$ **2.** $a - b$ **3.** $b - a$ **4.** $-a$ **5.** $a + 1$ **6.** $b - 1$ **7.** $2a + b$ **8.** $b + 3a$ **9.** $\frac{1}{5}(a + 2b)$ **10.** $\frac{1}{4}b + a$ **11.** $\frac{b}{a}$ **12.** $\frac{a}{b}$

13. $2 \ln x + \frac{1}{2}\ln(1 - x)$ **14.** $\ln x + \frac{1}{2}\ln(x^2 + 1)$ **15.** $3 \log_2 x - \log_2(x - 3)$ **16.** $\frac{1}{3}\log_5(x^2 + 1) - \log_5(x^2 - 1)$

17. $\log x + \log(x + 2) - 2\log(x + 3)$ **18.** $3 \log x + \frac{1}{2}\log(x + 1) - 2\log(x - 2)$ **19.** $\frac{1}{3}\ln(x - 2) + \frac{1}{3}\ln(x + 1) - \frac{2}{3}\ln(x + 4)$

20. $\frac{4}{3}\ln(x - 4) - \frac{2}{3}\ln(x^2 - 1)$ **21.** $\ln 5 + \ln x + \frac{1}{2}\ln(1 - 3x) - 3\ln(x - 4)$

22. $\ln 5 + 2\ln x + \frac{1}{3}\ln(1 - x) - \ln 4 - 2\ln(x + 1)$ **23.** $\log_5 u^3v^4$ **24.** $\log_3 \frac{u^2}{v}$ **25.** $-\frac{5}{2}\log_{1/2} x$ **26.** $-3\log_2 x$

27. $-2\ln(x - 1)$ **28.** $\log \frac{(x + 3)(x - 1)}{(x - 2)(x + 6)(x + 1)}$ **29.** $\log_2[x(3x - 2)^4]$ **30.** $9 \log_3 x$ **31.** $\log_a\left(\frac{25x^6}{\sqrt{2x + 3}}\right)$

32. $\log(\sqrt[3]{x^3 + 1}\sqrt{x^2 + 1})$ **33.** $\ln y = \ln a + (\ln b)x$ **34.** No; There is no slope or y-intercept. **35.** $\frac{5}{4}$ **36.** 42 **37.** 4 **38.** 3

39. 2.771 **40.** 1.796 **41.** −3.880 **42.** −3.907 **43.** 5.615 **44.** 2.584 **45.** 0.874 **46.** 0.303

47. $y = \left(\dfrac{\log x}{\log 4}\right)$

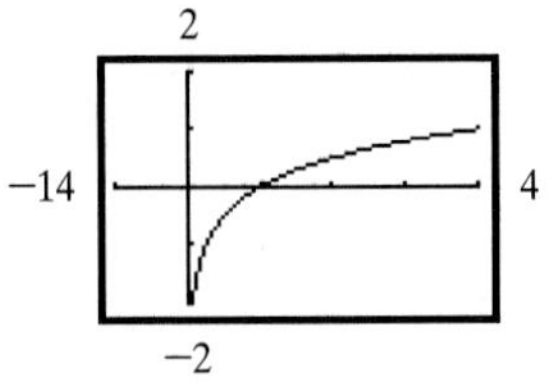

48. $y = \left(\dfrac{\log x}{\log 5}\right)$

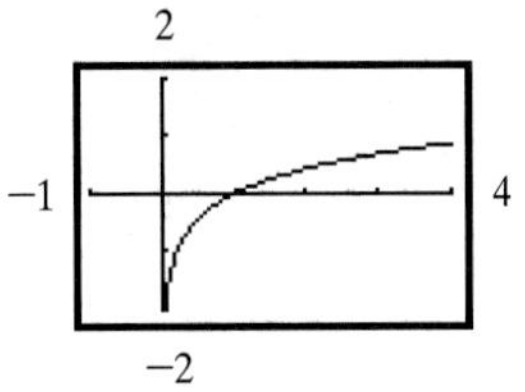

49. $y = \left(\dfrac{\log(x+2)}{\log 2}\right)$

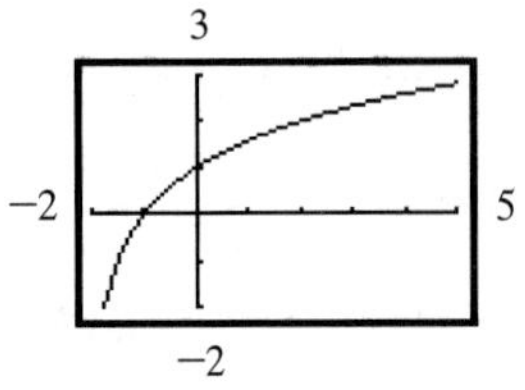

50. $y = \dfrac{\log(x-3)}{\log 4}$

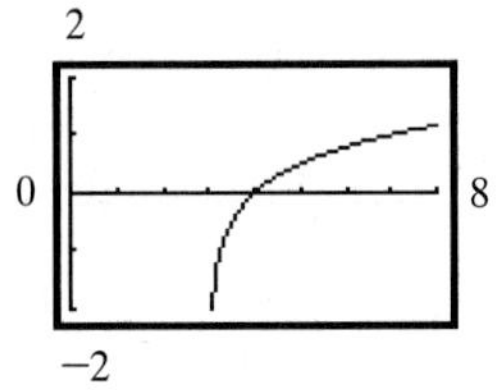

51. $y = \dfrac{\log(x+1)}{\log(x-1)}$

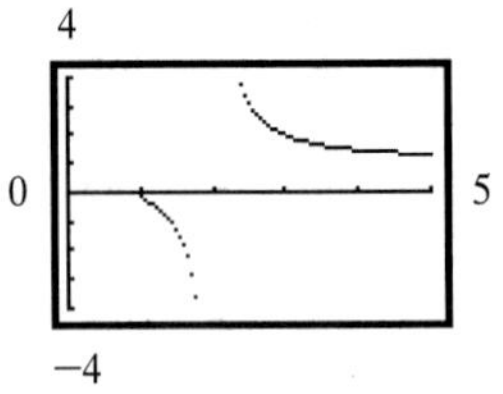

52. $y = \dfrac{\log(x-2)}{\log(x+2)}$

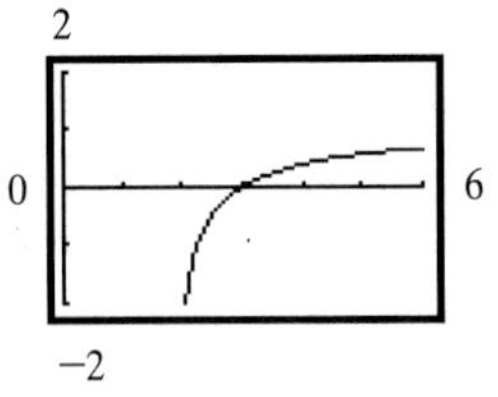

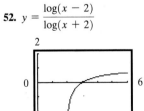

53. $y = Cx$ **54.** $y = x + C$ **55.** $y = Cx(x+1)$ **56.** $\dfrac{Cx^2}{x+1}$ **57.** $y = Ce^{3x}$ **58.** $y = Ce^{-2x}$ **59.** $y = Ce^{-4x} + 3$ **60.** $y = Ce^{5x} - 4$ **61.** $y = \dfrac{\sqrt[3]{C}(2x+1)^{1/6}}{(x+4)^{1/9}}$ **62.** $y = \sqrt{C}x^{-1/4}(x^2+1)^{1/6}$ **63.** 3 **64.** 3 **65.** 1 **66.** $n!$

67. If $A = \log_a M$ and $B = \log_a N$, then $a^A = M$ and $a^B = N$.

Then $\log_a\left(\dfrac{M}{N}\right) = \log_a\left(\dfrac{a^A}{a^B}\right) = \log_a a^{A-B} = A - B = \log_a M - \log_a N$.

68. $\log_a \dfrac{1}{N} = \log_a 1 - \log_a N = -\log_a N$

5.5 Exercises

1. $\frac{7}{2}$ **2.** $\frac{11}{3}$ **3.** $\{-2\sqrt{2}, 2\sqrt{2}\}$ **4.** $\left\{\frac{-1-\sqrt{85}}{2}, \frac{-1+\sqrt{85}}{2}\right\}$ **5.** 16 **6.** $\frac{1}{3}$ **7.** 8 **8.** $\frac{1}{3}$ **9.** 3 **10.** 5 **11.** 5 **12.** 4 **13.** 2 **14.** $\frac{1}{2}$ **15.** $\{-2, 4\}$ **16.** $\{-12, 4\}$ **17.** 21 **18.** −4 **19.** $\frac{1}{2}$ **20.** 1 **21.** $\{-\sqrt{2}, 0, \sqrt{2}\}$ **22.** $\left\{0, \frac{1}{2}\right\}$ **23.** $\left\{1 - \frac{\sqrt{6}}{3}, 1 + \frac{\sqrt{6}}{3}\right\}$ **24.** $\frac{1}{2}$ **25.** 0 **26.** 3 **27.** $\frac{\ln 3}{\ln 2} \approx 1.585$ **28.** 0 **29.** 0 **30.** 0 **31.** $\frac{3}{2}$ **32.** $\frac{3}{4}$ **33.** $\frac{\ln 10}{\ln 2} \approx 3.322$ **34.** $\frac{\ln 14}{\ln 3} \approx 2.402$ **35.** $-\frac{\ln 1.2}{\ln 8} \approx -0.088$ **36.** $-\frac{\ln 1.5}{\ln 2} \approx -0.585$ **37.** $\frac{\ln 3}{2\ln 3 + \ln 4} \approx 0.307$ **38.** $\frac{\ln 5 - \ln 2}{\ln 2 + 2\ln 5} \approx 0.234$ **39.** $\frac{\ln 7}{\ln 0.6 + \ln 7} \approx 1.356$ **40.** $\frac{\ln(4/3)}{\ln(4/3) + \ln 5} \approx 0.152$ **41.** 0 **42.** $\frac{\ln 0.3 + \ln 1.7}{2\ln 1.7 - \ln 0.3} \approx -0.297$ **43.** $\frac{\ln \pi}{1 + \ln \pi} \approx 0.534$ **44.** $\frac{-3}{1 - \ln \pi} \approx 20.728$ **45.** $\frac{\ln 1.6}{3\ln 2} \approx 0.226$ **46.** $\frac{5\ln(2/3)}{\ln 4} \approx -1.462$ **47.** $5\ln 1.5 \approx 2.027$ **48.** $\frac{\ln(6/5)}{0.3} \approx 0.608$ **49.** $\frac{9}{2}$ **50.** 4 **51.** 2 **52.** 67 **53.** −1 **54.** −1 **55.** 1 **56.** $\left\{\frac{26 - 8\sqrt{10}}{9}, \frac{26 + 8\sqrt{10}}{9}\right\}$ **57.** 16 **58.** 81 **59.** $\left\{-1, \frac{2}{3}\right\}$ **60.** $\left\{\frac{1}{4}, 4\right\}$ **61.** 1.92 **62.** 4.48 **63.** 2.79 **64.** 12.15 **65.** −0.57 **66.** {0.45, −1.98} **67.** −0.70 **68.** {1.86, 4.54} **69.** 0.57 **70.** 1.16 **71.** {0.39, 1.00} **72.** 0.65 **73.** 1.32 **74.** {0.05, 1.48} **75.** 1.31 **76.** 0.57

5.6 Exercises

1. \$108.29 **2.** \$59.83 **3.** \$609.50 **4.** \$358.84 **5.** \$697.09 **6.** \$789.24 **7.** \$12.46 **8.** \$49.35 **9.** \$125.23 **10.** \$156.83 **11.** \$88.72 **12.** \$59.14 **13.** \$860.72 **14.** \$626.61 **15.** \$554.09 **16.** \$266.08 **17.** \$59.71 **18.** \$654.98 **19.** \$361.93 **20.** \$886.92 **21.** 5.35% **22.** 6.82% **23.** 26% **24.** 7.18% **25.** $6\frac{1}{4}$% compounded annually **26.** 9% compounded quarterly **27.** 9% compounded monthly **28.** 7.9% compounded daily **29.** 104.32 months; 103.97 months **30.** 83.52 months; 83.18 months **31.** 61.02 months; 60.82 months **32.** 67.43 months; 67.15 months **33.** 15.27 years or 15 years, 4 months **34.** 16.62 years or 16 years, 8 months **35.** \$104,335 **36.** \$215.48 **37.** \$12,910.62 **38.** \$2955.39 **39.** About \$30.17 per share or \$3017 **40.** 15.47% **41.** 9.35% **42.** \$2315.97 **43.** Not quite. Jim will have \$1057.60. The second bank gives a better deal, since Jim will have \$1060.62 after 1 year. **44.** \$1021.60 **45.** Will has \$11,632.73; Henry has \$10,947.89. **46.** Take \$1000 now; it will be worth \$1349.86 in 3 years. **47. (a)** Interest is \$30,000 **(b)** Interest is \$38,613.59 **(c)** Interest is \$37,752.73. Simple interest at 12% is best. **48. (a)** 4.341344% **(b)** 4.341347% **49. (a)** \$1364.62 **(b)** \$1353.35 **50.** \$10,810.76 **51.** \$4631.93 **52.** 9.07%

59. **(a)** 6.1 years **(b)** 18.45 years

(c)
$$mP = P\left(1 + \frac{r}{n}\right)^{nt}$$
$$m = \left(1 + \frac{r}{n}\right)^{nt}$$
$$\ln m = \ln\left(1 + \frac{r}{n}\right)^{nt} = nt\ln\left(1 + \frac{r}{n}\right)$$
$$t = \frac{\ln m}{n\ln\left(1 + \frac{r}{n}\right)}$$

60. **(a)** 20.79 years **(b)** 7.74%

(c)
$$A = Pe^{rt}$$
$$\frac{A}{P} = e^{rt}$$
$$\ln\left(\frac{A}{P}\right) = rt$$
$$\ln A - \ln P = rt$$
$$t = \frac{\ln A - \ln P}{r}$$

5.7 Exercises

1. **(a)** 34.7 days; 69.3 days

(b)

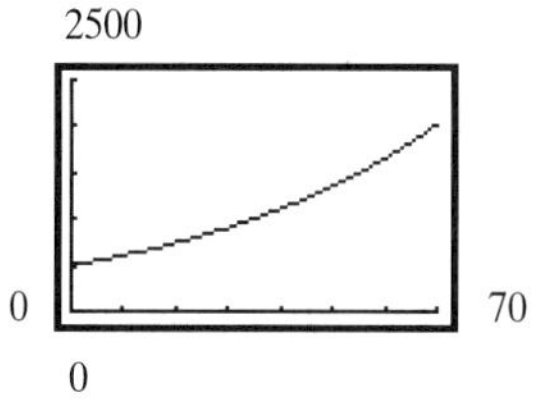

2. **(a)** 40.55 hours; 69.31 hours

(b)

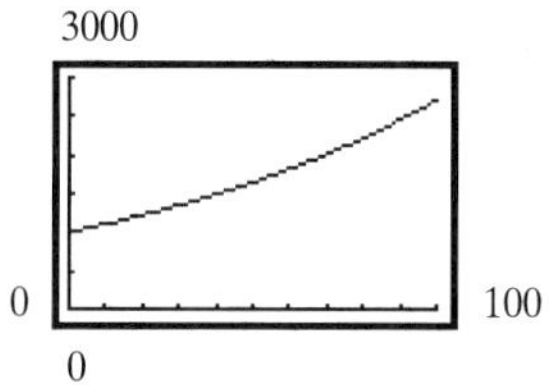

3. 28.4 years **4.** 7.97 days

5. **(a)** 94.4 years

(b)

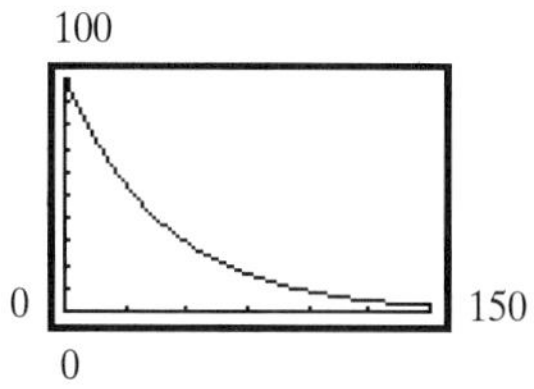

6. **(a)** 26.47 days

(b)

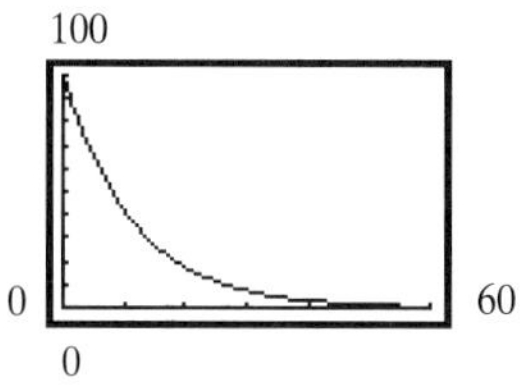

7. 5832; 3.9 days
8. 5243 bacteria; 7.85 hours
9. 25,198
10. 711,111
11. 9.797 grams
12. 9.99999947 grams; 9.9999947 grams

13. **(a)** 9727 years ago

(b)

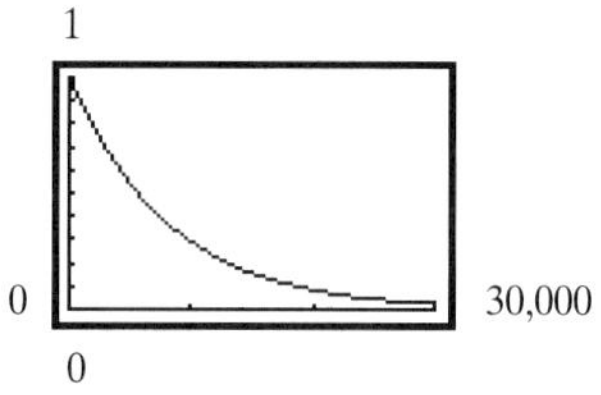

(c) 5600 years

14. **(a)** 2882 years old

(b)

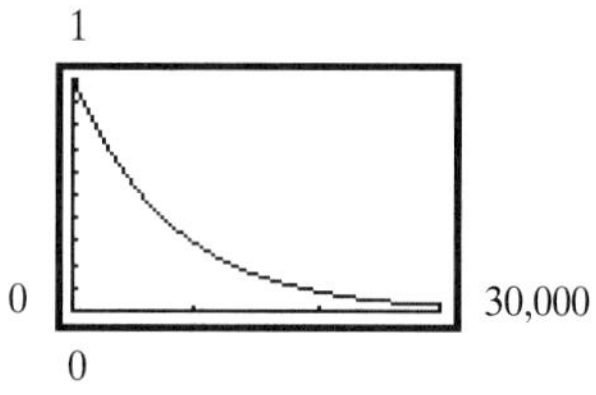

(c) 5600 years

15. **(a)** 5:18 PM

(b)

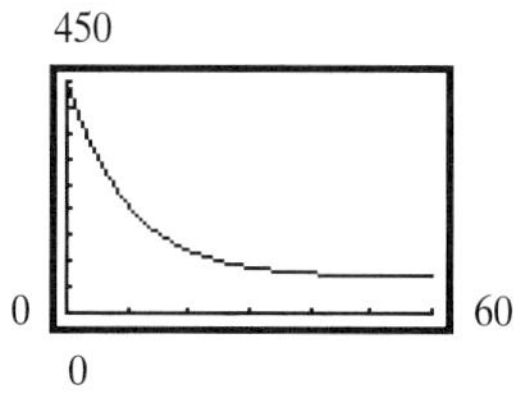

(c) After 14.3 minutes, the pizza will be 160°F.
(d) As time passes, the temperature of the pizza gets closer to 70°F.

16. 45.4°F

(a) 16.2 minutes

(b)

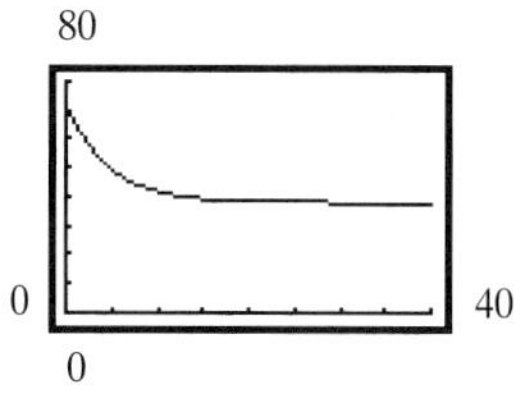

(c) 7.3 minutes
(d) As time passes, the temperature gets closer to 38°F.

17. **(a)** 18.63°C; 25.1°C
(b)

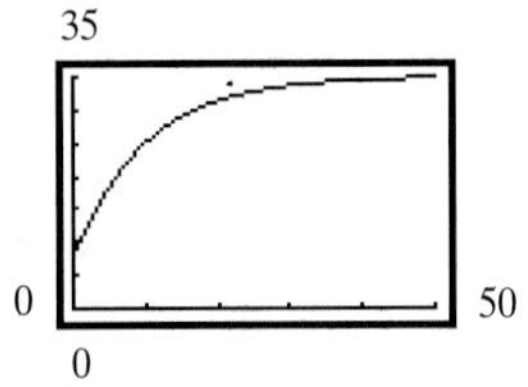

18. 45.7°F; 28.5 minutes

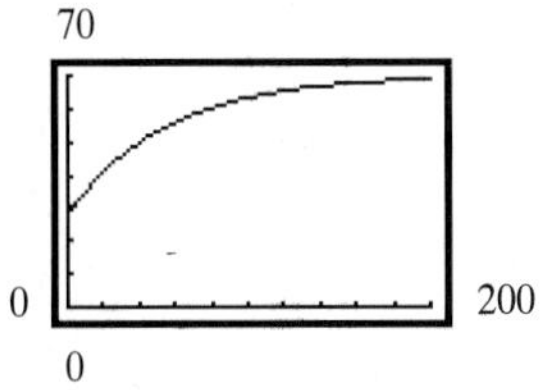

19. 7.34 kg; 76.6 hours
20. 1.25 volts
21. 26.5 days
22. 9:42 PM

23. **(a)** 0.1286 **(b)** 0.9
(c)

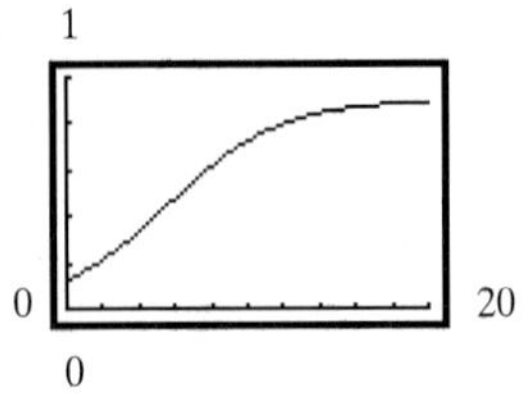

(d) 1996

24. **(a)** 0.2 **(b)** 0.9
(c)

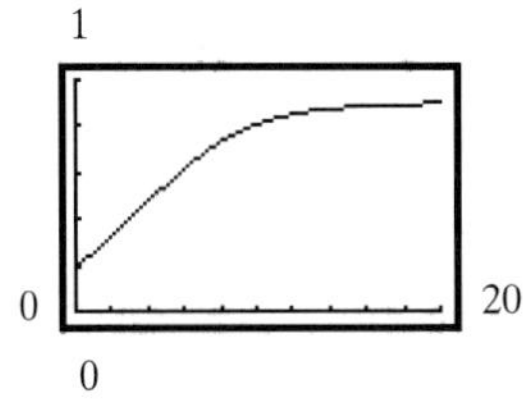

(d) 8.4 months after it has been introduced

25. **(a)**

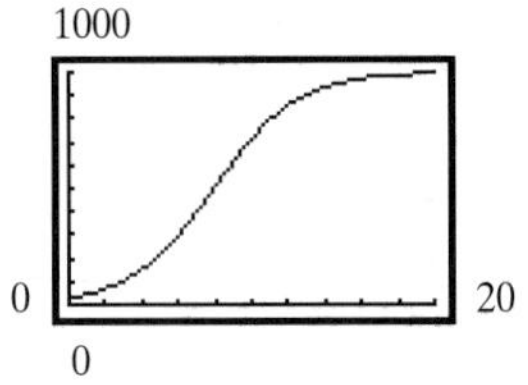

(b) 1000 **(c)** 30 **(d)** 11.076 hours

26. **(a)**

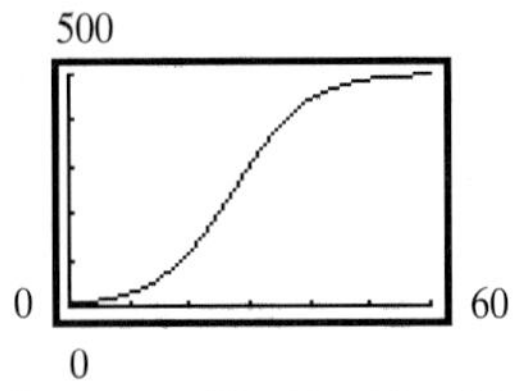

(b) 500 **(c)** 117 **(d)** After 29.8 years

5.8 Exercises

1. **(a)**

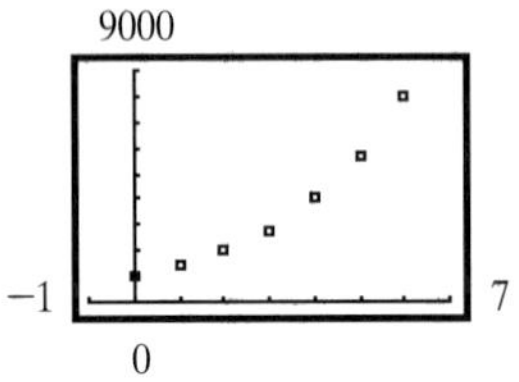

(b) $y = 1000.188(1.414)^x$ **(c)** $N = 1000.188e^{0.3465t}$ **(d)**

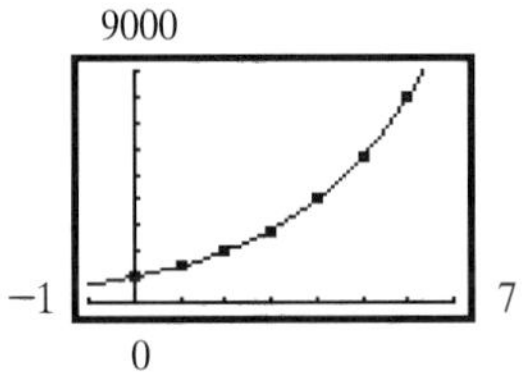

(e) 11,304

2. **(a)**

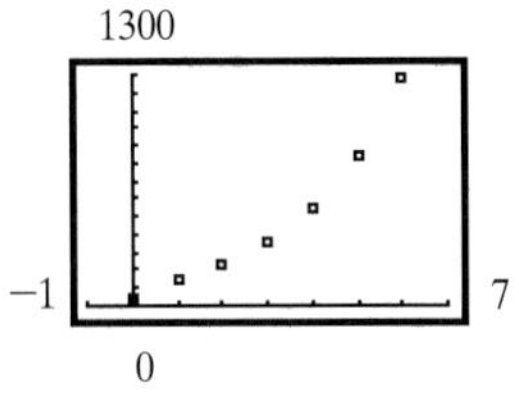

(b) $y = 72.46220(1.64757)^x$
(c) $N = 72.46220e^{0.49930t}$
(d)

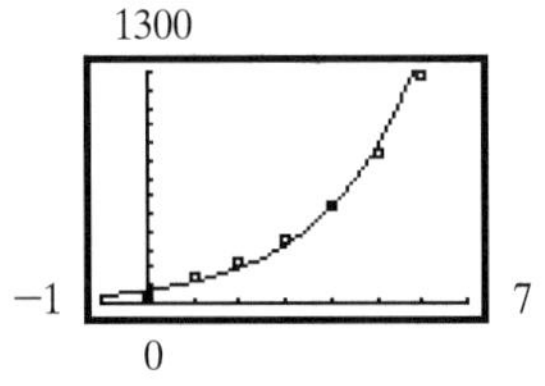

(e) 2388

3. **(a)**

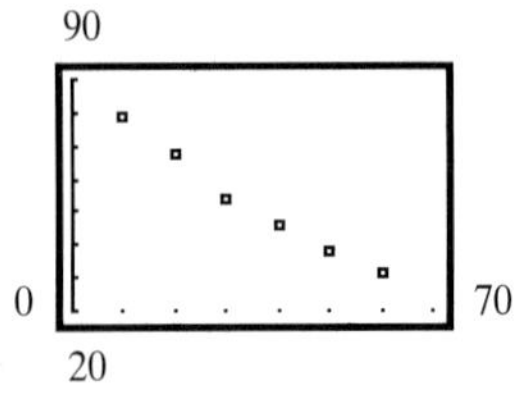

(b) $y = 96.01(0.9815)^x$ **(c)**

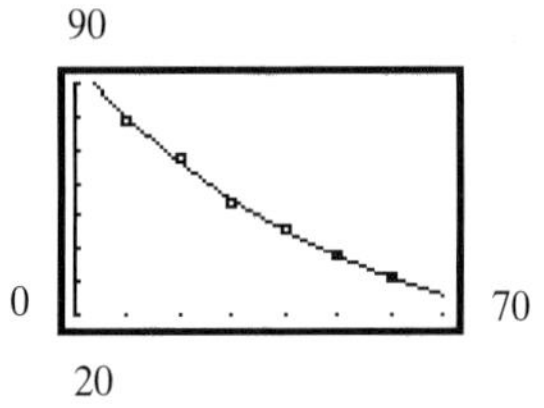

(d) $Q = 25$

4. (a)

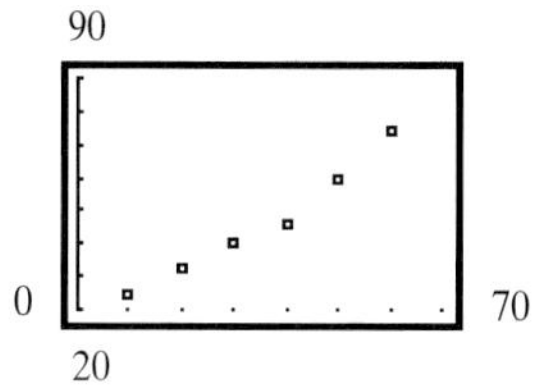

(b) $y = 20.52(1.0215)^x$ **(c)**

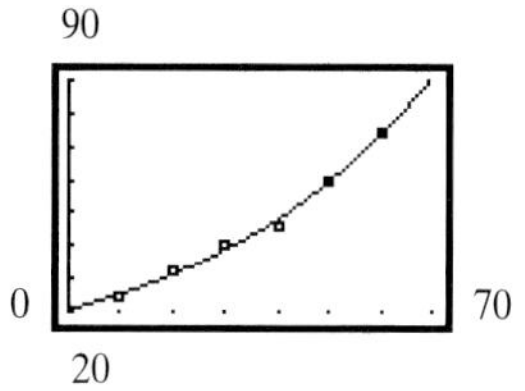

(d) $Q = 37$

5. (a)

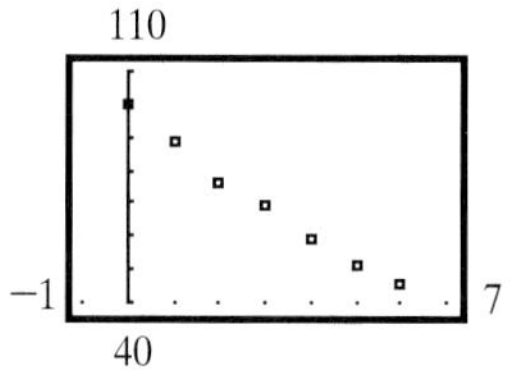

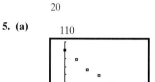

(b) $y = 100.326(0.877)^x$ **(c)** $A = 100.326e^{-0.1314t}$ **(d)**

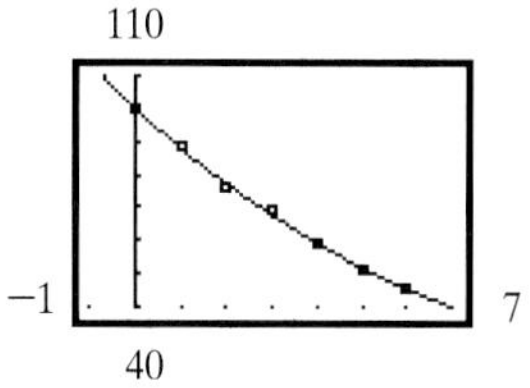

(e) 5.3 weeks **(f)** 0.14 grams

6. (a)

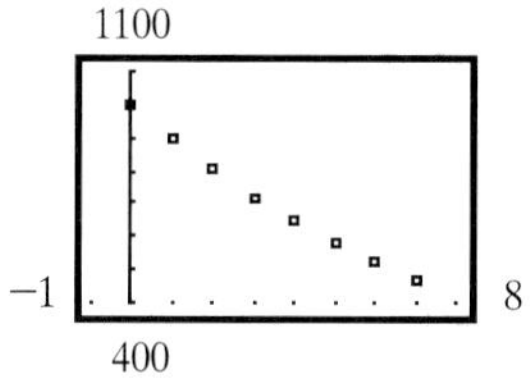

(b) $y = 998.90702(0.89757)^x$ **(c)** $A = 998.90702e^{-0.10807t}$ **(d)**

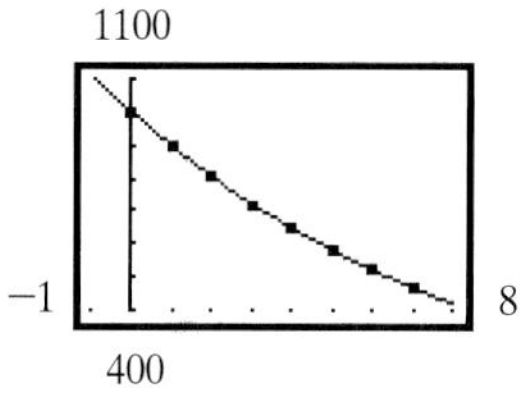

(e) 6.41 days **(f)** 115.0 grams

7. (a) x is the number of years after 1985

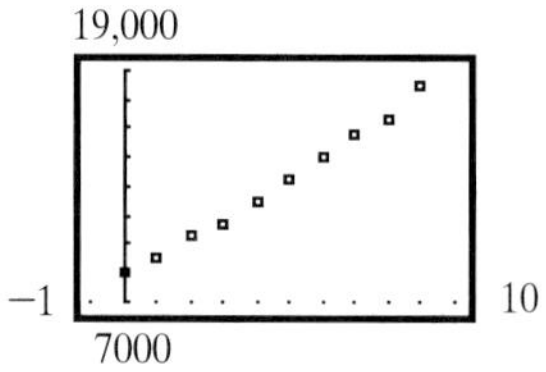

(b) $y = 10{,}014.4963(1.05655)^x$
(c) 5.655% **(d)** $68,672.21

8. (a) x is the number of years after 1988.

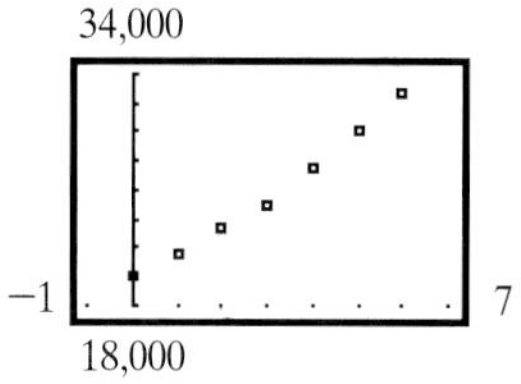

(b) $y = 19{,}819.1416(1.08550)^x$
(c) 8.55% **(d)** $273,677.65

9. (a) x is the number of months after the end of October, 1996.

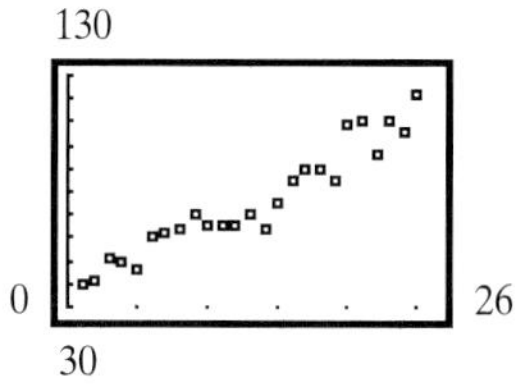

(b) $y = 41.8725(1.0427)^x$ **(c)** $A(t) = 41.8725e^{0.0418t}$

(d)

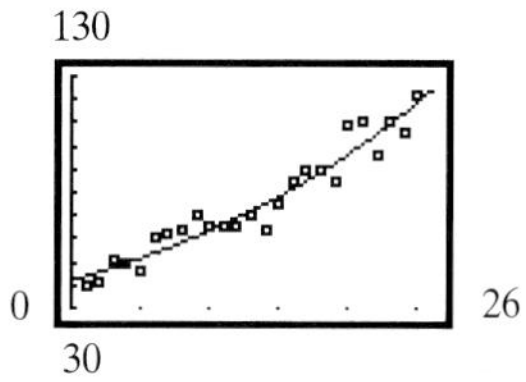

(e) $124.14

10. (a) x is the number of months after the end of October, 1996.

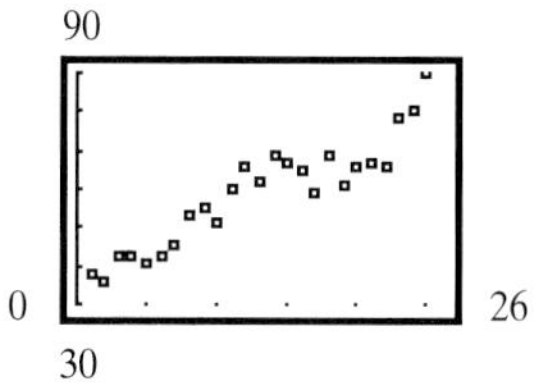

(b) $y = 38.0667(1.0319)^x$ **(c)** $A(t) = 86.0847e^{0.0314t}$

(d)

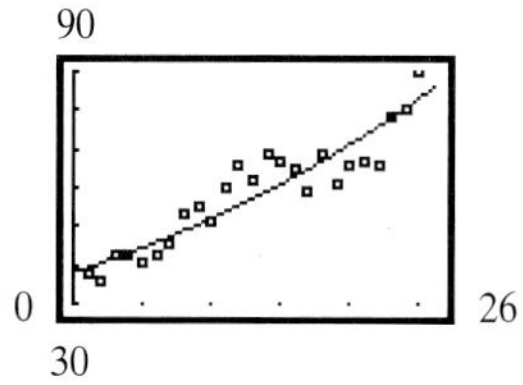

(e) $86.12

11. (a)

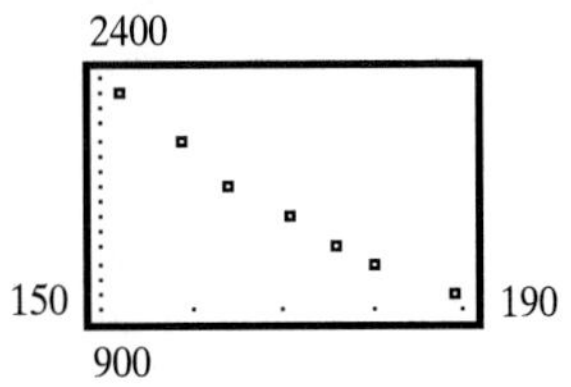

(b) $y = 32{,}741.02 - 6070.96 \ln x$

(c)

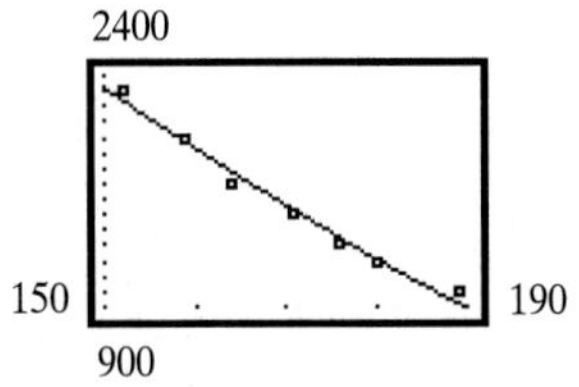

(d) Approximately 168 computers

12. (a)

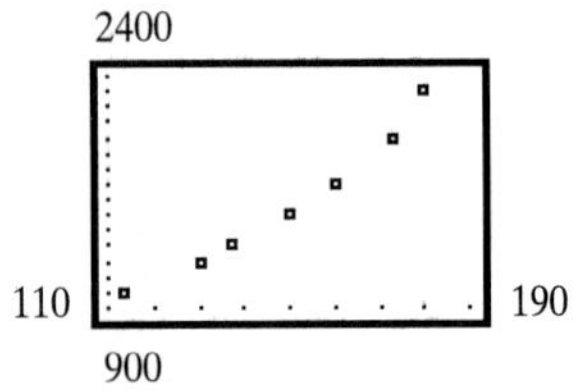

(b) $y = -11{,}850.72 + 2688.50 \ln x$

(c)

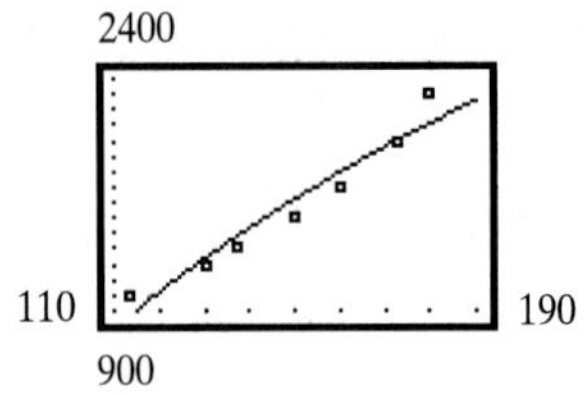

(d) Approximately 152 computers

13. (a)

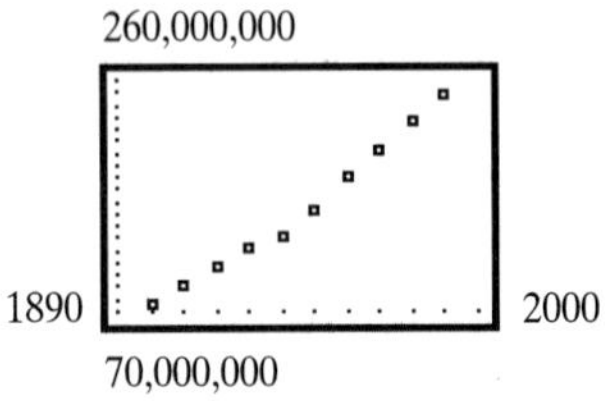

(b) $y = \dfrac{695{,}129{,}657.2}{1 + 4.1136 \times 10^{14} e^{-0.0166x}}$

(c)

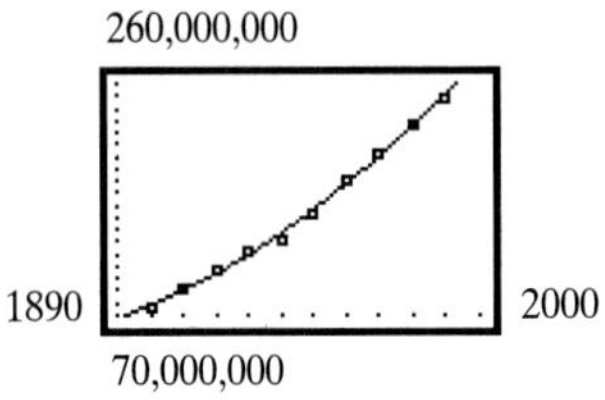

(d) 695,129,657 **(e)** 270,575,260

14. (a) $x = 1$ corresponds to 1981

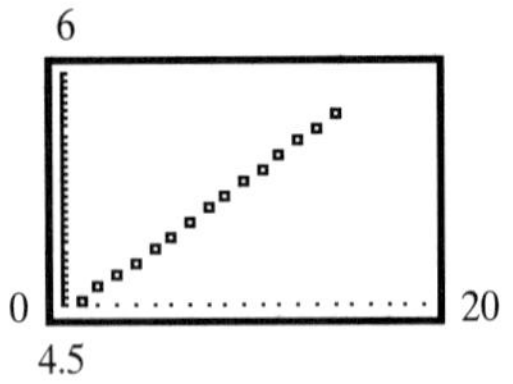

(b) $y = \dfrac{14.42090426}{1 + 2.2411 e^{-0.0261x}}$

(c)

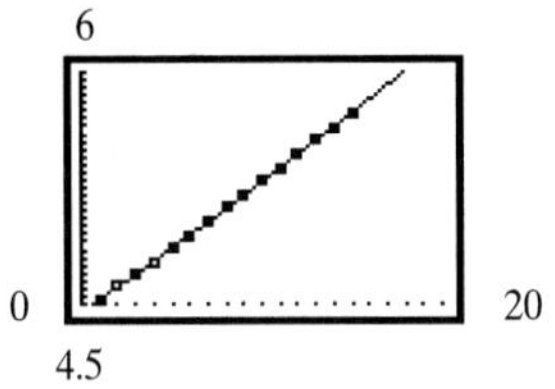

(d) 14.42090426 billion people
(e) 6.18998 billion people

15. (a)

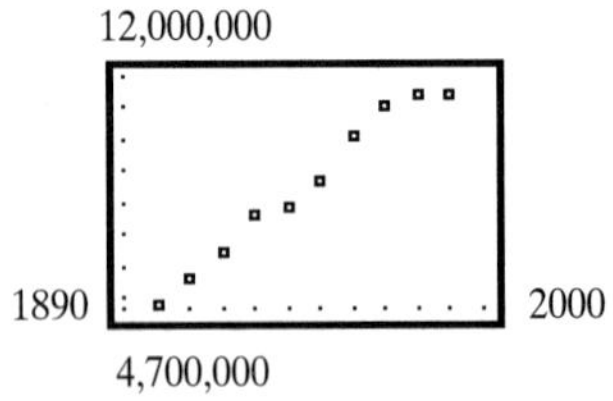

(b) $y = \dfrac{14{,}168{,}687.34}{1 + 1.5075745 \times 10^{21} e^{-0.02531x}}$

(c)

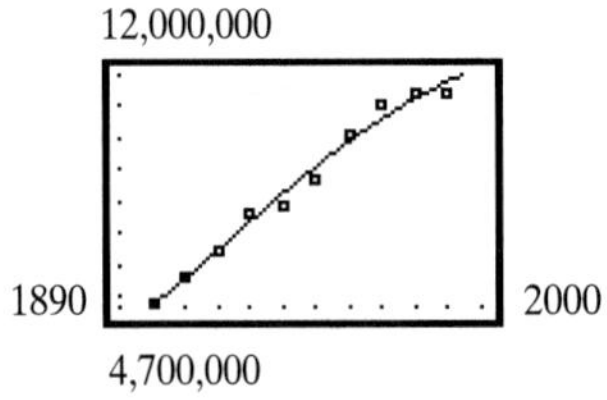

(d) 14,168,687 **(e)** Approximately 12,252,199

16. (a)

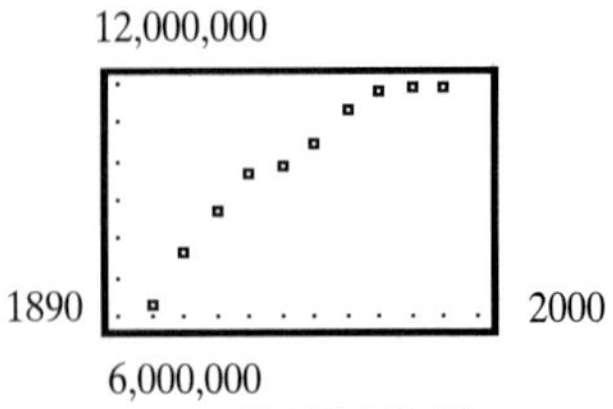

(b) $y = \dfrac{12{,}423{,}371.86}{1 + 4.5826685 \times 10^{29} e^{-0.03599x}}$

(c)

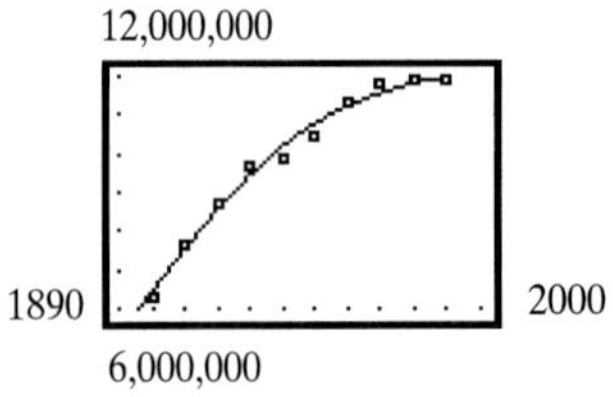

(d) 12,423,372 **(e)** Approximately 12,118,546

Fill-in-the Blank Items

1. One-to-one **2.** $y = x$ **3.** $(0, 1)$ and $(1, a)$ **4.** 1 **5.** 4 **6.** Sum **7.** 1 **8.** 7 **9.** All real numbers greater than 0 **10.** $(1, 0)$ and $(a, 1)$ **11.** 1 **12.** 7

True/False Items

1. False **2.** True **3.** True **4.** True **5.** False **6.** False **7.** True **8.** False **9.** False **10.** True

Review Exercises

1. $f^{-1}(x) = \dfrac{2x+3}{5x-2}$; $f(f^{-1}(x)) = \dfrac{2\left(\dfrac{2x+3}{5x-2}\right)+3}{5\left(\dfrac{2x+3}{5x-2}\right)-2} = x$; $f^{-1}(f(x)) = \dfrac{2\left(\dfrac{2x+3}{5x-2}\right)+3}{5\left(\dfrac{2x+3}{5x-2}\right)-2} = x$;

Domain f = Range f^{-1} = All real numbers except $\frac{2}{5}$; Range f = Domain f^{-1} = All real numbers except $\frac{2}{5}$

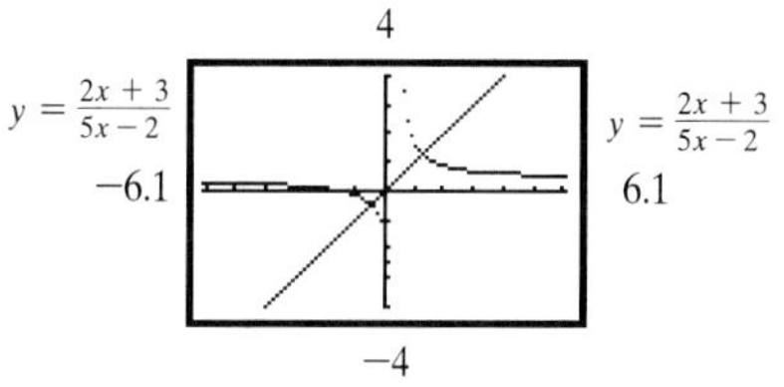

2. $f^{-1}(x) = \dfrac{2-3x}{x+1}$; $f(f^{-1}(x)) = \dfrac{2-\dfrac{2-3x}{x+1}}{3+\dfrac{2-3x}{x+1}} = x$; $f^{-1}(f(x)) = \dfrac{2-3\left(\dfrac{2-x}{3+x}\right)}{\dfrac{2-x}{3+x}+1} = x$;

Domain f = Range f^{-1} = All real numbers except -3; Range f = Domain f^{-1} = All real numbers except -1

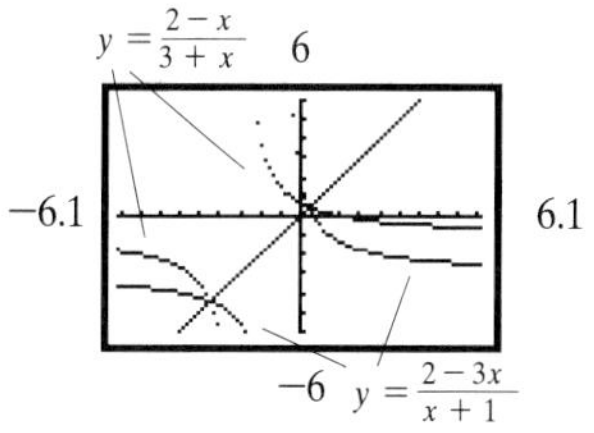

3. $f^{-1}(x) = \dfrac{x+1}{x}$; $f(f^{-1}(x)) = \dfrac{1}{\dfrac{x+1}{x}-1} = x$; $f^{-1}(f(x)) = \dfrac{\dfrac{1}{x-1}+1}{\dfrac{1}{x-1}} = x$;

Domain f = Range f^{-1} = All real numbers except 1; Range f = Domain f^{-1} = All real numbers except 0

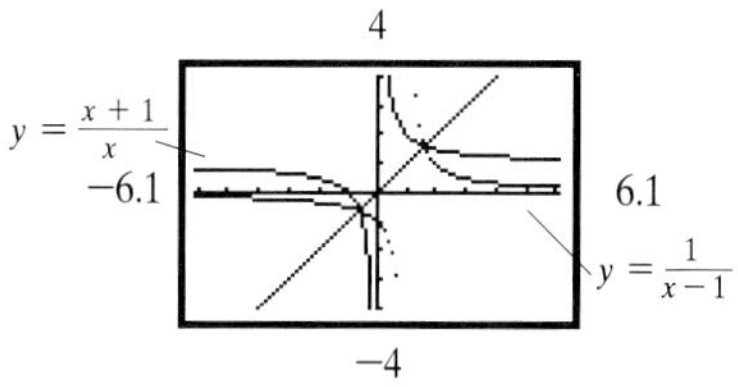

4. $f^{-1}(x) = x^2 + 2, x \geq 0; f(f^{-1}(x)) = \sqrt{(x^2 + 2) - 2} = |x| = x, x \geq 0; f^{-1}(f(x)) = (\sqrt{x - 2})^2 + 2 = x;$
Domain f = Range $f^{-1} = \{x | x \geq 2\}$; Range f = Domain $f^{-1} = \{x | x \geq 0\}$

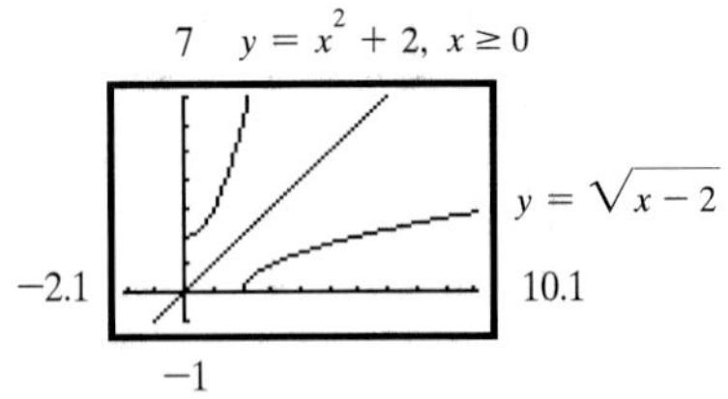

5. $f^{-1}(x) = \frac{27}{x^3}; f(f^{-1}(x)) = \frac{3}{\left(\frac{27}{x^3}\right)^{1/3}} = x; f^{-1}(f(x)) = \frac{27}{\left(\frac{3}{x^{1/3}}\right)^3} = x;$
Domain f = Range f^{-1} = All real numbers except 0; Range f = Domain f^{-1} = All real numbers except 0

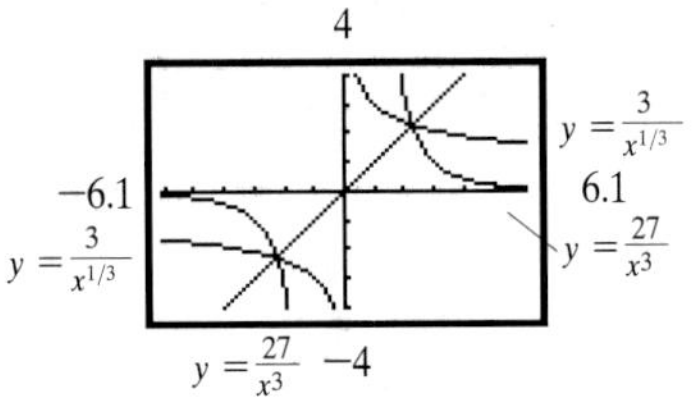

6. $f^{-1}(x) = (x - 1)^3; f(f^{-1}(x)) = [(x - 1)^3]^{1/3} + 1 = x; f^{-1}(f(x)) = [(x^{1/3} + 1) - 1]^3 = x;$
Domain f = Range f^{-1} = All real numbers; Range f = Domain f^{-1} = All real numbers

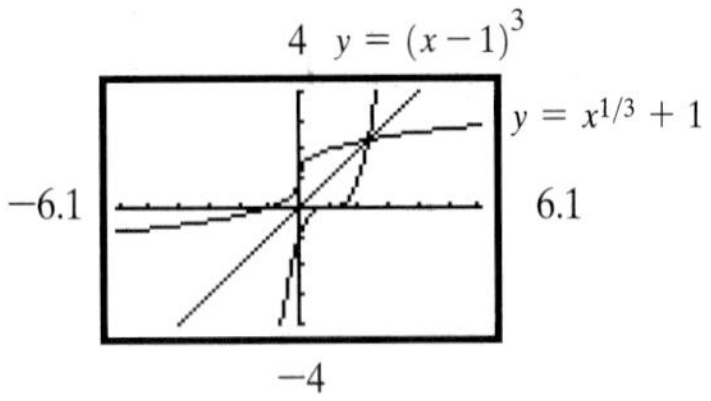

7. -3 **8.** 4 **9.** $\sqrt{2}$ **10.** 0.1 **11.** 0.4 **12.** $\sqrt{3}$ **13.** $\frac{25}{4}\log_4 x$ **14.** $\frac{13}{6}\log_3 x$ **15.** $-2\ln(x + 1)$ **16.** $\log\frac{x - 3}{x + 4}$
17. $\log\left(\frac{4x^3}{[(x + 3)(x - 2)]^{1/2}}\right)$ **18.** $\ln\left(16\sqrt{\frac{x^2 + 1}{x(x - 4)}}\right)$ **19.** $y = Ce^{2x^2}$ **20.** $y = 3 + 2Cx^2$ **21.** $y = (Ce^{3x^2})^2$
22. $y = \frac{C(x^2 + 3x + 2)}{2}$ **23.** $y = \sqrt{e^{x+C} + 9}$ **24.** $y = \sqrt{e^{-x+C} + 1}$ **25.** $y = \ln(x^2 + 4) - C$ **26.** $y = \frac{C + \ln(x + 4)^2}{3}$

27.
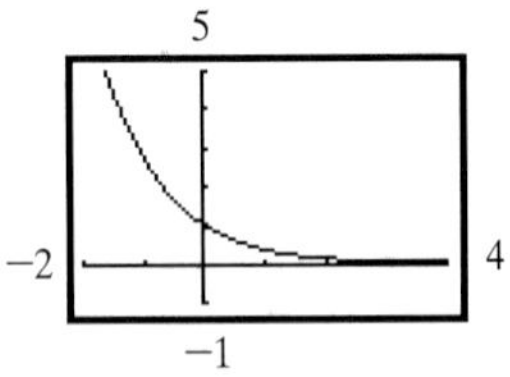

Domain: $(-\infty, \infty)$
Range: $(0, \infty)$
Asymptote: $y = 0$

28.
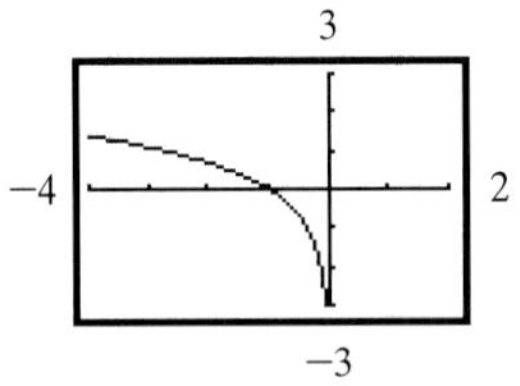

Domain: $(-\infty, 0)$
Range: $(-\infty, \infty)$
Asymptote: $x = 0$

29.
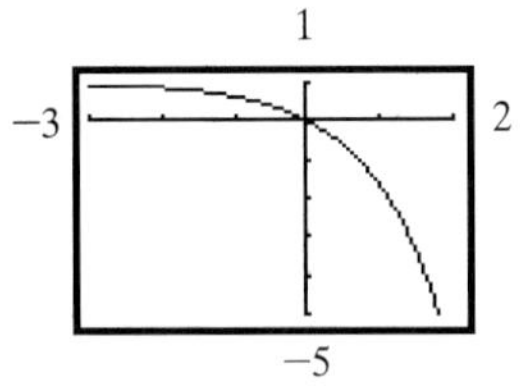

Domain: $(-\infty, \infty)$
Range: $(-\infty, 1)$
Asymptote: $y = 1$

30.
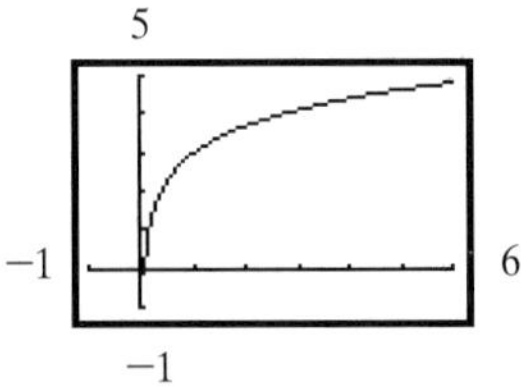

Domain: $(0, \infty)$
Range: $(-\infty, \infty)$
Asymptote: $x = 0$

31.
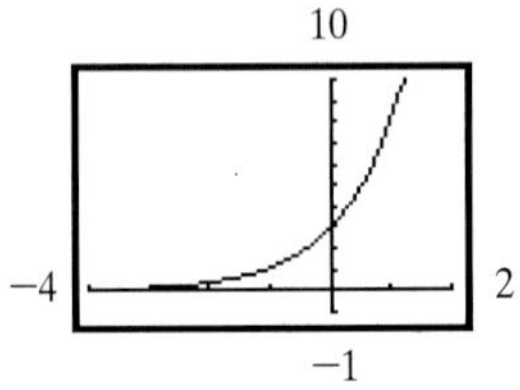

Domain: $(-\infty, \infty)$
Range: $(0, \infty)$
Asymptote: $y = 0$

32.
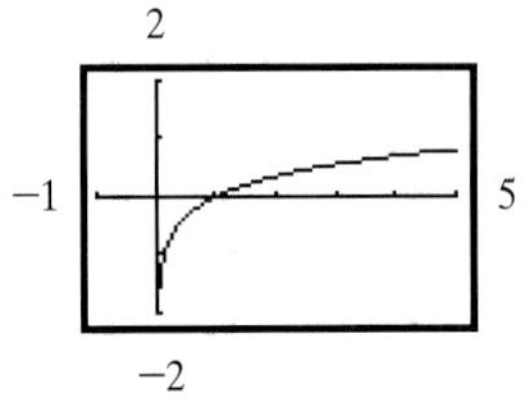

Domain: $(0, \infty)$
Range: $(-\infty, \infty)$
Asymptote: $x = 0$

33.
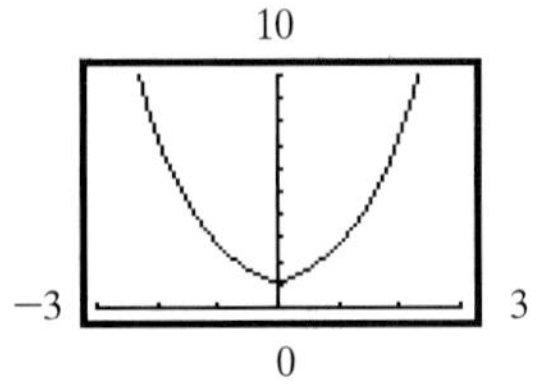

Domain: $(-\infty, \infty)$
Range: $[1, \infty)$
Asymptote: None

34.
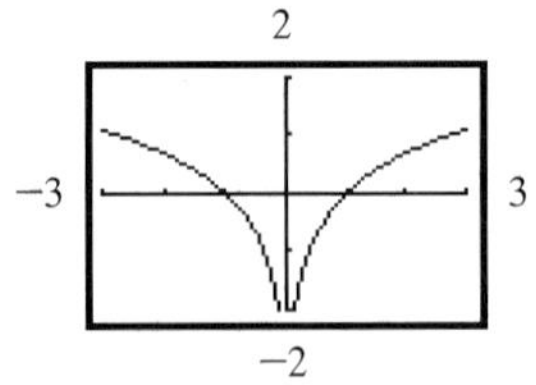

Domain: $\{x | x \neq 0\}$
Range: $(-\infty, \infty)$
Asymptote: $x = 0$

35.

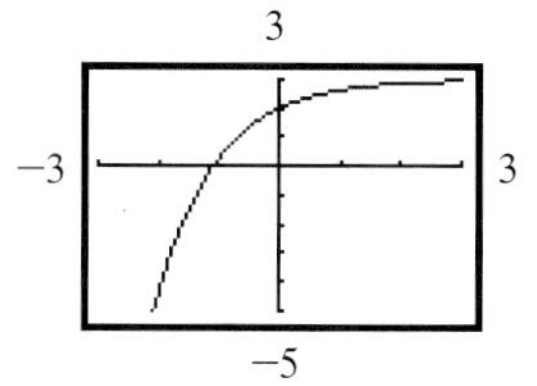

Domain: $(-\infty, \infty)$
Range: $(-\infty, 3)$
Asymptote: $y = 3$

36.

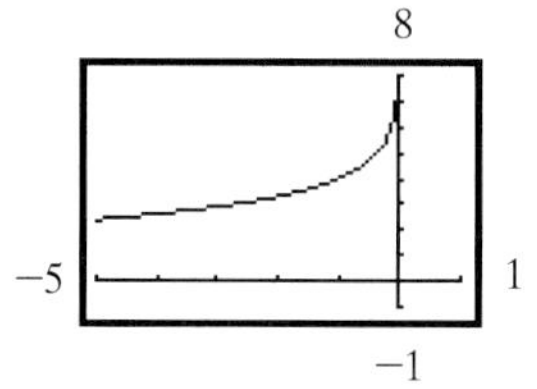

Domain: $(-\infty, 0)$
Range: $(-\infty, \infty)$
Asymptote: $x = 0$

37. $\frac{1}{4}$ **38.** $-\frac{16}{9}$
39. $\left\{\frac{-1-\sqrt{3}}{2}, \frac{-1+\sqrt{3}}{2}\right\}$
40. $\left\{\frac{1+\sqrt{3}}{2}, \frac{1-\sqrt{3}}{2}\right\}$ **41.** $\frac{1}{4}$ **42.** $\frac{1}{8}$
43. $\frac{2\ln 3}{\ln 5 - \ln 3} \approx 4.301$ **44.** $\frac{2(\ln 7 + \ln 5)}{\ln 7 - \ln 5} \approx 21.1331$

45. $\frac{12}{5}$ **46.** $\{-2, 6\}$ **47.** 83 **48.** $-\frac{1}{2}$ **49.** $\left\{-3, \frac{1}{2}\right\}$ **50.** 1 **51.** -1 **52.** $\{3, 4\}$ **53.** $1 - \ln 5 \approx -0.609$ **54.** $\frac{1}{2} - \ln 2 \approx -0.193$
55. $\frac{\ln 3}{3\ln 2 - 2\ln 3} \approx -9.327$ **56.** $\left\{0, \frac{\ln 3}{\ln 2}\right\}$ or $\{0, 1.585\}$ **57.** 3229.5 meters **58.** 1482 meters **59.** 7.6 mm of mercury
60. 61.5 mm Hg **61.** **(a)** 37.3 watts **(b)** 6.9 decibels **62.** **(a)** 11.8 **(b)** 9.56 inches **63.** **(a)** 71% **(b)** 85.5% **(c)** 90% **(d)** About 1.6 months **(e)** About 4.8 months **64.** About 4.2 months **65.** **(a)** 9.85 years **(b)** 4.27 years **66.** \$20,399 **67.** \$41,669 **68.** **(a)** $r = 10.436\%$ **(b)** \$32,249.24 **69.** 24,203 years ago **70.** 55.2 minutes

71. **(a)**

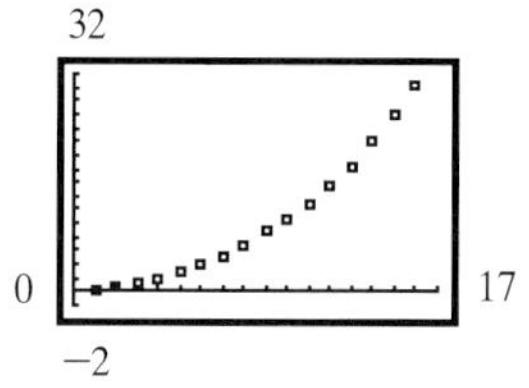

(b) $y = \frac{55.24286446}{1 + 69.63173349e^{-0.2760889967x}}$

(c)

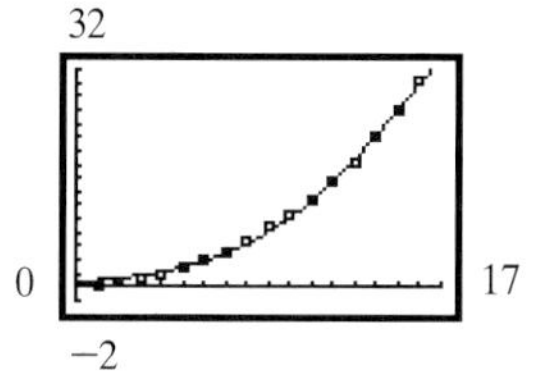

(d) Approximately 55.24 million

(e) $y = 0.4510626632(1.339709)^x$
(f)

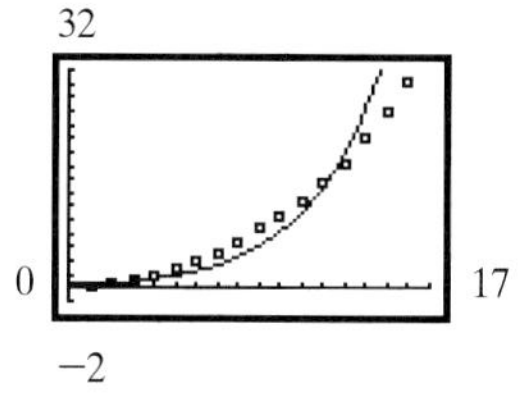

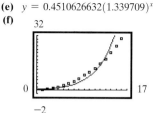

(g) Logistic model

CHAPTER 6 Systems of Equations and Inequalities

6.1 Exercises

1. $2(2) - (-1) = 5$ and $5(2) + 2(-1) = 8$ **2.** $3(-2) + 2(4) = 2$ and $(-2) - 7(4) = -30$
3. $3(2) - 4\left(\frac{1}{2}\right) = 4$ and $\frac{1}{2}(2) - 3\left(\frac{1}{2}\right) = -\frac{1}{2}$ **4.** $2\left(-\frac{1}{2}\right) + \frac{1}{2}(2) = 0$ and $3\left(-\frac{1}{2}\right) - 4(2) = -\frac{19}{2}$ **5.** $4 - 1 = 3$ and $\frac{1}{2}(4) + 1 = 3$
6. $-2 - (-5) = 3$ and $-3(-2) + (-5) = 1$ **7.** $3(1) + 3(-1) + 2(2) = 4$, $1 - (-1) - 2 = 0$, and $2(-1) - 3(2) = -8$
8. $4(2) - 1 = 7$, $8(2) + 5(-3) - 1 = 0$, and $-2 - (-3) + 5(1) = 6$ **9.** $x = 6, y = 2$ **10.** $x = 1, y = 2$ **11.** $x = 3, y = 2$
12. $x = -1, y = 2$ **13.** $x = 8, y = -4$ **14.** $x = -\frac{13}{4}, y = 2$ **15.** $x = \frac{1}{3}, y = -\frac{1}{6}$ **16.** $x = -\frac{5}{3}, y = 1$ **17.** Inconsistent
18. Inconsistent **19.** $x = 1, y = 2$ **20.** $x = 1, y = -\frac{4}{3}$ **21.** $x = 4 - 2y$, y is any real number **22.** $y = 3x - 7$, x is any real number
23. $x = 1, y = 1$ **24.** $x = \frac{1}{5}, y = \frac{3}{10}$ **25.** $x = \frac{3}{2}, y = 1$ **26.** $x = 2, y = -3$ **27.** $x = 4, y = 3$ **28.** $x = 12, y = 6$
29. $x = \frac{4}{3}, y = \frac{1}{5}$ **30.** $x = \frac{1}{2}, y = 2$ **31.** $x = \frac{1}{5}, y = \frac{1}{3}$ **32.** $x = 4, y = 3$ **33.** $x = 48.15, y = 15.18$ **34.** $x = 49.50, y = 14.26$
35. $x = -21.48, y = 16.12$ **36.** $x = -20.23, y = 6.03$ **37.** $x = 0.26, y = 0.07$ **38.** $x = -0.28, y = 0.16$
39. $p = \$16, q = 600$ T-shirts **40.** $p = \$3; q = 7000$ hot dogs **41.** Length 30 feet; width 15 feet **42.** 725 meters by 775 meters
43. Cheeseburger \$1.55; shakes \$0.85 **44.** 110 adults, 215 senior citizens **45.** 22.5 pounds
46. **(a)** \$90,000 in AA bonds and \$60,000 in a Bank Certificate **(b)** \$130,000 in AA bonds and \$20,000 in a Bank Certificate
47. Average wind speed 25 mph; average airspeed 175 mph **48.** Wind speed was 30 mph **49.** 80 \$25 sets and 120 \$45 sets
50. Hot dog \$1.00, soft drink \$0.50 **51.** \$5.56 **52.** Average speed of the swimmer is 4 mph; speed of current is 1 mph **53.** 5000
54. 7500 **55.** $b = -\frac{1}{2}, c = \frac{3}{2}$ **56.** $b = -3, c = 5$

6.2 Exercises

1. $x = 8, y = 2, z = 0$ **2.** $x = -3, y = 2, z = 1$ **3.** $x = 2, y = -1, z = 1$ **4.** $x = \frac{56}{13}, y = -\frac{7}{13}, z = \frac{35}{13}$ **5.** Inconsistent **6.** Inconsistent **7.** $x = 5z - 2, y = 4z - 3$ where z is any real number, or $x = \frac{5}{4}y + \frac{7}{4}, z = \frac{1}{4}y + \frac{3}{4}$, where y is any real number, or $y = \frac{4}{5}x - \frac{7}{5}, z = \frac{1}{5}x + \frac{2}{5}$ where x is any real number. **8.** $x = -\frac{4}{13}z + \frac{6}{13}, y = -\frac{7}{13}z + \frac{4}{13}$, where z is any real number **9.** Inconsistent **10.** Inconsistent **11.** $x = 1, y = 3, z = -2$ **12.** $x = 1, y = 3, z = -2$ **13.** $x = -3, y = \frac{1}{2}, z = 1$ **14.** $x = 3, y = -\frac{8}{3}, z = \frac{1}{9}$ **15.** $a = \frac{4}{3}, b = -\frac{5}{3}, c = 1$ **16.** $a = 3, b = -1, c = -6$ **17.** $I_1 = \frac{10}{71}, I_2 = \frac{65}{71}, I_3 = \frac{55}{71}$ **18.** $I_1 = \frac{11}{13}, I_2 = \frac{6}{13}, I_3 = \frac{17}{13}$ **19.** 100 orchestra, 210 main, and 190 balcony seats **20.** 110 adults, 220 children, 75 senior citizens **21.** 1.5 chicken, 1 corn, 2 milks **22.** \$8000 in Treasury bills, \$7000 in Treasury bonds, \$5000 in corporate bonds

6.3 Exercises

1. $\left[\begin{array}{cc|c} 1 & -5 & 5 \\ 4 & 3 & 6 \end{array}\right]$ **2.** $\left[\begin{array}{cc|c} 3 & 4 & 7 \\ 4 & -2 & 5 \end{array}\right]$ **3.** $\left[\begin{array}{cc|c} 2 & 3 & 6 \\ 4 & -6 & -2 \end{array}\right]$ **4.** $\left[\begin{array}{cc|c} 9 & -1 & 0 \\ 3 & -1 & 4 \end{array}\right]$ **5.** $\left[\begin{array}{cc|c} 0.01 & -0.03 & 0.06 \\ 0.13 & 0.10 & 0.20 \end{array}\right]$ **6.** $\left[\begin{array}{cc|c} \frac{4}{3} & -\frac{3}{2} & \frac{3}{4} \\ -\frac{1}{4} & \frac{1}{3} & \frac{2}{3} \end{array}\right]$

7. $\left[\begin{array}{ccc|c} 1 & -1 & 1 & 10 \\ 3 & 2 & 0 & 5 \\ 1 & 1 & 2 & 2 \end{array}\right]$ **8.** $\left[\begin{array}{ccc|c} 5 & -1 & -1 & 0 \\ 1 & 1 & 0 & 5 \\ 2 & 0 & -3 & 2 \end{array}\right]$ **9.** $\left[\begin{array}{ccc|c} 1 & 1 & -1 & 2 \\ 3 & -2 & 0 & 2 \\ 5 & 3 & -1 & 1 \end{array}\right]$ **10.** $\left[\begin{array}{ccc|c} 2 & 3 & -4 & 0 \\ 1 & 0 & -5 & -2 \\ 1 & 2 & -3 & -2 \end{array}\right]$ **11.** $\left[\begin{array}{ccc|c} 1 & -3 & -5 & -2 \\ 0 & 1 & 6 & 9 \\ 0 & 0 & 13 & 36 \end{array}\right]$

12. $\left[\begin{array}{ccc|c} 1 & -3 & -3 & -3 \\ 0 & 1 & 8 & 2 \\ 0 & 0 & 51 & 11 \end{array}\right]$ **13.** $\left[\begin{array}{ccc|c} 1 & -3 & 4 & 3 \\ 0 & 1 & -2 & 0 \\ 0 & 0 & 4 & 15 \end{array}\right]$ **14.** $\left[\begin{array}{ccc|c} 1 & -3 & -3 & -5 \\ 0 & 1 & 3 & 5 \\ 0 & 0 & 28 & 46 \end{array}\right]$ **15.** $\left[\begin{array}{ccc|c} 1 & -3 & 2 & -6 \\ 0 & 1 & -1 & 8 \\ 0 & 0 & -5 & 108 \end{array}\right]$ **16.** $\left[\begin{array}{ccc|c} 1 & -3 & -4 & -6 \\ 0 & 1 & 14 & 6 \\ 0 & 0 & 104 & 36 \end{array}\right]$

17. $\left[\begin{array}{ccc|c} 1 & -3 & 1 & -2 \\ 0 & 1 & 4 & 2 \\ 0 & 0 & 39 & 16 \end{array}\right]$ **18.** $\left[\begin{array}{ccc|c} 1 & -3 & -1 & 2 \\ 0 & 1 & 4 & 2 \\ 0 & 0 & 61 & 42 \end{array}\right]$ **19.** $\left[\begin{array}{ccc|c} 1 & -3 & -2 & 3 \\ 0 & 1 & 6 & -7 \\ 0 & 0 & 64 & -62 \end{array}\right]$ **20.** $\left[\begin{array}{ccc|c} 1 & -3 & 5 & -3 \\ 0 & 1 & -9 & 2 \\ 0 & 0 & -35 & 9 \end{array}\right]$

21. $\begin{cases} x = 5 \\ y = -1 \end{cases}$ consistent; $x = 5, y = -1$ **22.** $\begin{cases} x = -4 \\ y = 0 \end{cases}$ consistent; $x = -4, y = 0$ **23.** $\begin{cases} x = 1 \\ y = 2 \\ 0 = 3 \end{cases}$ inconsistent **24.** $\begin{cases} x = 0 \\ y = 0 \\ 0 = 2 \end{cases}$ inconsistent **25.** $\begin{cases} x + 2z = -1 \\ y - 4z = -2 \\ 0 = 0 \end{cases}$ consistent; $x = -1 - 2z$, $y = -2 + 4z$, z is any real number **26.** $\begin{cases} x + 4z = 4 \\ y + 3z = 2 \\ 0 = 0 \end{cases}$ consistent; $x = 4 - 4z$, $y = 2 - 3z$, z is any real number

27. $x = 6, y = 2$ **28.** $x = 1, y = 2$ **29.** $x = 2, y = 3$ **30.** $x = -1, y = 2$ **31.** $x = 4, y = -2$ **32.** $x = 2, y = 3$ **33.** Inconsistent **34.** Inconsistent **35.** $x = \frac{1}{2}, y = \frac{3}{4}$ **36.** $x = \frac{1}{3}, y = \frac{2}{3}$ **37.** $x = 4 - 2y$, y is any real number **38.** $y = 3x - 7$, x is any real number **39.** $x = \frac{3}{2}, y = 1$ **40.** $x = 2, y = -3$ **41.** $x = \frac{4}{3}, y = \frac{1}{5}$ **42.** $x = \frac{1}{2}, y = 2$ **43.** $x = 8, y = 2, z = 0$ **44.** $x = -3, y = 2, z = 1$ **45.** $x = 2, y = -1, z = 1$ **46.** $x = \frac{56}{13}, y = -\frac{7}{13}, z = \frac{35}{13}$ **47.** Inconsistent **48.** Inconsistent **49.** $x = 5z - 2, y = 4z - 3$ where z is any real number, or $x = \frac{5}{4}y + \frac{7}{4}, z = \frac{1}{4}y + \frac{3}{4}$, where y is any real number, or $y = \frac{4}{5}x - \frac{7}{5}, z = \frac{1}{5}x + \frac{2}{5}$ where x is any real number **50.** $x = -\frac{4}{13}z + \frac{6}{13}, y = -\frac{7}{13}z + \frac{4}{13}$ where z is any real number **51.** Inconsistent **52.** Inconsistent **53.** $x = 1, y = 3, z = -2$ **54.** $x = 1, y = 3, z = -2$ **55.** $x = 3, y = \frac{1}{2}, z = 1$ **56.** $x = 3, y = -\frac{8}{3}, z = \frac{1}{9}$ **57.** $x = \frac{1}{3}, y = \frac{2}{3}, z = 1$ **58.** $x = \frac{1}{3}, y = \frac{2}{3}, z = 1$ **59.** $y = 0, z = 1 - x$, x is any real number **60.** Inconsistent **61.** $x = 1, y = 2, z = 0, w = 1$ **62.** $x = 2, y = 1, z = 0, w = 1$ **63.** $y = -2x^2 + x + 3$ **64.** $y = x^2 - 4x + 2$ **65.** $f(x) = 3x^3 - 4x^2 + 5$ **66.** $f(x) = x^3 - 2x^2 + 6$ **67.** 1.5 salmon steak, 2 baked eggs, 1 acorn squash **68.** 1 serving pork chops, 2 servings corn, 2 glasses milk **69.** \$4000 in Treasury bills, \$4000 in Treasury bonds, \$2000 in corporate bonds **70.** \$12,000 in Treasury bills, \$2000 in Treasury bonds, and \$6000 in Corporate bonds. **71.** 8 Deltas, 5 Betas, 10 Sigmas **72.** 6 cases of orange juice, 20 cases grapefruit juice, 12 cases tomato juice **73.** $I_1 = \frac{44}{23}, I_2 = 2, I_3 = \frac{16}{23}, I_4 = \frac{28}{23}$ **74.** $I_1 = 3.5, I_2 = 2.5, I_3 = 1$

6.4 Exercises

1. 2 **2.** 7 **3.** 22 **4.** 28 **5.** −2 **6.** −2 **7.** 10 **8.** −169 **9.** −26 **10.** −119 **11.** $x = 6, y = 2$ **12.** $x = \frac{11}{3}, y = \frac{2}{3}$

13. $x = 3, y = 2$ **14.** $x = -1, y = 2$ **15.** $x = 8, y = -4$ **16.** $x = -\frac{13}{4}, y = 2$ **17.** $x = 4, y = -2$ **18.** $x = 2, y = 3$

19. Not applicable **20.** Not applicable **21.** $x = \frac{1}{2}, y = \frac{3}{4}$ **22.** $x = \frac{1}{3}, y = \frac{2}{3}$ **23.** $x = \frac{1}{10}, y = \frac{2}{5}$ **24.** $x = \frac{1}{5}, y = \frac{3}{10}$

25. $x = \frac{3}{2}, y = 1$ **26.** $x = 2, y = -3$ **27.** $x = \frac{4}{3}, y = \frac{1}{5}$ **28.** Not applicable **29.** $x = 1, y = 3, z = -2$ **30.** $x = 1, y = 3, z = -2$

31. $x = -3, y = \frac{1}{2}, z = 1$ **32.** $x = 3, y = -\frac{8}{3}, z = \frac{1}{9}$ **33.** Not applicable **34.** Not applicable **35.** $x = 0, y = 0, z = 0$

36. $x = 0, y = 0, z = 0$ **37.** Not applicable **38.** Not applicable **39.** $x = \frac{1}{5}, y = \frac{1}{3}$ **40.** $x = 4, y = 3$ **41.** −5 **42.** 1 or −1

43. $\frac{13}{11}$ **44.** $-\frac{7}{6}$ **45.** 0 or −9 **46.** 0 or $-\frac{1}{2}$ **47.** −4 **48.** 8 **49.** 12 **50.** 4 **51.** 8 **52.** 4 **53.** 8 **54.** 12

55.
$$\begin{aligned}(y_1 - y_2)x - (x_1 - x_2)y + (x_1y_2 - x_2y_1) &= 0\\ (y_1 - y_2)x + (x_2 - x_1)y &= x_2y_1 - x_1y_2\\ (x_2 - x_1)y - (x_2 - x_1)y_1 &= (y_2 - y_1)x + x_2y_1 - x_1y_2 - (x_2 - x_1)y_1\\ (x_2 - x_1)(y - y_1) &= (y_2 - y_1)x - (y_2 - y_1)x_1\\ y - y_1 &= \frac{y_2 - y_1}{x_2 - x_1}(x - x_1)\end{aligned}$$

56. This result follows from Problem 55 by exchanging rows.

57.
$$\begin{vmatrix} x^2 & x & 1\\ y^2 & y & 1\\ z^2 & z & 1\end{vmatrix} = x^2\begin{vmatrix} y & 1\\ z & 1\end{vmatrix} - x\begin{vmatrix} y^2 & 1\\ z^2 & 1\end{vmatrix} + \begin{vmatrix} y^2 & y\\ z^2 & z\end{vmatrix} = x^2(y - z) - x(y^2 - z^2) + yz(y - z)$$
$$= (y - z)[x^2 - x(y + z) + yz] = (y - z)[(x^2 - xy) - (xz - yz)] = (y - z)[x(x - y) - z(x - y)]$$
$$= (y - z)(x - y)(x - z)$$

58. Follow hint given in problem statement.

59.
$$\begin{vmatrix} a_{13} & a_{12} & a_{11}\\ a_{23} & a_{22} & a_{21}\\ a_{33} & a_{32} & a_{31}\end{vmatrix} = a_{13}(a_{22}a_{31} - a_{32}a_{21}) - a_{12}(a_{23}a_{31} - a_{33}a_{21}) + a_{11}(a_{23}a_{32} - a_{33}a_{22})$$
$$= [a_{11}(a_{22}a_{33} - a_{32}a_{23}) - a_{12}(a_{21}a_{33} - a_{31}a_{23}) + a_{13}(a_{21}a_{32} - a_{31}a_{22}) = -\begin{vmatrix} a_{11} & a_{12} & a_{13}\\ a_{21} & a_{22} & a_{23}\\ a_{31} & a_{32} & a_{33}\end{vmatrix}$$

60.
$$\begin{vmatrix} a_{11} & a_{12} & a_{13}\\ ka_{21} & ka_{22} & ka_{23}\\ a_{31} & a_{32} & a_{33}\end{vmatrix} = a_{11}\begin{vmatrix} ka_{22} & ka_{23}\\ a_{32} & a_{33}\end{vmatrix} - a_{12}\begin{vmatrix} ka_{21} & ka_{23}\\ a_{31} & a_{33}\end{vmatrix} + a_{13}\begin{vmatrix} ka_{21} & ka_{22}\\ a_{31} & a_{32}\end{vmatrix}$$
$$= a_{11}(ka_{23}a_{33} - ka_{23}a_{32}) - a_{12}(ka_{21}a_{33} - ka_{23}a_{31}) + a_{13}(ka_{21}a_{32} - ka_{22}a_{31})$$
$$= ka_{11}(a_{23}a_{33} - a_{23}a_{32}) - ka_{12}(a_{21}a_{33} - a_{23}a_{31}) + ka_{13}(a_{21}a_{32} - a_{22}a_{31})$$
$$= ka_{11}\begin{vmatrix} a_{22} & a_{23}\\ a_{32} & a_{33}\end{vmatrix} - ka_{12}\begin{vmatrix} a_{21} & a_{23}\\ a_{31} & a_{33}\end{vmatrix} - ka_{13}\begin{vmatrix} a_{21} & a_{22}\\ a_{31} & a_{32}\end{vmatrix} = k \cdot \begin{vmatrix} a_{11} & a_{12} & a_{13}\\ a_{21} & a_{22} & a_{23}\\ a_{31} & a_{32} & a_{33}\end{vmatrix}$$

61.
$$\begin{vmatrix} a_{11} & a_{12} & a_{11}\\ a_{21} & a_{22} & a_{21}\\ a_{31} & a_{32} & a_{31}\end{vmatrix} = a_{11}(a_{22}a_{31} - a_{32}a_{21}) - a_{12}(a_{21}a_{31} - a_{31}a_{21}) + a_{11}(a_{21}a_{32} - a_{31}a_{22})$$
$$= a_{11}a_{22}a_{31} - a_{11}a_{32}a_{21} - a_{12}(0) + a_{11}a_{21}a_{32} - a_{11}a_{31}a_{22} = 0$$

62. $\begin{vmatrix} ka_{21}+a_{11} & ka_{22}+a_{12} & ka_{23}+a_{13} \\ a_{21} & a_{22} & a_{23} \\ a_{31} & a_{32} & a_{33} \end{vmatrix} = (ka_{21}+a_{11})\begin{vmatrix} a_{22} & a_{23} \\ a_{32} & a_{33} \end{vmatrix} - (ka_{22}+a_{12})\begin{vmatrix} a_{21} & a_{23} \\ a_{31} & a_{33} \end{vmatrix} + (ka_{23}+a_{13})\begin{vmatrix} a_{21} & a_{22} \\ a_{31} & a_{32} \end{vmatrix}$

$= (ka_{21}+a_{11})(a_{22}a_{33}-a_{23}a_{32}) - (ka_{22}+a_{12})(a_{21}a_{33}-a_{23}a_{31})$
$+ (ka_{23}+a_{13})(a_{21}a_{32}-a_{22}a_{31})$
$= ka_{21}(a_{22}a_{33}-a_{23}a_{32}) - ka_{22}(a_{21}a_{33}-a_{23}a_{31}) + ka_{23}(a_{21}a_{32}-a_{22}a_{31})$
$+ a_{11}(a_{22}a_{33}-a_{23}a_{32}) - a_{12}(a_{21}a_{33}-a_{23}a_{31}) + a_{13}(a_{21}a_{32}-a_{22}a_{31})$
$= ka_{21}a_{22}a_{33} - ka_{21}a_{23}a_{32} - ka_{22}a_{21}a_{33} + ka_{22}a_{23}a_{31} + ka_{23}a_{21}a_{32} - ka_{23}a_{22}a_{31}$
$+ a_{11}\begin{vmatrix} a_{22} & a_{23} \\ a_{32} & a_{33} \end{vmatrix} - a_{12}\begin{vmatrix} a_{21} & a_{23} \\ a_{31} & a_{33} \end{vmatrix} + a_{13}\begin{vmatrix} a_{21} & a_{22} \\ a_{31} & a_{32} \end{vmatrix}$
$= 0 + \begin{vmatrix} a_{11} & a_{12} & a_{13} \\ a_{21} & a_{22} & a_{23} \\ a_{31} & a_{32} & a_{33} \end{vmatrix}$

Historical Problems

1. (a) $2-5i \longleftrightarrow \begin{bmatrix} 2 & -5 \\ 5 & 2 \end{bmatrix}, 1+3i \longleftrightarrow \begin{bmatrix} 1 & 3 \\ -3 & 1 \end{bmatrix}$ **(b)** $\begin{bmatrix} 2 & -5 \\ 5 & 2 \end{bmatrix}\begin{bmatrix} 1 & 3 \\ -3 & 1 \end{bmatrix} = \begin{bmatrix} 17 & 1 \\ -1 & 17 \end{bmatrix}$ **(c)** $17+i$ **(d)** $17+i$

2. (a) $x = (ka+lc)r + (kb+ld)s$; $y = (ma+nc)r + (mb+nd)s$ **(b)** $A = \begin{bmatrix} ka+lc & kb+ld \\ ma+nc & mb+nd \end{bmatrix}$

6.5 Exercises

1. $\begin{bmatrix} 4 & 4 & -5 \\ -1 & 5 & 4 \end{bmatrix}$ **2.** $\begin{bmatrix} -4 & 2 & -5 \\ 3 & -1 & 8 \end{bmatrix}$ **3.** $\begin{bmatrix} 0 & 12 & -20 \\ 4 & 8 & 24 \end{bmatrix}$ **4.** $\begin{bmatrix} -12 & -3 & 0 \\ 6 & -9 & 6 \end{bmatrix}$ **5.** $\begin{bmatrix} -8 & 7 & -15 \\ 7 & 0 & 22 \end{bmatrix}$ **6.** $\begin{bmatrix} 16 & 10 & -10 \\ -6 & 16 & 4 \end{bmatrix}$

7. $\begin{bmatrix} 28 & -9 \\ 4 & 23 \end{bmatrix}$ **8.** $\begin{bmatrix} 22 & 6 \\ 14 & -2 \end{bmatrix}$ **9.** $\begin{bmatrix} 1 & 14 & -14 \\ 2 & 22 & -18 \\ 3 & 0 & 28 \end{bmatrix}$ **10.** $\begin{bmatrix} 14 & 7 & -2 \\ 20 & 12 & -4 \\ -14 & 7 & -6 \end{bmatrix}$ **11.** $\begin{bmatrix} 15 & 21 & -16 \\ 22 & 34 & -22 \\ -11 & 7 & 22 \end{bmatrix}$ **12.** $\begin{bmatrix} 50 & -3 \\ 18 & 21 \end{bmatrix}$

13. $\begin{bmatrix} 25 & -9 \\ 4 & 20 \end{bmatrix}$ **14.** $\begin{bmatrix} 6 & 14 & -14 \\ 2 & 27 & -18 \\ 3 & 0 & 33 \end{bmatrix}$ **15.** $\begin{bmatrix} -13 & 7 & -12 \\ -18 & 10 & -14 \\ 17 & -7 & 34 \end{bmatrix}$ **16.** $\begin{bmatrix} 50 & -3 \\ 18 & 21 \end{bmatrix}$ **17.** $\begin{bmatrix} -2 & 4 & 2 & 8 \\ 2 & 1 & 4 & 6 \end{bmatrix}$ **18.** $\begin{bmatrix} -22 & 29 & 8 & -1 \\ -10 & 17 & 6 & -1 \end{bmatrix}$

19. $\begin{bmatrix} 9 & 2 \\ 34 & 13 \\ 47 & 20 \end{bmatrix}$ **20.** $\begin{bmatrix} 6 & 32 \\ 1 & 3 \\ -2 & -4 \end{bmatrix}$ **21.** $\begin{bmatrix} 1 & -1 \\ -1 & 2 \end{bmatrix}$ **22.** $\begin{bmatrix} 1 & 1 \\ 2 & 3 \end{bmatrix}$ **23.** $\begin{bmatrix} 1 & -\frac{5}{2} \\ -1 & 3 \end{bmatrix}$ **24.** $\begin{bmatrix} -1 & -\frac{1}{2} \\ -3 & -2 \end{bmatrix}$ **25.** $\begin{bmatrix} 1 & \frac{-1}{a} \\ -1 & \frac{2}{a} \end{bmatrix}$

26. $\begin{bmatrix} -\frac{2}{b} & \frac{3}{b} \\ 1 & -1 \end{bmatrix}$ **27.** $\begin{bmatrix} 3 & -3 & 1 \\ -2 & 2 & -1 \\ -4 & 5 & -2 \end{bmatrix}$ **28.** $\begin{bmatrix} 3 & -2 & -4 \\ 3 & -2 & -5 \\ -1 & 1 & 2 \end{bmatrix}$ **29.** $\begin{bmatrix} -\frac{5}{7} & \frac{1}{7} & \frac{3}{7} \\ \frac{9}{7} & \frac{1}{7} & -\frac{4}{7} \\ \frac{3}{7} & -\frac{2}{7} & \frac{1}{7} \end{bmatrix}$ **30.** $\begin{bmatrix} \frac{3}{7} & -\frac{4}{7} & \frac{1}{7} \\ \frac{1}{7} & \frac{1}{7} & -\frac{2}{7} \\ -\frac{5}{7} & \frac{9}{7} & \frac{3}{7} \end{bmatrix}$ **31.** $x=3, y=2$

32. $x=12, y=28$ **33.** $x=-5, y=10$ **34.** $x=9, y=23$ **35.** $x=2, y=-1$ **36.** $x=-7, y=-28$ **37.** $x=\frac{1}{2}, y=2$

38. $x=-\frac{1}{2}, y=3$ **39.** $x=-2, y=1$ **40.** $x=2, y=1$ **41.** $x=\frac{2}{a}, y=\frac{3}{a}$ **42.** $x=\frac{2}{b}, y=4$ **43.** $x=-2, y=3, z=5$

44. $x=4, y=-2, z=1$ **45.** $x=\frac{1}{2}, y=-\frac{1}{2}, z=1$ **46.** $x=1, y=-1, z=\frac{1}{2}$ **47.** $x=-\frac{34}{7}, y=\frac{85}{7}, z=\frac{12}{7}$

48. $x=\frac{8}{7}, y=\frac{5}{7}, z=\frac{17}{7}$ **49.** $x=\frac{1}{3}, y=1, z=\frac{2}{3}$ **50.** $x=1, y=-1, z=1$

51. $\left[\begin{array}{cc|cc} 4 & 2 & 1 & 0 \\ 2 & 1 & 0 & 1 \end{array}\right] \to \left[\begin{array}{cc|cc} 1 & \frac{1}{2} & \frac{1}{4} & 0 \\ 2 & 1 & 0 & 1 \end{array}\right] \to \left[\begin{array}{cc|cc} 1 & \frac{1}{2} & \frac{1}{4} & 0 \\ 0 & 0 & -\frac{1}{2} & 1 \end{array}\right]$ **52.** $\left[\begin{array}{cc|cc} -3 & \frac{1}{2} & 1 & 0 \\ 6 & -1 & 0 & 1 \end{array}\right] \to \left[\begin{array}{cc|cc} 1 & -\frac{1}{6} & -\frac{1}{3} & 0 \\ 6 & -1 & 0 & 1 \end{array}\right] \to \left[\begin{array}{cc|cc} 1 & -\frac{1}{6} & -\frac{1}{3} & 0 \\ 0 & 0 & 2 & 1 \end{array}\right]$

53. $\left[\begin{array}{cc|cc} 15 & 3 & 1 & 0 \\ 10 & 2 & 0 & 1 \end{array}\right] \to \left[\begin{array}{cc|cc} 1 & \frac{1}{5} & \frac{1}{15} & 0 \\ 10 & 2 & 0 & 1 \end{array}\right] \to \left[\begin{array}{cc|cc} 1 & \frac{1}{5} & \frac{1}{15} & 0 \\ 0 & 0 & -\frac{2}{3} & 1 \end{array}\right]$ **54.** $\left[\begin{array}{cc|cc} -3 & 0 & 1 & 0 \\ 4 & 0 & 0 & 1 \end{array}\right] \to \left[\begin{array}{cc|cc} 1 & 0 & -\frac{1}{3} & 0 \\ 4 & 0 & 0 & 1 \end{array}\right] \to \left[\begin{array}{cc|cc} 1 & 0 & -\frac{1}{3} & 0 \\ 0 & 0 & \frac{4}{3} & 1 \end{array}\right]$

55. $\left[\begin{array}{ccc|ccc} -3 & 1 & -1 & 1 & 0 & 0 \\ 1 & -4 & -7 & 0 & 1 & 0 \\ 1 & 2 & 5 & 0 & 0 & 1 \end{array}\right] \to \left[\begin{array}{ccc|ccc} 1 & 2 & 5 & 0 & 0 & 1 \\ 1 & -4 & -7 & 0 & 1 & 0 \\ -3 & 1 & -1 & 1 & 0 & 0 \end{array}\right] \to \left[\begin{array}{ccc|ccc} 1 & 2 & 5 & 0 & 0 & 1 \\ 0 & -6 & -12 & 0 & 1 & -1 \\ 0 & 7 & 14 & 1 & 0 & 3 \end{array}\right] \to \left[\begin{array}{ccc|ccc} 1 & 2 & 5 & 0 & 0 & 1 \\ 0 & 1 & 2 & 0 & -\frac{1}{6} & \frac{1}{6} \\ 0 & 1 & 2 & \frac{1}{7} & 0 & \frac{3}{7} \end{array}\right]$

$\to \left[\begin{array}{ccc|ccc} 1 & 2 & 5 & 0 & 0 & 1 \\ 0 & 1 & 2 & 0 & -\frac{1}{6} & \frac{1}{6} \\ 0 & 0 & 0 & \frac{1}{7} & \frac{1}{6} & \frac{11}{42} \end{array}\right]$

56. $\left[\begin{array}{rrr|rrr} 1 & 1 & -3 & 1 & 0 & 0 \\ 2 & -4 & 1 & 0 & 1 & 0 \\ -5 & 7 & 1 & 0 & 0 & 1 \end{array}\right] \to \left[\begin{array}{rrr|rrr} 1 & 1 & -3 & 1 & 0 & 0 \\ 0 & -6 & 7 & -2 & 1 & 0 \\ 0 & 12 & -14 & 5 & 0 & 1 \end{array}\right] \to \left[\begin{array}{rrr|rrr} 1 & 1 & -3 & 1 & 0 & 0 \\ 0 & 1 & -\frac{7}{6} & \frac{1}{3} & -\frac{1}{6} & 0 \\ 0 & 12 & -14 & 5 & 0 & 1 \end{array}\right] \to \left[\begin{array}{rrr|rrr} 1 & 0 & -\frac{11}{6} & \frac{2}{3} & \frac{1}{6} & 0 \\ 0 & 1 & -\frac{7}{16} & \frac{1}{3} & -\frac{1}{6} & 0 \\ 0 & 0 & 0 & 1 & 2 & 1 \end{array}\right]$

57. $\begin{bmatrix} 0.01 & 0.05 & -0.01 \\ 0.01 & -0.02 & 0.01 \\ -0.02 & 0.01 & 0.03 \end{bmatrix}$ **58.** $\begin{bmatrix} 0.26 & -0.29 & -0.20 \\ -1.21 & 1.63 & 1.20 \\ -1.84 & 2.53 & 1.80 \end{bmatrix}$ **59.** $\begin{bmatrix} 0.02 & -0.04 & -0.01 & 0.01 \\ -0.02 & 0.05 & 0.03 & -0.03 \\ 0.02 & 0.01 & -0.04 & 0.00 \\ -0.02 & 0.06 & 0.07 & 0.06 \end{bmatrix}$

60. $\begin{bmatrix} 0.01 & 0.04 & 0.00 & 0.03 \\ 0.02 & -0.02 & 0.01 & 0.01 \\ -0.04 & 0.02 & 0.04 & 0.06 \\ 0.05 & -0.02 & 0.00 & -0.09 \end{bmatrix}$ **61.** $x = 4.57, y = -6.44, z = -24.07$ **62.** $x = 4.56, y = -6.06, z = -22.55$ **63.** $x = -1.19, y = 2.46, z = 8.27$ **64.** $x = -2.05, y = 3.88, z = 13.36$

65. **(a)** $\begin{bmatrix} 500 & 350 & 400 \\ 700 & 500 & 850 \end{bmatrix}$; $\begin{bmatrix} 500 & 700 \\ 350 & 500 \\ 400 & 850 \end{bmatrix}$ **(b)** $\begin{bmatrix} 15 \\ 8 \\ 3 \end{bmatrix}$ **(c)** $\begin{bmatrix} 11{,}500 \\ 17{,}050 \end{bmatrix}$ **(d)** $[0.10 \quad 0.05]$ **(e)** \$2002.50

66. **(a)** January: $\begin{bmatrix} 400 & 250 & 50 \\ 450 & 200 & 140 \end{bmatrix}$ February: $\begin{bmatrix} 350 & 100 & 30 \\ 350 & 300 & 100 \end{bmatrix}$ **(b)** $\begin{bmatrix} 750 & 350 & 80 \\ 800 & 500 & 240 \end{bmatrix}$ **(c)** $\begin{bmatrix} \$100 \\ \$150 \\ \$200 \end{bmatrix}$ **(d)** $\begin{bmatrix} \$143{,}500 \\ \$203{,}000 \end{bmatrix}$

6.6 Exercises

1.
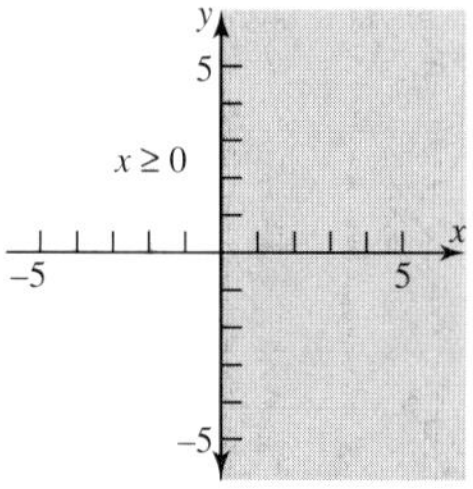

2.
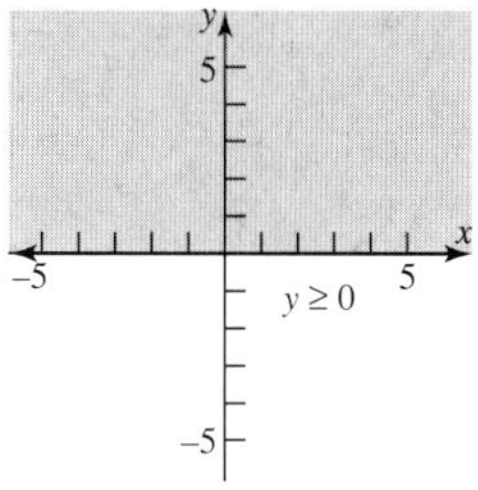

3.
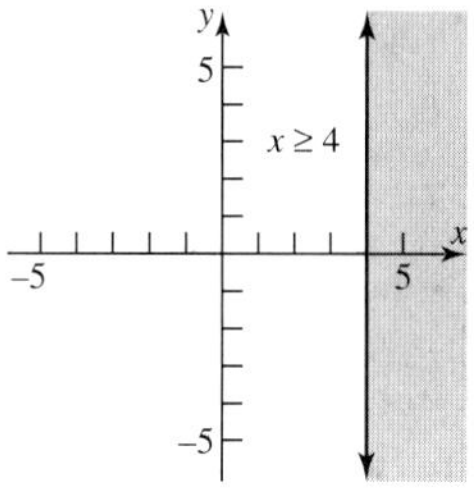

4.
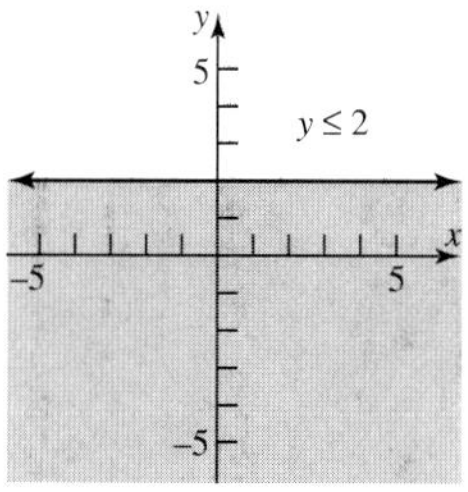

5.
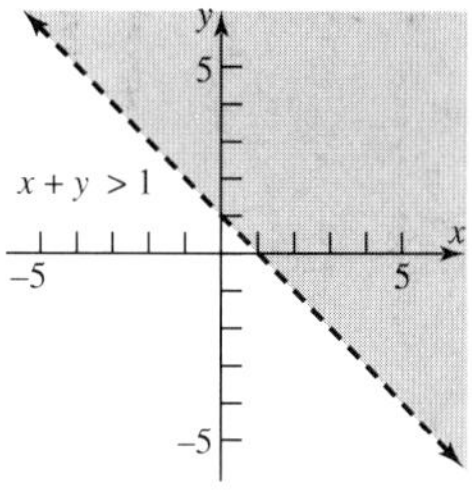

6.
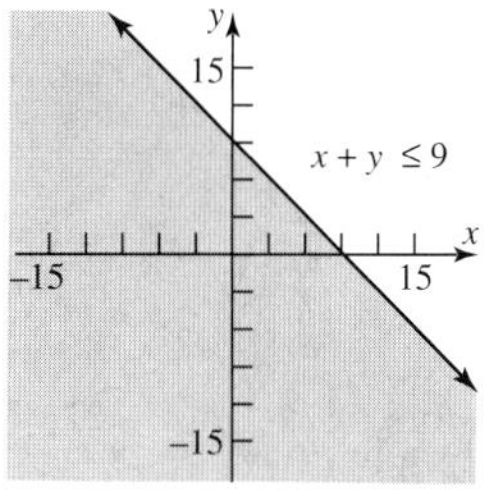

7.
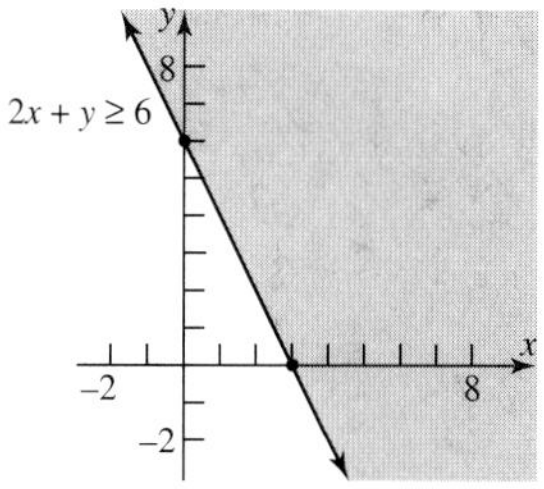

8.
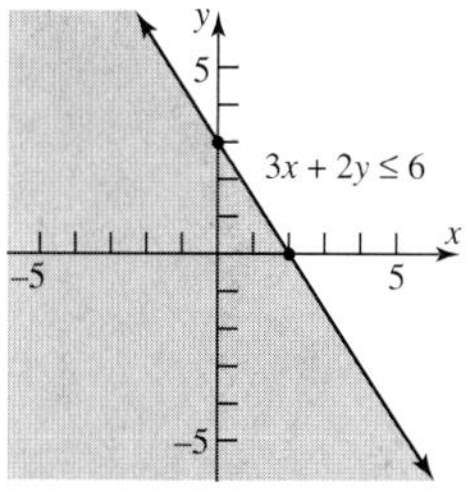

9.
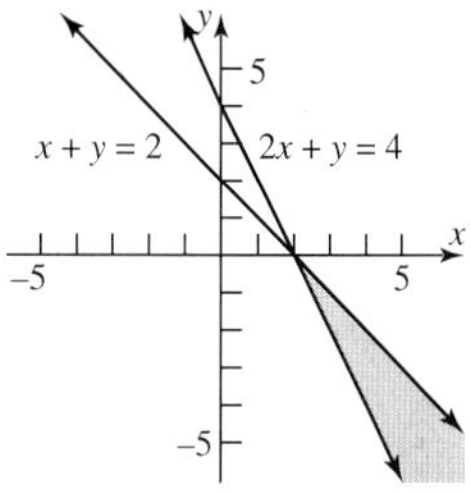

10.
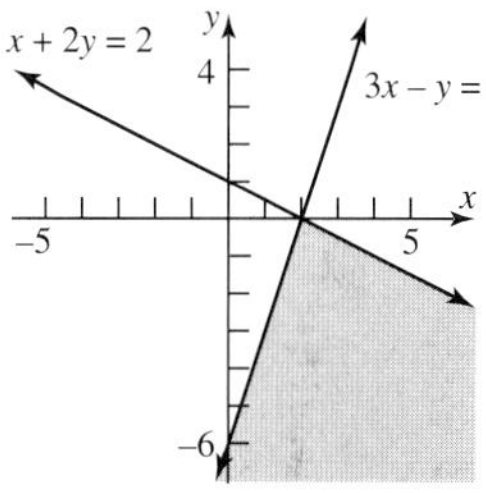

11.
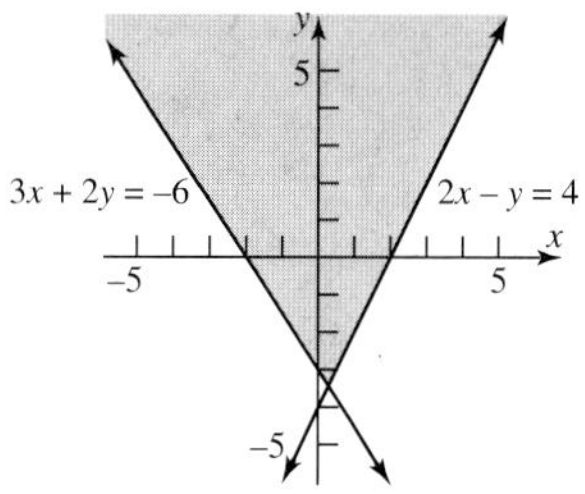

12.
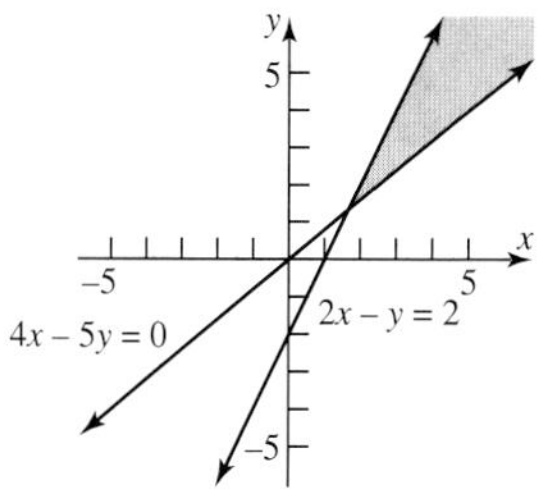

13.

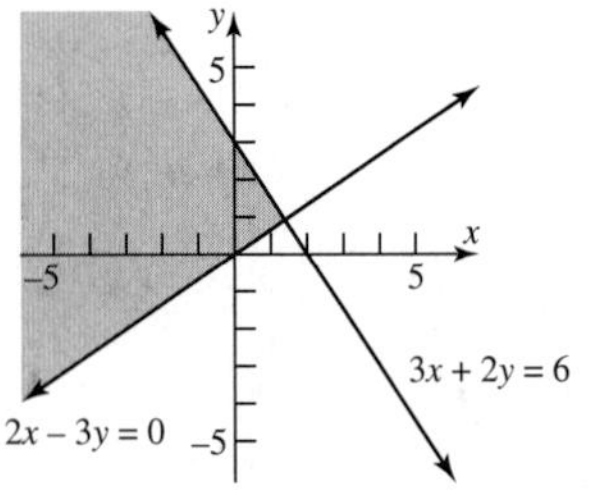

14.

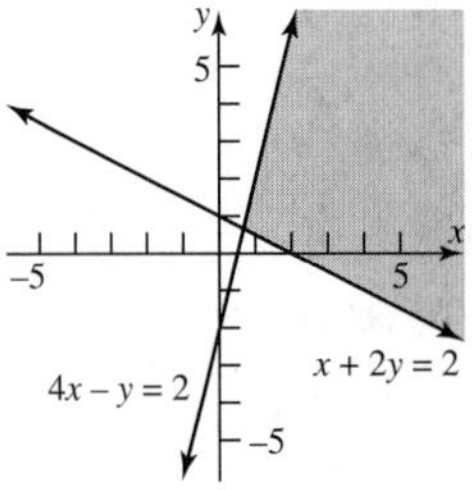

15.

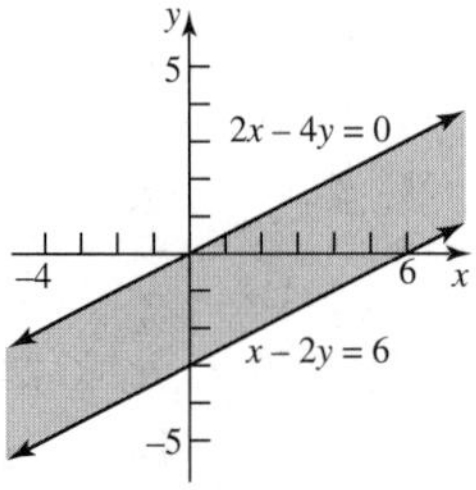

16.

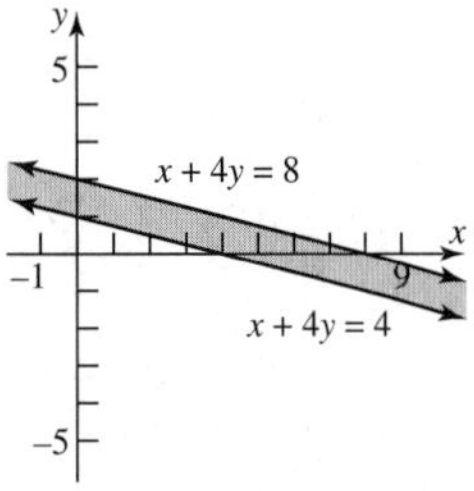

17.

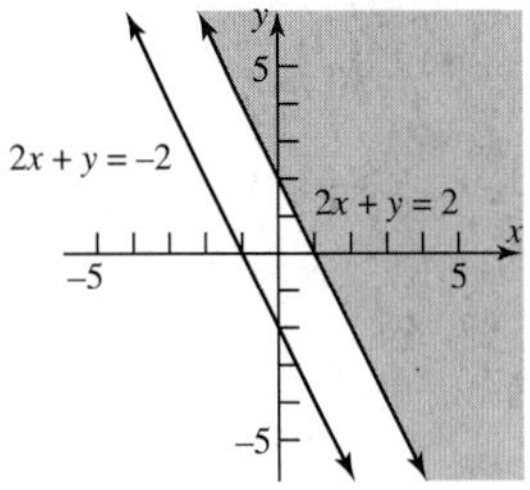

18.

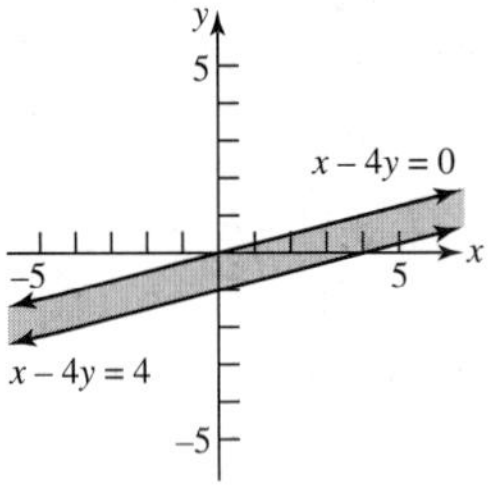

19. No solution

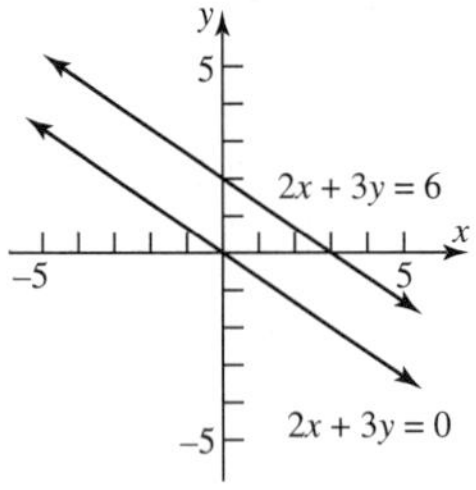

20.

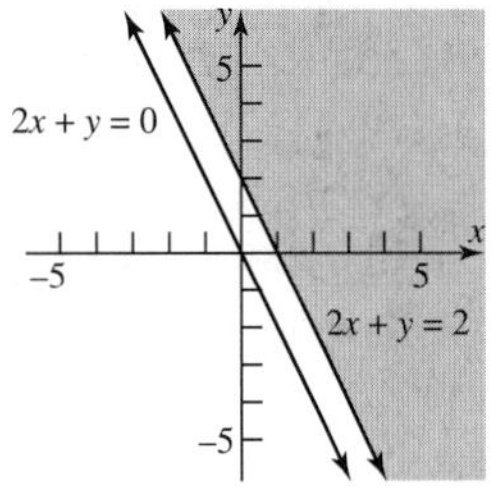

21. Bounded; corner points $(0, 0), (3, 0), (2, 2), (0, 3)$

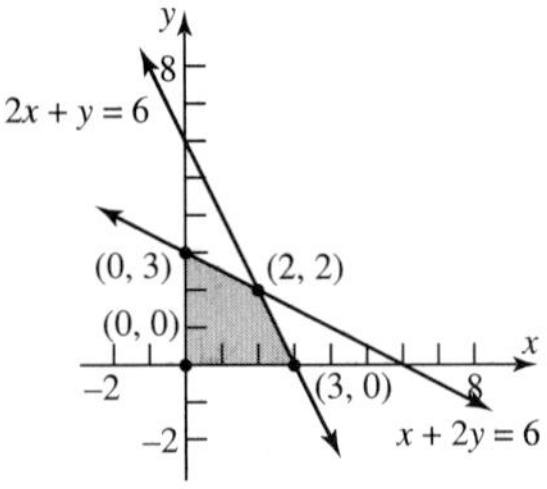

22. Unbounded; corner points $(0, 4), (4, 0)$

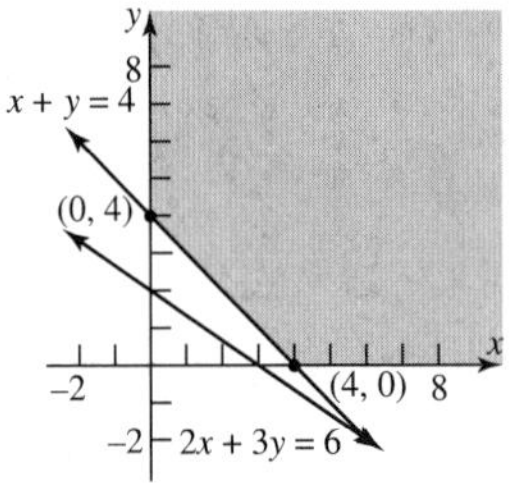

23. Unbounded; corner points $(2, 0), (0, 4)$

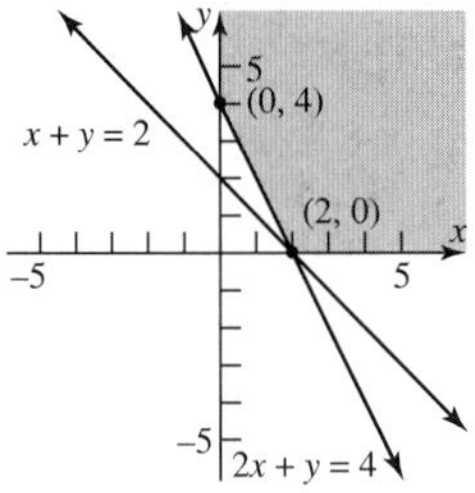

24. Bounded; corner points $(0, 0), (0, 2), (1, 0)$

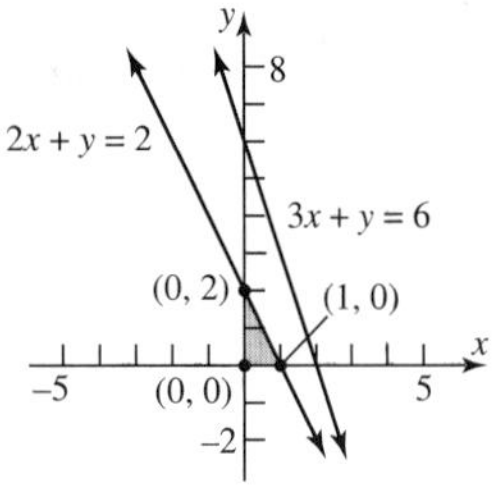

25. Bounded; corner points $(2, 0), (4, 0), \left(\frac{24}{7}, \frac{12}{7}\right), (0, 4), (0, 2)$

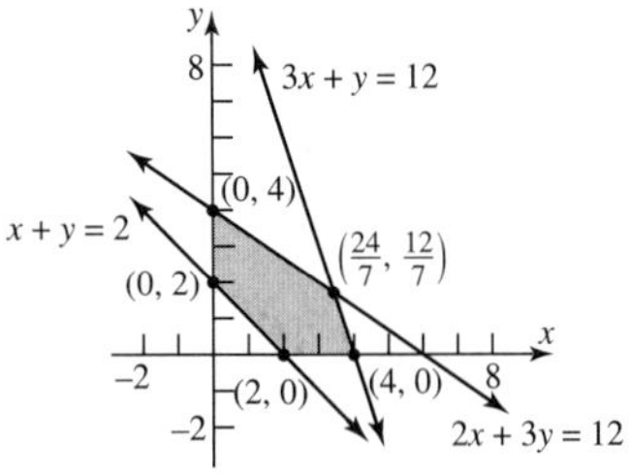

26. Bounded; corner points $(0, 2), (0, 3), (1, 1)$

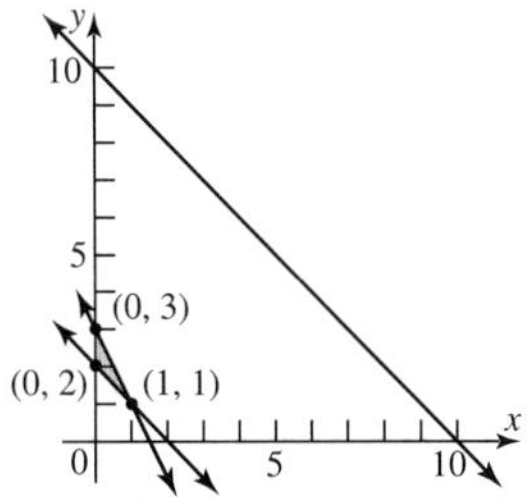

27. Bounded; corner points $(2, 0), (5, 0), (2, 6), (0, 8), (0, 2)$

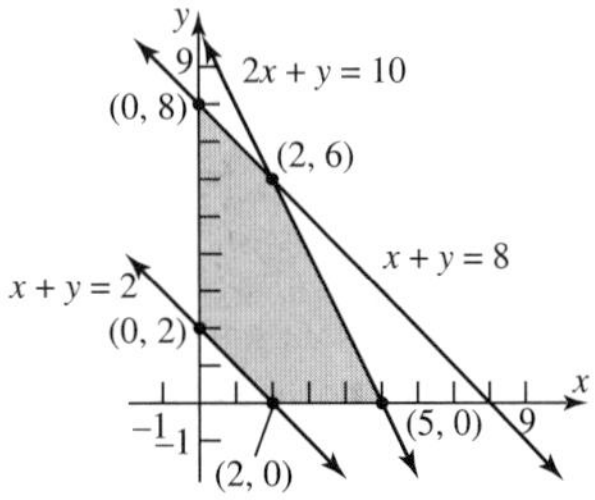

28. Bounded; corner points $(2, 0), (0, 2), (0, 8), (8, 0)$

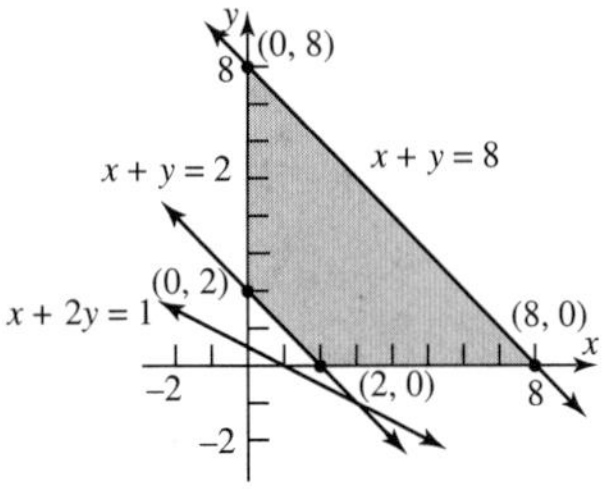

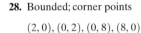

29. Bounded; corner points $(1, 0), (10, 0), (0, 5), \left(0, \frac{1}{2}\right)$

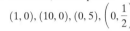

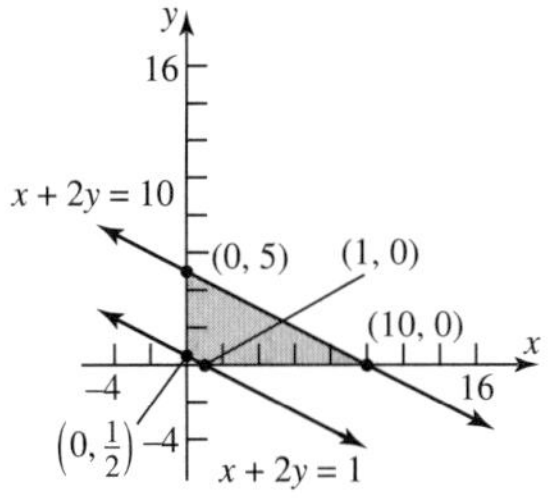

30. Bounded; corner points $(0, 2), (0, 5), (6, 2), (8, 0), (2, 0)$

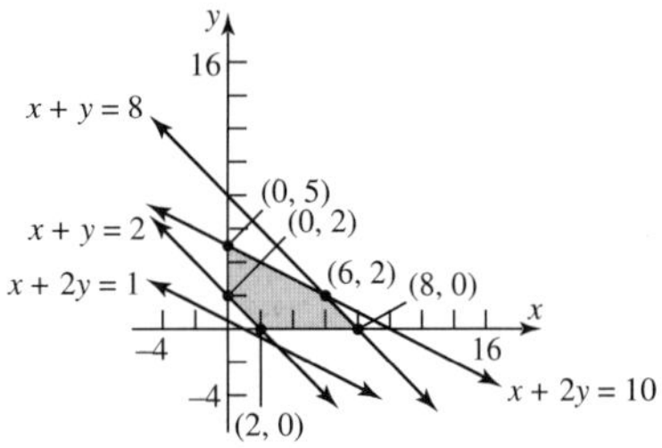

31. $\begin{cases} x \le 4 \\ x + y \le 6 \\ x \ge 0 \\ y \ge 0 \end{cases}$ **32.** $\begin{cases} y \le 5 \\ x + y \ge 2 \\ x \le 6 \\ x \ge 0,\ y \ge 0 \end{cases}$ **33.** $\begin{cases} x \le 20 \\ y \ge 15 \\ x + y \le 50 \\ x - y \le 0 \\ x \ge 0 \end{cases}$ **34.** $\begin{cases} y \le 6 \\ x \le 5 \\ 3x + 4y \ge 12 \\ y \ge 2x - 8 \\ x \ge 0,\ y \ge 0 \end{cases}$

35. Maximum value is 11; minimum value is 3 **36.** Maximum value is 28; minimum value is 8
37. Maximum value is 65; minimum value is 4 **38.** Maximum value is 56; minimum value is 3
39. Maximum value is 67; minimum value is 20 **40.** Maximum value is 65; minimum value is 15
41. The maximum value of z is 12, and it occurs at the point $(6, 0)$. **42.** Maximum value is 26, and it occurs at the point $(5, 7)$.
43. The minimum value of z is 4, and it occurs at the point $(2, 0)$. **44.** The minimum value of z is 8, and it occurs at the point $(0, 2)$.
45. The maximum value of z is 20, and it occurs at the point $(0, 4)$. **46.** The maximum value of z is 28, and it occurs at the point $(2, 6)$.
47. The minimum value of z is 8, and it occurs at the point $(0, 2)$. **48.** The minimum value of z is $\frac{15}{2}$, and it occurs at the point $\left(\frac{3}{2}, \frac{3}{2}\right)$.
49. The maximum value of z is 50, and it occurs at the point $(10, 0)$. **50.** The maximum value of z is 36, and it occurs at the point $(0. 9)$.
51. 8 downhill, 24 cross- country; \$1760; \$1920 **52.** 24 acres of soybeans and 12 acres of wheat for a maximum profit of \$5520; \$7200
53. 30 acres of soybeans and 10 acres of corn.
54. 15 ounces of supplement A and 0 ounces of supplement B or 7.5 ounces of supplement A and 11.25 ounces of supplement B, minimum cost \$22.50
55. $\frac{1}{2}$ hour on machine 1; $5\frac{1}{4}$ hours on machine 2 **56.** 15 newer trees and 10 older trees; minimum cost \$425
57. 100 pounds of ground beef and 50 pounds of pork **58. (a)** \$10,000 in a junk bond and \$10,000 in treasury bills **(b)** \$12,000 in junk bond and \$8,000 in treasury bills **59.** 10 racing skates, 15 figure skates
60. \$20,000 in AAA bond and \$30,000 in a CD **61.** 2 metal sample, 4 plastic samples; \$34
62. 15 cans of Gourmet Dog and 25 cans of Chow Hound **63. (a)** 10 first class, 120 coach **(b)** 15 first class, 120 coach
64. 5 ounces supplement A and 1 ounce of supplement B; minimum cost \$0.38

6.7 Exercises

1. Proper **2.** Proper **3.** Improper; $1 + \frac{9}{x^2 - 4}$ **4.** Improper; $3 + \frac{1}{x^2 - 1}$ **5.** Improper; $5x + \frac{22x - 1}{x^2 - 4}$
6. Improper; $3x + \frac{x^2 - 24x - 2}{x^3 + 8}$ **7.** Improper; $1 + \frac{-2(x - 6)}{(x + 4)(x - 3)}$ **8.** Improper; $2x + \frac{6x}{x^2 + 1}$ **9.** $\frac{-4}{x} + \frac{4}{x - 1}$
10. $\frac{2}{x + 2} + \frac{1}{x - 1}$ **11.** $\frac{1}{x} + \frac{-x}{x^2 + 1}$ **12.** $\frac{\frac{1}{5}}{x + 1} + \frac{-\frac{1}{5}x + \frac{1}{5}}{x^2 + 4}$ **13.** $\frac{-1}{x - 1} + \frac{2}{x - 2}$ **14.** $\frac{1}{x + 2} + \frac{2}{x - 4}$
15. $\frac{\frac{1}{4}}{x + 1} + \frac{\frac{3}{4}}{x - 1} + \frac{\frac{1}{2}}{(x - 1)^2}$ **16.** $\frac{-\frac{3}{4}}{x} + \frac{-\frac{1}{2}}{x^2} + \frac{\frac{3}{4}}{x - 2}$ **17.** $\frac{\frac{1}{12}}{x - 2} + \frac{-\frac{1}{12}(x + 4)}{x^2 + 2x + 4}$ **18.** $\frac{2}{x - 1} + \frac{-2x - 2}{x^2 + x + 1}$
19. $\frac{\frac{1}{4}}{x - 1} + \frac{\frac{1}{4}}{(x - 1)^2} + \frac{-\frac{1}{4}}{x + 1} + \frac{\frac{1}{4}}{(x + 1)^2}$ **20.** $\frac{\frac{1}{2}}{x} + \frac{\frac{1}{4}}{x^2} + \frac{-\frac{1}{2}}{x - 2} + \frac{\frac{3}{4}}{(x - 2)^2}$ **21.** $\frac{-5}{x + 2} + \frac{5}{x + 1} + \frac{-4}{(x + 1)^2}$
22. $\frac{\frac{2}{9}}{x + 2} + \frac{\frac{7}{9}}{x - 1} + \frac{\frac{2}{3}}{(x - 1)^2}$ **23.** $\frac{\frac{1}{4}}{x} + \frac{1}{x^2} + \frac{-\frac{1}{4}(x + 4)}{x^2 + 4}$ **24.** $\frac{\frac{14}{3}}{x - 1} + \frac{4}{(x - 1)^2} + \frac{-\frac{14}{3}x + \frac{4}{3}}{x^2 + 2}$ **25.** $\frac{\frac{2}{3}}{x + 1} + \frac{\frac{1}{3}(x + 1)}{x^2 + 2x + 4}$
26. $\frac{-6}{x} + \frac{7x + 7}{x^2 + 3x + 3}$ **27.** $\frac{\frac{2}{7}}{3x - 2} + \frac{\frac{1}{7}}{2x + 1}$ **28.** $\frac{-\frac{1}{7}}{2x + 3} + \frac{\frac{2}{7}}{4x - 1}$ **29.** $\frac{\frac{3}{4}}{x + 3} + \frac{\frac{1}{4}}{x - 1}$ **30.** $\frac{-3}{x + 1} + \frac{2}{x + 2} + \frac{2}{x + 3}$
31. $\frac{1}{x^2 + 4} + \frac{2x - 1}{(x^2 + 4)^2}$ **32.** $\frac{x}{x^2 + 16} + \frac{-16x + 1}{(x^2 + 16)^2}$ **33.** $\frac{-1}{x} + \frac{2}{x - 3} + \frac{-1}{x + 1}$ **34.** $\frac{-1}{x^2} - \frac{1}{x^4} + \frac{1}{x - 1}$
35. $\frac{4}{x - 2} + \frac{-3}{x - 1} + \frac{-1}{(x - 1)^2}$ **36.** $\frac{\frac{3}{8}}{x - 1} + \frac{\frac{1}{2}}{(x - 1)^2} + \frac{\frac{5}{8}}{x + 3}$ **37.** $\frac{x}{(x^2 + 16)^2} + \frac{-16x}{(x^2 + 16)^3}$ **38.** $\frac{1}{(x^2 + 4)^2} - \frac{4}{(x^2 + 4)^3}$
39. $\frac{-\frac{8}{7}}{2x + 1} + \frac{\frac{4}{7}}{x - 3}$ **40.** $\frac{\frac{4}{5}}{2x - 1} + \frac{\frac{8}{5}}{x + 2}$ **41.** $\frac{-\frac{2}{9}}{x} + \frac{-\frac{1}{3}}{x^2} + \frac{\frac{1}{6}}{x - 3} + \frac{\frac{1}{18}}{x + 3}$ **42.** $\frac{\frac{13}{24}}{x - 2} + \frac{-\frac{13}{24}}{x + 2} + \frac{-\frac{7}{6}}{x^2 + 2}$

True/False Items

1. False **2.** True **3.** False **4.** False **5.** False **6.** False **7.** False **8.** True **9.** True **10.** True **11.** True

Review Exercises

1. $x = 2, y = -1$ **2.** $x = \frac{11}{23}, y = \frac{8}{23}$ **3.** $x = 2, y = \frac{1}{2}$ **4.** $x = -\frac{1}{2}, y = 1$ **5.** $x = 2, y = -1$ **6.** $x = 0, y = \frac{5}{3}$
7. $x = \frac{11}{5}, y = -\frac{3}{5}$ **8.** $x = -\frac{1}{2}, y = -\frac{1}{2}$ **9.** $x = -\frac{8}{5}, y = \frac{12}{5}$ **10.** Inconsistent **11.** $x = 6, y = -1$ **12.** $x = 1, y = \frac{3}{2}$
13. $x = -4, y = 3$ **14.** $x = 0, y = -3$ **15.** $x = 2, y = 3$ **16.** $x = 84, y = -63$ **17.** Inconsistent **18.** Inconsistent
19. $x = -1, y = 2, z = -3$ **20.** $x = -1, y = 2, z = 7$ **21.** $\begin{bmatrix} 4 & -4 \\ 3 & 9 \\ 4 & 0 \end{bmatrix}$ **22.** $\begin{bmatrix} -2 & 4 \\ 1 & -1 \\ -6 & 4 \end{bmatrix}$ **23.** $\begin{bmatrix} 6 & 0 \\ 12 & 24 \\ -6 & 12 \end{bmatrix}$ **24.** $\begin{bmatrix} -16 & 12 & 0 \\ -4 & -4 & 8 \end{bmatrix}$
25. $\begin{bmatrix} 4 & -3 & 0 \\ 12 & -2 & -8 \\ -2 & 5 & -4 \end{bmatrix}$ **26.** $\begin{bmatrix} -2 & -12 \\ 5 & 0 \end{bmatrix}$ **27.** $\begin{bmatrix} 8 & -13 & 8 \\ 9 & 2 & -10 \\ 18 & -17 & 4 \end{bmatrix}$ **28.** $\begin{bmatrix} 9 & -31 \\ -6 & 5 \end{bmatrix}$ **29.** $\begin{bmatrix} \frac{1}{2} & -1 \\ -\frac{1}{6} & \frac{2}{3} \end{bmatrix}$ **30.** $\begin{bmatrix} -\frac{1}{2} & -\frac{1}{2} \\ -\frac{1}{4} & -\frac{3}{4} \end{bmatrix}$
31. $\begin{bmatrix} -\frac{5}{7} & \frac{9}{7} & \frac{3}{7} \\ \frac{1}{7} & \frac{1}{7} & -\frac{2}{7} \\ \frac{3}{7} & -\frac{4}{7} & \frac{1}{7} \end{bmatrix}$ **32.** $\begin{bmatrix} \frac{3}{7} & \frac{1}{7} & -\frac{5}{7} \\ -\frac{4}{7} & \frac{1}{7} & \frac{9}{7} \\ \frac{1}{7} & -\frac{2}{7} & \frac{3}{7} \end{bmatrix}$ **33.** Singular **34.** Singular **35.** $x = \frac{2}{5}, y = \frac{1}{10}$ **36.** $x = 1, y = \frac{3}{2}$
37. $x = \frac{1}{2}, y = \frac{2}{3}, z = \frac{1}{6}$ **38.** Inconsistent **39.** $x = -\frac{1}{2}, y = -\frac{2}{3}, z = -\frac{3}{4}$ **40.** $x = \frac{1}{3}, y = \frac{4}{5}, z = -\frac{1}{15}$
41. $z = -1, x = y + 1$, y is any real number **42.** x is any real number, $y = -2x, z = -2x$ **43.** $x = 1, y = 2, z = -3, t = 1$
44. $x = -\frac{17}{15}, y = -\frac{1}{15}, z = \frac{22}{15}, t = -\frac{2}{15}$ **45.** 5 **46.** −12 **47.** 108 **48.** −19 **49.** −100 **50.** 11 **51.** $x = 2, y = -1$
52. $x = 0, y = \frac{5}{3}$ **53.** $x = 2, y = 3$ **54.** $x = 0, y = -3$ **55.** $x = -1, y = 2, z = -3$ **56.** $x = \frac{37}{9}, y = -\frac{19}{6}, z = \frac{13}{18}$

57. Unbounded; corner point (0, 2)

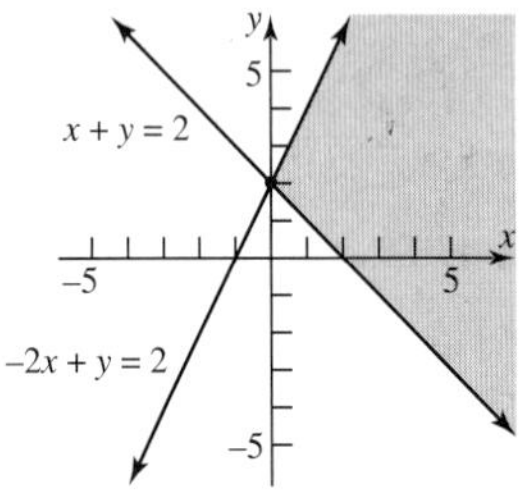

58. Unbounded; corner point (2, −2)

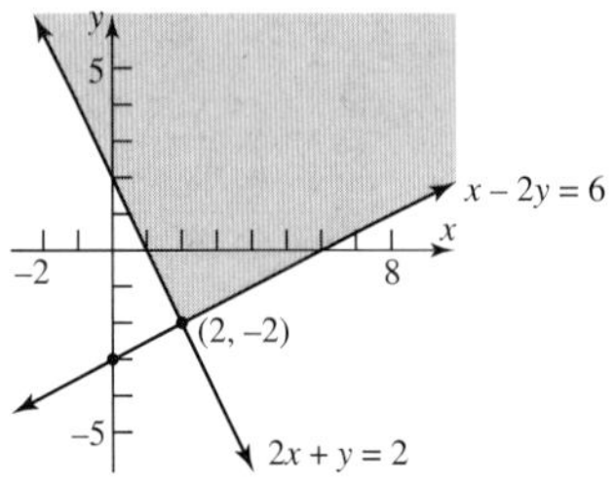

59. Bounded; corner points (0, 0), (0, 2), (3, 0)

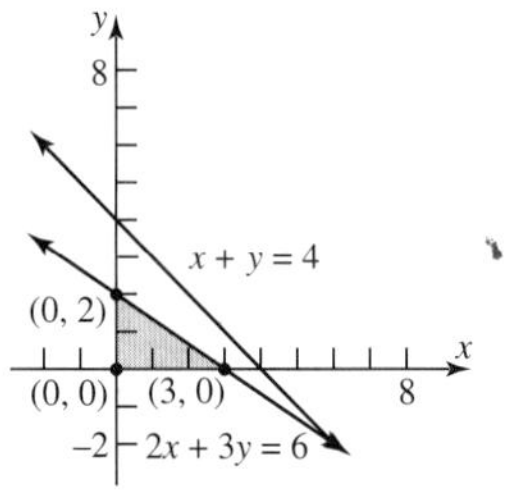

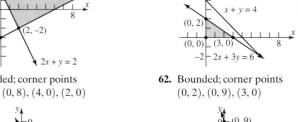

60. Unbounded; corner points (2, 0), (0, 6)

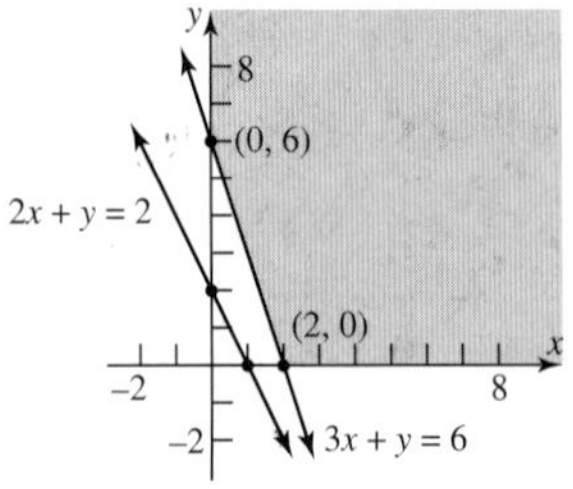

61. Bounded; corner points (0, 1), (0, 8), (4, 0), (2, 0)

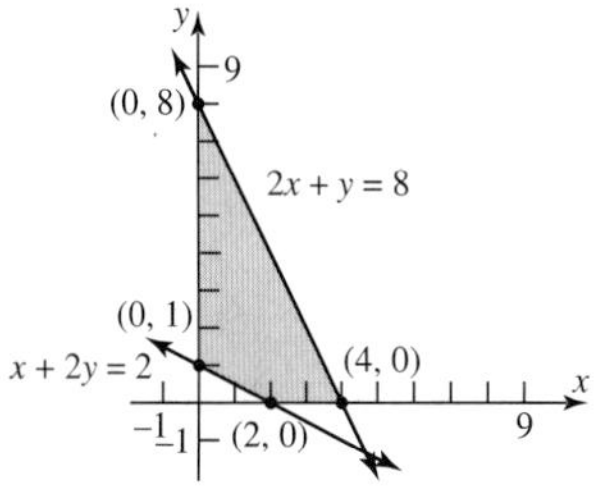

62. Bounded; corner points (0, 2), (0, 9), (3, 0)

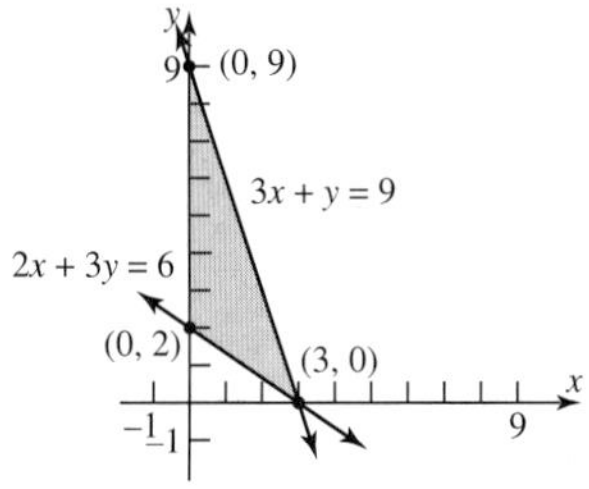

63. The maximum value is 32 when $x = 0$ and $y = 8$. **64.** The maximum value is 20, which occurs at (2, 4).
65. The minimum value is 3 when $x = 1$ and $y = 0$. **66.** The minimum value is 4, which occurs at (0, 4).

67. The minimum value is $\frac{108}{7}$ when $x = \frac{12}{7}$ and $y = \frac{12}{7}$. **68.** The maximum value is 36, which occurs at (4, 4). **69.** $\frac{-\frac{3}{2}}{x} + \frac{\frac{3}{2}}{x-4}$

70. $\frac{\frac{2}{5}}{x+2} + \frac{\frac{3}{5}}{x-3}$ **71.** $\frac{-3}{x-1} + \frac{3}{x} + \frac{4}{x^2}$ **72.** $\frac{4}{x-2} + \frac{-2}{(x-2)^2} + \frac{-4}{x-1}$ **73.** $\frac{-\frac{1}{10}}{x+1} + \frac{\frac{1}{10}x + \frac{9}{10}}{x^2+9}$ **74.** $\frac{\frac{6}{5}}{x-2} + \frac{-\frac{6}{5}x + \frac{3}{5}}{x^2+1}$

75. $\dfrac{x}{x^2+4}+\dfrac{-4x}{(x^2+4)^2}$ **76.** $\dfrac{x}{x^2+16}+\dfrac{-16x+1}{(x^2+16)^2}$ **77.** $\dfrac{\frac{1}{2}}{x^2+1}+\dfrac{\frac{1}{4}}{x-1}+\dfrac{-\frac{1}{4}}{x+1}$ **78.** $\dfrac{-\frac{4}{5}}{x^2+4}+\dfrac{\frac{2}{5}}{x-1}+\dfrac{-\frac{2}{5}}{x+1}$ **79.** 10

80. Inconsistent if $A \neq 10$ **81.** $y=-\frac{1}{3}x^2-\frac{2}{3}x+1$ **82.** $x^2+y^2+2x+2y-3=0$

83. 70 pounds of \$3 coffee and 30 pounds of \$6 coffee **84.** Corn—466.25 acres, soybeans—533.75 acres **85.** 1 small, 5 medium, 2 large

86. **(a)** $\begin{cases} \frac{1}{2}x+\frac{3}{8}y \le 120 \\ \frac{1}{4}x+\frac{3}{8}y \le 72 \\ x \ge 0 \\ y \ge 0 \end{cases}$

(b)

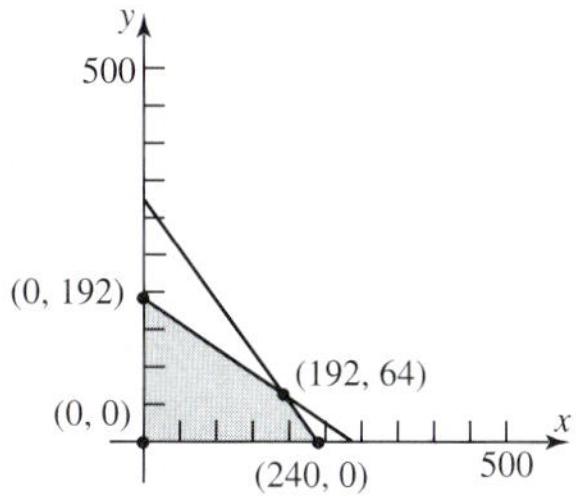

87. 24 feet by 10 feet
88. 2 feet by 2 feet
89. $4+\sqrt{2}$ inches and $4-\sqrt{2}$ inches
90. 5 inches
91. $100\sqrt{10}$ feet

92. x = amount of 10% HCl
y = amount of 25% HCl
z = amount of 40% HCl

x	y	z
0	66.7	33.3
16.7	33.3	50.0
33.3	0	66.7

93. Katy gets \$10, Mike gets \$20, Danny gets \$5, Colleen gets \$10
94. Jet-stream speed is 29.69 mph **95.** Bruce: 4 hours; Bryce: 2 hours; Marty: 8 hours
96. 12 dancing girls and 12 mermaids; maximum profit \$660. The process of molding has excess work-hours assigned.
97. 35 gasoline engines, 15 diesel engines; 15 gasoline engines, 0 diesel engines

CHAPTER 7 Sequences; Induction; The Binomial Theorem

7.1 Exercises

1. 1, 2, 3, 4, 5 **2.** 2, 5, 10, 17, 26 **3.** $\frac{1}{3}, \frac{2}{4}=\frac{1}{2}, \frac{3}{5}, \frac{4}{6}=\frac{2}{3}, \frac{5}{7}$ **4.** $\frac{3}{2}, \frac{5}{4}, \frac{7}{6}, \frac{9}{8}, \frac{11}{10}$ **5.** 1, −4, 9, −16, 25 **6.** $1, -\frac{2}{3}, \frac{3}{5}, -\frac{4}{7}, \frac{5}{9}$ **7.** $\frac{1}{2}, \frac{2}{5}, \frac{2}{7}, \frac{8}{41}, \frac{8}{61}$
8. $\frac{4}{3}, \frac{16}{9}, \frac{64}{27}, \frac{256}{81}, \frac{1024}{243}$ **9.** $-\frac{1}{6}, \frac{1}{12}, -\frac{1}{20}, \frac{1}{30}, -\frac{1}{42}$ **10.** $3, \frac{9}{2}, 9, \frac{81}{4}, \frac{243}{5}$ **11.** $\frac{1}{e}, \frac{2}{e^2}, \frac{3}{e^3}, \frac{4}{e^4}, \frac{5}{e^5}$ **12.** $\frac{1}{2}, 1, \frac{9}{8}, 1, \frac{25}{32}$ **13.** $\frac{n}{n+1}$
14. $\frac{1}{n(n+1)}$ **15.** $\frac{1}{2^{n-1}}$ **16.** $\frac{2^n}{3^n}$ **17.** $(-1)^{n+1}$ **18.** $n^{(-1)^{n+1}}$ **19.** $(-1)^{n+1}n$ **20.** $(-1)^{n+1}(2n)$
21. $a_1=2, a_2=5, a_3=8, a_4=11, a_5=14$ **22.** $a_1=3, a_2=1, a_3=3, a_4=1, a_5=3$ **23.** $a_1=-2, a_2=0, a_3=3, a_4=7, a_5=12$
24. $a_1=1, a_2=1, a_3=2, a_4=2, a_5=3$ **25.** $a_1=5, a_2=10, a_3=20, a_4=40, a_5=80$
26. $a_1=2, a_2=-2, a_3=2, a_4=-2, a_5=2$ **27.** $a_1=3, a_2=\frac{3}{2}, a_3=\frac{1}{2}, a_4=\frac{1}{8}, a_5=\frac{1}{40}$
28. $a_1=-2, a_2=-4, a_3=-9, a_4=-23, a_5=-64$ **29.** $a_1=1, a_2=2, a_3=2, a_4=4, a_5=8$
30. $a_1=-1, a_2=1, a_3=2, a_4=9, a_5=47$ **31.** $a_1=A, a_2=A+d, a_3=A+2d, a_4=A+3d, a_5=A+4d$
32. $a_1=A, a_2=Ar, a_3=Ar^2, a_4=Ar^3, a_5=Ar^4$
33. $a_1=\sqrt{2}, a_2=\sqrt{2+\sqrt{2}}, a_3=\sqrt{2+\sqrt{2+\sqrt{2}}}, a_4=\sqrt{2+\sqrt{2+\sqrt{2+\sqrt{2}}}}, a_5=\sqrt{2+\sqrt{2+\sqrt{2+\sqrt{2+\sqrt{2}}}}}$
34. $a_1=\sqrt{2}, a_2=2^{-1/4}, a_3=2^{-5/8}, a_4=2^{-13/16}, a_5=2^{-29/32}$ **35.** 50 **36.** 160 **37.** 21 **38.** −10 **39.** 90 **40.** 21 **41.** 26 **42.** 14
43. 42 **44.** 60 **45.** 96 **46.** 44 **47.** $3+4+\cdots+(n+2)$ **48.** $3+5+7+\cdots+(2n+1)$ **49.** $\frac{1}{2}+2+\frac{9}{2}+\cdots+\frac{n^2}{2}$
50. $4+9+16+\cdots+(n+1)^2$ **51.** $1+\frac{1}{3}+\frac{1}{9}+\cdots+\frac{1}{3^n}$ **52.** $1+\frac{3}{2}+\frac{9}{4}+\cdots+\left(\frac{3}{2}\right)^n$ **53.** $\frac{1}{3}+\frac{1}{9}+\cdots+\frac{1}{3^n}$
54. $1+3+5+\cdots+(2n-1)$ **55.** $\ln 2-\ln 3+\ln 4-\cdots+(-1)^n\ln n$ **56.** $8-16+32-\cdots+(-1)^{n+1}2^n$ **57.** $\sum_{k=1}^{20} k$ **58.** $\sum_{k=1}^{8} k^3$
59. $\sum_{k=1}^{13} \frac{k}{k+1}$ **60.** $\sum_{k=1}^{12} (2k-1)$ **61.** $\sum_{k=0}^{6} (-1)^k\left(\frac{1}{3^k}\right)$ **62.** $\sum_{k=0}^{11} (-1)^{k+1}\left(\frac{2}{3}\right)^k$ **63.** $\sum_{k=1}^{n} \frac{3^k}{k}$ **64.** $\sum_{k=1}^{n} \frac{k}{e^k}$
65. $\sum_{k=0}^{n} (a+kd)$ or $\sum_{k=1}^{n+1} [a+(k-1)d]$ **66.** $\sum_{k=1}^{n} ar^{k-1}$

67. **(a)** \$2930
(b)

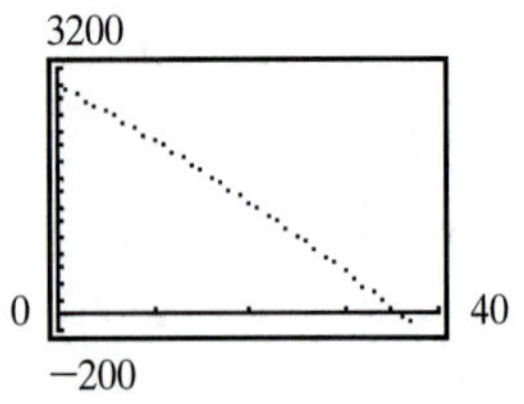

(c) At the beginning of the 15th month. 14 payments have been made.
(d) 36 payments. \$3583.78
(e) \$583.78

68. **(a)** \$18,058.03
(b)

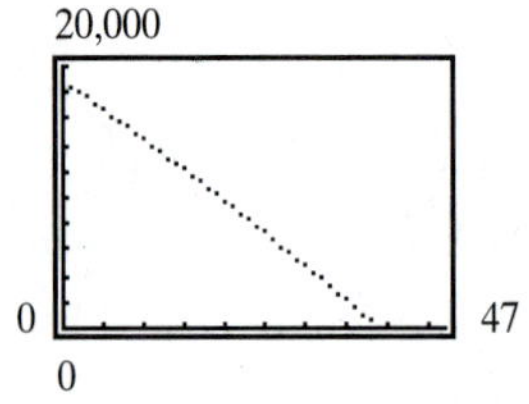

(c) At the beginning of the 20th month. 19 payments have been made
(d) 40 payments. \$20,898.51
(e) \$2398.51

69. **(a)** 2080
(b)

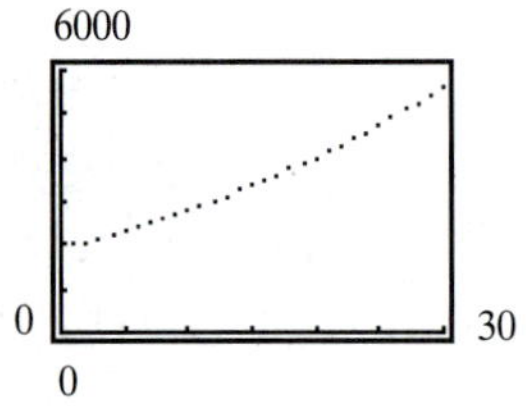

(c) At the beginning of the 27th month

70. **(a)** 240 tons
(b)

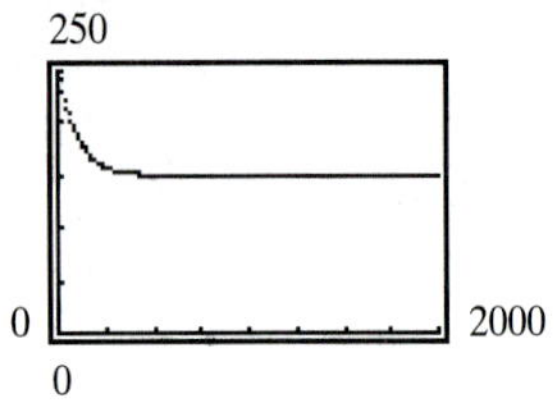

(c) 150 tons

71. 21 **73.** A Fibonacci sequence **74.** **(a)** 1, 1, 2, 3, 5, 8, 13, 21, 34, 55
(b) $\frac{1}{1} = 1, \frac{2}{1} = 2, \frac{3}{2} = 1.5, \frac{5}{3} \approx 1.667, \frac{8}{5} = 1.6, \frac{13}{8} = 1.625, \frac{21}{13} \approx 1.615, \frac{34}{21} \approx 1.619, \frac{55}{34} \approx 1.618$ **(c)** 1.618
(d) $\frac{1}{1} = 1, \frac{1}{2} = 0.5, \frac{2}{3} \approx 0.667, \frac{3}{5} = 0.6, \frac{5}{8} = 0.625, \frac{8}{13} \approx 0.615, \frac{13}{21} \approx 0.619, \frac{21}{34} \approx 0.618, \frac{34}{55} \approx 0.618$ **(e)** 0.618

7.2 Exercises

1. $d = 1; 5, 6, 7, 8$ **2.** $d = 1; -4, -3, -2, -1$ **3.** $d = 2; -3, -1, 1, 3$ **4.** $d = 3; 4, 7, 10, 13$ **5.** $d = -2; 4, 2, 0, -2$
6. $d = -2; 2, 0, -2, -4$ **7.** $d = -\frac{1}{3}; \frac{1}{6}, -\frac{1}{6}, -\frac{1}{2}, -\frac{5}{6}$ **8.** $d = \frac{1}{4}; \frac{11}{12}, \frac{7}{6}, \frac{17}{12}, \frac{5}{3}$ **9.** $d = \ln 3; \ln 3, 2\ln 3, 3\ln 3, 4\ln 3$ **10.** $d = 1; 1, 2, 3, 4$
11. $a_n = 3n - 1; a_5 = 14$ **12.** $a_n = 4n - 6; a_5 = 14$ **13.** $a_n = 8 - 3n; a_5 = -7$ **14.** $a_n = 8 - 2n; a_5 = -2$
15. $a_n = \frac{1}{2}(n - 1); a_5 = 2$ **16.** $a_n = \frac{4}{3} - \frac{1}{3}n; a_5 = -\frac{1}{3}$ **17.** $a_n = \sqrt{2}n; a_5 = 5\sqrt{2}$ **18.** $a_n = \pi(n - 1); a_5 = 4\pi$ **19.** $a_{12} = 24$
20. $a_8 = 13$ **21.** $a_{10} = -26$ **22.** $a_9 = -35$ **23.** $a_8 = a + 7b$ **24.** $a_7 = 14\sqrt{5}$ **25.** $a_1 = -13; d = 3; a_n = a_{n-1} + 3$
26. $a_1 = -3; d = 2; a_n = a_{n-1} + 2$ **27.** $a_1 = -53, d = 6; a_n = a_{n-1} + 6$ **28.** $a_1 = 74; d = -10; a_n = a_{n-1} - 10$
29. $a_1 = 28; d = -2; a_n = a_{n-1} - 2$ **30.** $a_1 = -18; d = 4; a_n = a_{n-1} + 4$ **31.** $a_1 = 25; d = -2; a_n = a_{n-1} - 2$
32. $a_1 = -40; d = 4; a_n = a_{n+1} + 4$ **33.** n^2 **34.** $n^2 + n$ **35.** $\frac{n}{2}(9 + 5n)$ **36.** $2n^2 - 3n$ **37.** 1260 **38.** 900 **39.** 324 **40.** 301

41.

42.

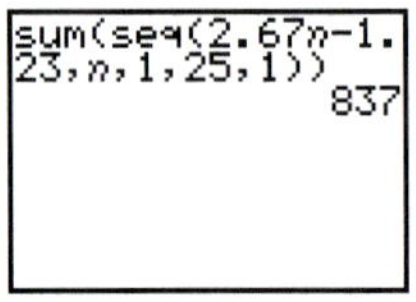

43.

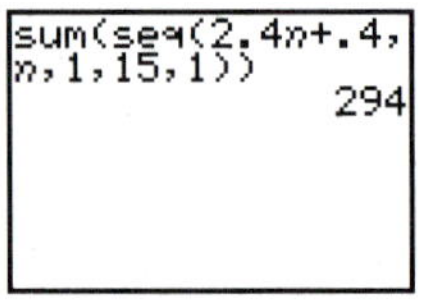

44.

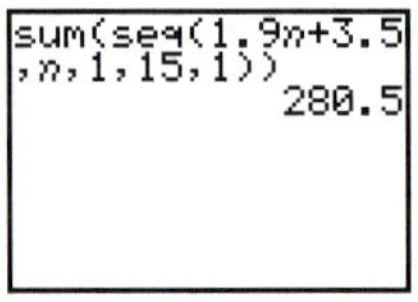

45.

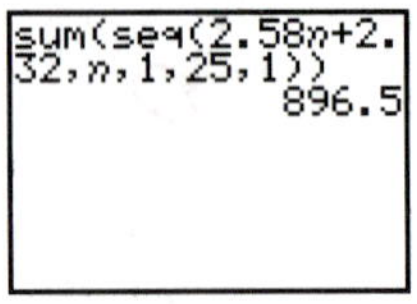

46.

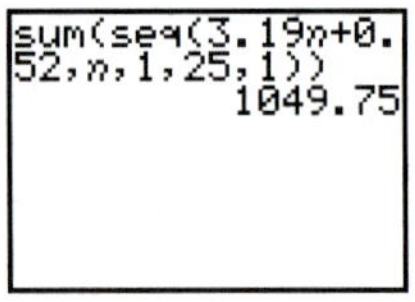

47. $-\frac{3}{2}$ **48.** 1 **49.** 1185 seats **50.** 2160 seats
51. 210 of beige and 190 blue
52. **(a)** 42 bricks **(b)** 2130 bricks **53.** 30 rows
54. 14 years

Historical Problems

1. $1\frac{2}{3}, 10\frac{5}{6}, 20, 29\frac{1}{6}, 38\frac{1}{3}$ **2.** **(a)** 1 **(b)** 2401 **(c)** 2800

7.3 Exercises

1. $r = 3; 3, 9, 27, 81$ **2.** $r = -5; -5, 25, -125, 625$ **3.** $r = \frac{1}{2}; -\frac{3}{2}, -\frac{3}{4}, -\frac{3}{8}, -\frac{3}{16}$ **4.** $r = \frac{5}{2}; \frac{5}{2}, \frac{25}{4}, \frac{125}{8}, \frac{625}{16}$ **5.** $r = 2; \frac{1}{4}, \frac{1}{2}, 1, 2$ **6.** $r = 3; \frac{1}{3}, 1, 3, 9$ **7.** $r = 2^{1/3}; 2^{1/3}, 2^{2/3}, 2, 2^{4/3}$ **8.** $r = 9; 9, 81, 729, 6561$ **9.** $r = \frac{3}{2}; \frac{1}{2}, \frac{3}{4}, \frac{9}{8}, \frac{27}{16}$ **10.** $r = \frac{2}{3}; 2, \frac{4}{3}, \frac{8}{9}, \frac{16}{27}$

11. Arithmetic; $d = 1$ **12.** Arithmetic; $d = 2$ **13.** Neither **14.** Neither **15.** Arithmetic; $d = -\frac{2}{3}$ **16.** Arithmetic; $d = -\frac{3}{4}$

17. Neither **18.** Arithmetic; $d = 2$ **19.** Geometric; $r = \frac{2}{3}$ **20.** Geometric; $r = \frac{5}{4}$ **21.** Geometric; $r = 2$ **22.** Neither

23. Geometric; $r = 3^{1/2}$ **24.** Geometric; $r = -1$ **25.** $a_5 = 162; a_n = 2 \cdot 3^{n-1}$ **26.** $a_5 = -512; a_n = -2 \cdot 4^{n-1}$

27. $a_5 = 5; a_n = 5 \cdot (-1)^{n-1}$ **28.** $a_5 = 96; a_n = 6 \cdot (-2)^{n-1}$ **29.** $a_5 = 0; a_n = 0$ **30.** $a_5 = \frac{1}{81}; a_n = \left(-\frac{1}{3}\right)^{n-1}$

31. $a_5 = 4\sqrt{2}; a_n = (\sqrt{2})^n$ **32.** $a_5 = 0; a_n = 0$ **33.** $a_7 = \frac{1}{64}$ **34.** $a_8 = 2187$ **35.** $a_9 = 1$ **36.** $a_{10} = 512$ **37.** $a_8 = 0.00000004$

38. $a_7 = 100{,}000.0$ **39.** $-\frac{1}{4}(1 - 2^n)$ **40.** $-\frac{1}{6}(1 - 3^n)$ **41.** $2\left[1 - \left(\frac{2}{3}\right)^n\right]$ **42.** $-2(1 - 3^n)$ **43.** $1 - 2^n$ **44.** $5\left[1 - \left(\frac{3}{5}\right)^{n+1}\right]$

45.
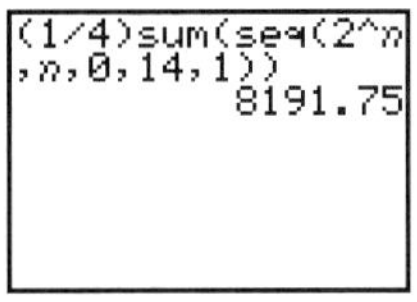
(1/4)sum(seq(2^n
,n,0,14,1))
8191.75

46.
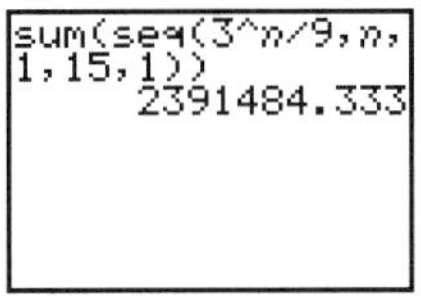
sum(seq(3^n/9,n,
1,15,1))
2391484.333

47.
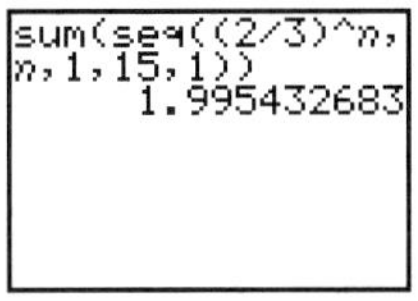
sum(seq((2/3)^n,
n,1,15,1))
1.995432683

48.
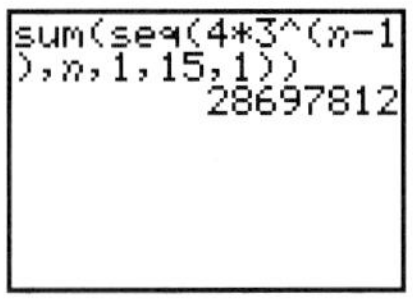
sum(seq(4*3^(n-1
),n,1,15,1))
28697812

49.
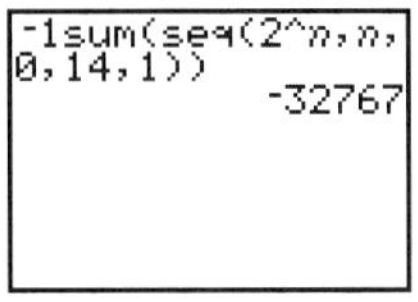
-1sum(seq(2^n,n,
0,14,1))
-32767

50.
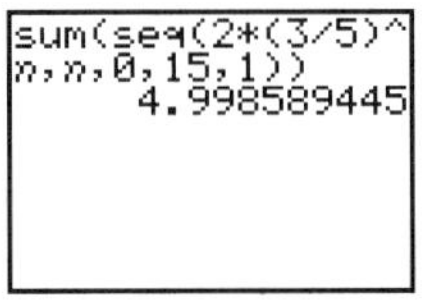
sum(seq(2*(3/5)^
n,n,0,15,1))
4.998589445

51. $\frac{3}{2}$ **52.** 6 **53.** 16 **54.** 9 **55.** $\frac{8}{5}$ **56.** $\frac{4}{7}$ **57.** $\frac{20}{3}$ **58.** 12 **59.** $\frac{18}{5}$ **60.** $\frac{8}{3}$ **61.** -4 **62.** 2 **63.** \$349,496.41 **64.** \$16,712.73 **65.** \$96,885.98 **66.** \$66,438.85 **67.** \$305.10 **68.** \$312.44

69. **(a)** $a_1 = 500; a_n = (1.02)a_{n-1} + 500$ **(b)** At the end of the 82nd quarter. **(c)** On December 31, 2023: \$156,116.15
70. **(a)** $a_1 = 45; a_n = 1.005a_{n-1} + 45$ **(b)** At the end of the 68th month [August 31, 2004] **(c)** On December 31, 2014: \$16,054.57
71. **(a)** $a_1 = 150{,}000; a_n = 1.005a_{n-1} - 899.33$
(b) \$149,850.67
(c)
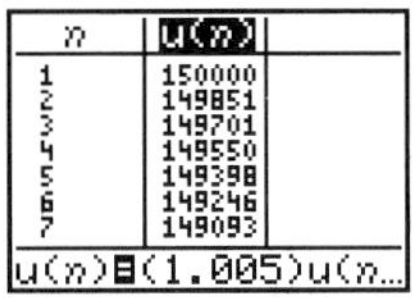
n
u(n)
1
2
3
4
5
6
7
150000
149851
149701
149550
149398
149246
149093
u(n)■(1.005)u(n...

(d) Beginning of the 59th month
(e) 30 years (360 months)
(f) \$173,758.80
(g) (a) $a_1 = 150{,}000; a_n = 1.005a_{n-1} - 999.33$
(b) \$149,750.67
(c)
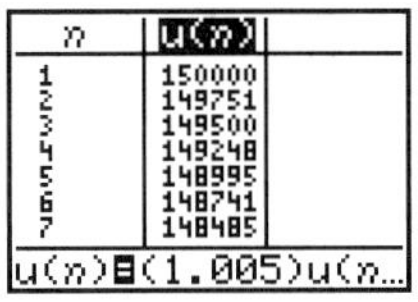
n
u(n)
1
2
3
4
5
6
7
150000
149751
149500
149248
148995
148741
148485
u(n)■(1.005)u(n...

(d) Beginning of 38th month
(e) Beginning of 280th month, last payment made.
(f) \$128,169.20

72. **(a)** $a_1 = 120{,}000; a_n = \left(1 + \frac{0.065}{12}\right)a_{n-1} - 758.48$
(b) \$119,891.52
(c)
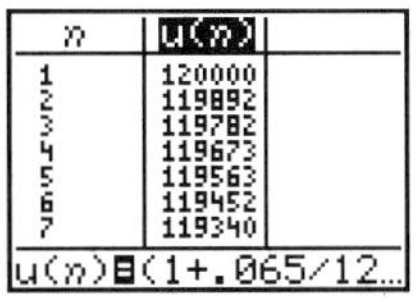
n
u(n)
1
2
3
4
5
6
7
120000
119892
119782
119673
119563
119452
119340
u(n)■(1+.065/12...

(d) Beginning of the 130th month [129 months]
(e) 360 months
(f) \$153,052.96
(g) (a) $a_1 = 120{,}000; a_n = \left(1 + \frac{0.065}{12}\right)a_{n-1} - 858.48$
(b) \$119,791.52
(c)
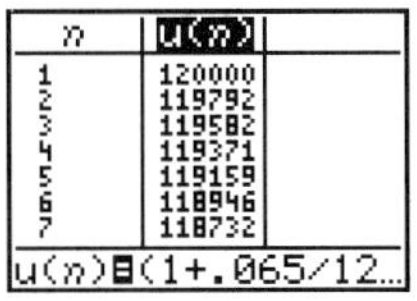
n
u(n)
1
2
3
4
5
6
7
120000
119792
119582
119371
119159
118946
118732
u(n)■(1+.065/12...

(d) Beginning of the 79th month [78 months]
(e) 262 months
(f) \$104,919.12

73. 10 **74.** 20 **75.** \$72.67 **76.** \$39.64 **77.** **(a)** 0.775 ft **(b)** 8th **(c)** 15.88 ft **(d)** 20 ft **78.** **(a)** 15.36 ft **(b)** $30(0.8)^n$ **(c)** 19 times **(d)** 270 ft **79.** \$21,879.11 **80.** \$6655.58 **81.** Option 2 results in the most: \$16,038,304; Option 1 results in the least: \$14,700,000
82. December 20, 1998 (111 days); \$9999.92 **83.** 1.845×10^{19}
84. Yes. A constant function is both arithmetic (with $d = 0$) and geometric (with $r = 1$).
87. 3, 5, 9, 24, 73 **88.** Option *B* results in more money (\$524,287, versus \$500,500 for option *A*)
89. *A*: \$25,250 per year in 5th year, \$112,742 total; *B*: \$24,761 per year in 5th year, \$116,801 total **90.** $\frac{1}{3}$

7.4 Exercises

1. (I) $n = 1: 2 \cdot 1 = 2$ and $1(1 + 1) = 2$
(II) If $2 + 4 + 6 + \cdots + 2k = k(k + 1)$, then $2 + 4 + 6 + \cdots + 2k + 2(k + 1) = (2 + 4 + 6 + \cdots + 2k) + 2(k + 1)$
$= k(k + 1) + 2(k + 1)$
$= k^2 + 3k + 2$
$= (k + 1)(k + 2)$.

2. (I) $n = 1: 4 \cdot 1 - 3 = 1$ and $1(2 \cdot 1 - 1) = 1$
(II) If $1 + 5 + 9 + \cdots + (4k - 3) = k(2k - 1)$, then $1 + 5 + 9 + \cdots + (4k - 3) + [4(k + 1) - 3]$
$= [1 + 5 + 9 + \cdots + (4k - 3)] + (4k + 1)$
$= k(2k - 1) + (4k + 1)$
$= 2k^2 + 3k + 1$
$= (k + 1)(2k + 1)$.

3. (I) $n = 1: 1 + 2 = 3$ and $\frac{1}{2}(1)(1 + 5) = \frac{1}{2}(6) = 3$
(II) If $3 + 4 + 5 + \cdots + (k + 2) = \frac{1}{2}k(k + 5)$, then $3 + 4 + 5 + \cdots + (k + 2) + [(k + 1) + 2]$
$= [3 + 4 + 5 + \cdots + (k + 2)] + (k + 3) = \frac{1}{2}k(k + 5) + k + 3 = \frac{1}{2}(k^2 + 7k + 6) = \frac{1}{2}(k + 1)(k + 6)$

4. (I) $n = 1: 2 \cdot 1 + 1 = 3$ and $1(1 + 2) = 3$
(II) If $3 + 5 + 7 \cdots + (2k + 1) = k(k + 2)$, then $3 + 5 + 7 + \cdots + (2k + 1) + [2(k + 1) + 1]$
$= [3 + 5 + 7 + \cdots + (2k + 1)] + (2k + 3)$
$= k(k + 2) + (2k + 3) = k^2 + 4k + 3$
$= (k + 1)(k + 3)$.

5. (I) $n = 1: 3 \cdot 1 - 1 = 2$ and $\frac{1}{2}(1)[3(1) + 1] = \frac{1}{2}(4) = 2$
(II) If $2 + 5 + 8 + \cdots + (3k - 1) = \frac{1}{2}k(3k + 1)$, then $2 + 5 + 8 + \cdots + (3k - 1) + [3(k + 1) - 1]$
$= [2 + 5 + 8 + \cdots + (3k - 1)] + 3k + 2 = \frac{1}{2}k(3k + 1) + (3k + 2) = \frac{1}{2}(3k^2 + 7k + 4) = \frac{1}{2}(k + 1)(3k + 4)$

6. (I) $n = 1: 3 \cdot 1 - 2 = 1$ and $\frac{1}{2} \cdot 1(3 \cdot 1 - 1) = \frac{1}{2}(2) = 1$
(II) If $1 + 4 + 7 + \cdots + (3k - 2) = \frac{1}{2}k(3k - 1)$, then $1 + 4 + 7 + \cdots + (3k - 2) + [3(k + 1) - 2]$
$= [1 + 4 + 7 + \cdots + (3k - 2)] + (3k + 1)$
$= \frac{1}{2}k(3k - 1) + (3k + 1) = \frac{1}{2}(3k^2 + 5k + 2)$
$= \frac{1}{2}(k + 1)(3k + 2)$.

7. (I) $n = 1: 2^{1-1} = 1$ and $2^1 - 1 = 1$
(II) If $1 + 2 + 2^2 + \cdots + 2^{k-1} = 2^k - 1$, then $1 + 2 + 2^2 + \cdots + 2^{k-1} + 2^{(k+1)-1} = (1 + 2 + 2^2 + \cdots + 2^{k-1}) + 2^k$
$= 2^k - 1 + 2^k = 2(2^k) - 1 = 2^{k+1} - 1$.

8. (I) $n = 1: 3^{1-1} = 1$ and $\frac{1}{2}(3^1 - 1) = \frac{1}{2}(2) = 1$
(II) If $1 + 3 + 3^2 + \cdots + 3^{k-1} = \frac{1}{2}(3^k - 1)$, then $1 + 3 + 3^2 + \cdots + 3^{k-1} + 3^{(k+1)-1} = [1 + 3 + 3^2 + \cdots + 3^{k-1}] + 3^k$
$= \frac{1}{2}(3^k - 1) + 3^k = \frac{1}{2}[3^k - 1 + 2(3^k)] = \frac{1}{2}[3(3^k) - 1]$
$= \frac{1}{2}(3^{k+1} - 1)$.

9. (I) $n = 1: 4^{1-1} = 1$ and $\frac{1}{3}(4^1 - 1) = \frac{1}{3}(3) = 1$
(II) If $1 + 4 + 4^2 + \cdots + 4^{k-1} = \frac{1}{3}(4^k - 1)$, then $1 + 4 + 4^2 + \cdots + 4^{k-1} + 4^{(k+1)-1} = (1 + 4 + 4^2 + \cdots + 4^{k-1}) + 4^k$
$= \frac{1}{3}(4^k - 1) + 4^k = \frac{1}{3}[4^k - 1 + 3(4^k)] = \frac{1}{3}[4(4^k) - 1] = \frac{1}{3}(4^{k+1} - 1)$.

10. (I) $n = 1: 5^{1-1} = 1$ and $\frac{1}{4}(5^1 - 1) = \frac{1}{4}(4) = 1$

(II) If $1 + 5 + 5^2 + \cdots + 5^{k-1} = \frac{1}{4}(5^k - 1)$, then $1 + 5 + 5^2 + \cdots + 5^{k-1} + 5^{(k+1)-1} = [1 + 5 + 5^2 + \cdots + 5^{k-1}] + 5^k$

$= \frac{1}{4}(5^k - 1) + 5^k = \frac{1}{4}[5^k - 1 + 4(5^k)] = \frac{1}{4}[5(5^k) - 1]$

$= \frac{1}{4}(5^{k+1} - 1)$

11. (I) $n = 1: \frac{1}{1 \cdot 2} = \frac{1}{2}$ and $\frac{1}{1+1} = \frac{1}{2}$

(II) If $\frac{1}{1 \cdot 2} + \frac{1}{2 \cdot 3} + \frac{1}{3 \cdot 4} + \cdots + \frac{1}{k(k+1)} = \frac{k}{k+1}$, then $\frac{1}{1 \cdot 2} + \frac{1}{2 \cdot 3} + \frac{1}{3 \cdot 4} + \cdots + \frac{1}{k(k+1)} + \frac{1}{(k+1)[(k+1)+1]}$

$= \left[\frac{1}{1 \cdot 2} + \frac{1}{2 \cdot 3} + \frac{1}{3 \cdot 4} + \cdots + \frac{1}{k(k+1)}\right] + \frac{1}{(k+1)(k+2)} = \frac{k}{k+1} + \frac{1}{(k+1)(k+2)} = \frac{k^2 + 2k + 1}{(k+1)(k+2)} = \frac{k+1}{k+2}.$

12. (I) $n = 1: \frac{1}{(2 \cdot 1 - 1)(2 \cdot 1 + 1)} = \frac{1}{3}$ and $\frac{1}{2 \cdot 1 + 1} = \frac{1}{3}$

(II) If $\frac{1}{1 \cdot 3} + \frac{1}{3 \cdot 5} + \frac{1}{5 \cdot 7} + \cdots + \frac{1}{(2k-1)(2k+1)} = \frac{k}{2k+1}$,

then $\frac{1}{1 \cdot 3} + \frac{1}{3 \cdot 5} + \frac{1}{5 \cdot 7} + \cdots + \frac{1}{(2k-1)(2k+1)} + \frac{1}{[2(k+1) - 1][2(k+1) + 1]}$

$= \left[\frac{1}{1 \cdot 3} + \frac{1}{3 \cdot 5} + \frac{1}{5 \cdot 7} + \cdots + \frac{1}{(2k-1)(2k+1)}\right] + \frac{1}{(2k+1)(2k+3)}$

$= \frac{k}{2k+1} + \frac{1}{(2k+1)(2k+3)}$

$= \frac{2k^2 + 3k + 1}{(2k+1)(2k+3)} = \frac{k+1}{2k+3} = \frac{k+1}{2(k+1)+1}.$

13. (I) $n = 1: 1^2 = 1$ and $\frac{1}{6} \cdot 1 \cdot 2 \cdot 3 = 1$

(II) If $1^2 + 2^2 + 3^2 + \cdots + k^2 = \frac{1}{6}k(k+1)(2k+1)$, then $1^2 + 2^2 + 3^2 + \cdots + k^2 + (k+1)^2$

$= (1^2 + 2^2 + 3^2 + \cdots + k^2) + (k+1)^2 = \frac{1}{6}k(k+1)(2k+1) + (k+1)^2 = \frac{1}{6}(2k^3 + 9k^2 + 13k + 6)$

$= \frac{1}{6}(k+1)(k+2)(2k+3).$

14. (I) $n = 1: 1^3 = 1$ and $\frac{1}{4} \cdot 1^2(1+1)^2 = \frac{1}{4}(4) = 1$

(II) If $1^3 + 2^3 + 3^3 + \cdots + k^3 = \frac{1}{4}k^2(k+1)^2$, then $1^3 + 2^3 + 3^3 + \cdots + k^3 + (k+1)^3$

$= [1^3 + 2^3 + 3^3 + \cdots + k^3] + (k+1)^3$

$= \frac{1}{4}k^2(k+1)^2 + (k+1)^3 = \frac{1}{4}(k+1)^2(k^2 + 4k + 4)$

$= \frac{1}{4}(k+1)^2(k+2)^2$

15. (I) $n = 1: 5 - 1 = 4$ and $\frac{1}{2}(1)(9 - 1) = \frac{1}{2} \cdot 8 = 4$

(II) If $4 + 3 + 2 + \cdots + (5 - k) = \frac{1}{2}k(9 - k)$, then $4 + 3 + 2 + \cdots + (5 - k) + [5 - (k+1)]$

$= [4 + 3 + 2 + \cdots + (5 - k)] + 4 - k = \frac{1}{2}k(9 - k) + 4 - k = \frac{1}{2}(-k^2 + 7k + 8) = \frac{1}{2}(k+1)(8 - k)$

$= \frac{1}{2}(k+1)[9 - (k+1)].$

16. (I) $n = 1: -(1 + 1) = -2$ and $-\frac{1}{2} \cdot 1(1 + 3) = -\frac{1}{2}(4) = -2$

(II) If $-2 - 3 - 4 - \cdots - (k+1) = -\frac{1}{2}k(k+3)$, then $-2 - 3 - 4 - \cdots - (k+1) - (k+2)$

$= [-2 - 3 - 4 - \cdots - (k+1)] - (k+2)$

$= -\frac{1}{2}k(k+3) - (k+2)$

$= -\frac{1}{2}(k^2 + 5k + 4) = -\frac{1}{2}(k+1)(k+4).$

17. (I) $n = 1: 1 \cdot (1 + 1) = 2$ and $\frac{1}{3} \cdot 1 \cdot 2 \cdot 3 = 2$

(II) If $1 \cdot 2 + 2 \cdot 3 + 3 \cdot 4 + \cdots + k(k + 1) = \frac{1}{3}k(k + 1)(k + 2)$, then

$$1 \cdot 2 + 2 \cdot 3 + 3 \cdot 4 + \cdots + k(k + 1) + (k + 1)(k + 2) = [1 \cdot 2 + 2 \cdot 3 + 3 \cdot 4 + \cdots + k(k + 1)] + (k + 1)(k + 2)$$
$$= \frac{1}{3}k(k + 1)(k + 2) + (k + 1)(k + 2) = \frac{1}{3}(k + 1)(k + 2)(k + 3)$$

18. (I) $n = 1: (2 \cdot 1 - 1)(2 \cdot 1) = 2$ and $\frac{1}{3} \cdot 1(1 + 1)(4 \cdot 1 - 1) = \frac{1}{3} \cdot 2 \cdot 3 = 2$

(II) If $1 \cdot 2 + 3 \cdot 4 + 5 \cdot 6 + \cdots + (2k - 1)(2k) = \frac{1}{3}k(k + 1)(4k - 1)$, then

$$1 \cdot 2 + 3 \cdot 4 + 5 \cdot 6 + \cdots + (2k - 1)(2k) + [2(k + 1) - 1][2(k + 1)]$$
$$= [1 \cdot 2 + 3 \cdot 4 + 5 \cdot 6 + \cdots + (2k - 1)(2k)] + (2k + 1)(2k + 2)$$
$$= \frac{1}{3}k(k + 1)(4k - 1) + 2(2k + 1)(k + 1)$$
$$= \frac{1}{3}(k + 1)[4k^2 - k + 6(2k + 1)]$$
$$= \frac{1}{3}(k + 1)(4k^2 + 11k + 6)$$
$$= \frac{1}{3}(k + 1)(k + 2)(4k + 3)$$

19. (I) $n = 1: 1^2 + 1 = 2$ is divisible by 2.
(II) If $k^2 + k$ is divisible by 2, then $(k + 1)^2 + (k + 1) = k^2 + 2k + 1 + k + 1 = (k^2 + k) + 2k + 2$. Since $k^2 + k$ is divisible by 2 and $2k + 2$ is divisible by 2, therefore, $(k + 1)^2 + k + 1$ is divisible by 2.

20. (I) $n = 1: 1^3 + 2 \cdot 1 = 3$ is divisible by 3.
(II) If $k^3 + 2k$ is divisible by 3, then $(k + 1)^3 + 2(k + 1) = k^3 + 3k^2 + 3k + 1 + 2k + 2 = k^3 + 2k + 3k^2 + 3k + 3$. Since $k^3 + 2k$ is divisible by 3 and $3k^2 + 3k + 3$ is divisible by 3, therefore, $(k + 1)^3 + 2(k + 1)$ is divisible by 3.

21. (I) $n = 1: 1^2 - 1 + 2 = 2$ is divisible by 2.
(II) If $k^2 - k + 2$ is divisible by 2, then $(k + 1)^2 - (k + 1) + 2 = k^2 + 2k + 1 - k - 1 + 2 = (k^2 - k + 2) + 2k$. Since $k^2 - k + 2$ is divisible by 2 and $2k$ is divisible by 2, therefore, $(k + 1)^2 - (k + 1) + 2$ is divisible by 2.

22. (I) $n = 1: 1(1 + 2)(1 + 2) = 6$ is divisible by 6.
(II) If $k(k + 1)(k + 2)$ is divisible by 6, then
$(k + 1)(k + 2)(k + 3) = k^3 + 6k^2 + 11k + 6 = k^3 + 3k^2 + 2k + 3k^2 + 9k + 6 = k(k^2 + 3k + 2) + 3(k^2 + 3k + 2)$
$= k(k + 1)(k + 2) + 3(k + 1)(k + 2)$. We know that $k(k + 1)(k + 2)$ is divisible by 6 and $3(k + 1)(k + 2)$ is also divisible by 6 since either $k + 1$ or $k + 2$ must be even (divisible by 2).

23. (I) $n = 1$: If $x > 1$, then $x^1 = x > 1$.
(II) Assume, for any natural number k, that if $x > 1$, then $x^k > 1$. Show that if $x > 1$, then $x^{k+1} > 1$:
$x^{k+1} = x^k \cdot x^1 > 1 \cdot x = x > 1$
↑
$x^k > 1$

24. (I) If $0 < x < 1$, then $0 < x^1 = x < 1$.
(II) Assume that if $0 < x < 1$, then $0 < x^k < 1$ for any natural number k.
Show that, if $0 < x < 1$, then $0 < x^{k+1} < 1$: $0 < x^{k+1} = x^k \cdot x < 1 \cdot x = x < 1$
↑
$x^k < 1$

25. (I) $n = 1$: $a - b$ is a factor of $a^1 - b^1 = a - b$.
(II) If $a - b$ is a factor of $a^k - b^k$, show that $a - b$ is a factor of $a^{k+1} - b^{k+1}$: $a^{k+1} - b^{k+1} = a(a^k - b^k) + b^k(a - b)$.
Since $a - b$ is a factor of $a^k - b^k$ and $a - b$ is a factor of $a - b$, therefore, $a - b$ is a factor of $a^{k+1} - b^{k+1}$.

26. (I) $n = 1$: $a + b$ is a factor of $a^{2\cdot1+1} + b^{2\cdot1+1} = a^3 + b^3 = (a + b)(a^2 - ab + b^2)$
(II) Suppose for some natural number k that $a + b$ is a factor of $a^{2k+1} + b^{2k+1}$. Show that $a + b$ is a factor of $a^{2k+3} + b^{2k+3}$:
$a^{2k+3} + b^{2k+3} = a^2(a^{2k+1} + b^{2k+1}) - b^{2k+1}(a^2 - b^2)$
Since $a + b$ divides $a^{2k+1} + b^{2k+1}$ and $a + b$ divides $a^2 - b^2 = (a + b)(a - b)$, then $a + b$ divides $a^{2k+3} + b^{2k+3}$.

27. $n = 1: 1^2 - 1 + 41 = 41$ is a prime number.
$n = 41: 41^2 - 41 + 41 = 1681 = 41^2$ is not prime.

28. (II) If $2 + 4 + 6 + \cdots + 2k = k^2 + k + 2$, then $2 + 4 + 6 + \cdots + 2k + 2(k + 1)$
$= [2 + 4 + 6 + \cdots + 2k] + 2k + 2 = k^2 + k + 2 + 2k + 2$
$= k^2 + 3k + 4 = (k + 1)^2 + (k + 1) + 2$
(I) $n = 1: 2 \cdot 1 = 2$ but $1^2 + 1 + 2 = 4 \neq 2$

29. (I) $n = 1$: $ar^{1-1} = a \cdot 1 = a$ and $a \cdot \dfrac{1 - r^1}{1 - r} = a$, because $r \neq 1$.

(II) If $a + ar + ar^2 + \cdots + ar^{k-1} = a\left(\dfrac{1 - r^k}{1 - r}\right)$, then $a + ar + ar^2 + \cdots + ar^{k-1} + ar^{(k+1)-1} = (a + ar + ar^2 + \cdots + ar^{k-1}) + ar^k$

$$= a\left(\frac{1 - r^k}{1 - r}\right) + ar^k = \frac{a(1 - r^k) + ar^k(1 - r)}{1 - r} = \frac{a - ar^k + ar^k - ar^{k+1}}{1 - r} = a\left(\frac{1 - r^{k+1}}{1 - r}\right)$$

30. (I) $n = 1$: $[a + (1 - 1)d] = a$ and $1 \cdot a + d\dfrac{1(1 - 1)}{2} = a$

(II) If $a + (a + d) + (a + 2d) + \cdots + [a + (k - 1)d] = ka + d\dfrac{k(k - 1)}{2}$, then

$$a + (a + d) + (a + 2d) + \cdots + [a + (k - 1)d] + (a + kd)$$
$$= \{a + (a + d) + (a + 2d) + \cdots + [a + (k - 1)d]\} + (a + kd)$$
$$= ka + d\frac{k(k - 1)}{2} + a + kd = (k + 1)a + d\frac{k^2 - k + 2k}{2}$$
$$= (k + 1)a + d\frac{k(k + 1)}{2} = (k + 1)a + d\frac{(k + 1)[(k + 1) - 1]}{2}$$

31. (I) $n = 3$: The sum of the angles of a triangle is $(3 - 2) \cdot 180° = 180°$.

(II) Assume for any k that the sum of the angles of a convex polygon of k sides is $(k - 2) \cdot 180°$. A convex polygon of $k + 1$ sides consists of a convex polygon of k sides plus a triangle (see the illustration). The sum of the angles is $(k - 2) \cdot 180° + 180° = (k - 1) \cdot 180°$.

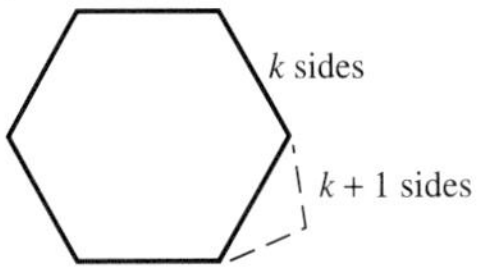

32. (I) $n = 4$: The number of diagonals of a quadrilateral is

$$\frac{1}{2} \cdot 4(4 - 3) = 2$$

(II) Assume that for any k the number of diagonals of a convex polygon of k sides (k vertices) is $\frac{1}{2}k(k - 3)$. A convex polygon of $k + 1$ sides ($k + 1$ vertices) consists of a convex polygon of k sides (k vertices) plus a triangle; see the illustration. The number of diagonals of this convex polygon consists of all the original ones plus $k - 1$ additional ones.

$$\frac{1}{2}k(k - 3) + (k - 1) = \frac{1}{2}(k^2 - k - 2)$$
$$= \frac{1}{2}(k + 1)(k - 2)$$

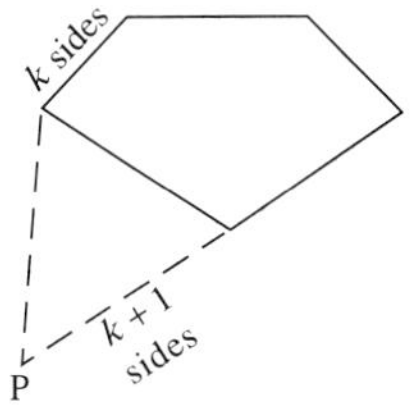

7.5 Exercises

1. 10 **2.** 35 **3.** 21 **4.** 36 **5.** 50 **6.** 4950 **7.** 1 **8.** 1 **9.** 1.866×10^{15} **10.** 4.192×10^{15} **11.** 1.483×10^{13} **12.** 1.767×10^{10}
13. $x^5 + 5x^4 + 10x^3 + 10x^2 + 5x + 1$ **14.** $x^5 - 5x^4 + 10x^3 - 10x^2 + 5x - 1$
15. $x^6 - 12x^5 + 60x^4 - 160x^3 + 240x^2 - 192x + 64$ **16.** $x^5 + 15x^4 + 90x^3 + 270x^2 + 405x + 243$
17. $81x^4 + 108x^3 + 54x^2 + 12x + 1$ **18.** $32x^5 + 240x^4 + 720x^3 + 1080x^2 + 810x + 243$
19. $x^{10} + 5y^2x^8 + 10y^4x^6 + 10y^6x^4 + 5y^8x^2 + y^{10}$ **20.** $x^{12} - 6y^2x^{10} + 15y^4x^8 - 20y^6x^6 + 15y^8x^4 - 6y^{10}x^2 + y^{12}$
21. $x^3 + 6\sqrt{2}x^{5/2} + 30x^2 + 40\sqrt{2}x^{3/2} + 60x + 24\sqrt{2}x^{1/2} + 8$ **22.** $x^2 - 4\sqrt{3}x^{3/2} + 18x - 12\sqrt{3}x^{1/2} + 9$
23. $(ax)^5 + 5by(ax)^4 + 10(by)^2(ax)^3 + 10(by)^3(ax)^2 + 5(by)^4(ax) + (by)^5$
24. $(ax)^4 - 4by(ax)^3 + 6(by)^2(ax)^2 - 4(by)^3(ax) + (by)^4$ **25.** 17,010 **26.** −262,440 **27.** −101,376 **28.** 1760 **29.** 41,472
30. −314,928 **31.** $2835x^3$ **32.** $189x^5$ **33.** $314{,}928x^7$ **34.** $48{,}384x^3$ **35.** 495 **36.** −84 **37.** 3360 **38.** 252 **39.** 1.00501
40. 0.98806 **41.** $\dbinom{n}{n} = \dfrac{n!}{n!(n - n)!} = \dfrac{n!}{n!0!} = \dfrac{n!}{n!} = 1$

42. If n and j are integers with $0 \le j \le n$, then

$$\binom{n}{n-j} = \left(\frac{n!}{(n-j)!(n-(n-j))!}\right) = \frac{n!}{(n-j)!j!} = \binom{n}{j}$$

43. $2^n = (1+1)^n = \binom{n}{0}1^n + \binom{n}{1}(1)(1)^{n-1} + \cdots + \binom{n}{n}1^n = \binom{n}{0} + \binom{n}{1} + \cdots + \binom{n}{n}$

44. Let $0 = (1-1)^n = \binom{n}{0}1^n + \binom{n}{1}(-1)1^{n-1} + \binom{n}{2}(-1)^2 1^{n-2} + \cdots + \binom{n}{n}(-1)^n$

Therefore, $\binom{n}{0} - \binom{n}{1} + \binom{n}{2} - \cdots + (-1)^n\binom{n}{n} = 0$

45. 1

46. $12! = 4.790016 \times 10^8$
$20! = 2.432902 \times 10^{18}$
$25! = 1.551121 \times 10^{25}$
Using Sterlings Formula
$12! \approx 4.79013972 \times 10^8$
$20! \approx 2.43292403 \times 10^{18}$
$25! \approx 1.55112992 \times 10^{25}$

Fill-in-the-Blank Items

1. sequence **2.** arithmetic **3.** geometric **4.** Pascal triangle **5.** 15

True/False Items

1. True **2.** True **3.** True **4.** True **5.** False **6.** False **7.** False

Review Exercises

1. $-\frac{4}{3}, \frac{5}{4}, -\frac{6}{5}, \frac{7}{6}, -\frac{8}{7}$ **2.** $5, -7, 9, -11, 13$ **3.** $2, 1, \frac{8}{9}, 1, \frac{32}{25}$ **4.** $e, \frac{e^2}{2}, \frac{e^3}{3}, \frac{e^4}{4}, \frac{e^5}{5}$ **5.** $3, 2, \frac{4}{3}, \frac{8}{9}, \frac{16}{27}$ **6.** $4, -1, \frac{1}{4}, -\frac{1}{16}, \frac{1}{64}$ **7.** $2, 0, 2, 0, 2$

8. $-3, 1, 5, 9, 13$ **9.** Arithmetic; $d = 1$; $\frac{n}{2}(n+11)$ **10.** Arithmetic; $d = 4$; $2n^2 + 5n$ **11.** Neither **12.** Neither

13. Geometric; $r = 8$; $\frac{8}{7}(8^n - 1)$ **14.** Geometric; $r = 9$; $\frac{9}{8}(9^n - 1)$ **15.** Arithmetic; $d = 4$; $2n(n-1)$

16. Arithmetic; $d = -4$; $3n - 2n^2$ **17.** Geometric; $r = \frac{1}{2}$; $6\left[1 - \left(\frac{1}{2}\right)^n\right]$ **18.** Geometric; $r = -\frac{1}{3}$; $\frac{15}{4}\left(1 - \left(-\frac{1}{3}\right)^n\right)$ **19.** Neither

20. Neither **21.** 115 **22.** 50 **23.** 75 **24.** -18 **25.** 0.49977 **26.** 682 **27.** 35 **28.** -13 **29.** $\frac{1}{10^{10}}$ **30.** 1024 **31.** $9\sqrt{2}$

32. $2^{9/2} = 16\sqrt{2}$ **33.** $5n - 4$ **34.** $4 - 3n$ **35.** $n - 10$ **36.** $6 + 2n$ **37.** $\frac{9}{2}$ **38.** 4 **39.** $\frac{4}{3}$ **40.** $\frac{18}{5}$ **41.** 8 **42.** $\frac{12}{7}$

43. (I) $n = 1: 3 \cdot 1 = 3$ and $\frac{3 \cdot 1}{2}(2) = 3$

(II) If $3 + 6 + 9 + \cdots + 3k = \frac{3k}{2}(k+1)$, then $3 + 6 + 9 + \cdots + 3k + 3(k+1) = (3 + 6 + 9 + \cdots + 3k) + (3k+3)$

$= \frac{3k}{2}(k+1) + (3k+3) = \frac{3k^2}{2} + \frac{9k}{2} + \frac{6}{2} = \frac{3}{2}(k^2 + 3k + 2) = \frac{3}{2}(k+1)(k+2) = \frac{3(k+1)}{2}[(k+1)+1].$

44. (I) $n = 1: 4 \cdot 1 - 2 = 2$ and $2 \cdot 1^2 = 2$

(II) If $2 + 6 + 10 + \cdots + (4k-2) = 2k^2$, then $2 + 6 + 10 + \cdots + (4k-2) + [4(k+1) - 2]$
$= [2 + 6 + 10 + \cdots + (4k-2)] + (4k+2)$
$= 2k^2 + (4k+2)$
$= 2(k^2 + 2k + 1) = 2(k+1)^2$

45. (I) $n = 1: 2 \cdot 3^{1-1} = 2$ and $3^1 - 1 = 2$

(II) If $2 + 6 + 18 + \cdots + 2 \cdot 3^{k-1} = 3^k - 1$, then $2 + 6 + 18 + \cdots + 2 \cdot 3^{k-1} + 2 \cdot 3^{(k+1)-1}$
$= (2 + 6 + 18 + \cdots + 2 \cdot 3^{k-1}) + 2 \cdot 3^k = 3^k - 1 + 2 \cdot 3^k = 3 \cdot 3^k - 1 = 3^{k+1} - 1.$

46. (I) $n = 1: 3 \cdot 2^{1-1} = 3$ and $3(2^1 - 1) = 3$

(II) If $3 + 6 + 12 + \cdots + 3 \cdot 2^{k-1} = 3(2^k - 1)$, then $3 + 6 + 12 + \cdots + 3 \cdot 2^{k-1} + 3 \cdot 2^k$
$= [3 + 6 + 12 + \cdots + 3 \cdot 2^{k-1}] + 3 \cdot 2^k$
$= 3(2^k - 1) + 3 \cdot 2^k$
$= 3 \cdot 2^k - 3 + 3 \cdot 2^k = 3(2^{k+1} - 1).$

47. (I) $n = 1: 1^2 = 1$ and $\frac{1}{2}(6 - 3 - 1) = \frac{1}{2}(2) = 1$

(II) If $1^2 + 4^2 + 7^2 + \cdots + (3k - 2)^2 = \frac{1}{2}k(6k^2 - 3k - 1)$, then

$1^2 + 4^2 + 7^2 + \cdots + (3k - 2)^2 + [3(k + 1) - 2]^2 = [1^2 + 4^2 + 7^2 + \cdots + (3k - 2)^2] + (3k + 1)^2$

$= \frac{1}{2}k(6k^2 - 3k - 1) + (3k + 1)^2$

$= \frac{1}{2}(6k^3 + 15k^2 + 11k + 2) = \frac{1}{2}(k + 1)(6k^2 + 9k + 2) = \frac{1}{2}(k + 1)[6(k + 1)^2 - 3(k + 1) - 1].$

48. (I) $n = 1: 1(1 + 2) = 3$ and $\frac{1}{6}(1 + 1)(2 \cdot 1 + 7) = 3$

(II) If $1 \cdot 3 + 2 \cdot 4 + 3 \cdot 5 + \cdots + k(k + 2) = \frac{k}{6}(k + 1)(2k + 7)$, then $1 \cdot 3 + 2 \cdot 4 + 3 \cdot 5 + \cdots + k(k + 2) + (k + 1)(k + 3)$

$= (1 \cdot 3 + 2 \cdot 4 + 3 \cdot 5 + \cdots + k(k + 2)] + (k + 1)(k + 3)$

$= \frac{k}{6}(k + 1)(2k + 7) + (k + 1)(k + 3)$

$= (k + 1)\left[\frac{k}{6}(2k + 7) + (k + 3)\right]$

$= (k + 1)\left[\frac{2k^2}{6} + \frac{7k}{6} + k + 3\right] = (k + 1)\left[\frac{2k^2 + 7k + 6k + 18}{6}\right]$

$= \frac{k + 1}{6}(2k^2 + 13k + 18) = \frac{k + 1}{6}(k + 2)(2k + 9) = \frac{k + 1}{6}[(k + 1) + 1][2(k + 1) + 7].$

49. $x^5 + 10x^4 + 40x^3 + 80x^2 + 80x + 32$ **50.** $x^4 - 12x^3 + 54x^2 - 108x + 81$ **51.** $32x^5 + 240x^4 + 720x^3 + 1080x^2 + 810x + 243$ **52.** $81x^4 - 432x^3 + 864x^2 - 768x + 256$ **53.** 144 **54.** −13,608 **55.** 84 **56.** 1792 **57.** **(a)** 8 **(b)** 1100 **58.** 360 **59.** $151,873.77

60. $244,129.08 **61.** **(a)** $\left(\frac{3}{4}\right)^3 \cdot 20 = \frac{135}{16}$ ft **(b)** $20\left(\frac{3}{4}\right)^n$ ft **(c)** after the 13th time **(d)** 140 ft **62.** $23,397.17

63. **(a)** $a_1 = 190{,}000$; $a_n = \left(1 + \frac{0.0675}{12}\right)a_{n-1} - 1232.34$

(b) $189,836.41

(c)

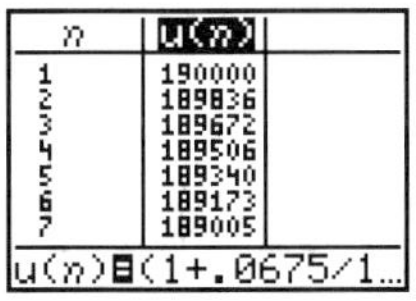

n
u(n)
1 190000
2 189836
3 189672
4 189506
5 189340
6 189173
7 189005
u(n)■(1+.0675/1...

(d) Beginning of 253rd month

(e) 360

(f) $253,642.40

(g) (a) $a_1 = 190{,}000$; $a_n = \left(1 + \frac{0.0675}{12}\right)a_{n-1} - 1332.34$

(b) $189,736.41

(c)

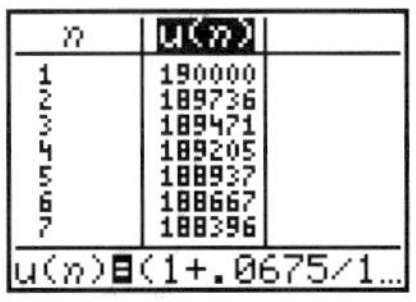

n
u(n)
1 190000
2 189736
3 189471
4 189205
5 188937
6 188667
7 188396
u(n)■(1+.0675/1...

(d) Beginning of 193rd month

(e) At the beginning of the 290th month

(f) $196,266.20

CHAPTER 8 Counting and Probability

8.1 Exercises

1. {1, 3, 5, 6, 7, 9} **2.** {1, 2, 3, 4, 5, 6, 7, 8, 9} **3.** {1, 5, 7} **4.** {1, 9} **5.** {1, 6, 9} **6.** {1, 6, 9} **7.** {1, 2, 4, 5, 6, 7, 8, 9} **8.** {1, 2, 3, 4, 5, 6, 7, 8, 9} **9.** {1, 2, 4, 5, 6, 7, 8, 9} **10.** {1} **11.** {0, 2, 6, 7, 8} **12.** {0, 2, 5, 7, 8, 9} **13.** {0, 1, 2, 3, 5, 6, 7, 8, 9} **14.** {0, 5, 9} **15.** {0, 1, 2, 3, 5, 6, 7, 8, 9} **16.** {0, 5, 9} **17.** {0, 1, 2, 3, 4, 6, 7, 8} **18.** {2, 7, 8} **19.** {0} **20.** {0, 1, 2, 3, 5, 6, 7, 8, 9} **21.** ∅, {a}, {b}, {c}, {d}, {a, b}, {a, c}, {a, d}, {b, c}, {b, d}, {c, d}, {a, b, c}, {b, c, d}, {a, c, d}, {a, b, d}, {a, b, c, d} **22.** ∅, {a}, {b}, {c}, {d}, {e}, {a, b}, {a, c}, {a, d}, {a, e}, {b, c}, {b, d}, {b, e}, {c, d}, {c, e}, {d, e}, {a, b, c}, {a, b, d}, {a, b, e}, {a, c, d}, {a, c, e}, {a, d, e}, {b, c, d}, {b, c, e}, {b, d, e}, {c, d, e}, {a, b, c, d}, {a, b, c, e}, {a, b, d, e}, {a, c, d, e}, {b, c, d, e}, {a, b, c, d, e} **23.** 25 **24.** 25 **25.** 40 **26.** 50 **27.** 25 **28.** 20 **29.** 37 **30.** 8 **31.** 18 **32.** 31 **33.** 5 **34.** 52 **35.** 175; 125 **36.** 550 **37.** **(a)** 15 **(b)** 15 **(c)** 15 **(d)** 25 **(e)** 40

38.

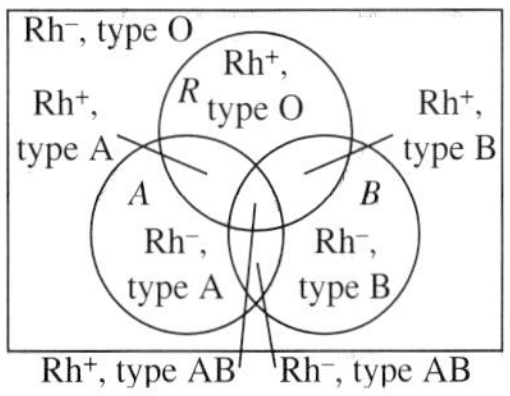

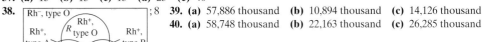

; 8 **39.** **(a)** 57,886 thousand **(b)** 10,894 thousand **(c)** 14,126 thousand **40.** **(a)** 58,748 thousand **(b)** 22,163 thousand **(c)** 26,285 thousand

8.2 Exercises

1. 30 **2.** 42 **3.** 120 **4.** 24 **5.** 1 **6.** 1 **7.** 336 **8.** 6720 **9.** 28 **10.** 28 **11.** 15 **12.** 15 **13.** 1 **14.** 18 **15.** 10,400,600 **16.** 48,620
17. {*abc, abd, abe, acb, acd, ace, adb, adc, ade, aeb, aec, aed*
bac, bad, bae, bca, bcd, bce, bda, bdc, bde, bea, bec, bed
cab, cad, cae, cba, cbd, cbe, cda, cdb, cde, cea, ceb, ced
dab, dac, dae, dba, dbc, dbe, dca, dcb, dce, dea, deb, dec
eab, eac, ead, eba, ebc, ebd, eca, ecb, ecd, eda, edb, edc}; 60
18. {*ab, ac, ad, ae, ba, bc, bd, be, ca, cb, cd, ce, da, db, dc, de, ea, eb, ec, ed*}; 20
19. {123, 124, 132, 134, 142, 143, 213, 214, 231, 234, 241, 243, 312, 314, 321, 324, 341, 342, 412, 413, 421, 423, 431, 432}; 24
20. {123, 124, 125, 126, 132, 134, 135, 136, 142, 143, 145, 146, 152, 153, 154, 156, 162, 163, 164, 165, 213, 214, 215, 216, 231, 234, 235, 236, 241, 243, 245, 246, 251, 253, 254, 256, 261, 263, 264, 265, 312, 314, 315, 316, 321, 324, 325, 326, 341, 342, 345, 346, 351, 352, 354, 356, 361, 362, 364, 365, 412, 413, 415, 416, 421, 423, 425, 426, 431, 432, 435, 436, 451, 452, 453, 456, 461, 462, 463, 465, 512, 513, 514, 516, 521, 523, 524, 526, 531, 532, 534, 536, 541, 542, 543, 546, 561, 562, 563, 564, 612, 613, 614, 615, 621, 623, 624, 625, 631, 632, 634, 635, 641, 642, 643, 645, 651, 652, 653, 654}; 120 **21.** {*abc, abd, abe, acd, ace, ade, bcd, bce, bde, cde*}; 10
22. {*ab, ac, ad, ae, bc, bd, be, cd, ce, de*}; 10 **23.** {123, 124, 134, 234}; 4
24. {123, 124, 125, 126, 134, 135, 136, 145, 146, 156, 234, 235, 236, 245, 246, 256, 345, 346, 356, 456}; 20 **25.** 15 **26.** 15 **27.** 16 **28.** 25
29. 8 **30.** 1000 **31.** 24 **32.** 120 **33.** 60 **34.** 360 **35.** 18,278 **36.** 12,338,352 **37.** 35 **38.** 56 **39.** 1024 **40.** 1024 **41.** 9000
42. 80,000 **43.** 120 **44. (a)** 6,760,000 **(b)** 3,407,040 **(c)** 3,276,000 **45.** 480 **46.** 125,000 **47.** 48,228,180
48. $365 \cdot 364 \cdot 363 \cdot 362 \cdot 361$ **49.** 336 **50.** 26,611,200 **51.** 5,209,344 **52.** 302,400 **53.** 362,880 **54.** 40,320 **55.** 90,720
56. 4,989,600 **57.** 1.156×10^{76} **58.** 70 **59.** 15 **60.** 126 **61. (a)** 63 **(b)** 35 **(c)** 1 **62. (a)** 3003 **(b)** 20,475 **(c)** 16,653

Historical Problems

1. (a) {*AA, ABA, BAA, ABBA, BBAA, BABA, BBB, ABBB, BABB, BBAB*}
(b) $P(A \text{ wins}) = \dfrac{C(4,2)+C(4,3)+C(4,4)}{2^4} = \dfrac{6+4+1}{16} = \dfrac{11}{16}$
$P(B \text{ wins}) = \dfrac{C(4,3)+C(4,4)}{2^4} = \dfrac{4+1}{16} = \dfrac{5}{16}$
The outcomes listed in part (a) are not equally likely.

2. (a) $E = \frac{1}{6}\cdot 0 + \frac{1}{6}\cdot 0 + \frac{1}{6}\cdot 0 + \frac{1}{6}\cdot 0 + \frac{1}{6}\cdot 6 + \frac{1}{6}\cdot 3 = \frac{6}{6} + \frac{3}{6} = 1 + \frac{1}{2} = \$1.50$
(b) $E = \frac{1}{6}(-1) + \frac{1}{6}(-1) + \frac{1}{6}(-1) + \frac{1}{6}(-1) + \frac{1}{6}(5) + \frac{1}{6}(2) = -\frac{1}{6} - \frac{1}{6} - \frac{1}{6} - \frac{1}{6} + \frac{5}{6} + \frac{2}{6} = \frac{3}{6} = \$0.50$

8.3 Exercises

1. 0, 0.01, 0.35, 1 **2.** $\frac{1}{2}, \frac{3}{4}, \frac{2}{3}, 0$ **3.** Probability model **4.** Probability model **5.** Not a probability model **6.** Not a probability model
7. $S = \{\text{HH, HT, TH, TT}\}$; $P(\text{HH}) = \frac{1}{4}, P(\text{HT}) = \frac{1}{4}, P(\text{TH}) = \frac{1}{4}, P(\text{TT}) = \frac{1}{4}$
8. $S = \{\text{HH, HT, TH, TT}\}$; $P(\text{HH}) = \frac{1}{4}, P(\text{HT}) = \frac{1}{4}, P(\text{TH}) = \frac{1}{4}, P(\text{TT}) = \frac{1}{4}$
9. S = {HH1, HH2, HH3, HH4, HH5, HH6, HT1, HT2, HT3, HT4, HT5, HT6, TH1, TH2, TH3, TH4, TH5, TH6, TT1, TT2, TT3, TT4, TT5, TT6}; each outcome has the probability of $\frac{1}{24}$.
10. S = {H1H, H1T, H2H, H2T, H3H, H3T, H4H, H4T, H5H, H5T, H6H, H6T, T1H, T1T, T2H, T2T, T3H, T3T, T4H, T4T, T5H, T5T, T6H, T6T}; each outcome has the probability of $\frac{1}{24}$.
11. S = {HHH, HHT, HTH, HTT, THH, THT, TTH, TTT}; each outcome has the probability of $\frac{1}{8}$.
12. S = {HHH, HHT, HTH, HTT, THH, THT, TTH, TTT}; each outcome has the probability of $\frac{1}{8}$.
13. S = {1 Yellow, 1 Red, 1 Green, 2 Yellow, 2 Red, 2 Green, 3 Yellow, 3 Red, 3 Green, 4 Yellow, 4 Red, 4 Green}; each outcome has the probability of $\frac{1}{12}$; thus, $P(\text{2 Red}) + P(\text{4 Red}) = \frac{1}{12} + \frac{1}{12} = \frac{1}{6}$.
14. S = {Forward Yellow, Forward Red, Forward Green, Backward Yellow, Backward Red, Backward Green}; each outcome has the probability of $\frac{1}{6}$; thus, $P(\text{Forward Yellow}) + P(\text{Forward Green}) = \frac{1}{6} + \frac{1}{6} = \frac{1}{3}$.

15. S = {1 Yellow Forward, 1 Yellow Backward, 1 Red Forward, 1 Red Backward, 1 Green Forward, 1 Green Backward, 2 Yellow Forward, 2 Yellow Backward, 2 Red Forward, 2 Red Backward, 2 Green Forward, 2 Green Backward, 3 Yellow Forward, 3 Yellow Backward, 3 Red Forward, 3 Red Backward, 3 Green Forward, 3 Green Backward, 4 Yellow Forward, 4 Yellow Backward, 4 Red Forward, 4 Red Backward, 4 Green Forward, 4 Green Backward}; each outcome has the probability of $\frac{1}{24}$; thus, $P(\text{1 Red Backward}) + P(\text{1 Green Backward}) = \frac{1}{24} + \frac{1}{24} = \frac{1}{12}$.
16. S = {Yellow 1 Forward, Yellow 1 Backward, Yellow 2 Forward, Yellow 2 Backward, Yellow 3 Forward, Yellow 3 Backward, Yellow 4 Forward, Yellow 4 Backward, Red 1 Forward, Red 1 Backward, Red 2 Forward, Red 2 Backward, Red 3 Forward, Red 3 Backward, Red 4 Forward, Red 4 Backward, Green 1 Forward, Green 1 Backward, Green 2 Forward, Green 2 Backward, Green 3 Forward, Green 3 Backward, Green 4 Forward, Green 4 Backward}; each outcome has the probability of $\frac{1}{24}$; thus, $P(\text{Yellow 2 Forward}) + P(\text{Yellow 4 Forward}) = \frac{1}{24} + \frac{1}{24} = \frac{1}{12}$.
17. S = {11 Red, 11 Yellow, 11 Green, 12 Red, 12 Yellow, 12 Green, 13 Red, 13 Yellow, 13 Green, 14 Red, 14 Yellow, 14 Green, 21 Red, 21 Yellow, 21 Green, 22 Red, 22 Yellow, 22 Green, 23 Red, 23 Yellow, 23 Green, 24 Red, 24 Yellow, 24 Green, 31 Red, 31 Yellow, 31 Green, 32 Red, 32 Yellow, 32 Green, 33 Red, 33 Yellow, 33 Green, 34 Red, 34 Yellow, 34 Green, 41 Red, 41 Yellow, 41 Green, 42 Red, 42 Yellow, 42 Green, 43 Red, 43 Yellow, 43 Green, 44 Red, 44 Yellow, 44 Green}; each outcome has the probability of $\frac{1}{48}$; thus, E = {22 Red, 22 Green, 24 Red, 24 Green}; $P(E) = \frac{n(E)}{n(S)} = \frac{4}{48} = \frac{1}{12}$.
18. S = {Forward 11, Forward 12, Forward 13, Forward 14, Forward 21, Forward 22, Forward 23, Forward 24, Forward 31, Forward 32, Forward 33, Forward 34, Forward 41, Forward 42, Forward 43, Forward 44, Backward 11, Backward 12, Backward 13, Backward 14, Backward 21, Backward 22, Backward 23, Backward 24, Backward 31, Backward 32, Backward 33, Backward 34, Backward 41, Backward 42, Backward 43, Backward 44}; each outcome has the probability of $\frac{1}{32}$; thus, E = {Forward 12, Forward 14, Forward 32, Forward 34}; $P(E) = \frac{n(E)}{n(S)} = \frac{4}{32} = \frac{1}{8}$.
19. A, B, C, F **20.** A **21.** B **22.** F **23.** $\frac{4}{5}; \frac{1}{5}$ **24.** $\frac{2}{3}, \frac{1}{3}$ **25.** $P(1) = P(3) = P(5) = \frac{2}{9}; P(2) = P(4) = P(6) = \frac{1}{9}$ **26.** $P(1) = P(2) = P(3) = P(4) = P(5) = \frac{1}{5}; P(6) = 0$ **27.** $\frac{3}{10}$ **28.** $\frac{2}{5}$ **29.** $\frac{1}{2}$ **30.** $\frac{1}{2}$ **31.** $\frac{1}{6}$ **32.** $\frac{7}{30}$ **33.** $\frac{1}{8}$ **34.** $\frac{1}{8}$ **35.** $\frac{1}{4}$ **36.** $\frac{3}{8}$ **37.** $\frac{1}{6}$ **38.** $\frac{1}{18}$ **39.** $\frac{1}{18}$ **40.** $\frac{1}{36}$ **41.** 0.55 **42.** 0.10 **43.** 0.70 **44.** 0 **45.** 0.30 **46.** 0.50 **47.** 0.747 **48.** 0.944 **49.** 0.7 **50.** 0.96 **51.** $\frac{17}{20}$ **52.** $\frac{3}{5}$ **53.** $\frac{11}{20}$ **54.** $\frac{3}{5}$ **55.** $\frac{1}{2}$ **56.** $\frac{1}{10}$ **57.** $\frac{3}{10}$ **58.** $\frac{13}{20}$ **59.** $\frac{2}{5}$ **60.** $\frac{3}{5}$ **61. (a)** 0.57 **(b)** 0.95 **(c)** 0.83 **(d)** 0.38 **(e)** 0.29 **(f)** 0.05 **(g)** 0.78 **(h)** 0.71 **62. (a)** 0.45 **(b)** 0.75 **(c)** 0.9 **63. (a)** $\frac{25}{33}$ **(b)** $\frac{25}{33}$ **64. (a)** $\frac{7}{13}$ **(b)** $\frac{11}{13}$ **65.** 0.167 **66.** 0.814 **67.** 0.000033069 **68.** 0.117 **69. (a)** $\frac{10}{32}$ **(b)** $\frac{1}{32}$ **70. (a)** $\frac{1}{4}$ **(b)** $\frac{5}{16}$ **71. (a)** 0.00463 **(b)** 0.049 **72. (a)** 0.869 **(b)** 0.402 **73.** 7.02×10^{-6}; 0.183 **74.** 0.000018 **75.** 0.1

8.4 Exercises

1. qualitative **2.** quantitative **3.** quantitative **4.** qualitative **5.** quantitative **6.** qualitative **7.** qualitative **8.** quantitative **9. (a)** Northwest **(b)** TWA **(c)** About 78% **(d)** About 78% **10. (a)** About $40,000 **(b)** 15-year fixed; about $50,000 **11. (a)** Housing, fuel, and utilities **(b)** Misc. goods and services **12. (a)** 80% **(b)** 20% **(c)** 48% **13. (a)** 13 **(b)** 20; 24 **(c)** 5 **(d)** About 145,000 **(e)** 30–34 **(f)** 80–84 **14. (a)** 8 **(b)** 85; 89 **(c)** 5 **(d)** 16 **(e)** 22 **(f)** 73

15. **(a)**

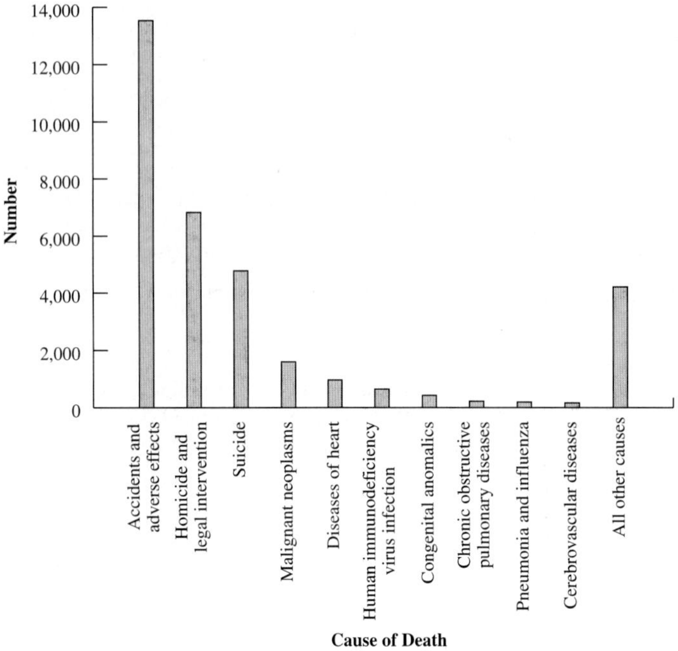

(b)

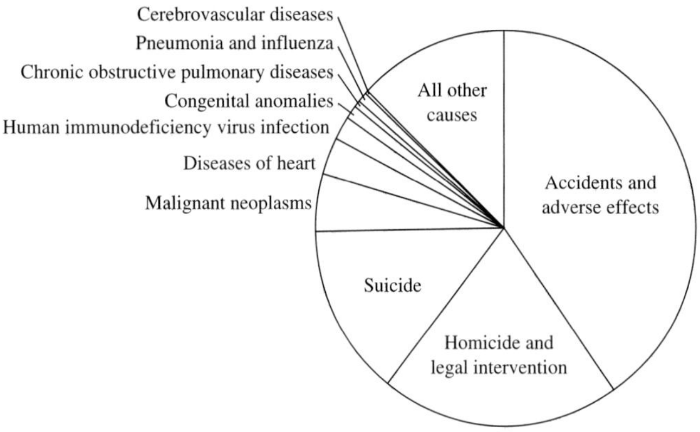

(d) Accidents and adverse effects.

(e)

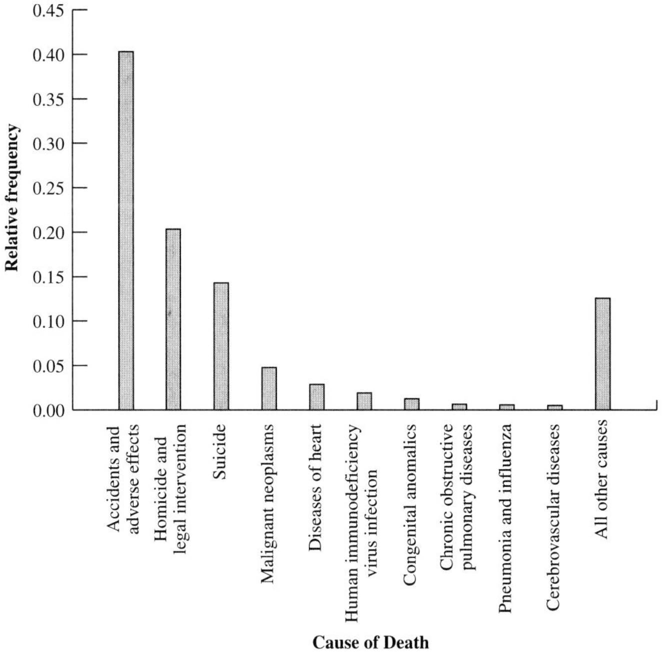

(f)

Cause of Death	Probability
Accidents and adverse effects	0.403
Homicide and legal intervention	0.203
Suicide	0.143
Malignant neoplasms	0.048
Diseases of heart	0.029
Human immunodeficiency virus infection	0.019
Congenital anomalies	0.013
Chronic obstructive pulmonary diseases	0.007
Pneumonia and influenza	0.006
Cerebrovascular diseases	0.005
All other causes	0.125

(g) 14.3%
(h) 19.1%
(i) 80.9%

16. (a)

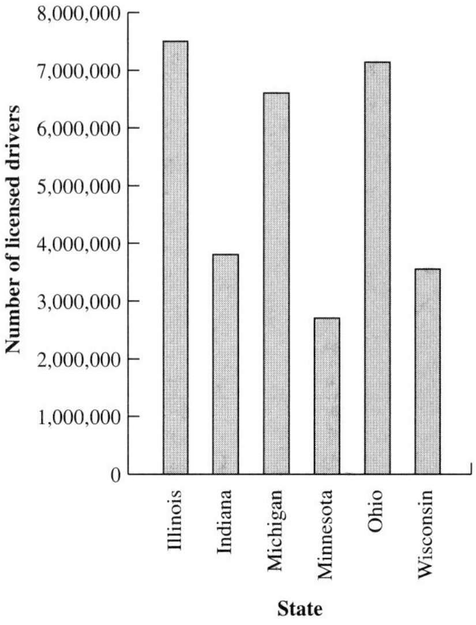

(b)

(c) Answers may vary.

(d) Illinois
(e) Minnesota

(f)

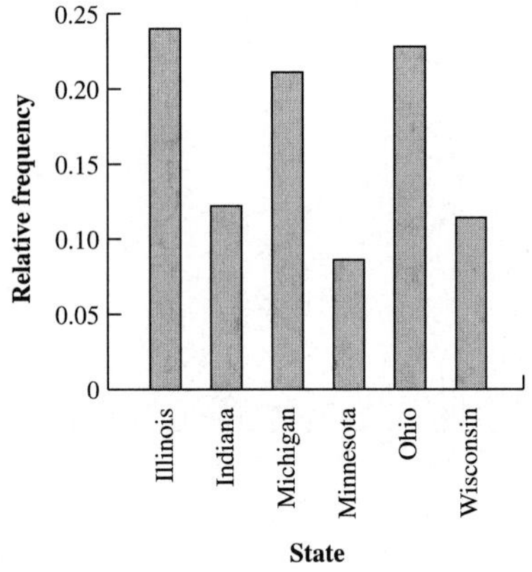

(g)

State	Probability
Illinois	0.2396
Indiana	0.1216
Michigan	0.2108
Minnesota	0.0864
Ohio	0.2281
Wisconsin	0.1135

(h) 21.08% **(i)** 36.12% **(j)** 63.88%

17. (a) 13 **(b)** 20; 24 **(c)** 5

(d)

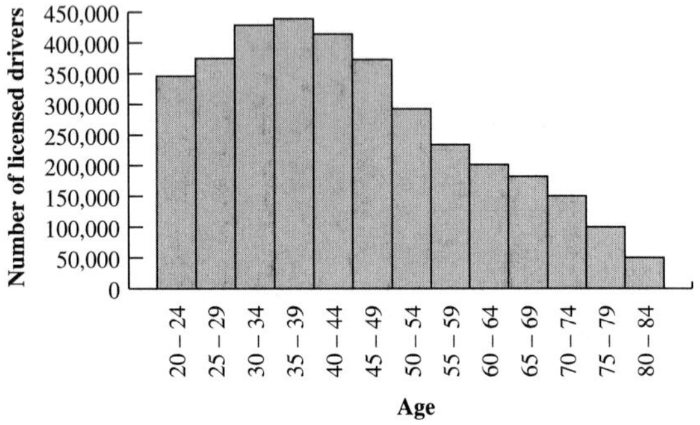

(e)

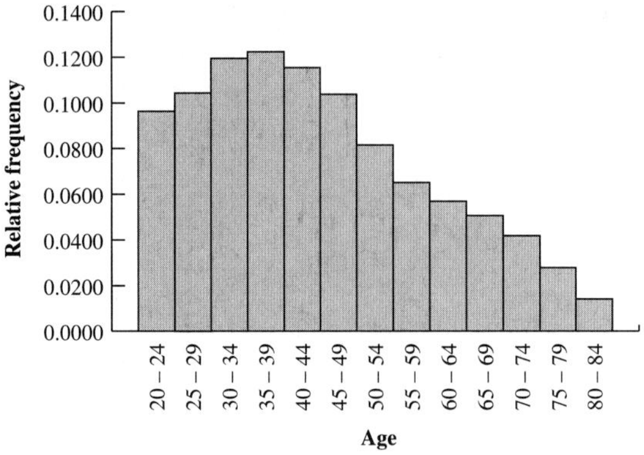

(f) 35–39 **(g)** 80–84 **(h)**

Age	Probability
20–24	0.0964
25–29	0.1044
30–34	0.1195
35–39	0.1224
40–44	0.1155
45–49	0.1039
50–54	0.0815
55–59	0.0651
60–64	0.0569
65–69	0.0507
70–74	0.0419
75–79	0.0279
80–84	0.0140

(i) 10.39%

18. **(a)** 13 **(b)** 20; 24 **(c)** 5

(d)

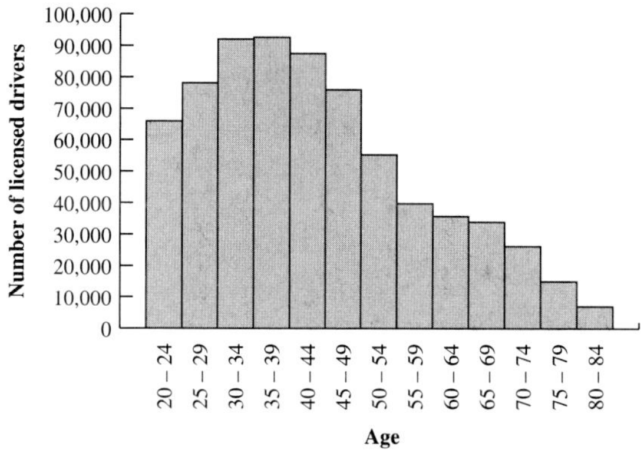

(e)

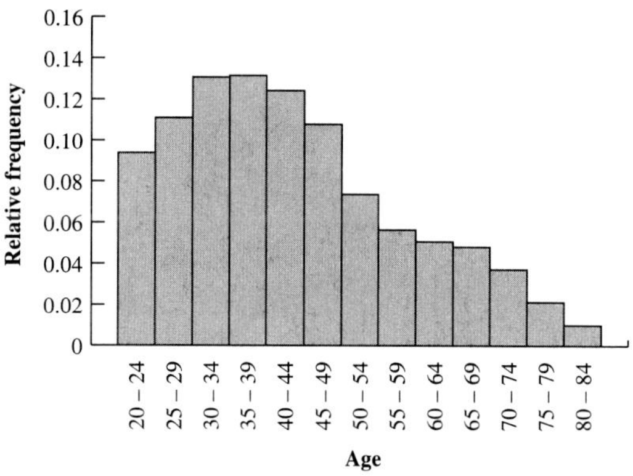

(f) 35–39 **(g)** 80–84 **(h)**

Age	Probability
20–24	0.0936
25–29	0.1109
30–34	0.1306
35–39	0.1314
40–44	0.1241
45–49	0.1078
50–54	0.0784
55–59	0.0563
60–64	0.0506
65–69	0.0481
70–74	0.0371
75–79	0.0213
80–84	0.0099

(i) 10.78%

19. **(a)** 15 **(b)** 0; 999 **(c)** 1000

(d)

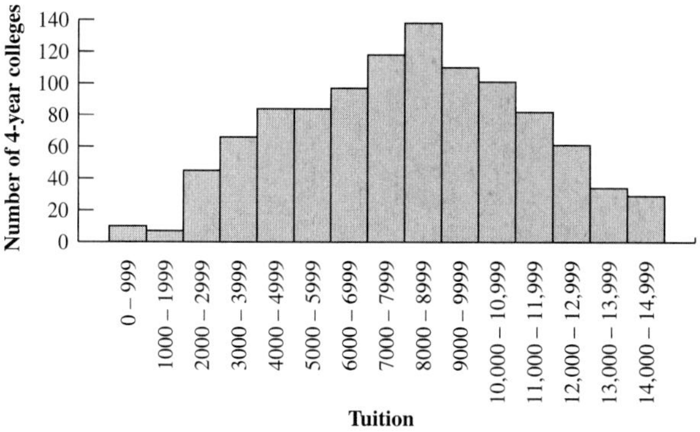

(e)

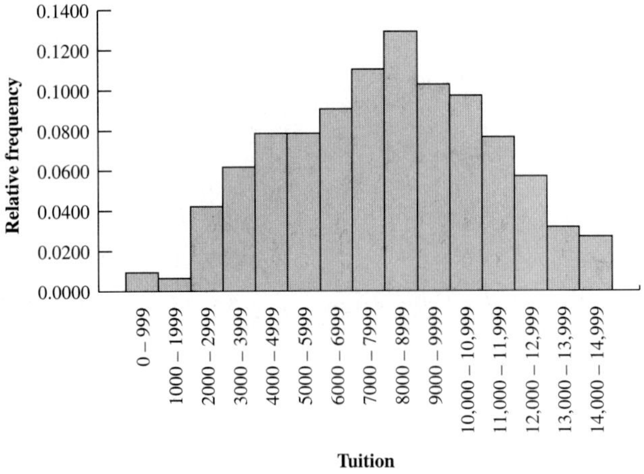

(f) 8000–8999

(g)

Tuition	Probability
0–999	0.0094
1000–1999	0.0065
2000–2999	0.0421
3000–3999	0.0617
4000–4999	0.0786
5000–5999	0.0786
6000–6999	0.0907
7000–7999	0.1104
8000–8999	0.1291
9000–9999	0.1029
10,000–10,999	0.0973
11,000–11,999	0.0767
12,000–12,999	0.0571
13,000–13,999	0.0318
14,000–14,999	0.0271

(h) 7.67%

20. (a) 15 **(b)** 0; 999 **(c)** 1000 **(d)**

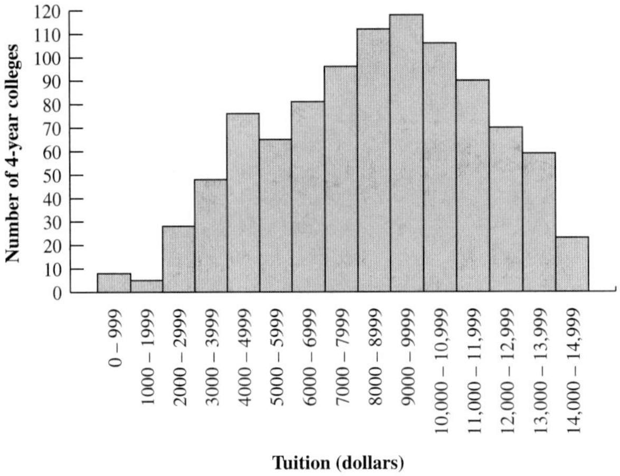

(e)

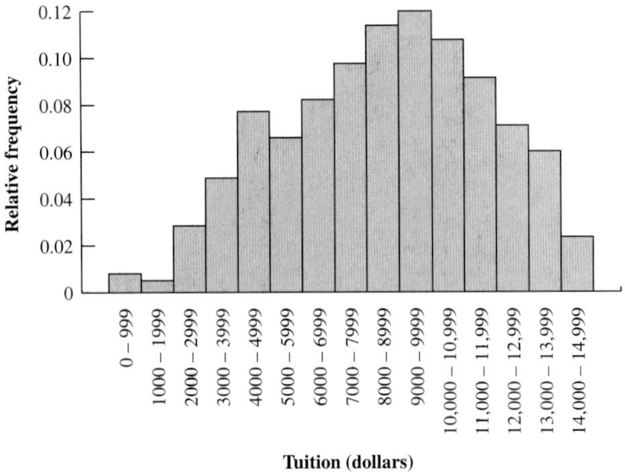

(f) 9000–9999

(g)

Tuition	Probability
0–999	0.0081
1000–1999	0.0051
2000–2999	0.0284
3000–3999	0.0487
4000–4999	0.0772
5000–5999	0.0660
6000–6999	0.0822
7000–7999	0.0975
8000–8999	0.1137
9000–9999	0.1198
10,000–10,999	0.1076
11,000–11,999	0.0914
12,000–12,999	0.0711
13,000–13,999	0.0599
14,000–14,999	0.0234

(h) 9.14%

21. (a)

Marital Status	Probability
Married, spouse present	0.5805
Married, spouse absent	0.0343
Widowed	0.0285
Divorced	0.0872
Never married	0.2695

(b) 58.05% **(c)** 26.95% **(d)** 61.48% **(e)** 41.95%

22. (a)

Marital Status	Probability
Married, spouse present	0.5386
Married, spouse absent	0.0406
Widowed	0.1090
Divorced	0.1095
Never married	0.2022

(b) 53.86% **(c)** 20.22% **(d)** 57.92% **(e)** 46.14%

Fill-in-the-Blank Items

1. union; intersection **2.** 20; 10 **3.** permutation **4.** combination **5.** equally likely **6.** complement

True/False Items

1. T **2.** F **3.** T **4.** T **5.** F **6.** T

Review Exercises

1. $\{1, 3, 5, 6, 7, 8\}$ **2.** $\{2, 3, 5, 6, 7, 8, 9\}$ **3.** $\{3, 7\}$ **4.** $\{3, 5, 7\}$ **5.** $\{1, 2, 4, 6, 8, 9\}$ **6.** $\{1, 4\}$ **7.** $\{1, 2, 4, 5, 6, 9\}$ **8.** $\{2, 4, 9\}$ **9.** 17 **10.** 24 **11.** 29 **12.** 34 **13.** 7 **14.** 45 **15.** 25 **16.** 7 **17.** 120 **18.** 720 **19.** 336 **20.** 210 **21.** 56 **22.** 35 **23.** 60 **24.** 120 **25.** 128 **26.** 64 **27.** 3024 **28.** 24 **29.** 70 **30.** 720 **31.** 91 **32. (a)** 14,400 **(b)** 14,400 **33.** 1,600,000 **34.** 60 **35.** 216,000 **36.** 256 **37.** 1260 **38.** 12,600 **39. (a)** 381,024 **(b)** 1260 **40. (a)** 280 **(b)** 280 **(c)** 640

41. (a)

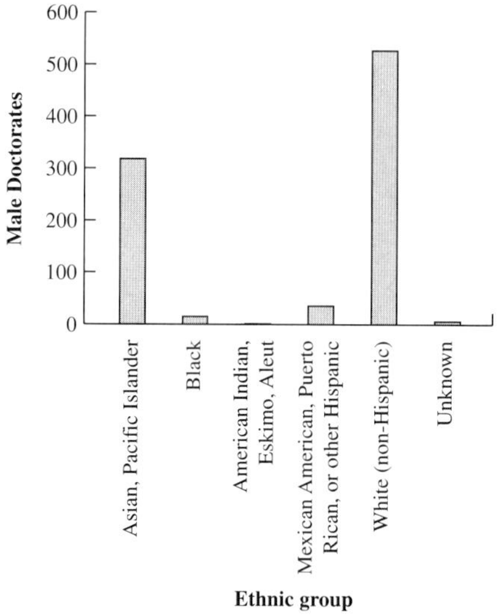

(b)

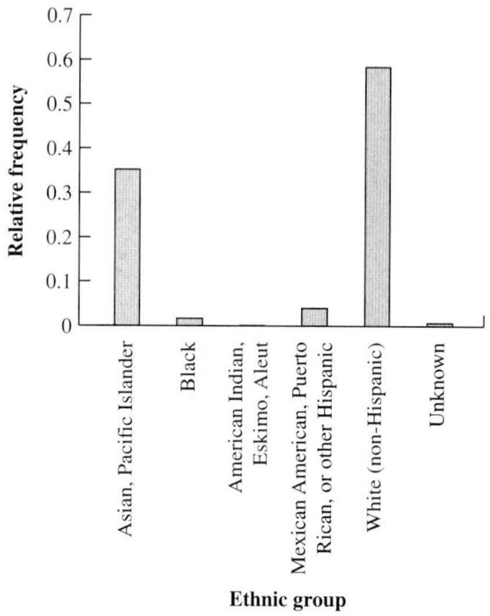

(c)

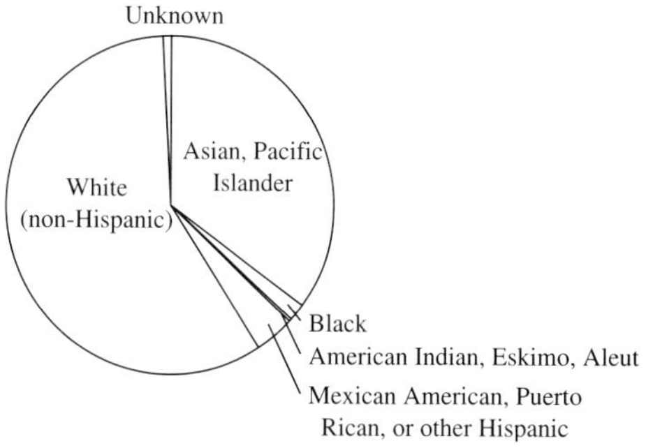

(d) White (non-Hispanic)

(e)

Ethnic Group	Probability
Asian, Pacific Islander	0.3522
Black	0.0166
American Indian, Eskimo, Aleut	0.0011
Mexican-American, Puerto Rican, or other Hispanic	0.0399
White (non-Hispanic)	0.5836
Unknown	0.0066

(f) 35.22%
(g) 39.21%
(h) 64.78%

42. (a)

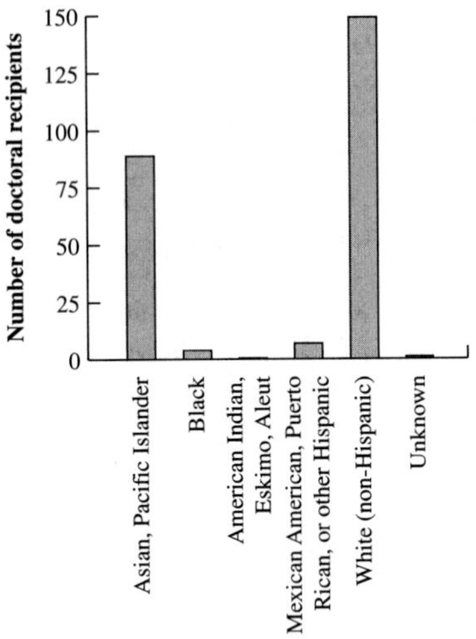

(b)

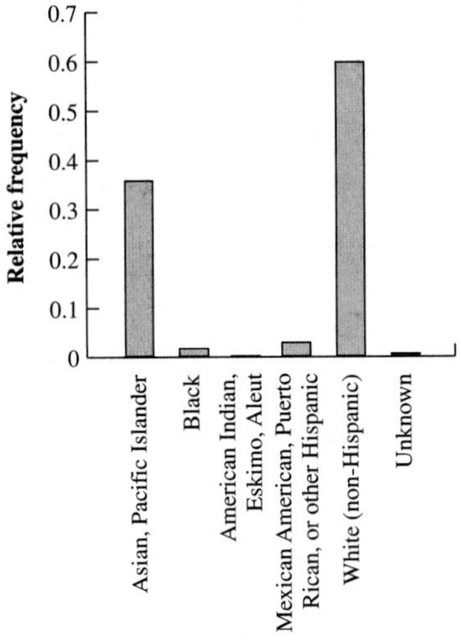

(c)

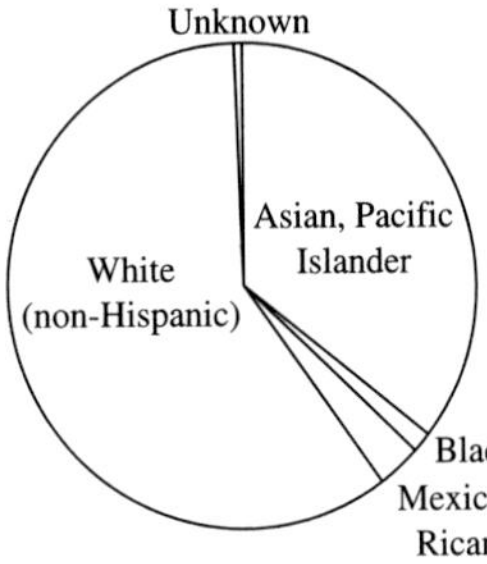

(d) White (non-Hispanic)

(e)

Ethnic Group	Probability
Asian, Pacific Islander	0.3560
Black	0.0160
Mexican-American, Puerto Rican, or other Hispanic	0.0280
White (non-Hispanic)	0.5960
Unknown	0.0040

(f) 35.60%
(g) 38.40%
(h) 64.40%

43. (a) 5 **(b)**

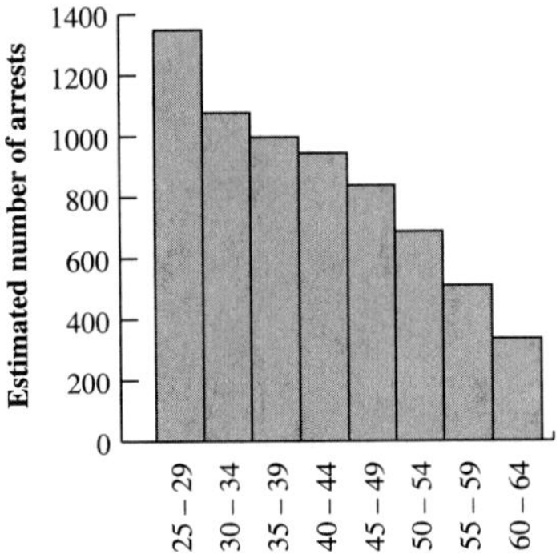

(c)

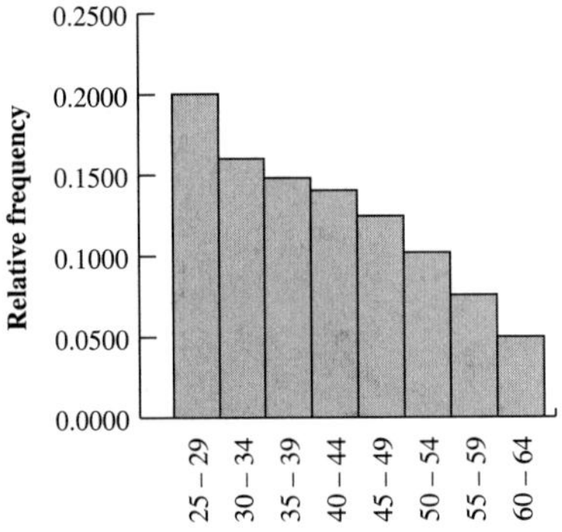

(d) 25–29

(e)

Age	Probability
25–29	0.2001
30–34	0.1599
35–39	0.1480
40–44	0.1403
45–49	0.1244
50–54	0.1019
55–59	0.0756
60–64	0.0498

(f) 14.80%
(g) 22.36%
(h) 85.20%

44. **(a)** 5 **(b)**

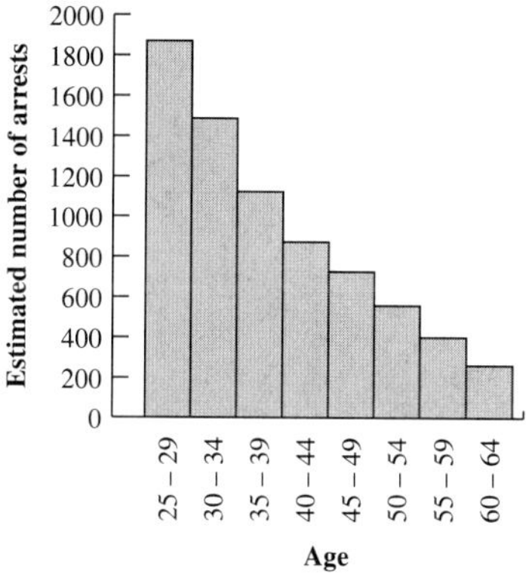

(c)

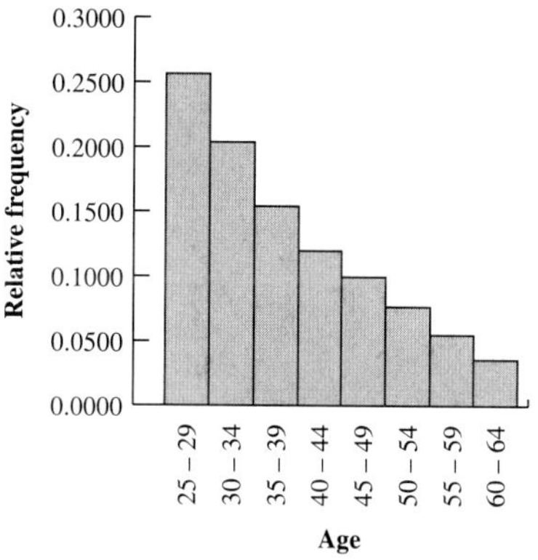

(d) 25–29
(e)

Age	Probability
25–29	0.2562
30–34	0.2037
35–39	0.1539
40–44	0.1195
45–49	0.0994
50–54	0.0765
55–59	0.0548
60–64	0.0359

(f) 15.39%
(g) 20.87%
(h) 84.61%

45. **(a)** $8.634628387 \times 10^{45}$ **(b)** 65.31% **(c)** 34.69% **46.** **(a)** 32.1% **(b)** 67.9% **47.** **(a)** 5.4% **(b)** 94.6% **48.** $\frac{3}{20}; \frac{9}{20}$ **49.** $\frac{4}{9}$ **50.** $\frac{1}{24}$ **51.** 0.2; 0.26 **52.** **(a)** $\frac{1}{22}$ **(b)** $\frac{7}{22}$ **(c)** $\frac{7}{44}$ **53.** **(a)** 0.246 **(b)** 9.8×10^{-4} **54.** **(a)** 0.68 **(b)** 0.58 **(c)** 0.32

CHAPTER 9 Conics

9.2 Exercises

1. B **2.** G **3.** E **4.** D **5.** H **6.** A **7.** C **8.** F **9.** F **10.** C **11.** G **12.** A **13.** D **14.** H **15.** B **16.** E

17. $y^2 = 16x$

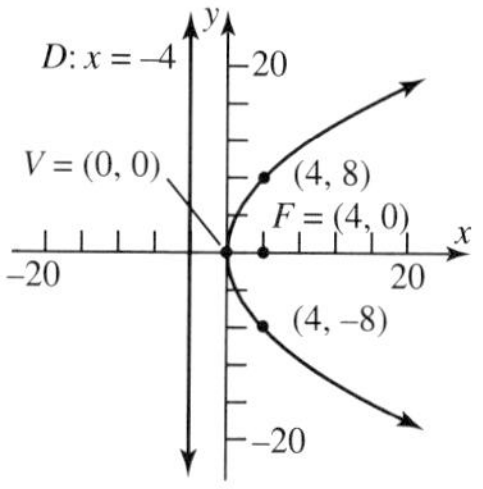

18. $x^2 = 8y$

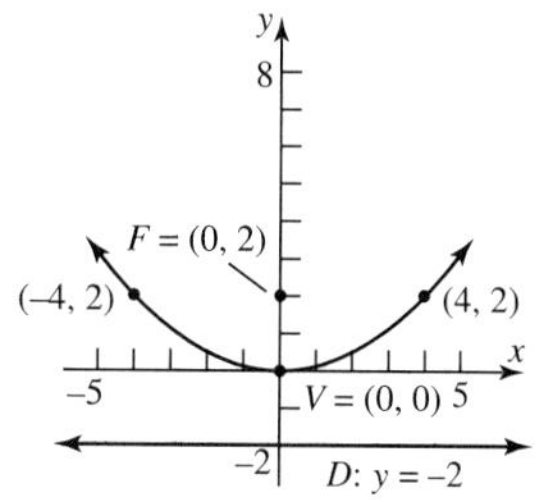

19. $x^2 = -12y$

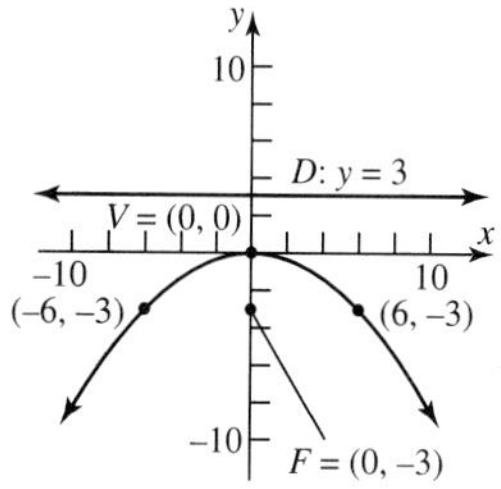

20. $y^2 = -16x$

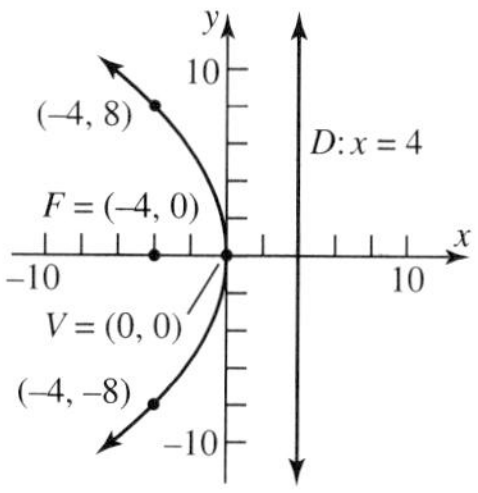

21. $y^2 = -8x$

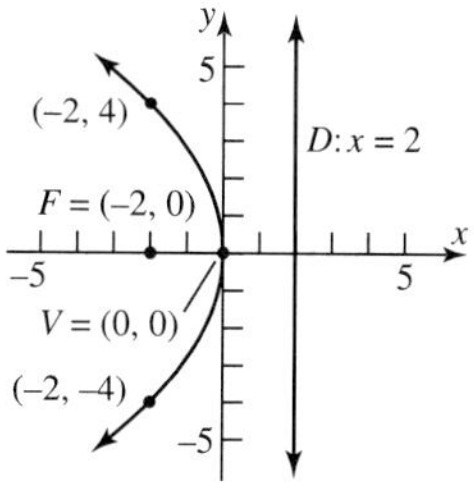

22. $x^2 = -4y$

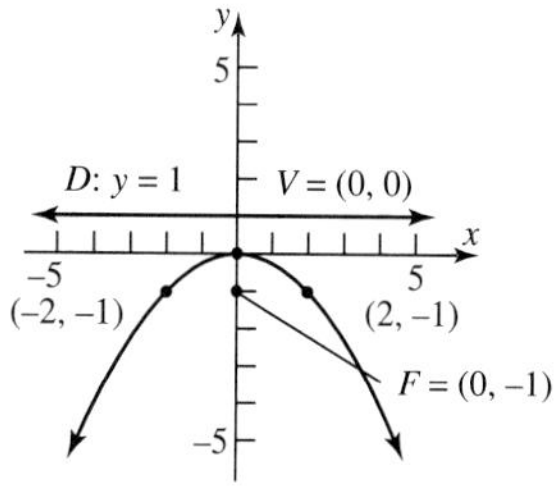

23. $x^2 = 2y$

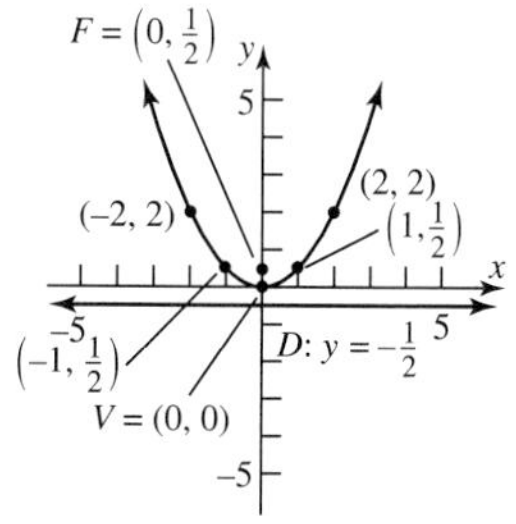

24. $y^2 = 2x$

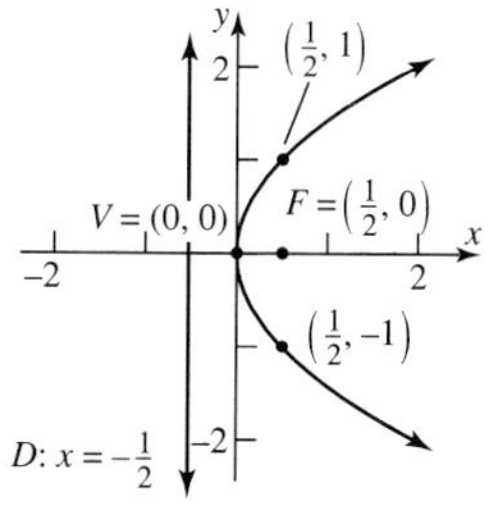

25. $(x - 2)^2 = -8(y + 3)$

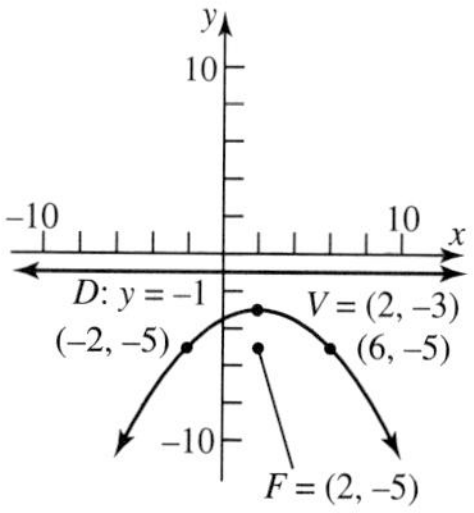

26. $(y+2)^2 = 8(x-4)$

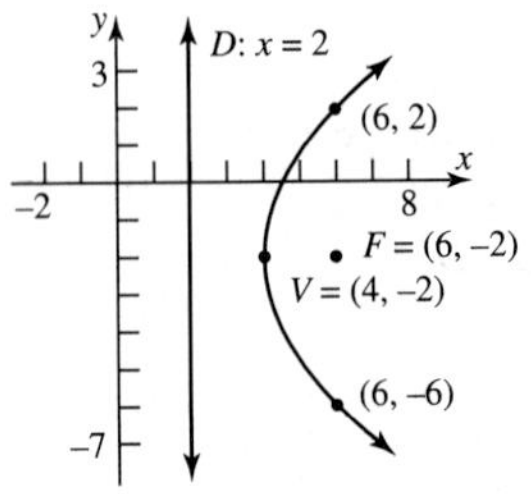

27. $x^2 = \frac{4}{3}y$

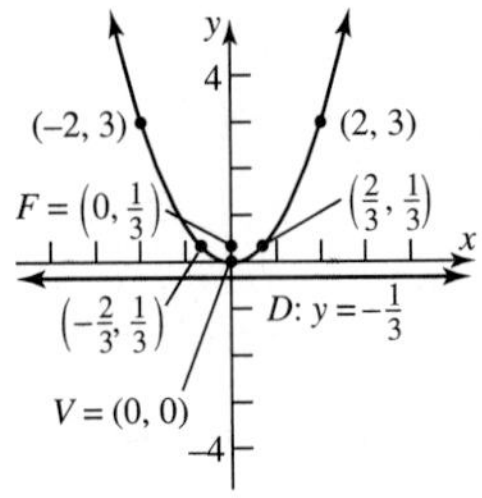

28. $y^2 = \frac{9}{2}x$

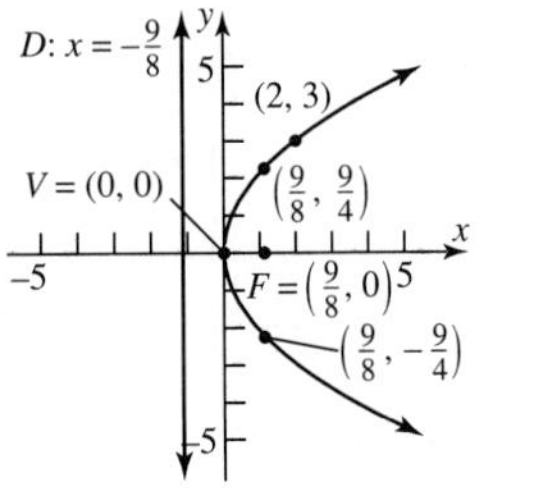

29. $(x+3)^2 = 4(y-3)$

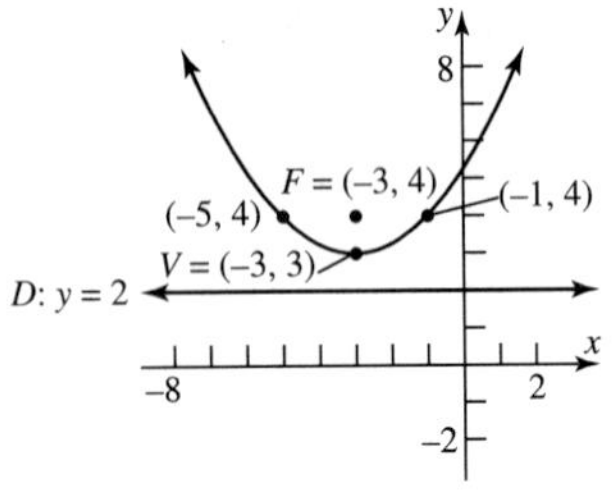

30. $(y-4)^2 = 12(x+1)$

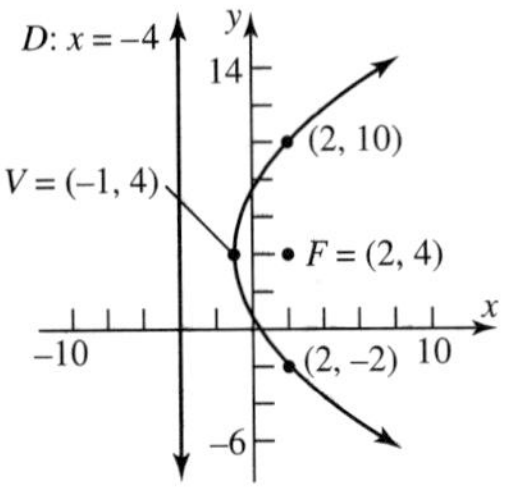

31. $(y+2)^2 = -8(x+1)$

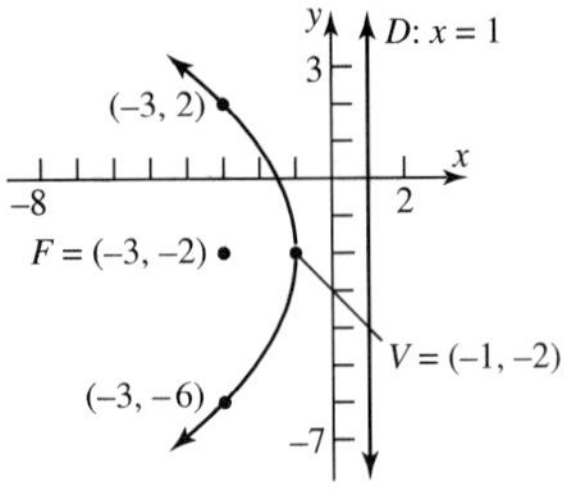

32. $(x+4)^2 = 12(y-1)$

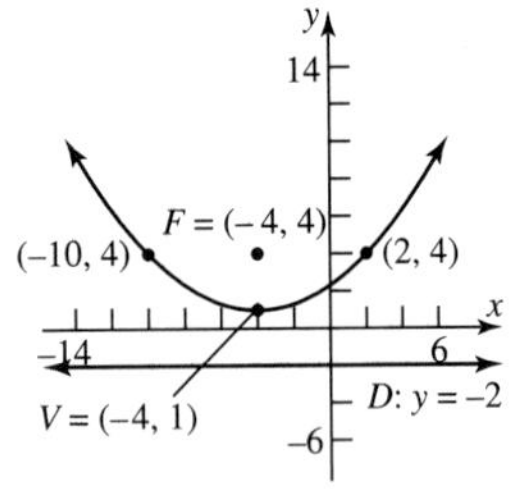

33. Vertex: $(0, 0)$; Focus: $(0, 1)$; Directrix: $y = -1$

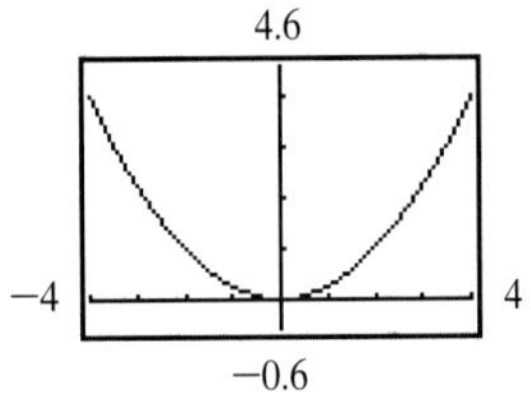

34. Vertex: $(0, 0)$; Focus: $(2, 0)$; Directrix: $x = -2$

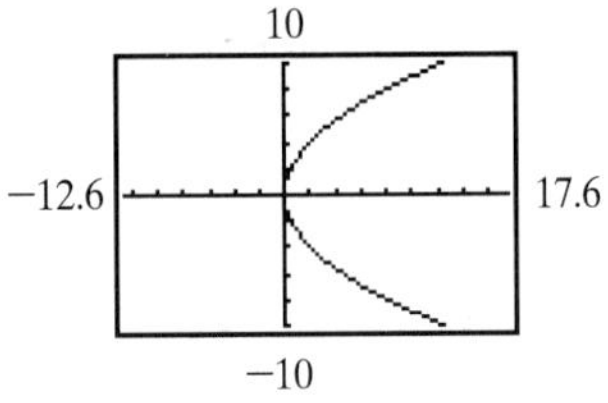

35. Vertex: $(0, 0)$; Focus: $(-4, 0)$; Directrix: $x = 4$

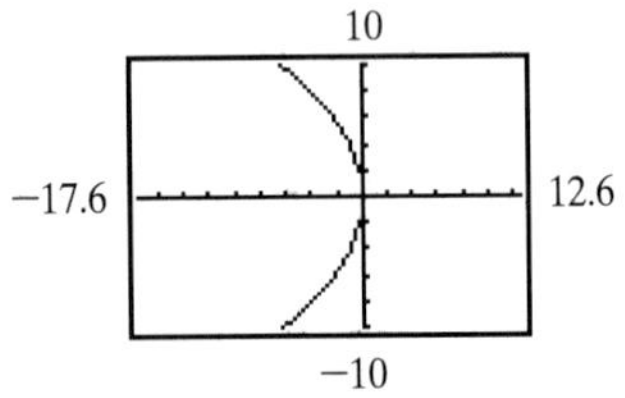

36. Vertex: $(0, 0)$; Focus: $(0, -1)$; Directrix: $y = 1$

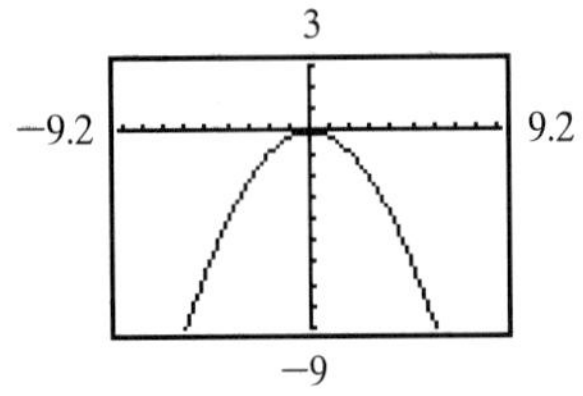

37. Vertex: $(-1, 2)$; Focus: $(1, 2)$; Directrix: $x = -3$

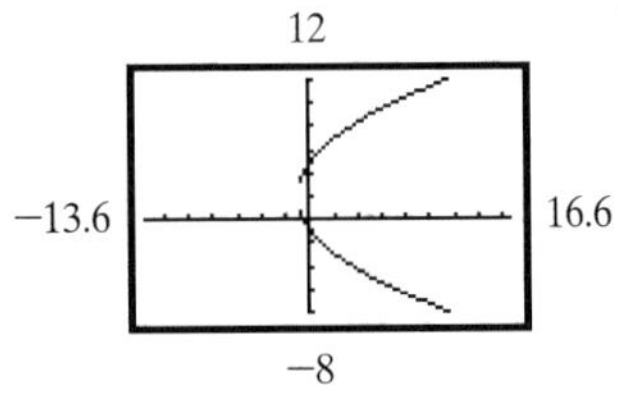

38. Vertex: $(-4, -2)$; Focus: $(-4, 2)$;

Directrix: $y = -6$

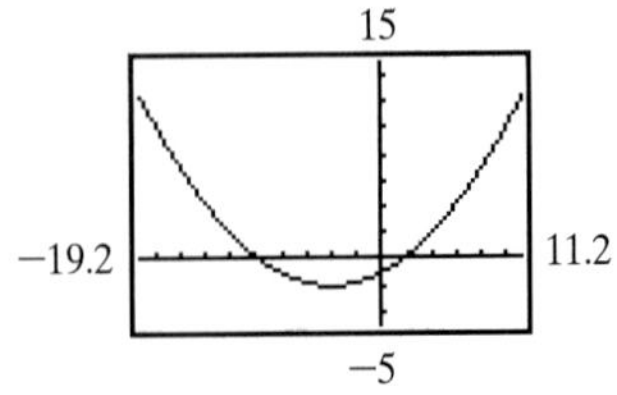

39. Vertex: $(3, -1)$; Focus: $\left(3, -\frac{5}{4}\right)$;

Directrix: $y = -\frac{3}{4}$

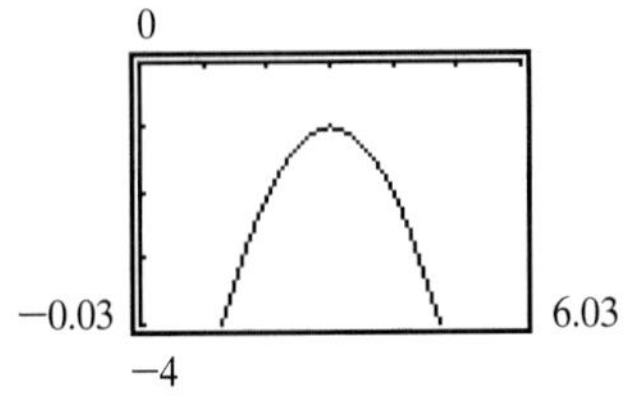

40. Vertex: $(2, -1)$; Focus: $(1, -1)$;

Directrix: $x = 3$

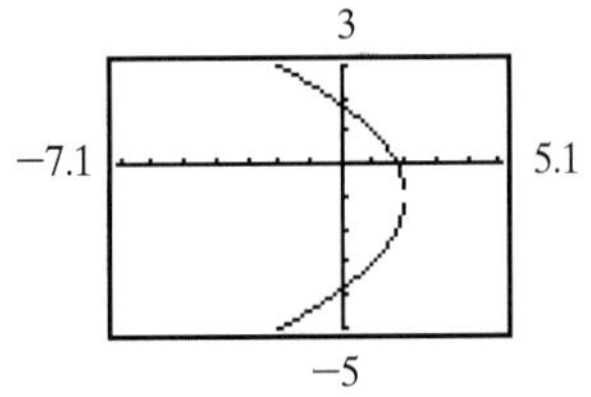

41. Vertex: $(2, -3)$; Focus: $(4, -3)$; Directrix: $x = 0$

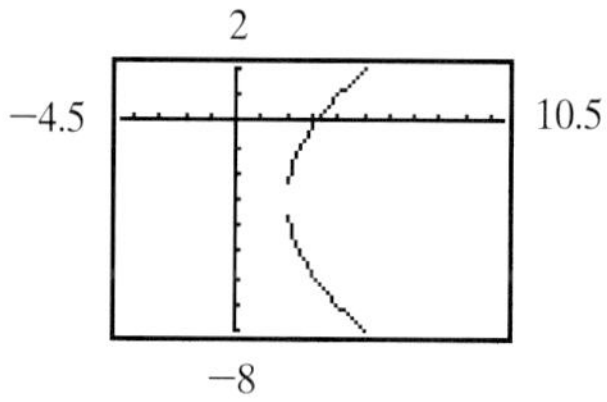

42. Vertex: $(2, 3)$; Focus: $(2, 4)$; Directrix: $y = 2$

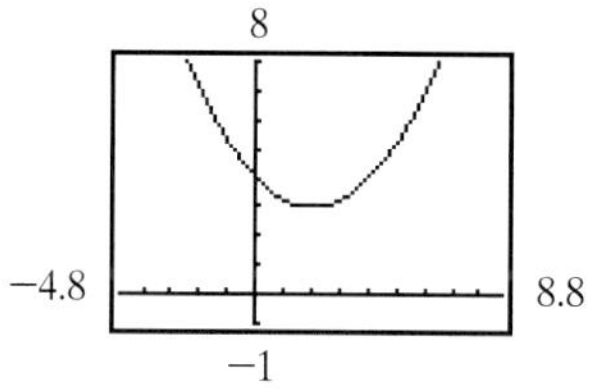

43. Vertex: $(0, 2)$; Focus: $(-1, 2)$; Directrix: $x = 1$

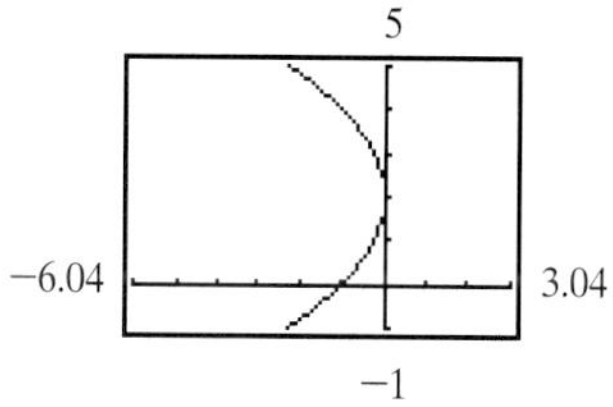

44. Vertex: $(-3, -2)$; Focus: $(-3, -1)$; Directrix: $y = -3$

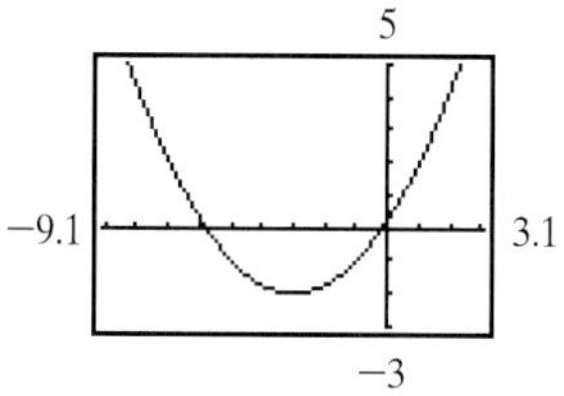

45. Vertex: $(-4, -2)$; Focus: $(-4, -1)$; Directrix: $y = -3$

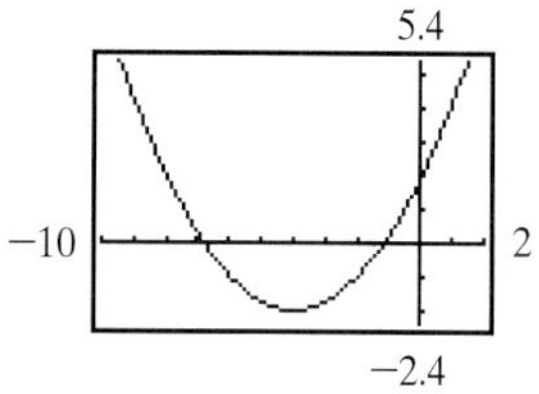

46. Vertex: $(0, 1)$; Focus: $(2, 1)$; Directrix: $x = -2$

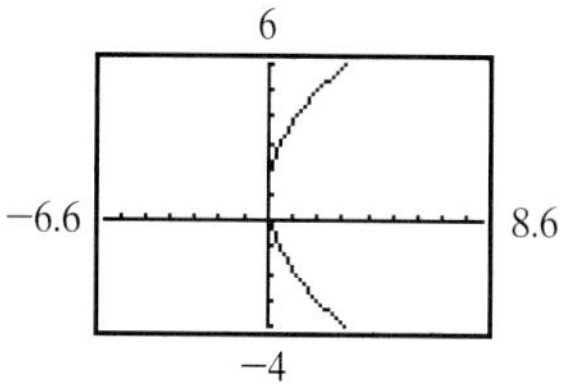

47. Vertex: $(-1, -1)$; Focus: $\left(-\frac{3}{4}, -1\right)$; Directrix: $x = -\frac{5}{4}$

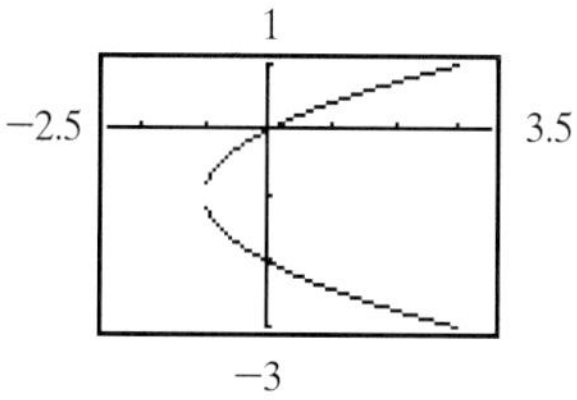

48. Vertex: $(2, -2)$; Focus: $\left(2, -\frac{3}{2}\right)$; Directrix: $y = -\frac{5}{2}$

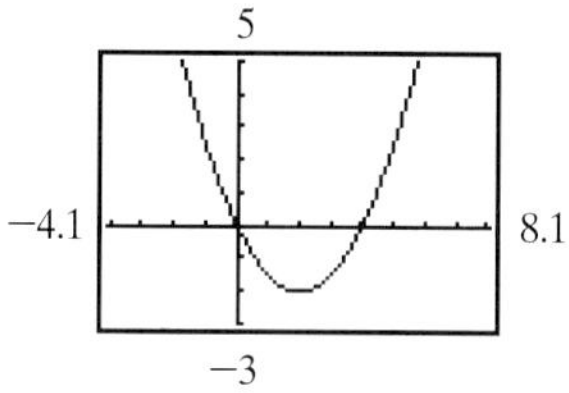

49. Vertex: $(2, -8)$; Focus: $\left(2, -\frac{31}{4}\right)$; Directrix: $y = -\frac{33}{4}$

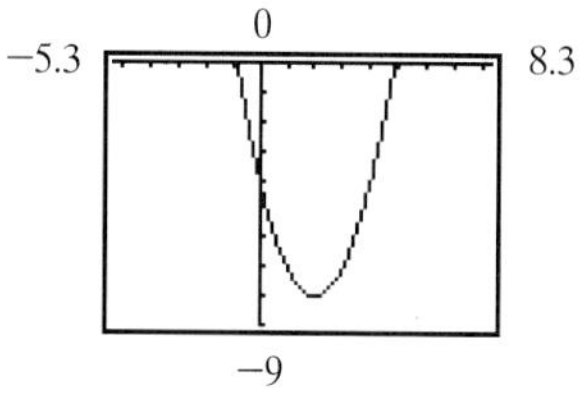

50. Vertex: $(37, -6)$; Focus: $\left(\frac{147}{4}, -6\right)$; Directrix: $x = \frac{149}{4}$

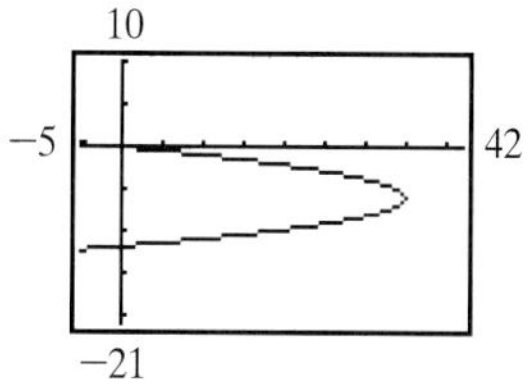

51. $(y - 1)^2 = x$ **52.** $(x - 1)^2 = -(y - 2)$ **53.** $(y - 1)^2 = -(x - 2)$

54. $x^2 = 4(y + 1)$ **55.** $x^2 = 4(y - 1)$ **56.** $(x - 1)^2 = \frac{1}{2}(y + 1)$ **57.** $y^2 = \frac{1}{2}(x + 2)$

58. $y^2 = -(x - 1)$ **59.** 1.5625 ft from the base of the dish, along the axis of symmetry
60. 1.125 ft or 13.5 in. from the base of the dish **61.** 1 in. from the vertex **62.** $4\sqrt{2}$ in.
63. 20 ft **64.** 15.625 ft **65.** 0.78125 ft **66.** $8\sqrt{2}$ ft
67. 4.17 ft from the base along the axis of symmetry
68. 0.0278 in. from the base of the mirror **69.** 24.31 ft, 18.75 ft, 7.64 ft **70.** 27.78 ft

71. $Ax^2 + Ey = 0$

$x^2 = -\frac{E}{A}y$

This is the equation of a parabola with vertex at $(0, 0)$ and axis of symmetry the y-axis. The focus is $\left(0, -\frac{E}{4A}\right)$; the directrix is the line $y = \frac{E}{4A}$.

72. $Cy^2 + Dx = 0$

$C \neq 0, D \neq 0$

$Cy^2 = -Dx$

$y^2 = -\frac{D}{C}x$

This is the equation of a parabola with vertex at $(0, 0)$ and axis of symmetry the x-axis. The focus is $\left(-\frac{D}{4C}, 0\right)$; the directrix is the line $x = \frac{D}{4C}$.

73. $Ax^2 + Dx + Ey + F = 0, A \neq 0$

$$Ax^2 + Dx = -Ey - F$$

$$x^2 + \frac{D}{A}x = -\frac{E}{A}y - \frac{F}{A}$$

$$\left(x + \frac{D}{2A}\right)^2 = -\frac{E}{A}y - \frac{F}{A} + \frac{D^2}{4A^2}$$

$$\left(x + \frac{D}{2A}\right)^2 = -\frac{E}{A}y + \frac{D^2 - 4AF}{4A^2}$$

(a) If $E \neq 0$, then the equation may be written as

$$\left(x + \frac{D}{2A}\right) = -\frac{E}{A}\left(y - \frac{D^2 - 4AF}{4AE}\right)$$

This is the equation of a parabola with vertex at

$\left(-\frac{D}{2A}, \frac{D^2 - 4AF}{4AE}\right)$ and axis of symmetry parallel to the y-axis.

(b)–(d) If $E = 0$, the graph of the equation contains no points if $D^2 - 4AF < 0$, is a single vertical line if $D^2 - 4AF = 0$, and is two vertical lines if $D^2 - 4AF > 0$.

74. $Cy^2 + Dx + Ey + F = 0, C \neq 0$

$$Cy^2 + Ey = -Dx - F$$

$$y^2 + \frac{E}{C}y = -\frac{D}{C}x - \frac{F}{C}$$

$$\left(y + \frac{E}{2C}\right)^2 = -\frac{D}{C}x - \frac{F}{C} + \frac{E^2}{4C^2}$$

$$\left(y + \frac{E}{2C}\right)^2 = -\frac{D}{C}x + \frac{E^2 - 4CF}{4C^2}$$

(a) If $D \neq 0$, then the equation may be written as

$$\left(y + \frac{E}{2C}\right)^2 = -\frac{D}{C}\left(x - \frac{E^2 - 4CF}{4CD}\right).$$

This is the equation of a parabola with vertex at

$\left(\frac{E^2 - 4CF}{4CD}, -\frac{E}{2C}\right)$ and axis of symmetry parallel to the x-axis.

(b)–(d) If $D \neq 0$, the graph of the equation contains no points if $E^2 - 4CF < 0$, is a single vertical line if $E^2 - 4CF = 0$, and is two horizontal lines if $E^2 - 4CF > 0$.

Exercises 9.3

1. C **2.** D **3.** B **4.** A **5.** C **6.** A **7.** D **8.** B

9. Vertices: $(-5, 0), (5, 0)$
Foci: $(-\sqrt{21}, 0), (\sqrt{21}, 0)$

3.2

−5 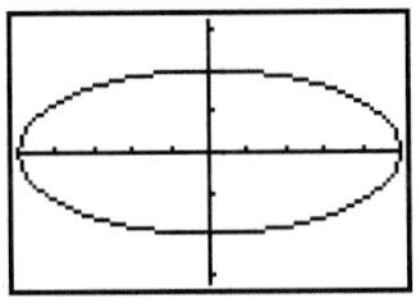 5

−3.2

10. Vertices: $(-3, 0), (3, 0)$
Foci: $(-\sqrt{5}, 0), (\sqrt{5}, 0)$

2

−3 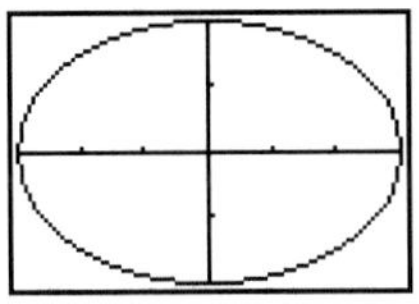 3

−2

11. Vertices: $(0, -5), (0, 5)$
Foci: $(0, -4), (0, 4)$

5

−7.5 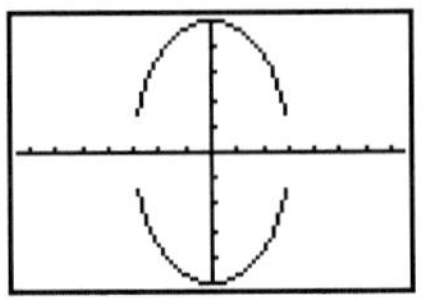 7.5

−5

12. Vertices: $(0, -4), (0, 4)$
Foci: $(0, -\sqrt{15}), (0, \sqrt{15})$

6

−4 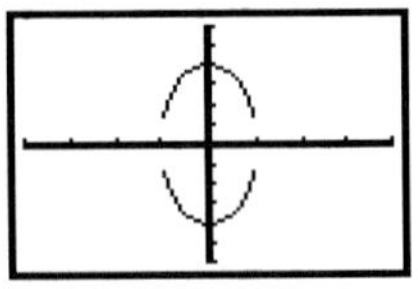 4

−6

13. Vertices: $(0, -4), (0, 4)$
Foci: $(0, -2\sqrt{3}), (0, 2\sqrt{3})$

4

−6.0 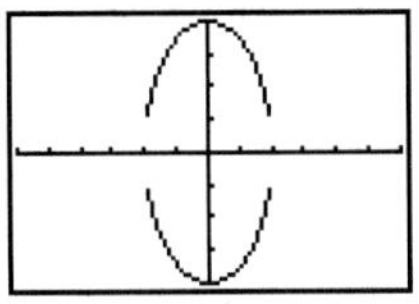 6.0

−4

14. Vertices: $(-3\sqrt{2}, 0), (3\sqrt{2}, 0)$
Foci: $(-4, 0), (4, 0)$

3

−4.5 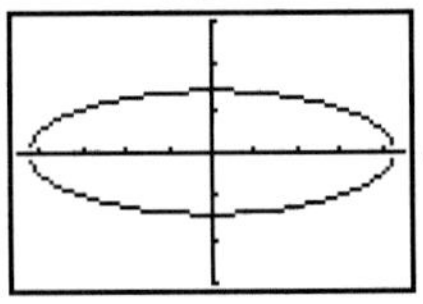 4.5

−3

15. Vertices: $(-2\sqrt{2}, 0), (2\sqrt{2}, 0)$
Foci: $(-\sqrt{6}, 0), (\sqrt{6}, 0)$

3.2

−5 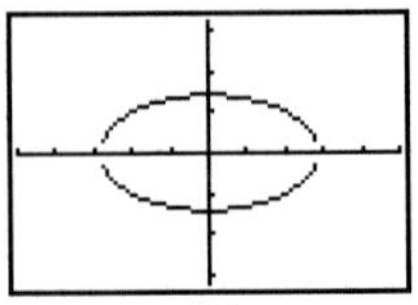 5

−3.2

16. Vertices: $(0, -3), (0, 3)$
Foci: $(0, -\sqrt{5}), (0, \sqrt{5})$

3

−4.5 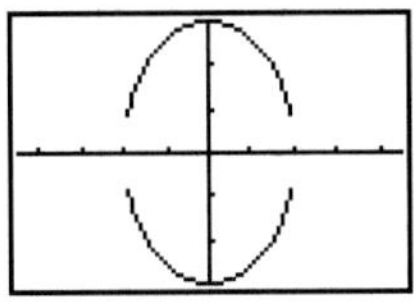 4.5

−3

17. Vertices: $(-4, 0), (4, 0), (0, -4), (0, 4)$
Focus: $(0, 0)$

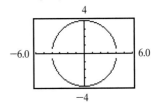

4

−6.0 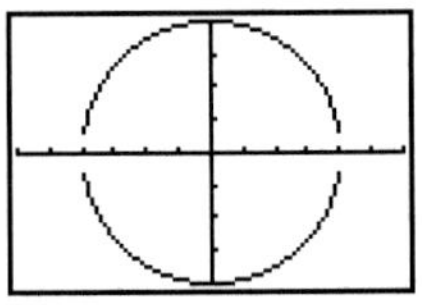 6.0

−4

18. Vertices: $(-2, 0), (2, 0), (0, -2), (0, 2)$
Focus: $(0, 0)$

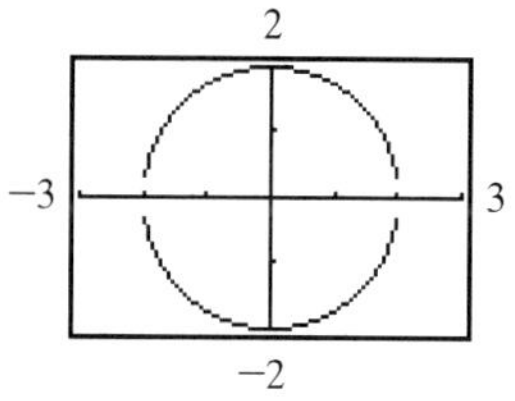

19. $\frac{x^2}{25} + \frac{y^2}{16} = 1$

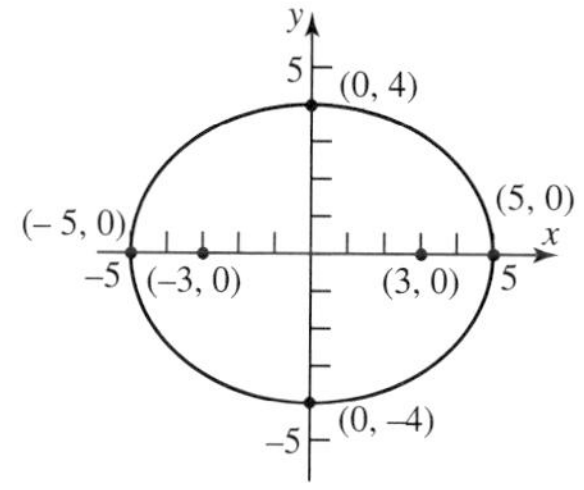

20. $\frac{x^2}{9} + \frac{y^2}{8} = 1$

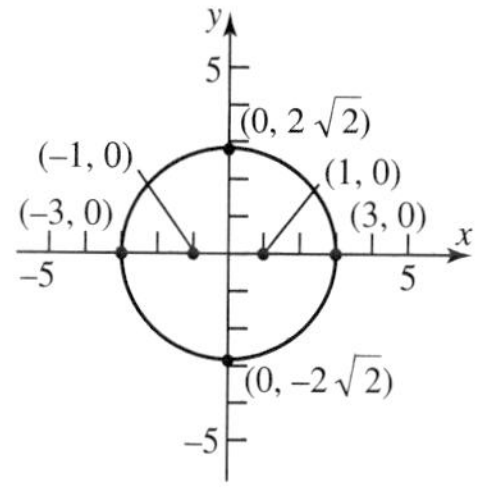

21. $\frac{x^2}{9} + \frac{y^2}{25} = 1$

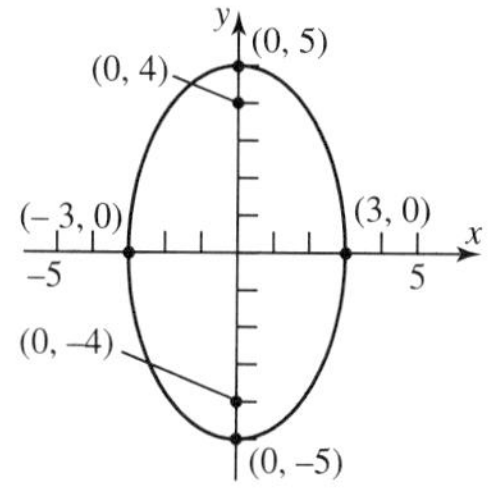

22. $\frac{x^2}{3} + \frac{y^2}{4} = 1$

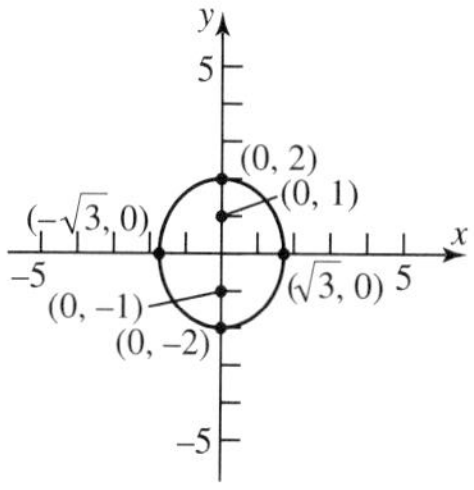

23. $\frac{x^2}{9} + \frac{y^2}{5} = 1$

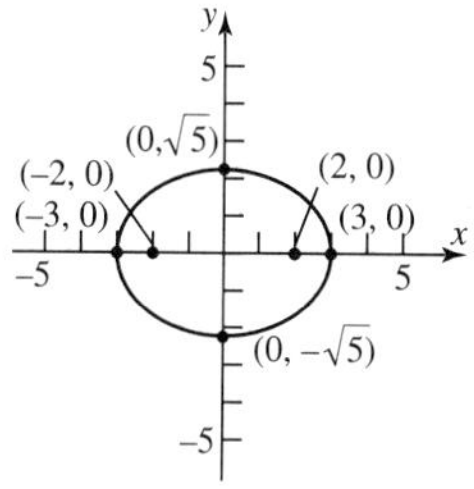

24. $\frac{x^2}{48} + \frac{y^2}{64} = 1$

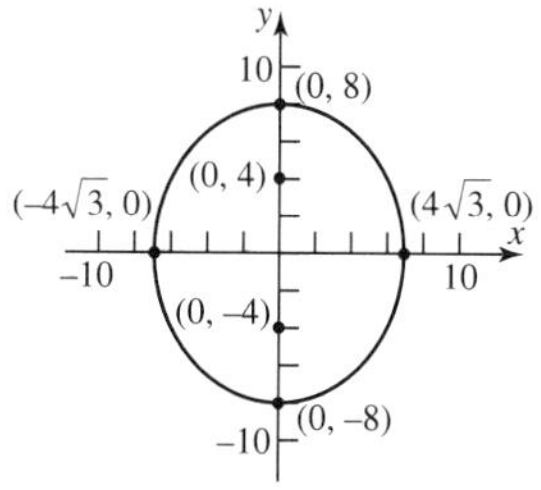

25. $\frac{x^2}{4} + \frac{y^2}{13} = 1$

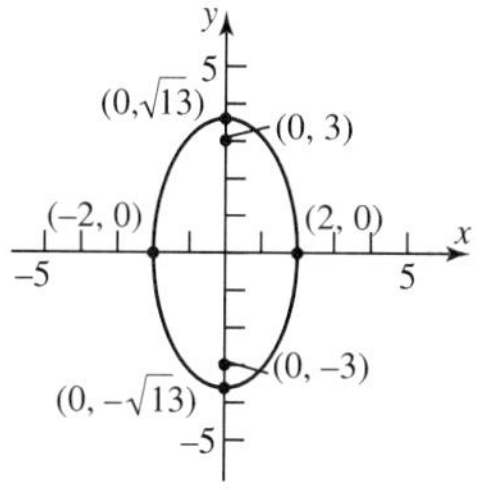

26. $\frac{x^2}{12} + \frac{y^2}{16} = 1$

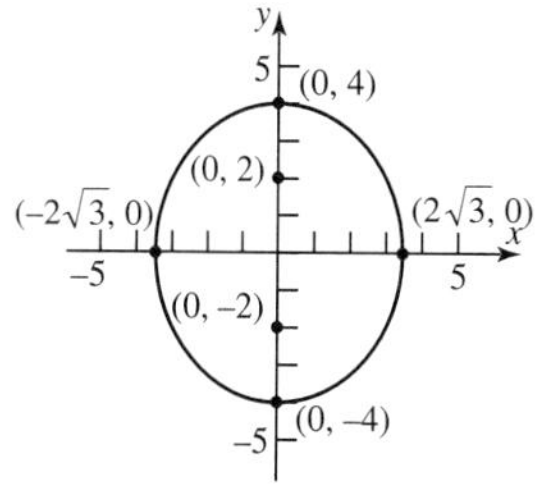

27. $x^2 + \frac{y^2}{16} = 1$

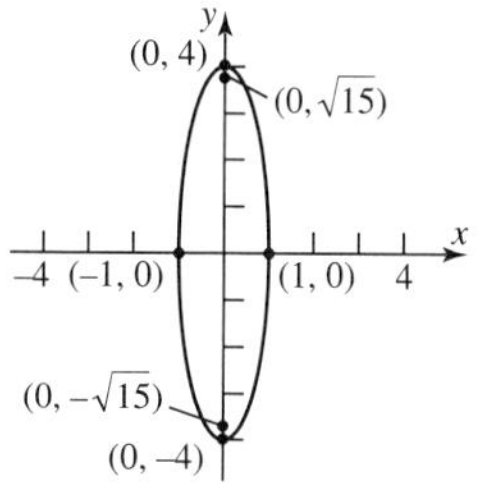

28. $\frac{x^2}{25} + \frac{y^2}{21} = 1$

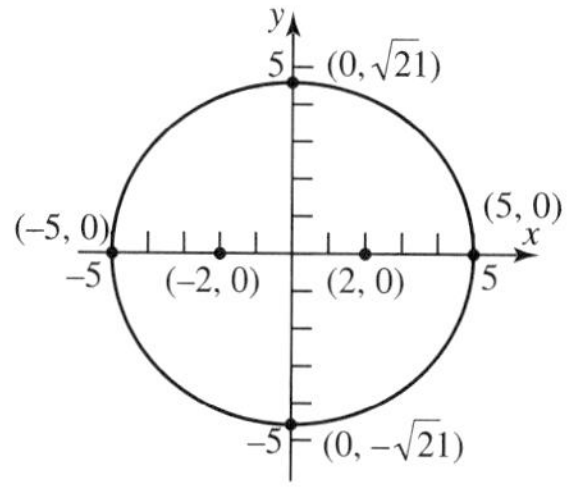

29. $\frac{(x + 1)^2}{4} + (y - 1)^2 = 1$

30. $(x + 1)^2 + \frac{(y + 1)^2}{4} = 1$

31. $(x - 1)^2 + \frac{y^2}{4} = 1$

32. $\frac{x^2}{4} + (y - 1)^2 = 1$

33. Center: $(3, -1)$
Vertices: $(3, -4), (3, 2)$
Foci: $(3, -1 - \sqrt{5}), (3, -1 + \sqrt{5})$

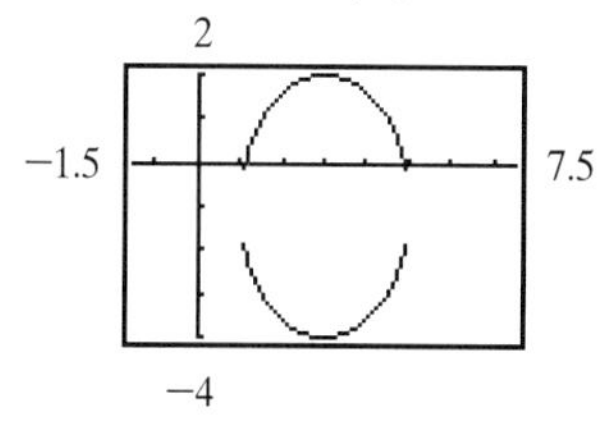

34. Center: $(-4, -2)$
Vertices: $(-7, -2), (-1, -2)$
Foci: $(-4 - \sqrt{5}, -2), (-4 + \sqrt{5}, -2)$

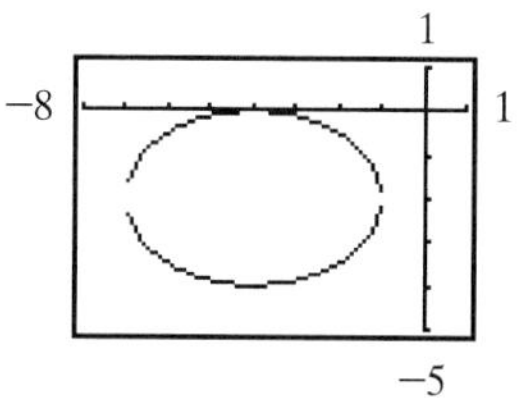

35. Center: $(-5, 4)$
Vertices: $(-9, 4), (-1, 4)$
Foci: $(-5 - 2\sqrt{3}, 4), (-5 + 2\sqrt{3}, 4)$

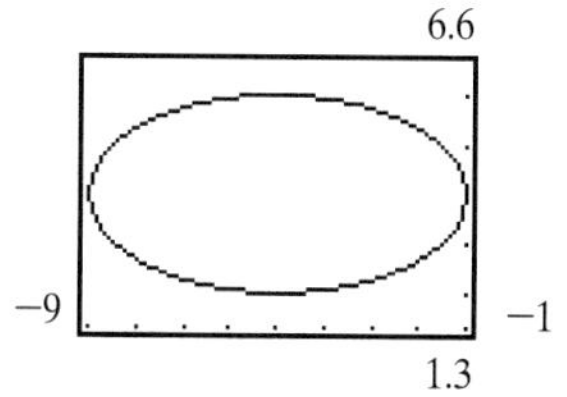

36. Center: $(3, -2)$
Vertices: $(3, -2 - 3\sqrt{2})$, $(3, -2 + 3\sqrt{2})$
Foci: $(3, -6)$, $(3, 2)$

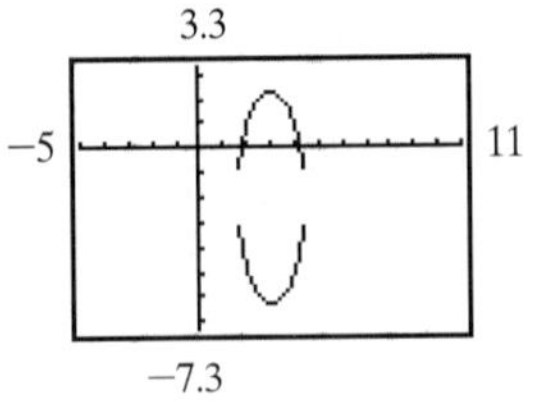

37. Center: $(-2, 1)$
Vertices: $(-4, 1)$, $(0, 1)$
Foci: $(-2 - \sqrt{3}, 1)$, $(-2 + \sqrt{3}, 1)$

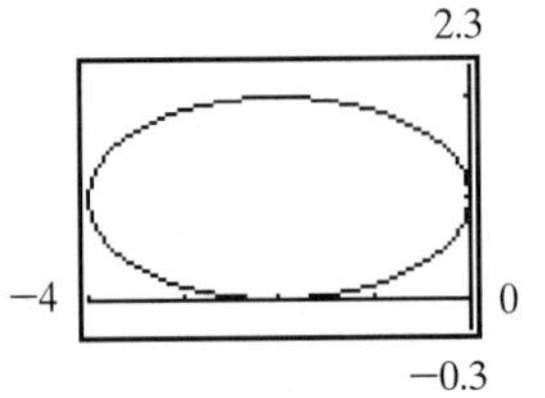

38. Center: $(0, 2)$
Vertices: $(-\sqrt{3}, 2)$, $(\sqrt{3}, 2)$
Foci: $(-\sqrt{2}, 2)$, $(\sqrt{2}, 2)$

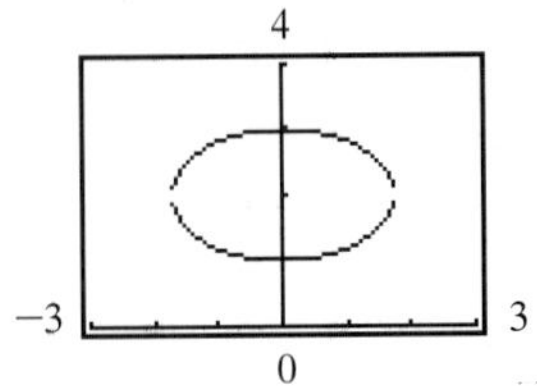

39. Center: $(2, -1)$
Vertices: $(2 - \sqrt{3}, -1)$, $(2 + \sqrt{3}, -1)$
Foci: $(1, -1)$, $(3, -1)$

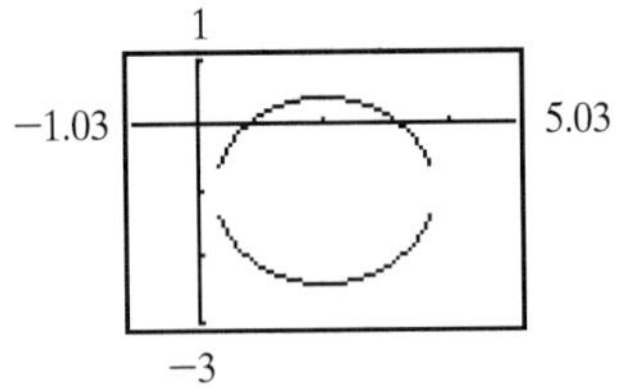

40. Center: $(-1, 1)$
Vertices: $(-1, -1)$, $(-1, 3)$
Foci: $(-1, 0)$, $(-1, 2)$

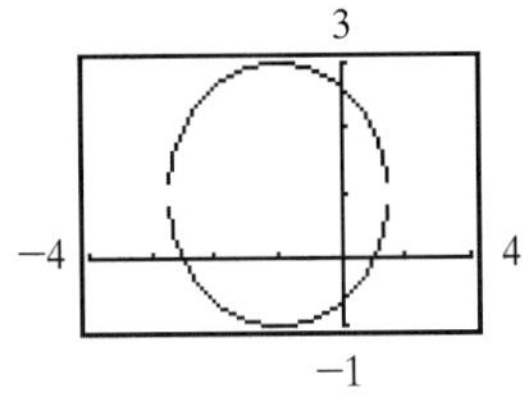

41. Center: $(1, -2)$
Vertices: $(1, -5)$, $(1, 1)$
Foci: $(1, -2 - \sqrt{5})$, $(1, -2 + \sqrt{5})$

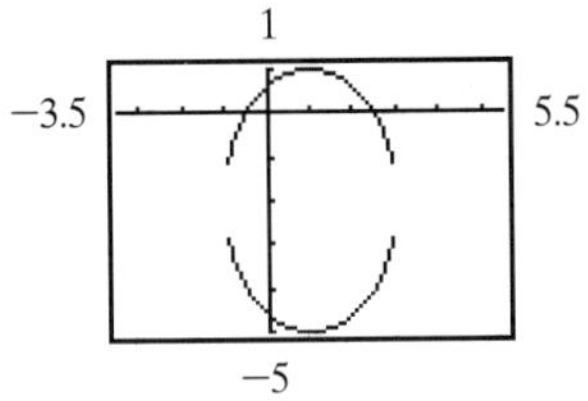

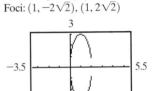

42. Center: $(-3, 1)$
Vertices: $(-6, 1)$, $(0, 1)$
Foci: $(-3 - 2\sqrt{2}, 1)$, $(-3 + 2\sqrt{2}, 1)$

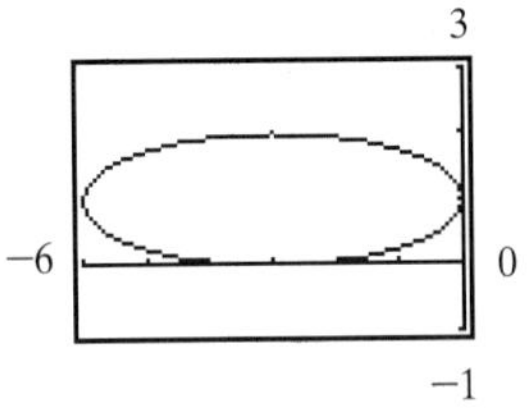

43. Center: $(0, -2)$
Vertices: $(0, -4)$, $(0, 0)$
Foci: $(0, -2 - \sqrt{3})$, $(0, -2 + \sqrt{3})$

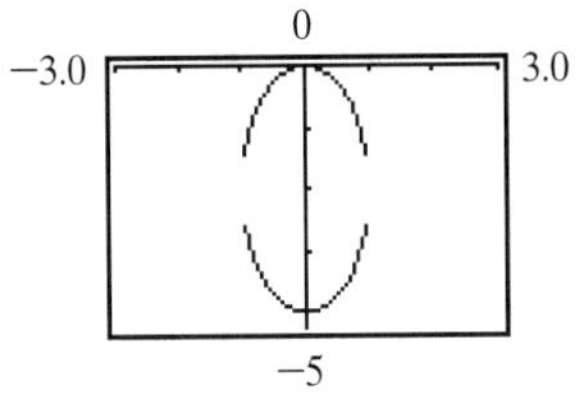

44. Center: $(1, 0)$
Vertices: $(1, -3)$, $(1, 3)$
Foci: $(1, -2\sqrt{2})$, $(1, 2\sqrt{2})$

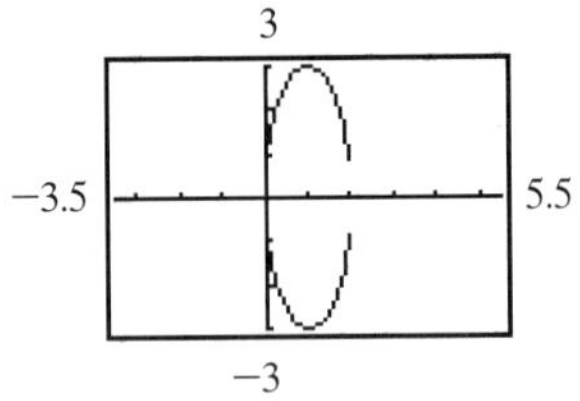

45. $\dfrac{(x-2)^2}{25} + \dfrac{(y+2)^2}{21} = 1$

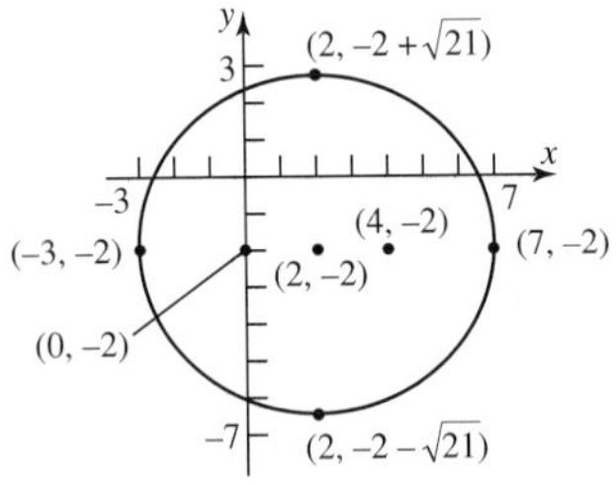

46. $\dfrac{(x+3)^2}{3} + \dfrac{(y-1)^2}{4} = 1$

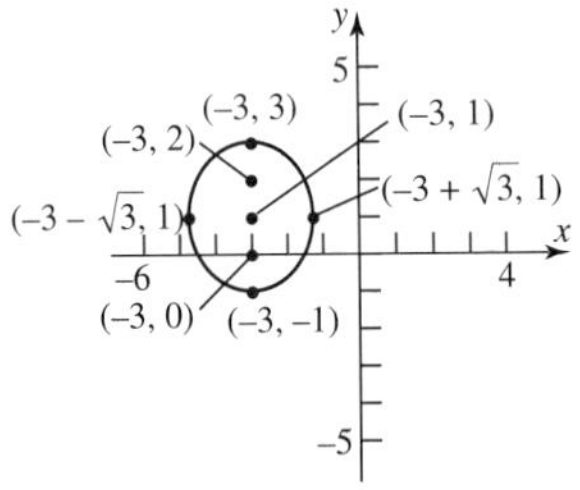

47. $\dfrac{(x-4)^2}{5} + \dfrac{(y-6)^2}{9} = 1$

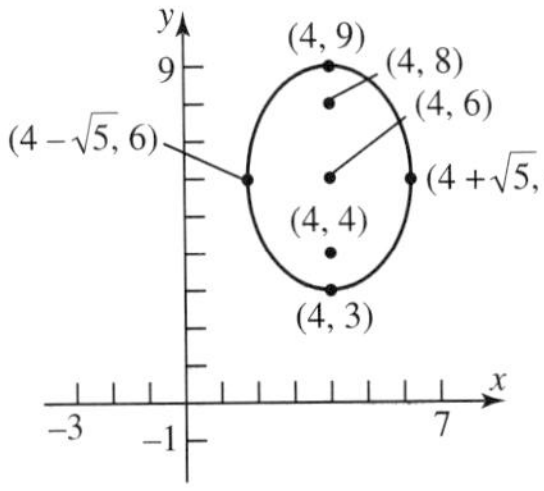

48. $\dfrac{(x+1)^2}{9} + \dfrac{(y-2)^2}{5} = 1$

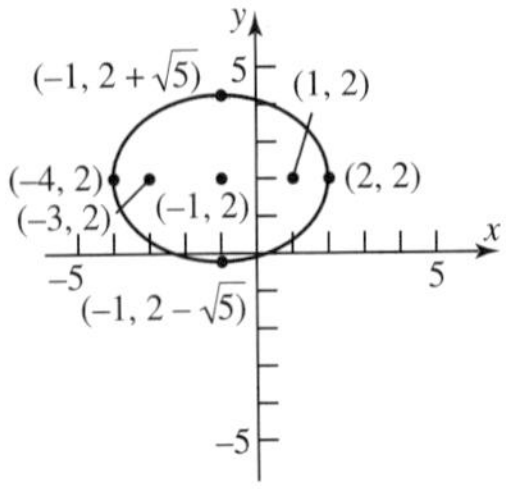

49. $\dfrac{(x-2)^2}{16} + \dfrac{(y-1)^2}{7} = 1$

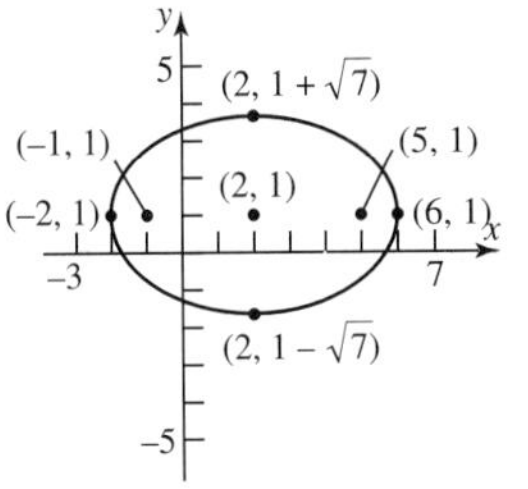

50. $\dfrac{(x-2)^2}{5} + \dfrac{(y-2)^2}{9} = 1$

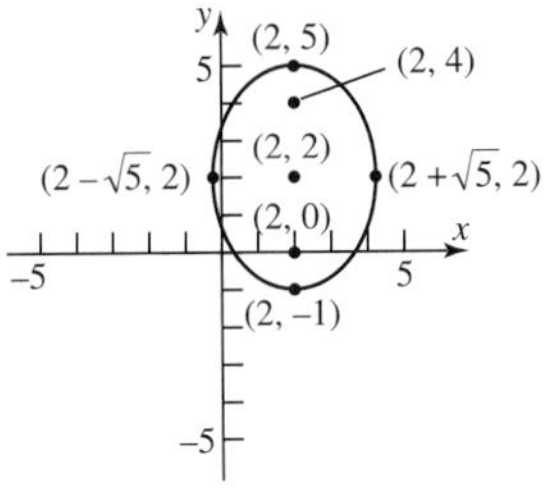

51. $\dfrac{(x-1)^2}{10} + (y-2)^2 = 1$

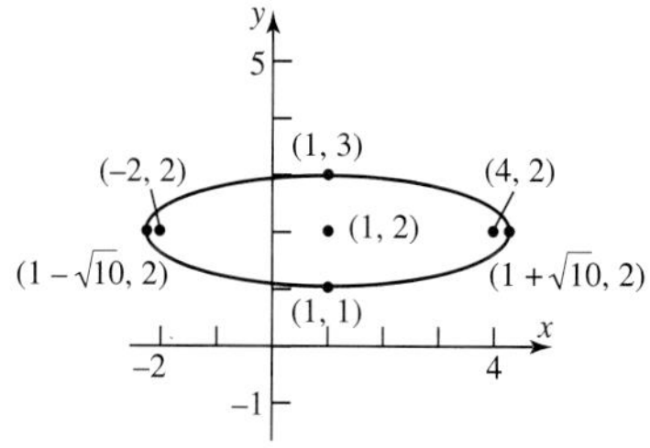

52. $\dfrac{(x-1)^2}{1} + \dfrac{(y-2)^2}{5} = 1$

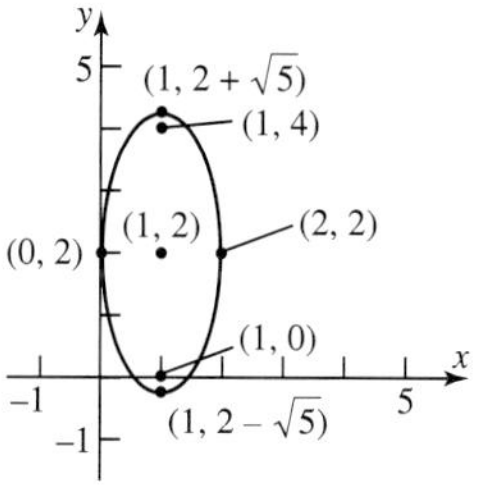

53. $\dfrac{(x-1)^2}{9} + (y-2)^2 = 1$

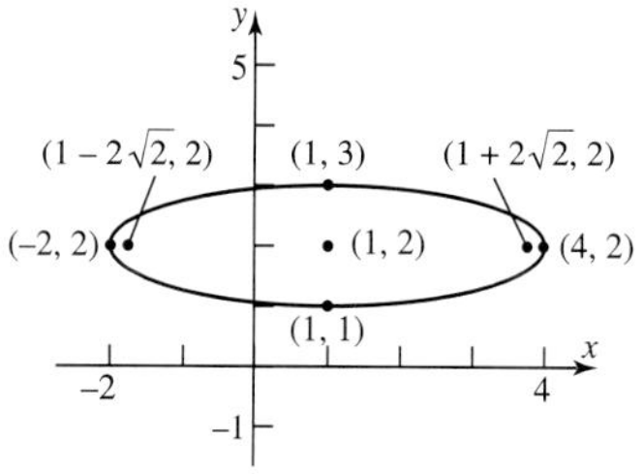

54. $\dfrac{(x-1)^2}{1} + \dfrac{(y-2)^2}{4} = 1$

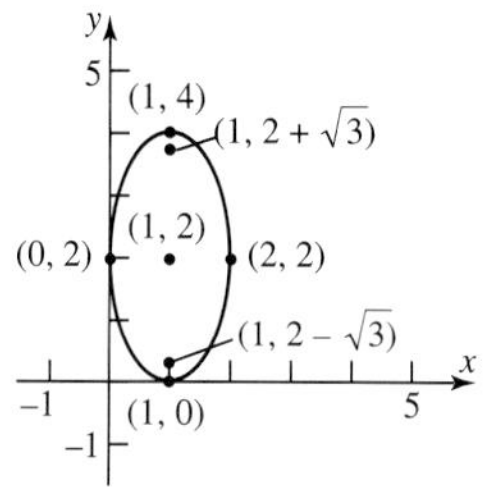

55.

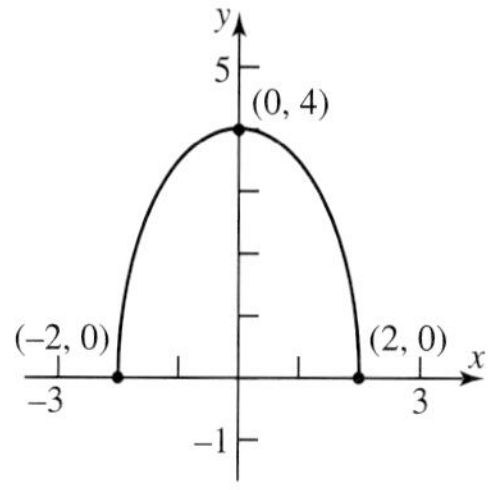

56.

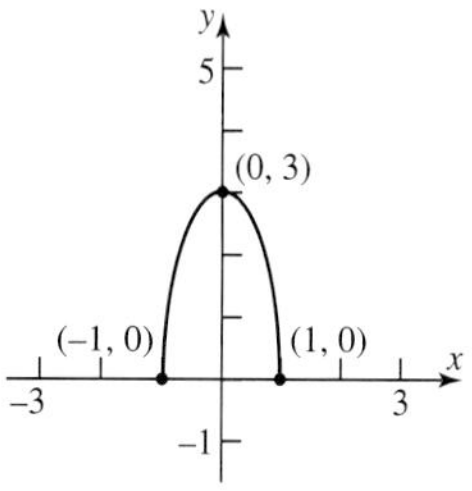

57.

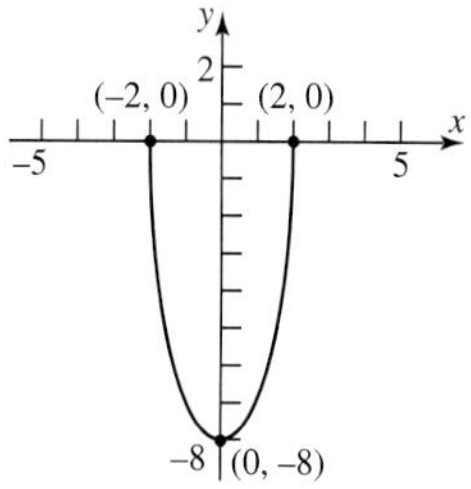

58.

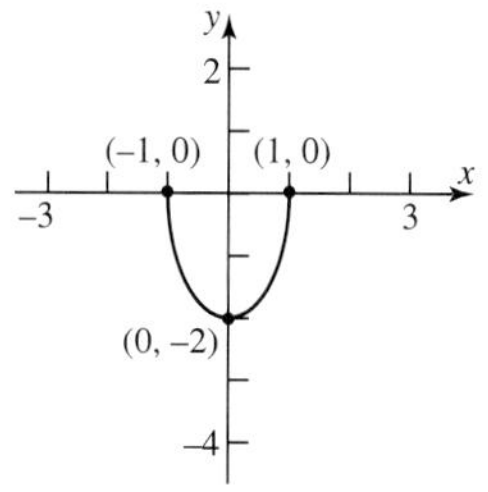

59. $\dfrac{x^2}{100} + \dfrac{y^2}{36} = 1$

60. at $x = 5$, distance is 2.57 ft; at $x = 10$, distance is 4.55 ft; at $x = 15$, distance is 12 ft

61. 43.3 ft **62.** 112 ft; 25.2 ft **63.** 24.65 ft, 21.65 ft, 13.82 ft **64.** 16.67 ft

65. 0 ft, 12.99 ft, 15 ft, 12.99 ft, 0 ft **66.** 73.69 ft

67. 91.5 million miles; $\dfrac{x^2}{(93)^2} + \dfrac{y^2}{8646.75} = 1$ **68.** 155.5 million miles; $\dfrac{x^2}{(142)^2} + \dfrac{y^2}{19{,}981.75} = 1$

69. perihelion: 460.6 million miles; mean distance: 483.8 million miles; $\dfrac{x^2}{(483.8)^2} + \dfrac{y^2}{233{,}524.2} = 1$

70. aphelion: 6346 million miles; mean distance: 5448.5 million miles; $\dfrac{x^2}{5448.5^2} + \dfrac{y^2}{28{,}880{,}646} = 1$ **71.** 30 ft **72.** $10\sqrt{7}$ ft

73. **(a)** $Ax^2 + Cy^2 + F = 0$

$Ax^2 + Cy^2 = -F$

If A and C are of the same and F is of opposite sign, then the equation takes the form $\dfrac{x^2}{\left(-\dfrac{F}{A}\right)} + \dfrac{y^2}{\left(-\dfrac{F}{C}\right)} = 1$, where $-\dfrac{F}{A}$ and $-\dfrac{F}{C}$ are positive. This is the equation of an ellipse with center at $(0, 0)$.

(b) If $A = C$, the equation may be written as $x^2 + y^2 = -\dfrac{F}{A}$. This is the equation of a circle with center at $(0, 0)$ and radius equal to $\sqrt{-\dfrac{F}{A}}$.

74.
$$Ax^2 + Cy^2 + Dx + Ey + F = 0 \qquad A \neq 0, C \neq 0$$
$$Ax^2 + Dx + Cy^2 + Ey = -F$$
$$A\left(x^2 + \frac{D}{A}x\right) + C\left(y^2 + \frac{E}{C}y\right) = -F$$
$$A\left(x + \frac{D}{2A}\right)^2 + C\left(y + \frac{E}{2C}\right)^2 = -F + \frac{D^2}{4A} + \frac{E^2}{4C}$$

(a) If $\dfrac{D^2}{4A} + \dfrac{E^2}{4C} - F$ is of the same sign as A (and C), this is the equation of an ellipse with center at $\left(\dfrac{-D}{2A}, \dfrac{-E}{2C}\right)$.

(b) If $\dfrac{D^2}{4A} + \dfrac{E^2}{4C} - F = 0$, the graph is the single point $\left(\dfrac{-D}{2A}, \dfrac{-E}{2C}\right)$.

(c) If $\dfrac{D^2}{4A} + \dfrac{E^2}{4C} - F$ is of the opposite sign to A (and C), the graph contains no points since, in this case, the left side has a sign opposite that of the right side.

Exercises 9.4

1. B **2.** C **3.** A **4.** D **5.** B **6.** D **7.** C **8.** A

9. $x^2 - \frac{y^2}{8} = 1$

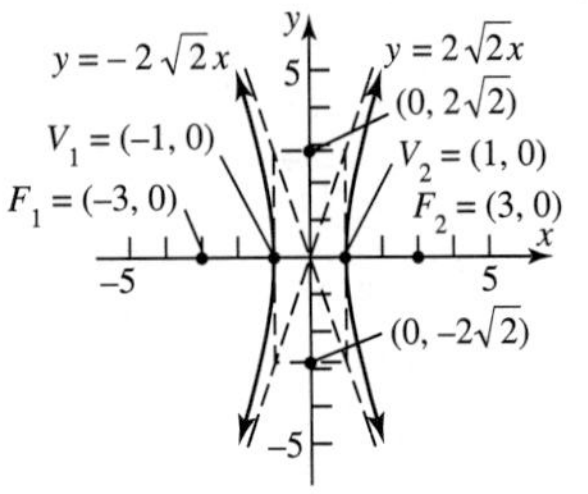

10. $\frac{y^2}{9} - \frac{x^2}{16} = 1$

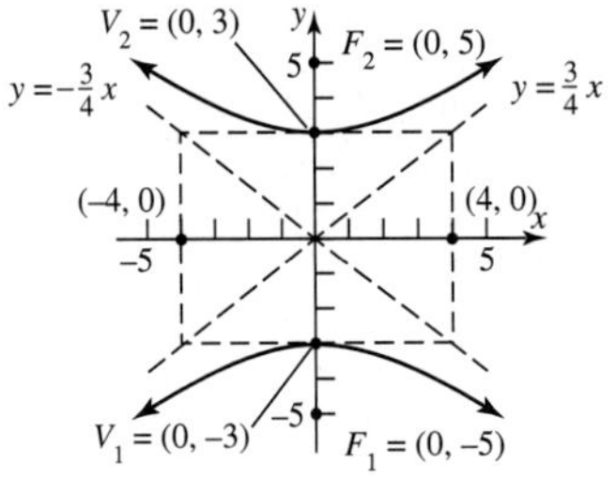

11. $\frac{y^2}{16} - \frac{x^2}{20} = 1$

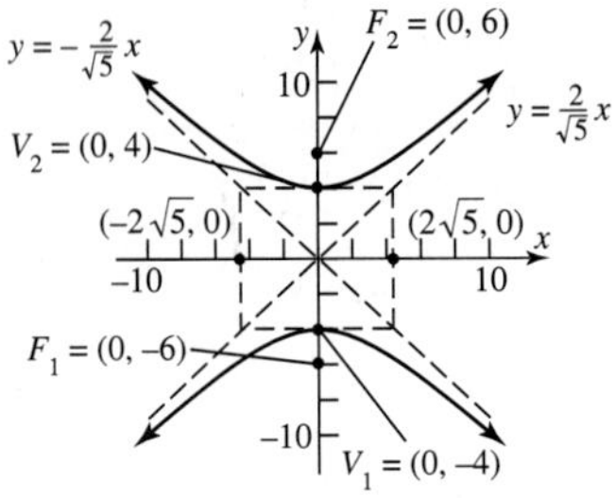

12. $\frac{x^2}{4} - \frac{y^2}{5} = 1$

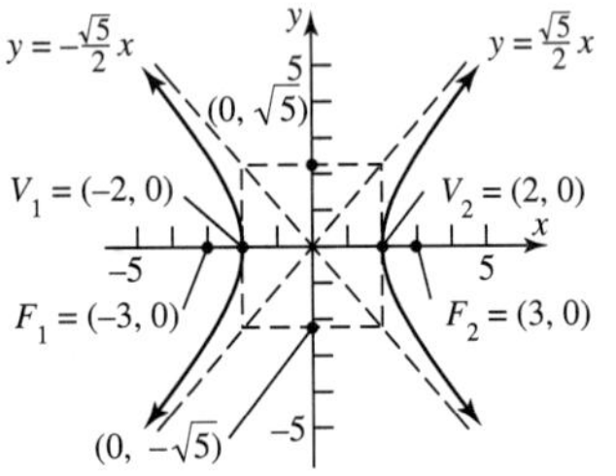

13. $\frac{x^2}{9} - \frac{y^2}{16} = 1$

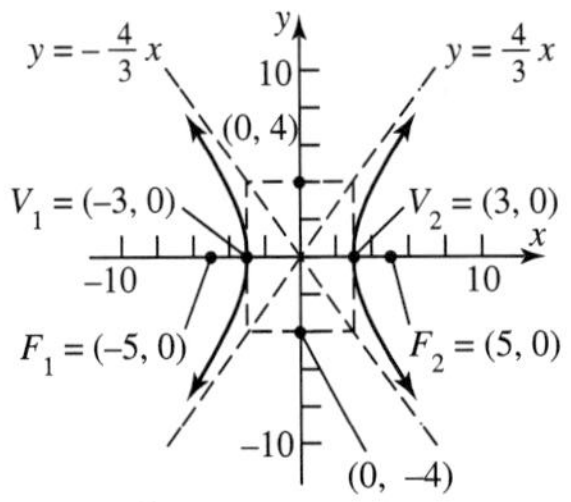

14. $\frac{y^2}{4} - \frac{x^2}{32} = 1$

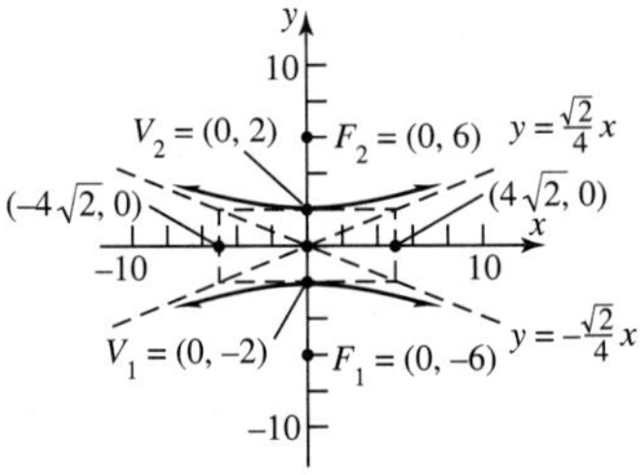

15. $\frac{y^2}{36} - \frac{x^2}{9} = 1$

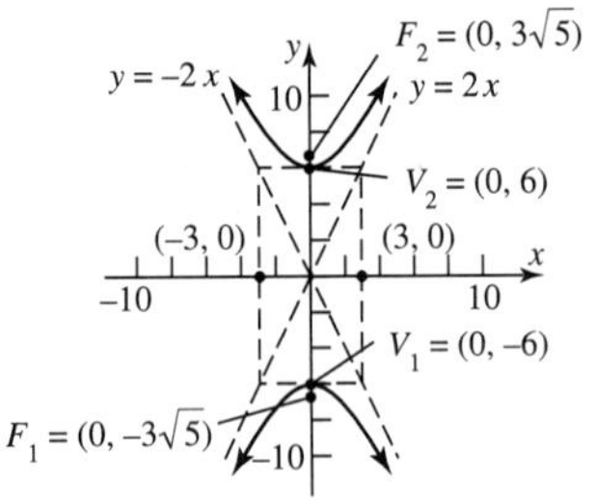

16. $\frac{x^2}{16} - \frac{y^2}{64} = 1$

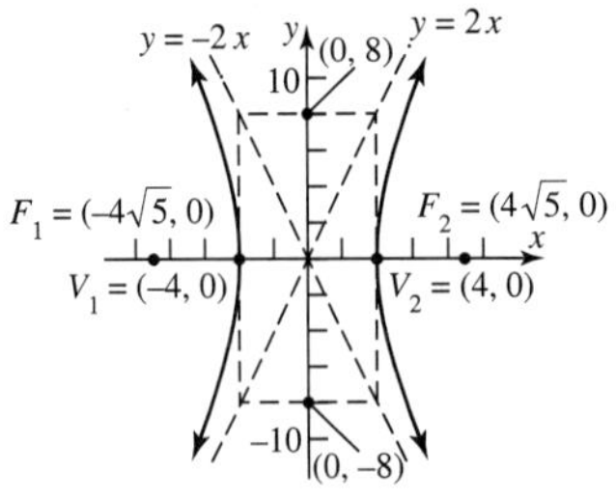

17. $\frac{x^2}{8} - \frac{y^2}{8} = 1$

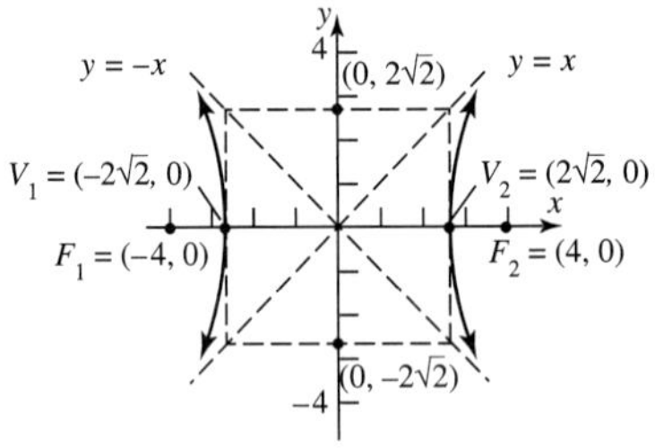

18. $\frac{y^2}{2} - \frac{x^2}{2} = 1$

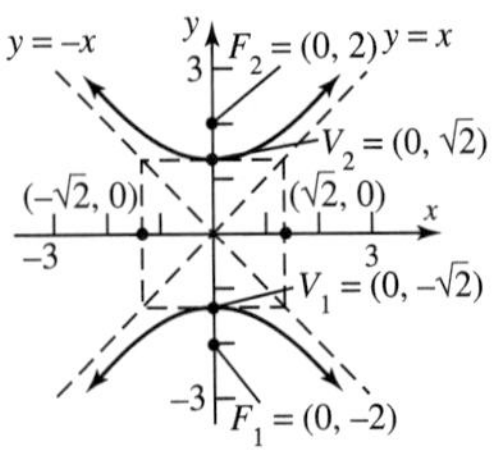

19. Center: (0, 0)
Transverse axis: x-axis
Vertices: $(-5, 0)$, $(5, 0)$
Foci: $(-\sqrt{34}, 0)$, $(\sqrt{34}, 0)$
Asymptotes: $y = \pm\dfrac{3}{5}x$

(a)

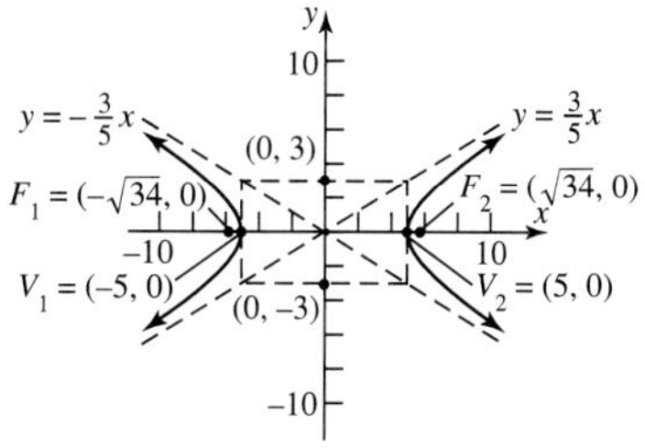

(b)

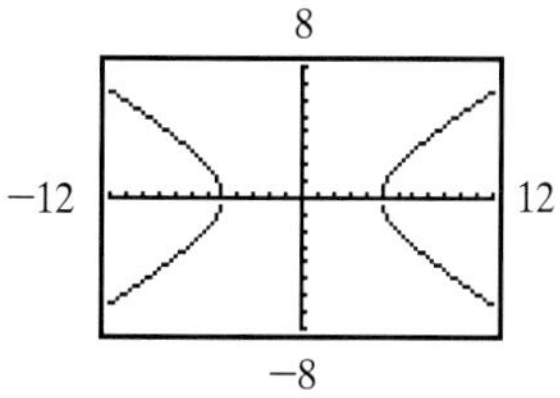

20. Center: (0, 0)
Transverse axis: y-axis
Vertices: $(0, -4)$, $(0, 4)$
Foci: $(0, -2\sqrt{5})$, $(0, 2\sqrt{5})$
Asymptotes: $y = \pm 2x$

(a)

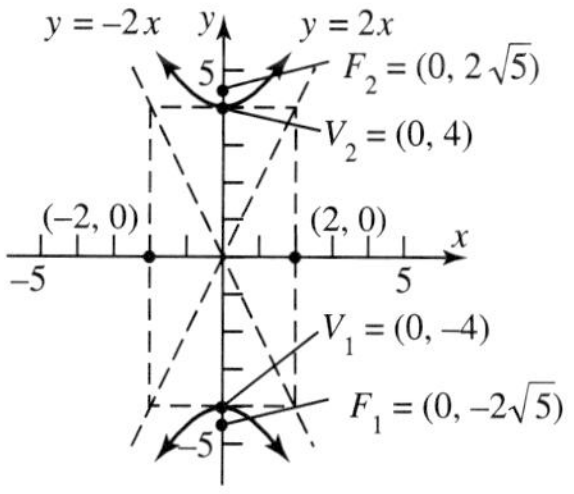

(b)

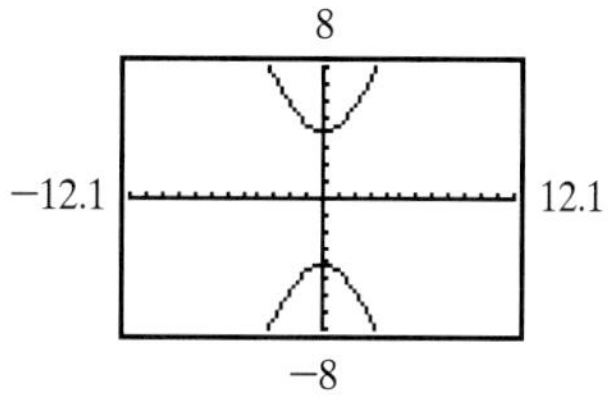

21. Center: (0, 0)
Transverse axis: x-axis
Vertices: $(-2, 0)$, $(2, 0)$
Foci: $(-2\sqrt{5}, 0)$, $(2\sqrt{5}, 0)$
Asymptotes: $y = \pm 2x$

(a)

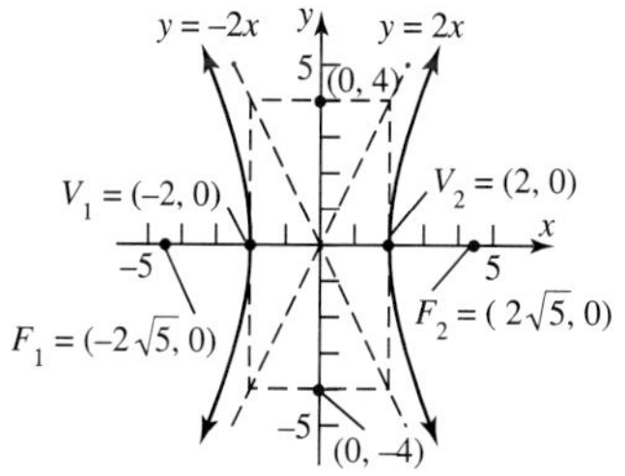

(b)

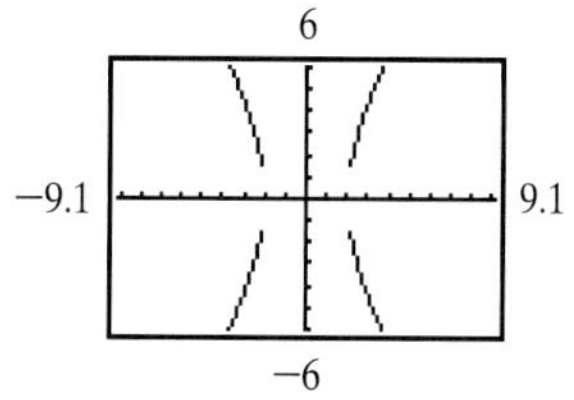

22. Center: (0, 0)
Transverse axis: y-axis
Vertices: $(0, -4)$, $(0, 4)$
Foci: $(0, -2\sqrt{5})$, $(0, 2\sqrt{5})$
Asymptotes: $y = \pm 2x$

(a)

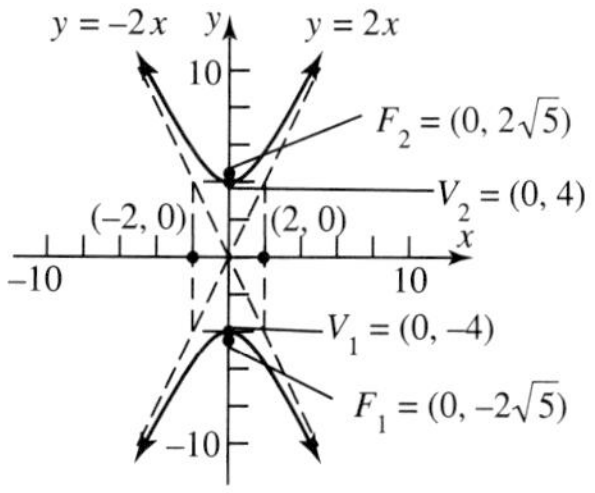

(b)

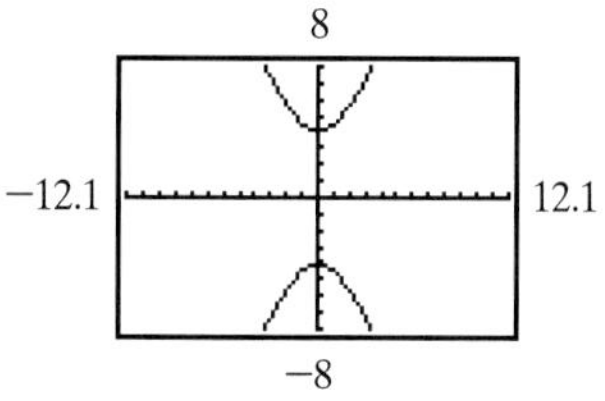

23. Center: (0, 0)
Transverse axis: y-axis
Vertices: $(0, -3)$, $(0, 3)$
Foci: $(0, -\sqrt{10})$, $(0, \sqrt{10})$
Asymptotes: $y = \pm 3x$

(a)

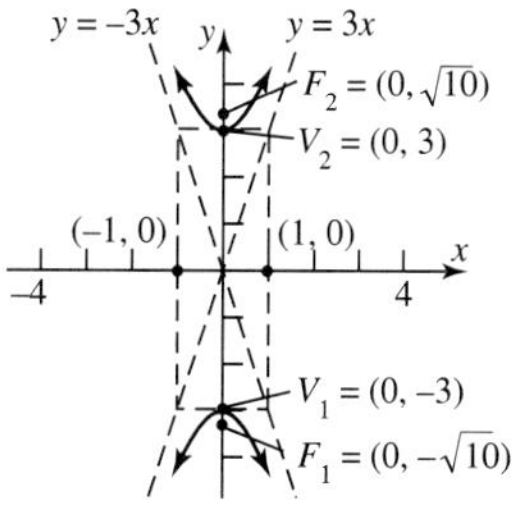

(b)

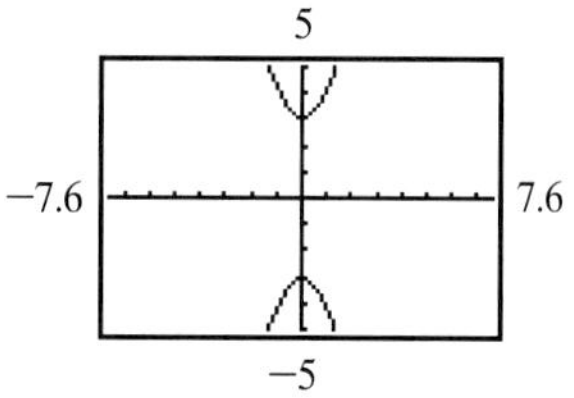

24. Center: (0, 0)
Transverse axis: x-axis
Vertices: $(-2, 0)$, $(2, 0)$
Foci: $(-2\sqrt{2}, 0)$, $(2\sqrt{2}, 0)$
Asymptotes: $y = \pm x$

(a)

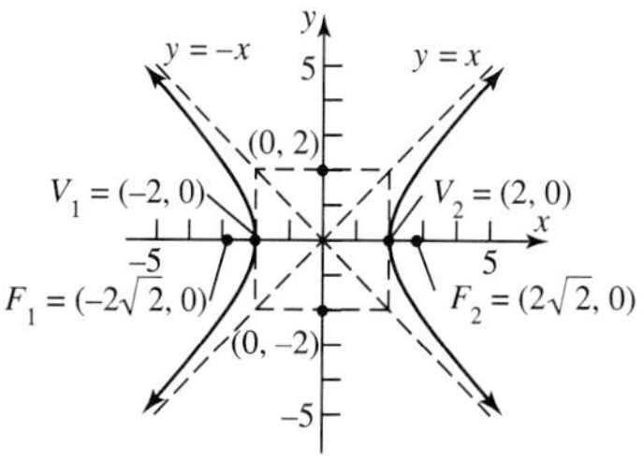

(b)

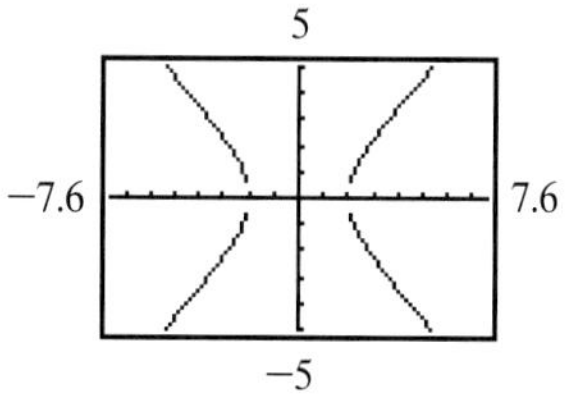

25. Center: (0, 0)
Transverse axis: y-axis
Vertices: $(0, -5)$, $(0, 5)$
Foci: $(0, -5\sqrt{2})$, $(0, 5\sqrt{2})$
Asymptotes: $y = \pm x$

(a)

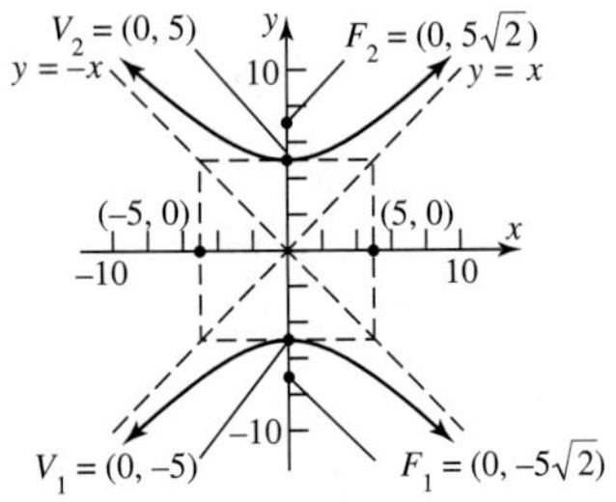

(b)

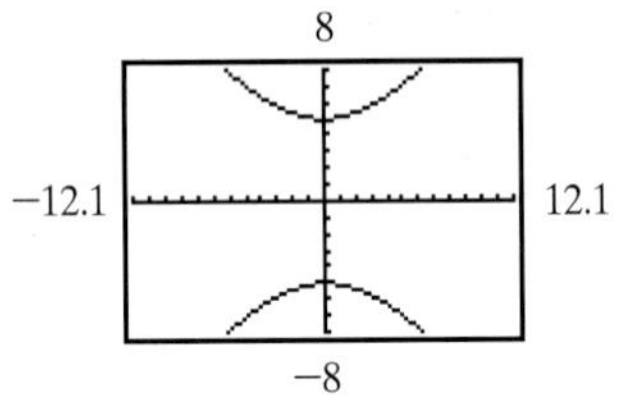

26. Center: (0, 0)
Transverse axis: x-axis
Vertices: $(-\sqrt{2}, 0)$, $(\sqrt{2}, 0)$
Foci: $(-\sqrt{6}, 0)$, $(\sqrt{6}, 0)$
Asymptotes: $y = \pm\sqrt{2}x$

(a)

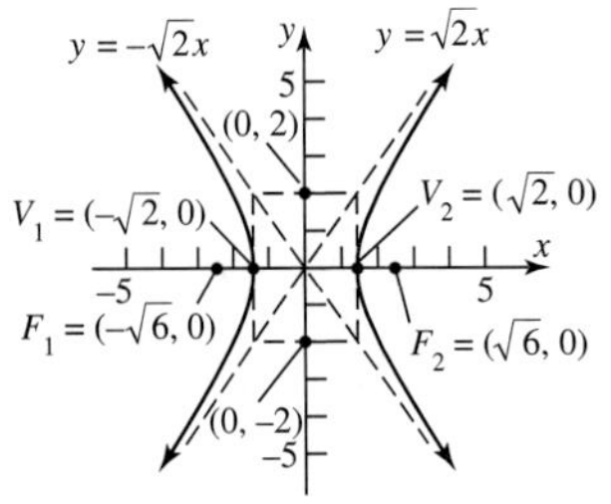

(b)

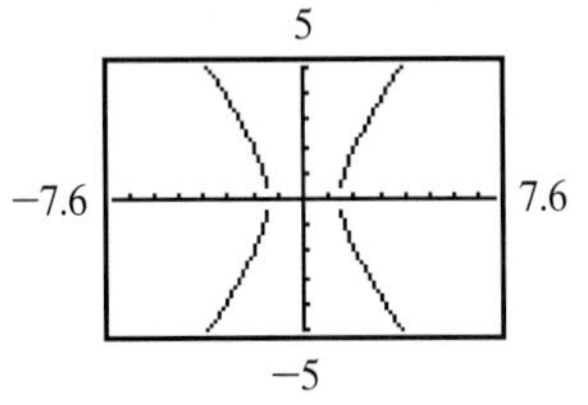

27. $x^2 - y^2 = 1$

28. $y^2 - x^2 = 1$

29. $\dfrac{y^2}{36} - \dfrac{x^2}{9} = 1$

30. $\dfrac{x^2}{4} - \dfrac{y^2}{16} = 1$

31. $\dfrac{(x-4)^2}{4} - \dfrac{(y+1)^2}{5} = 1$

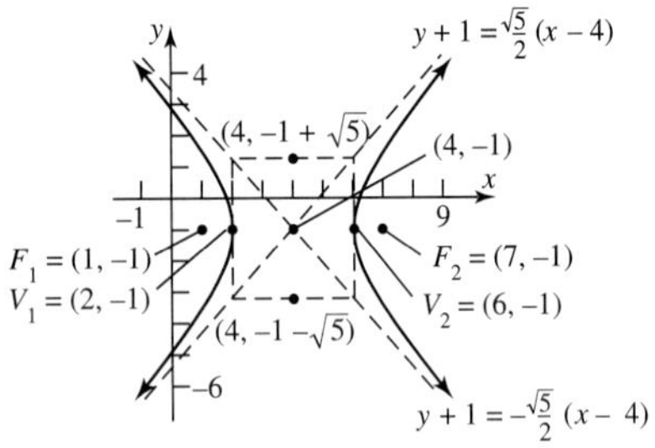

32. $\dfrac{(y-1)^2}{9} - \dfrac{(x+3)^2}{16} = 1$

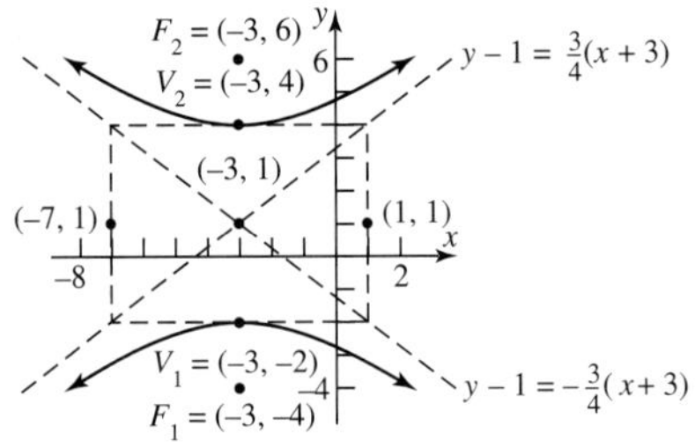

33. $\dfrac{(y+4)^2}{4} - \dfrac{(x+3)^2}{12} = 1$

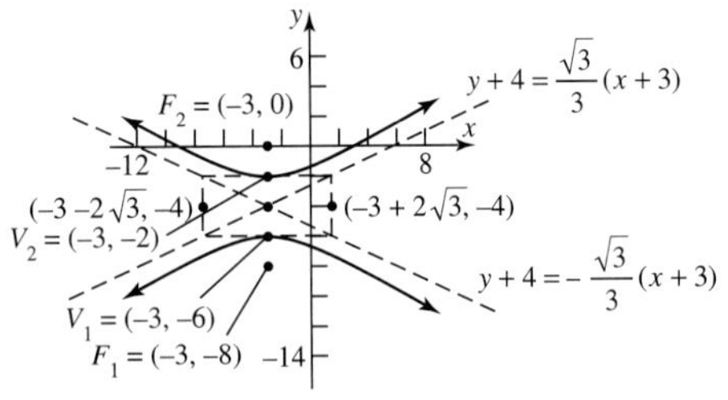

34. $\dfrac{(x-1)^2}{1} - \dfrac{(y-4)^2}{8} = 1$

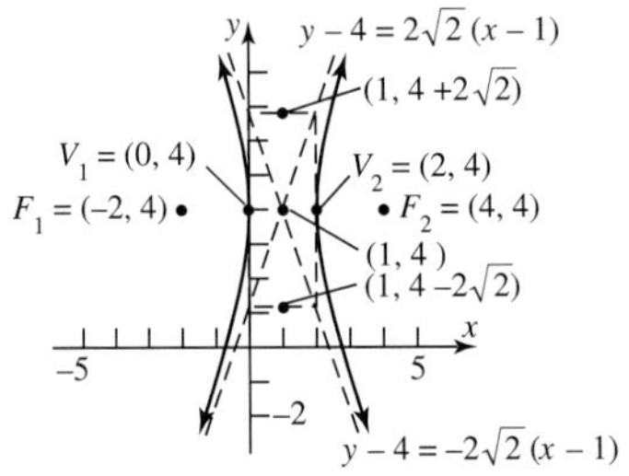

35. $(x-5)^2 - \dfrac{(y-7)^2}{3} = 1$

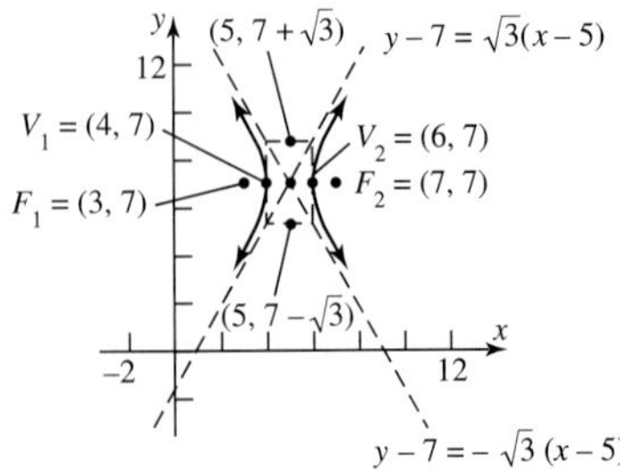

36. $\dfrac{(y-3)^2}{1} - \dfrac{(x+4)^2}{8} = 1$

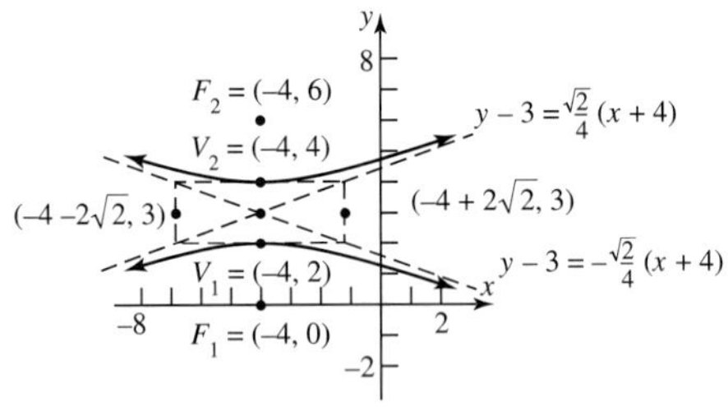

37. $\dfrac{(x-1)^2}{4} - \dfrac{(y+1)^2}{9} = 1$

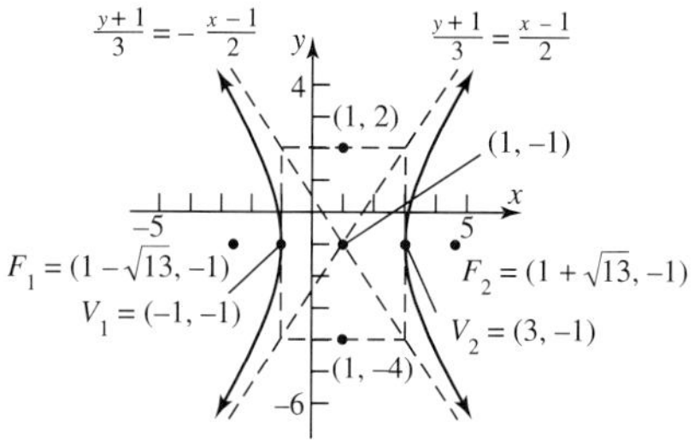

38. $\dfrac{(y+1)^2}{4} - \dfrac{9(x-1)^2}{16} = 1$

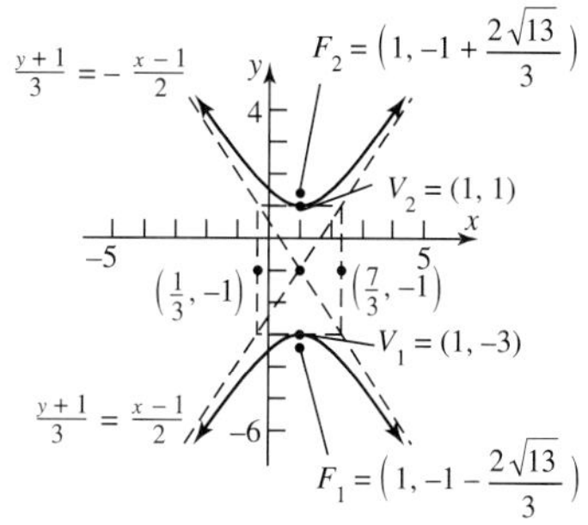

39. Center: $(2, -3)$
Transverse axis: Parallel to x-axis
Vertices: $(0, -3)$, $(4, -3)$
Foci: $(2 - \sqrt{13}, -3)$, $(2 + \sqrt{13}, -3)$
Asymptotes: $y + 3 = \pm\frac{3}{2}(x - 2)$

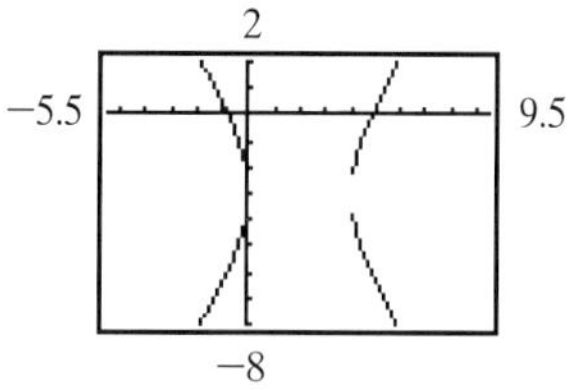

40. Center: $(2, -3)$
Transverse axis: Parallel to x-axis
Vertices: $(2, -5)$, $(2, -1)$
Foci: $(2, -3 + \sqrt{13})$, $(2, -3 - \sqrt{13})$
Asymptotes: $y + 3 = \pm\frac{2}{3}(x - 2)$

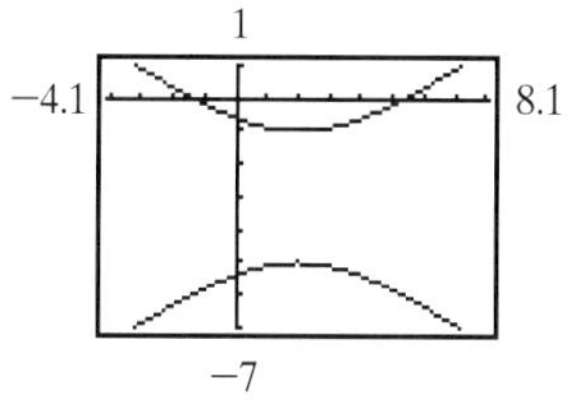

41. Center: $(-2, 2)$
Transverse axis: Parallel to y-axis
Vertices: $(-2, 0)$, $(-2, 4)$
Foci: $(-2, 2 - \sqrt{5})$, $(-2, 2 + \sqrt{5})$
Asymptotes: $y - 2 = \pm 2(x + 2)$

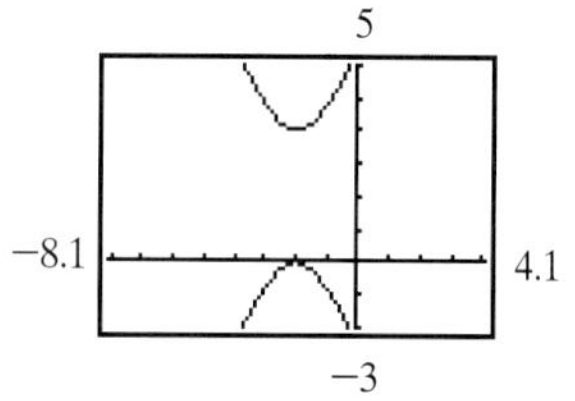

42. Center: $(-4, 3)$
Transverse axis: Parallel to x-axis
Vertices: $(-7, 3)$, $(-1, 3)$
Foci: $(-4 - \sqrt{10}, 3)$, $(-4 + \sqrt{10}, 3)$
Asymptotes: $y - 3 = \pm\frac{1}{3}(x + 4)$

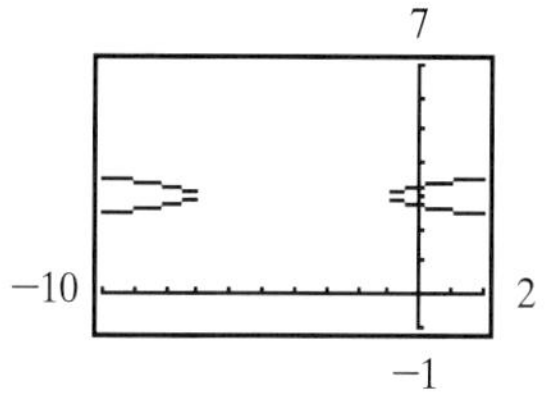

43. Center: $(-1, -2)$
Transverse axis: Parallel to x-axis
Vertices: $(-3, -2)$, $(1, -2)$
Foci: $(-1 - 2\sqrt{2}, -2)$, $(-1 + 2\sqrt{2}, -2)$
Asymptotes: $y + 2 = \pm(x + 1)$

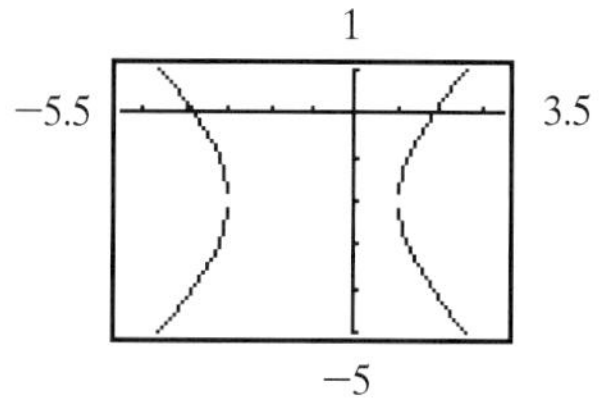

44. Center: $(-2, 3)$
Transverse axis: Parallel to y-axis
Vertices: $(-2, 5)$, $(-2, 1)$
Foci: $(-2, 3 + 2\sqrt{2})$, $(-2, 3 - 2\sqrt{2})$
Asymptotes: $y - 3 = \pm(x + 2)$

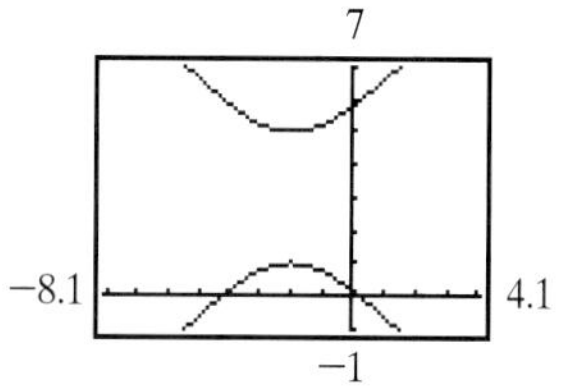

45. Center: $(1, -1)$
Transverse axis: Parallel to x-axis
Vertices: $(0, -1)$, $(2, -1)$
Foci: $(1 - \sqrt{2}, -1)$, $(1 + \sqrt{2}, -1)$
Asymptotes: $y + 1 = \pm(x - 1)$

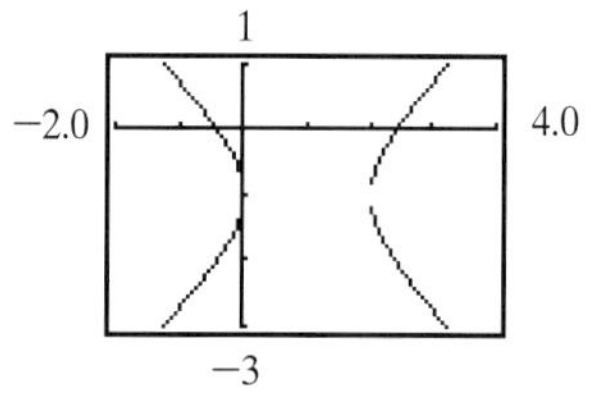

46. Center: $(2, 2)$
Transverse axis: Parallel to y-axis
Vertices: $(2, 1)$, $(2, 3)$
Foci: $(2, 2 - \sqrt{2})$, $(2, 2 + \sqrt{2})$
Asymptotes: $y - 2 = \pm(x - 2)$

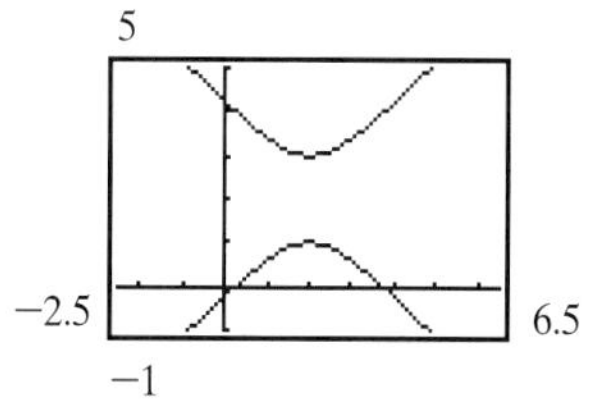

47. Center: $(-1, 2)$
Transverse axis: Parallel to y-axis
Vertices: $(-1, 0)$, $(-1, 4)$
Foci: $(-1, 2 - \sqrt{5})$, $(-1, 2 + \sqrt{5})$
Asymptotes: $y - 2 = \pm 2(x + 1)$

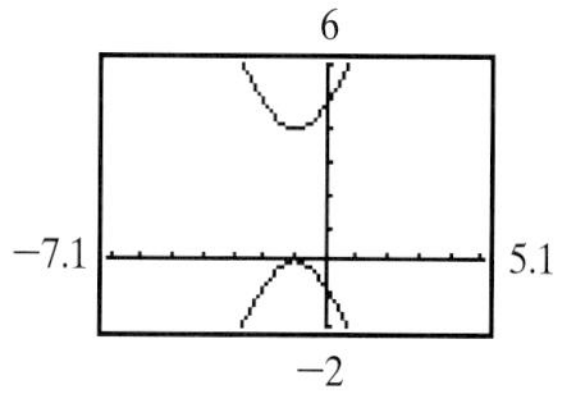

48. Center: $(-1, 2)$
Transverse axis: Parallel to x-axis
Vertices: $(-2, 2)$, $(0, 2)$
Foci: $(-1 - \sqrt{3}, 2)$, $(-1 + \sqrt{3}, 2)$
Asymptotes: $y - 2 = \pm\sqrt{2}(x + 1)$

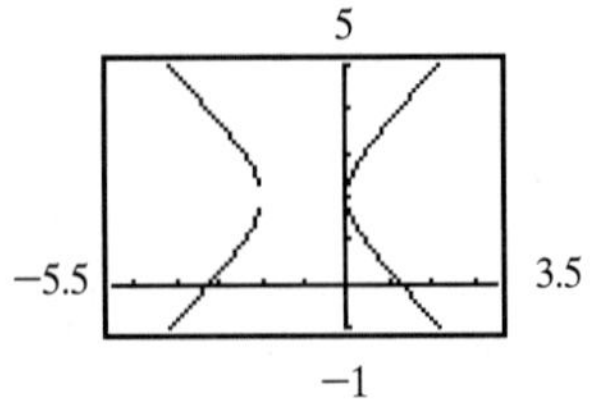

49. Center: $(3, -2)$
Transverse axis: Parallel to x-axis
Vertices: $(1, -2)$, $(5, -2)$
Foci: $(3 - 2\sqrt{5}, -2)$, $(3 + 2\sqrt{5}, -2)$
Asymptotes: $y + 2 = \pm 2(x - 3)$

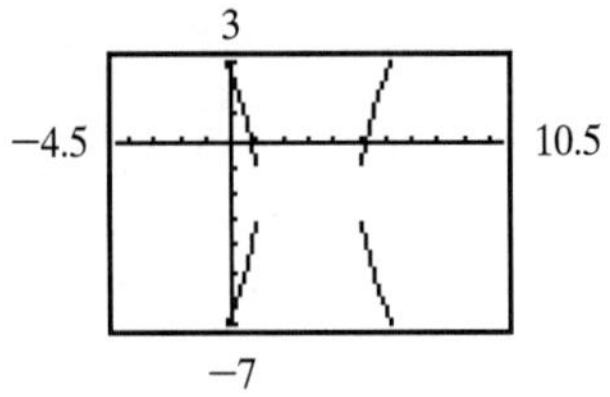

50. Center: $(1, -2)$
Transverse axis: Parallel to y-axis
Vertices: $(1, -2 - \sqrt{2})$, $(1, -2 + \sqrt{2})$
Foci: $(1, -2 - \sqrt{6})$, $(1, -2 + \sqrt{6})$
Asymptotes: $y + 2 = \pm\frac{\sqrt{2}}{2}(x - 1)$

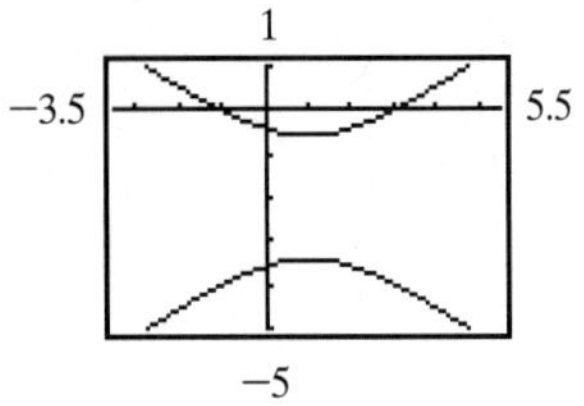

51. Center: $(-2, 1)$
Transverse axis: Parallel to y-axis
Vertices: $(-2, -1)$, $(-2, 3)$
Foci: $(-2, 1 - \sqrt{5})$, $(-2, 1 + \sqrt{5})$
Asymptotes: $y - 1 = \pm 2(x + 2)$

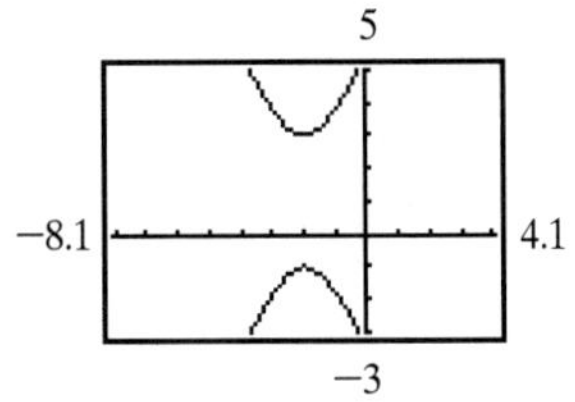

52. Center: $(-4, -1)$
Transverse axis: Parallel to x-axis
Vertices: $(-7, -1)$, $(-1, -1)$
Foci: $(-4 - 2\sqrt{3}, -1)$, $(-4 + 2\sqrt{3}, -1)$
Asymptotes: $y + 1 = \frac{\sqrt{3}}{3}(x + 4)$

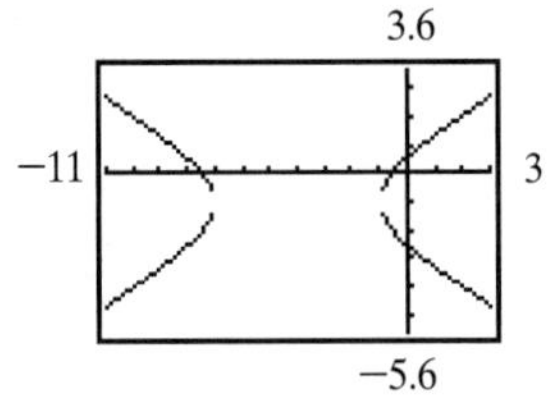

53.

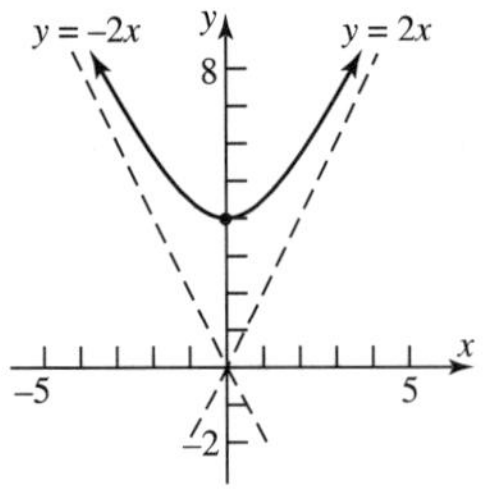

54.

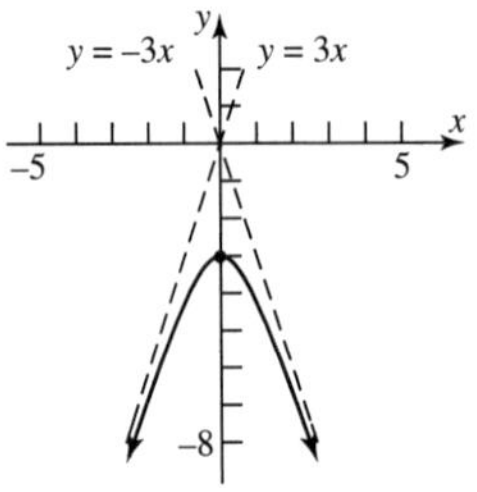

55.

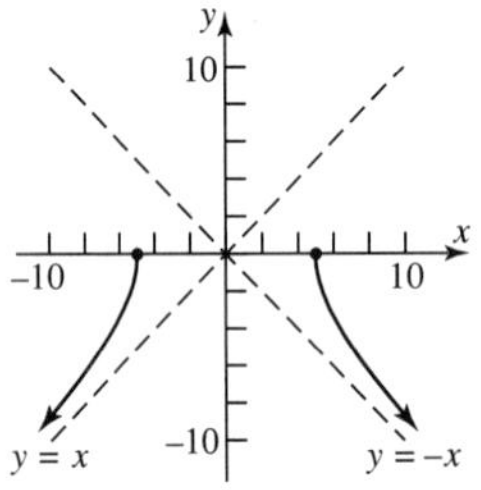

56.

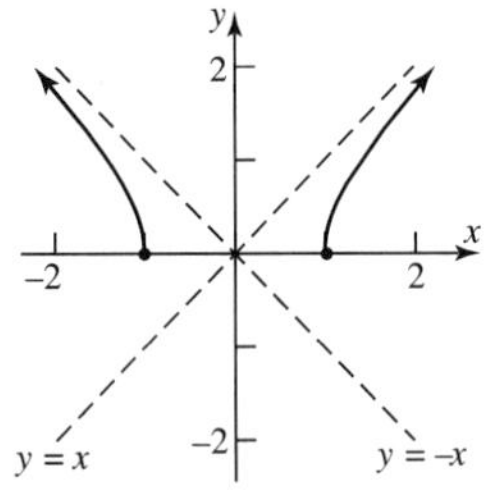

57. **(a)** The ship would reach shore 64.66 miles from the master station. **(b)** 0.00086 sec **(c)** (104, 50)
58. **(a)** The ship would reach shore 20.24 miles from the master station. **(b)** 0.0004301 sec **(c)** (48, 20)
59. **(a)** 450 ft **61.** If e is close to 1, narrow hyperbola; if e is very large, wide hyperbola **62.** $\sqrt{2}$
63. $\frac{x^2}{4} - y^2 = 1$; asymptotes $y = \pm\frac{1}{2}x$, $y^2 - \frac{x^2}{4} = 1$; asymptotes $y = \pm\frac{1}{2}x$

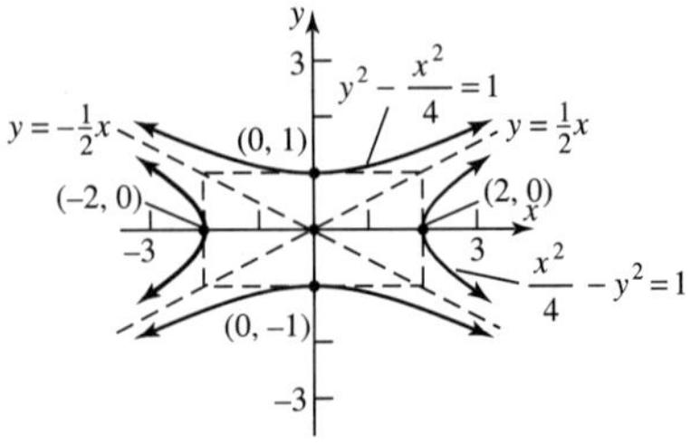

64. $\dfrac{y^2}{a^2} - \dfrac{x^2}{b^2} = 1$

$\dfrac{y^2}{a^2} = 1 + \dfrac{x^2}{b^2}$

$y^2 = a^2\left(1 + \dfrac{x^2}{b^2}\right)$

$y^2 = \dfrac{a^2x^2}{b^2}\left(\dfrac{b^2}{x^2} + 1\right)$

$y = \dfrac{ax}{b}\sqrt{\dfrac{b^2}{x^2} + 1}$

As $x \to -\infty$ or as $x \to \infty$, the term $\dfrac{b^2}{x^2}$ gets close to 0, so the expression under the radical gets closer to 1. Thus, the graph of the hyperbola gets closer to the lines

$y = -\dfrac{a}{b}x$ and $y = \dfrac{a}{b}x$

The lines are asymptotes of the hyperbola.

65. $Ax^2 + Cy^2 + F = 0$

$Ax^2 + Cy^2 = -F$

If A and C are of opposite sign and $F \neq 0$, this equation may be written as $\dfrac{x^2}{\left(-\dfrac{F}{A}\right)} + \dfrac{y^2}{\left(-\dfrac{F}{C}\right)} = 1$, where $-\dfrac{F}{A}$ and $-\dfrac{F}{C}$ are opposite in sign. This is the equation of a hyperbola with center $(0, 0)$. The transverse axis is the x-axis if $-\dfrac{F}{A} > 0$; the transverse axis is the y-axis if $-\dfrac{F}{A} < 0$.

66. $Ax^2 + Cy^2 + Dx + Ey + F = 0$ where A and C are of opposite sign

$Ax^2 + Dx + Cy^2 + Ey = -F$

$A\left(x^2 + \dfrac{D}{A}x\right) + C\left(y^2 + \dfrac{E}{C}y\right) = -F$

$A\left(x + \dfrac{D}{2A}\right)^2 + C\left(y + \dfrac{E}{2C}\right)^2 = -F + \dfrac{D^2}{4A} + \dfrac{E^2}{4C}$

(a) If $\dfrac{D^2}{4A} + \dfrac{E^2}{4C} - F \neq 0$, this is the equation of hyperbola with center at $\left(-\dfrac{D}{2A}, -\dfrac{E}{2C}\right)$.

(b) If $\dfrac{D^2}{4A} + \dfrac{E^2}{4C} - F = 0$, the graph is intersecting lines through the point $\left(-\dfrac{D}{2A}, -\dfrac{E}{2C}\right)$ with slopes $\sqrt{\left|\dfrac{A}{C}\right|}$.

Historical Problem

$x = 6, y = 8$

9.5 Exercises

1.

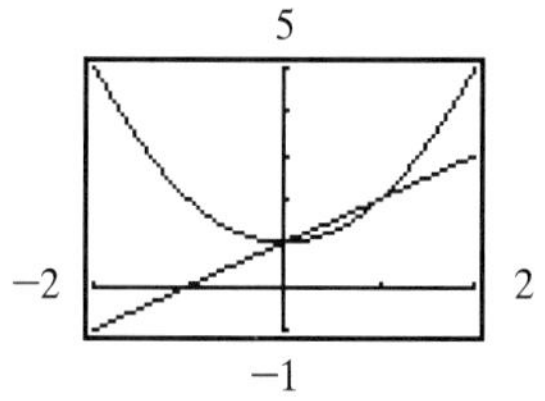

(0.00, 1.00); (1.00, 2.00)
(0, 1); (1, 2)

2.

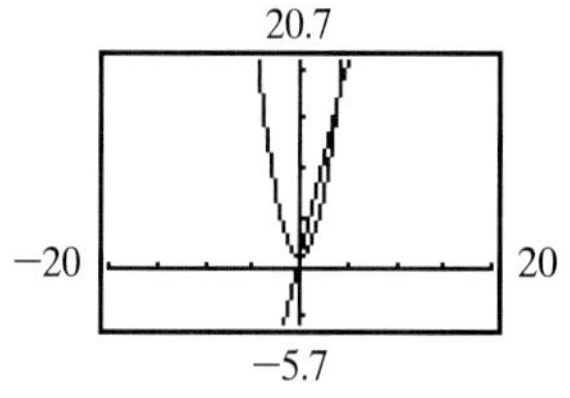

(0.00, 1.00); (4.00, 17.00)
(0, 1); (4, 17)

3.

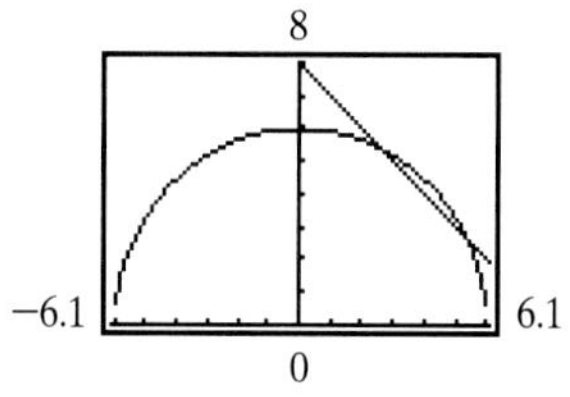

(2.59, 5.41); (5.41, 2.59)
$(4 - \sqrt{2}, 4 + \sqrt{2}); (4 + \sqrt{2}, 4 - \sqrt{2})$

4.

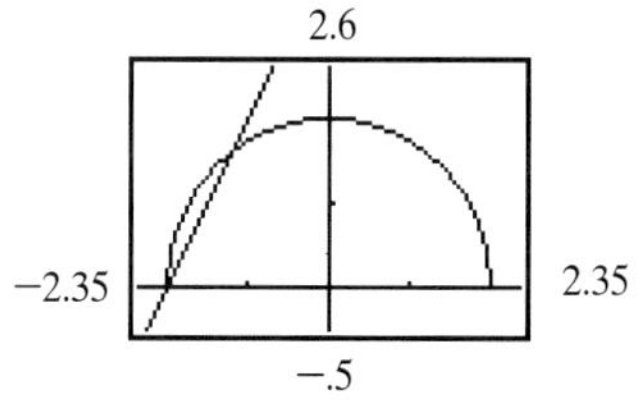

(−2.00, 0.00); (−1.20, 1.60)
$(-2, 0); \left(-\dfrac{6}{5}, \dfrac{8}{5}\right)$

5.

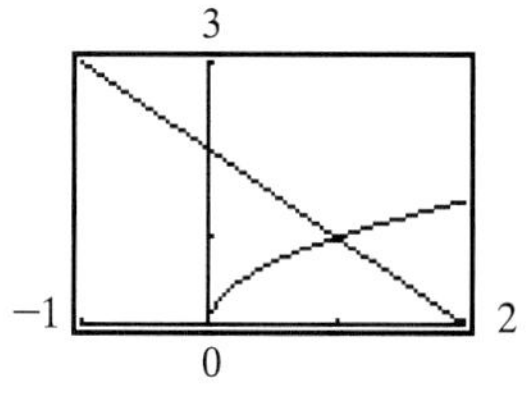

(1.00, 1.00)
(1, 1)

6.

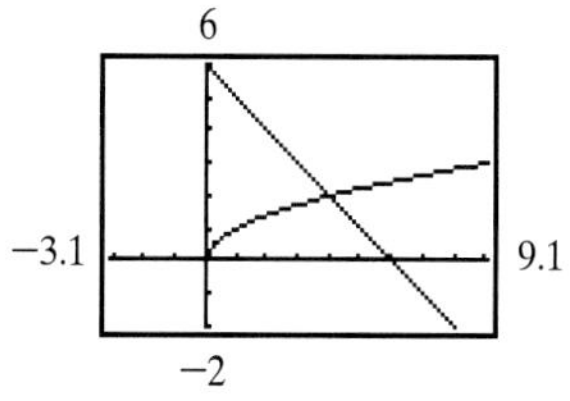

(4.00, 2.00)
(4, 2)

7.

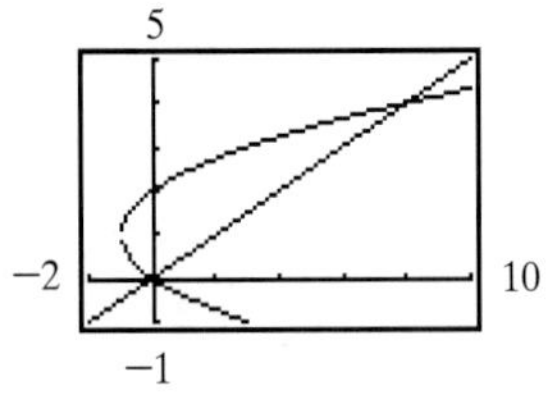

(0.00, 0.00); (8.00, 4.00)
(0, 0); (8, 4)

8.

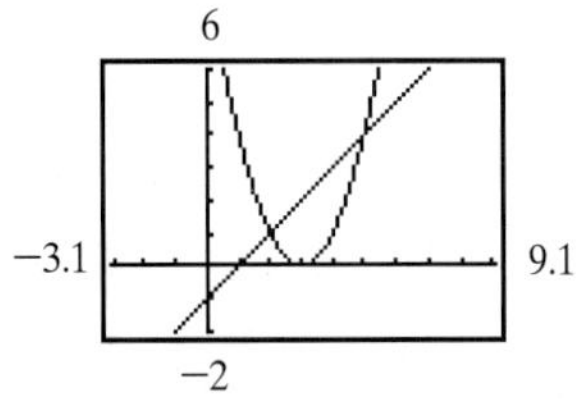

(2.00, 1.00); (5.00, 4.00)
(2, 1); (5, 4)

9.

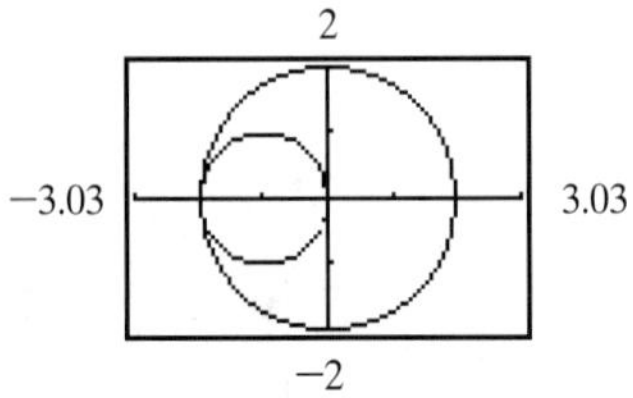

(−2.00, 0.00)
(−2, 0)

10.

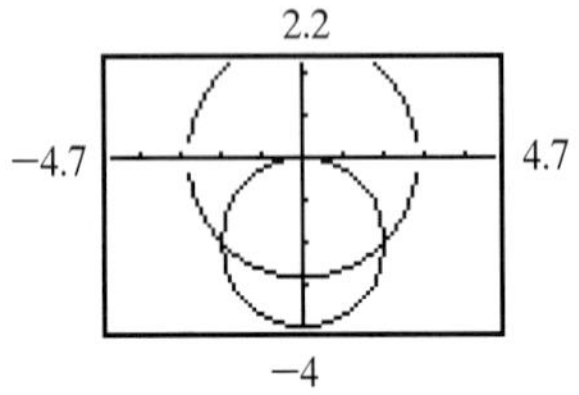

(−2.00, −2.00); (2.00, −2.00)
(−2, −2); (2, −2)

11.

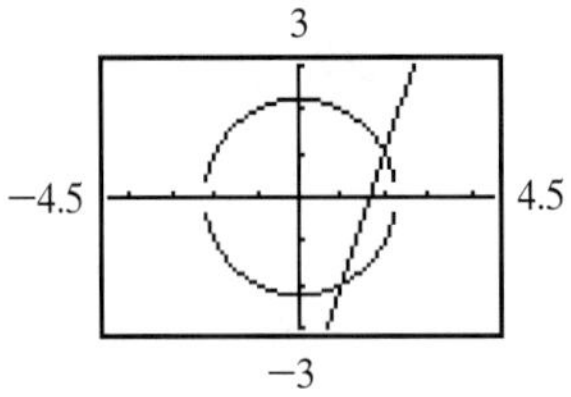

(1.00, −2.00); (2.00, 1.00)
(1, −2); (2, 1)

12.

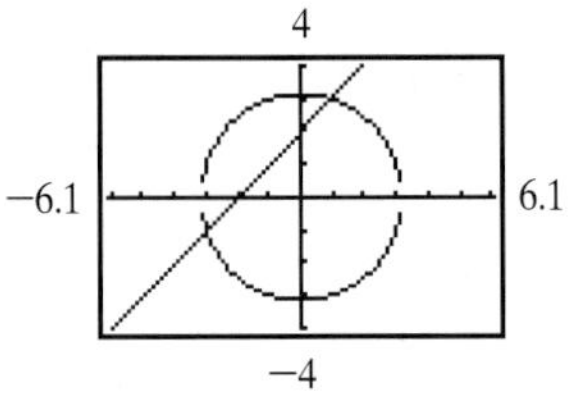

(−3.00, −1.00); (1.00, 3.00)
(−3, −1); (1, 3)

13.

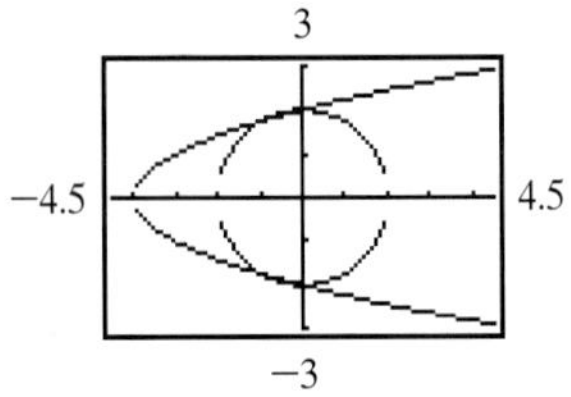

(0.00, 2.00); (0.00, −2.00);
(−1, −1.73); (−1, 1.73)
(0, 2); (0, −2); $(-1, -\sqrt{3})$; $(-1, \sqrt{3})$

14.

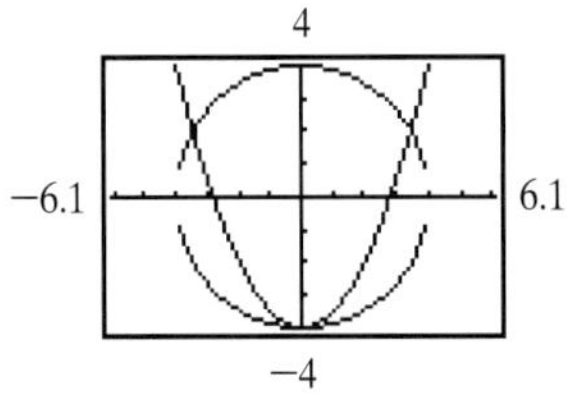

(−3.46, 2.00); (0.00, −4.00); (3.46, 2.00)
$(-2\sqrt{3}, 2)$; (0, −4); $(2\sqrt{3}, 2)$

15.

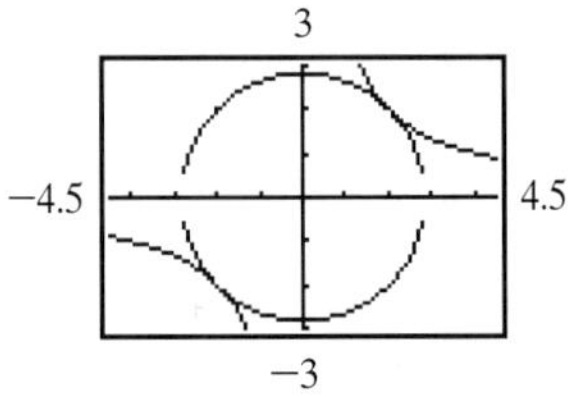

(2.00, 2.00); (−2.00, −2.00)
(2, 2); (−2, −2)

16.

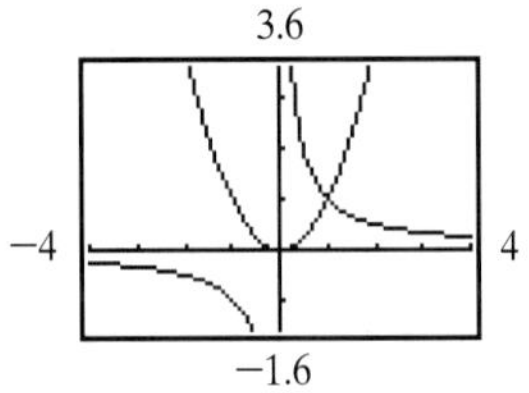

(1.00, 1.00)

(1, 1)

17.

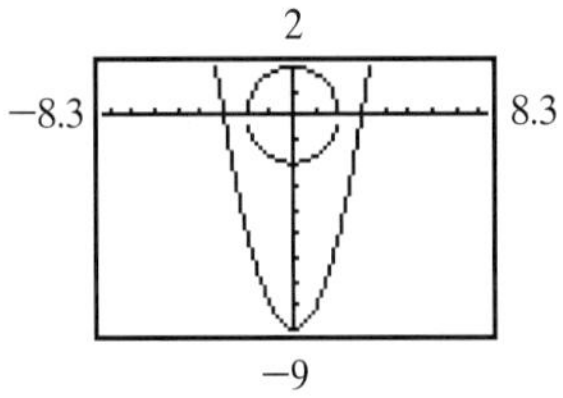

No solution; Inconsistent

18.

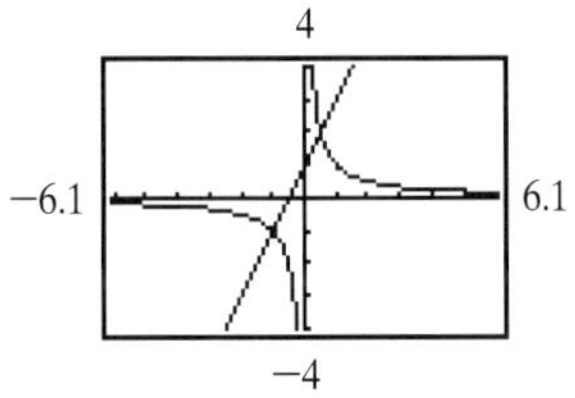

(−1.00, −1.00); (0.50, 2.00)
$(-1, -1)$; $\left(\frac{1}{2}, 2\right)$

19.

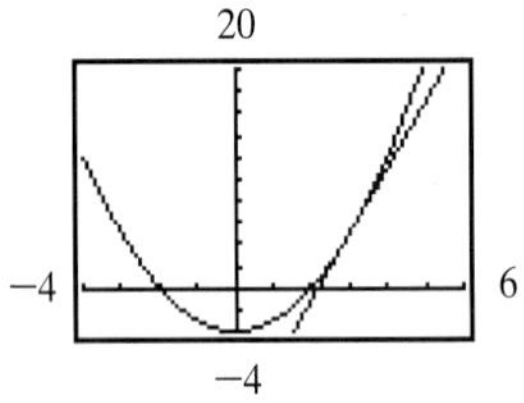

(3.00, 5.00)
(3, 5)

20.

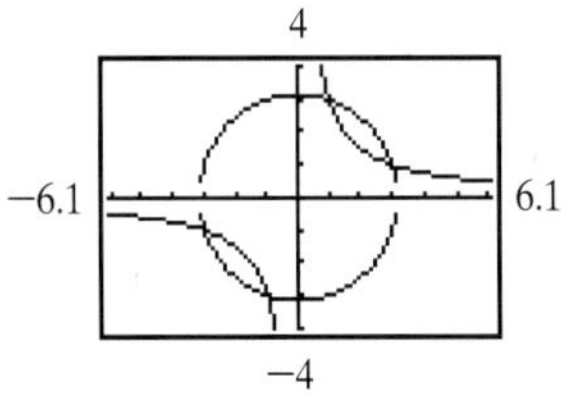

(−3.00, −1.00); (−1.00, −3.00);
(1.00, 3.00); (3.00, 1.00)
(−3, −1); (−1, −3); (1, 3); (3, 1)

21. $x = 1, y = 4; x = -1, y = -4; x = 2\sqrt{2}, y = \sqrt{2}; x = -2\sqrt{2}, y = -\sqrt{2}$ **22.** $x = 5, y = 2$ **23.** $x = 0, y = 1; x = -\frac{2}{3}, y = -\frac{1}{3}$

24. $x = -5, y = -\frac{3}{2}$ **25.** $x = 0, y = -1; x = \frac{5}{2}, y = -\frac{7}{2}$

26. $x = -2, y = -2; x = -\sqrt{2}, y = -2\sqrt{2}; x = \sqrt{2}, y = 2\sqrt{2}; x = 2, y = 2$ **27.** $x = 2, y = \frac{1}{3}; x = \frac{1}{2}, y = \frac{4}{3}$

28. $x = -2, y = 0; x = 7, y = 6$ **29.** $x = 3, y = 2; x = 3, y = -2; x = -3, y = 2; x = -3, y = -2$

30. $x = -1, y = -2; x = -1, y = 2; x = 1, y = -2; x = 1, y = 2$
31. $x = \frac{1}{2}, y = \frac{3}{2}; x = \frac{1}{2}, y = -\frac{3}{2}; x = -\frac{1}{2}, y = \frac{3}{2}; x = -\frac{1}{2}, y = -\frac{3}{2}$
32. $x = -2\sqrt{2}, y = \sqrt{3}; x = 2\sqrt{2}, y = \sqrt{3}; x = 2\sqrt{2}, y = -\sqrt{3}; x = -2\sqrt{2}, y = -\sqrt{3}$
33. $x = \sqrt{2}, y = 2\sqrt{2}; x = -\sqrt{2}, y = -2\sqrt{2}$ **34.** $x = -5\sqrt{3}, y = \sqrt{3}; x = 5\sqrt{3}, y = -\sqrt{3}$
35. No solution; system is inconsistent **36.** No solution; system is inconsistent
37. $x = \frac{8}{3}, y = \frac{2\sqrt{10}}{3}; x = -\frac{8}{3}, y = \frac{2\sqrt{10}}{3}; x = \frac{8}{3}, y = -\frac{2\sqrt{10}}{3}; x = -\frac{8}{3}, y = -\frac{2\sqrt{10}}{3}$
38. $x = -\frac{\sqrt{2}}{2}, y = -\frac{\sqrt{6}}{3}; x = -\frac{\sqrt{2}}{2}, y = \frac{\sqrt{6}}{3}; x = \frac{\sqrt{2}}{2}, y = -\frac{\sqrt{6}}{3}; x = \frac{\sqrt{2}}{2}, y = \frac{\sqrt{6}}{3}$
39. $x = 1, y = \frac{1}{2}; x = -1, y = \frac{1}{2}; x = 1, y = -\frac{1}{2}; x = -1, y = -\frac{1}{2}$
40. $x = -2, y = -\sqrt{2}; x = -2, y = \sqrt{2}; x = 2, y = -\sqrt{2}; x = 2, y = \sqrt{2}$ **41.** No solution; system is inconsistent
42. $x = \sqrt[4]{\frac{2}{5}}, y = \sqrt[4]{\frac{2}{3}}; x = -\sqrt[4]{\frac{2}{5}}, y = \sqrt[4]{\frac{2}{3}}; x = \sqrt[4]{\frac{2}{5}}, y = -\sqrt[4]{\frac{2}{3}}; x = -\sqrt[4]{\frac{2}{5}}, y = -\sqrt[4]{\frac{2}{3}}$
43. $x = \sqrt{3}, y = \sqrt{3}; x = -\sqrt{3}, y = -\sqrt{3}; x = 2, y = 1; x = -2, y = -1$ **44.** $x = -2, y = 2; x = 3, y = -3$
45. $x = 3, y = 2; x = -3, y = -2; x = 2, y = \frac{1}{2}; x = -2, y = -\frac{1}{2}$
46. $x = -3, y = 3; x = -\sqrt{3}, y = -\sqrt{3}; x = \sqrt{3}, y = \sqrt{3}; x = 3, y = -3$ **47.** $x = 3, y = 1; x = -1, y = -3$
48. $x = -1, y = 3; x = 3, y = -1$ **49.** $x = 0, y = -2; x = 0, y = 1; x = 2, y = -1$ **50.** $x = 2, y = 1$ **51.** $x = 2, y = 8$
52. $x = \frac{1}{2}, y = \frac{1}{16}$ **53.** $x = 0.48, y = 0.62$ **54.** $x = 0.65, y = 0.52$ **55.** $x = -1.65, y = -0.89$ **56.** $x = -1.37, y = 2.14$
57. $x = 0.58, y = 1.86; x = 1.81, y = 1.05; x = 0.58, y = -1.86; x = 1.81, y = -1.05$ **58.** $x = 0.64, y = 1.55; x = -0.64, y = -1.55$
59. $x = 2.35, y = 0.85$ **60.** $x = 1.90, y = 0.64; x = 0.14, y = -2.00$ **61.** 3 and 1; −3 and −1 **62.** 2 and 5 **63.** 2 and 2; −2 and −2
64. 2 and 5; −2 and −5 **65.** $\frac{1}{2}$ and $\frac{1}{3}$ **66.** $\frac{1}{2}$ and −1 **67.** 5 **68.** $\frac{1}{7}$

69.
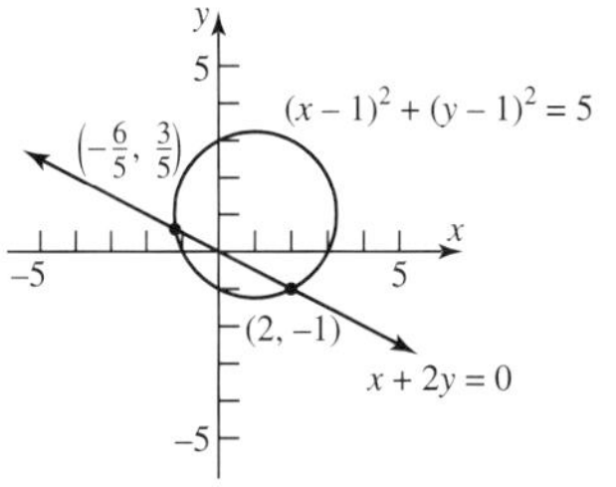

70.
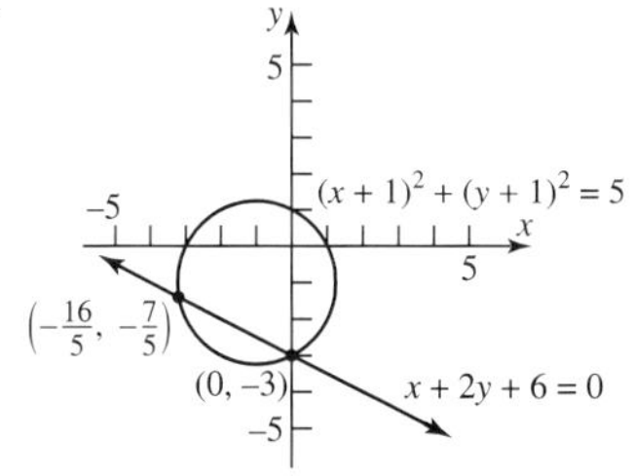

71.
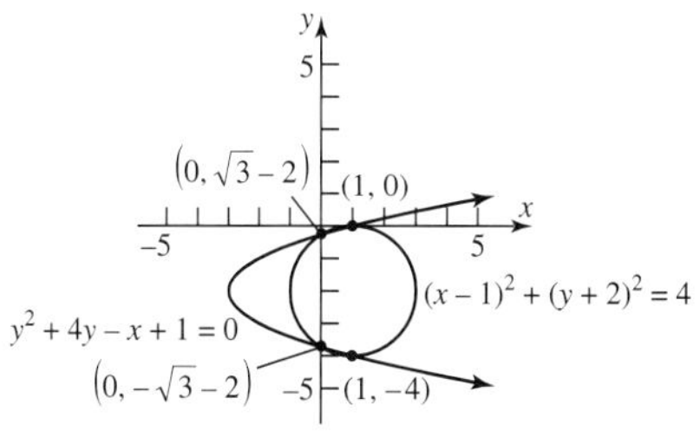

72.
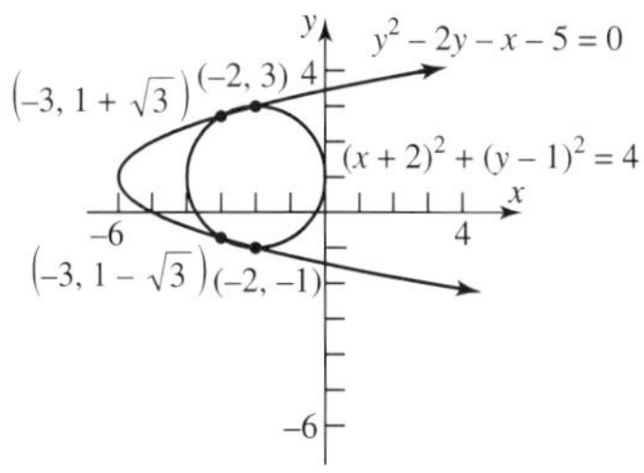

73.
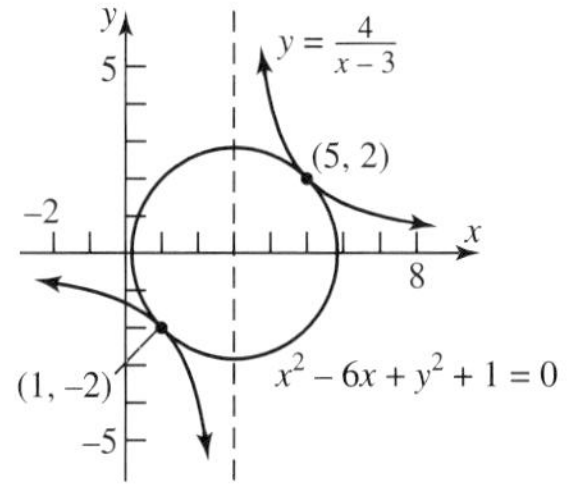

74.
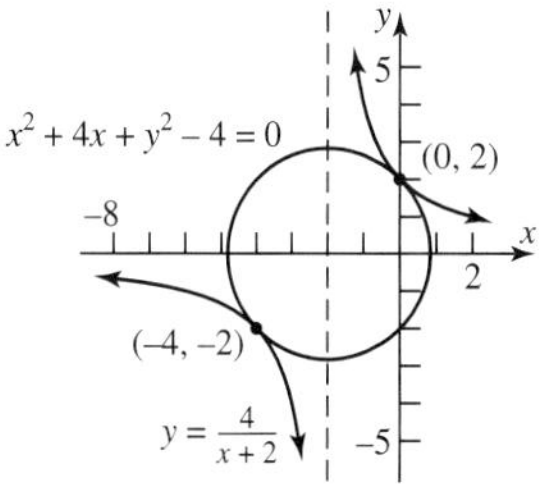

75. 5 in. by 3 in. **76.** 4 ft, 6 ft **77.** 2 cm and 4 cm **78.** 8 cm
79. tortoise: 7 m/hr, hare: $7\frac{1}{2}$ m/hr **80.** 10.02 ft **81.** 12 cm by 18 cm **82.** 16.57 cm by 13.03 cm **83.** $x = 60$ ft; $y = 30$ ft
84. It is not possible. **85.** $l = \frac{P + \sqrt{P^2 - 16A}}{4}; w = \frac{P - \sqrt{P^2 - 16A}}{4}$ **86.** $b = P - \frac{4h^2 + P^2}{2P}, l = \frac{4h^2 + P^2}{4P}$
87. $y = 4x - 4$ **88.** $y = -\frac{1}{3}x + \frac{10}{3}$ **89.** $y = 2x + 1$ **90.** $y = 4x + 9$ **91.** $y = -\frac{1}{3}x + \frac{7}{3}$ **92.** $y = \frac{3}{2}x + \frac{7}{2}$ **93.** $y = 2x - 3$
94. $y = \frac{1}{3}x + \frac{7}{3}$ **95.** $r_1 = \frac{-b + \sqrt{b^2 - 4ac}}{2a}; r_2 = \frac{-b - \sqrt{b^2 - 4ac}}{2a}$

Fill-in-the-Blank Items

1. parabola **2.** ellipse **3.** hyperbola **4.** major; transverse **5.** *y*-axis **6.** $\frac{y}{3} = \frac{x}{2}; \frac{y}{3} = -\frac{x}{2}$

True/False Items

1. True **2.** False **3.** True **4.** True **5.** True **6.** True

Review Exercises

1. Parabola; vertex $(0, 0)$, focus $(-4, 0)$, directrix $x = 4$ **2.** Parabola; vertex $(0, 0)$, focus $\left(0, \frac{1}{64}\right)$, directrix $y = -\frac{1}{64}$

3. Hyperbola; center $(0, 0)$, vertices $(5, 0)$ and $(-5, 0)$, foci $(\sqrt{26}, 0)$ and $(-\sqrt{26}, 0)$, asymptotes $y = \frac{1}{5}x$ and $y = -\frac{1}{5}x$

4. Hyperbola; center $(0, 0)$, vertices $(0, 5)$ and $(0, -5)$, foci $(0, \sqrt{26})$ and $(0, -\sqrt{26})$, asymptotes $y = 5x$ and $y = -5x$

5. Ellipse; center $(0, 0)$, vertices $(0, 5)$ and $(0, -5)$, foci $(0, 3)$ and $(0, -3)$

6. Ellipse; center $(0, 0)$, vertices $(0, 4)$ and $(0, -4)$, foci $(0, \sqrt{7})$ and $(0, -\sqrt{7})$

7. $x^2 = -4(y - 1)$: Parabola; vertex $(0, 1)$, focus $(0, 0)$, directrix $y = 2$

8. $\frac{y^2}{3} - \frac{x^2}{9} = 1$: Hyperbola; center $(0, 0)$, vertices $(0, \sqrt{3})$ and $(0, -\sqrt{3})$, foci $(0, 2\sqrt{3})$ and $(0, -2\sqrt{3})$; asymptotes $y = \frac{\sqrt{3}}{3}x$ and $y = -\frac{\sqrt{3}}{3}x$

9. $\frac{x^2}{2} - \frac{y^2}{8} = 1$: Hyperbola; center $(0, 0)$, vertices $(\sqrt{2}, 0)$ and $(-\sqrt{2}, 0)$, foci $(\sqrt{10}, 0)$ and $(-\sqrt{10}, 0)$, asymptotes $y = 2x$ and $y = -2x$

10. $\frac{x^2}{4} + \frac{y^2}{9} = 1$: Ellipse; center $(0, 0)$, vertices $(0, 3)$ and $(0, -3)$, foci $(0, \sqrt{5})$ and $(0, -\sqrt{5})$

11. $(x - 2)^2 = 2(y + 2)$: Parabola; vertex $(2, -2)$, focus $\left(2, -\frac{3}{2}\right)$, directrix $y = -\frac{5}{2}$

12. $(y - 1)^2 = \frac{1}{2}x$: Parabola; vertex $(0, 1)$, focus $\left(\frac{1}{8}, 1\right)$, directrix $y = -\frac{1}{8}$

13. $\frac{(y - 2)^2}{4} - (x - 1)^2 = 1$: Hyperbola; center $(1, 2)$, vertices $(1, 4)$ and $(1, 0)$, foci $(1, 2 + \sqrt{5})$ and $(1, 2 - \sqrt{5})$, asymptotes $y = -2x + 4$ and $y = 2x$

14. $(x + 1)^2 + \frac{(y - 2)^2}{4} = 1$: Ellipse; center $(-1, 2)$, vertices $(-1, 0)$ and $(-1, 4)$, foci $(-1, 2 + \sqrt{3})$ and $(-1, 2 - \sqrt{3})$

15. $\frac{(x - 2)^2}{9} + \frac{(y - 1)^2}{4} = 1$: Ellipse; center $(2, 1)$, vertices $(5, 1)$ and $(-1, 1)$, foci $(2 + \sqrt{5}, 1)$ and $(2 - \sqrt{5}, 1)$

16. $\frac{(x - 2)^2}{9} + \frac{(y + 1)^2}{4} = 1$: Ellipse; center $(2, -1)$; vertices $(5, -1)$ and $(-1, -1)$; foci $(2 + \sqrt{5}, 1)$ and $(2 - \sqrt{5}, 1)$

17. $(x - 2)^2 = -4(y + 1)$: Parabola; vertex $(2, -1)$, focus $(2, -2)$, directrix $y = 0$

18. $(y - 2)^2 = -\frac{3}{4}(x + 1)$: Parabola; vertex $(-1, 2)$; focus $\left(-\frac{19}{16}, 2\right)$; directrix $x = -\frac{13}{16}$

19. $\frac{(x - 1)^2}{4} + \frac{(y + 1)^2}{9} = 1$: Ellipse; center $(1, -1)$, vertices $(1, 2)$ and $(1, -4)$, foci $(1, -1 + \sqrt{5})$ and $(1, -1 - \sqrt{5})$

20. $(x - 1)^2 - (y + 1)^2 = 1$: Hyperbola; center $(1, -1)$; vertices $(0, -1)$ and $(2, -1)$; foci $(1 + \sqrt{2}, -1)$ and $(1 - \sqrt{2}, -1)$; asymptotes: $y = -x$ and $y = x - 2$

21. $y^2 = -8x$

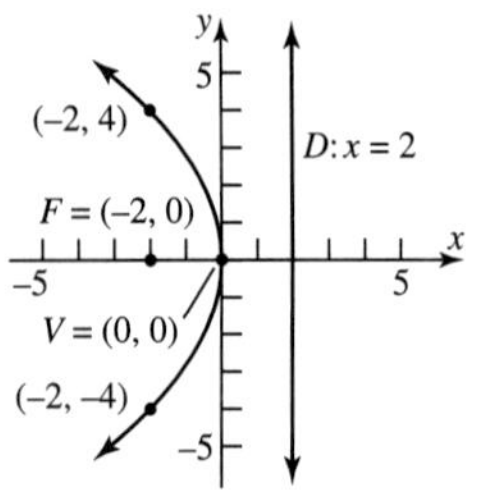

22. $\frac{x^2}{16} + \frac{y^2}{25} = 1$

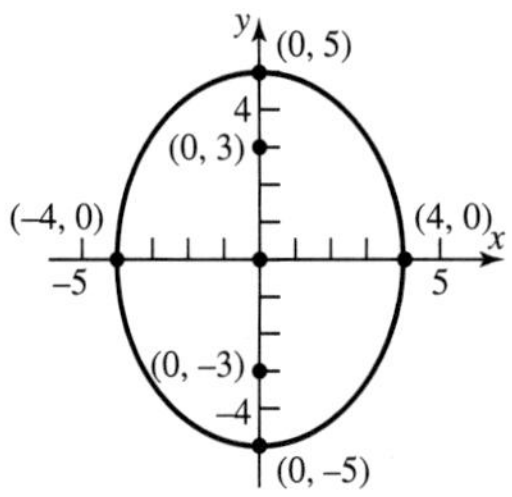

23. $\frac{y^2}{4} - \frac{x^2}{12} = 1$

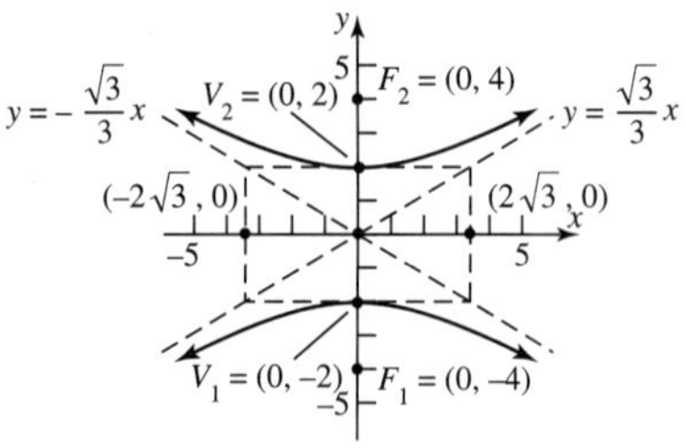

24. $x^2 = 12y$

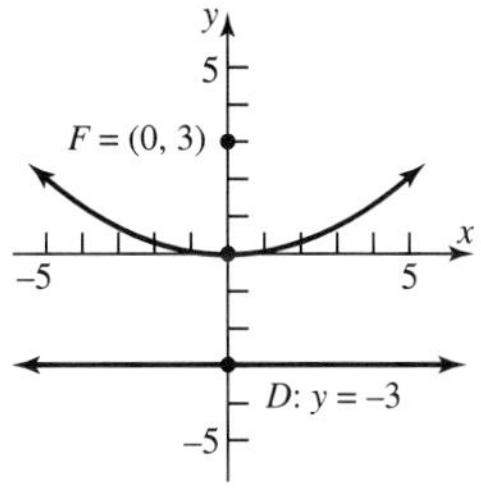

25. $\dfrac{x^2}{16} + \dfrac{y^2}{7} = 1$

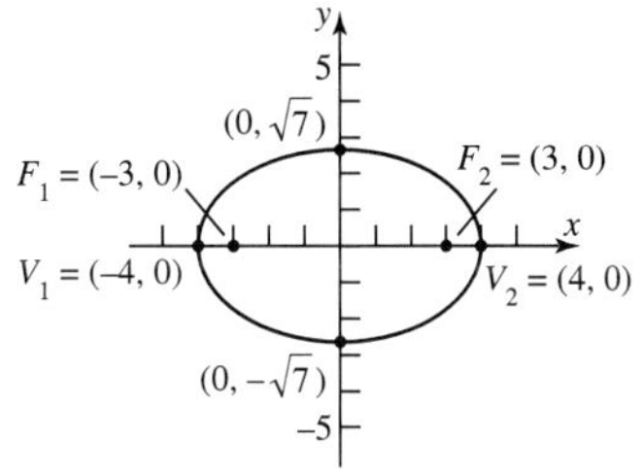

26. $\dfrac{x^2}{4} - \dfrac{y^2}{12} = 1$

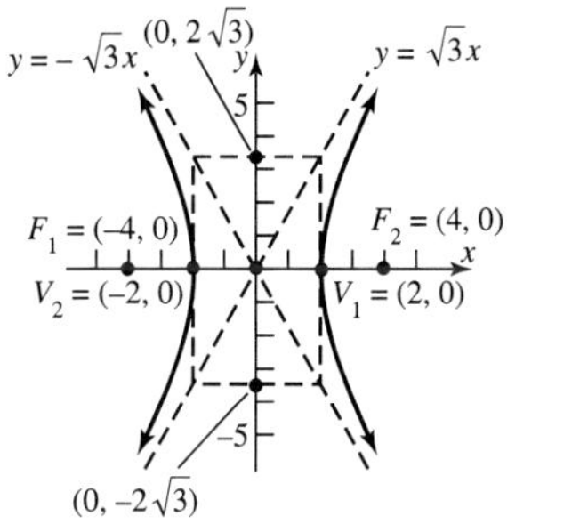

27. $(x - 2)^2 = -4(y + 3)$

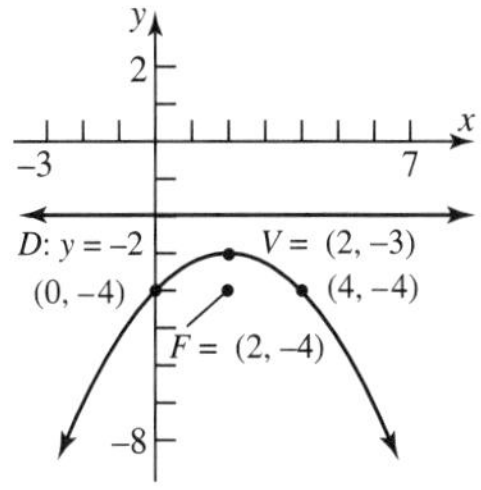

28. $\dfrac{(x + 1)^2}{9} + \dfrac{(y - 2)^2}{8} = 1$

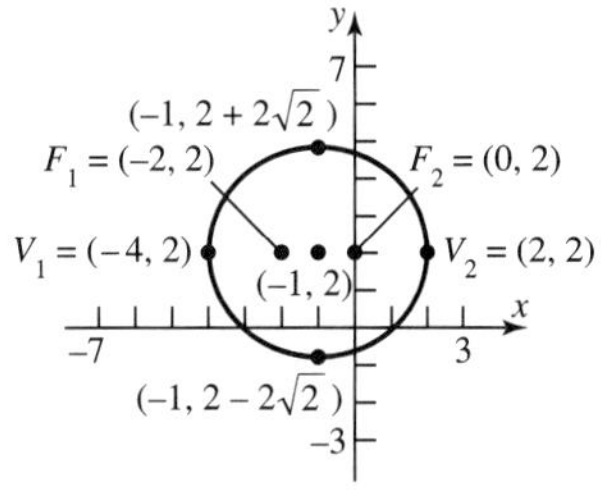

29. $(x + 2)^2 - \dfrac{(y + 3)^2}{3} = 1$

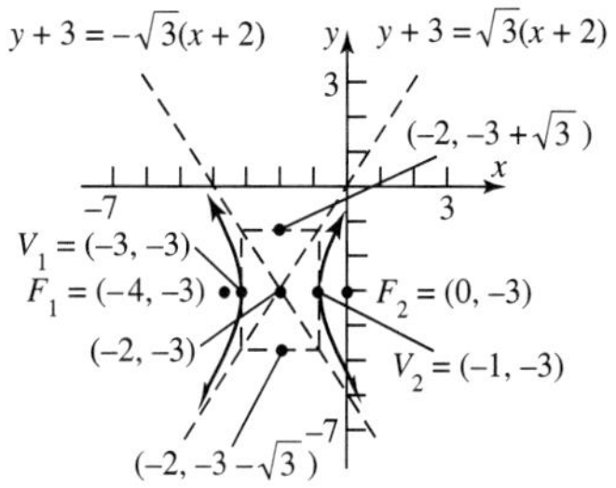

30. $(x - 3)^2 = -4(y - 7)$

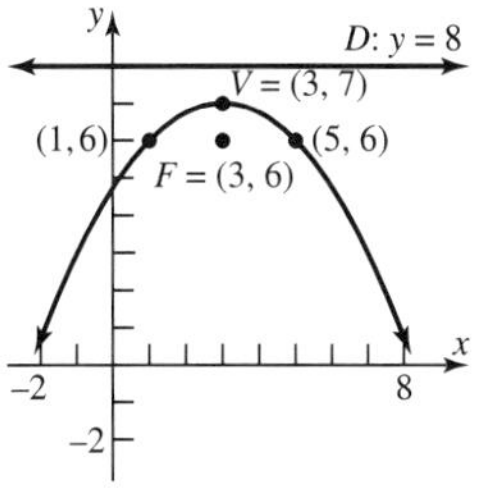

31. $\dfrac{(x + 4)^2}{16} + \dfrac{(y - 5)^2}{25} = 1$

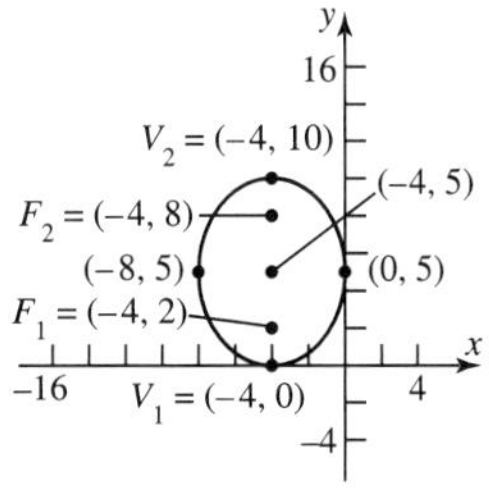

32. $\dfrac{(x - 1)^2}{16} - \dfrac{(y - 3)^2}{20} = 1$

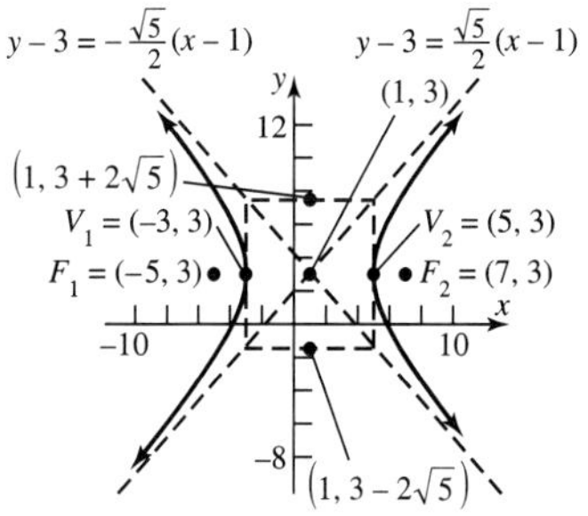

33. $\dfrac{(x + 1)^2}{9} - \dfrac{(y - 2)^2}{7} = 1$

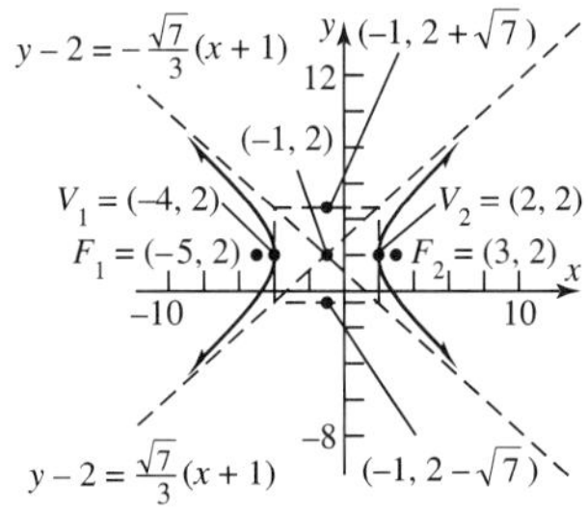

34. $(y + 2)^2 - \dfrac{(x - 4)^2}{15} = 1$

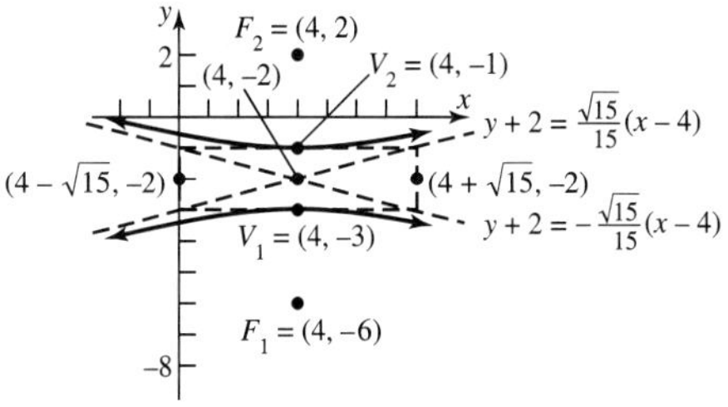

35. $\dfrac{(x-3)^2}{9} - \dfrac{(y-1)^2}{4} = 1$

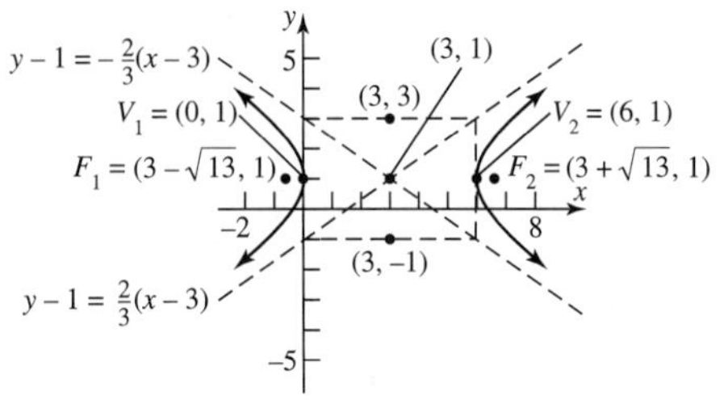

36. $\dfrac{(y-2)^2}{4} - \dfrac{(x-4)^2}{1} = 1$

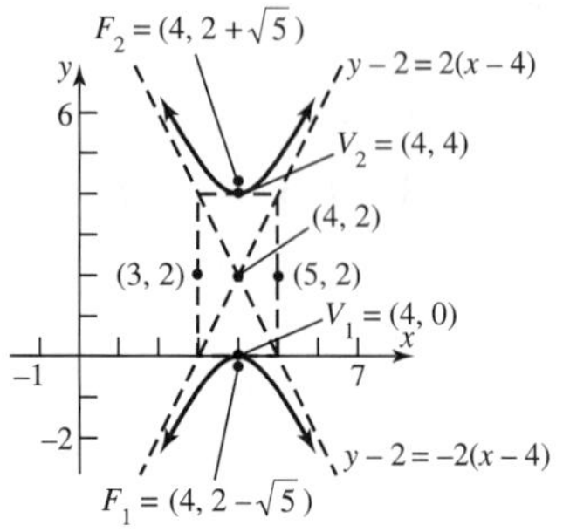

37. $\dfrac{x^2}{5} - \dfrac{y^2}{4} = 1$ **38.** $\dfrac{x^2}{20} + \dfrac{y^2}{4} = 1$ **39.** The ellipse $\dfrac{x^2}{16} + \dfrac{y^2}{7} = 1$ **40.** The hyperbola $\dfrac{x^2}{16} - \dfrac{y^2}{9} = 1$
41. $x = -\dfrac{2}{5}, y = -\dfrac{11}{5}; x = -2, y = 1$ **42.** $x = -4, y = 0; x = 2, y = -2\sqrt{3}; x = 2, y = 2\sqrt{3}$
43. $x = 2\sqrt{2}, y = \sqrt{2}; x = -2\sqrt{2}, y = -\sqrt{2}$ **44.** $x = -\sqrt{3}, y = -2\sqrt{2}; x = -\sqrt{3}, y = 2\sqrt{2}; x = \sqrt{3}, y = -2\sqrt{2}; x = \sqrt{3}, y = 2\sqrt{2}$
45. $x = 0, y = 0; x = -3, y = 3; x = 3, y = 3$ **46.** $x = 0, y = -3; x = 0, y = 3$
47. $x = \sqrt{2}, y = -\sqrt{2}; x = -\sqrt{2}, y = \sqrt{2}; x = \dfrac{4}{3}\sqrt{2}, y = -\dfrac{2}{3}\sqrt{2}; x = -\dfrac{4}{3}\sqrt{2}, y = \dfrac{2}{3}\sqrt{2}$
48. $x = -2, y = 1; x = 2, y = -1; x = -\dfrac{5}{21}\sqrt{42}, y = -\dfrac{2}{7}\sqrt{42}; x = \dfrac{5}{21}\sqrt{42}, y = \dfrac{2}{7}\sqrt{42}$ **49.** $x = 1, y = -1$
50. $x = -2, y = 0; x = -1, y = 2; x = 1, y = 0$ **51.** $\dfrac{1}{4}$ ft or 3 in. **52.** 19.44 ft, 17.78 ft, 11.11 ft **53.** 19.72 ft, 18.86 ft, 14.91 ft
54. 31.22 ft from the center of the ellipse **55. (a)** 45.24 mi from the Master Station **(b)** 0.000645 s **(c)** (66, 20)

APPENDIX Review

1 Exercises

1. (a) $\{2, 5\}$ **(b)** $\{-6, 2, 5\}$ **(c)** $\left\{-6, \dfrac{1}{2}, -1.333\ldots, 2, 5\right\}$ **(d)** $\{\pi\}$ **(e)** $\left\{-6, \dfrac{1}{2}, -1.333\ldots, \pi, 2, 5\right\}$ **2. (a)** $\{1\}$ **(b)** $\{0, 1\}$
(c) $\left\{-\dfrac{5}{3}, 2.060606\ldots, 1.25, 0, 1\right\}$ **(d)** $\{\sqrt{5}\}$ **(e)** $\left\{-\dfrac{5}{3}, 2.060606\ldots, 1.25, 0, 1, \sqrt{5}\right\}$ **3. (a)** $\{1\}$ **(b)** $\{0, 1\}$ **(c)** $\left\{0, 1, \dfrac{1}{2}, \dfrac{1}{3}, \dfrac{1}{4}\right\}$
(d) None **(e)** $\left\{0, 1, \dfrac{1}{2}, \dfrac{1}{3}, \dfrac{1}{4}\right\}$ **4. (a)** None **(b)** $\{-1\}$ **(c)** $\{-1, -1.1, -1.2, -1.3\}$ **(d)** None **(e)** $\{-1, -1.1, -1.2, -1.3\}$
5. (a) None **(b)** None **(c)** None **(d)** $\left\{\sqrt{2}, \pi, \sqrt{2} + 1, \pi + \dfrac{1}{2}\right\}$ **(e)** $\left\{\sqrt{2}, \pi, \sqrt{2} + 1, \pi + \dfrac{1}{2}\right\}$ **6. (a)** None **(b)** None
(c) $\left\{\dfrac{1}{2} + 10.3\right\}$ **(d)** $\left\{-\sqrt{2}, \pi + \sqrt{2}, \dfrac{1}{2} + 10.3\right\}$ **(e)** $\left\{-\sqrt{2}, \pi + \sqrt{2}, \dfrac{1}{2} + 10.3\right\}$ **7.** $3 + 2 = 5$ **8.** $5 \cdot 2 = 10$ **9.** $x + 2 = 3 \cdot 4$
10. $3 + y = 2 + 2$ **11.** $3 \cdot y = 1 + 2$ **12.** $2 \cdot x = 4 \cdot 6$ **13.** $x - 2 = 6$ **14.** $2 - y = 6$ **15.** $\dfrac{x}{2} = 6$ **16.** $\dfrac{2}{x} = 6$ **17.** 7 **18.** 5
19. 6 **20.** 0 **21.** 1 **22.** 1 **23.** $\dfrac{13}{3}$ **24.** $\dfrac{3}{2}$ **25.** −11 **26.** −23 **27.** 11 **28.** −11 **29.** −4 **30.** −12 **31.** 1 **32.** 3 **33.** 6 **34.** −1
35. $\dfrac{2}{7}$ **36.** $\dfrac{1}{6}$ **37.** $\dfrac{4}{45}$ **38.** 28 **39.** $\dfrac{23}{20}$ **40.** $\dfrac{11}{6}$ **41.** $\dfrac{79}{30}$ **42.** $\dfrac{151}{18}$ **43.** $\dfrac{13}{36}$ **44.** $\dfrac{46}{45}$ **45.** $\dfrac{-16}{45}$ **46.** $\dfrac{5}{42}$ **47.** $\dfrac{1}{60}$ **48.** $-\dfrac{3}{70}$ **49.** $\dfrac{15}{22}$
50. $\dfrac{25}{6}$ **51.** $6x + 24$ **52.** $8x - 4$ **53.** $x^2 - 4x$ **54.** $4x^2 + 12x$ **55.** $x^2 + 6x + 8$ **56.** $x^2 + 6x + 5$ **57.** $x^2 - x - 2$
58. $x^2 - 3x - 4$ **59.** $x^2 - 10x + 16$ **60.** $x^2 - 6x + 8$ **61.** $x^2 - 4$ **62.** $x^2 - 9$
63. $A = lw$; all the variables are positive real numbers **64.** $P = 2(l + w)$; all the variables are positive real numbers
65. $C = \pi d$; d and C are positive real numbers **66.** $A = \dfrac{1}{2}bh$; all the variables are positive real numbers
67. $A = \dfrac{\sqrt{3}}{4}x^2$; x and A are positive real numbers **68.** $P = 3x$; x and P are positive real numbers
69. $V = \dfrac{4}{3}\pi r^3$; r and V are positive real numbers **70.** $S = 4\pi r^2$; r and S are positive real numbers

71. $V = x^3$; x and V are positive real numbers **72.** $S = 6x^2$; x and S are positive real numbers **73.** $2x + 3x = x(2 + 3) = 5x$
74. $2 + 3 \cdot 4 = 2 + 12 = 14$; $(2 + 3) \cdot 4 = 5 \cdot 4 = 20$ **75.** $2(3 \cdot 4) = 2 \cdot 12 = 24$; $(2 \cdot 3) \cdot (2 \cdot 4) = 6 \cdot 8 = 48$
76. $\frac{4+3}{2+5} = \frac{7}{7} = 1$; $\frac{4}{2} + \frac{3}{5} = \frac{20}{10} + \frac{6}{10} = \frac{26}{10} = \frac{13}{5}$ **77.** No; $\frac{1}{3}$; 0.000333... **78.** No; $\frac{2}{3}$; 0.000666... **79.** No; $2 - 3 \neq 3 - 2$
80. No; $(5 - 2) - 1 \neq 5 - (2 - 1)$ **81.** No; $\frac{2}{3} \neq \frac{3}{2}$ **82.** No; $(12 \div 2) \div 2 \neq 12 \div (2 \div 2)$ **83.** Symmetric property
84. $x^2 + x = 5^2 + 5 = 25 + 5 = 30$ **85.** No; no **87.** 1

2 Exercises

1. 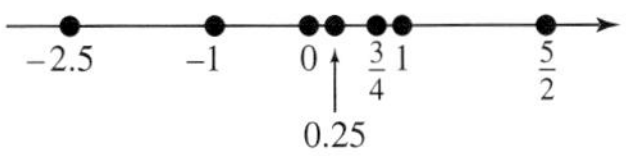 **2.** 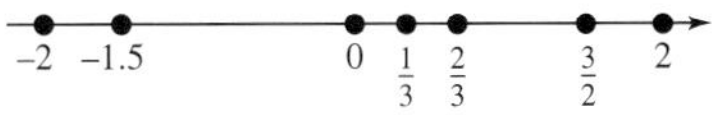 **3.** $>$ **4.** $<$ **5.** $>$ **6.** $<$ **7.** $>$ **8.** $>$ **9.** $=$ **10.** $>$ **11.** $<$ **12.** $=$ **13.** $x > 0$
14. $x < 0$ **15.** $x < 2$ **16.** $y > -5$ **17.** $x \leq 1$ **18.** $x \geq 2$ **19.** $2 < x < 5$ **20.** $0 < y \leq 2$
21. 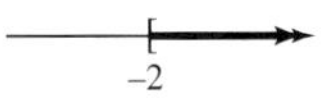 **22.**  **23.**  **24.** 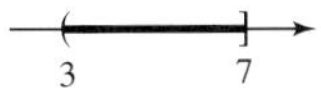 **25.** 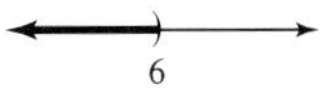
26. 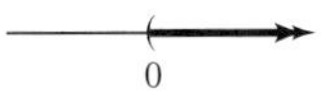 **27.** 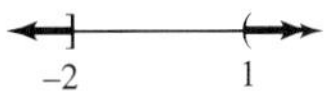 **28.**  **29.** 2 **30.** 3 **31.** 6 **32.** 2 **33.** 4 **34.** -3 **35.** -28 **36.** -2
37. $\frac{4}{5}$ **38.** $-\frac{1}{5}$ **39.** 0 **40.** $-\frac{7}{3}$ **41.** 1 **42.** 5 **43.** 5 **44.** 1 **45.** 1 **46.** -1 **47.** 22 **48.** 5 **49.** 2 **50.** 13 **51.** 3 **52.** 1 **53.** $-\frac{1}{24}$
54. $-\frac{1}{24}$ **55.** $-\frac{1}{6}$ **56.** $-\frac{1}{216}$ **57.** $-\frac{3}{2}$ **58.** $\frac{3}{4}$ **59.** $\frac{25}{36}$ **60.** $\frac{1}{144}$ **61.** -1 **62.** 7 **63.** 3 **64.** -4 **65.** -1 **66.** 1 **67.** $-\frac{4}{3}$ **68.** $\frac{2}{5}$
69. -18 **70.** $\frac{7}{5}$ **71.** -4 **72.** -1 **73.** $-\frac{3}{4}$ **74.** 2 **75.** -20 **76.** -10 **77.** 2 **78.** $-\frac{5}{2}$ **79.** $\{3, 4\}$ **80.** $\{-2, 3\}$ **81.** $\left\{-3, \frac{1}{2}\right\}$
82. $\left\{-2, \frac{1}{3}\right\}$ **83.** $\{-3, 0, 3\}$ **84.** $\{-1, 0, 1\}$ **85.** $\{-5, 0, 4\}$ **86.** $\{-7, 0, 1\}$ **87.** $\{-1, 1\}$ **88.** $\{-4, -1, 1\}$ **89.** $\{-2, 2, 3\}$
90. $\{-1, 1, 3\}$ **91.** 0°C **92.** 100°C **93.** 25°C **94.** -20°C **95. (a)** \$6000 **(b)** \$8000 **96.** \$98 **97. (a)** $2 \leq 5$ **(b)** $6 > 5$
98. (a) $6 \leq 8$ **(b)** $11 > 8$ **99. (a)** Acceptable **(b)** Not acceptable **100. (a)** $1.6 \geq 1.5$ **(b)** $1.4 < 1.5$ **101.** No **102.** 3.15 or 3.16

3 Exercises

1. 13 **2.** 10 **3.** 26 **4.** 5 **5.** 25 **6.** 50 **7.** Right triangles; 5 **8.** Right triangle; 10 **9.** Not a right triangle **10.** Not a right triangle
11. Right triangle; 25 **12.** Right triangle; 26 **13.** Not a right triangle **14.** Not a right triangle **15.** 8 in^2 **16.** 36 cm^2 **17.** 4 in^2
18. 18 cm^2 **19.** $A = 25\pi$ m^2; $C = 10\pi$ m **20.** $A = 4\pi$ ft^2; $C = 4\pi$ ft **21.** 224 ft^3 **22.** 288 in^2 **23.** $V = \frac{256}{3}\pi$ cm^3; $S = 64\pi$ cm^2
24. $V = 36\pi$ ft^3; $S = 36\pi$ ft^2 **25.** 648π in^3 **26.** 576π in^3 **27.** π square units **28.** $4 - \pi$ square units **29.** 2π square units
30. $2\pi - 4$ square units **31.** About 16.8 ft **32.** About 1.6 revolutions **33.** 64 square feet **34.** 12 ft^2
35. $24 + 2\pi \approx 30.28$ ft^2; $16 + 2\pi \approx 22.28$ ft^2 **36.** $69\pi \approx 216.77$ ft^2; $26\pi \approx 81.68$ ft **37.** About 5.477 mi **38.** About 3.000 mi
39. From 100 ft: 12.247 mi, From 150 ft: 15 mi

4 Exercises

1. 16 **2.** -16 **3.** $\frac{1}{16}$ **4.** 16 **5.** $-\frac{1}{16}$ **6.** $\frac{1}{16}$ **7.** $\frac{1}{8}$ **8.** $-\frac{1}{8}$ **9.** $\frac{1}{4}$ **10.** $\frac{2}{9}$ **11.** $\frac{1}{9}$ **12.** 4 **13.** $\frac{81}{64}$ **14.** 8 **15.** $\frac{27}{8}$ **16.** $\frac{4}{9}$ **17.** $\frac{81}{2}$
18. $\frac{25}{81}$ **19.** $\frac{4}{81}$ **20.** $\frac{125}{216}$ **21.** $\frac{1}{12}$ **22.** $\frac{1}{18}$ **23.** $-\frac{2}{3}$ **24.** $\frac{1}{8}$ **25.** y^2 **26.** $\frac{y}{x}$ **27.** $\frac{x}{y^2}$ **28.** y^4 **29.** $\frac{1}{64x^6}$ **30.** $\frac{1}{64x^6}$ **31.** $-\frac{4}{x}$ **32.** $-\frac{1}{4x}$
33. 3 **34.** 1 **35.** $\frac{1}{x^3y}$ **36.** $\frac{1}{x^3y}$ **37.** $\frac{1}{xy}$ **38.** $\frac{1}{x^3y^3}$ **39.** $\frac{y}{x}$ **40.** $\frac{3}{2x}$ **41.** $\frac{25x^2}{16y^2}$ **42.** $\frac{1}{x^4y^2}$ **43.** $\frac{1}{x^2y^2}$ **44.** $\frac{1}{xy}$ **45.** $\frac{1}{x^3y^3}$ **46.** $\frac{3y^4}{x^6}$
47. $-\frac{8x^3}{9yz^2}$ **48.** $\frac{4z}{25x^6y^3}$ **49.** $\frac{y^3}{x^8}$ **50.** y **51.** $\frac{16x^2}{9y^2}$ **52.** $\frac{216x^6}{125y^6}$ **53.** $\frac{1}{x^3y}$ **54.** $\frac{9x^5}{8y^5}$ **55.** $\frac{y^2}{x^2}$ **56.** y^5 **57.** 10; 0 **58.** 8; 44 **59.** 81 **60.** 8
61. 304,006.671 **62.** 693.440 **63.** 0.004 **64.** 0.019 **65.** 481.890 **66.** -481.890 **67.** 0.000 **68.** -0.000 **69.** 4.542×10^2
70. 3.214×10^1 **71.** 1.3×10^{-2} **72.** 4.21×10^{-3} **73.** 3.2155×10^4 **74.** 2.121×10^4 **75.** 4.23×10^{-4} **76.** 5.14×10^{-2} **77.** 61,500
78. 9700 **79.** 0.001214 **80.** 0.000988 **81.** 110,000,000 **82.** 411.2 **83.** 0.081 **84.** 0.6453 **85.** 400,000,000 m **86.** 8.872×10^3 m
87. 0.0000005 m **88.** 0.0000000001 m **89.** 5×10^{-4} in. **90.** 5×10^{-2} cm **91.** 1.92×10^9 barrels **92.** 7.4088×10^8 gal
93. 5.865696×10^{12} mi **94.** 5×10^2 s

5 Exercises

1. Monomial; Variable: x; Coefficient: 2; Degree: 3 **2.** Monomial; Variable: x; Coefficient: -4; Degree: 2 **3.** Not a monomial **4.** Not a monomial **5.** Monomial; Variables: x, y; Coefficient: -2; Degree: 3 **6.** Monomial; Variables: x, y; Coefficient: 5; Degree: 5 **7.** Not a monomial **8.** Not a monomial **9.** Not a monomial **10.** Not a monomial **11.** Yes; 2 **12.** Yes; 1 **13.** Yes; 0 **14.** Yes; 0 **15.** No **16.** No **17.** Yes; 3 **18.** Yes; 2 **19.** No **20.** No **21.** $x^2 + 7x + 2$ **22.** $x^3 + 4x^2 - 4x + 6$ **23.** $x^3 - 4x^2 + 9x + 7$ **24.** $-x^3 + 4x^2 - 4x - 9$ **25.** $6x^5 + 5x^4 + 3x^2 + x$ **26.** $10x^5 + 3x^3 - 10x^2 + 6$ **27.** $7x^2 - x - 7$ **28.** $-7x^2 - 3x$ **29.** $-2x^3 + 18x^2 - 18$ **30.** $8x^3 - 24x^2 - 48x + 4$ **31.** $2x^2 - 4x + 6$ **32.** $-2x^2 + x - 6$ **33.** $15y^2 - 27y + 30$ **34.** $-4y^3 + 4y^2 + 4y + 12$ **35.** $x^3 + x^2 - 4x$ **36.** $4x^5 - 4x^3 + 8x^2$ **37.** $-8x^5 - 10x^2$ **38.** $15x^4 - 20x^3$ **39.** $x^3 + 3x^2 - 2x - 4$ **40.** $2x^3 - x^2 - x - 3$ **41.** $x^2 + 6x + 8$ **42.** $x^2 + 8x + 15$ **43.** $2x^2 + 9x + 10$ **44.** $6x^2 + 5x + 1$ **45.** $x^2 - 2x - 8$ **46.** $x^2 + 2x - 8$ **47.** $x^2 - 5x + 6$ **48.** $x^2 - 6x + 5$ **49.** $2x^2 - x - 6$ **50.** $6x^2 - 10x - 4$ **51.** $-2x^2 + 11x - 12$ **52.** $-3x^2 - 4x - 1$ **53.** $2x^2 + 8x + 8$ **54.** $2x^2 - 3x - 9$ **55.** $x^2 - xy - 2y^2$ **56.** $2x^2 + xy - 3y^2$ **57.** $-6x^2 - 13xy - 6y^2$ **58.** $-2x^2 + 7xy - 3y^2$ **59.** $x^2 - 49$ **60.** $x^2 - 1$ **61.** $4x^2 - 9$ **62.** $9x^2 - 4$ **63.** $x^2 + 8x + 16$ **64.** $x^2 + 10x + 25$ **65.** $x^2 - 8x + 16$ **66.** $x^2 - 10x + 25$ **67.** $9x^2 - 16$ **68.** $25x^2 - 9$ **69.** $4x^2 - 12x + 9$ **70.** $9x^2 - 24x + 16$ **71.** $x^2 - y^2$ **72.** $x^2 - 9y^2$ **73.** $9x^2 - y^2$ **74.** $9x^2 - 16y^2$ **75.** $x^2 + 2xy + y^2$ **76.** $x^2 - 2xy + y^2$ **77.** $x^2 - 4xy + 4y^2$ **78.** $4x^2 + 12xy + 9y^2$ **79.** $x^3 - 6x^2 + 12x - 8$ **80.** $x^3 + 3x^2 + 3x + 1$ **81.** $8x^3 + 12x^2 + 6x + 1$ **82.** $27x^3 - 54x^2 + 36x - 8$ **83.** $x^3 + 6x^2y + 12xy^2 + 8y^3$ **84.** $x^3 - 6x^2y + 12xy^2 - 8y^3$ **85.** The degree of the product equals the degree of the product of the leading terms: $(a_nx^n + a_{n-1}x^{n-1} + \dots + a_1x + a_0)(b_mx^m + b_{m-1}x^{m-1} + \dots + b_1x + b_0) = a_nb_mx^{n+m} + (a_nb_{m-1} + a_{n-1}b_m)x^{n+m-1} + \dots + (a_1b_0 + a_0b_1)x + a_0b_0$, which is of degree $m + n$

6 Exercises

1. $3(x + 2)$ **2.** $7(x - 2)$ **3.** $a(x^2 + 1)$ **4.** $a(x - 1)$ **5.** $x(x^2 + x + 1)$ **6.** $x(x^2 - x + 1)$ **7.** $2x(x - 1)$ **8.** $3x(x - 1)$ **9.** $3xy(x - 2y + 4)$ **10.** $12xy(5x - 4y + 6x^2)$ **11.** $(x - 1)(x + 1)$ **12.** $(x + 2)(x - 2)$ **13.** $(2x + 1)(2x - 1)$ **14.** $(3x + 1)(3x - 1)$ **15.** $(x + 4)(x - 4)$ **16.** $(x + 5)(x - 5)$ **17.** $(5x + 2)(5x - 2)$ **18.** $9(2x + 1)(2x - 1)$ **19.** $(x + 1)^2$ **20.** $(x - 2)^2$ **21.** $(x + 2)^2$ **22.** $(x - 1)^2$ **23.** $(x - 5)^2$ **24.** $(x + 5)^2$ **25.** $(2x + 1)^2$ **26.** $(3x + 1)^2$ **27.** $(4x + 1)^2$ **28.** $(5x + 1)^2$ **29.** $(x - 3)(x^2 + 3x + 9)$ **30.** $(x + 5)(x^2 - 5x + 25)$ **31.** $(x + 3)(x^2 - 3x + 9)$ **32.** $-(2x - 3)(4x^2 + 6x + 9)$ **33.** $(2x + 3)(4x^2 - 6x + 9)$ **34.** $-(3x - 4)(9x^2 + 12x + 16)$ **35.** $(x + 2)(x + 3)$ **36.** $(x + 2)(x + 4)$ **37.** $(x + 6)(x + 1)$ **38.** $(x + 1)(x + 8)$ **39.** $(x + 5)(x + 2)$ **40.** $(x + 1)(x + 10)$ **41.** $(x - 8)(x - 2)$ **42.** $(x - 16)(x - 1)$ **43.** $(x - 8)(x + 1)$ **44.** $(x - 4)(x + 2)$ **45.** $(x + 8)(x - 1)$ **46.** $(x - 2)(x + 4)$ **47.** $(x + 2)(2x + 3)$ **48.** $(x - 1)(3x + 2)$ **49.** $(x - 2)(2x + 1)$ **50.** $(x + 2)(3x - 1)$ **51.** $(2x + 3)(3x + 2)$ **52.** Prime **53.** $(3x + 1)(x + 1)$ **54.** $(x + 1)(2x + 1)$ **55.** $(z + 1)(2z + 3)$ **56.** $(2z + 1)(3z + 1)$ **57.** $(x + 2)(3x - 4)$ **58.** $(x + 2)(3x + 4)$ **59.** $(x - 2)(3x + 4)$ **60.** $(x - 2)(3x - 4)$ **61.** $(x + 4)(3x + 2)$ **62.** $(x - 4)(3x - 2)$ **63.** $(x + 4)(3x - 2)$ **64.** $(x - 4)(3x + 2)$ **65.** $(x + 6)(x - 6)$ **66.** $(x - 3)(x + 3)$ **67.** $(1 + 2x)(1 - 2x)$ **68.** $(1 - 3x)(1 + 3x)$ **69.** $(x + 2)(x + 5)$ **70.** $(x + 1)(x + 4)$ **71.** $(x - 7)(x - 3)$ **72.** $(x - 4)(x - 2)$ **73.** Prime **74.** Prime **75.** Prime **76.** $(x + 6)^2$ **77.** $-(x - 5)(x + 3)$ **78.** Prime **79.** $3(x + 2)(x - 6)$ **80.** $x(x - 2)(x + 10)$ **81.** $y^2(y + 5)(y + 6)$ **82.** $3y(y - 8)(y + 2)$ **83.** $(2x + 3)^2$ **84.** $(3x - 2)^2$ **85.** $(3x + 1)(x + 1)$ **86.** $(x + 1)(4x - 1)$ **87.** $(x - 3)(x + 3)(x^2 + 9)$ **88.** $(x - 1)(x + 1)(x^2 + 1)$ **89.** $(x - 1)^2(x^2 + x + 1)^2$ **90.** $(x + 1)^2(x^2 - x + 1)^2$ **91.** $x^5(x - 1)(x + 1)$ **92.** $x^5(x - 1)(x^2 + x + 1)$ **93.** $(4x + 3)^2$ **94.** $(3x - 4)^2$ **95.** $-(4x - 5)(4x + 1)$ **96.** $-(x - 1)(16x + 5)$ **97.** $(2y - 5)(2y - 3)$ **98.** $(3y - 1)(3y + 4)$ **99.** $-(3x - 1)(3x + 1)(x^2 + 1)$ **100.** $-2(2x - 1)(2x + 1)(x^2 + 2)$ **101.** $(x + 3)(x - 6)$ **102.** $(x + 5)(3x - 7)$ **103.** $(x + 2)(x - 3)$ **104.** $(x - 3)(x - 1)$ **105.** $(3x - 5)(9x^2 - 3x + 7)$ **106.** $5x(25x^2 + 15x + 3)$ **107.** $(x + 5)(3x + 11)$ **108.** $(x - 3)(7x - 16)$ **109.** $(x - 1)(x + 1)(x + 2)$ **110.** $(x - 3)(x - 1)(x + 1)$ **111.** $(x - 1)(x + 1)(x^2 - x + 1)$ **112.** $(x + 1)^2(x^2 - x + 1)$ **113.** The possibilities are $(x \pm 1)(x \pm 4) = x^2 \pm 5x + 4$ or $(x \pm 2)(x \pm 2) = x^2 \pm 4x + 4$, none of which equal $x^2 + 4$

7 Exercises

1. 0 **2.** 0 **3.** 3 **4.** None **5.** None **6.** $1, -1$ **7.** $0, 1$ **8.** 0 **9.** $\frac{3}{x - 3}$ **10.** $\frac{x}{3}$ **11.** $\frac{x}{3}$ **12.** $\frac{5x + 8}{x}$ **13.** $\frac{4x}{2x - 1}$ **14.** $\frac{x^2 + 4x + 4}{x^2 - 16}$ **15.** $\frac{y + 5}{2(y + 1)}$ **16.** $\frac{y - 1}{y + 1}$ **17.** $\frac{x + 5}{x - 1}$ **18.** $-\frac{x}{x + 2}$ **19.** $\frac{x - 2}{x + 3}$ **20.** $\frac{2 - x}{x - 3}$ **21.** $-(x + 7)$ **22.** $-x - 3$ **23.** $\frac{3}{5x(x - 2)}$ **24.** $\frac{3x}{4(3x + 5)}$ **25.** $\frac{2x}{x + 4}$ **26.** $\frac{6(x + 1)}{x(2x - 1)}$ **27.** $\frac{8}{3x}$ **28.** $\frac{3}{5x}$ **29.** $\frac{x - 3}{x + 7}$ **30.** $\frac{(x - 5)(x - 2)(x + 3)}{(x - 3)(x - 1)(x + 5)}$ **31.** $\frac{4x}{(x - 2)(x - 3)}$ **32.** $\frac{3(x - 4)}{5x}$ **33.** $\frac{4}{5(x - 1)}$ **34.** $\frac{3}{x - 2}$ **35.** $\frac{(4 - x)(x - 4)}{4x}$ **36.** $-\frac{9x^3}{(x - 3)^2}$ **37.** $\frac{(x + 3)^2}{(x - 3)^2}$ **38.** $\frac{(x + 1)(x + 2)}{(x - 2)(x - 1)}$ **39.** $\frac{(x - 4)(x + 3)}{(x - 1)(2x + 1)}$ **40.** $\frac{(2x - 3)(4x + 1)}{(3x - 1)(4x - 1)}$ **41.** $\frac{x + 5}{2}$ **42.** $-\frac{3}{x}$ **43.** $\frac{(x - 2)(x + 2)}{2x - 3}$ **44.** $\frac{3(x^2 - 3)}{2x - 1}$ **45.** $\frac{3x - 2}{x - 3}$ **46.** $\frac{3x - 1}{3x + 2}$ **47.** $\frac{x + 9}{2x - 1}$ **48.** $\frac{4x - 5}{3x + 4}$ **49.** $\frac{4 - x}{x - 2}$ **50.** $\frac{x + 6}{x - 1}$ **51.** $\frac{2(x + 5)}{(x - 1)(x + 2)}$ **52.** $-\frac{3x + 35}{(x - 5)(x + 5)}$ **53.** $\frac{3x^2 - 2x - 3}{(x + 1)(x - 1)}$ **54.** $\frac{x(5x + 1)}{(x - 4)(x + 3)}$ **55.** $\frac{-11x - 2}{(x + 2)(x - 2)}$ **56.** $-\frac{2}{(x - 1)(x + 1)}$ **57.** $\frac{2(x^2 - 2)}{x(x - 2)(x + 2)}$ **58.** $\frac{x^4 + x^3 - x^2 + x - 1}{x^3(x^2 + 1)}$

59. $\frac{2x^3 - 2x^2 + 2x - 1}{x(x-1)^2}$ **60.** $\frac{x^2(3x^2 - 4x - 3)}{4(x-1)(x+1)}$ **61.** $\frac{x^3 - 7x^2 + 3x + 5}{(x+1)(x-1)(x-2)}$ **62.** $\frac{2x^3 + 4x^2 - 3x - 1}{x(x-1)(x+1)}$
63. $(x-2)(x+2)(x+1)$ **64.** $(x-4)^2(x+3)$ **65.** $x(x-1)(x+1)$ **66.** $3(x-3)(x+3)(2x+5)$ **67.** $x^3(2x-1)^2$
68. $x(x-3)(x+3)$ **69.** $x(x-1)^2(x+1)(x^2+x+1)$ **70.** $x^2(x+2)^4$ **71.** $\frac{5x}{(x-6)(x-1)(x+4)}$ **72.** $\frac{x^2 + 7x - 1}{(x-3)(x+8)}$
73. $\frac{2(2x^2 + 5x - 2)}{(x-2)(x+2)(x+3)}$ **74.** $\frac{3x^2 - 4x + 4}{(x-1)^2}$ **75.** $\frac{5x+1}{(x-1)^2(x+1)^2}$ **76.** $\frac{-2(2x+7)}{(x-1)^2(x+2)^2}$ **77.** $\frac{-x^2 + 3x + 13}{(x-2)(x+1)(x+4)}$
78. $\frac{x^2 - 6x + 11}{(x+7)(x+1)^2}$ **79.** $\frac{x^3 - 2x^2 + 4x + 3}{x^2(x+1)(x-1)}$ **80.** $\frac{3x^3 - 5x^2 + 2x + 1}{x^2(x-1)^2}$ **81.** $\frac{-1}{x(x+h)}$ **82.** $-\frac{2x+h}{x^2(x+h)^2}$ **83.** $\frac{x+1}{x-1}$ **84.** $\frac{4x^2+1}{3x^2-1}$
85. $\frac{(x-1)(x+1)}{x^2+1}$ **86.** $\frac{x}{(x+1)^2}$ **87.** $\frac{2(5x-1)}{(x-2)(x+1)^2}$ **88.** $\frac{-(5x-4)}{(x-2)(x+1)(x+3)}$ **89.** $\frac{-2x(x^2-2)}{(x+2)(x^2-x-3)}$
90. $\frac{(x+3)(x^2-x-15)}{x(4x^2+5x+3)}$ **91.** $\frac{-1}{x-1}$ **92.** $\frac{1}{x}$ **93.** $f = \frac{R_1 \cdot R_2}{(n-1)(R_1+R_2)}$; $\frac{2}{15}$ m **94.** $R = \frac{R_1R_2R_3}{R_1R_2 + R_1R_3 + R_2R_3}$; $\frac{20}{11}$ ohms

8 Exercises

1. 5 **2.** 9 **3.** 3 **4.** 5 **5.** −4 **6.** −2 **7.** $\frac{1}{3}$ **8.** $\frac{3}{2}$ **9.** $5x^2$ **10.** $4x^2$ **11.** $2(1+x)$ **12.** $2|x+4|$ **13.** $2\sqrt{2}$ **14.** $3\sqrt{3}$ **15.** $5\sqrt{2}$
16. $6\sqrt{2}$ **17.** $2\sqrt[3]{2}$ **18.** $2\sqrt[3]{3}$ **19.** $-2\sqrt[3]{2}$ **20.** $-2\sqrt[3]{2}$ **21.** $\frac{5}{3}|x|$ **22.** $\frac{1}{2x}$ **23.** $|x^3|y^2$ **24.** x^2y **25.** $6\sqrt{x}$ **26.** $3x^2\sqrt{x}$ **27.** $6x\sqrt{x}$
28. $10x^2$ **29.** 1 **30.** $\frac{5\sqrt[3]{x^2}}{2y}$ **31.** $\frac{4y^2}{3|x|}$ **32.** $\frac{3x^2}{4|y^3|}$ **33.** $15\sqrt[3]{3}$ **34.** $300\sqrt[3]{3}$ **35.** $\frac{1}{x(2x+3)}$ **36.** $\frac{x-1}{x+1}$ **37.** 1 **38.** $\frac{2\sqrt{x}}{x^2-4}$ **39.** $7\sqrt{2}$
40. $2\sqrt{5}$ **41.** $\sqrt{2}$ **42.** $-5\sqrt{3}$ **43.** $-\sqrt[3]{2}$ **44.** $15\sqrt[3]{3}$ **45.** $(2x-15)\sqrt{2x}$ **46.** $(9x+20)\sqrt{y}$ **47.** $(-x-5y)\sqrt[3]{2xy}$ **48.** $5xy$
49. $36\sqrt{2}$ **50.** $-60\sqrt{3}$ **51.** $3 - 4\sqrt{3}$ **52.** $5 + 6\sqrt{5}$ **53.** $42 + 9\sqrt{7}$ **54.** $36 + 9\sqrt{6}$ **55.** $3 - 2\sqrt{2}$ **56.** $8 + 2\sqrt{15}$
57. $1 - 3\sqrt[4]{4} + 3\sqrt[3]{2}$ **58.** $12(1 + \sqrt[3]{2} + \sqrt[3]{4})$ **59.** $4x + 4\sqrt{x} - 15$ **60.** $4x + 9\sqrt{x} - 9$ **61.** $\frac{-x^2}{\sqrt{1-x^2}}$ **62.** $\frac{1}{\sqrt{1-x^2}}$ **63.** $\frac{2\sqrt{5}}{5}$
64. $\frac{\sqrt{15}}{5}$ **65.** $\frac{4\sqrt{6}}{3}$ **66.** $\frac{\sqrt{10}}{2}$ **67.** $\frac{\sqrt{x}}{x}$ **68.** $\frac{x\sqrt{x^2+4}}{x^2+4}$ **69.** $\frac{15 - 3\sqrt{2}}{23}$ **70.** $\frac{2\sqrt{7}+4}{3}$ **71.** $\frac{4-\sqrt{7}}{3}$ **72.** $\frac{20 + 5\sqrt{2}}{7}$
73. $\frac{15 - 2\sqrt{5}}{41}$ **74.** $2 - \sqrt{3}$ **75.** $5 - 2\sqrt{6}$ **76.** $4 + \sqrt{15}$ **77.** $\frac{\sqrt{x}-2}{x-4}$ **78.** $\frac{\sqrt{x}+3}{x-9}$ **79.** 1.41 **80.** 2.65 **81.** 1.59 **82.** −1.71
83. 4.89 **84.** 0.04 **85.** 2.15 **86.** 1.33 **87.** **(a)** 15,660.4 gal **(b)** 390.7 gal **88.** **(a)** 16 ft/s **(b)** 32 ft/s **(c)** 12 ft/s **89.** $2\sqrt{2}\pi \approx 8.89$ s
90. $\pi\sqrt{2} \approx 4.44$ s **91.** $\pi \approx 3.14$ s **92.** $\frac{\pi\sqrt{2}}{2} \approx 2.22$ s

9 Exercises

1. 4 **2.** 8 **3.** 9 **4.** 16 **5.** $\frac{1}{8}$ **6.** $-\frac{1}{32}$ **7.** $\frac{1}{27}$ **8.** $\frac{1}{3125}$ **9.** $\frac{27}{8}$ **10.** $\frac{9}{4}$ **11.** $\frac{27}{8}$ **12.** $\frac{9}{4}$ **13.** 8 **14.** $\frac{1}{64}$ **15.** 8 **16.** $\frac{1}{27}$ **17.** 27
18. 16 **19.** $\frac{1}{5}$ **20.** $\frac{1}{9}$ **21.** 9 **22.** $5^{5/3}$ **23.** $\frac{1}{7}$ **24.** 6 **25.** 2 **26.** 3 **27.** 3 **28.** 2 **29.** $\frac{1}{2}$ **30.** $\frac{1}{3}$ **31.** $6^{2/3}$ **32.** $5^{3/4}$ **33.** $2^{5/6}$ **34.** $5^{5/6}$
35. $\sqrt{x}$ **36.** $\sqrt{x}$ **37.** $x^{7/4}$ **38.** $x^{7/6}$ **39.** x **40.** x **41.** x^2y^4 **42.** x^5y^{10} **43.** $x^{4/3}y^{5/3}$ **44.** $x^{5/4}y^{5/4}$ **45.** $\frac{8x^{3/2}}{y^{1/4}}$ **46.** $\frac{8y^{1/2}}{x^{3/2}}$ **47.** $\frac{x^{11}}{y^3}$
48. $\frac{x^4}{y^7}$

INDEX

Conics

Parabola

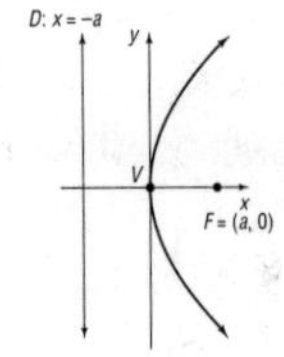

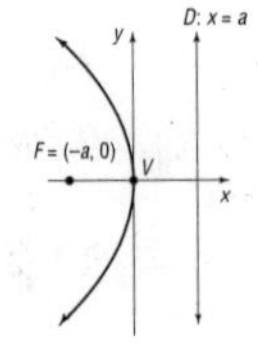

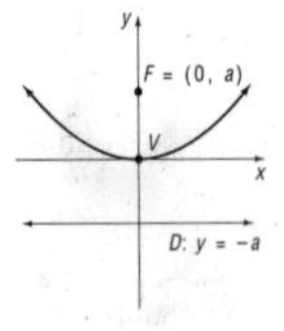

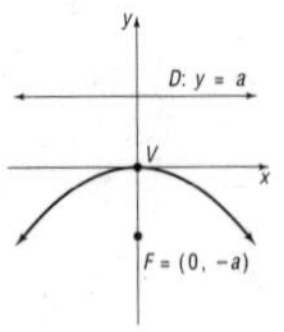

$y^2 = 4ax$ $\qquad y^2 = -4ax$ $\qquad x^2 = 4ay$ $\qquad x^2 = -4ay$

Ellipse

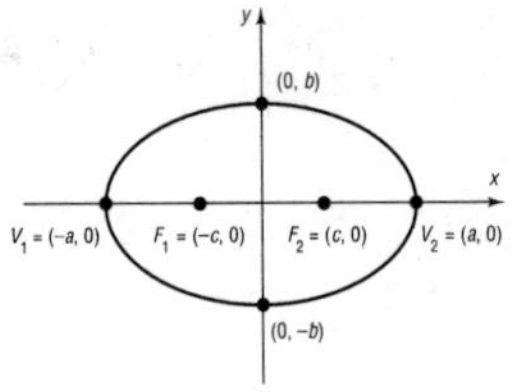

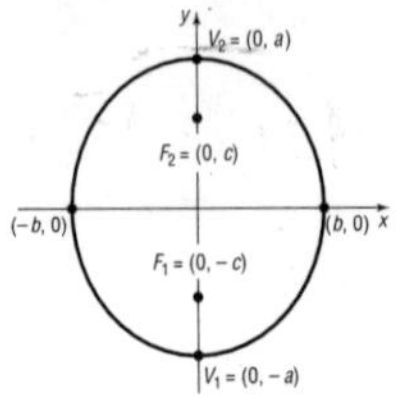

$$\frac{x^2}{a^2} + \frac{y^2}{b^2} = 1, \quad c^2 = a^2 - b^2$$

$$\frac{x^2}{b^2} + \frac{y^2}{a^2} = 1, \quad c^2 = a^2 - b^2$$

Hyperbola

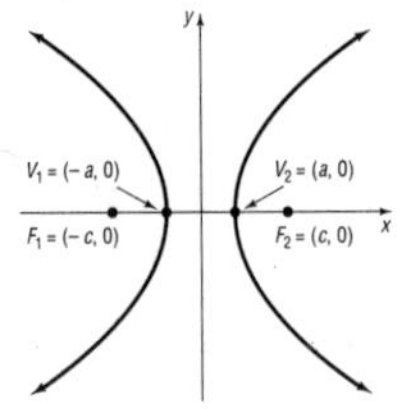

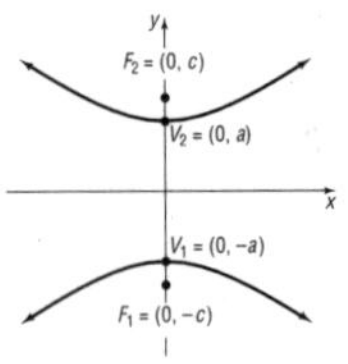

$$\frac{x^2}{a^2} - \frac{y^2}{b^2} = 1, \quad c^2 = a^2 + b^2$$

Asymptotes: $y = \frac{b}{a}x, \quad y = -\frac{b}{a}x$

$$\frac{y^2}{a^2} - \frac{x^2}{b^2} = 1, \quad c^2 = a^2 + b^2$$

Asymptotes: $y = \frac{a}{b}x, \quad y = -\frac{a}{b}x$